Prealgebra & Introductory Algebra

Third Edition

Margaret L. Lial
American River College

Diana L. Hestwood
Minneapolis Community and Technical College

John Hornsby
University of New Orleans

Terry McGinnis

Addison-Wesley

Boston • New York • San Francisco

London • Toronto • Sydney • Tokyo • Singapore • Madrid

Mexico City • Munich • Paris • Cape Town • Hong Kong • Montreal

Editorial Director	Christine Hoag
Editor in Chief	Maureen O'Connor
Executive Project Manager	Kari Heen
Project Editor	Courtney Slade
Editorial Assistant	Mary Gallagher
Senior Managing Editor	Karen Wernholm
Senior Production Supervisor	Kathleen A. Manley
Senior Designer	Barbara T. Atkinson
Photo Researcher	Beth Anderson
Supplements Production	Marianne Groth and Kayla Smith-Tarbox
Media Producers	Ceci Fleming and Lin Mahoney
Software Development	Eric Gregg, MathXL; Mary Durnwald, TestGen
Senior Marketing Manager	Michelle Renda
Marketing Assistant	Alicia Frankel
Senior Author Support/Technology Specialist	Joe Vetere
Senior Prepress Supervisor	Caroline Fell
Senior Media Buyer	Ginny Michaud
Rights and Permissions Advisor	Dana Weightman
Manufacturing Manager	Evelyn Beaton
Senior Manufacturing Buyer	Carol Melville
Composition/Production Coordination	Nesbitt Graphics, Inc.
Cover Image	Spring Interval Copyright © Lorraine Cota Manley

Library of Congress Cataloging-in-Publication Data

Prealgebra and Introductory Algebra / Margaret L. Lial, Diana L. Hestwood, John Hornsby, Terry McGinnis — 3rd ed.
   p. cm.
ISBN-13: 978-0-321-57873-0 (student edition)
ISBN-10: 0-321-57873-2 (student edition)
1. Algebra — Textbooks. I. Lial, Margaret L.
QA152.3.P74 2010
512—dc22   2008038400

1 2 3 4 5 6 7 8 9 10—RRDJC—12 11 10 09

Addison-Wesley
is an imprint of

www.pearsonhighered.com

ISBN 10: 0-321-57873-2
ISBN 13: 978-0-321-57873-0

CONTENTS

PREFACE

The third edition of *Prealgebra and Introductory Algebra* continues our ongoing commitment to provide the best possible text and supplements package that will help instructors teach and students succeed. To that end, we have addressed the diverse needs of today's students by integrating activities to help students improve their study skills, an attractive design, updated applications and graphs, helpful features, careful explanation of concepts, and an expanded package of supplements and study aids. We have also responded to the suggestions of users and reviewers and have added many new examples and exercises based on their feedback.

The text is designed for mathematics students who are new to algebra, are relearning the algebra they studied in the past, or are anxious about their ability to learn algebra. The text interweaves arithmetic review and geometry topics, as appropriate, into the algebraic themes of integers, variables and expressions, equations, solving application problems, positive and negative fractions and decimals, proportions, percents, measurement, graphing, and polynomials. The emphasis is on building a solid understanding of the foundations of algebra. This is accomplished by tying the content to students' experiences and previous knowledge, explaining important terminology in everyday English, showing *why* things work the way they do, and providing carefully sequenced exercises. This text is part of a series that also includes the following books:

- *Basic College Mathematics,* Eighth Edition, by Lial, Salzman, and Hestwood

- *Essential Mathematics,* Third Edition, by Lial and Salzman

- *Prealgebra*, Fourth Edition, by Lial and Hestwood

- *Introductory Algebra,* Ninth Edition, by Lial, Hornsby, and McGinnis

- *Intermediate Algebra,* Ninth Edition, by Lial, Hornsby, and McGinnis

- *Introductory and Intermediate Algebra,* Fourth Edition, by Lial, Hornsby, and McGinnis

- *Developmental Mathematics: Basic Mathematics and Algebra*, Second Edition, by Lial, Hornsby, McGinnis, Salzman, and Hestwood

Hallmark Features

We have retained the popular features of the previous edition that students and instructors have found helpful.

▶ *Chapter Openers* New and updated chapter openers feature real-world applications of mathematics that are relevant to students and tied to specific material within the chapters. Examples of topics include work/career applications, finding the best buy on cell phone service, home improvements, driving, weather, fishing, and discount and sales tax. (See Chapter 4, p. 215.)

▶ *Real-Life Applications* We are always on the lookout for interesting data to use in real-life applications. As a result, we have included many new or updated examples and exercises throughout the text that focus on real-life applications of mathematics. Students are often asked to find data in a table, chart, graph, or advertisement. (See pp. 297–98.) These applied problems provide an up-to-date flavor that will appeal to and motivate students.

▶ *Figures and Photos* Today's students are more visually oriented than ever. Thus, we have made a concerted effort to include mathematical figures, diagrams, tables, and graphs whenever possible. (See pp. 261–62.) Many of the graphs use a style similar to that seen by students in today's print and electronic media. Photos have been incorporated to enhance applications in examples and exercises. (See pp. 262 and 298.)

▶ *Emphasis on Problem Solving* Chapter 3 introduces students to our six-step process for solving application problems algebraically: *Read, Assign a Variable, Write an Equation,*

Solve, *State the Answer*, and *Check*. By devoting an entire chapter to this process, students build a strong foundation for problem solving, which is then reinforced through specific problem-solving sections in Chapters 4, 5, 6, 7, 8, 12, 14, and 15. (See p. 293.) The same six steps are also used throughout the other algebra titles in this textbook series.

▶ *Learning Objectives* Each section begins with clearly stated, numbered objectives, and the material within sections is keyed to these objectives so that students know exactly what concepts are covered. (See p. 265.)

▶ *Cautions and Notes* These color-coded and boxed comments, one of the most popular features of previous editions, warn students about common errors and emphasize important ideas throughout the exposition. (See pp. 272 and 301.) Cautions are highlighted in yellow and Notes are highlighted in purple.

▶ *Calculator Tips* These optional tips, marked with a red calculator icon, offer helpful information and instruction for students using calculators in the course. (See p. 241.)

▶ *Margin Problems* Margin problems, with answers immediately available on the bottom of the page, are found in every section of the text. (See p. 270.) This key feature allows students to immediately practice the material covered in the examples in preparation for the exercise sets.

▶ *Ample and Varied Exercise Sets* The text contains a wealth of exercises to provide students with opportunities to practice, apply, connect, and extend the skills they are learning. Numerous illustrations, tables, graphs, and photos help students visualize the problems they are solving. Problem types include skill building, writing, estimation, and calculator exercises, as well as applications and correct-the-error problems. In the Annotated Instructor's Edition of the text, the writing exercises are marked with an icon for writing so that instructors may assign these problems at their discretion. Exercises suitable for calculator work are marked in both the student and instructor editions with a calculator icon. (See pp. 305–06.) Students can watch an instructor work through the complete solution for each exercise marked with a DVD icon ◐ on the Videos on DVD.

▶ *Relating Concepts Exercises* These sets of exercises help students tie concepts together and develop higher level problem-solving skills as they compare and contrast ideas, identify and describe patterns, and extend concepts to new situations. (See p. 260.) These exercises make great collaborative activities for pairs or small groups of students.

▶ *Summary Exercises* Three more sets of these popular, in-chapter summary exercises have been added. These special exercises provide students with the all-important *mixed* practice they need at these critical points in their skill development. (See pp. 279–80.)

▶ *Ample Opportunity for Review* Each chapter concludes with a Chapter Summary that features: Key Terms with definitions and helpful graphics, New Formulas, New Symbols, Test Your Word Power, and a Quick Review of each section's content with additional examples. Also included is a comprehensive set of Chapter Review Exercises keyed to individual sections, a set of Mixed Review Exercises, and a Chapter Test. Students can watch an instructor work out the full solutions to the Chapter Test problems on the new Chapter Test Prep Video CD that comes with each new copy of the text.

▶ *Test Your Word Power* This feature, incorporated into each Chapter Summary, helps students understand and master mathematical vocabulary. Key terms from the chapter are presented along with four possible definitions in a multiple-choice format. Answers and examples illustrating each term are provided. (See p. 308.)

What's New in This Edition

The scope and sequence of topics in *Prealgebra and Introductory Algebra* has stood the test of time and rates highly with our reviewers. Therefore, you will find the table of contents intact, making the transition to the new edition easier.

NEW *Study Skills* There is an increased emphasis on study skills. Poor study skills are a major reason why students do not succeed in mathematics. A few generic tips sprinkled here and there are not enough to help students change their behavior. This text includes 13 carefully designed activities, integrated into the text material, that cover note taking, homework, study cards, math anxiety, test preparation, test taking, preparing for a final exam, and more. (See pp. 105–06, 163–64, and 213–14.) Most are found within the first few chapters so that students can use the skills throughout the course. (See the Table of Contents for titles and locations.) The first activity, "Your Brain *Can* Learn Mathematics," located before Chapter 1, explains *how* the brain actually learns and remembers so that students understand *why* the study skills will help them succeed in the course. Many of the other two-page activities may be used at any point in the course with students working individually or in small groups, or as source material for in-class discussions.

NEW *Math in the Media* Each one-page activity presents a relevant look at how mathematics is used in the media. Designed to help instructors answer the often-asked question, "When will I ever use this stuff?," these activities ask students to read and interpret data from newspaper articles, the Internet, and other familiar, real-world sources. (See pp. 258 and 284.) The activities are well suited to collaborative work or they can be completed by individuals or used for open-ended class discussions.

NEW *Solutions* Solutions to selected section exercises have been added to the back of the book (following the Answers section). This provides students with easily accessible step-by-step help in solving the exercises that are most commonly missed. Solutions are provided for the exercises marked with a square of blue color around the exercise number, for example, **15.**

NEW *Pointers* Pointers have been added to examples to provide students with important on-the-spot reminders and warnings about common pitfalls. (See pp. 282 and 290.)

NEW *Examples and Exercises* Throughout the text, examples and exercises have been adjusted or replaced to reflect current data and practices. Applications have been updated and cover a wider variety of topics such as the fields of technology, pop culture, ecology, and health sciences.

NEW *Chapter Test Prep Video CD* The Chapter Test Prep Video CD provides students with the opportunity to watch instructors work through step-by-step solutions to all the Chapter Test exercises from the textbook. The Chapter Test Prep Video CD is included with each new student text.

What Supplements Are Available?

For a comprehensive list of the supplements and study aids that accompany *Prealgebra and Introductory Algebra,* Third Edition, see pages x and xi.

STUDENT SUPPLEMENTS

Student's Solutions Manual
- By Jeffery A. Cole, *Anoka-Ramsey Community College*
- Provides detailed solutions to the odd-numbered section-level exercises and to all margin, Relating Concepts, Summary, Chapter Review, and Chapter Test exercises
  ISBNs: 0-321-59929-2, 978-0-321-59929-2

Worksheets for Classroom or Lab Practice
- Extra practice exercises for every section of the text with ample space for students to show their work
- These lab- and classroom-friendly workbooks also list the learning objectives and key vocabulary terms for every text section, along with vocabulary practice problems
  ISBNs: 0-321-60016-9, 978-0-321-60016-5

Videos on DVD
- Feature an engaging team of lecturers
- Complete set of lectures for each section of the text on DVD for student use at home or on campus
- Ideal for distance learning or supplemental instruction
- Include optional English and Spanish subtitles
- Students can watch an instructor work through the complete solutions for all exercises marked with a DVD icon ◉
  ISBNs: 0-321-59930-6, 978-0-321-59930-8

InterAct Math Tutorial Website *www.interactmath.com*
- Online practice and tutorial help
- Retry an exercise with new values each time for unlimited practice and mastery
- Every exercise is accompanied by an interactive guided solution that gives helpful feedback when an incorrect answer is entered
- View the steps of worked-out sample problems similar to those in the text

Chapter Test Prep Video CD
- Watch instructors work through step-by-step solutions to all the Chapter Test exercises from the textbook
- Included with each new student text
- Includes optional English subtitles

INSTRUCTOR SUPPLEMENTS

Annotated Instructor's Edition
- Provides answers to all text exercises in color next to the corresponding problems
- Icons identify writing ✐ and calculator ▦ exercises
  ISBNs: 0-321-59923-3, 978-0-321-59923-0

Instructor's Solutions Manual
- By Jeffery A. Cole, *Anoka-Ramsey Community College*
- Provides complete answers to all exercises in the text
  ISBNs: 0-321-59924-1, 978-0-321-59924-7

Additional Teaching Resources
- Includes resources to help both new and adjunct faculty with course preparation and classroom management by offering helpful teaching tips correlated to the sections of the text
  Available for download at *www.pearsonhighered.com*

Instructor's Resource Manual with Tests
- By James Ball, *Indiana State University*
- The resource manual contains a test bank with two diagnostic pretests, six free-response and two multiple-choice test forms per chapter, and two final exams.
- The manual also contains a mini-lecture for each section of the text with objectives, key examples, and teaching tips.
- A correlation guide from the second to the third edition and phonetic spellings for all key terms in the text are also included.
  ISBNs: 0-321-59925-X, 978-0-321-59925-4

PowerPoint Lecture Slides
- Present key concepts and definitions from the text
  Available for download at *www.pearsonhighered.com*

TestGen (*www.pearsonhighered.com/testgen*)
- Enables instructors to build, edit, print, and administer tests using a computerized bank of questions developed to cover all text objectives
- Algorithmically based, TestGen allows instructors to create multiple but equivalent versions of the same question or test with the click of a button
- Instructors can also modify test bank questions or add new questions
- Tests can be printed or administered online

Pearson Math Adjunct Support Center (*www. pearsontutor services.com/math-adjunct.html*) is staffed by qualified instructors with more than 50 years of combined experience at both the community college and university levels. Assistance is provided for faculty in the following areas:
- Suggested syllabus consultation
- Tips on using materials packed with your book
- Book-specific content assistance
- Teaching suggestions, including advice on classroom strategies

Available for Students and Instructors

MyMathLab® MyMathLab is a series of text-specific, easily customizable online courses for Pearson Education's textbooks in mathematics and statistics. Powered by CourseCompass™ (our online teaching and learning environment) and MathXL® (our online homework, tutorial, and assessment system), MyMathLab gives you the tools you need to deliver all or a portion of your course online, whether your students are in a lab setting or working from home. MyMathLab provides a rich and flexible set of course materials, featuring free-response exercises that are algorithmically generated for unlimited practice and mastery. Students can also use online tools, such as video lectures, animations, and a multimedia textbook, to independently improve their understanding and performance. Instructors can use MyMathLab's homework and test managers to select and assign online exercises correlated directly to the textbook, and they can also create and assign their own online exercises and import TestGen tests for added flexibility. MyMathLab's online gradebook—designed specifically for mathematics and statistics—automatically tracks students' homework and test results and gives the instructor control over how to calculate final grades. Instructors can also add offline (paper-and-pencil) grades to the gradebook. MyMathLab also includes access to the **Pearson Tutor Center** (*www.pearsontutorservices.com*). The Tutor Center is staffed by qualified mathematics instructors who provide textbook-specific tutoring for students via toll-free phone, fax, e-mail, and interactive Web sessions. MyMathLab is available to qualified adopters. For more information, visit our Web site at *www.mymathlab.com* or contact your sales representative.

MathXL® MathXL is a powerful online homework, tutorial, and assessment system that accompanies Pearson Education's textbooks in mathematics or statistics. With MathXL, instructors can create, edit, and assign online homework and tests using algorithmically generated exercises correlated at the objective level to the textbook. They can also create and assign their own online exercises and import TestGen tests for added flexibility. All student work is tracked in MathXL's online gradebook. Students can take chapter tests in MathXL and receive personalized study plans based on their test results. The study plan diagnoses weaknesses and links students directly to tutorial exercises for the objectives they need to study and retest. Students can also access supplemental animations and video clips directly from selected exercises. MathXL is available to qualified adopters. For more information, visit our Web site at *www.mathxl.com*, or contact your sales representative.

 MathXL® Tutorials on CD This interactive tutorial CD-ROM provides algorithmically generated practice exercises that are correlated at the objective level to the exercises in the textbook. Every practice exercise is accompanied by an example and a guided solution designed to involve students in the solution process. Selected exercises may also include a video clip to help students visualize concepts. The software provides helpful feedback for incorrect answers and can generate printed summaries of students' progress.
ISBNs:
0-321-59927-6
978-0-321-59927-8

Acknowledgments

The comments, criticisms, and suggestions of users, nonusers, instructors, and students have positively shaped this textbook over the years, and we are most grateful for the many responses we have received. The feedback gathered for this revision of the text was particularly helpful, and we especially wish to thank the following individuals who provided invaluable suggestions for this and the previous edition:

Carla Ainsworth, *Salt Lake Community College*
Randall Allbritton, *Daytona Beach Community College*
Jannette Avery, *Monroe Community College*
Pam Baenziger, *Kirkwood Community College*
Linda Beattie, *Western New Mexico University*
Jean Bolyard, *Fairmont State University*
Barbara Brown, *Anoka-Ramsey Community College*
Kim Brown, *Tarrant County College—Northeast Campus*
Hien Bui, *Hillsborough Community College*
Tim C. Caldwell, *Meridian Community College*
Russell Campbell, *Fairmont State University*
John Close, *Salt Lake Community College*
Jane Cuellar, *Taft College*
Ky Davis, *Muskingum Area Technical College*
Bill Dunn, *Las Positas College*
Lucy Edwards, *Las Positas College*
Randy Gallaher, *Lewis and Clark Community College*
Veronica Gold, *Assumption College*
Nancy Graham, *Rose State College*
Lynn Hargrove, *Sierra College*
J. Lloyd Harris, *Gulf Coast Community College*
Terry Haynes, *Eastern Oklahoma State College*
Edith Hays, *Texas Woman's University*
Karen Heavin, *Morehead State University*
Elizabeth Heston, *Monroe Community College*
Scott Higinbotham, *Middlesex Community College*
Lori Holdren, *Manatee Community College*
Sharon Jackson, *Brookhaven College*
Rosemary Karr, *Collin County Community College*
Harriet Kiser, *Floyd College*
Valerie Lazzara, *Palm Beach Community College*
Christine Heinecke Lehmann, *Purdue University—North Central*
Lou Ann Mahaney, *Tarrant County College—Northeast Campus*
Valerie H. Maley, *Cape Fear Community College*
Linda Marable, *Nashville State Community College*
Susan Martin, *Diablo Valley College*
Susan McClory, *San Jose State University*
Pam Miller, *Phoenix College*
Jeffrey Mills, *Ohio State University*
Michael Montano, *Riverside Community College*
Elizabeth Morrison, *Valencia Community College—West Campus*
Linda J. Murphy, *Northern Essex Community College*
Celia Nippert, *Western Oklahoma State College*
Elizabeth Olgilvie, *Horry-Georgetown Technical College*
Faith Peters, *Broward Community College*
Larry Pontaski, *Pueblo Community College*
Sara Pries, *Sierra College*
Brooke Quinlan, *Hillsborough Community College—Dale Marby Campus*
Manoj Raghunandanan, *Temple University*
Janalyn Richards, *Idaho State University*

Diann Robinson, *Ivy Tech State College—Lafayette*
Heather Roth, *Nova Southeastern University*
Rachael Schettenhelm, *Southern Connecticut State University*
Julia Simms, *Southern Illinois University—Edwardsville*
Dr. Yojana Sharma, *Stark State College of Technology*
Sounny Slitine, *Palo Alto College*
Lee Ann Spahr, *Durham Technical Community College*
Carol Stewart, *Fairmont State University*
Sharon Testone, *Onondaga Community College*
Sam Tinsley, *Richland College*
Shae Thompson, *Montana State University*
Cora S. West, *Florida Community College at Jacksonville*
Cheryl Wilcox, *Diablo Valley College*
Johanna Windmueller, *Seminole Community College*
Gabriel Yimesghen, *Community College of Philadelphia*
Kevin Yokoyama, *College of the Redwoods*
Carol A. Zavarella, Ph.D., *Hillsborough Community College*
Karl Zilm, *Lewis and Clark Community College*

Our sincere thanks go to the dedicated individuals at Pearson who have worked hard to make this revision a success: Greg Tobin, Maureen O'Connor, Kathy Manley, Barbara Atkinson, Michelle Renda, Beth Anderson, Kari Heen, Courtney Slade, Lin Mahoney, Ceci Fleming, Alicia Frankel, and Mary Gallagher. We are also grateful to Janette Krauss and Bonnie Boehme of Nesbitt Graphics for their excellent production work; Lucie Haskins for producing a useful Index; Jeff Cole for writing the Solutions manuals; and Janis Cimperman, Perian Herring, Paul Lorczak, Sarah Sponholz, Ann Ostberg, Sharon Testone, and Shannon d'Hemecourt for accuracy checking the manuscript.

Abby Tanenbaum did an outstanding job helping us with manuscript preparation. We are truly grateful for her contributions to so many of our books over the years. Special thanks go to Linda Russell, who wrote the Study Skills activities that appear throughout the text. In her roles as both an instructor and a specialist in reading and study skills at Minneapolis Community and Technical College, she created and revised these activities based on her work with hundreds of developmental-level students.

The ultimate measure of this textbook's success is whether it helps students master algebra skills, develop problem-solving techniques, and increase their confidence in learning and using mathematics. In order for us, as authors, to know what to keep and what to improve for the next edition, we need to hear from you, the instructor, and you, the student. Please tell us what you like and where you need additional help by sending an e-mail to math@pearson.com. We appreciate your feedback!

Margaret L. Lial
Diana L. Hestwood
John Hornsby
Terry McGinnis

This book is dedicated to my husband, Earl Orf, who is the wind beneath my wings, and to my students at Minneapolis Community and Technical College, from whom I have learned so much.

Diana L. Hestwood

Study Skills

▶▶▶ YOUR BRAIN *CAN* LEARN MATHEMATICS

Your brain knows how to learn, just as your lungs know how to breathe; however, there are important things you can do to maximize your brain's ability to do its work. This short introduction will help you choose effective strategies for learning mathematics. This is a simplified explanation of a complex process.

Your brain's outer layer is called the **neocortex,** which is where higher level thinking, language, reasoning, and purposeful behavior occur. The neocortex has about 100 billion (100,000,000,000) brain cells called **neurons.**

▶ As you learn something new, threadlike branches grow out of each neuron. These branches are called **dendrites.**

▶ When the dendrite from one neuron grows close enough to the dendrite from another neuron, a connection is made. There is a small gap at the connection point called a **synapse.** One dendrite sends an electrical signal across the gap to another dendrite.

▶ *Learning = growth and connecting of dendrites.*

▶ When you practice a skill just once or twice, the connections between neurons are very weak. If you do not practice the skill again, the dendrites at the connection points wither and die back. You have forgotten the new skill!

OBJECTIVES

1 Describe how practice fosters dendrite growth.

2 Explain the effect of anxiety on the brain.

Learning Something New

Remembering New Skills

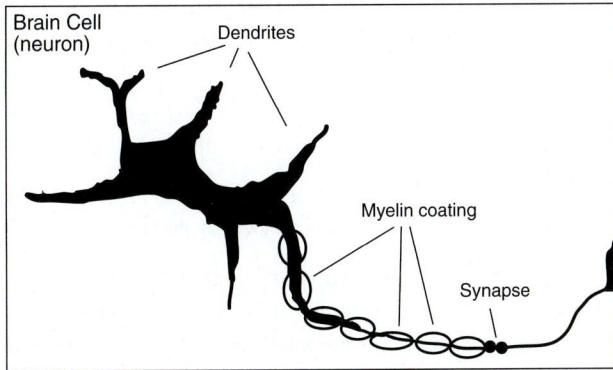

A neuron with several dendrites: one dendrite has developed a myelin coating through repeated practice.

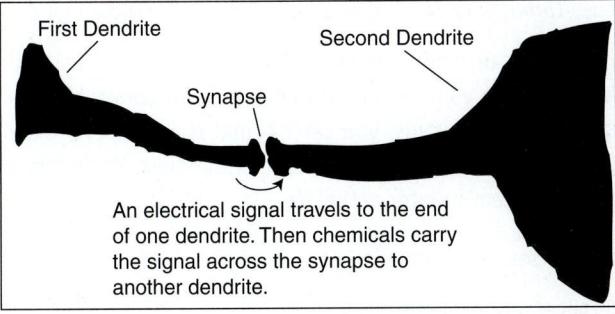

An electrical signal travels to the end of one dendrite. Then chemicals carry the signal across the synapse to another dendrite.

A close up view of the connection (synapse) between two dendrites.

- ▶ If you practice a new skill many times, the dendrites for that skill become coated with a fatty protein called **myelin.** Each time one dendrite sends a signal to another dendrite, the myelin coating becomes thicker and smoother, allowing the signals to move faster and with less interference. Thinking can now occur more quickly and easily, and *you will remember the skill for a long time* because the dendrite connections are very strong.

Become an Effective Student

- ▶ You grow dendrites specifically for the thing you are studying. If you practice dividing fractions, you will grow specialized dendrites just for dividing fractions. If you *watch other people* solve fraction problems, *you will grow dendrites for watching, not for solving.* So, be sure you are actively learning and practicing.

- ▶ If you practice something the *wrong* way, you will develop strong dendrite connections for doing it the wrong way! So, as you study, check frequently that you are getting correct answers.

- ▶ As you study a new topic that is related to things you already know, you will grow new dendrites, but your brain will also send signals throughout the network of dendrites for the related topics. In this way, you build a complex **neural network** that allows you to apply concepts, see differences and similarities between ideas, and understand relationships between concepts.

In the first few chapters of this textbook you will find "brain friendly" activities that are designed to help you grow and develop your own reliable neural networks for mathematics. Since you must grow your own dendrites (no one can grow them for you), these activities show you how to

- ▶ develop new dendrites,

- ▶ strengthen existing ones, and

- ▶ encourage the myelin coating to become thicker so signals are sent with less effort.

When you incorporate the activities into your regular study routine, you will discover that you understand better, remember longer, and forget less.

Also remember that *it does take time for dendrites to grow.* Trying to cram in several new concepts and skills at the last minute is not possible. Your dendrites simply can't grow that quickly. You can't expect to develop huge muscles by lifting weights for just one evening before a body building competition! In the same way, practice the study techniques *throughout the course* to facilitate strong growth of dendrites.

When Anxiety Strikes

If you are under stress or feeling anxious, such as during a test, your body secretes **adrenaline** into your system. Adrenaline in the brain blocks connections between neurons. In other words, you can't think! If you've ever experienced "blanking out" on a test, you know what adrenaline does. You'll learn several solutions to that problem in later activities.

Start Your Course Right!

- ▶ Attend all class sessions (especially the first one).

- ▶ Gather the necessary supplies.

- ▶ Carefully read the syllabus for the course, and ask questions if you don't understand.

1

Introduction to Algebra: Integers

T he weather—we all talk about it and often complain about it. When reporting temperatures, we need both *positive* and *negative* numbers. If you live in Chicago, the highest temperature ever recorded is 104 °F and the lowest is −27 °F, a difference of 131 degrees! In Atlanta, the record high and low are 105 °F and −8 °F. And in Barrow, Alaska, the range of extreme temperatures is 79 °F to −56 °F, a difference of 135 degrees. To make things even more uncomfortable, humidity can make hot temperatures feel hotter and wind can make cold temperatures feel colder. Learn more about "windchill" as you work Exercises 47–48 in **Section 1.4.** (*Source:* National Climatic Data Center.)

1.1 ▶▶▶ Place Value

It would be nice to earn millions of dollars like some of our favorite entertainers or sports stars. But how much is a million? If you received $1 every second, 24 hours a day, day after day, how many days would it take for you to receive a million dollars? How long to receive a billion dollars? Or a trillion dollars? Make some guesses and write them here.

It would take _____ to receive a million dollars.

It would take _____ to receive a billion dollars.

It would take _____ to receive a trillion dollars.

The answers are at the bottom left of the page. Later, in the exercises for **Section 1.7**, you'll find out how to calculate the answers.

OBJECTIVE 1 **Identify whole numbers.** First we have to be able to write the number that represents *one million*. We can write *one* as 1. How do we make it 1 *million?* Our number system is a **place value system.** That means that the location, or place, in which a number is written gives it a different value. Using money as an example, you can see that

$1 is one dollar.

$10 is ten dollars.

$100 is one hundred dollars.

$1000 is one thousand dollars.

Each time the 1 moved to the left one place, it was worth *ten times* as much. Can you keep moving it to the left? Yes, as many times as you like.

The chart below shows the *value* of each *place*. In other words, you write the 1 in the correct place to represent the number you want to express. It is important to memorize the place value names shown on the chart.

Whole Number Place Value Chart

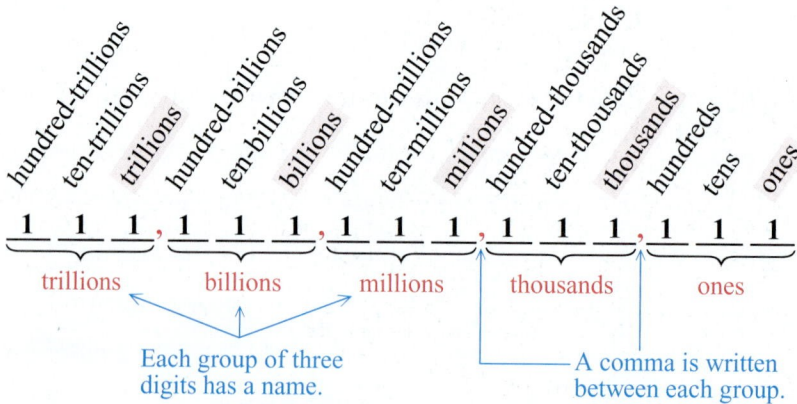

If there had been more room on this page, we could have continued to the left with quadrillions, quintillions, sextillions, septillions, octillions, and more.

Of course, we can use other *digits* besides 1. In our decimal system of writing numbers, we can use these ten **digits:** 0, 1, 2, 3, 4, 5, 6, 7, 8, and 9. In this section we will use the digits to write **whole numbers.**

These are whole numbers.

0   8   37   100   24,014

These are *not* whole numbers.

-6 $\dfrac{3}{4}$ 7.528   0.3 $5\dfrac{2}{3}$

Circle the whole numbers.

0.8	-14	502
$\dfrac{7}{9}$	3	$\dfrac{3}{2}$
14	0	$6\dfrac{4}{5}$
9.082	$-\dfrac{8}{3}$	60,005

EXAMPLE 1 **Identifying Whole Numbers**

Circle the whole numbers in this list.

75   -4 0   1.5 $\dfrac{5}{8}$ 300   0.666 $7\dfrac{1}{2}$ 2

Whole numbers *do* include zero. If we started a list of *all* the whole numbers, it would look like this: 0, 1, 2, 3, 4, 5, . . . with the three dots indicating that the list goes on and on. So the whole numbers in this example are: 75, 0, 300, and 2.

Work Problem **1** *at the Side.* ▶

OBJECTIVE 2 **Identify the place value of a digit through hundred-trillions.** The estimated world population in 2007 was 6,598,274,806 people. (*Source:* U.S. Census Bureau.) There are two 6s in this number but the value of each 6 is very different. The 6 on the right is in the ones place, so its value is simply 6. But the 6 on the left is worth a great deal more because of the *place* where it is written. Looking back at the place value chart, we see that this 6 is in the billions place, so its value is *6 billion*.

6 , 5 9 8 , 2 7 4 , 8 0 6 people in the world

Value of
6 *billion*

Value of
6 *ones*

2 Identify the place value of the digit 8 in each number.

(a) 45,628,665

(b) 800,503,622

EXAMPLE 2 **Identifying Place Value**

Identify the place value of each 8 in the number of people in the world.

6 , 5 9 **8** , 2 7 4 , **8** 0 6

Millions place

Hundreds place

Work Problem **2** *at the Side.* ▶

(c) 428,000,000,000

OBJECTIVE 3 **Write a whole number in words or digits.** To write a whole number in words, or to say it aloud, begin at the left. Write or say the number in each group of three, followed by the name for that group. When you get to the ones group, do *not* include the group name. Hyphens (dashes) are used whenever you write a number from 21 to 99, like twenty–one thousand, or thirty–seven million, or ninety–four billion.

EXAMPLE 3 **Writing Numbers in Words**

(a) Write 6,058,120 in words.
Start at the left.

6 , 0 5 8 , 1 2 0

Do **not** say "one hundred *and* twenty" here.

six million, fifty-eight thousand, one hundred twenty

Group name

Group name

Do not use group name "ones."

(d) 2,385,071

— **Continued on Next Page**

3 Write these numbers in words.

(a) 23,605

(b) 400,033,007

(c) 193,080,102,000,000

(b) Write 50,588,000,040,000 in words.
Start at the left.

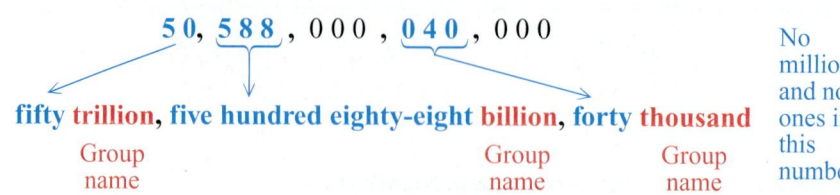

$$50, \underline{588}, 000, \underline{040}, 000$$

fifty trillion, five hundred eighty-eight billion, forty thousand

   Group   Group   Group
   name   name   name

No millions and no ones in this number

> **CAUTION**
> You often hear people say "and" when reading a group of three digits. For example, you may hear 120 as "one hundred *and* twenty," but this is *not* correct. The word *and* is used only when reading a *decimal point,* which we do not have here. The correct wording for 120 is "one hundred twenty." (See Example 3(a) at the bottom of the previous page.)

◀ *Work Problem* **3** *at the Side.*

---

4 Write each number using digits.

(a) Eighteen million, two thousand, three hundred five

(b) Two hundred billion, fifty million, six hundred sixteen

(c) Five trillion, forty-two billion, nine million

(d) Three hundred six million, seven hundred thousand, nine hundred fifty-nine

---

When you read or hear a number and want to write it in digits, look for the group names: **trillion, billion, million,** and **thousand.** Write the number in each group, followed by a comma. Do *not* put a comma at the end of the ones group.

EXAMPLE 4 **Writing Numbers in Digits**

Write each number using digits.

(a) Five hundred sixteen **thousand,** nine

The first group name is *thousand,* so you need to fill *two groups* of three digits: thousands and ones.

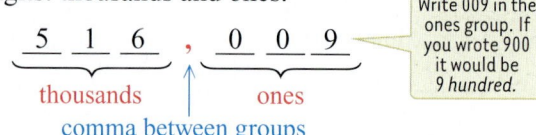

$$\underbrace{5\ \ 1\ \ 6}_{\text{thousands}}, \underbrace{0\ \ 0\ \ 9}_{\text{ones}}$$

comma between groups

Write 009 in the ones group. If you wrote 900 it would be *9 hundred.*

The number is 516,009.

(b) Seventy-seven **billion,** thirty **thousand,** five hundred

The first group name is *billion,* so you need to fill *four groups* of three digits: billions, millions, thousands, and ones.

$$\underbrace{0\ \ 7\ \ 7}_{\text{billions}}, \underbrace{0\ \ 0\ \ 0}_{\text{millions}}, \underbrace{0\ \ 3\ \ 0}_{\text{thousands}}, \underbrace{5\ \ 0\ \ 0}_{\text{ones}}$$

There are no millions, so fill the millions group with zeros.
When writing the number, you can omit the leading **0** in the billions group.
The number is 77,000,030,500.

◀ *Work Problem* **4** *at the Side.*

1.1 ▶▶▶ Exercises

Circle the whole numbers. See Example 1.

1. 15 $8\frac{3}{4}$ 0   3.781

83,001 -8 $\frac{7}{16}$ $\frac{9}{5}$

2. 33.7 -5 457 $\frac{8}{5}$

0   6 $1\frac{3}{4}$ -14.1

3. 5.8 -6 7 $\frac{5}{4}$

$\frac{1}{10}$ 362,049   0.1 $7\frac{7}{8}$

4. 75,039 $\frac{1}{3}$ -87 6.49

-0.5 $2\frac{7}{10}$ $\frac{15}{8}$ 4

Identify the place value of the digit 2 in each number. See Example 2.

5. 61,284

6. 82,110

7. 284,100

8. 823,415

9. 725,837,166

10. 442,653,199

11. 253,045,701,000

12. 823,000,419,567

13. From left to right, name the place value for each 0 in this number: 302,016,450,098,570.

14. From left to right, name the place value for each 0 in this number: 810,704,069,809,035.

In Exercises 15–26, write each number in words. See Example 3.

15. 8421

16. 1936

17. 46,205

18. 75,089

19. 3,064,801

20. 7,900,408

21. 840,111,003

22. 304,008,401

23. 51,006,888,321

24. 99,046,733,214

25. 3,000,712,000,000

26. 50,918,000,000,600

In Exercises 27–36, write each number using digits. See Example 4.

27. Forty-six thousand, eight hundred five

28. Seventy-nine thousand, forty-six

29. Five million, six hundred thousand, eighty-two

30. One million, thirty thousand, five

31. Two hundred seventy-one million, nine hundred thousand

32. Three hundred eleven million, four hundred

33. Twelve billion, four hundred seventeen million, six hundred twenty-five thousand, three hundred ten

34. Seventy-five billion, eight hundred sixty-nine million, four hundred eighty-eight thousand, five hundred six

35. Six hundred trillion, seventy-one million, four hundred

36. Four hundred forty trillion, thirty-six thousand, one hundred two

In Exercises 37–48, if the number is given in digits, write it in words. If the number is given in words, write it in digits. See Examples 3 and 4. (Source: The World Almanac, Statistical Abstract of the United States, Associated Press.)

37. In the United States, 3151 couples get married every day.

38. The New Year's Eve ball that drops at midnight on December 31 in Times Square, New York, has 9576 energy-efficient bulbs.

39. One hundred one million, two hundred eighty thousand adults are on diets at any one time in the United States.

40. Two million, five thousand Americans suffer from heartburn on any given day.

41. The average amount spent each day by individuals in the United States on computers and software is $173,523,700.

42. Americans spend an average of $131,703,860 daily on toys.

43. People eat fifty-five million, eight hundred hot dogs each day.

44. It costs about fifteen million, eight hundred forty thousand dollars to build a mile of 4-lane urban highway.

45. Dunkin' Donuts sells nearly 6,400,000 doughnuts every day. That's 2,336,000,000 doughnuts in one year!

46. The Hostess company bakeries are able to make 60,000 Twinkies every hour. That adds up to 525,600,000 Twinkies per year.

47. Americans sit in traffic jams for a total of four billion, two hundred million hours each year.

48. It is estimated that American drivers waste two billion, nine hundred million gallons of fuel yearly while sitting in traffic.

Relating Concepts (Exercises 49–52)   For Individual or Group Work

*Use your knowledge of place value to **work Exercises 49–52 in order.***

49. Here is a group of digits: 6, 0, 9, 1, 5, 0, 7, 1. Using each of these digits exactly once, arrange them to make the largest possible whole number and the smallest possible whole number. Then write each number in words.

50. Write these numbers in digits and in words.

   (a) Your house number or apartment building number
   (b) Your phone number, including the area code
   (c) The approximate cost of your tuition and books for this quarter or semester
   (d) Your zip code

   Now tell how you *usually* say each of the numbers in parts (a)–(d) above. Why do you think we ignore the rules when saying these numbers in everyday situations?

51. Look again at the Whole Number Place Value Chart at the beginning of this section. As you move to the left, each place is worth *ten* times as much as the previous place. Computers work on a *binary* system where each place is worth *two* times as much as the previous place. Complete this place value chart based on 2s.

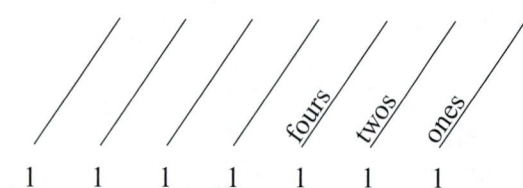

The *only* digits you may use in the binary system are 0 and 1. Here is an example.

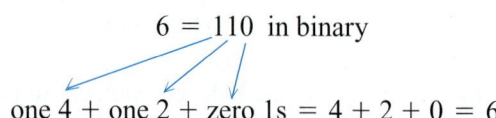

one 4 + one 2 + zero 1s = 4 + 2 + 0 = 6

Now try writing some numbers using the 2s (binary) place value chart.

   (a) 5 = _____
   (b) 10 = _____
   (c) 15 = _____

52. (a) Explain in your own words why our number system is called a *place value system*. Include an example as part of your explanation.

   (b) Find information on the Roman numeral system. Write these numbers using Roman numerals.

   8 = _____   38 = _____

   275 = _____   3322 = _____

   (c) Explain why the Roman system is *not* a place value system. What are the disadvantages of the Roman system?

Study Skills

▶▶▶ USING YOUR TEXTBOOK

Be sure to read *Your Brain Can Learn Mathematics* before this activity. You'll find out how your brain learns and remembers.

Your textbook can be very helpful. Find out what it has to offer. First, let's look at some general features that will help in all chapters.

Look on page iii in the very front of the book for the Table of Contents. Before you start Chapter 1, you'll want to look at Chapter R, Whole Numbers Review, near the end of the book.

On pages x–xi in the Preface is a list of supplementary resources for students. If you are interested in any of these, ask your instructor if they are available.

Each chapter is divided into sections, and each section has a number, such as 1.3 or 3.5. Your instructor will use these numbers to assign homework.

Chapter 3 ⟶ 3.5 ⟵ Section 5 within Chapter 3

There are four features to pay special attention to as you work in your book.

▶ **Objectives.** Each section lists the objectives in the upper corner of the first page. The objectives are listed again as each one is introduced. An objective tells you *what you will be able to do after you complete the section*. An excellent way to check your learning is to go back to the list of objectives when you are finished with a section and ask yourself if you can do them all.

▶ **Margin Exercises.** The exercises in the shaded margins of the pages in your textbook give you immediate practice. **This is a perfect way to get your dendrites growing right away!** The answers are given at the bottom, so you can check yourself easily.

▶ **Cautions, Pointers, Notes, and Calculator Tips.**
  • **Caution!** A bright yellow box is a comment about a common error that students make, or a common trouble spot you will want to avoid.
  • **Pointers** are little "clouds" next to worked examples. They point to specific places where common mistakes are made and give on-the-spot reminders.
  • Look for the specially marked purple **Note** boxes. They contain hints, explanations, or interesting side comments about a topic.
  • A small picture of a red calculator 🖩 appears several places. In the main part of the chapter, the icon means that there is a **Calculator Tip,** which helps you learn more about using your calculator. A calculator in an Exercise section is a recommendation to use your calculator to work that exercise.

Go back to the Table of Contents again. What is listed at the end of each chapter?

▶ **Chapter Summary** Turn to page 77 to find the Summary for Chapter 1. It lists the chapter's **Key Terms** (arranged in the order that they appear in the chapter) and **New Symbols** and/or **New Formulas.** Then, **Test Your Word Power** checks your understanding of the math vocabulary. Next is a **Quick Review** section. It lists each topic in the chapter and shows a worked example, with tips.

OBJECTIVES

1 Explain the meaning of text features such as section numbering, objectives, margin exercises.

2 Locate the Answers, Solutions, and Index sections.

Table of Contents

Section Numbering

Chapter Features

List a page number from Chapter 1 for each of these features:

A *Caution* appears on page _____.
A *Pointer* appears on page _____.
A *Note* appears on page _____.
A *Calculator Tip* appears on page _____.

End-of-Chapter Features

End of Chapter Features
(continued)

> How will you make good use of the features at the end of each chapter?
>
> _____
>
> _____
>
> _____

Answers

> *Flag the Answers section with a sticky note or other device, so that you can turn to it quickly.*

Solutions

Index

Why Are These Features Brain Friendly?

The textbook authors included text features that make it easier for you to understand the mathematics. **Your brain naturally seeks organization and predictability.** When you pay attention to the regular features of your textbook, you are allowing your brain to get familiar with all of the helpful tips, suggestions, and explanations that your book has to offer. You will make the best possible use of your textbook.

▶ **Review Exercises** Use these exercises as a way to check your understanding of all the concepts in the chapter. You can practice every type of problem. If you get stuck, the red numbers in brackets tell you which section of the chapter to go back to for more explanations. Make sure you do the **Mixed Review Exercises** to practice for tests.

▶ **Chapter Test** Plan to take the test as a practice exam. That way you can be sure you really know how to work all types of problems without looking back at the chapter.

▶ **Cumulative Review (starting with Chapter 2)** These exercises help you maintain the skills you've learned in all previous chapters. Practicing previous skills throughout the course will be a big help on the final exam.

How do you find out if you've worked the exercises correctly? Your textbook provides many of the answers. Throughout each chapter you should work the sample problems in the **margins**. The answers for those are at the **bottom of each page** in the margin area.

For homework, you can find the answers to all of the **odd-numbered section exercises** in the **Answers to Selected Exercises** section near the end of your textbook. Also, *all* of the answers are given for the Chapter Review Exercises and Chapter Tests. Check your textbook now, and find the page on which the Answers section begins.

The **Solutions** section near the end of the book shows how to solve some of the harder odd-numbered exercises step by step. Look for the problem numbers with a square of blue shading around them. These are the ones that have a solution.

The **Index** is the last thing in your textbook. All of the topics, vocabulary, and concepts are listed in alphabetical order in the Index. For example, look up the words below. Go to the page or pages listed and find each word. Write down the page that introduces or defines each one. There may be several subheadings listed under the main word, or several page numbers listed. Usually, the *first* place that a word appears in the textbook is where it is introduced and defined. So, the earliest page number is a good place to start.

> *Commutative property of multiplication* is defined on page _____.
>
> *Factors* of numbers are defined on page _____.
>
> *Rounding of mixed numbers* is explained on page _____.

1.2 ▶▶▶ Introduction to Signed Numbers

OBJECTIVE 1 Write positive and negative numbers used in everyday situations. The whole numbers in **Section 1.1** were either 0 or greater than 0. Numbers *greater* than 0 are called *positive numbers*. But many everyday situations involve numbers that are *less* than 0, called *negative numbers*. Here are a few examples.

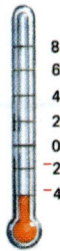

At midnight the temperature dropped to 4 degrees *below* zero.

−4 degrees

Jean had $30 in her checking account. She wrote a check for $40.75. We say she is now $10.75 "in the hole," or $10.75 "in the red," or *overdrawn* by $10.75.

−$10.75

The Packers football team *gained* 6 yards on the first play. On the second play they *lost* 9 yards. We can write the results using a positive number and a negative number.

+6 yards   and   **−9 yards**

A plane took off from the airport and climbed to 20,000 feet *above* sea level. We can write this using a positive number.

+20,000 feet

A scuba diver swam down to $25\frac{1}{2}$ feet *below* the surface. We can write this using a negative number.

−$25\frac{1}{2}$ feet

> **Note**
> To write a negative number, put a negative sign (a dash) in front of it: −10. Notice that the negative sign looks exactly like the subtraction sign, as in 5 − 3 = 2. The negative sign and subtraction sign do *not* mean the same thing (more on that in the next section). To avoid confusion for now, we will write negative signs in **red** and put them up higher than subtraction signs.
>
> ⁻10   means   **negative** 10   14 − 10 means 14 **minus** 10
> └── Raised dash
>
> Starting in **Chapter 4,** we will write negative signs in the traditional way. However, if you use a graphing calculator, it may show negative signs in the raised position.

Positive numbers can be written two ways:

1. Write a positive sign in front of the number: ⁺2 is *positive* 2. We will write the sign in the raised position to avoid confusion with the sign for addition, as in 6 + 3 = 9.

2. Do not write any sign. For example, 16 is assumed to be *positive* 16.

1 Write each negative number with a raised negative sign. Write each positive number in two ways.

(a) The temperature is $5\frac{1}{2}$ degrees below zero.

(b) Cameron lost 12 pounds on a diet.

(c) I deposited $210.35 in my checking account.

(d) I wrote too many checks, so my account is overdrawn by $65.

(e) The submarine dived to 100 feet below the surface of the sea.

(f) In this round of the card game, I won 50 points.

2 Graph each set of numbers.

(a) $^-2$ (b) 2 (c) 0

(d) $^-4$ (e) 4

$$\begin{array}{ccccccccccc} \\ ^-4 & ^-3 & ^-2 & ^-1 & 0 & 1 & 2 & 3 & 4 & 5 \end{array}$$

(f) $^-3\frac{1}{2}$ (g) $\frac{1}{2}$

(h) $^-1$ (i) 3

$$\begin{array}{ccccccccccc} \\ ^-4 & ^-3 & ^-2 & ^-1 & 0 & 1 & 2 & 3 & 4 & 5 \end{array}$$

ANSWERS

1. (a) $^-5\frac{1}{2}$ degrees (b) $^-12$ pounds
   (c) $210.35 or $^+\$210.35$ (d) $^-\$65$
   (e) $^-100$ feet (f) 50 points or $^+50$ points

2.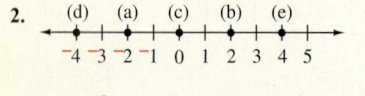
   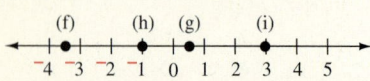

EXAMPLE 1 **Writing Positive and Negative Numbers**

Write each negative number with a raised negative sign. Write each positive number in two ways.

(a) The river rose to 8 feet above flood stage.

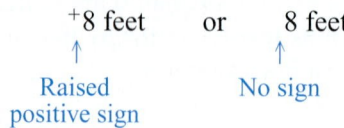

(b) Michael lost $500 in the stock market.

$$^-\$500$$
↑
Raised negative sign

◀ *Work Problem* **1** *at the Side.*

OBJECTIVE **2** **Graph signed numbers on a number line.** Mathematicians often use a **number line** to show how numbers relate to each other. A number line is like a thermometer turned sideways. *Zero* is the dividing point between the positive and negative numbers.

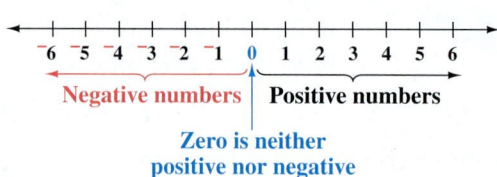

The number line could be shown with positive numbers on the left side of 0 instead of the right side. But it helps if everyone draws it the same way, as shown above. This method will also match what you do when graphing points and lines in **Chapter 9.**

EXAMPLE 2 **Graphing Numbers on a Number Line**

Graph each number on the number line.

(a) $^-5$ (b) 3 (c) $1\frac{1}{2}$ (d) 0 (e) $^-1$

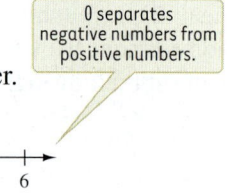

Draw a dot at the correct location for each number.

◀ *Work Problem* **2** *at the Side.*

OBJECTIVE **3** **Use the $<$ and $>$ symbols to compare integers.** In **Chapters 4 and 5** you will work with fractions and decimals. For the rest of this chapter, you will work only with *integers*. A list of **integers** can be written like this:

$$\dots,\ ^-6,\ ^-5,\ ^-4,\ ^-3,\ ^-2,\ ^-1,\ 0,\ 1,\ 2,\ 3,\ 4,\ 5,\ 6,\ \dots$$

The dots show that the list goes on forever in both directions.

We can use the number line to compare two integers.

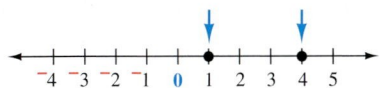 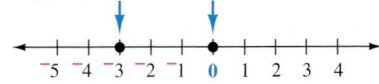

When comparing 1 and 4,
1 is to the *left* of 4.
1 is *less than* 4.
Use $<$ to mean "is less than."

$$1 \quad < \quad 4$$

1 **is less than** 4

When comparing 0 and $^-3$,
0 is to the *right* of $^-3$.
0 is *greater than* $^-3$.
Use $>$ to mean "is greater than."

$$0 \quad > \quad ^-3$$

0 **is greater than** $^-3$

> **Note**
> One way to remember which symbol to use is that the "smaller end of the symbol" points to the "smaller number" (the number that is less).
>
> $$1 < 4 \qquad\qquad 0 > ^-3$$
>
> Smaller number — └— Smaller end of symbol   Smaller end of symbol — └—Smaller number

EXAMPLE 3 **Comparing Integers, Using the $<$ and $>$ Symbols**

Write $<$ or $>$ between each pair of numbers to make a true statement.

(a) 0 ___ 2

   0 is to the *left* of 2 on the number line, so 0 is *less than* 2. Write 0 $<$ 2.

(b) 1 ___ $^-4$

   1 is to the *right* of $^-4$, so 1 is *greater than* $^-4$. Write 1 $>$ $^-4$.

(c) $^-4$ ___ $^-2$

   $^-4$ is to the *left* of $^-2$, so $^-4$ is *less than* $^-2$. Write $^-4$ $<$ $^-2$.

Work Problem **3** *at the Side.* ▶

OBJECTIVE 4 Find the absolute value of integers. In order to graph a number on the number line, you need to ask two things:

1. Which *direction* is it from 0? It can be in a positive direction or a negative direction. You can tell the direction by looking for a positive sign (or no sign, which means positive), or a negative sign.

2. How *far* is it from 0? The distance from 0 is the *absolute value* of a number.

> **Absolute Value**
> The **absolute value** of a number is its distance from 0 on the number line. *Absolute value* is indicated by two vertical bars. For example,
> $$|6| \quad \text{is read} \quad \text{"the absolute value of 6."}$$

3 Write $<$ or $>$ between each pair of numbers to make a true statement.

(a) 5 ___ 4

(b) 0 ___ 2

(c) $^-3$ ___ $^-2$

(d) $^-1$ ___ $^-4$

(e) 2 ___ $^-2$

(f) $^-5$ ___ 1

4 Find each absolute value.

(a) $|13|$

(b) $|{}^-7|$

(c) $|0|$

(d) $|{}^-350|$

(e) $|6000|$

The absolute value of a number will *always* be positive (or 0), because it is the *distance* from 0. A distance is never negative. (You wouldn't say that your living room is $^-16$ feet long.) So absolute value concerns only *how far away* the number is from 0; we don't care which direction it is from 0.

EXAMPLE 4 **Finding Absolute Values**

Find each absolute value.

(a) $|4|$ The distance from 0 to 4 on the number line is 4 spaces. So, $|4| = 4$.

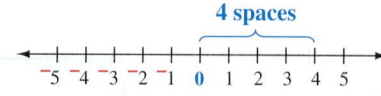

(b) $|{}^-4|$ The distance from 0 to $^-4$ on the number line is also 4 spaces. So, $|{}^-4| = 4$.

Absolute value is always positive (or 0), never negative.

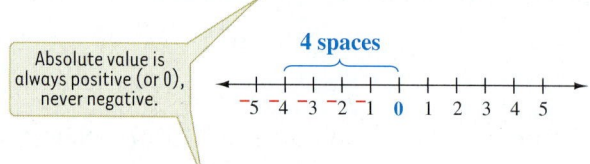

(c) $|0|$ $|0| = 0$ because the distance from 0 to 0 on the number line is 0 spaces.

◀ *Work Problem* **4** *at the Side.*

1.2 ▶▶▶ Exercises

Write each negative number with a raised negative sign. Write each positive number in two ways. See Example 1.

1. Mount Everest, the tallest mountain in the world, rises 29,035 feet above sea level. (*Source: World Almanac.*)

2. The bottom of Lake Baikal in Siberia, Russia, is 5371 feet below the surface of the water. (*Source: Guinness World Records.*)

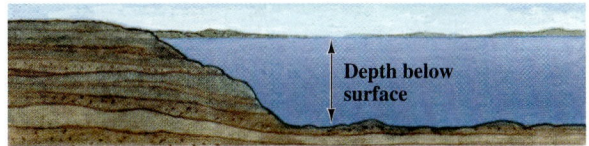

Depth below surface

◐ 3. The coldest temperature ever recorded on Earth is 128.6 degrees below zero in Antarctica. (*Source: Fact Finder.*)

4. Normal body temperature is 98.6 degrees Fahrenheit, although it varies slightly for some people. (*Source:* Mayo Clinic *Health Letter.*)

5. During the first three plays of the football game, the Trojans lost a total of 18 yards.

6. The Jets gained 25 yards on a pass play.

◐ 7. Angelique won $100 in a prize drawing at the shopping mall.

8. Derice overdrew his checking account by $37.

9. Keith lost $6\frac{1}{2}$ pounds while he was sick with the flu.

10. The mice in an experiment gained $2\frac{1}{2}$ ounces.

Graph each set of numbers. See Example 2.

◐ 11. $^-3, 3, 0, ^-5$

12. $^-2, 2, 0, 5$

13. $^-1, 4, ^-2, 5$

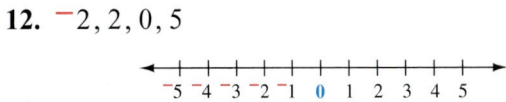

14. $3, ^-4, 1, ^-5$

15. $^-4\frac{1}{2}, \frac{1}{2}, 0, ^-8$

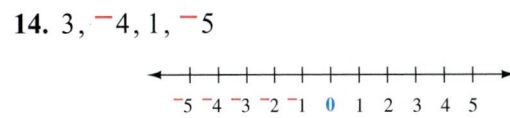

16. $^-7, 1\frac{1}{2}, -\frac{1}{2}, ^-9$

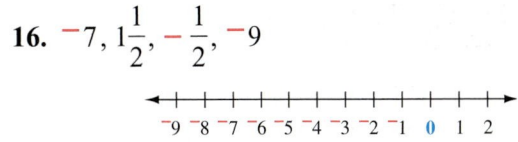

Write < or > between each pair of numbers to make a true statement. See Example 3.

17. 10 ___ 2

18. 6 ___ 0

19. ⁻1 ___ 0

20. ⁻3 ___ ⁻1

21. ⁻10 ___ 2

22. ⁻9 ___ 7

23. ⁻3 ___ ⁻6

24. 0 ___ ⁻1

25. ⁻10 ___ ⁻2

26. ⁻1 ___ ⁻5

27. 0 ___ ⁻8

28. 6 ___ ⁻4

29. 10 ___ ⁻2

30. ⁻2 ___ 1

31. ⁻4 ___ 4

32. 9 ___ ⁻9

Find each absolute value. See Example 4.

33. |15|

34. |10|

35. |⁻3|

36. |⁻8|

37. |0|

38. |100|

39. |200|

40. |⁻99|

41. |⁻75|

42. |⁻6320|

43. |⁻8042|

44. |0|

Relating Concepts (Exercises 45–48) For Individual or Group Work

A special type of X-ray, called a DEXA test, is a quick way to screen patients for osteoporosis (brittle bone disease). Use the information on the Patient Report Form to **work Exercises 45–48 in order.**

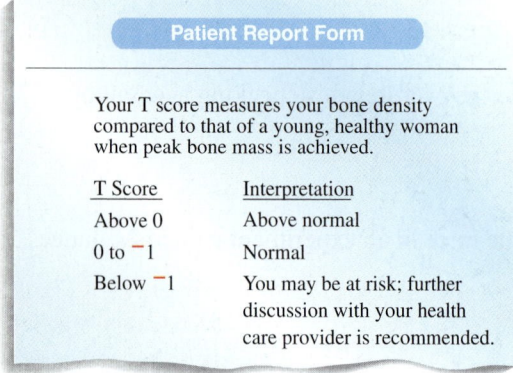

Patient Report Form

Your T score measures your bone density compared to that of a young, healthy woman when peak bone mass is achieved.

T Score	Interpretation
Above 0	Above normal
0 to ⁻1	Normal
Below ⁻1	You may be at risk; further discussion with your health care provider is recommended.

Source: Health Partners, Inc.

45. Here are the T scores for four patients. Draw a number line and graph the four scores.

    Patient A: ⁻1.5   Patient C: ⁻1

    Patient B: 0.5   Patient D: 0

46. List the patients' scores in order from lowest to highest.

47. What is the interpretation of each patient's score?

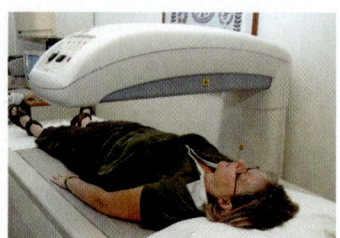

Taking a DEXA test of a patient's spine.

48. (a) What could happen if patient A did not understand the importance of a negative sign?

    (b) For which patient does the sign of the score make no difference? Explain your answer.

Study Skills

▶▶▶ HOMEWORK: HOW, WHY, AND WHEN

It is best for your brain if you keep up with the reading and homework in your math class. Remember that the more times you work with the information, the more dendrites you grow! So, give yourself every opportunity to read, work problems, and review your mathematics.

You have two choices for reading your math textbook. Read the short descriptions below and decide which will be best for you.

Maddy learns best by listening to her teacher explain things. She "gets it" when she sees the instructor work problems on the board. She likes to ask questions in class and put the information in her notes. She has learned that it helps if she has *previewed* the section before the lecture, so she knows generally what to expect in class. *But after the class instruction,* when Maddy gets home, she finds that she can understand the math textbook easily. She remembers what her teacher said, and she can double-check her notes if she gets confused. So, Maddy does her **careful** reading of the section in her text **after** hearing the classroom lecture on the topic.

De'Lore, on the other hand, feels he learns well by reading on his own. He prefers to read the section and try working the example problems before coming to class. That way, he already knows what the teacher is going to talk about. Then, he can follow the teacher's examples more easily. It is also easier for him to take notes in class. De'Lore likes to have his questions answered right away, which he can do if he has already read the chapter section. So, De'Lore **carefully** reads the section in his text **before** he hears the classroom lecture on the topic.

Notice that there is **no one right way** to work with your textbook. You always must figure out what works best for you. Note also that both Maddy and De'Lore work with one section at a time. **The key is that you read the textbook regularly!** The rest of this activity will give you some ideas of how to make the most of your reading.

Try the following steps as you **read** your math textbook.

▶ Read slowly. Read only one section—or even part of a section—at a time.

▶ Do the sample problems in the margins **as you go.** Check them right away. The answers are at the bottom of the page.

▶ If your mind wanders, work problems on separate paper and write explanations in your own words.

▶ Make study cards as you read each section. Pay special attention to the yellow and blue boxes in the book. Make cards for new vocabulary, rules, procedures, formulas, and sample problems.

▶ **NOW,** you are ready to do your homework assignment!

OBJECTIVES

1 Select an appropriate strategy for homework.

2 Use textbook features effectively.

Preview before Class; Read Carefully after Class

Read Carefully before Class

Why Are These Reading Techniques Brain Friendly?

The steps at the left encourage you to be **actively working with the material** in your text. Your brain **grows dendrites when it is doing something.**

These methods require you to **try several different techniques,** not just the same thing over and over. Your brain loves variety!

Also, the techniques allow you to take small breaks in your learning. Those **rest periods are crucial for good dendrite growth.**

Which steps for reading this book will be most helpful for you?

1. _____

2. _____

3. _____

Homework

Why Are These Homework Suggestions Brain Friendly?

Your brain will grow dendrites as you study the worked examples in the text and try doing them yourself on separate paper. So, when you see similar problems in the homework, you will already have dendrites to work from.

Giving yourself a practice test by trying to remember the steps (without looking at your card) is an excellent way to reinforce what you are learning.

Correcting errors right away is how you learn and reinforce the correct procedures. It is hard to unlearn a mistake, so always check to see that you are on the right track!

Teachers assign homework so you can grow your own dendrites (learn the material) and then coat the dendrites with myelin through practice (remember the material). Really! In learning, you get good at what you practice. So, completing homework every day will strengthen your neural network and prepare you for exams.

If you have read each section in your textbook according to the steps above, you will probably encounter few difficulties with the exercises in the homework. Here are some additional suggestions that will help you succeed with the homework.

▶ If you **have trouble with a problem,** find a similar worked example in the section. Pay attention to *every line* of the worked example to see how to get from step to step. Work it yourself too, on separate paper; don't just look at it.

▶ If it is **hard to remember the steps** to follow for certain procedures, write the steps on a separate card. Then write a short explanation of each step. Keep the card nearby while you do the exercises, but try *not* to look at it.

▶ If you **aren't sure you are working the assigned exercises correctly,** choose two or three odd-numbered problems that are a similar type and work them. Then check the answers in the Answers section at the back of your book to see if you are doing them correctly. If you aren't, go back to the section in the text and review the examples and find out how to correct your errors. When you are sure you understand, try the assigned problems again.

▶ **Make sure you do some homework every day,** even if the math class does not meet each day!

What are your biggest homework concerns?
List your two main concerns below. Then write a **brain friendly solution** for each one.

1. Concern: _____ Solution: _____

2. Concern: _____ Solution: _____

1.3 ▶▶▶ Adding Integers

OBJECTIVE 1 Add integers. Numbers that you are adding are called **addends,** and the result is called the **sum.** You can add integers while watching a football game. On each play, you can use a *positive* integer to stand for the yards *gained* by your team, and a *negative* integer for yards *lost*. Zero indicates no gain or loss. For example,

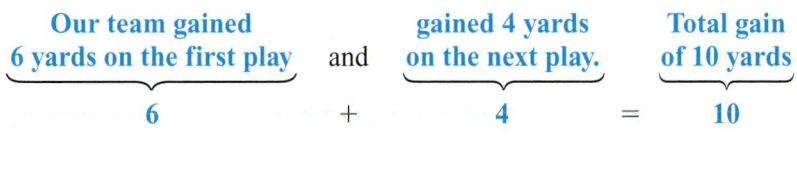

A drawing of a football field can help you add integers. Notice how similar the drawing is to a number line. Zero marks your team's starting point.

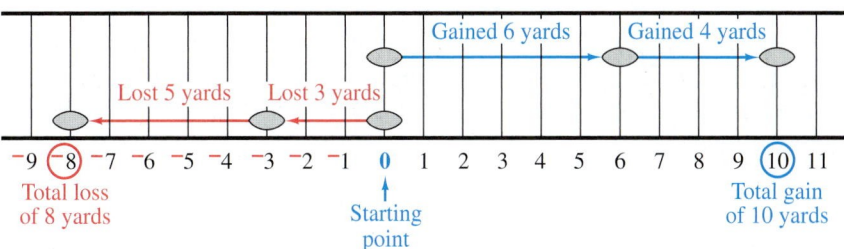

EXAMPLE 1 **Using a Number Line to Add Integers**

Use a number line to find $^-5 + {}^-4$.

Think of the number line as a football field. Your team starts at 0. On the first play it lost 5 yards. On the next play it lost 4 yards. The total loss is 9 yards.

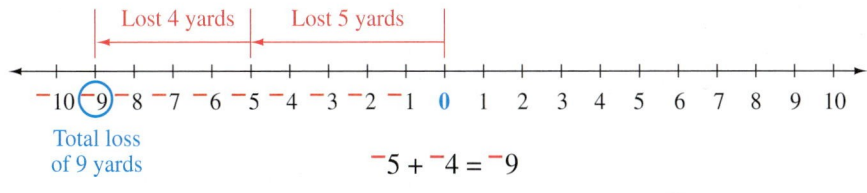

$^-5 + {}^-4 = {}^-9$

Work Problem **1** *at the Side.* ▶

Do you see a pattern in the margin problems you just did? The first two sums are $^-4$ and 4. They are the same, *except for the sign.* The same is true for the next two sums, $^-11$ and 11, and for the last two sums, $^-10$ and 10. This pattern leads to a rule for adding two integers when the signs are the same.

Adding Two Integers with the Same Sign

Step 1 *Add* the absolute values of the numbers.

Step 2 Use the *common sign* as the sign of the sum. If both numbers are positive, the sum is positive. If both numbers are negative, the sum is negative.

OBJECTIVES

1 Add integers.

2 Identify properties of addition.

---

1 Find each sum. Use the number line to help you.

(a) $^-2 + {}^-2$

(b) $2 + 2$

(c) $^-10 + {}^-1$

(d) $10 + 1$

(e) $^-3 + {}^-7$

(f) $3 + 7$

ANSWERS

1. **(a)** $^-4$ **(b)** 4 **(c)** $^-11$
   (d) 11 **(e)** $^-10$ **(f)** 10

2 Find each sum.

(a) $^-6 + {}^-6$

(b) $9 + 7$

(c) $^-5 + {}^-10$

(d) $^-12 + {}^-4$

(e) $13 + 2$

EXAMPLE 2 **Adding Two Integers with the Same Sign**

Add.

(a) $^-8 + {}^-7$

Step 1 *Add* the absolute values.

$$|{}^-8| = 8 \quad \text{and} \quad |{}^-7| = 7$$

Add $8 + 7$ to get 15.

Step 2 Use the *common sign* as the sign of the sum. Both numbers are negative, so the sum is negative.

$$^-8 + {}^-7 = {}^-15$$

Both negative　　Sum is negative.

(b) $3 + 6 = 9$

Both positive　　Sum is positive.

In *Step 1*, when both numbers are positive, their absolute values are also positive, so we only need to show *Step 2*.

◀ *Work Problem* **2** *at the Side.*

You can also use a number line (or drawing of a football field) to add integers with *different* signs. For example, suppose that your team gained 2 yards on the first play and then lost 7 yards on the next play. We can represent this as $2 + {}^-7$.

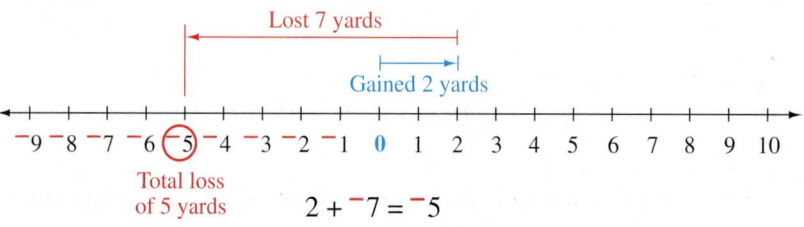

Total loss of 5 yards　　　$2 + {}^-7 = {}^-5$

Or, try this one. On the first play your team gained 10 yards, but then it lost 4 yards on the next play. We can represent this as $10 + {}^-4$.

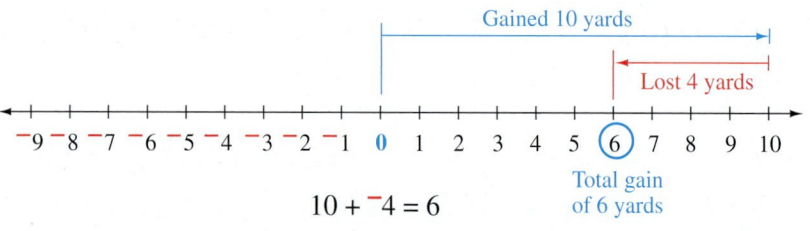

$10 + {}^-4 = 6$　　　Total gain of 6 yards

These examples illustrate the rule for adding two integers with unlike, or different, signs.

> **Adding Two Integers with Unlike Signs**
>
> **Step 1** *Subtract* the smaller absolute value from the larger absolute value.
>
> **Step 2** Use the sign of the number with the *larger absolute value* as the sign of the sum.

EXAMPLE 3 Adding Two Integers with Unlike Signs

Add.

(a) $^-8 + 3$

Step 1 $|^-8| = 8$ and $|3| = 3$

Subtract $8 - 3$ to get 5.

Step 2 $^-8$ has the *larger absolute value* and is negative, so the sum is also negative.

$$^-8 + 3 = ^-5$$

(b) $^-5 + 11$

Step 1 $|^-5| = 5$ and $|11| = 11$

Subtract $11 - 5$ to get 6.

Step 2 11 has the *larger absolute value* and is positive, so the sum is also positive.

$$^-5 + 11 = {}^+6 \quad \text{or} \quad 6$$

Work Problem **3** *at the Side.* ▶

EXAMPLE 4 Adding Several Integers

A football team has to gain at least 10 yards during four plays in order to keep the ball. Suppose that your college team lost 6 yards on the first play, gained 8 yards on the second play, lost 2 yards on the third play, and gained 7 yards on the fourth play. Did the team gain enough to keep the ball?

When you're adding several integers, work from left to right.

The team gained 7 yards, which is *not* enough yards to keep the ball.

Work Problem **4** *at the Side.* ▶

OBJECTIVE 2 Identify properties of addition. In our football model for adding integers, we said that 0 indicated no gain or loss, that is, no change in the position of the ball. This example illustrates one of the *properties of addition*. The properties of addition apply to all addition problems, regardless of the specific numbers you use.

Addition Property of 0
Adding 0 to any number leaves the number unchanged.

Some examples are shown below.

$$0 + 6 = 6 \qquad ^-25 + 0 = {}^-25 \qquad 72{,}399 + 0 = 72{,}399$$
$$0 + {}^-100 = {}^-100$$

3 Find each sum.

(a) $^-3 + 7$

(b) $6 + {}^-12$

(c) $12 + {}^-7$

(d) $^-10 + 2$

(e) $5 + {}^-9$

(f) $^-8 + 9$

4 Write an addition problem and solve it for each situation.

(a) The temperature was $^-15$ degrees this morning. It rose 21 degrees during the day, then dropped 10 degrees. What is the new temperature?

(b) Andrew had $60 in his checking account. He wrote a $20 check for gas and a $75 check for groceries. Later in the day he deposited an $85 tax refund in his account. What is the balance in his account?

5 Rewrite each sum using the commutative property of addition. Check that the sum is unchanged.

(a) $175 + 25 = \underline{} + \underline{}$

Both sums are ____.

(b) $7 + {}^{-}37 = \underline{} + \underline{}$

Both sums are ____.

(c) $^{-}16 + 16 = \underline{} + \underline{}$

Both sums are ____.

(d) $^{-}9 + {}^{-}41 = \underline{} + \underline{}$

Both sums are ____.

Another property of addition is that you can change the *order* of the addends and still get the same sum. For example,

Gaining 2 yards, then losing 7 yards, gives a result of $^{-}5$ yards.

Losing 7 yards, then gaining 2 yards, also gives a result of $^{-}5$ yards.

> **Commutative Property of Addition**
> Changing the *order* of two addends does *not* change the sum.
>
> Here are some examples.
>
> $$84 + 2 = 2 + 84 \qquad \text{Both sums are 86.}$$
>
> $$^{-}10 + 6 = 6 + {}^{-}10 \qquad \text{Both sums are } {}^{-}4.$$

EXAMPLE 5 **Using the Commutative Property of Addition**

Rewrite each sum, using the **commutative property of addition.** Check that the sum is unchanged.

(a) $65 + 35$

$$65 + 35 = 35 + 65$$
$$100 \quad = \quad 100$$

Both sums are 100, so the sum is unchanged.

Adding two numbers in a different order does not change the sum.

(b) $^{-}20 + {}^{-}30$

$$^{-}20 + {}^{-}30 = {}^{-}30 + {}^{-}20$$
$$^{-}50 \quad = \quad {}^{-}50$$

Both sums are $^{-}50$, so the sum is unchanged.

◀ *Work Problem* **5** *at the Side.*

When there are three addends, parentheses may be used to tell you which pair of numbers to add first, as shown below.

$(3 + 4) + 2$	First add $3 + 4$.		$3 + (4 + 2)$	First add $4 + 2$.
$7 \quad + 2$	Then add $7 + 2$.		$3 + \quad 6$	Then add $3 + 6$.
9			9	

Both sums are 9. This example illustrates another property of addition.

> **Associative Property of Addition**
> Changing the *grouping* of addends does *not* change the sum.
>
> Some examples are shown below.
>
> $$({}^{-}5 + 5) + 8 = {}^{-}5 + (5 + 8) \qquad 3 + ({}^{-}4 + {}^{-}6) = (3 + {}^{-}4) + {}^{-}6$$
> $$0 \quad + 8 = {}^{-}5 + \quad 13 \qquad\quad 3 + \quad {}^{-}10 \quad = \quad {}^{-}1 \quad + {}^{-}6$$
> $$8 \quad = \quad 8 \qquad\qquad\qquad {}^{-}7 \quad = \quad {}^{-}7$$
>
> Both sums are 8. | Both sums are $^{-}7$.

We can use the associative property to make addition problems easier. Notice in the first example in the blue box at the bottom of the previous page that it is easier to group $^-5 + 5$ (which is 0) and then add 8. In the second example in the box, it is helpful to group $^-4 + {}^-6$ because the sum is $^-10$, and it is easy to work with multiples of 10. (See **Section R.3** for more on multiples of 10.)

6 In each problem, write parentheses around the two addends that would be easiest to add. Then find the sum.

(a) $^-12 + 12 + {}^-19$

EXAMPLE 6 **Using the Associative Property of Addition**

In each addition problem, pick out the two addends that would be easiest to add. Write parentheses around those addends. Then find the sum.

(a) $6 + 9 + {}^-9$

Group $9 + {}^-9$ because the sum is 0.

> Use the associative property to add the easiest numbers first.

$$6 + \underbrace{(9 + {}^-9)}$$
$$\underbrace{6 + \quad 0}$$
$$6$$

(b) $17 + 3 + {}^-25$

Group $17 + 3$ because the sum is 20, which is a multiple of 10.

$$\underbrace{(17 + 3)} + {}^-25$$
$$\underbrace{20 \quad + {}^-25}$$
$$^-5$$

(b) $31 + 75 + {}^-75$

(c) $16 + {}^-1 + {}^-9$

Note

When using the associative property to make the addition of a group of numbers easier:

1. Look for two numbers whose sum is 0.

2. Look for two numbers whose sum is a multiple of 10 (the sum ends in 0, such as $10, 20, 30,$ or $^-100, {}^-200,$ etc.).

If neither of these occurs, look for two numbers that are easier for you to add. For example, in $98 + 43 + 5$, you may find that adding $43 + 5$ is easier than adding $98 + 43$.

(d) $^-8 + 5 + {}^-25$

Work Problem **6** *at the Side.* ▶

ANSWERS

6. (a) $\underbrace{(^-12 + 12)} + {}^-19$
$\underbrace{0 \quad + {}^-19}$
$^-19$

(b) $31 + \underbrace{(75 + {}^-75)}$
$\underbrace{31 + \quad 0}$
31

(c) $16 + \underbrace{(^-1 + {}^-9)}$
$\underbrace{16 + \quad {}^-10}$
6

(d) $^-8 + \underbrace{(5 + {}^-25)}$
$\underbrace{^-8 + \quad {}^-20}$
$^-28$

Math in the Media

GOLF SCOREBOARDS

At the 2007 PGA Championship golf tournament in Atlanta, Georgia, 72 strokes was the "par" score for each round of play. A negative score indicates that the player had less than 72 strokes for that round, and a positive score indicates that the player had more than 72 strokes. The scoreboard below shows the scores for several of the players.

Player	Round 1	Round 2	Round 3	Round 4
Tiger Woods	⁻6	⁻7	⁻6	⁻4
Zach Johnson	⁺1	⁻4	⁻10	⁻2
Sergio Garcia	⁻2	⁻6	⁻6	0

Source: www.pgatour.com

1. Identify the round and the player who had each score. There may be more than one correct answer.

   (a) Seven strokes less than par.

   (b) One stroke more than par.

   (c) Six strokes less than par.

2. What score did Garcia have in Round 4? Is the score positive or negative? What does that score tell you?

3. Graph Johnson's scores on the number line below. Label the round for each score that you graph.

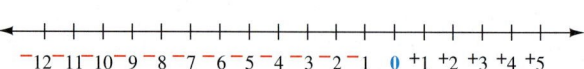

   ⁻12 ⁻11 ⁻10 ⁻9 ⁻8 ⁻7 ⁻6 ⁻5 ⁻4 ⁻3 ⁻2 ⁻1 0 ⁺1 ⁺2 ⁺3 ⁺4 ⁺5

4. (a) List Johnson's scores from least to greatest.

   (b) List Woods' scores from least to greatest.

   (c) List Garcia's scores from least to greatest.

5. To win at golf, you must get the least number of strokes. Which round was the best round for (a) Woods; (b) Johnson; (c) Garcia?

6. Find the total score for each player on the scoreboard.

Tiger Woods won the 2007 PGA Championship.

1.3 ▶▶▶ Exercises

Add by using the number line. See Example 1.

1. $^-2 + 5$

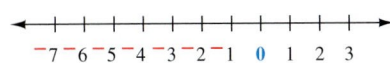

2. $^-3 + 4$

3. $^-5 + {}^-2$

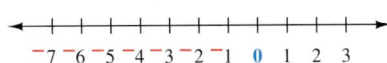

4. $^-2 + {}^-2$

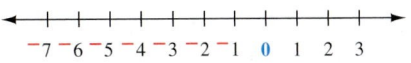

5. $3 + {}^-4$

6. $5 + {}^-1$

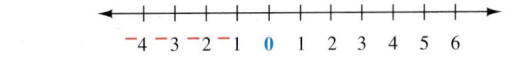

Add. See Example 2.

7. (a) $^-5 + {}^-5$

   (b) $5 + 5$

8. (a) $^-9 + {}^-9$

   (b) $9 + 9$

9. (a) $7 + 5$

   (b) $^-7 + {}^-5$

10. (a) $3 + 6$

   (b) $^-3 + {}^-6$

11. (a) $^-25 + {}^-25$

    (b) $25 + 25$

12. (a) $^-30 + {}^-30$

    (b) $30 + 30$

13. (a) $48 + 110$

    (b) $^-48 + {}^-110$

14. (a) $235 + 21$

    (b) $^-235 + {}^-21$

15. What pattern do you see in your answers to Exercises 7–14? Explain why this pattern occurs.

16. In your own words, explain how to add two integers that have the same sign.

Add. See Example 3.

17. (a) $^-6 + 8$ **18. (a)** $^-3 + 7$ **19. (a)** $^-9 + 2$ **20. (a)** $^-8 + 7$

(b) $6 + {}^-8$ **(b)** $3 + {}^-7$ **(b)** $9 + {}^-2$ **(b)** $8 + {}^-7$

21. (a) $20 + {}^-25$ **22. (a)** $30 + {}^-40$ **23. (a)** $200 + {}^-50$ **24. (a)** $150 + {}^-100$

(b) $^-20 + 25$ **(b)** $^-30 + 40$ **(b)** $^-200 + 50$ **(b)** $^-150 + 100$

25. What pattern do you see in your answers to Exercises 17–24? Explain why this pattern occurs.

26. In your own words, explain how to add two integers that have different signs.

Add. See Examples 2–4.

27. $^-8 + 5$ **28.** $^-3 + 2$ **29.** $^-1 + 8$ **30.** $^-4 + 10$

31. $^-2 + {}^-5$ **32.** $^-7 + {}^-3$ **33.** $6 + {}^-5$ **34.** $11 + {}^-3$

35. $4 + {}^-12$ **36.** $9 + {}^-10$ **37.** $^-10 + {}^-10$ **38.** $^-5 + {}^-20$

39. $^-17 + 0$ **40.** $0 + {}^-11$ **41.** $1 + {}^-23$ **42.** $13 + {}^-1$

43. $^-2 + {}^-12 + {}^-5$ **44.** $^-16 + {}^-1 + {}^-3$ **45.** $8 + 6 + {}^-8$

46. $^-5 + 2 + 5$

47. $^-7 + 6 + {^-4}$

48. $^-9 + 8 + {^-2}$

49. $^-3 + {^-11} + 14$

50. $15 + {^-7} + {^-8}$

51. $10 + {^-6} + {^-3} + 4$

52. $2 + {^-1} + {^-9} + 12$

53. $^-7 + 28 + {^-56} + 3$

54. $4 + {^-37} + 29 + {^-5}$

Write an addition problem for each situation and find the sum.

55. The football team gained 13 yards on the first play and lost 17 yards on the second play. How many yards did the team gain or lose in all?

56. At penguin breeding grounds on Antarctic islands, temperatures routinely drop to $^-15$ °C. Temperatures in the interior of the continent may drop another 60 °C below that. What is the temperature in the interior?

57. Nick's checking account was overdrawn by $62. He deposited $50 in his account. What is the balance in his account?

58. Cynthia had $100 in her checking account. She wrote a check for $83 and was charged $17 for overdrawing her account last month. What is her account balance?

59. $88 was stolen from Jay's car. He got $35 of it back. What was his net loss?

60. Marion lost 4 pounds in April, gained 2 pounds in May, and gained 3 pounds in June. How many pounds did she gain or lose in all?

61. Use the score sheet to find each player's point total after three rounds in a card game.

	Jeff	Terry
Round 1	Lost 20 pts	Won 42 pts
Round 2	Won 75 pts	Lost 15 pts
Round 3	Lost 55 pts	Won 20 pts

62. Use the information in the table on flood water depths to find the new flood level for each river.

	Red River	Mississippi
Monday	Rose 8 ft	Rose 4 ft
Tuesday	Fell 3 ft	Rose 7 ft
Wednesday	Fell 5 ft	Fell 13 ft

The table below shows the number of pounds gained or lost by several health club members during a four-month period. A negative number indicates a loss of pounds. A positive number indicates a gain. Find the total pounds gained or lost by each member.

	Member	Month 1	Month 2	Month 3	Month 4	Total
63.	Angela	⁻2	0	5	⁻5	
64.	Syshe	⁻1	2	⁻6	0	
65.	Brittany	3	⁻2	⁻2	3	
66.	Nicole	1	1	⁻4	2	

Rewrite each sum, using the commutative property of addition. Show that the sum is unchanged. See Example 5.

67. ⁻18 + ⁻5 = _____ + _____

Both sums are _____.

68. ⁻12 + 20 = _____ + _____

Both sums are _____.

69. ⁻4 + 15 = _____ + _____

Both sums are _____.

70. 17 + 1 = _____ + _____

Both sums are _____.

In each addition problem, write parentheses around the two addends that would be easiest to add. Then find the sum. See Example 6.

71. 6 + ⁻14 + 14

72. 9 + ⁻9 + ⁻8

73. ⁻14 + ⁻6 + ⁻7

74. ⁻18 + 3 + 7

75. Make up three of your own examples that illustrate the addition property of 0.

76. Make up three of your own examples that illustrate the associative property of addition. Show that the sum is unchanged.

Find each sum.

77. ⁻7081 + 2965

78. ⁻1398 + 3802

79. ⁻179 + ⁻61 + 8926

80. 36 + ⁻6215 + 428

81. 86 + ⁻99,000 + 0 + 2837

82. ⁻16,719 + 0 + 8878 + ⁻14

1.4 ▶▶▶ Subtracting Integers

OBJECTIVE 1 Find the opposite of a signed number.   Look at how the integers match up on this number line.

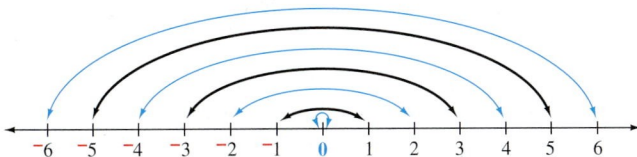

Each integer is matched with its *opposite*. **Opposites** are the same *distance* from 0 on the number line but are on *opposite sides* of 0.

$+2$ is the opposite of -2 and -2 is the opposite of $+2$

When you add opposites, the sum is always 0. The opposite of a number is also called its *additive inverse*.

$2 + {}^-2 = 0$ and ${}^-2 + 2 = 0$

EXAMPLE 1   Finding the Opposites of Signed Numbers

Find the opposite (additive inverse) of each number. Show that the sum of the number and its opposite is 0.

The sum of opposites is 0.

(a)   6   The opposite of 6 is -6 and $6 + {}^-6 = 0$

(b) -10 The opposite of -10 is 10   and $-10 + 10 = 0$

(c)   0   The opposite of 0 is 0   and $0 + 0 = 0$

———————————————————— *Work Problem* (**1**) *at the Side.* ▶

OBJECTIVE 2 Subtract integers.   Now that you know how to add integers and how to find opposites, you can subtract integers. Every subtraction problem has the same answer as a related addition problem. The examples below illustrate how to change subtraction problems into addition problems.

$6 - 2 = 4$ ⤦
↓ ↓   Same answer
$6 + {}^-2 = 4$ ⤦

$8 - 3 = 5$ ⤦
↓ ↓   Same answer
$8 + {}^-3 = 5$ ⤦

Subtracting Two Integers

To subtract two numbers, *add* the first number to the *opposite* of the second number. Remember to change *two* things:

Step 1   Make one pencil stroke to change the subtraction symbol to an addition symbol.

Step 2   Make a second pencil stroke to change the *second* number to its *opposite*. If the second number is positive, change it to negative. If the second number is negative, change it to positive.

CAUTION

When changing a subtraction problem to an addition problem, do *not* make any change in the *first* number. The pattern is

1st number − 2nd number = 1st number + opposite of 2nd number.

OBJECTIVES

1 Find the opposite of a signed number.

2 Subtract integers.

3 Combine adding and subtracting of integers.

1 Find the additive inverse (opposite) of each number. Show that the sum of the number and its additive inverse is 0.

(a) 5

(b) 48

(c) 0

(d) -1

(e) -24

ANSWERS

1. **(a)** $-5; 5 + {}^-5 = 0$
   (b) $-48; 48 + {}^-48 = 0$
   (c) $0; 0 + 0 = 0$
   (d) $1; {}^-1 + 1 = 0$
   (e) $24; {}^-24 + 24 = 0$

2 Subtract by changing subtraction to adding the opposite. (Make *two* pencil strokes.)

(a) $^-6 - 5$

(b) $3 - {}^-10$

(c) $^-8 - {}^-2$

(d) $0 - 10$

(e) $^-4 - {}^-12$

(f) $9 - 7$

3 Simplify.

(a) $6 - 7 + {}^-3$

(b) $^-2 + {}^-3 - {}^-5$

(c) $7 - 7 - 7$

(d) $^-3 - 9 + 4 - {}^-20$

EXAMPLE 2 **Subtracting Two Integers**

Make *two* pencil strokes to change each subtraction problem into an addition problem. Then find the sum.

Change 10 to $^-10$.

> Do **not** change 4.
> Change 10 to $^-10$.

(a) $4 - \mathbf{10} = 4 + {}^-\mathbf{10} = {}^-6$

Change subtraction to addition.

Change $^-6$ to $^+6$.

> Do **not** change $^-9$.
> Change $^-6$ to $^+6$.

(b) $^-9 - {}^-\mathbf{6} = {}^-9 + {}^+\mathbf{6} = {}^-3$

Change subtraction to addition.

Change $^-5$ to $^+5$.

(c) $3 - {}^-\mathbf{5} = 3 + {}^+\mathbf{5} = 8$ Make *two* pencil strokes.

Change subtraction to addition.

> Do **not** change the first number when subtracting.

Change 9 to $^-9$.

(d) $^-2 - \mathbf{9} = {}^-2 + {}^-\mathbf{9} = {}^-11$ Make *two* pencil strokes.

Change subtraction to addition.

◀ *Work Problem* **2** *at the Side.*

OBJECTIVE **3** **Combine adding and subtracting of integers.** When adding and subtracting more than two signed numbers, first change all subtractions to adding the opposite. Then add from left to right.

EXAMPLE 3 **Combining Addition and Subtraction**

Simplify by completing all the calculations.

$$
\begin{array}{llll}
^-5 - & \mathbf{10} - & \mathbf{12} + 1 & \text{Change all subtractions to adding the opposite.} \\
\downarrow \quad \downarrow & \downarrow \quad \downarrow & & \text{Change 10 to } ^-10. \text{ Change 12 to } ^-12. \\
^-5 + & {}^-\mathbf{10} + & {}^-\mathbf{12} + 1 & \text{Add from left to right. First add } ^-5 + {}^-10. \\
^-15 & + {}^-12 & + 1 & \text{Then add } ^-15 + {}^-12. \\
& ^-27 & + 1 & \text{Finally, add } ^-27 + 1. \\
& ^-26 & &
\end{array}
$$

◀ *Work Problem* **3** *at the Side.*

🖩 **Calculator Tip** You can use the *change of sign* key ⊕⁄⊖ or ⊕C⊖ on your *scientific* calculator to enter negative numbers. To enter $^-5$, press ⑤ ⊕⁄⊖. To enter $^+5$, just press ⑤. To enter Example 3 above, press the following keys.

5 ⊕⁄⊖ ⊖ 10 ⊖ 12 ⊕ 1 ⊜   The answer is $^-26$.

$^-5$ Subtract.

When using a calculator, you do *not* need to change subtraction to addition.

ANSWERS

2. (a) $^-11$ (b) 13 (c) $^-6$ (d) $^-10$
   (e) 8 (f) 2
3. (a) $^-4$ (b) 0 (c) $^-7$ (d) 12

1.4 ▶▶▶ Exercises

Find the opposite (additive inverse) of each number. Show that the sum of the number and its opposite is 0. See Example 1.

1. 6

2. 10

💿 **3.** ⁻13

4. ⁻3

5. 0

6. 1

Subtract by changing subtraction to addition. See Example 2.

7. 19 − 5

8. 24 − 11

💿 **9.** 10 − 12

10. 1 − 8

11. 7 − 19

12. 2 − 17

13. ⁻15 − 10

14. ⁻10 − 4

15. ⁻9 − 14

16. ⁻3 − 11

💿 **17.** ⁻3 − ⁻8

18. ⁻1 − ⁻4

19. 6 − ⁻14

20. 8 − ⁻1

21. 1 − ⁻10

22. 6 − ⁻1

23. ⁻30 − 30

24. ⁻25 − 25

25. ⁻16 − ⁻16

26. ⁻20 − ⁻20

27. 13 − 13

28. 19 − 19

29. 0 − 6

30. 0 − 12

31. (a) 3 − ⁻5

(b) 3 − 5

(c) ⁻3 − ⁻5

(d) ⁻3 − 5

32. (a) 9 − 6

(b) ⁻9 − 6

(c) 9 − ⁻6

(d) ⁻9 − ⁻6

33. (a) 4 − 7

(b) 4 − ⁻7

(c) ⁻4 − 7

(d) ⁻4 − ⁻7

34. (a) 8 − ⁻2

(b) ⁻8 − ⁻2

(c) 8 − 2

(d) ⁻8 − 2

Simplify. See Example 3.

35. ⁻2 − 2 − 2

36. ⁻8 − 4 − 8

37. 9 − 6 − 3 − 5

38. 12 − 7 − 5 − 4

39. 3 − ⁻3 − 10 − ⁻7

40. 1 − 9 − ⁻2 − ⁻6

💿 **41.** ⁻2 + ⁻11 − ⁻3

42. ⁻5 − ⁻2 + ⁻6

43. 4 − ⁻13 + ⁻5

44. 6 − ⁻1 + ⁻10

45. 6 + 0 − 12 + 1

46. ⁻10 − 4 + 0 + 18

WINDCHILL
Temperature (degrees Fahrenheit)

Calm	40	35	30	25	20	15	10	5	0	−5	−10	−15	−20	−25	−30
5	36	31	25	19	13	7	1	−6	−11	−16	−22	−28	−34	−40	−46
10	34	27	21	15	9	3	−4	−10	−16	−22	−28	−35	−41	−47	−53
15	32	25	19	13	6	0	−7	−13	−19	−26	−32	−39	−45	−51	−58
20	30	24	17	11	4	−2	−9	−15	−22	−29	−35	−42	−48	−55	−61
25	29	23	16	9	3	−4	−11	−17	−24	−31	−37	−44	−51	−58	−64
30	28	22	15	8	1	−5	−12	−19	−26	−33	−39	−46	−53	−60	−67
35	28	21	14	7	0	−7	−14	−21	−27	−34	−41	−48	−55	−62	−69
40	27	20	13	6	−1	−8	−15	−22	−29	−36	−43	−50	−57	−64	−71

Wind Speed (miles per hour)

Shaded area: Frostbite occurs in 15 minutes or less.

Source: National Weather Service

This windchill table shows how wind increases a person's heat loss. For example, find the temperature of 15 °F along the top of the table. Then find a wind speed of 20 mph along the left side of the table. This column and row intersect at $^-2$ °F, the "windchill temperature." The actual temperature is 15 °F but the wind makes it feel like $^-2$ °F. The difference between the actual temperature and the windchill temperature is $15 - {^-2} = 15 + {^+2} = 17$ degrees difference.

Use the table to find the windchill temperature under each set of conditions in Exercises 47–48. Then write and solve a subtraction problem to calculate actual temperature minus windchill temperature.

47. **(a)** 30 °F; 10 mph wind

(b) 15 °F; 15 mph wind

(c) 5 °F; 25 mph wind

(d) $^-10$ °F; 35 mph wind

48. **(a)** 40 °F; 20 mph wind

(b) 20 °F; 35 mph wind

(c) 10 °F; 15 mph wind

(d) $^-5$ °F; 30 mph wind

49. Find, correct, and explain the mistake made in this subtraction.

$$\begin{array}{c} {^-6} - 6 \\ \downarrow \quad \downarrow \\ {^-6} + 6 = 0 \end{array}$$

50. Find, correct, and explain the mistake made in this subtraction.

$$\begin{array}{c} {^-7} - \quad 5 \\ \downarrow \quad \downarrow \\ {^+7} + {^-5} = 2 \end{array}$$

Simplify. Begin each exercise by working inside the absolute value bars or the parentheses.

51. $^-2 + {^-11} + |{^-2}|$

52. $5 - |{^-3}| + 3$

53. $0 - |{^-7} + 2|$

54. $|1 - 8| - |0|$

55. $^-3 - ({^-2} + 4) + {^-5}$

56. $5 - 8 - (6 - 7) + 1$

Relating Concepts (Exercises 57–58) For Individual or Group Work

*Use your knowledge of the properties of addition to **work Exercises 57 and 58 in order.***

57. Look for a pattern in these pairs of subtractions.

$$^-3 - 5 = \underline{\hspace{1cm}} \qquad {^-4} - {^-3} = \underline{\hspace{1cm}}$$
$$5 - {^-3} = \underline{\hspace{1cm}} \qquad {^-3} - {^-4} = \underline{\hspace{1cm}}$$

Explain what happens when you try to apply the commutative property to subtraction.

58. Recall the addition property of 0. Can 0 be used in a subtraction problem without changing the other number? Explain what happens and give several examples. (*Hint:* Think about *order* in a subtraction problem.)

1.5 ▶▶▶ Problem Solving: Rounding and Estimating

One way to get a rough check on an answer is to *round* the numbers in the problem. **Rounding** a number means finding a number that is close to the original number, but easier to work with.

For example, a superintendent of schools in a large city might be discussing the need to build new schools. In making her point, it probably would not be necessary to say that the school district has 152,807 students—it probably would be sufficient to say that there are about 153,000 students, or even 150,000 students.

OBJECTIVE **1** **Locate the place to which a number is to be rounded.** The first step in rounding a number is to locate the *place to which the number is to be rounded*.

OBJECTIVES

1 Locate the place to which a number is to be rounded.

2 Round integers.

3 Use front end rounding to estimate answers in addition and subtraction.

> **EXAMPLE 1** **Finding the Place to Which a Number Is to Be Rounded**
>
> Locate and draw a line under the place to which each number is to be rounded. Then answer the question.
>
> **(a)** Round ⁻23 to the nearest ten. Is ⁻23 closer to ⁻2<u>0</u> or ⁻3<u>0</u>?
>
>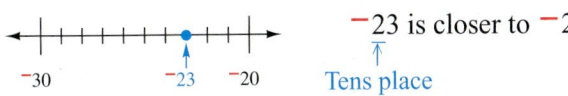
>
> ⁻23 is closer to ⁻20
>
> Tens place
>
> **(b)** Round $381 to the nearest hundred. Is it closer to $3<u>0</u>0 or $4<u>0</u>0?
>
>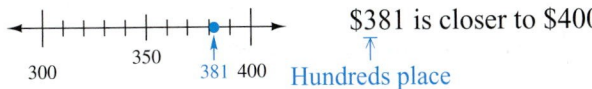
>
> $381 is closer to $400
>
> Hundreds place
>
> **(c)** Round ⁻54,702 to the nearest thousand. Is it closer to ⁻5<u>4</u>,000 or ⁻5<u>5</u>,000?
>
> ⁻54,702 is closer to ⁻55,000
>
> Thousands place ────────┘

Work Problem **1** *at the Side.* ▶

1 Locate and draw a line under the place to which the number is to be rounded. Then answer the question.

(a) ⁻746 (nearest ten)

Is it closer to ⁻740 or

⁻750? _____

(b) 2412 (nearest thousand)

Is it closer to 2000 or

3000? _____

(c) ⁻89,512 (nearest hundred)

Is it closer to ⁻89,500 or

⁻89,600? _____

(d) 546,325 (nearest ten-thousand)

Is it closer to 540,000 or

550,000? _____

OBJECTIVE **2** **Round integers.** Use these steps to round integers.

> **Rounding an Integer**
>
> **Step 1** Locate the *place* to which the number is to be rounded. Draw a line under that place.
>
> **Step 2** Look only at the next digit to the right of the one you underlined. If the next digit is *5 or more, increase* the underlined digit by 1. If the next digit is *4 or less,* do *not* change the digit in the underlined place.
>
> **Step 3** *Change* all digits to the right of the underlined place to zeros.

> **CAUTION**
> If you are rounding a negative number, be careful to write the negative sign in front of the rounded number. For example, ⁻79 rounds to ⁻80.

ANSWERS

1. **(a)** ⁻7<u>4</u>6 is closer to ⁻750
   (b) 2<u>4</u>12 is closer to 2000
   (c) ⁻89,<u>5</u>12 is closer to ⁻89,500
   (d) 5<u>4</u>6,325 is closer to 550,000

2 Round to the nearest ten.

(a) 34

(b) ⁻61

(c) ⁻683

(d) 1792

3 Round to the nearest thousand.

(a) 1725

(b) ⁻6511

(c) 58,829

(d) ⁻83,904

EXAMPLE 2   **Using the Rounding Rule for 4 or Less**

Round 349 to the nearest hundred.

Step 1   Locate the place to which the number is being rounded. Draw a line under that place.

$$349$$

Hundreds place

Step 2   Because the next digit to the right of the underlined place is 4, which is *4 or less,* do *not* change the digit in the underlined place.

Next digit is *4 or less.*

$$349$$

3 remains 3.

Step 3   Change all digits to the right of the underlined place to zeros.

Change to 0.

349 rounded to the nearest hundred is 300

Leave 3 as 3.

In other words, 349 is closer to 300 than to 400. 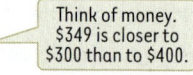 Think of money. $349 is closer to $300 than to $400.

◀ *Work Problem* **2** *at the Side.*

EXAMPLE 3   **Using the Rounding Rule for 5 or More**

Round 36,833 to the nearest thousand.

Step 1   Find the place to which the number is to be rounded. Draw a line under that place.

$$36,833$$

Thousands place

Step 2   Because the next digit to the right of the underlined place is 8, which is *5 or more,* add 1 to the underlined place.

Next digit is *5 or more.*

$$36,833$$

Change 6 to 7.

Step 3   Change all digits to the right of the underlined place to zeros.

Change to 0.

36,833 rounded to the nearest thousand is 37,000

Change 6 to 7.

In other words, 36,833 is closer to 37,000 than to 36,000.

◀ *Work Problem* **3** *at the Side.*

EXAMPLE 4 **Using the Rules for Rounding**

(a) Round $^-$2382 to the nearest ten.

Step 1 $^-$2382
⌐— Tens place

Step 2 The next digit to the right is 2, which is *4 or less.*

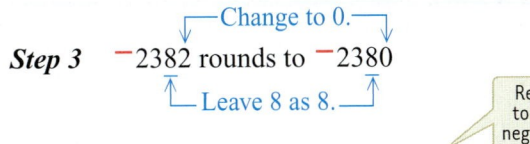

┌— Next digit is *4 or less.*

$^-$2382

└— Leave 8 as 8.

Step 3 ┌—Change to 0.—┐
$^-$2382 rounds to $^-$2380
└— Leave 8 as 8.—┘

> Remember to keep the negative sign.

$^-$2382 rounded to the nearest ten is $^-$2380.

(b) Round 13,961 to the nearest hundred.

Step 1 13,961
⌐— Hundreds place

Step 2 The next digit to the right is 6, which is *5 or more.*

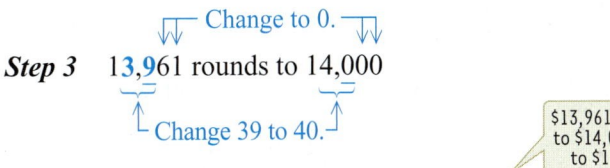

┌— Next digit is *5 or more.*

13,961

└— Change 9 to 10; write 0 and regroup 1 into thousands place.
└— 3 + regrouped 1 = 4

Step 3 ┌— Change to 0. —┐
13,961 rounds to 14,000
└ Change 39 to 40.┘

> $13,961 is closer to $14,000 than to $13,900.

13,961 rounded to the nearest hundred is 14,000.

> **Note**
> In Step 2 above, when you added 1 to the hundreds place, notice that the first three digits increased from 139 to 140.
>
> **13,9**61   rounded to   **14,0**00

Work Problem **4** *at the Side.* ▶

4 Round as indicated.

(a) $^-$6036 to the nearest ten

(b) 31,968 to the nearest hundred

(c) $^-$73,077 to the nearest thousand

(d) 4952 to the nearest thousand

(e) 85,949 to the nearest hundred

(f) 40,387 to the nearest thousand

5 Round as indicated.

(a) $^-$14,679 to the nearest ten-thousand

(b) 724,518,715 to the nearest million

(c) $^-$49,900,700 to the nearest million

(d) 306,779,000 to the nearest hundred-million

EXAMPLE 5   **Rounding Large Numbers**

(a) Round $^-$37,892 to the nearest ten-thousand.

Step 1 $^-$37,892
    └── Ten-thousands place

Step 2   The next digit to the right is 7, which is *5 or more*.

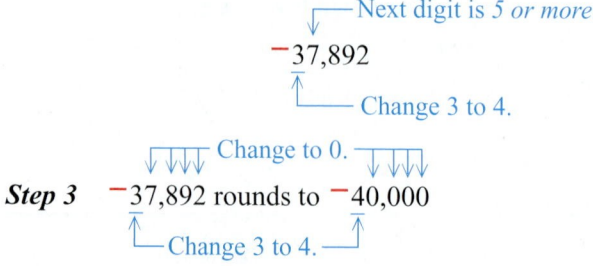

$^-$37,892
┌── Next digit is *5 or more*.
└── Change 3 to 4.

Step 3 $^-$37,892 rounds to $^-$40,000
    Change to 0. ── Change 3 to 4.

$^-$37,892 rounded to the nearest ten-thousand is $^-$40,000.

(b) Round 528,498,675 to the nearest million.

Step 1   528,498,675
    └── Millions place

Step 2   528,498,675
    ┌── Next digit is *4 or less*.
    └── Leave 8 as 8.

Step 3   528,498,675 rounds to 528,000,000
    Change to 0. ── Leave 8 as 8.

528,498,675 rounded to the nearest million is 528,000,000.

◀ *Work Problem* **5** *at the Side.*

OBJECTIVE **3** **Use front end rounding to estimate answers in addition and subtraction.**   In many everyday situations, we can round numbers and **estimate** the answer to a problem. For example, suppose that you're thinking about buying a sofa for $988 and a chair for $209. You can round the prices and estimate the total cost as $1000 + $200 ≈ $1200. The ≈ symbol means "approximately equal to." The estimated total of $1200 is close enough to help you decide whether you can afford both items. Of course, when it comes time to pay the bill, you'll want the *exact* total of $988 + $209 = $1197.

Sofa		**Chair**			
$988	+	$209	=	$1197	← Exact cost
Rounds to		Rounds to			
$1000	+	$200	=	$1200	← Estimated cost

In **front end rounding,** each number is rounded to the highest possible place, so all the digits become 0 except the first digit. Once the numbers have lots of zeros, working with them is easy. Front end rounding is often used to estimate answers.

EXAMPLE 6 **Using Front End Rounding**

Use front end rounding to round each number.

(a) $^-216$

Round to the highest possible place, that is, the leftmost digit. In this case, the leftmost digit, 2, is in the hundreds place, so round to the nearest hundred.

The rounded number is $^-200$. Notice that all the digits in the rounded number are 0, except the first digit. Also, remember to write the negative sign.

(b) 97,203

The leftmost digit, 9, is in the ten-thousands place, so round to the nearest ten-thousand.

The rounded number is 100,000. Notice that all the digits in the rounded number are 0, except the first digit.

Work Problem **6** *at the Side.* ▶

EXAMPLE 7 **Using Front End Rounding to Estimate an Answer**

Use front end rounding to estimate an answer. Then find the exact answer.

Meisha's paycheck showed gross pay of $823. It also listed deductions of $291. What is her net pay after deductions?

Estimate: Use front end rounding to round $823 and $291.

    ┌─ Next digit is *4 or less.*   ┌─ Next digit is *5 or more.*

$823 rounds to $800 $291 rounds to $300

    └─ Leave 8 as 8.┘   └─ Change 2 to 3.┘

Use the rounded numbers and subtract to *estimate* Meisha's net pay.

$$\$800 - \$300 = \$500 \leftarrow \text{Estimate}$$

Exact: Use the original numbers and subtract to find the *exact* amount.

$$\$823 - \$291 = \$532 \leftarrow \text{Exact}$$

Meisha's paycheck will show the *exact* amount of $532. Because $532 is fairly close to the *estimate* of $500, Meisha can quickly see that the amount shown on her paycheck probably is correct. She might also use the estimate when talking to a friend, saying, "My net pay is about $500."

Continued on Next Page

---

6 Use front end rounding to round each number.

(a) $^-94$

(b) 508

(c) $^-2522$

(d) 9700

(e) 61,888

(f) $^-963,369$

7 Use front end rounding to estimate an answer. Then find the exact answer.

Pao Xiong is a bookkeeper for a small business. The company checking account is overdrawn by $3881. He deposits a check for $2090. What is the balance in the account?

Estimate:

Exact:

CAUTION

Always *estimate* the answer first. Then, when you find the *exact* answer, check that it is close to the estimate. If your exact answer is very far off, rework the problem because you probably made an error.

Calculator Tip   It's easy to press the wrong key when using a calculator. If you use front end rounding and estimate the answer *before* entering the numbers, you can catch many such mistakes. For example, a student thought that he entered Example 7 from the previous page correctly.

$$823 \ominus 291 \circleq \qquad \boxed{1114}$$

Front end rounding gives an estimated answer of $800 - 300 = 500$, which is very different from 1114. Can you figure out which key the student pressed incorrectly?

Answer: The student pressed $\oplus$ instead of $\ominus$.

◀ *Work Problem* **7** *at the Side.*

1.5 ▶▶▶ Exercises

Round each number to the indicated place. See Examples 1–5.

1. 625 to the nearest ten

2. 206 to the nearest ten

◉ 3. ⁻1083 to the nearest ten

4. ⁻2439 to the nearest ten

5. 7862 to the nearest hundred

6. 6746 to the nearest hundred

7. ⁻86,813 to the nearest hundred

8. ⁻17,211 to the nearest hundred

9. 42,495 to the nearest hundred

10. 18,273 to the nearest hundred

◉ 11. ⁻5996 to the nearest hundred

12. ⁻8451 to the nearest hundred

13. ⁻78,499 to the nearest thousand

14. ⁻14,314 to the nearest thousand

15. 5847 to the nearest thousand

16. 49,706 to the nearest thousand

17. 595,008 to the nearest ten-thousand

18. 725,182 to the nearest ten-thousand

19. ⁻8,906,422 to the nearest million

20. ⁻13,713,409 to the nearest million

21. 139,610,000 to the nearest million

22. 609,845,500 to the nearest million

23. 19,951,880,500 to the nearest hundred-million

24. 5,993,505,000 to the nearest hundred-million

25. 8,608,200,000 to the nearest billion

26. 703,750,678,005 to the nearest billion

Use front end rounding to round each number. See Example 6.

27. Tyrone's truck shows this number on the odometer.

28. Ezra bought a used car with this odometer reading.

29. From summer to winter the average temperature drops 56 degrees.

30. The flood waters fell 42 inches yesterday.

31. Jan earned $9942 working part time.

32. Carol deposited $285 in her checking account.

33. 60,950,000 Americans go to a video store each week. (*Source:* Video Software Dealer's Assoc.)

34. 95,840,000 U.S. households have cable TV. (*Source:* Nielsen Media Research.)

35. The submarine will dive to 255 feet below the surface of the ocean.

36. DeAnne lost $1352 in the stock market.

37. The population of Alaska is 670,053 people, and the population of California is 36,457,549 people. (*Source:* U.S. Census Bureau.)

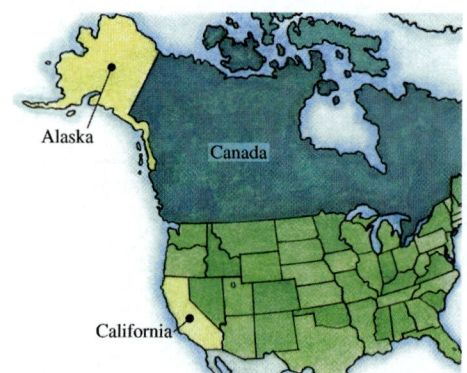

38. Within forty years, it is estimated that the U.S. population will be 420,080,587 people and Canada's population will be 41,429,579 people. (*Source:* U.S. Census Bureau.)

39. Explain in your own words how to do front end rounding. Also show two examples of numbers and how you round them.

40. Describe two situations in your own life when you might use rounded numbers. Describe two situations in which exact numbers are important.

First, use front end rounding to estimate each answer. Then find the exact answer.
In Exercises 49–54, change subtraction to adding the opposite. See Example 7.

🌐 **41.** $^-42 + 89$

Estimate: _____ + _____ = _____

Exact:

42. $^-66 + 25$

Estimate: _____ + _____ = _____

Exact:

43. $16 + {}^-97$

Estimate: _____ + _____ = _____

Exact:

44. $58 + {}^-19$

Estimate: _____ + _____ = _____

Exact:

45. $^-273 + {}^-399$

Estimate:

Exact:

46. $^-311 + {}^-582$

Estimate:

Exact:

47. $3081 + 6826$

Estimate:

Exact:

48. $4904 + 1181$

Estimate:

Exact:

49. $23 - 81$

Estimate:

Exact:

50. $72 - 84$

Estimate:

Exact:

51. $^-39 - 39$

Estimate:

Exact:

52. $^-91 - 91$

Estimate:

Exact:

53. $^-106 + 34 - {}^-72$

Estimate:

Exact:

54. $52 - {}^-87 - 139$

Estimate:

Exact:

First use front end rounding to estimate the answer to each application problem.
Then find the exact answer. See Example 7.

55. The community has raised $52,882 for the homeless shelter. If the goal is $78,650, how much more needs to be collected?

Estimate:

Exact:

56. A truck weighs 9250 pounds when empty. After being loaded with firewood, it weighs 21,375 pounds. What is the weight of the firewood?

Estimate:

Exact:

57. Dorene Cox decided to establish a monthly budget. She will spend $845 for rent, $325 for food, $365 for child care, $182 for transportation, $240 for other expenses, and put the remainder in savings. If her monthly take-home pay is $2120, find her monthly savings.

Estimate:

Exact:

58. Jared Ueda had $2874 in his checking account. He wrote checks for $308 for auto repairs, $580 for child support, and $778 for tuition. Find the amount remaining in his account.

Estimate:

Exact:

59. In a laboratory experiment, a mixture started at a temperature of $^-102$ degrees. First the temperature was raised 37 degrees and then raised 52 degrees. What was the final temperature?

Estimate:

Exact:

60. A scuba diver was photographing fish at 65 feet below the surface of the ocean. She swam up 24 feet and then swam down 49 feet. What was her final depth?

Estimate:

Exact:

61. The White House in Washington, D.C., has 132 rooms, 412 doors, and 147 windows. What is the total number of doors and windows? (*Source: Scholastic Book of World Records.*)

Estimate:

Exact:

62. There are 30,096 McDonald's restaurants through-out the world in 119 different countries. Burger King has 11,204 restaurants in 69 countries. How many restaurants do the two companies have together? (*Source:* www.mcdonalds.com and www.bk.com)

Estimate:

Exact:

1.6 ▶▶▶ Multiplying Integers

OBJECTIVES

1 Use a raised dot or parentheses to express multiplication.

2 Multiply integers.

3 Identify properties of multiplication.

4 Estimate answers to application problems involving multiplication.

OBJECTIVE 1 **Use a raised dot or parentheses to express multiplication.** In arithmetic we usually use "×" when writing multiplication problems. But in algebra, we use a raised dot or parentheses to show multiplication. The numbers being multiplied are called **factors** and the answer is called the **product.**

Arithmetic

$3 \times 5 = 15$
Factors Product

Algebra

$3 \cdot 5 = 15$ or $3(5) = 15$ or $(3)(5) = 15$
Factors Product   Factors Product   Factors Product

EXAMPLE 1 **Expressing Multiplication in Algebra**

Rewrite each multiplication in three different ways, using a dot or parentheses. Also identify the factors and the product.

(a) 10×7

Raised dot
↓
Rewrite it as $10 \cdot 7$ or $10(7)$ or $(10)(7)$

The factors are 10 and 7. The product is 70.

(b) 4×80

Rewrite it as $4 \cdot 80$ or $4(80)$ or $(4)(80)$

The factors are 4 and 80. The product is 320.

─────── Work Problem **1** at the Side. ▶

1 Rewrite each multiplication in three different ways using a dot or parentheses. Also identify the factors and the product.

(a) 100×6

(b) 7×12

Note

Parentheses are used to show several different things in algebra. When we discussed the associative property of addition earlier in this chapter, we used parentheses as shown below.

$6 + (9 + {}^-9)$ ← Parentheses show which numbers to add first.

$6 + \quad 0$

6

Now we are using parentheses to indicate multiplication, as in $3(5)$ or $(3)(5)$.

OBJECTIVE 2 **Multiply integers.** Suppose that our football team gained 5 yards on the first play, gained 5 yards again on the second play, and gained 5 yards again on the third play. We can add to find the result.

$$5 \text{ yards} + 5 \text{ yards} + 5 \text{ yards} = 15 \text{ yards}$$

A quick way to add the same number several times is to multiply.

Our team made 3 plays	and	gained 5 yards each time.	Our team gained a total of 15 yards.
3	•	5	= 15

When multiplying two integers, first multiply the absolute values. Then attach a positive sign or negative sign to the product according to the rules below.

> ### Multiplying Two Integers
>
> If two factors have *different signs,* the product is *negative.*
> For example,
>
> $$^-2 \cdot 6 = ^-12 \qquad \text{and} \qquad 4 \cdot ^-5 = ^-20$$
>
> If two factors have the *same sign,* the product is *positive.*
> For example,
>
> $$7 \cdot 3 = 21 \qquad \text{and} \qquad ^-3 \cdot ^-10 = 30$$

There are several ways to illustrate these rules. First we'll continue with football. Remember, you are interested in the results for *our* team. We will designate **our team** with a **positive sign** and **their team** with a **negative sign**.

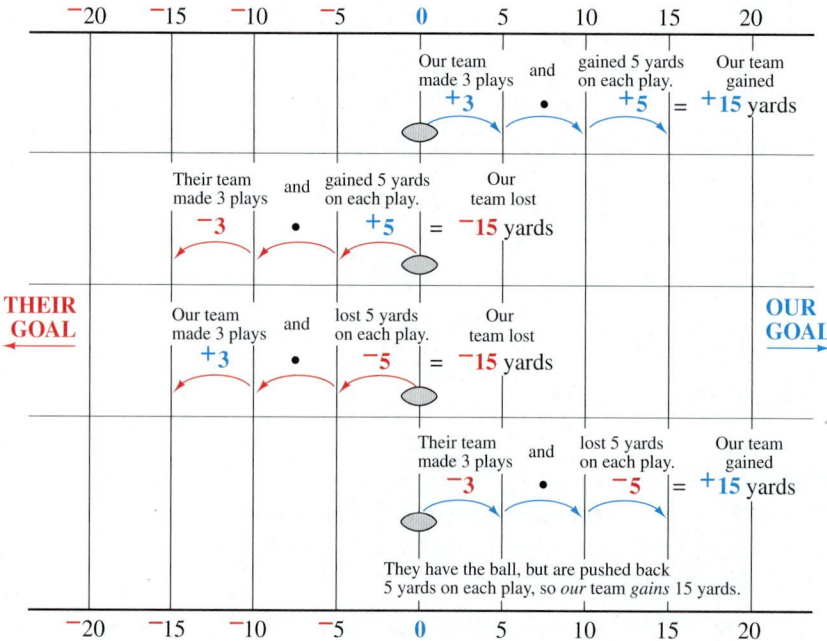

Here is a summary of the football examples.

When two factors have the *same* sign, the product is *positive.*

Both positive

$$3 \cdot 5 = 15$$

Product is positive.

Two factors with *matching* signs give a *positive* product.

$$^-3 \cdot ^-5 = 15$$

Both negative

When two factors have *different* signs, the product is *negative.*

$$^-3 \cdot 5 = ^-15$$

Product is negative.

Two factors with *different* signs give a *negative* product.

$$3 \cdot ^-5 = ^-15$$

There is another way to look at these multiplication rules. In mathematics, the rules or patterns must always be consistent.

Look for a pattern in this list of products.

$$4 \cdot 2 = \mathbf{8}$$
$$3 \cdot 2 = \mathbf{6}$$
$$2 \cdot 2 = \mathbf{4}$$
$$1 \cdot 2 = \mathbf{2}$$
$$0 \cdot 2 = \mathbf{0}$$
$$^-1 \cdot 2 = ?$$

Blue numbers decrease by 1.

Red numbers *decrease* by 2.

To keep the red pattern going, replace the **?** with a number that is 2 *less than* 0, which is $^-2$.

So, $^-1 \cdot 2 = {}^-2$. This pattern illustrates that the product of two numbers with *different* signs is *negative*.

Look for a pattern in this list of products.

$$4 \cdot {}^-2 = {}^-8$$
$$3 \cdot {}^-2 = {}^-6$$
$$2 \cdot {}^-2 = {}^-4$$
$$1 \cdot {}^-2 = {}^-2$$
$$0 \cdot {}^-2 = 0$$
$$^-1 \cdot {}^-2 = ?$$

Blue numbers decrease by 1.

Red numbers *increase* by 2.

To keep the red pattern going, replace the **?** with a number that is 2 *more than* 0, which is $^+2$.

So, $^-1 \cdot {}^-2 = {}^+2$. This pattern illustrates that the product of two numbers with the *same* sign is *positive*.

EXAMPLE 2 Multiplying Two Integers

(a) $^-2 \cdot 8 = {}^-16$ The factors have *different signs,* so the product is *negative*.

Positive
Negative

(b) $^-10\,(^-6) = 60$ The factors have the *same sign,* so the product is *positive*.

Both negative

(c) $(9)\,(^-11) = {}^-99$ The factors have *different signs,* so the product is *negative*.

Negative
Positive

Work Problem **2** *at the Side.* ▶

Sometimes there are more than two factors in a multiplication problem. If there are parentheses around two of the factors, multiply them first. If there aren't any parentheses, start at the left and work with two factors at a time.

EXAMPLE 3 Multiplying Several Factors

Multiply.

(a) $^-3 \cdot (4 \cdot {}^-5)$ Parentheses tell you to multiply $4 \cdot {}^-5$ first. The factors
$\ ^-3 \cdot {}^-20$ have *different* signs, so the product is *negative*.
$ 60$ Then multiply $^-3 \cdot {}^-20$. Both factors have the *same* sign, so the product is *positive*.

(b) $^-2 \cdot {}^-2 \cdot {}^-2$ There are no parentheses, so multiply $^-2 \cdot {}^-2$ first. The
$\ 4 \phantom{\cdot {}^-2\cdot} \cdot {}^-2$ factors have the *same* sign, so the product is *positive*.
${}^-8$ Then multiply $4 \cdot {}^-2$. The factors have *different* signs, so the product is *negative*.

Continued on Next Page

---

2 Multiply.

(a) $7(^-2)$

(b) $^-5 \cdot {}^-5$

(c) $^-1(14)$

(d) $10 \cdot 6$

(e) $(^-4)(^-9)$

3 Multiply.

(a) $5 \cdot (^-10 \cdot 2)$

(b) $^-1 \cdot 8 \cdot ^-5$

(c) $^-3 \cdot ^-2 \cdot ^-4$

(d) $^-2 \cdot (7 \cdot ^-3)$

(e) $(^-1)(^-1)(^-1)$

4 Multiply. Then name the property illustrated by each example.

(a) $819 \cdot 0$

(b) $1(^-90)$

(c) $25 \cdot 1$

(d) $(0)(^-75)$

CAUTION

In Example 3(b) you may be tempted to think that the final product will be *positive* because all the factors have the *same* sign. Be careful to **work with two factors at a time** and keep track of the sign at each step.

Calculator Tip   You can use the *change of sign* key for multiplication and division, just as you did for adding and subtracting. To enter Example 3(b) on your scientific calculator, press the following keys.

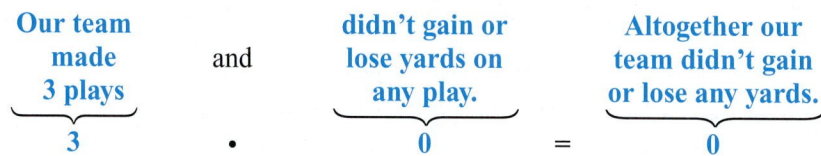

The answer is $^-8$.

$^-2$ $^-2$ $^-2$

◀ *Work Problem* **3** *at the Side.*

OBJECTIVE 3 Identify properties of multiplication.   Addition involving 0 is unusual because adding 0 does *not* change the number. For example, $7 + 0$ is still 7. (See **Section 1.3.**) But what happens in multiplication? Let's use our football team as an example.

Our team made **3 plays**   and   didn't gain or lose yards on any play.   Altogether our team didn't gain or lose any yards.

3   •   0   =   0

This example illustrates one of the properties of multiplication.

Multiplication Property of 0

Multiplying any number by 0 gives a product of 0.
Some examples are shown below.

$^-16 \cdot 0 = 0$ $(0)(5) = 0$ $32{,}977(0) = 0$

So, can you multiply a number by something that will *not* change the number?

$6 \cdot ? = 6$ $^-12(?) = ^-12$ $(?)(5876) = 5876$

The number 1 can replace the **?** in each example. This illustrates another property of multiplication.

Multiplication Property of 1

Multiplying a number by 1 leaves the number unchanged.
Some examples are shown below.

$6 \cdot 1 = 6$ $^-12(1) = ^-12$ $(1)(5876) = 5876$

EXAMPLE 4 Using Properties of Multiplication

Multiply. Then name the property illustrated by each example.

(a) $(0)(^-48) = 0$ Illustrates the **multiplication property of 0.**

(b) $615(1) = 615$ Illustrates the **multiplication property of 1.**

◀ *Work Problem* **4** *at the Side.*

ANSWERS

3. (a) $^-100$ (b) 40 (c) $^-24$
   (d) 42 (e) $^-1$
4. (a) 0; multiplication property of 0
   (b) $^-90$; multiplication property of 1
   (c) 25; multiplication property of 1
   (d) 0; multiplication property of 0

When adding, we said that changing the order of the addends did not change the sum (commutative property of addition). We also found that changing the *grouping* of addends did not change the sum (associative property of addition). These same ideas apply to multiplication.

> **Commutative Property of Multiplication**
>
> Changing the *order* of two factors does not change the product. For example,
> $$2 \cdot 5 = 5 \cdot 2 \quad \text{and} \quad {}^-4 \cdot 6 = 6 \cdot {}^-4$$

> **Associative Property of Multiplication**
>
> Changing the *grouping* of factors does not change the product. For example,
> $$9 \cdot (3 \cdot 2) = (9 \cdot 3) \cdot 2$$

EXAMPLE 5   Using the Commutative and Associative Properties

Show that the product is unchanged and name the property that is illustrated in each case.

(a) $\underbrace{{}^-7 \cdot {}^-4}_{28} = \underbrace{{}^-4 \cdot {}^-7}_{28}$ Multiplying two numbers in a different order does *not* change the product. Both products are 28.

This example illustrates the **commutative property of multiplication.**

(b) $5 \cdot (10 \cdot 2) = (5 \cdot 10) \cdot 2$ Use the associative property to multiply the easiest numbers first.

$5 \cdot \underbrace{20}_{100} = \underbrace{50}_{100} \cdot 2$ Both products are 100.

This example illustrates the **associative property of multiplication.**

Work Problem **5** *at the Side.* ▶

Now that you are familiar with multiplication and addition, we can look at a property that involves both operations.

> **Distributive Property**
>
> Multiplication *distributes* over addition. An example is shown below.
> $$3(6 + 2) = 3 \cdot 6 + 3 \cdot 2$$

What is the **distributive property** really saying? Notice that there is an understood multiplication symbol between the 3 and the parentheses. To "distribute" the 3 means to multiply 3 times each number inside the parentheses.

Understood to be *multiplying* by 3 $3(6 + 2)$
$$3 \cdot (6 + 2)$$

Using the distributive property,

$3 \cdot (6 + 2)$ can be rewritten as $3 \cdot 6 + 3 \cdot 2$

5 Show that the product is unchanged and name the property that is illustrated in each case.

(a) $(3 \cdot 3) \cdot 2 = 3 \cdot (3 \cdot 2)$

(b) $11 \cdot 8 = 8 \cdot 11$

(c) $2 \cdot {}^-15 = {}^-15 \cdot 2$

(d) ${}^-4 \cdot (2 \cdot 5) = ({}^-4 \cdot 2) \cdot 5$

6 Rewrite each product, using the distributive property. Show that the result is unchanged.

(a) $3(8 + 7)$

(b) $10(6 + {}^-9)$

(c) ${}^-6(4 + 4)$

EXAMPLE 6 **Using the Distributive Property**

Rewrite each product, using the distributive property. Show that the result is unchanged.

(a) $4(3 + 7)$

> Careful! Multiply *both* numbers by 4.

$$4(3 + 7) = 4 \cdot 3 + 4 \cdot 7$$
$$4(10) = 12 + 28$$
$$40 = 40 \qquad \text{Both results are 40.}$$

(b) ${}^-2({}^-5 + 1)$

> Multiple *both* numbers by ${}^-2$.

$${}^-2({}^-5 + 1) = {}^-2 \cdot {}^-5 + {}^-2 \cdot 1$$
$${}^-2({}^-4) = 10 + {}^-2$$
$$8 = 8 \qquad \text{Both results are 8.}$$

◀ *Work Problem* **6** *at the Side.*

OBJECTIVE 4 **Estimate answers to application problems involving multiplication.** Front end rounding can be used to estimate answers in multiplication, just as we did when adding and subtracting (see **Section 1.5**). Once the numbers have been rounded so that there are lots of zeros, we can use the multiplication shortcut described in the Review chapter (see **Section R.3**). As a brief review, look at the pattern in these examples.

$${}^-3 \cdot 2 \text{ is } {}^-6$$
$${}^-30 \cdot 200 = {}^-6000$$

Total of three zeros   Write three zeros after the ${}^-6$.

$${}^-2 \cdot {}^-5 \text{ is } 10$$
$${}^-2000 \cdot {}^-5000 = 10{,}000{,}000$$

Total of six zeros   Write six zeros after the 10.

---

7 Use front end rounding to estimate an answer. Then find the exact answer.

An average of 27,095 baseball fans attended each of the 81 home games during the season. What was the total home game attendance for the season?

Estimate:

Exact:

EXAMPLE 7 **Using Front End Rounding to Estimate an Answer**

Use front end rounding to estimate an answer. Then find the exact answer.

Last year the Video Land store had to replace 392 defective videos at a cost of \$19 each. How much money did the store lose on defective videos? (*Hint:* Because it's a loss, use a negative number for the cost.)

Estimate: Use front end rounding: 392 rounds to 400 and ${}^-\$19$ rounds to ${}^-\$20$. Use the rounded numbers and multiply to estimate the total amount of money lost.

$$4 \cdot {}^-2 \text{ is } {}^-8$$
$$400 \cdot {}^-\$20 = {}^-\$8000 \qquad \text{Estimate}$$

Total of three zeros   Write three zeros after the ${}^-8$.

Exact: $392 \cdot {}^-\$19 = {}^-\$7448$

Because the exact answer of ${}^-\$7448$ is fairly close to the estimate of ${}^-\$8000$, you can see that ${}^-\$7448$ probably is correct. The store manager could also use the estimate to say, "We lost about \$8000 on defective videos last year."

◀ *Work Problem* **7** *at the Side.*

ANSWERS

6. **(a)** $3 \cdot 8 + 3 \cdot 7$; both results are 45.
   (b) $10 \cdot 6 + 10 \cdot {}^-9$; both results are ${}^-30$.
   (c) ${}^-6 \cdot 4 + {}^-6 \cdot 4$; both results are ${}^-48$.
7. **(a)** *Estimate:* $30{,}000 \cdot 80 = 2{,}400{,}000$ fans
   (b) *Exact:* $27{,}095 \cdot 81 = 2{,}194{,}695$ fans

Multiply. See Examples 1–4.

1. **(a)** $9 \cdot 7$
   (b) $^-9 \cdot {}^-7$
   (c) $^-9 \cdot 7$
   (d) $9 \cdot {}^-7$

2. **(a)** $^-6 \cdot 9$
   (b) $6 \cdot {}^-9$
   (c) $^-6 \cdot {}^-9$
   (d) $6 \cdot 9$

3. **(a)** $7({}^-8)$
   (b) $^-7(8)$
   (c) $7(8)$
   (d) $^-7({}^-8)$

4. **(a)** $8(6)$
   (b) $^-8({}^-6)$
   (c) $^-8(6)$
   (d) $8({}^-6)$

5. $^-5 \cdot 7$

6. $^-10 \cdot 2$

7. $(^-5)(9)$

8. $(^-9)(4)$

9. $3({}^-6)$

10. $8({}^-9)$

11. $10({}^-5)$

12. $5({}^-11)$

13. $(^-1)(40)$

14. $(75)({}^-1)$

15. $^-56 \cdot 1$

16. $1 \cdot {}^-87$

17. $^-8({}^-4)$

18. $^-3({}^-9)$

19. $11 \cdot 7$

20. $4 \cdot 25$

21. $25 \cdot 0$

22. $0 \cdot 30$

23. $^-19({}^-7)$

24. $^-21({}^-3)$

25. $^-13({}^-1)$

26. $^-1({}^-31)$

27. $(0)({}^-25)$

28. $(^-50)(0)$

29. $^-4 \cdot {}^-6 \cdot 2$

30. $^-9 \cdot 3 \cdot {}^-3$

31. $(^-4)({}^-2)({}^-7)$

32. $(^-6)({}^-2)({}^-3)$

33. $5({}^-8)(4)$

34. $5(4)({}^-6)$

Write an integer in each blank to make a true statement.

35. $(^-3)(\underline{\quad}) = {}^-15$

36. $6 \cdot \underline{\quad} = {}^-24$

37. $\underline{\quad} \cdot 10 = {}^-30$

38. $(\underline{\quad})({}^-4) = 16$

39. $^-17 = 17(\underline{\quad})$

40. $29 = {}^-29(\underline{\quad})$

41. $(\underline{\quad})({}^-350) = 0$

42. $\underline{\quad} \cdot 99 = 99$

43. $5 \cdot {}^-4 \cdot \underline{\quad} = {}^-100$

44. $\underline{\quad} \cdot 2 \cdot {}^-2 = {}^-24$

45. $(\underline{\quad})({}^-5)({}^-2) = {}^-40$

46. $(^-3)(\underline{\quad})({}^-3) = {}^-27$

47. In your own words, explain the difference between the commutative and associative properties of multiplication. Show an example of each.

48. A student did this multiplication.

$$-3 \cdot -3 \cdot -3 = 27$$

He knew that $3 \cdot 3 \cdot 3$ is 27. Since all the factors have the same sign, he made the product positive. Do you agree with his reasoning? Explain.

Relating Concepts (Exercises 49–50) For Individual or Group Work

*Look for patterns as you **work Exercises 49 and 50 in order**.*

49. Write three numerical examples for each of these situations:

(a) A positive number multiplied by -1

(b) A negative number multiplied by -1

Now write a rule that explains what happens when you multiply a signed number by -1.

50. Do these multiplications.

$$-2 \cdot -2 = \underline{\quad}$$
$$-2 \cdot -2 \cdot -2 = \underline{\quad}$$
$$-2 \cdot -2 \cdot -2 \cdot -2 = \underline{\quad}$$
$$-2 \cdot -2 \cdot -2 \cdot -2 \cdot -2 = \underline{\quad}$$

Describe the pattern in the products. Then find the next three products without multiplying all the -2s.

Rewrite each multiplication, using the stated property. Show that the result is unchanged. See Examples 5 and 6.

51. Distributive property

$9(-3 + 5)$

52. Distributive property

$-6(4 + -5)$

53. Commutative property

$25 \cdot 8$

54. Commutative property

$-7 \cdot -11$

55. Associative property

$-3 \cdot (2 \cdot 5)$

56. Associative property

$(5 \cdot 5) \cdot 10$

First use front end rounding to estimate the answer to each application problem.
Then find the exact answer. See Example 7.

57. Alliette receives $324 per week for doing child care in her home. How much income will she have for an entire year? There are 52 weeks in a year.

Estimate:

Exact:

58. Enrollment at our community college has increased by 875 students each of the last four semesters. What is the total increase?

Estimate:

Exact:

59. A new computer software store had losses of $9950 during each month of its first year. What was the total loss for the year?

Estimate:

Exact:

60. A cell phone company estimates that it is losing 95 customers each week. How many customers will it lose in a year?

Estimate:

Exact:

61. Tuition at the state university is $182 per credit for undergraduates. How much tuition will Wei Chen pay for 13 credits?

Estimate:

Exact:

62. Pat ate a dozen crackers as a snack. Each cracker had 17 calories. How many calories did Pat eat?

Estimate:

Exact:

63. There are 24 hours in one day. How many hours are in one year (365 days)?

Estimate:

Exact:

64. There are 5280 feet in one mile. How many feet are in 17 miles?

Estimate:

Exact:

Simplify.

65. $^-8 \cdot |^-8 \cdot 8|$

66. $^-7 \cdot |7| \cdot |^-7|$

67. $(^-37)(^-1)(85)(0)$

68. $^-1(9732)(^-1)(^-1)$

69. $|6 - 7| \cdot {}^-355{,}299$

70. $987 \cdot {}^-65{,}432 \cdot |9 - 9|$

Each of these application problems requires several steps and may involve addition and subtraction as well as multiplication.

71. Each of Maurice's four cats needed a $24 rabies shot and a $29 shot to prevent respiratory infections. There was also one $35 office visit charge. What was the total amount of Maurice's bill?

72. Chantele has three children. Her older daughter had a throat culture taken at the clinic today. Her baby received three immunization shots and her son received two shots. The co-pay amounts were $8 for each shot, an $18 office charge for each child, and $12 for the throat culture. How much did Chantele pay?

73. There is a 3-degree drop in temperature for every thousand feet that an airplane climbs into the sky. If the temperature on the ground is 50 degrees, what will be the temperature when the plane reaches an altitude of 24,000 feet? (*Source:* Lands' End.)

74. An unmanned research submarine descends to 150 feet below the surface of the ocean. Then it continues to go deeper, taking a water sample every 25 feet. What is its depth when it takes the 15th sample?

75. In Ms. Zubero's algebra class, there are six tests of 100 points each, eight quizzes of six points each, and 20 homework assignments of five points each. There are also four "bonus points" on each test. What is the total number of possible points?

76. In Mr. Jackson's prealgebra class, there are three group projects worth 25 points each, five 100-point tests, a 150-point final exam, and seven quizzes worth 12 points each. Find the total number of possible points.

1.7 ▶▶▶ Dividing Integers

OBJECTIVE 1 Divide integers. In arithmetic, we usually use $\overline{)}$ to write division problems so that we can do them by hand. Calculator keys use the $\div$ symbol for division. In algebra, we usually show division by using a fraction bar, a slash mark, or the $\div$ symbol. The answer to a division problem is called the **quotient.**

Arithmetic

$$\text{Divisor} \to 2\overline{)16} \begin{array}{l} \leftarrow \text{Quotient} \\ \leftarrow \text{Dividend} \end{array} \quad 8$$

Calculator and Algebra

$$\underset{\uparrow\text{Quotient}}{16} \underset{\text{Dividend}}{\overset{\text{Divisor}}{\div}} 2 = 8$$

$$\text{Dividend} \to \frac{16}{2} = \underset{\uparrow\text{Quotient}}{8}$$
$$\text{Divisor} \to$$

For every division problem, we can write a related multiplication problem. (See **Section R.4.**) Because of this relationship, the sign rules for dividing integers are the same as the rules for multiplying integers. Examples:

$$\frac{16}{8} = 2 \qquad \text{because} \quad (2)(8) = 16$$

$$\frac{{}^-16}{{}^-8} = 2 \qquad \text{because} \quad (2)({}^-8) = {}^-16$$

$$\frac{{}^-16}{8} = {}^-2 \qquad \text{because} \quad ({}^-2)(8) = {}^-16$$

$$\frac{16}{{}^-8} = {}^-2 \qquad \text{because} \quad ({}^-2)({}^-8) = 16$$

Dividing Two Integers

If two numbers have *different signs,* the quotient is *negative.* Some examples are shown below.

$$\frac{{}^-18}{3} = {}^-6 \qquad \frac{40}{{}^-5} = {}^-8$$

If two numbers have the *same sign,* the quotient is *positive.* Some examples are shown below.

$$\frac{{}^-30}{{}^-6} = 5 \qquad \frac{48}{8} = 6$$

EXAMPLE 1 Dividing Two Integers

(a) $\dfrac{{}^-20}{5}$ Numbers have *different* signs, so the quotient is *negative.* $\dfrac{{}^-20}{5} = {}^-4$

(b) $\dfrac{{}^-24}{{}^-4}$ Numbers have the *same* sign, so the quotient is *positive.* $\dfrac{{}^-24}{{}^-4} = 6$

> The sign rules for dividing and multiplying are the same.

(c) $60 \div {}^-2$ Numbers have *different* signs, so the quotient is *negative.* $60 \div {}^-2 = {}^-30$

Work Problem **1** *at the Side.* ▶

OBJECTIVES

1 Divide integers.

2 Identify properties of division.

3 Combine multiplying and dividing of integers.

4 Estimate answers to application problems involving division.

5 Interpret remainders in division application problems.

---

1 Divide.

(a) $\dfrac{40}{{}^-8}$

(b) $\dfrac{49}{7}$

(c) $\dfrac{{}^-32}{4}$

(d) $\dfrac{{}^-10}{{}^-10}$

(e) ${}^-81 \div 9$

(f) ${}^-100 \div {}^-50$

ANSWERS

1. (a) ${}^-5$ **(b)** 7 **(c)** ${}^-8$ **(d)** 1
(e) ${}^-9$ **(f)** 2

2 Divide. Then state the property illustrated by each division.

(a) $\dfrac{^-12}{0}$

(b) $\dfrac{0}{39}$

(c) $\dfrac{^-9}{1}$

(d) $\dfrac{21}{21}$

OBJECTIVE **2** **Identify properties of division.** You have seen that 0 and 1 are used in special ways in addition and multiplication. This is also true in division.

Examples			Pattern (Division Property)
$\dfrac{5}{5} = 1$	$\dfrac{^-18}{^-18} = 1$	$\dfrac{^-793}{^-793} = 1$	When a nonzero number is divided by itself, the quotient is 1.
$\dfrac{5}{1} = 5$	$\dfrac{^-18}{1} = {^-18}$	$\dfrac{^-793}{1} = {^-793}$	When a number is divided by 1, the quotient is the number.
$\dfrac{0}{5} = 0$	$\dfrac{0}{^-18} = 0$	$\dfrac{0}{^-793} = 0$	When 0 is divided by any other number (except 0), the quotient is 0.
$\dfrac{5}{0}$ is undefined.		$\dfrac{^-18}{0}$ is undefined.	Division by 0 is *undefined.* There is no answer.

The most surprising property is that division by 0 *cannot be done.* Let's review the reason for that by rewriting this division problem as a related multiplication problem.

$$\dfrac{^-18}{0} = ? \quad \text{can be written as the multiplication} \quad ? \cdot 0 = {^-18}$$

If you thought the answer to $\frac{^-18}{0}$ should be 0, try replacing **?** with 0. It doesn't work in the related multiplication problem! Try replacing **?** with any number you like. The result in the related multiplication problem is always 0 instead of $^-18$. That is how we know that dividing by 0 cannot be done. Mathematicians say that it is *undefined* and have agreed never to divide by 0.

EXAMPLE 2 **Using the Properties of Division**

Divide. Then state the property illustrated by each example.

(a) $\dfrac{^-312}{^-312} = 1$ Any nonzero number divided by itself is 1.

(b) $\dfrac{75}{1} = 75$ Any number divided by 1 is the number.

(c) $\dfrac{0}{^-19} = 0$ Zero divided by any nonzero number is 0.

(d) $\dfrac{48}{0}$ is *undefined.*   Division by 0 is *undefined.*

> You *cannot* divide by 0. Write "undefined."

🖩 **Calculator Tip** Try Examples 2(c) and 2(d) above on your calculator. Use the change of sign key to enter $^-19$ on a *scientific* calculator.

0 ÷ 19 +/− =   Answer is 0

$^-19$

48 ÷ 0 =   Calculator shows "Error" or "ERR" or "E" for error because it cannot divide by 0.

◀ *Work Problem* **2** *at the Side.*

ANSWERS

2. (a) undefined; division by 0 is undefined.
  (b) 0; 0 divided by any nonzero number is 0.
  (c) $^-9$; any number divided by 1 is the number.
  (d) 1; any nonzero number divided by itself is 1.

OBJECTIVE **3** **Combine multiplying and dividing of integers.**
When a problem involves both multiplying and dividing, first check to see if
there are any parentheses. Do what is inside parentheses first. Then start at
the left and work toward the right, using two numbers at a time.

3 Simplify.

(a) $60 \div {}^-3({}^-5)$

EXAMPLE 3 **Combining Multiplication and Division of Integers**

Simplify.

(a) $6({}^-10) \div ({}^-3 \cdot 2)$ Do operations inside parentheses first: ${}^-3 \cdot 2$ is ${}^-6$.
The signs are *different*, so the product is *negative*.

$6({}^-10) \div {}^-6$ Start at the left: $6({}^-10)$ is ${}^-60$. The signs are
different, so the product is *negative*.

$ {}^-60 \div {}^-6 $ Finally, ${}^-60 \div {}^-6$ is 10. The signs are the *same*,
so the quotient is *positive*.

10

(b) ${}^-24 \div {}^-2(4) \div {}^-6$ No operations inside parentheses, so start
at the left: ${}^-24 \div {}^-2$ is 12. (*Same* signs,
positive quotient.)

$\mathbf{12}(4) \div {}^-6$ Next, 12(4) is 48. (*Same* signs, *positive* product.)

$48 \div {}^-6$ Finally, $48 \div {}^-6$ is ${}^-8$. (*Different* signs,
negative quotient.)

${}^-8$

(b) ${}^-6({}^-16 \div {}^-8) \cdot 2$

(c) ${}^-50 \div {}^-5 \div {}^-2$ There are no parentheses, so start at the left:
${}^-50 \div {}^-5$ is 10. (*Same* signs, *positive* quotient.)

$10 \div {}^-2$ Now, $10 \div {}^-2$ is ${}^-5$. (*Different* signs, *negative*
quotient.)

${}^-5$

(c) ${}^-8(10) \div 4({}^-3) \div {}^-6$

Work Problem **3** *at the Side.* ▶

OBJECTIVE **4** **Estimate answers to application problems
involving division.**   Front end rounding can be used to estimate answers
in division just as you did when multiplying (see **Section 1.6**). Once the
numbers have been rounded so that there are lots of zeros, you can use the
division shortcut described in the Review chapter (see **Section R.5**). As a
brief review, look at the pattern in these examples.

$$400\cancel{0} \div {}^-5\cancel{0} = 400 \div {}^-5 = {}^-80$$

Drop one 0 from both
dividend and divisor.

(d) $56 \div {}^-8 \div {}^-1$

$$ {}^-6\cancel{000} \div {}^-3\cancel{000} = {}^-6 \div {}^-3 = 2 $$

Drop three zeros from both
dividend and divisor.

4 First use front end rounding to estimate an answer. Then find the exact answer.

Laurie and Chuck Struthers lost $2724 on their stock investments last year. What was their average loss each month?

Estimate:

Exact:

EXAMPLE 4 **Using Front End Rounding to Estimate an Answer in Division**

First use front end rounding to estimate an answer. Then find the exact answer.

During a 24-hour laboratory experiment, the temperature of a solution dropped 96 degrees. What was the average drop in temperature each hour?

Estimate: Use front end rounding: ⁻96 degrees rounds to ⁻100 degrees and 24 hours rounds to 20 hours. To estimate the average, divide the rounded number of degrees by the rounded number of hours.

$$^-100 \text{ degrees} \div 20 \text{ hours} = {}^-5 \text{ degrees each hour} \leftarrow \text{Estimate}$$

Exact: $^-96 \text{ degrees} \div 24 \text{ hours} = {}^-4 \text{ degrees each hour} \leftarrow \text{Exact}$

Because the exact answer of ⁻4 degrees is close to the estimate of ⁻5 degrees, you can see that ⁻4 degrees probably is correct.

Calculator Tip The answer in Example 4 above "came out even." In other words, the quotient was an integer. Suppose that the drop in temperature had been 97 degrees. Do the division on your calculator.

97 (+/−) ÷ 24 (=) Calculator shows −4.041666667

⁻97

The quotient is *not* an integer. We will work with numbers like these in **Chapter 5,** Positive and Negative Decimals.

◄*Work Problem* **4** *at the Side.*

OBJECTIVE **5** **Interpret remainders in division application problems.** In arithmetic, division problems often have a remainder, as shown below.

```
        14 R10
   25) 360
        25
       110
       100
        10 ← Remainder
```

But what does **R10** really mean? Let's look at this same problem by using money amounts.

EXAMPLE 5 **Interpreting Remainders in Division Applications**

Divide; then interpret the remainder in each application.

(a) The math department at Lake Community College has $360 in its budget to buy scientific calculators for the math lab. If the calculators cost $25 each, how many can be purchased? How much money will be left over?

We can solve this problem by using the same division as shown above. But this time we can decide what the remainder really means.

```
                        14 ← Number of calculators purchased
Cost of one calculator → $25) $360 ← Budget
                        25
                       110
                       100
                       $10 ← Money left over
```

The remainder is the money that is left over.

The department can buy 14 calculators. There will be $10 left over.

— **Continued on Next Page**

▦ Calculator Tip You can use your calculator to solve Example 5(a) on the previous page. Recall that digits on the *right* side of the decimal point show *part* of one whole. You cannot order *part* of one calculator, so ignore those digits and use only the *whole number part* of the quotient.

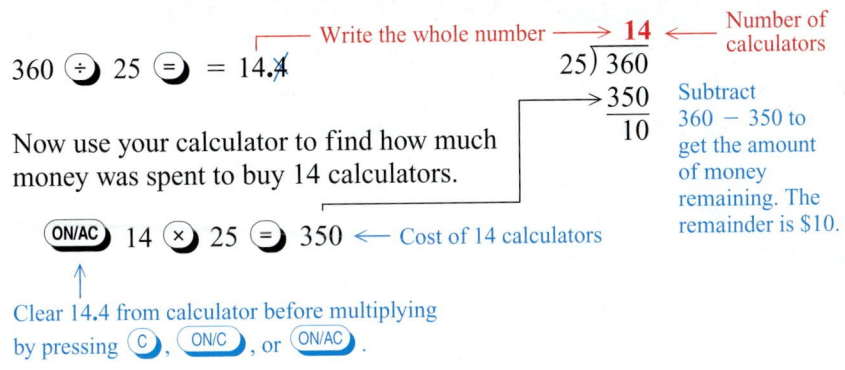

$$360 \div 25 = 14.4$$

Now use your calculator to find how much money was spent to buy 14 calculators.

(ON/AC) 14 (×) 25 (=) 350 ← Cost of 14 calculators

Clear 14.4 from calculator before multiplying by pressing (C), (ON/C), or (ON/AC).

Write the whole number → **14** ← Number of calculators

$$25\overline{)360}$$
$$\underline{350} \quad \text{Subtract } 360 - 350 \text{ to get the amount of money remaining. The remainder is \$10.}$$
$$10$$

(b) Luke's son is going on a Scout camping trip. There are 135 Scouts. Luke is renting tents that sleep 6 people each. How many tents should he rent?

We again use division to solve the problem. There is a remainder, but this time it must be interpreted differently than in the calculator example.

Each tent holds → $6\overline{)135}$ ← Total number of Scouts

22 ← Number of tents with 6 Scouts each

$$\begin{array}{r} 22 \\ 6\overline{)135} \\ \underline{12} \\ 15 \\ \underline{12} \\ \mathbf{3} \end{array}$$ ← Scouts left over

> 22 tents is *not* enough for all the Scouts.

If Luke rents 22 tents, 3 Scouts will have to sleep out in the rain. He must rent **23 tents** to accommodate all the Scouts. (One tent will have only 3 Scouts in it.)

Work Problem **5** *at the Side.* ▶

---

5 Divide; then interpret the remainder in each of these applications.

(a) Chad and Martha are baking cookies for a fund-raiser. They baked 116 cookies and are putting them into packages of a dozen each. How many packages will they have for the fund-raiser? How many cookies will be left over for them to eat?

(b) Coreen is a dispatcher for a bus company. A group of 249 senior citizens is going to a baseball game. If the buses will each hold 44 people, how many buses should she send to pick up the seniors?

ANSWERS

5. (a) 9 packages, with 8 cookies left over to eat
(b) 6 buses, because 5 buses would leave 29 seniors standing on the curb

Math in the Media

'TIL DEBT DO YOU PART!

Although prices can vary depending on where you live, average wedding costs in the United States increased from $15,208 in 1990 to the grand total you will calculate in the table below (in 2007). The average number of wedding guests is 165.

Category	Average Cost in 2007
Attire (bride's dress, men's formalwear, etc.)	$ 2710
Ceremony	2627
Favors and gifts	1166
Flowers	2048
Jewelry: engagement and wedding rings	4150
Music	991
Photography and video	3836
Reception	14,737
Stationery; invitations	881
Limousine	426
Grand Total	

Sources: www.costofwedding.com and *Bride's* magazine.

1. What is the grand total of expenses shown in the table?

2. How much more expensive was a wedding in 2007 compared to 1990?

3. If you budgeted $78 per person for the wedding reception and you invited 165 guests to a wedding in 2009, how much money would you have spent compared to the 2007 wedding reception costs?

4. If you budget $10,000 for the reception and the caterer charges $78 per person, how many guests can you invite? How much of your budget is left over?

5. If you budget $12,500 for the reception and the caterer charges $78 per person, how many guests can you invite? How much of your budget is left over?

6. What type of arithmetic problem did you work to get the answers to Problems 4 and 5? What is the mathematical term for the "left over" budget?

7. If you rent the limousine for six hours, what is the average cost per hour? When you solved this problem, which number was the dividend? the divisor? the quotient? Rewrite the division as a related multiplication.

1.7 ▶▶▶ Exercises

Divide. See Examples 1 and 2.

1. (a) $14 \div 2$

(b) $^-14 \div ^-2$

(c) $14 \div ^-2$

(d) $^-14 \div 2$

2. (a) $^-18 \div ^-3$

(b) $18 \div 3$

(c) $^-18 \div 3$

(d) $18 \div ^-3$

3. (a) $^-42 \div 6$

(b) $^-42 \div ^-6$

(c) $42 \div ^-6$

(d) $42 \div 6$

4. (a) $45 \div 5$

(b) $45 \div ^-5$

(c) $^-45 \div ^-5$

(d) $^-45 \div 5$

5. (a) $\dfrac{35}{35}$

(b) $\dfrac{35}{1}$

(c) $\dfrac{^-13}{1}$

(d) $\dfrac{^-13}{^-13}$

6. (a) $\dfrac{^-23}{1}$

(b) $\dfrac{^-23}{^-23}$

(c) $\dfrac{17}{1}$

(d) $\dfrac{17}{17}$

7. (a) $\dfrac{0}{50}$

(b) $\dfrac{50}{0}$

(c) $\dfrac{^-11}{0}$

(d) $\dfrac{0}{^-11}$

8. (a) $\dfrac{^-85}{0}$

(b) $\dfrac{0}{^-85}$

(c) $\dfrac{6}{0}$

(d) $\dfrac{0}{6}$

9. $\dfrac{^-8}{2}$

10. $\dfrac{^-14}{7}$

11. $\dfrac{21}{^-7}$

12. $\dfrac{30}{^-6}$

13. $\dfrac{^-54}{^-9}$

14. $\dfrac{^-48}{^-6}$

15. $\dfrac{55}{^-5}$

16. $\dfrac{70}{^-7}$

17. $\dfrac{^-28}{0}$

18. $\dfrac{^-40}{0}$

19. $\dfrac{14}{^-1}$

20. $\dfrac{25}{^-1}$

21. $\dfrac{^-20}{^-2}$

22. $\dfrac{^-80}{^-4}$

23. $\dfrac{^-48}{^-12}$

24. $\dfrac{^-30}{^-15}$

25. $\dfrac{^-18}{18}$

26. $\dfrac{50}{^-50}$

27. $\dfrac{0}{^-9}$

28. $\dfrac{0}{^-4}$

29. $\dfrac{^-573}{^-3}$

30. $\dfrac{^-580}{^-5}$

31. $\dfrac{163,672}{^-328}$

32. $\dfrac{^-69,496}{1022}$

Simplify. See Example 3.

33. $^-60 \div 10 \div {}^-3$

34. $36 \div {}^-4 \div 3$

35. $^-64 \div {}^-8 \div {}^-2$

36. $^-72 \div {}^-9 \div {}^-4$

37. $100 \div {}^-5({}^-2)$

38. $^-80 \div 4({}^-5)$

39. $48 \div 3 \bullet (12 \div {}^-4)$

40. $^-2 \bullet ({}^-3 \bullet {}^-7) \div 7$

41. $^-5 \div {}^-5({}^-10) \div {}^-2$

42. $^-9(4) \div {}^-36(50)$

43. $64 \bullet 0 \div {}^-8(10)$

44. $^-88 \div {}^-8 \div {}^-11(0)$

Relating Concepts (Exercises 45–50) For Individual or Group Work

Use your knowledge of the properties of multiplication as you **work Exercises 45–50 in order.**

45. Explain whether or not division is commutative like multiplication. Start by doing these two divisions on your calculator: $2 \div 1$ and $1 \div 2$.

46. Explain whether or not division is associative like multiplication. Start by doing these two divisions: $(12 \div 6) \div 2$ and $12 \div (6 \div 2)$.

47. Explain what is different and what is similar about multiplying and dividing two signed numbers.

48. In your own words, describe at least three division properties. Include examples to illustrate each property.

49. Write three numerical examples for each situation.

(a) A negative number divided by $^-1$

(b) A positive number divided by $^-1$

Now write a rule that explains what happens when you divide a signed number by $^-1$.

50. Explain why $\frac{0}{-3}$ and $\frac{-3}{0}$ do not give the same result.

Solve these application problems by using addition, subtraction, multiplication, or division. First use front end rounding to estimate the answer. Then find the exact answer. See Example 4.

51. The greatest ocean depth is 35,836 feet below sea level. If an unmanned research sub dives to that depth in 17 equal steps, how far does it dive in each step? (*Source: The Top 10 of Everything.*)

Estimate:

Exact:

52. Our college enrollment dropped by 3245 students over the last 11 years. What was the average drop in enrollment each year?

Estimate:

Exact:

53. When Ashwini discovered that her checking account was overdrawn by $238, she quickly transferred $450 from her savings to her checking account. What is the new balance in her checking account?

Estimate:

Exact:

54. The Tigers offensive team lost a total of 48 yards during the first half of the football game. During the second half they gained 191 yards. How many yards did they gain or lose during the entire game?

Estimate:

Exact:

55. The foggiest place in the United States is Cape Disappointment, Washington. It is foggy there an average of 106 days each year. How many days is it not foggy each year? (*Source:* National Weather Service.)

Estimate:

Exact:

56. The number of cell phone users in the United States in 1992 was 11 million. The number of users reached 233 million in 2007. What was the increase in the number of users during this 15-year period? (*Source: The World Almanac.*)

Estimate:

Exact:

57. A plane descended an average of 730 feet each minute during a 37-minute landing. How far did the plane descend during the landing?

Estimate:

Exact:

58. A discount store found that 174 items were lost to shoplifting last month. The average value of each item was $24. What was the total loss due to shoplifting?

Estimate:

Exact:

59. Mr. and Mrs. Martinez drove on the Interstate for five hours and traveled 315 miles. What was the average number of miles they drove each hour?

Estimate:

Exact:

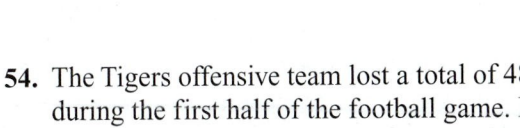

60. Rochelle has a 48-month car loan for $15,072. How much is her monthly payment?

Estimate:

Exact:

Find the exact answer in Exercises 61–66. Solving these problems requires more than one step.

61. Clarence bowled four games and had scores of 143, 190, 162, and 177. What was his average score? (*Hint:* To find the average, add all the scores and divide by the number of scores.)

62. Sheila kept track of her grocery expenses for six weeks. The amounts she spent were $184, $111, $136, $110, $98, and $153. What was the average weekly cost of her groceries?

63. On the back of an oatmeal box, it says that one serving weighs 40 grams and that there are 13 servings in the box. On the front of the box, it says that the weight of the contents is 510 grams. What is the difference in the total weight on the front and the back of the box? (*Source:* Quaker Oats.)

64. A 2000-calorie-per-day diet recommends that you eat no more than 65 grams of fat. If each gram of fat is 9 calories, how many calories can you consume in other types of food? (*Source:* U.S. Department of Agriculture.)

65. Stephanie had $302 in her checking account. She wrote a $116 check for day care and a $548 check for rent. She also deposited her $347 paycheck. What is the balance in her account?

66. Gary started a new checking account with a $500 deposit. The bank charged him $18 to print his checks. He also wrote a $193 check for car repairs and a $289 check to his credit card company. What is the balance in his account?

Divide; then interpret the remainder in each application. See Example 5.

67. A cellular phone company is offering 1000 free minutes of air time to new subscribers. How many hours of free time will a new subscriber receive?

68. Nikki is catering a large party. If one pie will serve eight guests, how many pies should she make for 100 guests?

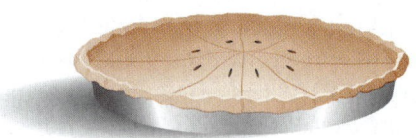

69. Hurricane victims are being given temporary shelter in a hotel. Each room can hold five people. How many rooms are needed for 163 people?

70. A college has received a $250,000 donation to be used for scholarships. How many $3500 scholarships can be given to students?

Simplify:

71. $|^-8| \div {}^-4 \cdot |^-5| \cdot |1|$

72. $^-6 \cdot |^-3| \div |9| \cdot {}^-2$

73. $^-6\,(^-8) \div (^-5 - {}^-5)$

74. $^-9 \div {}^-9\,(^-9 \div 9) \div (12 - 13)$

75. Look back at the opening page of **Section 1.1.** You guessed how many days it would take to receive a million dollars if you got $1 each second. Here's how to use your calculator to get the answer. If you get $1 per second, it would take 1,000,000 seconds to receive $1,000,000. Press these keys.

1000000 $\div$ 60 $=$ 16666.66667 $\div$ 60 $=$ 277.7777778 $\div$ 24 $=$ 11.57407407

There are 60 seconds in one minute.   About 16,667 minutes   There are 60 minutes in one hour.   About 278 hours   There are 24 hours in one day.   About $11\frac{1}{2}$ days (11.5 is equivalent to $11\frac{1}{2}$.)

Notice that you do *not* have to re-enter the intermediate answers. When the answer 16666.66667 appears on your calculator display, just go ahead and enter $\div$ 60.

Now use a *scientific* calculator to find how long it takes to receive a *billion* dollars. Start by entering 1000000000. Then follow the pattern shown above. You will need to do one more division step to get the number of years. (Assume that there are 365 days in one year.)

Summary Exercises on Operations with Integers

Simplify each expression.

1. $2 - 8$

2. $(^-16)(0)$

3. $^-14 - {}^-7$

4. $\dfrac{^-42}{6}$

5. $^-9(^-7)$

6. $\dfrac{^-12}{12}$

7. $(1)(^-56)$

8. $1 + {}^-23$

9. $5 - {}^-7$

10. $^-88 \div {}^-11$

11. $^-18 + 5$

12. $\dfrac{0}{^-10}$

13. $^-40 - {}^-40$

14. $^-17 + 0$

15. $8(^-6)$

16. $^-1 - 9$

17. $^-5(10)$

18. $\dfrac{30}{0}$

19. $0 - 14$

20. $\dfrac{18}{^-3}$

21. $^-13 + 13$

22. $\dfrac{^-16}{^-1}$

23. $20 - 50$

24. $\dfrac{^-7}{0}$

25. $(^-4)(^-6)(2)$

26. $^-2 + {}^-12 + {}^-5$

27. $^-60 \div 10 \div {}^-3$

28. $^-8 - 4 - 8$

29. $64(0) \div {}^-8$

30. $2 - {}^-5 + 9$

31. $^-9 + 8 + {}^-2$

32. $(^-6)(^-2)(^-3)$

33. $8 + 6 + {}^-8$

34. $9 - 0 - 16$

35. $^-25 \div {}^-1 \div {}^-5$

36. $1 - 32 + 0$

37. $^-72 \div {}^-9 \div {}^-4$

38. $^-7 + 28 + {}^-56 + 3$

39. $9 - 6 - 3 - 5$

40. $^-6({}^-8) \div ({}^-5 - 7)$

41. $^-1(9732)({}^-1)({}^-1)$

42. $^-80 \div 4({}^-5)$

43. $^-10 - 4 + 0 + 18$

44. $^-7 \cdot |7| \cdot |{}^-7|$

45. $5 - |{}^-3| + 3$

46. $^-2({}^-3)(7) \div {}^-7$

47. $^-3 - ({}^-2 + 4) - 5$

48. $0 - |{}^-7 + 2|$

49. Describe what happens in each situation.
   (a) Zero is divided by a nonzero number.

   (b) Any number is multiplied by 0.

   (c) A nonzero number is divided by itself.

50. Find, explain, and correct the errors in this student's work.

   (a) $\dfrac{15}{^-15} = 0$

   (b) $\dfrac{8}{0} = 8$

   (c) $^-10 \div {}^-2 \div {}^-5 = 1$

1.8 ▶▶▶ Exponents and Order of Operations

OBJECTIVE 1 Use exponents to write repeated factors. An **exponent** is a quick way to write repeated multiplication. Here is an example.

$$2 \cdot 2 \cdot 2 \cdot 2 \cdot 2 \quad \text{can be written} \quad 2^5 \leftarrow \text{Exponent}$$
$$\uparrow$$
$$\text{Base}$$

The *base* is the number being multiplied over and over, and the exponent tells how many times to use the number as a factor. This is called *exponential notation* or *exponential form*.

To simplify 2^5, actually do the multiplication.

$$2^5 = 2 \cdot 2 \cdot 2 \cdot 2 \cdot 2 = 32$$

Exponential form   Factored form   Simplified form

Here are some more examples, using 2 as the base.

$2 = 2^1$	is read	"2 to the **first power**."	
$2 \cdot 2 = 2^2$	is read	"2 to the **second power**" or, more commonly, "2 **squared**."	
$2 \cdot 2 \cdot 2 = 2^3$	is read	"2 to the **third power**" or, more commonly, "2 **cubed**."	
$2 \cdot 2 \cdot 2 \cdot 2 = 2^4$	is read	"2 to the **fourth power**."	
$2 \cdot 2 \cdot 2 \cdot 2 \cdot 2 = 2^5$	is read	"2 to the **fifth power**."	

and so on.

We usually don't write an exponent of 1, so if no exponent is shown, you can assume that it is 1. For example, 6 is actually 6^1, and 4 is actually 4^1.

> **Note**
> Exponents can also be negative numbers or 0, for example, 2^{-3} and 2^0. You will learn more about these exponents in **Chapter 13.**

EXAMPLE 1 Using Exponents

Complete this table

	Exponential Form	Factored Form	Simplified	Read as
(a)		$5 \cdot 5 \cdot 5$		
(b)		$(4)(4)$		
(c)		7		

Work Problem **1** *at the Side.* ▶

OBJECTIVE 2 Simplify expressions containing exponents.
Exponents are also used with signed numbers, as shown below.

$$(^-3)^2 = (^-3)(^-3) = 9 \qquad \text{The factors have the same sign, so the product is positive.}$$

$$(^-4)^3 = (^-4)(^-4)(^-4) \qquad \text{Multiply two numbers at a time.}$$
$$\underbrace{16} \quad (^-4) \qquad \text{First, } (^-4)(^-4) \text{ is positive 16.}$$
$$^-64 \qquad \text{Then, } 16(^-4) \text{ is } ^-64.$$

OBJECTIVES

1 Use exponents to write repeated factors.

2 Simplify expressions containing exponents.

3 Use the order of operations.

4 Simplify expressions with fraction bars.

---

1 Write each multiplication using exponents. Indicate how to read the exponential form.

(a) $3 \cdot 3 \cdot 3 \cdot 3$

(b) $6 \cdot 6$

(c) 9

(d) $(2)(2)(2)(2)(2)(2)$

ANSWERS

1. **(a)** 3^4 is read "3 to the fourth power."
   (b) 6^2 is read "6 squared" or "6 to the second power."
   (c) 9^1 is read "9 to the first power."
   (d) 2^6 is read "2 to the sixth power."

2 Simplify.

(a) $(^-2)^3$

(b) $(^-6)^2$

(c) $2^4(^-3)^2$

(d) $3^3(^-4)^2$

Simplify exponents before you do other multiplications, as shown below. Notice that the exponent applies only to the *first* thing to its *left*.

Exponent applies only to the 2. $2^3(5)(4^2)$ Exponent applies only to the 4.

(2)(2)(2) is 8. → $8(5)(16)$ ← (4)(4) is 16.

$40(16)$

640

EXAMPLE 2 Using Exponents with Negative Numbers

Simplify.

(a) $(^-5)^2 = (^-5)(^-5) = 25$

(b) $(^-5)^3 = (^-5)(^-5)(^-5)$

> Be careful! Work with two factors at a time and watch the signs.

$25(^-5)$

$^-125$

(c) $(^-2)^4 = (^-2)(^-2)(^-2)(^-2) = 16$

(d) $2^3(^-3)^2 = (2)(2)(2)(^-3)(^-3)$

$(8)(9)$

72

🖩 **Calculator Tip** On a *scientific* calculator, use the exponent key y^x to enter exponents. To enter 5^8, press the following keys.

5 y^x 8 = Answer is 390,625.

↑ Base   ↑ Exponent

Be careful when using your calculator's exponent key with a negative number, such as $(^-5)^3$. Different calculators use different keystrokes, so check the instruction manual, or experiment to see how your calculator works.

◀ *Work Problem* **2** *at the Side.*

OBJECTIVE 3 Use the order of operations. In **Sections 1.4** and **1.7,** you worked examples that mixed addition and subtraction or mixed multiplication and division. In those situations, you worked from left to right. Example 3 below is a review.

EXAMPLE 3 Working from Left to Right

Simplify.

(a) $^-8 - ^-6 + ^-11$ Do additions and subtractions from left to right.

$^-2 \quad + ^-11$

$^-13$

Continued on Next Page

ANSWERS

2. **(a)** $^-2 \cdot ^-2 \cdot ^-2 = ^-8$
   (b) $^-6 \cdot ^-6 = 36$
   (c) $16 \cdot 9 = 144$
   (d) $27 \cdot 16 = 432$

(b) $^-15 \div {}^-3\,(6)$ Do multiplications and divisions from left to right.

$$\to 5\,(6)$$

$$30$$

Work Problem **3** *at the Side.* ▶

Now we're ready to do problems that use a mix of the four operations, parentheses, and exponents. Let's start with a simple example: $4 + 2 \cdot 3$.

If we work from left to right	If we multiply first
$4 + \underbrace{2 \cdot 3}$	$4 + \underbrace{2 \cdot 3}$
$\underbrace{6 \quad \cdot 3}$	$4 + \quad 6$
18	**10**

← **Which answer is correct?** →

To be sure that everyone gets the same answer to a problem like this, mathematicians have agreed to do things in a certain order. The following order of operations shows that multiplying is done ahead of adding, so *the correct answer is 10.*

> ### Order of Operations
>
> *Step 1* Work inside *parentheses* or *other grouping symbols.*
>
> *Step 2* Simplify expressions with *exponents.*
>
> *Step 3* Do the remaining *multiplications and divisions* as they occur from left to right.
>
> *Step 4* Do the remaining *additions and subtractions* as they occur from left to right.

>  **Calculator Tip**   Enter the example above in your calculator.
>
> $$4 \;\oplus\; 2 \;\otimes\; 3 \;\ominus$$
>
> Which answer do you get? If you have a scientific calculator, it automatically uses the order of operations and multiplies first to get the correct answer of 10. Some standard, four-function calculators may *not* have the order of operations built into them and will give the *incorrect* answer of 18.

3 Simplify.

(a) $^-9 + {}^-15 - 3$

(b) $^-4 - 2 + {}^-6$

(c) $3\,(^-4) \div {}^-6$

(d) $^-18 \div 9\,(^-4)$

4 Simplify.

(a) $8 + 6(14 \div 2)$

(b) $4(1) + 8(9 - 2)$

(c) $3(5 + 1) + 20 \div 4$

5 Simplify.

(a) $2 + 40 \div (^-5 + 3)$

(b) $^-5(5) - (15 + 5)$

(c) $(^-24 \div 2) + (15 - 3)$

(d) $^-3(2 - 8) - 5(4 - 3)$

(e) $3(3) - (10 \cdot 3) \div 5$

(f) $6 - (2 + 7) \div (^-4 + 1)$

EXAMPLE 4 **Using the Order of Operations with Whole Numbers**

Simplify.

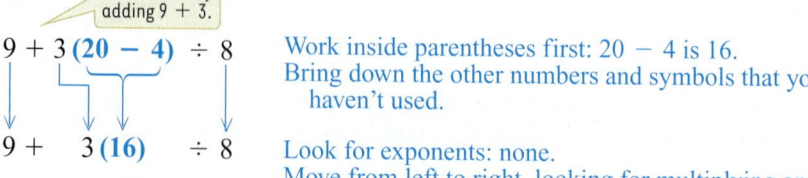

$9 + 3\,(20 - 4) \div 8$ Work inside parentheses first: $20 - 4$ is 16.
Bring down the other numbers and symbols that you haven't used.

$9 + \quad 3\,(16) \quad \div 8$ Look for exponents: none.
Move from left to right, looking for multiplying and dividing.

$9 + \quad 3\,(16) \quad \div 8$ Yes, here is multiplying: $3\,(16)$ is 48.

$9 + \quad 48 \quad \div 8$ Here is dividing: $48 \div 8$ is 6. There is no other multiplying or dividing, so look for adding and subtracting.

$9 + \qquad 6$ Add last: $9 + 6$ is 15.

15

◀ *Work Problem* **4** *at the Side.*

EXAMPLE 5 **Using the Order of Operations with Integers**

Simplify.

(a) $^-8 \div (7 - 5) - 9$ Work inside parentheses first: $7 - 5$ is 2.
Bring down the other numbers and symbols that you haven't used.
Look for exponents: none.

$^-8 \div \quad (2) \quad - 9$ Move from left to right, looking for multiplying and dividing.

$^-8 \div \quad (2) \quad - 9$ Here is dividing: $^-8 \div 2$ is $^-4$. No other multiplying or dividing, so look for adding and subtracting.

$^-4 \qquad - \quad 9$ Change subtracting to adding. Change 9 to its opposite.

$^-4 \qquad + \quad ^-9$ Add $^-4 + ^-9$.

$^-13$

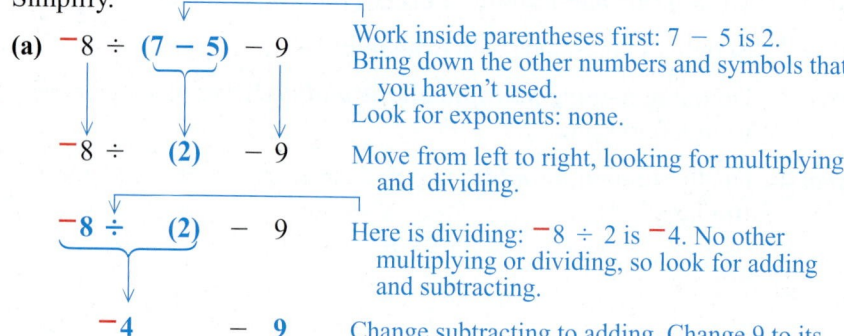

(b) $3 + 2\,(6 - 8) \cdot (15 \div 3)$ Work inside first set of parentheses.
Change $6 - 8$ to $6 + ^-8$ to get $^-2$.

$3 + 2\,(^-2) \cdot (15 \div 3)$ Work inside second set of parentheses: $15 \div 3$ is 5.

$3 + 2\,(^-2) \cdot 5$ Multiply and divide from left to right. First multiply $2\,(^-2)$ to get $^-4$.

$3 + ^-4 \cdot 5$ Then multiply $^-4 \cdot 5$ to get $^-20$.

$3 + \qquad ^-20$ Add last: $3 + ^-20$ is $^-17$.

$^-17$

◀ *Work Problem* **5** *at the Side.*

EXAMPLE 6 **Using the Order of Operations with Exponents**

Simplify.

(a) $4^2 - (^-3)^2$ The only parentheses are around $^-3$, but there
is no work to do inside these parentheses.

$4^2 - (^-3)^2$ Simplify the exponents: $4^2 = (4)(4) = 16$, and
$(^-3)^2 = (^-3)(^-3) = 9$.

$16 - 9$ There is no multiplying or dividing, so add and
subtract: $16 - 9$ is 7.

7

(b) $(^-4)^3 - (\mathbf{4 - 6})^2 (^-3)$ Work inside parentheses: $4 - 6$ becomes
$4 + {}^-6$, which is $^-2$.

$(^-4)^3 - (^-2)^2 (^-3)$ Simplify the exponents next:
$(^-4)^3$ is $(^-4)(^-4)(^-4) = {}^-64$, and
$(^-2)^2$ is $(^-2)(^-2) = 4$.

$^-64 - 4(^-3)$ Look for multiplying and dividing.
Multiply $4(^-3)$ to get $^-12$.

$^-64 - {}^-12$ Change subtraction to addition.
Change $^-12$ to its opposite.

$^-64 + {}^+12$ Add: $^-64 + 12$ is $^-52$.

$^-52$

CAUTION

To help in remembering the order of operations, you may have memorized the letters **PEMDAS,** or the phrase "Please Excuse My Dear Aunt Sally."

Please   **E**xcuse   **M**y   **D**ear   **A**unt   **S**ally

Parentheses; **E**xponents; **M**ultiply and **D**ivide; **A**dd and **S**ubtract

Be careful! Do **not** automatically do all multiplication before division. Multiplying and dividing are done *from left to right* (after parentheses and exponents).

Work Problem **6** *at the Side.* ▶

---

6 Simplify.

(a) $2^3 - 3^2$

(b) $6^2 \div (^-4)(^-3)$

(c) $(^-4)^2 - 3^2(5 - 2)$

(d) $(^-3)^3 + (3 - 9)^2$

7 Simplify.

(a) $\dfrac{-3(2^3)}{-10 - 6 + 8}$

(b) $\dfrac{(-10)(-5)}{-6 \div 3(5)}$

(c) $\dfrac{6 + 18 \div (-2)}{(1 - 10) \div 3}$

(d) $\dfrac{6^2 - 3^2(4)}{5 + (3 - 7)^2}$

OBJECTIVE 4 Simplify expressions with fraction bars. A fraction bar indicates division, as in $\dfrac{-6}{2}$, which means $-6 \div 2$. In the expression

$$\dfrac{-5 + 3^2}{16 - 7(2)}$$

the fraction bar also acts as a grouping symbol, like parentheses. It tells us to do the work in the numerator (above the bar) and then the work in the denominator (below the bar). The last step is to divide the results.

$$\dfrac{-5 + 3^2}{16 - 7(2)} \longrightarrow \dfrac{-5 + 9}{16 - 14} \longrightarrow \dfrac{4}{2} \longrightarrow 4 \div 2 = 2$$

The final result is 2.

EXAMPLE 7 Using the Order of Operations with Fraction Bars

Simplify $\dfrac{-8 + 5(4 - 6)}{4 - 4^2 \div 8}$.

First do the work in the numerator.

> Do **not** start by adding $-8 + 5$.

$-8 + 5(4 - 6)$ Work inside the parentheses.

$-8 + 5(-2)$ Multiply.

$-8 + (-10)$ Add.

Numerator $\longrightarrow -18$

Now do the work in the denominator.

$4 - 4^2 \div 8$ There are no parentheses; simplify the exponent.

$4 - 16 \div 8$ Divide.

$4 - 2$ Subtract.

Denominator $\longrightarrow 2$

The last step is the division.

$$\text{Numerator} \longrightarrow \dfrac{-18}{2} \longleftarrow \text{Denominator} = -9$$

◀ *Work Problem* **7** *at the Side.*

1.8 ▶▶▶ Exercises

Complete this table. See Example 1.

	Exponential Form	Factored Form	Simplified	Read as
1.	4^3		64	
2.	10^2		100	
3.		$2 \cdot 2 \cdot 2 \cdot 2 \cdot 2 \cdot 2 \cdot 2$		
4.		$3 \cdot 3 \cdot 3 \cdot 3 \cdot 3$		
5.		$5 \cdot 5 \cdot 5 \cdot 5$		
6.		$2 \cdot 2 \cdot 2 \cdot 2 \cdot 2 \cdot 2$		
7.				7 squared
8.				6 cubed
9.				10 to the first power
10.				4 to the fourth power

Simplify. See Examples 1 and 2.

11. (a) 10^1 **12. (a)** 5^1 **13. (a)** 4^1 **14. (a)** 3^1

    (b) 10^2 **(b)** 5^2 **(b)** 4^2 **(b)** 3^2

    (c) 10^3 **(c)** 5^3 **(c)** 4^3 **(c)** 3^3

    (d) 10^4 **(d)** 5^4 **(d)** 4^4 **(d)** 3^4

15. 5^{10} **16.** 4^9 **17.** 2^{12} **18.** 3^{10}

19. $(^-2)^2$ **20.** $(^-4)^2$ **21.** $(^-5)^2$ **22.** $(^-10)^2$

23. $(^-4)^3$ **24.** $(^-2)^3$ **25.** $(^-3)^4$ **26.** $(^-2)^4$

27. $(^-10)^3$ **28.** $(^-5)^3$ **29.** 1^4 **30.** 1^5

31. $3^3 \cdot 2^2$

32. $4^2 \cdot 5^2$

33. $2^3 (^-5)^2$

34. $3^2 (^-2)^2$

35. $6^1 (^-5)^3$

36. $7^1 (^-4)^3$

37. $(^-2)(^-2)^4$

38. $^-6(^-6)^2$

39. Simplify.

$(^-2)^2 =$ _____ $(^-2)^6 =$ _____

$(^-2)^3 =$ _____ $(^-2)^7 =$ _____

$(^-2)^4 =$ _____ $(^-2)^8 =$ _____

$(^-2)^5 =$ _____ $(^-2)^9 =$ _____

(a) Describe the pattern you see in the signs of the answers.

(b) What would be the sign of $(^-2)^{15}$ and the sign of $(^-2)^{24}$?

40. Explain why it is important to have rules for the order of operations. Why do you think our "natural instinct" is to just work from left to right?

Simplify. See Examples 3–7.

41. $12 \div 6(^-3)$

42. $10 - 30 \div 2$

43. $^-1 + 15 - 7 - 7$

44. $9 + {}^-5 + 2(^-2)$

45. $10 - 7^2$

46. $5 - 5^2$

47. $2 - {}^-5 + 3^2$

48. $6 - {}^-9 + 2^3$

49. $3 + 5(6 - 2)$

50. $4 + 3(8 - 3)$

◐ 51. $^-7 + 6(8 - 14)$

52. $^-3 + 5(9 - 12)$

53. $2(^-3 + 5) - (9 - 12)$

54. $3(2 - 7) - (^-5 + 1)$

55. $^-5(7 - 13) \div {}^-10$

56. $^-4(9 - 17) \div {}^-8$

57. $9 \div (^-3)^2 + {}^-1$

58. $^-48 \div (^-4)^2 + 3$

59. $2 - {}^-5(^-2)^3$

60. $1 - {}^-10(^-3)^3$

61. $^-2(^-7) + 3(9)$

62. $4(^-2) + {}^-3(^-5)$

63. $30 \div {}^-5 - 36 \div {}^-9$

64. $8 \div {}^-4 - 42 \div {}^-7$

65. $2(5) - 3(4) + 5(3)$

66. $9(3) - 6(4) + 3(7)$

67. $4(3^2) + 7(3 + 9) - {}^-6$

68. $5(4^2) - 6(1 + 4) - {}^-3$

69. $(^-4)^2 \cdot (7 - 9)^2 \div 2^3$

70. $(^-5)^2 \cdot (9 - 17)^2 \div (^-10)^2$

71. $\dfrac{^-1 + 5^2 - {}^-3}{^-6 - 9 + 12}$

72. $\dfrac{^-6 + 3^2 - {}^-7}{7 - 9 - 3}$

73. $\dfrac{^-2\,(4^2) - 4\,(6-2)}{^-4\,(8-13) \div \, ^-5}$

74. $\dfrac{3\,(3^2) - 5\,(9-2)}{8\,(6-9) \div \, ^-3}$

75. $\dfrac{2^3 \cdot (^-2 - 5) + 4\,(^-1)}{4 + 5\,(^-6 \cdot 2) + (5 \cdot 11)}$

76. $\dfrac{3^3 + 4\,(^-1 - 2) - 25}{^-4 + 4\,(3 \cdot 5) + (^-6 \cdot 9)}$

77. $5^2\,(9-11)\,(^-3)\,(^-3)^3$

78. $4^2\,(13-17)\,(^-2)\,(^-2)^3$

79. $|\,^-12\,| \div 4 + 2 \cdot |\,(^-2)^3\,| \div 4$

80. $6 - |\,2 - 3 \cdot 4\,| + (^-5)^2 \div 5^2$

81. $\dfrac{^-9 + 18 \div \, ^-3\,(^-6)}{32 - 4\,(12) \div 3\,(2)}$

82. $\dfrac{^-20 - 15\,(^-4) - \, ^-40}{14 + 27 \div 3\,(^-2) - \, ^-4}$

Study Skills

▶▶▶ TAKING LECTURE NOTES

S tudy the set of sample math notes in this section, and read the comments about them. Then try to incorporate the techniques into your own math note taking in class.

OBJECTIVES

1 Apply note taking strategies, such as writing problems as well as explanations.

2 Use appropriate abbreviations in notes.

▶ The **date and title** of the day's lecture topic are always at the top of every page. **Always begin a new day with a new page.**

▶ Note the **definitions** of base and exponent are written in parentheses—don't trust your memory!

▶ **Skipping lines** makes the notes easier to read.

▶ See how the **direction word** (*simplify*) is emphasized and explained.

▶ A **star marks an important concept.** This is a warning to avoid future mistakes. **Note the underlining**, too, which highlights the importance.

▶ Notice the two columns, which allow for the example and its explanation to be close together. **Whenever you know you'll be given a series of steps to follow, try the two-column method.**

▶ Note the **brackets and arrows**, which clearly show how the problem is set up to be simplified.

January 2 — *Exponents*

Exponents used to show repeated multiplication.

$3 \cdot 3 \cdot 3 \cdot 3$ can be written 3^4 ← exponent (how many times it's multiplied)

base (the number being multiplied)

Read 3^2 as 3 to the 2nd power or 3 squared

3^3 as 3 to the 3rd power or 3 cubed

3^4 as 3 to the 4th power

etc.

Simplifying an expression with exponents → actually do the repeated multiplication

2^3 means $2 \cdot 2 \cdot 2$ and $2 \cdot 2 \cdot 2 = 8$

★ Careful! [5^2 means $5 \cdot 5$ <u>NOT</u> $5 \cdot 2$

so $5^2 = 5 \cdot 5 = 25$ BUT $5^2 \neq 10$]

Example	*Explanation*
Simplify $(2^4) \cdot (3^2)$	Exponents mean <u>multiplication</u>.
$2 \cdot 2 \cdot 2 \cdot 2 \cdot 3 \cdot 3$	Use 2 as a factor 4 times. Use 3 as a factor 2 times. $2 \cdot 2 \cdot 2 \cdot 2$ is 16 $3 \cdot 3$ is 9 16 · 9 is 144
16 · 9	
144	simplified result is 144 (no exponents left)

Why Are These Notes Brain Friendly?

The notes are **easy to look at,** and you know that the brain responds to things that are visually pleasing. Other techniques that are visually memorable are the use of spacing (the two columns), stars, underlining, and circling. All of these methods **allow your brain to take note of important concepts and steps.**

The notes are also **systematic,** which means that they use certain techniques regularly. This way, your brain easily recognizes the topic of the day, the signals that show an important point, and the steps to follow for procedures. When you develop a system that you always use in your notes, your notes are easy to understand later when you are reviewing for a test.

Find one or two people in your math class to work with. Compare each other's lecture notes over a period of a week or so. Ask yourself the following questions as you examine the notes.

1. What are you doing in your notes to show the **main points** or larger concepts? (Such as underlining, boxing, using stars, capital letters, etc.)

2. In what ways do you **set off the explanations** for worked problems, examples, or smaller ideas (subpoints)? (Such as indenting, using arrows, circling or boxing)

3. What does **your instructor do** to show that he or she is moving from one idea to the next? (Such as saying "Next" or "Any questions," "Now," or erasing the board, etc.)

4. **How do you mark** that in your notes? (Such as skipping lines, using dashes or numbers, etc.)

5. What **explanations (in words) do you give yourself** in your notes, so when those new dendrites you grew in lecture are fading, you can read your notes and still remember the new concepts later when you try to do your homework?

6. What **did you learn** by examining your classmates' notes?

   - _____
   - _____
   - _____

7. What **will you try** in your own note taking? List **four** techniques that you will use next time you take notes in math class.

   - _____
   - _____
   - _____
   - _____

Chapter 1 ▶▶▶ Summary

▶ Key Terms

1.1 **place value system** — A place value system is a number system in which the location, or place, where a digit is written gives it a different value.

digits — The 10 digits in our number system are 0, 1, 2, 3, 4, 5, 6, 7, 8, and 9.

whole numbers — The whole numbers are 0, 1, 2, 3, and so on.

1.2 **number line** — A number line is like a thermometer turned sideways. It is used to show how numbers relate to each other.

integers — Integers are the whole numbers and their opposites.

absolute value — The absolute value of a number is its distance from 0 on the number line. Absolute value is indicated by two vertical bars and is always positive (or 0) but never negative.

1.3 **addends** — In an addition problem, the numbers being added are called addends.

sum — The answer to an addition problem is called the sum.

addition property of 0 — Adding 0 to any number leaves the number unchanged.

commutative property of addition — Changing the *order* of two addends does not change the sum.

associative property of addition — Changing the *grouping* of addends does not change the sum.

1.4 **opposite** — The opposite of a number is the same distance from 0 on the number line but on the opposite side of 0. It is also called the *additive inverse* because a number plus its opposite equals 0.

1.5 **rounding** — Rounding a number means finding a number that is close to the original number but easier to work with.

estimate — Use rounded numbers to get an approximate answer, or estimate.

front end rounding — Front end rounding is rounding numbers to the highest possible place, so all the digits become 0 except the first digit.

1.6 **factors** — In a multiplication problem, the numbers being multiplied are called factors.

product — The answer to a multiplication problem is called the product.

multiplication property of 0 — Multiplying any number by 0 gives a product of 0.

multiplication property of 1 — Multiplying a number by 1 leaves the number unchanged.

commutative property of multiplication — Changing the *order* of two factors does not change the product.

associative property of multiplication — Changing the *grouping* of factors does not change the product.

distributive property — Multiplication distributes over addition. For example, $3(6 + 2) = 3 \cdot 6 + 3 \cdot 2$.

1.7 **quotient** — The answer to a division problem is called the quotient.

1.8 **exponent** — An exponent tells how many times a number is used as a factor in repeated multiplication.

▶ New Symbols

$<$	is less than
$>$	is greater than
$\lvert 6 \rvert$ and $\lvert {}^-2 \rvert$	absolute value of 6; absolute value of ${}^-2$
$\approx$	approximately equal to

Exponent

2^5 and 10^3 exponential form

Base

▶ Test Your Word Power

See how well you have learned the vocabulary in this chapter. Answers follow the Quick Review.

1. In (4) ($^-$6) = $^-$24, the 4 and $^-$6 are called
  A. products
  B. factors
  C. addends
  D. opposites.

2. A list of all the **whole numbers** is
  A. 1, 2, 3, 4, ...
  B. ..., $^-$4, $^-$3, $^-$2, $^-$1, 0, 1, 2, 3, ...
  C. 0, 1, 2, 3, 4, 5, 6, 7, 8, 9
  D. 0, 1, 2, 3, 4, ...

3. The **absolute value** of a number is
  A. its distance from 0 on the number line
  B. never positive
  C. always negative
  D. equal to 0.

4. An **exponent**
  A. tells how many times a number is added
  B. is the number being multiplied
  C. tells how many times a number is used as a factor
  D. applies only to the first thing to its right.

5. The **opposite** of a number is
  A. never negative
  B. called the additive inverse
  C. called the absolute value
  D. at the same point on the number line.

6. **Front end rounding** is rounding numbers
  A. to the nearest thousand
  B. so there are no zeros
  C. to the highest possible place
  D. so they become integers.

7. A list of all the **integers** is
  A. 1, 2, 3, 4, ...
  B. ..., $^-$4, $^-$3, $^-$2, $^-$1, 0, 1, 2, 3, ...
  C. 0, 1, 2, 3, 4, 5, 6, 7, 8, 9
  D. 0, 1, 2, 3, 4, ...

8. The **associative property of multiplication** says that
  A. changing the order of two factors does not change the product
  B. multiplication distributes over addition
  C. multiplying a number by 0 is undefined
  D. changing the grouping of factors does not change the product.

▶ Quick Review

| Concepts | Examples |

1.1 Reading and Writing Whole Numbers

Do not use the word *and* when reading whole numbers. Commas separate groups of three digits. The first few group names are ones, thousands, millions, billions, trillions.

Write 3, 008, 160 in words.

three million, eight thousand, one hundred sixty

Write this number using digits:
twenty **billion**, sixty-five **thousand**, eighteen.

$$\underbrace{2\ 0}_{\text{billions}},\underbrace{0\ 0\ 0}_{\text{millions}},\underbrace{0\ 6\ 5}_{\text{thousands}},\underbrace{0\ 1\ 8}_{\text{ones}}$$

1.2 Graphing Signed Numbers

Place a dot at the correct location on the number line.

Graph: **(a)** $^-$4 **(b)** 0 **(c)** $^-$1 **(d)** $\frac{1}{2}$ **(e)** 2

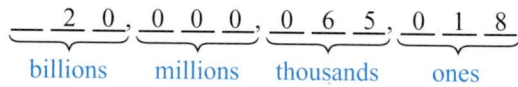

1.2 Comparing Integers

When comparing two integers, the one that is farther to the left on the number line is less than the other.
Use the $<$ symbol for "is less than" and the $>$ symbol for "is greater than."

Write $<$ or $>$ between each pair of numbers to make a true statement.

$$^-3 < {}^-2$$
$^-$3 is less than $^-$2
because $^-$3 is to the *left* of $^-$2 on the number line.

$$0 > {}^-4$$
0 is greater than $^-$4
because 0 is to the *right* of $^-$4 on the number line.

Concepts	Examples

1.2 Finding the Absolute Value of a Number

Find the distance on the number line from 0 to the number. The absolute value is always positive (or 0) but never negative.

Find each absolute value.

$|^-5| = 5$ because $^-5$ is 5 steps away from 0 on the number line.

$|3| = 3$ because 3 is 3 steps away from 0 on the number line.

1.3 Adding Two Integers

When both integers have the *same sign,* add the absolute values and use the common sign as the sign of the sum.

When the integers have *different signs,* subtract the smaller absolute value from the larger absolute value. Use the sign of the number with the larger absolute value as the sign of the sum.

Add.

(a) $^-6 + {}^-7$

Add the absolute values.

$|^-6| = 6$ and $|^-7| = 7$

Add $6 + 7 = 13$ and use the common sign as the sign of the sum: $^-6 + {}^-7 = {}^-13$.

(b) $^-10 + 4$

Subtract the smaller absolute value from the larger.

$|^-10| = 10$ and $|4| = 4$

Subtract $10 - 4 = 6$; the number with the larger absolute value is negative, so the sum is negative: $^-10 + 4 = {}^-6$.

1.3 Using Properties of Addition

Addition Property of 0: Adding 0 to any number leaves the number unchanged.

Commutative Property of Addition: Changing the *order* of two addends does not change the sum.

Associative Property of Addition: Changing the *grouping* of addends does not change the sum.

Name the property illustrated by each case.

(a) $^-16 + 0 = {}^-16$

(b) $4 + 10 = 10 + 4$ Both sums are 14.

(c) $2 + (^-6 + 1) = (2 + {}^-6) + 1$ Both sums are $^-3$.

(a) Addition property of 0

(b) Commutative property of addition

(c) Associative property of addition

1.4 Subtracting Two Integers

To subtract two numbers, add the first number to the opposite of the second number.

Step 1   Make one pencil stroke to change the subtraction symbol to an addition symbol.

Step 2   Make a second pencil stroke to change the *second number to its opposite.*

Subtract.

(a) $7 - {}^-2$ Change subtraction to addition.
↓ ↓ ↓   Change $^-2$ to its opposite, $^+2$.
$7 + {}^+2$
9

(b) $^-9 - 12$ Change subtraction to addition.
↓ ↓ ↓   Change 12 to its opposite, $^-12$.
$^-9 + {}^-12$
$^-21$

Concepts	Examples

1.5 Rounding Integers

Step 1   Draw a line under the place to which the number is to be rounded.

Step 2   Look only at the next digit to the right of the underlined place. If the next digit is *5 or more,* increase the underlined digit by 1. If the next digit is *4 or less,* do not change the digit in the underlined place.

Step 3   Change all digits to the right of the underlined place to zeros.

Round 36,833 to the nearest thousand.

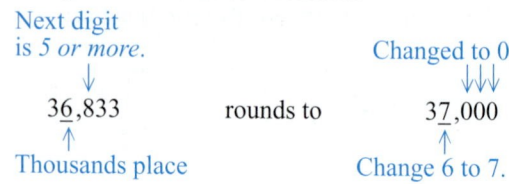

Round ⁻3582 to the nearest ten.

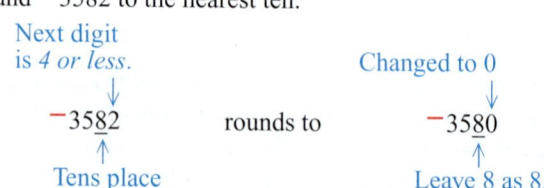

1.5 Front End Rounding

Round to the highest possible place so that all the digits become 0 except the first digit.

Use front end rounding.

Next digit
is *5 or more.*   Changed to 0

9̲7,203   rounds to   1̲00,000

Change 9 to 10. Regroup 1
into hundred-thousands place.

1.5 Estimating Answers in Addition and Subtraction

Use front end rounding to round the numbers in a problem. Then add or subtract the rounded numbers to estimate the answer.

First use front end rounding to estimate the answer. Then find the exact answer.

The temperature was 48 degrees below zero. During the morning it rose 21 degrees. What was the new temperature?

Estimate: ⁻50 + 20 = ⁻30 degrees

Exact: ⁻48 + 21 = ⁻27 degrees

1.6 Multiplying Two Integers

If two factors have *different signs,* the product is *negative.*

If two factors have the *same sign,* the product is *positive.*

Multiply.

(a)   ⁻5 (6) = ⁻30   The factors have *different* signs, so the product is *negative.*

(b)   (⁻10) (⁻2) = 20   The factors have the *same* sign, so the product is *positive.*

1.6 Using Properties of Multiplication

Multiplication property of 0: Multiplying any number by 0 gives a product of 0.

Multiplication property of 1: Multiplying any number by 1 leaves the number unchanged.

Commutative property of multiplication: Changing the *order* of two factors does not change the product.

Associative property of multiplication: Changing the *grouping* of factors does not change the product.

Distributive property: Multiplication distributes over addition.

Name the property illustrated by each case.

(a)   ⁻49 • 0 = 0

(b)   1 (675) = 675

(c)   (⁻8) (2) = (2) (⁻8)   Both products are ⁻16.

(d)   (⁻3 • ⁻2) • 4 = ⁻3 • (⁻2 • 4)   Both products are 24.

(e)   5 (2 + 4) = 5 • 2 + 5 • 4   Both results are 30.

(a)   Multiplication property of 0
(b)   Multiplication property of 1
(c)   Commutative property of multiplication
(d)   Associative property of multiplication
(e)   Distributive property

Concepts	Examples

1.6 Estimating Answers in Multiplication

First use front end rounding. Then multiply the rounded numbers using a shortcut: Multiply the nonzero digits in each factor; count the total number of zeros in the two factors and write that number of zeros in the product.

First use front end rounding to estimate the answer. Then find the exact answer.

At a PTA fund-raiser, Lionel sold 96 photo albums at $22 each. How much money did he take in?

Estimate: $100 \cdot \$20 = \$2000$

Exact: $96 \cdot \$22 = \$2112$

1.7 Dividing Two Integers

Use the same sign rules for dividing two integers as for multiplying two integers.

If two numbers have *different signs,* the quotient is *negative.*

If two numbers have the *same sign,* the quotient is *positive.*

Divide.

$\dfrac{^-24}{6} = {}^-4$ Numbers have *different* signs, so the quotient is *negative.*

$^-72 \div {}^-8 = 9$ Numbers have the *same* sign, so the quotient is *positive.*

$\dfrac{50}{^-5} = {}^-10$ Numbers have *different* signs, so the quotient is *negative.*

1.7 Using Properties of Division

(a) When a nonzero number is divided by itself, the quotient is 1.

(b) When a number is divided by 1, the quotient is the number.

(c) When 0 is divided by any other number (except 0), the quotient is 0.

(d) Division by 0 is *undefined.* There is no answer.

State the property illustrated by each case.

(a) $\dfrac{^-4}{^-4} = 1$ **(b)** $\dfrac{65}{1} = 65$

(c) $\dfrac{0}{9} = 0$ **(d)** $\dfrac{^-10}{0}$ is undefined.

The examples are in the same order as the properties listed at the left. Note that division is *not* commutative nor associative.

1.7 Estimating Answers in Division

First use front end rounding. Then divide the rounded numbers, using a shortcut: Drop the same number of zeros in both the divisor and the dividend.

First use front end rounding to estimate the answer. Then find the exact answer.

Joan has one year to pay off a $1020 loan. What is her monthly payment?

Estimate: $\$1000 \div 10 = \$100$

Exact: $\$1020 \div 12 = \$85$

1.7 Interpreting Remainders in Division

In some situations the remainder tells you how much is left over. In other situations, you must increase the quotient by 1 in order to accommodate the "left over."

Divide; then interpret the remainder.

Each chemistry student needs 35 milliliters of acid for an experiment. How many students can be served from a bottle holding 500 milliliters of acid?

$$
\begin{array}{r}
14 \\
35\overline{)500} \\
35 \\
\hline
150 \\
140 \\
\hline
10
\end{array}
$$

14 → 14 students served

10 → 10 milliliters of acid left over

Concepts	Examples

1.8 Using Exponents

An exponent tells how many times a number is used as a factor in repeated multiplication. An exponent applies only to its base (the first thing to the left of the exponent).

Simplify.

Exponent
↓

(a) $2^5 = 2 \cdot 2 \cdot 2 \cdot 2 \cdot 2 = 32$

(b) $(^-3)^2 = (^-3)(^-3) = 9$

1.8 Order of Operations

Mathematicians have agreed to follow this order.

Step 1 Work inside parentheses or other grouping symbols.

Step 2 Simplify expressions with exponents.

Step 3 Do the remaining multiplications and divisions as they occur from left to right.

Step 4 Do the remaining additions and subtractions as they occur from left to right.

Simplify.

$$(^-2)^4 + 3(^-4 - ^-2) \qquad \text{Work inside parentheses.}$$
$$(^-2)^4 + \quad 3(^-2) \qquad \text{Simplify exponents.}$$
$$16 + \quad 3(^-2) \qquad \text{Multiply.}$$
$$16 + \quad ^-6 \qquad \text{Add.}$$
$$10$$

1.8 Using the Order of Operations with Fraction Bars

When there is a fraction bar, do all the work in the numerator. Then do all the work in the denominator. Finally, divide numerator by denominator.

Simplify.

$$\frac{^-10 + 4^2 - 6}{2 + 3(1 - 4)} = \frac{^-10 + 16 - 6}{2 + 3(-3)} = \frac{0}{^-7} = 0$$

ANSWERS TO TEST YOUR WORD POWER

1. B; *Example:* In $7(10) = 70$, the factors are 7 and 10.

2. D; *Example:* 12, 0, 710, and 89,475 are all whole numbers.

3. A; *Example:* $|^-3| = 3$ and $|3| = 3$ because both $^-3$ and 3 are 3 steps away from 0 on the number line.

4. C; *Example:* In 2^5, the exponent is 5, which indicates that 2 is used as a factor 5 times, so $2^5 = 2 \cdot 2 \cdot 2 \cdot 2 \cdot 2 = 32$.

5. B; *Example:* The opposite of 5 is $^-5$; it is the additive inverse because $5 + (^-5) = 0$.

6. C; *Example:* Using front end rounding, round 48,299 to the highest possible place, which is ten-thousands. So, 48,299 rounds to 50,000. All digits are 0 except the first digit.

7. B; *Example:* $^-10$, 0, and 6 are all integers.

8. D; *Example:* $4 \cdot (^-2 \cdot 3)$ can be rewritten as $(4 \cdot ^-2) \cdot 3$; both products are $^-24$.

Study Skills

REVIEWING A CHAPTER

This activity is really about **preparing for tests.** Some of the suggestions are ideas that you will learn to use a little later in the term, but get started trying them out now. Often, the first chapter in your math textbook will be review, so it is good to practice some of the study techniques on material that is not too challenging.

Use these **chapter reviewing techniques.**

▶ **Make a study card for each vocabulary word and concept.** Include a definition, an example, a sketch, and a page reference. Include the symbol or formula if there is one. See the *Using Study Cards* activity for a quick look at some sample study cards.

▶ **Go back to the section** to find more explanations or information about any new vocabulary, formulas, or symbols.

▶ **Use the Chapter Summary** to practice each type of problem. Do not expect the Summary to substitute for reading and working through the whole chapter! First, take the "Test Your Word Power" quiz to check your understanding of new vocabulary. The answers are at the end of the Quick Review. Then read the Quick Review. **Pay special attention to the red headings.** Check the explanations for the solutions to problems given. Try to think about the **whole chapter.**

▶ **Study your lecture notes** to see what your instructor has emphasized in class. Then review that material in your text.

▶ **Do the Review Exercises.**
  ✓ Check your answers **after** you're done with each **section of exercises.**
  ✓ If you get stuck on a problem, **first** check the Chapter Summary. If that doesn't clear up your confusion, then check the section and your lecture notes.
  ✓ Pay attention to **direction words** for the problems, such as *simplify, round, solve,* and *estimate.*
  ✓ Make **study cards for especially difficult problems.**

▶ **Do the Mixed Review exercises.** This is a good check to see if you can still do the problems when they are in mixed-up order. **Check your answers carefully** in the Answers section in the back of your book. Are your answers **exact** and **complete?** Make sure you are **labeling** answers correctly, using the right **units.** For example, does your answer need to include *$, cm², ft,* and so on?

OBJECTIVES

1 Use the Chapter Summary to practice every type of problem.

2 Create study cards for vocabulary.

3 Practice by doing review and mixed review exercises.

4 Take the Chapter Test as a practice test.

Chapter Reviewing Techniques

You have already become familiar with the features of your textbook. This activity requires you to make good use of them. Your **brain needs repetition** to strengthen dendrites and the connections between them. By following the steps outlined here, you will be reinforcing the concepts, procedures, and skills you need to use for tests (and for the next chapters).

The combination of techniques provides repetition in different ways. That **promotes good branching of dendrites** instead of just relying on one branch or route to connect to the other dendrites. A thorough review of each chapter will **solidify your dendrite connections.** It will help you be sure that you understand the concepts **completely and accurately.** Also, taking the Chapter Test will **simulate the testing situation**, which gives you practice in test taking conditions.

Now Try This ▶▶▶

▶ **Take the Chapter Test as if it is a real test.** If your instructor has skipped sections in the chapter, figure out which problems to skip on the test before you start.

✓ **Time yourself** just as you would for a real test.

✓ **Use a calculator or notes** just as you would be permitted to (or not) on a real test.

✓ **Take the test in one sitting,** just like a real test is given in one sitting.

✓ **Show all your work.** Practice showing your work just the way your instructor has asked you to show it.

✓ **Practice neatness.** Can someone else follow your steps?

✓ **Check your answers** in the back of the book.

Notice that reviewing a chapter will take some time. Remember that it takes time for dendrites to grow! You cannot grow a good network of dendrites by rushing through a review in one night. But if you use the suggestions over a few days or evenings, you will notice that you understand the material more thoroughly and remember it longer.

Follow the reviewing techniques listed above for your next test. For each technique, write a comment *about how it worked for you in the spaces below.*

1. **Make a study card for each vocabulary word and concept.**

2. **Go back to the section** to find more explanations or information.

3. **Take the Test Your Word Power quiz and use the Quick Review** to review each concept in the chapter.

4. **Study your lecture notes** to see what your instructor has emphasized in class.

5. **Do the Review Exercises,** following the specific suggestions on the previous page.

6. **Do the Mixed Review exercises.**

7. **Take the Chapter Test** as if it is a real test.

Chapter 1 ▶▶▶ Review Exercises

If you need help with any of these Review Exercises, look in the section indicated in the red brackets.

[1.1] **1.** Circle the whole numbers: 86 2.831 $^-4$ 0 $\frac{2}{3}$ 35,600

Write these numbers in words.

2. 806

3. 319,012

4. 60,003,200

5. 15,749,000,000,006

Write these numbers using digits.

6. Five hundred four thousand, one hundred

7. Six hundred twenty million, eighty thousand

8. Ninety-nine billion, seven million, three hundred fifty-six

[1.2] **9.** Graph these numbers: $^-3\frac{1}{2}, 2, ^-5, 0.$

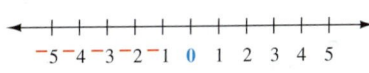

Write < or > between each pair of numbers to make a true statement.

10. 0 ___ $^-4$

11. $^-3$ ___ $^-1$

12. 2 ___ $^-2$

13. $^-2$ ___ 1

Find each absolute value.

14. $|^-5|$

15. $|9|$

16. $|0|$

17. $|^-125|$

[1.3] *Add.*

18. $^-9 + 8$

19. $^-8 + ^-5$

20. $16 + ^-19$

21. $^-4 + 4$

22. $6 + ^-5$

23. $^-12 + ^-12$

24. $0 + ^-7$

25. $^-16 + 19$

26. $9 + ^-4 + ^-8 + 3$

27. $^-11 + ^-7 + 5 + ^-4$

[1.4] *Find the opposite (additive inverse) of each number. Show that the sum of the number and its opposite is 0.*

28. $^-5$

29. 18

Subtract by changing subtraction to addition.

30. $5 - 12$

31. $24 - 7$

32. $^-12 - 4$

33. $4 - ^-9$

34. $^-12 - ^-30$

35. $^-8 - 14$

36. $^-6 - ^-6$

37. $^-10 - 10$

38. $^-8 - ^-7$

39. $0 - 3$

40. $1 - ^-13$

41. $15 - 0$

Simplify.

42. $3 - 12 - 7$

43. $^-7 - ^-3 + 7$

44. $4 + ^-2 - 0 - 10$

45. $^-12 - 12 + 20 - ^-4$

[1.5] *Round each number as indicated.*

46. 205 to the nearest ten

47. 59,499 to the nearest thousand

48. 85,066,000 to the nearest million

49. $^-2963$ to the nearest hundred

50. $^-7,063,885$ to the nearest ten-thousand

51. 399,712 to the nearest thousand

Use front end rounding to round each number.

52. The combined weight loss of ten dieters was 197 pounds.

53. The land on the shore of the Dead Sea in the Middle East is 1312 feet below sea level. (*Source: Goode's World Atlas.*)

54. There are 362,000,000 Yellow Pages directories published in the United States each year. (*Source: Yellow Pages Publishers Association.*)

55. Within forty years, the Census Bureau predicts a world population of 9,402,000,000 people. (*Source: U.S. Census Bureau.*)

[1.6] *Multiply.*

56. $^-6\,(9)$

57. $(^-7)\,(^-8)$

58. $10\,(^-10)$

59. $^-45 \cdot 0$

60. $^-1\,(^-24)$

61. $17 \cdot 1$

62. $4\,(^-12)$

63. $(^-5)\,(^-25)$

64. $^-3\,(^-4)\,(^-3)$

65. $^-5\,(2)\,(^-5)$

66. $(^-8)\,(^-1)\,(^-9)$

[1.7] *Simplify.*

67. $\dfrac{^-63}{^-7}$

68. $\dfrac{70}{^-10}$

69. $\dfrac{^-15}{0}$

70. $^-100 \div {}^-20$

71. $18 \div {}^-1$

72. $\dfrac{0}{12}$

73. $\dfrac{^-30}{^-2}$

74. $\dfrac{^-35}{35}$

75. $^-40 \div {}^-4 \div {}^-2$

76. $^-18 \div 3\,(^-3)$

77. $0 \div {}^-10\,(5) \div 5$

78. Divide; then interpret the remainder. It took 1250 hours to build the set for a new play. How many working days of eight hours each did it take to build the set?

[1.8] *Simplify.*

79. 10^4

80. 2^5

81. 3^3

82. $(^-4)^2$

83. $(^-5)^3$

84. 8^1

85. $6^2 \cdot 3^2$

86. $5^2\,(^-2)^3$

87. $^-30 \div 6 - 4\,(5)$

88. $6 + 8\,(2 - 3)$

89. $16 \div 4^2 + (^-6 + 9)^2$

90. $^-3\,(4) - 2\,(5) + 3\,(^-2)$

91. $\dfrac{^-10 + 3^2 - {}^-9}{3 - 10 - 1}$

92. $\dfrac{^-1\,(1 - 3)^3 + 12 \div 4}{^-5 + 24 \div 8 \cdot 2\,(6 - 6) + 5}$

▶▶▶ Mixed Review Exercises

Name the property illustrated by each case.

93. $^-3 + (5 + 1) = (^-3 + 5) + 1$　**94.** $^-7(2) = 2(^-7)$　　　**95.** $0 + 19 = 19$

96. $^-42 \cdot 0 = 0$　　　　**97.** $2(^-6 + 4) = 2 \cdot {}^-6 + 2 \cdot 4$　**98.** $(^-6 \cdot 3) \cdot {}^-1 = {}^-6 \cdot (3 \cdot {}^-1)$

First use front end rounding to estimate each answer. Then find the exact answer.

99. Last year, 192 Elvis jukeboxes were sold at a price of $11,900 each. What was the total value of the jukeboxes?

Estimate:

Exact:

100. Chad had $185 in his checking account. He deposited his $428 paycheck and then wrote a $706 check for car repairs. What is the balance in his account?

Estimate:

Exact:

101. Georgia's hybrid car used 22 gallons of gas on her 880-mile vacation trip. What was the average number of miles she drove on each gallon of gas?

Estimate:

Exact:

102. When inventory was taken at Mathtronic Company, 19 calculators and 12 computer modems were missing. Each calculator is worth $39 and each modem is worth $85. What is the total value of the missing items?

Estimate:

Exact:

Elena Sanchez opened a shop that does alterations and designs custom clothing. Use the table of her income and expenses to answer Exercises 103–106.

Month	Income	Expenses	Profit or Loss
Jan.	$2400	$^-$3100	
Feb.	$1900	$^-$2000	
Mar.	$2500	$^-$1800	
Apr.	$2300	$^-$1400	
May	$1600	$^-$1600	
June	$1900	$^-$1200	

103. Complete the table by finding Elena's profit or loss for each month.

104. Which month had the greatest loss? Which month had the greatest profit?

105. What was Elena's average monthly income?

106. What was the average monthly amount of expenses?

1. Write this number in words: 20,008,307

2. Write this number using digits:
thirty billion, seven hundred thousand, five

3. Graph the numbers 3, $^-2$, 0, $-\dfrac{1}{2}$ on the number line at the right.

4. Write $<$ or $>$ between each pair of numbers to make a true statement.

$$0 \rule{1cm}{0.4pt} {}^-3 \qquad\qquad {}^-2 \rule{1cm}{0.4pt} {}^-1$$

5. Find each absolute value: $|10|$ and $|{}^-14|$.

Add, subtract, multiply, or divide.

6. $3 - 9$

7. $^-12 + 7$

8. $\dfrac{-28}{-4}$

9. $^-1(40)$

10. $^-5 - {}^-15$

11. $({}^-8)({}^-8)$

12. $^-25 + {}^-25$

13. $\dfrac{17}{0}$

14. $^-30 - 30$

15. $\dfrac{50}{-10}$

16. $5({}^-9)$

17. $0 - {}^-6$

Simplify.

18. $^-35 \div 7({}^-5)$

19. $^-15 - {}^-8 + 7$

20. $3 - 7({}^-2) - 8$

21. $({}^-4)^2 \cdot 2^3$

22. $\dfrac{5^2 - 3^2}{(4)({}^-2)}$

23. $^-2({}^-4 + 10) + 5(4)$

24. $^-3 + ({}^-7 - {}^-10) + 4(6 - 10)$

25. Explain how an exponent is used. Include two examples, one using a positive number as the base and one using a negative number as the base.

1. _____

2. _____

3.

   <-- + + + + + + + -->
   $^-3$ $^-2$ $^-1$ **0** 1 2 3

4. _____

5. _____

6. _____

7. _____

8. _____

9. _____

10. _____

11. _____

12. _____

13. _____

14. _____

15. _____

16. _____

17. _____

18. _____

19. _____

20. _____

21. _____

22. _____

23. _____

24. _____

25. _____

26. _____

26. Explain the commutative property of addition and the associative property of addition. Also give an example to illustrate each property.

Round each number as indicated.

27. _____

27. 851 to the nearest hundred

28. _____

28. 36,420,498,725 to the nearest million

29. _____

29. 349,812 to the nearest thousand

First use front end rounding to estimate the answer to each application problem. Then find the exact answer.

30. *Estimate:* _____

   Exact: _____

30. Lorene had $184 in her checking account. She deposited her $293 paycheck and then wrote a $506 check for tuition. What is the balance in her account?

31. *Estimate:* _____

   Exact: _____

31. The Cardinals football team had a bad year. It lost a total of 1140 yards in 12 games. What was the average loss in each game?

32. *Estimate:* _____

   Exact: _____

32. One kind of cereal has 220 calories in each serving. Another kind has 110 calories in each serving. During a month with 31 days, how many calories would you save by eating the second kind of cereal each morning for breakfast? (*Source:* General Mills and Post Cereals.)

Write and solve a subtraction problem to answer this question.

33. _____

33. On the planet Mars the average high temperature is ⁻10 °F and the average low is ⁻100 °F. What is the difference between the high and low temperatures? (*Source:* NASA.)

Divide; then interpret the remainder.

34. _____

34. Anthony has a part-time job as a shipping clerk. He is sending 1276 mathematics books to a college bookstore. Each shipping carton can safely hold 48 books. What is the minimum number of cartons he will need?

Study Skills

▶▶▶ MANAGING YOUR TIME

Many college students find themselves juggling a difficult schedule and multiple responsibilities. Perhaps you are going to school, working part time, and managing family demands. Here are some tips to help you develop good time management skills and habits.

▶ **Read the syllabus for each class.** Check on class policies, such as attendance, late homework, and make-up tests. Find out how you are graded. Keep the syllabus in your notebook.

▶ **Make a semester or quarter calendar.** Put test dates and major due dates for *all* your classes on the same calendar. That way you will see which weeks are the really busy ones. Try using a different color pen for each class. Your brain responds well to the use of color. A semester calendar is on the next page.

▶ **Make a weekly schedule.** After you fill in your classes and other regular responsibilities (such as work, picking up kids from school, etc.), block off some study periods during the day that you can guarantee you will use for studying. Aim for 2 hours of study for each 1 hour you are in class.

▶ **Make "To Do" lists.** Then use them by crossing off the tasks as you complete them. You might even number them in the order they need to be done (most important ones first).

▶ **Break big assignments into smaller chunks.** They won't seem so big that way. Make deadlines for each small part so you stay on schedule.

▶ **Give yourself small breaks in your studying.** Do not try to study for hours at a time! Your brain needs rest between periods of learning. Try to give yourself a 10 minute break each hour or so. You will learn more and remember it longer.

▶ **If you get off schedule, just try to get back on schedule tomorrow.** We all slip from time to time. All is not lost! Make a new "To Do" list and start doing the most important things first.

▶ **Get help when you need it.** Talk with your instructor during office hours. Also, most colleges have some kind of Learning Center, tutoring center, or counseling office. If you feel lost and overwhelmed, ask for help. Someone can help you decide what to do first and what to spend your time on right away.

> **What two or three of the suggestions above will you try this week? How do you think they will help you?**
>
> 1. _____
>
> 2. _____
>
> 3. _____

OBJECTIVES

1 Create a semester schedule.

2 Create a "to do" list.

Why Are These Techniques Brain Friendly?

Your brain appreciates some order. It enjoys a little routine, for example, choosing the same study time and place each day. You will find that you quickly settle in to your reading or homework.

Also, your brain **functions better when you are calm.** Too much rushing around at the last minute to get your homework and studying done sends hostile chemicals to your brain and makes it more difficult for you to learn and remember. So, a little planning can really pay off.

Building rest into your schedule is good for your brain. Remember, it takes time for dendrites to grow.

We've suggested using color on your calendars. This too, is brain friendly. Remember, your brain **likes pleasant colors and visual material** that are nice to look at. Messy and hard to read calendars will not be helpful, and you probably won't look at them often.

SEMESTER CALENDAR

WEEK	MON	TUES	WED	THUR	FRI	SAT	SUN
1							
2							
3							
4							
5							
6							
7							
8							
9							
10							
11							
12							
13							
14							
15							
16							

Understanding Variables and Solving Equations

As many as 50,000 thunderstorms occur worldwide in a single day. Thunderstorms always include lightning, and in the United States an average of 300 people are struck by lightning each year. (*Source: AccuWeather.com*) But you don't need a weather forecaster to tell you how far away a thunderstorm is. You can use the algebraic expression in **Section 2.1,** Exercises 59–60, to estimate your distance from the storm and, if it is close, move to shelter.

There are other algebraic expressions related to weather. For example, in summertime, you can find the temperature by counting the chirps of a cricket and using the expression in **Section 2.1,** Exercise 14.

2

2.1 ▶▶▶ Introduction to Variables

OBJECTIVE 1 Identify variables, constants, and expressions.
You probably know that algebra uses letters, especially the letter x. But why use letters when numbers are easier to understand? Here is an example.

Suppose that you run your college bookstore. When deciding how many books to order for a certain class, you first find out the class limit, that is, the maximum number of students allowed in the class. You will need at least that many books. But you decide to order 5 extra copies for emergencies.

Rule for ordering books: Order the class limit + 5 extra

How many books would you order for a prealgebra class with a limit of 25 students?

Class limit ⌐   ⌐ Extra

$$25 + 5 \qquad \text{You would order 30 prealgebra books.}$$

How many books would you order for a geometry class that allows 40 students to register?

Class limit ⌐   ⌐ Extra

$$40 + 5 \qquad \text{You would order 45 geometry books.}$$

You could set up a table to keep track of the number of books to order for various classes.

Class	Rule for Ordering Books: Class Limit + 5 Extra	Number of Books to Order
Prealgebra	25 + 5	30
Geometry	40 + 5	45
College algebra	35 + 5	40
Calculus 1	50 + 5	55

A shorthand way to write your rule is shown below.

$$c + 5$$
↑
The c stands for class limit.

You can't write your rule by using just numbers because the class limit varies, or changes, depending on which class you're talking about. So you use a letter, called a **variable,** to represent the part of the rule that varies. Notice the similarity in the words *varies* and *variable*. When part of a rule does *not* change, it is called a **constant.**

The variable, or the part of
the rule that varies or changes
↓
$$c + 5$$
↑
The constant, or the part of
the rule that does *not* change

$c + 5$ is called an **expression.** It expresses (tells) the rule for ordering books. You could use any letter you like for the variable part of the expression, such as $x + 5$, or $n + 5$, and so on. But one suggestion is to use a letter that reminds you of what it stands for. In this situation, the letter c reminds us of "class limit."

EXAMPLE 1   Writing an Expression and Identifying the Variable and Constant

Write an expression for this rule. Identify the variable and the constant.

Order the class limit minus 10 books because some students will buy used books.

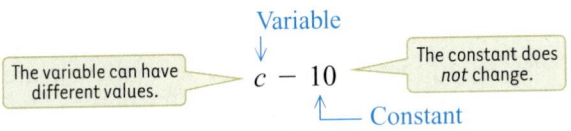

Variable

The variable can have different values. → $c - 10$ ← The constant does *not* change.

Constant

Work Problem **1** *at the Side.* ▶

1 Write an expression for this rule. Identify the variable and the constant.

Order the class limit plus 15 extra books because it is a very large class.

OBJECTIVE 2 Evaluate variable expressions for given replacement values.   When you need to figure out how many books to order for a particular class, you use a specific value for the class limit, like 25 students in prealgebra. Then you **evaluate the expression,** that is, you follow the rule.

Ordering Books for a Prealgebra Class

$c + 5$ Expression (rule for ordering books) is $c + 5$.
↓ Replace c with 25, the class limit for prealgebra.

$25 + 5$ Follow the rule. Add $25 + 5$.

30 Order 30 prealgebra books.

EXAMPLE 2   Evaluating an Expression

Use this rule for ordering books: Order the class limit minus 10. The expression is $c - 10$.

(a) Evaluate the expression when the class limit is 32.

$c - 10$ Replace c with 32.
↓

$32 - 10$ Follow the rule. Subtract to find $32 - 10$.

22 Order 22 books.

(b) Evaluate the expression when the class limit is 48.

$c - 10$ Replace c with 48.
↓

$48 - 10$ Follow the rule. Subtract to find $48 - 10$.

38 Order 38 books.

Work Problem **2** *at the Side.* ▶

2 Use this expression for ordering books: $c + 3$.

(a) Evaluate the expression when the class limit is 25.

(b) Evaluate the expression when the class limit is 60.

In any career you choose, there will be many useful "rules" that need to be written using variables (letters) because part of the rule changes depending on the situation. This is one reason algebra is such a powerful tool. Here is another example.

Suppose that you work in a landscaping business. You are putting a fence around a square-shaped garden. Each side of the garden is 6 feet long. How much fencing material should you bring to finish the job? You could add the lengths of the four sides.

←— 6 feet —→

6 feet 6 feet

←— 6 feet —→

$$6 \text{ feet} + 6 \text{ feet} + 6 \text{ feet} + 6 \text{ feet} = 24 \text{ feet of fencing}$$

3 **(a)** Evaluate the expression 4*s* when the length of one side of a square table is 3 feet.

Or, recall that multiplication is a quick way to do repeated addition. The square garden has 4 sides, so multiply by 4.

$$4 \cdot 6 \text{ feet} = 24 \text{ feet of fencing}$$

So the rule for calculating the amount of fencing for a square garden is:

$$4 \cdot \text{length of one side}$$

Other jobs may require fencing for larger or smaller square shapes. The following table shows how much fencing you will need.

Length of One Side of Square Shape	Expression (Rule) to Find Total Amount of Fencing: 4 • Length of One Side	Total Amount of Fencing Needed
6 feet	4 • 6 feet	24 feet
9 feet	4 • 9 feet	36 feet
10 feet	4 • 10 feet	40 feet
3 feet	4 • 3 feet	12 feet

The expression (rule) can be written in shorthand form as shown below.

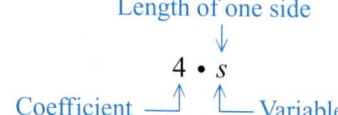

Length of one side

$$4 \cdot s$$

Coefficient ⎯ Variable

The number part in a *multiplication* expression is called the numerical coefficient, or just the **coefficient.** We usually don't write multiplication dots in expressions, so we do the following.

$$4 \cdot s \quad \text{is written as} \quad 4s$$

You can use the expression 4*s* any time you need to know the *perimeter* of a square shape, that is, the total distance around all four sides of the square.

(b) Evaluate the expression 4*s* when the length of one side of a square park is 7 miles.

> **CAUTION**
> If an expression involves adding, subtracting, or dividing, then you *do* have to write +, −, or ÷ . It is *only* multiplication that is understood without writing an operation symbol.
>
> $4 + s$ $4 − s$ $4 ÷ s$ $4s$
>
> Add *s*.   Subtract *s*.   Divide by *s*.   Multiply by *s*.

EXAMPLE 3 **Evaluating an Expression with Multiplication**

The expression (rule) for finding the perimeter of a square shape is 4*s*. Evaluate the expression when the length of one side of a square parking lot is 30 yards.

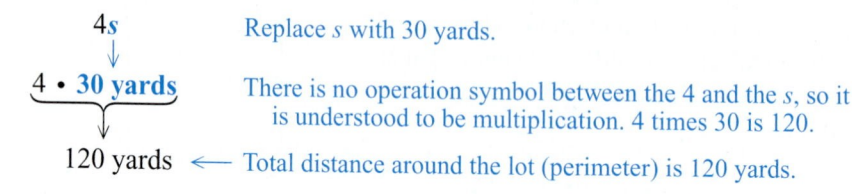

4*s*   Replace *s* with 30 yards.

4 • **30 yards**   There is no operation symbol between the 4 and the *s*, so it is understood to be multiplication. 4 times 30 is 120.

120 yards ⟵ Total distance around the lot (perimeter) is 120 yards.

◀ Work Problem **3** at the Side.

Some expressions (rules) involve several different steps. An expression for finding the approximate systolic blood pressure of a person of a certain age is shown below.

$$100 + \frac{a}{2} \quad \leftarrow \text{Age of person (the variable)}$$

Remember that a fraction bar means division, so $\frac{a}{2}$ is the person's age divided by 2. You also need to remember the order of operations, which means doing division before addition.

EXAMPLE 4 **Evaluating an Expression with Several Steps**

Evaluate the expression $100 + \frac{a}{2}$ when the age of the person is 24.

$$100 + \frac{a}{2} \quad \text{Replace } a \text{ with 24, the age of the person.}$$

$$100 + \frac{24}{2} \quad \text{Follow the rule using the order of operations. First divide: } 24 \div 2 \text{ is 12.}$$

$$100 + 12 \quad \text{Now add: } 100 + 12 \text{ is 112.}$$

$$112 \quad \leftarrow \text{The approximate systolic blood pressure is 112.}$$

Work Problem **4** *at the Side.* ▶

🖩 **Calculator Tip** If you like to fish, you can use an expression (rule) like the one below to find the approximate weight (in pounds) of a fish you catch. Measure the length of the fish (in inches) and then use the correct expression for that type of fish. For a northern pike, the weight expression is shown below.

Variable (length of fish) ⟶

$$\frac{l^3}{3600}$$

where l is the length of the fish in inches. (*Source: InFisherman.*)

To evaluate this expression for a fish that is 43 inches long, follow the rule by calculating as follows.

$$\frac{43^3}{3600} \quad \text{Replace } l \text{ with 43, the length of the fish in inches.}$$

In the numerator, you can multiply 43 • 43 • 43 or use the y^x key on your calculator. Then divide by 3600.

Enter 43 y^x 3 ÷ 3600 =   Calculator shows 22.08527778

Base Exponent

The fish weighs about 22 pounds.

Now use the expression to find the approximate weight of a northern pike that is 37 inches long. (Answer: about 14 pounds.)

(continued)

4 Evaluate the expression $100 + \frac{a}{2}$ when the age of the person is 40.

ANSWER

4. $100 + \frac{40}{2}$ ← Replace a with 40.

$100 + 20$

120 ← Approximate systolic blood pressure is 120.

5 **(a)** Use the expression for finding your average bowling score. Evaluate the expression if your total score for 4 games is 532.

Notice that variables are used on your calculator keys. On the y^x key, y represents the base and x represents the exponent. You first evaluated y^x by entering 43 as the base and 3 as the exponent for the first fish. Then you evaluated y^x again by entering 37 as the base and 3 as the exponent for the second fish.

Some expressions (rules) involve several variables. For example, if you go bowling and want to know your average score, you can use this expression.

$$\underline{t} \;\leftarrow\; \text{Total score for all games (variable)}$$
$$\underline{g} \;\leftarrow\; \text{Number of games (variable)}$$

EXAMPLE 5   **Evaluating Expressions with Two Variables**

(a) Find your average score if you bowl three games and your total score for all three games is 378.
  Use the expression (rule) for finding your average score.

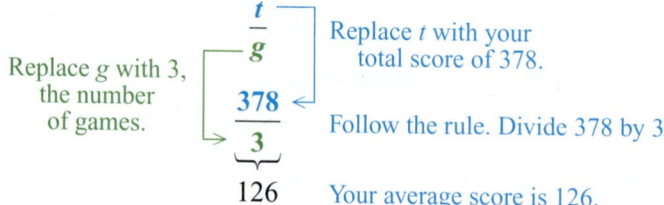

Replace g with 3, the number of games.

$\dfrac{t}{g}$ Replace t with your total score of 378.

$\dfrac{378}{3}$ Follow the rule. Divide 378 by 3.

126   Your average score is 126.

(b) Complete this table.

Value of x	Value of y	Expression $x - y$
16	10	16 − 10 is 6
100	5	
3	7	
8	0	

(b) Complete these tables to show how to evaluate each expression.

Value of x	Value of y	Expression (Rule) $x + y$
2	5	2 + 5 is 7
−6	4	__ + __ is __
0	16	__ + __ is __

Value of x	Value of y	Expression (Rule) xy
2	5	2 • 5 is 10
−6	4	__ • __ is __
0	16	__ • __ is __

The expression (rule) is to *add* the two variables. So the completed table is:

Value of x	Value of y	Expression (Rule) $x + y$
2	5	2 + 5 is 7
−6	4	−6 + 4 is −2
0	16	0 + 16 is 16

The expression (rule) is to *multiply* the two variables. We know that it's multiplication because there is no operation symbol between the x and y. So the completed table is:

Value of x	Value of y	Expression (Rule) xy
2	5	2 • 5 is 10
−6	4	−6 • 4 is −24
0	16	0 • 16 is 0

ANSWERS
5. **(a)** $\dfrac{532}{4}$; average score is 133.
   (b) 100 − 5 is 95
      3 − 7 is −4
      8 − 0 is 8

◀ *Work Problem* **5** *at the Side.*

OBJECTIVE **3** **Write properties of operations using variables.**
Now you can use variables as a shorthand way to express the properties you learned about in **Sections 1.3, 1.6,** and **1.7.** We'll use the letters a and b to represent any two numbers.

Commutative Property of Addition

$$a + b = b + a$$

Commutative Property of Multiplication

$$a \cdot b = b \cdot a$$

$$\text{or,} \quad ab = ba$$

To get specific examples, you can pick values for a and b. For example, if a is $^-3$, replace every a with $^-3$. If b is 5, replace every b with 5.

$a + b = b + a$

$^-3 + 5 = 5 + ^-3$

$2 = 2$

Both sums are 2

$ab = ba$

$^-3 \cdot 5 = 5 \cdot ^-3$

$^-15 = ^-15$

Both products are $^-15$

Of course, you could pick many different values for a and b, because the commutative "rule" will always work for adding *any* two numbers or multiplying *any* two numbers.

EXAMPLE 6 **Writing Properties of Operations Using Variables**

Use the variable b to state this property: When any number is divided by 1, the quotient is the number.

Use the letter b to represent any number.

$$\frac{b}{1} = b$$

Work Problem **6** *at the Side.* ▶

OBJECTIVE **4** **Use exponents with variables.** In **Section 1.8** we used an exponent as a quick way to write repeated multiplication. For example,

$3 \cdot 3 \cdot 3 \cdot 3 \cdot 3$ can be written $3^5 \leftarrow$ Exponent

↑ — Base

3 is used as a factor 5 times.

The meaning of an exponent is the same when a variable is the base.

$c \cdot c \cdot c \cdot c \cdot c$ can be written $c^5 \leftarrow$ Exponent

↑ — Base

c is used as a factor 5 times.

m^2 means $m \cdot m$ Here m is used as a factor 2 times.

$x^4 y^3$ means $x^4 \cdot y^3$ or

$$\underbrace{x \cdot x \cdot x \cdot x}_{x^4} \cdot \underbrace{y \cdot y \cdot y}_{y^3}$$

$7b^2$ means $7 \cdot b \cdot b$ The exponent applies *only* to b.

$^-4xy^2z$ means $^-4 \cdot x \cdot y \cdot y \cdot z$ The exponent applies *only* to y.

---

6 **(a)** Use the variable a to state this property: Multiplying any number by 0 gives a product of 0.

(b) Use the variables a, b, and c to state the associative property of addition: Changing the grouping of addends does not change the sum.

7 Rewrite each expression without exponents.

(a) x^5

(b) $4a^2b^2$

(c) $-10xy^3$

(d) s^4tu^2

EXAMPLE 7 Understanding Exponents Used with Variables

Rewrite each expression without exponents.

(a) y^6 can be written as $y \cdot y \cdot y \cdot y \cdot y \cdot y$
y is used as a factor 6 times.

(b) $12bc^3$ can be written as $12 \cdot b \cdot c \cdot c \cdot c$
Coefficient is 12. The exponent applies *only* to c. c^3

(c) $-2m^2n^4$ can be written as $-2 \cdot m \cdot m \cdot n \cdot n \cdot n \cdot n$
Coefficient is -2. m^2 n^4

◀ Work Problem **7** at the Side.

To evaluate an expression with exponents, multiply all the factors.

EXAMPLE 8 Evaluating Expressions with Exponents

Evaluate each expression.

(a) x^2 when x is -3

x^2 means $x \cdot x$ Replace each x with -3.
$-3 \cdot -3$ Multiply -3 times -3.
9

So x^2 becomes $(-3)^2$, which is $(-3)(-3)$, or 9.

8 Evaluate each expression.

(a) y^3 when y is -5

(b) r^2s^2 when r is 6 and s is 3

(c) $10xy^2$ when x is 4 and y is -3

(d) $-3c^4$ when c is 2

(b) x^3y when x is -4 and y is -10

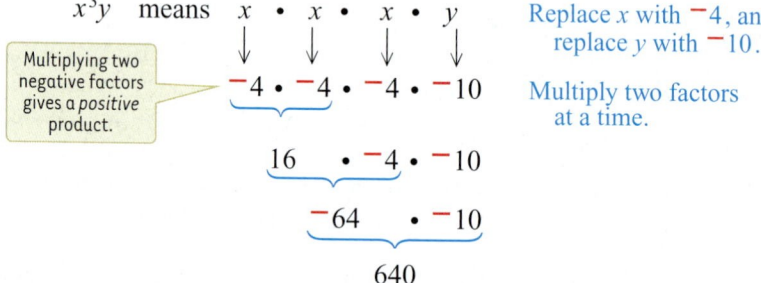

x^3y means $x \cdot x \cdot x \cdot y$ Replace x with -4, and replace y with -10.

Multiplying two negative factors gives a *positive* product.

$-4 \cdot -4 \cdot -4 \cdot -10$ Multiply two factors at a time.

$16 \cdot -4 \cdot -10$

$-64 \cdot -10$

640

So x^3y becomes $(-4)^3(-10)$, which is $(-4)(-4)(-4)(-10)$, or 640.

(c) $-5ab^2$ when a is 5 and b is 3

$-5ab^2$ means $-5 \cdot a \cdot b \cdot b$ Replace a with 5, and replace b with 3.

Multiplying two factors with *different* signs gives a *negative* product.

$-5 \cdot 5 \cdot 3 \cdot 3$ Multiply two factors at a time.

$-25 \cdot 3 \cdot 3$

$-75 \cdot 3$

-225

So $-5ab^2$ becomes $-5(5)(3)^2$, which is $-5(5)(3)(3)$, or -225.

Coefficient is -5.

◀ Work Problem **8** at the Side.

2.1   Exercises

FOR EXTRA HELP

Identify the variable and the constant in each expression. See Example 1.

1. $c + 4$ **2.** $d + 6$ **3.** $^-3 + m$ **4.** $^-4 + n$

Identify the parts of each expression. Choose from these labels: variable, constant, *and* coefficient.

5. $5h$ **6.** $3s$ **7.** $2c - 10$ **8.** $6b - 1$

9. $x - y$ **10.** xy **11.** $^-6g + 9$ **12.** $^-10k + 15$

Evaluate each expression. See Examples 2–5.

13. The expression (rule) for ordering robes for the graduation ceremony at West Community College is $g + 10$, where g is the number of graduates. Evaluate the expression when

    (a) there are 654 graduates.

    (b) there are 208 graduates.

    (c) there are 95 graduates.

14. Crickets chirp faster as the weather gets hotter. The expression (rule) for finding the approximate temperature (in degrees Fahrenheit) is $c + 37$ where c is the number of chirps made in 15 seconds by a field cricket. (*Source:* AccuWeather.com) Evaluate the expression when the cricket

    (a) chirps 45 times.

    (b) chirps 33 times.

    (c) chirps 58 times.

15. The expression for finding the perimeter of a triangle with sides of equal length is $3s$, where s is the length of one side. Evaluate the expression when

    (a) the length of one side is 11 inches.

    (b) the length of one side is 3 feet.

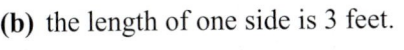

16. The expression for finding the perimeter of a pentagon with sides of equal length is $5s$, where s is the length of one side. Evaluate the expression when

    (a) the length of one side is 25 meters.

    (b) the length of one side is 8 inches.

17. The expression for ordering brushes for an art class is $3c - 5$, where c is the class limit. Evaluate the expression when

(a) the class limit is 12.

(b) the class limit is 16.

18. The expression for ordering doughnuts for the office staff is $2n - 4$, where n is the number of people at work. Evaluate the expression when

(a) there are 13 people at work.

(b) there are 18 people at work.

19. The expression for figuring a student's average test score is $\frac{p}{t}$, where p is the total points earned on all the tests and t is the number of tests. Evaluate the expression when

(a) 332 points were earned on 4 tests.

(b) there were 7 tests and 637 points were earned.

20. The expression for deciding how many buses are needed for a group trip is $\frac{p}{b}$, where p is the total number of people and b is the number of people that one bus will hold. Evaluate the expression when

(a) 176 people are going on a trip and one bus holds 44 people.

(b) a bus holds 36 people and 72 people are going on a trip.

Complete each table by evaluating the expressions. See Example 5.

21.

Value of x	Expression $x + x + x + x$	Expression $4x$
$^{-}2$	$^{-}2 + {}^{-}2 + {}^{-}2 + {}^{-}2$ is $^{-}8$	$4 \cdot {}^{-}2$ is $^{-}8$
12		
0		
$^{-}5$		

22.

Value of y	Expression $3y$	Expression $y + 2y$
$^{-}6$	$3(^{-}6)$ is $^{-}18$	$^{-}6 + 2(^{-}6)$ is $^{-}6 + {}^{-}12$, or $^{-}18$
10		
$^{-}3$		
0		

23.

Value of x	Value of y	Expression $^{-}2x + y$
3	7	$^{-}2(3) + 7$ is $^{-}6 + 7$, or 1
$^{-}4$	5	
$^{-}6$	$^{-}2$	
0	$^{-}8$	

24.

Value of x	Value of y	Expression $^{-}2xy$
3	7	$^{-}2 \cdot 3 \cdot 7$ is $^{-}42$
$^{-}4$	5	
$^{-}6$	$^{-}2$	
0	$^{-}8$	

25. Explain the words *variable* and *expression*.

26. Explain the words *coefficient* and *constant*.

Use the variable b to express each of these properties. See Example 6.

27. Multiplying a number by 1 leaves the number unchanged.

28. Adding 0 to any number leaves the number unchanged.

29. Any number divided by 0 is undefined.

30. Multiplication distributes over addition. (Use a, b, and c as the variables.)

Rewrite each expression without exponents. See Example 7.

31. c^6

32. d^7

33. x^4y^3

34. c^2d^5

35. $-3a^3b$

36. $-8m^2n$

37. $9xy^2$

38. $5ab^4$

39. $-2c^5d$

40. $-4x^3y$

41. a^3bc^2

42. x^2yz^6

Evaluate each expression when r is -3, s is 2, and t is -4. See Example 8.

43. t^2

44. r^2

45. rs^3

46. s^4t

47. $3rs$

48. $6st$

49. $-2s^2t^2$

50. $-4rs^4$

51. $r^2s^5t^3$

52. $r^3s^4t^2$

53. $-10r^5s^7$

54. $-5s^6t^5$

Evaluate each expression when x is 4, y is -2, and z is -6.

55. $|xy| + |xyz|$

56. $x + |y^2| + |xz|$

57. $\dfrac{z^2}{-3y + z}$

58. $\dfrac{y^2}{x + 2y}$

Relating Concepts (Exercises 59–60) For Individual or Group Work

At the beginning of this chapter, you read that there is an expression that tells your approximate distance from a thunderstorm. First, count the number of seconds from the time you see a lightning flash until you start to hear the thunder. Then, to estimate the distance (in miles), use the expression $\dfrac{s}{5}$, where s is the number of seconds. Use this expression as you **work Exercises 59 and 60 in order.** *(This expression is based on the fact that the light from the lightning travels faster than the sound from the thunder.)*

59. Evaluate the thunderstorm expression for each number of seconds. How far away is the storm?

   (a) 15 seconds

   (b) 10 seconds

   (c) 5 seconds

60. Explain how you can use your answers from Exercise 59 to:

   (a) Estimate the distance when the time is $2\frac{1}{2}$ seconds.

   (b) Find the number of seconds when the distance is $1\frac{1}{2}$ miles.

   (c) Find the number of seconds when the distance is $2\frac{1}{2}$ miles.

Study Skills

▶▶▶ USING STUDY CARDS

You may have used "flash cards" in other classes before. In math, study cards can be helpful, too. However, they are different because the main things to remember in math are *not* necessarily terms and definitions; they are *sets of steps to follow* to solve problems (and how to know which set of steps to follow) and *concepts about how math works* (principles). So, the cards will look different but will be just as useful.

In this two-part activity, you will find four types of study cards to use in math. Look carefully at what kinds of information to put on them and where to put it. Then use them the way you would any flash card:

▶ to quickly review when you have a few minutes,

▶ to do daily reviews,

▶ to review before a test.

Remember, the most helpful thing about study cards is making them. It is in the making of them that you have to do the kind of thinking that is most brain friendly and will improve your neural network of dendrites. After each card description you will find an assignment to try. It is marked "**NOW TRY THIS.**"

For **new vocabulary cards**, put the word (spelled correctly) and the page number where it is found on the front of the card. On the back, write:

▶ the definition (in your own words if possible),

▶ an example, an exception (if there are any),

▶ any related words, and

▶ a sample problem (if appropriate).

Front of Card

> variable   p. 94

Back of Card

> Definition: A letter representing a number that varies (changes) depending on the situation.
>
> Example: $c + 5$ where c represents the class limit which varies for different classes.
>   variable / constant   varies ⟷ variable
>
> – Use any letter you like: a, b, c, d, h, k, m, n, o, p, x, y, etc.
> – Related words: constant, coefficient

Now Try This ▶▶▶

List four new vocabulary words/concepts you need to learn right now. Make a card for each one.

_____   _____

_____   _____

Procedure ("Steps") Cards

For **procedure cards,** write the name of the procedure at the top on the front of the card. Then write each step *in words*. If you need to know abbreviations for some words, include them along with the whole words written out. On the back, put an example of the procedure, showing each step you need to take. You can review by looking at the front and practicing a new worked example, or by looking at the back and remembering what the procedure is called and what the steps are.

Evaluating an expression

Step 1: *Replace each variable with the value you are given.*

Step 2: *Do the calculations, following the order of operations.*

Front of Card

Example: Evaluate 3c – 5 when c is 20

must be given the replacement value

3 c – 5
↓
3(20)–5   *Replace c with 20.*
           Do the calculations.

60 – 5   *Follow order of operations; multiply first, then subtract.*

55   *No variables in final answer.*

Back of Card

Now Try This ▶▶▶

What procedure are you learning right now? Make a "steps" card for it.

Procedure: _____

2.2 ▶▶▶ Simplifying Expressions

OBJECTIVE **1** **Combine like terms, using the distributive property.**   In **Section 2.1**, the expression for ordering math textbooks was $c + 5$. This expression was simple and easy to use. Sometimes expressions are *not* written in the simplest possible way. For example:

Evaluate this expression when c is 20.

$c + 5$	Replace c with 20.
↓	
$\underbrace{20 + 5}$	Add $20 + 5$.
25	

Evaluate this expression when c is 20.

$2c - 10 - c + 15$	Replace c with 20.
↓ ↓	
$2 \cdot 20 - 10 - 20 + 15$	Multiply $2 \cdot 20$.
$40 \quad - 10 - 20 + 15$	Change subtraction to adding the opposite.
↓ ↓ ↓ ↓	
$40 \; + {}^-10 + {}^-20 + 15$	Add from left to right.
$\underbrace{30} \qquad + {}^-20 + 15$	
$\underbrace{10} \qquad\quad + 15$	
25	

These two expressions are actually equivalent. When you evaluate them, the final result is the same, but it takes a lot more work when you use the right-hand expression. To save a lot of work, you need to learn how to *simplify expressions*. Then you can rewrite $2c - 10 - c + 15$ in the simplest way possible, which is $c + 5$.

The basic idea in **simplifying expressions** is to *combine*, or *add*, like terms. Each addend in an expression is a **term.** Here are two examples.

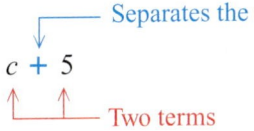

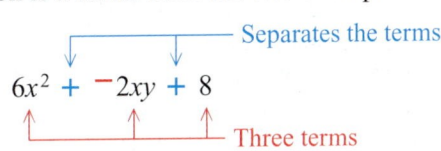

In $6x^2 + {}^-2xy + 8$, the 8 is the *constant term*. There are also two *variable terms* in the expression: $6x^2$ is a variable term, and ${}^-2xy$ is a variable term. A **variable term** has a number part (coefficient) and a letter part (variable).

If no coefficient is shown, it is assumed to be 1. Remember from **Section 1.6** that multiplying any number by 1 does *not* change the number.

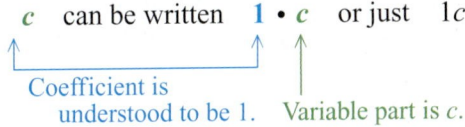

Also, ${}^-c$ can be written ${}^-1 \cdot c$. The coefficient of ${}^-c$ is understood to be ${}^-1$.

Like Terms

Like terms are terms with exactly the same variable parts (the same letters and exponents). The coefficients do *not* have to match.

OBJECTIVES

1 Combine like terms, using the distributive property.

2 Simplify expressions.

3 Use the distributive property to multiply.

1 List the like terms in each expression. Then identify the coefficients of the like terms.

(a) $3b^2 + {}^-3b + 3 + b^3 + b$

Like Terms		Unlike Terms	
$5x$ and $3x$	Variable parts match; both are x.	$3x$ and $3x^2$	Variable parts do *not* match; exponents are different.
${}^-6y^3$ and y^3	Variable parts match; both are y^3.	${}^-2x$ and ${}^-2y$	Variable parts do *not* match; letters are different.
$4a^2b$ and $5a^2b$	Variable parts match; both are a^2b.	a^3b and a^2b	Variable parts do *not* match; exponents are different.
${}^-8$ and 4	There are no variable parts; numbers are like terms.	${}^-8c$ and 4	Variable parts do *not* match; one term has a variable part, but the other term does not.

(b) ${}^-4xy + 4x^2y + {}^-4xy^2 + {}^-4 + 4$

EXAMPLE 1 **Identifying Like Terms and Their Coefficients**

List the like terms in each expression. Then identify the coefficients of the like terms.

(a) ${}^-5x + {}^-5x^2 + 3xy + x + {}^-5$
The like terms are ${}^-5x$ and x.
The coefficient of ${}^-5x$ is ${}^-5$, and the coefficient of x is understood to be 1.

(b) $2yz^2 + 2y^2z + {}^-3y^2z + 2 + {}^-6yz$
The like terms are $2y^2z$ and ${}^-3y^2z$.
The coefficients are 2 and ${}^-3$.

(c) $5r^2 + 2r + {}^-2r^2 + 5 + 5r^3$

(c) $10ab + 12 + {}^-10a + 12b + {}^-6$
The like terms are 12 and ${}^-6$.
The like terms are constants (there are no variable parts).

> Numbers alone (no variable parts) are like terms.

◀ *Work Problem* **1** *at the Side.*

The distributive property (see **Section 1.6**) can be used "in reverse" to combine like terms. Here is an example.

(d) ${}^-10 + {}^-x + {}^-10x + {}^-x^2 + {}^-10y$

$$\underbrace{\mathbf{3} \cdot x}_{3x} + \underbrace{\mathbf{4} \cdot x}_{4x} \quad \text{can be written as} \quad \underbrace{(\mathbf{3 + 4})}_{\underbrace{7}} \cdot x$$
$$7x$$

Thus, $3x + 4x$ can be written in *simplified form* as $7x$. To check that $3x + 4x$ is the same as $7x$, evaluate each expression when x is 2.

$$\begin{array}{cc} \mathbf{3}x + \mathbf{4}x & \mathbf{7}x \\ \downarrow \quad \downarrow & \downarrow \\ 3 \cdot \mathbf{2} + 4 \cdot \mathbf{2} & 7 \cdot \mathbf{2} \\ \underbrace{6}\ +\ \underbrace{8} & 14 \\ 14 \end{array}$$

Both results are 14, so the expressions are equivalent. But you can see how much easier it is to work with $7x$, the simplified expression.

CAUTION
Notice that $3x + 4x$ is simplified to $7x$, **not** to $7x^2$.

Variable part is unchanged.   **Do *not* change x to x^2.**

Combining Like Terms

Step 1 If there are any variable terms with no coefficient, write in the understood 1.

Step 2 If there are any subtractions, change each one to adding the opposite.

Step 3 Find *like* terms (the variable parts match).

Step 4 Add the coefficients (number parts) of like terms. *The variable part stays the same.*

EXAMPLE 2 **Combining Like Terms**

Combine like terms.

(a) $2x + 4x + x$

$2x + 4x + x$ No coefficient; write understood 1.
There are no subtractions to change.

$2x + 4x + 1x$ Find like terms: $2x$, $4x$, and $1x$ are like terms, so add the coefficients, $2 + 4 + 1$.

$(2 + 4 + 1)x$ The variable part, x, stays the same.

$7x$

Therefore, $2x + 4x + x$ can be written as $7x$.

*Do **not** write $7x^3$. Keep x as x.*

(b) $^-3y^2 - 8y^2$

$^-3y^2 - 8y^2$ Both coefficients are shown.
Change subtraction to adding the opposite.

$^-3y^2 + {}^-8y^2$ Find like terms: $^-3y^2$ and $^-8y^2$ are like terms, so add the coefficients, $^-3 + {}^-8$.

$(^-3 + {}^-8)y^2$ The variable part, y^2, stays the same.

$^-11y^2$

Therefore, $^-3y^2 - 8y^2$ can be written as $^-11y^2$.

Work Problem **2** *at the Side.* ▶

OBJECTIVE 2 Simplify expressions. When simplifying expressions, be careful to combine only *like* terms—those having variable parts that match. You *cannot* combine terms if the variable parts are different.

---

2 Combine like terms.

(a) $10b + 4b + 10b$

(b) $y^3 + 8y^3$

(c) $^-7n - n$

(d) $3c - 5c - 4c$

(e) $^-9xy + xy$

(f) $^-4p^2 - 3p^2 + 8p^2$

(g) $ab - ab$

ANSWERS
2. (a) $24b$ (b) $9y^3$ (c) ^-8n (d) ^-6c
(e) ^-8xy (f) $1p^2$, or just p^2
(g) 0, because $1ab + {}^-1ab$ is $(1 + {}^-1)\,ab$, or $0ab$, and 0 times anything is 0.

3 Simplify each expression by combining like terms.

(a) $3b^2 + 4d^2 + 7b^2$

(b) $4a + b - 6a + b$

(c) $^-6x + 5 + 6x + 2$

(d) $2y - 7 - y + 7$

(e) $^-3x - 5 + 12 + 10x$

EXAMPLE 3 Simplifying Expressions

Simplify each expression by combining like terms.

(a) $6xy + 2y + 3xy$

The *like* terms are $6xy$ and $3xy$. We can use the commutative property to rewrite the expression so that the like terms are next to each other. This helps to organize our work.

$6xy + 3xy + 2y$ Combine *like* terms only.

$(6 + 3)xy + 2y$ Add the coefficients, $6 + 3$.
         The variable part, xy, stays the same.

$9xy \quad + 2y$ Keep writing $2y$, the term that was not combined; it is still part of the expression.

The simplified expression is $9xy + 2y$.

(b) Here is the expression from the first page in this section.

$2c - 10 - c + 15$ Write the understood 1 as the coefficient of c.

$2c - 10 - 1c + 15$ Change subtractions to adding the opposite.

$2c + {^-10} + {^-1c} + 15$ We can add in any order, so rewrite the expression so that like terms are next to each other.

$2c + {^-1c} \; + \; {^-10} + 15$ Combine $2c + {^-1c}$. Also combine $^-10 + 15$.

$(2 + {^-1})c + \quad 5$

$1c \quad + \quad 5$

The simplified expression is $1c + 5$ or just $c + 5$

$1c$ is the same as c.

> **Note**
>
> In this book, when combining like terms we will usually write the variable terms in *alphabetical order*. A constant term (number only) will be written last. So, in Examples 3(a) and 3(b) above, the preferred and alternative ways of writing the expressions are as follows.
>
> The simplified expression is $9xy + 2y$ (alphabetical order). However, by the commutative property of addition, $2y + 9xy$ is also correct.
>
> The simplified expression is $c + 5$ (constant written last). However, by the commutative property of addition, $5 + c$ is also correct.

◀ **Work Problem** **3** **at the Side.**

We can use the associative property of multiplication to simplify an expression such as $4(3x)$.

$4(3x)$ can be written as $4 \cdot (3 \cdot x)$

Understood multiplications

Using the associative property, we can regroup the factors.

$4 \cdot (3 \cdot x)$ can be written as $(4 \cdot 3) \cdot x$ To simplify, multiply $4 \cdot 3$.

$\underbrace{12}\ \cdot x$ Write $12 \cdot x$ without the multiplication dot.

$12x$

The simplified expression is $12x$.

EXAMPLE 4 **Simplifying Multiplication Expressions**

Simplify.

(a) $5(10y)$

Use the associative property.

$5 \cdot (10 \cdot y)$ can be written as $(5 \cdot 10) \cdot y$ Multiply $5 \cdot 10$.

$\underbrace{50}\ \cdot y$ Write $50 \cdot y$ without the multiplication dot.

$50y$

So, $5(10y)$ simplifies to $50y$.

(b) $^-6(3b)$

Use the associative property.

$^-6(3b)$ can be written as $(^-6 \cdot 3)b$

^-18b

So, $^-6(3b)$ simplifies to ^-18b.

(c) $^-4(^-2x^2)$

Use the associative property.

$^-4(^-2x^2)$ can be written as $(^-4 \cdot {}^-2)x^2$

$8x^2$

So, $^-4(^-2x^2)$ simplifies to $8x^2$.

Work Problem **4** *at the Side.* ▶

OBJECTIVE 3 Use the distributive property to multiply.   The distributive property can also be used to simplify expressions such as $3(x + 5)$. You *cannot* add the terms inside the parentheses because x and 5 are *not* like terms. But notice the understood multiplication dot between the 3 and the parentheses.

$$3(x + 5)$$
$$\downarrow$$
$$3 \cdot (x + 5)$$

Thus you can *distribute* multiplication over addition, as you did in **Section 1.6.** That is, multiply 3 times each term inside the parentheses.

$3 \cdot (x + 5)$ can be written as $\underbrace{3 \cdot x} + \underbrace{3 \cdot 5}$

$3x\ +\ 15$

So, $3(x + 5)$ simplifies to $3x + 15$.

└─ Stays as addition ─┘

4 Simplify.

(a) $7(4c)$

(b) $^-3(5y^3)$

(c) $20(^-2a)$

(d) $^-10(^-x)$

5 Simplify.

(a) $7(a + 10)$

(b) $3(x - 3)$

(c) $4(2y + 6)$

(d) $^-5(3b + 2)$

(e) $^-8(c + 4)$

Multiplication also distributes over subtraction.

$4(y - 2)$ means $4 \cdot (y - 2)$ and can be written as $\underbrace{4 \cdot y}_{4y} - \underbrace{4 \cdot 2}_{8}$

So, $4(y - 2)$ simplifies to $4y - 8$.
Stays as subtraction

Notice that we did *not* need to change subtraction to adding the opposite.

EXAMPLE 5 **Using the Distributive Property**

Simplify.

(a) $6(y - 4)$ can be written as $6 \cdot y - 6 \cdot 4$
$6y - 24$
Stays as subtraction

Remember to multiply $6 \cdot y$ **and** $6 \cdot 4$

So, $6(y - 4)$ simplifies to $6y - 24$

(b) $5(3x + 2)$ can be written as $5 \cdot 3x + 5 \cdot 2$
$5 \cdot 3 \cdot x + 10$
$\underbrace{15 \cdot x}_{} + 10$
$15x + 10$

So, $5(3x + 2)$ simplifies to $15x + 10$

(c) $^-2(4a + 3)$ can be written as $^-2 \cdot 4a + {}^-2 \cdot 3$
$^-2 \cdot 4 \cdot a + {}^-6$
$\underbrace{^-8}_{} \cdot a + {}^-6$
$^-8a + {}^-6$

Now we will use the definition of subtraction "in reverse" to rewrite $^-8a + {}^-6$.

Change $^-6$ to its opposite, $^+6$.

Write $^-8a + {}^-6$ as $^-8a - 6$

Change addition to subtraction.

Think back to the way we changed subtraction to adding the opposite. Here we are "working backward." From now on, whenever addition is followed by a negative number, we will change it to subtracting a positive number.

$^-8a + {}^-6$
$\updownarrow \quad \updownarrow$
$^-8a - 6$

Equivalent expressions

So, $^-2(4a + 3)$ simplifies to $^-8a - 6$.

◀ *Work Problem* **5** *at the Side.*

Sometimes you need to do several steps to simplify an expression.

EXAMPLE 6 **Simplifying a More Complex Expression**

Simplify: $8 + 3(x - 2)$

$8 + 3(x - 2)$	Do *not* add $8 + 3$. Use the distributive property first because multiplying is done *before* adding.
$8 + 3 \cdot x - 3 \cdot 2$	Do the multiplications.
$8 + 3x \quad - \quad 6$	Rewrite so that like terms are next to each other.
$3x + 8 \quad - \quad 6$	Subtract to find $8 - 6$ or change to adding $8 + {}^-6$.
$3x + \quad 2$	

The simplified expression is $3x + 2$.

> **CAUTION**
> Do *not* add $8 + 3$ as the first step in Example 6 above. Remember that the order of operations tells you to do multiplying *before* adding.

Work Problem **6** *at the Side.* ▶

6 Simplify.

(a) $^-4 + 5(y + 1)$

(b) $2(3w + 4) - 5$

(c) $5(6x - 2) + 3x$

(d) $21 + 7(a^2 - 3)$

(e) $^-y + 3(2y + 5) - 18$

Math in the Media

EXPRESSIONS AND TRAFFIC SIGNALS

Traffic engineers have to decide how long to have the green, yellow, and red lights showing on a traffic signal.

To find the number of seconds for the green light, the engineers use this expression.

$$2.1n + 3.7$$

The variable n stands for the average number of vehicles traveling in each lane of the street during one complete light cycle (green to yellow to red, and back to green again).

1. How many seconds should the green light be on if a street averages 5 vehicles in each lane per light cycle? Round to the nearest whole number.

2. How many seconds for the green light, to the nearest whole number, if the average is 10 vehicles per light cycle? If the average is 15 vehicles per light cycle?

3. As the average number of vehicles increases, what is happening to the number of seconds for the green light?

4. Based on the answers to Questions 1 and 2, make a guess on the number of seconds for the green light when the average is 20 vehicles per light cycle. Then calculate the number of seconds using the expression.

To decide the number of seconds that a yellow light should be on, traffic engineers use this expression.

$$\frac{5v}{100} + 1 \qquad \text{where } v \text{ is the speed limit in miles per hour (mph).}$$

5. How many seconds should the yellow light be on if the speed limit is 20 mph? 40 mph? 60 mph?

6. Based on the answers you just calculated, how could you estimate the time for a yellow light if the speed limit is 30 mph? 50 mph?

7. Use the given expression to find the number of seconds that the yellow light should be on if the speed limit is 30 mph and 50 mph. Did you get the same result as in Question 6?

Source: Applying Mathematics: A Consumer/Career Approach.

2.2 ▶▶▶ **Exercises**

 FOR EXTRA HELP      Math XL PRACTICE   WATCH   DOWNLOAD   READ   REVIEW

Circle the like terms in each expression. Then identify the coefficients of the like terms.
See Example 1.

1. $2b^2 + 2b + 2b^3 + b^2 + 6$

2. $3x + x^3 + 3x^2 + 3 + 2x^3$

3. $^-x^2y + ^-xy + 2xy + ^-2xy^2$

4. $ab^2 + ^-a^2b + 2ab + ^-3a^2b$

5. $7 + 7c + 3 + 7c^3 + ^-4$

6. $4d + ^-5 + 1 + ^-5d^2 + 4$

Simplify each expression. See Example 2.

7. $6r + 6r$

8. $4t + 10t$

9. $x^2 + 5x^2$

10. $9y^3 + y^3$

11. $p - 5p$

12. $n - 3n$

13. $^-2a^3 - a^3$

14. $^-10x^2 - x^2$

15. $c - c$

16. $b^2 - b^2$

17. $9xy + xy - 9xy$

18. $r^2s - 7r^2s + 7r^2s$

19. $5t^4 + 7t^4 - 6t^4$

20. $10mn - 9mn + 3mn$

21. $y^2 + y^2 + y^2 + y^2$

22. $a + a + a$

23. $^-x - 6x - x$

24. $^-y - y - 3y$

Simplify by combining like terms. Write each answer with the variables in alphabetical order and any constant term last. See Example 3.

25. $8a + 4b + 4a$

26. $6x + 5y + 4y$

27. $6 + 8 + 7rs$

28. $10 + 2c^2 + 15$

29. $a + ab^2 + ab^2$

30. $n + mn + n$

31. $6x + y - 8x + y$

32. $d + 3c - 7c + 3d$

33. $8b^2 - a^2 - b^2 + a^2$

34. $5ab - ab + 3a^2b - 4ab$

35. $^-x^3 + 3x - 3x^2 + 2$

36. $a^2b - 2ab - ab^3 + 3a^3b$

37. $^-9r + 6t - s - 5r + s + t - 6t + 5s - r$

38. $^-x - 3y + 4z + x - z + 5y - 8x - y$

Simplify by using the associative property of multiplication. See Example 4.

39. $3(10a)$

40. $8(4b)$

41. $^-4(2x^2)$

42. $^-7(3b^3)$

43. $5(^-4y^3)$

44. $2(^-6x)$

45. $^-9(^-2cd)$

46. $^-6(^-4rs)$

47. $7(3a^2bc)$

48. $4(2xy^2z^2)$

49. $^-12(^-w)$

50. $^-10(^-k)$

Use the distributive property to simplify each expression. See Example 5.

51. $6(b + 6)$

52. $5(a + 3)$

53. $7(x - 1)$

54. $4(y - 4)$

55. $3(7t + 1)$

56. $8(2c + 5)$

57. $^-2(5r + 3)$

58. $^-5(6z + 2)$

59. $^-9(k + 4)$

60. $^-3(p + 7)$

61. $50(m - 6)$

62. $25(n - 1)$

Simplify each expression. See Example 6.

63. $10 + 2(4y + 3)$

64. $4 + 7(x^2 + 3)$

65. $6(a^2 - 2) + 15$

66. $5(b - 4) + 25$

67. $2 + 9(m - 4)$

68. $6 + 3(n - 8)$

69. $^-5(k + 5) + 5k$

70. $^-7(p + 2) + 7p$

71. $4(6x - 3) + 12$

72. $6(3y - 3) + 18$

73. $5 + 2(3n + 4) - n$

74. $8 + 8(4z + 5) - z$

75. $^-p + 6(2p - 1) + 5$

76. $^-k + 3(4k - 1) + 2$

77. Explain the difference between *simplifying* an expression and *evaluating* an expression.

78. Simplify each expression. Are the answers equivalent? Explain why or why not.

$$5(3x + 2) \qquad 5(2 + 3x)$$

79. Explain what makes two terms *like* terms. Include several examples in your explanation.

80. Explain how to combine like terms. Include an example in your explanation.

81. Explain and correct the error made by a student who simplified this expression.

$$\underbrace{{}^-2x + 7x}_{5x^2} + 8$$

$$5x^2 \quad + 8$$

82. Explain and correct the error made by a student who simplified this expression.

$$\underbrace{{}^-10a + 6a}_{{}^-4a} \quad \underbrace{- \quad 7 + 2}_{+ \quad {}^-7 + 2}$$

$${}^-4a \quad + \quad {}^-7 + 2$$

$${}^-4a \quad + \quad {}^-5$$

$$4a - 5$$

Simplify.

83. ${}^-4(3y) - 5 + 2(5y + 7)$

84. $6({}^-3x) - 9 + 3({}^-2x + 6)$

85. ${}^-10 + 4({}^-3b + 3) + 2(6b - 1)$

86. $12 + 2(4a - 4) + 4({}^-2a - 1)$

87. ${}^-5({}^-x + 2) + 8({}^-x) + 3({}^-2x - 2) + 16$

88. ${}^-7({}^-y) + 6(y - 1) + 3({}^-2y) + 6 - y$

Study Skills

This is the second part of the Study Cards activity. As you get further into a chapter, you can choose particular problems that will serve as a good test review. Here are two more types of study cards that will help you.

When you are doing your homework and find yourself saying, "This is really hard," or "I'm worried I'll make a mistake," make a **tough problem** study card! On the front, write out the procedure to work the type of problem *in words*. If there are special notes (like what *not* to do), include them. On the back, work at least one example; make sure you label what you are doing.

OBJECTIVES

1. Create study cards for difficult problems.
2. Create study cards of quiz problems.

Tough Problems Card

Simplifying Expressions

Combine (add) <u>like</u> terms only.

Like terms have variable parts that match.

Use the distributive property if possible.

Front of Card

Example	Simplify this expression.
$7 + 3(2n - 4)$	Do <u>not</u> add $7 + 3$!
$7 + 3(2n - 4)$	Use the distributive property.
$7 + 3 \cdot 2n - 3 \cdot 4$	Do the multiplications.
$7 + 6n - 12$	Rewrite so like terms are next to each other.
$6n + 7 + {}^-12$	Change subtraction to adding the opposite.
$6n + {}^-5$	Do <u>not</u> stop here!
$6n - 5$	Use definition of subtraction "in reverse."

Back of Card

Choose three types of difficult problems or problems that have given you trouble, and work them out on study cards. Be sure to put the words for solving the problem on one side and the worked problem on the other side.

◀◀◀ **Now Try This**

Practice Quiz Cards

Make up a few **quiz cards** for each type of problem you learn, and use them to prepare for a test. Choose two or three problems from the different sections of the chapter. Be sure you don't just choose the easiest problems! Put the problem **with the direction words** (like *solve, simplify, evaluate*) on the front, and work the problem on the back. If you like, put the page number from the text there, too. When you review, you work the problem on a separate paper, and check it by looking at the back.

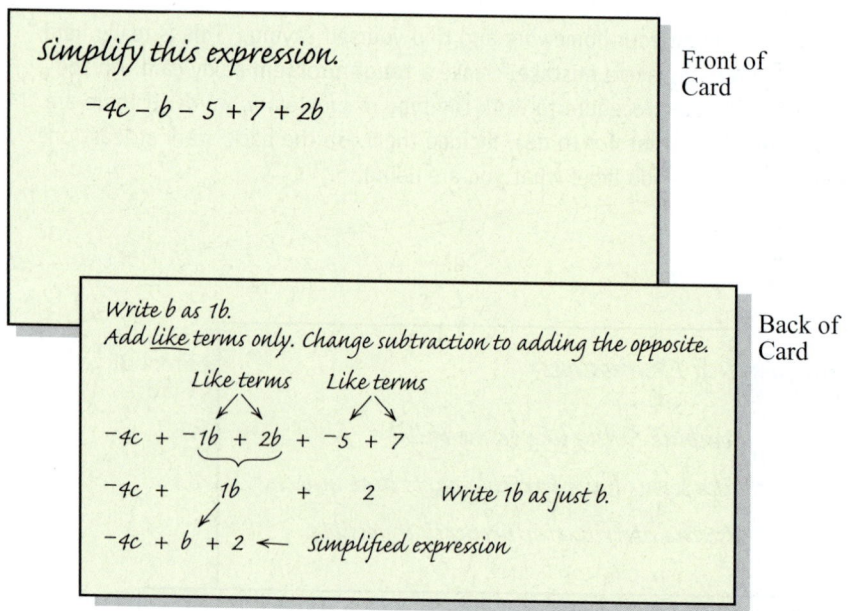

Front of Card

Back of Card

Why Are Study Cards Brain Friendly?

First, making the study cards is an **active technique** that really gets your dendrites growing. You have to make decisions about what is most important and how to put it on the card. This kind of thinking is more in depth than just memorizing, and as a result, you will understand the concepts better and remember them longer.

Second, the cards are **visually appealing** (if you write neatly and try some color). Your brain responds to pleasant visual images, and again, you will remember longer and may even be able to "picture in your mind" how your cards look. This will help you during tests.

Third, because study cards are small and portable, you can review them easily whenever you have a few minutes. Even while you're waiting for a bus or have a few minutes between classes you can take out your cards and read them to yourself. Your **brain really benefits from repetition;** each time you review your cards your dendrites are growing thicker and stronger. After a while, the information will become automatic and you will remember it for a long time.

Summary Exercises on Variables and Expressions

Identify the parts of each expression. Choose from these labels: variable, constant, coefficient.

1. $^-10 - m$

2. ^-8cd

3. $6 + 4x$

4. The expression for finding the perimeter of an octagon with sides of equal length is $8s$. Evaluate the expression when

   (a) the length of one side is 4 yards.

   (b) the length of one side is 15 inches.

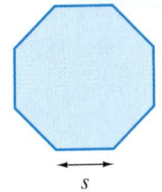

5. The expression for the total cost of a car is $d + mt$ where d is the down payment, m is the monthly payment, and t is the number of months you must make payments. Evaluate the expression when

   (a) the down payment is \$3000 and you make 36 monthly payments of \$280.

   (b) the down payment is \$1750 and you make 48 monthly payments of \$429.

Rewrite each expression without exponents.

6. ad^4

7. b^3cd

8. $^-7ab^5c^2$

Evaluate each expression when w is 5, x is $^-2$, y is $^-6$, and z is 0.

9. w^4

10. $5xz$

11. yz^2

12. wxy

13. x^3

14. ^-4wy

15. $3xy^2$

16. w^2x^5

17. $^-7wx^4y^3$

Simplify each expression.

18. $10b + 4b + 10b$

19. $^-3x - 5 + 12 + 10x$

20. $^-8(c + 4)$

21. $^-9xy + 9xy$

22. $^-4(^-3c^2d)$

23. $3f - 5f - 4f$

24. $2(3w + 4)$

25. $^-a - 6b - a$

26. $^-10(^-5x^3y^2)$

27. $5r^2 + 2r - 2r^2 + 5r^3$

28. $21 + 7(h^2 - 3)$

29. $^-3(m + 3) + 3m$

30. $^-4(8y - 5) + 5$

31. $2 + 12(3x - 1)$

32. $^-n + 5(4n - 2) + 11$

33. Explain and correct each error made by the student who simplified these expressions.

(a) $6(n + 2)$
$6n + 2$

(b) $^-5(^-4a)$
^-20a

(c) $\underbrace{3y + 2y} - 10$
$5y^2 \quad - 10$

34. Explain and correct the error made by this student.

$\underbrace{^-8x + x} \overset{\downarrow}{-} \underbrace{\overset{\downarrow}{12} + 3}$
$^-7x \quad + \quad ^-12 + 3$
$^-7x \quad + \quad ^-9$
$7x - 9$

2.3 ▷▷▷ Solving Equations Using Addition

Now you are ready for a look at the "heart" of algebra, writing and solving **equations.** *Writing* an equation is a way to show the relationship between what you *know* about a problem and what you *don't* know. Then, *solving* the equation is a way to figure out the part that you didn't know and answer your question.

The questions you can answer by writing and solving equations are as varied as the careers people choose. A zookeeper can solve an equation that answers the question of how long to incubate the egg of a particular tropical bird. An aerobics instructor can solve an equation that answers the question of how hard a certain person should exercise for maximum benefit.

OBJECTIVE 1 Determine whether a given number is a solution of an equation.   Let's start with the example from the beginning of this chapter: ordering textbooks for math classes. The expression we used to order books was $c + 5$, where c was the class limit (the maximum number of students allowed in the class). Suppose that 30 prealgebra books were ordered. What is the class limit for prealgebra? To answer this question, write an equation showing the relationship between what you know and what you don't know.

You don't know the class limit.   You do know the total number of books ordered.

$$c + 5 = 30$$

You do know that 5 extra books were ordered.

> **Note**
> An equation has an equal sign. Notice the similarity in the words **equa**tion and **equa**l. An *expression* does *not* have an equal sign.

The equal sign in an equation is like the balance point on a playground teeter-totter, or seesaw. To have a true equation, the two sides must balance.

These equations balance, so we can use the $=$ sign.

These equations do *not* balance, so we write $\neq$ to mean "is not equal to."

$$6 + 8 = 14$$

$$6 + 8 \neq 15$$

$$10 = 5 \cdot 2$$

$$10 \neq 4 \cdot 2$$

$$3 \cdot 2 = 5 + 1$$

$$4 + 5 \neq 5 \cdot 4$$

OBJECTIVES

1 Determine whether a given number is a solution of an equation.

2 Solve equations using the addition property of equality.

3 Simplify equations before using the addition property of equality.

When an equation has a variable, we **solve the equation** by finding a number that can replace the variable and make the equation balance. For the example about ordering prealgebra textbooks:

$$c + 5 = 30$$

What number can replace c so that the equation balances?

1 **(a)** Which of these numbers, 95, 65, or 70, is the solution of the equation $c + 15 = 80$?

Try replacing c with **15**. $\quad 15 + 5 \neq 30$ $\qquad$ Does *not* balance: $\quad 15 + 5$ is only 20.

Try replacing c with **40**. $\quad 40 + 5 \neq 30$ $\qquad$ Does *not* balance: $\quad 40 + 5$ is more than 30.

Try replacing c with **25**. $\quad 25 + 5 = 30$ $\qquad$ Balances: $25 + 5$ is 30.

The **solution** is **25** because **25** is the *only* number that makes the equation balance. By solving the equation, you have answered the question about the class limit for prealgebra. The class limit is 25.

> **Note**
>
> Most of the equations that you will solve in **Chapters 2–8** have only one solution, that is, one number that makes the equation balance. Later in this book, you will solve equations that have two or more solutions.

(b) Which of these numbers, 20, 24, or 32, is the solution of the equation $28 = c - 4$?

EXAMPLE 1 **Identifying the Solution of an Equation**

Which of these numbers, 70, 40, or 60, is the solution of the equation $c - 10 = 50$?

Replace c with each of the numbers. The one that makes the equation balance is the solution.

$$70 - 10 \neq 50 \qquad 40 - 10 \neq 50 \qquad 60 - 10 = 50$$

Does *not* balance: $70 - 10$ is more than 50.

Does *not* balance: $40 - 10$ is only 30.

Balances: $60 - 10$ is 50.

The solution is 60 because, when c is 60, the equation balances.

 ◀ *Work Problem* **1** *at the Side.*

OBJECTIVE **2** **Solve equations using the addition property of equality.** When solving the book ordering equation, $c + 5 = 30$, you could just look at the equation and think, "What number, plus 5, would balance with 30?" You could easily see that c had to be 25. Not all equations can be solved this easily, so you'll need some tools for the harder ones. The first tool, called the **addition property of equality,** allows us to add the *same* number to *both* sides of an equation.

Addition Property of Equality

If $a = b$, then $a + c = b + c$.

In other words, you may add the same number to both sides of an equation and still keep it balanced.

Think of the teeter-totter. If there are 3 children of the same size on each side, it will balance. If 2 more children climb onto the left side, the only way to keep the balance is to have 2 more children of the same size climb onto the right side as well.

$$3 = 3$$

$$3 + 2 = 3 + 2$$

All the tools you will learn to use with equations have one goal.

Goal in Solving an Equation

The goal is to end up with the variable (letter) on one side of the equal sign balancing a number on the other side.

We work on the original equation until we get:

variable = number   or   number = variable

Once we have arrived at that point, the number balancing the variable is the solution to the original equation.

> **EXAMPLE 2** **Using the Addition Property of Equality**

Solve each equation and check the solution.

(a) $c + 5 = 30$

We want to get the variable, c, by itself on the left side of the equal sign. To do that, we add the *opposite* of 5, which is $^-5$. Then $5 + ^-5$ will be 0.

$$c + 5 = 30$$

Add $^-5$ to the left side. $\longrightarrow$ $\underline{^-5 \quad ^-5}$ $\longleftarrow$ To keep the balance, add $^-5$ to the right side also.

$$c + \mathbf{0} = 25 \longleftarrow 30 + ^-5 \text{ is } 25.$$

$5 + ^-5$ is 0.

Recall that adding 0 to any number leaves the number unchanged, so $c + 0$ is c.

$$\underbrace{c + 0}_{} = 25$$
$$c \quad\; = 25$$

Because c *balances* with 25, the *solution* is 25.

Check the solution by replacing c with 25 in the *original equation*.

$$c + 5 = 30 \qquad \text{Original equation}$$
$$\underbrace{25 + 5}_{} = 30 \qquad \text{Replace } c \text{ with 25.}$$
$$\mathbf{30} \quad = \mathbf{30} \qquad \text{Balances}$$

> The solution is 25, **not** 30.

Because the equation balances when we use 25 to replace the variable, we know that **25 is the correct solution**. If it had *not* balanced, we would need to rework the problem, find our error, and correct it.

(b) $^-5 = x - 3$

We want the variable, x, by itself on the right side of the equal sign. (Remember, it doesn't matter which side of the equal sign the variable is on, just so it ends up by itself.) To see what number to add, we change the subtraction to adding the opposite.

$$^-5 = x - \quad 3 \qquad \text{Change subtraction to adding}$$
$$\qquad\qquad \downarrow \quad \downarrow \qquad\qquad \text{the opposite.}$$
$$^-5 = x + {}^-3 \qquad \text{To get } x \text{ by itself on the right side,}$$

To keep the balance, add 3 to the left side also. $^-5 + 3$ is $^-2$.

$$\underline{3 \qquad\qquad 3} \qquad \text{add the opposite of } ^-3, \text{ which is 3. Then } ^-3 + 3 \text{ is 0.}$$
$$^-2 = \underbrace{x + \quad 0}_{} \qquad \text{Adding 0 to } x \text{ leaves } x \text{ unchanged.}$$
$$^-2 = \quad x \qquad x \text{ balances with } ^-2, \text{ so } ^-2 \text{ is the solution.}$$

We check the solution by replacing x with $^-2$ in the *original equation*. If the equation balances when we use $^-2$, we know that it is the correct solution. If the equation does *not* balance when we use $^-2$, we made an error and need to try solving the equation again.

Continued on Next Page

Check

$-5 = x - 3$		Original equation
$-5 = -2 - 3$		Replace x with -2.
$-5 = -2 + -3$		Change subtraction to adding the opposite.
$-5 = -5$		Balances; this shows -2 is the correct solution.

When x is replaced with -2, the equation balances, so **-2 is the correct solution**.

> **CAUTION**
> When checking the solution to Example 2(b) above, we ended up with $-5 = -5$. Notice that -5 is ***not*** the solution. The solution is -2, the number used to replace x in the original equation.

Work Problem **2** *at the Side.* ▶

OBJECTIVE 3 Simplify equations before using the addition property of equality. Sometimes you can simplify the expression on one or both sides of the equal sign. Doing so will make it easier to solve the equation.

EXAMPLE 3 **Simplifying before Solving Equations**

Solve each equation and check each solution.

(a) $y + 8 = 3 - 7$

You cannot simplify the left side because y and 8 are *not* like terms.

$$y + 8 = 3 - 7$$

Simplify the right side by changing subtraction to adding the opposite.

$$y + 8 = 3 + -7$$

Add $3 + -7$.

To get y by itself on the left side, add the opposite of 8, which is -8.

$$y + 8 = -4$$

To keep the balance, add -8 to the right side also.

$8 + -8$ is 0.

$$\underline{\quad -8 \quad \quad -8 \quad}$$

$$y + 0 = -12$$

$-4 + -8$ is -12.

$$y = -12$$

The solution is -12. Now check the solution.

Check

$y + 8 = 3 - 7$		Go back to the *original* equation and replace y with -12.
Add $-12 + 8$. $\quad -12 + 8 = 3 - 7$		Change $3 - 7$ to $3 + -7$.
$-4 = 3 + -7$		Add $3 + -7$.
$-4 = -4$		Balances; so -12 is the correct solution.

When y is replaced with -12, the equation balances, so **-12 is the correct solution** (not -4).

Continued on Next Page

---

2 Solve each equation and check each solution.

(a) $12 = y + 5$

Check

(b) $b - 2 = -6$

Check

(b) $^-2 + 2 = {}^-4b - 6 + 5b$

3 Simplify each side of the equation when possible. Then solve the equation and check the solution.

(a) $2 - 8 = k - 2$

Simplify the left side by adding $^-2 + 2$.

To keep the balance, add 6 to the left side also. ⟶

$$\underbrace{{}^-2 + 2}_{} = {}^-4b \underset{\downarrow}{-} \underset{\downarrow}{6} + 5b$$

Simplify the right side by changing subtraction to adding the opposite.

$0 = {}^-4b + {}^-6 + 5b$ Find like terms.

$0 = \underbrace{{}^-4b + 5b}_{} + {}^-6$ Combine $^-4b + 5b$.

$0 = 1b + {}^-6$ To get $1b$ by itself, add the opposite of $^-6$, which is 6.

$\dfrac{6}{6} = 1b \qquad \dfrac{6}{}$ ⟵

$6 = 1b + 0$

$6 = \underset{\downarrow}{1b}$ $1b$ is equivalent to b.

$6 = b$

Check

The solution is 6.

Check $^-2 + 2 = {}^-4b - 6 + 5b$ Go back to the *original* equation and replace each b with 6.

Add $^-2 + 2$. $\underbrace{{}^-2 + 2}_{} = \underbrace{{}^-4 \cdot 6}_{} - 6 + \underbrace{5 \cdot 6}_{}$ On the right side, do multiplications first.

$0 = \underbrace{{}^-24 + {}^-6}_{} + 30$ Change subtraction to adding the opposite.

(b) $4r + 1 - 3r = {}^-8 + 11$

$0 = \underbrace{{}^-30 + 30}_{}$ Add from left to right.

$0 = 0$ Balances

When b is replaced with 6, the equation balances, so **6 is the correct solution (not 0).**

CAUTION
When checking a solution, always go back to the *original* equation. That way, you will catch any errors you made when simplifying each side of the equation.

Check

◀ Work Problem **3** at the Side.

ANSWERS

3. **(a)** $k = {}^-4$
   Check $2 - 8 = \underset{\downarrow}{k} - 2$

   $\underbrace{2 + {}^-8}_{} = \underbrace{{}^-4 + {}^-2}_{}$

   Balances $^-6 = {}^-6$

(b) $r = 2$
   Check $4r + 1 - \underset{\downarrow}{3r} = {}^-8 + 11$

   $\underbrace{4 \cdot 2}_{} + 1 - \underbrace{3 \cdot 2}_{} = \underbrace{{}^-8 + 11}_{}$

   $\underbrace{8 + 1}_{} - 6 = 3$

   $\underbrace{9 + {}^-6}_{} = 3$

   Balances $3 = 3$

2.3 ▶▶▶ Exercises

In each list of numbers, find the one that is a solution of the given equation. See Example 1.

1. $n - 50 = 8$

58, 42, 60

2. $r - 20 = 5$

15, 30, 25

🌐 **3.** $^-6 = y + 10$

$^-4, \,^-16, \, 16$

4. $^-4 = x + 13$

$17, \,^-17, \,^-9$

5. $t + 12 = 0$

$0, \,^-12, \,^-24$

6. $b - 8 = 0$

$8, 0, \,^-8$

Solve each equation and check each solution. See Example 2.

7. $p + 5 = 9$ **Check** $p + 5 = 9$

8. $a + 3 = 12$ **Check** $a + 3 = 12$

9. $8 = r - 2$ **Check** $8 = r - 2$

10. $3 = b - 5$ **Check** $3 = b - 5$

🌐 **11.** $^-5 = n + 3$ **Check**

12. $^-1 = a + 8$ **Check**

13. $^-4 + k = 14$ **Check**

14. $^-9 + y = 7$ **Check**

15. $y - 6 = 0$ **Check**

16. $k - 15 = 0$ **Check**

17. $7 = r + 13$ **Check**

18. $12 = z + 19$ **Check**

19. $x - 12 = {}^{-}1$ **Check**

20. $m - 3 = {}^{-}9$ **Check**

21. ${}^{-}5 = {}^{-}2 + t$ **Check**

22. ${}^{-}1 = {}^{-}10 + w$ **Check**

A solution is given for each equation. Show how to check the solution. If the solution is correct, leave it. If the solution is not correct, solve the equation and check your new solution. See Example 2.

23. $z - 5 = 3$
The solution is ${}^{-}2$.   **Check** $z - 5 = 3$
                   $\downarrow$

24. $x - 9 = 4$
The solution is 13.   **Check** $x - 9 = 4$
                   $\downarrow$

25. $7 + x = {}^{-}11$
The solution is ${}^{-}18$.   **Check**

26. $2 + k = {}^{-}7$
The solution is ${}^{-}5$.   **Check**

27. ${}^{-}10 = {}^{-}10 + b$
The solution is 10.   **Check**

28. $0 = {}^{-}14 + a$
The solution is 0.   **Check**

Simplify each side of the equation when possible. Then solve the equation and check the solution. Show your work. See Example 3.

29. $c - 4 = {}^-8 + 10$ **Check**

30. $b - 8 = 10 - 6$ **Check**

31. ${}^-1 + 4 = y - 2$ **Check**

32. $2 + 3 = k - 4$ **Check**

33. $10 + b = {}^-14 - 6$ **Check**

34. $1 + w = {}^-8 - 8$ **Check**

35. $t - 2 = 3 - 5$ **Check**

36. $p - 8 = {}^-10 + 2$ **Check**

37. $10z - 9z = {}^-15 + 8$ **Check**

38. $2r - r = 5 - 10$ **Check**

39. ${}^-5w + 2 + 6w = {}^-4 + 9$ **Check**

40. ${}^-2t + 4 + 3t = 6 - 7$ **Check**

Solve each equation. Show your work. See Examples 2 and 3.

41. $^-3 - 3 = 4 - 3x + 4x$

42. $^-5 - 5 = {}^-2 - 6b + 7b$

43. $^-3 + 7 - 4 = {}^-2a + 3a$

44. $6 - 11 + 5 = {}^-8c + 9c$

45. $y - 75 = {}^-100$

46. $a - 200 = {}^-100$

47. $^-x + 3 + 2x = 18$

48. $^-s + 2s - 4 = 13$

49. $82 = {}^-31 + k$

50. $^-5 = 72 + w$

51. $^-2 + 11 = 2b - 9 - b$

52. $^-6 + 7 = 2h - 1 - h$

53. $r - 6 = 7 - 10 - 8$

54. $m - 5 = 2 - 9 + 1$

55. $^-14 = n + 91$

56. $66 = x - 28$

57. $^-9 + 9 = 5 + h$

58. $18 - 18 = 6 + p$

59. A student did this work when solving an equation. Do you agree that the solution is $^-7$? Explain why or why not.

$$\underbrace{^-8 + 1} = x + 7$$
$$^-7 = x + 7$$
$$\underline{^-7 \qquad ^-7}$$
$$^-14 = \underbrace{x + 0}$$
$$^-14 = x$$

Check

$$^-8 + 1 = x + 7$$
$$\underbrace{^-8 + 1} = \underbrace{^-14 + 7}$$
$$^-7 = ^-7$$

Balances, so $^-7$ is the solution.

60. A student did this work when solving an equation. Show how to check the solution. If the solution does not check, find and correct the errors.

$$^-3 - 6 = n - 5$$
$$\underbrace{^-3 + 6} = n - 5$$
$$3 = n - 5$$
$$\underline{^-5 \qquad ^-5}$$
$$^-2 = \underbrace{n + 0}$$
$$^-2 = n$$

61. West Community College always orders 10 extra robes for the graduation ceremony. The college ordered 305 robes this year. Solving the equation $g + 10 = 305$ will give the number of graduates (g) this year. Solve the equation.

62. Refer to Exercise 61. The college ordered 278 robes last year. Solve the equation $g + 10 = 278$ to find the number of graduates last year.

63. The warmer the temperature, the faster a field cricket chirps. Solving the equation $92 = c + 37$ will give you the number of chirps (in 15 seconds) when the temperature is 92 degrees. Solve the equation.

64. Refer to Exercise 63. Solve the equation $77 = c + 37$ to find the number of times a field cricket chirps (in 15 seconds) when the temperature is 77 degrees.

65. During the summer months, Ernesto spends an average of only \$45 per month on parking fees by riding his bike to work on nice days. This is \$65 less per month than what he spends for parking in the winter. Solving the equation $p - 65 = 45$ will give you his monthly parking fees in the winter. Solve the equation.

66. By walking to work several times a week in the summer, Aimee spends an average of \$56 less per month on parking fees. If she spends \$98 per month on parking in the summer, solve the equation $p - 56 = 98$ to find her monthly parking fees in the winter.

Solve each equation. Show your work.

67. $^-17 - 1 + 26 - 38 = {}^-3 - m - 8 + 2m$

68. $19 - 38 - 9 + 11 = {}^-t - 6 + 2t - 6$

69. $^-6x + 2x + 6 + 5x = |0 - 9| - |^-6 + 5|$

70. $^-h - |^-9 - 9| + 8h - 6h = {}^-12 - |^-5 + 0|$

Relating Concepts (Exercises 71–72) For Individual or Group Work

Use what you have learned about solving equations to **work Exercises 71 and 72 in order.**

71. (a) Write two *different* equations that have $^-2$ as the solution. Be sure that you have to use the *addition property of equality* to solve the equations. Show how to solve each equation. Use Exercises 7 to 22 as models.

(b) Follow the directions in part (a), but this time write two equations that have 0 as the solution.

72. Not all equations have solutions that are integers. Try solving these equations.

(a) $x + 1 = 1\dfrac{1}{2}$

(b) $\dfrac{1}{4} = y - 1$

(c) $\$2.50 + n = \$3.35$

(d) Write two more equations that have fraction or decimal solutions.

2.4 ▶▶▶ Solving Equations Using Division

OBJECTIVE 1 Solve equations, using the division property of equality. In **Section 2.1** you worked with the expression for finding the perimeter of a square-shaped garden, that is, finding the total distance around all four sides of the garden:

$$4s, \text{ where } s \text{ is the length of one side of the square}$$

Suppose you know that 24 feet of fencing was used around a square-shaped garden. What was the length of one side of the garden? To answer this question, write an equation showing the relationship between what you know and what you don't know.

You don't know the length of one side.

You do know that there are 4 sides. → You do know the perimeter.

$$4\,s = 24$$

To solve the equation, what number can replace s so that the equation balances? You can see that s is 6 feet.

$$4 \cdot 6 = 24 \qquad \text{Balances:} \\ 4 \cdot 6 \text{ is exactly } 24.$$

The **solution** is **6 feet** because **6** is the *only* number that makes the equation balance. You have answered the question about the length of one side: The length is 6 feet.

There is a tool that you can use to solve equations such as $4s = 24$. Called the **division property of equality,** it allows you to *divide* both sides of an equation by the *same* number. (The only exception is that you cannot divide by 0.)

> **Division Property of Equality**
>
> If $a = b$, then $\dfrac{a}{c} = \dfrac{b}{c}$ as long as c is *not* 0.
>
> In other words, you may divide both sides of an equation by the same nonzero number and still keep it balanced.

In **Section 2.3,** you saw that *adding* the same number to both sides of an equation kept it balanced. We could also have *subtracted* the same number from both sides because subtraction is defined as adding the opposite. Now we're saying that you can *divide* both sides by the same number. In **Chapter 4** we'll *multiply* both sides by the same number.

> **Equality Principle for Solving an Equation**
>
> As long as you do the *same* thing to *both* sides of an equation, the balance is maintained. (The only exception is that you cannot divide by 0.)

OBJECTIVES

1 Solve equations using the division property of equality.

2 Simplify equations before using the division property of equality.

3 Solve equations such as $^-x = 5$.

1 Solve each equation and check each solution.

(a) $4s = 44$

Check

(b) $27 = {}^-9p$

Check

(c) ${}^-40 = {}^-5x$

Check

(d) $7t = {}^-70$

Check

EXAMPLE 1 Using the Division Property of Equality

Solve each equation and check each solution.

(a) $4s = 24$

As with any equation, the goal is to get the variable by itself on one side of the equal sign. On the left side we have $4s$, which means $4 \cdot s$. The variable is multiplied by 4. Division is the opposite of multiplication, so dividing by 4 can be used to "undo" multiplying by 4.

Divide $4s$ by 4. The fraction bar indicates division: $4s \div 4$ is s. $\quad \dfrac{4s}{4} = \dfrac{24}{4} \quad$ To keep the balance, divide the right side by 4 also: $24 \div 4$ is 6

$$s = 6$$

So, as we already knew, 6 is the solution. We check the solution by replacing s with 6 in the original equation.

Check $\quad 4s = 24 \quad$ Original equation

$\quad 4 \cdot 6 = 24 \quad$ Replace s with 6.

$\quad 24 = 24 \quad$ Balances [The solution is 6, **not** 24.]

When s is replaced with 6, the equation balances, so **6 is the correct solution**.

(b) $42 = {}^-6w$

On the right side of the equation, the variable is *multiplied* by ${}^-6$. To undo the multiplication, *divide* by ${}^-6$.

To keep the balance, divide by ${}^-6$ on the left side also. $\quad \dfrac{42}{{}^-6} = \dfrac{{}^-6w}{{}^-6} \quad$ Use division to undo multiplication: ${}^-6w \div {}^-6$ is w.

$${}^-7 = w$$

The solution is ${}^-7$.

Check $\quad 42 = {}^-6w \quad$ Original equation

$\quad 42 = {}^-6 \cdot {}^-7 \quad$ Replace w with ${}^-7$. [The solution is ${}^-7$, **not** 42.]

$\quad 42 = 42 \quad$ Balances

When w is replaced with ${}^-7$, the equation balances, so ${}^-7$ **is the correct solution**.

> **CAUTION**
> Be careful to divide both sides by the *same* number as the coefficient of the variable term. In Example 1(b) above, the coefficient of ${}^-6w$ is ${}^-6$, so divide both sides by ${}^-6$. (Do **not** divide by the *opposite* of ${}^-6$, which is 6. Use the opposite only when you're *adding* the same number to both sides.)

◀ **Work Problem** **1** at the Side.

OBJECTIVE 2 Simplify equations before using the division property of equality. You can sometimes simplify the expression on one or both sides of the equal sign, as you did in **Section 2.3.**

EXAMPLE 2 Simplifying before Solving Equations

Solve each equation and check each solution.

(a) $4y - 7y = {}^-12$

Simplify the left side by combining like terms.

Change subtraction to adding the opposite.

Divide by the coefficient, which is $^-3$.

$$4y - 7y = {}^-12$$ The right side cannot be simplified.

$$4y + {}^-7y = {}^-12$$

$$\frac{^-3y}{^-3} = \frac{^-12}{^-3}$$ To keep the balance, divide by $^-3$ on the right side.

$$y = 4$$ $^-12 \div {}^-3$ is 4.

The solution is 4.

Check

Do multiplications first.

Change subtraction to adding the opposite.

$$4y - 7y = {}^-12$$ Go back to the *original* equation and replace each y with 4.

$$4 \cdot 4 - 7 \cdot 4 = {}^-12$$

$$16 - 28 = {}^-12$$

$$16 + {}^-28 = {}^-12$$

$${}^-12 = {}^-12$$ Balances

When y is replaced with 4, the equation balances, so **4 is the correct solution**.

(b) $3 - 10 + 7 = h + 7h$

Change subtraction to adding the opposite.

Add from left to right.

$$3 - 10 + 7 = h + 7h$$ Write the understood 1 as the coefficient of h.

$$3 + {}^-10 + 7 = 1h + 7h$$ Combine like terms.

$${}^-7 + 7 = 8h$$

To keep the balance, divide by 8 on the left side also.

$$\frac{0}{8} = \frac{8h}{8}$$ Divide by the coefficient, which is 8.

$0 \div 8$ is 0. $\qquad 0 = h$

The solution is 0.

Check

$$3 - 10 + 7 = h + 7h$$ Go back to the *original* equation and replace each h with 0.

$$3 + {}^-10 + 7 = 0 + 7 \cdot 0$$

$${}^-7 + 7 = 0 + 0$$

$$0 = 0$$ Balances

When h is replaced with 0, the equation balances, so **0 is the correct solution**.

Work Problem **2** *at the Side.* ▶

2 Simplify each side of the equation when possible. Then solve the equation and check the solution.

(a) $^-28 = {}^-6n + 10n$

Check

(b) $p - 14p = {}^-2 + 18 - 3$

Check

ANSWERS

2. (a) $n = {}^-7$

Check $^-28 = {}^-6n + 10n$

$^-28 = {}^-6 \cdot {}^-7 + 10 \cdot {}^-7$

$^-28 = 42 + {}^-70$

Balances $^-28 = {}^-28$

(b) $p = {}^-1$

Check $p - 14p = {}^-2 + 18 - 3$

$^-1 - 14(^-1) = 16 - 3$

$^-1 - {}^-14 = 13$

$^-1 + {}^+14 = 13$

Balances $13 = 13$

3 Solve each equation and check each solution.

(a) $^-k = ^-12$

Check

(b) $7 = ^-t$

Check

(c) $^-m = ^-20$

Check

OBJECTIVE **3** **Solve equations such as $^-x = 5$.** When solving equations, do **not** leave a negative sign in front of the variable.

EXAMPLE 3 **Solving an Equation of the Type $^-x = 5$**

Solve $^-x = 5$ and check the solution.

It may look as if there is nothing more we can do to the equation $^-x = 5$, but ^-x is *not* the same as x. To see this, we write in the understood $^-1$ as the coefficient of ^-x.

Coefficient is understood to be $^-1$.

$^-x = 5$ can be written $^-1x = 5$

We want the coefficient of x to be $^+1$, not $^-1$. To accomplish that, we can divide both sides by the coefficient of x, which is $^-1$.

$$\frac{^-1x}{^-1} = \frac{5}{^-1} \quad \text{Divide both sides by } ^-1.$$

On the left side, $^-1 \div ^-1$ is 1. $\quad 1x = ^-5 \quad$ On the right side, $5 \div ^-1$ is $^-5$.

Now x is by itself on one side of the equal sign and has a coefficient of $^+1$. The solution is $^-5$.

Check $\quad ^-x = 5 \quad$ Go back to the *original* equation.

$^-1x = 5 \quad$ Write in the understood $^-1$ as the coefficient of ^-x.

$^-1 \cdot ^-5 = 5 \quad$ Replace x with $^-5$. ⟵ The solution is $^-5$, **not** 5.

$5 = 5 \quad$ Balances

When x is replaced with $^-5$, the equation balances, so **$^-5$ is the correct solution**.

> **CAUTION**
> As the last step in solving an equation, do **not** leave a negative sign in front of a variable. For example, do *not* leave $^-y = ^-8$. Write in the understood $^-1$ as the coefficient, so that
>
> $^-y = ^-8$ is written as $^-1y = ^-8$
>
> Then divide both sides by $^-1$ to get $y = 8$. The solution is 8.

◀ *Work Problem* **3** *at the Side.*

Solve each equation and check each solution. See Example 1.

1. $6z = 12$ **Check** $6z = 12$ **2.** $8k = 24$ **Check** $8k = 24$

3. $48 = 12r$ **Check**   **4.** $99 = 11m$ **Check**

5. $3y = 0$ **Check**   **6.** $5a = 0$ **Check**

7. $^-7k = 70$ **Check**   **8.** $^-6y = 36$ **Check**

9. $^-54 = ^-9r$ **Check**   **10.** $^-36 = ^-4p$ **Check**

11. $^-25 = 5b$ **Check**   **12.** $^-70 = 10x$ **Check**

Simplify where possible. Then solve each equation and check each solution. See Example 2.

13. $2r = ^-7 + 13$ **Check** $2r = \underbrace{^-7 + 13}$ **14.** $6y = 28 - 4$ **Check** $6y = \underbrace{28 - 4}$

15. $^-12 = 5p - p$ **Check**   **16.** $20 = z - 11z$ **Check**

Solve each equation. Show your work. See Examples 1 and 2.

17. $3 - 28 = 5a$

18. $^-55 + 7 = 8n$

19. $x - 9x = 80$

20. $4c - c = {}^-27$

21. $13 - 13 = 2w - w$

22. $^-11 + 11 = 8t - 7t$

23. $3t + 9t = 20 - 10 + 26$

24. $6m + 6m = 40 + 20 - 12$

25. $0 = {}^-9t$

26. $^-10 = 10b$

27. $^-14m + 8m = 6 - 60$

28. $7w - 14w = 1 - 50$

29. $100 - 96 = 31y - 35y$

30. $150 - 139 = 20x - 9x$

Use multiplication to simplify the side of the equation with the variable. Then solve each equation.

31. $3(2z) = {}^-30$

32. $2(4k) = 16$

33. $50 = {}^-5(5p)$

34. $60 = 4({}^-3a)$

35. ${}^-2({}^-4k) = 56$

36. ${}^-5(4r) = {}^-80$

37. ${}^-90 = {}^-10({}^-3b)$

38. ${}^-90 = {}^-5({}^-2y)$

Solve each equation. See Example 3.

39. ${}^-x = 32$

40. ${}^-c = 23$

41. ${}^-2 = {}^-w$

42. ${}^-75 = {}^-t$

43. ${}^-n = {}^-50$

44. ${}^-x = {}^-1$

45. $10 = {}^-p$

46. $100 = {}^-k$

47. Look again at the solutions to Exercises 39–46. Describe the pattern you see. Then write a rule for solving equations with a negative sign in front of the variable, such as $^-x = 5$.

48. Explain the division property of equality in your own words.

49. Explain and correct the error made by a student who solved this equation.

$$3x = \underbrace{16 - 1}$$

$$\frac{3x}{-3} = \frac{15}{-3}$$

$$x = {}^-5$$

50. Write two *different* equations that have $^-4$ as the solution. Be sure that you have to use the division property of equality to solve the equations. Show how to solve each equation. Use Exercises 1–14 as models.

51. The perimeter of a triangle with sides of equal length is 3 times the length of one side (s). If the perimeter is 45 ft, solving the equation $3s = 45$ will give the length of one side. Solve the equation.

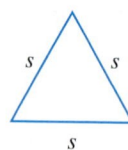

52. Refer to Exercise 51. If the perimeter of the triangle is 63 inches, solve the equation $3s = 63$ to find the length of one side.

53. The perimeter of a pentagon with sides of equal length is 5 times the length of one side (s). If the perimeter is 120 meters, solve the equation $120 = 5s$ to find the length of one side.

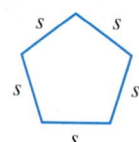

54. Refer to Exercise 53. If the pentagon has a perimeter of 335 yards, solving the equation $335 = 5s$ will give the length of one side. Solve the equation.

Solve each equation. Show your work.

55. $89 - 116 = {}^-4({}^-4y) - 9(2y) + y$

56. $58 - 208 = {}^-b + 8({}^-3b) + 5({}^-5b)$

57. $^-37(14x) + 28(21x) = |72 - 72| + |{}^-166 + 96|$

58. $6a - 10a - 3(2a) = |{}^-25 - 25| - 5(8)$

2.5 ▶▶▶ Solving Equations with Several Steps

OBJECTIVE 1 Solve equations using the addition and division properties of equality. To solve some equations, you need to use both the addition property of equality (see **Section 2.3**) and the division property of equality (see **Section 2.4**). Here are the steps.

> ### Solving an Equation Using the Addition and Division Properties
>
> **Step 1** *Add* the same amount to both sides of the equation so that the variable term (the variable and its coefficient) ends up by itself on one side of the equal sign.
>
> **Step 2** *Divide* both sides by the coefficient of the variable term to find the solution.
>
> **Step 3** *Check* the solution by going back to the *original* equation.

EXAMPLE 1 Solving an Equation with Several Steps

Solve this equation and check the solution: $5m + 1 = 16$.

Step 1 Get the variable term by itself on one side of the equal sign. The variable term is $5m$. Adding $^-1$ to the left side of the equation will leave $5m$ by itself. To keep the balance, add $^-1$ to the right side also.

$$5m + 1 = 16$$
$$\underline{\quad ^-1 \quad ^-1 \quad}$$
$$\underline{5m + 0} = 15$$
$$5m \quad = 15$$

First add the same thing to both sides.

Step 2 Divide both sides by the coefficient of the variable term. In $5m$, the coefficient is 5, so divide both sides by 5.

Divide both sides by the coefficient.

$$\frac{5m}{5} = \frac{15}{5}$$
$$m = 3$$

Step 3 Check the solution by going back to the *original* equation.

$$5m + 1 = 16 \quad \text{Use the } original \text{ equation}$$
$$\text{and replace } m \text{ with 3.}$$
$$5(3) + 1 = 16$$
$$15 + 1 = 16$$
$$16 = 16 \quad \text{Balances}$$

*The solution is 3, **not** 16.*

When m is replaced with 3, the equation balances, so **3 is the correct solution**.

Work Problem ① at the Side. ▶

So far, variable terms have appeared on just one side of the equal sign. But some equations start with variable terms on both sides. In that case, you can use the addition property of equality to add the same *variable term* to both sides of the equation, just as you have added the same *number* to both sides.

OBJECTIVES

1 Solve equations using the addition and division properties of equality.

2 Solve equations using the distributive, addition, and division properties.

---

① Solve each equation and check each solution.

(a) $2r + 7 = 13$

Check

(b) $20 = 6y - 4$

Check

(c) $^-10z - 9 = 11$

Check

ANSWERS

1. (a) $r = 3$
Check $2(3) + 7 = 13$
$6 + 7 = 13$
Balances $13 = 13$

(b) $y = 4$
Check $20 = 6(4) - 4$
$20 = 24 - 4$
Balances $20 = 20$

(c) $z = ^-2$
Check $^-10(^-2) - 9 = 11$
$20 - 9 = 11$
Balances $11 = 11$

2 Solve each equation *two ways*. First keep the variable term on the *left* side when you solve. Then solve again, keeping the variable term on the *right* side. Compare the solutions.

(a) $3y - 1 = 2y + 7$

$3y - 1 = 2y + 7$

(b) $3p - 2 = p - 6$

$3p - 2 = p - 6$

First decide whether to keep the variable term on the left side, or to keep the variable term on the right side. It doesn't matter which one you keep; just pick one side or the other. Then use the addition property to "get rid of" the variable term on the *other* side by adding its opposite.

EXAMPLE 2   **Solving an Equation with Variable Terms on Both Sides**

Solve this equation and check the solution: $2x - 2 = 5x - 11$.

First let's keep $2x$, the variable term on the *left* side. That means we need to "get rid of" $5x$ on the *right* side. We can do that by adding the opposite of $5x$, which is ^-5x.

To keep the balance, add ^-5x to the left side also. Write ^-5x under $2x$, *not* under 2.

Write ^-5x under $5x$, *not* under 11. $5x + {}^-5x$ is $0x$, or 0.

Change subtractions to adding the opposite.

$$
\begin{array}{rclcrcr}
2x & - & 2 & = & 5x & - & 11 \\
{}^-5x & & & & {}^-5x & & \\
\hline
{}^-3x & - & 2 & = & 0 & - & 11 \\
& \downarrow & \downarrow & & & \downarrow & \downarrow \\
{}^-3x & + & {}^-2 & = & 0 & + & {}^-11 \\
& & 2 & & & & 2 \\
\hline
{}^-3x & + & 0 & = & 0 & + & {}^-9 \\
\end{array}
$$

To get ^-3x by itself, add 2 to both sides.

Divide both sides by $^-3$, the coefficient of the variable term.

$$
\frac{{}^-3x}{{}^-3} = \frac{{}^-9}{{}^-3}
$$

$$
x = 3
$$

Suppose that, in the first step, we decided to keep $5x$ on the *right* side and "get rid of" $2x$ on the *left* side. Let's see what happens.

Add ^-2x to both sides.

Change subtractions to adding the opposite.

To get $3x$ by itself, add 11 to both sides.

$$
\begin{array}{rclcrcr}
2x & - & 2 & = & 5x & - & 11 \\
{}^-2x & & & & {}^-2x & & \\
\hline
0 & - & 2 & = & 3x & - & 11 \\
& \downarrow & \downarrow & & & \downarrow & \downarrow \\
0 & + & {}^-2 & = & 3x & + & {}^-11 \\
& & 11 & & & & 11 \\
\hline
0 & + & 9 & = & 3x & + & 0 \\
\end{array}
$$

Divide both sides by 3.

$$
\frac{9}{3} = \frac{3x}{3}
$$

$$
3 = x
$$

The two solutions are the same. In both cases, x balances with 3.

Notice that we used the addition principle *twice:* once to "get rid of" the variable term $2x$ and once to "get rid of" the number $^-11$. We could have done those steps in the reverse order without changing the result.

> **Note**
>
> More than one sequence of steps will work to solve complicated equations. The basic approach is the following:
>
> • Simplify each side of the equation, if possible.
> • Get the variable term by itself on one side of the equal sign and a number by itself on the other side.
> • Divide both sides by the coefficient of the variable term.

ANSWERS

2. **(a)** $y = 8$ and $8 = y$
   (b) $p = {}^-2$ and $^-2 = p$

◀ **Work Problem** **2** at the Side.

OBJECTIVE 2 Solve equations using the distributive, addition, and division properties. If an equation contains parentheses, check to see whether you can use the distributive property to remove them.

EXAMPLE 3 Solving an Equation Using the Distributive Property

Solve this equation and check the solution: $^-6 = 3(y - 2)$.

We can use the distributive property to simplify the right side of the equation. Recall from **Section 2.2** that

$$3\,(y - 2) \quad \text{can be written as} \quad \underbrace{3 \cdot y}_{3y} - \underbrace{3 \cdot 2}_{6}$$

So the original equation $^-6 = 3(y - 2)$ becomes $^-6 = 3y - 6$.

$$
\begin{array}{ll}
^-6 = 3y - 6 & \text{Change subtraction to} \\
\quad\quad\downarrow\quad\downarrow & \quad\text{adding the opposite.} \\
^-6 = 3y + {}^-6 & \text{To get } 3y \text{ by itself,} \\
\underline{6 \qquad\quad 6} & \quad\text{add 6 to both sides.} \\
\dfrac{0}{3} = \dfrac{3y}{3} & \text{Divide both sides by 3,} \\
& \quad\text{the coefficient of } 3y. \\
0 = y
\end{array}
$$

The solution is 0.

Check
$$
\begin{array}{ll}
^-6 = 3(y - 2) & \text{Go back to the } \textit{original} \text{ equation and replace} \\
\quad\quad\downarrow & \quad y \text{ with 0.} \\
^-6 = 3(0 - 2) & \text{Follow the order of operations: work inside} \\
\quad\quad\downarrow\quad\downarrow & \quad\text{parentheses first.} \\
^-6 = 3(0 + {}^-2) & \text{Change subtraction to addition.} \\
^-6 = \quad 3({}^-2) & \\
^-6 = \quad\quad ^-6 & \text{Balances}
\end{array}
$$

When y is replaced with 0, the equation balances, so **0 is the correct solution**.

——————————— *Work Problem* **3** *at the Side.* ▶

Here is a summary of all the steps you can use to solve an equation. Sometimes you will use only two or three steps, and sometimes you will need all five steps, as in the example on the next page.

Solving an Equation

Step 1 If possible, use the *distributive property* to remove parentheses.

Step 2 **Combine** any like terms on the left side of the equation. Combine any like terms on the right side of the equation.

Step 3 **Add** the same amount to both sides of the equation so that the variable term ends up by itself on one side of the equal sign and a number is by itself on the other side. You may have to do this step more than once.

Step 4 **Divide** both sides by the coefficient of the variable term to find the solution.

Step 5 **Check** your solution by going back to the *original* equation. Replace the variable with your solution. Follow the order of operations to complete the calculations. If the two sides of the equation balance, your solution is correct.

---

3 Solve each equation and check each solution.

(a) $^-12 = 4(y - 1)$

(b) $5(m + 4) = 20$

(c) $6(t - 2) = 18$

ANSWERS

3. (a) $y = {}^-2$ (b) $m = 0$ (c) $t = 5$

4 Solve each equation and check each solution.

(a) $3(b + 7) = 2b - 1$

Solving an Equation

Solve this equation and check the solution: $8 + 5(m + 2) = 6 + 2m$

Step 1 Use the distributive property on the left side.

$$8 + 5(m + 2) = 6 + 2m$$

Step 2 Combine like terms on the left side.

$$8 + 5m + 10 = 6 + 2m$$ No like terms on the right side.

$$5m + 18 = 6 + 2m$$

Step 3 Add ^-2m to both sides.

$$\begin{array}{r} {}^-2m \qquad\qquad {}^-2m \\ \overline{} \\ 3m + 18 = 6 + 0 \\ 3m + 18 = 6 \end{array}$$

Step 3 To get $3m$ by itself, add $^-18$ to both sides.

$$\begin{array}{r} {}^-18 \qquad {}^-18 \\ \overline{} \\ 3m + 0 = {}^-12 \end{array}$$

Step 4 Divide both sides by 3, the coefficient of the variable term $3m$.

$$\frac{3m}{3} = \frac{^-12}{3}$$

$$m = {}^-4$$

The solution is $^-4$.

Step 5 **Check**

$$8 + 5(m + 2) = 6 + 2m$$ Replace each m with $^-4$.

$$8 + 5(^-4 + 2) = 6 + 2(^-4)$$

$$8 + 5(^-2) = 6 + ^-8$$

$$8 + ^-10 = ^-2$$

$$^-2 = ^-2$$ Balances

When m is replaced with $^-4$, the equation balances, so **$^-4$ is the correct solution (not $^-2$).**

(b) $6 - 2n = 14 + 4(n - 5)$

◀ *Work Problem* **4** *at the Side.*

Solve each equation and check each solution. See Example 1.

 1. $7p + 5 = 12$ **Check** $7p + 5 = 12$
↓

2. $6k + 3 = 15$ **Check** $6k + 3 = 15$
↓

3. $2 = 8y - 6$ **Check**

4. $10 = 11p - 12$ **Check**

5. $^-3m + 1 = 1$ **Check**

6. $^-4k + 5 = 5$ **Check**

7. $28 = ^-9a + 10$ **Check**

8. $75 = ^-10w + 25$ **Check**

9. $^-5x - 4 = 16$ **Check**

10. $^-12b - 3 = 21$ **Check**

*In Exercises 11–16, solve each equation **two** ways. First keep the variable term on the left side when you solve it. Then solve it again, keeping the variable term on the right side. Finally, check your solution. See Example 2.*

11. $6p - 2 = 4p + 6$ $6p - 2 = 4p + 6$ **Check** $6p - 2 = 4p + 6$
                                                                                              $\downarrow$ $\downarrow$

12. $5y - 5 = 2y + 10$ $5y - 5 = 2y + 10$ **Check** $5y - 5 = 2y + 10$
                                                                                              $\downarrow$ $\downarrow$

13. $^{-}2k - 6 = 6k + 10$ $^{-}2k - 6 = 6k + 10$ **Check**

14. $5x + 4 = {}^{-}3x - 4$ $5x + 4 = {}^{-}3x - 4$ **Check**

15. $^{-}18 + 7a = 2a + 7$ $^{-}18 + 7a = 2a + 7$ **Check**

16. $^-9 + 2z = 9z + 12$ $\qquad$ $^-9 + 2z = 9z + 12$ $\qquad$ **Check**

Use the distributive property to help you solve each equation. Show your work.
See Example 3.

17. $8(w - 2) = 32$ $\qquad$ **18.** $9(b - 4) = 27$ $\qquad$ **19.** $^-10 = 2(y + 4)$

20. $^-3 = 3(x + 6)$ $\qquad$ **◉ 21.** $^-4(t + 2) = 12$ $\qquad$ **22.** $^-5(k + 3) = 25$

23. $6(x - 5) = ^-30$ $\qquad$ **24.** $7(r - 7) = ^-49$ $\qquad$ **25.** $^-12 = 12(h - 2)$

26. $^-11 = 11(c - 3)$ $\qquad$ **27.** $0 = ^-2(y + 2)$ $\qquad$ **28.** $0 = ^-9(b + 1)$

Solve each equation. Show your work. See Example 4.

29. $6m + 18 = 0$

30. $8p - 40 = 0$

31. $6 = 9w - 12$

32. $8 = 8h + 24$

33. $5x = 3x + 10$

34. $7n = {}^{-}2n - 36$

35. $2a + 11 = 8a - 7$

36. $r - 10 = 10r + 8$

37. $7 - 5b = 28 + 2b$

38. $1 - 8t = {}^{-}9 - 3t$

39. ${}^{-}20 + 2k = k - 4k$

40. $6y - y = {}^{-}16 + y$

41. $10(c - 6) + 4 = 2 + c - 58$

42. $8(z + 7) - 6 = z + 60 - 10$

43. ${}^{-}18 + 13y + 3 = 3(5y - 1) - 2$

44. $3 + 5h - 9 = 4(3h + 4) - 1$

45. $6 - 4n + 3n = 20 - 35$

46. $^-19 + 8 = 6p - 7p - 5$

47. $6(c - 2) = 7(c - 6)$

48. $^-3(5 + x) = 4(x - 2)$

49. $^-5(2p + 2) - 7 = 3(2p + 5)$

50. $4(3m - 6) = 72 + 3(m - 8)$

51. $^-6b - 4b + 7b = 10 - b + 3b$

52. $w + 8 - 5w = {}^-w - 15w + 11w$

53. Solve $^-2t - 10 = 3t + 5$. Show each step you take while solving it. Next to each step, write a sentence that explains what you did in that step. Be sure to tell when you used the addition property of equality and when you used the division property of equality.

54. Explain the distributive property in your own words. Show two examples of using the distributive property to remove parentheses in an expression.

55. Here is one student's solution to an equation. Show how to check the solution. If the solution doesn't check, explain the error and correct it.

$$
\begin{aligned}
-8 + 4a &= 2a + 2 \\
\underline{-2a} \quad\;\; &\quad \underline{-2a} \\
-10 + 4a &= 0 + 2 \\
-10 + 4a &= 2 \\
\underline{10} \quad\;\; &\quad \underline{10} \\
0 + 4a &= 12 \\
\frac{4a}{4} &= \frac{12}{4} \\
a &= 3
\end{aligned}
$$

56. Here is one student's solution to an equation. Show how to check the solution. If the solution doesn't check, explain the error and correct it.

$$
\begin{aligned}
2(x + 4) &= -16 \\
2x + 4 &= -16 \\
\underline{-4} \quad\;\; &\quad \underline{-4} \\
2x + 0 &= -20 \\
\frac{2x}{2} &= \frac{-20}{2} \\
x &= -10
\end{aligned}
$$

Relating Concepts (Exercises 57–60) For Individual or Group Work

Work Exercises 57–60 in order.

57. (a) Suppose that the sum of two numbers is negative and you know that one of the numbers is positive. What can you conclude about the other number?

(b) How can you tell, just by looking, that the solution to $x + 5 = -7$ must be a negative number? Recall your answer from part (a).

58. (a) Suppose that the sum of two numbers is positive and you know that one of the numbers is negative. What can you conclude about the other number?

(b) How can you tell, just by looking, that the solution to $-8 + d = 2$ must be a positive number? Recall your answer from part (a).

59. (a) Suppose the product of two numbers is negative and you know that one of the numbers is negative. What can you conclude about the other number?

(b) How can you tell, just by looking, that the solution to $-15n = -255$ must be positive?

60. (a) Suppose the product of two numbers is positive and you know that one of the numbers is negative. What can you conclude about the other number?

(b) How can you tell, just by looking, that the solution to $437 = -23y$ must be negative?

Chapter 2 ▶▶▶ Summary

▶ Key Terms

2.1	**variable**	A variable is a letter that represents a number that varies or changes, depending on the situation.
	constant	A constant is a number that is added or subtracted in an expression. It does not vary. For example, 5 is the constant in the expression $c + 5$.
	expression	An expression expresses, or tells, the rule for doing something. It is a combination of operations on variables and numbers. An expression does *not* have an equal sign.
	evaluate the expression	To evaluate an expression, replace each variable with specific values (numbers) and follow the order of operations.
	coefficient	The number part in a multiplication expression is the coefficient. For example, 4 is the coefficient in the expression $4s$.
2.2	**simplifying expressions**	To simplify an expression, write it in a simpler way by combining all the like terms.
	term	Each addend in an expression is a term.
	variable term	A variable term has a number part (called the coefficient) multiplied by a variable part (a letter). An example is $4s$.
	like terms	Like terms are terms with exactly the same variable parts (the same letters and exponents). The coefficients may be different.
2.3	**equations**	An equation has an equal sign. It shows the relationship between what is known about a problem and what isn't known.
	solve the equation	To solve an equation, find a number that can replace the variable and make the equation balance.
	solution	A solution of an equation is a number that can replace the variable and make the equation balance.
	addition property of equality	The addition property of equality states that adding the same quantity to both sides of an equation will keep it balanced.
	check the solution	To check the solution of an equation, go back to the *original* equation and replace the variable with the solution. If the equation balances, the solution is correct.
2.4	**division property of equality**	The division property of equality states that dividing both sides of an equation by the same nonzero number will keep it balanced.

▶ Test Your Word Power

See how well you have learned the vocabulary in this chapter. Answers follow the Quick Review.

1. A **variable**
   A. can only be the letter x
   B. is never an addend in an expression
   C. is the solution of an equation
   D. represents a number that varies.

2. Which expression has 2 as a **coefficient?**
   A. x^2
   B. $2x$
   C. $x + 2$
   D. $2 - x$

3. Which expression has 4 as a **constant** term?
   A. $4y$
   B. y^4
   C. $4 + y$
   D. $\dfrac{y}{4}$

4. Which expression has four **terms?**
   A. $2 + 3x = {}^{-}6 + x$
   B. $2 + 3x - 6 + x$
   C. $(2)(3x)({}^{-}6)(x)$
   D. $2(3x) = {}^{-}6(x)$

5. Like terms
   A. can be multiplied but not added
   B. have the same coefficients
   C. have the same solutions
   D. have the same variable parts.

6. To **simplify an expression,**
   A. combine all the like terms
   B. multiply the exponents
   C. add all the numbers in the expression
   D. add the same quantity to both sides.

▶ Quick Review

Concepts	Examples

(2.1) Evaluating Expressions

Replace each variable with the specified value. Then follow the order of operations to simplify the expression.

The expression for ordering textbooks for two prealgebra classes is $2c + 10$, where c is the class limit. Evaluate the expression when the class limit is 24.

$$2c + 10 \qquad \text{Replace } c \text{ with 24.}$$

$$2 \cdot 24 + 10 \qquad \text{Multiply first.}$$

$$48 + 10 \qquad \text{Add last.}$$

$$58 \qquad \text{Order 58 books.}$$

(2.1) Using Exponents with Variables

An exponent next to a variable tells how many times to use the variable as a factor in multiplication.

Rewrite $^-6x^4$ without exponents.

$^-6x^4$ can be written as $^-6 \cdot x \cdot x \cdot x \cdot x$

Coefficient is $^-6$ — x is used as a factor 4 times.

(2.1) Evaluating Expressions with Exponents

Rewrite the expression without exponents, replace each variable with the specified value, and multiply all the factors.

Evaluate x^3y when x is $^-4$ and y is 5.

x^3y means $x \cdot x \cdot x \cdot y$ Replace x with $^-4$ and y with 5.

$^-4 \cdot ^-4 \cdot ^-4 \cdot 5$ Multiply two factors at a time.

$16 \quad \cdot ^-4 \cdot 5$

$^-64 \quad \cdot 5$

$^-320$

(2.2) Identifying Like Terms

Like terms have *exactly* the same letters and exponents. The coefficients may be different.

List the like terms in this expression. Then identify the coefficients of the like terms.

$$^-3b + ^-3b^2 + 3ab + b + 3$$

The like terms are ^-3b and b. The coefficient of ^-3b is $^-3$, and the coefficient of b is understood to be 1.

(2.2) Combining Like Terms

Step 1 If there are any variable terms with no coefficient, write in the understood 1.

Step 2 Change any subtractions to adding the opposite.

Step 3 Find like terms.

Step 4 Add the coefficients of like terms, keeping the variable part the same.

Simplify $4x^2 - 10 + x^2 + 15$.

Write understood 1.

$$4x^2 - 10 + 1x^2 + 15$$

Change subtraction to adding the opposite.

$$4x^2 + ^-10 + 1x^2 + 15$$

$$4x^2 + 1x^2 + ^-10 + 15$$

Combine $4x^2 + 1x^2$. The variable part stays the same. Also combine $^-10 + 15$.

$$(4 + 1)x^2 + 5$$

$$5x^2 + 5$$

The simplified expression is $5x^2 + 5$.

Concepts	Examples

2.2 Simplifying Multiplication Expressions

Use the associative property to rewrite the expression so that the two number parts can be multiplied. The variable part stays the same.

Simplify: $^{-}7(5k)$
Use the associative property of multiplication.

$$^{-}7 \cdot (5 \cdot k) \text{ can be written as } \underbrace{(^{-}7 \cdot 5)} \cdot k$$
$$\underbrace{^{-}35} \cdot k$$
$$^{-}35k$$

The simplified expression is $^{-}35k$.

2.2 Using the Distributive Property

Multiplication distributes over addition and over subtraction. Be careful to multiply *every* term inside the parentheses by the number outside the parentheses.

Simplify.

(a) $6(w - 4)$ can be written as $\underbrace{6 \cdot w}_{6w} - \underbrace{6 \cdot 4}_{24}$

The simplified expression is $6w - 24$.

(b) $^{-}3(2b + 5)$ can be written as $\underbrace{^{-}3 \cdot 2b}_{^{-}6b} + \underbrace{^{-}3 \cdot 5}_{+ \, ^{-}15}$

Use the definition of subtraction "in reverse" to write $^{-}6b + \, ^{-}15$ as $^{-}6b - 15$.

The simplified expression is $^{-}6b - 15$.

2.3 Solving and Checking Equations Using the Addition Property of Equality

If possible, *simplify* the expression on one or both sides of the equal sign.

Next, to get the variable by itself on one side of the equal sign, *add* the same number to both sides.

Finally, *check* the solution by going back to the original equation and replacing the variable with the solution. If the equation balances, the solution is correct.

Solve this equation and check the solution.

$$
\begin{array}{rcll}
\underbrace{^{-}5 + 8} & = & 9 + r & \text{Simplify the left side by} \\
 & & & \text{adding } ^{-}5 + 8. \\
3 & = & 9 + r & \text{To get } r \text{ by itself, add} \\
\underline{^{-}9} & & \underline{^{-}9} & \text{the opposite of 9,} \\
 & & & \text{which is } ^{-}9, \text{ to both sides.} \\
^{-}6 & = & 0 + r & \\
^{-}6 & = & r &
\end{array}
$$

The solution is $^{-}6$.

Check $^{-}5 + 8 = 9 + \underset{\downarrow}{r}$ Use the original equation and replace r with $^{-}6$.

$$\underbrace{^{-}5 + 8} = \underbrace{9 + \, ^{-}6}$$
$$3 = 3 \quad \text{Balances}$$

When r is replaced with $^{-}6$, the equation balances, so $^{-}6$ is the correct solution (**not** 3).

Concepts

Examples

(2.4) Solving and Checking Equations Using the Division Property of Equality

If possible, *simplify* the expression on one or both sides of the equal sign.

Next, to get the variable by itself on one side of the equal sign, *divide* both sides by the coefficient of the variable term.

Finally, *check* the solution by going back to the original equation and replacing the variable with the solution. If the equation balances, the solution is correct.

Solve this equation and check the solution.

Simplify the left side. Change subtraction to adding the opposite.

$$2h - 6h = 18 + 22$$

Simplify the right side; add $18 + 22$.

$$2h + {}^{-}6h = 40$$

Divide by $^{-}4$, the coefficient of $^{-}4h$.

$$\frac{{}^{-}4h}{{}^{-}4} = \frac{40}{{}^{-}4}$$

Also divide 40 by $^{-}4$ to keep the balance.

$$h = {}^{-}10$$

The solution is $^{-}10$.

Check

$$2h - 6h = 18 + 22$$

Original equation; replace h with $^{-}10$.

$$2({}^{-}10) - 6({}^{-}10) = 40$$

$${}^{-}20 - {}^{-}60 = 40$$

$${}^{-}20 + {}^{+}60$$

$$40 = 40$$ Balances

When h is replaced with $^{-}10$, the equation balances, so $^{-}10$ is the correct solution (**not** 40).

(2.4) Solving Equations Such as $^{-}x = 5$

As the last step in solving an equation, do **not** leave a negative sign in front of the variable, such as $^{-}x = 5$, because ^{-}x is **not** the same as x. Divide both sides by $^{-}1$, the understood coefficient of ^{-}x.

Solve this equation and check the solution.

$$9 = {}^{-}n$$

Write the understood $^{-}1$ as the coefficient of n.

$$9 = {}^{-}n \quad \text{can be written as} \quad 9 = {}^{-}1n$$

Now divide both sides by $^{-}1$.

$$\frac{9}{{}^{-}1} = \frac{{}^{-}1n}{{}^{-}1}$$

$${}^{-}9 = n$$

The solution is $^{-}9$.

Check

$$9 = {}^{-}n$$ Original equation

$$9 = {}^{-}1n$$ Write understood $^{-}1$.

$$9 = {}^{-}1({}^{-}9)$$ Replace n with $^{-}9$.

$$9 = 9$$ Balances

When n is replaced with $^{-}9$, the equation balances, so $^{-}9$ is the correct solution (**not** 9).

Concepts	Examples

2.5 Solving Equations with Several Steps

Solve this equation and check the solution.

Step 1 If possible, use the distributive property to remove parentheses.

Step 1 $\quad 3 + 2(y + 8) = 5y + 4$

Step 2 Combine any like terms on the left side of the equal sign. Combine any like terms on the right side of the equal sign.

Step 2 $\quad 3 + 2y + 16 = 5y + 4$

$$2y + 19 = 5y + 4$$

Step 3 Add the same amount to both sides of the equation so that the variable term ends up by itself on one side of the equal sign, and a number is by itself on the other side. You may have to do this step more than once.

Step 3
$$\begin{array}{rcl} -2y & & -2y \\ \hline 0 + 19 &=& 3y + 4 \end{array}$$

Step 3
$$\begin{array}{rcl} & -4 & & -4 \\ \hline 0 + 15 &=& 3y + 0 \end{array}$$

Step 4 Divide both sides by the coefficient of the variable term to find the solution.

Step 4
$$\frac{15}{3} = \frac{3y}{3}$$

$$5 = y$$

The solution is 5.

Step 5 Check the solution by going back to the original equation. Replace the variable with the solution. If the equation balances, the solution is correct.

Check Step 5

$$3 + 2(y + 8) = 5y + 4 \quad \text{Original equation}$$

$$3 + 2(5 + 8) = 5(5) + 4$$

$$3 + 2(13) = 25 + 4$$

$$3 + 26 = 29$$

$$29 = 29 \quad \text{Balances}$$

When y is replaced with 5, the equation balances, so 5 is the correct solution (**not** 29).

ANSWERS TO TEST YOUR WORD POWER

1. D; *Example:* In $c + 5$, the variable is c.
2. B; *Example:* $2y^3$ and $2n$ also have 2 as a coefficient.
3. C; *Example:* $-5a^2 + 4$ also has 4 as a constant term.
4. B; *Example:* $3y^2 - 6y + 2y - 5$ also has four terms. Choices (A) and (D) are equations, not expressions; choice (C) has four *factors*.
5. D; *Example:* $7n^2$ and $-3n^2$ are like terms.
6. A; *Example:* To simplify $4a - 9 + 6a$, combine $4a$ and $6a$ by adding the coefficients. The simplified expression is $10a - 9$.

Math in the Media

ALGEBRAIC EXPRESSIONS AND TUITION COSTS

Algebraic expressions are useful in the real world whenever the same set of instructions is repeated for different choices of numbers. Calculating college tuition and fees is an example. The information below is for "resident of the district" students at North Harris Montgomery Community College District (NHMCCD) in Texas.

Fees Required at NHMCCD, Spring 2008

[*Residents of the district pay*] tuition at the rate of $36 per credit hour, a $6 per credit hour technology fee, a $2 per credit hour student activity fee, a $2 per credit hour general use fee, and a registration fee of $12.

1. Calculate the tuition and fees for a student who is a resident of the district and who enrolls at NHMCCD for the number of credit hours listed in parts **(a)**, **(b)**, and **(c)** below. Then, for part **(d)**, let x represent the number of credit hours. Pay attention to the *process* you used in your calculations so that you can write the algebraic expression for x credit hours.

   (a) 3 credit hours: _____   **(b)** 9 credit hours: _____

   (c) 12 credit hours: _____   **(d)** x credit hours: _____ dollars

Write the algebraic expression that represents the tuition and fees for each institution for one semester. Let x represent the number of credit hours.

Institution/Web Site	Description of Tuition and Fees	Algebraic Expression
2. American River College, California Nonresident student www.arc.losrios.edu	Enrollment: $20 per credit hour Parking: $30 per semester Additional nonresident enrollment: $193 per credit hour Student representation fee: $1 per semester	
3. Austin Community College, Texas Out-of-district student www.austincc.edu	Tuition: $39 per credit hour General fee: $13 per credit hour Parking: $5 per semester Additional out-of-district tuition: $79 per credit hour Student activity fee: $2 per credit hour	
4. Valdosta State University, Georgia In-state student www.valdosta.edu	Tuition: $124 per credit hour Health fee: $87 per semester Student activity fee: $209 per semester Athletics fee: $116 per semester Technology fee: $48 per semester Parking fee: $50 per semester	

5. List the tuition and fees for your college. Then write an algebraic expression using x to represent the number of credit hours.

Source: www.nhmdd.edu

Chapter 2 ▶▶▶ Review Exercises

[2.1]

1. (a) Identify the variable, the coefficient, and the constant in this expression.

$$^-3 + 4k$$

(b) Circle the expression that has 20 as the constant term and $^-9$ as the coefficient.

$20m - 9$ $^-9 + 20x$

$^-9y + 20$ $20 + 9n$

2. The expression for ordering test tubes for a chemistry lab is $4c + 10$, where c is the class limit. Evaluate the expression when

(a) the class limit is 15.

(b) the class limit is 24.

3. Rewrite each expression without exponents.

(a) x^2y^4

(b) $5ab^3$

4. Evaluate each expression when m is 2, n is $^-3$, and p is 4.

(a) n^2 **(b)** n^3

(c) $^-4mp^2$ **(d)** $5m^4n^2$

[2.2] *Simplify.*

5. $ab + ab^2 + 2ab$

6. $^-3x + 2y - x - 7$

7. $^-8(^-2g^3)$

8. $4(3r^2t)$

9. $5(k + 2)$

10. $^-2(3b + 4)$

11. $3(2y - 4) + 12$

12. $^-4 + 6(4x + 1) - 4x$

13. Write an expression with four terms that *cannot* be simplified.

[2.3] *Solve each equation and check each solution.*

14. $16 + n = 5$ **Check** $16 + \quad n = 5$

15. $^-4 + 2 = 2a - 6 - a$ **Check**

$^-4 + 2 = 2a \quad - 6 - 4$

[2.4] *Solve each equation. Show your work.*

16. $48 = ^-6m$

17. $k - 5k = ^-40$

18. $^-17 + 11 + 6 = 7t$

19. $^-2p + 5p = 3 - 21$ **20.** $^-30 = 3(^-5r)$ **21.** $12 = {}^-h$

[2.5] *Solve each equation. Show your work.*

22. $12w - 4 = 8w + 12$ **23.** $0 = {}^-4(c + 2)$

24. Every Wednesday is "treat day" at the office. The person who brings the treats buys two treat items for each employee plus four extras. If Roger brings 34 doughnuts, solve the equation $34 = 2n + 4$ to find the number of employees.

▶▶▶ Mixed Review Exercises

Solve each equation. Show your work.

25. $12 + 7a = 4a - 3$ **26.** $^-2(p - 3) = {}^-14$ **27.** $10y = 6y + 20$

28. $2m - 7m = 5 - 20$ **29.** $20 = 3x - 7$ **30.** $b + 6 = 3b - 8$

31. $z + 3 = 0$ **32.** $3(2n - 1) = 3(n + 3)$ **33.** $^-4 + 46 = 7(^-3t + 6)$

34. $6 + 10d - 19 = 2(3d + 4) - 1$ **35.** $^-4(3b + 9) = 24 + 3(2b - 8)$

Chapter 2 ▸▸▸ Test  Use the Chapter Test Prep Video CD to see fully worked-out solutions to any of the exercises you want to review.

1. Identify the parts of this expression: $-7w + 6$
Choose from these labels: variable, constant, coefficient.

1. _____

2. The expression for buying hot dogs for the company picnic is $3a + 2c$, where a is the number of adults and c is the number of children. Evaluate the expression when there are 45 adults and 21 children.

2. _____

Rewrite each expression without exponents.

3. $x^5 y^3$

4. $4ab^4$

3. _____

4. _____

5. Evaluate $-2s^2 t$ when s is -5 and t is 4.

5. _____

Simplify each expression.

6. $3w^3 - 8w^3 + w^3$

7. $xy - xy$

6. _____

7. _____

8. $-6c - 5 + 7c + 5$

9. $3m^2 - 3m + 3mn$

8. _____

9. _____

10. $-10(4b^2)$

11. $-5(-3k)$

10. _____

11. _____

12. $7(3t + 4)$

13. $-4(a + 6)$

12. _____

13. _____

14. $-8 + 6(x - 2) + 5$

15. $-9b - c - 3 + 9 + 2c$

14. _____

15. _____

Solve each equation and check each solution.

16. $^-4 = x - 9$ **Check**

17. $^-7w = 77$ **Check**

18. $^-p = 14$ **Check**

19. $^-15 = ^-3(a + 2)$ **Check**

Solve each equation. Show your work.

20. $6n + 8 - 5n = ^-4 + 4$

21. $5 - 20 = 2m - 3m$

22. $^-2x + 2 = 5x + 9$

23. $3m - 5 = 7m - 13$

24. $2 + 7b - 44 = ^-3b + 12 + 9b$

25. $3c - 24 = 6(c - 4)$

26. Write an equation that requires the *addition* property of equality to solve it and has $^-4$ as its solution. Then write a different equation that requires the *division* property of equality to solve it and has $^-4$ as its solution. Show how to solve each equation.

16. _____

17. _____

18. _____

19. _____

20. _____

21. _____

22. _____

23. _____

24. _____

25. _____

26. _____

Study Skills

▶▶▶ PREPARING FOR TESTS

Many things besides studying can improve your test scores. You may not realize that eating the right foods, and getting enough exercise and sleep can also improve your scores. Your brain (and therefore your ability to think) is affected by the condition of your whole body. So, part of your preparation for tests includes keeping yourself in good physical shape as well as spending time on the actual course material. Try these suggestions and see the difference.

OBJECTIVES

1 Restate the importance of sleep and good nutrition as it affects learning.

2 Explain the effect of anxiety and stress on learning.

Performance Health Tips

Performance Health Tips to Improve Your Test Score	Explanation
Get **seven to eight hours of sleep** the night before the exam. (It's helpful to get that much sleep *every* night.)	**Fatigue and exhaustion** reduce efficiency. They also cause poor memory and recall. If you didn't sleep much the night before a test, 20 minutes of relaxation or meditation can help. (Also see the comments below about eating carbohydrates to help you sleep.)
Eat a **small, high-energy meal** about two hours before the test. Start the meal with a small amount of protein such as fish, chicken, or nonfat yogurt. Include carbohydrates if you like, but no high-fat foods.	Just 3 to 4 ounces of protein increases the amount of a chemical in the brain called tyrosine, which **improves your alertness, accuracy, and motivation.** High-fat foods dull your mind and slow down your brain.
Drink plenty of water. Don't wait until you feel thirsty; your body is already dehydrated by the time you feel it.	Research suggests that staying well hydrated improves the electrochemical communications in your brain.
Give your brain the time it needs to grow dendrites!	**Cramming doesn't work;** your brain cannot grow dendrites that quickly. **Studying every day** using these study skills techniques is the way to give your brain the time it needs.

Anxiety Prevention Tips

To Prevent Anxiety	Explanation
Practice slow, deep breathing for five minutes each day. Then do a minute or two of deep breathing right before the test. Also, if you feel your anxiety building during the test, stop for a minute, close your eyes, and do some deep breathing.	When **test anxiety** hits, you breathe more quickly and shallowly, which causes hyperventilation. Symptoms may be confusion, inability to concentrate, shaking, dizziness, and more. Slow, deep breathing will **calm you and prevent panic.**

To Prevent Anxiety	Explanation
Do 15 to 20 minutes of *moderate exercise* (like walking) shortly before the test. Daily exercise is even better!	*Exercise reduces stress* and will help prevent "blanking out" on a test. Exercise also increases your alertness, clear thinking, and energy.
To help you sleep the night before the test, or any time you need to calm down, *eat high-carbohydrate foods* such as popcorn, bread, rice, crackers, muffins, bagels, pasta, corn, baked potatoes (not fries or chips), and cereals.	Carbohydrates increase the level of a chemical in the brain called serotonin, which has a *calming effect on the mind*. It reduces feelings of tension and stress and improves your ability to concentrate. You only need to eat a small amount, like half a bagel, to get this effect.
Before the test, *go easy on caffeinated beverages* such as coffee, tea, and soft drinks. Do not eat candy bars or other sugary snacks.	Extra caffeine can *make you jittery,* "hyper," and shaky for the test. It can increase the tendency to panic. Too much sugar causes negative emotional reactions in some people.

Now Try This ▶▶▶

What will you do to improve your next test score? List the three or four tips you think will help you the most.

1. _____

2. _____

3. _____

4. _____

What changes will you have to make in order to try the tips you chose?

_____

_____

_____

_____

See *Tips for Taking Math Tests* **and** *Preparing for Your Final Exam* **for more ideas about managing anxiety.** (Check the Table of Contents to find their locations.)

3

Solving Application Problems

A century ago there were only 8000 cars and 144 miles of paved roads in the United States. But now there are 137,000,000 registered cars and 4,000,000 miles of paved roads. One 12-lane segment of Interstate Highway 5 in California carries over 365,000 vehicles every day! (*Source:* Federal Highway Administration.)

A handy formula for drivers to use is the distance formula, $d = rt$. In **Section 3.1,** Exercises 51–54, you'll see how to use the formula to find your driving time, rate (speed), or distance traveled on long trips.

3.1 ▶▶▶ Problem Solving: Perimeter

OBJECTIVES

1 Use the formula for perimeter of a square to find the perimeter or the length of one side.

2 Use the formula for perimeter of a rectangle to find the perimeter, the length, or the width.

3 Find the perimeter of parallelograms, triangles, and irregular shapes.

OBJECTIVE 1 Use the formula for perimeter of a square to find the perimeter or the length of one side. If you have ever studied geometry, you probably used several different formulas such as $P = 2l + 2w$ and $A = lw$. A **formula** is just a shorthand way of writing a rule for solving a particular type of problem. A formula uses variables (letters) and it has an equal sign, so it is an equation. That means you can use the equation-solving techniques you learned in **Chapter 2** to work with formulas.

But let's start at the beginning. Geometry was developed centuries ago when people needed a way to measure land. The name *geometry* comes from the Greek words *ge,* meaning earth, and *metron,* meaning measure. Today we still use geometry to measure land. It is also important in architecture, construction, navigation, art and design, physics, chemistry, and astronomy. You can use geometry at home when you buy carpet or wallpaper, hang a picture, or do home repairs. In this chapter you'll learn about two basic ideas, perimeter and area. Other geometry concepts will appear in later chapters.

In **Section 2.1,** you found the *perimeter* of a square garden.

> ### Perimeter
> The distance around the outside edges of any flat shape is called the **perimeter** of the shape.

To review, a **square** has four sides that are all the same length. Also, the sides meet to form *right angles,* which measure 90° (90 degrees). This means that the sides form "square corners." Two examples of squares are shown below.

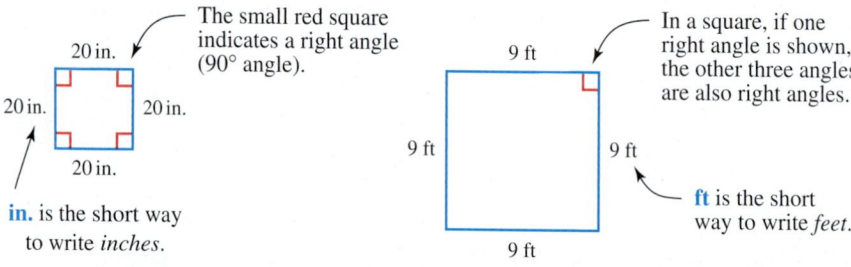

To find the *perimeter* of a square, we can "unfold" the shape so the four sides lie end-to-end, as shown below.

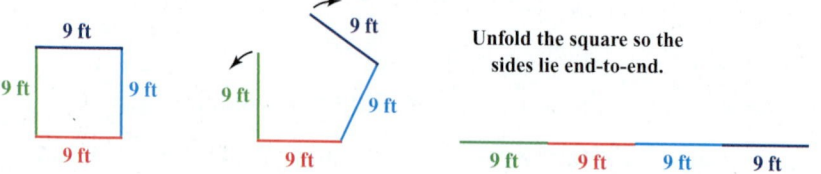

Now we can see the total length of the four sides. The total length is the *perimeter* of the square. We can find the perimeter by adding.

$$\text{Perimeter} = 9 \text{ ft} + 9 \text{ ft} + 9 \text{ ft} + 9 \text{ ft} = 36 \text{ ft}$$

A shorter way is to multiply the length of one side times 4, because all 4 sides are the same length.

Finding the Perimeter of a Square

Perimeter of a square = side + side + side + side

or $P = 4 \cdot$ side

$P = 4s$

1 Find the perimeter of each square, using the appropriate formula.

(a) The 20 in. square shown on the previous page

EXAMPLE 1 **Finding the Perimeter of a Square**

Find the perimeter of the square on the previous page that measures 9 ft on each side.

Use the formula for perimeter of a square, $P = 4s$. You know that for this particular square, the value of s is 9 ft.

$P = 4s$ Formula for perimeter of a square

$P = 4 \cdot 9 \text{ ft}$ Replace s with 9 ft. Multiply 4 times 9 ft.

$P = 36 \text{ ft}$ Write 36 ft; ft is the unit of measure.

The perimeter of the square is 36 ft. Notice that this answer matches the result obtained from adding the four sides.

Work Problem **1** *at the Side.* ▶

(b) A square measuring 14 miles (14 mi) on each side (*Hint:* Draw a sketch of the square and label each side with its length.)

2 Use the perimeter of each square and the appropriate formula to find the length of one side. Then check your solution by drawing a square, labeling each side, and finding the perimeter.

(a) Perimeter is 28 in.

EXAMPLE 2 **Finding the Length of One Side of a Square**

If the perimeter of a square is 40 cm, find the length of one side. (Note: **cm** is the short way to write *centimeters*.)

Use the formula for perimeter of a square, $P = 4s$. This time you know that the value of P (the perimeter) is 40 cm.

$P = 4s$ Formula for perimeter of a square

$40 \text{ cm} = 4s$ Replace P with 40 cm

$\dfrac{40 \text{ cm}}{4} = \dfrac{4s}{4}$ To get the variable by itself on the right side, divide both sides by 4.

$10 \text{ cm} = s$

The length of one side of the square is 10 cm.

> Write **cm** as part of your answer.

Check Check the solution by drawing a square and labeling the length of each side as 10 cm. The perimeter is $10 \text{ cm} + 10 \text{ cm} + 10 \text{ cm} + 10 \text{ cm} = 40 \text{ cm}$. This result matches the perimeter given in the problem.

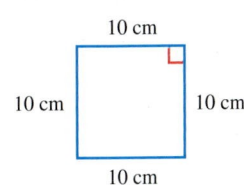

Work Problem **2** *at the Side.* ▶

(b) Perimeter is 100 ft.

(c) Perimeter is 64 cm.

ANSWERS

1. (a) $P = 80$ in.
   (b) $P = 56$ miles

   14 mi, 14 mi, 14 mi, 14 mi

2. (a) $s = 7$ in.

   7 in., 7 in., 7 in., 7 in.

   Check
   $P = 7 \text{ in.} + 7 \text{ in.} + 7 \text{ in.} + 7 \text{ in.} = 28 \text{ in.}$

   (b) $s = 25$ ft

   25 ft, 25 ft, 25 ft, 25 ft

   Check
   $P = 25 \text{ ft} + 25 \text{ ft} + 25 \text{ ft} + 25 \text{ ft} = 100 \text{ ft}$

   (c) $s = 16$ cm

   16 cm, 16 cm, 16 cm, 16 cm

   Check
   $P = 16 \text{ cm} + 16 \text{ cm} + 16 \text{ cm} + 16 \text{ cm} = 64 \text{ cm}$

3 Find the perimeter of each rectangle using the appropriate formula. Check your solutions by adding the lengths of the four sides.

(a)

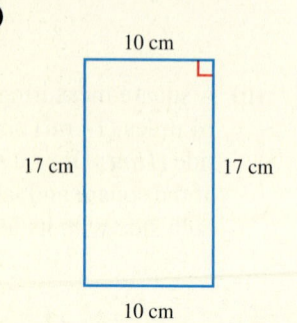

(b) A rectangle 6 m wide and 11 m long (*Hint:* First draw a sketch of the rectangle and label the length of each side.)

(c)

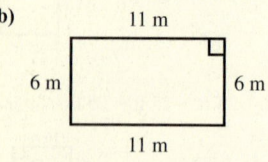

ANSWERS

3. **(a)** $P = 54$ cm
   Check 17 cm + 17 cm + 10 cm + 10 cm = 54 cm

(b)

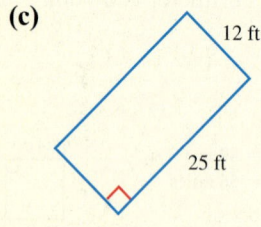

$P = 34$ m
   Check
   11 m + 11 m + 6 m + 6 m = 34 m

(c) $P = 74$ ft
   Check
   25 ft + 25 ft + 12 ft + 12 ft = 74 ft

OBJECTIVE 2 Use the formula for perimeter of a rectangle to find the perimeter, the length, or the width. A rectangle is a figure with four sides that meet to form right angles, which measure 90°. Each set of opposite sides is parallel and congruent (has the same length). Three examples of rectangles are shown below.

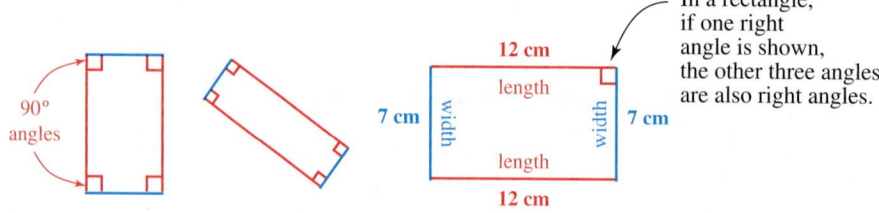

Each longer side of a rectangle is called the length (*l*) and each shorter side is called the width (*w*).

Look at the rectangle above with the lengths of the sides labeled. To find the perimeter (distance around), you could unfold the shape so the sides lie end-to-end. Then add the lengths of the sides.

$$P = 12 \text{ cm} + 7 \text{ cm} + 12 \text{ cm} + 7 \text{ cm} = 38 \text{ cm}$$

or $P = 12 \text{ cm} + 12 \text{ cm} + 7 \text{ cm} + 7 \text{ cm} = 38 \text{ cm}$ Commutative property

Because the two long sides are both 12 cm, and the two short sides are both 7 cm, you can also use the formula below.

Finding the Perimeter of a Rectangle
Perimeter of a rectangle = length + length + width + width
$$P = (2 \cdot \text{length}) + (2 \cdot \text{width})$$
$$P = 2l + 2w$$

EXAMPLE 3 Finding the Perimeter of a Rectangle

Find the perimeter of this rectangle.

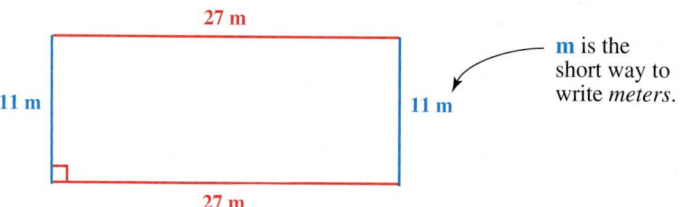

The length is **27 m**, and the width is **11 m**.

$P = 2l + 2w$ Replace *l* with 27 m and *w* with 11 m.
$P = 2 \cdot 27 \text{ m} + 2 \cdot 11 \text{ m}$ Do the multiplications first.
$P = 54 \text{ m} + 22 \text{ m}$ Add last.
$P = 76 \text{ m}$

The perimeter of the rectangle (the distance you would walk around the outside edges of the rectangle) is 76 m.

Check To check the solution, add the lengths of the four sides.
$$P = 27 \text{ m} + 27 \text{ m} + 11 \text{ m} + 11 \text{ m}$$
$$P = 76 \text{ m}$$ Matches the solution above

◄ *Work Problem* **3** *at the Side.*

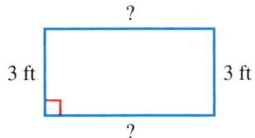 **EXAMPLE 4** **Finding the Length or Width of a Rectangle**

If the perimeter of a rectangle is 20 ft and the width is 3 ft, find the length.

First draw a sketch of the rectangle and label the widths as 3 ft.

Then use the formula for perimeter of a rectangle, $P = 2l + 2w$. The value of P is 20 ft and the value of w is 3 ft.

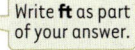

$$P = 2l + 2w \qquad \text{Formula for perimeter of a rectangle}$$

$$20 \text{ ft} = 2l + \underline{2 \cdot 3 \text{ ft}} \qquad \text{Replace } P \text{ with 20 ft and } w \text{ with 3 ft.}$$
$$\text{Simplify the right side by multiplying } 2 \cdot 3 \text{ ft.}$$

$$20 \text{ ft} = 2l + 6 \text{ ft}$$
$$\underline{{}^-6 \text{ ft} \qquad \qquad {}^-6 \text{ ft}} \qquad \text{To get } 2l \text{ by itself, add } {}^-6 \text{ ft to both sides:}$$
$$14 \text{ ft} = 2l + 0 \qquad \qquad 6 + {}^-6 \text{ is 0.}$$

$$\frac{14 \text{ ft}}{2} = \frac{2l}{2} \qquad \text{To get } l \text{ by itself, divide both sides by 2.}$$

$$7 \text{ ft} = l$$

The length is 7 ft. ⟵ Write **ft** as part of your answer.

Check To check the solution, put the length measurements on your sketch. Then add the four measurements.

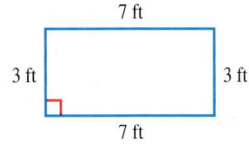

$$P = 7 \text{ ft} + 7 \text{ ft} + 3 \text{ ft} + 3 \text{ ft}$$
$$P = 20 \text{ ft}$$

A perimeter of 20 ft matches the information in the original problem, so 7 ft is the correct length of the rectangle.

Work Problem **4** *at the Side.* ▶

OBJECTIVE **3** **Find the perimeter of parallelograms, triangles, and irregular shapes.** A **parallelogram** is a four-sided figure in which opposite sides are both parallel and equal in length. Some examples are shown below. Notice that opposite sides have the same length.

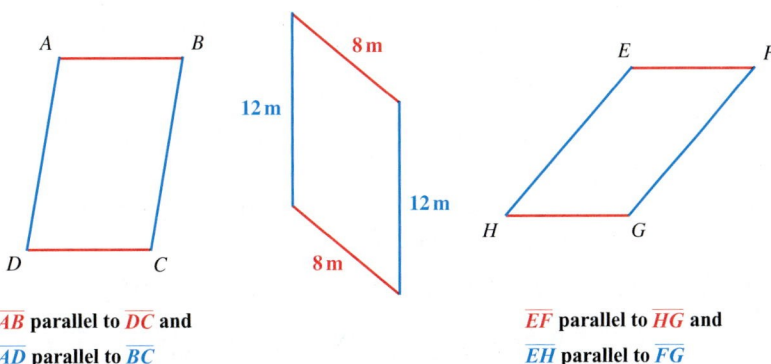

$\overline{AB}$ parallel to $\overline{DC}$ and
$\overline{AD}$ parallel to $\overline{BC}$

$\overline{EF}$ parallel to $\overline{HG}$ and
$\overline{EH}$ parallel to $\overline{FG}$

Perimeter is the distance around a flat shape, so the easiest way to find the perimeter of a parallelogram is to add the lengths of the four sides.

4 Use the perimeter of each rectangle and the appropriate formula to find the length or width. Draw a sketch of each rectangle and use it to check your solution.

(a) The perimeter of a rectangle is 36 in. and the width is 8 in. Find the length.

(b) A rectangle has a width of 4 cm. The perimeter is 32 cm. Find the length.

(c) A rectangle with a perimeter of 14 ft has a length of 6 ft. Find the width.

ANSWERS

4. (a) $l = 10$ in.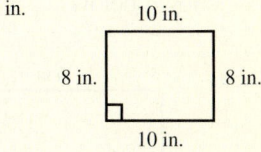

Check 10 in. + 10 in. + 8 in. + 8 in. = 36 in.

(b) $l = 12$ cm

Check 12 cm + 12 cm + 4 cm + 4 cm = 32 cm

(c) $w = 1$ ft

Check 6 ft + 6 ft + 1 ft + 1 ft = 14 ft

5 Find the perimeter of each parallelogram.

(a)

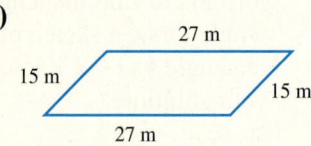

(b)

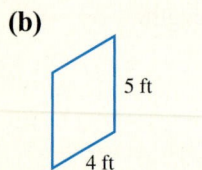

6 Find the perimeter of each triangle.

(a)

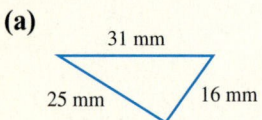

(b) A triangle with sides that each measure 5 in. Draw a sketch of the triangle and label the length of each side.

7 How much fencing will be needed to go around a flower bed with the measurements shown below?

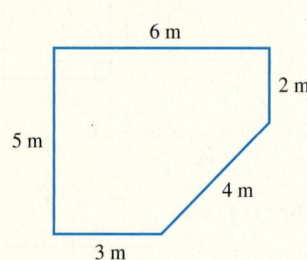

 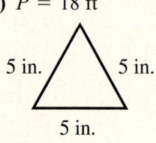
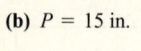
EXAMPLE 5 **Finding the Perimeter of a Parallelogram**

Find the perimeter of the middle parallelogram on the previous page.

$$P = \textbf{12 m} + \textbf{12 m} + \textbf{8 m} + \textbf{8 m}$$
$$P = 40 \text{ m}$$

◀ *Work Problem* **5** *at the Side.*

A **triangle** is a figure with exactly three sides. Some examples are shown below.

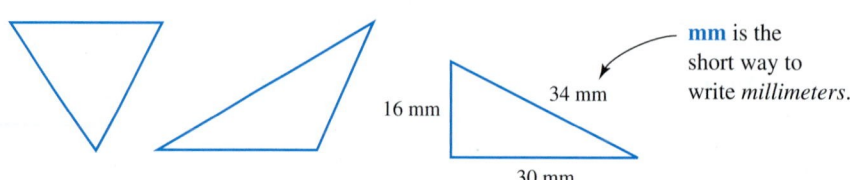

To find the perimeter of a triangle (the distance around the edges), add the lengths of the three sides.

EXAMPLE 6 **Finding the Perimeter of a Triangle**

Find the perimeter of the triangle above on the right.

To find the perimeter, add the lengths of the sides.

$$P = 16 \text{ mm} + 30 \text{ mm} + 34 \text{ mm}$$
$$P = 80 \text{ mm}$$

◀ *Work Problem* **6** *at the Side.*

As with any other shape, you can find the perimeter of (distance around) an irregular shape by adding the lengths of the sides.

EXAMPLE 7 **Finding the Perimeter of an Irregular Shape**

The floor of a room has the shape shown below.

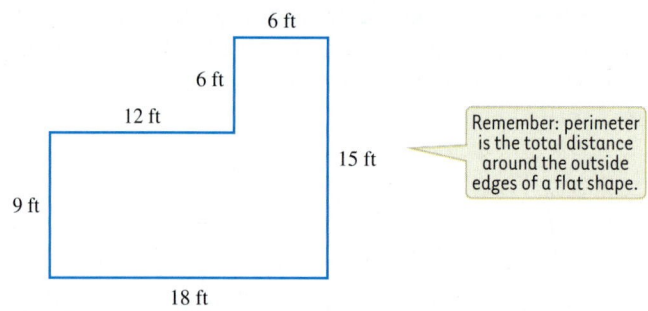

Remember: perimeter is the total distance around the outside edges of a flat shape.

Suppose you want to put a new wallpaper border along the top of all the walls. How much material do you need?

Find the perimeter of the room by adding the lengths of the sides.

$$P = 9 \text{ ft} + 12 \text{ ft} + 6 \text{ ft} + 6 \text{ ft} + 15 \text{ ft} + 18 \text{ ft}$$
$$P = 66 \text{ ft}$$

You need 66 ft of wallpaper border.

◀ *Work Problem* **7** *at the Side.*

3.1 ▶▶▶ Exercises

Find the perimeter of each square, using the appropriate formula. See Example 1.

1.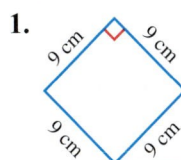

2.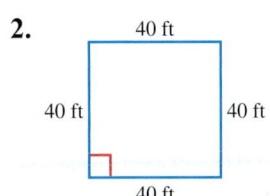

3.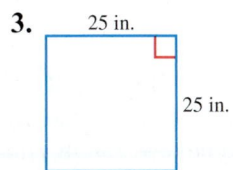

4.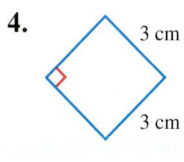

Draw a sketch of each square and label the lengths of the sides. Then find the perimeter. (Sketches may vary; show your sketches to your instructor.)

5. A square park measuring 1 mile on each side

6. A square garden measuring 4 meters on each side

7. A 22 mm square postage stamp

8. A 10 in. square piece of cardboard

For the given perimeter of each square, find the length of one side using the appropriate formula. See Example 2.

9. The perimeter is 120 ft.

10. The perimeter is 52 cm.

11. The perimeter is 4 mm.

12. The perimeter is 20 miles.

13. A square parking lot with a perimeter of 92 yards

14. A square building with a perimeter of 144 meters

15. A square closet with a perimeter of 8 ft

16. A square bedroom with a perimeter of 44 ft

Find the perimeter of each rectangle, using the appropriate formula. Check your solutions by adding the lengths of the four sides. See Example 3.

17.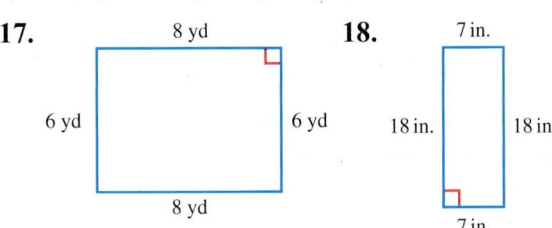

18.

19.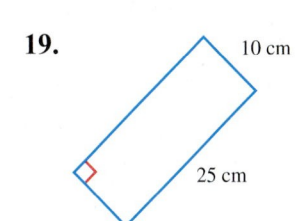

20.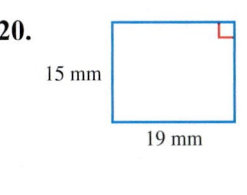

Draw a sketch of each rectangle and label the lengths of the sides. Then find the perimeter by using the appropriate formula. (Sketches may vary; show your sketches to your instructor.)

21. A rectangular living room 20 ft long by 16 ft wide

22. A rectangular placemat 45 cm long by 30 cm wide

23. An 8 in. by 5 in. rectangular piece of paper

24. A 2 ft by 3 ft rectangular window

For each rectangle, you are given the perimeter and either the length or width. Find the unknown measurement using the appropriate formula. Draw a sketch of each rectangle and use it to check your solution. See Example 4. (Show your sketches to your instructor.)

25. The perimeter is 30 cm and the width is 6 cm.

26. The perimeter is 48 yards and the length is 14 yards.

27. The length is 4 miles and the perimeter is 10 miles.

28. The width is 8 meters and the perimeter is 34 meters.

29. A 6 ft long rectangular table has a perimeter of 16 ft.

30. A 13 in. wide rectangular picture frame has a perimeter of 56 in.

31. A rectangular door 1 meter wide has a perimeter of 6 meters.

|←1 m→|

32. A rectangular house 33 ft long has a perimeter of 118 ft.

|←——— 33 ft ———→|

In exercises 33–44, find the perimeter of each shape. The figures in Exercises 33–36 are parallelograms. See Examples 5–7.

33.

58 m

46 m   46 m

58 m

34.

11 in.

10 in.

10 in.

11 in.

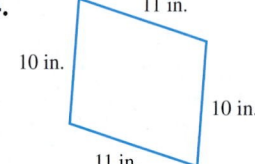

35.

100 ft

60 ft

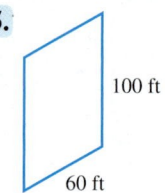

36.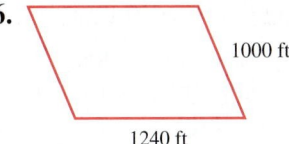

37.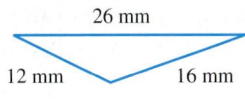

38.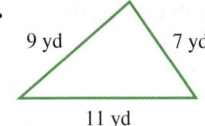

39.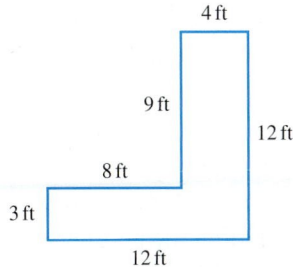

40.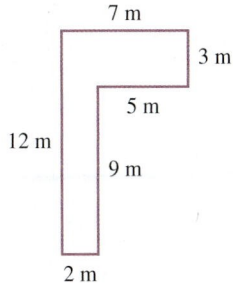

41.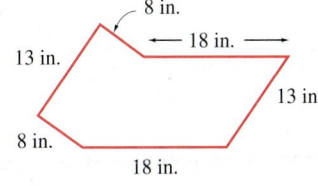

42.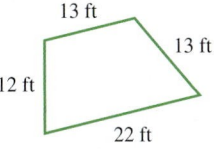

43.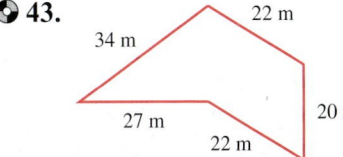

44.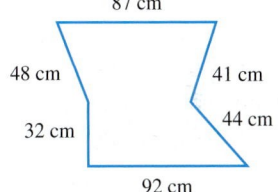

For each shape you are given the perimeter and the lengths of all sides except one. Find the length of the unlabeled side.

45. The perimeter is 115 cm.

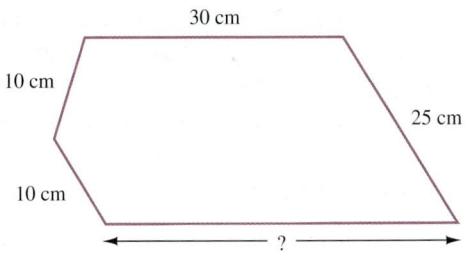

46. The perimeter is 63 in.

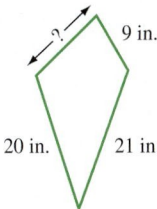

47. The perimeter is 78 in.

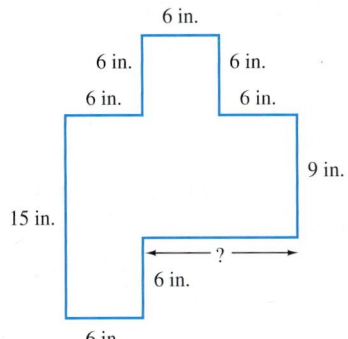

48. The perimeter is 196 ft.

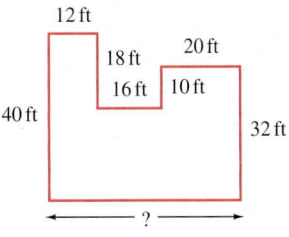

49. In an *equilateral* triangle, all sides have the same length.

  (a) Draw sketches of four different equilateral triangles, label the lengths of the sides, and find the perimeters.

  (b) Write a "shortcut" rule (a formula) for finding the perimeter of an equilateral triangle.

  (c) Will your formula work for other kinds of triangles that are not equilateral? Explain why or why not.

50. Be sure that you have done Exercise 49 first.

  (a) Draw a sketch of a figure with five sides of equal length. Write a "shortcut" rule (a formula) for finding the perimeter of this shape.

  (b) Draw a sketch of a figure with six sides of equal length. Write a formula for finding the perimeter of the shape.

  (c) Write a formula for finding the perimeter of a shape with 10 sides of equal length.

  (d) Write a formula for finding the perimeter of a shape with n sides of equal length.

Relating Concepts (Exercises 51–54) For Individual or Group Work

A formula that has many uses for drivers is $d = rt$, called the distance formula. If you are driving a car, then

  d is the *distance* you travel (how many miles)
  r is the *rate* (how fast you are driving in miles per hour)
  t is the *time* (how many hours you drive).

*Use the distance formula as you **work Exercises 51–54 in order.***

51. Suppose you are driving on Interstate highways at an average rate of 70 miles per hour. Use the distance formula to find out how far you will travel in **(a)** 2 hours; **(b)** 5 hours; **(c)** 8 hours.

52. If an ice storm slows your driving rate to 35 miles per hour, how far will you travel in **(a)** 2 hours, **(b)** 5 hours, **(c)** 8 hours? Show how to find each answer using the formula. **(d)** Explain how to find each answer using the results from Exercise 51 instead of the formula.

53. Use the distance formula to find out how many hours you would have to drive to travel the 3000 miles from Boston to San Francisco if your average rate is **(a)** 60 miles per hour; **(b)** 50 miles per hour; **(c)** 20 miles per hour (which was the speed limit 100 years ago).

54. Use the distance formula to find the average driving rate (speed) on each of these trips. (Distances are from *World Almanac.*)

  (a) It took 11 hours to drive 671 miles from Atlanta to Chicago.

  (b) Sam drove 1539 miles from New York City to Dallas in 27 hours.

  (c) Carlita drove 16 hours to travel 1040 miles from Memphis to Denver.

3.2 ▶▶▶ Problem Solving: Area

OBJECTIVE 1 Use the formula for area of a rectangle to find the area, the length, or the width.

> **Understanding the Difference between Perimeter and Area**
> **Perimeter** is the *distance around the outside edges* of a flat shape.
> **Area** is the amount of *surface inside* a flat shape.

The *perimeter* of a rectangle is the distance around the *outside edges*. Recall that we unfolded a shape and laid the sides end-to-end so we could see the total distance. The *area* of a rectangle is the amount of surface *inside* the rectangle. We measure area by finding the number of squares of a certain size needed to cover the surface inside the rectangle. Think of covering the floor of a living room with carpet. Carpet is measured in square yards, that is, square pieces that measure 1 yard along each side. Here is a drawing of a rectangular living room floor.

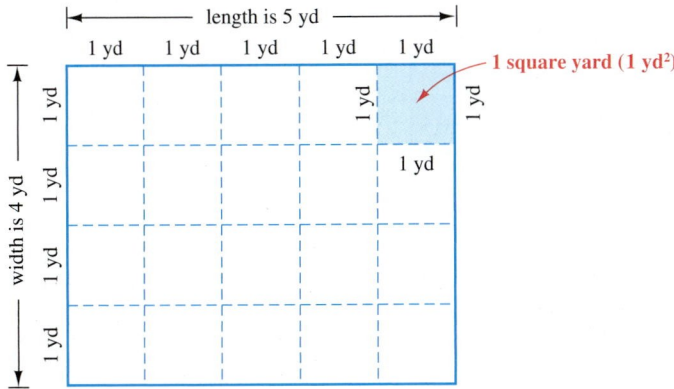

You can see from the drawing that it takes 20 squares to cover the floor. We say that the area of the floor is 20 *square yards*. A short way to write square yards is yd².

20 square yards   can be written as   20 yd²

To find the number of squares, you can count them, or you can multiply the number of squares in the length (5) times the number of squares in the width (4) to get 20. The formula is given below.

> **Finding the Area of a Rectangle**
>
> Area of a rectangle = length • width
> $$A = lw$$
> Remember to use *square units* when measuring area.

Squares of many sizes can be used to measure area. For smaller areas, you might use the ones shown at the right.

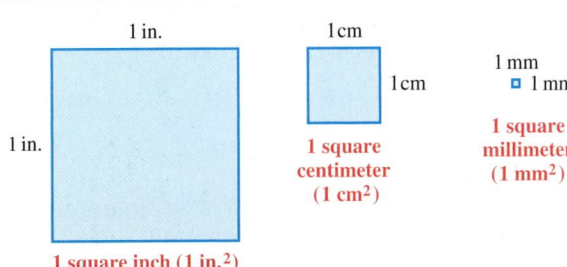

Actual-size drawings

OBJECTIVES

1 Use the formula for area of a rectangle to find the area, the length, or the width.

2 Use the formula for area of a square to find the area or the length of one side.

3 Use the formula for area of a parallelogram to find the area, the base, or the height.

4 Solve application problems involving perimeter and area of rectangles, squares, or parallelograms.

1 Find the area of each rectangle, using the appropriate formula.

(a)

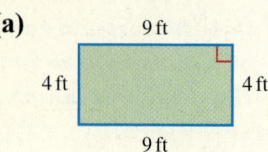

9 ft

4 ft   4 ft

9 ft

(b) A rectangle is 35 yd long and 20 yd wide. (First make a sketch of the rectangle and label the lengths of the sides.)

(c) A rectangular patio measures 3 m by 2 m. (First make a sketch of the patio and label the lengths of the sides.)

Other sizes of squares that are often used to measure area are listed here, but they are too large to draw on this page.

1 square meter (1 m^2)   1 square foot (1 ft^2)
1 square kilometer (1 km^2)   1 square yard (1 yd^2)
                                   1 square mile (1 mi^2)

> **CAUTION**
> The raised 2 in 4^2 means that you multiply $4 \cdot 4$ to get 16. The raised 2 in cm^2 or yd^2 is a short way to write the word "square." It means that you multiplied cm times cm to get cm^2, or yd times yd to get yd^2. Recall that a short way to write $x \cdot x$ is x^2. Similarly, cm $\cdot$ cm is cm^2. When you see 5 cm^2, say "five square centimeters." Do *not* multiply $5 \cdot 5$, because the exponent applies only to the *first* thing to its immediate left. The exponent applies to cm, *not* to the number.

EXAMPLE 1 **Finding the Area of Rectangles**

Find the area of each rectangle.

(a)

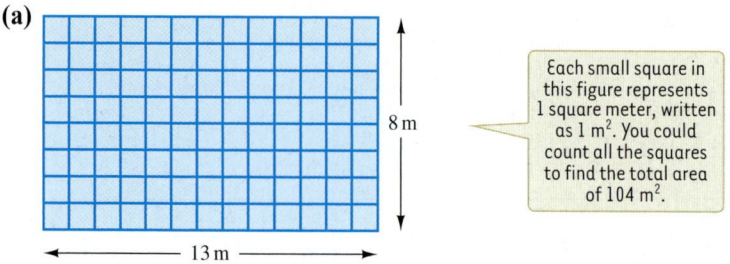

8 m

13 m

Each small square in this figure represents 1 square meter, written as 1 m^2. You could count all the squares to find the total area of 104 m^2.

The length of this rectangle is 13 m and the width is 8 m. Use the formula $A = lw$, which means $A = l \cdot w$.

$A = \quad l \quad \cdot \quad w$ Replace *l* with 13 m and *w* with 8 m.

$A = 13 \text{ m} \cdot 8 \text{ m}$ Multiply 13 times 8 to get 104.

$A = 104 \text{ m}^2$ Multiply m times m to get m^2.

The area of the rectangle is 104 m^2. If you count the number of squares in the sketch, you will also get 104 m^2. Each square in the sketch represents 1 m by 1 m, which is 1 square meter (1 m^2).

(b) A rectangle measuring 7 cm by 21 cm

First make a sketch of the rectangle. The length is 21 cm (the longer measurement) and the width is 7 cm. Then use the formula for area of a rectangle, $A = lw$.

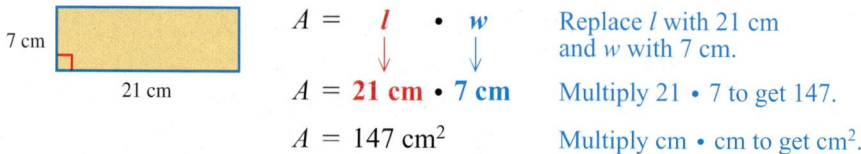

7 cm

21 cm

$A = \quad l \quad \cdot \quad w$ Replace *l* with 21 cm and *w* with 7 cm.

$A = 21 \text{ cm} \cdot 7 \text{ cm}$ Multiply $21 \cdot 7$ to get 147.

$A = 147 \text{ cm}^2$ Multiply cm $\cdot$ cm to get cm^2.

The area of the rectangle is 147 cm^2.

◀ **Work Problem** **1** at the Side.

ANSWERS

1. (a) $A = 36$ ft^2

(b)

20 yd

35 yd

$A = 700$ yd^2

(c)

2 m

3 m

$A = 6$ m^2

> **CAUTION**
> The units for *area* will always be *square* units (cm^2, m^2, yd^2, mi^2, and so on). The units for *perimeter* will always be *linear* units (cm, m, yd, mi, and so on), *not* square units.

EXAMPLE 2 **Finding the Length or Width of a Rectangle**

If the area of a rectangular rug is 12 yd² and the length is 4 yd, find the width.

First draw a sketch of the rug and label the length as 4 yd.

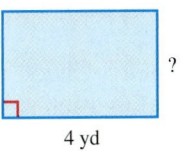

Use the formula for area of a rectangle, $A = lw$.
The value of A is 12 yd² and the value of l is 4 yd.

A	$=$	$l \cdot w$	Replace A with 12 yd² and replace l with 4 yd.
12 yd^2	$=$	$4 \text{ yd} \cdot w$	To get w by itself, divide both sides by 4 yd.

$$\frac{12 \text{ yd} \cdot yd}{4 \ yd} = \frac{4 \text{ yd} \cdot w}{4 \text{ yd}}$$ On the left side, rewrite yd² as yd • yd. Then $\frac{yd}{yd}$ is 1, so they "cancel out."

$$3 \text{ yd} = w$$ On the left side, 12 yd ÷ 4 is 3 yd.

The width of the rug is 3 yd.

Check   To check the solution, put the width measurement on your sketch. Then use the area formula.

$$A = l \cdot w$$
$$A = 4 \text{ yd} \cdot 3 \text{ yd}$$
$$A = 12 \text{ yd}^2$$

An area of 12 yd² matches the information in the original problem. So 3 yd is the correct width of the rug.

Work Problem **2** *at the Side.* ▶

OBJECTIVE 2 Use the formula for area of a square to find the area or the length of one side.   As with a rectangle, you can multiply length times width to find the area (surface inside) of a square. Because the length and the width are the same in a square, the formula is written as shown below.

Finding the Area of a Square

Area of a square = side • side
$$A = s \cdot s$$
$$A = s^2$$

Remember to use *square units* when measuring area.

EXAMPLE 3 **Finding the Area of a Square**

Find the area of a square highway sign that is 4 ft on each side.
Use the formula for area of a square, $A = s^2$.

$A = s^2$	Remember that s^2 means $s \cdot s$.
$A = s \cdot s$	Replace s with 4 ft.
$A = 4 \text{ ft} \cdot 4 \text{ ft}$	Multiply 4 • 4 to get 16.
$A = 16 \text{ ft}^2$	Multiply ft • ft to get ft².

> **Be careful!**
> s^2 is 4 • 4 or 16
> (**not** 2 • 4).

The area of the sign is 16 ft².

——— Continued on Next Page

---

2 Use the area of each rectangle and the appropriate formula to find the length or width. Draw a sketch of each rectangle and use it to check your solution.

(a) The area of a microscope slide is 12 cm², and the length is 6 cm. Find the width.

(b) A child's play lot is 10 ft wide and has an area of 160 ft². Find the length.

(c) A hallway floor is 31 m long and has an area of 93 m². Find the width of the floor.

ANSWERS

2. (a) $w = 2$ cm

	2 cm
6 cm	

Check $A = 6 \text{ cm} \cdot 2 \text{ cm}$
$A = 12 \text{ cm}^2$
Matches original problem

(b) $l = 16$ ft

	10 ft
16 ft	

Check $A = 16 \text{ ft} \cdot 10 \text{ ft}$
$A = 160 \text{ ft}^2$
Matches original problem

(c) $w = 3$ m

	3 m
31 m	

Check $A = 31 \text{ m} \cdot 3 \text{ m}$
$A = 93 \text{ m}^2$
Matches original problem

3 Find the area of each square, using the appropriate formula. Make a sketch of each square.

(a) A 12 in. square piece of fabric

(b) A square township 7 miles on a side

(c) A square earring measuring 20 mm on each side

4 Given the area of each square, find the length of one side by inspection.

(a) The area of a square-shaped nature center is 16 mi².

(b) A square floor with an area of 100 m²

(c) A square clock face with an area of 81 in.²

CAUTION

Be careful! s^2 means $s \cdot s$. It does **not** mean $2 \cdot s$. In this example s is 4 ft, so $(4 \text{ ft})^2$ is 4 ft $\cdot$ 4 ft = 16 ft². It is **not** $2 \cdot 4$ ft = 8 ft.

Check Check the solution by drawing a square and labeling each side as 4 ft. You can multiply length (4 ft) times width (4 ft), as you did for a rectangle. So the area is 4 ft $\cdot$ 4 ft, or 16 ft². This result matches the solution you got by using the formula $A = s^2$.

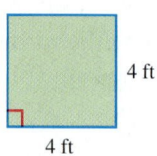

4 ft

4 ft

◀ *Work Problem* **3** *at the Side.*

EXAMPLE 4 **Finding the Length of One Side of a Square**

If the area of a square township is 49 mi², what is the length of one side of the township?

Use the formula for area of a square, $A = s^2$. The value of A is 49 mi².

$$A = s^2$$ Replace A with 49 mi².

$$49 \text{ mi}^2 = s^2$$ To get s by itself, we have to "undo" the squaring of s. This is called *finding the square root* (more on square roots in **Chapter 5**).

$$49 \text{ mi}^2 = s \cdot s$$ For now, solve by inspection. Ask, what number times itself gives 49?

$$49 \text{ mi}^2 = 7 \text{ mi} \cdot 7 \text{ mi}$$ $7 \cdot 7$ is 49, so 7 mi $\cdot$ 7 mi is 49 mi².

The value of s is 7 mi, so the length of one side of the township is 7 mi. Notice how this result matches the information about the township in Margin Problem 3(b) at the left.

◀ *Work Problem* **4** *at the Side.*

OBJECTIVE **3** **Use the formula for area of a parallelogram to find the area, the base, or the height.** To find the area of a parallelogram, first draw a dashed line inside the figure, as shown here.

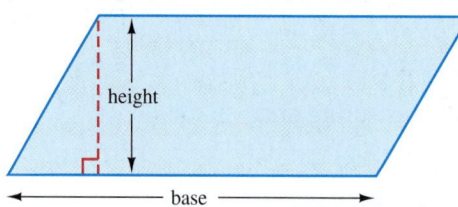

As an experiment, try this yourself by tracing this parallelogram onto a piece of paper.

height

base

The length of the dashed line is the *height* of the parallelogram. It forms a 90° angle (a right angle) with the base. The height is the shortest distance between the base and the opposite side.

(continued)

ANSWERS

3. (a) $A = 144$ in.²

12 in.

12 in.

(b) $A = 49$ mi²

7 mi

7 mi

(c) $A = 400$ mm²

20 mm

20 mm

4. (a) $s = 4$ mi **(b)** $s = 10$ m **(c)** $s = 9$ in.

Now cut off the triangle created on the left side of the parallelogram and move it to the right side, as shown below.

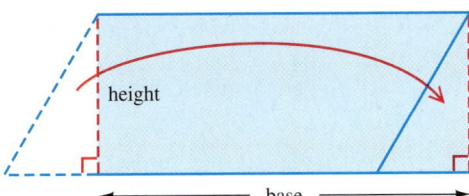

You have made the parallelogram into a rectangle. You can see that the area of the parallelogram and the rectangle are the same.

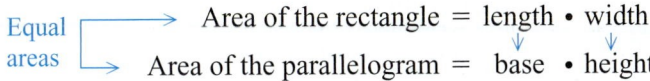

Equal areas
→ Area of the rectangle = length • width
→ Area of the parallelogram = base • height

> **Finding the Area of a Parallelogram**
>
> Area of a parallelogram = base • height
>
> $$A = bh$$
>
> Remember to use *square units* when measuring area.

EXAMPLE 5 **Finding the Area of Parallelograms**

Find the area of each parallelogram.

(a)
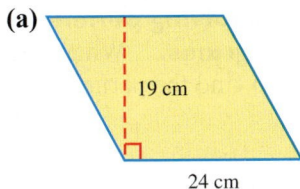

(b)
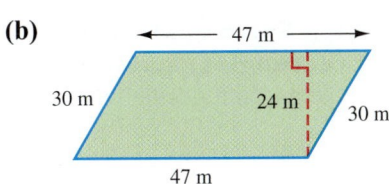

(a) The base is 24 cm and the height is 19 cm. The formula for the area of a parallelogram is $A = bh$.

$A = b \cdot h$	Replace b with 24 cm and h with 19 cm.
$A = $ **24 cm • 19 cm**	Multiply 24 • 19 to get 456.
$A = 456 \text{ cm}^2$	Multiply cm • cm to get cm².

The area of the parallelogram is 456 cm².

> Write **cm²** as part of your answer.

(b) Use the formula for area of a parallelogram, $A = bh$.

$A = b \cdot h$	Replace b with 47 m and h with 24 m.
$A = $ **47 m • 24 m**	Multiply 47 • 24 to get 1128.
$A = 1128 \text{ m}^2$	Multiply m • m to get m².

Notice that the 30 m sides are *not* used in finding the area. But you would use them when finding the *perimeter* of the parallelogram.

Work Problem **5** *at the Side.* ▶

EXAMPLE 6 **Finding the Base or Height of a Parallelogram**

The area of a parallelogram is 24 ft² and the base is 6 ft. Find the height.

First draw a sketch of the parallelogram and label the base as 6 ft.

Continued on Next Page

---

5 Find the area of each parallelogram.

(a)
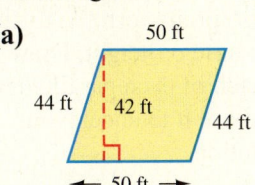

(b)
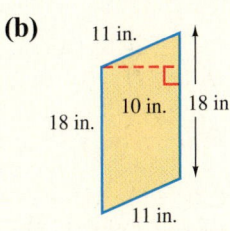

(c) A parallelogram with base 8 cm and height 1 cm.

6 Use the area of each parallelogram and the appropriate formula to find the base or height. Draw a sketch of each parallelogram and use it to check your solution.

(a) The area of a parallelogram is 140 in.² and the base is 14 in. Find the height.

(b) A parallelogram has an area of 4 yd². The height is 1 yd. Find the base.

7 If sod costs $3 per square yard, how much will the neighbors in Example 7 spend to cover the playground with grass?

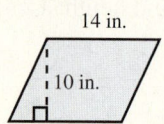
Use the formula for the area of a parallelogram, $A = bh$. The value of A is 24 ft², and the value of b is 6 ft.

$A \quad = \quad b \quad \cdot \quad h$ Replace A with 24 ft² and b with 6 ft.

$24 \text{ ft}^2 \quad = \quad 6 \text{ ft} \cdot h$ To get h by itself, divide both sides by 6 ft.

$$\frac{24 \text{ ft} \cdot \text{ft}}{6 \text{ ft}} = \frac{6 \text{ ft} \cdot h}{6 \text{ ft}}$$ On the left side, rewrite ft² as ft $\cdot$ ft. Then $\frac{\text{ft}}{\text{ft}}$ is 1, so they "cancel out."

$4 \text{ ft} = \quad h$ On the left side, 24 ft ÷ 6 is 4 ft.

The height of the parallelogram is 4 ft.

Check   To check the solution, put the height measurement on your sketch. Then use the area formula.

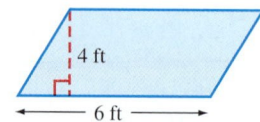
4 ft
6 ft

$A = \quad b \quad \cdot \quad h$

$A = 6 \text{ ft} \cdot 4 \text{ ft}$

$A = 24 \text{ ft}^2$

An area of 24 ft² matches the information in the original problem. So 4 ft is the correct height of the parallelogram.

◀ Work Problem **6** at the Side.

OBJECTIVE 4 Solve application problems involving perimeter and area of rectangles, squares, or parallelograms.   When you are solving problems, first decide whether you need to find the perimeter or the area.

EXAMPLE 7   **Solving an Application Problem Involving Perimeter or Area**

A group of neighbors is fixing up a playground for their children. The rectangular lot is 22 yd by 16 yd. If chain-link fencing costs $6 per yard, how much will they spend to put a fence around the lot?

First draw a sketch of the rectangular lot and label the lengths of the sides. The fence will go around the edges of the lot, so you need to find the *perimeter* of the lot.

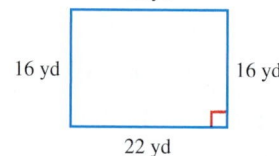
22 yd
16 yd   16 yd
22 yd

$P = \quad 2l \quad + \quad 2w$ Formula for perimeter of a rectangle

$P = 2 \cdot 22 \text{ yd} + 2 \cdot 16 \text{ yd}$ Replace l with 22 yd and w with 16 yd.

$P = \quad 44 \text{ yd} \quad + \quad 32 \text{ yd}$

$P = 76 \text{ yd}$

The perimeter of the lot is 76 yd, so the neighbors need to buy 76 yd of fencing. The cost of the fencing is $6 *per yard,* which means $6 *for 1 yard.* To find the cost for 76 yd, multiply $6 $\cdot$ 76. The neighbors will spend $456 on the fence.

An application involving the *area* of the playground is found in Margin Problem 7 at the left.

◀ Work Problem **7** at the Side.

3.2 ▶▶▶ Exercises

FOR EXTRA HELP

Find the area of each rectangle, square, or parallelogram using the appropriate formula. See Examples 1, 3, and 5.

1.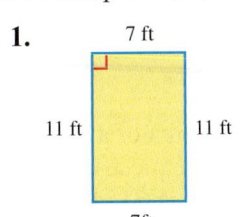
7 ft
11 ft   11 ft
7ft

2.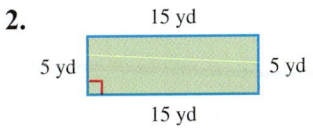
15 yd
5 yd   5 yd
15 yd

3.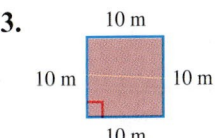
10 m
10 m   10 m
10 m

4.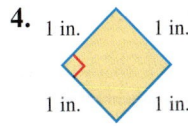
1 in.   1 in.
1 in.   1 in.

5.
31 mm
31 mm
25 mm   31 mm
31 mm

6.
← 21 m →
20 m   13 m   20 m
21 m

7.
6 in.

8.
20 cm

In Exercises 9–16, first draw a sketch of the shape and label the lengths of the sides or base and height. Then find the area. (Sketches may vary; show your sketches to your instructor.)

9. A rectangular calculator that measures 15 cm by 7 cm

10. A rectangular piece of plywood that is 8 ft long and 2 ft wide

11. A parallelogram with height of 9 ft and base of 8 ft

12. A parallelogram measuring 18 mm on the base and 3 mm on the height

13. A fire burned a square-shaped forest 25 mi on a side.

14. An 11 in. square pillow

15. A square piece of window glass 1 m on each side

16. A table 12 ft long by 3 ft wide

In Exercises 17–22, use the area of each rectangle and either its length or width, and the appropriate formula, to find the other measurement. Draw a sketch of each rectangle and use it to check your solution. See Example 2. (Sketches may vary; show your sketches to your instructor.)

17. The area of a desk is 18 ft², and the width is 3 ft. Find its length.

18. The area of a classroom is 630 ft², and the length is 30 ft. Find its width.

19. A parking lot is 90 yd long and has an area of 7200 yd². Find its width.

20. A playground is 60 yd wide and has an area of 6000 yd². Find its length.

◑ 21. A 154 in.² photo has a width of 11 in. Find its length.

22. A 15 in.² note card has a width of 3 in. Find its length.

Given the area of each square, find the length of one side by inspection. See Example 4.

23. A square floor has an area of 36 m².

24. A square stamp has an area of 9 cm².

25. The area of a square sign is 4 ft².

26. The area of a square piece of metal is 64 in.².

Use the area of each parallelogram and either its base or height, and the appropriate formula, to find the other measurement. Draw a sketch of each parallelogram and use it to check your solution. See Example 6. (Sketches may vary; show your sketches to your instructor.)

27. The area is 500 cm², and the base is 25 cm. Find the height.

28. The area is 1500 m², and the height is 30 m. Find the base.

29. The height is 13 in. and the area is 221 in.². Find the base.

30. The base is 19 cm, and the area is 114 cm². Find the height.

31. The base is 9 m, and the area is 9 m². Find the height.

32. The area is 25 mm², and the height is 5 mm. Find the base.

*Explain and correct the **two** errors made by students in Exercises 33 and 34.*

33.

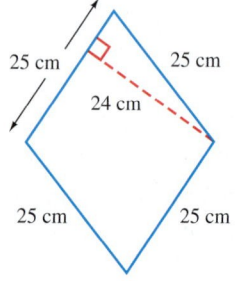

25 cm   25 cm

24 cm

25 cm   25 cm

$$P = 25 \text{ cm} + 24 \text{ cm} + 25 \text{ cm} + 25 \text{ cm} + 25 \text{ cm}$$
$$P = 124 \text{ cm}^2$$

34.

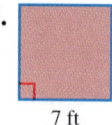

7 ft

$$A = s^2$$
$$A = 2 \cdot 7 \text{ ft}$$
$$A = 14 \text{ ft}$$

Name each figure and find its perimeter and area.

35.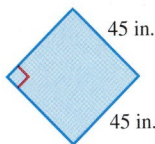
45 in.
45 in.

36.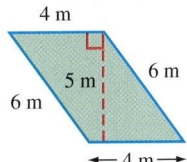
4 m
5 m
6 m
6 m
4 m

37.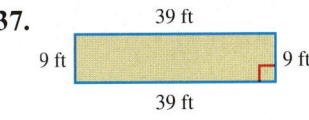
39 ft
9 ft
9 ft
39 ft

38.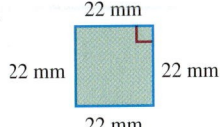
22 mm
22 mm
22 mm
22 mm

39.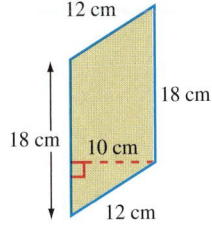
12 cm
18 cm
18 cm
10 cm
12 cm

40.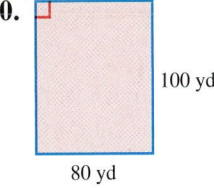
100 yd
80 yd

Solve each application problem. You may need to find the perimeter, the area, or one of the side measurements. In Exercises 41–46, draw a sketch for each problem and label the sketch with the appropriate measurements. See Example 7. (Sketches may vary; show your sketches to your instructor.)

41. Gymnastic floor exercises are performed on a square mat that is 12 meters on a side. Find the perimeter and area of a mat. (*Source:* www.nist.gov)

42. A regulation volleyball court is 18 meters by 9 meters. Find the perimeter and area of a regulation court. (*Source:* www.nist.gov)

43. Tyra's kitchen is 4 m wide and 5 m long. She is pasting a decorative strip that costs $6 per meter around the top edge of all the walls. How much will she spend?

44. The Wang's family room measures 20 ft by 25 ft. They are covering the floor with square tiles that measure 1 ft on a side and cost $1 each. How much will they spend on tile?

45. Mr. and Mrs. Gomez are buying carpet for their square-shaped bedroom, which measures 5 yd along each wall. The carpet is $23 per square yard and padding and installation is another $6 per square yard. How much will they spend in all?

46. A page in this book measures about 27 cm from top to bottom and 21 cm from side to side. Find the perimeter and the area of the page.

47. A regulation football field is 100 yd long (excluding end zones) and has an area of 5300 yd². Find the width of the field. (*Source:* NFL.)

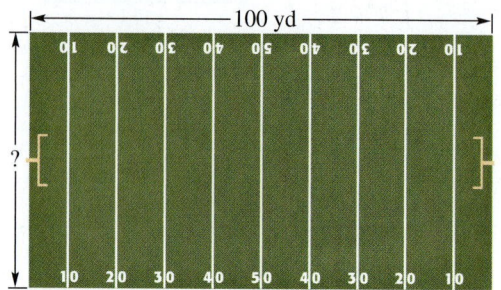

48. There are 14,790 ft² of ice in the rectangular playing area for a major league hockey game (excluding the area behind the goal lines). If the playing area is 85 ft wide, how long is it? (*Source:* NHL.)

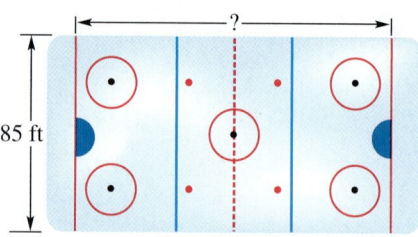

49. The table below shows information on two rectangular tents for camping.

Tents	Coleman Family Dome	Eddie Bauer Dome Tent
Dimensions	13 ft × 13 ft	12 ft × 12 ft
Sleeps	8 campers	6 campers
Sale price	$127	$99

Source: target.com

For the Coleman tent, find the perimeter, area, and number of square feet of floor space for each camper. Round to the nearest whole number if necessary.

50. Look at the table in Exercise 49. For the Eddie Bauer tent, find the perimeter, area, and number of square feet of floor space for each camper. Round to the nearest whole number if necessary.

Relating Concepts (Exercises 51–54) For Individual or Group Work

*Use your knowledge of perimeter and area to **work Exercises 51–54 in order.***

51. Suppose you have 12 ft of fencing to make a square or rectangular garden plot. Draw sketches of *all* the possible plots that use exactly 12 ft of fencing and label the lengths of the sides. Use only *whole number* lengths. (*Hint:* There are three possibilities.)

52. (a) Find the area of each plot in Exercise 51.

   (b) Which plot has the greatest area?

53. Repeat Exercise 51 using 16 ft of fencing. Be sure to draw *all* possible square or rectangular plots that have whole number lengths for the sides.

54. (a) Find the area of each plot in Exercise 53.

   (b) Compare your results to those from Exercise 52. What do you notice about the plots with the greatest area?

Summary Exercises on Perimeter and Area

Name each figure and find its perimeter and area using the appropriate formulas.

1.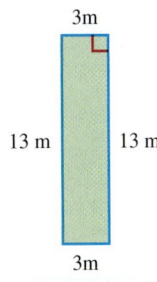

3m

13 m   13 m

3m

2.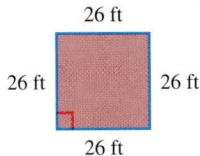

26 ft

26 ft   26 ft

26 ft

3.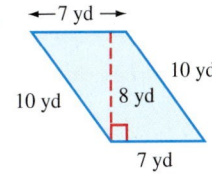

←7 yd→

10 yd

10 yd   8 yd

7 yd

4.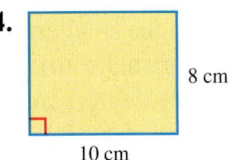

8 cm

10 cm

5.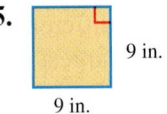

9 in.

9 in.

6.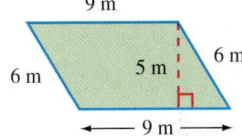

9 m

6 m   5 m   6 m

←9 m→

7.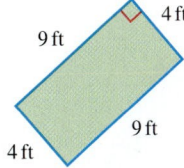

4 ft

9 ft

9 ft

4 ft

8.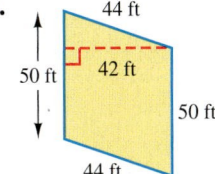

44 ft

50 ft   42 ft

50 ft

44 ft

In Exercises 9–14, use the appropriate formula to find the unknown measurement. Draw a sketch of the figure and use it to check your solution.

9. A rectangle has a length of 7 ft and a perimeter of 16 ft. Find the width.

10. A parallelogram with an area of 5 yd^2 has a height of 1 yd. Find the base.

11. A square photograph has an area of 36 in.^{2} Find the length on one side.

12. A sidewalk is 42 m long and has an area of 84 m². Find the width of the sidewalk.

13. The perimeter of a square map is 64 cm. What is the length of one side?

14. A rectangular patio is 9 ft wide and has a perimeter of 48 ft. What is the patio's length?

Solve each application problem using the appropriate formula. Show your work.

15. How much fencing is needed to enclose a triangular building site that measures 168 meters on each side?

16. Regulation soccer fields can measure 50 to 100 yards wide and 100 to 130 yards long, depending upon the age and skill level of the players. Find the difference in playing room between the smallest and largest fields. (*Source:* Womenssportsnet.com)

17. A Toshiba laptop computer has a rectangular screen with a perimeter of 42 inches. The width of the screen is 12 inches. Find the height of the screen. (*Source:* Toshiba.)

18. Kari is decorating the cover of a square-shaped photo album that is 22 cm along each side. She is gluing braid along the edges of the front and back covers. How much braid will she need?

19. A school flag measures 3 feet by 5 feet. Find the amount of fabric needed to make seven flags, and the amount of binding needed to go around all the edges of all the flags.

20. The lobby floor of a new skyscraper is in the shape of a parallelogram, with an area of 3024 ft². If the base of the parallelogram measures 63 ft, what is its height?

3.3 ▶▶▶ Solving Application Problems with One Unknown Quantity

OBJECTIVE **1** **Translate word phrases into algebraic expressions.** In **Sections 3.1** and **3.2** you worked with applications involving perimeter and area. You were able to use well-known rules (formulas) to set up equations that could be solved. However, you will encounter many problems for which no formula is available. Then you need to analyze the problem and translate the words into an equation that fits the particular situation. We'll start by translating word phrases into algebraic expressions.

EXAMPLE 1 **Translating Word Phrases into Algebraic Expressions**

Write each phrase as an algebraic expression. Use x as the variable.

Words	Algebraic Expression	
A number **plus** 2	$x + 2$ or $2 + x$	
The **sum** of 8 and a number	$8 + x$ or $x + 8$	*Two* correct ways to write each *addition* expression
5 **more than** a number	$x + 5$ or $5 + x$	
$^-35$ **added to** a number	$^-35 + x$ or $x + ^-35$	
A number **increased by** 6	$x + 6$ or $6 + x$	
9 **less than** a number	$x - 9$	
A number **subtracted from** 3	$3 - x$	*Only one* correct way to write each *subtraction* expression
3 **subtracted from** a number	$x - 3$	
A number **decreased by** 4	$x - 4$	
10 **minus** a number	$10 - x$	

> **CAUTION**
> Recall that addition can be done in any order, so $x + 2$ gives the same result as $2 + x$. This is *not* true in subtraction, so be careful. $10 - x$ does *not* give the same result as $x - 10$.

Work Problem **1** *at the Side.* ▶

EXAMPLE 2 **Translating Word Phrases into Algebraic Expressions**

Write each phrase as an algebraic expression. Use x as the variable.

Words	Algebraic Expression
8 **times** a number	$8x$
The **product** of 12 and a number	$12x$
Double a number (meaning "2 times")	$2x$
The **quotient** of $^-6$ and a number	$\dfrac{^-6}{x}$
A number **divided by** 10	$\dfrac{x}{10}$
15 **subtracted from** 4 **times** a number	$4x - 15$
The result **is**	$=$

Work Problem **2** *at the Side.* ▶

OBJECTIVES

1 Translate word phrases into algebraic expressions.

2 Translate sentences into equations.

3 Solve application problems with one unknown quantity.

---

1 Write each phrase as an algebraic expression. Use x as the variable.

(a) 15 less than a number

(b) 12 more than a number

(c) A number increased by 13

(d) A number minus 8

(e) 10 plus a number

(f) A number subtracted from 6

(g) 6 subtracted from a number

2 Write each phrase as an algebraic expression. Use x as the variable.

(a) Double a number

(b) The product of $^-8$ and a number

(c) The quotient of 15 and a number

(d) 5 times a number subtracted from 30

ANSWERS

1. (a) $x - 15$ (b) $x + 12$ or $12 + x$
   (c) $x + 13$ or $13 + x$ (d) $x - 8$
   (e) $10 + x$ or $x + 10$ (f) $6 - x$
   (g) $x - 6$

2. (a) $2x$ (b) ^-8x (c) $\dfrac{15}{x}$
   (d) $30 - 5x$ (*not* $5x - 30$)

3 Translate each sentence into an equation and solve it. Check your solution by going back to the words in the original problem.

(a) If 3 times a number is added to 4, the result is 19. Find the number.

(b) If 7 is subtracted from 6 times a number, the result is $^-25$. Find the number.

OBJECTIVE 2 Translate sentences into equations.   The next example shows you how to translate a sentence into an equation that you can solve.

EXAMPLE 3   **Translating a Sentence into an Equation**

If 5 times a number is added to 11, the result is 26. Find the number.

Let x represent the unknown number. Use the information in the problem to write an equation.

$$\underbrace{5 \text{ times a number}}\ \underbrace{\text{added to}}\ \underbrace{11}\ \underbrace{\text{is}}\ \underbrace{26}$$
$$5x \qquad + \qquad 11 = 26$$

Next, solve the equation.

$$\begin{array}{rcl} 5x + 11 &=& 26 \\ \underline{-11} & & \underline{-11} \\ 5x + 0 &=& 15 \end{array}$$ To get $5x$ by itself, add $^-11$ to both sides.

$$\frac{5x}{5} = \frac{15}{5}$$ To get x by itself, divide both sides by 5.

$$x = 3$$

The number is 3.

Check   Go back to the words of the *original* problem.

$$\text{If } \underbrace{5 \text{ times}}\ \underbrace{\text{a number}}\ \underbrace{\text{is added to}}\ \underbrace{11,}\ \text{the result is } 26.$$
$$5 \quad \cdot \quad 3 \quad + \quad 11 \quad = \quad 26$$

Does $5 \cdot 3 + 11$ really equal 26? Yes, $5 \cdot 3 + 11 = 15 + 11 = 26$.
   So 3 is the correct solution because it "works" when you put it back into the original problem.

◄ *Work Problem* **3** *at the Side.*

OBJECTIVE 3 Solve application problems with one unknown quantity.   Now you are ready to tackle application problems. The steps we will use are summarized below.

Solving an Application Problem

Step 1   **Read** the problem once to see what it is about. Read it carefully a second time. As you read, make a sketch or write word phrases that identify the known and the unknown parts of the problem.

Step 2(a)   If there is one unknown quantity, **assign a variable** to represent it. Write down what your variable represents.

Step 2(b)   If there is more than one unknown quantity, **assign a variable** to represent "the thing you know the least about." Then write variable expression(s), using the same variable, to show the relationship of the other unknown quantities to the first one.

Step 3   **Write an equation,** using your sketch or word phrases as the guide.

Step 4   **Solve** the equation.

Step 5   **State the answer** to the question in the problem and label your answer.

Step 6   **Check** whether your answer fits all the facts given in the *original* statement of the problem. If it does, you are done. If it doesn't, start again at Step 1.

ANSWERS

3. (a) $3x + 4 = 19$
        $x = 5$
   Check $\underbrace{3 \cdot 5} + 4$ does equal 19
              $\underbrace{15} + 4$
                  19

(b) $6x - 7 = {}^-25$
        $x = {}^-3$
   Check $\underbrace{6 \cdot {}^-3} - 7$ does equal $^-25$
              $\underbrace{{}^-18} + {}^-7$
                  $^-25$

EXAMPLE 4   **Solving an Application Problem with One Unknown Quantity**

Heather put some money aside in an envelope for small household expenses. Yesterday she took out $20 for groceries. Today a friend paid back a loan and Heather put the $34 in the envelope. Now she has $43 in the envelope. How much was in the envelope at the start?

Step 1   **Read** the problem once. It is about money in an envelope. Read it a second time and write word phrases.

Unknown: amount of money in the envelope at the start
Known: took out $20; put in $34; ended up with $43

Step 2(a)   There is only one unknown quantity, so **assign a variable** to represent it. Let *m* represent the money at the start.

Step 3   **Write an equation,** using the phrases you wrote as a guide.

Money at the start	Took out $20	Put in $34	Ended up with $43
m	$- \$20$	$+ \$34$	$= \$43$

Step 4   **Solve** the equation.

$$m - 20 + 34 = 43 \qquad \text{Change subtraction to adding the opposite.}$$
$$m + {}^-20 + 34 = 43 \qquad \text{Simplify the left side.}$$
$$m + 14 = 43$$
$$\qquad {}^-14 \qquad {}^-14 \qquad \begin{array}{l}\text{To get } m \text{ by itself.}\\ \text{add } {}^-14 \text{ to both sides.}\end{array}$$
$$m + 0 = 29$$
$$m = 29$$

Step 5   **State the answer** to the question, "How much was in the envelope at the start?" There was $29 in the envelope.

Step 6   **Check** the solution by going back to the *original* problem and inserting the solution.

Started with $29 in the envelope

Took out $20, so $29 − $20 = $9 in the envelope

Put in $34, so $9 + $34 = $43

Now has $43 ⟵ Matches

Because $29 "works" when you put it back into the original problem, you know it is the correct solution.

> If your answer does **not** work in the original problem, start again at *Step 1*.

Work Problem **4** *at the Side.* ▶

4 Some people got on an empty bus at its first stop. At the second stop, 3 people got on. At the third stop, 5 more people got on. At the fourth stop, 10 people got off, but 4 people were still on the bus. How many people got on at the first stop? Show your work for each of the six problem-solving steps.

ANSWER

4. *Step 1*   **Read.**
Unknown: number of people who got on at first stop
Known: 3 got on; 5 got on; 10 got off; 4 people still on bus

Step 2(a)   **Assign a variable.**
Let *p* be people who got on at first stop. (You may use any letter you like as the variable.)

Step 3   **Write an equation.**
$$p + 3 + 5 - 10 = 4$$

Step 4   **Solve.**
$$p + 3 + 5 + {}^-10 = 4$$
$$p + {}^-2 = 4$$
$$\quad +2 \qquad +2$$
$$p + 0 = 6$$
$$p = 6$$

Step 5   **State the answer.**
6 people got on at the first stop.

Step 6   **Check.**
6 got on at first stop
3 got on at 2nd stop: $6 + 3 = 9$
5 got on at 3rd stop: $9 + 5 = 14$
10 got off at 4th stop: $14 - 10 = 4$
4 people are left. ⟵ Matches

5 Five donors each gave the same amount of money to a college to use for scholarships. From the money, scholarships of $1250, $900, and $850 were given to students; $250 was left. How much money did each donor give to the college? Show your work for each of the six problem-solving steps.

ANSWER

5. *Step 1*   Read.
Unknown: money given by each donor
Known: 5 donors; gave out $1250, $900, $850; $250 left

Step 2(a)   Assign a variable.
Let m be each donor's money.

Step 3   Write an equation.
$5 \cdot m - \$1250 - \$900 - \$850 = \$250$

Step 4   Solve.

$5m + {}^-1250 + {}^-900 + {}^-850 = 250$

$$
\begin{array}{rcl}
5m + \quad {}^-3000 & = & 250 \\
\quad\quad {}^+3000 & & {}^+3000 \\
\hline
5m + \quad\quad 0 & = & 3250 \\
\dfrac{5m}{5} & = & \dfrac{3250}{5} \\
m & = & 650
\end{array}
$$

Step 5   State the answer.
Each donor gave $650.

Step 6   Check.
5 donors each gave $650, so
$5 \cdot \$650 = \$3250$
Gave out $1250, $900, $850, so
$\$3250 - 1250 - 900 - 850 = \$250$
Had $250 left ←— Matches —↗

EXAMPLE 5 **Solving an Application Problem with One Unknown Quantity**

Three friends each put in the same amount of money to buy a gift. After they spent $2 for a card and $31 for the gift, they had $6 left. How much money had each friend put in originally?

Step 1   **Read** the problem. It is about 3 friends buying a gift.

Unknown: amount of money each friend contributed
Known: 3 friends put in money; spent $2 and $31; had $6 left

Step 2(a)   There is only one unknown quantity. **Assign a variable,** m, to represent the amount of money each friend contributed.

Step 3   **Write an equation.**

Number of friends	Amount each friend put in	Spent on card	Spent on gift	Left over
3 •	m	$2	$31	$6

$$3 \cdot m - \$2 - \$31 = \$6$$

To see why this is multiplication, think of an example. If each friend put in $10, how much money would there be? $3 \cdot \$10$, or $30

Step 4   **Solve.**

$$
\begin{array}{rcll}
3m - 2 - 31 & = & 6 & \text{Change subtractions to adding the opposite.} \\
3m + {}^-2 + {}^-31 & = & 6 & \text{Simplify the left side.} \\
3m + \quad {}^-33 & = & 6 & \\
\quad\quad {}^+33 & & {}^+33 & \text{To get } 3m \text{ by itself, add 33 to both sides.} \\
\hline
3m + \quad 0 & = & 39 & \\
\dfrac{3m}{3} & = & \dfrac{39}{3} & \text{To get } m \text{ by itself, divide both sides by 3.} \\
m & = & 13 &
\end{array}
$$

Step 5   **State the answer.** Each friend put in $13.   ⟵ Write a **$** as part of your answer.

Step 6   **Check** the solution by putting it back into the *original* problem.

3 friends each put in $13, so $3 \cdot \$13 = \$39$.
Spent $2, spent $31, so $\$39 - \$2 - \$31 = \$6$
Had $6 left ⟵——— Matches ———↗

$13 is the correct solution because it "works."

◀ Work Problem **5** at the Side.

EXAMPLE 6 **Solving a More Complex Application Problem with One Unknown Quantity**

Michael has completed 5 less than three times as many lab experiments as David. If Michael has completed 13 experiments, how many experiments has David completed?

Step 1   **Read** the problem. It is about the number of experiments done by two students.

Unknown: number of experiments David did
Known: Michael did 5 less than 3 times the number David did; Michael did 13.

Step 2(a)   **Assign a variable.** Let n represent the number of experiments David did.

Step 3   **Write an equation.**

The number Michael did	is	5 less than 3 times David's number.
13	=	$3n - 5$

> Be careful with subtraction! $5 - 3n$ is **not** the same as $3n - 5$.

Step 4   **Solve.**

$13 = 3n - 5$ Change subtraction to adding the opposite.

$13 = 3n + {}^-5$

$$\begin{array}{c} {}^+5 \qquad\quad {}^+5 \\ \hline 18 = \underline{3n + \;\;0} \end{array}$$ To get $3n$ by itself, add 5 to both sides.

$\dfrac{18}{3} = \dfrac{3n}{3}$ To get n by itself, divide both sides by 3.

$6 = n$

Step 5   **State the answer.** David did 6 experiments.

Step 6   **Check** the solution by putting it back into the *original* problem.

3 times David's number $3 \cdot 6 = 18$

Less 5 $18 - 5 = 13$

Michael did 13. $\longleftarrow$ Matches $\longrightarrow$

The correct solution is: David did 6 experiments.

> Label your answer. Write *6 experiments*, **not** just 6.

Work Problem **6** *at the Side.* ▶

6 Susan donated $10 more than twice what LuAnn donated. If Susan donated $22, how much did LuAnn donate?

Show your work for each of the six problem-solving steps.

ANSWER

6. *Step 1*   **Read.**
Unknown: LuAnn's donation
Known: Susan donated $10 more than twice what LuAnn donated; Susan donated $22.

Step 2(a)   **Assign a variable.**
Let d be LuAnn's donation.

Step 3   **Write an equation.**
$2d + \$10 = \$22$
(or $\$10 + 2d = \$22$)

Step 4   **Solve.**

$$\begin{array}{c} 2d + \;\;10 = \;\;22 \\ \underline{\quad\;\; {}^-10 = {}^-10} \\ 2d + \;\;\;\;0 = \;\;12 \end{array}$$

$\dfrac{2d}{2} = \dfrac{12}{2}$

$d = 6$

Step 5   **State the answer.**
LuAnn donated $6.

Step 6   **Check.**
$10 more than twice $6 is
$\$10 + (2 \cdot \$6) = \$10 + \$12 = \$22$
Susan donated $22. ← Matches

Math in the Media

FORMULAS

To estimate the number of words in a child's vocabulary, a pediatrician may use this formula.

$$V = 60A - 900$$

where V is the number of vocabulary words and A is the age of the child in months.

Source: Pediatrics.

Use the formula to find the *age* of each child.

1. The child's vocabulary is 180 words.

2. The child's vocabulary is 1440 words.

3. The child's vocabulary is 60 words.

4. Describe *in words* what you were doing as you found each child's age.

5. Find the child's age if the child's vocabulary is 0 words.

6. Refer to Problem 5 above. *Without solving an equation,* are there any other ages at which children would have a vocabulary of 0 words?

7. Is the formula useful for all ages of children? Explain your answer.

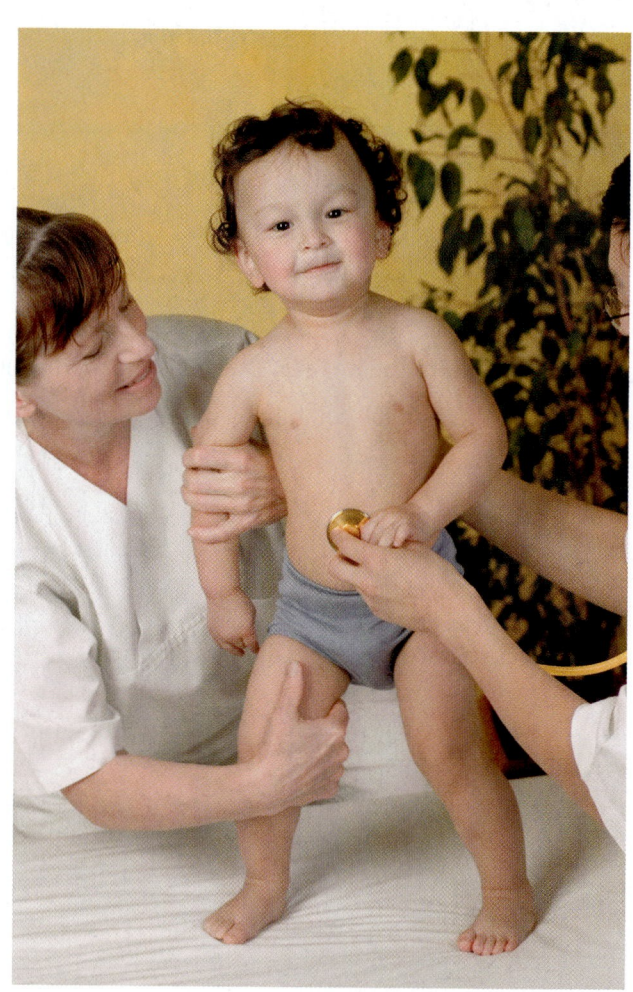

Write an algebraic expression, using x as the variable. See Examples 1 and 2.

1. 14 plus a number

2. The sum of a number and $^-8$

3. $^-5$ added to a number

4. 16 more than a number

5. 20 minus a number

6. A number decreased by 25

7. 9 less than a number

8. A number subtracted from $^-7$

9. Subtract 4 from a number.

10. 3 fewer than a number

11. $^-6$ times a number

12. The product of $^-3$ and a number

13. Double a number

14. A number times 10

15. A number divided by 2

16. 4 divided by a number

17. Twice a number added to 8

18. Five times a number plus 5

19. 10 fewer than seven times a number

20. 12 less than six times a number

21. The sum of twice a number and the number

22. Triple a number subtracted from the number

Translate each sentence into an equation and solve it. Check your solution by going back to the words in the original problem. See Example 3.

23. If four times a number is decreased by 2, the result is 26. Find the number.

24. The sum of 8 and five times a number is 53. Find the number.

25. If a number is added to twice the number, the result is ⁻15. What is the number?

26. If a number is subtracted from three times the number, the result is ⁻8. What is the number?

27. If the product of some number and 5 is increased by 12, the result is seven times the number. Find the number.

28. If eight times a number is subtracted from eleven times the number, the result is ⁻9. Find the number.

29. When three times a number is subtracted from 30, the result is 2 plus the number. What is the number?

30. When twice a number is decreased by 8, the result is the number increased by 7. Find the number.

Solve each application problem. Use the six problem-solving steps you learned in this section. See Examples 4–6.

31. Ricardo gained 15 pounds over the winter. He went on a diet and lost 28 pounds. Then he regained 5 pounds and weighed 177 pounds. How much did he weigh originally?

32. Mr. Chee deposited $80 into his checking account. Then, after writing a $23 check for gas and a $90 check for his child's day care, the balance in his account was $67. How much was in his account before he made the deposit?

33. There were 18 cookies in Magan's cookie jar. While she was busy in another room, her children ate some of the cookies. Magan bought three dozen cookies and added them to the jar. At that point she had 49 cookies in the jar. How many cookies did her children eat?

34. The Greens had a 20-pound bag of bird seed in their garage. Mice got into the bag and ate some of it. The Greens then bought an 8-pound bag of seed and put all the seed in a metal container. They now have 24 pounds of seed. How much did the mice eat?

35. A college bookstore ordered six boxes of red pens. The store sold 32 red pens last week and 35 red pens this week. Five pens were left on the shelf. How many pens were in each box?

36. A local charity received a donation of eight cartons filled with cans of soup. The charity gave out 100 cans of soup yesterday and 92 cans today before running out. How many cans were in each carton?

37. The 14 music club members each paid the same amount for dues. The club also earned $340 selling magazine subscriptions. They spent $575 to organize a jazz festival. Now their bank account, which started at $0, is overdrawn by $25. How much did each member pay in dues?

38. The manager of an apartment complex had 11 packages of lightbulbs on hand. He replaced 29 burned out bulbs in hallway lights and 7 bulbs in the party room. Eight bulbs were left. How many bulbs were in each package?

39. When 75 is subtracted from four times Tamu's age, the result is Tamu's age. How old is Tamu?

40. If three times Linda's age is decreased by 36, the result is twice Linda's age. How old is Linda?

41. While shopping for clothes, Consuelo spent $3 less than twice what Brenda spent. Consuelo spent $81. How much did Brenda spend?

42. Dennis weighs 184 pounds. His weight is 2 pounds less than six times his child's weight. How much does his child weigh?

43. Paige bought five bags of candy for Halloween. Forty-eight children visited her home and she gave each child three pieces of candy. At the end of the night she still had one bag of candy. How many pieces of candy were in each bag?

44. A restaurant ordered four packages of paper napkins. Yesterday they used up one package, and today they used up 140 napkins. Two packages plus 60 napkins remain. How many napkins are in each package?

45. The recommended daily intake of iron for an adult female is 4 mg less than twice the recommended amount for an infant. The amount for an adult female is 18 mg. How much should the infant receive? (*Source:* Food and Nutrition Board.)

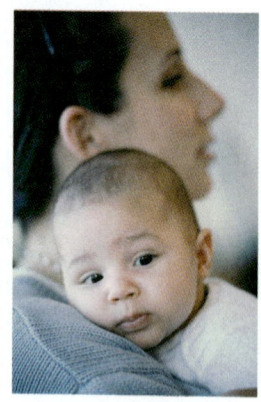

46. A cheetah's sprinting speed is 61 miles per hour less than three times a zebra's running speed. A cheetah can sprint 68 miles per hour. Find the zebra's running speed. (*Source: Grolier Multimedia Encyclopedia.*)

3.4 ▶▶▶ Solving Application Problems with Two Unknown Quantities

OBJECTIVE 1 **Solve application problems with two unknown quantities.** In the preceding section, the problems had only one unknown quantity. As a result, we used Step 2(a) rather than Step 2(b) in the problem-solving steps. For easy reference, we repeat the steps here.

OBJECTIVE

1 Solve application problems with two unknown quantities.

Solving an Application Problem

Step 1 **Read** the problem once to see what it is about. Read it carefully a second time. As you read, make a sketch or write word phrases that identify the known and the unknown parts of the problem.

Step 2(a) If there is one unknown quantity, **assign a variable** to represent it. Write down what your variable represents.

Step 2(b) If there is more than one unknown quantity, **assign a variable** to represent "the thing you know the least about." Then write variable expression(s), using the same variable, to show the relationship of the other unknown quantities to the first one.

Step 3 **Write an equation,** using your sketch or word phrases as the guide.

Step 4 **Solve** the equation.

Step 5 **State the answer** to the question in the problem and label your answer.

Step 6 **Check** whether your answer fits all the facts given in the *original* statement of the problem. If it does, you are done. If it doesn't, start again at Step 1.

Now you are ready to solve problems with two unknown quantities.

EXAMPLE 1 **Solving an Application Problem with Two Unknown Quantities**

Last month, Sheila worked 72 hours more than Russell. Together they worked a total of 232 hours. Find the number of hours each person worked last month.

Step 1 **Read** the problem. It is about the number of hours worked by Sheila and by Russell.

Unknowns: hours worked by Sheila;
　　　　　hours worked by Russell
Known: Sheila worked 72 hours more than Russell;
　　　　232 hours total for Sheila and Russell

Step 2(b) There are *two* unknowns so **assign a variable** to represent "the thing you know the least about." You know the *least* about the hours worked by Russell, so let h represent Russell's hours.

Sheila worked 72 hours more than Russell, so her hours are $h + 72$, that is, Russell's hours (h) plus 72 more.

Step 3 **Write an equation.**

Hours worked by Russell		Hours worked by Sheila		Total hours worked
h	$+$	$h + 72$	$=$	232

Continued on Next Page

1 In a day of work, Keonda made \$12 more than her daughter. Together they made \$182. Find the amount that each person made. (*Hint:* Which amount do you know the *least* about, Keonda's or her daughter's? Let *m* be that amount.) Use the six problem-solving steps.

Step 4 **Solve.**

$$\underbrace{h + h}_{2h} + 72 = 232 \qquad \text{Simplify the left side by combining like terms.}$$

$$\underline{ \; ^-72 \quad ^-72} \qquad \text{To get } 2h \text{ by itself, add } ^-72 \text{ to both sides.}$$

$$\underbrace{2h + 0}_{} = 160$$

$$\frac{2h}{2} = \frac{160}{2} \qquad \text{To get } h \text{ by itself, divide both sides by 2.}$$

$$h = 80$$

Step 5 **State the answer.**

Because *h* represents Russell's hours, and the solution of the equation is $h = 80$, Russell worked 80 hours.

$\underset{\downarrow}{h} + 72$ represents Sheila's hours. Replace *h* with 80.

$\mathbf{80} + 72 = 152$, so Sheila worked 152 hours.

The final answer is:
Russell worked 80 hours and
Sheila worked 152 hours.

> Write **hours** as part of both answers.

Step 6 **Check** the solution by putting both numbers back into the *original* problem.

"Sheila worked 72 hours more than Russell."
Sheila's 152 hours are 72 more than Russell's 80 hours, so the solution checks.

$$\begin{array}{r} 152 \\ -72 \\ \hline 80 \end{array}$$

"Together they worked a total of 232 hours."
Sheila's 152 hours + Russell's 80 hours = 232 hours, so the solution checks.

$$\begin{array}{r} 152 \\ +80 \\ \hline 232 \end{array}$$

You've answered the question correctly because 80 hours and 152 hours fit all the facts given in the problem.

> **CAUTION**
> Check the solution to an application problem by putting the numbers back in the *original* problem. If they do *not* work, recheck your work or try solving the problem in a different way.

◀ *Work Problem* **1** *at the Side.*

EXAMPLE 2 **Solving an Application Problem with Two Unknown Quantities**

Riley cut a cord that was 83 ft long into two pieces. One piece was 19 ft shorter than the other. How long was each piece?

Step 1 **Read** the problem. It is about a cord that is cut into two pieces.

> **Unknowns:** length of first piece;
> length of second piece
> **Known:** Second piece is 19 ft shorter than first piece;
> total length of both pieces is 83 ft

Step 2(b) There are *two* unknowns so **assign a variable** to represent "the thing you know the least about." You know the *least* about the first piece, so let *p* represent the length of the first piece.

ANSWER

1. Daughter made *m*.
Keonda made $m + 12$.
$m + m + 12 = 182$
Daughter made \$85.
Keonda made \$97.

Check $\$97 - \$85 = \$12$
and $\$97 + \$85 = \$182$

Continued on Next Page

The second piece is 19 ft shorter than the first piece, so its length is $p - 19$.

Step 3 **Write an equation.**

Length of first piece		Length of second piece		Total length
p	$+$	$p - 19$	$=$	83

Step 4 **Solve.**

$$p + p - 19 = 83 \quad \text{Simplify the left side by combining like terms.}$$

$$2p - 19 = 83 \quad \text{Change subtraction to adding the opposite.}$$

$$2p + {}^{-}19 = 83$$

$$\underline{\quad\quad +19 \quad +19\quad} \quad \text{To get } 2p \text{ by itself, add 19 to both sides.}$$

$$2p + 0 = 102$$

$$\frac{2p}{2} = \frac{102}{2} \quad \text{To get } p \text{ by itself, divide both sides by 2.}$$

$$p = 51$$

Step 5 **State the answer.**

p represents the length of the first piece, and $p = 51$, so the first piece is 51 ft long.

$p - 19$ represents the second piece. Replace p with 51.

$51 - 19 = 32$, so the second piece is 32 ft long.

The final answer is: One piece was 51 ft long and the other piece was 32 ft long.

Step 6 **Check** the solution by putting both numbers back into the *original* problem.

"Riley cut a cord that was 83 ft long."

  $51 \text{ ft} + 32 \text{ ft} = 83 \text{ ft}$, so the solution checks.

"One piece was 19 ft shorter than the other."

> If your answers do **not** fit all the given facts, start over at *Step 1*.

32 ft is 19 ft shorter than 51 ft, so the solution checks.

You've answered the question correctly because 51 ft and 32 ft fit all the facts given in the problem.

Work Problem **2** *at the Side.* ▶

2 Charles had 175 yd of fishing line to put on two fishing reels. He put 25 yd less of line on one reel than on the other. How much line did he put on each reel? Use the six problem-solving steps.

EXAMPLE 3 **Solving a Geometry Application with Two Unknown Quantities**

The length of a rectangle is 2 cm more than the width. The perimeter is 68 cm. Find the length and width.

Step 1 **Read** the problem. It is about a rectangle. Make a sketch of a rectangle.

  Unknowns: length of the rectangle; width of the rectangle
  Known: The length is 2 cm more than the width; the perimeter is 68 cm.

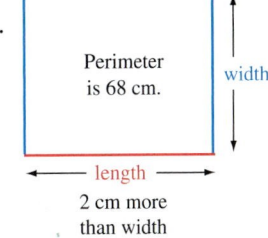

Perimeter is 68 cm. width length 2 cm more than width

ANSWER

2. Length on first reel is r.
   Length on second reel is $r - 25$.
   $r + r - 25 = 175$
   First reel had 100 yd of line.
   Second reel had 75 yd of line.

   Check $100 \text{ yd} + 75 \text{ yd} = 175 \text{ yd}$ and $100 \text{ yd} - 75 \text{ yd} = 25 \text{ yd}$

Continued on Next Page

3 Make a sketch to help solve this problem.

The length of Ann's rectangular garden plot is 3 yd more than the width. She used 22 yd of fencing to go around the entire garden. Find the length and the width of the garden, using the six problem-solving steps.

Step 2(b)   There are *two* unknowns so **assign a variable** to represent "the thing you know the least about." You know the *least* about the width, so let w represent the **width**.

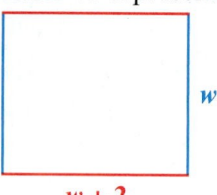

The length is 2 cm more than the width, so the **length** is $w + 2$.

Step 3   **Write an equation.**
Use the formula for perimeter of a rectangle, $P = 2l + 2w$, to help you write the equation.

$$P = 2 \quad l \quad + 2 \quad w$$

Replace P with 68.
Replace l with $(w + 2)$.

$$68 = 2(w + 2) + 2 \cdot w$$

Step 4   **Solve.**

$$68 = 2(w + 2) + 2w \qquad \text{Use the distributive property.}$$

$$68 = 2w + 4 + 2w \qquad \text{Combine like terms.}$$

$$68 = 4w + 4$$

$$\underline{-4 \qquad\qquad -4} \qquad \text{To get } 4w \text{ by itself, add } -4 \text{ to both sides.}$$

$$64 = 4w + 0$$

$$\frac{64}{4} = \frac{4w}{4} \qquad \text{To get } w \text{ by itself, divide both sides by 4.}$$

$$16 = w$$

Step 5   **State the answer.**

w represents the width, and $w = 16$, so the width is 16 cm.

$w + 2$ represents the length. Replace w with 16.

$16 + 2 = 18$, so the length is 18 cm.

Write **cm** as the label for both width and length.

The final answer is: The width is 16 cm and the length is 18 cm.

Step 6   **Check** the solution by putting the measurements on your sketch and going back to the original problem.

"The length of a rectangle is 2 cm more than the width."

18 cm is 2 cm more than 16 cm, so the solution checks.

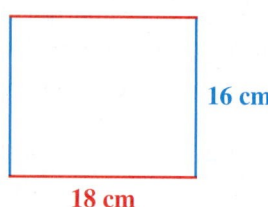

"The perimeter is 68 cm."

$$P = 2 \cdot 18 \text{ cm} + 2 \cdot 16 \text{ cm}$$

$$P = 36 \text{ cm} + 32 \text{ cm}$$

$$P = 68 \text{ cm} \leftarrow \text{This matches the perimeter given in the original problem, so the solution checks.}$$

◀ *Work Problem* **3** *at the Side.*

ANSWER

3.

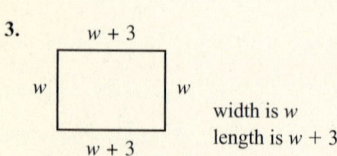

width is w
length is $w + 3$

$22 = 2(w + 3) + 2 \cdot w$
width is 4 yd
length is 7 yd

Check   7 yd is 3 yd more than 4 yd.
$P = 2 \cdot 7 \text{ yd} + 2 \cdot 4 \text{ yd} = 22 \text{ yd}$
Matches perimeter given in the original problem

3.4 ▶▶▶ Exercises

FOR EXTRA HELP · **MyMathLab** · Math XL PRACTICE · WATCH · DOWNLOAD · READ · REVIEW

Solve each application problem using the six problem-solving steps you learned in this section. See Examples 1 and 2.

1. My sister is 9 years older than I am. The sum of our ages is 51. Find our ages.

2. Ed and Marge were candidates for city council. Marge won with 93 more votes than Ed. The total number of votes cast in the election was 587. Find the number of votes received by each candidate.

3. Last year, Lien earned $1500 more than her husband. Together they earned $37,500. How much did each of them earn?

4. A $149,000 estate is to be divided between two charities so that one charity receives $18,000 less than the other. How much will each charity receive?

5. Jason paid five times as much for his computer as he did for his printer. He paid a total of $1320 for both items. What did each item cost?

6. The attendance at the Saturday night baseball game was three times the attendance at Sunday's game. In all, 56,000 fans attended the games. How many fans were at each game?

7. A board is 78 cm long. Rosa cut the board into two pieces, with one piece 10 cm longer than the other. Find the length of both pieces. (*Hint:* Make a sketch of the board. Which piece do you know the least about? Let *x* represent the length of that piece.)

Longer piece ┆ Shorter piece

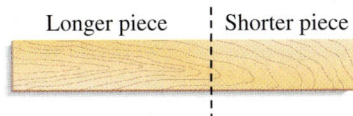

8. A rope is 21 yd long. Marcos cut it into two pieces so that one piece is 3 yd longer than the other. Find the length of each piece.

Longer piece ┆ Shorter piece

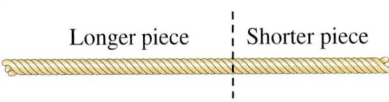

9. A wire is cut into two pieces, with one piece 7 ft shorter than the other. The wire was 31 ft long before it was cut. How long was each piece?

10. A 90 cm pipe is cut into two pieces so that one piece is 6 cm shorter than the other. Find the length of each piece.

11. In the U.S. Congress, the number of Representatives is 65 less than five times the number of Senators. There are a total of 535 members of Congress. Find the number of Senators and the number of Representatives. (*Source: World Almanac.*)

12. Florida's record low temperature is 68 degrees higher than Montana's record low. The sum of the two record lows is ⁻72 degrees. What is the record low for each state? (*Source*: National Climatic Data Center.)

13. A fence is 706 m long. It is to be cut into three parts. Two parts are the same length, and the third part is 25 m longer than either of the other two. Find the length of each part.

14. A wooden railing is 82 m long. It is to be divided into four pieces. Three pieces will be the same length, and the fourth piece will be 2 m longer than each of the other three. Find the length of each piece.

In Exercises 15–20, use the formula for the perimeter of a rectangle, $P = 2l + 2w$. Make a sketch to help you solve each problem. See Example 3. (Sketches may vary; show your sketches to your instructor.)

15. The perimeter of a rectangle is 48 yd. The width is 5 yd. Find the length.

16. The length of a rectangle is 27 cm, and the perimeter is 74 cm. Find the width of the rectangle.

17. A rectangular dog pen is twice as long as it is wide. The perimeter of the pen is 36 ft. Find the length and the width of the pen.

18. A new city park is a rectangular shape. The length is triple the width. It will take 240 meters of fencing to go around the park. Find the length and width of the park.

19. The length of a rectangular jewelry box is 3 in. more than twice the width. The perimeter is 36 in. Find the length and the width.

20. The perimeter of a rectangular house is 122 ft. The width is 5 ft less than the length. Find the length and the width.

21. A photograph measures 8 in. by 10 in. Earl put it in a frame that added 2 in. to every side. Find the outside perimeter and total area of the photograph and frame.

22. Barb had a 16 in. by 20 in. photograph. She cropped 3 in. off every side of the photo. What are the perimeter and the area of the cropped photo?

Chapter 3 ▶▶▶ Summary

▶ Key Terms

3.1 **formula** Formulas are well-known rules for solving common types of problems. They are written in a shorthand form that uses variables.

perimeter Perimeter is the distance around the outside edges of a flat shape. It is measured in linear units such as in., ft, yd, mm, cm, m, and so on.

square A square is a figure with four sides that are all the same length and meet to form right angles, which measure 90°. See example at the right.

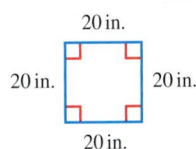

rectangle A rectangle is a four-sided figure in which all sides meet to form right angles, which measure 90°. The opposite sides are the same length. See example at the right.

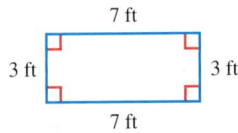

parallelogram A parallelogram is a four-sided figure in which opposite sides are both parallel and equal in length. See example at the right.

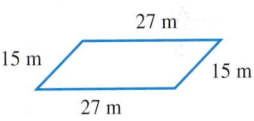

triangle A triangle is a figure with exactly three sides. See example at the right.

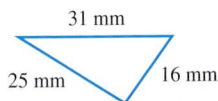

3.2 **area** Area is the surface inside a two-dimensional (flat) shape. It is measured by determining the number of squares of a certain size needed to cover the surface inside the shape. Some of the commonly used units for measuring area are square inches (in.^{2}), square feet (ft^2), square yards (yd^2), square centimeters (cm^2), and square meters (m^2).

▶ New Symbols

Right angle:

(90° angle)

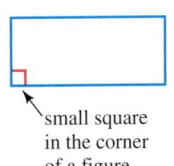

small square in the corner of a figure

Square units: in.^{2} ft^2 yd^2 mi^2

(for measuring area) mm^2 cm^2 m^2 km^2

▶ New Formulas

Perimeter of a square: $P = 4s$

Perimeter of a rectangle: $P = 2l + 2w$

Area of a square: $A = s^2$

Area of a rectangle: $A = lw$

Area of a parallelogram: $A = bh$

▶ Test Your Word Power

See how well you have learned the vocabulary in this chapter. Answers follow the Quick Review.

1. The **perimeter** of a flat shape is
   A. measured in square units
   B. found by adding the lengths of the sides
   C. measured in cubic units
   D. found by multiplying length times width.

2. The **area** of a flat shape is
   A. found by adding length plus width
   B. measured in linear units
   C. found by squaring the height
   D. measured in square units.

3. When working with a **square shape,**
   A. the perimeter formula is $P = s^2$
   B. the area formula is $A = \frac{1}{2}bh$
   C. all sides have the same length
   D. all angles measure 180°.

4. When working with a **rectangular shape,**
   A. the area formula is $A = lw$
   B. opposite sides have different lengths
   C. the perimeter formula is $P = bh$
   D. the width is the longer measurement.

5. In *all* **triangles**
   A. the sides meet at 90° angles
   B. the sides have the same length
   C. there are exactly three sides
   D. the area formula is $A = s^2$.

6. In *all* **parallelograms**
   A. there are exactly six sides
   B. the height is parallel to the base
   C. all sides have the same length
   D. opposite sides are parallel and equal in length.

▶ Quick Review

Concepts	Examples

(3.1) Finding Perimeter

To find the perimeter of *any* shape, add the lengths of the sides. Perimeter is measured in linear units (cm, m, ft, yd, and so on).

 Or, for squares and rectangles, you can use the formulas shown below.

Find the perimeter of each figure.

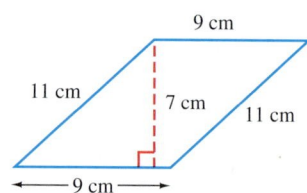

$P = 9 \text{ cm} + 11 \text{ cm} + 9 \text{ cm} + 11 \text{ cm}$

$P = 40 \text{ cm}$

Perimeter of a square: $P = 4s$,
 where s is the length of one side.

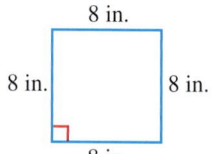

$P = 4s$
$P = 4 \cdot 8 \text{ in.}$
$P = 32 \text{ in.}$

Perimeter of a rectangle: $P = 2l + 2w$,
 where l is the length and w is the width.

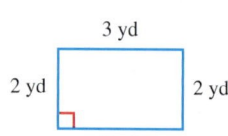

$P = 2 \cdot l + 2 \cdot w$
$P = 2 \cdot 3 \text{ yd} + 2 \cdot 2 \text{ yd}$
$P = 6 \text{ yd} + 4 \text{ yd}$
$P = 10 \text{ yd}$

Concepts	Examples

3.1 Finding the Length of One Side of a Square

If you know the perimeter of a square, use the formula $P = 4s$. Replace P with the value for the perimeter and solve the equation for s.

If the perimeter of a square room is 44 ft, find the length of one side.

$$P = 4s \qquad \text{Replace } P \text{ with 44 ft.}$$

$$44 \text{ ft} = 4s$$

$$\frac{44 \text{ ft}}{4} = \frac{4s}{4} \qquad \text{Divide both sides by 4.}$$

$$11 \text{ ft} = s$$

The length of one side is 11 ft.

Check Check the solution by drawing a sketch of the room. The perimeter is $11 \text{ ft} + 11 \text{ ft} + 11 \text{ ft} + 11 \text{ ft} = 44 \text{ ft}$. The result matches the perimeter given in the problem.

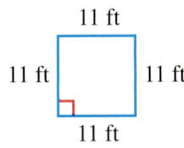

3.1 Finding the Length or Width of a Rectangle

If you know the perimeter of a rectangle and either its width or length, use the formula $P = 2l + 2w$. Replace P and either l or w with the values that you know. Then solve the equation.

The width of a rectangular rug is 8 ft. The perimeter is 36 ft. Find the length.

$$P = 2l + 2w \qquad \text{Replace } P \text{ with 36 ft and } w \text{ with 8 ft.}$$

$$36 \text{ ft} = 2l + 2 \cdot 8 \text{ ft}$$

$$36 \text{ ft} = 2l + 16 \text{ ft}$$

$$\underline{{}^{-}16 \text{ ft} \qquad {}^{-}16 \text{ ft}} \qquad \text{Add } {}^{-}16 \text{ ft to both sides.}$$

$$20 \text{ ft} = 2l + 0$$

$$\frac{20 \text{ ft}}{2} = \frac{2l}{2} \qquad \text{Divide both sides by 2.}$$

$$10 \text{ ft} = l \qquad \text{The length is 10 ft.}$$

Check To check the solution, draw a sketch of the rectangle and label the lengths of the sides. Then add the four measurements.

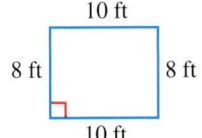

$$10 \text{ ft} + 10 \text{ ft} + 8 \text{ ft} + 8 \text{ ft} = 36 \text{ ft}$$

The result matches the perimeter given in the problem.

Concepts	Examples

[3.2] Finding Area

Use the appropriate formula. Remember to measure area in *square* units (cm^2, m^2, ft^2, yd^2, and so on).

Area of a rectangle: $A = lw$,
  where l is the length and w is the width.

Find the area of each figure.

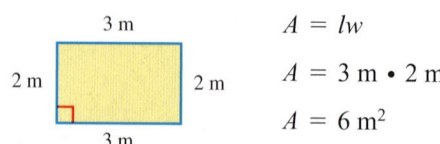

$A = lw$

$A = 3 \text{ m} \cdot 2 \text{ m}$

$A = 6 \text{ m}^2$

Area of a square: $A = s^2$, which means $s \cdot s$,
  where s is the length of one side.

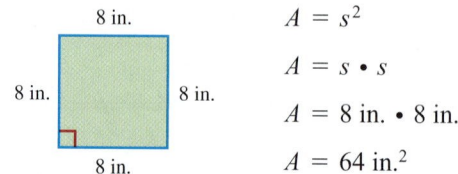

$A = s^2$

$A = s \cdot s$

$A = 8 \text{ in.} \cdot 8 \text{ in.}$

$A = 64 \text{ in.}^2$

Area of a parallelogram: $A = bh$,
  where b is the base and h is the height.

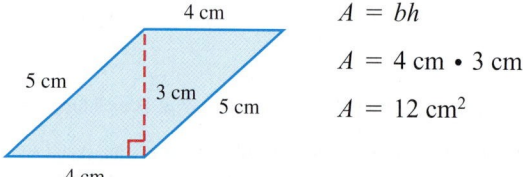

$A = bh$

$A = 4 \text{ cm} \cdot 3 \text{ cm}$

$A = 12 \text{ cm}^2$

[3.2] Finding the Unknown Length in a Rectangle or Parallelogram

If you know the area of a rectangle or parallelogram and one of the other measurements, use the appropriate area formula (see above). Replace A and one of the other variables with the values that you know. Then solve the equation.

The area of a parallelogram is 72 yd^2, and its height is 9 yd. Find the base.

$$A = bh \qquad \text{Replace } A \text{ with } 72 \text{ yd}^2 \text{ and } h \text{ with 9 yd.}$$

$$72 \text{ yd}^2 = b \cdot 9 \text{ yd}$$

$$\frac{72 \text{ yd} \cdot \text{yd}}{9 \text{ yd}} = \frac{b \cdot 9 \text{ yd}}{9 \text{ yd}} \qquad \text{Divide both sides by 9 yd.}$$

$$8 \text{ yd} = b$$

The base of the parallelogram is 8 yd.

[3.2] Finding the Length of One Side of a Square

If you know the area of a square, use the formula $A = s^2$. Replace A with the value that you know. Then solve the equation by asking, "What number, times itself, gives the value of A?"

A square ceiling has an area of 100 ft^2. What is the length of each side of the ceiling?

$$A = s^2 \qquad \text{Replace } A \text{ with } 100 \text{ ft}^2.$$

$$100 \text{ ft}^2 = s^2 \qquad \text{Rewrite } s^2 \text{ as } s \cdot s.$$

$$100 \text{ ft}^2 = s \cdot s \qquad \text{Ask, "What number times itself gives 100?"}$$

$$100 \text{ ft}^2 = 10 \text{ ft} \cdot 10 \text{ ft}$$

The value of s is 10 ft, so each side of the ceiling is 10 ft long.

Concepts	Examples

3.3 Translating Sentences into Equations

Translate word phrases into symbols using x (or any other letter) as the variable. Then solve the equation. Check the solution by putting it back in the original problem.

If 10 is subtracted from three times a number, the result is 14. Find the number.

Let x represent the unknown number.

$$3x - 10 = 14 \qquad \text{Change subtraction to adding the opposite.}$$
$$3x + {}^-10 = 14$$
$$\underline{\quad +10 \qquad +10} \qquad \text{Add 10 to both sides.}$$
$$3x + \quad 0 = 24$$
$$\frac{3x}{3} = \frac{24}{3} \qquad \text{Divide both sides by 3.}$$
$$x = 8$$

The number is 8.

Check   If 10 is subtracted from three times 8, do you get 14? Yes, $3 \cdot 8 - 10 = 24 - 10 = 14$.

3.3 Solving Application Problems with One Unknown Quantity

Use the six problem-solving steps outlined in **Section 3.3.** They are listed below in abbreviated form.

Denise had some money in her purse this morning. She gave $15 to her daughter and paid $4 to park in the lot at work. At that point she still had $27. How much was in her purse this morning?

Step 1   **Read** the problem and identify what is known and what is unknown.

Step 1   **Unknown:** money in purse this morning
Known: took out $15; took out $4; still had $27.

Step 2   **Assign a variable** to represent the unknown quantity.

Step 2   Let m represent the money in her purse this morning.

Step 3   **Write an equation.**

Step 3 $m - \$15 - \$4 = \$27$

Step 4   **Solve** the equation.

Step 4
$$m - 15 - 4 = 27$$
$$m + {}^-15 + {}^-4 = 27 \qquad \text{Change subtractions.}$$
$$m + \quad {}^-19 = 27 \qquad \text{Combine like terms.}$$
$$\underline{\quad +19 \qquad +19} \qquad \text{Add 19 to both sides.}$$
$$m + \quad 0 = 46$$
$$m = 46$$

Step 5   **State the answer.**

Step 5   She had $46 in her purse this morning.

Step 6   **Check** whether your answer fits all the facts given in the *original* statement of the problem.

Step 6   Started with $46.
Took out $15, so $46 - \$15 = \$31$
Took out $4, so $31 - \$4 = \$27$
Had $27 left ⟵ Matches

Concepts	Examples

(3.4) Solving Application Problems with Two Unknown Quantities

Use the six problem-solving steps outlined in **Section 3.3.**

In *Step 2*, there are *two* unknown quantities, so assign a variable to represent "the thing you know the least about." Then write a variable expression, using the same variable, to show the relationship of the other unknown quantity to the first one.

Last week, Brian earned $50 more than twice what Dan earned. How much did each person earn if the total for both of them was $254?

Step 1 **Unknowns :** Brian's earnings;
Dan's earnings

Known: Brian earned $50 more than twice what Dan earned;
the sum of their earnings was $254.

Step 2 You know the *least* about Dan's earnings, so let m represent Dan's earnings.

Brian earned $50 more than twice what Dan earned, so Brian's earnings are $2m + \$50$.

Step 3 $\underbrace{m + 2m}_{} + 50 = 254$

Step 4 $\underbrace{3m}_{} + 50 = 254$

$$\frac{\; ^-50 \quad ^-50}{3m \;+\; 0 = 204} \qquad \text{Add } ^-50 \text{ to both sides.}$$

$$\frac{3m}{3} = \frac{204}{3} \qquad \text{Divide both sides by 3.}$$

$$m = 68$$

Step 5 m represents Dan's earnings, so Dan earned $68.

$2m + \$50$ represents Brian's earnings,

and $2 \cdot \$68 + \$50$ is $\$136 + \$50 = \$186$.

Dan earned $68; Brian earned $186.

Step 6 Is $186 actually $50 more than twice $68?
Yes, the solution checks.

Does $68 + \$186 = \$254$?
Yes, the solution checks.

ANSWERS TO TEST YOUR WORD POWER

1. B; *Example:* If a triangle has sides measuring 4 ft, 10 ft, and 8 ft, then $P = 4 \text{ ft} + 10 \text{ ft} + 8 \text{ ft} = 22 \text{ ft}$.
2. D; *Example:* Area is measured in square units such as in.^{2}, ft^2, yd^2, cm^2, and km^2.
3. C; *Example:* If one side of a square is 5 in. long, all the other sides will also be 5 in. long.
4. A; *Example:* If the length of a rectangle is 12 cm and the width is 8 cm, then $A = 12 \text{ cm} \cdot 8 \text{ cm} = 96 \text{ cm}^2$.
5. C; *Examples:*

6. D; *Example:* In parallelogram *ABCD*, sides *AB* and *DC* are parallel and equal in length. Also, sides *AD* and *BC* are parallel and equal in length.

Chapter 3 ▷▷▷ Review Exercises

[3.1] *In Exercises 1–4, find the perimeter of each figure. Also, name each figure in Exercises 1–3.*

1.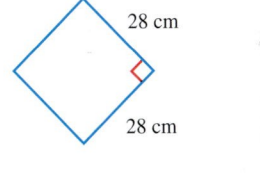
28 cm
28 cm

2.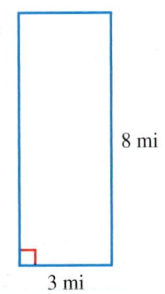
8 mi
3 mi

3.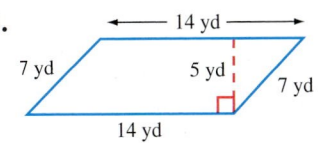
14 yd
7 yd   5 yd   7 yd
14 yd

4.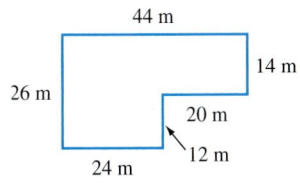
44 m
26 m   14 m
20 m
24 m   12 m

In Exercises 5–7, use the appropriate formula to find the unknown measurement.

5. A square card table has a perimeter of 12 ft. Find the length of one side of the table.

6. A rectangular playground has a perimeter of 128 yd. Find its length if it is 31 yd wide.

7. A rectangular watercolor painting that is 21 in. long has a perimeter of 72 in. What is the width of the painting?

[3.2] *In Exercises 8–10, draw a sketch of each shape and label the lengths of the sides or the base and height. Then find the area, using the appropriate formula. (Sketches may vary; show your sketches to your instructor.)*

8. A tablecloth that measures 5 ft by 8 ft

9. A 25 m square dance floor

10. A parallelogram-shaped lot with a base of 16 yd and a height of 13 yd

In Exercises 11–13, use the appropriate formula to find the unknown measurement.

11. A rectangular patio that is 14 ft long has an area of 126 ft^2. Find its width.

12. A parallelogram has an area of 88 cm^2. If the base is 11 cm, what is the height?

13. The area of a square piece of land is 100 mi^2. What is the length of one side?

[3.3] *Write each phrase as an algebraic expression. Use x as the variable.*

14. A number subtracted from 57

15. The sum of 15 and twice a number

16. The product of $^-9$ and a number

Translate each sentence into an equation and solve it. Show your work.

17. The sum of four times a number and 6 is $^-30$. What is the number?

18. When twice a number is subtracted from 10, the result is 4 plus the number. Find the number.

[3.3–3.4] *Use the six problem-solving steps to solve each problem. Show your work.*

19. Grace wrote a $600 check for her rent. Then she deposited her $750 paycheck and a $75 tax refund into her account. The new balance was $309. How much was in her account before she wrote the rent check?

20. Yoku ordered four boxes of candles for his restaurant. One candle was put on each of the 25 tables. There were 23 candles left. How many candles were originally in each box?

21. $1000 in prize money in an essay contest is being split between Reggie and Donald. Donald should get $300 more than Reggie. How much will each man receive?

22. A rectangular photograph is twice as long as it is wide. The perimeter of the photograph is 84 cm. Find the length and the width of the photograph.

▶▶▶ **Mixed Review Exercises**

Use the information in the advertisement and the appropriate formulas to solve Exercises 23–26. Show your work.

NOW ON SALE

Build Your Own Dog Pen

Kit #1 includes 20 feet of fencing.

Kit #2 includes 36 feet of fencing.

23. Anthony made a square dog pen using Kit #2.

    (a) What was the length of each side of the pen?

    (b) What was the area of the pen?

24. First draw sketches of two *different* rectangular dog pens that you could build using all the fencing in Kit #1. Label the lengths of the sides. Then find the area of each pen.

25. Timotha bought Kit #2. But she used some of the fencing around her garden, so she went back and bought Kit #1. Now she has 41 ft of fencing for a dog pen. How much fencing did she use around her garden? Use the six problem-solving steps.

26. Diana bought Kit #2. The pen she built had a length that was 2 ft more than the width. Find the length and width of the pen. Use the six problem-solving steps.

Find the perimeter of each shape.

1.

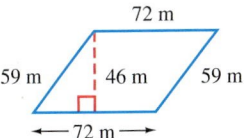
72 m

59 m   46 m   59 m

72 m

2.

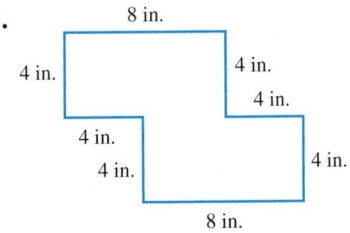
8 in.

4 in.   4 in.

4 in.

4 in.

4 in.   4 in.

8 in.

3. A square wetland 3 miles on a side

4. A rectangular mirror that measures 2 ft by 4 ft

5.

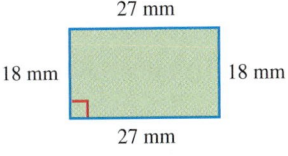
50 cm

15 cm

45 cm

Find the area of each shape.

6.

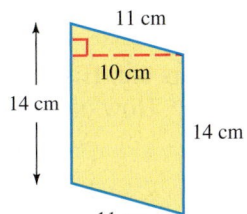
27 mm

18 mm   18 mm

27 mm

7.

11 cm

10 cm

14 cm

14 cm

11 cm

8. A rectangular animal preserve is 55 mi wide and 68 mi long.

9. The floor of a square room measures 6 m on each side.

Solve Exercises 10–14 using the appropriate formulas. Show your work.

10. A square table has a perimeter of 12 ft. Find the length of one side.

11. The Mercado family has 34 ft of fencing to put around a garden plot. The plot is rectangular in shape. If it is 6 ft wide, find the length of the plot.

12. The area of a parallelogram is 65 in.^{2}, and the base is 13 in. What is the height of the parallelogram?

1. _____

2. _____

3. _____

4. _____

5. _____

6. _____

7. _____

8. _____

9. _____

10. _____

11. _____

12. _____

13. _____

13. A rectangular postage stamp has a length of 4 cm and an area of 12 cm². Find its width.

14. _____

14. A square bulletin board has an area of 16 ft². How long is each side of the bulletin board?

15. _____

15. Explain the difference between ft and ft². For which types of problems might you use each of these units?

Translate each sentence into an equation and solve it. Show your work.

16. _____

16. If 40 is added to four times a number, the result is zero. Find the number.

17. _____

17. When 7 times a number is decreased by 23, the result is the number plus 7. What is the number?

Solve each application problem, using the six problem-solving steps.

18. _____

18. Josephine had $43 in her wallet. Her son used some of it when he bought groceries. Josephine found $16 in her desk drawer and put it in her wallet. She counted $44 in the wallet. How much money did her son spend on groceries?

19. _____

19. Ray is 39 years old. His age is 4 years more than five times his daughter's age. How old is his daughter?

20. _____

20. A board is 118 cm long. Karin cut it into two pieces, with one piece 4 cm longer than the other. Find the length of both pieces.

21. _____

21. The perimeter of a rectangular building is 420 ft. The length is four times as long as the width. Find the length and the width. Draw a sketch to help you solve this problem.

22. _____

22. Marcella and her husband, Tim, spent a total of 19 hours redecorating their living room. Tim spent 3 hours less time than Marcella. How long did each person work on the room?

Study Skills

TIPS FOR TAKING MATH TESTS

OBJECTIVES

1 Apply suggestions to tests and quizzes.

2 Develop a set of "best practices" to apply while testing.

Improving Your Test Score

To Improve Your Test Score	Comments
Come prepared with a pencil, eraser, and calculator, if allowed. If you are easily distracted, sit in the corner farthest from the door.	**Working in pencil lets you erase,** keeping your work neat and readable.
Scan the entire test, note the point value of different problems, and plan your time accordingly. Allow at least five minutes to check your work at the end of the testing time.	If you have 50 minutes to do 20 problems, $50 \div 20 = 2.5$ minutes per problem. **Spend less time on easy ones,** more time on problems with higher point values.
Read directions carefully, and circle any significant words. When you finish a problem, read the directions again to make sure you did what was asked.	**Pay attention to announcements** written on the board or made by your instructor. Ask if you don't understand. You don't want to get problems wrong because you misread the directions!
Show your work. Most math teachers give partial credit if some of the steps in your work are correct, even if the final answer is wrong. **Write neatly.** If you like to scribble when first working or checking a problem, do it on scratch paper.	**If your teacher can't read your writing, you won't get credit for it.** If you need more space to work, ask if you can use extra pieces of paper that you hand in with your test paper.
Check that the **answer to an application problem is reasonable** and makes sense. Read the problem again to make sure you've answered the question.	**Use common sense.** Can the father really be seven years old? Would a month's rent be $32,140? Label your answer: $, years, inches, etc.
To check for careless errors, you need to **rework the problem again, without looking at your previous work.** Cover up your work with a piece of scratch paper, and pretend you are doing the problem for the first time. Then compare the two answers.	If you just "look over" your work, your mind can make the same mistake again without noticing it. Reworking the problem from the beginning **forces you to rethink it.** If possible, use a different method to solve the problem the second time.

Reducing Anxiety

To Reduce Anxiety	Comments
Do not try to review up until the last minute before the test. Instead, go for a walk, do some deep breathing, and arrive just in time for the test. Ignore other students.	Listening to anxious classmates before the test **may cause you to panic.** Moderate exercise and deep breathing will calm your mind.

(continued)

Reducing Anxiety
(continued)

To Reduce Anxiety	Comments
Do a "knowledge dump" as soon as you get the test. Write important notes to yourself in a corner of the test paper: formulas, or common errors you want to watch out for.	Writing down tips and things that you've memorized **lets you relax;** you won't have to worry about forgetting those things and can refer to them as needed.
Do the easy problems first to build confidence. If you feel your anxiety starting to build, *immediately* stop for a minute, close your eyes, and take several slow, deep breaths.	Greater confidence helps you **get the easier problems correct.** Anxiety causes shallow breathing, which leads to confusion and reduced concentration. Deep breathing calms you.
As you work on more difficult problems, **notice your "inner voice."** You may have negative thoughts such as, "I can't do it," or "who cares about this test anyway." In your mind, yell "STOP" and take several deep, slow breaths. Or, replace the negative thought with a positive one.	Here are **examples of positive statements.** Try writing one of them on the top of your test paper. • I know I can do it. • I can do this one step at a time. • I've studied hard, and I'll do the best I can.
If you still can't solve a difficult problem when you come back to it the second time, **make a guess and do not change it.** In this situation, your first guess is your best bet. Do not change the answer just because you're a little unsure. **Change it only if you find an obvious mistake.**	If you are thinking about changing an answer, be sure you have a good reason for changing it. If you cannot find a specific error, leave your first answer alone. **When the tests are returned, check to see if changing answers helped or hurt you.**
Read the harder problems twice. Write down *anything* that might help solve the problem: a formula, a picture, etc. If you still can't get it, circle the problem and **come back to it later.** Do *not* erase any of the things you wrote down.	If you know even a *little* bit about the problem, write it down. The **answer may come to you** as you work on it, or you may get partial credit. Don't spend too long on any one problem. Your subconscious mind will work on the tough problem while you go on with the test.
Ignore students who finish early. Use the entire test time. *You do not get extra credit for finishing early.* Use the extra time to rework problems and correct careless errors.	Students who leave early are often the ones who didn't study or who are too anxious to continue working. If they bother you, *sit as far from the door as possible.*

Why Are These Suggestions Brain Friendly?

Several suggestions address anxiety. **Reducing anxiety** allows your brain to make the connections between dendrites; in other words, you can think clearly.

Remember that **your brain continues to work** on a difficult problem even if you skip it and go on to the next one. Your subconscious mind will come through for you if you are open to the idea!

Some of the suggestions ask you to **use your common sense.** Follow the directions, show your work, write neatly, and pay attention to whether your answers really make sense.

Rational Numbers: Positive and Negative Fractions

Americans spend over $290 billion each year buying materials for their home repair, construction, or remodeling projects. Both men and women take on home improvement projects in about equal numbers. One of the small but essential items for many projects is nails. There are many kinds and sizes of nails, and you'll need to buy the right size nail for each project. Nails are sized according to the "penny" system, which is a number followed by the abbreviation "d." (The "d" is the traditional British abbreviation for "penny.") An algebraic expression, and the ability to solve equations with fractions, will help you find the relationship between a nail's "penny size" and its length in inches. See **Section 4.7,** Exercises 33–36. (*Sources:* U.S. Census Bureau; Opinion Research Corp.; *Season by Season Home Maintenance*.)

4.1 ▶▶▶ Introduction to Signed Fractions

OBJECTIVE 1 Use a fraction to name part of a whole. In Chapters 1–3 you worked with integers. Recall that a list of the integers can be written as follows.

$$\ldots, \ ^-6, \ ^-5, \ ^-4, \ ^-3, \ ^-2, \ ^-1, 0, 1, 2, 3, 4, 5, 6, \ldots$$

The dots show that the list goes on forever in both directions.

Now we will work with *fractions*.

> **Fractions**
>
> A **fraction** is a number of the form $\dfrac{a}{b}$ where a and b are integers and b is not 0.

One use for fractions is situations in which we need a number that is between two integers. Here is an example.

$\frac{2}{3}$ cup

A recipe uses $\frac{2}{3}$ cup of milk.

$\frac{2}{3}$ is between 0 and 1.

$\frac{2}{3}$ is a fraction because it is of the form $\dfrac{a}{b}$ and 2 and 3 are integers.

The number $\frac{2}{3}$ is a fraction that represents 2 of 3 equal parts. In this example, the cup is divided into 3 equal parts and we use enough milk to fill 2 of the parts.

We read $\frac{2}{3}$ as "two-thirds."

1 Write fractions for the shaded portion and the unshaded portion of each figure.

(a)

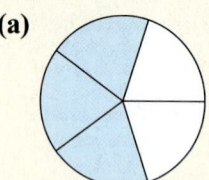

(b)

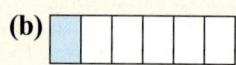

(c)

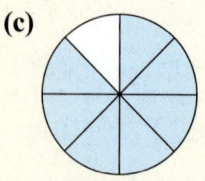

> **EXAMPLE 1 Using Fractions to Represent Part of One Whole**
>
> Use fractions to represent the shaded portion and the unshaded portion of each figure.
>
> **(a)**
>
>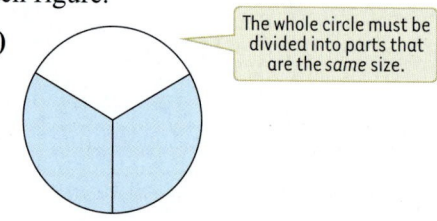
>
> The whole circle must be divided into parts that are the *same* size.
>
> The figure has 3 equal parts. The 2 shaded parts are represented by the fraction $\frac{2}{3}$. The *un*shaded part is $\frac{1}{3}$ of the figure.
>
> **(b)**
>
>
>
> The figure has 7 equal parts. The 4 shaded parts are represented by the fraction $\frac{4}{7}$. The *un*shaded part is $\frac{3}{7}$ of the figure.

◀ *Work Problem* **1** *at the Side.*

Fractions can also be used to represent more than one whole object.

EXAMPLE 2   Using Fractions to Represent More Than One Whole

Use a fraction to represent the shaded parts.

(a)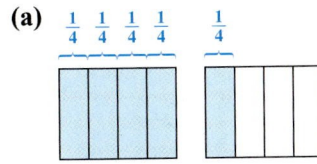

(b)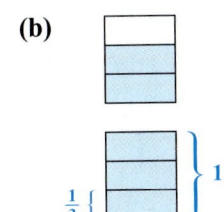

An area equal to 5 of the $\frac{1}{4}$ parts is shaded, so $\frac{5}{4}$ is shaded.

An area equal to 5 of the $\frac{1}{3}$ parts is shaded, so $\frac{5}{3}$ is shaded.

Work Problem **2** *at the Side.* ▶

OBJECTIVE 2   Identify numerators, denominators, proper fractions, and improper fractions. In the fraction $\frac{2}{3}$, the number 2 is the *numerator* and 3 is the *denominator*. The bar between the numerator and the denominator is the *fraction bar*.

Fraction bar → $\dfrac{2}{3}$ ← Numerator ← Denominator

Numerator and Denominator

The **denominator** of a fraction shows the number of equal parts in the whole, and the **numerator** shows how many parts are being considered.

Note

Recall that a fraction bar, —, is a symbol for division and division by 0 is *undefined*. Therefore a fraction with a denominator of 0 is also *undefined*.

EXAMPLE 3   Identifying Numerators and Denominators

Identify the numerator and denominator in each fraction. Then state the number of equal parts in the whole.

(a) $\dfrac{5}{9}$ $\dfrac{5}{9}$ ← Numerator ← Denominator

9 equal parts in the whole

(b) $\dfrac{11}{7}$ $\dfrac{11}{7}$ ← Numerator ← Denominator

7 equal parts in the whole

Work Problem **3** *at the Side.* ▶

Fractions are sometimes called *proper* or *improper* fractions.

Proper and Improper Fractions

If the numerator of a fraction is *less* than the denominator, the fraction is a **proper fraction.** A proper fraction is less than 1.

If the numerator is *greater than or equal to* the denominator, the fraction is an **improper fraction.** An improper fraction is greater than or equal to 1.

2 Write fractions for the shaded portions.

(a) $\frac{1}{7}$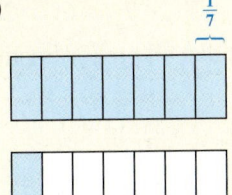

(b) $\frac{1}{4}$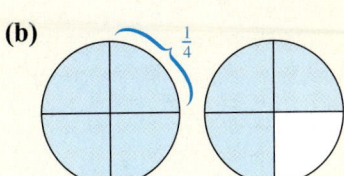

3 Identify the numerator and the denominator. Draw a picture with shaded parts to show each fraction. Your drawings may vary, but they should have the correct number of shaded parts.

(a) $\dfrac{2}{3}$

(b) $\dfrac{1}{4}$

(c) $\dfrac{8}{5}$

(d) $\dfrac{5}{2}$

ANSWERS

2. (a) $\frac{8}{7}$ (b) $\frac{7}{4}$

3. (a) N: 2; D: 3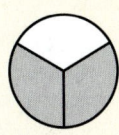

(b) N: 1; D: 4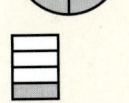

(c) N: 8; D: 5

(d) N: 5; D: 2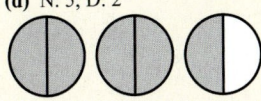

4 From this group of fractions:

$$\frac{3}{4} \quad \frac{8}{7} \quad \frac{5}{7} \quad \frac{6}{6} \quad \frac{1}{2} \quad \frac{2}{1}$$

(a) list all proper fractions.

Proper Fractions	Improper Fractions
$\dfrac{1}{2} \quad \dfrac{5}{11} \quad \dfrac{35}{36}$	$\dfrac{9}{7} \quad \dfrac{126}{125} \quad \dfrac{7}{7}$

EXAMPLE 4 **Classifying Types of Fractions**

(a) Identify all proper fractions in this list.

$$\frac{3}{4} \quad \frac{5}{9} \quad \frac{17}{5} \quad \frac{9}{7} \quad \frac{3}{3} \quad \frac{12}{25} \quad \frac{1}{9} \quad \frac{5}{3}$$

Proper fractions have a numerator that is *less* than the denominator. The proper fractions in the list are shown below.

$$\frac{3}{4} \;\leftarrow 3 \text{ is less than 4.} \qquad \frac{5}{9} \quad \frac{12}{25} \quad \frac{1}{9}$$

(b) Identify all improper fractions in the list in part (a).

Improper fractions have a numerator that is *equal to or greater* than the denominator. The improper fractions in the list are shown below.

$$\frac{17}{5} \;\leftarrow 17 \text{ is greater than 5.} \quad \frac{9}{7} \quad \frac{3}{3} \quad \frac{5}{3}$$

◀ *Work Problem* **4** *at the Side.*

(b) list all improper fractions.

> **CAUTION**
>
> In **Chapters 1–3** we used a raised negative sign to help you avoid confusion between negative numbers and subtraction. Now you are ready to start writing the negative sign in the more traditional way. In this chapter, the negative sign will still be red, but it will be centered on the number instead of raised: for example, -2 instead of $^-2$. When the negative sign might be confused with the sign for subtraction, we will write parentheses around the negative number. Here is an example.
>
> $$3 - (-2) \quad \text{means} \quad 3 \text{ minus } (\text{negative } 2)$$
>
> For fractions, the negative sign will be written in front of the fraction bar: for example, $-\frac{3}{4}$. As with integers, the negative sign tells you that a fraction is *less than 0;* it is to the *left* of 0 on the number line. When there is *no* sign in front of a fraction, the fraction is assumed to be positive. For example, $\frac{3}{4}$ is assumed to be $+\frac{3}{4}$. It is to the *right* of 0 on the number line.

OBJECTIVE **3** **Graph positive and negative fractions on a number line.** Sometimes we need *negative* numbers that are between two integers. For example, $-\frac{3}{4}$ is between 0 and -1. Graphing numbers on a number line helps us see the difference between $\frac{3}{4}$ and $-\frac{3}{4}$. Both represent 3 out of 4 equal parts, but they are in opposite directions from 0 on the number line. For $\frac{3}{4}$, divide the distance from 0 to 1 into 4 equal parts. Then start at 0, count over 3 parts, and make a dot. For $-\frac{3}{4}$, repeat the same process between 0 and -1.

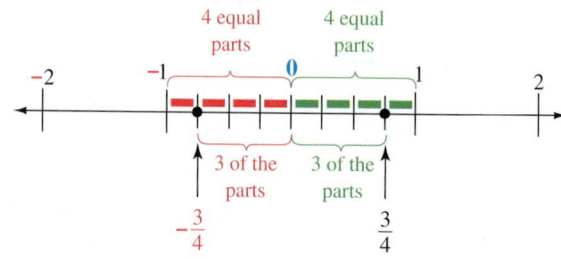

EXAMPLE 5 **Graphing Positive and Negative Fractions**

Graph each fraction on the number line.

(a) $\dfrac{2}{5}$

There is *no* sign in front of $\frac{2}{5}$, so it is *positive*. Because $\frac{2}{5}$ is between 0 and 1, we divide that space into 5 equal parts. Then we start at 0 and count to the right 2 parts.

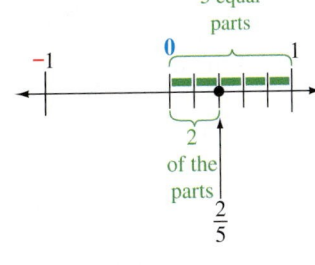

(b) $-\dfrac{4}{5}$

The fraction is *negative,* so it is between 0 and -1. We divide that space into 5 equal parts. Then we start at 0 and count to the left 4 parts.

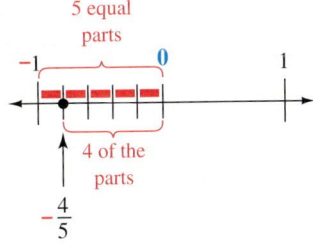

Work Problem **5** *at the Side.* ▶

OBJECTIVE **4** **Find the absolute value of a fraction.** In **Section 1.2** we said that the *absolute value* of a number was its distance from 0 on the number line. Two vertical bars indicate absolute value, as shown below.

$$\left| -\frac{3}{4} \right| \text{ is read "the absolute value of negative three-fourths."}$$

As with integers, the absolute value of fractions will *always* be positive (or 0) because it is the *distance* from 0 on the number line.

EXAMPLE 6 **Finding the Absolute Value of Fractions**

Find each absolute value: $\left| \dfrac{1}{2} \right|$ and $\left| -\dfrac{1}{2} \right|$.

The distance from 0 to $\dfrac{1}{2}$ on the number line is $\dfrac{1}{2}$ space, so $\left| \dfrac{1}{2} \right| = \dfrac{1}{2}$.

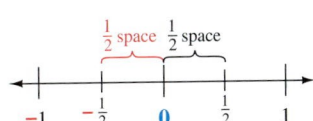

The distance from 0 to $-\dfrac{1}{2}$ is also $\dfrac{1}{2}$ space, so $\left| -\dfrac{1}{2} \right| = \dfrac{1}{2}$.

Work Problem **6** *at the Side.* ▶

5 Graph each fraction on the number line.

(a) $\dfrac{2}{4}$

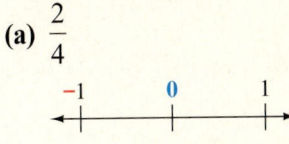

(b) $\dfrac{1}{2}$

(c) $-\dfrac{2}{3}$

6 Find each absolute value.

(a) $\left| -\dfrac{3}{4} \right|$

(b) $\left| \dfrac{5}{8} \right|$

(c) $\left| 0 \right|$

ANSWERS

5. **(a)**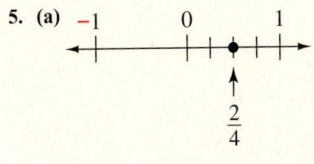

(b)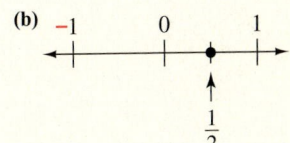

(c)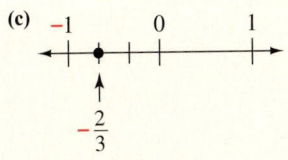

6. **(a)** $\dfrac{3}{4}$ **(b)** $\dfrac{5}{8}$ **(c)** 0

OBJECTIVE **5** **Write equivalent fractions.** You may have noticed in Margin Problems 5(a) and 5(b) on the previous page that $\frac{2}{4}$ and $\frac{1}{2}$ were at the same point on the number line. Both of them were halfway between 0 and 1. There are actually *many* different names for this point. We illustrate some of them below.

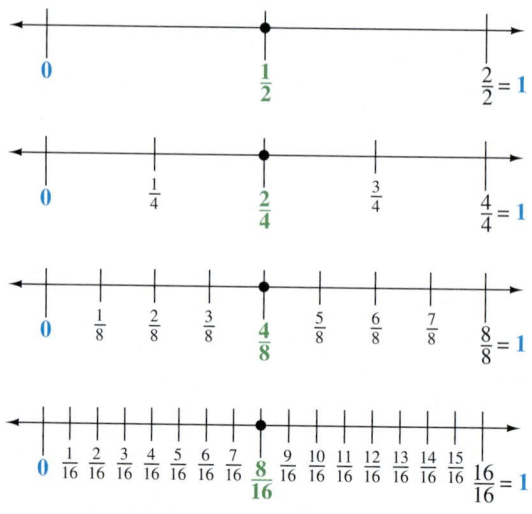

That is, $\frac{8}{16} = \frac{4}{8} = \frac{2}{4} = \frac{1}{2}$. If you have used a standard ruler with inches divided into sixteenths, you probably already noticed that these distances are the same. Although the fractions look different, they all name the same point that is halfway between 0 and 1. In other words, they have the same value. We say that they are *equivalent fractions*.

> **Equivalent Fractions**
>
> Fractions that represent the same number (the same point on a number line) are **equivalent fractions.**

Drawing number lines is tedious, so we usually find equivalent fractions by multiplying or dividing both the numerator and denominator by the same number. We can use some of the fractions that we just graphed to illustrate this method.

$$\frac{1}{2} = \frac{1 \cdot 2}{2 \cdot 2} = \frac{2}{4} \qquad\qquad \frac{8}{16} = \frac{8 \div 4}{16 \div 4} = \frac{2}{4}$$

Multiply both numerator and denominator by 2. Divide both numerator and denominator by 4.

> **Writing Equivalent Fractions**
>
> If *a*, *b*, and *c* are numbers (and *b* and *c* are not 0), then:
>
> $$\frac{a}{b} = \frac{a \cdot c}{b \cdot c} \qquad \text{or} \qquad \frac{a}{b} = \frac{a \div c}{b \div c}$$
>
> In other words, if the numerator and denominator of a fraction are multiplied or divided by the *same* nonzero number, the result is an *equivalent* fraction.

EXAMPLE 7 **Writing Equivalent Fractions**

(a) Write $-\dfrac{1}{2}$ as an equivalent fraction with a denominator of 16.

In other words, $-\dfrac{1}{2} = -\dfrac{?}{16}$.

The original denominator is 2. *Multiplying* 2 times 8 gives 16, the new denominator. To write an equivalent fraction, multiply *both* the numerator and denominator by 8.

$$-\dfrac{1}{2} = -\dfrac{1 \cdot 8}{2 \cdot 8} = -\dfrac{8}{16}$$

Multiply numerator and denominator by the *same* number.

Keep the negative sign.

So, $-\dfrac{1}{2}$ is equivalent to $-\dfrac{8}{16}$.

(b) Write $\dfrac{12}{15}$ as an equivalent fraction with a denominator of 5.

In other words, $\dfrac{12}{15} = \dfrac{?}{5}$.

The original denominator is 15. *Dividing* 15 by 3 gives 5, the new denominator. To write an equivalent fraction, divide *both* the numerator and denominator by 3.

$$\dfrac{12}{15} = \dfrac{12 \div 3}{15 \div 3} = \dfrac{4}{5}$$

Divide numerator and denominator by the *same* number.

So, $\dfrac{12}{15}$ is equivalent to $\dfrac{4}{5}$.

Work Problem **7** *at the Side.* ▶

Look back at the set of four number lines on the previous page. Notice that there are many different names for 1.

$$\dfrac{2}{2} = 1 \qquad \dfrac{4}{4} = 1 \qquad \dfrac{8}{8} = 1 \qquad \dfrac{16}{16} = 1$$

Because a fraction bar is a symbol for division, you can think of $\frac{2}{2}$ as $2 \div 2$, which equals 1. Similarly, $\frac{4}{4}$ is $4 \div 4$, which also is 1, and so on. These examples illustrate one of the division properties from **Section 1.7.**

Division Properties

If a is any number (except 0), then $\dfrac{a}{a} = 1$. In other words, when a nonzero number is divided by itself, the result is 1.

For example: $\dfrac{6}{6} = 1$ and $\dfrac{-4}{-4} = 1$

Also recall that when any number is divided by 1, the result is the number. That is, $\dfrac{a}{1} = a$.

For example: $\dfrac{6}{1} = 6$ and $-\dfrac{12}{1} = -12$

---

7 **(a)** Write $\frac{2}{5}$ as an equivalent fraction with a denominator of 20.

(b) Write $-\frac{21}{28}$ as an equivalent fraction with a denominator of 4.

8 Simplify each fraction by dividing the numerator by the denominator.

(a) $\dfrac{10}{10}$

EXAMPLE 8   Using Division to Simplify Fractions

Simplify each fraction by dividing the numerator by the denominator.

(a) $\dfrac{5}{5}$ Think of $\dfrac{5}{5}$ as $5 \div 5$. The result is 1, so $\dfrac{5}{5} = 1$.

> A fraction bar is a symbol for division.

(b) $-\dfrac{12}{4}$ Think of $-\dfrac{12}{4}$ as $-12 \div 4$. The result is -3, so $-\dfrac{12}{4} = -3$.

Keep the negative sign.

(c) $\dfrac{6}{1}$ Think of $\dfrac{6}{1}$ as $6 \div 1$. The result is 6, so $\dfrac{6}{1} = 6$.

◀ *Work Problem* **8** *at the Side.*

(b) $-\dfrac{3}{1}$

Note

The title of this chapter is "Rational Numbers: Positive and Negative Fractions." *Rational numbers* are numbers that can be written in the form $\dfrac{a}{b}$, where a and b are integers and b is not 0. In Example 8(c) above, you saw that an integer can be written in the form $\dfrac{a}{b}$ (6 can be written as $\frac{6}{1}$).

So rational numbers include all the integers and all the fractions. In **Chapter 5** you'll work with rational numbers that are in decimal form.

(c) $\dfrac{8}{2}$

(d) $-\dfrac{25}{5}$

Write the fractions that represent the shaded and unshaded portions of each figure. See Examples 1 and 2.

1. 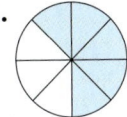   **2.** 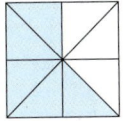   **3.** 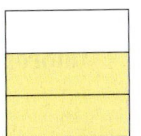   **4.**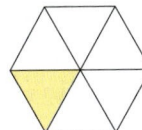

5. 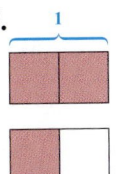   **6.** 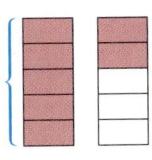   **7.** 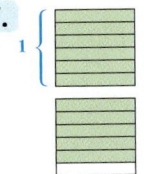   **8.**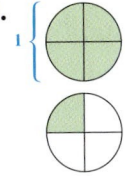

9. What fraction of these coins are dimes? What fraction are pennies? What fraction are nickels?

10. What fraction of these recording artists are men? What fraction are women? What fraction are wearing something white?

11. Of the 71 computers in the lab, 58 are laptops. What fraction of the computers are *not* laptops? What fraction are laptops?

12. A community college basketball team has 12 members. If five of the players are sophomores and the rest are freshmen, find the fraction of the members that are sophomores and the fraction that are freshmen.

The circle graph shows the results of a survey on where women would like to have flowers delivered on Valentine's Day. Use the graph to answer Exercises 13–14.

13. (a) What fraction of the women would like flowers delivered at work? **(b)** Delivered at home or at work?

14. (a) What fraction picked a location other than home or work? **(b)** What fraction picked at home or other?

Delivering Flowers On Valentine's Day

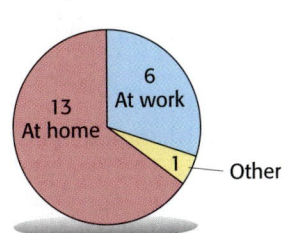

Out of every 20 women surveyed, the number who would like flowers delivered at home, at work, or elsewhere.

Source: FTD, Inc.

Identify the numerator and denominator in each fraction. Then state the number of equal parts in the whole. See Example 3.

15. $\dfrac{3}{4}$

16. $\dfrac{5}{8}$

17. $\dfrac{12}{7}$

18. $\dfrac{8}{3}$

List the proper and improper fractions in each group of numbers. See Example 4.

	Proper	Improper
19. $\dfrac{8}{5}, \dfrac{1}{3}, \dfrac{5}{8}, \dfrac{6}{6}, \dfrac{12}{2}, \dfrac{7}{16}$	_____	_____

	Proper	Improper
20. $\dfrac{1}{6}, \dfrac{5}{8}, \dfrac{15}{14}, \dfrac{11}{9}, \dfrac{7}{7}, \dfrac{3}{4}$	_____	_____

Graph each pair of fractions on the number line. See Example 5.

21. $\dfrac{1}{4}, -\dfrac{1}{4}$

22. $-\dfrac{1}{3}, \dfrac{1}{3}$

23. $-\dfrac{3}{5}, \dfrac{3}{5}$

24. $\dfrac{5}{6}, -\dfrac{5}{6}$

25. $\dfrac{7}{8}, -\dfrac{7}{8}$

26. $-\dfrac{3}{4}, \dfrac{3}{4}$

Write a positive or negative fraction to describe each situation.

27. The baby lost $\frac{3}{4}$ pound in weight while she was sick.

28. Greta needed $\frac{1}{3}$ cup of brown sugar for the cookie recipe.

29. Barb Brown's driveway is $\frac{3}{10}$ mile long.

30. The oil level in my car is $\frac{1}{2}$ quart below normal.

Find each absolute value. See Example 6.

31. $\left| -\dfrac{2}{5} \right|$

32. $\left| -\dfrac{4}{4} \right|$

33. $|0|$

34. $\left| \dfrac{9}{10} \right|$

35. Rewrite each fraction as an equivalent fraction with a denominator of 24. See Example 7.

   (a) $\dfrac{1}{2} = \dfrac{}{24}$ **(b)** $\dfrac{1}{3} = ---$ **(c)** $\dfrac{2}{3} = ---$ **(d)** $\dfrac{1}{4} = ---$ **(e)** $\dfrac{3}{4} = ---$

   (f) $\dfrac{1}{6} = ---$ **(g)** $\dfrac{5}{6} = ---$ **(h)** $\dfrac{1}{8} = ---$ **(i)** $\dfrac{3}{8} = ---$ **(j)** $\dfrac{5}{8} = ---$

36. Rewrite each fraction as an equivalent fraction with a denominator of 36. See Example 7.

(a) $\dfrac{1}{2} = \dfrac{}{36}$ **(b)** $\dfrac{1}{3} = \underline{\quad}$ **(c)** $\dfrac{2}{3} = \underline{\quad}$ **(d)** $\dfrac{1}{4} = \underline{\quad}$ **(e)** $\dfrac{3}{4} = \underline{\quad}$

(f) $\dfrac{1}{6} = \underline{\quad}$ **(g)** $\dfrac{5}{6} = \underline{\quad}$ **(h)** $\dfrac{1}{9} = \underline{\quad}$ **(i)** $\dfrac{4}{9} = \underline{\quad}$ **(j)** $\dfrac{8}{9} = \underline{\quad}$

37. Rewrite each fraction as an equivalent fraction with a denominator of 3. See Example 7.

(a) $-\dfrac{2}{6} = -\dfrac{}{3}$ **(b)** $-\dfrac{4}{6} = \underline{\quad}$ **(c)** $-\dfrac{12}{18} = \underline{\quad}$ **(d)** $-\dfrac{6}{18} = \underline{\quad}$ **(e)** $-\dfrac{200}{300} = \underline{\quad}$

(f) Write two more fractions that are equivalent to $-\frac{1}{3}$ and two more fractions equivalent to $-\frac{2}{3}$.

38. Rewrite each fraction as an equivalent fraction with a denominator of 4. See Example 7.

(a) $-\dfrac{2}{8} = -\dfrac{}{4}$ **(b)** $-\dfrac{6}{8} = \underline{\quad}$ **(c)** $-\dfrac{15}{20} = \underline{\quad}$ **(d)** $-\dfrac{50}{200} = \underline{\quad}$ **(e)** $-\dfrac{150}{200} = \underline{\quad}$

(f) Write two more fractions that are equivalent to $-\frac{1}{4}$ and two more fractions equivalent to $-\frac{3}{4}$.

Relating Concepts (Exercises 39–42)   For Individual or Group Work

Use your calculator as you **work Exercises 39–42 in order.**

39. (a) Write $\frac{3}{8}$ as an equivalent fraction with a denominator of 3912.

(b) Explain how you solved part (a).

40. (a) Write $\frac{7}{9}$ as an equivalent fraction with a denominator of 5472.

(b) Explain how you solved part (a).

41. (a) Is $-\dfrac{697}{3485}$ equivalent to $-\dfrac{1}{2}$, $-\dfrac{1}{3}$, or $-\dfrac{1}{5}$?

(b) Explain how you solved part (a).

42. (a) Is $-\dfrac{817}{4902}$ equivalent to $-\dfrac{1}{4}$, $-\dfrac{1}{6}$, or $-\dfrac{1}{8}$?

(b) Explain how you solved part (a).

---

43. Can you write $\frac{3}{5}$ as an equivalent fraction with a denominator of 18? Explain why or why not. If not, what denominators could you use instead of 18?

44. Can you write $\frac{3}{4}$ as an equivalent fraction with a denominator of 0? Explain why or why not.

Simplify each fraction by dividing the numerator by the denominator. See Example 8.

45. $\dfrac{10}{1}$

46. $\dfrac{9}{9}$

47. $-\dfrac{16}{16}$

48. $-\dfrac{7}{1}$

49. $-\dfrac{18}{3}$

50. $-\dfrac{40}{4}$

51. $\dfrac{24}{8}$

52. $\dfrac{42}{6}$

53. $\dfrac{14}{7}$

54. $\dfrac{8}{2}$

55. $-\dfrac{90}{10}$

56. $-\dfrac{45}{9}$

57. $\dfrac{150}{150}$

58. $\dfrac{55}{5}$

59. $-\dfrac{32}{4}$

60. $-\dfrac{200}{200}$

There are many correct ways to draw the answers for Exercises 61–68, so ask your instructor to check your work.

61. Shade $\frac{3}{5}$ of this figure. What fraction is unshaded?

62. Shade $\frac{5}{6}$ of this figure. What fraction is unshaded?

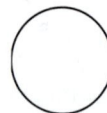

63. Shade $\frac{3}{8}$ of this figure. What fraction is unshaded?

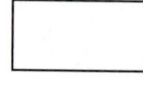

64. Shade $\frac{1}{3}$ of this figure. What fraction is unshaded?

65. Draw a group of figures. Make $\frac{1}{10}$ of the figures circles, $\frac{6}{10}$ of the figures squares, and $\frac{3}{10}$ of the figures triangles. Then shade $\frac{1}{6}$ of the squares and $\frac{2}{3}$ of the triangles.

66. Write a group of capital letters. Make $\frac{4}{9}$ of the letters A's, $\frac{2}{9}$ of the letters B's, and $\frac{3}{9}$ of the letters C's. Then draw a line under $\frac{3}{4}$ of the A's and $\frac{1}{2}$ of the B's.

67. Draw a group of punctuation marks. Make $\frac{5}{12}$ of them exclamation points, $\frac{1}{12}$ of them commas, $\frac{3}{12}$ of them periods, and add enough question marks to make a full $\frac{12}{12}$ in all. Then circle $\frac{2}{5}$ of the exclamation points.

68. Draw a group of symbols. Make $\frac{2}{15}$ of them addition signs, $\frac{4}{15}$ of them subtraction signs, $\frac{2}{15}$ of them division signs, and add enough equal signs to make a full $\frac{15}{15}$ in all. Then circle $\frac{3}{4}$ of the subtraction signs.

4.2 ▶▶▶ Writing Fractions in Lowest Terms

OBJECTIVE 1 Identify fractions written in lowest terms. You can see from these drawings that $\frac{1}{2}$ and $\frac{4}{8}$ are different names for the same amount of pizza.

$\frac{1}{2}$ of the pizza has pepperoni on it.

$\frac{4}{8}$ of the pizza has pepperoni on it.

You saw in the last section that $\frac{1}{2}$ and $\frac{4}{8}$ are equivalent fractions. But we say that the fraction $\frac{1}{2}$ is in *lowest terms* because the numerator and denominator have no *common factor* other than 1. That means that 1 is the only number that divides evenly into both 1 and 2. However, the fraction $\frac{4}{8}$ is *not* in lowest terms because its numerator and denominator have a common factor of 4. That means 4 will divide evenly into both 4 and 8.

> **Note**
> Recall that *factors* are numbers being multiplied to give a product. For example,
>
> $1 \cdot 4 = 4$, so 1 and 4 are factors of 4.
>
> $2 \cdot 4 = 8$, so 2 and 4 are factors of 8.
>
> 4 is a factor of both 4 and 8, so 4 is a *common factor* of those numbers.

> **Writing a Fraction in Lowest Terms**
> A fraction is written in **lowest terms** when the numerator and denominator have no common factor other than 1. Examples are $\frac{1}{3}, \frac{3}{4}, \frac{2}{5}$, and $\frac{7}{10}$.
>
> When you work with fractions, always write the final answer in lowest terms.

EXAMPLE 1 Identifying Fractions Written in Lowest Terms

Are the following fractions in lowest terms? If not, find a common factor of the numerator and denominator (other than 1).

(a) $\frac{3}{8}$

The numerator and denominator have no common factor other than 1, so the fraction is in lowest terms.

(b) $\frac{21}{36}$

The numerator and denominator have a common factor of 3, so the fraction is *not* in lowest terms.

Work Problem ① *at the Side.* ▶

OBJECTIVES

1. Identify fractions written in lowest terms.

2. Write a fraction in lowest terms using common factors.

3. Write a number as a product of prime factors.

4. Write a fraction in lowest terms using prime factorization.

5. Write a fraction with variables in lowest terms.

① Are the following fractions in lowest terms? If not, find a common factor of the numerator and denominator (other than 1).

(a) $\dfrac{2}{3}$

(b) $-\dfrac{8}{10}$

(c) $-\dfrac{9}{11}$

(d) $\dfrac{15}{20}$

2 Divide by a common factor to write each fraction in lowest terms.

(a) $\dfrac{5}{10}$

(b) $\dfrac{9}{12}$

(c) $-\dfrac{24}{30}$

(d) $\dfrac{15}{40}$

(e) $-\dfrac{50}{90}$

ANSWERS

2. (a) $\dfrac{1}{2}$ (b) $\dfrac{3}{4}$ (c) $-\dfrac{4}{5}$ (d) $\dfrac{3}{8}$ (e) $-\dfrac{5}{9}$

OBJECTIVE **2** **Write a fraction in lowest terms using common factors.** We will show you two methods for writing a fraction in lowest terms. The first method, dividing by a common factor, works best when the numerator and denominator are small numbers.

EXAMPLE 2 Using Common Factors to Write Fractions in Lowest Terms

Divide by a common factor to write each fraction in lowest terms.

(a) $\dfrac{20}{24}$

The *greatest* common factor of 20 and 24 is 4. Divide both numerator and denominator by 4.

$$\dfrac{20}{24} = \dfrac{20 \div 4}{24 \div 4} = \dfrac{5}{6}$$

(b) $\dfrac{30}{50} = \dfrac{30 \div 10}{50 \div 10} = \dfrac{3}{5}$ Divide both numerator and denominator by 10.

(c) $-\dfrac{24}{42} = -\dfrac{24 \div 6}{42 \div 6} = -\dfrac{4}{7}$ Divide both numerator and denominator by 6. Keep the negative sign.

(d) $\dfrac{60}{72}$

Suppose we made an error and thought that 4 was the greatest common factor of 60 and 72. Dividing by 4 gives the following.

$$\dfrac{60}{72} = \dfrac{60 \div 4}{72 \div 4} = \dfrac{15}{18}$$

But $\frac{15}{18}$ is *not* in lowest terms because 15 and 18 have a common factor of 3. Therefore, divide the numerator and denominator by 3.

$$\dfrac{15}{18} = \dfrac{15 \div 3}{18 \div 3} = \dfrac{5}{6} \leftarrow \text{Lowest terms}$$

The fraction $\frac{60}{72}$ could have been written in lowest terms in one step by dividing by 12, the *greatest* common factor of 60 and 72.

$$\dfrac{60}{72} = \dfrac{60 \div 12}{72 \div 12} = \dfrac{5}{6} \begin{cases} \text{Same answer} \\ \text{as above} \end{cases}$$

Either way works. Just keep dividing until the fraction is in lowest terms.

This method of writing a fraction in lowest terms by dividing by a common factor is summarized below.

> **Dividing by a Common Factor to Write a Fraction in Lowest Terms**
>
> **Step 1** Find the *greatest* number that will divide evenly into both the numerator and denominator. This number is a **common factor.**
>
> **Step 2** **Divide** both numerator and denominator by the common factor.
>
> **Step 3** **Check** to see if the new numerator and denominator have any common factors (besides 1). If they do, repeat Steps 2 and 3. If the only common factor is 1, the fraction is in lowest terms.

◀ *Work Problem* **2** *at the Side.*

OBJECTIVE **3** **Write a number as a product of prime factors.**
In Example 2(d) on the previous page, the greatest common factor of 60 and 72 was difficult to see quickly. You can handle a problem like that by writing the numerator and denominator as a product of *prime numbers*.

3 Label each number as *prime* or *composite* or *neither*.
1, 2, 3, 4, 7, 9, 13, 19, 25, 29

> **Prime Numbers**
> A **prime number** is a whole number that has exactly *two different* factors, itself and 1.

The number 3 is a prime number because it can be divided evenly only by itself and 1. The number 8 is *not* a prime number. The number 8 is a *composite number* because it can be divided evenly by 2 and 4, as well as by itself and 1.

> **Composite Numbers**
> A number with a factor other than itself or 1 is called a **composite number.**

> **CAUTION**
> A prime number has *only two* different factors, itself and 1. The number 1 is *not* a prime number because it does not have *two different* factors; the only factor of 1 is 1. Also, 0 is *not* a prime number. Therefore, 0 and 1 are *neither* prime nor composite numbers.

EXAMPLE 3 **Finding Prime Numbers**

Label each number as *prime* or *composite* or *neither*.

0   2   5   10   11   15

First, 0 is *neither* prime nor composite. Next, 2, 5, and 11, are *prime*. Each of these numbers is divisible only by itself and 1. The number 10 can be divided by 5 and 2, so it is *composite*. Also, 15 is a *composite* number because 15 can be divided by 5 and 3.

Work Problem **3** *at the Side.* ▶

For reference, here are the prime numbers smaller than 100.

2	3	5	7	11
13	17	19	23	29
31	37	41	43	47
53	59	61	67	71
73	79	83	89	97

> **CAUTION**
> All prime numbers are odd numbers except the number 2. Be careful though, because *not all odd numbers are prime numbers*. For example, 9, 15, and 21 are odd numbers but they are *not* prime numbers.

The *prime factorization* of a number can be especially useful when working with fractions.

> **Prime Factorization**
> A **prime factorization** of a number is a factorization in which every factor is a prime number.

4 Find the prime factorization of each number.

(a) 8

(b) 42

(c) 90

(d) 100

(e) 81

EXAMPLE 4 **Factoring Using the Division Method**

(a) Find the prime factorization of 48.

$2\overline{)48}$ ← Divide 48 by 2 (the first prime number); quotient is 24

$2\overline{)24}$ ← Divide 24 by 2; quotient is 12

$2\overline{)12}$ ← Divide 12 by 2; quotient is 6

$2\overline{)6}$ ← Divide 6 by 2; quotient is 3

$3\overline{)3}$ ← Divide 3 by 3; quotient is 1

1 ← Continue to divide until the quotient is 1

All the divisors are prime factors.

Because all the factors (divisors) are prime, the prime factorization of 48 is

2 • 2 • 2 • 2 • 3

Check by multiplying the factors to see if the product is 48.
Yes, 2 • 2 • 2 • 2 • 3 does equal 48.

> **Note**
>
> You may write the factors in any order because multiplication is commutative. So you could write the factorization of 48 as 3 • 2 • 2 • 2 • 2. We will show the factors from least to greatest in our examples.

(b) Find the prime factorization of 225.

$3\overline{)225}$ ← 225 is not divisible by 2 (first prime) so use 3 (next prime)

$3\overline{)75}$ ← Divide 75 by 3

$5\overline{)25}$ ← 25 is not divisible by 3; use 5

$5\overline{)5}$ ← Divide 5 by 5

1 ← Quotient is 1

All the divisors are prime factors.

225 = **3 • 3 • 5 • 5**

CHECK: **3 • 3** is 9;
9 • **5** is 45; 45 • **5** is **225**

> **CAUTION**
>
> When you're using the division method of factoring, the last quotient is 1. Do **not** list 1 as a prime factor because 1 is not a prime number.

◀ *Work Problem* **4** *at the Side.*

Another method of factoring uses what is called a *factor tree*.

EXAMPLE 5 **Factoring Using a Factor Tree**

Find the prime factorization of each number.

(a) 60

Try to divide 60 by the first prime number, 2. The quotient is 30. Write the factors 2 and 30 under the 60. Circle the 2, because it is a prime.

60

2   30

Continued on Next Page

Try dividing 30 by 2. The quotient is 15. Write the factors 2 and 15 under the 30.

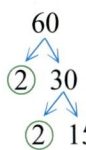

60
② 30
② 15

Because 15 cannot be divided evenly by 2, try dividing 15 by the next prime number, 3. The quotient is 5. Write the factors 3 and 5 under the 15.

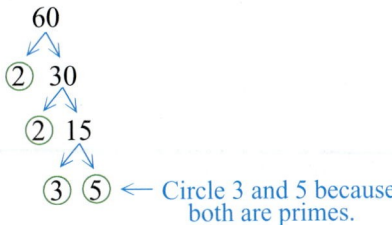

60
② 30
② 15
③ ⑤ ← Circle 3 and 5 because both are primes.

No uncircled factors remain, so you have found the prime factorization (the circled factors).

$$60 = 2 \cdot 2 \cdot 3 \cdot 5$$

CHECK: **2** • **2** is 4; 4 • **3** is 12; 12 • **5** is **60**

(b) 72

Divide 72 by 2, the first prime number.

72
② 36 ← Divide by 2 again; 36 = 2 • 18
② 18 ← Divide by 2 again; 18 = 2 • 9
② 9 ← Divide by 3; 9 = 3 • 3
③ ③

$$72 = 2 \cdot 2 \cdot 2 \cdot 3 \cdot 3$$

(c) 45

Because 45 cannot be divided evenly by 2, try dividing by the next prime, 3.

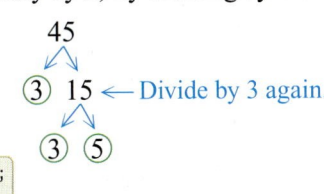

45
③ 15 ← Divide by 3 again.
③ ⑤

$$45 = 3 \cdot 3 \cdot 5$$

CHECK: **3** • **3** is 9; 9 • **5** is **45**.

Note

Here is a reminder about the quick way to see whether a number is *divisible* by 2, 3, or 5; in other words, there is no remainder when you do the division.

A number is divisible by 2 if the ones digit is 0, 2, 4, 6, or 8. For example, 30, 512, 76, and 3018 are all divisible by 2.

A number is divisible by 3 if the *sum* of the digits is divisible by 3. For example, 129 is divisible by 3 because 1 + 2 + 9 = 12 and 12 is divisible by 3.

A number is divisible by 5 if the ones digit is 0 or 5. For example, 85, 610, and 1725 are all divisible by 5.

See **Section R.4** for more information.

Work Problem **5** *at the Side.* ▶

5 Complete each factor tree and write the prime factorization.

(a) 28
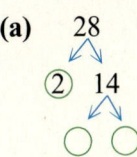
② 14
○ ○

(b) 35
5

(c) 90

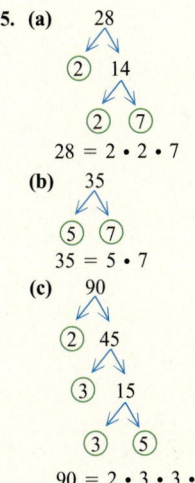

▦ **Calculator Tip**   You can use your calculator to find the prime factorization of a number. Here is an example that uses 539.

Try dividing 539 by the first prime number, 2.

$$539 \div 2 = 269.5 \quad \text{Does not divide evenly}$$

Try dividing 539 by the next prime number, 3.

$$539 \div 3 = 179.6666667 \quad \text{Does not divide evenly}$$

Keep trying the next prime numbers until you find one that divides evenly.

$$539 \div 5 = 107.8 \quad \text{Does not divide evenly}$$

$$539 \div 7 = 77 \quad \text{Divides evenly}$$

Once you have found that 7 works, try using it again on the quotient 77.

$$77 \div 7 = 11 \quad \text{Divides evenly}$$

Because 11 is prime, you're finished. The prime factorization of 539 is $7 \cdot 7 \cdot 11$.

Now try factoring 2431 using your calculator. (The answer is at the bottom left of this page.)

OBJECTIVE 4 **Write a fraction in lowest terms using prime factorization.**   Now you can use the second method for writing fractions in lowest terms: prime factorization. This is a good method to use when the numerator and denominator are larger numbers or include variables.

EXAMPLE 6   **Using Prime Factorization to Write Fractions in Lowest Terms**

(a) Write $\frac{20}{35}$ in lowest terms.

20 can be written as $2 \cdot 2 \cdot 5$ ← Prime factors
35 can be written as $5 \cdot 7$ ← Prime factors

$$\frac{20}{35} = \frac{2 \cdot 2 \cdot 5}{5 \cdot 7}$$

The numerator and denominator have 5 as a common factor. Dividing both numerator and denominator by 5 will give an equivalent fraction.

Any number divided by itself is 1

$$\frac{20}{35} = \frac{2 \cdot 2 \cdot 5}{5 \cdot 7} = \frac{2 \cdot 2 \cdot 5}{7 \cdot 5} = \frac{2 \cdot 2 \cdot \boxed{5 \div 5}}{7 \cdot \boxed{5 \div 5}} = \frac{2 \cdot 2 \cdot 1}{7 \cdot 1} = \frac{4}{7}$$

Multiplication is commutative.

$\frac{20}{35}$ is written in lowest terms as $\frac{4}{7}$.

To shorten the work, you may use slashes to indicate the divisions. For example, the work on $\frac{20}{35}$ can be shown as follows.

$$\frac{20}{35} = \frac{2 \cdot 2 \cdot \overset{1}{\cancel{5}}}{\underset{1}{\cancel{5}} \cdot 7}$$

Slashes indicate $5 \div 5$, and the result is 1

Continued on Next Page

(b) Write $\frac{60}{72}$ in lowest terms.

Use the prime factorizations of 60 and 72 from Examples 5(a) and 5(b) on page 235.

$$\frac{60}{72} = \frac{2 \cdot 2 \cdot 3 \cdot 5}{2 \cdot 2 \cdot 2 \cdot 3 \cdot 3}$$

This time there are three common factors. Use slashes to show the three divisions.

$$\frac{60}{72} = \frac{\overset{1}{\cancel{2}} \cdot \overset{1}{\cancel{2}} \cdot \overset{1}{\cancel{3}} \cdot 5}{\underset{1}{\cancel{2}} \cdot \underset{1}{\cancel{2}} \cdot 2 \cdot \underset{1}{\cancel{3}} \cdot 3} = \frac{5}{6}$$

2 ÷ 2 is 1   2 ÷ 2 is 1   3 ÷ 3 is 1

← Multiply 1 · 1 · 1 · 5 to get 5
← Multiply 1 · 1 · 2 · 1 · 3 to get 6

(c) $\dfrac{18}{90}$

$$\frac{18}{90} = \frac{\overset{1}{\cancel{2}} \cdot \overset{1}{\cancel{3}} \cdot \overset{1}{\cancel{3}}}{\underset{1}{\cancel{2}} \cdot \underset{1}{\cancel{3}} \cdot \underset{1}{\cancel{3}} \cdot 5} = \frac{1}{5}$$

← Multiply 1 · 1 · 1 to get 1
← Multiply 1 · 1 · 1 · 5 to get 5

> **CAUTION**
> In Example 6(c) above, all factors of the numerator divided out. But $1 \cdot 1 \cdot 1$ is still 1, so the final answer is $\frac{1}{5}$ (**not** 5).

This method of using prime factorization to write a fraction in lowest terms is summarized below.

> **Using Prime Factorization to Write a Fraction in Lowest Terms**
>
> **Step 1** Write the **prime factorization** of both numerator and denominator.
>
> **Step 2** Use slashes to show where you are **dividing** the numerator and denominator by any common factors.
>
> **Step 3** **Multiply** the remaining factors in the numerator and in the denominator.

_____ *Work Problem* **6** *at the Side.* ▶

OBJECTIVE 5 Write a fraction with variables in lowest terms.
Fractions may have variables in the numerator or denominator. Examples are shown below.

$$\frac{6}{2x} \qquad \frac{3xy}{9xy} \qquad \frac{4b^3}{8ab} \qquad \frac{7ab^2}{n^2}$$

You can use prime factorization to write these fractions in lowest terms.

6 Use the method of prime factorization to write each fraction in lowest terms.

(a) $\dfrac{16}{48}$

(b) $\dfrac{28}{60}$

(c) $\dfrac{74}{111}$

(d) $\dfrac{124}{340}$

ANSWERS

6. (a) $\dfrac{\overset{1}{\cancel{2}} \cdot \overset{1}{\cancel{2}} \cdot \overset{1}{\cancel{2}} \cdot \overset{1}{\cancel{2}}}{\underset{1}{\cancel{2}} \cdot \underset{1}{\cancel{2}} \cdot \underset{1}{\cancel{2}} \cdot \underset{1}{\cancel{2}} \cdot 3} = \dfrac{1}{3}$

(b) $\dfrac{\overset{1}{\cancel{2}} \cdot \overset{1}{\cancel{2}} \cdot 7}{\underset{1}{\cancel{2}} \cdot \underset{1}{\cancel{2}} \cdot 3 \cdot 5} = \dfrac{7}{15}$

(c) $\dfrac{2 \cdot \overset{1}{\cancel{37}}}{3 \cdot \underset{1}{\cancel{37}}} = \dfrac{2}{3}$

(d) $\dfrac{\overset{1}{\cancel{2}} \cdot \overset{1}{\cancel{2}} \cdot 31}{\underset{1}{\cancel{2}} \cdot \underset{1}{\cancel{2}} \cdot 5 \cdot 17} = \dfrac{31}{85}$

7 Write each fraction in lowest terms.

(a) $\dfrac{5c}{15}$

(b) $\dfrac{10x^2}{8x^2}$

(c) $\dfrac{9a^3}{11b^3}$

(d) $\dfrac{6m^2n}{9n^2}$

EXAMPLE 7 Writing Fractions with Variables in Lowest Terms

Write each fraction in lowest terms.

(a) $\dfrac{6}{2x}$ ← Prime factors of 6 are 2 • 3
← 2x means 2 • x.

$$\dfrac{6}{2x} = \dfrac{\overset{1}{\cancel{2}} \cdot 3}{\underset{1}{\cancel{2}} \cdot x} = \dfrac{3}{x} \quad \begin{array}{l} \leftarrow 1 \cdot 3 \text{ is } 3 \\ \leftarrow 1 \cdot x \text{ is } x \end{array}$$

(b) $\dfrac{3xy}{9xy} = \dfrac{3 \cdot x \cdot y}{3 \cdot 3 \cdot x \cdot y} = \dfrac{\overset{1}{\cancel{3}} \cdot \overset{1}{\cancel{x}} \cdot \overset{1}{\cancel{y}}}{\underset{1}{\cancel{3}} \cdot 3 \cdot \underset{1}{\cancel{x}} \cdot \underset{1}{\cancel{y}}} = \dfrac{1}{3}$

3xy means $3 \cdot x \cdot y$

The prime factors of 9 are 3 • 3

> **Be careful!** In the numerator, 1 • 1 • 1 is 1. The answer is $\frac{1}{3}$ (**not** 3).

(c) $\dfrac{4b^3}{8ab} = \dfrac{2 \cdot 2 \cdot b \cdot b \cdot b}{2 \cdot 2 \cdot 2 \cdot a \cdot b} = \dfrac{\overset{1}{\cancel{2}} \cdot \overset{1}{\cancel{2}} \cdot \overset{1}{\cancel{b}} \cdot b \cdot b}{\underset{1}{\cancel{2}} \cdot \underset{1}{\cancel{2}} \cdot 2 \cdot a \cdot \underset{1}{\cancel{b}}} = \dfrac{b^2}{2a} \quad \begin{array}{l} \leftarrow b \cdot b \text{ is } b^2 \\ \leftarrow 2 \cdot a \text{ is } 2a \end{array}$

b^3 means $b \cdot b \cdot b$

The prime factors of 8 are 2 • 2 • 2

(d) $\dfrac{7ab^2}{n^2} = \dfrac{7 \cdot a \cdot b \cdot b}{n \cdot n}$ There are no common factors.

$\dfrac{7ab^2}{n^2}$ is already in lowest terms.

◀ *Work Problem* **7** *at the Side.*

4.2 ▶▶▶ Exercises

Are the following fractions in lowest terms? If not, find a common factor of the numerator and denominator (other than 1). See Example 1.

1. (a) $-\dfrac{3}{10}$ (b) $\dfrac{10}{15}$ (c) $\dfrac{9}{16}$ (d) $-\dfrac{4}{21}$ (e) $\dfrac{6}{9}$ (f) $-\dfrac{7}{28}$

2. (a) $\dfrac{10}{12}$ (b) $\dfrac{3}{18}$ (c) $-\dfrac{8}{15}$ (d) $-\dfrac{22}{33}$ (e) $\dfrac{2}{25}$ (f) $-\dfrac{14}{15}$

Write each fraction in lowest terms. Use the method of dividing by a common factor. See Example 2.

🌐 **3.** (a) $\dfrac{10}{15}$ (b) $\dfrac{6}{9}$ (c) $-\dfrac{7}{28}$ (d) $-\dfrac{25}{50}$ (e) $\dfrac{16}{18}$ (f) $-\dfrac{8}{20}$

4. (a) $\dfrac{10}{12}$ (b) $\dfrac{3}{18}$ (c) $-\dfrac{22}{33}$ (d) $\dfrac{12}{16}$ (e) $-\dfrac{15}{20}$ (f) $-\dfrac{9}{15}$

Label each number as prime *or* composite *or* neither. *See Example 3.*

5.   9   2   8   1   5   11   10   21

6.   12   3   7   6   0   15   13   25

Find the prime factorization of each number. See Examples 4 and 5.

7. 6   **8.** 12   **9.** 20   **10.** 30

11. 25   **12.** 18   🌐 **13.** 36   **14.** 56

15. (a) 44   **16.** (a) 45   **17.** (a) 75   **18.** (a) 80

(b) 88   (b) 135   (b) 68   (b) 64

(c) 189   (c) 385

Write each numerator and denominator as a product of prime factors. Then use the prime factorization to write the fraction in lowest terms. See Example 6.

19. $\dfrac{8}{16}$

20. $\dfrac{6}{8}$

21. $\dfrac{32}{48}$

22. $\dfrac{9}{27}$

23. $\dfrac{14}{21}$

24. $\dfrac{20}{32}$

25. $\dfrac{36}{42}$

26. $\dfrac{22}{33}$

27. $\dfrac{50}{63}$

28. $\dfrac{72}{80}$

29. $\dfrac{27}{45}$

30. $\dfrac{36}{63}$

31. $\dfrac{12}{18}$

32. $\dfrac{63}{90}$

33. $\dfrac{35}{40}$

34. $\dfrac{36}{48}$

35. $\dfrac{90}{180}$

36. $\dfrac{16}{64}$

37. $\dfrac{210}{315}$

38. $\dfrac{96}{192}$

39. $\dfrac{429}{495}$

40. $\dfrac{135}{182}$

Write your answers to Exercises 41–46 in lowest terms.

41. There are 60 minutes in an hour. What fraction of an hour is

(a) 15 minutes? (b) 30 minutes?

(c) 6 minutes? (d) 60 minutes?

42. There are 24 hours in a day. What fraction of a day is

(a) 8 hours? (b) 18 hours?

(c) 12 hours? (d) 3 hours?

43. SueLynn's monthly income is $2400.

(a) She spends $800 on rent. What fraction of her income is spent on rent?

(b) She spends $400 on food. What fraction of her income is spent on food?

(c) What fraction of her income is left for other expenses?

44. There are 10,000 students at Minneapolis Community and Technical College.

(a) 7500 of the students receive some form of financial aid. What fraction of the students receive financial aid?

(b) 6000 of the students are women. What fraction are women?

(c) What fraction of the students are men?

45. What fraction of the time spent on household chores is done by (a) husbands, (b) wives, (c) children?

Of the typical 48 hours spent weekly on household chores, here is the amount of time spent by each family member.

| Husbands 10 hours | Wives 32 hours | Children 6 hours |

Source: *Journal of Marriage and the Family.*

46. A survey asked people whether certain types of advertising were believable. Out of every 100 people in the survey, here is the number who said the advertising was believable.

Type of Advertising	Number Who Said Advertising Was Believable
Computer software	35
Pharmaceutical companies	28
Auto manufacturers	18
Insurance companies	15

Source: Porter Novella.

What fraction of the people said each type of advertising was believable?

47. Explain the error in each of these problems and correct it.

(a) $\dfrac{9}{36} = \dfrac{\not 3 \cdot \not 3}{2 \cdot 2 \cdot \not 3 \cdot \not 3} = 4$

(b) $\dfrac{9}{16} = \dfrac{9 \div 3}{16 \div 4} = \dfrac{3}{4}$

48. (a) Explain how you could use your calculator to find the prime factorization of 437. Then find the prime factorization.

(b) The text lists all the prime numbers less than 100. Use the divisibility rules and your calculator to find at least five prime numbers between 100 and 150.

Write each fraction in lowest terms. See Example 7.

49. $\dfrac{16c}{40}$

50. $\dfrac{36}{54a}$

51. $\dfrac{20x}{35x}$

52. $\dfrac{21n}{28n}$

53. $\dfrac{18r^2}{15rs}$

54. $\dfrac{18ab}{48b^2}$

55. $\dfrac{6m}{42mn^2}$

56. $\dfrac{10g^2}{90g^2h}$

57. $\dfrac{9x^2}{16y^2}$

58. $\dfrac{5rst}{8st}$

59. $\dfrac{7xz}{9xyz}$

60. $\dfrac{6a^3}{23b^3}$

61. $\dfrac{21k^3}{6k^2}$

62. $\dfrac{16x^3}{12x^4}$

63. $\dfrac{13a^2bc^3}{39a^2bc^3}$

64. $\dfrac{22m^3n^4}{55m^3n^4}$

65. $\dfrac{14c^2d}{14cd^2}$

66. $\dfrac{19rs}{19s^3}$

67. $\dfrac{210ab^3c}{35b^2c^2}$

68. $\dfrac{81w^4xy^2}{300wy^4}$

69. $\dfrac{25m^3rt^2}{36n^2s^3w^2}$

70. $\dfrac{42a^5b^4c^3}{7a^4b^3c^2}$

71. $\dfrac{33e^2fg^3}{11efg}$

72. $\dfrac{21xy^2z^3}{17ab^2c^3}$

4.3 ▶▶▶ Multiplying and Dividing Signed Fractions

OBJECTIVE 1 Multiply signed fractions. Suppose that you give $\frac{1}{3}$ of your candy bar to your friend Ann. Then Ann gives $\frac{1}{2}$ of her share to Tim. How much of the bar does Tim get to eat?

OBJECTIVES

1. **Multiply signed fractions.**

2. **Multiply fractions that involve variables.**

3. **Divide signed fractions.**

4. **Divide fractions that involve variables.**

5. **Solve application problems involving multiplying and dividing fractions.**

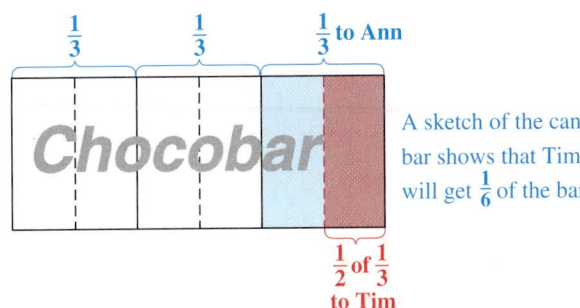

$\frac{1}{3}$ $\frac{1}{3}$ $\frac{1}{3}$ to Ann

A sketch of the candy bar shows that Tim will get $\frac{1}{6}$ of the bar.

$\frac{1}{2}$ of $\frac{1}{3}$ to Tim

Tim's share is $\frac{1}{2}$ **of** $\frac{1}{3}$ candy bar. When used with fractions, the word **of** indicates multiplication.

$$\frac{1}{2}\text{ of }\frac{1}{3}\quad\text{means}\quad\frac{1}{2}\cdot\frac{1}{3}$$

Tim's share is $\frac{1}{6}$ bar, so $\frac{1}{2}\cdot\frac{1}{3}=\frac{1}{6}$.

This example illustrates the rule for multiplying fractions.

> **Multiplying Fractions**
>
> If a, b, c, and d are numbers (but b and d are not 0), then
> $$\frac{a}{b}\cdot\frac{c}{d}=\frac{a\cdot c}{b\cdot d}$$
> In other words, multiply the numerators and multiply the denominators.

When we apply this rule to find Tim's part of the candy bar, we get

$$\frac{1}{2}\cdot\frac{1}{3}=\frac{1\cdot 1}{2\cdot 3}=\frac{1}{6}\quad\begin{array}{l}\leftarrow\text{Multiply numerators.}\\\leftarrow\text{Multiply denominators.}\end{array}$$

EXAMPLE 1 **Multiplying Signed Fractions**

Find each product.

(a) $-\dfrac{5}{8}\cdot-\dfrac{3}{4}$ Recall that the product of two negative numbers is a positive number.

Multiply the numerators and multiply the denominators.

$$-\frac{5}{8}\cdot-\frac{3}{4}=\frac{5\cdot 3}{8\cdot 4}=\frac{15}{32}\quad\leftarrow\text{Lowest terms}$$

The product of two negative numbers is positive.

The answer is in lowest terms because 15 and 32 have no common factor other than 1.

Continued on Next Page

1 Find each product.

(a) $-\dfrac{3}{4} \cdot \dfrac{1}{2}$

(b) $\left(-\dfrac{2}{5}\right)\left(-\dfrac{2}{3}\right)$

(c) $\dfrac{3}{4}\left(\dfrac{3}{8}\right)$

(b) $\left(\dfrac{4}{7}\right)\left(-\dfrac{2}{5}\right) = -\dfrac{4 \cdot 2}{7 \cdot 5} = -\dfrac{8}{35}$ Recall that the product of a negative number and a positive number is negative.

◀ **Work Problem** **1** **at the Side.**

Sometimes the result won't be in lowest terms. For example, find $\frac{3}{10}$ of $\frac{5}{6}$.

$$\dfrac{3}{10} \text{ of } \dfrac{5}{6} \quad \text{means} \quad \dfrac{3}{10} \cdot \dfrac{5}{6} = \dfrac{3 \cdot 5}{10 \cdot 6} = \dfrac{15}{60} \quad \left\{\begin{array}{l}\text{Not in lowest} \\ \text{terms}\end{array}\right.$$

Now write $\frac{15}{60}$ in lowest terms.

$$\dfrac{15}{16} = \dfrac{\overset{1}{\cancel{3}} \cdot \overset{1}{\cancel{5}}}{2 \cdot 2 \cdot \underset{1}{\cancel{3}} \cdot \underset{1}{\cancel{5}}} = \dfrac{1}{4} \quad \leftarrow \text{Lowest terms}$$

You used prime factorization in **Section 4.2** to write fractions in lowest terms. You can also use it when multiplying fractions. Writing the prime factors of the original fractions and dividing out common factors *before* multiplying usually saves time. If you divide out *all* the common factors, the result will automatically be in lowest terms. Let's see how that works when finding $\frac{3}{10}$ of $\frac{5}{6}$.

3 and 5 are already prime.

Write 10 as 2 • 5

Write 6 as 2 • 3

$$\dfrac{3}{10} \cdot \dfrac{5}{6} = \dfrac{3 \cdot 5}{2 \cdot 5 \cdot 2 \cdot 3} = \dfrac{\overset{1}{\cancel{3}} \cdot \overset{1}{\cancel{5}}}{2 \cdot \underset{1}{\cancel{5}} \cdot 2 \cdot \underset{1}{\cancel{3}}} = \dfrac{1}{4} \quad \left\{\begin{array}{l}\text{Same result} \\ \text{as above}\end{array}\right.$$

Divide out the common factors.

> **CAUTION**
>
> When you are working with fractions, always write the final result in lowest terms. Visualizing $\frac{15}{60}$ is hard to do. But when $\frac{15}{60}$ is written as $\frac{1}{4}$, working with it is much easier. A fraction is *simplified* when it is written in lowest terms.

EXAMPLE 2   **Using Prime Factorization to Multiply Fractions**

(a) $-\dfrac{8}{5}\left(\dfrac{5}{12}\right)$

Multiplying a negative number times a positive number gives a negative product.

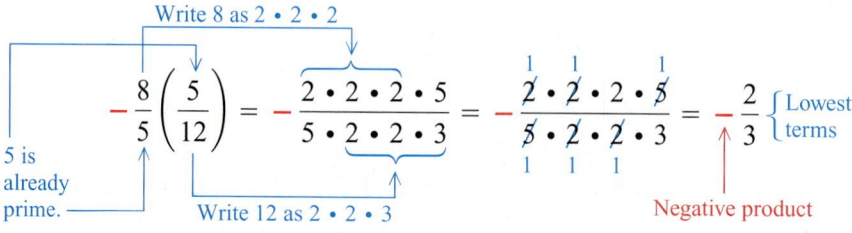

Continued on Next Page

(b) Find $\dfrac{2}{9}$ of $\dfrac{15}{16}$.

Recall that, when used with fractions, *of* indicates multiplication.

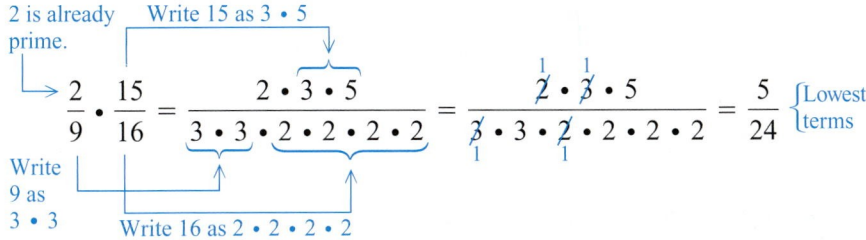

2 is already prime.

Write 15 as 3 • 5

Write 9 as 3 • 3

Write 16 as 2 • 2 • 2 • 2

$$\frac{2}{9} \cdot \frac{15}{16} = \frac{2 \cdot 3 \cdot 5}{3 \cdot 3 \cdot 2 \cdot 2 \cdot 2 \cdot 2} = \frac{\overset{1}{\cancel{2}} \cdot \overset{1}{\cancel{3}} \cdot 5}{\cancel{3} \cdot 3 \cdot \cancel{2} \cdot 2 \cdot 2 \cdot 2} = \frac{5}{24} \quad \{\text{Lowest terms}$$

Work Problem **2** *at the Side.* ▶

🖩 **Calculator Tip** If your calculator has a fraction key $\boxed{a^{b/c}}$, you can do calculations with fractions. You'll also need the change of sign key $\boxed{+\!/\!-}$ to enter negative fractions, as you did to enter negative integers in **Section 1.4.**

Start by entering several different fractions. Clear your calculator after each one.

To enter $\frac{3}{4}$, press 3 $\boxed{a^{b/c}}$ 4. The display will show $\boxed{3 \lrcorner 4}$

Fraction bar

To enter $-\frac{9}{10}$, press 9 $\boxed{a^{b/c}}$ 10 $\boxed{+\!/\!-}$. The display will show $\boxed{-9 \lrcorner 10}$.

Try entering a fraction that is *not* in lowest terms. As soon as you press an operation key, such as $\boxed{\times}$ or $\boxed{\div}$, most calculators will automatically show the fraction in lowest terms. Suppose that you start to enter the multiplication problem $\frac{4}{16} \cdot \frac{2}{3}$. Press 4 $\boxed{a^{b/c}}$ 16 $\boxed{\times}$. The display shows $\boxed{1 \lrcorner 4}$, or $\frac{1}{4}$, which is $\frac{4}{16}$ in lowest terms. The calculator will always show fractions in lowest terms.

Let's check the result of Example 2(a) on the previous page: Multiply $-\frac{8}{5}\left(\frac{5}{12}\right)$ by pressing the following keys.

8 $\boxed{a^{b/c}}$ 5 $\boxed{+\!/\!-}$ $\boxed{\times}$ 5 $\boxed{a^{b/c}}$ 12 $\boxed{=}$ The display shows $\boxed{-2 \lrcorner 3}$.

$\underbrace{\qquad\qquad}_{-\dfrac{8}{5}}$ $\underbrace{\qquad\qquad}_{\dfrac{5}{12}}$ $\underbrace{\qquad\qquad}_{-\dfrac{2}{3}}$

Now try Example 2(b) above: Find $\frac{2}{9}$ of $\frac{15}{16}$. (Did you get $\frac{5}{24}$?)

There are some limitations to the calculations that you can do using the fraction key.

Try entering the fraction $\frac{9}{1000}$. What happens?

(You can't enter denominators >999.)

Try doing this multiplication: $\frac{7}{10} \cdot \frac{3}{100}$. The result should be $\frac{21}{1000}$. What happens?

(The answer is given in decimal form because the denominator is >999.)

2 Use prime factorization to multiply these fractions.

(a) $\dfrac{15}{28}\left(-\dfrac{6}{5}\right)$

(b) $\dfrac{12}{7} \cdot \dfrac{7}{24}$

(c) $\left(-\dfrac{11}{18}\right)\left(-\dfrac{9}{20}\right)$

ANSWERS

2. **(a)** $-\dfrac{3 \cdot \overset{1}{\cancel{5}} \cdot \overset{1}{\cancel{2}} \cdot 3}{2 \cdot \underset{1}{\cancel{2}} \cdot 7 \cdot \underset{1}{\cancel{5}}} = -\dfrac{9}{14}$

(b) $\dfrac{\overset{1}{\cancel{2}} \cdot \overset{1}{\cancel{2}} \cdot \overset{1}{\cancel{3}} \cdot \overset{1}{\cancel{7}}}{\underset{1}{\cancel{7}} \cdot \underset{1}{\cancel{2}} \cdot \underset{1}{\cancel{2}} \cdot 2 \cdot \underset{1}{\cancel{3}}} = \dfrac{1}{2}$

(c) $\dfrac{11 \cdot \overset{1}{\cancel{3}} \cdot \overset{1}{\cancel{3}}}{2 \cdot \underset{1}{\cancel{3}} \cdot \underset{1}{\cancel{3}} \cdot 2 \cdot 2 \cdot 5} = \dfrac{11}{40}$

3 Use prime factorization to find these products.

(a) $\dfrac{3}{4}$ of 36

(b) $-10 \cdot \dfrac{2}{5}$

(c) $\left(-\dfrac{7}{8}\right)(-24)$

4 Use prime factorization to find these products.

(a) $\dfrac{2c}{5} \cdot \dfrac{c}{4}$

(b) $\left(\dfrac{m}{6}\right)\left(\dfrac{9}{m^2}\right)$

(c) $\left(\dfrac{w^2}{y}\right)\left(\dfrac{x^2 y}{w}\right)$

ANSWERS

3. (a) $\dfrac{3 \cdot \overset{1}{\cancel{2}} \cdot \overset{1}{\cancel{2}} \cdot 3 \cdot 3}{\underset{1}{\cancel{2}} \cdot \underset{1}{\cancel{2}} \cdot 1} = \dfrac{27}{1} = 27$

(b) $-\dfrac{2 \cdot \overset{1}{\cancel{5}} \cdot 2}{1 \cdot \underset{1}{\cancel{5}}} = -\dfrac{4}{1} = -4$

(c) $\dfrac{7 \cdot \overset{1}{\cancel{2}} \cdot \overset{1}{\cancel{2}} \cdot \overset{1}{\cancel{2}} \cdot 3}{\underset{1}{\cancel{2}} \cdot \underset{1}{\cancel{2}} \cdot \underset{1}{\cancel{2}}} = \dfrac{21}{1} = 21$

4. (a) $\dfrac{\overset{1}{\cancel{2}} \cdot c \cdot c}{5 \cdot \underset{1}{\cancel{2}} \cdot 2} = \dfrac{c^2}{10}$

(b) $\dfrac{\overset{1}{\cancel{m}} \cdot \overset{1}{\cancel{3}} \cdot 3}{2 \cdot \underset{1}{\cancel{3}} \cdot \underset{1}{\cancel{m}} \cdot m} = \dfrac{3}{2m}$

(c) $\dfrac{\overset{1}{\cancel{y}} \cdot w \cdot x \cdot x \cdot \overset{1}{\cancel{y}}}{\underset{1}{\cancel{y}} \cdot \underset{1}{\cancel{w}}} = \dfrac{wx^2}{1} = wx^2$

CAUTION

The fraction key on a calculator is useful for *checking* your work. But knowing the rules for fraction computation is important because you'll need them when fractions involve variables. You *cannot* enter fractions such as $\dfrac{3x}{5}$ or $\dfrac{9}{m^2}$ on your calculator (see Example 4 below).

EXAMPLE 3 **Multiplying a Fraction and an Integer**

Find $\frac{2}{3}$ of 6.

We can write 6 in fraction form as $\frac{6}{1}$. Recall that $\frac{6}{1}$ means $6 \div 1$, which is 6. So we can write any integer a as $\frac{a}{1}$.

$$\underset{\frac{2}{3}\ \textbf{of}\ 6\ \text{means}}{} \ \dfrac{2}{3} \cdot \dfrac{6}{1} = \dfrac{2 \cdot 2 \cdot \overset{1}{\cancel{3}}}{\underset{1}{\cancel{3}} \cdot 1} = \dfrac{4}{1} = 4$$

$4 \div 1$ is 4

◀ *Work Problem* **3** *at the Side.*

OBJECTIVE **2** **Multiply fractions that involve variables.** The multiplication method that uses prime factors also works when there are variables in the numerators and/or denominators of the fractions.

EXAMPLE 4 **Multiplying Fractions with Variables**

Find each product.

(a) $\dfrac{3x}{5} \cdot \dfrac{2}{9x}$

$3x$ means $3 \cdot x$. $\dfrac{x}{x}$ is 1

$$\dfrac{3x}{5} \cdot \dfrac{2}{9x} = \dfrac{3 \cdot x \cdot 2}{5 \cdot 3 \cdot 3 \cdot x} = \dfrac{\overset{1}{\cancel{3}} \cdot \overset{1}{\cancel{x}} \cdot 2}{5 \cdot \underset{1}{\cancel{3}} \cdot 3 \cdot \underset{1}{\cancel{x}}} = \dfrac{2}{15}$$

The prime factors of 9 are $3 \cdot 3$, so $9x$ is $3 \cdot 3 \cdot x$.

(b) $\left(\dfrac{3y}{4x}\right)\left(\dfrac{2x^2}{y}\right)$

$2x^2$ means $2 \cdot x \cdot x$.

$$\left(\dfrac{3y}{4x}\right)\left(\dfrac{2x^2}{y}\right) = \dfrac{3 \cdot y \cdot 2 \cdot x \cdot x}{2 \cdot 2 \cdot x \cdot y} = \dfrac{3 \cdot \overset{1}{\cancel{y}} \cdot \overset{1}{\cancel{2}} \cdot \overset{1}{\cancel{x}} \cdot x}{2 \cdot \underset{1}{\cancel{2}} \cdot \underset{1}{\cancel{x}} \cdot \underset{1}{\cancel{y}}} = \dfrac{3x}{2}$$

The prime factors of 4 are $2 \cdot 2$, so $4x$ is $2 \cdot 2 \cdot x$.

◀ *Work Problem* **4** *at the Side.*

OBJECTIVE **3** **Divide signed fractions.** To divide fractions, you will rewrite division problems as multiplication problems. For division, you will leave the first number (the dividend) as it is, but change the second number (the divisor) to its *reciprocal*.

Reciprocal of a Fraction

Two numbers are **reciprocals** of each other if their product is 1.

The reciprocal of the fraction $\dfrac{a}{b}$ is $\dfrac{b}{a}$ because

$$\frac{a}{b} \cdot \frac{b}{a} = \frac{\overset{1}{\cancel{a}} \cdot \overset{1}{\cancel{b}}}{\underset{1}{\cancel{b}} \cdot \underset{1}{\cancel{a}}} = \frac{1}{1} = 1$$

Notice that you "flip" or "invert" a fraction to find its reciprocal. Here are some examples.

Number	Reciprocal	Reason
$\dfrac{1}{6}$	$\dfrac{6}{1}$	Because $\dfrac{1}{6} \cdot \dfrac{6}{1} = \dfrac{6}{6} = 1$
$-\dfrac{2}{5}$	$-\dfrac{5}{2}$	Because $\left(-\dfrac{2}{5}\right)\left(-\dfrac{5}{2}\right) = \dfrac{10}{10} = 1$
4 Think of 4 as $\frac{4}{1}$	$\dfrac{1}{4}$	Because $4 \cdot \dfrac{1}{4} = \dfrac{4}{1} \cdot \dfrac{1}{4} = \dfrac{4}{4} = 1$

Note

Every number has a reciprocal except 0. Why not 0? Recall that a number times its reciprocal equals 1. But that doesn't work for 0.

$$0 \cdot (\text{reciprocal}) \neq 1$$

      Put any number here. When you multiply it by 0, you get 0, never 1

Dividing Fractions

If a, b, c, and d are numbers (but b, c, and d are not 0), then we have the following.

$$\frac{a}{b} \div \frac{c}{d} = \frac{a}{b} \cdot \frac{d}{c}$$

     Reciprocals

In other words, change division to multiplying by the reciprocal of the divisor.

Use this method to find the quotient for $\frac{2}{3} \div \frac{1}{6}$. Rewrite it as a multiplication problem and then use the steps for multiplying fractions.

Change division to multiplication.

$$\frac{2}{3} \div \frac{1}{6} = \frac{2}{3} \cdot \frac{6}{1} = \frac{2 \cdot 2 \cdot \overset{1}{\cancel{3}}}{\underset{1}{\cancel{3}} \cdot 1} = \frac{4}{1} = 4$$

The reciprocal of $\frac{1}{6}$ is $\frac{6}{1}$

Does it make sense that $\frac{2}{3} \div \frac{1}{6} = 4$? Let's compare dividing fractions to dividing whole numbers.

$$15 \div 3 \text{ is asking, "How many 3s are in 15?"}$$
$$\frac{2}{3} \div \frac{1}{6} \text{ is asking, "How many } \frac{1}{6}\text{s are in } \frac{2}{3}\text{?"}$$

The figure below illustrates $\frac{2}{3} \div \frac{1}{6}$.

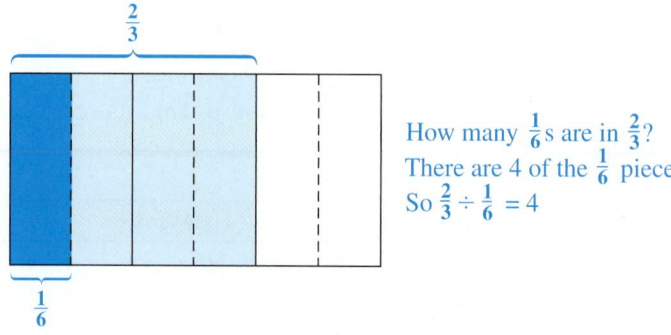

How many $\frac{1}{6}$s are in $\frac{2}{3}$?
There are 4 of the $\frac{1}{6}$ pieces in $\frac{2}{3}$.
So $\frac{2}{3} \div \frac{1}{6} = 4$

As a final check on this method of dividing, try changing $15 \div 3$ into a multiplication problem. You know that the quotient should be 5.

$$15 \div 3 = 15 \cdot \frac{1}{3} = \frac{15}{1} \cdot \frac{1}{3} = \frac{\overset{1}{\cancel{3}} \cdot 5 \cdot 1}{1 \cdot \underset{1}{\cancel{3}}} = \frac{5}{1} = 5 \quad \left\{ \begin{array}{l} \text{The quotient} \\ \text{we expected} \end{array} \right.$$

Reciprocals

So, you can see that *dividing* by a fraction is the same as *multiplying* by the *reciprocal* of the fraction.

EXAMPLE 5 **Dividing Signed Fractions**

Rewrite each division problem as a multiplication problem. Then multiply.

Change the *divisor* $\left(\frac{4}{5}\right)$ to its reciprocal, $\frac{5}{4}$. Do **not** change $\frac{3}{10}$.

(a) $\dfrac{3}{10} \div \dfrac{4}{5} = \dfrac{3}{10} \cdot \dfrac{5}{4} = \dfrac{3 \cdot \overset{1}{\cancel{5}}}{2 \cdot \underset{1}{\cancel{5}} \cdot 2 \cdot 2} = \dfrac{3}{8}$

Reciprocals

> **Note**
>
> When multiplying fractions, you don't always have to factor the numerator and denominator completely into prime numbers. In part (a) above, if you notice that 5 is a common factor of the numerator and denominator, you can write
>
> $$\frac{3}{10} \cdot \frac{5}{4} = \frac{3 \cdot \overset{1}{\cancel{5}}}{2 \cdot \underset{1}{\cancel{5}} \cdot 4} = \frac{3}{8}$$
>
> Factor 10 into $2 \cdot 5$
>
> Leave 4 as it is.
>
> If no common factors are obvious to you, then write out the complete prime factorization to help find the common factors.

Continued on Next Page

(b) $2 \div \left(-\dfrac{1}{3}\right)$

First notice that the numbers have different signs. In a division problem, different signs mean that the quotient is negative. Then write 2 in fraction form as $\frac{2}{1}$.

$$2 \div \left(-\frac{1}{3}\right) = \frac{2}{1} \cdot \left(-\frac{3}{1}\right) = -\frac{2 \cdot 3}{1 \cdot 1} = -\frac{6}{1} = -6$$

Reciprocals

Negative product

(c) $-\dfrac{3}{4} \div (-8) = -\dfrac{3}{4} \cdot \left(-\dfrac{1}{8}\right) = \dfrac{3 \cdot 1}{4 \cdot 8} = \dfrac{3}{32}$

Reciprocals

No common factor to divide out

> Both numbers in the problem were negative. When signs *match*, the quotient is *positive*.

(d) $\dfrac{9}{16} \div 0$

> Dividing by 0 **cannot** be done.

This is *undefined*, just as dividing by 0 was undefined for integers (see **Section 1.7**). Recall that 0 does *not* have a reciprocal, so you can't change the division to multiplying by the reciprocal of the divisor.

(e) $0 \div \dfrac{9}{16} = 0 \cdot \dfrac{16}{9} = 0$

Reciprocals

Recall that 0 divided by any nonzero number gives a result of 0.

Work Problem **5** *at the Side.* ▶

OBJECTIVE 4 Divide fractions that involve variables. The method for dividing fractions also works when there are variables in the numerators and/or denominators of the fractions.

EXAMPLE 6 **Dividing Fractions with Variables**

Divide. (Assume that none of the variables represent zero.)

(a) $\dfrac{x^2}{y} \div \dfrac{x}{3y} = \dfrac{x^2}{y} \cdot \dfrac{3y}{x} = \dfrac{x \cdot x \cdot 3 \cdot y}{y \cdot x} = \dfrac{3x}{1} = 3x$

Reciprocals

(b) $\dfrac{8b}{5} \div b^2 = \dfrac{8b}{5} \cdot \dfrac{1}{b^2} = \dfrac{8 \cdot b \cdot 1}{5 \cdot b \cdot b} = \dfrac{8}{5b}$

Think of b^2 as $\dfrac{b^2}{1}$ so the reciprocal is $\dfrac{1}{b^2}$.

Work Problem **6** *at the Side.* ▶

---

5 Rewrite each division problem as a multiplication problem. Then multiply.

(a) $-\dfrac{3}{4} \div \dfrac{5}{8}$

(b) $0 \div \left(-\dfrac{7}{12}\right)$

(c) $\dfrac{5}{6} \div 10$

(d) $-9 \div \left(-\dfrac{9}{16}\right)$

(e) $\dfrac{2}{5} \div 0$

6 Divide. (Assume none of the variables represent zero.)

(a) $\dfrac{c^2 d^2}{4} \div \dfrac{c^2 d}{4}$

(b) $\dfrac{20}{7h} \div \dfrac{5h}{7}$

(c) $\dfrac{n}{8} \div mn$

ANSWERS

5. (a) $-\dfrac{3}{4} \cdot \dfrac{8}{5} = -\dfrac{3 \cdot 2 \cdot 4}{4 \cdot 5} = -\dfrac{6}{5}$

(b) $0 \cdot \left(-\dfrac{12}{7}\right) = 0$

(c) $\dfrac{5}{6} \cdot \dfrac{1}{10} = \dfrac{5 \cdot 1}{6 \cdot 2 \cdot 5} = \dfrac{1}{12}$

(d) $-\dfrac{9}{1} \cdot \left(-\dfrac{16}{9}\right) = \dfrac{9 \cdot 16}{1 \cdot 9} = \dfrac{16}{1} = 16$

(e) undefined; can't be written as multiplication because 0 doesn't have a reciprocal

6. (a) d **(b)** $\dfrac{4}{h^2}$ **(c)** $\dfrac{1}{8m}$

7 Look for indicator words or draw sketches to help you with these problems.

(a) How many times can a $\frac{2}{3}$ quart spray bottle be filled before 18 quarts of window cleaner are used up?

OBJECTIVE **5** **Solve application problems involving multiplying and dividing fractions.** When you're solving application problems, some indicator words are used to suggest multiplication and some are used to suggest division.

INDICATOR WORDS FOR MULTIPLICATION	INDICATOR WORDS FOR DIVISION
product	per
double	each
triple	goes into
times	divided by
twice	divided into
of (when *of* follows a fraction)	divided equally

Look for these indicator words in the following examples. However, *you won't always find an indicator word.* Then, you need to think through the problem to decide what to do. Sometimes, drawing a sketch of the situation described in the problem will help you decide which operation to use.

(b) A retiring police officer will receive $\frac{5}{8}$ of her highest annual salary as retirement income. If her highest annual salary is $64,000, how much will she receive as retirement income?

> **EXAMPLE 7** **Using Indicator Words and Sketches to Solve Application Problems**

(a) Lois gives $\frac{1}{10}$ **of** her income to her church. Last month she earned $1980. How much of that did she give to her church?

Notice the word **of**. Because the word **of** *follows the fraction* $\frac{1}{10}$, it indicates multiplication.

$$\frac{1}{10} \text{ of } 1980 = \frac{1}{10} \cdot \frac{1980}{1} = \frac{1 \cdot \overset{1}{\cancel{10}} \cdot 198}{\underset{1}{\cancel{10}} \cdot 1} = \frac{198}{1} = 198$$

Lois gave $198 to her church.

(b) The apparel design class is making infant snowsuits to give to a local shelter. A fabric store donated a 12 yd length of fabric for the project. If one snowsuit needs $\frac{2}{3}$ yd of fabric, how many suits can the class make?

The word **of** appears in the second sentence: "A fabric store donated a 12 yd length **of** fabric." But the word **of** does *not follow a fraction,* so it is *not* an indicator to multiply. Let's try a sketch. There is a piece of fabric 12 yd long. One snowsuit will use $\frac{2}{3}$ yd. The question is, how many $\frac{2}{3}$ yd pieces can be cut from the 12 yards?

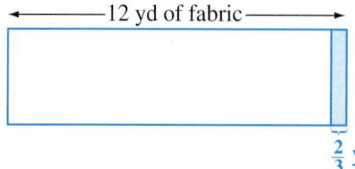

⟵————12 yd of fabric————⟶

$\frac{2}{3}$ yd

Cutting 12 yards into equal size pieces indicates division. How many $\frac{2}{3}$s are in 12?

$$12 \div \frac{2}{3} = \frac{12}{1} \cdot \frac{3}{2} = \frac{\overset{1}{\cancel{2}} \cdot 6 \cdot 3}{1 \cdot \underset{1}{\cancel{2}}} = \frac{18}{1} = 18$$

Reciprocals

The class can make 18 snowsuits.

◀ *Work Problem* **7** *at the Side.*

4.3 ▶▶▶ Exercises

Multiply. Write the products in lowest terms. See Examples 1–4.

1. $-\dfrac{3}{8} \cdot \dfrac{1}{2}$

2. $\left(\dfrac{2}{3}\right)\left(-\dfrac{5}{7}\right)$

3. $\left(-\dfrac{3}{8}\right)\left(-\dfrac{12}{5}\right)$

4. $\dfrac{4}{9} \cdot \dfrac{12}{7}$

5. $\dfrac{21}{30}\left(\dfrac{5}{7}\right)$

6. $\left(-\dfrac{6}{11}\right)\left(-\dfrac{22}{15}\right)$

7. $10\left(-\dfrac{3}{5}\right)$

8. $-20\left(\dfrac{3}{4}\right)$

9. $\dfrac{4}{9}$ of 81

10. $\dfrac{2}{3}$ of 48

11. $\left(\dfrac{3x}{4}\right)\left(\dfrac{5}{xy}\right)$

12. $\left(\dfrac{2}{5a^2}\right)\left(\dfrac{a}{8}\right)$

Divide. Write the quotients in lowest terms. (Assume that none of the variables represent zero.) See Examples 5 and 6.

13. $\dfrac{1}{6} \div \dfrac{1}{3}$

14. $-\dfrac{1}{2} \div \dfrac{2}{3}$

15. $-\dfrac{3}{4} \div \left(-\dfrac{5}{8}\right)$

16. $\dfrac{7}{10} \div \dfrac{2}{5}$

17. $6 \div \left(-\dfrac{2}{3}\right)$

18. $-7 \div \left(-\dfrac{1}{4}\right)$

19. $-\dfrac{2}{3} \div 4$

20. $\dfrac{5}{6} \div (-15)$

21. $\dfrac{11c}{5d} \div 3c$

22. $8x^2 \div \dfrac{4x}{7}$

23. $\dfrac{ab^2}{c} \div \dfrac{ab}{c}$

24. $\dfrac{mn}{6} \div \dfrac{n}{3m}$

25. Explain and correct the error in each of these calculations.

(a) $\dfrac{3}{14} \cdot \dfrac{7}{9} = \dfrac{\cancel{3} \cdot \cancel{7}}{2 \cdot \cancel{7} \cdot \cancel{3} \cdot 3} = 6$

(b) $8 \cdot \dfrac{2}{3} = \dfrac{8}{1} \cdot \dfrac{3}{2} = \dfrac{\cancel{2} \cdot 4 \cdot 3}{1 \cdot \cancel{2}} = \dfrac{12}{1} = 12$

26. Explain and correct the error in each of these calculations.

(a) $\dfrac{3}{4} \cdot \dfrac{8}{9} = \dfrac{3 \cdot 8}{4 \cdot 9} = \dfrac{24}{36}$

(b) $\dfrac{2}{5} \cdot \dfrac{3}{8} = \dfrac{\overset{1}{\cancel{2}}}{5} \cdot \dfrac{3}{\underset{2}{\cancel{8}}} = \dfrac{3}{10}$

27. Explain and correct the error in each of these calculations.

(a) $\dfrac{2}{3} \div 4 = \dfrac{2}{3} \cdot \dfrac{4}{1} = \dfrac{2 \cdot 4}{3 \cdot 1} = \dfrac{8}{3}$

(b) $\dfrac{5}{6} \div \dfrac{10}{9} = \dfrac{6}{5} \cdot \dfrac{10}{9} = \dfrac{2 \cdot \cancel{3}^{1} \cdot 2 \cdot \cancel{5}^{1}}{\cancel{5} \cdot \cancel{3} \cdot 3} = \dfrac{4}{3}$

28. Explain and correct the error in each of these calculations.

(a) $\dfrac{1}{2} \div 0 = 0$

(b) $\dfrac{\cancel{5}^{1}}{10} \div \dfrac{1}{\cancel{5}_{1}} = \dfrac{1}{10}$

29. Your friend missed class and is confused about how to divide fractions. Write a short explanation for your friend.

30. Mary spilled coffee on her math homework, and part of one problem is covered up.

$$\dfrac{3}{} \div \dfrac{4}{5}$$

She knows the answer given in the back of the book is $\frac{3}{4}$. Describe how to find the missing number.

Find each product or quotient. Write all answers in lowest terms. See Examples 1–6.

31. $\dfrac{4}{5} \div 3$

32. $\left(-\dfrac{20}{21} \right)\left(-\dfrac{14}{15} \right)$

33. $-\dfrac{3}{8}\left(\dfrac{3}{4} \right)$

34. $-\dfrac{8}{17} \div \dfrac{4}{5}$

35. $\dfrac{3}{5}$ of 35

36. $\dfrac{2}{3} \div (-6)$

37. $-9 \div \left(-\dfrac{3}{5} \right)$

38. $\dfrac{7}{8} \cdot \dfrac{25}{21}$

39. $\dfrac{12}{7} \div 0$

40. $\dfrac{5}{8}$ of (-48)

41. $\left(\dfrac{11}{2} \right)\left(-\dfrac{5}{6} \right)$

42. $\dfrac{3}{4} \div \dfrac{3}{16}$

43. $\dfrac{4}{7}$ of $14b$

44. $\dfrac{ab}{6} \div \dfrac{b}{9}$

45. $\dfrac{12}{5} \div 4d$

46. $\dfrac{18}{7} \div 2t$

47. $\dfrac{x^2}{y} \div \dfrac{w}{2y}$

48. $\dfrac{5}{6}$ of $18w$

Solve each application problem. See Example 7.

49. Al is helping Tim make a mahogany lamp table for Jill's birthday. Find the area of the rectangular top of the table if it is $\frac{4}{5}$ yd long by $\frac{3}{8}$ yd wide.

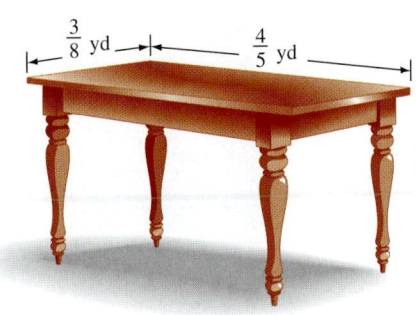

50. A rectangular dog bed is $\frac{7}{8}$ yd by $\frac{10}{9}$ yd. Find its area.

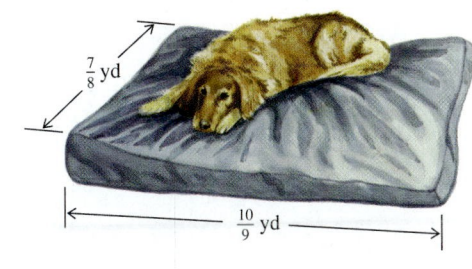

51. How many $\frac{1}{8}$-ounce eyedrop dispensers can be filled from a container that holds 10 ounces of eyedrops?

52. Ms. Shaffer has a piece of property with an area that is $\frac{9}{10}$ acre. She wishes to divide it into three equal parts for her children. How many acres of land will each child get?

53. Todd estimates that it will cost him $12,400 to attend a community college for one year. He thinks he can earn $\frac{3}{4}$ of the cost and borrow the balance. Find the amount he must earn and the amount he must borrow.

54. Joyce Chen wants to make vests to sell at a craft fair. Each vest requires $\frac{3}{4}$ yd of material. She has 36 yd of material. Find the number of vests she can make.

55. Pam Trizlia has a small pickup truck that can carry $\frac{2}{3}$ cord of firewood. Find the number of trips needed to deliver 6 cords of wood.

56. At the Garlic Festival Fun Run, $\frac{5}{12}$ of the runners are women. If there are 780 runners, how many are women? How many are men?

57. There are 234 baseball players in the Baseball Hall of Fame. About one-third of these men played at infield positions (1st base, 2nd base, 3rd base, or short stop). About how many infield players are in the Hall of Fame? (*Source:* National Baseball Hall of Fame.)

58. Parking lot A is $\frac{1}{4}$ mile long and $\frac{3}{16}$ mile wide, and parking lot B is $\frac{3}{8}$ mile long and $\frac{1}{8}$ mile wide. Which parking lot has the larger area?

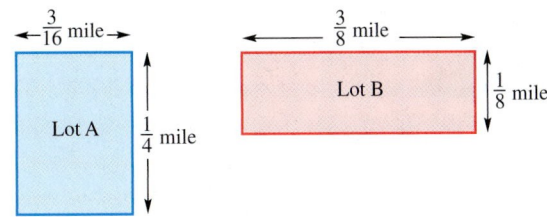

59. An adult male alligator may weigh 400 pounds. A newly hatched alligator weighs only $\frac{1}{8}$ pound. The adult weighs how many times the hatchling? (*Source:* St. Marks NWR.)

60. A female alligator lays about 35 eggs in a nest of marsh grass and mud. But $\frac{4}{5}$ of the eggs or hatchlings will fall prey to raccoons, wading birds, or larger alligators. How many of the 35 eggs will hatch and survive? (*Source:* St. Marks NWR.)

The table below shows the income for the Gomez family last year and the circle graph shows how they spent their income. Use this information to work Exercises 61–64.

Month	Income	Month	Income
January	$4575	July	$5540
February	$4312	August	$3732
March	$4988	September	$4170
April	$4530	October	$5512
May	$4320	November	$4965
June	$4898	December	$6458

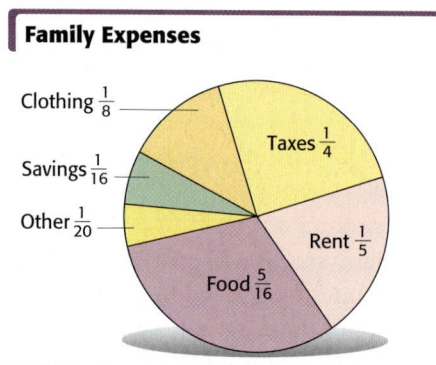

Family Expenses

Clothing $\frac{1}{8}$

Savings $\frac{1}{16}$

Other $\frac{1}{20}$

Taxes $\frac{1}{4}$

Rent $\frac{1}{5}$

Food $\frac{5}{16}$

61. (a) What was the family's total income for the year?

(b) Find the amount of the family's rent for the year.

62. (a) How much did the family pay in taxes during the year?

(b) How much more did the family spend on taxes than on rent?

63. How much did the family spend for food and clothing last year?

64. Find the amount the family saved during the year.

There are a total of about 175 million companion pets in the United States. This table shows the fraction of pets that are dogs, cats, birds, and horses. Use the table to answer Exercises 65–68.

COMPANION PETS IN THE UNITED STATES

Type of Pet	Fraction of All Pets
Dog	$\frac{2}{5}$
Cat	$\frac{12}{25}$
Bird	$\frac{3}{50}$
Horse	$\frac{1}{25}$

Source: American Veterinary Medical Association.

65. How many U.S. pets are horses?

66. How many U.S. pets are dogs?

67. How many dogs and cats are pets?

68. How many birds are pets?

4.4 ▶▶▶ Adding and Subtracting Signed Fractions

OBJECTIVE 1 Add and subtract like fractions. You probably remember learning something about "common denominators" in other math classes. When fractions have the *same* denominator, we say that they have a *common* denominator, which makes them **like fractions.** When fractions have different denominators, they are called **unlike fractions.** Here are some examples.

Like Fractions

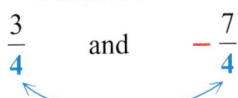

$\dfrac{3}{4}$ and $-\dfrac{7}{4}$

Common denominator

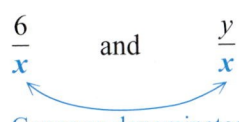

$\dfrac{6}{x}$ and $\dfrac{y}{x}$

Common denominator

Unlike Fractions

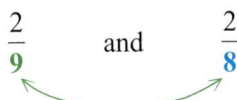

$\dfrac{2}{9}$ and $\dfrac{2}{8}$

Different denominators

$\dfrac{a^2}{3}$ and $\dfrac{3}{a}$

Different denominators

You can add or subtract fractions *only* when they have a common denominator. To see why, let's look at more pizzas.

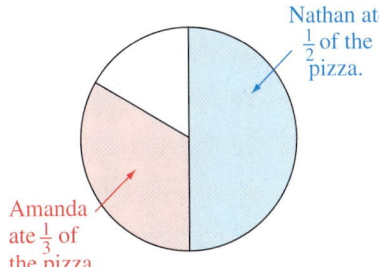

Nathan ate $\frac{1}{2}$ of the pizza.

Amanda ate $\frac{1}{3}$ of the pizza.

What fraction of the pizza has been eaten? We can't write a fraction until the pizza is cut into pieces of *equal* size. That's what the denominator of a fraction tell us: the number of *equal* size pieces in the pizza.

Now the pizza is cut into 6 *equal* pieces, and we can find out how much was eaten.

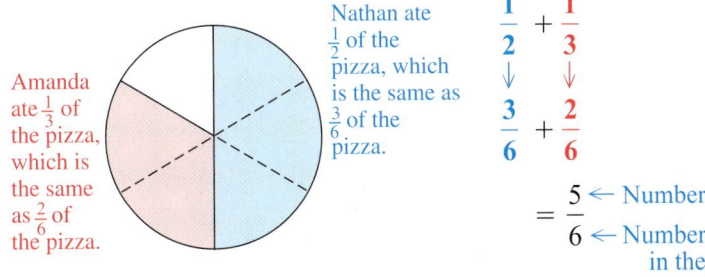

Amanda ate $\frac{1}{3}$ of the pizza, which is the same as $\frac{2}{6}$ of the pizza.

Nathan ate $\frac{1}{2}$ of the pizza, which is the same as $\frac{3}{6}$ of the pizza.

$$\dfrac{1}{2} + \dfrac{1}{3}$$
$$\downarrow \qquad \downarrow$$
$$\dfrac{3}{6} + \dfrac{2}{6}$$

$$= \dfrac{5}{6} \leftarrow \text{Number of pieces eaten}$$
$$\phantom{= \dfrac{5}{6}} \leftarrow \text{Number of equal pieces in the pizza}$$

Adding and Subtracting Like Fractions

You can add or subtract fractions *only* when they have a common denominator. If a, b, and c are numbers (and b is not 0), then

$$\dfrac{a}{b} + \dfrac{c}{b} = \dfrac{a+c}{b} \qquad \text{and} \qquad \dfrac{a}{b} - \dfrac{c}{b} = \dfrac{a-c}{b}$$

In other words, add or subtract the numerators and write the result over the common denominator. Then check to be sure that the answer is in lowest terms.

1 Write each sum or difference in lowest terms.

(a) $\dfrac{1}{6} + \dfrac{5}{6}$

(b) $-\dfrac{11}{12} + \dfrac{5}{12}$

(c) $-\dfrac{2}{9} - \dfrac{3}{9}$

(d) $\dfrac{8}{ab} + \dfrac{3}{ab}$

ANSWERS

1. (a) $\dfrac{6}{6} = 1$ (b) $-\dfrac{1}{2}$

(c) $-\dfrac{5}{9}$ (d) $\dfrac{11}{ab}$

EXAMPLE 1 **Adding and Subtracting Like Fractions**

Find each sum or difference.

(a) $\dfrac{1}{8} + \dfrac{3}{8}$

These are *like* fractions because they have a *common denominator of 8.* So they are ready to be added. Add the numerators and write the sum over the common denominator.

$$\underset{\uparrow}{\dfrac{1}{8}} + \underset{\uparrow}{\dfrac{3}{8}} = \underset{\uparrow}{\dfrac{1+3}{8}} = \dfrac{4}{8}$$

Common denominator

Now write $\frac{4}{8}$ in lowest terms.

$$\dfrac{4}{8} = \dfrac{\overset{1}{\cancel{2}} \cdot \overset{1}{\cancel{2}}}{\underset{1}{\cancel{2}} \cdot \underset{1}{\cancel{2}} \cdot 2} = \dfrac{1}{2}$$

The sum, in lowest terms, is $\frac{1}{2}$.

> **CAUTION**
> Add *only* the numerators. **Do not add the denominators.** In part (a) above we *kept the common denominator.*
>
> $$\dfrac{1}{8} + \dfrac{3}{8} = \dfrac{1+3}{8} \qquad \textbf{not} \qquad \dfrac{1}{8} + \dfrac{3}{8} = \dfrac{1+3}{8+8} = \dfrac{4}{16}$$
>
> Incorrect
>
> To help you understand why we add *only* the numerators, think of $\frac{1}{8}$ as $1(\frac{1}{8})$ and $\frac{3}{8}$ as $3(\frac{1}{8})$. Then we use the distributive property.
>
> $$\dfrac{1}{8} + \dfrac{3}{8} = 1\left(\dfrac{1}{8}\right) + 3\left(\dfrac{1}{8}\right) = (1+3)\left(\dfrac{1}{8}\right) = 4\left(\dfrac{1}{8}\right) = \dfrac{4}{8} = \dfrac{1}{2}$$
>
> Use the distributive property.
>
> Or think about a pie cut into 8 equal pieces. If you eat 1 piece and your friend eats 3 pieces, together you've eaten 4 of the 8 pieces or $\frac{4}{8}$ of the pie (*not* $\frac{4}{16}$ of the pie, which would be 4 out of 16 pieces).

(b) $-\underset{\uparrow}{\dfrac{3}{5}} + \underset{\uparrow}{\dfrac{4}{5}} = \underset{\uparrow}{\dfrac{-3+4}{5}} = \dfrac{1}{5}$ ← Lowest terms

Common denominator

Rewrite subtraction as adding the opposite.

(c) $\underset{\uparrow}{\dfrac{3}{10}} - \underset{\uparrow}{\dfrac{7}{10}} = \underset{\uparrow}{\dfrac{3-7}{10}} = \dfrac{3+(-7)}{10} = \dfrac{-4}{10}$ or $-\dfrac{4}{10}$

Common denominator

Always write fraction answers in lowest terms.

Now write $-\frac{4}{10}$ in lowest terms.

$$-\dfrac{4}{10} = -\dfrac{\overset{1}{\cancel{2}} \cdot 2}{\underset{1}{\cancel{2}} \cdot 5} = -\dfrac{2}{5}$$

(d) $\underset{\uparrow}{\dfrac{5}{x^2}} - \underset{\uparrow}{\dfrac{2}{x^2}} = \underset{\uparrow}{\dfrac{5-2}{x^2}} = \dfrac{3}{x^2}$

Common denominator

◄ **Work Problem** **1** at the Side.

OBJECTIVE **2** **Find the lowest common denominator for unlike fractions.** When we first tried to add the pizza eaten by Nathan and Amanda, we could *not* do so because the pizza was *not cut into pieces of the same size*. So we rewrote $\frac{1}{2}$ and $\frac{1}{3}$ as equivalent fractions that both had 6 as the common denominator.

$$\frac{1}{2} = \frac{1 \cdot 3}{2 \cdot 3} = \frac{3}{6} \left.\begin{array}{l} \\ \end{array}\right\} \text{Common}$$
$$\frac{1}{3} = \frac{1 \cdot 2}{3 \cdot 2} = \frac{2}{6} \quad \text{denominator of 6}$$

Then we could add the fractions, because the pizza was cut into pieces of *equal* size.

$$\frac{1}{2} + \frac{1}{3} = \frac{3}{6} + \frac{2}{6} = \frac{3+2}{6} = \frac{5}{6} \leftarrow \text{Lowest terms}$$

In other words, when you want to add or subtract unlike fractions, the first thing you must do is rewrite them so that they have a common denominator.

> **A Common Denominator for Unlike Fractions**
>
> To find a common denominator for two unlike fractions, find a number that is divisible by *both* of the original denominators.
>
> For example, a common denominator for $\frac{1}{2}$ and $\frac{1}{3}$ is 6 because 2 goes into 6 evenly and 3 goes into 6 evenly.

Notice that 12 is also a common denominator for $\frac{1}{2}$ and $\frac{1}{3}$ because 2 and 3 both go into 12 evenly.

$$\frac{1}{2} = \frac{1 \cdot 6}{2 \cdot 6} = \frac{6}{12} \left.\begin{array}{l} \\ \end{array}\right\} \text{Common}$$
$$\frac{1}{3} = \frac{1 \cdot 4}{3 \cdot 4} = \frac{4}{12} \quad \text{denominator of 12}$$

Now that the fractions have a common denominator, we can add them.

$$\frac{1}{2} + \frac{1}{3} = \frac{6}{12} + \frac{4}{12} = \frac{6+4}{12} = \frac{10}{12} \left\{\begin{array}{l}\text{Not in}\\ \text{lowest}\\ \text{terms}\end{array}\right. \quad \text{but} \quad \frac{10}{12} = \frac{\overset{1}{\cancel{2}} \cdot 5}{\underset{1}{\cancel{2}} \cdot 6} = \frac{5}{6} \left\{\begin{array}{l}\text{Same}\\ \text{result as}\\ \text{above}\end{array}\right.$$

Both 6 and 12 worked as common denominators for adding $\frac{1}{2}$ and $\frac{1}{3}$, but using the smaller number saved some work. You should always try to find the smallest common denominator. If you don't for some reason, you can still work the problem—but it may take you longer. You'll have to divide out some common factors at the end in order to write the answer in lowest terms.

> **Least Common Denominator (LCD)**
>
> The **least common denominator** (LCD) for two fractions is the *smallest* positive number divisible by both denominators of the original fractions. For example, both 6 and 12 are common denominators for $\frac{1}{2}$ and $\frac{1}{3}$, but 6 is smaller, so it is the LCD.

2 Find the LCD for each pair of fractions by inspection.

(a) $\dfrac{3}{5}$ and $\dfrac{3}{10}$

(b) $\dfrac{1}{2}$ and $\dfrac{2}{5}$

(c) $\dfrac{3}{4}$ and $\dfrac{1}{6}$

(d) $\dfrac{5}{6}$ and $\dfrac{7}{18}$

3 Use prime factorization to find the LCD for each pair of fractions.

(a) $\dfrac{1}{10}$ and $\dfrac{13}{14}$

(b) $\dfrac{5}{12}$ and $\dfrac{17}{20}$

(c) $\dfrac{7}{15}$ and $\dfrac{7}{9}$

ANSWERS

2. (a) 10 (b) 10 (c) 12 (d) 18

3. (a) $10 = 2 \cdot 5$
$14 = 2 \cdot 7$ $\quad$ $LCD = 2 \cdot 5 \cdot 7 = 70$

(b) $12 = 2 \cdot 2 \cdot 3$
$20 = 2 \cdot 2 \cdot 5$ $\quad$ $LCD = 2 \cdot 2 \cdot 3 \cdot 5 = 60$

(c) $15 = 3 \cdot 5$
$9 = 3 \cdot 3$ $\quad$ $LCD = 3 \cdot 3 \cdot 5 = 45$

There are several ways to find the LCD. When the original denominators are small numbers, you can often find the LCD by inspection. *Hint:* **Always check to see if the larger denominator will work as the LCD.**

EXAMPLE 2 **Finding the LCD by Inspection**

(a) Find the LCD for $\frac{2}{3}$ and $\frac{1}{9}$ by inspection.

Check to see if 9 (the larger denominator) will work as the LCD. Is 9 divisible by 3 (the other denominator)? Yes, so 9 is the LCD for $\frac{2}{3}$ and $\frac{1}{9}$.

(b) Find the LCD for $\frac{5}{8}$ and $\frac{5}{6}$ by inspection.

Check to see if 8 (the larger denominator) will work. No, 8 is not divisible by 6. So start checking numbers that are multiples of 8, that is, 16, 24, and 32. Notice that 24 will work because it is divisible by 8 and by 6.

The LCD for $\frac{5}{8}$ and $\frac{5}{6}$ is 24.

◀ *Work Problem* **2** *at the Side.*

For larger denominators, you can use prime factorization to find the LCD. Factor each denominator completely into prime numbers. Then use the factors to build the LCD.

EXAMPLE 3 **Using Prime Factors to Find the LCD**

(a) What is the LCD for $\frac{7}{12}$ and $\frac{13}{18}$?

Write 12 and 18 as the product of prime factors. Then use prime factors in the LCD that "cover" both 12 and 18.

$$12 = 2 \cdot 2 \cdot 3$$
$$18 = 2 \cdot 3 \cdot 3$$

Factors of 12

$$LCD = 2 \cdot 2 \cdot 3 \cdot 3 = 36$$

Factors of 18

Check whether 36 is divisible by 12 (yes) and by 18 (yes). So 36 is the LCD for $\frac{7}{12}$ and $\frac{13}{18}$.

CAUTION

When finding the LCD, notice that we did *not* have to repeat the factors that 12 and 18 have in common. If we had used *all* the 2s and 3s, we would get a common denominator, but not the *smallest* one.

(b) What is the LCD for $\frac{11}{15}$ and $\frac{9}{70}$?

$$15 = 3 \cdot 5$$
$$70 = 2 \cdot 5 \cdot 7$$

Factors of 15

$$LCD = 3 \cdot 5 \cdot 2 \cdot 7 = 210$$

Factors of 70

Check whether 210 is divisible by 15 (yes) and divisible by 70 (yes). So 210 is the LCD for $\frac{11}{15}$ and $\frac{9}{70}$.

◀ *Work Problem* **3** *at the Side.*

OBJECTIVE **3** **Add and subtract unlike fractions.** Here are the steps for adding or subtracting unlike fractions. The key idea is that you must rewrite the fractions so that they have a common denominator before you can add or subtract them.

> **Adding and Subtracting Unlike Fractions**
>
> **Step 1** Find the LCD, the smallest number divisible by both denominators in the problem.
>
> **Step 2** Rewrite each original fraction as an equivalent fraction whose denominator is the LCD.
>
> **Step 3** Add or subtract the numerators of the like fractions. Keep the common denominator.
>
> **Step 4** Write the sum or difference in lowest terms.

EXAMPLE 4 **Adding and Subtracting Unlike Fractions**

Find each sum or difference.

(a) $\dfrac{1}{5} + \dfrac{3}{10}$ *First check to see if the larger denominator is the LCD.*

Step 1 The larger denominator (10) is the LCD.

Step 2 $\dfrac{1}{5} = \dfrac{1 \cdot 2}{5 \cdot 2} = \dfrac{2}{10} \leftarrow \text{LCD}$ and $\dfrac{3}{10}$ already has the LCD.

Step 3 Add the numerators. Write the sum over the common denominator.

$$\frac{1}{5} + \frac{3}{10} = \frac{2}{10} + \frac{3}{10} = \frac{2+3}{10} = \frac{5}{10}$$

Step 4 Write $\frac{5}{10}$ in lowest terms.

$$\frac{5}{10} = \frac{\overset{1}{\cancel{5}}}{2 \cdot \underset{1}{\cancel{5}}} = \frac{1}{2} \leftarrow \text{Lowest terms}$$

(b) $\dfrac{3}{4} - \dfrac{5}{6}$

Step 1 The LCD is 12.

Step 2 $\dfrac{3}{4} = \dfrac{3 \cdot 3}{4 \cdot 3} = \dfrac{9}{12} \leftarrow \text{LCD}$ and $\dfrac{5}{6} = \dfrac{5 \cdot 2}{6 \cdot 2} = \dfrac{10}{12} \leftarrow \text{LCD}$

Step 3 Subtract the numerators. Write the difference over the common denominator.

$$9 + (-10) \text{ is } -1$$

$$\frac{3}{4} - \frac{5}{6} = \frac{9}{12} - \frac{10}{12} = \frac{9-10}{12} = \frac{-1}{12} \quad \text{or} \quad -\frac{1}{12}$$

Step 4 $-\frac{1}{12}$ is in lowest terms.

Continued on Next Page

4 Find each sum or difference. Write all answers in lowest terms.

(a) $\dfrac{2}{3} + \dfrac{1}{6}$

(b) $\dfrac{1}{12} - \dfrac{5}{6}$

(c) $3 - \dfrac{4}{5}$

(d) $-\dfrac{5}{12} + \dfrac{9}{16}$

(c) $-\dfrac{5}{12} + \dfrac{5}{9}$

Step 1 Use prime factorization to find the LCD.

$$12 = 2 \cdot 2 \cdot 3$$
$$9 = 3 \cdot 3$$

Factors of 12

$$LCD = 2 \cdot 2 \cdot 3 \cdot 3 = 36$$

Factors of 9

Step 2 $-\dfrac{5}{12} = -\dfrac{5 \cdot 3}{12 \cdot 3} = -\dfrac{15}{36}$ and $\dfrac{5}{9} = \dfrac{5 \cdot 4}{9 \cdot 4} = \dfrac{20}{36}$

Step 3 Add the numerators. Keep the common denominator.

$$-\dfrac{5}{12} + \dfrac{5}{9} = -\dfrac{15}{36} + \dfrac{20}{36} = \dfrac{-15 + 20}{36} = \dfrac{5}{36}$$

Step 4 $\frac{5}{36}$ is in lowest terms.

(d) $4 - \dfrac{2}{3}$

Step 1 Think of 4 as $\frac{4}{1}$. The LCD for $\frac{4}{1}$ and $\frac{2}{3}$ is 3, the larger denominator.

Step 2 $\dfrac{4}{1} = \dfrac{4 \cdot 3}{1 \cdot 3} = \dfrac{12}{3}$ and $\dfrac{2}{3}$ already has the LCD.

Step 3 Subtract the numerators. Keep the common denominator.

$$\dfrac{4}{1} - \dfrac{2}{3} = \dfrac{12}{3} - \dfrac{2}{3} = \dfrac{12 - 2}{3} = \dfrac{10}{3}$$

Step 4 $\frac{10}{3}$ is in lowest terms.

◀ *Work Problem* **4** *at the Side.*

OBJECTIVE 4 Add and subtract unlike fractions that contain variables. We use the same steps to add or subtract unlike fractions with variables in the numerators or denominators.

EXAMPLE 5 Adding and Subtracting Unlike Fractions with Variables

Find each sum or difference.

(a) $\dfrac{1}{4} + \dfrac{b}{5}$

Step 1 The LCD is 20.

Step 2 $\dfrac{1}{4} = \dfrac{1 \cdot 5}{4 \cdot 5} = \dfrac{5}{20}$ and $\dfrac{b}{5} = \dfrac{b \cdot 4}{5 \cdot 4} = \dfrac{4b}{20}$

Continued on Next Page

Step 3 $\dfrac{1}{4} + \dfrac{b}{5} = \dfrac{5}{20} + \dfrac{4b}{20} = \dfrac{5+4b}{20}$ ← Add the numerators.
← Keep the common denominator.

Step 4 $\dfrac{5+4b}{20}$ is in lowest terms.

> **CAUTION**
> In *Step 4* above, we could *not* add $5 + 4b$ in the numerator of the answer because 5 and $4b$ are *not* like terms. We *could* add $5b + 4b$ but *not* $5 + 4b$.
>
> Variable parts match.

(b) $\dfrac{2}{3} - \dfrac{6}{x}$

Step 1 The LCD is $3 \cdot x$, or $3x$.

Step 2 $\dfrac{2}{3} = \dfrac{2 \cdot x}{3 \cdot x} = \dfrac{2x}{3x}$ and $\dfrac{6}{x} = \dfrac{6 \cdot 3}{x \cdot 3} = \dfrac{18}{3x}$

Step 3 $\dfrac{2}{3} - \dfrac{6}{x} = \dfrac{2x}{3x} - \dfrac{18}{3x} = \dfrac{2x - 18}{3x}$ ← Subtract the numerators.
← Keep the common denominator.

Step 4 $\dfrac{2x - 18}{3x}$ is in lowest terms.

> **Note**
> Notice in part (b) above that we found the LCD for $\frac{2}{3} - \frac{6}{x}$ by multiplying the two denominators. The LCD is $3 \cdot x$ or $3x$.
>
> Multiplying the two denominators will *always* give you a common denominator, but it may not be the *least* common denominator. Here are more examples.
>
> $\dfrac{1}{3} - \dfrac{2}{5}$ If you multiply the denominators, $3 \cdot 5 = 15$ and 15 is the LCD.
>
> $\dfrac{5}{6} + \dfrac{3}{4}$ If you multiply the denominators, $6 \cdot 4 = 24$ and 24 will work. But you'll save some time by using the least common denominator, which is 12.
>
> $\dfrac{7}{y} + \dfrac{a}{4}$ If you multiply the denominators, $y \cdot 4 = 4y$ and 4y is the LCD.

Work Problem **5** *at the Side.* ▶

---

5 Find each sum or difference.

(a) $\dfrac{5}{6} - \dfrac{h}{2}$

(b) $\dfrac{7}{t} + \dfrac{3}{5}$

(c) $\dfrac{4}{x} - \dfrac{8}{3}$

Math in the Media

MUSIC

The time signature at the beginning of a piece of music looks like a fraction. Commonly used time signatures are $\frac{2}{4}$, $\frac{3}{4}$, $\frac{4}{4}$, and $\frac{6}{8}$. Musicians use the time signature to tell how long to hold each note. The values of different notes can be written as fractions:

$$\mathbf{o} = 1 \qquad \text{♩} = \frac{1}{2} \qquad \text{♩} = \frac{1}{4} \qquad \text{♪} = \frac{1}{8} \qquad \text{♬} = \frac{1}{16}$$

Music is divided into measures. In $\frac{4}{4}$ time, each measure contains notes that add up to $\frac{4}{4}$ (or 1). In $\frac{2}{4}$ time the notes in each measure add up to $\frac{2}{4}$ $\left(\text{or } \frac{1}{2}\right)$, and so on for $\frac{3}{4}$ time and $\frac{6}{8}$ time.

Write one or more notes in each measure to make it add up to its time signature. Use as many different kinds of notes as possible. Write the addition problem underneath the notes that shows how the notes add up to the time signature.

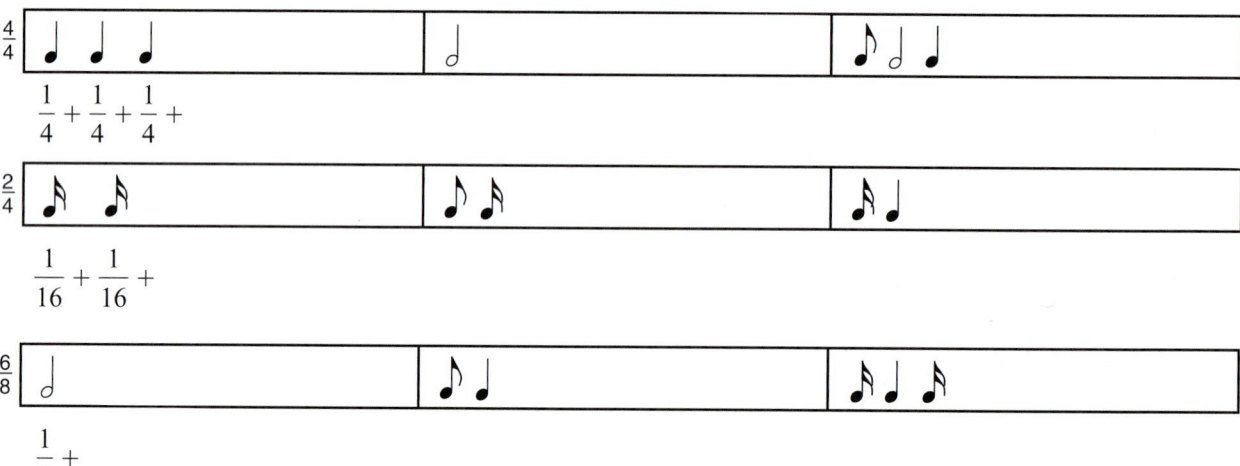

$$\frac{1}{4} + \frac{1}{4} + \frac{1}{4} +$$

$$\frac{1}{16} + \frac{1}{16} +$$

$$\frac{1}{2} +$$

Below is an excerpt from "The Star-Spangled Banner." Divide the music into measures based on the time signature.

say   does   that   Star-span-gled   Ban-ner   yet   wave   O'er   the

258

4.4 ▶▶▶ Exercises

Find each sum or difference. Write all answers in lowest terms. See Examples 1–5.

1. $\dfrac{3}{4} + \dfrac{1}{8}$

2. $\dfrac{1}{3} + \dfrac{1}{2}$

3. $-\dfrac{1}{14} + \left(-\dfrac{3}{7}\right)$

4. $-\dfrac{2}{9} + \dfrac{2}{3}$

5. $\dfrac{2}{3} - \dfrac{1}{6}$

6. $\dfrac{5}{12} - \dfrac{1}{4}$

7. $\dfrac{3}{8} - \dfrac{3}{5}$

8. $\dfrac{1}{3} - \dfrac{3}{5}$

9. $-\dfrac{5}{8} + \dfrac{1}{12}$

10. $-\dfrac{13}{16} + \dfrac{13}{16}$

11. $-\dfrac{7}{20} - \dfrac{5}{20}$

12. $-\dfrac{7}{9} - \dfrac{5}{6}$

13. $0 - \dfrac{7}{18}$

14. $-\dfrac{7}{8} + 3$

15. $2 - \dfrac{6}{7}$

16. $5 - \dfrac{2}{5}$

17. $-\dfrac{1}{2} + \dfrac{3}{24}$

18. $\dfrac{7}{10} + \dfrac{7}{15}$

19. $\dfrac{1}{5} + \dfrac{c}{3}$

20. $\dfrac{x}{4} + \dfrac{2}{3}$

21. $\dfrac{5}{m} - \dfrac{1}{2}$

22. $\dfrac{2}{9} - \dfrac{4}{y}$

23. $\dfrac{3}{b^2} + \dfrac{5}{b^2}$

24. $\dfrac{10}{xy} - \dfrac{7}{xy}$

25. $\dfrac{c}{7} + \dfrac{3}{b}$

26. $\dfrac{2}{x} - \dfrac{y}{5}$

27. $-\dfrac{4}{c^2} - \dfrac{d}{c}$

28. $-\dfrac{1}{n} + \dfrac{m}{n^2}$

29. $-\dfrac{11}{42} - \dfrac{11}{70}$

30. $\dfrac{7}{45} - \dfrac{7}{20}$

31. A key step in adding or subtracting unlike fractions is to rewrite the fractions so that they have a common denominator. Explain why this step is necessary.

32. Explain how to write a fraction with an indicated greater denominator. As part of your explanation, show how to rewrite $\frac{3}{4}$ as an equivalent fraction with a denominator of 12.

33. Explain the error in each calculation and correct it.

(a) $\dfrac{3}{4} + \dfrac{2}{5} = \dfrac{3+2}{4+5} = \dfrac{5}{9}$

(b) $\dfrac{5}{6} - \dfrac{4}{9} = \dfrac{5}{18} - \dfrac{4}{18} = \dfrac{5-4}{18} = \dfrac{1}{18}$

34. Explain the error in each calculation and correct it.

(a) $-\dfrac{1}{4} + \dfrac{7}{12} = -\dfrac{3}{12} + \dfrac{7}{12} = \dfrac{-3+7}{12} = \dfrac{4}{12}$

(b) $\dfrac{3}{10} - \dfrac{1}{4} = \dfrac{3-1}{10-4} = \dfrac{2}{6} = \dfrac{1}{3}$

Relating Concepts (Exercises 35–36)   For Individual or Group Work

*As you **work Exercises 35 and 36 in order,** think about the properties you learned when working with integers. Then explain what each pair of problems illustrates. (For a review of the properties, see the **Chapter 1 Summary.**)*

35. (a) $-\dfrac{2}{3} + \dfrac{3}{4} = $ _____ $\dfrac{3}{4} + \left(-\dfrac{2}{3}\right) = $ _____

(b) $\dfrac{5}{6} - \dfrac{1}{2} = $ _____ $\dfrac{1}{2} - \dfrac{5}{6} = $ _____

(c) $\left(-\dfrac{2}{3}\right)\left(\dfrac{9}{10}\right) = $ _____ $\left(\dfrac{9}{10}\right)\left(-\dfrac{2}{3}\right) = $ _____

(d) $\dfrac{2}{5} \div \dfrac{1}{15} = $ _____ $\dfrac{1}{15} \div \dfrac{2}{5} = $ _____

36. (a) $-\dfrac{7}{12} + \dfrac{7}{12} = $ _____ $\dfrac{3}{5} + \left(-\dfrac{3}{5}\right) = $ _____

(b) $-\dfrac{13}{16} \div \left(-\dfrac{13}{16}\right) = $ _____ $\dfrac{1}{8} \div \dfrac{1}{8} = $ _____

(c) $\dfrac{5}{6} \cdot 1 = $ _____ $1\left(-\dfrac{17}{20}\right) = $ _____

(d) $\left(-\dfrac{4}{5}\right)\left(-\dfrac{5}{4}\right) = $ _____ $7 \cdot \dfrac{1}{7} = $ _____

Solve each application problem. Write all answers in lowest terms.

37. When assembling shelving units, Pam Phelps must use the proper type and size of hardware. Find the total length of the bolt shown.

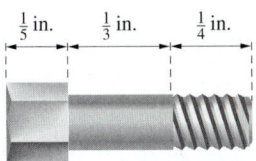

$\frac{1}{5}$ in. $\frac{1}{3}$ in. $\frac{1}{4}$ in.

38. How much fencing will be needed to enclose this rectangular wildflower preserve?

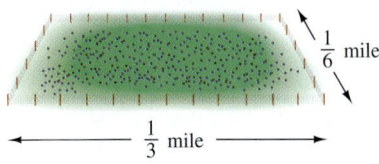

$\frac{1}{6}$ mile

$\frac{1}{3}$ mile

39. The owner of Racy's Feed Store ordered $\frac{1}{3}$ cubic yard of corn, $\frac{3}{8}$ cubic yard of oats, and $\frac{1}{4}$ cubic yard of washed medium mesh gravel. How many cubic yards of material were ordered?

40. A flower grower purchased $\frac{9}{10}$ acre of land one year and $\frac{3}{10}$ acre the next year. She then sold $\frac{7}{10}$ acre of land. How much land does she now have?

41. A forester planted $\frac{5}{12}$ acre of seedlings in the morning and $\frac{11}{12}$ acre in the afternoon. The next day, $\frac{7}{12}$ acre of seedlings were destroyed by a brush fire. How many acres of seedlings remained?

42. Adrian Ortega drives a tanker for the British Petroleum Company. He leaves the refinery with his tanker filled to $\frac{7}{8}$ of capacity. If he delivers $\frac{1}{4}$ of the tanker's capacity at the first stop and $\frac{1}{3}$ of the tanker's capacity at the second stop, find the fraction of the tanker's capacity remaining.

The circle graph shows the fraction of U.S. workers who used various methods to learn about computers. Use the graph to answer Exercises 43–46.

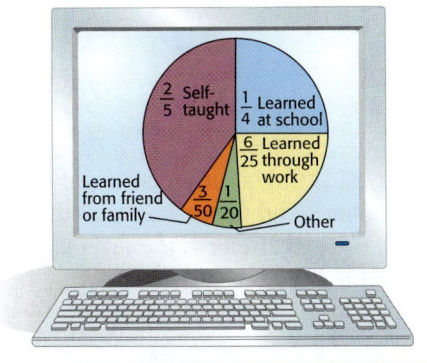

Ways To Learn Computers

$\frac{2}{5}$ Self-taught

$\frac{1}{4}$ Learned at school

$\frac{6}{25}$ Learned through work

Learned from friend or family $\frac{3}{50}$ $\frac{1}{20}$ Other

Source: John J. Heldrich Center for Workforce Development.

43. What fraction of workers are self-taught or learned from friends or family?

44. What fraction of workers learned at school or through work?

45. What is the difference in the fraction of workers who are self-taught and those who learned at work?

46. What is the difference in the fraction of workers who are self-taught and those who learned at school?

Refer to the circle graph to answer Exercises 47–50.

47. What fraction of the day was spent in class and study? How many hours is that?

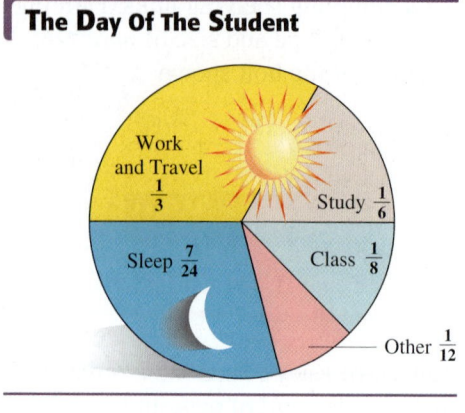

The Day Of The Student

48. What fraction of the day was spent in work and travel and sleep? How many hours is that?

49. How much more of the day was spent sleeping than studying? Write your answer as a fraction of the day.

50. How much more of the day was spent working and traveling than in class? Write your answer as a fraction of the day.

Use the photo to answer Exercises 51–52. A nut driver is like a screwdriver but is used to tighten nuts instead of screws. The ends of the drivers are sized from $\frac{3}{16}$ inch to $\frac{1}{2}$ inch to fit smaller or larger nuts. (The symbol " is for inches.)

Source: Author's tool collection.

51. The rightmost driver fits a nut that is how much larger than the nut for the leftmost driver?

52. The nut size for the yellow-handled driver is how much less than the nut size for the blue-handled driver?

53. A hazardous waste dump site will require $\frac{7}{8}$ mile of security fencing. The site has four sides with three of the sides measuring $\frac{1}{4}$ mile, $\frac{1}{6}$ mile, and $\frac{3}{8}$ mile. Find the length of the fourth side.

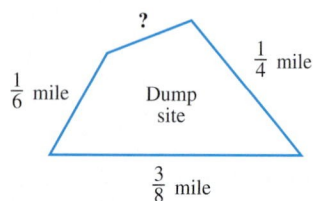

54. Chakotay is fitting a turquoise stone into a bear claw pendant. Find the diameter of the hole in the pendant. (The diameter is the distance across the center of the hole.)

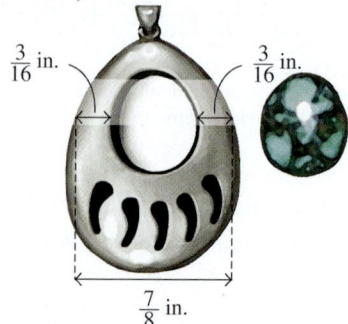

Study Skills

▶▶▶ MAKING A MIND MAP

Mind mapping is a visual way to show information that you have learned. It is an excellent way to review. Mapping is flexible and can be personalized, which is helpful for your memory. Your brain likes to see things that are **pleasing** to look at, **colorful,** and that **show connections** between ideas. Take advantage of that by creating maps that

▶ are easy to read,

▶ use color in a systematic way, and

▶ clearly show you how different concepts are related (using arrows or dotted lines, for example).

Below are some general directions for making a map. After you read them, go to the next page and work on completing the map that has been started for you. It is from **Sections 4.3** and **4.4.**

▶ To begin a mind map, write the concept in the center of a piece of paper and either circle it or draw a box around it.

▶ Make a line out from the center concept, and draw a box large enough to write the definition of the concept.

▶ Think of the other aspects (subpoints) of the concept that you have learned, such as procedures to follow or formulas. Make a separate line and box connecting each subpoint to the center.

▶ From each of the new boxes, add the information you've learned. You can continue making new lines and boxes and circles, or you can list items below the new information.

▶ Use color to highlight the major points. For example, everything related to one subpoint might be the same color. That way you can easily see related ideas.

▶ You may also use arrows, underlining, or small drawings to help yourself remember.

Directions for Making a Mind Map

> **Why Is Mapping Brain Friendly?**
> Remember that your brain grows dendrites when you are **actively thinking** about and working with information. Making a map requires you to think hard about **how to place the information, how to show connections** between parts of the map, and **how color will be useful.** It also takes a lot of thinking to fill in all related details and **show how those details connect to the larger concept.** All that thinking will let your brain grow a complex, many-branched neural network of interconnected dendrites. It is time well spent.

Try this Fractions Mind Map Using Sections 4.3 and 4.4

On a separate paper, make a map that summarizes Computations with Fractions. Follow the directions below. Use the starter map below.

▶ The longest rectangles are instructions for all four operations. (The first one starts "Rewrite all numbers as fractions..." and the second one is at the bottom of the map.)

▶ Notice the wavy dividing lines that separate the map into two sides.

▶ Your job is to complete the map by writing the steps used in multiplying fractions (from **Section 4.3**) and the steps used in adding and subtracting fractions (from **Section 4.4**).

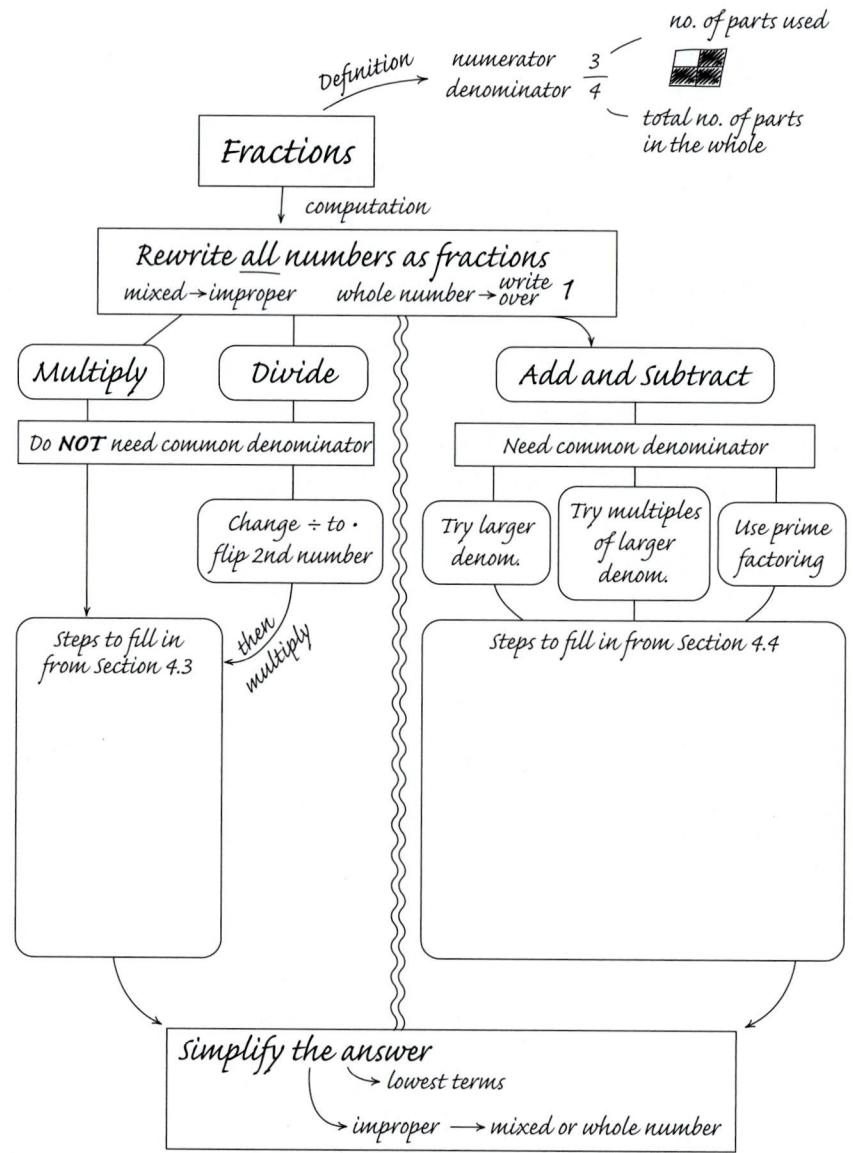

4.5 ▶▶▶ Problem Solving: Mixed Numbers and Estimating

OBJECTIVE **1** **Identify mixed numbers and graph them on a number line.** When a fraction and a whole number are written together, the result is a **mixed number.** For example, the mixed number

$$3\frac{1}{2} \quad \text{represents} \quad 3 + \frac{1}{2}$$

or 3 wholes and $\frac{1}{2}$ of a whole. Read $3\frac{1}{2}$ as "three and one half."

One common use of mixed numbers is to measure things. Examples are shown below.

Juan worked $5\frac{1}{2}$ hours.　　　The box weighs $2\frac{3}{4}$ pounds.

The park is $1\frac{7}{10}$ miles long.　　Add $1\frac{2}{3}$ cups of flour.

> **EXAMPLE 1** **Illustrating a Mixed Number with a Diagram and a Number Line**

As this diagram shows, the mixed number $3\frac{1}{2}$ is equivalent to the improper fraction $\frac{7}{2}$.

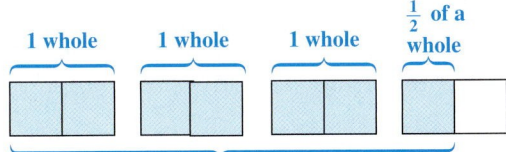

$$\frac{7}{2} \quad \leftarrow \text{Number of shaded parts}$$
$$\quad \leftarrow \text{Number of equal parts in each whole}$$

We can also use a number line to show mixed numbers, as in this graph of $3\frac{1}{2}$ and $-3\frac{1}{2}$.

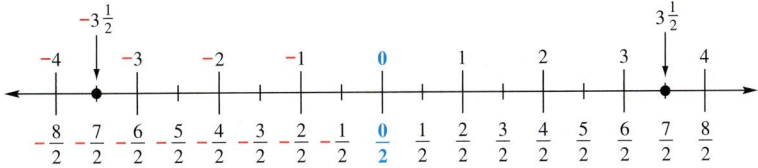

The number line shows the following.

$$3\frac{1}{2} \quad \text{is equivalent to} \quad \frac{7}{2}$$

$$-3\frac{1}{2} \quad \text{is equivalent to} \quad -\frac{7}{2}$$

> **Note**
> $3\frac{1}{2}$ represents $3 + \frac{1}{2}$.
> $-3\frac{1}{2}$ represents $-3 + (-\frac{1}{2})$, which can also be written as $-3 - \frac{1}{2}$.
> In algebra we usually work with the improper fraction form of mixed numbers, especially for negative mixed numbers. However, positive mixed numbers are frequently used in daily life, so it's important to know how to work with them. For example, we usually say $3\frac{1}{2}$ inches rather than $\frac{7}{2}$ inches.

Work Problem **1** *at the Side.* ▶

OBJECTIVES

1 Identify mixed numbers and graph them on a number line.

2 Rewrite mixed numbers as improper fractions, or the reverse.

3 Estimate the answer and multiply or divide mixed numbers.

4 Estimate the answer and add or subtract mixed numbers.

5 Solve application problems containing mixed numbers.

1 **(a)** Use this diagram to write $1\frac{2}{3}$ as an improper fraction.

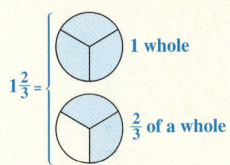

(b) Now graph $1\frac{2}{3}$ and $-1\frac{2}{3}$ on this number line.

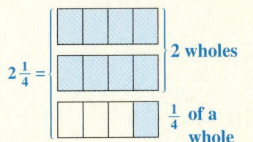

(c) Use this diagram to write $2\frac{1}{4}$ as an improper fraction.

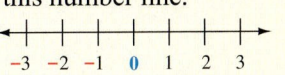

(d) Now graph $2\frac{1}{4}$ and $-2\frac{1}{4}$ on this number line.

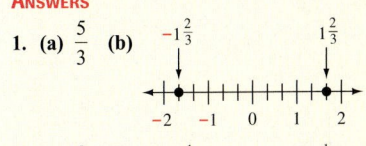

ANSWERS

1. (a) $\frac{5}{3}$ **(b)**

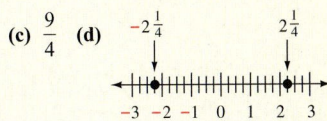

(c) $\frac{9}{4}$ **(d)**

2 Write each mixed number as an equivalent improper fraction.

(a) $3\dfrac{2}{3}$

(b) $4\dfrac{7}{10}$

(c) $5\dfrac{3}{4}$

(d) $8\dfrac{5}{6}$

OBJECTIVE **2** **Rewrite mixed numbers as improper fractions, or the reverse.** You can use the following steps to write $3\frac{1}{2}$ as an improper fraction without drawing a diagram or a number line.

Step 1 Multiply 2 times 3 and add 1 to the product.

$$3\dfrac{1}{2} \qquad 2 \cdot 3 = 6 \qquad \text{Then } 6 + 1 = 7$$

Step 2 Use 7 (from Step 1) as the numerator and 2 as the denominator.

$$3\dfrac{1}{2} = \dfrac{7}{2} \leftarrow (2 \cdot 3) + 1$$

Same denominator

To see why this method works, recall that $3\frac{1}{2}$ represents $3 + \frac{1}{2}$. Let's add $3 + \frac{1}{2}$.

$$3 + \dfrac{1}{2} = \dfrac{3}{1} + \dfrac{1}{2} = \dfrac{6}{2} + \dfrac{1}{2} = \dfrac{6+1}{2} = \dfrac{7}{2} \left\{ \begin{array}{l} \text{Same result} \\ \text{as above} \end{array} \right.$$

Common denominator

In summary, use the following steps to *write a mixed number as an improper fraction*.

Writing a Mixed Number as an Improper Fraction

Step 1 **Multiply** the denominator of the fraction times the whole number and **add** the numerator of the fraction to the product.

Step 2 Write the result of *Step 1* as the **numerator** and keep the original **denominator.**

EXAMPLE 2 **Writing a Mixed Number as an Improper Fraction**

Write $7\frac{2}{3}$ as an improper fraction (numerator greater than denominator).

Step 1 $7\dfrac{2}{3} \qquad 3 \cdot 7 = 21 \qquad \text{Then } 21 + 2 = 23$

Step 2 $7\dfrac{2}{3} = \dfrac{23}{3} \leftarrow (3 \cdot 7) + 2$

> Keep the same denominator.

Same denominator

◀ *Work Problem* **2** *at the Side.*

We used *multiplication* for the first step in writing a mixed number as an improper fraction. To work in *reverse*, writing an improper fraction as a mixed number, we will use *division*. Recall that the fraction bar is a symbol for division.

> ### Writing an Improper Fraction as a Mixed Number
>
> To write an *improper fraction* as a mixed number, divide the numerator by the denominator. The quotient is the whole number part (of the mixed number), the remainder is the numerator of the fraction part, and the denominator remains the same.

Always check to be sure that the fraction part of the mixed number is in lowest terms. Then the mixed number is in *simplest form*.

EXAMPLE 3   **Writing Improper Fractions as Mixed Numbers**

Write each improper fraction as an equivalent mixed number in simplest form.

(a) $\dfrac{17}{5}$

Divide 17 by 5.

$$\begin{array}{r} 3 \\ 5\overline{)17} \\ 15 \\ \hline 2 \end{array}$$

$3 \leftarrow$ Whole number part
$2 \leftarrow$ Remainder

The quotient **3** is the whole number part of the mixed number. The remainder **2** is the numerator of the fraction, and the denominator stays as **5**.

$$\frac{17}{5} = 3\frac{2}{5} \leftarrow \text{Remainder}$$

Same denominator

Let's look at a drawing of $\frac{17}{5}$ to check our work.

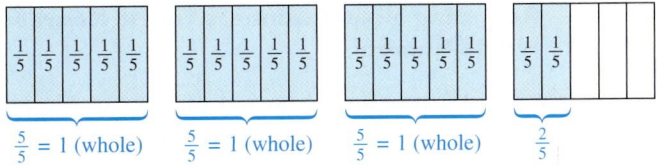

$\frac{5}{5} = 1$ (whole) $\frac{5}{5} = 1$ (whole) $\frac{5}{5} = 1$ (whole) $\frac{2}{5}$

Continued on Next Page

3 Write each improper fraction as an equivalent mixed number in simplest form.

(a) $\dfrac{5}{2}$

(b) $\dfrac{14}{4}$

(c) $\dfrac{33}{5}$

(d) $\dfrac{58}{10}$

4 Round each mixed number to the nearest whole number.

(a) $2\dfrac{3}{4}$

(b) $6\dfrac{3}{8}$

(c) $4\dfrac{2}{3}$

(d) $1\dfrac{7}{10}$

(e) $3\dfrac{1}{2}$

(f) $5\dfrac{4}{9}$

In other words,

$$\frac{17}{5} = \frac{5}{5} + \frac{5}{5} + \frac{5}{5} + \frac{2}{5}$$
$$= 1 + 1 + 1 + \frac{2}{5}$$
$$= 3 + \frac{2}{5}$$
$$= 3\frac{2}{5}$$

(b) $\dfrac{26}{4}$

Divide 26 by 4.

$$\begin{array}{r} 6 \\ 4\overline{)26} \\ 24 \\ \hline 2 \end{array} \quad \text{so} \quad \frac{26}{4} = 6\frac{2}{4} = 6\frac{1}{2} \quad \begin{cases} \text{Simplest} \\ \text{form} \end{cases}$$

Write $\frac{2}{4}$ in lowest terms.

You could write $\frac{26}{4}$ in lowest terms first.

$$\frac{26}{4} = \frac{\cancel{2} \cdot 13}{\cancel{2} \cdot 2} = \frac{13}{2} \quad \text{Then} \quad \begin{array}{r} 6 \\ 2\overline{)13} \\ 12 \\ \hline 1 \end{array} \quad \text{so} \quad \frac{13}{2} = 6\frac{1}{2} \quad \begin{cases} \text{Same result} \\ \text{as above} \end{cases}$$

◀ *Work Problem* **3** *at the Side.*

OBJECTIVE 3 Estimate the answer and multiply or divide mixed numbers. Once you have rewritten mixed numbers as improper fractions, you can use the steps you learned in **Section 4.3** to multiply and divide. However, it's a good idea to estimate the answer before you start any other work.

EXAMPLE 4 **Rounding Mixed Numbers to the Nearest Whole Number**

To estimate answers, first round each mixed number to the *nearest whole number*. If the numerator is *half* of the denominator or *more*, round up the whole number part. If the numerator is *less* than half the denominator, leave the whole number as it is.

(a) Round $1\dfrac{5}{8}$ ← 5 is more than 4 ← Half of 8 is 4 $1\dfrac{5}{8}$ rounds up to 2

(b) Round $3\dfrac{2}{5}$ ← 2 is less than $2\frac{1}{2}$ ← Half of 5 is $2\frac{1}{2}$ $3\dfrac{2}{5}$ rounds to 3

◀ *Work Problem* **4** *at the Side.*

ANSWERS

3. (a) $2\dfrac{1}{2}$ (b) $3\dfrac{1}{2}$ (c) $6\dfrac{3}{5}$ (d) $5\dfrac{4}{5}$

4. (a) 3 (b) 6 (c) 5 (d) 2 (e) 4 (f) 5

Multiplying and Dividing Mixed Numbers

Step 1 *Rewrite* each mixed number as an improper fraction.

Step 2 *Multiply* or *divide* the improper fractions.

Step 3 Write the answer in lowest terms and change it to a mixed number or whole number where possible. This step gives you an answer that is in *simplest form.*

EXAMPLE 5 **Estimating the Answer and Multiplying Mixed Numbers**

First, round the numbers and estimate each answer. Then find the exact answer. Write exact answers in simplest form.

(a) $2\frac{1}{2} \cdot 3\frac{1}{5}$

Estimate the answer by rounding the mixed numbers.

$2\frac{1}{2}$ rounds to 3   and $3\frac{1}{5}$ rounds to 3

$$3 \cdot 3 = 9 \leftarrow \text{Estimated answer}$$

To find the exact answer, first rewrite each mixed number as an improper fraction.

Step 1 $2\frac{1}{2} = \frac{5}{2}$ and $3\frac{1}{5} = \frac{16}{5}$

Next, multiply.

|Step 2 | Step 3|

$$2\frac{1}{2} \cdot 3\frac{1}{5} = \frac{5}{2} \cdot \frac{16}{5} = \frac{\overset{1}{\cancel{5}} \cdot \overset{1}{\cancel{2}} \cdot 8}{\underset{1}{\cancel{2}} \cdot \underset{1}{\cancel{5}}} = \frac{8}{1} = 8 \begin{cases}\text{Simplest} \\ \text{form}\end{cases}$$

The estimate was 9, so an exact answer of 8 is reasonable.

> Estimating helps you catch errors.

(b) $\left(3\frac{5}{8}\right)\left(4\frac{4}{5}\right)$

First, round each mixed number and estimate the answer.

$3\frac{5}{8}$ rounds to 4   and $4\frac{4}{5}$ rounds to 5

$$4 \cdot 5 = 20 \leftarrow \text{Estimated answer}$$

Now find the exact answer.

|Step 1 | Step 2 | Step 3|

$$\left(3\frac{5}{8}\right)\left(4\frac{4}{5}\right) = \left(\frac{29}{8}\right)\left(\frac{24}{5}\right) = \frac{29 \cdot 3 \cdot \overset{1}{\cancel{8}}}{\underset{1}{\cancel{8}} \cdot 5} = \frac{87}{5} = 17\frac{2}{5} \begin{cases}\text{Simplest} \\ \text{form}\end{cases}$$

The estimate was 20, so an exact answer of $17\frac{2}{5}$ is reasonable.

——————————— *Work Problem* **5** *at the Side.* ▶

5 First, round the numbers and estimate each answer. Then find the exact answer. Write exact answers in simplest form.

(a) $2\frac{1}{4} \cdot 7\frac{1}{3}$

$$\underline{\quad} \cdot \underline{\quad} = \underline{\quad} \text{ estimate}$$

(b) $\left(4\frac{1}{2}\right)\left(1\frac{2}{3}\right)$

$$(\underline{\quad})(\underline{\quad}) = \underline{\quad} \text{ estimate}$$

(c) $3\frac{3}{5} \cdot 4\frac{4}{9}$

$$\underline{\quad} \cdot \underline{\quad} = \underline{\quad} \text{ estimate}$$

(d) $\left(3\frac{1}{5}\right)\left(5\frac{3}{8}\right)$

$$(\underline{\quad})(\underline{\quad}) = \underline{\quad} \text{ estimate}$$

ANSWERS

5. **(a)** *Estimate:* $2 \cdot 7 = 14$; *Exact:* $16\frac{1}{2}$

(b) *Estimate:* $(5)(2) = 10$; *Exact:* $7\frac{1}{2}$

(c) *Estimate:* $4 \cdot 4 = 16$; *Exact:* 16

(d) *Estimate:* $(3)(5) = 15$; *Exact:* $17\frac{1}{5}$

6 First, round the numbers and estimate each answer. Then find the exact answer. Write exact answers in simplest form.

(a) $6\frac{1}{4} \div 3\frac{1}{3}$

$\downarrow$ $\qquad$ $\downarrow$

_____ ÷ _____ = _____ *estimate*

(b) $3\frac{3}{8} \div 2\frac{4}{7}$

$\downarrow$ $\qquad$ $\downarrow$

_____ ÷ _____ = _____ *estimate*

(c) $8 \div 5\frac{1}{3}$

$\downarrow$ $\qquad$ $\downarrow$

_____ ÷ _____ = _____ *estimate*

(d) $4\frac{1}{2} \div 6$

$\downarrow$ $\qquad$ $\downarrow$

_____ ÷ _____ = _____ *estimate*

ANSWERS

6. **(a)** *Estimate:* $6 \div 3 = 2$; *Exact:* $1\frac{7}{8}$

(b) *Estimate:* $3 \div 3 = 1$; *Exact:* $1\frac{5}{16}$

(c) *Estimate:* $8 \div 5 = 1\frac{3}{5}$; *Exact:* $1\frac{1}{2}$

(d) *Estimate:* $5 \div 6 = \frac{5}{6}$; *Exact:* $\frac{3}{4}$

EXAMPLE 6 **Estimating the Answer and Dividing Mixed Numbers**

First, round the numbers and estimate each answer. Then find the exact answer. Write exact answers in simplest form.

(a) $3\frac{3}{5} \div 1\frac{1}{2}$

To estimate the answer, round each mixed number to the nearest whole number.

$$3\frac{3}{5} \qquad \div \qquad 1\frac{1}{2}$$
$$\downarrow \quad \text{Rounded} \quad \downarrow$$
$$4 \qquad \div \qquad 2 \quad = \quad 2 \leftarrow \text{Estimate}$$

To find the exact answer, first rewrite each mixed number as an improper fraction.

$$3\frac{3}{5} \div 1\frac{1}{2} = \frac{18}{5} \div \frac{3}{2}$$

> You do **not** need a common denominator when multiplying or dividing fractions.

Now rewrite the problem as multiplying by the reciprocal of $\frac{3}{2}$.

$$\frac{18}{5} \div \frac{3}{2} = \frac{18}{5} \cdot \frac{2}{3} = \frac{\overset{1}{\cancel{3}} \cdot 6 \cdot 2}{5 \cdot \underset{1}{\cancel{3}}} = \frac{12}{5} = 2\frac{2}{5} \quad \left\{ \begin{array}{l}\text{Simplest} \\ \text{form}\end{array}\right.$$

Reciprocals

The estimate was 2, so an exact answer of $2\frac{2}{5}$ is reasonable.

(b) $4\frac{3}{8} \div 5$

First, round the numbers and estimate the answer.

$$4\frac{3}{8} \qquad \div \qquad 5$$
$$\downarrow \quad \text{Rounded} \quad \downarrow$$
$$4 \qquad \div \qquad 5$$

Write $4 \div 5$ using a fraction bar. $\left. \right\} \frac{4}{5} \leftarrow \text{Estimate}$

Now find the exact answer.

Write 5 as $\frac{5}{1}$

$$4\frac{3}{8} \div 5 = \frac{35}{8} \div \frac{5}{1} = \frac{35}{8} \cdot \frac{1}{5} = \frac{\overset{1}{\cancel{5}} \cdot 7 \cdot 1}{8 \cdot \underset{1}{\cancel{5}}} = \frac{7}{8} \quad \left\{ \begin{array}{l}\text{Simplest} \\ \text{form}\end{array}\right.$$

Reciprocals

The estimate was $\frac{4}{5}$, so an exact answer of $\frac{7}{8}$ is reasonable. They are both less than 1.

◀ *Work Problem* **6** *at the Side.*

OBJECTIVE **4** **Estimate the answer and add or subtract mixed numbers.**   The steps you learned for adding and subtracting fractions in **Section 4.4** will also work for mixed numbers: Just rewrite the mixed numbers as equivalent improper fractions. Again, it is a good idea to estimate the answer before you start any other work.

EXAMPLE 7 **Estimating the Answer and Adding or Subtracting Mixed Numbers**

First, estimate each answer. Then add or subtract to find the exact answer. Write exact answers in simplest form.

(a) $2\dfrac{3}{8} + 3\dfrac{3}{4}$

To estimate the answer, round each mixed number to the nearest whole number.

$$2\dfrac{3}{8} + 3\dfrac{3}{4}$$
$$\downarrow \qquad \downarrow$$
$$2 + 4 \; = \; 6 \longleftarrow \text{Estimate}$$

To find the exact answer, first rewrite each mixed number as an equivalent improper fraction.

$$2\dfrac{3}{8} + 3\dfrac{3}{4} = \dfrac{19}{8} + \dfrac{15}{4}$$

You **do** need a common denominator to add or subtract fractions.

You can't add fractions until they have a common denominator. The LCD for $\frac{19}{8}$ and $\frac{15}{4}$ is 8. Rewrite $\frac{15}{4}$ as an equivalent fraction with a denominator of 8.

$$\dfrac{19}{8} + \dfrac{15}{4} = \dfrac{19}{8} + \dfrac{30}{8} = \dfrac{19 + 30}{8} = \dfrac{49}{8} = 6\dfrac{1}{8} \; \begin{cases} \text{Simplest} \\ \text{form} \end{cases}$$

Common denominator

The estimate was 6, so an exact answer of $6\frac{1}{8}$ is reasonable.

(b) $4\dfrac{2}{3} - 2\dfrac{4}{5}$

Round each number and estimate the answer.

$$4\dfrac{2}{3} - 2\dfrac{4}{5}$$
$$\downarrow \qquad \downarrow$$
$$5 - 3 \; = \; 2 \longleftarrow \text{Estimate}$$

To find the exact answer, rewrite the mixed numbers as improper fractions and subtract.

$$4\dfrac{2}{3} - 2\dfrac{4}{5} = \dfrac{14}{3} - \dfrac{14}{5} = \dfrac{70}{15} - \dfrac{42}{15} = \dfrac{70 - 42}{15} = \dfrac{28}{15} = 1\dfrac{13}{15} \; \begin{cases} \text{Simplest} \\ \text{form} \end{cases}$$

LCD is 15

The estimate was 2, so an exact answer of $1\frac{13}{15}$ is reasonable.

Continued on Next Page

7 First, round the numbers and estimate each answer. Then add or subtract to find the exact answer.

(a) $5\dfrac{1}{3} \;-\; 2\dfrac{5}{6}$

$\downarrow \qquad\quad \downarrow$

$\rule{1cm}{0.4pt} - \rule{1cm}{0.4pt} = \rule{1cm}{0.4pt}$ *estimate*

(b) $\dfrac{3}{4} \;+\; 3\dfrac{1}{8}$

$\downarrow \qquad\quad \downarrow$

$\rule{1cm}{0.4pt} + \rule{1cm}{0.4pt} = \rule{1cm}{0.4pt}$ *estimate*

(c) $6 \;-\; 3\dfrac{4}{5}$

$\downarrow \qquad\quad \downarrow$

$\rule{1cm}{0.4pt} - \rule{1cm}{0.4pt} = \rule{1cm}{0.4pt}$ *estimate*

(c) $5 - 1\dfrac{3}{8}$

$$5 - 1\dfrac{3}{8}$$
$$\downarrow \qquad \downarrow$$
$$5 - 1 = 4 \;\leftarrow \text{Estimate}$$

Write 5 as $\dfrac{5}{1}$

$$5 - 1\dfrac{3}{8} = \dfrac{5}{1} - \dfrac{11}{8} = \dfrac{40}{8} - \dfrac{11}{8} = \dfrac{29}{8} = 3\dfrac{5}{8} \quad \begin{cases}\text{Simplest}\\ \text{form}\end{cases}$$

LCD is 8

The estimate was 4, so an exact answer of $3\frac{5}{8}$ is reasonable.

◀ Work Problem **7** at the Side.

Note

In some situations the method of rewriting mixed numbers as improper fractions may result in very large numerators. Consider this example.

Last year Hue's child was $48\frac{3}{8}$ in. tall. This year the child is $51\frac{1}{4}$ in. tall. How much has the child grown?

First, estimate the answer by rounding each mixed number to the nearest whole number.

$$51\dfrac{1}{4} - 48\dfrac{3}{8}$$
$$\downarrow \qquad\quad \downarrow$$
$$51 - 48 = 3 \text{ in.} \;\leftarrow \text{Estimate}$$

To find the exact answer, rewrite the mixed numbers as improper fractions.

Rewrite $\frac{205}{4}$ as $\frac{410}{8}$

$$51\dfrac{1}{4} - 48\dfrac{3}{8} = \dfrac{205}{4} - \dfrac{387}{8} = \dfrac{410}{8} - \dfrac{387}{8} = \dfrac{410 - 387}{8} = \dfrac{23}{8} = 2\dfrac{7}{8} \text{ in.}$$

LCD is 8

You can also use the fraction key $\boxed{a^{b/c}}$ on your *scientific* calculator to solve this problem.

To enter $51\frac{1}{4}$, press

$$51 \;\boxed{a^{b/c}}\; 1 \;\boxed{a^{b/c}}\; 4 \;\boxed{a^{b/c}}\; . \text{ The display shows } \boxed{51\_1\lrcorner4}.$$

$\uparrow$ Whole number $\uparrow$ Numerator $\uparrow$ Denominator

Then press $\boxed{-}$ $48 \;\boxed{a^{b/c}}\; 3 \;\boxed{a^{b/c}}\; 8 \;\boxed{=}$. The display shows $\boxed{2\_7\lrcorner8}$.

Subtract. $48\dfrac{3}{8}$ $2\dfrac{7}{8}$

Either way the exact answer is $2\frac{7}{8}$ in., which is close to the estimate of 3 in.

Another efficient method for handling large mixed numbers is to rewrite them in decimal form. You will learn how to do that in **Chapter 5.**

ANSWERS

7. **(a)** *Estimate:* $5 - 3 = 2$; *Exact:* $2\dfrac{1}{2}$

(b) *Estimate:* $1 + 3 = 4$; *Exact:* $3\dfrac{7}{8}$

(c) *Estimate:* $6 - 4 = 2$; *Exact:* $2\dfrac{1}{5}$

OBJECTIVE 5 Solve application problems containing mixed numbers. Rounding mixed numbers to the nearest whole number can also help you decide whether to solve an application problem by adding, subtracting, multiplying, or dividing.

EXAMPLE 8 Solving Application Problems with Mixed Numbers

First, estimate the answer to each application problem. Then find the exact answer.

(a) Gary needs to haul **$15\frac{3}{4}$ tons** of sand to a construction site. His truck can carry **$2\frac{1}{4}$ tons**. How many trips will he need to make?

First, round each mixed number to the nearest whole number.

$$15\frac{3}{4} \quad \text{rounds to} \quad 16 \quad \text{and} \quad 2\frac{1}{4} \quad \text{rounds to} \quad 2$$

Now read the problem again, *using the rounded numbers.*

Gary needs to haul **16 tons** of sand to a construction site. His truck can carry **2 tons**. How many trips will he need to make?

Using the rounded numbers in the problem makes it easier to see that you need to *divide.*

$$16 \div 2 = 8 \text{ trips} \leftarrow \text{Estimate}$$

To find the exact answer, use the original mixed numbers and divide.

$$15\frac{3}{4} \div 2\frac{1}{4} = \frac{63}{4} \div \frac{9}{4} = \frac{63}{4} \cdot \frac{4}{9} = \frac{7 \cdot \cancel{9} \cdot \cancel{4}}{\cancel{4} \cdot \cancel{9}} = \frac{7}{1} = 7 \quad \left\{ \begin{array}{l}\text{Simplest} \\ \text{form}\end{array}\right.$$

Reciprocals

Gary needs to make 7 trips to haul all the sand. This result is close to the estimate of 8 trips.

(b) Zenitia worked **$3\frac{5}{6}$ hours** on Monday and **$6\frac{1}{2}$ hours** on Tuesday. How much longer did she work on Tuesday than on Monday?

First, round each mixed number to the nearest whole number.

$$3\frac{5}{6} \quad \text{rounds to} \quad 4 \quad \text{and} \quad 6\frac{1}{2} \quad \text{rounds to} \quad 7$$

Now read the problem again, *using the rounded numbers.*

Zenitia worked **4 hours** on Monday and **7 hours** on Tuesday. How much longer did she work on Tuesday than on Monday?

Using the rounded numbers in the problem makes it easier to see that you need to *subtract.*

$$7 - 4 = 3 \text{ hours} \leftarrow \text{Estimate}$$

To find the exact answer, use the original mixed numbers and subtract.

Write answer in simplest form.

$$6\frac{1}{2} - 3\frac{5}{6} = \frac{13}{2} - \frac{23}{6} = \frac{39}{6} - \frac{23}{6} = \frac{39 - 23}{6} = \frac{16}{6} = 2\frac{4}{6} = 2\frac{2}{3}$$

LCD is 6

Zenitia worked $2\frac{2}{3}$ hours longer on Tuesday. This result is close to the estimate of 3 hours.

Work Problem **8** *at the Side.* ▶

8 First, round the numbers and estimate the answer to each problem. Then find the exact answer.

(a) Richard's son grew $3\frac{5}{8}$ inches last year and $2\frac{1}{4}$ inches this year. How much has his height increased over the two years?

Estimate:

Exact:

(b) Ernestine used $2\frac{1}{2}$ packages of chocolate chips in her cookie recipe. Each package has $5\frac{1}{2}$ ounces of chips. How many ounces of chips did she use in the recipe?

Estimate:

Exact:

ANSWERS

8. (a) *Estimate:* $4 + 2 = 6$ inches;

Exact: $5\frac{7}{8}$ inches

(b) *Estimate:* $3 \cdot 6 = 18$ ounces;

Exact: $13\frac{3}{4}$ ounces

Math in the Media

RECIPES

Guilt-free Gravy

Makes 2 cups.

- 3 tablespoons corn starch
- $\frac{1}{4}$ cup apple cider
- 2 cups fat-free broth

- $\frac{1}{2}$ teaspoon rubbed sage
- $\frac{3}{4}$ teaspoon salt
- $\frac{1}{8}$ teaspoon pepper

To make in the microwave:

Combine corn starch, apple cider, broth, and sage in a 2-quart, microwave-safe bowl. With a whisk, stir mixture until corn starch is completely dissolved. Microwave on high power for 7 to 9 minutes or until mixture boils, stirring every minute. Boil for 1 minute. Add salt and pepper.

Source: www.argostarch.com

You decide to make gravy for Thanksgiving dinner and find this recipe through an Internet search.

1. (a) How much gravy does the recipe make?

   (b) You need 5 cups of gravy. By what factor should you multiply each ingredient in the recipe?

   (c) Find the amount of each ingredient needed to make 5 cups of gravy.

   corn starch: _____   sage: _____

   apple cider: _____   salt: _____

   fat-free broth: _____   pepper: _____

2. Most sets of measuring spoons have a spoon for each of these amounts:

   1 tablespoon, $\frac{1}{2}$ tablespoon, 1 teaspoon, $\frac{1}{2}$ teaspoon, $\frac{1}{4}$ teaspoon, $\frac{1}{8}$ teaspoon

   (a) How could you measure $1\frac{7}{8}$ teaspoons of salt in the fewest steps?

   (b) How might you measure the pepper from Question 1?

4.5 ▶▶▶ Exercises

Graph the mixed numbers or improper fractions on the number line. See Example 1.

1. Graph $2\frac{1}{3}$ and $-2\frac{1}{3}$.

2. Graph $1\frac{3}{4}$ and $-1\frac{3}{4}$.

3. Graph $\frac{3}{2}$ and $-\frac{3}{2}$.

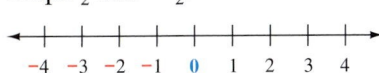

4. Graph $\frac{11}{3}$ and $-\frac{11}{3}$.

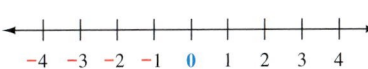

Write each mixed number as an improper fraction. See Example 2.

5. $4\frac{1}{2}$

6. $2\frac{1}{4}$

7. $-1\frac{3}{5}$

8. $-1\frac{5}{6}$

9. $2\frac{3}{8}$

10. $3\frac{4}{9}$

11. $-5\frac{7}{10}$

12. $-4\frac{5}{7}$

13. $10\frac{11}{15}$

14. $12\frac{9}{11}$

Write each improper fraction as a mixed number in simplest form. See Example 3.

15. $\frac{13}{3}$

16. $\frac{11}{2}$

17. $-\frac{10}{4}$

18. $-\frac{14}{5}$

19. $\frac{22}{6}$

20. $\frac{28}{8}$

21. $-\frac{51}{9}$

22. $-\frac{44}{10}$

◑ 23. $\frac{188}{16}$

24. $\frac{200}{15}$

First, round each mixed number to the nearest whole number and estimate the answer.
Then find the exact answer. Write exact answers in simplest form. See Examples 4–7.

◑ 25. *Exact:*

$2\frac{1}{4} \cdot 3\frac{1}{2}$

Estimate:

___ • ___ = ___

26. *Exact:*

$\left(1\frac{1}{2}\right)\left(3\frac{3}{4}\right)$

Estimate:

(___)(___) = ___

◑ 27. *Exact:*

$3\frac{1}{4} \div 2\frac{5}{8}$

Estimate:

___ ÷ ___ = ___

28. *Exact:*

$2\frac{1}{4} \div 1\frac{1}{8}$

Estimate:

___ ÷ ___ = ___

◑ 29. *Exact:*

$3\frac{2}{3} + 1\frac{5}{6}$

Estimate:

___ + ___ = ___

30. *Exact:*

$4\frac{4}{5} + 2\frac{1}{3}$

Estimate:

___ + ___ = ___

31. *Exact:*

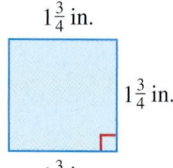

$$4\frac{1}{4} - \frac{7}{12}$$

Estimate:

___ − ___ = ___

32. *Exact:*

$$10\frac{1}{3} - 6\frac{5}{6}$$

Estimate:

___ − ___ = ___

33. *Exact:*

$$5\frac{2}{3} \div 6$$

Estimate:

___ ÷ ___ = ___

34. *Exact:*

$$1\frac{7}{8} \div 6\frac{1}{4}$$

Estimate:

___ ÷ ___ = ___

35. *Exact:*

$$8 - 1\frac{4}{5}$$

Estimate:

___ − ___ = ___

36. *Exact:*

$$7 - 3\frac{3}{10}$$

Estimate:

___ − ___ = ___

Find the perimeter and the area of each square or rectangle. Write all answers in simplest form.

37. $1\frac{3}{4}$ in.

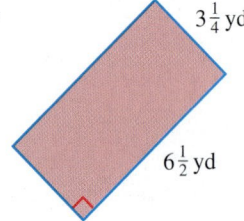

$1\frac{3}{4}$ in.

$1\frac{3}{4}$ in.

38.

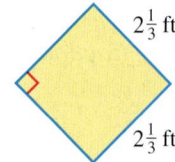

$2\frac{1}{3}$ ft

$2\frac{1}{3}$ ft

39.

$3\frac{1}{4}$ yd

$6\frac{1}{2}$ yd

40.

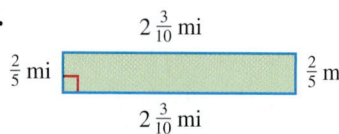

$2\frac{3}{10}$ mi

$\frac{2}{5}$ mi $\frac{2}{5}$ mi

$2\frac{3}{10}$ mi

First, estimate the answer to each application problem. Then find the exact answer. Write all answers in simplest form. See Example 8.

41. A carpenter has two pieces of oak trim. One piece of trim is $12\frac{1}{2}$ ft long and the other is $8\frac{2}{3}$ ft long. How many feet of oak trim does he have in all?

Estimate:

Exact:

42. On Monday, $5\frac{3}{4}$ tons of cans were recycled, and on Tuesday, $9\frac{3}{5}$ tons were recycled. How many tons were recycled on these two days?

Estimate:

Exact:

43. The directions for mixing an insect spray say to use $1\frac{3}{4}$ ounces of chemical in each gallon of water. How many ounces of chemical should be mixed with $5\frac{1}{2}$ gallons of water?

Estimate:

Exact:

44. Shirley Cicero wants to make 16 holiday wreaths to sell at a craft fair. Each wreath requires $2\frac{1}{4}$ yd of ribbon. How many yards does she need in all?

Estimate:

Exact:

45. The Boy Scout troop has volunteered to pick up trash along a 4-mile stretch of highway. So far they have done $1\frac{7}{10}$ miles. How much do they have left to do?

Estimate:

Exact:

46. Marv bought a 10 yd length of Italian silk fabric. He used $3\frac{7}{8}$ yd to make a jacket. What length of fabric is left for other sewing projects?

Estimate:

Exact:

47. Suppose that a bridesmaid's floor-length dress requires a piece of material $3\frac{3}{4}$ yd in length. What length of material is needed to make dresses for five bridesmaids?

Estimate:

Exact:

48. A cookie recipe uses $\frac{2}{3}$ cup brown sugar. How much brown sugar is needed to make $2\frac{1}{2}$ times the original recipe?

Estimate:

Exact:

Solve each application problem. Write all answers in simplest form. Use the information on the Oregon state park sign for Exercises 49–52. The distances on the sign are in miles.

FACE ROCK ¼ MI.
DEVILS KITCHEN 1 ¾ MI.
PICNIC PARKING 2 ¼ MI.
BEACH PARKING 2 ⅜ MI.

49. What is the distance from Devils Kitchen to the beach parking?

50. How far is it from Face Rock to the beach parking?

51. Suppose you bicycle from the park sign to the picnic parking, then bicycle back to Face Rock, and finally bicycle to the beach parking. How far will you have traveled?

52. If you drive back and forth from the sign to the picnic parking each day for a week, how many miles will you drive?

53. Find the length of the arrow shaft.

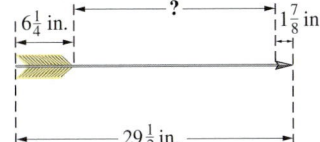

$6\frac{1}{4}$ in. ? $1\frac{7}{8}$ in.

$29\frac{1}{2}$ in.

54. Find the length of the indented section on this board.

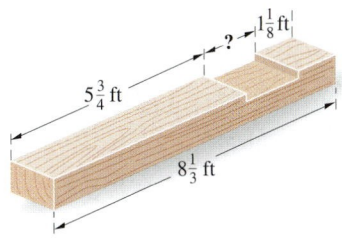

$1\frac{1}{8}$ ft

?

$5\frac{3}{4}$ ft

$8\frac{1}{3}$ ft

First, estimate the answer to each application problem. Then use your calculator to find the exact answer.

55. A craftsperson must attach a lead strip around all four sides of a rectangular stained glass window before it is installed. Find the length of lead stripping needed for the window shown.

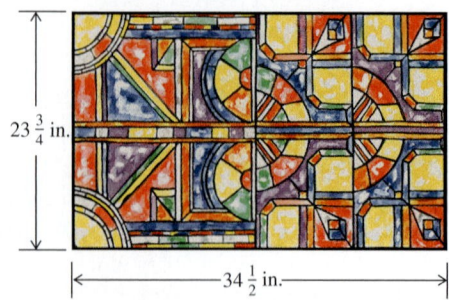

$23\frac{3}{4}$ in.

$34\frac{1}{2}$ in.

Estimate:

Exact:

56. To complete a custom order, Zak Morten of Home Depot must find the number of inches of brass trim needed to go around the four sides of the lamp base plate shown. Find the length of brass trim needed.

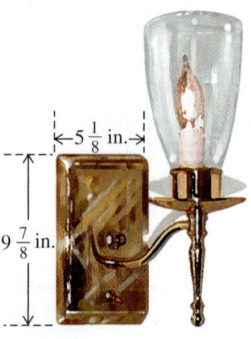

$5\frac{1}{8}$ in.

$9\frac{7}{8}$ in.

Estimate:

Exact:

57. A fishing boat anchor requires $10\frac{3}{8}$ pounds of steel. Find the number of anchors that can be manufactured with 25,730 pounds of steel.

Estimate:

Exact:

58. Each apartment requires $62\frac{1}{2}$ square yards of carpet. Find the number of apartments that can be carpeted with 6750 square yards of carpet.

Estimate:

Exact:

59. Claire and Deb create custom hat bands that people can put around the crowns of their hats. The finished bands are the lengths shown in the table. The strip of fabric for each band must include the finished length plus an extra $\frac{3}{4}$ in. for the seam.

Band Size	Finished Length
Small	$21\frac{7}{8}$ in.
Medium	$22\frac{5}{8}$ in.
Large	$23\frac{1}{2}$ in.

What length of fabric strip is needed to make 4 small bands, 5 medium bands, and 3 large bands, including the seam allowance?

Estimate:

Exact:

60. Three sides of a parking lot are $108\frac{1}{4}$ ft, $162\frac{3}{8}$ ft, and $143\frac{1}{2}$ ft. If the distance around the lot is $518\frac{3}{4}$ ft, find the length of the fourth side.

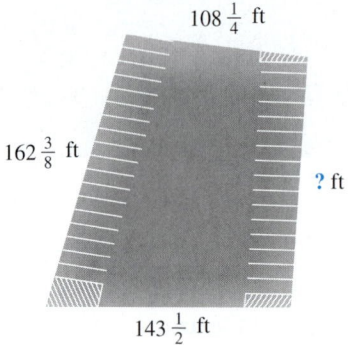

$108\frac{1}{4}$ ft

$162\frac{3}{8}$ ft

? ft

$143\frac{1}{2}$ ft

Estimate:

Exact:

Summary Exercises on Fractions

1. Write fractions that represent the shaded and unshaded portion of each figure.

(a)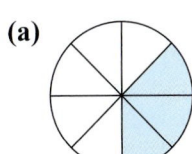

(b)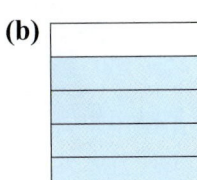

2. Graph $\frac{2}{3}$ and $-\frac{2}{3}$ on the number line.

3. Rewrite each fraction with the indicated denominator.

(a) $-\frac{4}{5} = -\frac{}{30}$

(b) $\frac{2}{7} = \frac{}{14}$

4. Simplify.

(a) $\frac{15}{15}$

(b) $-\frac{24}{6}$

(c) $\frac{9}{1}$

5. Write the prime factorization of each number.

(a) 72

(b) 105

6. Write each fraction in lowest terms.

(a) $\frac{24}{30}$

(b) $\frac{175}{200}$

Simplify.

7. $\left(-\frac{3}{4}\right)\left(-\frac{2}{3}\right)$

8. $-\frac{7}{8} + \frac{2}{3}$

9. $\frac{7}{16} + \frac{5}{8}$

10. $\frac{5}{8} \div \frac{3}{4}$

11. $\frac{2}{3} - \frac{4}{5}$

12. $\frac{7}{12}\left(-\frac{9}{14}\right)$

13. $-21 \div \left(-\frac{3}{8}\right)$

14. $\frac{7}{8} - \frac{5}{12}$

15. $-\frac{35}{45} \div \frac{10}{15}$

16. $-\frac{5}{6} - \frac{3}{4}$

17. $\frac{7}{12} + \frac{5}{6} + \frac{2}{3}$

18. $\frac{5}{8}$ of 56

First round the numbers and estimate each answer. Then find the exact answer.

19. *Exact:*

$4\frac{3}{4} + 2\frac{5}{6}$

Estimate:

___ + ___ = ___

20. *Exact:*

$2\frac{2}{9} \cdot 5\frac{1}{7}$

Estimate:

___ • ___ = ___

21. *Exact:*

$6 - 2\frac{7}{10}$

Estimate:

___ − ___ = ___

22. *Exact:*

$1\frac{3}{5} \div 3\frac{1}{2}$

Estimate:

___ ÷ ___ = ___

23. *Exact:*

$4\frac{2}{3} \div 1\frac{1}{6}$

Estimate:

___ ÷ ___ = ___

24. *Exact:*

$3\frac{5}{12} - \frac{3}{4}$

Estimate:

___ − ___ = ___

Solve each application problem. Write all answers in simplest form.

25. When installing cabinets, Cecil Feathers must be certain that the proper type and size of mounting screw is used.

(a) Find the total length of the screw shown.

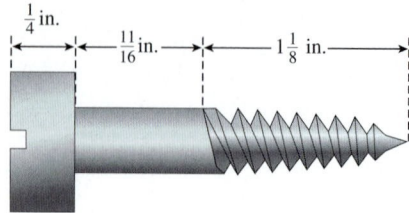

(b) If the screw is put into a board that is $1\frac{3}{4}$ in. thick, how much of the screw will stick out the back of the board?

26. Find the perimeter and the area of this postage stamp.

27. A batch of cookies requires $\frac{3}{4}$ pound of chocolate chips. If you have nine pounds of chocolate chips, how many batches of cookies can you make?

28. The Municipal Utility District says that the cost of operating a hair dryer is $\frac{1}{5}$¢ per minute. Find the cost of operating the hair dryer for a half hour.

29. A survey asked 1500 adults if UFOs (unidentified flying objects) are real. The circle graph shows the fraction of the adults who gave each answer. How many adults gave each answer?

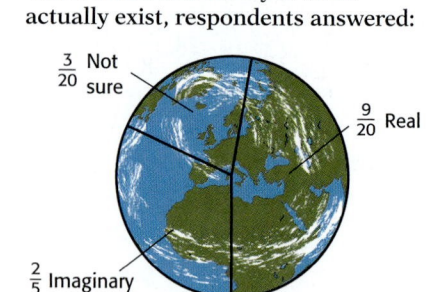

Out Of This World?

When asked in a survey if UFOs actually exist, respondents answered:

$\frac{3}{20}$ Not sure

$\frac{9}{20}$ Real

$\frac{2}{5}$ Imaginary

Source: Yankelovich Partners for *Life* magazine.

30. Find the diameter of the hole in the rectangular mounting bracket shown. (The diameter is the distance across the center of the hole.) Then find the perimeter of the bracket.

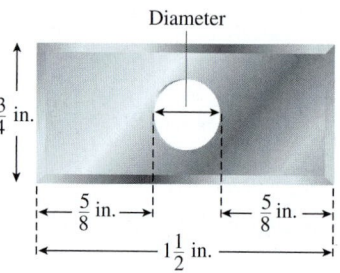

31. A bottle of contact lens daily cleaning solution holds $\frac{2}{3}$ fluid ounce. How many bottles can be filled with $15\frac{1}{3}$ fluid ounces? (*Source:* Alcon Laboratories, Inc.)

32. A home builder bought two parcels of land that were $5\frac{7}{8}$ acres and $10\frac{3}{4}$ acres. She is setting aside $2\frac{7}{8}$ acres for a park and using the rest for $1\frac{1}{4}$ acre home lots. How many lots will be in the development?

4.6 ▶▶▶ Exponents, Order of Operations, and Complex Fractions

OBJECTIVE 1 Simplify fractions with exponents. We have used exponents as a quick way to write repeated multiplication of integers and variables. Here are two examples as a review.

$$(-3)^2 = \underbrace{(-3)(-3)}_{\text{Two factors of } -3} = 9 \quad \text{and} \quad x^3 = \underbrace{x \cdot x \cdot x}_{\text{Three factors of } x}$$

Exponent ↓ (−3)²   Base ↑

Exponent ↓ x³   Base ↑

The meaning of an exponent remains the same when a fraction is the base.

OBJECTIVES

1 Simplify fractions with exponents.

2 Use the order of operations with fractions.

3 Simplify complex fractions.

EXAMPLE 1 Simplifying Fractions with Exponents

Simplify.

(a) $\left(-\dfrac{1}{2}\right)^3$

The base is $-\frac{1}{2}$. The exponent indicates that there are three factors of $-\frac{1}{2}$.

$$\left(-\frac{1}{2}\right)^3 = \underbrace{\left(-\frac{1}{2}\right)\left(-\frac{1}{2}\right)\left(-\frac{1}{2}\right)}_{\text{Three factors of } -\frac{1}{2}} \quad \text{Multiply } \left(-\frac{1}{2}\right)\left(-\frac{1}{2}\right) \text{ to get } \frac{1}{4}$$

> Watch the signs! Multiply *two* factors at a time.

$$= \frac{1}{4}\underbrace{\left(-\frac{1}{2}\right)}_{} \quad \text{Now multiply } \frac{1}{4}\left(-\frac{1}{2}\right) \text{ to get } -\frac{1}{8}$$

$$= -\frac{1}{8} \quad \text{The product is negative.}$$

(b) $\left(\dfrac{3}{4}\right)^2 \left(\dfrac{2}{3}\right)^3$

Two factors of $\frac{3}{4}$

Three factors of $\frac{2}{3}$

$$\left(\frac{3}{4}\right)^2\left(\frac{2}{3}\right)^3 = \left(\frac{3}{4}\right)\left(\frac{3}{4}\right)\left(\frac{2}{3}\right)\left(\frac{2}{3}\right)\left(\frac{2}{3}\right)$$

$$= \frac{\overset{1}{\cancel{3}} \cdot \overset{1}{\cancel{3}} \cdot \overset{1}{\cancel{2}} \cdot \overset{1}{\cancel{2}} \cdot \overset{1}{\cancel{2}}}{\underset{1}{\cancel{2}} \cdot \underset{1}{\cancel{2}} \cdot \underset{1}{\cancel{2}} \cdot 2 \cdot \underset{1}{\cancel{3}} \cdot \underset{1}{\cancel{3}} \cdot 3} \quad \begin{array}{l}\text{Divide out all}\\\text{common factors.}\end{array}$$

$$= \frac{1}{6}$$

Work Problem **1** *at the Side.* ▶

OBJECTIVE 2 Use the order of operations with fractions. The order of operations that you used in **Section 1.8** for integers also applies to fractions.

---

1 Simplify.

(a) $\left(-\dfrac{3}{5}\right)^2$

(b) $\left(\dfrac{1}{3}\right)^4$

(c) $\left(-\dfrac{2}{3}\right)^3\left(\dfrac{1}{2}\right)^2$

(d) $\left(-\dfrac{1}{2}\right)^2\left(\dfrac{1}{4}\right)^2$

2 Simplify.

(a) $\dfrac{1}{3} - \dfrac{5}{9}\left(\dfrac{3}{4}\right)$

(b) $-\dfrac{3}{4} + \left(-\dfrac{1}{2}\right)^2 \div \dfrac{2}{3}$

(c) $\dfrac{12}{5} - \dfrac{1}{6}\left(3 - \dfrac{3}{5}\right)$

Order of Operations

Step 1 Work inside *parentheses* or *other grouping symbols.*

Step 2 Simplify expressions with *exponents.*

Step 3 Do the remaining *multiplications and divisions* as they occur from left to right.

Step 4 Do the remaining *additions and subtractions* as they occur from left to right.

EXAMPLE 2 Using the Order of Operations with Fractions

Simplify.

(a) $-\dfrac{1}{3} + \dfrac{1}{2}\left(\dfrac{4}{5}\right)$ There is no work to be done inside the parentheses. There are no exponents, so start with Step 3, multiplying and dividing.

$-\dfrac{1}{3} + \dfrac{1 \cdot 4}{2 \cdot 5}$ Multiply.

> Multiply $\frac{1}{2} \cdot \frac{4}{5}$ first. Do **not** start by adding $-\frac{1}{3} + \frac{1}{2}$.

$-\dfrac{1}{3} + \dfrac{4}{10}$ Now add. The LCD is 30. $-\dfrac{1}{3} = -\dfrac{10}{30}$ and $\dfrac{4}{10} = \dfrac{12}{30}$

$\dfrac{-10 + 12}{30}$ Add the numerators; keep the common denominator.

$\dfrac{2}{30}$ Write $\dfrac{2}{30}$ in lowest terms: $\dfrac{\overset{1}{\cancel{2}}}{\underset{1}{\cancel{2}} \cdot 15} = \dfrac{1}{15}$

$\dfrac{1}{15}$ The answer is in lowest terms.

(b) $-2 + \left(\dfrac{1}{4} - \dfrac{3}{2}\right)^2$ Work inside parentheses. The LCD for $\dfrac{1}{4}$ and $\dfrac{3}{2}$ is 4

$-2 + \left(\dfrac{1 - 6}{4}\right)^2$ Rewrite $\dfrac{3}{2}$ as $\dfrac{6}{4}$ and subtract.

$-2 + \left(\dfrac{-5}{4}\right)^2$ Simplify the term with the exponent. Multiply $\left(-\dfrac{5}{4}\right)\left(-\dfrac{5}{4}\right)$. *Signs match,* so the product is *positive.*

$-2 + \left(\dfrac{25}{16}\right)$ Add last. Write -2 as $-\dfrac{2}{1}$

$-\dfrac{2}{1} + \dfrac{25}{16}$ The LCD is 16. Rewrite $-\dfrac{2}{1}$ as $-\dfrac{32}{16}$

$\dfrac{-32 + 25}{16}$ Add the numerators; keep the common denominator.

$-\dfrac{7}{16}$ ⟵ The answer is in lowest terms.

ANSWERS

2. (a) $-\dfrac{1}{12}$ (b) $-\dfrac{3}{8}$ (c) 2

◀ *Work Problem* **2** *at the Side.*

OBJECTIVE **3** **Simplify complex fractions.** We have used both the ÷ symbol and a fraction bar to indicate division. For example,

Indicates division → $\dfrac{6}{2}$ can be written as $6 \div 2$
Indicates division

That means we could write $-\frac{4}{5} \div \left(-\frac{3}{10}\right)$ using a fraction bar instead of ÷.

$$-\frac{4}{5} \div \left(-\frac{3}{10}\right) \quad \text{can be written as} \quad \dfrac{-\dfrac{4}{5}}{-\dfrac{3}{10}} \leftarrow \text{Indicates division}$$

Indicates division

The result looks a bit complicated, and its name reflects that fact. We call it a *complex fraction.*

Complex Fractions

A **complex fraction** is a fraction in which the numerator and/or denominator contain one or more fractions.

To simplify a complex fraction, rewrite it in horizontal format using the ÷ symbol for division.

EXAMPLE 3 **Simplifying a Complex Fraction**

Simplify: **(a)** $\dfrac{-\dfrac{4}{5}}{-\dfrac{3}{10}}$ **(b)** $\dfrac{\left(\dfrac{2}{3}\right)^2}{6}$

(a) Rewrite the complex fraction using the ÷ symbol for division. Then follow the steps for dividing fractions.

$$\dfrac{-\dfrac{4}{5}}{-\dfrac{3}{10}} = -\frac{4}{5} \div -\frac{3}{10} = -\frac{4}{5} \cdot -\frac{10}{3} = \frac{4 \cdot 2 \cdot \overset{1}{\cancel{5}}}{\cancel{5} \cdot 3} = \frac{8}{3} \quad \text{or} \quad 2\frac{2}{3}$$

Reciprocals

The quotient is positive because the numbers in the problem had matching signs (both were negative).

(b) Rewrite the complex fraction using the ÷ symbol for division. Then follow the order of operations.

$$\dfrac{\left(\dfrac{2}{3}\right)^2}{6} = \left(\frac{2}{3}\right)^2 \div 6 = \frac{4}{9} \div 6 = \frac{4}{9} \cdot \frac{1}{6} = \frac{\overset{1}{\cancel{2}} \cdot 2 \cdot 1}{9 \cdot \cancel{2} \cdot 3} = \frac{2}{27}$$

Reciprocals

Work Problem **3** *at the Side.* ▶

3 Simplify.

(a) $\dfrac{-\dfrac{3}{5}}{\dfrac{9}{10}}$

(b) $\dfrac{6}{\dfrac{3}{4}}$ ← *Hint:* Write 6 as $\frac{6}{1}$

(c) $\dfrac{-\dfrac{15}{16}}{-5}$

(d) $\dfrac{-3}{\left(-\dfrac{3}{4}\right)^2}$

ANSWERS

3. **(a)** $-\dfrac{2}{3}$ **(b)** 8 **(c)** $\dfrac{3}{16}$

(d) $-\dfrac{16}{3}$ or $-5\dfrac{1}{3}$

Math in the Media

HEART-RATE TRAINING ZONE

Performing aerobic exercise benefits your aerobic fitness and helps burn fat. For best results, you should keep your heart rate within the training zone for a minimum of 12 minutes.

Example: The Training Zone (TZ) is based on your Heart Rate (HR) for one minute. To see if you are in the training zone, measure your heart rate for 15 seconds. Compare it to the 15-second training zone. Find the exact answer, then round to the nearest whole number.

Source: www.runningforfitness.org

Instruction	Calculation	Example (age 22)
Calculate maximum heart rate (MHR)	$220 - \text{your age}$	$220 - 22 = 198$
Calculate lower limit of training zone (TZ)	$\frac{3}{5} \times (\text{MHR})$	$\frac{3}{5} \times (198) = \frac{594}{5} = 118\frac{4}{5}$
Calculate upper limit of training zone (TZ)	$\frac{4}{5} \times (\text{MHR})$	$\frac{4}{5} \times (198) = \frac{792}{5} = 158\frac{2}{5}$
Calculate the exact 15-second training zone; then round to the nearest whole number.	$\left(\frac{1}{4} \times \text{lower TZ}, \frac{1}{4} \times \text{Upper TZ}\right)$	$\frac{1}{4} \times \frac{594}{5} = 29\frac{7}{10}; \frac{1}{4} \times \frac{792}{5} = 39\frac{3}{5}$ $29\frac{7}{10} < \text{HR} < 39\frac{3}{5}$ (exact) $30 < \text{HR} < 40$ (rounded)

1. Suppose you work in a physical fitness center and decide to design a poster to remind the clients of the training zone for their age. Compute the exact 15-second training zone for people of each age below. Write fractions in lowest terms. Then round the answers to the nearest whole number.

2. Explain why the lower and upper training zones (TZ) are multiplied by $\frac{1}{4}$.

Age	MHR	Lower Limit of TZ	Upper Limit of TZ	15-Second TZ (exact)	15-Second TZ (rounded)
20					
30					
40					
50					
60					

4.6 ▶▶▶ Exercises

Simplify. See Example 1.

1. $\left(-\dfrac{3}{4}\right)^2$

2. $\left(-\dfrac{4}{5}\right)^2$

3. $\left(\dfrac{2}{5}\right)^3$

4. $\left(\dfrac{1}{4}\right)^3$

5. $\left(-\dfrac{1}{3}\right)^3$

6. $\left(-\dfrac{3}{5}\right)^3$

7. $\left(\dfrac{1}{2}\right)^5$

8. $\left(\dfrac{1}{3}\right)^4$

9. $\left(\dfrac{7}{10}\right)^2$

10. $\left(\dfrac{8}{9}\right)^2$

11. $\left(-\dfrac{6}{5}\right)^2$

12. $\left(-\dfrac{8}{7}\right)^2$

13. $\dfrac{15}{16}\left(\dfrac{4}{5}\right)^3$

14. $-8\left(-\dfrac{3}{8}\right)^2$

15. $\left(\dfrac{1}{3}\right)^4\left(\dfrac{9}{10}\right)^2$

16. $\left(\dfrac{4}{5}\right)^2\left(\dfrac{1}{2}\right)^6$

17. $\left(-\dfrac{3}{2}\right)^3\left(-\dfrac{2}{3}\right)^2$

18. $\left(\dfrac{5}{6}\right)^2\left(-\dfrac{2}{5}\right)^3$

Relating Concepts (Exercises 19–20) For Individual or Group Work

Use your knowledge of exponents as you **work Exercises 19 and 20 in order.**

19. (a) Evaluate this series of examples.

$$\left(-\frac{1}{2}\right)^2 = \underline{\hspace{1cm}} \qquad \left(-\frac{1}{2}\right)^6 = \underline{\hspace{1cm}}$$

$$\left(-\frac{1}{2}\right)^3 = \underline{\hspace{1cm}} \qquad \left(-\frac{1}{2}\right)^7 = \underline{\hspace{1cm}}$$

$$\left(-\frac{1}{2}\right)^4 = \underline{\hspace{1cm}} \qquad \left(-\frac{1}{2}\right)^8 = \underline{\hspace{1cm}}$$

$$\left(-\frac{1}{2}\right)^5 = \underline{\hspace{1cm}} \qquad \left(-\frac{1}{2}\right)^9 = \underline{\hspace{1cm}}$$

20. Several drops of ketchup fell on Ron's homework. Explain how he can figure out what real number is covered by each drop. Be careful. More than one number may work, or there may not be any real number that works.

(a) $\left(\blacksquare\right)^2 = \dfrac{4}{9}$ (b) $\left(\bullet\right)^3 = -\dfrac{1}{27}$

(c) $\left(\bullet\right)^4 = \dfrac{1}{16}$ (d) $\left(\blacklozenge\right)^2 = -\dfrac{9}{16}$

(e) $\left(\blacklozenge\right)^2\left(\blacksquare\right)^2 = \dfrac{1}{36}$

(b) Explain the pattern in the sign of the answers.

Simplify. See Example 2.

21. $\dfrac{1}{5} - 6\left(\dfrac{7}{10}\right)$

22. $\dfrac{2}{9} - 4\left(\dfrac{5}{6}\right)$

23. $\left(\dfrac{4}{3} \div \dfrac{8}{3}\right) + \left(-\dfrac{3}{4} \cdot \dfrac{1}{4}\right)$

24. $\left(-\dfrac{1}{3} \cdot \dfrac{3}{5}\right) + \left(\dfrac{3}{4} \div \dfrac{1}{4}\right)$

25. $-\dfrac{3}{10} \div \dfrac{3}{5}\left(-\dfrac{2}{3}\right)$

26. $5 \div \left(-\dfrac{10}{3}\right)\left(-\dfrac{4}{9}\right)$

27. $\dfrac{8}{3}\left(\dfrac{1}{4}-\dfrac{1}{2}\right)^2$

28. $\dfrac{1}{3}\left(\dfrac{4}{5}-\dfrac{3}{10}\right)^3$

29. $-\dfrac{3}{8}+\dfrac{2}{3}\left(-\dfrac{2}{3}+\dfrac{1}{6}\right)$

30. $\dfrac{1}{6}+4\left(\dfrac{2}{5}-\dfrac{7}{10}\right)$

31. $2\left(\dfrac{1}{3}\right)^3-\dfrac{2}{9}$

32. $8\left(-\dfrac{3}{4}\right)^2+\dfrac{3}{2}$

33. $\left(-\dfrac{2}{3}\right)^3\left(\dfrac{1}{8}-\dfrac{1}{2}\right)-\dfrac{2}{3}\left(\dfrac{1}{8}\right)$

34. $\left(\dfrac{3}{5}\right)^2\left(\dfrac{5}{9}-\dfrac{2}{3}\right)\div\left(-\dfrac{1}{5}\right)^2$

35. A square operation key on a calculator is $\frac{3}{8}$ inch on each side. What is the area of the key? Use the formula $A = s^2$. (*Source:* Texas Instruments.)

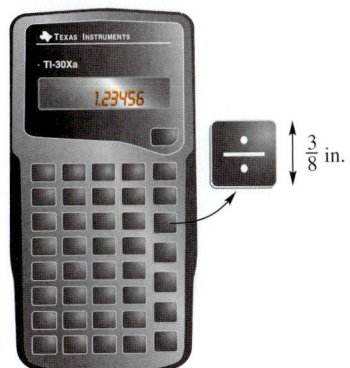 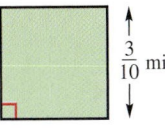

36. A square lot for sale in the country is $\frac{3}{10}$ mile on a side. Find the area of the lot by using the formula $A = s^2$.

$\frac{3}{10}$ mi

37. A rectangular parking lot at the megamall is $\frac{7}{10}$ mile long and $\frac{1}{4}$ mile wide. How much fencing is needed to enclose the lot? Use the formula $P = 2l + 2w$.

38. A computer chip in a rectangular shape is $\frac{7}{8}$ inch long and $\frac{5}{16}$ inch wide. An insulating strip must be put around all sides of the chip. Find the length of the strip. Use the formula $P = 2l + 2w$.

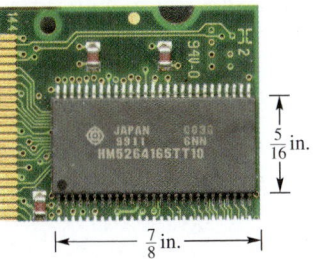

$\frac{5}{16}$ in.

$\frac{7}{8}$ in.

Simplify. See Example 3.

39. $\dfrac{-\dfrac{7}{9}}{-\dfrac{7}{36}}$

40. $\dfrac{\dfrac{15}{32}}{-\dfrac{5}{64}}$

41. $\dfrac{-15}{\dfrac{6}{5}}$

42. $\dfrac{-6}{-\dfrac{5}{8}}$

43. $\dfrac{\dfrac{4}{7}}{8}$

44. $\dfrac{-\dfrac{11}{5}}{3}$

45. $\dfrac{-\dfrac{2}{3}}{-2\dfrac{2}{5}}$

46. $\dfrac{\dfrac{1}{2}}{3\dfrac{1}{3}}$

47. $\dfrac{-4\dfrac{1}{2}}{\left(\dfrac{3}{4}\right)^2}$

48. $\dfrac{1\dfrac{2}{3}}{\left(-\dfrac{1}{6}\right)^2}$

49. $\dfrac{\left(\dfrac{2}{5}\right)^2}{\left(-\dfrac{4}{3}\right)^2}$

50. $\dfrac{\left(-\dfrac{5}{6}\right)^2}{\left(\dfrac{1}{2}\right)^3}$

4.7 ▶▶▶ Problem Solving: Equations Containing Fractions

OBJECTIVE 1   Use the multiplication property of equality to solve equations containing fractions. In **Section 2.4** you used the division property of equality to solve an equation such as $4s = 24$. The division property says that you may divide both sides of an equation by the same nonzero number and the equation will still be balanced. Now that you have some experience with fractions, let's look again at how the division property works.

$$4s \text{ means } 4 \cdot s \quad \begin{array}{l} \longrightarrow 4s = 24 \\ \longrightarrow \dfrac{4 \cdot s}{4} = \dfrac{24}{4} \end{array} \quad \text{Divide both sides by 4}$$

$$\text{Divide out the common factor of 4} \quad \dfrac{\overset{1}{\cancel{4}} \cdot s}{\underset{1}{\cancel{4}}} = 6$$

$$s = 6$$

Because multiplication and division are related to each other, we can also *multiply* both sides of an equation by the same nonzero number and keep it balanced.

> **Multiplication and Division Properties of Equality**
>
> The **multiplication property of equality** says, if $a = b$, then $a \cdot c = b \cdot c$
>
> The **division property of equality** says, if $a = b$, then $\dfrac{a}{c} = \dfrac{b}{c}$ as long as
>
> c is not 0. In other words, you may multiply or divide both sides of an equation by the same nonzero number and it will still be balanced.

EXAMPLE 1   Using the Multiplication Property of Equality

Solve each equation and check each solution.

(a) $\dfrac{1}{2}b = 5$

As in **Chapter 2**, you want the variable by itself on one side of the equal sign. In this example, you want $1b$, not $\frac{1}{2}b$, on the left side. (Recall that $1b$ is equivalent to b.)

In **Section 4.3** you learned that the product of a number and its reciprocal is 1. Thus, multiplying $\frac{1}{2}$ by $\frac{2}{1}$ will give the desired coefficient of 1.

$$\dfrac{1}{2}b = 5$$

$$\dfrac{2}{1}\left(\dfrac{1}{2}b\right) = \dfrac{2}{1}(5) \quad \text{Multiply both sides by } \tfrac{2}{1} \text{ (the reciprocal of } \tfrac{1}{2}\text{).}$$

On the left side, use the associative property to regroup the factors.
$$\left(\dfrac{2}{1} \cdot \dfrac{1}{2}\right)b = \dfrac{2}{1}\left(\dfrac{5}{1}\right) \quad \text{On the right side, 5 is equivalent to } \tfrac{5}{1}$$

$$\left(\dfrac{\overset{1}{\cancel{2}}}{1} \cdot \dfrac{1}{\underset{1}{\cancel{2}}}\right)b = \dfrac{10}{1}$$

$1b$ is equivalent to b.
$$\begin{array}{l} \longrightarrow 1b = 10 \\ \longrightarrow b = 10 \leftarrow \text{ The solution is 10.} \end{array}$$

Continued on Next Page

OBJECTIVES

1 Use the multiplication property of equality to solve equations containing fractions.

2 Use both the addition and multiplication properties of equality to solve equations containing fractions.

3 Solve application problems using equations containing fractions.

Once you understand the process, you don't have to show every step. Here is a shorthand solution of the same problem.

$$\frac{1}{2}b = 5$$

$$\frac{\overset{1}{\cancel{2}}}{1}\left(\frac{1}{\underset{1}{\cancel{2}}}b\right) = \frac{2}{1}\left(\frac{5}{1}\right)$$

$$b = 10 \quad \leftarrow \text{Solution}$$

The solution is 10. Check the solution by going back to the *original* equation.

Check $\quad \dfrac{1}{2}\,b = 5 \qquad$ Replace b with 10 in the *original* equation.

$$\frac{1}{2}(\mathbf{10}) = 5 \qquad \text{Multiply on the left side: } \tfrac{1}{2}(10) \text{ is}$$
$$\tfrac{1}{2} \cdot \tfrac{10}{1} \quad \text{or} \quad \tfrac{1}{2} \cdot \tfrac{2\,\cdot\,5}{1}$$

$$\frac{1 \cdot \overset{1}{\cancel{2}} \cdot 5}{\underset{1}{\cancel{2}} \cdot 1} = 5$$

$$5 = 5 \qquad \text{Balances}$$

When b is 10, the equation balances, so 10 is the correct solution (***not*** 5).

(b) $12 = -\dfrac{3}{4}x$

$$12 = -\frac{3}{4}x \qquad \begin{array}{l}\text{Multiply both sides by } -\tfrac{4}{3} \text{ (the reciprocal of } -\tfrac{3}{4}\text{).} \\ \text{The reciprocal of a negative number} \\ \text{is also negative.}\end{array}$$

$$-\frac{4}{3}(12) = -\frac{\overset{1}{\cancel{4}}}{\cancel{3}}\left(-\frac{\overset{1}{\cancel{3}}}{\cancel{4}}x\right) \qquad \begin{array}{l}\text{On the left side } -\tfrac{4}{3}(12) \text{ is } -\tfrac{4}{3} \cdot \tfrac{12}{1} \\ \text{or} \quad -\frac{4 \cdot \overset{4}{\cancel{3}} \cdot 4}{\underset{1}{\cancel{3}} \cdot 1} = -16\end{array}$$

$$-16 = x$$

The solution is -16. Check the solution by going back to the *original* equation.

Check $\quad 12 = -\dfrac{3}{4}x \qquad$ Replace x with -16 in the *original* equation.

$$12 = -\frac{3}{4}(\mathbf{-16}) \qquad \begin{array}{l}\text{The product of two negative} \\ \text{numbers is positive.}\end{array}$$
$$-\tfrac{3}{4}(-16) \text{ is } \tfrac{3}{4} \cdot \tfrac{16}{1} \quad \text{or} \quad \tfrac{3}{4} \cdot \tfrac{4\,\cdot\,4}{1}$$

$$12 = \frac{3 \cdot \overset{1}{\cancel{4}} \cdot 4}{\underset{1}{\cancel{4}} \cdot 1}$$

$$12 = 12 \qquad \text{Balances} \quad \boxed{\text{If the equation does \textbf{not} balance, your solution is wrong. Start over.}}$$

When x is -16, the equation balances, so -16 is the correct solution (***not*** 12).

Continued on Next Page

(c) $-\dfrac{2}{5}n = -\dfrac{1}{3}$

$$-\dfrac{\cancel{5}^{1}}{\cancel{2}_{1}}\left(-\dfrac{\cancel{2}^{1}}{\cancel{5}_{1}}n\right) = \left(-\dfrac{5}{2}\right)\left(-\dfrac{1}{3}\right)$$

Multiply both sides by $-\dfrac{5}{2}$
(the reciprocal of $-\dfrac{2}{5}$).

$$n = \dfrac{5 \cdot 1}{2 \cdot 3}$$

The product of two negative numbers is positive.

$$n = \dfrac{5}{6} \longleftarrow \text{The solution is } \tfrac{5}{6}.$$

Check

$$-\dfrac{2}{5}n = -\dfrac{1}{3} \qquad \text{Original equation}$$

$$\left(-\dfrac{2}{5}\right)\left(\dfrac{5}{6}\right) = -\dfrac{1}{3} \qquad \text{Replace } n \text{ with } \tfrac{5}{6}$$

$$-\dfrac{\cancel{2}^{1} \cdot \cancel{5}^{1}}{\cancel{5}_{1} \cdot \cancel{2}_{1} \cdot 3} = -\dfrac{1}{3} \qquad \text{Multiply on the left side.}$$

$$-\dfrac{1}{3} = -\dfrac{1}{3} \qquad \text{Balances}$$

When n is $\tfrac{5}{6}$, the equation balances, so $\tfrac{5}{6}$ is the correct solution (***not*** $-\tfrac{1}{3}$).

Work Problem **1** *at the Side.* ▶

OBJECTIVE **2** **Use both the addition and multiplication properties of equality to solve equations containing fractions.** In **Section 2.5** you used both the addition and *division* properties of equality to solve equations. Now you can use both the addition and *multiplication* properties.

EXAMPLE 2 **Using the Addition and Multiplication Properties of Equality**

Solve each equation and check each solution.

(a) $\dfrac{1}{3}c + 5 = 7$

The first step is to get the variable term $\tfrac{1}{3}c$ by itself on the left side of the equal sign. Recall that to "get rid of" the 5 on the left side, add the opposite of 5, which is -5, to both sides.

$$\dfrac{1}{3}c + 5 = 7$$

$$\underline{\qquad -5 \quad -5} \qquad \text{Add } -5 \text{ to both sides.}$$

$$\dfrac{1}{3}c + 0 = 2$$

$$\dfrac{1}{3}c = 2$$

$$\dfrac{\cancel{3}^{1}}{1}\left(\dfrac{1}{\cancel{3}_{1}}c\right) = \dfrac{3}{1}\left(\dfrac{2}{1}\right) \qquad \begin{array}{l}\text{Multiply both sides by } \tfrac{3}{1}\\ \text{(the reciprocal of } \tfrac{1}{3}).\end{array}$$

$$c = 6$$

The solution is 6. Check the solution by going back to the *original* equation.

Continued on Next Page

---

1 Solve each equation. Check each solution.

(a) $\dfrac{1}{6}m = 3$

Check

(b) $\dfrac{3}{2}a = -9$

Check

(c) $\dfrac{3}{14} = -\dfrac{2}{7}x$

Check

ANSWERS

1. (a) $m = 18$ **Check** $\dfrac{1}{6}m = 3$
$$\dfrac{1}{6}(18) = 3$$
Balances $3 = 3$

(b) $a = -6$ **Check** $\dfrac{3}{2}a = -9$
$$\dfrac{3}{2}(-6) = -9$$
Balances $-9 = -9$

(c) $x = -\dfrac{3}{4}$ **Check** $\dfrac{3}{14} = -\dfrac{2}{7}x$
$$\dfrac{3}{14} = -\dfrac{2}{7}\left(-\dfrac{3}{4}\right)$$
Balances $\dfrac{3}{14} = \dfrac{3}{14}$

2 Solve each equation. Check each solution.

(a) $18 = \frac{4}{5}x + 2$

Check

(b) $\frac{1}{4}h - 5 = 1$

Check

(c) $\frac{4}{3}r + 4 = -8$

Check

Check $\quad \frac{1}{3}c + 5 = 7 \qquad$ Original equation

$\quad \frac{1}{3}(6) + 5 = 7 \qquad$ Replace c with 6.

$\quad \underbrace{2} + 5 = 7$

$\quad 7 = 7 \qquad$ Balances

When c is 6, the equation balances, so 6 is the correct solution (**not** 7).

(b) $-3 = \frac{2}{3}y + 7$

To get the variable term $\frac{2}{3}y$ by itself on the right side, add -7 to both sides.

$$-3 = \frac{2}{3}y + 7$$

$$\underline{-7} \qquad \underline{-7} \qquad \text{Add } -7 \text{ to both sides.}$$

$$-10 = \frac{2}{3}y + 0$$

$$\frac{3}{2}(-10) = \frac{\cancel{3}}{\cancel{2}}\left(\frac{\cancel{2}}{\cancel{3}}y\right) \qquad \begin{array}{l}\text{Multiply both sides by } \frac{3}{2} \\ \text{(the reciprocal of } \frac{2}{3}\text{).}\end{array}$$

$$-15 = y \qquad \longleftarrow \quad \text{The solution is } -15.$$

Check $\quad -3 = \frac{2}{3}y + 7 \qquad$ Original equation

$\quad -3 = \frac{2}{3}(-15) + 7 \qquad$ Replace y with -15

$\quad -3 = \underbrace{-10} + 7$

$\quad -3 = -3 \qquad$ Balances

When y is -15, the equation balances, so -15 is the correct solution (**not** -3).

◀ *Work Problem* **2** *at the Side.*

ANSWERS

2. (a) $x = 20$ **Check** $18 = \frac{4}{5}x + 2$

$18 = \frac{4}{5}(20) + 2$

$18 = 16 + 2$

Balances $18 = 18$

(b) $h = 24$ **Check** $\frac{1}{4}h - 5 = 1$

$\frac{1}{4}(24) - 5 = 1$

$6 - 5 = 1$

Balances $\quad 1 = 1$

(c) $r = -9$ **Check** $\frac{4}{3}r + 4 = -8$

$\frac{4}{3}(-9) + 4 = -8$

$-12 + 4 = -8$

Balances $\quad -8 = -8$

OBJECTIVE **3** **Solve application problems using equations containing fractions.** Use the six problem-solving steps from **Section 3.3** to solve application problems.

EXAMPLE 3 **Solving an Application Problem Using an Equation with Fractions**

The expression for finding a person's approximate systolic blood pressure is $100 + \dfrac{age}{2}$. Suppose your friend's systolic blood pressure is 116 (and he has normal blood pressure). Find his age.

Step 1 **Read** the problem. It is about blood pressure and age.

Unknown: friend's age

Known: Blood pressure expression is $100 + \dfrac{age}{2}$; friend's pressure is 116.

Step 2 **Assign a variable.** Let a represent the friend's age.

Step 3 **Write an equation.**

$$100 + \frac{\textbf{age}}{2} \text{ is blood pressure}$$

$$100 + \frac{a}{2} = 116$$

Step 4 **Solve.**

$$100 + \frac{a}{2} = 116$$

$$\underline{-100 \qquad\qquad -100} \qquad \text{Add } -100 \text{ to both sides.}$$

$$0 + \frac{a}{2} = 16$$

$$\frac{a}{2} = 16$$

$\frac{1}{2}a$ is equivalent to $\frac{a}{2}$ because $\frac{1}{2}a$ is $\frac{1}{2} \cdot \frac{a}{1}$

$$\frac{1}{2}a = 16$$

$$\overset{1}{\cancel{\frac{2}{1}}}\left(\overset{1}{\cancel{\frac{1}{2}}}a\right) = \frac{2}{1}(16) \qquad \begin{array}{l}\text{Multiply both sides by } \frac{2}{1}\\ \text{(the reciprocal of } \frac{1}{2}\text{).}\end{array}$$

$$a = 32$$

Step 5 **State the answer.** Your friend is 32 years old.

Step 6 **Check** the solution by putting it back into the *original* problem.

Approximate systolic blood pressure is $100 + \dfrac{age}{2}$.

If the age is 32, then $100 + \frac{32}{2} = 100 + 16 = 116$.

Friend's blood pressure is 116. ← Matches ——

> Your answer must fit all the information in the original problem.

Age 32 is the correct solution because it "works" when you put it back into the original problem.

Work Problem **3** *at the Side.* ▶

3 A woman's systolic blood pressure is 111. Find her age, using the expression for systolic blood pressure from Example 3 and the six problem-solving steps. (Assume that the woman has normal blood pressure.)

ANSWER

3. Age is a.

$100 + \dfrac{a}{2} = 111$ The woman is 22 years old.

Check $100 + \dfrac{22}{2} = 100 + 11 = 111$

Math in the Media

HOTEL EXPENSES

Mathematics teachers attending a national conference in Salt Lake City, Utah, in April, 2008 found the following information about hotel rates on their organization's Web site. The rates do not include taxes, so ignore taxes when answering the problems.

Hotel (Salt Lake City)	Single	Double	Triple	Quad
Grand America (executive suite rooms)	$208	$219	$235	$248
Embassy Suites	$146	$166	$181	$196
Little America (Tower rooms)	$172	$172	$182	$202
Super 8 Airport	$105	$115	$125	$135

Source: www.nctm.org

1. **(a)** The double rate is for two people sharing a room. What fractional part does each person pay?

   (b) **Multiply** the double rate at the Super 8 Airport by $\frac{1}{2}$. What is the result?

   (c) **Divide** the double rate at the Super 8 Airport by 2. What is the result?

   (d) Explain what happened. How much money would one person owe if he or she shared a double room at the Super 8 Airport hotel?

2. **(a)** The triple rate is for three people sharing a room. What fractional part does each person pay?

   (b) How much money would one person owe if he or she shared a triple room at the Little America Hotel? Find your answer using *two different methods,* based on your observations in Problem 1.

3. **(a)** Suppose you have a travel allotment of $800 that can be spent on transportation, hotel, food, and registration fees. You and a colleague decide to attend the 3-day conference in Salt Lake City and plan to share a room for three nights at the Grand America. Conference registration is $225; the flight costs $334 round-trip; the shuttle between the airport and the hotel costs $20 per person each way; and you budget $35 per day for meals. How much out-of-pocket expense will you have to pay?

   (b) How much would you save if you could recruit a third person to share the room?

294

4.7 ▶▶▶ **Exercises**

FOR
EXTRA
HELP
PRACTICE   WATCH   DOWNLOAD   READ   REVIEW

Solve each equation and check each solution. See Examples 1 and 2.

1. $\frac{1}{3}a = 10$ **Check**

2. $7 = \frac{1}{5}y$ **Check**

3. $-20 = \frac{5}{6}b$ **Check**

4. $-\frac{4}{9}w = 16$ **Check**

 5. $-\frac{7}{2}c = -21$ **Check**

6. $-25 = \frac{5}{3}x$ **Check**

7. $\frac{9}{16} = \frac{3}{4}m$ **Check**

8. $\frac{5}{12}k = \frac{15}{16}$ **Check**

9. $\frac{3}{10} = -\frac{1}{4}d$ **Check**

10. $-\frac{7}{8}h = -\frac{1}{6}$ **Check**

11. $\frac{1}{6}n + 7 = 9$ **Check**

12. $3 + \frac{1}{4}p = 5$ **Check**

13. $-10 = \dfrac{5}{3}r + 5$ **Check**

14. $0 = 6 + \dfrac{3}{2}t$ **Check**

15. $\dfrac{3}{8}x - 9 = 0$ **Check**

16. $\dfrac{1}{3}s - 10 = -5$ **Check**

Solve each equation. Show your work.

17. $7 - 2 = \dfrac{1}{5}y - 4$

18. $0 - 8 = \dfrac{1}{10}k - 3$

19. $4 + \dfrac{2}{3}n = -10 + 2$

20. $-\dfrac{2}{5}m - 3 = -9 + 0$

21. $3x + \dfrac{1}{2} = \dfrac{3}{4}$

22. $4y + \dfrac{1}{3} = \dfrac{7}{9}$

23. $\dfrac{3}{10} = -4b - \dfrac{1}{5}$

24. $\dfrac{5}{6} = -3c - \dfrac{2}{3}$

25. Check the solution given for each equation. If a solution doesn't check, show how to find the correct solution.

    (a) $\dfrac{1}{6}x + 1 = -2$ **(b)** $-\dfrac{3}{2} = \dfrac{9}{4}k$

            $x = 18$ $k = -\dfrac{2}{3}$

26. Check the solution given for each equation. If a solution doesn't check, show how to find the correct solution.

    (a) $-\dfrac{3}{4}y = -\dfrac{5}{8}$ **(b)** $16 = -\dfrac{7}{3}w + 2$

            $y = \dfrac{5}{6}$ $w = 6$

27. Write two different equations that have 8 as a solution. Write your equations with a fraction as the coefficient of the variable term. Use Exercises 1–10 as models.

28. Write two different equations that have -12 as the solution. Write your equations with a fraction as the coefficient of the variable term. Use Exercises 1–10 as models.

In Exercises 29–32, find each person's age using the six problem-solving steps and this expression for approximate systolic blood pressure: $100 + \dfrac{age}{2}$. *Assume that all the people have normal blood pressure. See Example 3.*

29. A man has systolic blood pressure of 109. How old is he?

30. A man has systolic blood pressure of 118. How old is he?

31. A woman has systolic blood pressure of 122. How old is she?

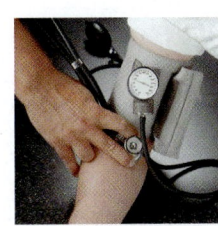

32. A woman has systolic blood pressure of 113. How old is she?

As you read at the start of this chapter, nails for home repair projects are classified by the "penny" system, which is a number that indicates the nail's length. The expression for finding the length of a nail, in inches, is $\dfrac{\text{penny size}}{4} + \dfrac{1}{2}$ inch. In Exercises 33–36, find the penny size for each nail using this expression and the six problem-solving steps. (Source: Season by Season Home Maintenance.)

33. The length of a common nail is 3 inches. What is its penny size?

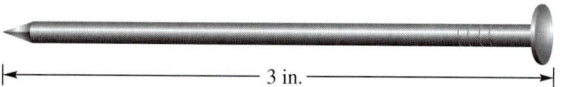

34. The length of a drywall nail is 2 inches. Find the nail's penny size.

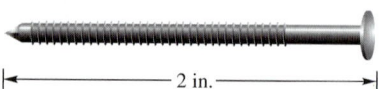

35. The length of a box nail is $2\frac{1}{2}$ inches. What penny size would you ask for when buying these nails?

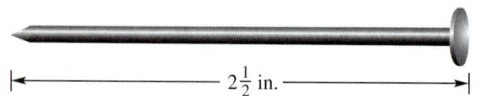

36. The length of a finishing nail is $1\frac{1}{2}$ inches. What is its penny size?

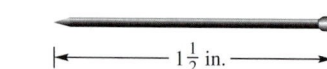

An expression for the recommended weight of an adult is $\frac{11}{2}$ (height in inches) − 220. In Exercises 37–40, find each person's height using this expression and the six problem-solving steps. Assume that all the people are at their recommended weight.

37. A man weighs 209 pounds. What is his height in inches?

38. A woman weighs 165 pounds. What is her height in inches?

39. A woman weighs 132 pounds. What is her height in inches?

40. A man weighs 176 pounds. What is his height in inches?

4.8 ▶▶▶ Geometry Applications: Area and Volume

OBJECTIVES

1 Find the area of a triangle.

2 Find the volume of a rectangular solid.

3 Find the volume of a pyramid.

OBJECTIVE 1 Find the area of a triangle. In **Section 3.1** you worked with triangles, which are flat shapes that have exactly three sides. You found the perimeter of a triangle by adding the lengths of the three sides. Now you are ready to find the area of a triangle (the amount of surface inside the triangle).

You can find the *height* of a triangle by measuring the distance from one vertex of the triangle to the opposite side (the base). The height must be *perpendicular* to the base, that is, it must form a right angle with the base. Sometimes you have to extend the base before you can draw the height perpendicular to it, as shown on the right-hand figure below.

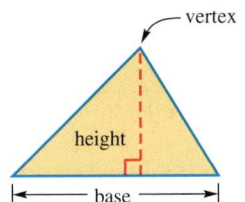 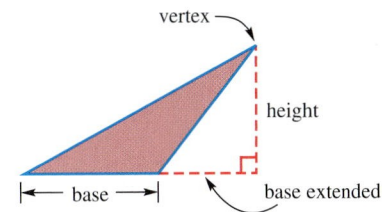

If you cut out two identical triangles and turn one upside down, you can fit them together to form a parallelogram, as shown below.

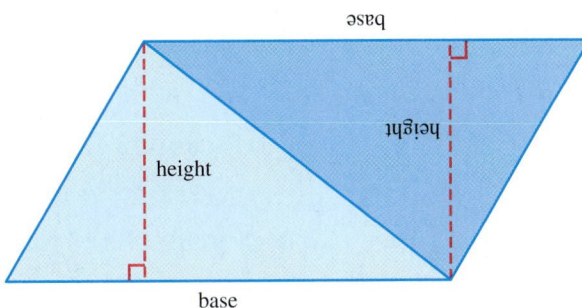

Recall from **Section 3.2** that the area of the parallelogram is *base* times *height*. Because each triangle is *half* of the parallelogram, the area of one triangle is

$$\frac{1}{2} \text{ of base times height}$$

Use the following formula to find the area of a triangle.

Finding the Area of a Triangle

$$\text{Area of triangle} = \frac{1}{2} \cdot \text{base} \cdot \text{height}$$

$$A = \frac{1}{2} bh$$

Remember to use *square units* when measuring area.

---

1 Find the area of each triangle.

(a)

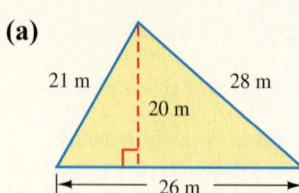

(b) |⟵ 6 yd ⟶|

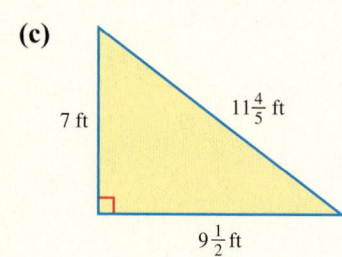

(c)

EXAMPLE 1 **Finding the Area of Triangles**

Find the area of each triangle.

(a)

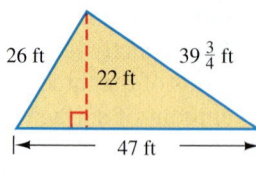

The base is 47 ft and the height is 22 ft. You do *not* need the 26 ft or $39\frac{3}{4}$ ft sides to find the area.

$$A = \frac{1}{2} \cdot \boldsymbol{b} \cdot \boldsymbol{h}$$ Replace b with 47 ft and h with 22 ft.

$$A = \frac{1}{2} \cdot \mathbf{47\ ft} \cdot \mathbf{22\ ft}$$

$$A = \frac{1}{2} \cdot \frac{47\ ft}{1} \cdot \frac{22\ ft}{1}$$ Divide out the common factor of 2

$$A = \frac{1 \cdot 47\ ft \cdot \overset{1}{\cancel{2}} \cdot 11\ ft}{\underset{1}{\cancel{2}} \cdot 1 \cdot 1}$$ Multiply $47 \cdot 11$ to get 517

This is *area*, so write **ft²** in the answer.

$$A = 517\ ft^2$$ Multiply ft • ft to get ft².

(b)

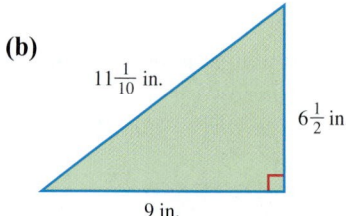

Because two sides of the triangle are perpendicular to each other, use those sides as the base and the height. (Remember that the height must be perpendicular to the base.)

$$A = \frac{1}{2}\boldsymbol{bh}$$ Formula for area of a triangle

$$A = \frac{1}{2} \cdot \mathbf{9\ in.} \cdot \mathbf{6\frac{1}{2}\ in.}$$ Replace b with 9 in. and h with $6\frac{1}{2}$ in.

$$A = \frac{1}{2} \cdot \frac{9\ in.}{1} \cdot \frac{13\ in.}{2}$$ Write 9 in. and $6\frac{1}{2}$ in. as improper fractions.

$$A = \frac{1 \cdot 9\ in. \cdot 13\ in.}{2 \cdot 1 \cdot 2}$$ ⟵ Multiply $9 \cdot 13$ to get 117 and in. • in. to get in.²

$$A = \frac{117}{4}\ in.^2 \quad \text{or} \quad 29\frac{1}{4}\ in.^2$$

◀ *Work Problem* **1** *at the Side.*

EXAMPLE 2 **Using the Concept of Area**

Find the area of the shaded part in this figure.

The *entire* figure is a rectangle.

$$A = lw$$
$$A = 30\ cm \cdot 40\ cm$$
$$A = 1200\ cm^2$$

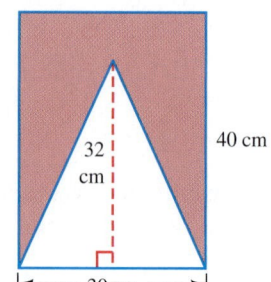

ANSWERS

1. **(a)** 260 m² **(b)** 15 yd²
   (c) $\frac{133}{4}$ ft² or $33\frac{1}{4}$ ft²

Continued on Next Page

The *un*shaded part of the figure is a triangle.

$$A = \frac{1}{2} bh$$

$$A = \frac{1}{2} \cdot \frac{30 \text{ cm}}{1} \cdot \frac{32 \text{ cm}}{1}$$

$$A = \frac{1 \cdot \overset{1}{2} \cdot 15 \text{ cm} \cdot 32 \text{ cm}}{\underset{1}{2} \cdot 1 \cdot 1}$$

$$A = 480 \text{ cm}^2$$

Subtract to find the area of the shaded part.

Entire area Unshaded part Shaded part

$$A = \overbrace{1200 \text{ cm}^2} - \overbrace{480 \text{ cm}^2} = \overbrace{720 \text{ cm}^2}$$

> This is *area*, so write **cm²** in the answer.

Work Problem **2** *at the Side.* ▶

OBJECTIVE **2** **Find the volume of a rectangular solid.** A shoe box and a cereal box are examples of three-dimensional (or solid) figures. The three dimensions are length, width, and height. (A rectangle or square is a two-dimensional figure. The two dimensions are length and width.) If we want to know how much a shoe box will hold, we find its *volume*. We measure volume by seeing how many cubes of a certain size will fill the space inside the box. Three sizes of *cubic units* are shown here. Notice that all the edges of a cube have the same length and all sides meet at right angles.

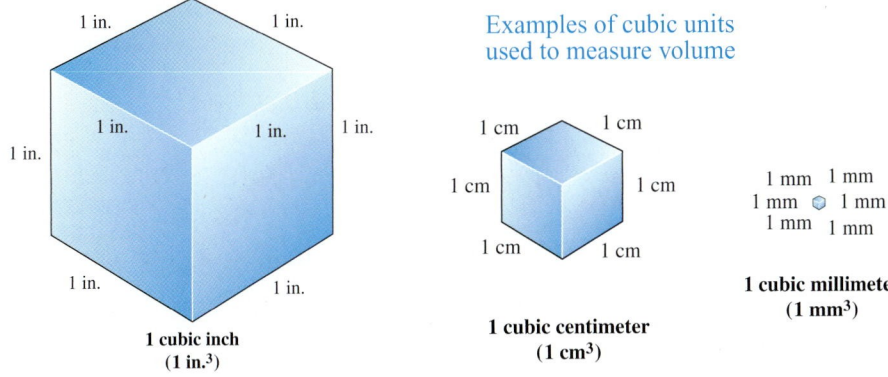

Examples of cubic units used to measure volume

1 cubic inch (1 in.³)

1 cubic centimeter (1 cm³)

1 cubic millimeter (1 mm³)

Some other sizes of cubes that are used to measure volume are 1 cubic foot (1 ft³), 1 cubic yard (1 yd³), and 1 cubic meter (1 m³).

CAUTION
The raised 3 in 4^3 means that you multiply $4 \cdot 4 \cdot 4$ to get 64. The raised 3 in cm³ or ft³ is a short way to write the word "cubic." It means that you multiplied cm times cm times cm to get cm³, or ft times ft times ft to get ft³. Recall that a short way to write $x \cdot x \cdot x$ is x^3. Similarly, cm $\cdot$ cm $\cdot$ cm is cm³. When you see 5 cm³, say "five cubic centimeters." Do *not* multiply $5 \cdot 5 \cdot 5$ because the exponent applies only to the *first* thing to its left. The exponent applies to cm, *not* to the 5.

Volume

Volume is a measure of the space inside a solid shape. The volume of a solid is the number of cubic units it takes to fill the solid.

2 Find the area of the shaded part in this figure.

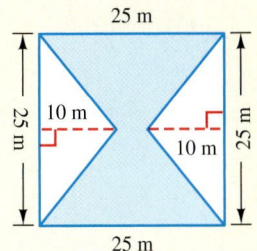

3 Find the volume of each rectangular solid.

(a)

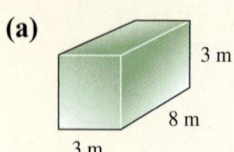

3 m
8 m
3 m

(b) Length $6\frac{1}{4}$ ft, width $3\frac{1}{2}$ ft, height 2 ft

Use the formula below to find the volume of *rectangular solids* (box-like shapes).

> **Finding the Volume of Rectangular Solids**
>
> Volume of a rectangular solid = length • width • height
> $$V = lwh$$
> Remember to use *cubic units* when measuring volume.

EXAMPLE 3 **Finding the Volume of Rectangular Solids**

Find the volume of each box.

(a)

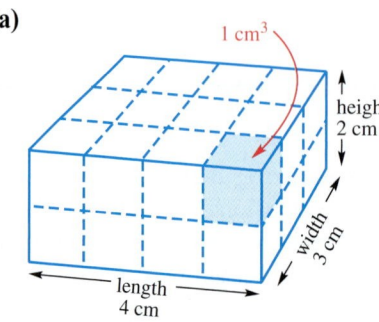

1 cm³
height 2 cm
width 3 cm
length 4 cm

Each cube that fits in the box is 1 cubic centimeter (1 cm³). To find the volume, you can count the number of cubes.

Bottom layer has 12 cubes. ⎫
Top layer has 12 cubes. ⎭ Total of 24 cubes (24 cm³)

Or you can use the formula for rectangular solids.

$$V = \quad l \quad • \quad w \quad • \quad h$$
$$V = \textbf{4 cm} • \textbf{3 cm} • \textbf{2 cm} \qquad \text{Multiply } 4 • 3 • 2 \text{ to get } 24$$
$$V = 24 \text{ cm}^3 \qquad \text{Multiply cm • cm • cm to get cm}^3.$$

(b)

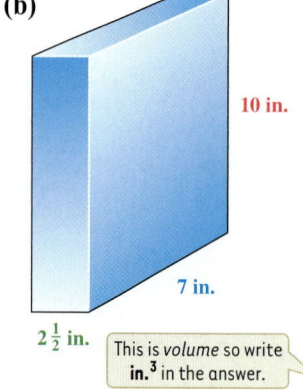

10 in.
7 in.
$2\frac{1}{2}$ in.

This is *volume* so write **in.³** in the answer.

Use the formula $V = lwh$.

$$V = 7 \text{ in.} • 2\frac{1}{2} \text{ in.} • 10 \text{ in.} \qquad \begin{array}{l}\text{Write each measurement as an improper fraction.}\end{array}$$

$$V = \frac{7 \text{ in.}}{1} • \frac{5 \text{ in.}}{2} • \frac{10 \text{ in.}}{1}$$

$$V = \frac{7 \text{ in.} • 5 \text{ in.} • \overset{1}{2} • 5 \text{ in.}}{1 • \underset{1}{2} • 1} \qquad \begin{array}{l}\text{Divide out the common factor of 2}\end{array}$$

$$V = 175 \text{ in.}^3 \qquad \text{Cubic units for volume}$$

◄ *Work Problem* **3** *at the Side.*

OBJECTIVE **3** **Find the volume of a pyramid.** A pyramid is a solid shape like the one shown below. We will study pyramids with square or rectangular bases.

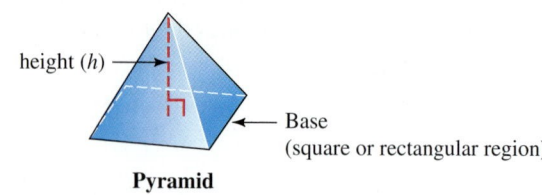

height (h)
Base (square or rectangular region)

Pyramid

The height is the distance from the base to the highest point of the pyramid. The height must be perpendicular to the base.

Note

In this book we will work only with pyramids that have a base with four sides that is square or rectangular. In later math courses you may work with pyramids that have a base with three sides (triangle), five sides (pentagon), six sides (hexagon), and so on.

Use this formula to find the volume of a pyramid.

Finding the Volume of a Pyramid

$$\text{Volume of a pyramid} = \frac{1}{3} \cdot B \cdot h$$

$$V = \frac{1}{3} Bh$$

where B is the area of the base of the pyramid and h is the height of the pyramid.

Remember to use *cubic units* when measuring volume.

EXAMPLE 4 **Finding the Volume of a Pyramid**

Find the volume of this pyramid with a rectangular base.

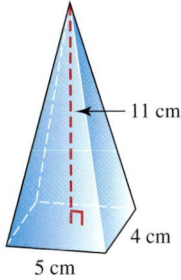

11 cm

4 cm

5 cm

First find the value of B in the formula, which is the *area of the rectangular base*. Recall that the area of a rectangle is found by multiplying length times width.

$$B = 5 \text{ cm} \cdot 4 \text{ cm}$$
$$\mathbf{B = 20 \text{ cm}^2}$$

Now find the volume.

$$V = \frac{1}{3} \boldsymbol{Bh} \qquad \text{Formula for volume of pyramid}$$

$$V = \frac{1}{3} \cdot \mathbf{20 \text{ cm}^2 \cdot 11 \text{ cm}} \qquad \text{Replace } B \text{ with 20 cm}^2 \text{ and } h \text{ with 11 cm.}$$

$$V = \frac{1}{3} \cdot \frac{20 \text{ cm}^2}{1} \cdot \frac{11 \text{ cm}}{1} \qquad \text{There are no common factors to divide out.}$$

$$V = \frac{1 \cdot 20 \text{ cm}^2 \cdot 11 \text{ cm}}{3 \cdot 1 \cdot 1}$$

> This is *volume*, so write **cm³** in the answer.

$$V = \frac{220}{3} \text{ cm}^3 \quad \text{or} \quad 73\frac{1}{3} \text{ cm}^3 \qquad \text{Cubic units for volume}$$

Work Problem **4** *at the Side.* ▶

4 Find the volume of a pyramid with a square base measuring 10 ft by 10 ft. The height of the pyramid is 6 ft.

Math in the Media

QUILT PATTERNS

People who make quilts often base their designs on a block that is cut into a grid of 4, 9, 16, or 25 squares. The quilter chooses various colors for the pieces. Each quilt design shown below was selected from an Archive of American Quilt Designs.

Source: Patchwork Persuasion: Quilts from Traditional Designs.

1. Identify the makeup of the block as 4, 9, 16, etc. Each color is what fractional part of the block?

(a)

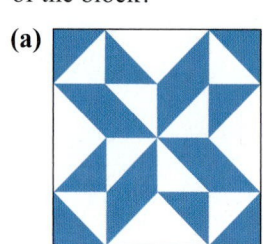

Barbara Fritchie Star

Block size: ____

Blue: ____ White: ____

(b)

Handy Andy

Block size: ____

Blue: ____ White: ____

(c)

Peace and Plenty

Block size: ____

Blue: ____ Yellow: ____

Whitish: ____

(d)

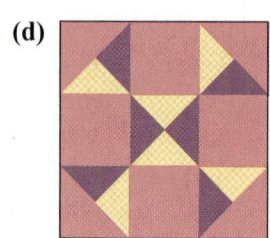

Cobwebs

Block size: ____

Mauve: ____ Purple: ____

Yellow: ____

2. Use the empty block to design and color your own quilt pattern. Tell the fractional part of the block that is represented by each color.

4.8 ▶▶▶ **Exercises**

FOR EXTRA HELP

MyMathLab Math XL PRACTICE WATCH DOWNLOAD READ REVIEW

Find the perimeter and area of each triangle. See Example 1.

1.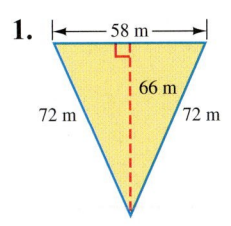

2.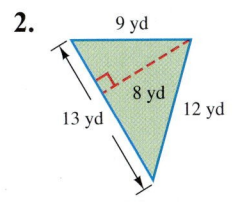

🌐 **3.**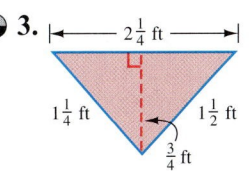

4.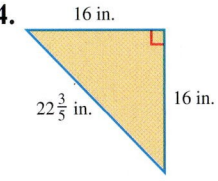

5.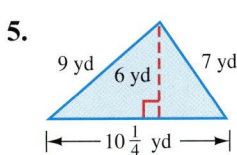

6.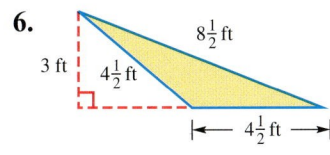

7.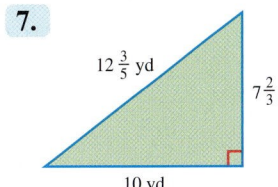

8.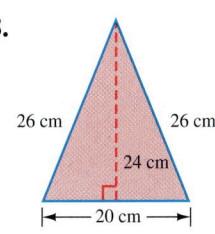

Find the shaded area in each figure. See Example 2.

9.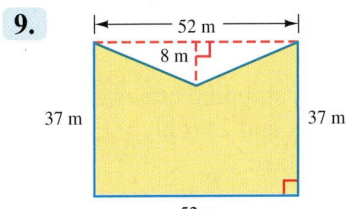

10.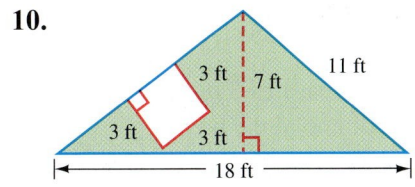

Solve each application problem.

11. A triangular tent flap measures $3\frac{1}{2}$ ft along the base and has a height of $4\frac{1}{2}$ ft. How much canvas is needed to make the flap?

12. A wooden sign in the shape of a right triangle has perpendicular sides measuring $1\frac{1}{2}$ yd and 1 yd. How much surface area does the front of the sign have?

13. A triangular space between three streets has the measurements shown below. How much new curbing is needed to go around the space? How much sod is needed to cover the space?

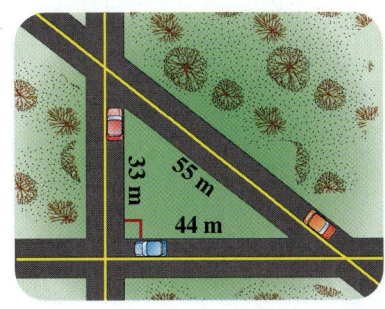

14. A city lot with an unusual shape is shown below.
  (a) How much frontage (distance along streets) does the lot have?
  (b) What is the area of the lot?

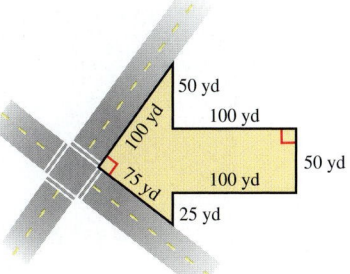

In Exercises 15–20, name each solid and find its volume. See Examples 3 and 4.

15.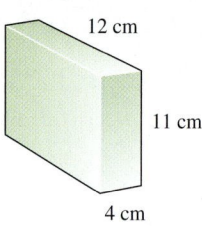
12 cm
11 cm
4 cm

16.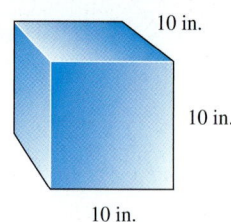
10 in.
10 in.
10 in.

17.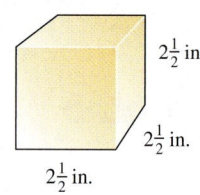
$2\frac{1}{2}$ in.
$2\frac{1}{2}$ in.
$2\frac{1}{2}$ in.

18.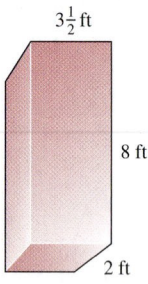
$3\frac{1}{2}$ ft
8 ft
2 ft

19.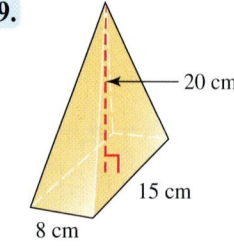
20 cm
15 cm
8 cm

20.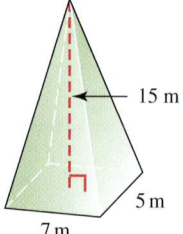
15 m
5 m
7 m

21. Find the volume of a pyramid with a height of 5 ft and a square base that is 8 ft on each side.

22. A pyramid has a height of 4 yd. The rectangular base measures 10 yd by 4 yd. What is the volume of the pyramid?

23. A box to hold pencils measures 3 in. by 8 in. by $\frac{3}{4}$ in. high. Find the volume of the box. (*Source:* Faber Castell.)

No. 2 Pencils
One Dozen
$\frac{3}{4}$ in.
8 in.
3 in.

24. A train is being loaded with shipping crates. Each crate is 12 ft long, 8 ft wide, and $2\frac{1}{4}$ ft high. How much space will each crate take?

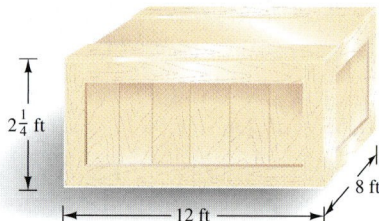
$2\frac{1}{4}$ ft
8 ft
12 ft

25. One of the ancient stone pyramids in Egypt has a square base that measures 145 m on each side. The height is 93 m. What is the volume of the pyramid? (*Source: Columbia Encyclopedia.*)

26. A cardboard model of an ancient stone pyramid has a square base that is $10\frac{3}{8}$ in. on each side. The height is $6\frac{1}{2}$ in. Find the volume of the model.

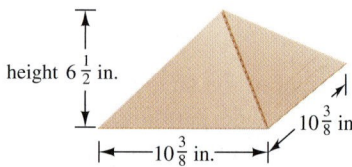
height $6\frac{1}{2}$ in.
$10\frac{3}{8}$ in
$10\frac{3}{8}$ in.

Chapter 4 ▶▶▶ Summary

▶ **Key Terms**

4.1 **fraction**   A fraction is a number of the form $\frac{a}{b}$, where a and b are integers and b is not 0.

     numerator   The top number in a fraction is the numerator. It shows how many of the equal parts are being considered.

     denominator   The bottom number in a fraction is the denominator. It shows how many equal parts are in the whole.

     proper fraction   In a proper fraction, the numerator is less than the denominator. The fraction is less than 1.

     improper fraction   In an improper fraction, the numerator is greater than or equal to the denominator. The fraction is greater than or equal to 1.

     equivalent fractions   Equivalent fractions have the same value even though they look different. When graphed on a number line, they are names for the same point.

4.2 **lowest terms**   A fraction is written in lowest terms when its numerator and denominator have no common factor other than 1.

     prime number   A prime number is a whole number that has exactly two different factors, itself and 1. The first few prime numbers are 2, 3, 5, 7, 11, 13, and 17.

     composite number   A composite number has at least one factor other than itself and 1. Examples are 4, 6, 9, and 10. The numbers 0 and 1 are neither prime nor composite.

     prime factorization   In a prime factorization, every factor is a prime number. For example, the prime factorization of 24 is 2 • 2 • 2 • 3.

4.3 **reciprocal**   Two numbers are reciprocals of each other if their product is 1. The reciprocal of $\frac{a}{b}$ is $\frac{b}{a}$ because $\frac{a}{b} \cdot \frac{b}{a} = 1$.

4.4 **like fractions**   Like fractions have the same denominator.

     unlike fractions   Unlike fractions have different denominators.

     least common denominator   The least common denominator (LCD) for two fractions is the smallest positive number that can be divided evenly by both denominators.

4.5 **mixed number**   A mixed number is a fraction and a whole number written together. It represents the sum of the whole number and the fraction. For example, $5\frac{1}{3}$ represents $5 + \frac{1}{3}$.

4.6 **complex fraction**   A complex fraction is a fraction in which the numerator and/or denominator contain one or more fractions.

4.7 **multiplication property of equality**   The multiplication property of equality states that you may multiply both sides of an equation by the same nonzero number and it will still be balanced.

     division property of equality   The division property of equality states that you may divide both sides of an equation by the same nonzero number and it will still be balanced.

4.8 **volume**   Volume is a measure of the space inside a solid shape. Volume is measured in cubic units, such as in.³, ft³, yd³, mm³, cm³, and so on.

▶ **New Symbols**

Cubic units (for measuring volume)   in.³   ft³   yd³   mm³   cm³   m³

▶ **New Formulas**

Area of a triangle: $A = \frac{1}{2}bh$

Volume of a rectangular solid: $V = lwh$

Volume of a pyramid: $V = \frac{1}{3}Bh$

▶ Test Your Word Power

See how well you have learned the vocabulary in this chapter. Answers follow the Quick Review.

1. A **fraction** is in lowest terms if
   A. it has a value less than 1
   B. its numerator and denominator have no common factor other than 1
   C. its numerator and denominator are composite numbers
   D. it is rewritten as a mixed number.

2. The **denominator** of a fraction
   A. is written above the fraction bar
   B. is a prime number
   C. shows how many equal parts are in the whole
   D. is the smallest number divisible by the numerator.

3. The **LCD** of two fractions is the
   A. smallest number divisible by both denominators
   B. largest factor common to both denominators
   C. smallest number divisible by both numerators
   D. smallest prime number that divides evenly into both denominators.

4. Two numbers are **reciprocals** of each other if
   A. they have the same prime factorizations
   B. their sum is 0
   C. they are written in lowest terms
   D. their product is 1.

5. **Volume** is
   A. measured in square units
   B. the space inside a solid shape
   C. the sum of the lengths of the sides of a shape
   D. found by multiplying base times height.

6. A **mixed number**
   A. has a value equal to 1
   B. is the reciprocal of an improper fraction
   C. is the sum of a whole number and a fraction
   D. has a value less than 1.

7. A whole number is **prime** if
   A. it is divisible by itself and 1
   B. it has only composite factors
   C. it cannot be divided
   D. it has exactly two factors, itself and 1.

8. **Equivalent fractions**
   A. have the same denominators
   B. are written in lowest terms
   C. name the same point on a number line
   D. are reciprocals of each other.

▶ Quick Review

Concepts	Examples

4.1 Understanding Fraction Terminology

The *numerator* is the top number. The *denominator* is the bottom number.

Proper fractions

$$\frac{2}{3}, \frac{3}{4}, \frac{15}{16}, \frac{1}{8} \quad \leftarrow \text{Numerator} \atop \leftarrow \text{Denominator}$$

In a *proper fraction,* the numerator is less than the denominator, so the fraction is less than 1. In an *improper fraction,* the numerator is greater than or equal to the denominator, so the fraction is greater than or equal to 1.

Improper fractions

$$\frac{17}{8}, \frac{19}{12}, \frac{11}{2}, \frac{5}{3}, \frac{7}{7}$$

4.1 Writing Equivalent Fractions

Multiply or divide the numerator and denominator by the same nonzero number. The result is an equivalent fraction.

$$\frac{1}{2} = \frac{1 \cdot 8}{2 \cdot 8} = \frac{8}{16} \quad \leftarrow \text{Equivalent to } \tfrac{1}{2}$$

$$-\frac{12}{15} = -\frac{12 \div 3}{15 \div 3} = -\frac{4}{5} \quad \leftarrow \text{Equivalent to } -\tfrac{12}{15}$$

Concepts	Examples

4.2 Finding Prime Factorizations

A prime factorization of a number shows the number as the product of prime numbers. The first few prime numbers are 2, 3, 5, 7, 11, 13, and 17. You can use a division method or a factor tree to find the prime factorization.

Find the prime factorization of 24.

Division Method:

$2\overline{)24}$ ← Divide 24 by 2, the first prime; quotient is 12

$2\overline{)12}$ ← Divide 12 by 2; quotient is 6

$2\overline{)6}$ ← Divide 6 by 2; quotient is 3

$3\overline{)3}$ ← Divide 3 by 3; quotient is 1

1 ← Continue to divide until the quotient is 1

$24 = 2 \cdot 2 \cdot 2 \cdot 3$

Factor Tree Method:

Circle each prime number.

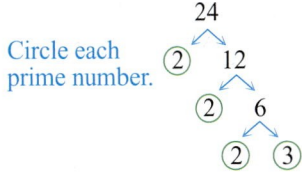

$24 = 2 \cdot 2 \cdot 2 \cdot 3$

4.2 Writing Fractions in Lowest Terms

Write the prime factorization of both numerator and denominator. Divide out all common factors, using slashes to show the division. Multiply any remaining factors in the numerator and in the denominator.

$$\frac{18}{90} = \frac{\overset{1}{\cancel{2}} \cdot \overset{1}{\cancel{3}} \cdot \overset{1}{\cancel{3}}}{\underset{1}{\cancel{2}} \cdot \underset{1}{\cancel{3}} \cdot \underset{1}{\cancel{3}} \cdot 5} = \frac{1}{5}$$

$$\frac{2b^3}{8ab} = \frac{\overset{1}{\cancel{2}} \cdot \overset{1}{\cancel{b}} \cdot b \cdot b}{\underset{1}{\cancel{2}} \cdot 2 \cdot 2 \cdot a \cdot \underset{1}{\cancel{b}}} = \frac{b^2}{4a}$$

4.3 Multiplying Fractions

Multiply the numerators and multiply the denominators. The product must be written in lowest terms. One way to do this is to write each original numerator and each original denominator as the product of primes and divide out any common factors before multiplying.

$$\left(-\frac{7}{10}\right)\left(\frac{5}{6}\right) = -\frac{7 \cdot \overset{1}{\cancel{5}}}{2 \cdot \underset{1}{\cancel{5}} \cdot 2 \cdot 3} = -\frac{7}{12}$$

$$\frac{3x^2}{5} \cdot \frac{2}{9x} = \frac{\overset{1}{\cancel{3}} \cdot \overset{1}{\cancel{x}} \cdot x \cdot 2}{5 \cdot \underset{1}{\cancel{3}} \cdot 3 \cdot \underset{1}{\cancel{x}}} = \frac{2x}{15}$$

4.3 Dividing Fractions

Rewrite the division problem as multiplying by the reciprocal of the divisor. In other words, the first number (dividend) stays the same and the second number (divisor) is changed to its reciprocal. Then use the steps for multiplying fractions. The quotient must be in lowest terms.

Division by 0 is undefined.

$$2 \div \left(-\frac{1}{3}\right) = \frac{2}{1} \cdot \left(-\frac{3}{1}\right) = -\frac{2 \cdot 3}{1 \cdot 1} = -6$$

Reciprocals

$$\frac{x^2}{y^2} \div \frac{x}{3y} = \frac{x^2}{y^2} \cdot \frac{3y}{x} = \frac{\overset{1}{\cancel{x}} \cdot x \cdot 3 \cdot \overset{1}{\cancel{y}}}{\underset{1}{\cancel{y}} \cdot y \cdot \underset{1}{\cancel{x}}} = \frac{3x}{y}$$

Reciprocals

Concepts	Examples

4.4 Adding and Subtracting Like Fractions

You can add or subtract fractions *only* when they have the *same* denominator. Add or subtract the numerators and write the result over the common denominator. Be sure that the final result is in lowest terms.

$$\frac{3}{10} - \frac{7}{10} = \frac{3-7}{10} = \frac{-4}{10} \quad \text{or} \quad -\frac{4}{10}$$

Write $-\frac{4}{10}$ in lowest terms.

$$-\frac{4}{10} = -\frac{\overset{1}{\cancel{2}} \cdot 2}{\underset{1}{\cancel{2}} \cdot 5} = -\frac{2}{5} \left\{ \begin{array}{l} \text{Lowest} \\ \text{terms} \end{array} \right.$$

$$\frac{5}{a} + \frac{7}{a} = \frac{5+7}{a} = \frac{12}{a} \left\{ \begin{array}{l} \text{Lowest} \\ \text{terms} \end{array} \right.$$

4.4 Finding the Lowest Common Denominator (LCD)

Write the prime factorization of each denominator. Then use enough prime factors in the LCD to "cover" both denominators.

What is the LCD for $\frac{5}{12}$ and $\frac{5}{18}$?

$$12 = 2 \cdot 2 \cdot 3$$
$$18 = 2 \cdot 3 \cdot 3$$

Factors of 12

$$\text{LCD} = 2 \cdot 2 \cdot 3 \cdot 3 = 36$$

Factors of 18

The LCD for $\frac{5}{12}$ and $\frac{5}{18}$ is 36.

4.4 Adding and Subtracting Unlike Fractions

Find the LCD. Rewrite each original fraction as an equivalent fraction whose denominator is the LCD. Then add or subtract the numerators and keep the common denominator. Be sure that the final result is in lowest terms.

$$-\frac{5}{12} + \frac{7}{9}$$

The LCD is 36.

Rewrite: $-\dfrac{5}{12} = -\dfrac{5 \cdot 3}{12 \cdot 3} = -\dfrac{15}{36}$

Rewrite: $\dfrac{7}{9} = \dfrac{7 \cdot 4}{9 \cdot 4} = \dfrac{28}{36}$

Add: $-\dfrac{15}{36} + \dfrac{28}{36} = \dfrac{-15+28}{36} = \dfrac{13}{36} \left\{ \begin{array}{l} \text{Lowest} \\ \text{terms} \end{array} \right.$

$$\frac{2}{3} - \frac{6}{x}$$

The LCD is $3 \cdot x$ or $3x$.

Rewrite: $\dfrac{2}{3} = \dfrac{2 \cdot x}{3 \cdot x} = \dfrac{2x}{3x}$

Rewrite: $\dfrac{6}{x} = \dfrac{6 \cdot 3}{x \cdot 3} = \dfrac{18}{3x}$

Subtract: $\dfrac{2x}{3x} - \dfrac{18}{3x} = \dfrac{2x - 18}{3x} \left\{ \begin{array}{l} \text{Lowest} \\ \text{terms} \end{array} \right.$

Concepts	Examples

4.5 Mixed Numbers and Improper Fractions

Changing Mixed Numbers to Improper Fractions
Multiply the denominator by the whole number, add the numerator, and place the result over the original denominator.

Changing Improper Fractions to Mixed Numbers
Divide the numerator by the denominator and place the remainder over the original denominator.

Mixed to improper

$$7\frac{2}{3} = \frac{23}{3} \leftarrow \frac{(3 \cdot 7) + 2}{\text{Same denominator}}$$

Improper to mixed

$$\frac{17}{5} = 3\frac{2}{5}$$

Same denominator

4.5 Multiplying Mixed Numbers

First, round the numbers and estimate the answer. Then follow these steps to find the exact answer.

Step 1 Rewrite each mixed number as an improper fraction.

Step 2 Multiply.

Step 3 Write the answer in lowest terms and change the answer to a mixed number if desired. Then the answer is in simplest form.

Estimate:

$$1\frac{3}{5} \quad \cdot \quad 3\frac{1}{3}$$

$$\downarrow \quad \text{Rounded} \quad \downarrow$$

$$2 \quad \cdot \quad 3 = 6$$

Exact:

$$1\frac{3}{5} \cdot 3\frac{1}{3} = \frac{8}{5} \cdot \frac{10}{3}$$

$$= \frac{8 \cdot 2 \cdot \overset{1}{\cancel{5}}}{\underset{1}{\cancel{5}} \cdot 3}$$

$$= \frac{16}{3} = 5\frac{1}{3} \leftarrow$$

Close to estimate

4.5 Dividing Mixed Numbers

First, round the numbers and estimate the answer. Then follow these steps to find the exact answer.

Step 1 Rewrite each mixed number as an improper fraction.

Step 2 Divide. (Rewrite as multiplication using the reciprocal of the divisor.)

Step 3 Write the answer in lowest terms and change the answer to a mixed number if desired. Then the answer is in simplest form.

Estimate:

$$3\frac{3}{4} \quad \div \quad 2\frac{2}{5}$$

$$\downarrow \quad \text{Rounded} \quad \downarrow$$

$$4 \quad \div \quad 2 = 2$$

Exact:

$$3\frac{3}{4} \div 2\frac{2}{5} = \frac{15}{4} \div \frac{12}{5} \leftarrow$$

Reciprocal of $\frac{12}{5}$ is $\frac{5}{12}$

$$= \frac{15}{4} \cdot \frac{5}{12}$$

$$= \frac{\overset{1}{\cancel{3}} \cdot 5 \cdot 5}{4 \cdot \underset{1}{\cancel{3}} \cdot 4}$$

$$= \frac{25}{16} = 1\frac{9}{16} \leftarrow$$

Close to estimate

4.5 Adding and Subtracting Mixed Numbers

First round the numbers and estimate the answer. Then rewrite the mixed numbers as improper fractions and follow the steps for adding and subtracting fractions. Write the answer in simplest form.

Estimate:

$$2\frac{3}{8} + 3\frac{3}{4}$$

$$\downarrow \qquad \downarrow$$

$$2 + 4 = 6 \leftarrow \text{Estimate}$$

Exact:

$$2\frac{3}{8} + 3\frac{3}{4} = \frac{19}{8} + \frac{15}{4} = \frac{19}{8} + \frac{30}{8} = \frac{19 + 30}{8}$$

$$= \frac{49}{8} = 6\frac{1}{8} \leftarrow \text{Close to estimate}$$

Concepts	Examples

(4.6) Simplifying Fractions with Exponents

The meaning of an exponent is the same for fractions as it is for integers. An exponent is a way to write repeated multiplication.

$$\left(-\frac{2}{3}\right)^2 \quad \text{means} \quad \left(-\frac{2}{3}\right)\left(-\frac{2}{3}\right) = \frac{2 \cdot 2}{3 \cdot 3} = \frac{4}{9}$$

The product of two negative numbers is positive.

(4.6) Order of Operations

The order of operations is the same for fractions as for integers.

1. Work inside *parentheses* or *other grouping symbols*.

2. Simplify expressions with *exponents*.

3. Do the remaining *multiplications and divisions* as they occur from left to right.

4. Do the remaining *additions and subtractions* as they occur from left to right.

Simplify.

$$-\frac{2}{3} + 3\left(\frac{1}{4}\right)^2$$

Cannot work inside parentheses. Apply the exponent: $\frac{1}{4} \cdot \frac{1}{4}$ is $\frac{1}{16}$

$$-\frac{2}{3} + 3\left(\frac{1}{16}\right)$$

Multiply next: $3\left(\frac{1}{16}\right)$ is $\frac{3}{1} \cdot \frac{1}{16} = \frac{3}{16}$

$$-\frac{2}{3} + \frac{3}{16}$$

Add last. The LCD is 48

$$-\frac{32}{48} + \frac{9}{48}$$

Rewrite $-\frac{2}{3}$ as $-\frac{32}{48}$
Rewrite $\frac{3}{16}$ as $\frac{9}{48}$

$$\frac{-32 + 9}{48}$$

Add the numerators. Keep the common denominator.

$$-\frac{23}{48}$$

The answer is in lowest terms.

(4.6) Simplifying Complex Fractions

Recall that the fraction bar indicates division. Rewrite the complex fraction using the ÷ symbol for division. Then follow the steps for dividing fractions.

Simplify. $\dfrac{-\dfrac{4}{5}}{10}$

Rewrite as $-\dfrac{4}{5} \div 10$.

$$-\frac{4}{5} \div 10 = -\frac{4}{5} \cdot \frac{1}{10} = -\frac{\overset{1}{2} \cdot 2 \cdot 1}{5 \cdot \underset{1}{2} \cdot 5} = -\frac{2}{25}$$

Reciprocals

Concepts	Examples

(4.7) Solving Equations Containing Fractions

Solve the equation. Check the solution.

$$\frac{1}{3}b + 6 = 10$$

Step 1 If necessary, add the same number to both sides of the equation so that the variable term is by itself on one side of the equal sign.

$$\frac{-6 \quad -6}{\frac{1}{3}b + 0 = 4} \quad \text{Add } -6 \text{ to both sides.}$$

$$\frac{1}{3}b = 4$$

Step 2 Multiply both sides by the reciprocal of the coefficient of the variable term.

$$\frac{\cancel{3}}{1}\left(\frac{1}{\cancel{3}}b\right) = \frac{3}{1}(4) \quad \begin{array}{l}\text{Multiply both sides by } \frac{3}{1}\\ \text{(the reciprocal of } \frac{1}{3}\text{).}\end{array}$$

$$b = 12$$

The solution is 12.

Step 3 To check your solution, go back to the *original* equation and replace the variable with your solution. If the equation balances, your solution is correct. If it does not balance, rework the problem.

Check $\quad \frac{1}{3} b + 6 = 10 \quad$ Original equation

$$\frac{1}{3}(12) + 6 = 10 \quad \text{Replace } b \text{ with 12}$$

$$4 + 6 = 10$$

$$10 = 10 \quad \text{Balances}$$

When b is 12, the equation balances, so 12 is the correct solution (**not** 10).

(4.8) Finding the Area of a Triangle

Use this formula to find the area of a triangle.

$$\text{Area} = \frac{1}{2} \cdot \text{base} \cdot \text{height}$$

$$A = \frac{1}{2}bh$$

Remember that area is measured in **square units**.

Find the area of this triangle.

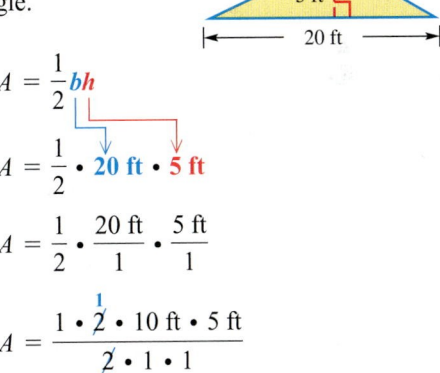

$$A = \frac{1}{2}bh$$

$$A = \frac{1}{2} \cdot 20 \text{ ft} \cdot 5 \text{ ft}$$

$$A = \frac{1}{2} \cdot \frac{20 \text{ ft}}{1} \cdot \frac{5 \text{ ft}}{1}$$

$$A = \frac{1 \cdot \cancel{2} \cdot 10 \text{ ft} \cdot 5 \text{ ft}}{\cancel{2} \cdot 1 \cdot 1}$$

$$A = 50 \text{ ft}^2 \quad \text{Measure area in square units.}$$

(4.8) Finding the Volume of a Rectangular Solid

Use this formula to find the volume of box-like solids.

$$\text{Volume} = \text{length} \cdot \text{width} \cdot \text{height}$$

$$V = lwh$$

Volume is measured in **cubic units**.

Find the volume of this box.

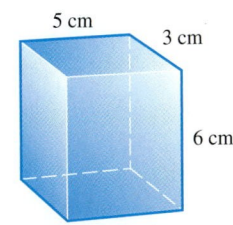

$$V = l \cdot w \cdot h$$

$$V = 5 \text{ cm} \cdot 3 \text{ cm} \cdot 6 \text{ cm}$$

$$V = 90 \text{ cm}^3$$

Measure volume in cubic units.

Concepts	Examples

4.8 Finding the Volume of a Pyramid

Use this formula to find the volume of a pyramid.

$$\text{Volume} = \frac{1}{3} \cdot B \cdot h$$

$$V = \frac{1}{3}Bh$$

where B is the area of the base and h is the height of the pyramid.

Volume is measured in **cubic units**.

Find the volume of a pyramid with a square base 2 cm by 2 cm and a height of 6 cm.

$$\text{Area of square base} = 2\text{ cm} \cdot 2\text{ cm}$$
$$B = \textbf{4 cm}^2$$

$$V = \frac{1}{3} \cdot \textbf{B} \cdot \textbf{h}$$

$$V = \frac{1}{3} \cdot \textbf{4 cm}^2 \cdot \textbf{6 cm}$$

$$V = \frac{1}{3} \cdot \frac{4\text{ cm}^2}{1} \cdot \frac{6\text{ cm}}{1}$$

$$V = \frac{1 \cdot 4\text{ cm}^2 \cdot \overset{1}{\cancel{3}} \cdot 2\text{ cm}}{\underset{1}{\cancel{3}} \cdot 1 \cdot 1}$$

$$V = \textbf{8 cm}^3 \qquad \text{Measure volume in cubic units.}$$

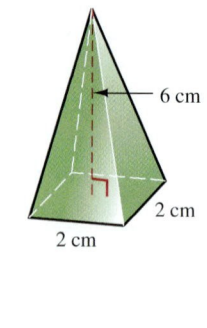

ANSWERS TO TEST YOUR WORD POWER

1. B; *Example:* $\frac{2}{5}$ is in lowest terms, but $\frac{4}{10}$ is not.

2. C; *Example:* In $\frac{3}{4}$ the denominator, 4, shows that the whole is divided into 4 equal parts.

3. A; *Example:* The LCD of $\frac{1}{4}$ and $\frac{5}{6}$ is 12, because 12 is the smallest number that can be divided evenly by 4 and by 6.

4. D; *Example:* The reciprocal of $\frac{3}{8}$ is $\frac{8}{3}$ because $\frac{3}{8} \cdot \frac{8}{3} = \frac{24}{24} = 1$.

5. B; *Example:* The volume of a rectangular solid (box-like shape) is the space inside the box, measured in cubic units.

6. C; *Example:* The mixed number $2\frac{3}{4}$ represents $2 + \frac{3}{4}$.

7. D; *Example:* 7 is prime because it has exactly two factors, 7 and 1.

8. C; *Example:* $\frac{1}{2}$ and $\frac{2}{4}$ are equivalent fractions because they both name the point halfway between 0 and 1.

Chapter 4 ▶▶▶ Review Exercises

[4.1] **1.** What fraction of these figures are squares? What fraction are circles?

   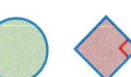

2. Write fractions to represent the shaded and unshaded portions of this figure.

3. Graph $-\frac{1}{2}$ and $1\frac{1}{2}$ on the number line.

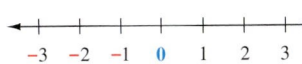

4. Simplify each fraction.

(a) $-\dfrac{20}{5}$ (b) $\dfrac{8}{1}$ (c) $-\dfrac{3}{3}$

[4.2] *Write each fraction in lowest terms.*

5. $\dfrac{28}{32}$

6. $\dfrac{54}{90}$

7. $\dfrac{16}{25}$

8. $\dfrac{15x^2}{40x}$

9. $\dfrac{7a^3}{35a^3b}$

10. $\dfrac{12mn^2}{21m^3n}$

[4.3] *Multiply or divide. Write all answers in lowest terms.*

11. $-\dfrac{3}{8} \div (-6)$

12. $\dfrac{2}{5}$ of (-30)

13. $\dfrac{4}{9}\left(\dfrac{2}{3}\right)$

14. $\left(\dfrac{7}{3x^3}\right)\left(\dfrac{x^2}{14}\right)$

15. $\dfrac{ab}{5} \div \dfrac{b}{10a}$

16. $\dfrac{18}{7} \div 3k$

[4.4] *Add or subtract. Write all answers in lowest terms.*

17. $-\dfrac{5}{12} + \dfrac{5}{8}$

18. $\dfrac{2}{3} - \dfrac{4}{5}$

19. $4 - \dfrac{5}{6}$

20. $\dfrac{7}{9} + \dfrac{13}{18}$

21. $\dfrac{n}{5} + \dfrac{3}{4}$

22. $\dfrac{3}{10} - \dfrac{7}{y}$

[4.5] *First, round each mixed number to the nearest whole number and estimate the answer. Then find the exact answer.*

23. *Exact:*

$$2\frac{1}{4} \div 1\frac{5}{8}$$

Estimate:

_____ ÷ _____ = _____

24. *Exact:*

$$7\frac{1}{3} - 4\frac{5}{6}$$

Estimate:

_____ − _____ = _____

25. *Exact:*

$$1\frac{3}{4} + 2\frac{3}{10}$$

Estimate:

_____ + _____ = _____

[4.6] *Simplify.*

26. $\left(-\dfrac{3}{4}\right)^3$

27. $\left(\dfrac{2}{3}\right)^2\left(-\dfrac{1}{2}\right)^4$

28. $\dfrac{2}{5} + \dfrac{3}{10}(-4)$

29. $-\dfrac{5}{8} \div \left(-\dfrac{1}{2}\right)\left(\dfrac{14}{15}\right)$

30. $\dfrac{\frac{5}{8}}{\frac{1}{16}}$

31. $\dfrac{\frac{8}{9}}{-6}$

[4.7] *Solve each equation. Show your work.*

32. $-12 = -\dfrac{3}{5}w$

33. $18 + \dfrac{6}{5}r = 0$

34. $3x - \dfrac{2}{3} = \dfrac{5}{6}$

[4.8] *Find the area of the triangle. Name each solid and find its volume.*

35.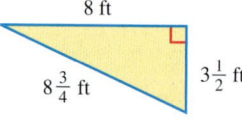

36.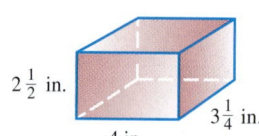

37.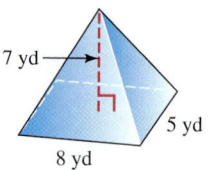

Mixed Review Exercises

Solve each application problem.

38. A chili recipe that makes 10 servings uses $2\frac{1}{2}$ pounds of meat. How much meat will be in each serving? How much meat would be needed to make 30 servings?

39. Yanli worked as a math tutor for $4\frac{1}{2}$ hours on Monday, $2\frac{3}{4}$ hours on Tuesday, and $3\frac{2}{3}$ hours on Friday. How much longer did she work on Monday than on Friday? How many hours did she work in all?

40. There are 60 children in the day care center. If $\frac{1}{5}$ of the children are preschoolers, $\frac{2}{3}$ of the children are toddlers, and the rest are infants, find the number of children in each age group.

41. A rectangular city park is $\frac{3}{4}$ mile long and $\frac{3}{10}$ mile wide. Find the perimeter and area of the park.

1. Write fractions to represent the shaded and unshaded portions of the figure.

2. Graph $-\frac{2}{3}$ and $2\frac{1}{3}$ on the number line at the right.

Write each fraction in lowest terms.

3. $\dfrac{21}{84}$

4. $\dfrac{25}{54}$

5. $\dfrac{6a^2b}{9b^2}$

Add, subtract, multiply, or divide, as indicated. Write all answers in lowest terms.

6. $\dfrac{1}{6} + \dfrac{7}{10}$

7. $-\dfrac{3}{4} \div \dfrac{3}{8}$

8. $\dfrac{5}{8} - \dfrac{4}{5}$

9. $(-20)\left(-\dfrac{7}{10}\right)$

10. $\dfrac{\frac{4}{9}}{-6}$

11. $4 - \dfrac{7}{8}$

12. $-\dfrac{2}{9} + \dfrac{2}{3}$

13. $\dfrac{21}{24}\left(\dfrac{9}{14}\right)$

14. $\dfrac{12x}{7y} \div 3x$

15. $\dfrac{6}{n} - \dfrac{1}{4}$

16. $\dfrac{2}{3} + \dfrac{a}{5}$

17. $\left(\dfrac{5}{9b^2}\right)\left(\dfrac{b}{10}\right)$

18. Simplify.

$\left(-\dfrac{1}{2}\right)^3 \left(\dfrac{2}{3}\right)^2$

19. Simplify.

$\dfrac{1}{6} + 4\left(\dfrac{2}{5} - \dfrac{7}{10}\right)$

---

1. _____

2.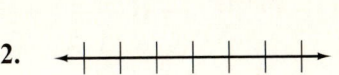
   -3 -2 -1 0 1 2 3

3. _____

4. _____

5. _____

6. _____

7. _____

8. _____

9. _____

10. _____

11. _____

12. _____

13. _____

14. _____

15. _____

16. _____

17. _____

18. _____

19. _____

20. Estimate: _____

    Exact: _____

21. Estimate: _____

    Exact: _____

22. _____

23. _____

24. _____

25. _____

26. _____

27. _____

28. _____

29. _____

30. _____

31. _____

32. _____

First, round the numbers and estimate each answer. Then find the exact answer. Write exact answers in simplest form.

20. $4\dfrac{4}{5} \div 1\dfrac{1}{8}$

21. $3\dfrac{2}{5} - 1\dfrac{9}{10}$

Solve each equation. Show your work.

22. $7 = \dfrac{1}{5}d$

23. $-\dfrac{3}{10}t = \dfrac{9}{14}$

24. $0 = \dfrac{1}{4}b - 2$

25. $\dfrac{4}{3}x + 7 = -13$

Find the area of each triangle.

26.

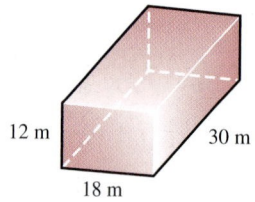

12 m   8 m   9 m
13 m

27.

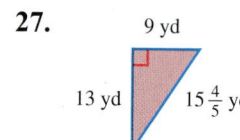

9 yd
13 yd $15\dfrac{4}{5}$ yd

Name each solid and find its volume.

28.

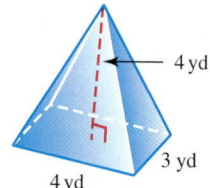

12 m   30 m
18 m

29.

4 yd
3 yd
4 yd

Solve each application problem.

30. Ann-Marie Sargent is training for an upcoming wheelchair race. She rides $4\dfrac{5}{6}$ hours on Monday, $6\dfrac{2}{3}$ hours on Tuesday, and $3\dfrac{1}{4}$ hours on Wednesday. How many hours did she spend in all? How many more hours did she train on Tuesday than on Monday?

31. A new vaccine is synthesized at the rate of $2\dfrac{1}{2}$ ounces per day. How long will it take to synthesize $8\dfrac{3}{4}$ ounces?

32. There are 8448 students at the Metro Community College campus. If $\dfrac{7}{8}$ of the students work either full time or part time, find the total number of students who work.

Study Skills

After taking a test, many students heave a big sigh of relief and try to forget it ever happened. Don't fall into this trap! An exam is a learning opportunity. It gives you clues about *what your instructor thinks is important,* what *concepts and skills are valued* in mathematics, and *if you are on the right track*.

Jot down problems that caused you trouble. Find out how to solve them by checking your textbook, notes, or asking your instructor or tutor (if available). You might see those same problems again on a final exam.

Find out what you got wrong and why you had points deducted. Write down the problem so you can learn how to do it correctly. Sometimes you only have a short time in class to review your test. *If you need more time,* ask your instructor if you can look at the test in his or her office.

Here is a list of typical reasons for making errors on math tests.

1. You read the directions wrong.
2. You read the question wrong or skipped over something.
3. You made a computation error (maybe even an easy one).
4. Your answer is not accurate.
5. Your answer is not complete.
6. You labeled your answer wrong. For example, you labeled it "feet" and it should have been "feet^2."
7. You didn't show your work.
8. *You didn't understand the concept.
9. *You were unable to go from words (in a word problem) to setting up the problem.
10. *You were unable to apply a procedure to a new situation.
11. You were so anxious that you made errors even when you knew the material.

The first seven errors are **test-taking errors.** They are easy to correct if you decide to carefully read test questions and directions, proofread or rework your problems, show all your work, and double check units and labels every time.

The three starred errors (*) are **test preparation errors.** Remember that to grow a complex neural network, you need to practice the kinds of problems that you will see on the tests. So, for example, if application problems are difficult for you, you must *do more application problems!* If you have practiced the study skills techniques, however, you are less likely to make these kinds of errors on tests because you will have a deeper understanding of course concepts and you will be able to remember them better.

The last error isn't really an error. **Anxiety** can play a big part in your test results. Go back to the *Preparing for Tests* activity and read the suggestions about exercise and deep breathing. Recall from the *Your Brain Can Learn Mathematics* activity that when you are anxious, your body produces adrenaline. The presence

OBJECTIVES

1 Determine the reason for errors.

2 Develop a plan to avoid test-taking errors.

3 Review material to correct misunderstandings.

Immediately After the Test

After the Test Is Returned

Find Out Why You Made the Errors You Made

of *adrenaline in the brain blocks connections* between dendrites. If you can *reduce the adrenaline* in your system, you will be able to *think more clearly* during your test. Just five minutes of brisk walking right before your test can help do that. Also, *practicing a relaxation technique while you do your homework* will make it more likely that you can benefit from using the technique during a test. *Deep breathing* is helpful because it gets *oxygen into your brain*. When you are anxious you tend to breathe more shallowly, which can make you feel confused and easily distracted.

Make a Plan for the Next Test

Make a plan for your next test based on your results from this test. You might review the Chapter Summary and work the problems in the Chapter Review Exercises or the Chapter Test. Ask your instructor or a tutor (if available) for more help if you are confused about any of the problems.

Now Try This ▶▶▶

Below is a record sheet to track your progress in test taking. Use it to find out if you make particular kinds of errors. Then you can work specifically on correcting them. Just check in the box when you made one of the errors. If you take more than five tests, make your own grid on separate paper.

Test Taking Errors

Test #	Read directions wrong	Read question wrong	Computation error	Not exact or accurate	Not complete	Labeled wrong	Didn't show work
1							
2							
3							
4							
5							

Test Preparation Errors

Test #	Didn't understand concept	Didn't set up problem correctly	Couldn't apply concept to new situation
1			
2			
3			
4			
5			

Anxiety

Test #	Felt anxious *before* the exam	Felt anxious *during* the exam	Blanked out on questions	Got questions wrong that I knew how to do
1				
2				
3				
4				
5				

What will you do to avoid Test Taking Errors?

_____

_____

_____

What will you do to avoid Test Preparation Errors?

_____

_____

_____

What will you do to reduce anxiety?

_____

_____

_____

5

Rational Numbers: Positive and Negative Decimals

Over 41 million Americans go fishing at least once a year, making it America's sixth most popular recreational activity. (*Source:* National Sporting Goods Association.) Record-size fish caught include a 67.5-pound muskie in Wisconsin, a 58-pound channel catfish in South Carolina, and a 97.25 pound chinook salmon in Alaska.

In **Section 5.3**, Exercises 57–60, this father and daughter will use decimal numbers when paying for new fishing equipment. But will decimals help them catch their limit? (See **Section 5.1**, Exercises 59–62, and **Section 5.6**, Exercises 67–68.)

5.1 ▶▶▶ Reading and Writing Decimal Numbers

In **Chapter 4,** you worked with rational numbers written in fraction form to represent parts of a whole. In this chapter, we will use rational numbers written as **decimals** to show parts of a whole. For example, our money system is based on decimals. One dollar is divided into 100 equivalent parts. One cent ($0.01) is one of the parts, and a dime ($0.10) is 10 of the parts. Metric measurement (see **Chapter 8**) is also based on decimals.

OBJECTIVE 1 Write parts of a whole using decimals. Decimals are used when a whole is divided into 10 equivalent parts or into 100 or 1000 or 10,000 equivalent parts. In other words, decimals are fractions with denominators that are a power of 10. For example, the square at the right is cut into 10 equivalent parts. Written as a fraction, each part is $\frac{1}{10}$ of the whole. Written as a decimal, each part is **0.1**. Both $\frac{1}{10}$ and 0.1 are read as "*one tenth*."

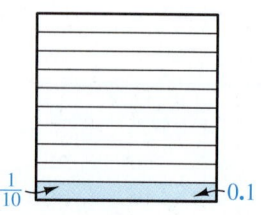

One tenth of the square is shaded.

The dot in 0.1 is called the **decimal point.**

$$0.1$$
↑
Decimal point

1 There are 10 dimes in one dollar. Each dime is $\frac{1}{10}$ of a dollar. Write a fraction, a decimal, and the words that name the yellow shaded portion of each dollar.

(a)

The square at the right has **7** of its 10 parts shaded.
Written as a *fraction,* $\frac{7}{10}$ of the square is shaded.
Written as a *decimal,* **0.7** of the square is shaded.
Both $\frac{7}{10}$ and 0.7 are read as "*seven tenths*."

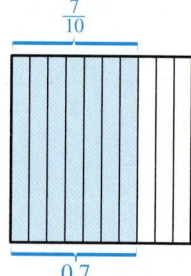

Seven tenths of the square is shaded.

(b)

◀ *Work Problem* **1** *at the Side.*

Each square below is cut into 100 equivalent parts.
Written as a fraction, each part is $\frac{1}{100}$ of the whole.
Written as a decimal, each part is **0.01** of the whole.
Both $\frac{1}{100}$ and 0.01 are read as "*one hundredth*."

(c)

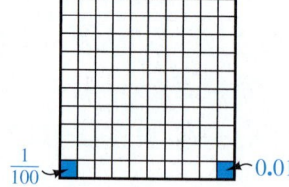

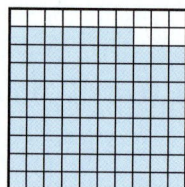

Eighty-seven hundredths of the square is shaded.

The square above on the right has 87 of its 100 parts shaded.
Written as a fraction, $\frac{87}{100}$ of the total area is shaded.
Written as a decimal, **0.87** of the total area is shaded.
Both $\frac{87}{100}$ and 0.87 are read as "*eighty-seven hundredths*."

Work Problem **2** *at the Side.* ▶

The example below shows several numbers written as fractions, as decimals, and in words.

> **EXAMPLE 1** **Using the Decimal Forms of Fractions**
>
> | | Fraction | Decimal | Read As |
> |---|---|---|---|
> | **(a)** | $\frac{4}{10}$ | 0.4 | four tenths |
> | **(b)** | $-\frac{9}{100}$ | -0.09 | negative nine hundredths |
> | **(c)** | $\frac{71}{100}$ | 0.71 | seventy-one hundredths |
> | **(d)** | $\frac{8}{1000}$ | 0.008 | eight thousandths |
> | **(e)** | $-\frac{45}{1000}$ | -0.045 | negative forty-five thousandths |
> | **(f)** | $\frac{832}{1000}$ | 0.832 | eight hundred thirty-two thousandths |

Work Problem **3** *at the Side.* ▶

OBJECTIVE 2 Identify the place value of a digit. The decimal point separates the *whole number part* from the *fractional part* in a decimal number. In the chart below, you see that the **place value** names for fractional parts are similar to those on the whole number side but end in "***ths***."

Decimal Place Value Chart

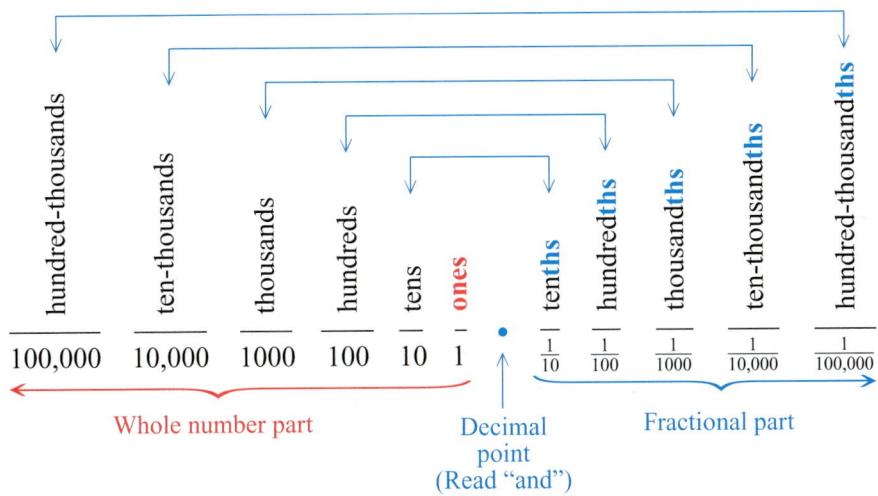

hundred-thousands	ten-thousands	thousands	hundreds	tens	ones	tenths	hundredths	thousandths	ten-thousandths	hundred-thousandths
100,000	10,000	1000	100	10	1	$\frac{1}{10}$	$\frac{1}{100}$	$\frac{1}{1000}$	$\frac{1}{10,000}$	$\frac{1}{100,000}$

← Whole number part

Decimal point (Read "and")

Fractional part

> **Note**
>
> Notice that the **ones** place is at the center of the place value chart above. There is no "oneths" place.
>
> Also notice that each place is 10 times the value of the place to its right.
>
> Finally, be sure to write a hyphen (dash) in ten-thousand**ths** and hundred-thousand**ths**.

2 Write the portion of each square that is shaded as a fraction, as a decimal, and in words.

(a)

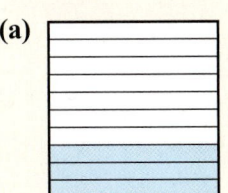

(b)

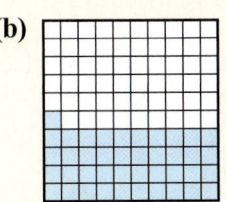

3 Write each decimal as a fraction.

(a) -0.7

(b) 0.2

(c) -0.03

(d) 0.69

(e) 0.047

(f) -0.351

4 Identify the place value of each digit.

(a) 971.54

(b) 0.4

(c) 5.60

(d) 0.0835

5 Tell how to read each decimal in words.

(a) 0.6

(b) 0.46

(c) 0.05

(d) 0.409

(e) 0.0003

(f) 0.2703

(g) 0.088

4. (a)
hundreds | tens | ones | . | tenths | hundredths
9 7 1 . 5 4   (b) 0 . 4 (ones, tenths)

(c) 5 . 6 0 (ones, tenths, hundredths)   (d) 0 . 0 8 3 5 (ones, tenths, hundredths, thousandths, ten-thousandths)

5. (a) six tenths
(b) forty-six hundredths
(c) five hundredths
(d) four hundred nine thousandths
(e) three ten-thousandths
(f) two thousand seven hundred three ten-thousandths
(g) eighty-eight thousandths

CAUTION

If a number does *not* have a decimal point, it is an *integer*. An integer has no fractional part. If you want to show the decimal point in an integer, it is just to the ***right*** of the digit in the ones place. Here are three examples.

$$8 = 8. \qquad 306 = 306. \qquad -42 = -42.$$

↑ Decimal point ↑ Decimal point ↑ Decimal point

EXAMPLE 2 **Identifying the Place Value of a Digit**

Identify the place value of each digit.

(a) 178.36

hundreds | tens | ones | tenths | hundredths
1 7 8 . 3 6

(b) 0.00935

ones | tenths | hundredths | thousandths | ten-thousandths | hundred-thousandths
0 . 0 0 9 3 5

Notice in Example 2(b) that we do *not* use commas on the right side of the decimal point.

◄ Work Problem **4** at the Side.

OBJECTIVE **3** **Read decimal numbers.** A decimal number is read according to its form as a fraction.

ones | tenths
0.**9**

We read 0.9 as "nine tenths" because 0.9 is the same as $\frac{9}{10}$. Notice that 0.9 ends in the tenths place.

ones | tenths | hundredths
0.0 **2**

We read 0.02 as "two hundredths" because 0.02 is the same as $\frac{2}{100}$. Notice that 0.02 ends in the hundredths place.

EXAMPLE 3 **Reading Decimal Numbers**

Tell how to read each decimal in words.

(a) 0.3

Because $0.3 = \frac{3}{10}$, read the decimal as: three ten**ths**.

(b) 0.49   Read it as: forty-nine hundred**ths**.

(c) 0.08   Read it as: eight hundred**ths**.

 Think: $0.08 = \frac{8}{100}$ so write *hundredths*.

(d) 0.918   Read it as: nine hundred eighteen thousand**ths**.

(e) 0.0106   Read it as: one hundred six ten-thousand**ths**.

 Think: $0.0106 = \frac{106}{10,000}$

◄ Work Problem **5** at the Side.

Reading a Decimal Number

Step 1 Read any whole number part to the *left* of the decimal point as you normally would.

Step 2 Read the decimal point as "*and*."

Step 3 Read the part of the number to the *right* of the decimal point as if it were an ordinary whole number.

Step 4 Finish with the place value name of the rightmost digit; these names all end in "*ths*."

Note

If there is *no whole number part,* you will use only Steps 3 and 4.

EXAMPLE 4 Reading Decimal Numbers

Read each decimal.

(a)

→ 9 is in tenths place.

16.9

Remember to say or write "and" **only** when you see a decimal point.

sixteen **and** nine **tenths** ←

16.9 is read "sixteen and nine tenths."

(b)

→ 5 is in hundredths place.

482.35

four hundred eighty-two **and** thirty-five **hundredths** ←

482.35 is read "four hundred eighty-two and thirty-five hundredths."

→ 3 is in thousandths place.

(c) 0.063 is "sixty-three **thousandths**." (No whole number part)

(d) 11.1085 is "eleven **and** one thousand eighty-five **ten-thousandths**."

CAUTION
Use "and" *only* when reading a decimal point. A common mistake is to read the whole number 405 as "four hundred *and* five." But there is *no decimal point* shown in 405, so it is read "four hundred five."

Work Problem **6** *at the Side.* ▶

OBJECTIVE 4 Write decimals as fractions or mixed numbers.
Knowing how to read decimals will help you when writing decimals as fractions or mixed numbers.

Writing a Decimal as a Fraction or Mixed Number

Step 1 The digits to the right of the decimal point are the numerator of the fraction.

Step 2 The denominator is 10 for tenths, 100 for hundredths, 1000 for thousandths, 10,000 for ten-thousandths, and so on.

Step 3 If the decimal has a whole number part, it is written as a mixed number with the same whole number part.

6 Tell how to read each decimal in words.

(a) 3.8

(b) 15.001

(c) 0.0073

(d) 764.309

7 Write each decimal as a fraction or mixed number.

(a) 0.7

(b) 12.21

(c) 0.101

(d) 0.007

(e) 1.3717

8 Write each decimal as a fraction or mixed number in lowest terms.

(a) 0.5

(b) 12.6

(c) 0.85

(d) 3.05

(e) 0.225

(f) 420.0802

EXAMPLE 5 **Writing Decimals as Fractions or Mixed Numbers**

Write each decimal as a fraction or mixed number.

(a) 0.19

The digits to the right of the decimal point, 19, are the numerator of the fraction. The denominator is 100 for hundredths because the rightmost digit is in the hundredths place.

$$0.19 = \frac{19}{100} \leftarrow 100 \text{ for hundredths}$$

$\uparrow$ Hundredths place

(b) 0.863

$$0.863 = \frac{863}{1000} \leftarrow 1000 \text{ for thousandths}$$

$\uparrow$ Thousandths place

(c) 4.0099

The whole number part stays the same.

$$4.0099 = 4\frac{99}{10,000} \leftarrow 10,000 \text{ for ten-thousandths}$$

$\uparrow$ Ten-thousandths place

◄ *Work Problem* **7** *at the Side.*

EXAMPLE 6 **Writing Decimals as Fractions or Mixed Numbers**

Write each decimal as a fraction or mixed number in lowest terms.

(a) $0.4 = \dfrac{4}{10} \leftarrow 10$ for tenths

Write $\dfrac{4}{10}$ in lowest terms. $\quad \dfrac{4}{10} = \dfrac{4 \div 2}{10 \div 2} = \dfrac{2}{5} \leftarrow$ Lowest terms

(b) $0.75 = \dfrac{75}{100} = \dfrac{75 \div 25}{100 \div 25} = \dfrac{3}{4} \leftarrow$ Lowest terms

The whole number part stays the same.

(c) $18.105 = 18\dfrac{105}{1000} = 18\dfrac{105 \div 5}{1000 \div 5} = 18\dfrac{21}{200} \leftarrow$ Lowest terms

(d) $42.8085 = 42\dfrac{8085}{10,000} = 42\dfrac{8085 \div 5}{10,000 \div 5} = 42\dfrac{1617}{2000} \leftarrow$ Lowest terms

CAUTION
Always check that your fraction answers are in lowest terms.

◄ *Work Problem* **8** *at the Side.*

Calculator Tip In this book, we will write **0.**45 instead of just .45, to emphasize that there is no whole number. Your *scientific* calculator shows these zeros also. Enter ⊙ ④ ⑤ and notice that the display automatically shows 0.45 even though you did not press 0. For comparison, enter the whole number 45 by pressing ④ ⑤ ⊕ and notice that the decimal point automatically appears to the *right* of the 5. (*Graphing* calculators may not automatically show a 0 in the ones place.)

ANSWERS

7. (a) $\dfrac{7}{10}$ (b) $12\dfrac{21}{100}$ (c) $\dfrac{101}{1000}$

  (d) $\dfrac{7}{1000}$ (e) $1\dfrac{3717}{10,000}$

8. (a) $\dfrac{1}{2}$ (b) $12\dfrac{3}{5}$ (c) $\dfrac{17}{20}$ (d) $3\dfrac{1}{20}$

  (e) $\dfrac{9}{40}$ (f) $420\dfrac{401}{5000}$

5.1 ▶▶▶ ▶ **Exercises**

FOR EXTRA HELP

MyMathLab

Math XL PRACTICE   WATCH   DOWNLOAD   READ   REVIEW

Identify the digit that has the given place value. See Example 2.

1. 70.489
  tens
  ones
  tenths

2. 135.296
  ones
  tenths
  tens

3. 0.2518
  hundredths
  thousandths
  ten-thousandths

4. 0.9347
  hundredths
  thousandths
  ten-thousandths

5. 93.01472
  thousandths
  ten-thousandths
  tenths

6. 0.51968
  tenths
  ten-thousandths
  hundredths

7. 314.658
  tens
  tenths
  hundreds

8. 51.325
  tens
  tenths
  hundredths

9. 149.0832
  hundreds
  hundredths
  ones

10. 3458.712
  hundreds
  hundredths
  tenths

11. 6285.7125
  thousands
  thousandths
  hundredths

12. 5417.6832
  thousands
  thousandths
  ones

Write the decimal number that has the specified place values. See Example 2.

13. 0 ones, 5 hundredths, 1 ten, 4 hundreds, 2 tenths

14. 7 tens, 9 tenths, 3 ones, 6 hundredths, 8 hundreds

15. 3 thousandths, 4 hundredths, 6 ones, 2 ten-thousandths, 5 tenths

16. 8 ten-thousandths, 4 hundredths, 0 ones, 2 tenths, 6 thousandths

17. 4 hundredths, 4 hundreds, 0 tens, 0 tenths, 5 thousandths, 5 thousands, 6 ones

18. 7 tens, 7 tenths, 6 thousands, 6 thousandths, 3 hundreds, 3 hundredths, 2 ones

Write each decimal as a fraction or mixed number in lowest terms. See Examples 5 and 6.

19. 0.7   **20.** 0.1   ⊙ **21.** 13.4   **22.** 9.8   **23.** 0.35

24. 0.85   **25.** 0.66   **26.** 0.33   **27.** 10.17   **28.** 31.99

29. 0.06   **30.** 0.08   **31.** 0.205   **32.** 0.805

33. 5.002   **34.** 4.008   **35.** 0.686   **36.** 0.492

Tell how to read each decimal in words. See Examples 1, 3, and 4.

37. 0.5   **38.** 0.2

39. 0.78   **40.** 0.55

41. 0.105   **42.** 0.609
⊙

43. 12.04   **44.** 86.09
⊙

45. 1.075   **46.** 4.025

Write each decimal in numbers. See Examples 3 and 4.

47. Six and seven tenths

48. Eight and twelve hundredths

49. Thirty-two hundredths

50. One hundred eleven thousandths

51. Four hundred twenty and eight thousandths

52. Two hundred and twenty-four thousandths

53. Seven hundred three ten-thousandths

54. Eight hundred and six hundredths

55. Seventy-five and thirty thousandths

56. Sixty and fifty hundredths

57. Anne read the number 4302 as "four thousand three hundred and two." Explain what is wrong with the way Anne read the number.

58. Jerry read the number 9.0106 as "nine and one hundred and six ten-thousandths." Explain the error he made.

The father on the first page of this chapter needs to select the correct fishing line for his daughter's reel. Fishing line is sold according to how many pounds of "pull" the line can withstand before breaking. Use the table to answer Exercises 59–62. Write all fractions in lowest terms. (Note: The diameter of the fishing line is its thickness.)

FISHING LINE

Test Strength (pounds)	Average Diameter (inches)
4	0.008
8	0.010
12	0.013
14	0.014
17	0.015
20	0.016

Source: Berkley Outdoor Technologies Group.

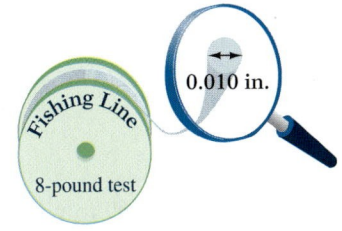

The diameter is the distance across the end of the line (or its thickness).

59. Write the diameter of 8-pound test line in words and as a fraction.

60. Write the diameter of 17-pound test line in words and as a fraction.

61. What is the test strength of the line with a diameter of $\frac{13}{1000}$ inch?

62. What is the test strength of the line with a diameter of sixteen thousandths inch?

Suppose your job is to take phone orders for precision parts. Use the table below. In Exercises 63–66, write the correct part number that matches what you hear the customer say over the phone. In Exercises 67–68, write the words you would say to the customer.

Part Number	Size in Centimeters
3-A	0.06
3-B	0.26
3-C	0.6
3-D	0.86
4-A	1.006
4-B	1.026
4-C	1.06
4-D	1.6
4-E	1.602

63. "Please send the six tenths centimeter bolt."

Part number _____

64. "The part missing from our order was the one and six hundredths size."

Part number _____

65. "The size we need is one and six thousandths centimeters."

Part number _____

66. "Do you still stock the twenty-six hundredths centimeter bolt?"

Part number _____

67. "What size is part number 4-E?" Write your answer in words.

68. "What size is part number 4-B?" Write your answer in words.

Relating Concepts (Exercises 69—76) For Individual or Group Work

*Use your knowledge of place value to **work Exercises 69–76 in order.***

69. Look back at the decimal place value chart on page 329 of this section. What do you think would be the names of the next four places to the *right* of hundred-thousandths? What information did you use to come up with these names?

70. A common mistake is to think that the first place to the right of the decimal point is "oneths" and the second place is "tenths." Why might someone make that mistake? How would you explain why there is no "oneths" place?

71. Use your answer from Exercise 69 to write 0.72436955 in words.

72. Use your answer from Exercise 69 to write 0.000678554 in words.

73. Write 8006.500001 in words.

74. Write 20,060.000505 in words.

75. Write this decimal using digits.
three hundred two thousand forty ten-millionths

76. Write this decimal using digits.
nine billion, eight hundred seventy-six million, five hundred forty-three thousand, two hundred ten and one hundred million two hundred thousand three hundred billionths

5.2 ▶▶▶ Rounding Decimal Numbers

Section 1.5 showed how to round integers. For example, 89 rounded to the nearest ten is 90, and 8512 rounded to the nearest hundred is 8500.

OBJECTIVE 1 Learn the rules for rounding decimals. It is also important to be able to **round** decimals. For example, a store is selling 2 candy mints for $0.75 but you want only one mint. The price of each mint is $0.75 ÷ 2, which is $0.375, but you cannot pay part of a cent. Is $0.375 closer to $0.37 or to $0.38? Actually, it's exactly halfway between. When this happens in everyday situations, the rule is to round *up*. The store will charge you $0.38 for the mint.

OBJECTIVES

1 **Learn the rules for rounding decimals.**

2 **Round decimals to any given place.**

3 **Round money amounts to the nearest cent or nearest dollar.**

Rounding a Decimal Number

Step 1 Find the place to which the rounding is being done. Draw a "cut-off" line *after* that place to show that you are cutting off and dropping the rest of the digits.

Step 2 Look *only* at the *first* digit you are cutting off.

Step 3A If this digit is *4 or less,* the part of the number you are keeping *stays the same.*

Step 3B If this digit is *5 or more,* you must *round up* the part of the number you are keeping.

Step 4 You can use the ≈ symbol or the ≐ symbol to indicate that the rounded number is now an approximation (close, but *not exact*). Both symbols mean "is approximately equal to." (In this book we will use the ≈ symbol.)

CAUTION
Do *not* move the decimal point when rounding.

OBJECTIVE 2 Round decimals to any given place. These examples show you how to round decimals.

EXAMPLE 1 **Rounding a Decimal Number**

Round 14.39652 to the nearest thousandth. (Is it closer to 14.396 or to 14.397?)

Step 1 Draw a "cut-off" line after the thousandths place.

$$1\ 4\ .\ 3\ 9\ 6\ \big|\ 5\ 2 \quad \text{You are cutting off the 5 and 2}$$
$$\text{Thousandths} \qquad \text{They will be dropped.}$$

Step 2 Look *only* at the *first* digit you are cutting off. Ignore the other digits you are cutting off.

$$1\ 4\ .\ 3\ 9\ 6\ \big|\ 5\ 2 \quad \begin{array}{l}\text{Look } only \text{ at the 5}\\ \text{Ignore the 2}\end{array}$$

Continued on Next Page

1 Round to the nearest thousandth.

(a) 0.33492

(b) 8.00851

(c) 265.42068

(d) 10.70180

Step 3   If the first digit you are cutting off is *5 or more,* round up the part of the number you are keeping.

$$1\ 4\ .\ 3\ 9\ 6 \quad 5\ 2$$
$$+\ \ 0\ .\ 0\ 0\ 1$$
$$\overline{1\ 4\ .\ 3\ 9\ 7}$$

First digit cut is 5 *or more,* so round up by adding 1 thousandth to the part you are keeping.

So, 14.39652 rounded to the nearest thousandth is 14.397
You can write 14.39652 ≈ 14.397

Rounding to **thousandths** means **three** decimal places.

> **CAUTION**
> When rounding integers in **Section 1.5,** you kept all the digits but changed some to zeros. With decimals, you cut off and *drop the extra digits.* In the example above, 14.39652 rounds to 14.397 (***not*** 14.39700).

◀ *Work Problem* **1** *at the Side.*

In Example 1 above, the rounded number 14.397 had *three decimal places.* **Decimal places** are the number of digits to the *right* of the decimal point. The first decimal place is tenths, the second is hundredths, the third is thousandths, and so on.

EXAMPLE 2   **Rounding Decimals to Different Places**

Round to the place indicated.

(a) Round 5.3496 to the nearest tenth. (Is it closer to 5.3 or to 5.4?)

Step 1   Draw a cut-off line after the tenths place.

Tenths is **one** decimal place.

$$5\ .\ 3 \quad 4\ 9\ 6$$

You are cutting off the 4, 9, and 6

Tenths

Step 2 $5\ .\ 3 \quad \underline{4}\ 9\ 6$

Look *only* at the 4

Ignore these digits.

Step 3 $5\ .\ 3 \quad 4\ 9\ 6$

First digit cut is 4 *or less,* so the part you are keeping stays the same.

$5\ .\ 3$ ← Stays the same

5.3496 rounded to the nearest tenth is 5.3 (*one decimal place for* *tenths*).
You can write 5.3496 ≈ 5.3
Notice: 5.3496 does *not* round to 5.3000 (which would be ten-thousandths instead of tenths).

(b) Round 0.69738 to the nearest hundredth. (Is it closer to 0.69 or to 0.70?)

Step 1 $0\ .\ 6\ 9\ |\ 7\ 3\ 8$ Draw a cut-off line after the hundredths place.

Hundredths

Look *only* at the 7

Step 2 $0\ .\ 6\ 9\ |\ 7\ 3\ 8$

Continued on Next Page

Step 3   0 . 6 9 | 7 3 8   ┌─ First digit cut is *5 or more,* so round up
                              by adding 1 hundredth to the part you
                              are keeping.

     1

    0 . 6 9   ← Keep this part.
 + 0 . 0 1   ← To round up, add 1 hundredth.
  0 . 7 **0**   ← 9 + 1 is 10; write 0 and regroup 1 to the tenths place.

So, 0.69738 rounded to the nearest hundredth is 0.70. Hundredths is *two*
decimal places so you *must* write the 0 in the hundredths place.
    You can write $0.69738 \approx 0.70$

> **CAUTION**
> If a *rounded* number has a 0 in the rightmost place, you *must* keep the 0.
> As shown above, 0.69738 rounded to the nearest hundredth is 0.7**0**. Do
> *not* write 0.7, which is rounded to tenths instead of hundredths.

(c) Round 0.01806 to the nearest thousandth. (Is it closer to 0.018 or
to 0.019?)

    0 . 0 1 8 | 0 6   ┌─ First digit cut is *4 or less,* so the part
                                           you are keeping stays the same.

    0 . 0 1 8   ← Stays the same

So, 0.01806 rounded to the nearest thousandth is 0.018 (three decimal places
for thousandths).
    You can write $0.01806 \approx 0.018$

(d) Round 57.976 to the nearest tenth. (Is it closer to 57.9 or to 58.0?)

    57.9 | 76   ┌─ First digit cut is *5 or more,* so round up by
                                     adding 1 tenth to the part you are keeping.

    1
   57.9
 + 0.1
  58.0   ← 9 + 1 is 10; write 0 and
      regroup 1 to the ones place.

> Be sure to write
> the 0 in the
> *tenths* place.

So, 57.976 rounded to the nearest tenth is 58.0. You can write $57.976 \approx 58.\mathbf{0}$
You *must* write the 0 in the tenths place to show that the number was rounded
to the nearest tenth.

> **CAUTION**
> Check that your rounded answer shows *exactly* the number of decimal
> places asked for in the problem. Be sure your answer shows *one decimal
> place* if you rounded to *tenths,* *two decimal places* for *hundredths,* *three
> decimal places* for *thousandths,* and so on.

─────────────── *Work Problem* ② *at the Side.* ▶

**OBJECTIVE 3   Round money amounts to the nearest cent or
nearest dollar.**   When you are shopping in a store, money amounts are
usually rounded to the nearest cent. There are 100 cents in a dollar.

$$\text{Each cent is } \frac{1}{100} \text{ of a dollar.}$$

Another way to write $\frac{1}{100}$ is 0.01. So rounding to the *nearest cent* is the
same as rounding to the *nearest hundredth of a dollar.*

---

2 Round to the place indicated.

  (a) 0.8988 to the nearest
     hundredth

  (b) 5.8903 to the nearest
     hundredth

  (c) 11.0299 to the nearest
     thousandth

  (d) 0.545 to the nearest tenth

ANSWERS

2. **(a)** 0.90   **(b)** 5.89
  (c) 11.030   **(d)** 0.5

3 Round each money amount to the nearest cent.

(a) $14.595

You pay ———.

(b) $578.0663

You pay ———.

(c) $0.849

You pay ———.

(d) $0.0548

You pay ———.

EXAMPLE 3 **Rounding to the Nearest Cent**

Round each money amount to the nearest cent.

(a) $2.4238 (Is it closer to $2.42 or to $2.43?)

$$\$2.42\,|\,38$$

First digit cut is *4 or less,* so the part you are keeping stays the same.

$2.42 ← You pay $2.42

$2.4238 rounded to the nearest cent is $2.42.

Rounding to the *nearest cent* is rounding to hundredths.

(b) $0.695 (Is it closer to $0.69 or to $0.70?)

5 or more; round up.

$$\$0.69\,|\,5$$

$$\begin{array}{r} \$0.69 \\ + \ \$0.01 \\ \hline \$0.70 \end{array}$$

To round up, add 1 hundredth (1 cent).

You pay $0.70

$0.695 rounded to the nearest cent is $0.70.

◀ *Work Problem* **3** *at the Side.*

Note

Some stores round *all* money amounts up to the next higher cent, even if the next digit is *4 or less.* In Example 3(a) above, some stores would round $2.4238 *up* to $2.43, even though it is closer to $2.42.

It is also common to round money amounts to the nearest dollar. For example, you can do that on your federal and state income tax returns to make the calculations easier.

EXAMPLE 4 **Rounding to the Nearest Dollar**

Round to the nearest dollar.

(a) $48.69 (Is it closer to $48 or to $49?)

$$\$48.\,|\,69$$

First digit cut is *5 or more,* so round up by adding $1

$$\begin{array}{r} \$48 \\ + \ 1 \\ \hline \$49 \end{array}$$

Write $49 **not** $49.00

$48.69 rounded to the nearest dollar is $49

CAUTION

$48.69 rounded to the nearest dollar is $49. Be careful to write the answer as **$49** to show that the rounding is to the *nearest dollar*. Writing $49.00 would show rounding to the *nearest cent.*

Continued on Next Page

ANSWERS

3. **(a)** $14.60 **(b)** $578.07
   (c) $0.85 **(d)** $0.05

(b) $594.36 (Is it closer to $594 or to $595?)

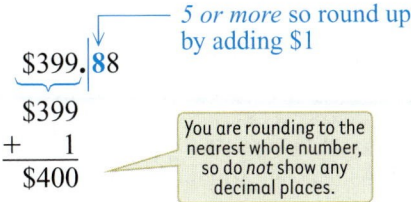

First digit cut is *4 or less,* so the part you are keeping stays the same.

$594.|36

$594

$594.36 rounded to the nearest dollar is $594 — Write $594 **not** $594.00

(c) $399.88 (Is it closer to $399 or to $400?)

$399.|88

5 or more so round up by adding $1

$399
+   1
$400

You are rounding to the nearest whole number, so do *not* show any decimal places.

$399.88 rounded to the nearest dollar is $400

(d) $2689.50 (Is it closer to $2689 or to $2690?)

$2689.|50

5 or more, so round up by adding $1

$2689
+   1
$2690

$2689.50 rounded to the nearest dollar is $2690

> **Note**
>
> When rounding $2689.50 to the nearest dollar, above, notice that it is exactly halfway between $2689 and $2690. When this happens in everyday situations, the rule is to round *up.* (Scientists working with technical data may use a more complicated rule when rounding numbers that are exactly in the middle.)

(e) $0.61 (Is it closer to $0 or to $1?)

$0.|61

5 or more, so round up.

$0.61 rounded to the nearest dollar is $1 — Write $1 **not** $1.00

> 🖩 **Calculator Tip** Accountants and other people who work with money amounts often set their calculators to automatically round to two decimal places (nearest cent) or to round to zero decimal places (nearest dollar). Your calculator may have this feature.

Work Problem **4** *at the Side.* ▶

---

4 Round to the nearest dollar.

  (a) $29.10

  (b) $136.49

  (c) $990.91

  (d) $5999.88

  (e) $49.60

  (f) $0.55

  (g) $1.08

ANSWERS

4. **(a)** $29   **(b)** $136   **(c)** $991
  (d) $6000   **(e)** $50   **(f)** $1   **(g)** $1

Math in the Media

LAWN FERTILIZER

Gotta Be Green

A lot's being written about personal responsibility these days, and the idea seems to be ending up on the front lawn—literally! Each spring, homeowners across the country gear up to green up their lawns, and the increased use of fertilizer has a lot of environmentalists concerned about the potential effects of chemical runoff into nearby rivers and streams.

Every year, according to a study conducted by the University of Minnesota's Department of Agriculture, each household in the Minneapolis/St. Paul metro area uses an average of 36 pounds of lawn fertilizer. That adds up to 25,529,295 pounds, or 12,765 tons. Add to that another 193,000 pounds of weed killer and you're looking at the total picture for keeping it green in the Twin Cities.

Source: Minneapolis Star Tribune.

1. According to the article,

    (a) How many pounds of lawn fertilizer are used each year in the *entire metro area?*

    (b) Do a division on your calculator to find the number of *households* in the metro area.

    (c) Why does it make sense to round your answer to part (b)? If so, how would you round it?

2. There are 2000 pounds in one ton.

    (a) Find the number of tons equivalent to 25,529,295 pounds of fertilizer.

    (b) Does your answer match the figure given in the article? If not, what did the author of the article do to get 12,765 tons?

    (c) Is the author's figure accurate? Why or why not?

3. According to the article, "each household in the Minneapolis/St. Paul metro area uses an average of 36 pounds of lawn fertilizer" each year. When the calculations were done to find the average, the answer was probably not *exactly* 36 pounds.

    (a) List three different values that are *less than* 36 that would round to 36. List a value with one decimal place, a value with two decimal places, and a value with three decimal places.

    (b) List three different values that are *greater than* 36 that would round to 36. List one value each with one, two, and three decimal places.

5.2 ▶▶▶ Exercises

Round each number to the place indicated. See Examples 1 and 2.

1. 16.8974 to the nearest tenth

2. 193.845 to the nearest hundredth

3. 0.95647 to the nearest thousandth

4. 96.81584 to the nearest ten-thousandth

5. 0.799 to the nearest hundredth

6. 0.952 to the nearest tenth

7. 3.66062 to the nearest thousandth

8. 1.5074 to the nearest hundredth

9. 793.988 to the nearest tenth

10. 476.1196 to the nearest thousandth

11. 0.09804 to the nearest ten-thousandth

12. 176.004 to the nearest tenth

13. 48.512 to the nearest one

14. 3.385 to the nearest one

15. 9.0906 to the nearest hundredth

16. 30.1290 to the nearest thousandth

17. 82.000151 to the nearest ten-thousandth

18. 0.400594 to the nearest ten-thousandth

Nardos is grocery shopping. The store will round the amount she pays for each item to the nearest cent. Write the rounded amounts. See Example 3.

19. Soup is three cans for $2.45, so one can is $0.81666. Nardos pays _____.

20. Orange juice is two cartons for $3.89, so one carton is $1.945. Nardos pays _____.

21. Facial tissue is four boxes for $4.89, so one box is $1.2225. Nardos pays _____.

22. Muffin mix is three packages for $1.75, so one package is $0.58333. Nardos pays _____.

23. Candy bars are six for $2.99, so one bar is $0.4983. Nardos pays _____.

24. Boxes of spaghetti are four for $4.39, so one box is $1.0975. Nardos pays _____.

As she gets ready to do her income tax return, Ms. Chen rounds each amount to the nearest dollar. Write the rounded amounts. See Example 4.

25. Income from job, $48,649.60

26. Income from interest on bank account, $69.58

27. Union dues, $310.08

28. Federal withholding, $6064.49

29. Donations to charity, $848.91

30. Medical expenses, $609.38

Round each money amount as indicated.

31. $499.98 to the nearest dollar

32. $9899.59 to the nearest dollar

33. $0.996 to the nearest cent

34. $0.09929 to the nearest cent

35. $999.73 to the nearest dollar

36. $9999.80 to the nearest dollar

The table lists speed records for various types of transportation. Use the table to answer Exercises 37–40.

Record	Speed (miles per hour)
Land speed record (specially built car)	763.04
Motorcycle speed record (specially adapted motorcycle)	322.16
Fastest rollercoaster	106.9
Fastest military jet	2193.167
Boeing 737-300 airplane (regular passenger service)	495
Indianapolis 500 auto race (fastest average winning speed)	185.981
Daytona 500 auto race (fastest average winning speed)	177.602

Sources: Guinness World Records and The World Almanac.

37. Round these speed records to the nearest whole number.

   (a) Motorcycle

   (b) Rollercoaster

38. Round these speed records to the nearest hundredth.

   (a) Daytona 500 average winning speed

   (b) Indianapolis 500 average winning speed

39. Round these speed records to the nearest tenth.

   (a) Indianapolis 500 average winning speed   **(b)** Land speed record

40. Round these speed records to the nearest hundred.

   (a) military jet   **(b)** Boeing 737-300 airplane

Relating Concepts (Exercises 41–44) For Individual or Group Work

Use your knowledge about rounding money amounts to **work Exercises 41–44** *in order.*

41. Explain what happens when you round $0.499 to the nearest dollar. Why does this happen?

42. Look again at Exercise 41. How else could you round $0.499 that would be more helpful? What kind of guideline does this suggest about rounding to the nearest dollar?

43. Explain what happens when you round $0.0015 to the nearest cent. Why does this happen?

44. Suppose you want to know which of these amounts is less, so you round them both to the nearest cent.

     $0.5968 $0.6014

Explain what happens. Describe what you could do instead of rounding to the nearest cent.

5.3 ▶▶▶ Adding and Subtracting Signed Decimal Numbers

OBJECTIVE **1** **Add and subtract positive decimals.** When adding or subtracting *whole* numbers, you line up the numbers in columns so that you are adding ones to ones, tens to tens, and so on. A similar idea applies to adding or subtracting *decimal* numbers. With decimals, you line up the decimal points to be sure that you are adding tenths to tenths, hundredths to hundredths, and so on.

> **Adding and Subtracting Decimal Numbers**
>
> *Step 1* Write the numbers in columns with the decimal points lined up.
>
> *Step 2* If necessary, write in zeros so both numbers have the same number of decimal places. Then add or subtract as if they were whole numbers.
>
> *Step 3* Line up the decimal point in the answer directly below the decimal points in the problem.

OBJECTIVES

1 Add and subtract positive decimals.

2 Add and subtract negative decimals.

3 Estimate the answer when adding or subtracting decimals.

EXAMPLE 1 Adding Decimal Numbers

Find each sum.

(a) 16.92 and 48.34

Step 1 Write the numbers in columns with the decimal points lined up.

$$
\begin{array}{r}
\text{tens}\ \text{ones}\ .\ \text{tenths}\ \text{hundredths} \\
1\,6\,.\,9\,2 \\
+\,4\,8\,.\,3\,4 \\
\hline
\end{array}
$$

Decimal points are lined up.

Step 2 Add as if these were whole numbers.

$$
\begin{array}{r}
1\,1\quad\ \\
16\,.\,92 \\
+\ 48\,.\,34 \\
\hline
65\,.\,26
\end{array}
$$

Step 3

Decimal point in answer is lined up under decimal points in problem.

(b) 5.897 + 4.632 + 12.174

Write the numbers in columns with the decimal points lined up. Then add.

$$
\begin{array}{r}
11\quad 21\quad \\
5\,.\,897 \\
4\,.\,632 \\
+\ 12\,.\,174 \\
\hline
22\,.\,703
\end{array}
$$

Decimal points are lined up.

Work Problem **1** *at the Side.* ▶

In Example 1(a) above, both numbers had *two* decimal places (two digits to the right of the decimal point). In Example 1(b), all the numbers had *three decimal places* (three digits to the right of the decimal point). That made it easy to add tenths to tenths, hundredths to hundredths, and so on.

1 Find each sum.

(a) 2.86 + 7.09

(b) 13.761 + 8.325

(c) 0.319 + 56.007 + 8.252

(d) 39.4 + 0.4 + 177.2

ANSWERS

1. (a) 9.95 **(b)** 22.086 **(c)** 64.578
  (d) 217.0

2 Find each sum.

(a) $6.54 + 9.8$

(b) $0.831 + 222.2 + 10$

(c) $8.64 + 39.115 + 3.0076$

(d) $5 + 429.823 + 0.76$

If the numbers of decimal places do *not* match, you can write in zeros as placeholders to make them match. This is shown in Example 2 below.

EXAMPLE 2   **Writing Zeros as Placeholders before Adding**

Find each sum.

(a) $7.3 + 0.85$

There are two decimal places in 0.85 (tenths and hundredths), so write a 0 in the hundredths place in 7.3 so that it has two decimal places also.

$$
\begin{array}{r}
7.3\mathbf{0} \;\leftarrow \text{One 0 is}\\
+\,0.85 \quad \text{written in.}\\
\hline
8.15
\end{array}
$$

7.30   is equivalent to   7.3   because

$$7\frac{30}{100} \text{ in lowest terms is } 7\frac{3}{10}$$

(b) $6.42 + 9 + 2.576$

Write in zeros so that all the addends have three decimal places. Notice how the whole number 9 is written with the decimal point on the *right* side. (If you put the decimal point on the *left* side of the 9, you would turn it into the decimal fraction 0.9.)

Write the decimal point on the *right* side of the 9.

$$
\begin{array}{r}
6.42\,\mathbf{0} \;\leftarrow \text{One 0 is written in.}\\
9.\mathbf{000} \;\leftarrow \text{9 is a whole number; decimal point}\\
\text{and three zeros are written in.}\\
+\,2.576 \;\leftarrow \text{No zeros are needed.}\\
\hline
17.996
\end{array}
$$

Note

Writing zeros to the right of a *decimal* number does *not* change the value of the number, as shown in Example 2(a) above.

◀ *Work Problem* **2** *at the Side.*

EXAMPLE 3   **Subtracting Decimal Numbers**

Find each difference.

(a) 15.82 from 28.93

Step 1

$$
\begin{array}{r}
28.93\\
-\,15.82
\end{array}
$$

Line up decimal points. Then you will be subtracting hundredths from hundredths and tenths from tenths.

Step 2

$$
\begin{array}{r}
28.93\\
-\,15.82\\
\hline
13\ 11
\end{array}
$$

Both numbers have two decimal places; no need to write in zeros.

⟵ Subtract as if they were whole numbers.

Step 3

$$
\begin{array}{r}
28.93\\
-\,15.82\\
\hline
13.11
\end{array}
$$

⟵ Decimal point in answer is lined up.

Continued on Next Page

(b) 146.35 minus 58.98

Regrouping is needed here.

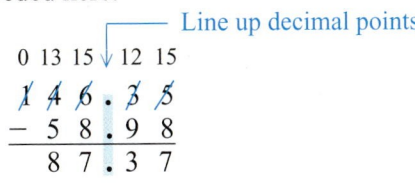

                — Line up decimal points.

$$\begin{array}{r} 0\ 13\ 15\ |\ 12\ 15 \\ 1\,4\,6\,.\,3\,5 \\ -\ \ 5\,8\,.\,9\,8 \\ \hline 8\,7\,.\,3\,7 \end{array}$$

Work Problem **3** *at the Side.* ▶

EXAMPLE 4   **Writing Zeros as Placeholders before Subtracting**

Find each difference.

(a) 16.5 from 28.362

Use the same steps as in Example 3 above. Remember to write in zeros so both numbers have three decimal places.

               — Line up decimal points.

16.500 is equivalent to 16.5	28.362
	$-\ 16.\mathbf{500}$ ← Write two zeros.
	$\overline{11.862}$ ← Subtract as usual.

(b) $59.7 - 38.914$

$$\begin{array}{r} 59.\mathbf{700} \leftarrow \text{Write two zeros.} \\ -\ 38.914 \\ \hline 20.786 \leftarrow \text{Subtract as usual.} \end{array}$$

(c) 12 less 5.83

$$12.\mathbf{00} \leftarrow \text{Write a decimal point and two zeros.}$$

12.00 is equivalent to 12	$-\ 5.83$
	$\overline{6.17}$ ← Subtract as usual.

Work Problem **4** *at the Side.* ▶

🖩 **Calculator Tip**   If you are *adding* decimal numbers, you can enter them in any order on your calculator. Try these; jot down the answers.

     9.82 ⊕ 1.86 ⊜ _____ 1.86 ⊕ 9.82 ⊜ _____

The answers are the same because addition is *commutative* (see **Section 1.3**). But subtraction is *not* commutative. It *does* matter which number you enter first. Try these:

     9.82 ⊖ 1.86 ⊜ _____ 1.86 ⊖ 9.82 ⊜ _____

The answers are 7.96 and -7.96. As you know, positive numbers are *greater* than 0, but negative numbers are *less* than 0. So it is important to do subtraction in the correct order, particularly if it is in your checkbook!

OBJECTIVE 2 **Add and subtract negative decimals.**   The rules that you used to add integers in **Section 1.3** will also work for positive and negative decimal numbers.

Adding Signed Numbers

To add two numbers with the *same* sign, add the absolute values of the numbers. Use the common sign as the sign of the sum.

To add two numbers with *unlike* signs, subtract the smaller absolute value from the larger absolute value. Use the sign of the number with the larger absolute value as the sign of the sum.

---

3 Find each difference.

(a) 22.7 from 72.9

(b) 6.425 from 11.813

(c) $\$20.15 - \$19.67$

4 Find each difference.

(a) 18.651 from 25.3

(b) $5.816 - 4.98$

(c) 40 less 3.66

(d) $1 - 0.325$

5 Find each sum.

(a) $13.245 + (-18)$

(b) $-0.7 + (-0.33)$

(c) $-6.02 + 100.5$

ANSWERS

5. **(a)** -4.755 **(b)** -1.03 **(c)** 94.48

EXAMPLE 5 **Adding Positive and Negative Decimal Numbers**

Find each sum.

(a) $-3.7 + (-16)$

Both addends are negative, so the sum will be negative. To begin, $|-3.7|$ is 3.7 and $|-16|$ is 16. Then add the absolute values.

$$
\begin{array}{r}
3.7 \\
+\ 16.\mathbf{0} \leftarrow \text{Write a decimal point and one 0} \\
\hline
19.7
\end{array}
$$

$$-3.7 + (-16) = -19.7$$

Both negative   Negative sum

Note

In **Chapters 1–4**, the negative sign was **red** to help you distinguish it from the subtraction symbol. From now on it will be **black.** We will continue to write parentheses around negative numbers when the negative sign might be confused with other symbols. Thus in part (a) above,

$$-3.7 + (-16) \quad \text{means} \quad \textbf{negative } 3.7 \quad \textbf{plus} \quad \textbf{negative } 16$$

(b) $-5.23 + 0.792$

The addends have different signs. To begin, $|-5.23|$ is 5.23 and $|0.792|$ is 0.792. Then subtract the smaller absolute value from the larger.

$$
\begin{array}{r}
5.23\mathbf{0} \leftarrow \text{Write one 0} \\
-\ 0.792 \\
\hline
4.438
\end{array}
$$

$$-5.23 + 0.792 = -4.438$$

Number with larger absolute value is negative.   Answer is negative.

◄ *Work Problem* **5** *at the Side.*

In **Section 1.4** you rewrote subtraction of integers as addition of the first number to the opposite of the second number. This same strategy works with positive and negative decimal numbers.

EXAMPLE 6 **Subtracting Positive and Negative Decimal Numbers**

Find each difference.

(a) $4.3 - 12.73$

Rewrite subtraction as adding the opposite.

$$4.3 \ \mathbf{-}\ \mathbf{12.73} \qquad \text{The opposite of 12.73 is } -12.73$$
$$4.3 \ \mathbf{+}\ (\mathbf{-12.73})$$

-12.73 has the larger absolute value and is negative, so the answer will be negative.

$$4.3 + (-12.73) = -8.43 \qquad \text{Subtract the absolute values:}$$

Answer is negative.
$$
\begin{array}{r}
12.73 \\
-\ 4.30 \\
\hline
8.43
\end{array}
$$

Continued on Next Page

(b) $-3.65 - (-4.8)$

Rewrite subtraction as adding the opposite.

$$-3.65 \ -\ (-\mathbf{4.8}) \qquad \text{The opposite of } -4.8 \text{ is } 4.8$$
$$\downarrow \quad \downarrow$$
$$-3.65 \ +\ \mathbf{4.8}$$

4.8 has the larger absolute value and is positive, so the answer will be positive.

$$-3.65 + 4.8 = 1.15 \qquad \text{Subtract the absolute values:}$$

$$\begin{array}{c} \uparrow \\ \text{Answer is} \\ \text{positive.} \end{array} \qquad \begin{array}{r} 4.80 \\ -3.65 \\ \hline 1.15 \end{array}$$

(c) $14.2 - \underbrace{(1.69 + 0.48)}$ Work inside parentheses first.

$14.2 - \quad (2.17)$ Change subtraction to adding the opposite.

$\underbrace{14.2 + \quad (-2.17)}$ 14.2 has the larger absolute value and is positive, so the answer will be positive.

12.03

Work Problem 6 *at the Side.* ▶

OBJECTIVE 3 **Estimate the answer when adding or subtracting decimals.** A common error when working decimal problems by hand is to misplace the decimal point in the answer. Or, when using a calculator, you may accidentally press the wrong key. Using *front end rounding* to estimate the answer will help you avoid these mistakes. Start by rounding each number to the highest possible place (as you did in **Section 1.5**). Here are several examples. Notice that in the rounded numbers, only the leftmost digit is something other than 0.

3.25	rounds to	3	6.812	rounds to	7
532.6	rounds to	500	26.397	rounds to	30
7094.2	rounds to	7000	351.24	rounds to	400

EXAMPLE 7 **Estimating Decimal Answers**

First, use front end rounding to round each number. Then add or subtract the rounded numbers to get an estimated answer. Finally, find the exact answer.

(a) Find the sum of 194.2 and 6.825

Estimate: *Exact:*

$$\begin{array}{r} 200 \\ + \quad 7 \\ \hline 207 \end{array} \begin{array}{l} \xleftarrow{\text{Rounds to}} \\ \xleftarrow{\text{Rounds to}} \end{array} \begin{array}{r} 194.200 \\ + \quad 6.825 \\ \hline 201.025 \end{array}$$

The estimate goes out to the hundreds place (three places to the *left* of the decimal point), and so does the exact answer. Therefore, the decimal point is probably in the correct place in the exact answer.

(b) Subtract $13.78 from $69.42

Estimate: *Exact:*

$$\begin{array}{r} \$70 \\ - \quad 10 \\ \hline \$60 \end{array} \begin{array}{l} \xleftarrow{\text{Rounds to}} \\ \xleftarrow{\text{Rounds to}} \end{array} \begin{array}{r} \$69.42 \\ - \quad 13.78 \\ \hline \$55.64 \end{array} \xleftarrow{\substack{\text{Exact answer is close to estimate, so} \\ \text{it is probably correct.}}}$$

Continued on Next Page

---

6 Find each difference.

(a) $-0.37 - (-6)$

(b) $5.8 - 10.03$

(c) $-312.72 - 65.7$

(d) $0.8 - (6 - 7.2)$

7 Use front end rounding to estimate each answer. Then find the exact answer.

(a) $2.83 + 5.009 + 76.1$

Estimate:

Exact:

(b) $19.28 less 1.53

Estimate:

Exact:

(c) $11.365 - 38$

Estimate:

Exact:

(d) $-214.6 + 300.72$

Estimate:

Exact:

(c) $-1.861 - 7.3$

Rewrite subtraction as adding the opposite. Then use front end rounding to get an estimated answer.

$$-1.861 \quad - \quad 7.3$$
$$\downarrow \quad\quad \downarrow$$
$$-1.861 \quad + \quad (-7.3)$$
$$\downarrow \quad\quad \downarrow$$
Rounded $\quad -2 \quad + \quad (-7) \quad = -9 \leftarrow$ Estimate

To find the exact answer, add the absolute values.

$$1.861 \leftarrow \text{Absolute value of } -1.861$$
$$+ \, 7.300 \leftarrow \text{Absolute value of } -7.3 \text{ with two zeros written in}$$
$$\overline{9.161}$$

The answer will be negative because both numbers are negative.

$$-1.861 + (-7.3) = -9.161 \leftarrow \text{Exact}$$

The exact answer of -9.161 is reasonable because it is close to the estimated answer of -9.

◀ *Work Problem* **7** *at the Side.*

5.3 ▶▶▶ Exercises

Find each sum. See Examples 1 and 2.

1. 5.69
    0.24
    + 11.79

2. 372.1
    33.7
    + 42.3

3. 0.38
    7
    + 4.6

4. 3.7
    0.812
    + 55

5. $14.23 + 8 + 74.63 + 18.715 + 0.286$

6. $197.4 + 0.72 + 17.43 + 25 + 1.4$

7. $27.65 + 18.714 + 9.749 + 3.21$

8. $58.546 + 19.2 + 8.735 + 14.58$

9. Explain and correct the error that a student made when he added $0.72 + 6 + 39.5$ this way:

    0.72
    6
    + 39.50
    40.28

10. Explain and correct the error that a student made when she added $7.21 + 65 + 13.15$ this way:

    7.21
    .65
    + 13.15
    21.01

11. Show why 0.3 is equivalent to 0.3000.

12. Explain why 7 may be written as 7.0 but not as 0.7.

Find each difference. See Examples 3 and 4.

13. $90.5 - 0.8$

14. $303.72 - 0.68$

15. 0.4 less 0.291

16. 0.35 less 0.088

17. 6 minus 5.09

18. 80 minus 16.3

19. Subtract 8.339 from 15

20. Subtract 0.08 from 44

21. Explain and correct the error that a student made when he subtracted 7.45 from 15.32 this way:

    7.45
    − 15.32
    12.13

22. Explain the difference between saying "subtract 2.9 from 8" and saying "2.9 minus 8."

This drawing of a human skeleton shows the average length of the longest bones, in inches.
Use the drawing to answer Exercises 23–26.

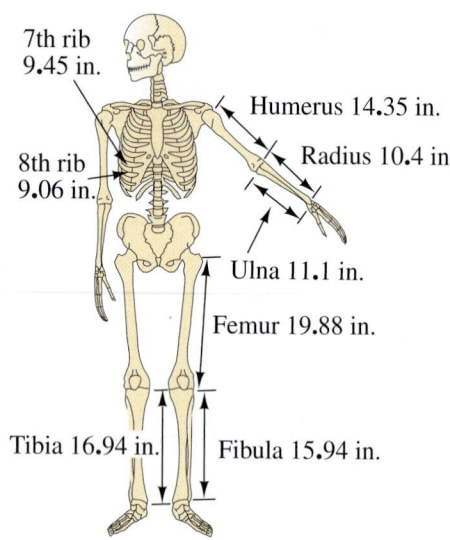

7th rib
9.45 in.

Humerus 14.35 in.

Radius 10.4 in.

8th rib
9.06 in.

Ulna 11.1 in.

Femur 19.88 in.

Tibia 16.94 in.   Fibula 15.94 in.

Source: *The Top 10 of Everything.*

23. (a) What is the combined length of the humerus and radius bones?

(b) What is the difference in the lengths of these two bones?

24. (a) What is the total length of the femur and tibia bones?

(b) How much longer is the femur than the tibia?

25. (a) Find the sum of the lengths of the humerus, ulna, femur, and tibia.

(b) How much shorter is the 8th rib than the 7th rib?

26. (a) What is the difference in the lengths of the two bones in the lower arm?

(b) What is the difference in the lengths of the two bones in the lower leg?

Find each sum or difference. See Examples 5 and 6.

27. $24.008 + (-0.995)$

28. $0.77 - 3.06$

29. $-6.05 + (-39.7)$

30. $-6.409 + 8.224$

31. $0.9 - 7.59$

32. $-489.7 - 38$

33. $-2 - 4.99$

34. $2.068 - (-32.7)$

35. $-5.009 + 0.73$

36. $-0.33 - 65$

37. $-1.7035 - (5 - 6.7)$

38. $60 + (-0.9345 + 1.4)$

39. $8000 - (8002.63 - 8)$

40. $-210 - (-0.7306 + 0.5)$

Use front end rounding to estimate each answer. Then use your estimate to select the correct exact answer. Circle your choice. See Example 7.

41. 18 − 11.725

Estimate:

Exact: 29.725   6.275   −11.545

42. 20 − 1.37

Estimate:

Exact: −21.37   −1.863   18.63

43. −6.5 + 0.7

Estimate:

Exact: −5.8   7.2   −0.58

44. −9.67 + 3.09

Estimate:

Exact: −12.76   −6.58   6.58

45. −42.671 − 194.9

Estimate:

Exact: −152.229   −23.7571
        −237.571

46. −803.25 − 0.6

Estimate:

Exact: −803.85   80.385
        803.85

47. 8.4 − (−50.83)

Estimate:

Exact: −42.43   −59.23   59.23

48. 14.98 − (−6.506)

Estimate:

Exact: −8.474   21.486   8.004

Use front end rounding to estimate each sum or difference. Then find the exact answer to each application problem. Use the information in the table for Exercises 49–52.

INTERNET USERS IN SELECTED COUNTRIES

Country	Number of Users
United States	197.8 million
China	119.5 million
Japan	86.3 million
India	50.6 million
South Korea	34 million
Canada	21.9 million
Mexico	16.9 million
World total	**1081.1 million**

Source: Computer Industry Almanac.

49. How many fewer Internet users are there in Canada compared to South Korea?

Estimate:

Exact:

50. How many more users are there in China compared to India?

Estimate:

Exact:

51. How many Internet users are there in all the countries listed in the table?

Estimate:

Exact:

52. Using the answer from Exercise 51, calculate the number of worldwide Internet users in countries other than the ones in the table.

Estimate:

Exact:

53. The tallest known land mammal is a prehistoric ancestor of the rhino, measuring 6.4 m. Compare the rhino's height to the combined heights of these NBA basketball stars: Kevin Garnett at 2.1 m, Allen Iverson at 1.83 m, and Shaquille O'Neal at 2.16 m. Is their combined height greater or less than the prehistoric rhino's? By how much? (*Source:* www.NBA.com/players)

6.4 m

Estimate:

Exact:

54. Sammy works in a veterinarian's office. He weighed two kittens. One was 3.9 ounces and the other was 4.05 ounces. What was the difference in the weight of the two kittens?

Estimate:

Exact:

55. Steven One Feather gave the cashier a $20 bill to pay for $9.12 worth of groceries. How much change did he get?

Estimate:

Exact:

56. The cost of Julie's tennis racket, with tax, was $41.09. She gave the clerk two $20 bills and a $10 bill. What amount of change did Julie receive?

Estimate:

Exact:

*When buying fishing equipment, the father and daughter on the first page of this chapter brought along the store's sale insert from the Sunday paper. Use the information below on sale prices to answer Exercises 57–60. **When estimating, round to the nearest whole number.***

Fishing Opener Sale
Catch your limit of savings!

Bobbers 3 for 87¢

Environmentally safe tin split shot
$2.07
Leaded split shot
94¢

8-pound test fishing line
regular $4.84
invisible $7.47
fluorescent $5.14
No-See Line

Tackle boxes
Two trays $7.96   Three trays $9.96

Spinning reels: $9.88, $12.54, $18.84, $24.96
Spinning rods: $9.97, $18.97, $22.96, $28.94

Source: Wal-Mart.

57. What is the difference in price between the fluorescent and regular fishing line?

Estimate:

Exact:

58. How much more does the least expensive spinning rod cost than the least expensive spinning reel?

Estimate:

Exact:

59. Find the total cost of the second-highest-priced spinning reel, two packages of tin split shot, and a three-tray tackle box. Sales tax for all the items is $2.31.

Estimate:

Exact:

60. The father bought three bobbers on sale. He also bought some SPF15 sunscreen for $7.53 and a flotation vest for $44.96. Sales tax was $3.74. How much did he spend in all?

Estimate:

Exact:

Olivia Sanchez kept track of her expenses for one month. Use her list to answer Exercises 61–64.

61. What were Olivia's total expenses for the month?

62. How much did Olivia pay for her cell phone, cable TV, and Internet access?

63. What was the difference in the amounts spent for groceries and for the car payment?

64. How much more did Olivia spend on rent than on all her car expenses?

Monthly Expenses	
Rent	$994
Car payment	$190.78
Car repairs, gas	$205
Cable TV	$39.95
Internet access	$19.95
Electricity	$40.80
Cell phone	$57.32
Groceries	$186.81
Entertainment	$97.75
Clothing, laundry	$107

Find the length of the dashed line in each rectangle or circle.

65.

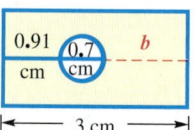

0.91 cm   0.7 cm b
3 cm

66.

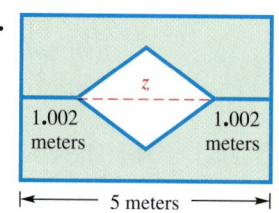

z
1.002 meters   1.002 meters
5 meters

67.

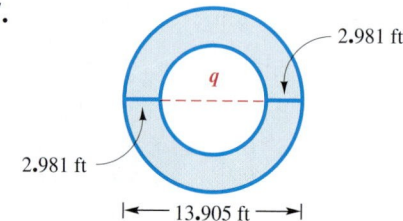

2.981 ft
q
2.981 ft
13.905 ft

5.4 ▶▶▶ Multiplying Signed Decimal Numbers

OBJECTIVE 1 Multiply positive and negative decimals. The decimals 0.3 and 0.07 can be multiplied by writing them as fractions.

$$0.3 \times 0.07 = \frac{3}{10} \times \frac{7}{100} = \frac{3 \times 7}{10 \times 100} = \frac{21}{1000} = 0.021$$

1 decimal place + 2 decimal places → 3 decimal places

Can you see a way to multiply decimals without writing them as fractions? Try these steps. Remember that each number in a multiplication problem is called a *factor*, and the answer is called the *product*.

OBJECTIVES

1 Multiply positive and negative decimals.

2 Estimate the answer when multiplying decimals.

Multiplying Two Decimal Numbers

Step 1 Multiply the factors (the numbers being multiplied) as if they were whole numbers.

Step 2 Find the *total* number of decimal places in *both* factors.

Step 3 Write the decimal point in the product (the answer) so it has the same number of decimal places as the total from Step 2. You may need to write in extra zeros on the left side of the product in order to get the correct number of decimal places.

Step 4 If two factors have the *same sign,* the product is *positive*. If two factors have *different signs,* the product is *negative*.

Note

When multiplying decimals, you do **not** need to line up decimal points. (You *do* need to line up decimal points when adding or subtracting.)

EXAMPLE 1 Multiplying Decimal Numbers

Find the product of 8.34 and (−4.2)

Step 1 Multiply the numbers as if they were whole numbers.

$$\begin{array}{r} 8.34 \\ \times\ 4.2 \\ \hline 1668 \\ 3336 \\ \hline 35028 \end{array}$$

You do *not* have to line up decimal points when multiplying.

Step 2 Count the total number of decimal places in both factors.

$$\begin{array}{r} 8.34 \leftarrow 2\text{ decimal places} \\ \times\ 4.2 \leftarrow 1\text{ decimal place} \\ \hline 1668 \quad 3\text{ total decimal places} \\ 3336 \\ \hline 35028 \end{array}$$

Continued on Next Page

1 Find each product.

(a) $-2.6\,(0.4)$

(b) $(45.2)\,(0.25)$

(c)
$$
\begin{array}{r}
0.104 \leftarrow \text{3 decimal places} \\
\times \qquad 7 \leftarrow \text{0 decimal places} \\
\hline
\leftarrow \text{3 decimal places} \\
\text{in the product}
\end{array}
$$

(d) $(-3.18)^2$
Hint: Recall that squaring a number means multiplying the number by itself, so this is $(-3.18)\,(-3.18)$.

2 Find each product.

(a) $0.04\,(-0.09)$

(b) $(0.2)\,(0.008)$

(c) $(-0.063)\,(-0.04)$

(d) $(0.003)^2$

Step 3 Count over 3 places in the product and write the decimal point. Count from *right to left*.

$$
\begin{array}{r}
8.3\,4 \leftarrow \text{2 decimal places} \\
\times \quad 4.2 \leftarrow \text{1 decimal place} \\
\hline
1\,6\,6\,8 \qquad \text{3 total decimal places} \\
3\,3\,3\,6 \qquad \\
\hline
3\,5.0\,2\,8 \leftarrow \text{3 decimal places in product}
\end{array}
$$

Count over 3 places from right to left to position the decimal point.

Step 4 The factors have *different* signs, so the product is *negative:*
8.34 times (-4.2) is -35.028

◀ *Work Problem* **1** *at the Side.*

EXAMPLE 2 **Writing Zeros as Placeholders in the Product**

Find the product: $(-0.042)\,(-0.03)$

Start by multiplying, then count decimal places.

$$
\begin{array}{r}
0.0\,4\,2 \leftarrow \text{3 decimal places} \\
\times \quad 0.0\,3 \leftarrow \text{2 decimal places} \\
\hline
1\,2\,6 \leftarrow \text{5 decimal places needed in product}
\end{array}
$$

After multiplying, the answer has only three decimal places, but five are needed. So write two zeros on the *left* side of the answer.

$$
\begin{array}{r}
0.0\,4\,2 \\
\times \quad 0.0\,3 \\
\hline
\mathbf{0\,0}\,1\,2\,6 \\
\uparrow \uparrow
\end{array}
\qquad
\begin{array}{r}
0.0\,4\,2 \leftarrow \text{3 decimal places} \\
\times \quad 0.0\,3 \leftarrow \text{2 decimal places} \\
\hline
.0\,0\,1\,2\,6 \leftarrow \text{5 decimal places}
\end{array}
$$

Write two zeros on *left* side of answer.

Now count over 5 places and write in the decimal point.

The final product is 0.00126, which has five decimal places. The product is *positive* because the factors have the *same* sign (both factors are negative).

◀ *Work Problem* **2** *at the Side.*

⊞ **Calculator Tip** When working with money amounts, you may need to write a 0 in your answer. For example, try multiplying $\$3.54 \times 5$ on your calculator. Write down the result.

$$
3.54 \; \otimes \; 5 \; \ominus \; \underline{\qquad}
$$

Notice that the result is 17.7, which is *not* the way to write a money amount. You have to write the 0 in the hundredths place: $\$17.7\mathbf{0}$ is correct. The calculator does not show the "extra" 0 because:

$$
17.70 \text{ or } 17\frac{70}{100} \quad \text{simplifies to} \quad 17\frac{7}{10} \text{ or } 17.7
$$

So keep an eye on your calculator—it doesn't know when you're working with money amounts.

ANSWERS

1. **(a)** -1.04 **(b)** 11.300 or 11.3 **(c)** 0.728
   (d) 10.1124
2. **(a)** -0.0036 **(b)** 0.0016 **(c)** 0.00252
   (d) 0.000009

OBJECTIVE 2 Estimate the answer when multiplying decimals.
If you are doing multiplication problems by hand, estimating the answer helps you check that the decimal point is in the right place. When you are using a calculator, estimating helps you catch an error like pressing the ÷ key instead of the × key.

EXAMPLE 3   Estimating before Multiplying

First, use front end rounding to estimate (76.34) (12.5). Then find the exact answer.

Estimate: *Exact:*

$$
\begin{array}{r}
80 \\
\times\ 10 \\
\hline
800
\end{array}
$$
Rounds to

$$
\begin{array}{r}
7\,6.3\,4 \quad \leftarrow \text{2 decimal places}\\
\times\ \ 1\,2.5 \quad \leftarrow \text{1 decimal place}\\
\hline
3\,8\,1\,7\,0 \\
1\,5\,2\,6\,8 \\
7\,6\,3\,4 \\
\hline
9\,5\,4.2\,5\,0
\end{array}
$$
3 decimal places are in the product.

Both the estimate and the exact answer go out to the hundreds place, so the decimal point in 954.250 is probably in the correct place.

*Work Problem **3** at the Side.* ▶

3 First, use front end rounding to estimate the answer. Then find the exact answer.

(a) (11.62) (4.01)

Estimate:

Exact:

(b) (−5.986) (−33)

Estimate:

Exact:

(c) 8 ($4.35)

Estimate:

Exact:

(d) 58.6 (−17.4)

Estimate:

Exact:

ANSWERS

3. **(a)** *Estimate:* (10) (4) = 40; *Exact:* 46.5962
   (b) *Estimate:* (−6) (−30) = 180; *Exact:* 197.538
   (c) *Estimate:* 8 ($4) = $32; *Exact:* $34.80
   (d) *Estimate:* 60 (−20) = −1200; *Exact:* −1019.64

Math in the Media

Many Americans complain about being "too busy." Yet, according to surveys, they find time to watch television. Use the information in the table to answer the questions on TV viewing.

Time Spent by American Men Watching TV
2.5 hours on an average weekday
3.59 hours on an average weekend day
Time Spent by American Women Watching TV
2.22 hours on an average weekday
2.7 hours on an average weekend day

Sources: American Time Use Survey, Bureau of
Labor Statistics and Americans' Use of Time Project.

1. **(a)** How many more hours do men spend than women watching TV on an average weekend day? How many minutes is this, to the nearest whole minute?

   (b) Who spends less time watching TV on an average weekday, men or women? How much less time in hours? How many minutes is this, to the nearest whole minute?

2. **(a)** How many hours would a man spend watching TV during one week with five weekdays and two weekend days? How many hours in one year, to the nearest whole hour? (Use 1 year = 52 weeks.)

   (b) How many more or fewer hours would a woman spend than a man watching TV in the same week as part (a)? How many more or fewer hours in one year, to the nearest whole hour?

3. **(a)** How many hours do you spend watching TV on an average weekday? On an average weekend day?

   (b) Using your answers from part (a), how many hours do you watch TV during an average week? During one year?

5.4 ▶▶▶ Exercises

Find each product. See Example 1.

1. $\begin{array}{r} 0.042 \\ \times\ \ 3.2 \\ \hline \end{array}$

2. $\begin{array}{r} 0.571 \\ \times\ \ 2.9 \\ \hline \end{array}$

3. $-21.5(7.4)$

4. $-85.4(-3.5)$

5. $(-23.4)(-0.66)$

6. $0.896(-0.7)$

7. $\begin{array}{r} \$51.88 \\ \times\ \ \ 665 \\ \hline \end{array}$

8. $\begin{array}{r} \$736.75 \\ \times\ \ \ \ 118 \\ \hline \end{array}$

Use the fact that $(72)(6) = 432$ *to solve Exercises 9–16 by simply counting decimal places and writing the decimal point in the correct location. Be sure to indicate the sign of the product.*

9. $72(-0.6) = $ ⠀4 3 2

10. $7.2(-6) = $ ⠀4 3 2

11. $(7.2)(0.06) = $ ⠀4 3 2

12. $(0.72)(0.6) = $ ⠀4 3 2

13. $-0.72(-0.06) = $ ⠀4 3 2

14. $-72(-0.0006) = $ ⠀4 3 2

15. $(0.0072)(0.6) = $ ⠀4 3 2

16. $(0.072)(0.006) = $ ⠀4 3 2

Find each product. See Example 2.

17. $(0.006)(0.0052)$

18. $(0.0052)(0.009)$

19. $(-0.003)^2$

20. $(0.0004)^2$

Relating Concepts (Exercises 21–22) For Individual or Group Work

Look for patterns as you **work Exercises 21 and 22 in order.**

21. Do these multiplications:

$(5.96)(10) = $ _____ $(3.2)(10) = $ _____

$(0.476)(10) = $ _____ $(80.35)(10) = $ _____

$(722.6)(10) = $ _____ $(0.9)(10) = $ _____

What pattern do you see? Write a "rule" for multiplying by 10. What do you think the rule is for multiplying by 100? by 1000? Write the rules and try them out on the numbers above.

22. Do these multiplications:

$(59.6)(0.1) = $ _____ $(3.2)(0.1) = $ _____

$(0.476)(0.1) = $ _____ $(80.35)(0.1) = $ _____

$(65)(0.1) = $ _____ $(523)(0.1) = $ _____

What pattern do you see? Write a "rule" for multiplying by 0.1. What do you think the rule is for multiplying by 0.01? by 0.001? Write the rules and try them out on the numbers above.

First, use front end rounding to estimate the answer. Then find the exact answer.
See Example 3.

23. *Estimate:*   *Exact:*
⊕   ← Rounds to   39.6
                ← Rounds to
   × ____   × 4.8

24. *Estimate:*   *Exact:*
                                 18.7
   × ____   × 2.3

25. *Estimate:*   *Exact:*
                                 37.1
   × ____   × 42

26. *Estimate:*   *Exact:*
                                 5.08
   × ____   × 71

27. *Estimate:*   *Exact:*
                                 6.53
   × ____   × 4.6

28. *Estimate:*   *Exact:*
                                 7.51
   × ____   × 8.2

29. *Estimate:*   *Exact:*
                                 2.809
   × ____   × 6.85

30. *Estimate:*   *Exact:*
                                 73.52
   × ____   × 22.34

Even with most of the problem missing, you can tell whether or not these answers are
reasonable. Circle reasonable or unreasonable. If the answer is unreasonable, move the
decimal point or insert a decimal point to make the answer reasonable.

31. How much was his car payment? $28.90

reasonable

unreasonable, should be _____

32. How many hours did she work today? 25 hours

reasonable

unreasonable, should be _____

33. How tall is her son? 60.5 inches

reasonable

unreasonable, should be _____

34. How much does he pay for rent now? $6.92

reasonable

unreasonable, should be _____

35. What is the price of one gallon of milk? $419

reasonable

unreasonable, should be _____

36. How long is the living room? 16.8 feet

reasonable

unreasonable, should be _____

37. How much did the baby weigh? 0.095 pound

reasonable

unreasonable, should be _____

38. What was the sale price of the jacket? $1.49

reasonable

unreasonable, should be _____

Solve each application problem. Round money answers to the nearest cent when necessary.

39. LaTasha worked 50.5 hours over the last two weeks.
She earns $18.73 per hour. How much did she make?

40. Michael's time card shows 42.2 hours at $10.03 per
hour. What are his earnings?

41. Sid needs a piece of canvas material 0.6 meter in length to make a carry-all bag that fits on his wheelchair. If canvas is $4.09 per meter, how much will Sid spend? (*Note:* $4.09 *per* meter means $4.09 for *one* meter.)

42. How much will Mrs. Nguyen pay for 3.5 yards of lace trim that costs $0.87 per yard?

43. Michelle filled the tank of her SUV with regular unleaded gas. Use the information shown on the pump to find how much she paid for gas.

GALLONS	PRICE PER GALLON	GALLONS	PRICE PER GALLON
10.329	$ 4.289	20.510	$ 3.979

SUPRA UNLEADED
Minimum Octane Rating 90

UNLEADED REGULAR
Minimum Octane Rating 87

Source: Holiday.

44. Ground beef and chicken legs are on sale. Juma bought 1.7 pounds of legs. Use the information in the ad to find the amount she paid.

BIG ONE FOODS Sale

Ground Beef
$2.09
per pound
For juicy burgers

Chicken Legs
$0.98
per pound

PRICES GOOD THROUGH SUNDAY!

45. Ms. Rolack is a real estate broker who helps people sell their homes. Her fee is 0.07 times the price of the home. What was her fee for selling a $289,500 home?

46. Manny Ramirez of the Boston Red Sox had a batting average of 0.296 in the 2007 season. He went to bat 483 times. How many hits did he make? (*Hint:* Multiply the number of times at bat by his batting average.) Round to the nearest whole number. (*Source: World Almanac.*)

Paper money in the United States has not always been the same size. Shown below are the measurements of bills printed before 1929 and the measurements from 1929 on. Use this information to answer Exercises 47–50. Recall that perimeter is the total distance around the edges of a figure (see **Section 3.1**). *The area of a rectangle is found by multiplying the length by the width (see* **Section 3.2**). *(Source: www.moneyfactory.com)*

Before 1929

3.125 in.

7.4218 in.

From 1929 on

2.61 in.

6.14 in.

47. (a) Find the area of each bill, rounded to the nearest tenth.

(b) What is the difference in the rounded areas?

48. (a) Find the perimeter of each bill, to the nearest hundredth.

(b) How much less is the perimeter of today's bills than the bills printed before 1929?

49. The thickness of one piece of today's paper money is 0.0043 inch.

(a) If you had a pile of 100 bills, how high would it be?

(b) How high would a pile of 1000 bills be?

50. (a) Use your answers from Exercise 49 to find the number of bills in a pile that is 43 inches high.

(b) How much money would you have if the pile is all $20 bills?

51. Judy Lewis pays $38.96 per month for basic cable TV. The one-time installation fee was $49. How much will she pay for cable over two years? How much would she pay in two years for the deluxe cable package that costs $89.95 per month?

52. Chuck's car payment is $420.27 per month for four years. He also made a down payment of $5000 at the time he bought the car. How much will he pay altogether?

53. Paper for the copy machine at the library costs $0.015 per sheet. How much will the library pay for 5100 sheets?

54. A student group collected 2200 pounds of plastic as a fund-raiser. How much will they make if the recycling center pays $0.142 per pound?

55. Barry bought 16.5 meters of rope at $0.47 per meter and three meters of wire at $1.05 per meter. How much change did he get from three $5 bills?

56. Susan bought a 42-inch plasma HDTV that cost $1999.99. She paid $68.83 per month for 36 months. How much could she have saved by paying for the HDTV when she bought it?

Use the information below from the Look Smart mail order catalog to answer Exercises 57–60.

Knit Shirt Ordering Information		
43–2A	short sleeved, solid colors	$14.75 each
43–2B	short sleeved, stripes	$16.75 each
43–3A	long sleeved, solid colors	$18.95 each
43–3B	long sleeved, stripes	$21.95 each
XXL size, add $2 per shirt.		
Monogram, $4.95 each. Gift box, $5 each.		

Total Price of Items (excluding monograms and gift boxes)	Shipping, Packing, and Handling
$0–25.00	$3.50
$25.01–75.00	$5.95
$75.01–125.00	$7.95
$125.01+	$9.95
Shipping to each additional address, add $4.25.	

57. Find the total cost of ordering four long-sleeved, solid-color shirts and two short-sleeved, striped shirts, all in the XXL size, and all shipped to your home.

58. What is the total cost of eight long-sleeved shirts, five in solid colors and three striped? None of the shirts are the XXL size. Include the cost of shipping the solid shirts to your home and the striped shirts to your brother's home.

59. (a) What is the total cost, including shipping, of sending three short-sleeved, solid-color shirts, size M, with monograms, in a gift box to your aunt for her birthday?

(b) How much did the monograms, gift box, and shipping add to the cost of your gift?

60. (a) Suppose you order one of each type of shirt for yourself, adding a monogram on each of the solid-color shirts. At the same time, you order three long-sleeved striped shirts, in the XXL size, shipped to your dad in a gift box. Find the total cost of your order.

(b) What is the difference in total cost (excluding shipping) between the shirts for yourself and the gift for your dad?

5.5 ▶▶▶ Dividing Signed Decimal Numbers

There are two kinds of decimal division problems: those in which a decimal is divided by an integer, and those in which a number is divided by a decimal. First recall the parts of a division problem.

$$\begin{array}{c} 8 \leftarrow \text{Quotient} \\ \text{Divisor} \rightarrow 2\overline{)16} \leftarrow \text{Dividend} \end{array}$$

Dividend
↓   ⌐ Divisor
$16 \div 2 = 8$
         ↑
     Quotient

Dividend → $\dfrac{16}{2}$ = 8
Divisor →   ↑
              Quotient

OBJECTIVES

1. Divide a decimal by an integer.

2. Divide a number by a decimal.

3. Estimate the answer when dividing decimals.

4. Use the order of operations with decimals.

OBJECTIVE **1** **Divide a decimal by an integer.** When the divisor is an integer, use these steps.

> **Dividing a Decimal Number by an Integer**
>
> *Step 1* Write the decimal point in the quotient (answer) directly above the decimal point in the dividend.
>
> *Step 2* Divide as if both numbers were whole numbers.
>
> *Step 3* If both numbers have the *same sign,* the quotient is *positive*. If they have *different signs,* the quotient is *negative*.

EXAMPLE 1 **Dividing Decimals by Integers**

Find each quotient. Check the quotients by multiplying.

(a) $\underbrace{21.93}_{\text{Dividend}} \div \underbrace{(-3)}_{\text{Divisor}}$

First consider $21.93 \div 3$.

$3\overline{)21.93}$

Step 1 Write the decimal point in the quotient directly above the decimal point in the dividend.

Decimal points lined up

$3\overline{)21\overset{\bullet}{.}93}$

Step 2 Divide as if the numbers were whole numbers.

Check by multiplying the quotient times the divisor.

$\begin{array}{r} 7.31 \\ 3\overline{)21.93} \end{array}$

Matches, so 7.31 is correct.

Check

$\begin{array}{r} 7.31 \\ \times\ \ \ 3 \\ \hline 21.93 \end{array}$

Step 3 The quotient is -7.31 because the numbers have *different* signs.

$21.93 \div (-3) = -7.31$

Different signs, negative quotient.

Different signs   Negative quotient

Continued on Next Page

1 Divide. Check the quotients by multiplying.

(a) $4\overline{)93.6}$

(b) $6\overline{)6.804}$

(c) $\dfrac{278.3}{11}$

(d) $-0.51835 \div 5$

(e) $-213.45 \div (-15)$

(b)

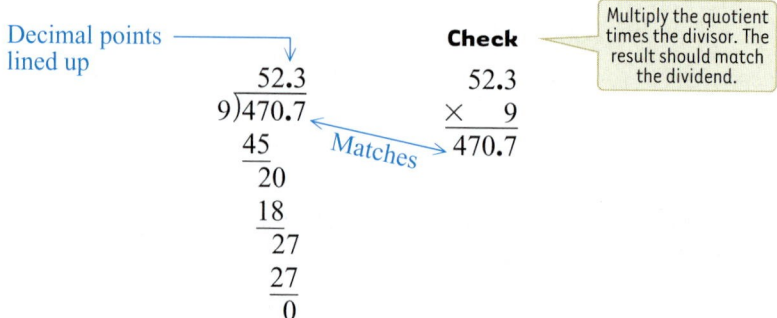

Write the decimal point in the quotient directly above the decimal point in the dividend. Then divide as if they were whole numbers.

Decimal points lined up

$$
\begin{array}{r}
52.3 \\
9\overline{)470.7} \\
\underline{45} \\
20 \\
\underline{18} \\
27 \\
\underline{27} \\
0
\end{array}
$$

Check
$$
\begin{array}{r}
52.3 \\
\times\ \ 9 \\
\hline
470.7
\end{array}
$$

Matches

Multiply the quotient times the divisor. The result should match the dividend.

The quotient is 52.3 and is *positive* because both numbers have the *same* sign.

◀ Work Problem **1** at the Side.

EXAMPLE 2 Writing Extra Zeros to Complete a Division

Divide 1.5 by 8. Use multiplying to check the quotient.

Keep dividing until the remainder is 0, or until the digits in the quotient begin to repeat in a pattern. In Example 1(b) above, you ended up with a remainder of 0. But sometimes you run out of digits in the dividend before that happens. If so, write extra zeros on the right side of the dividend so you can continue dividing.

$$
\begin{array}{r}
0.1 \\
8\overline{)1.5} \\
\underline{8} \\
7
\end{array}
$$
← All digits have been used.

← Remainder is not yet 0.

Write a 0 after the 5 in the dividend so you can continue dividing. Keep writing more zeros in the dividend if needed. Recall that writing zeros to the *right* of a decimal number does **not** change its value.

$$
\begin{array}{r}
0.1\ 8\ 7\ 5 \\
8\overline{)1.5\ 0\ 0\ 0} \\
\underline{8} \\
7\ 0 \\
\underline{6\ 4} \\
6\ 0 \\
\underline{5\ 6} \\
4\ 0 \\
\underline{4\ 0} \\
0
\end{array}
$$
← Three zeros needed to complete the division

← Stop dividing when the remainder is 0.

Check
$$
\begin{array}{r}
0.1875 \\
\times\ \ \ \ 8 \\
\hline
1.5000
\end{array}
$$

Matches dividend, so 0.1875 is correct.

CAUTION
Notice that in decimals the dividend might *not* be the larger number. In Example 2 above, the dividend is 1.5, which is *smaller* than 8.

———— **Continued on Next Page**

📱 **Calculator Tip** When *multiplying* numbers, you can enter them in any order because multiplication is commutative (see **Section 1.6**). But division is *not* commutative. It *does* matter which number you enter first. Try Example 2 on the previous page both ways; jot down your answers.

1.5 ÷ 8 = _____   8 ÷ 1.5 = _____

Notice that the first answer, 0.1875, matches the result from Example 2. But the second answer is much different: 5.333333333. Be careful to enter the dividend first.

Work Problem **2** *at the Side.* ▶

In the next example the remainder is never 0, even if we keep dividing.

EXAMPLE 3   **Rounding a Decimal Quotient**

Divide 4.7 by 3. Round the quotient to the nearest thousandth.

Write extra zeros in the dividend so that you can continue dividing.

```
      1.5 6 6 6
   3)4.7 0 0 0   ← Three zeros added so far
     3
     ───
     1 7
     1 5 ↓
     ───
       2 0
       1 8 ↓
       ───
         2 0
         1 8 ↓
         ───
           2 0
           1 8
           ───
             2   ← Remainder is still not 0.
```

Notice that the digit 6 in the quotient is repeating. It will continue to do so. The remainder will *never be 0*. There are two ways to show that an answer is a **repeating decimal** that goes on forever. You can write three dots after the answer, or you can write a bar above the digits that repeat (in this case, the 6).

$$1.5\underline{666}\ldots \quad \text{or} \quad 1.5\overline{6}$$
            Three dots   ← Bar above repeating digit

CAUTION
Do not use *both* the dots *and* the bar at the same time. Use three dots *or* the bar.

When repeating decimals occur, round the quotient according to the directions in the problem. In this example, to round to thousandths, divide out one *more* place, to ten-thousandths.

$4.7 \div 3 = 1.5666\ldots$ rounds to 1.567 ⟵ Nearest **thousandth** is **three** decimal places.

Check the answer by multiplying 1.567 by 3. Because 1.567 is a rounded answer, the check will *not* give exactly 4.7, but it should be very close.

$(1.567)(3) = 4.701$ ← Does not equal exactly 4.7 because 1.567 was rounded

Continued on Next Page

2 Divide. Check the quotients by multiplying.

(a) $\dfrac{6.4}{5}$

(b) $30.87 \div (-14)$

(c) $\dfrac{-259.5}{-30}$

(d) $0.3 \div 8$

ANSWERS

2. (a) 1.28;
   Check (1.28)(5) = 6.40 or 6.4
   (b) −2.205;
   Check (−2.205)(−14) = 30.870 or 30.87
   (c) 8.65;
   Check (8.65)(−30) = −259.50 or −259.5
   (d) 0.0375;
   Check (0.0375)(8) = 0.3000 or 0.3

3 Divide. Round quotients to the nearest thousandth. If it is a repeating decimal, also write the answer using a bar. Check your answers by multiplying.

(a) $13\overline{)267.01}$

(b) $6\overline{)20.5}$

(c) $\dfrac{10.22}{9}$

(d) $16.15 \div 3$

(e) $116.3 \div 11$

CAUTION
When checking quotients that you've rounded, the check will *not* match the dividend exactly, but it should be very close.

◀ *Work Problem* **3** *at the Side.*

OBJECTIVE **2** **Divide a number by a decimal.** To divide by a *decimal* divisor, first change the divisor to a whole number. Then divide as before. To see how this is done, write the problem in fraction form. Here is an example.

$$1.2\overline{)6.36} \quad \text{can be written} \quad \frac{6.36}{1.2}$$

In **Section 4.1** you learned that multiplying the numerator and denominator by the same number gives an equivalent fraction. We want the divisor (1.2) to be a whole number. Multiplying by 10 will accomplish that.

$$\text{Decimal divisor} \rightarrow \frac{6.36}{1.2} = \frac{(6.36)\,(10)}{(1.2)\,(10)} = \frac{63.6}{12} \leftarrow \text{Whole number divisor}$$

The short way to multiply by 10 is to move the decimal point *one place* to the *right* in both the divisor and the dividend.

$$1.2\overline{)6.36} \quad \text{is equivalent to} \quad 12\overline{)63.6}$$

Note

Moving the decimal points the *same* number of places to the right in *both* the divisor and dividend will *not* change the answer.

Dividing by a Decimal Number

Step 1 Count the number of decimal places in the divisor and move the decimal point that many places to the *right*. (This changes the divisor to a whole number.)

Step 2 Move the decimal point in the dividend the *same* number of places to the *right*. (Write in extra zeros if needed.)

Step 3 Write the decimal point in the quotient directly above the decimal point in the dividend. Then divide as usual.

Step 4 If both numbers have the *same sign,* the quotient is *positive.* If they have *different signs,* the quotient is *negative.*

ANSWERS

3. **(a)** 20.539; no repeating digits visible on calculator;
   Check (20.539) (13) = 267.007
   (b) 3.417; 3.41$\overline{6}$;
   Check (3.417) (6) = 20.502
   (c) 1.136; 1.13$\overline{5}$;
   Check (1.136) (9) = 10.224
   (d) 5.383; 5.38$\overline{3}$;
   Check (5.383) (3) = 16.149
   (e) 10.573; 10.5$\overline{72}$;
   Check (10.573) (11) = 116.303

| EXAMPLE 4 | **Dividing by Decimal Numbers** |

(a) $\dfrac{27.69}{0.003}$

Move the decimal point in the divisor *three* places to the *right* so that 0.003 becomes the whole number 3. In order to move the decimal point in the dividend the same number of places, write in an extra 0.

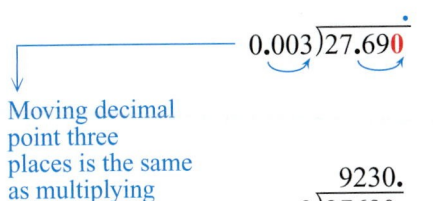

Move decimal points in divisor and dividend. Then line up decimal point in the quotient.

Moving decimal point three places is the same as multiplying by 1000 ⟶ $3\overline{)27690.}$ quotient 9230.

Divide as usual.

(b) Divide -5 by -4.2 and round the quotient to the nearest hundredth.

First consider $5 \div 4.2$. Move the decimal point in the divisor one place to the right so that 4.2 becomes the whole number 42. The decimal point in the dividend starts on the right side of 5 and is also moved one place to the right.

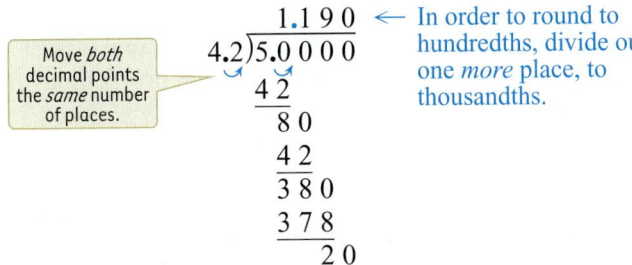

Move *both* decimal points the *same* number of places.

In order to round to hundredths, divide out one *more* place, to thousandths.

Rounding the quotient to the nearest hundredth gives 1.19. The quotient is *positive* because both the divisor and dividend have the *same* sign (both are negative).

$$-5 \div (-4.2) \approx 1.19$$

Same sign → Positive quotient

Work Problem ④ *at the Side.* ▶

OBJECTIVE 3 Estimate the answer when dividing decimals.
Estimating the answer to a division problem helps you catch errors. Compare the estimate to your exact answer. If they are very different, do the division again.

④ Divide. If the quotient does not come out even, round to the nearest hundredth.

(a) $0.2\overline{)1.04}$

(b) $0.06\overline{)1.8072}$

(c) $0.005\overline{)32}$

(d) $-8.1 \div 0.025$

(e) $\dfrac{7}{1.3}$

(f) $-5.3091 \div (-6.2)$

5 Decide whether each answer is reasonable by using front end rounding to estimate the answer. If the exact answer is *not* reasonable, find and correct the error.

(a) $42.75 \div 3.8 = 1.125$

Estimate:

(b) $807.1 \div 1.76 = 458.580$
to nearest thousandth

Estimate:

(c) $48.63 \div 52 = 93.519$
to nearest thousandth

Estimate:

(d) $9.0584 \div 2.68 = 0.338$

Estimate:

ANSWERS

5. (a) Estimate is $40 \div 4 = 10$; exact answer is not reasonable, should be 11.25

(b) Estimate is $800 \div 2 = 400$; exact answer is reasonable.

(c) Estimate is $50 \div 50 = 1$; exact answer is not reasonable, should be 0.935

(d) Estimate is $9 \div 3 = 3$; exact answer is not reasonable, should be 3.38

EXAMPLE 5 **Estimating before Dividing**

First, use front end rounding to estimate the answer. Then divide to find the exact answer.

$$580.44 \div 2.8$$

Here is how one student solved this problem. She rounded 580.44 to 600 and 2.8 to 3 to estimate the answer.

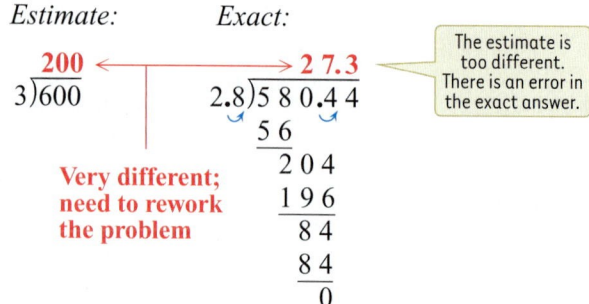

Estimate: *Exact:*

The estimate is too different. There is an error in the exact answer.

Very different; need to rework the problem

Notice that the estimate, which is in the hundreds, is very different from the exact answer, which is only in the tens. This tells the student that she needs to rework the problem. Can you find the error?
(The exact answer should be 207.3, which fits with the estimate of 200.)

◀ *Work Problem* **5** *at the Side.*

OBJECTIVE 4 Use the order of operations with decimals. Use the order of operations from **Section 1.8** when a decimal problem involves more than one operation.

Order of Operations

Step 1 Work inside *parentheses* or *other grouping symbols.*

Step 2 Simplify expressions with *exponents.*

Step 3 Do the remaining *multiplications and divisions* as they occur from left to right.

Step 4 Do the remaining *additions and subtractions* as they occur from left to right.

EXAMPLE 6 **Using the Order of Operations**

Simplify by using the order of operations.

(a) $2.5 + (-6.3)^2 + 9.62$ Apply the exponent: $(-6.3)(-6.3)$ is 39.69

$2.5 + 39.69 + 9.62$ Add from left to right.

$42.19 + 9.62$

51.81

(b) $1.82 + (5.2 - 6.7)(5.8)$ Work inside parentheses.

$1.82 + (-1.5)(5.8)$ Multiply next.

$1.82 + (-8.7)$ Add last.

-6.88

Continued on Next Page

(c) $3.7^2 - 1.8 \div 5(1.5)$ Apply the exponent.

$13.69 - 1.8 \div 5(1.5)$ Multiply and divide from *left to right,*
                              so first divide 1.8 by 5

$13.69 - 0.36\,(1.5)$ Multiply 0.36 by 1.5

$13.69 - 0.54$ Subtract last.

13.15

Work Problem **6** *at the Side.* ▶

🖩 **Calculator Tip**   Most calculators that have parentheses keys ⟮ ⟯ can handle calculations like those in Example 6(b) on the previous page just by entering the numbers in the order given. For example, the keystrokes for Example 6(b) are:

Parentheses

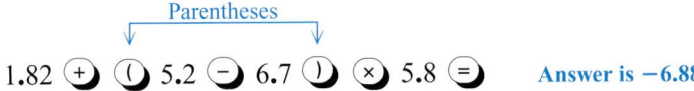

1.82 ⊕ ⟮ 5.2 ⊖ 6.7 ⟯ ⊗ 5.8 ⊜   **Answer is −6.88**

Standard, four-function calculators generally do *not* give the correct answer if you enter the numbers in the order given. Check the instruction manual that came with your calculator for information on "order of calculations" to see if your model has the rules for order of operations built into it. For a quick check, try entering this problem:

2 ⊕ 2 ⊗ 2 ⊜

If the result is 6, the calculator follows the order of operations. If the result is 8, it does *not* have the rules built into it. Use the space below to explain how this test works.

Answer: The test works because a calculator that follows the order of operations will automatically do the multiplication first. If the calculator does *not* have the rules built into it, it will work from left to right.

Following Order of Operations | **Working from Left to Right**

$2 + 2 \times 2$ Multiply $2 + 2 \times 2$
                    before adding.

$2 + \quad 4$ $4 \quad \times 2$

$6 \leftarrow$ **Correct** $8 \leftarrow$ **Incorrect**

---

6 Simplify.

(a) $-4.6 - 0.79 + 1.5^2$

(b) $3.64 \div 1.3\,(3.6)$

(c) $0.08 + 0.6\,(2.99 - 3)$

(d) $10.85 - 2.3\,(5.2) \div 3.2$

Math in the Media

DOLLAR-COST AVERAGING

Making money in the stock market can be difficult. You want to buy shares of stock when the price is low and sell them when the price is high. Predicting the right time to buy is tricky. One strategy is dollar-cost averaging. You invest the same amount at regular intervals, like the first of each month. Over time you will usually buy more shares at lower prices, though there is no guarantee of making a profit.

The table below shows Microsoft's closing share price on the first day of each month in 2007. See what happens if you invest $100 each month rather than buying a lot of shares at one time.

Microsoft Corp. (MSFT)	Amount Invested	Price Per Share	Number of Shares Bought
January	$100	30.86	($100 ÷ 30.86) ≈ 3.2404 shares
February	$100	28.17	
March	$100	27.87	
April	$100	29.94	
May	$100	30.69	
June	$100	29.47	
July	$100	28.99	
August	$100	28.73	
September	$100	29.46	
October	$100	36.81	
November	$100	34.74	
December	$100	33.60	
Total investment			

Source: http://finance.yahoo.com

Use a calculator to help answer these questions.

1. Calculate the total amount invested and enter the value in the table.

2. Calculate the number of shares bought each month. Round the number of shares to the nearest ten-thousandth. The calculation for January is shown.

3. Calculate the average market price per share. (*Hint:* Add the monthly prices per share and divide by 12.)

4. Calculate the average price based on *dollar-cost averaging*. (*Hint:* Divide the total amount invested by the total number of shares.)

5. Rank the following scenarios in order of which was the best investment (most profit or least loss). Show the value of each investment in December 2007 as a basis for your answer.

   (a) $1200 invested in Microsoft in January 2007

   (b) $1200 invested in Microsoft using *dollar-cost averaging*

   (c) $1200 invested in Microsoft in October 2007

5.5 ▶▶▶ Exercises

Find each quotient. See Examples 1, 2, and 4.

 1. $27.3 \div (-7)$

2. $-50.4 \div 8$

3. $\dfrac{4.23}{9}$

4. $\dfrac{1.62}{6}$

5. $-20.01 \div (-0.05)$

6. $-16.04 \div (-0.08)$

7. $1.5\overline{)54}$

8. $2.4\overline{)132}$

Use the fact that $108 \div 18 = 6$ to work Exercises 9–16 simply by moving decimal points. See Examples 1, 2, and 4.

9. $1.8\overline{)0.108}$

10. $18\overline{)10.8}$

11. $0.018\overline{)108}$

12. $0.18\overline{)1.08}$

13. $0.18\overline{)10.8}$

14. $0.18\overline{)108}$

15. $18\overline{)0.0108}$

16. $1.8\overline{)0.0108}$

Divide. Round quotients to the nearest hundredth when necessary. See Examples 3 and 4.

17. $4.6\overline{)116.38}$

18. $2.6\overline{)4.992}$

19. $\dfrac{-3.1}{-0.006}$

20. $\dfrac{-1.7}{0.09}$

Divide. Round quotients to the nearest thousandth. See Examples 3 and 4.

21. $-240.8 \div 9$

22. $-76.43 \div (-7)$

23. $0.034\overline{)342.81}$

24. $0.043\overline{)1748.4}$

Relating Concepts (Exercises 25–26) For Individual or Group Work

*First, look back at your work in **Section 5.4**, Exercises 21 and 22. Then look for patterns as you **work Exercises 25 and 26 in order.***

25. Do these division problems.

$3.77 \div 10 = $ _____ $9.1 \div 10 = $ _____

$0.886 \div 10 = $ _____ $30.19 \div 10 = $ _____

$406.5 \div 10 = $ _____ $6625.7 \div 10 = $ _____

(a) What pattern do you see? Write a "rule" for dividing by 10. What do you think the rule is for dividing by 100? by 1000? Write the rules and try them out on the numbers above.

(b) Compare your rules to the ones you wrote in **Section 5.4**, Exercise 21.

26. Do these division problems.

$40.2 \div 0.1 = $ _____ $7.1 \div 0.1 = $ _____

$0.339 \div 0.1 = $ _____ $15.77 \div 0.1 = $ _____

$46 \div 0.1 = $ _____ $873 \div 0.1 = $ _____

(a) What pattern do you see? Write a "rule" for dividing by 0.1. What do you think the rule is for dividing by 0.01? by 0.001? Write the rules and try them out on the numbers above.

(b) Compare your rules to the ones you wrote in **Section 5.4**, Exercise 22.

Decide whether each answer is reasonable *or* unreasonable *by using front end rounding to estimate the answer. If the exact answer is not reasonable, find and correct the error.* See Example 5.

27. $37.8 \div 8 = 47.25$

Estimate:

28. $345.6 \div 3 = 11.52$

Estimate:

29. $54.6 \div 48.1 \approx 1.135$

Estimate:

30. $2428.8 \div 4.8 = 56$

Estimate:

31. $307.02 \div 5.1 = 6.2$

Estimate:

32. $395.415 \div 5.05 = 78.3$

Estimate:

33. $9.3 \div 1.25 = 0.744$

Estimate:

34. $78 \div 14.2 = 0.182$

Estimate:

Solve each application problem. Round money answers to the nearest cent when necessary.

35. Rob has discovered that his daughter's favorite brand of tights are on sale. He decided to buy one pair as a surprise for her. How much did he pay?

Special Purchase!
Girls'
Tights
6 pairs for $23.98
Stock up now!

36. The bookstore has a special price on notepads. How much did Randall pay for one notepad?

Notepads 4 for $1.69

37. It will take 21 months for Aimee to pay off her credit card balance of $1408.68. How much is she paying each month?

38. Marcella Anderson bought a 2.6 meter length of microfiber woven suede fabric for $33.77. How much did she pay per meter?

39. Adrian Webb bought 619 bricks to build a barbecue pit, paying $185.70. Find the cost per brick. (*Hint:* Cost *per* brick means the cost for *one* brick.)

40. Lupe Wilson is a newspaper distributor. Last week she paid the newspaper $130.51 for 842 copies. Find the cost per copy.

41. Darren Jackson earned $476.80 for 40 hours of work. Find his earnings per hour.

42. At a CD manufacturing company, 400 CDs cost $289. Find the cost per CD.

43. It took 16.35 gallons of gas to fill the gas tank of Kim's car. She had driven 346.2 miles since her last fill-up. How many miles per gallon did she get? Round to the nearest tenth.

44. Mr. Rodriquez pays $53.19 each month to House-hold Finance. How many months will it take him to pay off $1436.13?

Use the table of women's longest long jumps (through the year 2007) to answer Exercises 45–50. To find an average, add up the values you are interested in and then divide the sum by the number of values. Round your answers to the nearest hundredth.

Athlete	Country	Year	Length (meters)
Galina Christyakova	USSR	1988	7.52
Jackie Joyner-Kersee	U.S.	1994	7.49
Heike Drechsler	Germany	1992	7.48
Jackie Joyner-Kersee	U.S.	1987	7.45
Jackie Joyner-Kersee	U.S.	1988	7.40
Jackie Joyner-Kersee	U.S.	1991	7.32
Jackie Joyner-Kersee	U.S.	1996	7.20
Chioma Ajunwa	Nigeria	1996	7.12
Fiona May	Italy	2000	7.09
Tatyana Lebedeva	Russia	2004	7.07

Source: CNNSI.com

45. Find the average length of the long jumps made by Jackie Joyner-Kersee.

46. Find the average length of all the long jumps listed in the table.

47. How much longer was the fifth-longest jump than the sixth-longest jump?

48. If the first-place athlete made five jumps of the same length, what would be the total distance jumped?

49. What was the total length jumped by the top three athletes in the table?

50. How much less was the last-place jump than the next-to-last-place jump?

Simplify. See Example 6.

51. $7.2 - 5.2 + 3.5^2$

52. $6.2 + 4.3^2 - 9.72$

53. $38.6 + 11.6(10.4 - 13.4)$

54. $2.25 - 1.06(0.85 - 3.95)$

55. $-8.68 - 4.6(10.4) \div 6.4$

56. $25.1 + 11.4 \div 7.5(-3.75)$

57. $33 - 3.2(0.68 + 9) + (-1.3)^2$

58. $0.6 + (-1.89 + 0.11) \div 0.004(0.5)$

Solve each application problem.

59. Soup is on sale at six cans for $3.25, or you can purchase individual cans for $0.57. How much will you save per can if you buy six cans? Round to the nearest cent.

60. Nadia's diet says she can eat 3.5 ounces of chicken nuggets. The package weighs 10.5 ounces and contains 15 nuggets. How many nuggets can Nadia eat?

61. The U.S. Treasury prints about 38,000,000 pieces of paper money each day. The printing presses run 24 hours a day. How many pieces of money are printed, to the nearest whole number:

38,000,000 pieces of paper money are printed each day

(a) each hour

(b) each minute

(c) each second

(*Source:* www.moneyfactory.com)

62. Mach 1 is the speed of sound. Dividing a vehicle's speed by the speed of sound gives its speed on the Mach scale. In 1997, a specially built car with two 110,000-horsepower engines broke the world land speed record by traveling 763.035 miles per hour. The speed of sound changes slightly with the weather. That day it was 748.11 miles per hour. What was the car's Mach speed, to the nearest hundredth? (*Source:* Associated Press.)

General Mills will give a school 10¢ for each box top logo from its cereals and other products. A school can earn up to $10,000 per year. Use this information to answer Exercises 63–66. Round answers to the nearest whole number. (Source: General Mills.)

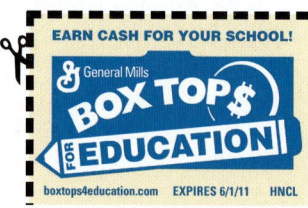

63. How many box tops would a school need to collect in one year to earn the maximum amount?

64. (Complete Exercise 63 first.) If a school has 550 children, how many box tops would each child need to collect to reach the maximum?

65. How many box tops would need to be collected during each of the 38 weeks in the school year to reach the maximum amount?

66. How many box tops would each of the 550 children need to collect during each of the 38 weeks of school to reach the maximum amount?

Summary Exercises on Decimals

Write each decimal as a fraction or mixed number in lowest terms.

1. 0.8

2. 6.004

3. 0.35

Write each decimal in words.

4. 94.5

5. 2.0003

6. 0.706

Write each decimal in numbers.

7. five hundredths

8. three hundred nine ten-thousandths

9. ten and seven tenths

Round to the place indicated.

10. 6.1873 to the nearest hundredth

11. 0.95 to the nearest tenth

12. 0.42025 to the nearest thousandth

13. $0.893 to the nearest cent

14. $3.0017 to the nearest cent

15. $99.64 to the nearest dollar

Simplify.

16. $-0.27(3.5)$

17. $50 - 0.3801$

18. $0.35 \div (-0.007)$

19. $\dfrac{-90.18}{-6}$

20. $(0.004)(1.22)$

21. $1.55 - 3.7$

22. $-0.95 + 10.005$

23. $3.6 + 0.718 + 9 + 5.0829$

24. $32.305 - 40 + 0.7$

25. $-8.9 + 4^2 \div (-0.02)$

26. $(-0.18 + 2.5) + 4(-0.05)$

27. $0.64 \div 16.3$ Round your answer to the nearest hundredth.

Find the perimeter of each figure.

28.

19.75 in.

6.3 in. | | 6.3 in.

19.75 in.

29.

2 meters   1 meter

0.9 meter

1.7 meters

1.18 meters

0.86 meter

2.095 meters

Solve each application problem.

30. At a bakery, Sue Chee bought $7.42 worth of muffins and $10.09 worth of croissants for a staff party and a $0.69 cookie for herself. How much change did she receive from two $10 bills?

31. A craft cooperative paid $40.32 for enough fabric to make eight baby blankets for their store. They sold the blankets for $15.95 each. How much profit was made on the blankets?

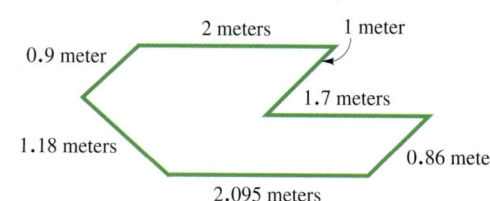

The table below shows information on two tents for camping. Use the table to answer Exercises 32 and 33.

Tents	Coleman Family Dome	Eddie Bauer Dome Tent
Dimensions	13 ft × 13 ft	12 ft × 12 ft
Sleeps	8 campers	6 campers
Sale price	$127	$99

Source: target.com

32. For the Coleman tent, find:

(a) the area of the tent floor

(b) the cost per square foot of floor space, to the nearest cent.

33. Find the same information for the Eddie Bauer tent as you did for the Coleman tent.

Use the information in the table to answer Exercises 34–36.

Animal	Average Weight of Animal (ounces)	Average Weight of Food Eaten Each Day (ounces)
Hamster	3.5	0.4
Queen bee	0.004	?
Hummingbird	?	0.07

Source: NCTM News Bulletin.

34. (a) In how many days will a hamster eat enough food to equal its body weight? Round to the nearest whole number of days.

(b) If a 140-pound woman ate her body weight of food in the same number of days as the hamster, how much would she eat each day? Round to the nearest tenth.

35. While laying eggs, a queen bee eats eighty times her weight each day. Use this information to fill in one of the missing values in the table.

36. A hummingbird's body weight is about 1.6 times the weight of its daily food intake. Find its body weight.

5.6 ▶▶▶   Fractions and Decimals

Writing fractions as equivalent decimals can help you do calculations or compare the size of two numbers more easily.

OBJECTIVE 1 Write fractions as equivalent decimals.   Recall that a fraction is one way to show division (see **Section 1.7**). For example, $\frac{3}{4}$ means $3 \div 4$. If you are doing the division by hand, write it as $4\overline{)3}$. When you do the division, the result is 0.75, the decimal equivalent of $\frac{3}{4}$.

> **Writing a Fraction as an Equivalent Decimal**
>
> **Step 1**   Divide the numerator of the fraction by the denominator.
>
> **Step 2**   If necessary, round the answer to the place indicated.

Work Problem **1** *at the Side.* ▶

EXAMPLE 1   **Writing Fractions or Mixed Numbers as Decimals**

(a) Write $\frac{1}{8}$ as a decimal.

$\frac{1}{8}$ means $1 \div 8$. Write it as $8\overline{)1}$. The decimal point in the dividend is on the *right* side of the 1. Write extra zeros in the dividend so you can continue dividing until the remainder is 0.

$$\frac{1}{8} \;\Rightarrow\; 1 \div 8 \;\Rightarrow\; 8\overline{)1} \;\Rightarrow\;$$

Be careful to divide $8\overline{)1}$, **not** $1\overline{)8}$

Decimal points lined up

$$\begin{array}{r} 0.125 \\ 8\overline{)1.000} \\ \underline{8} \\ 20 \\ \underline{16} \\ 40 \\ \underline{40} \\ 0 \end{array}$$

← Three extra zeros needed

← Remainder is 0

Therefore, $\frac{1}{8} = 0.125$

To check, write 0.125 as a fraction, then write it in lowest terms.

$$0.125 = \frac{125}{1000} \quad \text{In lowest terms:} \quad \frac{125 \div 125}{1000 \div 125} = \frac{1}{8} \quad \left\{ \begin{array}{l} \text{Original} \\ \text{fraction} \end{array} \right.$$

> ▦ **Calculator Tip**   When using your calculator to write fractions as decimals, enter the numbers from the top down. Remember that the *order* in which you enter the numbers *does* matter in division. Example 1(a) above works like this:
>
> $\frac{1}{8}$ ↓ Top down   Enter 1 ⊙ 8 ⊜   Answer is 0.125
>
> What happens if you enter 8 ⊙ 1 ⊜? Do you see why that cannot possibly be correct? (Answer: $8 \div 1 = 8$. A proper fraction like $\frac{1}{8}$ *cannot* be equivalent to a whole number.)

Continued on Next Page

OBJECTIVES

1 Write fractions as equivalent decimals.

2 Compare the size of fractions and decimals.

1 Rewrite each fraction so you could do the division by hand. Do *not* complete the division.

(a) $\frac{1}{9}$ is written $9\overline{)}$

(b) $\frac{2}{3}$ is written $\overline{)}$

(c) $\frac{5}{4}$ is written $\overline{)}$

(d) $\frac{3}{10}$ is written $\overline{)}$

(e) $\frac{21}{16}$ is written $\overline{)}$

(f) $\frac{1}{50}$ is written $\overline{)}$

ANSWERS

1. (a) $9\overline{)1}$ **(b)** $3\overline{)2}$ **(c)** $4\overline{)5}$
(d) $10\overline{)3}$ **(e)** $16\overline{)21}$ **(f)** $50\overline{)1}$

2 Write each fraction or mixed number as a decimal.

(a) $\dfrac{1}{4}$

(b) $2\dfrac{1}{2}$

(c) $\dfrac{5}{8}$

(d) $4\dfrac{3}{5}$

(e) $\dfrac{7}{8}$

(b) Write $2\dfrac{3}{4}$ as a decimal.

One method is to divide 3 by 4 to get 0.75 for the fraction part. Then add the whole number part to 0.75.

$$\dfrac{3}{4} \longrightarrow \begin{array}{r} 0.75 \\ 4\overline{)3.00} \\ 2\,8 \\ \hline 20 \\ 20 \\ \hline 0 \end{array}$$

Fraction part

$$\begin{array}{r} 2.00 \leftarrow \text{Whole number part} \\ +\;0.75 \\ \hline 2.75 \end{array}$$

So, $2\dfrac{3}{4} = 2.75$ **Check** $2.75 = 2\dfrac{75}{100} = 2\dfrac{3}{4}$ ←Lowest terms

Whole number parts match.

A second method is to write $2\dfrac{3}{4}$ as an improper fraction before dividing numerator by denominator.

$$2\dfrac{3}{4} = \dfrac{11}{4}$$

$$\dfrac{11}{4} \longrightarrow 11 \div 4 \longrightarrow 4\overline{)11} \longrightarrow \begin{array}{r} 2.7\,5 \\ 4\overline{)11.0\,0} \\ 8 \\ \hline 3\,0 \\ 2\,8 \\ \hline 2\,0 \\ 2\,0 \\ \hline 0 \end{array} \leftarrow \text{Two extra zeros needed}$$

Whole number parts match.

So, $2\dfrac{3}{4} = 2.75$

$\dfrac{3}{4}$ is equivalent to $\dfrac{75}{100}$ or 0.75

◀ **Work Problem** **2** **at the Side.**

EXAMPLE 2 **Writing a Fraction as a Decimal with Rounding**

Write $\dfrac{2}{3}$ as a decimal and round to the nearest thousandth.

$\dfrac{2}{3}$ means $2 \div 3$. To round to thousandths, divide out one *more* place, to ten-thousandths.

$$\dfrac{2}{3} \longrightarrow 2 \div 3 \longrightarrow 3\overline{)2} \longrightarrow \begin{array}{r} 0.6666 \\ 3\overline{)2.0000} \\ 1\,8 \\ \hline 20 \\ 18 \\ \hline 20 \\ 18 \\ \hline 20 \\ 18 \\ \hline 2 \end{array} \leftarrow \begin{cases} \text{Four zeros needed} \\ \text{for ten-thousandths} \end{cases}$$

Be careful to divide $3\overline{)2}$, **not** $2\overline{)3}$

Written as a repeating decimal, $\dfrac{2}{3} = 0.\overline{6}$ ← Bar above repeating digit

Rounded to the nearest thousandth, $\dfrac{2}{3} \approx 0.667$

Continued on Next Page

▦ **Calculator Tip**   Try Example 2 on the previous page on your calculator. Enter 2 ÷ 3. Which answer do you get?

> **0.666666667**   or   **0.666666666**

Most *scientific* and *graphing* calculators will show a 7 as the last digit. Because the 6s keep on repeating forever, the calculator automatically rounds in the last decimal place it has room to show. If you have a 10-digit display space, the calculator is rounding as shown below.

> 0.6666666666 (11 digits)   rounds to   0.66666666**7**
>
> Next digit is 5 *or more,* so 6 rounds to 7

Other calculators, especially standard, four-function ones, may *not* round. They just cut off, or *truncate,* the extra digits. Such a calculator would show 0.6666666 in the display.

Would this difference in calculators show up when changing $\frac{1}{3}$ to a decimal? Why not? (Answer: The repeating digit is a 3, which is *4 or less,* so it stays as a 3 whether it's rounded or not.)

Work Problem ③ *at the Side.* ▶

OBJECTIVE 2 **Compare the size of fractions and decimals.** You can use a number line to compare fractions and decimals. For example, the number line below shows the space between 0 and 1. The locations of some commonly used fractions are marked, along with their decimal equivalents.

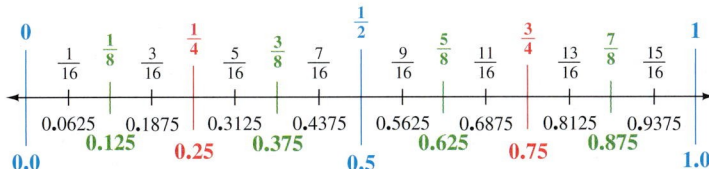

The next number line shows the locations of some commonly used fractions between 0 and 1 that are equivalent to repeating decimals. The decimal equivalents use a bar above repeating digits.

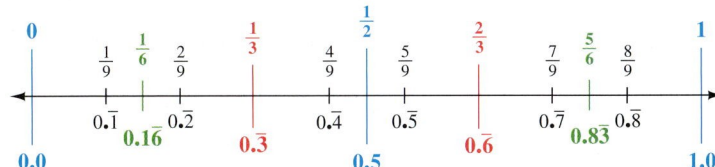

EXAMPLE 3   **Using a Number Line to Compare Numbers**

Use the number lines above to decide whether to write >, <, or = in the blank between each pair of numbers.

(a) 0.6875 _____ 0.625

You learned in **Section 1.2** that the number farther to the right on the number line is the greater number. On the first number line, 0.6875 is to the *right* of 0.625, so use the > symbol.

> 0.6875   **is greater than**   0.625   0.6875 > 0.625

Continued on Next Page

③ Write as decimals. Round ▦ to the nearest thousandth.

(a) $\frac{1}{3}$

(b) $2\frac{7}{9}$

(c) $\frac{10}{11}$

(d) $\frac{3}{7}$

(e) $3\frac{5}{6}$

ANSWERS

3. **(a)** $\frac{1}{3} \approx 0.333$ **(b)** $2\frac{7}{9} \approx 2.778$

 (c) $\frac{10}{11} \approx 0.909$ **(d)** $\frac{3}{7} \approx 0.429$

 (e) $3\frac{5}{6} \approx 3.833$

4 Use the number lines on the previous page to help you decide whether to write $<$, $>$, or $=$ in each blank.

(a) 0.4375 _____ 0.5

(b) 0.75 _____ 0.6875

(c) 0.625 _____ 0.0625

(d) $\dfrac{2}{8}$ _____ 0.375

(e) 0.8$\overline{3}$ _____ $\dfrac{5}{6}$

(f) $\dfrac{1}{2}$ _____ 0.$\overline{5}$

(g) 0.$\overline{1}$ _____ 0.1$\overline{6}$

(h) $\dfrac{8}{9}$ _____ 0.$\overline{8}$

(i) 0.$\overline{7}$ _____ $\dfrac{4}{6}$

(j) $\dfrac{1}{4}$ _____ 0.25

5 Arrange each group in order from least to greatest.

(a) 0.7, 0.703, 0.7029

(b) 6.39, 6.309, 6.401, 6.4

(c) 1.085, $1\dfrac{3}{4}$, 0.9

(d) $\dfrac{1}{4}, \dfrac{2}{5}, \dfrac{3}{7}$, 0.428

ANSWERS

4. (a) $<$ (b) $>$ (c) $>$ (d) $<$ (e) $=$
   (f) $<$ (g) $<$ (h) $=$ (i) $>$ (j) $=$
5. (a) 0.7, 0.7029, 0.703
   (b) 6.309, 6.39, 6.4, 6.401
   (c) 0.9, 1.085, $1\dfrac{3}{4}$
   (d) $\dfrac{1}{4}, \dfrac{2}{5}$, 0.428, $\dfrac{3}{7}$

(b) $\dfrac{3}{4}$ _____ 0.75

On the first number line, $\dfrac{3}{4}$ and 0.75 are at the same point on the number line. They are equivalent, so use the $=$ symbol.

$$\dfrac{3}{4} = 0.75$$

(c) 0.5 _____ 0.$\overline{5}$

On the second number line, 0.5 is to the *left* of 0.$\overline{5}$ (which is actually 0.555 . . .), so use the $<$ symbol.

0.5 **is less than** 0.$\overline{5}$ 0.5 $<$ 0.$\overline{5}$

(d) $\dfrac{2}{6}$ _____ 0.$\overline{3}$

Write $\dfrac{2}{6}$ in lowest terms as $\dfrac{1}{3}$.
On the second number line you can see that $\dfrac{1}{3}$ and 0.$\overline{3}$ are equivalent.

$$\dfrac{1}{3} = 0.\overline{3}$$

◀ Work Problem **4** at the Side.

You can also compare fractions by first writing each one as a decimal. You can then compare the decimals by writing each one with the same number of decimal places.

EXAMPLE 4   **Arranging Numbers in Order**

Write each group of numbers in order, from least to greatest.

(a) 0.49   0.487   0.4903

It is easier to compare decimals if they are all tenths, or all hundredths, and so on. Because 0.4903 has four decimal places (ten-thousandths), write zeros to the right of 0.49 and 0.487 so they also have four decimal places. Writing zeros to the right of a decimal number does *not* change its value (see **Section 5.3**). Then find the least and greatest number of ten-thousandths.

0.49 = 0.4900 = **4900** ten-thousandths   ← 4900 is in the middle.
0.487 = 0.4870 = **4870** ten-thousandths   ← 4870 is the least.
0.4903 = **4903** ten-thousandths   ← 4903 is the greatest.

From least to greatest, the correct order is shown below.

0.487   0.49   0.4903

(b) $2\dfrac{5}{8}$ 2.63   2.6

Write $2\dfrac{5}{8}$ as $\dfrac{21}{8}$ and divide $8\overline{)21}$ to get the decimal form, 2.625. Then, because 2.625 has three decimal places, write zeros so all the numbers have three decimal places.

$2\dfrac{5}{8} = 2.625 = 2$ and **625** thousandths   ← 625 is in the middle.

2.63 = 2.630 = 2 and **630** thousandths   ← 630 is the greatest.

2.6 = 2.600 = 2 and **600** thousandths   ← 600 is the least.

From least to greatest, the correct order is shown below.

2.6 $2\dfrac{5}{8}$ 2.63

◀ Work Problem **5** at the Side.

5.6 ▶▶▶ Exercises

Write each fraction or mixed number as a decimal. Round to the nearest thousandth when necessary. See Examples 1 and 2.

1. $\dfrac{1}{2}$

2. $\dfrac{1}{4}$

3. $\dfrac{3}{4}$

4. $\dfrac{1}{10}$

5. $\dfrac{3}{10}$

6. $\dfrac{7}{10}$

7. $\dfrac{9}{10}$

8. $\dfrac{4}{5}$

9. $\dfrac{3}{5}$

10. $\dfrac{2}{5}$

11. $\dfrac{7}{8}$

12. $\dfrac{3}{8}$

13. $2\dfrac{1}{4}$

14. $1\dfrac{1}{2}$

15. $14\dfrac{7}{10}$

16. $23\dfrac{3}{5}$

17. $3\dfrac{5}{8}$

18. $2\dfrac{7}{8}$

19. $6\dfrac{1}{3}$

20. $5\dfrac{2}{3}$

21. $\dfrac{5}{6}$

22. $\dfrac{1}{6}$

23. $1\dfrac{8}{9}$

24. $5\dfrac{4}{7}$

Relating Concepts (Exercises 25–28) For Individual or Group Work

*Use your knowledge of fractions and decimals as you **work Exercises 25–28 in order.***

25. (a) Explain how you can tell that Keith made an error *just by looking at his final answer*. Here is his work.

$$\dfrac{5}{9} \quad 5\overline{)9.0}^{\,1.8} \quad \text{so} \quad \dfrac{5}{9} = 1.8$$

(b) Show the correct way to change $\dfrac{5}{9}$ to a decimal. Explain why your answer makes sense.

26. (a) How can you prove to Sandra that $2\dfrac{7}{20}$ is *not* equivalent to 2.035? Here is her work.

$$2\dfrac{7}{20} \quad 20\overline{)7.00}^{\,0.35} \quad \text{so} \quad 2\dfrac{7}{20} = 2.035$$

(b) What is the correct answer? Show how to prove that it is correct.

27. Ving knows that $\dfrac{3}{8} = 0.375$. How can he write $1\dfrac{3}{8}$ as a decimal *without* having to do a division? How can he write $3\dfrac{3}{8}$ as a decimal? $295\dfrac{3}{8}$ as a decimal? Explain your answer.

28. Iris has found a shortcut for writing mixed numbers as decimals.

$$2\dfrac{7}{10} = 2.7 \qquad 1\dfrac{13}{100} = 1.13$$

Does her shortcut work for all mixed numbers? Explain when it works and why it works.

Find the decimal or fraction equivalent for each number. Write fractions in lowest terms.

Fraction	Decimal	Fraction	Decimal
29. _____	0.4	**30.** _____	0.75
31. _____	0.625	**32.** _____	0.111
33. _____	0.35	**34.** _____	0.9
35. $\frac{7}{20}$	_____	**36.** $\frac{1}{40}$	_____
37. _____	0.04	**38.** _____	0.52
39. _____	0.15	**40.** _____	0.85
41. $\frac{1}{5}$	_____	**42.** $\frac{1}{8}$	_____
43. _____	0.09	**44.** _____	0.02

Solve each application problem.

45. The average length of a newborn baby is 20.8 inches. Charlene's baby is 20.08 inches long. Is her baby longer or shorter than the average? By how much?

46. The patient in room 830 is supposed to get 8.3 milligrams of medicine. She was actually given 8.03 milligrams. Did she get too much or too little medicine? What was the difference?

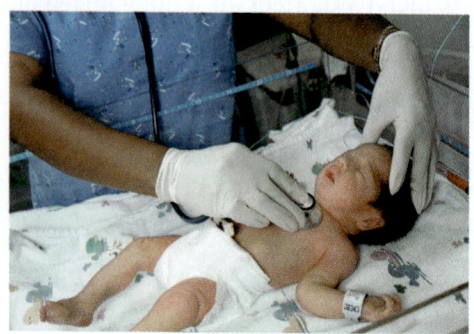

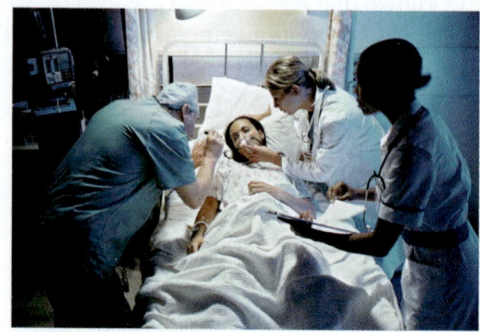

47. The label on the bottle of vitamins says that each capsule contains 0.5 gram of calcium. When checked, each capsule had 0.505 gram of calcium. Was there too much or too little calcium? What was the difference?

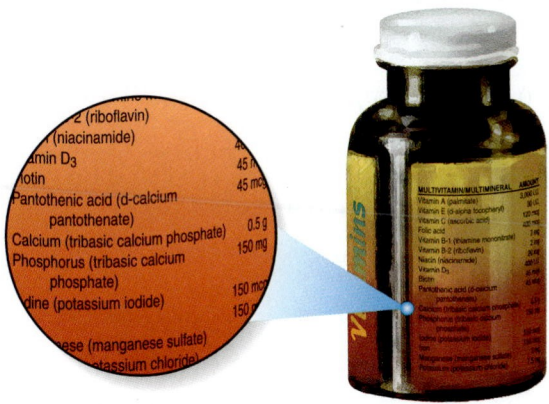

48. The glass mirror of the Hubble telescope had to be repaired in space because it would not focus properly. The problem was that the mirror's outer edge had a thickness of 0.6248 cm when it was supposed to be 0.625 cm. Was the edge too thick or too thin? By how much? (*Source:* NASA.)

49. Precision Medical Parts makes an artificial heart valve that must measure between 0.998 centimeter and 1.002 centimeters. Circle the lengths that are acceptable.

1.01 cm   0.9991 cm   1.0007 cm   0.99 cm

50. The white rats in a medical experiment must start out weighing between 2.95 ounces and 3.05 ounces. Circle the weights that can be used.

3.0 ounces   2.995 ounces   3.055 ounces

3.005 ounces

51. Ginny Brown had hoped her crops would get $3\frac{3}{4}$ inches of rain this month. The newspaper said the area received 3.8 inches of rain. Was that more or less than Ginny had hoped for? By how much?

52. The rats in the experiment in Exercise 50 gained $\frac{3}{8}$ ounce. They were expected to gain 0.3 ounce. Was their actual gain more or less than expected? By how much?

Use the number lines near Example 3 in the text to decide whether to write $>$, $<$, or $=$ between each pair of numbers. See Example 3.

53. **(a)** 0.3125 ____ 0.375   **(b)** $\frac{6}{8}$ ____ 0.75   **(c)** $0.\overline{8}$ ____ $0.8\overline{3}$ **(d)** 0.5 ____ $\frac{5}{9}$

54. **(a)** 0.125 ____ 0.0625   **(b)** $\frac{4}{9}$ ____ $\frac{3}{6}$ **(c)** 1.0 ____ $\frac{4}{4}$ **(d)** $\frac{1}{6}$ ____ $0.\overline{1}$

Arrange each group of numbers in order from least to greatest. See Example 4.

55. 0.54, 0.5455, 0.5399

56. 0.76, 0.7, 0.7006

57. 5.8, 5.79, 5.0079, 5.804

58. 12.99, 12.5, 13.0001, 12.77

59. 0.628, 0.62812, 0.609, 0.6009

60. 0.27, 0.281, 0.296, 0.3

61. 5.8751, 4.876, 2.8902, 3.88

62. 0.98, 0.89, 0.904, 0.9

63. 0.043, 0.051, 0.006, $\dfrac{1}{20}$

64. 0.629, $\dfrac{5}{8}$, 0.65, $\dfrac{7}{10}$

65. $\dfrac{3}{8}$, $\dfrac{2}{5}$, 0.37, 0.4001

66. 0.1501, 0.25, $\dfrac{1}{10}$, $\dfrac{1}{5}$

Four boxes of fishing line are in a sale bin. The thicker the line, the stronger it is. The diameter of the fishing line is its thickness. Use the information on the boxes to answer Exercises 67–68.

67. (a) Which color box has the strongest line? The weakest line?

(b) What is the difference in line diameter between the weakest and strongest lines?

68. (a) Which color box has the line that is $\frac{1}{125}$ inch in diameter?

(b) What is the difference in line diameter between the blue and purple boxes?

Some rulers for technical drawings show each inch divided into tenths. Use this scale drawing for Exercises 69–74. Change the measurements on the drawing to decimals and round them to the nearest tenth of an inch.

69. Length **(a)** is _____

70. Length **(b)** is _____

71. Length **(c)** is _____

72. Length **(d)** is _____

73. Length **(e)** is _____

74. Length **(f)** is _____

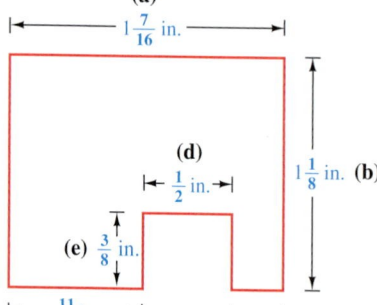

5.7 ▶▶▶ Problem Solving with Statistics: Mean, Median, and Mode

The word *statistics* originally came from words that mean *state numbers*. State numbers refer to numerical information, or *data,* gathered by the government such as the number of births, deaths, or marriages in a population. Today the word *statistics* has a much broader meaning; data from the fields of economics, social science, science, and business can all be organized and studied under the branch of mathematics called *statistics*.

OBJECTIVE 1 Find the mean of a list of numbers. Making sense of a long list of numbers can be difficult. So when you analyze data, one of the first things to look for is a *measure of central tendency*—a single number that you can use to represent the entire list of numbers. One such measure is the *average* or **mean.** The mean can be found with the following formula.

> ### Finding the Mean (Average)
>
> $$\text{mean} = \frac{\text{sum of all values}}{\text{number of values}}$$

EXAMPLE 1 Finding the Mean (Average)

David had test scores of 84, 90, 95, 98, and 88. Find his mean (average) score.
  Use the formula for finding the mean. Add up all the test scores and then divide the sum by the number of tests.

$$\text{mean} = \frac{84 + 90 + 95 + 98 + 88}{5} \quad \text{◀ Add all the test scores.}$$

$$\text{mean} = \frac{455}{5} \quad \text{◀ Divide by the number of tests.}$$

$$\text{mean} = 91$$

David has a mean (average) score of 91.

────────────── *Work Problem* ① *at the Side.* ▶

EXAMPLE 2 Applying the Mean (Average)

The sales of photo albums at Sarah's Card Shop for each day last week were $86, $103, $118, $117, $126, $158, and $149. Find the mean daily sales of photo albums.
  To find the mean, add all the daily sales amounts and then divide the sum by the number of days (7).

$$\text{mean} = \frac{\$86 + \$103 + \$118 + \$117 + \$126 + \$158 + \$149}{7} \quad \begin{array}{l}\leftarrow \text{Sum of sales}\\ \leftarrow \text{Number of days}\end{array}$$

$$\text{mean} = \frac{\$857}{7}$$

$$\text{mean} \approx \$122.43 \quad \text{Rounded to nearest cent}$$

────────────── *Work Problem* ② *at the Side.* ▶

---

① Tanya had test scores of 95, 91, 81, 78, 81, and 90. Find her mean (average) score.

② Find the mean for each list of numbers.

(a) Jorge's monthly cell phone bills last year were $25.12, $42.58, $76.19, $32, $81.11, $26.41, $49.76, $59.32, $71.18, $30.09, $60.50, and $79.84. Find his average monthly bill to the nearest cent.

(b) The sales for one year at eight different office supply stores: $749,820; $765,480; $643,744; $824,222; $485,886; $668,178; $702,294; $525,800

ANSWERS

1. $\dfrac{516}{6} = 86$ average score

2. (a) $\dfrac{\$634.10}{12} \approx \$52.84$

   (b) $\dfrac{\$5,365,424}{8} = \$670,678$

3 Alison Nakano works downtown. Some days she can park in cheaper lots that charge $6, $7, or $8. Other days, she has to park in lots that charge $9 or $10. Last month she kept track of the amount she spent each day for parking and the number of days she spent that amount. Find her average daily parking cost.

Parking Fee	Frequency
$ 6	2
$ 7	6
$ 8	3
$ 9	4
$10	6

OBJECTIVE **2** **Find a weighted mean.** Some items in a list of data might appear more than once. In this case, we find a **weighted mean,** in which each value is "weighted" by multiplying it by the number of times it occurs.

EXAMPLE 3 **Finding a Weighted Mean**

The table below shows the amount of contribution and the number of times the amount was given (frequency) to a food pantry. Find the weighted mean.

Contribution Value	Frequency
$ 3	4
$ 5	2
$ 7	1
$ 8	5
$ 9	3
$10	2
$12	1
$13	2

4 people each contributed $3.

The same amount was given by more than one person: for example, $3 was given by four people, and $8 was given by five people. Other amounts, such as $12, were given by only one person.

To find the mean, multiply each contribution value by its frequency. Then add the products. Next, add the numbers in the *frequency* column to find the total number of values, that is, the total number of people who contributed money.

Value	Frequency	Product
$ 3	4	($3 • 4) = $12
$ 5	2	($5 • 2) = $10
$ 7	1	($7 • 1) = $ 7
$ 8	5	($8 • 5) = $40
$ 9	3	($9 • 3) = $27
$10	2	($10 • 2) = $20
$12	1	($12 • 1) = $12
$13	2	($13 • 2) = $26
Totals	**20**	**$154**

Finally, divide the totals.

$$\text{mean} = \frac{\$154}{20} = \$7.70$$

The mean contribution to the food pantry was $7.70.

◀ *Work Problem* **3** *at the Side.*

ANSWER
3. $\frac{\$174}{21} \approx \$8.29$ (to nearest cent)

A common use of the weighted mean is to find a student's *grade point average (GPA),* as shown in the next example.

EXAMPLE 4 **Applying the Weighted Mean**

Find the GPA (grade point average) for a student who earned the following grades last semester. Assume A = 4, B = 3, C = 2, D = 1, and F = 0. The number of credits determines how many times the grade is counted (the frequency).

Course	Credits	Grade	Credits · Grade
Mathematics	4	A (= 4)	4 · 4 = 16
Speech	3	C (= 2)	3 · 2 = 6
English	3	B (= 3)	3 · 3 = 9
Computer science	2	A (= 4)	2 · 4 = 8
Theater	2	D (= 1)	2 · 1 = 2
Totals	**14**		**41**

It is common to round grade point averages to the nearest hundredth. So the grade point average for this student is rounded to 2.93.

$$\text{GPA} = \frac{41}{14} \approx 2.93$$ This GPA is close to a B.

Work Problem **4** *at the Side.* ▶

4 Find the GPA (grade point average) for a student who earned the following grades. Round to the nearest hundredth.

Course	Credits	Grade
Mathematics	5	A (= 4)
English	3	C (= 2)
Biology	4	B (= 3)
History	3	B (= 3)

OBJECTIVE 3 Find the median. Because it can be affected by extremely high or low numbers, the mean is often a poor indicator of central tendency for a list of numbers. In cases like this, another measure of central tendency, called the *median,* can be used. The **median** divides a group of numbers in half; half the numbers lie above the median, and half lie below the median.

Find the median by listing the numbers *in order* from *least* to *greatest.* If the list contains an *odd* number of items, the median is the *middle number.*

5 Find the median for the numbers of customers helped each hour at the order desk.

35, 33, 27, 30, 39, 50, 59, 25, 30

EXAMPLE 5 **Finding the Median (Odd Number of Items)**

Find the median for this list of prices.

$7, $23, $15, $6, $18, $12, $24

First, arrange the numbers in numerical order from least to greatest.

Least ⟶ 6, 7, 12, 15, 18, 23, 24 ⟵ Greatest

Next, find the *middle* number in the list.

6, 7, 12, **15**, 18, 23, 24   List the numbers from least to greatest **before** finding the middle.

Three are below. ↓ Three are above.
Middle number

The median price is $15.

Work Problem **5** *at the Side.* ▶

If a list contains an *even* number of items, there is no single middle number. In this case, the median is defined as the mean (average) of the *middle two* numbers.

ANSWERS

4. GPA = $\frac{47}{15} \approx 3.13$

5. 33 customers (the middle number when the numbers are arranged from least to greatest)

6 Find the median for this list of measurements.

178 ft, 261 ft, 126 ft, 189 ft, 121 ft, 195 ft, 121 ft, 200 ft

7 Find the mode for each list of numbers.

(a) Ages of part-time employees (in years): 28, 21, 16, 22, 28, 34, 22, 28, 19, 18

(b) Total points on a screening exam: 312, 219, 782, 312, 219, 426, 507, 600

(c) Monthly commissions of salespeople: $1706, $1289, $1653, $1892, $1301, $1782, $1450, $1566

EXAMPLE 6 **Finding the Median (Even Number of Items)**

Find the median for this list of ages, in years.

$$74, 7, 15, 13, 25, 28, 47, 59, 33, 68$$

First, arrange the numbers in numerical order from least to greatest. Then, because the list has an even number of ages, find the middle *two* numbers.

Least $\longrightarrow$ 7, 13, 15, 25, **28, 33**, 47, 59, 68, 74 $\longleftarrow$ Greatest

Middle two numbers

The median age is the mean (average) of the two middle numbers.

$$\text{median} = \frac{28 + 33}{2} = \frac{61}{2} = 30.5 \text{ years}$$

◀ Work Problem **6** at the Side.

OBJECTIVE **4** **Find the mode.** Another statistical measure is the **mode**, which is the number that occurs *most often* in a list of numbers. For example, the test scores for ten students are shown below.

$$\downarrow \qquad \downarrow \qquad \downarrow$$
74, 81, 39, **74**, 82, 80, 100, 92, **74**, 85

The mode is 74. Three students earned a score of 74, so 74 appears more times on the list than any other score. It is *not* necessary to place the numbers in numerical order when looking for the mode, although that may help you find it more easily.

A list can have two modes; such a list is sometimes called *bimodal*. If no number occurs more frequently than any other number in a list, the list has *no mode*.

EXAMPLE 7 **Finding the Mode**

Find the mode for each list of numbers.

(a) 51, 32, 49, 51, 49, 90, 49, 60, 17, 60

The number 49 occurs three times, which is more often than any other number. Therefore, **49** is the mode.

(b) 482, 485, 483, 485, 487, 487, 489, 486

Because both **485** and **487** occur twice, each is a mode. This list is *bimodal*.

(c) $10,708; $11,519; $10,972; $12,546; $13,905; $12,182

No price occurs more than once. This list has *no mode*.

Measures of Central Tendency

The **mean** is the sum of all the values divided by the number of values. It is the mathematical *average*.

The **median** is the *middle number* (or the average of the two middle numbers) in a group of values that are listed from least to greatest. It divides a group of numbers in half.

The **mode** is the value that occurs *most often* in a group of values.

ANSWERS

6. $\frac{178 + 189}{2} = 183.5$ ft

7. **(a)** 28 years
   (b) bimodal, 219 points and 312 points (this list has two modes)
   (c) no mode (no number occurs more than once)

◀ Work Problem **7** at the Side.

5.7 ▶▶▶ Exercises

FOR EXTRA HELP | MyMathLab | Math XL PRACTICE | WATCH | DOWNLOAD | READ | REVIEW

Find the mean for each list of numbers. Round answers to the nearest tenth when necessary. See Examples 1 and 2.

1. Final exam scores: 92, 51, 59, 86, 68, 73, 49, 80

2. Quiz scores: 18, 25, 21, 8, 16, 13, 23, 19

3. Annual salaries: $31,900; $32,850; $34,930; $39,712; $38,340, $60,000

4. Numbers of people attending baseball games: 27,500; 18,250; 17,357; 14,298; 33,110

5. The Athletic Shoe Store sold shoes at the following prices: $75.52, $36.15, $58.24, $21.86, $47.68, $106.57, $82.72, $52.14, $28.60, $72.92.

6. In one evening, a server collected the following checks from her dinner customers: $30.10, $42.80, $91.60, $51.20, $88.30, $21.90, $43.70, $51.20.

Find the weighted mean. Round answers to the nearest tenth when necessary. See Example 3.

7.

Quiz Score	Frequency
3	4
5	2
6	5
8	5
9	2

8.

Credits per Student	Frequency
9	3
12	5
13	2
15	6
18	1

Find the GPA (grade point average) for students earning the following grades. Assume $A = 4$, $B = 3$, $C = 2$, $D = 1$, and $F = 0$. Round answers to the nearest hundredth. See Example 4.

9.

Course	Credits	Grade
Biology	4	B
Biology lab	2	A
Mathematics	5	C
Health	1	F
Psychology	3	B

10.

Course	Credits	Grade
Chemistry	3	A
English	3	B
Mathematics	4	B
Theater	2	C
Astronomy	3	C

11. Look again at the grades in Exercise 9. Find the student's GPA in each of these situations.

   (a) The student earned a B instead of an F in the 1-credit class.

   (b) The student earned a B instead of a C in the 5-credit class.

   (c) Both (a) and (b) happened.

12. List the credits for the courses you're taking at this time. List the lowest grades you think you will earn in each class and find your GPA. Then list the highest grades you think you will earn and find your GPA.

Find the median for each list of numbers. See Examples 5 and 6.

13. Number of e-mail messages received:
9, 15, 23, 12, 14, 24, 28

14. Patients seen each day at a clinic:
99, 108, 123, 109, 126, 146, 129, 168, 170

15. Students enrolled in algebra each semester:
328, 549, 420, 592, 715, 483

16. Number of cars in the parking lot each day:
520, 523, 513, 1283, 338, 509, 290, 420

17. Pounds of shrimp sold each day: 51, 48, 96, 40, 47,
40, 95, 56, 34, 49

18. Number of gallons of paint sold per week:
1072, 1068, 1093, 1042, 1056, 205, 1009, 1081

The table lists the cruising speed and distance flown without refueling for several types of larger airplanes used to carry passengers. Use the table to answer Exercises 19–22.

Type of Airplane	Cruising Speed (miles per hour)	Distance without Refueling (miles)
747-400	565	7650
747-200	558	6450
DC-9	505	1100
DC-10	550	5225
727	530	1550
757	530	2875

Source: Northwest Airlines World Traveler.

21. Find the median distance flown.

19. What is the average distance flown without refueling, to the nearest mile?

20. Find the average cruising speed to the nearest mile per hour.

22. Find the median cruising speed.

Find the mode or modes for each list of numbers. See Example 7.

23. Number of samples taken each hour:
3, 8, 5, 1, 7, 6, 8, 4, 5, 8

24. Monthly water bills:
$21, $32, $46, $32, $49, $32, $49, $25, $32

25. Ages of senior residents (in years):
74, 68, 68, 68, 75, 75, 74, 74, 70, 77

26. Patients admitted to the hospital each week:
30, 19, 25, 78, 36, 20, 45, 85, 38

27. The number of boxes of candy sold by each child:
5, 9, 17, 3, 2, 8, 19, 1, 4, 20, 10, 6

28. The weights of soccer players (in pounds):
158, 161, 165, 162, 165, 157, 163, 162

The table lists monthly normal temperatures from November through April for two of the coldest U.S. cities. Use the table to answer Exercises 29–30. Round answers to the nearest whole degree.

Normal Monthly Temperatures (in Degrees Fahrenheit)						
City	Nov.	Dec.	Jan.	Feb.	Mar.	Apr.
Barrow, Alaska	−2	−11	−13	−18	−15	−2
Fairbanks, Alaska	3	−7	−10	−4	11	31

Source: National Climatic Data Center.

29. Find Barrow's mean temperature and Fairbanks' mean temperature for the six-month period. How much warmer is Fairbanks' mean than Barrow's?

30. Find Barrow's median temperature and Fairbanks' median temperature for the six-month period. How much cooler is Barrow's median than Fairbanks'?

5.8 ▷▷▷ Geometry Applications: Pythagorean Theorem and Square Roots

In **Section 3.2,** you used this formula for area of a square, $A = s^2$. The blue square below has an area of 25 cm^2 because $(5 \text{ cm})(5 \text{ cm}) = 25 \text{ cm}^2$.

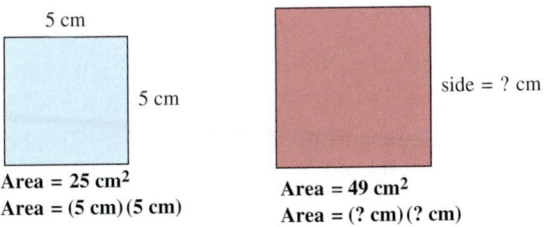

5 cm

5 cm

Area = 25 cm^2
Area = (5 cm)(5 cm)

side = ? cm

Area = 49 cm^2
Area = (? cm)(? cm)

The red square above has an area of 49 cm^2. To find the length of a side, ask yourself, "What number can be multiplied by itself to give 49?" Because $(7)(7) = 49$, the length of each side is 7 cm. Also, because $(7)(7) = 49$ we say that 7 is the *square root* of 49, or $\sqrt{49} = 7$.

> **Square Root**
>
> The positive **square root** of a positive number is one of two identical positive factors of that number.
>   For example, $\sqrt{36} = 6$ because $(6)(6) = 36$.

> **Note**
> There is another square root for 36. We know that
> $$(6)(6) = 36 \quad \text{and} \quad (-6)(-6) = 36$$
> so the *positive* square root of 36 is 6 and the *negative* square root of 36 is -6. In this section we will work only with positive square roots. You will learn more about negative square roots in **Chapter 16**.

Work Problem ① *at the Side.* ▶

A number that has a whole number as its square root is called a *perfect square*. For example, 9 is a perfect square because $\sqrt{9} = 3$, and 3 is a whole number. The first few perfect squares are listed below.

> **The First Twelve Perfect Squares**
>
> $\sqrt{1} = 1$ $\sqrt{16} = 4$ $\sqrt{49} = 7$ $\sqrt{100} = 10$
>
> $\sqrt{4} = 2$ $\sqrt{25} = 5$ $\sqrt{64} = 8$ $\sqrt{121} = 11$
>
> $\sqrt{9} = 3$ $\sqrt{36} = 6$ $\sqrt{81} = 9$ $\sqrt{144} = 12$

OBJECTIVE **1** **Find square roots using the square root key on a calculator.** If a number is *not* a perfect square, then you can find its *approximate* square root by using a calculator with a square root key.

OBJECTIVES

1 Find square roots using the square root key on a calculator.

2 Find the unknown length in a right triangle.

3 Solve application problems involving right triangles.

1 Find each square root.

(a) $\sqrt{36}$

(b) $\sqrt{25}$

(c) $\sqrt{9}$

(d) $\sqrt{100}$

(e) $\sqrt{121}$

ANSWERS

1. **(a)** 6 **(b)** 5 **(c)** 3 **(d)** 10 **(e)** 11

2 Use a calculator with a square root key to find each square root. Round to the nearest thousandth when necessary.

(a) $\sqrt{11}$

(b) $\sqrt{40}$

(c) $\sqrt{56}$

(d) $\sqrt{196}$

(e) $\sqrt{147}$

Calculator Tip To find a square root on a *scientific* calculator, use the (√) or the (√x) key. (On some models you may have to press the (2nd) key to access the square root function.) You do *not* need to use the (=) key. Try these.

To find $\sqrt{16}$ press:	16 (√x)	Answer is 4
To find $\sqrt{7}$ press:	7 (√x)	Answer is 2.645751311

For $\sqrt{7}$, your calculator shows 2.645751311, which is an *approximate* answer. (Some calculators may show more or fewer digits.) We will round to the nearest thousandth, so $\sqrt{7} \approx 2.646$. To check, multiply 2.646 times 2.646. Do you get 7 as the result? No, you get 7.001316, which is very close to 7. The difference is due to rounding.

EXAMPLE 1 **Finding the Square Root of Numbers**

Use a calculator to find each square root. Round to the nearest thousandth.

(a) $\sqrt{35}$ Calculator shows 5.916079783; round to 5.916

(b) $\sqrt{124}$ Calculator shows 11.13552873; round to 11.136

◀ Work Problem **2** at the Side.

OBJECTIVE 2 **Find the unknown length in a right triangle.** One place you will use square roots is when working with the *Pythagorean Theorem*. This theorem applies only to *right* triangles (triangles with a 90° angle). The longest side of a right triangle is called the **hypotenuse.** It is opposite the right angle. The other two sides are called *legs*. The legs form the right angle. Here are some right triangles.

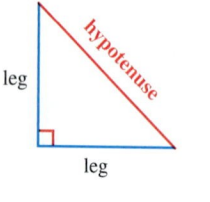

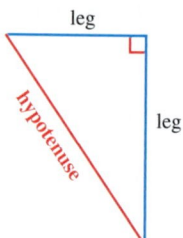

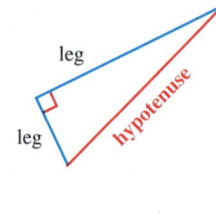

Examples of right triangles

Pythagorean Theorem

$$(\text{hypotenuse})^2 = (\text{leg})^2 + (\text{leg})^2$$

In other words, square the length of each side. After you have squared all the sides, the sum of the squares of the two legs will equal the square of the hypotenuse. An example is shown below.

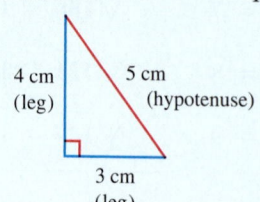

$$(\text{hypotenuse})^2 = (\text{leg})^2 + (\text{leg})^2$$
$$5^2 = 4^2 + 3^2$$
$$25 = 16 + 9$$
$$25 = 25$$

The theorem is named after Pythagoras, a Greek mathematician who lived about 2500 years ago. He and his followers may have used floor tiles to prove the theorem, as shown on the next page.

ANSWERS

2. (a) $\sqrt{11} \approx 3.317$ (b) $\sqrt{40} \approx 6.325$ (c) $\sqrt{56} \approx 7.483$ (d) $\sqrt{196} = 14$ (e) $\sqrt{147} \approx 12.124$

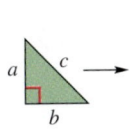
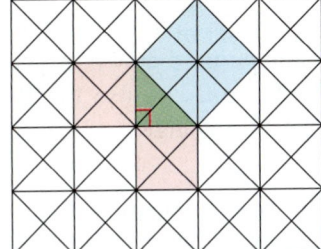

Right triangle

3 Find the unknown length in each right triangle. Round your answers to the nearest tenth when necessary.

(a)

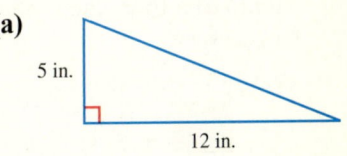

5 in.

12 in.

The green right triangle in the center of the floor tiles has sides a, b, and c. The pink square drawn on side a contains four triangular tiles. The pink square on side b contains four tiles. The blue square on side c contains eight tiles. The number of tiles in the square on side c equals the sum of the number of tiles in the squares on sides a and b, that is, 8 tiles = 4 tiles + 4 tiles. As a result, you often see the Pythagorean Theorem written as $c^2 = a^2 + b^2$.

If you know the lengths of any two sides in a right triangle, you can use the Pythagorean Theorem to find the length of the third side.

(b)

7 cm   90°   25 cm

> **Formulas Based on the Pythagorean Theorem**
>
> To find the hypotenuse: **hypotenuse = $\sqrt{(\text{leg})^2 + (\text{leg})^2}$**
>
> To find a leg: **leg = $\sqrt{(\text{hypotenuse})^2 - (\text{leg})^2}$**

(c)

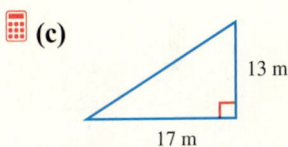

13 m

17 m

EXAMPLE 2 **Finding the Unknown Length in Right Triangles**

Find the unknown length in each right triangle. Round your answers to the nearest tenth when necessary.

(a)

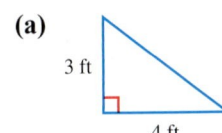

3 ft

4 ft

The unknown length is the side opposite the right angle, which is the hypotenuse. Use the formula for finding the hypotenuse.

$$\text{hypotenuse} = \sqrt{(\text{leg})^2 + (\text{leg})^2} \qquad \text{Find the hypotenuse.}$$
$$\text{hypotenuse} = \sqrt{(3)^2 + (4)^2} \qquad \text{Legs are 3 and 4}$$
$$= \sqrt{9 + 16} \qquad (3)(3) \text{ is 9} \quad \text{and} \quad (4)(4) \text{ is 16}$$
$$= \sqrt{25}$$
$$= 5$$

The hypotenuse is 5 ft long.   *Write **ft** in the answer (**not** ft².)*

(d)

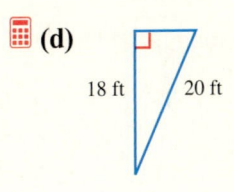

18 ft   20 ft

(b)

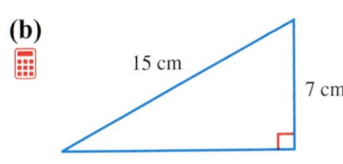

15 cm   7 cm

You *do* know the length of the hypotenuse (15 cm), so it is the length of one of the legs that is unknown. Use the formula for finding a leg.

$$\text{leg} = \sqrt{(\text{hypotenuse})^2 - (\text{leg})^2} \qquad \text{Find a leg.}$$
$$\text{leg} = \sqrt{(15)^2 - (7)^2} \qquad \text{Hypotenuse is 15; one leg is 7}$$
$$= \sqrt{225 - 49} \qquad (15)(15) \text{ is 225} \quad \text{and} \quad (7)(7) \text{ is 49}$$
$$= \sqrt{176} \qquad \text{Use a calculator to find } \sqrt{176}$$
$$\approx 13.3 \qquad \text{Round 13.26649916 to 13.3}$$

The length of the leg is approximately 13.3 cm.

(e)

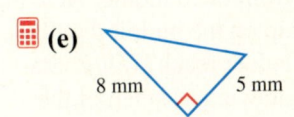

8 mm   5 mm

ANSWERS

3. **(a)** $\sqrt{169} = 13$ in. **(b)** $\sqrt{576} = 24$ cm
   (c) $\sqrt{458} \approx 21.4$ m **(d)** $\sqrt{76} \approx 8.7$ ft
   (e) $\sqrt{89} \approx 9.4$ mm

Work Problem **3** *at the Side.* ▶

4 These problems show ladders leaning against buildings. Find the unknown lengths. Round answers to the nearest tenth of a foot when necessary.

(a)

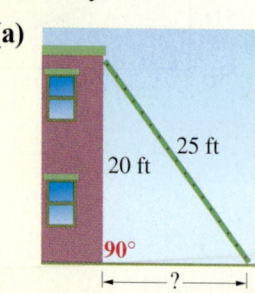

How far away from the building is the bottom of the ladder?

(b)

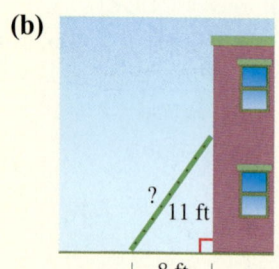

How long is the ladder?

(c) A 17 ft ladder is leaning against a building. The bottom of the ladder is 10 ft from the building. How high up on the building will the ladder reach? (*Hint:* Start by drawing a sketch of the building and the ladder.)

> **CAUTION**
>
> *Remember:* A small square drawn in one angle of a triangle indicates a right angle (90°). You can use the Pythagorean Theorem *only* on triangles that have a right angle.

OBJECTIVE 3 **Solve application problems involving right triangles.** This example is an application of the Pythagorean Theorem.

EXAMPLE 3 **Using the Pythagorean Theorem**

A television antenna is on the roof of a house, as shown. Find the length of the support wire. Round your answer to the nearest tenth of a meter if necessary.

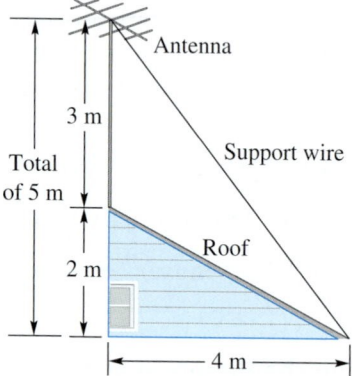

A right triangle is formed. The total length of the leg on the left is 3 m + 2 m = 5 m.

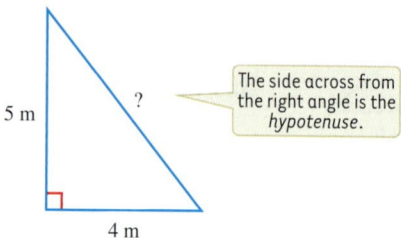

The side across from the right angle is the *hypotenuse.*

Notice that the support wire is opposite the right angle, so it is the *hypotenuse* of the right triangle.

$$\text{hypotenuse} = \sqrt{(\text{leg})^2 + (\text{leg})^2} \quad \text{Find the hypotenuse.}$$

$$\text{hypotenuse} = \sqrt{(5)^2 + (4)^2} \quad \text{Legs are 5 and 4}$$

$$= \sqrt{25 + 16} \quad 5^2 \text{ is 25 and } 4^2 \text{ is 16}$$

$$= \sqrt{41} \quad \text{Use a calculator to find } \sqrt{41}$$

$$\approx 6.4 \quad \longleftarrow \text{ Rounded}$$

The length of the support wire is approximately **6.4** m. 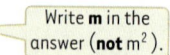 Write **m** in the answer (**not** m²).

> **CAUTION**
>
> You use the Pythagorean Theorem to find the *length* of one side, *not* the area of the triangle. Your answer will be in linear units, such as ft, yd, cm, m, and so on (*not* ft², yd², cm², m²).

◀ *Work Problem* **4** *at the Side.*

ANSWERS

4. **(a)** leg = $\sqrt{225}$ = 15 ft

   (b) hypotenuse = $\sqrt{185}$ ≈ 13.6 ft

   (c) leg = $\sqrt{189}$ ≈ 13.7 ft

5.8 ▶▶▶ Exercises

 FOR EXTRA HELP    **MyMathLab**   Math **XL** PRACTICE   WATCH   DOWNLOAD   READ   REVIEW

Find each square root. Starting with Exercise 5, find the square root using a calculator.
Round your answers to the nearest thousandth when necessary. See Example 1.

1. $\sqrt{16}$ **2.** $\sqrt{4}$ **3.** $\sqrt{64}$ **4.** $\sqrt{81}$

5. $\sqrt{11}$ **6.** $\sqrt{23}$ **7.** $\sqrt{5}$ **8.** $\sqrt{2}$

9. $\sqrt{73}$ **10.** $\sqrt{80}$ **11.** $\sqrt{101}$ **12.** $\sqrt{125}$

13. $\sqrt{361}$ **14.** $\sqrt{729}$ **15.** $\sqrt{1000}$ **16.** $\sqrt{2000}$

17. You know that $\sqrt{25} = 5$ and $\sqrt{36} = 6$. Using just that information (no calculator), describe how you could *estimate* $\sqrt{30}$. How would you estimate $\sqrt{26}$ or $\sqrt{35}$? Now check your estimates using a calculator.

18. Explain the relationship between *squaring* a number and finding the *square root* of a number. Include two examples to illustrate your explanation.

Find the unknown length in each right triangle. Use a calculator to find square roots.
Round your answers to the nearest tenth when necessary. See Example 2.

19.

 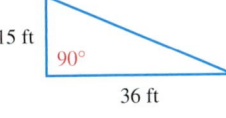

15 ft, 90°, 36 ft

20.

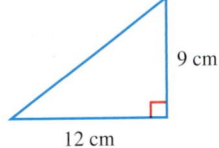

9 cm, 12 cm

21.

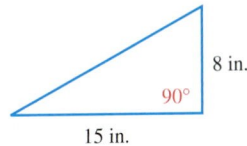

8 in., 90°, 15 in.

22.

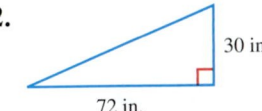

30 in., 72 in.

23.

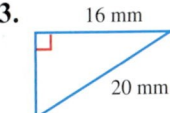

16 mm, 20 mm

24.

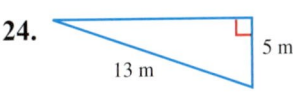

5 m, 13 m

25.
3 in.
8 in.

26.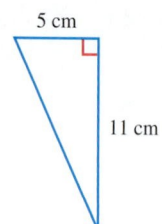
5 cm
11 cm

27.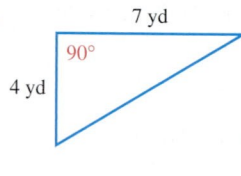
7 yd
90°
4 yd

28.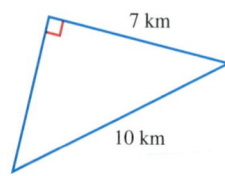
7 km
10 km

29.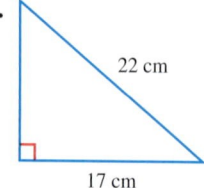
22 cm
17 cm

30.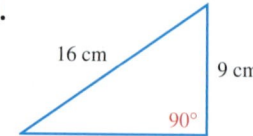
16 cm
9 cm
90°

31.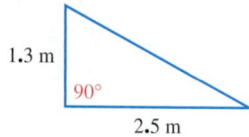
1.3 m
90°
2.5 m

32.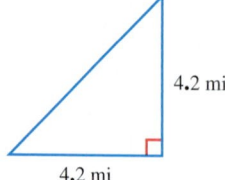
4.2 mi
4.2 mi

33.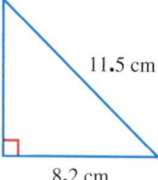
11.5 cm
8.2 cm

34.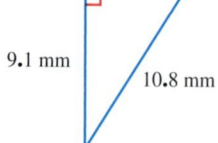
9.1 mm
10.8 mm

35.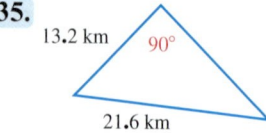
13.2 km
90°
21.6 km

36.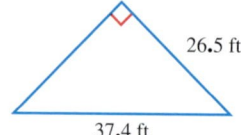
26.5 ft
37.4 ft

Solve each application problem. Round your answers to the nearest tenth when necessary. See Example 3.

37. Find the length of this loading ramp.

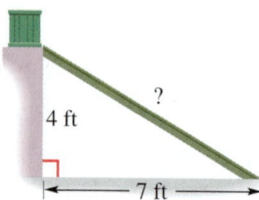
?
4 ft
7 ft

38. Find the unknown length in this window frame.

?
4 ft
3 ft

39. How high is the airplane above the ground?

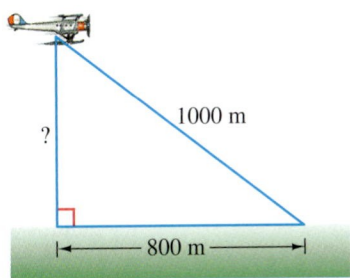

40. Find the height of this farm silo.

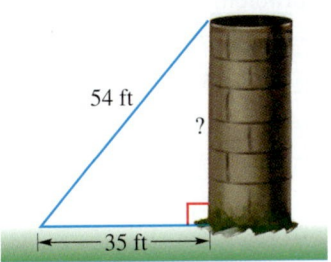

41. How long is the diagonal brace on this storage shed door?

42. Find the height of this rectangular television screen. (*Source:* Sears.)

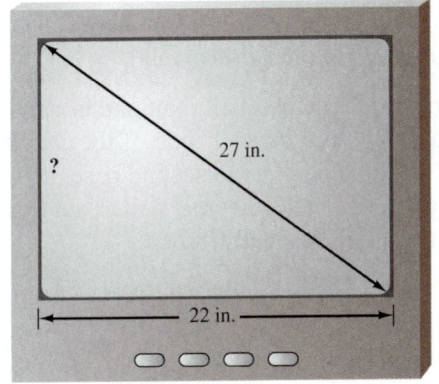

43. To reach his ladylove, a knight placed a 12 ft ladder against the castle wall. If the base of the ladder is 3 ft from the building, how high on the castle will the top of the ladder reach? Draw a sketch of the castle and ladder and solve the problem.

44. William drove his car 15 miles north, then made a right turn and drove 7 miles east. How far is he, in a straight line, from his starting point? Draw a sketch to illustrate the problem and then solve it.

45. Explain the *two* errors made by a student in solving this problem. Also find the correct answer. Round to the nearest tenth.

$$? = \sqrt{(13)^2 + (20)^2}$$

$$= \sqrt{169 + 400}$$

$$= \sqrt{569} \approx 23.9 \text{ m}^2$$

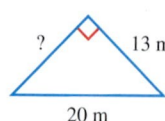
? 13 m

20 m

46. Explain the *two* errors made by a student in solving this problem. Also find the correct answer. Round to the nearest tenth.

$$? = \sqrt{(9)^2 + (7)^2}$$

$$= \sqrt{18 + 14}$$

$$= \sqrt{32} \approx 5.657 \text{ in.}$$

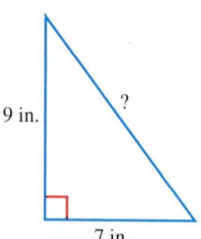
9 in.   ?

7 in.

Relating Concepts (Exercises 47–50) For Individual or Group Work

Use your knowledge of the Pythagorean Theorem to **work Exercises 47–50** *in order.* *Round answers to the nearest tenth.*

47. A major league baseball diamond is a square shape measuring 90 ft on each side. If the catcher throws a ball from home plate to second base, how far is he throwing the ball? (*Source:* American League of Professional Baseball Clubs.)

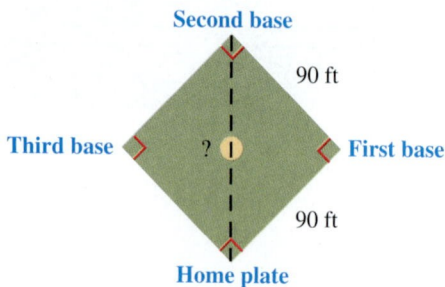
Second base

90 ft

Third base   ?   First base

90 ft

Home plate

48. A softball diamond is only 60 ft on each side. (*Source:* Amateur Softball Association.)

(a) Draw a sketch of the softball diamond and label the bases and the lengths of the sides.

(b) How far is it to throw a ball from home plate to second base?

49. Look back at your answer to Exercise 47. Explain how you can tell the distance from third base to first base without doing any further calculations.

50. (a) Look back at your answer to Exercise 48. Suppose you measured the distance from home plate to second base on a softball diamond and found it was 80 ft. What would this tell you about the length of each side of the diamond? (Assume the diamond is still a square.)

(b) Bonus question: For the diamond in part (a), find the length of each side, to the nearest tenth.

5.9 ▶▶▶ Problem Solving: Equations Containing Decimals

OBJECTIVE 1 Solve equations containing decimals using the addition property of equality. In **Section 2.3** you used the addition property of equality to solve an equation like $c + 5 = 30$. The addition property says that you can add the *same* number to *both* sides of an equation and still keep it balanced. You can also use this property when an equation contains decimal numbers.

EXAMPLE 1 **Using the Addition Property of Equality**

Solve each equation and check each solution.

(a) $w + 2.9 = -0.6$

The first step is to get the variable term (w) by itself on the left side of the equal sign. Use the addition property to "get rid of" the 2.9 on the left side by adding its opposite, -2.9, to both sides.

$$w + 2.9 = -0.6$$
$$\underline{\quad -2.9 \qquad -2.9} \qquad \text{Add } -2.9 \text{ to both sides.}$$
$$\underbrace{w + 0}_{w} = -3.5$$
$$w = -3.5$$

The solution is -3.5. To check the solution, go back to the *original* equation.

Check $\quad w + 2.9 = -0.6 \qquad$ Original equation
$$\underbrace{-3.5 + 2.9} = -0.6 \qquad \text{Replace } w \text{ with } -3.5$$
$$-0.6 = -0.6 \qquad \text{Balances}$$

When w is replaced with -3.5, the equation balances, so -3.5 is the correct solution (**not** -0.6).

(b) $7 = -4.3 + x$

To get x by itself on the right side of the equal sign, add 4.3 to both sides.

$$7 = -4.3 + x$$
$$\underline{+\,4.3 \qquad +4.3} \qquad \text{Add 4.3 to both sides.}$$
$$11.3 = \underbrace{0 + x}$$
$$11.3 = x$$

The solution is 11.3. To check the solution, go back to the *original* equation.

Check $\quad 7 = -4.3 + x \qquad$ Original equation
$$7 = \underbrace{-4.3 + 11.3} \qquad \text{Replace } x \text{ with } 11.3 \quad \boxed{\text{The solution is } 11.3 \text{ (not 7).}}$$
$$7 = 7 \qquad \text{Balances}$$

When x is replaced with 11.3, the equation balances, so 11.3 is the correct solution (**not** 7).

Work Problem **1** *at the Side.* ▶

OBJECTIVES

1 Solve equations containing decimals using the addition property of equality.

2 Solve equations containing decimals using the division property of equality.

3 Solve equations containing decimals using both properties of equality.

4 Solve application problems involving equations with decimals.

---

1 Solve each equation and check each solution.

(a) $8.1 = h + 9 \qquad$ **Check**

(b) $-0.75 + y = 0 \qquad$ **Check**

(c) $c - 6.8 = -4.8 \qquad$ **Check**

ANSWERS

1. (a) $h = -0.9 \qquad$ **Check** $\quad 8.1 = h + 9$
$$8.1 = -0.9 + 9$$
$$\text{Balances} \quad 8.1 = 8.1$$

(b) $y = 0.75 \qquad$ **Check** $\quad -0.75 + y = 0$
$$-0.75 + 0.75 = 0$$
$$\text{Balances} \quad 0 = 0$$

(c) $c = 2 \qquad$ **Check** $\quad c - 6.8 = -4.8$
$$2 - 6.8 = -4.8$$
$$\text{Balances} \quad -4.8 = -4.8$$

2 Solve each equation and check each solution.

(a) $-3y = -0.63$

Check

(b) $2.25r = -18$

Check

(c) $1.7 = 0.5n$

Check

2. **(a)** $y = 0.21$ **Check** $-3y = -0.63$
   $$-3(0.21) = -0.63$$
   Balances $-0.63 = -0.63$

(b) $r = -8$ **Check** $2.25r = -18$
   $$2.25(-8) = -18$$
   Balances $-18 = -18$

(c) $n = 3.4$ **Check** $1.7 = 0.5n$
   $$1.7 = 0.5(3.4)$$
   Balances $1.7 = 1.7$

OBJECTIVE 2 Solve equations containing decimals using the division property of equality. You can also use the division property of equality (from **Section 2.4**) when an equation contains decimals.

EXAMPLE 2 **Using the Division Property of Equality**

Solve each equation and check each solution.

(a) $5x = 12.4$

On the left side of the equation, the variable is multiplied by 5. To undo the multiplication, divide both sides by 5.

$5x$ means $5 \cdot x$.

$$5x = 12.4$$

$$\frac{5 \cdot x}{5} = \frac{12.4}{5} \quad \text{Divide both sides by 5}$$

On the left side, divide out the common factor of 5

$$\frac{\overset{1}{5} \cdot x}{\underset{1}{5}} = 2.48 \quad \text{On the right side, } 12.4 \div 5 \text{ is } 2.48$$

$$x = 2.48$$

The solution is 2.48. To check the solution, go back to the original equation.

Check $5x = 12.4$ Original equation

$$5(2.48) = 12.4 \quad \text{Replace } x \text{ with } 2.48$$

The solution is 2.48 (**not** 12.4)

$$12.4 = 12.4 \quad \text{Balances}$$

When x is replaced with 2.48, the equation balances, so 2.48 is the correct solution (**not** 12.4).

(b) $-9.3 = 1.5t$

Signs are different, so quotient is negative.

$$\frac{-9.3}{1.5} = \frac{\overset{1}{\cancel{1.5}}t}{\underset{1}{\cancel{1.5}}} \quad \text{Divide both sides by the coefficient of the variable term, } 1.5$$

$$-6.2 = t$$

The solution is -6.2. To check the solution, go back to the original equation.

Check $-9.3 = 1.5t$ Original equation

$$-9.3 = 1.5(-6.2) \quad \text{Replace } t \text{ with } -6.2$$

$$-9.3 = -9.3 \quad \text{Balances}$$

When t is replaced with -6.2, the equation balances, so -6.2 is the correct solution (**not** -9.3).

◀ Work Problem **2** at the Side.

OBJECTIVE 3 Solve equations containing decimals using both properties of equality. Sometimes you need to use both the addition and division properties to solve an equation, as shown in Example 3.

EXAMPLE 3 **Solving Equations with Several Steps**

(a) $2.5b + 0.35 = -2.65$

The first step is to get the variable term, $2.5b$, by itself on the left side of the equal sign.

$$
\begin{array}{rcl}
2.5b + 0.35 &=& -2.65 \\
\underline{-0.35 \qquad -0.35} & & \qquad \text{Add } -0.35 \text{ to both sides.} \\
2.5b + \;\; 0 &=& -3.00 \\
2.5b &=& -3
\end{array}
$$

The next step is to divide both sides by the coefficient of the variable term. In $2.5b$, the coefficient is 2.5.

$$
\frac{\overset{1}{\cancel{2.5}}b}{\underset{1}{\cancel{2.5}}} = \frac{-3}{2.5} \qquad
\begin{array}{l}
\text{On the right side, signs do} \\
\textit{not} \text{ match, so the quotient} \\
\text{is } \textit{negative.}
\end{array}
$$

$$
b = -1.2
$$

The solution is -1.2. To check the solution, go back to the original equation.

Check

$$
\begin{array}{rcll}
2.5\boldsymbol{b} + 0.35 &=& -2.65 & \text{Original equation} \\
2.5\,(\boldsymbol{-1.2}) + 0.35 &=& -2.65 & \text{Replace } b \text{ with } -1.2 \\
-3 \quad + 0.35 &=& -2.65 & \\
-2.65 &=& -2.65 & \text{Balances, so } -1.2 \text{ is the} \\
& & & \text{correct solution.}
\end{array}
$$

(b) $5x - 0.98 = 2x + 0.4$

There is a variable term on both sides of the equation. You can choose to keep the variable term on the left side, or to keep the variable term on the right side. Either way will work. Just pick the left side or the right side.

Suppose that you decide to keep the variable term, $5x$, on the left side. Use the addition property to "get rid of" $2x$ on the right side by adding its opposite, $-2x$, to both sides.

$$
\begin{array}{rcll}
5x - 0.98 &=& 2x + 0.4 & \\
\underline{-2x \qquad\quad} & & \underline{-2x \qquad\quad} & \text{Add } -2x \text{ to both sides.} \\
3x - 0.98 &=& 0 \;+ 0.4 & \\
3x + (-0.98) &=& 0.4 & \\
\underline{+0.98} & & \underline{+0.98} & \text{Add } 0.98 \text{ to both sides.} \\
3x + \quad 0 &=& 1.38 &
\end{array}
$$

Change subtraction to adding the opposite.

$$
\frac{\overset{1}{\cancel{3}}x}{\underset{1}{\cancel{3}}} = \frac{1.38}{3} \qquad \text{Divide both sides by 3}
$$

$$
x = 0.46
$$

The solution is 0.46. To check the solution, go back to the original equation.

Check

$$
\begin{array}{rcll}
5x - 0.98 &=& 2x + 0.4 & \text{Original equation} \\
5\,(\boldsymbol{0.46}) - 0.98 &=& 2\,(\boldsymbol{0.46}) + 0.4 & \text{Replace } x \text{ with } 0.46 \\
2.3 \;- 0.98 &=& 0.92 + 0.4 & \\
1.32 &=& 1.32 & \text{Balances, so } 0.46 \text{ is the} \\
& & & \text{correct solution.}
\end{array}
$$

Work Problem **3** *at the Side.* ▶

---

3 Solve each equation and check each solution.

(a) $4 = 0.2c - 2.6$

Check

(b) $3.1k - 4 = 0.5k + 13.42$

Check

(c) $-2y + 3 = 3y - 6$

Check

Answers

3. (a) $c = 33$ **Check**
$$
\begin{array}{rcl}
4 &=& 0.2c - 2.6 \\
4 &=& 0.2\,(33) - 2.6 \\
4 &=& 6.6 \; - 2.6
\end{array}
$$
Balances $\;4 = 4$

(b) $k = 6.7$ **Check**
$$
\begin{array}{rcl}
3.1k - 4 &=& 0.5k + 13.42 \\
3.1\,(6.7) - 4 &=& 0.5\,(6.7) + 13.42 \\
20.77 - 4 &=& 3.35 + 13.42
\end{array}
$$
Balances $16.77 = 16.77$

(c) $y = 1.8$ **Check**
$$
\begin{array}{rcl}
-2y + 3 &=& 3y - 6 \\
-2(1.8) + 3 &=& 3(1.8) - 6 \\
-3.6 + 3 &=& 5.4 - 6
\end{array}
$$
Balances $-0.6 = -0.6$

4 During April, a special rate was offered on air-to-ground cell phone calls. The connection fee was $1.34 and the cost per minute was $2.69. Maureen made a call that cost $39. How long did the call last? Use the six problem-solving steps.

OBJECTIVE **4** **Solve application problems involving equations with decimals.**   Use the six problem-solving steps from **Section 3.3.**

EXAMPLE 4 **Solving an Application Problem**

Many larger airplanes have phones that can be used to call people on the ground. In January 2004, the cost of using the air-to-ground cell phone was $3.28 per minute plus a $2.99 connection charge. (*Source:* AT&T.) Hernando was billed $19.39 for one call. How many minutes did the call last?

Step 1   **Read** the problem. It is about the cost of a telephone call.

> **Unknown:** number of minutes the call lasted
> **Known:** Costs are $3.28 per minute plus $2.99; total cost was $19.39.

Step 2   **Assign a variable:** There is only one unknown, so let m be the number of minutes.

Step 3   **Write an equation.**

Cost per minute		Number of minutes		Connection charge		Total cost
3.28	•	m	+	2.99	=	19.39

Step 4   **Solve** the equation.

$$3.28m + 2.99 = 19.39$$
$$\underline{-2.99 \quad -2.99} \qquad \text{Add } -2.99 \text{ to both sides.}$$
$$3.28m + \quad 0 = 16.40$$

$$\frac{3.28m}{3.28} = \frac{16.40}{3.28} \qquad \text{Divide both sides by 3.28}$$

$$m = 5$$

> Write **minutes** as part of your answer.

Step 5   **State the answer.** The call lasted 5 minutes.

Step 6   **Check** the solution by putting it back into the original problem.

> $3.28 per minute times 5 minutes = $16.40
> $16.40 plus $2.99 connection charge = $19.39
> Hernando was billed $19.39. ← Matches ⟶

Because 5 minutes "works" when you put it back into the *original* problem, it is the correct solution.

> If your answer does **not** work in the original problem, start again at Step 1.

◀ *Work Problem* **4** *at the Side.*

5.9 ▶▶▶ Exercises

Solve each equation and check each solution. See Examples 1 and 2.

1. $h + 0.63 = 5.1$ **Check**

2. $-0.2 = k - 0.7$ **Check**

3. $-20.6 + n = -22$ **Check**

4. $g - 5 = 6.03$ **Check**

5. $0 = b - 0.008$ **Check**

6. $0.18 + m = -4.5$ **Check**

7. $2.03 = 7a$ **Check**

8. $-6.2c = 0$ **Check**

9. $0.8p = -96$ **Check**

10. $-10.16 = -4r$ **Check**

11. $-3.3t = -2.31$ **Check**

12. $8.3w = -49.8$ **Check**

Solve each equation. Show your work. See Example 3.

13. $7.5x + 0.15 = -6$

14. $0.8 = 0.2y + 3.4$

15. $-7.38 = 2.05z - 7.38$

16. $6.2h - 0.4 = 2.7$

17. $3c + 10 = 6c + 8.65$

18. $2.1b + 5 = 1.6b + 10$

19. $0.8w - 0.4 = -6 + w$

20. $7r + 9.64 = -2.32 + 5r$ **21.** $-10.9 + 0.5p = 0.9p + 5.3$ **22.** $0.7x - 4.38 = x - 2.16$

Solve each application problem using the six problem-solving steps. See Example 4.

23. Most adult medication doses are for a person weighing 150 pounds. For a 45-pound child, the adult dose should be multiplied by 0.3. If the child's dose of a decongestant is 9 milligrams, what is the adult dose?

24. For a 30-pound child, an adult dose of medication should be multiplied by 0.2. If the child's dose of a cough suppressant is 3 milliliters, find the adult dose.

25. A storm blew down many trees. Several neighbors rented a chain saw for $275.80 and helped each other cut up and stack the wood. The rental company charges $65.95 per day plus a $12 sharpening fee. How many days was the saw rented? (*Source:* Central Rental.)

26. A 20-inch chain saw can be rented for $29.95 for the first two hours, and $9 for each additional hour. Steve's rental charge was $56.95. How many hours did he rent the saw? (*Source:* Central Rental.)

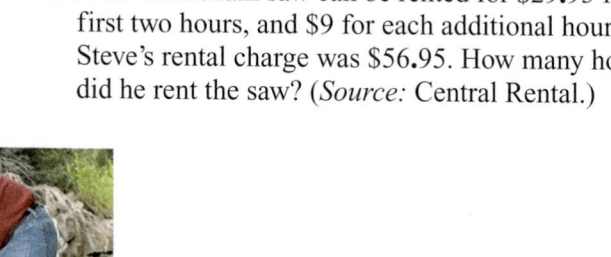

Relating Concepts (Exercises 27–30) For Individual or Group Work

When doing aerobic exercises, it is important to increase your heart rate (the number of beats per minute) so that you get the maximum benefit from the exercise. But you don't want your heart rate to be so fast that it is dangerous. Here is an expression for finding a safe maximum heart rate for a healthy person with no heart disease: $0.7(220 - a)$ *where a is the person's age.* **Work Exercises 27–30** *in order: Write an equation and solve it to find each person's age. Assume all the people are healthy.*

27. How old is a person who has a maximum safe heart rate of 140 beats per minute? *Hint:* Use the distributive property to simplify $0.7(220 - a)$.

28. How old is a person who has a maximum safe heart rate of 126 beats per minute?

29. If a person's maximum safe heart rate is 134 (rounded to the nearest whole number), how old is the person, to the nearest whole year?

30. If a person's maximum safe heart rate is 117 (rounded to the nearest whole number), how old is the person, to the nearest whole year?

5.10 ▶▶▶ Geometry Applications: Circles, Cylinders, and Surface Area

OBJECTIVE 1 Find the radius and diameter of a circle. Suppose you start with one dot on a piece of paper. Then you draw many dots that are each 2 cm away from the first dot. If you draw enough dots (points) you'll end up with a circle, as shown below on the left. Each point on the circle is exactly 2 cm away from the *center* of the circle. The 2 cm distance is called the *radius, r*, of the circle. The distance across the circle (passing through the center) is called the *diameter, d*, of the circle. In this circle, the diameter is 4 cm.

OBJECTIVES

1 Find the radius and diameter of a circle.

2 Find the circumference of a circle.

3 Find the area of a circle.

4 Find the volume of a cylinder.

5 Find the surface area of a rectangular solid.

6 Find the surface area of a cylinder.

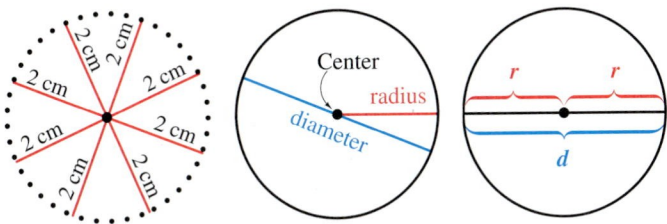

Circle, Radius, and Diameter

A **circle** is a two-dimensional (flat) figure with all points the same distance from a fixed center point.

The **radius** (*r*) is the distance from the center of the circle to any point on the circle.

The **diameter** (*d*) is the distance across the circle passing through the center.

Using the circle above on the right as a model, you can see some relationships between the radius and diameter.

Finding the Diameter and Radius of a Circle

$$\text{diameter} = 2 \cdot \text{radius}$$

$$d = 2r$$

$$\text{and} \quad r = \frac{d}{2}$$

EXAMPLE 1 Finding the Diameter and Radius of a Circle

Find the unknown length of the diameter or radius in each circle.

(a)

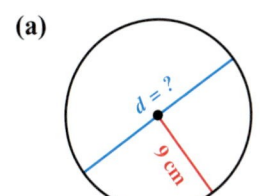

Because the radius is 9 cm, the diameter is twice as long.

$$d = 2 \cdot r$$

$$d = 2 \cdot 9 \text{ cm}$$

Multiply the radius by 2 to get the diameter.

$$d = 18 \text{ cm}$$

Continued on Next Page

1 Find the unknown length of the diameter or radius in each circle.

(a)

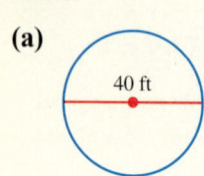

40 ft

(b)

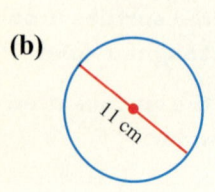

11 cm

(c)

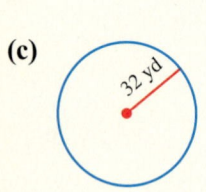

32 yd

(d)

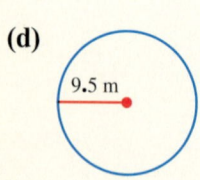

9.5 m

(b)

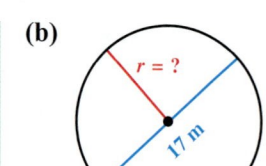

r = ?

17 m

The radius is half the diameter.

$r = \dfrac{d}{2}$ so $r = \dfrac{17 \text{ m}}{2}$

$r = 8.5$ m   or $8\dfrac{1}{2}$ m

> $8.5 = 8\frac{1}{2}$ because $8\frac{5}{10}$ simplifies to $8\frac{1}{2}$.

◀ Work Problem **1** at the Side.

OBJECTIVE 2 Find the circumference of a circle. The perimeter of a circle is called its **circumference.** Circumference is the distance around the edge of a circle.

The diameter of the can in the drawing is about 10.6 cm, and the circumference of the can is about 33.3 cm. Dividing the circumference of the circle by the diameter gives an interesting result.

$$\frac{\text{Circumference}}{\text{diameter}} = \frac{33.3}{10.6} \approx 3.14 \qquad \textcolor{blue}{\text{Rounded to the nearest hundredth}}$$

Dividing the circumference of *any* circle by its diameter *always* gives an answer close to 3.14. This means that going around the edge of any circle is a little more than 3 times as far as going straight across the circle.

This ratio of circumference to diameter is called **π** (the Greek letter **pi**, pronounced PIE). There is no decimal that is exactly equal to π, but here is the *approximate* value.

$$\pi \approx 3.14159265359$$

> **Rounding the Value of Pi (π)**
>
> We usually round π to 3.14. Therefore, calculations involving π will give approximate answers and should be written using the ≈ symbol.

Use the following formulas to find the *circumference* of a circle.

> **Finding the Circumference (Distance Around a Circle)**
>
> Circumference = π • diameter
>
> $C = \pi d$
>
> or, because $d = 2r$ then $C = \pi • 2r$ usually written $C = 2\pi r$
>
> Remember to use linear units such as ft, yd, m, and cm when measuring circumference (**not** square units).

ANSWERS

1. **(a)** $r = 20$ ft   **(b)** $r = 5.5$ cm
   (c) $d = 64$ yd   **(d)** $d = 19$ m

EXAMPLE 2 **Finding the Circumference of Circles**

Find the circumference of each circle. Use 3.14 as the approximate value for π. Round answers to the nearest tenth.

(a)

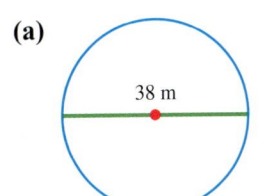

38 m

The *diameter* is 38 m, so use the formula with *d* in it.

$$C = \pi \cdot d$$
$$C \approx 3.14 \cdot 38 \text{ m}$$

Write **m** in your answer.

$$C \approx 119.3 \text{ m} \leftarrow \text{Rounded}$$

(b)

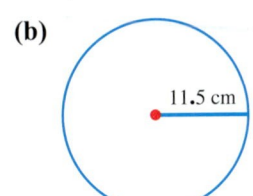

11.5 cm

In this example, the length of the *radius* is labeled, so it is easier to use the formula with *r* in it.

$$C = 2 \cdot \pi \cdot r$$
$$C \approx 2 \cdot 3.14 \cdot 11.5 \text{ cm}$$

Write **cm** in your answer.

$$C \approx 72.2 \text{ cm} \leftarrow \text{Rounded}$$

▦ **Calculator Tip** Most *scientific* calculators have a ⓟ key. Try pressing it. With a 10-digit display, you'll see the value of π to the nearest billionth. (Some calculators may show more or fewer digits.)

> **3.141592654**

But this is still an approximate value, although it is more precise than rounding π to 3.14. Try finding the circumference in Example 2(a) above using the ⓟ key.

ⓟ ⓧ 38 ⓪ **119.3805208**   Rounds to 119.4

When you used 3.14 as the approximate value of π, the result rounded to 119.3, so the answers are slightly different. In this book we will use 3.14. Our measurements of radius and diameter are given as whole numbers or with tenths, so it is acceptable to round π to hundredths. Also, some students may be using a calculator without a ⓟ key.

Work Problem **2** *at the Side.* ▶

OBJECTIVE **3** **Find the area of a circle.** To find the formula for the area of a circle, start by cutting two circles into many pie-shaped pieces.

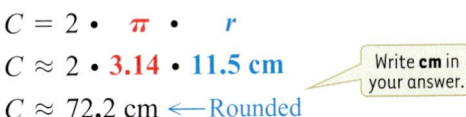

Circumference (distance around) is $2 \cdot \pi \cdot r$ $2 \cdot \pi \cdot r$ $2 \cdot \pi \cdot r$

Unfold the circles, much as you might "unfold" a peeled orange, and put them together as shown here.

$2 \cdot \pi \cdot r$

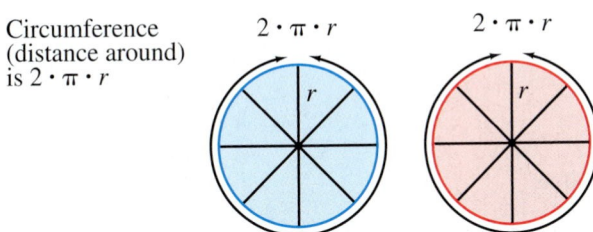

r r

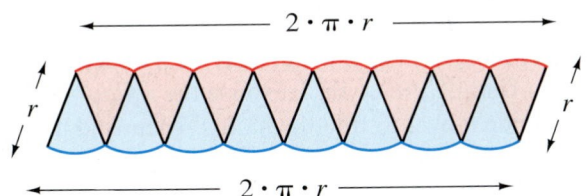

$2 \cdot \pi \cdot r$

Continued on Next Page

2 Find the circumference of each circle. Use 3.14 as the approximate value for π. Round answers to the nearest tenth.

(a)

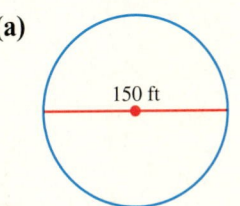

150 ft

(b)

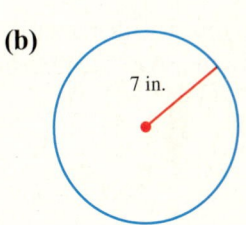

7 in.

(c) diameter 0.9 km

(d) radius 4.6 m

3 Find the area of each circle. Use 3.14 for π. Round your answers to the nearest tenth.

(a)

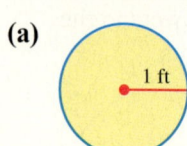

1 ft

(b)

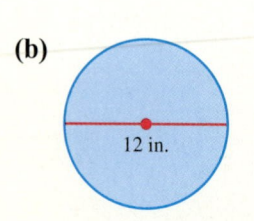

12 in.

(*Hint*: The diameter is 12 in., so $r =$ _____ in.)

(c)

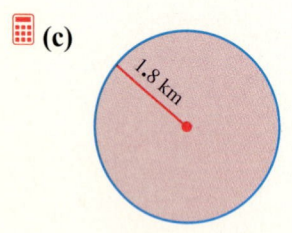

1.8 km

(d)

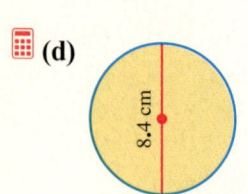

8.4 cm

ANSWERS

3. (a) $A \approx 3.1$ ft^2
   (b) $r = 6$ in.; $A \approx 113.0$ in.^{2}
   (c) $A \approx 10.2$ km^2
   (d) $A \approx 55.4$ cm^2

The figure is approximately a parallelogram with height r (the radius of the original circle) and base $2 \cdot \pi \cdot r$ (the circumference of the original circle). The area of the "parallelogram" is base times height.

Area of "parallelogram" $= \overbrace{b \quad \cdot h}$
Area of "parallelogram" $= 2 \cdot \pi \cdot r \cdot r$
Area of "parallelogram" $= 2 \cdot \pi \cdot r^2$ ← Recall that $r \cdot r$ is r^2

Because the "parallelogram" was formed from *two* circles, the area of *one* circle is half as much.

$$\frac{1}{2} \cdot 2 \cdot \pi \cdot r^2 = 1 \cdot \pi \cdot r^2 \quad \text{or simply} \quad \pi r^2$$

Finding the Area of a Circle

Area of a circle $= \pi \cdot$ radius $\cdot$ radius
$$A = \pi r^2$$
Remember to use **square units** when measuring area.

EXAMPLE 3 **Finding the Area of Circles**

Find the area of each circle. Use 3.14 for π. Round your answers to the nearest tenth.

(a) A circle with a radius of 8.2 cm
Use the formula $A = \pi r^2$, which means $A = \pi \cdot r \cdot r$.

$$A = \pi \cdot r \cdot r$$
$$A \approx 3.14 \cdot 8.2 \text{ cm} \cdot 8.2 \text{ cm}$$
$$A \approx 211.1 \text{ cm}^2 \leftarrow \text{Rounded}$$

This is **area**, so write **cm²** in the answer.

(b)

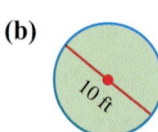

10 ft

To use the area formula $A = \pi r^2$, you need to know the radius (r). In this circle, the *diameter* is 10 ft. First find the radius.

$$r = \frac{d}{2}$$

You *cannot* use the diameter in the area formula, so **find the radius first.**

$$r = \frac{10 \text{ ft}}{2} = 5 \text{ ft}$$

Now find the area.

$$A \approx 3.14 \cdot 5 \text{ ft} \cdot 5 \text{ ft}$$
$$A \approx 78.5 \text{ ft}^2 \leftarrow \text{Square units for area}$$

CAUTION
When finding *circumference*, you can start with either the radius or the diameter. When finding *area*, you must use the *radius*. If you are given the diameter, divide it by 2 to find the radius. Then find the area.

◀ **Work Problem** **3** at the Side.

> ▦ **Calculator Tip**   You can find the area of the circle in Example 3(a) on the previous page using your calculator. The first method works on all types of calculators.
>
> $$3.14 \;\text{⊗}\; 8.2 \;\text{⊗}\; 8.2 \;\text{⊜}\; \mathbf{211.1336}$$
>
> You round the answer to 211.1 (nearest tenth).
>
> On a *scientific* or *graphing* calculator you can also use the ⬭x^2⬭ key, which automatically squares the number you enter (that is, multiplies the number times itself).
>
> $$3.14 \;\text{⊗}\; 8.2 \;\text{⬭}x^2\text{⬭}\; \text{⊜}\; \mathbf{211.1336}$$

In the next example we will find the area of a *semicircle,* which is half the area of a circle.

EXAMPLE 4   **Finding the Area of a Semicircle**

Find the area of the semicircle below. Use 3.14 for π. Round your answer to the nearest tenth.

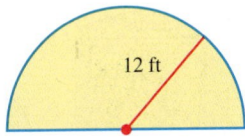

First, find the area of a whole circle with a radius of 12 ft.

$$A = \pi \cdot r \cdot r$$
$$A \approx 3.14 \cdot 12\text{ ft} \cdot 12\text{ ft}$$
$$A \approx 452.16\text{ ft}^2 \qquad \text{Do not round yet.}$$

Divide the area of the whole circle by 2 to find the area of the semicircle.

$$\frac{452.16\text{ ft}^2}{2} = 226.08\text{ ft}^2$$

The *last* step is rounding 226.08 to the nearest tenth.

> Write **ft²** in your answer.

$$\text{Area of semicircle} \approx 226.1\text{ ft}^2 \qquad \text{Rounded}$$

——————— *Work Problem* ④ *at the Side.* ▶

EXAMPLE 5   **Applying the Concept of Circumference**

A circular rug is 8 feet in diameter. The cost of fringe for the edge is $2.25 per foot. What will it cost to add fringe to the rug? Use 3.14 for π.

$$\text{Circumference} = \pi \cdot d$$
$$C \approx 3.14 \cdot 8\text{ ft}$$
$$C \approx 25.12\text{ ft}$$

$$\text{cost} = \text{cost per foot} \cdot \text{Circumference}$$
$$\text{cost} = \frac{\$2.25}{1\text{ ft}} \cdot \frac{25.12\text{ ft}}{1}$$
$$\text{cost} = \$56.52$$

The cost of adding fringe to the rug is $56.52.

——————— *Work Problem* ⑤ *at the Side.* ▶

④ Find the area of each semicircle. Use 3.14 for π. Round your answers to the nearest tenth.

(a)

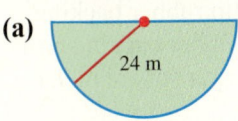

24 m

(b)

35.4 ft

(c)

9.8 m

⑤ Find the cost of binding around the edge of a circular rug that is 3 meters in diameter. The binder charges $4.50 per meter. Use 3.14 for π.

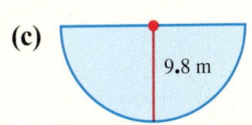

ANSWERS

4. **(a)** $A \approx 904.3\text{ m}^2$ **(b)** $A \approx 491.9\text{ ft}^2$
   (c) $A \approx 150.8\text{ m}^2$
5. $42.39

6 Find the cost of covering the underside of the rug in Margin Problem 5 on the previous page with a nonslip rubber backing. The rubber backing costs $3.89 per square meter.

EXAMPLE 6 Applying the Concept of Area

Find the cost of covering the underside of the rug in Example 5 (on the previous page) with a nonslip rubber backing. The rubber backing costs $1.50 per square foot. Use 3.14 for π.

First find the radius.

$$r = \frac{d}{2} = \frac{8 \text{ ft}}{2} = 4 \text{ ft}$$

Then find the area.

$$A = \pi \cdot r^2$$
$$A \approx 3.14 \cdot 4 \text{ ft} \cdot 4 \text{ ft}$$
$$A \approx 50.24 \text{ ft}^2$$

> Write a **$** in your answer.

$$\text{cost} = \frac{\$1.50}{1 \text{ ft}^2} \cdot \frac{50.24 \text{ ft}^2}{1} = \$75.36$$

◀ **Work Problem 6** at the Side.

---

7 Find the volume of each cylinder. Use 3.14 for π. Round your answers to the nearest tenth.

(a)

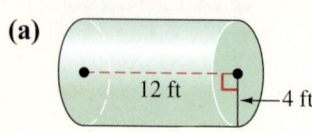

12 ft, 4 ft

(b)

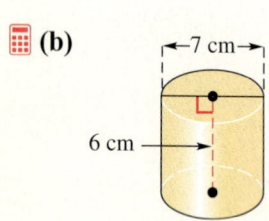

7 cm, 6 cm

(c) radius 14.5 yd, height 3.2 yd

OBJECTIVE 4 Find the volume of a cylinder. Several *cylinders* are shown below.

The height must be perpendicular to the circular top and bottom of the cylinder.

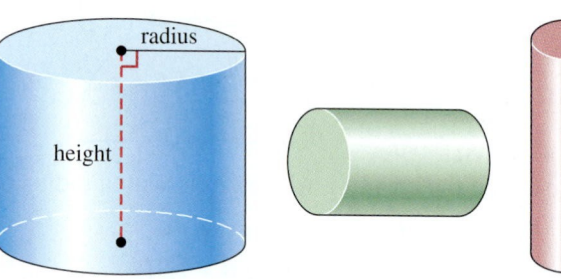

radius, height

These are called *right circular cylinders* because the top and bottom are circles, and the side makes a right angle with the top and bottom. Examples of cylinders are a soup can, a home water heater, and a piece of pipe.

Use the following formula to find the *volume* of a *cylinder*. Notice that the first part of the formula, $\pi \cdot r \cdot r$, is the *area* of the circular base.

Finding the Volume of a Cylinder

$$\text{Volume of a cylinder} = \pi \cdot r \cdot r \cdot h$$
$$V = \pi r^2 h$$

Remember to use **cubic units** when measuring volume.

EXAMPLE 7 Finding the Volume of Cylinders

Find the volume of each cylinder. Use 3.14 as the approximate value of π. Round your answers to the nearest tenth if necessary.

(a)

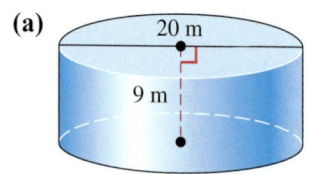

20 m, 9 m

The diameter is 20 m, so the radius is 20 m ÷ 2 = 10 m. The height is 9 m. Use the formula.

$$V = \pi \cdot r \cdot r \cdot h$$
$$V \approx 3.14 \cdot 10 \text{ m} \cdot 10 \text{ m} \cdot 9 \text{ m}$$
$$V \approx 2826 \text{ m}^3 \leftarrow \text{Cubic units for volume}$$

(b)

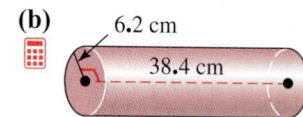

6.2 cm, 38.4 cm

$$V \approx 3.14 \cdot 6.2 \text{ cm} \cdot 6.2 \text{ cm} \cdot 38.4 \text{ cm}$$
$$V \approx 4634.94144 \qquad \text{Now round to tenths.}$$
$$V \approx 4634.9 \text{ cm}^3 \leftarrow \text{Cubic units for volume}$$

◀ **Work Problem 7** at the Side.

ANSWERS

6. $27.48

7. **(a)** $V \approx 602.9 \text{ ft}^3$ **(b)** $V \approx 230.8 \text{ cm}^3$
   (c) $V \approx 2112.6 \text{ yd}^3$

OBJECTIVE 5 Find the surface area of a rectangular solid.   You have just learned how to find the *volume* of a cylinder. In **Section 4.8** you found the *volume* of a rectangular solid. For example, the volume of the cereal box shown below is $V = lwh = (7 \text{ in.}) (2 \text{ in.}) (10 \text{ in.}) = 140 \text{ in.}^3$ But if your company makes cereal boxes, you also need to know how much cardboard is needed for each box. You need to find the *surface area* of the box.

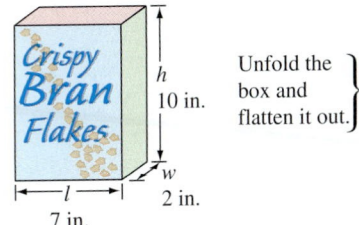

Unfold the box and flatten it out.

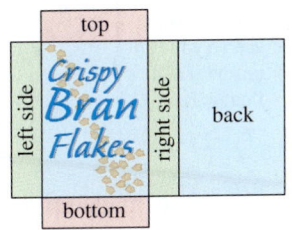

The unfolded box is made up of six rectangles: front, back, top, bottom, left side, right side.

Surface area is the area on the surface of a three-dimensional object (a solid). For a rectangular solid like the cereal box, the surface area is the sum of the areas of the six rectangular sides. Notice that the top and bottom have the same area, the front and back have the same area, and the left and right sides have the same area.

Surface Area = | top | w + | bottom | w + | Crispy Bran Flakes | h + | back | h + | left side | h + | right side | h

$SA = \underbrace{l \cdot w \ + \ l \cdot w}_{2lw} \ + \ \underbrace{l \cdot h \ + \ l \cdot h}_{2lh} \ + \underbrace{w \cdot h + w \cdot h}_{2wh}$

$SA = \qquad\quad 2lw \qquad\qquad + \qquad 2lh \qquad\quad + \qquad 2wh$

> **Finding the Surface Area of a Rectangular Solid**
>
> $\text{Surface Area} = (2 \cdot l \cdot w) + (2 \cdot l \cdot h) + (2 \cdot w \cdot h)$
> $\qquad SA = \quad\; 2lw \quad + \quad 2lh \quad + \quad 2wh$
>
> Remember that area is measured in square units, so use **square units** when measuring *surface* area.

EXAMPLE 8   Finding the Volume and Surface Area of a Rectangular Solid

Find the volume and surface area of this shipping carton.

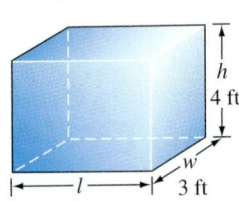

First find the volume.

$V = lwh$

$V = 5 \text{ ft} \cdot 3 \text{ ft} \cdot 4 \text{ ft}$  Write **ft³** for **volume.**

$V = 60 \text{ ft}^3$ ← **Cubic** units for **volume**

Next find the surface area.

$SA = \qquad\quad 2lw \qquad\quad + \qquad 2lh \qquad\quad + \qquad 2wh$
$SA = (2 \cdot 5 \text{ ft} \cdot 3 \text{ ft}) + (2 \cdot 5 \text{ ft} \cdot 4 \text{ ft}) + (2 \cdot 3 \text{ ft} \cdot 4 \text{ ft})$
$SA = \qquad\quad 30 \text{ ft}^2 \qquad + \qquad 40 \text{ ft}^2 \qquad + \qquad 24 \text{ ft}^2$ Write **ft²** for **area.**
$SA = 94 \text{ ft}^2$ ← **Square** units for **area**

Work Problem **8** *at the Side.* ▶

8 Find the volume and surface area of each rectangular solid.

(a)

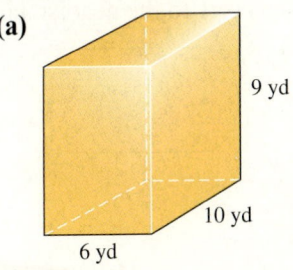

9 yd

10 yd

6 yd

(b)

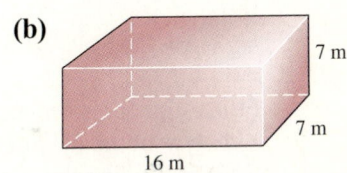

7 m

7 m

16 m

ANSWERS

8. (a) $V = 540 \text{ yd}^3$ (*cubic* yd for *volume*)
$SA = 408 \text{ yd}^2$ (*square* yd for *area*)
(b) $V = 784 \text{ m}^3$
$SA = 546 \text{ m}^2$

9 Find the volume and surface area of each cylinder. Use 3.14 for π. Round your answers to the nearest tenth.

(a)

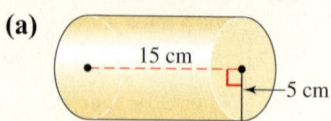
15 cm — 5 cm

(b)

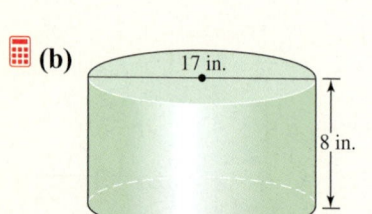
17 in. — 8 in.

(*Hint:* You are given the *diameter* of the cylinder. Start by finding the *radius*.)

OBJECTIVE 6 Find the surface area of a cylinder. You can use the same idea of "unfolding" a shape to find the surface area of a cylinder, such as the soup can shown below. Finding the surface area will tell you how much aluminum you need to make the can.

The unfolded soup can is made up of a rectangular side, a circular top, and a circular bottom.

Remember that the formula for the area of a circle is πr^2.

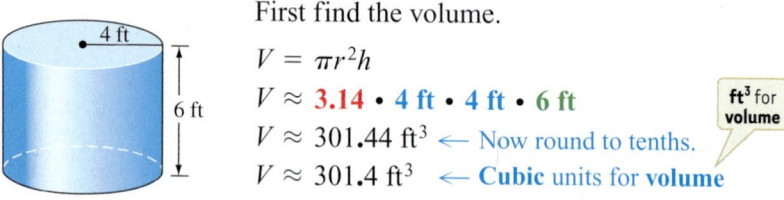

$$SA = \underbrace{2\pi r \cdot h}_{} + \underbrace{\pi r^2 + \pi r^2}_{}$$
$$SA = 2\pi rh + 2\pi r^2$$

Finding the Surface Area of a Right Circular Cylinder

Surface Area $= (2 \cdot \pi \cdot r \cdot h) + (2 \cdot \pi \cdot r \cdot r)$
$$SA = 2\pi rh + 2\pi r^2$$
Remember that area is measured in square units, so use **square units** when measuring *surface* area.

EXAMPLE 9 Finding the Volume and Surface Area of a Right Circular Cylinder

Find the volume and surface area of this water tank. Use 3.14 as the approximate value for π. Round your answers to the nearest tenth when necessary.

4 ft — 6 ft

First find the volume.
$V = \pi r^2 h$
$V \approx 3.14 \cdot 4\text{ ft} \cdot 4\text{ ft} \cdot 6\text{ ft}$ **ft³ for volume**
$V \approx 301.44\text{ ft}^3$ ← Now round to tenths.
$V \approx 301.4\text{ ft}^3$ ← **Cubic** units for **volume**

Now find the surface area.
$SA = 2\pi rh + 2\pi r^2$
$SA \approx (2 \cdot 3.14 \cdot 4\text{ ft} \cdot 6\text{ ft}) + (2 \cdot 3.14 \cdot 4\text{ ft} \cdot 4\text{ ft})$
$SA \approx 150.72\text{ ft}^2 + 100.48\text{ ft}^2$
$SA \approx 251.2\text{ ft}^2$ ← **Square** units for **area** **ft² for area**

◀ Work Problem **9** at the Side.

Find the unknown length in each circle. See Example 1.

1.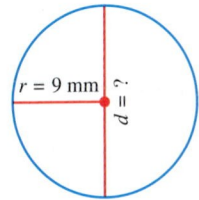
$r = 9$ mm $d = ?$

2.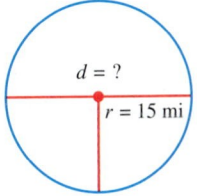
$d = ?$ $r = 15$ mi

3.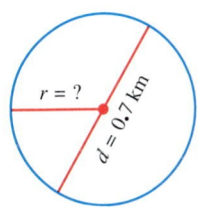
$r = ?$ $d = 0.7$ km

4.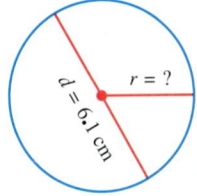
$d = 6.1$ cm $r = ?$

Find the circumference and area of each circle. Use 3.14 as the approximate value for π.
Round your answers to the nearest tenth. See Examples 2 and 3.

5.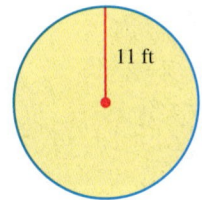
11 ft

6.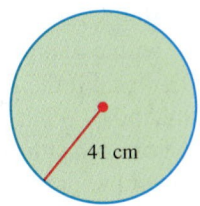
41 cm

7.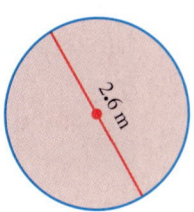
2.6 m

8.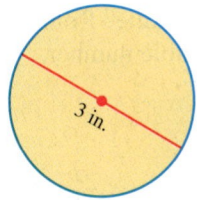
3 in.

Find the circumference and area of circles having the following diameters. Use 3.14 for π.
Round your answers to the nearest tenth. See Examples 2 and 3.

9. $d = 15$ cm

10. $d = 39$ ft

11. $d = 7\frac{1}{2}$ ft

12. $d = 4\frac{1}{2}$ yd

13. $d = 8.65$ km

14. $d = 19.5$ mm

Find each shaded area. Note that Exercises 15–18 all contain semicircles. Use 3.14 as the
approximate value of π. Round your answers to the nearest tenth when necessary. See Example 4.

15.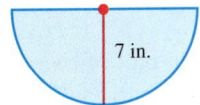
7 in.

16.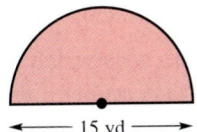
15 yd

17.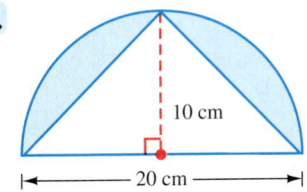
10 cm
20 cm

18.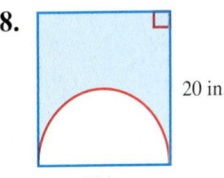
20 in.
18 in.

Solve each application problem. Use 3.14 as the approximate value of π. Round answers to the nearest tenth. See Examples 5 and 6.

19. An irrigation system moves around a center point to water a circular area for crops. If the irrigation system is 50 yd long, how large is the watered area?

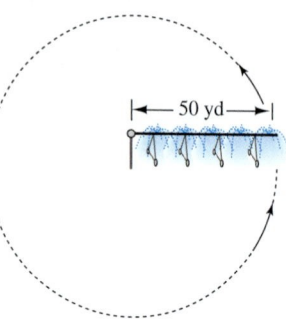

50 yd

20. If you swing a ball held at the end of a string 20 cm long, how far will the ball travel on each turn?

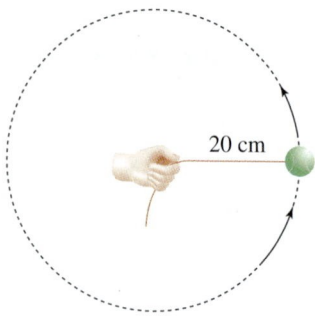

20 cm

21. A Michelin Cross Terrain SUV tire has an overall diameter of 29.10 inches. How far will a point on the tire tread move in one complete turn? (*Source: Michelin.*)

Bonus question: How many revolutions does the tire make per mile? Round to the nearest whole number.

22. In August 2005, hurricane Katrina slammed into coastal Mississippi and flooded New Orleans. The diameter of the circular storm was 210 miles. In July 2005, hurricane Dennis, with a diameter of only 80 miles, hit the Florida Panhandle. What was the area of each storm, to the nearest hundred square miles? (*Source: Associated Press.*)

For Exercises 23–28, first draw a circle and label the radius or diameter. Then solve the problem. For all Exercises 23–34, use 3.14 for π and round answers to the nearest tenth.

23. A radio station can be heard 150 miles in all directions during evening hours. How many square miles are in the station's broadcast area?

24. An earthquake was felt by people 900 km away in all directions from the epicenter (the source of the earthquake). How much area was affected by the quake?

25. The diameter of Diana Hestwood's wristwatch is 1 in. and the radius of the clock face on her kitchen wall is 3 in. Find the circumference and the area of each clock face.

26. The diameter of the largest known ball of twine is 12 ft 9 in. The sign posted near the ball says it has a circumference of 40 ft. Is the sign correct? *Hint:* First change 9 in. to feet and add it to 12 ft. (*Source: Guinness World Records.*)

27. Blaine Fenstad wants to buy a pair of two-way radios. Some models have a range of 2 miles under ideal conditions. More expensive models have a range of 5 miles. What is the difference in the area covered by the 2-mile and 5-mile models? (*Source: Best Buy.*)

28. The National Audubon Society holds an end-of-year bird count. Volunteers count all the birds they see in a circular area during a 24-hour period. Each circle has a diameter of 15 miles. About 1700 circular areas are counted across the United States each December. What is the total area covered by the count?

29. A forester measures the circumference of a living tree at chest height, then calculates the diameter.

(a) If the circumference of one tree is 144 cm, what is the diameter?

(b) Explain how you solved part (a).

30. In Atlanta, Interstate 285 circles the city and is known as the "perimeter." If the circumference of the circle made by the highway is 62.8 miles, find:

(a) the diameter of the circle

(b) the area inside the circle.

(*Source: Greater Atlanta Newcomer's Guide.*)

31. The Mormons traveled west to Utah by covered wagon in 1847. They tied a rag to a wagon wheel to keep track of the distance they traveled. The radius of the wheel was 2.33 ft. How far did the rag travel each time the wheel made a complete revolution? (*Source: Trail of Hope.*)

32. First work Exercise 31. Then find how many wheel revolutions equaled one mile. There are 5280 ft in one mile.

33. Find the shaded area.

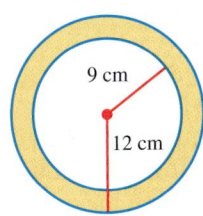

9 cm

12 cm

34. Find the area of this skating rink.

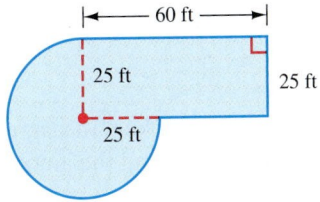

60 ft

25 ft

25 ft

25 ft

Find the volume and surface area of each cylinder or rectangular solid. Use 3.14 as the approximate value of π. Round your answers to the nearest tenth when necessary. See Examples 7–9.

35.

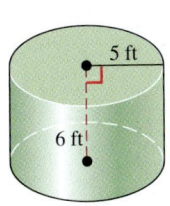

5 ft

6 ft

36.

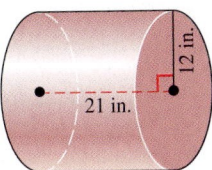

21 in.

12 in.

37.

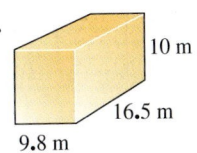

10 m

16.5 m

9.8 m

38.

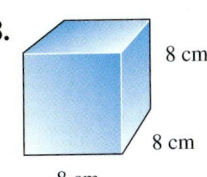

8 cm

8 cm

8 cm

39.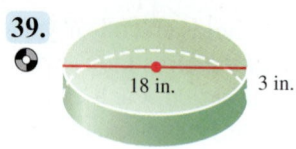

18 in. 3 in.

40.

2 ft

8 ft

41.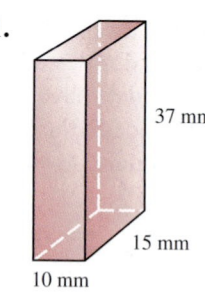

37 mm

15 mm

10 mm

42.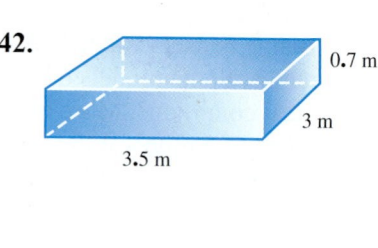

0.7 m

3 m

3.5 m

43. Explain the *two* errors made by a student in finding the volume of a cylinder with a diameter of 7 cm and a height of 5 cm. Find the correct answer.

$$V \approx 3.14 \cdot 7 \cdot 7 \cdot 5$$
$$V \approx 769.3 \text{ cm}^2$$

44. (a) Look again at Exercise 38 on the previous page. The figure is a *cube*. What is special about the measurements of a cube?

(b) Find a shortcut you can use to calculate the surface area of a cube.

Solve each application problem. Use 3.14 as the approximate value of π. Round your answers to the nearest tenth when necessary.

45. A city sewer pipe has a diameter of 5 ft and a length of 200 ft. Find the volume of the pipe.

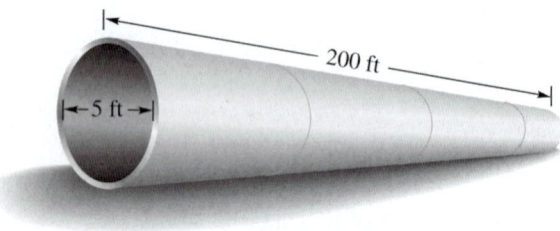

200 ft

5 ft

46. A cylindrical woven basket made by a Northwest Coast tribe is 8 cm high and has a diameter of 11 cm. What is the volume of the basket?

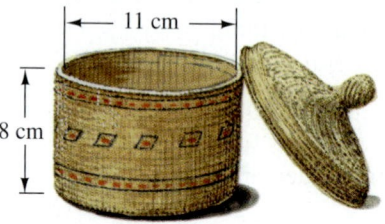

11 cm

8 cm

47. A box for graham crackers measures 5.5 in. by 2.8 in. by 8 in. high. Find the amount of cardboard needed to make the box.

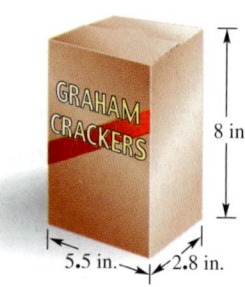

GRAHAM CRACKERS

8 in.

5.5 in. 2.8 in.

48. A soda can is 12.5 cm tall and the round top has a diameter of 6.7 cm. How much aluminum is needed to make the can? (Assume the can has a flat top and bottom.)

6.7 cm

LEMON BURST

12.5 cm

Chapter 5 ▶▶▶ Summary

▶ Key Terms

5.1 **decimals** Decimals, like fractions, are used to show parts of a whole.

decimal point A decimal point is the dot that is used to separate the whole number part from the fractional part of a decimal number.

place value Place value is the value assigned to each place to the right or left of the decimal point. Whole numbers, such as ones and tens, are to the *left* of the decimal point. Fractional parts, such as tenths and hundredths, are to the *right* of the decimal point.

5.2 **round** To round is to "cut off" a number after a certain place, such as to round to the nearest hundredth. The rounded number is less accurate than the original number. You can use the symbol " ≈ " to mean "is approximately equal to."

decimal places Decimal places are the number of digits to the *right* of the decimal point; for example, 6.37 has two decimal places, 4.706 has three decimal places.

5.5 **repeating decimal** A repeating decimal (like the 6 in 0.166…) is a decimal number with one or more digits that repeat forever; it never ends. Use three dots to indicate that it is a repeating decimal. Or write the number with a bar above the repeating digits, as in $0.1\overline{6}$. (Use the dots or the bar, but not both.)

5.7 **mean** The mean is the sum of all the values divided by the number of values. It is often called the *average*.

weighted mean The weighted mean is a mean calculated so that each value is multiplied by its frequency.

median The median is the middle number in a group of values that are listed from smallest to largest. It divides a group of values in half. If there is an even number of values, the median is the mean (average) of the two middle values.

mode The mode is the value that occurs most often in a group of values.

5.8 **square root** A positive square root of a positive number is one of two equal positive factors of the number.

hypotenuse The hypotenuse is the side of a right triangle opposite the 90° angle; it is the longest side.
Example: See the red side in the triangle at the right.

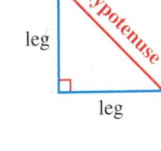

5.10 **circle** A circle is a two-dimensional (flat) figure with all points the same distance from a fixed center point.
Example: See figure at the right.

radius Radius is the distance from the center of a circle to any point on the circle.
Example: See the red radius in the circle at the right.

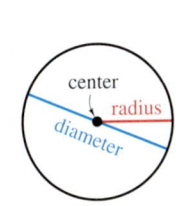

diameter Diameter is the distance across a circle, passing through the center.
Example: See the blue diameter in the circle at the right.

circumference Circumference is the distance around a circle.

π (pi) π is the ratio of the circumference to the diameter of any circle. It is approximately equal to 3.14.

surface area Surface area is the area on the surface of a three-dimensional object (a solid). Surface area is measured in square units.

▶ New Symbols

$3.8\overline{6}$ ◀—— Bar above repeating digit(s) in a decimal $\approx$ is approximately equal to

$\sqrt{}$ square root

$3.8666\ldots$ Three dots indicate a repeating decimal π Greek letter pi (pronounced PIE); ratio of the circumference of a circle to its diameter

▶ New Formulas

$$\mathbf{mean} = \frac{\text{sum of all values}}{\text{number of values}}$$

Circumference of a circle: $C = \pi d$
or $C = 2\pi r$

$$\mathbf{hypotenuse} = \sqrt{(\text{leg})^2 + (\text{leg})^2}$$

Area of a circle: $A = \pi r^2$

$$\mathbf{leg} = \sqrt{(\text{hypotenuse})^2 - (\text{leg})^2}$$

Volume of a cylinder: $V = \pi r^2 h$

diameter of a circle: $d = 2r$

Surface Area of a rectangular solid:
$SA = 2lw + 2lh + 2wh$

radius of a circle: $r = \dfrac{d}{2}$

Surface Area of a cylinder: $SA = 2\pi rh + 2\pi r^2$

▶ Test Your Word Power

See how well you have learned the vocabulary in this chapter. Answers follow the Quick Review.

1. **Decimal numbers** are like fractions in that they both
   A. must be written in lowest terms
   B. need common denominators
   C. have decimal points
   D. represent parts of a whole.

2. **Decimal places** refer to
   A. the digits from 0 to 9
   B. digits to the left of the decimal point
   C. digits to the right of the decimal point
   D. the number of zeros in a decimal number.

3. The **hypotenuse** is
   A. the long base in a rectangle
   B. the height in a parallelogram
   C. the longest side in a right triangle
   D. the distance across a circle, passing through the center.

4. The **decimal point**
   A. separates the whole number part from the fractional part
   B. is always moved when finding a quotient
   C. separates tenths from hundredths
   D. is at the far left side of a whole number.

5. The number $0.\overline{3}$ is an example of
   A. an estimate
   B. a repeating decimal
   C. a rounded number
   D. a truncated number.

6. π is the ratio of
   A. the diameter to the radius of a circle
   B. the circumference to the diameter of a circle
   C. the circumference to the radius of a circle
   D. the diameter to the circumference of a circle.

7. The **median** for a set of values is
   A. the mathematical average
   B. the value that occurs most often
   C. the sum of each value times its frequency
   D. the middle value when the values are listed from least to greatest.

8. The **circumference** of a circle is
   A. the perimeter of the circle
   B. found using the expression πr^2
   C. the distance across the circle
   D. measured in square units.

▶ **Quick Review**

Concepts	Examples

5.1 Reading and Writing Decimals

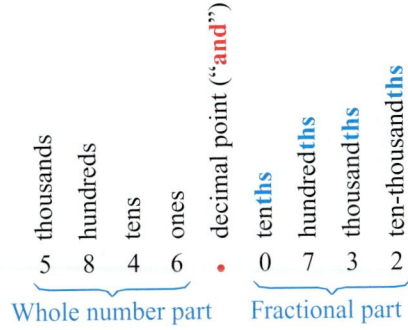

Write each decimal in words.

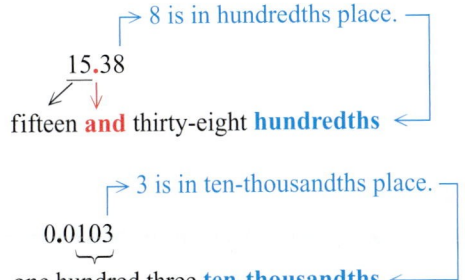

15.38

fifteen **and** thirty-eight **hundredths**

0.0103

one hundred three **ten-thousandths**

5.1 Writing Decimals as Fractions

The digits to the right of the decimal point are the numerator. The place value of the rightmost digit determines the denominator.

Always write the fraction in lowest terms.

Write 0.45 as a fraction in lowest terms.

The numerator is 45. The rightmost digit, 5, is in the hundredths place, so the denominator is 100. Then write the fraction in lowest terms.

$$\frac{45}{100} = \frac{45 \div 5}{100 \div 5} = \frac{9}{20} \leftarrow \text{Lowest terms}$$

5.2 Rounding Decimals

Find the place to which you are rounding. Draw a cut-off line to the right of that place; the rest of the digits will be dropped. Look *only* at the first digit being cut. If it is *4 or less,* the part you are keeping stays the same. If it is *5 or more,* the part you are keeping rounds up. Do not move the decimal point when rounding. Use the sign "≈" to mean "is approximately equal to."

Round 0.17952 to the nearest thousandth.

First digit cut is *5 or more* so round up.

0.179|52

0.179 ← Keep this part.
+ 0.001 ← To round up, add 1 thousandth.
0.180

0.17952 rounds to 0.180.
Write 0.17952 ≈ 0.180

5.3 Adding Positive and Negative Decimals

Estimate the answer by using front end rounding: round each number to the highest possible place.

   To find the exact answer, line up the decimal points. If needed, write in zeros as placeholders. Add or subtract the absolute values as if they were whole numbers. Line up the decimal point in the answer.

   If the numbers have the same sign, use the common sign as the sign of the sum. If the numbers have different signs, the sign of the sum is the sign of the number with the greater absolute value.

Find the sum.
5.68 + 785.3 + 12 + 2.007

Estimate: *Exact:*

6 ←	5.**680**	Use zeros as
800 ←	785.**300**	placeholders so
10 ←	12.**000**	all numbers have
+ 2 ←	+ 2.007	three decimal
818	804.987	places.

Line up decimal points.

The estimate and exact answer are both in hundreds, so the decimal point is probably in the correct place.

Concepts	Examples

5.3 Subtracting Positive and Negative Decimals

Rewrite subtracting as adding the opposite of the second number. Then follow the rules for adding positive and negative decimals.

Subtract.

$$4.2 - 12.91$$
$$4.2 + (-12.91)$$

$|4.2|$ is 4.2 and $|-12.91|$ is 12.91.
Subtract $12.91 - 4.2$ to get 8.71.
Because -12.91 has the larger absolute value and is negative, the answer will be negative.

$$4.2 + (-12.91) = -8.71$$

5.4 Multiplying Positive and Negative Decimals

Step 1 Multiply as you would for whole numbers.
Step 2 Count the total number of decimal places in both factors.
Step 3 Write the decimal point in the answer so it has the same number of decimal places as the total from Step 2. You may need to write extra zeros on the left side of the product in order to get enough decimal places in the answer.
Step 4 If two factors have the *same sign,* the product is *positive.* If two factors have *different signs,* the product is negative.

Multiply $(0.169)(-0.21)$

$$
\begin{array}{r}
0.169 \leftarrow \text{3 decimal places} \\
\times \quad 0.21 \leftarrow \text{2 decimal places} \\
\hline
169 \qquad \text{5 total decimal places} \\
338 \qquad \\
\hline
.03549 \leftarrow \text{5 decimal places in product}
\end{array}
$$

Write in a 0 so you can count over 5 decimal places.
The factors have different signs, so the product is negative.

$$(0.169)(-0.21) = -0.03549$$

5.5 Dividing by a Decimal

Step 1 Change the divisor to a whole number by moving the decimal point to the right.
Step 2 Move the decimal point in the dividend the same number of places to the right.
Step 3 Write the decimal point in the quotient directly above the decimal point in the dividend. Then divide as with whole numbers.
Step 4 If the numbers have the *same sign,* the quotient is *positive.* If they have *different signs,* the quotient is *negative.*

Divide -52.8 by -0.75.
First consider $52.8 \div 0.75$.

$$
\begin{array}{r}
70.4 \\
0.75\overline{)52.800} \\
525 \\
\hline
300 \\
300 \\
\hline
0
\end{array}
$$

Move decimal point two places to the right in divisor and dividend.
Write zeros in the dividend so you can move the decimal point and continue dividing until the remainder is 0.

The quotient is positive because both the divisor and dividend were negative (same signs means positive quotient).

5.6 Writing Fractions as Decimals

Divide the numerator by the denominator. If necessary, round to the place indicated.

Write $\frac{1}{8}$ as a decimal.

$\frac{1}{8}$ means $1 \div 8$. Write it as $8\overline{)1}$.
The decimal point is on the right side of 1.

$$
\begin{array}{r}
0.125 \\
8\overline{)1.000} \leftarrow \\
8 \\
\hline
20 \\
16 \\
\hline
40 \\
40 \\
\hline
0
\end{array}
$$

Write the decimal point and three zeros so you can continue dividing.

Therefore, $\frac{1}{8}$ is equivalent to 0.125.

Concepts

Examples

5.6 Comparing the Size of Fractions and Decimals

Step 1 Write any fractions as decimals.
Step 2 Write zeros so that all the numbers being compared have the same number of decimal places.
Step 3 Use $<$ to mean "is less than," $>$ to mean "is greater than," or list the numbers from least to greatest.

Arrange in order from least to greatest.

$$0.505 \quad \frac{1}{2} \quad 0.55$$

$0.505 = 505$ thousandths $\leftarrow$ 505 is in the middle.

$\frac{1}{2} = 0.5 = 0.500 = 500$ thousandths $\leftarrow$ 500 is least.

$0.55 = 0.550 = 550$ thousandths $\leftarrow$ 550 is greatest.

$$(\text{least}) \frac{1}{2} \quad 0.505 \quad 0.55 \ (\text{greatest})$$

5.7 Finding the Mean (Average) of a Set of Values

Step 1 Add all values to obtain a total.
Step 2 Divide the total by the number of values.

Here are Heather Hall's test scores in her math course.

$$93 \quad 76 \quad 83 \quad 93$$
$$78 \quad 82 \quad 87 \quad 85$$

Find Heather's mean score to the nearest tenth.

$$\text{mean} = \frac{93 + 76 + 83 + 93 + 78 + 82 + 87 + 85}{8}$$

$$= \frac{677}{8} \approx 84.6 \leftarrow \text{Mean test score}$$

5.7 Finding the Median of a Set of Values

Step 1 Arrange the data from least to greatest.
Step 2 If there is an odd number of values, select the middle value. If there is an even number of values, find the average of the two middle values.

Find the median for Heather Hall's scores from the previous example.

List the scores from least to greatest.

$$76 \quad 78 \quad 82 \quad \textbf{83} \quad \textbf{85} \quad 87 \quad 93 \quad 93$$

Middle values

The middle two values are 83 and 85. Find the average of these two values.

$$\frac{83 + 85}{2} = 84 \leftarrow \text{Median test score}$$

5.7 Finding the Mode of a Set of Values

Find the value that appears most often in the list of values. This is the mode.

If no value appears more than once, there is no mode. If two different values appear the same number of times, the list is bimodal.

Find the mode for Heather's scores in the previous example.

The most frequently occurring score is 93 (it occurs twice). Therefore, the mode is 93.

Concepts	Examples

5.8 Finding the Square Root of a Number

Use the square root key on a calculator, .
Round to the nearest thousandth when
necessary.

$$\sqrt{64} = 8$$

$$\sqrt{43} \approx 6.557 \quad \text{6.557438524 is rounded to the nearest thousandth.}$$

5.8 Finding the Unknown Length in a Right Triangle

To find the hypotenuse, use:

$$\text{hypotenuse} = \sqrt{(\text{leg})^2 + (\text{leg})^2}$$

The hypotenuse is the side opposite the right angle.
It is the longest side in a right triangle.

Find the unknown length in this right triangle. Round to the nearest tenth.

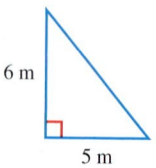

6 m

5 m

$$\text{hypotenuse} = \sqrt{(6)^2 + (5)^2}$$
$$= \sqrt{36 + 25}$$
$$= \sqrt{61} \approx 7.8$$

The hypotenuse is about 7.8 m long.

To find a leg, use:

$$\text{leg} = \sqrt{(\text{hypotenuse})^2 - (\text{leg})^2}$$

The legs are the sides that form the right angle.

Find the unknown length in this right triangle. Round to the nearest tenth.

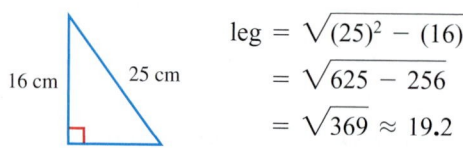

16 cm   25 cm

$$\text{leg} = \sqrt{(25)^2 - (16)^2}$$
$$= \sqrt{625 - 256}$$
$$= \sqrt{369} \approx 19.2$$

The leg is about 19.2 cm long.

5.9 Solving Equations Containing Decimals

Solve the equation and check the solution.

Step 1 Use the addition property of equality to get the variable term by itself on one side of the equal sign.

$$4.5x + 0.7 = -5.15$$
$$\underline{-0.7 \qquad -0.7} \qquad \text{Add } -0.7 \text{ to both sides.}$$
$$4.5x + 0 \qquad -5.85$$

Step 2 Divide both sides by the coefficient of the variable term to find the solution.

$$\frac{\overset{1}{4.5x}}{\underset{1}{4.5}} = \frac{-5.85}{4.5} \qquad \text{Divide both sides by 4.5}$$

$$x = -1.3$$

The solution is -1.3. To check the solution, go back to the original equation.

Step 3 Check the solution by going back to the *original* equation. Replace the variable with the solution. If the equation balances, the solution is correct.

Check $\quad 4.5x + 0.7 = -5.15 \qquad$ Original equation

$$4.5(-1.3) + 0.7 = -5.15 \qquad \text{Replace } x \text{ with } -1.3$$

$$-5.85 \quad + 0.7 = -5.15$$

$$-5.15 \quad = -5.15 \qquad \text{Balances}$$

When x is replaced with -1.3, the equation balances, so -1.3 is the correct solution (***not*** -5.15).

Concepts	Examples

5.10 Circles

Use this formula to find the *diameter* of a circle when you are given the radius.

$$\text{diameter} = 2 \cdot \text{radius} \quad \text{or} \quad d = 2r$$

Find the diameter of a circle if the radius is 3 ft.

$$d = 2 \cdot r$$

$$d = 2 \cdot \textbf{3 ft} = 6 \text{ ft}$$

Use this formula to find the *radius* of a circle when you are given the diameter.

$$\text{radius} = \frac{\text{diameter}}{2} \quad \text{or} \quad r = \frac{d}{2}$$

Find the radius of a circle if the diameter is 5 cm.

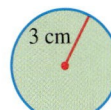

$$r = \frac{d}{2}$$

$$r = \frac{\textbf{5 cm}}{2} = 2.5 \text{ cm}$$

Use these formulas to find the *circumference* of a circle.

When you know the *radius,* use this formula.

$$C = 2 \cdot \pi \cdot \text{radius}$$

$$\text{or} \quad C = 2\pi r$$

When you know the *diameter,* use this formula.

$$C = \pi \cdot \text{diameter}$$

$$\text{or} \quad C = \pi d$$

Use 3.14 as the approximate value for π.

Find the circumference of a circle with a radius of 7 yd. Round your answer to the nearest tenth.

$$\text{Circumference} = 2 \cdot \pi \cdot r$$

$$C \approx 2 \cdot \textbf{3.14} \cdot \textbf{7 yd}$$

$$C \approx 44.0 \text{ yd} \leftarrow \text{Rounded}$$

Use this formula to find the *area* of a circle.

$$A = \pi \cdot r \cdot r$$

$$\text{or} \quad A = \pi r^2$$

Use 3.14 as the approximate value of π.
Area is measured in **square units.**

Find the area of the circle. Round your answer to the nearest tenth.

$$\text{Area} = \pi \cdot r \cdot r$$

$$A \approx \textbf{3.14} \cdot \textbf{3 cm} \cdot \textbf{3 cm}$$

$$A \approx 28.3 \text{ cm}^2 \leftarrow \text{Rounded;}$$
$$\text{square units for area}$$

5.10 Volume of a Cylinder

Use this formula to find the volume of a cylinder.

$$\text{Volume} = \pi \cdot r \cdot r \cdot h$$

$$\text{or} \quad V = \pi r^2 h$$

where r is the radius of the circular base and h is the height of the cylinder.

Volume is measured in **cubic units.**

Find the volume of this cylinder.

First, find the radius. $\quad r = \dfrac{8 \text{ m}}{2} = 4 \text{ m}$

$$V = \pi \cdot r \cdot r \cdot h$$

$$V \approx \textbf{3.14} \cdot \textbf{4 m} \cdot \textbf{4 m} \cdot \textbf{10 m}$$

$$V \approx 502.4 \text{ m}^3 \leftarrow \text{Cubic units for volume}$$

Concepts	Examples

5.10 Surface Area of a Rectangular Solid

Use this formula to find the surface area of a rectangular solid.

Surface Area $= (2 \cdot l \cdot w) + (2 \cdot l \cdot h) + (2 \cdot w \cdot h)$

or $\quad SA = 2lw + 2lh + 2wh$

where l is the length, w is the width, and h is the height of the solid.

Surface area is measured in **square units**.

Find the surface area of this packing crate.

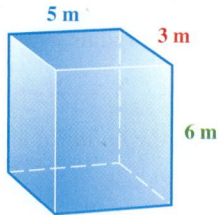

$SA = 2lw + 2lh + 2wh$

$SA = (2 \cdot 5 \text{ m} \cdot 3 \text{ m}) + (2 \cdot 5 \text{ m} \cdot 6 \text{ m}) + (2 \cdot 3 \text{ m} \cdot 6 \text{ m})$

$SA = 30 \text{ m}^2 + 60 \text{ m}^2 + 36 \text{ m}^2$

$SA = 126 \text{ m}^2 \leftarrow$ Square units for surface area

---

5.10 Surface Area of a Cylinder

Use this formula to find the surface area of a cylinder.

Surface Area $= (2 \cdot \pi \cdot r \cdot h) + (2 \cdot \pi \cdot r \cdot r)$

or $\quad SA = 2\pi rh + 2\pi r^2$

where r is the radius of the circular base and h is the height of the cylinder. Use 3.14 as the approximate value of π.

Surface area is measured in **square units**.

Find the surface area of a hot water tank with a height of 4.5 ft and a diameter of 1.8 ft. Round your answer to the nearest tenth.

First, find the radius. $r = \dfrac{1.8 \text{ ft}}{2} = 0.9 \text{ ft}$

$SA = 2\pi rh + 2\pi r^2$

$SA \approx (2 \cdot 3.14 \cdot 0.9 \text{ ft} \cdot 4.5 \text{ ft}) + (2 \cdot 3.14 \cdot 0.9 \text{ ft} \cdot 0.9 \text{ ft})$

$SA \approx 25.434 \text{ ft}^2 + 5.0868 \text{ ft}^2$

$SA \approx 30.5208 \text{ ft}^2 \quad$ Now round to tenths.

$SA \approx 30.5 \text{ ft}^2 \leftarrow$ Square units for surface area

---

ANSWERS TO TEST YOUR WORD POWER

1. D; *Example:* For 0.7, the whole is cut into ten parts, and you are interested in 7 of the parts.
2. C; *Examples:* The number 6.87 has two decimal places; 0.309 has three decimal places.
3. C; *Example:* In triangle *ABC*, side *AC* (the red side) is the hypotenuse.

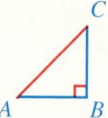

4. A; *Example:* In 5.42, the decimal point separates the whole number part, 5 ones, from the decimal part, 42 hundredths.
5. B; *Example:* The bar above the 3 in $0.\overline{3}$ indicates that the 3 repeats forever.

6. B; *Example:* The ratio of a circumference of 12.57 cm to a diameter of 4 cm is $\dfrac{12.57}{4} \approx 3.14$ (rounded).

7. D; *Example:* For this set of prices—$4, $3, $7, $2, $5, $6, $7—the median is $5.
8. A; *Example:* If the radius of a circle is 4 ft, then $C \approx 2 \cdot 3.14 \cdot 4 \approx 25.1$ ft. So the distance around the circle is about 25.1 ft.

Chapter 5 ▶▶▶ Review Exercises

[5.1] *Name the digit that has the given place value.*

1. 243.059

   tenths

   hundredths

2. 0.6817

   ones

   tenths

3. $5824.39

   hundreds

   hundredths

4. 896.503

   tenths

   tens

5. 20.73861

   tenths

   ten-thousandths

Write each decimal as a fraction or mixed number in lowest terms.

6. 0.5

7. 0.75

8. 4.05

9. 0.875

10. 0.027

11. 27.8

Write each decimal in words.

12. 0.8

13. 400.29

14. 12.007

15. 0.0306

Write each decimal in numbers.

16. Eight and three tenths

17. Two hundred five thousandths

18. Seventy and sixty-six ten-thousandths

19. Thirty hundredths

[5.2] *Round to the place indicated.*

20. 275.635 to the nearest tenth

21. 72.789 to the nearest hundredth

22. 0.1604 to the nearest thousandth

23. 0.0905 to the nearest thousandth

24. 0.98 to the nearest tenth

Round each money amount to the nearest cent.

25. $15.8333

26. $0.698

27. $17,625.7906

Round each income or expense item to the nearest dollar.

28. The income from the pancake breakfast was $350.48.

29. Each member paid $129.50 in dues.

30. The refreshments cost $99.61.

31. The bank charges were $29.37.

[5.3] *Find each sum or difference.*

32. $0.4 - 6.07$

33. $-20 + 19.97$

34. $-1.35 + 7.229$

35. $0.005 + (3 - 9.44)$

First, use front end rounding to estimate each answer. Then find the exact answer.

36. Americans' favorite recreational activity is exercise walking. About 81.3 million people go walking at least once during the year. Swimming is second with 56.3 million people, and camping is third with 49.9 million people. How many more people go walking than camping? (*Source:* National Sporting Goods Association.)

Estimate:

Exact:

37. Today, Jasmin had $306 in her checking account. She wrote a check to the day care center for $215.53 and a check for $44.67 at the grocery store. What is the new balance in her account?

Estimate:

Exact:

38. Joey spent $1.59 for toothpaste, $5.33 for a gift, and $18.94 for a toaster. He gave the clerk three $10 bills. How much change did he get?

Estimate:

Exact:

39. Roseanne is training for a wheelchair race. She raced 2.3 kilometers on Monday, 4 kilometers on Wednesday, and 5.25 kilometers on Friday. How far did she race altogether?

Estimate:

Exact:

[5.4] *First, use front end rounding to estimate each answer. Then find the exact answer.*

40. *Estimate:*   *Exact:*

$$\begin{array}{r} 6.138 \\ \times\ \underline{} \qquad \times\ \underline{\ 3.7} \end{array}$$

41. *Estimate:*   *Exact:*

$$\begin{array}{r} 42.9 \\ \times\ \underline{} \qquad \times\ \underline{\ 3.3} \end{array}$$

Find each product.

42. $(-5.6)(-0.002)$

43. $(0.071)(-0.005)$

[5.5] *Decide if each answer is reasonable by rounding the numbers and estimating the answer. If the exact answer is not reasonable, find and correct the error.*

44. $706.2 \div 12 = 58.85$

Estimate:

45. $26.6 \div 2.8 = 0.95$

Estimate:

Divide. Round quotients to the nearest thousandth when necessary.

46. $3\overline{)43.4}$

47. $\dfrac{-72}{-0.06}$

48. $-0.00048 \div 0.0012$

[5.4–5.5] *Solve each application problem.*

49. Adrienne worked 46.5 hours this week. Her hourly wage is $14.24 for the first 40 hours and 1.5 times that rate over 40 hours. Find her total earnings to the nearest dollar.

50. A book of 12 tickets costs $35.89 at the State Fair midway. What is the cost per ticket, to the nearest cent?

51. Stock in Math Tronic sells for $3.75 per share. Kenneth is thinking of investing $500. How many whole shares could he buy?

52. Grapes are on sale at $0.99 per pound. How much will Ms. Lee pay for 3.5 pounds of grapes, to the nearest cent?

Simplify.

53. $3.5^2 + 8.7(-1.95)$

54. $11 - 3.06 \div (3.95 - 0.35)$

[5.6] *Write each fraction as a decimal. Round to the nearest thousandth when necessary.*

55. $3\dfrac{4}{5}$

56. $\dfrac{16}{25}$

57. $1\dfrac{7}{8}$

58. $\dfrac{1}{9}$

Arrange each group of numbers in order from least to greatest.

59. $3.68, 3.806, 3.6008$

60. $0.215, 0.22, 0.209, 0.2102$

61. $0.17, \dfrac{3}{20}, \dfrac{1}{8}, 0.159$

[5.7] *Find the mean and the median for each set of data.*

62. Digital cameras sold each day:
18, 12, 15, 24, 9, 42, 54, 87, 21, 3

63. Number of insurance claims processed:
54, 28, 35, 43, 17, 37, 68, 75, 39

64. Find the weighted mean.

Dollar Value	Frequency
$42	3
$47	7
$53	2
$55	3
$59	5

65. Find the mode or modes for each set of data.

Hiking boots at store J priced at $107, $69, $139, $107, $160, $84, $160

Hiking boots at store K priced at $119, $136, $99, $119, $139, $119, $95

[5.8] *Find the unknown length in each right triangle. Use a calculator to find square roots. Round your answers to the nearest tenth when necessary.*

66.

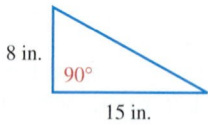

8 in. 90° 15 in.

67.

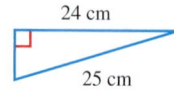

24 cm, 25 cm

68.

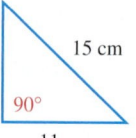

15 cm, 90°, 11 cm

69.

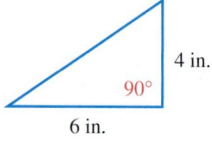

4 in. 90° 6 in.

70.

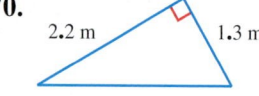

2.2 m, 1.3 m

71.

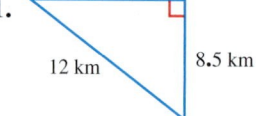

12 km, 8.5 km

[5.9] *Solve each equation. Show your work.*

72. $-0.1 = b - 0.35$

73. $-3.8x = 0$

74. $6.8 + 0.4n = 1.6$

75. $-0.375 + 1.75a = 2a$

76. $0.3y - 5.4 = 2.7 + 0.8y$

[5.10] *Find the unknown radius or diameter.*

77. The radius of a circular irrigation field is 68.9 m. What is the diameter of the field?

78. The diameter of a juice can is 3 in. What is the radius of the can?

Find the circumference and area of each circle. Use **3.14** *as the approximate value for π. Round your answers to the nearest tenth.*

79.

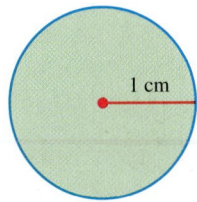

1 cm

80.

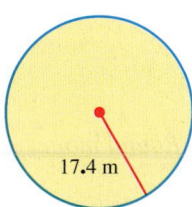

17.4 m

81.

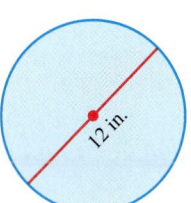

12 in.

Find the volume and surface area of each solid. Use **3.14** *as the approximate value for π. Round your answers to the nearest tenth when necessary.*

82.

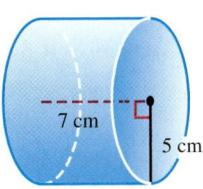

7 cm   5 cm

83.

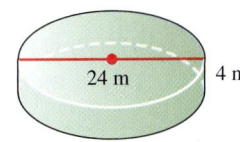

24 m   4 m

84. A rectangular cooler that measures 3.5 ft by 1.5 ft and is 1.5 ft high

▶▶▶ Mixed Review Exercises

Simplify.

85. $89.19 + 0.075 + 310.6 + 5$

86. $72.8(-3.5)$

87. $1648.3 \div 0.46$ Round to thousandths.

88. $30 - 0.9102$

89. $(4.38)(0.007)$

90. $0.005\overline{)0.047}$

91. $72.105 + 8.2 - 95.37$

92. $\dfrac{-81.36}{9}$

93. $(0.6 - 1.22) + 4.8(-3.15)$

94. $0.455(18)$

95. $(-1.6)(-0.58)$

96. $0.218\overline{)7.63}$

97. $-21.059 - 20.8$

98. $18.3 - 3^2 \div 0.5$

Use the information in the ad to solve Exercises 99–103. Round money answers to the nearest cent. (Disregard any sales tax.)

Grand Opening Sale!
Save on Clothing for the Entire Family

Jeans for Teens
only $19.95 each
women's sizes $24.99

Athletic Shoes
regularly priced
$89.99 to $149.50
NOW just $71 to $119.60

Men's socks NOW 3 pairs for $8.99
Children's socks 6 pairs for $5

Hurry in — TWO DAYS ONLY

99. How much would one pair of men's socks cost?

100. How much more would one pair of men's socks cost than one pair of children's socks?

101. How much would Fernando pay for a dozen pairs of men's socks?

102. How much would Akiko pay for five pairs of teen jeans and four pairs of women's jeans?

103. What is the difference between the cheapest sale price for athletic shoes and the highest regular price?

Solve each equation.

104. $4.62 = -6.6y$

105. $1.05x - 2.5 = 0.8x + 5$

Solve each application problem. Round answers to the nearest tenth when necessary.

106. A circular table has a diameter of 5 ft. What length of rubber strip is needed to go around the edge of the table? What is the area of the tabletop?

107. Jerry missed one math test, so his test scores are 82, 0, 78, 93, 85. Find his mean and median scores.

108. LaRae drove 16 miles south, then made a 90° right turn and drove 12 miles west. How far is she, in a straight line, from her starting point?

109. A juice can that is 7 in. tall has a diameter of 3 in. What is the volume of the can?

Chapter 5 ▷▷▷ Test

 **Test Prep** VIDEO CD   Use the Chapter Test Prep Video CD to see fully worked-out solutions to any of the exercises you want to review.

Write each decimal as a fraction or mixed number in lowest terms.

1. 18.4

2. 0.075

Write each decimal in words.

3. 60.007

4. 0.0208

First, use front end rounding to round each number and estimate the answer. Then find the exact answer.

5. $7.6 + 82.0128 + 39.59$

6. $-5.79(1.2)$

7. $-79.1 - 3.602$

8. $-20.04 \div (-4.8)$

Find the exact answer.

9. $670 - 0.996$

10. $0.15\overline{)72}$

11. $(-0.006)(-0.007)$

Solve each application problem.

12. Pat bought 6.5 ft of decorative gold chain to hang a light over her dining table. She paid $11.64. What was the cost per foot, to the nearest cent?

13. Davida ran a race in 3.059 minutes. Angela ran the race in 3.5 minutes. Who won? By how much?

14. Mr. Yamamoto bought 1.85 pounds of cheese at $2.89 per pound. What was the total amount he paid, to the nearest cent?

Solve each equation. Show your work.

15. $-5.9 = y + 0.25$

16. $-4.2x = 1.47$

17. $3a - 22.7 = 10$

18. $-0.8n + 1.88 = 2n - 6.1$

1. _____

2. _____

3. _____

4. _____

5. *Estimate:* _____
   Exact: _____

6. *Estimate:* _____
   Exact: _____

7. *Estimate:* _____
   Exact: _____

8. *Estimate:* _____
   Exact: _____

9. _____

10. _____

11. _____

12. _____

13. _____

14. _____

15. _____

16. _____

17. _____

18. _____

19. _____

19. Arrange in order from least to greatest: $0.44, 0.451, \dfrac{9}{20}, 0.4506$

20. _____

20. Simplify: $6.3^2 - 5.9 + 3.4(-0.5)$

21. _____

21. Find the mean number of books loaned: 52, 61, 68, 69, 73, 75, 79, 84, 91, 98.

22. _____

22. Find the mode for hot tub temperatures (Fahrenheit) of 96°, 104°, 103°, 104°, 103°, 104°, 91°, 74°, 103°.

23. _____

23. Find the weighted mean.

Cost	Frequency
$ 6	7
$10	3
$11	4
$14	2
$19	3
$24	1

24. _____

24. Find the median cost of an online textbook: $54.50, $48, $39.75, $89, $56.25, $49.30, $46.90, $51.80.

Find the unknown lengths. Round your answers to the nearest tenth when necessary.

25. _____

25.
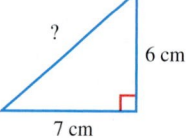

26. _____

26.
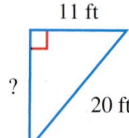

Use 3.14 as the approximate value for π and round your answers to the nearest tenth when necessary.

27. _____

27. Find the radius.
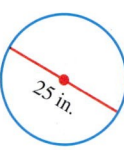

28. _____

28. Find the circumference.
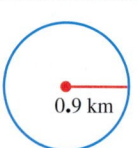

29. _____

29. Find the area.
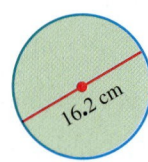

30. _____

30. Find the volume.
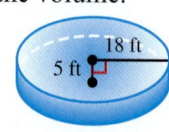

31. _____

31. Find the surface area of the solid in Problem 30.

6

Ratio, Proportion, and Line/Angle/Triangle Relationships

Three out of every four Americans have cellular phones. Worldwide, about 1 out of 3 people use a cell phone! (*Source:* International Telecommunications Union.)

You can use unit rates to find the best deal on cell phone service plans. But most plans do not cover long-distance calls to certain U.S. locations or to other countries. Then you can use unit rates to find the cheapest long-distance calling card. (See **Section 6.2,** Exercises 37–38 for calling cards and Exercises 47–50 for cell phone plans; and **Cumulative Review,** Exercises 30–33 for international calling cards.)

6.1 ▶▶▶ Ratios

OBJECTIVES

1. Write ratios as fractions.
2. Solve ratio problems involving decimals or mixed numbers.
3. Solve ratio problems after converting units.

A **ratio** compares two quantities. You can compare two numbers, such as 8 and 4, or two measurements that have the *same* type of units, such as 3 days and 12 days. (*Rates* compare measurements with different types of units and are covered in the next section.)

Ratios can help you see important relationships. For example, if the ratio of your monthly expenses to your monthly income is 10 to 9, then you are spending $10 for every $9 you earn and going deeper into debt.

OBJECTIVE 1 Write ratios as fractions. A ratio can be written in three ways.

Writing a Ratio

The ratio of $7 **to** $3 can be written as follows.

$$7 \text{ to } 3 \quad \text{or} \quad 7{:}3 \quad \text{or} \quad \frac{7}{3} \leftarrow \text{Fraction bar indicates “to.”}$$

“:” indicates “**to.**”

Writing a ratio as a fraction is the most common method, and the one we will use here. All three ways are read, “the ratio of 7 **to** 3.” The word **to** separates the quantities being compared.

Writing a Ratio as a Fraction

Order is important when you're writing a ratio. The quantity mentioned **first** is the **numerator**. The quantity mentioned **second** is the **denominator**. For example:

$$\text{The ratio of 5 to 12} \quad \text{is written} \quad \frac{5}{12}$$

EXAMPLE 1 Writing Ratios

Ancestors of the Pueblo Indians built multistory apartment towns in New Mexico about 1100 years ago. A room might measure 14 feet long, 11 feet wide, and 15 feet high.

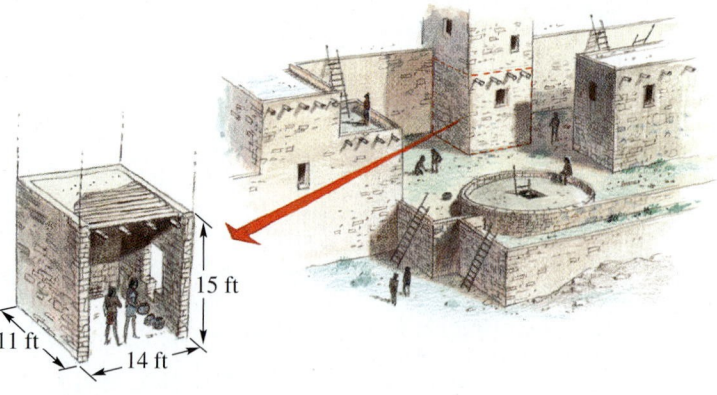

Continued on Next Page

Write each ratio as a fraction, using the room measurements.

(a) Ratio of length to width

> Do **not** rewrite the ratio as $1\frac{3}{11}$.

The ratio of **length to width** is $\dfrac{14 \text{ feet}}{11 \text{ feet}} = \dfrac{14}{11}$

Numerator (mentioned first) Denominator (mentioned second)

You can divide out common *units* just as you divided out common *factors* when writing fractions in lowest terms. (See **Section 4.2.**) However, do *not* rewrite the fraction as a mixed number. Keep it as the ratio of 14 to 11.

(b) Ratio of width to height

> Divide out the common units (ft).

The ratio of width **to** height is $\dfrac{11 \text{ feet}}{15 \text{ feet}} = \dfrac{11}{15}$

CAUTION

Remember, the *order* of the numbers is important in a ratio. Look for the words "ratio of *a* to *b*." Write the ratio as $\dfrac{a}{b}$, *not* $\dfrac{b}{a}$. The quantity mentioned first is the numerator.

Work Problem **1** *at the Side.* ▶

Any ratio can be written as a fraction. Therefore, you can write a ratio in *lowest terms,* just as you do with any fraction.

EXAMPLE 2 **Writing Ratios in Lowest Terms**

Write each ratio as a fraction in lowest terms.

(a) 60 days to 20 days

The ratio is $\frac{60}{20}$. Write this ratio in lowest terms by dividing the numerator and the denominator by 20.

$$\frac{60}{20} = \frac{60 \div 20}{20 \div 20} = \frac{3}{1} \quad \left\{ \begin{array}{l} \text{Ratio in} \\ \text{lowest terms} \end{array} \right.$$

So, the ratio of 60 days to 20 days is 3 to 1, or, written as a fraction, $\frac{3}{1}$.

CAUTION

In the fractions chapter (**Chapter 4**), you would have rewritten $\frac{3}{1}$ as 3. But a *ratio* compares *two* quantities, so you need to keep both parts of the ratio and write it as $\frac{3}{1}$.

(b) 50 ounces of medicine to 120 ounces of medicine

The ratio is $\frac{50}{120}$. Divide the numerator and the denominator by 10.

$$\frac{50}{120} = \frac{50 \div 10}{120 \div 10} = \frac{5}{12} \quad \left\{ \begin{array}{l} \text{Ratio in} \\ \text{lowest terms} \end{array} \right.$$

So, the ratio of 50 ounces to 120 ounces is $\frac{5}{12}$.

Continued on Next Page

1 Shane spent $14 on meat, $5 on milk, and $7 on fresh fruit. Write the following ratios as fractions.

(a) The ratio of amount spent on fruit to amount spent on milk

(b) The ratio of amount spent on milk to amount spent on meat

(c) The ratio of amount spent on meat to amount spent on milk

ANSWERS

1. **(a)** $\frac{7}{5}$ **(b)** $\frac{5}{14}$ **(c)** $\frac{14}{5}$

2 Write each ratio as a fraction in lowest terms.

(a) 9 hours to 12 hours

(b) 100 meters to 50 meters

(c) The ratio of width to length for this rectangle

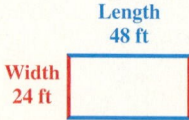

Length
48 ft

Width
24 ft

3 Write each ratio as a ratio of whole numbers in lowest terms.

(a) The price of Tamar's favorite brand of lipstick increased from $5.50 to $7.00. Find the ratio of the increase in price to the original price.

(b) Last week, Lance worked 4.5 hours each day. This week he cut back to 3 hours each day. Find the ratio of the decrease in hours to the original number of hours.

(c) 15 people in a large van to 6 people in a small van

$$\text{The ratio is } \frac{15}{6} = \frac{15 \div 3}{6 \div 3} = \frac{5}{2} \leftarrow \begin{cases} \text{Ratio in} \\ \text{lowest terms} \end{cases}$$

> **Note**
> Although $\frac{5}{2} = 2\frac{1}{2}$, ratios are *not* written as mixed numbers. Nevertheless, in Example 2(c) above, the ratio $\frac{5}{2}$ does mean the large van holds $2\frac{1}{2}$ times as many people as the small van.

◀ Work Problem **2** at the Side.

OBJECTIVE 2 Solve ratio problems involving decimals or mixed numbers. Sometimes a ratio compares two decimal numbers or two fractions. It is easier to understand if we rewrite the ratio as a ratio of two whole numbers.

EXAMPLE 3 Using Decimal Numbers in a Ratio

The price of a Sunday newspaper increased from $1.50 to $1.75. Find the ratio of the <u>increase in price</u> **to** the <u>original price</u>.

The words <u>increase in price</u> are mentioned first, so the increase will be the numerator. How much did the price go up? Use subtraction.

$$\text{new price} - \text{original price} = \text{increase}$$
$$\$1.75 - \$1.50 = \$0.25$$

The words <u>original price</u> are mentioned second, so the original price of $1.50 is the denominator.

The ratio of <u>increase in price</u> **to** <u>original price</u> is shown below.

$$\frac{0.25}{1.50} \begin{array}{l} \leftarrow \text{increase in price} \\ \leftarrow \text{original price} \end{array}$$

Now rewrite the ratio as a ratio of whole numbers. Recall that if you multiply both the numerator and denominator of a fraction by the same number, you get an equivalent fraction. The decimals in this example are hundredths, so multiply by 100 to get whole numbers. (If the decimals are tenths, multiply by 10. If thousandths, multiply by 1000.) Then write the ratio in lowest terms.

$$\frac{0.25}{1.50} = \underbrace{\frac{(0.25)(100)}{(1.50)(100)} = \frac{25}{150}}_{\substack{\text{Ratio as two} \\ \text{whole numbers}}} = \frac{25 \div 25}{150 \div 25} = \frac{1}{6} \leftarrow \begin{cases} \text{Ratio in} \\ \text{lowest terms} \end{cases}$$

◀ Work Problem **3** at the Side.

EXAMPLE 4 Using Mixed Numbers in Ratios

Write each ratio as a comparison of whole numbers in lowest terms.

(a) 2 days to $2\frac{1}{4}$ days

Write the ratio as follows. Divide out the common units.

$$\frac{2 \text{ days}}{2\frac{1}{4} \text{ days}} = \frac{2}{2\frac{1}{4}}$$

Continued on Next Page

Next, write 2 as $\frac{2}{1}$ and $2\frac{1}{4}$ as the improper fraction $\frac{9}{4}$.

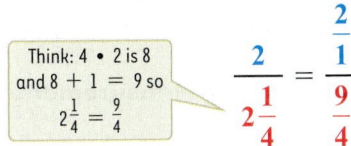

Think: 4 • 2 is 8 and 8 + 1 = 9 so $2\frac{1}{4} = \frac{9}{4}$

$$\frac{2}{2\frac{1}{4}} = \frac{\frac{2}{1}}{\frac{9}{4}}$$

Now rewrite the problem in horizontal format, using the "÷" symbol for division. Finally, multiply by the reciprocal of the divisor, as you did in **Section 4.3.**

$$\frac{\frac{2}{1}}{\frac{9}{4}} = \frac{2}{1} \div \frac{9}{4} = \frac{2}{1} \cdot \frac{4}{9} = \frac{8}{9}$$

Reciprocals

The ratio, in lowest terms, is $\frac{8}{9}$.

(b) $3\frac{1}{4}$ to $1\frac{1}{2}$

Write the ratio as $\dfrac{3\frac{1}{4}}{1\frac{1}{2}}$. Then write $3\frac{1}{4}$ and $1\frac{1}{2}$ as improper fractions.

$$3\frac{1}{4} = \frac{13}{4} \quad \text{and} \quad 1\frac{1}{2} = \frac{3}{2}$$

The ratio is shown below.

$$\frac{3\frac{1}{4}}{1\frac{1}{2}} = \frac{\frac{13}{4}}{\frac{3}{2}}$$

Rewrite as a division problem in horizontal format, using the "÷" symbol. Then multiply by the reciprocal of the divisor.

$$\frac{13}{4} \div \frac{3}{2} = \frac{13}{4} \cdot \frac{2}{3} = \frac{13 \cdot \overset{1}{2}}{\underset{1}{2} \cdot 2 \cdot 3} = \frac{13}{6} \quad \leftarrow \left\{ \begin{array}{l} \text{Ratio in} \\ \text{lowest terms} \end{array} \right.$$

Reciprocals

> **Note**
>
> We can also work Examples 4(a) and 4(b) above by using decimals.
>
> **(a)** $2\frac{1}{4}$ is equivalent to 2.25, so we have the ratio shown below.
>
> $$\frac{2}{2\frac{1}{4}} = \frac{2}{2.25} = \frac{(2)(100)}{(2.25)(100)} = \frac{200}{225} = \frac{200 \div 25}{225 \div 25} = \frac{8}{9} \quad \leftarrow \text{Same result}$$
>
> **(b)** $3\frac{1}{4}$ is equivalent to 3.25 and $1\frac{1}{2}$ is equivalent to 1.5.
>
> $$\frac{3\frac{1}{4}}{1\frac{1}{2}} = \frac{3.25}{1.5} = \frac{(3.25)(100)}{(1.5)(100)} = \frac{325}{150} = \frac{325 \div 25}{150 \div 25} = \frac{13}{6} \quad \leftarrow \text{Same result}$$
>
> This method would *not* work for fractions that are repeating decimals, such as $\frac{1}{3}$ or $\frac{5}{6}$.

Work Problem ④ *at the Side.* ▶

④ Write each ratio as a ratio of whole numbers in lowest terms.

(a) $3\frac{1}{2}$ to 4

(b) $5\frac{5}{8}$ pounds to $3\frac{3}{4}$ pounds

(c) $3\frac{1}{3}$ inches to $\frac{5}{6}$ inch

5 Write each ratio as a fraction in lowest terms. (*Hint:* Recall that it is usually easier to write the ratio using the smaller measurement unit.)

(a) 9 inches to 6 feet

(b) 2 days to 8 hours

(c) 7 yards to 14 feet

(d) 3 quarts to 3 gallons

(e) 25 minutes to 2 hours

(f) 4 pounds to 12 ounces

OBJECTIVE 3 Solve ratio problems after converting units. When a ratio compares measurements, both measurements must be in the *same* units. For example, *feet* must be compared to *feet, hours* to *hours,* and so on.

EXAMPLE 5 Ratio Applications Using Measurement

(a) Write the ratio of the length of the shorter board on the left to the length of the longer board on the right. Compare in inches.

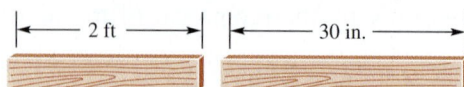

First, express 2 feet in inches. Because 1 foot has 12 inches, 2 feet is

$$2 \cdot \textbf{12 inches} = 24 \text{ inches}$$

The length of the board on the left is 24 inches, so the ratio of the lengths is shown below. The common units divide out.

> Once the units match, you can divide them out.

$$\frac{2 \text{ ft}}{30 \text{ in.}} = \frac{24 \text{ inches}}{30 \text{ inches}} = \frac{24}{30}$$

Write the ratio in lowest terms.

$$\frac{24}{30} = \frac{24 \div 6}{30 \div 6} = \frac{4}{5} \begin{cases} \text{Ratio in} \\ \text{lowest terms} \end{cases}$$

The shorter board on the left is $\frac{4}{5}$ the length of the longer board on the right.

> **Note**
>
> Notice in the example above that we wrote the ratio using the smaller unit (inches are smaller than feet). Using the smaller unit will help you avoid working with fractions.

(b) Write the ratio of 28 days to 3 weeks.

Since it is usually easier to write the ratio using the smaller measurement, compare in *days* because days are shorter than weeks.

First express 3 weeks in days. Because 1 week has 7 days, 3 weeks is

$$3 \cdot \textbf{7 days} = 21 \text{ days}$$

So the ratio in days is shown below.

$$\frac{28 \text{ days}}{3 \text{ weeks}} = \frac{28 \text{ days}}{21 \text{ days}} = \frac{28}{21} = \frac{28 \div 7}{21 \div 7} = \frac{4}{3} \begin{cases} \text{Ratio in} \\ \text{lowest terms} \end{cases}$$

Use the table below to help set up ratios that compare measurements.

Measurement Comparisons

Length	Weight
12 inches = 1 foot	16 ounces = 1 pound
3 feet = 1 yard	2000 pounds = 1 ton
5280 feet = 1 mile	**Time**
Capacity (Volume)	60 seconds = 1 minute
2 cups = 1 pint	60 minutes = 1 hour
2 pints = 1 quart	24 hours = 1 day
4 quarts = 1 gallon	7 days = 1 week

◀ *Work Problem* **5** *at the Side.*

6.1 ▶▶▶   **Exercises**

FOR EXTRA HELP    MyMathLab   Math XL PRACTICE   WATCH   DOWNLOAD   READ   REVIEW

Write each ratio as a fraction in lowest terms. See Examples 1 and 2.

1. 8 days to 9 days

2. $11 to $15

3. $100 to $50

4. 35¢ to 7¢

5. 30 minutes to 90 minutes

6. 9 pounds to 36 pounds

Write each ratio as a ratio of whole numbers in lowest terms. See Examples 3 and 4.

7. $4.50 to $3.50

8. $0.08 to $0.06

9. $1\frac{1}{4}$ to $1\frac{1}{2}$

10. $2\frac{1}{3}$ to $2\frac{2}{3}$

Write each ratio as a fraction in lowest terms. For help, use the table of measurement relationships on page 442. See Example 5.

11. 4 feet to 30 inches

12. 8 feet to 4 yards

13. 5 minutes to 1 hour

14. 3 pounds to 6 ounces

15. 5 gallons to 5 quarts

16. 3 cups to 3 pints

The table shows the number of greeting cards that Americans buy for various occasions. Use the information to answer Exercises 17–22. Write each ratio as a fraction in lowest terms.

Holiday/Event	Cards Sold
Valentine's Day	900 million
Mother's Day	150 million
Father's Day	95 million
Graduation	60 million
Thanksgiving	30 million
Halloween	25 million

Source: Hallmark Cards.

17. Find the ratio of Thanksgiving cards to graduation cards.

18. Find the ratio of Halloween cards to Mother's Day cards.

19. Find the ratio of Valentine's Day cards to Halloween cards.

20. Find the ratio of Mother's Day cards to Father's Day cards.

21. Explain how you might use the information in the table if you owned a shop selling gifts and greeting cards.

22. Why is the ratio of Valentine's Day cards to graduation cards $\frac{15}{1}$? Give two possible reasons.

The bar graph shows worldwide sales of the most popular songs of all time. Use the graph to complete Exercises 23–24. Write each ratio as a fraction in lowest terms.

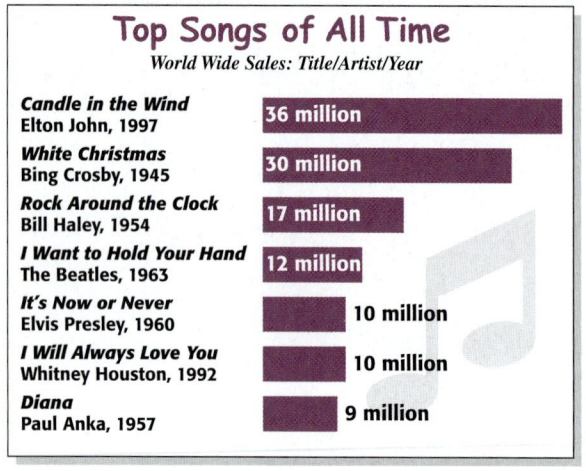

Top Songs of All Time
World Wide Sales: Title/Artist/Year

Candle in the Wind Elton John, 1997	36 million
White Christmas Bing Crosby, 1945	30 million
Rock Around the Clock Bill Haley, 1954	17 million
I Want to Hold Your Hand The Beatles, 1963	12 million
It's Now or Never Elvis Presley, 1960	10 million
I Will Always Love You Whitney Houston, 1992	10 million
Diana Paul Anka, 1957	9 million

Sources: The Music Information Database, www.songfacts.com

23. Sales of which two songs give a ratio of $\frac{3}{1}$? There may be more than one correct answer.

24. Sales of which two songs give a ratio of $\frac{5}{6}$? There may be more than one correct answer.

For each figure in Exercises 25–28, find the ratio of the length of the longest side to the length of the shortest side. Write each ratio as a fraction in lowest terms. See Examples 2–4.

25.
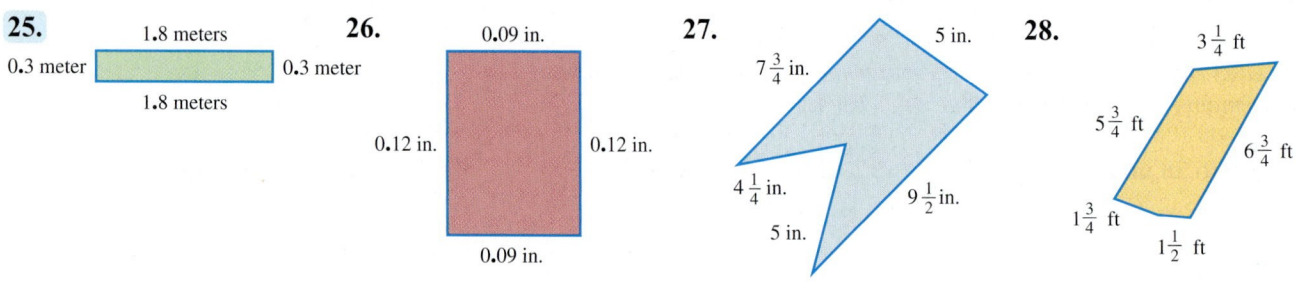
1.8 meters
0.3 meter ▭ 0.3 meter
1.8 meters

26.
0.09 in.
0.12 in. ▭ 0.12 in.
0.09 in.

27.
5 in.
$7\frac{3}{4}$ in.
$4\frac{1}{4}$ in. $9\frac{1}{2}$ in.
5 in.

28.
$3\frac{1}{4}$ ft
$5\frac{3}{4}$ ft $6\frac{3}{4}$ ft
$1\frac{3}{4}$ ft
$1\frac{1}{2}$ ft

Write each ratio as a fraction in lowest terms.

29. The price of automobile engine oil has gone from $10 to $12.50 for the 5 quarts needed for an oil change. Find the ratio of the increase in price to the original price.

30. The price of an antibiotic decreased from $8.80 to $5.60 for 10 tablets. Find the ratio of the decrease in price to the original price.

31. The first time a movie was made in Minnesota, the cast and crew spent $59\frac{1}{2}$ days filming winter scenes. The next year, another movie was filmed in $8\frac{3}{4}$ weeks. Find the ratio of the first movie's filming time to the second movie's time. Compare in weeks.

32. The percheron, a large draft horse, measures about $5\frac{3}{4}$ feet at the shoulder. The prehistoric ancestor of the horse measured only $15\frac{3}{4}$ inches at the shoulder. Find the ratio of the percheron's height to its prehistoric ancestor's height. Compare in inches.
(*Source: Eyewitness Books: Horse.*)

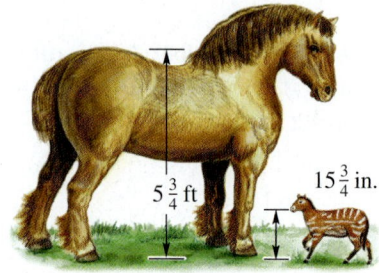

$5\frac{3}{4}$ ft $15\frac{3}{4}$ in.

6.2 ▷▷▷ Rates

A *ratio* compares two measurements with the same type of units, such as 9 <u>feet</u> **to** 12 <u>feet</u> (both length measurements). But many of the comparisons we make use measurements with different types of units, such as shown here.

160 <u>dollars</u> **for** 8 <u>hours</u> (money to time)

450 <u>miles</u> **on** 15 <u>gallons</u> (distance to capacity)

This type of comparison is called a **rate.**

OBJECTIVE 1 Write rates as fractions. Suppose that yesterday you hiked 18 <u>miles</u> **in** 4 <u>hours</u>. The *rate* at which you hiked can be written as a fraction in lowest terms.

$$\frac{18 \text{ miles}}{4 \text{ hours}} = \frac{18 \text{ miles} \div 2}{4 \text{ hours} \div 2} = \frac{9 \text{ miles}}{2 \text{ hours}} \leftarrow \text{Rate in lowest terms}$$

In a rate, you often find one of these words separating the quantities you are comparing.

in   for   on   per   from

> **CAUTION**
> When writing a rate, always include the units, such as miles, hours, dollars, and so on. Because the units in a rate are different, the units do *not* divide out.

EXAMPLE 1 Writing Rates in Lowest Terms

Write each rate as a fraction in lowest terms.

(a) 5 gallons of chemical **for** $60

$$\frac{5 \text{ gallons} \div 5}{60 \text{ dollars} \div 5} = \frac{1 \text{ gallon}}{12 \text{ dollars}} \searrow \text{Write the units: gallons and dollars.}$$

(b) $1500 wages **in** 10 weeks

$$\frac{1500 \text{ dollars} \div 10}{10 \text{ weeks} \div 10} = \frac{150 \text{ dollars}}{1 \text{ week}}$$

Be sure to write the units in a *rate*: dollars, miles, gallons, etc.

(c) 2225 miles **on** 75 gallons of gas

$$\frac{2225 \text{ miles} \div 25}{75 \text{ gallons} \div 25} = \frac{89 \text{ miles}}{3 \text{ gallons}}$$

Work Problem **1** *at the Side.* ▶

OBJECTIVE 2 Find unit rates. When the *denominator* of a rate is 1, it is called a **unit rate.** We use unit rates frequently. For example, you earn $12.75 for *1 hour* of work. This unit rate is written:

$12.75 **per** hour   or $12.75 / hour.

You drive 28 miles on *1 gallon* of gas. This unit rate is written

28 miles **per** gallon   or   28 miles / gallon.

Use **per** or a / mark when writing unit rates.

OBJECTIVES

1 **Write rates as fractions.**

2 **Find unit rates.**

3 **Find the best buy based on cost per unit.**

1 Write each rate as a fraction in lowest terms.

(a) $6 for 30 packages

(b) 500 miles in 10 hours

(c) 4 teachers for 90 students

(d) 1270 bushels from 30 acres

ANSWERS

1. **(a)** $\dfrac{1 \text{ dollar}}{5 \text{ packages}}$ **(b)** $\dfrac{50 \text{ miles}}{1 \text{ hour}}$

   (c) $\dfrac{2 \text{ teachers}}{45 \text{ students}}$ **(d)** $\dfrac{127 \text{ bushels}}{3 \text{ acres}}$

2 Find each unit rate.

(a) $4.35 for 3 pounds of cheese

(b) 304 miles on 9.5 gallons of gas

(c) $850 in 5 days

(d) 24-pound turkey for 15 people

EXAMPLE 2   **Finding Unit Rates**

Find each unit rate.

(a) 337.5 miles on 13.5 gallons of gas
Write the rate as a fraction.

$$\frac{337.5 \text{ miles}}{13.5 \text{ gallons}} \longleftarrow \text{The fraction bar indicates division.}$$

Divide 337.5 by 13.5 to find the unit rate.

$$13.5 \overline{)337.5}^{2\,5.}$$

$$\frac{337.5 \text{ miles} \div \textbf{13.5}}{13.5 \text{ gallons} \div \textbf{13.5}} = \frac{25 \text{ miles}}{1 \text{ gallon}}$$

The unit rate is 25 miles **per** gallon, or 25 miles/gallon.

(b) 549 miles in 18 hours

$$\frac{549 \text{ miles}}{18 \text{ hours}} \qquad \text{Divide.} \qquad 18 \overline{)549.0}^{30.5}$$

The unit rate is 30.5 miles **per** hour, or 30.5 miles/hour.

(c) $810 in 6 days

$$\frac{810 \text{ dollars}}{6 \text{ days}} \qquad \text{Divide.} \qquad 6 \overline{)810}^{135}$$

> Use *per* or a slash mark to write unit rates.

The unit rate is $135 **per** day, or $135/day.

◄ *Work Problem* **2** *at the Side.*

OBJECTIVE **3** **Find the best buy based on cost per unit.**   When shopping for groceries, household supplies, and health and beauty items, you will find many different brands and package sizes. You can save money by finding the lowest *cost per unit*.

> **Cost per Unit**
>
> **Cost per unit** is a rate that tells how much you pay for *one* item or *one* unit. Examples are $3.25 per gallon, $47 per shirt, and $2.98 per pound.

EXAMPLE 3   **Determining the Best Buy**

The local store charges the following prices for pancake syrup. Find the best buy.

Continued on Next Page

The best buy is the container with the *lowest* cost per unit. All the containers are measured in *ounces* (oz), so you first need to find the *cost per ounce* for each one. Divide the price of the container by the number of ounces in it. Round to the nearest thousandth if necessary.

Let the *order* of the *words* help you set up the rate.

cost $\longrightarrow$ **$1.28**

per (means divide) $\rightarrow$ ───────

ounce $\longrightarrow$ **12 ounces**

Size	Cost per Unit (Rounded)
12 ounces	$\dfrac{\$1.28}{12 \text{ ounces}} \approx \$0.107$ per ounce (highest)
24 ounces	$\dfrac{\$1.81}{24 \text{ ounces}} \approx \$0.075$ per ounce (lowest)
36 ounces	$\dfrac{\$2.73}{36 \text{ ounces}} \approx \$0.076$ per ounce

The lowest cost per ounce is $0.075, so the 24-ounce container is the best buy.

> **Note**
>
> Earlier we rounded money amounts to the nearest hundredth (nearest cent). But when comparing unit costs, rounding to the nearest thousandth will help you see the difference between very similar unit costs. Notice that the 24-ounce and 36-ounce syrup containers above would both have rounded to $0.08 per ounce if we had rounded to hundredths.

Work Problem **3** *at the Side.* ▶

🖩 **Calculator Tip**   When using a calculator to find unit prices, remember that division is *not* commutative. In Example 3 above you wanted to find cost per ounce. Let the *order* of the *words* help you enter the numbers in the correct order.

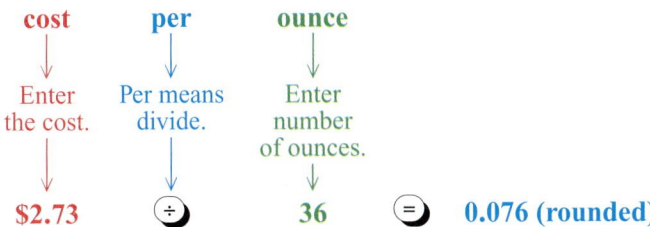

cost	**per**	**ounce**
↓	↓	↓
Enter the cost.	Per means divide.	Enter number of ounces.
↓	↓	↓
$2.73	÷	**36** = **0.076 (rounded)**

If you reversed the order and entered 36 ÷ 2.73 = , the result is the number of *ounces* per *dollar*. How could you use that information to find the best buy? (*Answer:* The best buy would be the greatest number of ounces per dollar.)

Finding the best buy is sometimes a complicated process. Things that affect the cost per unit can include "cents off" coupons and differences in how much use you'll get out of each unit.

---

3 Find the best buy (lowest cost per unit) for each purchase.

(a) 2 quarts for $3.25
3 quarts for $4.95
4 quarts for $6.48

(b) 6 cans of cola for $1.99
12 cans of cola for $3.49
24 cans of cola for $7

4 Solve each problem.

(a) Some batteries claim to last longer than others. If you believe these claims, which brand is the best buy?

Four-pack of AA-size batteries for $2.79

One AA-size battery for $1.19; lasts twice as long

(b) Which tube of toothpaste is the best buy? You have a coupon for 85¢ off Brand C and a coupon for 20¢ off Brand D.

Brand C is $3.89 for 6 ounces.

Brand D is $1.59 for 2.5 ounces.

> **EXAMPLE 4** **Solving Best Buy Applications**

Solve each application problem.

(a) There are many brands of liquid laundry detergent. If you feel they all do a good job of cleaning your clothes, you can base your purchase on cost per unit. But some brands are "concentrated" so you can use less detergent for each load of clothes. Which of the choices shown below is the best buy?

try **SUDZY** to clean your clothes!

50 fluid ounces for $3.99
Does same number of washloads as the old 64-ounce bottle!

WHITE-O gets out **ALL** the stains!

One gallon (128 ounces) for $9.89

Does twice the washloads of the old gallon bottle!

To find Sudzy's unit cost, divide $3.99 by 64 ounces, not 50 ounces. You're getting as many clothes washed as if you bought 64 ounces. Similarly, to find White-O's unit cost, divide $9.89 by 256 ounces (twice 128 ounces, or 2 • 128 ounces = 256 ounces).

$$\text{Sudzy} \quad \frac{\$3.99}{64 \text{ ounces}} \approx \$0.062 \text{ per ounce}$$

> The best buy is the *lower* cost per ounce.

$$\text{White-O} \quad \frac{\$9.89}{256 \text{ ounces}} \approx \$0.039 \text{ per ounce}$$

White-O has the lower cost per ounce and is the best buy. (However, if you try it and it doesn't get out all the stains, Sudzy may be worth the extra cost.)

(b) "Cents-off" coupons also affect the best buy. Suppose you are looking at these choices for "extra-strength" pain reliever. Both brands have the same amount of pain reliever in each tablet.

Brand X is $2.29 for 50 tablets.

Brand Y is $10.75 for 200 tablets.

You have a 40¢ coupon for Brand X and a 75¢ coupon for Brand Y. Which choice is the best buy?

To find the best buy, first subtract the coupon amounts, then divide to find the lowest cost per ounce.

Brand X costs $2.29 − $0.40 = $1.89

$$\frac{\$1.89}{50 \text{ tablets}} \approx \$0.038 \text{ per tablet}$$

> Look for the *lower* cost per tablet.

Brand Y costs $10.75 − $0.75 = $10.00

$$\frac{\$10.00}{200 \text{ tablets}} = \$0.05 \text{ per tablet}$$

Brand X has the lower cost per tablet and is the best buy.

◀ *Work Problem* **4** *at the Side.*

6.2 ▶▶▶ Exercises

Write each rate as a fraction in lowest terms. See Example 1.

1. 10 cups for 6 people

2. $12 for 30 pens

3. 15 feet in 35 seconds

4. 100 miles in 30 hours

5. 14 people for 28 dresses

6. 12 wagons for 48 horses

7. 25 letters in 5 minutes

8. 68 pills for 17 people

9. $63 for 6 visits

10. 25 doctors for 310 patients

11. 72 miles on 4 gallons

12. 132 miles on 8 gallons

Find each unit rate. See Example 2.

13. $60 in 5 hours

14. $2500 in 20 days

15. 50 eggs from 10 chickens

16. 36 children from 12 families

17. 7.5 pounds for 6 people

18. 44 bushels from 8 trees

19. $413.20 for 4 days

20. $74.25 for 9 hours

Earl kept the following record of the gas he bought for his car. For each entry, find the number of miles he traveled and the unit rate. Round your answers to the nearest tenth.

	Date	Odometer at Start	Odometer at End	Miles Traveled	Gallons Purchased	Miles per Gallon
21.	2/4	27,432.3	27,758.2		15.5	
22.	2/9	27,758.2	28,058.1		13.4	
23.	2/16	28,058.1	28,396.7		16.2	
24.	2/20	28,396.7	28,704.5		13.3	

Source: Author's car records.

Find the best buy (based on the cost per unit) for each item. See Example 3.
(Sources: Cub Foods, Target, Rainbow Foods.)

25. Black pepper

26. Shampoo

27. Cereal
12 ounces for $2.49
14 ounces for $2.89
18 ounces for $3.96

28. Soup (same size cans)
2 cans for $2.18
3 cans for $3.57
5 cans for $5.29

29. Chunky peanut butter
12 ounces for $1.29
18 ounces for $1.79
28 ounces for $3.39
40 ounces for $4.39

30. Baked beans
8 ounces for $0.59
16 ounces for $0.99
21 ounces for $1.29
28 ounces for $1.89

31. Suppose you are choosing between two brands of chicken noodle soup. Brand A is $0.88 per can and Brand B is $0.98 per can. The cans are the same size, but Brand B has more chunks of chicken in it. Which soup is the best buy? Explain your choice.

32. A small bag of potatoes costs $0.19 per pound. A large bag costs $0.15 per pound. But there are only two people in your family, so half the large bag would probably spoil before you used it up. Which bag is the best buy? Explain.

Solve each application problem. See Examples 2–4.

33. Makesha lost 10.5 pounds in six weeks. What was her rate of loss in pounds per week?

34. Enrique's taco recipe uses three pounds of meat to feed 10 people. Give the rate in pounds per person.

35. Russ works 7 hours to earn $85.82. What is his pay rate per hour?

36. Find the cost of 1 gallon of Hawaiian Punch beverage if 18 gallons for a graduation party cost $55.62.

The table lists information about three long-distance calling cards. The connection fee is charged each time you make a call, no matter how long the call lasts. Use the information in the table to answer Exercises 37–40. Round answers to the nearest thousandth when necessary.

LONG-DISTANCE CALLING CARDS (U.S.)

Card Name	Cost per Minute	Connection Fee
Radiant Penny	$0.01	$0.39
IDT Special	$0.022	$0.14
Access America	$0.047	$0.00

Source: www.1callcard.com

37. (a) Find the *actual* total cost, including the connection charge, for a five-minute call using each card.

(b) Find the cost per minute for this call using each card and select the best buy.

38. (a) Find the *actual* total cost, including the connection charge, for a 30-minute call using each card.

(b) Find the cost per minute for this call using each card and select the best buy.

39. Find the *actual* total cost per minute for a 15-minute call and a 20-minute call using each card. Then select the best buy for each call.

40. All the cards round calls up to the next full minute.

(a) Suppose you call the wrong number. How much would you pay for this 40-*second* call on each card?

(b) How much would you save on this call by using Access America instead of Radiant Penny?

41. In the 2000 Olympics, Michael Johnson ran the 400-meter event in a record time of approximately 44 seconds (actually 43.84 seconds). Give his rate in seconds per meter and in meters per second. Use 44 seconds as the time. (*Source:* www.Olympics.com)

42. Sofia can clean and adjust five hearing aids in four hours. Give her rate in hearing aids per hour and in hours per hearing aid.

43. If you believe the claims that some batteries last longer, which is the best buy?

44. Which is the best buy, assuming these laundry detergents both clean equally well?

45. Three brands of cornflakes are available. Brand G is priced at \$2.39 for 10 ounces. Brand K is \$3.99 for 20.3 ounces and Brand P is \$3.39 for 16.5 ounces. You have a coupon for 50¢ off Brand P and a coupon for 60¢ off Brand G. Which cereal is the best buy based on cost per unit?

46. Two brands of facial tissue are available. Brand K has a special price of \$5 for three boxes of 175 tissues each. Brand S is priced at \$1.29 per box of 125 tissues. You have a coupon for 20¢ off one box of Brand S and a coupon for 45¢ off one box of Brand K. How can you get the best buy on one box of tissue?

Relating Concepts (Exercises 47–50) For Individual or Group Work

On the first page of this chapter, we said that unit rates can help you get the best deal on cell phone service. Use the information in the table to **work Exercises 47–50 in order.**

CELL PHONE SERVICE PLANS

Company	Anytime Minutes	One-Time Activation Fee	Monthly Charge	Termination Fee
Verizon	400	\$35	\$59.99	\$175
T-Mobile	600	\$35	\$39.99	\$200
Nextel	500	\$35	\$45.99	\$200
Sprint	500	\$36	\$55	\$150

Source: Advertisements appearing in *Minneapolis Star Tribune*.

Notes:

1. All companies require that you sign a contract for one year of service and charge a one-time termination fee if you quit early.

2. Unused minutes cannot be carried over to the next month.

47. How much will each company's activation fee cost you on a monthly basis during the one-year contract?

48. All the plans allow unlimited calls on nights and weekends, so you would be using the "anytime minutes" on weekdays. Figure out the average number of weekdays per month. Then, for each plan, how many minutes could you use per weekday? Round to the nearest whole minute.

49. Find the actual average cost per "anytime minute" during the one-year contract for each company, including the activation fee. Assume you use all the minutes and no more. Decide how to round your answers so you can find the best buy.

50. Suppose that after two months you canceled your service because you found that you only used 100 "anytime minutes" per month. Under those conditions, find the actual cost per "anytime minute" for each company, to the nearest cent.

6.3 ▶▶▶ Proportions

OBJECTIVE 1 Write proportions. A **proportion** states that two ratios (or rates) are equivalent. For example,

$$\frac{\$20}{4 \text{ hours}} = \frac{\$40}{8 \text{ hours}}$$

is a proportion that says the rate $\dfrac{\$20}{4 \text{ hours}}$ is equivalent to the rate $\dfrac{\$40}{8 \text{ hours}}$.
As the amount of money doubles, the number of hours also doubles. This proportion is read:

20 dollars **is to** 4 hours   **as**   40 dollars **is to** 8 hours

---

EXAMPLE 1 Writing Proportions

Write each proportion.

(a) 6 feet is to 11 feet **as** 18 feet is to 33 feet.

$$\frac{6 \text{ feet}}{11 \text{ feet}} = \frac{18 \text{ feet}}{33 \text{ feet}} \quad \text{so} \quad \frac{6}{11} = \frac{18}{33}$$ The common units (feet) divide out and are not written.

(b) $9 is to 6 liters **as** $3 is to 2 liters.

$$\frac{\$9}{6 \text{ liters}} = \frac{\$3}{2 \text{ liters}}$$ The units do *not* match so you must write them in the proportion.

Work Problem **1** *at the Side.* ▶

OBJECTIVE 2 Determine whether proportions are true or false. There are two ways to see whether a proportion is true. One way is to *write both of the ratios in lowest terms*.

EXAMPLE 2 Writing Both Ratios in Lowest Terms

Determine whether each proportion is true or false by writing both ratios in lowest terms.

(a) $\dfrac{5}{9} = \dfrac{18}{27}$

Write each ratio in lowest terms.

$$\frac{5}{9} \leftarrow \begin{array}{l}\text{Already in} \\ \text{lowest terms}\end{array} \qquad \frac{18 \div 9}{27 \div 9} = \frac{2}{3} \leftarrow \text{Lowest terms}$$

Because $\frac{5}{9}$ is *not* equivalent to $\frac{2}{3}$, the proportion is *false*. The ratios are *not* proportional.

(b) $\dfrac{16}{12} = \dfrac{28}{21}$

Write each ratio in lowest terms.

$$\frac{16 \div 4}{12 \div 4} = \frac{4}{3} \quad \text{and} \quad \frac{28 \div 7}{21 \div 7} = \frac{4}{3}$$

Both ratios are equivalent to $\frac{4}{3}$, so the proportion is *true*. The ratios are proportional.

Work Problem **2** *at the Side.* ▶

---

1 Write each proportion.

(a) $7 is to 3 cans as $28 is to 12 cans.

(b) 9 meters is to 16 meters as 18 meters is to 32 meters.

(c) 5 is to 7 as 35 is to 49.

(d) 10 is to 30 as 60 is to 180.

2 Determine whether each proportion is true or false by writing both ratios in lowest terms.

(a) $\dfrac{6}{12} = \dfrac{15}{30}$

(b) $\dfrac{20}{24} = \dfrac{3}{4}$

(c) $\dfrac{25}{40} = \dfrac{30}{48}$

(d) $\dfrac{35}{45} = \dfrac{12}{18}$

ANSWERS

1. **(a)** $\dfrac{\$7}{3 \text{ cans}} = \dfrac{\$28}{12 \text{ cans}}$ **(b)** $\dfrac{9}{16} = \dfrac{18}{32}$

   (c) $\dfrac{5}{7} = \dfrac{35}{49}$ **(d)** $\dfrac{10}{30} = \dfrac{60}{180}$

2. **(a)** $\dfrac{1}{2} = \dfrac{1}{2}$; true **(b)** $\dfrac{5}{6} \neq \dfrac{3}{4}$; false

   (c) $\dfrac{5}{8} = \dfrac{5}{8}$; true **(d)** $\dfrac{7}{9} \neq \dfrac{2}{3}$; false

A second way to see whether a proportion is true is to find *cross products*.

Using Cross Products to Determine Whether a Proportion Is True

To see whether a proportion is true, first multiply along one diagonal, then multiply along the other diagonal, as shown here.

$$5 \cdot 4 = 20$$

$$\frac{2}{5} = \frac{4}{10}$$

Cross products are equal.

$$2 \cdot 10 = 20$$

In this case the **cross products** are both 20. When cross products are *equal,* the proportion is *true.* If the cross products are *unequal,* the proportion is *false.*

Note

Why does the cross products test work? It is based on rewriting both fractions with a common denominator of $5 \cdot 10$ or 50. (We do not search for the *lowest* common denominator. We simply use the product of the two given denominators.)

$$\frac{2 \cdot 10}{5 \cdot 10} = \frac{20}{50} \quad \text{and} \quad \frac{4 \cdot 5}{10 \cdot 5} = \frac{20}{50}$$

We see that $\frac{2}{5}$ and $\frac{4}{10}$ are equivalent because both can be rewritten as $\frac{20}{50}$. The cross products test takes a shortcut by comparing only the two numerators ($20 = 20$).

EXAMPLE 3 **Using Cross Products**

Use cross products to see whether each proportion is true or false.

(a) $\dfrac{3}{5} = \dfrac{12}{20}$ Multiply along one diagonal and then multiply along the other diagonal.

$$5 \cdot 12 = 60$$

$$\frac{3}{5} = \frac{12}{20}$$

Equal — When cross products are equal, the proportion is true.

$$3 \cdot 20 = 60$$

The cross products are *equal,* so the proportion is *true.*

CAUTION

Use cross products *only* when working with *proportions.* Do **not** use cross products when multiplying fractions, adding fractions, or writing fractions in lowest terms.

Continued on Next Page

(b) $\dfrac{2\frac{1}{3}}{3\frac{1}{3}} = \dfrac{9}{16}$ Find the cross products.

> Write $3\frac{1}{3}$ as $\frac{10}{3}$ and write 9 as $\frac{9}{1}$.

$$3\frac{1}{3} \cdot 9 = \dfrac{10}{\cancel{3}} \cdot \dfrac{\cancel{9}^{3}}{1} = \dfrac{30}{1} = 30$$

$$\dfrac{2\frac{1}{3}}{3\frac{1}{3}} = \dfrac{9}{16}$$

Unequal; proportion is false.

$$2\frac{1}{3} \cdot 16 = \dfrac{7}{3} \cdot \dfrac{16}{1} = \dfrac{112}{3} = 37\frac{1}{3}$$

The cross products are *unequal,* so the proportion is *false.*

> **Note**
> The numbers in a proportion do *not* have to be whole numbers. They may be fractions, mixed numbers, decimal numbers, and so on.

Work Problem **3** *at the Side.* ▶

OBJECTIVE 3 Find the unknown number in a proportion. Four numbers are used in a proportion. If any three of these numbers are known, the fourth can be found. For example, find the unknown number that will make this proportion true.

$$\dfrac{3}{5} = \dfrac{x}{40}$$

The variable x represents the unknown number. First find the cross products.

$$\dfrac{3}{5} = \dfrac{x}{40}$$

$5 \cdot x$
$3 \cdot 40$ Cross products

To make the proportion true, the cross products must be equal. This gives us the following equation.

$$\underbrace{5 \cdot x}_{} = \underbrace{3 \cdot 40}_{}$$
$$5x = 120$$

Recall from **Section 2.4** that we can solve an equation of this type by dividing both sides by the coefficient of the variable term. In this case, the coefficient of $5x$ is 5.

$$\dfrac{5x}{5} = \dfrac{120}{5} \quad \leftarrow \text{Divide both sides by 5}$$

On the left side, divide out the common factor of 5; slashes indicate the division.

$$\dfrac{\cancel{5} \cdot x}{\cancel{5}} = 24$$

On the right side, divide 120 by 5 to get 24

$$x = 24$$

3 Find the cross products to see whether each proportion is true or false.

(a) $\dfrac{5}{9} = \dfrac{10}{18}$

(b) $\dfrac{32}{15} = \dfrac{16}{8}$

(c) $\dfrac{10}{17} = \dfrac{20}{34}$

(d) $\dfrac{2.4}{6} = \dfrac{5}{12}$

$(6)(5) =$

$(2.4)(12) =$

(e) $\dfrac{3}{4.25} = \dfrac{24}{34}$

(f) $\dfrac{1\frac{1}{6}}{2\frac{1}{3}} = \dfrac{4}{8}$

ANSWERS

3. **(a)** $90 = 90$; true
   (b) $240 \neq 256$; false
   (c) $340 = 340$; true
   (d) $(6)(5) = 30$; $(2.4)(12) = 28.8$; false
   (e) $102 = 102$; true
   (f) $9\frac{1}{3} = 9\frac{1}{3}$; true

The unknown number in the proportion is 24. The complete proportion is shown below.

$$\frac{3}{5} = \frac{24}{40} \quad \leftarrow x \text{ is } 24$$

Check by finding the cross products. If they are equal, you solved the problem correctly. If they are unequal, rework the problem.

$$5 \cdot 24 = \mathbf{120} \quad \leftarrow$$

$$\frac{3}{5} = \frac{24}{40}$$

Equal; proportion is true.

$$3 \cdot 40 = \mathbf{120} \quad \leftarrow$$

The solution is 24, **not** 120.

The cross products are equal, so the solution, $x = 24$, is correct.

> **CAUTION**
> The solution is 24, which is the unknown number in the proportion. 120 is *not* the solution; it is the cross product you get when *checking* the solution.

Solve a proportion for an unknown number by using the following steps.

> **Solving a Proportion to Find an Unknown Number**
>
> **Step 1** Find the cross products.
>
> **Step 2** Show that the cross products are equivalent.
>
> **Step 3** Divide both sides of the equation by the coefficient of the variable term.
>
> **Step 4** Check by writing the solution in the *original* proportion and finding the cross products.

EXAMPLE 4 **Solving Proportions**

Find the unknown number in each proportion. Round answers to the nearest hundredth when necessary.

(a) $\dfrac{16}{x} = \dfrac{32}{20}$ You can write $\frac{32}{20}$ in lowest terms as $\frac{8}{5}$

Recall that ratios can be rewritten in lowest terms. If desired, you can do that *before* finding the cross products. In this example, write $\frac{32}{20}$ in lowest terms as $\frac{8}{5}$, which gives the proportion $\dfrac{16}{x} = \dfrac{8}{5}$.

Step 1

$$x \cdot 8 \quad \leftarrow$$

$$\frac{16}{x} = \frac{8}{5}$$

Find the cross products.

$$16 \cdot 5 \quad \leftarrow$$

Continued on Next Page

Step 2 $x \cdot 8 = 16 \cdot 5$ Show that the cross products are equivalent.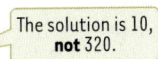

$x \cdot 8 = 80$

Step 3 $\dfrac{x \cdot \overset{1}{\cancel{8}}}{\underset{1}{\cancel{8}}} = \dfrac{80}{8}$ ← Divide both sides by 8

$x = 10$ ← Find x. (No rounding is necessary.)

Step 4 Write the solution in the *original* proportion and check by finding the cross products.

$10 \cdot 32 = \textbf{320}$ ←

x is 10 → $\dfrac{16}{10} = \dfrac{32}{20}$ Equal; proportion is true.

$16 \cdot 20 = \textbf{320}$ ←

The cross products are equal, so 10 is the correct solution.

> The solution is 10, **not** 320.

> **Note**
> It is not necessary to write the ratios in lowest terms before solving. However, if you do, you will have smaller numbers to work with.

(b) $\dfrac{7}{12} = \dfrac{15}{x}$

$12 \cdot 15 = 180$ ←

Step 1 $\dfrac{7}{12} = \dfrac{15}{x}$ Find the cross products.

$7 \cdot x$ ←

Step 2 $7 \cdot x = 180$ ← Show that cross products are equivalent.

Step 3 $\dfrac{\overset{1}{\cancel{7}} \cdot x}{\underset{1}{\cancel{7}}} = \dfrac{180}{7}$ ← Divide both sides by 7

$x \approx 25.71$ ← Rounded to nearest hundredth

When the division does not come out even, check for directions on how to round your answer. Divide out one more place, then round.

$\begin{array}{r} 25.714 \\ 7\overline{)180.000} \end{array}$ ← Divide out to thousandths so you can round to hundredths.

Step 4 Write the solution in the original proportion and check by finding the cross products.

$(12)(15) = \textbf{180}$ ←

$\dfrac{7}{12} \approx \dfrac{15}{25.71}$ Very close, but *not* equal due to rounding the solution

$(7)(25.71) = \textbf{179.97}$ ←

The cross products are slightly different because you rounded the value of x. However, they are close enough to see that the problem was done correctly and 25.71 is the approximate solution (**not** 179.97).

Work Problem **4** at the Side. ▶

4 Find the unknown numbers. Round to hundredths when necessary. Check your answers by finding the cross products.

(a) $\dfrac{1}{2} = \dfrac{x}{12}$

(b) $\dfrac{6}{10} = \dfrac{15}{x}$

(c) $\dfrac{28}{x} = \dfrac{21}{9}$

(d) $\dfrac{x}{8} = \dfrac{3}{5}$

(e) $\dfrac{14}{11} = \dfrac{x}{3}$

ANSWERS

4. **(a)** $x = 6$ **(b)** $x = 25$
   (c) $x = 12$ **(d)** $x = 4.8$
   (e) $x \approx 3.82$ (rounded to nearest hundredth)

The next example shows how to solve for the unknown number in a proportion with fractions or decimals.

EXAMPLE 5 **Solving Proportions with Mixed Numbers and Decimals**

Find the unknown number in each proportion.

(a) $\dfrac{2\frac{1}{5}}{6} = \dfrac{x}{10}$

Step 1
$$\dfrac{2\frac{1}{5}}{6} = \dfrac{x}{10}$$

$6 \cdot x$

$2\frac{1}{5} \cdot 10$

Find the cross products.

Find $2\frac{1}{5} \cdot 10$.

$$2\frac{1}{5} \cdot 10 = \dfrac{11}{5} \cdot \dfrac{10}{1} = \dfrac{11 \cdot 2 \cdot \overset{1}{\cancel{5}}}{\underset{1}{\cancel{5}} \cdot 1} = \dfrac{22}{1} = 22$$

Changed to improper fraction

Step 2 $6 \cdot x = 22$ ← Show that cross products are equivalent.

Step 3 $\dfrac{\overset{1}{\cancel{6}} \cdot x}{\underset{1}{\cancel{6}}} = \dfrac{22}{6}$ ← Divide both sides by 6.

Write the solution as a mixed number in lowest terms.

$$x = \dfrac{22 \div 2}{6 \div 2} = \dfrac{11}{3} = 3\frac{2}{3}$$

Step 4 Write the solution in the original proportion and check by finding the cross products.

$$6 \cdot 3\frac{2}{3} = \dfrac{2 \cdot \overset{1}{\cancel{3}}}{1} \cdot \dfrac{11}{\underset{1}{\cancel{3}}} = \dfrac{22}{1} = \mathbf{22}$$

$$\dfrac{2\frac{1}{5}}{6} = \dfrac{3\frac{2}{3}}{10}$$

$$2\frac{1}{5} \cdot 10 = \dfrac{11}{\underset{1}{\cancel{5}}} \cdot \dfrac{2 \cdot \overset{1}{\cancel{5}}}{1} = \dfrac{22}{1} = \mathbf{22}$$

Equal

The cross products are equal, so $3\frac{2}{3}$ is the correct solution.

> The solution is $3\frac{2}{3}$, **not** 22.

Continued on Next Page

Note

You can use decimal numbers and your calculator to solve Example 5(a) on the previous page. $2\frac{1}{5}$ is equivalent to 2.2, so the cross products are

$$6 \cdot x = (2.2)(10)$$

$$\frac{\overset{1}{\cancel{6}} \cdot x}{\underset{1}{\cancel{6}}} = \frac{22}{6}$$

When you divide 22 by 6 on your calculator, it shows 3.666666667. Write the answer using a bar to show the repeating digit: $3.\overline{6}$. Or round the answer to 3.67 (nearest hundredth).

(b) $\dfrac{1.5}{0.6} = \dfrac{2}{x}$

$(1.5)(x) = (0.6)(2)$ ← Show that cross products are equivalent.

$(1.5)(x) = 1.2$

$$\frac{(\cancel{1.5})(x)}{\cancel{1.5}} = \frac{1.2}{1.5} \quad \longleftarrow \text{ Divide both sides by 1.5.}$$

$$x = \frac{1.2}{1.5} \qquad \text{Complete the division.} \quad 1.5\overline{)1.20}\,\,^{.8}$$

$$x = 0.8$$

So the solution is 0.8.

Write the solution in the original equation, and check by finding the cross products.

$$(0.6)(2) = \mathbf{1.2}$$

$$\frac{\mathbf{1.5}}{\mathbf{0.6}} = \frac{\mathbf{2}}{\mathbf{0.8}} \qquad \text{Equal}$$

$$(1.5)(0.8) = \mathbf{1.2}$$

The cross products are equal, so 0.8 is the correct solution. The solution is 0.8 (**not** 1.2)

Work Problem **5** *at the Side.* ▶

5 Find the unknown numbers. Round to hundredths on the decimal problems when necessary. Check your solutions by finding the cross products.

(a) $\dfrac{3\frac{1}{4}}{2} = \dfrac{x}{8}$

(b) $\dfrac{x}{3} = \dfrac{1\frac{2}{3}}{5}$

(c) $\dfrac{0.06}{x} = \dfrac{0.3}{0.4}$

(d) $\dfrac{2.2}{5} = \dfrac{13}{x}$

(e) $\dfrac{x}{6} = \dfrac{0.5}{1.2}$

(f) $\dfrac{0}{2} = \dfrac{x}{7.092}$

ANSWERS

5. **(a)** $x = 13$ **(b)** $x = 1$ **(c)** $x = 0.08$
 (d) $x \approx 29.55$ (rounded to nearest hundredth)
 (e) $x = 2.5$ **(f)** $x = 0$

Math in the Media

FEEDING HUMMINGBIRDS

Filling Your Feeder

After getting a hummingbird feeder, the next step is to fill it! You have two choices at this point: you can either buy one of the commercial mixtures or you can make your own solution, shown at the right.

The concentration of the sugar is important. The 1 to 4 ratio of sugar to water is recommended because it approximates the ratio of sugar to water found in the nectar of many hummingbird flowers.

Boiling the solution helps retard fermentation. Sugar-and-water solutions are subject to rapid spoiling, especially in hot weather.

Source: The Hummingbird Book.

> **Recipe for Homemade Mixture:**
> **1 part sugar (not honey)**
> **4 parts water**
> **Boil for 1 to 2 minutes. Cool.**
> **Store extra in refrigerator.**

A recipe can be used to make as much of a mixture as you need as long as the ingredients are kept proportional. Use the recipe for a homemade mixture of sugar water for hummingbird feeders to answer these problems.

1. What is the ratio of sugar to water in the recipe?

   What is the ratio of water to sugar in the recipe?

2. Complete each table.

Sugar	Water
1 cup	4 cups
	5 cups
	6 cups
	7 cups
2 cups	8 cups

Sugar	Water
1 cup	4 cups
	3 cups
	2 cups
	1 cup

3. How much water would you need if you used
   (a) 3 cups of sugar?
   (b) 4 cups of sugar?
   (c) $\frac{1}{3}$ cup of sugar?

4. As you change the amounts of water and sugar, should you change the length of time that you boil the mixture? Explain your answer.

6.3 ▶▶▶ Exercises

Write each proportion. See Example 1.

1. $9 is to 12 cans as $18 is to 24 cans.

2. 28 people is to 7 cars as 16 people is to 4 cars.

3. 200 adults is to 450 children as 4 adults is to 9 children.

4. 150 trees is to 1 acre as 1500 trees is to 10 acres.

5. 120 feet is to 150 feet as 8 feet is to 10 feet.

6. $6 is to $9 as $10 is to $15.

7. 2.2 hours is to 3.3 hours as 3.2 hours is to 4.8 hours.

8. 4 meters is to 4.75 meters as 6 meters is to 7.125 meters.

Determine whether each proportion is true *or* false *by writing the ratios in lowest terms. Show the simplified ratios and then write* true *or* false. *See Example 2.*

9. $\dfrac{6}{10} = \dfrac{3}{5}$

10. $\dfrac{1}{4} = \dfrac{9}{36}$

11. $\dfrac{5}{8} = \dfrac{25}{40}$

12. $\dfrac{2}{3} = \dfrac{20}{27}$

13. $\dfrac{150}{200} = \dfrac{200}{300}$

14. $\dfrac{100}{120} = \dfrac{75}{100}$

In Exercises 15–26, use cross products to determine whether each proportion is true *or* false. *Show the cross products and then circle* True *or* False. *See Example 3.*

15. $\dfrac{2}{9} = \dfrac{6}{27}$

True   False

16. $\dfrac{20}{25} = \dfrac{4}{5}$

True   False

17. $\dfrac{20}{28} = \dfrac{12}{16}$

True   False

18. $\dfrac{16}{40} = \dfrac{22}{55}$

True   False

19. $\dfrac{110}{18} = \dfrac{160}{27}$

True   False

20. $\dfrac{600}{420} = \dfrac{20}{14}$

True   False

21. $\dfrac{3.5}{4} = \dfrac{7}{8}$

True   False

22. $\dfrac{36}{23} = \dfrac{9}{5.75}$

True   False

23. $\dfrac{18}{15} = \dfrac{2\frac{5}{6}}{2\frac{1}{2}}$

True   False

24. $\dfrac{1\frac{3}{10}}{3\frac{1}{5}} = \dfrac{4}{9}$

True   False

25. $\dfrac{6}{3\frac{2}{3}} = \dfrac{18}{11}$

True   False

26. $\dfrac{16}{13} = \dfrac{2}{1\frac{5}{8}}$

True   False

27. Suppose Joe Mauer of the Minnesota Twins had 17 hits in 50 times at bat, and Freddy Sanchez of the Pittsburgh Pirates was at bat 450 times and got 153 hits. Paul is trying to convince Jamie that the two men hit equally well. Show how you could use a proportion and cross products to see whether Paul is correct.

28. Jay worked 3.5 hours and packed 91 cartons. Craig packed 126 cartons in 5.25 hours. To see if the men worked equally fast, Barry set up this proportion:

$$\dfrac{3.5}{91} = \dfrac{126}{5.25}$$

Explain what is wrong with Barry's proportion and write a correct one. Is the correct proportion true or false?

Solve each proportion to find the unknown number. Round your answers to hundredths when necessary. Check your answers by finding cross products. See Examples 4 and 5.

29. $\dfrac{1}{3} = \dfrac{x}{12}$

30. $\dfrac{x}{6} = \dfrac{15}{18}$

● **31.** $\dfrac{15}{10} = \dfrac{3}{x}$

32. $\dfrac{5}{x} = \dfrac{20}{8}$

33. $\dfrac{x}{11} = \dfrac{32}{4}$

34. $\dfrac{12}{9} = \dfrac{8}{x}$

35. $\dfrac{42}{x} = \dfrac{18}{39}$

36. $\dfrac{49}{x} = \dfrac{14}{18}$

37. $\dfrac{x}{25} = \dfrac{4}{20}$

38. $\dfrac{6}{x} = \dfrac{4}{8}$

● **39.** $\dfrac{8}{x} = \dfrac{24}{30}$

40. $\dfrac{32}{5} = \dfrac{x}{10}$

41. $\dfrac{99}{55} = \dfrac{44}{x}$

42. $\dfrac{x}{12} = \dfrac{101}{147}$

43. $\dfrac{0.7}{9.8} = \dfrac{3.6}{x}$

44. $\dfrac{x}{3.6} = \dfrac{4.5}{6}$

45. $\dfrac{250}{24.8} = \dfrac{x}{1.75}$

46. $\dfrac{4.75}{17} = \dfrac{43}{x}$

Find the unknown number in each proportion. Write your answers as whole or mixed numbers when possible. See Example 5.

47. $\dfrac{15}{1\frac{2}{3}} = \dfrac{9}{x}$

48. $\dfrac{x}{\frac{3}{10}} = \dfrac{2\frac{2}{9}}{1}$

49. $\dfrac{2\frac{1}{3}}{1\frac{1}{2}} = \dfrac{x}{2\frac{1}{4}}$

50. $\dfrac{1\frac{5}{6}}{x} = \dfrac{\frac{3}{14}}{\frac{6}{7}}$

Solve each proportion two different ways. First change all the numbers to decimal form and solve. Then change all the numbers to fraction form and solve; write your answers in lowest terms.

51. $\dfrac{\frac{1}{2}}{x} = \dfrac{2}{0.8}$

52. $\dfrac{\frac{3}{20}}{0.1} = \dfrac{0.03}{x}$

53. $\dfrac{x}{\frac{3}{50}} = \dfrac{0.15}{1\frac{4}{5}}$

54. $\dfrac{8\frac{4}{5}}{1\frac{1}{10}} = \dfrac{x}{0.4}$

Relating Concepts (Exercises 55–56)   For Individual or Group Work

Work Exercises 55–56 in order. First prove that the proportions are **not** true. Then create four true proportions for each exercise by changing only one number at a time.

55. $\dfrac{10}{4} = \dfrac{5}{3}$

56. $\dfrac{6}{8} = \dfrac{24}{30}$

Summary Exercises on Ratios, Rates, and Proportions

Use the circle graph of one college's enrollment to complete Exercises 1–4. Write each ratio as a fraction in lowest terms.

1. Write the ratio of freshmen to juniors.

2. What is the ratio of freshmen to the total college enrollment?

3. Find the ratio of seniors and sophomores to juniors.

4. Write the ratio of freshmen and sophomores to juniors and seniors.

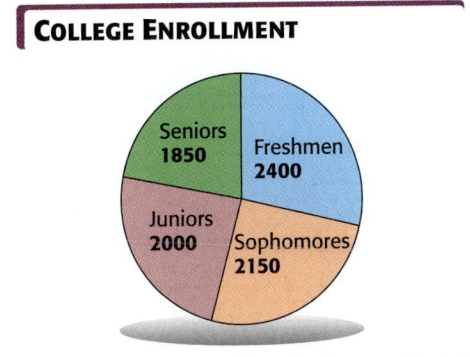

COLLEGE ENROLLMENT

Seniors 1850 | Freshmen 2400 | Juniors 2000 | Sophomores 2150

The bar graph shows the number of Americans who play various instruments. Use the graph to complete Exercises 5 and 6. Write each ratio as a fraction in lowest terms.

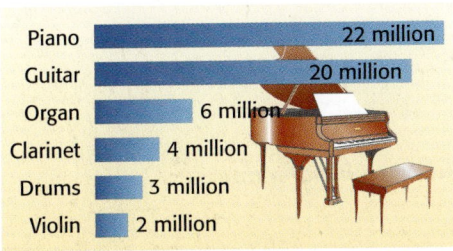

AMERICANS MAKE MUSIC BY THE MILLIONS

How many people play each type of instrument?

Piano	22 million
Guitar	20 million
Organ	6 million
Clarinet	4 million
Drums	3 million
Violin	2 million

Source: *America by the Numbers.*

5. Write six ratios that compare the least popular instrument to each of the other instruments.

6. Which two instruments give each of these ratios:
   (a) $\frac{5}{1}$; **(b)** $\frac{2}{1}$? There may be more than one correct answer.

The table lists data on the top three individual scoring NBA basketball games of all time. Use the data to answer Exercises 7 and 8. Round answers to the nearest tenth.

7. What was Wilt Chamberlain's scoring rate in points per minute and in minutes per point?

Player	Date	Points	Min
Wilt Chamberlain	3/2/62	100	48
David Thompson	4/9/78	73	43
David Robinson	4/24/94	71	44

Source: www.NBA.com

8. Find David Robinson's scoring rate in points per minute and minutes per point.

Wilt Chamberlain (#13)

9. Lucinda's paycheck showed gross pay of $652.80 for 40 hours of work and $195.84 for 8 hours of overtime work. Find her regular hourly pay rate and her overtime rate.

10. Satellite TV is being offered to new subscribers at $25 per month for 50 channels, $39 per month for 100 channels, or $48 per month for 150 channels. What is the monthly cost per channel under each plan? (*Source:* Dish1Up Satellites.)

11. Find the best buy on gourmet coffee beans.

11 ounces for $6.79

12 ounces for $7.24

16 ounces for $10.99

(*Source:* Cub Foods.)

12. Which brand of cat food is the best buy? You have a coupon for $2 off on Brand P, and another for $1 off on Brand N.

Brand N is $3.75 for 3.5 pounds

Brand P is $5.99 for 7 pounds

Brand R is $10.79 for 18 pounds

Use either the method of writing in lowest terms or the method of finding cross products to decide whether each proportion is true *or* false. *Show your work and then write* true *or* false.

13. $\dfrac{28}{21} = \dfrac{44}{33}$

14. $\dfrac{2.3}{8.05} = \dfrac{0.25}{0.9}$

15. $\dfrac{2\frac{5}{8}}{3\frac{1}{4}} = \dfrac{21}{26}$

Solve each proportion to find the unknown number. Round your answers to hundredths when necessary.

16. $\dfrac{7}{x} = \dfrac{25}{100}$

17. $\dfrac{15}{8} = \dfrac{6}{x}$

18. $\dfrac{x}{84} = \dfrac{78}{36}$

19. $\dfrac{10}{11} = \dfrac{x}{4}$

20. $\dfrac{x}{17} = \dfrac{3}{55}$

21. $\dfrac{2.6}{x} = \dfrac{13}{7.8}$

22. $\dfrac{0.14}{1.8} = \dfrac{x}{0.63}$

23. $\dfrac{\frac{1}{3}}{8} = \dfrac{x}{24}$

24. $\dfrac{6\frac{2}{3}}{4\frac{1}{6}} = \dfrac{6\frac{5}{}}{x}$

6.4 ▷▷▷ Problem Solving with Proportions

OBJECTIVE 1 Use proportions to solve application problems.
Proportions can be used to solve a wide variety of problems. Watch for problems in which you are given a ratio or rate and then asked to find part of a corresponding ratio or rate. Remember that a ratio or rate compares two quantities and often includes one of these indicator words.

in   for   on   per   from   to

Use the six problem-solving steps from **Section 3.3.** When setting up the proportion, use a variable to represent the unknown number. We have used the letter *x*, but you may use any letter you like.

EXAMPLE 1 Solving a Proportion Application

Mike's car can travel 163 **miles** on 6.4 **gallons** of gas. How far can it travel on a full tank of 14 **gallons** of gas? Round to the nearest mile.

Step 1 **Read** the problem. It is about how far a car can travel on a certain amount of gas.

> Unknown: miles traveled on 14 gallons of gas
> Known: 163 miles traveled on 6.4 gallons of gas

Step 2 **Assign a variable.** There is only one unknown, so let *x* be the number of miles traveled on 14 gallons.

Step 3 **Write an equation.** The equation is in the form of a proportion. Decide what is being compared. This problem compares **miles** to **gallons.** Write the two rates described in the problem. Be sure that *both* rates compare miles to gallons in the same order. In other words, miles is in both numerators and gallons is in both denominators.

This rate compares **miles** to **gallons**.
$$\frac{163 \text{ miles}}{6.4 \text{ gallons}} = \frac{x \text{ miles}}{14 \text{ gallons}}$$
This rate compares **miles** to **gallons**. (Matching units)

Step 4 **Solve** the equation. Ignore the units while finding the cross products.

$$\frac{163 \text{ miles}}{6.4 \text{ gallons}} = \frac{x \text{ miles}}{14 \text{ gallons}}$$

$(6.4)(x) = (163)(14)$ Show that cross products are equivalent.

$(6.4)(x) = 2282$

$\frac{(6.4)(x)}{6.4} = \frac{2282}{6.4}$ Divide both sides by 6.4

$x = 356.5625$ Round to 357 *Always check the problem for rounding directions.*

Step 5 **State the answer.** Mike's car can travel 357 miles, rounded to the nearest mile, on a full tank of gas.

Continued on Next Page

1 Set up and solve a proportion for each problem using the six problem-solving steps.

(a) If 2 pounds of fertilizer will cover 50 square feet of garden, how many pounds are needed for 225 square feet?

(b) A U.S. map has a scale of 1 inch to 75 miles. Lake Superior is 4.75 inches long on the map. What is the lake's actual length, to the nearest whole mile?

(c) Cough syrup is to be given at the rate of 30 milliliters for each 100 pounds of body weight. How much should be given to a 34-pound child? Round to the nearest whole milliliter.

Step 6 **Check** that the solution is reasonable and fits the facts given in the original statement of the problem. The car traveled 163 miles on 6.4 gallons of gas; 14 gallons is a little more than *twice as much* gas, so the car should travel a little more than *twice as far.*

$$(2)(163 \text{ miles}) = 326 \text{ miles} \leftarrow \text{Estimate}$$

The solution, 357 miles, is a little more than the estimate of 326 miles, so it is reasonable.

CAUTION

When setting up the proportion, do *not* mix up the units in the rates.

$$\left. \begin{array}{c} \text{compares } \textbf{miles} \\ \text{to } \textbf{gallons} \end{array} \right\} \quad \dfrac{163 \textbf{ miles}}{6.4 \textbf{ gallons}} \neq \dfrac{14 \textbf{ gallons}}{x \textbf{ miles}} \quad \left\{ \begin{array}{c} \text{compares } \textbf{gallons} \\ \text{to } \textbf{miles} \end{array} \right.$$

These rates do *not* compare things in the same order and *cannot* be set up as a proportion.

◀ *Work Problem* **1** *at the Side.*

EXAMPLE 2 **Solving a Proportion Application**

A newspaper report says that 7 out of 10 people surveyed watch the news on TV. At that rate, how many of the 3200 people in town would you expect to watch the news?

Step 1 **Read** the problem. It is about people watching the news on TV.

> Unknown: how many people in town are expected to watch the news on TV
> Known: 7 out of 10 people surveyed watched the news on TV.

Step 2 **Assign a variable.** There is only one unknown, so let x be the number of people in town who watch the news on TV.

Step 3 **Write an equation.** Set up the two rates as a proportion. You are comparing people who watch the news to people surveyed. Write the two rates described in the example. Be sure that both rates make the same comparison. "People who watch the news" is mentioned first, so it should be in the numerator of *both* rates.

$$\text{People who watch the news} \rightarrow \dfrac{7}{10} = \dfrac{x}{3200} \leftarrow \text{People who watch the news}$$
$$\begin{array}{c} \text{Total group} \rightarrow \\ \text{(people surveyed)} \end{array} \qquad \qquad \begin{array}{c} \leftarrow \text{Total group} \\ \text{(people in town)} \end{array}$$

Step 4 **Solve** the equation.

$$\dfrac{7}{10} = \dfrac{x}{3200}$$

$$(10)(x) = (7)(3200) \qquad \text{Show that cross products are equivalent.}$$

$$(10)(x) = 22{,}400$$

$$\dfrac{\cancel{10}^{1}(x)}{\cancel{10}_{1}} = \dfrac{22{,}400}{10} \qquad \text{Divide both sides by 10}$$

$$x = 2240 \quad \boxed{\text{No rounding is needed here.}}$$

Continued on Next Page

Step 5   **State the answer.** You would expect 2240 people in town to watch the news on TV.

Step 6   **Check** that the solution is reasonable by putting it back into the original statement of the problem.

Notice that 7 out of 10 people is more than half the people, but less than all the people. Half of the 3200 people in town is $3200 \div 2 = 1600$, so between 1600 and 3200 people would be expected to watch the news on TV. The solution, 2240 people, is between 1600 and 3200, so it is reasonable.

CAUTION
Always check that your answer is reasonable. If it isn't, look at the way your proportion is set up. Be sure you have matching units in the numerators and matching units in the denominators.

For example, suppose you set up the last proportion *incorrectly,* as shown below.

$$\frac{7}{10} = \frac{3200}{x} \quad \leftarrow \textbf{Incorrect setup}$$

$$(7)(x) = (10)(3200)$$

$$\frac{\overset{1}{\cancel{(7)}}(x)}{\underset{1}{\cancel{7}}} = \frac{32{,}000}{7}$$

$$x \approx 4571 \text{ people} \quad \leftarrow \textbf{Unreasonable answer}$$

This answer is ***unreasonable*** because there are only 3200 people in the town; it is ***not*** possible for 4571 people to watch the news.

Work Problem **2** *at the Side.* ▶

2 Solve each problem to find a reasonable answer. Then flip one side of your proportion to see what answer you get with an **incorrect** setup. Explain why the second answer is **unreasonable.**

(a) A survey showed that 2 out of 3 people would like to lose weight. At this rate, how many people in a group of 150 want to lose weight?

(b) In one state, 3 out of 5 college students receive financial aid. At this rate, how many of the 4500 students at Central Community College receive financial aid?

(c) An advertisement says that 9 out of 10 dentists recommend sugarless gum. If the ad is true, how many of the 60 dentists in our city would recommend sugarless gum?

ANSWERS

2. **(a)** 100 people (reasonable); incorrect setup gives 225 people (only 150 people in the group).
   (b) 2700 students (reasonable); incorrect setup gives 7500 students (only 4500 students at the college).
   (c) 54 dentists (reasonable); incorrect setup gives ≈ 67 dentists (only 60 dentists in the city).

Math in the Media

CURRENCY EXCHANGE

When you travel between countries, you will exchange U.S. dollars for the local currency. The exchange rate between currencies changes daily, and you can easily find the updated rates using the Internet or any major newspaper. The table below has been extracted from the Oanda Web page, www.oanda.com. It shows how much of each country's currency was equivalent to 1 U.S. dollar on April 16, 2008.

NORTH AMERICA/CARIBBEAN CURRENCY RATES
(APRIL 16, 2008)

Currency	Symbol	Value
Canadian dollar	CAD	1.0198
Cayman Islands dollar	KYD	0.833
Jamaican dollar	JMD	74.75
Mexican peso	MXN	10.5
United States dollar	USD	1.00

From the table, $1.00 U.S. was equivalent to 10.5 Mexican pesos. You can set up a proportion to convert dollars to pesos. For example, suppose you want to determine the number of pesos that is equivalent to $50.00.

$$\frac{\$1}{10.5 \text{ pesos}} = \frac{\$50}{x \text{ pesos}} \quad \text{or} \quad \frac{1}{10.5} = \frac{50}{x}$$
$$(1)(x) = (10.5)(50)$$
$$x = 525.0 \text{ pesos}$$

So $50 buys 525 pesos.

1. Based on the currency exchange rates for April 16, 2008, find the amount of each local currency that is equivalent to $50 U.S. and find the number of U.S. dollars that is equivalent to 200 units of each local currency. Round your answers to the nearest hundredth.

   (a) $50 = _____ Canadian dollars, and

   200 Canadian dollars = _____ U.S. dollars.

   (b) $50 = _____ Cayman Islands dollars, and

   200 Cayman Islands dollars = _____ U.S. dollars.

   (c) $50 = _____ Jamaican dollars, and

   200 Jamaican dollars = _____ U.S. dollars.

2. Set up a proportion to find the number of U.S. dollars that was equivalent to 1 Mexican Peso. Round your answer to the nearest cent.
   1 Mexican peso was equivalent to $_____ (U.S.).

3. From Problem 2, you should recognize the conversion rate based on 1 Mexican peso as the expression $\frac{1}{10.5}$. What is the mathematical word that describes the relationship between the conversion rates 10.5 and $\frac{1}{10.5}$?

6.4 ▶▶▶   **Exercises**

FOR
EXTRA
HELP

MyMathLab   Math XL
PRACTICE   WATCH   DOWNLOAD   READ   REVIEW

Set up and solve a proportion for each application problem. See Example 1.

1. Caroline can sketch four cartoon strips in five hours. How long will it take her to sketch 18 strips?

2. The Cosmic Toads recorded eight songs on their first CD in 26 hours. How long will it take them to record 14 songs for their second CD?

🌐 **3.** Sixty newspapers cost $27. Find the cost of 16 newspapers.

4. Twenty-two guitar lessons cost $528. Find the cost of 12 lessons.

▦ **5.** If three pounds of fescue grass seed cover about 350 square feet of ground, how many pounds are needed for 4900 square feet?

▦ **6.** Anna earns $1242.08 in 14 days. How much does she earn in 260 days?

7. Tom makes $672.80 in five days. How much does he make in three days?

8. If 5 ounces of a medicine must be mixed with 8 ounces of water, how many ounces of medicine would be mixed with 20 ounces of water?

🌐 **9.** The bag of rice noodles shown below makes 7 servings. At that rate, how many ounces of noodles do you need for 12 servings, to the nearest ounce?

10. This can of sweet potatoes is enough for four servings. How many ounces are needed for nine servings, to the nearest ounce?

11. Three quarts of a latex enamel paint will cover about 270 square feet of wall surface. How many quarts will you need to cover 350 square feet of wall surface in your kitchen and 100 square feet of wall surface in your bathroom?

12. One gallon of clear gloss wood finish covers about 550 square feet of surface. If you need to apply three coats of finish to 400 square feet of surface, how many gallons do you need, to the nearest tenth?

Use the floor plan shown to answer Exercises 13–16. On the plan, one inch represents four feet.

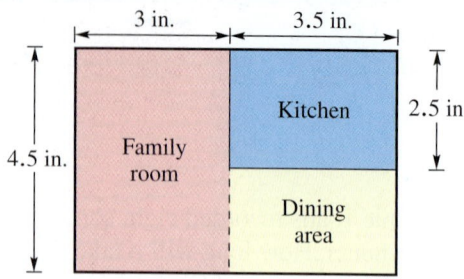

13. What is the actual length and width of the kitchen?

14. What is the actual length and width of the family room?

15. What is the actual length and width of the dining area?

16. What is the actual length and width of the entire floor plan?

The table below lists recommended amounts of food to order for 25 party guests. Use the table to answer Exercises 17 and 18. (Source: Cub Foods.)

FOOD FOR 25 GUESTS

Item	Amount
Fried chicken	40 pieces
Lasagna	14 pounds
Deli meats	4.5 pounds
Sliced cheese	$2\frac{1}{3}$ pounds
Bakery buns	3 dozen
Potato salad	6 pounds

17. How much of each food item should Nathan and Amanda order for a graduation party with 60 guests?

18. Taisha is having 20 neighbors over for a Fourth of July picnic. How much food should she buy?

In Exercises 19–24, set up a proportion to solve each problem. Check to see whether your answer is reasonable. Then flip one side of your proportion to see what answer you get with an incorrect setup. Explain why the second answer is unreasonable. See Example 2.

19. About 7 out of 10 people entering our community college need to take a refresher math course. If we have 2950 entering students, how many will probably need refresher math? (*Source:* Minneapolis Community and Technical College.)

20. In a survey, only 3 out of 100 people like their eggs poached. At that rate, how many of the 60 customers who ordered eggs at Soon-Won's restaurant this morning asked to have them poached? Round to the nearest whole person.

21. About 1 out of 3 people choose vanilla as their favorite ice cream flavor. If 238 people attend an ice cream social, how many would you expect to choose vanilla? Round to the nearest whole person.

22. In a test of 200 sewing machines, only one had a defect. At that rate, how many of the 5600 machines shipped from the factory have defects?

23. About 98 out of 100 U.S. households have at least one TV set. There were 113,100,000 U.S. households in 2005. How many households had one or more TVs? (*Source:* Nielsen Media Research.)

24. In a survey, 3 out of 100 dog owners washed their pets by having the dogs go into the shower with them. If the survey is accurate, how many of the 31,200,000 dog owners in the United States use this method? (*Source:* Teledyne Water Pik; American Veterinary Medical Association.)

Set up and solve a proportion for each problem.

25. The stock market report says that five stocks went up for every six stocks that went down. If 750 stocks went down yesterday, how many went up?

26. The human body contains 90 pounds of water for every 100 pounds of body weight. How many pounds of water are in a child who weighs 80 pounds?

27. The ratio of the length of an airplane wing to its width is 8 to 1. If the length of a wing is 32.5 meters, how wide must it be? Round to the nearest hundredth.

28. The Rosebud School District wants a student-to-teacher ratio of 19 to 1. How many teachers are needed for 1850 students? Round to the nearest whole number.

29. The number of calories you burn is proportional to your weight. A 150-pound person burns 222 calories during 30 minutes of tennis. How many calories would a 210-pound person burn, to the nearest whole number? (*Source: Wellness Encyclopedia.*)

30. (Complete Exercise 29 first.) A 150-pound person burns 189 calories during 45 minutes of grocery shopping. How many calories would a 115-pound person burn, to the nearest whole number? (*Source: Wellness Encyclopedia.*)

31. At 3 P.M., Coretta's shadow is 1.05 meters long. Her height is 1.68 meters. At the same time, a tree's shadow is 6.58 meters long. How tall is the tree? Round to the nearest hundredth.

32. (Complete Exercise 31 first.) Later in the day, Coretta's shadow was 2.95 meters long. How long a shadow did the tree have at that time? Round to the nearest hundredth.

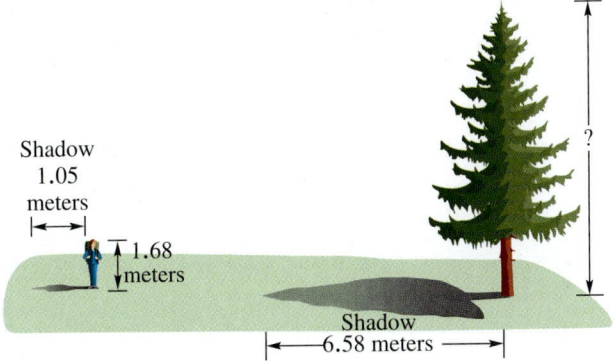

Shadow 1.05 meters
1.68 meters
Shadow 6.58 meters

Shadow 2.95 meters
Shadow ?

33. Can you set up a proportion to solve this problem? Explain why or why not. Jim is 25 years old and weighs 180 pounds. How much will he weigh when he is 50 years old?

34. Write your own application problem that can be solved by setting up a proportion. Also show the proportion and the steps needed to solve your problem.

35. A survey of college students shows that 4 out of 5 drink coffee. Of the students who drink coffee, 1 out of 8 adds cream to it. How many of the 50,500 students at Ohio State University would be expected to use cream in their coffee?

36. About 9 out of 10 adults think it is a good idea to exercise regularly. But of the ones who think it is a good idea, only 1 in 6 actually exercises at least three times a week. At this rate, how many of the 300 employees in our company exercise regularly?

37. The nutrition information on a bran cereal box says that a $\frac{1}{3}$ cup serving provides 80 calories and 8 grams of dietary fiber. At that rate, how many calories and grams of fiber are in a $\frac{1}{2}$ cup serving? (*Source:* Kraft Foods, Inc.)

38. A $\frac{2}{3}$ cup serving of penne pasta has 210 calories and 2 grams of dietary fiber. How many calories and grams of fiber would be in a 1 cup serving? (*Source:* Borden Foods.)

Relating Concepts (Exercises 39–42) For Individual or Group Work

A box of instant mashed potatoes has the list of ingredients shown in the table. Use this information to **work Exercises 39–42 in order.**

Ingredient	For 12 Servings
Water	$3\frac{1}{2}$ cups
Margarine	6 tablespoons
Milk	$1\frac{1}{2}$ cups
Potato flakes	4 cups

Source: General Mills.

39. Find the amount of each ingredient needed for six servings. Show *two* different methods for finding the amounts. One method should use proportions.

40. Find the amount of each ingredient needed for 18 servings. Show *two* different methods for finding the amounts, one using proportions and one using your answers from Exercise 39.

41. Find the amount of each ingredient needed for three servings, using your answers from either Exercise 39 or Exercise 40.

42. Find the amount of each ingredient needed for nine servings, using your answers from either Exercise 40 or Exercise 41.

6.5 ▶▶▶ Geometry: Lines and Angles

Geometry starts with the idea of a point. A **point** can be described as a location in space. It has no length or width. A point is represented by a dot and is named by writing a capital letter next to the dot.

Point *P*

OBJECTIVE ① Identify and name lines, line segments, and rays.
A **line** is a straight row of points that goes on forever in both directions. A line is drawn by using arrowheads to show that it never ends. The line is named by using the letters of any two points on the line.

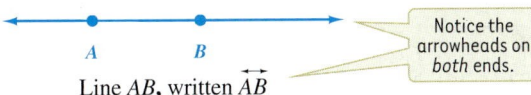

Notice the arrowheads on *both* ends.

Line *AB*, written $\overleftrightarrow{AB}$

A piece of a line that has two endpoints is called a **line segment.** A line segment is named for its endpoints. The segment with endpoints *P* and *Q* is shown below. It can be named $\overline{PQ}$ or $\overline{QP}$.

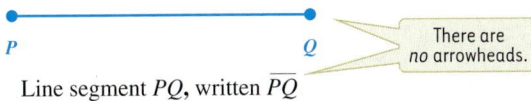

There are *no* arrowheads.

Line segment *PQ*, written $\overline{PQ}$

A **ray** is a part of a line that has only one endpoint and goes on forever in one direction. A ray is named by using the endpoint and some other point on the ray. The endpoint is always mentioned first.

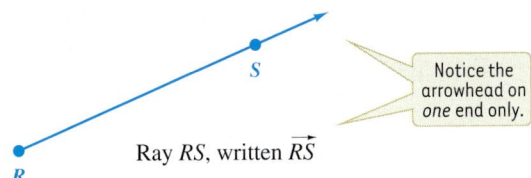

Notice the arrowhead on *one* end only.

Ray *RS*, written $\overrightarrow{RS}$

EXAMPLE 1 Identifying and Naming Lines, Rays, and Line Segments

Identify each figure below as a line, line segment, or ray and name it using the appropriate symbol.

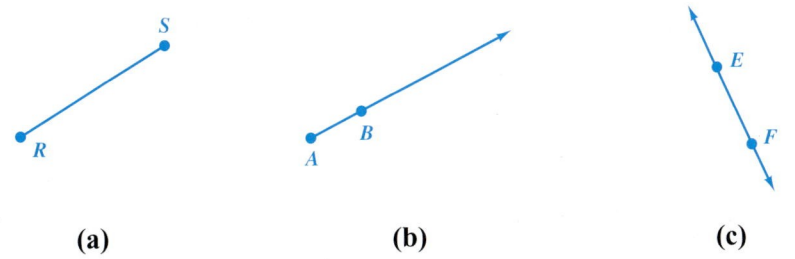

(a)   (b)   (c)

Figure **(a)** has two endpoints, so it is a *line segment* named $\overline{RS}$ or $\overline{SR}$.
Figure **(b)** starts at point *A* and goes on forever in one direction, so it is a *ray* named $\overrightarrow{AB}$.
Figure **(c)** goes on forever in both directions, so it is a *line* named $\overleftrightarrow{EF}$ or $\overleftrightarrow{FE}$.

Work Problem ① *at the Side.* ▶

① Identify each figure as a line, line segment, or ray, and name it.

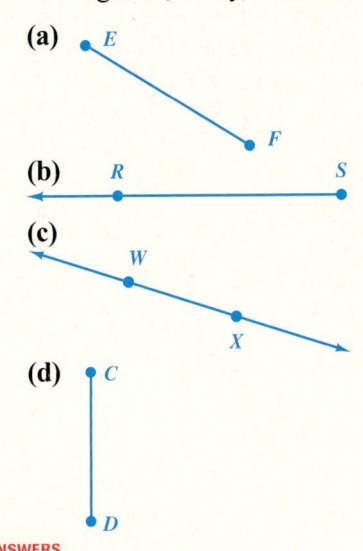

(a)

(b)

(c)

(d)

ANSWERS

1. **(a)** line segment named $\overline{EF}$ or $\overline{FE}$
   (b) ray named $\overrightarrow{SR}$ **(c)** line named $\overleftrightarrow{WX}$ or $\overleftrightarrow{XW}$
   (d) line segment named $\overline{CD}$ or $\overline{DC}$

2 Label each pair of lines as appearing to be parallel or as intersecting.

(a)

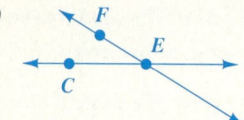

(b)

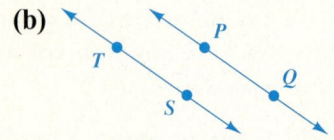

(c)

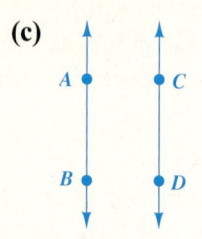

OBJECTIVE **2** **Identify parallel and intersecting lines.** A *plane* is an infinitely large, flat surface. A floor or a wall is part of a plane. Lines that are in the *same plane,* but that never intersect (never cross), are called **parallel lines,** while lines that cross are called **intersecting lines.** (Think of an intersection, where two streets cross each other.)

EXAMPLE 2 **Identifying Parallel and Intersecting Lines**

Label each pair of lines as appearing to be parallel or as intersecting.

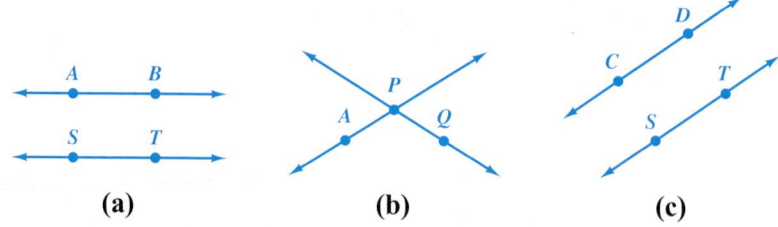

(a)   **(b)**   **(c)**

The lines in Figures **(a)** and **(c)** do not intersect; they appear to be *parallel lines*.

The lines in Figure **(b)** cross at *P*, so they are *intersecting lines*.

CAUTION
Appearances may be deceiving! Do not assume that lines are parallel unless it is stated that they are parallel.

◀ *Work Problem* **2** *at the Side.*

OBJECTIVE **3** **Identify and name angles.** An **angle** is made up of two rays that start at a common endpoint. This common endpoint is called the *vertex.*

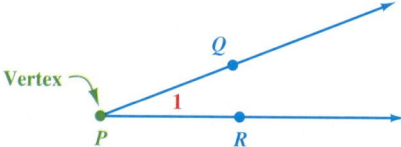

$\overrightarrow{PQ}$ and $\overrightarrow{PR}$ are called the *sides* of the angle. The angle can be named in four different ways, as shown below.

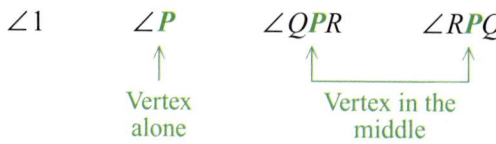

Naming an Angle
To name an angle, write the vertex alone or write the vertex in the middle of two other points, one from each side. If two or more angles have the *same vertex,* as in Example 3 on the next page, do *not* use the vertex alone to name an angle.

EXAMPLE 3 **Identifying and Naming an Angle**

Name the highlighted angle in three different ways.

In this situation, do **not** use the *vertex alone* to name the angle.

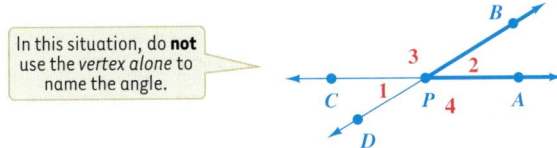

The angle can be named $\angle BPA$, $\angle APB$, or $\angle 2$. It *cannot* be named $\angle P$, using the vertex alone, because four different angles have P as their vertex.

Work Problem **3** *at the Side.* ▶

OBJECTIVE 4 Classify angles as right, acute, straight, or obtuse. Angles can be measured in **degrees.** The symbol for degrees is a small, raised circle °. Think of the minute hand on a clock as a ray of an angle. Suppose it is at 12:00. During one hour of time, the minute hand moves around in a complete circle. It moves 360 *degrees*, or 360°. In half an hour, at 12:30, the minute hand has moved halfway around the circle, or 180°. An angle of 180° is called a **straight angle.** When two rays go in opposite directions and form a straight line, then the rays form a straight angle.

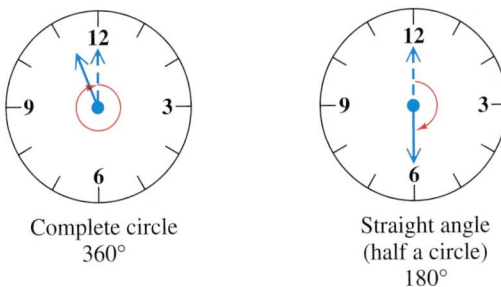

Complete circle
360°

Straight angle
(half a circle)
180°

In a quarter of an hour, at 12:15, the minute hand has moved $\frac{1}{4}$ of the way around the circle, or 90°. An angle of 90° is called a **right angle.** The rays of a right angle form one corner of a square. So, to show that an angle is a **right angle**, we draw a **small square** at the vertex.

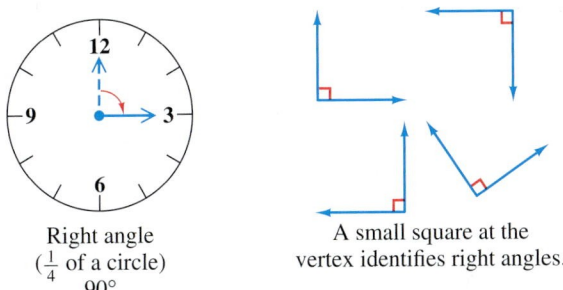

Right angle
($\frac{1}{4}$ of a circle)
90°

A small square at the vertex identifies right angles.

An angle that measures 1° is shown below. You can see that an angle of 1° is very small.

1° angle

3 (a) Name the highlighted angle in three different ways.

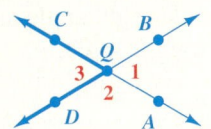

(b) Darken the rays that make up $\angle ZTW$.

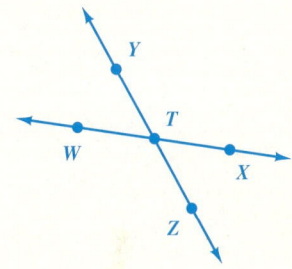

(c) Name this angle in four different ways.

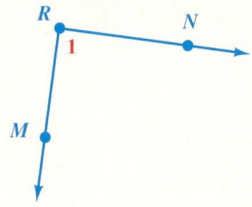

4 Label each angle as acute, right, obtuse, or straight. State the number of degrees in the right angle and in the straight angle.

(a)

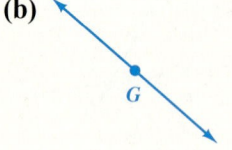

(b)

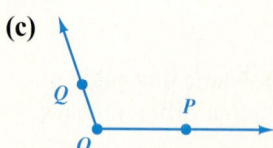

(c)

(d)

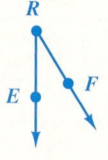

Some other terms used to describe angles are shown below.

Acute angles measure less than 90°.

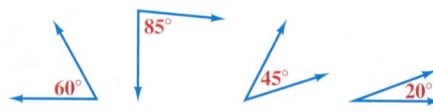

Examples of acute angles

Obtuse angles measure more than 90° but less than 180°.

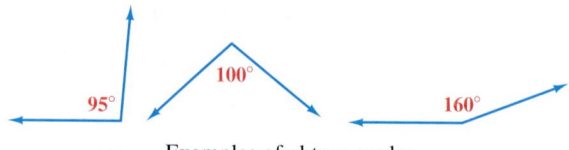

Examples of obtuse angles

Section 9.2 shows you how to use a tool called a *protractor* to measure the number of degrees in an angle.

Classifying Angles

Acute angles measure less than 90°.

Right angles measure *exactly* 90°.

Obtuse angles measure more than 90° but less than 180°.

Straight angles measure *exactly* 180°.

Note

Angles can also be measured in radians, which you will learn about in a later math course.

EXAMPLE 4 Classifying an Angle

Label each angle as acute, right, obtuse, or straight.

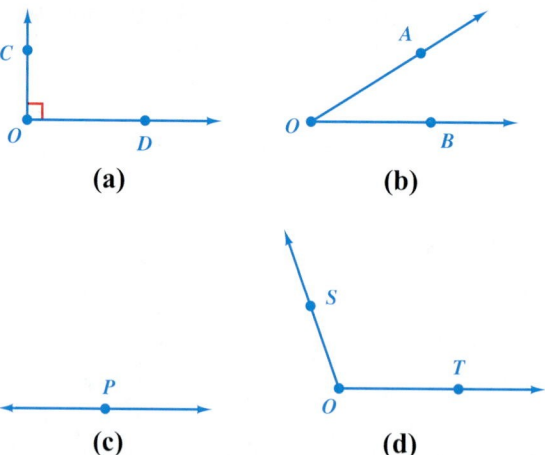

Figure **(a)** shows a *right angle* (exactly 90° and identified by a small square at the vertex).

Figure **(b)** shows an *acute angle* (less than 90°).

Figure **(c)** shows a *straight angle* (exactly 180°).

Figure **(d)** shows an *obtuse angle* (more than 90° but less than 180°).

◀ *Work Problem* **4** *at the Side.*

OBJECTIVE **5** **Identify perpendicular lines.** Two lines are called **perpendicular lines** if they intersect to form a right angle.

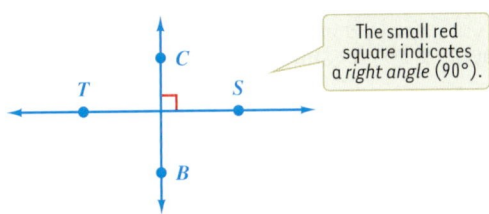

The small red square indicates a *right angle* (90°).

$\overleftrightarrow{CB}$ and $\overleftrightarrow{ST}$ are **perpendicular** lines because they intersect at right angles, as indicated by the small red square in the figure.

Perpendicular lines can be written in the following way: $\overleftrightarrow{CB} \perp \overleftrightarrow{ST}$.

EXAMPLE 5 **Identifying Perpendicular Lines**

Which pairs of lines are perpendicular?

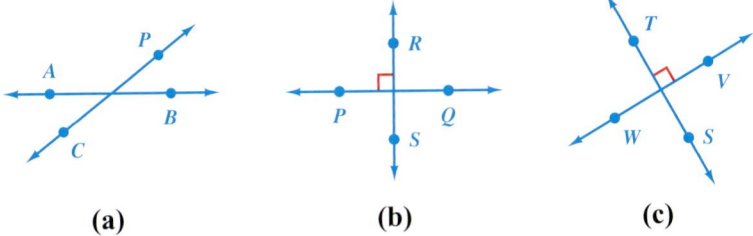

(a)   (b)   (c)

The lines in Figures **(b)** and **(c)** are *perpendicular* to each other, because they intersect at right angles.

The lines in Figure **(a)** are *intersecting lines,* but they are *not* perpendicular because they do *not* form a right angle.

Work Problem **5** *at the Side.* ▶

OBJECTIVE **6** **Identify complementary angles and supplementary angles and find the measure of a complement or supplement of a given angle.** Two angles are called **complementary angles** if the sum of their measures is 90°. If two angles are complementary, each angle is the *complement* of the other.

EXAMPLE 6 **Identifying Complementary Angles**

Identify each pair of complementary angles.

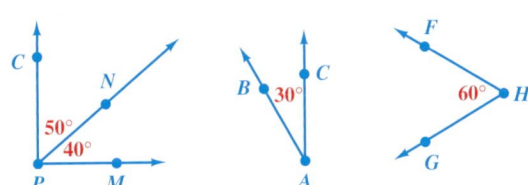

$\angle MPN$ (40°) and $\angle NPC$ (50°) are complementary angles because

$$40° + 50° = \mathbf{90°}$$

$\angle CAB$ (30°) and $\angle FHG$ (60°) are complementary angles because

$$30° + 60° = \mathbf{90°}$$

Work Problem **6** *at the Side.* ▶

---

5 Which pair of lines is perpendicular? How can you describe the other pair of lines?

(a)

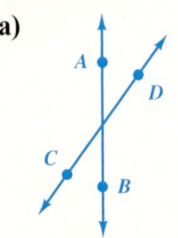

(b)

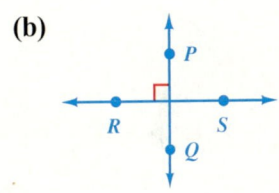

6 Identify each pair of complementary angles.

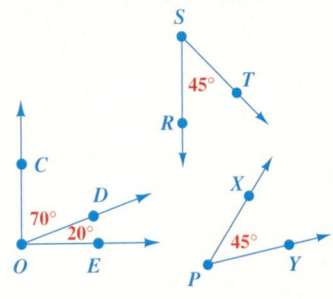

---

7 Find the complement of each angle.

(a) 35°

(b) 80°

EXAMPLE 7 Finding the Complement of Angles

Find the complement of each angle.

(a) 30°
Find the complement of 30° by subtracting. **90° − 30° = 60°** ← Complement

(b) 75°
Find the complement of 75° by subtracting. **90° − 75° = 15°** ← Complement

◀ Work Problem **7** at the Side.

Two angles are called **supplementary angles** if the sum of their measures is 180°. If two angles are supplementary, each angle is the *supplement* of the other.

EXAMPLE 8 Identifying Supplementary Angles

Identify each pair of supplementary angles.

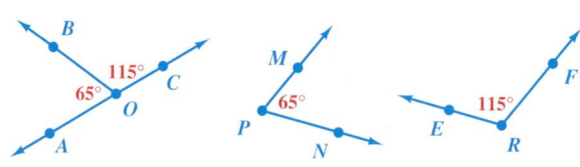

∠*BOA* and ∠*BOC*, because 65° + 115° = **180°**
∠*BOA* and ∠*ERF*, because 65° + 115° = **180°**
∠*BOC* and ∠*MPN*, because 115° + 65° = **180°**
∠*MPN* and ∠*ERF*, because 65° + 115° = **180°**

◀ Work Problem **8** at the Side.

---

8 Identify each pair of supplementary angles. (*Hint:* There are four pairs.)

---

9 Find the supplement of each angle.

(a) 175°

(b) 30°

EXAMPLE 9 Finding the Supplement of Angles

Find the supplement of each angle.

(a) 70°
Find the supplement of 70° by subtracting. **180° − 70° = 110°** ← Supplement

(b) 140°
Find the supplement of 140° by subtracting. **180° − 140° = 40°** ← Supplement

◀ Work Problem **9** at the Side.

OBJECTIVE 7 Identify congruent angles and vertical angles and use this knowledge to find the measures of angles. Two angles are called **congruent angles** if they measure the same number of degrees. If two angles are congruent, this is written as ∠*A* ≅ ∠*B* and read as, "angle *A* **is congruent to** angle *B*." Here is an example.

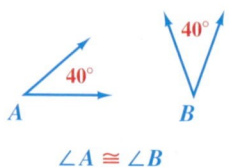

∠*A* ≅ ∠*B*

Example of congruent angles

EXAMPLE 10 **Identifying Congruent Angles**

Identify the angles that are congruent.

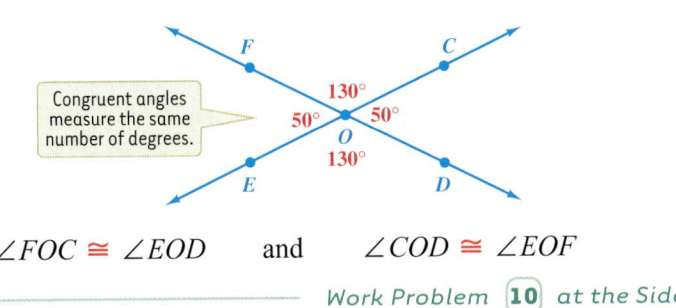

$\angle FOC \cong \angle EOD$ and $\angle COD \cong \angle EOF$

Work Problem **10** *at the Side.* ▶

Angles that share a common side and a common vertex are called *adjacent* angles, such as $\angle FOC$ and $\angle COD$ in Example 10 above. Angles that do *not* share a common side are called *nonadjacent* angles. Two nonadjacent angles formed by two intersecting lines are called **vertical angles.**

EXAMPLE 11 **Identifying Vertical Angles**

Identify the vertical angles in this figure.

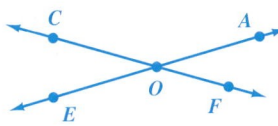

$\angle AOF$ and $\angle COE$ are vertical angles because they do *not* share a common side and they are formed by two intersecting lines ($\overleftrightarrow{CF}$ and $\overleftrightarrow{EA}$).

$\angle COA$ and $\angle EOF$ are also vertical angles.

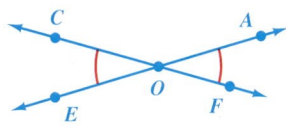 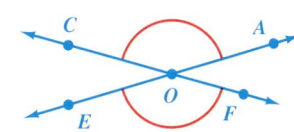

Work Problem **11** *at the Side.* ▶

Look back at Example 10 at the top of the page. Notice that the two *congruent* angles that measure 130° are also *vertical* angles. Also, the two congruent angles that measure 50° are vertical angles. This illustrates the following property.

> **Vertical Angles Are Congruent**
> If two angles are *vertical* angles, they are *congruent;* that is, they measure the same number of degrees.

10 Identify the angles that are congruent.

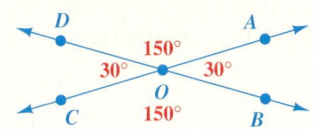

11 Identify the vertical angles. What is special about vertical angles?

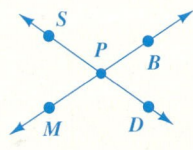

ANSWERS

10. $\angle BOC \cong \angle AOD$; $\angle AOB \cong \angle DOC$
11. $\angle SPB$ and $\angle MPD$; $\angle BPD$ and $\angle SPM$; vertical angles are congruent (they measure the same number of degrees).

12 In the figure below, find the measure of each unlabeled angle. Write the angle measures on the figure.

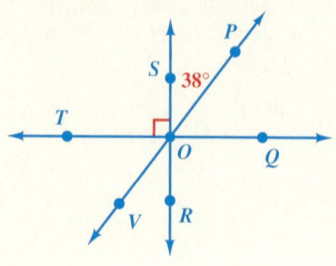

(a) $\angle TOS$

(b) $\angle QOR$

(c) $\angle VOR$

(d) $\angle POQ$

(e) $\angle TOV$

ANSWERS

12. (a) 90° (b) 90° (c) 38° (d) 52°
    (e) 52°

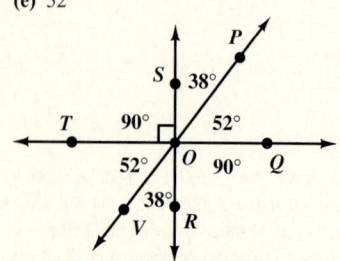

EXAMPLE 12 Finding the Measures of Vertical Angles

In the figure below, find the measure of each unlabeled angle.

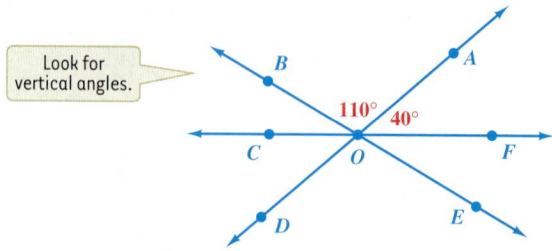

Look for vertical angles.

(a) $\angle COD$
   $\angle COD$ and $\angle AOF$ are vertical angles, so they are congruent. This means they measure the same number of degrees.

   The measure of $\angle AOF$ is 40°   so   the measure of $\angle COD$ is **40°** also.

(b) $\angle DOE$
   $\angle DOE$ and $\angle BOA$ are vertical angles, so they are congruent.

   The measure of $\angle BOA$ is 110°   so   the measure of $\angle DOE$ is **110°** also.

(c) $\angle COB$
   Look at $\angle COB$, $\angle BOA$, and $\angle AOF$. Notice that $\overrightarrow{OC}$ and $\overrightarrow{OF}$ go in opposite directions. Therefore, $\angle COF$ is a straight angle and measures 180°. To find the measure of $\angle COB$, subtract the sum of the other two angles from 180°.

$$180° - (110° + 40°) = 180° - (150°) = 30°$$

   The measure of $\angle COB$ is **30°.**

(d) $\angle EOF$
   $\angle EOF$ and $\angle COB$ are vertical angles, so they are congruent. We know from part (c) above that the measure of $\angle COB$ is 30° so the measure of $\angle EOF$ is **30°** also.

◀ *Work Problem* **12** *at the Side.*

OBJECTIVE **8** **Identify corresponding angles and alternate interior angles and use this knowledge to find the measures of angles.** We can also find congruent angles (angles with the same measure) when two *parallel lines* are crossed by a third line, called a *transversal*. When a transversal crosses two *parallel* lines, eight angles are formed, as shown below. There are special names for certain pairs of angles.

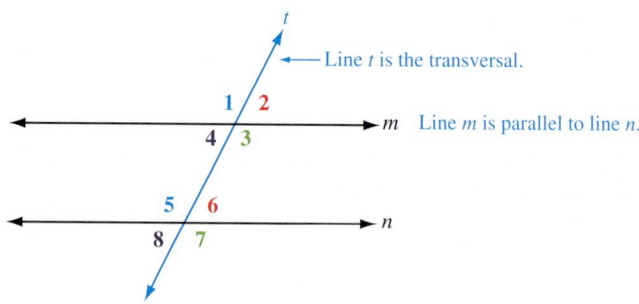

Line t is the transversal.

Line m is parallel to line n.

$\angle 1$ and $\angle 5$ are called **corresponding angles.** Notice that they are both on the same side of the transversal (line t) and in the same relative position. *Corresponding angles are congruent,* so $\angle 1$ and $\angle 5$ measure the same number of degrees. There are four pairs of corresponding angles.

∠1 and ∠5 are corresponding angles, so ∠1 ≅ ∠5.

∠2 and ∠6 are corresponding angles, so ∠2 ≅ ∠6.

∠3 and ∠7 are corresponding angles, so ∠3 ≅ ∠7.

∠4 and ∠8 are corresponding angles, so ∠4 ≅ ∠8.

When a transversal crosses two parallel lines, angles 3, 4, 5, and 6 are called *interior angles*. You can see that they are "inside" the *parallel* lines.

∠3 and ∠5 are alternate interior angles.

∠4 and ∠6 are alternate interior angles.

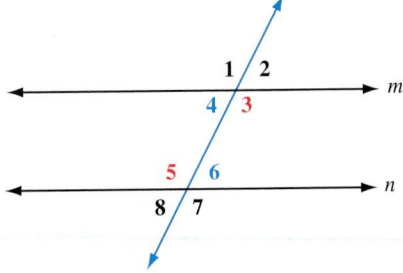

When two lines are *parallel,* then **alternate interior angles** *are congruent* (they have the same measure). Notice that alternate interior angles are on opposite sides of the transversal.

$$\angle 3 \cong \angle 5 \quad \text{and} \quad \angle 4 \cong \angle 6$$

> **Angles Formed by Parallel Lines and a Transversal**
>
> When two parallel lines are crossed by a transversal:
> 1. Corresponding angles are congruent, and
> 2. Alternate interior angles are congruent.

EXAMPLE 13 **Identifying Corresponding Angles and Alternate Interior Angles**

In each figure, line *m* is parallel to line *n*. Identify all pairs of corresponding angles and all pairs of alternate interior angles.

(a)

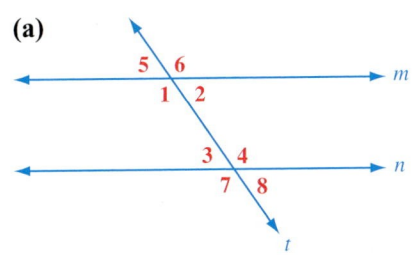

There are four pairs of corresponding angles:

∠5 and ∠3   ∠6 and ∠4
∠1 and ∠7   ∠2 and ∠8

Alternate interior angles:
∠1 and ∠4   ∠2 and ∠3

(b)

Corresponding angles:
∠1 and ∠5   ∠3 and ∠7
∠2 and ∠6   ∠4 and ∠8

Alternate interior angles:
∠3 and ∠6   ∠4 and ∠5

Work Problem **13** *at the Side.* ▶

13 In each figure below, line *m* is parallel to line *n*. Identify all pairs of corresponding angles and all pairs of alternate interior angles.

(a)

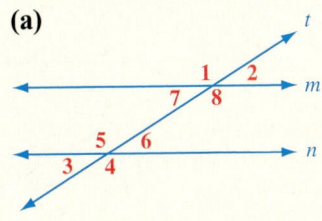

(b)

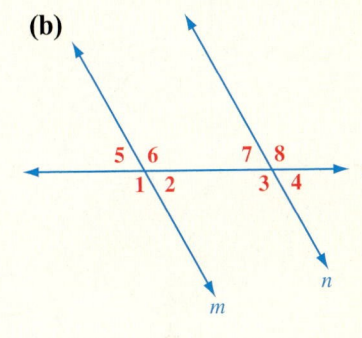

14 In each figure below, line *m* is parallel to line *n*.

(a) The measure of ∠6 is 150°. Find the measures of the other angles.

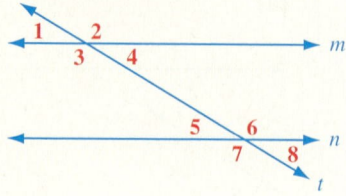

(b) The measure of ∠1 is 45°. Find the measures of the other angles.

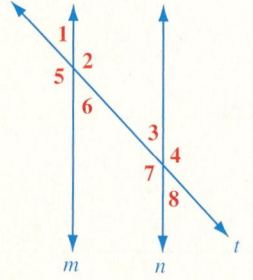

Recall that two angles are supplementary angles if the sum of their measures is 180°. Also remember that two rays that form a 180° angle form a straight line. Now you can combine your knowledge about supplementary angles with the information on parallel lines.

EXAMPLE 14   **Working with Parallel Lines**

In the figure at the right, line *m* is parallel to line *n* and the measure of ∠4 is 70°. Find the measures of the other angles.

Look for corresponding angles and for alternate interior angles.

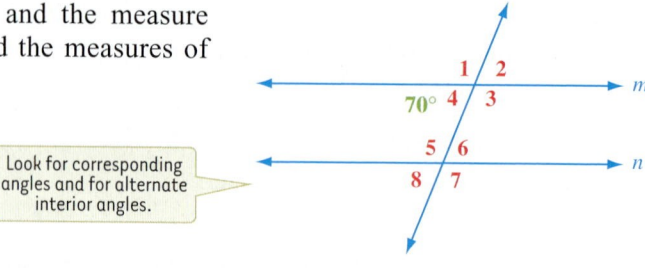

As you find the measure of each angle, write it on the figure.

∠4 ≅ ∠8 (corresponding angles), so the measure of ∠8 is also 70°.
∠4 ≅ ∠6 (alternate interior angles), so the measure of ∠6 is also 70°.
∠6 ≅ ∠2 (corresponding angles), so the measure of ∠2 is also 70°.

Notice that the exterior sides of ∠4 and ∠3 form a straight line, that is, a straight angle of 180°. Therefore, ∠4 and ∠3 are supplementary angles and the sum of their measures is 180°. If ∠4 is 70° then ∠3 must be 110° because 180° − 70° = 110°. So the measure of ∠3 is 110°.

∠3 ≅ ∠7 (corresponding angles), so the measure of ∠7 is also 110°.
∠3 ≅ ∠5 (alternate interior angles), so the measure of ∠5 is also 110°.
∠5 ≅ ∠1 (corresponding angles), so the measure of ∠1 is also 110°.

With the measures of all the angles labeled, you can double-check that each pair of angles that forms a straight angle also adds up to 180°.

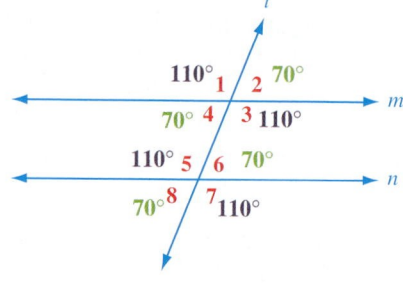

◀ *Work Problem* **14** *at the Side.*

ANSWERS

14. (a)

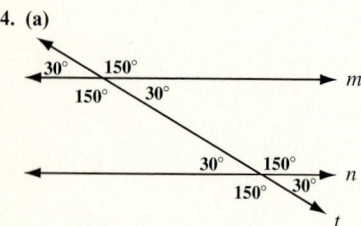

(b)

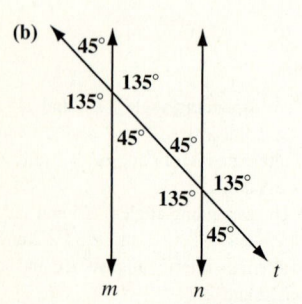

6.5 ▶▶▶ Exercises

Identify each figure as a line, line segment, *or* ray *and name it using the appropriate symbol. See Example 1.*

1.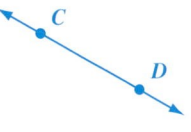

2.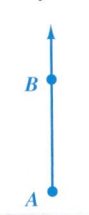

3.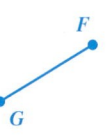

4.

5.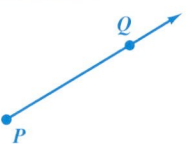

6.

Label each pair of lines as appearing to be parallel, *as* perpendicular, *or as* intersecting. *See Examples 2 and 5.*

7.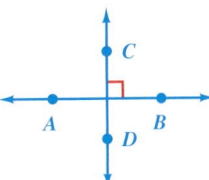

8.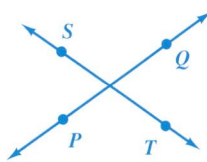

9.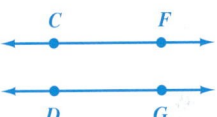

10.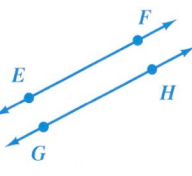

11.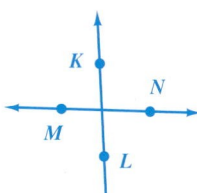

12.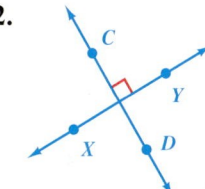

Name each highlighted angle by using the three-letter form of identification. See Example 3.

13.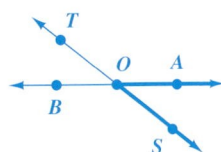

14.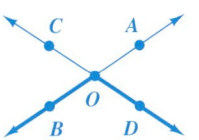

15.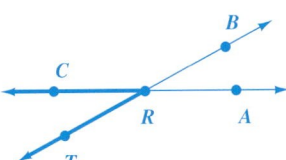

16.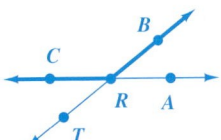

17.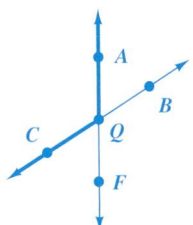

18.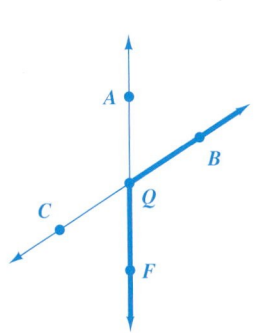

Label each angle as acute, right, obtuse, *or* straight. *For right angles and straight angles, indicate the number of degrees in the angle. See Example 4.*

19.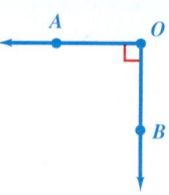

20.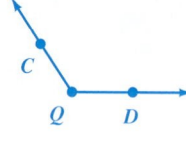

21.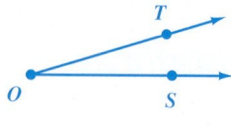

22.

23.

24.

Identify each pair of complementary angles. See Example 6.

25.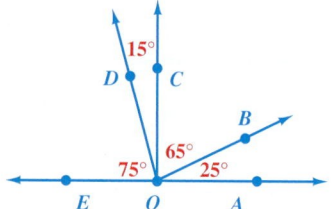

26.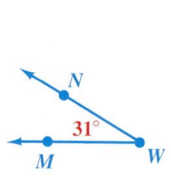

Identify each pair of supplementary angles. See Example 8.

27.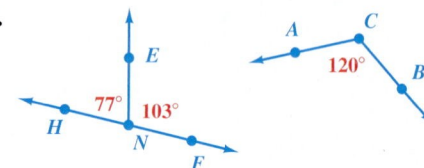

28.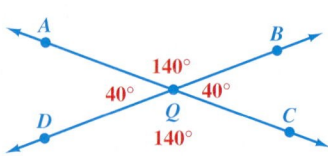

Find the complement of each angle. See Example 7.

29. 40°　　　　**30.** 35°　　　　**31.** 86°　　　　**32.** 59°

Find the supplement of each angle. See Example 9.

33. 130°　　　　**34.** 75°　　　　**35.** 90°　　　　**36.** 5°

In Exercises 37 and 38, identify the angles that are congruent. See Examples 10 and 11.

37.

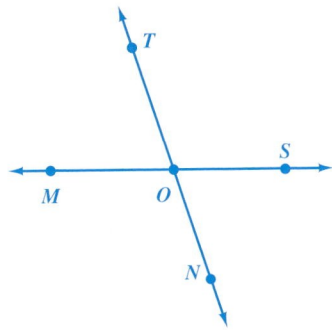

38.

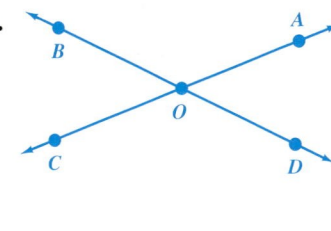

In Exercises 39 and 40, find the measure of each of the angles. See Example 12.

39. In the figure below, $\angle AOH$ measures 37° and $\angle COE$ measures 63°.

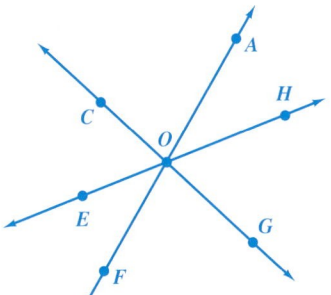

40. In the figure below, $\angle POU$ measures 105° and $\angle UOT$ measures 40°.

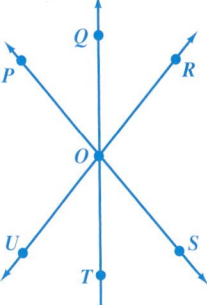

Relating Concepts (Exercises 41–46)   For Individual or Group Work

Use the figure to **work Exercises 41–46 in order.** *Decide whether each statement is* true *or* false.
If it is true, explain why. If it is false, rewrite it to make a true statement.

41. $\angle UST$ is 90°.

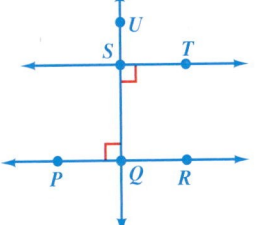

42. $\overleftrightarrow{SQ}$ and $\overleftrightarrow{PQ}$ are perpendicular.

43. The measure of $\angle USQ$ is less than the measure of $\angle PQR$.

44. $\overleftrightarrow{ST}$ and $\overleftrightarrow{PR}$ are intersecting.

45. $\overleftrightarrow{QU}$ and $\overleftrightarrow{TS}$ are parallel.

46. $\angle UST$ and $\angle UQR$ measure the same number of degrees.

In each figure, line m is parallel to line n. Identify all pairs of corresponding angles and all pairs of alternate interior angles. See Example 13.

47.

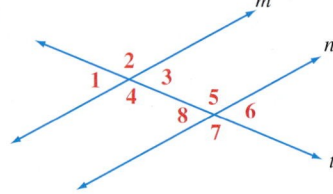

48.

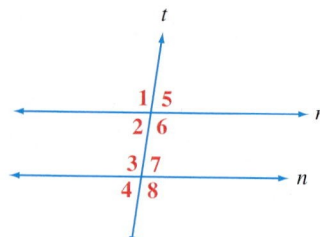

In each figure, line m is parallel to line n. Find the measure of each angle. See Example 14.

49. ∠8 measures 130°.

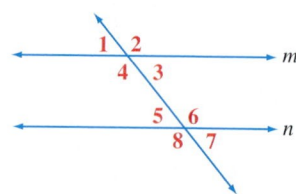

50. ∠2 measures 80°.

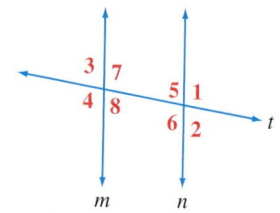

51. ∠6 measures 47°.

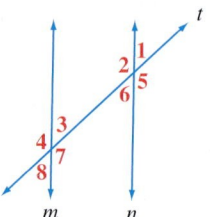

52. ∠2 measures 108°.

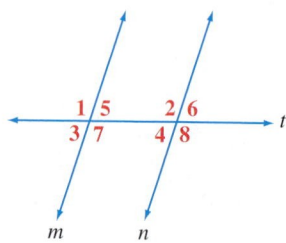

53. ∠6 measures 114°.

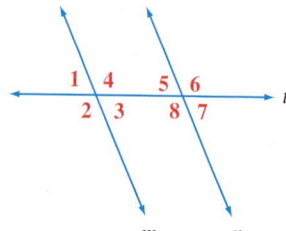

54. ∠3 measures 59°.

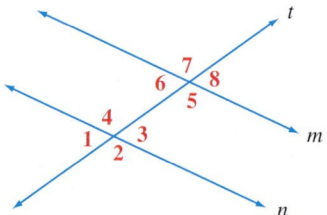

In each figure, $\overrightarrow{BA}$ is parallel to $\overrightarrow{CD}$. Find the measure of each numbered angle.

55.

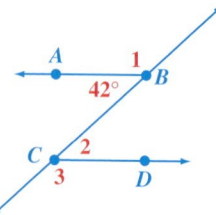

56.

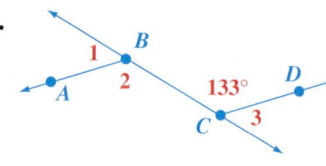

6.6 ▶▶▶ Geometry Applications: Congruent and Similar Triangles

Two useful concepts in geometry are *congruence* and *similarity*. If two figures are *identical,* both in *shape* and in *size,* we say the figures are **congruent.** In other words, the figures are perfect duplicates of each other, like getting two identical prints made from your digital camera. If two figures have the *same shape* but are *different sizes,* we say the figures are **similar,** like getting a print made from your digital camera and then an enlargement of the same print. We'll explore the ideas of congruence and similarity using triangles.

OBJECTIVES

❶ **Identify corresponding parts of congruent triangles.**

❷ **Prove that triangles are congruent using SAS, SSS, and ASA.**

❸ **Identify corresponding parts of similar triangles.**

❹ **Find the unknown lengths of sides in similar triangles.**

❺ **Solve application problems involving similar triangles.**

OBJECTIVE 1 Identify corresponding parts of congruent triangles. The two triangles shown below are *congruent* because they are the *same shape* and the *same size.*

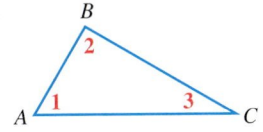
One way to name this triangle is
$\triangle ABC$.

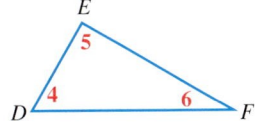
One way to name this triangle is
$\triangle DEF$.

Suppose you picked up $\triangle ABC$ and slid it over on top of $\triangle DEF$. You would see that the two triangles are a perfect match. $\angle 1$ would be on top of $\angle 4$, so they are called *corresponding angles*. Similarly, $\angle 2$ and $\angle 5$ are corresponding angles, and $\angle 3$ and $\angle 6$ are corresponding angles. You would see that corresponding angles have the same measure, as indicated below.

$$m\angle 1 = m\angle 4 \qquad m\angle 2 = m\angle 5 \qquad m\angle 3 = m\angle 6$$

The abbreviation for measure is m, so $m\angle 1$ is read, "the measure of angle 1."

When you put $\triangle ABC$ on top of $\triangle DEF$, you would also see that side AB is on top of side DE. We say that $\overline{AB}$ and $\overline{DE}$ are *corresponding sides*. Similarly, $\overline{BC}$ and $\overline{EF}$ are corresponding sides, and $\overline{AC}$ and $\overline{DF}$ are corresponding sides. You would see that corresponding sides have the same length.

$$AB = DE \qquad BC = EF \qquad AC = DF$$

Because corresponding angles have the same measure, and corresponding sides have the same length, we know that $\triangle ABC$ **is congruent to** $\triangle DEF$. We can write this as $\triangle ABC \cong \triangle DEF$.

> **Congruent Triangles**
>
> If two triangles are congruent, then
> 1. Corresponding angles have the same measure, and
> 2. Corresponding sides have the same length.

EXAMPLE 1 Identifying Corresponding Parts in Congruent Triangles

Each pair of triangles is congruent. List the corresponding angles and corresponding sides.

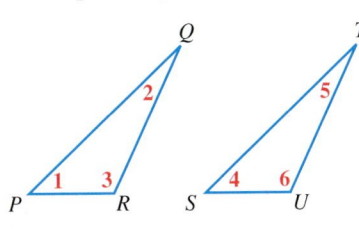

(a) If you picked up $\triangle PQR$ and slid it over on top of $\triangle STU$, the two triangles would match.
The corresponding parts are:

$\angle 1$ and $\angle 4$	$\overline{PQ}$ and $\overline{ST}$
$\angle 2$ and $\angle 5$	$\overline{PR}$ and $\overline{SU}$
$\angle 3$ and $\angle 6$	$\overline{QR}$ and $\overline{TU}$

Continued on Next Page

1 Each pair of triangles is congruent. List the corresponding angles and the corresponding sides.

(a)

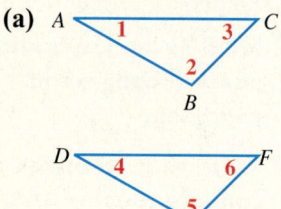

(b)

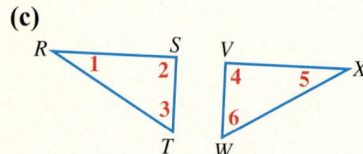

(*Hint:* Rotate $\triangle FGH$, then slide it on top of $\triangle JLK$.)

(c)

(*Hint:* Flip $\triangle RST$ over, then slide it on top of $\triangle VWX$.)

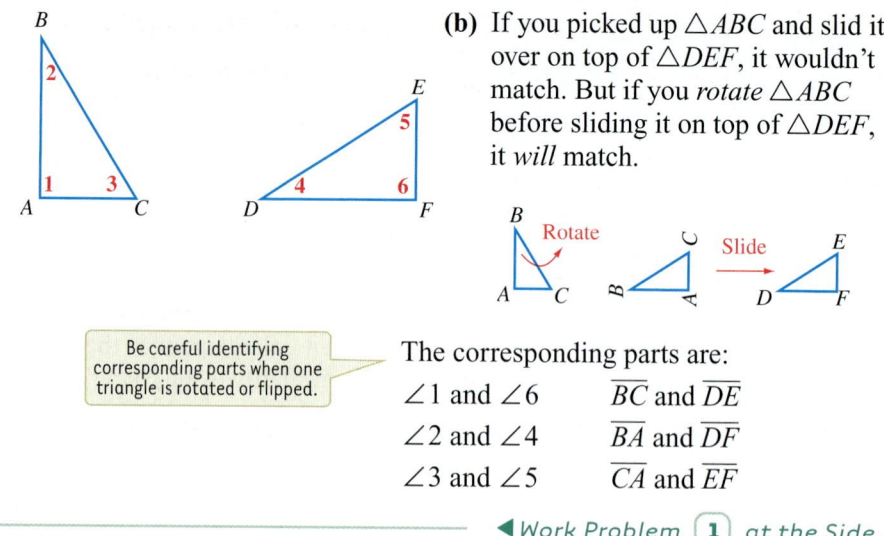

(b) If you picked up $\triangle ABC$ and slid it over on top of $\triangle DEF$, it wouldn't match. But if you *rotate* $\triangle ABC$ before sliding it on top of $\triangle DEF$, it *will* match.

> Be careful identifying corresponding parts when one triangle is rotated or flipped.

The corresponding parts are:

∠1 and ∠6 $\quad$ $\overline{BC}$ and $\overline{DE}$

∠2 and ∠4 $\quad$ $\overline{BA}$ and $\overline{DF}$

∠3 and ∠5 $\quad$ $\overline{CA}$ and $\overline{EF}$

◀ Work Problem **1** at the Side.

OBJECTIVE 2 Prove that triangles are congruent using SAS, SSS, and ASA. One way to prove that two triangles are congruent would be to measure all the angles and all the sides. If the measures of the corresponding angles and sides are equal, then the triangles are congruent. But here are three quicker methods to prove that two triangles are congruent.

Proving That Two Triangles Are Congruent

1. Angle–Side–Angle (ASA) Method
If two angles and the side between them on one triangle measure the same as the corresponding parts on another triangle, the triangles are congruent.

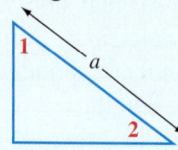 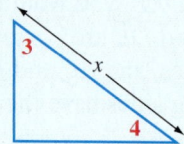

If $m\angle 1 = m\angle 3$ and $m\angle 2 = m\angle 4$ and $a = x$ then the two triangles are congruent.

2. Side–Side–Side (SSS) Method
If three sides of one triangle measure the same as the corresponding sides of another triangle, the triangles are congruent.

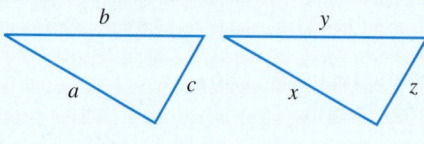

If $a = x$ and $b = y$ and $c = z$ then the two triangles are congruent.

3. Side–Angle–Side (SAS) Method
If two sides and the angle between them on one triangle measure the same as the corresponding parts on another triangle, the triangles are congruent.

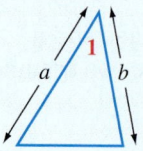 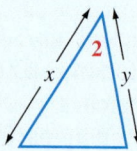

If $a = x$ and $b = y$ and $m\angle 1 = m\angle 2$ then the two triangles are congruent.

ANSWERS

1. **(a)** ∠1 and ∠4, ∠2 and ∠5, ∠3 and ∠6;
$\overline{AC}$ and $\overline{DF}$, $\overline{AB}$ and $\overline{DE}$, $\overline{BC}$ and $\overline{EF}$,
(b) ∠1 and ∠6, ∠2 and ∠5, ∠3 and ∠4;
$\overline{GF}$ and $\overline{KL}$, $\overline{FH}$ and $\overline{LJ}$, $\overline{GH}$ and $\overline{KJ}$,
(c) ∠1 and ∠5, ∠2 and ∠4, ∠3 and ∠6;
$\overline{RS}$ and $\overline{XV}$, $\overline{RT}$ and $\overline{XW}$, $\overline{ST}$ and $\overline{VW}$

EXAMPLE 2 **Proving That Two Triangles Are Congruent**

Explain which method can be used to prove that each pair of triangles is congruent. Choose from ASA, SSS, and SAS.

(a)
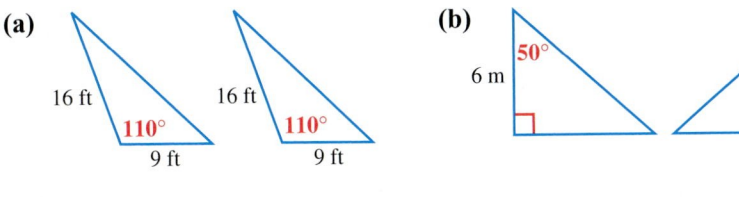

(b)

(c)
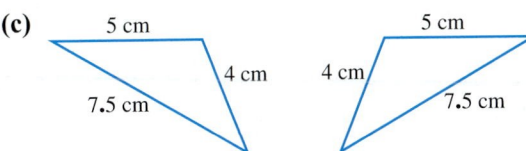

(a) On both triangles, two corresponding sides and the angle between them measure the same, so the Side–Angle–Side (SAS) method can be used to prove that the triangles are congruent.

(b) On both triangles, two corresponding angles and the side between them measure the same, so the Angle–Side–Angle (ASA) method can be used to prove that the triangles are congruent.

(c) Each pair of corresponding sides has the same length, so the Side–Side–Side (SSS) method can be used to prove that the triangles are congruent.

Work Problem **2** *at the Side.* ▶

OBJECTIVE 3 Identify corresponding parts of similar triangles.
Now that you've worked with *congruent* triangles, let's look at *similar* triangles. Remember that congruent triangles match exactly, both in shape and in size. Similar triangles, on the other hand, have the same shape but are *different sizes*. Three pairs of similar triangles are shown here.

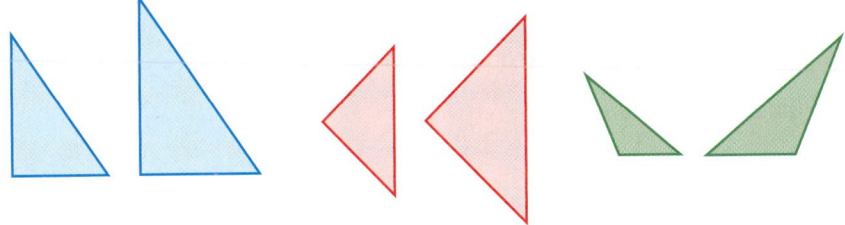

Each pair of triangles has the same shape because the corresponding angles have the same measure. But the corresponding sides are *not* the same length, so the triangles are of *different sizes*.

Two similar triangles are shown to the right. Notice that corresponding angles have the same measure, but corresponding sides have different lengths.

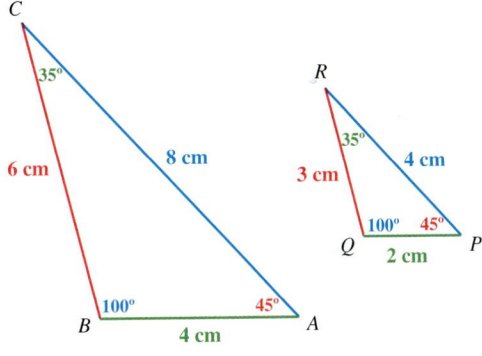

$\overline{CB}$ corresponds to $\overline{RQ}$. Similarly, $\overline{CA}$ corresponds to $\overline{RP}$, and $\overline{BA}$ corresponds to $\overline{QP}$. Notice that each side in the larger triangle is *twice* the length of the corresponding side in the smaller triangle.

Work Problem **3** *at the Side.* ▶

2 Determine which method can be used to prove that each pair of triangles is congruent.

(a)
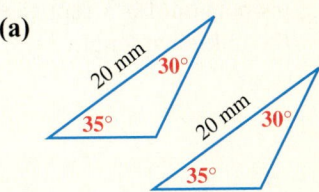

(b)
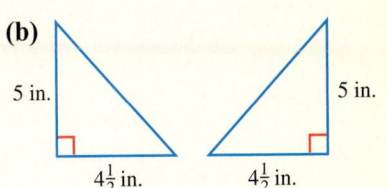

(c)
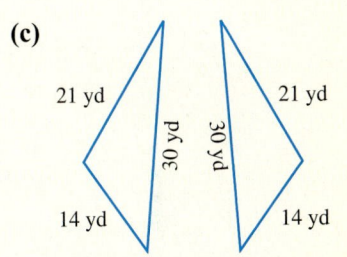

3 Identify corresponding angles and sides in these similar triangles.

(a)
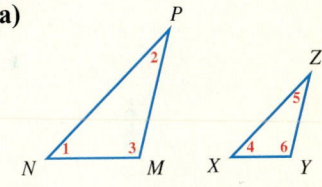

Angles: Sides:

1 and _____ $\overline{PN}$ and _____
2 and _____ $\overline{PM}$ and _____
3 and _____ $\overline{NM}$ and _____

(b)
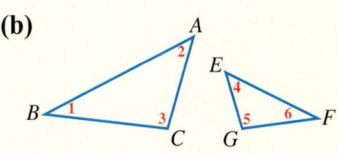

Angles: Sides:

1 and _____ $\overline{AB}$ and _____
2 and _____ $\overline{BC}$ and _____
3 and _____ $\overline{AC}$ and _____

ANSWERS

2. **(a)** ASA **(b)** SAS **(c)** SSS
3. **(a)** 4; 5; 6; $\overline{ZX}$; $\overline{ZY}$; $\overline{XY}$
   (b) 6; 4; 5; $\overline{EF}$; $\overline{FG}$; $\overline{EG}$

4 Find the length of $\overline{EF}$ in Example 3 at the right by setting up and solving a proportion. Let x represent the unknown length.

OBJECTIVE **4** **Find the unknown lengths of sides in similar triangles.** Similar triangles are useful because of the following definition.

> **Similar Triangles**
>
> If two triangles are similar, then
> 1. Corresponding angles have the same measure, and
> 2. The ratios of the lengths of corresponding sides are equal.

EXAMPLE 3   **Finding the Unknown Lengths of Sides in Similar Triangles**

Find the length of $\overline{DF}$ in the smaller triangle. Assume the triangles are similar.

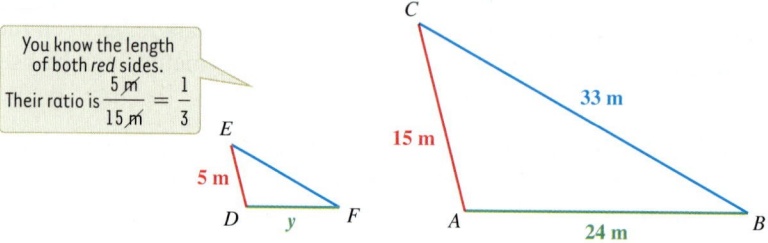

You know the length of both *red* sides. Their ratio is $\dfrac{5 \text{ m}}{15 \text{ m}} = \dfrac{1}{3}$

The length you want to find in the smaller triangle is $\overline{DF}$, and it corresponds to $\overline{AB}$ in the *larger* triangle. Then, notice that $\overline{ED}$ in the smaller triangle corresponds to $\overline{CA}$ in the larger triangle, and you know both of their lengths. Since the *ratios* of the lengths of corresponding sides are equal, you can set up a proportion. (Recall that a proportion states that two ratios are equal.)

$$\text{Corresponding sides} \begin{cases} DF \to \mathbf{y} \\ AB \to \mathbf{24} \end{cases} = \frac{\mathbf{5} \leftarrow ED}{\mathbf{15} \leftarrow CA} \begin{cases} \text{Corresponding} \\ \text{sides} \end{cases}$$

$$\frac{y}{24} = \frac{1}{3} \quad \text{Write } \tfrac{5}{15} \text{ in lowest terms as } \tfrac{1}{3}$$

Find the cross products.

$$24 \cdot 1 = 24$$

$$\frac{y}{24} \bowtie \frac{1}{3}$$

$$y \cdot 3$$

$$y \cdot 3 = 24 \qquad \text{Show that the cross products are equivalent.}$$

$$\frac{y \cdot \overset{1}{\cancel{3}}}{\cancel{3}_{1}} = \frac{24}{3} \qquad \text{Divide both sides by 3.}$$

Write **m** in the answer.

$$y = 8$$

$\overline{DF}$ has a length of 8 m.

◀ *Work Problem* **4** *at the Side.*

EXAMPLE 4   **Finding an Unknown Length and the Perimeter**

Find the perimeter of the smaller triangle. Assume the triangles are similar.

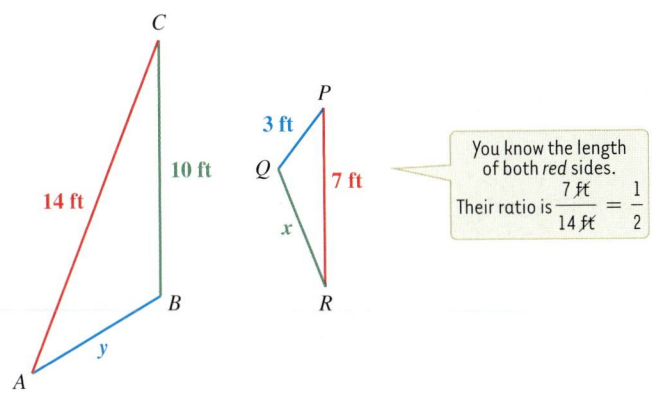

5 **(a)** Find the perimeter of triangle ABC in Example 4 at the left.

First find x, the length of $\overline{QR}$ in the smaller triangle, then add the lengths of all three sides to find the perimeter.

The smaller triangle is turned "upside down" compared to the larger triangle, so be careful when identifying corresponding sides. $\overline{PR}$ is the longest side in the smaller triangle, and $\overline{AC}$ is the longest side in the larger triangle. So $\overline{PR}$ and $\overline{AC}$ are corresponding sides and you know both of their lengths. $\overline{QR}$, the length you want to find in the smaller triangle, corresponds to $\overline{BC}$ in the larger triangle. The ratios of the lengths of corresponding sides are equal, so you can set up a proportion.

$$\begin{array}{l} QR \rightarrow \\ BC \rightarrow \end{array} \frac{x}{10} = \frac{7}{14} \begin{array}{l} \leftarrow PR \\ \leftarrow AC \end{array}$$

$$\frac{x}{10} = \frac{1}{2} \quad \text{Write } \tfrac{7}{14} \text{ in lowest terms as } \tfrac{1}{2}$$

Find the cross products.

$$10 \cdot 1 = 10$$

$$\frac{x}{10} = \frac{1}{2}$$

$$x \cdot 2$$

$$x \cdot 2 = 10 \quad \text{Show that the cross products are equivalent.}$$

$$\frac{x \cdot \overset{1}{\cancel{2}}}{\cancel{2}} = \frac{10}{2} \quad \text{Divide both sides by 2.}$$

Write **ft** for the length.

$$x = 5$$

$\overline{QR}$ has a length of 5 ft.

Now add the lengths of all three sides to find the perimeter of the smaller triangle.

$$\text{Perimeter} = 5 \text{ ft} + 3 \text{ ft} + 7 \text{ ft} = 15 \text{ ft}$$

(b) Find the perimeter of each triangle. Assume the triangles are similar.

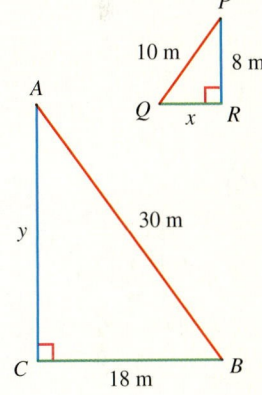

Work Problem **5** *at the Side.* ▶

6 Find the height of each flagpole.

(a)

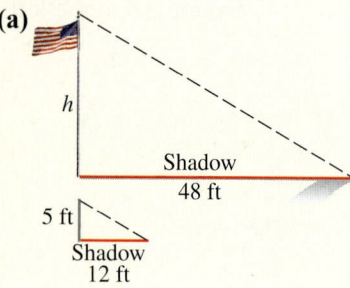

Shadow 48 ft

5 ft

Shadow 12 ft

(b)

7.2 m

5 m

h

12.5 m

OBJECTIVE **5** **Solve application problems involving similar triangles.** The next example shows an application of similar triangles.

EXAMPLE 5 **Using Similar Triangles in an Application**

A flagpole casts a shadow 99 m long at the same time that a pole 10 m tall casts a shadow 18 m long. Find the height of the flagpole.

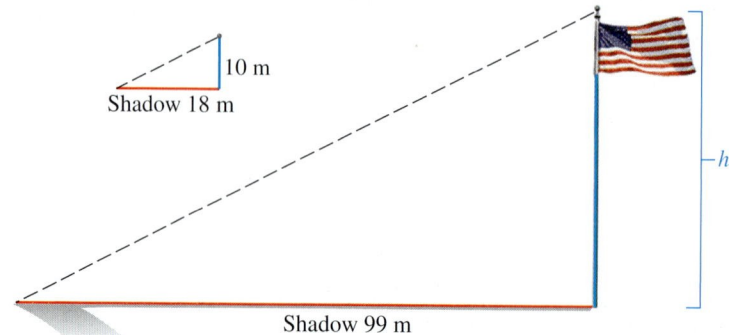

10 m
Shadow 18 m

Shadow 99 m

The triangles shown are similar, so write a proportion to find h.

Height in larger triangle → $\dfrac{h}{10} = \dfrac{99}{18}$ ← Shadow in larger triangle
Height in smaller triangle → ← Shadow in smaller triangle

Find the cross products and show that they are equivalent.

$$h \cdot 18 = 10 \cdot 99$$
$$h \cdot 18 = 990$$

$$\frac{h \cdot \overset{1}{\cancel{18}}}{\underset{1}{\cancel{18}}} = \frac{990}{18} \qquad \text{Divide both sides by 18.}$$

$$h = 55$$

The flagpole is 55 m high.

> **Note**
>
> There are several other correct ways to set up the proportion in Example 5 above. One way is to simply flip the ratios on *both* sides of the equal sign.
>
> $$\frac{10}{h} = \frac{18}{99}$$
>
> But there is another option, shown below.
>
> Height in larger triangle → $\dfrac{h}{99} = \dfrac{10}{18}$ ← Height in smaller triangle
> Shadow in larger triangle → ← Shadow in smaller triangle
>
> Notice that both ratios compare *height* to *shadow* in the same order. The ratio on the left describes the larger triangle, and the ratio on the right describes the smaller triangle.

ANSWERS

6. (a) $h = 20$ ft   **(b)** $h = 18$ m

6.6 ▶▶▶ **Exercises**

Each pair of triangles is congruent. List the corresponding angles and the corresponding sides. See Example 1.

1.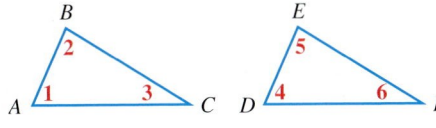

2.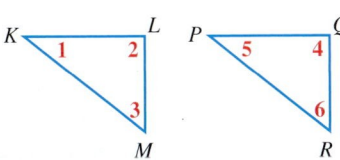

3.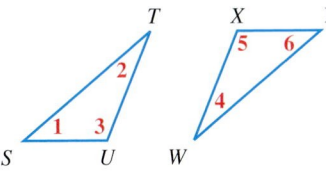

4.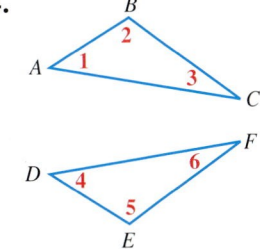

5.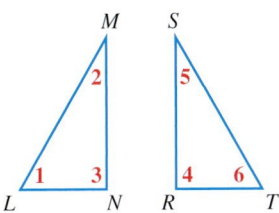

6.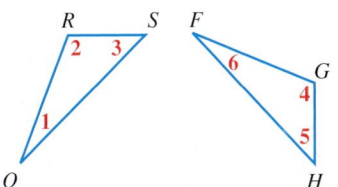

Determine which of these methods can be used to prove that each pair of triangles is congruent;
Angle–Side–Angle (ASA), Side–Side–Side (SSS), or Side–Angle–Side (SAS). See Example 2.

7.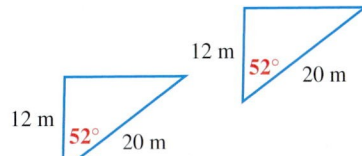

8.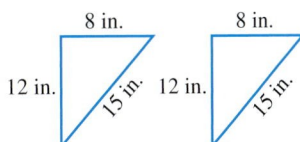

9.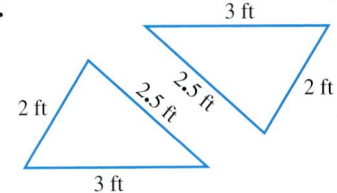

10.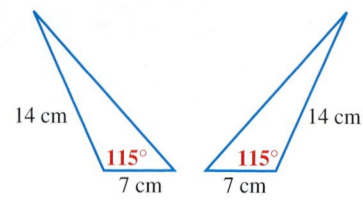

11.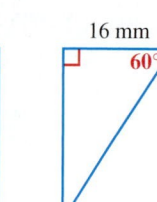

12.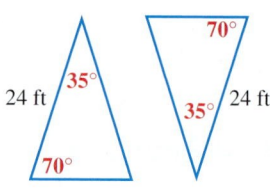

Relating Concepts (Exercises 13–16) For Individual or Group Work

Work Exercises 13–16 in order. Given the information in each exercise, explain how you can prove that the indicated triangles are congruent. Note: A midpoint divides a segment into two congruent parts.

13.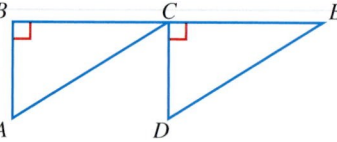

C is the midpoint of $\overline{BE}$ and $CD = BA$. Prove that $\triangle ABC \cong \triangle DCE$.

14.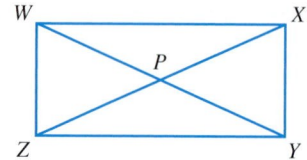

P is the midpoint of both $\overline{WY}$ and $\overline{XZ}$; $WZ = XY$. Prove that $\triangle WPZ \cong \triangle YPX$.

15.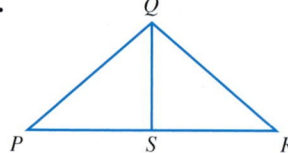

$\overline{QS} \perp \overline{PR}$ and S is the midpoint of $\overline{PR}$. Prove that $\triangle PQS \cong \triangle RQS$.

16.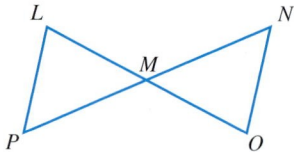

M is the midpoint of both $\overline{LO}$ and $\overline{PN}$. Prove that $\triangle PLM \cong \triangle NOM$.

Find the unknown lengths in each pair of similar triangles. See Example 3.

17.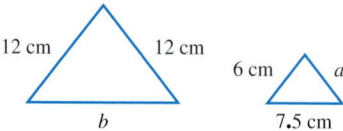

18.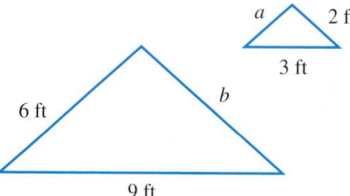

19.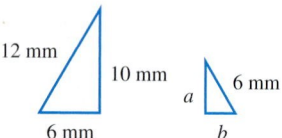

20.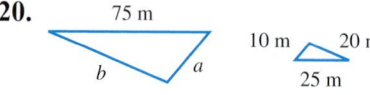

21.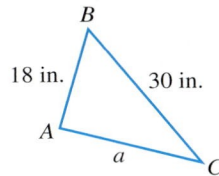

22.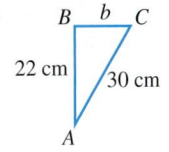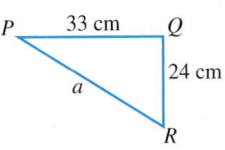

In Exercises 23 and 24, find the perimeter of each triangle. Assume the triangles are similar. See Example 4.

23.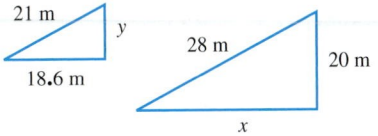

24.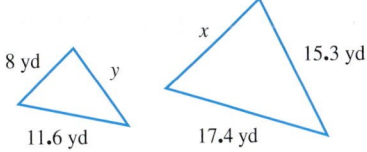

25. Triangles *CDE* and *FGH* are similar. Find the perimeter and area of triangle *FGH*. *Note:* The heights of similar triangles have the same ratio as corresponding sides. Round to the nearest tenth when necessary.

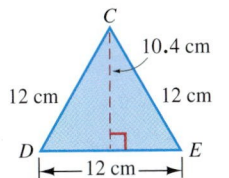 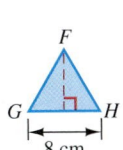

26. Triangles *JKL* and *MNO* are similar. Find the perimeter and area of triangle *MNO*.

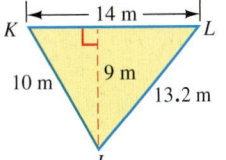 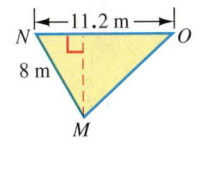

Solve each application problem. See Example 5.

27. The height of the house shown here can be found by comparing its shadow to the shadow cast by a 3-foot stick. Find the height of the house by writing a proportion and solving it.

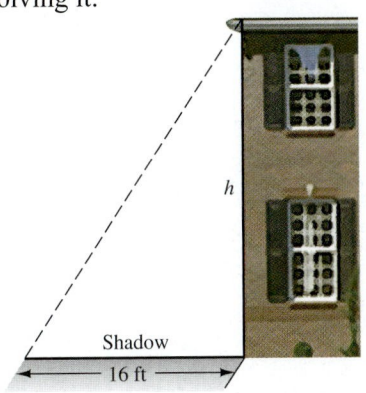

28. A fire lookout tower provides an excellent view of the surrounding countryside. The height of the tower can be found by lining up the top of the tower with the top of a 2-meter stick. Use similar triangles to find the height of the tower.

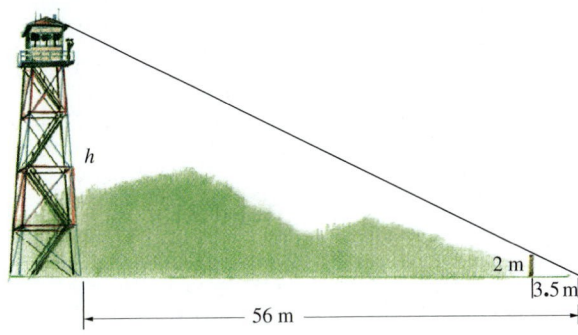

29. Look up the word *similar* in a dictionary. What is the nonmathematical definition of this word? Describe two examples of similar objects at home, school, or work.

30. Look up the word *congruent* in a dictionary. What is the nonmathematical definition of this word? Describe two examples of congruent objects at home, school, or work.

Find the unknown length in Exercises 31–34. Round your answers to the nearest tenth. Note: When a line is drawn parallel to one side of a triangle, the smaller triangle that is formed will be similar to the original triangle. In Exercises 31–32, the red segments are parallel.

31.

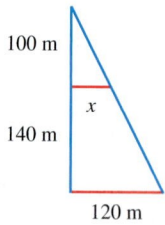

Hint: Redraw the two triangles and label the sides.

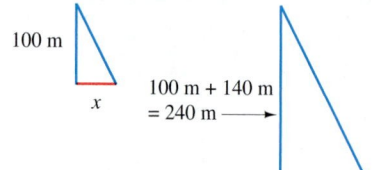

32.

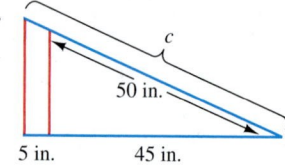

Hint: Redraw the two triangles and label the sides.

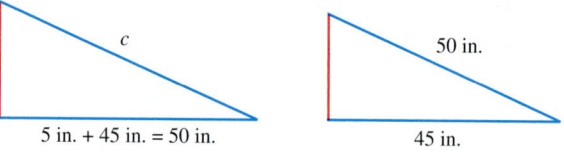

33. Use similar triangles and a proportion to find the length of the lake shown here. (*Hint:* The side 100 m long in the smaller triangle corresponds to the side of 100 m + 120 m = 220 m in the larger triangle.)

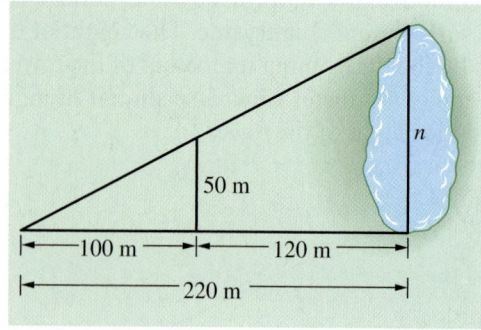

34. To find the height of the tree, find y and then add $5\frac{1}{2}$ ft for the distance from the ground to the eye level of the person.

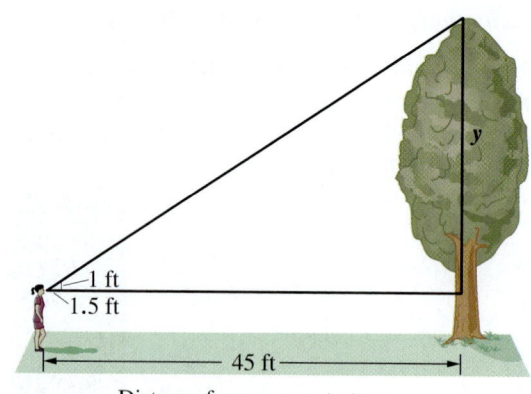

Distance from person to tree

Chapter 6 ▶▶▶ Summary

▶ Key Terms

6.1 **ratio**

A ratio compares two quantities that have the same type of units. For example, the ratio of 6 apples to 11 apples is written in fraction form as $\frac{6}{11}$. The common units (apples) divide out.

6.2 **rate**

A rate compares two measurements with different types of units. Examples are 96 dollars for 8 hours, or 450 miles on 18 gallons.

unit rate

A unit rate has 1 in the denominator.

cost per unit

Cost per unit is a rate that tells how much you pay for one item or one unit. The lowest cost per unit is the best buy.

6.3 **proportion**

A proportion states that two ratios or rates are equivalent.

cross products

Multiply along one diagonal and then along the other diagonal to find the cross products of a proportion. If the cross products are equal, the proportion is true.

6.5 **point**

A point is a location in space. *Example:* Point P at the right.

line

A line is a straight row of points that goes on forever in both directions. *Example:* Line AB, written $\overleftrightarrow{AB}$, at the right.

line segment

A line segment is a piece of a line with two endpoints. *Example:* Line segment PQ, written $\overline{PQ}$, at the right.

ray

A ray is a part of a line that has one endpoint and extends forever in one direction. *Example:* Ray RS, written $\overrightarrow{RS}$, at the right.

parallel lines

Parallel lines are two lines in the same plane that never intersect (never cross). *Example:* $\overleftrightarrow{AB}$ is parallel to $\overleftrightarrow{ST}$ at the right.

intersecting lines

Intersecting lines cross. *Example:* $\overleftrightarrow{RQ}$ intersects $\overleftrightarrow{AB}$ at point P at the right.

angle

An angle is made up of two rays that have a common endpoint called the vertex. *Example:* Angle 1 at the right.

degrees

Degrees are used to measure angles; a complete circle is 360 degrees, written 360°.

straight angle

A straight angle is an angle that measures *exactly* 180°; its sides form a straight line. *Example:* Angle G at the right.

right angle

A right angle is an angle that measures *exactly* 90°. *Example:* Angle AOB at the right.

acute angle

An acute angle is an angle that measures less than 90°. *Example:* Angle E at the right.

obtuse angle

An obtuse angle is an angle that measures more than 90° but less than 180°. *Example:* Angle F at the right.

perpendicular lines

Perpendicular lines are two lines that intersect to form a right angle. *Example:* $\overleftrightarrow{PQ}$ is perpendicular to $\overleftrightarrow{RS}$ at the right.

(continued)

▶ Key Terms

complementary angles	Complementary angles are two angles whose measures add up to 90°.
supplementary angles	Supplementary angles are two angles whose measures add up to 180°.
congruent angles	Congruent angles are angles that measure the same number of degrees.
vertical angles	Vertical angles are two nonadjacent congruent angles formed by two intersecting lines. *Example*: ∠COA and ∠EOF are vertical angles at the right.
corresponding angles	Corresponding angles are formed when two parallel lines are crossed by a transversal; corresponding angles are congruent and are on the same side of the transversal and in the same relative position. *Example*: In the figure at the right, line *m* is parallel to line *n*. The pairs of corresponding angles are ∠1 and ∠5, ∠2 and ∠6, ∠3 and ∠7, ∠4 and ∠8.
alternate interior angles	When two parallel lines are crossed by a transversal, there are two pairs of alternate interior angles and each pair is congruent. They are on opposite sides of the transversal. *Example*: In the figure at the right, line *m* is parallel to line *n*. The pairs of alternate interior angles are ∠3 and ∠5, ∠4 and ∠6.

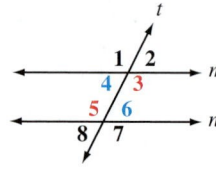

6.6	**congruent figures**	Congruent figures are identical both in shape and in size.
	similar figures	Similar figures have the same shape but are different sizes.
	congruent triangles	Congruent triangles are triangles with the same shape and the same size; corresponding angles measure the same number of degrees and corresponding sides have the same length.
	similar triangles	Similar triangles are triangles with the same shape but not necessarily the same size; corresponding angles measure the same number of degrees and the *ratios* of the lengths of corresponding sides are equal.

▶ New Symbols

$\overleftrightarrow{AB}$	Line *AB*	⊥	is perpendicular to	≅	is congruent to
$\overline{EF}$	Line segment *EF*	∠**MRN**	angle *MRN*	**m∠2**	the measure of angle 2
$\overrightarrow{RS}$	ray *RS*	1°	one degree	△**ABC**	triangle *ABC*

▶ Test Your Word Power

See how well you have learned the vocabulary in this chapter. Answers follow the Quick Review.

1. A **rate**
   A. can be written only as a decimal
   B. compares two quantities that have the same type of units
   C. compares two quantities that have different types of units
   D. is the reciprocal of a ratio.

2. **Cost per unit** is
   A. the best buy
   B. a ratio written in lowest terms
   C. found by comparing cross products
   D. the price of one item or one unit.

3. A **proportion**
   A. shows that two ratios or rates are equivalent
   B. contains only whole numbers or decimals
   C. always has one unknown number
   D. states that two improper fractions are equivalent.

4. An **obtuse angle**
   A. is formed by perpendicular lines
   B. is congruent to a right angle
   C. measures more than 90° but less than 180°
   D. measures less than 90°.

5. Two angles that are **complementary**
   A. have measures that add up to 180°
   B. are always congruent
   C. form a straight angle
   D. have measures that add up to 90°.

6. **Perpendicular lines**
   A. intersect to form a right angle
   B. intersect to form an acute angle
   C. never intersect
   D. have a common endpoint called the vertex.

▶ Quick Review

Concepts	Examples

6.1 Writing a Ratio

A ratio compares two quantities that have the same type of units. A ratio is usually written as a fraction with the number that is mentioned first in the numerator. The common units divide out and are not written in the answer. Check that the fraction is in lowest terms.

Write this ratio as a fraction in lowest terms.

60 ounces of medicine **to** 160 ounces of medicine

$$\frac{60 \text{ ounces}}{160 \text{ ounces}} = \frac{60 \div 20}{160 \div 20} = \frac{3}{8} \quad \left\{ \begin{array}{l} \text{Ratio in} \\ \text{lowest terms} \end{array} \right.$$

↑ Divide out common units.

6.1 Using Mixed Numbers in a Ratio

If a ratio has mixed numbers, change the mixed numbers to improper fractions. Rewrite the problem in horizontal form, using the "÷" symbol for division. Finally, multiply by the reciprocal of the divisor.

Write as a ratio of whole numbers in lowest terms.

$$2\frac{1}{2} \quad \text{to} \quad 3\frac{3}{4}$$

$$\frac{2\frac{1}{2}}{3\frac{3}{4}} \qquad \text{Ratio with mixed numbers}$$

$$= \frac{\frac{5}{2}}{\frac{15}{4}} \qquad \text{Ratio with improper fractions}$$

$$= \frac{5}{2} \div \frac{15}{4} = \frac{5}{2} \cdot \frac{4}{15} = \frac{\cancel{5} \cdot \cancel{2} \cdot 2}{2 \cdot 3 \cdot \cancel{5}} = \frac{2}{3} \quad \left\{ \begin{array}{l} \text{Ratio in} \\ \text{lowest} \\ \text{terms} \end{array} \right.$$

↑ Reciprocals ↑

6.1 Using Measurements in Ratios

When a ratio compares measurements, both measurements must be in the *same* units. It is usually easier to compare the measurements using the smaller unit, for example, inches instead of feet.

Write 8 inches to 6 feet as a ratio in lowest terms.

Compare using the smaller unit, inches. Because 1 foot has 12 inches, 6 feet is

$$6 \cdot \textbf{12 inches} = 72 \text{ inches}$$

The ratio is shown below.

$$\frac{8 \text{ inches}}{6 \text{ feet}} = \frac{8 \text{ inches}}{72 \text{ inches}} = \frac{8 \div 8}{72 \div 8} = \frac{1}{9} \quad \left\{ \begin{array}{l} \text{Ratio in} \\ \text{lowest terms} \end{array} \right.$$

6.2 Writing Rates

A rate compares two measurements with different types of units. The units do *not* divide out, so you must write them as part of the rate.

Write the rate as a fraction in lowest terms.

475 miles in 10 hours

$$\frac{475 \text{ miles} \div 5}{10 \text{ hours} \div 5} = \frac{95 \text{ miles}}{2 \text{ hours}} \quad \begin{array}{l} \text{Must write units:} \\ \text{miles and hours} \end{array}$$

Concepts	Examples

6.2 Finding a Unit Rate

A unit rate has 1 in the denominator. To find the unit rate, divide the numerator by the denominator. Write unit rates using the word **per** or a / mark.

Write as a unit rate: $1278 in 9 days.

$$\frac{\$127}{9\text{ day}} \quad \leftarrow \text{Fraction bar indicates division.}$$

$$9\overline{)1278}^{\,142} \quad \text{so} \quad \frac{\$1278 \div 9}{9 \text{ days} \div 9} = \frac{\$142}{1 \text{ day}}$$

Write the answer as $142 **per** day or $142/day.

6.2 Finding the Best Buy

The best buy is the item with the lowest cost per unit. Divide the price by the number of units. Round to thousandths when necessary. Then compare to find the lowest cost per unit.

Find the best buy on cheese.

2 pounds for $2.25

3 pounds for $3.40

Find cost per unit (cost per pound).

$$\frac{\$2.25}{2} = \$1.125 \text{ per pound}$$

$$\frac{\$3.40}{3} \approx \$1.133 \text{ per pound}$$

The lower cost per pound is $1.125, so 2 pounds of cheese is the best buy.

6.3 Writing Proportions

A proportion states that two ratios or rates are equivalent. The proportion "5 is to 6 as 25 is to 30" is written as shown below.

$$\frac{5}{6} = \frac{25}{30}$$

To see whether a proportion is true or false, multiply along one diagonal, then multiply along the other diagonal. If the two cross products are equal, the proportion is true. If the two cross products are unequal, the proportion is false.

Write as a proportion: 8 is to 40 as 32 is to 160

$$\frac{8}{40} = \frac{32}{160}$$

Is this proportion true or false?

$$\frac{6}{8\frac{1}{2}} = \frac{24}{34}$$

Find the cross products.

$$8\frac{1}{2} \cdot 24 = \frac{17}{\cancel{2}} \cdot \frac{\cancel{2} \cdot 12}{1} = \mathbf{204}$$

$$\frac{6}{8\frac{1}{2}} \bowtie \frac{24}{34} \qquad 6 \cdot 34 = \mathbf{204} \quad \leftarrow$$

Equal

The cross products are equal, so the proportion is true.

Concepts	Examples

6.3 Solving Proportions

Solve for an unknown number in a proportion by using these steps.

Find the unknown number.

$$\frac{12}{x} = \frac{6}{8} \quad \rbrack \text{Write } \tfrac{6}{8} \text{ in lowest terms.}$$

$$\frac{12}{x} = \frac{3}{4} \leftarrow$$

Step 1 Find the cross products. (If desired, you can rewrite the ratios in lowest terms before finding the cross products.)

Step 1
$$\frac{12}{x} = \frac{3}{4} \qquad \begin{array}{l} x \cdot 3 \leftarrow \\ 12 \cdot 4 \leftarrow \end{array} \rbrack \begin{array}{l} \text{Find the cross} \\ \text{products.} \end{array}$$

Step 2 Show that the cross products are equivalent.

Step 2
$$x \cdot 3 = \underbrace{12 \cdot 4} \qquad \begin{array}{l}\text{Show that cross products are} \\ \text{equivalent.}\end{array}$$
$$x \cdot 3 = 48$$

Step 3 Divide both sides of the equation by the coefficient of the variable term.

Step 3
$$\frac{x \cdot \overset{1}{\cancel{3}}}{\underset{1}{\cancel{3}}} = \frac{48}{3} \qquad \text{Divide both sides by 3}$$
$$x = 16$$

Step 4 Check by writing the solution in the *original* proportion and finding the cross products.

Step 4

$$x \text{ is } 16 \rightarrow \frac{12}{16} = \frac{6}{8} \qquad \begin{array}{l} 16 \cdot 6 = \mathbf{96} \leftarrow \\ 12 \cdot 8 = \mathbf{96} \leftarrow \end{array} \rbrack \text{Equal}$$

The cross products are equal, so 16 is the correct solution (**not** 96).

6.4 Applications of Proportions

Use the six problem-solving steps from **Section 3.3.**

If 3 pounds of grass seed cover 450 square feet of lawn, how much seed is needed for 1500 square feet of lawn?

Step 1 **Read** the problem.

Step 1 The problem is about the amount of grass seed needed for a lawn.

Unknown: pounds of seed needed for 1500 square feet of lawn
Known: 3 pounds cover 450 square feet of lawn

Step 2 **Assign a variable.**

Step 2 There is only one unknown, so let x be the pounds of seed needed for 1500 square feet of lawn.

Step 3 **Write an equation.**

Step 3 The equation is in the form of a proportion. Be sure that both rates in the proportion compare pounds to square feet in the same order.

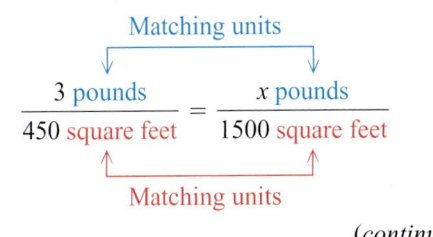

$$\frac{3 \text{ pounds}}{450 \text{ square feet}} = \frac{x \text{ pounds}}{1500 \text{ square feet}}$$

Matching units

(continued)

Concepts	Examples

6.4 Applications of Proportions *(continued)*

Step 4 **Solve** the equation.

Step 4 Ignore the units while finding the cross products and solving for x.

$$450 \cdot x = 3 \cdot 1500 \qquad \text{Show that cross products}$$
$$450 \cdot x = 4500 \qquad \text{are equivalent.}$$

$$\frac{\cancel{450} \cdot x}{\cancel{450}} = \frac{4500}{450} \qquad \text{Divide both sides by 450}$$

$$x = 10$$

Step 5 **State the answer.**

Step 5 10 pounds of grass seed are needed for 1500 square feet of lawn.

Step 6 **Check** that the solution is reasonable by putting it back into the original statement of the problem.

Step 6 450 square feet of lawn needs 3 pounds of seed; 1500 square feet is about *three times as much* lawn, so about *three times as much* seed is needed.

$$(3)(3 \text{ pounds}) = 9 \text{ pounds} \leftarrow \text{Estimate}$$

The solution, 10 pounds, is close to the estimate of 9 pounds, so it is reasonable.

6.5 Lines

A *line* is a straight row of points that goes on forever in both directions. If part of a line has one endpoint, it is a *ray*. If it has two endpoints, it is a *line segment*.

Identify each figure as a line, line segment, or ray and name it using the appropriate symbol.

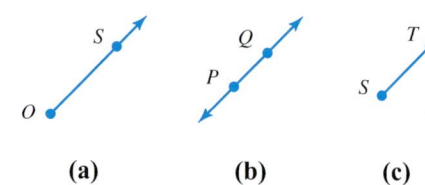

(a)	(b)	(c)

Figure **(a)** shows a ray named $\overrightarrow{OS}$.

Figure **(b)** shows a line named $\overleftrightarrow{PQ}$ or $\overleftrightarrow{QP}$.

Figure **(c)** shows a line segment named $\overline{ST}$ or $\overline{TS}$.

If two lines intersect at right angles, they are *perpendicular*. If two lines in the same plane never intersect, they are *parallel*.

Label each pair of lines as appearing to be parallel or as perpendicular.

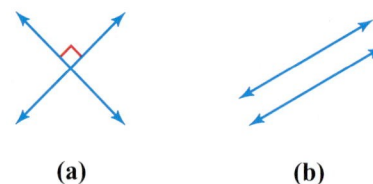

(a)	(b)

Figure **(a)** shows perpendicular lines (they intersect at 90°).

Figure **(b)** shows lines that appear to be parallel (they never intersect).

Concepts	Examples

(6.5) Angles

If the sum of the measures of two angles is 90°, they are *complementary*.
If the sum of the measures of two angles is 180°, they are *supplementary*.

If two angles measure the same number of degrees, the angles are *congruent*. The symbol for congruent is ≅.

Two nonadjacent angles formed by two intersecting lines are called *vertical angles*. Vertical angles are congruent.

Find the complement and supplement of a 35° angle.

$$90° - 35° = 55° \quad \text{(the complement)}$$
$$180° - 35° = 145° \text{ (the supplement)}$$

Identify the vertical angles in this figure. Which angles are congruent?

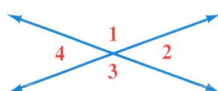

$\angle 1$ and $\angle 3$ are vertical angles.

$\angle 2$ and $\angle 4$ are vertical angles.

Vertical angles are congruent, so $\angle 1 \cong \angle 3$ and $\angle 2 \cong \angle 4$.

(6.5) Parallel Lines

When two parallel lines are crossed by a transversal, corresponding angles are congruent, and alternate interior angles are congruent. Use this information to find the measures of the other angles.

Read $m\angle 1$ as "the measure of angle 1."

Line m is parallel to line n and the measure of $\angle 4$ is 125°. Find the measures of the other angles.

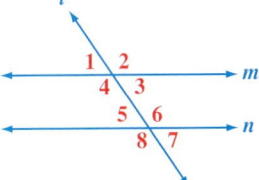

$\angle 4 \cong \angle 8$ (corresponding angles), so $m\angle 8 = 125°$.

$\angle 4 \cong \angle 6$ (alternate interior angles), so $m\angle 6 = 125°$.

$\angle 6 \cong \angle 2$ (corresponding angles), so $m\angle 2 = 125°$.

$\angle 4$ and $\angle 3$ are supplements, so $m\angle 3 = 180° - 125° = 55°$.

$\angle 3 \cong \angle 7$ (corresponding angles), so $m\angle 7 = 55°$.

$\angle 3 \cong \angle 5$ (alternate interior angles), so $m\angle 5 = 55°$.

$\angle 5 \cong \angle 1$ (corresponding angles), so $m\angle 1 = 55°$.

(6.6) Proving That Two Triangles Are Congruent

Congruent triangles are identical both in shape and in size. This means that corresponding angles have the same measure and corresponding sides have the same length.

Here are three ways to prove that two triangles are congruent.

1. Angle–Side–Angle (ASA) method: If two angles and the side between them on one triangle measure the same as the corresponding parts on another triangle, the triangles are congruent.

2. Side–Side–Side (SSS) method: If three sides of one triangle measure the same as the corresponding sides of another triangle, the triangles are congruent.

3. Side–Angle–Side (SAS) method: If two sides and the angle between them on one triangle measure the same as the corresponding parts on another triangle, the triangles are congruent.

Determine which method can be used to prove that each pair of triangles is congruent.

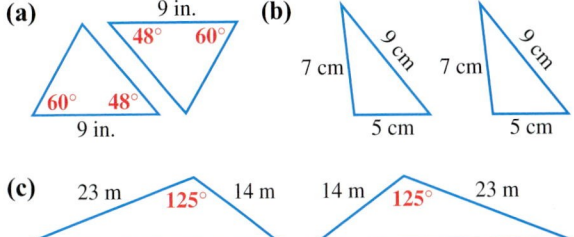

(a) On both triangles, two corresponding angles and the side between them measure the same, so use ASA.

(b) Each pair of corresponding sides has the same length, so use SSS.

(c) On both triangles, two corresponding sides and the angle between them measure the same, so use SAS.

Concepts	Examples

(6.6) Finding the Unknown Lengths in Similar Triangles

Use the fact that in similar triangles, the ratios of the lengths of corresponding sides are equal.
Write a proportion. Then find the cross products and show that they are equivalent. Finish solving for the unknown length.

Find the unknown lengths in this pair of similar triangles.

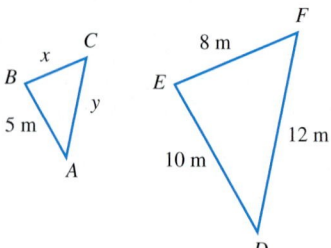

$$\frac{x}{8} = \frac{5}{10}$$

$$x \cdot 10 = 8 \cdot 5$$

$$\frac{x \cdot \overset{1}{\cancel{10}}}{\underset{1}{\cancel{10}}} = \frac{40}{10}$$

$$x = 4 \text{ m}$$

$$\frac{y}{12} = \frac{5}{10}$$

$$y \cdot 10 = 12 \cdot 5$$

$$\frac{y \cdot \overset{1}{\cancel{10}}}{\underset{1}{\cancel{10}}} = \frac{60}{10}$$

$$y = 6 \text{ m}$$

ANSWERS TO TEST YOUR WORD POWER

1. C; *Example:* $4.50 for 3 pounds is a rate comparing dollars to pounds.

2. D; *Example:* $1.95 per gallon tells the price of one gallon (one unit).

3. A; *Example:* The proportion $\frac{5}{6} = \frac{25}{30}$ says that $\frac{5}{6}$ is equivalent to $\frac{25}{30}$.

4. C; *Examples:* Angles that measure 91°, 120°, and 175° are all obtuse angles.

5. D; *Example:* If $\angle 1$ measures 35° and $\angle 2$ measures 55°, the angles are complementary because 35° + 55° = 90°.

6. A; *Example:* $\overleftrightarrow{EF}$ is perpendicular to $\overleftrightarrow{GH}$.

Chapter 6 ▶▶▶ Review Exercises

[6.1] *Write each ratio as a fraction in lowest terms. Change to the same units when necessary, using the table of measurement comparisons in **Section 6.1.** Use the information in the graph to answer Exercises 1–3.*

AVERAGE LENGTH OF SHARKS AND WHALES

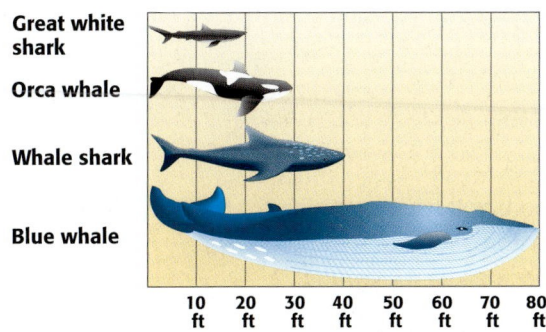

Great white shark

Orca whale

Whale shark

Blue whale

10 ft 20 ft 30 ft 40 ft 50 ft 60 ft 70 ft 80 ft

Source: Grolier Multimedia Encyclopedia.

1. Orca whale's length to whale shark's length

2. Blue whale's length to great white shark's length

3. Which two animals' lengths give a ratio of $\frac{1}{2}$? There is more than one correct answer.

4. $2.50 to $1.25

5. $0.30 to $0.45

6. $1\frac{2}{3}$ cups to $\frac{2}{3}$ cup

7. $2\frac{3}{4}$ miles to $16\frac{1}{2}$ miles

8. 5 hours to 100 minutes

9. 9 in. to 2 ft

10. 1 ton to 1500 pounds

11. 8 hours to 3 days

12. Jake sold $350 worth of his kachina figures. Ramona sold $500 worth of her pottery. What is the ratio of Ramona's sales to Jake's sales?

13. Ms. Wei's new car gets 35 miles per gallon. Her old car got 25 miles per gallon. Find the ratio of the new car's mileage to the old car's mileage.

14. This fall, 6000 students are taking math courses and 7200 students are taking English courses. Find the ratio of math students to English students.

[6.2] *Write each rate as a fraction in lowest terms.*

15. $88 for 8 dozen

16. 96 children in 40 families

17. When entering data into his computer, Patrick can type four pages in 20 minutes. Give his rate in pages per minute and minutes per page.

18. Elena made $24 in three hours. Give her earnings in dollars per hour and hours per dollar.

Find the best buy.

19. Minced onion

     8 ounces for $4.98

     3 ounces for $2.49

     2 ounces for $1.89

20. Dog food; you have a coupon for $1 off on 8 pounds or more.

     35.2 pounds for $36.96

     17.6 pounds for $18.69

     3.5 pounds for $4.25

[6.3] *Use either the method of writing in lowest terms or of finding cross products to decide whether each proportion is* true *or* false. *Show your work and then write* true *or* false.

21. $\dfrac{6}{10} = \dfrac{9}{15}$

22. $\dfrac{6}{48} = \dfrac{9}{36}$

23. $\dfrac{47}{10} = \dfrac{98}{20}$

24. $\dfrac{1.5}{2.4} = \dfrac{2}{3.2}$

25. $\dfrac{3\frac{1}{2}}{2\frac{1}{3}} = \dfrac{6}{4}$

Find the unknown number in each proportion. Round answers to the nearest hundredth when necessary.

26. $\dfrac{4}{42} = \dfrac{150}{x}$

27. $\dfrac{16}{x} = \dfrac{12}{15}$

28. $\dfrac{100}{14} = \dfrac{x}{56}$

29. $\dfrac{5}{8} = \dfrac{x}{20}$

30. $\dfrac{x}{24} = \dfrac{11}{18}$

31. $\dfrac{7}{x} = \dfrac{18}{21}$

32. $\dfrac{x}{3.6} = \dfrac{9.8}{0.7}$

33. $\dfrac{13.5}{1.7} = \dfrac{4.5}{x}$

34. $\dfrac{0.82}{1.89} = \dfrac{x}{5.7}$

[6.4] *Set up and solve a proportion for each application problem.*

35. The ratio of cats to dogs at the animal shelter is 3 to 5. If there are 45 dogs, how many cats are there?

36. Danielle had 8 hits in 28 times at bat during last week's games. If she continues to hit at the same rate, how many hits will she get in 161 times at bat?

37. If 3.5 pounds of ground beef cost $9.77, what will 5.6 pounds cost? Round to the nearest cent.

38. About 4 out of 10 students are expected to vote in campus elections. There are 8247 students. How many are expected to vote? Round to the nearest whole number.

39. The scale on Brian's model railroad is 1 inch to 16 feet. One of the scale model boxcars is 4.25 inches long. What is the length of a real boxcar in feet?

40. Marvette makes necklaces to sell at a local gift shop. She made 2 dozen necklaces in $16\frac{1}{2}$ hours. How long will it take her to make 40 necklaces?

41. A 180-pound person burns 284 calories playing basketball for 25 minutes. At this rate, how many calories would the person burn in 45 minutes, to the nearest whole number? (*Source: Wellness Encyclopedia.*)

42. In the hospital pharmacy, Michiko sees that a medicine is to be given at the rate of 3.5 milligrams for every 50 pounds of body weight. How much medicine should be given to a patient who weighs 210 pounds?

[6.5] *Identify each figure as a* line, line segment, *or* ray *and name it using the appropriate symbol.*

43.

44.

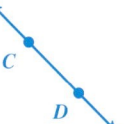

45.

Label each pair of lines as appearing to be parallel, *as* perpendicular, *or as* intersecting.

46.

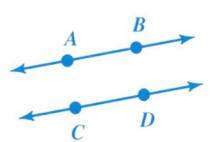

47.

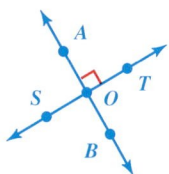

48.

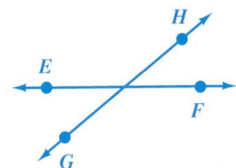

Label each angle as an acute, right, obtuse, or straight angle. For right and straight angles, indicate the number of degrees in the angle.

49.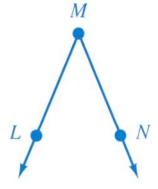

50.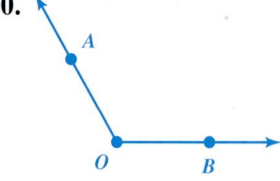

51.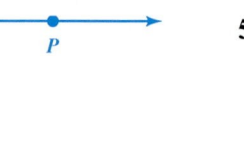

52.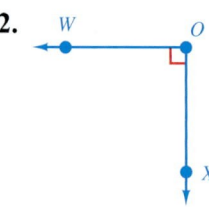

Find the complement or supplement of each angle.

53. Find each complement.

   (a) 80°

   (b) 45°

   (c) 7°

54. Find each supplement.

   (a) 155°

   (b) 90°

   (c) 33°

55. In the figure below, $\angle 2$ measures 60°. Find the measure of each of the other angles.

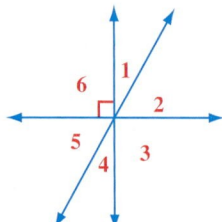

56. Line m is parallel to line n and $\angle 8$ measures 160°. Find the measures of the other angles.

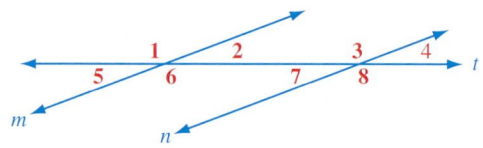

[6.6] *Determine which method can be used to prove that each pair of triangles is congruent.*

57.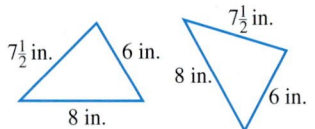

58.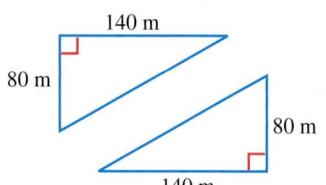

59.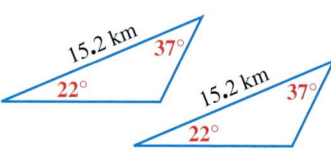

Find the unknown lengths in each pair of similar triangles. Then find the perimeter of the larger triangle in each pair.

60.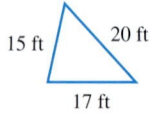

61.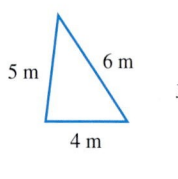

62.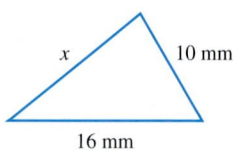

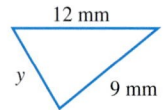

▶▶▶ Mixed Review Exercises

Find the unknown number in each proportion. Round to hundredths when necessary.

63. $\dfrac{x}{45} = \dfrac{70}{30}$

64. $\dfrac{x}{52} = \dfrac{0}{20}$

65. $\dfrac{64}{10} = \dfrac{x}{20}$

66. $\dfrac{15}{x} = \dfrac{65}{100}$

67. $\dfrac{7.8}{3.9} = \dfrac{13}{x}$

68. $\dfrac{34.1}{x} = \dfrac{0.77}{2.65}$

Write each ratio as a fraction in lowest terms. Change to the same units when necessary.

69. 4 dollars to 10 quarters

70. $4\frac{1}{8}$ inches to 10 inches

71. 10 yards to 8 feet

72. \$3.60 to \$0.90

73. 12 eggs to 15 eggs

74. 37 meters to 7 meters

75. 3 pints to 4 quarts

76. 15 minutes to 3 hours

77. $4\frac{1}{2}$ miles to $1\frac{3}{10}$ miles

Set up and solve a proportion for each application problem.

78. Nearly 7 out of 8 fans buy something to drink at rock concerts. How many of the 28,500 fans at today's concert would be expected to buy a beverage? Round to the nearest hundred fans.

79. Emily spent \$150 on car repairs and \$400 on car insurance. What is the ratio of the amount spent on insurance to the amount spent on repairs?

80. Antonio is choosing among three packages of plastic wrap. Is the best buy 25 feet for \$0.78; 75 feet for \$1.99; or 100 feet for \$2.59? He has a coupon for 50¢ off that is good for either of the larger two packages.

81. On this scale drawing of a backyard patio, 0.5 inch represents 6 feet. If the patio measures 1.75 inches long and 1.25 inches wide on the drawing, what will be the actual length and width of the patio when it is built?

0.5 in. = 6 ft

82. A vitamin supplement for cats is to be given at the rate of 1000 milligrams for a 5-pound cat. (*Source:* St. Jon Pet Care Products.)

   (a) How much should be given to a 7-pound cat?

   (b) How much should be given to an 8-ounce kitten?

83. Charles made 251 points during 169 minutes of playing time last year. At that same rate, how many points would you expect him to make if he plays 14 minutes in tonight's game? Round to the nearest whole number.

84. An antibiotic is to be given at the rate of $1\frac{1}{2}$ teaspoons for every 24 pounds of body weight. How much should be given to an infant who weighs 8 pounds?

Label each figure. Choose from these labels: line segment, ray, parallel lines, perpendicular lines, intersecting lines, acute angle, right angle, straight angle, obtuse angle. *Indicate the number of degrees in the right angle and the straight angle.*

85.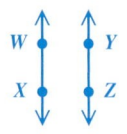

86.

87.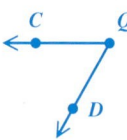

88.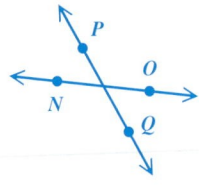

89.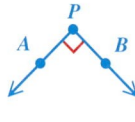

90.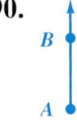

91.

92.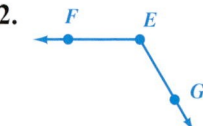

93.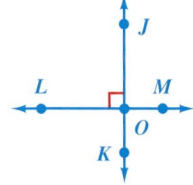

94. Explain what is happening in each sentence.

    (a) The road was so slippery that my car did a 360.

    (b) After the election, the governor's view on new taxes took a 180° turn.

95. (a) Can two obtuse angles be supplementary? Explain why or why not.

    (b) Can two acute angles be complementary? Explain why or why not.

96. In the figure below, ∠2 measures 45° and ∠7 measures 55°. Find the measure of each of the other angles.

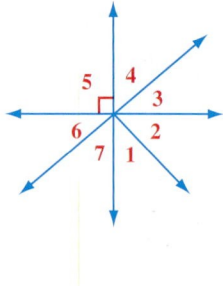

97. In the figure below, line *m* is parallel to line *n* and ∠5 measures 75°. Find the measures of the other angles.

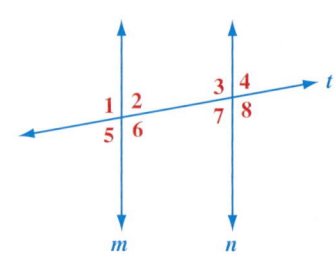

Chapter 6 ▶▶▶ Test

Use the Chapter Test Prep Video CD to see fully worked-out solutions to any of the exercises you want to review.

Write each rate or ratio as a fraction in lowest terms. Change to the same units when necessary.

1. $15 for 75 minutes

2. 3 hours to 40 minutes

3. The little theater at our college has 320 seats. The auditorium has 1200 seats. Find the ratio of auditorium seats to theater seats.

4. Find the best buy on spaghetti sauce. You have a coupon for 75¢ off Brand X and a coupon for 50¢ off Brand Y.

      26 ounces of Brand X for $3.89
      16 ounces of Brand Y for $1.89
      14 ounces of Brand Z for $1.29

Find the unknown number in each proportion. In Problems 5–7, round answers to the nearest hundredth when necessary.

5. $\dfrac{5}{9} = \dfrac{x}{45}$

6. $\dfrac{3}{1} = \dfrac{8}{x}$

7. $\dfrac{x}{20} = \dfrac{6.5}{0.4}$

8. $\dfrac{2\frac{1}{3}}{x} = \dfrac{\frac{8}{9}}{4}$

Set up and solve a proportion for each application problem.

9. Pedro entered 18 orders into his computer in 30 minutes at his job. At that rate, how many orders could he enter in forty minutes?

10. About 2 out of every 15 people are left-handed. How many of the 650 students in our school would you expect to be left-handed? Round to the nearest whole number.

11. A medication is given at the rate of 8.2 grams for every 50 pounds of body weight. How much should be given to a 145-pound person? Round to the nearest tenth.

12. On a scale model, 1 inch represents 8 feet. If a building in the model is 7.5 inches tall, what is the actual height of the building in feet?

1. _____

2. _____

3. _____

4. _____

5. _____

6. _____

7. _____

8. _____

9. _____

10. _____

11. _____

12. _____

13. _____

14. _____

15. _____

16. _____

17. _____

Choose the figure that matches each label. For right and straight angles, indicate the number of degrees in the angle.

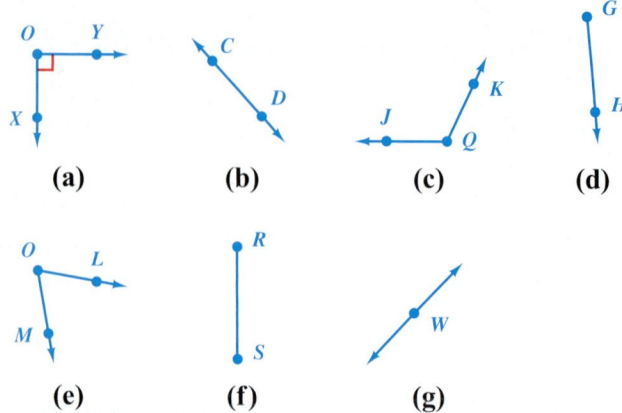

(a)   (b)   (c)   (d)

(e)   (f)   (g)

13. Acute angle is figure _____.

14. Right angle is figure _____ and its measure is _____.

15. Ray is figure _____.

16. Straight angle is figure _____ and its measure is _____.

17. Write a definition of parallel lines and a definition of perpendicular lines. Make a sketch to illustrate each definition.

18. Find the complement of an 81° angle.

19. Find the supplement of a 20° angle.

18. _____

19. _____

20. In the figure below, ∠4 measures 50° and ∠6 measures 95°. Find the measures of the other angles.

21. In the figure below, line *m* is parallel to line *n* and ∠3 measures 65°. Find the measures of the other angles.

20. _____

21. _____

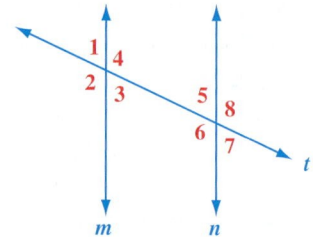

Determine which method can be used to prove that each pair of triangles is congruent.

22. _____

22.

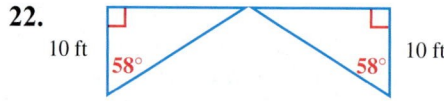

23.

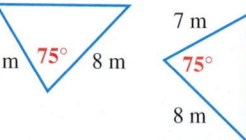

23. _____

24. Find the unknown lengths in these similar triangles.

25. Find the perimeter of each of these similar triangles.

24. _____

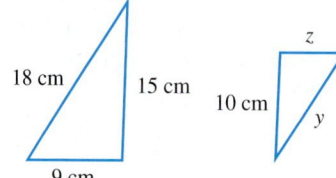

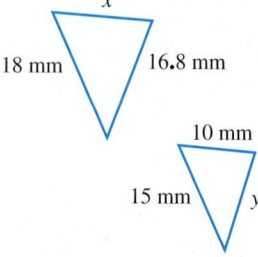

25. _____

7

Percent

Everyone likes to buy things that are "on sale." It helps if you can estimate the sale price *before* deciding to buy. Most states add on a sales tax, so that is part of your final cost as well. Find out how to calculate discounts and sales tax in **Section 7.5**, Examples 1 and 4. Then get some practice in Exercises 1–10, 17–24, and 51–54.

In **Section 7.5** you'll also learn how to use percents to estimate restaurant tips and calculate the interest on loans.

7.1 ▶▶▶ The Basics of Percent

1 Write a percent to describe each situation.

(a) You leave a $20 tip for a restaurant bill of $100. What percent tip did you leave?

(b) The tax on a $100 graphing calculator is $5. What is the tax rate?

(c) You earn 94 points on a 100 point test. What percent of the points did you earn?

OBJECTIVE **1** **Learn the meaning of percent.** You have probably seen percents frequently in daily life. The symbol for percent is **%**. For example, during one day you may leave a 15% tip for the waitress at dinner, pay 7% sales tax on a CD player, and buy shoes at 25% off the regular price. The next day your score on a math test may be 89% correct.

> **The Meaning of Percent**
>
> A **percent** is a ratio with a denominator of 100. So percent means "per 100" or "how many out of 100." The symbol for percent is **%**. Read 15% as "fifteen percent."

EXAMPLE 1 **Understanding Percent**

Write a percent to describe each situation.

(a) If you left a $15 tip when the restaurant bill was $100, then you left $15 per $100 or $\frac{15}{100}$ or 15%.

(b) If you pay $7 in tax on a $100 CD player, then the tax rate is $7 per $100 or $\frac{7}{100}$ or 7%.

(c) If you earn 89 points on a 100-point math test, then your score is 89 out of 100 or $\frac{89}{100}$ or 89%.

◀ *Work Problem* **1** *at the Side.*

OBJECTIVE **2** **Write percents as decimals.** In order to work with percents, you will need to write them as decimals or as fractions. We'll start by writing percents as equivalent decimal numbers. Twenty-five percent, or 25%, means 25 parts out of 100 parts, or $\frac{25}{100}$. Remember that the fraction bar indicates division. So we can write $\frac{25}{100}$ as $25 \div 100$. When you do the division, $25 \div 100$ is 0.25.

Indicates division $\longrightarrow \dfrac{25}{100}$ can be written as $25 \div 100 = 0.25$
↑
Indicates division

Another way to remember how to write a percent as a decimal is to use the meaning of the word *percent*. The first part of the word, *per*, is an indicator word for *division*. The last part of the word, *cent*, comes from the Latin word for *hundred*. (Recall that there are 100 *cent*s in a dollar, and 100 years in a *cent*ury.)

25% is 25 **percent**
$$25 \div 100 = 0.25$$

> **Writing a Percent as an Equivalent Decimal**
>
> To write a percent as a decimal, drop the % symbol and divide by 100.

EXAMPLE 2 **Writing Percents as Decimals**

Write each percent as a decimal.

(a) 47% $47\% = 47 \div 100 = 0.47$ Decimal form

(b) 3% $3\% = 3 \div 100 = 0.03$ Decimal form

(c) 28.2% $28.2\% = 28.2 \div 100 = 0.282$ Decimal form

(d) 100% $100\% = 100 \div 100 = 1.00$ Decimal form

(e) 135% $135\% = 135 \div 100 = 1.35$ Decimal form

> **CAUTION**
> In Example 2(d) above, notice that 100% is 1.00, or 1, which is a whole number. Whenever you have a percent that is *100% or greater*, the equivalent decimal number will be *1 or greater*. Notice in Example 2(e) above that 135% is 1.35 (greater than 1).

Work Problem 2 *at the Side.* ▶

In the exercise set for **Section 5.5,** you discovered a shortcut for *dividing* by 100: Move the decimal point *two* places to the *left*. You can use this shortcut when writing percents as decimals.

EXAMPLE 3 **Changing Percents to Decimals by Moving the Decimal Point**

Write each percent as a decimal by dropping the percent symbol and moving the decimal point two places to the left.

(a) 17%

$17\% = 17.\%$ ⟵ Decimal point starts at far right side.

.17 ⟵ Percent symbol is dropped.

⟶ Decimal point is moved *two places* to the *left*.

> Moving the decimal point *two places* to the *left* is a quick way to *divide by 100*.

$17\% = 0.17$

(b) 160%

$160\% = 160.\% = 1.60$ or 1.6 Decimal point starts at far right side.
1.60 is equivalent to 1.6

(c) 4.9%

04.9% 0 is attached so the decimal point can be moved *two places* to the *left*.

$4.9\% = 0.049$

(d) 0.6%

$00.6\% = 0.006$ 0 is attached so the decimal point can be moved *two places* to the *left*.

Continued on Next Page

2 Write each percent as a decimal.

(a) 68%

(b) 5%

(c) 40.6%

(d) 200%

(e) 350%

3 Write each percent as a decimal by moving the decimal point.

(a) 90%

(b) 9%

(c) 900%

(d) 9.9%

(e) 0.9%

> **Note**
>
> In Example 3(d) on the previous page, notice that 0.6% is less than 1%. Because 1% is equivalent to 0.01 or $\frac{1}{100}$, any fraction of a percent smaller than 1% is less than 0.01. The decimal equivalent of 0.6% is 0.006, which is less than 0.01.

◀ *Work Problem* **3** *at the Side.*

OBJECTIVE **3** **Write decimals as percents.** You can write a decimal as a percent. For example, the decimal 0.25 is the same as the fraction $\frac{25}{100}$.

This fraction means 25 out of 100 parts, or 25%. Notice that multiplying 0.25 by 100 gives the same result.

$$(0.25)(100) = 25 \quad \text{so} \quad 0.25 = 25\%$$

This result makes sense because we are doing the opposite of what we did to change a percent to a decimal.

To change a percent to a decimal, we *drop* the % symbol and *divide* by 100. So, to *reverse* the process and change a decimal to a percent, we *multiply* by 100 and *attach* a % symbol.

> **Writing a Decimal as a Percent**
>
> To write a decimal as a percent, multiply by 100 and attach a % symbol.

> **Note**
>
> A quick way to *multiply* a number by 100 is to move the decimal point *two* places to the *right*. Notice that this is the opposite of the shortcut for *dividing* by 100.
>
> Drop % symbol; *divide* by 100.
> (Move decimal point two places *left*.)
>
> **Decimal** ⟵ ⟶ **Percent**
>
> *Multiply* by 100; attach % symbol.
> (Move decimal point two places *right*.)

EXAMPLE 4 **Changing Decimals to Percents by Moving the Decimal Point**

Write each decimal as a percent.

(a) 0.21

0.21 — Moving the decimal point *two places to the right* is a quick way to *multiply by 100*.

0.21 = 21% ⟵ Percent symbol is attached after decimal point is moved.

⟵ Decimal point is *not* written with whole number percents.

(b) 0.529 = 52.9% ⟵ Percent symbol is attached after decimal point is moved.

(c) 1.92 = 192% ⟵ Percent symbol is attached after decimal point is moved.

Continued on Next Page

(d) 2.5

    2.50   0 is attached so the decimal point can be moved
                    two places to the right.

    2.5 = 250%

(e) 3

    3. = 3.**00** so 3.**00** = 300%

Hint:
1 = 100%
2 = 200%
3 = 300%
and so on.

CAUTION
In Examples 4(c), 4(d), and 4(e) above, notice that 1.92, 2.5, and 3 are greater than 1. Because the number 1 is equivalent to 100%, all *numbers greater than 1* will be equivalent to *percents greater than 100%.*

Work Problem **4** *at the Side.* ▶

OBJECTIVE **4** **Write percents as fractions.** Percents can also be written as fractions. Recall that a percent is a ratio with a denominator of 100. For example, 89% is $\frac{89}{100}$. Because the fraction bar indicates division, we are dividing by 100, just as we did when writing a percent as a decimal.

Writing a Percent as a Fraction
To write a percent as a fraction, drop the % symbol and write the number over 100. Then write the fraction in lowest terms.

EXAMPLE 5 **Writing Percents as Fractions**

Write each percent as a fraction or mixed number in lowest terms or as a whole number.

(a) 25% Drop the % symbol and write 25 over 100.

$$25\% = \frac{25}{100} \leftarrow 25 \text{ per } 100$$

$$= \frac{25 \div 25}{100 \div 25} = \frac{1}{4} \leftarrow \text{Lowest terms}$$

As a check, write 25% as a decimal.

$$25\% = 25 \div 100 = 0.25 \leftarrow \text{Percent sign dropped}$$

Recall that 0.25 means 25 hundredths.

$$0.25 = \frac{25}{100} = \frac{25 \div 25}{100 \div 25} = \frac{1}{4} \leftarrow \text{Same result as above}$$

(b) 76% Drop the % symbol and write 76 over 100.

        This number becomes the numerator.

$$76\% = \frac{76}{100} \leftarrow \text{The } denominator \text{ is always 100 because percent means } parts\ per\ 100$$

Write $\frac{76}{100}$ in lowest terms. $\frac{76 \div 4}{100 \div 4} = \frac{19}{25} \leftarrow$ Lowest terms

Continued on Next Page

4 Write each number as a percent.

(a) 0.95

(b) 0.16

(c) 0.09

(d) 0.617

(e) 0.4

(f) 5.34

(g) 2.8

(h) 4

5 Write each percent as a fraction or mixed number in lowest terms or as a whole number.

(a) 50%

(b) 19%

(c) 80%

(d) 6%

(e) 125%

(f) 300%

(c) 150%

$$150\% = \frac{150}{100} = \frac{150 \div 50}{100 \div 50} = \frac{3}{2} = 1\frac{1}{2} \leftarrow \text{Mixed number}$$

(d) $100\% = \dfrac{100}{100} = 1 \leftarrow$ Whole number

> Always simplify fraction answers.

> **Note**
> Remember that percent means *per 100*.

◀ *Work Problem* **5** *at the Side.*

Example 6 below shows how to write decimal percents and fraction percents as fractions.

EXAMPLE 6 **Writing Decimal Percents or Fraction Percents as Fractions**

Write each percent as a fraction in lowest terms.

(a) 15.5%

Drop the % symbol and write 15.5 over 100.

$$15.5\% = \frac{15.5}{100}$$

To get a whole number in the numerator, multiply the numerator and denominator by 10. (Multiplying by $\frac{10}{10}$ is the same as multiplying by 1.)

$$\frac{15.5}{100} = \frac{(15.5)(10)}{(100)(10)} = \frac{155}{1000}$$

Write the fraction in lowest terms.

$$\frac{155 \div 5}{1000 \div 5} = \frac{31}{200} \leftarrow \text{Lowest terms}$$

(b) $33\frac{1}{3}\%$

Drop the % symbol and write $33\frac{1}{3}$ over 100.

$$33\frac{1}{3}\% = \frac{33\frac{1}{3}}{100}$$

When there is a mixed number in the numerator, write it as an improper fraction. So $33\frac{1}{3}$ is $\frac{100}{3}$.

> Think:
> 3 • 33 is 99
> and 99 + 1 is 100
> so $33\frac{1}{3}$ is $\frac{100}{3}$

$$\frac{33\frac{1}{3}}{100} = \frac{\frac{100}{3}}{100}$$

Continued on Next Page

Now you have a complex fraction (see **Section 4.6**). Rewrite the complex fraction using the ÷ symbol for division. Then follow the steps for dividing fractions.

$$\frac{\frac{100}{3}}{100} = \frac{100}{3} \div 100 = \frac{100}{3} \div \frac{100}{1} = \frac{100}{3} \cdot \frac{1}{100} = \frac{100 \cdot 1}{3 \cdot 100} = \frac{1}{3}$$

Reciprocals

> **Note**
>
> In Example 6(a) on the previous page, we could have changed 15.5% to $15\frac{1}{2}\%$ and then written it as the improper fraction $\frac{31}{2}$ over 100. But it is usually easier to work with decimal percents as they are.

Work Problem **6** *at the Side.* ▶

OBJECTIVE **5** **Write fractions as percents.** Recall that to write a percent as a fraction, you *drop* the percent symbol and *divide* by 100. So, to reverse the process and change a fraction to a percent, you *multiply* by 100 and *attach* a percent symbol.

> **Writing a Fraction as a Percent**
>
> To write a fraction as a percent, multiply by 100 and attach a % symbol. This is the same as multiplying by 100%.

> **Note**
>
> Look back at Example 5(d) near the top of the previous page to see that 100% = 1. Recall that multiplying a number by 1 does *not* change the value of the number. So multiplying by 100% does not change the value of a number; it just gives us an *equivalent percent.*

EXAMPLE 7 **Writing Fractions as Percents**

Write each fraction as a percent. Round to the nearest tenth of a percent when necessary.

(a) $\frac{2}{5}$ Multiply $\frac{2}{5}$ by 100%.

$$\frac{2}{5} = \left(\frac{2}{5}\right)(100\%) = \left(\frac{2}{5}\right)\left(\frac{100}{1}\%\right) = \left(\frac{2}{5}\right)\left(\frac{5 \cdot 20}{1}\%\right) = \frac{2 \cdot 5 \cdot 20}{5 \cdot 1}\%$$

$$= \frac{40}{1}\% = 40\%$$

To check the result, write 40% as $\frac{40}{100}$ and simplify the fraction.

$$40\% = \frac{40}{100} = \frac{40 \div 20}{100 \div 20} = \frac{2}{5} \leftarrow \text{Original fraction}$$

Continued on Next Page

6 Write each percent as a fraction in lowest terms.

(a) 18.5%

(b) 87.5%

(c) 6.5%

(d) $66\frac{2}{3}\%$

(e) $12\frac{1}{3}\%$

(f) $62\frac{1}{2}\%$

7 Write each fraction as a percent. If you're using a calculator, first work each one by hand. Then use your calculator and round to the nearest tenth of a percent when necessary.

(a) $\dfrac{1}{2}$

(b) $\dfrac{3}{4}$

(c) $\dfrac{1}{10}$

(d) $\dfrac{7}{8}$

(e) $\dfrac{5}{6}$

(f) $\dfrac{2}{3}$

(b) $\dfrac{5}{8}$ Multiply $\frac{5}{8}$ by 100%.

$$\frac{5}{8} = \left(\frac{5}{8}\right)(100\%) = \left(\frac{5}{8}\right)\left(\frac{100}{1}\%\right) = \left(\frac{5}{2\cdot 4}\right)\left(\frac{\overset{1}{\overbrace{4}}\cdot 25}{1}\%\right) = \frac{5\cdot \overset{1}{\cancel{4}}\cdot 25}{2\cdot 4\cdot 1}\%$$

$$= \frac{125}{2}\% = 62\frac{1}{2}\%$$

You can also do the last step of simplifying $\frac{125}{2}$ on your calculator. Enter $\frac{125}{2}$ as 125 ÷ 2 =. The result is 62.5, so $\frac{5}{8} = 62\frac{1}{2}\%$ or $\frac{5}{8} = 62.5\%$.

(c) $\dfrac{1}{6}$ Multiply $\frac{1}{6}$ by 100%.

$$\frac{1}{6} = \left(\frac{1}{6}\right)(100\%) = \left(\frac{1}{6}\right)\left(\frac{100}{1}\%\right) = \left(\frac{1}{2\cdot 3}\right)\left(\frac{2\cdot 50}{1}\%\right) = \frac{1\cdot \overset{1}{\cancel{2}}\cdot 50}{\underset{1}{\cancel{2}}\cdot 3\cdot 1}\%$$

$$= \frac{50}{3}\% = 16\frac{2}{3}\%$$

To simplify $\frac{50}{3}$ on your calculator, enter 50 ÷ 3 =. The result is 16.66666666, with the 6 continuing to repeat. The directions say to round to the nearest *tenth* of a percent.

┌ Next digit is *5 or more.*

16.**6**6666666%   rounds to   16.**7**%

↑
Tenths place

> When the next digit is *5 or more*, round up.

So $\frac{1}{6} = 16\frac{2}{3}\%$ (exact answer)   or $\frac{1}{6} \approx 16.7\%$ (rounded answer).

🖩 **Calculator Tip**

In Example 7(a) on the previous page, you can use your calculator to write $\frac{2}{5}$ as a percent.

Step 1   Enter $\frac{2}{5}$ as 2 ÷ 5 =. Your calculator shows   **0.4**

↑
Decimal equivalent of $\dfrac{2}{5}$

Step 2   Change the decimal number 0.4 to a percent by moving the decimal point two places to the right (multiply by 100%).

0.4**0** = 40% ← Attach % symbol.

Try this technique on Examples 7(b) and 7(c) on this page.

For $\frac{5}{8}$, enter 5 ÷ 8 =. Your calculator shows   **0.625**

Move the decimal point in 0.625 two places to the right.

0.625 = 62.5% ← Attach % symbol.

Continued on Next Page

ANSWERS

7. **(a)** 50%   **(b)** 75%   **(c)** 10%

   (d) $87\frac{1}{2}$% or 87.5% (Both are exact answers.)

   (e) exactly $83\frac{1}{3}$%, or 83.3% (rounded)

   (f) exactly $66\frac{2}{3}$%, or 66.7% (rounded)

For $\frac{1}{6}$, enter 1 ÷ 6 =. Your calculator shows | **0.166666666** |

Some calculators show 7 in the last place. ──→

Move the decimal point two places to the right. Then round to the nearest tenth.

$$0.1\underset{\frown}{66666666} = 16.66666666\% \approx 16.7\% \leftarrow \text{Attach \% symbol.}$$

With a fraction such as $\frac{1}{6}$, your calculator gives only an *approximate* answer. You can't get the exact answer of $16\frac{2}{3}\%$ by using your calculator.

Work Problem **7** *on the Previous Page at the Side.* ▶

OBJECTIVE 6 Use 100% and 50%. When working with percents, it is helpful to have several reference points. 100% and 50% are two helpful reference points.

100% means 100 parts out of 100 parts. That's *all* of the parts. If you pay 100% of a $245 dentist bill, you pay $245 (*all* of it).

EXAMPLE 8 **Finding 100% of a Number**

Fill in the blanks.

(a) 100% of $42 is _____.

100% is *all* of the money.

So 100% of $42 is $42.

(b) 100% of 9 miles is _____.

100% is *all* of the miles.

So 100% of 9 miles is 9 miles.

Work Problem **8** *at the Side.* ▶

50% means 50 parts out of 100 parts, which is *half* of the parts because $\frac{50}{100} = \frac{1}{2}$. So 50% of $12 is $6 (*half* of the money).

EXAMPLE 9 **Finding 50% of a Number**

Fill in the blanks.

(a) 50% of $42 is _____.

50% is *half* of the money.

So 50% of $42 is $21.

(b) 50% of 9 miles is _____.

50% is *half* of the miles.

So 50% of 9 miles is $4\frac{1}{2}$ miles.

Work Problem **9** *at the Side.* ▶

---

8 Fill in the blanks.

(a) 100% of $4.60 is _____.

(b) 100% of 3000 students is _____.

(c) 100% of 7 pages is _____.

(d) 100% of 272 miles is _____.

(e) 100% of $10\frac{1}{2}$ hours is _____.

9 Fill in the blanks.

(a) 50% of $4.60 is _____.

(b) 50% of 3000 students is _____.

(c) 50% of 7 pages is _____.

(d) 50% of 272 miles is _____.

(e) 50% of $10\frac{1}{2}$ hours is _____.

ANSWERS

8. (a) $4.60 **(b)** 3000 students **(c)** 7 pages
(d) 272 miles **(e)** $10\frac{1}{2}$ hours

9. (a) $2.30 **(b)** 1500 students
(c) $3\frac{1}{2}$ pages **(d)** 136 miles
(e) $5\frac{1}{4}$ hours

Math in the Media

DECIMALS, PERCENTS, AND QUILT PATTERNS

Please complete the Quilt Patterns activity in Section 4.8 before starting this activity.

Two of the quilt patterns you worked with in **Section 4.8** are repeated here. For each pattern, complete the table by copying the fractions you found and then changing the fractions to decimals and to percents. Use your calculator and round decimals to the nearest thousandth and percents to the nearest tenth when necessary. In the last row of the table, add the fractions. Then add the decimals, and finally, add the percents.

Source: Patchwork Persuasions: Quilts from Traditional Designs.

Peace and Plenty

Color	Fraction	Decimal	Percent
Blue			
Yellow			
Whitish			
Total			

Cobwebs

Color	Fraction	Decimal	Percent
Mauve			
Purple			
Yellow			
Total			

3. What pattern do you notice in the Total row in each table?

4. Explain why some of the totals in the Cobwebs' table don't quite fit the pattern.

5. If you created a quilt pattern for the quilt activity in **Section 4.8**, make a table like the ones in Questions 1 and 2 for your design.

7.1 ▶▶▶ Exercises

Write each percent as a decimal. See Examples 2 and 3.

1. 25%   **2.** 35%   **3.** 30%   **4.** 20%

5. 6%   **6.** 3%   **7.** 140%   **8.** 250%

9. 7.8%   **10.** 6.7%   **11.** 100%   **12.** 600%

13. 0.5%   **14.** 0.2%   **15.** 0.35%   **16.** 0.076%

Write each decimal as a percent. See Example 4.

17. 0.5   **18.** 0.6   **19.** 0.62   **20.** 0.18

21. 0.03   **22.** 0.07   **23.** 0.125   **24.** 0.875

25. 0.629   **26.** 0.494   **27.** 2   **28.** 5

29. 2.6   **30.** 1.8   **31.** 0.0312   **32.** 0.0625

Write each percent as a fraction or mixed number in lowest terms. See Examples 5 and 6.

33. 20%   **34.** 40%   **35.** 50%   **36.** 75%

37. 55%   **38.** 35%   **39.** 37.5%   **40.** 87.5%

41. 6.25%   **42.** 43.75%   **43.** $16\frac{2}{3}\%$ **44.** $83\frac{1}{3}\%$

45. 130%   **46.** 175%   **47.** 250%   **48.** 325%

Write each fraction as a percent. If you're using a calculator, first work each one by hand. Then use your calculator and round to the nearest tenth of a percent when necessary. See Example 7.

49. $\dfrac{1}{4}$ **50.** $\dfrac{1}{5}$ **51.** $\dfrac{3}{10}$ **52.** $\dfrac{9}{10}$

53. $\dfrac{3}{5}$ **54.** $\dfrac{3}{4}$ **55.** $\dfrac{37}{100}$ **56.** $\dfrac{63}{100}$

57. $\dfrac{3}{8}$ **58.** $\dfrac{1}{8}$ **59.** $\dfrac{1}{20}$ **60.** $\dfrac{1}{50}$

61. $\dfrac{5}{9}$ **62.** $\dfrac{7}{9}$ **63.** $\dfrac{1}{7}$ **64.** $\dfrac{5}{7}$

In each statement, write percents as decimals and decimals as percents. See Examples 2–4.

65. In 1900, only 8% of U.S. homes had a telephone. (*Source: Harper's Index.*)

66. In 1900, only 14% of homes in the United States had a bathtub. (*Source: Harper's Index.*)

67. Tornadoes can occur on any day of the year, but 42% of them appear in May and June. (*Source:* National Severe Storms Laboratory.)

68. Only 2.1% of tornadoes occur in December, making it the least likely month for twisters. (*Source:* National Severe Storms Laboratory.)

69. The property tax rate in Alpine County is 0.035.

70. A church building fund has 0.49 of the money needed.

71. The number of people taking CPR training this session is 2 times that of the last session.

72. Attendance at this year's company picnic is 3 times last year's attendance.

Write a fraction and a percent for the shaded part of each figure. Then write a fraction and a percent for the unshaded part of each figure.

73.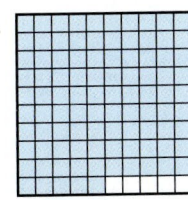

74.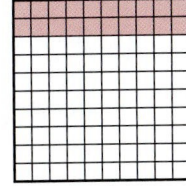

75.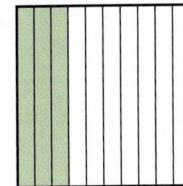

76.

77.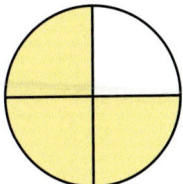

78.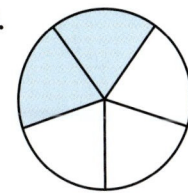

Complete this table. Write fractions and mixed numbers in lowest terms.

	Fraction	Decimal	Percent
79.	$\frac{1}{100}$	_____	_____
80.	$\frac{1}{10}$	_____	_____
81.	_____	0.2	_____
82.	_____	0.25	_____
83.	_____	_____	30%
84.	_____	_____	40%
85.	$\frac{1}{2}$	_____	_____
86.	$\frac{3}{4}$	_____	_____
87.	_____	_____	90%
88.	_____	_____	100%
89.	_____	1.5	_____
90.	_____	2.25	_____

The road signs in Exercises 91–92 tell drivers that they are approaching a steep, downward hill. For example, the 8% sign means that for every 100 ft of roadway length, the road will drop a total of 8 ft in elevation.

91. Write the percent on the sign as a decimal and as a fraction in lowest terms.

92. Write the percent on the sign as a decimal and as a fraction in lowest terms.

93. Suppose the roadway dropped 5 ft in elevation for every 100 ft of length. Write a percent, a decimal, and a fraction in lowest terms to describe this situation.

94. Suppose the roadway dropped 7 ft in elevation for every 100 ft of length. Write a percent, a decimal, and a fraction in lowest terms to describe this situation.

The diagram shows the different types of teeth in an adult's mouth. Use the diagram to answer Exercises 95–100. Write each answer in Exercises 95–98 as a fraction in lowest terms, as a decimal, and as a percent.

95. What portion of an adult's teeth are incisors, designed to bite and cut?

96. The pointy canine teeth tear and rip food. They are what portion of an adult's teeth?

97. The molars, which grind up food, are what portion of an adult's teeth?

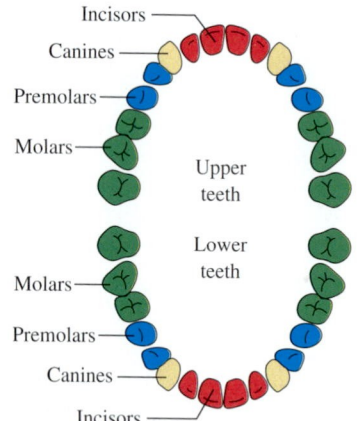

98. What portion of an adult's teeth are premolars?

Source: Time-Life *Human Body*.

99. Some people have four fewer molars than shown in the tooth diagram. For these people, the canines are what portion of their teeth? Write your answer as a fraction in lowest terms and as a percent.

100. For the adults who have four fewer molars than shown, the incisors are what portion of their teeth? Write your answer as a fraction in lowest terms and as a percent.

101. Explain and correct the errors made by students when they used their calculators on these problems.

(a) Write $\frac{7}{20}$ as a percent.

Student entered 7 ÷ 20 =, and the result was 0.35, so $\frac{7}{20} = 0.35\%$

(b) Write $\frac{16}{25}$ as a percent.

Student entered 25 ÷ 16 =, and the result was 1.5625, so $\frac{16}{25} = 156.25\%$

102. Explain and correct the errors made by students who moved decimal points to solve these problems.

(a) Write 3.2 as a percent.

$03.2 = 0.032$ so $3.2 = 0.032\%$

(b) Write 60% as a decimal.

$00.60 = 0.0060$ so $60\% = 0.0060$

Fill in the blanks in Exercises 103–118. Remember that 100% is all of something and 50% is half of it. See Examples 8 and 9.

103. (a) 100% of \$78 is _____.

(b) 50% of \$78 is _____.

104. (a) 100% of 5 hours is _____.

(b) 50% of 5 hours is _____.

105. (a) 100% of 15 inches is _____.

(b) 50% of 15 inches is _____.

106. (a) 100% of \$6000 is _____.

(b) 50% of \$6000 is _____.

107. (a) 100% of 2.8 miles is _____.

(b) 50% of 2.8 miles is _____.

108. (a) 100% of \$2.50 is _____.

(b) 50% of \$2.50 is _____.

109. There are 20 children in the preschool class. 100% of the children are served breakfast and lunch. How many children are served both meals?

_____

110. The Speedy Delivery company owns 345 vans. 100% of the vans are painted white with blue lettering. How many company vans are painted white with blue lettering?

_____

111. Alyssa needs 120 credits to graduate. She has earned 50% of the credits. How many credits has she earned?

_____

112. Ian has a limit of \$1500 on his credit card. He has used 50% of the limit. How many dollars of credit has he used?

_____

113. (a) John owes $285 for tuition. Financial aid will pay 50% of the cost. Financial aid will pay

_____.

(b) What *percent* of the tuition will John have to pay?

_____

(c) How much *money* will John have to pay?

_____

115. (a) About 50% of the 8200 students at our college work more than 20 hours per week. How many *students* is this?

_____

(b) What *percent* of the students work 20 hours or less per week?

_____

117. (a) Dylan's test score was 100%. How many of the 35 problems did he work correctly?

_____

(b) How many *problems* did Dylan miss?

_____

119. Describe a shortcut way to find 50% of a number. Include two examples to illustrate the shortcut.

114. (a) The Animal Humane Society took in 20,000 animals last year. About 50% of them were dogs. The number of dogs taken in was about how many?

_____

(b) What *percent* of the animals were *not* dogs?

_____

(c) How many *animals* were not dogs?

_____

116. (a) Shalayna's cell phone plan includes 1600 minutes of "off-peak" calling time per month. Last month she used 50% of her "off-peak" minutes. How many *minutes* did she use?

_____

(b) What *percent* of her "off-peak" minutes were *not* used?

_____

118. (a) Latrell made 100% of his 12 free throws during a practice game today. How many free throws did he make?

_____

(b) How many *free throws* did Latrell miss?

_____

120. Describe a shortcut way to find 100% of a number. Include two examples to illustrate the shortcut.

7.2 ▷▷▷ The Percent Proportion

We will show you two ways to solve percent problems. One is the proportion method, which we discuss in this section. The other is the percent equation method, which we explain in **Section 7.3**.

OBJECTIVE 1 Identify the percent, whole, and part. You have learned that a statement of two equivalent ratios is called a proportion (see **Section 6.3**). For example, the fraction $\frac{3}{5}$ is the same as the ratio 3 to 5, and 60% is the ratio 60 to 100. As the figure below shows, these two ratios are equivalent and make a proportion.

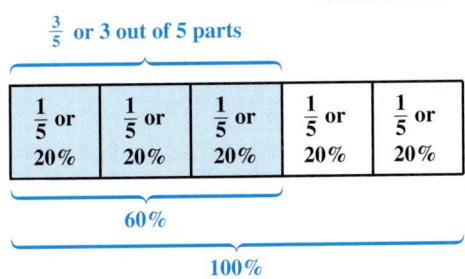

The **percent proportion** can be used to solve percent problems.

The Percent Proportion

Percent is to *100*   **as**   *part* is to *whole*.

$$\frac{\text{percent}}{100} = \frac{\text{part}}{\text{whole}}$$

Always 100 ⟶ because percent means "per 100"

In some textbooks the percent proportion is written using the words *amount* and *base*.

$$\frac{\text{percent}}{100} = \frac{\text{amount}}{\text{base}}$$

Here is the proportion for the figure at the top of the page.

60% means 60 parts out of 100 parts. $\left.\right\}$ $\frac{60}{100} = \frac{3}{5}$ ← Shaded (3 parts)
← Whole (5 parts)

If we write $\frac{60}{100}$ in lowest terms, it is equal to $\frac{3}{5}$, so the proportion is true.

$$\frac{60}{100} = \frac{60 \div 20}{100 \div 20} = \frac{3}{5} \leftarrow \left\{ \begin{array}{l} \text{Matches ratio on} \\ \text{right side of proportion} \end{array} \right.$$

In order to use the percent proportion to solve problems, you must be able to pick out the *percent,* the *whole,* and the *part.* Look for the percent first, because it is the easiest to identify.

Identifying the Percent

The **percent** is a ratio of a part to a whole, with 100 as the denominator. In a problem, the percent appears with the word *percent* or with the symbol **%** after it.

---

1 Identify the percent.

(a) Of the $2000, 15% will be spent on a washing machine.

(b) 60 employees is what percent of 750 employees?

(c) The state sales tax is $6\frac{1}{2}$ percent of the $590 price.

(d) $30 is 48% of what amount of money?

(e) 75 of the 110 rental cars were rented today. What percent were rented?

2 Identify the whole.

(a) Of the $2000, 15% will be spent on a washing machine.

(b) 60 employees is what percent of 750 employees?

(c) The state sales tax is $6\frac{1}{2}$ percent of the $590 price.

(d) $30 is 48% of what amount of money?

(e) 75 of the 110 rental cars were rented today. What percent were rented?

EXAMPLE 1 **Identifying the Percent in Percent Problems**

Identify the percent in each problem.

(a) 32% of the 900 women were retired. How many were retired?

Percent

The percent is 32. The number 32 appears with the symbol %.

(b) $150 is 25 percent of what number?

Percent

The percent is 25 because 25 appears with the word *percent*.

(c) If 7 students failed, what percent of the 350 students failed?

Percent (unknown)

The word *percent* has no number with it, so the percent is the unknown part of the problem.

◄ *Work Problem* **1** *at the Side.*

The second thing to look for in a percent problem is the *whole* (sometimes called the *base*).

Identifying the Whole

The **whole** is the entire quantity. In a problem, the *whole* often appears after the word **of**.

EXAMPLE 2 **Identifying the Whole in Percent Problems**

These problems are the same as those in Example 1 above. Now identify the *whole*.

(a) 32% **of** the 900 women were retired. How many were retired?

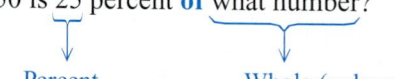

Percent   Whole

The whole is 900 women. The number 900 appears after the word *of*.

(b) $150 is 25 percent **of** what number?

Percent   Whole (unknown; follows *of*)

The whole is unknown.

(c) If 7 students failed, what percent **of** the 350 students failed?

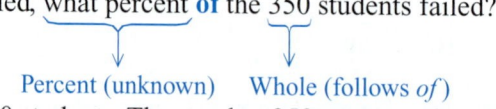

Percent (unknown)   Whole (follows *of*)

The whole is 350 students. The number 350 appears after the word *of*.

◄ *Work Problem* **2** *at the Side.*

The third and final thing to identify in a percent problem is the *part* (sometimes called the *amount*).

Identifying the Part

The **part** is the number being compared to the whole.

> **Note**
> If you have trouble identifying the *part,* find the *whole* and the *percent* first. The remaining number is the *part.*

EXAMPLE 3 | **Identifying the Part in Percent Problems**

These problems are the same as those in Examples 1 and 2 on the previous page. Identify the *part.*

(a) 32% **of** the 900 women were retired. How many were retired?

   Percent   Whole   Part (unknown)

The part of the women who were retired is unknown. In other words, some part of 900 women were retired.

(b) $150 is 25 percent **of** what number?

   Part Percent   Whole (unknown)

$150 is the remaining number, so $150 is the part.

(c) If 7 students failed, what percent **of** 350 students failed?

   Part   Percent (unknown) Whole

The part of the students who failed is 7 students.

Work Problem **3** *at the Side.* ▶

OBJECTIVE **2** **Solve percent problems using the percent proportion.**

EXAMPLE 4 | **Using the Percent Proportion to Find the Part**

Use the percent proportion to answer this question.

15% of $165 is how much money?

Recall that the percent proportion is $\dfrac{\text{percent}}{100} = \dfrac{\text{part}}{\text{whole}}$. First identify the percent by looking for the % symbol or the word *percent.* Then look for the *whole* (usually follows the word *of*). Finally, identify the *part.*

15% **of** $165 is how much money?

   Percent   Whole   Part (unknown)
         (follows *of*)

Set up the percent proportion. Here we use *n* as the variable representing the unknown part. You may use any letter you like.

$$\text{Percent} \rightarrow \frac{15}{100} = \frac{n}{165} \begin{matrix} \leftarrow \text{Part (unknown)} \\ \leftarrow \text{Whole} \end{matrix}$$
$$\text{Always 100} \rightarrow$$

Recall from **Section 6.3** that the first step in solving a proportion is to find the cross products.

Step 1 $\dfrac{15}{100} = \dfrac{n}{165}$ $\begin{matrix} 100 \cdot n \\ 15 \cdot 165 \end{matrix}$ ← Find the cross products.

Continued on Next Page

---

3 Identify the part.

(a) Of the $2000, 15% will be spent on a washing machine.

(b) 60 employees is what percent of 750 employees?

(c) The state sales tax is $6\frac{1}{2}$ percent of the $590 price.

(d) $30 is 48% of what amount of money?

(e) 75 of the 110 rental cars were rented today. What percent were rented?

ANSWERS

3. **(a)** unknown   **(b)** 60 employees
   (c) unknown   **(d)** $30   **(e)** 75 cars

4 Use the percent proportion to answer these questions.

  (a) 9% of 3250 miles is how many miles?

  (b) What is 20% of 180 calories?

  (c) 78% of $5.50 is how much?

  (d) What is $12\frac{1}{2}\%$ of 400 homes? (*Hint:* Write $12\frac{1}{2}\%$ as 12.5%.)

5 Use the percent proportion to answer these questions.

  (a) 1200 books is what percent of 5000 books?

  (b) What percent of $6.50 is $0.52?

  (c) 20 athletes is what percent of 32 athletes?

Step 2 $100 \cdot n = \underline{15 \cdot 165}$ Show that the cross products are equivalent.

    $100 \cdot n = 2475$

Step 3 $\dfrac{\overset{1}{\cancel{100}} \cdot n}{\underset{1}{\cancel{100}}} = \dfrac{2475}{100}$ Divide both sides by 100, the coefficient of the variable term. On the left side, divide out the common factor of 100.

    $n = 24.75$ On the right side, $2475 \div 100$ is 24.75

The part is **$24.75**, so 15% of $165 is **$24.75**.   *Write a $ in your answer.*

> **CAUTION**
> When you use the percent proportion, do **not** move the decimal point in the percent or in the answer.

◀ Work Problem **4** at the Side.

EXAMPLE 5 **Using the Percent Proportion to Find the Percent**

Use the percent proportion to answer this question.

8 pounds is what percent of 160 pounds?

Part   Percent   Whole
         (unknown)   (follows of)

The percent proportion is $\dfrac{\text{percent}}{100} = \dfrac{\text{part}}{\text{whole}}$. Set up the proportion using p as the variable representing the unknown percent. Then find the cross products.

Percent (unknown) → $\dfrac{p}{100} = \dfrac{8}{160}$ ← Part
    Always 100 →   ← Whole

$\dfrac{p}{100} = \dfrac{8}{160}$ $100 \cdot 8 = 800$ Cross products
            $p \cdot 160$

$p \cdot 160 = 800$ Show that the cross products are equivalent.

$\dfrac{p \cdot \overset{1}{\cancel{160}}}{\underset{1}{\cancel{160}}} = \dfrac{800}{160}$ Divide both sides by 160

$p = 5$ *The percent was unknown. Write % in your answer.*

The percent is **5%**. So 8 pounds is **5%** of 160 pounds.

> **CAUTION**
> When you're finding an unknown percent, as in Example 5 above, be careful to label your answer with the % symbol. Do **not** add a decimal point or move the decimal point in your answer.

◀ Work Problem **5** at the Side.

ANSWERS

4. **(a)** $\dfrac{9}{100} = \dfrac{n}{3250}$ The part is 292.5 miles.

   (b) $\dfrac{20}{100} = \dfrac{n}{180}$ The part is 36 calories.

   (c) $\dfrac{78}{100} = \dfrac{n}{5.50}$ The part is $4.29.

   (d) $\dfrac{12.5}{100} = \dfrac{n}{400}$ The part is 50 homes.

5. **(a)** $\dfrac{p}{100} = \dfrac{1200}{5000}$;   24%

   (b) $\dfrac{p}{100} = \dfrac{0.52}{6.50}$;   8%

   (c) $\dfrac{p}{100} = \dfrac{20}{32}$;   62.5% or $62\frac{1}{2}\%$

EXAMPLE 6 **Using the Percent Proportion to Find the Whole**

Use the percent proportion to answer this question.

162 credits is 90% **of** how many credits?

Part   Percent   Whole (unknown);
                         (follows *of*)

$$\text{Percent} \to \frac{90}{100} \gets \text{Always 100} = \frac{162}{n} \gets \text{Part}$$
$$\gets \text{Whole (unknown)}$$

$$\frac{90}{100} = \frac{162}{n} \qquad \begin{array}{l} 100 \cdot 162 = 16{,}200 \\ 90 \cdot n \end{array} \quad \text{Cross products}$$

$$90 \cdot n = 16{,}200 \qquad \text{Show that the cross products are equivalent.}$$

$$\frac{\overset{1}{\cancel{90}} \cdot n}{\underset{1}{\cancel{90}}} = \frac{16{,}200}{90} \qquad \text{Divide both sides by 90}$$

$$n = 180$$

> Write **credits** in your answer.

The whole is **180 credits**. So 162 credits is 90% of **180 credits.**

——————— *Work Problem* **6** *at the Side.* ▶

So far in all the examples, the part has been *less* than the whole. This is because all the percents have been less than 100%. Recall that 100% of something is *all* of it. When the percent is *less* than 100%, you have *less* than all of it.

Now let's look at percents *greater* than 100%. For example,

100% of $20 is all of the money, or $20.

150% of $20 is *greater* than $20.

100%	+	**50%**	=	**150%**
of the money is		of the money is		of the money is
$20	+	**$10**	=	**$30**

So 150% of $20 is $30.

When the percent is *greater* than 100%, the part is *greater* than the whole, as you'll see in Example 7 on the next page.

---

6 Use the percent proportion to answer these questions.

(a) 37 cars is 74% of how many cars?

(b) 45% of how much money is $139.59?

(c) 1.2 tons is $2\frac{1}{2}$% of how many tons?

7 Use the percent proportion to answer each question.

(a) 350% of $6 is how much?

(b) 23 hours is what percent of 20 hours?

(c) What percent of $47.32 is $106.47?

EXAMPLE 7 **Working with Percents Greater Than 100%**

Use the percent proportion to answer each question.

(a) How many students is 210% **of** 40 students?

Part (unknown)   Percent   Whole (follows *of*)

$$\text{Percent} \rightarrow \frac{210}{100} = \frac{n}{40} \begin{array}{l} \leftarrow \text{Part (unknown)} \\ \leftarrow \text{Whole} \end{array}$$
Always 100 →

$$\frac{210}{100} = \frac{n}{40}$$

$100 \cdot n$ ← Cross products
$210 \cdot 40 = 8400$ ←

$100 \cdot n = 8400$ Show that the cross products are equivalent.

$$\frac{\overset{1}{\cancel{100}} \cdot n}{\cancel{100}_{1}} = \frac{8400}{100}$$ Divide both sides by 100

$$n = 84$$

The part is **84 students**, which is *greater* than the whole of 40 students. This result makes sense because the percent is 210%. If it was exactly 200%, we would have *2 times the whole,* and 2 times 40 students is 80 students. So 210% should be even a little more than 80 students. Our answer of 84 students is reasonable.

(b) What percent **of** $50 is $68?

Percent   Whole   Part
(unknown) (follows *of*)

$$\text{Percent (unknown)} \rightarrow \frac{p}{100} = \frac{68}{50} \begin{array}{l} \leftarrow \text{Part} \\ \leftarrow \text{Whole} \end{array}$$
Always 100 →

$$\frac{p}{100} = \frac{68}{50}$$

$100 \cdot 68 = 6800$ ← Cross products
$p \cdot 50$ ←

$p \cdot 50 = 6800$ Show that the cross products are equivalent.

$$\frac{p \cdot \overset{1}{\cancel{50}}}{\cancel{50}_{1}} = \frac{6800}{50}$$ Divide both sides by 50

Write % in your answer.

$$p = 136$$

The percent is **136%**. This result makes sense because $68 is *greater* than $50, so $68 has to be *greater than 100% of $50.*

◀ *Work Problem* **7** *at the Side.*

7.2 ▶▶▶ Exercises

In Exercises 1–12, (a) identify the percent; (b) identify the whole; (c) identify the part; (d) write a percent proportion, and solve it to answer the question. See Examples 1–6.

1. What is 10% of 3000 runners?

2. What is 35% of 2340 volunteers?

3. 4% of 120 feet is how many feet?

4. 9% of $150 is how much money?

5. 16 pepperoni pizzas is what percent of 32 pizzas?

6. 35 hours is what percent of 140 hours?

7. What percent of 200 calories is 16 calories?

8. What percent of 350 parking spaces is 7 handicapped parking spaces?

9. 495 successful students is 90% of what number of students?

10. 84 e-mails is 28% of what number of e-mails?

11. $12\frac{1}{2}$% of what amount is $3.50?

12. $5\frac{1}{2}$% of what amount is $17.60?

Write a percent proportion and solve it to answer Exercises 13–24. If necessary, round money answers to the nearest cent and percent answers to the nearest tenth of a percent. See Examples 4–7.

13. 250% of 7 hours is how long?

14. What is 130% of 60 trees?

15. What percent of $172 is $32?

16. $14 is what percent of $398?

17. 748 books is 110% of what number of books?

18. 145% of what number of inches is 11.6 inches?

19. What is 14.7% of $274?

20. 8.3% of $43 is how much?

21. 105 employees is what percent of 54 employees?

22. What percent of 46 animals is 100 animals?

23. $0.33 is 4% of what amount?

24. 6% of what amount is $0.03?

25. A student turned in the following answers on a test. You can see that *two* of the answers are incorrect *without working the problems*. Find the incorrect answers and explain how you identified them (without actually solving the problems).

$$50\% \text{ of } \$84 \quad \text{is} \quad \underline{\$42}$$
$$150\% \text{ of } \$30 \quad \text{is} \quad \underline{\$20}$$
$$25\% \text{ of } \$16 \quad \text{is} \quad \underline{\$32}$$
$$100\% \text{ of } \$217 \text{ is} \quad \underline{\$217}$$

26. Name the three parts in a percent problem. For each of these three parts, write a sentence telling how you identify it.

27. Explain and correct the *two* errors that a student made when solving this problem:
$14 is what percent of $8?

$$\frac{p}{100} = \frac{8}{14} \qquad p \cdot 14 = 100 \cdot 8$$

$$\frac{p \cdot \overset{1}{\cancel{14}}}{\underset{1}{\cancel{14}}} = \frac{800}{14}$$

$$p \approx 57.1$$

The answer is 57.1 (rounded).

28. Explain and correct the *two* errors that a student made when solving this problem: 9 children is 30% of what number of children?

$$\frac{30}{100} = \frac{n}{9} \qquad 100 \cdot n = 30 \cdot 9$$

$$\frac{\overset{1}{\cancel{100}} \cdot n}{\underset{1}{\cancel{100}}} = \frac{270}{100}$$

$$n = 2.7$$

The answer is 2.7%.

7.3 ▶▶▶ The Percent Equation

OBJECTIVE 1 **Estimate answers to percent problems involving 25%.** Before showing you the percent equation, we need to do some more estimation. As you have learned when working with integers, fractions, and decimals, it is always a good idea to estimate the answer. Doing so helps you catch mistakes. Also, when you're out shopping or eating in a restaurant, you will be able to estimate the sales tax, discount, or tip.

In **Section 7.1** we used shortcuts for 100% of a number (all of the number) and 50% of a number (divide the number by 2). Now let's look at a quick way to work with 25%.

25% means 25 parts out of 100 parts, or $\frac{25}{100}$, which is the same as $\frac{1}{4}$

25% of $40 would be $\frac{1}{4}$ of $40, or $10

A quick way to find $\frac{1}{4}$ of a number is to *divide it by 4*. Recall that the denominator, 4, tells you that the whole is divided into 4 equal parts.

EXAMPLE 1 **Estimating 25% of a Number**

Estimate the answer to each question.

(a) What is 25% of $817?
Use front end rounding to round $817 to $800.
Then divide $800 by 4. The estimate is **$200**.

> 25% is $\frac{1}{4}$, so to find 25% of a number, divide it by 4.

(b) Find 25% of 19.7 miles.
Use front end rounding to round 19.7 miles to 20 miles.
Then divide 20 miles by 4. The estimate is **5 miles**.

(c) 25% of 49 days is how long?
You could round 49 days to 50 days, using front end rounding.
Then divide 50 by 4 to get an estimate of **12.5 days**.

However, the division step is simpler if you notice that 48 is a multiple of 4. You can round 49 days to 48 days and divide by 4 to get an estimate of **12 days**. Either way gives you a fairly good idea of the correct answer.

Work Problem ① *at the Side.* ▶

OBJECTIVE 2 **Find 10% and 1% of a number by moving the decimal point.** There are also helpful shortcuts for finding 10% or 1% of a number.

Ten percent, or 10%, means 10 parts out of 100 parts or $\frac{10}{100}$, which is the same as $\frac{1}{10}$. A quick way to find $\frac{1}{10}$ of a number is to *divide it by 10*. The denominator, 10, tells you that the whole is divided into 10 equal parts. The shortcut for dividing by 10 is to move the decimal point *one* place to the *left*.

EXAMPLE 2 **Finding 10% of a Number by Moving the Decimal Point**

Find the *exact* answer to each question by moving the decimal point.

(a) What is 10% of $817?
To find 10% of $817, divide $817 by 10. Do the division by moving the decimal point *one* place to the *left*. The decimal point starts at the far right side of $817.

$$10\% \text{ of } \$817. = \$81.70 \leftarrow \text{Exact answer}$$

↑
Write this 0 because it's money.

> Do **not** leave it as $81.7

So 10% of $817 is **$81.70**.

Continued on Next Page

── **Continued on Next Page**

OBJECTIVES

1 Estimate answers to percent problems involving 25%.

2 Find 10% and 1% of a number by moving the decimal point.

3 Solve basic percent problems using the percent equation.

---

1 Estimate the answer to each question.

(a) What is 25% of $110.38?

(b) Find 25% of 7.6 hours.

(c) 25% of 34 pounds is how many pounds?

ANSWERS

1. **(a)** $100 ÷ 4 gives an estimate of $25.
   (b) 8 hours ÷ 4 gives an estimate of 2 hours.
   (c) 30 pounds ÷ 4 gives an estimate of 7.5 pounds, or 32 pounds ÷ 4 gives an estimate of 8 pounds.

2 Find the *exact* answer to each question by moving the decimal point.

(a) What is 10% of $110.38?

(b) Find 10% of 7.6 hours.

(c) 10% of 34 pounds is how many pounds?

3 Find the *exact* answer to each question by moving the decimal point.

(a) What is 1% of $110.38?

(b) Find 1% of 7.6 hours.

(c) 1% of 34 pounds is how many pounds?

(b) Find 10% of 19.7 miles.
To find 10% of 19.7 miles, divide 19.7 by 10.
Move the decimal point *one* place to the *left*.

$$10\% \text{ of } 19.7 \text{ miles} = 1.97 \text{ miles} \longleftarrow \text{Exact answer}$$

So 10% of 19.7 miles is **1.97 miles**.

◀ *Work Problem* **2** *at the Side.*

One percent, or 1%, is 1 part out of 100 parts or $\frac{1}{100}$. This time the denominator of 100 tells you that the whole is divided into 100 parts. Recall that a quick way to divide by 100 is to move the decimal point *two* places to the *left*.

EXAMPLE 3 **Finding 1% of a Number by Moving the Decimal Point**

Find the *exact* answer to each question by moving the decimal point.

(a) What is 1% of $817?
To find 1% of $817, divide $817 by 100. Do the division by moving the decimal point *two* places to the *left*.

$$1\% \text{ of } \$817. = \$8.17 \longleftarrow \text{Exact answer}$$

So 1% of $817 is **$8.17**.

(b) Find 1% of 19.7 miles.
To find 1% of 19.7 miles, divide 19.7 by 100.
Move the decimal point *two* places to the *left*.

$$1\% \text{ of } 19.7 \text{ miles} = 0.197 \text{ mile} \longleftarrow \text{Exact answer}$$

So 1% of 19.7 miles is **0.197 mile**.

◀ *Work Problem* **3** *at the Side.*

Here is a summary of some of the shortcuts you can use with percents.

Percent Shortcuts

200% of a number is 2 times the number; **300% of a number** is 3 times the number; and so on.

100% of a number is the entire number.

To find **50% of a number,** divide the number by 2.

To find **25% of a number,** divide the number by 4.

To find **10% of a number,** divide the number by 10. To do the division, move the decimal point in the number *one* place to the *left*.

To find **1% of a number,** divide the number by 100. To do the division, move the decimal point in the number *two* places to the *left*.

ANSWERS

2. **(a)** $110.38 = $11.038 or $11.04 to the nearest cent
   (b) 7.6 hours = 0.76 hour
   (c) 34. pounds = 3.4 pounds
3. **(a)** $110.38 = $1.1038 or $1.10 to the nearest cent
   (b) 07.6 hours = 0.076 hour
   (c) 34. pounds = 0.34 pound

OBJECTIVE 3 **Solve basic percent problems using the percent equation.** In **Section 7.2** you used a proportion to solve percent problems. Now you will learn how to solve these problems using the percent equation.

Percent Equation

$$\text{percent } \textbf{of } \text{whole} = \text{part}$$

The word **of** indicates multiplication, so the **percent equation** becomes

$$\text{percent} \cdot \text{whole} = \text{part}$$

Be sure to write the percent as a decimal or fraction before using the equation.

The percent equation is just a rearrangement of the percent proportion. Recall that in the proportion you wrote the percent over 100. Because there is no 100 in the equation, you have to change the percent to a decimal or fraction by dividing by 100 *before* using the equation.

Note

Once you have set up a percent equation, we encourage you to use your calculator to do the multiplying or dividing needed to solve the equation. For this reason, we will always write the percent as a decimal. If you're doing the problems by hand, changing the percent to a fraction may be easier at times. Either method will work.

Examples 4, 5, and 6 below are the same percent questions that were in the examples in **Section 7.2.** There we used a proportion to answer each question. Now we will use an equation to answer them. You can then compare the equation method with the proportion method.

EXAMPLE 4 **Using the Percent Equation to Find the Part**

Write and solve a percent equation to answer each question.

(a) 15% of $165 is how much money?

Translate the sentence into an equation. Recall that *of* indicates multiplication and *is* translates to the equal sign. The percent must be written in decimal form. Use any letter you like to represent the unknown quantity. (We will use n for an unknown number and p for an unknown percent.)

Write 15% as the decimal 0.15

$$15.\% \text{ of } \$165 \text{ is how much money?}$$
$$0.15 \quad \cdot \quad 165 \quad = \quad n$$

To solve the equation, simplify the left side, multiplying 0.15 by 165.

$$\underline{(0.15)(165)} = n$$
$$24.75 \quad = n$$

So 15% of $165 is **$24.75**, which matches the answer obtained using a proportion (see Example 4 in **Section 7.2**).

Continued on Next Page

4 Write and solve an equation to answer each question.

(a) 9% of 3250 miles is how many miles?

(b) 78% of $5.50 is how much?

(c) What is $12\frac{1}{2}\%$ of 400 homes? (*Hint:* Write $12\frac{1}{2}\%$ as 12.5%. Then move the decimal point two places to the left.)

(d) How much is 350% of $6?

Check Use estimation to check that the solution is reasonable. First find 10% of $165 by moving the decimal point.

10% of 165. is $16.50   and

5% of $165 would be half as much, that is, half of $16.50 or about $8.

So the *estimate* for 15% of $165 is $16.50 + $8 = $24.50. The exact answer of $24.75 is very close to this estimate, so it is reasonable.

(b) How many students is 210% of 40 students?
Translate the sentence into an equation. Write the percent in decimal form.

How many students is 210.% of 40 students?

$$n \quad = \quad 2.10 \quad \cdot \quad 40$$

Write 210% as the decimal 2.10

This time the two sides of the percent equation are reversed, so

$$\text{part} = \text{percent} \cdot \text{whole}$$

Recall that the variable may be on either side of the equal sign. To solve the equation, simplify the right side, multiplying 2.10 by 40.

$$n = (2.10)(40)$$
$$n = 84$$

So **84 students** is 210% of 40 students. This matches the answer obtained by using a proportion (see Example 7(a) in **Section 7.2**).

Check Use estimation to check that the solution is reasonable. 210% is close to 200%.

200% of 40 students is 2 times 40 students = 80 students ← Estimate

The exact answer of 84 students is close to the estimate, so it is reasonable.

◀ Work Problem **4** at the Side.

EXAMPLE 5 **Using the Percent Equation to Find the Percent**

Write and solve a percent equation to answer each question.

(a) 8 pounds is what percent of 160 pounds?
Translate the sentence into an equation. This time the percent is unknown. Do **not** move the decimal point in the other numbers.

8 pounds is what percent of 160 pounds?

$$8 \quad = \quad p \quad \cdot \quad 160$$

To solve the equation, divide both sides by 160.

On the left side, divide 8 by 160

$$\frac{8}{160} = \frac{p \cdot \overset{1}{\cancel{160}}}{\underset{1}{\cancel{160}}}$$

On the right side, divide out the common factor of 160

Solution in *decimal* form

$$0.05 = p$$

Multiply the solution by **100%** to change it from a *decimal* to a *percent*.

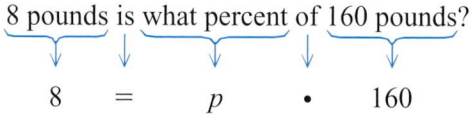

$$0.05 = 5\%$$

So 8 pounds is **5%** of 160 pounds. This matches the answer obtained by using a proportion (see Example 5 in **Section 7.2**).

Continued on Next Page

Check The solution makes sense because 10% of 160 pounds would be 16 pounds.

$$10\% \text{ of } 160 \text{ pounds is } 16 \text{ pounds}\quad \text{so}$$

5% of 160 pounds is half as much, that is, half of 16 pounds, or 8 pounds.

8 pounds matches the number given in the original problem, so 5% is the correct solution.

(b) What percent of $50 is $68?
Translate the sentence into an equation and solve it.

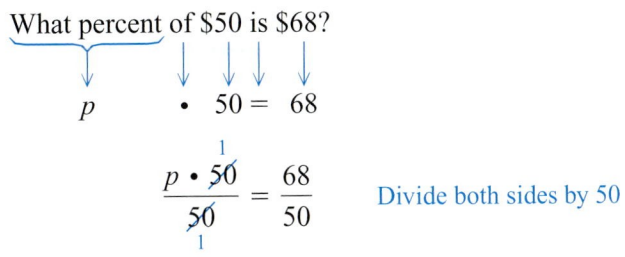

$$\frac{p \cdot \overset{1}{\cancel{50}}}{\underset{1}{\cancel{50}}} = \frac{68}{50}\qquad \text{Divide both sides by 50}$$

$$p = 1.36 \leftarrow \text{Solution in } decimal \text{ form}$$

> **Multiply the solution by 100%** to change it from a *decimal* to a *percent.*

$$1.36 = 136\%$$

So **136%** of $50 is $68. This matches the answer obtained by using a proportion (see Example 7(b) in **Section 7.2**).

Check The solution makes sense because 100% of $50 would be $50 (all of it), and 200% of $50 would be 2 times $50, or $100. So $68 has to be between 100% and 200%.

136% is between 100% and 200%. → $$\begin{array}{l} 100\% \text{ of } \$50 = \$50 \\ 200\% \text{ of } \$50 = \$100 \end{array}$$ ← $68 is between $50 and $100

The solution of 136% fits the conditions.

> **CAUTION**
> When you use an equation to solve for an unknown percent, *the solution will be in decimal form.* Remember to ***multiply the solution by 100%*** to change it from decimal form to a percent. The shortcut is to move the decimal point in the solution *two* places to the *right* and attach the % symbol.
>
> Recall that 100% = 1, and multiplying a number by 1 does *not* change the value of the number; it just gives us an *equivalent percent.*

Work Problem **5** *at the Side.* ▶

---

5 Write and solve an equation to answer each question.

(a) 1200 books is what percent of 5000 books?

(b) 23 hours is what percent of 20 hours?

(c) What percent of $6.50 is $0.52?

6 Write and solve an equation to answer each question.

(a) 74% of how many cars is 37 cars?

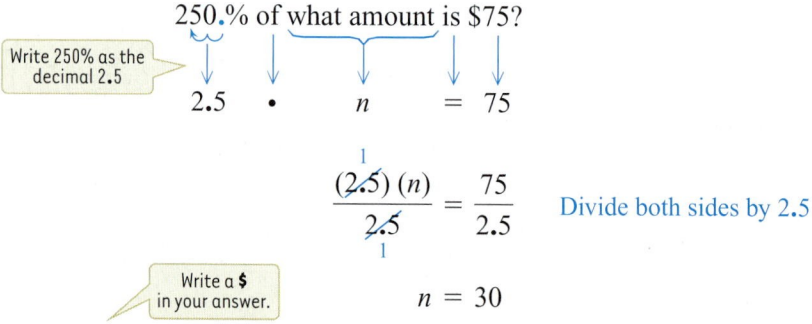

EXAMPLE 6   **Using the Percent Equation to Find the Whole**

Write and solve a percent equation to answer each question.

(a) 162 credits is 90% of how many credits?

Translate the sentence into an equation. Write the percent in decimal form.

162 credits is   90.% of how many credits?

$$162 = 0.90 \cdot n$$

Write 90% as the decimal 0.90

Recall that 0.90 is equivalent to 0.9, so use 0.9 in the equation.

$$\frac{162}{0.9} = \frac{(0.9)(n)}{0.9}$$ Divide both sides by 0.9

$$180 = n$$ Write **credits** in your answer.

So 162 credits is 90% of **180 credits**.

Check   The solution makes sense because 90% of 180 credits should be 10% less than 100% of the credits, and 10% of 180. credits is 18 credits.

100% of 180 credits	−	10% of 180. credits	=	90% of 180 credits
180 credits	−	18 credits	=	162 credits ← Matches the number given in the original problem

(b) 1.2 tons is $2\frac{1}{2}$% of how many tons?

(b) 250% of what amount is $75?

Translate the sentence into an equation. Write the percent in decimal form.

250.% of what amount is $75?

Write 250% as the decimal 2.5

$$2.5 \cdot n = 75$$

$$\frac{(2.5)(n)}{2.5} = \frac{75}{2.5}$$ Divide both sides by 2.5

Write a **$** in your answer.

$$n = 30$$

So 250% of **$30** is $75.

(c) 216 calculators is 160% of how many calculators?

Check   The solution makes sense because 200% of $30 is 2 times $30 = $60, and 50% of $30 is $30 ÷ 2 = $15.

200% of $30	+	50% of $30	=	250% of $30
(2)($30)		$30 ÷ 2		
$60	+	$15	=	$75 ← Matches the number given in the original problem

◄ Work Problem **6** at the Side.

ANSWERS

6. **(a)** $0.74 \cdot n = 37$;   50 cars
   (b) $1.2 = 0.025 \cdot n$;   48 tons
   (c) $216 = 1.6 \cdot n$;   135 calculators

7.3 ▷▷▷ Exercises

Use your estimation skills and the percent shortcuts to select the most reasonable answers.
*Circle your choices. Do **not** write an equation or proportion and solve it. See Examples 1–3.*

1. Find 50% of 3000 patients.

150 patients   1500 patients   300 patients

2. What is 50% of 192 pages?

48 pages   384 pages   96 pages

3. 25% of $60 is how much?

$15 $6 $30

4. Find 25% of $2840.

$28.40 $710 $284

5. What is 10% of 45 pounds?

0.45 pound   22.5 pounds   4.5 pounds

6. 10% of 7 feet is how many feet?

0.7 foot   3.5 feet   14 feet

7. Find 200% of $3.50.

$0.35 $1.75 $7.00

8. What is 300% of $12?

$4 $36 $1.20

9. 1% of 5200 students is how many students?

520 students   52 students   2600 students

10. Find 1% of 460 miles.

0.46 mile   46 miles   4.6 miles

11. Find 10% of 8700 cell phones.

8700 phones   4350 phones   870 phones

12. 25% of 128 CDs is how many CDs?

64 CDs   32 CDs   1280 CDs

13. What is 25% of 19 hours?

4.75 hours   1.9 hours   2.5 hours

14. What is 1% of $37?

$370 $3.70 $0.37

15. (a) Describe a shortcut for finding 10% of a number and explain *why* your shortcut works.

(b) Once you know 10% of a certain number, explain how you could use that information to find 20% and 30% of the same number.

16. (a) Describe a shortcut for finding 1% of a number and explain *why* it works.

(b) Once you know 1% of a certain number, explain how you could use that information to find 2% and 3% of the same number.

Write and solve an equation to answer each question in Exercises 17–48. See Examples 4–6.

17. 35% of 660 programs is how many programs?

18. 55% of 740 canisters is how many canisters?

19. 70 truckloads is what percent of 140 truckloads?

20. 30 crew members is what percent of 75 crew members?

21. 476 circuits is 70% of what number of circuits?

22. 621 tons is 45% of what number of tons?

23. $12\frac{1}{2}$% of what number of people is 135 people?

24. $6\frac{1}{2}$% of what number of bottles is 130 bottles?

25. What is 65% of 1300 species?

26. What is 75% of 360 dosages?

27. 4% of $520 is how much?

28. 7% of $480 is how much?

29. 38 styles is what percent of 50 styles?

30. 75 offices is what percent of 125 offices?

31. What percent of $264 is $330?

32. What percent of $480 is $696?

33. 141 employees is 3% of what number of employees?

34. 16 books is 8% of what number of books?

35. 32% of 260 quarts is how many quarts?

36. 44% of 430 liters is how many liters?

37. $1.48 is what percent of $74?

38. $0.51 is what percent of $8.50?

39. How many tablets is 140% of 500 tablets?

40. How many patients is 175% of 540 patients?

41. 40% of what number of salads is 130 salads?

42. 75% of what number of wrenches is 675 wrenches?

43. What percent of 160 liters is 2.4 liters?

44. What percent of 600 miles is 7.5 miles?

45. 225% of what number of gallons is 11.25 gallons?

46. 180% of what number of ounces is 6.3 ounces?

47. What is 12.4% of 8300 meters?

48. What is 13.2% of 9400 acres?

49. Explain and correct the error in each of these solutions.

(a) 3 hours is what percent of 15 hours?

$$3 = p \cdot 15$$

$$\frac{3}{15} = \frac{p \cdot \overset{1}{\cancel{15}}}{\underset{1}{\cancel{15}}}$$

$$0.2 = p$$

The answer is 0.2%.

(b) $50 is what percent of $20?

$$50 \cdot p = 20$$

$$\frac{\overset{1}{\cancel{50}} \cdot p}{\underset{1}{\cancel{50}}} = \frac{20}{50}$$

$$p = 0.40 = 40\%$$

The answer is 40%.

50. Explain and correct the error in each of these solutions.

(a) 12 inches is 5% of what number of inches?

$$(12)(0.05) = n$$

$$0.6 = n$$

The answer is 0.6 inch.

(b) What is 4% of 30 pounds?

$$n = (4)(30)$$

$$n = 120$$

The answer is 120 pounds.

Relating Concepts (Exercises 51–52) For Individual or Group Work

*Use your knowledge of fractions to **work Exercises 51 and 52 in order.***

51. Suppose that you have this problem: $33\frac{1}{3}\%$ of $162 is how much?

(a) First, change $33\frac{1}{3}\%$ to a fraction. (See **Section 7.1** for help.) Then, write an equation and solve it.

(b) Now solve the problem by changing $33\frac{1}{3}\%$ to a decimal. (*Hint:* Look at part (a) to see what fraction is equivalent to $33\frac{1}{3}\%$. Change the fraction to a decimal, using your calculator. Keep *all* the decimal places shown in the calculator's display window. Now write the equation and solve it, using your calculator.)

(c) Compare your answers from part (a) and part (b). How different are they?

52. Now suppose that you have this problem: 22 cans is $66\frac{2}{3}\%$ of what number of cans?

(a) First, change $66\frac{2}{3}\%$ to a fraction. (See **Section 7.1** for help.) Then, write an equation and solve it. (See **Section 4.7** for help.)

(b) Now solve the problem by changing $66\frac{2}{3}\%$ to a decimal. Use your calculator and keep all the decimal places shown. Now write the equation and solve it.

(c) Compare your answers from part (a) and part (b). How different are they?

Summary Exercises on Percent

1. Complete this table. Write fractions in lowest terms and as whole or mixed numbers when possible.

	Fraction	Decimal	Percent
(a)	$\dfrac{3}{100}$		
(b)			30%
(c)		0.375	
(d)			160%
(e)	$\dfrac{1}{16}$		
(f)			5%
(g)		2.0	
(h)	$\dfrac{4}{5}$		
(i)		0.072	

2. Use percent shortcuts to answer these questions.

(a) 10% of 35 ft is _____.

(b) 100% of 19 miles is _____.

(c) 50% of 210 cows is _____.

(d) 1% of $8 is _____.

(e) 25% of 2000 women is _____.

(f) 300% of $15 is _____.

(g) 10% of $875 is _____.

(h) 25% of 48 pounds is _____.

(i) 1% of 9500 students is _____.

Use the percent proportion or percent equation to answer each question. If necessary, round money answers to the nearest cent and percent answers to the nearest tenth of a percent.

3. 9 Web sites is what percent of 72 Web sites?

4. 30 DVDs is 40% of what number of DVDs?

5. 6% of $8.79 is how much?

6. 945 students is what percent of 540 students?

7. $3\frac{1}{2}$% of 168 pounds is how much, to the nearest tenth of a pound?

8. 1.25% of what number of hours is 7.5 hours?

9. What percent of 80,000 deer is 40,000 deer?

10. 465 camp sites is 93% of what number of camp sites?

11. What number of golf balls is 280% of 35 golf balls?

12. What percent of $66 is $1.80?

13. 9% of what number of apartments is 207 apartments?

14. What weight is 84% of 0.75 ounce?

15. $1160 is what percent of $800?

16. Find 3.75% of 6500 voters, to the nearest whole number.

17. What is 300% of 0.007 inch?

18. 24 minutes is what percent of 6 minutes?

19. What percent of 60 yards is 4.8 yards?

20. $0.17 is 25% of what amount?

The circle graph shows the average costs for various wedding expenses. Use the graph to answer Exercises 21–24. Round money answers to the nearest cent, if necessary.

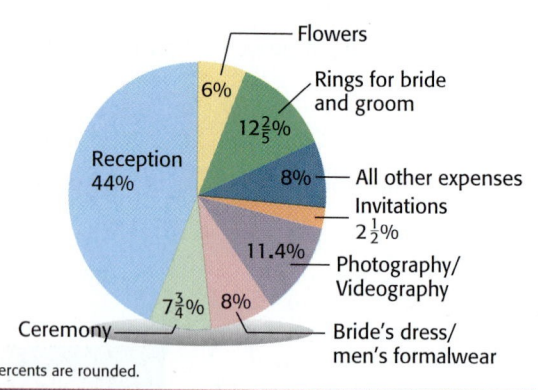

WHAT IT COSTS TO RING THE WEDDING BELLS
The average wedding has 165 guests and costs $33,600. Here is how couples are spending the money.

Flowers 6%
Rings for bride and groom 12$\frac{2}{5}$%
All other expenses 8%
Invitations 2$\frac{1}{2}$%
Photography/Videography 11.4%
Bride's dress/men's formalwear 8%
Ceremony 7$\frac{3}{4}$%
Reception 44%

Percents are rounded.

Source: *Bride's* magazine.

21. Which item in the graph is least expensive, and how much is spent on it?

22. What amount is spent on the ceremony?

23. **(a)** Find the cost of photography and videography.

   (b) What amount is spent on rings for the bride and groom?

24. **(a)** How much is spent on the reception?

   (b) On average, what is the cost for each guest at the reception?

7.4 ▷▷▷ Problem Solving with Percent

OBJECTIVE 1 Solve percent application problems. Solving percent problems involves identifying three items: the *percent*, the *whole*, and the *part*. Then you can write a percent equation or percent proportion and solve it to answer the question in the problem. Use the six problem-solving steps from **Section 3.3**.

OBJECTIVES

1 Solve percent application problems.

2 Solve problems involving percent of increase or decrease.

EXAMPLE 1 Finding the Part

A new low-income housing project charges 30% of a family's income as rent. The Smiths' family income is $1260 per month. How much will the Smiths pay for rent?

Step 1 **Read the problem.** It is about a family paying part of its income for rent.

  Unknown: amount of rent

  Known: 30% of income paid for rent; $1260 monthly income

Step 2 **Assign a variable.** There is only one unknown, so let n be the amount paid for rent.

Step 3 **Write an equation.** Use the percent equation.

$$\text{percent} \cdot \text{whole} = \text{part}$$

Recall that the *whole* often follows the word *of*.

  The percent is given in the problem: 30%. The key word **of** appears *right after* 30%, which means that you can use the phrase "30% **of** a family's income" to help you write one side of the equation. Write 30% as the decimal 0.30.

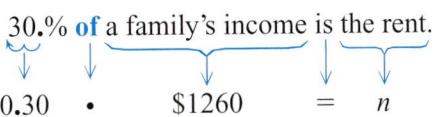

30.% **of** a family's income is the rent.

$$0.30 \quad \cdot \quad \$1260 \quad = \quad n$$

Step 4 **Solve the equation.** Simplify the left side, multiplying 0.30 by 1260.

$$\underline{(0.30)(1260)} = n$$
$$378 \qquad = n$$

> Write a **$** in your answer.

Step 5 **State the answer.** The Smith family will pay **$378** for rent.

Step 6 **Check** the solution. The solution of $378 makes sense because 10% of $1260 is $126, so 30% would be 3 times $126, or $378.

> **Note**
>
> You could also use the percent proportion in *Step 3* above.
>
> $$\text{Percent} \rightarrow \frac{30}{100} = \frac{n}{1260} \begin{array}{l}\leftarrow \text{Part (unknown)} \\ \leftarrow \text{Whole}\end{array}$$
> $$\text{Always 100} \rightarrow$$
>
> The answer will be the same, $378.
>
> Throughout the rest of this chapter, we will use the percent equation. You may, if you wish, use the percent proportion instead. The final answers will be the same.

Work Problem **1** *at the Side.* ▶

1 About 65% of the students at City Center College receive some form of financial aid. How many of the 9280 students enrolled this year are receiving aid? Use the six problem-solving steps.

2 There were 50 points on the first math test. Hue's score was 83% correct. How many points did Hue earn? Use the six problem-solving steps.

EXAMPLE 2 **Finding the Part**

When Britta received her first \$180 paycheck as a math tutor, $12\frac{1}{2}$% was withheld for federal income tax. How much was withheld?

Step 1 **Read** the problem. It is about part of Britta's pay being withheld for taxes.

> **Unknown:** amount withheld for taxes
>
> **Known:** $12\frac{1}{2}$% of earnings withheld; \$180 in pay

Step 2 **Assign a variable.** There is only one unknown, so let n be the amount withheld for taxes.

Step 3 **Write an equation.** Use the percent equation. The *percent* is given: $12\frac{1}{2}$%. Write $12\frac{1}{2}$% as 12.5% and then move the decimal point two places to the left. So 12.5% becomes the decimal 0.125.

The key word *of* doesn't appear after $12\frac{1}{2}$%. Instead, think about whether you know the *whole* or the *part*. You know Britta's *whole* paycheck is \$180, but you do *not* know what *part* of it was withheld.

> Be careful!
> $12\frac{1}{2}$% is 12.5%, but remember to *move the decimal point two places to the left* to get 0.125

$$\text{percent} \cdot \text{whole} = \text{part}$$
$$12.5\% \cdot \$180 = n$$

Step 4 **Solve.**
$$(0.125)(180) = n$$
$$22.5 = n$$

Step 5 **State the answer.** **\$22.50** was withheld from Britta's paycheck.

Step 6 **Check the solution.** The solution of \$22.50 makes sense because 10% of \$180 is \$18, so a little more than \$18 should be withheld.

◀ *Work Problem* **2** *at the Side.*

EXAMPLE 3 **Finding the Percent**

On a 15-point quiz, Zenitia earned 13 points. What percent correct is this, to the nearest whole percent?

Step 1 **Read** the problem. It is about points earned on a quiz.

> **Unknown:** percent correct
>
> **Known:** earned 13 out of 15 points

Step 2 **Assign a variable.** Let p be the unknown percent.

Step 3 **Write an equation.** Use the percent equation. There is no number with a % symbol in the problem. The question, "What percent is this?" tells you that the *percent* is unknown. The *whole* is all the points on the quiz (15 points), and 13 points is the *part* of the quiz that Zenitia did correctly.

$$\text{percent} \cdot \text{whole} = \text{part}$$
$$p \cdot 15 \text{ points} = 13 \text{ points}$$

Continued on Next Page

ANSWER

2. Let n be the points earned.
   $(0.83)(50) = n$
   Hue earned 41.5 points.
   Check: 10% of 50 points is 5 points, so 80% would be 8 times 5 points = 40 points. Hue earned a little more than 80%, so 41.5 points is reasonable.

Step 4 **Solve** the equation.

$$\frac{p \cdot \cancel{15}^{1}}{\cancel{15}_{1}} = \frac{13}{15} \qquad \text{Divide both sides by 15}$$

> CAUTION! **Multiply the solution by 100%** to change it from a *decimal* to a *percent*.

$$p = 0.8\overline{6} \longleftarrow \text{The solution is a repeating decimal.}$$

$$0.8\underset{\smile}{66666667} \approx 86.6666667\% \approx 87\% \longleftarrow \text{Rounded}$$

Step 5 **State the answer.** Zenitia had **87%** correct, rounded to the nearest whole percent.

Step 6 **Check the solution.** The solution of 87% makes sense because she earned most of the possible points, so the percent should be fairly close to 100%.

Work Problem ③ *at the Side.* ▶

EXAMPLE 4 **Finding the Percent**

The rainfall in the Red River Valley was 33 inches this year. The average rainfall is 30 inches. This year's rainfall is what percent of the average rainfall?

Step 1 **Read the problem.** It is about comparing this year's rainfall to the average rainfall.

> **Unknown:** This year's rain is what percent of the average?
>
> **Known:** 33 inches this year; 30 inches is average

Step 2 **Assign a variable.** Let p be the unknown percent.

Step 3 **Write an equation.** The percent is unknown. The key word **of** appears *right after* the word *percent,* so you can use that sentence to help you write the equation.

$$\underbrace{\text{This year's rainfall}}_{\text{33 inches}} \; \underbrace{\text{is}}_{=} \; \underbrace{\text{what percent}}_{p} \; \underbrace{\textbf{of}}_{\cdot} \; \underbrace{\text{the average rainfall?}}_{\text{30 inches}}$$

Step 4 **Solve the equation.**

$$33 = p \cdot 30$$

$$\frac{33}{30} = \frac{p \cdot \cancel{30}^{1}}{\cancel{30}_{1}} \qquad \text{Divide both sides by 30}$$

Solution in *decimal* form $\rightarrow$ $1.1 = p$

> **Multiply the solution by 100%** to change it from a *decimal* to a *percent*.

$$1.\underset{\smile}{10} = 110\%$$

Step 5 **State the answer.** This year's rainfall is **110%** of the average rainfall.

Step 6 **Check the solution.** The solution of 110% makes sense because 33 inches is *more* than 30 inches (more than 100% of the average rainfall), so 33 inches must be *more* than 100% of 30 inches.

Work Problem ④ *at the Side.* ▶

---

③ The Los Angeles Lakers made 47 of 80 field goal attempts in one game. What percent is this, to the nearest whole percent? Use the six problem-solving steps.

④ Valley College predicted that 1200 new students would enroll in the fall. It actually had 1620 new students enroll. The actual enrollment is what percent of the predicted number? Use the six problem-solving steps.

ANSWERS

3. Let p be the unknown percent.
$$p \cdot 80 = 47$$
$$p = 0.5875$$
$$0.5\underset{\smile}{875} = 58.75\% \approx 59\%$$
The Lakers made 59% of their field goals, to the nearest whole percent.
Check: The Lakers made a little more than half of their field goals. $\frac{1}{2} = 50\%$, so the solution of 59% is reasonable.

4. Let p be the unknown percent.
$$p \cdot 1200 = 1620$$
$$p = 1.35$$
$$1.\underset{\smile}{35} = 135\%$$
Enrollment is 135% of the predicted number.
Check: More than 1200 students enrolled (more than 100%), so 1620 students must be more than 100% of the predicted number. The solution of 135% is reasonable.

5 Use the six problem-solving steps to answer each question.

(a) Ezra did 15 problems correctly on a test, giving him a score of $62\frac{1}{2}\%$. How many problems were on the test?

(b) A frozen dinner advertises that only 18% of its calories are from fat. If the dinner contains 55 calories from fat, what is the total number of calories in the dinner? Round to the nearest whole number.

EXAMPLE 5 Finding the Whole

A newspaper article stated that 648 pints of blood were donated at the blood bank last month, which was only 72% of the number of pints needed. How many pints of blood were needed?

Step 1 **Read the problem.** It is about blood donations.

> **Unknown:** number of pints needed
>
> **Known:** 648 pints were donated;
> 648 pints is 72% of the number needed.

Step 2 **Assign a variable.** Let n be the number of pints needed.

Step 3 **Write an equation.** The percent is given in the problem: 72%. The key word **of** appears *right after* 72%, so you can use the phrase "72% **of** the number of pints needed" to help you write one side of the equation.

72.% **of** the number of pints needed is 648 pints.

$$0.72 \quad \cdot \quad n \quad = 648$$

Step 4 **Solve.**

$$\frac{(0.72)(n)}{0.72} = \frac{648}{0.72} \qquad \text{Divide both sides by 0.72}$$

$$n = 900$$

> Write **pints** in your answer.

Step 5 **State the answer. 900 pints** of blood were needed.

Step 6 **Check the solution.** The solution of 900 pints makes sense because 10% of 900 pints is 90 pints, so 70% would be 7 times 90 pints, or 630 pints, which is close to the number given in the problem (648 pints).

◀ *Work Problem* **5** *at the Side.*

OBJECTIVE 2 Solve problems involving percent of increase or decrease. We are often interested in looking at increases or decreases in prices, earnings, population, and many other numbers. This type of problem involves finding the percent of change. Use the following steps to find the **percent of increase.**

Finding the Percent of Increase

Step 1 Use subtraction to find the *amount* of increase.

Step 2 Use a form of the percent equation to find the *percent* of increase.

percent **of** whole = part

percent **of** original value = amount of increase

EXAMPLE 6 **Finding the Percent of Increase**

Brad's hourly wage as assistant manager of a fast-food restaurant was raised from $9.40 to $9.87. What was the percent of increase?

Step 1   **Read the problem.** It is about an increase in wages.

> **Unknown:** percent of increase
> **Known:** Original hourly wage was $9.40;
> new hourly wage is $9.87.

Step 2   **Assign a variable.** Let p be the percent of increase.

Step 3   **Write an equation.** First subtract $9.87 − $9.40 to find how much Brad's wage went up. That is the *amount* of increase. Then write an equation to find the unknown *percent* of increase. Be sure to use his *original* wage ($9.40) in the equation because we are looking for the change from his *original* wage. Do **not** use the new wage of $9.87 in the equation.

$$\$9.87 - 9.40 = \$0.47 \leftarrow \textit{Amount of increase}$$

$$\underbrace{\text{percent } \textbf{of} \text{ original wage}} = \underbrace{\text{amount of increase}}$$

$$p \quad \bullet \quad \$9.40 \quad = \quad \$0.47$$

> Do **not** use $9.87 in the equation. Use the **original** wage of $9.40

Step 4   **Solve.**
$$\frac{(p)(9.40)}{9.40} = \frac{0.47}{9.40} \quad \begin{array}{l}\text{Divide both sides} \\ \text{by 9.40}\end{array}$$

$$p = 0.05 \leftarrow \text{Solution in } \textit{decimal} \text{ form}$$

Multiply the solution by 100% to change it from a *decimal* to a *percent*.

$$0.05 = 5\%$$

Step 5   **State the answer.** Brad's hourly wage increased **5%**.

Step 6   **Check the solution.** 10% of $9.40 would be a $0.94 raise. A raise of $0.47 is half as much, and half of 10% is 5%, so the solution checks.

> **CAUTION**
> When writing the percent equation in *Step 3* above, be sure to multiply the percent (p) by the **original** wage. Do **not** use the *new* wage in the percent equation.

Work Problem **6** *at the Side.* ▶

Use a similar procedure to find the **percent of decrease.**

> **Finding the Percent of Decrease**
>
> **Step 1**   Use subtraction to find the *amount* of decrease.
>
> **Step 2**   Use a form of the percent equation to find the *percent* of decrease.
>
> $$\underset{\downarrow}{\text{percent}} \; \underset{\downarrow}{\textbf{of}} \; \underset{\downarrow}{\text{whole}} \; = \; \underset{\downarrow}{\text{part}}$$
>
> $$\underbrace{\text{percent } \textbf{of} \text{ original value}} = \underbrace{\text{amount of decrease}}$$

---

6 Use the six problem-solving steps to answer each question.

(a) Over the last two years, Duyen's rent has increased from $650 per month to $767. What is the percent increase?

(b) A shopping mall increased the number of handicapped parking spaces from 8 to 20. What is the percent increase?

EXAMPLE 7 **Finding the Percent of Decrease**

7 Use the six problem-solving steps to answer each question. Round answers to the nearest whole percent.

(a) During a severe winter storm, average daily attendance at an elementary school fell from 425 students to 200 students. What was the percent decrease?

Rozenia trained for six months to run in a marathon. Her weight dropped from 137 pounds to 122 pounds. Find the percent of decrease. Round to the nearest whole percent.

Step 1   **Read the problem.** It is about a decrease in weight.

> **Unknown:** percent of decrease
>
> **Known:** Original weight was 137 pounds; new weight is 122 pounds.

Step 2   **Assign a variable.** Let p be the percent of decrease.

Step 3   **Write an equation.** First subtract 137 pounds − 122 pounds to find how much Rozenia's weight went down. That is the *amount* of decrease. Then write an equation to find the unknown *percent* of decrease. Be sure to use her *original* weight (137 pounds) in the equation because we are looking for the change from her *original* weight. Do ***not*** use the new weight of 122 pounds in the equation.

137 pounds − 122 pounds = 15 pounds ← *Amount* of decrease

(b) The makers of a brand of spaghetti sauce claim that the number of calories from fat in each serving has been reduced by 20%. Is the claim correct if the number of calories from fat dropped from 70 calories to 60 calories per serving? Explain your answer.

percent **of** original weight = amount of decrease

$$p \cdot 137 \text{ pounds} = 15 \text{ pounds}$$

> Do **not** use 122 pounds in the equation. Use the **original** weight of 137 pounds.

Step 4   **Solve.**

$$\frac{p \cdot \cancel{137}}{\cancel{137}} = \frac{15}{137} \quad \text{Divide both sides by 137}$$

$$p = 0.109489051 \leftarrow \text{Solution in } decimal \text{ form}$$

Multiply the solution by 100% to change it from a *decimal* to a *percent*.

$$0.109489051 = 10.9489051\% \approx 11\%$$

Rounded to nearest whole percent

Step 5   **State the answer.** Rozenia's weight decreased approximately **11%**.

Step 6   **Check the solution.** A 10% decrease would be 13.7 = 13.7 pounds, so an 11% decrease is a reasonable solution.

ANSWERS

7. **(a)** 425 − 200 = 225 student decrease
Let p be percent of decrease.
$$p \cdot 425 = 225$$
$$p \approx 0.529 \approx 53\% \text{ (rounded)}$$
Daily attendance decreased 53%.
Check: A 50% decrease would be
425 ÷ 2 ≈ 212, so a 53% decrease is reasonable.

(b) 70 − 60 = 10 calorie decrease
Let p be percent of decrease.
$$p \cdot 70 = 10$$
$$p \approx 0.143 \approx 14\% \text{ (rounded)}$$
The claim of a 20% decrease is not true; the decrease in calories is about 14%.
Check: A 10% decrease would be
70. = 7 calories; a 20% decrease would be 2 · 7 = 14 calories, so a 14% decrease is reasonable.

> **CAUTION**
> When writing the percent equation in *Step 3* above, be sure to multiply the percent (p) by the ***original*** weight. Do ***not*** use the *new* weight in the percent equation.

◀ *Work Problem* **7** *at the Side.*

7.4 ▷▷▷ Exercises

Use the six problem-solving steps in Exercises 1–22. Round percent answers to the nearest tenth of a percent when necessary. See Examples 1–5.

1. Robert Garrett, who works part-time, earns $210 per week and has 18% of this amount withheld for taxes, Social Security, and Medicare. Find the amount withheld.

2. Most shampoos contain 75% to 90% water. If a 16-ounce bottle of shampoo contains 78% water, find the number of ounces of water in the bottle, to the nearest tenth. (*Source: Consumer Reports.*)

3. An ATM machine charges $2 for any size cash withdrawal. The $2 fee is what percent of a

  (a) $20 withdrawal

  (b) $40 withdrawal

  (c) $100 withdrawal

  (d) $200 withdrawal.

4. An Internet (on-line) company charges $8 for shipping and handling on any order of $50 or less. The $8 shipping charge is what percent of a

  (a) $10 order

  (b) $16 order

  (c) $25 order

  (d) $50 order.

The figure below shows, on average, what portion of the human body is made up of water, protein, fat, and so on. Use the figure to answer Exercises 5 and 6. Round answers to the nearest tenth.

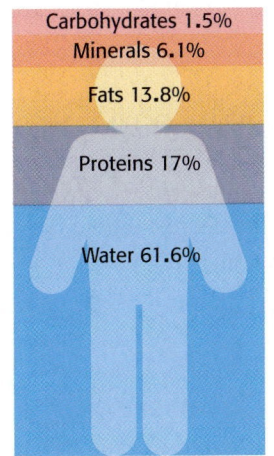

Carbohydrates 1.5%
Minerals 6.1%
Fats 13.8%
Proteins 17%
Water 61.6%

Source: Beakman & Jax, Universal Press Syndicate.

5. If an adult weighs 165 pounds, how much of that weight is

  (a) water

  (b) minerals.

6. If a teenager weighs 92 pounds, how much of that weight is

  (a) fat

  (b) carbohydrates.

7. The guided-missile destroyer USS *Sullivans* has a 335-person crew of which 44 are female. What percent of the crew is female? What percent of the crew is male? (*Source:* U.S. Navy.)

8. In a test by *Consumer Reports,* 6 of the 123 cans of tuna analyzed contained more than the 30-microgram intake limit of mercury. What percent of the cans contained an excessive level of mercury? What percent of the cans contained less than or equal to 30 micrograms of mercury?

9. The U.S. Census Bureau reported that Americans who are 65 years of age or older made up 14.2% of the total population in 2006. It said that there were 42.6 million Americans in this group. Find the total U.S. population for that year.

10. Julie Ward has 8.5% of her earnings deposited into the credit union. If this amounts to $263.50 per month, find her monthly and annual earnings.

11. The campus honor society hoped to raise $50,000 in donations from businesses for scholarships. It actually raised $69,000. This amount was what percent of the goal?

12. Doug had budgeted $220 for textbooks but ended up spending $316.80. The amount he spent was what percent of his budget?

13. Alfonso earned a score of 85% on his test. He did 34 problems correctly. How many problems were on the test?

14. In a telephone survey, 467 U.S. women said they had done a home improvement project in the past two years. This was 45% of the women in the survey. How many women were surveyed, to the nearest whole number? (*Source:* Opinion Research Corp.)

15. During the 2006–2007 NBA season, Kevin Garnett made 638 field goals, which was 47.6% of the shots he tried. How many shots did he try, to the nearest whole number? (*Source:* www.NBA.com)

16. Ray Allen, who plays basketball in the NBA, attempted 309 free throws during the 2006–2007 season. He made 90.3% of his shots. How many free throws did Ray make, to the nearest whole number? (*Source:* www.NBA.com)

17. An ad for steel-belted radial tires promises 15% better mileage. If Sheera's SUV has gotten 20.6 miles per gallon in the past, what mileage can she expect after the new tires are installed? (Round to the nearest tenth of a mile.)

18. Chris Chike set a world record for the highest score for the video game "Guitar Hero III: Legends of Rock." Chris earned 870,647 points, which is 97% of the possible points for a single song on the Expert Level. How many points were possible, to the nearest whole number? (*Source: Rochester Post-Bulletin.*)

The graph (pictograph) shows the percent of chicken noodle soup sold during the cold and flu season. Use this information to answer Exercises 19–22.

SOUP'S ON

350 million cans of chicken noodle soup are sold each year. More than half are bought during cold-and-flu season, with January being the number one month. The percent sold during each flu-season month is shown.

October November December January February March

Source: *USA Today.*

19. Which of the flu season months had the lowest sales of chicken noodle soup? How many cans were bought that month?

20. What percent of the chicken noodle soup sales take place during the flu months of October through March? What percent of sales take place in the *non-flu* season months?

21. Find the number of cans of soup sold in the highest sales month and in the second-highest sales month.

22. How many more cans of soup were sold in October than in November? How many more were sold in November than December?

Use the six problem-solving steps to find the percent increase or decrease. Round your answers to the nearest tenth of a percent. See Examples 6 and 7.

23. Henry Ford started the Ford Motor Company. The Model T first appeared in 1908 and cost $825. In 1913, Ford started using an assembly line and the price of the Model T dropped to $290. What was the percent of decrease? (*Source:* Kenneth C. Davis.)

24. In 1914, Henry Ford became a hero to his workers because he doubled the daily minimum wage. He also cut the workday from nine hours to eight hours. Find the percent of decrease in the number of hours in the workday. (*Source:* Kenneth C. Davis.)

25. Students at Lane College were charged $1449 as tuition this semester. If the tuition was $1328 last semester, find the percent of increase.

26. Americans are eating more fish. This year the average American will eat 16.1 pounds compared to only 11.7 pounds per year in 1970. Find the percent of increase. (*Source:* USDA/Economic Research Service.)

27. Jordan's part-time work schedule has been reduced to 18 hours per week. He had been working 30 hours per week. What is the percent decrease?

28. Janis works as a hair stylist. During January, she cut her price on haircuts from $28 to $25.50 to try to get more customers. By what percent did she decrease the price?

29. In 1967 there were 78 animal species listed as threatened with extinction in the United States. In 2007 there were 607 animals on the list. What was the percent increase, to the nearest whole percent? (*Source:* U.S. Fish and Wildlife Service.)

Condor with wing tag   Florida panther   Bighorn sheep

30. The world population was estimated at 6,602,275,000 people in 2007. It is projected to reach 7,985,550,000 people by 2025. By what percent will the world's population increase in those 18 years? (*Source:* United Nations.)

31. You can have an *increase* of 150% in the price of something. Could there be a 150% *decrease* in its price? Explain why or why not.

32. Show how to use a shortcut to find 25% of $80. Then explain how to use the result to find 75% of $80 and 125% of $80 *without* solving a proportion or equation.

Relating Concepts (Exercises 33–36) For Individual or Group Work

*As you **work Exercises 33–36 in order,** explain why each solution does **not** make sense. Then find and correct the error.*

33. The recommended maximum daily amount of dietary fat is 65 grams. George ate 78 grams of fat today. He ate what percent of the recommended amount?

$$p \cdot 78 = 65$$

$$\frac{p \cdot \overset{1}{\cancel{78}}}{\underset{1}{\cancel{78}}} = \frac{65}{78}$$

$$p = 0.833 = 83.3\%$$

34. The Goblers soccer team won 18 of its 25 games this season. What percent did the team win?

$$p \cdot 25 = 18$$

$$\frac{p \cdot \overset{1}{\cancel{25}}}{\underset{1}{\cancel{25}}} = \frac{18}{25}$$

$$p = 0.72\%$$

35. The human brain is $2\frac{1}{2}\%$ of total body weight. How much would the brain of a 150-pound person weigh?

$$2\tfrac{1}{2}\% \text{ of } 150 = n$$

$$(2.5)(150) = n$$

$$375 = n$$

36. Yesterday, because of an ice storm, 80% of the students were absent. How many of the 800 students made it to class?

$$80\% \text{ of } 800 = n$$

$$(0.80)(800) = n$$

$$640 = n$$

The answer is 640 students.

7.5 ▶▶▶ Consumer Applications: Sales Tax, Tips, Discounts, and Simple Interest

Four of the more common uses of percent in daily life are sales taxes, tips, discounts, and simple interest on loans.

OBJECTIVE 1 Find sales tax and total cost. Most states collect **sales taxes** on the purchases you make in stores. Your county or city may also add on a small amount of sales tax. For example, your state may charge $6\frac{1}{2}\%$ on purchases and your city may add on another $\frac{1}{2}\%$ for a total of 7%. The exact percent varies from place to place but is usually from 4% to 8%. The stores collect the tax and send it to the city or state government where it is used to pay for things like road repair, public schools, parks, police and fire protection, and so on.

You can use a form of the percent equation to calculate sales tax. The **tax rate** is the *percent*. The cost of the item(s) you are buying is the *whole*. The amount of tax you pay is the *part*.

> **Finding Sales Tax and Total Cost**
>
> Use a form of the percent equation to find sales tax.
>
> $$\text{percent} \quad \cdot \quad \text{whole} \quad = \quad \text{part}$$
> $$\underbrace{\text{tax rate}} \quad \cdot \quad \underbrace{\text{cost of item}} \quad = \quad \underbrace{\text{amount you pay in sales tax}}$$
>
> Then add to find how much you will pay in all.
>
> $$\text{cost of item} + \text{sales tax} = \text{total cost paid by you}$$

EXAMPLE 1 Finding Sales Tax and Total Cost

Suppose that you buy a DVD player for $289 from A-1 Electronics. The sales tax rate in your state is $6\frac{1}{2}\%$. How much is the tax? What is the total cost of the DVD player?

Step 1 **Read the problem.** It asks for the sales tax on a DVD player and the total cost.

Step 2 **Assign a variable.** Let n be the amount of tax.

Step 3 **Write an equation.** Use the sales tax equation. Write $6\frac{1}{2}\%$ as 6.5% and then move the decimal point two places to the left.

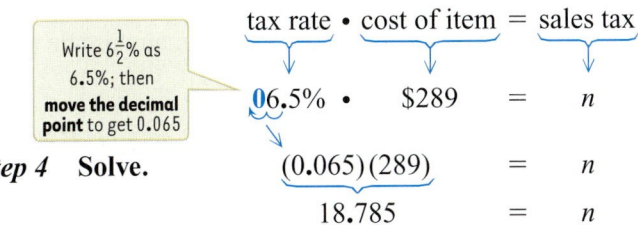

$$\text{tax rate} \cdot \text{cost of item} = \text{sales tax}$$
$$06.5\% \cdot \$289 = n$$

Write $6\frac{1}{2}\%$ as 6.5%; then **move the decimal point** to get 0.065

Step 4 **Solve.**
$$(0.065)(289) = n$$
$$18.785 = n$$

The store will round the tax to the nearest cent, so $18.785 rounds to **$18.79**.

Now add the sales tax to the cost of the DVD player to find your total cost.

$$\underbrace{\text{cost of item}} + \underbrace{\text{sales tax}} = \underbrace{\text{total cost}}$$
$$\$289 \quad + \quad \$18.79 = \textbf{\$307.79}$$

Continued on Next Page

1 Find the sales tax and the total cost. Round the sales tax to the nearest cent when necessary. Check your answer by estimating the sales tax.

(a) $495 camcorder; $5\frac{1}{2}$% sales tax

(b) $29.98 watch; 7% sales tax

(c) $1.19 candy bar; 4% sales tax

2 Find the sales tax rate on each purchase. Then use estimation to check your solution.

(a) The tax on a $57 textbook is $3.42.

(b) The tax on a $4 notebook is $0.18.

(c) The tax on a $998 sofa is $49.90.

ANSWERS

1. **(a)** sales tax = $27.23 (rounded)
   total cost = $522.23
   Check: 1% of $500. is $5;

   6% is 6 times $5 = $30
   (b) sales tax = $2.10 (rounded)
   total cost = $32.08
   Check: 1% of $30. is $0.30;

   7% is 7 times $0.30 = $2.10
   (c) Sales tax = $0.05 (rounded)
   total cost = $1.24
   Check: 1% of $1 is $0.01;
   4% is 4 times $0.01 = $0.04
2. **(a)** 6%   *Check:* 1% of $60. is $0.60;
   6% would be 6 times $0.60, or $3.60
   (which is close to $3.42 given in the problem).
   (b) 4.5%   *Check:* 1% of $4 is $0.04; round
   4.5% to 5%; 5% of $4 is 5 times $0.04,
   or $0.20 (which is close to $0.18 given
   in the problem).
   (c) 5%   *Check:* 10% of $1000. is $100;

   5% is half of that, or $50 (which is close
   to $49.90 given in the problem).

Step 5   **State the answer.** The tax is **$18.79** and the total cost of the DVD player, including tax, is **$307.79**.

Step 6   **Check:** Use estimation to check that the amount of sales tax is reasonable. Round $289 to $300. Then 1% of $300. is $3. Round $6\frac{1}{2}$% to 7%. Then 7% would be 7 times $3 or $21 for sales tax. Our solution of $18.79 is close to $21, so it is reasonable.

◀ *Work Problem* **1** *at the Side.*

EXAMPLE 2   **Finding the Sales Tax Rate**

Ms. Ortiz bought a $21,950 hybrid car. She paid an additional $1646.25 in sales tax. What was the sales tax rate?

Step 1   **Read** the problem. It asks for the sales tax rate on a car purchase.

Step 2   **Assign a variable.** Let p be the tax rate (the percent).

Step 3   **Write an equation.** Use the sales tax equation.

tax rate • cost of item = sales tax

$$p \quad • \quad \$21,950 \quad = \$1646.25$$

Step 4   **Solve.** $$\frac{(p)(21,950)}{21,950} = \frac{1646.25}{21,950}$$ Divide both sides by 21,950

Multiply the solution by 100% $p = 0.075$ ⟵ Solution in *decimal* form

Change the solution from a decimal to a percent: $0.075 = 7.5\%$.

Step 5   **State the answer.** The sales tax rate is **7.5%** (or $7\frac{1}{2}$%).

Step 6   **Check.** Use estimation to check that the solution is reasonable. If the tax rate was 1%, then 1% of $21950. = $219.50, or about $200.

Round 7.5% to 8%. Then 8% would be 8 times $200, or $1600.

The tax amount given in the original problem, $1646.25, is close to the estimate, so our solution of 7.5% is reasonable.

◀ *Work Problem* **2** *at the Side.*

OBJECTIVE **2** **Estimate and calculate restaurant tips.** Waiters and waitresses rely on tips as a major part of their income. The general rule of thumb is to leave 15% of your bill for food and beverages as a tip for the server. If you receive exceptional service or are eating in an upscale restaurant, consider leaving a 20% tip.

EXAMPLE 3   **Estimating 15% and 20% Tips**

First estimate each tip. Then calculate the exact amount.

(a) Kirby took his wife to dinner at a nice restaurant to celebrate her promotion at work. The bill came to $77.85. How much should he leave for a 20% tip?

Estimate: Round $77.85 to $80. Then 10% of $80. is $8.

20% would be 2 times $8, or **$16**. ⟵ Estimate

Continued on Next Page

Exact: Use the percent equation. Write 20% as a decimal. The bill for food and beverages is the *whole* and the tip is the *part.*

$$\text{percent} \cdot \text{whole} = \text{part}$$
$$20.\% \cdot \$77.85 = n$$

20.% = 0.20 ► $(0.20)(77.85) = n$
$$15.57 = n$$

20% of $77.85 is **$15.57**, which is close to the estimate of $16.

A tip is usually rounded off to a convenient amount, such as the nearest quarter or nearest dollar, so Kirby left $16.

(b) Linda, Peggy, and Mary ordered similarly priced lunches and agreed to split the bill plus a 15% tip. How much should each woman pay if the bill is $21.63?

Estimate: Round $21.63 to $20. Then 10% of $20. is $2.

5% of $20 would be half as much, that is, half of $2, or $1.

So an estimate of the 15% tip is $2 + $1 = **$3**. ← Estimates ┐

An estimate of the amount each woman should pay is ($20 + $3) ÷ 3 ≈ **$8**.

Exact: Use the percent equation to calculate the 15% tip. Add the tip to the bill. Then divide the total by 3 to find the amount each woman should pay.

$$\text{percent} \cdot \text{whole} = \text{part}$$
$$15.\% \cdot \$21.63 = n$$

15.% = 0.15 ► $(0.15)(21.63) = n$
$$3.2445 = n$$

Round $3.2445 to **$3.24** (nearest cent), which is close to the estimate of $3 for the tip.

Add: $21.63 + $3.24 = $24.87 ← Total cost of lunch and tip

Divide: $24.87 ÷ 3 = **$8.29** ← Amount paid by each woman

Work Problem **3** *at the Side.* ▶

OBJECTIVE 3 **Find the discount and sale price.** Most people prefer buying things when they are on sale. A store will reduce prices, or **discount,** to attract additional customers. You can use a form of the percent equation to calculate the discount. The *rate* of discount is the *percent.* The original price is the *whole.* The amount that will be discounted (subtracted from the original price) is the *part.*

┌───┐
Finding the Discount and Sale Price

Use a form of the percent equation to find the discount.

$$\text{percent} \cdot \text{whole} = \text{part}$$
$$\text{rate of discount} \cdot \text{original price} = \text{amount of discount}$$

Then subtract to find the sale price.

$$\text{original price} - \text{amount of discount} = \text{sale price}$$
└───┘

3 First estimate each tip. Then calculate the exact tip.

(a) 20% tip on a bill of $58.37

(b) 15% tip on a bill of $11.93

(c) A bill of $89.02 plus a 15% tip shared equally by four friends. How much will each friend pay?

4 Find the amount of the discount and the sale price for each item.

(a) An Easy-Boy leather recliner originally priced at $950 is offered at a 35% discount.

EXAMPLE 4 Finding the Discount and Sale Price

The Oak Mill Furniture Store has an oak dining room set with an original price of $840 on sale at 15% off. Find the sale price.

Step 1 **Read** the problem. It asks for the sale price on a dining room set.

Step 2 **Assign a variable.** Let n be the amount of discount.

Step 3 **Write an equation.** Use a form of the percent equation to find the discount. Write 15% as the decimal 0.15.

rate of discount • original price = amount of discount

Step 4 **Solve.**

$$(0.15)(840) = n$$
$$126 = n$$

The amount of discount is **$126**. Find the sale price by subtracting the amount of the discount ($126) from the original price.

original price − amount of discount = sale price

$$\$840 - \$126 = \$714$$

Remember to subtract the discount.

Step 5 **State the answer.** During the sale, you can buy the dining room set for **$714**.

Step 6 **Check.** Round $840 to $800. Then 10% of $800, is $80 and 5% is half as much, or $40. So 15% of $800 is $80 + $40 = $120. An estimate of the sale price is $800 − $120 = $680, so the exact answer of $714 is reasonable.

◀ *Work Problem* **4** *at the Side.*

(b) In August, Eastside Boutique has women's swimsuits on sale at 40% off. One swimsuit was originally priced at $68. Another suit was originally priced at $97.

⊞ **Calculator Tip** In Example 4 above, you can use a *scientific* calculator to find the amount of discount and subtract the discount from the original price.

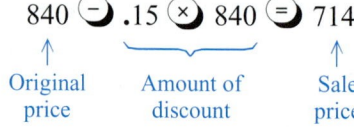

840 ⊖ .15 ⊗ 840 ⊜ 714

Original price — Amount of discount — Sale price

Your *scientific* calculator observes the order of operations, so it will automatically do the multiplication before the subtraction. (Recall that simple, four-function calculators *may not* follow the order of operations; they might give an incorrect result.)

OBJECTIVE 4 Calculate simple interest and the total amount due on a loan. **Interest** is a fee paid, or a charge made, for lending or borrowing money. The amount of money borrowed is called the **principal**. The charge for interest is usually given as a percent, called the **interest rate.** The interest rate is assumed to be *per year* (for *one* year) unless stated otherwise.

In most cases, interest is calculated on the original principal and is called **simple interest.** Use the following **interest formula** to find simple interest.

Formula for Simple Interest

$$\text{Interest} = \text{principal} \cdot \text{rate} \cdot \text{time}$$

The formula is usually written using variables.

$$I = p \cdot r \cdot t \quad \text{or} \quad I = prt$$

When you repay a loan, the interest is added to the original principal to find the total amount due.

Finding the Total Amount Due

$$\text{amount due} = \text{principal} + \text{interest}$$

Note

Simple interest calculations are used for most short-term business loans, automobile loans, and consumer loans.

EXAMPLE 5 **Finding Simple Interest and Total Amount Due for 1 Year**

Find the simple interest and total amount due on a $2000 loan at 6% for 1 year.

The amount borrowed, or principal (p), is $2000. The interest rate (r) is 6%, which is 0.06 as a decimal, and the time of the loan (t) is 1 year. Use the interest formula.

$$I = p \cdot r \cdot t$$
$$I = (2000)(0.06)(1)$$
$$I = 120$$

Don't stop here. Add the interest to the principal.

The interest is **$120**.

Now add the principal and the interest to find the total amount due.

$$\begin{aligned} \text{amount due} &= \text{principal} + \text{interest} \\ &= \$2000 + \$120 \\ &= \$2120 \end{aligned}$$

The total amount due is **$2120**.

Work Problem **5** *at the Side.* ▶

EXAMPLE 6 **Finding Simple Interest and Total Amount Due for More Than 1 Year**

Find the simple interest and total amount due on a $4200 loan at $8\frac{1}{2}\%$ for $3\frac{1}{2}$ years.

The principal (p) is $4200. The rate ($r$) is $8\frac{1}{2}\%$, which is the same as 8.5%. Move the decimal point two places to the left to change 8.5% to a decimal.

$$8\frac{1}{2}\% = 8.5\% = 08.5 = 0.085$$

*Remember to **move the decimal point** in 8.5% to get 0.085*

The time (t) is $3\frac{1}{2}$ or 3.5 years. Use the formula.

$$I = p \cdot r \cdot t$$
$$I = (4200)(0.085)(3.5)$$
$$I = 1249.50$$

The interest is **$1249.50**.

Continued on Next Page

---

5 Find the simple interest and total amount due on each loan.

(a) $500 at 4% for 1 year

(b) $1850 at $9\frac{1}{2}\%$ for 1 year (*Hint:* Write $9\frac{1}{2}\%$ as 9.5%. Then *move the decimal point* two places to the left to change 9.5% to a decimal.)

ANSWERS

5. (a) $20; $520 **(b)** $175.75; $2025.75

6 Find the simple interest and total amount due for each loan.

(a) $340 at 5% for $3\frac{1}{2}$ years

(b) $2450 at 8% for $3\frac{1}{4}$ years (*Hint:* Write $3\frac{1}{4}$ years as 3.25 years.)

(c) $14,200 at $7\frac{1}{2}$% for $2\frac{3}{4}$ years

7 Find the simple interest and total amount due for each loan.

(a) $1600 at 7% for 4 months

(b) $25,000 at $10\frac{1}{2}$% for 3 months

(c) $4350 at $12\frac{1}{4}$% for 9 months

Now add the principal and the interest to find the total amount due.

$$\textbf{amount due} = \text{principal} + \textbf{interest}$$
$$= \$4200 + \$1249.50$$
$$= \$5449.50$$

The total amount due is **$5449.50**.

> **CAUTION**
> Be careful when changing a mixed number percent, like $8\frac{1}{2}$%, to a decimal. Writing $8\frac{1}{2}$% as 8.5% is only the first step. There is a decimal point in 8.5% but there is still a **%** sign. You must divide by 100 before dropping the % sign. ***Remember to move the decimal point two places to the left,*** as shown in Example 6 on the previous page.
> $$8\tfrac{1}{2}\% = 8.5\% = 0.085$$

◀ *Work Problem* **6** *at the Side.*

Interest rates are given *per year*. For loan periods of less than one year, be careful to express the time as a fraction of a year.

If the time is given in months, use a denominator of 12, because there are 12 months in a year. A loan of 9 months would be for $\frac{9}{12}$ of a year, a loan of 7 months would be for $\frac{7}{12}$ of a year, and so on.

EXAMPLE 7 **Finding Simple Interest and Total Amount Due for Less Than 1 Year**

Find the simple interest and total amount due on $840 at $9\frac{3}{4}$% for 7 months.
The principal is $840. The rate is $9\frac{3}{4}$% or 0.0975.

$$9\frac{3}{4}\% = 9.75\% = 09.75 = 0.0975$$

The time is $\frac{7}{12}$ of a year. Use the formula ***I = prt***.

$$I = (840)(0.0975)\left(\frac{7}{12}\right) \qquad \text{7 months is } \tfrac{7}{12} \text{ of a year.}$$
$$= (81.9)\left(\frac{7}{12}\right)$$
$$= \left(\frac{81.9}{1}\right)\left(\frac{7}{12}\right) \qquad \begin{array}{l}\text{Multiply numerators.}\\\text{Multiply denominators.}\end{array}$$
$$= \frac{573.3}{12} = 47.775 \qquad \text{Divide 573.3 by 12}$$

The interest is **$47.78**, rounded to the nearest cent.

The total amount due is $840 + **$47.78** = **$887.78**

> 🖩 **Calculator Tip** The calculator solution for finding the interest in Example 7 above uses chain calculations.
>
> $840 \otimes .0975 \otimes 7 \div 12 \ominus 47.775$ ← Round to $47.78

◀ *Work Problem* **7** *at the Side.*

Find the amount of the sales tax or the tax rate and the total cost. Round money answers to the nearest cent. See Examples 1 and 2.

	Cost of Item	Tax Rate	Amount of Tax	Total Cost
1.	$100	6%	_____	_____
2.	$200	4%	_____	_____
3.	$68	_____	$2.04	_____
4.	$185	_____	$9.25	_____
5.	$365.98	8%	_____	_____
6.	$28.49	7%	_____	_____
7.	$2.10	$5\frac{1}{2}\%$	_____	_____
8.	$7.00	$7\frac{1}{2}\%$	_____	_____
9.	$12,600	_____	$567	_____
10.	$21,800	_____	$1417	_____

For each restaurant bill, estimate a 15% tip and a 20% tip. Then find the exact amounts for a 15% tip and a 20% tip. Round exact amounts to the nearest cent when necessary. See Example 3.

Bill	Estimate of 15% Tip	Exact 15% Tip	Estimate of 20% Tip	Exact 20% Tip
11. $32.17	_____	_____	_____	_____
12. $21.94	_____	_____	_____	_____
13. $78.33	_____	_____	_____	_____
14. $67.85	_____	_____	_____	_____
15. $9.55	_____	_____	_____	_____
16. $52.61	_____	_____	_____	_____

Find the amount or rate of discount and the sale price. Round money answers to the nearest cent when necessary. See Example 4.

Original Price	Rate of Discount	Amount of Discount	Sale Price
17. $100	15%	_____	_____
18. $200	20%	_____	_____
19. $180	_____	$54	_____
20. $38	_____	$9.50	_____
21. $17.50	25%	_____	_____
22. $76	60%	_____	_____
23. $37.88	10%	_____	_____
24. $59.99	40%	_____	_____

Find the simple interest and total amount due on each loan. See Examples 5–7.
Round answers to the nearest cent when necessary.

	Principal	Rate	Time	Interest	Total Amount Due
25.	$300	14%	1 year	_____	_____
26.	$600	11%	6 months	_____	_____
27.	$740	6%	9 months	_____	_____
28.	$1180	9%	2 years	_____	_____
29.	$1500	$9\frac{1}{2}\%$	$1\frac{1}{2}$ years	_____	_____
30.	$3000	$6\frac{1}{2}\%$	$2\frac{1}{2}$ years	_____	_____
31.	$17,800	$7\frac{3}{4}\%$	8 months	_____	_____
32.	$20,500	$8\frac{1}{4}\%$	5 months	_____	_____

Solve each application problem. Round money answers to the nearest cent when necessary.

33. Diamonds at Discounts sells diamond engagement rings at 40% off the regular price. Find the sale price of a $\frac{1}{2}$-carat diamond ring normally priced at $1950.

34. An 80 GB iPOD classic originally priced at $249 is marked down 8%. Find the price of the iPOD after the markdown.

35. Evelina Jones lent $7500 to her son Rick, the owner of Rick's Limousine Service. He repaid the loan at the end of 9 months at $8\frac{1}{2}\%$ simple interest. What total amount did Rick pay his mother?

36. The owners of Delta Trucking purchased four diesel-powered tractors for cross-country hauling at a cost of $87,500 per tractor. If they borrowed the purchase price for $1\frac{1}{2}$ years at 11% simple interest, find the total amount due.

37. A Motorola H350 Bluetooth headset is sale priced at $24.99. The sales tax rate is $6\frac{1}{2}\%$. Find the total cost of the headset. (*Source:* www.BestBuy.com)

38. A weekday "golf/breakfast special" includes breakfast, 18 holes of golf, and use of a cart for $47.95 plus tax per person. If the sales tax rate is $7\frac{1}{2}\%$, find the total cost per person. How much will three friends pay to play golf?

39. An Anderson wood frame French door is priced at $1980 with a sales tax of $99. Find the sales tax rate.

40. Textbooks for two classes cost $185 plus sales tax of $11.10. Find the sales tax rate.

41. A "super 45% off sale" begins today. What is the sale price of a ski parka normally priced at $135?

42. A discontinued Whirlpool model side-by-side refrigerator with in-door icemaker originally sold for $1197. What is the sale price with a 35% discount?

43. Ricia and Seitu split a $43.70 dinner bill plus 15% tip. How much did each person pay?

44. Marvette took her brother out to dinner for his birthday. The bill for food was $58.36 and for wine was $15.44. How much was her 20% tip, rounded to the nearest dollar?

45. A 50" slim depth projection HDTV normally priced at $1199.99 is on sale for 18% off. Find the discount and the sale price.

46. This week, Honda CRVs are offered at 15% off the manufacturer's suggested price. Find the discount and the sale price of a CRV originally priced at $23,500.

47. Ms. Henderson owes $1900 in taxes. She is charged a penalty of $12\frac{1}{4}\%$ annual interest and pays the taxes and penalty after 6 months. Find the total amount she must pay.

48. Norell Di Loreto, owner of Sunset Realtors, borrowed $27,000 to update her office computer system. If the loan is for 24 months at $7\frac{3}{4}\%$, find the total amount due on the loan.

49. Vincente and Samuel ordered a large deep-dish pizza for $17.98. How much did they give the delivery person to pay for the pizza and a 15% tip, rounded to the nearest dollar?

50. Cher, Maya, and Adara shared a $28.50 bill for a buffet lunch. Because the server only brought their beverages, they left a 10% tip instead of the usual 15%. How much did each person pay?

Use the information in the store ad to answer Exercises 51–54. Round sale prices and sales tax to the nearest cent when necessary.

STORE CLOSE-OUT!

All clothing is now 45% off!

All jewelry is now 30% off!

All electronics are now 65% off!

6% sales tax added to jewelry and electronics purchases.

CASH ONLY! ALL SALES ARE FINAL!

51. Danika bought a computer modem originally priced at $129 and a $60 pair of earrings. What was her bill for the two items?

52. Find David's total bill for a $189 jacket and a $75 graphing calculator.

53. Sergei purchased a camcorder originally priced at $287.95, two pairs of $48 jeans, and a $95 ring. Find his total bill.

54. Richard picked out three pairs of $15 running shorts, two $28 shirts, and a digital camera originally priced at $99.99. How much did he pay in all?

Relating Concepts (Exercises 55–56) For Individual or Group Work

*Use your knowledge of percent to **work Exercises 55 and 56 in order.***

55. (a) College students are offered a 6% discount on a dictionary that sells for $18.50. If the sales tax is 6%, find the cost of the dictionary, including the sales tax, to the nearest cent.

(b) In part (a) the rate of discount and the sales tax rate are the same percent. Explain why the answer did *not* end up back at $18.50.

56. (a) A combination printer, scanner, copier, and FAX machine priced at $398 is marked down 7% to promote the new model. If the sales tax is also 7%, find the cost of the machine, including sales tax, to the nearest cent.

(b) What rate of sales tax would have made the final answer in part (a) end up back at $398? Round to the nearest hundredth of a percent.

Math in the Media

MAKE YOUR INVESTMENTS GROW— COMPOUND INTEREST

Simple interest is paid only on the original principal. But savings accounts and most investments earn *compound interest*. In that case, interest is paid on the principal *and* the interest earned. Let's see the difference made by compounding the interest.

If you invest $10,000 at 8% simple interest for 3 years, here is what you get.

$$I = (\$10,000)(0.08)(3) = \textbf{\$2400} \leftarrow \text{Interest earned}$$

$$\text{Total amount you have} = \$10,000 + \textbf{\$2400} = \textbf{\$12,400}$$

But suppose you invest $10,000 in an account that earns 8% *compounded annually,* and you leave the initial investment and the interest in the account for 3 years. The diagram below shows the compounded amount in your account at the end of each year, to the nearest dollar.

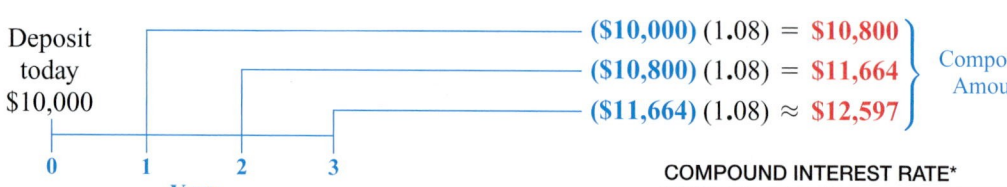

Deposit today $10,000

($10,000) (1.08) = **$10,800**
($10,800) (1.08) = **$11,664** } Compound Amount
($11,664) (1.08) ≈ **$12,597**

| 0 | 1 | 2 | 3 |
Year

You made an extra $197 with compounding ($12,597 − $12,400 = $197).

However, compound interest *really* makes a tremendous difference when you invest money over 10 or 20 or more years. The table at the right shows how a $10,000 invest-ment, plus the interest it earns, grows. Amounts are rounded to the nearest dollar.

COMPOUND INTEREST RATE*

Years	5%	8%	10%
10	$16,289	$21,589	$25,937
20	26,533	46,610	67,275
30	43,219	100,627	174,494
40	70,400	217,245	452,593
50	114,674	469,016	1,173,909

*Interest compounded annually
Source: www.moneychimp.com/calculator

Each investment in Questions 1–3 is $10,000.

1. **(a)** Calculate the amount you would have after 10 years with a 5% *simple* interest investment.

   (b) How much more would you have with *compound* interest?

2. **(a)** Calculate the amount you would have after 30 years with an 8% simple interest investment.

   (b) How much more would you have with compound interest?

3. **(a)** Repeat the calculations you did in Questions 1 and 2 for a 10% simple interest account after 50 years.

   (b) The compounded amount is how many *times* greater than the simple interest amount?

Chapter 7 ▶▶▶ Summary

▶ Key Terms

7.1	**percent**	Percent means "per one hundred." A percent is a ratio with a denominator of 100.

7.2   **percent proportion**   The proportion used to solve percent problems is $\dfrac{\text{percent}}{100} = \dfrac{\text{part}}{\text{whole}}$

    whole   The *whole* in a percent problem is the entire quantity or the total. It is sometimes called the *base*.

    part   The *part* in a percent problem is the number being compared to the *whole*.

7.3   **percent equation**   The percent equation is: percent • whole = part. The equation can be used instead of the percent proportion to solve percent problems.

7.4   **percent of increase or decrease**   Percent of increase or decrease is the amount of change (increase or decrease) expressed as a percent of the original value.

7.5   **sales tax**   Sales tax is a percent of the total sales charged as a tax.

    tax rate   The tax rate is the percent used when calculating the amount of tax.

    discount   Discount is often expressed as a percent of the original price; it is then deducted from the original price, resulting in the sale price.

    interest   Interest is a fee paid or a charge made for lending or borrowing money.

    principal   Principal is the amount of money on which interest is charged.

    interest rate   Often referred to as *rate,* it is the charge for interest and is given as a percent.

    simple interest   When interest is calculated on the original principal, it is called simple interest.

    interest formula   The interest formula is used to calculate simple interest. It is
Interest = principal • rate • time   or $I = prt$.

▶ New Formulas

Finding simple interest: $I = prt$

▶ New Symbols

%   percent symbol (meaning "per 100")

▶ Test Your Word Power

See how well you have learned the vocabulary in this chapter. Answers follow the Quick Review.

1. Percent means
  A. per one thousand
  B. part divided by whole
  C. per one hundred
  D. part times whole.

2. The **whole** in a percent problem is
  A. the entire quantity or total
  B. a ratio with a denominator of 100
  C. the amount of change
  D. always 100.

3. When calculating sales tax, the **rate** is
  A. the part
  B. the whole
  C. the total cost
  D. the percent.

4. The **interest formula** is
  A. tax rate • cost of item = sales tax
  B. $I = prt$
  C. percent • whole = part
  D. $\dfrac{\text{percent}}{100} = \dfrac{\text{part}}{\text{whole}}$.

5. When calculating interest on a loan, the **principal** is the
  A. amount of money borrowed
  B. total amount due
  C. rate charged for borrowing money
  D. amount of simple interest.

6. The **percent of increase or decrease** compares the amount of change to
  A. 100
  B. the percent

  C. the original value before the change
  D. the new value after the change.

7. The **percent equation** is
  A. tax rate • cost of item = sales tax
  B. $I = prt$
  C. percent • whole = part
  D. $\dfrac{\text{percent}}{100} = \dfrac{\text{part}}{\text{whole}}$.

8. A **discount** is
  A. added to the original price
  B. divided by 100
  C. multiplied by the percent
  D. subtracted from the original price.

▶ **Quick Review**

Concepts	Examples

7.1 **Basics of Percent**

Writing a Percent as a Decimal
To write a percent as a decimal, move the decimal point *two* places to the *left* and drop the % sign.

$$50\% = 50.\% = 0.50 \text{ or } 0.5$$
$$3\% = 03.\% = 0.03$$

Writing a Decimal as a Percent
To write a decimal as a percent, move the decimal point *two* places to the *right* and attach a % sign.

$$0.75 = 0.75 = 75\%$$
$$3.6 = 3.60 = 360\%$$

Writing a Percent as a Fraction
To write a percent as a fraction, drop the % symbol and write the number over 100. Then write the fraction in lowest terms.

$$35\% = \frac{35}{100} = \frac{35 \div 5}{100 \div 5} = \frac{7}{20} \leftarrow \text{Lowest terms}$$

$$125\% = \frac{125}{100} = \frac{125 \div 25}{100 \div 25} = \frac{5}{4} = 1\frac{1}{4}$$

Writing a Fraction as a Percent
To write a fraction as a percent, multiply by 100 and attach a % symbol. This is the same as multiplying by 100%.

$$\frac{3}{5} = \frac{3}{5} \cdot \frac{100}{1}\% = \frac{3 \cdot 5 \cdot 20}{5 \cdot 1}\% = \frac{60}{1}\% = 60\%$$

$$\frac{7}{8} = \frac{7}{8} \cdot \frac{100}{1}\% = \frac{7 \cdot 4 \cdot 25}{2 \cdot 4 \cdot 1}\% = \frac{175}{2}\% = 87\frac{1}{2}\%$$

or 87.5%

7.2 **Using the Percent Proportion**

The percent proportion is shown below.

$$\text{Always 100} \rightarrow \frac{\text{percent}}{100} = \frac{\text{part}}{\text{whole}}$$

Identify the percent first. It appears with the word *percent* or the % symbol.

The *whole* is the entire quantity or total. It often appears after the word **of**.

The *part* is the number being compared to the whole.

30 children is what percent **of** 75 children?

$$\text{Percent (unknown)} \rightarrow \frac{p}{100} = \frac{30}{75} \leftarrow \text{Part}$$
$$\text{Always 100} \rightarrow \qquad\qquad \leftarrow \text{Whole (follows } of\,)$$

To solve the proportion, find the cross products.

$$\frac{P}{100} = \frac{30}{75}$$

$$100 \cdot 30 = 3000 \leftarrow$$
$$p \cdot 75 \leftarrow \qquad \text{Cross products}$$

$$p \cdot 75 = 3000 \qquad \text{Show that the cross products are equivalent.}$$

$$\frac{p \cdot 75}{75} = \frac{3000}{75} \qquad \text{Divide both sides by 75}$$

$$p = 40$$

30 children is **40%** of 75 children.

7.3 **Percent Shortcuts**

200% of a number is 2 times the number.

100% of a number is the entire number.

To find 50% of a number, divide the number by 2.

To find 25% of a number, divide the number by 4.

To find 10% of a number, move the decimal point *one* place to the *left*.

To find 1% of a number, move the decimal point *two* places to the *left*.

200% of $35 is 2 times $35 = $70.

100% of 600 women is *all* the women (600 women).

50% of $8000 is $8000 ÷ 2 = $4000.

25% of 40 pens is 40 ÷ 4 = 10 pens.

10% of $92.40 is $9.24.

1% of 62. miles is 0.62 mile.

Concepts	Examples

7.3–7.4 **Using the Percent Equation**

Use the six problem-solving steps from **Chapter 3**.

Step 1 **Read** the problem.

Step 2 **Assign a variable.**

Step 3 **Write an equation.**

Step 4 **Solve the equation.**

Step 5 **State the answer.**

Step 6 **Check** the solution.

Todd's regular pay is \$540 per week but \$43.20 is taken out of each paycheck for medical insurance. What percent is that?

Step 1 The problem asks for the percent of Todd's pay that is taken out for insurance.

     Unknown: percent taken out
     Known: Whole paycheck is \$540;
              \$43.20 taken out.

Step 2 Let p be the unknown percent.

Step 3 Use the percent equation. The *whole* is Todd's entire pay of \$540 and the *part* is \$43.20 (the part taken out of his paycheck).

$$\text{percent} \cdot \text{whole} = \text{part}$$
$$p \quad \cdot \quad 540 \quad = 43.20$$

Step 4
$$\frac{p \cdot 540}{540} = \frac{43.20}{540} \quad \text{Divide both sides by 540}$$

$$p = 0.08 \leftarrow \text{Decimal form}$$

Multiply 0.08 by 100% so that $0.08 = 8\%$.

Step 5 8% of Todd's pay is taken out for medical insurance.

Step 6 Use estimation. If 10% of Todd's pay were withheld, then 10% of \$540 = \$54, which is a little more than the \$43.20 actually withheld. So 8% is a reasonable solution.

7.4 **Finding Percent of Increase or Decrease**

Use subtraction to find the *amount* of increase or decrease. When writing the equation, be careful to use the *original* value as the whole.

Enrollment rose from 3820 students to 5157 students. Find the percent of increase.

$$5157 \text{ students} - 3820 \text{ students} = 1337 \text{ students}$$
$$\uparrow$$
$$\text{Amount of increase}$$

$$\text{percent} \underline{\text{ of }} \underline{\text{original value}} = \underline{\text{amount of increase}}$$
$$p \quad \cdot \quad 3820 \quad = \quad 1337$$

$$p \cdot 3820 = 1337$$

$$\frac{p \cdot 3820}{3820} = \frac{1337}{3820} \quad \text{Divide both sides by 3820}$$

$$p = 0.35 \leftarrow \text{Decimal form}$$
$$0.35 = 35\% \leftarrow \text{Percent increase}$$

The enrollment increased **35%**.

Concepts	Examples

(7.5) Consumer Applications

Finding Sales Tax

To find sales tax, use this equation.

$$\text{tax rate} \cdot \text{cost of item} = \text{sales tax}$$

Write the percent as a decimal.

Find the sales tax on an \$89 pair of binoculars if the sales tax rate is 6%.

$$\text{tax rate} \cdot \text{cost of item} = \text{sales tax}$$

$$06.\% \cdot \$89 = n$$

$$(0.06)(89) = n$$

$$5.34 = n$$

The sales tax is **\$5.34**.

Estimating and Calculating Restaurant Tips

To estimate a 15% tip, first round the food bill up to a convenient whole dollar amount. Then find 10% of the rounded amount by moving the decimal point *one* place to the *left*. Then add half of that amount for the other 5%.

To estimate a 20% tip, first find 10% by moving the decimal point *one* place to the *left*. Then double the amount.

To find the exact tip, write the percent as a decimal. The bill for food and beverages is the *whole* and the tip is the *part*.

Estimate a 15% tip on a restaurant bill of \$38.72. Then find the exact tip.

Estimate: Round \$38.72 to \$40.

   10% of \$40. is \$4.

   Half of \$4 is \$2.

So an estimate of the 15% tip is \$4 + \$2 = \$6.

Exact:

$$\text{percent} \cdot \text{whole} = \text{part}$$

$$15.\% \cdot 38.72 = n$$

$$(0.15)(38.72) = n$$

$$5.808 = n$$

The exact tip is **\$5.81** (rounded to nearest cent).

Finding a Discount

To find a discount, use this formula.

$$\text{rate of discount} \cdot \text{original price} = \text{amount of discount}$$

Then subtract to find the sale price.

$$\text{original price} - \text{amount of discount} = \text{sale price}$$

All calculators are on sale at 20% off. Find the sale price of a calculator originally marked \$35.

$$20\% \cdot \$35 = n$$

$$(0.20)(35) = n$$

$$7 = n$$

The amount of discount is \$7.

Then \$35 − \$7 = **\$28** ← Sale price

Finding Simple Interest and Total Amount Due

To find the simple interest on a loan, use the formula $I = prt$.

$$\text{Interest} = \text{principal} \cdot \text{rate} \cdot \text{time}$$

Time (t) is in years. When the time is given in months, use a fraction with 12 in the denominator because there are 12 months in a year.

Write the rate (the percent) as a decimal.

To find the total amount due on the loan, add the interest to the original principal.

\$2800 is borrowed at 8% for 5 months. Find the amount of interest and the total amount due.

$$I = p \cdot r \cdot t$$

$$= (2800)(0.08)\left(\frac{5}{12}\right)$$

$$= (224)\left(\frac{5}{12}\right) = \frac{(224)(5)}{12} \approx \textbf{\$93.33}$$

Total amount due = \$2800 + **\$93.33** = **\$2893.33**

ANSWERS TO TEST YOUR WORD POWER

1. C; *Example:*   7% means 7 per 100, or, 7 out of 100.

2. A; *Example:*   In the problem "15 computers is what percent of 75 computers," the whole is 75 computers, which is the total group of computers.

3. D; *Example:*   Houston has a sales tax rate of 8.25%; Minneapolis has a sales tax rate of 7%.

4. B; *Example:*   The interest formula is Interest = principal • rate • time. If you borrow $4000 for 2 years at a rate of 9%, then $I = (4000)(0.09)(2) = \$720$.

5. A; *Example:*   If you borrow $4000 for 2 years at a rate of 9%, the principal is $4000.

6. C; *Example:*   If your rent increased from $800 to $850, the *percent of increase* compares the amount of change ($50) to the *original* rent ($800).

7. C; *Example:*   To answer the question, "What percent of 40 Web pages is 12 Web pages," let p be the unknown percent and write the equation as: $p \cdot 40 = 12$.

8. D; *Example:*   If sunglasses are on sale at 10% off, then a pair of sunglasses regularly priced at $25 will have a discount of $2.50 subtracted from the price; you will pay $22.50. To find the amount of discount, multiply $(0.10)(\$25)$ to get $2.50.

Math in the Media

MOVIES MAKE BIG BUCKS

Americans love movies and spend millions of dollars each year to see them. The table below lists some of the movies that have made the most money (gross earnings).

SELECTED ALL-TIME TOP GROSSING AMERICAN MOVIES

Title (Year Released)	Gross Earnings (nearest $10 million)
Titanic (1997)	$600 million
Star Wars: Episode IV (1977)	
Spider-Man (2002)	$400 million
Jurassic Park (1993)	
Harry Potter and the Sorcerer's Stone (2001)	$320 million
Pirates of the Caribbean (2003)	$300 million
X-Men: The Last Stand (2006)	$230 million

Write ratios as fractions in lowest terms. Round percent answers to the nearest whole percent when necessary.

1. **(a)** Write a ratio that compares the earnings of *Harry Potter and the Sorcerer's Stone* to *Spider-Man,* then change the fraction to a decimal and to a percent.

   (b) Repeat the steps in part (a) but compare the earnings of *Titanic* to *Pirates of the Caribbean.*

2. **(a)** The earnings of *Jurassic Park* are 90% of the earnings of *Spider-Man.* Use the percent proportion or equation to find *Jurassic Park*'s earnings. Write your answer in the table.

   (b) What are *Star War*'s earnings if they are 115% of the earnings for *Spider-Man*? Write your answer in the table.

3. **(a)** The earnings for *X-Men* are what percent of the earnings for *Titanic*?

   (b) What percent of *Harry Potter*'s earnings are *SpiderMan*'s earnings?

   (c) The earnings for *Star Wars* are what percent of the earnings for *X-Men*?

Source: From *Titanic* (1997)

Chapter 7  Review Exercises

[7.1] *Write each percent as a decimal and each decimal as a percent.*

1. 25% **2.** 180% **3.** 12.5% **4.** 7%

5. 2.65 **6.** 0.02 **7.** 0.3 **8.** 0.002

Write each percent as a fraction or mixed number in lowest terms. Write each fraction as a percent.

9. 12% **10.** 37.5% **11.** 250% **12.** 5%

13. $\dfrac{3}{4}$ **14.** $\dfrac{5}{8}$ **15.** $3\dfrac{1}{4}$ **16.** $\dfrac{3}{50}$

Complete this table.

Fraction	Decimal	Percent
$\dfrac{1}{8}$	**17.** _____	**18.** _____
19. _____	0.15	**20.** _____
21. _____	**22.** _____	180%

Use percent shortcuts to fill in the blanks.

23. 100% of $46 is _____.

24. 50% of $46 is _____.

25. 100% of 9 hours is _____.

26. 50% of 9 hours is _____.

[7.2] *Use a percent proportion to answer each question. If necessary, round percent answers to the nearest tenth of a percent.*

27. 338.8 meters is 140% of what number of meters?

28. 2.5% of what number of cases is 425 cases?

29. What is 6% of 450 cellular phones?

30. 60% of 1450 reference books is how many books?

31. What percent of 380 pairs is 36 pairs?

32. 1440 cans is what percent of 640 cans?

[7.3] *Use the percent equation to answer each question. Round money answers to the nearest cent, if necessary.*

33. 11% of $23.60 is how much?

34. What is 125% of 64 days?

35. 1.28 ounces is what percent of 32 ounces?

36. $46 is 8% of what number of dollars?

37. 8 people is 40% of what number of people?

38. What percent of 174 ft is 304.5 ft?

[7.4] *Use the six problem-solving steps to answer each question. If necessary, round percent answers to the nearest tenth of a percent.*

39. (a) A medical clinic found that 16.8% of the patients were late for their appointments in January. The number of patients who were late was 504. Find the total number of patients in January.

   (b) In February, only 345 patients were late. Find the percent of decrease in late patients from January to February.

40. (a) Coreen budgeted $280 for food on her vacation. She actually spent 130% of that amount. How much did she spend on food?

   (b) Coreen's vacation budget included $50 for gifts and souvenirs, but she spent $112. What percent of the budgeted amount did she spend?

41. (a) In the first part of a tree-planting project, 640 of the 800 trees planted were still living one year later. What percent of the trees planted were still living?

   (b) In the second part of the project, 850 trees were planted. What was the percent of increase in trees planted?

42. Scientists tell us that worldwide there are 9600 species of birds and that 1000 of these species are in danger of extinction. What percent of the bird species are in danger of extinction, to the nearest tenth of a percent?

The ivory-billed woodpecker, thought to be extinct, was recently found again in the southeastern United States.

[7.5] *Find the amount of sales tax or the tax rate and the total cost. Round to the nearest cent, if necessary.*

Amount of Sale	Tax Rate	Amount of Tax	Total Cost
43. $2.79	4%	_____	_____
44. $780	_____	$58.50	_____

For each restaurant bill, estimate a 15% tip and a 20% tip. Then find the exact amount for a 15% tip and a 20% tip. Round to the nearest cent, if necessary.

Bill	Estimated 15%	Exact 15%	Estimated 20%	Exact 20%
45. $42.73	_____	_____	_____	_____
46. $8.05	_____	_____	_____	_____

Find the amount or rate of discount and the sale price.

Original Price	Rate of Discount	Amount of Discount	Sale Price
47. $37.50	10%	_____	_____
48. $252	_____	$63	_____

Find the simple interest and total amount due on each loan.

Principal	Rate	Time	Interest	Total Amount Due
49. $350	$6\frac{1}{2}\%$	3 years	_____	_____
50. $1530	16%	9 months	_____	_____

> ▶▶▶ **Mixed Review Exercises**

The bar graph shows the types of electronic/computer games that adults like to play. Use the information in the graph to answer Exercises 51–54.

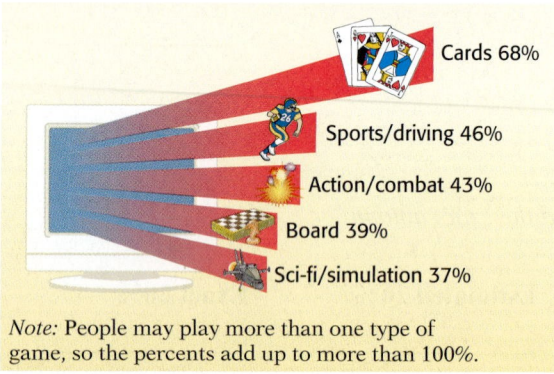

CARE FOR A GAME OF CARDS?
One in three adults say they play electronic or computer games. Types of games they play:

Cards 68%
Sports/driving 46%
Action/combat 43%
Board 39%
Sci-fi/simulation 37%

Note: People may play more than one type of game, so the percents add up to more than 100%.

Source: Cable & Telecommunications Association for Marketing.

51. Write the portion of adults who like to play electronic/computer games as a fraction and as a percent.

52. What type of game is most popular? Write the portion of players who picked this type as a percent, a decimal, and a fraction in lowest terms.

53. If 830 adults were surveyed, how many of them play electronic/computer games, to the nearest whole number?

54. Using your answer from Exercise 53, how many of the playing adults chose the most popular and least popular types of game?

The table shows the number of animals received during the first nine months of the year by the local Animal Humane Society and the number placed in new homes. Use the table to answer Exercises 55–60. Round percent answers to the nearest tenth of a percent.

Animals Received	Placed in New Homes
5371 dogs	2599 dogs
6447 cats	2346 cats
2223 other*	406 other*

*"Other" includes birds, rabbits, guinea pigs, gerbils, etc.

55. What percent of the dogs were placed in new homes?

56. The number of dogs received so far this year is 75% of the total number expected. How many dogs are expected, to the nearest whole number?

57. What percent of the cats were placed in new homes?

58. During the first nine months of last year, 2300 cats were received. What is the percent increase in cats received from last year to this year?

59. What percent of all the animals received were placed in new homes?

60. The Society's goal was to place 40% of all animals received. So far they have missed their goal by how many animals? Round to the nearest whole number.

Chapter 7 ▶▶▶ Test

Use the Chapter Test Prep Video CD to see fully worked-out solutions to any of the exercises you want to review.

Write each percent as a decimal and each decimal as a percent.

1. 75%   **2.** 0.6   **3.** 1.8

4. 0.075   **5.** 300%   **6.** 2%

Write each percent as a fraction or mixed number in lowest terms.

7. 62.5%   **8.** 240%

Write each fraction or mixed number as a percent.

9. $\dfrac{1}{20}$ **10.** $\dfrac{7}{8}$ **11.** $1\dfrac{3}{4}$

Write and solve a proportion or an equation to answer each question. Show your work.

12. 16 laptops is 5% of what number of laptops?

13. $192 is what percent of $48?

14. Erica Green has saved 75% of the amount needed for a down payment on a condominium. If she has saved $14,625, find the total down payment needed.

15. The price of a used car is $7950 plus sales tax of $6\frac{1}{2}$%. Find the total cost of the car including sales tax.

16. Enrollment in mathematics courses increased from 1440 students last semester to 1925 students this semester. Find the percent of increase, to the nearest whole percent.

1. _____

2. _____

3. _____

4. _____

5. _____

6. _____

7. _____

8. _____

9. _____

10. _____

11. _____

12. _____

13. _____

14. _____

15. _____

16. _____

17. _____

17. Explain a shortcut for finding 50% of a number and a shortcut for finding 25% of a number. Show an example of how to use each shortcut.

18. _____

18. Explain how you would *estimate* a 15% tip on a restaurant bill of $31.94. Then explain how you would *estimate* a 20% tip on the same bill.

19. _____

19. Find the exact 15% tip, to the nearest cent, for the restaurant bill in Problem 18. If you and two friends are sharing the bill and the exact tip, how much will each person pay?

Find the amount of discount and the sale price of each item in Problems 20 and 21. Round answers to the nearest cent when necessary. Show your work.

20. _____

20. Jeremy plans to use his 8% employee discount to buy a $48 clock radio so he can get to work on time.

21. _____

21. A Samsung DVD/VCR combination player regularly priced at $229.95 is on sale at 18% off.

22. _____

22. Jamal found a $1089 computer on sale at 30% off because it was a "discontinued" model. The store will let him pay for it over 6 months with no interest charge. How much will each monthly payment need to be to cover the discounted price plus 7% sales tax?

23. _____

23. What is the simple interest and total amount due on a four-year loan of $5000 at $8\frac{1}{4}\%$?

24. _____

24. Kendra borrowed $860 to pay medical expenses. The loan is for 6 months at 12% simple interest. Find the simple interest and total amount due on the loan.

Study Skills

Your math final exam is likely to be a **comprehensive exam.** This means that it will cover material from the **entire term.** The end of the term will be less stressful if you **make a plan** for how you will prepare for each of your exams.

First, figure out the **score you need to earn on the final exam** to get the course grade you are aiming for. Check your course syllabus for grading policies, or ask your instructor if you are not sure of them. This allows you to set a goal for yourself.

> How many points do you need to earn on your mathematics final exam to get the grade you want? _____
>
> _____

Second, create a **final exam week plan for your work and personal life.** If you need to make an adjustment in your work schedule, do it in advance, so you aren't scrambling at the last minute. If you have family members to care for, you might want to enlist some help from others so you can spend extra time studying. Try to plan in advance so you don't create additional stress for yourself. You will have to set some priorities, and studying has to be at the top of the list! Although life doesn't stop for finals, some things can be ignored for a short time. You don't want to "burn out" during final exam week; **get enough sleep and healthy food so you can perform your best.**

> What adjustments in your personal life do you need to make for final exam week? _____
>
> _____

Third, use the following suggestions to guide your studying and reviewing.

- **Know exactly which chapters and sections will be on the final exam.**

- **Divide up the chapters,** and decide how much you will review each day.

- Begin your reviewing **several days** before the exam.

- **Use returned quizzes and tests** to review earlier material (if you have them).

- **Practice all types of problems,** but emphasize the types that are most difficult for you. Use the **Cumulative Reviews** that are at the end of each chapter in your textbook.

- **Rewrite your notes or make mind maps** to create summaries.

- **Make study cards for all types of problems.** Be sure to use the same **direction words** (such as *simplify, solve, estimate*) that your exam will use. Carry the cards with you and review them whenever you have a few spare minutes.

OBJECTIVES

1. **Create a final exam week plan.**

2. **Break studying into chunks and study over several days.**

3. **Practice all types of problems.**

Create a Plan

Study and Review

Managing Stress

Reducing Physical Stress

> Which techniques will you try?
>
> _____
>
> _____
>
> _____

Reducing Mental Stress

> Which techniques will you try?
>
> _____
>
> _____
>
> _____

Of course, a week of final exams produces stress. **Students who develop skills for reducing and managing stress do better on their final exams and are less likely to "bomb" an exam.** You already know the damaging effect of adrenaline on your ability to think clearly. But several days (or weeks) of elevated stress is also harmful to your brain and your body. You will feel better if you make a conscious effort to reduce your stress level. Even if it takes you away from studying for a little while each day, the time will be well spent.

Examples of ways to reduce **physical stress** are listed below. Can you add any of your own ideas to the list?

- *Laugh until your eyes water.* Watching your favorite funny movie, exchanging a joke with a friend, or viewing a comedy bit on the Internet are all ways to generate a healthy laugh. Laughing raises the level of *calming* chemicals (endorphins) in your brain.

- *Exercise for 20 to 30 minutes.* If you normally exercise regularly, do NOT stop during final exam week! Exercising helps relax muscles, diffuses adrenaline, and raises the level of endorphins in your body. If you don't exercise much, get some gentle exercise, such as a daily walk, to help you relax.

- *Practice deep breathing.* Several minutes of deep, smooth breathing will calm you. Close your eyes too.

- *Visualize a relaxing scene.* Choose something that you find peaceful and picture it. Imagine what it feels like and sounds like. Try to put yourself in the picture.

- If you feel stress in your muscles, such as your shoulders or back, *slowly squeeze the muscles as much as you can, and then release them.* Sometimes we don't realize we are clenching our teeth or holding tension in our shoulders until we consciously work with them. Try to notice what it feels like when they are relaxed and loose. Squeezing and then releasing muscles is also something you can do during an exam if you feel yourself tightening up.

Mental stress reduction is also a powerful tool both before and during an exam. In addition to these suggestions, do you have any of your own techniques?

- *Talk positively to yourself.* Tell yourself you will get through it.

- *Reward yourself.* Give yourself small breaks, a little treat—something that makes you happy—every day of final exam week.

- *Make a list of things to do* and feel the sense of accomplishment when you cross each item off.

- When you take time to relax or exercise, *make sure you are relaxing your mind too.* Use your mind for something *completely* different from the kind of thinking you do when you study. Plan your garden, play your favorite music, walk your dog, read a good book.

- *Visualize.* Picture yourself completing exams and projects successfully. Picture yourself taking the test calmly and confidently.

8

◀ The world's best-preserved crater made by a meteor crashing into Earth is located in northern Arizona. Notice the buildings at the upper left edge of the crater for size comparison.

Measurement

W e are constantly measuring things, from very large to very small.

For example, a meteor crashed into Earth in northern Arizona nearly 50,000 years ago. The bowl-shaped crater from that crash is sixty stories deep and 4150 feet across. (See **Section 8.1,** Exercises 63–66.)

On the other hand, computer microchips often measure a tiny 5 mm long by 1 mm wide. (Learn about millimeters in **Section 8.2** and then try Exercise 39.)

A greatly enlarged photo of a computer microchip.

8.1 Problem Solving with U.S. Customary Measurements

8.2 The Metric System— Length

8.3 The Metric System— Capacity and Weight (Mass)

Summary Exercises on U.S. Customary and Metric Units

8.4 Problem Solving with Metric Measurement

8.5 Metric—U.S. Customary Conversions and Temperature

577

8.1 ▶▶▶ Problem Solving with U.S. Customary Measurements

OBJECTIVES

1. **Learn the basic U.S. customary measurement units.**

2. **Convert among measurement units using multiplication or division.**

3. **Convert among measurement units using unit fractions.**

4. **Solve application problems using U.S. customary measurement units.**

We measure things all the time: the distance traveled on vacation, the floor area we want to cover with carpet, the amount of milk in a recipe, the weight of the bananas we buy at the store, the number of hours we work, and many more.

In the United States, we still use **U.S. customary measurement units** for many everyday activities. Examples are inches, feet, quarts, ounces, and pounds. However, science, medicine, sports, and manufacturing use the **metric system** (meters, liters, and grams). And, because the rest of the world uses *only* the metric system, U.S. businesses have been changing to the metric system in order to compete internationally.

OBJECTIVE 1 Learn the basic U.S. customary measurement units. Until the switch to the metric system is complete, we still need to know how to use U.S. customary measurement units. The table below lists the relationships you should memorize. The time relationships are used in both the U.S. customary and metric systems.

1 After memorizing the measurement conversions, answer these questions.

(a) 1 c = _____ fl oz

(b) _____ qt = 1 gal

(c) 1 wk = _____ days

(d) _____ ft = 1 yd

(e) 1 ft = _____ in.

(f) _____ oz = 1 lb

(g) 1 ton = _____ lb

(h) _____ min = 1 hr

(i) 1 pt = _____ c

(j) _____ hr = 1 day

(k) 1 min = _____ sec

(l) 1 qt = _____ pt

(m) _____ ft = 1 mi

U.S. Customary Measurement Relationships

Length	Weight
12 inches (in.) = 1 foot (ft)	16 ounces (oz) = 1 pound (lb)
3 feet (ft) = 1 yard (yd)	2000 pounds (lb) = 1 ton
5280 feet (ft) = 1 mile (mi)	

Capacity	Time
8 fluid ounces (fl oz) = 1 cup (c)	60 seconds (sec) = 1 minute (min)
2 cups (c) = 1 pint (pt)	60 minutes (min) = 1 hour (hr)
2 pints (pt) = 1 quart (qt)	24 hours (hr) = 1 day
4 quarts (qt) = 1 gallon (gal)	7 days = 1 week (wk)

As you can see, there is no simple way to convert among these various measures. The units evolved over hundreds of years and were based on a variety of "standards." For example, one yard was the distance from the tip of a king's nose to his thumb when his arm was outstretched. An inch was three dried barleycorns laid end to end.

EXAMPLE 1 Knowing U.S. Customary Measurement Units

Memorize the U.S. customary measurement conversions shown above. Then answer these questions.

(a) 24 hr = _____ day   Answer: 1 day

(b) 1 yd = _____ ft   Answer: 3 ft

◀ *Work Problem* **1** *at the Side.*

OBJECTIVE 2 Convert among measurement units using multiplication or division. You often need to convert from one unit of measure to another. Two methods of converting measurements are shown here. Study each way and use the method you prefer. The first method involves deciding whether to multiply or divide.

ANSWERS

1. (a) 8 (b) 4 (c) 7 (d) 3 (e) 12
   (f) 16 (g) 2000 (h) 60 (i) 2 (j) 24
   (k) 60 (l) 2 (m) 5280

> **Converting among Measurement Units**
> 1. *Multiply* when converting from a larger unit to a smaller unit.
> 2. *Divide* when converting from a smaller unit to a larger unit.

EXAMPLE 2 **Converting from One Unit of Measure to Another**

Convert each measurement.

(a) 7 ft to inches
   You are converting from a *larger* unit to a *smaller* unit (a *foot* is longer than an *inch*), so multiply.
   Because *1 ft* = **12** *in.*, multiply by 12.

$$7 \text{ ft} = 7 \cdot \mathbf{12} = 84 \text{ in.}$$

(b) $3\frac{1}{2}$ lb to ounces
   You are converting from a *larger* unit to a *smaller* unit (a *pound* is heavier than an *ounce*), so multiply.
   Because *1 lb* = **16** *oz*, multiply by 16.

> Divide 16 and 2 by their common factor of 2.
> 16 ÷ 2 is 8 and 2 ÷ 2 is 1.

$$3\frac{1}{2} \text{ lb} = 3\frac{1}{2} \cdot \mathbf{16} = \frac{7}{\cancel{2}_{1}} \cdot \frac{\cancel{16}^{8}}{1} = \frac{56}{1} = 56 \text{ oz}$$

(c) 20 qt to gallons
   You are converting from a *smaller* unit to a *larger* unit (a *quart* is smaller than a *gallon*) so divide.
   Because **4** *qt* = *1 gal*, divide by 4.

$$20 \text{ qt} = \frac{20}{4} = 5 \text{ gal}$$

Divide by 4. ⤴

(d) 45 min to hours
   You are converting from a *smaller* unit to a *larger* unit (a *minute* is less than an *hour*), so divide.
   Because **60** *min* = *1 hr*, divide by 60 and write the fraction in lowest terms.

$$45 \text{ min} = \frac{45}{\mathbf{60}} = \frac{45 \div \mathbf{15}}{60 \div \mathbf{15}} = \frac{3}{4} \text{ hr} \quad \leftarrow \text{Lowest terms}$$

Divide by 60. ⤴

Work Problem **2** *at the Side.* ▶

OBJECTIVE 3 Convert among measurement units using unit fractions.   If you have trouble deciding whether to multiply or divide when converting measurements, use *unit fractions* to solve the problem. You'll also use this method in science courses. A **unit fraction** is equivalent to 1. Here is an example.

$$\frac{12 \text{ in.}}{12 \text{ in.}} = \frac{\cancel{12} \text{ in.}^{1}}{\cancel{12} \text{ in.}_{1}} = 1$$

Use the table of measurement relationships on the previous page to find that 12 in. is the same as 1 ft. So you can substitute 1 ft for 12 in. in the numerator, or you can substitute 1 ft for 12 in. in the denominator. This makes two useful unit fractions.

$$\frac{\mathbf{1 \ ft}}{12 \text{ in.}} = 1 \quad \text{or} \quad \frac{12 \text{ in.}}{\mathbf{1 \ ft}} = 1$$

---

2 Convert each measurement using multiplication or division.

(a) $5\frac{1}{2}$ ft to inches

(b) 64 oz to pounds

(c) 6 yd to feet

(d) 2 tons to pounds

(e) 35 pt to quarts

(f) 20 min to hours

(g) 4 wk to days

ANSWERS

2. **(a)** 66 in. **(b)** 4 lb **(c)** 18 ft
   (d) 4000 lb **(e)** $17\frac{1}{2}$ qt **(f)** $\frac{1}{3}$ hr
   (g) 28 days

To convert from one measurement unit to another, just multiply by the appropriate unit fraction. Remember, a unit fraction is equivalent to 1. Multiplying something by 1 does *not* change its value.

Use these guidelines to choose the correct unit fraction.

3 First write the unit fraction needed to make each conversion. Then complete the conversion.

(a) 36 in. to feet

unit fraction $\left.\right\}$ $\dfrac{1 \text{ ft}}{12 \text{ in.}}$

> **Choosing a Unit Fraction**
>
> The *numerator* should use the measurement unit you want in the *answer*.
> The *denominator* should use the measurement unit you want to *change*.

EXAMPLE 3 Using Unit Fractions with Length Measurements

(a) Convert 60 in. to feet.

Use a unit fraction with feet (the unit for your answer) in the numerator, and inches (the unit being changed) in the denominator. Because *1 ft = 12 in.*, the necessary unit fraction is

$$\dfrac{1 \text{ ft}}{12 \text{ in.}} \quad \leftarrow \text{Unit for your answer is feet.}$$
$$\leftarrow \text{Unit being changed is inches.}$$

(b) 14 ft to inches

unit fraction $\left.\right\}$ $\dfrac{\text{in.}}{\text{ft}}$

Next, multiply 60 in. times this unit fraction. Write 60 in. as the fraction $\dfrac{60 \text{ in.}}{1}$ and divide out common units and factors wherever possible.

$$60 \text{ in.} \cdot \dfrac{1 \text{ ft}}{12 \text{ in.}} = \dfrac{\overset{5}{\cancel{60 \text{ in.}}}}{1} \cdot \dfrac{1 \text{ ft}}{\underset{1}{\cancel{12 \text{ in.}}}} = \dfrac{5 \cdot 1 \text{ ft}}{1} = 5 \text{ ft}$$

(c) 60 in. to feet

unit fraction $\left.\right\}$ _____

These units should match.

Divide out inches.

Divide 60 and 12 by 12.

(b) Convert 9 ft to inches.

Select the correct unit fraction to change 9 ft to inches.

$$\dfrac{12 \text{ in.}}{1 \text{ ft}} \quad \leftarrow \text{Unit for your answer is inches.}$$
$$\leftarrow \text{Unit being changed is feet.}$$

(d) 4 yd to feet

unit fraction $\left.\right\}$ _____

Multiply 9 ft times the unit fraction.

$$9 \text{ ft} \cdot \dfrac{12 \text{ in.}}{1 \text{ ft}} = \dfrac{9 \cancel{\text{ ft}}}{1} \cdot \dfrac{12 \text{ in.}}{1 \cancel{\text{ ft}}} = \dfrac{9 \cdot 12 \text{ in.}}{1} = 108 \text{ in.}$$

These units should match.

Divide out feet.

(e) 39 ft to yards

unit fraction $\left.\right\}$ _____

> **CAUTION**
> If no units will divide out, you made a mistake in choosing the unit fraction.

(f) 2 mi to feet

unit fraction $\left.\right\}$ _____

◀ *Work Problem* **3** *at the Side.*

EXAMPLE 4 Using Unit Fractions with Capacity and Weight Measurements

(a) Convert 9 pt to quarts.

First select the correct unit fraction.

$$\dfrac{1 \text{ qt}}{2 \text{ pt}} \quad \leftarrow \text{Unit for your answer is quarts.}$$
$$\leftarrow \text{Unit being changed is pints.}$$

Continued on Next Page

Next multiply.

Write as a mixed number.

$$9 \text{ pt} \cdot \frac{1 \text{ qt}}{2 \text{ pt}} = \frac{9 \text{ pt}}{1} \cdot \frac{1 \text{ qt}}{2 \text{ pt}} = \frac{9}{2} \text{ qt} = 4\frac{1}{2} \text{ qt}$$

These units should match.

Divide out pints.

4 Convert using unit fractions.

(a) 16 qt to gallons

(b) Convert $7\frac{1}{2}$ gal to quarts.

Write as an improper fraction.

$$\frac{7\frac{1}{2} \text{ gal}}{1} \cdot \frac{4 \text{ qt}}{1 \text{ gal}} = \frac{15}{2} \cdot \frac{4}{1} \text{ qt}$$

Divide out gallons.

Divide 4 and 2 by their common factor of 2.
$4 \div 2$ is 2.
$2 \div 2$ is 1.

$$= \frac{15}{\overset{}{2}_{1}} \cdot \frac{\overset{2}{4}}{1} \text{ qt}$$

$$= 30 \text{ qt}$$

(b) 3 c to pints

(c) Convert 36 oz to pounds.

Notice that **oz** divides out, leaving **lb**, the unit you want for the answer.

$$\frac{\overset{9}{36} \text{ oz}}{1} \cdot \frac{1 \text{ lb}}{\underset{4}{16} \text{ oz}} = \frac{9}{4} \text{ lb} = 2\frac{1}{4} \text{ lb}$$

(c) $3\frac{1}{2}$ tons to pounds

Note

In Example 4(c) above you get $\frac{9}{4}$ lb. Recall that $\frac{9}{4}$ means $9 \div 4$. If you do $9 \div 4$ on your calculator, you get 2.25 lb. U.S. customary measurements usually use fractions or mixed numbers, like $2\frac{1}{4}$ lb. However, 2.25 lb is also correct and is the way grocery stores often show weights of produce, meat, and cheese.

(d) $1\frac{3}{4}$ lb to ounces

Work Problem **4** *at the Side.* ▶

EXAMPLE 5 **Using Several Unit Fractions**

Sometimes you may need to use two or three unit fractions to complete a conversion.

(a) Convert 63 in. to yards.

Use the unit fraction $\frac{1 \text{ ft}}{12 \text{ in.}}$ to change inches to feet and the unit fraction $\frac{1 \text{ yd}}{3 \text{ ft}}$ to change feet to yards. Notice how all the units divide out except yards, which is the unit you want in the answer.

$$\frac{63 \text{ in.}}{1} \cdot \frac{1 \text{ ft}}{12 \text{ in.}} \cdot \frac{1 \text{ yd}}{3 \text{ ft}} = \frac{63}{36} \text{ yd} = \frac{63 \div 9}{36 \div 9} \text{ yd} = \frac{7}{4} \text{ yd} = 1\frac{3}{4} \text{ yd}$$

(e) 4 oz to pounds

Continued on Next Page

5 Convert using two or three unit fractions.

(a) 4 tons to ounces

(b) 3 mi to inches

(c) 36 pt to gallons

(d) 2 wk to minutes

You can also divide out common factors in the numbers.

$$\frac{\overset{7}{\cancel{63}}}{1} \cdot \frac{1}{\underset{4}{\cancel{12}}} \cdot \frac{1}{\underset{1}{\cancel{3}}} = \frac{7}{4} = 1\frac{3}{4} \text{ yd}$$

Instead of changing $\frac{7}{4}$ to $1\frac{3}{4}$, you can enter $7 \div 4$ on your calculator to get 1.75 yd. Both answers are correct because 1.75 is equivalent to $1\frac{3}{4}$.

(b) Convert 2 days to seconds.

Use three unit fractions. The first one changes days to hours, the next one changes hours to minutes, and the last one changes minutes to seconds. All the units divide out except seconds, which is the unit you want in your answer.

$$\frac{2 \text{ days}}{1} \cdot \frac{24 \text{ hr}}{1 \text{ day}} \cdot \frac{60 \text{ min}}{1 \text{ hr}} \cdot \frac{60 \text{ sec}}{1 \text{ min}} = 172{,}800 \text{ sec}$$

Divide out **days**. —⌐
    Divide out **hr**. —⌐
        Divide out **min**. —⌐

◀ *Work Problem* **5** *at the Side.*

OBJECTIVE **4** **Solve application problems using U.S. customary measurement units.** To solve measurement application problems, we will use the steps summarized here.

Step 1 **Read** the problem.

Step 2 **Work out a plan.**

Step 3 **Estimate** a reasonable answer.

Step 4 **Solve** the problem.

Step 5 **State the answer.**

Step 6 **Check** your work.

Because measurement applications often involve conversions, writing an equation may not be the best way to solve the problem. Therefore, Steps 2 and 3 are different from the ones you learned in **Chapter 3;** the rest are the same.

EXAMPLE 6 **Solving U.S. Customary Measurement Applications**

(a) A 36-oz can of coffee is on sale at Jerry's Foods for $7.89. What is the cost per pound, to the nearest cent? (*Source:* Jerry's Foods.)

Step 1 **Read** the problem. The problem asks for the cost per *pound* of coffee.

Step 2 **Work out a plan.** The weight of the coffee is given in *ounces,* but the answer must be cost *per pound.* Convert ounces to pounds. The word *per* indicates division. You need to divide the cost by the number of pounds.

Step 3 **Estimate** a reasonable answer. To estimate, round $7.89 to $8. Then, there are 16 oz in a pound, so 36 oz are a little more than 2 pounds. So, $8 ÷ 2 = $4 per pound as our estimate.

Continued on Next Page

ANSWERS

5. (a) 128,000 oz **(b)** 190,080 in.

(c) $4\frac{1}{2}$ gal or 4.5 gal **(d)** 20,160 min

Step 4   **Solve** the problem. Use a unit fraction to convert 36 oz to pounds.

Notice that **oz** divides out, leaving **lb** for the answer.

On your calculator, $9 \div 4 = 2.25$

$$\frac{36 \text{ oz}}{1} \cdot \frac{1 \text{ lb}}{16 \text{ oz}} = \frac{9}{4} \text{ lb} = 2.25 \text{ lb}$$

Then divide to find the *cost* per *pound*.

Cost per pound → $\dfrac{\$7.89}{2.25 \text{ lb}} = 3.50\overline{6} \approx 3.51$ (rounded)

Step 5   **State the answer.** The coffee costs $3.51 per pound (to the nearest cent).

Step 6   **Check** your work. The exact answer of $3.51 is close to our estimate of $4.

(b) Bilal's favorite cake recipe uses $1\frac{2}{3}$ cups of milk. If he makes six cakes for a bake sale at his son's school, how many quarts of milk will he need?

Step 1   **Read** the problem. The problem asks for the number of *quarts* of milk needed for six cakes.

Step 2   **Work out a plan.** Multiply to find the number of *cups* of milk for six cakes. Then convert *cups* to *quarts* (the unit required in the answer).

Step 3   **Estimate** a reasonable answer. To estimate, round $1\frac{2}{3}$ cups to 2 cups. Then, 2 cups times $6 = 12$ cups. There are 4 cups in a quart, so 12 cups $\div 4 = 3$ quarts as our estimate.

Step 4   **Solve** the problem. First multiply. Then use unit fractions to convert.

$$1\frac{2}{3} \cdot 6 = \frac{5}{3} \cdot \frac{6}{1} = \frac{10}{1} = 10 \text{ cups}$$ { Milk needed for six cakes

$$\frac{10 \text{ cups}}{1} \cdot \frac{1 \text{ pt}}{2 \text{ cups}} \cdot \frac{1 \text{ qt}}{2 \text{ pt}} = \frac{5}{2} \text{ qt} = 2\frac{1}{2} \text{ qt}$$

Both **cups** and **pt** divide out, leaving **qt**, the unit you want for the answer.

Step 5   **State the answer.** Bilal needs $2\frac{1}{2}$ qt (or 2.5 qt) of milk.

Step 6   **Check** your work. The exact answer of $2\frac{1}{2}$ qt is close to our estimate of 3 qt.

Note

In Step 2 above, we *first multiplied* $1\frac{2}{3}$ cups by 6 to find the number of cups needed, then *converted* 10 cups to $2\frac{1}{2}$ quarts. It would also work to *first convert* $1\frac{2}{3}$ cups to $\frac{5}{12}$ qt, then *multiply* $\frac{5}{12}$ qt by 6 to get $2\frac{1}{2}$ qt.

Work Problem **6** *at the Side.* ▶

6 Solve each application problem using the six problem-solving steps.

(a) Kristin paid $3.29 for 12 oz of extra sharp cheddar cheese. What is the price per pound, to the nearest cent?

(b) A moving company estimates 11,000 lb of furnishings for an average 3-bedroom house. If the company made five such moves last week, how many tons of furnishings did they move? (*Source:* North American Van Lines.)

Math in the Media

GROWING SUNFLOWERS

The front and back of a seed packet for sunflowers are shown at the right. Look at the top of the packet first. (Ignore sales tax in Questions 1 and 2.)

1. There were 42 seeds in the packet. If 40 of the seeds sprouted, what was the cost per sprout, to the nearest cent?

2. If vegetable and flower seeds were on sale at 30% off, what was the cost per sprout, to the nearest cent?

3. What percent of the seeds sprouted, to the nearest whole percent?

4. How many seeds would add up to a weight of 1 gram?

5. The table at the bottom of the packet uses the symbol (′) for feet, and the symbol (″) for inches.

   (a) How tall will the plants grow, in feet?

   (b) How tall will they grow in inches?

   (c) How tall will they grow in yards?

6. If you plant all 42 seeds in one long row, using the spacing given on the package, how long will your row be in feet?

7. How many inches tall should the plants be when you thin them (remove less vigorous plants to give others room to grow)? How tall is that in feet?

8. What is the range in the diameter of the flowers, in inches, and in feet? Diameter is the distance across the circular flower.

Sunflower, Mammoth Grey Stripe

Tall Plants, Huge Flowers

Net Wt. 3.5 g $1.19

Sunflower, Mammoth Grey Stripe

The stalk of this sunflower will grow to 12'/4 m. Flowers will range from 6"/15 cm to 15"/38 cm in diameter. Sunflowers can thrive in poor soil with little moisture.

Type	Height	Planting Depth	Seed Spacing	Thinning Height	Spacing After Thinning	Days to Germination
Annual	8-10' 2.4-3 m	1/2" 13 mm	6" 15 cm	3" 8 cm	2' 61 cm	10-20

Select a sunny or lightly shaded location and plant outdoors, where plants are to remain, after all danger of frost is past. For tallest plants, sow in good soil with moderate moisture.

Stock #1185

7 18964 98119 7

Source: Olds Seed Solutions.

8.1 ▶▶▶ Exercises

Fill in the blanks with the measurement relationships you have memorized. See Example 1.

1. 1 yd = _____ ft;   _____ in. = 1 ft

2. 1 ft = _____ in.;   _____ ft = 1 mi

3. _____ fl oz = 1 c;   1 qt = _____ pt

4. _____ qt = 1 gal;   1 pt = _____ c

5. 1 mi = _____ ft;   _____ ft = 1 yd

6. 1 wk = _____ days;   _____ sec = 1 min

7. _____ lb = 1 ton;   1 lb = _____ oz

8. _____ oz = 1 lb;   1 ton = _____ lb

9. 1 min = _____ sec;   _____ min = 1 hr

10. 1 day = _____ hr;   _____ sec = 1 min

Convert each measurement by multiplying or dividing. See Example 2.

11. (a) 120 sec = _____ min

  (b) 4 hr = _____ min

12. (a) 180 min = _____ hr

  (b) 5 min = _____ sec

13. (a) 2 qt = _____ gal

  (b) $6\frac{1}{2}$ ft = _____ in.

14. (a) $4\frac{1}{2}$ gal = _____ qt

  (b) 12 oz = _____ lb

15. An adult African elephant could weigh 7 to 8 tons. How many pounds could it weigh? (*Source: The Top 10 of Everything.*)

16. A reticulated python snake is the world's longest snake. It grows to a length of 18 to 33 feet. How many yards long can the snake be? (*Source: The Top 10 of Everything.*)

Convert each measurement in Exercises 17–38 using unit fractions. See Examples 3 and 4.

17. 9 yd = _____ ft

18. 20,000 lb = _____ tons

19. 7 lb = _____ oz

20. 96 oz = _____ lb

21. 5 qt = _____ pt

22. 26 pt = _____ qt

23. 90 min = _____ hr

24. 45 sec = _____ min

25. 3 in. = _____ ft

26. 30 in. = _____ ft

27. 24 oz = _____ lb

28. 36 oz = _____ lb

29. 5 c = _____ pt

30. 15 qt = _____ gal

Use the information in the bar graph below to answer Exercises 31–32.

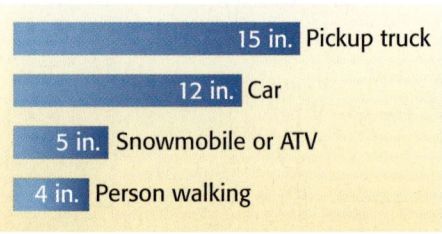

Thickness of Lake Ice Needed for Safe Walking/Driving

- 15 in. Pickup truck
- 12 in. Car
- 5 in. Snowmobile or ATV
- 4 in. Person walking

Source: Wisconsin DNR.

31. If the ice on a lake is $\frac{1}{2}$ ft thick, what will it safely support?

32. How many feet of ice are needed to safely drive a pickup truck on a lake?

33. $2\frac{1}{2}$ tons = _____ lb

34. $4\frac{1}{2}$ pt = _____ c

35. $4\frac{1}{4}$ gal = _____ qt

36. $2\frac{1}{4}$ hr = _____ min

37. After 15 years, a saguaro cactus is still only one-third to two-thirds of a foot tall, depending upon rainfall. How tall could the cactus be in inches? (*Source: Ecology of the Saguaro III.*)

38. Yao Ming, an NBA basketball player from China, is $7\frac{1}{2}$ ft tall. What is his height in inches? (*Source:* www.NBA.com)

Use two or three unit fractions to make each conversion. See Example 5.

39. 6 yd = _____ in.

40. 2 tons = _____ oz

41. 112 c = _____ qt

42. 336 hr = _____ wk

43. 6 days = _____ sec

44. 5 gal = _____ c

45. $1\frac{1}{2}$ tons = _____ oz

46. $3\frac{1}{3}$ yd = _____ in.

47. The statement 8 = 2 is *not* true. But with appropriate measurement units, it *is* true.

$$8 \text{ quarts} = 2 \text{ gallons}$$

Attach measurement units to these numbers to make the statement true.

(a) 1 _____ = 16 _____

(b) 10 _____ = 20 _____

(c) 120 _____ = 2 _____

(d) 2 _____ = 24 _____

(e) 6000 _____ = 3 _____

(f) 35 _____ = 5 _____

48. Explain in your own words why you can add 2 feet + 12 inches to get 3 feet, but you cannot add 2 feet + 12 pounds.

Convert each measurement. See Example 5.

49. $2\frac{3}{4}$ mi = _____ in.

50. $5\frac{3}{4}$ tons = _____ oz

51. $6\frac{1}{4}$ gal = _____ fl oz

52. $3\frac{1}{2}$ days = _____ sec

53. 24,000 oz = _____ ton

54. 57,024 in. = _____ mi

Solve each application problem. Show your work. See Example 6.

55. Geralyn bought 20 oz of strawberries for $2.29. What was the price per pound for the strawberries, to the nearest cent?

56. Zach paid $0.90 for a 0.8 oz candy bar. What was the cost per pound?

57. Dan orders supplies for the science labs. Each of the 24 stations in the chemistry lab needs 2 ft of rubber tubing. If rubber tubing sells for $8.75 per yard, how much will it cost to equip all the stations?

58. In 2006, Marquette, Michigan, had 170 inches of snowfall, while Detroit, Michigan, had 15 inches. What was the difference in snowfall between the two cities, in feet? Round to the nearest tenth. (*Source: World Almanac.*)

59. Tropical cockroaches are the fastest land insects. They can run about 5 feet per second. At this rate, how long would it take the cockroach to travel one mile? (*Source: Guinness World Records.*)

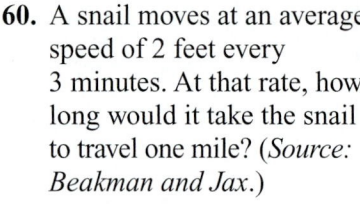

Give your answer

(a) in seconds;

(b) in minutes.

60. A snail moves at an average speed of 2 feet every 3 minutes. At that rate, how long would it take the snail to travel one mile? (*Source: Beakman and Jax.*)

Give your answer

(a) in hours;

(b) in days.

61. At the day care center, each of the 15 toddlers drinks about $\frac{2}{3}$ cup of milk with lunch. The center is open 5 days a week.

(a) How many quarts of milk will the center need for one week of lunches?

(b) If the center buys milk in gallon containers, how many containers should be ordered for one week?

62. Bob's Candies in Albany, Georgia, makes 135,000 pounds of candy canes each day. (*Source: Bob's Candies, Inc.*)

(a) How many tons of candy canes are produced during a 5-day workweek?

(b) The plant operates 24 hours per day. How many tons of candy canes are produced each hour, to the nearest tenth?

Relating Concepts (Exercises 63–66) For Individual or Group Work

*On the first page of this chapter, we said that the bowl-shaped crater in northern Arizona made by a meteor crash was sixty stories deep and 4150 ft across. Use this information as you **work Exercises 63–66 in order.** (Source: The Meteor Crater Story.)*

63. (a) The distance across the crater is what part of a mile, to the nearest tenth?

(b) The crater is nearly circular. In a circle, the distance around the outside edge is about 3.14 times the distance across the circle. How far is it to walk around the edge of the crater in feet?

(c) How far is it to walk around the edge of the crater in miles, to the nearest tenth?

64. (a) The crater is 550 ft deep. The depth is how many yards, to the nearest whole number?

(b) How many inches deep is the crater?

(c) The depth of the crater is what part of a mile, to the nearest tenth?

(d) When we say that the crater is as deep as a sixty-story building, we are assuming that each story is how many feet tall, to the nearest foot?

65. On one side of the crater there are a few small juniper trees. The trees are 700 years old but only 18 inches to 30 inches tall because of the strong winds and lack of rain.

(a) How tall are the trees in feet?

(b) How many months old are the trees?

(c) At this rate of growth, how long would it take a 30-inch tree to reach a height of three feet?

66. Evidence of two huge meteor crashes has been found on the floor of the Caribbean Sea. One giant circular crater is 90 miles across and the other is 120 miles across.

(a) Using the information about circles in Exercise 63, what is the approximate distance around the edge of the smaller crater, to the nearest mile?

(b) Around the larger crater?

8.2 ▶▶▶ The Metric System—Length

Around 1790, a group of French scientists developed the metric system of measurement. It is an organized system based on multiples of 10, like our number system and our money. After you are familiar with metric units, you will see that they are easier to use than the hodgepodge of U.S. customary measurement relationships you used in **Section 8.1.**

> **Note**
> The metric system information in this text is consistent with usage guidelines from the National Institute of Standards and Technology, www.nist.gov/metric.

OBJECTIVE 1 Learn the basic metric units of length. The basic unit of length in the metric system is the **meter** (also spelled *metre*). Use the symbol **m** for meter; do not put a period after it. If you put five of the pages from this textbook side by side, they would measure about 1 meter. Or, look at a yardstick—a meter is just a little longer. A yard is 36 inches long; a meter is about 39 inches long.

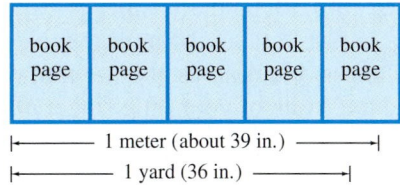

book page	book page	book page	book page	book page

|← 1 meter (about 39 in.) →|
|← 1 yard (36 in.) →|

In the metric system, you use meters for things like measuring the length of your living room, talking about heights of buildings, or describing track and field athletic events.

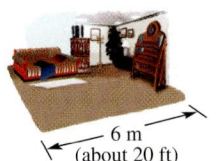
6 m
(about 20 ft)

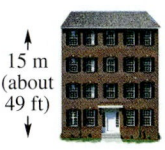
15 m
(about 49 ft)

START
Run the
100 m dash.

Work Problem **1** *at the Side.* ▶

To make longer or shorter length units in the metric system, **prefixes** are written in front of the word *meter*. For example, the prefix *kilo* means **1000**, so a *kilo*meter is **1000** meters. The table below shows how to use the prefixes for length measurements. It is helpful to memorize the prefixes because they are also used with weight and capacity measurements. The purple boxes are the units you will use most often in daily life.

Prefix	kilo- meter	hecto- meter	deka- meter	meter	deci- meter	centi- meter	milli- meter
Meaning	1000 meters	100 meters	10 meters	1 meter	$\frac{1}{10}$ of a meter	$\frac{1}{100}$ of a meter	$\frac{1}{1000}$ of a meter
Symbol	**km**	**hm**	**dam**	m	**dm**	**cm**	**mm**

↑ ↑ ↑ ↑
Length units that are used most often

OBJECTIVES

1 Learn the basic metric units of length.

2 Use unit fractions to convert among units.

3 Move the decimal point to convert among units.

1 Circle the items that measure about 1 meter.

Length of a pencil

Length of a baseball bat

Height of doorknob from the floor

Height of a house

Basketball player's arm length

Length of a paper clip

ANSWER

1. baseball bat, height of doorknob, basketball player's arm length

Here are some comparisons to help you get acquainted with the commonly used length units: km, m, cm, mm.

Kilometers are used instead of miles. A kilometer is **1000** meters. It is about 0.6 mile (a little more than half a mile) or about 5 to 6 city blocks. If you participate in a 10 km run, you'll run about 6 miles.

A meter is divided into 100 smaller pieces called ***centi***meters. Each centimeter is $\frac{1}{100}$ of a meter. Centimeters are used instead of inches. A centimeter is a little shorter than $\frac{1}{2}$ inch. The cover of this textbook is about 21 cm wide. A nickel is about 2 cm across. Measure the width and length of your little finger on this centimeter ruler. The width of your little finger is probably about 1 cm, or a little more.

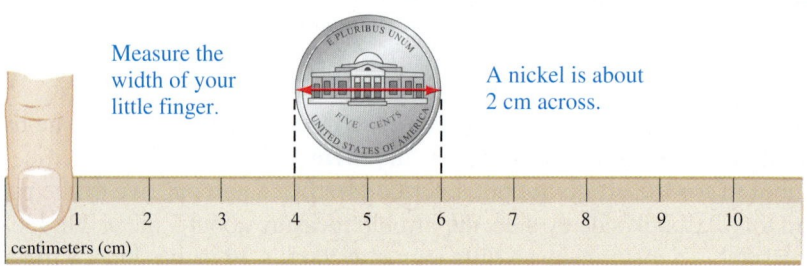

A meter is divided into 1000 smaller pieces called ***milli***meters. Each millimeter is $\frac{1}{1000}$ of a meter. It takes 10 mm to equal 1 cm, so it is a very small length. The thickness of a dime is about 1 mm. Measure the width of your pen or pencil and the width of your little finger on this millimeter ruler.

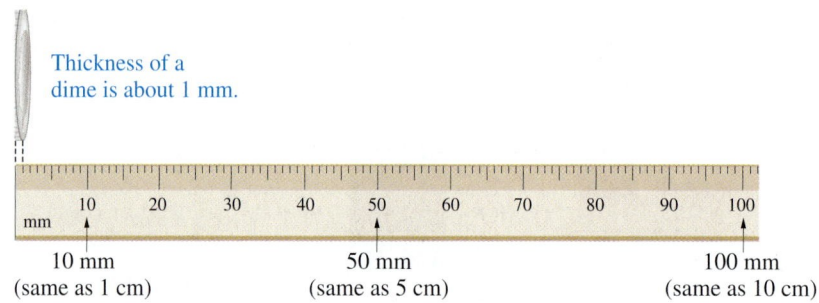

2 Write the most reasonable metric unit in each blank. Choose from km, m, cm, and mm.

(a) The woman's height is 168 _____ .

(b) The man's waist is 90 _____ around.

(c) Louise ran the 100 _____ dash in the track meet.

(d) A postage stamp is 22 _____ wide.

(e) Michael paddled his canoe 2 _____ down the river.

(f) The pencil lead is 1 _____ thick.

(g) A stick of gum is 7 _____ long.

(h) The highway speed limit is 90 _____ per hour.

(i) The classroom was 12 _____ long.

(j) A penny is about 18 _____ across.

EXAMPLE 1 **Using Metric Length Units**

Write the most reasonable metric unit in each blank. Choose from km, m, cm, and mm.

(a) The distance from home to work is 20 _____.

20 **km** because kilometers are used instead of miles. 20 km is about 12 miles.

(b) My wedding ring is 4 _____ wide.

4 **mm** because the width of a ring is very small.

(c) The newborn baby is 50 _____ long.

50 **cm**; which is half of a meter; a meter is about 39 inches so half a meter is around 20 inches.

ANSWERS

2. **(a)** cm **(b)** cm **(c)** m **(d)** mm **(e)** km **(f)** mm **(g)** cm **(h)** km **(i)** m **(j)** mm

◀ *Work Problem* **2** *at the Side.*

OBJECTIVE 2 Use unit fractions to convert among units. You can convert among metric length units using unit fractions. Keep these relationships in mind when setting up the unit fractions.

Metric Length Relationships

1 km = 1000 m so the unit fractions are:	**1 m = 1000 mm** so the unit fractions are:
$\dfrac{1 \text{ km}}{1000 \text{ m}}$ or $\dfrac{1000 \text{ m}}{1 \text{ km}}$	$\dfrac{1 \text{ m}}{1000 \text{ mm}}$ or $\dfrac{1000 \text{ mm}}{1 \text{ m}}$
1 m = 100 cm so the unit fractions are:	**1 cm = 10 mm** so the unit fractions are:
$\dfrac{1 \text{ m}}{100 \text{ cm}}$ or $\dfrac{100 \text{ cm}}{1 \text{ m}}$	$\dfrac{1 \text{ cm}}{10 \text{ mm}}$ or $\dfrac{10 \text{ mm}}{1 \text{ cm}}$

EXAMPLE 2 **Using Unit Fractions to Convert Length Measurements**

Convert each measurement using unit fractions.

(a) 5 km to m

Put the unit for the answer (meters) in the numerator of the unit fraction; put the unit you want to change (km) in the denominator.

Unit fraction equivalent to 1 $\begin{cases} 1000 \text{ m} & \leftarrow \text{Unit for answer} \\ 1 \text{ km} & \leftarrow \text{Unit being changed} \end{cases}$

Multiply. Divide out common units where possible.

$$5 \text{ km} \cdot \frac{1000 \text{ m}}{1 \text{ km}} = \frac{5 \text{ km}}{1} \cdot \frac{1000 \text{ m}}{1 \text{ km}} = \frac{5 \cdot 1000 \text{ m}}{1} = 5000 \text{ m}$$

These units should match.

Here, **km** divides out leaving **m**, the unit you want for your answer.

5 km = 5000 m

The answer makes sense because a kilometer is much longer than a meter, so 5 km will contain many meters.

(b) 18.6 cm to m

Multiply by a unit fraction that allows you to divide out centimeters.

Unit fraction

$$\frac{18.6 \text{ cm}}{1} \cdot \frac{1 \text{ m}}{100 \text{ cm}} = \frac{18.6}{100} \text{ m} = 0.186 \text{ m}$$

Do **not** write a period here.

18.6 cm = 0.186 m

There are 100 cm in a meter, so 18.6 cm will be a small part of a meter. The answer makes sense.

Work Problem **3** *at the Side.* ▶

---

3 First write the unit fraction needed to make each conversion. Then complete the conversion.

(a) 3.67 m to cm

unit fraction $\Big\}$ $\dfrac{100 \text{ cm}}{1 \text{ m}}$

(b) 92 cm to m

unit fraction $\Big\}$ $\dfrac{\text{m}}{\text{cm}}$

(c) 432.7 cm to m

unit fraction $\Big\}$ _____

(d) 65 mm to cm

unit fraction $\Big\}$ _____

(e) 0.9 m to mm

unit fraction $\Big\}$ _____

(f) 2.5 cm to mm

unit fraction $\Big\}$ _____

ANSWERS

3. **(a)** 367 cm   **(b)** $\dfrac{1 \text{ m}}{100 \text{ cm}}$; 0.92 m

(c) $\dfrac{1 \text{ m}}{100 \text{ cm}}$; 4.327 m

(d) $\dfrac{1 \text{ cm}}{10 \text{ mm}}$; 6.5 cm

(e) $\dfrac{1000 \text{ mm}}{1 \text{ m}}$; 900 mm

(f) $\dfrac{10 \text{ mm}}{1 \text{ cm}}$; 25 mm

4 Do each multiplication or division by hand or on a calculator. Compare your answer to the one you get by moving the decimal point.

(a) $(43.5)(10) = $ _____

43.5 gives 435

(b) $43.5 \div 10 = $ _____

43.5 gives _____

(c) $(28)(100) = $ _____

28.00 gives _____

(d) $28 \div 100 = $ _____

28. gives _____

(e) $(0.7)(1000) = $ _____

0.700 gives _____

(f) $0.7 \div 1000 = $ _____

000.7 gives _____

OBJECTIVE 3 **Move the decimal point to convert among units.**
By now you have probably noticed that conversions among metric units are made by multiplying or dividing by 10, by 100, or by 1000. A quick way to *multiply* by 10 is to move the decimal point one place to the *right*. Move it two places to the right to multiply by 100, three places to multiply by 1000. *Dividing* is done by moving the decimal point to the *left* in the same manner.

◀ *Work Problem* **4** *at the Side.*

An alternate conversion method to unit fractions is moving the decimal point using this **metric conversion line.**

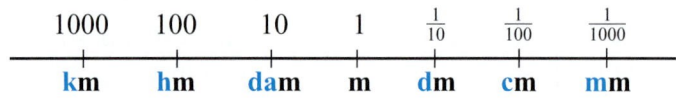

Here are the steps for using the conversion line.

Using the Metric Conversion Line

Step 1 Find the unit you are given on the metric conversion line.

Step 2 Count the number of places to get from the unit you are given to the unit you want in the answer.

Step 3 Move the decimal point the **same number of places** and in the **same direction** as you did on the conversion line.

EXAMPLE 3 **Using the Metric Conversion Line**

Use the metric conversion line to make the following conversions.

(a) 5.702 km to m

Find **km** on the metric conversion line. To get to **m**, you move *three places* to the *right*. So move the decimal point in 5.702 *three places* to the *right*.

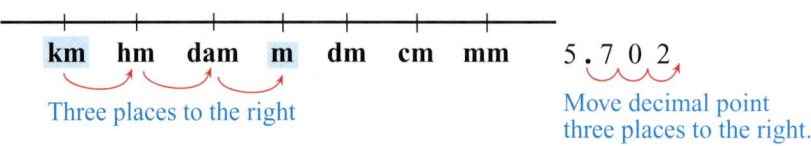

5.702 km = 5702 m

(b) 69.5 cm to m

Find **cm** on the conversion line. To get to **m**, move *two places* to the *left*.

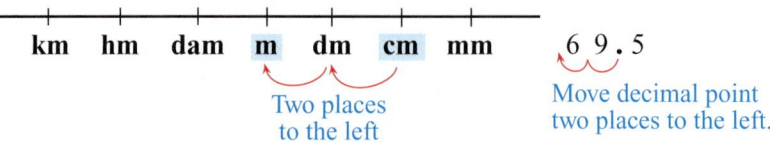

69.5 cm = 0.695 m

Continued on Next Page

(c) 8.1 cm to mm

From **cm** to **mm** is *one place* to the *right*.

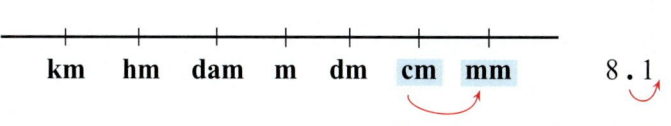

km hm dam m dm **cm mm**   8 . 1

One place   Move decimal point
to the right   one place to the right.

8.1 cm = 81 mm

Work Problem **5** *at the Side.* ▶

EXAMPLE 4 **Practicing Length Conversions**

Convert using the metric conversion line.

(a) 1.28 m to mm

Moving from **m** to **mm** is going *three places to the right.* In order to move the decimal point in 1.28 three places to the right, you must write a 0 as a placeholder.

1.28**0**   Zero is written in as a placeholder.

Move decimal point
three places to the right.

> The answer is a *whole number,* so you do not need to write the decimal point.

1.28 m = 1280 mm

(b) 60 cm to m

From **cm** to **m** is two places to the left. The decimal point in 60 starts at the *far right side* because 60 is a whole number. Then move it two places to the left.

60.   60.

↑
Decimal point starts here.   Move decimal point
                              two places to the left.

60 cm = 0.60 m, and 0.60 m is equivalent to 0.6 m.

(c) 8 m to km

From **m** to **km** is three places to the left. The decimal point in 8 starts at the far right side. In order to move it three places to the left, you must write two zeros as placeholders.

                           Two zeros are written
                           in as placeholders.

8.   008.

↑
Decimal point starts here.   Move decimal point
                              three places to the left.

8 m = 0.008 km

Work Problem **6** *at the Side.* ▶

---

5 Convert using the metric conversion line.

(a) 12.008 km to m

(b) 561.4 m to km

(c) 20.7 cm to m

(d) 20.7 cm to mm

(e) 4.66 m to cm

(f) 85.6 mm to cm

6 Convert using the metric conversion line.

(a) 9 m to mm

(b) 3 cm to m

(c) 14.6 km to m

(d) 5 mm to cm

(e) 70 m to km

(f) 0.8 m to cm

ANSWERS

5. (a) 12,008 m **(b)** 0.5614 km
   (c) 0.207 m **(d)** 207 mm
   (e) 466 cm **(f)** 8.56 cm
6. (a) 9000 mm **(b)** 0.03 m
   (c) 14,600 m **(d)** 0.5 cm
   (e) 0.07 km **(f)** 80 cm

Math in the Media

MEASURING UP

Hair and Nail Growth

Q: How fast do hair and nails grow? Do they grow faster in the summer?

A: Fingernails grow, on average, about one-tenth of a millimeter per day, although there is considerable variation among individuals. Fingernails grow faster than toenails, and nails on the longest fingers appear to grow the fastest.

Fingernails, as well as hair and skin, grow faster in the summer, presumably under the influence of sunlight, which expands blood vessels, bringing more oxygen and nutrients to the area and allowing for faster growth.

The rate the scalp hair grows is 0.3 to 0.4 millimeter per day, or about 6 inches a year.

Source: Minneapolis Star Tribune.

1. **(a)** How much do fingernails grow in one week?

   (b) In one month?

   (c) In one year?

2. **(a)** Using metric units, how much does hair grow in one week?

   (b) In one month?

   (c) In one year?

3. When you have finished **Section 8.5,** come back to this article. Is the statement about hair growing 6 inches a year accurate? Explain your answer.

4. **(a)** According to *Guinness World Records,* the longest toenails measure 15.2 cm. How many millimeters is that length? How many meters?

   (b) The longest eyelash is listed as 5.08 cm. How long is the eyelash in millimeters? In meters?

8.2 ▶▶▶ Éxercises

Use your knowledge of the meaning of metric prefixes to fill in the blanks.

1. *kilo* means _____ so

   1 km = _____ m

2. *deka* means _____ so

   1 dam = _____ m

3. *milli* means _____ so

   1 mm = _____ m

4. *deci* means _____ so

   1 dm = _____ m

5. *centi* means _____ so

   1 cm = _____ m

6. *hecto* means _____ so

   1 hm = _____ m

Use this ruler to measure the width of your thumb and hand for Exercises 7–10.

7. The width of your hand in centimeters

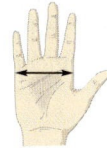

8. The width of your hand in millimeters

9. The width of your thumb in millimeters

10. The width of your thumb in centimeters

Write the most reasonable metric length unit in each blank. Choose from km, m, cm, and mm.
See Example 1.

11. The child was 91 _____ tall.

12. The cardboard was 3 _____ thick.

13. Ming-Na swam in the 200 _____ backstroke race.

14. The bookcase is 75 _____ wide.

15. Adriana drove 400 _____ on her vacation.

16. The door is 2 _____ high.

17. An aspirin tablet is 10 _____ across.

18. Lamard jogs 4 _____ every morning.

19. A paper clip is about 3 _____ long.

20. My pen is 145 _____ long.

21. Dave's truck is 5 _____ long.

22. Wheelchairs need doorways that are at least 80 _____ wide.

23. Describe at least three examples of metric length units that you have come across in your daily life.

24. Explain one reason the metric system would be easier for a child to learn than the U.S. customary system.

Convert each measurement. Use unit fractions or the metric conversion line. See Examples 2–4.

25. 7 m to cm

26. 18 m to cm

27. 40 mm to m

28. 6 mm to m

⊙ **29.** 9.4 km to m

30. 0.7 km to m

31. 509 cm to m
⊙

32. 30 cm to m

33. 400 mm to cm

34. 25 mm to cm

35. 0.91 m to mm

36. 4 m to mm

37. Is 82 cm greater than or less than 1 m? What is the difference in the lengths?

38. Is 1022 m greater than or less than 1 km? What is the difference in the lengths?

39. On the first page of this chapter, we said that computer microchips may be only 5 mm long and 1 mm wide. Using the ruler on the previous page, draw a rectangle that measures 5 mm by 1 mm. Then convert each measurement to centimeters.

A greatly enlarged photo of a computer microchip.

40. The world's smallest butter-fly has a wingspan of 15 mm. The smallest mouse is 50 mm long. Using the ruler on the previous page, draw a line that is 15 mm long and a line 50 mm long. Then convert each measure-ment to centimeters. (*Source: Top 10 of Everything.*)

41. The Roe River near Great Falls, Montana, is the shortest river in the world, with a north fork that is just under 18 m long. How many kilometers long is the north fork of the river? (*Source: Guinness Book of Amazing Nature.*)

42. There are 60,000 km of blood vessels in the human body. How many meters of blood vessels are in the body? (*Source: Big Book of Knowledge.*)

43. The median height for U.S. females who are 20 to 29 years old is about 1.64 m. Convert this height to centimeters and to millimeters. (*Source: U.S. National Center for Health Statistics.*)

44. The median height for 20- to 29-year-old males in the United States is about 177 cm. Convert this height to meters and to millimeters. (*Source: U.S. National Center for Health Statistics.*)

45. Use two unit fractions to convert 5.6 mm to km.

46. Use two unit fractions to convert 16.5 km to mm.

8.3 ▷▷▷ The Metric System—Capacity and Weight (Mass)

We use capacity units to measure liquids, such as the amount of milk in a recipe, the gasoline in our car tank, and the water in an aquarium. (The capacity units in the U.S. customary system are cups, pints, quarts, and gallons.) The basic metric unit for capacity is the **liter** (also spelled *litre*). The capital letter L is the symbol for liter, to avoid confusion with the numeral 1.

OBJECTIVE 1 **Learn the basic metric units of capacity.** The liter is related to metric length in this way: a box that measures 10 cm on every side holds exactly one liter. (The volume of the box is 10 cm • 10 cm • 10 cm = 1000 cubic centimeters. Volume was discussed in **Section 4.8.**) A liter is just a little more than 1 quart.

Holds exactly 1 liter (L)

About 1 liter of milk

About 1 liter of oil for your car

A liter is a little more than one quart (just $\frac{1}{4}$ cup more).

In the metric system you use liters for things like buying shampoo and soda at the store, filling a pail with water, and describing the size of your home aquarium.

Buy a 2 L bottle of soda

Use a 12 L pail to wash floors

Watch the fish in your 40 L aquarium

Work Problem **1** *at the Side.* ▶

To make larger or smaller capacity units, we use the same **prefixes** as we did with length units. For example, *kilo* means 1000 so a *kilo*meter is 1000 meters. In the same way, a *kilo*liter is 1000 liters.

Prefix	*kilo-* liter	*hecto-* liter	*deka-* liter	liter	*deci-* liter	*centi-* liter	*milli-* liter
Meaning	1000 liters	100 liters	10 liters	1 liter	$\frac{1}{10}$ of a liter	$\frac{1}{100}$ of a liter	$\frac{1}{1000}$ of a liter
Symbol	**k**L	**h**L	**da**L	L	**d**L	**c**L	**m**L

↑ ↑
Capacity units used most often

1 Which things can be measured in liters?

Amount of water in the bathtub

Length of the bathtub

Width of your car

Amount of gasoline you buy for your car

Weight of your car

Height of a pail

Amount of water in a pail

ANSWER

1. water in bathtub, gasoline, water in a pail

The capacity units you will use most often in daily life are liters (L) and *milli*liters (mL). A tiny box that measures 1 cm on every side holds exactly one milliliter. (In medicine, this small amount is also called 1 cubic centimeter, or 1 cc for short.) It takes 1000 mL to make 1 L. Here are some useful comparisons.

Holds exactly 1 milliliter (mL)   Teaspoon holds 5 mL   One cup holds about 250 mL

② Write the most reasonable metric unit in each blank. Choose from L and mL.

(a) I bought 8 _____ of soda at the store.

(b) The nurse gave me 10 _____ of cough syrup.

(c) This is a 100 _____ garbage can.

(d) It took 10 _____ of paint to cover the bedroom walls.

(e) My car's gas tank holds 50 _____.

(f) I added 15 _____ of oil to the pancake mix.

(g) The can of orange soda holds 350 _____.

(h) My friend gave me a 30 _____ bottle of expensive perfume.

EXAMPLE 1 **Using Metric Capacity Units**

Write the most reasonable metric unit in each blank. Choose from L and mL.

(a) The bottle of shampoo held 500 _____.
500 **mL** because 500 L would be about 500 quarts, which is too much.

(b) I bought a 2 _____ carton of orange juice.
2 **L** because 2 mL would be less than a teaspoon.

◀ *Work Problem* ② *at the Side.*

OBJECTIVE 2 Convert among metric capacity units. Just as with length units, you can convert between milliliters and liters using unit fractions.

Metric Capacity Relationships

1 L = 1000 mL, so the unit fractions are:

$$\frac{1\text{ L}}{1000\text{ mL}} \quad \text{or} \quad \frac{1000\text{ mL}}{1\text{ L}}$$

Or you can use a metric conversion line to decide how to move the decimal point.

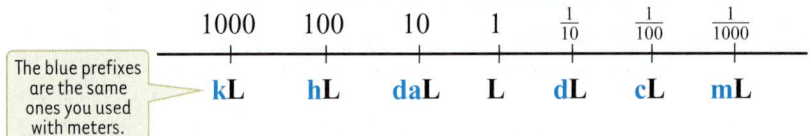

The blue prefixes are the same ones you used with meters.

1000	100	10	1	$\frac{1}{10}$	$\frac{1}{100}$	$\frac{1}{1000}$
kL	hL	daL	L	dL	cL	mL

EXAMPLE 2 **Converting among Metric Capacity Units**

Convert using the metric conversion line or unit fractions.

(a) 2.5 L to mL
Using the metric conversion line:
From **L** to **mL** is *three places* to the *right*.

2.500   Write two zeros as placeholders.

2.5 L = 2500 mL

Using unit fractions:

Multiply by a unit fraction that allows you to divide out liters.

$$\frac{2.5\text{ L}}{1} \cdot \frac{1000\text{ mL}}{1\text{ L}} = 2500\text{ mL}$$

L divides out, leaving mL for your answer.

Continued on Next Page

(b) 80 mL to L

Using the metric conversion line:

From **mL** to **L** is *three places* to the *left*.

80.   080.

↑ Decimal point starts here.

Move decimal point three places to the left.

80 mL = 0.080 L or 0.08 L

Using unit fractions:

Multiply by a unit fraction that allows you to divide out mL.

$$\frac{80 \ \text{mL}}{1} \cdot \frac{1 \ \text{L}}{1000 \ \text{mL}}$$

$$= \frac{80}{1000} \ \text{L} = 0.08 \ \text{L}$$

Do **not** write a period here.

Work Problem **3** *at the Side.* ▶

OBJECTIVE 3 Learn the basic metric units of weight (mass).
The **gram** is the basic metric unit for *mass*. Although we often call it "weight," there is a difference. Weight is a measure of the pull of gravity; the farther you are from the center of Earth, the less you weigh. In outer space you become weightless, but your mass, the amount of matter in your body, stays the same regardless of where you are. In science courses, it will be important to distinguish between the weight of an object and its mass. But for everyday purposes, we will use the word *weight*.

The gram is related to metric length in this way: The weight of the water in a box measuring 1 cm on every side is 1 gram. This is a very tiny amount of water (1 mL) and a very small weight. One gram is also the weight of a dollar bill or a single raisin. A nickel weighs 5 grams. A plain, regular-sized hamburger and bun weighs from 175 to 200 grams.

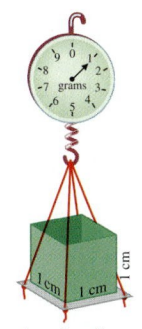

The 1 mL of water in this tiny box weighs 1 gram.

A nickel weighs 5 grams.

A dollar bill weighs 1 gram.

A plain hamburger weighs 175 to 200 grams.

Work Problem **4** *at the Side.* ▶

---

3 Convert.

(a) 9 L to mL

(b) 0.75 L to mL

(c) 500 mL to L

(d) 5 mL to L

(e) 2.07 L to mL

(f) 3275 mL to L

4 Which things would weigh about 1 gram?

A small paper clip

A pair of scissors

One playing card from a deck of cards

A calculator

An average-sized apple

The check you wrote to the cable company

ANSWERS

3. **(a)** 9000 mL   **(b)** 750 mL   **(c)** 0.5 L
(d) 0.005 L   **(e)** 2070 mL   **(f)** 3.275 L
4. paper clip, playing card, check

5 Write the most reasonable metric unit in each blank. Choose from kg, g, and mg.

(a) A thumbtack weighs

800 _____.

(b) A teenager weighs

50 _____.

(c) This large cast-iron

frying pan weighs

1 _____.

(d) Jerry's basketball

weighed 600 _____.

(e) Tamlyn takes a

500 _____ calcium

tablet every morning.

(f) On his diet, Greg can eat

90 _____ of meat for

lunch.

(g) One strand of hair weighs

2 _____.

(h) One banana might weigh

150 _____.

To make larger or smaller weight units, we use the same **prefixes** as we did with length and capacity units. For example, *kilo* means 1000 so a *kilo*meter is 1000 meters, a *kilo*liter is 1000 liters, and a *kilo*gram is 1000 grams.

Prefix	kilo-gram	hecto-gram	deka-gram	gram	deci-gram	centi-gram	milli-gram
Meaning	1000 grams	100 grams	10 grams	1 gram	$\frac{1}{10}$ of a gram	$\frac{1}{100}$ of a gram	$\frac{1}{1000}$ of a gram
Symbol	**kg**	**hg**	**dag**	g	**dg**	**cg**	**mg**

Weight (mass) units that are used most often

The units you will use most often in daily life are kilograms (kg), grams (g), and milligrams (mg). *Kilo*grams are used instead of pounds. A kilogram is 1000 grams. It is about 2.2 pounds. Two packages of butter plus one stick of butter weigh about 1 kg. An average newborn baby weighs 3 to 4 kg; a college football player might weigh 100 to 130 kg.

1 kilogram is
about 2.2 pounds   100 to 130 kg   3 to 4 kg

Extremely small weights are measured in *milli*grams. It takes 1000 mg to make 1 g. Recall that a dollar bill weighs about 1 g. Imagine cutting it into 1000 pieces; the weight of one tiny piece would be 1 mg. Dosages of medicine and vitamins are given in milligrams. You will also use milligrams in science classes.

Cut a dollar bill into 1000 pieces.
One tiny piece weighs 1 milligram.

EXAMPLE 3 **Using Metric Weight Units**

Write the most reasonable metric unit in each blank.
Choose from kg, g, and mg.

(a) Ramon's suitcase weighed 20 _____.
20 **kg** because kilograms are used instead of pounds.
20 kg is about 44 pounds.

(b) LeTia took a 350 _____ aspirin tablet.
350 **mg** because 350 g would be more than the weight of a hamburger, which is too much.

(c) Jenny mailed a letter that weighed 30 _____.
30 **g** because 30 kg would be much too heavy and 30 mg is less than the weight of a dollar bill.

◀ *Work Problem* **5** *at the Side.*

OBJECTIVE **4** **Convert among metric weight (mass) units.** As with length and capacity, you can convert among metric weight units by using unit fractions. The unit fractions you need are shown here.

Metric Weight (Mass) Relationships

1 kg = 1000 g so the unit fractions are:	**1 g = 1000 mg** so the unit fractions are:
$\dfrac{1 \text{ kg}}{1000 \text{ g}}$ or $\dfrac{1000 \text{ g}}{1 \text{ kg}}$	$\dfrac{1 \text{ g}}{1000 \text{ mg}}$ or $\dfrac{1000 \text{ mg}}{1 \text{ g}}$

Or you can use a metric conversion line to decide how to move the decimal point.

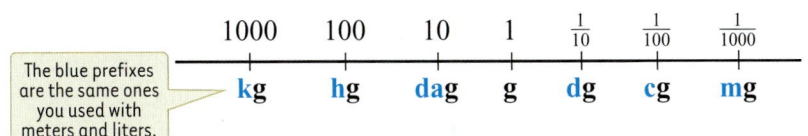

The blue prefixes are the same ones you used with meters and liters.

1000	100	10	1	$\frac{1}{10}$	$\frac{1}{100}$	$\frac{1}{1000}$
kg	**hg**	**dag**	**g**	**dg**	**cg**	**mg**

EXAMPLE 4 **Converting among Metric Weight Units**

Convert using the metric conversion line or unit fractions.

(a) 7 mg to g

Using the metric conversion line:

From **mg** to **g** is *three places* to the *left*.

7.   007.

↑ Decimal point starts here.

Move decimal point three places to the left.

7 mg = 0.007 g

Using unit fractions:

Multiply by a unit fraction that allows you to divide out mg.

$$\frac{7 \text{ mg}}{1} \cdot \frac{1 \text{ g}}{1000 \text{ mg}} = \frac{7}{1000} \text{ g}$$

$$= 0.007 \text{ g}$$

Three decimal places for thousandths.

(b) 13.72 kg to g

Using the metric conversion line:

From **kg** to **g** is *three places* to the *right*.

13.720   Decimal point moves three places to the right.

13.72 kg = 13,720 g

↑ A comma (not a decimal point)

Using unit fractions:

Multiply by a unit fraction that allows you to divide out kg.

$$\frac{13.72 \text{ kg}}{1} \cdot \frac{1000 \text{ g}}{1 \text{ kg}} = 13,720 \text{ g}$$

↑ A comma (not a decimal point)

Work Problem **6** *at the Side.* ▶

---

6 Convert.

(a) 10 kg to g

(b) 45 mg to g

(c) 6.3 kg to g

(d) 0.077 g to mg

(e) 5630 g to kg

(f) 90 g to kg

ANSWERS

6. (a) 10,000 g   **(b)** 0.045 g   **(c)** 6300 g
(d) 77 mg   **(e)** 5.63 kg   **(f)** 0.09 kg

7 First decide which type of units are needed: length, capacity, or weight. Then write the most appropriate unit in the blank. Choose from km, m, cm, mm, L, mL, kg, g, and mg.

(a) Gail bought a 4 _____ can of paint.

Use _____ units.

(b) The bag of chips weighed 450 _____.

Use _____ units.

(c) Give the child 5 _____ of cough syrup.

Use _____ units.

(d) The width of the window is 55 _____.

Use _____ units.

(e) Akbar drives 18 _____ to work.

Use _____ units.

(f) The laptop computer weighs 2 _____.

Use _____ units.

(g) A credit card is 55 _____ wide.

Use _____ units.

OBJECTIVE 5 Distinguish among basic metric units of length, capacity, and weight (mass). As you encounter things to be measured at home, on the job, or in your classes at school, be careful to use the correct type of measurement unit.

Use *length units* (kilometers, meters, centimeters, millimeters) to measure:

how long	how high	how far away
how wide	how tall	how far around (perimeter)
how deep	distance	

Use *capacity units* (liters, milliliters) to measure liquids (things that can be poured) such as:

water	shampoo	gasoline
milk	perfume	oil
soft drinks	cough syrup	paint

Also use liters and milliliters to describe how much liquid something can hold, such as an eyedropper, measuring cup, pail, or bathtub.

Use *weight units* (kilograms, grams, milligrams) to measure:

the weight of something   how heavy something is

In **Chapters 3–5** you used square units (such as cm^2 and m^2) to measure area, and cubic units (such as cm^3 and m^3) to measure volume.

EXAMPLE 5 Using a Variety of Metric Units

First decide which type of units are needed: length, capacity, or weight. Then write the most appropriate metric unit in the blank. Choose from km, m, cm, mm, L, mL, kg, g, and mg.

(a) The letter needs another stamp because it weighs 40 _____.

Use _____ units.

The letter weighs 40 **g** because 40 mg is less than the weight of a dollar bill and 40 kg would be about 88 pounds.

Use **weight** units because of the word "weighs."

(b) The swimming pool is 3 _____ deep at the deep end.

Use _____ units.

The pool is 3 **m** deep because 3 cm is only about an inch and 3 km is more than a mile.

Use **length** units because of the word "deep."

(c) This is a 340 _____ can of juice.

Use _____ units.

It is a 340 **mL** can because 340 liters would be more than 340 quarts.

Use **capacity** units because juice is a liquid.

◄ *Work Problem* **7** *at the Side.*

8.3 ▶▶▶ Exercises

Write the most reasonable metric unit in each blank. Choose from L, mL, kg, g, and mg.
See Examples 1 and 3.

1. The glass held

250 _____ of water.

2. Hiromi used 12 _____ of water to wash the kitchen floor.

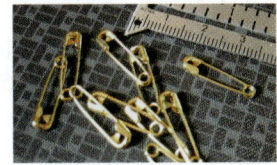

3. Dolores can make 10 _____ of soup in that pot.

4. Jay gave 2 _____ of vitamin drops to the baby.

5. Our yellow Labrador dog grew up to weigh 40 _____.

6. A small safety pin weighs 750 _____.

7. Lori caught a small sunfish weighing 150 _____.

8. One dime weighs 2 _____.

9. Andre donated 500 _____ of blood today.

10. Barbara bought the 2 _____ bottle of cola.

11. The patient received a 250 _____ tablet of medication each hour.

12. The 8 people on the elevator weighed a total of 500 _____.

13. The gas can for the lawn mower holds 4 _____.

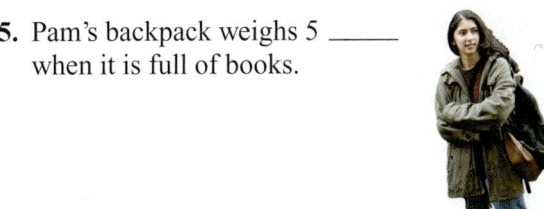

14. Kevin poured 10 _____ of vanilla into the mixing bowl.

15. Pam's backpack weighs 5 _____ when it is full of books.

16. One grain of salt weighs 2 _____.

Today, medical measurements are usually given in the metric system. Since we convert among metric units of measure by moving the decimal point, it is possible that mistakes can be made. Examine the following dosages and indicate whether they are reasonable or unreasonable. If a dose is unreasonable, indicate whether it is too much or too little.

17. Drink 4.1 L of Kaopectate after each meal.

18. Drop 1 mL of solution into the eye twice a day.

19. Soak your feet in 5 kg of Epsom salts per liter of water.

20. Inject 0.5 L of insulin each morning.

21. Take 15 mL of cough syrup every four hours.

22. Take 200 mg of vitamin C each day.

23. Take 350 mg of aspirin three times a day.

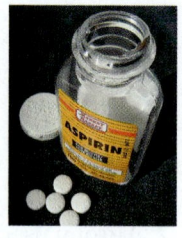

24. Buy a tube of ointment weighing 0.002 g.

25. Describe at least two examples of metric capacity units and two examples of metric weight units that you have come across in your daily life.

26. Explain in your own words how the meter, liter, and gram are related.

27. Describe how you decide which unit fraction to use when converting 6.5 kg to grams.

28. Write an explanation of each step you would use to convert 20 mg to grams using the metric conversion line.

Convert each measurement. Use unit fractions or the metric conversion line. See Examples 2 and 4.

29. 15 L to mL

30. 6 L to mL

31. 3000 mL to L

32. 18,000 mL to L

33. 925 mL to L

34. 200 mL to L

35. 8 mL to L

36. 25 mL to L

37. 4.15 L to mL

38. 11.7 L to mL

39. 8000 g to kg

40. 25,000 g to kg

41. 5.2 kg to g

42. 12.42 kg to g

43. 0.85 g to mg

44. 0.2 g to mg

45. 30,000 mg to g

46. 7500 mg to g

47. 598 mg to g

48. 900 mg to g

49. 60 mL to L

50. 6.007 kg to g

51. 3 g to kg

52. 12 mg to g

53. 0.99 L to mL

54. 13,700 mL to L

Write the most appropriate metric unit in each blank. Choose from km, m, cm, mm, L, mL, kg, g, and mg. See Example 5.

55. The masking tape is 19 _____ wide.

56. The roll has 55 _____ of tape on it.

57. Buy a 60 _____ jar of acrylic paint for art class.

58. One onion weighs 200 _____.

59. My waist measurement is 65 _____.

60. Add 2 _____ of windshield washer fluid to your car.

61. A single postage stamp weighs 90 _____.

62. The hallway is 10 _____ long.

Solve each application problem. Show your work. (Source for Exercises 63–68: Top 10 of Everything.)

63. Human skin has about 3 million sweat glands, which release an average of 300 mL of sweat per day. How many liters of sweat are released each day?

64. In hot climates, the sweat glands in a person's skin may release up to 3.5 L of sweat in one day. How many milliliters is that?

65. The average weight of a human brain is 1.34 kg. How many grams is that?

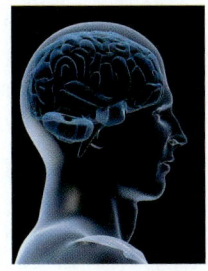

66. A healthy human heart pumps about 70 mL of blood per beat. How many liters of blood does it pump per beat?

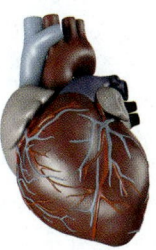

67. On average, we breathe in and out roughly 900 mL of air every 10 seconds. How many liters of air is that?

68. In the Victorian era, people believed that heavier brains meant greater intelligence. They were impressed that Otto von Bismarck's brain weighed 1907 g, which is how many kilograms?

69. A small adult cat weighs from 3000 g to 4000 g. How many kilograms is that? (*Source: Lyndale Animal Hospital.*)

70. If the letter you are mailing weighs 29 g, you must put additional postage on it. How many kilograms does the letter weigh? (*Source:* U.S. Postal Service.)

71. Is 1005 mg greater than or less than 1 g? What is the difference in the weights?

72. Is 990 mL greater than or less than 1 L? What is the difference in the amounts?

73. One nickel weighs 5 g. How many nickels are in 1 kg of nickels?

74. The ratio of the total length of all the fish to the amount of water in an aquarium can be 3 cm of fish for every 4 L of water. What is the total length of all fish you can put in a 40 L aquarium? (*Source: Tropical Aquarium Fish.*)

Relating Concepts (Exercises 75–78) For Individual or Group Work

*Recall that the prefix **kilo** means 1000, so a **kilo**meter is 1000 meters. You'll learn about other prefixes for numbers greater than 1000 as you **work Exercises 75–78 in order.***

75. (a) The prefix *mega* means one million. Use the symbol M (capitalized) for *mega*. So a megameter (Mm) is how many meters?

1 Mm = _____ m

(b) Figure out a unit fraction that you can use to convert megameters to meters. Then use it to convert 3.5 Mm to meters.

76. (a) The prefix *giga* means one billion. Use the symbol G (capitalized) for *giga*. So a gigameter (Gm) is how many meters?

1 Gm = _____ m

(b) Figure out a unit fraction you can use to convert meters to gigameters. Then use it to convert 2500 m to gigameters.

77. (a) The prefix *tera* means one trillion. Use the symbol T (capitalized) for *tera*. So a *tera*meter (Tm) is how many meters?

1 Tm = _____ m.

(b) Think carefully before you fill in the blanks:

1 Tm = _____ Gm

1 Tm = _____ Mm

78. A computer's memory is measured in *bytes*. A byte can represent a single letter, a digit, or a punctuation mark. The memory for a desktop computer may be measured in megabytes (abbreviated MB) or gigabytes (abbreviated GB). Using the meanings of *mega* and *giga*, it would seem that

1 MB = _____ bytes and

1 GB = _____ bytes.

However, because computers use a base 2 or binary system, 1 MB is actually 2^{20} and 1 GB is 2^{30}. Use your calculator to find the actual values.

2^{20} = _____ 2^{30} = _____

Summary Exercises on U.S. Customary and Metric Units

The most commonly used U.S. customary and metric system units are listed in mixed-up order.
Write each unit in the correct box in the table below. Within each box, write the units in order from
smallest to largest.

pound	yard	liter	kilogram	ton	gallon	pint
millimeter	gram	quart	meter	inch	cup	centimeter
milliliter	mile	foot	milligram	kilometer	ounce	fluid ounce

1. U.S. Customary Units			2. Metric System Units		
Length	Weight	Capacity	Length	Weight	Capacity

Fill in the blanks with the measurement relationships that you memorized.

3. (a) 1 _____ = 12 in.

   (b) 3 ft = 1 _____

   (c) 1 mi = _____ ft

4. (a) 60 sec = 1 _____

   (b) 1 hr = _____ min

   (c) _____ hr = 1 day

5. (a) 1 cup = _____ fl oz

   (b) 4 qt = 1 _____

   (c) _____ pt = 1 qt

6. (a) 16 oz = 1 _____

   (b) _____ lb = 1 ton

   (c) 1 lb = _____ oz

Write the most reasonable metric unit in each blank. Choose from km, m, cm, mm, L, mL, kg, g, and mg.

7. My water bottle holds 450 _____.

8. Michael won the 200 _____ race today.

9. The child weighed 23 _____.

10. Jifar took a 375 _____ aspirin tablet.

11. The red pen is 14 _____ long.

12. A quarter is about 25 _____ across.

13. Merlene made 12 _____ of fruit punch for her daughter's birthday party.

14. This cereal has 4 _____ of protein in each serving.

Convert each measurement using unit fractions or the metric conversion line. Show your work.

15. 45 cm to meters

16. $\frac{3}{4}$ min to seconds

17. 0.6 L to milliliters

18. 8 g to milligrams

19. 300 mm to centimeters

20. 45 in. to feet

21. 50 mL to liters

22. 18 qt to gallons

23. 7.28 kg to grams

24. $2\frac{1}{4}$ lb to ounces

25. 9 g to kilograms

26. 5 yd to inches

Solve Exercises 27–30 using the data in the table at the right on some of the world's tallest people. (Source: Top 10 of Everything.)

27. What is the height of the tallest person in centimeters? In millimeters?

Robert Wadlow was 2.72 m tall.

28. Find the 4th tallest person's height in meters and in kilometers.

WORLD'S TALLEST PEOPLE

Rank	Name/Dates/Country	Height
1st	Robert Wadlow (1918–1940) USA	2.72 m
2nd	John Rogan (1868–1905) USA	268 cm
4th	John Carroll (1932–1969) USA	264 cm
7th	Edouard Beaupré (1881–1904) Canada	2.5 m
10th	Jeng Jinlian (1964–1982) China	248 cm

29. How much taller is the tallest person than the second tallest, in meters? Convert this difference to centimeters and millimeters.

30. What is the difference in height between the 7th tallest and 4th tallest people, in centimeters? Convert the height difference to meters and millimeters.

Solve each application problem. Show your work.

31. Vernice bought a 12 oz bag of chips for $3.49 today. What was the price per pound to the nearest cent?

32. In 2006, Miami, Florida, received about 64 in. of rain. In 2005, about 59 in. of rain fell. How many feet of rain did Miami get in all during the two years, to the nearest tenth?

8.4 ▶▶▶ Problem Solving with Metric Measurement

OBJECTIVE 1 Solve application problems involving metric measurements. One advantage of the metric system is the ease of comparing measurements in application situations. Just be sure that you are comparing similar units: mg to mg, km to km, and so on.

We will use the same problem-solving steps that you used for U.S. customary measurement applications in **Section 8.1.**

EXAMPLE 1 Solving a Metric Application

Cheddar cheese is on sale at $8.99 per kilogram. Jake bought 350 g of the cheese. How much did he pay, to the nearest cent?

Step 1 Read the problem. The problem asks for the cost of 350 g of cheese.

Step 2 Work out a plan. The price is $8.99 per *kilogram,* but the amount Jake bought is given in *grams.* Convert grams to kilograms (the unit in the price). Then multiply the weight by the cost per kilogram.

Step 3 Estimate a reasonable answer. Round the cost of 1 kg from $8.99 to $9. There are 1000 g in a kilogram, so 350 g is about $\frac{1}{3}$ of a kilogram. Jake is buying about $\frac{1}{3}$ of a kilogram, so $\frac{1}{3}$ of $9 = $3 as our estimate.

Step 4 Solve the problem. Use a unit fraction to convert 350 g to kilograms.

> **g** divides out, leaving **kg** for your answer.

$$\frac{350 \text{ g}}{1} \cdot \frac{1 \text{ kg}}{1000 \text{ g}} = \frac{350}{1000} \text{ kg} = 0.35 \text{ kg}$$

Now multiply 0.35 kg times the cost per kilogram.

> Nearest cent is the nearest hundredth.

$$\frac{\$8.99}{1 \text{ kg}} \cdot \frac{0.35 \text{ kg}}{1} = \$3.1465 \approx \$3.15 \quad \text{(rounded)}$$

Step 5 State the answer. Jake paid $3.15, rounded to the nearest cent.

Step 6 Check your work. The exact answer of $3.15 is close to our estimate of $3.

Work Problem **1** *at the Side.* ▶

1 Solve this problem using the six problem-solving steps.

Satin ribbon is on sale at $0.89 per meter. How much will 75 cm cost, to the nearest cent?

EXAMPLE 2 Solving a Metric Application

Olivia has 2.6 m of lace. How many centimeters of lace can she use to trim each of six hair ornaments? Round to the nearest tenth of a centimeter.

Step 1 Read the problem. The problem asks for the number of centimeters of lace for each of six hair ornaments.

Step 2 Work out a plan. The given amount of lace is in *meters,* but the answer must be in *centimeters.* Convert meters to centimeters, then divide by 6 (the number of hair ornaments).

Continued on Next Page

2 Lucinda's doctor wants her to take 1.2 g of medication each day in three equal doses. How many milligrams should be in each dose? Use the six problem-solving steps.

Step 3 **Estimate** a reasonable answer. To estimate, round 2.6 m of lace to 3 m. Then, 3 m = 300 cm, and 300 cm ÷ 6 = 50 cm as our estimate.

Step 4 **Solve** the problem. On the metric conversion line, moving from **m** to **cm** is two places to the right, so move the decimal point in 2.6 m two places to the right. Then divide by 6.

$$2.60\,\text{m} = 260\,\text{cm} \qquad \frac{260\ \text{cm}}{6\ \text{ornaments}} \approx 43.3\,\text{cm per ornament}$$

Step 5 **State the answer.** Olivia can use about 43.3 cm of lace on each ornament.

Step 6 **Check** your work. The exact answer of 43.3 cm is close to our estimate of 50 cm.

◀ *Work Problem* **2** *at the Side.*

> **Note**
>
> In Example 1 we used a unit fraction to convert the measurement, and in Example 2 we moved the decimal point. Use whichever method you prefer. Also, there is more than one way to solve an application problem. Another way to solve Example 2 is to divide 2.6 m by 6 to get 0.4333 m of lace for each ornament. Then convert 0.4333 m to 43.3 cm (rounded to the nearest tenth).

3 Andrea has two pieces of fabric. One measures 2 m 35 cm and the other measures 1 m 85 cm. How many meters of fabric does she have in all? Use the six problem-solving steps.

EXAMPLE 3 **Solving a Metric Application**

Rubin measured a board and found that the length was 3 m plus an additional 5 cm. He cut off a piece measuring 1 m 40 cm for a shelf. Find the length in meters of the remaining piece of board.

Step 1 **Read** the problem. Part of a board is cut off. The problem asks what length of board, in meters, is left over. It may help to make a drawing of the board and label the lengths given in the problem.

Step 2 **Work out a plan.** The lengths involve two units, m and cm. Rewrite both lengths in meters (the unit called for in the answer), and then subtract.

Step 3 **Estimate** a reasonable answer. To estimate, 3 m 5 cm can be rounded to 3 m, because 5 cm is less than half of a meter (less than 50 cm). Round 1 m 40 cm down to 1 m. Then, 3 m − 1 m = 2 m as our estimate.

Step 4 **Solve** the problem. Rewrite the lengths in meters. Then subtract.

$$
\begin{array}{ll}
3\ \text{m} \longrightarrow & 3.00\ \text{m} \\
\text{plus 5 cm} \longrightarrow & + 0.05\ \text{m} \\
\hline
& 3.05\ \text{m}
\end{array}
\qquad
\begin{array}{ll}
1\ \text{m} \longrightarrow & 1.0\ \text{m} \\
\text{plus 40 cm} \longrightarrow & + 0.4\ \text{m} \\
\hline
& 1.4\ \text{m}
\end{array}
$$

Subtract to find leftover length.

$$
\begin{array}{ll}
3.05\ \text{m} & \longleftarrow \text{Board} \\
- 1.40\ \text{m} & \longleftarrow \text{Shelf} \\
\hline
1.65\ \text{m} & \longleftarrow \text{Leftover piece}
\end{array}
$$

Step 5 **State the answer.** The length of the remaining piece is 1.65 m.

Step 6 **Check** your work. The exact answer of 1.65 m is close to our estimate of 2 m.

◀ *Work Problem* **3** *at the Side.*

Solve each application problem. Show your work. Round money answers to the nearest cent. See Examples 1–3.

 **1.** Bulk rice at the food co-op is on special at $0.98 per kilogram. Pam scooped some rice into a bag and put it on the scale. How much will she pay for 850 g of rice?

2. Lanh is buying a piece of plastic tubing measuring 315 cm for the science lab. The price is $4.75 per meter. How much will Lanh pay?

3. A miniature Yorkshire terrier, one of the smallest dogs, may weigh only 500 g. But a St. Bernard, the heaviest dog, could easily weigh 90 kg. What is the difference in the weights of the two dogs, in kilograms? (*Source: Big Book of Knowledge.*)

4. The world's longest insect is the giant stick insect of Indonesia, measuring 33 cm. The fairy fly, the smallest insect, is just 0.2 mm long. How much longer is the giant stick insect, in millimeters? (*Source: Big Book of Knowledge.*)

5. An adult human body contains about 5 L of blood. If each beat of the heart pumps 70 mL of blood, how many times must the heart beat to pass all the blood through the heart? Round to the nearest whole number of beats. (*Source: Harper's Index.*)

6. A floor tile measures 30 cm by 30 cm and weighs 185 g. How many kilograms would a stack of 24 tiles weigh? How much would five stacks of tiles weigh? (*Source: The Tile Shop.*)

 **7.** Each piece of lead for a mechanical pencil has a thickness of 0.5 mm and is 60 mm long. Find the total length in centimeters of the lead in a package with 30 pieces. If the price of the package is $3.29, find the cost per centimeter for the lead. (*Source: Pentel.*)

60 mm

8. The apartment building caretaker puts 750 mL of chlorine into the swimming pool every day. How many liters should he order to have a one-month (30-day) supply on hand? If chlorine is sold in containers that hold 4 L, how many containers should be ordered for one month? How much chlorine will be left over at the end of the month?

9. Rosa is building a bookcase. She has one board that is 2 m 8 cm long and another that is 2 m 95 cm long. What is the total length of the two boards in meters?

10. Janet has a piece of fabric that is 10 m 30 cm in length. She wants to make curtains for three windows that are all the same size. What length of fabric is available for each window, to the nearest tenth of a meter?

11. In a chemistry lab, each of the 45 students needs 85 mL of acid. How many 1 L bottles of acid need to be ordered? How much acid will be left over?

12. James needs two 1.3 m pieces and two 85 cm pieces of wood molding to frame a picture. The price is $5.89 per meter plus 7% sales tax. How much will James pay?

Use the bar graph below to answer Exercises 13 and 14.

Caffeine Meter
Average milligrams of caffeine per 8 oz cup or equivalent

Double espresso 160 mg

Drip coffee 90 mg

Cola 45 mg

25 mg Chocolate bar

5 mg Decaffeinated coffee

Source: Celestial Seasonings.

13. If Agnete usually drinks three 8 oz cups of drip coffee each day, how many grams of caffeine will she consume in one week?

14. Lorenzo's doctor has suggested that he cut down on caffeine. So Lorenzo switched from drinking four 8 oz cups of cola every day to drinking two 8 oz cups of decaffeinated coffee. How many fewer grams of caffeine is he consuming each week?

15. During August 2003, Mars moved closer to Earth at a rate of about 10,000 meters per second. How much closer, in kilometers, did Mars get to Earth:

   (a) in one second,

   (b) in one minute,

   (c) in one hour?

   (*Source:* NASA.)

16. Some of the newest football stadiums have Field Turf instead of grass. Use the drawing below to find the total thickness in centimeters of the top two layers of Field Turf.

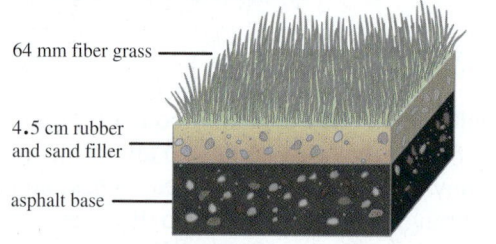

64 mm fiber grass

4.5 cm rubber and sand filler

asphalt base

Source: Sports Facilities Commission.

Relating Concepts (Exercises 17–20) For Individual or Group Work

It is difficult to weigh very light objects, such as a single sheet of paper or a single staple (unless you have an expensive scientific scale). But you can weigh a large number of the items and then divide to find the weight of one item. Before dividing, subtract the weight of the box or wrapper that the items are packaged in to find the net weight. **Work Exercises 17–20 in order** *and complete the table.*

	Item	Total Weight in Grams	Weight of Packaging	Net Weight	Weight of One Item in Grams	Weight of One Item in Milligrams
17.	Box of 50 envelopes	255 g	40 g	_____	_____	_____
18.	Box of 1000 staples	350 g	20 g	_____	_____	_____
19.	Ream of paper (500 sheets)	_____	50 g	_____		3000 mg
20.	Box of 100 small paper clips	_____	5 g	_____		500 mg

8.5 ▶▶▶ Metric–U.S. Customary Conversions and Temperature

OBJECTIVE **1** **Use unit fractions to convert between metric and U.S. customary units.** Until the United States has switched completely from customary units to the metric system, it will be necessary to make conversions from one system to the other. *Approximate* conversions can be made with the help of the table below, in which the values have been rounded to the nearest hundredth or thousandth. (The only value that is exact, not rounded, is 1 inch = 2.54 cm.)

Metric to U.S. Customary		U.S. Customary to Metric	
1 kilometer	$\approx$ 0.62 mile	1 mile	$\approx$ 1.61 kilometers
1 meter	$\approx$ 1.09 yards	1 yard	$\approx$ 0.91 meter
1 meter	$\approx$ 3.28 feet	1 foot	$\approx$ 0.30 meter
1 centimeter	$\approx$ 0.39 inch	1 inch	= 2.54 centimeters
1 liter	$\approx$ 0.26 gallon	1 gallon	$\approx$ 3.79 liters
1 liter	$\approx$ 1.06 quarts	1 quart	$\approx$ 0.95 liter
1 kilogram	$\approx$ 2.20 pounds	1 pound	$\approx$ 0.45 kilogram
1 gram	$\approx$ 0.035 ounce	1 ounce	$\approx$ 28.35 grams

EXAMPLE 1 **Converting between Metric and U.S. Customary Length Units**

Convert 10 m to yards using unit fractions. Round your answer to the nearest tenth if necessary.

We're changing from a *metric* unit to a *U.S. customary* unit. In the "Metric to U.S. Customary" side of the table above, you see that 1 meter $\approx$ 1.09 yards. Two unit fractions can be written using that information.

$$\frac{1 \text{ m}}{1.09 \text{ yd}} \quad \text{or} \quad \frac{1.09 \text{ yd}}{1 \text{ m}}$$

Multiply by the unit fraction that allows you to divide out meters (that is, meters is in the denominator).

$$10 \text{ m} \cdot \frac{1.09 \text{ yd}}{1 \text{ m}} = \frac{10 \text{ m}}{1} \cdot \frac{1.09 \text{ yd}}{1 \text{ m}} = \frac{(10)(1.09 \text{ yd})}{1} = 10.9 \text{ yd}$$

These units should match.

Meters (**m**) divide out leaving **yd**, the unit you want for the answer.

10 m $\approx$ 10.9 yd

Note

In Example 1 above, you could also use the numbers from the "U.S. Customary to Metric" side of the table that involve meters and yards: 1 yard $\approx$ 0.91 meter.

$$\frac{10 \text{ m}}{1} \cdot \frac{1 \text{ yd}}{0.91 \text{ m}} = \frac{10}{0.91} \text{ yd} \approx 10.99 \text{ yd}$$

The answer is slightly different because the values in the table are rounded. Also, you have to divide instead of multiply, which is usually more difficult to do without a calculator. We will use the first method in this chapter.

OBJECTIVES

1 Use unit fractions to convert between metric and U.S. customary units.

2 Learn common temperatures on the Celsius scale.

3 Use formulas to convert between Celsius and Fahrenheit temperatures.

1 Convert using unit fractions. Round your answers to the nearest tenth.

(a) 23 m to yards

(b) 40 cm to inches

(c) 5 mi to kilometers (Look at the "U.S. Customary to Metric" side of the table.)

(d) 12 in. to centimeters

ANSWERS

1. **(a)** 23 m $\approx$ 25.1 yd **(b)** 40 cm $\approx$ 15.6 in.
(c) 5 mi $\approx$ 8.1 km **(d)** 12 in. $\approx$ 30.5 cm

Work Problem **1** *at the Side.* ▶

2 Convert. Use the values from the table on the previous page to make unit fractions. Round answers to the nearest tenth.

(a) 17 kg to pounds

(b) 5 L to quarts

(c) 90 g to ounces

(d) 3.5 gal to liters

(e) 145 lb to kilograms

(f) 8 oz to grams

ANSWERS

2. **(a)** 17 kg ≈ 37.4 lb **(b)** 5 L ≈ 5.3 qt
   (c) 90 g ≈ 3.2 oz **(d)** 3.5 gal ≈ 13.3 L
   (e) 145 lb ≈ 65.3 kg **(f)** 8 oz ≈ 226.8 g

EXAMPLE 2 **Converting between Metric and U.S. Customary Weight and Capacity Units**

Convert using unit fractions. Round your answers to the nearest tenth.

(a) 3.5 kg to pounds

Look in the "Metric to U.S. Customary" side of the table on the previous page to see that 1 kilogram ≈ 2.20 pounds. Use this information to write a unit fraction that allows you to divide out kilograms.

$$\frac{3.5 \text{ kg}}{1} \cdot \frac{2.20 \text{ lb}}{1 \text{ kg}} = \frac{(3.5)(2.20 \text{ lb})}{1} = 7.7 \text{ lb}$$

3.5 kg ≈ 7.7 lb

The conversion value is approximate, so use the ≈ symbol in your answer.

(b) 18 gal to liters

Look in the "U.S. Customary to Metric" side of the table to see that 1 gallon ≈ 3.79 liters. Write a unit fraction that allows you to divide out gallons.

gal divides out, leaving L for your answer.

$$\frac{18 \text{ gal}}{1} \cdot \frac{3.79 \text{ L}}{1 \text{ gal}} = \frac{(18)(3.79 \text{ L})}{1} = 68.22 \text{ L}$$

68.22 rounded to the nearest tenth is 68.2
18 gal ≈ 68.2 L

(c) 300 g to ounces

In the "Metric to U.S. Customary" side of the table, 1 gram ≈ 0.035 ounce.

$$\frac{300 \text{ g}}{1} \cdot \frac{0.035 \text{ oz}}{1 \text{ g}} = \frac{(300)(0.035 \text{ oz})}{1} = 10.5 \text{ oz}$$

300 g ≈ 10.5 oz

CAUTION
Because the metric and U.S. customary systems were developed independently, almost all comparisons are approximate. Your answers should be written with the "≈" symbol to show they are approximate.

◀ *Work Problem* **2** *at the Side.*

OBJECTIVE 2 Learn common temperatures on the Celsius scale. In the metric system, temperature is measured on the **Celsius** scale. On the Celsius scale, water freezes at 0 °C and boils at 100 °C. The small raised circle stands for "degrees" and the capital **C** is for Celsius. Read the temperatures like this:

Water freezes at 0 degrees Celsius (0 °C).

Water boils at 100 degrees Celsius (100 °C).

The U.S. customary temperature system, used only in the United States, is measured on the **Fahrenheit** scale. On this scale:

Water freezes at 32 degrees Fahrenheit (32 °F).

Water boils at 212 degrees Fahrenheit (212 °F).

The thermometer below shows some typical temperatures in both Celsius and Fahrenheit. For example, comfortable room temperature is about 20 °C or 68 °F, and normal body temperature is about 37 °C or 98.6 °F.

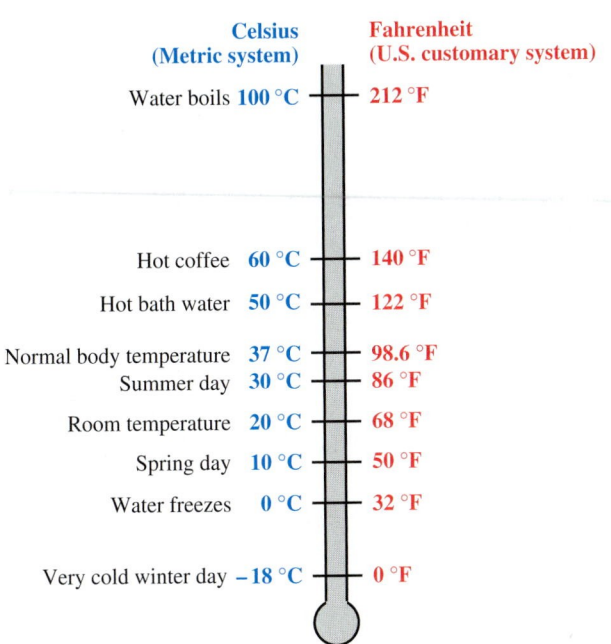

> **Note**
> The freezing and boiling temperatures are exact. The other temperatures are approximate. Even normal body temperature varies slightly from person to person.

EXAMPLE 3 **Using Celsius Temperatures**

Circle the Celsius temperature that is most reasonable for each situation.

(a) Warm summer day   29 °C   64 °C   90 °C

29 °C is reasonable. 64 °C and 90 °C are too hot; they're both above the temperature of hot bath water (above 122 °F).

(b) Inside a freezer   −10 °C   3 °C   25 °C

−10 °C is the reasonable temperature because it is the only one below the freezing point of water (0 °C). Your frozen foods would start thawing at 3 °C or 25 °C.

Work Problem **3** *at the Side.* ▶

OBJECTIVE **3** **Use formulas to convert between Celsius and Fahrenheit temperatures.**   You can use these formulas to convert between Celsius and Fahrenheit temperatures.

> **Celsius–Fahrenheit Conversion Formulas**
>
> Converting from Fahrenheit (F)   Converting from Celsius (C)
> to Celsius (C)   to Fahrenheit (F)
>
> $$C = \frac{5(F - 32)}{9}$$ $$F = \frac{9C}{5} + 32$$

4 Convert to Celsius. Round your answers to the nearest degree if necessary.

(a) 72 °F

(b) 20 °F

(c) 212 °F

(d) 98.6 °F

---

5 Convert to Fahrenheit. Round your answers to the nearest degree if necessary.

(a) 100 °C

(b) −25 °C

(c) 32 °C

(d) −18 °C

As you use these formulas, be sure to follow the order of operations from **Section 1.8**.

> **Order of Operations**
> **Step 1** Work inside *parentheses* or *other grouping symbols.*
> **Step 2** Simplify any expressions with *exponents.*
> **Step 3** Do the remaining *multiplications and divisions* as they occur from left to right.
> **Step 4** Do the remaining *additions and subtractions* as they occur from left to right.

EXAMPLE 4 **Converting Fahrenheit to Celsius**

Convert 10 °F to Celsius. Round your answer to the nearest degree.

Use the correct formula and follow the order of operations.

$$C = \frac{5(F - 32)}{9}$$ Fahrenheit to Celsius formula.

$$= \frac{5(10 - 32)}{9}$$ Replace F with 10.
Work inside parentheses first.
$10 - 32$ becomes $10 + (-32)$.

$$= \frac{5(-22)}{9}$$ Multiply in the numerator; positive times negative gives a negative product.

$$= \frac{-110}{9}$$ Divide; negative divided by positive gives a negative quotient.

$$= -12.\overline{2}$$ Round to −12 (nearest degree).

Thus, 10 °C ≈ −12 °F.

◀ *Work Problem* **4** *at the Side.*

EXAMPLE 5 **Converting Celsius to Fahrenheit**

Convert 15 °C to Fahrenheit.

Use the correct formula and follow the order of operations.

$$F = \frac{9C}{5} + 32$$ Celsius to Fahrenheit formula.

$$= \frac{9 \cdot 15}{5} + 32$$ Replace C with 15.

$$= \frac{9 \cdot 3 \cdot \overset{1}{\cancel{5}}}{\underset{1}{\cancel{5}}} + 32$$ Divide out the common factor.
Multiply in the numerator.

$$= 27 + 32$$ Add.

$$= 59$$

Thus, 15 °C = 59 °F.

◀ *Work Problem* **5** *at the Side.*

8.5 ▶▶▶ **Exercises**

FOR
EXTRA
HELP

MyMathLab

Math XL
PRACTICE

WATCH

DOWNLOAD

READ

REVIEW

Use the table on the first page of this section and unit fractions to make approximate conversions from metric to U.S. customary or U.S. customary to metric. Round your answers to the nearest tenth. See Examples 1 and 2.

1. 20 m to yards

2. 8 km to miles

3. 80 m to feet

4. 85 cm to inches

5. 16 ft to meters

6. 3.2 yd to meters

7. 150 g to ounces

8. 2.5 oz to grams

9. 248 lb to kilograms

10. 7.68 kg to pounds

11. 28.6 L to quarts

12. 15.75 L to gallons

13. For the 2000 Olympics, the 3M Company used 5 g of pure gold to coat Michael Johnson's track shoes. (*Source:* 3M Company.)

   (a) How many ounces of gold were used, to the nearest tenth?

   (b) Was this enough extra weight to slow him down?

14. The label on a Van Ness auto feeder for cats and dogs says it holds 1.4 kg of dry food. How many pounds of food does it hold, to the nearest tenth? (*Source:* Van Ness Plastics.)

15. The heavy-duty wash cycle in a dishwater uses 8.4 gal of water. How many liters does it use, to the nearest tenth? (*Source:* Frigidaire.)

16. The rinse-and-hold cycle in a dishwasher uses only 4.5 L of water. How many gallons does it use, to the nearest tenth? (*Source:* Frigidaire.)

17. The smallest pet fish are dwarf gobies, which are half an inch long. How many centimeters long is a dwarf gobie, to the nearest tenth? (*Hint:* Write half an inch in decimal form.)

18. The fastest nerve signals in the human body travel 120 meters per second. How many feet per second do the signals travel? (*Source: Big Book of Knowledge.*)

The BabyBjörn is a popular baby carrier imported from Sweden. Use the information from the instruction sheet that comes with the carrier to answer Exercises 19–20. (Source: BabyBjörn, Sweden.)

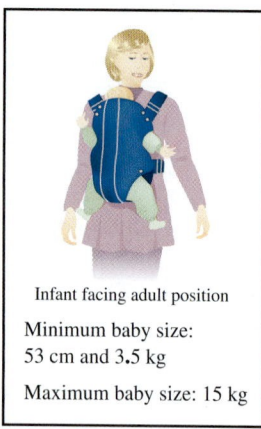

Infant facing adult position

Minimum baby size:
53 cm and 3.5 kg

Maximum baby size: 15 kg

19. Can the carrier be safely used for a newborn infant who weighs 8 lb and is 19.5 in. long? Explain your answer.

20. Can the carrier be safely used for a baby who weighs 30 lb? Explain your answer.

Circle the most reasonable Celsius temperature for each situation. See Example 3.

21. A snowy day

   12 °C   28 °C   −8 °C

22. Brewing coffee

   80 °C   180 °C   15 °C

23. A high fever

   21 °C   40 °C   103 °C

24. Swimming pool water

   90 °C   78 °C   25 °C

25. Oven temperature

   150 °C   50 °C   30 °C

26. Light jacket weather

   0 °C   10 °C   −10 °C

Use the conversion formulas from this section to convert Fahrenheit temperatures to Celsius and Celsius temperatures to Fahrenheit. Round your answers to the nearest degree if necessary. See Examples 4 and 5.

27. 60 °F

28. 80 °F

29. −4 °F

30. 15 °F

31. 8 °C

32. 18 °C

33. −5 °C

34. 0 °C

Solve each application problem. Round your answers to the nearest degree, if necessary.

35. The highest temperature ever recorded on Earth was 136 °F at El Azizia, Libya, in 1922. The lowest was −129 °F in Antarctica. Convert these temperatures to Celsius. (*Source: The World Almanac.*)

36. Hummingbirds have a normal body temperature of 107 °F. But on cold nights they go into a state of torpor where their body temperature drops to 39 °F. What are these temperatures in Celsius? (*Source: Wildbird.*)

37. The directions for a self-stick clothes hook with adhesive on the back are as follows: "Apply to surfaces above 50 °F. Adhesive could soften and lose adhesion above 105 °F." What are these temperatures in the metric system? (*Source: 3M Company.*)

38. A box of imported Belgian chocolates carries a warning to keep the box dry and at <18 °C. Translate this warning into the U.S. Customary system. (*Source: Chocolaterie Guylian N.V.*)

39. The tag on a pair of hiking boots is shown below.

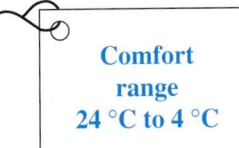

Comfort range
24 °C to 4 °C

Source: Sorel.

   (a) In what kind of weather would you be most comfortable wearing these boots?

   (b) For what Fahrenheit temperatures are the boots designed?

   (c) What range of metric temperatures do you have in January where you live? Would you be comfortable in these boots?

40. Sleeping bags made by Eddie Bauer are sold around the world. Each type of sleeping bag is designed for outdoor camping in certain temperatures.

OUTDOOR SLEEPING BAGS	
Junior bag	5 °C or warmer
Removable liner bag	0 °C to 15 °C
Conversion bag	−7 °C to 0 °C

Source: Eddie Bauer.

   (a) At what Fahrenheit temperatures should you use the Junior bag?

   (b) What Fahrenheit temperatures is the Removable liner bag designed for?

   (c) What are the Fahrenheit temperatures for the Conversion bag?

Relating Concepts (Exercises 41–48) For Individual or Group Work

The article below appeared in American newspapers. However, both Newfoundland (part of Canada) and Ireland use the metric system. Their newspapers would have reported all the measurements in metric units. Complete the conversions to metric, rounding answers to the nearest tenth when necessary.

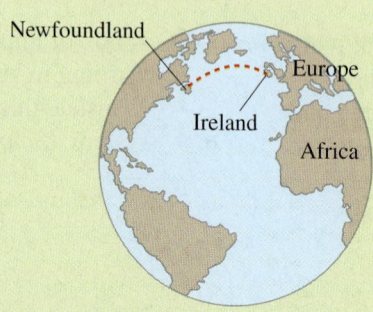

Q.: A recent news brief reported on some men who flew a model airplane from Newfoundland to Ireland. Can you provide some details of the flight?

A.: The model plane is 6 feet long and weighs 11 pounds. Made of balsa wood and mylar, it crossed the Atlantic—the flight path took it 1,888.3 miles—in 38 hours, 23 minutes. It soared at a cruising altitude of 1,000 feet. The plane used a souped-up piston engine and carried less than a gallon of fuel, as mandated by rules of the Federation Aeronautique Internationale, the governing body of model airplane building. When it landed in County Galway, Ireland, it had less than 2 fluid ounces of fuel left. The plane was built by Maynard Hill of Silver Spring, Maryland.

(*Source*: *New York Times.*)

41. Length of model plane

42. Weight of plane

43. Length of flight path

44. Time of flight

45. Cruising altitude

46. Fuel at the start, in milliliters

47. Fuel left after landing, in milliliters
(*Hint:* First convert 2 fl oz to quarts.)

48. What *percent* of the fuel was left at the end of the flight?

Chapter 8 ▷▷▷ Summary

▶ **Key Terms**

8.1	**U.S. customary measurement units**	The U.S. customary measurement units are used for many daily activities only in the United States. Commonly used units include quarts, pounds, feet, miles, and degrees Fahrenheit.
	metric system	The metric system of measurement is an international system used in manufacturing, science, medicine, sports, and other fields. Commonly used units in this system include meters, liters, grams, and degrees Celsius.
	unit fraction	A unit fraction involves measurement units and is equivalent to 1. Unit fractions are used to convert among different measurements.
8.2	**meter**	The meter is the basic unit of length in the metric system. The symbol **m** is used for meter. One meter is a little longer than a yard.
	prefixes	Attaching a prefix such as *kilo-* or *milli-* to the words meter, liter, or gram gives names of larger or smaller units. For example, the prefix *kilo* means 1000 so a *kilo*meter is 1000 meters.
	metric conversion line	The metric conversion line is a line showing the various metric measurement prefixes and their size relationship to each other.
8.3	**liter**	The liter is the basic unit of capacity in the metric system. The symbol **L** is used for liter. One liter is a little more than one quart.
	gram	The gram is the basic unit of weight (mass) in the metric system. The symbol **g** is used for gram. One gram is the weight of 1 milliliter of water or one dollar bill.
8.5	**Celsius**	The Celsius scale is used to measure temperature in the metric system. Water boils at 100 °C and freezes at 0 °C.
	Fahrenheit	The Fahrenheit scale is used to measure temperature in the U.S. customary system. Water boils at 212 °F and freezes at 32 °F.

▶ **New Symbols**

Frequently used metric length units

- **km** kilometer
- **m** meter
- **cm** centimeter
- **mm** millimeter

Frequently used metric capacity units

- **L** liter
- **mL** milliliter

Frequently used metric weight (mass) units

- **kg** kilogram
- **g** gram
- **mg** milligram

°C degrees Celsius
(metric temperature unit)

°F degrees Fahrenheit
(U.S. customary temperature unit)

▶ **New Formulas**

Converting from Celsius to Fahrenheit: $F = \dfrac{9C}{5} + 32$

Converting from Fahrenheit to Celsius: $C = \dfrac{5(F - 32)}{9}$

▶ Test Your Word Power

See how well you have learned the vocabulary in this chapter. Answers follow the Quick Review.

1. The **metric system**
   A. uses meters, liters, and degrees Fahrenheit
   B. is based on multiples of 10
   C. is used only in the United States
   D. has evolved over centuries.

2. The **U.S. customary measurement units**
   A. are used throughout the world
   B. are based on multiples of 12
   C. include feet, inches, quarts, and pounds
   D. were developed by a group of scientists in 1790.

3. A **unit fraction**
   A. has the unit you want to change in the numerator
   B. has a denominator of 1
   C. must be written in lowest terms
   D. is equivalent to 1.

4. A **gram** is
   A. the weight of 1 mL of water
   B. abbreviated gm
   C. equivalent to 1000 kg
   D. approximately equal to 2.2 pounds.

5. A **meter** is
   A. equivalent to 1000 cm
   B. approximately equal to $\frac{1}{2}$ inch
   C. abbreviated m with no period after it
   D. the basic unit of capacity in the metric system.

6. The **Celsius** temperature scale
   A. shows water freezing at 32°
   B. is used in the U.S. customary system of measurement
   C. shows water boiling at 100°
   D. cannot be converted to the Fahrenheit temperature scale.

▶ Quick Review

Concepts	Examples

8.1 The U.S. Customary Measurement Units

Memorize the basic measurement relationships. Then, to convert units, multiply when changing from a larger unit to a smaller unit; divide when changing from a smaller unit to a larger unit.

Convert each measurement.
(a) 5 ft to inches
$$5 \text{ ft} = 5 \cdot 12 = 60 \text{ in.}$$

(b) 3 lb to ounces
$$3 \text{ lb} = 3 \cdot 16 = 48 \text{ oz}$$

(c) 15 qt to gallons
$$15 \text{ qt} = \frac{15}{4} = 3\frac{3}{4} \text{ gal}$$

8.1 Using Unit Fractions

Another, more useful, conversion method is multiplying by a unit fraction. The unit you want in the answer should be in the numerator. The unit you want to change should be in the denominator.

Convert 32 oz to pounds.

$$32 \text{ oz} \cdot \frac{1 \text{ lb}}{16 \text{ oz}} \Big\} \text{ Unit fraction}$$

These units should match.

$$= \frac{\overset{2}{\cancel{32} \cancel{oz}}}{1} \cdot \frac{1 \text{ lb}}{\underset{1}{\cancel{16} \cancel{oz}}} \quad \begin{array}{l} \text{Divide out ounces.} \\ \text{Divide out common factors.} \end{array}$$

$$= 2 \text{ lb}$$

Concepts	Examples

8.1 Solving U.S. Customary Measurement Application Problems

To solve application problems, use these problem-solving steps.

Use the six steps to solve this problem.

Mr. Green has 10 yd of rope. He is cutting it into eight pieces so his sailing class can practice knot tying. How many feet of rope will each of his eight students get?

Step 1   Read the problem.

Step 1   The problem asks how many feet of rope can be given to each of eight students.

Step 2   Work out a plan.

Step 2   Convert 10 yd to feet (the unit required in the answer). Then divide by eight students.

Step 3   Estimate a reasonable answer.

Step 3   There are 3 ft in one yard, so there are 30 ft in 10 yd. Then 30 ft ÷ 8 ≈ 4 ft as our estimate.

Step 4   Solve the problem.

Step 4   Use a unit fraction to convert 10 yd to feet, then divide.

$$\frac{10 \text{ yd}}{1} \cdot \frac{3 \text{ ft}}{1 \text{ yd}} = 30 \text{ ft}$$

$$\frac{30 \text{ ft}}{8 \text{ students}} = 3\frac{3}{4} \text{ ft or } 3.75 \text{ ft per student}$$

Step 5   State the answer.

Step 5   Each student gets $3\frac{3}{4}$ ft or 3.75 ft of rope.

Step 6   Check your work.

Step 6   The exact answer of 3.75 ft is close to our estimate of 4 ft.

8.2 Basic Metric Length Units

Use approximate comparisons to judge which length units are appropriate:

> 1 mm is the thickness of a dime.
>
> 1 cm is about $\frac{1}{2}$ inch.
>
> 1 m is a little more than 1 yard.
>
> 1 km is about 0.6 mile.

Write the most reasonable metric unit in each blank. Choose from km, m, cm, and mm.

The room is 6 __m__ long.

A paper clip is 30 __mm__ long.

He drove 20 __km__ to work.

8.2 and 8.3 Converting within the Metric System

Using Unit Fractions
One conversion method is to multiply by a unit fraction. Use a fraction with the unit you want in the answer in the numerator and the unit you want to change in the denominator.

Convert.
(a) 9 g to kg

$$\frac{9 \text{ g}}{1} \cdot \frac{1 \text{ kg}}{1000 \text{ g}} = \frac{9}{1000} \text{ kg} = 0.009 \text{ kg}$$

9 g = 0.009 kg

(b) 3.6 m to cm

$$\frac{3.6 \text{ m}}{1} \cdot \frac{100 \text{ cm}}{1 \text{ m}} = 360 \text{ cm}$$

3.6 m = 360 cm

(continued)

Concepts	Example

8.2 and 8.3 Converting within the Metric System
(*continued*)

Using the Metric Conversion Line

Another conversion method is to find the unit you are given on the metric conversion line. Count the number of places to get from the unit you are given to the unit you want. Move the decimal point the same number of places and in the same direction.

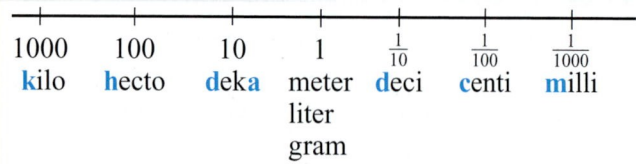

1000	100	10	1	$\frac{1}{10}$	$\frac{1}{100}$	$\frac{1}{1000}$
kilo	**hecto**	**deka**	meter	**deci**	centi	**milli**
			liter			
			gram			

Convert.

(a) 68.2 kg to g

From **kg** to **g** is three places to the right.

6 8.2 0 0   Decimal point is moved three places to the right.

68.2 kg = 68,200 g

(b) 300 mL to L

From **mL** to **L** is three places to the left.

3 0 0.   Decimal point is moved three places to the left.

300 mL = 0.3 L

(c) 825 cm to m

From **cm** to **m** is two places to the left.

8 2 5.   Decimal point is moved two places to the left.

825 cm = 8.25 m

8.3 Basic Metric Capacity Units

Use approximate comparisons to judge which capacity units are appropriate:

> 1 L is a little more than 1 quart.
>
> 1 mL is the amount of water in a cube 1 cm on each side.
>
> 5 mL is about one teaspoon.
>
> 250 mL is about one cup.

Write the most reasonable metric unit in each blank. Choose from L and mL.

The pail holds 12 __L__.

The milk carton from the vending machine holds 250 __mL__.

8.3 Basic Metric Weight (Mass) Units

Use approximate comparisons to judge which weight units are appropriate:

> 1 kg is about 2.2 pounds.
>
> 1 g is the weight of 1 mL of water or one dollar bill.
>
> 1 mg is $\frac{1}{1000}$ of a gram; very tiny!

Write the most reasonable metric unit in each blank. Choose from kg, g, and mg.

The wrestler weighed 95 __kg__.

She took a 500 __mg__ aspirin tablet.

One banana weighs 150 __g__.

Concepts	Examples

8.4 Solving Metric Measurement Application Problems

Convert units so you are comparing kg to kg, cm to cm, and so on. When a measurement involves two units, such as 6 m 20 cm, write it in terms of the unit called for in the answer (6.2 m or 620 cm).

Use these problem-solving steps.

Step 1 **Read** the problem.

Step 2 **Work out a plan.**

Step 3 **Estimate** a reasonable answer.

Step 4 **Solve** the problem.

Step 5 **State the answer.**

Step 6 **Check** your work.

Use the six steps to solve this problem.

George cut 1 m 35 cm off of a 3 m board. How long was the leftover piece, in meters?

Step 1 The problem asks for the length of the leftover piece in meters.

Step 2 Convert the cut-off measurement to meters (the unit required in the answer), and then subtract to find the "leftover."

Step 3 To estimate, round 1 m 35 cm to 1 m, because 35 cm is less than half a meter. Then, 3 m − 1 m = 2 m as our estimate.

Step 4 Convert the cut-off measurement to meters, then subtract.

$$
\begin{array}{ll}
1\ \text{m} \rightarrow \quad 1.00\ \text{m} & 3.00\ \text{m} \leftarrow \text{Board} \\
\text{plus } 35\ \text{cm} \rightarrow \underline{+\ 0.35\ \text{m}} & \underline{-\ 1.35\ \text{m}} \leftarrow \text{Cut off} \\
\quad\quad\quad\quad 1.35\ \text{m} & 1.65\ \text{m} \leftarrow \text{Left}
\end{array}
$$

Step 5 The leftover piece is 1.65 m long.

Step 6 The exact answer of 1.65 m is close to our estimate of 2 m.

8.5 Converting between Metric and U.S. Customary Units

Write a unit fraction using the values in the table of conversion factors on the first page of **Section 8.5.** Because the values in the table are rounded, your answers will be approximate.

Convert. Round answers to the nearest tenth.

(a) 23 m to yards
From the table, 1 meter $\approx$ 1.09 yards.

$$\frac{23\ \cancel{\text{m}}}{1} \cdot \frac{1.09\ \text{yd}}{1\ \cancel{\text{m}}} = 25.07\ \text{yd}$$

25.07 rounds to 25.1, so 23 m $\approx$ 25.1 yd.

(b) 4 oz to grams
From the table, 1 ounce $\approx$ 28.35 grams.

$$\frac{4\ \cancel{\text{oz}}}{1} \cdot \frac{28.35\ \text{g}}{1\ \cancel{\text{oz}}} = 113.4\ \text{g}$$

4 oz $\approx$ 113.4 g

Concepts	Examples

8.5 Common Celsius Temperatures

Use approximate and exact comparisons to judge which temperatures are appropriate:

Exact comparisons:
0 °C is the freezing point of water (32 °F).

100 °C is the boiling point of water (212 °F).

Approximate comparisons:
10 °C for a spring day (50 °F)

20 °C for room temperature (68 °F)

30 °C for summer day (86 °F)

37 °C for normal body temperature (98.6 °F)

Circle the Celsius temperature that is most reasonable.

(a) Hot summer day:
　(35 °C)　　90 °C　　110 °C

(b) The first snowy day in winter:
　−20 °C　(0 °C)　　15 °C

8.5 Converting between Fahrenheit and Celsius Temperatures

Use this formula to convert from Fahrenheit (F) to Celsius (C).

$$C = \frac{5(F - 32)}{9}$$

Convert 100 °F to Celsius. Round your answer to the nearest degree if necessary.

$$C = \frac{5(\mathbf{100} - 32)}{9} \qquad \text{Replace F with 100.}$$

$$= \frac{5(68)}{9} = \frac{340}{9}$$

$$= 37.\overline{7} \qquad \text{Round to 38}$$

100 °F ≈ 38 °C

Use this formula to convert from Celsius (C) to Fahrenheit (F).

$$F = \frac{9C}{5} + 32$$

Convert −8 °C to Fahrenheit. Round your answer to the nearest degree if necessary.

$$F = \frac{9(\mathbf{-8})}{5} + 32 \qquad \text{Replace C with } -8$$

$$= \frac{-72}{5} + 32$$

$$= -14.4 + 32$$

$$= 17.6 \qquad \text{Round to 18}$$

−8 °C ≈ 18 °F

ANSWERS TO TEST YOUR WORD POWER

1. B; *Examples:* 10 meters = 1 dekameter; 100 meters = 1 hectometer; 1000 meters = 1 kilometer.

2. C; *Examples:* Feet and inches are used to measure length, quarts to measure capacity, pounds to measure weight.

3. D; *Example:* Because 12 in. = 1 ft, the unit fraction $\frac{12 \text{ in.}}{1 \text{ ft}}$ is equivalent to $\frac{12 \text{ in.}}{12 \text{ in.}} = 1$.

4. A; *Example:* A small box measuring 1 cm on every edge holds exactly 1 mL of water, and the water weighs 1 g.

5. C; *Example:* A measurement of 16 meters is written 16 m (without a period).

6. C; *Example:* In the metric system, water freezes at 0 °C and boils at 100 °C. The U.S. customary system uses the Fahrenheit temperature scale where water freezes at 32 °F and boils at 212 °F.

Chapter 8 ▶▶▶ Review Exercises

[8.1] *Fill in the blanks with the measurement relationships you have memorized.*

1. 1 lb = _____ oz

2. _____ ft = 1 yd

3. 1 ton = _____ lb

4. _____ qt = 1 gal

5. 1 hr = _____ min

6. 1 c = _____ fl oz

7. _____ sec = 1 min

8. _____ ft = 1 mi

9. _____ in. = 1 ft

Convert using unit fractions.

10. 4 ft = _____ in.

11. 6000 lb = _____ tons

12. 64 oz = _____ lb

13. 18 hr = _____ day

14. 150 min = _____ hr

15. $1\frac{3}{4}$ lb = _____ oz

16. $6\frac{1}{2}$ ft = _____ in.

17. 7 gal = _____ c

18. 4 days = _____ sec

19. The average depth of the world's oceans is 12,460 ft. (*Source: Handy Ocean Answer Book.*)

    (a) What is the average depth in yards?

    (b) What is the average depth in miles, to the nearest tenth?

20. During the first year of a program to recycle office paper, a company recycled 123,260 pounds of paper. The company received $40 per ton for the paper. How much money did the company make? Use the six problem-solving steps. (*Source:* I. C. System.)

[8.2] *Write the most reasonable metric length unit in each blank. Choose from km, m, cm, and mm.*

21. My thumb is 20 _____ wide.

22. Her waist measurement is 66 _____.

23. The two towns are 40 _____ apart.

24. A basketball court is 30 _____ long.

25. The height of the picnic bench is 45 _____.

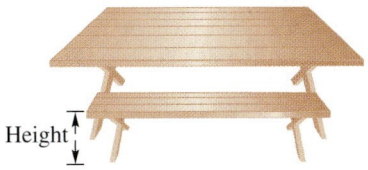

Height

26. The eraser on the end of my pencil is 5 _____ long.

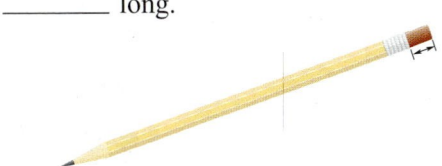

Convert using unit fractions or the metric conversion line.

27. 5 m to cm

28. 8.5 km to m

29. 85 mm to cm

30. 370 cm to m

31. 70 m to km

32. 0.93 m to mm

[8.3] *Write the most reasonable metric unit in each blank. Choose from L, mL, kg, g, and mg.*

33. The eyedropper holds 1 _____.

34. I can heat 3 _____ of water in this saucepan.

35. Loretta's hammer weighed 650 _____.

36. Yongshu's suitcase weighed 20 _____ when it was packed.

37. My fish tank holds 80 _____ of water.

38. I'll buy the 500 _____ bottle of mouthwash.

39. Mara took a 200 _____ antibiotic pill.

40. This piece of chicken weighs 100 _____

Convert using unit fractions or the metric conversion line.

41. 5000 mL to L

42. 8 L to mL

43. 4.58 g to mg

44. 0.7 kg to g

45. 6 mg to g

46. 35 mL to L

[8.4] *Solve each application problem. Show your work.*

47. Each serving of punch at the wedding reception will be 180 mL. How many liters of punch are needed for 175 servings?

48. Jason is serving a 10 kg turkey to 28 people. How many grams of meat is he allowing for each person? Round to the nearest whole gram.

49. Yerald weighed 92 kg. Then he lost 4 kg 750 g. What is his weight now, in kilograms?

50. Young-Mi bought 950 g of onions. The price was $1.49 per kilogram. How much did she pay, to the nearest cent?

[8.5] *Use the table on the first page of **Section 8.5** and unit fractions to make approximate conversions. Round your answers to the nearest tenth if necessary.*

51. 6 m to yards

52. 30 cm to inches

53. 108 km to miles

54. 800 mi to kilometers

55. 23 qt to liters

56. 41.5 L to quarts

*Write the appropriate **metric** temperature in each blank.*

57. Water freezes at _____.

58. Water boils at _____.

59. Normal body temperature is about _____.

60. Comfortable room temperature is about _____.

*Use the conversion formulas in **Section 8.5** to convert each temperature to Fahrenheit or to Celsius. Round to the nearest degree if necessary.*

61. 77 °F

62. 5 °F

63. −2 °C

64. Water coming into a dishwasher should be at least 49 °C to clean the dishes properly. What Fahrenheit temperature is that? (*Source:* Frigidaire.)

▶▶▶ Mixed Review Exercises

Write the most reasonable metric unit in each blank. Choose from km, m, cm, mm, L, mL, kg, g, and mg.

65. I added 1 _____ of oil to my car.

66. The box of books weighed 15 _____.

67. Larry's shoe is 30 _____ long.

68. Jan used 15 _____ of shampoo on her hair.

69. My fingernail is 10 _____ wide.

70. I walked 2 _____ to school.

71. The tiny bird weighed 15 _____.

72. The new library building is 18 _____ wide.

73. The cookie recipe uses 250 _____ of milk.

74. Renee's pet mouse weighs 30 _____.

75. One postage stamp weighs 90 _____.

76. I bought 30 _____ of gas for my car.

Convert the following using unit fractions, the metric conversion line, or the temperature conversion formulas.

77. 10.5 cm to millimeters

78. 45 min to hours

79. 90 in. to feet

80. 1.3 m to centimeters

81. 25 °C to Fahrenheit

82. $3\frac{1}{2}$ gal to quarts

83. 700 mg to grams   **84.** 0.81 L to milliliters   **85.** 5 lb to ounces

86. 60 kg to grams   **87.** 1.8 L to milliliters   **88.** 86 °F to Celsius

89. 0.36 m to centimeters   **90.** 55 mL to liters

Solve each application problem. Show your work.

91. Peggy had a board measuring 2 m 4 cm. She cut off 78 cm. How long is the board now, in meters?

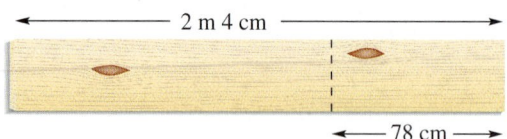

92. During the 12-day Minnesota State Fair, one of the biggest in the United States, Sweet Martha's booth sells an average of 3000 pounds of cookies per day. How many tons of cookies are sold in all? (*Source: Minneapolis Star Tribune.*)

93. Olivia is sending a recipe to her mother in Mexico. Among other things, the recipe calls for 4 oz of rice and a baking temperature of 350 °F. Convert these measurements to metric, rounding to the nearest gram and nearest degree.

94. While on vacation in Canada, Jalo became ill and went to a health clinic. They said he weighed 80.9 kg and was 1.83 m tall. Find his weight in pounds and height in feet. Round to the nearest tenth.

The largest two-axle trucks in the world are mining trucks used to haul huge loads of rock and iron ore. Some information about these $2 million trucks is given below. Fill in the blank spaces in the table. Then use the information to answer Exercises 101–102. Round answers to the nearest tenth.

WORLD'S LARGEST TWO-AXLE TRUCK

	Mining Truck	Measurements	
		Metric Units	Customary Units
95.	*Length of truck*	_____ m	44 ft
96.	*Height of truck*	6.7 m	_____ ft
97.	*Height of tire*	_____ m	4 yd
98.	*Width of tire tread*	102 cm	_____ in.
99.	*Weight of load truck can carry*	220,000 kg	_____ lb
100.	*Fuel needed to travel 1 mile*	_____ to _____ L	5 to 6 gal

Source: Hull Rust Mahoning Mine.

101. Using U.S. customary measurements:

   (a) The truck can carry a load weighing how many tons?

   (b) How many inches high is each tire?

102. Using metric measurements:

   (a) What is the width of the tire tread in meters?

   (b) How tall is the truck in centimeters?

Chapter 8 ▶▶▶ Test  Use the Chapter Test Prep Video CD to see fully worked-out solutions to any of the exercises you want to review.

Convert each measurement.

1. 9 gal = _____ qt

2. 45 ft = _____ yd

3. 135 min = _____ hr

4. 9 in. = _____ ft

5. $3\frac{1}{2}$ lb = _____ oz

6. 5 days = _____ min

Write the most reasonable metric unit in each blank. Choose from km, m, cm, mm, L, mL, kg, g, and mg.

7. My husband weighs 75 _____ .

8. I hiked 5 _____ this morning.

9. She bought 125 _____ of cough syrup.

10. This apple weighs 180 _____ .

11. This page is about 21 _____ wide.

12. My watch band is 10 _____ wide.

13. I bought 10 _____ of soda for the picnic.

14. The bracelet is 16 _____ long.

Convert the following measurements. Show your work.

15. 250 cm to meters

16. 4.6 km to meters

17. 5 mm to centimeters

18. 325 mg to grams

19. 16 L to milliliters

20. 0.4 kg to grams

21. 10.55 m to centimeters

22. 95 mL to liters

1. _____

2. _____

3. _____

4. _____

5. _____

6. _____

7. _____

8. _____

9. _____

10. _____

11. _____

12. _____

13. _____

14. _____

15. _____

16. _____

17. _____

18. _____

19. _____

20. _____

21. _____

22. _____

23. _____

23. The rainiest place in the world is Mount Waialeale in Hawaii, which receives 460 inches of rain each year. What is the average rainfall per month, in feet, to the nearest tenth? (*Source:* National Geographic Society.)

24. **(a)** _____

    (b) _____

24. A 6-inch Subway Veggie Delite sandwich has 590 mg of sodium. A 6-inch Super Subway Melt sandwich has 2.9 g of sodium. (*Source:* Subway.)

   (a) How much more sodium is in the Super Melt sandwich, in milligrams, than in the Veggie Delite?

   (b) The recommended amount of sodium is less than 2400 mg daily. How much more or less sodium does the Super Melt have than the recommended daily amount?

Pick the metric temperature that is most reasonable in each situation.

25. _____

25. The water is almost boiling.

    210 °C   155 °C   95 °C

26. _____

26. The tomato plants may freeze tonight.

    30 °C   20 °C   0 °C

27. _____

*Use the table from **Section 8.5** and unit fractions to convert each measurement. Round your answers to the nearest tenth if necessary.*

27. 6 ft to meters

28. _____

28. 125 lb to kilograms

29. _____

29. 50 L to gallons

30. _____

30. 8.1 km to miles

31. _____

Use the conversion formulas to convert each temperature. Round your answers to the nearest degree if necessary.

31. 74 °F to Celsius

32. _____

32. -12 °C to Fahrenheit

Solve this application problem. Show your work.

33. _____

33. Denise is making five matching pillows. She needs 120 cm of braid to trim each pillow. If the braid costs $3.98 per yard, how much will she spend to trim the pillows, to the nearest cent? (First find the number of meters of braid Denise needs.)

34. _____

34. Describe two benefits the United States would achieve by switching entirely to the metric system.

Graphs

The Internet can be a great place to do research. Whether you're writing term papers and speeches for college courses, or planning on a career as a writer or reporter, you'll need to find information at various Web sites, interpret the data, and communicate it to other people. The data will often be presented in the form of a table (see **Section 9.1**, Examples 1 and 2) or a graph (see **Sections 9.2 and 9.3**). Interpreting the data accurately is critical to the success of your report, article, or speech. (See **Section 9.3**, Exercises 37–44.)

9.1 Problem Solving with Tables and Pictographs

9.2 Reading and Constructing Circle Graphs

9.3 Bar Graphs and Line Graphs

9.4 The Rectangular Coordinate System

9.5 Introduction to Graphing Linear Equations

9.1 ▶▶▶ Problem Solving with Tables and Pictographs

OBJECTIVES

1. **Read and interpret data presented in a table.**

2. **Read and interpret data from a pictograph.**

Throughout this book you have used numbers, expressions, formulas, and equations to communicate rules or information. In this chapter you'll see how *tables, pictographs, circle graphs, bar graphs,* and *line graphs* are also used to communicate information.

OBJECTIVE 1 Read and interpret data presented in a table.
A **table** presents data organized into rows and columns. The advantage of a table is that you can find very specific, exact values. The disadvantages are that you may have to spend some time searching through the table to find what you want, and it may not be easy to see trends or patterns.

The table below shows information about the performance of eight U.S. airlines during the year 2007 (January–December).

PERFORMANCE DATA FOR SELECTED U.S. AIRLINES
JANUARY–DECEMBER 2007

Airline	On-Time Performance	Luggage Handling*
Airtran	77%	4.1
American	69%	7.3
Continental	74%	5.3
Delta	77%	7.6
Northwest	70%	5.0
Southwest	80%	5.9
United	70%	5.8
US Airways	69%	8.5

*Luggage problems per 1000 passengers.
Source: Department of Transportation Air Travel Consumer Report.

For example, by reading from left to right along the row marked United, you first see that 70% of United's flights were on time during 2007. The next number is 5.8 and the heading at the top of that column is *Luggage Handling**. The little star (asterisk) tells you to look below the table for more information. Next to the asterisk below the table it says "Luggage problems per 1000 passengers." So, almost 6 passengers out of every 1000 passengers on United flights had some sort of problem with their luggage.

EXAMPLE 1 Reading and Interpreting Data from a Table

Use the table above on airline performance to answer these questions.

(a) What percent of American's flights were on time?
Look across the row labeled American to see that 69% of its flights were on time.

(b) Which airline had the worst luggage handling record?
Look down the column headed Luggage Handling. To find the *worst* record, look for the *highest* number of luggage problems. The highest number is 8.5. Then look to the left to find the airline, which is US Airways.

(c) What was the average percent of on-time flights for the three airlines with the best performance, to the nearest whole percent?
Look down the column headed On-Time Performance to find the three highest numbers: 80%, 77%, and 77%. To find the average, add the values and divide by 3.

Continued on Next Page

$$\frac{80 + 77 + 77}{3} = \frac{234}{3} = 78$$

The average on-time performance for the three best airlines was 78%.

—— *Work Problem* (**1**) *at the Side.* ▶

EXAMPLE 2   Interpreting Data from a Table

The table below shows the maximum cab fares in five different cities in 2006. The "flag drop" charge is made when the driver starts the meter. "Wait time" is the charge for having to wait in the middle of a ride.

MAXIMUM TAXICAB FARES ALLOWED IN SELECTED CITIES IN 2006

City	Flag Drop	Price per Mile	Wait Time (per Hour)
Chicago	$2.25	$1.80	$20
Denver	$1.60	$2	$22.50
Miami	$2.50	$2.40	$24
New York	$2.50	$2	$12
San Francisco	$2.85	$2.25	$27

Source: www.schallerconsult.com

Use the table to answer these questions.

(a) What is the maximum fare for a 9-mile ride in New York that includes having the cab wait 15 minutes while you pick up a package at a store?

The price per mile in New York is $2, so the cost for 9 miles is $9\,(\$2) = \$18$. Then, add the flag drop charge of $2.50. Finally, figure out the cost of the wait time. One way to do that is to set up a proportion. Recall that 1 hour is 60 minutes, so each side of the proportion compares the cost to the number of minutes of wait time.

$$\text{Cost} \rightarrow \frac{\$12}{60 \text{ min}} = \frac{\$x}{15 \text{ min}} \leftarrow \text{Cost}$$
$$\text{Wait time} \rightarrow \quad\quad\quad\quad \leftarrow \text{Wait time}$$

Show that cross products are equivalent. → $60 \cdot x = 12 \cdot 15$

$$\frac{\overset{1}{\cancel{60}}\,x}{\underset{1}{\cancel{60}}} = \frac{180}{60} \quad\quad \text{Divide both sides by 60}$$

$$x = \$3 \longleftarrow \text{Charge for waiting 15 minutes}$$

Total fare $= \$18 + \$2.50 + \$3 = \$23.50$

(b) It is customary to give the cab driver a tip. Find the total cost of the cab ride in part (a) above if the passenger added a 15% tip, rounded to the nearest quarter (nearest $0.25).

Use the percent equation to find the exact tip.

$$\text{percent} \cdot \text{whole} = \text{part} \quad \text{Percent equation}$$
$$(15.\%)\,(\$23.50) = n \quad \text{Write 15\% as a decimal.}$$

Write 15% as 0.15 → $(0.15)\,(\$23.50) = n$

$$\$3.525 = n \quad \text{Round to \$3.50 (nearest \$0.25).}$$

The total cost of the cab ride is $\$23.50 + \$3.50 \text{ tip} = \$27.00$.

—— *Work Problem* (**2**) *at the Side.* ▶

1 Use the table of airline performance on the previous page to answer these questions.

(a) What percent of Continental's flights were on time?

(b) Which airline had the best on-time performance?

(c) Which airline had the best record for luggage handling?

(d) Which airline(s) had 5 or fewer luggage handling problems per 1000 passengers?

(e) What was the average number of luggage problems for all eight airlines, to the nearest tenth?

2 Use the table of cab fares at the left to answer these questions.

(a) What is the difference in the maximum fare for a cab ride of 6.5 miles in Chicago compared to New York? Assume there is no wait time.

(b) What is the maximum fare for a 12-mile cab ride in Miami that includes 10 minutes of wait time?

(c) Find the total cost for a cab ride of 4.5 miles in Denver, including 30 minutes of wait time and a 15% tip. Round the tip to the nearest $0.25.

ANSWERS

1. **(a)** 74%   **(b)** Southwest
   (c) Airtran   **(d)** Airtran, Northwest
   (e) $49.5 \div 8 \approx 6.2$ (rounded)
2. **(a)** Chicago $13.95; New York $15.50;
   New York fare is $1.55 higher.   **(b)** $35.30
   (c) $21.85 + $3.25 tip = $25.10

3 Use the pictograph in Example 3 to answer these questions.

(a) What is the approximate population of Chicago?

(b) What is the approximate population of Atlanta?

(c) Approximately how much less is the population of Los Angeles than New York?

(d) The population of Dallas is approximately how much greater than Atlanta?

ANSWERS

3. **(a)** 10 million people (or 10,000,000)
   (b) 5 million people (or 5,000,000)
   (c) 6 million people (or 6,000,000)
   (d) 1 million people (or 1,000,000)

OBJECTIVE 2 Read and interpret data from a pictograph. Tables show numbers in rows and columns. Graphs, on the other hand, are a *visual* way to communicate data; that is, they show a *picture* of the information rather than a list of numbers. In this section you'll work with one type of graph, *pictographs,* and in the next two sections you'll learn about circle graphs, bar graphs, and line graphs.

The advantage of a graph is that you can easily make comparisons or see trends just by looking. The disadvantage is that the graph may not give you the specific, more exact numbers you need in some situations.

EXAMPLE 3 Reading and Interpreting a Pictograph

A **pictograph** uses symbols or pictures to represent various amounts. The pictograph below shows the population of five U.S. metropolitan areas (cities with their surrounding suburbs) in 2005. The *key* at the bottom of the graph tells you that each symbol of a person represents 2 million people. Half of a symbol represents half of 2 million people.

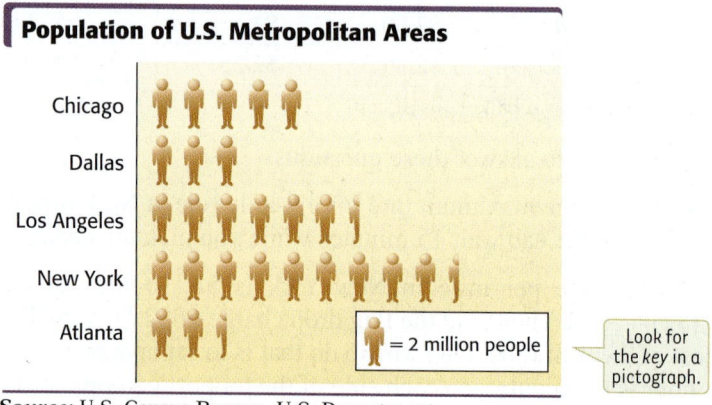

Population of U.S. Metropolitan Areas

= 2 million people

Look for the *key* in a pictograph.

Source: U.S. Census Bureau, U.S. Department of Commerce.

Use the pictograph to answer these questions.

(a) What is the approximate population of Los Angeles?

The population of Los Angeles is represented by 6 whole symbols ($6 \cdot 2$ million $= 12$ million) plus half of a symbol ($\frac{1}{2}$ of 2 million is 1 million) for a total of 13 million people.

Recall that 2 million can be written as 2,000,000, so another way to get the answer is to multiply 6.5 (the number of symbols) times 2,000,000 people for a total of 13,000,000 people.

(b) How much greater is the population of New York than Atlanta?

New York shows $9\frac{1}{2}$ symbols and Atlanta shows $2\frac{1}{2}$ symbols. So New York has 7 more symbols than Atlanta, and $7 \cdot 2$ million $= 14$ million. Thus, New York has about 14 million (or 14,000,000) more people than Atlanta.

> **Note**
> One disadvantage of a pictograph is that you often have to round numbers a great deal; in this case, to the nearest *million* people. The actual population of the Dallas area is 5,819,000 but the graph shows 3 symbols (6,000,000 people) so it is off by about 181,000 people.

◀ *Work Problem* **3** *at the Side.*

This table lists the basketball players in the NBA with the highest scoring average at the end of the 2006–2007 season. (Players must have a minimum of 10,000 points or 400 games.) Use the table to answer Exercises 1–10. See Examples 1 and 2.

ALL-TIME NBA STATISTICAL LEADERS—SCORING AVERAGE
(AT THE END OF THE 2006–2007 SEASON)

Player	Games	Points	Average Points Per Game
Michael Jordan	1072	32,292	30.1
Wilt Chamberlain	1045	31,419	30.1
Allen Iverson*	747	20,824	27.9
Elgin Baylor	846	23,149	27.4
Jerry West	932	25,192	
Bob Pettit	792	20,880	
Shaquille O'Neal*	981	25,454	

*Player still actively playing in the NBA.
Source: National Basketball Association.

1. (a) How many points did Wilt Chamberlain score during his NBA career?

   (b) Which player(s) scored more points than Chamberlain?

2. (a) How many games did Elgin Baylor play in during his career in the NBA?

   (b) Which player is closest to Baylor in number of games?

3. (a) Which player has been in the greatest number of games?

   (b) Which player has been in the fewest number of games?

4. Which players have scored more than 25,000 points? List them in order, starting with the player with the greatest number of points.

5. What is the difference in points scored between the player with the greatest number of points and the player with the least number of points?

6. How many fewer games has Shaquille O'Neal played in than Michael Jordan?

7. Complete the table by finding the average number of points scored per game by Jerry West, by Bob Pettit, and by Shaquille O'Neal. Look at the other averages in the table to decide how to round your answers.

8. Find the overall scoring average (points per game) for all seven players listed in the table. Use the numbers in the *Games* column and the *Points* column to calculate your answer.

9. According to the table, what is different about Iverson and O'Neal from the rest of the players? (Hint: Look at the note below the table.)

10. Michael Jordan and Wilt Chamberlain are tied in first place for average points per game. Explain what happens if you round their average points per game to the nearest hundredth instead of the nearest tenth.

This table shows the number of calories burned during 30 minutes of various types of exercise. The table also shows how the number of calories burned varies according to the weight of the person doing the exercise. Use the table to answer Exercises 11–20. See Examples 1 and 2.

CALORIES BURNED DURING 30 MINUTES OF EXERCISE BY PEOPLE OF DIFFERENT WEIGHTS

Activity	Calories Burned in 30 Minutes		
	110 Pounds	140 Pounds	170 Pounds
Moderate jogging	322	410	495
Moderate walking	110	140	170
Moderate bicycling	140	180	220
Aerobic dance	200	255	310
Racquetball	210	268	325
Tennis	160	205	250

Source: Fairview Health Services.

11. A person weighing 140 pounds is looking at the table.

   (a) How many calories will be burned during 30 minutes of aerobic dance?

   (b) Which activity burns the most calories?

12. A person weighing 170 pounds is looking at the table.

   (a) How many calories are burned during 30 minutes of tennis?

   (b) Which activity burns the fewest calories?

13. (a) Which activities can a 110-pound person do to burn at least 200 calories in 30 minutes?

   (b) Which activities can a 170-pound person do to burn at least 200 calories in 30 minutes?

14. (a) Which activities would burn fewer than 200 calories in 30 minutes for a 140-pound person?

   (b) Which activities would burn fewer than 300 calories in 30 minutes for a 170-pound person?

15. How many total calories will a 140-pound person burn during 15 minutes of bicycling and 60 minutes of moderate walking?

16. How many total calories will a 110-pound person burn during 90 minutes of tennis and 15 minutes of aerobic dance?

Set up and solve proportions to answer Exercises 17–20.

17. How many calories would you expect a 125-pound person to burn

   (a) during 30 minutes of moderate jogging?

   (b) during 30 minutes of racquetball? Round to the nearest whole number.

18. How many calories would you expect a 185-pound person to burn

   (a) during 30 minutes of walking?

   (b) during 30 minutes of bicycling? Round to the nearest ten.

19. How many more calories would a 158-pound person burn during 15 minutes of aerobic dance than during 20 minutes of walking? Round to the nearest whole number.

20. How many fewer calories would a 196-pound person burn during 25 minutes of walking than during 20 minutes of tennis? Round to the nearest whole number.

This pictograph shows the approximate number of passenger arrivals and departures at selected U.S. airports in 2006. Use the pictograph to answer Exercises 21–28. See Example 3.

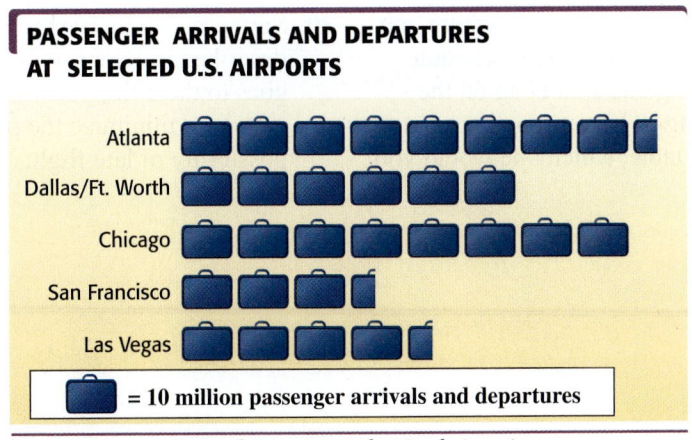

PASSENGER ARRIVALS AND DEPARTURES AT SELECTED U.S. AIRPORTS

Atlanta
Dallas/Ft. Worth
Chicago
San Francisco
Las Vegas

= 10 million passenger arrivals and departures

Source: Airports Council International—North America.

21. Approximately how many passenger arrivals and departures took place at the

   (a) Chicago airport?

   (b) San Francisco airport?

22. Approximately how many passenger arrivals and departures took place at the

   (a) Dallas airport?

   (b) Las Vegas airport?

23. What is the approximate total number of arrivals and departures at the two busiest airports?

24. What is the difference in the number of arrivals and departures at the busiest airport and the least-busy airport?

25. How many fewer arrivals and departures did Chicago's airport have compared to Atlanta's airport?

26. Find the approximate total number of arrivals and departures for the three least-busy airports.

27. What is the approximate total number of arrivals and departures for all five airports?

28. Find the average number of arrivals and departures for the five airports.

Relating Concepts (Exercises 29—34) For Individual or Group Work

Look back at the first table in this section, Performance Data for Selected U.S. Airlines.
*Use the table as you **work Exercises 29–34 in order.***

29. Suppose you are planning a business trip where you will fly to a new city each day on a tight schedule. You'll travel light, carrying one small bag on the plane rather than checking it. If you can choose any one of the airlines in the table, which one would you pick? Explain why.

30. Suppose you are planning the business trip described in Exercise 29 and the only airline that goes to the cities you want is American. What could you do to minimize the problems caused by the possibility of late flights?

31. Now you are planning a two-week vacation trip to a beachfront resort. You'll be checking several bags and your expensive golf clubs. If you can choose any one of the airlines in the table, which one would you pick? Explain why.

32. Suppose you are planning the vacation trip described in Exercise 31 and US Airways is the only airline that goes to the city you want. What could you do to minimize possible luggage handling problems?

33. Think of three possible reasons an airline might have a lower percentage of on-time flights during a particular month than they usually do. What, if anything, could the airline do to resolve each of the problems you listed?

34. Describe three other factors you might consider when selecting an airline, other than on-time performance and luggage handling problems.

9.2 ▶▶▶ Reading and Constructing Circle Graphs

A *circle graph* is another way to show a *picture* of a set of data. This picture can often be understood faster and more easily than a formula or a list of numbers.

OBJECTIVE 1 Read a circle graph. A **circle graph** is used to show how a total amount is divided into parts. The circle graph below shows you how 24 hours in the life of a college student are divided among different activities.

THE DAY OF A COLLEGE STUDENT

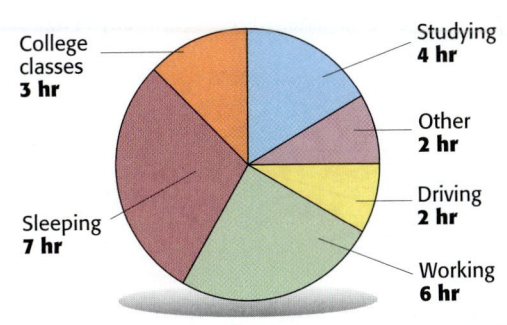

College classes 3 hr
Studying 4 hr
Other 2 hr
Driving 2 hr
Working 6 hr
Sleeping 7 hr

Work Problem **1** *at the Side.* ▶

OBJECTIVE 2 Use a circle graph. The circle graph above uses pie-shaped pieces called *sectors* to show the amount of time spent on each activity (the total must be 24 hours). The circle graph can therefore be used to compare the time spent on one activity to the total number of hours in the day.

EXAMPLE 1 Using a Circle Graph

Find the ratio of time spent in college classes to the total number of hours in a day. Write the ratio as a fraction in lowest terms. (See **Section 6.1.**)

The circle graph shows that 3 of the 24 hours in a day are spent in class. The ratio of class time to the hours in a day is shown below.

$$\frac{3 \text{ hours (college classes)}}{24 \text{ hours (whole day)}} = \frac{3 \text{ hours}}{24 \text{ hours}} = \frac{\overset{1}{\cancel{3}}}{\underset{1}{\cancel{3}} \cdot 8} = \frac{1}{8} \left\{ \begin{array}{l} \text{Ratio in} \\ \text{lowest terms} \end{array} \right.$$

Work Problem **2** *at the Side.* ▶

The circle graph above can also be used to find the ratio of the time spent on one activity to the time spent on any other activity. See the next example.

EXAMPLE 2 Finding a Ratio from a Circle Graph

Find the ratio of working time to class time.

The circle graph shows 6 hours spent working and 3 hours spent in class. The ratio of working time to class time is shown below.

$$\frac{6 \text{ hours (working)}}{3 \text{ hours (class)}} = \frac{6 \text{ hours}}{3 \text{ hours}} = \frac{\overset{1}{\cancel{3}} \cdot 2}{\underset{1}{\cancel{3}}} = \frac{2}{1} \leftarrow \text{Ratio in lowest terms}$$

Work Problem **3** *at the Side.* ▶

OBJECTIVES

1 Read a circle graph.
2 Use a circle graph.
3 Use a protractor to draw a circle graph.

1 Use the circle graph at the left to answer each question.

(a) The greatest number of hours is spent in which activity?

(b) How many more hours are spent working than studying?

(c) Find the total number of hours spent studying, working, and attending classes.

2 Use the circle graph to find each ratio. Write the ratios as fractions in lowest terms.

(a) Hours spent driving to whole day

(b) Hours spent studying to whole day

(c) Hours spent sleeping and doing other to whole day

3 Use the circle graph to find each ratio. Write the ratios as fractions in lowest terms.

(a) Hours spent studying to hours spent working

(b) Hours spent working to hours spent sleeping

(c) Hours spent studying to hours spent driving

ANSWERS

1. (a) sleeping (b) 2 hours (c) 13 hours
2. (a) $\frac{1}{12}$ (b) $\frac{1}{6}$ (c) $\frac{3}{8}$
3. (a) $\frac{2}{3}$ (b) $\frac{6}{7}$ (c) $\frac{2}{1}$

4 Use the circle graph on frozen pizza sales to find the amount of sales for each company.

(a) Kraft

(b) Van De Kamps

(c) Tony's Pizza Service

(d) Pillsbury Corp.

A circle graph often shows data as percents. For example, total U.S. sales of frozen pizza are $2 billion each year. The circle graph below shows how sales are divided among various companies that make frozen pizza. The entire circle represents $2 billion in sales. Each sector represents the sales of one company as a percent of the total sales. The total in a circle graph must be 100%, although it may be slightly more or less due to rounding the percent for each sector.

HOT SALES OF FROZEN PIZZA

Americans eat $2 billion worth of frozen pizzas each year. The percent of total sales for each company is rounded to the nearest whole percent.

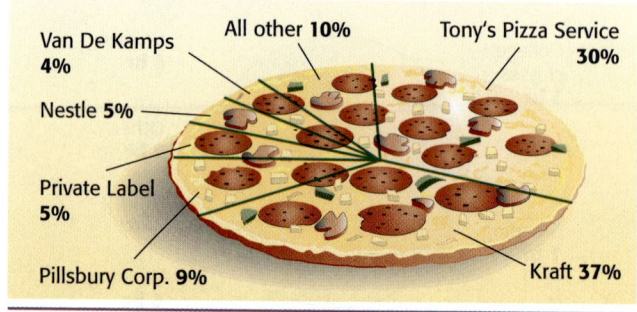

All other **10%**

Van De Kamps **4%**

Tony's Pizza Service **30%**

Nestle **5%**

Private Label **5%**

Pillsbury Corp. **9%**

Kraft **37%**

Source: Information Resources, Inc.

EXAMPLE 3 **Calculating an Amount Using a Circle Graph**

Use the circle graph above on frozen pizza sales to find the amount of sales for Nestle.

Recall the percent equation.

$$\text{percent} \cdot \text{whole} = \text{part}$$

The percent for Nestle is 5%. Rewrite 5% as the decimal 0.05. The *whole* is the total sales of $2 billion (the entire circle).

$$
\begin{array}{ccc}
\text{percent} \cdot & \text{whole} & = \text{part} \\
\downarrow & \downarrow & \downarrow \\
05.\% \cdot & \$2 \text{ billion} & = n \\
\downarrow & \downarrow & \\
(0.05)(\$2{,}000{,}000{,}000) & = n \\
\$100{,}000{,}000 & = n
\end{array}
$$

Write 5% as 0.05

Write $2 billion as $2,000,000,000

The sales for Nestle are $100,000,000.

◀ *Work Problem* **4** *at the Side.*

OBJECTIVE **3** **Use a protractor to draw a circle graph.** The coordinator of the Fair Oaks Youth Soccer League organizes teams in five age groups. She counts the number of registered players in each age group as shown in the table on the next page. Then she calculates what percent of the total each group represents. For example, there are 59 players in the "Under 8" group, out of 298 total players.

$$
\begin{array}{c}
\text{percent} \cdot \text{whole} = \text{part} \\
p \cdot 298 = 59 \\
\dfrac{p \cdot 298}{298} = \dfrac{59}{298} \\
p \approx 0.198 \approx 19.8\% \quad \text{rounds to 20\%}
\end{array}
$$

Multiply the decimal answer by 100% to get the percent.

ANSWERS

4. (a) $740,000,000 **(b)** $80,000,000
(c) $600,000,000 **(d)** $180,000,000

FAIR OAKS YOUTH SOCCER LEAGUE

Age Group	Number of Players	Percent of Total (rounded to nearest whole percent)
Under 8 years	59	20% ← 59 players ≈ 20% of 298
Ages 8–9	46	15%
Ages 10–11	75	25%
Ages 12–13	74	25%
Ages 14–15	44	15%
Total	**298**	**100%**

You can show these percents in a circle graph. Recall that a circle has 360 degrees (written 360°). The 360° represents the entire league, or 100% of the soccer players.

EXAMPLE 4 **Drawing a Circle Graph**

Using the data on *age groups,* find the number of degrees in the sector that would represent the "Under 8" group, and begin constructing a circle graph.

A complete circle has 360°. Because the "Under 8" group makes up 20% of the total number of players, the number of degrees needed for the "Under 8" sector of the circle graph is 20% of 360°.

$$20.\% \text{ of } 360° = n$$

Write 20% as 0.20

$$(0.20)(360°) = n$$

$$72° = n$$

Use the circle drawn below and a tool called a **protractor** to make a circle graph. First, using a ruler or straightedge, draw a line from the center of the circle to the left edge. Place the hole in the protractor over the center of the circle, making sure that the 0 mark and the black line on the protractor are right over the line you drew. Find 72° and make a mark as shown in the illustration. Then remove the protractor and use the straightedge to draw a line from the 72° mark to the center of the circle. This sector is 72° and represents the "Under 8" group.

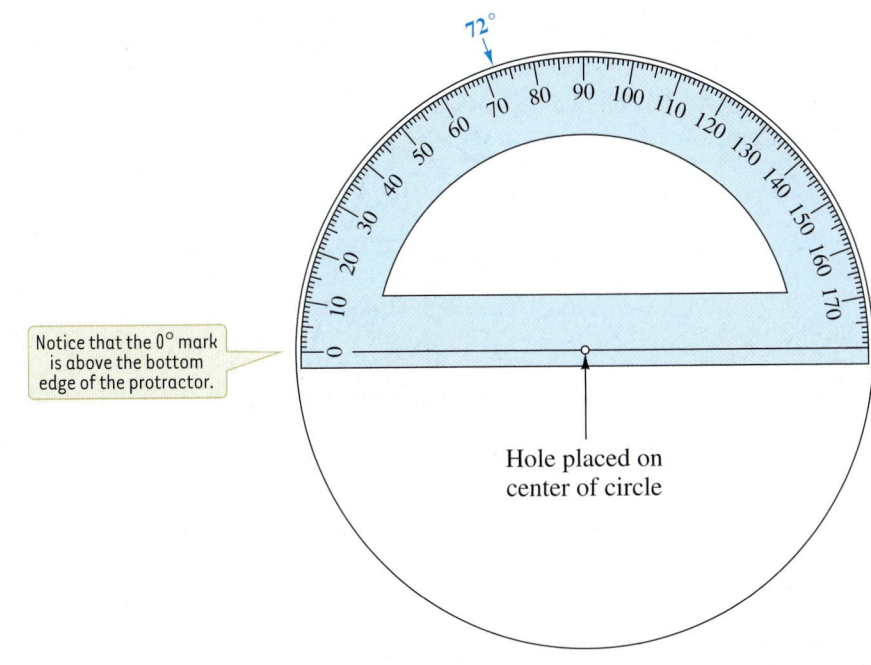

Notice that the 0° mark is above the bottom edge of the protractor.

Hole placed on center of circle

Continued on Next Page

5 Using the information on the soccer age groups in the table on the previous page, find the number of degrees needed for each sector. Then complete the circle graph at the bottom right on this page.

(a) "Ages 10–11" sector

(b) "Ages 12–13" sector

(c) "Ages 14–15" sector

To draw the "Ages 8–9" sector, begin by finding the number of degrees in the sector, which is 15% of the total circle, or 15% of 360°.

$$15.\% \text{ of } 360° = n$$
$$(0.15)(360°) = n$$
$$54° = n$$

Again, place the hole of the protractor over the center of the circle, but this time align 0 with the previous 72° mark. Make a mark at 54° and draw a line from the mark to the center of the circle. This sector is 54° and represents the "Ages 8–9" group.

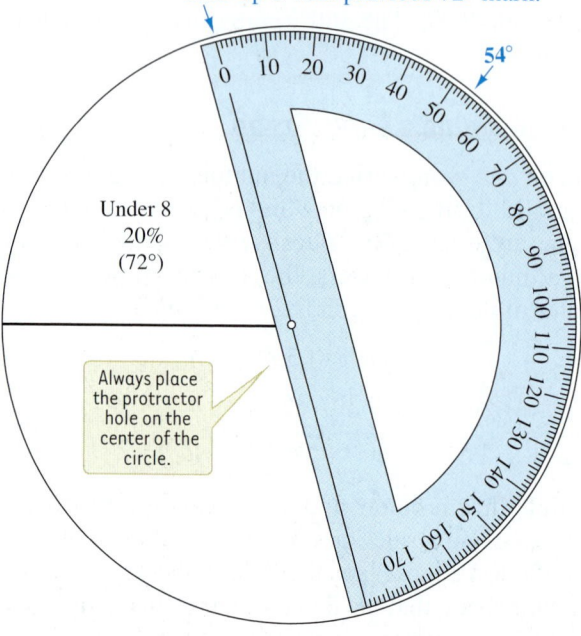

Line up 0 with previous 72° mark.

Always place the protractor hole on the center of the circle.

CAUTION

You must be certain that the hole in the protractor is placed on the exact center of the circle each time you measure the size of a sector.

◀ *Work Problem* **5** *at the Side.*

Use this circle for Problem 5 in the margin.

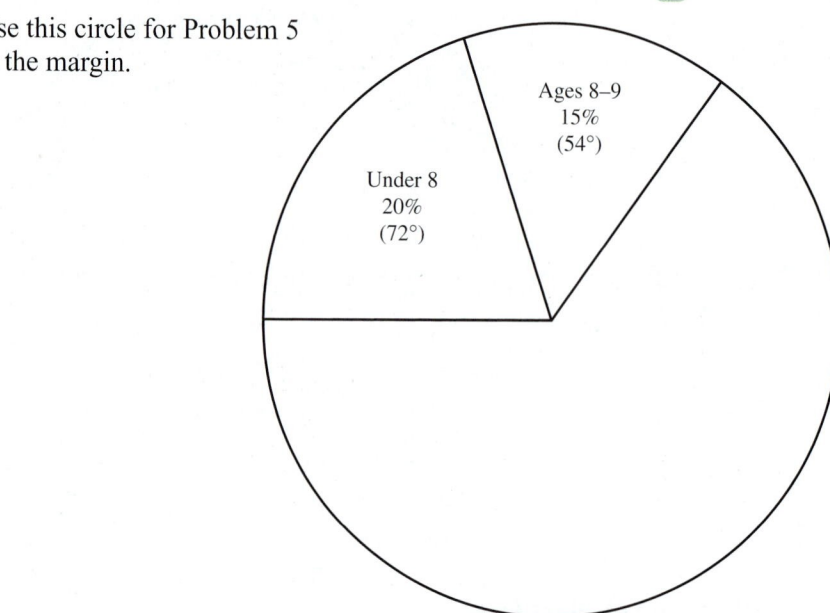

Ages 8–9
15%
(54°)

Under 8
20%
(72°)

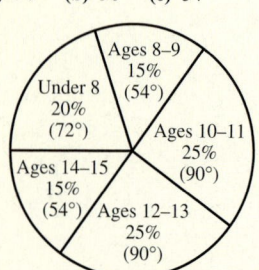

This circle graph shows the budget for adding a family room to an existing home. Use the graph to answer Exercises 1–6. Write ratios as fractions in lowest terms. See Examples 1 and 2.

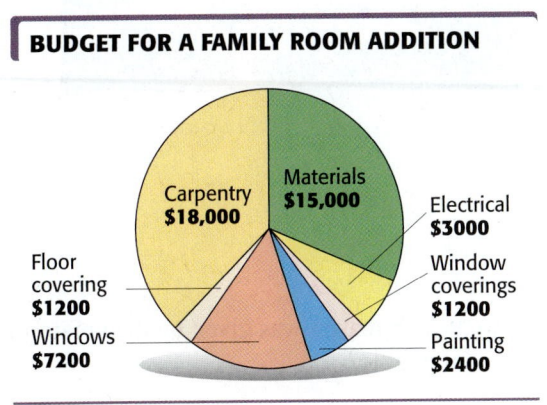

BUDGET FOR A FAMILY ROOM ADDITION

Carpentry $18,000
Materials $15,000
Electrical $3000
Floor covering $1200
Window coverings $1200
Windows $7200
Painting $2400

1. **(a)** Find the total budget for adding the family room.

   (b) What is the greatest single expense, and how much is it?

2. **(a)** What is the second-greatest expense in adding the family room, and how much is it?

   (b) What is least expensive?

3. **(a)** Find the ratio of the amount for carpentry to the total remodeling budget.

   (b) What is the ratio of the amount for materials to the amount for electrical.

4. **(a)** Find the ratio of the amount for painting to the total remodeling budget.

   (b) Find the ratio of the amount for windows to the amount for window coverings.

5. **(a)** Find the ratio of the amount for floor covering, painting, and window coverings to the total remodeling budget.

   (b) If the actual expenses for adding the family room amounted to $57,600, what is the ratio of the actual total to the budget total?

6. **(a)** Find the ratio of the amount for carpentry and electrical to the amount for materials.

   (b) The homeowner found a sale on carpet and spent only $900 to cover the floor. What is the ratio of the carpet cost to the budget amount for floor covering?

This circle graph, adapted from USA Today, *shows the number of people in a survey who gave various reasons for eating dinner at restaurants. Each person could pick only one reason. Use the graph to answer Exercises 7–14. See Examples 1 and 2.*

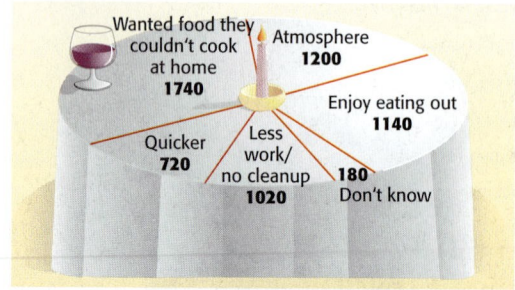

ON THE TOWN

When asked in a survey why they ate dinner in restaurants, a group of people gave these reasons.

Wanted food they couldn't cook at home **1740**

Atmosphere **1200**

Enjoy eating out **1140**

Quicker **720**

Less work/ no cleanup **1020**

180 Don't know

Source: Market Facts for Tyson Foods.

7. (a) Which reason was given by the least number of people?

(b) Which reason was given by the second-fewest number of people?

8. (a) Which reason was given by the greatest number of people?

(b) Which reason was given by the second-greatest number of people?

Find each ratio in Exercises 9–14. Write the ratios as fractions in lowest terms.

9. Those who said dining out is "Quicker" to total people in the survey

10. Those who said "Enjoy eating out" to the total people in the survey

11. Those who said "Less work/no cleanup" to those who said "Atmosphere"

12. Those who said "Don't know" to those who said "Quicker"

13. Those who said "Wanted food they couldn't cook at home" to those who said "Don't know"

14. Those who said "Atmosphere" to those who said "Enjoy eating out"

This circle graph shows the results of an on-line survey of 400 people who have a cat as a pet. They answered the question, "How does your cat react when you entertain?" The responses are expressed as percents of the 400 people in the survey. Use the graph to answer Exercises 15–20. See Example 3.

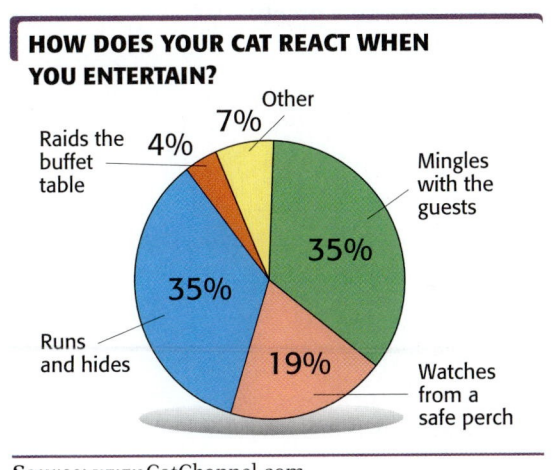

HOW DOES YOUR CAT REACT WHEN YOU ENTERTAIN?

Other 7%
Raids the buffet table 4%
Mingles with the guests 35%
Runs and hides 35%
Watches from a safe perch 19%

Source: www.CatChannel.com

15. How many people said their cat mingles with the guests?

16. How many cat owners said their cats watch from a safe perch?

17. Which response was given least often and by how many people?

18. Which response category represented 7% of the people surveyed? How many people were in this category?

19. How many fewer people said "watches from a safe perch" than said "runs and hides"?

20. How many more people said "mingles with the guests" than said "raids the buffet table"?

This circle graph shows the results of a survey on favorite hot dog toppings in the United States. Each topping is expressed as a percent of the 3200 people in the survey. Use the graph to find the number of people favoring each of the toppings in Exercises 21–26.

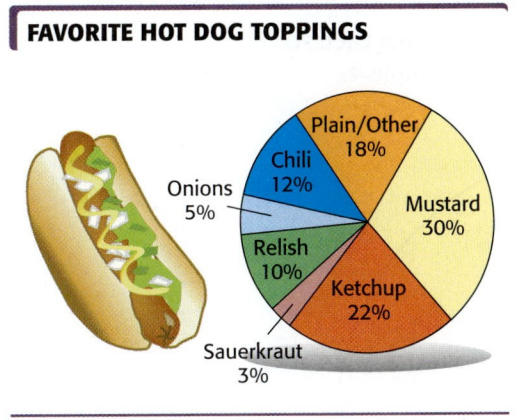

FAVORITE HOT DOG TOPPINGS

Plain/Other 18%
Chili 12%
Onions 5%
Mustard 30%
Relish 10%
Ketchup 22%
Sauerkraut 3%

Source: National Hot Dog and Sausage Council.

21. Onions

22. Sauerkraut

23. The most popular topping

24. The second-most-popular topping

25. How many more people chose chili than chose relish?

26. How many fewer people chose onions than chose mustard?

27. Describe the steps for calculating the size of the sector for each item in a circle graph.

28. A protractor is the tool used to draw a circle graph. Give a brief explanation of what the protractor does and how you would use it to measure and draw each sector in the circle graph.

29. During one semester Kara Diano spent $10,920 for school expenses as shown in this table. Find all numbers missing from the table.

Item	Dollar Amount	Percent of Total	Degrees of a Circle
(a) Rent	$2730	25%	_____
(b) Food	$2184	_____	72°
(c) Clothing	$1092	_____	_____
(d) Books and supplies	$1092	_____	_____
(e) Tuition and fees	$1638	_____	_____
(f) Savings	$ 546	_____	_____
(g) Entertainment	$1638	_____	_____

(h) Draw a circle graph using the budget information from the table. Label each sector with the budget item and the percent of total for that item. See Example 4.

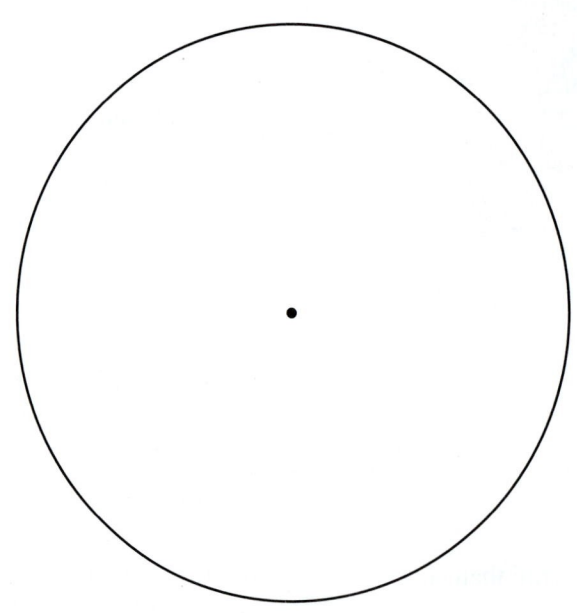

30. The Pathfinder Research Group asked 4488 Americans how they fall asleep. The results are shown in the figure on the right.

Use this information to complete the table. Round to the nearest whole percent and to the nearest degree.

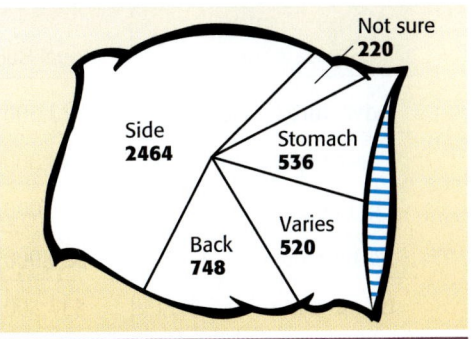

SET TO SLEEP

Number of Americans surveyed who fall asleep on their:

Source: Pathfinder Research Group.

	Sleeping Position	Number of Americans	Percent of Total	Number of Degrees
(a)	Side	_____	_____	_____
(b)	Back	_____	_____	_____
(c)	Stomach	_____	_____	_____
(d)	Varies	_____	_____	_____
(e)	Not sure	_____	_____	_____

(f) Add up the percents. Is the total 100%? Explain why or why not.

(g) Add up the degrees. Is the total 360°? Explain why or why not.

(h) Draw a circle graph using the information from the table above. Label each sector with the sleeping position and the percent of total for that position.

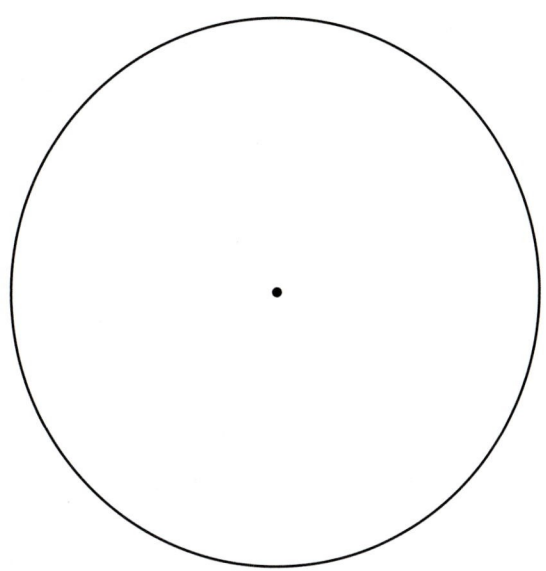

31. White Water Rafting Company divides its annual sales into five categories as shown in the table below.

Category	Annual Sales	Percent of Total
Adventure classes	$12,500	_____
Grocery/provision sales	$40,000	_____
Equipment rentals	$60,000	_____
Rafting tours	$50,000	_____
Equipment sales	$37,500	_____

(a) Find the total sales for the year.

(b) Find the percent of the total for each category and write it in the table.

(c) Find the number of degrees in a circle graph for each item.

(d) Make a circle graph showing the sales information. Label each sector with the category and the percent of total for that category.

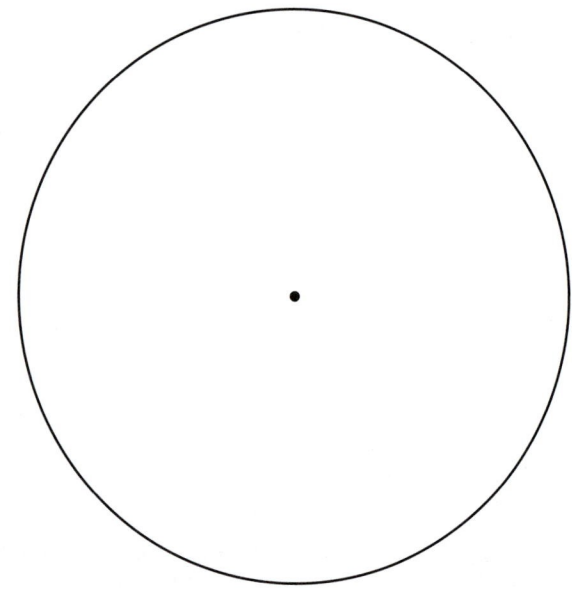

32. A book publisher had 25% of total sales in mysteries, 10% in biographies, 15% in cookbooks, 15% in romance novels, 20% in science, and the rest in travel books.

(a) Find the number of degrees in a circle graph for each type of book.

(b) Draw a circle graph, using the information given. Label each sector with the type of book and the percent of total for that type.

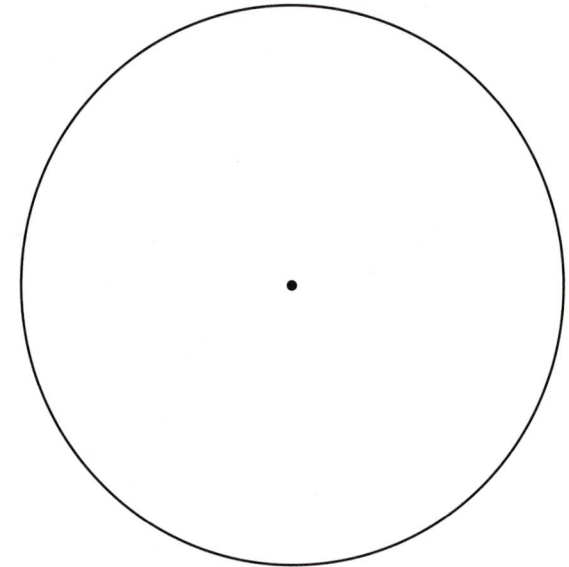

9.3 ▶▶▶ Bar Graphs and Line Graphs

OBJECTIVE 1 Read and understand a bar graph. A bar graph is useful for showing comparisons. For example, the bar graph below compares the number of college graduates who continued taking advanced courses in their major field during each of five years.

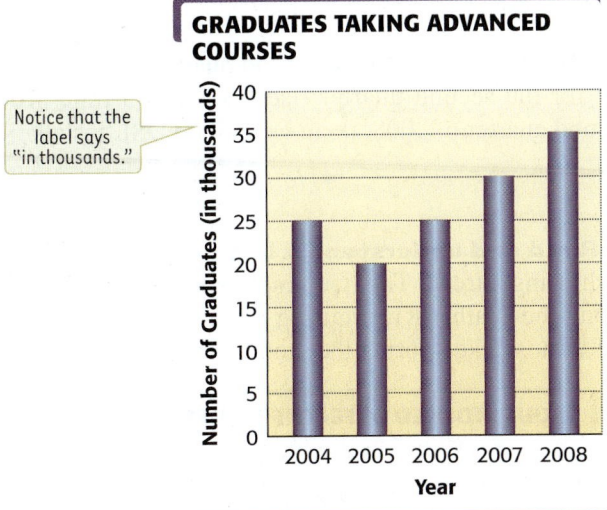

GRADUATES TAKING ADVANCED COURSES

Notice that the label says "in thousands."

OBJECTIVES

Read and understand

1 a bar graph;

2 a double-bar graph;

3 a line graph;

4 a comparison line graph.

EXAMPLE 1 Using a Bar Graph

How many college graduates took advanced courses in their major field in 2006?

The bar for 2006 rises to 25. Notice the label along the left side of the graph that says "Number of Graduates (in thousands)." The phrase *in thousands* means you have to **multiply 25 by 1000** to get 25,000. So, 25,000 (**not** 25) graduates took advanced courses in their major field in 2006.

Work Problem **1** *at the Side.* ▶

OBJECTIVE 2 Read and understand a double-bar graph. A **double-bar graph** can be used to compare two sets of data. The graph below shows the number of DSL (digital subscriber line) installations each quarter for two different years.

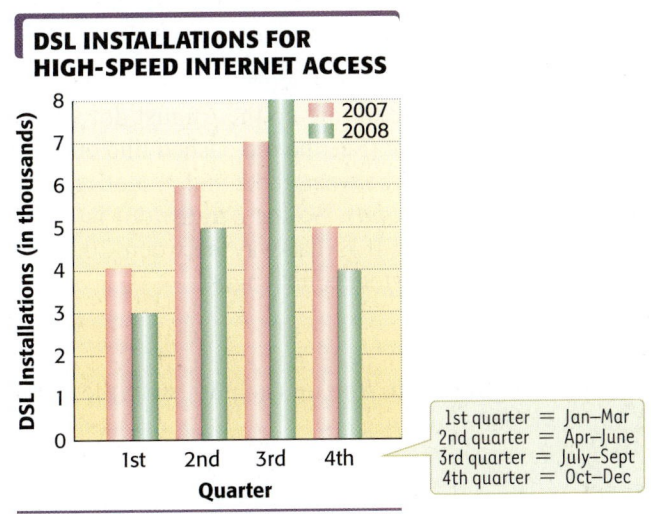

DSL INSTALLATIONS FOR HIGH-SPEED INTERNET ACCESS

1st quarter = Jan–Mar
2nd quarter = Apr–June
3rd quarter = July–Sept
4th quarter = Oct–Dec

1 Use the bar graph at the left to find the number of college graduates who took advanced courses in their major field in each of these years.

(a) 2004

(b) 2005

(c) 2007

(d) 2008

ANSWERS

1. **(a)** 25,000 graduates **(b)** 20,000 graduates
   (c) 30,000 graduates **(d)** 35,000 graduates

2 Use the double-bar graph on the previous page to find the number of DSL installations in 2007 and 2008 for each quarter.

(a) 1st quarter

(b) 3rd quarter

(c) 4th quarter

(d) Identify the quarter and year with the greatest number of installations. How many installations were made?

3 Use the line graph at the right to answer these questions.

(a) How many trout were stocked in June?

(b) How many fewer trout were stocked in May than in April?

(c) Find the total number of trout stocked during the five months.

(d) In which month were the most trout stocked? How many were stocked?

EXAMPLE 2 **Reading a Double-Bar Graph**

Use the double-bar graph on the previous page to find the following.

(a) The number of DSL installations in the second quarter of 2007.

There are two bars for the second quarter. The color code in the upper right-hand corner of the graph tells you that the **red bars** represent 2007. So the **red bar** on the *left* is for the 2nd quarter of 2007. It rises to 6. Multiply 6 by 1000 because the label on the left side of the graph says *in thousands*. So there were 6000 DSL installations for the second quarter in 2007.

(b) The number of DSL installations in the second quarter of 2008.

The **green bar** for the second quarter rises to 5 and 5 times 1000 is 5000. So, in the second quarter of 2008, there were 5000 DSL installations.

◀ *Work Problem* **2** *at the Side.*

OBJECTIVE **3** **Read and understand a line graph.** A **line graph** is often useful for showing a trend. The line graph below shows the number of trout stocked along the Feather River during five months. Each dot indicates the number of trout stocked during the month directly below that dot.

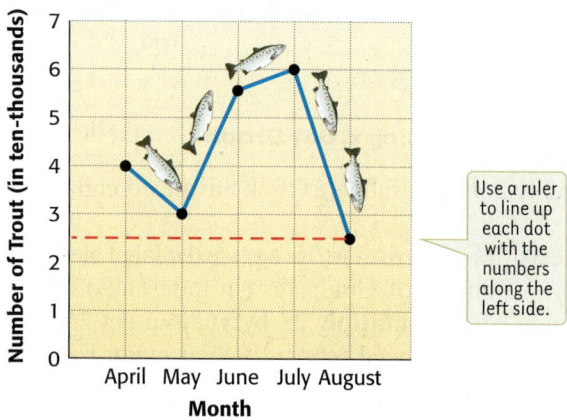

TROUT STOCKED IN FEATHER RIVER

Use a ruler to line up each dot with the numbers along the left side.

EXAMPLE 3 **Understanding a Line Graph**

Use the line graph above to answer each question.

(a) In which month were the least number of trout stocked?

The lowest point on the graph is the dot directly above August, so the least number of trout were stocked in August.

(b) How many trout were stocked in August?

Use a ruler or straightedge to line up the August dot with the numbers along the left edge of the graph. (See the red dashed line on the graph.)

The August dot is halfway between the 2 and 3. Notice that the label on the left side says *in ten-thousands*. So the August dot is halfway between (2 • 10,000) and (3 • 10,000). It is halfway between 20,000 and 30,000. That means 25,000 trout were stocked in August.

> **CAUTION**
> Use a ruler or straightedge to line up each dot with the number on the left side of the graph.

◀ *Work Problem* **3** *at the Side.*

ANSWERS

2. **(a)** 4000; 3000 installations
   (b) 7000; 8000 installations
   (c) 5000; 4000 installations
   (d) 3rd quarter of 2008; 8000
3. **(a)** 55,000 trout   **(b)** 10,000 fewer trout
   (c) 210,000 trout   **(d)** July; 60,000 trout

OBJECTIVE **4** **Read and understand a comparison line graph.**
Two sets of data can also be compared by drawing two line graphs together as a **comparison line graph.** For example, the line graph below compares the number of minivans sold and the number of Sport Utility Vehicles (SUVs) sold during each of five years.

SALES OF MINIVANS AND SUVS

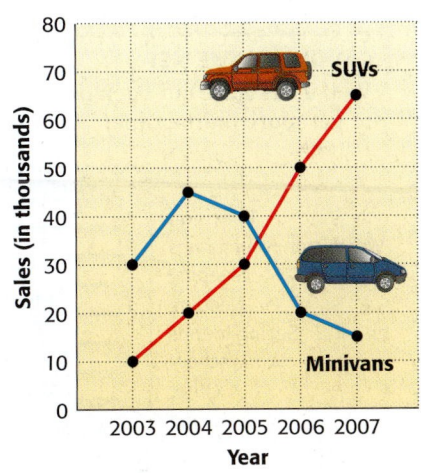

EXAMPLE 4 **Interpreting a Comparison Line Graph**

Use the comparison line graph above to find the following.

(a) The number of minivans sold in 2004
Find the dot on the **blue line** above 2004. Use a ruler or straightedge to line up the dot with the numbers along the left edge. The dot is halfway between 40 and 50, which is 45. Then, 45 times 1000 is 45,000 minivans sold in 2004.

(b) The number of SUVs sold in 2007
The **red line** on the graph shows that 65,000 SUVs were sold in 2007.

(c) The amount of decrease and the percent of decrease in minivan sales from 2004 to 2005. Round to the nearest whole percent.

Use subtraction to find the *amount* of decrease.

minivan sales in 2004	−	**minivan sales in 2005**	=	**amount of decrease**
45,000	−	40,000	=	5000

Now use the percent equation to find the *percent* of decrease.

percent **of** original sales = amount of decrease

$$p \cdot 45{,}000 = 5000$$

$$\frac{p \cdot \overset{1}{\cancel{45{,}000}}}{\underset{1}{\cancel{45{,}000}}} = \frac{5000}{45{,}000} \quad \text{Divide both sides by 45,000}$$

$$p = 0.11\overline{1} \leftarrow \text{Solution in } decimal \text{ form}$$

> **Multiply by 100%** to change the decimal to a percent.

$$p = 0.11\overline{1} = 11.\overline{1}\% \approx 11\% \text{ to nearest whole percent}$$

The *amount* of decrease is 5000 minivans.
The **percent** of decrease is 11% (rounded).

Work Problem **4** *at the Side.* ▶

4 Use the comparison line graph at the left to find the following.

(a) The number of minivans sold in 2003, 2005, 2006, and 2007

(b) The number of SUVs sold in 2003, 2004, 2005, and 2006

(c) The first full year in which the number of SUVs sold was greater than the number of minivans sold

(d) The amount of increase and percent of increase in SUV sales from 2005 to 2006; round to the nearest whole percent

Math in the Media

SURFING THE NET

1. Look at the "Source" information at the bottom of the graph. How were the numbers in the graph obtained?

2. (a) How many people in the poll said they cut back on television viewing to find time to use the Internet?

   (b) Find the number of people in the poll who cut back on each of the other activities listed in the graph.

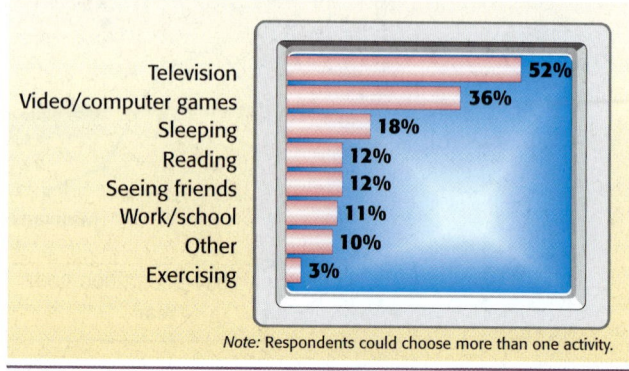

CAUGHT IN THE NET
Web users have cut back on the following activities to get more online time:

Activity	Percent
Television	52%
Video/computer games	36%
Sleeping	18%
Reading	12%
Seeing friends	12%
Work/school	11%
Other	10%
Exercising	3%

Note: Respondents could choose more than one activity.

Source: NUKE InterNETWORK poll of 500 regular users.

3. Add up all the responses to the poll from Problem 2. Why is the total more than the 500 people that were in the poll?

4. Suppose you took a poll on Internet use among 500 students at your school. Would you expect the results to be similar to those shown in the graph? Why or why not?

5. Conduct a survey of your class members. First find out if they regularly use the Internet, then ask which of the activities they cut back on to have more surfing time. Each person polled can select more than one activity.

   (a) How many students are in your class poll? _____

   (b) Complete the table using the responses from those who regularly use the Internet.

Activity	Number Who Cut Back on the Activity	Percent Who Cut Back on the Activity
Television		
Video/computer games		
Sleeping		
Reading		
Seeing friends		
Work/school		
Other		
Exercising		

   (c) Make a bar graph showing your survey data. How is your data similar to the graph shown above? How is it different?

FOR EXTRA HELP

This bar graph shows the top seven reasons people say they shop on-line. Use the graph to answer Exercises 1–6. See Example 1.

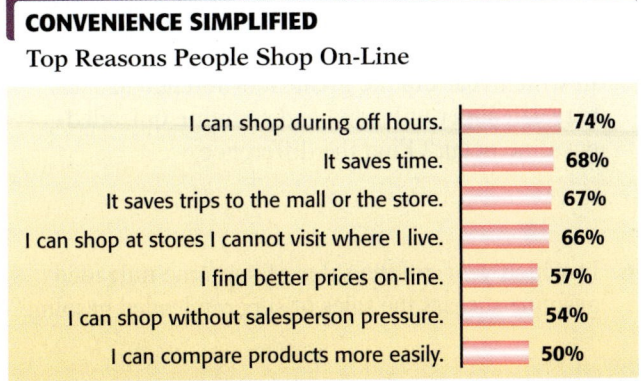

CONVENIENCE SIMPLIFIED

Top Reasons People Shop On-Line

I can shop during off hours.	74%
It saves time.	68%
It saves trips to the mall or the store.	67%
I can shop at stores I cannot visit where I live.	66%
I find better prices on-line.	57%
I can shop without salesperson pressure.	54%
I can compare products more easily.	50%

Source: EMARKETER.

1. What is the top reason people shop on-line? What percent gave this reason?

2. What is the second most popular reason for shopping on-line? What percent gave this reason?

3. What percent of the people say they find better prices on-line? If 600 people were surveyed, how many gave this answer?

4. What percent say it saves trips to the mall or store? If 600 people were surveyed, how many gave this answer?

5. Which reason(s) were given by $\frac{1}{2}$ of the people? Which reason(s) were given by nearly $\frac{3}{4}$ of the people?

6. Which reason(s) were given by about $\frac{2}{3}$ of the people?

This double-bar graph shows the number of workers who were unemployed in a city during the first six months of 2007 and 2008. Use this graph to answer Exercises 7–12. Round percents to the nearest whole number. See Example 2.

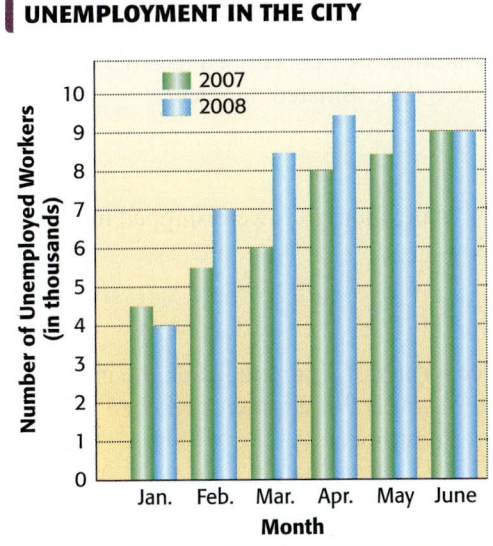

UNEMPLOYMENT IN THE CITY

7. In the first half of 2008, which month had the greatest number of unemployed workers? What was the total number unemployed in that month?

8. How many workers were unemployed in January of 2007?

9. How many more workers were unemployed in February of 2008 than in February of 2007?

10. How many fewer workers were unemployed in March of 2007 than in March of 2008?

11. Find the amount of increase and the percent of increase in the number of unemployed workers from February 2007 to April 2007.

12. Find the amount of increase and the percent of increase in the number of unemployed workers from January 2008 to June 2008.

This double-bar graph shows sales of super unleaded and supreme unleaded gasoline at a service station for each of five years. Use this graph to answer Exercises 13–18. See Example 2.

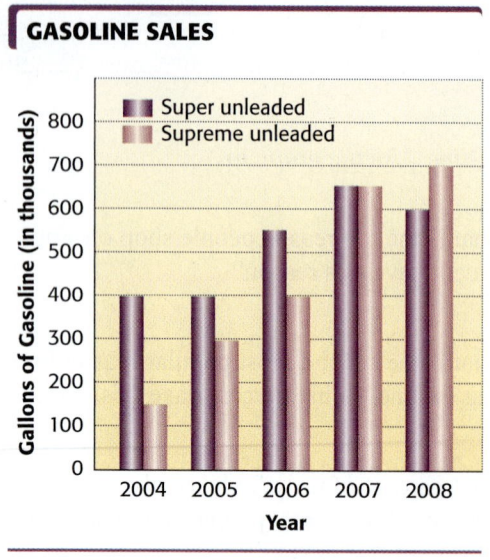

GASOLINE SALES

13. How many gallons of supreme unleaded gasoline were sold in 2004?

14. How many gallons of super unleaded gasoline were sold in 2007?

15. In which year did the greatest difference in sales between super unleaded and supreme unleaded gasoline occur? Find the difference.

16. In which year did the sales of supreme unleaded gasoline surpass the sales of super unleaded gasoline?

17. Find the amount of increase and percent of increase in supreme unleaded gasoline sales from 2004 to 2008. Round to the nearest whole percent.

18. Find the amount of increase and percent of increase in super unleaded gasoline sales from 2004 to 2008.

This line graph shows how sales of personal computers (PCs) have increased since they first became widely available in 1985. Use the line graph to answer Exercises 19–24. Round percents to the nearest whole percent. See Example 3.

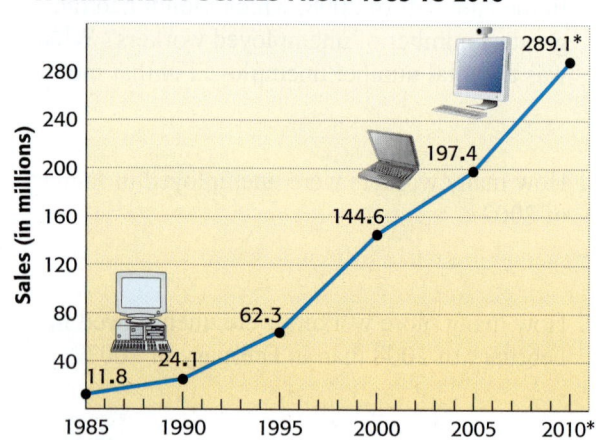

A QUARTER CENTURY OF COMPUTING
WORLDWIDE PC SALES FROM 1985 TO 2010

*Estimated

Source: Gartner Dataquest and ARS Technica.

19. How many PCs were shipped in 1990?

20. The number of PCs shipped in 2010 is an estimate. What is the estimated number?

21. How many more PCs were shipped in 2005 than in 1985?

22. Place a ruler or straightedge on the graph to find the year in which PC shipments reached **(a)** 40 million; **(b)** 120 million.

23. Find the amount of increase in shipments from 1995 to 2000. What was the percent of increase, to the nearest whole percent?

24. What was the amount of increase and the percent of increase in PC shipments from 1985 to 1990?

This comparison line graph shows the number of DVDs sold by two different chain stores during each of five years. Use this graph to find the annual number of DVDs sold in each year listed in Exercises 25–28. See Example 4.

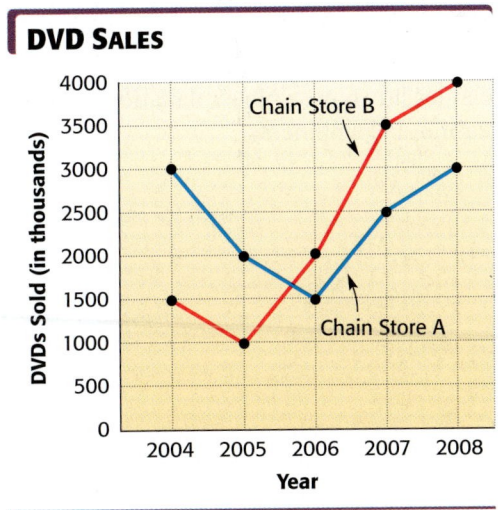

DVD SALES

25. **(a)** Chain Store A in 2004
    (b) Chain Store B in 2004

26. **(a)** Chain Store A in 2005
    (b) Chain Store B in 2005

27. **(a)** Chain Store A in 2007
    (b) Chain Store A in 2008

28. **(a)** Chain Store B in 2006
    (b) Chain Store B in 2007

29. Describe the pattern(s) or trend(s) you see in the graph.

30. Store B used to have lower sales than Store A. What might have happened to cause this change? Give four possible explanations.

This comparison line graph shows the sales and profits of Tacos-To-Go for each of four years. Use the graph to answer Exercises 31–36.

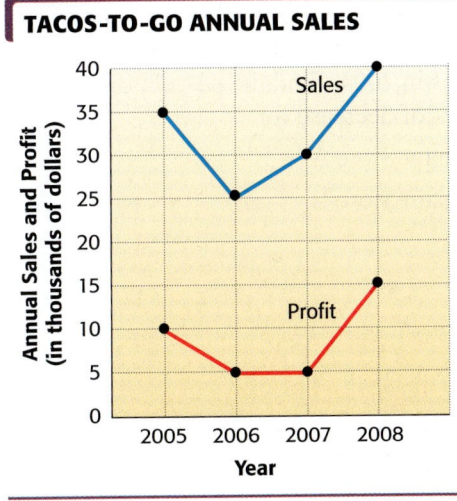

TACOS-TO-GO ANNUAL SALES

31. Find the year of lowest sales and the amount of sales.

32. Find the year of greatest profit and the amount of profit.

33. For each of the four years, the profit is what percent of sales? Round to the nearest whole percent.

34. Find the total sales for all four years and the total profit for all four years. Total profit is what percent of total sales? Round to the nearest whole percent.

35. Give two possible explanations for the decrease in sales from 2005 to 2006 and two possible explanations for the increase in sales from 2006 to 2008.

36. *Based on the graph,* what conclusion can you make about the relationship between sales and profits for Tacos-To-Go?

Relating Concepts (Exercises 37–44) For Individual or Group Work

*Find the line graph on Worldwide Shipments of PCs and the double-line graph on DVD Sales earlier in this exercise set. Use the graphs as you **work Exercises 37–44 in order.** Round percent answers to the nearest whole percent.*

37. What overall trend do you see in the line graph on worldwide shipments of Personal Computers (PCs)?

38. Give at least two possible explanations for the increase in PC shipments.

39. Look back at your earlier answers or find the percent of increase in PC shipments from

   (a) 1985 to 1990 (see Exercise 24)

   (b) 1990 to 1995

   (c) 1995 to 2000 (see Exercise 23)

   (d) 2000 to 2005

   (e) 2005 to 2010.

40. Look at the percents of increase in Exercise 39. What trend do you see?

41. What future conditions could result in a decrease in PC shipments? Give at least two possibilities.

42. Look back at the double-line graph of DVD sales. For each store, compare the sales of DVDs in 2008 to sales in 2004. Find the percent of increase or the percent of decrease for:

   (a) Chain Store A

   (b) Chain Store B.

43. *Based on the graph,* what amount of DVD sales would you predict for

   (a) Store A in 2009?

   (b) Store B in 2009?

   (c) Explain how you arrived at your predictions.

44. (a) *Based on the graph,* which store would you like to own? Explain why.

   (b) Name at least three other things you would want to know before deciding which store to buy.

9.4 ▶▶▶ The Rectangular Coordinate System

OBJECTIVE **1** **Plot a point, given the coordinates, and find the coordinates, given a point.** A bar graph or line graph shows the relationship between two things. The line graph below is from Example 3 in **Section 9.3.** It shows the relationship between the month of the year and the number of trout stocked in the Feather River.

OBJECTIVES

1 Plot a point, given the coordinates, and find the coordinates, given a point.

2 Identify the four quadrants and determine which points lie within each one.

TROUT STOCKED IN FEATHER RIVER

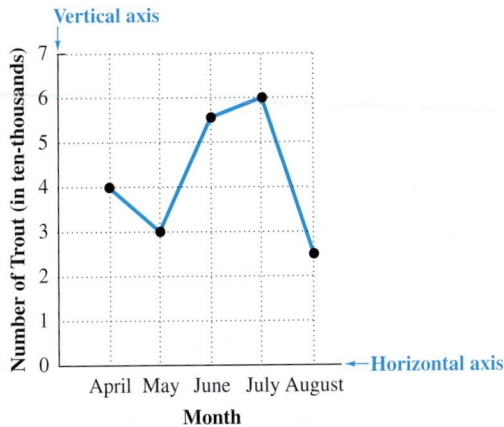

Each black dot on the graph represents a particular month paired with a particular number of trout. This is an example of **paired data.** We write each pair inside parentheses, with a comma separating the two items. To be consistent, we will always list the item on the *horizontal axis* first. In this case, the months are shown on the **horizontal axis** (the line that goes "left and right"), and the number of trout is shown along the **vertical axis** (the line that goes "up and down").

Paired Data from Line Graph on Trout Stocked in Feather River

(Apr, 40,000) (May, 30,000) (June, 55,000) (July, 60,000) (Aug, 25,000)

Each data pair gives you the location of a particular spot on the graph, and that spot is marked with a dot. This idea of paired data can be used to locate particular places on any flat surface.

Think of a small town laid out in a grid of square blocks, as shown below. To tell a taxi driver where to go, you could say, "the corner of 4th Avenue and 2nd Street" or just "4th and 2nd." As an *ordered pair,* it would be (4, 2). Of course, both you and the taxi driver need to know that the avenue is mentioned first (the number on the horizontal axis) and that the street is mentioned second (the number on the vertical axis). If the driver goes to (2, 4) instead, you'll be at the wrong corner.

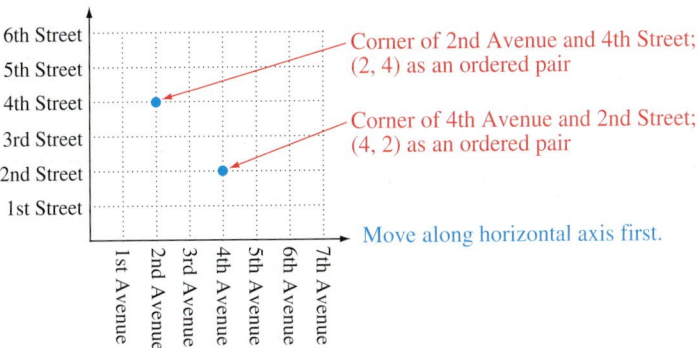

1 Plot each point on the grid. Write the ordered pair next to each point.

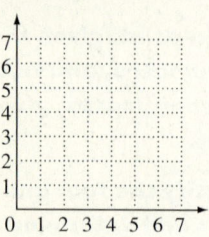

(a) (1, 4)

(b) (5, 2)

(c) (4, 1)

(d) (3, 3)

EXAMPLE 1 **Plotting Points on a Grid**

Use the grid at the right to plot each point.

(a) (3, 5)

Start at 0. Move *to the right* along the horizontal axis until you reach 3. Then move *up* 5 units so that you are aligned with 5 on the vertical axis. Make a dot. This is the plot, or graph, of the point (3, 5).

(b) (5, 3)

Start at 0. Move *to the right* along the horizontal axis until you reach 5. Then move *up* 3 units so that you are aligned with 3 on the vertical axis. Make a dot. This is the plot, or graph, of the point (5, 3).

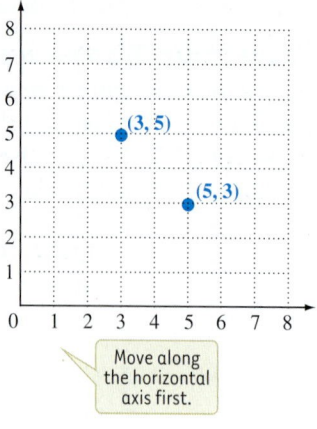

Move along the horizontal axis first.

CAUTION

In Example 1, notice that the points (3, 5) and (5, 3) are ***not*** the same. The "address" of a point is called an **ordered pair** because the *order* within the pair is important. Always move along the horizontal axis first.

◀ *Work Problem* **1** *at the Side.*

You have been using both positive and negative numbers throughout this book. We can extend our grid system to include negative numbers, as shown below. The horizontal axis is now a number line with 0 at the center, positive numbers extending to the right and negative numbers to the left. This horizontal number line is called the ***x*-axis.**

The vertical axis is also a number line, with positive numbers extending upward from 0 and negative numbers extending downward from 0. The vertical axis is called the ***y*-axis.** Together, the *x*-axis and the *y*-axis form a rectangular **coordinate system.** The point (0, 0) is where the *x*-axis crosses the *y*-axis; it is called the **origin.**

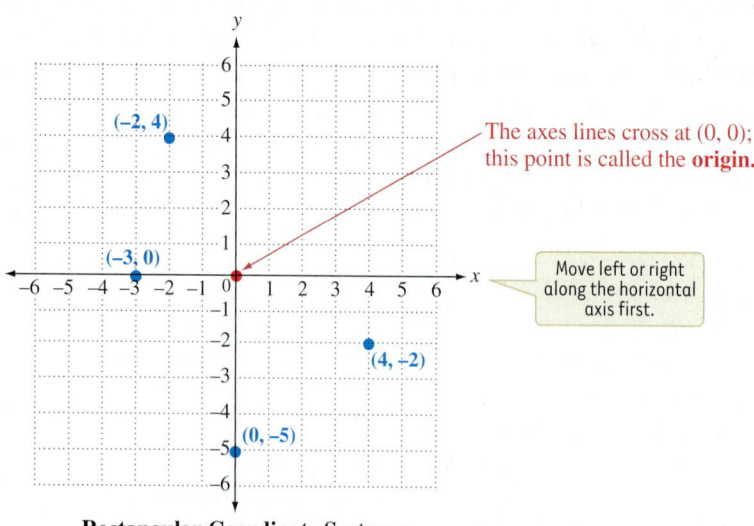

The axes lines cross at (0, 0); this point is called the **origin.**

Move left or right along the horizontal axis first.

Rectangular Coordinate System

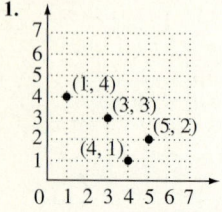

EXAMPLE 2 Plotting Points on a Rectangular Coordinate System

Plot each point on the rectangular coordinate system shown at the bottom of the previous page.

(a) (4, −2)

Start at 0. Then move left or right along the horizontal *x*-axis first. Because 4 is *positive,* move *to the right* until you reach 4. Now, because the 2 is *negative,* move *down* 2 units so that you are aligned with −2 on the *y*-axis. Make a dot and label it (4, −2).

(b) (−2, 4)

Starting at 0, move left or right along the horizontal *x*-axis first. In this case, move *to the left* until you reach −2. Then move *up* 4 units. Make a dot and label it (−2, 4). Notice that (−2, 4) is **not** the same as (4, −2).

(c) (0, −5)

Start at 0. Move left or right along the horizontal *x*-axis first. However, because the first number is 0, do not move left or right. Then move *down* 5 units. Make a dot and label it (0, −5).

(d) (−3, 0)

Starting at 0, move *to the left* along the horizontal *x*-axis to −3. Then, because the second number is 0, stay on −3. Do *not* move up or down. Make a dot and label it (−3, 0).

Note

When the *first* number in an ordered pair is 0, the point is on the *y*-axis, as in Example 2(c) above. When the *second* number in an ordered pair is 0, the point is on the *x*-axis, as in Example 2(d) above.

Work Problem **2** *at the Side.* ▶

We can use a coordinate system and an ordered pair to show the location of any point. The numbers in the ordered pair are called the **coordinates** of the point.

EXAMPLE 3 Finding the Coordinates of Points

Find the coordinates of points *A, B, C,* and *D*.

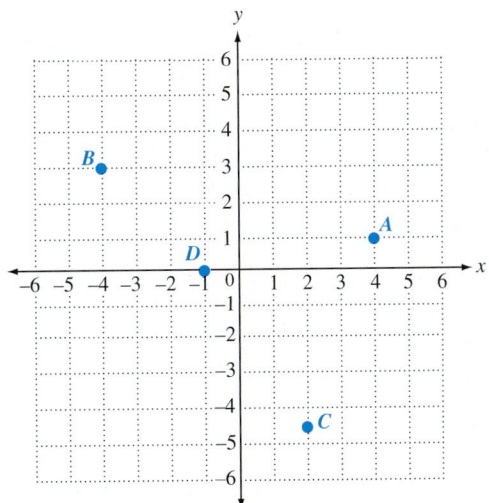

To reach point *A* from the origin, move 4 units *to the right;* then move *up* 1 unit. The coordinates are (4, 1).

Continued on Next Page

2 Plot each point on the coordinate system below. Write the ordered pair next to each point.

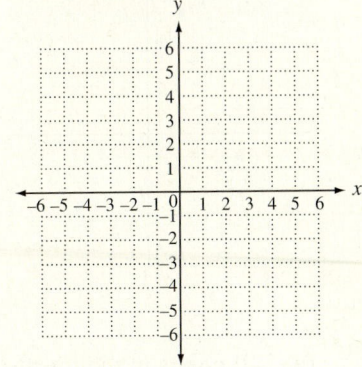

(a) (5, −3)

(b) (−5, 3)

(c) (0, 3)

(d) (−4, −4)

(e) (−2, 0)

ANSWERS

2.

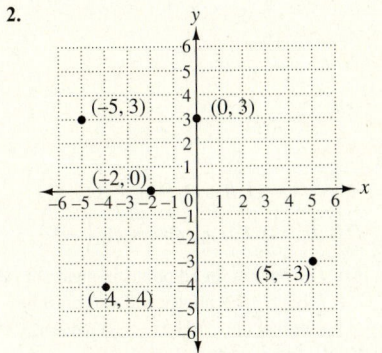

3 Find the coordinates of points *A*, *B*, *C*, *D*, and *E*.

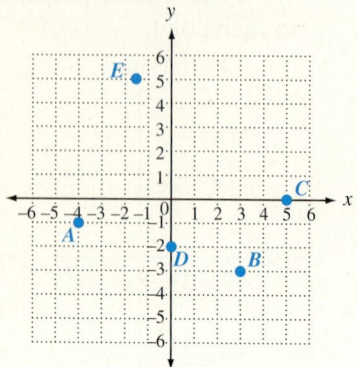

4 **(a)** All points in the fourth quadrant are similar in what way? Give two examples of points in the fourth quadrant.

(b) In which quadrant is each point located: $(-2, -6)$; $(0, 5)$; $(-3, 1)$; $(4, -1)$?

To reach point *B* from the origin, move 4 units *to the left;* then move *up* 3 units. The coordinates are $(-4, 3)$.

To reach point *C* from the origin, move 2 units *to the right;* then move *down* approximately $4\frac{1}{2}$ units. The approximate coordinates are $(2, -4\frac{1}{2})$.

To reach point *D* from the origin, move 1 unit *to the left;* then do *not* move either up or down. The coordinates are $(-1, 0)$.

> **Note**
>
> If a point is between the lines on the coordinate system, you can use fractions to give the approximate coordinates. For example, the approximate coordinates of point *C* above are $(2, -4\frac{1}{2})$.

◀ *Work Problem* **3** *at the Side.*

OBJECTIVE 2 Identify the four quadrants and determine which points lie within each one. The *x*-axis and *y*-axis divide the coordinate system into four regions, called **quadrants.** These quadrants are numbered with Roman numerals, as shown below. Points on the axes lines are not in any quadrant.

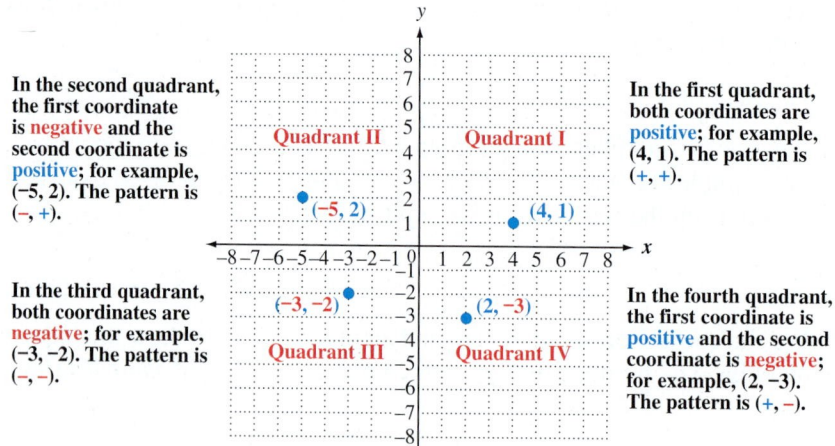

In the second quadrant, the first coordinate is **negative** and the second coordinate is **positive**; for example, $(-5, 2)$. The pattern is $(-, +)$.

In the first quadrant, both coordinates are **positive**; for example, $(4, 1)$. The pattern is $(+, +)$.

In the third quadrant, both coordinates are **negative**; for example, $(-3, -2)$. The pattern is $(-, -)$.

In the fourth quadrant, the first coordinate is **positive** and the second coordinate is **negative**; for example, $(2, -3)$. The pattern is $(+, -)$.

EXAMPLE 4 Working with Quadrants

(a) All points in the third quadrant are similar in what way? Give two examples of points in the third quadrant.

For all points in quadrant III, both coordinates are negative. The pattern is $(-, -)$. There are many possible examples, such as $(-2, -5)$ and $(-4, -4)$. Just be sure that both numbers are negative.

(b) In which quadrant is each point located: $(3, 5)$; $(1, -6)$; $(-4, 0)$?

For $(3, 5)$ the pattern is $(+, +)$, so the point is in **quadrant I.**

For $(1, -6)$ the pattern is $(+, -)$, so the point is in **quadrant IV.**

The point corresponding to $(-4, 0)$ is on the *x*-axis, so it isn't in any quadrant.

> When one of the numbers is 0, the point is on an axis line.

◀ *Work Problem* **4** *at the Side.*

9.4 ▶▶▶ Exercises

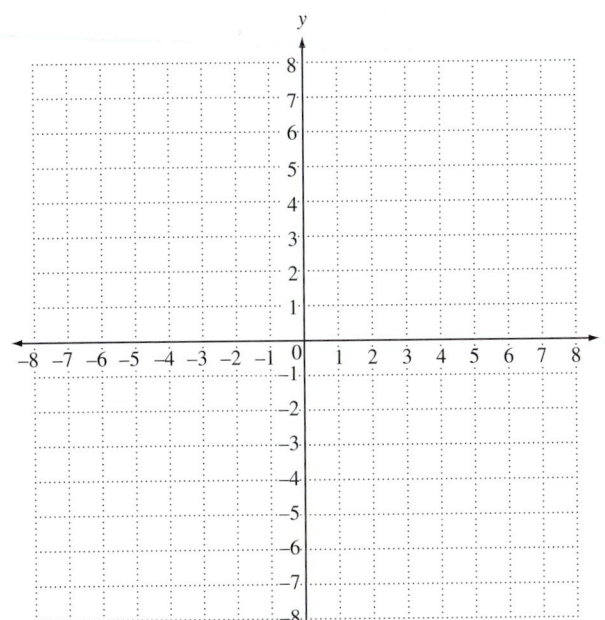

Plot each point on the rectangular coordinate system. Label each point with its coordinates. See Examples 1 and 2.

1. $(3, 7)$ $(-2, 2)$ $(-3, -7)$ $(2, -2)$ $(0, 6)$
 $(6, 0)$ $(0, -4)$ $(-4, 0)$

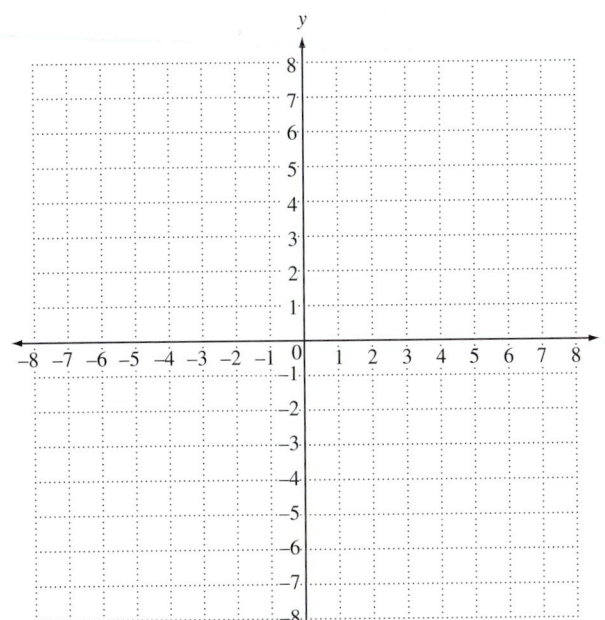

2. $(5, 2)$ $(-3, -3)$ $(4, -1)$ $(-4, 1)$ $(-1, 0)$
 $(0, 3)$ $(2, 0)$ $(0, -5)$

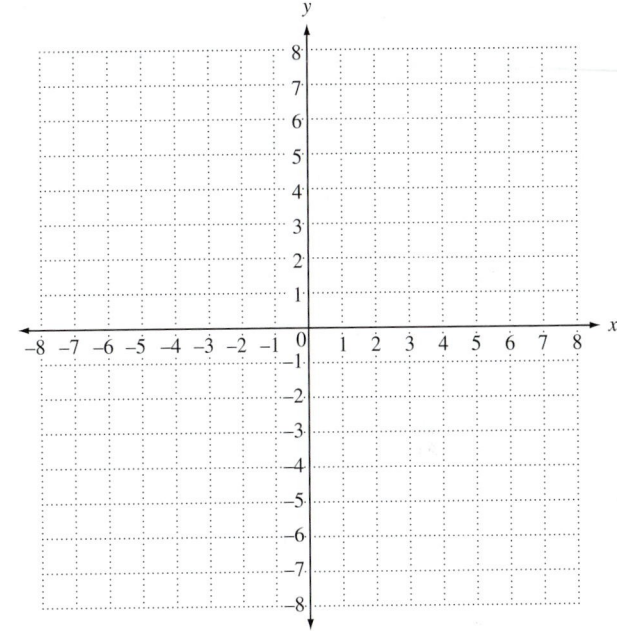

3. $(-5, 3)$ $(4, 4)$ $(-2\frac{1}{2}, 0)$ $(3, -5)$ $(0, 0)$
 $(2, \frac{1}{2})$ $(-7, -5)$ $(-1, -6)$

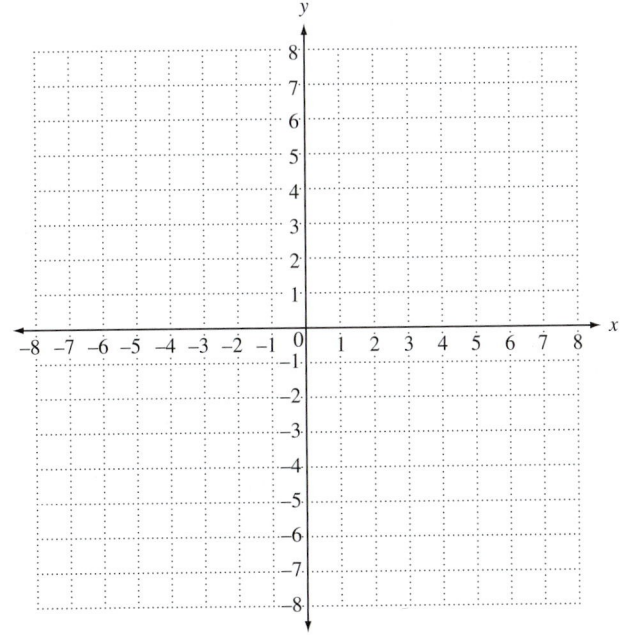

4. $(1, 7)$ $(0, 3\frac{1}{2})$ $(-5, -1)$ $(6, -2)$ $(-2, 6)$
 $(0, 0)$ $(-3, 3)$ $(-\frac{1}{2}, -2)$

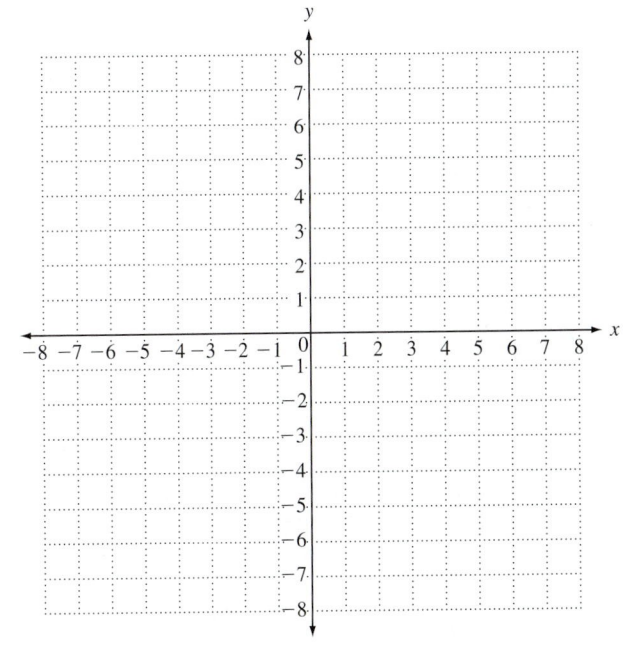

Give the coordinates of each point. See Example 3.

5.

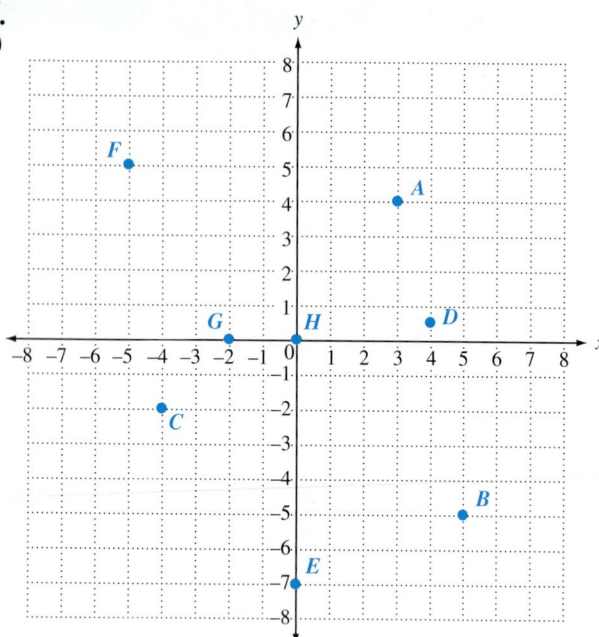

6.

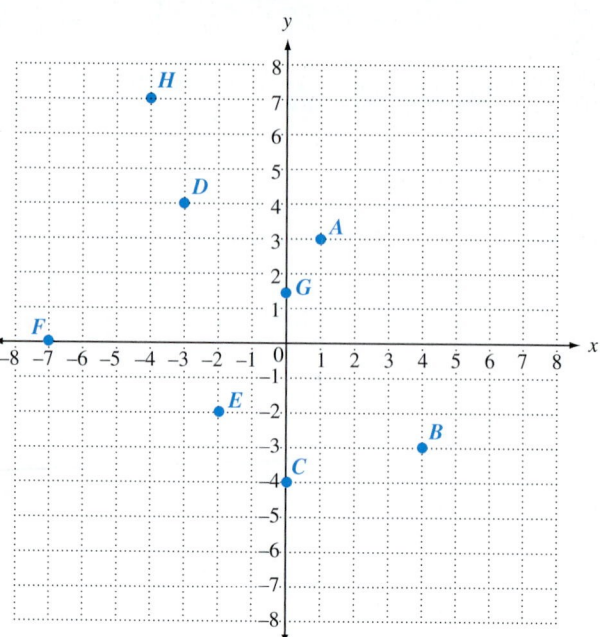

Identify the quadrant in which each point is located. See Example 4.

7. In which quadrant is each point located?
$(-3, -7)$ $(0, 4)$ $(10, -16)$ $(-9, 5)$

8. In which quadrant is each point located?
$(1, 12)$ $(20, -8)$ $(-5, 0)$ $(-14, 14)$

Complete each ordered pair with a number that will make the point fall in the specified quadrant.

9. (a) Quadrant II $(-4, \underline{\quad})$

   (b) Quadrant IV $(7, \underline{\quad})$

   (c) No quadrant $(\underline{\quad}, -2)$

   (d) Quadrant III $(\underline{\quad}, -1\frac{1}{2})$

   (e) Quadrant I $(3\frac{1}{4}, \underline{\quad})$

10. (a) Quadrant III $(-5, \underline{\quad})$

   (b) Quadrant I $(\underline{\quad}, 3)$

   (c) Quadrant IV $(\underline{\quad}, -\frac{1}{2})$

   (d) No quadrant $(6, \underline{\quad})$

   (e) Quadrant II $(\underline{\quad}, 1\frac{3}{4})$

11. Explain how to graph the ordered pair (a, b), where a and b are positive or negative integers.

12. Explain how to graph the ordered pair (a, b) where a is 0 and b is an integer. Explain how to graph (a, b) where a is an integer and b is 0.

9.5 ▶▶▶   Introduction to Graphing Linear Equations

In **Chapters 2–7** you solved equations that had only one variable, such as $2n - 3 = 7$ or $\frac{1}{3}x = 10$. Each of these equations had exactly one solution; n is 5 in the first equation, and x is 30 in the second equation. In other words, there was only *one* number that could replace the variable and make the equation balance. Later in this book, you will work with equations that have two variables and many different numbers that will make the equation balance. This section will get you started.

OBJECTIVE 1 Graph linear equations in two variables.   Suppose that you have 6 hours of study time available during a weekend. You plan to study math and psychology. For example, you could spend 4 hours on math and then 2 hours on psychology, for a total of 6 hours. Or you could spend $1\frac{1}{2}$ hours on math and then $4\frac{1}{2}$ hours on psychology, for a total of 6 hours. Here is a list of *some* of the possible combinations.

Hours on Math	+	Hours on Psychology	=	Total Hours Studying
0	+	6	=	6
1	+	5	=	6
$1\frac{1}{2}$	+	$4\frac{1}{2}$	=	6
3	+	3	=	6
4	+	2	=	6
$5\frac{1}{2}$	+	$\frac{1}{2}$	=	6
6	+	0	=	6

We can write an equation to represent this situation.

$$\underset{m}{\overset{\textbf{hours studying math}}{}} \quad + \quad \underset{p}{\overset{\textbf{hours studying psychology}}{}} \quad = \quad \overset{\text{total of 6 hours}}{6}$$

This equation, $m + p = 6$, has *two* variables. The hours spent on math (m) can vary, and the hours spent on psychology (p) can vary.

As you can see, there is more than one solution for this equation. We can list possible solutions as *ordered pairs*. The first number in the pair is the value of m (math), and the second number in the pair is the corresponding value of p (psychology).

$$(m, p) \quad (m, p) \quad (m, p) \quad (m, p) \quad (m, p) \quad (m, p) \quad (m, p)$$
$$\downarrow\downarrow \quad\quad \downarrow\downarrow \quad\quad \downarrow\downarrow \quad\quad \downarrow\downarrow \quad\quad \downarrow\downarrow \quad\quad \downarrow\downarrow \quad\quad \downarrow\downarrow$$
$$(0, 6) \quad (1, 5) \quad \left(1\frac{1}{2}, 4\frac{1}{2}\right) \quad (3, 3) \quad (4, 2) \quad \left(5\frac{1}{2}, \frac{1}{2}\right) \quad (6, 0)$$

Another way to show the solutions is to plot the ordered pairs, as you learned to do in **Section 9.4.** This method will give us a "picture" of the solutions that we listed on the previous page.

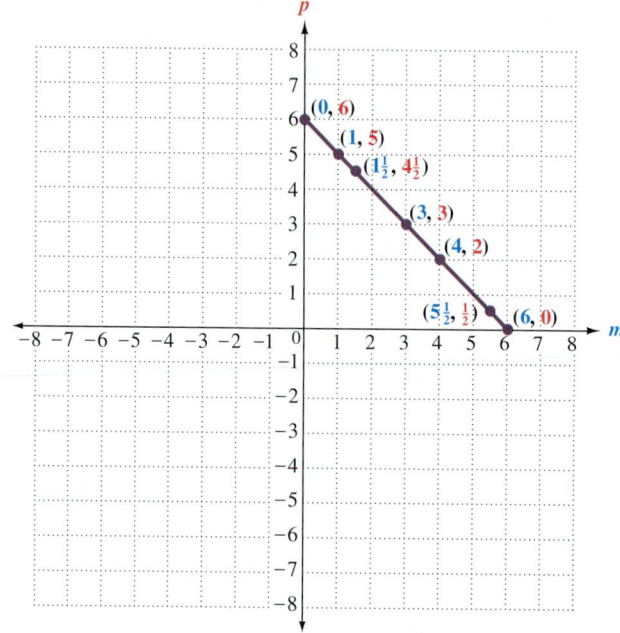

$m + p = 6$

The variables m and p represent hours, and hours can be 0 or positive numbers, but *not* negative numbers.

Notice that all the solutions (all the ordered pairs) lie on a straight line. When you draw a line connecting the ordered pairs, you have graphed the solutions. **Every point on the line is a solution.** You can use the line to find additional solutions besides the ones that we listed. For example, the point $(5, 1)$ is on the line. This point tells you that another solution is 5 hours on math and 1 hour on psychology. The fact that the line is a *straight* line tells you that $m + p = 6$ is a *linear equation.* (The word *line* is part of the word *line*ar.) Later on in algebra you will work with equations whose solutions form a curve rather than a straight line when you graph them.

To draw the line for $m + p = 6$, we really needed only two solutions (two ordered pairs). But it's a good idea to use a third ordered pair as a check. If the three ordered pairs are *not* in a straight line, there is an error in your work.

> **Graphing a Linear Equation**
>
> To **graph a linear equation,** find at least three ordered pairs that satisfy the equation. Then plot the ordered pairs on a coordinate system and connect them with a straight line. *Every* point on the line is a solution of the equation.

EXAMPLE 1 **Graphing a Linear Equation**

Graph $x + y = 3$ by finding three solutions and plotting the ordered pairs. Then use the graph to find a fourth solution of the equation.

There are many possible solutions. Start by picking three different values for x. You can choose any numbers you like, but 0 and small numbers usually are easy to use. Then find the value of y that will make the sum equal to 3. Set up a table to organize the information.

Continued on Next Page

	x	y	Check that $x + y = 3$	Ordered Pair (x, y)
Start by picking easy numbers for x.	0	3	$0 + 3 = 3$	$(0, 3)$
	1	2	$1 + 2 = 3$	$(1, 2)$
	2	1	$2 + 1 = 3$	$(2, 1)$

Plot the ordered pairs and draw a line through the points, extending it in both directions as shown below.

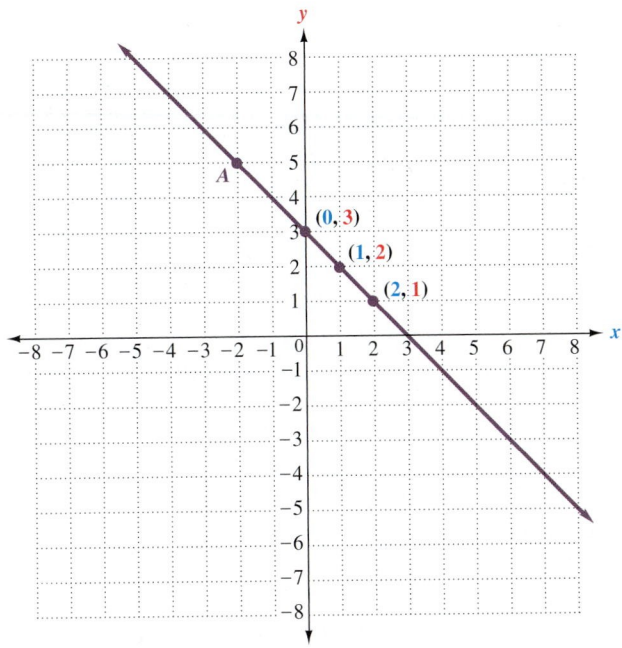

Now you can use the graph to find more solutions of $x + y = 3$. *Every* point on the line is a solution. Suppose that you pick **point A**. The coordinates are $(-2, 5)$.

To check that $(-2, 5)$ is a solution, substitute -2 for x and 5 for y in the original equation.

$$x + y = 3 \quad \text{Original equation}$$
$$-2 + 5 = 3$$
$$3 = 3 \quad \text{Balances} \quad \boxed{(-2, 5) \text{ is a solution, \textbf{not} 3.}}$$

The equation balances, so $(-2, 5)$ is another solution of $x + y = 3$.

> **Note**
>
> The line in Example 1 above was extended in both directions because *every* point on the line is a solution of $x + y = 3$. However, when we graphed the line for the hours spent studying, $m + p = 6$, we did *not* extend the line. That is because the variables m and p represented hours, and hours can only be 0 or positive numbers; all the solutions had to be in the first quadrant.

Work Problem **1** at the Side. ▶

---

1 Graph $x + y = 5$ by finding three solutions and plotting the ordered pairs. Then use the graph to find *two* other solutions of the equation.

x	y	Check that $x + y = 5$	Ordered Pair (x, y)
0			
1			
2			

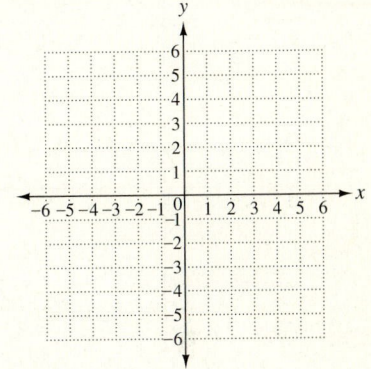

Two other solutions are (___, ___) and (___, ___).

1. Plot $(0, 5)$, $(1, 4)$, and $(2, 3)$.

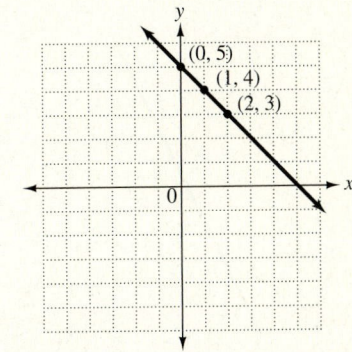

There are many other solutions. Some possibilities are $(-1, 6)$; $(3, 2)$; $(4, 1)$; $(5, 0)$; $(6, -1)$.

2 Graph $y = 2x$ by finding three solutions and plotting the ordered pairs. Then use the graph to find *two* other solutions of the equation.

x	y = 2 • x	Ordered Pair (x, y)
0		
1		
2		

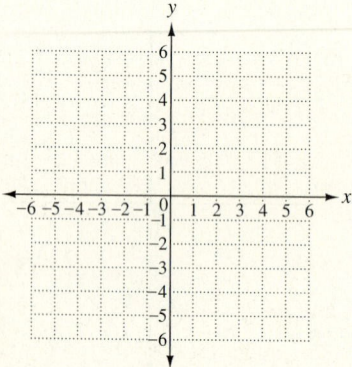

Two other solutions are (__, __) and (__, __).

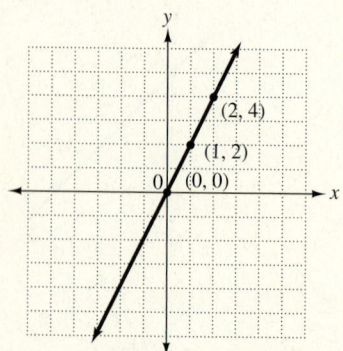

EXAMPLE 2 **Graphing a Linear Equation**

Graph $y = -3x$ by finding three solutions and plotting the ordered pairs. Then use the graph to find a fourth solution of the equation.

You can choose any three values for x, but small numbers such as 0, 1, and 2 are easy to use. Then $y = -3x$ tells you that y is -3 times the value of x.

$$y = -3x$$

y is -3 times x

First set up a table.

> Start by picking easy numbers for x.

x	y = −3 • x	Ordered Pair (x, y)
0	−3 • 0 is 0	(0, 0)
1	−3 • 1 is −3	(1, −3)
2	−3 • 2 is −6	(2, −6)

Plot the ordered pairs and draw a line through the points. Be sure to draw arrows on both ends of the line to show that it continues in both directions.

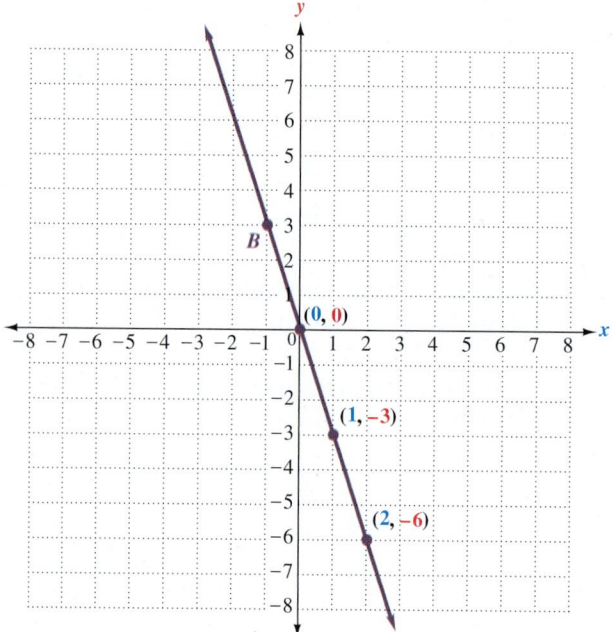

Now use the graph to find more solutions. *Every* point on the line is a solution.

Suppose that you pick **point B**. The coordinates are $(-1, 3)$. To check that $(-1, 3)$ is a solution, substitute -1 for x and 3 for y in the original equation.

$$y = -3x \quad \text{Original equation}$$

$$3 = -3(-1)$$

$$3 = 3 \quad \text{Balances}$$

The equation balances, so $(-1, 3)$ is another solution of $y = -3x$.

◀ Work Problem **2** at the Side.

EXAMPLE 3 **Graphing a Linear Equation**

Graph $y = \dfrac{1}{2}x$ by finding three solutions and plotting the ordered pairs. Then use the graph to find a fourth solution of the equation.

Complete the table. The coefficient of x is $\frac{1}{2}$, so choose even numbers like 2, 4, and 6 as values for x because they are easy to divide in half. The equation $y = \frac{1}{2}x$ tells you that y is $\frac{1}{2}$ *times* the value of x.

> Pick *even* numbers for *x*; they are easy to multiply by $\frac{1}{2}$

x	$y = \frac{1}{2} \cdot x$	Ordered Pair (x, y)
2	$\frac{1}{2} \cdot 2$ is **1**	**(2, 1)**
4	$\frac{1}{2} \cdot 4$ is **2**	**(4, 2)**
6	$\frac{1}{2} \cdot 6$ is **3**	**(6, 3)**

Plot the ordered pairs and draw a line through the points.

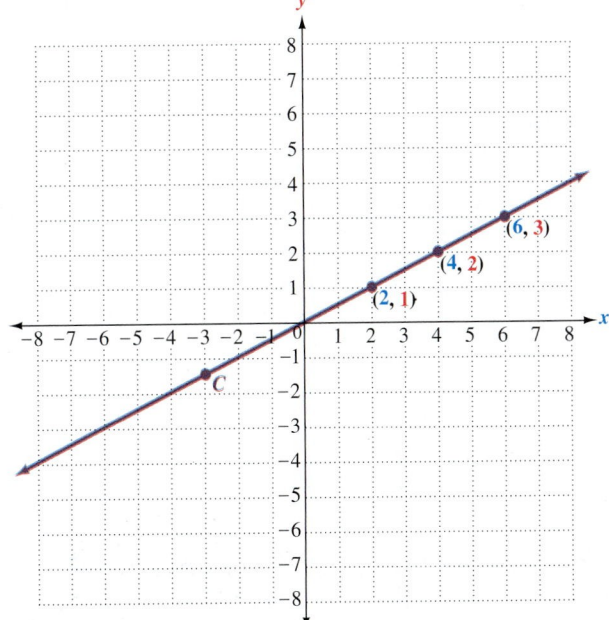

Now use the graph to find more solutions. *Every* point on the line is a solution.
Suppose that you pick **point C**. The coordinates are $(-3, -1\frac{1}{2})$. Check that $(-3, -1\frac{1}{2})$ is a solution by substituting -3 for x and $-1\frac{1}{2}$ for y.

$$y = \frac{1}{2}x \qquad \text{Original equation}$$

$$-1\frac{1}{2} = \frac{1}{2}(-3)$$

$$-1\frac{1}{2} = -\frac{3}{2} \qquad \text{Balances}$$

> $-1\frac{1}{2}$ is equivalent to $-\frac{3}{2}$

The equation balances, so $(-3, -1\frac{1}{2})$ is another solution of $y = \frac{1}{2}x$.

Work Problem **3** *at the Side.* ▶

---

3 Graph $y = -\frac{1}{2}x$ by finding three solutions and plotting the ordered pairs. Then use the graph to find *two* more solutions.

x	$y = -\frac{1}{2} \cdot x$	Ordered Pair (x, y)
2		
4		
6		

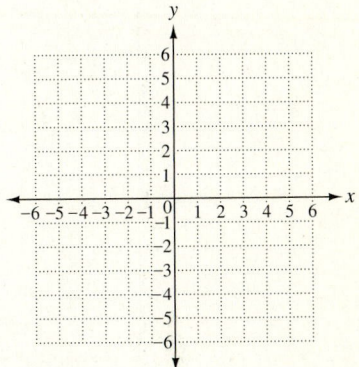

Two other solutions are (__, __) and (__, __).

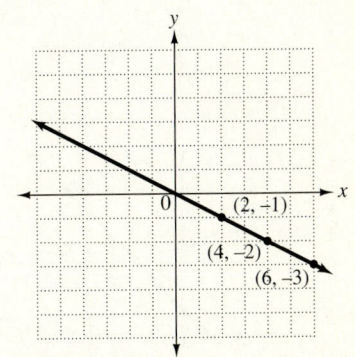

4 Graph the equation $y = x - 5$ by finding three solutions and plotting the ordered pairs. Then use the graph to find *two* more solutions.

x	y = x − 5	Ordered Pair (x, y)
1		
2		
3		

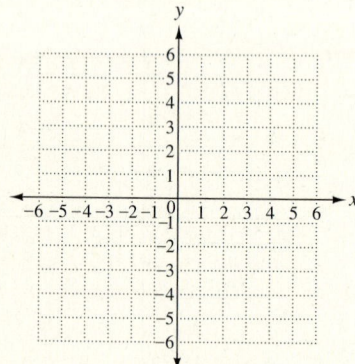

Two other solutions are (___, ___) and (___, ___).

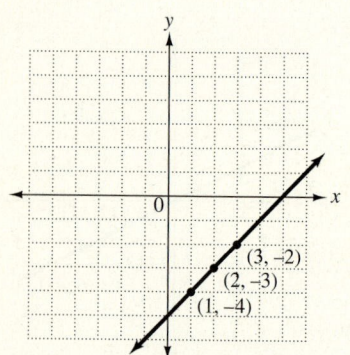
EXAMPLE 4 **Graphing a Linear Equation**

Graph the equation $y = x + 4$ by finding three solutions and plotting the ordered pairs. Then use the graph to find two more solutions of the equation.

First set up a table. The equation $y = x + 4$ tells you that y must be 4 more than the value of x.

x	y = x + 4	Ordered Pair (x, y)
0	0 + 4 is 4	(0, 4)
1	1 + 4 is 5	(1, 5)
2	2 + 4 is 6	(2, 6)

Plot the ordered pairs and draw a line through the points.

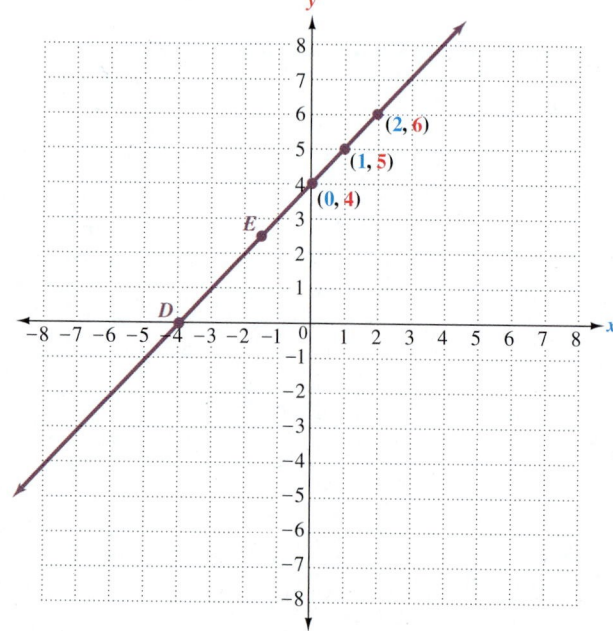

Now use the graph to find two more solutions. *Every* point on the line is a solution. Suppose that you pick **point D** at $(-4, 0)$ and **point E** at $\left(-1\frac{1}{2}, 2\frac{1}{2}\right)$. Check that both ordered pairs are solutions.

Check $(-4, 0)$

$$y = x + 4$$
$$0 = -4 + 4$$
$$0 = 0 \quad \text{Balances}$$

$(-4, 0)$ is a solution, **not** 0.

Check $\left(-1\frac{1}{2}, 2\frac{1}{2}\right)$

$$y = x + 4$$
$$2\frac{1}{2} = -1\frac{1}{2} + 4$$
$$\frac{5}{2} = -\frac{3}{2} + \frac{8}{2}$$
$$\frac{5}{2} = \frac{5}{2} \quad \text{Balances}$$

$2\frac{1}{2}$ can be written as $\frac{5}{2}$ and $-1\frac{1}{2}$ can be written as $-\frac{3}{2}$

Both equations balance, so $(-4, 0)$ and $\left(-1\frac{1}{2}, 2\frac{1}{2}\right)$ are also solutions of $y = x + 4$.

◀ Work Problem **4** at the Side.

OBJECTIVE 2 Identify the slope of a line as positive or negative. Let's look again at some of the lines that we graphed for various equations. All are straight lines, but some are almost flat and some tilt steeply upward or downward.

Graph from Example 1: $x + y = 3$

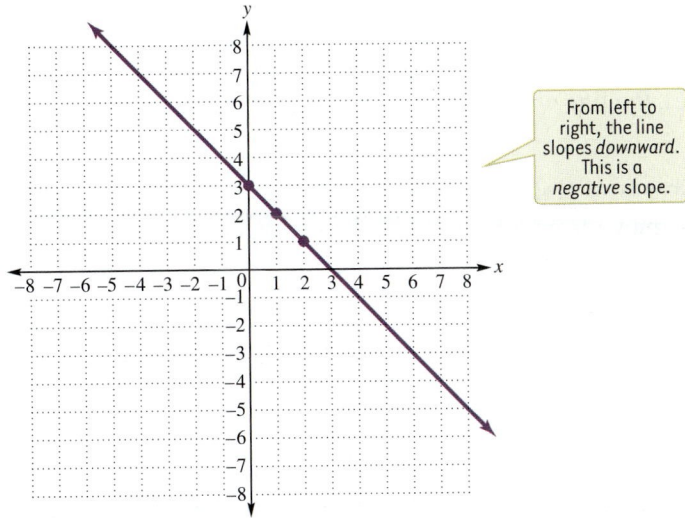

As you move from *left to right,* the line slopes downward, as if you were walking down a hill. When a line tilts downward, we say that it has a *negative slope.*

Now look at the table of solutions we used to draw the line.

The value of *x* is *increasing* from 0 to 1 to 2

x	y
0	**3**
1	**2**
2	**1**

The value of *y* is *decreasing* from 3 to 2 to 1

As the value of *x increases,* the value of *y* does the *opposite*—it *decreases.* Whenever one variable increases while the other variable decreases, the line will have a negative slope.

Graph from Example 3: $y = \dfrac{1}{2}x$

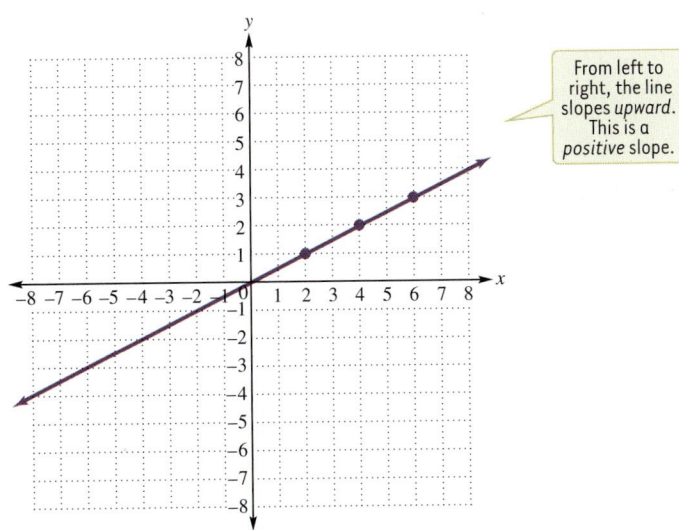

As you move from *left to right,* this line slopes upward, as if you were walking up a hill. When a line tilts upward, we say that it has a *positive slope.*

5 Look back at the graphs in Margin Problems 2 and 3. Then complete these sentences.

(a) The graph of $y = 2x$ has a _____ slope. As the value of x increases, the value of y _____.

Now look at the table of solutions we used to draw the line.

The value of x is *increasing* from 2 to 4 to 6

x	y
2	1
4	2
6	3

The value of y is *increasing* from 1 to 2 to 3

As the value of x *increases,* the value of y does the *same* thing—it also *increases*. Whenever both variables do the same thing (both increase or both decrease), the line will have a positive slope.

> **Positive and Negative Slopes**
>
> As you move from left to right, a line with a *positive* slope tilts *upward* or rises. As the value of one variable increases, the value of the other variable also increases (does the same).
>
> As you move from left to right, a line with a *negative* slope tilts *downward* or falls. As the value of one variable increases, the value of the other variable decreases (does the opposite).

EXAMPLE 5 Identifying Positive or Negative Slope in a Line

Look back at the graph of $y = -3x$ in Example 2. Then complete these sentences.

The graph of $y = -3x$ has a _____ slope.

As the value of x increases, the value of y _____.

(b) The graph of $y = -\frac{1}{2}x$ has a _____ slope. As the value of x increases, the value of y _____.

The graph of $y = -3x$ has a <u>negative</u> slope (because it tilts downward).
As the value of x increases, the value of y <u>decreases</u> (does the opposite).

◀ Work Problem 5 at the Side.

FOR
EXTRA
HELP

Graph each equation by completing the table to find three solutions and plotting the ordered pairs. Then use the graph to find two *other solutions. See Example 1.*

1. $x + y = 4$

x	y	Ordered Pair (x, y)
0		
1		
2		

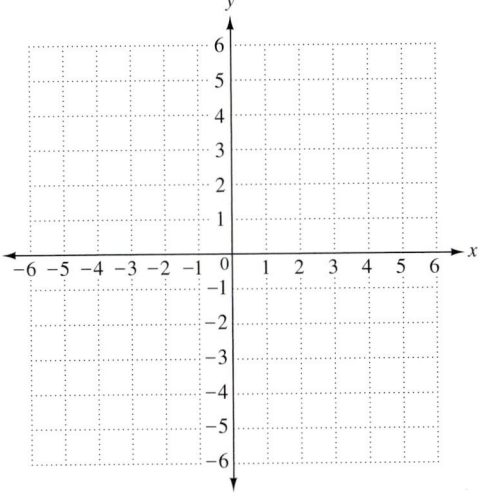

Two other solutions are (__, __) and (__, __).

2. $x + y = -4$

x	y	Ordered Pair (x, y)
0		
1		
2		

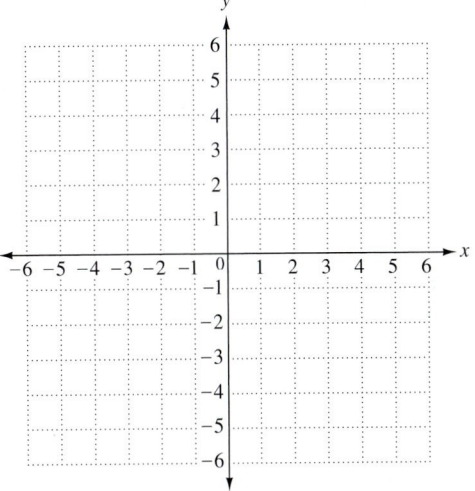

Two other solutions are (__, __) and (__, __).

3. $x + y = -1$

x	y	Ordered Pair (x, y)
0		
1		
2		

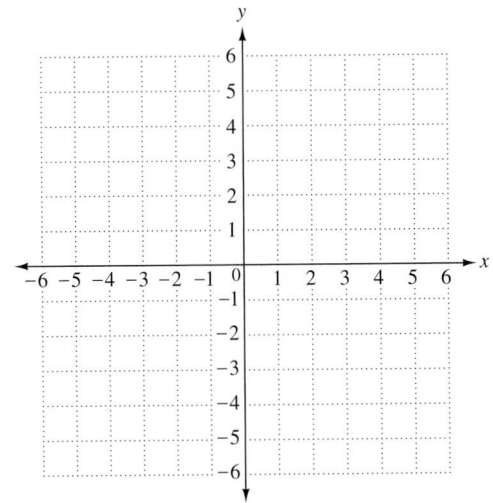

Two other solutions are (__, __) and (__, __).

4. $x + y = 1$

x	y	Ordered Pair (x, y)
0		
1		
2		

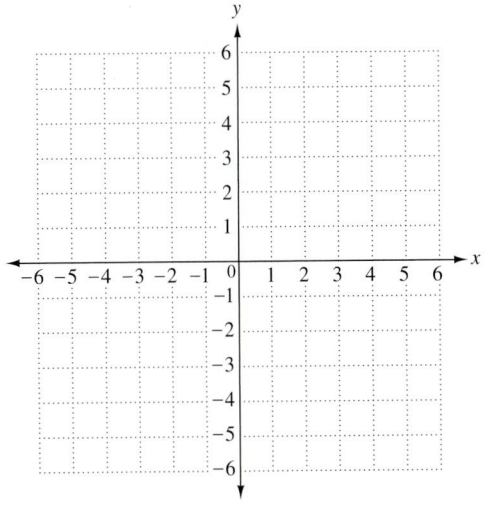

Two other solutions are (__, __) and (__, __).

5. The line in Exercise 1 crosses the y-axis at what point? _____ The line in Exercise 3 crosses the y-axis at what point? _____
Based on these examples, where would the graph of $x + y = -6$ cross the y-axis? _____
Where would the graph of $x + y = 99$ cross the y-axis? _____

6. Look at where the line crosses the x-axis and where it crosses the y-axis in Exercises 2 and 4. What pattern do you see?

Graph each equation. Make your own table using the listed values of x. See Examples 2 and 4.

7. $y = x - 2$

Use 1, 2, and 3 as the values of x.

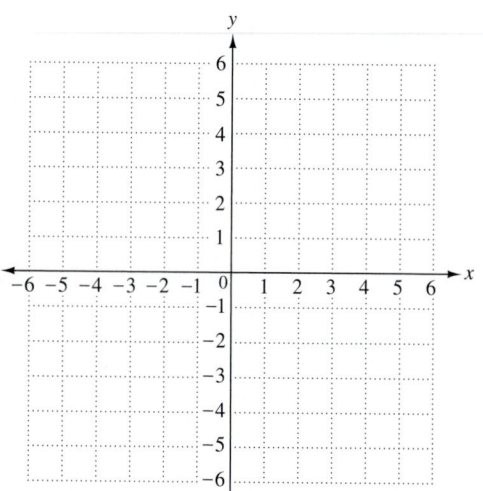

8. $y = x + 1$

Use 1, 2, and 3 as the values of x.

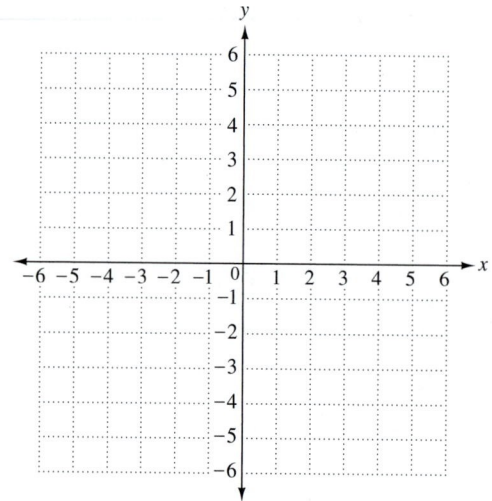

9. $y = x + 2$

Use 0, -1, and -2 as the values of x.

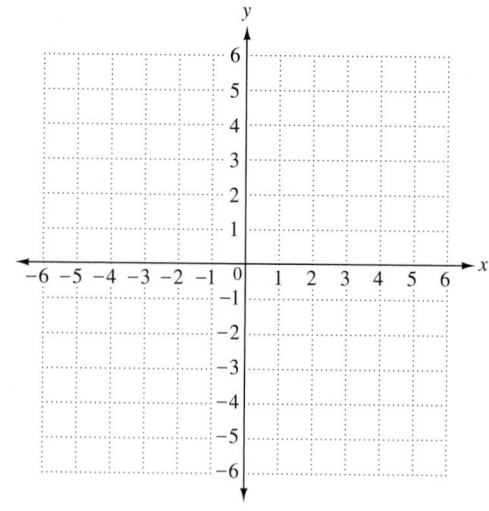

10. $y = x - 1$

Use 0, -1, and -2 as the values of x.

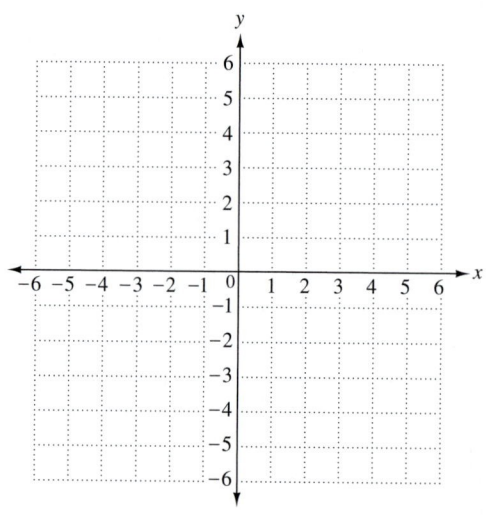

11. $y = -3x$

Use 0, 1, and 2 as the values of x.

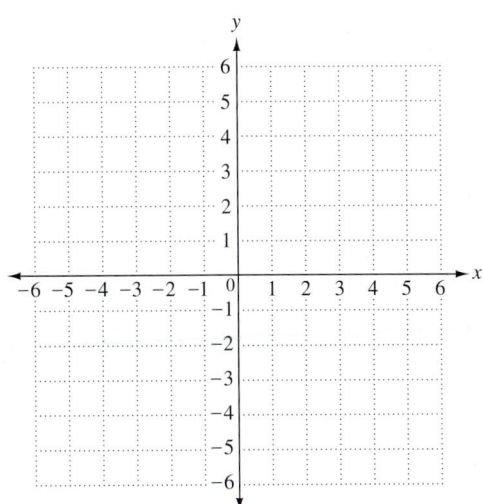

12. $y = -2x$

Use 0, 1, and 2 as the values of x.

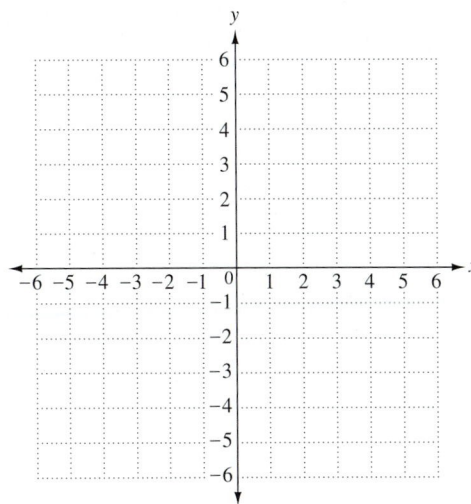

13. Look back at the graphs in Exercises 1, 3, 7, 9, and 11. Which lines have a positive slope? Which lines have a negative slope?

14. Look back at the graphs in Exercises 2, 4, 8, 10, and 12. Which lines have a positive slope? Which lines have a negative slope?

Graph each equation. Make your own table using the listed values of x. See Examples 1–4.

15. $y = \dfrac{1}{3}x$

Use 0, 3, and 6 as the values of x.

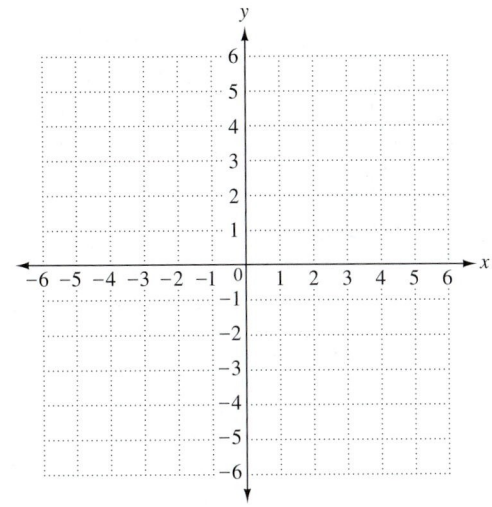

16. $y = \dfrac{1}{2}x$

Use 0, 2, and 4 as the values of x.

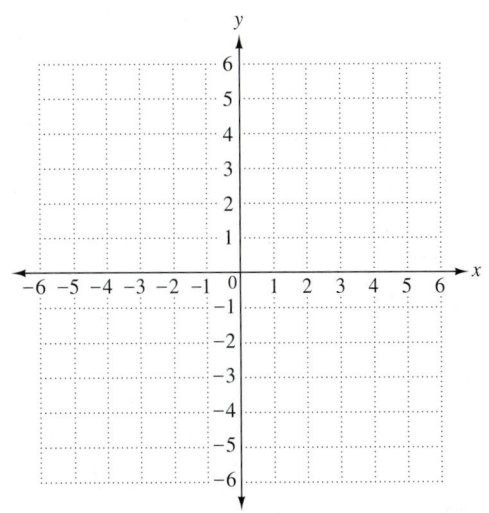

17. $y = x$

Use -1, -2, and -3 as the values of x.

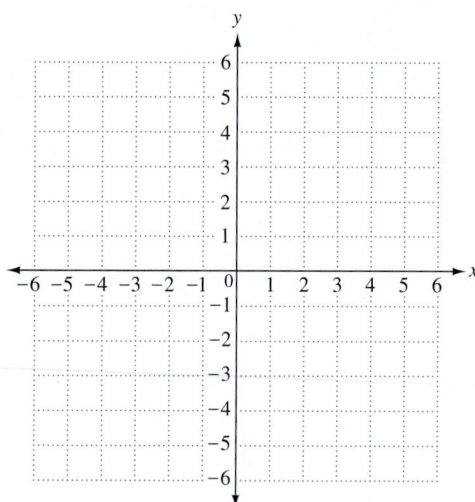

18. $x + y = 0$

Use 1, 2, and 3 as the values of x.

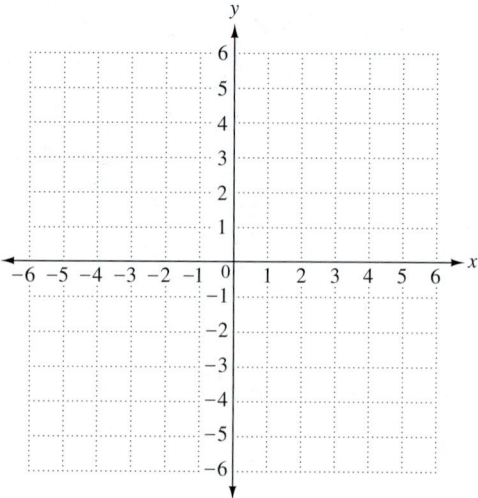

19. $y = -2x + 3$

Use 0, 1, and 2 as the values of x.

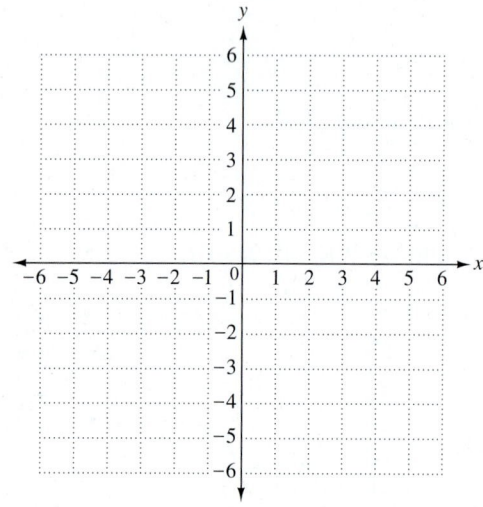

20. $y = 3x - 4$

Use 0, 1, and 2 as the values of x.

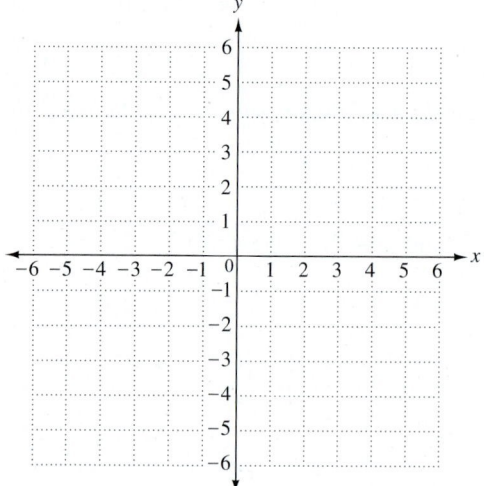

Graph each equation. Choose three values for x. Make a table showing your x values and the corresponding y values. After graphing the equation, state whether the line has a positive or negative slope. See Examples 1–5.

21. $x + y = -3$

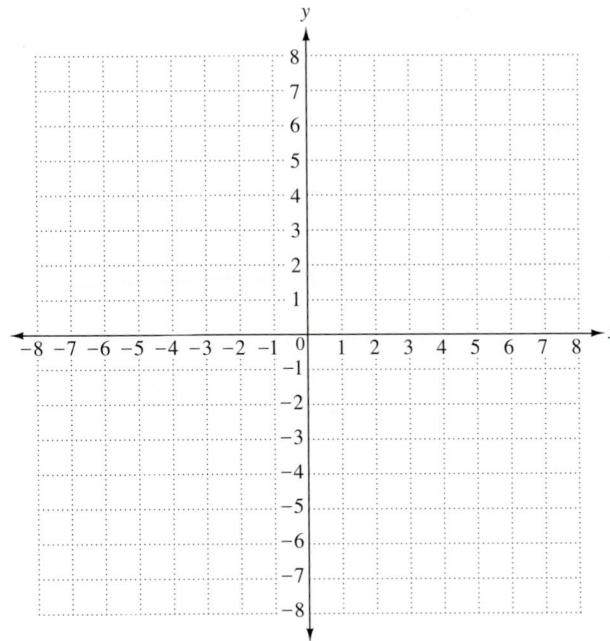

22. $x + y = 2$

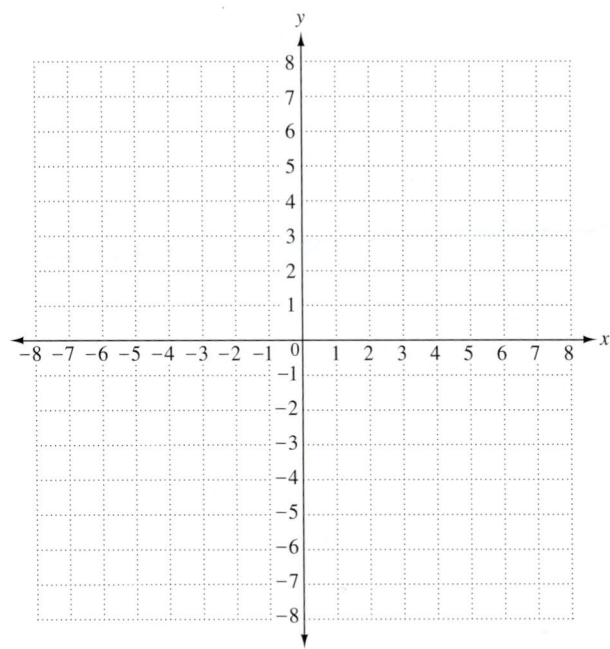

23. $y = \dfrac{1}{4}x$ (*Hint:* Try using 0 and multiples of 4 as the values of x.)

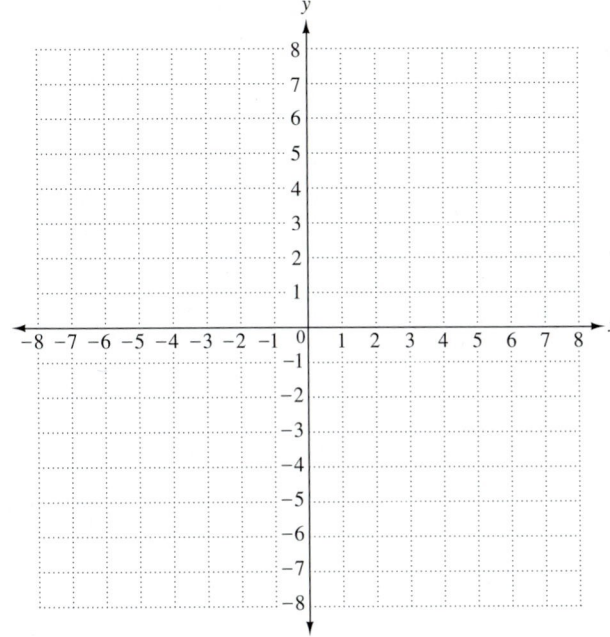

24. $y = -\dfrac{1}{3}x$ (*Hint:* Try using 0 and multiples of 3 as the values of x.)

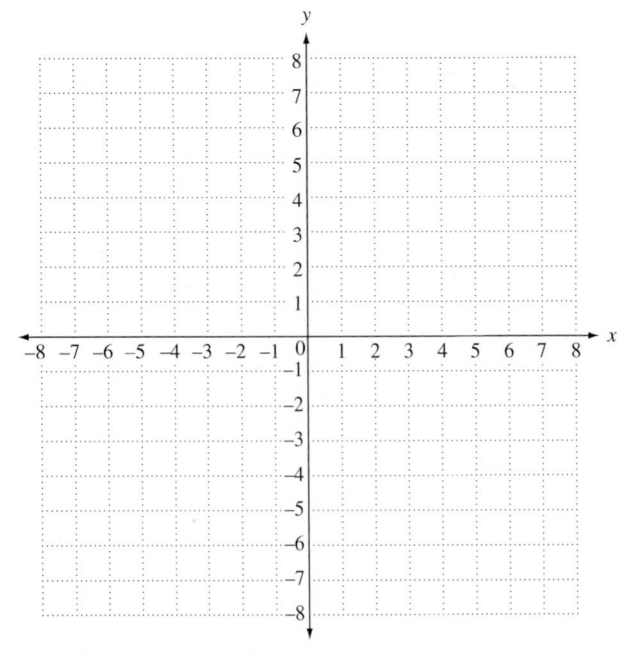

25. $y = x - 5$

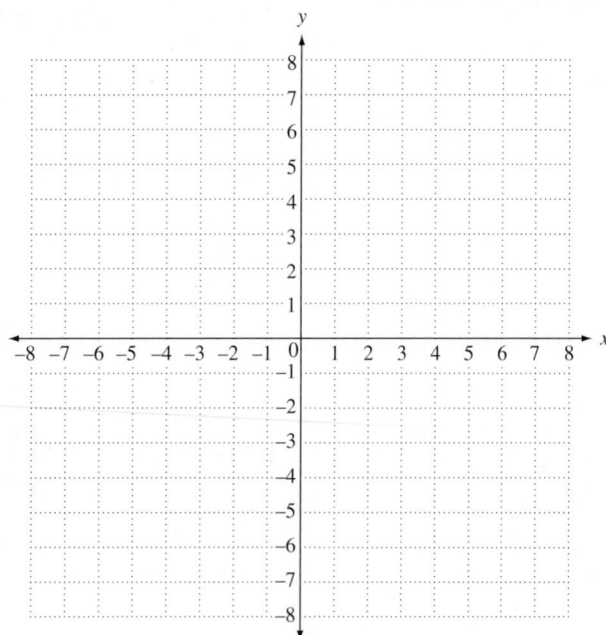

26. $y = x + 4$

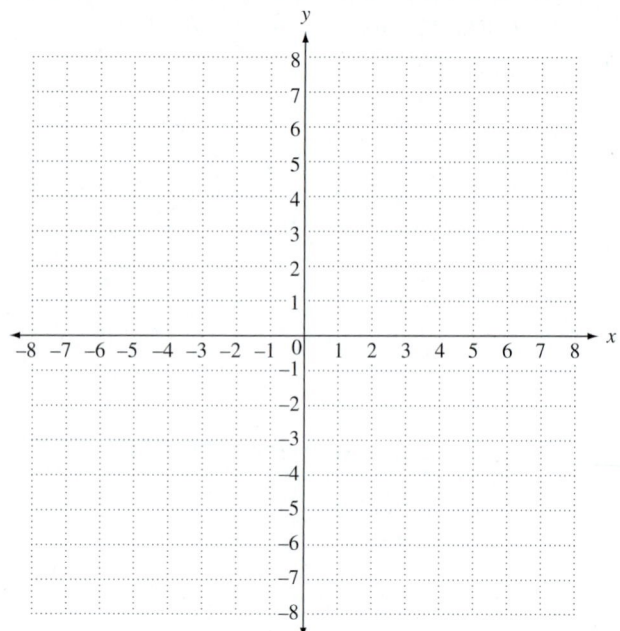

27. $y = -3x + 1$

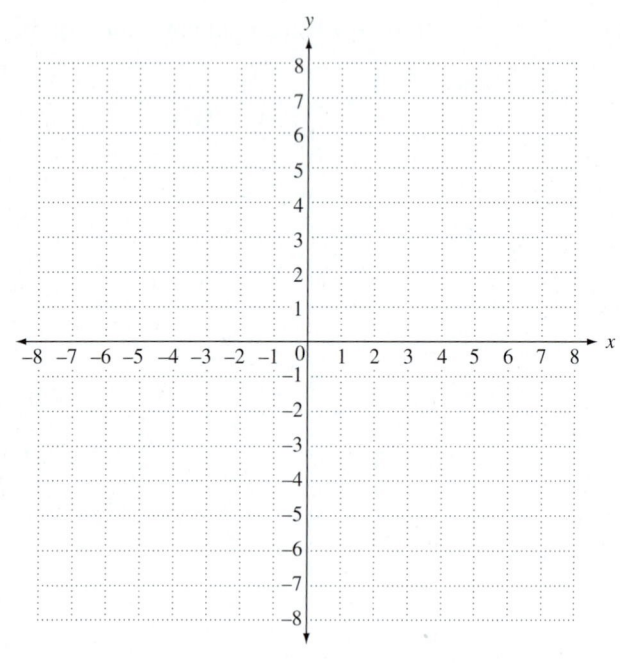

28. $y = 2x - 2$

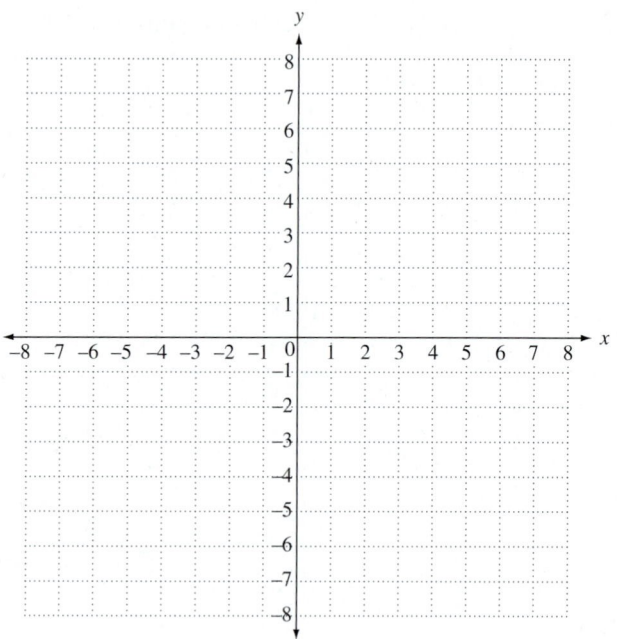

Chapter 9 ▶▶▶ Summary

▶ Key Terms

9.1	**table**	A table presents data organized into rows and columns.
	pictograph	A pictograph uses symbols or pictures to represent various amounts.

9.2	**circle graph**	A circle graph shows how a total amount is divided into parts or sectors. It is based on percents of 360°.
	protractor	A protractor is a tool (usually in the shape of a half-circle) used to measure the number of degrees in an angle or part of a circle.

9.3	**bar graph**	A bar graph uses bars of various heights to show quantity or frequency.
	double-bar graph	A double-bar graph compares two sets of data by showing two sets of bars.
	line graph	A line graph uses dots connected by line segments to show trends.
	comparison line graph	A comparison line graph shows how two or more sets of data relate to each other by showing a line graph for each set of data.

9.4	**paired data**	When each number in a set of data is matched with another number by some rule of association, we call it paired data.
	horizontal axis	The horizontal axis is the number line in a coordinate system that goes "left and right."
	vertical axis	The vertical axis is the number line in a coordinate system that goes "up and down."
	ordered pair	An ordered pair is the "address" of a point in a coordinate system. The *order* of the numbers is important. The first number tells how far to move left or right from 0 along the horizontal axis. The second number tells how far to move up or down from 0 along the vertical axis.
	***x*-axis**	The horizontal axis is called the *x*-axis.
	***y*-axis**	The vertical axis is called the *y*-axis.
	coordinate system	Together, the *x*-axis and the *y*-axis form a rectangular coordinate system. *Example:* See figure at the right.
	origin	The axes lines in a rectangular coordinate system cross at (0, 0); this point is called the origin. *Example:* See red dot marking (0, 0) in figure at the right.
	coordinates	Coordinates are the numbers in the ordered pair that specify the location of a point on a rectangular coordinate system. *Example:* See the point (−4, 2) at the right.
	quadrants	The *x*-axis and the *y*-axis divide the coordinate system into four regions called quadrants; they are designated with Roman numerals. *Example:* The point (−4, 2) at the right is in quadrant II.

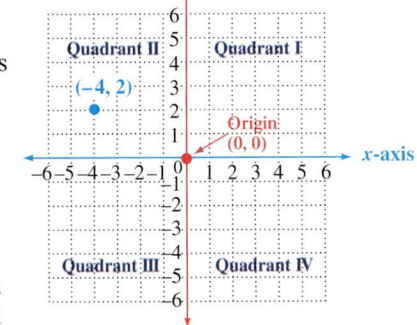

9.5	**graph a linear equation**	All the solutions of a linear equation (all the ordered pairs that satisfy the equation) lie along a straight line. When you draw the line, you have graphed the equation.

▶ New Symbols

(x, y)   ordered pair

▶ Test Your Word Power

See how well you have learned the vocabulary in this chapter. Answers follow the Quick Review.

1. A **circle graph**
   A. uses symbols to represent various amounts
   B. is useful for showing trends
   C. shows how a total amount is divided into parts
   D. compares two sets of data.

2. A **line graph**
   A. uses symbols to represent various amounts
   B. is useful for showing trends
   C. shows how a total amount is divided into parts
   D. compares two sets of data.

3. A **protractor** is used to
   A. measure degrees in an angle
   B. draw circles of various sizes
   C. measure lengths
   D. calculate circumference.

4. Two sets of data can be compared using a
   A. circle graph
   B. pictograph
   C. line graph
   D. double-bar graph.

5. The **x-axis** in a rectangular coordinate system is the
   A. number line that goes "up and down"
   B. number line that goes "left and right"
   C. center point of the grid
   D. vertical axis.

6. **Quadrants** are the
   A. numbers in an ordered pair
   B. solutions of a linear equation
   C. four regions in a coordinate system
   D. paired data shown on a graph.

7. **Coordinates** are
   A. points in a straight line on a coordinate system
   B. designated with Roman numerals
   C. the solutions of a linear equation
   D. numbers used to locate a point in a coordinate system.

8. A **rectangular coordinate system**
   A. is formed by the x-axis and y-axis
   B. is divided into eight quadrants
   C. is the solution of a linear equation
   D. has only positive numbers.

▶ Quick Review

Concepts	Examples

9.1 Reading a Table

The data in a table is organized into rows and columns. As you read from left to right along each row, check the heading at the top of each column.

Use the table below to answer the question.

PER CAPITA CONSUMPTION OF SELECTED BEVERAGES IN GALLONS

	1990	1995	2000	2005
Milk	25.3	23.6	22.2	21.0
Coffee	26.9	20.2	26.3	24.6
Bottled water	8.8	11.6	16.7	23.2
Soft drinks	47.1	50.6	53.2	52.3

Source: U.S. Dept. of Agriculture Economic Research Service.

In which years was the consumption of milk greater than the consumption of coffee?

Read across the rows labeled "milk" and "coffee" from left to right, comparing the numbers in each column. In 1995 the figure for milk (23.6 gallons) is greater than for coffee (20.2 gallons).

Concepts	Examples

9.1 Reading a Pictograph

A pictograph uses symbols or pictures to represent various amounts. The *key* tells you how much each symbol represents. A fractional part of a symbol represents a fractional part of the symbol's value.

Use the pictograph to answer these questions.

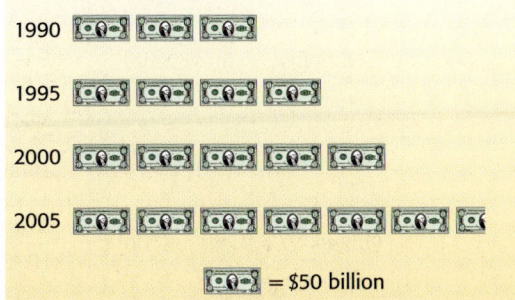

SALES OF FOOD AND DRINK AT U.S. RESTAURANTS

1990	
1995	
2000	
2005	

= $50 billion

Source: National Restaurant Association.

(a) How much was spent at U.S. restaurants in 2005?

Sales for 2005 are represented by 6 whole symbols (6 • $50 billion = $300 billion) plus half of a symbol ($\frac{1}{2}$ of $50 billion = $25 billion) for a total of $325 billion.

(b) How much did restaurant expenditures increase from 1990 to 2000?

The year 2000 shows two more symbols than 1990, so the increase is 2 • $50 billion = $100 billion.

9.2 Constructing a Circle Graph

Step 1 Determine the percent of the total for each item.

Step 2 Find the number of degrees out of 360° that each percent represents.

Step 3 Use a protractor to measure the number of degrees for each item in the circle.

Step 4 Label each sector in the circle with the item name and percent of total for that item.

Construct a circle graph for these expenses from a business trip.

Item	Amount
Transportation	$350
Lodging	$300
Food	$250
Other	$100
Total	**$1000**

(continued)

Concepts	Examples

9.2 Constructing a Circle Graph *(continued)*

Item	Amount	Percent of Total		Sector Size
Transportation	$350	$\frac{\$350}{\$1000} = \frac{7}{20} = $ **35%**	so 35% of 360° = (0.35)(360)	= 126°
Lodging	$300	$\frac{\$300}{\$1000} = \frac{3}{10} = $ **30%**	so 30% of 360° = (0.30)(360)	= 108°
Food	$250	$\frac{\$250}{\$1000} = \frac{1}{4} = $ **25%**	so 25% of 360° = (0.25)(360)	= 90°
Other	$100	$\frac{\$100}{\$1000} = \frac{1}{10} = $ **10%**	so 10% of 360° = (0.10)(360)	= 36°

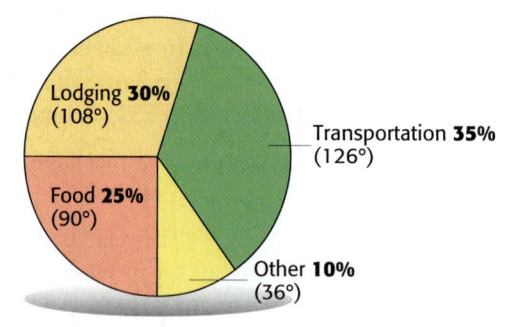

BUSINESS TRIP EXPENSES

Lodging **30%** (108°)
Transportation **35%** (126°)
Food **25%** (90°)
Other **10%** (36°)

9.3 Reading a Bar Graph

The height of the bar is used to show the quantity or frequency (number) in a specific category. Use a ruler or straightedge to line up the top of each bar with the numbers on the left side of the graph.

Use the bar graph below to determine the number of students who earned each letter grade.

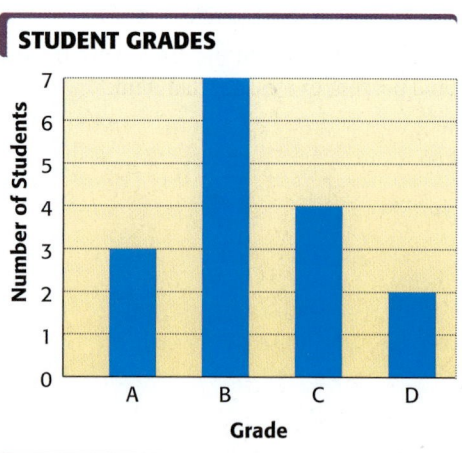

STUDENT GRADES

The number of students earning each letter grade were A: 3 students; B: 7 students; C: 4 students; D: 2 students.

Concepts	Examples

9.3 Reading a Line Graph

A dot is used to show the number or quantity in a specific class. The dots are connected with line segments. This kind of graph is used to show a trend.

The line graph below shows the annual sales for the Fabric Supply Center for each of four years.

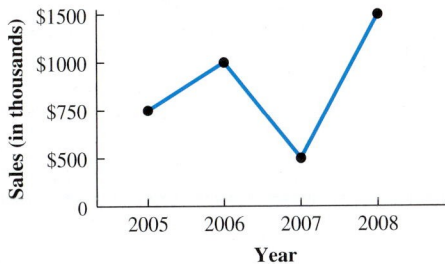

Which year had the lowest sales? What was the amount of sales that year?

2007; sales were $500 • 1000 = $500,000

9.4 Plotting Points

Start at the center of the coordinate system (the origin). The first number in an ordered pair tells you how far to move *left* or *right* along the horizontal axis; *positive* numbers are to the *right,* and *negative* numbers are to the *left.* The second number in an ordered pair tells you how far to move *up* or *down; positive* numbers are *up,* and *negative* numbers are *down.*

To plot $(3, -2)$, start at 0, move to the *right* 3 units, and then move *down* 2 units.

To plot $(-2, 3)$, start at 0, move to the *left* 2 units, and then move *up* 3 units.

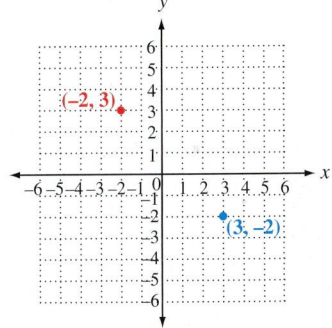

9.4 Identifying Quadrants

The *x*-axis and *y*-axis divide the coordinate system into four regions called quadrants. The quadrants are designated with Roman numerals.

Points in the first quadrant fit the pattern $(+, +)$.

Points in the second quadrant fit the pattern $(-, +)$.

Points in the third quadrant fit the pattern $(-, -)$.

Points in the fourth quadrant fit the pattern $(+, -)$.

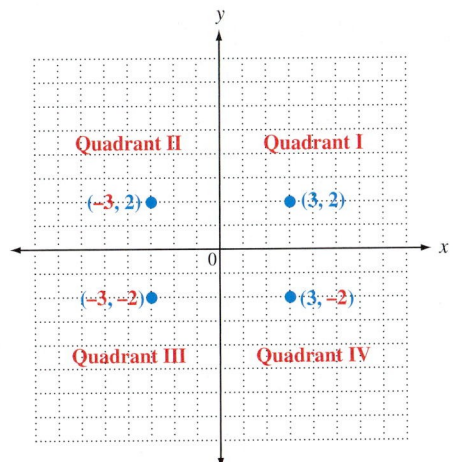

Concepts	Examples

9.5 Graphing Linear Equations

Choose any three values for x. Then find the corresponding values of y. Plot the three ordered pairs on a coordinate system. Draw a line through the points, extending it in both directions. If the three points do *not* lie on a straight line, there is an error in your work.

Every point on the line is a solution of the given equation.

Graph $y = -2x$. Then use the graph to find *two* other solutions.

x	$y = -2 \cdot x$	Ordered Pair (x, y)
1	$-2 \cdot 1$ is -2	$(1, -2)$
2	$-2 \cdot 2$ is -4	$(2, -4)$
3	$-2 \cdot 3$ is -6	$(3, -6)$

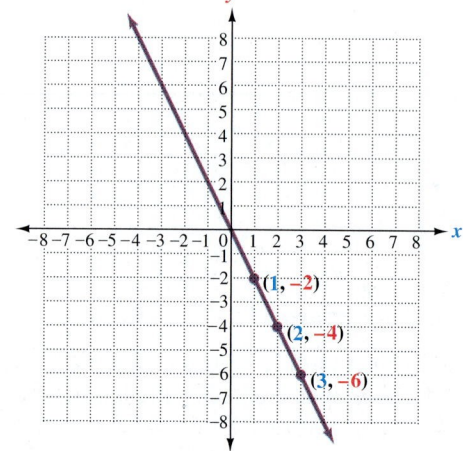

Every point on the line is a solution. Some of the other solutions are $(0, 0)$; $(-1, 2)$; $(-2, 4)$; and $(-2\frac{1}{2}, 5)$.

ANSWERS TO TEST YOUR WORD POWER

1. C; *Example:* A circle graph can show how a 24-hour day is divided among various activities.
2. B; *Example:* A line graph can show changes in the amount of sales over several months or several years.
3. A; *Example:* A protractor is a tool, usually in the shape of a half-circle, used to measure or draw angles of a certain number of degrees. A tool called a compass is used to draw circles of various sizes.
4. D; *Example:* Two sets of bars in different colors can compare monthly unemployment figures for two different years.
5. B; *Example:* The x-axis is the horizontal number line with 0 at the center, negative numbers extending to the left, and positive numbers extending to the right.
6. C; *Example:* The x-axis and the y-axis divide the coordinate system into four regions; each region is designated by a Roman numeral.
7. D; *Example:* To locate the point $(2, -3)$, start at the origin and move 2 units to the right along the x-axis, then move down 3 units.
8. A; *Example:* Together, a horizontal number line (x-axis) and vertical number line (y-axis) form a rectangular coordinate system.

Chapter 9 ▷▷▷ Review Exercises

[9.1] *Use the table at the right to answer Exercises 1–4.*

1. Which sport had the
   (a) fewest men's teams?

   (b) second-greatest number of women's teams?

PARTICIPATION IN SELECTED NCAA SPORTS 2004–2005

Sport	Males			Females		
	Teams	Athletes	Average Squad	Teams	Athletes	Average Squad
Basketball	1000	16,291	16.3	1025	14,686	14.3
Cross country	865	11,638	13.5	940	12,901	13.7
Golf	762	7953	10.4	483	3828	7.9
Gymnastics	19	329	17.3	85	1402	16.5
Volleyball	79	1161		982	13,634	

Source: The National Collegiate Athletic Association (NCAA).

2. Which sport had
   (a) about 13,000 female athletes?

   (b) about 1000 male athletes?

3. **(a)** How many more men participated in basketball than cross country?

   (b) How many fewer women participated in gymnastics than basketball?

4. Find the average squad size for men's and women's volleyball teams and complete the table. Round to the nearest tenth.

Use the pictograph on average yearly snowfall to answer Exercises 5–8.

5. What is the average yearly snowfall in
   (a) Juneau?
   (b) Washington, D.C.?

6. What is the average yearly snowfall in
   (a) Minneapolis?
   (b) Cleveland?

7. Find the difference in the average yearly snowfall between
   (a) Buffalo and Cleveland.
   (b) Memphis and Minneapolis.

8. Find the difference in the average yearly snowfall between the city with the greatest amount and the city with the least amount.

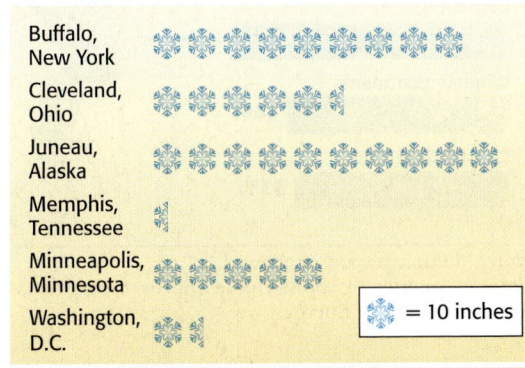

AVERAGE YEARLY SNOWFALL IN SELECTED U.S. CITIES

Source: National Climatic Data Center.

[9.2] *Use this circle graph for Exercises 9–13.*

9. **(a)** What was the largest single expense of the vacation? How much was that item?

   (b) What was the second-most-expensive item? How much was that item?

   (c) What was the total cost of the vacation?

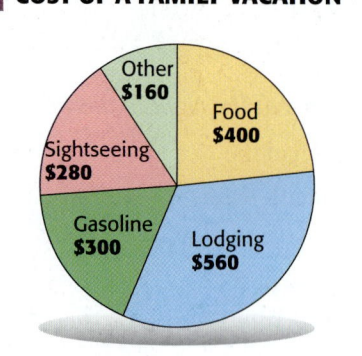

COST OF A FAMILY VACATION

Other $160
Food $400
Sightseeing $280
Gasoline $300
Lodging $560

Use the family vacation circle graph above to find each ratio. Write the ratios as fractions in lowest terms.

10. Cost of lodging to the cost of food

11. Cost of gasoline to the total cost of the vacation

12. Cost of sightseeing to the total cost of the vacation

13. Cost of gasoline to the cost of the Other category

[9.3] *This bar graph shows the top six home improvement projects that people do themselves. Use the graph to answer Exercises 14–19.*

CAN YOU FIX IT?

Projects homeowners take on themselves according to a recent survey:

Painting, wallpapering — 63%
Landscaping, gardening — 54%
Interior decorating — 51%
Carpentry — 43%
Window treatments — 34%
Construction work — 33%

Note: Multiple responses allowed.
Source: American Express Home Improvement Telephone Survey.

14. What is the most popular project? What percent of the homeowners in the survey gave that answer?

15. Which project was selected the least? What percent of the homeowners in the survey gave that answer?

16. What percent of the homeowners in the survey selected carpentry projects? There were 341 people surveyed. How many of them selected carpentry, to the nearest whole number?

17. Of the 341 homeowners surveyed, how many selected landscaping or gardening projects? Round to the nearest whole number.

18. Name the project selected by about **(a)** $\frac{1}{2}$ of the homeowners **(b)** $\frac{1}{3}$ of the homeowners.

19. Give two reasons painting and wallpapering might be selected so much more than construction work.

This double-bar graph shows the number of acre-feet of water in the Lake Natoma reservoir for each of the first six months of 2007 and 2008. Use this graph to answer Exercises 20–25.

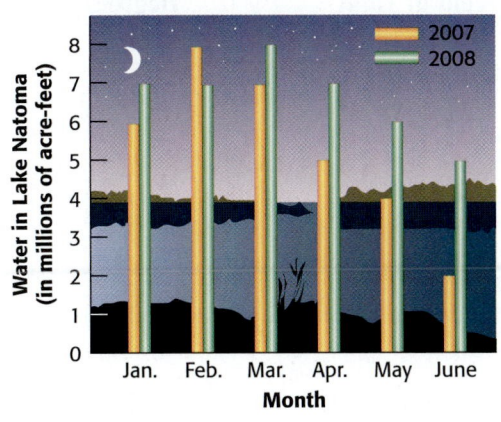

WATER IN LAKE NATOMA

20. From January through June 2008, which month had the greatest amount of water in the lake? How much was there?

21. During which month in the first half of 2007 was the least amount of water in the lake? How much was there?

22. How many acre-feet of water were in the lake in June of 2008?

23. How many acre-feet of water were in the lake in January of 2007?

24. Find the amount of decrease and the percent of decrease in the acre-feet of water in the lake from March 2008 to June 2008.

25. Find the amount of decrease and the percent of decrease in the acre-feet of water in the lake from April 2007 to June 2007.

This comparison line graph shows the annual floor-covering sales of two different home improvement centers during each of five years. Use this graph to find the amount of sales in each year listed in Exercises 26–29 and to answer Exercises 30 and 31.

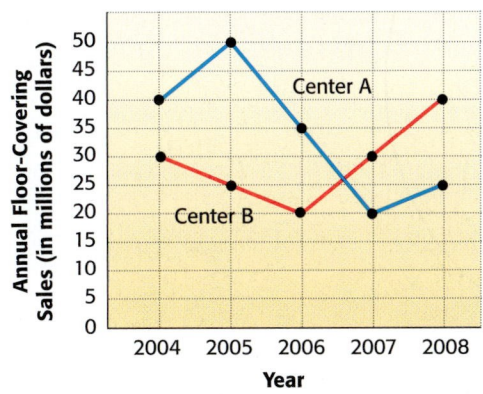

ANNUAL FLOOR-COVERING SALES

26. Center A in 2005

27. Center A in 2007

28. Center B in 2006

29. Center B in 2008

30. What trend do you see in Center A's sales from 2005 to 2008? Why might this have happened?

31. What trend do you see in Center B's sales starting in 2006? Why might this have happened?

[9.2] *The Broadway Hair Salon spent a total of $22,400 to remodel their salon. The breakdown of expenditures for various items is shown. Find all the missing numbers in Exercises 32–36.*

Item	Dollar Amount	Percent of Total	Degrees of Circle
32. Plumbing and electrical changes	$2240	10%	_____
33. Work stations	$7840	_____	_____
34. Small appliances	$4480	_____	_____
35. Interior decoration	$5600	_____	_____
36. Supplies	_____	_____	36°

37. Draw a circle graph using the information in Exercises 32–36. Label each sector with the item and the percent of total for that item.

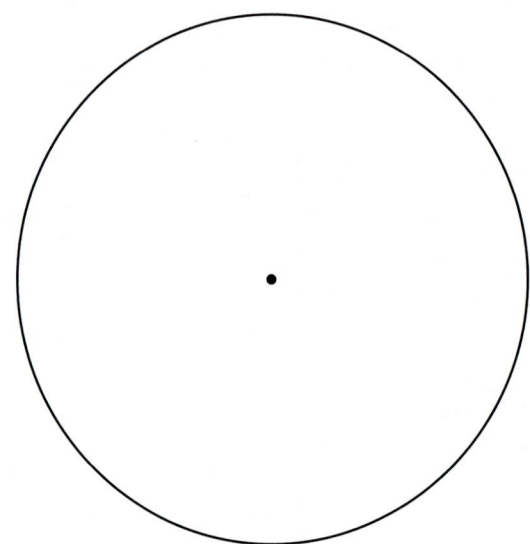

[9.4] *In Exercise 38, plot each point on the rectangular coordinate system and label it with its coordinates. In Exercise 39, give the coordinates of each point.*

38. $(1, 7)$ $(-1\frac{1}{2}, 0)$ $(-4, -2)$ $(0, 3)$ $(2, -\frac{1}{2})$

$(-7, 1)$ $(5, -5)$ $(0, -4)$

39.

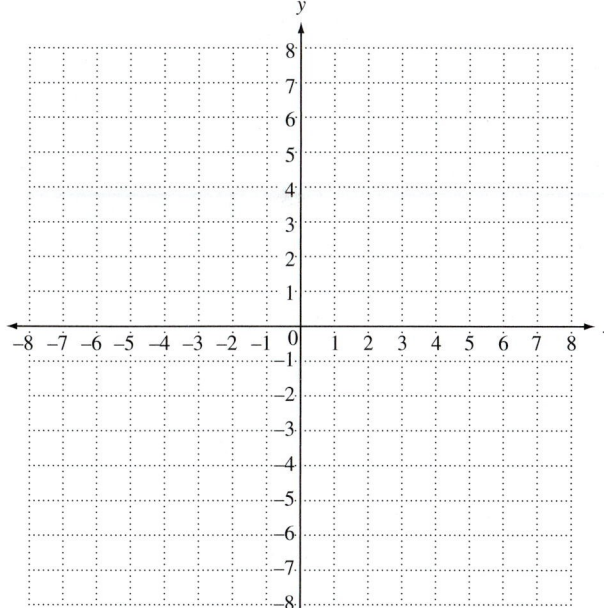 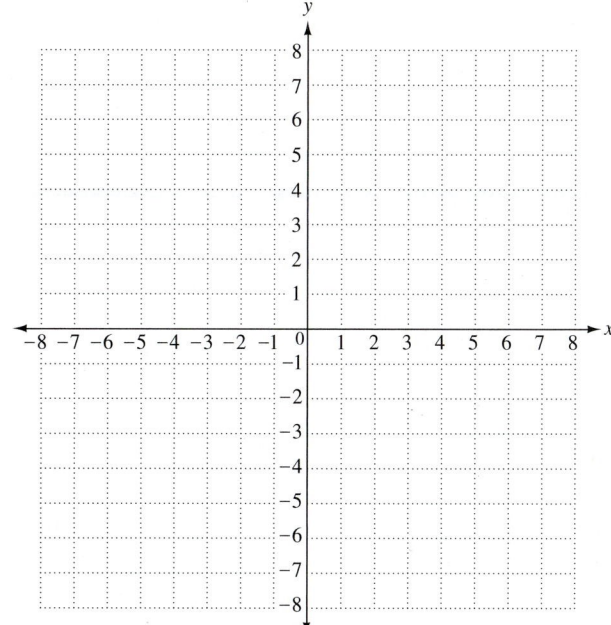

[9.5] *Graph each equation. Complete the table using the listed values of x. State whether each line has a positive or negative slope. Use the graph to find two other solutions of the equation.*

40. $x + y = -2$

Use 0, 1, and 2 as the values of x.

x	y	(x, y)

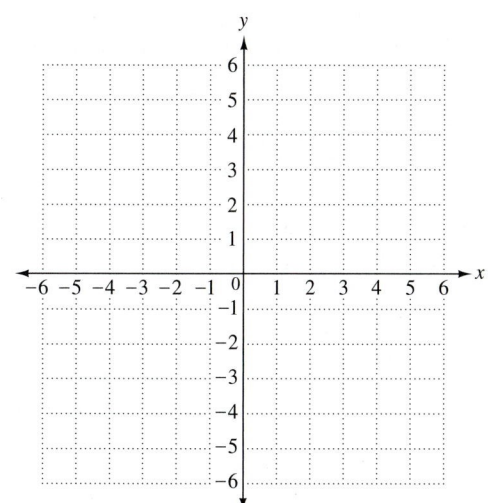

The graph of $x + y = -2$ has a _____ slope.

Two other solutions of $x + y = -2$ are

(____ , ____) and (____ , ____).

41. $y = x + 3$

Use 0, 1, and 2 as the values of x.

x	y	(x, y)

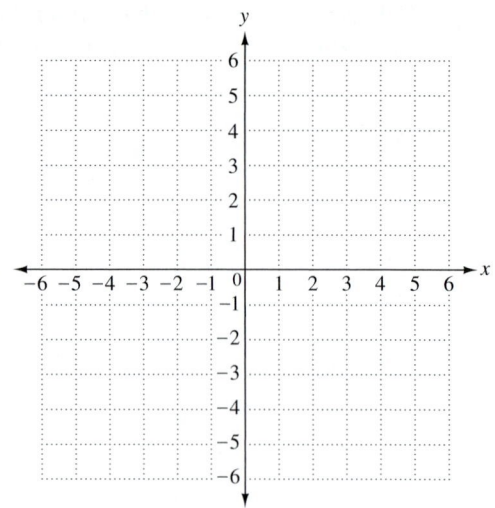

The graph of $y = x + 3$ has a _____ slope.

Two other solutions for $y = x + 3$ are

(____ , ____) and (____ , ____).

42. $y = -4x$

Use -1, 0, and 1 as the values of x.

x	y	(x, y)

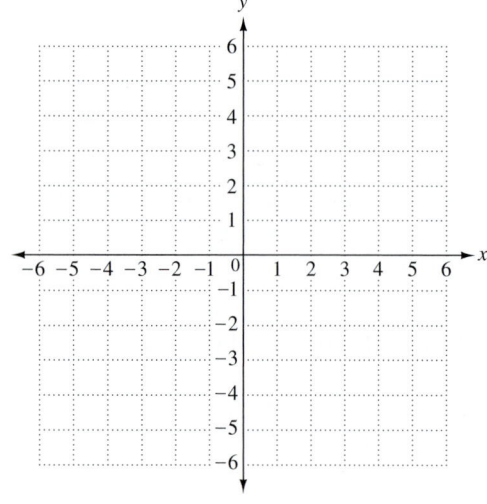

The graph of $y = -4x$ has a _____ slope.

Two other solutions of $y = -4x$ are

(____ , ____) and (____ , ____).

Chapter 9 ▶▶▶ Test

 Use the Chapter Test Prep Video CD to see fully worked-out solutions to any of the exercises you want to review.

Use this table to answer Problems 1–3.

AMOUNT OF CALCIUM IN SELECTED FOODS

Food	Measure	Calories	Calcium (mg)
Swiss cheese	1 oz	95	219
Cream cheese	1 oz	100	23
Yogurt, fruit flavor	8 oz	230	345
Skim milk	1 cup	85	302
Sardines	3 oz	175	371

Source: U.S. Department of Agriculture.

1. Which food has **(a)** the greatest amount of calcium? **(b)** the least amount of calcium?

   1. (a) _____

   (b) _____

2. A packet of Swiss cheese contains 1.75 oz. How many calories are in that amount of Swiss cheese? Round to the nearest whole number.

   2. _____

3. Find the amount of calcium in a 6 oz container of fruit-flavored yogurt. Round to the nearest whole number.

   3. _____

Use this pictograph to answer Problems 4–6.

NUMBER OF U.S. ENDANGERED WILDLIFE SPECIES

Mammals	🐏 🐏 🐏 🐏 🐏 🐏 🐏
Birds	🐏 🐏 🐏 🐏 🐏 🐏 🐏 🐏
Reptiles	🐏 🐏
Fish	🐏 🐏 🐏 🐏 🐏 🐏 🐏 🐏
Insects	🐏 🐏 🐏 🐏

🐏 = 10 Species

Source: U.S. Fish and Wildlife Service.

4. **(a)** Which group (or groups) has the greatest number of endangered species?

   (b) How many endangered species are in the top group(s)?

   4. (a) _____

   (b) _____

5. How many more species of mammals are endangered than reptiles?

   5. _____

6. What is the total number of endangered species shown in the pictograph?

   6. _____

This circle graph shows the advertising budget for Lakeland Amusement Park. Find the dollar amount budgeted for each category listed in Problems 7–10. The total advertising budget is $2,800,000.

7. _____

7. Which form of advertising has the largest budget? What amount is this?

8. _____

8. Which form of advertising has the smallest budget? How much is this?

9. _____

9. How much is budgeted for Internet advertising?

10. _____

10. How much is budgeted for newspaper ads?

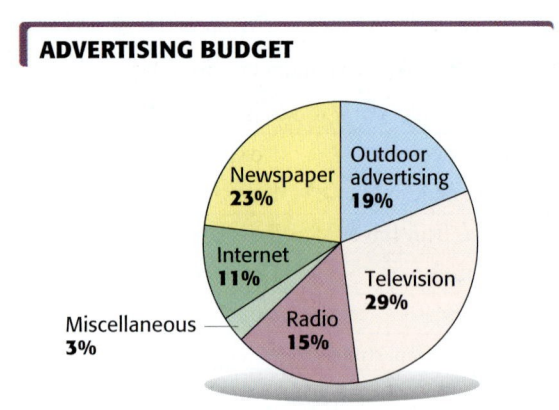

ADVERTISING BUDGET

Newspaper 23%
Outdoor advertising 19%
Internet 11%
Television 29%
Miscellaneous 3%
Radio 15%

This graph shows one student's income and expenses for four years.

11. _____

11. In what year did the student's expenses exceed income? By how much?

12. _____

12. Find the amount of increase and the percent of increase in the student's expenses from 2005 to 2006. Round to the nearest whole percent.

13. _____

13. In what year did the student's income decline? Give two possible explanations for the decline.

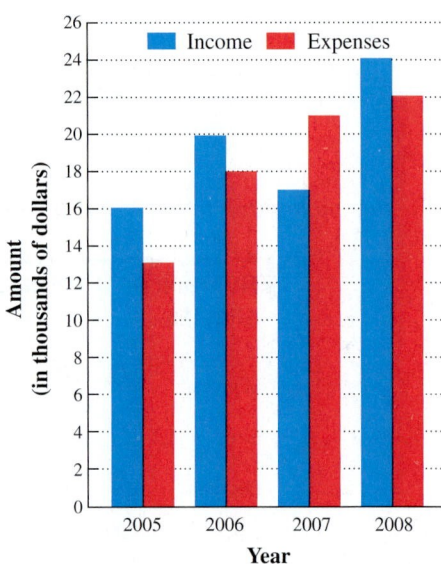

This graph shows enrollment at two community colleges.

14. _____

14. What was College A's enrollment in 2005? What was College B's enrollment in 2006?

15. _____

15. Which college had higher enrollment in 2008? How much higher?

16. _____

16. Give two possible explanations for the fact that College B's enrollment passed College A's.

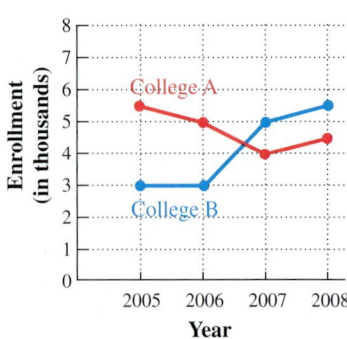

*During a one-year period, Oak Mill Furniture Sales had a total of $480,000
in expenses. Find all the numbers missing from the table.*

Item	Dollar Amount	Percent of Total	Degrees of a Circle	
17. Salaries	$168,000	_____	_____	17. _____
18. Delivery expense	$24,000	_____	_____	18. _____
19. Advertising	$96,000	_____	_____	19. _____
20. Rent	$144,000	_____	_____	20. _____
21. Other	_____	_____	36°	21. _____

22. Draw a circle graph using a protractor and the information in
Problems 17–21. Label each sector with the item and the percent of
total for that item.

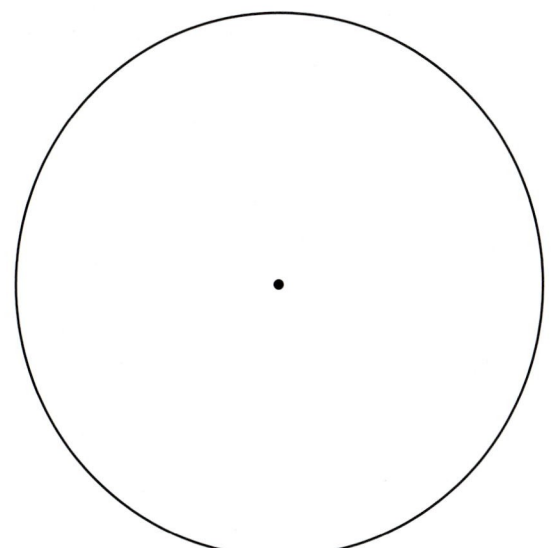

Plot each point on the coordinate system below. Label each point with its coordinates.

23. $(-5, 3)$

24. $(1, -4)$

25. $(0, 6)$

26. $(2, 0)$

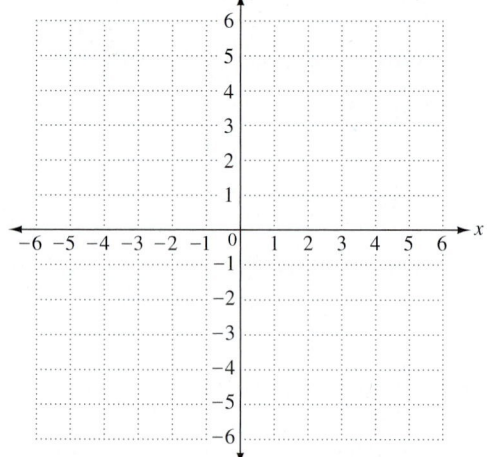

Give the coordinates of each lettered point shown below, and state which quadrant the point is in.

27. _____

28. _____

29. _____

30. _____

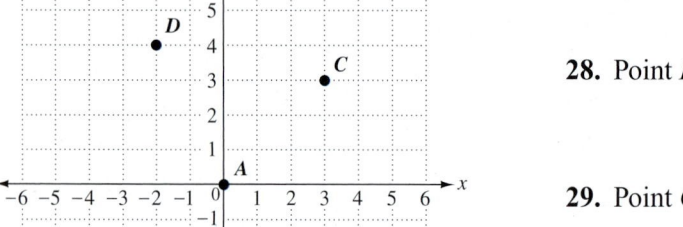

27. Point A

28. Point B

29. Point C

30. Point D

31. Graph $y = x - 4$ on the coordinate system below. Make your own table using 0, 1, and 2 as the values of x.

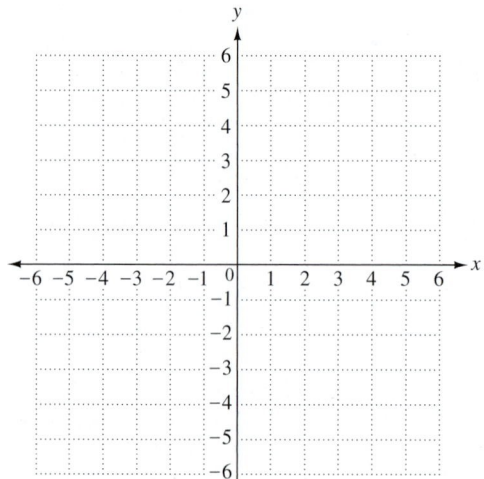

32. (a) _____

(b) _____

32. (a) Use the graph in Problem 31 to find *two* other solutions of $y = x - 4$.

(b) State whether the graph of $y = x - 4$ has a positive or negative slope.

Real Numbers, Equations, and Inequalities

Americans have enjoyed "going to the movies" since the first movie theater opened in 1905. The highest number of theater tickets ever sold in one year was more than 4.49 billion tickets back in 1929. The U.S. population then was about 123 million, so, on average, each American went to a movie theater at least 36 times that year.

(*Source: World Almanac, Guinness World Records.*)

We can use inequality symbols to show these comparisons.

1929 U.S. ticket sales $>$ 4.49 billion

1929 average theater visits per person $\geq$ 36

In **Section 10.1,** Exercises 35–38, you'll use these symbols to show comparisons between the all-time top money-making films.

10.1 ▶▶▶ Real Numbers and Expressions

Throughout Chapters 1–9, you have been developing basic concepts and skills to help you understand and use algebra. In this chapter we will introduce some more advanced terminology and techniques.

OBJECTIVE 1 Identify rational numbers, irrational numbers, and real numbers.   You have worked with different kinds of numbers: integers in Chapter 1, fractions in Chapter 4, decimals in Chapter 5, and so on. Now let's be more specific about these different groups, or *sets*, of numbers.

The first numbers a young child learns are the ones used for counting, called the set of **natural numbers.** The numbers in the set are written between braces.

$$\{1, 2, 3, 4, 5, \ldots\}$$

> Three dots indicate that the list of *natural numbers* goes on forever.

Once zero is included with the natural numbers, we have the set of **whole numbers.**

$$\{0, 1, 2, 3, 4, 5, \ldots\}$$

> The list of *whole numbers* goes on forever.

So far we have only zero and positive numbers. The next set of numbers, the **integers,** includes the natural numbers, their *opposites,* and zero. You worked with integers in Chapter 1.

$$\{\ldots -5, -4, -3, -2, -1, 0, 1, 2, 3, 4, 5, \ldots\}$$

> The list of *integers* goes on forever in *both* directions.

Recall how we used a number line to show the position of each integer.

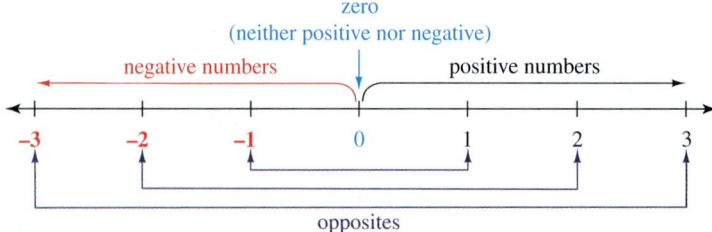

In Chapters 4 and 5 you used positive and negative fractions and decimals which are also located on the number line. All integers and fractions and some of the decimal numbers are examples of *rational numbers.*

> **Rational Numbers**
>
> **Rational numbers** are numbers that can be written as quotients of integers, with the denominator not equal to zero. In other words, rational numbers can be written in the form $\frac{a}{b}$ where a and b are integers and $b \neq 0$.

The name "*ratio*nal numbers" comes from the word *ratio*, which indicates a quotient. Some examples of rational numbers are shown here, including how to write them as a quotient of integers.

$$5 \qquad \text{can be written as } \frac{5}{1}$$
$$-18 \qquad \text{can be written as } \frac{-18}{1}$$

All integers are rational numbers.

$$3\frac{1}{2} \qquad \text{can be written as } \frac{7}{2}$$

$$-\frac{5}{8} \qquad \text{can be written as } \frac{-5}{8} \text{ or } \frac{5}{-8}$$

All fractions are rational numbers.

$$0.19 \qquad \text{can be written as } \frac{19}{100}$$

$$2.7 \qquad \text{can be written as } 2\frac{7}{10} = \frac{27}{10}$$

$$0.333\ldots \qquad \text{can be written as } \frac{1}{3}$$

$$0.\overline{72} \qquad \text{can be written as } \frac{8}{11}$$

These decimals are rational numbers, but *not all decimals* are rational numbers.

Any integer can be written as the quotient of itself and 1, so all integers are rational numbers. A decimal number that comes to an end (terminates), such as 0.19 or 2.7, is a rational number. Decimal numbers that repeat in a *fixed block* of digits, such as 0.333... and 0.$\overline{72}$, are also rational numbers. (Recall that the bar above the digits 7 and 2 means that they repeat indefinitely, so 0.727272... can be written as 0.$\overline{72}$.)

Recall that to *graph* a number, we place a dot on the number line at the point corresponding to the number.

> **EXAMPLE 1** | **Graphing Rational Numbers**
>
> Graph these rational numbers on the number line.
>
> $$-\frac{3}{2}, \quad -\frac{2}{3}, \quad \frac{1}{2}, \quad 1\frac{1}{3}, \quad \frac{23}{8}, \quad 3\frac{1}{4}$$
>
> To locate the improper fractions on the number line, write them in the form of mixed numbers or decimals.
>
>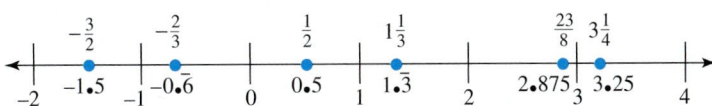

Work Problem **1** *at the Side.* ▶

There are also numbers that are *not* rational. For example, in **Section 5.10** you worked with π. Recall that we found π by dividing the circumference of a circle by its diameter. In other words, π is a quotient, but it never ends and the digits never repeat in a fixed block, no matter how far you carry out the division.

$$\frac{\text{circumference of circle}}{\text{diameter of circle}} = 3.14159265\ldots$$

Powerful computers have calculated π out to millions of decimal places, but the pattern of digits never repeats and never ends. So π is an example of an *irrational number.*

> **Irrational Numbers**
>
> **Irrational numbers** are nonrational numbers represented by points on the number line. The decimal form of an irrational number does not terminate and does not repeat in a fixed block of digits.

---

1 Graph these rational numbers on the number line.

$$-3, \; -2.75, \; -\frac{3}{4}, \; 1\frac{1}{2}, \; \frac{17}{8}$$

ANSWER

1.

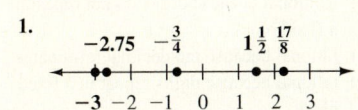

2 Identify each number as rational or irrational, and explain why.

(a) $\sqrt{36}$

(b) $0.6666\ldots$

(c) $\sqrt{13}$

(d) $0.454545\ldots$

(e) $0.131131113\ldots$

(f) 9.4375

(g) $0.\overline{27}$

The decimal form of an irrational number never terminates (never ends) and never repeats in a fixed block of digits. An example is the number $0.10110111011110\ldots$ Another example is $\sqrt{7}$. In **Section 5.8** we said that the *approximate* value for $\sqrt{7}$ is 2.646. It's approximate because $(2.646)^2$ is 7.001316, not 7. The actual value of $\sqrt{7}$ in decimal form never terminates and never repeats, so it is an irrational number. These numbers lie between rational numbers on the number line. You will work with irrational numbers in Chapter 16.

> **CAUTION**
>
> Some square roots are irrational. Examples are $\sqrt{2}$, $\sqrt{3}$, and $\sqrt{7}$. However, *not all* square roots are irrational. For example, $\sqrt{9}$ is *rational*, because $\sqrt{9} = 3$.

EXAMPLE 2 Identifying Rational and Irrational Numbers

Identify each number as *rational* or *irrational,* and explain why. Use your calculator to find square roots.

(a) $0.181818\ldots$ (b) 3.125 (c) $0.20220222022220\ldots$

(d) $\sqrt{11}$ (e) $\sqrt{16}$ (f) $0.\overline{36}$

(a) Rational, because the digits repeat in a fixed block.

(b) Rational, because the decimal terminates (comes to an end).

(c) Irrational, because the digits do *not* repeat in a fixed block.

(d) Irrational, because the decimal value of $\sqrt{11}$ never terminates or repeats.

(e) Rational, because $\sqrt{16} = 4$.

(f) Rational, because the digits repeat in a fixed block.

◀ *Work Problem* **2** *at the Side.*

Finally, *all* numbers that can be represented by points on the number line are called *real numbers.*

> **Real Numbers**
>
> The set of **real numbers** includes all the rational numbers *and* all the irrational numbers. All the real numbers can be represented by points on the number line.

All the numbers mentioned so far in this section are *real numbers*. The relationships between the various types of numbers are shown in the drawing on the next page. Notice that every real number is either a rational number or an irrational number.

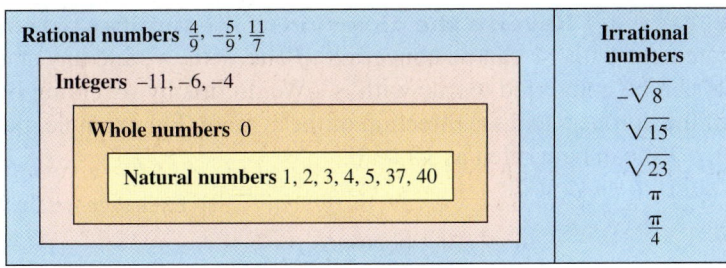

All numbers shown are real numbers.

OBJECTIVE 2 Use the symbols $\neq$, $<$, $\leq$, $>$, and $\geq$ to compare real numbers. The symbols you used in earlier chapters when comparing numbers are shown below.

$=$	$<$	$>$
is equal to	is less than	is greater than
$3(7) = 7(3)$	$-2 < 0$	$0.65 > 0.6$

Keep the meanings of $<$ and $>$ clear by remembering that the symbol always points to the lesser number. (See **Section 1.2** for more review.) Here are three other useful symbols.

$\neq$	$\leq$	$\geq$
is not equal to	is less than or equal to	is greater than or equal to
$3 - 7 \neq 7 - 3$	$0 \leq 4$	$1.25 \geq 1.2$
$\frac{1}{2} \neq 0.3$	$\frac{3}{4} \leq \frac{3}{4}$	$-8 \geq -8$

Use $\neq$ when two numbers or quantities are *not* equal.

We read the inequality $0 \leq 4$ as "zero is less than or equal to 4." If either the $<$ part *or* the $=$ part is true, then the inequality is true. So $0 \leq 4$ is true because $0 < 4$. Then $\frac{3}{4} \leq \frac{3}{4}$ is true because $\frac{3}{4} = \frac{3}{4}$.

In a similar manner, $1.25 \geq 1.2$ is true because $1.25 > 1.2$. Also, $-8 \geq -8$ is true because $-8 = -8$.

EXAMPLE 3   Using the Symbols $\neq$, $\leq$, $\geq$

Label each statement as true or false and explain why.

(a) $-4 \leq -3$ **(b)** $\frac{1}{10} \geq \frac{1}{2}$ **(c)** $-6.2 \geq -6.2$ **(d)** $2 \div 1 \neq 1 \div 2$

(a) True, because $-4 < -3$. Recall that -4 is to the left of -3 on the number line.

(b) False, because $\frac{1}{10} < \frac{1}{2}$ and $\frac{1}{10} \neq \frac{1}{2}$.

(c) True, because $-6.2 = -6.2$.

(d) True, because they are not equal. $2 \div 1$ is 2, but $1 \div 2$ is $\frac{1}{2}$ or 0.5.

Work Problem **3** *at the Side.* ▶

---

3 Label each statement as true or false and explain why.

(a) $0 \geq -\dfrac{3}{4}$

(b) $\sqrt{25} \leq \sqrt{25}$

(c) $3.06 \geq 3.6$

(d) $\dfrac{2}{3}(10) \neq \dfrac{3}{2}(10)$

(e) $-15 \leq -16$

(f) $\dfrac{1}{4} \neq \dfrac{12}{48}$

(g) $0.5 \geq \dfrac{1}{2}$

ANSWERS

3. (a) True because $0 > -\dfrac{3}{4}$

(b) True because $\sqrt{25} = \sqrt{25}$

(c) False because $3.06 < 3.6$ and $3.06 \neq 3.6$

(d) True because $6\dfrac{2}{3} \neq 15$

(e) False because $-15 > -16$ and $-15 \neq -16$

(f) False because $\dfrac{12}{48}$ in lowest terms is $\dfrac{1}{4}$

(g) True because $0.5 = \dfrac{1}{2}$

4 Rewrite each statement with the inequality symbol reversed.

(a) $0.3 \leq 0.33$

(b) $\dfrac{2}{3} > \dfrac{2}{9}$

(c) $-5 < -1$

(d) $\dfrac{9}{10} \geq 0.7$

OBJECTIVE **3** **Reverse the direction of inequality statements.**
Any statement with $<$ can be converted to one with $>$, and any statement with $>$ can be converted to one with $<$. We do this by reversing both the order of the numbers and the direction of the symbol. For example, the statement $6 < 10$ can be written as $10 > 6$.

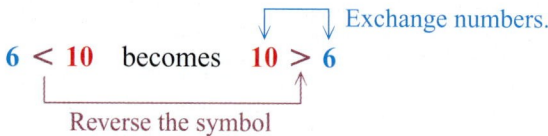

Exchange numbers.

$6 < 10 \quad$ becomes $\quad 10 > 6$

Reverse the symbol

You can verify that both statements are true: 6 is less than 10, and 10 is greater than 6.

EXAMPLE 4 **Converting Between $<$ and $>$**

The list below shows each statement written in two equally correct ways.

(a) $9 < 16 \quad$ becomes $\quad 16 > 9$

(b) $0 > -2 \quad$ becomes $\quad -2 < 0$

(c) $\dfrac{1}{2} \leq \dfrac{3}{4} \quad$ becomes $\quad \dfrac{3}{4} \geq \dfrac{1}{2}$

> Reverse the direction of the symbol **and** reverse the order of the numbers.

(d) $-7 \geq -8 \quad$ becomes $\quad -8 \leq -7$

◀ Work Problem **4** at the Side.

OBJECTIVE **4** **Use the order of operations to simplify expressions with brackets.** We have been using parentheses to show several different things.

Parentheses used to indicate multiplication:	Parentheses used to indicate a negative number:	Parentheses used to indicate order of operations:
$3\,(5)$	$6 - (-4)$	$6 + (3 + 4)$
means multiply 3 times 5.	means 6 minus negative 4.	means add $3 + 4$ first.

An expression with double parentheses, such as the expression $2\,(8 + 3\,(6 + 5))$, can be confusing. We can avoid confusion by using square brackets, [ ], in place of one pair of parentheses.

5 Simplify each expression.

(a) $4[7 + 3\,(6 + 1)]$

(b) $3[(-20 \div 5) - 7]$

EXAMPLE 5 **Using Brackets**

Simplify. $2[8 + 3\,(6 + 5)]$

Begin inside the parentheses. Then follow the order of operations as you complete the work inside the brackets.

$2[8 + 3\,(6 + 5)]$ Work inside parentheses: add $6 + 5$.

$2[8 + 3\,(11)]$ Multiply $3\,(11)$.

$2[8 + 33]$ Add $8 + 33$.

$2[41]$ Multiply 2 times 41.

82

◀ Work Problem **5** at the Side.

ANSWERS

4. **(a)** $0.33 \geq 0.3$ **(b)** $\dfrac{2}{9} < \dfrac{2}{3}$

   (c) $-1 > -5$ **(d)** $0.7 \leq \dfrac{9}{10}$

5. **(a)** 112   **(b)** -33

OBJECTIVE 5 Remove parentheses and simplify expressions using the distributive property. In **Section 2.2** you used the *distributive property* to simplify expressions. Here are two examples.

$$6(y + 4) \qquad 5(3x - 2)$$

$$6 \cdot y + 6 \cdot 4 \qquad 5 \cdot 3x - 5 \cdot 2$$

$$6y + 24 \qquad 15x - 10$$

We can also use the distributive property to remove parentheses when there is a *negative* number in front of them.

EXAMPLE 6 Using the Distributive Property to Remove Parentheses

Write without parentheses.

(a) $-2(2y + 3)$

$$-2(2y + 3)$$

> Multiply **every** term inside the parentheses by -2.

$$(-2 \cdot 2y) + (-2 \cdot 3)$$

$$-4y \quad + \quad -6$$
$$\downarrow \qquad \downarrow$$
$$-4y \quad - \quad 6$$

Use the definition of subtraction "in reverse" as in Section 2.2; change adding to subtracting; change -6 to its opposite, $+6$.

The simplified expression is $-4y - 6$.

(b) $-5(3a - 4)$

$$-5(3a - 4)$$

$$(-5 \cdot 3a) - (-5 \cdot 4)$$

Multiply every term inside the parentheses by -5.

$$-15a \quad - \quad (-20)$$
$$\downarrow \qquad \downarrow$$
$$-15a \quad + \quad 20$$

Use the definition of subtraction: change subtracting to adding; change -20 to its opposite, $+20$.

The simplified expression is $-15a + 20$.

> **CAUTION**
> Watch the signs carefully when there is a *negative* number in front of the parentheses, as in Example 6. Notice that every term in the simplified expression has the *opposite sign* from the original expression.

Work Problem **6** *at the Side.* ▶

Sometimes an expression may have just a negative sign in front of parentheses, such as $-(3x + 5)$. We can rewrite this as $-1(3x + 5)$ and then use the distributive property to multiply every term inside the parentheses by -1.

---

6 Write without parentheses.

(a) $-3(4b + 1)$

(b) $-7(2x - 3)$

(c) $-4(h - 5)$

(d) $-6(-2y + 4)$

7 Write without parentheses.

(a) $-(6k - 5)$

(b) $-(-2 - r)$

(c) $-(-5y + 8)$

(d) $-(z + 4)$

> **EXAMPLE 7** Using the Distributive Property to Remove Parentheses

Write without parentheses.

(a) $-(3x + 5)$ can be written $-1(3x + 5)$ Multiply **every** term inside the parentheses by -1.

$$(-1 \cdot 3x) + (-1 \cdot 5)$$

$$-3x \; + \; (-5) \qquad \text{Use the definition of subtraction "in reverse."}$$

$$-3x \; - \; 5$$

The simplified expression is $-3x - 5$.

(b) $-(-7r - 8)$ can be written as $-1(-7r - 8)$ Multiply every term inside the parentheses by -1.

$$(-1 \cdot -7r) - (-1 \cdot 8)$$

$$7r \; - \; (-8) \qquad \text{Use the definition of subtraction.}$$

$$7r \; + \; 8$$

The simplified expression is $7r + 8$.

◀ Work Problem **7** at the Side.

---

8 Simplify.

(a) $10p + 3(5 + 2p)$

(b) $7x - 2 - (1 + x)$

(c) $-(3k^2 + 5k) + 7(k^2 - 4k)$

(d) $-2(4b - 3) - 5(2b^2 + 1)$

> **EXAMPLE 8** Simplifying Expressions Involving Like Terms

Simplify each expression.

(a) $14y + 2(6 + 3y)$

$$14y + 2(6 + 3y) \qquad \text{Multiply every term inside the parentheses by 2.}$$

$$14y + 12 + 6y \qquad \text{Combine like terms.}$$

$$20y + 12$$

(b) $-(2 - r) + 10r$ is written $-1(2 - r) + 10r$ Rewrite $-(2 - r)$ as $-1(2 - r)$. Then use the distributive property.

$$-2 + r + 10r$$

$$-2 + r + 10r \qquad \text{Combine like terms.}$$

$$-2 + 11r$$

(c) $5(2a^2 - 6a) - 3(4a^2 - 9)$

Multiply every term inside these parentheses by 5. $5(2a^2 - 6a) -3(4a^2 - 9)$ Multiply every term inside these parentheses by -3.

$$10a^2 - 30a - 12a^2 + 27 \qquad \text{Combine like terms.}$$

$$-2a^2 - 30a + 27$$

◀ Work Problem **8** at the Side.

ANSWERS

7. (a) $-6k + 5$ (b) $2 + r$ (c) $5y - 8$
    (d) $-z - 4$
8. (a) $16p + 15$ (b) $6x - 3$ (c) $4k^2 - 33k$
    (d) $-10b^2 - 8b + 1$

10.1 ▸▸▸ Exercises

*Identify each number as **rational** or **irrational** and explain why. See Example 2.*

1. -0.0625

2. $\sqrt{5}$

3. π

4. 0

5. $\dfrac{3}{4}$

6. $-5\dfrac{2}{3}$

7. $0.636363\ldots$

8. $0.3233233323333\ldots$

9. $\sqrt{2}$

10. $0.416666\ldots$

*List all numbers from each set that are **(a)** natural numbers, **(b)** whole numbers, **(c)** integers, **(d)** rational numbers, **(e)** irrational numbers, **(f)** real numbers.*

11. $\left\{-9,\ -\sqrt{7},\ -1\dfrac{1}{4},\ -\dfrac{3}{5},\ 0,\ \sqrt{5},\ 3,\ 5.9,\ 7\right\}$

12. $\left\{-5.3,\ -5,\ -\sqrt{3},\ -1,\ -\dfrac{1}{9},\ 0,\ 1.2,\ 4,\ \sqrt{12}\right\}$

Write each statement in words. Then label the statement as true or false, and explain why. See Example 3.

13. $0.75 \neq \dfrac{3}{4}$

14. $0 \geq -1$

15. $-4 \leq -5$

16. $\dfrac{1}{3} \neq 0.3$

17. $0 \geq 0$

18. $4 \leq 5$

Write each statement with the inequality symbol reversed. See Example 4.

19. $12 < 19$

20. $0.55 > 0.5$

21. $\dfrac{4}{5} \geq \dfrac{1}{2}$

22. $-40 \leq -30$

First simplify each statement wherever possible. Then label the statement as true or false. See Examples 3 and 5.

23. $-17 \leq 1 - 18$

24. $-12 \geq -10 - 2$

25. $-6(8) + 9(5) \geq 0$

26. $-4(-20) - 15(5) \geq 0$

27. $6[5 - 3(4 - 2)] \neq 6$

28. $-3[(0 - 5) + 2] \neq 9$

29. $3^2[(-10 \div 5) + 7] \le 40$

30. $-(5^2)[-3 + 2(-6 \div 2)] \le -225$

31. $[6 - 4(4)] \div (-10) \ge 1$

32. $-1 \ne 4 \div [8(2) - 20]$

33. $0 \ne 12 \div [3(2) - 6]$

34. $[12 - 2(36 \div 6)] \div 9 \le 0$

Relating Concepts (Exercises 35–38) For Individual or Group Work

Use the table on movie gross receipts (all the income, before paying expenses) to work
Exercises 35–38 in order.

ALL-TIME TOP AMERICAN MOVIES THROUGH OCTOBER 2008

Rank	Title (Year)	Gross Receipts (millions)
1	*Titanic (1997)*	$600.8
2	*The Dark Knight (2008)*	527.4
3	*Star Wars (1977)*	461.0
4	*Shrek 2 (2004)*	436.5
5	*E.T.: The Extra-Terrestrial (1982)*	435.1
6	*The Phantom Menace (1999)*	431.1

Source: www.boxofficereport.com

35. (a) Which films had gross receipts greater than 461 million dollars?

  (b) Which films had gross receipts greater than or equal to 461 million dollars?

36. (a) Which films had gross receipts less than 435.1 million dollars?

  (b) Which films had gross receipts less than or equal to 435.1 million dollars?

37. Write a statement using the $\le$ symbol that describes the gross receipts for the films ranked 4, 5, and 6.

38. Write a statement using the $\ge$ symbol that describes the gross receipts for the top four films.

Use the distributive property to simplify each expression. See Example 7.

39. $-(4t + 5m)$

40. $-(9x + 12y)$

41. $-(-5c - 4d)$

42. $-(-13x - 15y)$

43. $-(6h - n)$

44. $-(a - 7b)$

45. $-(-3q + 5r - 8s)$

46. $-(-4z + 5w - 9y)$

Simplify each expression. See Examples 6–8.

47. $13p + 4(4 - 8p)$

48. $5x + 3(7 - 2x)$

49. $-4(y - 7) - 6$

50. $-5(t - 13) - 4$

51. $-(6 - y) + y^2 - 6$

52. $-(w + 5) - w^2 + 5$

53. $2(3b^2 - b) - 4(b - 2)$

54. $7(x^2 - 3) - 5(x + 6)$

55. $-3(-a + 1) - (2a - 4)$

56. $-8(-5 + c^2) - (10 - 7c^2)$

57. $-10(-3k - 2) + 6(-4 - k)$

58. $-7(-4 + 3x^2) + 5(x^2 - 6)$

10.2 ▶▶▶ More on Solving Linear Equations

The equations you solved in Chapters 2–7 were all *linear equations in one variable.* Such equations contain only one variable, and that variable is always raised to the first power and never appears in a denominator. Some examples of linear equations are shown below.

$$5m + 1 = 16 \qquad 2k - 2 = 5k - 11 \qquad 100 + \frac{a}{2} = 116$$

OBJECTIVE 1 Solve more difficult linear equations. We developed a 5-step process in **Sections 2.5** and **4.7** to solve linear equations. As a review, here are the steps.

Solving Linear Equations

Step 1 If possible, use the **distributive property** to remove parentheses.

Step 2 **Combine** any like terms on the left side of the equation. Combine any like terms on the right side of the equation.

Step 3 **Add** or **subtract** the same amount on both sides of the equation so that the variable term ends up by itself on one side of the equal sign and a number is by itself on the other side. You may have to do this step more than once.

Step 4 **Multiply** or **divide** both sides by the same number to find the solution.

Step 5 **Check** your solution by going back to the *original* equation. Replace the variable with your solution. Follow the order of operations to complete the calculations. If the two sides of the equation balance, your solution is correct.

EXAMPLE 1 Review of Solving Linear Equations

Solve this equation and check the solution: $-2 + 3(2x + 7) = 7 + 3x$

Step 1 Use the distributive property. $-2 + 3(2x + 7) = 7 + 3x$

Step 2 Combine like terms. $-2 + 6x + 21 = 7 + 3x$

$$6x + 19 = 7 + 3x$$

Step 3 Subtract 19 from both sides.

$$\underline{\qquad -19 \qquad -19 \qquad}$$

$$6x + 0 = -12 + 3x$$

$$6x = -12 + 3x$$

Subtract $3x$ from both sides. $\underline{-3x \qquad\qquad -3x}$

$$3x = -12 + 0$$

Step 4 Divide both sides by 3. $\dfrac{3x}{3} = \dfrac{-12}{3}$

The solution is -4. $x = -4 \leftarrow$ Solution

Continued on Next Page

1 Solve each equation.

(a) $2p + 4 = 7(p - 2) + p$

Step 5 Check by replacing each x with -4 in the *original* equation.

$$-2 + 3(2x + 7) = 7 + 3x$$

$$-2 + 3(2 \cdot -4 + 7) = 7 + 3(-4)$$

$$-2 + 3(-8 + 7) = 7 + -12$$

$$-2 + 3(-1) = -5$$

$$-2 + -3 = -5$$

$$-5 = -5 \quad \text{Balances}$$

When x is replaced with -4, the equation balances, so **-4 is the correct solution** (***not*** -5).

> **Note**
>
> Recall from **Section 2.5** that more than one sequence of steps will work. For example, in *Step 3* above, we could first subtract $3x$ from both sides, then subtract 19 from both sides. The solution would be the same.

◀ Work Problem **1** at the Side.

(b)
$-5y + 7y - 6y - 9 = 3 + 2y$

EXAMPLE 2 Removing Parentheses Before Solving an Equation

Solve this equation: $8a - (3 + 2a) = 3a + 1$

Start by removing the parentheses on the left side of the equation. Remember that the $-$ sign in this situation acts like a factor of -1, changing the sign of *every* term inside the parentheses.

Step 1 Use distributive property to multiply every term inside the parentheses by -1.

$$8a - (3 + 2a) = 3a + 1$$
$$8a - 1(3 + 2a) = 3a + 1$$

Step 2 Combine like terms.

$$8a - 3 - 2a = 3a + 1$$
$$6a - 3 = 3a + 1$$

Step 3 Add 3 to both sides.

$$\underline{\quad +3 \qquad +3}$$
$$6a = 3a + 4$$

Subtract $3a$ from both sides.

$$\underline{-3a \qquad -3a}$$

Step 4 Divide both sides by 3.

$$\frac{3a}{3} = \frac{4}{3}$$
$$a = \frac{4}{3}$$

The solution is $\frac{4}{3}$. You can *check* the solution by going back to the *original* equation and replacing each a with $\frac{4}{3}$.

Continued on Next Page

ANSWERS

1. (a) $p = 3$ **(b)** $y = -2$

CAUTION
When there is a subtraction sign in front of parentheses, be very careful to *change* the sign of *every* term inside the parentheses. In Example 2, the signs on the left side of the equation changed as shown below.

$$8a - (3 + 2a) \qquad \text{Another} \atop \text{example} \qquad -6x - (2x - 5)$$

$$8a - 3 - 2a \qquad\qquad\qquad -6x - 2x + 5$$

Work Problem **2** *at the Side.* ▶

2 Solve each equation.

(a) $2m - (7m + 6) = 39$

EXAMPLE 3 **Solving Linear Equations**

Solve: $4(8 - 3t) = 32 - 8(t + 2)$

$$4(8 - 3t) = 32 - 8(t + 2) \qquad \text{Distributive property; notice change in signs on right side.}$$

$$32 - 12t = 32 - 8t - 16 \qquad \text{Combine like terms.}$$

$$32 - 12t = 16 - 8t$$

$$\underline{-32 \qquad\qquad -32} \qquad \text{Subtract 32 from both sides.}$$

$$-12t = -16 - 8t$$

$$\underline{+ 8t \qquad\qquad + 8t} \qquad \text{Add } 8t \text{ to both sides.}$$

$$\frac{-4t}{-4} = \frac{-16}{-4} \qquad \text{Divide both sides by } -4.$$

$$t = 4$$

The solution is 4.

Check

$$4(8 - 3t) = 32 - 8(t + 2) \qquad \text{Original equation.}$$

$$4(8 - 3 \cdot 4) = 32 - 8(4 + 2) \qquad \text{Replace } t \text{ with 4.}$$

$$4(8 - 12) = 32 - 8(6)$$

$$4(-4) = 32 - 48$$

$$-16 = -16 \qquad \text{Balances}$$

The solution is 4 (**not** -16).

When t is replaced with 4, the equation balances, so **4 is the correct solution** (**not** -16).

Work Problem **3** *at the Side.* ▶

(b) $6 = 4x - (3 - 2x)$

3 Solve.

(a) $2(4 - 3r) = 3(r + 1) + 14$

(b) $2 - 3(2 + 6z) = 4(z + 1)$

OBJECTIVE 2 **Solve equations that have no solution or infinitely many solutions.** The equations solved so far have had exactly *one* solution. However, as you'll see in the next examples, some equations have *many* solutions or *no* solution.

4 Solve each equation.

(a) $2(x - 6) = 3 + 2x - 15$

(b) $4b + 2(3 - 2b) = 6$

EXAMPLE 4 Solving an Equation That Has Infinitely Many Solutions

Solve: $5x - 15 = 5(x - 3)$

$5x - 15 = 5(x - 3)$ Use the distributive property.

$5x - 15 = 5x - 15$

$\underline{\;+\,15 +\,15}$ Add 15 to both sides.

$5x = 5x$

$\underline{-5x -5x}$ Subtract $5x$ from both sides.

$0 = 0$ True statement; no variable.

The variable has "disappeared." It is true that $0 = 0$, but what is the solution? Actually there are an *infinite* number of solutions because *any real number* is a solution. Let's try a few different possibilities.

Try 7 as a solution. Replace each x with 7.	Try -2 as a solution. Replace each x with -2.	Try $\frac{1}{2}$ as a solution. Replace each x with $\frac{1}{2}$.
$5x - 15 = 5(x - 3)$	$5x - 15 = 5(x - 3)$	$5x - 15 = 5(x - 3)$
$5 \cdot 7 - 15 = 5(7 - 3)$	$5(-2) - 15 = 5(-2 - 3)$	$5(\frac{1}{2}) - 15 = 5(\frac{1}{2} - 3)$
$35 - 15 = 5(4)$	$-10 - 15 = 5(-5)$	$\frac{5}{2} - 15 = 5(-\frac{5}{2})$
$20 = 20$	$-25 = -25$	$-\frac{25}{2} = -\frac{25}{2}$
Balances	Balances	Balances

Any real number you tried would balance. Let's see why that is so. Look again at the first step in solving the equation. After using the distributive property to remove the parentheses, both sides are exactly the same.

$5x - 15 = 5(x - 3)$ Use the distributive property.

$5x - 15 = 5x - 15$ Both sides are exactly the same.

When both sides of an equation are exactly the same, the equation is called an **identity.** An identity is true for all replacements of the variables.

The solution for $5x - 15 = 5(x - 3)$ is **all real numbers**.   Write **all real numbers** as the solution (**not** 0).

> **CAUTION**
> When solving an equation like the one in Example 4, do ***not*** write "0" as the solution. While 0 is *one* of the solutions, there are *infinitely many other solutions.* So, you need to write "the solution is all real numbers."

◄ *Work Problem* **4** *at the Side.*

EXAMPLE 5 **Solving an Equation That Has No Solution**

Solve: $2w - 7(w + 1) = -5w + 4$

$2w - 7(w + 1) = -5w + 4$ Use the distributive property.

$2w - 7w - 7 = -5w + 4$ Combine like terms.

$-5w - 7 = -5w + 4$

$\underline{+5w \qquad\quad +5w}$ Add $5w$ to both sides.

$-7 = 4$ False statement; no variable. — In this situation, write **no solution.**

As in Example 4, the variable has "disappeared." This time, however, we are left with a *false* statement: $-7 \neq 4$. Whenever this happens in solving an equation, it is a signal that the equation has *no solution*. So, you write "no solution."

Work Problem **5** *at the Side.* ▶

OBJECTIVE **3** **Solve equations by first clearing fractions and decimals.** In **Section 4.7** we solved equations containing fractions such as $\frac{3}{4}x = 12$. We multiplied both sides by $\frac{4}{3}$ (the reciprocal of $\frac{3}{4}$).

$$\frac{\cancel{4}}{\cancel{3}}\left(\frac{\cancel{3}}{\cancel{4}}x\right) = 12\left(\frac{4}{3}\right)$$

$$x = 16$$

Another method is to first clear the equation of fractions. You can do that by multiplying both sides by the LCD (lowest common denominator) of all the fractions in the equation. (See **Section 4.4** for ways to find the LCD.)

EXAMPLE 6 **Clearing an Equation of Fractions**

Solve: $\frac{2}{3}x - \frac{1}{2}x = -\frac{1}{6}x + 2$

Remember that 2 can be written as $\frac{2}{1}$. The least common denominator for $\frac{2}{3}, \frac{1}{2}, -\frac{1}{6}$, and $\frac{2}{1}$ is 6. Start by multiplying both sides of the equation by 6.

$$\frac{2}{3}x - \frac{1}{2}x = -\frac{1}{6}x + 2$$

$$6\left(\frac{2}{3}x - \frac{1}{2}x\right) = 6\left(-\frac{1}{6}x + 2\right)$$ — Multiply **both** sides by the LCD, which is 6.

$$\cancel{6}\left(\frac{2}{\cancel{3}}x\right) - \cancel{6}\left(\frac{1}{\cancel{2}}x\right) = \cancel{6}\left(-\frac{1}{\cancel{6}}x\right) + 6(2)$$ Use the distributive property to multiply each term by 6.

$$4x - 3x = -1x + 12$$ Combine like terms.

$$1x = -1x + 12$$

$$\underline{+1x \qquad\quad +1x}$$ Add $1x$ to both sides.

$$\frac{2x}{2} = \frac{12}{2}$$ Divide both sides by 2.

$$x = 6$$

The solution is 6. Check the solution by going back to the *original* equation and replacing each x with 6.

——— **Continued on Next Page**

5 Solve each equation.

(a) $9y - 4 = 3y + 6(y + 1)$

(b)
$2m - 5(m + 2) = -3(m - 5)$

6 Solve: $\dfrac{1}{4}x - 4 = \dfrac{3}{2}x + \dfrac{3}{4}x$

> **CAUTION**
> When clearing an equation of fractions, be careful to multiply *every* term on both sides by the LCD.

◀ *Work Problem* **6** *at the Side.*

You can also clear an equation of decimals by multiplying both sides by a power of 10.

> **EXAMPLE 7** **Clearing an Equation of Decimals**
>
> Solve: $0.1t + 0.05(20 - t) = 0.09(20)$
>
> Our goal is to multiply by some number that will change the coefficients from decimals to whole numbers. Since decimal numbers involve tenths, hundredths, thousandths, and so on, we will multiply by 10, or by 100, or by 1000, and so on. In this example there are at most two decimal places (hundredths) so we multiply by 100. A quick way to multiply a number by 100 is to move the decimal point two places to the right.
>
> $$0.10t + 0.05(20 - t) = 0.09(20) \qquad \text{To multiply by 100, move the decimal point two places to the right.}$$
> $$10t + 5(20 - t) = 9(20) \qquad \text{Use the distributive property.}$$
> $$10t + 100 - 5t = 180 \qquad \text{Combine like terms.}$$
> $$5t + 100 = 180$$
> $$\underline{-100 \qquad -100} \qquad \text{Subtract 100 from both sides.}$$
> $$\frac{5t}{5} = \frac{80}{5} \qquad \text{Divide both sides by 5.}$$
> $$t = 16$$
>
> The solution is 16. Check the solution by going back to the *original* equation and replacing each t with 16.

◀ *Work Problem* **7** *at the Side.*

7 Solve:

$0.06(100 - y) + 0.4y = 0.05(86)$

10.2 ▶▶▶ Exercises

Solve each equation. See Examples 1–3.

1. $5m + 8 = 7 + 4m$

2. $2(2r + 1) = 3r - 6$

3. $10p + 6 = 4(3p - 1)$

4. $-5x + 8 = -3x + 10$

5. $7r - 5r + 2 = 5r - r$

6. $9p - 4p + 6 = 7p - 3p$

7. $x + 3 = -(2x + 2)$

8. $2x + 1 = -(x + 3)$

9. $4(2x - 1) = -6(x + 3)$

10. $6(3w + 5) = -2(-10w + 10)$

11. $(5y + 6) - (3 + 4y) = 10$

12. $(8r - 3) - (7r + 1) = -6$

13. $2(p + 5) - (9 + p) = -3$

14. $4(k - 6) - (3k + 2) = -5$

15. $-6(2b + 1) + (13b - 7) = 0$

16. $-5(3w - 3) + (1 + 16w) = 0$

17. $-2(8p + 2) - 3(2 - 7p) = 2(4 + 2p)$

18. $-5(1 - 2z) + 4(3 - z) = 7(3 + z)$

19. $4(7x - 1) + 3(2 - 5x) = 4(3x + 5) - 6$

20. $9(2m - 3) - 4(5 + 3m) = 5(4 + m) - 3$

Solve each equation. See Examples 4 and 5.

21. $3x + 9 = 3x + 8$

22. $-2x + 5 = -2x$

23. $8x + 1 = 1 + 8x$

24. $4w - 5 = -5 + 4w$

25. $6x + 5 - 7x + 3 = 5x - 6x - 4$

26. $4x - 3 - 8x + 1 = 5x - 9x + 7$

27. $6(4x - 1) = 12(2x + 3)$

28. $6(2x + 8) = 4(3x - 6)$

29. $3(2x - 4) = 6(x - 2)$

30. $3(6 - 4x) = 2(-6x + 9)$

31. $10(-2x + 1) = -14(x + 2) + 38 - 6x$

32. $2(2 - 3r) = 5(1 - r) - r - 1$

33. After working correctly through several steps of the solution of a linear equation, a student obtains the equation $7x = 3x$. Then the student divides both sides by x to get $7 = 3$ and gives "no solution" as the answer. Is this correct? If not, explain why.

34. If the final step in solving a linear equation leads to the statement $0 = 0$, explain why it is incorrect to say that 0 is the solution of the equation. What are the solutions of the equation?

Solve each equation by first clearing it of fractions or decimals. See Examples 6 and 7.

35. $\dfrac{3}{5}t - \dfrac{1}{10}t = t - \dfrac{5}{2}$

36. $-\dfrac{2}{7}r + 2r = \dfrac{1}{2}r + \dfrac{17}{2}$

37. $-\dfrac{1}{4}(x - 12) + \dfrac{1}{2}(x + 2) = x + 4$

38. $\dfrac{1}{9}(y + 18) + \dfrac{1}{3}(2y + 3) = y + 3$

39. $\dfrac{2}{3}k - \left(k + \dfrac{1}{4}\right) = \dfrac{1}{12}(k + 4)$

40. $-\dfrac{5}{6}q - \left(q - \dfrac{1}{2}\right) = \dfrac{1}{4}(q + 1)$

41. $5.6x + 2 = 4.6x$

42. $9.1x - 5 = 8.1x$

43. $5.2q - 4.6 - 7.1q = -2.1 - 1.9q - 2.5$

44. $-4.0x + 2.7 - 1.6x = 1.3 - 5.6x + 1.4$

45. $0.2(60) + 0.05x = 0.1(60 + x)$

46. $0.3(30) + 0.15x = 0.2(30 + x)$

47. $x + 0.05(12 - x) = 0.1(63)$

48. $0.92x + 0.98(12 - x) = 0.96(12)$

49. $0.06(10,000) + 0.08x = 0.072(10,000 + x)$

50. $0.02(5000) + 0.03x = 0.025(5000 + x)$

Solve each equation, and check your solution. See Examples 1–7.

51. $9(v + 1) - 3v = 2(3v + 1) - 8$

52. $-3(5z + 24) + 2 = 2(3 - 2z) - 4$

53. $\frac{1}{2}(x + 2) + \frac{3}{4}(x + 4) = x + 5$

54. $-(6k - 5) - (-5k + 8) = -3$

55. $-(4y + 2) - (-3y - 5) = 3$

56. $\frac{1}{3}(x + 3) + \frac{1}{6}(x - 6) = x + 3$

57. $0.10(x + 80) + 0.20x = 14$

58. $0.30(x + 15) + 0.40(x + 25) = 25$

59. $4(x + 8) = 2(2x + 6) + 20$

60. $4(x + 3) = 2(2x + 8) - 4$

61. $-2(2s - 4) - 8 = -3(4s + 4) - 1$

62. $8(t - 3) + 4t = 6(2t + 1) - 10$

10.3 ▷▷▷ Formulas and Solving for a Specified Variable

Many application problems can be solved with a formula. For example, in Chapters 3–8 you used formulas for perimeter and area of geometric figures, for money earned on bank savings, and for converting between Celsius and Fahrenheit temperatures. The formulas used in this book are shown on the inside back cover.

OBJECTIVE 1 Solve a formula for one variable given the values of the other variables. Given the values of all but one of the variables in a formula, we can find the value of the remaining variable by using the methods introduced in Chapter 3.

OBJECTIVES

1 Solve a formula for one variable, given the values of the other variables.

2 Solve a formula for a specified variable.

EXAMPLE 1 Using a Formula to Evaluate a Variable

Find the value of the remaining variable.

(a) $A = lw$; $A = 64$, $l = 10$

Recall that this formula gives the area of a rectangle with length l and width w. Substitute the given values into the formula and then solve for w.

$$A = lw$$
$$64 = 10w \qquad \text{Replace } A \text{ with 64 and } l \text{ with 10.}$$
$$\frac{64}{10} = \frac{10w}{10} \qquad \text{Divide both sides by 10.}$$
$$6.4 = w$$

Check that the width of the rectangle is 6.4.

Check

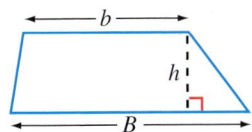

$$A = l \quad w$$
$$64 = 10\,(6.4)$$
$$64 = \quad 64 \qquad \text{Balances}$$

(b) $A = \frac{1}{2}h(b + B)$; $A = 210$, $B = 27$, $h = 10$

This formula gives the area of a *trapezoid* with parallel sides of lengths b and B and height h between the parallel sides.

Again, begin by substituting the given values into the formula.

$$A = \frac{1}{2}h(b + B)$$

$$210 = \frac{1}{2}(10)(b + 27) \qquad \text{Replace } A \text{ with 210, } h \text{ with 10, and } B \text{ with 27.}$$

Continued on Next Page

1 Find the value of the remaining variable in each formula.

(a) $I = prt$; $I = \$246$, $r = 0.06$, $t = 2$

Now solve for b.

$$210 = \frac{1}{2}(10)(b + 27) \qquad \text{Multiply } \frac{1}{2}(10) \text{ to get } 5.$$

$$210 = 5(b + 27) \qquad \text{Use the distributive property.}$$

$$210 = 5b + 135$$

$$\underline{-135 \qquad\qquad -135} \qquad \text{Subtract 135 from both sides.}$$

$$75 = 5b$$

$$\frac{75}{5} = \frac{5b}{5} \qquad \text{Divide both sides by 5.}$$

$$15 = b$$

Check that the length of the shorter parallel side, b, is 15.

◀ *Work Problem* **1** *at the Side.*

Formulas are often used to solve application problems. *It is a good idea to draw a sketch when a problem involves a geometric figure.*

(b) $P = 2l + 2w$; $P = 126$, $w = 25$

EXAMPLE 2 **Finding the Width of a Rectangular Lot**

A rectangular garden plot has a perimeter of 8 meters and a length of 2.5 meters. Find the width of the lot.

Begin by drawing a rectangle and labeling the sides.

We chose w to represent the width of the garden plot in meters.

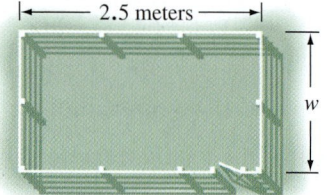

The formula for the perimeter of a rectangle is $P = 2l + 2w$.

Find the width by substituting 8 for P and 2.5 for l in the formula.

$$8 = 2(\mathbf{2.5}) + 2w \qquad \text{Replace } P \text{ with } \mathbf{8} \text{ and } l \text{ with } \mathbf{2.5}$$

Now solve the equation.

$$8 = \underbrace{2(2.5)}_{} + 2w \qquad \text{Simplify.}$$

$$8 = \quad 5 \quad + 2w$$

$$\underline{-5 \quad -5} \qquad \text{Subtract 5 from both sides.}$$

$$\frac{3}{2} = \frac{2w}{2} \qquad \text{Divide both sides by 2.}$$

$$1.5 = w$$

The width is 1.5 meters.

2 A farmer has 800 meters of fencing material to enclose a rectangular field. The width of the field is 175 meters. Find the length of the field.

Check If the length is 2.5 meters and the width is 1.5 meters, then $P = 2(2.5) + 2(1.5) = 5 + 3 = 8$ meters. This matches the perimeter given in the problem.

◀ *Work Problem* **2** *at the Side.*

ANSWERS

1. (a) $p = \$2050$ (b) $l = 38$
2. $l = 225$ meters

OBJECTIVE 2 Solve a formula for a specified variable. Sometimes it is necessary to solve many problems that use the same formula. For example, a surveying class might need to solve several problems that involve the formula for the area of a rectangle, $A = lw$. Suppose that in each problem the area (A) and the length (l) of a rectangle are given, and the width (w) must be found. Rather than solving for w each time the formula is used, it would be simpler to rewrite the *formula* so that it is solved for w. This process is called *solving for a specified variable.*

In solving a formula for a specified variable, we treat the specified variable as if it were the *only* variable in the equation, and *treat the other variables as if they were numbers.* Then we use the same steps to solve the equation for the specified variable that we used earlier to solve equations with just one variable.

3 **(a)** Solve $I = prt$ for t.

EXAMPLE 3 **Solving for a Specified Variable**

Solve $A = lw$ for w.

Think of "undoing" what has been done to w. Since w is *multiplied* by l, undo the multiplication by *dividing* both sides by l. (Recall that division is the opposite of multiplication.)

$$A = lw$$
$$\frac{A}{l} = \frac{lw}{l} \qquad \text{Divide both sides by } l. \text{ On the right side, divide out the common factor of } l.$$
$$\frac{A}{l} = w$$

The formula is now solved for w.

(b) Solve $P = a + b + c$ for a.

> **Note**
>
> Look back at Example 1(a) in this section. It is a specific example where we knew that A was 64 and l was 10. Notice that we divided 64 by 10 to find w. In other words, we divided the area by the length to find the width. Or, $\frac{A}{l} = w$.

Work Problem **3** *at the Side.* ▶

EXAMPLE 4 **Solving for a Specified Variable**

(a) Solve $P = 2l + 2w$ for l.

We want to get l by itself on one side of the equation. We begin by subtracting $2w$ from both sides.

$$P = 2l + 2w$$
$$\underline{ -2w \qquad\qquad -2w} \qquad \text{Subtract } 2w \text{ from both sides.}$$
$$P - 2w = 2l$$
$$\frac{P - 2w}{2} = \frac{2l}{2} \qquad \text{Divide both sides by 2.}$$
$$\frac{P - 2w}{2} = l$$

The last step gives the formula solved for l. It is in simplest form, so you need not do anything else to it.

Continued on Next Page

4 **(a)** Solve $A = p + prt$ for t.

(b) Solve $F = \frac{9}{5}C + 32$ for C.

We need to get C by itself on one side of the equation.

First "undo" the *addition* of 32 to $\frac{9}{5}$ C by *subtracting* 32 from both sides.

$$F = \frac{9}{5}C + 32$$

$$\underline{-32 \qquad\qquad -32} \qquad \text{Subtract 32 from both sides.}$$

$$F - 32 = \frac{9}{5}C$$

Now multiply both sides by $\frac{5}{9}$ (the reciprocal of $\frac{9}{5}$).

$$\frac{5}{9}(F - 32) = \frac{1}{\cancel{9}} \cdot \frac{\cancel{9}}{\cancel{5}}C \qquad \text{Multiply both sides by } \frac{5}{9}.$$

$$\frac{5}{9}(F - 32) = C$$

The result is the formula for converting temperatures from Fahrenheit to Celsius that you used in **Section 8.5.**

◀ *Work Problem* **4** *at the Side.*

(b) Solve $Ax + By = C$ for y.

FOR EXTRA HELP

Find the value of the remaining variable in each formula. See Example 1.

1. $P = 2l + 2w$ (perimeter of a rectangle)
$P = 20, w = 4$

2. $P = 2l + 2w$
$P = 26, l = 8$

3. $A = \frac{1}{2}bh$ (area of triangle); $A = 70, b = 10$

4. $A = \frac{1}{2}bh$; $A = 64, h = 16$

5. $P = a + b + c$ (perimeter of a triangle)
$P = 15, a = 3, b = 7$

6. $P = a + b + c$
$P = 12, a = 3, c = 5$

7. $d = rt$ (distance formula); $d = 100, t = 2.5$

8. $d = rt$; $d = 252, r = 45$

9. $I = prt$ (simple interest)
$I = 875, p = 5000, r = 0.025$

10. $I = prt$
$I = 1575, r = 0.035, t = 6$

11. $C = 2\pi r$ (circumference of a circle)
$C = 8.164, \pi \approx 3.14$

12. $C = 2\pi r$, $C = 16.328, \pi \approx 3.14$

13. $V = lwh$ (volume of a rectangular solid)
$V = 384, l = 12, h = 4$

14. $V = lwh$
$V = 150, w = 5, h = 3$

Use a formula to write an equation for each problem, then solve it. See Example 2.

15. The newspaper *The Constellation,* printed in 1859 in New York City, had a page length of 51 inches and a perimeter of 172 inches. **(a)** What was the width of the page? **(b)** What was the area of the page?

16. The *Daily Banner,* published in Roseberg, Oregon, in the nineteenth century, had a page width of 3 inches and a perimeter of 13 inches. **(a)** What was the page length? **(b)** What was the area of the page?

17. The Skydome in Toronto, Canada, is the first stadium with a hard-shell, retractable roof. The steel dome has a circumference of 1978 feet. To the nearest foot, what is the diameter of this dome? Use 3.14 as the approximate value of π.

18. The largest drum ever constructed was built in Japan in 2001. The circular face of the drum had a circumference of 49.42 ft. What was the diameter of the drum face, to the nearest hundredth of a foot? Use 3.14 as the approximate value of π.

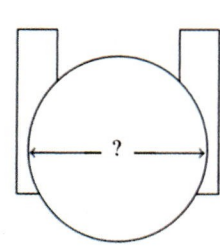

Solve the formula for the specified variable. See Examples 3–5.

19. $d = rt$ for r

20. $d = rt$ for t

21. $A = lw$ for l

22. $V = lwh$ for h

23. $P = a + b + c$ for a

24. $P = a + b + c$ for b

25. $I = prt$ for p

26. $I = prt$ for r

27. $A = \frac{1}{2}bh$ for b

28. $A = \frac{1}{2}bh$ for h

29. $A = p + prt$ for r

30. $P = 2l + 2w$ for w

31. $V = \pi r^2 h$ for h

32. $V = \frac{1}{3}\pi r^2 h$ for h

33. $F = \frac{9}{5}C + 32$ for C

Relating Concepts (Exercises 34–36) For Individual or Group Work

In many cases there are equally acceptable equivalent answers for a problem. ***Work Exercises 34–36 in order*** *to see how two seemingly different answers to a problem can both be correct.*

34. Solve the formula $P = 2l + 2w$ for w using the following steps.
 (a) Subtract $2l$ from both sides.
 (b) Divide both sides by 2.

35. Referring to the formula in Exercise 34, solve for w again using the following steps.
 (a) Divide each term on both sides by 2.
 (b) Subtract l from both sides.

36. Compare the results in Exercises 34(b) and 35(b). Both are acceptable answers. To show that they are equivalent, give a justification for each step below.

 (a) $\dfrac{P}{2} - l = \dfrac{P}{2} - \dfrac{2}{2} \cdot l$

 (b) $\phantom{\dfrac{P}{2} - l} = \dfrac{P}{2} - \dfrac{2}{2} \cdot \dfrac{l}{1}$

 (c) $\phantom{\dfrac{P}{2} - l} = \dfrac{P}{2} - \dfrac{2l}{2}$

 (d) $\phantom{\dfrac{P}{2} - l} = \dfrac{P - 2l}{2}$

10.4 ▶▶▶ Solving Linear Inequalities

The addition and multiplication properties can be extended to inequalities. **Inequalities** are statements with algebraic expressions related in the following ways:

$<$ "is less than"

$\le$ "is less than or equal to"

$>$ "is greater than"

$\ge$ "is greater than or equal to."

We solve an inequality by finding all real number solutions for it. For example, the solutions of $x \le 2$ include all *real numbers* that are less than or equal to 2, and not just the *integers* less than or equal to 2.

> **OBJECTIVE 1 Graph the solutions of inequalities on a number line.** Graphing is a good way to show the solutions of an inequality. To graph all real numbers satisfying $x \le 2$, we place a solid circle at 2 on a number line and draw an arrow extending from the circle to the left (to represent the fact that all numbers less than 2 are also part of the graph). The graph is shown below.

> **Closed** circle shows that **2 is included.**

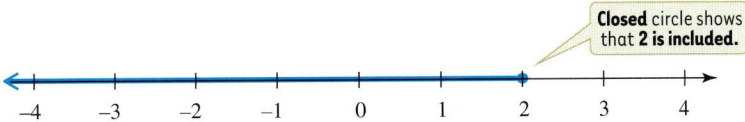

EXAMPLE 1 Graphing the Solutions of an Inequality

(a) Graph $x > -5$.

The statement $x > -5$ says that x can take any value greater than -5, but x *cannot equal* -5 itself. We show this on a graph by placing an *open* circle at -5 and drawing an arrow to the right. The open circle at -5 shows that -5 is *not* part of the graph.

> **Open** circle shows that -5 is **not** included.

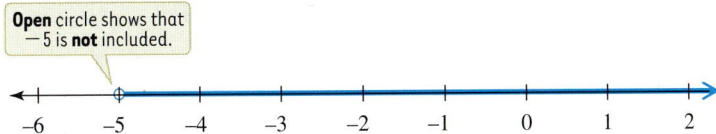

(b) Graph $3 > x$.

Recall from **Section 10.1** that $3 > x$ can be rewritten as $x < 3$. We do this by reversing the direction of the symbol from $>$ to $<$ and reversing the order of 3 and x. The graph of $x < 3$ is shown below.

> **Open** circle shows that 3 is **not** included.

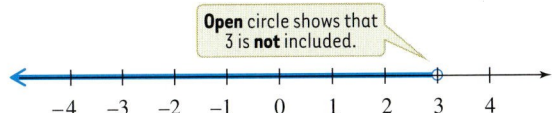

Note

To graph an inequality like the one in part (b) above, first rewrite it with the variable on the left. Fewer errors occur this way.

Work Problem **1** *at the Side.* ▶

1 Graph each inequality.

(a) $x \le 3$

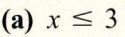

(b) $x > -4$

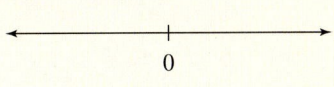

(c) $-4 \ge x$

(d) $0 < x$

ANSWERS

1. **(a)**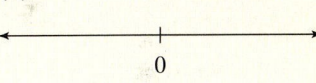

(b)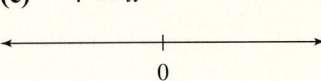

(c)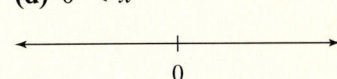

(d)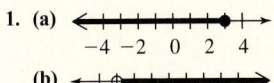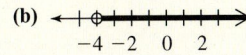

2 Graph each inequality.

(a) $-7 < x < -2$

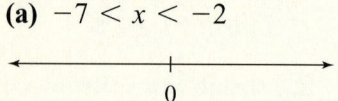

EXAMPLE 2 Graphing the Solutions of an Inequality

Graph $-3 \leq x < 2$.

The statement $-3 \leq x < 2$ is read "-3 is less than or equal to x *and* x is less than 2." We graph the solutions of this inequality by placing a *closed* circle at -3 (because -3 is part of the graph) and an *open* circle at 2 (because 2 is *not* part of the graph), then drawing a line segment between the two circles. Notice that the graph includes all points *between* -3 and 2, and includes -3 as well.

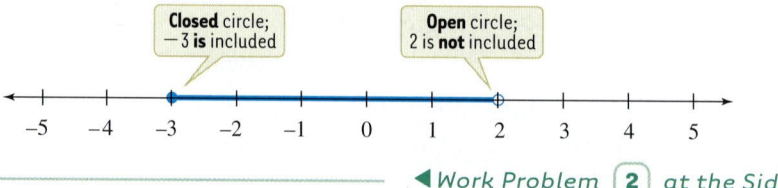

Closed circle; -3 **is** included

Open circle; 2 is **not** included

◀ Work Problem **2** at the Side.

OBJECTIVE 2 Use the addition property of inequality. We can solve inequalities such as $x + 4 \leq 9$ in much the same way as equations.

> **Linear Inequality in One Variable**
>
> A **linear inequality in one variable** can be written in the form
> $$Ax + B < C$$
> where A, B, and C are real numbers, with $A \neq 0$.

All definitions and rules are also valid for $>$, $\leq$, and $\geq$.

Examples of linear inequalities in one variable are shown below.

$$x + 5 < 2 \qquad t - 3 \geq 5 \qquad 2k + 5 \leq 10$$

Consider the inequality $2 < 5$. If 4 is added to both sides of this inequality, the result is still a true statement, as shown below.

$$2 + 4 < 5 + 4$$
$$6 < 9 \qquad \text{True}$$

(b) $-6 < x \leq 4$

Now try subtracting 8 from both sides.

$$2 - 8 < 5 - 8$$
$$-6 < -3 \qquad \text{True}$$

The result is again a true statement. These examples suggest the **addition property of inequality.**

> **Addition Property of Inequality**
>
> For any real numbers A, B, and C, the inequalities
> $$A < B \quad \text{and} \quad A + C < B + C$$
> have exactly the same solutions. In other words, the same number may be added to each side of an inequality without changing the solutions.

We can replace $<$ in the addition property of inequality with $>$, $\leq$, or $\geq$. Also, as with the addition property of equality, the same number may be *subtracted* on each side of an inequality.

> **EXAMPLE 3** **Using the Addition Property of Inequality**

Solve $7 + 3k > 2k - 5$.

$$7 + 3k > 2k - 5$$

$$\underline{-2k > -2k} \qquad \text{Subtract } 2k \text{ from both sides.}$$

$$7 + k > -5$$

$$\underline{-7 > -7} \qquad \text{Subtract } 7 \text{ from both sides.}$$

$$k > -12$$

A graph of the solutions, $k > -12$, is shown below.

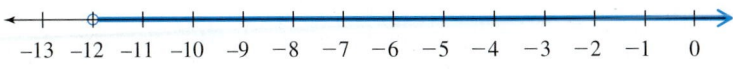

Work Problem **3** *at the Side.* ▶

OBJECTIVE 3 Use the multiplication property of inequality.
The addition property of inequality cannot be used to solve inequalities such as $4y \geq 28$. These inequalities require the *multiplication property of inequality.* To see how this property works, it is helpful to look at some examples.

First, write the inequality $3 < 7$ and then multiply both sides by the positive number 2.

$$3 < 7$$

$$\mathbf{2}(3) < \mathbf{2}(7) \qquad \text{Multiply both sides by 2.}$$

$$6 < 14 \qquad \text{True}$$

Now multiply both sides of $3 < 7$ by the negative number -5.

$$3 < 7$$

$$\mathbf{-5}(3) < \mathbf{-5}(7) \qquad \text{Multiply both sides by } -5.$$

$$-15 < -35 \qquad \text{False}$$

To get a true statement when multiplying both sides by -5, we must *reverse the direction of the inequality symbol.*

$$3 < 7$$

$$\mathbf{-5}(3) > \mathbf{-5}(7) \qquad \text{Multiply by } -5 \text{ and reverse the symbol.}$$

$$-15 > -35 \qquad \text{True}$$

Take the inequality $-6 < 2$ as another example. Multiply both sides by the positive number 4.

$$-6 < 2$$

$$\mathbf{4}(-6) < \mathbf{4}(2) \qquad \text{Multiply both sides by 4.}$$

$$-24 < 8 \qquad \text{True}$$

Multiply both sides of $-6 < 2$ by -5 **and at the same time reverse the direction of the inequality symbol.**

$$-6 < 2$$

$$\mathbf{-5}(-6) > \mathbf{-5}(2) \qquad \text{Multiply by } -5 \text{ and reverse the symbol.}$$

$$30 > -10 \qquad \text{True}$$

Work Problem **4** *at the Side.* ▶

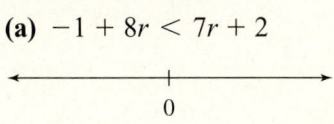

3 Solve each inequality, and graph the solutions.

(a) $-1 + 8r < 7r + 2$

<!-- number line with 0 marked -->

(b) $4m \geq 3m - 1$

<!-- number line with 0 marked -->

4 (a) Multiply both sides of $-2 < 8$ by 6 and then by -5. Reverse the direction of the inequality symbol when necessary to make a true statement.

(b) Multiply both sides of $-4 > -9$ by 2 and then by -8. Reverse the direction of the inequality symbol when necessary to make a true statement.

ANSWERS

3. **(a)** $r < 3$

<!-- number line from -4 to 4, open circle at 3 -->
$$-4 \ -2 \quad 0 \quad 2\ 3\ 4$$

(b) $m \geq -1$

<!-- number line, closed dot at -1 -->
$$-2 \ -1 \quad 0 \quad 1 \quad 2$$

4. **(a)** $-12 < 48; 10 > -40$
(b) $-8 > -18; 32 < 72$

In summary, the multiplication property of inequality has two parts.

> **Multiplication Property of Inequality**
>
> For any real numbers A, B, and C (where $C \neq 0$),
> 1. if C is *positive,* then the inequalities
>
> $$A < B \qquad \text{and} \qquad AC < BC$$
>
> have exactly the same solutions;
>
> 2. if C is *negative,* then the inequalities
>
> $$A < B \qquad \text{and} \qquad AC > BC$$
>
> have exactly the same solutions.
>
> In other words, both sides of an inequality may be multiplied by the same *positive* number without changing the solutions. ***If the multiplier is negative, we must reverse the direction of the inequality symbol.***

We can replace $<$ in the multiplication property of inequality with $>$, $\leq$, or $\geq$. As with the multiplication property of equality, the same nonzero number may be *divided* into both sides.

> **CAUTION**
>
> It is important to remember the differences in the multiplication property of inequality for positive and negative numbers.
> 1. When both sides of an inequality are multiplied or divided by a positive number, the direction of the inequality symbol *does not change.* Adding or subtracting terms on both sides also *does not change* the symbol.
> 2. When both sides of an inequality are multiplied or divided by a negative number, the direction of the symbol *does change.* ***Reverse the direction of the symbol of inequality only when multiplying or dividing by a negative number.***

EXAMPLE 4 **Using the Multiplication Property of Inequality**

Solve $3r < -18$.

Using the multiplication property of inequality, we divide both sides by 3. Since 3 is a *positive* number, the direction of the inequality symbol *does not* change. ***It does not matter that the number on the right side of the inequality is negative.***

$$3r < -18$$

$$\frac{3r}{3} < \frac{-18}{3}$$

> Divide by 3, a **positive** number, so the direction of the inequality symbol does **not** change.

$$r < -6$$

The graph of the solutions is shown below.

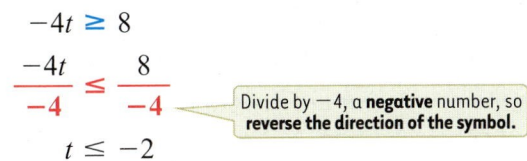 **EXAMPLE 5** **Using the Multiplication Property of Inequality**

Solve $-4t \geq 8$.

Here both sides of the inequality must be divided by -4, a *negative* number, which *does* change the direction of the inequality symbol.

$$-4t \geq 8$$

$$\frac{-4t}{-4} \leq \frac{8}{-4}$$ Divide by -4, a **negative** number, so **reverse the direction of the symbol.**

$$t \leq -2$$

The solutions are graphed below.

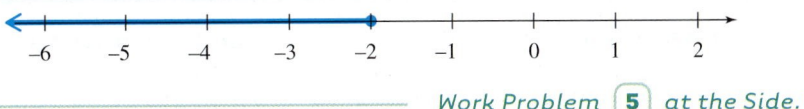

Work Problem **5** *at the Side.* ▶

5 Solve each inequality. Graph the solutions.

(a) $9y < -18$

(b) $-2r > -12$

(c) $-5p \leq 0$

OBJECTIVE **4** **Solve inequalities using both properties of inequality.** The steps in solving an inequality are summarized below. (Remember that $<$ can be replaced with $>$, $\leq$, or $\geq$ in this summary.)

> **Solving Inequalities**
>
> **Step 1** Simplify each side separately. If possible, use the distributive property to remove any parentheses. Then combine like terms on each side.
>
> **Step 2** Use the addition property. Add or subtract the same amount on both sides of the inequality so that the variable term ends up by itself on one side of the inequality sign and a number is by itself on the other side. You may have to do this step more than once.
>
> **Step 3** Use the multiplication property to write the inequality in the form $x < c$ or $x > c$.

EXAMPLE 6 **Solving an Inequality**

Solve $5(k - 3) - 7k \geq 4(k - 3) + 9$.

Step 1 Use the distributive property to remove the parentheses; then combine like terms.

$5(k - 3) - 7k \geq 4(k - 3) + 9$ Use distributive property.

$5k - 15 - 7k \geq 4k - 12 + 9$ Combine like terms.

$-2k - 15 \geq 4k - 3$

Step 2 Use the addition property.

$$-2k - 15 \geq \quad 4k - 3$$
$$\underline{-4k} \qquad \underline{-4k} \qquad \text{Subtract } 4k \text{ from both sides.}$$
$$-6k - 15 \geq \qquad -3$$
$$\underline{+ 15} \qquad \underline{+ 15} \qquad \text{Add 15 to both sides.}$$
$$-6k \qquad \geq \qquad 12$$

Continued on Next Page

ANSWERS

5. (a) $y < -2$

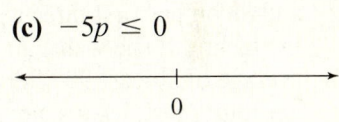

(b) $r < 6$

(c) $p \geq 0$

6 Solve each inequality and graph the solutions.

(a) $5r - r + 2 < 7r - 5$

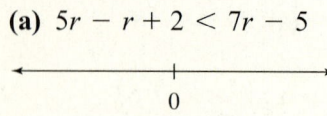

(b) $4(y - 1) - 3y >$
$-15 - (2y + 1)$

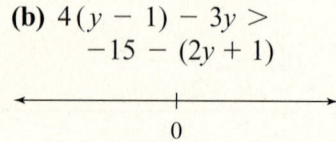

Step 3 Divide both sides by -6, a *negative* number, and *reverse the direction* of the inequality symbol.

$$\frac{-6k}{-6} \leq \frac{12}{-6}$$

Divide by -6 and **reverse the symbol.**

$$k \leq -2$$

A graph of the solutions is shown below.

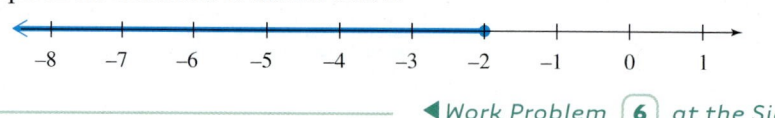

◀ *Work Problem* **6** *at the Side.*

OBJECTIVE 5 Use inequalities to solve application problems.
Inequalities can be used to solve application problems involving phrases that suggest inequality. The table below gives some of the more common phrases, along with examples and translations.

Phrase	Example	Inequality
Is greater than	A number *is greater than* 4	$x > 4$
Is less than	A number *is less than* -12	$x < -12$
Is at least	A number *is at least* 6	$x \geq 6$
Is at most	A number *is at most* 8	$x \leq 8$

CAUTION

Do not confuse phrases like "5 less than a number" and statements like "5 **is** less than a number." The first of these is expressed as "$x - 5$" while the second is expressed at "$5 < x$."

7 Maggie has scores of 98, 86, and 88 on her first three tests in algebra. If she wants an average of at least 90 after her fourth test, what score must she make on her fourth test?

EXAMPLE 7 Finding an Average Test Score

Brent has scores of 86, 88, and 78 on his first three tests in geometry. If he wants an average of at least 80 after his fourth test, what score must he make on his fourth test?

Let x represent Brent's score on his fourth test. To find the average of the four scores, add them and divide the sum by 4.

$$\underbrace{\frac{86 + 88 + 78 + x}{4}}_{\text{Average}} \overset{\text{is at}}{\underset{\text{least 80.}}{\geq}} 80$$

$$4\left(\frac{252 + x}{4}\right) \geq 4(80) \qquad \text{Multiply both sides by 4.}$$

$$252 + x \geq 320$$
$$\underline{-252 \qquad\qquad -252} \qquad \text{Subtract 252 from both sides.}$$
$$x \geq 68$$

He must score 68 or more on the fourth test to have an average of *at least* 80.

◀ *Work Problem* **7** *at the Side.*

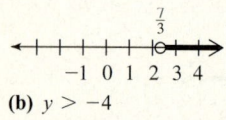

10.4 ▶▶▶ Exercises

MyMathLab Math XL PRACTICE WATCH DOWNLOAD READ REVIEW

1. Explain how you can determine whether to use an open circle or a closed circle at an endpoint when graphing an inequality on a number line.

2. Explain how the graph of $t \geq -7$ differs from the graph of $t > -7$.

3. Explain why we can't list the solutions of an inequality such as $x < 2$.

4. What inequality symbol is used to express the statement "y is at least 7"?

Graph each inequality on the given number line. See Examples 1 and 2.

5. $k \leq 4$

6. $r \leq -11$

7. $x < -3$

8. $y < 3$

9. $8 \leq x \leq 10$

10. $3 \leq x \leq 5$

11. $0 < y \leq 10$

12. $-3 \leq x < 5$

13. Why is it *wrong* to write $3 < x < -2$ to indicate that x is between -2 and 3?

14. If $p < q$ and $r < 0$, which one of the following statements is false?

   (a) $pr < qr$ **(b)** $pr > qr$ **(c)** $p + r < q + r$ **(d)** $p - r < q - r$

Solve each inequality and graph the solutions. See Example 3.

15. $z - 8 \geq -7$

16. $p - 3 \geq -11$

17. $2k + 3 \geq k + 8$

18. $3x + 7 \geq 2x + 11$

19. $3n + 5 < 2n - 6$

20. $5x - 2 < 4x - 5$

21. Under what conditions must the inequality symbol be reversed when using the multiplication property of inequality?

22. Explain the steps you would use to solve the inequality $-5x > 20$.

23. Your friend tells you that when solving the inequality $6x < -42$, he reversed the direction of the inequality because of the presence of -42. How would you respond?

24. By what number must you *multiply* both sides of $0.2x > 6$ to get just x on the left side?

Solve each inequality and graph the solutions. See Examples 4 and 5.

25. $3x < 18$

26. $5x < 35$

27. $2y \geq -20$

28. $6m \geq -24$

29. $-8t > 24$

30. $-7x > 49$

31. $-x \geq 0$

32. $-k < 0$

33. $-\dfrac{3}{4}r < -15$

34. $-\dfrac{7}{8}t < -14$

35. $-0.02x \leq 0.06$

36. $-0.03v \geq -0.12$

Solve each inequality and graph the solutions. See Example 6.

37. $5r + 1 \geq 3r - 9$

38. $6t + 3 < 3t + 12$

39. $6x + 3 + x < 2 + 4x + 4$

40. $-4w + 12 + 9w \geq w + 9 + w$

41. $-x + 4 + 7x \leq -2 + 3x + 6$

42. $14y - 6 + 7y > 4 + 10y - 10$

43. $5(x + 3) - 6x \leq 3(2x + 1) - 4x$

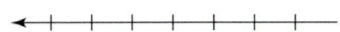

44. $2(x - 5) + 3x < 4(x - 6) + 1$

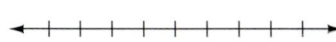

45. $\frac{2}{3}(p + 3) > \frac{5}{6}(p - 4)$

46. $\frac{7}{9}(y - 5) \leq \frac{4}{3}(y + 5)$

47. $4x - (6x + 1) \leq 8x + 2(x - 3)$

48. $2y - (4y + 3) < 6y + 3(y + 4)$

49. $5(2k + 3) - 2(k - 8) > 3(2k + 4) + k - 2$

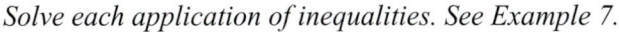

50. $2(3z - 5) + 4(z + 6) \geq 2(3z + 2) + 3z - 15$

Solve each application of inequalities. See Example 7.

51. When 8 is subtracted from the sum of three times a number and 6, the result is less than 4 more than the number. Find all such numbers.

52. When 2 is added to the difference between six times a number and 5, the result is greater than 13 added to 5 times the number. Find all such numbers.

53. Twylene Johnson has scores of 89, 78, 73, and 81 on her first four algebra tests. If she wants an average of at least 80 after her fifth test, what score must she make on her fifth test?

54. Mabimi Pampo has scores of 96 and 86 on his first two geometry tests. What must he score on his third test so that his average is at least 90?

55. The formula for converting Fahrenheit temperature to Celsius is shown below.

$$C = \frac{5}{9}(F - 32)$$

If the Celsius temperature on a certain summer day in Houston is never more than 30 degrees, how would you describe the corresponding Fahrenheit temperatures?

56. The formula for converting Celsius temperature to Fahrenheit is shown below.

$$F = \frac{9}{5}C + 32$$

The Fahrenheit temperature of Key West, Florida, has never exceeded 95 degrees. How would you describe this using Celsius temperatures?

57. For what values of x would the rectangle have perimeter of at least 400?

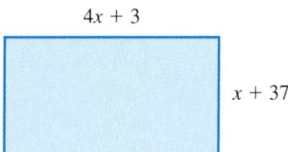

4x + 3

x + 37

58. For what values of x would the triangle have perimeter of at least 72?

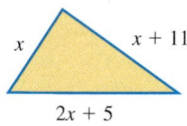

x

x + 11

2x + 5

59. Audrey earned $200 at odd jobs during July, $300 during August, and $225 during September. If her average salary for the four months from July through October is to be at least $250, how much must she earn during October?

60. In order to qualify for a company pension plan, an employee must average at least $1000 per month in earnings. During the first four months of the year, an employee made $900, $1200, $1040, and $760. What amount of earnings during the fifth month will qualify the employee for the pension plan?

61. An international phone call costs $2.00 for the first three minutes plus $0.30 per minute for each minute or fractional part of a minute after the first three minutes. If x represents the number of minutes of the length of the call after the first three minutes, then $2 + 0.30x$ represents the cost of the call. If Jorge has $5.60 to spend on a call, what is the maximum total time he can use the phone?

62. At the Speedy Gas 'n Go, a car wash costs $5.00, and gasoline is selling for $3.50 per gal. Terri Hoelker has $38.25 to spend, and her car is so dirty that she must have it washed. What is the maximum number of gallons of gasoline that she can purchase?

Relating Concepts (Exercises 63–67) For Individual or Group Work

Work Exercises 63–67 in order, to see the connection between the solution of an equation and the solutions of the corresponding inequalities. *Graph the solutions in Exercises 63–65.*

63. $3x + 2 = 14$

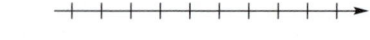

64. $3x + 2 < 14$

65. $3x + 2 > 14$

66. Now graph all the solutions together on the following number line.

How would you describe the graph?

67. Based on your results from Exercises 63–66, if you were to graph the solutions of

$$-4x + 3 = -1, \quad -4x + 3 > -1,$$
$$\text{and} \quad -4x + 3 < -1$$

on the same number line, what do you think the graph would be?

Chapter 10 ▷▷▷ Summary

▶ Key Terms

10.1	**natural numbers**	The set of natural numbers is $\{1, 2, 3, 4, \dots\}$.
	whole numbers	The set of whole numbers is $\{0, 1, 2, 3, 4, \dots\}$.
	integers	The set of integers is $\{\dots, -3, -2, -1, 0, 1, 2, 3, \dots\}$.
	rational numbers	Rational numbers can be written as quotients of two integers, with denominator not equal to 0.
	irrational numbers	Irrational numbers are nonrational numbers represented by points on the number line. The decimal form of an irrational number does not terminate (end) and does not repeat in a fixed block of digits.
	real numbers	Real numbers include all numbers that can be represented by points on the number line, that is, all rational and irrational numbers.
10.2	**identity**	An identity is an equation that is true for all replacements of the variable. The solution of an identity is "all real numbers."
10.4	**inequality**	An inequality is a statement with algebraic expressions related by $<$, $\leq$, $>$, or $\geq$.
	linear inequality in one variable	A linear inequality in one variable can be written in the form $Ax + B < C$, $Ax + B \leq C$, $Ax + B > C$, or $Ax + B \geq C$, where A, B, and C are real numbers, with $A \neq 0$.

▶ New Symbols

$\{0, 1, 2, 3, \dots\}$	The items in a *set* are written between *set braces*.
$\neq$ "is not equal to"	$\leq$ "is less than or equal to" $\geq$ "is greater than or equal to"

▶ Test Your Word Power

See how well you have learned the vocabulary in this chapter. Answers, with examples, follow the Quick Review.

1. Which number is the decimal form of an **irrational number?**
   A. $0.63636363\dots$
   B. -0.524
   C. $0.02022022202222\dots$
   D. $\sqrt{0.16}$

2. All **real numbers** are
   A. quotients of two integers, with the denominator not equal to zero.
   B. rational
   C. decimals that repeat in a fixed block
   D. represented by points on the number line.

3. A **rational number** can be written as
   A. the quotient of two integers, with the denominator not equal to zero
   B. a decimal that does not repeat in a fixed block
   C. a decimal that does not terminate
   D. the product of two integers, with neither factor equal to zero.

4. An **inequality** is
   A. a statement that two algebraic expressions are equal
   B. a point on a number line
   C. an equation with no solutions
   D. a statement with algebraic expressions related by $<$, $\leq$, $>$, or $\geq$.

5. An **identity** is an equation that
   A. has no solution
   B. has all real numbers as solutions
   C. cannot be solved
   D. has 0 as the only solution.

6. Which statement is a **linear inequality**?
   A. $2x^2 + 6 > 9$
   B. $0x + 25 \geq -1$
   C. $3x + 10 \leq -5$
   D. $7x + 1 = 15$

▶ Quick Review

Concepts	Examples

10.1 Using the Symbols ≠, ≤, ≥

Read the inequality $-2 \leq 0$ as "negative 2 is less than or equal to 0." If either the $<$ part or the $=$ part is true, then the inequality is true. Use the symbol $\geq$ in a similar way to mean "is greater than or equal to."

Tell whether each statement is true or false.

(a) $-3 \leq -1$ **(b)** $6.7 \geq 6.7$

(a) True because $-3 < -1$, that is, -3 is to the left of -1 on the number line.

(b) True because $6.7 = 6.7$.

To convert between $<$ and $>$ (or between $\leq$ and $\geq$) reverse both the order of the numbers and the direction of the symbol.

Rewrite each statement with the inequality symbol reversed.

(a) $0.1 > 0.01$ **(b)** $\dfrac{1}{2} \leq 0.7$

$0.01 < 0.1$ $0.7 \geq \dfrac{1}{2}$

Exchange numbers   Exchange numbers

10.1 Using Brackets

Brackets are used instead of double parentheses. Use the order of operations to complete all work inside the brackets.

Simplify.

$-3[-4 + 6(\mathbf{-3 + 5})]$ Work inside parentheses.

$-3[-4 + 6(\mathbf{2})]$ Multiply $6(2)$.

$-3[-4 + \mathbf{12}]$ Add $-4 + 12$.

$-3[\mathbf{8}]$ Multiply $-3 \cdot 8$.

-24

10.1 Using the Distributive Property to Remove Parentheses

Multiply *every term* inside the parentheses by the number in front of the parentheses.

Watch the signs carefully if there is a *negative* number in front of the parentheses; the sign of every term inside the parentheses will change to its opposite.

Simplify each expression.

(a) $-(4a - 6)$ is written $\mathbf{-1}(4a - 6)$

$-4a + 6$ ⟵ Simplified expression.

(b) $\mathbf{-6}(3 + x) - \mathbf{2}(4x + 5)$ Use distributive property

$-18 - 6x - 8x - 10$ Combine like terms.

$-14x - 28$ ⟵ Simplified expression.

Concepts	Examples

10.2 Solve Equations That Have No Solution or Infinitely Many Solutions

If, when solving an equation, the result is an *identity* (both sides of the equation are exactly the same), then there are an infinite number of solutions. Write "all real numbers" as the solution because *any* real number will work.

Solve each equation.

(a) $3(a - 4) = -12 + 3a$ Use distributive property.

$$3a - 12 = -12 + 3a$$

$$\underline{+12 \qquad +12}$$ Add 12 to both sides.

$$3a \quad = \quad 3a$$

$$\underline{-3a \qquad\qquad -3a}$$ Subtract $3a$ from both sides.

$$0 \quad = \quad 0$$ Identity

The solution is *all real numbers*.

If, when solving an equation, the result is a *false* statement, such as $-7 = 4$, it is a signal that the equation has *no* solution.

(b) $x - 6(x + 1) = 4 - 5x$ Use distributive property.

$$x - 6x - 6 = 4 - 5x$$ Combine like terms.

$$-5x - 6 = 4 - 5x$$

$$\underline{+5x \qquad\qquad +5x}$$ Add $5x$ to both sides.

$$-6 = 4$$ False statement

There is *no solution*.

10.2 Clearing an Equation of Fractions or Decimals

To clear an equation of fractions, multiply *every term* on both sides of the equation by the lowest common denominator (LCD) of all the fractions.

To clear an equation of decimals, multiply *every term* on both sides of the equation by a power of 10 that will change the coefficients from decimals to whole numbers.

Solve.

$$-\frac{5}{6}b = \frac{2}{3}b - \frac{1}{2}$$

$$6\left(-\frac{5}{6}b\right) = 6\left(\frac{2}{3}b - \frac{1}{2}\right)$$ Multiply both sides by 6 (the LCD).

$$\overset{1}{\cancel{6}}\left(-\frac{5}{\cancel{6}}b\right) = \overset{2}{\cancel{6}}\left(\frac{2}{\cancel{3}}b\right) - \overset{3}{\cancel{6}}\left(\frac{1}{\cancel{2}}\right)$$ Use the distributive property to multiply every term by 6.

$$-5b = 4b - 3$$ Subtract $4b$ from both sides.

$$\underline{-4b \qquad -4b}$$

$$\frac{-9b}{-9} = \frac{-3}{-9}$$ Divide both sides by -9.

$$b = \frac{1}{3}$$

10.3 Solving Formulas

To find the value of one of the variables in a formula, given values for the others, substitute the known values into the formula.

Find l if $A = lw$, given that $A = 24$ and $w = 3$.

$$24 = l \cdot 3$$ Replace A with 24 and w with 3

$$\frac{24}{3} = \frac{l \cdot 3}{3}$$ Divide both sides by 3.

$$8 = l$$

Concepts	Examples

10.3 Solving for a Specified Variable

To solve a formula for one of the variables, isolate that variable by treating the other variables as numbers, and then using the steps for solving equations.

Solve $P = 2l + 2w$ for w.

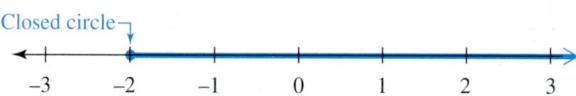

$$P = 2l + 2w$$

$$\underline{\ -2l \qquad -2l}$$ Subtract $2l$ from both sides.

$$P - 2l = \qquad 2w$$

$$\frac{P - 2l}{2} = \frac{2w}{2}$$ Divide both sides by 2.

$$\frac{P - 2l}{2} = w$$ ← Formula solved for w.

10.4 Graphing the Solutions of an Inequality

On the number line, use a closed circle to indicate that a number is part of the graph. Use an open circle to indicate that a number is *not* part of the graph.

(a) Graph $x \geq -2$.

(b) Graph $0 \leq x < 4$

10.4 Solving Linear Inequalities

To solve an inequality:

1. Simplify each side separately: clear parentheses and combine like terms.

2. Add or subtract the same number on both sides to get the variable term by itself on one side and a number on the other side.

3. Multiply or divide by the same number on both sides to get the form $x > a$ or $x < a$.

   When multiplying or dividing by a negative number, reverse the direction of the inequality symbol.

Solve $3(1 - x) + 5 - 2x > 9 - 6$ and graph the solutions.

$3(1 - x) + 5 - 2x > 9 - 6$ Use distributive property.

$3 - 3x + 5 - 2x > 9 - 6$ Combine like terms.

$8 - 5x > 3$

$\underline{-8 \qquad\qquad -8}$ Subtract 8 from both sides.

$-5x > -5$

$\dfrac{-5x}{-5} < \dfrac{-5}{-5}$ Divide both sides by -5 and change $>$ to $<$.

$x < 1$

ANSWERS TO TEST YOUR WORD POWER

1. C; *Examples:* $0.1213141516\ldots$, $\sqrt{13}$
2. D; *Examples:* Real numbers include all the rational numbers and all the irrational numbers.

3. A; *Examples:* $3, -12, 4.1, \dfrac{5}{6}$

4. D; *Examples:* $x < 5$ and $7 + 2y \leq 11$
5. B; *Example:* $4x + 12 = 4(x + 3)$ because when the distributive property is used on the right side, the equation becomes
   $4x + 12 = 4x + 12$ (both sides are exactly the same)
6. C; *Example:* $2x + 5 \geq 7$

Chapter 10 ▶▶▶ Review Exercises

[10.1] *Identify each number as rational or irrational and explain why.*

1. 24.625

2. 0.363636 …

 **3.** $\sqrt{144}$

4. $\sqrt{23}$

Label each statement as true or false, and explain why.

5. $\dfrac{2}{3} \neq 0.6$

6. $-\dfrac{5}{8} \leq -\dfrac{5}{8}$

7. $2 - 10 \geq 10 - 2$

8. $-75 \geq -50$

Simplify.

9. $6[2 + 8(3^3)]$

10. $3^2[(11 + 3) - 4]$

11. $-8 + [(-4 + 17) - (-3 - 3)]$

12. $[-9 - (2 \cdot 1) - (-3)] + [8 + (-13 + 13)]$

13. $-5(2x - 4)$

14. $-(-5 + 3p)$

15. $-(-17c - 6)$

16. $-2 - (8 - 10y)$

17. $-10 - (7 + 14r)$

18. $-8 - (-3r - 6)$

19. $-8(5k - 6) + 3(7k + 2)$

20. $-7(2t - 4) - 4(3t + 8) + 27t$

[10.2] *Solve each equation.*

21. $5x + 8 = -2(-2x - 1)$

22. $8t = 7t + \dfrac{3}{2}$

23. $(4r - 8) - (3r + 12) = 0$

24. $7(2x + 1) = 6(2x - 9)$

25. $-\dfrac{6}{5}y = -18$

26. $\dfrac{1}{2}r - \dfrac{1}{6}r + 3 = 2 + \dfrac{1}{6}r + 1$

27. $3x - (-2x + 6) = 4(x - 4) + x$

28. $0.10(x + 80) + 0.20x = 14$

[10.3] *Find the value of the remaining variable in each formula.*

29. $A = \dfrac{1}{2}bh;\ A = 44,\ b = 8$

30. $C = 2\pi r;\ C = 29.83,\ \pi = 3.14$

Solve the formula for the specified variable.

31. $V = lwh$ for w

32. $A = \dfrac{1}{2}h(b + B)$ for h

Solve each application problem using the appropriate formula.

33. A cinema screen in Indonesia has a length of 92.75 feet and a perimeter of 326.5 feet. What is the screen's width?

34. The largest box of popcorn was filled by students in Jacksonville, Florida. The box was approximately 40 feet long, 20.7 feet wide, and had a volume of 6624 cubic feet. Find the height of the box.

20.7 ft

WORLD'S LARGEST BOX OF POPCORN

40 ft

[10.4] *Graph each inequality on the number line provided.*

35. $p \geq -4$

36. $x < 7$

37. $-5 \leq y < 6$

38. $r \geq \dfrac{1}{2}$

Solve each inequality. Graph the solutions.

39. $y + 6 \geq 3$

40. $5t < 4t + 2$

41. $-6x \leq -18$

42. $8(k - 5) - (2 + 7k) \geq 4$

43. $4x - 3x > 10 - 4x + 7x$

44. $3(2w + 5) + 4(8 + 3w) < 5(3w + 2) + 2w$

45. Carlotta Valdez has scores of 81, 77, and 88 on her first three calculus tests. What scores on a fourth test will give her an average of at least 85?

46. If nine times a number is added to 6, the result is at most 3. Find all such numbers.

▶▶▶   **Mixed Review Exercises**

Solve.

47. $\dfrac{y}{7} = \dfrac{y - 5}{2}$

48. $I = prt$ for r

49. $-2x > -4$

50. $2k - 5 = 4k + 13$

51. $0.05x + 0.02x = 4.9$

52. $2 - 3(y - 5) = 4 + y$

53. $9x - (7x + 2) = 3x + (2 - x)$

54. $\dfrac{1}{3}s + \dfrac{1}{2}s + 7 = \dfrac{5}{6}s + 5 + 2$

55. $4 - 5(a + 2) = 3(a + 1) - 1$

56. $P = a + b + c$ for a

57. $\dfrac{2}{3}y + \dfrac{3}{4}y = -17$

58. $2 - 6(z + 1) = 4(z - 2) + 10$

59. The area of a triangle is 182 square inches. The height is 14 inches. Find the length of the base.

60. The perimeter of a rectangle is 75 inches. The width is 17 inches. What is the length?

61. On the first two days of their vacation to Florida, Wally and Nikolas drove 430 miles and 470 miles. If they must average at least 450 miles per day, what distance must they drive on the third day?

62. Nalima has grades of 82, 91, 97, and 96 on her first four English tests. What must she make on her fifth test so that her average will be at least 90?

63. Gina used 4.7 meters of fringe around the edge of a circular tablecloth that she is making. What is the diameter of the cloth, to the nearest tenth of a meter? Use 3.14 as the approximate value of π.

64. The area of a rectangular oriental rug is 93.5 square feet. If the width is $8\frac{1}{2}$ feet, find the length.

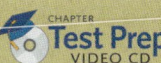

Identify each number as rational or irrational and explain why.

1. π **2.** $3.909090\ldots$

Label each statement as true or false and explain why.

3. $-6 \leq -8$ **4.** $\dfrac{3}{4} \neq 0.75$ **5.** $1\dfrac{3}{10} \geq 0.95$

Simplify.

6. $4[-20 + 7(-2)]$ **7.** $-6 - [-7 + (2 - 3)]$

8. $-(-5a + 14)$ **9.** $-2(3x^2 + 4) - 3(x^2 + 2x)$

Solve.

10. $5x + 9 = 7x + 21$

11. $2 - 3(y - 5) = 3 + (y + 1)$

12. $2.3x + 13.7 = 1.3x + 2.9$

13. $7 - (m - 4) = -3m + 2(m + 1)$

14. $-\dfrac{4}{7}x = -1 - \dfrac{2}{3}x$

15. $0.06(x + 20) + 0.08(x - 10) = 4.6$

16. $-8(2x + 4) = -4(4x + 8)$

1. _____

2. _____

3. _____

4. _____

5. _____

6. _____

7. _____

8. _____

9. _____

10. _____

11. _____

12. _____

13. _____

14. _____

15. _____

16. _____

17. _____

17. The formula for simple interest is $I = prt$. If $I = 600$, $r = 0.08$, and $t = 3$, find p.

18. _____

18. The Ziegfield Room in Reno, Nevada has a circular turntable for performers. The circumference of the turntable is 62.5 feet. What is the diameter of the turntable, to the nearest tenth of a foot? Use 3.14 as the approximate value of π.

19. _____

19. Solve $d = 2r$ for r.

20. _____

20. Solve $A = \dfrac{1}{2}bh$ for h.

21. _____

21. Solve $A = p + prt$ for r.

Solve each inequality and graph the solutions.

22. _____

⊢—+—+—+—+—+—+—▶

22. $-3x > -33$

23. _____

⊢—+—+—+—+—+—+—▶

23. $-4x + 2(x - 3) \geq 4x - (3 + 5x) - 7$

24. _____

24. Paula Story has scores of 95, 84, and 100 on her first three college algebra tests. What must she score on her fourth test so that her average is at least 90?

Graphs of Linear Equations and Inequalities in Two Variables

While U.S. consumers in general continue to pile up credit card debt, fewer undergraduate college students are carrying credit cards, and those with cards are using them less. In 2004, 76% of undergraduates carried at least one credit card, down from a peak of 83% in 2001. The average outstanding balance also dropped to $2169, from a high of $2748 in 2000. These declines are attributed to increased financial education aimed specifically at high school and college students. (*Source:* Nellie Mae.)

In Example 7 of **Section 11.2,** we examine a *linear equation in two variables* that models credit card debt in the United States.

741

11.1 ▶▶▶ Reading Graphs; Linear Equations in Two Variables

OBJECTIVES

1. Interpret graphs.
2. Write a solution as an ordered pair.
3. Decide whether a given ordered pair is a solution of a given equation.
4. Complete ordered pairs for a given equation.
5. Complete a table of values.
6. Plot ordered pairs.

As we saw in **Chapter 9,** circle graphs (pie charts) provide a convenient way to organize and communicate information. Along with *bar graphs* and *line graphs,* they can be used to analyze data, make predictions, or simply to entertain us.

OBJECTIVE 1 Interpret graphs. A **bar graph** is used to show comparisons. It consists of a series of bars (or simulations of bars) arranged either vertically or horizontally. In a bar graph, values from two categories are paired with each other.

EXAMPLE 1 Interpreting a Bar Graph

The bar graph in Figure 1 shows U.S. sales of motor scooters, which have gained popularity due to their fuel efficiency. The graph compares sales in thousands.

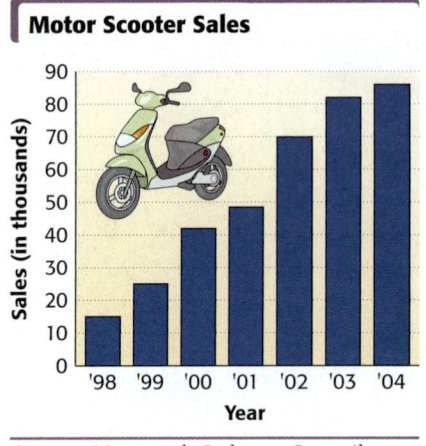

Motor Scooter Sales

Source: Motorcycle Industry Council.

Figure 1

1. Refer to the bar graph in Figure 1.

   (a) Which years had sales less than 50 thousand?

   (b) Estimate sales of motor scooters in 1999 and 2001.

   (c) Describe the change in sales of motor scooters from 1999 to 2001.

(a) In what years were sales greater than 50 thousand?

Locate 50 on the vertical scale and follow the line across to the right. Three years—2002, 2003, and 2004—have bars that extend above the line for 50, so sales were greater than 50 thousand in those years.

(b) Estimate sales in 2000 and 2004.

Locate the top of the bar for 2000, and move horizontally across to the vertical scale to see that it is about 40. Sales in 2000 were about 40 thousand. Follow the top of the bar for 2004 across to the vertical scale to see that it lies about halfway between 80 and 90 thousand, so sales in 2004 were about 85,000.

(c) Describe the change in sales as the years progressed.

As the years progressed, sales increased steadily, from about 15 thousand in 1998 to about 85 thousand in 2004.

◀ *Work Problem* 1 *at the Side.*

ANSWERS

1. (a) 1998, 1999, 2000, 2001
   (b) 1999: about 25 thousand; 2001: about 50 thousand
   (c) Sales approximately doubled from 1999 to 2001.

A **line graph** is used to show changes or trends in data over time. To form a line graph, we connect a series of points representing data with line segments.

┌─ **EXAMPLE 2** | **Interpreting a Line Graph**

Current projections indicate that funding for Medicare will not cover its costs unless the program changes. The line graph in Figure 2 shows Medicare funds in billions of dollars for the years 2004 through 2013.

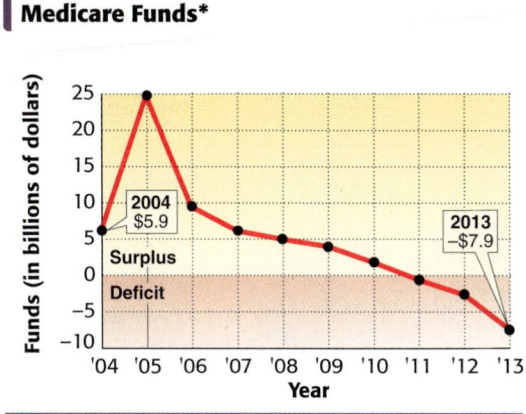

Medicare Funds*

Source: Centers for Medicare and Medicaid Services.
*Projected

Figure 2

(a) Which is the only period in which Medicare funds increased?

Because the graph *rises* from 2004 to 2005 and falls in every other case, funds increased between these two years.

(b) What is the projected trend from 2005 to 2013?

Funds will decrease, since the graph *falls* during this period.

(c) In which year is it projected that funds will first show a deficit?

From 2004 to 2010, the graph is always above 0, but in 2011, it falls slightly below 0 for the first time, indicating a deficit.

(d) Based on the figures shown in the graph, what is the difference in Medicare funds from 2004 to 2013?

$$\underbrace{-\$7.9 \text{ billion}}_{\text{2013 amount}} - \underbrace{\$5.9 \text{ billion}}_{\text{2004 amount}} = \underbrace{-\$13.8 \text{ billion}}_{\text{Difference}}$$

The fund amount will have *decreased* $13.8 billion (as indicated by the negative sign in −$13.8).

─────────────── *Work Problem* **2** *at the Side.* ▶

The line graph in Figure 2 relates years to Medicare funds. We can also represent these two related quantities using a table of data, as shown at the side. Notice that in table form, we can see specific data rather than estimating it. Trends in the data are easier to see from the graph, however, which gives us a "picture" of the data.

We can extend these ideas to the subject of this chapter, *linear equations in two variables*. A linear equation in two variables, one for each of the quantities being related, can be used to represent the data in the table or graph. We introduced this concept in **Section 9.5** and extend the topic in this chapter.

---

2 Refer to the line graph in Figure 2.

(a) Which year has the greatest amount of Medicare funds?

(b) Estimate projected Medicare funds for 2010. Is there a surplus or a deficit in 2010?

(c) About how much is it projected that funds will decrease from 2006 to 2011?

Year	Medicare Funds (in billions of dollars)*
2004	5.9
2005	25.0
2006	9.5
2007	6.0
2008	5.0
2009	4.0
2010	2.0
2011	−0.5
2012	−2.5
2013	−7.9

*Projected

ANSWERS

2. (a) 2005
　　(b) $2 billion; surplus
　　(c) $10 billion

3 Write each solution as an ordered pair.

(a) $x = 5$ and $y = 7$

(b) $y = 6$ and $x = -1$

(c) $y = 4$ and $x = -3$

(d) $x = \dfrac{2}{3}$ and $y = -12$

(e) $y = 1.5$ and $x = -2.4$

(f) $x = 0$ and $y = 0$

The graph of a linear equation in two variables is a line.

> **Linear Equation in Two Variables**
>
> A **linear equation in two variables** is an equation that can be written in the form
>
> $$Ax + By = C,$$
>
> where A, B, and C are real numbers and A and B are not both 0.

Some examples of linear equations in two variables in this form, called *standard form,* are

$$3x + 4y = 9, \quad x - y = 0, \quad \text{and} \quad x + 2y = -8.$$ Linear equations in two variables

> **Note**
>
> Other linear equations in two variables, such as
>
> $$y = 4x + 5 \quad \text{and} \quad 3x = 7 - 2y,$$
>
> are not written in standard form but could be. We discuss the forms of linear equations in more detail in **Section 11.4.**

OBJECTIVE 2 Write a solution as an ordered pair. Recall from **Section 2.3** that a *solution* of an equation is a number that makes the equation true when it replaces the variable. For example, the linear equation in one variable $x - 2 = 5$ has solution 7, since replacing x with 7 gives a true statement.

A solution of a linear equation in two variables requires two numbers, one for each variable. For example, a true statement results when we replace x with 2 and y with 13 in the equation $y = 4x + 5$ since

$$13 = 4(2) + 5. \quad \text{Let } x = 2, y = 13.$$

The pair of numbers $x = 2$ and $y = 13$ gives one solution of the equation $y = 4x + 5$. The phrase "$x = 2$ and $y = 13$" is abbreviated

x-value ⟶ ⟵ y-value

$$(2, 13)$$

Ordered pair

with the x-value, 2, and the y-value, 13, given as a pair of numbers written inside parentheses. *The x-value is always given first.* A pair of numbers such as (2, 13) is called an **ordered pair.** As the name indicates, the order in which the numbers are written is important. The ordered pairs (2, 13) and (13, 2) are *not* the same. The second pair indicates that $x = 13$ and $y = 2$. *For two ordered pairs to be equal, their x-values must be equal and their y-values must be equal.*

◀ *Work Problem* **3** *at the Side.*

OBJECTIVE 3 Decide whether a given ordered pair is a solution of a given equation. We substitute the x- and y-values of an ordered pair into a linear equation in two variables to see whether the ordered pair is a solution.

ANSWERS

3. (a) $(5, 7)$ (b) $(-1, 6)$ (c) $(-3, 4)$
   (d) $\left(\dfrac{2}{3}, -12\right)$ (e) $(-2.4, 1.5)$ (f) $(0, 0)$

EXAMPLE 3 **Deciding Whether Ordered Pairs Are Solutions of an Equation**

Decide whether each ordered pair is a solution of the equation $2x + 3y = 12$.

(a) $(3, 2)$

To see whether $(3, 2)$ is a solution of the given equation $2x + 3y = 12$, substitute 3 for x and 2 for y in the equation.

$$2x + 3y = 12$$
$$2(3) + 3(2) \stackrel{?}{=} 12 \qquad \text{Let } x = 3; \text{ let } y = 2.$$
$$6 + 6 \stackrel{?}{=} 12 \qquad \text{Multiply.}$$
$$12 = 12 \qquad \text{True}$$

This result is true, so $(3, 2)$ is a solution of $2x + 3y = 12$.

(b) $(-2, -7)$

$$2x + 3y = 12$$
$$2(-2) + 3(-7) \stackrel{?}{=} 12 \qquad \text{Let } x = -2; \text{ let } y = -7.$$

> Use parentheses to avoid errors.

$$-4 + (-21) \stackrel{?}{=} 12 \qquad \text{Multiply.}$$
$$-25 = 12 \qquad \text{False}$$

This result is false, so $(-2, -7)$ is *not* a solution of $2x + 3y = 12$.

Work Problem **4** *at the Side.* ▶

OBJECTIVE **4** **Complete ordered pairs for a given equation.** Choosing a number for one variable in a linear equation makes it possible to find the value of the other variable.

EXAMPLE 4 **Completing Ordered Pairs**

Complete each ordered pair for the equation $y = 4x + 5$.

(a) $(7, \_\_)$ | The x-value always comes first.

In this ordered pair, $x = 7$. To find the corresponding value of y, replace x with 7 in the equation.

$$y = 4x + 5$$
$$y = 4(7) + 5 \qquad \text{Let } x = 7.$$
$$y = 28 + 5 \qquad \text{Multiply.}$$
$$y = 33 \qquad \text{Add.}$$

The ordered pair is $(7, 33)$.

(b) $(\_\_, -3)$

In this ordered pair, $y = -3$. Find the value of x by replacing y with -3 in the equation; then solve for x.

$$y = 4x + 5$$
$$-3 = 4x + 5 \qquad \text{Let } y = -3.$$
$$-8 = 4x \qquad \text{Subtract 5 from each side.}$$
$$-2 = x \qquad \text{Divide each side by 4.}$$

The ordered pair is $(-2, -3)$.

Work Problem **5** *at the Side.* ▶

---

4 Decide whether each ordered pair is a solution of the equation $5x + 2y = 20$.

(a) $(0, 10)$

$$5x + 2y = 20$$
$$5(\_\_) + 2(\_\_) \stackrel{?}{=} 20$$
$$\_\_\_\_ + 20 \stackrel{?}{=} 20$$
$$\_\_\_\_ = 20$$

Is $(0, 10)$ a solution?

(b) $(2, -5)$

(c) $(3, 2)$

(d) $(-4, 20)$

5 Complete each ordered pair for the equation $y = 2x - 9$.

(a) $(5, \_\_)$

$$y = 2(\_\_) - 9$$
$$y = \_\_\_\_ - 9$$
$$y = \_\_\_\_$$

The ordered pair is _____.

(b) $(2, \_\_)$

(c) $(\_\_, 7)$

(d) $(\_\_, -13)$

ANSWERS

4. **(a)** 0; 10; 0; 20; yes **(b)** no **(c)** no
   (d) yes
5. **(a)** 5; 10; 1; (5, 1) **(b)** (2, −5) **(c)** (8, 7)
   (d) (−2, −13)

6 Complete the table of values for each equation.

(a) $2x - 3y = 12$

x	y
0	
	0
3	
	-3

(b) $x = -1$

x	y
	-4
	0
	2

(c) $y = 4$

x	y
-3	
2	
5	

ANSWERS

6. **(a)**

x	y
0	-4
6	0
3	-2
$\frac{3}{2}$	-3

(b)

x	y
-1	-4
-1	0
-1	2

(c)

x	y
-3	4
2	4
5	4

OBJECTIVE 5 Complete a table of values. Ordered pairs are often displayed in a **table of values.** The table may be written either vertically or horizontally.

EXAMPLE 5 Completing Tables of Values

Complete the table of values for each equation. Then write the results as ordered pairs.

(a) $x - 2y = 8$

x	y
2	
10	
	0
	-2

To complete the first two ordered pairs of the table, let $x = 2$ and $x = 10$, respectively.

If $\quad x = 2,$	If $\quad x = 10,$
then $\quad x - 2y = 8$	then $\quad x - 2y = 8$
becomes $\quad 2 - 2y = 8$	becomes $\quad 10 - 2y = 8$
$-2y = 6$	$-2y = -2$
$y = -3.$	$y = 1.$

Now complete the last two ordered pairs by letting $y = 0$ and $y = -2$, respectively.

If $\quad y = 0,$	If $\quad y = -2,$
then $\quad x - 2y = 8$	then $\quad x - 2y = 8$
becomes $\quad x - 2(0) = 8$	becomes $\quad x - 2(-2) = 8$
$x - 0 = 8$	$x + 4 = 8$
$x = 8.$	$x = 4.$

The completed table of values follows.

x	y
2	-3
10	1
8	0
4	-2

Write *y*-values here.

Write *x*-values here.

The corresponding ordered pairs are $(2, -3)$, $(10, 1)$, $(8, 0)$, and $(4, -2)$. Each ordered pair is a solution of the given equation $x - 2y = 8$.

(b) $x = 5$

x	y
	-2
	6
	3

The given equation is $x = 5$. No matter which value of y is chosen, the value of x is *always* 5.

x	y
5	-2
5	6
5	3

The corresponding ordered pairs are $(5, -2)$, $(5, 6)$, and $(5, 3)$.

◄ *Work Problem* **6** *at the Side.*

Note

We can think of $x = 5$ in Example 5(b) as an equation in two variables
by rewriting $x = 5$ as $x + 0y = 5$. This form of the equation shows that
for any value of y, the value of x is 5. Similarly, $y = 4$ in Problem 6(c)
in the margin on the preceding page is the same as $0x + y = 4$.

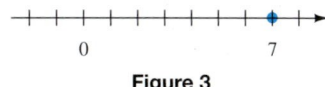 **OBJECTIVE 6 Plot ordered pairs.** In **Section 10.2,** we saw that lin-
ear equations in *one* variable had either one, zero, or an infinite number of
real number solutions. These solutions could be graphed on *one* number line.
For example, the linear equation in one variable $x - 2 = 5$ has solution 7,
which is graphed on the number line in Figure 3.

+——+——+——+——+——+——+——◆—+——→
       0   7

Figure 3

Every linear equation in *two* variables has an infinite number of ordered
pairs as solutions. Each choice of a number for one variable leads to a particular real number for the other variable.

To graph these solutions, represented as the ordered pairs (x, y), we need
two number lines, one for each variable, as drawn in Figure 4. The horizontal number line is called the **x-axis,** and the vertical line is called the **y-axis.**
Together, the x-axis and y-axis form a **rectangular coordinate system,** also
called the **Cartesian coordinate system,** in honor of René Descartes, the
French mathematician who is credited with its invention.

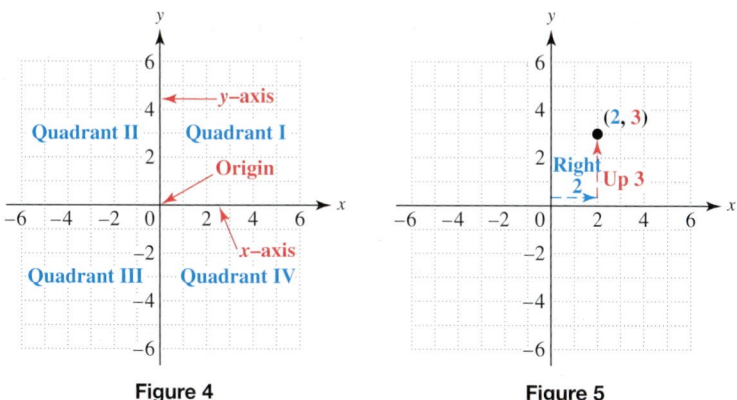

Figure 4   **Figure 5**

The coordinate system is divided into four regions, called **quadrants.**
These quadrants are numbered counterclockwise, as shown in Figure 4.
Points on the axes themselves are not in any quadrant. The point at which
the x-axis and y-axis cross is called the **origin.** The origin, which is labeled 0
in Figure 4, is the point corresponding to $(0, 0)$.

Work Problem **7** *at the Side.* ▶

The x-axis and y-axis determine a **plane**—a flat surface illustrated by a
sheet of paper. By referring to the two axes, every point in the plane can be
associated with an ordered pair. The numbers in the ordered pair are called
the **coordinates** of the point.

For example, we locate the point associated with the ordered pair $(2, 3)$
by starting at the origin. Since the x-coordinate is 2, we go 2 units to the right
along the x-axis. Then, since the y-coordinate is 3, we turn and go up 3 units
on a line parallel to the y-axis. The point $(2, 3)$ is **plotted** in Figure 5. From
now on, we will refer to the point with x-coordinate 2 and y-coordinate 3 as
the point $(2, 3)$.

7 Name the quadrant in which
each point in the figure is
located.

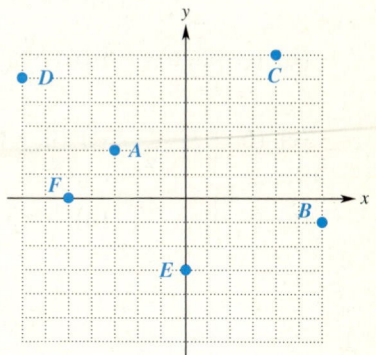

**René Descartes
(1596–1650)**

ANSWER

7. *A:* II; *B:* IV; *C:* I; *D:* II; *E:* no quadrant;
   F: no quadrant

8 Plot each ordered pair on a coordinate system.

(a) $(3, 5)$ **(b)** $(-2, 6)$

(b) $(-4.5, 0)$ **(d)** $(-5, -2)$

(e) $(6, -2)$ **(f)** $(0, -6)$

(g) $(0, 0)$ **(h)** $\left(-3, \dfrac{5}{2}\right)$

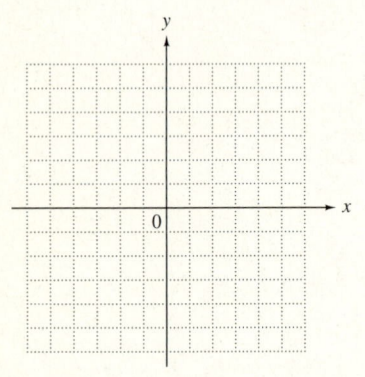

> **Note**
>
> When we graph on a number line (as in Figure 3), one number corresponds to each point. On a plane, however, *both* numbers in an ordered pair are needed to locate a point (as in Figure 5). The ordered pair is a name for the point.
>
> We mentioned that René Descartes is credited with inventing the Cartesian coordinate system. Legend has it that Descartes, who was lying in bed ill, was watching a fly crawl about on the ceiling near a corner of the room. It occurred to him that the location of the fly could be described by determining its distances from the two adjacent walls. See the figure.
>
>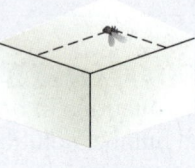
>
> Locating a fly on a ceiling

EXAMPLE 6 **Plotting Ordered Pairs**

Plot each ordered pair on a coordinate system.

(a) $(1, 5)$ **(b)** $(-2, 3)$ **(c)** $(-1, -4)$ **(d)** $(3, -2)$

(e) $\left(\dfrac{3}{2}, 2\right)$ **(f)** $(5, 0)$ **(g)** $(0, -3)$ **(h)** $(4, -3.75)$

See Figure 6. In each case, begin at the origin. Move right or left the number of units that corresponds to the x-coordinate in the ordered pair— *right if the x-coordinate is positive or left if it is negative.* Then turn and move up or down the number of units that corresponds to the y-coordinate— *up if the y-coordinate is positive or down if it is negative.* So in part (c), locate the point $(-1, -4)$ by first going 1 unit to the *left* along the x-axis. Then turn and go 4 units *down,* parallel to the y-axis.

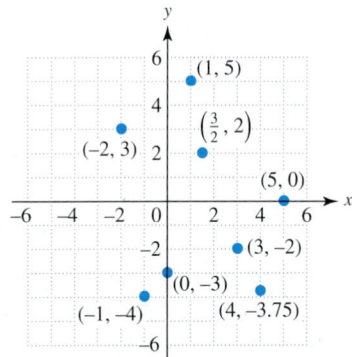

Figure 6

Notice the difference in the locations of the points $(-2, 3)$ and $(3, -2)$ in parts (b) and (d). The point $(-2, 3)$ is in quadrant II, whereas the point $(3, -2)$ is in quadrant IV. *The order of the coordinates is important. Remember that the x-coordinate is always given first in an ordered pair.*

To plot the point $\left(\frac{3}{2}, 2\right)$ in part (e), think of the improper fraction $\frac{3}{2}$ as the mixed number $1\frac{1}{2}$ and move $\frac{3}{2}$ (or $1\frac{1}{2}$) units to the right along the x-axis. Then turn and go 2 units up, parallel to the y-axis. The point $(4, -3.75)$ in part (h) is plotted similarly, by approximating the location of the decimal y-coordinate.

In part (f), the point $(5, 0)$ lies on the x-axis since the y-coordinate is 0. In part (g), the point $(0, -3)$ lies on the y-axis since the x-coordinate is 0.

◀ *Work Problem* **8** *at the Side.*

ANSWERS

8.

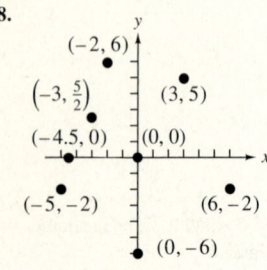

Sometimes we can use a linear equation in two variables to mathematically describe, or *model*, a real-life situation, as shown in the next example.

EXAMPLE 7 **Completing Ordered Pairs to Estimate the Number of Twin Births**

The number of twin births in the United States has increased steadily in recent years. The annual number of twin births from 2000 through 2005 can be closely approximated by the linear equation

Number of twin births ——┐   ┌— Year

$$y = 3.074x - 6029.7,$$

which relates x, the year, and y, the number of twin births in thousands. (*Source: National Vital Statistics Reports,* Vol. 56, No. 6, December 5, 2007.)

(a) Complete the table of values for the given linear equation.

x (Year)	y (Number of Twin Births, in thousands)
2000	
2002	
2005	

To find y when $x = 2000$, substitute into the equation.

$$y = 3.074x - 6029.7$$

$$y = 3.074(2000) - 6029.7 \quad \text{Let } x = 2000.$$

≈ means "is approximately equal to."

$$y \approx 118 \qquad \text{Use a calculator.}$$

This means that in 2000, there were about 118 thousand (or 118,000) twin births in the United States.

Work Problem **9** *at the Side.* ▶

Including the results from Problem 9 at the side gives the completed table that follows.

x (Year)	y (Number of Twin Births, in thousands)
2000	118
2002	124
2005	134

We can write the results from the table of values as ordered pairs (x, y). Each year x is paired with its number of twin births y (in thousands):

$$(2000, 118), \quad (2002, 124), \quad \text{and} \quad (2005, 134).$$

—— **Continued on Next Page**

9 Refer to the linear equation in Example 7.

(a) Find the y-value for $x = 2002$. Round to the nearest whole number.

(b) Find the y-value for $x = 2005$. Interpret your result.

(b) Graph the ordered pairs found in part (a).

The ordered pairs (2000, 118), (2002, 124), and (2005, 134) are graphed in Figure 7. This graph of ordered pairs of data is called a **scatter diagram.** Notice how the axes are labeled: x represents the year, and y represents the number of twin births in thousands. Different scales are used on the two axes. Here, each square represents one unit in the horizontal direction and 5 units in the vertical direction. Because the numbers in the first ordered pair are large, we show a break in the axes near the origin.

x (Year)	y (Number of Twin Births, in thousands)
2000	118
2002	124
2005	134

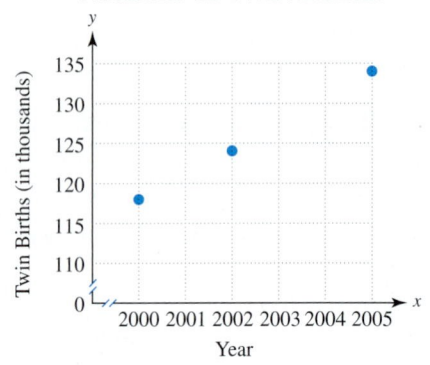

NUMBER OF TWIN BIRTHS

Figure 7

A scatter diagram enables us to tell whether two quantities are related to each other. In Figure 7, the plotted points could be connected to closely approximate a straight *line,* so the variables x (year) and y (number of twin births) have a *line*ar relationship. The increase in the number of twin births is also reflected.

CAUTION
The equation in Example 7 is valid only for the years 2000 through 2005 because it was based on data for those years. *Do not assume that this equation would provide reliable data for other years since the data for those years may not follow the same pattern.*

FOR EXTRA HELP

The bar graph shows total U.S. milk production in billions of pounds for the years 2001 through 2007. Use the bar graph to work Exercises 1–4. See Example 1.

1. In what years was U.S. milk production greater than 175 billion pounds?

2. In what years was U.S. milk production about the same?

3. Estimate U.S. milk production in 2001 and 2007.

4. Describe the change in U.S. milk production from 2001 to 2007.

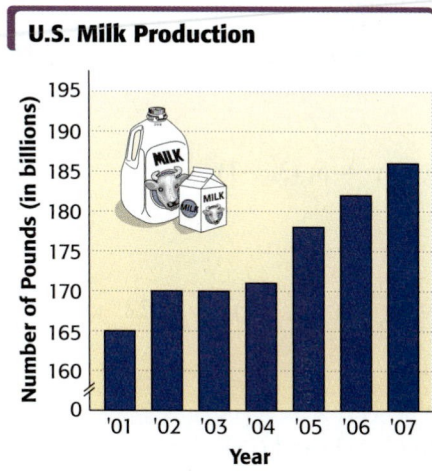

U.S. Milk Production

Source: U.S. Department of Agriculture.

The line graph shows the average price, adjusted for inflation, that Americans have paid for a gallon of gasoline for selected years since 1970. Use the line graph to work Exercises 5–8. See Example 2.

5. Over which period of years did the greatest increase in the price of a gallon of gas occur? About how much was this increase?

6. Estimate the price of a gallon of gas during 1985, 1990, 1995, and 2000.

7. Describe the trend in gas prices from 1980 to 1995.

8. During which year(s) did a gallon of gas cost approximately $1.50?

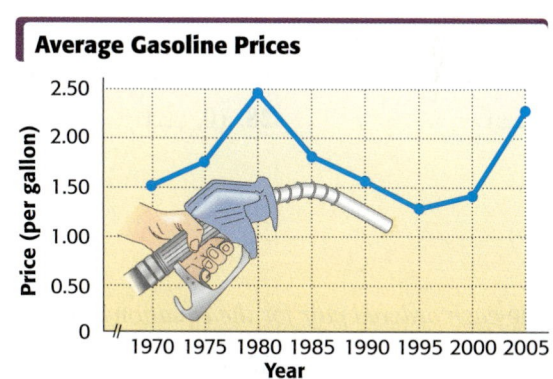

Average Gasoline Prices

Source: Energy Information Administration.

Use the concepts of this section to fill in each blank with the correct response.

9. The symbol (x, y) _____ represent an ordered pair, while the
   (does/does not)

   symbols $[x, y]$ and $\{x, y\}$ _____ represent ordered pairs.
   (do/do not)

10. The point whose graph has coordinates $(-4, 2)$ is in quadrant _____.

11. The point whose graph has coordinates $(0, 5)$ lies on the _____-axis.

12. The ordered pair $(4, \underline{\hspace{1cm}})$ is a solution of the equation $y = 3$.

13. The ordered pair $(\underline{\hspace{1cm}}, -2)$ is a solution of the equation $x = 6$.

14. The ordered pair $(3, 2)$ is a solution of the equation $2x - 5y = \underline{\hspace{0.7cm}}$.

Decide whether each ordered pair is a solution of the given equation. See Example 3.

15. $x + y = 9; (0, 9)$ **16.** $x + y = 8; (0, 8)$ **17.** $2x - y = 6; (4, 2)$

18. $2x + y = 5; (3, -1)$ **19.** $4x - 3y = 6; (2, 1)$ **20.** $5x - 3y = 15; (5, 2)$

21. $y = \dfrac{2}{3}x; (-6, -4)$ **22.** $y = -\dfrac{1}{4}x; (-8, 2)$ **23.** $x = -6; (5, -6)$ **24.** $y = 2; (2, 4)$

25. Do $(4, -1)$ and $(-1, 4)$ represent the same ordered pair? Explain.

26. Explain why it would be easier to find the corresponding y-value for $x = \frac{1}{3}$ in the equation $y = 6x + 2$ than it would be for $x = \frac{1}{7}$.

Complete each ordered pair for the equation $y = 2x + 7$. See Example 4.

27. $(2, \underline{\hspace{0.5cm}})$ **28.** $(0, \underline{\hspace{0.5cm}})$ **29.** $(\underline{\hspace{0.5cm}}, 0)$ **30.** $(\underline{\hspace{0.5cm}}, -3)$

Complete each ordered pair for the equation $y = -4x - 4$. See Example 4.

31. $(0, \underline{\hspace{0.5cm}})$ **32.** $(\underline{\hspace{0.5cm}}, 0)$ **33.** $(\underline{\hspace{0.5cm}}, 16)$ **34.** $(\underline{\hspace{0.5cm}}, 24)$

Complete each table of values. In Exercises 35–38, write the results as ordered pairs. See Example 5.

35. $2x + 3y = 12$

x	y
0	
	0
	8

36. $4x + 3y = 24$

x	y
0	
	0
	4

37. $3x - 5y = -15$

x	y
0	
	0
	-6

38. $4x - 9y = -36$

x	y
	0
0	
	8

39. $x = -9$

x	y
	6
	2
	-3

40. $x = 12$

x	y
	3
	8
	0

41. $y = -6$

x	y
8	
4	
-2	

42. $y = -10$

x	y
4	
0	
-4	

43. $x - 8 = 0$

x	y
	8
	3
	0

44. $y + 2 = 0$

x	y
9	
2	
0	

Give the ordered pairs for the points labeled A–F in the figure. Tell the quadrant in which each point is located.

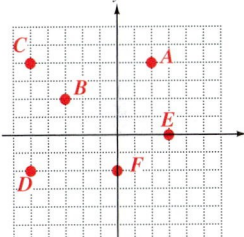

45. A

46. B

47. C

48. D

49. E

50. F

Fill in each blank with the word positive *or the word* negative.

The point with coordinates (x, y) is in

51. quadrant III if x is _____ and y is _____.

52. quadrant II if x is _____ and y is _____.

53. quadrant IV if x is _____ and y is _____.

54. quadrant I if x is _____ and y is _____.

55. A point (x, y) has the property that $xy < 0$. In which quadrant(s) must the point lie? Explain.

56. A point (x, y) has the property that $xy > 0$. In which quadrant(s) must the point lie? Explain.

Plot each ordered pair on the rectangular coordinate system provided. See Example 6.

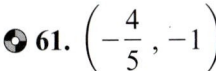 **57.** $(6, 2)$

58. $(5, 3)$

59. $(-4, 2)$

60. $(-3, 5)$

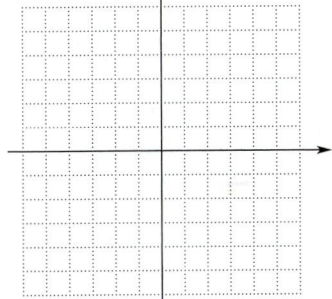

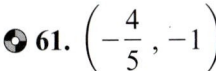 **61.** $\left(-\dfrac{4}{5}, -1\right)$

62. $\left(-\dfrac{3}{2}, -4\right)$

63. $(3, -1.75)$

64. $(5, -4.25)$

65. $(0, 4)$

66. $(0, -3)$

67. $(4, 0)$

68. $(-3, 0)$

Complete each table of values, and then plot the ordered pairs. See Examples 5 and 6.

69. $x - 2y = 6$

x	y
0	
	0
2	
	−1

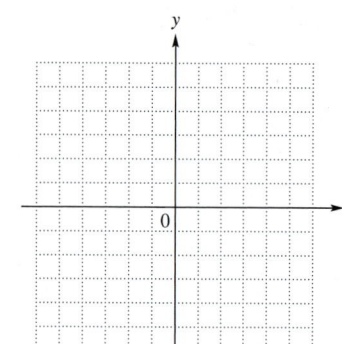

70. $2x - y = 4$

x	y
0	
	0
1	
	−6

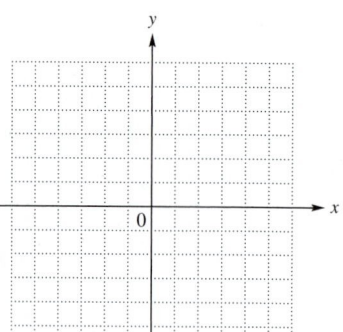

71. $3x - 4y = 12$

x	y
0	
	0
−4	
	−4

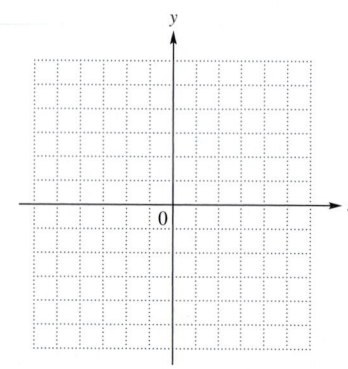

72. $2x - 5y = 10$

x	y
0	
	0
−5	
	−3

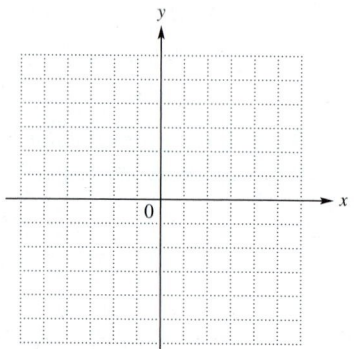

73. $y + 4 = 0$

x	y
0	
5	
−2	
−3	

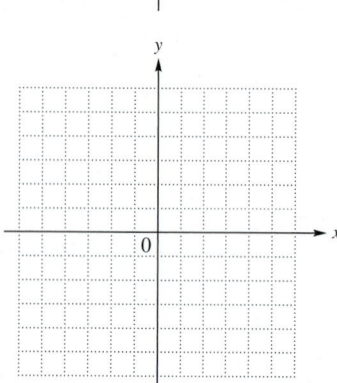

74. $x - 5 = 0$

x	y
	1
	0
	6
	−4

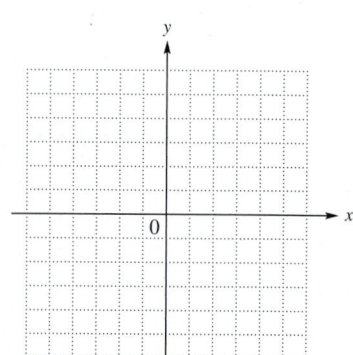

75. Look at the graphs of the ordered pairs in Exercises 69–74. Describe the pattern indicated by the plotted points.

Work each problem. See Example 7.

76. Suppose that it costs $5000 to start up a business selling snow cones. Furthermore, it costs $0.50 per cone in labor, ice, syrup, and overhead. Then the cost to make *x* snow cones is given by *y* dollars, where

$$y = 0.50x + 5000.$$

Express each of the following as an ordered pair.

(a) When 100 snow cones are made, the cost is $5050. (*Hint:* What does *x* represent? What does *y* represent?)

(b) When the cost is $6000, the number of snow cones made is 2000.

77. It costs a flat fee of $20 plus $5 per day to rent a pressure washer. Therefore, the cost to rent the pressure washer for *x* days is given by

$$y = 5x + 20,$$

where *y* is in dollars. Express each of the following as an ordered pair.

(a) When the washer is rented for 5 days, the cost is $45. (*Hint:* What does *x* represent? What does *y* represent?)

(b) I paid $50 when I returned the washer, so I must have rented it for 6 days.

78. The table shows the number of U.S. students studying abroad (in thousands) for several academic years.

Academic Year	Number of Students (in thousands)
2000	154
2001	161
2002	175
2003	191
2004	206
2005	224

Source: Institute of International Education.

(a) Write the data from the table as ordered pairs (x, y), where x represents the year and y represents the number of U.S. students studying abroad.

(b) What does the ordered pair (2004, 206) mean in the context of this problem?

(c) Make a scatter diagram of the data using the ordered pairs from part (a).

U.S. STUDENTS STUDYING ABROAD

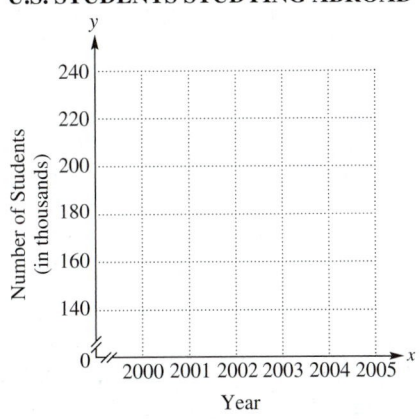

(d) Describe the pattern indicated by the points on the scatter diagram. What is the trend in the number of U.S. students studying abroad?

79. The table shows the rate (in percent) at which 2-year college students (public) complete a degree within 3 years.

Year	Percent
2000	32.4
2001	31.6
2002	31.6
2003	30.1
2004	29.0
2005	27.5

Source: ACT.

(a) Write the data from the table as ordered pairs (x, y), where x represents the year and y represents the percent.

(b) What would the ordered pair (2007, 27.1) mean in the context of this problem?

(c) Make a scatter diagram of the data using the ordered pairs from part (a).

2-YEAR COLLEGE STUDENTS COMPLETING A DEGREE WITHIN 3 YEARS

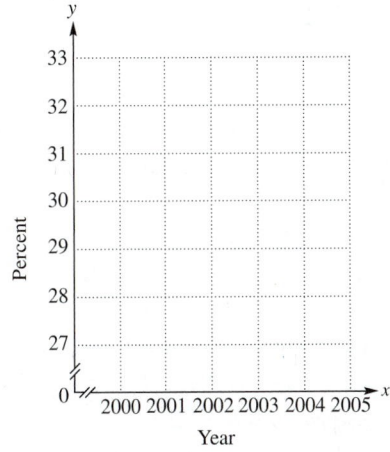

(d) Describe the pattern indicated by the points on the scatter diagram. What is happening to the rates at which 2-year college students complete a degree within 3 years?

80. The maximum benefit for the heart from exercising occurs if the heart rate is in the target heart rate zone. The lower limit of this target zone can be approximated by the linear equation

$$y = -0.5x + 108,$$

where x represents age and y represents heartbeats per minute. (*Source:* www.fitresource.com)

(a) Complete the table of values for this linear equation.

Age	Heartbeats (per minute)
20	
40	
60	
80	

(b) Write the data from the table of values as ordered pairs.

(c) Make a scatter diagram of the data. Do the points lie in a linear pattern?

TARGET HEART RATE ZONE
(Lower Limit)

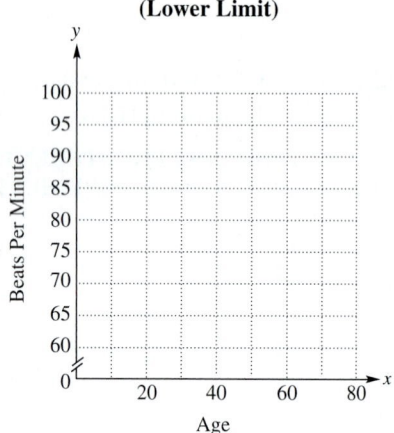

81. (See Exercise 80.) The upper limit of the target heart rate zone can be approximated by the linear equation

$$y = -0.8x + 173,$$

where x represents age and y represents heartbeats per minute. (*Source:* www.fitresource.com)

(a) Complete the table of values for this linear equation.

Age	Heartbeats (per minute)
20	
40	
60	
80	

(b) Write the data from the table of values as ordered pairs.

(c) Make a scatter diagram of the data. Describe the pattern indicated by the data.

TARGET HEART RATE ZONE
(Upper Limit)

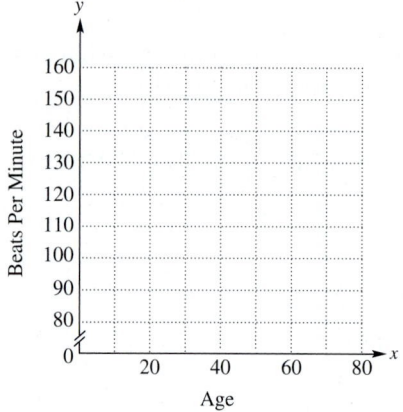

82. Refer to Exercises 80 and 81. What is the target heart rate zone for age 20? age 40?

11.2 ▷▷▷ Graphing Linear Equations in Two Variables

OBJECTIVE 1 Graph linear equations by plotting ordered pairs. There are infinitely many ordered pairs that satisfy an equation in two variables. We find these ordered-pair solutions by choosing as many values of x (or y) as we wish and then completing each ordered pair.

For example, consider the equation $x + 2y = 7$. If we choose $x = 1$, then $y = 3$, so the ordered pair $(1, 3)$ is a solution of the equation $x + 2y = 7$.

$$1 + 2(3) = 7$$

Work Problem ⃝**1** *at the Side.* ▶

Figure 8 shows a graph of all the ordered-pair solutions found above and in Problem 1 at the side for $x + 2y = 7$.

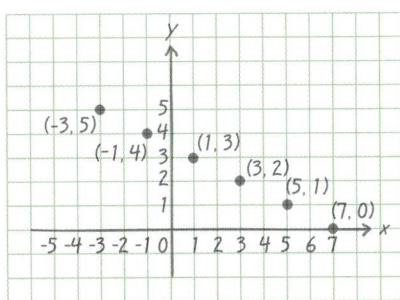

Figure 8

Notice that the points plotted in Figure 8 all appear to lie on a straight line, as shown in Figure 9. In fact, the following is true.

Every point on the line represents a solution of the equation $x + 2y = 7$, and every solution of the equation corresponds to a point on the line.

The line gives a "picture" of all the solutions of the equation $x + 2y = 7$. Only a portion of the line is shown here, but it extends indefinitely in both directions, as suggested by the arrowhead on each end.

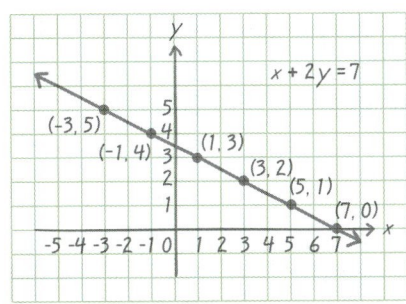

Figure 9

The line in Figure 9 is called the **graph** of the equation $x + 2y = 7$, and the process of plotting the ordered pairs and drawing the line through the corresponding points is called **graphing.**

OBJECTIVES

1 Graph linear equations by plotting ordered pairs.

2 Find intercepts.

3 Graph linear equations of the form $Ax + By = 0$.

4 Graph linear equations of the form $y = k$ or $x = k$.

5 Use a linear equation to model data.

⃝**1** Complete each ordered pair for the equation $x + 2y = 7$.

(a) $(-3, \underline{\quad})$

(b) $(-1, \underline{\quad})$

(c) $(3, \underline{\quad})$

(d) $(5, \underline{\quad})$

(e) $(7, \underline{\quad})$

ANSWERS

1. **(a)** $(-3, 5)$ **(b)** $(-1, 4)$ **(c)** $(3, 2)$
(d) $(5, 1)$ **(e)** $(7, 0)$

The preceding discussion can be generalized.

2 Complete the table of values, and graph the linear equation.

$x + y = 6$

x	y
0	
	0
2	

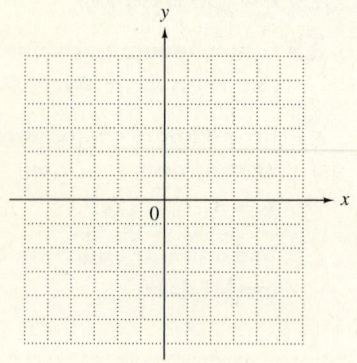

> **Graph of a Linear Equation**
>
> The graph of any linear equation in two variables is a straight line.

(Notice that the word *line* appears in the term *"line*ar equation.")

Because two distinct points determine a line, a straight line can be graphed by finding any two different points on the line. However, it is a good idea to plot a third point as a check.

EXAMPLE 1 **Graphing a Linear Equation**

Graph the linear equation $4x - 5y = 20$.

At least two different points are needed to draw the graph. First let $x = 0$ and then let $y = 0$ to complete two ordered pairs.

$$4x - 5y = 20$$
$$4(0) - 5y = 20 \quad \text{Let } x = 0.$$
$$0 - 5y = 20$$
$$-5y = 20$$
$$y = -4$$

$$4x - 5y = 20$$
$$4x - 5(0) = 20 \quad \text{Let } y = 0.$$
$$4x - 0 = 20$$
$$4x = 20$$
$$x = 5$$

> Write each x-value first.

The ordered pairs are $(0, -4)$ and $(5, 0)$. Find a third ordered pair (as a check) by choosing a number other than 0 for x or y. We choose $y = 2$.

$$4x - 5y = 20$$
$$4x - 5(2) = 20 \quad \text{Let } y = 2.$$
$$4x - 10 = 20$$
$$4x = 30 \quad \text{Add 10.}$$
$$x = \frac{30}{4}, \quad \text{or} \quad \frac{15}{2} \quad \text{Divide by 4; lowest terms}$$

This gives the ordered pair $(\frac{15}{2}, 2)$, or $(7\frac{1}{2}, 2)$. Plot the three ordered pairs $(0, -4)$, $(5, 0)$, and $(7\frac{1}{2}, 2)$, and draw a line through them. This line, shown in Figure 10, is the graph of $4x - 5y = 20$.

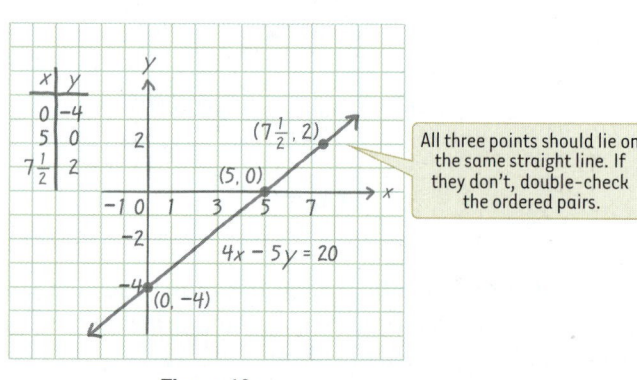

All three points should lie on the same straight line. If they don't, double-check the ordered pairs.

Figure 10

ANSWER

2.

x	y
0	6
6	0
2	4

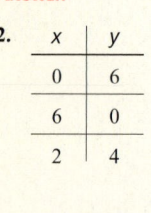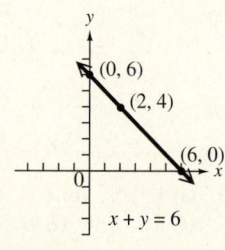

◀ *Work Problem* **2** *at the Side.*

EXAMPLE 2 **Graphing a Linear Equation**

Graph the linear equation $y = -\frac{3}{2}x + 3$.

Although this equation is not in the form $Ax + By = C$, it *could* be written in that form, so it is a linear equation. Two different points on the graph can be found by first letting $x = 0$ and then letting $y = 0$.

If $x = 0$, then

$$y = -\frac{3}{2}x + 3$$

$$y = -\frac{3}{2}(0) + 3 \qquad \text{Let } x = 0.$$

$$y = 0 + 3 \qquad \text{Multiply.}$$

$$y = 3. \qquad \text{Add.}$$

If $y = 0$, then

$$y = -\frac{3}{2}x + 3$$

$$0 = -\frac{3}{2}x + 3 \qquad \text{Let } y = 0.$$

$$\frac{3}{2}x = 3 \qquad \text{Add } \tfrac{3}{2}x.$$

$$x = 2. \qquad \text{Multiply by } \tfrac{2}{3}.$$

This gives the ordered pairs $(0, 3)$ and $(2, 0)$. We find a third point (as a check) by letting x or y equal some other number. For example, let $x = -2$.

$$y = -\frac{3}{2}x + 3$$

> Choosing a multiple of 2 makes multiplying by $-\frac{3}{2}$ easier.

$$y = -\frac{3}{2}(-2) + 3 \qquad \text{Let } x = -2.$$

$$y = 3 + 3 \qquad \text{Multiply.}$$

$$y = 6 \qquad \text{Add.}$$

This gives the ordered pair $(-2, 6)$. These three ordered pairs are shown in the table with Figure 11. Plot the corresponding points, and then draw a line through them. This line, shown in Figure 11, is the graph of $y = -\frac{3}{2}x + 3$.

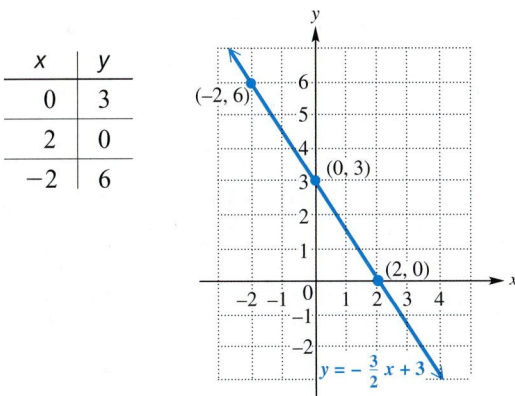

x	y
0	3
2	0
−2	6

Figure 11

Work Problem **3** *at the Side.* ▶

3 Make a table of values, and graph the linear equation.

$$y = \frac{2}{3}x - 2$$

x	y

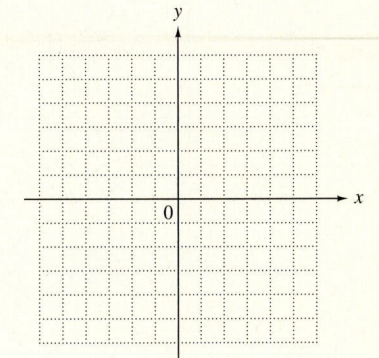

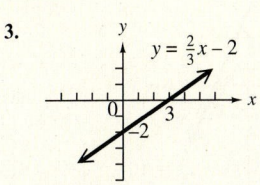

4 Find the intercepts for the graph of $5x + 2y = 10$. Then draw the graph. (Be sure to get a third point as a check.)

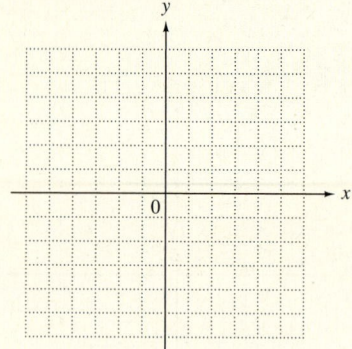

OBJECTIVE 2 Find intercepts. In Figure 11, the graph crosses, or intersects, the y-axis at $(0, 3)$ and the x-axis at $(2, 0)$. For this reason, $(0, 3)$ is called the **y-intercept,** and $(2, 0)$ is called the **x-intercept** of the graph.

The intercepts are particularly useful for graphing linear equations. The intercepts are found by replacing, in turn, each variable with 0 in the equation and solving for the value of the other variable.

Finding Intercepts

To find the *x*-intercept, let $y = 0$ in the given equation and solve for x. Then $(x, 0)$ is the *x*-intercept.

To find the *y*-intercept, let $x = 0$ in the given equation and solve for y. Then $(0, y)$ is the *y*-intercept.

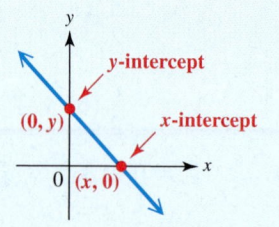

EXAMPLE 3 **Finding Intercepts**

Find the intercepts for the graph of $2x + y = 4$. Then draw the graph.

To find the y-intercept, let $x = 0$; to find the x-intercept, let $y = 0$.

$2x + y = 4$	$2x + y = 4$
$2(0) + y = 4$ Let $x = 0$.	$2x + 0 = 4$ Let $y = 0$.
$0 + y = 4$	$2x = 4$
$y = 4$	$x = 2$

The y-intercept is $(0, 4)$. The x-intercept is $(2, 0)$. Find a third point as a check. For example, choosing $x = 1$ gives $y = 2$. Plot $(0, 4)$, $(2, 0)$, and $(1, 2)$ and draw the line through them. This line, shown in Figure 12, is the graph.

x	y
0	4
2	0
1	2

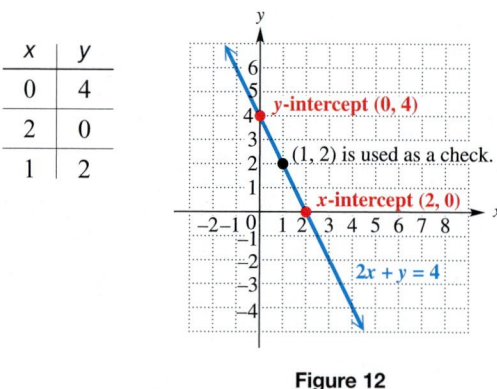

Figure 12

◀ *Work Problem* **4** *at the Side.*

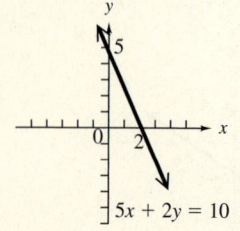
OBJECTIVE 3 Graph linear equations of the form $Ax + By = 0$. In the preceding examples, the x- and y-intercepts were used to help draw the graphs. This is not always possible. Example 4 shows what to do when the x- and y-intercepts are the same point (that is, coincide).

EXAMPLE 4 **Graphing an Equation of the Form $Ax + By = 0$**

Graph the linear equation $x - 3y = 0$.

If we let $x = 0$, then $y = 0$, giving the ordered pair $(0, 0)$. Letting $y = 0$ also gives $(0, 0)$. This is the same ordered pair, so we choose two *other* values for x or y. Choosing 2 for y gives $x - 3 \cdot 2 = 0$, leading to $x = 6$, so another ordered pair is $(6, 2)$. Choosing -2 for y gives $x - 3(-2) = 0$, leading to $x = -6$, so a third ordered pair is $(-6, -2)$. We use the ordered pairs $(-6, -2)$, $(0, 0)$, and $(6, 2)$ to sketch the graph in Figure 13.

x	y
0	0
6	2
-6	-2

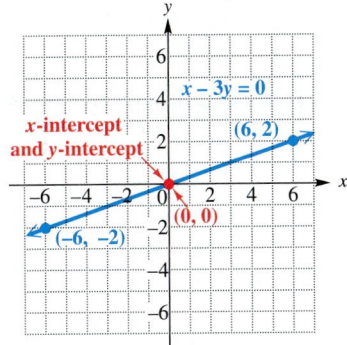

Figure 13

Work Problem **5** *at the Side.* ▶

Line through the Origin

If A and B are nonzero real numbers, the graph of a linear equation of the form

$$Ax + By = 0$$

passes through the origin $(0, 0)$.

OBJECTIVE 4 **Graph linear equations of the form $y = k$ or $x = k$.** The equation $y = -4$ is a linear equation in which the coefficient of x is 0. (To see this, write $y = -4$ as $0x + y = -4$.) Also, $x = 3$ is a linear equation in which the coefficient of y is 0. These equations lead to horizontal or vertical straight lines, as the next examples show.

EXAMPLE 5 **Graphing an Equation of the Form $y = k$**

Graph $y = -4$.

As the equation states, for any value of x, y is always equal to -4. Three ordered pairs that satisfy the equation are shown. The graph is the horizontal line in Figure 14. The y-intercept is $(0, -4)$; there is no x-intercept.

x	y
-2	-4
0	-4
3	-4

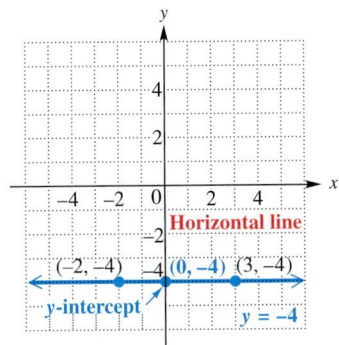

Figure 14

Work Problem **6** *at the Side.* ▶

5 Graph $2x - y = 0$.

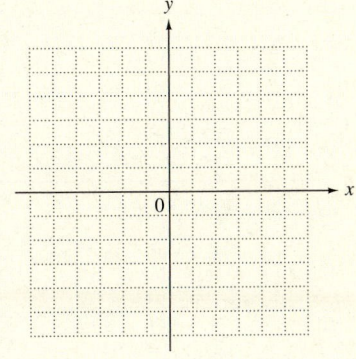

6 Graph $y = -5$.

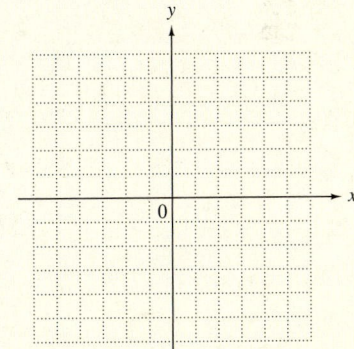

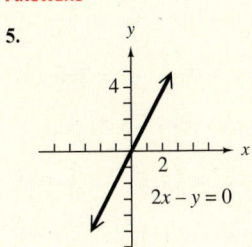
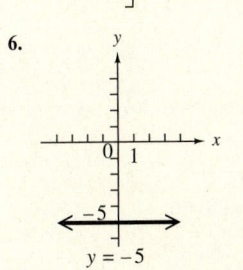

7 Graph $x + 4 = 6$.

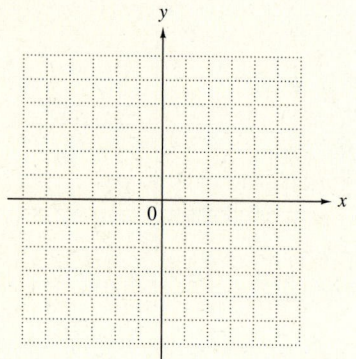

EXAMPLE 6 **Graphing an Equation of the Form $x = k$**

Graph $x - 3 = 0$.

First add 3 to each side of $x - 3 = 0$ to get $x = 3$. All the ordered pairs that satisfy this equation have x-coordinate 3. Any number can be used for y. See Figure 15 for the graph of this vertical line, along with a table of values. The x-intercept is $(3, 0)$; there is no y-intercept.

x	y
3	3
3	0
3	−2

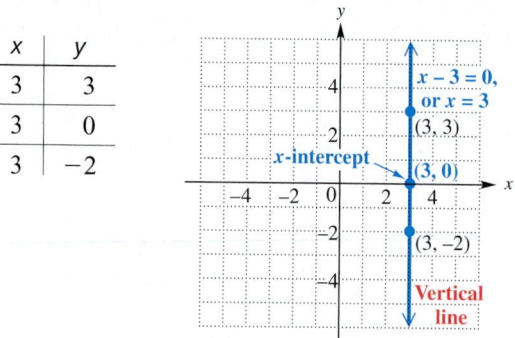

Figure 15

◀ *Work Problem* **7** *at the Side.*

From the results in Examples 5 and 6, we make the following observations.

Horizontal and Vertical Lines

The graph of the linear equation $y = k$, where k is a real number, is the horizontal line with y-intercept $(0, k)$ and no x-intercept.

The graph of the linear equation $x = k$, where k is a real number, is the vertical line with x-intercept $(k, 0)$ and no y-intercept.

In particular, notice that the horizontal line $y = 0$ is the x-axis and the vertical line $x = 0$ is the y-axis. The different forms of linear equations from this section and the methods of graphing them are summarized below.

Graphing a Linear Equation

Equation	*Graphing Method*	*Example*
$y = k$	Draw a horizontal line through $(0, k)$.	
$x = k$	Draw a vertical line through $(k, 0)$.	

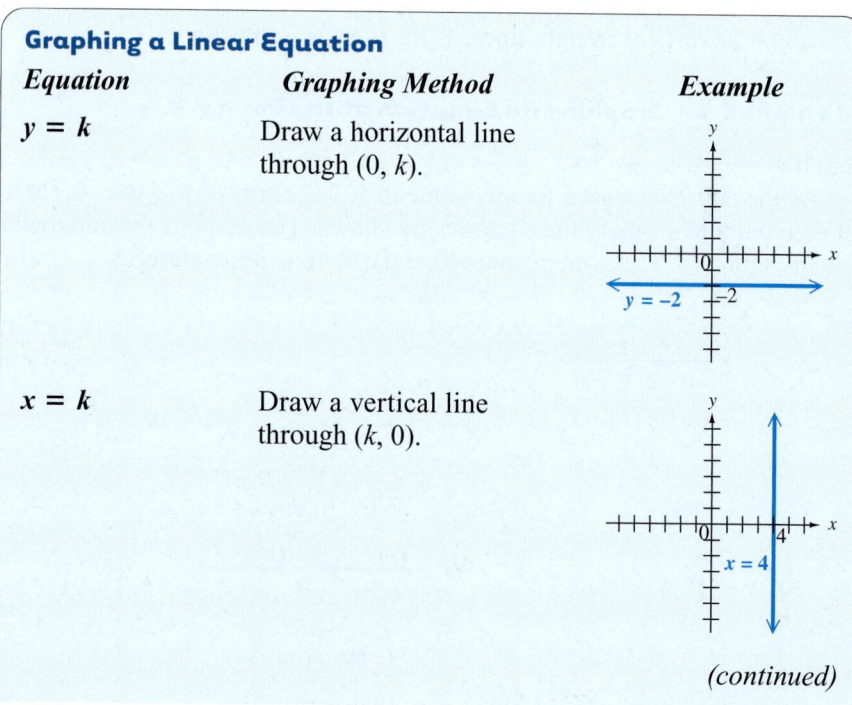

(continued)

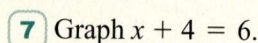

Equation	Graphing Method	Example
$Ax + By = 0$	Graph passes through $(0, 0)$. To get additional points that lie on the graph, choose any values for x or y, except 0.	
$Ax + By = C$ (but not of the types above)	Find any two points on the line. A good choice is to find the intercepts. Let $x = 0$, and find the corresponding value of y; then let $y = 0$, and find x. As a check, get a third point by choosing a value of x or y that has not yet been used.	

Work Problem **8** **at the Side.** ▶

Note

Another method of graphing linear equations, using the concepts of slope and y-intercept, will be covered in Objective 2 of **Section 11.4.**

OBJECTIVE **5** **Use a linear equation to model data.**

EXAMPLE 7 **Using a Linear Equation to Model Credit Card Debt**

Credit card debt in the United States has increased steadily during recent years. The amount of debt y in billions of dollars can be modeled by the linear equation

$$y = 38.7x + 450,$$

where $x = 0$ represents the year 1995, $x = 1$ represents 1996, and so on. (*Source:* Board of Governors of the Federal Reserve System.)

(a) Use the equation to approximate credit card debt in the years 1995, 2000, and 2003.

Substitute the appropriate value for each year x to find credit card debt in that year.

For 1995: $\quad y = 38.7(0) + 450 \qquad$ Replace x with 0.
$\qquad\qquad\;\; y = 450$ billion dollars

For 2000: $\quad y = 38.7(5) + 450 \qquad$ $2000 - 1995 = 5$;
$\qquad\qquad\;\; y = 643.5$ billion dollars $\quad$ Replace x with 5.

For 2003: $\quad y = 38.7(8) + 450 \qquad$ $2003 - 1995 = 8$;
$\qquad\qquad\;\; y = 759.6$ billion dollars $\quad$ Replace x with 8.

Continued on Next Page

8 Match the information about the graphs in parts (a)–(d) with the linear equations in A–D.

A. $x = 5$
B. $2x - 5y = 8$
C. $y - 2 = 3$
D. $x + 4y = 0$

(a) The graph of the equation is a horizontal line.

(b) The graph of the equation passes through the origin.

(c) The graph of the equation is a vertical line.

(d) The graph of the equation passes through $(9, 2)$.

ANSWERS

8. **(a)** C **(b)** D **(c)** A **(d)** B

9 Use the graph and then the equation in Example 7 to approximate credit card debt in 1997.

(b) Write the information from part (a) as three ordered pairs, and use them to graph the given linear equation.

Since x represents the year and y represents the debt, the ordered pairs are $(0, 450)$, $(5, 643.5)$, and $(8, 759.6)$. See Figure 16. (Arrowheads are not included with the graphed line, since the data are for the years 1995 to 2003 only—that is, from $x = 0$ to $x = 8$.)

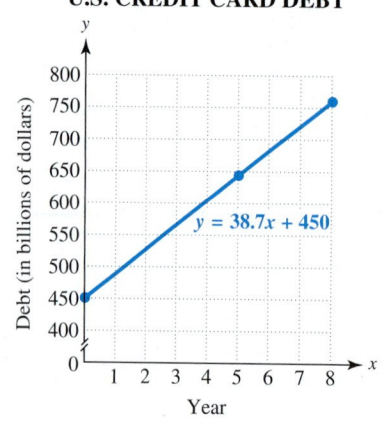

U.S. CREDIT CARD DEBT

$y = 38.7x + 450$

Debt (in billions of dollars)

Year

Figure 16

(c) Use the graph and then the equation to approximate credit card debt in 2002.

For 2002, $x = 7$. On the graph, find 7 on the horizontal axis, move up to the graphed line and then across to the vertical axis. It appears that credit card debt in 2002 was about 725 billion dollars. To use the equation, substitute 7 for x.

$$y = 38.7x + 450$$

$$y = 38.7(7) + 450 \qquad \text{Let } x = 7.$$

$$y = 720.9 \text{ billion dollars}$$

This result for 2002 is close to our estimate of 725 billion dollars from the graph.

◀ *Work Problem* **9** *at the Side.*

11.2 ▶▶▶ **Exercises**

Complete the given ordered pairs for each equation. Then graph each equation by plotting the points and drawing the line through them. See Examples 1 and 2.

1. $x + y = 5$

$(0, \underline{\quad}), (\underline{\quad}, 0), (2, \underline{\quad})$

2. $x - y = 2$

$(0, \underline{\quad}), (\underline{\quad}, 0), (5, \underline{\quad})$

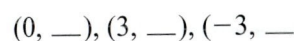 **3.** $y = \dfrac{2}{3}x + 1$

$(0, \underline{\quad}), (3, \underline{\quad}), (-3, \underline{\quad})$

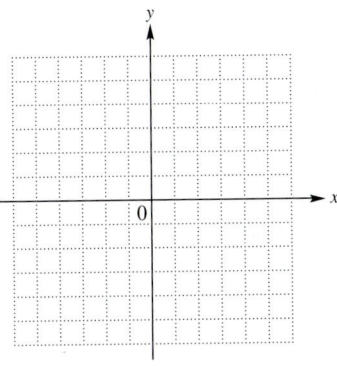 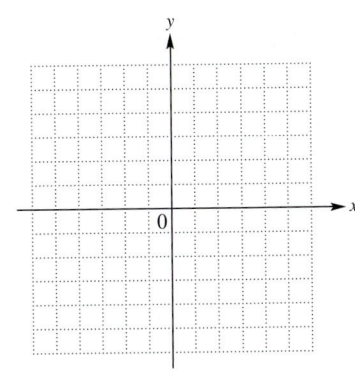 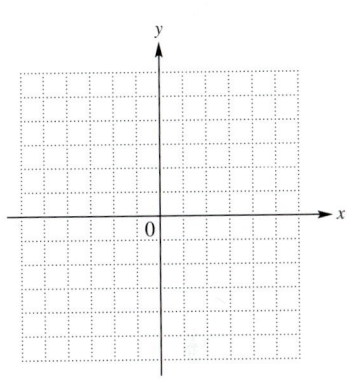

4. $y = -\dfrac{3}{4}x + 2$

$(0, \underline{\quad}), (4, \underline{\quad}), (-4, \underline{\quad})$

5. $3x = -y - 6$

$(0, \underline{\quad}), (\underline{\quad}, 0), \left(-\dfrac{1}{3}, \underline{\quad}\right)$

6. $x = 2y + 3$

$(\underline{\quad}, 0), (0, \underline{\quad}), \left(\underline{\quad}, \dfrac{1}{2}\right)$

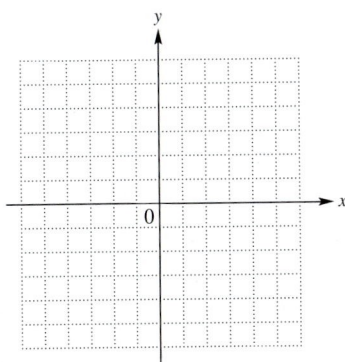 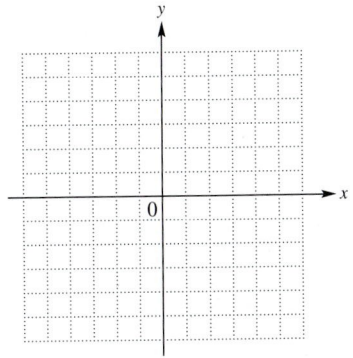 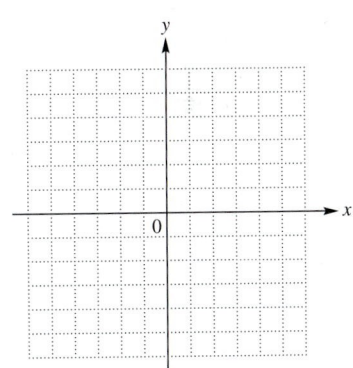

7. Match the information about each graph in Column I with the correct linear equation in Column II.

I

(a) The graph of the equation has y-intercept $(0, -4)$.

(b) The graph of the equation has $(0, 0)$ as x-intercept and y-intercept.

(c) The graph of the equation does not have an x-intercept.

(d) The graph of the equation has x-intercept $(4, 0)$.

II

A. $3x + y = -4$

B. $x - 4 = 0$

C. $y = 4x$

D. $y = 4$

8. Write a few sentences summarizing how to graph a linear equation in two variables.

Find the intercepts for the graph of each equation. See Example 3.

9. $2x - 3y = 24$

   x-intercept:

   y-intercept:

10. $-3x + 8y = 48$

   x-intercept:

   y-intercept:

11. $x + 6y = 0$

   x-intercept:

   y-intercept:

12. $3x - y = 0$

   x-intercept:

   y-intercept:

Graph each linear equation. See Examples 1–6.

13. $y = x - 2$

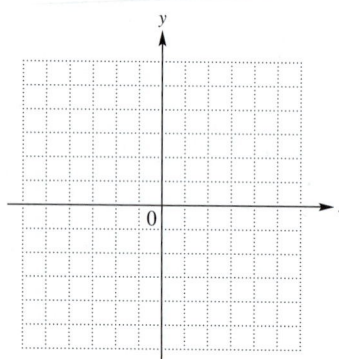

14. $y = -x + 6$

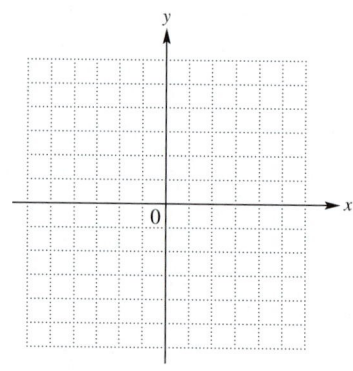

15. $x - y = 4$

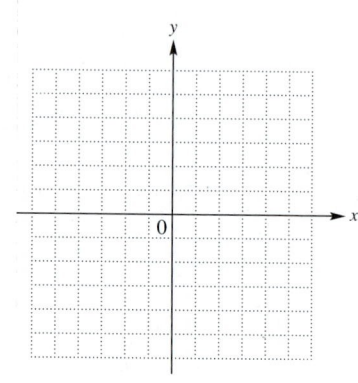

16. $x - y = 5$

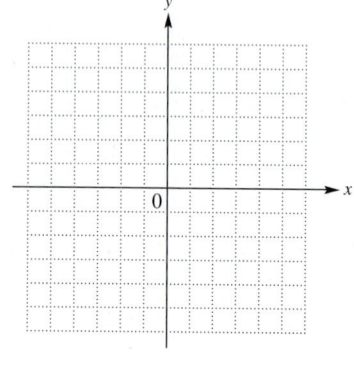

17. $2x + y = 6$

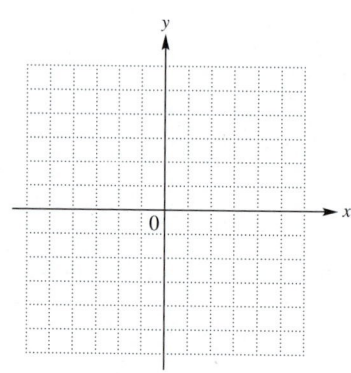

18. $-3x + y = -6$

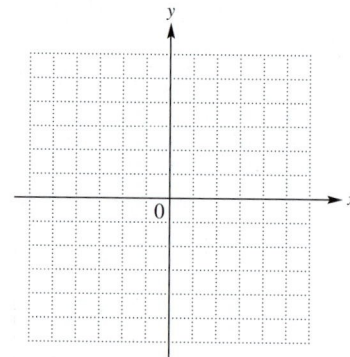

19. $3x + 7y = 14$

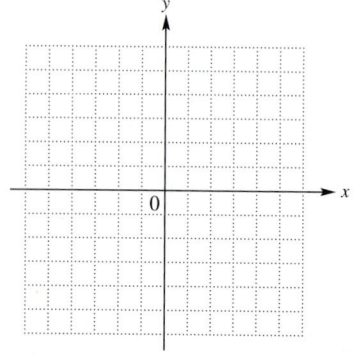

20. $6x - 5y = 18$

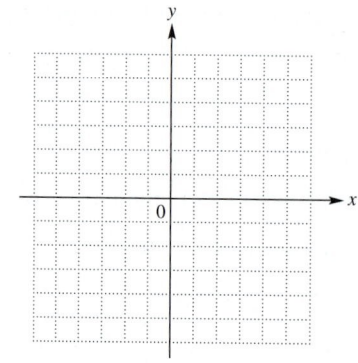

21. $y - 2x = 0$

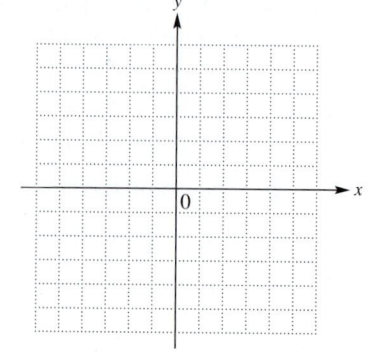

22. $y + 3x = 0$

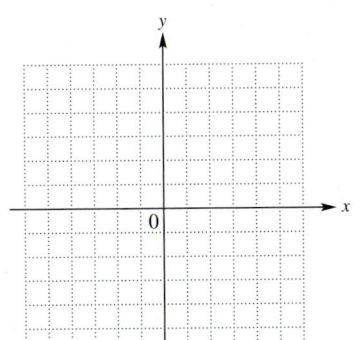

23. $y = -6x$

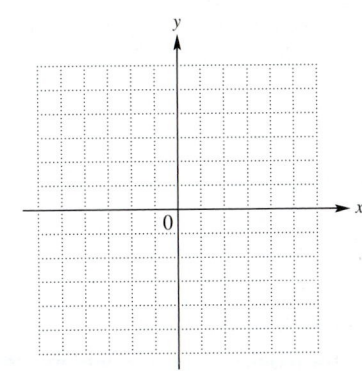

24. $y = 4x$

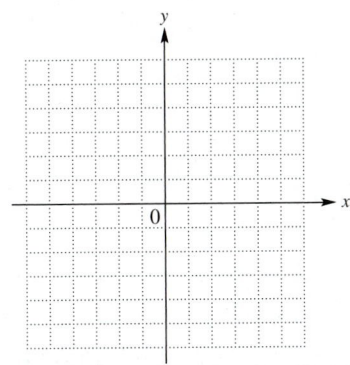

25. $x = -2$

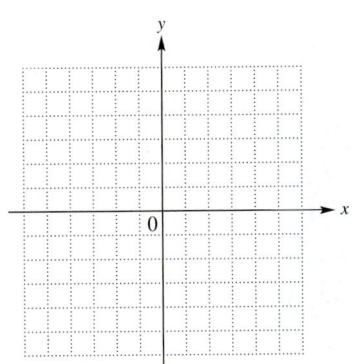

26. $x = 4$

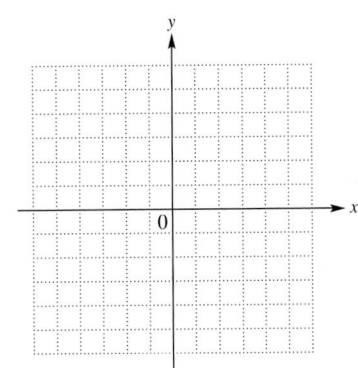

27. $y - 3 = 0$

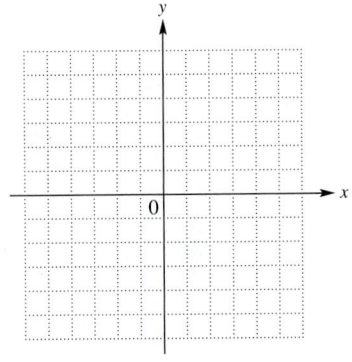

28. $y + 1 = 0$

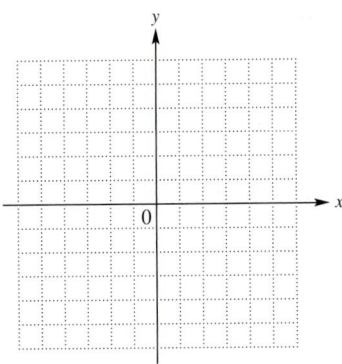

29. $-3y = 15$

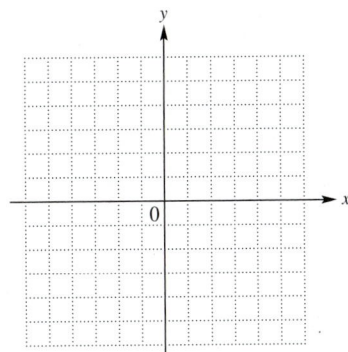

30. $-2y = 12$

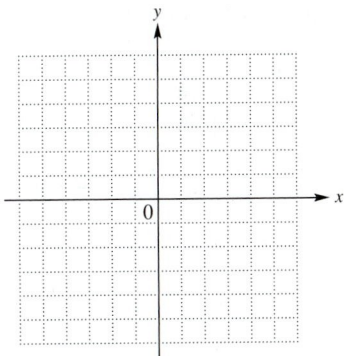

In Exercises 31–34, describe what the graph of each linear equation will look like on the coordinate plane. (Hint: Rewrite the equation if necessary so that it is in a more recognizable form.)

31. $3x = y - 9$
32. $x - 10 = 1$
33. $3y = -6$
34. $2x = 4y$

35. A student attempted to graph $4x + 5y = 0$ by finding intercepts. She first let $x = 0$ and found y; then she let $y = 0$ and found x. In both cases, the resulting point was $(0, 0)$. She knew that she needed at least two points to graph the line, but was unsure what to do next because finding intercepts gave her only one point. How would you explain to her what to do next?

36. What is the equation of the x-axis? What is the equation of the y-axis?

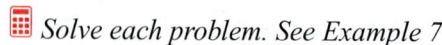 *Solve each problem. See Example 7.*

37. The height y (in centimeters) of a woman is related to the length of her radius bone x (from the wrist to the elbow) and is approximated by the linear equation

$$y = 3.9x + 73.5.$$

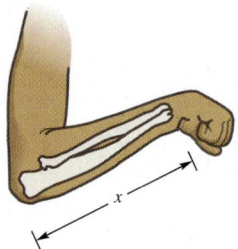

(a) Use the equation to find the approximate heights of women with radius bones of lengths 20 cm, 22 cm, and 26 cm.

(b) Write the information from part (a) as three ordered pairs.

(c) Graph the equation using the data from part (b).

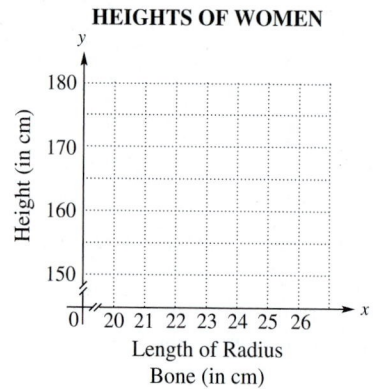

HEIGHTS OF WOMEN

Length of Radius Bone (in cm)

(d) Use the graph to estimate the length of the radius bone in a woman who is 167 cm tall. Then use the equation to find the length of this radius bone to the nearest centimeter. (*Hint:* Substitute for y in the equation.)

38. The weight y (in pounds) of a man taller than 60 in. can be roughly approximated by the linear equation

$$y = 5.5x - 220,$$

where x is the height of the man in inches.

(a) Use the equation to approximate the weights of men whose heights are 62 in., 66 in., and 72 in.

(b) Write the information from part (a) as three ordered pairs.

(c) Graph the equation using the data from part (b).

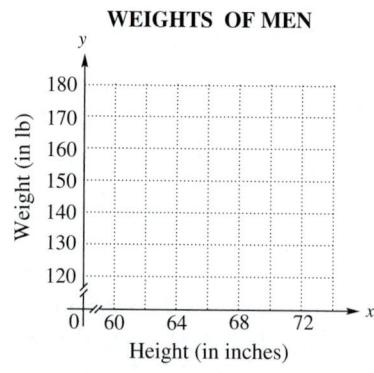

WEIGHTS OF MEN

Height (in inches)

(d) Use the graph to estimate the height of a man who weighs 155 lb. Then use the equation to find the height of this man to the nearest inch. (*Hint:* Substitute for y in the equation.)

39. As a fundraiser, a school club is selling posters. The printer charges a \$25 set-up fee, plus \$0.75 for each poster. Then the cost y in dollars to print x posters is given by the linear equation

$$y = 0.75x + 25.$$

(a) What is the cost y in dollars to print 50 posters? to print 100 posters?

(b) Find the number of posters x if the printer billed the club for costs of \$175.

(c) Write the information from parts (a) and (b) as three ordered pairs.

(d) Use the data from part (c) to graph the equation.

POSTER COSTS

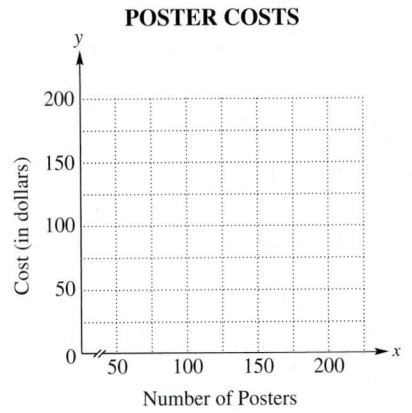

Number of Posters

40. A gas station is selling gasoline for \$4.50 per gallon and charges \$7 for a car wash. Then the cost y in dollars for x gallons of gasoline and a car wash is given by the linear equation

$$y = 4.50x + 7.$$

(a) What is the cost y in dollars for 9 gallons of gasoline and a car wash? for 4 gallons of gasoline and a car wash?

(b) Find the number of gallons of gasoline x if the cost for the gasoline and a car wash is \$43.00.

(c) Write the information from parts (a) and (b) as three ordered pairs.

(d) Use the data from part (c) to graph the equation.

GASOLINE AND CAR WASH COSTS

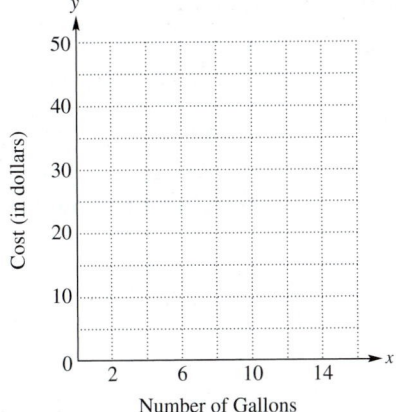

Number of Gallons

41. The graph shows the value of a certain sport-utility vehicle over the first five years of ownership. Use the graph to do the following.

(a) Determine the initial value of the SUV.

(b) Find the **depreciation** (loss in value) from the original value after the first three years.

(c) What is the annual or yearly depreciation in each of the first five years?

SUV VALUE

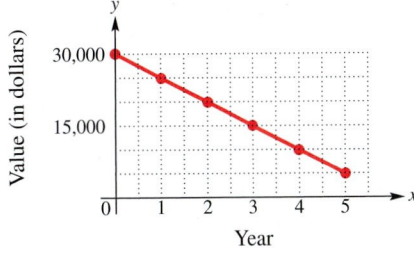

Year

(d) What does the ordered pair (5, 5000) mean in the context of this problem?

42. Demand for an item is often closely related to its price. As price increases, demand decreases, and as price decreases, demand increases. Suppose demand for a video game is 2000 units when the price is $40, and demand is 2500 units when the price is $30.

(a) Let x be the price and y be the demand for the game. Graph the two given pairs of prices and demands.

VIDEO GAME PRICE/DEMAND

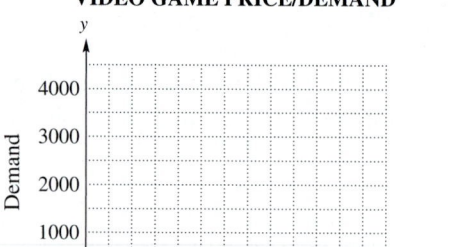

(b) Assume the relationship is linear. Draw a line through the two points from part (a). From your graph, estimate the demand if the price drops to $20.

(c) Use the graph to estimate the price if the demand is 3500 units.

(d) Write the prices and demands from parts (b) and (c) as ordered pairs.

43. U.S. per capita consumption of cheese increased for the years 1980 through 2005 as shown in the graph. If $x = 0$ represents 1980, $x = 5$ represents 1985, and so on, per capita consumption y in pounds can be modeled by the linear equation

$$y = 0.5383x + 18.74.$$

Cheese Consumption

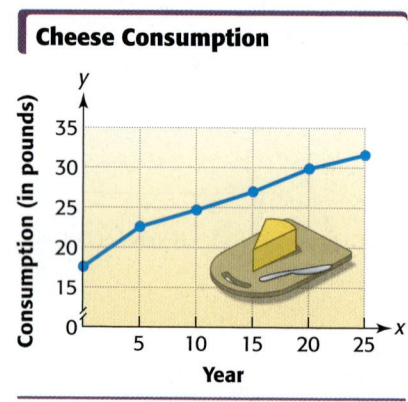

Source: U.S. Department of Agriculture.

(a) Use the equation to approximate consumption in 1990, 2000, and 2005 to the nearest tenth.

(b) Use the graph to estimate consumption for the same years.

(c) How do the approximations using the equation compare to the estimates from the graph?

44. In the United States, sporting goods sales y (in billions of dollars) from 2000 through 2006 are shown in the graph and modeled by the linear equation

$$y = 3.018x + 72.52,$$

where $x = 0$ corresponds to 2000, $x = 1$ corresponds to 2001, and so on.

Sporting Goods Sales

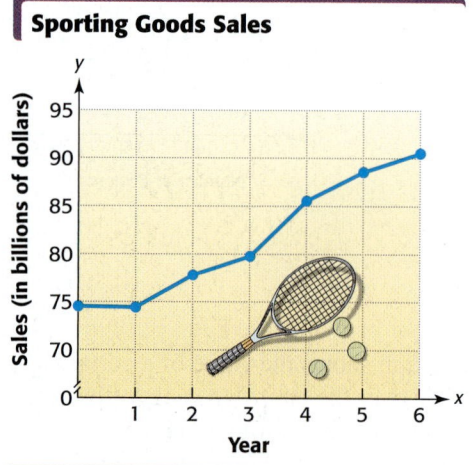

Source: National Sporting Goods Association.

(a) Use the equation to approximate sporting goods sales in 2000, 2004, and 2006. Round your answers to the nearest billion dollars.

(b) Use the graph to estimate sales for the same years.

(c) How do the approximations using the equation compare to the estimates using the graph?

11.3 ▶▶▶ Slope of a Line

An important characteristic of the lines we graphed in the previous section is their slant or "steepness", as viewed from *left to right*. See Figure 17.

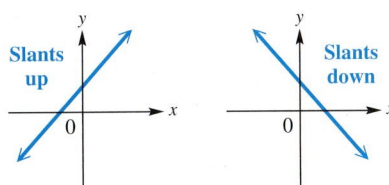

Figure 17

OBJECTIVES

1 Find the slope of a line given two points.

2 Find the slope from the equation of a line.

3 Use slope to determine whether two lines are parallel, perpendicular, or neither.

One way to measure the steepness of a line is to compare the vertical change in the line to the horizontal change while moving along the line from one fixed point to another. This measure of steepness is called the *slope* of the line.

OBJECTIVE 1 Find the slope of a line given two points.   To find the steepness, or slope, of the line in Figure 18, we begin at point Q and move to point P. The vertical change, or **rise**, is the change in the y-values, which is the difference $6 - 1 = 5$ units. The horizontal change, or **run**, from Q to P is the change in the x-values, which is the difference $5 - 2 = 3$ units.

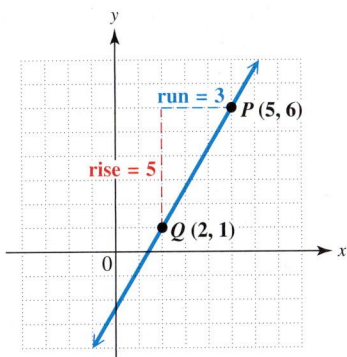

Figure 18

Remember from **Section 6.1** that one way to compare two numbers is by using a ratio. The **slope** is the ratio of the vertical change in y to the horizontal change in x. The line in Figure 18 has

$$\text{slope} = \frac{\text{vertical change in } y \text{ (rise)}}{\text{horizontal change in } x \text{ (run)}} = \frac{5}{3}.$$

To confirm this ratio, we can count grid squares. We start at point Q in Figure 18 and count *up* 5 grid squares to find the vertical change (rise). To find the horizontal change (run) and arrive at point P, we count to the *right* 3 grid squares. The slope is $\frac{5}{3}$, as found above. ***Slope is a single number that allows us to determine the direction in which a line is slanting from left to right, as well as how much slant there is to the line.***

1 Find the slope ratio of each line.

(a)

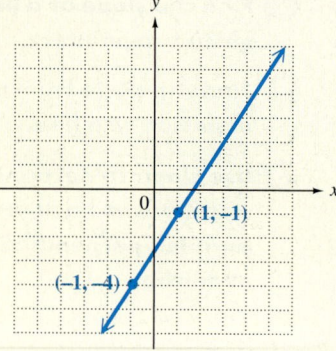

(b)

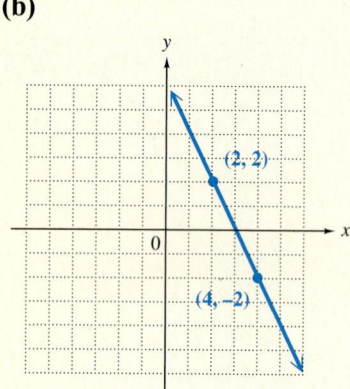

EXAMPLE 1 Finding the Slope of a Line

Find the slope of the line in Figure 19.

We use the two points shown on the line. The vertical change is the difference in the y-values, or $-1 - 3 = -4$, and the horizontal change is the difference in the x-values, or $6 - 2 = 4$. Thus, the line has

$$\text{slope} = \frac{\text{change in } y \text{ (rise)}}{\text{change in } x \text{ (run)}} = \frac{-4}{4}, \quad \text{or} \quad -1.$$

Counting grid squares, we begin at point P and count *down* 4 grid squares. Because we counted down, we write the vertical change as a negative number, -4 here. Then we count to the *right* 4 grid squares to reach point Q. The slope is $\frac{-4}{4}$, or -1.

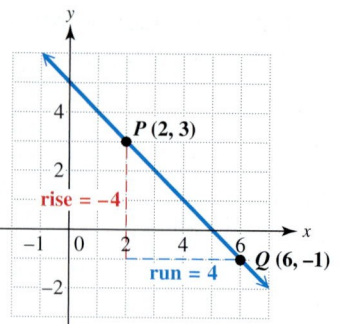

Figure 19

◀ Work Problem **1** at the Side.

> **Note**
>
> **The slope of a line is the same for any two points on the line.** To see this, refer to Figure 19. Find the points $(3, 2)$ and $(5, 0)$, which also lie on the line. If we start at $(3, 2)$ and count *down* 2 units and then to the *right* 2 units, we arrive at $(5, 0)$. The slope is $\frac{-2}{2}$, or -1, the same slope we found in Example 1.

The concept of slope is used in many everyday situations. See Figure 20. For example, a highway with a 10%, or $\frac{1}{10}$, grade (or slope) rises 1 m for every 10 m horizontally. Architects specify the pitch of a roof by using slope; a $\frac{5}{12}$ roof means that the roof rises 5 ft for every 12 ft that it runs in the horizontal direction. The slope of a stairwell also indicates the ratio of the vertical rise to the horizontal run. In the figure, the slope of the stairwell is $\frac{8}{12}$, or $\frac{2}{3}$.

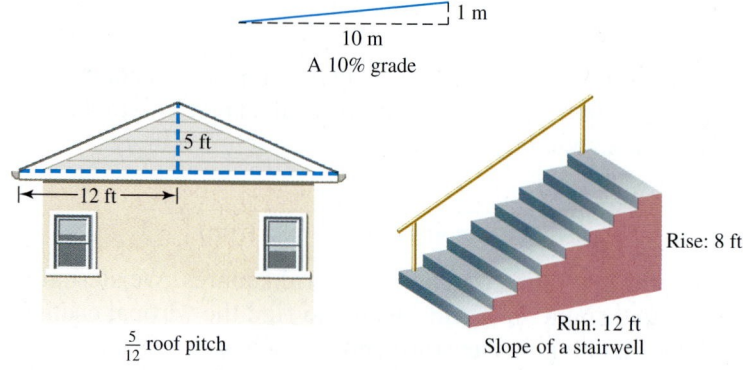

Figure 20

We can generalize the preceding discussion and find the slope of a line through two nonspecific points (x_1, y_1) and (x_2, y_2). (This notation is called **subscript notation.** Read x_1 as "*x*-sub-one" and x_2 as "*x*-sub-two.") See Figure 21.

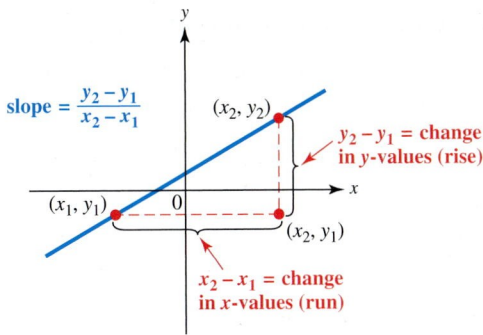

Figure 21

2 Find $\dfrac{y_2 - y_1}{x_2 - x_1}$ for the following values.

(a) $y_2 = 4, y_1 = -1,$
$\quad x_2 = 3, x_1 = 4$

Moving along the line from the point (x_1, y_1) to the point (x_2, y_2), we see that y changes by $y_2 - y_1$ units. This is the vertical change (rise). Similarly, x changes by $x_2 - x_1$ units, which is the horizontal change (run). The slope of the line is the ratio of $y_2 - y_1$ to $x_2 - x_1$.

> **Note**
> Subscript notation is used to identify a point. It does *not* indicate any operation. Note the difference between x_2, a nonspecific value, and x^2, which means $x \cdot x$. Read x_2 as "x-sub-two," *not* "x squared."

(b) $x_1 = 3, x_2 = -5,$
$\quad y_1 = 7, y_2 = -9$

Traditionally, the letter m represents slope. The slope m of a line is defined as follows.

> **Slope Formula**
> The **slope** of the line through the points (x_1, y_1) and (x_2, y_2) is
> $$m = \frac{\textbf{change in } y}{\textbf{change in } x} = \frac{y_2 - y_1}{x_2 - x_1} \qquad (x_1 \neq x_2).$$

The slope gives the change in y for each unit of change in x.

Work Problem **2** at the Side. ▶

(c) $x_1 = 2, x_2 = 7,$
$\quad y_1 = 4, y_2 = 9$

EXAMPLE 2 **Finding Slopes of Lines**

Find the slope of each line.

(a) The line through $(-4, 7)$ and $(1, -2)$

Use the slope formula. Let $(-4, 7) = (x_1, y_1)$ and $(1, -2) = (x_2, y_2)$.

$$\text{slope } m = \frac{\text{change in } y}{\text{change in } x} = \frac{y_2 - y_1}{x_2 - x_1} = \frac{-2 - 7}{1 - (-4)} = \frac{-9}{5} = -\frac{9}{5}$$

> Substitute carefully here.

Continued on Next Page

ANSWERS

2. **(a)** -5 **(b)** 2 **(c)** 1

3 Find the slope of each line.

(a) The line through $(6, -2)$ and $(5, 4)$

Count grid squares in Figure 22 to confirm that the slope is $-\frac{9}{5}$.

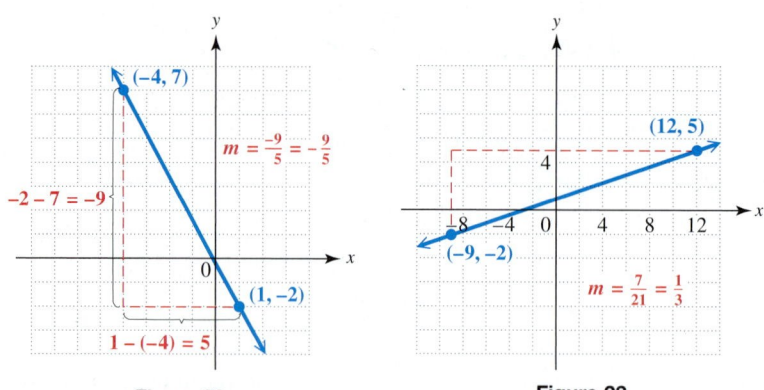

Figure 22   **Figure 23**

(b) The line through $(-3, 5)$ and $(-4, -7)$

(b) The line through $(-9, -2)$ and $(12, 5)$

$$\text{slope } m = \frac{\overset{y\text{-value}}{\mathbf{5 - (-2)}}}{\underset{x\text{-value from the } same \text{ ordered pair}}{\mathbf{12 - (-9)}}} = \frac{7}{21} = \frac{1}{3}$$

See Figure 23. Note that the same slope is obtained by subtracting in reverse order.

$$\text{slope } m = \frac{\overset{y\text{-value}}{\mathbf{-2 - 5}}}{\underset{x\text{-value from the } same \text{ ordered pair}}{\mathbf{-9 - 12}}} = \frac{-7}{-21} = \frac{1}{3}$$

(c) The line through $(6, -8)$ and $(-2, 4)$

(Find this slope in two different ways as in Example 2(b).)

> **CAUTION**
> *It makes no difference which point is (x_1, y_1) or (x_2, y_2); however, be consistent.* Start with the x- and y-values of one point (either one), and subtract the corresponding values of the other point.

◀ *Work Problem* **3** *at the Side.*

The slopes we found for the lines in Figures 22 and 23 suggest the following generalization.

> **Positive and Negative Slopes**
> A line with positive slope rises (slants up) from left to right.
> A line with negative slope falls (slants down) from left to right.

ANSWERS

3. (a) -6 **(b)** 12 **(c)** $-\frac{3}{2}; -\frac{3}{2}$

EXAMPLE 3 **Showing that the Slope of a Horizontal Line Is Zero**

Find the slope of the line through $(-8, 4)$ and $(2, 4)$.

$$m = \frac{y_2 - y_1}{x_2 - x_1} = \frac{4 - 4}{-8 - 2} = \frac{0}{-10} = 0 \qquad \text{Zero slope}$$

As shown in Figure 24, the line through the given points is horizontal, with equation $y = 4$. *All horizontal lines have slope 0* since the difference in their y-values is always 0.

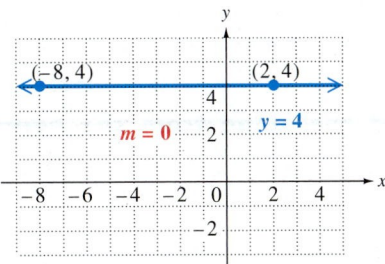

Figure 24

EXAMPLE 4 **Showing that a Vertical Line Has Undefined Slope**

Find the slope of the line through $(6, 2)$ and $(6, -4)$.

$$m = \frac{y_2 - y_1}{x_2 - x_1} = \frac{2 - (-4)}{6 - 6} = \frac{6}{0} \qquad \text{Undefined slope}$$

Because division by 0 is undefined, this line has undefined slope. (This is why the slope formula at the beginning of this section had the restriction $x_1 \neq x_2$.) The graph in Figure 25 shows that this line is vertical, with equation $x = 6$. All points on a vertical line have the same x-value, so *all vertical lines have undefined slope.*

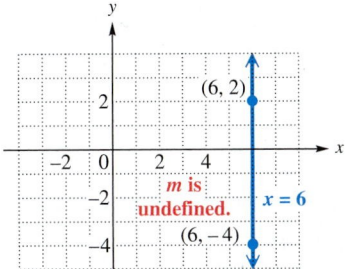

Figure 25

> **Slopes of Horizontal and Vertical Lines**
>
> **Horizontal lines,** which have equations of the form $y = k$, have **slope 0.**
>
> **Vertical lines,** which have equations of the form $x = k$, have **undefined slope.**

Work Problem **4** *at the Side.* ▶

---

4 Find the slope of each line.

(a) The line through $(2, 5)$ and $(-1, 5)$

(b) The line through $(3, 1)$ and $(3, -4)$

(c) The line with equation $y = -1$

(d) The line with equation $x - 4 = 0$

Figure 26 summarizes the four cases for slopes of lines.

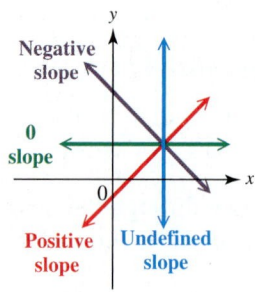

Slopes of lines
Figure 26

OBJECTIVE 2 Find the slope from the equation of a line. Consider the equation

$$y = -3x + 5.$$

We can find its slope using any two points on the line. We get these two points by first choosing two different values of x and then finding the corresponding values of y. For example, choose $x = -2$ and $x = 4$.

$y = -3x + 5$	$y = -3x + 5$
$y = -3(-2) + 5$ Let $x = -2$.	$y = -3(4) + 5$ Let $x = 4$.
$y = 6 + 5$	$y = -12 + 5$
$y = 11$	$y = -7$

The ordered pairs are $(-2, 11)$ and $(4, -7)$. Now use the slope formula.

$$m = \frac{11 - (-7)}{-2 - 4} = \frac{18}{-6} = -3$$

The slope, $m = -3$, is the same number as the coefficient of x in the equation $y = -3x + 5$. It can be shown that this always happens, *as long as the equation is solved for y.* This fact is used to find the slope of a line from its equation.

Finding the Slope of a Line from Its Equation

Step 1 Solve the equation for y.

Step 2 The slope is given by the coefficient of x.

Note

We will see in **Section 11.4** that the equation $y = -3x + 5$ is written using a special form of the equation of a line, called *slope-intercept form,*

$$y = mx + b.$$

EXAMPLE 5 **Finding Slopes from Equations**

Find the slope of each line.

(a) $2x - 5y = 4$

    Step 1 Solve the equation for y.

$$2x - 5y = 4 \quad \boxed{\text{Isolate } y \text{ on one side.}}$$

$$-5y = -2x + 4 \qquad \text{Subtract } 2x.$$

$$y = \frac{2}{5}x - \frac{4}{5} \qquad \text{Divide by } -5.$$

    Step 2 The slope is given by the coefficient of x, so the slope is $\frac{2}{5}$.

(b)

$$8x + 4y = 1$$

$$\boxed{\text{Solve for } y.} \quad 4y = -8x + 1 \qquad \text{Subtract } 8x.$$

$$y = -2x + \frac{1}{4} \qquad \text{Divide by } 4.$$

The slope of this line is given by the coefficient of x, which is -2.

Work Problem **5** *at the Side.* ▶

OBJECTIVE 3 Use slope to determine whether two lines are parallel, perpendicular, or neither. Two lines in a plane that never intersect are **parallel.** We use slopes to tell whether two lines are parallel. For example, Figure 27 shows the graphs of $x + 2y = 4$ and $x + 2y = -6$. These lines appear to be parallel. Solving $x + 2y = 4$ for y gives $y = -\frac{1}{2}x + 2$. Solving $x + 2y = -6$ for y gives $y = -\frac{1}{2}x - 3$. Both lines have slope $-\frac{1}{2}$. **Nonvertical parallel lines always have equal slopes.**

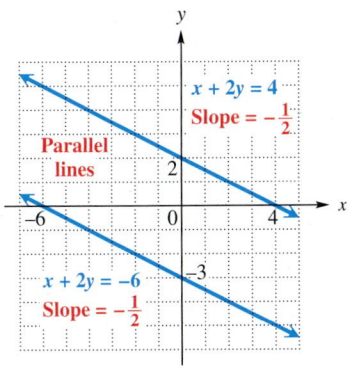

Figure 27

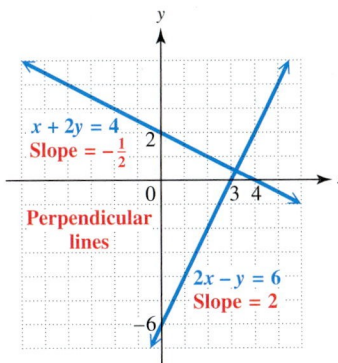

Figure 28

    Figure 28 shows the graphs of $x + 2y = 4$ and $2x - y = 6$. These lines appear to be **perpendicular** (that is, they intersect at a 90° angle). Solving $x + 2y = 4$ for y gives $y = -\frac{1}{2}x + 2$, with slope $-\frac{1}{2}$. Solving $2x - y = 6$ for y gives $y = 2x - 6$, with slope 2. The product of $-\frac{1}{2}$ and 2 is

$$-\frac{1}{2}(2) = -1.$$

This condition is true in general. **The product of the slopes of two perpendicular lines, neither of which is vertical, is always -1.** This means that the slopes of perpendicular lines are negative (or opposite) reciprocals—if one slope is the nonzero number a, then the other is $-\frac{1}{a}$. The table in the margin shows several examples.

---

5 Find the slope of each line.

(a) $y = -\dfrac{7}{2}x + 1$

(b) $3x + 2y = 9$

(c) $y + 4 = 0$

(d) $x + 3 = 7$

Number	Negative Reciprocal
$\frac{3}{4}$	$-\frac{4}{3}$
$\frac{1}{2}$	$-\frac{2}{1}$, or -2
-6, or $-\frac{6}{1}$	$\frac{1}{6}$
-0.4, or $-\frac{4}{10}$	$\frac{10}{4}$, or 2.5

The product of each number and its negative reciprocal is -1.

ANSWERS

5. **(a)** $-\dfrac{7}{2}$ **(b)** $-\dfrac{3}{2}$ **(c)** 0 **(d)** undefined

6 Decide whether each pair of lines is *parallel, perpendicular,* or *neither.*

(a) $x + y = 6$

$x + y = 1$

(b) $3x - y = 4$

$x + 3y = 9$

(c) $2x - y = 5$

$2x + y = 3$

(d) $3x - 7y = 35$

$7x - 3y = -6$

> **Slopes of Parallel and Perpendicular Lines**
>
> Two lines with the same slope are parallel.
>
> Two lines whose slopes have a product of -1 are perpendicular.

EXAMPLE 6 Deciding Whether Lines Are Parallel, Perpendicular, or Neither

Decide whether each pair of lines is *parallel, perpendicular,* or *neither.*

(a) $x + 2y = 7$

$-2x + y = 3$

Find the slope of each line by first solving each equation for y.

$x + 2y = 7$	$-2x + y = 3$
$2y = -x + 7$ Subtract x.	$y = 2x + 3$ Add $2x$.
$y = -\dfrac{1}{2}x + \dfrac{7}{2}$ Divide by 2.	
Slope is $-\frac{1}{2}$.	Slope is 2.

Because the slopes are not equal, the lines are not parallel. Check the product of the slopes: $-\frac{1}{2}(2) = -1$. The two lines are perpendicular because the product of their slopes is -1. See Figure 29.

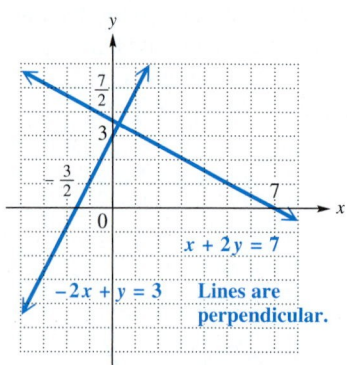 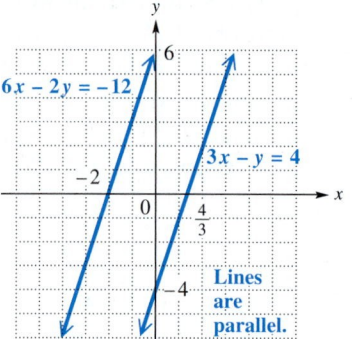

Figure 29	**Figure 30**

(b) $3x - y = 4$ Solve for y. $y = 3x - 4$

$6x - 2y = -12$ $y = 3x + 6$

Both lines have slope 3, so the lines are parallel. See Figure 30.

(c) $4x + 3y = 6$ $y = -\dfrac{4}{3}x + 2$

       Solve for y.

$2x - y = 5$ $y = 2x - 5$

Here the slopes are $-\frac{4}{3}$ and 2. Because $-\frac{4}{3} \neq 2$ and $-\frac{4}{3}(2) \neq -1$, these lines are neither parallel nor perpendicular.

(d) $5x - y = 1$ $y = 5x - 1$

       Solve for y.

$x - 5y = -10$ $y = \dfrac{1}{5}x + 2$

The slopes are 5 and $\frac{1}{5}$. The lines are not parallel, nor are they perpendicular. (***Be careful!*** $5\left(\frac{1}{5}\right) = 1$, ***not*** -1.)

◀ *Work Problem* **6** *at the Side.*

11.3 ▶▶▶ Exercises

Use the coordinates of the indicated points to find the slope of each line. See Example 1.

1.

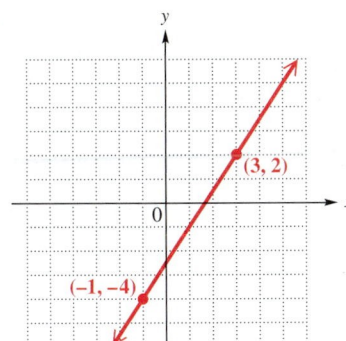

2.

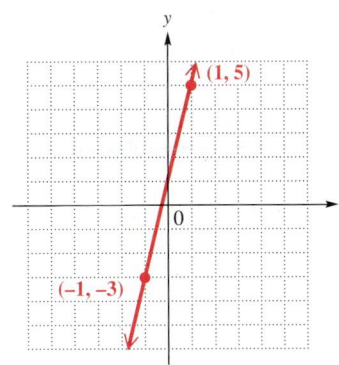

3.

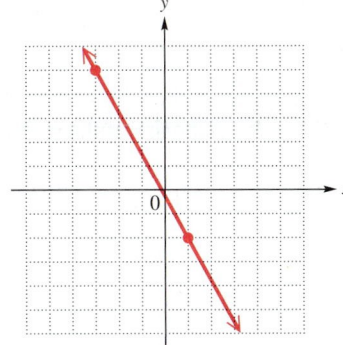

4.

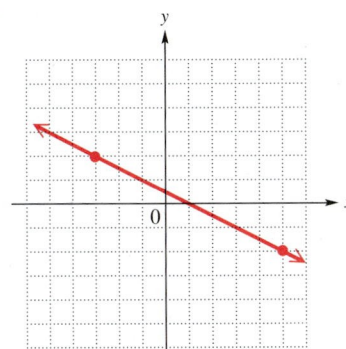

5.

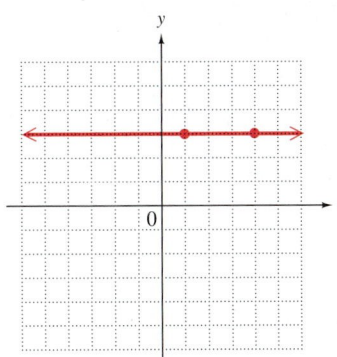

6.

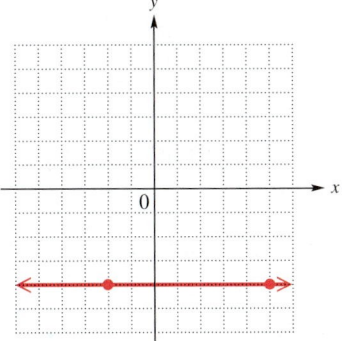

7. In the context of the graph of a straight line, what is meant by "rise"? What is meant by "run"?

8. Look at the graph in Exercise 1, and answer the following.

   (a) Start at the point $(-1, -4)$ and count vertically up to the horizontal line that goes through the other plotted point. What is this vertical change? (Remember: "up" means positive, "down" means negative.) _____

   (b) From this new position, count horizontally to the other plotted point. What is this horizontal change? (Remember: "right" means positive, "left" means negative.) _____

   (c) What is the quotient of the numbers found in parts (a) and (b)? _____
   What do we call this number? _____

   (d) If we were to *start* at the point $(3, 2)$ and *end* at the point $(-1, -4)$, would the answer to part (c) be the same? Explain why or why not.

On the given coordinate system, sketch the graph of a straight line with the indicated slope.

9. Negative

10. Positive

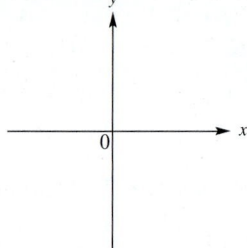

11. Undefined

12. Zero

13. Decide whether the line with the given slope rises from left to right, falls from left to right, is horizontal, or is vertical.

(a) $m = -4$ **(b)** $m = 0$ **(c)** m is undefined.   **(d)** $m = \dfrac{3}{7}$

14. Explain in your own words what is meant by the *slope* of a line.

15. A student found the slope of the line through the points $(2, 5)$ and $(-1, 3)$ and got $-\frac{2}{3}$ as his answer. He showed his work as

$$\frac{3 - 5}{2 - (-1)} = \frac{-2}{3} = -\frac{2}{3}.$$

WHAT WENT WRONG? Give the correct slope.

Find the slope of the line through each pair of points. See Examples 2–4.

16. $(4, -1)$ and $(-2, -8)$ **17.** $(1, -2)$ and $(-3, -7)$ **18.** $(-8, 0)$ and $(0, -5)$

19. $(0, 3)$ and $(-2, 0)$ **20.** $(-4, -5)$ and $(-5, -8)$ **21.** $(-2, 4)$ and $(-3, 7)$

22. $(6, -5)$ and $(-12, -5)$ **23.** $(4, 3)$ and $(-6, 3)$ **24.** $(-8, 6)$ and $(-8, -1)$

25. $(-12, 3)$ and $(-12, -7)$ **26.** $(3.1, 2.6)$ and $(1.6, 2.1)$ **27.** $\left(-\dfrac{7}{5}, \dfrac{3}{10}\right)$ and $\left(\dfrac{1}{5}, -\dfrac{1}{2}\right)$

Find the slope of each line. See Example 5.

28. $y = 2x - 3$ **29.** $y = 5x + 12$ **30.** $2y = -x + 4$ **31.** $4y = x + 1$

32. $-6x + 4y = 4$ **33.** $3x - 2y = 3$ **34.** $y = 4$ **35.** $y = 6$

36. $x = 5$ **37.** $x = -2$ **38.** $x + y = 0$ **39.** $x - y = 0$

The figure at the right shows a line that has a positive slope (because it rises from left to right) and a positive y-value for the y-intercept (because it intersects the y-axis above the origin).

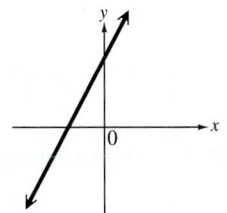

*For each figure in Exercises 40–45, decide whether (**a**) the slope is positive, negative, or 0 and whether (**b**) the y-value of the y-intercept is positive, negative, or 0.*

40. (a) _____   **41. (a)** _____   **42. (a)** _____

    (b) _____   **(b)** _____   **(b)** _____

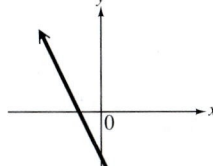   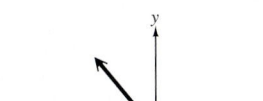   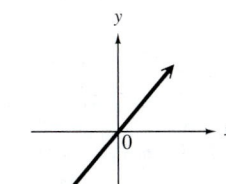

43. (a) _____   **44. (a)** _____   **45. (a)** _____

    (b) _____   **(b)** _____   **(b)** _____

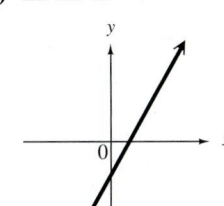   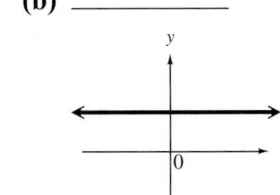   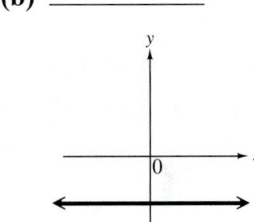

In each pair of equations, give the slope of each line, and then determine whether the two lines are parallel, perpendicular, *or* neither parallel nor perpendicular. *See Example 6.*

46. $2x + 5y = 4$ 🌐 **47.** $-4x + 3y = 4$ **48.** $8x - 9y = 6$

    $4x + 10y = 1$ $-8x + 6y = 0$ $8x + 6y = -5$

49. $5x - 3y = -2$ **50.** $3x - 2y = 6$ 🌐 **51.** $3x - 5y = -1$

    $3x - 5y = -8$ $2x + 3y = 3$ $5x + 3y = 2$

52. What is the slope (or pitch) of this roof?

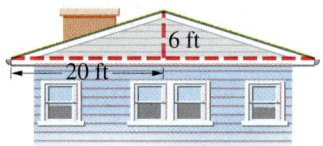

53. What is the slope (or grade) of this hill?

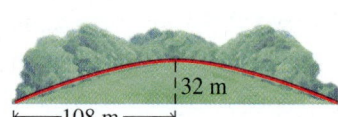

54. What is the slope (or grade) of this ski slope?

Relating Concepts (Exercises 55—60) For Individual or Group Work

Figure A gives public school enrollment (in thousands) in grades 9–12 in the United States. Figure B gives the (average) number of public school students per computer.

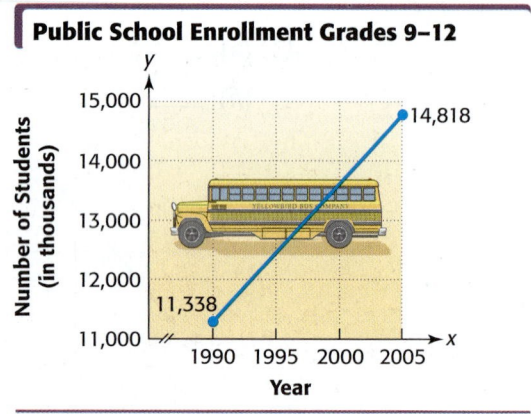

Source: U.S. Department of Education.

Figure A

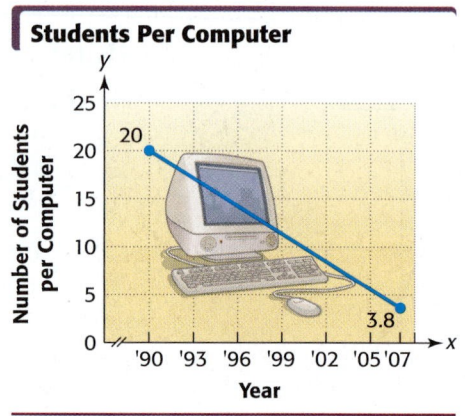

Source: Quality Education Data, Inc.

Figure B

Work Exercises 55–60 in order.

55. Use the ordered pairs (1990, 11,338) and (2005, 14,818) to find the slope of the line in Figure A.

56. The slope of the line in Figure A is _____. This means that during
(positive/negative)

the period represented, enrollment _____.
(increased/decreased)

57. The slope of a line represents its *rate of change*. Based on Figure A, what was the increase in students *per year* during the period shown?

58. Use the given information to find the slope, to the nearest hundredth, of the line in Figure B.

59. The slope of the line in Figure B is _____. This means that during
(positive/negative)

the period represented, the number of students per computer _____.
(increased/decreased)

60. Based on Figure B, what was the decrease in students per computer *per year* during the period shown?

11.4 ▶▶▶ Equations of Lines

In **Section 11.3,** we found the slope (steepness) of a line from the equation of the line by solving the equation for y. In that form, the slope is the coefficient of x. For example, the slope of the line with equation $y = 2x + 3$ is 2, the coefficient of x. What does the number **3** represent? If $x = 0$, the equation becomes

$$y = 2(0) + 3$$
$$y = 3.$$

Since $y = 3$ corresponds to $x = 0$, $(0, 3)$ is the y-intercept of the graph of $y = 2x + 3$. An equation like $y = 2x + 3$ that is solved for y is said to be in **slope-intercept form** because both the slope and the y-intercept of the line can be read directly from the equation.

Slope-Intercept Form

The slope-intercept form of the equation of a line with slope m and y-intercept $(0, b)$ is

$$y = mx + b.$$

Slope ⟶↑   ↑⟶ $(0, b)$ is the y-intercept.

REMEMBER: The intercept in slope-intercept form is the y-intercept.

Note

The slope-intercept form is the most useful form for a linear equation because of the information we can determine from it. It is also the form used by graphing calculators and the one that describes a *linear function,* an important concept in mathematics.

OBJECTIVE 1 Write an equation of a line given its slope and y-intercept. Given the slope and y-intercept of a line, we can use the slope-intercept form to write an equation of the line.

EXAMPLE 1 Writing an Equation of a Line

Write an equation of the line with slope $\frac{2}{3}$ and y-intercept $(0, -1)$.
Here $m = \frac{2}{3}$ and $b = -1$, so an equation is

Slope ⟶↓   ↓⟶ y-intercept $(0, b)$
$$y = mx + b$$
$$y = \frac{2}{3}x + (-1), \quad \text{or} \quad y = \frac{2}{3}x - 1.$$

Work Problem **1** *at the Side.* ▶

OBJECTIVES

1 Write an equation of a line given its slope and y-intercept.

2 Graph a line given its slope and a point on the line.

3 Write an equation of a line given its slope and any point on the line.

4 Write an equation of a line given two points on the line.

5 Find an equation of a line that fits a data set.

---

1 Write an equation of the line with the given slope and y-intercept.

(a) slope $\frac{1}{2}$; y-intercept $(0, -4)$

(b) slope -1; y-intercept $(0, 8)$

(c) slope 3; y-intercept $(0, 0)$

(d) slope 0; y-intercept $(0, 2)$

(e) slope 1; y-intercept $(0, 0.75)$

ANSWERS

1. (a) $y = \frac{1}{2}x - 4$ (b) $y = -x + 8$
   (c) $y = 3x$ (d) $y = 2$
   (e) $y = x + 0.75$

2 Graph $3x - 4y = 8$ by using the slope and y-intercept.

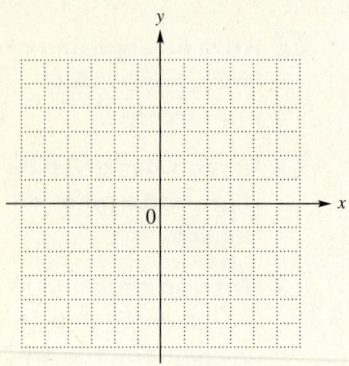

OBJECTIVE 2 Graph a line given its slope and a point on the line. We can use the slope and y-intercept to graph a line.

> **Graphing a Line by Using the Slope and y-Intercept**
>
> **Step 1** Write the equation in slope-intercept form, if necessary, by solving for y.
>
> **Step 2** Identify the y-intercept. Graph the point $(0, b)$.
>
> **Step 3** Identify slope m of the line. Use the geometric interpretation of slope ("rise over run") to find another point on the graph by counting from the y-intercept.
>
> **Step 4** Join the two points with a line to obtain the graph.

EXAMPLE 2 Graphing a Line by Using the Slope and y-Intercept

Graph $2x - 3y = 3$ by using the slope and y-intercept.

Step 1 Solve for y to write the equation in slope-intercept form.

$$2x - 3y = 3 \qquad \text{Given equation}$$

Isolate y on one side. $\qquad -3y = -2x + 3 \qquad$ Subtract $2x$.

$$y = \frac{2}{3}x - 1 \qquad \text{Divide by } -3.$$

Slope m ⎯⎯↑   ⎿⎯⎯ y-intercept $(0, b)$

Step 2 The y-intercept is $(0, -1)$. Graph this point. See Figure 31.

Step 3 The slope is $\frac{2}{3}$. By the definition of slope,

$$m = \frac{\textbf{change in } y}{\textbf{change in } x} = \frac{2}{3}.$$

Counting from the y-intercept 2 units up and 3 units to the right, we obtain another point on the graph, $(3, 1)$.

Step 4 Draw the line through the points $(0, -1)$ and $(3, 1)$ to obtain the graph of the given equation $2x - 3y = 3$. See Figure 31.

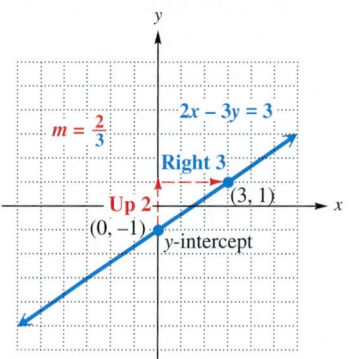

Figure 31

◀ Work Problem **2** at the Side.

ANSWER

2.

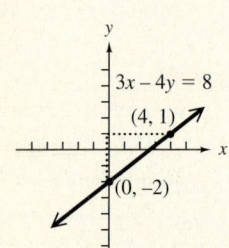

The method of Example 2 can be extended to graph a line given its slope and *any* point on the line.

EXAMPLE 3 Graphing a Line by Using the Slope and a Point

Graph the line passing through the point $(-2, 3)$, with slope -4.

First, locate the point $(-2, 3)$. See Figure 32. Then write the slope -4 as

$$\text{slope } m = \frac{\text{change in } y}{\text{change in } x} = -4 = \frac{-4}{1}.$$

Locate another point on the line by counting 4 units *down* (because of the negative sign) from $(-2, 3)$ and then 1 unit to the right. Finally, draw the line through this new point P and the given point $(-2, 3)$. See Figure 32.

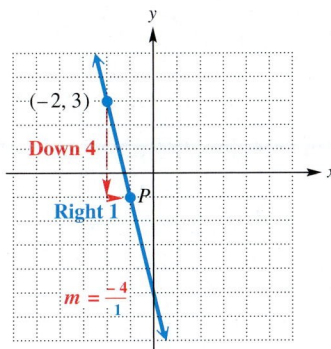

Figure 32

Note

In Example 3, we could have written the slope as $\frac{4}{-1}$ instead. In this case, we would move 4 units up from $(-2, 3)$ and then 1 unit to the *left* (because of the negative sign). Verify that this produces the same line.

Work Problem **3** *at the Side.* ▶

OBJECTIVE 3 Write an equation of a line given its slope and any point on the line. We can use the slope-intercept form to write the equation of a line if we know the slope and any point on the line.

EXAMPLE 4 Using the Slope-Intercept Form to Write an Equation of a Line

Write an equation, in slope-intercept form, of the line having slope 4 passing through the point $(2, 5)$.

Since the line passes through the point $(2, 5)$, we can substitute $x = 2$, $y = 5$, and the given slope $m = 4$ into $y = mx + b$ and solve for b.

$y = mx + b$	Slope-intercept form
$5 = 4(2) + b$	Let $x = 2$, $y = 5$, and $m = 4$.
$5 = 8 + b$	Multiply.
$-3 = b$	Subtract 8.

Remember: $(0, b)$ is the y-intercept. Don't stop here.

The y-intercept is $(0, -3)$. Using the given slope, 4, an equation of the line is

$$y = 4x - 3. \qquad \text{Slope-intercept form}$$

Work Problem **4** *at the Side.* ▶

3 Graph the line passing through the point $(2, -3)$, with slope $-\frac{1}{3}$.

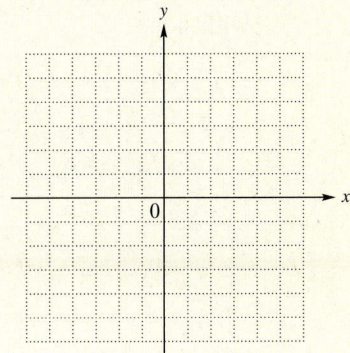

4 Write an equation, in slope-intercept form, of the line having slope -2 and passing through the point $(-1, 4)$.

ANSWERS

3.

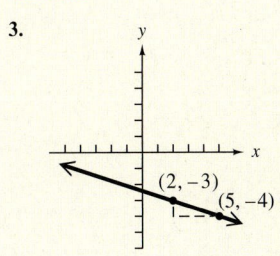

4. $y = -2x + 2$

5 Write an equation of each line. Give the final answer in slope-intercept form.

(a) The line through $(-1, 3)$, with slope -2

$$y - y_1 = m(x - x_1)$$
$$y - \underline{\quad} = \underline{\quad} [x - (\underline{\quad})]$$
$$y - 3 = -2(x + \underline{\quad})$$
$$y - 3 = -2x - \underline{\quad}$$
$$y = \underline{\qquad}$$

(b) The line through $(5, 2)$, with slope $-\frac{1}{3}$

There is another form that can be used to write the equation of a line. To develop this form, let m represent the slope of a line and let (x_1, y_1) represent a given point on the line. Let (x, y) represent any other point on the line. See Figure 33. Then,

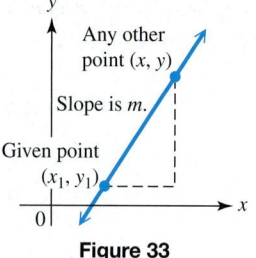

Figure 33

$$m = \frac{y - y_1}{x - x_1} \qquad \text{Definition of slope}$$
$$m(x - x_1) = y - y_1 \qquad \text{Multiply each side by } x - x_1.$$
$$y - y_1 = m(x - x_1). \qquad \text{Rewrite.}$$

This result is the **point-slope form** of the equation of a line.

> **Point-Slope Form**
>
> The point-slope form of the equation of a line with slope m passing through the point (x_1, y_1) is
>
> $$y - y_1 = m(x - x_1).$$
>
> Slope ↓   Given point ↑

EXAMPLE 5 **Using the Point-Slope Form to Write Equations**

Write an equation of each line. Give the final answer in slope-intercept form.

(a) The line through $(-2, 4)$, with slope -3

The given point is $(-2, 4)$ so $x_1 = -2$ and $y_1 = 4$. Also, $m = -3$. Substitute these values into the point-slope form.

$$y - y_1 = m(x - x_1) \qquad \text{Point-slope form}$$
$$y - 4 = -3[x - (-2)] \qquad \text{Let } x_1 = -2, y_1 = 4, m = -3.$$
$$y - 4 = -3(x + 2) \qquad \boxed{\text{Be careful substituting.}}$$
$$y - 4 = -3x - 6 \qquad \text{Distributive property}$$
$$y = -3x - 2 \qquad \text{Add 4.}$$

(b) The line through $(4, 2)$, with slope $\frac{3}{5}$

$$y - y_1 = m(x - x_1) \qquad \text{Point-slope form}$$
$$y - 2 = \frac{3}{5}(x - 4) \qquad \text{Let } x_1 = 4, y_1 = 2, m = \frac{3}{5}.$$
$$y - 2 = \frac{3}{5}x - \frac{12}{5} \qquad \text{Distributive property}$$
$$y = \frac{3}{5}x - \frac{12}{5} + \frac{10}{5} \qquad \text{Add } 2 = \frac{10}{5}.$$
$$y = \frac{3}{5}x - \frac{2}{5} \qquad \text{Combine like terms.}$$

ANSWERS

5. **(a)** $3; -2; -1; 1; 2; -2x + 1$

    (b) $y = -\frac{1}{3}x + \frac{11}{3}$

We did not clear fractions after the substitution step because we want the equation in slope-intercept form—that is, solved for y.

◀ **Work Problem** **5** at the Side.

OBJECTIVE 4 Write an equation of a line given two points on the line. We can also use the point-slope form to find an equation of a line when two points on the line are known.

EXAMPLE 6 Writing an Equation of a Line Given Two Points

Write an equation of the line through the points $(-2, 5)$ and $(3, 4)$. Give the final answer in slope-intercept form.

First, find the slope of the line, using the slope formula.

$$\text{slope } m = \frac{y_2 - y_1}{x_2 - x_1} = \frac{5 - 4}{-2 - 3} = \frac{1}{-5} = -\frac{1}{5}$$

Now use either $(-2, 5)$ or $(3, 4)$ and the point-slope form. Using $(3, 4)$ gives

$$y - y_1 = m(x - x_1) \qquad \text{Point-slope form}$$

$$y - 4 = -\frac{1}{5}(x - 3) \qquad \text{Let } x_1 = 3, y_1 = 4, m = -\tfrac{1}{5}.$$

$$y - 4 = -\frac{1}{5}x + \frac{3}{5} \qquad \text{Distributive property}$$

$$y = -\frac{1}{5}x + \frac{3}{5} + \frac{20}{5} \qquad \text{Add } 4 = \tfrac{20}{5}.$$

$$y = -\frac{1}{5}x + \frac{23}{5}. \qquad \text{Combine like terms.}$$

The same result would be found using $(-2, 5)$ for (x_1, y_1).

Work Problem **6** *at the Side.* ▶

Note

In Example 6, the same result would also be found by substituting the slope and either given point in slope-intercept form $y = mx + b$ and then solving for b, as in Example 4. Try this.

Many of the linear equations in **Sections 11.1–11.3** were given in the form

$$Ax + By = C,$$

called **standard form,** where A, B, and C are real numbers and A and B are not both 0. In most cases, A, B, and C are rational numbers. For consistency in this book, we give answers so that A, B, and C are integers with greatest common factor 1 and $A \geq 0$.

Note

The definition of standard form is not the same in all texts. A linear equation can be written in many different, equally correct, ways. For example,

$$3x + 4y = 12, \quad 6x + 8y = 24, \quad \text{and} \quad -9x - 12y = -36$$

all represent the same set of ordered pairs. When giving answers, let us agree that $3x + 4y = 12$ is preferable to the other forms because the greatest common factor of 3, 4, and 12 is 1 and $A \geq 0$.

6 Write an equation in slope-intercept form of the line through each pair of points.

(a) $(-3, 1)$ and $(2, 4)$

(b) $(2, 5)$ and $(-1, 6)$

A summary of the forms of linear equations follows.

Forms of Linear Equations		
Equation	*Description*	*Example*
$x = k$	**Vertical line** Slope is undefined; x-intercept is $(k, 0)$.	$x = 3$
$y = k$	**Horizontal line** Slope is 0; y-intercept is $(0, k)$.	$y = 3$
$y = mx + b$	**Slope-intercept form** Slope is m; y-intercept is $(0, b)$.	$y = \dfrac{3}{2}x - 6$
$y - y_1 = m(x - x_1)$	**Point-slope form** Slope is m; line passes through (x_1, y_1).	$y + 3 = \dfrac{3}{2}(x - 2)$
$Ax + By = C$	**Standard form** Slope is $-\frac{A}{B}$; x-intercept is $\left(\frac{C}{A}, 0\right)$; y-intercept is $\left(0, \frac{C}{B}\right)$.	$3x - 2y = 12$

OBJECTIVE 5 Find an equation of a line that fits a data set.
Earlier in this chapter, we gave linear equations that modeled real data, such as number of twin births and amounts of credit card debt, and then used these equations to estimate or predict values. We now develop a procedure to find such an equation if the given set of data fits a linear pattern—that is, its graph consists of points lying close to a straight line.

EXAMPLE 7 Finding an Equation of a Line That Describes Data

The table lists the average annual cost (in dollars) of tuition and fees for in-state students at public 4-year colleges and universities for selected years. Year 1 represents 2001, year 3 represents 2003, and so on.

Year	Cost (in dollars)
1	3766
3	4645
5	5491
7	6185

Source: The College Board.

Plot the data and find an equation that approximates it.

Letting y represent the cost in year x, we plot the data as shown in Figure 34 on the next page.

Continued on Next Page

AVERAGE ANNUAL COSTS AT PUBLIC 4-YEAR COLLEGES

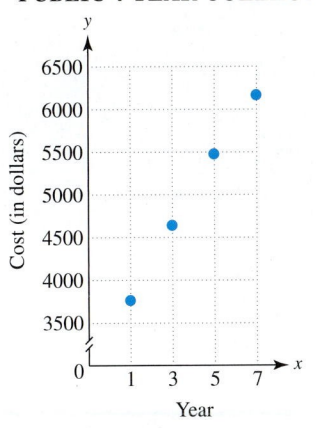

Figure 34

The points appear to lie approximately in a straight line. We can use two of the data pairs and the slope-intercept form of the equation of a line to get an equation that describes the relationship between the year and the cost. We choose the ordered pairs $(5, 5491)$ and $(7, 6185)$ from the table and find the slope of the line through these points.

$$m = \frac{y_2 - y_1}{x_2 - x_1} = \frac{6185 - 5491}{7 - 5} = 347 \qquad \text{Let } (7, 6185) = (x_2, y_2) \text{ and } (5, 5491) = (x_1, y_1).$$

As we might expect, the slope, 347, is positive, indicating that tuition and fees *increased* \$347 each year. Now use this slope and the point $(5, 5491)$ in the slope-intercept form to find an equation of the line.

$$y = mx + b \qquad \text{Slope-intercept form}$$

Solve for b, the y-value of the y-intercept.

$$5491 = 347(5) + b \qquad \text{Substitute for } x, y, \text{ and } m.$$
$$5491 = 1735 + b \qquad \text{Multiply.}$$
$$3756 = b \qquad \text{Subtract 1735.}$$

Thus, $m = 347$ and $b = 3756$, so an equation of the line is

$$y = 347x + 3756.$$

To see how well this equation approximates the ordered pairs in the data table, let $x = 3$ (for 2003) and find y.

$$y = 347x + 3756 \qquad \text{Equation of the line}$$
$$y = 347(3) + 3756 \qquad \text{Substitute 3 for } x.$$
$$y = 4797 \qquad \text{Multiply; add.}$$

The corresponding value in the table for $x = 3$ is 4645, so the equation approximates the data reasonably well. With caution, the equation could be used to predict values for years that are not included in the table.

> **Note**
> In Example 7, if we had chosen two different data points, we would have gotten a slightly different equation.

Work Problem **7** *at the Side.* ▶

7 Use the points $(3, 4645)$ and $(7, 6185)$ to find an equation in slope-intercept form that approximates the data of Example 7. How well does this equation approximate the cost in 2005?

Math in the Media

INTERPRETING (AND MISINTERPRETING) LINE GRAPHS

The graph shown here is typical of many graphs that appear in magazines and newspapers. This one shows that between 1996 and 2006, the number of McDonald's restaurants worldwide rose from 20,000 to 31,000. This is depicted by the line segment joining the two points labeled *A* and *B*.

Use the graph to answer each of the following.

1. To represent point *A*, write an ordered pair in the form

   (year, number of restaurants in thousands).

   Do this for point *B* also.

Number of McDonald's Restaurants Worldwide (in thousands)

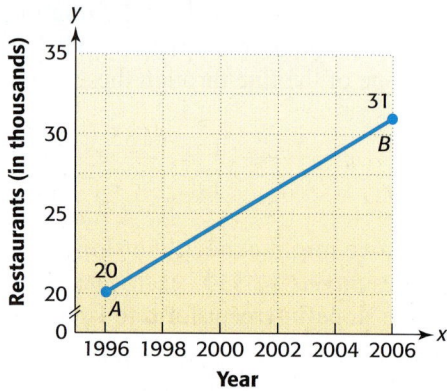

Source: McDonald's Corp.; Hoovers.

2. The points *A* and *B* as well as the points on line *AB* between them make up line segment *AB*. We can find the coordinates of a point *M* on line segment *AB* that is exactly halfway between *A* and *B*. (This point is called the **midpoint** of the segment.)

   (a) To find the *x*-coordinate of *M*, we find the average of the *x*-coordinates of *A* and *B* by adding them and dividing by 2. What is the *x*-coordinate of *M*?

   (b) To find the *y*-coordinate of *M*, we find the average of the *y*-coordinates of *A* and *B*. What is the *y*-coordinate of *M*?

3. Fill in the blanks with the appropriate responses: The ordered pair that represents *M*, the midpoint of segment *AB*, is (_____ , _____). This suggests that in the year _____, there were _____ thousand McDonald's restaurants worldwide.

4. Use the points (1996, 20) and (2006, 31) to find the $y = mx + b$ form of the equation of the line containing *A* and *B*.

5. Use the result of Exercise 4 to find the value of *y* when $x = 2001$. Does this correspond to the result you found in Exercise 2(b)?

6. The actual number of McDonald's restaurants worldwide was 30 thousand in 2001. How does this compare to your answers in Exercises 3 and 5? Explain how a line graph such as this one can be misleading. Use the concept of *slope* in your explanation.

11.4 ▶▶▶ Exercises

1. Match the correct equation in Column II with the description given in Column I.

I	**II**
(a) Slope -2, the line through the point $(4, 1)$	**A.** $y = 4x$
(b) Slope -2, y-intercept $(0, 1)$	**B.** $y = \dfrac{1}{4}x$
(c) The line through the points $(0, 0)$ and $(4, 1)$	**C.** $y = -2x + 1$
(d) The line through the points $(0, 0)$ and $(1, 4)$	**D.** $y - 1 = -2(x - 4)$

2. In the summary box on page 266, we give the equations $y = \frac{3}{2}x - 6$ and $y + 3 = \frac{3}{2}(x - 2)$ as examples of equations in slope-intercept form and point-slope form, respectively. Write each of these equations in standard form. What do you notice?

*Use the geometric interpretation of slope (rise divided by run, from **Section 11.3**) to find the slope of each line. Then, by identifying the y-intercept from the graph, write the slope-intercept form of the equation of the line.*

3.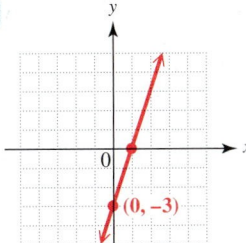

4.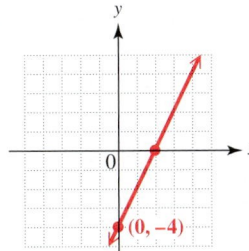

5.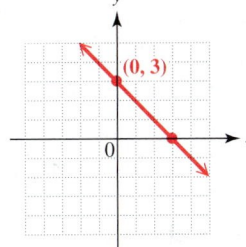

6.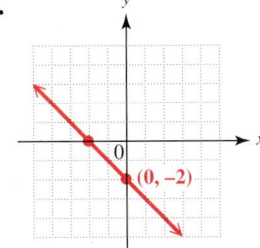

7.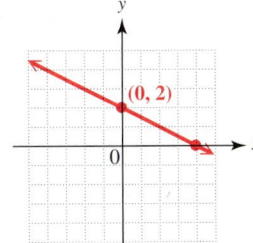

8.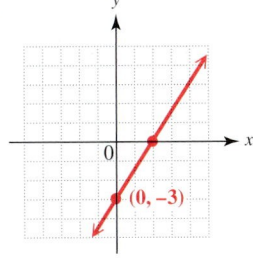

Write the equation of the line with the given slope and y-intercept. See Example 1.

9. slope 4;
y-intercept $(0, -3)$

10. slope -5;
y-intercept $(0, 6)$

11. slope 0;
y-intercept $(0, 3)$

12. slope 3;
y-intercept $(0, 0)$

13. Match each equation with the graph that would most closely resemble its graph.

(a) $y = x + 3$

(b) $y = -x + 3$

(c) $y = x - 3$

(d) $y = -x - 3$

A.

B.
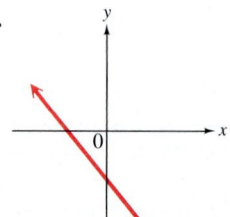

C.

D.

14. Explain why the equation of a vertical line cannot be written in the form $y = mx + b$.

Graph each equation by finding the slope and y-intercept, and using their definitions to find two points on the line. See Example 2.

15. $y = 3x + 2$

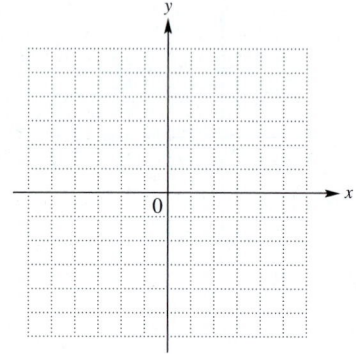

16. $y = 4x - 4$

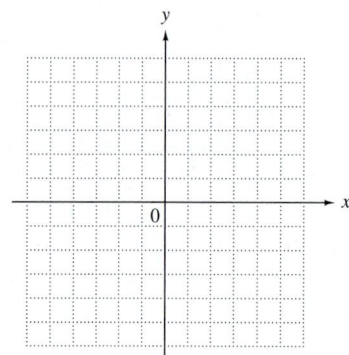

17. $2x + y = -5$

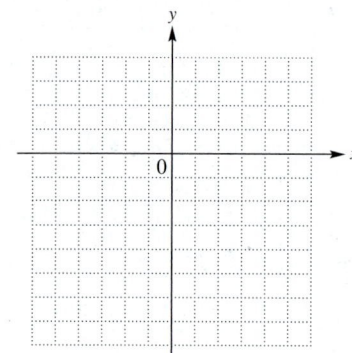

18. $3x + y = -2$

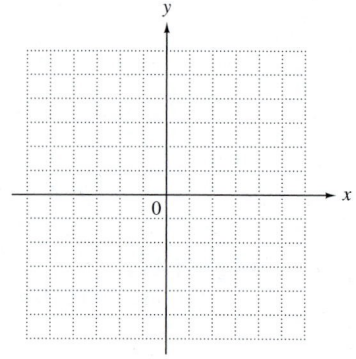

19. $x + 2y = 4$

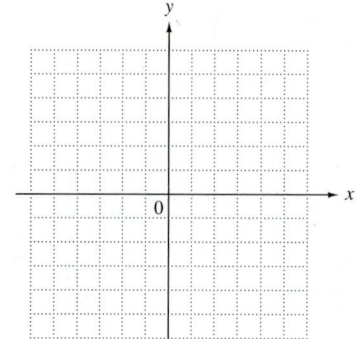

20. $x + 3y = 12$

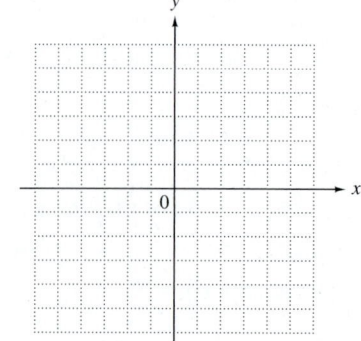

Graph each line passing through the given point and having the given slope. (In Exercises 25–28, recall the types of lines having slope 0 and undefined slope.) Give the slope-intercept form of the equation of the line if possible. See Example 3.

21. $(-2, 3)$, $m = \dfrac{1}{2}$

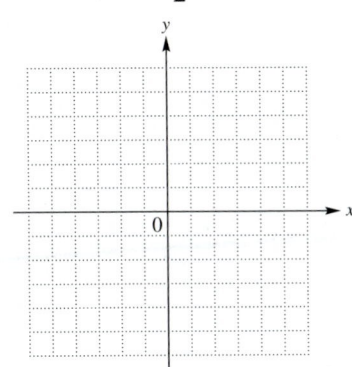

22. $(-4, -1)$, $m = \dfrac{3}{4}$

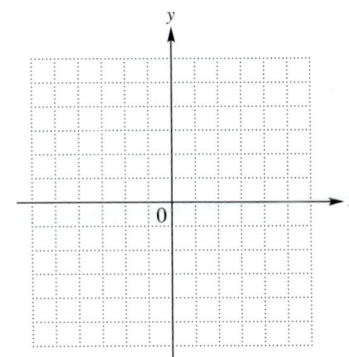

23. $(1, -5)$, $m = -\dfrac{2}{5}$

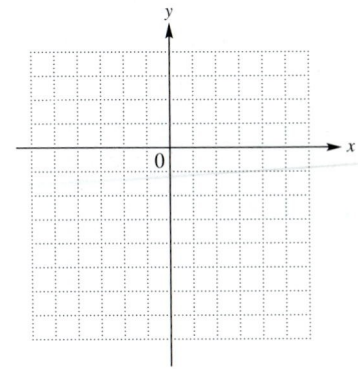

24. $(2, -1)$, $m = -\dfrac{1}{3}$

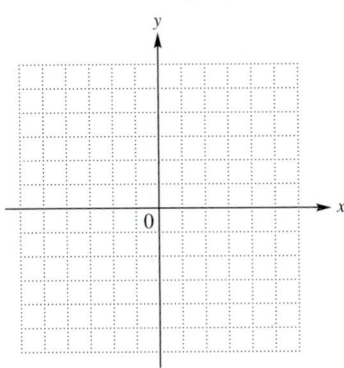

25. $(3, 2)$, $m = 0$

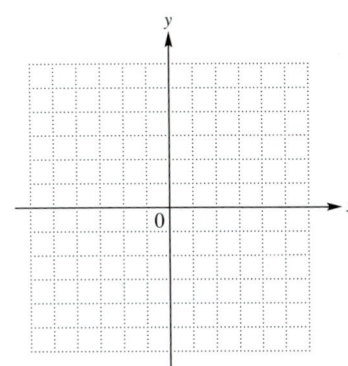

26. $(-2, 3)$, $m = 0$

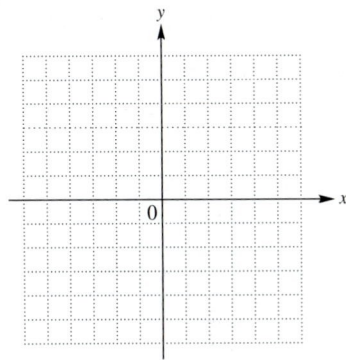

27. $(3, -2)$, undefined slope

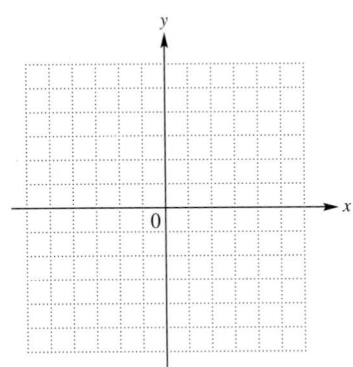

28. $(2, 4)$, undefined slope

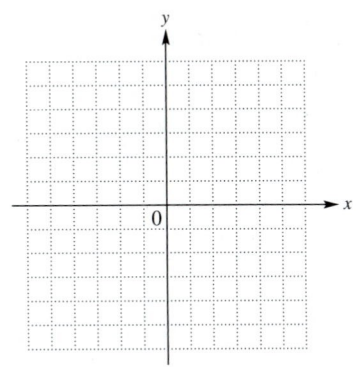

29. $(0, 0)$, $m = \dfrac{2}{3}$

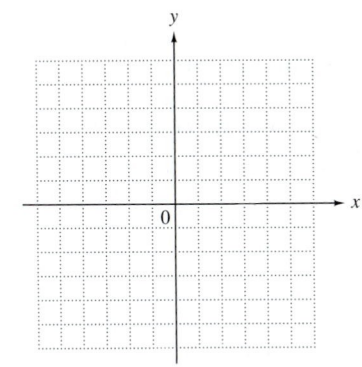

30. (a) What is the common name given to the vertical line whose x-intercept is the origin?

(b) What is the common name given to the line with slope 0 whose y-intercept is the origin?

Write an equation of the line passing through the given point and having the given slope. Give the final answer in slope-intercept form. See Examples 4 and 5.

31. $(4, 1)$, $m = 2$

32. $(2, 7)$, $m = 3$

33. $(3, -10)$, $m = -2$

34. $(2, -5)$, $m = -4$

35. $(-2, 5)$, $m = \dfrac{2}{3}$

36. $(-4, 1)$, $m = \dfrac{3}{4}$

Write an equation of the line passing through each pair of points. Give the final answer in slope-intercept form, if possible. See Example 6.

37. $(8, 5)$ and $(9, 6)$

38. $(4, 10)$ and $(6, 12)$

39. $(-1, -7)$ and $(-8, -2)$

40. $(-2, -1)$ and $(3, -4)$

41. $(0, -2)$ and $(-3, 0)$

42. $(-4, 0)$ and $(0, 2)$

43. $(3, 5)$ and $(3, -2)$

44. $(3, -5)$ and $(-1, -5)$

45. $\left(\dfrac{1}{2}, \dfrac{3}{2}\right)$ and $\left(-\dfrac{1}{4}, \dfrac{5}{4}\right)$

46. $\left(-\dfrac{2}{3}, \dfrac{8}{3}\right)$ and $\left(\dfrac{1}{3}, \dfrac{7}{3}\right)$

Write an equation of the line satisfying the given conditions. Give the final answer in slope-intercept form.

47. The line through $(2, -3)$, parallel to $3x = 4y + 5$

48. The line through $(-1, 4)$, perpendicular to $2x + 3y = 8$

49. The line perpendicular to $x - 2y = 7$, y-intercept $(0, -3)$

50. The line parallel to $5x = 2y + 10$, y-intercept $(0, 4)$

Relating Concepts (Exercises 51–58) For Individual or Group Work

If we think of ordered pairs of the form (C, F), then the two most common methods of measuring temperature, Celsius and Fahrenheit, can be related as follows: When C = 0, F = 32, and when C = 100, F = 212. **Work Exercises 51–58 in order.**

51. Write two ordered pairs relating these two temperature scales.

52. Find the slope of the line through the two points.

53. Use the point-slope form to find an equation of the line. (Your variables should be C and F rather than x and y.)

54. Write an equation for F in terms of C.

55. Use the equation from Exercise 54 to write an equation for C in terms of F.

56. Use the equation from Exercise 54 to find the Fahrenheit temperature when $C = 30$.

57. Use the equation from Exercise 55 to find the Celsius temperature when $F = 50$.

58. For what temperature is $F = C$?

*The cost to produce x items is, in some cases, expressed as y = mx + b. The number b gives the **fixed cost** (the cost that is the same no matter how many items are produced), and the number m is the **variable cost** (the cost to produce an additional item). Use this information to work Exercises 59 and 60.*

59. It costs $400 to start up a business selling campaign buttons. Each button costs $0.25 to produce.

   (a) What is the fixed cost?

   (b) What is the variable cost?

   (c) Write the cost equation.

   (d) What will be the cost to produce 100 campaign buttons, based on the cost equation?

   (e) How many campaign buttons will be produced if total cost is $775?

60. It costs $2000 to purchase a copier, and each copy costs $0.02 to make.

   (a) What is the fixed cost?

   (b) What is the variable cost?

   (c) Write the cost equation.

   (d) What will be the cost to produce 10,000 copies, based on the cost equation?

   (e) How many copies will be produced if total cost is $2600?

Solve each problem. See Example 7.

61. The table lists the average annual cost (in dollars) of tuition and fees at 2-year colleges for selected years, where year 1 represents 2003, year 2 represents 2004, and so on.

Year	Cost (in dollars)
1	1909
2	2079
3	2182
4	2272
5	2361

Source: The College Board.

(a) Write five ordered pairs for the data.

(b) Plot the ordered pairs. Do the points lie approximately in a straight line?

AVERAGE ANNUAL COSTS AT 2-YEAR COLLEGES

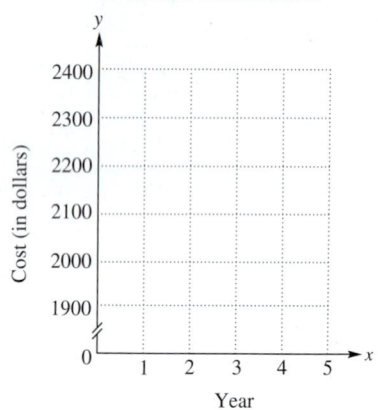

Year

(c) Use the ordered pairs (2, 2079) and (5, 2361) to find the equation of a line that approximates the data. Write the final equation in slope-intercept form.

(d) Use the equation from part (c) to estimate the average annual cost at 2-year colleges in 2008 to the nearest dollar. (*Hint:* What is the value of x for 2008?)

62. The table gives heavy-metal nuclear waste (in thousands of metric tons) from spent reactor fuel now stored temporarily at reactor sites, awaiting permanent storage. (*Source:* "Burial of Radioactive Nuclear Waste Under the Seabed," *Scientific American*, January 1998.)

Year x	Waste y
1995	32
2000	42
2010*	61
2020*	76

*Estimates by the U.S. Department of Energy.

Let $x = 0$ represent 1995, $x = 5$ represent 2000 (since $2000 - 1995 = 5$), and so on.

(a) For 1995, the ordered pair is (0, 32). Write ordered pairs for the data for the other years given in the table.

(b) Plot the ordered pairs (x, y). Do the points lie approximately in a straight line?

HEAVY-METAL NUCLEAR WASTE AWAITING STORAGE

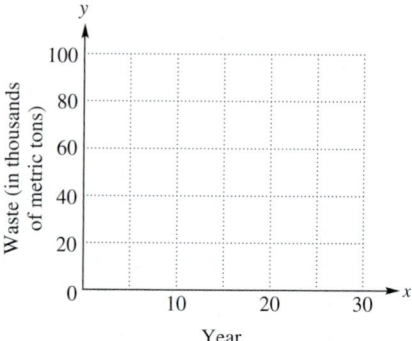

Year

(c) Use the ordered pairs (0, 32) and (25, 76) to find the equation of a line that approximates the data. Write the equation in slope-intercept form.

(d) Use the equation from part (c) to estimate the amount of nuclear waste in 2015. (*Hint:* What is the value of x for 2015?)

Summary Exercises on Linear Equations and Graphs

Identify the slope and the y-intercept of the graph of each equation.

1. $3x + y = -6$

2. $2x + y = -4$

3. $-4x - y = 3$

4. $-5x - y = 8$

5. $-3x + 2y = 12$

6. $-5x + 3y = 15$

Graph each line, using the given information or equation.

7. $m = 1, b = -2$

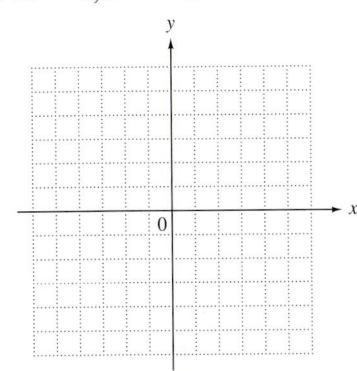

8. $m = 1$, y-intercept $(0, -4)$

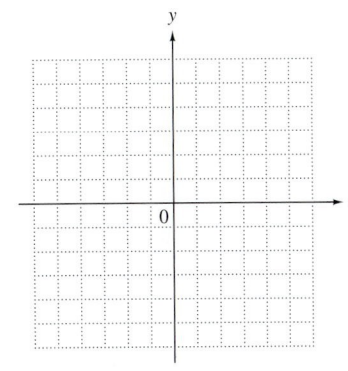

9. $y = -2x + 6$

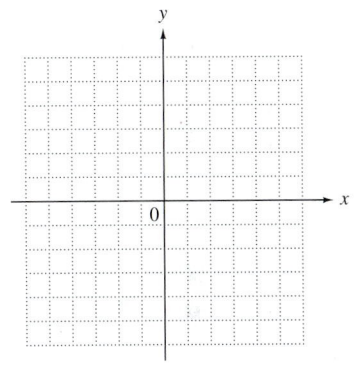

10. $x + 4 = 0$

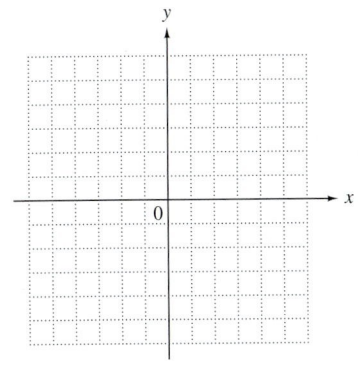

11. $m = -\dfrac{2}{3}$, passes through $(3, -4)$

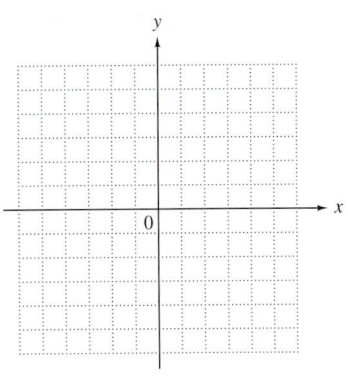

12. $y = -\dfrac{1}{2}x + 2$

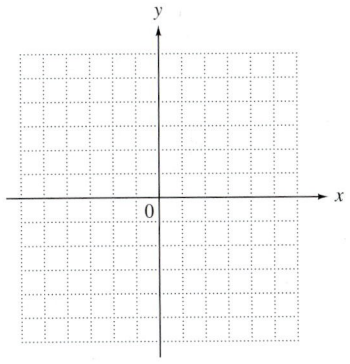

13. $y - 4 = -9$

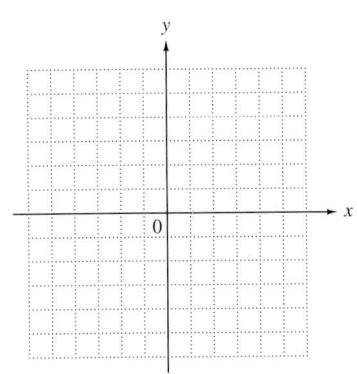

14. $m = -\dfrac{3}{4}$, passes through $(4, -4)$

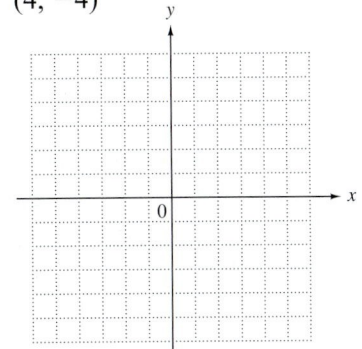

15. Undefined slope, passes through $(3.5, 0)$

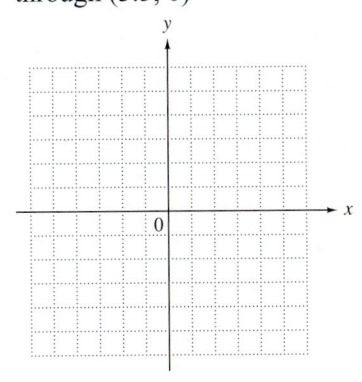

16. Slope $-\dfrac{1}{5}$, passes through $(0, 0)$ **17.** $4x - 5y = 20$ **18.** $6x - 5y = 30$

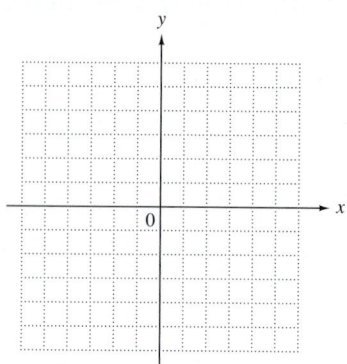 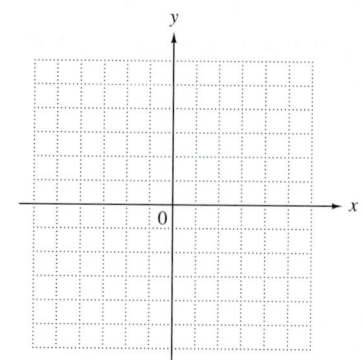

19. $x - 4y = 0$ **20.** $m = 0$, passes through $\left(0, \dfrac{3}{2}\right)$ **21.** $3y = 12 - 2x$

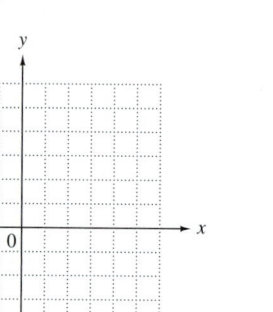 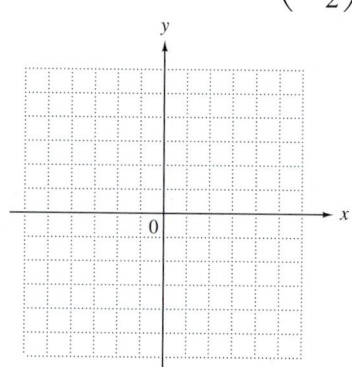 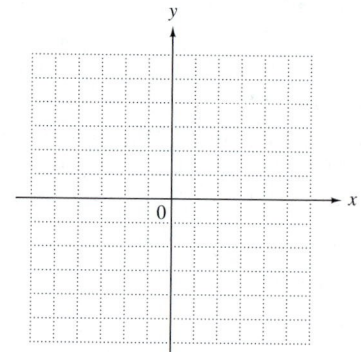

Write an equation of each line. Give the final answer in slope-intercept form if possible.

22. $m = -3, b = -6$ **23.** The line through $(1, -7)$ and $(-2, 5)$

24. The line through $(0, 0)$ and $(5, 3)$ **25.** The line through $(0, 0)$, undefined slope

26. The line through $(0, 0)$, $m = 0$ **27.** $m = -2$, y-intercept $(0, -4)$ **28.** $m = \dfrac{5}{3}$, through $(-3, 0)$

11.5 ▷▷▷ Graphing Linear Inequalities in Two Variables

In **Section 11.2** we graphed linear equations, such as $2x + 3y = 6$. Now we extend this work to include *linear inequalities in two variables,* such as $2x + 3y \leq 6$. (Recall that $\leq$ is read "is less than or equal to.")

> **Linear Inequality in Two Variables**
>
> An inequality that can be written as
>
> $$Ax + By < C \quad \text{or} \quad Ax + By > C,$$
>
> where A, B, and C are real numbers and A and B are not both 0, is a **linear inequality in two variables.**

The symbols $\leq$ and $\geq$ may replace $<$ and $>$ in the definition.

OBJECTIVE 1 Graph linear inequalities. The linear inequality $2x + 3y \leq 6$ means that

$$2x + 3y < 6 \quad \textbf{or} \quad 2x + 3y = 6.$$

As we found earlier, the graph of $2x + 3y = 6$ is a line. This **boundary line** divides the plane into two regions. The graph of the solutions of the inequality $2x + 3y < 6$ will include only *one* of these regions. We find the required region by solving the original inequality for y.

$$2x + 3y \leq 6 \quad \boxed{\text{Isolate } y \text{ on one side.}}$$

$$3y \leq -2x + 6 \qquad \text{Subtract } 2x.$$

$$y \leq -\frac{2}{3}x + 2 \qquad \text{Divide by 3.}$$

All ordered pairs in which y is *less than or equal to* $-\frac{2}{3}x + 2$ will be solutions of the inequality. The ordered pairs in which y is equal to $-\frac{2}{3}x + 2$ are on the boundary line, so the ordered pairs in which y *is less than* $-\frac{2}{3}x + 2$ will be *below* that line. (As we move *down* vertically, the y-values *decrease.*) To indicate the solutions, we shade the region below the line, as in Figure 35. The shaded region, along with the boundary line, is the desired graph.

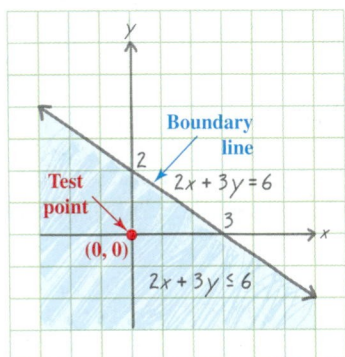

Figure 35

Work Problem ① *at the Side.* ▶

Alternatively, a test point gives a quick way to find the correct region to shade. We choose any point *not* on the boundary line. Because $(0, 0)$ is easy to substitute, we often use it. We substitute 0 for x and 0 for y in the original inequality to see whether the resulting statement is true or false.

OBJECTIVES

① **Graph linear inequalities.**

② **Graph an inequality with boundary through the origin.**

---

① Shade the appropriate region for each linear inequality.

(a) $x + 2y \geq 6$

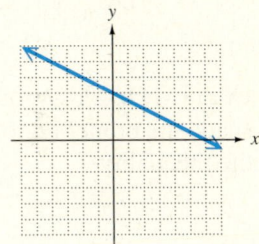

(b) $3x + 4y \leq 12$

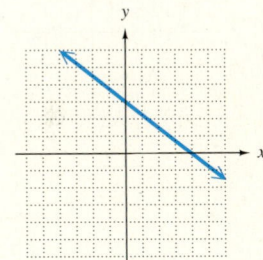

ANSWERS

1. (a)

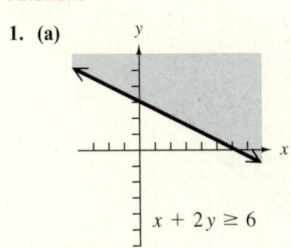

$x + 2y \geq 6$

(b)

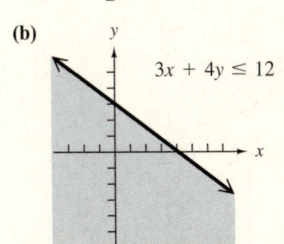

$3x + 4y \leq 12$

2 Use (0, 0) as a test point to shade the proper region for the inequality

$$4x - 5y \le 20.$$

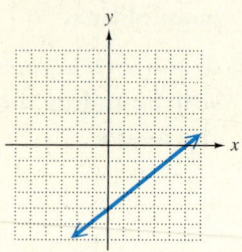

$$2x + 3y \le 6 \qquad \text{Original inequality}$$

$$2(0) + 3(0) \overset{?}{\le} 6 \qquad \text{Let } x = 0 \text{ and } y = 0.$$

Use (0, 0) as a test point.

$$0 + 0 \overset{?}{\le} 6$$

$$0 \le 6 \qquad \text{True}$$

Since the last statement is true, we shade the region that includes the test point (0, 0). This agrees with the result shown in Figure 35 on the preceding page.

◀ *Work Problem* **2** *at the Side.*

EXAMPLE 1 **Graphing a Linear Inequality**

Graph the inequality $x - y > 5$.

This inequality does *not* include the equals sign. Therefore, the points on the line $x - y = 5$ do *not* belong to the graph. However, the line still serves as a boundary for two regions, one of which satisfies the inequality. To graph the inequality, first graph the equation $x - y = 5$. Use a *dashed line* to show that the points on the line are *not* solutions of the inequality $x - y > 5$. See Figure 36.

Now choose a test point to see which side of the line satisfies the inequality. Again, (0, 0) is a convenient choice.

$$x - y > 5 \qquad \text{Original inequality}$$

$$0 - 0 \overset{?}{>} 5 \qquad \text{Let } x = 0 \text{ and } y = 0.$$

$$0 > 5 \qquad \text{False}$$

Since $0 > 5$ is false, the graph of the inequality is the region that *does not* contain (0, 0). Shade the *other* region, as shown in Figure 36, to obtain the required graph.

3 Use (1, 1) as a test point to shade the proper region for the inequality

$$3x + 5y > 15.$$

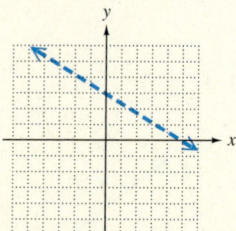

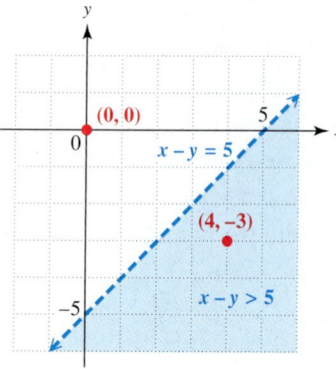

Figure 36

To check that the proper region is shaded, we test a point in the shaded region. For example, we use (4, −3) from the shaded region as follows.

$$x - y > 5$$

$$4 - (-3) \overset{?}{>} 5 \qquad \text{Let } x = 4 \text{ and } y = -3.$$

Use parentheses to avoid errors.

$$7 > 5 \qquad \text{True}$$

This verifies that the correct region is shaded in Figure 36.

◀ *Work Problem* **3** *at the Side.*

ANSWERS

2.

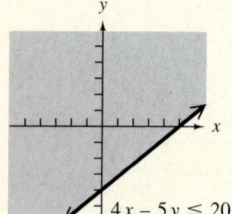

$4x - 5y \le 20$

3.

$3x + 5y > 15$

A summary of the steps used to graph a linear inequality in two variables follows.

> ### Graphing a Linear Inequality
>
> *Step 1* **Graph the boundary.** Graph the line that is the boundary of the region. Use the methods of **Section 11.2.** Draw a solid line if the inequality has $\leq$ or $\geq$; draw a dashed line if the inequality has $<$ or $>$.
>
> *Step 2* **Shade the appropriate side.** Use any point not on the line as a test point. Substitute for x and y in the *inequality*. If a true statement results, shade the side containing the test point. If a false statement results, shade the other side.

4 Graph $2x - y \geq -4$.

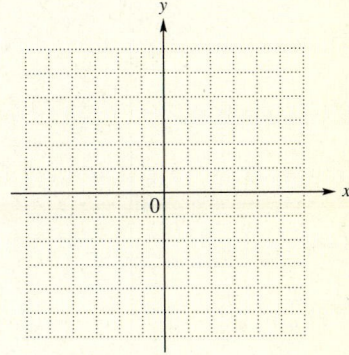

EXAMPLE 2 Graphing a Linear Inequality

Graph the inequality $2x - 5y \geq 10$.

Start by graphing the equation $2x - 5y = 10$. Use a solid line to show that the points on the line are solutions of the inequality $2x - 5y \geq 10$. Choose any test point not on the line. Again, we choose $(0, 0)$.

$$2x - 5y \geq 10$$
$$2(0) - 5(0) \overset{?}{\geq} 10 \qquad \text{Let } x = 0 \text{ and } y = 0.$$
$$0 - 0 \overset{?}{\geq} 10$$
$$0 \geq 10 \qquad \text{False}$$

Because $0 \geq 10$ is false, shade the region *not* containing $(0, 0)$. See Figure 37. Verify that a point in the shaded region satisfies the inequality.

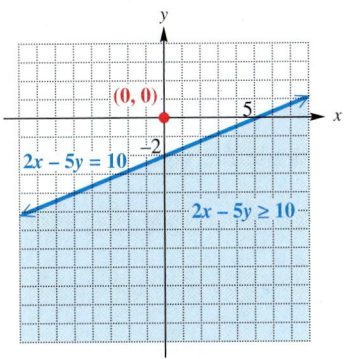

Figure 37

Work Problem **4** at the Side. ▶

EXAMPLE 3 Graphing a Linear Inequality with a Vertical Boundary Line

Graph the inequality $x < 3$.

First graph $x = 3$, a vertical line through the point $(3, 0)$. Use a dashed line (why?) and choose $(0, 0)$ as a test point.

$$x < 3 \qquad \text{Original inequality}$$
$$0 \overset{?}{<} 3 \qquad \text{Let } x = 0.$$
$$0 < 3 \qquad \text{True}$$

Continued on Next Page

ANSWER

4.

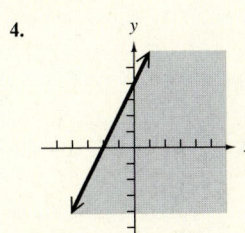

$2x - y \geq -4$

5 Graph $y < 4$.

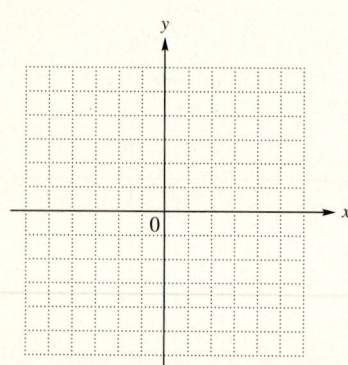

Because $0 < 3$ is true, we shade the region containing $(0, 0)$, as in Figure 38.

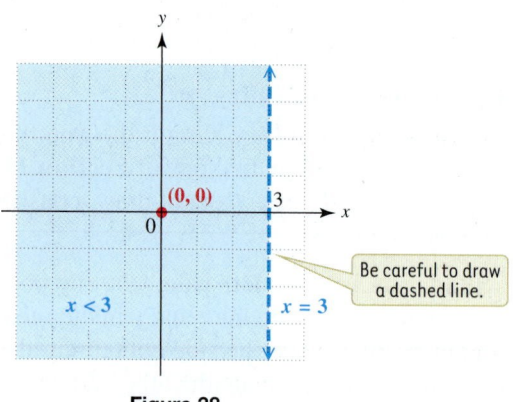

Figure 38

◀ Work Problem **5** at the Side.

OBJECTIVE 2 Graph an inequality with boundary through the origin. *If the graph of an inequality has a boundary line through the origin, (0, 0) cannot be used as a test point.*

EXAMPLE 4 **Graphing a Linear Inequality with a Boundary Line through the Origin**

Graph the inequality $x \le 2y$.

We begin by graphing $x = 2y$, using a solid line. Some ordered pairs that can be used to graph this line are $(0, 0)$, $(6, 3)$, and $(4, 2)$. We cannot use $(0, 0)$ as a test point because $(0, 0)$ is *on* the line $x = 2y$. Instead, we choose a test point *off* the line, say $(1, 3)$.

$$x \le 2y \qquad \text{Original inequality}$$
$$1 \stackrel{?}{\le} 2(3) \qquad \text{Let } x = 1 \text{ and } y = 3.$$
$$1 \le 6 \qquad \text{True}$$

Because $1 \le 6$ is true, we shade the side of the graph containing the test point $(1, 3)$. See Figure 39.

6 Graph $x \ge -3y$.

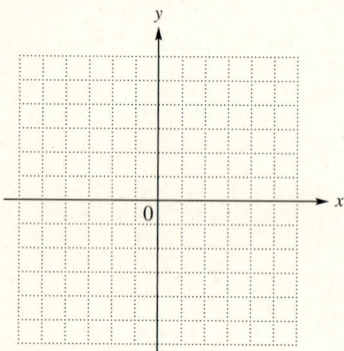

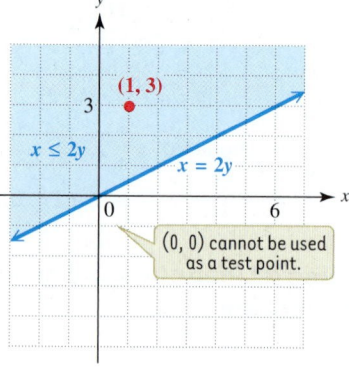

Figure 39

◀ Work Problem **6** at the Side.

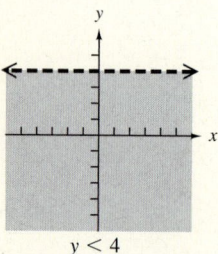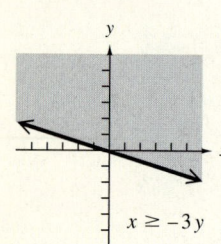

11.5 ▶▶▶ **Exercises**

FOR EXTRA HELP

MyMathLab

Math XL
PRACTICE

WATCH

DOWNLOAD

READ

REVIEW

Decide whether each statement is true *or* false. *If false, explain why.*

1. The point $(4, 0)$ lies on the graph of $3x - 4y < 12$.

2. The point $(4, 0)$ lies on the graph of $3x - 4y \le 12$.

3. Both points $(4, 1)$ and $(0, 0)$ lie on the graph of $3x - 2y \ge 0$.

4. The graph of $y > x$ does not contain points in quadrant IV.

The following statements were taken from various media. Each includes a phrase that can be symbolized with one of the inequality symbols $<, \le, >,$ *or* $\ge$. *In Exercises 5–8, give the inequality symbol for the bold faced words.*

5. Since it was recognized in 1981, HIV/AIDS has killed **more than** 25 million people worldwide and infected **more than** 60 million, about two-thirds of whom live in Africa. (*Source:* The President's Emergency Plan for AIDS Relief, February, 2008.)

6. The average national automobile insurance premium of $1896 in 2007 was $20 **less than** the 2006 average premium. (*Source:* 2007 Mid-Year Auto Insurance Pricing Report.)

7. As of December 2007, airline passengers were allowed one carry-on bag, with dimensions totaling **at most** 45 in. (*Source:* The Gazette.)

8. As of February 2008, all major airlines except US Airways award **at least** 500 frequent flier miles per flight. (*Source:* USA Today.)

In Exercises 9–16, the straight-line boundary has been drawn. Complete each graph by shading the correct region. See Examples 1–4.

9. $x + y \geq 4$

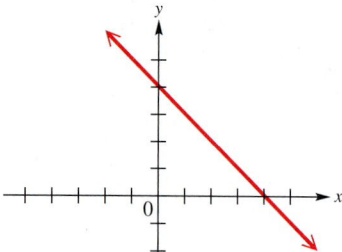

10. $x + y \leq 2$

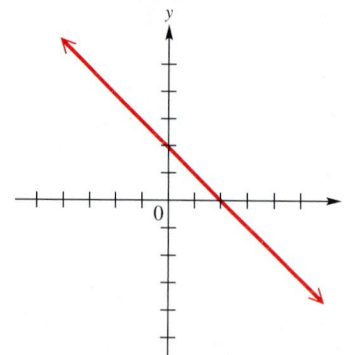

11. $x + 2y \geq 7$

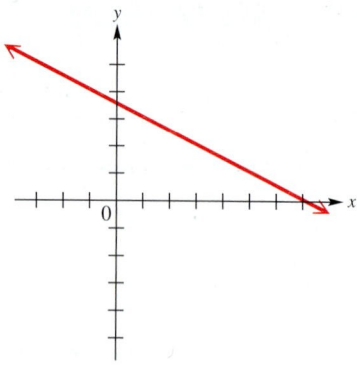

12. $2x + y \geq 5$

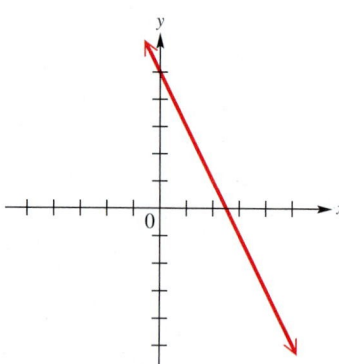

13. $-3x + 4y > 12$

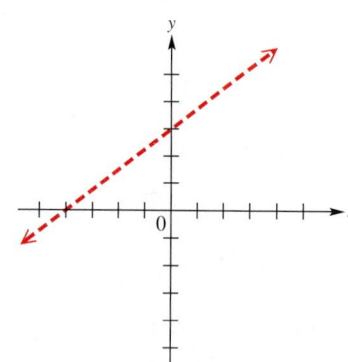

14. $4x - 5y < 20$

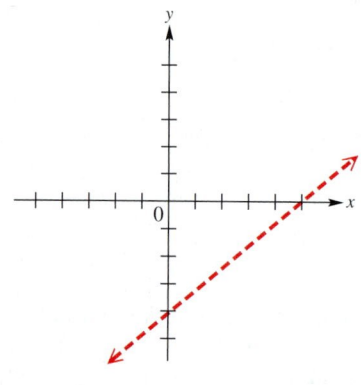

15. $x > 4$

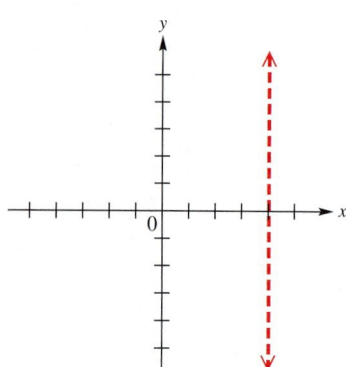

16. $y < -1$

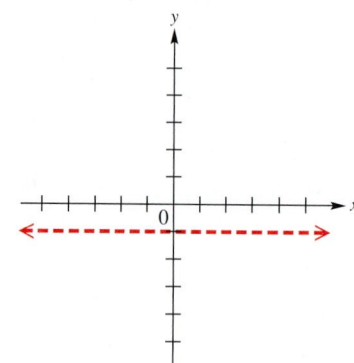

17. $x \geq -y$

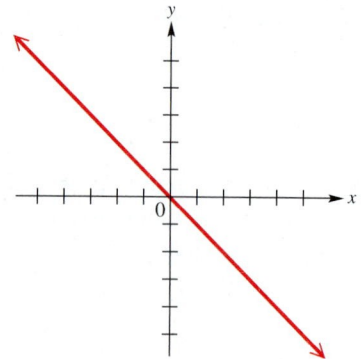

18. Explain how to determine whether to use a dashed line or a solid line when graphing a linear inequality in two variables.

19. Explain why the point $(0, 0)$ is not an appropriate choice for a test point when graphing an inequality whose boundary goes through the origin.

Graph each linear inequality. See Examples 1–4.

20. $x + y \geq 3$

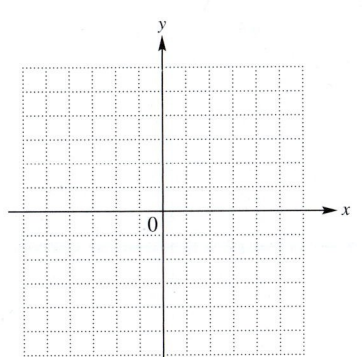

21. $x + y \leq 5$

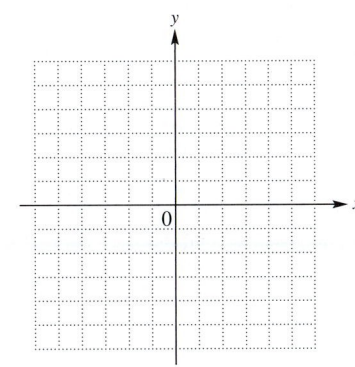

22. $x + 3y > 6$

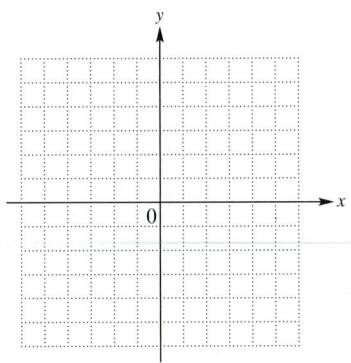

23. $x + 2y < 4$

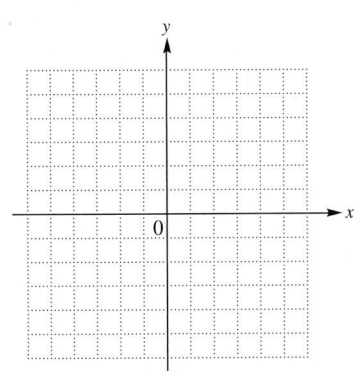

24. $-4y > 3x - 12$

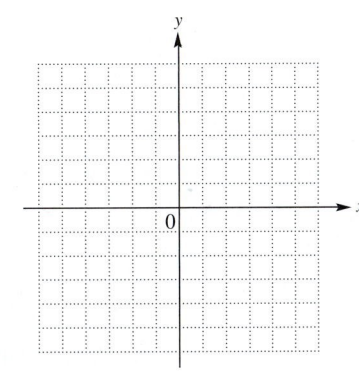

25. $2x + 6 > -3y$

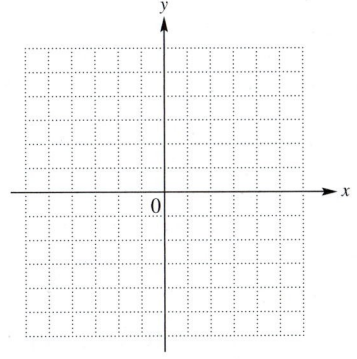

26. $y < -3x + 1$

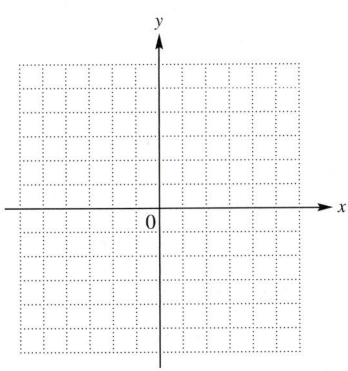

27. $y \geq 2x + 1$

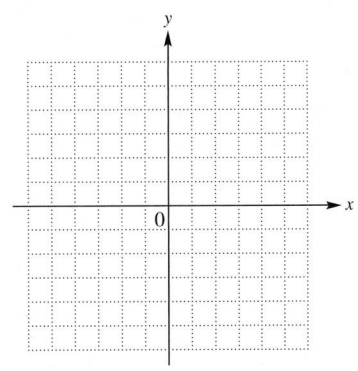

28. $x \geq 1$

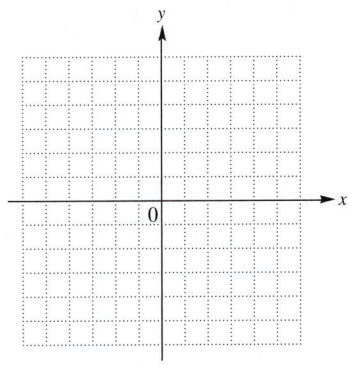

29. $x \leq -2$

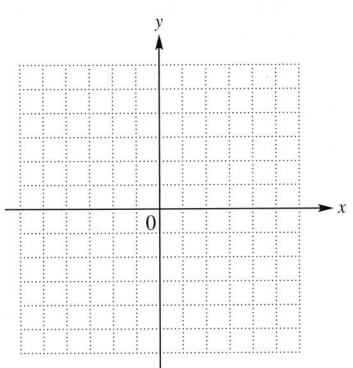

30. $y < -3$

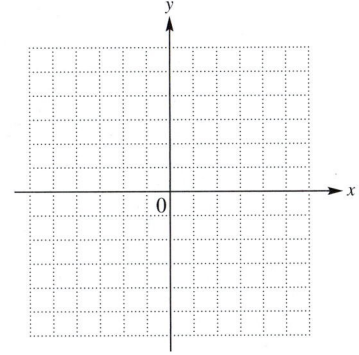

31. $y < 5$

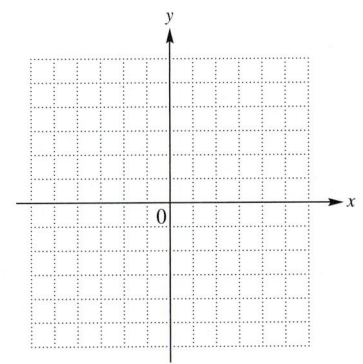

32. $y \leq 2x$

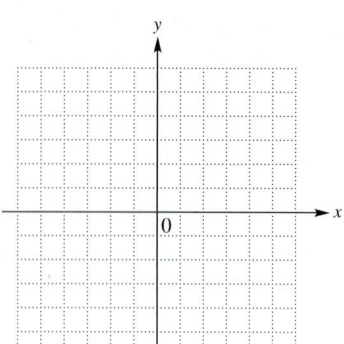

🌑 **33.** $y \geq 4x$

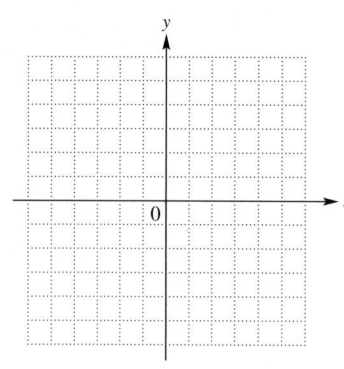

34. $x > -5y$

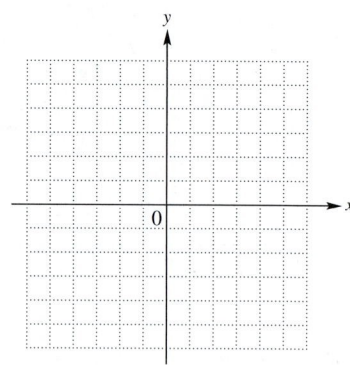

35. Explain why the graph of $y > x$ cannot lie in quadrant IV.

36. Explain why the graph of $y < x$ cannot lie in quadrant II.

Solve each problem. In part (a), $x \geq 0$ and $y \geq 0$, so graph only the part of the inequality in quadrant I.

37. A company will ship x units of merchandise to outlet I and y units of merchandise to outlet II. The company must ship a total of at least 500 units to these two outlets. This can be expressed by writing

$$x + y \geq 500.$$

(a) Graph the inequality.

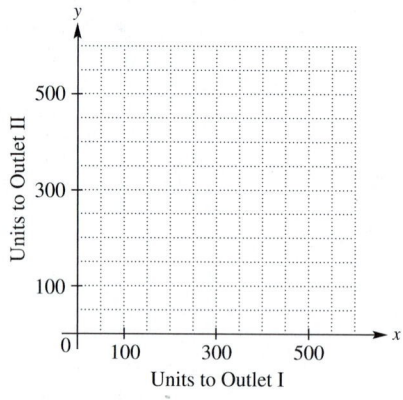

(b) Give two ordered pairs that satisfy the inequality.

38. A toy manufacturer makes stuffed bears and geese. It takes 20 min to sew a bear and 30 min to sew a goose. There is a total of 480 min of sewing time available to make x bears and y geese. These restrictions lead to the inequality

$$20x + 30y \leq 480.$$

(a) Graph the inequality.

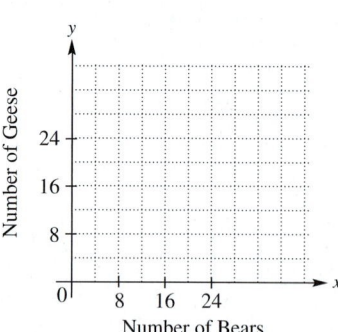

(b) Give two ordered pairs that satisfy the inequality.

Chapter 11 ▷▷▷ Summary

▶ Key Terms

11.1 **bar graph**   A bar graph is a series of bars used to show comparisons between two categories of data.

line graph   A line graph consists of a series of points that are connected with line segments and is used to show changes or trends in data.

linear equation in two variables   An equation that can be written in the form $Ax + By = C$ is a linear equation in two variables. (A and B are real numbers that cannot both be 0.)

ordered pair   A pair of numbers written between parentheses in which order is important is called an ordered pair.

x	y
0	4
2	0
1	2

Table of values
for $2x + y = 4$

table of values   A table showing selected ordered pairs of numbers that satisfy an equation is called a table of values.

***x*-axis**   The horizontal axis in a coordinate system is called the x-axis.

***y*-axis**   The vertical axis in a coordinate system is called the y-axis.

rectangular (Cartesian) coordinate system   An x-axis and y-axis at right angles form a coordinate system.

quadrants   A coordinate system divides the plane into four regions called quadrants.

origin   The point at which the x-axis and y-axis intersect is called the origin.

plane   A flat surface determined by two intersecting lines is a plane.

coordinates   The numbers in an ordered pair are called the coordinates of the corresponding point.

plot   To plot an ordered pair is to find the corresponding point on a coordinate system.

scatter diagram   A graph of ordered pairs of data is a scatter diagram.

11.2 **graph**   The graph of an equation is the set of all points that correspond to the ordered pairs that satisfy the equation.

graphing   The process of plotting the ordered pairs that satisfy a linear equation and drawing a line through them is called graphing.

y-intercept   If a graph intersects the y-axis at k, then the y-intercept is $(0, k)$.

x-intercept   If a graph intersects the x-axis at k, then the x-intercept is $(k, 0)$.

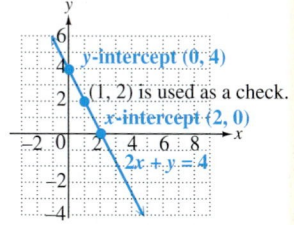

Graph of $2x + y = 4$

11.3 **rise**   Rise is the vertical change between two different points on a line.

run   Run is the horizontal change between two different points on a line.

slope   The slope of a line is the ratio of the change in y compared to the change in x when moving along the line from one point to another.

parallel lines   Two lines in a plane that never intersect are parallel.

perpendicular lines   Perpendicular lines intersect at a 90° angle.

11.5 **linear inequality in two variables**   An inequality that can be written in the form $Ax + By < C$, $Ax + By > C$, $Ax + By \leq C$, or $Ax + By \geq C$ is a linear inequality in two variables.

boundary line   In the graph of a linear inequality, the boundary line separates the region that satisfies the inequality from the region that does not satisfy the inequality.

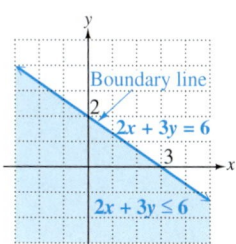

▶ New Symbols

(x, y) ordered pair

(x_1, y_1) subscript notation; x-sub-one, y-sub-one

m slope

▶ Test Your Word Power

See how well you have learned the vocabulary in this chapter. Answers, with examples, follow the Quick Review.

1. An **ordered pair** is a pair of numbers written
   A. in numerical order between brackets
   B. between parentheses or brackets
   C. between parentheses in which order is important
   D. between parentheses in which order does not matter.

2. The **coordinates** of a point are
   A. the numbers in the corresponding ordered pair
   B. the solution of an equation
   C. the values of the x- and y-intercepts
   D. the graph of the point.

3. An **intercept** is
   A. the point where the x-axis and y-axis intersect
   B. a pair of numbers written in parentheses in which order is important
   C. one of the four regions determined by a rectangular coordinate system
   D. the point where a graph intersects the x-axis or the y-axis.

4. The **slope** of a line is
   A. the measure of the run over the rise of the line
   B. the distance between two points on the line
   C. the ratio of the change in y to the change in x along the line
   D. the horizontal change compared to the vertical change of two points on the line.

5. Two lines in a plane are **parallel** if
   A. they represent the same line
   B. they never intersect
   C. they intersect at a 90° angle
   D. one has a positive slope and one has a negative slope.

6. Two lines in a plane are **perpendicular** if
   A. they represent the same line
   B. they never intersect
   C. they intersect at a 90° angle
   D. one has a positive slope and one has a negative slope.

▶ Quick Review

Concepts	Examples

11.1 Reading Graphs; Linear Equations in Two Variables

Bar graphs and line graphs are ways to "picture", or represent, the relationship between two variables.

U.S. Marathon Finishers

Source: Running USA.

The line graph illustrates the number of U.S. runners in thousands who finished marathons in the years 2001–2005.

Concepts	Examples

11.1 Reading Graphs; Linear Equations in Two Variables *(continued)*

An ordered pair is a solution of an equation if it makes the equation a true statement.

Is $(2, -5)$ or $(0, -6)$ a solution of $4x - 3y = 18$?

$$4(2) - 3(-5) = 23 \neq 18 \quad \bigg| \quad 4(0) - 3(-6) = 18$$

$(2, -5)$ is not a solution. $\quad \bigg| \quad (0, -6)$ is a solution.

If a value of either variable in an equation is given, the value of the other variable can be found by substitution.

Complete the ordered pair $(0, \underline{\quad})$ for $3x = y + 4$.

$$3(0) = y + 4 \qquad \text{Let } x = 0.$$
$$0 = y + 4 \qquad \text{Multiply.}$$
$$-4 = y \qquad \text{Subtract 4.}$$

The ordered pair is $(0, -4)$.

To plot the ordered pair $(-3, 4)$, start at the origin, go 3 units to the left, and from there go 4 units up.

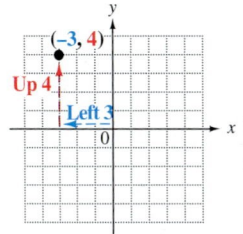

11.2 Graphing Linear Equations in Two Variables

To graph a linear equation:

Step 1 Find at least two ordered pairs that are solutions of the equation.

Step 2 Plot the corresponding points.

Step 3 Draw a straight line through the points.

The graph of $y = k$ is a horizontal line through $(0, k)$.

The graph of $x = k$ is a vertical line through $(k, 0)$.

Graph $x - 2y = 4$.

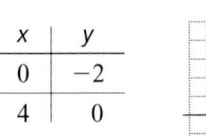

x	y
0	-2
4	0

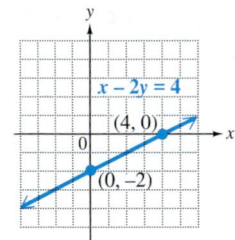

11.3 Slope of a Line

The slope of the line through (x_1, y_1) and (x_2, y_2) is

$$m = \frac{\text{change in } y}{\text{change in } x} = \frac{y_2 - y_1}{x_2 - x_1} \quad (x_1 \neq x_2).$$

Horizontal lines have slope 0.

Vertical lines have undefined slope.

To find the slope of a line from its equation, solve for y. The slope is the coefficient of x.

The line through $(-2, 3)$ and $(4, -5)$ has slope

$$m = \frac{-5 - 3}{4 - (-2)} = \frac{-8}{6} = -\frac{4}{3}.$$

The line $y = -2$ has slope 0.

The line $x = 4$ has undefined slope.

Find the slope of the graph of $3x - 4y = 12$.

$$-4y = -3x + 12 \qquad \text{Add } -3x.$$
$$y = \frac{3}{4}x - 3 \qquad \text{Divide by } -4.$$
$$\uparrow$$
$$\text{Slope}$$

Concepts	Examples

11.4 Equations of Lines

Slope-Intercept Form

$y = mx + b$

m is the slope.

$(0, b)$ is the y-intercept.

Write an equation of the line with slope **2** and y-intercept $(0, -5)$.

$$y = 2x - 5$$

Point-Slope Form

$y - y_1 = m(x - x_1)$

m is the slope.

(x_1, y_1) is a point on the line.

Write an equation of the line with slope $-\frac{1}{2}$ through $(-4, 5)$.

$$y - 5 = -\frac{1}{2}[x - (-4)] \qquad \text{Substitute.}$$

$$y - 5 = -\frac{1}{2}(x + 4)$$

$$y - 5 = -\frac{1}{2}x - 2 \qquad \text{Distributive property}$$

$$y = -\frac{1}{2}x + 3 \qquad \text{Add 5.}$$

Standard Form

$Ax + By = C$

This equation is written in standard form as

$$x + 2y = 6,$$

with $A = 1$, $B = 2$, and $C = 6$.

11.5 Graphing Linear Inequalities in Two Variables

Step 1 Graph the line that is the boundary of the region. Make it solid if the inequality is $\leq$ or $\geq$; make it dashed if the inequality is $<$ or $>$.

Step 2 Use any point not on the line as a test point. Substitute for x and y in the inequality. If the result is true, shade the side of the line containing the test point; if the result is false, shade the other side.

Graph $2x + y \leq 5$.

Graph the line $2x + y = 5$. Make it solid because the symbol $\leq$ includes equality.

Use $(0, 0)$ as a test point.

$$2(0) + 0 \overset{?}{\leq} 5$$

$$0 \leq 5 \qquad \text{True}$$

Shade the side of the line containing $(0, 0)$.

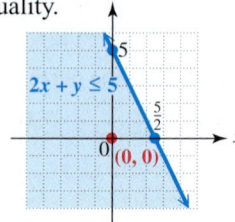

ANSWERS TO TEST YOUR WORD POWER

1. C; *Examples:* $(0, 3)$, $(3, 8)$, $(4, 0)$
2. A; *Example:* The point associated with the ordered pair $(1, 2)$ has x-coordinate 1 and y-coordinate 2.
3. D; *Example:* The graph of the equation $4x - 3y = 12$ has x-intercept at $(3, 0)$ and y-intercept at $(0, -4)$.
4. C; *Example:* The line through $(3, 6)$ and $(5, 4)$ has slope $\dfrac{4 - 6}{5 - 3} = \dfrac{-2}{2} = -1$.
5. B; *Example:* See Figure 27 in **Section 11.3**.
6. C; *Example:* See Figure 28 in **Section 11.3**.

Chapter 11 ▶▶▶ Review Exercises

[11.1] *The percent of first-year college students at two-year public institutions who returned for a second year for the years 2001 through 2007 are shown in the graph.*

1. Write ordered pairs of the form (year, percent) for the data shown in the graph.

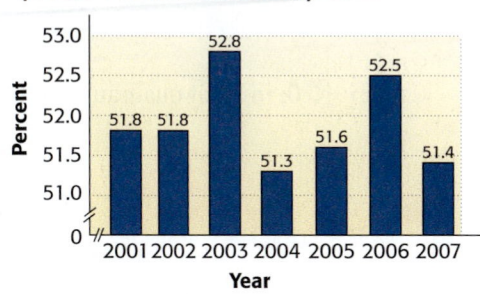

Percent of Students Who Return for Second Year (2-Year Public Institutions)

Source: ACT.

2. What does the ordered pair (2006, 52.5) mean in the context of these problems?

3. In what year did the percent show the greatest decrease from the previous year? What was this decrease?

4. In what year did the percent show the greatest increase from the previous year? What was this increase?

Complete the given ordered pairs for each equation.

5. $y = 3x + 2$ $(-1, \_\_), (0, \_\_), (\_\_, 5)$

6. $4x + 3y = 6$ $(0, \_\_), (\_\_, 0), (-2, \_\_)$

7. $x = 3y$ $(0, \_\_), (8, \_\_), (\_\_, -3)$

8. $x - 7 = 0$ $(\_\_, -3)\ (\_\_, 0), (\_\_, 5)$

Decide whether each ordered pair is a solution of the given equation.

9. $x + y = 7; (2, 5)$

10. $2x + y = 5; (-1, 3)$

11. $3x - y = 4; \left(\dfrac{1}{3}, -3\right)$

Plot each ordered pair on the given coordinate system.

12. $(2, 3)$ **13.** $(-4, 2)$

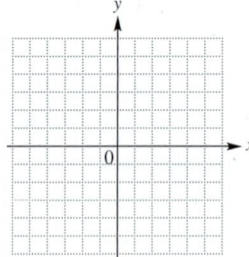

14. $(3, 0)$ **15.** $(0, -6)$

16. If $x > 0$ and $y < 0$, in what quadrant(s) must (x, y) lie? Explain.

17. On what axis does the point $(k, 0)$ lie for any real value of k? the point $(0, k)$? Explain.

Without plotting the given point, name the quadrant in which each point lies.

18. $(-2, 3)$ **19.** $(-1, -4)$ **20.** $\left(0, -5\dfrac{1}{2}\right)$

[11.2] *Find the intercepts for the graph of each equation.*

21. $y = 2x + 5$ **22.** $2x + y = -7$ **23.** $3x + 2y = 8$

   x-intercept: x-intercept: x-intercept:

   y-intercept: y-intercept: y-intercept:

Graph each linear equation.

24. $2x - y = 3$ **25.** $x + 2y = -4$ **26.** $x + y = 0$

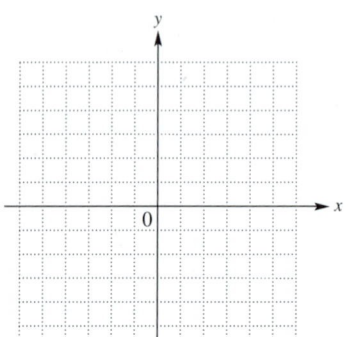   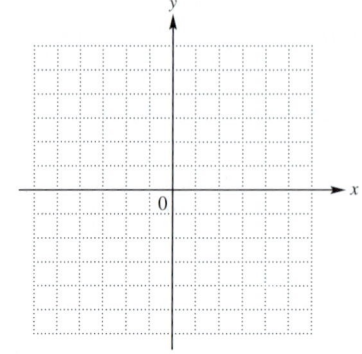   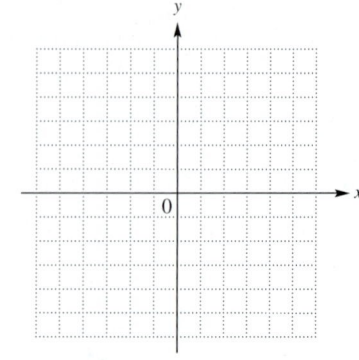

[11.3] *Find the slope of each line.*

27. The line through $(2, 3)$ and $(-4, 6)$

28. The line through $(0, 0)$ and $(-3, 2)$

29. The line through $(0, 6)$ and $(1, 6)$

30. The line through $(2, 5)$ and $(2, 8)$

31. $y = 3x - 4$

32. $y = \dfrac{2}{3}x + 1$

33.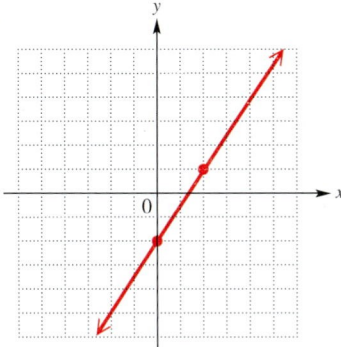

34.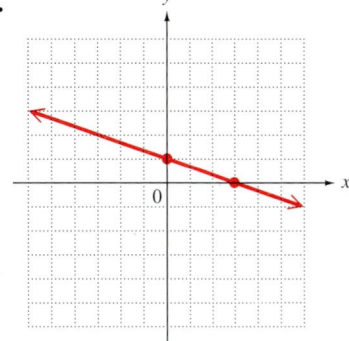

35. $x = 0$

36. $y = 4$

37. The line having these points

x	y
0	1
2	4
6	10

38. **(a)** A line parallel to the graph of $y = 2x + 3$

(b) A line perpendicular to the graph of $y = -3x + 3$

Decide whether each pair of lines is parallel, perpendicular, *or* neither.

39. $3x + 2y = 6$
  $6x + 4y = 8$

40. $x - 3y = 1$
  $3x + y = 4$

41. $x - 2y = 8$
  $x + 2y = 8$

42. What is the slope of a line perpendicular to a line with undefined slope?

[11.4] *Write an equation of each line. Give the final answer in slope-intercept form (if possible).*

43. $m = -1, b = \dfrac{2}{3}$

44. The line in Exercise 34

45. The line through $(4, -3)$, $m = 1$

46. The line through $(-1, 4)$, $m = \dfrac{2}{3}$

47. The line through $(1, -1)$, $m = -\dfrac{3}{4}$

48. The line through $(2, 1)$ and $(-2, 2)$

49. The line through $(-4, 1)$, slope 0

50. The line through $\left(\dfrac{1}{3}, -\dfrac{3}{4}\right)$, undefined slope

51. Consider the equation $x + 3y = 15$.

  (a) Write it in the form $y = mx + b$.

  (b) What is the slope? What is the y-intercept?

  (c) Use the slope and y-intercept to graph the line. Indicate two points on the graph.

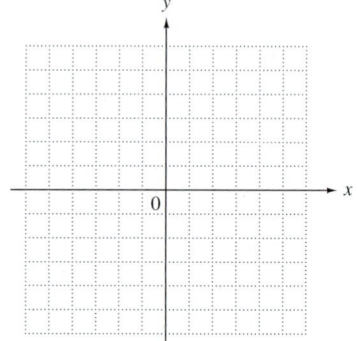

52. Match the description in Column I with the correct equation in Column II.

I	**II**
(a) Slope -0.5, $b = -2$	**A.** $y = -\dfrac{1}{2}x$
(b) x-intercept $(4, 0)$, y-intercept $(0, 2)$	**B.** $y = -\dfrac{1}{2}x - 2$
(c) The line through $(4, -2)$ and $(0, 0)$	**C.** $x - 2y = 2$
(d) $m = \dfrac{1}{2}$, passes through $(-2, -2)$	**D.** $x + 2y = 4$
	E. $x = 2y$

[11.5] *Graph each linear inequality.*

53. $3x + 5y > 9$

54. $2x - 3y > -6$

55. $x \geq -4$

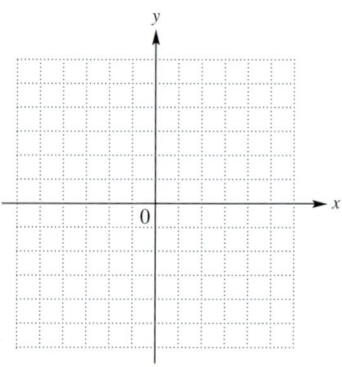
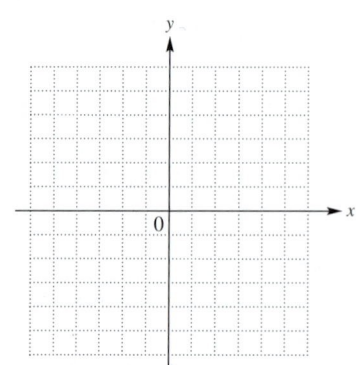
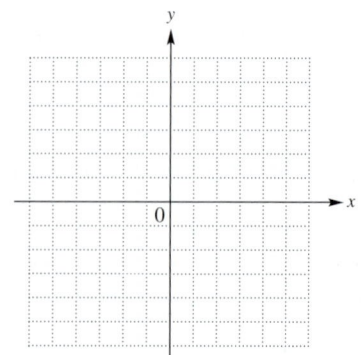

Mixed Review Exercises

*In Exercises 56–61, match each statement to the appropriate graph or graphs in
A–D. Graphs may be used more than once.*

A. 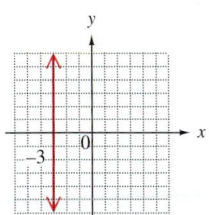 **B.** 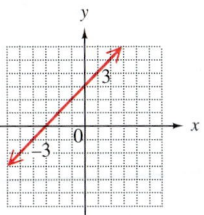 **C.** 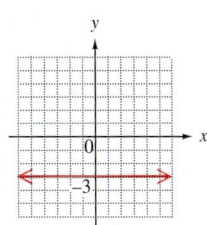 **D.**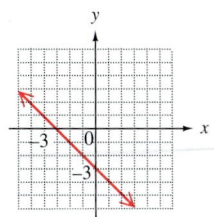

56. The line shown in the graph has undefined slope.

57. The graph of the equation has y-intercept $(0, -3)$.

58. The graph of the equation has x-intercept $(-3, 0)$.

59. The line shown in the graph has negative slope.

60. The graph is that of the equation $y = -3$.

61. The line shown in the graph has slope 1.

Find the intercepts and the slope of each line. Then graph the line.

62. $y = -2x - 5$

x-intercept:

y-intercept:

slope:

63. $x + 3y = 0$

x-intercept:

y-intercept:

slope:

64. $y - 5 = 0$

x-intercept:

y-intercept:

slope:

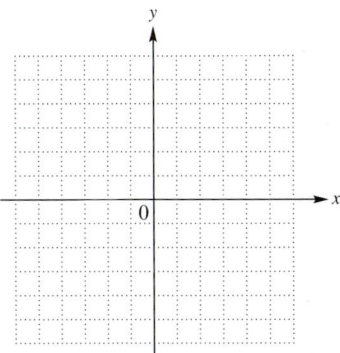 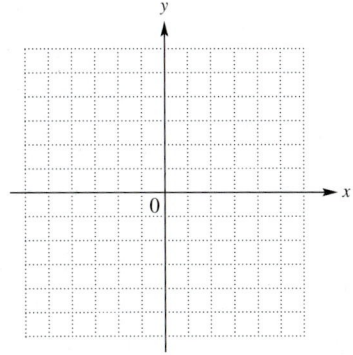 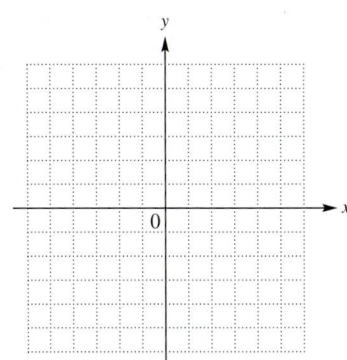

Write an equation of each line. Give the final answer in slope-intercept form.

65. $m = -\dfrac{1}{4}, b = -\dfrac{5}{4}$

66. The line through $(8, 6)$, $m = -3$

67. The line through $(3, -5)$ and $(-4, -1)$

Graph each inequality.

68. $y < -4x$

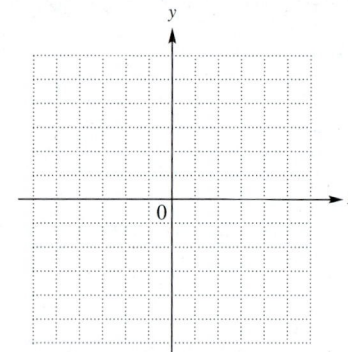

69. $x - 2y \leq 6$

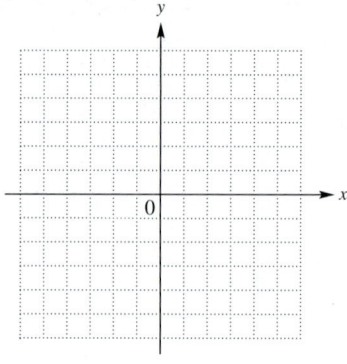

Relating Concepts (Exercises 70–76) For Individual or Group Work

*The percents of four-year college students in public schools who earned a degree within five years of entry between 2002 and 2007 are shown in the graph. Use the graph to **work Exercises 70–76 in order.***

70. What was the percent increase from 2002 to 2007?

71. Since the points of the graph lie approximately in a linear pattern, a straight line can be used to model the data. Will this line have positive or negative slope? Explain.

72. Write two ordered pairs for the data for 2002 and 2007.

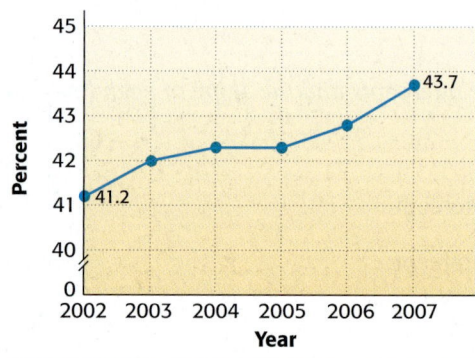

Percents of Students Graduating Within 5 Years (Public Institutions)

Source: ACT.

73. Use the ordered pairs from Exercise 72 to find the equation of a line that models the data. Write the equation in slope-intercept form.

74. Based on the equation you found in Exercise 73, what is the slope of the line? Does it agree with your answer in Exercise 71?

75. Use the equation from Exercise 73 to approximate the percents for 2003 through 2006, and complete the table.

Year	Percent
2003	
2004	
2005	
2006	

76. Use the equation from Exercise 73 to estimate the percent for 2008. Can we be sure that this estimate is accurate?

Chapter 11 ▶▶▶ **Test** Test Prep VIDEO CD

Use the Chapter Test Prep Video CD to see fully worked-out solutions to any of the exercises you want to review.

The line graph shows the overall unemployment rate in the U.S. civilian labor force for the years 1998 through 2005. Use the graph to work Exercises 1–3.

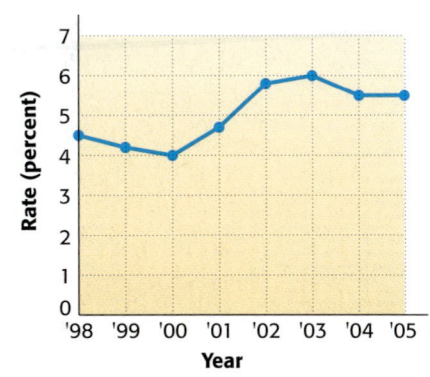

Unemployment Rate

Rate (percent)

Year

Source: U.S. Department of Labor.

1. Between which pairs of consecutive years did the unemployment rate decrease?

2. What was the general trend in the unemployment rate between 2000 and 2003?

3. Estimate the overall unemployment rate in 2003 and 2004. About how much did the unemployment rate decline between 2003 and 2004?

Graph each linear equation. Give the x- and y-intercepts.

4. $3x + y = 6$

5. $y - 2x = 0$

1. _____

2. _____

3. _____

4. *x*-intercept: _____

   y-intercept: _____

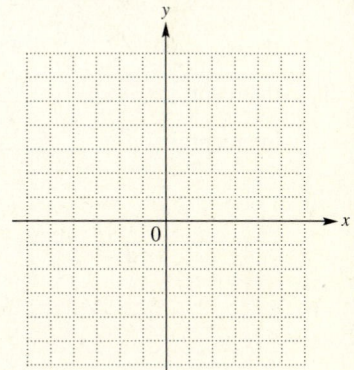

5. *x*-intercept: _____

   y-intercept: _____

6. *x*-intercept: _____

y-intercept: _____

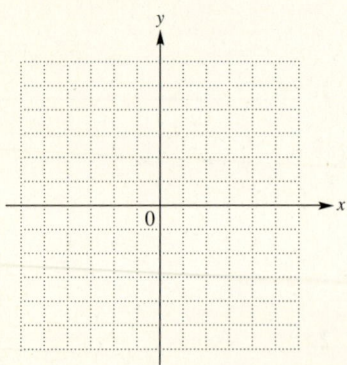

6. $x + 3 = 0$

7. *x*-intercept: _____

y-intercept: _____

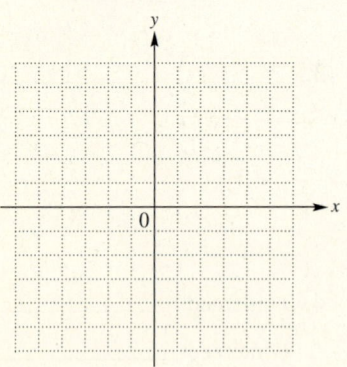

7. $y = 1$

8. *x*-intercept: _____

y-intercept: _____

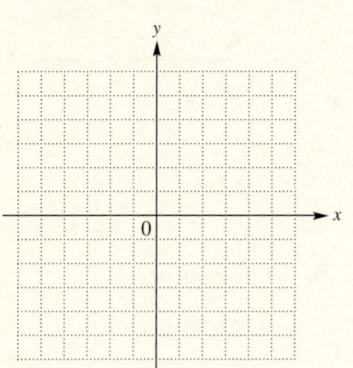

8. $x - y = 4$

Find the slope of each line.

9. _____

9. The line through $(-4, 6)$ and $(-1, -2)$

10. _____

10. $2x + y = 10$

11. _____

11. $x + 12 = 0$

12.

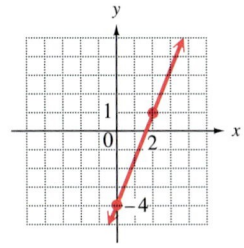

13. A line parallel to the graph of $y - 4 = 6$

13. _____

Write an equation for each line. Give the final answer in slope-intercept form.

14. The line through $(-1, 4)$, $m = 2$

14. _____

15. The line in Exercise 12

15. _____

16. The line through $(2, -6)$ and $(1, 3)$

16. _____

17. x-intercept: $(3, 0)$; y-intercept: $\left(0, \dfrac{9}{2}\right)$

17. _____

Graph each linear inequality.

18. $x + y \leq 3$

18.

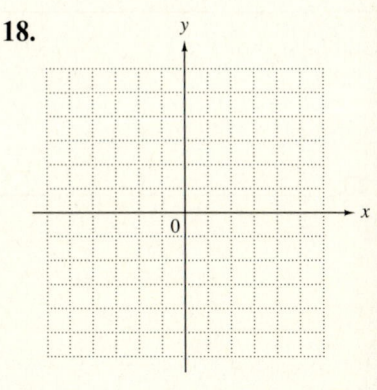

19.

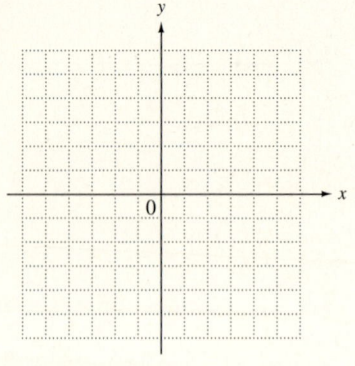

19. $3x - y > 0$

The graph shows worldwide snowmobile sales from 2000 through 2007, where 2000 corresponds to $x = 0$. Use the graph to work Exercises 20–24.

Worldwide Snowmobile Sales

Source: www.snowmobile.org

20. _____

20. Is the slope of the line in the graph positive or negative? Explain.

21. _____

21. Write two ordered pairs for the data points shown in the graph. Use them to find the slope of the line.

22. _____

22. Use the ordered pairs and slope from Exercise 21 to find an equation of a line that models the data. Write the equation in slope-intercept form.

23. _____

23. Use the equation from Exercise 22 to approximate worldwide snowmobile sales for 2005. How does your answer compare to the actual sales of 173.7 thousand?

24. _____

24. What does the ordered pair (7, 160) mean in the context of this problem?

Systems of Linear Equations and Inequalities

Over the years, Americans have continued their fascination with Hollywood and the movies. Although many people now prefer to watch movies at home on DVD, the number of tickets sold at domestic movie theaters increased in 2007 to about 1.4 billion, with a total gross of over $9.6 billion. The top box office draws of that year—*Spider-Man 3* and *Shrek the Third*—attracted millions of adults and children wishing to get away from it all for a few hours. (*Source:* www.boxofficemojo.com)

In Exercises 13 and 14 of **Section 12.4,** we use *systems of equations* to find out how much money these top films earned in total and on their opening weekends.

12.1 ▶▶▶ Solving Systems of Linear Equations by Graphing

OBJECTIVES

1 Decide whether a given ordered pair is a solution of a system.

2 Solve linear systems by graphing.

3 Solve special systems by graphing.

4 Identify special systems without graphing.

A **system of linear equations,** often called a **linear system,** consists of two or more linear equations with the same variables.

$$
\begin{array}{lll}
2x + 3y = 4 & x + 3y = 1 & x - y = 1 \\
3x - y = -5 & -y = 4 - 2x & y = 3
\end{array}
\quad
\begin{array}{l}
\text{Linear} \\
\text{systems}
\end{array}
$$

In the system on the right, think of $y = 3$ as an equation in two variables by writing it as $0x + y = 3$.

OBJECTIVE 1 Decide whether a given ordered pair is a solution of a system. A **solution of a system** of linear equations is an ordered pair that makes both equations true at the same time. A solution of an equation is said to *satisfy* the equation.

EXAMPLE 1 Determining Whether an Ordered Pair Is a Solution

Is $(4, -3)$ a solution of each system?

(a) $x + 4y = -8$
$3x + 2y = 6$
To decide whether $(4, -3)$ is a solution of the system, substitute 4 for x and -3 for y in each equation.

$$
\begin{array}{ll}
x + 4y = -8 & \qquad 3x + 2y = 6 \\
4 + 4(-3) \stackrel{?}{=} -8 & \qquad 3(4) + 2(-3) \stackrel{?}{=} 6 \\
4 + (-12) \stackrel{?}{=} -8 \quad \text{Multiply.} & \qquad 12 + (-6) \stackrel{?}{=} 6 \quad \text{Multiply.} \\
-8 = -8 \quad \text{True} & \qquad 6 = 6 \quad \text{True}
\end{array}
$$

Because $(4, -3)$ satisfies both equations, it is a solution of the system.

(b) $2x + 5y = -7$
$3x + 4y = 2$
Again, substitute 4 for x and -3 for y in both equations.

$$
\begin{array}{ll}
2x + 5y = -7 & \qquad 3x + 4y = 2 \\
2(4) + 5(-3) \stackrel{?}{=} -7 & \qquad 3(4) + 4(-3) \stackrel{?}{=} 2 \\
8 + (-15) \stackrel{?}{=} -7 \quad \text{Multiply.} & \qquad 12 + (-12) \stackrel{?}{=} 2 \quad \text{Multiply.} \\
-7 = -7 \quad \text{True} & \qquad 0 = 2 \quad \text{False}
\end{array}
$$

The ordered pair $(4, -3)$ is not a solution of this system because it does not satisfy the second equation.

◀ *Work Problem* **1** *at the Side.*

---

1 Fill in the blanks, and decide whether the given ordered pair is a solution of the system.

(a) $(2, 5)$

$3x - 2y = -4$
$5x + y = 15$

$$3x - 2y = -4$$
$$3(\underline{\quad}) - 2(\underline{\quad}) \stackrel{?}{=} -4$$

$$5x + y = 15$$
$$5(2) + \underline{\quad} \stackrel{?}{=} \underline{\quad}$$

$(2, 5)$ _____ a solution.
(is/is not)

(b) $(1, -2)$

$x - 3y = 7$
$4x + y = 5$

$(1, -2)$ _____ a solution.
(is/is not)

OBJECTIVE 2 Solve linear systems by graphing. The set of all ordered pairs that are solutions of a system is its **solution set.** One way to find the solution set of a system of two linear equations is to graph both equations on the same axes. The graph of each line shows points whose coordinates satisfy the equation of that line. Any intersection point would be on both lines and would therefore be a solution of *both* equations. ***Thus, the coordinates of any point where the lines intersect give a solution of the system.***

ANSWERS

1. **(a)** 2; 5; 5; 15; is **(b)** is not

The graph in Figure 1 shows that the solution of the system in Example 1(a) is the intersection point $(4, -3)$. Because *two different* straight lines can intersect at no more than one point, there can never be more than one solution for such a system.

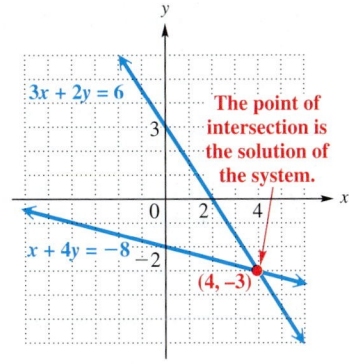

Figure 1

EXAMPLE 2 **Solving a System by Graphing**

Solve the system of equations by graphing both equations on the same axes.

$$2x + 3y = 4$$

$$3x - y = -5$$

We graph these two equations by plotting several points for each line. Recall from **Section 11.2** that the intercepts are often convenient choices.

$2x + 3y = 4$

x	y
0	$\frac{4}{3}$
2	0
-2	$\frac{8}{3}$

Find a third ordered pair as a check.

$3x - y = -5$

x	y
0	5
$-\frac{5}{3}$	0
-2	-1

The lines in Figure 2 suggest that the graphs intersect at the point $(-1, 2)$. We check this by substituting -1 for x and 2 for y in both equations.

Check $2x + 3y = 4$

$2(-1) + 3(2) \overset{?}{=} 4$

$\qquad\qquad 4 = 4$ True

$\qquad 3x - y = -5$

$3(-1) - 2 \overset{?}{=} -5$

$\qquad -5 = -5$ True

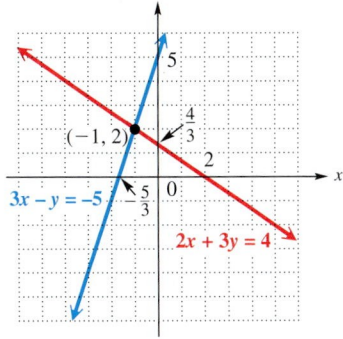

Figure 2

Because $(-1, 2)$ satisfies both equations, the *solution set* of this system is $\{(-1, 2)\}$. We write the ordered pair in the solution set between set braces.

Work Problem **2** *at the Side.* ▶

Note

We can also graph a linear system by writing each equation in the system in slope-intercept form and using the slope and y-intercept to graph each line. For Example 2,

$2x + 3y = 4$ becomes $y = -\frac{2}{3}x + \frac{4}{3}$ y-intercept $(0, \frac{4}{3})$; slope $-\frac{2}{3}$

$3x - y = -5$ becomes $y = 3x + 5$. y-intercept $(0, 5)$; slope 3, or $\frac{3}{1}$

Confirm that graphing these equations results in the same lines and the same solution shown in Figure 2.

---

2 Solve each system of equations by graphing both equations on the same axes. Check your solutions.

(a) $5x - 3y = 9$

$\qquad x + 2y = 7$

(One of the lines is already graphed.)

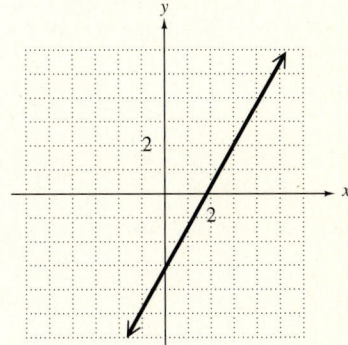

(b) $x + y = 4$

$\qquad 2x - y = -1$

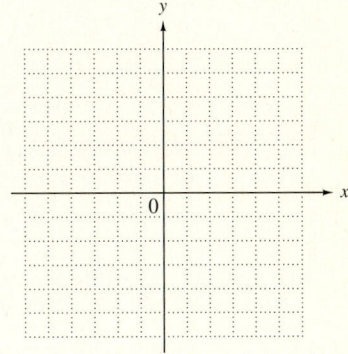

To solve a linear system by graphing, follow these steps.

> **Solving a Linear System by Graphing**
>
> *Step 1* **Graph each equation** of the system on the same coordinate axes.
>
> *Step 2* **Find the coordinates of the point of intersection** of the graphs if possible. This is the solution of the system.
>
> *Step 3* **Check** the solution in *both* of the original equations. Then write the solution set.

> **CAUTION**
> A difficulty with the graphing method of solution is that it may not be possible to determine from the graph the exact coordinates of the point that represents the solution, particularly if these coordinates are not integers. For this reason, algebraic methods of solution are explained later in this chapter. The graphing method does, however, show geometrically how solutions are found and is useful when approximate answers will do.

OBJECTIVE 3 Solve special systems by graphing. Sometimes the graphs of the two equations in a system either do not intersect at all or are the same line, as in the systems in Example 3.

EXAMPLE 3 **Solving Special Systems**

Solve each system by graphing.

(a) $2x + y = 2$
$2x + y = 8$

The graphs of these lines are shown in Figure 3. The two lines are parallel and have no points in common. For such a system, **there is no solution.** Its solution set is the **empty set,** or **null set,** symbolized $\emptyset$.

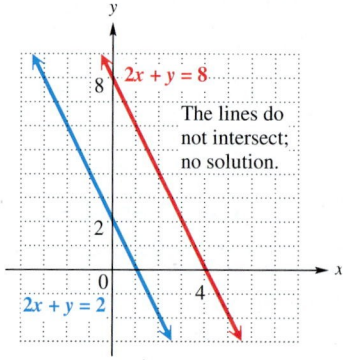

The lines do not intersect; no solution.

Figure 3

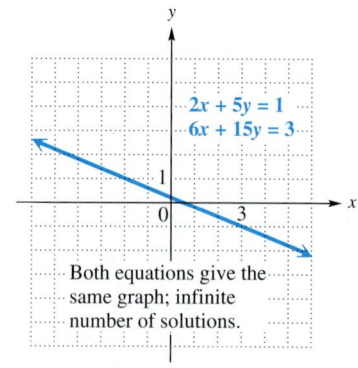

Both equations give the same graph; infinite number of solutions.

Figure 4

(b) $2x + 5y = 1$
$6x + 15y = 3$

The graphs of these two equations are the same line. See Figure 4. The second equation can be obtained by multiplying each side of the first equation by 3. In this case, every point on the line is a solution of the system, and *the solution set contains an infinite number of ordered pairs* that satisfy the equations.

Continued on Next Page

We write the solution set as

$$\{(x, y) \mid 2x + 5y = 1\},$$

> This is the first equation in the system. See the Note below.

read "the set of ordered pairs (x, y) such that $2x + 5y = 1$." This notation is called **set-builder notation.** It is convenient to use this notation when it is not possible to list all the elements of a set

Note

When a system has an infinite number of solutions, as in Example 3(b), either equation of the system could be used to write the solution set. *We prefer to use the equation in standard form with coefficients that are integers having greatest common factor 1.* If neither of the given equations of the system is in this form, use an *equivalent* equation that is in standard form with coefficients that are integers having greatest common factor 1 to write the solution set with set-builder notation.

Work Problem **3** *at the Side.* ▶

The system in Example 2 has exactly one solution. A system with at least one solution is called a **consistent system.** A system of equations with no solution, such as the one in Example 3(a), is called an **inconsistent system.** The equations in Example 2 are **independent equations** with different graphs. The equations of the system in Example 3(b) have the same graph and are equivalent. Because they are different forms of the same equation, these equations are called **dependent equations.**

Examples 2 and 3 show the three cases that may occur when solving a system of two equations with two variables.

Three Cases for Solutions of Systems

1. The graphs intersect at exactly one point, which gives the (single) ordered-pair solution of the system. The **system is consistent** and the **equations are independent.** See Figure 5(a).
2. The graphs are parallel lines, so there is no solution and the solution set is ∅. The **system is inconsistent** and the **equations are independent.** See Figure 5(b).
3. The graphs are the same line. There is an infinite number of solutions, and the solution set is written in set-builder notation as $\{(x, y) \mid$ _____ $\}$, where one of the equations is written after the | symbol. The **system is consistent** and the **equations are dependent.** See Figure 5(c).

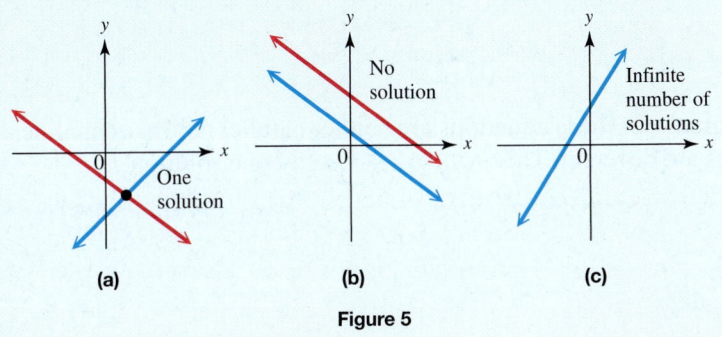

Figure 5

---

3 Solve each system of equations by graphing both equations on the same axes.

(a) $3x - y = 4$
$6x - 2y = 12$

(One of the lines is already graphed.)

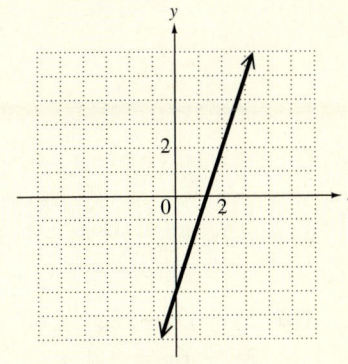

(b) $x - 3y = -2$
$2x - 6y = -4$

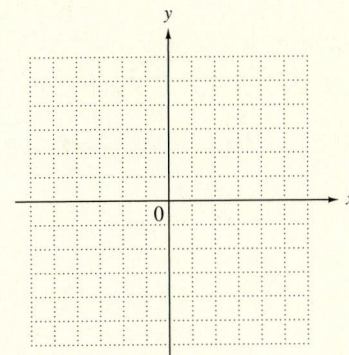

ANSWERS

3. **(a)** There is *no solution*, so write the solution set as ∅.
   (b) There is an *infinite number of solutions*, so use set-builder notation.
   $\{(x, y) \mid x - 3y = -2\}$

4 Describe each system without graphing. State the number of solutions.

(a) $2x - 3y = 5$
$3y = 2x - 7$

(b) $-x + 3y = 2$
$2x - 6y = -4$

(c) $6x + y = 3$
$2x - y = -11$

OBJECTIVE 4 Identify special systems without graphing. Example 3 showed that the graphs of an inconsistent system are parallel lines and the graphs of a system of dependent equations are the same line. We can recognize these special kinds of systems without graphing by using slopes.

EXAMPLE 4 Identifying the Three Cases by Using Slopes

Describe each system without graphing. State the number of solutions.

(a) $3x + 2y = 6$
$-2y = 3x - 5$
Write each equation in slope-intercept form, $y = mx + b$, by solving for y.

$3x + 2y = 6$	$-2y = 3x - 5$
$2y = -3x + 6$ Subtract $3x$.	
$y = -\dfrac{3}{2}x + 3$ Divide by 2.	$y = -\dfrac{3}{2}x + \dfrac{5}{2}$ Divide by -2.

Both equations have slope $-\frac{3}{2}$ but they have different y-intercepts, 3 and $\frac{5}{2}$. In **Section 11.3,** we found that lines with the same slope are parallel, so these equations have graphs that are parallel lines. Thus, the system has no solution.

(b) $2x - y = 4$
$x = \dfrac{y}{2} + 2$

Again, write the equations in slope-intercept form.

$2x - y = 4$	$x = \dfrac{y}{2} + 2$
$-y = -2x + 4$	$\dfrac{y}{2} + 2 = x$
$y = 2x - 4$	$\dfrac{y}{2} = x - 2$
	$y = 2x - 4$

The equations are exactly the same—their graphs are the same line. Thus, the system has an infinite number of solutions.

(c) $x - 3y = 5$
$2x + y = 8$
In slope-intercept form, the equations are as follows.

$x - 3y = 5$	$2x + y = 8$
$-3y = -x + 5$	$y = -2x + 8$
$y = \dfrac{1}{3}x - \dfrac{5}{3}$	

The graphs of these equations are neither parallel nor the same line, since the slopes are different. This system has exactly one solution.

◀ *Work Problem* **4** *at the Side.*

12.1 ▶▶▶ Exercises

1. Which ordered pair could be a solution of the system graphed? Why is it the only valid choice?

   A. $(2, 2)$

   B. $(-2, 2)$

   C. $(-2, -2)$

   D. $(2, -2)$

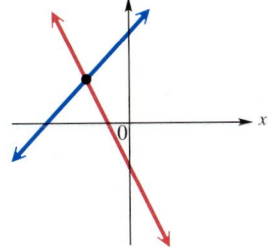

2. Which ordered pair could be a solution of the system graphed? Why is it the only valid choice?

   A. $(2, 0)$

   B. $(0, 2)$

   C. $(-2, 0)$

   D. $(0, -2)$

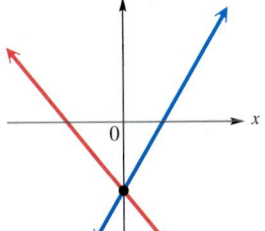

3. How can you tell without graphing that this system has no solution?

$$x + y = 2$$
$$x + y = 4$$

4. Explain why a system of two linear equations cannot have exactly two solutions.

Decide whether the given ordered pair is a solution of the given system. See Example 1.

5. $(2, -3)$
$$x + y = -1$$
$$2x + 5y = 19$$

6. $(4, 3)$
$$x + 2y = 10$$
$$3x + 5y = 3$$

7. $(-1, -3)$
$$3x + 5y = -18$$
$$4x + 2y = -10$$

8. $(-9, -2)$
$$2x - 5y = -8$$
$$3x + 6y = -39$$

9. $(7, -2)$
$$4x = 26 - y$$
$$3x = 29 + 4y$$

10. $(9, 1)$
$$2x = 23 - 5y$$
$$3x = 24 + 3y$$

11. $(6, -8)$
$$-2y = x + 10$$
$$3y = 2x + 30$$

12. $(-5, 2)$
$$5y = 3x + 20$$
$$3y = -2x - 4$$

Solve each system of equations by graphing. If the system is inconsistent or the equations are dependent, say so. See Examples 2 and 3.

13. $x - y = 2$
$x + y = 6$

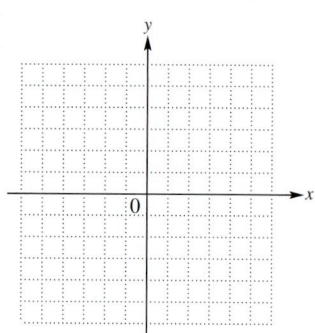

14. $x - y = 3$
$x + y = -1$

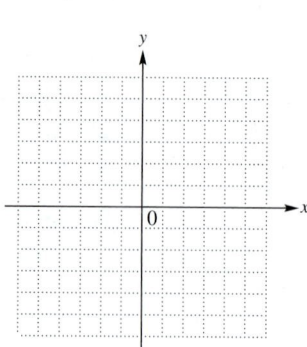

15. $x + y = 4$
$y - x = 4$

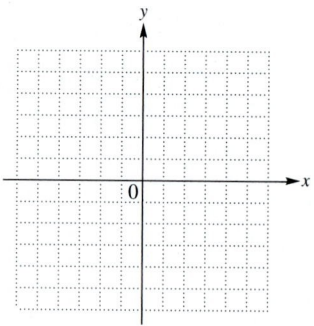

16. $x + y = -5$
$x - y = 5$

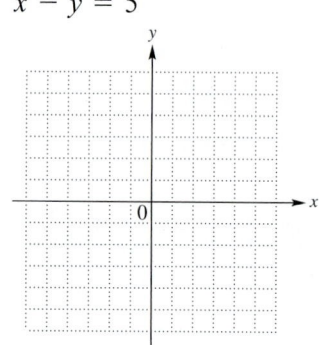

17. $x - 2y = 6$
$x + 2y = 2$

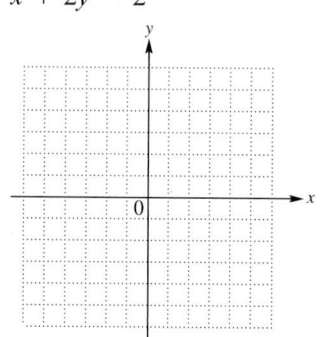

18. $2x - y = 4$
$4x + y = 2$

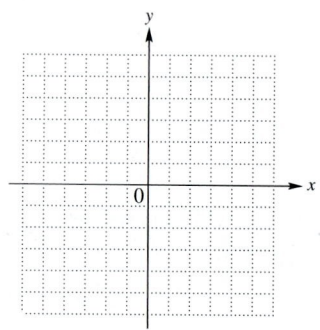

19. $3x - 2y = -3$
$-3x - y = -6$

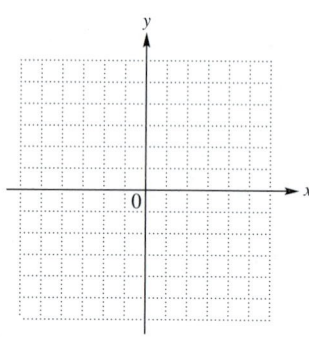

20. $2x - y = 4$
$2x + 3y = 12$

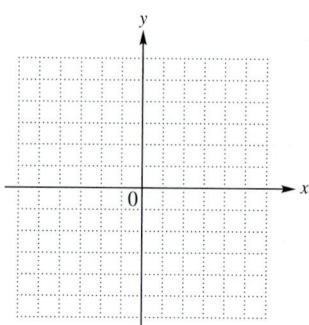

21. $2x - 3y = -6$
$y = -3x + 2$

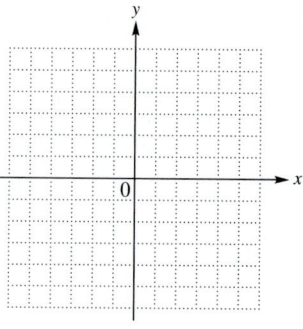

22. $-3x + y = -3$
$y = x - 3$

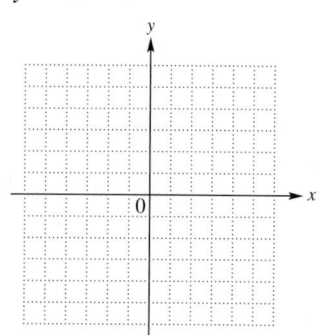

23. $x + 2y = 6$
$2x + 4y = 8$

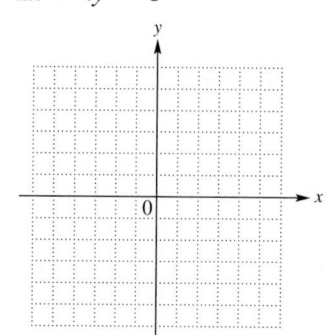

24. $2x - y = 6$
$6x - 3y = 12$

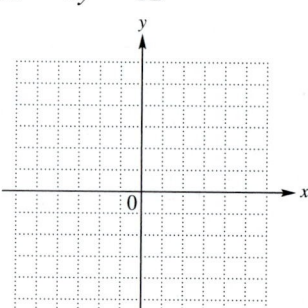

25. $4x - 2y = 8$
$2x = y + 4$

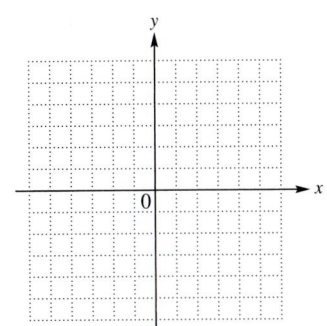

26. $3x = 5 - y$
$6x + 2y = 10$

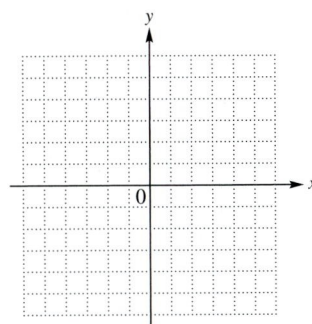

27. $3x - 4y = 24$
$y = -\dfrac{3}{2}x + 3$

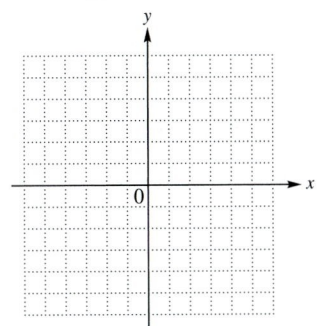

28. $3x - 2y = 12$
$y = -4x + 5$

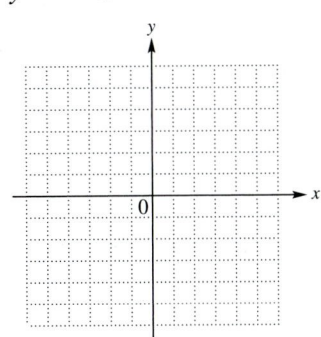

29. $3x = y + 5$
$6x - 5 = 2y$

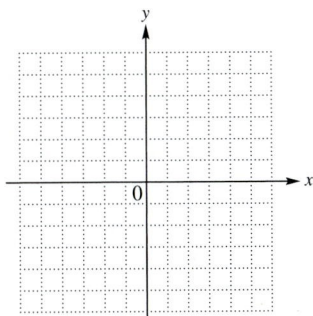

30. $2x = y - 4$
$4x - 2y = -4$

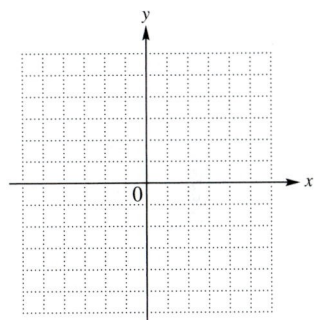

Without graphing, answer the following questions for each linear system. See Example 4.

(a) *Is the system inconsistent, are the equations dependent, or neither?*
(b) *Is the graph a pair of intersecting lines, a pair of parallel lines, or one line?*
(c) *Does the system have one solution, no solution, or an infinite number of solutions?*

31. $y - x = -5$
$x + y = 1$

32. $2x + y = 6$
$x - 3y = -4$

33. $x + 2y = 0$
$4y = -2x$

34. $y = 3x$
$y + 3 = 3x$

35. $5x + 4y = 7$
$10x + 8y = 4$

36. $4x - 6y = 10$
$-6x + 9y = -15$

The numbers of daily morning and evening newspapers in the United States in selected years are shown in the graph. Use the graph to work Exercises 37 and 38.

37. For which years were there more evening dailies than morning dailies?

38. Estimate the year in which the number of evening and morning dailies was closest to the same. About how many newspapers of each type were there in that year?

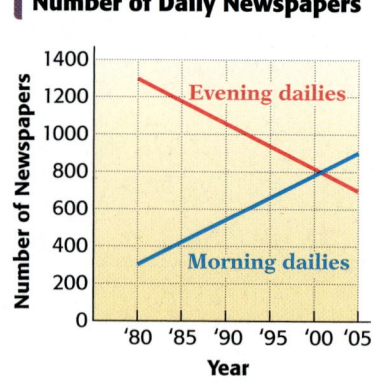

Number of Daily Newspapers

Source: Editor & Publisher International Year Book.

Work Exercises 39 and 40 using the graphs provided.

39. The graph shows how college students managed their money during the period 1997–2004.

   (a) During what period did ATM use dominate both credit card *and* debit card use?

   (b) In what year did debit card use overtake credit card use?

   (c) In what year did debit card use overtake ATM use?

   (d) Write an ordered pair for the debit card use data in the year 2004.

   (e) Describe the trend in debit card use over this period.

How College Students Manage Their Money

Source: Georgetown University Credit Research Center.

40. The graph shows how the average viewing hours for broadcast TV and cable/satellite TV in the United States has changed during the period 1998–2004.

   (a) In approximately what year did Americans spend almost the same number of hours watching broadcast and cable/satellite TV? How many hours per year was this?

   (b) Express the point of intersection of the two graphs as an ordered pair of the form (year, hours).

   (c) During what period was the time spent watching broadcast TV almost constant?

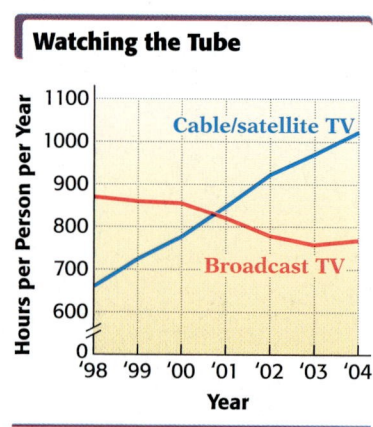

Watching the Tube

Source: Veronis Suhler Stevenson.

12.2 ▶▶▶ Solving Systems of Linear Equations by Substitution

OBJECTIVE 1 **Solve linear systems by substitution.**

Work Problem (**1**) *at the Side.* ▶

As we saw in Problem 1 at the side, graphing to solve a system of equations has a serious drawback: It is difficult to accurately find a solution such as $\left(\frac{11}{3}, -\frac{4}{9}\right)$ from a graph. One algebraic method for solving a system of equations is the **substitution method.**

OBJECTIVES

1 **Solve linear systems by substitution.**

2 **Solve special systems by substitution.**

3 **Solve linear systems with fractions and decimals by substitution.**

EXAMPLE 1 **Using the Substitution Method**

Solve the system by the substitution method.

$$3x + 5y = 26$$
$$y = 2x$$

The second equation is already solved for y. This equation says that $y = 2x$. Substituting $2x$ for y in the first equation gives

$$3x + 5y = 26$$
$$3x + 5(2x) = 26 \qquad \text{Let } y = 2x.$$
$$3x + 10x = 26 \qquad \text{Multiply.}$$
$$13x = 26 \qquad \text{Combine like terms.}$$
$$x = 2. \qquad \text{Divide by 13.}$$

Because $x = 2$, we find y from the equation $y = 2x$ by substituting 2 for x.

$$y = 2(2) = 4 \qquad \text{Let } x = 2.$$

Check that the solution set of the given system is $\{(2, 4)\}$ by substituting 2 for x and 4 for y in *both* equations.

Work Problem (**2**) *at the Side.* ▶

EXAMPLE 2 **Using the Substitution Method**

Solve the system by the substitution method.

$$2x + 5y = 7$$
$$x = -1 - y$$

The second equation gives x in terms of y. Substitute $-1 - y$ for x in the first equation.

$$2x + 5y = 7$$
$$2(-1 - y) + 5y = 7 \qquad \text{Let } x = -1 - y.$$
$$-2 - 2y + 5y = 7 \qquad \text{Distributive property}$$
$$-2 + 3y = 7 \qquad \text{Combine like terms.}$$
$$3y = 9 \qquad \text{Add 2.}$$
$$y = 3 \qquad \text{Divide by 3.}$$

> Distribute 2 to *both* −1 and −*y*.

To find x, substitute 3 for y in the equation $x = -1 - y$ to get

$$x = -1 - 3 = -4.$$

> Write the *x*-coordinate first.

Check that the solution set of the given system is $\{(-4, 3)\}$.

Work Problem (**3**) *at the Side.* ▶

---

1 Solve the system by graphing.

$$2x + 3y = 6$$
$$x - 3y = 5$$

Can you determine the answer? Why or why not?

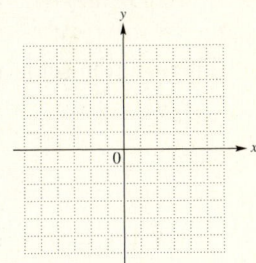

2 Fill in the blanks to solve by the substitution method. Check your solution.

$$3x + 5y = 69$$
$$y = 4x$$
$$3x + 5(\underline{}) = 69$$
$$\underline{} = 69$$
$$x = \underline{}$$
$$y = 4(\underline{}) = \underline{}$$

The solution set is _____.

3 Solve by the substitution method. Check your solution.

$$2x + 7y = -12$$
$$x = 3 - 2y$$

ANSWERS

1. The answer cannot be determined from the graph because it is too difficult to read the exact coordinates.
2. $4x$; $23x$; 3; 3; 12; $\{(3, 12)\}$
3. $\{(15, -6)\}$

4 Solve each system by substitution. Check each solution.

(a) Fill in the blanks to solve

$$x + 4y = -1$$
$$2x - 5y = 11.$$

Solve the first equation for x.

$$x = -1 - \underline{\quad}$$

Substitute into the second equation to find y.

$$2(\underline{\quad}) - 5y = 11$$
$$-2 - 8y - 5y = 11$$
$$-2 - \underline{\quad}y = 11$$
$$\underline{\quad}y = 13$$
$$y = \underline{\quad}$$

Find x.

$$x = -1 - \underline{\quad}$$
$$x = \underline{\quad}$$

The solution set is $\underline{\quad}$.

(b) $2x + 5y = 4$
    $x + y = -1$

CAUTION

Even though we found y first in Example 2, *the x-coordinate is always written first in the ordered-pair solution of a system.*

To solve a system by substitution, follow these steps.

Solving a Linear System by Substitution

Step 1 **Solve one equation for either variable.** If one of the variables has coefficient 1 or −1, choose it, since it usually makes the substitution easier.

Step 2 **Substitute** for that variable in the other equation. The result should be an equation with just one variable.

Step 3 **Solve** the equation from Step 2.

Step 4 **Substitute** the result from Step 3 into the equation from Step 1 to find the value of the other variable.

Step 5 **Check** the solution in *both* of the original equations. Then write the solution set.

EXAMPLE 3   Using the Substitution Method

Use substitution to solve the system.

$$2x = 4 - y \qquad (1)$$
$$5x + 3y = 10 \qquad (2)$$

Step 1 For the substitution method, we must solve one of the equations for either x or y. Because the coefficient of y in equation (1) is −1, we choose equation (1) and solve for y.

$$2x = 4 - y \qquad (1)$$
$$y + 2x = 4 \qquad \text{Add } y.$$
$$y = -2x + 4 \qquad \text{Subtract } 2x.$$

Step 2 Now substitute $-2x + 4$ for y in equation (2).

$$5x + 3y = 10 \qquad (2)$$
$$5x + 3(-2x + 4) = 10 \qquad \text{Let } y = -2x + 4.$$

Step 3 Now solve the equation from Step 2.

$$5x - 6x + 12 = 10 \qquad \text{Distributive property}$$
$$-x + 12 = 10 \qquad \text{Combine like terms.}$$
$$-x = -2 \qquad \text{Subtract } 12.$$
$$x = 2 \qquad \text{Multiply by } -1.$$

Distribute 3 to both −2x and 4.

Step 4 Since $y = -2x + 4$ and $x = 2$, $y = -2(2) + 4 = 0$.

Step 5 Check that $(2, 0)$ is the solution.

Check

$$2x = 4 - y \qquad (1) \qquad\qquad 5x + 3y = 10 \qquad (2)$$
$$2(2) \overset{?}{=} 4 - 0 \qquad\qquad\qquad 5(2) + 3(0) \overset{?}{=} 10$$
$$4 = 4 \qquad \text{True} \qquad\qquad\qquad 10 = 10 \qquad \text{True}$$

Since both results are true, the solution set of the system is $\{(2, 0)\}$.

◀ *Work Problem* **4** *at the Side.*

EXAMPLE 4 **Using the Substitution Method**

Use substitution to solve the system.

$$2x + 3y = 10 \quad (1)$$
$$-3x - 2y = 0 \quad (2)$$

Step 1 To use the substitution method, we must solve one of the equations for one of the variables. We choose equation (1) and solve for x.

$$2x + 3y = 10 \qquad (1)$$
$$2x = 10 - 3y \qquad \text{Subtract } 3y.$$
$$x = 5 - \frac{3}{2}y \qquad \text{Divide by 2.}$$

Step 2 Substitute this expression for x in equation (2).

$$-3x - 2y = 0 \qquad (2)$$
$$-3\left(5 - \frac{3}{2}y\right) - 2y = 0 \qquad \text{Let } x = 5 - \frac{3}{2}y.$$

Step 3
$$-15 + \frac{9}{2}y - 2y = 0 \qquad \text{Distributive property}$$

$$\boxed{\begin{aligned}-3\left(-\tfrac{3}{2}\right) &= \left(-\tfrac{3}{1}\right)\left(-\tfrac{3}{2}\right) \\ &= \tfrac{9}{2}\end{aligned}}$$

$$-15 + \frac{5}{2}y = 0 \qquad \text{Combine like terms.}$$
$$\frac{5}{2}y = 15 \qquad \text{Add 15.}$$

$$\boxed{\begin{aligned}\tfrac{2}{5}\left(\tfrac{5}{2}y\right) &= \left(\tfrac{2}{5}\cdot\tfrac{5}{2}\right)y \\ &= 1\cdot y = y\end{aligned}} \qquad y = \frac{30}{5} = 6 \qquad \text{Multiply by } \tfrac{2}{5}.$$

Step 4 Find x by substituting **6** for y in $x = 5 - \frac{3}{2}y$.

$$x = 5 - \frac{3}{2}(6) = -4$$

Step 5 Check that $(-4, 6)$ is the solution.

Check
$$\begin{array}{ll|ll}
2x + 3y = 10 & (1) & -3x - 2y = 0 & (2) \\
2(-4) + 3(6) \stackrel{?}{=} 10 & & -3(-4) - 2(6) \stackrel{?}{=} 0 & \\
-8 + 18 \stackrel{?}{=} 10 & & 12 - 12 \stackrel{?}{=} 0 & \\
10 = 10 & \text{True} & 0 = 0 & \text{True}
\end{array}$$

Both results are true, so the solution set of the system is $\{(-4, 6)\}$.

> **Note**
>
> In Example 4, we could have started the solution by solving the second equation for either x or y and then substituting the result into the first equation. The solution would be the same.

Work Problem **5** *at the Side.*

---

5 Solve the system by substitution. Check your solution.

$$3x + 2y = 1$$
$$3x - 4y = -11$$

OBJECTIVE **2** **Solve special systems by substitution.** We can solve inconsistent systems with graphs that are parallel lines and systems of dependent equations with graphs that are the same line using the substitution method.

EXAMPLE 5 **Solving an Inconsistent System by Substitution**

Use substitution to solve the system.

$$x = 5 - 2y \quad (1)$$
$$2x + 4y = 6 \quad (2)$$

Substitute $5 - 2y$ for x in equation (2).

$$2x + 4y = 6 \quad (2)$$
$$2(5 - 2y) + 4y = 6 \qquad \text{Let } x = 5 - 2y.$$
$$10 - 4y + 4y = 6 \qquad \text{Distributive property}$$
$$10 = 6 \qquad \text{False}$$

This false result means that the equations in the system have graphs that are parallel lines. The system is inconsistent, and the solution set is ∅. See Figure 6.

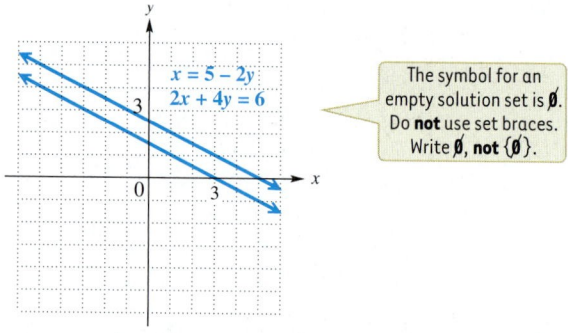

The symbol for an empty solution set is ∅. Do **not** use set braces. Write ∅, **not** {∅}.

Figure 6

CAUTION
It is a common error to give "false" as the answer to an inconsistent system. The correct response is ∅.

EXAMPLE 6 **Solving a System with Dependent Equations by Substitution**

Solve the system by the substitution method.

$$3x - y = 4 \quad (1)$$
$$-9x + 3y = -12 \quad (2)$$

Begin by solving equation (1) for y to get $y = 3x - 4$. Substitute $3x - 4$ for y in equation (2) and solve the resulting equation.

$$-9x + 3y = -12 \quad (2)$$
$$-9x + 3(3x - 4) = -12 \qquad \text{Let } y = 3x - 4.$$
$$-9x + 9x - 12 = -12 \qquad \text{Distributive property}$$
$$0 = 0 \qquad \text{Add 12; combine like terms.}$$

Continued on Next Page

This true result means that every solution of one equation is also a solution of the other, so the system has an infinite number of solutions—all the ordered pairs corresponding to points that lie on the common graph. The solution set is $\{(x, y) \mid 3x - y = 4\}$. A graph of the equations of this system is shown in Figure 7.

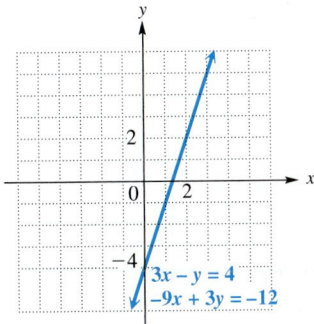

Figure 7

CAUTION

It is a common error to give "true" as the solution of a system of dependent equations. Remember that we give the solution set in set-builder notation using an equation that is in standard form, with integer coefficients having greatest common factor 1.

Work Problem **6** *at the Side.* ▶

OBJECTIVE **3** **Solve linear systems with fractions and decimals by substitution.** When a system includes an equation with fractions as coefficients, eliminate the fractions by multiplying each side of the equation by a common denominator. Then solve the resulting system.

EXAMPLE 7 **Using the Substitution Method with Fractions as Coefficients**

Solve the system by the substitution method.

$$3x + \frac{1}{4}y = 2 \qquad (1)$$

$$\frac{1}{2}x + \frac{3}{4}y = -\frac{5}{2} \qquad (2)$$

Clear equation (1) of fractions by multiplying each side by 4.

$$4\left(3x + \frac{1}{4}y\right) = 4(2) \qquad \text{Multiply by 4.}$$

$$4(3x) + 4\left(\frac{1}{4}y\right) = 4(2) \qquad \text{Distributive property}$$

$$12x + y = 8 \qquad (3)$$

Continued on Next Page

---

6 Solve each system by substitution.

(a) $8x - y = 4$
$y = 8x + 4$

(b) $\quad 7x - 6y = 10$
$\quad -14x + 20 = -12y$

---

7 Solve the system by substitution. First clear all fractions.

$$\frac{2}{3}x + \frac{1}{2}y = 6$$

$$\frac{1}{2}x - \frac{3}{4}y = 0$$

Now clear equation (2) of fractions by multiplying each side by the common denominator 4.

$$\frac{1}{2}x + \frac{3}{4}y = -\frac{5}{2} \qquad (2)$$

$$4\left(\frac{1}{2}x + \frac{3}{4}y\right) = 4\left(-\frac{5}{2}\right) \qquad \text{Multiply by 4.}$$

$$4\left(\frac{1}{2}x\right) + 4\left(\frac{3}{4}y\right) = 4\left(-\frac{5}{2}\right) \qquad \text{Distributive property}$$

$$2x + 3y = -10 \qquad (4)$$

The given system of equations has been simplified to the equivalent system

$$12x + y = 8 \qquad (3)$$

$$2x + 3y = -10. \qquad (4)$$

To solve this system by substitution, equation (3) can be solved for y.

$$12x + y = 8 \qquad (3)$$

$$y = -12x + 8 \qquad \text{Subtract } 12x.$$

Now substitute this result for y in equation (4).

$$2x + 3y = -10 \qquad (4)$$

$$2x + 3(-12x + 8) = -10 \qquad \text{Let } y = -12x + 8.$$

$$2x - 36x + 24 = -10 \qquad \text{Distributive property}$$

> Distribute 3 to both $-12x$ and 8.

$$-34x = -34 \qquad \text{Combine like terms; subtract 24.}$$

$$x = 1 \qquad \text{Divide by } -34.$$

8 Complete the process of solving the system given in Example 8 by substitution. (*Hint:* Solve equation (4) for x. Then substitute this result for x in equation (3) to find y.)

Substitute 1 for x in $y = -12x + 8$ to get

$$y = -12(1) + 8 = -4.$$

Check by substituting 1 for x and -4 for y in both of the original equations. The solution set is $\{(1, -4)\}$.

◀ *Work Problem* **7** *at the Side.*

If any of the coefficients in the equations of a system are decimals, we can eliminate the decimals by multiplying by a power of 10, as we did when solving linear equations with decimal coefficients in **Section 10.2.**

EXAMPLE 8 **Using the Substitution Method with Decimals as Coefficients**

Solve the system by the substitution method.

$$0.5x + 2.4y = 4.2 \qquad (1)$$

$$-0.1x + 1.5y = 5.1 \qquad (2)$$

To eliminate (or "clear") decimals, multiply each equation by 10.

$$5x + 24y = 42 \qquad (3)$$

$$-x + 15y = 51 \qquad (4) \qquad (-0.1x) \cdot 10 = -1x = -x$$

Now we can solve this equivalent system by substitution.

◀ *Work problem* **8** *at the Side.*

12.2 ▶▶▶ Exercises

1. A student solves the system

$$5x - y = 15$$
$$7x + y = 21$$

and finds that $x = 3$, which is the correct value for x. The student gives the solution set as $\{3\}$. ***WHAT WENT WRONG?***

2. When you use the substitution method, how can you tell that a system has

(a) no solution? **(b)** an infinite number of solutions?

Solve each system by the substitution method. Check each solution. See Examples 1–8.

3. $x + y = 12$
   $y = 3x$

4. $x + 3y = -28$
   $y = -5x$

5. $3x + 2y = 27$
   $x = y + 4$

6. $4x + 3y = -5$
   $x = y - 3$

7. $3x + 5y = 14$
   $x - 2y = -10$

8. $5x + 2y = -1$
   $2x - y = -13$

9. $3x + 4 = -y$
   $2x + y = 0$

10. $2x - 5 = -y$
   $x + 3y = 0$

11. $7x + 4y = 13$
   $x + y = 1$

12. $3x - 2y = 19$
   $x + y = 8$

13. $3x - y = 5$
   $y = 3x - 5$

14. $4x - y = -3$
   $y = 4x + 3$

15. $6x - 8y = 6$
   $2y = -2 + 3x$

16. $3x + 2y = 6$
   $6x = 8 + 4y$

17. $2x + 8y = 3$
   $x = 8 - 4y$

18. $2x + 10y = 3$
$x = 1 - 5y$

19. $12x - 16y = 8$
$3x = 4y + 2$

20. $6x + 9y = 6$
$2x = 2 - 3y$

21. $\dfrac{1}{5}x + \dfrac{2}{3}y = -\dfrac{8}{5}$
$3x - y = 9$

22. $\dfrac{1}{3}x - \dfrac{1}{2}y = \dfrac{1}{6}$
$3x - 2y = 9$

23. $\dfrac{x}{2} - \dfrac{y}{3} = 9$
$\dfrac{x}{5} - \dfrac{y}{4} = 5$

24. $\dfrac{1}{6}x + \dfrac{1}{6}y = 2$
$-\dfrac{1}{2}x - \dfrac{1}{3}y = -8$

25. $\dfrac{x}{5} + 2y = \dfrac{16}{5}$
$\dfrac{3x}{5} + \dfrac{y}{2} = -\dfrac{7}{5}$

26. $\dfrac{x}{3} - \dfrac{3y}{4} = -\dfrac{1}{2}$
$\dfrac{x}{6} + \dfrac{y}{8} = \dfrac{3}{4}$

27. $0.1x + 0.9y = -2$
$0.5x - 0.2y = 4.1$

28. $0.2x - 1.3y = -3.2$
$-0.1x + 2.7y = 9.8$

29. $0.08x - 0.01y = 1.3$
$0.22x + 0.15y = 8.9$

Relating Concepts (Exercises 30–33)   For Individual or Group Work

A system of linear equations can be used to model the cost and the revenue of a business. **Work Exercises 30–33 in order.**

30. Suppose that you start a business manufacturing and selling bicycles, and it costs you \$5000 to get started. You determine that each bicycle will cost \$400 to manufacture. Explain why the linear equation $y_1 = 400x + 5000$ gives your *total* cost to manufacture x bicycles (y_1 in dollars).

31. You decide to sell each bike for \$600. What expression in x represents the revenue you will take in if you sell x bikes? Write an equation using y_2 to express your revenue when you sell x bikes (y_2 in dollars).

32. Form a system from the two equations in Exercises 30 and 31, and then solve the system, assuming $y_1 = y_2$, that is, cost = revenue.

33. The value of x from Exercise 32 is the number of bikes it takes to *break even*. Fill in the blanks: When _____ bikes are sold, the break-even point is reached. At that point, you have spent _____ dollars and taken in _____ dollars.

12.3 ▶▶▶ Solving Systems of Linear Equations by Elimination

OBJECTIVE 1 Solve linear systems by elimination. An algebraic method that depends on the addition property of equality can be used to solve systems. As mentioned earlier, adding the same quantity to each side of an equation results in equal sums.

$$\text{If} \quad A = B, \quad \text{then} \quad A + C = B + C.$$

This addition can be taken a step further. Adding *equal* quantities, rather than the *same* quantity, to both sides of an equation also results in equal sums.

$$\text{If} \quad A = B \quad \text{and} \quad C = D, \quad \text{then} \quad A + C = B + D.$$

Using the addition property to solve systems is called the **elimination method.** When using this method, the idea is to *eliminate* one of the variables. *To do this, one of the variables in the two equations must have coefficients that are opposites.*

OBJECTIVES

1 Solve linear systems by elimination.

2 Multiply when using the elimination method.

3 Use an alternative method to find the second value in a solution.

4 Use the elimination method to solve special systems.

EXAMPLE 1 Using the Elimination Method

Use the elimination method to solve the system.

$$x + y = 5$$
$$x - y = 3$$

Each equation in this system is a statement of equality, so the sum of the left sides equals the sum of the right sides. Adding in this way gives

$$(x + y) + (x - y) = 5 + 3.$$

Combine like terms and simplify to get

$$2x = 8$$
$$x = 4. \quad \text{Divide by 2.}$$

Notice that y has been eliminated. The result, $x = 4$, gives the x-value of the solution of the given system. To find the y-value of the solution, substitute 4 for x in either of the two equations of the system.

Work Problem **1** *at the Side.* ▶

Check the solution set found at the side, $\{(4, 1)\}$, by substituting 4 for x and 1 for y in both equations of the given system.

Check

$x + y = 5$	$x - y = 3$
$4 + 1 \overset{?}{=} 5$	$4 - 1 \overset{?}{=} 3$
$5 = 5$ True	$3 = 3$ True

Since both results are true, the solution set of the system is $\{(4, 1)\}$.

1 (a) Substitute 4 for x in the equation $x + y = 5$ to find the value of y.

(b) Give the solution set of the system.

CAUTION

A system is not completely solved until values for both x and y are found. Do not stop after finding the value of only one variable. Remember to write the solution set as a set containing an ordered pair.

◀ Work Problem **2** at the Side.

2 Solve each system by the elimination method. Check each solution.

(a) Fill in the blanks to solve the following system.

$$x + y = 8$$
$$x - y = 2$$

Add.

$$(x + y) + (x - y) = 8 + \underline{\quad}$$
$$2 \underline{\quad} = \underline{\quad}$$
$$x = \underline{\quad}$$

Find y.

$$x - y = 2$$
$$\underline{\quad} - y = 2$$
$$-y = \underline{\quad}$$
$$y = \underline{\quad}$$

The solution set is ___.

(b) $3x - y = 7$
$2x + y = 3$

In general, to solve a system by elimination, follow these steps.

Solving a Linear System by Elimination

Step 1 **Write both equations in standard form** $Ax + By = C$.

Step 2 **Transform so that the coefficients of one pair of variable terms are opposites.** Multiply one or both equations by appropriate numbers so that the sum of the coefficients of either the x- or y-terms is 0.

Step 3 **Add** the new equations to eliminate a variable. The sum should be an equation with just one variable.

Step 4 **Solve** the equation from Step 3 for the remaining variable.

Step 5 **Substitute** the result from Step 4 into *either* of the original equations and solve for the other variable.

Step 6 **Check** the solution in *both* of the original equations. Then write the solution set.

It does not matter which variable is eliminated first. Usually we choose the one that is more convenient to work with.

EXAMPLE 2 Using the Elimination Method

Solve the system.

$$y + 11 = 2x$$
$$5x = y + 26$$

Step 1 Rewrite both equations in the form $Ax + By = C$ to get the system

$$-2x + y = -11 \qquad \text{Subtract } 2x \text{ and } 11.$$
$$5x - y = 26. \qquad \text{Subtract } y.$$

Step 2 Because the coefficients of y are 1 and -1, adding will eliminate y. It is not necessary to multiply either equation by a number.

Step 3 Add the two equations. This time we use vertical addition.

$$\begin{array}{rcr} -2x + y &=& -11 \\ 5x - y &=& 26 \\ \hline 3x &=& 15 \end{array} \qquad \text{Add in columns.}$$

Step 4 Solve the equation.

$$3x = 15$$

Don't stop here. → $x = 5$ Divide by 3.

Step 5 Find the value of y by substituting 5 for x in either of the original equations. Choosing the first equation gives

$$y + 11 = 2x$$
$$y + 11 = 2(5) \qquad \text{Let } x = 5.$$
$$y + 11 = 10$$
$$y = -1. \qquad \text{Subtract 11.}$$

Continued on Next Page

Step 6 Check the solution by substituting $x = 5$ and $y = -1$ into both of the original equations.

Check

$$y + 11 = 2x \qquad\qquad 5x = y + 26$$
$$-1 + 11 \stackrel{?}{=} 2(5) \qquad\qquad 5(5) \stackrel{?}{=} -1 + 26$$
$$10 = 10 \quad \text{True} \qquad\qquad 25 = 25 \quad \text{True}$$

Since $(5, -1)$ is a solution of *both* equations, the solution set is $\{(5, -1)\}$.

Work Problem **3** *at the Side.* ▶

OBJECTIVE 2 Multiply when using the elimination method. Sometimes we need to multiply each side of one or both equations in a system by some number before adding the equations will eliminate a variable.

EXAMPLE 3 Multiplying Both Equations When Using the Elimination Method

Solve the system.

$$2x + 3y = -15 \qquad (1)$$
$$5x + 2y = 1 \qquad (2)$$

Adding the two equations gives $7x + 5y = -14$, which does not eliminate either variable. However, we can multiply each equation by a suitable number so that the coefficients of one of the two variables are opposites. For example, to eliminate x, multiply each side of equation (1) by 5, and each side of equation (2) by -2.

$$
\begin{array}{ll}
10x + 15y = -75 & \text{Multiply equation (1) by 5.} \\
-10x - 4y = -2 & \text{Multiply equation (2) by } -2. \\
\hline
11y = -77 & \text{Add.} \\
y = -7 & \text{Divide by 11.}
\end{array}
$$

Substituting -7 for y in either equation (1) or (2) gives $x = 3$. Check that the solution set of the system is $\{(3, -7)\}$.

Work Problem **4** *at the Side.* ▶

OBJECTIVE 3 Use an alternative method to find the second value in a solution. Sometimes it is easier to find the value of the second variable in a solution by using the elimination method twice.

EXAMPLE 4 Finding the Second Value Using an Alternative Method

Solve the system.

$$4x = 9 - 3y \qquad (1)$$
$$5x - 2y = 8 \qquad (2)$$

Rearrange the terms in equation (1) so that like terms are aligned in columns. To do this, add $3y$ to each side to get the following system.

$$4x + 3y = 9 \qquad (3)$$
$$5x - 2y = 8 \qquad (2)$$

One way to proceed is to eliminate y by multiplying each side of equation (3) by 2 and each side of equation (2) by 3, and then adding.

— **Continued on Next Page**

---

3 Solve each system by the elimination method. Check each solution.

(a) $2x - y = 2$
$4x + y = 10$

(b) $8x - 5y = 32$
$4x + 5y = 4$

4 (a) Solve the system in Example 3 by first eliminating the variable y. Check your solution.

(b) Solve

$$6x + 7y = 4$$
$$5x + 8y = -1,$$

and check your solution.

ANSWERS

3. (a) $\{(2, 2)\}$ **(b)** $\left\{\left(3, -\dfrac{8}{5}\right)\right\}$

4. (a) $\{(3, -7)\}$ **(b)** $\{(3, -2)\}$

5 Solve each system of equations.

(a) $5x = 7 + 2y$
$5y = 5 - 3x$

(b) $3y = 8 + 4x$
$6x = 9 - 2y$

$$\begin{array}{ll} 8x + 6y = 18 & \text{Multiply equation (3) by 2.} \\ \underline{15x - 6y = 24} & \text{Multiply equation (2) by 3.} \\ 23x \qquad = 42 & \text{Add.} \\ \qquad x = \dfrac{42}{23} & \text{Divide by 23.} \end{array}$$

Substituting $\frac{42}{23}$ for x in one of the given equations would give y, but the arithmetic involved would be messy. Instead, solve for y by starting again with the original equations and eliminating x. Multiply each side of equation (3) by 5 and each side of equation (2) by -4, and then add.

$$\begin{array}{ll} \mathbf{20x} + 15y = 45 & \text{Multiply equation (3) by 5.} \\ \underline{\mathbf{-20x} + 8y = -32} & \text{Multiply equation (2) by } -4. \\ 23y = 13 & \text{Add.} \\ \qquad y = \dfrac{13}{23} & \text{Divide by 23.} \end{array}$$

Check that the solution set is $\{ (\frac{42}{23}, \frac{13}{23}) \}$.

◀ *Work Problem* **5** *at the Side.*

When the value of the first variable is a fraction, the method used in Example 4 helps avoid arithmetic errors. Of course, this method could be used to solve any system of equations.

6 Solve each system by the elimination method.

(a) $4x + 3y = 10$

$2x + \dfrac{3}{2}y = 12$

(b) $4x - 6y = 10$
$-10x + 15y = -25$

OBJECTIVE **4** **Use the elimination method to solve special systems.**

EXAMPLE 5 **Using the Elimination Method for an Inconsistent System or Dependent Equations**

Solve each system by the elimination method.

(a) $2x + 4y = 5$

$4x + 8y = -9$

Multiply each side of $2x + 4y = 5$ by -2; then add to $4x + 8y = -9$.

$$\begin{array}{l} -4x - 8y = -10 \\ \underline{4x + 8y = -9} \\ \qquad 0 = -19 \qquad \text{False} \end{array}$$

Write $\emptyset$, not $\{\emptyset\}$.

The false statement $0 = -19$ indicates that the solution set is $\emptyset$.

(b) $3x - y = 4$

$-9x + 3y = -12$

Multiply each side of the first equation by 3; then add the two equations.

$$\begin{array}{l} 9x - 3y = 12 \\ \underline{-9x + 3y = -12} \\ \qquad 0 = 0 \qquad \text{True} \end{array}$$

A true statement occurs when the equations are equivalent. As before, this indicates that every solution of one equation is also a solution of the other. The solution set is $\{(x, y) \mid 3x - y = 4\}$. (See **Section 12.2**, Example 6, where the same system was solved using substitution.)

◀ *Work Problem* **6** *at the Side.*

ANSWERS

5. (a) $\left\{ \left(\dfrac{45}{31}, \dfrac{4}{31} \right) \right\}$ (b) $\left\{ \left(\dfrac{11}{26}, \dfrac{42}{13} \right) \right\}$

6. (a) $\emptyset$ (b) $\{(x, y) \mid 2x - 3y = 5\}$

In Exercises 1–4, answer true *or* false *for each statement. If* false, *tell why.*

1. The ordered pair $(0, 0)$ *must* be a solution of a system of the form

$$Ax + By = 0$$
$$Cx + Dy = 0.$$

2. To eliminate the y-terms in the system

$$2x + 12y = 7$$
$$3x + 4y = 1,$$

we should multiply the bottom equation by 3 and then add.

3. The system

$$x + y = 1$$
$$x + y = 2$$

has $\emptyset$ as its solution set.

4. The ordered pair $(4, -5)$ cannot be a solution of a system that contains the equation $5x - 4y = 0$.

Solve each system by the elimination method. Check each solution. See Examples 1 and 2.

5. $x + y = 2$
   $2x - y = -5$

6. $3x - y = -12$
   $x + y = 4$

7. $2x + y = -5$
   $x - y = 2$

8. $2x + y = -15$
   $-x - y = 10$

9. $3x + 2y = 0$
   $-3x - y = 3$

10. $5x - y = 5$
   $-5x + 2y = 0$

11. $6x - y = -1$
   $5y = 17 + 6x$

12. $y = 9 - 6x$
   $-6x + 3y = 15$

Solve each system by the elimination method. Check each solution. See Examples 3–5.

13. $2x - y = 12$
$3x + 2y = -3$

14. $x + y = 3$
$-3x + 2y = -19$

15. $x + 3y = 19$
$2x - y = 10$

16. $4x - 3y = -19$
$2x + y = 13$

17. $x + 4y = 16$
$3x + 5y = 20$

18. $2x + y = 8$
$5x - 2y = -16$

19. $5x - 3y = -20$
$-3x + 6y = 12$

20. $4x + 3y = -28$
$5x - 6y = -35$

21. $2x - 8y = 0$
$4x + 5y = 0$

22. $3x - 15y = 0$
$6x + 10y = 0$

23. $x + y = 7$
$x + y = -3$

24. $x - y = 4$
$x - y = -3$

25. $-x + 3y = 4$
$-2x + 6y = 8$

26. $6x - 2y = 24$
$-3x + y = -12$

27. $4x - 3y = -19$
$3x + 2y = 24$

28. $5x + 4y = 12$
$3x + 5y = 15$

29. $3x - 7 = -5y$
$5x + 4y = -10$

30. $2x + 3y = 13$
$6 + 2y = -5x$

31. $2x + 3y = 0$
$4x + 12 = 9y$

32. $-4x + 3y = 2$
$5x + 3 = -2y$

33. $24x + 12y = -7$
$16x - 17 = 18y$

34. $9x + 4y = -3$
$6x + 7 = -6y$

35. $3x = 3 + 2y$
$-\dfrac{4}{3}x + y = \dfrac{1}{3}$

36. $3x = 27 + 2y$
$x - \dfrac{7}{2}y = -25$

37. $5x - 2y = 3$
$10x - 4y = 5$

38. $3x - 5y = 1$
$6x - 10y = 4$

39. $6x + 3y = 0$
$-18x - 9y = 0$

40. $3x - 5y = 0$
$9x - 15y = 0$

Relating Concepts (Exercises 41–46) For Individual or Group Work

Attending the movies is one of America's favorite forms of entertainment. The graph shows U.S. movie attendance from 1996 through 2004. In 1996, attendance was 1339 million, as represented by the point P(1996, 1339). In 2004, attendance was 1536 million, as represented by the point Q(2004, 1536). We can find an equation of line segment PQ by using a system of equations. Then we use the equation we found to approximate the attendance in any of the years between 1996 and 2004. **Work Exercises 41–46 in order.**

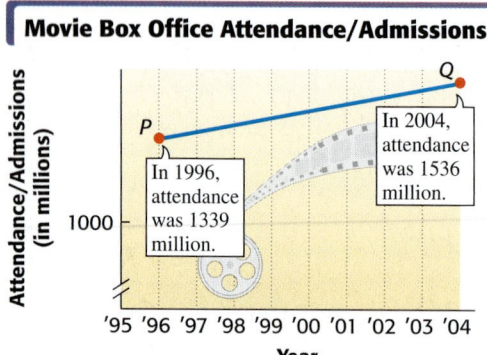

Movie Box Office Attendance/Admissions

In 1996, attendance was 1339 million.

In 2004, attendance was 1536 million.

Source: Motion Picture Association of America.

41. The line segment has an equation that can be written in the form $y = ax + b$. Using the coordinates of point P with $x = 1996$ and $y = 1339$, write an equation in the variables a and b.

42. Using the coordinates of point Q with $x = 2004$ and $y = 1536$, write a second equation in the variables a and b.

43. Write the system of equations formed from the two equations in Exercises 41 and 42, and solve the system using the elimination method.

44. What is the equation of the segment PQ?

45. Let $x = 2002$ in the equation of Exercise 44, and solve for y to the nearest tenth. How does the result compare with the actual figure of 1639 million?

46. The actual data points for the years 1996 through 2004 do not lie in a perfectly straight line. Explain the pitfalls of relying too heavily on using the equation in Exercise 44 to predict attendance.

Summary Exercises on Solving Systems of Linear Equations

The exercises in this summary include a variety of problems on solving systems of linear equations. Since we do not usually specify the method of solution, use the following guidelines to help you decide whether to use substitution or elimination.

> **Guidelines for Choosing a Method to Solve a System of Linear Equations**
>
> 1. If one of the equations of the system is already solved for one of the variables, as in the systems
>
> $$3x + 4y = 9 \qquad -5x + 3y = 9$$
> $$\text{or}$$
> $$y = 2x - 6 \qquad x = 3y - 7,$$
>
> the substitution method is the better choice.
>
> 2. If both equations are in standard $Ax + By = C$ form, as in
>
> $$4x - 11y = 3$$
> $$-2x + 3y = 4,$$
>
> and none of the variables has coefficient -1 or 1, the elimination method is the better choice.
>
> 3. If one or both of the equations are in standard form and the coefficient of one of the variables is -1 or 1, as in the systems
>
> $$3x + y = -2 \qquad -x + 3y = -4$$
> $$\text{or}$$
> $$-5x + 2y = 4 \qquad 3x - 2y = 8,$$
>
> either method is appropriate.

Use the preceding guidelines to solve each problem.

1. Assuming you want to minimize the amount of work required, tell whether you would use the substitution or elimination method to solve each system. Explain your answers. *Do not actually solve.*

   (a) $3x + 2y = 18$
   $y = 3x$

   (b) $3x + y = -7$
   $x - y = -5$

   (c) $3x - 2y = 0$
   $9x + 8y = 7$

2. Which one of the following systems would be easier to solve using the substitution method? Why?

   $$5x - 3y = 7 \qquad 7x + 2y = 4$$
   $$2x + 8y = 3 \qquad y = -3x + 1$$

In Exercises 3 and 4, (a) solve the system by the elimination method, (b) solve the system by the substitution method, and (c) tell which method you prefer for that particular system and why.

3. $4x - 3y = -8$
$\quad x + 3y = 13$

4. $2x + 5y = 0$
$\quad x = -3y + 1$

Solve each system by the method of your choice. (For Exercises 5–7, see your answers for Exercise 1.)

5. $3x + 2y = 18$
$\quad y = 3x$

6. $3x + y = -7$
$\quad x - y = -5$

7. $3x - 2y = 0$
$\quad 9x + 8y = 7$

8. $x + y = 7$
$\quad x = -3 - y$

9. $5x - 4y = 15$
$\quad -3x + 6y = -9$

10. $4x + 2y = 3$
$\quad y = -x$

11. $3x = 7 - y$
$\quad 2y = 14 - 6x$

12. $3x - 5y = 7$
$\quad 2x + 3y = 30$

13. $3y = 4x + 2$
$\quad 5x - 2y = -3$

14. $4x + 3y = 1$
$\quad 3x + 2y = 2$

15. $2x - 3y = 7$
$\quad -4x + 6y = 14$

16. $0.2x + 0.3y = 1.0$
$\quad -0.3x + 0.1y = 1.8$

17. $6x + 5y = 13$
$\quad 3x + 3y = 4$

18. $x - 3y = 7$
$\quad 4x + y = 5$

19. $\dfrac{1}{4}x - \dfrac{1}{5}y = 9$
$\quad y = 5x$

20. $\dfrac{1}{2}x + \dfrac{1}{3}y = -\dfrac{1}{3}$
$\quad \dfrac{1}{2}x + 2y = -7$

21. $-\dfrac{1}{2}x - \dfrac{1}{3}y = -5$
$\quad -\dfrac{1}{2}x - \dfrac{1}{3}y = -5$

22. $\dfrac{x}{5} + 2y = \dfrac{8}{5}$
$\quad \dfrac{3x}{5} + \dfrac{y}{2} = -\dfrac{7}{10}$

23. $\dfrac{x}{5} + y = \dfrac{6}{5}$
$\quad \dfrac{x}{10} + \dfrac{y}{3} = \dfrac{5}{6}$

24. $\dfrac{2}{5}x + \dfrac{4}{3}y = -8$
$\quad \dfrac{7}{10}x - \dfrac{2}{9}y = 9$

25. $0.5x + 0.2y = 0.2$
$\quad x - 0.6y = -0.5$

12.4 ▶▶▶ Applications of Linear Systems

You have used a six-step method for solving applied problems throughout this text. We modify those steps slightly to allow for two variables and two equations.

> **Solving an Applied Problem with Two Variables**
>
> *Step 1* **Read** the problem, several times if necessary, until you understand what is given and what is to be found.
>
> *Step 2* **Assign variables** to represent the unknown values, using diagrams or tables as needed. Write down what each variable represents.
>
> *Step 3* **Write two equations** using both variables.
>
> *Step 4* **Solve** the system of two equations.
>
> *Step 5* **State the answer** to the problem. Is the answer reasonable?
>
> *Step 6* **Check** the answer in the words of the original problem.

OBJECTIVE 1 Solve problems about unknown numbers.

1 Solve the system.
$$x = 3551 + y$$
$$x + y = 25{,}953$$

EXAMPLE 1 Solving a Problem about Two Unknown Numbers

In 2004, sales of athletic/sports footwear were $3551 million more than sales of sports clothing. Together, total sales for these items were $25,953 million. (*Source:* National Sporting Goods Association.) What were the sales for each?

Step 1 **Read** the problem carefully. We must find the 2004 sales (in millions of dollars) for athletic/sports footwear and clothing. We know how much more footwear sales were than clothing sales. Also, we know the total sales.

Step 2 **Assign variables.**

Let x = sales of footwear in millions of dollars,
and y = sales of clothing in millions of dollars.

Step 3 **Write two equations.**

$x = 3551 + y$ — Sales of footwear were $3551 million more than sales of clothing.

$x + y = 25{,}953$ — Total sales were $25,953 million.

Step 4 **Solve** the system for x and y from Step 3. The substitution method works well here since the first equation is already solved for x.

Work Problem **1** at the Side. ▶

Step 5 **State the answer.** Footwear sales were $14,752 million, and clothing sales were $11,201 million.

Step 6 **Check** the answer in the original problem. Since

$$14{,}752 - 11{,}201 = 3551 \quad \text{and} \quad 14{,}752 + 11{,}201 = 25{,}953,$$

the answer satisfies the information in the problem.

ANSWER

1. $x = 14{,}752, y = 11{,}201$

2 Set up a system of equations for the following problem. Do not solve the system.

Two of the most popular movies of 2007 were *Ratatouille* and *The Simpsons Movie*. Together, their domestic gross was $389.5 million. *The Simpsons Movie* grossed $23.3 million less than *Ratatouille*. How much did each movie gross? (*Source:* www.boxofficemojo.com)

Let x = the amount (in millions) that *Ratatouille* grossed, and y = the amount (in millions) that _____ grossed.

> **CAUTION**
> If an applied problem asks for *two* values as in Example 1, be sure to give both of them in your answer.

◀ *Work Problem* **2** *at the Side.*

OBJECTIVE 2 **Solve problems about quantities and their costs.** We can also use a linear system to solve an applied problem involving two quantities and their costs.

EXAMPLE 2 **Solving a Problem about Quantities and Costs**

Musicals have long been the most popular shows on Broadway. The musical *Wicked,* based on a "re-imagining" of *The Wizard of Oz*, has played to sold-out houses in cities around the world. (*Source:* www.broadway.com)

For the production playing at the Ford Center in Chicago, orchestra (main floor) seats cost $148, while the best balcony tickets cost $65. (*Source*: www.ticketmaster.com) Suppose that the members of a club spent a total of $2614 for 30 tickets to *Wicked*. How many tickets of each kind did they buy?

Step 1 **Read** the problem several times.

Step 2 **Assign variables.**

$$\text{Let } x = \text{ the number of orchestra seats,}$$
$$\text{and } y = \text{ the number of balcony seats.}$$

Summarize the information given in the problem in a table. The entries in the first two rows of the Total Value column were found by multiplying the number of tickets sold by the price per ticket.

	Number of Tickets	Price per Ticket (in dollars)	Total Value
Orchestra	x	148	$148x$
Balcony	y	65	$65y$
Total	30	✗✗✗✗✗	2614

Step 3 **Write two equations.** The total number of tickets was 30, so

$$x + y = 30. \qquad \text{Total number of tickets}$$

Since the total value was $2614, the final column leads to

$$148x + 65y = 2614. \qquad \text{Total value of tickets}$$

These two equations form the system

$$x + y = 30 \qquad (1)$$
$$148x + 65y = 2614. \qquad (2)$$

Step 4 **Solve** the system using the elimination method. To eliminate the y-terms, multiply each side of equation (1) by -65 to get

$$-65x - 65y = -1950.$$

Continued on Next Page

Then add this result to equation (2).

$$\begin{aligned} -65x - 65y &= -1950 \\ \underline{148x + 65y} &= \underline{2614} \qquad (2) \\ 83x &= 664 \qquad \text{Add.} \\ x &= \mathbf{8} \qquad \text{Divide by 83.} \end{aligned}$$

Substitute 8 for x in equation (1) to get

$$\begin{aligned} \mathbf{x} + y &= 30 \qquad (1) \\ \mathbf{8} + y &= 30 \qquad \text{Let } x = 8. \\ y &= 22 \end{aligned}$$

Step 5 **State the answer.** The club members bought 8 orchestra tickets and 22 balcony tickets.

Step 6 **Check.** The sum of 8 and 22 is 30, so the total number of tickets is correct. Since 8 tickets were purchased at $148 each and 22 at $65 each, the total of all the ticket prices is

$$\$148(8) + \$65(22) = \$2614,$$

which agrees with the total amount stated in the problem.

Work Problem ③ *at the Side.* ▶

OBJECTIVE ③ **Solve problems about mixtures.** In **Section 7.3** we solved percent equations using one variable. Many problems about mixtures that involve percent can be solved using a system of two equations in two variables.

EXAMPLE 3 **Solving a Mixture Problem Involving Percent**

A pharmacist needs 100 L of 50% alcohol solution. She has on hand 30% alcohol solution and 80% alcohol solution, which she can mix. How many liters of each will be required to make the 100 L of 50% alcohol solution?

Step 1 **Read** the problem. Note the percent of each solution and of the mixture.

Step 2 **Assign variables.**

Let x = the number of liters of 30% alcohol needed,

and y = the number of liters of 80% alcohol needed.

Summarize the information given in the problem in a table. Percents are written as decimals.

Liters of Mixture	Percent	Liters of Pure Alcohol
x	0.30	$0.30x$
y	0.80	$0.80y$
100	0.50	$0.50(100)$

Continued on Next Page

---

③ For the production of *Wicked* playing at the Pantages Theatre in Los Angeles, orchestra seats cost $96 and mid-priced mezzanine tickets cost $58. (*Source:* www.ticketmaster.com) If a group of 18 people attended the show and spent a total of $1234 for their tickets, how many of each kind of ticket did they buy?

(a) Complete the table.

	Number of Tickets Sold	Price (in dollars)	Total Value
Orchestra	x		
Mezzanine	y		
Total		XXXXXX	

(b) Write a system of equations.

(c) Solve the system and check your answer in the words of the original problem.

4 How many liters of 25% alcohol solution must be mixed with 12% solution to get 13 L of 15% solution?

(a) Complete the table.

Liters	Percent	Liters of Pure Alcohol
x	0.25	0.25x
y	0.12	
13	0.15	

(b) Write a system of equations, and solve it.

5 Solve the problem.
Joe needs 60 milliliters (mL) of 20% acid solution for a chemistry experiment. The lab has on hand only 10% and 25% solutions. How much of each should he mix to get the desired amount of 20% solution?

6 Solve using the formula $d = rt$.
A small plane traveled from Stockholm, Sweden, to Oslo, Norway, averaging 244 km per hr. The trip took 1.7 hr. To the nearest kilometer, what is the distance between the two cities?

Figure 8 gives an idea of what is actually happening in this problem.

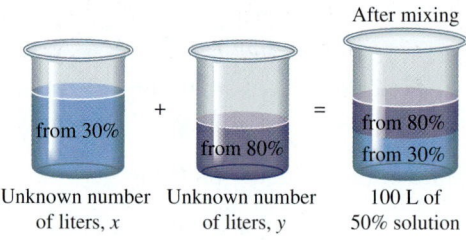

After mixing

from 30% + from 80% = from 80%
from 30%

Unknown number of liters, x Unknown number of liters, y 100 L of 50% solution

Figure 8

Step 3 **Write two equations.** Since the total number of liters in the final mixture will be 100, the first equation is

$$x + y = 100.$$

To find the amount of pure alcohol in each mixture, multiply the number of liters by the concentration. The amount of pure alcohol in the 30% solution added to the amount of pure alcohol in the 80% solution will equal the amount of pure alcohol in the final 50% solution. This gives the second equation,

$$0.30x + 0.80y = 0.50(100).$$

These two equations form the system

> Be sure to write two equations.

$$x + \quad y = 100$$
$$0.30x + 0.80y = 50. \qquad 0.50(100) = 50$$

Step 4 **Solve** this system by the substitution method. Solving the first equation of the system for x gives $x = 100 - y$. Substitute $100 - y$ for x in the second equation.

$$0.30x + 0.80y = 50$$
$$0.30(\mathbf{100 - y}) + 0.80y = 50 \qquad \text{Let } x = 100 - y.$$
$$30 - 0.30y + 0.80y = 50 \qquad \text{Distributive property}$$
$$30 + 0.50y = 50 \qquad \text{Combine like terms.}$$
$$0.50y = 20 \qquad \text{Subtract 30.}$$
$$y = 40 \qquad \text{Divide by 0.50.}$$

Then $x = 100 - \mathbf{y} = 100 - \mathbf{40} = 60.$

Step 5 **State the answer.** The pharmacist should use 60 L of the 30% solution and 40 L of the 80% solution.

Step 6 Since $60 + 40 = 100$ and $0.30(60) + 0.80(40) = 50$, this mixture will give the 100 L of 50% solution, as required in the original problem.

 ◀ *Work Problems* **4** *and* **5** *at the Side.*

OBJECTIVE 4 Solve problems about distance, rate (or speed), and time. If an automobile travels at an average rate of 50 mph for 2 hr, then it travels $50 \times 2 = 100$ mi. This is an example of the basic relationship between distance, rate, and time:

> **distance = rate × time,** given by the formula $d = rt.$

◀ *Work Problem* **6** *at the Side.*

EXAMPLE 4 **Solving a Problem about Distance, Rate, and Time**

Two executives in cities 400 mi apart drive to a business meeting at a location on the line between their cities. They meet after 4 hr. Find the speed of each car if one car travels 20 mph faster than the other.

Step 1 **Read** the problem carefully.

Step 2 **Assign variables.** Let x = the speed of the faster car,

and y = the speed of the slower car.

We use the formula $d = rt$. Since each car travels for 4 hr, the time, t, for each car is 4. The distance is found by using the formula $d = rt$ and the expressions already entered in the table.

	r	t	d
Faster Car	x	4	$4x$
Slower Car	y	4	$4y$

Find d from $d = rt$.

Figure 9 shows what is happening in the problem.

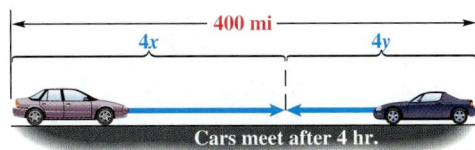

Figure 9

Step 3 **Write two equations.** As shown in the figure, since the total distance traveled by both cars is 400 mi, one equation is

$$4x + 4y = 400.$$

Because the faster car goes 20 mph faster than the slower car, the second equation is

$$x = 20 + y.$$

Step 4 **Solve** the system of equations,

$$4x + 4y = 400 \quad (1)$$
$$x = 20 + y, \quad (2)$$

by substitution. Replace x with $20 + y$ in equation (1) and then solve for y.

$$4(\mathbf{20 + y}) + 4y = 400 \quad \text{Let } x = 20 + y.$$
$$80 + 4y + 4y = 400 \quad \text{Distributive property}$$
$$80 + 8y = 400 \quad \text{Combine like terms.}$$
$$8y = 320 \quad \text{Subtract 80.}$$
$$y = 40 \quad \text{Divide by 8.}$$

Since $x = 20 + y$ and $y = \mathbf{40}$,

$$x = 20 + \mathbf{40} = 60.$$

Step 5 **State the answer.** The speeds of the cars are 40 mph and 60 mph.

Step 6 **Check** the answer. Since each car travels for 4 hr, total distance is

$$4(60) + 4(40) = 240 + 160 = 400 \text{ mi}, \quad \text{as required.}$$

Work Problem **7** *at the Side.* ▶

7 Two cars that were 450 mi apart traveled toward each other. They met after 5 hr. If one car traveled twice as fast as the other, what were their speeds?

(a) Complete this table.

	r	t	d
Faster Car	x	5	
Slower Car	y	5	

(b) Write a system, and solve it.

ANSWERS

7. (a)

	r	t	d
Faster Car	x	5	$5x$
Slower Car	y	5	$5y$

(b) $5x + 5y = 450$
$x = 2y$
faster car: $x = 60$ mph;
slower car: $y = 30$ mph

8 Solve the system.

$$x + y = 320$$
$$x - y = 280$$

CAUTION
Be careful! *When you use two variables to solve a problem, you must write two equations.*

EXAMPLE 5 **Solving a Problem about Distance, Rate, and Time**

A plane flies 560 mi in 1.75 hr traveling with the wind. The return trip against the same wind takes the plane 2 hr. Find the speed of the plane and the speed of the wind.

Step 1 **Read** the problem several times.

Step 2 **Assign variables.**

$$\text{Let } x = \text{the speed of the plane,}$$
$$\text{and } y = \text{the speed of the wind.}$$

The speed (rate) of the plane *with* the wind is $(x + y)$ mph, and the speed (rate) of the plane *against* the wind is $(x - y)$ mph. See Figure 10.

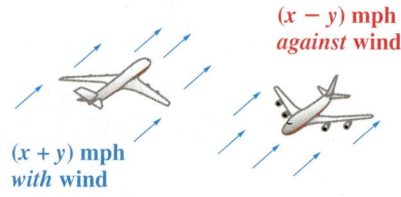

$(x - y)$ **mph**
against **wind**

$(x + y)$ **mph**
with **wind**

Figure 10

9 Solve the problem.

In 1 hr, Gigi can row 2 mi against the current or 10 mi with the current. Find the speed of the current and Gigi's speed in still water. (*Hint:* Let $x =$ the speed of the current and $y =$ Gigi's speed in still water. Then her rate against the current is $(y - x)$ mph, and her rate with the current is $(y + x)$ mph.)

We use this information and the formula $d = rt$ (or $rt = d$) to complete a table.

	r	t	d
With Wind	$x + y$	1.75	560
Against Wind	$x - y$	2	560

Step 3 **Write two equations.** From the table,

$$1.75\,(x + y) = 560 \quad \xrightarrow{\text{Divide by 1.75}} \quad x + y = 320 \quad (1)$$
$$2\,(x - y) = 560 \quad \xrightarrow{\text{Divide by 2}} \quad x - y = 280. \quad (2)$$

Step 4 **Solve** the system of equations (1) and (2).

◀ *Work Problem* **8** *at the Side.*

Step 5 **State the answer.** From Problem 8 at the side, the speed of the plane is 300 mph and the speed of the wind is 20 mph.

Step 6 **Check.** The answer seems reasonable, and true statements result when the values are substituted into the equations of the system.

◀ *Work Problem* **9** *at the Side.*

12.4 ▶▶▶ **Exercises**

Choose the correct response in Exercises 1–7.

1. Which expression represents the monetary value of x 20-dollar bills?

   A. $\dfrac{x}{20}$ dollars   **B.** $\dfrac{20}{x}$ dollars   **C.** $(20 + x)$ dollars   **D.** $20x$ dollars

2. Which expression represents the cost of t pounds of candy that sells for \$1.95 per lb?

   A. \$1.95t **B.** $\dfrac{\$1.95}{t}$ **C.** $\dfrac{t}{\$1.95}$ **D.** \$1.95 + t

3. Which expression represents the amount of interest earned on d dollars at an interest rate of 2%?

   A. $2d$ dollars   **B.** $0.02d$ dollars   **C.** $0.2d$ dollars   **D.** $200d$ dollars

4. Suppose that x liters of a 40% acid solution are mixed with y liters of a 35% solution to obtain 100 L of a 38% solution. One equation in a system for solving this problem is $x + y = 100$. Which one of the following is the other equation?

   A. $0.35x + 0.40y = 0.38(100)$ **B.** $0.40x + 0.35y = 0.38(100)$

   C. $35x + 40y = 38$ **D.** $40x + 35y = 0.38(100)$

5. According to *Natural History* magazine, the speed of a cheetah is 70 mph. If a cheetah runs for x hours, how many miles does the cheetah cover?

   A. $(70 + x)$ miles   **B.** $(70 - x)$ miles   **C.** $\dfrac{70}{x}$ miles   **D.** $70x$ miles

6. What is the speed of a plane that travels at a rate of 560 mph *against* a wind of r mph?

   A. $(560 + r)$ mph   **B.** $\dfrac{560}{r}$ mph   **C.** $(560 - r)$ mph   **D.** $(r - 560)$ mph

7. What is the speed of a plane that travels at a rate of 560 mph *with* a wind of r mph?

   A. $\dfrac{r}{560}$ mph   **B.** $(560 - r)$ mph   **C.** $(560 + r)$ mph   **D.** $(r - 560)$ mph

8. Using the list of steps for solving an applied problem with two variables, describe the general procedure you will use to solve the problems that follow in this exercise set.

Exercises 9 and 10 are good warm-up problems. In each case, refer to the six-step problem-solving method, fill in the blanks for Steps 2 and 3, and then complete the solution by applying Steps 4–6.

9. The sum of two numbers is 98 and the difference between them is 48. Find the two numbers.

Step 1 **Read** the problem carefully.

Step 2 **Assign variables.**

Let $x =$ the first number and let

$y =$ _____.

Step 3 **Write two equations.**

First equation: $x + y = 98$

Second equation: _____

10. The sum of two numbers is 201 and the difference between them is 11. Find the two numbers.

Step 1 **Read** the problem carefully.

Step 2 **Assign variables.**

Let $x =$ the first number and let

$y =$ _____.

Step 3 **Write two equations.**

First equation: $x + y = 201$

Second equation: _____

Write a system of equations for each problem, and then solve the problem. See Example 1.

11. As of 2008, the two longest-running shows in Broadway history were *The Phantom of the Opera* and *Cats.* As of September 26, 2007, there had been a total of 15,682 Broadway performances of the two shows, with 712 more performances of *The Phantom of the Opera* than *Cats.* How many performances were there of each show? (*Source:* The Broadway League.)

12. Two other musicals that had very long Broadway runs were *A Chorus Line* and *Beauty and the Beast.* During their runs, there were 676 fewer performances of *Beauty and the Beast* than of *A Chorus Line,* and a total of 11,598 performances of the two shows. How many performances were there of each show? (*Source:* The Broadway League.)

13. The two domestic top-grossing movies of 2007 were *Spider-Man 3* and *Shrek the Third. Shrek the Third* grossed $13.8 million less than *Spider-Man 3*, and together the two films took in $659.2 million. How much did each of these movies earn? (*Source:* www.boxofficemojo.com)

14. During their opening weekends, *Spider-Man 3* and *Shrek the Third* grossed a total of $272.7 million, with *Spider-Man 3* grossing $29.5 million more than *Shrek the Third.* How much did each of these movies earn during their opening weekends? (*Source:* www.boxofficemojo.com)

15. The Terminal Tower in Cleveland, Ohio, is 242 ft shorter than the Key Tower, also in Cleveland. The total of the heights of the two buildings is 1658 ft. Find the heights of the buildings. (*Source: World Almanac and Book of Facts.*)

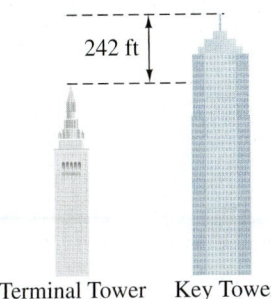

242 ft

Terminal Tower   Key Tower

16. In 2006, a total of 646.3 thousand people lived in the Twin Cities of Minneapolis and St. Paul, Minnesota. Minneapolis had 99.3 thousand more residents than St. Paul. What was the population of each city? (*Source:* U.S. Census Bureau.)

If x units of a product cost C dollars to manufacture and earn revenue of R dollars, the value of x where the expressions for C and R are equal is called the **break-even quantity,** *the number of units that produce* 0 *profit. In Exercises 17 and 18,* **(a)** *find the break-even quantity, and* **(b)** *decide whether the product should be produced based on whether it will earn a profit. (Profit equals revenue minus cost.)*

17. $C = 85x + 900$; $R = 105x$; no more than 38 units can be sold.

18. $C = 105x + 6000$; $R = 255x$; no more than 400 units can be sold.

Write a system of equations for each problem, and then solve the system. See Example 2.

19. A motel clerk counts his $1 and $10 bills at the end of a day. He finds that he has a total of 74 bills having a combined monetary value of $326. Find the number of bills of each denomination that he has.

Number of Bills	Denomination of Bill	Total Value
x	$1	
y	$10	
74	✗✗✗	$326

20. Carly is a bank teller. At the end of a day, she has a total of 69 $5 and $10 bills. The total value of the money is $590. How many of each denomination does she have?

Number of Bills	Denomination of Bill	Total Value
x	$5	$5x
y	$10	
✗✗✗		

21. A newspaper advertised DVDs and CDs. Tracy Sudak went shopping and bought each of her seven nephews a gift, either a DVD of the movie *Night at the Museum* or the latest Linkin Park CD. The DVD cost $14.95 and the CD cost $16.88, and she spent a total of $114.30. How many DVDs and how many CDs did she buy?

22. Terry Wong saw the ad (see Exercise 21) and he, too, went shopping. He bought each of his five nieces a gift, either a DVD of *Hairspray* or the CD soundtrack to *High School Musical 2*. The DVD cost $14.99 and the soundtrack cost $13.88, and he spent a total of $70.51. How many DVDs and CDs did he buy?

23. Maria Lopez has twice as much money invested at 5% simple annual interest as she does at 4%. If her yearly income from these two investments is $350, how much does she have invested at each rate?

24. Charles Miller invested his textbook royalty income in two accounts, one paying 3% annual simple interest and the other paying 2% interest. He earned a total of $11 interest. If he invested three times as much in the 3% account as he did in the 2% account, how much did he invest at each rate?

25. The two top-grossing North American concert tours in 2007 were The Police and Van Halen. Based on the average ticket prices for these tours, it cost a total of $1217 to buy six tickets for The Police and five tickets to a Van Halen concert. Three tickets for The Police and four tickets for Van Halen cost a total of $781. How much did an average ticket cost for each tour? (*Source:* Pollstar.)

26. Two other popular North American concert tours in 2007 were Billy Joel and Neil Young. Based on the average ticket prices for these tours, it cost a total of $986 to buy eight tickets for Billy Joel and three tickets to a Neil Young concert. Four tickets for Billy Joel and five tickets for Neil Young cost a total of $878. How much did an average ticket cost for each tour? (*Source:* Pollstar.)

Write a system of equations for each problem, and then solve the system. See Example 3.

27. A 40% dye solution is to be mixed with a 70% dye solution to get 120 L of a 50% solution. How many liters of the 40% and 70% solutions will be needed?

Liters of Solution	Percent (as a Decimal)	Liters of Pure Dye
x	0.40	
y	0.70	
120	0.50	

28. A 90% antifreeze solution is to be mixed with a 75% solution to make 120 L of a 78% solution. How many liters of the 90% and 75% solutions will be used?

Liters of Solution	Percent (as a Decimal)	Liters of Pure Antifreeze
x	0.90	
y	0.75	
120	0.78	

29. Ahmad Hashemi wishes to mix coffee worth $6 per lb with coffee worth $3 per lb to get 90 lb of a mixture worth $4 per lb. How many pounds of the $6 and the $3 coffees will be needed?

Pounds	Dollars per Pound	Cost
x	6	
y		
90		

30. Mariana Coanda wishes to blend candy selling for $1.20 per lb with candy selling for $1.80 per lb to get a mixture that will be sold for $1.40 per lb. How many pounds of the $1.20 and the $1.80 candies should be used to get 45 lb of the mixture?

Pounds	Dollars per Pound	Cost
x		
y	1.80	
45		

31. How many pounds of nuts selling for $6 per lb and raisins selling for $3 per lb should Kelli Hammer combine to obtain 60 lb of a trail mix selling for $5 per lb?

32. Avis Proctor works at a gourmet delicatessen. She is preparing cheese trays for a large reception. She is using some cheeses that sell for $8 per lb and others that sell for $12 per lb. How many pounds of cheese at each price should she use in order for the mixed cheeses on the trays to weigh a total of 56 lb and sell for $10.50 per lb?

Write a system of equations for each problem, and then solve the system. See Examples 4 and 5.

33. RAGBRAI®, the Des Moines **R**egister's **A**nnual **G**reat **B**icycle **R**ide **A**cross **I**owa, is the longest and oldest touring bicycle ride in the world. Suppose a cyclist began the 471 mi ride on July 20, 2008, in western Iowa at the same time that a car traveling toward it left eastern Iowa. If the bicycle and the car met after 7.5 hr and the car traveled 35.8 mph faster than the bicycle, find the average speed of each. (*Source:* www.ragbrai.org)

34. In 2006, Atlanta's Hartsfield Airport was the nation's busiest. Suppose two planes leave the airport at the same time, one traveling east and the other traveling west. If the planes are 2100 mi apart after 2 hr and one plane travels 50 mph faster than the other, find the speed of each plane. (*Source:* Airports Council International.)

35. Toledo and Cincinnati are 200 mi apart. A car leaves Toledo traveling toward Cincinnati, and another car leaves Cincinnati at the same time, traveling toward Toledo. The car leaving Toledo averages 15 mph faster than the other, and they meet after 1 hr and 36 min. What are the rates of the cars?

36. Kansas City and Denver are 600 mi apart. Two cars start from these cities, traveling toward each other. They meet after 6 hr. Find the rate of each car if one travels 30 mph slower than the other.

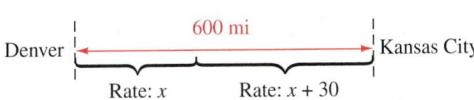

37. At the beginning of a bicycle ride for charity, Roberto and Juana are 30 mi apart. If they leave at the same time and ride in the same direction, Roberto overtakes Juana in 6 hr. If they ride toward each other, they meet in 1 hr. What are their speeds?

38. Mr. Abbot left Farmersville in a plane at noon to travel to Exeter. Mr. Baker left Exeter in his automobile at 2 P.M. to travel to Farmersville. It is 400 mi from Exeter to Farmersville. If the sum of their speeds was 120 mph, and if they crossed paths at 4 P.M., find the speed of each.

39. A boat takes 3 hr to go 24 mi upstream. It can go 36 mi downstream in the same time. Find the speed of the current and the speed of the boat in still water if x = the speed of the boat in still water and y = the speed of the current.

	r	t	d
Downstream	$x + y$		36
Upstream	$x - y$		24

40. It takes a boat $1\frac{1}{2}$ hr to go 12 mi downstream, and 6 hr to return. Find the speed of the boat in still water and the speed of the current. Let x = the speed of the boat in still water and y = the speed of the current.

	r	t	d
Downstream	$x + y$	$\frac{3}{2}$	12
Upstream		6	

41. If a plane can travel 440 mph against the wind and 500 mph with the wind, find the speed of the wind and the speed of the plane in still air.

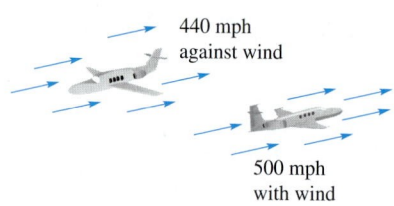

440 mph
against wind

500 mph
with wind

42. A small plane travels 200 mph with the wind and 120 mph against it. Find the speed of the wind and the speed of the plane in still air.

12.5 ▷▷▷ Solving Systems of Linear Inequalities

We graphed the solutions of a linear inequality in **Section 11.5.** Recall that to graph the solutions of $x + 3y > 12$, for example, we first graph $x + 3y = 12$ by finding and plotting a few ordered pairs that satisfy the equation. Because the points on the line do *not* satisfy the inequality, we use a dashed line. To decide which side of the line includes the points that are solutions, we choose a test point not on the line, such as $(0, 0)$. Substituting these values for x and y in the inequality gives

$$x + 3y > 12$$
$$0 + 3(0) \overset{?}{>} 12$$
$$0 > 12. \quad \text{False}$$

This false result indicates that the solutions are those points on the side of the line that does not include $(0, 0)$, as shown in Figure 11.

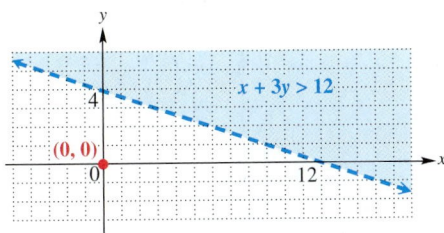

Figure 11

Now we use the same techniques to solve systems of linear inequalities.

OBJECTIVE 1 **Solve systems of linear inequalities by graphing.** A **system of linear inequalities** consists of two or more linear inequalities. The **solution set of a system of linear inequalities** includes all points that make all inequalities of the system true at the same time. To solve a system of linear inequalities, use the following steps.

> **Solving a System of Linear Inequalities**
>
> *Step 1* **Graph the inequalities.** Graph each inequality using the method of **Section 11.5.**
>
> *Step 2* **Choose the intersection.** Indicate the solution set of the system by shading the intersection of the graphs (the region where the graphs overlap).

OBJECTIVE

1 Solve systems of linear inequalities by graphing.

EXAMPLE 1 **Solving a System of Two Linear Inequalities**

Graph the solution set of the system.

$$3x + 2y \leq 6$$
$$2x - 5y \geq 10$$

To graph $3x + 2y \leq 6$, graph the solid boundary line $3x + 2y = 6$ and shade the region containing $(0, 0)$, as shown in Figure 12(a) on the next page. Then graph $2x - 5y \geq 10$ with the solid boundary line $2x - 5y = 10$. The test point $(0, 0)$ makes this inequality false, so shade the region on the other side of the boundary line. See Figure 12(b).

Continued on Next Page

1 Graph the solution set of the system.

$$x - 2y \le 8$$
$$3x + y \ge 6$$

To get you started, the graphs of $x - 2y = 8$ and $3x + y = 6$ are shown.

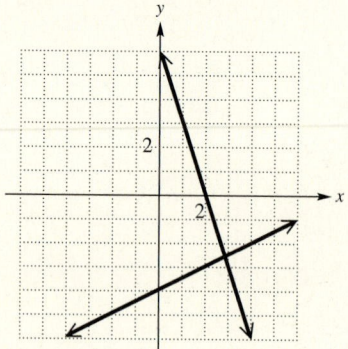

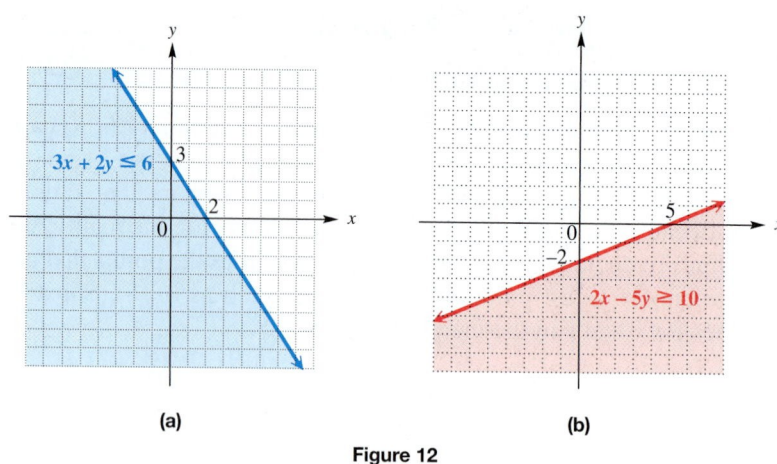

(a)   **(b)**

Figure 12

The solution set of this system includes all points in the intersection (overlap) of the graphs of the two inequalities. It includes the shaded region and portions of the two boundary lines shown in Figure 13.

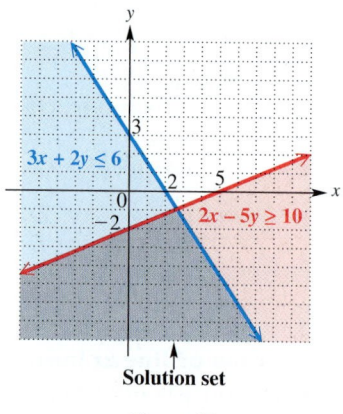

Solution set

Figure 13

◀ *Work Problem* **1** *at the Side.*

> **Note**
>
> We usually do all the work on one set of axes. In the following examples, only one graph is shown. Be sure that the region of the final solution set is clearly indicated.

EXAMPLE 2 **Solving a System of Two Linear Inequalities**

Graph the solution set of the system.

$$x - y > 5$$
$$2x + y < 2$$

Figure 14 shows the graphs of both $x - y > 5$ and $2x + y < 2$. Dashed lines show that the graphs of the inequalities do not include their boundary lines. The solution set of the system is the region with the darkest shading. The solution set does not include either boundary line.

Continued on Next Page

ANSWER

1.

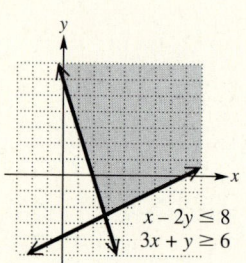

$x - 2y \le 8$
$3x + y \ge 6$

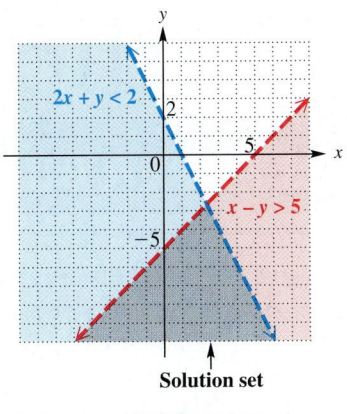

Figure 14

EXAMPLE 3 Solving a System of Three Linear Inequalities

Graph the solution set of the system.

$$4x - 3y \leq 8$$
$$x \geq 2$$
$$y \leq 4$$

Recall that $x = 2$ is a vertical line through the point $(2, 0)$, and $y = 4$ is a horizontal line through $(0, 4)$. The graph of the solution set is the shaded region in Figure 15, including all boundary lines.

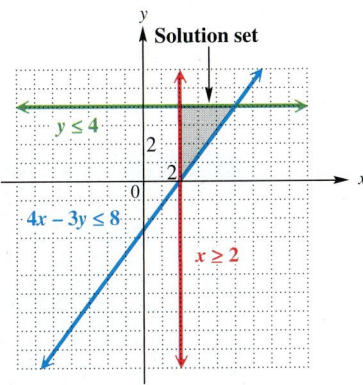

Figure 15

Work Problem **2** *at the Side.* ▶

2 Graph the solution set of each system.

(a) $x + 2y < 0$
$3x - 4y < 12$

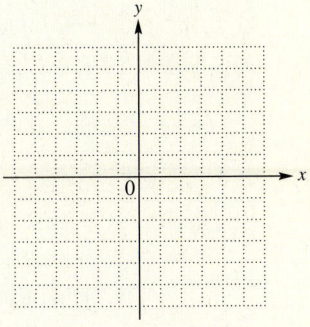

(b) $3x + 2y \leq 12$
$x \leq 2$
$y \leq 4$

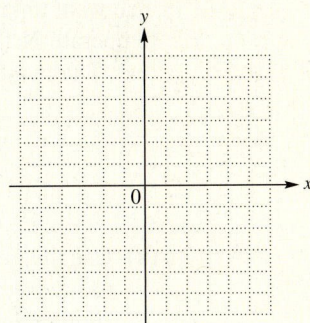

ANSWERS

2. **(a)**

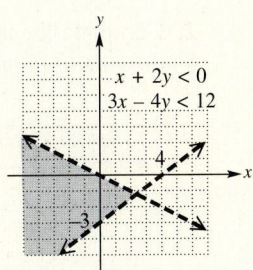

(b)

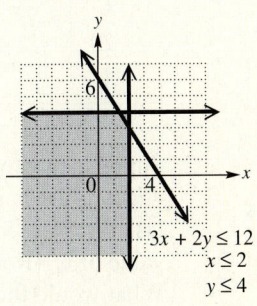

Math in the Media

The 1966 film *Fahrenheit 451*, directed by François Truffaut and based on the Ray Bradbury book of the same name, is a science fiction classic. In a future dominated by oppression, a fireman is assigned the task of burning books to discourage independent thinking. He eventually begins to read them and questions the motive of the government.

The Fahrenheit scale for temperature is used in the United States, but in all other countries temperature is routinely reported in Celsius. The formula for converting Celsius to Fahrenheit is $F = \frac{9}{5}C + 32$, but a quick "rule of thumb" given by travel books is *"Double the Celsius temperature and add 30,"* which is mathematically stated by the approximation formula

$$F = 2C + 30.$$

1. Suppose you are interested in knowing for what temperature the rule of thumb and the actual formulas give the same result. You also want to know if the rule of thumb formula is predicting temperatures that are lower or higher than the actual temperature. The two formulas can be written as the system of equations

$$F = \frac{9}{5}C + 32$$

$$F = 2C + 30.$$

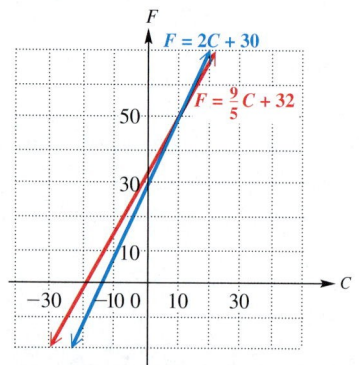

(a) Use the graph of the system of equations to find the point of intersection. (*Hint:* To check your answer, use substitution to see if it satisfies both formulas.)

(b) For what temperature in degrees Celsius do the two formulas agree?

(c) For what temperature in degrees Fahrenheit do the two formulas agree?

2. Complete the table of values to compare the *actual* and the *rule of thumb* formulas for temperature conversion.

°C	°F (*Actual*)	°F (*Rule of Thumb*)
0		
5		
10		
15		
20		
30		

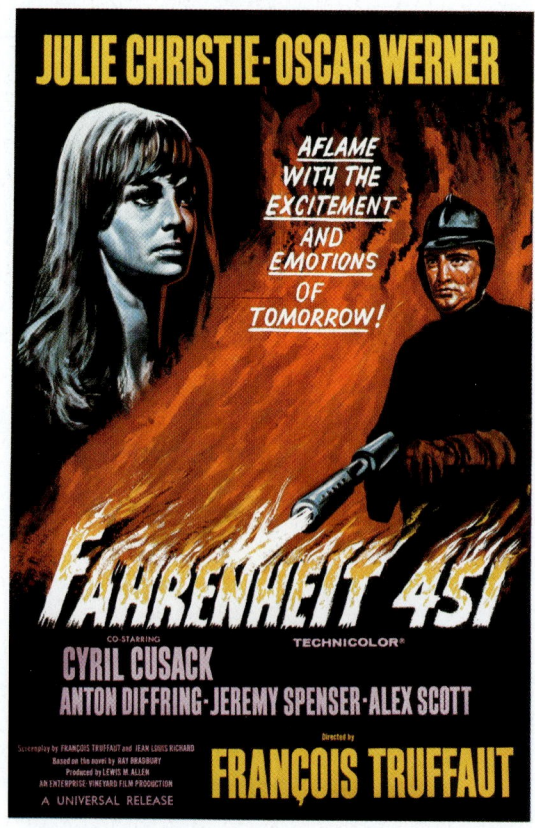

3. Suppose that the movie and book title *Fahrenheit 451* was given in Celsius rather than Fahrenheit. Use the exact formula $F = \frac{9}{5}C + 32$ to solve for this value of C. (Round to the nearest whole number.) What would the title then be?

12.5 ▶▶▶ **Exercises**

 PRACTICE   WATCH   DOWNLOAD   READ REVIEW

Match each system of inequalities with the correct graph from choices A–D.

1. $x \geq 5$
   $y \leq -3$

2. $x \leq 5$
   $y \geq -3$

3. $x > 5$
   $y < -3$

4. $x < 5$
   $y > -3$

A.

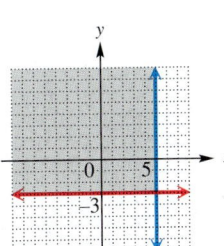

B.

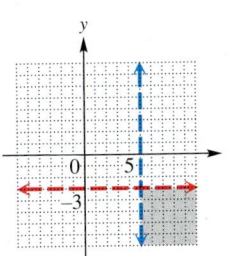

C.

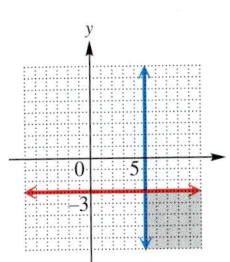

D.

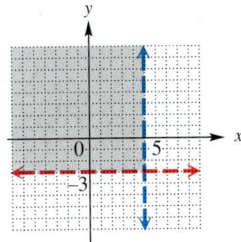

Graph the solution set of each system of linear inequalities. See Examples 1–3.

5. $x + y \leq 6$
   $x - y \geq 1$

6. $x + y \leq 2$
   $x - y \geq 3$

7. $4x + 5y \geq 20$
   $x - 2y \leq 5$

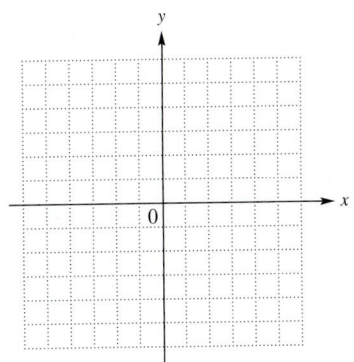

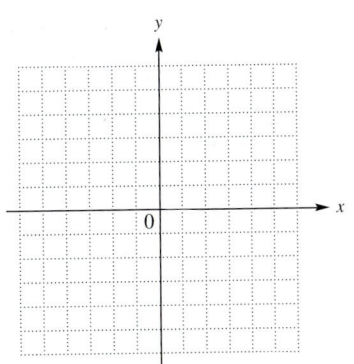

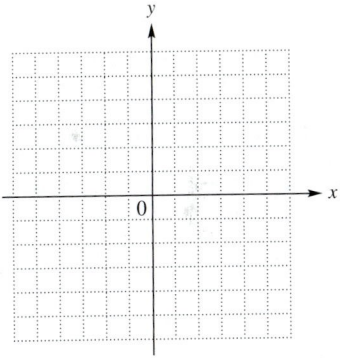

8. $x + 4y \leq 8$
   $2x - y \geq 4$

9. $2x + 3y < 6$
   $x - y < 5$

10. $x + 2y < 4$
   $x - y < -1$

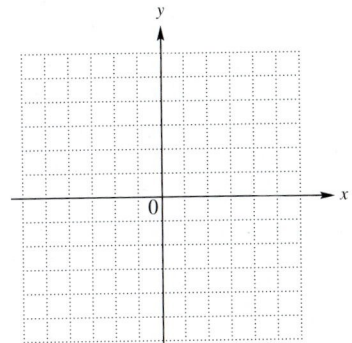

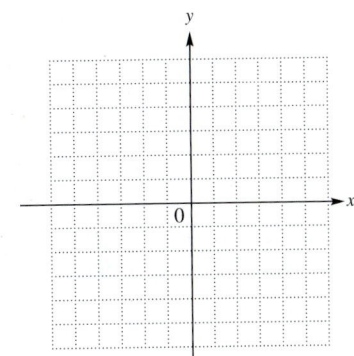

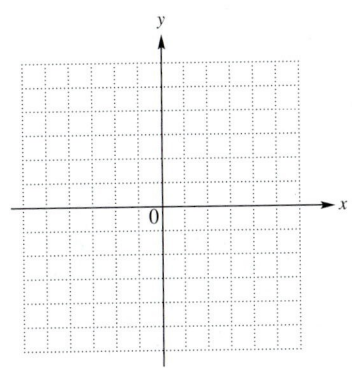

11. $y \leq 2x - 5$
   $x < 3y + 2$

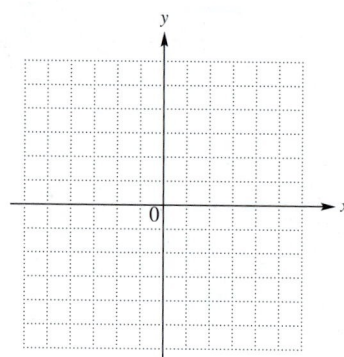

12. $x \geq 2y + 6$
   $y > -2x + 4$

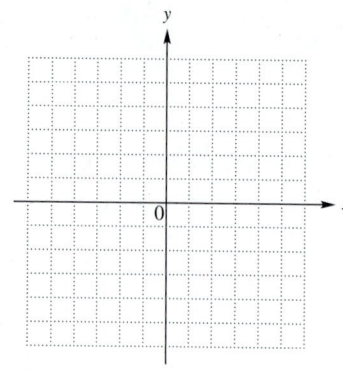

13. $4x + 3y < 6$
   $x - 2y > 4$

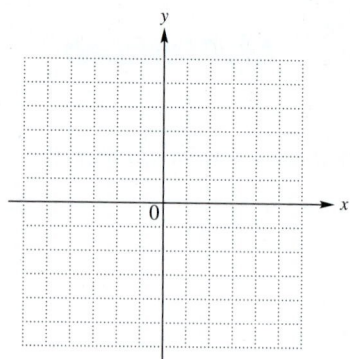

14. $3x + y > 4$
   $x + 2y < 2$

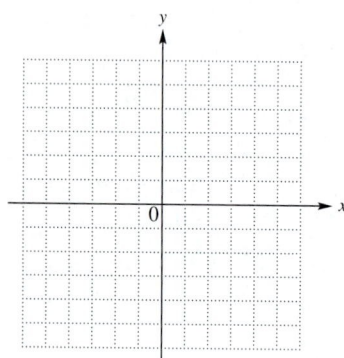

15. $x \leq 2y + 3$
   $x + y < 0$

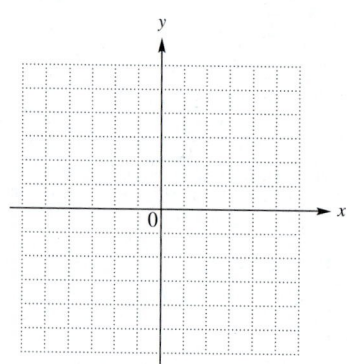

16. $x \leq 4y + 3$
   $x + y > 0$

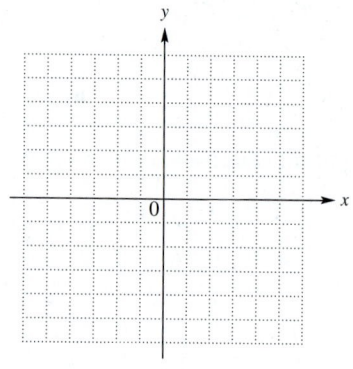

17. $4x + 5y < 8$
   $y > -2$
   $x > -4$

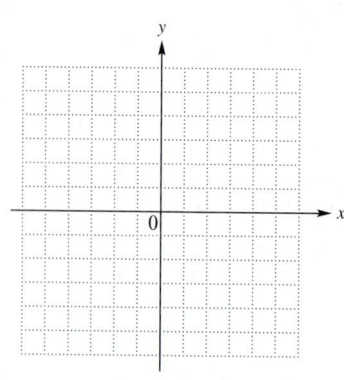

18. $x + y \geq -3$
   $x - y \leq 3$
   $y \leq 3$

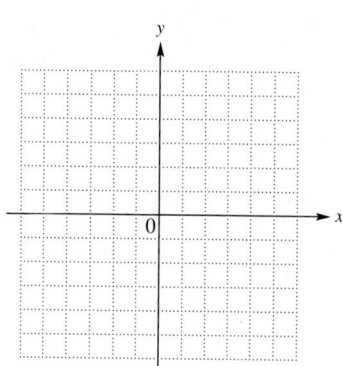

19. $3x - 2y \geq 6$
   $x + y \leq 4$
   $x \geq 0$
   $y \geq -4$

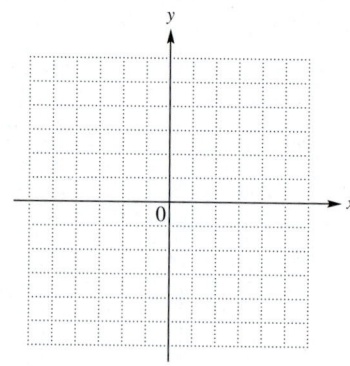

20. Every system of inequalities illustrated in the examples of this section has infinitely many solutions. Explain why this is so. Does this mean that *any* ordered pair is a solution?

Chapter 12 ▶▶▶ Summary

▶ Key Terms

12.1 **system of linear equations**
A system of linear equations (or **linear system**) consists of two or more linear equations with the same variables.

solution of a system
The solution of a system of linear equations includes all the ordered pairs that make all the equations of the system true at the same time.

solution set of a system
The set of all ordered pairs that are solutions of a system is its solution set.

set-builder notation
Set builder notation uses a variable and a description to describe a set. It is often used to describe sets whose elements cannot easily be listed.

consistent system
A system of equations with at least one solution is a consistent system.

inconsistent system
An inconsistent system of equations is a system with no solution.

independent equations
Equations of a system that have different graphs are called independent equations.

dependent equations
Equations of a system that have the same graph (because they are different forms of the same equation) are called dependent equations.

12.5 **system of linear inequalities**
A system of linear inequalities contains two or more linear inequalities (and no other kinds of inequalities).

solution set of a system of linear inequalities
The solution set of a system of linear inequalities includes all points that make all inequalities of the system true at the same time.

▶ New Symbols

$\emptyset$ empty (null) set

$\{x \mid x \text{ has a certain property}\}$ set-builder notation

▶ Test Your Word Power

See how well you have learned the vocabulary in this chapter. Answers, with examples, follow the Quick Review.

1. A **system of linear equations** consists of
   A. at least two linear equations with different variables
   B. two or more linear equations that have an infinite number of solutions
   C. two or more linear equations with the same variables
   D. two or more linear inequalities.

2. A **solution of a system** of linear equations is
   A. an ordered pair that makes one equation of the system true

   B. an ordered pair that makes all the equations of the system true at the same time
   C. any ordered pair that makes one or the other or both equations of the system true
   D. the set of values that make all the equations of the system false.

3. A **consistent system** is a system of equations
   A. with at least one solution
   B. with no solution
   C. with an infinite number of solutions
   D. that have the same graph.

4. An **inconsistent system** is a system of equations
   A. with one solution
   B. with no solution
   C. with an infinite number of solutions
   D. that have the same graph.

5. **Dependent equations**
   A. have different graphs
   B. have no solution
   C. have one solution
   D. are different forms of the same equation.

▶ Quick Review

Concepts

Examples

12.1 Solving Systems of Linear Equations by Graphing

An ordered pair is a solution of a system if it makes all equations of the system true at the same time.

Is $(4, -1)$ a solution of the system $\begin{array}{l} x + y = 3 \\ 2x - y = 9 \end{array}$?

Because $4 + (-1) = 3$ and $2(4) - (-1) = 9$ are both true, $(4, -1)$ is a solution.

If the graphs of the equations of a system are both sketched on the same axes, then the points of intersection, if any, are solutions of the system.

Solve by graphing.

$$x + y = 5$$
$$2x - y = 4$$

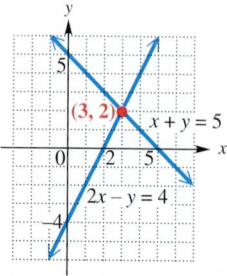

If the graphs of the equations do not intersect (that is, the lines are parallel), then the system has no solution and the solution set is ∅.

If the graphs of the equations are the same line, then the system has an infinite number of solutions. Use set-builder notation to write the solution set: $\{(x, y) \mid \underline{\hspace{2cm}}\}$.

The ordered pair $(3, 2)$ satisfies both equations, so $\{(3, 2)\}$ is the solution set.

12.2 Solving Systems of Linear Equations by Substitution

Step 1 Solve one equation for either variable.

Solve by substitution.

$$x + 2y = -5 \quad (1)$$
$$y = -2x - 1 \quad (2)$$

Equation (2) is already solved for y.

Step 2 Substitute for that variable in the other equation to get an equation in one variable.

Substitute $-2x - 1$ for y in equation (1).

Step 3 Solve the equation from Step 2.

$$x + 2(-2x - 1) = -5$$
$$x - 4x - 2 = -5$$
$$-3x - 2 = -5$$
$$-3x = -3$$
$$x = 1$$

Step 4 Substitute the result into the equation from Step 1 to get the value of the other variable.

To find y, let $x = 1$ in equation (2):

$$y = -2(1) - 1 = -3.$$

Step 5 Check. Write the solution set.

The solution $(1, -3)$ checks, so $\{(1, -3)\}$ is the solution set.

12.3 Solving Systems of Linear Equations by Elimination

Step 1 Write both equations in standard form $Ax + By = C$.

Solve by elimination.

$$x + 3y = 7 \quad (1)$$
$$3x - y = 1 \quad (2)$$

Step 2 If necessary, multiply one or both equations by appropriate numbers so that the sum of the coefficients of either the x- or y-terms is 0.

Multiply equation (1) by -3 to eliminate the x-terms.

(continued)

Concepts	Examples

12.3 Solving Systems of Linear Equations by Elimination *(continued)*

Step 3 Add the equations to get an equation with only one variable (or no variable).

$$-3x - 9y = -21$$
$$\underline{3x - y = 1}$$
$$-10y = -20 \quad \text{Add.}$$
$$y = 2 \quad \text{Divide by } -10.$$

Step 4 Solve the equation from Step 3.

Step 5 Substitute the solution from Step 4 into either of the original equations to find the value of the remaining variable.

Substitute to get the value of x.

$$x + 3(2) = 7 \quad (1)$$
$$x + 6 = 7$$
$$x = 1$$

Step 6 Check. Write the solution set.

Since $1 + 3(2) = 7$ and $3(1) - 2 = 1$, the solution $(1, 2)$ checks, so the solution set is $\{(1, 2)\}$.

12.4 Applications of Linear Systems

Use the modified six-step method.

Step 1 **Read** the problem carefully.

Step 2 **Assign variables** for each unknown value. Use diagrams or tables as needed.

Step 3 **Write two equations** using both variables.

Step 4 **Solve** the system.

The sum of two numbers is 30. Their difference is 6. Find the numbers.

Let x represent one number.

Let y represent the other number.

$$x + y = 30$$
$$\underline{x - y = 6}$$
$$2x = 36 \quad \text{Add.}$$
$$x = 18 \quad \text{Divide by 2.}$$

Step 5 **State the answer.**

Let $x = 18$ in the first equation: $18 + y = 30$. Solve to get $y = 12$. The numbers are 18 and 12.

Step 6 **Check** the answer in the words of the original problem.

The sum of 18 and 12 is 30, and the difference between 18 and 12 is 6, so the answer checks.

12.5 Solving Systems of Linear Inequalities

To solve a system of two or more linear inequalities, graph the inequalities on the same axes. (This was explained in **Section 11.5.**) The solution of the system is the intersection (overlap) of the regions of the graphs. The portions of the boundary lines that bound the region of solutions are included for a $\leq$ or $\geq$ inequality and excluded for a $<$ or $>$ inequality.

The shaded region is the solution of the system

$$2x + 4y \geq 5$$
$$x \geq 1.$$

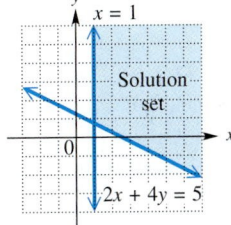

ANSWERS TO TEST YOUR WORD POWER

1. C; *Example:* $2x + y = 7$, $3x - y = 3$
2. B; *Example:* The ordered pair $(2, 3)$ satisfies both equations of the system in the Answer 1 example, so it is a solution of the system.
3. A; *Example:* The system in the Answer 1 example is consistent. The graphs of the equations intersect at exactly one point, in this case the solution $(2, 3)$.
4. B; *Example:* The equations of two parallel lines make up an inconsistent system; their graphs never intersect, so there is no solution to the system.
5. D; *Example:* The equations $4x - y = 8$ and $8x - 2y = 16$ are dependent because their graphs are the same line.

Math in the Media

CONNECTING GRAPHS IN THE MEDIA WITH SYSTEMS OF EQUATIONS

The March 16, 2008, headline and accompanying graph shown here indicate the Port of New Orleans cargo volume declined since 1998, after a period of growth between 1991 and 1998. The graph shows year-to-year fluctuations, as it consists of line segments describing the data from each year to the next.

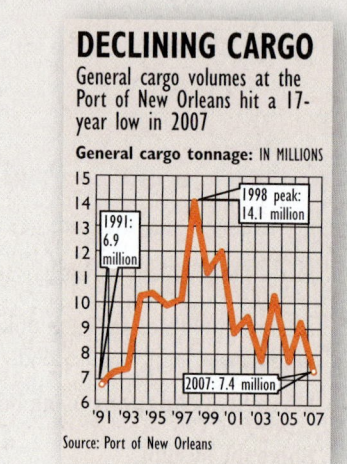

N.O. port fights to grow as cargo lags

DECLINING CARGO

General cargo volumes at the Port of New Orleans hit a 17-year low in 2007

General cargo tonnage: IN MILLIONS

1991: 6.9 million
1998 peak: 14.1 million
2007: 7.4 million

'91 '93 '95 '97 '99 '01 '03 '05 '07

Source: Port of New Orleans

1. Suppose that we wish to depict the basic idea but provide a less detailed graph of the data consisting of only three data points: the point for the year 1991, represented by $A(1991, 6.9)$; the point for the year 1998, represented by $B(1998, 14.1)$; and the point for the year 2007, represented by $C(2007, 7.4)$. We could simply graph the segment AB and then graph the segment BC. Do this on the axes provided.

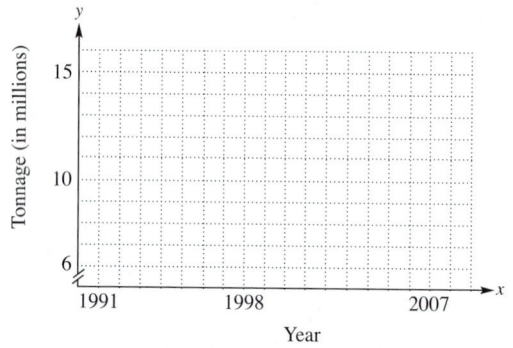

2. Use the coordinates for points A and B to find the equation of the line on which segment AB lies. Express it in slope-intercept form.

3. Use the coordinates for points B and C to find the equation of the line on which segment BC lies. Express it in slope-intercept form.

4. Consider the two equations you found in Exercises 3 and 4 as a system of linear equations. Solve the system, and confirm that the solution of the system is (1998, 14.1).

Chapter 12 ▶▶▶ Review Exercises

[12.1] *Decide whether the given ordered pair is a solution of the given system.*

1. $(3, 4)$
$$4x - 2y = 4$$
$$5x + y = 19$$

2. $(-5, 2)$
$$x - 4y = -13$$
$$2x + 3y = 4$$

Solve each system by graphing.

3. $x + y = 4$
$$2x - y = 5$$

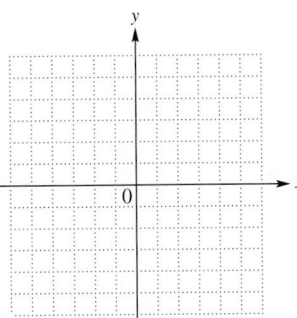

4. $x - 2y = 4$
$$2x + y = -2$$

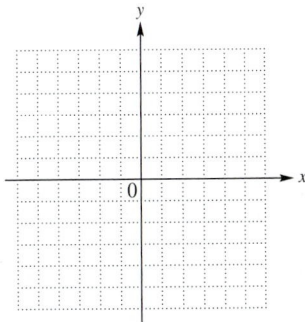

5. $x - 2 = 2y$
$$2x - 4y = 4$$

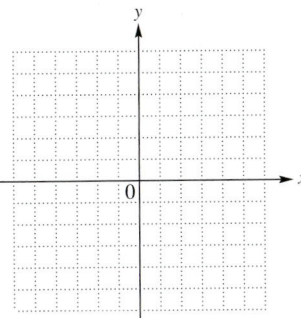

6. $2x + 4 = 2y$
$$y - x = -3$$

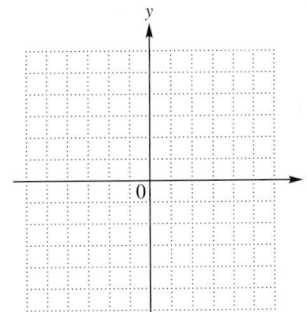

7. When a student was asked to determine whether the ordered pair $(1, -2)$ is a solution of the system

$$x + y = -1$$
$$2x + y = 4,$$

he answered "yes." His reasoning was that the ordered pair satisfies the equation $x + y = -1$; that is, $1 + (-2) = -1$ is true. Why is his answer wrong?

[12.2] *Solve each system by the substitution method.*

8. $3x + y = 7$
$$x = 2y$$

9. $2x - 5y = -19$
$$y = x + 2$$

10. $4x + 5y = 44$

$x + 2 = 2y$

11. $5x + 15y = 3$

$x + 3y = 2$

[12.3] *Solve each system by the elimination method.*

12. $2x - y = 13$

$x + y = 8$

13. $3x - y = -13$

$x - 2y = -1$

14. $-4x + 3y = 25$

$6x - 5y = -39$

15. $3x - 4y = 9$

$6x - 8y = 18$

16. For the system

$$2x + 12y = 7$$
$$3x + 4y = 1,$$

if we were to multiply the first (top) equation by -3, by what number would we have to multiply the second (bottom) equation in order to

(a) eliminate the x-terms when solving by the elimination method?

(b) eliminate the y-terms when solving by the elimination method?

Solve each system by any method.

17. $x - 2y = 5$

$y = x - 7$

18. $5x - 3y = 11$

$2y = x - 4$

19. $\dfrac{x}{2} + \dfrac{y}{3} = 7$

$\dfrac{x}{4} + \dfrac{2y}{3} = 8$

20. $\dfrac{3x}{4} - \dfrac{y}{3} = \dfrac{7}{6}$

$\dfrac{x}{2} + \dfrac{2y}{3} = \dfrac{5}{3}$

21. $2.4x + 1.7y = 7.6$

$1.2x - 0.5y = 9.2$

22. $0.5x + 3.4y = 13$

$1.5x - 2.6y = -25$

[12.4] *Solve each problem by using a system of equations.*

23. At the end of 2006, Subway topped McDonald's as the largest restaurant chain in the United States. Subway operated 6981 more restaurants than McDonald's, and together the two chains had 34,529 restaurants. How many restaurants did each company operate? (*Source:* Technomic.)

24. In 2006, the two magazines with the largest circulations in the United States were *AARP The Magazine* and *Reader's Digest*. Together, the average total circulation of these two magazines was 33.5 million copies. The circulation of *Reader's Digest* was 13.3 million less than that of *AARP The Magazine*. What were the circulation figures for each magazine? (*Source:* www.myjobsource.com)

25. The perimeter of a rectangle is 90 m. Its length is $1\frac{1}{2}$ times its width. Find the length and width of the rectangle.

26. A cashier has 20 bills, all of which are $10 or $20 bills. The total value of the money is $330. How many of each type does the cashier have?

Number of Bills	Denomination of Bills	Total Value
x	$10	$10x$
	$20	
✕✕✕✕✕✕		$330

27. Candy that sells for $1.30 per lb is to be mixed with candy selling for $0.90 per lb to get 100 lb of a mix that will sell for $1 per lb. How much of each type should be used?

28. A certain plane flying with the wind travels 540 mi in 2 hr. Later, flying against the same wind, the plane travels 690 mi in 3 hr. Find the speed of the plane in still air and the speed of the wind.

29. After taxes, Ms. Cesar's game show winnings were $18,000. She invested part of it at 3% annual simple interest and the rest at 4%. Her interest income for the first year was $650. How much did she invest at each rate?

Amount of Principal	Percent (as a Decimal)	Interest
x	0.03	
y	0.04	
$18,000	✕✕✕✕✕✕	

30. A 40% antifreeze solution is to be mixed with a 70% solution to get 90 L of a 50% solution. How many liters of the 40% and 70% solutions will be needed?

Number of Liters	Percent (as a Decimal)	Amount of Pure Antifreeze
x	0.40	
y	0.70	
90	0.50	✕✕✕✕✕✕

[12.5] *Graph the solution set for each system of linear inequalities.*

31. $x + y \geq 2$
$x - y \leq 4$

32. $y \geq 2x$
$2x + 3y \leq 6$

33. $x + y < 3$
$2x > y$

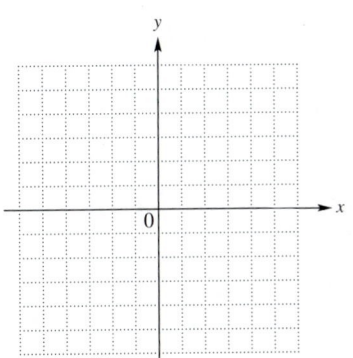

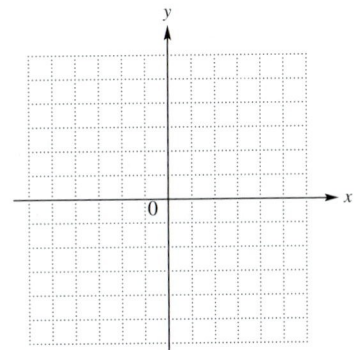

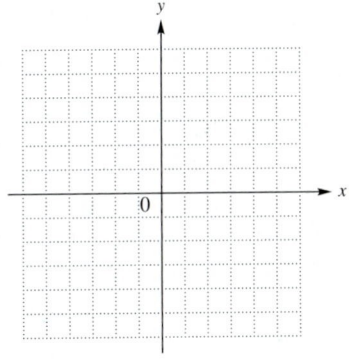

34. Which system of linear inequalities is graphed in the figure?

  A. $x \le 3$ **B.** $x \le 3$ **C.** $x \ge 3$ **D.** $x \ge 3$

    $y \le 1$ $y \ge 1$ $y \le 1$ $y \ge 1$

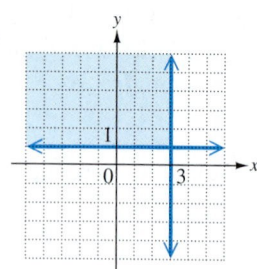

35. Without actually graphing, determine which system of inequalities has no solution.

  A. $x \ge 4$ **B.** $x + y > 4$ **C.** $x > 2$ **D.** $x + y < 4$

    $y \le 3$ $x + y < 3$ $y < 1$ $x - y < 3$

▶▶▶ Mixed Review Exercises

Solve each system.

36. $3x + 4y = 6$
  $4x - 5y = 8$

37. $\dfrac{3x}{2} + \dfrac{y}{5} = -3$

  $4x + \dfrac{y}{3} = -11$

38. $x + 6y = 3$
   $2x + 12y = 2$

39. $x + y < 5$
  $x - y \ge 2$

40. $y \le 2x$
  $x + 2y > 4$

41. $y < -4x$
  $y < -2$

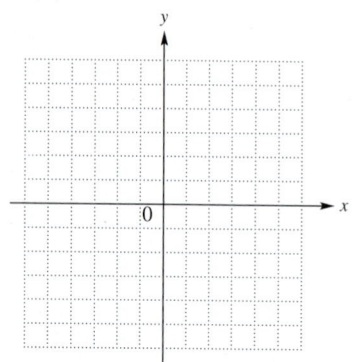

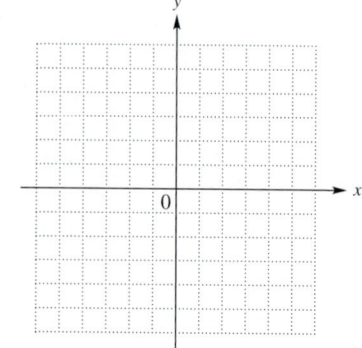

42. The perimeter of an isosceles triangle is 29 in. One side of the triangle is 5 in. longer than each of the two equal sides. Find the lengths of the sides of the triangle.

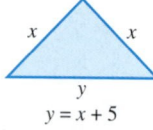

    $y = x + 5$

43. Super Bowl XLII was played in Glendale, Arizona, on February 3, 2008. The New York Giants beat the New England Patriots by 3 points, and the winning score was 11 points less than twice the losing score. What was the final score of the game? (*Source:* NFL.)

44. Eboni Perkins compared the monthly payments she would incur for two types of mortgages: fixed-rate and variable-rate. Her observations led to the following graph.

 (a) For which years would the monthly payment be more for the fixed-rate mortgage than for the variable-rate mortgage?

 (b) In what year would the payments be the same, and what would those payments be?

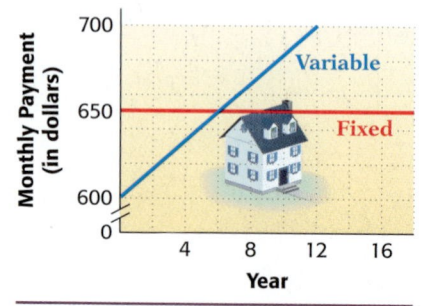

Chapter 12 ▶▶▶ Test

Use the Chapter Test Prep Video CD to see fully worked-out solutions to any of the exercises you want to review.

1. Solve the system by graphing.

$$2x + y = 1$$
$$3x - y = 9$$

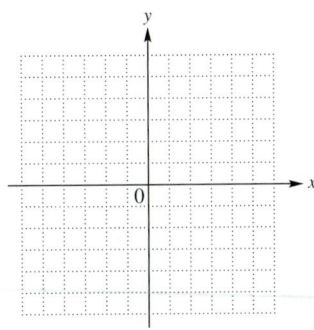

1. _____

2. Suppose that the graph of a system of two linear equations consists of lines that have the same slope but different y-intercepts. How many solutions does the system have?

2. _____

3. _____

4. _____

Solve each system by the substitution method.

3. $2x + y = -4$
$\quad x = y + 7$

4. $4x + 3y = -35$
$\quad x + y = 0$

5. _____

6. _____

Solve each system by the elimination method.

5. $2x - y = 4$
$\quad 3x + y = 21$

6. $4x + 2y = 2$
$\quad 5x + 4y = 7$

7. _____

7. $6x - 5y = 0$
$\quad -2x + 3y = 0$

8. $4x + 5y = 2$
$\quad -8x - 10y = 6$

8. _____

Solve each system by any method.

9. _____

9. $3x = 6 + y$
$\quad 6x - 2y = 12$

10. $\dfrac{x}{2} - \dfrac{y}{4} = 7$

$\quad \dfrac{2x}{3} + \dfrac{5y}{4} = 3$

10. _____

Solve each problem.

11. The distance between Memphis and Atlanta is 782 mi less than the distance between Minneapolis and Houston. Together, the two distances total 1570 mi. How far is it between Memphis and Atlanta? How far is it between Minneapolis and Houston? (*Source: Rand McNally Road Atlas.*)

12. In 2004, a total of 9.0 million people visited the Statue of Liberty and the National World War II Memorial, two popular tourist attractions. The Statue of Liberty had 1.8 million fewer visitors than the National World War II Memorial. How many visitors did each of these attractions have? (*Source:* National Park Service, Department of the Interior.)

13. A 15% solution of alcohol is to be mixed with a 40% solution to get 50 L of a final mixture that is 30% alcohol. How much of each of the original solutions should be used?

14. Two cars leave from Perham, Minnesota, and travel in the same direction. One car travels $1\frac{1}{3}$ times as fast as the other. After 3 hr they are 45 mi apart. What are the speeds of the cars?

Graph the solution set of each system of inequalities.

15. $2x + 7y \le 14$
$x - y \ge 1$

16. $2x - y > 6$
$4y + 12 \ge -3x$

11. _____

12. _____

13. _____

14. _____

15.

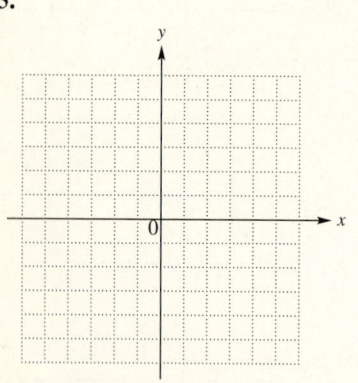

16.

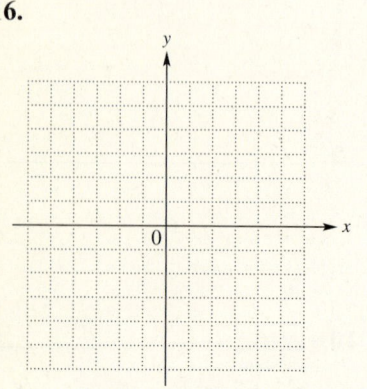

13

Exponents and Polynomials

Just how much is a *trillion*? A trillion, written 1,000,000,000,000, is a million million, or a thousand billion. A trillion seconds would last more than 31,000 years—that is, 310 centuries. By 2017, the U.S. government projects that consumers and taxpayers will spend more than $4 trillion on health care, accounting for $1 of every $5 spent. (*Source:* Centers for Medicare and Medicaid Services.)

In **Section 13.8**, we use *exponents* and *scientific notation* to write and calculate with large numbers, such as the national debt, tax revenue, and the distances of a double-helix nebula and the star Pollux from Earth.

13.1 ▶▶▶ Adding and Subtracting Polynomials

OBJECTIVES

1 Review combining like terms.

2 Know the vocabulary for polynomials.

3 Evaluate polynomials.

4 Add polynomials.

5 Subtract polynomials.

6 Add and subtract polynomials with more than one variable.

Term	Numerical Coefficient
$-7y$	-7
$34r^3$	34
$-26x^5yz^4$	-26
$-k = -1k$	-1
$r = 1r$	1
$\frac{3x}{8} = \frac{3}{8}x$	$\frac{3}{8}$
$\frac{x}{3} = \frac{1x}{3} = \frac{1}{3}x$	$\frac{1}{3}$

1 Add like terms.

(a) $5x^4 + 7x^4$

(b) $9pq + 3pq - 2pq$

(c) $r^2 + 3r + 5r^2$

(d) $x + \dfrac{1}{2}x$

(e) $8t + 6w$

(f) $3x^4 - 3x^2$

ANSWERS

1. **(a)** $12x^4$ **(b)** $10pq$ **(c)** $6r^2 + 3r$ **(d)** $\dfrac{3}{2}x$

   (e) These are unlike terms. They cannot be added.

   (f) These are unlike terms. They cannot be added.

Recall from **Section 2.2** that in an expression such as

$$4x^3 + 6x^2 + 5x + 8,$$

the quantities that are added, $4x^3$, $6x^2$, $5x$, and 8, are called **terms.** In the term $4x^3$, the number 4 is called the **numerical coefficient,** or simply the **coefficient,** of x^3. In the same way, 6 is the coefficient of x^2 in the term $6x^2$, 5 is the coefficient of x in the term $5x$, and 8 is the **constant** term. Other examples are given in the table at the side.

OBJECTIVE 1 Review combining like terms. Recall from **Section 2.2** that **like terms** have exactly the same combination of variables, with the same exponents on the variables. *Only the coefficients may differ.*

$$\left.\begin{array}{ll} 19m^5 & \text{and} \quad 14m^5 \\ -37y^9 & \text{and} \quad y^9 \\ 3pq & \text{and} \quad -2pq \\ 2xy^2 & \text{and} \quad -xy^2 \end{array}\right\} \begin{array}{c} \text{Examples} \\ \text{of} \\ \text{like terms} \end{array} \qquad \left.\begin{array}{ll} 7x & \text{and} \quad 7y \\ z^4 & \text{and} \quad z \\ 2pq & \text{and} \quad 2p \\ -4xy^2 & \text{and} \quad 5x^2y \end{array}\right\} \begin{array}{c} \text{Examples} \\ \text{of} \\ \text{unlike terms} \end{array}$$

Using the distributive property, we combine, or add, like terms by adding their coefficients.

EXAMPLE 1 Adding Like Terms

Simplify each expression by adding like terms.

(a) $-4x^3 + 6x^3$

$= (-4 + 6)x^3$ Distributive property

$= 2x^3$

(b) $9x^6 - 14x^6 + x^6$

$= (9 - 14 + 1)x^6$ $x^6 = 1x^6$

$= -4x^6$

(c) $12m^2 + 5m + 4m^2$ **(d)** $3x^2y + 4x^2y - x^2y$

$= (12 + 4)m^2 + 5m$ $= (3 + 4 - 1)x^2y$

$= 16m^2 + 5m$ $= 6x^2y$

In Example 1(c), we cannot combine $16m^2$ and $5m$. These two terms are unlike because the exponents on the variables are different. *Unlike terms have different variables or different exponents on the same variables.*

◀ Work Problem **1** at the Side.

OBJECTIVE 2 Know the vocabulary for polynomials. A **polynomial in x** is a term or the sum of a finite number of terms of the form ax^n, for any real number a and any whole number n. For example,

$$16x^8 - 7x^6 + 5x^4 - 3x^2 + 4 \qquad \text{Polynomial}$$

is a polynomial in x. This polynomial is written in **descending powers,** because the exponents on x decrease from left to right.

On the other hand, $2x^3 - x^2 + \frac{4}{x}$ is not a polynomial, since a variable appears in a denominator. We can define a *polynomial* using any variable, not just x, as in Example 1(c). Polynomials may have terms with more than one variable, as in Example 1(d).

Work Problem **2** *at the Side.* ▶

The **degree of a term** is the sum of the exponents on the variables. A constant term has degree 0. For example, $3x^4$ has degree **4**, while $6x^{17}$ has degree **17**. The term $5x$ (or $5x^1$) has degree **1**, -7 has degree 0, and $2x^2y$ has degree $2 + 1 = 3$ (y has an exponent of 1).

The **degree of a polynomial** is the greatest degree of any nonzero term of the polynomial. For example, $3x^4 - 5x^2 + 6$ is of degree **4**, the polynomial $5x + 7$ is of degree 1, 3 is of degree 0, and $x^2y + xy - 5xy^2$ is of degree 3.

Three types of polynomials are very common and are given special names. A polynomial with only one term is called a **monomial.** (*Mono-* means "one," as in *mono*rail.) Examples are

$$9m, \quad -6y^5, \quad a^2, \quad \text{and} \quad 6. \qquad \text{Monomials}$$

A polynomial with exactly two terms is called a **binomial.** (*Bi-* means "two," as in *bi*cycle.) Examples are

$$-9x^4 + 9x^3, \quad 8m^2 + 6m, \quad \text{and} \quad 3m^5 - 9m^2. \qquad \text{Binomials}$$

A polynomial with exactly three terms is called a **trinomial.** (*Tri-* means "three," as in *tri*angle.) Examples are

$$9m^3 - 4m^2 + 6, \quad \frac{19}{3}y^2 + \frac{8}{3}y + 5, \quad \text{and} \quad -3m^5 - 9m^2 + 2. \qquad \text{Trinomials}$$

EXAMPLE 2 Classifying Polynomials

Simplify each polynomial if possible. Then give the degree and tell whether the polynomial is a *monomial*, a *binomial*, a *trinomial*, or *none of these*.

(a) $2x^3 + 5$ — We cannot simplify further. This is a binomial of degree 3.

(b) $4x - 5x + 2x$

Add like terms to simplify: $4x - 5x + 2x = x$. The degree is 1 (since $x = x^1$). The simplified polynomial is a monomial.

Work Problem **3** *at the Side.* ▶

OBJECTIVE 3 Evaluate polynomials. A polynomial usually represents different numbers for different values of the variable.

EXAMPLE 3 Evaluating a Polynomial

Find the value of $3x^4 + 5x^3 - 4x - 4$ when $x = -2$ and when $x = 3$.

First, substitute -2 for x.

$$3x^4 + 5x^3 - 4x - 4$$

Use parentheses to avoid errors.

$$= 3(-2)^4 + 5(-2)^3 - 4(-2) - 4 \qquad \text{Let } x = -2.$$
$$= 3(16) + 5(-8) - 4(-2) - 4 \qquad \text{Apply the exponents.}$$
$$= 48 - 40 + 8 - 4 \qquad \text{Multiply.}$$
$$= 12 \qquad \text{Add and subtract.}$$

Continued on Next Page

2 Choose all descriptions that apply for each of the expressions in parts (a)–(d).

A. Polynomial
B. Polynomial written in descending powers
C. Not a polynomial

(a) $3m^5 + 5m^2 - 2m + 1$

(b) $2p^4 + p^6$

(c) $\frac{1}{x} + 2x^2 + 3$

(d) $x - 3$

3 Simplify each polynomial if possible. Then give the degree and tell whether the polynomial is a *monomial, binomial, trinomial,* or *none of these*.

(a) $3x^2 + 2x - 4$

(b) $x^3 + 4x^3$

(c) $x^8 - x^7 + 2x^8$

ANSWERS

2. **(a)** A and B **(b)** A **(c)** C **(d)** A and B
3. **(a)** degree 2; trinomial
  (b) degree 3; monomial (simplify to $5x^3$)
  (c) degree 8; binomial (simplify to $3x^8 - x^7$)

4 Find the value of $2x^3 + 8x - 6$ in each case.

(a) When $x = -1$

(b) When $x = 4$

Next, replace x with 3.

$$3x^4 + 5x^3 - 4x - 4$$
$$= 3(3)^4 + 5(3)^3 - 4(3) - 4 \qquad \text{Let } x = 3.$$
$$= 3(81) + 5(27) - 4(3) - 4 \qquad \text{Apply the exponents.}$$
$$= 243 + 135 - 12 - 4 \qquad \text{Multiply.}$$
$$= 362 \qquad \text{Add and subtract.}$$

CAUTION
Use parentheses around the numbers that are substituted for the variable in Example 3, particularly when substituting a negative number for a variable that is raised to a power. Otherwise, a sign error may result.

◀ *Work Problem* **4** *at the Side.*

OBJECTIVE **4** **Add polynomials.** Polynomials may be added, subtracted, multiplied, and divided.

Adding Polynomials
To add two polynomials, add like terms.

5 Add each pair of polynomials.

(a) $4x^3 - 3x^2 + 2x$ and $6x^3 + 2x^2 - 3x$

(b) $x^2 - 2x + 5$ and $4x^2 - 2$

EXAMPLE 4 **Adding Polynomials Vertically**

(a) Add $6x^3 - 4x^2 + 3$ and $-2x^3 + 7x^2 - 5$.
Write like terms in columns.

$$6x^3 - 4x^2 + 3$$
$$\underline{-2x^3 + 7x^2 - 5}$$

Now add, column by column.

$$
\begin{array}{rrr}
6x^3 & -4x^2 & 3 \\
-2x^3 & 7x^2 & -5 \\
\hline
4x^3 & 3x^2 & -2
\end{array}
$$

Add the three sums together.

$$4x^3 + 3x^2 + (-2) = 4x^3 + 3x^2 - 2$$

(b) Add $2x^2 - 4x + 3$ and $x^3 + 5x$.
Write like terms in columns and add column by column.

$$
\begin{array}{l}
\quad\ 2x^2 - 4x + 3 \\
x^3 \qquad\ + 5x \\
\hline
x^3 + 2x^2 +\ x + 3
\end{array}
$$

Leave spaces for missing terms.

◀ *Work Problem* **5** *at the Side.*

The polynomials in Example 4 also could be added horizontally.

EXAMPLE 5 **Adding Polynomials Horizontally**

(a) Add $6x^3 - 4x^2 + 3$ and $-2x^3 + 7x^2 - 5$.
Combine like terms.

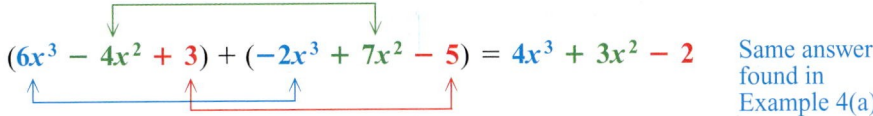

$(6x^3 - 4x^2 + 3) + (-2x^3 + 7x^2 - 5) = 4x^3 + 3x^2 - 2$ Same answer found in Example 4(a)

(b) Add $2x^2 - 4x + 3$ and $x^3 + 5x$.

$$(2x^2 - 4x + 3) + (x^3 + 5x)$$
$$= x^3 + 2x^2 - 4x + 5x + 3 \quad \text{Commutative property}$$
$$= x^3 + 2x^2 + x + 3 \quad \text{Combine like terms.}$$

Work Problem **6** *at the Side.* ▶

OBJECTIVE 5 **Subtract polynomials.**   In **Section 1.4**, the difference $x - y$ was defined as $x + (-y)$. (We find the difference $x - y$ by adding x and the opposite of y.) For example,

$$7 - 2 = 7 + (-2) = 5 \quad \text{and} \quad -8 - (-2) = -8 + 2 = -6.$$

A similar method is used to subtract polynomials.

> **Subtracting Polynomials**
>
> To subtract two polynomials, change all the signs of the second polynomial and add the result to the first polynomial.

EXAMPLE 6 **Subtracting Polynomials**

(a) Perform the subtraction $(5x - 2) - (3x - 8)$.
Change the signs in the second polynomial and add.

$$(5x - 2) - (3x - 8)$$
$$= (5x - 2) + (-3x + 8)$$
$$= 2x + 6$$

(b) Subtract $6x^3 - 4x^2 + 2$ from $11x^3 + 2x^2 - 8$.

$$(11x^3 + 2x^2 - 8) - (6x^3 - 4x^2 + 2)$$ *Write the problem in the correct order.*
$$= (11x^3 + 2x^2 - 8) + (-6x^3 + 4x^2 - 2)$$
$$= 5x^3 + 6x^2 - 10$$

To check a subtraction problem, use the following fact:

$$\text{If} \quad a - b = c, \quad \text{then} \quad a = b + c.$$

For example, $6 - 2 = 4$, so we check by writing $6 = 2 + 4$, which is correct. We check the polynomial subtraction above as follows:

$$(6x^3 - 4x^2 + 2) + (5x^3 + 6x^2 - 10)$$
$$= 11x^3 + 2x^2 - 8.$$

Since the sum is $11x^3 + 2x^2 - 8$, the subtraction was performed correctly.

Work Problem **7** *at the Side.* ▶

6 Find each sum.

(a) $(2x^4 - 6x^2 + 7)$
  $+ (-3x^4 + 5x^2 + 2)$

(b) $(3x^2 + 4x + 2)$
  $+ (6x^3 - 5x - 7)$

7 Subtract, and check your answers by addition.

(a) $(14y^3 - 6y^2 + 2y - 5)$
  $- (2y^3 - 7y^2 - 4y + 6)$

(b) Subtract

$$\left(-\frac{3}{2}y^2 + \frac{4}{3}y + 6\right)$$

from $\left(\frac{7}{2}y^2 - \frac{11}{3}y + 8\right)$.

8 Subtract by columns.

$(4y^3 - 16y^2 + 2y)$
$- (12y^3 - 9y^2 + 16)$

Subtraction also can be done in columns. We use vertical subtraction in **Section 13.7** when we study polynomial division.

EXAMPLE 7 **Subtracting Polynomials Vertically**

Subtract by columns: $(14y^3 - 6y^2 + 2y - 5) - (2y^3 - 7y^2 - 4y + 6)$.

$$14y^3 - 6y^2 + 2y - 5$$
$$\underline{2y^3 - 7y^2 - 4y + 6}$$ Arrange like terms in columns.

Change all signs in the second row, and then add.

$$14y^3 - 6y^2 + 2y - 5$$
$$\underline{-2y^3 + 7y^2 + 4y - 6}$$ Change signs.
$$12y^3 + y^2 + 6y - 11$$ Add.

◀ *Work Problem* **8** *at the Side.*

---

9 Perform the indicated operations.

$(6p^4 - 8p^3 + 2p - 1)$
$- (-7p^4 + 6p^2 - 12)$
$+ (p^4 - 3p + 8)$

EXAMPLE 8 **Adding and Subtracting More Than Two Polynomials**

Perform the indicated operations to simplify the expression

$$(4 - x + 3x^2) - (2 - 3x + 5x^2) + (8 + 2x - 4x^2).$$

Rewrite, changing the subtraction to adding the opposite.

$$(4 - x + 3x^2) - (2 - 3x + 5x^2) + (8 + 2x - 4x^2)$$
$$= (4 - x + 3x^2) + (-2 + 3x - 5x^2) + (8 + 2x - 4x^2)$$
$$= (2 + 2x - 2x^2) + (8 + 2x - 4x^2)$$ Combine like terms.
$$= 10 + 4x - 6x^2$$ Combine like terms.

◀ *Work Problem* **9** *at the Side.*

---

10 Add or subtract.

(a) $(3mn + 2m - 4n)$
$+ (-mn + 4m + n)$

OBJECTIVE **6** **Add and subtract polynomials with more than one variable.**   Polynomials in more than one variable are added and subtracted by combining like terms, just as with single-variable polynomials.

EXAMPLE 9 **Adding and Subtracting Multivariable Polynomials**

Add or subtract as indicated.

(a) $(4a + 2ab - b) + (3a - ab + b)$

$$= 4a + 2ab - b + 3a - ab + b$$

$$= 7a + ab$$ Combine like terms.

(b) $(2x^2y + 3xy + y^2) - (3x^2y - xy - 2y^2)$

$$= 2x^2y + 3xy + y^2 - 3x^2y + xy + 2y^2$$

$$= -x^2y + 4xy + 3y^2$$ Be careful with signs.

(b) $(5p^2q^2 - 4p^2 + 2q)$
$- (2p^2q^2 - p^2 - 3q)$

◀ *Work Problem* **10** *at the Side.*

13.1 ▶▶▶ Exercises

Fill in each blank with the correct response.

1. In the term $7x^5$, the coefficient is _____ and the exponent is _____.

2. The expression $5x^3 - 4x^2$ has _____ term(s).
   (how many?)

3. The degree of the term $-4x^8$ is _____.

4. The polynomial $4x^2 - y^2$ _____ an example of a trinomial.
   (is/is not)

5. When $x^2 + 10$ is evaluated for $x = 4$, the result is _____.

6. _____ is an example of a monomial with coefficient 5, in the variable x, having degree 9.

For each polynomial, determine the number of terms, and name the coefficient of each term.

7. $6x^4$

8. $-9y^5$

9. t^4

10. s^7

11. $\dfrac{x}{5}$

12. $\dfrac{z}{8}$

13. $-19r^2 - r$

14. $2y^3 - y$

15. $x - 8x^2 + \dfrac{2}{3}x^3$

16. $v - 2v^3 + \dfrac{3}{4}v^2$

In each polynomial, combine like terms whenever possible. Write the result with descending powers. See Example 1.

17. $-3m^5 + 5m^5$

18. $-4y^3 + 3y^3$

19. $2r^5 + (-3r^5)$

20. $-19y^2 + 9y^2$

21. $\dfrac{1}{2}x^4 + \dfrac{1}{6}x^4$

22. $\dfrac{3}{10}x^6 + \dfrac{1}{5}x^6$

23. $0.2m^5 - 0.5m^2$

24. $-0.9y + 0.9y^2$

25. $-3x^5 + 2x^5 - 4x^5$

26. $6x^3 - 8x^3 + 9x^3$

27. $-4p^7 + 8p^7 + 5p^9$

28. $-3a^8 + 4a^8 - 3a^2$

29. $-4y^2 + 3y^2 - 2y^2 + y^2$

30. $3r^5 - 8r^5 + r^5 + 2r^5$

For each polynomial, first simplify, if possible, and write it with descending powers. Then give the degree of the resulting polynomial, and tell whether it is a monomial, a binomial, a trinomial, or none of these. See Example 2.

31. $6x^4 - 9x$

32. $7t^3 - 3t$

33. $5m^4 - 3m^2 + 6m^5 - 7m^3$

34. $6p^5 + 4p^3 - 8p^4 + 10p^2$

35. $\dfrac{5}{3}x^4 - \dfrac{2}{3}x^4 + \dfrac{1}{3}x^2 - 4$

36. $\dfrac{4}{5}r^6 + \dfrac{1}{5}r^6 - r^4 + \dfrac{2}{5}r$

37. $0.8x^4 - 0.3x^4 - 0.5x^4 + 7$

38. $1.2t^3 - 0.9t^3 - 0.3t^3 + 9$

39. $2.5x^2 + 0.5x + x^2 - x - 2x^2$

Find the value of each polynomial (a) when $x = 2$ and (b) when $x = -1$. See Example 3.

40. $5x - 4$

41. $-2x + 3$

42. $-3x^2 + 14x - 2$

43. $2x^2 + 5x + 1$

44. $x^4 - 6x^3 + x^2 + 1$

45. $2x^5 - 4x^4 + 5x^3 - x^2$ 🌐

46. $2x^6 - 4x$

47. $-4x^5 + x^2$

Relating Concepts (Exercises 48–52) For Individual or Group Work

A polynomial can model the distance in feet that a car going approximately 68 mph will skid in t seconds. If we let D represent this distance, then

$$D = 100t - 13t^2.$$

*Each time we evaluate this polynomial for a value of t, we get one and only one output value D. This idea is basic to the concept of a **function**, an important concept in mathematics. Exercises 48–52 illustrate this idea with this polynomial and three others. **Work them in order.***

48. Evaluate the given polynomial when $t = 5$. Use the result to fill in the blanks: In _____ seconds, the car will skid _____ feet.

49. Use the polynomial equation $D = 100t - 13t^2$ to find the distance the car will skid in 1 sec. Write an ordered pair of the form (t, D).

50. If gasoline costs \$4.00 per gal, then the monomial $4.00x$ gives the cost, in dollars, of x gallons. How much would 4 gal cost?

51. If it costs \$15 plus \$2 per day to rent a chain saw, the binomial $2x + 15$ gives the cost in dollars to rent the chain saw for x days. How much would it cost to rent the saw for 6 days?

52. If an object is projected upward under certain conditions, its height in feet is given by the trinomial $-16t^2 + 60t + 80$, where t is in seconds. Evaluate this trinomial for $t = 2.5$, and then use the result to fill in the blanks: If _____ seconds have elapsed, the height of the object is _____ feet.

Add or subtract as indicated. See Examples 4 and 7.

53. Add.
$$3m^2 + 5m$$
$$2m^2 - 2m$$

54. Add.
$$4a^3 - 4a^2$$
$$6a^3 + 5a^2$$

55. Subtract.
$$12x^4 - x^2$$
$$8x^4 + 3x^2$$

56. Subtract.
$$13y^5 - y^3$$
$$7y^5 + 5y^3$$

57. Add.
$$\frac{2}{3}x^2 + \frac{1}{5}x + \frac{1}{6}$$
$$\frac{1}{2}x^2 - \frac{1}{3}x + \frac{2}{3}$$

58. Add.
$$\frac{4}{7}y^2 - \frac{1}{5}y + \frac{7}{9}$$
$$\frac{1}{3}y^2 - \frac{1}{3}y + \frac{2}{5}$$

59. Subtract.
$$12m^3 - 8m^2 + 6m + 7$$
$$5m^2 \qquad - 4$$

60. Subtract.
$$5a^4 - 3a^3 + 2a^2 - a + 6$$
$$-6a^4 \qquad - a^2 + a - 1$$

61. Subtract.
$$4.3x^3 - 6.1x^2 - 3.0x - 5$$
$$1.4x^3 - 2.6x^2 - 1.5x + 4$$

Perform the indicated operations. See Examples 5, 6, and 8.

62. $(3r^2 + 5r - 6) + (2r - 5r^2)$

63. $(2r^2 + 3r - 12) + (6r^2 + 2r)$

64. $(x^2 + x) - (3x^2 + 2x - 1)$

65. $(8m^2 - 7m) - (3m^2 + 7m - 6)$

66. $(-2b^6 + 3b^4 - b^2) + (b^6 + 2b^4 + 2b^2)$

67. $(16x^3 - x^2 + 3x) + (-12x^3 + 3x^2 + 2x)$

68. $(8t^5 + 3t^3 + 5t) - (19t^4 - 6t^2 + t)$

69. $(7y^4 + 3y^2 + 2y) - (18y^5 - 5y^3 + y)$

70. $[(9b^3 - 4b^2 + 3b + 2) - (-2b^3 + b)] - (8b^3 + 6b + 4)$

71. $[(8m^2 + 4m - 7) - (2m^3 - 5m + 2)] - (m^2 + m)$

72. Subtract $-5w^3 + 5w^2 - 7$ from $6w^3 + 8w + 5$.

73. Subtract $9x^2 - 3x + 7$ from $-2x^2 - 6x + 4$.

74. Find the difference when $9x^4 + 3x^2 + 5$ is subtracted from $8x^4 - 2x^3 + x - 1$.

Find a polynomial that represents the perimeter of each square, rectangle, or triangle.

75.
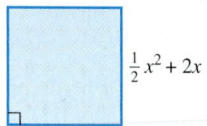
$\frac{1}{2}x^2 + 2x$

76.
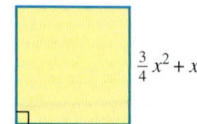
$\frac{3}{4}x^2 + x$

77.
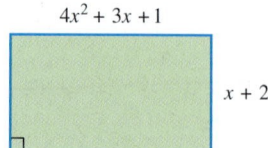
$4x^2 + 3x + 1$
$x + 2$

78.
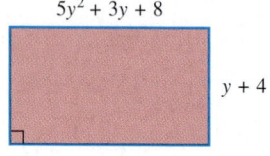
$5y^2 + 3y + 8$
$y + 4$

79.
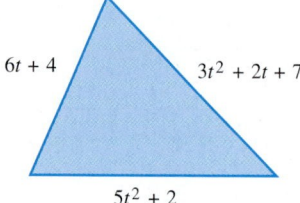
$6t + 4$ $3t^2 + 2t + 7$
$5t^2 + 2$

80.
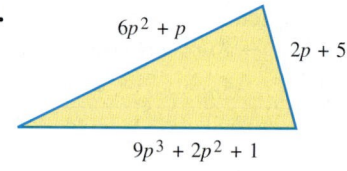
$6p^2 + p$ $2p + 5$
$9p^3 + 2p^2 + 1$

Add or subtract as indicated. See Example 9.

81. $(9a^2b - 3a^2 + 2b) + (4a^2b - 4a^2 - 3b)$

82. $(4xy^3 - 3x + y) + (5xy^3 + 13x - 4y)$

83. $(2c^4d + 3c^2d^2 - 4d^2) - (c^4d + 8c^2d^2 - 5d^2)$

84. $(3k^2h^3 + 5kh + 6k^3h^2) - (2k^2h^3 - 9kh + k^3h^2)$

85. Subtract.
$$9m^3n - 5m^2n^2 + 4mn^2$$
$$-3m^3n + 6m^2n^2 + 8mn^2$$

86. Subtract.
$$12r^5t + 11r^4t^2 - 7r^3t^3$$
$$-8r^5t + 10r^4t^2 + 3r^3t^3$$

*Find **(a)** a polynomial that represents the perimeter of each triangle and **(b)** the measures of the angles of the triangle. (Hint: In part (b), the sum of the measures of the angles of any triangle is 180°.)*

87.
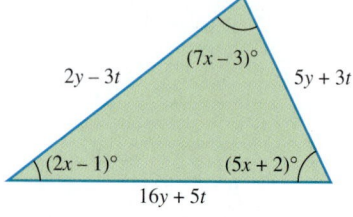
$(7x - 3)°$
$2y - 3t$ $5y + 3t$
$(2x - 1)°$ $(5x + 2)°$
$16y + 5t$

88.
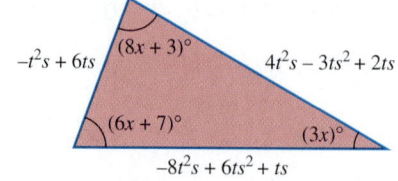
$(8x + 3)°$
$-t^2s + 6ts$ $4t^2s - 3ts^2 + 2ts$
$(6x + 7)°$ $(3x)°$
$-8t^2s + 6ts^2 + ts$

13.2 ▶▶▶ The Product Rule and Power Rules for Exponents

OBJECTIVE 1 Use exponents. In **Section 1.8,** we used exponents to write repeated products. Recall that in the expression 5^2, the number 5 is called the **base** and 2 is called the **exponent,** or **power.** The expression 5^2 is called an **exponential expression.** Although we do not usually write a quantity with an exponent of 1, in general, for any quantity a, $a = a^1$.

EXAMPLE 1 Using Exponents

Write $3 \cdot 3 \cdot 3 \cdot 3 \cdot 3$ in exponential form and evaluate.
    Since 3 occurs as a factor five times, the base is **3** and the exponent is **5.** The exponential expression is 3^5, read "3 to the fifth power," or simply "3 to the fifth."

$$\underbrace{3 \cdot 3 \cdot 3 \cdot 3 \cdot 3}_{\text{5 factors of 3}} \quad \text{means} \quad \mathbf{3^5}, \quad \text{or} \quad 243.$$

Work Problem **1** *at the Side.* ▶

EXAMPLE 2 Evaluating Exponential Expressions

Evaluate. Name the base and the exponent.

		Base	**Exponent**
(a) $5^4 = 5 \cdot 5 \cdot 5 \cdot 5 = 625$		5	4
(b) $-5^4 = -1 \cdot 5^4 = -1 \cdot (5 \cdot 5 \cdot 5 \cdot 5) = -625$		5	4
(c) $(-5)^4 = (-5)(-5)(-5)(-5) = 625$		-5	4

CAUTION
Look at Examples 2(b) and (c). In -5^4, the absence of parentheses shows that the exponent 4 applies only to the base 5, and not -5. In $(-5)^4$, the parentheses show that the exponent 4 applies to the base -5. In summary, $-a^n$ and $(-a)^n$ are not necessarily the same.

Expression	Base	Exponent	Example
$-a^n$	a	n	$-3^2 = -(3 \cdot 3) = -9$
$(-a)^n$	$-a$	n	$(-3)^2 = (-3)(-3) = 9$

Work Problem **2** *at the Side.* ▶

OBJECTIVE 2 Use the product rule for exponents. To develop the product rule, we use the definition of an exponent.

$$2^4 \cdot 2^3 = \underbrace{(2 \cdot 2 \cdot 2 \cdot 2)}_{\text{4 factors}}\underbrace{(2 \cdot 2 \cdot 2)}_{\text{3 factors}}$$
$$= \underbrace{2 \cdot 2 \cdot 2 \cdot 2 \cdot 2 \cdot 2 \cdot 2}_{4 + 3 = 7 \text{ factors}}$$
$$= 2^7$$

OBJECTIVES

1 Use exponents.

2 Use the product rule for exponents.

3 Use the rule $(a^m)^n = a^{mn}$.

4 Use the rule $(ab)^m = a^m b^m$.

5 Use the rule $\left(\dfrac{a}{b}\right)^m = \dfrac{a^m}{b^m}$.

6 Use combinations of the rules for exponents.

7 Use the rules for exponents in a geometry application.

1 Write $2 \cdot 2 \cdot 2 \cdot 2$ in exponential form and evaluate.

2 Evaluate. Name the base and the exponent.

(a) $(-2)^5$ **(b)** -2^5

(c) -4^2 **(d)** $(-4)^2$

ANSWERS

1. 2^4, or 16
2. **(a)** -32; -2; 5   **(b)** -32; 2; 5
   (c) -16; 4; 2   **(d)** 16; -4; 2

3 Simplify by using the product rule, if possible.

(a) $8^2 \cdot 8^5$

(b) $(-7)^5(-7)^3$

(c) $y^3 \cdot y$

(d) $z^2 z^5 z^6$

(e) $4^2 \cdot 3^5$

(f) $6^4 + 6^2$

Also,

$$6^2 \cdot 6^3 = (6 \cdot 6)(6 \cdot 6 \cdot 6)$$
$$= 6 \cdot 6 \cdot 6 \cdot 6 \cdot 6$$
$$= 6^5.$$

Generalizing from these examples,

$$2^4 \cdot 2^3 = 2^{4+3} = 2^7 \quad \text{and} \quad 6^2 \cdot 6^3 = 6^{2+3} = 6^5.$$

In each case, adding the exponents gives the exponent of the product, suggesting the **product rule for exponents.**

Product Rule for Exponents

For any positive integers m and n, $\quad a^m \cdot a^n = a^{m+n}$.
(Keep the same base and add the exponents.)

Example: $\quad 6^2 \cdot 6^5 = 6^{2+5} = 6^7$

CAUTION

Do not multiply the bases when using the product rule. *Keep the same base and add the exponents.* For example,

$$6^2 \cdot 6^5 = 6^7, \quad \textbf{\textit{not}} \quad 36^7.$$

EXAMPLE 3 **Using the Product Rule**

Use the product rule for exponents to simplify, if possible.

(a) $6^3 \cdot 6^5 = 6^{3+5} = 6^8$ — Keep the same base.

(b) $(-4)^7(-4)^2 = (-4)^{7+2} = (-4)^9$

(c) $x^2 \cdot x = x^2 \cdot x^1 = x^{2+1} = x^3$

(d) $m^4 m^3 m^5 = m^{4+3+5} = m^{12}$

(e) $2^3 \cdot 3^2$

The product rule does not apply to the product $2^3 \cdot 3^2$ because the bases are different.

$$2^3 \cdot 3^2 = 8 \cdot 9 = 72 \qquad \text{Evaluate } 2^3 \text{ and } 3^2; \text{ then multiply.}$$

Think: 2^3 means $2 \cdot 2 \cdot 2$.

Think: 3^2 means $3 \cdot 3$.

(f) $2^3 + 2^4$

The product rule does not apply to $2^3 + 2^4$ because it is a *sum,* not a *product.*

$$2^3 + 2^4 = 8 + 16 = 24 \qquad \text{Evaluate } 2^3 \text{ and } 2^4; \text{ then add.}$$

CAUTION

The bases of the factors must be the same before we can apply the product rule for exponents.

ANSWERS

3. **(a)** 8^7 **(b)** $(-7)^8$ **(c)** y^4
   (d) z^{13} **(e)** The product rule does not apply.
   (product: 3888) **(f)** The product rule does
   not apply. (sum: 1332)

◀ *Work Problem* **3** *at the Side.*

EXAMPLE 4 Using the Product Rule

Multiply $2x^3$ and $3x^7$.

$2x^3 \cdot 3x^7$ $\quad$ $2x^3 = 2 \cdot x^3; \ 3x^7 = 3 \cdot x^7$

$= (2 \cdot 3) \cdot (x^3 \cdot x^7)$ $\quad$ Commutative and associative properties

$= 6x^{3+7}$ $\quad$ Multiply; product rule

$= 6x^{10}$ $\quad$ Add the exponents.

CAUTION

Be sure you understand the difference between *adding* and *multiplying* exponential expressions. For example,

$$8x^3 + 5x^3 \quad \text{means} \quad (8 + 5)x^3, \quad \text{or} \quad 13x^3,$$

but $\quad (8x^3)(5x^3) \quad$ means $\quad (8 \cdot 5)x^{3+3}, \quad$ or $\quad 40x^6.$

Work Problem **4** *at the Side.* ▶

OBJECTIVE 3 Use the rule $(a^m)^n = a^{mn}$. We can simplify an expression such as $(8^3)^2$ with the product rule for exponents, as follows.

$$(8^3)^2 = (8^3)(8^3) = 8^{3+3} = 8^6$$

The product of the exponents in $(8^3)^2$, **3 · 2**, gives the exponent in 8^6. Also,

$(5^2)^4 = 5^2 \cdot 5^2 \cdot 5^2 \cdot 5^2$ $\quad$ Definition of exponent

$= 5^{2+2+2+2}$ $\quad$ Product rule

$= 5^8,$ $\quad$ Add the exponents.

and $2 \cdot 4 = 8$. These examples suggest **power rule (a) for exponents.**

Power Rule (a) for Exponents

For any positive integers m and n, $\quad (a^m)^n = a^{mn}$.
(Raise a power to a power by multiplying exponents.)

Example: $\quad (3^2)^4 = 3^{2 \cdot 4} = 3^8$

EXAMPLE 5 Using Power Rule (a)

Use power rule (a) for exponents to simplify.

(a) $(2^5)^3 = 2^{5 \cdot 3} = 2^{15}$ $\quad$ **(b)** $(5^7)^2 = 5^{7 \cdot 2} = 5^{14}$ $\quad$ **(c)** $(x^2)^5 = x^{2 \cdot 5} = x^{10}$

Work Problem **5** *at the Side.* ▶

OBJECTIVE 4 Use the rule $(ab)^m = a^m b^m$. We can rewrite the expression $(4x)^3$ as shown below.

$(4x)^3 = (4x)(4x)(4x)$ $\quad$ Definition of exponent

$= 4 \cdot 4 \cdot 4 \cdot x \cdot x \cdot x$ $\quad$ Commutative and associative properties

$= 4^3 x^3$ $\quad$ Definition of exponent

This example suggests **power rule (b) for exponents.**

---

4 Multiply.

(a) $5m^2 \cdot 2m^6$

(b) $3p^5 \cdot 9p^4$

(c) $-7p^5 \cdot (3p^8)$

5 Simplify.

(a) $(5^3)^4$

(b) $(6^2)^5$

(c) $(3^2)^4$

(d) $(a^6)^5$

ANSWERS

4. (a) $10m^8$ **(b)** $27p^9$ **(c)** $-21p^{13}$

5. (a) 5^{12} **(b)** 6^{10} **(c)** 3^8 **(d)** a^{30}

6 Simplify.

(a) $(2ab)^4$

(b) $5(mn)^3$

(c) $(3a^2b^4)^5$

(d) $(-5m^2)^3$

Power Rule (b) for Exponents

For any positive integer m, $(ab)^m = a^m b^m$.
(Raise a product to a power by raising each factor to the power.)

Example: $(2p)^5 = 2^5 p^5$

EXAMPLE 6 Using Power Rule (b)

Use power rule (b) for exponents to simplify.

(a) $(3xy)^2$

$= 3^2 x^2 y^2$ Power rule (b)

$= 9x^2 y^2$ $3^2 = 3 \cdot 3 = 9$

(b) $9(pq)^2$

$= 9(p^2 q^2)$ Power rule (b)

$= 9p^2 q^2$ Multiply.

(c) $5(2m^2 p^3)^4$

$= 5[2^4 (m^2)^4 (p^3)^4]$ Power rule (b)

$= 5(2^4 m^8 p^{12})$ Power rule (a)

$= 5 \cdot 2^4 m^8 p^{12}$

$= 80m^8 p^{12}$ $5 \cdot 2^4 = 5 \cdot 16 = 80$

(d) $(-5^6)^3$

$= (-1 \cdot 5^6)^3$ $-a = -1 \cdot a$

$= (-1)^3 (5^6)^3$ Power rule (b)

$= -1 \cdot 5^{18}$ Power rule (a)

$= -5^{18}$

Raise -1 to the designated power.

CAUTION

Power rule (b) does not apply to a sum:

$$(4x)^2 = 4^2 x^2, \quad \text{but} \quad (4 + x)^2 \neq 4^2 + x^2.$$

◀ **Work Problem** **6** **at the Side.**

OBJECTIVE **5** **Use the rule** $\left(\frac{a}{b}\right)^m = \frac{a^m}{b^m}$**.** Since the quotient $\frac{a}{b}$ can be written as $a \cdot \frac{1}{b}$, we can use power rule (b), together with some of the properties of real numbers, to get **power rule (c) for exponents.**

Power Rule (c) for Exponents

For any positive integer m, $\left(\dfrac{a}{b}\right)^m = \dfrac{a^m}{b^m}$ $(b \neq 0)$.

(Raise a quotient to a power by raising both the numerator and the denominator to the power.)

Example: $\left(\dfrac{5}{3}\right)^2 = \dfrac{5^2}{3^2}$

EXAMPLE 7 **Using Power Rule (c)**

Use power rule (c) for exponents to simplify.

(a) $\left(\dfrac{2}{3}\right)^5 = \dfrac{2^5}{3^5} = \dfrac{32}{243}$

(b) $\left(\dfrac{m}{n}\right)^4 = \dfrac{m^4}{n^4}, \quad n \neq 0$

(c) $\left(\dfrac{1}{5}\right)^4 = \dfrac{1^4}{5^4} = \dfrac{1}{5^4} = \dfrac{1}{625}$ $\quad 1^4 = 1 \cdot 1 \cdot 1 \cdot 1 = 1$

Note

In Example 7(c), we used the fact that $1^4 = 1$.

> *In general, $1^n = 1$, for any integer n.*

Work Problem **7** *at the Side.* ▶

The rules for exponents discussed in this section are basic to the study of algebra and should be *memorized*.

Rules for Exponents

For positive integers m and n: *Examples*

Product rule $a^m \cdot a^n = a^{m+n}$ $6^2 \cdot 6^5 = 6^{2+5} = 6^7$

Power rules **(a)** $(a^m)^n = a^{mn}$ $(3^2)^4 = 3^{2 \cdot 4} = 3^8$

(b) $(ab)^m = a^m b^m$ $(2p)^5 = 2^5 p^5$

(c) $\left(\dfrac{a}{b}\right)^m = \dfrac{a^m}{b^m}$ *(b ≠ 0).* $\left(\dfrac{5}{3}\right)^2 = \dfrac{5^2}{3^2}$

OBJECTIVE **6** **Use combinations of the rules for exponents.**
More than one rule may be needed to simplify an exponential expression.

EXAMPLE 8 **Using Combinations of Rules**

Simplify each expression.

(a) $\left(\dfrac{2}{3}\right)^2 \cdot 2^3$

$= \dfrac{2^2}{3^2} \cdot \dfrac{2^3}{1}$ Power rule (c)

$= \dfrac{2^2 \cdot 2^3}{3^2 \cdot 1}$ Multiply fractions.

$= \dfrac{2^{2+3}}{3^2}$ Product rule

$= \dfrac{2^5}{3^2}$

$= \dfrac{32}{9}$

(b) $(5x)^3 (5x)^4$

$= (5x)^7$ Product rule

$= 5^7 x^7$ Power rule (b)

Continued on Next Page

---

7 Simplify. Assume that all variables represent nonzero real numbers.

(a) $\left(\dfrac{5}{2}\right)^4$

(b) $\left(\dfrac{p}{q}\right)^2$

(c) $\left(\dfrac{r}{t}\right)^3$

(d) $\left(\dfrac{1}{3}\right)^5$

(e) $\left(\dfrac{1}{x}\right)^{10}$

ANSWERS

7. **(a)** $\dfrac{625}{16}$ **(b)** $\dfrac{p^2}{q^2}$ **(c)** $\dfrac{r^3}{t^3}$ **(d)** $\dfrac{1}{243}$ **(e)** $\dfrac{1}{x^{10}}$

8 Simplify.

(a) $(2m)^3 (2m)^4$

(b) $\left(\dfrac{5k^3}{3}\right)^2$

(c) $\left(\dfrac{1}{5}\right)^4 (2x)^2$

(d) $(-3xy^2)^3 (x^2y)^4$

9 Find a polynomial that represents the area of the figure.

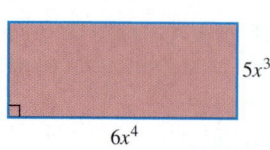

$4x^2$

$8x^4$

(c) $(2x^2y^3)^4 (3xy^2)^3$

$= 2^4 (x^2)^4 (y^3)^4 \cdot 3^3x^3 (y^2)^3$ Power rule (b)

$= 2^4x^8y^{12} \cdot 3^3x^3y^6$ Power rule (a)

$= 2^4 \cdot 3^3 x^8x^3y^{12}y^6$ Commutative and associative properties

$= 16 \cdot 27x^{11}y^{18}$ Product rule

$= 432x^{11}y^{18}$ Multiply.

Notice that $(2x^2y^3)^4$ means $2^4x^{2\cdot4}y^{3\cdot4}$, **not** $(2\cdot4)x^{2\cdot4}y^{3\cdot4}$.

> Do *not* multiply the coefficient 2 and the exponent 4.

(d) $(-x^3y)^2 (-x^5y^4)^3$

> Think of the negative sign in each factor as -1.

$= (-1x^3y)^2 (-1x^5y^4)^3$ $-a = -1 \cdot a$

$= (-1)^2 (x^3)^2 (y^2) \cdot (-1)^3 (x^5)^3 (y^4)^3$ Power rule (b)

$= (-1)^2 (x^6) (y^2) \cdot (-1)^3 (x^{15}) (y^{12})$ Power rule (a)

$= (-1)^5 (x^{6+15}) (y^{2+12})$ Product rule

$= -1x^{21}y^{14}$

$= -x^{21}y^{14}$

◀ *Work Problem* **8** *at the Side.*

OBJECTIVE **7** **Use the rules for exponents in a geometry application.**

EXAMPLE 9 **Using Area Formulas**

Find a polynomial that represents the area of each geometric figure.

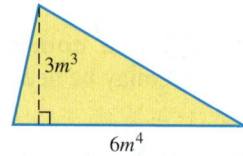

$5x^3$

$6x^4$

Figure 1

$3m^3$

$6m^4$

Figure 2

For Figure 1, use the formula for the area of a rectangle, $A = lw$.

$A = (6x^4)(5x^3)$ Area formula

$A = 6 \cdot 5 \cdot x^{4+3}$ Product rule

$A = 30x^7$

Figure 2 is a triangle with base $6m^4$ and height $3m^3$. Substitute into the formula for the area of a triangle and simplify.

$A = \dfrac{1}{2}bh$ Area formula

$A = \dfrac{1}{2}(6m^4)(3m^3)$ Substitute.

$A = \dfrac{1}{2}(18m^7)$,   or $9m^7$ Product rule; multiply.

◀ *Work Problem* **9** *at the Side.*

13.2 ▶▶▶ Exercises

1. What exponent is understood on the base x in the expression xy^2?

2. How are the expressions 3^2, 5^3, and 7^4 read?

Decide whether each statement is true *or* false.

3. $3^3 = 9$

4. $(-2)^4 = 2^4$

5. $(a^2)^3 = a^5$

6. $\left(\dfrac{1}{4}\right)^2 = \dfrac{1}{4^2}$

Write each expression using exponents. See Example 1.

7. $t \cdot t \cdot t \cdot t \cdot t \cdot t \cdot t \cdot t$

8. $w \cdot w \cdot w \cdot w \cdot w \cdot w$

9. $\left(\dfrac{1}{2}\right)\left(\dfrac{1}{2}\right)\left(\dfrac{1}{2}\right)\left(\dfrac{1}{2}\right)\left(\dfrac{1}{2}\right)$

10. $\left(-\dfrac{1}{4}\right)\left(-\dfrac{1}{4}\right)\left(-\dfrac{1}{4}\right)\left(-\dfrac{1}{4}\right)$

11. $(-8p)(-8p)$

12. $(-7x)(-7x)(-7x)$

13. Explain how the expressions $(-3)^4$ and -3^4 are different.

14. Explain how the expressions $(5x)^3$ and $5x^3$ are different.

Identify the base and the exponent for each exponential expression. In Exercises 15–18, also evaluate the expression. See Example 2.

15. 3^5

16. 2^7

17. $(-3)^5$

18. $(-2)^7$

19. $(-6x)^4$

20. $(-8x)^4$

21. $-6x^4$

22. $-8x^4$

23. Explain why the product rule does not apply to the expression $5^2 + 5^3$. Then evaluate the expression.

24. Explain why the product rule does not apply to the expression $3^2 \cdot 4^3$. Then evaluate the expression.

Use the product rule for exponents to simplify each expression, if possible. Write each answer in exponential form. See Examples 3 and 4.

25. $5^2 \cdot 5^6$

26. $3^6 \cdot 3^7$

27. $4^2 \cdot 4^7 \cdot 4^3$

28. $5^3 \cdot 5^8 \cdot 5^2$

29. $(-7)^3(-7)^6$

30. $(-9)^8(-9)^5$

31. $t^3 t^8 t^{13}$

32. $n^5 n^6 n^9$

33. $(-8r^4)(7r^3)$

34. $(10a^7)(-4a^3)$

35. $(-6p^5)(-7p^5)$

36. $(-5w^8)(-9w^8)$

37. $3^8 + 3^9$

38. $4^{12} + 4^5$

39. $5^8 \cdot 3^8$

40. $6^3 \cdot 8^3$

Use the power rules for exponents to simplify each expression. See Examples 5–7.

41. $(4^3)^2$

42. $(8^3)^6$

43. $(t^4)^5$

44. $(y^6)^5$

45. $(7r)^3$

46. $(11x)^4$

47. $(-5^2)^6$

48. $(-9^4)^8$

49. $(-8^3)^5$

50. $(-7^5)^7$

51. $(5xy)^5$

52. $(9pq)^6$

53. $8(qr)^3$

54. $4(vw)^5$

55. $\left(\dfrac{1}{2}\right)^3$

56. $\left(\dfrac{1}{3}\right)^5$

57. $\left(\dfrac{a}{b}\right)^3$, $b \neq 0$

58. $\left(\dfrac{r}{t}\right)^4$, $t \neq 0$

59. $\left(\dfrac{9}{5}\right)^8$

60. $\left(\dfrac{12}{7}\right)^6$

61. $(-2x^2y)^3$

62. $(-5m^4p^2)^3$

63. $(3a^3b^2)^2$

64. $(4x^3y^5)^4$

Simplify each expression. See Example 8.

65. $\left(\dfrac{5}{2}\right)^3 \cdot \left(\dfrac{5}{2}\right)^2$

66. $\left(\dfrac{3}{4}\right)^5 \cdot \left(\dfrac{3}{4}\right)^6$

67. $\left(\dfrac{9}{8}\right)^3 \cdot 9^2$

68. $\left(\dfrac{8}{5}\right)^4 \cdot 8^3$

69. $(2x)^9 (2x)^3$

70. $(6y)^5 (6y)^8$

71. $(-6p)^4 (-6p)$

72. $(-13q)^3 (-13q)$

73. $(6x^2y^3)^5$

74. $(5r^5t^6)^7$

75. $(x^2)^3 (x^3)^5$

76. $(y^4)^5 (y^3)^5$

77. $(2w^2x^3y)^2 (x^4y)^5$

78. $(3x^4y^2z)^3 (yz^4)^5$

79. $(-r^4s)^2 (-r^2s^3)^5$

80. $(-ts^6)^4 (-t^3s^5)^3$

81. $\left(\dfrac{5a^2b^5}{c^6}\right)^3$, $c \neq 0$

82. $\left(\dfrac{6x^3y^9}{z^5}\right)^4$, $z \neq 0$

83. $(-5m^3p^4q)^2 (p^2q)^3$

84. $(-a^4b^5)(-6a^3b^3)^2$

85. $(2x^2y^3z)^4 (xy^2z^3)^2$

Find a polynomial that represents the area of each figure. See Example 9.

86.

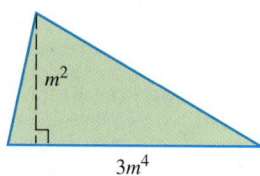

87.

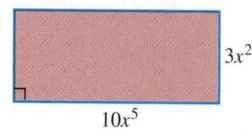

88.

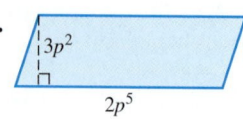

13.3 ▷▷▷ Multiplying Polynomials

OBJECTIVE **1** **Multiply a monomial and a polynomial.** As shown in **Section 13.2,** we find the product of two monomials by using the rules for exponents and the commutative and associative properties. For example,

$$(-8m^6)(-9n^6)$$
$$= (-8)(-9)(m^6)(n^6)$$
$$= 72m^6n^6.$$

> **CAUTION**
> *Do not confuse addition of terms with multiplication of terms.*
> $$7q^5 + 2q^5 = 9q^5, \quad \text{but} \quad (7q^5)(2q^5) = 7 \cdot 2q^{5+5} = 14q^{10}.$$

To find the product of a monomial and a polynomial with more than one term, we use the distributive property and multiplication of monomials.

EXAMPLE 1 **Multiplying Monomials and Polynomials**

Find each product.

(a) $4x^2(3x + 5)$

$$4x^2(3x + 5) = 4x^2(3x) + 4x^2(5) \qquad \text{Distributive property}$$
$$= 12x^3 + 20x^2 \qquad \text{Multiply monomials.}$$

(b) $-8m^3(4m^3 + 3m^2 + 2m - 1)$
$$= -8m^3(4m^3) + (-8m^3)(3m^2)$$
$$+ (-8m^3)(2m) + (-8m^3)(-1) \qquad \text{Distributive property}$$
$$= -32m^6 - 24m^5 - 16m^4 + 8m^3 \qquad \text{Multiply monomials.}$$

Work Problem ① *at the Side.* ▶

OBJECTIVE **2** **Multiply two polynomials.** We can use the distributive property repeatedly to find the product of any two polynomials. For example, to find the product of the polynomials $x^2 + 3x + 5$ and $x - 4$, think of $x - 4$ as a single quantity and use the distributive property as follows.

$$(x^2 + 3x + 5)(x - 4)$$
$$= x^2(x - 4) + 3x(x - 4) + 5(x - 4) \qquad \text{Distributive property}$$
$$= x^2(x) + x^2(-4) + 3x(x) + 3x(-4) + 5(x) + 5(-4)$$
$$\qquad\qquad \text{Distributive property again}$$
$$= x^3 - 4x^2 + 3x^2 - 12x + 5x - 20 \qquad \text{Multiply monomials.}$$
$$= x^3 - x^2 - 7x - 20 \qquad \text{Combine like terms.}$$

This example suggests the following rule.

> **Multiplying Polynomials**
> To multiply two polynomials, multiply each term of the second polynomial by each term of the first polynomial and add the products.

OBJECTIVES

1 Multiply a monomial and a polynomial.

2 Multiply two polynomials.

3 Multiply binomials by the FOIL method.

① Find each product.

(a) $5m^3(2m + 7)$

(b) $2x^4(3x^2 + 2x - 5)$

(c) $-4y^2(3y^3 + 2y^2 - 4y + 8)$

ANSWERS

1. **(a)** $10m^4 + 35m^3$
   (b) $6x^6 + 4x^5 - 10x^4$
   (c) $-12y^5 - 8y^4 + 16y^3 - 32y^2$

2 Multiply.

(a) $(m + 3)(m^2 - 2m + 1)$

(b) $(6p^2 + 2p - 4)(3p^2 - 5)$

3 Find the product.

$$3x^2 + 4x - 5$$
$$\underline{x + 4}$$

4 Use the rectangle method to find each product.

(a) $(4x + 3)(x + 2)$

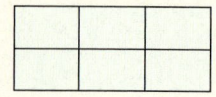

(b) $(x + 5)(x^2 + 3x + 1)$

EXAMPLE 2 Multiplying Two Polynomials

Multiply $(m^2 + 5)(4m^3 - 2m^2 + 4m)$.

Multiply each term of the second polynomial by each term of the first.

$(m^2 + 5)(4m^3 - 2m^2 + 4m)$

$= m^2(4m^3) + m^2(-2m^2) + m^2(4m) + 5(4m^3) + 5(-2m^2) + 5(4m)$

$= 4m^5 - 2m^4 + 4m^3 + 20m^3 - 10m^2 + 20m$

$= 4m^5 - 2m^4 + 24m^3 - 10m^2 + 20m$ Combine like terms.

◀ Work Problem **2** at the Side.

EXAMPLE 3 Multiplying Polynomials Vertically

Multiply $(x^3 + 2x^2 + 4x + 1)(3x + 5)$ vertically.

Write the polynomials as follows.

$$x^3 + 2x^2 + 4x + 1$$
$$\underline{3x + 5}$$

Begin by multiplying each of the terms in the top row by 5.

$$x^3 + 2x^2 + 4x + 1$$
$$\underline{3x + 5}$$
$$5x^3 + 10x^2 + 20x + 5 \quad 5(x^3 + 2x^2 + 4x + 1)$$

Notice how this process is similar to multiplication of whole numbers. Now multiply each term in the top row by $3x$. Then add like terms.

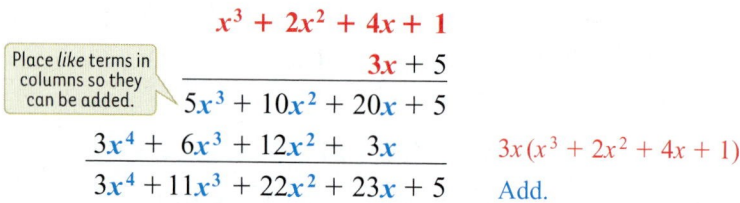

The product is $3x^4 + 11x^3 + 22x^2 + 23x + 5$.

◀ Work Problem **3** at the Side.

We can use a rectangle to model polynomial multiplication. For example, to find the product

$$(2x + 1)(3x + 2),$$

label a rectangle with each term as shown below on the left. Then put the product of each pair of monomials in the appropriate box as shown on the right.

The product of the binomials is the sum of these four monomial products.

$(2x + 1)(3x + 2)$
$= 6x^2 + 4x + 3x + 2$
$= 6x^2 + 7x + 2$

◀ Work Problem **4** at the Side.

ANSWERS

2. (a) $m^3 + m^2 - 5m + 3$
   (b) $18p^4 + 6p^3 - 42p^2 - 10p + 20$
3. $3x^3 + 16x^2 + 11x - 20$
4. (a) $4x^2 + 11x + 6$
   (b) $x^3 + 8x^2 + 16x + 5$

OBJECTIVE 3 Multiply binomials by the FOIL method. In algebra, many of the polynomials to be multiplied are both binomials (with just two terms). For these products, the **FOIL method** reduces the rectangle method to a systematic approach without the rectangle. To develop the FOIL method, we use the distributive property to find $(x + 3)(x + 5)$.

$$(x + 3)(x + 5)$$

$$= (x + 3)x + (x + 3)5 \qquad \text{Distributive property}$$

$$= x(x) + 3(x) + x(5) + 3(5) \qquad \text{Distributive property again}$$

$$= x^2 + 3x + 5x + 15 \qquad \text{Multiply.}$$

$$= x^2 + 8x + 15 \qquad \text{Combine like terms.}$$

Here is where the letters of the word FOIL originate.

$(x + 3)(x + 5)$ Multiply the **First terms**: $x(x)$.   **F**

$(x + 3)(x + 5)$ Multiply the **Outer terms**: $x(5)$.   **O**
This is the **outer product.**

$(x + 3)(x + 5)$ Multiply the **Inner terms**: $3(x)$.   **I**
This is the **inner product.**

$(x + 3)(x + 5)$ Multiply the **Last terms**: $3(5)$.   **L**

The outer product, $5x$, and the inner product, $3x$, should be added mentally so that the three terms of the answer can be written without extra steps.

$$(x + 3)(x + 5)$$

$$= x^2 + 8x + 15$$

A summary of the steps in the FOIL method follows.

Multiplying Binomials by the FOIL Method

Step 1   Multiply the two **F**irst terms of the binomials to get the first term of the answer.

Step 2   Find the **O**uter product and the **I**nner product and add them (when possible) to get the middle term of the answer.

Step 3   Multiply the two **L**ast terms of the binomials to get the last term of the answer.

$$\mathbf{F} = x^2 \qquad \mathbf{L} = 15$$

$$(x + 3)(x + 5)$$

$$\mathbf{I} = 3x$$
$$\underline{\mathbf{O} = 5x}$$
$$8x \qquad \text{Add.}$$

Work Problem **5** *at the Side.* ▶

5 For the product

$$(2p - 5)(3p + 7),$$

find the following.

(a) Product of first terms

(b) Outer product

(c) Inner product

(d) Product of last terms

(e) Complete product in simplified form

6 Use the FOIL method to find each product.

(a) $(m + 4)(m - 3)$

(b) $(y + 7)(y + 2)$

(c) $(r - 8)(r - 5)$

EXAMPLE 4 Using the FOIL Method

Use the FOIL method to find the product $(x + 8)(x - 6)$.

Step 1 **F** Multiply the **first** terms: $x(x) = x^2$.

Step 2 **O** Find the **outer** product: $x(-6) = -6x$.

   I Find the **inner** product: $8(x) = 8x$.

   Add the outer and inner products mentally: $-6x + 8x = \mathbf{2x}$.

Step 3 **L** Multiply the **last** terms: $8(-6) = \mathbf{-48}$.

The product $(x + 8)(x - 6)$ is $x^2 + \mathbf{2x} - \mathbf{48}$, the sum of the terms found in Steps 1–3. As a shortcut, this product can be found as follows.

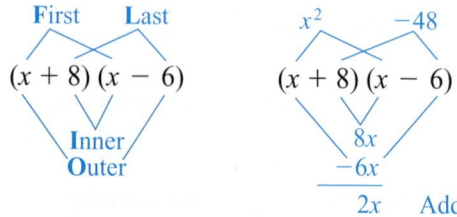

◀ *Work Problem* **6** *at the Side.*

EXAMPLE 5 Using the FOIL Method

Multiply $(9x - 2)(3y + 1)$.

First $(9x - 2)(3y + 1)$ $27xy$

Outer $(9x - 2)(3y + 1)$ $9x$ — These unlike terms cannot be added.

Inner $(9x - 2)(3y + 1)$ $-6y$

Last $(9x - 2)(3y + 1)$ -2

   F **O** **I** **L**

The product $(9x - 2)(3y + 1)$ is $27xy + 9x - 6y - 2$.

◀ *Work Problem* **7** *at the Side.*

---

7 Find the product.

$(4x - 3)(2y + 5)$

---

8 Find each product.

(a) $(6m + 5)(m - 4)$

(b) $(3r + 2t)(3r + 4t)$

(c) $y^2(8y + 3)(2y + 1)$

EXAMPLE 6 Using the FOIL Method

Find each product.

(a) $(2k + 5y)(k + 3y)$

   F **O** **I** **L**
$$= 2k(k) + 2k(3y) + 5y(k) + 5y(3y)$$
$$= 2k^2 + 6ky + 5ky + 15y^2$$
$$= 2k^2 + 11ky + 15y^2$$

(b) $(7p + 2q)(3p - q)$
$$= 21p^2 - 7pq + 6pq - 2q^2$$
$$= 21p^2 - pq - 2q^2$$

(c) $2x^2(x - 3)(3x + 4)$
$$= 2x^2(3x^2 - 5x - 12)$$
$$= 6x^4 - 10x^3 - 24x^2$$

◀ *Work Problem* **8** *at the Side.*

Note

In Example 6(c), we could have multiplied $2x^2$ and $x - 3$ first.

$$2x^2(x - 3)(3x + 4)$$
$$= (2x^3 - 6x^2)(3x + 4)$$
$$= 6x^4 - 10x^3 - 24x^2 \quad \text{Same answer}$$

13.3 ▶▶▶ Exercises

Find each product using the rectangle method shown in the text.

1. $(x + 3)(x + 4)$ **2.** $(x + 5)(x + 2)$ **3.** $(2x + 1)(x^2 + 3x + 2)$ **4.** $(x + 4)(3x^2 + 2x + 1)$

5. In multiplying a monomial by a polynomial, such as in $4x(3x^2 + 7x^3) = 4x(3x^2) + 4x(7x^3)$, the first property that is used is the _____ property.

6. Match each product in parts (a)–(d) with the correct polynomial in choices A–D.

   (a) $(x - 5)(x + 3)$ **(b)** $(x + 5)(x + 3)$ **(c)** $(x - 5)(x - 3)$ **(d)** $(x + 5)(x - 3)$

   A. $x^2 + 8x + 15$ **B.** $x^2 - 8x + 15$ **C.** $x^2 - 2x - 15$ **D.** $x^2 + 2x - 15$

Find each product. See Example 1.

7. $-2m(3m + 2)$ **8.** $-5p(6 + 3p)$ **9.** $\dfrac{3}{4}p(8 - 6p + 12p^3)$

10. $\dfrac{4}{3}x(3 + 2x + 5x^3)$ **11.** $2y^5(3 + 2y + 5y^4)$ **12.** $2m^4(3m^2 + 5m + 6)$

Find each product. See Examples 2 and 3.

13. $(6x + 1)(2x^2 + 4x + 1)$ **14.** $(9y - 2)(8y^2 - 6y + 1)$

15. $(2r - 1)(3r^2 + 4r - 4)$ **16.** $(9a + 2)(9a^2 + a + 1)$

17. $(4m + 3)(5m^3 - 4m^2 + m - 5)$ **18.** $(y + 4)(3y^3 - 2y^2 + y + 3)$

19. $(5x^2 + 2x + 1)(x^2 - 3x + 5)$ **20.** $(2m^2 + m - 3)(m^2 - 4m + 5)$

Find each product. See Examples 4–6.

21. $(m + 7)(m + 5)$ **22.** $(x + 4)(x + 7)$ **23.** $(n - 2)(n + 3)$ **24.** $(r - 6)(r + 8)$

25. $(4r + 1)(2r - 3)$ **26.** $(5x + 2)(2x - 7)$ **27.** $(3x + 2)(3x - 2)$ **28.** $(7x + 3)(7x - 3)$

29. $(3q + 1)(3q + 1)$

30. $(4w + 7)(4w + 7)$

31. $(5x + 7)(3y - 8)$

32. $(4x + 3)(2y - 1)$

33. $(3t + 4s)(2t + 5s)$

34. $(8v + 5w)(2v + 3w)$

35. $(-0.3t + 0.4)(t + 0.6)$

36. $(-0.5x + 0.9)(x - 0.2)$

37. $\left(x - \dfrac{2}{3}\right)\left(x + \dfrac{1}{4}\right)$

38. $\left(-\dfrac{8}{3} + 3k\right)\left(-\dfrac{2}{3} - k\right)$

39. $\left(-\dfrac{5}{4} + 2r\right)\left(-\dfrac{3}{4} - r\right)$

40. $2m^3(4m - 1)(2m + 3)$

41. $x(2x - 5)(x + 3)$

42. $5t^4(t + 3)(3t - 1)$

43. $3y^3(2y + 3)(y - 5)$

Relating Concepts (Exercises 44–48) For Individual or Group Work

Work Exercises 44–48 in order. (All units are in yards.)

44. Find a polynomial that represents the area of the rectangle.

$3x + 6$

10

45. Suppose you know that the area of the rectangle is 600 yd². Use this information and the polynomial from Exercise 44 to write an equation in x, and solve it.

46. (a) What are the dimensions of the rectangle?

(b) Use the result of part (a) to find the perimeter of the lawn.

47. Suppose the rectangle represents a lawn and it costs \$3.50 per square yard to lay sod on the lawn. How much will it cost to sod the entire lawn?

48. Again, suppose the rectangle represents a lawn and it costs \$9.00 per yard to fence the lawn. How much will it cost to fence the lawn?

---

49. Perform the following multiplications: $(x + 4)(x - 4)$; $(y + 2)(y - 2)$; $(r + 7)(r - 7)$. Observe your answers, and explain the pattern that can be found in the answers.

50. Repeat Exercise 49 for the following: $(x + 4)(x + 4)$; $(y - 2)(y - 2)$; $(r + 7)(r + 7)$.

13.4 ▶▶▶ Special Products

In this section, we develop shortcuts to find certain binomial products.

OBJECTIVE 1 Square binomials. The square of a binomial can be found quickly by using the method shown in Example 1.

EXAMPLE 1 Squaring a Binomial

Find $(m + 3)^2$.

$(m + 3)^2$ means $(m + 3)(m + 3)$.

$$(m + 3)(m + 3)$$
$$= m^2 + 3m + 3m + 9 \quad \text{FOIL}$$
$$= m^2 + 6m + 9 \quad \text{Combine like terms.}$$

This result has the squares of the first and the last terms of the binomial:

$$m^2 = m^2 \quad \text{and} \quad 3^2 = 9.$$

The middle term, 6m, is twice the product of the two terms of the binomial, since the outer and inner products are $m(3)$ and $3(m)$, and

$$m(3) + 3(m) = 2(m)(3) = 6m.$$

Work Problem **1** *at the Side.* ▶

Example 1 suggests the following rules.

> **Square of a Binomial**
>
> The square of a binomial is a trinomial consisting of the square of the first term, plus twice the product of the two terms, plus the square of the last term of the binomial. For a and b,
>
> $$(a + b)^2 = a^2 + 2ab + b^2.$$
>
> Also, $\qquad (a - b)^2 = a^2 - 2ab + b^2.$

EXAMPLE 2 Squaring Binomials

Square each binomial.

$$(a - b)^2 = a^2 - 2 \cdot a \cdot b + b^2$$

(a) $(5z - 1)^2 = (5z)^2 - 2(5z)(1) + (1)^2$
$$= 25z^2 - 10z + 1 \qquad (5z)^2 = 5^2z^2 = 25z^2$$

(b) $(3b + 5r)^2$
$$= (3b)^2 + 2(3b)(5r) + (5r)^2$$
$$= 9b^2 + 30br + 25r^2$$

(c) $(2a - 9x)^2$
$$= (2a)^2 - 2(2a)(9x) + (9x)^2$$
$$= 4a^2 - 36ax + 81x^2$$

Continued on Next Page

OBJECTIVES

1. Square binomials.
2. Find the product of the sum and difference of two terms.
3. Find greater powers of binomials.

1 Consider the binomial $x + 4$.

  (a) What is the first term of the binomial? Square it.

  (b) What is the last term of the binomial? Square it.

  (c) Find twice the product of the two terms of the binomial.

  (d) Find $(x + 4)^2$.

ANSWERS

1. **(a)** $x; x^2$ **(b)** $4; 16$ **(c)** $8x$
   (d) $x^2 + 8x + 16$

2 Square each binomial.

(a) $(t - 6)^2$

(b) $(2m - p)^2$

(c) $(4p + 3q)^2$

(d) $(5r - 6s)^2$

(e) $\left(3k - \dfrac{1}{2}\right)^2$

(f) $x(2x + 7)^2$

(d) $\left(4m + \dfrac{1}{2}\right)^2$

$$= (4m)^2 + 2(4m)\left(\dfrac{1}{2}\right) + \left(\dfrac{1}{2}\right)^2 \qquad (a+b)^2 = a^2 + 2ab + b^2$$

$$= 16m^2 + 4m + \dfrac{1}{4}$$

(e) $x(4x - 3)^2$

> Remember the middle term.

$$= x(16x^2 - 24x + 9) \qquad \text{Square the binomial.}$$

$$= 16x^3 - 24x^2 + 9x \qquad \text{Distributive property}$$

Notice that in the square of a sum, all of the terms are positive, as in Examples 2(b) and (d). *In the square of a difference, the middle term is negative,* as in Examples 2(a) and (c).

> **CAUTION**
> A common error when squaring a binomial is to forget the middle term of the product. In general,
> $$(a + b)^2 = a^2 + 2ab + b^2, \quad not \quad a^2 + b^2,$$
> and $$(a - b)^2 = a^2 - 2ab + b^2, \quad not \quad a^2 - b^2.$$

◀ *Work Problem* **2** *at the Side.*

OBJECTIVE **2** **Find the product of the sum and difference of two terms.** In binomial products of the form $(a + b)(a - b)$, one binomial is the sum of two terms, and the other is the difference of the *same* two terms. For example, the product of $x + 2$ and $x - 2$ is

$$(x + 2)(x - 2)$$

$$= x^2 - 2x + 2x - 4 \qquad \text{FOIL}$$

$$= x^2 - 4. \qquad \text{Combine like terms.}$$

As the above example suggests, the product of $a + b$ and $a - b$ is the difference of two squares.

> **Product of the Sum and Difference of Two Terms**
> $$(a + b)(a - b) = a^2 - b^2$$

> **Note**
> The expressions $a + b$ and $a - b$, the sum and difference of the *same* two terms, are called **conjugates.** In the example above, $x + 2$ and $x - 2$ are conjugates.

EXAMPLE 3   **Finding the Product of the Sum and Difference of Two Terms**

Find each product.

(a) $(x + 4)(x - 4)$

Use the rule for the product of the sum and difference of two terms.

$$(x + 4)(x - 4)$$
$$= x^2 - 4^2$$
$$= x^2 - 16$$

(b) $\left(\dfrac{2}{3} - w\right)\left(\dfrac{2}{3} + w\right)$

$$= \left(\dfrac{2}{3} + w\right)\left(\dfrac{2}{3} - w\right) \qquad \text{Commutative property}$$

$$= \left(\dfrac{2}{3}\right)^2 - w^2 \qquad \text{Multiply.}$$

$$= \dfrac{4}{9} - w^2 \qquad \text{Square } \tfrac{2}{3}.$$

(c) $x(x + 2)(x - 2)$

$$= x(x^2 - 4) \qquad \text{Find the product of the sum and difference of two terms.}$$

$$= x^3 - 4x \qquad \text{Distributive property}$$

EXAMPLE 4   **Finding the Product of the Sum and Difference of Two Terms**

Find each product.

$$(a \;+\; b)\,(a \;-\; b)$$

(a) $(5m + 3)(5m - 3)$

Use the rule for the product of the sum and difference of two terms.

$$(5m + 3)(5m - 3)$$
$$= (5m)^2 - 3^2 \qquad (a + b)(a - b) = a^2 - b^2$$
$$= 25m^2 - 9 \qquad \text{Apply the exponents.}$$

(b) $(4x + y)(4x - y)$

$$= (4x)^2 - y^2$$
$$= 16x^2 - y^2$$

(c) $\left(z - \dfrac{1}{4}\right)\left(z + \dfrac{1}{4}\right)$

$$= z^2 - \left(\dfrac{1}{4}\right)^2$$

$$= z^2 - \dfrac{1}{16}$$

(d) $2p(p^2 + 3)(p^2 - 3)$

$$= 2p(p^4 - 9) \qquad \text{Multiply the conjugates.}$$

$$= 2p^5 - 18p \qquad \text{Distributive property}$$

Work Problem **3** at the Side. ▶

---

3 Find each product.

(a) $(y + 3)(y - 3)$

(b) $(10m + 7)(10m - 7)$

(c) $(7p + 2q)(7p - 2q)$

(d) $\left(3r - \dfrac{1}{2}\right)\left(3r + \dfrac{1}{2}\right)$

(e) $3x(x^3 - 4)(x^3 + 4)$

ANSWERS

3. (a) $y^2 - 9$ **(b)** $100m^2 - 49$
  (c) $49p^2 - 4q^2$ **(d)** $9r^2 - \dfrac{1}{4}$
  (e) $3x^7 - 48x$

The product rules of this section will be important in **Chapters 14** and **15** and should be *memorized*.

4 Find each product.

(a) $(m + 1)^3$

OBJECTIVE **3** **Find greater powers of binomials.** The methods used in the previous section and this section can be combined to find greater powers of binomials.

EXAMPLE 5 **Finding Greater Powers of Binomials**

Find each product.

(a) $(x + 5)^3$

$$= (x + 5)^2 (x + 5) \qquad a^3 = a^2 \cdot a$$

$$= (x^2 + 10x + 25)(x + 5) \qquad \text{Square the binomial.}$$

$$= x^3 + 10x^2 + 25x + 5x^2 + 50x + 125 \qquad \text{Multiply polynomials.}$$

$$= x^3 + 15x^2 + 75x + 125 \qquad \text{Combine like terms.}$$

(b) $(2y - 3)^4$

$$= (2y - 3)^2 (2y - 3)^2 \qquad a^4 = a^2 \cdot a^2$$

$$= (4y^2 - 12y + 9)(4y^2 - 12y + 9) \qquad \text{Square each binomial.}$$

$$= 16y^4 - 48y^3 + 36y^2 - 48y^3 + 144y^2 \qquad \text{Multiply polynomials.}$$

$$\qquad - 108y + 36y^2 - 108y + 81$$

$$= 16y^4 - 96y^3 + 216y^2 - 216y + 81 \qquad \text{Combine like terms.}$$

(b) $(3k - 2)^4$

(c) $-2r (r + 2)^3$

$$= -2r (r + 2)(r + 2)^2$$

$$= -2r (r + 2)(r^2 + 4r + 4)$$

$$= -2r (r^3 + 4r^2 + 4r + 2r^2 + 8r + 8)$$

$$= -2r (r^3 + 6r^2 + 12r + 8)$$

$$= -2r^4 - 12r^3 - 24r^2 - 16r$$

(c) $-3x (x - 4)^3$

◀ *Work Problem* **4** *at the Side.*

13.4 ▶▶▶ Exercises

1. Consider the square $(2x + 3)^2$.

   (a) What is the square of the first term, $(2x)^2$?

   (b) What is twice the product of the two terms, $2(2x)(3)$?

   (c) What is the square of the last term, 3^2?

   (d) Write the final product, which is a trinomial, using your results from parts (a)–(c).

2. Repeat Exercise 1 for the square $(3x - 2)^2$.

Find each square. See Examples 1 and 2.

3. $(p + 2)^2$ **4.** $(r + 5)^2$ **5.** $(z - 5)^2$ **6.** $(x - 3)^2$

7. $(4x - 3)^2$ **8.** $(5y + 2)^2$ **9.** $(2p + 5q)^2$ **10.** $(8a - 3b)^2$

11. $(0.8t + 0.7s)^2$ **12.** $(0.7z - 0.3w)^2$ **13.** $\left(5x + \dfrac{2}{5}y\right)^2$ **14.** $\left(6m - \dfrac{4}{5}n\right)^2$

15. $t(3t - 1)^2$ **16.** $x(2x + 5)^2$ **17.** $-(4r - 2)^2$ **18.** $-(3y - 8)^2$

19. Consider the product $(7x + 3y)(7x - 3y)$.

   (a) What is the product of the first terms, $7x(7x)$?

   (b) Multiply the outer terms, $7x(-3y)$. Then multiply the inner terms, $3y(7x)$. Add the results. What is this sum?

   (c) What is the product of the last terms, $3y(-3y)$?

   (d) Write the complete product using your answers in parts (a) and (c). Why is the sum found in part (b) omitted here?

20. Repeat Exercise 19 for the product $(5x + 7y)(5x - 7y)$.

Find each product. See Examples 3 and 4.

21. $(q + 2)(q - 2)$ **22.** $(x + 8)(x - 8)$ **23.** $(2w + 5)(2w - 5)$ **24.** $(3z + 8)(3z - 8)$

25. $(10x + 3y)(10x - 3y)$ **26.** $(13r + 2z)(13r - 2z)$ **27.** $(2x^2 - 5)(2x^2 + 5)$ **28.** $(9y^2 - 2)(9y^2 + 2)$

29. $\left(7x + \dfrac{3}{7}\right)\left(7x - \dfrac{3}{7}\right)$ **30.** $\left(9y + \dfrac{2}{3}\right)\left(9y - \dfrac{2}{3}\right)$ **31.** $p(3p + 7)(3p - 7)$ **32.** $q(5q - 1)(5q + 1)$

Relating Concepts (Exercises 33–42) For Individual or Group Work

Special products can be illustrated by using areas of rectangles. Use the figure and **work Exercises 33–38 in order,** *to justify the special product* $(a + b)^2 = a^2 + 2ab + b^2$.

33. Express the area of the entire square figure as the square of a binomial.

34. Give the monomial that represents the area of the red square.

35. Give the monomial that represents the sum of the areas of the blue rectangles.

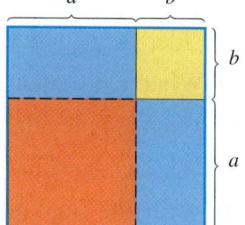

36. Give the monomial that represents the area of the yellow square.

37. What is the sum of the monomials you obtained in Exercises 34–36?

38. Explain why the binomial square you found in Exercise 33 must equal the polynomial you found in Exercise 37.

To understand how the special product $(a + b)^2 = a^2 + 2ab + b^2$ *can be applied to a purely numerical problem,* **work Exercises 39–42 in order.**

39. Evaluate 35^2 using either traditional paper-and-pencil methods or a calculator.

40. The number 35 can be written as $30 + 5$. Therefore, $35^2 = (30 + 5)^2$. Use the special product for squaring a binomial with $a = 30$ and $b = 5$ to write an expression for $(30 + 5)^2$. Do not simplify at this time.

41. Use the order of operations to simplify the expression you found in Exercise 40.

42. How do the answers in Exercises 39 and 41 compare?

Find each product. See Example 5.

43. $(m - 5)^3$

44. $(p + 3)^3$

45. $(y + 2)^3$

46. $(x - 7)^3$

47. $(2a + 1)^3$

48. $(3m - 1)^3$

49. $(3r - 2t)^4$

50. $(2z + 5y)^4$

51. $3x^2(x - 3)^3$

52. $4p^3(p + 4)^3$

53. $-8x^2y(x + y)^4$

In Exercises 54 and 55, refer to the figure shown here.

54. Find a polynomial that represents the volume of the cube.

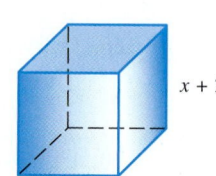

$x + 2$

55. If the value of x is 6, what is the volume of the cube?

13.5 ▷▷▷ Integer Exponents and the Quotient Rule

In all our earlier work, exponents were positive integers. Now we want to develop meaning for exponents that are *not* positive integers.

Consider the following list.

$$2^4 = 16$$
$$2^3 = 8$$
$$2^2 = 4$$

Do you see the pattern in the values? Each time we reduce the exponent by 1, the value is divided by 2 (the base). Using this pattern, we can continue the list to smaller and smaller integer exponents.

$$2^1 = 2$$
$$2^0 = 1$$
$$2^{-1} = \frac{1}{2}$$

Work Problem **1** *at the Side.* ▶

From the preceding list and the answers to Problem 1 at the side, it appears that we should define 2^0 as 1 and negative exponents as reciprocals.

OBJECTIVE **1** **Use 0 as an exponent.** We want the definitions of 0 and negative exponents to satisfy the rules for exponents from **Section 13.2.** For example, if $6^0 = 1$,

$$6^0 \cdot 6^2 = 1 \cdot 6^2 = 6^2 \quad \text{and} \quad 6^0 \cdot 6^2 = 6^{0+2} = 6^2,$$

so the product rule is satisfied. Check that the power rules are also valid for a 0 exponent. Thus, we define a 0 exponent as follows.

> **Zero Exponent**
>
> For any nonzero real number a, $\quad a^0 = 1.$
>
> *Example:* $\quad 17^0 = 1$

EXAMPLE 1 **Using Zero Exponents**

Evaluate.

(a) $60^0 = 1$

(b) $(-60)^0 = 1$

(c) $-60^0 = -(1) = -1$

(d) $y^0 = 1, \quad y \neq 0$

(e) $6y^0 = 6(1) = 6, \quad y \neq 0$

(f) $(6y)^0 = 1, \quad y \neq 0$

> **CAUTION**
> Look again at Examples 1(b) and (c). In $(-60)^0$, the base is -60 and the exponent is 0. Any nonzero base raised to the exponent 0 is 1. In -60^0, the base is 60. Then $60^0 = 1$, and $-60^0 = -1$.

Work Problem **2** *at the Side.* ▶

1 Continue the list of exponentials using -2, -3, and -4 as exponents.

$$2^{-2} = \underline{\quad\quad}$$
$$2^{-3} = \underline{\quad\quad}$$
$$2^{-4} = \underline{\quad\quad}$$

2 Evaluate.

(a) 28^0

(b) $(-16)^0$

(c) -7^0

(d) $m^0, \quad m \neq 0$

(e) $-p^0, \quad p \neq 0$

ANSWERS

1. $2^{-2} = \dfrac{1}{4}; 2^{-3} = \dfrac{1}{8}; 2^{-4} = \dfrac{1}{16}$

2. **(a)** 1 **(b)** 1 **(c)** -1 **(d)** 1 **(e)** -1

OBJECTIVE 2 Use negative numbers as exponents. From the lists at the beginning of this section and margin Problem 1, since $2^{-2} = \frac{1}{4}$ and $2^{-3} = \frac{1}{8}$, we can deduce that 2^{-n} should equal $\frac{1}{2^n}$. Is the product rule valid in such cases? For example, if we multiply 6^{-2} by 6^2, we get

$$6^{-2} \cdot 6^2 = 6^{-2+2} = 6^0 = 1.$$

The expression 6^{-2} behaves as if it were the reciprocal of 6^2, because their product is 1. The reciprocal of 6^2 may be written $\frac{1}{6^2}$, leading us to define 6^{-2} as $\frac{1}{6^2}$. This is a particular case of the definition of negative exponents.

Negative Exponents

For any nonzero real number a and any integer n, $\quad a^{-n} = \dfrac{1}{a^n}.$

Example: $\quad 3^{-2} = \dfrac{1}{3^2}$

By definition, a^{-n} and a^n are reciprocals, since

$$a^n \cdot a^{-n} = a^n \cdot \frac{1}{a^n} = 1.$$

Since $1^n = 1$, the definition of a^{-n} can also be written

$$a^{-n} = \frac{1}{a^n} = \frac{1^n}{a^n} = \left(\frac{1}{a}\right)^n.$$

For example, $\quad 6^{-3} = \left(\dfrac{1}{6}\right)^3 \quad$ and $\quad \left(\dfrac{1}{3}\right)^{-2} = 3^2.$

EXAMPLE 2 Using Negative Exponents

Simplify by writing with positive exponents. Assume that all variables represent nonzero real numbers.

(a) $3^{-2} = \dfrac{1}{3^2} = \dfrac{1}{9} \qquad a^{-n} = \frac{1}{a^n}$

(b) $5^{-3} = \dfrac{1}{5^3} = \dfrac{1}{125}$

(c) $\left(\dfrac{1}{2}\right)^{-3} = 2^3 = 8 \qquad \frac{1}{2}$ and 2 are reciprocals.

  Notice that we can change the base to its reciprocal if we also change the sign of the exponent.

(d) $\left(\dfrac{2}{5}\right)^{-4} = \left(\dfrac{5}{2}\right)^4 = \dfrac{5^4}{2^4} = \dfrac{625}{16} \qquad$ **(e)** $\left(\dfrac{4}{3}\right)^{-5} = \left(\dfrac{3}{4}\right)^5 = \dfrac{3^5}{4^5} = \dfrac{243}{1024}$

  $\frac{2}{5}$ and $\frac{5}{2}$ are reciprocals.

(f) $4^{-1} - 2^{-1}$

$\quad = \dfrac{1}{4} - \dfrac{1}{2} = \dfrac{1}{4} - \dfrac{2}{4} = -\dfrac{1}{4} \qquad$ Apply the exponents first; then subtract.

Continued on Next Page

(g) $p^{-2} = \dfrac{1}{p^2}$

(h) $\dfrac{1}{x^{-4}} = \dfrac{1^{-4}}{x^{-4}}$　　$1^n = 1$, for any integer n

　　　　$= \left(\dfrac{1}{x}\right)^{-4}$　　Power rule (c)

　　　　$= x^4$　　$\dfrac{1}{x}$ and x are reciprocals.

Notice that, in general, $\dfrac{1}{a^{-n}} = a^n$.

(i) $x^3 y^{-4} = \dfrac{x^3}{1} \cdot \dfrac{1}{y^4} = \dfrac{x^3}{y^4}$

CAUTION

A negative exponent does not indicate a negative number. Negative exponents lead to reciprocals.

Expression	Example	
a^{-n}	$3^{-2} = \dfrac{1}{3^2} = \dfrac{1}{9}$	Not negative
$-a^{-n}$	$-3^{-2} = -\dfrac{1}{3^2} = -\dfrac{1}{9}$	Negative

Work Problem **3** *at the Side.* ▶

Consider the following:

$$\frac{2^{-3}}{3^{-4}} = \frac{\dfrac{1}{2^3}}{\dfrac{1}{3^4}} = \frac{1}{2^3} \div \frac{1}{3^4} = \frac{1}{2^3} \cdot \frac{3^4}{1} = \frac{3^4}{2^3}.$$　　To divide by a fraction, multiply by its reciprocal.

Therefore,

$$\frac{2^{-3}}{3^{-4}} = \frac{3^4}{2^3}.$$

Changing from Negative to Positive Exponents

For any nonzero numbers a and b, and any integers m and n,

$$\frac{a^{-m}}{b^{-n}} = \frac{b^n}{a^m} \quad \text{and} \quad \left(\frac{a}{b}\right)^{-m} = \left(\frac{b}{a}\right)^m.$$

Examples: $\dfrac{3^{-5}}{2^{-4}} = \dfrac{2^4}{3^5}$ and $\left(\dfrac{4}{5}\right)^{-3} = \left(\dfrac{5}{4}\right)^3$

---

3 Simplify by writing with positive exponents. Assume that all variables represent nonzero real numbers.

(a) 4^{-3}

(b) 6^{-2}

(c) $\left(\dfrac{1}{4}\right)^{-2}$

(d) $\left(\dfrac{2}{3}\right)^{-2}$

(e) $2^{-1} + 5^{-1}$

(f) m^{-5}

(g) $\dfrac{1}{z^{-4}}$

(h) $p^2 q^{-5}$

ANSWERS

3. **(a)** $\dfrac{1}{4^3} = \dfrac{1}{64}$ **(b)** $\dfrac{1}{6^2} = \dfrac{1}{36}$ **(c)** $4^2 = 16$

(d) $\left(\dfrac{3}{2}\right)^2 = \dfrac{9}{4}$ **(e)** $\dfrac{1}{2} + \dfrac{1}{5} = \dfrac{7}{10}$

(f) $\dfrac{1}{m^5}$ **(g)** z^4 **(h)** $\dfrac{p^2}{q^5}$

4 Simplify. Assume that all variables represent nonzero real numbers.

(a) $\dfrac{7^{-1}}{5^{-4}}$

(b) $\dfrac{x^{-3}}{y^{-2}}$

(c) $\dfrac{4h^{-5}}{m^{-2}k}$

(d) $\left(\dfrac{3m}{p}\right)^{-2}$

EXAMPLE 3 Changing from Negative to Positive Exponents

Simplify. Assume that all variables represent nonzero real numbers.

(a) $\dfrac{4^{-2}}{5^{-3}} = \dfrac{5^3}{4^2} = \dfrac{125}{16}$

(b) $\dfrac{m^{-5}}{p^{-1}} = \dfrac{p^1}{m^5} = \dfrac{p}{m^5}$

(c) $\dfrac{a^{-2}b}{3d^{-3}} = \dfrac{bd^3}{3a^2}$ Notice that b in the numerator and the coefficient 3 in the denominator are not affected.

(d) $\left(\dfrac{x}{2y}\right)^{-4}$

$= \left(\dfrac{2y}{x}\right)^4$ Negative-to-positive rule

$= \dfrac{2^4 y^4}{x^4}$ Power rule (c)

$= \dfrac{16y^4}{x^4}$

◀ Work Problem **4** at the Side.

CAUTION

Be careful. We cannot use the rule $\dfrac{a^{-m}}{b^{-n}} = \dfrac{b^n}{a^m}$ to change negative exponents to positive exponents if the exponents occur in a *sum* or *difference* of terms. For example,

$\dfrac{5^{-2} + 3^{-1}}{7 - 2^{-3}}$ would be written with positive exponents as $\dfrac{\dfrac{1}{5^2} + \dfrac{1}{3}}{7 - \dfrac{1}{2^3}}$.

OBJECTIVE **3** **Use the quotient rule for exponents.** Consider a quotient of two exponential expressions with the same base.

$$\dfrac{6^5}{6^3} = \dfrac{6 \cdot 6 \cdot 6 \cdot 6 \cdot 6}{6 \cdot 6 \cdot 6} = 6^2$$

Notice that the difference between the exponents, $5 - 3 = 2$, is the exponent in the quotient. Also,

$$\dfrac{6^2}{6^4} = \dfrac{6 \cdot 6}{6 \cdot 6 \cdot 6 \cdot 6} = \dfrac{1}{6^2} = 6^{-2}.$$

Here, $2 - 4 = -2$. These examples suggest the **quotient rule for exponents.**

Quotient Rule for Exponents

For any nonzero real number a and any integers m and n,

$$\dfrac{a^m}{a^n} = a^{m-n}.$$

(Keep the same base and subtract the exponents.)

Example: $\dfrac{5^8}{5^4} = 5^{8-4} = 5^4$

CAUTION

A common **error** is to write $\dfrac{5^8}{5^4} = 1^{8-4} = 1^4$. **This is incorrect.** By the quotient rule, the quotient must have the *same base*, 5, so

$$\frac{5^8}{5^4} = 5^{8-4} = 5^4.$$

We can confirm this by using the definition of exponents to write out the factors:

$$\frac{5^8}{5^4} = \frac{5 \cdot 5 \cdot 5 \cdot 5 \cdot 5 \cdot 5 \cdot 5 \cdot 5}{5 \cdot 5 \cdot 5 \cdot 5} = 5^4.$$

EXAMPLE 4 **Using the Quotient Rule**

Simplify. Assume that all variables represent nonzero real numbers.

(a) $\dfrac{5^8}{5^6} = 5^{8-6} = 5^2 = 25$

> Keep the same base.

(b) $\dfrac{4^2}{4^9} = 4^{2-9} = 4^{-7} = \dfrac{1}{4^7}$

(c) $\dfrac{5^{-3}}{5^{-7}} = 5^{-3-(-7)} = 5^4 = 625$

> Be careful with signs.

(d) $\dfrac{q^5}{q^{-3}} = q^{5-(-3)} = q^8$

(e) $\dfrac{3^2 x^5}{3^4 x^3}$

$= \dfrac{3^2}{3^4} \cdot \dfrac{x^5}{x^3}$

$= 3^{2-4} \cdot x^{5-3}$

$= 3^{-2} x^2$

$= \dfrac{x^2}{3^2}$

$= \dfrac{x^2}{9}$

(f) $\dfrac{(m+n)^{-2}}{(m+n)^{-4}}$

$= (m+n)^{-2-(-4)}$

$= (m+n)^{-2+4}$

$= (m+n)^2, \quad m \neq -n$

The restriction $m \neq -n$ is necessary to prevent a denominator of 0 in the original expression. Division by 0 is undefined.

(g) $\dfrac{7x^{-3}y^2}{2^{-1}x^2y^{-5}}$

$= \dfrac{7 \cdot 2^1 y^2 y^5}{x^2 x^3}$ Definition of negative exponent

$= \dfrac{14y^7}{x^5}$ Multiply; product rule

Work Problem **5** *at the Side.* ▶

The definitions and rules for exponents given in this section and **Section 13.2** are summarized on the next page.

---

5 Simplify. Assume that all variables represent nonzero real numbers.

(a) $\dfrac{5^{11}}{5^8}$

(b) $\dfrac{4^7}{4^{10}}$

(c) $\dfrac{6^{-5}}{6^{-2}}$

(d) $\dfrac{8^4 m^9}{8^5 m^{10}}$

(e) $\dfrac{3^{-1}(x+y)^{-3}}{2^{-2}(x+y)^{-4}}, \quad x \neq -y$

ANSWERS

5. (a) 125   **(b)** $\dfrac{1}{64}$ **(c)** $\dfrac{1}{216}$ **(d)** $\dfrac{1}{8m}$

    (e) $\dfrac{4}{3}(x+y)$

Definitions and Rules for Exponents

For any integers m and n: **Examples**

Product rule	$a^m \cdot a^n = a^{m+n}$		$7^4 \cdot 7^5 = 7^{4+5} = 7^9$

Zero exponent $a^0 = 1$ $(a \neq 0)$ $(-3)^0 = 1$

Negative exponent $a^{-n} = \dfrac{1}{a^n}$ $(a \neq 0)$ $5^{-3} = \dfrac{1}{5^3}$

Quotient rule $\dfrac{a^m}{a^n} = a^{m-n}$ $(a \neq 0)$ $\dfrac{2^2}{2^5} = 2^{2-5} = 2^{-3} = \dfrac{1}{2^3}$

Power rule (a) $(a^m)^n = a^{mn}$ $(4^2)^3 = 4^{2\cdot3} = 4^6$

Power rule (b) $(ab)^m = a^m b^m$ $(3k)^4 = 3^4 k^4$

Power rule (c) $\left(\dfrac{a}{b}\right)^m = \dfrac{a^m}{b^m}$ $(b \neq 0)$ $\left(\dfrac{2}{3}\right)^2 = \dfrac{2^2}{3^2}$

Negative-to-positive rules $\dfrac{a^{-m}}{b^{-n}} = \dfrac{b^n}{a^m}$ $(a, b \neq 0)$ $\dfrac{2^{-4}}{5^{-3}} = \dfrac{5^3}{2^4}$

$$\left(\dfrac{a}{b}\right)^{-m} = \left(\dfrac{b}{a}\right)^m. \qquad \left(\dfrac{4}{7}\right)^{-2} = \left(\dfrac{7}{4}\right)^2$$

OBJECTIVE 4 Use combinations of rules.   We sometimes need to use more than one rule to simplify an expression.

EXAMPLE 5   **Using a Combination of Rules**

Simplify each expression. Assume that all variables represent nonzero real numbers.

(a) $\dfrac{(4^2)^3}{4^5}$

$= \dfrac{4^6}{4^5}$ Power rule (a)

$= 4^{6-5}$ Quotient rule

$= 4^1$

$= 4$

(b) $(2x)^3 (2x)^2$

$= (2x)^5$ Product rule

$= 2^5 x^5$ Power rule (b)

$= 32x^5$

(c) $\left(\dfrac{2x^3}{5}\right)^{-4}$

$= \left(\dfrac{5}{2x^3}\right)^4$ Negative-to-positive rule

$= \dfrac{5^4}{2^4 x^{12}}$ Power rules (a)–(c)

$= \dfrac{625}{16x^{12}}$

(d) $\left(\dfrac{3x^{-2}}{4^{-1}y^3}\right)^{-3}$

$= \dfrac{3^{-3}x^6}{4^3 y^{-9}}$ Power rules (a)–(c)

$= \dfrac{x^6 y^9}{4^3 \cdot 3^3}$ Negative-to-positive rule

$= \dfrac{x^6 y^9}{1728}$ $4^3 \cdot 3^3 = 64 \cdot 27 = 1728$

Continued on Next Page

(e) $\dfrac{(4m)^{-3}}{(3m)^{-4}}$

$\quad = \dfrac{4^{-3}m^{-3}}{3^{-4}m^{-4}}$ Power rule (b)

$\quad = \dfrac{3^4 m^4}{4^3 m^3}$ Negative-to-positive rule

$\quad = \dfrac{3^4 m^{4-3}}{4^3}$ Quotient rule

$\quad = \dfrac{3^4 m}{4^3}$

$\quad = \dfrac{81m}{64}$

> **Note**
> Since the steps can be done in several different orders, there are many equally correct ways to simplify expressions like those in Examples 5(c) through 5(e).

Work Problem **6** *at the Side.* ▶

6 Simplify each expression. Assume that all variables represent nonzero real numbers.

(a) $\dfrac{(3^4)^2}{3^3}$

(b) $(4x)^2 (4x)^4$

(c) $\dfrac{(6x)^{-1}}{(3x^2)^{-2}}$

(d) $\dfrac{3^9 \cdot (x^2 y)^{-2}}{3^3 \cdot x^{-4} y}$

ANSWERS

6. **(a)** 243 **(b)** $4^6 x^6$, or $4096 x^6$ **(c)** $\dfrac{3x^3}{2}$

(d) $\dfrac{729}{y^3}$

Math in the Media

MORE POWER TO YOU, CAPTAIN KIRK

The original *Star Trek* series first aired during the 1966 to 1967 television season and started the phenomenon that continues today. There have been five different television series and 10 feature movies with the *Star Trek* theme.

Captain James T. Kirk, portrayed by William Shatner, led the Starship Enterprise during its first three seasons. During the first season, the February 2, 1967, episode "Court Martial" told the story of Kirk being put on trial. He was accused of negligence in the death of a crewmember, because the computer records of the ship contradicted Kirk's logs. As the trial begins, Kirk explains how the sounds on the ship can be recorded and magnified:

Kirk: *Gentlemen, this computer has an auditory sensor. It can, in effect, hear sounds. By installing a booster we can increase that capability on an order of one to the fourth power. The computer should be able to bring us every sound occurring on the ship.*

1. Read Captain Kirk's statement carefully. What error did he make?

2. What is the result if we raise the number 1 to any whole number power?

3. It is possible that Kirk meant "10 to the fourth power." Express 10^4 in expanded form.

4. The word **googol** was invented to express a very large power of 10. The search engine Google was named in honor of it. Look up the meaning of googol, and write it in exponential form.

5. Investigate the meaning of the word **googolplex**.

13.5 ▶▶▶ Exercises

Decide whether each expression is positive, negative, or 0.

1. $(-2)^{-3}$ **2.** $(-3)^{-2}$ **3.** -2^4 **4.** -3^6

5. $\left(\dfrac{1}{4}\right)^{-2}$ **6.** $\left(\dfrac{1}{5}\right)^{-2}$ **7.** $1 - 5^0$ **8.** $1 - 7^0$

Decide whether each expression is equal to either 0, 1, or −1. See Example 1.

9. 9^0 **10.** 5^0 **11.** $(-4)^0$ **12.** $(-10)^0$ **13.** -9^0

14. -5^0 **15.** $(-2)^0 - 2^0$ **16.** $(-8)^0 - 8^0$ **17.** $\dfrac{0^{10}}{10^0}$ **18.** $\dfrac{0^5}{5^0}$

Evaluate each expression. See Examples 1 and 2.

19. $7^0 + 9^0$ **20.** $8^0 + 6^0$ 🌐 **21.** 4^{-3} **22.** 5^{-4} **23.** $\left(\dfrac{1}{2}\right)^{-4}$

24. $\left(\dfrac{1}{3}\right)^{-3}$ 🌐 **25.** $\left(\dfrac{6}{7}\right)^{-2}$ **26.** $\left(\dfrac{2}{3}\right)^{-3}$ **27.** $(-3)^{-4}$ **28.** $(-4)^{-3}$

29. $5^{-1} + 3^{-1}$ **30.** $6^{-1} + 2^{-1}$ **31.** $-2^{-1} + 3^{-2}$ **32.** $(-3)^{-2} + (-4)^{-1}$

Relating Concepts (Exercises 33–36) For Individual or Group Work

In Objective 1, we used the product rule to motivate the definition of a 0 exponent. We can also use the quotient rule. To see this, **work Exercises 33–36 in order.**

33. Consider the expression $\frac{25}{25}$. What is its simplest form?

34. Write the quotient in Exercise 33 using the fact that $25 = 5^2$.

35. Apply the quotient rule for exponents to your answer for Exercise 34. Give the answer as a power of 5.

36. Because your answers for Exercises 33 and 35 both represent $\frac{25}{25}$, they must be equal. Write this equality. What definition does it support?

Simplify by writing each expression with positive exponents. Assume that all variables represent nonzero real numbers. See Examples 2–4.

37. $\dfrac{9^4}{9^5}$ **38.** $\dfrac{7^3}{7^4}$ **39.** $\dfrac{6^{-3}}{6^2}$ **40.** $\dfrac{4^{-2}}{4^3}$ **41.** $\dfrac{1}{6^{-3}}$

42. $\dfrac{1}{5^{-2}}$ **43.** $\dfrac{2}{r^{-4}}$ **44.** $\dfrac{3}{s^{-8}}$ **45.** $\dfrac{4^{-3}}{5^{-2}}$ **46.** $\dfrac{6^{-2}}{5^{-4}}$

47. $p^5 q^{-8}$ **48.** $x^{-8} y^4$ **49.** $\dfrac{r^5}{r^{-4}}$ **50.** $\dfrac{a^6}{a^{-4}}$ **51.** $\dfrac{6^4 x^8}{6^5 x^3}$

52. $\dfrac{3^8 y^5}{3^{10} y^2}$ **53.** $\dfrac{6y^3}{2y}$ **54.** $\dfrac{5m^2}{m}$ **55.** $\dfrac{3x^5}{3x^2}$ **56.** $\dfrac{10p^8}{2p^4}$

57. $\dfrac{x^{-3} y}{4z^{-2}}$ **58.** $\dfrac{p^{-5} q^{-4}}{9r^{-3}}$ **59.** $\dfrac{(a+b)^{-3}}{(a+b)^{-4}}$ **60.** $\dfrac{(x+y)^{-8}}{(x+y)^{-9}}$

Simplify by writing each expression with positive exponents. Assume that all variables represent nonzero real numbers. See Example 5.

61. $\dfrac{(7^4)^3}{7^9}$ **62.** $\dfrac{(5^3)^2}{5^2}$ **63.** $x^{-3} \cdot x^5 \cdot x^{-4}$ **64.** $y^{-8} \cdot y^5 \cdot y^{-2}$

65. $\dfrac{(3x)^{-2}}{(4x)^{-3}}$ **66.** $\dfrac{(2y)^{-3}}{(5y)^{-4}}$ **67.** $\left(\dfrac{x^{-1} y}{z^2}\right)^{-2}$ **68.** $\left(\dfrac{p^{-4} q}{r^{-3}}\right)^{-3}$

69. $(6x)^4 (6x)^{-3}$ **70.** $(10y)^9 (10y)^{-8}$ **71.** $\dfrac{(m^7 n)^{-2}}{m^{-4} n^3}$ **72.** $\dfrac{(m^8 n^{-4})^2}{m^{-2} n^5}$

73. $\dfrac{5x^{-3}}{(4x)^2}$ **74.** $\dfrac{-3k^5}{(2k)^2}$ **75.** $\left(\dfrac{2p^{-1} q}{3^{-1} m^2}\right)^2$ **76.** $\left(\dfrac{4xy^2}{x^{-1} y}\right)^{-2}$

Summary Exercises on the Rules for Exponents

Simplify each expression. Assume that all variables represent nonzero real numbers.

1. $\left(\dfrac{6x^2}{5}\right)^{12}$

2. $\left(\dfrac{rs^2t^3}{3t^4}\right)^6$

3. $(10x^2y^4)^2(10xy^2)^3$

4. $(-2ab^3c)^4(-2a^2b)^3$

5. $\left(\dfrac{9wx^3}{y^4}\right)^3$

6. $(4x^{-2}y^{-3})^{-2}$

7. $\dfrac{c^{11}(c^2)^4}{(c^3)^3(c^2)^{-6}}$

8. $\left(\dfrac{k^4t^2}{k^2t^{-4}}\right)^{-2}$

9. $5^{-1}+6^{-1}$

10. $\dfrac{(3y^{-1}z^3)^{-1}(3y^2)}{(y^3z^2)^{-3}}$

11. $\dfrac{(2xy^{-1})^3}{2^3x^{-3}y^2}$

12. $-8^0+(-8)^0$

13. $(z^4)^{-3}(z^{-2})^{-5}$

14. $\left(\dfrac{r^2st^5}{3r}\right)^{-2}$

15. $\dfrac{(3^{-1}x^{-3}y)^{-1}(2x^2y^{-3})^2}{(5x^{-2}y^2)^{-2}}$

16. $\left(\dfrac{5x^2}{3x^{-4}}\right)^{-1}$

17. $\left(\dfrac{-2x^{-2}}{2x^2}\right)^{-2}$

18. $\dfrac{(x^{-4}y^2)^3(x^2y)^{-1}}{(xy^2)^{-3}}$

19. $\dfrac{(a^{-2}b^3)^{-4}}{(a^{-3}b^2)^{-2}(ab)^{-4}}$

20. $(2a^{-30}b^{-29})(3a^{31}b^{30})$

21. $5^{-2}+6^{-2}$

22. $\left(\dfrac{(x^{47}y^{23})^2}{x^{-26}y^{-42}}\right)^0$

23. $\left(\dfrac{7a^2b^3}{2}\right)^3$

24. $-(-12^0)$

25. $-(-12)^0$

26. $\dfrac{0^{12}}{12^0}$

27. $\dfrac{(2xy^{-3})^{-2}}{(3x^{-2}y^4)^{-3}}$

28. $\left(\dfrac{a^2b^3c^4}{a^{-2}b^{-3}c^{-4}}\right)^{-2}$

29. $(6x^{-5}z^3)^{-3}$

30. $(2p^{-2}qr^{-3})(2p)^{-4}$

31. $\dfrac{(xy)^{-3}(xy)^5}{(xy)^{-4}}$

32. $42^0 - (-12)^0$

33. $\dfrac{(7^{-1}x^{-3})^{-2}(x^4)^{-6}}{7^{-1}x^{-3}}$

34. $\left(\dfrac{3^{-4}x^{-3}}{3^{-3}x^{-6}}\right)^{-2}$

35. $(5p^{-2}q)^{-3}(5pq^3)^4$

36. $8^{-1} + 6^{-1}$

37. $\left(\dfrac{4r^{-6}s^{-2}t}{2r^8s^{-4}t^2}\right)^{-1}$

38. $(13x^{-6}y)(13x^{-6}y)^{-1}$

39. $\dfrac{(8pq^{-2})^4}{(8p^{-2}q^{-3})^3}$

40. $\left(\dfrac{mn^{-2}p}{m^2np^4}\right)^{-2}\left(\dfrac{mn^{-2}p}{m^2np^4}\right)^3$

41. $-(-3^0)^0$

42. $5^{-1} - 8^{-1}$

43. A student simplified $(10^2)^3$ as 1000^6. ***WHAT WENT WRONG?*** Give the correct answer.

44. A student simplified -5^4 as shown:
$$-5^4 = (-5)^4 = 625.$$
WHAT WENT WRONG? Give the correct answer.

13.6 ▶▶▶ Dividing a Polynomial by a Monomial

OBJECTIVE **1** **Divide a polynomial by a monomial.** We add two fractions with a common denominator as follows.

$$\frac{a}{c} + \frac{b}{c} = \frac{a + b}{c}$$

In reverse, this statement gives a rule for dividing a polynomial by a monomial.

> **Dividing a Polynomial by a Monomial**
>
> To divide a polynomial by a monomial, divide each term of the polynomial by the monomial:
>
> $$\frac{a + b}{c} = \frac{a}{c} + \frac{b}{c} \quad (c \neq 0).$$
>
> *Examples:* $\quad \frac{2 + 5}{3} = \frac{2}{3} + \frac{5}{3} \quad$ and $\quad \frac{x + 3z}{2y} = \frac{x}{2y} + \frac{3z}{2y} \quad (y \neq 0)$

The parts of a division problem are named here.

$$\text{Dividend} \rightarrow \frac{12x^2 + 6x}{6x} = 2x + 1 \leftarrow \text{Quotient}$$
$$\text{Divisor} \rightarrow$$

EXAMPLE 1 Dividing a Polynomial by a Monomial

Divide $5m^5 - 10m^3$ by $5m^2$.

$$\frac{5m^5 - 10m^3}{5m^2}$$

$$= \frac{5m^5}{5m^2} - \frac{10m^3}{5m^2} \qquad \text{Use the preceding rule, with } + \text{ replaced by } -.$$

$$= m^3 - 2m \qquad \text{Quotient rule}$$

Check Multiply: $5m^2 \cdot (m^3 - 2m) = 5m^5 - 10m^3.$

$\qquad\qquad\qquad$ ↑ $\qquad$ ↑ $\qquad\qquad\qquad$ Original polynomial
$\qquad\qquad$ Divisor $\quad$ Quotient $\qquad\qquad$ (Dividend)

Because division by 0 is undefined, the quotient $\frac{5m^5 - 10m^3}{5m^2}$ is undefined if $m = 0$. From now on, we assume that no denominators are 0.

Work Problem **1** *at the Side.* ▶

EXAMPLE 2 Dividing a Polynomial by a Monomial

Divide $\dfrac{16a^5 - 12a^4 + 8a^2}{4a^3}$.

$$\frac{16a^5 - 12a^4 + 8a^2}{4a^3}$$

$$= \frac{16a^5}{4a^3} - \frac{12a^4}{4a^3} + \frac{8a^2}{4a^3} \qquad \text{Divide each term by } 4a^3.$$

$$= 4a^2 - 3a + \frac{2}{a} \qquad \text{Quotient rule}$$

Continued on Next Page

OBJECTIVE

1 Divide a polynomial by a monomial.

1 Divide.

(a) $\dfrac{6p^4 + 18p^7}{3p^2}$

(b) $\dfrac{12m^6 + 18m^5 + 30m^4}{6m^2}$

(c) $(18r^7 - 9r^2) \div (3r)$

ANSWERS

1. (a) $2p^2 + 6p^5$ (b) $2m^4 + 3m^3 + 5m^2$
   (c) $6r^6 - 3r$

2 Divide.

(a) $\dfrac{20x^4 - 25x^3 + 5x}{5x^2}$

(b) $\dfrac{50m^4 - 30m^3 + 20m}{10m^3}$

3 Divide.

(a) $\dfrac{-9y^6 + 8y^7 - 11y - 4}{y^2}$

(b) $\dfrac{-8p^4 - 6p^3 - 12p^5}{-3p^3}$

4 Divide.

$\dfrac{45x^4y^3 + 30x^3y^2 - 60x^2y}{-15x^2y}$

The quotient $4a^2 - 3a + \frac{2}{a}$ is not a polynomial because of the expression $\frac{2}{a}$, which has a variable in the denominator. While the sum, difference, and product of two polynomials are always polynomials, the quotient of two polynomials may not be.

Check $4a^3\left(4a^2 - 3a + \dfrac{2}{a}\right)$ *Divisor × Quotient should equal Dividend.*

$$= 4a^3(4a^2) + 4a^3(-3a) + 4a^3\left(\dfrac{2}{a}\right) \quad \text{Distributive property}$$

$$= 16a^5 - 12a^4 + 8a^2 \quad \text{Dividend}$$

◀ *Work Problem* **2** *at the Side.*

EXAMPLE 3 **Dividing a Polynomial by a Monomial with a Negative Coefficient**

Divide $-7x^3 + 12x^4 - 4x$ by $-4x$.

Write the polynomial in descending powers as $12x^4 - 7x^3 - 4x$ before dividing.

Write in descending powers. → $\dfrac{12x^4 - 7x^3 - 4x}{-4x}$

$$= \dfrac{12x^4}{-4x} - \dfrac{7x^3}{-4x} - \dfrac{4x}{-4x} \quad \begin{array}{l}\text{Divide each term}\\ \text{by } -4x.\end{array}$$

$$= -3x^3 - \dfrac{7x^2}{-4} - (-1) \quad \text{Quotient rule}$$

$$= -3x^3 + \dfrac{7}{4}x^2 + \mathbf{1} \quad \boxed{\begin{array}{l}\text{Be sure to include}\\ \text{the 1 in the answer.}\end{array}}$$

Check by multiplying.

◀ *Work Problem* **3** *at the Side.*

EXAMPLE 4 **Dividing a Polynomial by a Monomial**

Divide $180x^4y^{10} - 150x^3y^8 + 120x^2y^6 - 90xy^4 + 100y$ by $-30xy^2$.

$$\dfrac{180x^4y^{10} - 150x^3y^8 + 120x^2y^6 - 90xy^4 + 100y}{-30xy^2}$$

$$= \dfrac{180x^4y^{10}}{-30xy^2} - \dfrac{150x^3y^8}{-30xy^2} + \dfrac{120x^2y^6}{-30xy^2} - \dfrac{90xy^4}{-30xy^2} + \dfrac{100y}{-30xy^2}$$

$$= -6x^3y^8 + 5x^2y^6 - 4xy^4 + 3y^2 - \dfrac{10}{3xy}$$

◀ *Work Problem* **4** *at the Side.*

ANSWERS

2. (a) $4x^2 - 5x + \dfrac{1}{x}$ (b) $5m - 3 + \dfrac{2}{m^2}$

3. (a) $8y^5 - 9y^4 - \dfrac{11}{y} - \dfrac{4}{y^2}$

   (b) $4p^2 + \dfrac{8p}{3} + 2$

4. $-3x^2y^2 - 2xy + 4$

13.6 ▶▶▶ Exercises

FOR EXTRA HELP

MyMathLab Math XP PRACTICE WATCH DOWNLOAD READ REVIEW

Fill in each blank with the correct response.

1. In the statement $\dfrac{6x^2 + 8}{2} = 3x^2 + 4$, _____ is the dividend, _____ is the divisor, and _____ is the quotient.

2. The expression $\dfrac{3x + 12}{x}$ is undefined if $x =$ _____.

3. To check the division shown in Exercise 1, multiply _____ by _____ and show that the product is _____.

4. The expression $5x^2 - 3x + 6 + \frac{2}{x}$ _____ a polynomial.
(is/is not)

5. Explain why the division problem $\dfrac{16m^3 - 12m^2}{4m}$ can be performed using the method of this section, while the division problem $\dfrac{4m}{16m^3 - 12m^2}$ cannot.

6. Evaluate $\dfrac{5y + 6}{2}$ when $y = 2$. Evaluate $5y + 3$ when $y = 2$. Does $\dfrac{5y + 6}{2}$ equal $5y + 3$?

Perform each division. See Examples 1–4.

7. $\dfrac{60x^4 - 20x^2 + 10x}{2x}$

8. $\dfrac{120x^6 - 60x^3 + 80x^2}{2x}$

9. $\dfrac{20m^5 - 10m^4 + 5m^2}{-5m^2}$

10. $\dfrac{12t^5 - 6t^3 + 6t^2}{-6t^2}$

11. $\dfrac{8t^5 - 4t^3 + 4t^2}{2t}$

12. $\dfrac{8r^4 - 4r^3 + 6r^2}{2r}$

13. $\dfrac{4a^5 - 4a^2 + 8}{4a}$

14. $\dfrac{5t^8 + 5t^7 + 15}{5t}$

15. $\dfrac{12x^5 - 4x^4 + 6x^3}{-6x^2}$

16. $\dfrac{24x^6 - 12x^5 + 30x^4}{-6x^2}$

17. $\dfrac{4x^2 + 20x^3 - 36x^4}{4x^2}$

18. $\dfrac{5x^2 - 30x^4 + 30x^5}{5x^2}$

19. $\dfrac{-3x^3 - 4x^4 + 2x}{-3x^2}$

20. $\dfrac{-8x + 6x^3 - 5x^4}{-3x^2}$

21. $\dfrac{27r^4 - 36r^3 - 6r^2 + 3r - 2}{3r}$

22. $\dfrac{8k^4 - 12k^3 - 2k^2 - 2k - 3}{2k}$

23. $\dfrac{2m^5 - 6m^4 + 8m^2}{-2m^3}$

24. $\dfrac{6r^5 - 8r^4 + 10r^2}{-2r^4}$

25. $(120x^{11} - 60x^{10} + 140x^9 - 100x^8) \div (10x^{12})$

26. $(120x^{12} - 84x^9 + 60x^8 - 36x^7) \div (12x^9)$

27. $(20a^4b^3 - 15a^5b^2 + 25a^3b) \div (-5a^4b)$

28. $(16y^5z - 8y^2z^2 + 12yz^3) \div (-4y^2z^2)$

29. What polynomial represents the length of the rectangle?

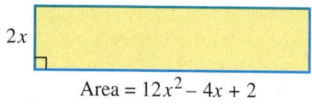

2x

Area $= 12x^2 - 4x + 2$

30. What polynomial represents the length of the base of the triangle?

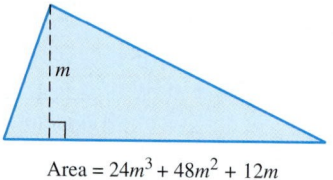

m

Area $= 24m^3 + 48m^2 + 12m$

31. What polynomial, when divided by $5x^3$, yields $3x^2 - 7x + 7$ as a quotient?

32. The quotient of a certain polynomial and $-12y^3$ is $6y^3 - 5y^2 + 2y - 3 + \frac{7}{y}$. Find the polynomial.

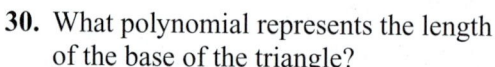

Relating Concepts (Exercises 33—36) For Individual or Group Work

Our system of numeration is called a decimal system. It is based on powers of ten. In a whole number such as 2846, each digit is understood to represent the number of powers of ten for its place value. The 2 represents two thousands (2×10^3), *the 8 represents eight hundreds* (8×10^2), *the 4 represents four tens* (4×10^1), *and the 6 represents six ones (or units)* (6×10^0). *In expanded form we write*

$$2846 = (2 \times 10^3) + (8 \times 10^2) + (4 \times 10^1) + (6 \times 10^0).$$

Keeping this information in mind, **work Exercises 33–36 in order.**

33. Divide 2846 by 2, using paper-and-pencil methods: $2\overline{)2846}$.

34. Write your answer in Exercise 33 in expanded form.

35. Use the methods of this section to divide the polynomial $2x^3 + 8x^2 + 4x + 6$ by 2.

36. Compare your answers in Exercises 34 and 35. How are they similar? How are they different? For what value of x does the answer in Exercise 35 equal the answer in Exercise 34?

13.7 ▸▸▸ Dividing a Polynomial by a Polynomial

OBJECTIVE 1 Divide a polynomial by a polynomial. We use a method of "long division" to divide a polynomial by a polynomial (other than a monomial). *Both polynomials must be written in descending powers.*

OBJECTIVES

1 Divide a polynomial by a polynomial.

2 Apply division to a geometry problem.

Dividing Whole Numbers

Step 1

Divide 6696 by 27.

$$27\overline{)6696}$$

Step 2

66 divided by $27 = 2$;
$2 \cdot 27 = 54$.

$$\begin{array}{r} 2 \\ 27\overline{)6696} \\ 54 \end{array}$$

Step 3

Subtract; then bring down the next digit.

$$\begin{array}{r} 2 \\ 27\overline{)6696} \\ 54\downarrow \\ \hline 129 \end{array}$$

Step 4

129 divided by $27 = 4$;
$4 \cdot 27 = 108$.

$$\begin{array}{r} 24 \\ 27\overline{)6696} \\ 54 \\ \hline 129 \\ 108 \end{array}$$

Step 5

Subtract; then bring down the next digit.

$$\begin{array}{r} 24 \\ 27\overline{)6696} \\ 54 \\ \hline 129 \\ 108\downarrow \\ \hline 216 \end{array}$$

Dividing Polynomials

Step 1

Divide $8x^3 - 4x^2 - 14x + 15$ by $2x + 3$.

$$2x + 3\overline{)8x^3 - 4x^2 - 14x + 15}$$

Step 2

$8x^3$ divided by $2x = 4x^2$;
$4x^2(2x + 3) = 8x^3 + 12x^2$.

$$\begin{array}{r} 4x^2 \\ 2x + 3\overline{)8x^3 - 4x^2 - 14x + 15} \\ 8x^3 + 12x^2 \end{array}$$

Step 3

Subtract; then bring down the next term.

$$\begin{array}{r} 4x^2 \\ 2x + 3\overline{)8x^3 - 4x^2 - 14x + 15} \\ 8x^3 + 12x^2 \downarrow \\ \hline -16x^2 - 14x \end{array}$$

(To subtract two polynomials, change the signs of the second and then add.)

Step 4

$-16x^2$ divided by $2x = -8x$;
$-8x(2x + 3) = -16x^2 - 24x$.

$$\begin{array}{r} 4x^2 - 8x \\ 2x + 3\overline{)8x^3 - 4x^2 - 14x + 15} \\ 8x^3 + 12x^2 \\ \hline -16x^2 - 14x \\ -16x^2 - 24x \end{array}$$

Step 5

Subtract; then bring down the next term.

$$\begin{array}{r} 4x^2 - 8x \\ 2x + 3\overline{)8x^3 - 4x^2 - 14x + 15} \\ 8x^3 + 12x^2 \\ \hline -16x^2 - 14x \\ -16x^2 - 24x\downarrow \\ \hline 10x + 15 \end{array}$$

(continued)

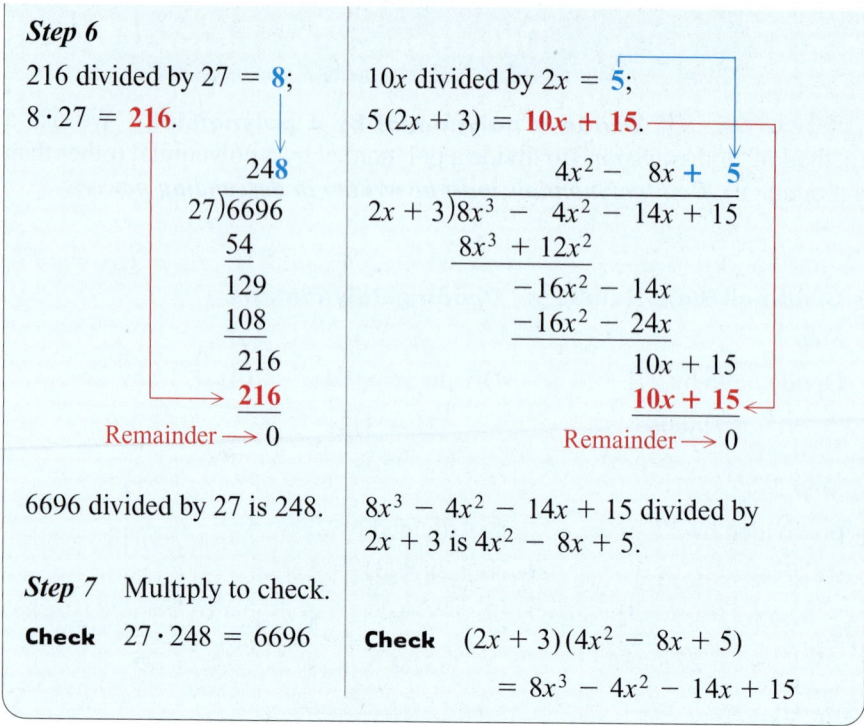

Step 6

216 divided by 27 = **8**; $8 \cdot 27 = \mathbf{216}$.

$$\begin{array}{r} 24\mathbf{8} \\ 27\overline{)6696} \\ 54 \\ \hline 129 \\ 108 \\ \hline 216 \\ \mathbf{216} \\ \hline \end{array}$$

Remainder $\longrightarrow 0$

10x divided by 2x = **5**; $5(2x + 3) = \mathbf{10x + 15}$.

$$\begin{array}{r} 4x^2 - 8x + \mathbf{5} \\ 2x + 3\overline{)8x^3 - 4x^2 - 14x + 15} \\ 8x^3 + 12x^2 \\ \hline -16x^2 - 14x \\ -16x^2 - 24x \\ \hline 10x + 15 \\ \mathbf{10x + 15} \\ \hline \end{array}$$

Remainder $\longrightarrow 0$

6696 divided by 27 is 248.

$8x^3 - 4x^2 - 14x + 15$ divided by $2x + 3$ is $4x^2 - 8x + 5$.

Step 7   Multiply to check.

Check $27 \cdot 248 = 6696$

Check $(2x + 3)(4x^2 - 8x + 5)$
$= 8x^3 - 4x^2 - 14x + 15$

EXAMPLE 1   **Dividing a Polynomial by a Polynomial**

Divide $5x + 4x^3 - 8 - 4x^2$ by $2x - 1$.

The first polynomial must be written with the exponents in descending powers as $4x^3 - 4x^2 + 5x - 8$. Then divide by $2x - 1$.

$$\begin{array}{r} \mathbf{2x^2 - x + 2} \\ 2x - 1\overline{)4x^3 - 4x^2 + 5x - 8} \\ 4x^3 - 2x^2 \\ \hline -2x^2 + 5x \\ -2x^2 + x \\ \hline 4x - 8 \\ 4x - 2 \\ \hline \mathbf{-6} \leftarrow \text{Remainder} \end{array}$$

To subtract, add the opposite.

Write in descending powers.

Step 1 $4x^3$ divided by $2x = \mathbf{2x^2}$; $2x^2(2x - 1) = 4x^3 - 2x^2$.

Step 2   Subtract; bring down the next term.

Step 3 $-2x^2$ divided by $2x = \mathbf{-x}$; $-x(2x - 1) = -2x^2 + x$.

Step 4   Subtract; bring down the next term.

Step 5 $4x$ divided by $2x = \mathbf{2}$; $2(2x - 1) = 4x - 2$.

Step 6   Subtract. The remainder is $\mathbf{-6}$. Write the remainder as the numerator of a fraction that has $2x - 1$ as its denominator. The answer is not a polynomial because of the nonzero remainder.

$$\underset{\text{Divisor} \rightarrow}{\overset{\text{Dividend} \rightarrow}{\frac{4x^3 - 4x^2 + 5x - 8}{2x - 1}}} = \underbrace{2x^2 - x + 2}_{\substack{\text{Quotient} \\ \text{polynomial}}} + \mathbf{\frac{-6}{2x - 1}}$$

Continued on Next Page

Step 7 Multiply to check.

Check $(2x - 1)\left(2x^2 - x + 2 + \dfrac{-6}{2x - 1} \right)$

$$= (2x - 1)(2x^2) + (2x - 1)(-x) + (2x - 1)(2)$$

$$+ (2x - 1)\left(\dfrac{-6}{2x - 1} \right)$$

$$= 4x^3 - 2x^2 - 2x^2 + x + 4x - 2 - 6$$

$$= 4x^3 - 4x^2 + 5x - 8$$

—————— Work Problem ① at the Side. ▶

1 Divide.

(a) $(x^3 + x^2 + 4x - 6)$
$\div (x - 1)$

(b) $\dfrac{p^3 - 2p^2 - 5p + 9}{p + 2}$

EXAMPLE 2 **Dividing into a Polynomial with Missing Terms**

Divide $x^3 - 1$ by $x - 1$.

Here the polynomial $x^3 - 1$ is missing the x^2-term and the x-term. When terms are missing, use **0** as the coefficient for each missing term. (Zero acts as a placeholder here, just as it does in our numeration system.) Thus, $x^3 - 1 = x^3 + 0x^2 + 0x - 1$. Now divide.

$$
\begin{array}{r}
x^2 + x + 1 \\
x - 1 \overline{) x^3 + 0x^2 + 0x - 1} \\
\underline{x^3 - x^2} \\
x^2 + 0x \\
\underline{x^2 - x} \\
x - 1 \\
\underline{x - 1} \\
0
\end{array}
$$

Insert placeholders for the missing terms.

The remainder is 0. The quotient is $x^2 + x + 1$.

Check $(x - 1)(x^2 + x + 1)$

$$= x^3 + x^2 + x - x^2 - x - 1$$

$$= x^3 - 1$$

—————— Work Problem ② at the Side. ▶

2 Divide.

(a) $\dfrac{r^2 - 5}{r + 4}$

(b) $(x^3 - 8) \div (x - 2)$

EXAMPLE 3 **Dividing by a Polynomial with Missing Terms**

Divide $x^4 + 2x^3 + 2x^2 - x - 1$ by $x^2 + 1$.

Since $x^2 + 1$ has a missing x-term, write it as $x^2 + 0x + 1$.

$$
\begin{array}{r}
x^2 + 2x + 1 \\
x^2 + 0x + 1 \overline{) x^4 + 2x^3 + 2x^2 - x - 1} \\
\underline{x^4 + 0x^3 + x^2} \\
2x^3 + x^2 - x \\
\underline{2x^3 + 0x^2 + 2x} \\
x^2 - 3x - 1 \\
\underline{x^2 + 0x + 1} \\
-3x - 2 \leftarrow \text{Remainder}
\end{array}
$$

Insert a placeholder for the missing term.

Continued on Next Page

ANSWERS

1. (a) $x^2 + 2x + 6$

   (b) $p^2 - 4p + 3 + \dfrac{3}{p + 2}$

2. (a) $r - 4 + \dfrac{11}{r + 4}$

   (b) $x^2 + 2x + 4$

3 Divide.

(a)

$(2x^4 + 3x^3 - x^2 + 6x + 5)$
$\div (x^2 - 1)$

(b)

$$\dfrac{2m^5 + m^4 + 6m^3 - 3m^2 - 18}{m^2 + 3}$$

4 Divide $3x^3 + 7x^2 + 7x + 10$ by $3x + 6$.

5 Divide $x^3 + 4x^2 + 8x + 8$ by $x + 2$.

When the result of subtracting ($-3x - 2$, in this case) is a constant or a polynomial of degree less than the divisor ($x^2 + 0x + 1$), that constant or polynomial is the remainder. We write the answer as

$$x^2 + 2x + 1 + \dfrac{-3x - 2}{x^2 + 1}.$$

> Remember to include "$+\ \frac{\text{remainder}}{\text{divisor}}$."

Multiply to check that this is the correct quotient.

◀ *Work Problem* **3** *at the Side.*

EXAMPLE 4 **Dividing a Polynomial when the Quotient Has Fractional Coefficients**

Divide $4x^3 + 2x^2 + 3x + 2$ by $4x - 4$.

$$\dfrac{6x^2}{4x} = \dfrac{3}{2}x$$
$$\dfrac{9x}{4x} = \dfrac{9}{4}$$

$$
\begin{array}{r}
x^2 + \frac{3}{2}x + \frac{9}{4} \\
4x - 4\overline{)4x^3 + 2x^2 + 3x + 2} \\
\underline{4x^3 - 4x^2} \\
6x^2 + 3x \\
\underline{6x^2 - 6x} \\
9x + 2 \\
\underline{9x - 9} \\
11
\end{array}
$$

The answer is $x^2 + \dfrac{3}{2}x + \dfrac{9}{4} + \dfrac{11}{4x - 4}$.

◀ *Work Problem* **4** *at the Side.*

OBJECTIVE **2** **Apply division to a geometry problem.**

EXAMPLE 5 **Using an Area Formula**

The area of the rectangle in Figure 3 is $x^3 + 4x^2 + 8x + 8$ square units and the width is $x + 2$ units. What is its length?

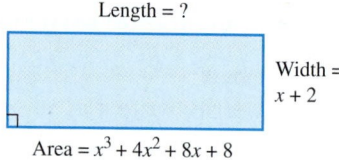

Length = ?

Width = $x + 2$

Area = $x^3 + 4x^2 + 8x + 8$

Figure 3

Since $A = lw$, solving for l gives $l = \frac{A}{w}$. Divide $x^3 + 4x^2 + 8x + 8$ by the width, $x + 2$.

◀ *Work Problem* **5** *at the Side.*

The quotient from Problem 5 at the side, $x^2 + 2x + 4$, represents the length of the rectangle in units.

13.7 ▶▶▶ Exercises

1. In the division problem $(4x^4 + 2x^3 - 14x^2 + 19x + 10) \div (2x + 5) = 2x^3 - 4x^2 + 3x + 2$, which polynomial is the divisor? Which is the quotient?

2. When dividing one polynomial by another, how do you know when to stop dividing?

3. In dividing $12m^2 - 20m + 3$ by $2m - 3$, what is the first step?

4. In the division in Exercise 3, what is the second step?

Perform each division. See Example 1.

5. $\dfrac{x^2 - x - 6}{x - 3}$

6. $\dfrac{m^2 - 2m - 24}{m - 6}$

7. $\dfrac{2y^2 + 9y - 35}{y + 7}$

8. $\dfrac{2y^2 + 9y + 7}{y + 1}$

9. $\dfrac{p^2 + 2p + 20}{p + 6}$

10. $\dfrac{x^2 + 11x + 16}{x + 8}$

11. $(r^2 - 8r + 15) \div (r - 3)$

12. $(t^2 + 2t - 35) \div (t - 5)$

13. $\dfrac{4a^2 - 22a + 32}{2a + 3}$

14. $\dfrac{9w^2 + 6w + 10}{3w - 2}$

15. $\dfrac{8x^3 - 10x^2 - x + 3}{2x + 1}$

16. $\dfrac{12t^3 - 11t^2 + 9t + 18}{4t + 3}$

Perform each division. See Examples 2–4.

17. $\dfrac{3y^3 + y^2 + 2}{y + 1}$

18. $\dfrac{2r^3 - 6r - 36}{r - 3}$

19. $\dfrac{2x^3 + x + 2}{x + 1}$

20. $\dfrac{3x^3 + x + 5}{x + 1}$

21. $\dfrac{3k^3 - 4k^2 - 6k + 10}{k^2 - 2}$

22. $\dfrac{5z^3 - z^2 + 10z + 2}{z^2 + 2}$

23. $(x^4 - x^2 - 2) \div (x^2 - 2)$

24. $(r^4 + 2r^2 - 3) \div (r^2 - 1)$

25. $\dfrac{x^4 - 1}{x^2 - 1}$

26. $\dfrac{y^3 + 1}{y + 1}$

27. $\dfrac{6p^4 - 15p^3 + 14p^2 - 5p + 10}{3p^2 + 1}$

28. $\dfrac{6r^4 - 10r^3 - r^2 + 15r - 8}{2r^2 - 3}$

29. $\dfrac{2x^5 + x^4 + 11x^3 - 8x^2 - 13x + 7}{2x^2 + x - 1}$

30. $\dfrac{4t^5 - 11t^4 - 6t^3 + 5t^2 - t + 3}{4t^2 + t - 3}$

31. $(10x^3 + 13x^2 + 4x + 1) \div (5x + 5)$

32. $(6x^3 - 19x^2 - 19x - 4) \div (2x - 8)$

Work each problem. See Example 5.

33. What is the length of the rectangle if the area is $5x^3 + 7x^2 - 13x - 6$ square units?

$5x + 2$

34. Find the measure of the base of the parallelogram if the area is $2x^3 + 2x^2 - 3x + 1$ square units.

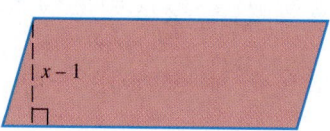

$x - 1$

Relating Concepts (Exercises 35–38) For Individual or Group Work

We can find the value of a polynomial in x for a given value of x by substituting that number for x. Surprisingly, we can accomplish the same thing by division. For example, to find the value of $2x^2 - 4x + 3$ for $x = -3$, we would divide $2x^2 - 4x + 3$ by $x - (-3)$. The remainder will give the value of the polynomial for $x = -3$. **Work Exercises 35–38 in order.**

35. Find the value of $2x^2 - 4x + 3$ for $x = -3$ by substitution.

36. Divide $2x^2 - 4x + 3$ by $x + 3$. Give the remainder.

37. Compare your answers to Exercises 35 and 36. What do you notice?

38. Choose another polynomial and evaluate it both ways for some value of the variable. Do the answers agree?

13.8 ▷▷▷ An Application of Exponents: Scientific Notation

OBJECTIVE 1 Express numbers in scientific notation. Numbers occurring in science are often extremely large (such as the distance from Earth to the sun, 93,000,000 mi) or extremely small (the wavelength of yellow-green light, approximately 0.0000006 m). Because of the difficulty of working with many zeros, scientists often express such numbers with exponents, using a form called *scientific notation*.

OBJECTIVES

1 Express numbers in scientific notation.

2 Convert numbers in scientific notation to numbers without exponents.

3 Use scientific notation in calculations.

> **Scientific Notation**
>
> A number is written in **scientific notation** when it is expressed in the form
>
> $$a \times 10^n,$$
>
> where $1 \leq |a| < 10$ and n is an integer.

In **scientific notation,** there is always one nonzero digit before the decimal point. This is shown in the following examples.

$3.19 \times 10^1 = 3.19 \times 10 = 31.9$	Decimal point moves 1 place to the right.
$3.19 \times 10^2 = 3.19 \times 100 = 319.$	Decimal point moves 2 places to the right.
$3.19 \times 10^3 = 3.19 \times 1000 = 3190.$	Decimal point moves 3 places to the right.
$3.19 \times 10^{-1} = 3.19 \times 0.1 = 0.319$	Decimal point moves 1 place to the left.
$3.19 \times 10^{-2} = 3.19 \times 0.01 = 0.0319$	Decimal point moves 2 places to the left.
$3.19 \times 10^{-3} = 3.19 \times 0.001 = 0.00319$	Decimal point moves 3 places to the left.

> **Note**
>
> In scientific notation, the times symbol, $\times$, is commonly used.

A number in scientific notation is always written with the decimal point after the first nonzero digit and then multiplied by the appropriate power of 10. For example, 56,200 is written 5.62×10^4, since

$$56,200 = 5.62 \times \mathbf{10,000} = 5.62 \times \mathbf{10^4}.$$

Other examples include

42,000,000	written	4.2×10^7,
0.000586	written	5.86×10^{-4},
and 2,000,000,000	written	2×10^9.

> It is not necessary to write 2.0.

To write a number in scientific notation, follow the steps given on the next page. (For a negative number, follow these steps using the *absolute value* of the number; then make the result negative.)

1 Write each number in scientific notation.

(a) 63,000

(b) 5,870,000

(c) 7.0065

(d) 0.0571

(e) −0.00062

Writing a Number in Scientific Notation

Step 1 Move the decimal point to the right of the first nonzero digit.

Step 2 Count the number of places you moved the decimal point.

Step 3 The number of places in Step 2 is the absolute value of the exponent on 10.

Step 4 The exponent on 10 is positive if the original number is greater than the number in Step 1; the exponent is negative if the original number is less than the number in Step 1. If the decimal point is not moved, the exponent is 0.

EXAMPLE 1 Using Scientific Notation

Write each number in scientific notation.

(a) 93,000,000

Move the decimal point to follow the first nonzero digit (the 9). Count the number of places the decimal point was moved.

$$93{,}000{,}000. \quad \longleftarrow \text{Decimal point}$$
7 places

The number will be written in scientific notation as 9.3×10^n. To find the value of n, first compare the original number, 93,000,000, with 9.3. Since 93,000,000 is *greater* than 9.3, we must multiply by a *positive* power of 10 so that the product 9.3×10^n will equal the larger number.

Since the decimal point was moved 7 places, and since n is positive,

$$93{,}000{,}000 = 9.3 \times 10^7.$$

(b) $63{,}200{,}000{,}000 = 6.3200000000 = 6.32 \times 10^{10}$
10 places

(c) $3.021 = 3.021 \times 10^0$

(d) 0.00462

Move the decimal point to the right of the first nonzero digit and count the number of places the decimal point was moved.

$$0.00462 \quad 3 \text{ places}$$

Since 0.00462 is *less* than 4.62, the exponent must be *negative*.

$$0.00462 = 4.62 \times 10^{-3}$$

(e) $-0.0000762 = -7.62 \times 10^{-5}$
5 places

Remember the negative sign.

◀ Work Problem **1** at the Side.

Note

To choose the exponent when you write a number in scientific notation, think: If the original number is "large," like 93,000,000, use a *positive* exponent on 10, since positive is greater than negative. However, if the original number is "small," like 0.00462, use a *negative* exponent on 10, since negative is less than positive.

OBJECTIVE 2 **Convert numbers in scientific notation to numbers without exponents.** To convert a number written in scientific notation to a number without exponents, work in reverse. *Multiplying a number by a positive power of 10 will make the number greater; multiplying by a negative power of 10 will make the number less.*

EXAMPLE 2 **Writing Numbers without Exponents**

Write each number without exponents.

(a) 6.2×10^3
Since the exponent is positive, make 6.2 greater by moving the decimal point 3 places to the right. It is necessary to attach two 0s.

$$6.2 \times \mathbf{10^3} = 6.2\underset{\frown}{\mathbf{00}} = 6200$$

(b) $4.283 \times 10^5 = 4.283\underset{\frown}{\mathbf{00}} = 428{,}300$ Move 5 places to the right; attach 0s as necessary.

(c) $-9.73 \times 10^{-2} = -\underset{\frown}{\mathbf{0}}9.73 = -0.0973$ Move 2 places to the left.

The exponent tells the number of places and the direction that the decimal point is moved.

───────── *Work Problem* 2 *at the Side.* ▶

OBJECTIVE 3 **Use scientific notation in calculations.** The next example uses scientific notation with products and quotients.

EXAMPLE 3 **Multiplying and Dividing with Scientific Notation**

Perform each calculation. Write answers in scientific notation and also without exponents.

(a) $(7 \times 10^3)(5 \times 10^4)$

$= (7 \times 5)(10^3 \times 10^4)$ Commutative and associative properties

$= \mathbf{35} \times 10^7$ Multiply; product rule

> Don't stop! This number is *not* in scientific notation, since 35 is not between 1 and 10.

$= (\mathbf{3.5 \times 10^1}) \times 10^7$ Write 35 in scientific notation.

$= 3.5 \times (\mathbf{10^1 \times 10^7})$ Associative property

$= 3.5 \times \mathbf{10^8}$ Product rule

$= 350{,}000{,}000$ Write without exponents.

(b) $\dfrac{4 \times 10^{-5}}{2 \times 10^3} = \dfrac{4}{2} \times \dfrac{10^{-5}}{10^3} = 2 \times 10^{-8} = 0.00000002$

───────── *Work Problem* 3 *at the Side.* ▶

Note

Multiplying or dividing numbers written in scientific notation may produce an answer in the form $a \times \mathbf{10^0}$. Since $10^0 = 1$, $a \times 10^0 = a$. For example,

$$(8 \times 10^{-4})(5 \times 10^4) = 40 \times \mathbf{10^0} = 40. \qquad 10^0 = 1$$

Also, if $a = 1$, then $a \times 10^n = 10^n$. For example, we could write $1{,}000{,}000$ as 10^6 instead of 1×10^6.

---

2 Write without exponents.

(a) 4.2×10^3

(b) 8.7×10^5

(c) 6.42×10^{-3}

3 Perform each calculation. Write answers in scientific notation and also without exponents.

(a) $(2.6 \times 10^4)(2 \times 10^{-6})$

(b) $(3 \times 10^5)(5 \times 10^{-2})$

(c) $\dfrac{4.8 \times 10^2}{2.4 \times 10^{-3}}$

ANSWERS

2. **(a)** 4200 **(b)** 870,000 **(c)** 0.00642
3. **(a)** 5.2×10^{-2}; 0.052
   (b) 1.5×10^4; 15,000
   (c) 2×10^5; 200,000

4 The speed of light is approximately 3.0×10^5 km per sec. How far does light travel in 6.0×10^1 sec? (*Source: World Almanac and Book of Facts.*)

> 🖩 **Calculator Tip** Calculators usually have a key labeled EE or EXP for scientific notation. See your owner's manual for more information.

EXAMPLE 4 Using Scientific Notation to Solve an Application

A *nanometer* is a very small unit of measure that is equivalent to about 0.00000003937 in. About how much would 700,000 nanometers measure in inches? (*Source: World Almanac and Book of Facts.*)

Write each number in scientific notation, and then multiply.

$$700,000\,(0.00000003937)$$

$$= (7 \times 10^5)(3.937 \times 10^{-8}) \qquad \text{Write in scientific notation.}$$

$$= (7 \times 3.937)(10^5 \times 10^{-8}) \qquad \text{Properties of real numbers}$$

$$= \mathbf{27.559 \times 10^{-3}} \qquad \text{Multiply; product rule}$$

> Don't stop here.

$$= \mathbf{(2.7559 \times 10^1)} \times 10^{-3} \qquad \begin{array}{l}\text{Write 27.559 in}\\\text{scientific notation.}\end{array}$$

$$= 2.7559 \times 10^{-2} \qquad \text{Product rule}$$

$$= 0.027559 \qquad \text{Write without exponents.}$$

Thus, 700,000 nanometers would measure

$$2.7559 \times 10^{-2} \text{ in.,} \quad \text{or} \quad 0.027559 \text{ in.}$$

◀ *Work Problem* **4** *at the Side.*

5 If the speed of light is approximately 3.0×10^5 km per sec, how many seconds does it take light to travel approximately 1.5×10^8 km from the sun to Earth? (*Source: World Almanac and Book of Facts.*)

EXAMPLE 5 Using Scientific Notation to Solve an Application

In 2003, the national debt was $\$3.9136 \times 10^{12}$ (which is more than \$3 trillion). The population of the United States was approximately 290 million that year. About how much would each person have had to contribute in order to pay off the national debt? (*Source: U.S. Office of Management and Budget; U.S. Census Bureau.*)

Write the population in scientific notation. Then divide to obtain the per person contribution.

$$\frac{3.9136 \times 10^{12}}{\mathbf{290,000,000}}$$

$$= \frac{3.9136 \times 10^{12}}{\mathbf{2.9 \times 10^8}} \qquad \begin{array}{l}\text{Write 290 million in}\\\text{scientific notation.}\end{array}$$

$$= \frac{3.9136}{2.9} \times 10^4 \qquad \text{Quotient rule}$$

$$\approx 1.3495 \times 10^4 \qquad \text{Divide; round to 4 decimal places.}$$

$$\approx 13,495 \qquad \text{Write without exponents.}$$

Each person would have to pay about \$13,495.

◀ *Work Problem* **5** *at the Side.*

ANSWERS

4. 1.8×10^7 km, or 18,000,000 km
5. 5×10^2 sec, or 500 sec

13.8 ▶▶▶ **Exercises**

FOR
EXTRA
HELP

MyMathLab

Math XL
PRACTICE

WATCH

DOWNLOAD

READ

REVIEW

Write the numbers (other than dates) mentioned in the following statements in scientific notation.

1. NASA has budgeted $6,130,900,000 for 2003 and $5,868,900,000 for 2004 for the international space station. (*Source:* U.S. National Aeronautics and Space Administration.)

2. The mass of Pluto is 0.0021 times that of Earth; the mass of Jupiter is 317.83 times that of Earth. (*Source: World Almanac and Book of Facts.*)

Determine whether or not the given number is written in scientific notation as defined in Objective 1. If it is not, write it as such.

3. 4.56×10^3

4. 7.34×10^5

5. 5,600,000

6. 34,000

7. 0.004

8. 0.0007

9. 0.8×10^2

10. 0.9×10^3

11. Explain in your own words what it means for a number to be written in scientific notation.

12. Explain how to multiply a number by a positive power of ten. Then explain how to multiply a number by a negative power of ten.

Write each number in scientific notation. See Example 1.

13. 5,876,000,000

14. 9,994,000,000

15. 82,350

16. 78,330

17. 0.000007

18. 0.0000004

19. −0.00203

20. −0.0000578

Write each number without exponents. See Example 2.

21. 7.5×10^5

22. 8.8×10^6

23. 5.677×10^{12}

24. 8.766×10^9

25. 1×10^{12}

26. 1×10^7

27. -6.21×10^0

28. -8.56×10^0

29. 7.8×10^{-4}

30. 8.9×10^{-5}

31. 5.134×10^{-9}

32. 7.123×10^{-10}

Perform the indicated operations. Write the answers in scientific notation and then without exponents. See Example 3.

33. $(2 \times 10^8)(3 \times 10^3)$

34. $(3 \times 10^7)(3 \times 10^3)$

35. $(5 \times 10^4)(3 \times 10^2)$

36. $(8 \times 10^5)(2 \times 10^3)$

37. $(4 \times 10^{-6})(2 \times 10^3)$

38. $(3 \times 10^{-7})(2 \times 10^2)$

39. $(6 \times 10^3)(4 \times 10^{-2})$

40. $(7 \times 10^5)(3 \times 10^{-4})$

41. $(9 \times 10^4)(7 \times 10^{-7})$

42. $(6 \times 10^4)(8 \times 10^{-8})$

43. $(3.15 \times 10^{-4})(2.04 \times 10^8)$

44. $(4.92 \times 10^{-3})(2.25 \times 10^7)$

45. $\dfrac{9 \times 10^{-5}}{3 \times 10^{-1}}$

46. $\dfrac{12 \times 10^{-4}}{4 \times 10^{-3}}$

47. $\dfrac{8 \times 10^3}{2 \times 10^2}$

48. $\dfrac{15 \times 10^4}{3 \times 10^3}$

49. $\dfrac{2.6 \times 10^{-3}}{2 \times 10^2}$

50. $\dfrac{9.5 \times 10^{-1}}{5 \times 10^3}$

51. $\dfrac{4 \times 10^5}{8 \times 10^2}$

52. $\dfrac{3 \times 10^9}{6 \times 10^5}$

53. $\dfrac{2.6 \times 10^{-3} \times 7.0 \times 10^{-1}}{2 \times 10^2 \times 3.5 \times 10^{-3}}$

54. $\dfrac{9.5 \times 10^{-1} \times 2.4 \times 10^4}{5 \times 10^3 \times 1.2 \times 10^{-2}}$

55. $\dfrac{(1.65 \times 10^8)(5.24 \times 10^{-2})}{(6 \times 10^4)(2 \times 10^7)}$

Work each problem. In Exercises 58–60, give answers without exponents. See Examples 4 and 5.

56. Pollux, one of the brightest stars in the night sky, is 33.7 light-years from Earth. If one light-year is about 6,000,000,000,000 mi (that is, 6 trillion mi), about how many miles is Pollux from Earth? (*Source: World Almanac and Book of Facts.*)

57. In March 2006, astronomers using the Spitzer Space Telescope discovered a twisted double-helix nebula, a conglomeration of dust and gas stretching across the center of the Milky Way galaxy. This nebula is 25,000 light-years from Earth. If one light-year is about 6,000,000,000,000 mi, about how many miles is the twisted double-helix nebula from Earth? (*Source*: http://articles.news.aol.com)

58. In 2003, the U.S. government collected about $6730 per person in taxes. If the population at that time was 290,000,000, how much did the government collect in taxes for 2003? (*Source*: U.S. Internal Revenue Service.)

59. In 2000, the population of the United States was about 281.4 million. To the nearest dollar, calculate how much each person in the United States would have had to contribute in order to make one lucky person a trillionaire (that is, to give that person $1,000,000,000,000). (*Source*: U.S. Census Bureau.)

60. In 2006, Congress raised the government's debt limit to 9×10^{12}. When this national debt limit is reached, about how much is it for every man, woman, and child in the country? Use 300 million as the population of the United States. (*Source: The Gazette*, Cedar Rapids, Iowa, March 17, 2006.)

Chapter 13 ▶▶▶ Summary

▶ Key Terms

13.1	**term**	A term is a number, a variable, or a product or quotient of a number and one or more variables raised to powers.
	like terms	Terms with exactly the same variables (including the same exponents) are called like terms.
	polynomial	A polynomial is a term or the sum of a finite number of terms with whole number exponents.
	descending powers	A polynomial in x is written in descending powers if the exponents on x in its terms are in decreasing order.
	degree of a term	The degree of a term is the sum of the exponents on the variables.
	degree of a polynomial	The degree of a polynomial is the greatest degree of any term of the polynomial.
	monomial	A monomial is a polynomial with exactly one term.
	binomial	A binomial is a polynomial with exactly two terms.
	trinomial	A trinomial is a polynomial with exactly three terms.

13.2 exponential expression — A number written with an exponent is an exponential expression.

3^4 ← Exponent ⎤ Exponential
⎰ expression
Base ⎦

13.3 FOIL — FOIL is a shortcut method for finding the product of two binomials. The letters of the word **FOIL** originate as follows: Multiply the **F**irst terms, multiply the **O**uter terms (to get the outer product), multiply the **I**nner terms (to get the inner product), and multiply the **L**ast terms.

outer product — The outer product of $(2x + 3)(x - 5)$ is $2x(-5)$.

inner product — The inner product of $(2x + 3)(x - 5)$ is $3x$.

13.4 conjugate — The conjugate of $a + b$ is $a - b$.

13.8 scientific notation — A number written as $a \times 10^n$, where $1 \le |a| < 10$ and n is an integer, is in scientific notation.

▶ New Symbols

x^{-n} x to the negative n power $a \times 10^n$ scientific notation

▶ Test Your Word Power

See how well you have learned the vocabulary in this chapter. Answers, with examples, follow the Quick Review.

1. A **polynomial** is an algebraic expression made up of
   A. a term or a finite product of terms with positive coefficients and exponents
   B. a term or a finite sum of terms with real coefficients and whole number exponents
   C. the product of two or more terms with positive exponents
   D. the sum of two or more terms with whole number coefficients and exponents.

2. The **degree of a term** is
   A. the number of variables in the term
   B. the product of the exponents on the variables
   C. the least exponent on the variables
   D. the sum of the exponents on the variables.

3. A **trinomial** is a polynomial with
   A. only one term
   B. exactly two terms
   C. exactly three terms
   D. more than three terms.

4. A **binomial** is a polynomial with
   A. only one term
   B. exactly two terms
   C. exactly three terms
   D. more than three terms.

5. A **monomial** is a polynomial with
   A. only one term
   B. exactly two terms
   C. exactly three terms
   D. more than three terms.

6. **FOIL** is a method for
   A. adding two binomials
   B. adding two trinomials
   C. multiplying two binomials
   D. multiplying two trinomials.

▶ Quick Review

Concepts	Examples

13.1 Adding and Subtracting Polynomials

Addition
Add like terms.

Add.
$$\begin{array}{r} 2x^2 + 5x - 3 \\ 5x^2 - 2x + 7 \\ \hline 7x^2 + 3x + 4 \end{array}$$

Subtraction
Change the signs of the terms in the second polynomial and add to the first polynomial.

Subtract. $(2x^2 + 5x - 3) - (5x^2 - 2x + 7)$
$$= (2x^2 + 5x - 3) + (-5x^2 + 2x - 7)$$
$$= -3x^2 + 7x - 10$$

13.2 The Product Rule and Power Rules for Exponents

For any integers m and n:

Product rule $\quad a^m \cdot a^n = a^{m+n}$

Power rules (a) $\quad (a^m)^n = a^{mn}$

$\qquad$ **(b)** $\quad (ab)^m = a^m b^m$

$\qquad$ **(c)** $\quad \left(\dfrac{a}{b}\right)^m = \dfrac{a^m}{b^m} \quad (b \neq 0)$.

Simplify.

$$2^4 \cdot 2^5 = 2^{4+5} = 2^9$$

$$(3^4)^2 = 3^{4 \cdot 2} = 3^8$$

$$(6a)^5 = 6^5 a^5$$

$$\left(\frac{2}{3}\right)^4 = \frac{2^4}{3^4}$$

13.3 Multiplying Polynomials

Multiply each term of the first polynomial by each term of the second polynomial. Then add like terms.

Multiply.
$$\begin{array}{r} 3x^3 - 4x^2 + 2x - 7 \\ 4x + 3 \\ \hline 9x^3 - 12x^2 + 6x - 21 \\ 12x^4 - 16x^3 + 8x^2 - 28x \\ \hline 12x^4 - 7x^3 - 4x^2 - 22x - 21 \end{array}$$

FOIL Method

Step 1 Multiply the two **First** terms to get the first term of the answer.

Step 2 Find the **Outer** product and the **Inner** product and mentally add them, when possible, to get the middle term of the answer.

Step 3 Multiply the two **Last** terms to get the last term of the answer.

Add the terms found in Steps 1–3.

Multiply $(2x + 3)(5x - 4)$.
$$2x(5x) = 10x^2$$

$$2x(-4) + 3(5x) = 7x$$

$$3(-4) = -12$$

The product is $10x^2 + 7x - 12$.

13.4 Special Products

Square of a Binomial

$$(a + b)^2 = a^2 + 2ab + b^2$$

$$(a - b)^2 = a^2 - 2ab + b^2$$

Product of the Sum and Difference of Two Terms

$$(a + b)(a - b) = a^2 - b^2$$

Multiply.

$$(3x + 1)^2$$
$$= (3x)^2 + 2(3x)(1) + 1^2$$
$$= 9x^2 + 6x + 1$$

$$(2m - 5n)^2$$
$$= (2m)^2 - 2(2m)(5n) + (5n)^2$$
$$= 4m^2 - 20mn + 25n^2$$

$$(4a + 3)(4a - 3)$$
$$= (4a)^2 - 3^2$$
$$= 16a^2 - 9$$

Concepts	Examples

13.5 Integer Exponents and the Quotient Rule

If $a, b \neq 0$, for integers m and n:

Zero exponent $\quad a^0 = 1$

Negative exponent $\quad a^{-n} = \dfrac{1}{a^n}$

Quotient rule $\quad \dfrac{a^m}{a^n} = a^{m-n}$

Negative-to-positive rules $\quad \dfrac{a^{-m}}{b^{-n}} = \dfrac{b^n}{a^m} \quad \left(\dfrac{a}{b}\right)^{-m} = \left(\dfrac{b}{a}\right)^{m}.$

Simplify.

$$15^0 = 1$$

$$5^{-2} = \frac{1}{5^2} = \frac{1}{25}$$

$$\frac{4^8}{4^3} = 4^{8-3} = 4^5$$

$$\frac{6^{-2}}{7^{-3}} = \frac{7^3}{6^2} \qquad \left(\frac{5}{3}\right)^{-4} = \left(\frac{3}{5}\right)^{4}$$

13.6 Dividing a Polynomial by a Monomial

Divide each term of the polynomial by the monomial:

$$\frac{a + b}{c} = \frac{a}{c} + \frac{b}{c}.$$

Divide. $\dfrac{4x^3 - 2x^2 + 6x - 8}{2x}$

$$= \frac{4x^3}{2x} - \frac{2x^2}{2x} + \frac{6x}{2x} - \frac{8}{2x}$$

$$= 2x^2 - x + 3 - \frac{4}{x}$$

13.7 Dividing a Polynomial by a Polynomial

Use "long division."

Divide.
$$
\begin{array}{r}
2x - 5 + \dfrac{-1}{3x + 4} \\[4pt]
3x + 4 \overline{) 6x^2 - 7x - 21} \\
\underline{6x^2 + 8x } \\
-15x - 21 \\
\underline{-15x - 20} \\
-1 \leftarrow \text{Remainder}
\end{array}
$$

13.8 An Application of Exponents: Scientific Notation

To write a number in scientific notation (as $a \times 10^n$, where $1 \leq |a| < 10$), move the decimal point to the right of the first nonzero digit. If the decimal point is moved n places, and this makes the number smaller, n is positive; if it makes the number larger, n is negative. If the decimal point is not moved, n is 0.

Write in scientific notation.

$$247 = 2.47 \times 10^2$$

$$0.0051 = 5.1 \times 10^{-3}$$

Write without exponents.

$$3.25 \times 10^5 = 325{,}000$$

$$8.44 \times 10^{-6} = 0.00000844$$

ANSWERS TO TEST YOUR WORD POWER

1. B; *Example:* $5x^3 + 2x^2 - 7$
2. D; *Examples:* The term 6 has degree 0, $3x$ has degree 1, $-2x^8$ has degree 8, and $5x^2y^4$ has degree 6.
3. C; *Example:* $2a^2 - 3ab + b^2$
4. B; *Example:* $3t^3 + 5t$
5. A; *Examples:* -5 and $4xy^5$
6. C; *Example:* $(m + 4)(m - 3)$

$$
\begin{array}{cccc}
\mathbf{F} & \mathbf{O} & \mathbf{I} & \mathbf{L} \\
\end{array}
$$
$$= m(m) - 3m + 4m + 4(-3)$$
$$= m^2 + m - 12$$

Math in the Media

FLOODS, HURRICANES, AND EARTHQUAKES, OH MY!

In recent years, the number of natural disasters seems to be on the increase. Charles F. Richter devised a scale in 1935 to compare the intensities, or relative power, of earthquakes. The **intensity** of an earthquake (often mentioned in media reports) is measured relative to the intensity of a standard **zero-level** earthquake of intensity I_0. The relationship is equivalent to $I = I_0 \times 10^R$, where R is the **Richter scale** measure. For example, if an earthquake has magnitude 5.0 on the Richter scale, then its intensity is calculated as $I = I_0 \times 10^{5.0} = I_0 \times 100{,}000$, which is 100,000 times as intense as a zero-level earthquake.

To compare an earthquake that measures 8.0 on the Richter scale to one that measures 5.0, find the ratio of the intensities:

$$\frac{\text{intensity } 8.0}{\text{intensity } 5.0} = \frac{I_0 \times 10^{8.0}}{I_0 \times 10^{5.0}} = \frac{10^8}{10^5} = 10^{8-5} = 10^3 = 1000.$$

Therefore, an earthquake that measures 8.0 is 1000 times as intense as one that measures 5.0.

The Gazette TUESDAY April 19, 2008

Quake rattles Iowans

Chances of 'big one' happening here remote UNI professor says

Source: ESRI; USGS.

The table gives Richter scale measurements for several earthquakes.

Year	Earthquake	Richter Scale Measurement
2008	West Salem, IL	5.2
2005	Northern Sumatra, Indonesia	8.6
2004	West coast of Northern Sumatra	9.1
2003	Southeastern Iran	6.6
1998	Balleny Islands region	8.1
1906	San Francisco, CA	7.7

Source: U.S. Geological Survey.

1. Compare the intensity of the 2004 west coast of northern Sumatra earthquake to that of the 1998 Balleny Islands region earthquake.

2. Compare the intensity of the 2005 northern Sumatra, Indonesia earthquake to that of the 2003 southeastern Iran earthquake.

3. Compare the intensity of the 1906 San Francisco earthquake, the most powerful to strike the United States, to that of the 2008 West Salem earthquake. (*Hint*: Use the exponential key of a scientific calculator to compute the required power of 10.)

4. Suppose an earthquake measures a value of x on the Richter scale. How would the intensity of a second earthquake compare if its Richter scale measure is $x + 4.0$? How would it compare if its Richter scale measure is $x - 1.0$?

Chapter 13 ▷▷▷ Review Exercises

[13.1] *Combine terms where possible in each polynomial. Write the answer in descending powers of the variable. Give the degree of the answer. Identify the polynomial as a* monomial, binomial, trinomial, *or* none of these.

1. $9m^2 + 11m^2 + 2m^2$

2. $-4p + p^3 - p^2 + 8p + 2$

3. $12a^5 - 9a^4 + 8a^3 + 2a^2 - a + 3$

4. $-7y^5 - 8y^4 - y^5 + y^4 + 9y$

Add or subtract as indicated.

5. Add.
$$\begin{array}{r} -2a^3 + 5a^2 \\ -3a^3 - a^2 \\ \hline \end{array}$$

6. Add.
$$\begin{array}{r} 4r^3 - 8r^2 + 6r \\ -2r^3 + 5r^2 + 3r \\ \hline \end{array}$$

7. Subtract.
$$\begin{array}{r} 6y^2 - 8y + 2 \\ -5y^2 + 2y - 7 \\ \hline \end{array}$$

8. Subtract.
$$\begin{array}{r} -12k^4 - 8k^2 + 7k - 5 \\ k^4 + 7k^2 + 11k + 1 \\ \hline \end{array}$$

9. $(2m^3 - 8m^2 + 4) + (8m^3 + 2m^2 - 7)$

10. $(-5y^2 + 3y + 11) + (4y^2 - 7y + 15)$

11. $(6p^2 - p - 8) - (-4p^2 + 2p + 3)$

12. $(12r^4 - 7r^3 + 2r^2) - (5r^4 - 3r^3 + 2r^2 + 1)$

[13.2] *Simplify each expression.*

13. $4^3 \cdot 4^8$

14. $(-5)^6(-5)^5$

15. $(-8x^4)(9x^3)$

16. $(2x^2)(5x^3)(x^9)$

17. $(19x)^5$

18. $(-4y)^7$

19. $5(pt)^4$

20. $\left(\dfrac{7}{5}\right)^6$

21. $(3x^2y^3)^3$

22. $(t^4)^8(t^2)^5$

23. $(6x^2z^4)^2(x^3yz^2)^4$

24. $\left(\dfrac{2m^3n}{p^2}\right)^3$

25. Find a polynomial that represents the volume of the figure. (If necessary, refer to the formulas on the inside back cover.)

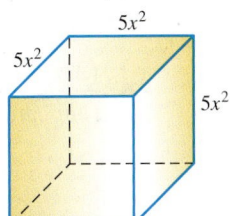

26. Explain why the product rule for exponents does not apply to the expression $7^2 + 7^4$.

[13.3] *Find each product.*

27. $5x(2x + 14)$

28. $-3p^3(2p^2 - 5p)$

29. $(3r - 2)(2r^2 + 4r - 3)$

30. $(2y + 3)(4y^2 - 6y + 9)$

31. $(5p^2 + 3p)(p^3 - p^2 + 5)$

32. $(x + 6)(x - 3)$

33. $(3k - 6)(2k + 1)$

34. $(6p - 3q)(2p - 7q)$

35. $(m^2 + m - 9)(2m^2 + 3m - 1)$

[13.4] *Find each product.*

36. $(a + 4)^2$

37. $(3p - 2)^2$

38. $(2r + 5s)^2$

39. $(r + 2)^3$

40. $(2x - 1)^3$

41. $(2z + 7)(2z - 7)$

42. $(6m - 5)(6m + 5)$

43. $(5a + 6b)(5a - 6b)$

44. $(2x^2 + 5)(2x^2 - 5)$

45. The square of a binomial leads to a polynomial with how many terms? The product of the sum and difference of two terms leads to a polynomial with how many terms?

46. Explain why $(a + b)^2$ is not equal to $a^2 + b^2$.

[13.5] *Evaluate each expression.*

47. $5^0 + 8^0$

48. 2^{-5}

49. $\left(\dfrac{6}{5}\right)^{-2}$

50. $4^{-2} - 4^{-1}$

Simplify each expression. Assume that all variables represent nonzero numbers.

51. $\dfrac{6^{-3}}{6^{-5}}$

52. $\dfrac{x^{-7}}{x^{-9}}$

53. $\dfrac{p^{-8}}{p^4}$

54. $\dfrac{r^{-2}}{r^{-6}}$

55. $(2^4)^2$

56. $(9^3)^{-2}$

57. $(5^{-2})^{-4}$

58. $(8^{-3})^4$

59. $\dfrac{(m^2)^3}{(m^4)^2}$

60. $\dfrac{y^4 \cdot y^{-2}}{y^{-5}}$

61. $\dfrac{r^9 \cdot r^{-5}}{r^{-2} \cdot r^{-7}}$

62. $(-5m^3)^2$

63. $(2y^{-4})^{-3}$

64. $\dfrac{ab^{-3}}{a^4b^2}$

65. $\dfrac{(6r^{-1})^2 \cdot (2r^{-4})}{r^{-5}(r^2)^{-3}}$

66. $\dfrac{(2m^{-5}n^2)^3(3m^2)^{-1}}{m^{-2}n^{-4}(m^{-1})^2}$

[13.6] *Perform each division.*

67. $\dfrac{-15y^4}{-9y^2}$

68. $\dfrac{-12x^3y^2}{6xy}$

69. $\dfrac{6y^4 - 12y^2 + 18y}{-6y}$

70. $\dfrac{2p^3 - 6p^2 + 5p}{2p^2}$

71. $(5x^{13} - 10x^{12} + 20x^7 - 35x^5) \div (-5x^4)$

72. $(-10m^4n^2 + 5m^3n^3 + 6m^2n^4) \div (5m^2n)$

[13.7] *Perform each division.*

73. $(2r^2 + 3r - 14) \div (r - 2)$

74. $\dfrac{12m^2 - 11m - 10}{3m - 5}$

75. $\dfrac{10a^3 + 5a^2 - 14a + 9}{5a^2 - 3}$

76. $\dfrac{2k^4 + 4k^3 + 9k^2 - 8}{2k^2 + 1}$

[13.8] *Write each number in scientific notation.*

77. 48,000,000

78. 28,988,000,000

79. 0.000065

80. 0.0000000824

Write each number without exponents.

81. 2.4×10^4

82. 7.83×10^7

83. 8.97×10^{-7}

84. 9.95×10^{-12}

Perform the indicated operations. Write the answers in scientific notation and then without exponents.

85. $(2 \times 10^{-3})(4 \times 10^5)$

86. $\dfrac{8 \times 10^4}{2 \times 10^{-2}}$

87. $\dfrac{12 \times 10^{-5} \times 5 \times 10^4}{4 \times 10^3 \times 6 \times 10^{-2}}$

88. $\dfrac{2.5 \times 10^5 \times 4.8 \times 10^{-4}}{7.5 \times 10^8 \times 1.6 \times 10^{-5}}$

89. A computer can perform 466,000,000 calculations per second. How many calculations can it perform per minute? Per hour?

90. There are 1×10^9 Social Security numbers. The population of the United States is about 3×10^8. How many Social Security numbers are available for each person? (*Source*: U.S. Census Bureau.)

▶▶▶ Mixed Review Exercises

Perform each indicated operation. Assume that all variables represent nonzero real numbers.

91. $19^0 - 3^0$

92. $(3p)^4 (3p^{-7})$

93. 7^{-2}

94. $(-7 + 2k)^2$

95. $\dfrac{2y^3 + 17y^2 + 37y + 7}{2y + 7}$

96. $\left(\dfrac{6r^2s}{5}\right)^4$

97. $-m^5(8m^2 + 10m + 6)$

98. $\left(\dfrac{1}{2}\right)^{-5}$

99. $(25x^2y^3 - 8xy^2 + 15x^3y) \div (5x)$

100. $(6r^{-2})^{-1}$

101. $(2x + y)^3$

102. $2^{-1} + 4^{-1}$

103. $(a + 2)(a^2 - 4a + 1)$

104. $(5y^3 - 8y^2 + 7) - (-3y^3 + y^2 + 2)$

105. $(2r + 5)(5r - 2)$

106. $(12a + 1)(12a - 1)$

107. What polynomial represents the area of this rectangle?

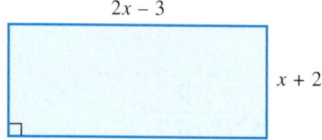

$2x - 3$

$x + 2$

108. What polynomial represents the perimeter of this square? The area?

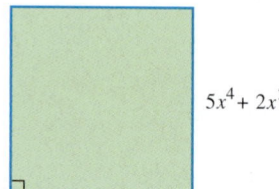

$5x^4 + 2x^2$

Perform the indicated operations.

1. $(5t^4 - 3t^2 + 7t + 3) - (t^4 - t^3 + 3t^2 + 8t + 3)$

1. _____

2. $(2y^2 - 8y + 8) + (-3y^2 + 2y + 3) - (y^2 + 3y - 6)$

2. _____

3. Subtract.

$$9t^3 - 4t^2 + 2t + 2$$
$$9t^3 + 8t^2 - 3t - 6$$

3. _____

Simplify.

4. $(-2)^3(-2)^2$

4. _____

5. $\left(\dfrac{6}{m^2}\right)^3, \quad m \neq 0$

5. _____

6. $3x^2(-9x^3 + 6x^2 - 2x + 1)$

6. _____

7. $(2r - 3)(r^2 + 2r - 5)$

7. _____

8. $(t - 8)(t + 3)$

8. _____

9. $(4x + 3y)(2x - y)$

9. _____

10. $(5x - 2y)^2$

10. _____

11. $(10v + 3w)(10v - 3w)$

11. _____

12. $(x + 1)^3$

12. _____

13. What polynomial represents the perimeter of this square?   The area?

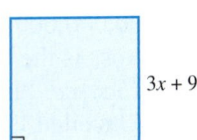
$3x + 9$

13. _____

14. _____

15. _____

Evaluate each expression.

14. 5^{-4} **15.** $(-3)^0 + 4^0$ **16.** $4^{-1} + 3^{-1}$

16. _____

Perform the indicated operations. In Exercises 17 and 18, write each answer using only positive exponents. Assume that variables represent nonzero numbers.

17. _____

17. $\dfrac{8^{-1} \cdot 8^4}{8^{-2}}$

18. _____

18. $\dfrac{(x^{-3})^{-2}\,(x^{-1}y)^2}{(xy^{-2})^2}$

19. _____

19. $\dfrac{8y^3 - 6y^2 + 4y + 10}{2y}$

20. _____

20. $(-9x^2y^3 + 6x^4y^3 + 12xy^3) \div (3xy)$

21. _____

21. $\dfrac{2x^2 + x - 36}{x - 4}$

22. _____

22. $(3x^3 - x + 4) \div (x - 2)$

Write each number in scientific notation.

23. (a) _____

    (b) _____

23. (a) 344,000,000,000

    (b) 0.00000557

Write each number without exponents.

24. (a) _____

    (b) _____

24. (a) 2.96×10^7

    (b) 6.07×10^{-8}

25. _____

25. A satellite galaxy of our own Milky Way, known as the Large Magellanic Cloud, is 1000 light-years across. If a light-year is equal to 5,890,000,000,000 mi, how many miles across is the Large Magellanic Cloud? (*Source:* "Images of Brightest Nebula Unveiled," *USA Today*, June 12, 2002.)

Factoring and Applications

Wireless communication uses radio waves to carry signals and messages across distances. Cellular phones, one of the most popular forms of wireless communication, have become an invaluable tool for people to stay connected to family, friends, and work while on the go. In 2007, there were about 243 million cell phone subscribers in the United States, with 81% of the population having cell phone service. Total revenue from this service was about $133 billion. (*Source:* CITA–The Wireless Association.)

In Exercise 31 of **Section 14.7,** we use a *quadratic equation* to model the number of cell phone subscribers in the United States in recent years.

14.1 ▶▶▶ Factors; The Greatest Common Factor

OBJECTIVES

1 Find the greatest common factor of a list of numbers.

2 Find the greatest common factor of a list of variable terms.

3 Factor out the greatest common factor.

4 Factor by grouping.

Recall from **Section 4.2** that to **factor** a number means to write it as the product of two or more numbers. The product is called the **factored form** of the number. Here is an example.

$$12 = \underbrace{6 \cdot 2}_{\text{Factored form}}^{\text{Factors}}$$

Factoring is a process that "undoes" multiplying. We multiply $6 \cdot 2$ to get 12, but we factor 12 by writing it as $6 \cdot 2$.

OBJECTIVE 1 Find the greatest common factor of a list of numbers. An integer that is a factor of two or more integers is a **common factor** of those integers. For example, 6 is a common factor of 18 and 24 because 6 is a factor of both 18 and 24. Other common factors of 18 and 24 are 1, 2, and 3. The **greatest common factor (GCF)** of a list of integers is the largest common factor of those integers. This means 6 is the greatest common factor of 18 and 24, since it is the largest of their common factors.

> **Note**
>
> Factors of a number are also divisors of the number. The greatest common factor is the same as the greatest common divisor.

EXAMPLE 1 **Finding the Greatest Common Factor for Numbers**

Find the greatest common factor for each list of numbers.

(a) 30, 45

First write each number in prime factored form.

$$30 = 2 \cdot \mathbf{3} \cdot \mathbf{5}$$
$$45 = \mathbf{3} \cdot 3 \cdot \mathbf{5}$$

Use each prime the *least* number of times it appears in *all* the factored forms. There is no 2 in the prime factored form of 45, so there will be no 2 in the greatest common factor. The least number of times 3 appears in all the factored forms is 1. The least number of times 5 appears is also 1. From this, the

$$\text{GCF} = \mathbf{3}^1 \cdot \mathbf{5}^1 = 3 \cdot 5 = 15.$$

(b) 72, 120, 432

Find the prime factored form of each number.

$$72 = \mathbf{2 \cdot 2 \cdot 2 \cdot 3} \cdot 3$$
$$120 = \mathbf{2 \cdot 2 \cdot 2 \cdot 3} \cdot 5$$
$$432 = \mathbf{2 \cdot 2 \cdot 2} \cdot 2 \cdot \mathbf{3} \cdot 3 \cdot 3$$

The least number of times 2 appears in all the factored forms is 3, and the least number of times 3 appears is 1. There is no 5 in the prime factored form of either 72 or 432, so the

$$\text{GCF} = \mathbf{2}^3 \cdot \mathbf{3}^1 = 24.$$

Continued on Next Page

(c) 10, 11, 14

Write the prime factored form of each number.

$$10 = 2 \cdot 5$$
$$11 = 11$$
$$14 = 2 \cdot 7$$

There are no primes common to all three numbers, so the GCF is 1.

Work Problem **1** *at the Side.* ▶

OBJECTIVE **2** **Find the greatest common factor of a list of variable terms.** The terms x^4, x^5, x^6, and x^7 have x^4 as the greatest common factor because the least exponent on the variable x is 4.

$$x^4 = 1 \cdot \boldsymbol{x^4}, \quad x^5 = x \cdot \boldsymbol{x^4}, \quad x^6 = x^2 \cdot \boldsymbol{x^4}, \quad x^7 = x^3 \cdot \boldsymbol{x^4}$$

Note

The exponent on a variable in the GCF is the *least* exponent that appears on that variable in *all* the terms.

EXAMPLE 2 **Finding the Greatest Common Factor for Variable Terms**

Find the greatest common factor for each list of terms.

(a) $21m^7, -18m^6, 45m^8$

$$21m^7 = \boldsymbol{3} \cdot 7 \cdot \boldsymbol{m^7}$$
$$-18m^6 = -1 \cdot 2 \cdot \boldsymbol{3} \cdot 3 \cdot \boldsymbol{m^6}$$
$$45m^8 = \boldsymbol{3} \cdot 3 \cdot 5 \cdot \boldsymbol{m^8}$$

First, 3 is the greatest common factor of the coefficients 21, -18, and 45. The least exponent on m is 6, so the

$$GCF = 3m^6.$$

(b) $x^4y^2, x^7y^5, x^3y^7, y^{15}$

$$\boldsymbol{x^4 y^2}, \quad x^7 \boldsymbol{y^5}, \quad x^3 \boldsymbol{y^7}, \quad \boldsymbol{y^{15}}$$

There is no x in the last term, y^{15}, so x will not appear in the greatest common factor. There is a y in each term, however, and 2 is the least exponent on y. The GCF is y^2.

(c) $-a^2b, -ab^2$

$$-a^2b = -1a^2b = -1 \cdot 1 \cdot a^2b$$
$$-ab^2 = -1ab^2 = -1 \cdot 1 \cdot ab^2$$

The factors of -1 are -1 and 1. Since $1 > -1$, the GCF is $1ab$, or ab.

Note

In a list of negative terms, sometimes a negative common factor is preferable (even though it is not the greatest common factor). In Example 2(c), for instance, we might prefer $-ab$ as the common factor. In factoring exercises like this, either answer will be acceptable.

---

1 Find the greatest common factor for each list of numbers.

(a) 30, 20, 15

$$30 = 2 \cdot 3 \cdot 5$$
$$20 = 2 \cdot \underline{\quad} \cdot \underline{\quad}$$
$$15 = 3 \cdot \underline{\quad}$$

GCF = $\underline{\quad}$

(b) 42, 28, 35

(c) 12, 18, 26, 32

(d) 10, 15, 21

2 Find the greatest common factor for each list of terms.

(a) $6m^4, 9m^2, 12m^5$

$$6m^4 = 2 \cdot \underline{\hspace{1cm}} \cdot m^4$$

$$9m^2 = 3 \cdot \underline{\hspace{1cm}} \cdot \underline{\hspace{1cm}}$$

$$12m^5 = 2 \cdot 2 \cdot \underline{\hspace{1cm}} \cdot \underline{\hspace{1cm}}$$

$$\text{GCF} = \underline{\hspace{1cm}}$$

(b) $-12p^5, -18q^4$

(c) y^4z^2, y^6z^8, z^9

(d) $12p^{11}, 17q^5$

Finding the Greatest Common Factor (GCF)

Step 1 **Factor.** Write each number in prime factored form.

Step 2 **List common factors.** List each prime number or each variable that is a factor of every term in the list. (If a prime does not appear in one of the prime factored forms, it cannot appear in the greatest common factor.)

Step 3 **Choose least exponents.** Use as exponents on the common prime factors the *least* exponents from the prime factored forms.

Step 4 **Multiply.** Multiply the primes from Step 3. If there are no primes left after Step 3, the greatest common factor is 1.

◄ *Work Problem* **2** *at the Side.*

OBJECTIVE 3 Factor out the greatest common factor. The polynomial

$$3m + 12$$

has two terms, $3m$ and 12. The greatest common factor of these two terms is 3. We can write $3m + 12$ so that each term is a product with 3 as one factor.

$$3m + 12$$
$$= \mathbf{3} \cdot m + \mathbf{3} \cdot 4$$
$$= \mathbf{3}(m + 4) \qquad \text{Distributive property}$$

The factored form of $3m + 12$ is $3(m + 4)$. This process is called **factoring out the greatest common factor.**

CAUTION

The polynomial $3m + 12$ is *not* in factored form when written as the *sum*

$$3 \cdot m + 3 \cdot 4. \qquad \text{Not in factored form}$$

The terms are factored, but the polynomial is not. The factored form of $3m + 12$ is the *product*

$$3(m + 4). \qquad \text{In factored form}$$

Writing a polynomial as a product, that is, in factored form, is called **factoring** the polynomial.

EXAMPLE 3 **Factoring Out the Greatest Common Factor**

Factor out the greatest common factor.

(a) $5y^2 + 10y$

$$= \mathbf{5y}(y) + \mathbf{5y}(2) \qquad \text{GCF} = 5y$$
$$= \mathbf{5y}(y + 2) \qquad \text{Distributive property}$$

Check   Multiply the factored form.

$$5y(y + 2)$$
$$= 5y(y) + 5y(2) \qquad \text{Distributive property}$$
$$= 5y^2 + 10y \qquad \text{Original polynomial}$$

Continued on Next Page

(b) $20m^5 + 10m^4 - 15m^3$

$$= \mathbf{5m^3}(4m^2) + \mathbf{5m^3}(2m) - \mathbf{5m^3}(3) \qquad \text{GCF} = 5m^3$$

$$= \mathbf{5m^3}(4m^2 + 2m - 3) \qquad \text{Factor out } 5m^3.$$

Check $5m^3(4m^2 + 2m - 3)$

$$= 20m^5 + 10m^4 - 15m^3 \qquad \text{Original polynomial}$$

(c) $x^5 + x^3$

$$= \mathbf{x^3}(x^2) + \mathbf{x^3}(\mathbf{1})$$

$$= \mathbf{x^3}(x^2 + \mathbf{1}) \overset{\boxed{\text{Don't forget the 1.}}}{\longleftarrow}$$

(d) $20m^7p^2 - 36m^3p^4$

$$= 4m^3p^2(5m^4) - 4m^3p^2(9p^2) \qquad \text{GCF} = 4m^3p^2$$

$$= 4m^3p^2(5m^4 - 9p^2) \qquad \text{Factor out } 4m^3p^2.$$

(e) $\dfrac{1}{6}n^2 + \dfrac{5}{6}n$

$$= \frac{1}{6}n(n) + \frac{1}{6}n(5) \qquad \text{GCF} = \tfrac{1}{6}n$$

$$= \frac{1}{6}n(n + 5)$$

> **CAUTION**
> Be sure to include the **1** in a problem like Example 3(c). *Check that the factored form can be multiplied out to give the original polynomial.*

Work Problem ③ *at the Side.* ▶

┌─ **EXAMPLE 4**   **Factoring Out a Negative Common Factor**

Factor $-8x^4 + 16x^3 - 4x^2$.

We can factor out either $4x^2$ or $-4x^2$ here. We factor out $-4x^2$ so that the coefficient of the first term in the trinomial factor will be positive.

$-8x^4 + 16x^3 - 4x^2$ $\overset{\boxed{\text{Be careful with signs.}}}{}$

$$= -4x^2(2x^2) - 4x^2(-4x) - 4x^2(1) \qquad -4x^2 \text{ is a common factor.}$$

$$= -4x^2(2x^2 - 4x + 1) \qquad \text{Factor out } -4x^2.$$

Check $-4x^2(2x^2 - 4x + 1)$

$$= -4x^2(2x^2) - 4x^2(-4x) - 4x^2(1) \qquad \text{Distributive property}$$

$$= -8x^4 + 16x^3 - 4x^2 \qquad \text{Original polynomial}$$

Work Problem ④ *at the Side.* ▶

> **Note**
> Whenever we factor a polynomial in which the coefficient of the first term of a polynomial is negative, we will factor out the negative common factor, even if it is just -1. However, it would also be correct to factor out $4x^2$ in Example 4 to obtain $4x^2(-2x^2 + 4x - 1)$.

---

③ Factor out the greatest common factor.

(a) $4x^2 + 6x$

(b) $10y^5 - 8y^4 + 6y^2$

(c) $m^7 + m^9$

(d) $8p^5q^2 + 16p^6q^3 - 12p^4q^7$

(e) $\dfrac{1}{3}b^2 - \dfrac{2}{3}b$

④ Factor

$$-14a^3b^2 - 21a^2b^3 + 7ab$$

by factoring out a negative common factor.

ANSWERS

3. **(a)** $2x(2x + 3)$
   (b) $2y^2(5y^3 - 4y^2 + 3)$
   (c) $m^7(1 + m^2)$
   (d) $4p^4q^2(2p + 4p^2q - 3q^5)$
   (e) $\dfrac{1}{3}b(b - 2)$

4. $-7ab(2a^2b + 3ab^2 - 1)$

5 Factor out the greatest common factor.

(a) $r(t - 4) + 5(t - 4)$

(b) $y^2(y + 2) - 3(y + 2)$

(c) $x(x - 1) - 5(x - 1)$

EXAMPLE 5   **Factoring Out a Common Binomial Factor**

Factor out the greatest common factor.

(a) $a(a + 3) + 4(a + 3)$

Sometimes the GCF has a factor with more than one term. The binomial $a + 3$ is the greatest common factor here.

Same

$$a(a + 3) + 4(a + 3)$$

$$= (a + 3)(a + 4)$$

(b) $x^2(x + 1) - 5(x + 1)$

$$= (x + 1)(x^2 - 5) \qquad \text{Factor out } x + 1.$$

◀ Work Problem **5** at the Side.

OBJECTIVE 4 Factor by grouping. *When a polynomial has four terms, common factors can sometimes be used to factor by grouping.*

EXAMPLE 6   **Factoring by Grouping**

Factor by grouping.

(a) $2x + 6 + ax + 3a$

Group the first two terms and the last two terms, since the first two terms have a common factor of 2 and the last two terms have a common factor of a.

$$2x + 6 + ax + 3a$$

$$= (2x + 6) + (ax + 3a)$$

$$= 2(x + 3) + a(x + 3)$$

The expression is still not in factored form because it is the **sum** of two terms. Now, however, $x + 3$ is a common factor and can be factored out.

$$2x + 6 + ax + 3a$$

$$= (2x + 6) + (ax + 3a) \qquad \text{Group the terms.}$$

$$= 2(x + 3) + a(x + 3) \qquad \text{Factor each group.}$$

$$= (x + 3)(2 + a) \qquad \text{Factor out } x + 3.$$

The final result is in factored form because it is a **product.** Note that the goal in factoring by grouping is to get a common factor, $x + 3$ here, so that the last step is possible. Check by multiplying the binomials using the FOIL method from **Section 13.3.**

Check $(x + 3)(2 + a)$

$$= 2x + ax + 6 + 3a \qquad \text{FOIL}$$

$$= 2x + 6 + ax + 3a, \qquad \text{Rearrange terms.}$$

which is the original polynomial.

Continued on Next Page

(b) $6ax + 24x + a + 4$

$\quad = (6ax + 24x) + (a + 4)$ Group the terms.

$\quad = 6x\,(a + 4) + 1\,(a + 4)$ Factor each group.

> Remember the 1.

$\quad = (a + 4)(6x + 1)$ Factor out $a + 4$.

Check $(a + 4)(6x + 1)$

$\quad = 6ax + a + 24x + 4$ FOIL

$\quad = 6ax + 24x + a + 4,$ Rearrange terms.

which is the original polynomial.

(c) $2x^2 - 10x + 3xy - 15y$

$\quad = (2x^2 - 10x) + (3xy - 15y)$ Group the terms.

$\quad = 2x\,(x - 5) + 3y\,(x - 5)$ Factor each group.

$\quad = (x - 5)(2x + 3y)$ Factor out the common factor, $x - 5$.

Check $(x - 5)(2x + 3y)$

$\quad = 2x^2 + 3xy - 10x - 15y$ FOIL

$\quad = 2x^2 - 10x + 3xy - 15y$ Original polynomial

(d) $t^3 + 2t^2 - 3t - 6$

> Be sure to write a $+$ sign between the groups.

$\quad = (t^3 + 2t^2) + (-3t - 6)$ Group the terms.

$\quad = t^2\,(t + 2) - 3\,(t + 2)$ Factor out -3 so there is a common factor,

> Be careful with signs.

$\quad\quad\quad\quad\quad\quad\quad\quad\quad\quad\quad\quad$ $t + 2;\ -3\,(t + 2) = -3t - 6.$

$\quad = (t + 2)(t^2 - 3)$ Factor out $t + 2$.

Check by multiplying.

> **CAUTION**
> *Be careful with signs when grouping* in a problem like Example 6(d). It is wise to check the factoring in the second step, as shown in the example side comment, before continuing.

Work Problem **6** *at the Side.* ▶

> **Factoring a Polynomial with Four Terms by Grouping**
>
> *Step 1*   **Group terms.** Collect the terms into two groups so that each group has a common factor.
>
> *Step 2*   **Factor within groups.** Factor out the greatest common factor from each group.
>
> *Step 3*   **Factor the entire polynomial.** Factor a common binomial factor from the results of Step 2.
>
> *Step 4*   **If necessary, rearrange terms.** If Step 2 does not result in a common binomial factor, try a different grouping.

---

6 Factor by grouping.

    (a) $pq + 5q + 2p + 10$

    (b) $2xy + 3y + 2x + 3$

    (c) $2a^2 - 4a + 3ab - 6b$

    (d) $x^3 + 3x^2 - 5x - 15$

Answers

6. **(a)** $(p + 5)(q + 2)$
    (b) $(2x + 3)(y + 1)$
    (c) $(a - 2)(2a + 3b)$
    (d) $(x + 3)(x^2 - 5)$

7 Factor by grouping.

(a) $6y^2 - 20w + 15y - 8yw$

EXAMPLE 7 **Rearranging Terms Before Factoring by Grouping**

Factor by grouping.

(a) $10x^2 - 12y + 15x - 8xy$

Factoring out the common factor of 2 from the first two terms and the common factor of x from the last two terms gives

$$10x^2 - 12y + 15x - 8xy$$
$$= 2(5x^2 - 6y) + x(15 - 8y).$$

This does not lead to a common factor, so we try rearranging the terms. There is usually more than one way to do this. We try the following.

$$10x^2 - 12y + 15x - \mathbf{8xy}$$

$= 10x^2 - \mathbf{8xy} - 12y + 15x$		Commutative property
$= (10x^2 - 8xy) + (-12y + 15x)$		Group the terms.
$= 2x(5x - 4y) + 3\,\mathbf{(-4y + 5x)}$		Factor each group.
$= 2x\,\mathbf{(5x - 4y)} + 3\,\mathbf{(5x - 4y)}$		Rewrite $-4y + 5x$.
$= \mathbf{(5x - 4y)}(2x + 3)$		Factor out $5x - 4y$.

Check $(5x - 4y)(2x + 3)$

$= 10x^2 + 15x - 8xy - 12y$		FOIL
$= 10x^2 - 12y + 15x - 8xy$		Original polynomial

(b) $2xy + 12 - 3y - 8x$

We need to rearrange these terms to get two groups that each have a common factor. Trial and error suggests the following grouping.

$$2xy + 12 - 3y - 8x \quad \boxed{\text{Always write a + sign between the two groups.}}$$

$= (2xy - 3y) + (-8x + 12)$		Group the terms.
$= y(2x - 3) - 4(2x - 3)$	$\boxed{\text{Be careful with signs.}}$	Factor each group.
$= (2x - 3)(y - 4)$		Factor out $2x - 3$.

Since the quantities in parentheses in the second step must be the same, we factored out -4 rather than 4. *Check* by multiplying.

(b) $9mn - 4 + 12m - 3n$

CAUTION

Use negative signs carefully when grouping, as in Example 7(b), or a sign error will occur. ***Always check by multiplying.***

◀ *Work Problem* **7** *at the Side.*

ANSWERS

7. **(a)** $(2y + 5)(3y - 4w)$
   (b) $(3m - 1)(3n + 4)$

14.1 ▶▶▶ Exercises

Find the greatest common factor for each list of numbers. See Example 1.

1. 12, 16

2. 18, 24

3. 40, 20, 4

4. 50, 30, 5

5. 18, 24, 36, 48

6. 15, 30, 45, 75

7. 4, 9, 12

8. 9, 16, 24

Find the greatest common factor for each list of terms. See Example 2.

9. $16y$, 24

10. $18w$, 27

11. $30x^3$, $40x^6$, $50x^7$

12. $60z^4$, $70z^8$, $90z^9$

13. $-x^4y^3$, $-xy^2$

14. $-a^4b^5$, $-a^3b$

15. $42ab^3$, $-36a$, $90b$, $-48ab$

16. $45c^3d$, $75c$, $90d$, $-105cd$

Complete each factoring.

17. $9m^4 = 3m^2 (\quad)$

18. $12p^5 = 6p^3 (\quad)$

19. $-8z^9 = -4z^5 (\quad)$

20. $-15k^{11} = -5k^8 (\quad)$

21. $6m^4n^5 = 3m^3n (\quad)$

22. $27a^3b^2 = 9a^2b (\quad)$

23. $12y + 24 = 12 (\quad)$

24. $18p + 36 = 18 (\quad)$

25. $10a^2 - 20a = 10a (\quad)$

26. $15x^2 - 30x = 15x (\quad)$

27. $8x^2y + 12x^3y^2 = 4x^2y (\quad)$

28. $18s^3t^2 + 10st = 2st (\quad)$

Factor out the greatest common factor, or a negative common factor if the coefficient of the term of greatest degree is negative. See Examples 3–5.

29. $x^2 - 4x$

30. $m^2 - 7m$

31. $6t^2 + 15t$

32. $8x^2 + 6x$

33. $\frac{1}{4}d^2 - \frac{3}{4}d$

34. $\frac{1}{5}z^2 + \frac{3}{5}z$

35. $-12x^3 - 6x^2$

36. $-21b^3 + 7b^2$

37. $65y^{10} + 35y^6$

38. $100a^5 + 16a^3$

39. $11w^3 - 100$

40. $13z^5 - 80$

41. $8m^2n^3 + 24m^2n^2$

42. $19p^2y - 38p^2y^3$

43. $-4x^3 + 10x^2 - 6x$

44. $-9z^3 + 6z^2 - 12z$

45. $13y^8 + 26y^4 - 39y^2$

46. $5x^5 + 25x^4 - 20x^3$

47. $45q^4p^5 + 36qp^6 + 81q^2p^3$

48. $125a^3z^5 + 60a^4z^4 - 85a^5z^2$

49. $c(x + 2) + d(x + 2)$

50. $r(5 - x) + t(5 - x)$

51. $a^2(2a + b) - b(2a + b)$

52. $3x(x^2 + 5) - y(x^2 + 5)$

53. $q(p + 4) - 1(p + 4)$

54. $y^2(x - 4) + 1(x - 4)$

Factor by grouping. See Examples 6 and 7.

55. $5m + mn + 20 + 4n$

56. $ts + 5t + 2s + 10$

⬤ **57.** $6xy - 21x + 8y - 28$

58. $2mn - 8n + 3m - 12$

59. $3xy + 9x + y + 3$

60. $6n + 4mn + 3 + 2m$

61. $7z^2 + 14z - az - 2a$

62. $2b^2 + 3b - 8ab - 12a$

63. $18r^2 + 12ry - 3xr - 2xy$

64. $5m^2 + 15mp - 2mr - 6pr$

65. $w^3 + w^2 + 9w + 9$

66. $y^3 + y^2 + 6y + 6$

67. $3a^3 + 6a^2 - 2a - 4$

68. $10x^3 + 15x^2 - 8x - 12$

69. $16m^3 - 4m^2p^2 - 4mp + p^3$

70. $10t^3 - 2t^2s^2 - 5ts + s^3$

71. $y^2 + 3x + 3y + xy$

72. $m^2 + 14p + 7m + 2mp$

73. $2z^2 + 6w - 4z - 3wz$

74. $2a^2 + 20b - 8a - 5ab$

Relating Concepts (Exercises 75–78) For Individual or Group Work

In many cases, the choice of which pairs of terms to group when factoring by grouping can be made in different ways. To see this for Example 7(b), **work Exercises 75–78 in order.**

75. Start with the polynomial from Example 7(b), $2xy + 12 - 3y - 8x$, and rearrange the terms as follows: $2xy - 8x - 3y + 12$. What property from **Section 1.3** allows this?

76. Group the first two terms and the last two terms of the rearranged polynomial in Exercise 75. Then factor each group.

77. Is your result from Exercise 76 in factored form? Explain your answer.

78. If your answer to Exercise 77 is *no*, factor the polynomial. Is the result the same as the one shown for Example 7(b)?

14.2 ▶▶▶ Factoring Trinomials

OBJECTIVES

1 Factor trinomials with a coefficient of 1 for the squared term.

2 Factor trinomials after factoring out the greatest common factor.

Using FOIL, the product of the binomials $k - 3$ and $k + 1$ is

$$(k - 3)(k + 1) = k^2 - 2k - 3. \qquad \text{Multiplying}$$

Suppose instead that we are given the polynomial $k^2 - 2k - 3$ and want to rewrite it as the product $(k - 3)(k + 1)$. That is,

$$k^2 - 2k - 3 = (k - 3)(k + 1). \qquad \text{Factoring}$$

Recall from **Section 14.1** that this process is called *factoring* the polynomial. Factoring reverses or, "undoes," multiplying.

OBJECTIVE 1 Factor trinomials with a coefficient of 1 for the squared term. When factoring polynomials with integer coefficients, we use only integers in the factors. For example, we can factor $x^2 + 5x + 6$ by finding integers m and n such that

$$x^2 + 5x + 6 \quad \text{is written as} \quad (x + m)(x + n).$$

To find these integers m and n, we first use FOIL to multiply the two binomials on the right above:

$$(x + m)(x + n)$$
$$= x^2 + nx + mx + mn$$
$$= x^2 + (n + m)x + mn. \qquad \text{Distributive property}$$

Comparing this result with $x^2 + 5x + 6$ shows that we must find integers m and n having a sum of 5 and a product of 6.

Product of m and n is 6.

$$x^2 + 5x + 6 = x^2 + (n + m)x + mn$$

Sum of m and n is 5.

Because many pairs of integers have a sum of 5, it is best to begin by listing those pairs of integers whose product is 6. Both 5 and 6 are positive, so we consider only pairs in which both integers are positive.

Work Problem **1** *at the Side.* ▶

1 (a) List all pairs of positive integers whose product is 6.

(b) Find the pair from part (a) whose sum is 5.

From Problem 1 at the side, we see that the numbers 1 and 6 and the numbers 2 and 3 both have a product of 6, but only the pair 2 and 3 has a sum of 5. So 2 and 3 are the required integers, and

$$x^2 + 5x + 6 \quad \text{is factored as} \quad (x + 2)(x + 3).$$

Check by multiplying the binomials using FOIL. *Make sure that the sum of the outer and inner products produces the correct middle term.*

Check $(x + 2)(x + 3) = x^2 + 5x + 6$ Correct
$$\frac{\begin{array}{c} 2x \\ 3x \end{array}}{5x} \quad \text{Add.}$$

This method of factoring can be used only for trinomials that have 1 as the coefficient of the squared term.

2 Factor each trinomial.

(a) $y^2 + 12y + 20$

First complete the given list of numbers.

Factors of 20	Sums of Factors
20, 1	20 + 1 = 21
10, ___	10 + ___ = ___
5, ___	5 + ___ = ___

(b) $x^2 + 9x + 18$

---

3 Factor each trinomial.

(a) $t^2 - 12t + 32$

First complete the given list of numbers.

Factors of 32	Sums of Factors
−32, −1	−32 + (−1) = −33
−16, ___	−16 + (___) = ___
−8, ___	−8 + (___) = ___

(b) $y^2 - 10y + 24$

EXAMPLE 1   Factoring a Trinomial with All Positive Terms

Factor $m^2 + 9m + 14$.

Look for two integers whose product is **14** and whose sum is **9**. List the pairs of integers whose products are 14. Then examine the sums. Only positive integers are needed since all signs in $m^2 + 9m + 14$ are positive.

Factors of 14	Sums of Factors	
14, 1	14 + 1 = 15	
7, 2	7 + 2 = **9**	Sum is 9.

From the list, 7 and 2 are the required integers, since $7 \cdot 2 = $ **14** and $7 + 2 = $ **9**. Thus,

$$m^2 + 9m + 14 \quad \text{factors as} \quad (m + 2)(m + 7).$$

Check $(m + 2)(m + 7)$

$$= m^2 + 7m + 2m + 14 \qquad \text{FOIL}$$

$$= m^2 + 9m + 14 \qquad \text{Original polynomial}$$

Note

In Example 1, the answer $(m + 2)(m + 7)$ also could have been written

$$(m + 7)(m + 2).$$

Because of the commutative property of multiplication, the order of the factors does not matter. *Always check by multiplying.*

◀ Work Problem **2** at the Side.

EXAMPLE 2   Factoring a Trinomial with a Negative Middle Term

Factor $x^2 - 9x + 20$.

Find two integers whose product is **20** and whose sum is **−9**. Since the numbers we are looking for have a *positive product* and a *negative sum*, we consider only pairs of negative integers.

Factors of 20	Sums of Factors	
−20, −1	−20 + (−1) = −21	
−10, −2	−10 + (−2) = −12	
−5, −4	−5 + (−4) = **−9**	Sum is −9.

The required integers are −5 and −4, so

$$x^2 - 9x + 20 \quad \text{factors as} \quad (x - 5)(x - 4).$$

Check $(x - 5)(x - 4)$

$$= x^2 - 4x - 5x + 20 \qquad \text{FOIL}$$

$$= x^2 - 9x + 20 \qquad \text{Original polynomial}$$

◀ Work Problem **3** at the Side.

EXAMPLE 3   **Factoring a Trinomial with Two Negative Terms**

Factor $p^2 - 2p - 15$.

Find two integers whose product is -15 and whose sum is -2. If these numbers do not come to mind right away, find them (if they exist) by listing all the pairs of integers whose product is -15. Because the last term, -15, is negative, we need pairs of integers with different signs.

Factors of -15	Sums of Factors
$15, -1$	$15 + (-1) = 14$
$-15, 1$	$-15 + 1 = -14$
$5, -3$	$5 + (-3) = 2$
$\mathbf{-5, 3}$	$-5 + 3 = \mathbf{-2}$ Sum is -2.

The required integers are -5 and 3, so

$$p^2 - 2p - 15 \quad \text{factors as} \quad (p - 5)(p + 3).$$

Check   Multiply $(p - 5)(p + 3)$ to obtain $p^2 - 2p - 15$.

Note

In Examples 1–3, notice that we listed factors in descending order (disregarding sign) when we were looking for the required pair of integers. This helps avoid skipping the correct combination.

Work Problem **4** *at the Side.* ▶

As shown in the next example, some trinomials cannot be factored using only integers. We call such trinomials **prime polynomials.**

EXAMPLE 4   **Deciding Whether Polynomials Are Prime**

Factor each trinomial.

(a) $x^2 - 5x + 12$

As in Example 2, both factors must be negative to give a positive product and a negative sum. First, list all the pairs of negative integers whose product is 12. Then examine the sums.

Factors of 12	Sums of Factors
$-12, -1$	$-12 + (-1) = -13$
$-6, -2$	$-6 + (-2) = -8$
$-4, -3$	$-4 + (-3) = -7$

None of the pairs of integers has a sum of -5. Therefore, the trinomial $x^2 - 5x + 12$ *cannot be factored using only integers; it is a prime polynomial.*

(b) $k^2 - 8k + 11$

There is no pair of integers whose product is 11 and whose sum is -8, so $k^2 - 8k + 11$ is a prime polynomial.

Work Problem **5** *at the Side.* ▶

---

4 Factor each trinomial.

(a) $a^2 - 9a - 22$

(b) $r^2 - 6r - 16$

5 Factor each trinomial, if possible.

(a) $r^2 - 3r - 4$

(b) $m^2 - 2m + 5$

6 Factor each trinomial.

(a) $b^2 - 3ab - 4a^2$

(b) $r^2 - 6rs + 8s^2$

Guidelines for factoring a trinomial of the form $x^2 + bx + c$ are summarized here.

> **Factoring $x^2 + bx + c$**
>
> Find two integers whose product is c and whose sum is b.
>
> **1.** Both integers must be positive if b and c are positive.
>
> **2.** Both integers must be negative if c is positive and b is negative.
>
> **3.** One integer must be positive and one must be negative if c is negative.

EXAMPLE 5 **Factoring a Trinomial with Two Variables**

Factor $z^2 - 2bz - 3b^2$.

Here, the coefficient of z in the middle term is $-2b$, so we need to find two expressions whose product is $-3b^2$ and whose sum is $-2b$. The expressions are $-3b$ and b, so

$$z^2 - 2bz - 3b^2 \quad \text{factors as} \quad (z - 3b)(z + b).$$

Check $(z - 3b)(z + b)$

$$= z^2 + zb - 3bz - 3b^2 \qquad \text{FOIL}$$

$$= z^2 + 1bz - 3bz - 3b^2 \qquad \text{Identity and commutative properties}$$

$$= z^2 - 2bz - 3b^2 \qquad \text{Combine like terms.}$$

◀ *Work Problem* **6** *at the Side.*

---

7 Factor each trinomial completely.

(a) $2p^3 + 6p^2 - 8p$

OBJECTIVE **2** **Factor trinomials after factoring out the greatest common factor.** The trinomial in the next example does not have a coefficient of 1 for the squared term. (In fact, there is no squared term.) However, there may be a common factor.

EXAMPLE 6 **Factoring a Trinomial with a Common Factor**

Factor $4x^5 - 28x^4 + 40x^3$.

First, factor out the greatest common factor, $4x^3$.

$$4x^5 - 28x^4 + 40x^3$$

$$= 4x^3(x^2 - 7x + 10)$$

Now factor $x^2 - 7x + 10$. The integers -5 and -2 have a product of 10 and a sum of -7. The complete factored form is

Include $4x^3$. ⟶ $4x^3(x - 5)(x - 2)$.

(b) $-3x^4 + 15x^3 - 18x^2$

Check $4x^3(x - 5)(x - 2)$

$$= 4x^3(x^2 - 7x + 10) \qquad \text{FOIL}$$

$$= 4x^5 - 28x^4 + 40x^3 \qquad \text{Distributive property}$$

◀ *Work Problem* **7** *at the Side.*

> **CAUTION**
> When factoring, *always look for a common factor first.* Remember to include the common factor as part of the answer. As a check, multiplying out the complete factored form should give the original polynomial.

14.2 ▶▶▶ Exercises

1. When factoring a trinomial in x as $(x + a)(x + b)$, what must be true of a and b, if the last term of the trinomial is negative?

2. In Exercise 1, what must be true of a and b if the last term is positive?

3. What is meant by a *prime polynomial*?

4. How can you check your work when factoring a trinomial? Does the check ensure that the trinomial is *completely* factored?

In Exercises 5–8, list all pairs of integers with the given product. Then find the pair whose sum is given. See the tables in Examples 1–4.

5. Product: 12; Sum: 7

6. Product: 18; Sum: 9

7. Product: -24; Sum: -5

8. Product: -36; Sum: -16

9. Which one of the following is the correct factored form of $x^2 - 12x + 32$?

A. $(x - 8)(x + 4)$ **B.** $(x + 8)(x - 4)$
C. $(x - 8)(x - 4)$ **D.** $(x + 8)(x + 4)$

10. What would be the first step in factoring
$$2x^3 + 8x^2 - 10x?$$

Complete each factoring.

11. $x^2 + 15x + 44 = (x + 4)(\quad)$

12. $r^2 + 15r + 56 = (r + 7)(\quad)$

13. $x^2 - 9x + 8 = (x - 1)(\quad)$

14. $t^2 - 14t + 24 = (t - 2)(\quad)$

15. $y^2 - 2y - 15 = (y + 3)(\quad)$

16. $t^2 - t - 42 = (t + 6)(\quad)$

17. $x^2 + 9x - 22 = (x - 2)(\quad)$

18. $x^2 + 6x - 27 = (x - 3)(\quad)$

19. $y^2 - 7y - 18 = (y + 2)(\quad)$

20. $y^2 - 2y - 24 = (y + 4)(\quad)$

Factor completely. If a polynomial cannot be factored, write prime. See Examples 1–4.

21. $y^2 + 9y + 8$

22. $a^2 + 9a + 20$

◑ 23. $b^2 + 8b + 15$

24. $x^2 + 6x + 8$

25. $m^2 + m - 20$

26. $p^2 + 4p - 5$

27. $x^2 + 3x - 40$

28. $d^2 + 4d - 45$

◑ 29. $y^2 - 8y + 15$

30. $y^2 - 6y + 8$

31. $z^2 - 15z + 56$

32. $x^2 - 13x + 36$

33. $r^2 - r - 30$
◑

34. $q^2 - q - 42$

35. $a^2 - 8a - 48$

36. $m^2 - 10m - 24$

◑ 37. $x^2 + 4x + 5$

38. $t^2 + 11t + 12$

Factor completely. See Examples 5 and 6.

39. $r^2 + 3ra + 2a^2$

40. $x^2 + 5xa + 4a^2$

41. $x^2 + 4xy + 3y^2$

42. $p^2 + 9pq + 8q^2$

◑ 43. $t^2 - tz - 6z^2$

44. $a^2 - ab - 12b^2$

45. $v^2 - 11vw + 30w^2$

46. $v^2 - 11vx + 24x^2$

47. $4x^2 + 12x - 40$

48. $5y^2 - 5y - 30$

◑ 49. $2t^3 + 8t^2 + 6t$

50. $3t^3 + 27t^2 + 24t$

51. $-2x^6 - 8x^5 + 42x^4$

52. $-4y^5 - 12y^4 + 40y^3$

53. $a^5 + 3a^4b - 4a^3b^2$

54. $z^{10} - 4z^9y - 21z^8y^2$

55. $m^3n - 10m^2n^2 + 24mn^3$

56. $y^3z + 3y^2z^2 - 54yz^3$

57. Use the FOIL method from **Section 13.3** to show that $(2x + 4)(x - 3) = 2x^2 - 2x - 12$. Why, then, is it incorrect to completely factor $2x^2 - 2x - 12$ as $(2x + 4)(x - 3)$?

58. Why is it incorrect to completely factor $3x^2 + 9x - 12$ as the product $(x - 1)(3x + 12)$?

14.3 ▶▶▶ Factoring Trinomials by Grouping

Trinomials like $2x^2 + 7x + 6$, in which the coefficient of the squared term is *not* 1, are factored with extensions of the methods from the previous sections. One such method uses factoring by grouping from **Section 14.1.**

OBJECTIVE 1 Factor trinomials by grouping when the coefficient of the squared term is not 1. Recall that a trinomial such as $m^2 + 3m + 2$ is factored by finding two integers whose product is 2 and whose sum is 3. To factor $2x^2 + 7x + 6$, we look for two integers whose product is $2 \cdot 6 = 12$ and whose sum is 7.

$$\text{Sum is 7.}$$
$$2x^2 + 7x + 6$$
$$\text{Product is } 2 \cdot 6 = 12.$$

By considering pairs of positive integers whose product is 12, the necessary integers are found to be 3 and 4. We use these integers to write the middle term, $7x$, as $7x = 3x + 4x$. The trinomial $2x^2 + 7x + 6$ becomes

$$2x^2 + 7x + 6$$
$$= 2x^2 + \underbrace{3x + 4x}_{7x} + 6$$
$$= (2x^2 + 3x) + (4x + 6) \qquad \text{Group the terms.}$$
$$= x(2x + 3) + 2(2x + 3) \qquad \text{Factor each group.}$$
$$\qquad\qquad \text{Must be the same}$$
$$= (2x + 3)(x + 2). \qquad \text{Factor out } 2x + 3.$$

Check Multiply $(2x + 3)(x + 2)$ to obtain $2x^2 + 7x + 6$.

In the preceding example, we could have written $7x$ as $4x + 3x$. Factoring by grouping this way would give the same answer.

Work Problem **1** *at the Side.* ▶

EXAMPLE 1 Factoring Trinomials by Grouping

Factor each trinomial.

(a) $6r^2 + r - 1$

We must find two integers with a product of $6(-1) = -6$ and a sum of 1. The integers are -2 and 3. We write the middle term, r, as $-2r + 3r$.

$$6r^2 + r - 1$$
$$= 6r^2 - 2r + 3r - 1 \qquad r = -2r + 3r$$
$$= (6r^2 - 2r) + (3r - 1) \qquad \text{Group the terms.}$$
$$= 2r(3r - 1) + 1(3r - 1) \qquad \text{The binomials must be the same.}$$
$$\qquad\qquad \text{Remember the 1.}$$
$$= (3r - 1)(2r + 1) \qquad \text{Factor out } 3r - 1.$$

Check Multiply $(3r - 1)(2r + 1)$ to obtain $6r^2 + r - 1$.

Continued on Next Page

OBJECTIVE

1 Factor trinomials by grouping when the coefficient of the squared term is not 1.

1 **(a)** Factor $2x^2 + 7x + 6$ by writing $7x$ as $4x + 3x$. Complete the following.

$$2x^2 + 7x + 6$$
$$= 2x^2 + 4x + 3x + 6$$
$$= (2x^2 + \underline{\quad}) + (3x + \underline{\quad})$$
$$= 2x(x + \underline{\quad}) + 3(x + \underline{\quad})$$
$$= (\underline{\quad\quad})(2x + 3)$$

(b) Is the answer in part (a) the same as in the example? (Remember that the order of the factors does not matter.)

ANSWERS

1. (a) $4x$; 6; 2; 2; $x + 2$ **(b)** yes

2 Factor each trinomial by grouping.

(a) $2m^2 + 7m + 3$

(b) $5p^2 - 2p - 3$

(c) $15k^2 - km - 2m^2$

(b) $12z^2 - 5z - 2$

Look for two integers whose product is $12(-2) = -24$ and whose sum is -5. The required integers are 3 and -8, so

$$12z^2 - 5z - 2$$
$$= 12z^2 + 3z - 8z - 2 \qquad -5z = 3z - 8z$$
$$= (12z^2 + 3z) + (-8z - 2) \qquad \text{Group the terms.}$$
$$= 3z(4z + 1) - 2(4z + 1) \qquad \text{Factor each group.}$$

> Be careful with signs.

$$= (4z + 1)(3z - 2). \qquad \text{Factor out } 4z + 1.$$

Check Multiply $(4z + 1)(3z - 2)$ to obtain $12z^2 - 5z - 2$.

(c) $10m^2 + mn - 3n^2$

Two integers whose product is $10(-3) = -30$ and whose sum is 1 are -5 and 6. Rewrite the trinomial with four terms.

$$10m^2 + mn - 3n^2$$
$$= 10m^2 - 5mn + 6mn - 3n^2 \qquad mn = -5mn + 6mn$$
$$= 5m(2m - n) + 3n(2m - n) \qquad \text{Group the terms; factor each group.}$$
$$= (2m - n)(5m + 3n) \qquad \text{Factor out } 2m - n.$$

> Check by multiplying.

◀ *Work Problem* **2** *at the Side.*

3 Factor each trinomial completely.

(a) $-4x^2 + 2x + 30$

(b) $18p^4 + 63p^3 + 27p^2$

(c) $6a^2 + 3ab - 18b^2$

EXAMPLE 2 **Factoring a Trinomial with a Common Factor by Grouping**

Factor $28x^5 - 58x^4 - 30x^3$.

First factor out the greatest common factor, $2x^3$.

$$28x^5 - 58x^4 - 30x^3$$
$$= 2x^3(14x^2 - 29x - 15)$$

To factor $14x^2 - 29x - 15$, find two integers whose product is $14(-15) = -210$ and whose sum is -29. Factoring 210 into prime factors gives

$$210 = 2 \cdot 3 \cdot 5 \cdot 7.$$

Combine these prime factors in pairs in different ways, using one positive factor and one negative factor to get -210. The factors 6 and -35 have the correct sum, -29. Now rewrite the given trinomial and factor it.

$$28x^5 - 58x^4 - 30x^3$$

> Remember the common factor.

$$= 2x^3(14x^2 - 29x - 15)$$
$$= 2x^3(14x^2 + 6x - 35x - 15)$$
$$= 2x^3[(14x^2 + 6x) + (-35x - 15)]$$
$$= 2x^3[2x(7x + 3) - 5(7x + 3)]$$
$$= 2x^3[(7x + 3)(2x - 5)]$$
$$= 2x^3(7x + 3)(2x - 5)$$

◀ *Work Problem* **3** *at the Side.*

ANSWERS

2. **(a)** $(2m + 1)(m + 3)$
   (b) $(5p + 3)(p - 1)$
   (c) $(5k - 2m)(3k + m)$
3. **(a)** $-2(2x + 5)(x - 3)$
   (b) $9p^2(2p + 1)(p + 3)$
   (c) $3(2a - 3b)(a + 2b)$

14.3 ▶▶▶ Exercises

FOR
EXTRA
HELP

MathXL
PRACTICE   WATCH   DOWNLOAD   READ   REVIEW

The middle term of each trinomial has been rewritten. Now factor by grouping.
See Example 1.

1. $m^2 + 8m + 12$
$= m^2 + 6m + 2m + 12$

2. $x^2 + 9x + 14$
$= x^2 + 7x + 2x + 14$

3. $a^2 + 3a - 10$
$= a^2 + 5a - 2a - 10$

4. $y^2 - 2y - 24$
$= y^2 + 4y - 6y - 24$

5. $10t^2 + 9t + 2$
$= 10t^2 + 5t + 4t + 2$

6. $6x^2 + 13x + 6$
$= 6x^2 + 9x + 4x + 6$

7. $15z^2 - 19z + 6$
$= 15z^2 - 10z - 9z + 6$

8. $12p^2 - 17p + 6$
$= 12p^2 - 9p - 8p + 6$

9. $8s^2 + 2st - 3t^2$
$= 8s^2 - 4st + 6st - 3t^2$

10. $3x^2 - xy - 14y^2$
$= 3x^2 - 7xy + 6xy - 14y^2$

11. $15a^2 + 22ab + 8b^2$
$= 15a^2 + 10ab + 12ab + 8b^2$

12. $25m^2 + 25mn + 6n^2$
$= 25m^2 + 15mn + 10mn + 6n^2$

13. Which pair of integers would be used to rewrite the middle term when factoring $12y^2 + 5y - 2$ by grouping?

A. $-8, 3$ **B.** $8, -3$ **C.** $-6, 4$ **D.** $6, -4$

14. Which pair of integers would be used to rewrite the middle term when factoring $20b^2 - 13b + 2$ by grouping?

A. $10, 3$ **B.** $-10, -3$ **C.** $8, 5$ **D.** $-8, -5$

Complete the steps to factor each trinomial by grouping.

15. $2m^2 + 11m + 12$

  (a) Find two integers whose product is

    _____ · _____ = _____ and whose

    sum is _____.

  (b) The required integers are _____ and

    _____.

  (c) Write the middle term $11m$ as _____ +

    _____.

  (d) Rewrite the given trinomial using four terms.

  (e) Factor the polynomial in part (d) by grouping.

  (f) Check by multiplying.

16. $6y^2 - 19y + 10$

  (a) Find two integers whose product is

    _____ · _____ = _____ and whose

    sum is _____.

  (b) The required integers are _____ and

    _____.

  (c) Write the middle term $-19y$ as _____ +

    _____.

  (d) Rewrite the given trinomial using four terms.

  (e) Factor the polynomial in part (d) by grouping.

  (f) Check by multiplying.

Factor each trinomial by grouping. See Examples 1 and 2.

17. $2x^2 + 7x + 3$

18. $3y^2 + 13y + 4$

19. $4r^2 + r - 3$

20. $4r^2 + 3r - 10$

21. $8m^2 - 10m - 3$

22. $20x^2 - 28x - 3$

23. $21m^2 + 13m + 2$

24. $38x^2 + 23x + 2$

25. $6b^2 + 7b + 2$

26. $6w^2 + 19w + 10$

27. $12y^2 - 13y + 3$

28. $15a^2 - 16a + 4$

29. $24x^2 - 42x + 9$

30. $48b^2 - 74b - 10$

31. $2m^3 + 2m^2 - 40m$

32. $3x^3 + 12x^2 - 36x$

33. $-32z^5 + 20z^4 + 12z^3$

34. $-18x^5 - 15x^4 + 75x^3$

35. $12p^2 + 7pq - 12q^2$

36. $6m^2 - 5mn - 6n^2$

37. $6a^2 - 7ab - 5b^2$

38. $25g^2 - 5gh - 2h^2$

39. $5 - 6x + x^2$

40. $7 + 8x + x^2$

41. On a quiz, a student factored $16x^2 - 24x + 5$ by grouping as follows.

$$16x^2 - 24x + 5$$
$$= 16x^2 - 4x - 20x + 5$$
$$= 4x(4x - 1) - 5(4x - 1) \qquad \text{His answer}$$

He thought his answer was correct since it checked by multiplying. Why was his answer marked wrong? What is the correct factored form?

42. On the same quiz, another student factored $3k^3 - 12k^2 - 15k$ by first factoring out the common factor $3k$ to get $3k(k^2 - 4k - 5)$. Then she wrote

$$k^2 - 4k - 5$$
$$= k^2 - 5k + k - 5$$
$$= k(k - 5) + 1(k - 5)$$
$$= (k - 5)(k + 1). \qquad \text{Her answer}$$

Why was her answer marked wrong? What is the correct factored form?

14.4 ▶▶▶ Factoring Trinomials Using FOIL

OBJECTIVE 1 Factor trinomials using FOIL. This section shows an alternative method of factoring trinomials in which the coefficient of the squared term is not 1. This method uses trial and error.

To factor $2x^2 + 7x + 6$ (the same trinomial factored at the beginning of **Section 14.3**) by trial and error, we use FOIL backwards. We want to write $2x^2 + 7x + 6$ as the product of two binomials.

$$(\quad)(\quad)$$

The product of the two first terms of the binomials is $2x^2$. The possible factors of $2x^2$ are $2x$ and x or $-2x$ and $-x$. Since all terms of the trinomial are positive, we consider only positive factors. Thus, we have

$$(2x\quad)(x\quad).$$

The product of the two last terms, 6, can be factored as $1 \cdot 6, 6 \cdot 1, 2 \cdot 3$, or $3 \cdot 2$. Try each pair to find the pair that gives the correct middle term, $7x$.

Work Problem 1 at the Side. ▶

In part (b) at the side, since $2x + 6 = 2(x + 3)$, the binomial $2x + 6$ has a common factor of 2, while $2x^2 + 7x + 6$ has no common factor other than 1. The product $(2x + 6)(x + 1)$ cannot be correct. (Part (c) also has one binomial factor with a common factor.)

Note

If the original polynomial has no common factor, then none of its binomial factors will either.

Now try the remaining numbers 3 and 2 as factors of 6.

$$(2x + 3)(x + 2) = 2x^2 + 7x + 6 \quad \text{Correct}$$

$$3x$$
$$4x$$
$$7x \quad \text{Add.}$$

Finally, we see that $2x^2 + 7x + 6$ factors as $(2x + 3)(x + 2)$.

Check Multiply $(2x + 3)(x + 2)$ to obtain $2x^2 + 7x + 6$.

EXAMPLE 1 Factoring a Trinomial with All Positive Terms Using FOIL

Factor $8p^2 + 14p + 5$.

The number 8 has several possible pairs of factors, but 5 has only 1 and 5 or -1 and -5. For this reason, it is easier to begin by considering the factors of 5. Ignore the negative factors, since all coefficients in the trinomial are positive. If $8p^2 + 14p + 5$ can be factored, the factors will have the form

$$(\quad + 5)(\quad + 1).$$

Continued on Next Page

OBJECTIVE

1 Factor trinomials using FOIL.

1 Multiply to decide whether each factored form is correct or incorrect for
$$2x^2 + 7x + 6.$$
(a) $(2x + 1)(x + 6)$

(b) $(2x + 6)(x + 1)$

(c) $(2x + 2)(x + 3)$

2 Factor each trinomial.

(a) $2p^2 + 9p + 9$

(b) $6p^2 + 19p + 10$

(c) $8x^2 + 14x + 3$

3 Factor each trinomial.

(a) $4y^2 - 11y + 6$

(b) $9x^2 - 21x + 10$

When factoring $8p^2 + 14p + 5$, the possible pairs of factors of $8p^2$ are $8p$ and p, or $4p$ and $2p$. Try various combinations, checking to see if the middle term is $14p$ in each case.

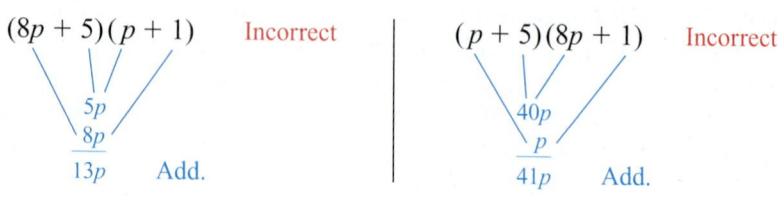

Since $14p$ is the correct middle term,

$$8p^2 + 14p + 5 \quad \text{factors as} \quad (4p + 5)(2p + 1).$$

Check Multiply $(4p + 5)(2p + 1)$ to obtain $8p^2 + 14p + 5$.

◀ *Work Problem* **2** *at the Side.*

EXAMPLE 2 **Factoring a Trinomial with a Negative Middle Term Using FOIL**

Factor $6x^2 - 11x + 3$.

Since 3 has only 1 and 3 or -1 and -3 as factors, it is better here to begin by factoring 3. The last term of the trinomial $6x^2 - 11x + 3$ is positive and the middle term has a negative coefficient, so we consider only negative factors. We need two negative factors because the *product* of two negative factors is positive and their *sum* is negative, as required.

Try -3 and -1 as factors of 3:

$$(\quad - 3)(\quad - 1).$$

The factors of $6x^2$ may be either $6x$ and x, or $2x$ and $3x$.

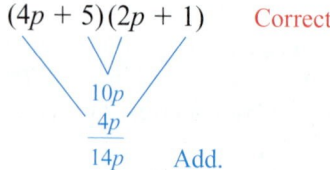

The factors $2x$ and $3x$ produce $-11x$, the correct middle term, so

$$6x^2 - 11x + 3 \quad \text{factors as} \quad (2x - 3)(3x - 1).$$

Check Multiply $(2x - 3)(3x - 1)$ to obtain $6x^2 - 11x + 3$.

Note

In Example 2, we might also realize that our initial attempt to factor $6x^2 - 11x + 3$ as $(6x - 3)(x - 1)$ *cannot* be correct since $6x - 3$ has a common factor of 3 and the original polynomial does not.

◀ *Work Problem* **3** *at the Side.*

> **EXAMPLE 3** **Factoring a Trinomial with a Negative Last Term Using FOIL**

Factor $8x^2 + 6x - 9$.

The integer 8 has several possible pairs of factors, as does -9. Since the last term is negative, one positive factor and one negative factor of -9 are needed. Since the coefficient of the middle term is small, it is wise to avoid large factors such as 8 or 9. We try 4 and 2 as factors of 8, and 3 and -3 as factors of -9, and check the middle term.

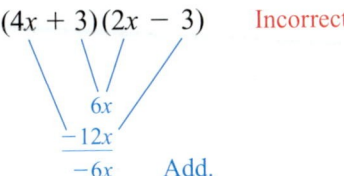

$(4x + 3)(2x - 3)$ Incorrect

Now we try interchanging 3 and -3, since only the sign of the middle term is incorrect.

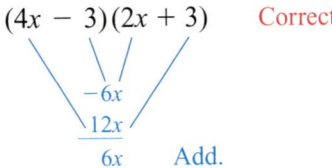

$(4x - 3)(2x + 3)$ Correct

This combination produces $6x$, the correct middle term, so

$$8x^2 + 6x - 9 \quad \text{factors as} \quad (4x - 3)(2x + 3).$$

Work Problem **4** *at the Side.* ▶

> **EXAMPLE 4** **Factoring a Trinomial with Two Variables**

Factor $12a^2 - ab - 20b^2$.

There are several pairs of factors of $12a^2$, including

$$12a \text{ and } a, \quad 6a \text{ and } 2a, \quad \text{and} \quad 4a \text{ and } 3a,$$

just as there are many possible pairs of factors of $-20b^2$, including

$$20b \text{ and } -b, \quad -20b \text{ and } b, \quad 10b \text{ and } -2b,$$
$$-10b \text{ and } 2b, \quad 4b \text{ and } -5b, \quad \text{and} \quad -4b \text{ and } 5b.$$

Once again, since the coefficient of the middle term is small, avoid the larger factors. Try the factors $6a$ and $2a$ and $4b$ and $-5b$.

$$(6a + 4b)(2a - 5b)$$

This cannot be correct, as mentioned before, since $6a + 4b$ has 2 as a common factor, while the given trinomial does not. Try $3a$ and $4a$ with $4b$ and $-5b$.

$$(3a + 4b)(4a - 5b)$$
$$= 12a^2 + ab - 20b^2 \quad \text{Incorrect}$$

Here the middle term is ab, rather than $-ab$. Interchange the signs of the last two terms in the factors.

$$(3a - 4b)(4a + 5b)$$
$$= 12a^2 - ab - 20b^2 \quad \text{Correct}$$

Work Problem **5** *at the Side.* ▶

---

4 Factor each trinomial, if possible.

(a) $6x^2 + 5x - 4$

(b) $6m^2 - 11m - 10$

(c) $4x^2 - 3x - 7$

(d) $3y^2 + 8y - 6$

5 Factor each trinomial.

(a) $2x^2 - 5xy - 3y^2$

(b) $8a^2 + 2ab - 3b^2$

ANSWERS

4. (a) $(3x + 4)(2x - 1)$
 (b) $(2m - 5)(3m + 2)$
 (c) $(4x - 7)(x + 1)$
 (d) prime
5. (a) $(2x + y)(x - 3y)$
 (b) $(4a + 3b)(2a - b)$

6 Factor each trinomial.

(a) $36z^3 - 6z^2 - 72z$

(b) $-24x^3 + 32x^2y + 6xy^2$

EXAMPLE 5 **Factoring Trinomials with Common Factors**

Factor each trinomial.

(a) $15y^3 + 55y^2 + 30y$

First factor out the greatest common factor, $5y$.

$$15y^3 + 55y^2 + 30y$$
$$= 5y(3y^2 + 11y + 6)$$

Now factor $3y^2 + 11y + 6$. Try $3y$ and y as factors of $3y^2$ and 2 and 3 as factors of 6.

$$(3y + 2)(y + 3)$$
$$= 3y^2 + 11y + 6 \qquad \text{Correct}$$

The complete factored form of $15y^3 + 55y^2 + 30y$ is

$$5y(3y + 2)(y + 3).$$

Remember the common factor.

Check $5y(3y + 2)(y + 3)$

$$= 5y(3y^2 + 11y + 6) \qquad \text{FOIL}$$
$$= 15y^3 + 55y^2 + 30y \qquad \text{Distributive property}$$

(b) $-24a^3 - 42a^2 + 45a$

The common factor could be $3a$ or $-3a$. If we factor out $-3a$, the first term of the trinomial will be positive, which makes it easier to factor.

$$-24a^3 - 42a^2 + 45a$$
$$= -3a(8a^2 + 14a - 15) \qquad \text{Factor out } -3a.$$
$$= -3a(4a - 3)(2a + 5) \qquad \text{Use trial and error.}$$

Check $-3a(4a - 3)(2a + 5)$

$$= -3a(8a^2 + 14a - 15)$$
$$= -24a^3 - 42a^2 + 45a$$

CAUTION
This caution bears repeating: *Remember to include the common factor in the final factored form.*

◀ *Work Problem* **6** *at the Side.*

14.4 ▶▶▶ Exercises

Decide which is the correct factored form of the given polynomial.

1. $2x^2 - x - 1$
  A. $(2x - 1)(x + 1)$ **B.** $(2x + 1)(x - 1)$

2. $3a^2 - 5a - 2$
  A. $(3a + 1)(a - 2)$ **B.** $(3a - 1)(a + 2)$

3. $4y^2 + 17y - 15$
  A. $(y + 5)(4y - 3)$ **B.** $(2y - 5)(2y + 3)$

4. $12c^2 - 7c - 12$
  A. $(6c - 2)(2c + 6)$ **B.** $(4c + 3)(3c - 4)$

5. $4k^2 + 13mk + 3m^2$
  A. $(4k + m)(k + 3m)$ **B.** $(4k + 3m)(k + m)$

6. $2x^2 + 11x + 12$
  A. $(2x + 3)(x + 4)$ **B.** $(2x + 4)(x + 3)$

Complete each factoring.

7. $6a^2 + 7ab - 20b^2$
  $= (3a - 4b)(\qquad)$

8. $9m^2 - 3mn - 2n^2$
  $= (3m + n)(\qquad)$

9. $2x^2 + 6x - 8$
  $= 2(\qquad)$
  $= 2(\qquad)(\qquad)$

10. $3x^2 - 9x - 30$
  $= 3(\qquad)$
  $= 3(\qquad)(\qquad)$

11. $4z^3 - 10z^2 - 6z$
  $= 2z(\qquad)$
  $= 2z(\qquad)(\qquad)$

12. $15r^3 - 39r^2 - 18r$
  $= 3r(\qquad)$
  $= 3r(\qquad)(\qquad)$

13. For the polynomial $12x^2 + 7x - 12$, 2 is not a common factor. Explain why the binomial $2x - 6$, then, cannot be a factor of the polynomial.

14. How are the signs of the last terms of the two binomial factors of a trinomial determined?

Factor each trinomial completely. See Examples 1–5.

◐ 15. $3a^2 + 10a + 7$

16. $7r^2 + 8r + 1$

17. $2y^2 + 7y + 6$

18. $5z^2 + 12z + 4$

19. $15m^2 + m - 2$

20. $6x^2 + x - 1$

21. $12s^2 + 11s - 5$

22. $20x^2 + 11x - 3$

◐ 23. $10m^2 - 23m + 12$

24. $6x^2 - 17x + 12$

25. $8w^2 - 14w + 3$

26. $9p^2 - 18p + 8$

27. $20y^2 - 39y - 11$

28. $10x^2 - 11x - 6$

29. $3x^2 - 15x + 16$

30. $2t^2 + 13t - 18$

31. $20x^2 + 22x + 6$

32. $36y^2 + 81y + 45$

33. $-40m^2q - mq + 6q$

34. $-15a^2b - 22ab - 8b$

35. $15n^4 - 39n^3 + 18n^2$

36. $24a^4 + 10a^3 - 4a^2$

37. $-15x^2y^2 + 7xy^2 + 4y^2$

38. $-14a^2b^3 - 15ab^3 + 9b^3$

39. $5a^2 - 7ab - 6b^2$

40. $6x^2 - 5xy - y^2$

41. $12s^2 + 11st - 5t^2$

42. $25a^2 + 25ab + 6b^2$

☢ **43.** $6m^6n + 7m^5n^2 + 2m^4n^3$

44. $12k^3q^4 - 4k^2q^5 - kq^6$

If a trinomial has a negative coefficient for the squared term, such as $-2x^2 + 11x - 12$, *it may be easier to factor by first factoring out the common factor* -1:

$$-2x^2 + 11x - 12$$
$$= -1(2x^2 - 11x + 12)$$
$$= -1(2x - 3)(x - 4).$$

Use this method to factor the trinomials in Exercises 45–50.

45. $-x^2 - 4x + 21$

46. $-x^2 + x + 72$

47. $-3x^2 - x + 4$

48. $-5x^2 + 2x + 16$

49. $-2a^2 - 5ab - 2b^2$

50. $-3p^2 + 13pq - 4q^2$

Relating Concepts (Exercises 51–56) For Individual or Group Work

One of the most common problems that beginning algebra students face is this: If an answer obtained doesn't look exactly like the one given in the back of the book, is it necessarily incorrect? Often there are several different equivalent forms of an answer that are all correct. **Work Exercises 51–56 in order,** *to see how and why this is possible for factoring problems.*

51. Factor the integer 35 as the product of two prime numbers.

52. Factor the integer 35 as the product of the negatives of two prime numbers.

53. Verify that $6x^2 - 11x + 4$ factors as $(3x - 4)(2x - 1)$.

54. Verify that $6x^2 - 11x + 4$ factors as $(4 - 3x)(1 - 2x)$.

55. Compare the two valid factored forms in Exercises 53 and 54. How do the factors in each case compare?

56. Suppose you know that the correct factored form of a particular trinomial is $(7t - 3)(2t - 5)$. Based on your observations in Exercises 51–55, what is another valid factored form?

14.5 ▶▶▶ Special Factoring Techniques

By reversing the rules for multiplication of binomials from **Section 13.4,** we obtain rules for factoring polynomials in certain forms.

OBJECTIVE 1 Factor a difference of squares.   The formula for the product of the sum and difference of the same two terms is

$$(a + b)(a - b) = a^2 - b^2.$$

Reversing this rule leads to the following special factoring rule.

> **Factoring a Difference of Squares**
> $$a^2 - b^2 = (a + b)(a - b)$$

For example, $m^2 - 16$

$$= m^2 - 4^2$$

$$= (m + 4)(m - 4).$$

As the next examples show, the following conditions must be true for a binomial to be a difference of squares.

1. Both terms of the binomial must be squares, such as

$$x^2, \quad 9y^2, \quad 25, \quad 1, \quad m^4.$$

2. The terms of the binomial must have different signs (one positive and one negative).

EXAMPLE 1   **Factoring Differences of Squares**

Factor each binomial, if possible. (In part (c), use fractions.)

$$a^2 - b^2 = (a + b)(a - b)$$

(a) $x^2 - 49 = x^2 - 7^2 = (x + 7)(x - 7)$

(b) $y^2 - m^2$

$$= (y + m)(y - m)$$

(c) $z^2 - \dfrac{9}{16}$

$$= \left(z + \frac{3}{4}\right)\left(z - \frac{3}{4}\right) \qquad \tfrac{9}{16} = \left(\tfrac{3}{4}\right)^2$$

(d) $x^2 - 8$
Because 8 is not the square of an integer, this binomial is not a difference of squares. It is a prime polynomial.

(e) $p^2 + 16$
Since $p^2 + 16$ is a *sum* of squares, it is not equal to $(p + 4)(p - 4)$. Also, using FOIL,

$$(p - 4)(p - 4)$$
$$= p^2 - 8p + 16, \quad \text{not} \quad p^2 + 16,$$

and

$$(p + 4)(p + 4)$$
$$= p^2 + 8p + 16, \quad \text{not} \quad p^2 + 16,$$

so $p^2 + 16$ is a prime polynomial.

OBJECTIVES

1 Factor a difference of squares.

2 Factor a perfect square trinomial.

1 Factor, if possible. (In part (b), use fractions.)

(a) $p^2 - 100$

(b) $x^2 - \dfrac{25}{36}$

(c) $x^2 + y^2$

(d) $9m^2 - 49$

(e) $64a^2 - 25$

2 Factor completely.

(a) $50r^2 - 32$

(b) $27y^2 - 75$

(c) $25a^2 - 64b^2$

(d) $k^4 - 49$

(e) $81r^4 - 16$

EXAMPLE 2 **Factoring Differences of Squares**

Factor each difference of squares.

$$a^2 \;-\; b^2 \;=\; (a \;+\; b)\,(a \;-\; b)$$

(a) $25m^2 - 16 = (5m)^2 - 4^2 = (5m + 4)(5m - 4)$

(b) $49z^2 - 64$

$= (7z)^2 - \mathbf{8}^2$

$= (7z + \mathbf{8})(7z - \mathbf{8})$

◀ Work Problem **1** at the Side.

Note

Always check a factored form by multiplying.

EXAMPLE 3 **Factoring More Complex Differences of Squares**

Factor completely.

(a) $81y^2 - 36$

$= \mathbf{9}(9y^2 - 4)$ Factor out the GCF, 9.

$= 9[(3y)^2 - 2^2]$

$= 9(3y + 2)(3y - 2)$ Difference of squares

(b) $9x^2 - 4z^2$

$= (\mathbf{3}x)^2 - (\mathbf{2}z)^2$

$= (\mathbf{3}x + \mathbf{2}z)(\mathbf{3}x - \mathbf{2}z)$

(c) $p^4 - 36$

$= (p^2)^2 - 6^2$

$= (p^2 + 6)(p^2 - 6)$

(d) $m^4 - 16$

$= (m^2)^2 - 4^2$

Don't stop here.

$= (m^2 + 4)(m^2 - 4)$ Difference of squares

$= (m^2 + 4)(m + 2)(m - 2)$ Difference of squares again

◀ Work Problem **2** at the Side.

OBJECTIVE 2 Factor a perfect square trinomial. The expressions 144, $4x^2$, and $81m^6$ are called *perfect squares* because

$$144 = \mathbf{12}^2, \quad 4x^2 = (\mathbf{2x})^2, \quad \text{and} \quad 81m^6 = (\mathbf{9m^3})^2.$$

ANSWERS

1. **(a)** $(p + 10)(p - 10)$

(b) $\left(x + \dfrac{5}{6}\right)\left(x - \dfrac{5}{6}\right)$

(c) prime

(d) $(3m + 7)(3m - 7)$

(e) $(8a + 5)(8a - 5)$

2. **(a)** $2(5r + 4)(5r - 4)$

(b) $3(3y + 5)(3y - 5)$

(c) $(5a + 8b)(5a - 8b)$

(d) $(k^2 + 7)(k^2 - 7)$

(e) $(9r^2 + 4)(3r + 2)(3r - 2)$

A **perfect square trinomial** is a trinomial that is the square of a binomial. For example, $x^2 + 8x + 16$ is a perfect square trinomial because it is the square of the binomial $x + 4$:

$$x^2 + 8x + 16$$
$$= (x + 4)(x + 4)$$
$$= (x + 4)^2.$$

On the one hand, a necessary condition for a trinomial to be a perfect square is that *two of its terms must be perfect squares*. For this reason, $16x^2 + 4x + 15$ is not a perfect square trinomial because only the term $16x^2$ is a perfect square.

On the other hand, even if two of the terms are perfect squares, the trinomial may not be a perfect square trinomial. For example, $x^2 + 6x + 36$ has two perfect square terms, but it is not a perfect square trinomial. (Try to find a binomial that can be squared to give $x^2 + 6x + 36$.)

We can multiply to see that the square of a binomial gives one of the following perfect square trinomials.

> **Factoring Perfect Square Trinomials**
> $$a^2 + 2ab + b^2 = (a + b)^2$$
> $$a^2 - 2ab + b^2 = (a - b)^2$$

The middle term of a perfect square trinomial is always twice the product of the two terms in the squared binomial. (See **Section 13.4.**) Use this to check any attempt to factor a trinomial that appears to be a perfect square.

EXAMPLE 4 **Factoring a Perfect Square Trinomial**

Factor $x^2 + 10x + 25$.
   The term x^2 is a perfect square, and so is 25. Try to factor the trinomial
$$x^2 + 10x + 25 \quad \text{as} \quad (x + 5)^2.$$

To check, take twice the product of the two terms in the squared binomial.

$$2 \cdot x \cdot 5 = 10x$$

Twice First term — of binomial Last term of binomial

Since $10x$ is the middle term of the trinomial, the trinomial is a perfect square and can be factored as $(x + 5)^2$. Thus,
$$x^2 + 10x + 25 \quad \text{factors as} \quad (x + 5)^2.$$

Work Problem ③ *at the Side.* ▶

EXAMPLE 5 **Factoring Perfect Square Trinomials**

Factor each trinomial.

(a) $x^2 - 22x + 121$
   The first and last terms are perfect squares ($121 = 11^2$ or $(-11)^2$). Check to see whether the middle term of $x^2 - 22x + 121$ is twice the product of the first and last terms of the binomial $x - 11$.

— **Continued on Next Page**

③ Factor each trinomial.

(a) $p^2 + 14p + 49$

(b) $m^2 + 8m + 16$

(c) $x^2 + 2x + 1$

4 Factor each trinomial.

(a) $p^2 - 18p + 81$

(b) $16a^2 + 56a + 49$

(c) $121p^2 + 110p + 100$

(d) $64x^2 - 48x + 9$

(e) $27y^3 + 72y^2 + 48y$

$$2 \cdot x \cdot (-11) = -22x$$

Twice — First term — Last term

Since twice the product of the first and last terms of the binomial is the middle term, $x^2 - 22x + 121$ is a perfect square trinomial and

$$x^2 - 22x + 121 \quad \text{factors as} \quad (x - 11)^2.$$

Same sign

Notice that the sign of the second term in the squared binomial is the same as the sign of the middle term in the trinomial.

(b) $9m^2 - 24m + 16 = (3m)^2 + 2(3m)(-4) + (-4)^2 = (3m - 4)^2$

Twice — First term — Last Term

(c) $25y^2 + 20y + 16$

The first and last terms are perfect squares.

$$25y^2 = (5y)^2 \quad \text{and} \quad 16 = 4^2$$

However, twice the product of the first and last terms of the binomial $5y + 4$ is $2 \cdot 5y \cdot 4 = 40y$, which is not the middle term of $25y^2 + 20y + 16$. This trinomial is not a perfect square. In fact, the trinomial cannot be factored, even with the methods of the previous sections. It is a prime polynomial.

(d) $12z^3 + 60z^2 + 75z$

$$= 3z(4z^2 + 20z + 25) \qquad \text{Factor out } 3z.$$
$$= 3z[(2z)^2 + 2(2z)(5) + 5^2] \qquad 4z^2 + 20z + 25 \text{ is a perfect square trinomial.}$$
$$= 3z(2z + 5)^2 \qquad \text{Factor.}$$

> **Note**
>
> 1. The sign of the second term in the squared binomial is always the same as the sign of the middle term in the trinomial.
> 2. The first and last terms of a perfect square trinomial must be *positive*, because they are squares. For example, the polynomial $x^2 - 2x - 1$ cannot be a perfect square because the last term is negative.
> 3. Perfect square trinomials can also be factored using grouping or FOIL, although using the method of this section is often easier.

◀ Work Problem **4** at the Side.

The methods of factoring discussed in this section are summarized here.

Special Factoring Rules

Difference of squares $\quad a^2 - b^2 = (a + b)(a - b)$

Perfect square trinomials $\quad a^2 + 2ab + b^2 = (a + b)^2$

$$a^2 - 2ab + b^2 = (a - b)^2$$

14.5 ▶▶▶ Exercises

1. To help you factor a difference of squares, complete the following list of squares.

$1^2 =$ _____ $2^2 =$ _____ $3^2 =$ _____ $4^2 =$ _____ $5^2 =$ _____

$6^2 =$ _____ $7^2 =$ _____ $8^2 =$ _____ $9^2 =$ _____ $10^2 =$ _____

$11^2 =$ _____ $12^2 =$ _____ $13^2 =$ _____ $14^2 =$ _____ $15^2 =$ _____

$16^2 =$ _____ $17^2 =$ _____ $18^2 =$ _____ $19^2 =$ _____ $20^2 =$ _____

2. To use the factoring techniques described in this section, you will sometimes need to recognize fourth powers of integers. Complete the following list of fourth powers.

$1^4 =$ _____ $2^4 =$ _____ $3^4 =$ _____ $4^4 =$ _____ $5^4 =$ _____

3. The following powers of x are all perfect squares: $x^2, x^4, x^6, x^8, x^{10}$. Based on this observation, we may make a conjecture (an educated guess) that if the power of a variable is divisible by _____ (with 0 remainder), then it is a perfect square.

4. Which of the following are differences of squares?

  A. $x^2 - 4$ **B.** $y^2 + 9$ **C.** $2a^2 - 25$ **D.** $9m^2 - 1$

Factor each binomial completely. In Exercises 7, 8, 13, and 14, use fractions. See Examples 1–3.

◐ 5. $y^2 - 25$

6. $t^2 - 16$

7. $p^2 - \dfrac{1}{9}$

8. $q^2 - \dfrac{1}{4}$

9. $m^2 - 12$

10. $k^2 - 18$

◐ 11. $9r^2 - 4$

12. $4x^2 - 9$

13. $4m^2 - \dfrac{9}{25}$

14. $100b^2 - \dfrac{49}{81}$

◐ 15. $36x^2 - 16$

16. $32a^2 - 8$

17. $196p^2 - 225$

18. $361q^2 - 400$

19. $16r^2 - 25a^2$

20. $49m^2 - 100p^2$

21. $100x^2 + 49$

22. $81w^2 + 16$

23. $p^4 - 49$

24. $r^4 - 25$

25. $x^4 - 1$

26. $y^4 - 10,000$

27. $p^4 - 256$

28. $16k^4 - 1$

29. When a student was directed to factor $x^4 - 81$ completely, his teacher did not give him full credit for the answer $(x^2 + 9)(x^2 - 9)$. The student argued that because his answer does indeed give $x^4 - 81$ when multiplied out, he should be given full credit. *WHAT WENT WRONG?* Give the correct factored form.

30. The binomial $4x^2 + 16$ is a sum of squares that *can* be factored. How is this binomial factored? When can a sum of squares be factored?

31. In the polynomial $9y^2 + 14y + 25$, the first and last terms are perfect squares. Can the polynomial be factored? If it can, factor it. If it cannot, explain why it is not a perfect square trinomial.

32. Which of the following are perfect square trinomials?

    A. $y^2 - 13y + 36$ **B.** $x^2 + 6x + 9$ **C.** $4z^2 - 4z + 1$ **D.** $16m^2 + 10m + 1$

Factor each trinomial completely. It may be necessary to factor out the greatest common factor first. In Exercises 37–40, use fractions or decimals, as appropriate. See Examples 4 and 5.

33. $w^2 + 2w + 1$ **34.** $p^2 + 4p + 4$ **35.** $x^2 - 8x + 16$

36. $x^2 - 10x + 25$ **37.** $t^2 + t + \dfrac{1}{4}$ **38.** $m^2 + \dfrac{2}{3}m + \dfrac{1}{9}$

39. $x^2 - 1.0x + 0.25$ **40.** $y^2 - 1.4y + 0.49$ **41.** $2x^2 + 24x + 72$

42. $3y^2 - 48y + 192$ **43.** $16x^2 - 40x + 25$ **44.** $36y^2 - 60y + 25$

45. $49x^2 - 28xy + 4y^2$ **46.** $4z^2 - 12zw + 9w^2$ **47.** $64x^2 + 48xy + 9y^2$

48. $9t^2 + 24tr + 16r^2$ **49.** $-50h^3 + 40h^2y - 8hy^2$ **50.** $-18x^3 - 48x^2y - 32xy^2$

Relating Concepts (Exercises 51–54) For Individual or Group Work

We have seen that multiplication and factoring are reverse processes. We know that multiplication and division are also related. To check a division problem, we multiply the quotient by the divisor to get the dividend. To see how factoring and division are related, **work Exercises 51–54 in order.**

51. Factor $10x^2 + 11x - 6$.

52. Use long division from **Section 13.7** to divide $10x^2 + 11x - 6$ by $2x + 3$.

53. Could we have predicted the result in Exercise 52 from the result in Exercise 51? Explain.

54. Divide $x^3 - 1$ by $x - 1$. Use your answer to factor $x^3 - 1$.

Summary Exercises on Factoring

As you factor a polynomial, we suggest asking yourself these questions to decide on a suitable factoring technique.

> **Factoring a Polynomial**
>
> 1. **Is there a common factor other than 1?** If so, factor out the greatest common factor (GCF) of all terms of the given polynomial.
>
> 2. **How many terms are in the polynomial?**
>
>   *Two terms:* Check to see whether it is a difference of squares.
>
>   *Three terms:* Is it a perfect square trinomial? If the trinomial is not a perfect square, check to see whether the coefficient of the second-degree term is 1. If so, use the method of **Section 14.2.** If the coefficient of the squared term of the trinomial is not 1, use the general factoring methods of **Sections 14.3** and **14.4.**
>
>   *Four terms:* Try to factor the polynomial by grouping.
>
> 3. **Can any factors be factored further?** If so, factor them.

> **CAUTION**
> Be careful when checking your answer to a factoring problem.
>
> 1. *Check* that the product of all the factors does indeed yield the original polynomial.
>
> 2. *Check* that the original polynomial has been factored *completely.* (See Question 3 above.)
>
> *Checking by multiplication alone will not ensure that you have factored the original polynomial completely.*

Suppose we are asked to completely factor the trinomial

$$6x^2 + 24xy + 24y^2$$

and give

$$(2x + 4y)(3x + 6y)$$

as the answer. If we only check by multiplying, we might conclude that our answer is correct because, by FOIL,

$$(2x + 4y)(3x + 6y)$$
$$= 6x^2 + 24xy + 24y^2.$$

However, we would not have factored the given polynomial *completely* because the binomial $2x + 4y$ has a common factor of 2 and the binomial $3x + 6y$ has a common factor of 3. Rather than factoring out these common factors at the *end* of our work, it is more efficient to factor out the GCF, 6 (which is $2 \cdot 3$), at the beginning, as follows.

$$6x^2 + 24xy + 24y^2$$
$$= 6(x^2 + 4xy + 4y^2) \qquad \text{Factor out the GCF, 6.}$$
$$= 6(x + 2y)^2 \qquad \text{Factor the perfect square trinomial.}$$

To avoid leaving out common factors in your final answer, *remember to always factor out the GCF first.*

We now use the factoring guidelines given on the preceding page to factor $12x^2 + 26xy + 12y^2$ completely.

Question 1: *Is there a common factor?* Yes, there is a common factor. The GCF is 2, so factor it out.

$$12x^2 + 26xy + 12y^2$$
$$= 2(6x^2 + 13xy + 6y^2)$$

Question 2: *How many terms are in the polynomial?* The polynomial $6x^2 + 13xy + 6y^2$ has three terms, but it is not a perfect square. To factor the trinomial by grouping, as in **Section 14.3,** begin by finding two integers with a product of $6 \cdot 6$, or 36, and a sum of 13. These integers are 4 and 9.

$12x^2 + 26xy + 12y^2$

$= \mathbf{2}(6x^2 + \mathbf{13xy} + 6y^2)$	Factor out the GCF, 2.
$= 2(6x^2 + \mathbf{4xy} + \mathbf{9xy} + 6y^2)$	$4 \cdot 9 = 36; 4 + 9 = 13$
$= 2[(6x^2 + 4xy) + (9xy + 6y^2)]$	Group the terms.
$= 2[2x(\mathbf{3x + 2y}) + 3y(\mathbf{3x + 2y})]$	Factor each group.
$= 2(\mathbf{3x + 2y})(2x + 3y)$	Factor out the common factor, $3x + 2y$.

(The trinomial $6x^2 + 13xy + 6y^2$ could also be factored by trial and error, using FOIL backwards, as in **Section 14.4.**)

Question 3: *Can any factors be factored further?* None of the factors can be factored further, so the original polynomial has been factored completely.

Match each polynomial in Column I with the method you would use to factor it in Column II. The choices in Column II may be used once, more than once, or not at all.

I	**II**
1. $12x^2 + 20x + 8$	**A.** Factor out the GCF; no further factoring is possible.
2. $x^2 - 17x + 72$	**B.** Factor a difference of squares once.
3. $-16m^2n + 24mn - 40mn^2$	**C.** Factor a difference of squares twice.
4. $64a^2 - 121b^2$	**D.** Factor a perfect square trinomial.
5. $36p^2 - 60pq + 25q^2$	**E.** Factor by grouping.
6. $z^2 - 4z + 6$	**F.** Factor out the GCF; then factor a trinomial by grouping or trial and error.
7. $625 - r^4$	**G.** Factor into two binomials by finding two integers whose product is the constant in the trinomial and whose sum is the coefficient of the middle term.
8. $x^6 + 4x^4 - 3x^2 - 12$	
9. $4w^2 + 49$	**H.** The polynomial is prime.
10. $144 - 24z + z^2$	

Factor each polynomial completely. In Exercises 26, 32, 64, and 81, use fractions or decimals, as appropriate. Remember to check by multiplying.

11. $32m^9 + 16m^5 + 24m^3$

12. $2m^2 - 10m - 48$

13. $14k^3 + 7k^2 - 70k$

14. $9z^2 + 64$

15. $6z^2 + 31z + 5$

16. $m^2 - 3mn - 4n^2$

17. $49z^2 - 16y^2$

18. $100n^2r^2 + 30nr^3 - 50n^2r$

19. $16x^2 + 20x$

20. $20 + 5m + 12n + 3mn$

21. $10y^2 - 7yz - 6z^2$

22. $y^4 - 81$

23. $m^2 + 2m - 15$

24. $6y^2 - 5y - 4$

25. $32z^3 + 56z^2 - 16z$

26. $p^2 - 2.4p + 1.44$

27. $z^2 - 12z + 36$

28. $9m^2 - 64$

29. $y^2 - 4yk - 12k^2$

30. $16z^2 - 8z + 1$

31. $6y^2 - 6y - 12$

32. $x^2 + \dfrac{1}{2}x + \dfrac{1}{16}$

33. $p^2 - 17p + 66$

34. $a^2 + 17a + 72$

35. $k^2 + 100$

36. $108m^2 - 36m + 3$

37. $z^2 - 3za - 10a^2$

38. $2a^3 + a^2 - 14a - 7$

39. $4k^2 - 12k + 9$

40. $a^2 - 3ab - 28b^2$

41. $16r^2 + 24rm + 9m^2$

42. $3k^2 + 4k - 4$

43. $n^2 - 12n - 35$

44. $a^4 - 625$

45. $16k^2 - 48k + 36$

46. $8k^2 - 10k - 3$

47. $36y^6 - 42y^5 - 120y^4$

48. $5z^3 - 45z^2 + 70z$

49. $8p^2 + 23p - 3$

50. $8k^2 - 2kh - 3h^2$

51. $54m^2 - 24z^2$

52. $4k^2 - 20kz + 25z^2$

53. $6a^2 + 10a - 4$

54. $15h^2 + 11hg - 14g^2$

55. $28a^2 - 63b^2$

56. $10z^2 - 7z - 6$

57. $125m^4 - 400m^3n + 195m^2n^2$

58. $9y^2 + 12y - 5$

59. $9u^2 + 66uv + 121v^2$

60. $36x^2 + 32x + 9$

61. $27p^{10} - 45p^9 - 252p^8$

62. $10m^2 + 25m - 60$

63. $4 - 2q - 6p + 3pq$

64. $k^2 - \dfrac{64}{121}$

65. $64p^2 - 100m^2$

66. $m^3 + 4m^2 - 6m - 24$

67. $100a^2 - 81y^2$

68. $8a^2 + 23ab - 3b^2$

69. $a^2 + 8a + 16$

70. $4y^2 - 25$

71. $2x^2 + 5x + 6$

72. $-3x^3 + 12xy^2$

73. $25a^2 - 70ab + 49b^2$

74. $8t^4 - 8$

75. $-4x^2 + 24xy - 36y^2$

76. $100a^2 - 25b^2$

77. $-2x^2 + 26x - 72$

78. $2m^2 - 15n - 5mn + 6m$

79. $12x^2 + 22x - 20$

80. $y^6 + 5y^4 - 3y^2 - 15$

81. $y^2 - 0.64$

82. $12p^3 - 54p^2 - 30p$

14.6 ▶▶▶ Solving Quadratic Equations by Factoring

Galileo Galilei developed theories to explain physical phenomena and set up experiments to test his ideas. According to legend, Galileo dropped objects of different weights from the Leaning Tower of Pisa to disprove the belief that heavier objects fall faster than lighter objects. He developed a formula for freely falling objects described by $d = 16t^2$, where d is the distance in feet that an object falls (disregarding air resistance) in t seconds, regardless of weight.

The equation $d = 16t^2$ is a *quadratic equation*. A quadratic equation contains a squared term and no terms of higher degree.

OBJECTIVES

1 Solve quadratic equations by factoring.

2 Solve other equations by factoring.

> **Quadratic Equation**
>
> A **quadratic equation** is an equation that can be written in the form
>
> $$ax^2 + bx + c = 0,$$
>
> where a, b, and c are real numbers, with $a \neq 0$. The given form is called **standard form**.

$$x^2 + 5x + 6 = 0, \quad 2t^2 - 5t = 3, \quad y^2 = 4 \qquad \text{Quadratic equations}$$

In these examples, only $x^2 + 5x + 6 = 0$ is in standard form.

Work Problems **1** *and* **2** *at the Side.* ▶

Galileo Galilei (1564–1642)

Up to now, we have factored *expressions,* including many quadratic expressions of the form $ax^2 + bx + c$. In this section, we use factored quadratic expressions to solve quadratic *equations.*

OBJECTIVE 1 **Solve quadratic equations by factoring.** We use the **zero-factor property** to solve a quadratic equation by factoring.

> **Zero-Factor Property**
>
> **If a and b are real numbers and $ab = 0$, then $a = 0$ or $b = 0$.**
>
> In words, if the product of two numbers is 0, then at least one of the numbers must be 0. One number *must* be 0, but both *may* be 0.

EXAMPLE 1 **Using the Zero-Factor Property**

Solve each equation.

(a) $(x + 3)(2x - 1) = 0$

The product $(x + 3)(2x - 1)$ is equal to 0. By the zero-factor property, the only way that the product of these two factors can be 0 is if at least one of the factors equals 0. Therefore, either $x + 3 = 0$ or $2x - 1 = 0$.

$$x + 3 = 0 \quad \text{or} \quad 2x - 1 = 0 \qquad \text{Zero-factor property}$$
$$x = -3 \quad \text{or} \qquad 2x = 1 \qquad \text{Solve each equation.}$$
$$x = \frac{1}{2}$$

Continued on Next Page

1 Which of the following equations are quadratic equations?

  A. $y^2 - 4y - 5 = 0$

  B. $x^3 - x^2 + 16 = 0$

  C. $2z^2 + 7z = -3$

  D. $x + 2y = -4$

2 Write each quadratic equation in standard form.

  (a) $x^2 - 3x = 4$

  (b) $y^2 = 9y - 8$

ANSWERS
1. A, C
2. **(a)** $x^2 - 3x - 4 = 0$
   (b) $y^2 - 9y + 8 = 0$

3 Solve each equation. Check your solutions.

(a) $(x - 5)(x + 2) = 0$

(b) $(3x - 2)(x + 6) = 0$

(c) $z(2z + 5) = 0$

The given equation, $(x + 3)(2x - 1) = 0$, has two solutions, -3 and $\frac{1}{2}$. *Check* these solutions by substituting -3 for x in the original equation, $(x + 3)(2x - 1) = 0$. Then start over and substitute $\frac{1}{2}$ for x.

Check If $x = -3$, then

$$(x + 3)(2x - 1) = 0$$

$$(-3 + 3)[2(-3) - 1] \overset{?}{=} 0$$

$$0(-7) = 0. \quad \text{True}$$

If $x = \dfrac{1}{2}$, then

$$(x + 3)(2x - 1) = 0$$

$$\left(\frac{1}{2} + 3\right)\left(2 \cdot \frac{1}{2} - 1\right) \overset{?}{=} 0$$

$$\frac{7}{2}(1 - 1) \overset{?}{=} 0$$

$$\frac{7}{2} \cdot 0 = 0. \quad \text{True}$$

Both -3 and $\frac{1}{2}$ result in true equations, so the solution set is $\{-3, \frac{1}{2}\}$.

(b)
$$y(3y - 4) = 0$$

$y = 0$ or $3y - 4 = 0$ Zero-factor property

> *Don't forget that 0 is a solution.*

$$3y = 4$$

$$y = \frac{4}{3}$$

Check these solutions by substituting each one in the original equation. The solution set is $\{0, \frac{4}{3}\}$.

◀ *Work Problem* **3** *at the Side.*

> **Note**
>
> The word *or* as used in Example 1 means "one or the other or both."

In Example 1, each equation to be solved was given with the polynomial in factored form. If the polynomial in an equation is not already factored, first make sure that the equation is in standard form. Then factor and solve.

EXAMPLE 2 **Solving Quadratic Equations**

Solve each equation.

(a) $x^2 - 5x = -6$

First, write the equation in standard form by adding 6 to each side.

> *Don't factor x out at this step.*

$$x^2 - 5x = -6$$

$$x^2 - 5x + 6 = 0 \quad \text{Add 6.}$$

Now factor $x^2 - 5x + 6$. Find two numbers whose product is 6 and whose sum is -5. These two numbers are -2 and -3, so the equation becomes

$$(x - 2)(x - 3) = 0. \quad \text{Factor the trinomial.}$$

$x - 2 = 0$ or $x - 3 = 0$ Zero-factor property

$x = 2$ or $x = 3$ Solve each equation.

Continued on Next Page

ANSWERS

2. (a) $\{-2, 5\}$ **(b)** $\left\{-6, \frac{2}{3}\right\}$ **(c)** $\left\{-\frac{5}{2}, 0\right\}$

Check   If $x = 2$, then

$$2^2 - 5(2) \stackrel{?}{=} -6$$

$$4 - 10 \stackrel{?}{=} -6$$

$$-6 = -6. \quad \text{True}$$

If $x = 3$, then

$$3^2 - 5(3) \stackrel{?}{=} -6$$

$$9 - 15 \stackrel{?}{=} -6$$

$$-6 = -6. \quad \text{True}$$

Both solutions check, so the solution set is $\{2, 3\}$.

(b)

$$y^2 = y + 20$$

$$y^2 - y - 20 = 0 \qquad \text{Write in standard form.}$$

$$(y - 5)(y + 4) = 0 \qquad \text{Factor the trinomial.}$$

$$y - 5 = 0 \quad \text{or} \quad y + 4 = 0 \qquad \text{Zero-factor property}$$

$$y = 5 \quad \text{or} \qquad y = -4 \qquad \text{Solve each equation.}$$

Check by substituting in the original equation. The solution set is $\{-4, 5\}$.

Work Problem **4** *at the Side.* ▶

Solving a Quadratic Equation by Factoring

Step 1   **Write the equation in standard form,** that is, with all terms on one side of the equals sign in descending powers of the variable and 0 on the other side.

Step 2   **Factor** completely.

Step 3   **Use the zero-factor property** to set each factor with a variable equal to 0.

Step 4   **Solve** the resulting equations.

Step 5   **Check** each solution in the original equation.

EXAMPLE 3   **Solving a Quadratic Equation (Common Factor)**

Solve $4p^2 + 40 = 26p$.

$$4p^2 - 26p + 40 = 0 \qquad \text{Standard form}$$

$$2(2p^2 - 13p + 20) = 0 \qquad \text{Factor out 2.}$$

$$2p^2 - 13p + 20 = 0 \qquad \text{Divide each side by 2.}$$

$$(2p - 5)(p - 4) = 0 \qquad \text{Factor.}$$

$$2p - 5 = 0 \quad \text{or} \quad p - 4 = 0 \qquad \text{Zero-factor property}$$

$$2p = 5 \quad \text{or} \qquad p = 4 \qquad \text{Solve each equation.}$$

$$p = \frac{5}{2}$$

Check that the solution set is $\{\frac{5}{2}, 4\}$ by substituting in the original equation.

CAUTION

A common error is to include the common factor **2** as a solution in Example 3. *Only factors containing variables lead to solutions.*

Work Problem **5** *at the Side.* ▶

---

4 Solve each equation. Check your solutions.

(a) $m^2 - 3m - 10 = 0$

(b) $r^2 + 2r = 8$

5 Solve each equation. Check your solutions.

(a) $10a^2 - 5a - 15 = 0$

(b) $4x^2 - 2x = 42$

ANSWERS

4.   (a) $\{-2, 5\}$ (b) $\{-4, 2\}$

5.   (a) $\left\{-1, \dfrac{3}{2}\right\}$ (b) $\left\{-3, \dfrac{7}{2}\right\}$

6 Solve each equation. Check your solutions.

(a) $49m^2 - 9 = 0$

(b) $p(4p + 7) = 2$

(c) $m^2 = 3m$

EXAMPLE 4 Solving Quadratic Equations

Solve each equation.

(a) $16m^2 - 25 = 0$

We can factor the left side of the equation as the difference of squares (**Section 14.5**).

$$16m^2 - 25 = 0$$

$$(4m + 5)(4m - 5) = 0 \qquad \text{Factor the difference of squares.}$$

$$4m + 5 = 0 \quad \text{or} \quad 4m - 5 = 0 \qquad \text{Zero-factor property}$$

$$4m = -5 \quad \text{or} \qquad 4m = 5 \qquad \text{Solve each equation.}$$

$$m = -\frac{5}{4} \quad \text{or} \qquad m = \frac{5}{4}$$

Check the solutions $-\frac{5}{4}$ and $\frac{5}{4}$ in the original equation. The solution set is $\{-\frac{5}{4}, \frac{5}{4}\}$.

(b) $k(2k + 5) = 3$

We need to write this equation in standard form.

$$k(2k + 5) = 3 \qquad \text{To be in standard form, 0 must be on one side.}$$

$$2k^2 + 5k = 3 \qquad \text{Multiply.}$$

$$2k^2 + 5k - 3 = 0 \qquad \text{Subtract 3.}$$

$$(2k - 1)(k + 3) = 0 \qquad \text{Factor.}$$

$$2k - 1 = 0 \quad \text{or} \quad k + 3 = 0 \qquad \text{Zero-factor property}$$

$$2k = 1 \quad \text{or} \qquad k = -3$$

$$k = \frac{1}{2}$$

The solution set is $\{-3, \frac{1}{2}\}$.

(c) $y^2 = 2y$

$$y^2 - 2y = 0 \qquad \text{Standard form}$$

$$y(y - 2) = 0 \qquad \text{Factor.} \qquad \text{(Don't forget to set the variable factor } y \text{ equal to 0.)}$$

$$y = 0 \quad \text{or} \quad y - 2 = 0 \qquad \text{Zero-factor property}$$

$$y = 2$$

The solution set is $\{0, 2\}$.

CAUTION

In Example 4(b), the zero-factor property could not be used to solve the equation $k(2k + 5) = 3$ in its given form because of the 3 on the right. *The zero-factor property applies only to a product that equals 0.*

In Example 4(c), it is tempting to begin by dividing each side of the equation $y^2 = 2y$ by y to get $y = 2$. Note that we do not get the other solution, 0, if we divide by a variable. (We *may* divide each side of an equation by a *nonzero* real number, however. For instance, in Example 3 we divided each side by 2.)

ANSWERS

6. **(a)** $\left\{-\frac{3}{7}, \frac{3}{7}\right\}$ **(b)** $\left\{-2, \frac{1}{4}\right\}$ **(c)** $\{0, 3\}$

◀ *Work Problem* **6** *at the Side.*

EXAMPLE 5 Solving a Quadratic Equation with a Double Solution

Solve each equation.

(a)
$$z^2 + 121 = 22z$$
$$z^2 - 22z + 121 = 0 \qquad \text{Standard form}$$

Because $121 = 11^2$ and $22z = 2 \cdot z \cdot 11$, the trinomial on the left is a perfect square.

$$(z - 11)^2 = 0 \qquad \text{Factor.}$$

To apply the zero-product property, write $(z - 11)^2$ as two separate factors.

$$(z - 11)(z - 11) = 0 \qquad a^2 = a \cdot a$$
$$z - 11 = 0 \quad \text{or} \quad z - 11 = 0 \qquad \text{Zero-factor property}$$

Because the two factors are identical, they both lead to the same solution. (This is called a **double solution.**) Thus,

$$z - 11 = 0$$
$$z = 11. \qquad \text{Add 11.}$$

Check
$$z^2 + 121 = 22z$$
$$11^2 + 121 \stackrel{?}{=} 22(11) \qquad \text{Let } z = 11.$$
$$121 + 121 \stackrel{?}{=} 242$$
$$242 = 242 \qquad \text{True}$$

The solution set is $\{11\}$.

(b)
$$9t^2 - 30t = -25$$
$$9t^2 - 30t + 25 = 0 \qquad \text{Standard form}$$
$$(3t - 5)^2 = 0 \qquad \text{Factor the perfect square trinomial.}$$
$$3t - 5 = 0 \quad \text{or} \quad 3t - 5 = 0 \qquad \text{Zero-factor property}$$
$$3t = 5 \qquad \text{Solve the equation.}$$
$$t = \frac{5}{3}$$

Check the double solution by substituting $\frac{5}{3}$ in the original equation. The solution set is $\left\{\frac{5}{3}\right\}$.

———————————————— *Work Problem* **7** *at the Side.* ▶

> **7** Solve each equation. Check your solutions.
>
> **(a)** $x^2 + 16x = -64$
>
> **(b)** $4x^2 - 4x + 1 = 0$

CAUTION
When a trinomial has two identical factors (a perfect square trinomial), as in Examples 5(a) and (b), it is common for students to write the solution of the corresponding quadratic equation twice in the solution set. Each of these equations has only *one* distinct solution. ***There is no need to write the same number more than once in a solution set.***

Note
Not all quadratic equations can be solved by factoring. A more general method for solving such equations is given in **Chapter 17.**

8 Solve each equation. Check your solutions.

(a) $r^3 - 16r = 0$

(b) $x^3 - 3x^2 - 18x = 0$

OBJECTIVE **2** **Solve other equations by factoring.** We can extend the zero-factor property to solve equations that involve more than two factors with variables, as shown in Examples 6 and 7. (These equations are *not* quadratic equations. Why not?)

EXAMPLE 6 **Solving an Equation with More Than Two Variable Factors**

Solve $6z^3 - 6z = 0$.

$$6z^3 - 6z = 0$$
$$6z(z^2 - 1) = 0 \qquad \text{Factor out } 6z.$$
$$6z(z + 1)(z - 1) = 0 \qquad \text{Factor } z^2 - 1.$$

By an extension of the zero-factor property, this product can equal 0 only if at least one of the factors equals 0. Write and solve three equations, one for each factor with a variable.

$$6z = 0 \quad \text{or} \quad z + 1 = 0 \quad \text{or} \quad z - 1 = 0$$
$$z = 0 \quad \text{or} \qquad z = -1 \quad \text{or} \qquad z = 1$$

Check by substituting, in turn, 0, -1, and 1 in the original equation. The solution set is $\{-1, 0, 1\}$.

◀ Work Problem **8** at the Side.

9 Solve each equation. Check your solutions.

(a) $(m + 3)(m^2 - 11m + 10) = 0$

(b) $(2x + 5)(4x^2 - 9) = 0$

EXAMPLE 7 **Solving an Equation with a Quadratic Factor**

Solve $(2x - 1)(x^2 - 9x + 20) = 0$.

$$(2x - 1)(x^2 - 9x + 20) = 0$$
$$(2x - 1)(x - 5)(x - 4) = 0 \qquad \text{Factor } x^2 - 9x + 20.$$
$$2x - 1 = 0 \quad \text{or} \quad x - 5 = 0 \quad \text{or} \quad x - 4 = 0 \qquad \text{Zero-factor property}$$
$$x = \frac{1}{2} \quad \text{or} \qquad x = 5 \quad \text{or} \qquad x = 4$$

Check to verify that the solution set is $\{\frac{1}{2}, 4, 5\}$.

◀ Work Problem **9** at the Side.

CAUTION
In Example 7, it would be unproductive to begin by multiplying the two factors together. Keep in mind that the zero-factor property and its extension requires the product of two or more factors to equal 0. *Always consider first whether an equation is given in the appropriate form to apply the zero-factor property.*

ANSWERS

8. (a) $\{-4, 0, 4\}$ (b) $\{-3, 0, 6\}$

9. (a) $\{-3, 1, 10\}$ (b) $\left\{-\frac{5}{2}, -\frac{3}{2}, \frac{3}{2}\right\}$

14.6 ▶▶▶ Exercises

Solve each equation, and check your solutions. See Example 1.

1. $(x + 5)(x - 2) = 0$

2. $(x - 1)(x + 8) = 0$

3. $(2m - 7)(m - 3) = 0$

4. $(6k + 5)(k + 4) = 0$

5. $t(6t + 5) = 0$

6. $w(4w + 1) = 0$

7. $2x(3x - 4) = 0$

8. $6x(4x + 9) = 0$

9. $\left(x + \dfrac{1}{2}\right)\left(2x - \dfrac{1}{3}\right) = 0$

10. $\left(a + \dfrac{2}{3}\right)\left(5a - \dfrac{1}{2}\right) = 0$

11. $(x - 9)(x - 9) = 0$

12. $(2x + 1)(2x + 1) = 0$

13. Look at this "solution." ***WHAT WENT WRONG?***

$$2x(3x - 4) = 0$$
$$x = 2 \quad \text{or} \quad x = 0 \quad \text{or} \quad 3x - 4 = 0$$
$$x = \frac{4}{3}$$

The solution set is $\{2, 0, \frac{4}{3}\}$.

14. Look at this "solution." ***WHAT WENT WRONG?***

$$x(7x - 1) = 0$$
$$7x - 1 = 0 \quad \text{Zero-factor property}$$
$$x = \frac{1}{7}$$

The solution set is $\{\frac{1}{7}\}$.

Solve each equation, and check your solutions. See Examples 2–7.

15. $y^2 + 3y + 2 = 0$

16. $p^2 + 8p + 7 = 0$

17. $y^2 - 3y + 2 = 0$

18. $r^2 - 4r + 3 = 0$

19. $x^2 = 24 - 5x$

20. $t^2 = 2t + 15$

21. $x^2 = 3 + 2x$

22. $m^2 = 4 + 3m$

23. $z^2 + 3z = -2$

24. $p^2 - 2p = 3$

25. $m^2 + 8m + 16 = 0$

26. $b^2 - 6b + 9 = 0$

27. $3x^2 + 5x - 2 = 0$

28. $6r^2 - r - 2 = 0$

29. $6p^2 = 4 - 5p$

30. $6x^2 = 4 + 5x$

31. $9s^2 + 12s = -4$

32. $36x^2 + 60x = -25$

33. $y^2 - 9 = 0$

34. $m^2 - 100 = 0$

35. $16k^2 - 49 = 0$

36. $4w^2 - 9 = 0$

37. $n^2 = 121$

38. $x^2 = 400$

39. $x^2 = 7x$

40. $t^2 = 9t$

41. $6r^2 = 3r$

42. $10y^2 = -5y$

43. $g(g - 7) = -10$

44. $r(r - 5) = -6$

45. $z(2z + 7) = 4$

46. $b(2b + 3) = 9$

47. $2(y^2 - 66) = -13y$

48. $3(t^2 + 4) = 20t$

49. $5x^3 - 20x = 0$

50. $3x^3 - 48x = 0$

51. $9y^3 - 49y = 0$

52. $16r^3 - 9r = 0$

53. $(2r + 5)(3r^2 - 16r + 5) = 0$

54. $(3m + 4)(6m^2 + m - 2) = 0$

55. $(2x + 7)(x^2 + 2x - 3) = 0$

56. $(x + 1)(6x^2 + x - 12) = 0$

57. Galileo's formula for freely falling objects, $d = 16t^2$, was given at the beginning of this section. The distance d in feet an object falls depends on the time elapsed t in seconds. (This is an example of an important mathematical concept, a **function.**)

(a) Use Galileo's formula and complete the following table. (*Hint:* Substitute each given value into the formula and solve for the unknown value.)

t in seconds	0	1	2	3		
d in feet	0	16			256	576

(b) When $t = 0$, $d = 0$. Explain this in the context of the problem.

58. In Exercise 57, when you substituted 256 for d and solved for t, you should have found two solutions: 4 and -4. Why doesn't -4 make sense as an answer?

14.7 ▶▶▶ Applications of Quadratic Equations

We can use factoring to solve quadratic equations that arise in applications. We follow the same basic problem-solving steps used throughout this text.

OBJECTIVES

1 Solve problems about geometric figures.

2 Solve problems about consecutive integers.

3 Solve problems using the Pythagorean formula.

4 Solve problems using given quadratic models.

Solving an Applied Problem

Step 1 **Read** the problem, several times if necessary, until you *understand* what is given and what is to be found.

Step 2 **Assign a variable** to represent the unknown value, using diagrams or tables as needed. Write a statement that tells what the variable represents. Express any other unknown values in terms of the variable.

Step 3 **Write an equation** using the variable expression(s).

Step 4 **Solve** the equation.

Step 5 **State the answer.** Does it seem reasonable?

Step 6 **Check** the answer in the words of the original problem.

OBJECTIVE 1 Solve problems about geometric figures.   Refer to the formulas given on the inside back cover of this text, if necessary.

EXAMPLE 1 **Solving an Area Problem**

The Monroes want to plant a rectangular garden in their yard. The width of the garden will be 4 ft less than its length, and they want it to have an area of 96 ft^2. (ft^2 means square feet.) Find the length and width of the garden.

Step 1 **Read** the problem carefully. We need to find the dimensions of a garden with area 96 ft^2.

Step 2 **Assign a variable.**

Let $x =$ the length of the garden.

Then $x - 4 =$ the width. (The width is 4 ft less than the length.)

See Figure 1.

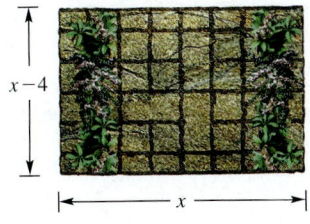

Figure 1

Step 3 **Write an equation.** The area of a rectangle is given by

$$\text{Area} = lw = \text{length} \times \text{width}. \qquad \text{Area formula}$$

Substitute 96 for area, x for length, and $x - 4$ for width.

$$A = lw$$

$$96 = x(x - 4) \qquad \text{Let } A = 96, l = x, w = x - 4.$$

Continued on Next Page

1 Solve each problem.

(a) The length of a rectangular room is 2 m more than the width. The area of the floor is 48 m². Find the length and width of the room.

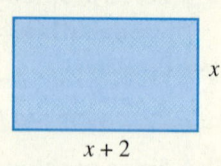

$x + 2$

x

Step 4 **Solve.**

$$96 = x(x - 4)$$
$$96 = x^2 - 4x \qquad \text{Distributive property}$$
$$x^2 - 4x - 96 = 0 \qquad \text{Standard form}$$
$$(x - 12)(x + 8) = 0 \qquad \text{Factor.}$$
$$x - 12 = 0 \quad \text{or} \quad x + 8 = 0 \qquad \text{Zero-factor property}$$
$$x = 12 \quad \text{or} \qquad x = -8$$

Step 5 **State the answer.** The solutions are 12 and -8. A rectangle cannot have a side of negative length, so discard -8. The length of the garden will be 12 ft. The width will be $12 - 4 = 8$ ft.

Step 6 **Check.** The width of the garden is 4 ft less than the length; the area is $12 \cdot 8 = 96$ ft².

Problem-Solving Hint

When solving applied problems, ***always check solutions against physical facts*** and discard any answers that are not appropriate.

◄ *Work Problem* **1** *at the Side.*

OBJECTIVE **2** **Solve problems about consecutive integers.** Two **consecutive integers** are integers that are next to each other on a number line, such as 5 and 6, or -11 and -10. **Consecutive odd integers** are *odd* integers that are next to each other, such as 5 and 7, or -13 and -11. **Consecutive even integers** are defined similarly; 4 and 6 are consecutive even integers, as are -10 and -8. (In this book, we will list consecutive integers in increasing order from left to right.)

(b) The length of each side of a square is increased by 4 in. The sum of the areas of the original square and the larger square is 106 in². What is the length of a side of the original square?

Problem-Solving Hint

In consecutive integer problems, if x represents the first integer, then for

two consecutive integers, use	$x, \quad x + 1;$
three consecutive integers, use	$x, \quad x + 1, \quad x + 2;$
two consecutive even or odd integers, use	$x, \quad x + 2;$
three consecutive even or odd integers, use	$x, \quad x + 2, \quad x + 4.$

EXAMPLE 2 **Solving a Consecutive Integer Problem**

The product of the numbers on two consecutive post-office boxes is 210. Find the box numbers.

Step 1 **Read** the problem. Note that the boxes are numbered consecutively.

Step 2 **Assign a variable.**

Let $x =$ the first box number.

Then $x + 1 =$ the next consecutive box number.

See Figure 2.

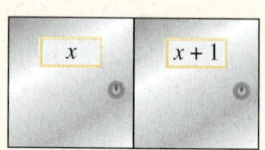

Figure 2

x $\qquad$ $x + 1$

Step 3 **Write an equation.** The product of the box numbers is 210, so

$$x(x + 1) = 210.$$

Continued on Next Page

Step 4 **Solve.**
$$x(x + 1) = 210$$
$$x^2 + x = 210 \qquad \text{Distributive property}$$
$$x^2 + x - 210 = 0 \qquad \text{Standard form}$$
$$(x + 15)(x - 14) = 0 \qquad \text{Factor.}$$
$$x + 15 = 0 \quad \text{or} \quad x - 14 = 0 \qquad \text{Zero-factor property}$$
$$x = -15 \quad \text{or} \qquad x = 14$$

Step 5 **State the answer.** The solutions are -15 and 14. Discard the solution -15 since a box number cannot be negative. When $x = 14$, then $x + 1 = 15$, so the post-office boxes have the numbers 14 and 15.

Step 6 **Check.** The numbers 14 and 15 are consecutive and their product is $14 \cdot 15 = 210$, as required.

——————————————— *Work Problem* **2** *at the Side.* ▶

EXAMPLE 3 **Solving a Consecutive Integer Problem**

The product of two consecutive odd integers is 1 less than five times their sum. Find the integers.

Step 1 **Read** carefully. This problem is a little more complicated.

Step 2 **Assign a variable.** We must find two consecutive *odd* integers.

Let $x =$ the lesser integer.

Then $x + 2 =$ the next greater odd integer.

Step 3 **Write an equation.** According to the problem, the product is 1 less than five times the sum.

The product	is	five times the sum	less 1.
↓	↓	↓	↓
$x(x + 2)$	$=$	$5(x + x + 2)$	$- \quad 1$

Step 4 **Solve.**
$$x^2 + 2x = 5x + 5x + 10 - 1 \qquad \text{Distributive property}$$
$$x^2 + 2x = 10x + 9 \qquad \text{Combine like terms.}$$
$$x^2 - 8x - 9 = 0 \qquad \text{Standard form}$$
$$(x - 9)(x + 1) = 0 \qquad \text{Factor.}$$
$$x - 9 = 0 \quad \text{or} \quad x + 1 = 0 \qquad \text{Zero-factor property}$$
$$x = 9 \quad \text{or} \qquad x = -1$$

Step 5 **State the answer.** We need to find two consecutive odd integers.

If $x = 9$ is the lesser, then $x + 2 = 9 + 2 = 11$ is the greater.

If $x = -1$ is the lesser, then $x + 2 = -1 + 2 = 1$ is the greater.

There are two sets of answers here since integers can be positive or negative.

Step 6 **Check.** The product of the first pair of integers is $9 \cdot 11 = 99$. One less than five times their sum is $5(9 + 11) - 1 = 99$. Thus 9 and 11 satisfy the problem. Repeat the check with -1 and 1.

——————————————— *Work Problem* **3** *at the Side.* ▶

---

2 Solve the problem.

The product of the numbers on two consecutive lockers at a health club is 132. Find the locker numbers.

3 Solve each problem.

(a) The product of two consecutive even integers is 4 more than two times their sum. Find the integers.

(b) Find three consecutive odd integers such that the product of the least and greatest is 16 more than the middle integer.

CAUTION
Do *not* use $x, x + 1, x + 3$, and so on to represent consecutive odd integers. To see why, let $x = 3$. Then $x + 1 = 3 + 1 = 4$ and $x + 3 = 3 + 3 = 6$, and 3, 4, and 6 are not consecutive odd integers.

OBJECTIVE **3** **Solve problems using the Pythagorean formula.**
The next example uses the Pythagorean formula from **Section 5.8.**

Pythagorean Formula

If a right triangle (a triangle with a 90° angle) has longest side of length c and two other sides of lengths a and b, then

$$a^2 + b^2 = c^2.$$

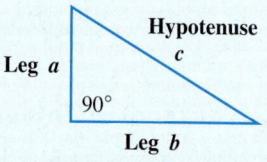

Recall that the longest side, the **hypotenuse,** is opposite the right angle. The two shorter sides are the **legs** of the triangle.

EXAMPLE 4 **Using the Pythagorean Formula**

Amy and Kevin leave their office, with Amy traveling north and Kevin traveling east. When Kevin is 1 mi farther than Amy from the office, the distance between them is 2 mi more than Amy's distance from the office. Find their distances from the office and the distance between them.

Step 1 **Read** the problem again. We must find three distances.

Step 2 **Assign a variable.** Let x represent Amy's distance from the office, $x + 1$ represent Kevin's distance from the office, and $x + 2$ represent the distance between them. Place these on a right triangle, as in Figure 3.

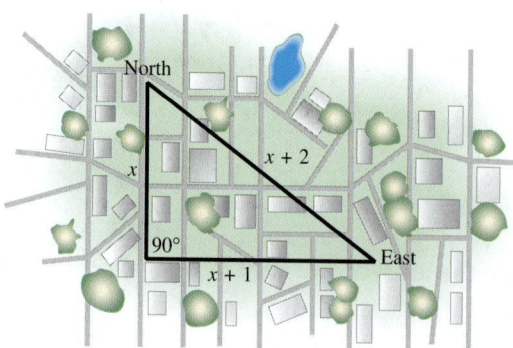

Figure 3

Step 3 **Write an equation.** Substitute into the Pythagorean formula.

$$a^2 + b^2 = c^2$$

$$x^2 + (x + 1)^2 = (x + 2)^2$$

> Be careful to substitute properly.

Continued on Next Page

Step 4 **Solve.** $x^2 + x^2 + 2x + 1 = x^2 + 4x + 4$

$$x^2 - 2x - 3 = 0 \quad \text{Standard form}$$
$$(x - 3)(x + 1) = 0 \quad \text{Factor.}$$
$$x - 3 = 0 \quad \text{or} \quad x + 1 = 0 \quad \text{Zero-factor property}$$
$$x = 3 \quad \text{or} \quad x = -1$$

Step 5 **State the answer.** Since -1 cannot represent a distance, 3 is the only possible answer. Amy's distance is 3 mi, Kevin's distance is $3 + 1 = 4$ mi, and the distance between them is $3 + 2 = 5$ mi.

Step 6 **Check.** Since $3^2 + 4^2 = 5^2$, the answer is correct.

> **CAUTION**
> When solving a problem involving the Pythagorean formula, be sure that the expressions for the sides are properly placed.
>
> $$\textbf{leg}^2 + \textbf{leg}^2 = \textbf{hypotenuse}^2$$

Work Problem **4** *at the Side.* ▶

OBJECTIVE 4 Solve problems using given quadratic models.
In Examples 1–4, we wrote quadratic equations to model, or mathematically describe, various situations and then solved the equations. Now we are given the quadratic models and must use them to determine data.

EXAMPLE 5 **Finding the Height of a Ball**

A tennis player can hit a ball 180 ft per sec. If she hits a ball directly upward, the height h of the ball in feet at time t in seconds is modeled by the quadratic equation

$$h = -16t^2 + 180t + 6.$$

When will the ball be 206 ft above the ground?

A height of 206 ft means $h = 206$, so we substitute 206 for h in the equation and then solve for t.

$$\mathbf{206} = -16t^2 + 180t + 6 \quad \text{Let } h = 206.$$
$$-16t^2 + 180t + 6 = 206 \quad \text{Interchange sides.}$$
$$-16t^2 + 180t - 200 = 0 \quad \text{Standard form}$$
$$4t^2 - 45t + 50 = 0 \quad \text{Divide by } -4.$$
$$(4t - 5)(t - 10) = 0 \quad \text{Factor.}$$
$$4t - 5 = 0 \quad \text{or} \quad t - 10 = 0 \quad \text{Zero-factor property}$$
$$t = \frac{5}{4} \quad \text{or} \quad t = 10$$

Since we found two acceptable answers, the ball will be 206 ft above the ground twice (once on its way up and once on its way down)—at $\frac{5}{4}$ sec and at 10 sec after it is hit. See Figure 4.

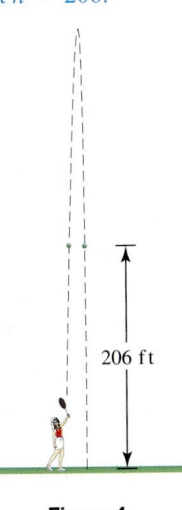

206 ft

Figure 4

Work Problem **5** *at the Side.* ▶

---

4 Solve the problem.

The hypotenuse of a right triangle is 3 in. longer than the longer leg. The shorter leg is 3 in. shorter than the longer leg. Find the lengths of the sides of the triangle.

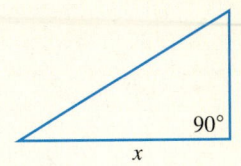

5 Solve the problem.

The number of impulses fired after a nerve has been stimulated is modeled by

$$I = -x^2 + 2x + 60,$$

where x is in milliseconds (ms) after the stimulation. When will 45 impulses occur? Do you get two solutions? Why is only one answer given?

6 Use the model in Example 6 to find the foreign-born population of the United States in 1990. Give your answer to the nearest tenth of a million. How does it compare to the actual value from the table?

EXAMPLE 6 **Modeling the Foreign–Born Population of the United States**

After decreasing in the middle of the 20th century, the foreign-born population of the United States started to increase in the later part of the century and is now increasing rapidly. The foreign-born population over the years 1930–2004 can be modeled by the quadratic equation

$$y = 0.01036x^2 - 0.5316x + 15.36,$$

where $x = 0$ represents 1930, $x = 10$ represents 1940, and so on, and y is the number of people in millions. (*Source:* U.S. Census Bureau.)

(a) Use the model to find the foreign-born population in 1980 to the nearest tenth of a million.

Since $x = 0$ represents 1930, $x = 50$ represents 1980. Substitute 50 for x in the equation.

$y = 0.01036(\mathbf{50})^2 - 0.5316(\mathbf{50}) + 15.36$ Let $x = 50$.

$y = 14.7$ Round to the nearest tenth.

In 1980, the foreign-born population of the United States was about 14.7 million.

(b) Repeat part (a) for 2004.

$y = 0.01036(\mathbf{74})^2 - 0.5316(\mathbf{74}) + 15.36$ For 2004, let $x = 74$.

$y = 32.8$ Round to the nearest tenth.

In 2004, the foreign-born population of the United States was about 32.8 million.

(c) The model used in parts (a) and (b) was developed using the data in the table below. How do the results in parts (a) and (b) compare to the actual data from the table?

Year	Foreign-Born Population (millions)
1930	14.2
1940	11.6
1950	10.3
1960	9.7
1970	9.6
1980	14.1
1990	19.8
2000	28.4
2004	34.2

From the table, the actual value for 1980 is 14.1 million. Our answer in part (a), 14.7 million, is slightly high. For 2004, the actual value is 34.2 million, so our answer of 32.8 million in part (b) is somewhat low.

◀ *Work Problem* **6** *at the Side.*

ANSWER

6. 20.8 million; The actual value is 19.8 million, so our answer using the model is somewhat high.

1. To review the six problem-solving steps, complete each statement.

   Step 1: _____ the problem, several times if necessary, until you understand what is given and what must be found.

   Step 2: Assign a _____ to represent the unknown value.

   Step 3: Write a(n) _____ using the variable expression(s).

   Step 4: _____ the equation.

   Step 5: State the _____.

   Step 6: _____ the answer in the words of the _____ problem.

2. A student solves an applied problem and gets 6 or -3 for the length of the side of a square. Which of these answers is reasonable? Explain.

In Exercises 3–6, a figure and a corresponding geometric formula are given. Using x as the variable, complete Steps 3–6 for each problem. (Refer to the steps in Exercise 1 as needed.)

3.

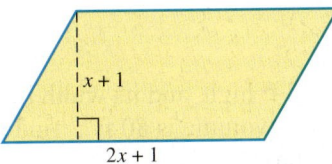

Area of a parallelogram: $A = bh$

The area of this parallelogram is 45 sq. units. Find its base and height.

4.

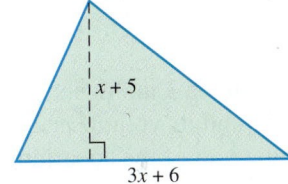

Area of a triangle: $A = \frac{1}{2}bh$

The area of this triangle is 60 sq. units. Find its base and height.

5.

Volume of a rectangular Chinese box: $V = lwh$

The volume of this box is 192 cu. units. Find its length and width.

6.

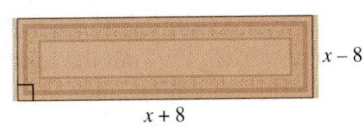

Area of a rectangular rug: $A = lw$

The area of this rug is 80 sq. units. Find its length and width.

Solve each problem. Check your answers to be sure they are reasonable. Refer to the formulas on the inside back cover. See Example 1.

7. The length of a standard jewel case is 2 cm more than its width. The area of the rectangular top of the case is 168 cm². Find the length and width of the jewel case.

8. A standard DVD case is 6 cm longer than it is wide. The area of the rectangular top of the case is 247 cm². Find the length and width of the case.

9. The dimensions of an HPf1905 flat-panel monitor are such that its length is 3 in. more than its width. If the length were doubled and if the width were decreased by 1 in., the area would be increased by 150 in.². What are the length and width of the flat panel?

10. The keyboard of the computer in Exercise 9 is 11 in. longer than it is wide. If both its length and width are increased by 2 in., the area of the top of the keyboard is increased by 54 in.². Find the length and width of the keyboard. (*Source:* Author's computer.)

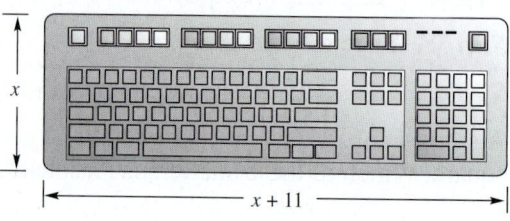

$x + 3$

x

11. A 10-gal aquarium is 3 in. higher than it is wide. Its length is 21 in., and its volume is 2730 in.³. What are the height and width of the aquarium?

12. A toolbox is 2 ft high, and its width is 3 ft less than its length. If its volume is 80 ft³, find the length and width of the box.

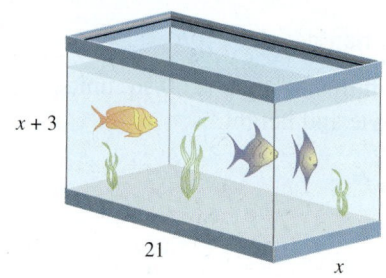

$x + 3$

21

x

x $x - 3$ 2

13. A square mirror has sides measuring 2 ft less than the sides of a square painting. If the difference between their areas is 32 ft², find the lengths of the sides of the mirror and the painting.

14. The sides of one square have length 3 m more than the sides of a second square. If the area of the larger square is subtracted from 4 times the area of the smaller square, the result is 36 m². What are the lengths of the sides of each square?

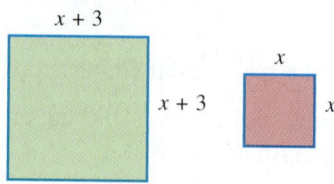

$x + 3$

$x + 3$

x

x

Solve each problem about consecutive integers. See Examples 2 and 3.

15. The product of the numbers on two consecutive volumes of research data is 420. Find the volume numbers.

16. The product of the page numbers on two facing pages of a book is 600. Find the page numbers.

17. The product of two consecutive integers is 11 more than their sum. Find the integers.

18. The product of two consecutive integers is 4 less than four times their sum. Find the integers.

19. Find two consecutive odd integers such that their product is 15 more than three times their sum.

20. Find two consecutive odd integers such that five times their sum is 23 less than their product.

21. Find three consecutive even integers such that the sum of the squares of the lesser two is equal to the square of the greatest.

22. Find three consecutive even integers such that the square of the sum of the lesser two is equal to twice the greatest.

Use the Pythagorean formula to solve each problem. See Example 4.

23. The hypotenuse of a right triangle is 1 cm longer than the longer leg. The shorter leg is 7 cm shorter than the longer leg. Find the length of the longer leg of the triangle.

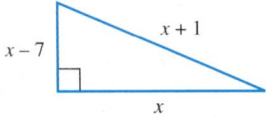

24. The longer leg of a right triangle is 1 m longer than the shorter leg. The hypotenuse is 1 m shorter than twice the shorter leg. Find the length of the shorter leg of the triangle.

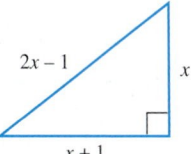

25. Terri works due north of home. Her husband Denny works due east. They leave for work at the same time. By the time Terri is 5 mi from home, the distance between them is 1 mi more than Denny's distance from home. How far from home is Denny?

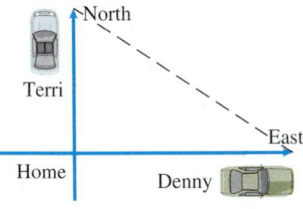

26. Two cars left an intersection at the same time. One traveled north. The other traveled 14 mi farther, but to the east. How far apart were they then, if the distance between them was 4 mi more than the distance traveled east?

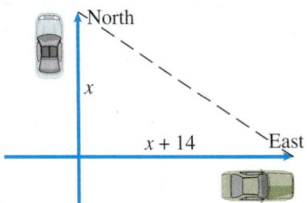

27. A ladder is leaning against a building. The distance from the bottom of the ladder to the building is 4 ft less than the length of the ladder. How high up the side of the building is the top of the ladder if that distance is 2 ft less than the length of the ladder?

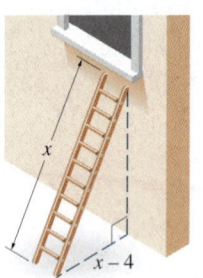

28. A lot has the shape of a right triangle with one leg 2 m longer than the other. The hypotenuse is 2 m less than twice the length of the shorter leg. Find the length of the shorter leg.

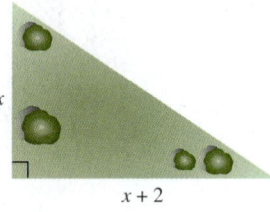

Solve each problem. See Examples 5 and 6.

29. An object projected from a height of 48 ft with an initial velocity of 32 ft per sec after t seconds has height

$$h = -16t^2 + 32t + 48.$$

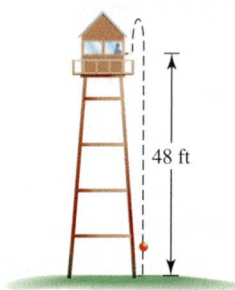

(a) After how many seconds is the height 64 ft? (*Hint:* Let $h = 64$ and solve.)

(b) After how many seconds is the height 60 ft?

(c) After how many seconds does the object hit the ground? (*Hint:* When the object hits the ground, $h = 0$.)

(d) The quadratic equation from part (c) has two solutions, yet only one of them is appropriate for answering the question. Why is this so?

30. If an object is projected upward from ground level with an initial velocity of 64 ft per sec, its height h in feet t seconds later is

$$h = -16t^2 + 64t.$$

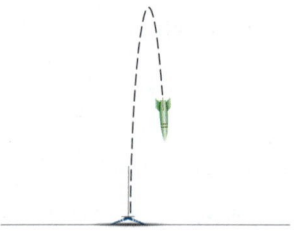

(a) After how many seconds is the height 48 ft?

(b) The object reaches its maximum height 2 sec after it is projected. What is this maximum height?

(c) After how many seconds does the object hit the ground?

(d) The quadratic equation from part (c) has two solutions, yet only one of them is appropriate for answering the question. Why is this so?

31. The table shows the number of cellular phone subscribers (in millions) in the United States.

Year	Subscribers (in millions)
1990	5
1992	11
1994	24
1996	44
1998	69
2000	109
2002	141
2004	182
2006	233

Source: CTIA-The Wireless Association.

We used the data to develop the quadratic equation

$$y = 0.734x^2 + 2.62x + 3.37,$$

which models the number of cellular phone subscribers y (in millions) in the year x, where $x = 0$ represents 1990, $x = 2$ represents 1992, and so on.

(a) Use the model to find the number of cellular phones in 1996 to the nearest million. How does the result compare to the actual data in the table?

(b) What value of x corresponds to 2004?

(c) Use the model to find the number of cellular phones in 2004 to the nearest million. How does the result compare to the actual data in the table?

(d) Assuming that the trend in the data continues, use the quadratic equation to estimate the number of cellular phones in 2009 to the nearest million.

Relating Concepts (Exercises 32–40) For Individual or Group Work

The U.S. trade deficit represents the amount by which exports are less than imports. It provides not only a sign of economic prosperity but also a warning of potential decline. The data in the table shows the U.S. trade deficit in goods and services for 2001 through 2005.

Year	Deficit (in billions of dollars)
2001	365.1
2002	423.7
2003	496.9
2004	612.1
2005	714.4

Source: U.S. Census Bureau.

Use the data to **work Exercises 32–40 in order.**

32. How much did the trade deficit in goods and services increase from 2001 to 2002? What percent increase is this (to the nearest percent)?

33. The U.S. trade deficit might be approximated by the linear equation

$$y = 88.7x + 256,$$

where y is the deficit in billions of dollars. Here $x = 1$ represents 2001, $x = 2$ represents 2002, and so on. Use this equation to approximate the trade deficits in 2003, 2004, and 2005.

34. How do your answers from Exercise 33 compare to the actual data in the table?

35. The trade deficit y (in billions of dollars) might also be approximated by the quadratic equation

$$y = 9.24x^2 + 33.24x + 321,$$

where $x = 1$ again represents 2001, $x = 2$ represents 2002, and so on. Use this equation to approximate the trade deficits in 2003, 2004, and 2005.

36. Compare your answers from Exercise 35 to the actual data in the table. Which equation, the linear one in Exercise 33 or the quadratic one in Exercise 35, models the data better?

37. Write the data from the table as a set of ordered pairs (x, y), where x represents the years starting with 2001, such that $x = 1$ for 2001, $x = 2$ for 2002, and so on, and y represents the trade deficit in billions of dollars.

Year	Deficit (in billions of dollars)
2001	365.1
2002	423.7
2003	496.9
2004	612.1
2005	714.4

Source: U.S. Census Bureau.

38. Plot the ordered pairs from Exercise 37 on the graph.

U.S. TRADE DEFICIT
(Goods and Services)

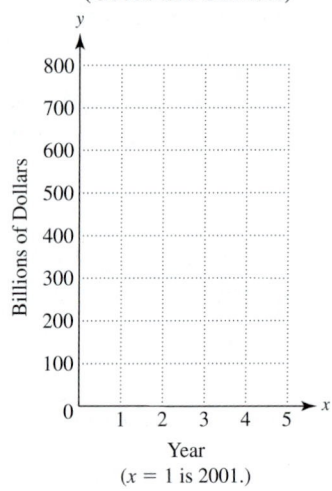

Year
($x = 1$ is 2001.)

39. Assuming that the trend in the data continues, use the quadratic equation from Exercise 35 to estimate the trade deficit for the year 2006.

40. The actual trade deficit for 2006 was 758.2 billion dollars.

   (a) How does the actual deficit for 2006 compare to your estimate from Exercise 39?

   (b) Should the quadratic equation be used to estimate the U.S. trade deficit for years after 2005? Explain.

Chapter 14 ▶▶▶ Summary

▶ Key Terms

14.1 **factor**   An expression A is a factor of an expression B if B can be divided by A with 0 remainder.

factored form   An expression is in factored form when it is written as a product.

greatest common factor (GCF)   The greatest common factor is the largest quantity that is a factor of each of a group of quantities.

factoring   The process of writing a polynomial as a product is called factoring.

14.2 **prime polynomial**   A prime polynomial is a polynomial that cannot be factored using only integers.

14.5 **perfect square trinomial**   A perfect square trinomial is a trinomial that can be factored as the square of a binomial.

14.6 **quadratic equation**   A quadratic equation is an equation that can be written in the form $ax^2 + bx + c = 0$, with $a \neq 0$.

standard form   The form $ax^2 + bx + c = 0$ is the standard form of a quadratic equation.

14.7 **hypotenuse**   The longest side of a right triangle, opposite the right angle, is the hypotenuse.

legs   The two shorter sides of a right triangle are the legs.

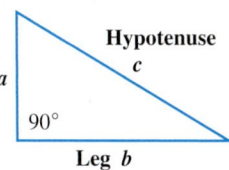

▶ Test Your Word Power

See how well you have learned the vocabulary in this chapter. Answers, with examples, follow the Quick Review.

1. Factoring is
A. a method of multiplying polynomials
B. the process of writing a polynomial as a product
C. the answer in a multiplication problem
D. a way to add the terms of a polynomial.

2. A polynomial is in **factored form** when
A. it is prime
B. it is written as a sum
C. the squared term has a coefficient of 1
D. it is written as a product.

3. The **greatest common factor** of a polynomial is
A. the least integer that divides evenly into all the terms of the polynomial
B. the least term that is a factor of all the terms in the polynomial
C. the greatest term that is a factor of all the terms in the polynomial
D. the variable that is common to all the terms in the polynomial.

4. A **perfect square trinomial** is a trinomial
A. that can be factored as the square of a binomial
B. that cannot be factored
C. that is multiplied by a binomial
D. where all terms are perfect squares.

5. A **quadratic equation** is an equation that can be written in the form
A. $y = mx + b$
B. $ax^2 + bx + c = 0 \, (a \neq 0)$
C. $Ax + By = C$
D. $x = k$.

6. A **hypotenuse** is
A. either of the two shorter sides of a triangle
B. the shortest side of a right triangle
C. the side opposite the right angle in a right triangle
D. the longest side in any triangle.

▶ Quick Review

Concepts	Examples

14.1 Factors; The Greatest Common Factor

Finding the Greatest Common Factor (GCF)

Step 1 Write each number in prime factored form.

Step 2 List each prime number or each variable that is a factor of every term in the list.

Step 3 Use as exponents on the common prime factors the least exponents from the prime factored forms.

Step 4 Multiply the primes from Step 3.

Find the greatest common factor of $4x^2y$, $-6x^2y^3$, and $2xy^2$.

$$4x^2y = \mathbf{2} \cdot 2 \cdot \mathbf{x^2} \cdot \mathbf{y}$$
$$-6x^2y^3 = -1 \cdot \mathbf{2} \cdot 3 \cdot \mathbf{x^2} \cdot \mathbf{y^3}$$
$$2xy^2 = \mathbf{2} \cdot \mathbf{x} \cdot \mathbf{y^2}$$

The greatest common factor is $2xy$.

Factoring by Grouping

Step 1 Group the terms.

Step 2 Factor out the greatest common factor from each group.

Step 3 Factor a common binomial factor from the results of Step 2.

Step 4 If necessary, rearrange terms and try a different grouping.

Factor by grouping.

$$2a^2 + 2ab + a + b$$
$$= (2a^2 + 2ab) + (a + b)$$
$$= 2a\,\mathbf{(a + b)} + 1\,\mathbf{(a + b)}$$
$$= \mathbf{(a + b)}(2a + 1)$$

14.2 Factoring Trinomials

To factor $x^2 + bx + c$, find m and n such that $mn = c$ and $m + n = b$.

$$x^2 + bx + c$$

Then $x^2 + bx + c$ factors as $(x + m)(x + n)$.

Check by multiplying.

Factor $x^2 + 6x + 8$.

$$x^2 + 6x + 8$$

$m = 2$ and $n = 4$

$x^2 + 6x + 8$ factors as $(x + 2)(x + 4)$.

Check $(x + 2)(x + 4)$
$$= x^2 + 4x + 2x + 8$$
$$= x^2 + 6x + 8$$

14.3 Factoring Trinomials by Grouping

To factor $ax^2 + bx + c$, find m and n such that $mn = ac$.

$$ax^2 + bx + c$$

Then factor $ax^2 + mx + nx + b$ by grouping.

Factor $3x^2 + 14x - 5$.

-15

Find two integers with a product of $3(-5) = -15$ and a sum of 14. The integers are -1 and 15.

$$3x^2 + 14x - 5$$
$$= 3x^2 - x + 15x - 5$$
$$= (3x^2 - x) + (15x - 5)$$
$$= x(3x - 1) + 5(3x - 1)$$
$$= (3x - 1)(x + 5)$$

Concepts	Examples

14.4 Factoring Trinomials Using FOIL

To factor $ax^2 + bx + c$ by trial and error, use FOIL backwards.

By trial and error,

$$3x^2 + 14x - 5 \quad \text{factors as} \quad (3x - 1)(x + 5).$$

14.5 Special Factoring Techniques

Difference of Squares

$$a^2 - b^2 = (a + b)(a - b)$$

Perfect Square Trinomials

$$a^2 + 2ab + b^2 = (a + b)^2$$

$$a^2 - 2ab + b^2 = (a - b)^2$$

Factor.

$$4x^2 - 9$$
$$= (2x + 3)(2x - 3)$$

$$9x^2 + 6x + 1 \qquad\qquad 4x^2 - 20x + 25$$
$$= (3x + 1)^2 \qquad\qquad = (2x - 5)^2$$

14.6 Solving Quadratic Equations by Factoring

Zero-Factor Property

If a and b are real numbers and $ab = 0$, then $a = 0$ or $b = 0$.

If $(x - 2)(x + 3) = 0$, then $x - 2 = 0$ or $x + 3 = 0$.

Solving a Quadratic Equation by Factoring

Solve $2x^2 = 7x + 15$.

Step 1 Write the equation in standard form.

$$2x^2 - 7x - 15 = 0$$

Step 2 Factor.

$$(2x + 3)(x - 5) = 0$$

Step 3 Use the zero-factor property.

$$2x + 3 = 0 \quad \text{or} \quad x - 5 = 0$$

Step 4 Solve the resulting equations.

$$2x = -3 \qquad\qquad x = 5$$
$$x = -\frac{3}{2}$$

Step 5 Check.

The solutions $-\frac{3}{2}$ and 5 satisfy the original equation. The solution set is $\{-\frac{3}{2}, 5\}$.

14.7 Applications of Quadratic Equations

Pythagorean Formula

In a right triangle, the square of the hypotenuse equals the sum of the squares of the legs.

$$a^2 + b^2 = c^2$$

In a right triangle, one leg measures 2 ft longer than the other. The hypotenuse measures 4 ft longer than the shorter leg. Find the lengths of the three sides of the triangle.

Let x = the length of the shorter leg. Then

$$x^2 + (x + 2)^2 = (x + 4)^2.$$

Solve this equation to get $x = 6$ or $x = -2$. Discard -2 as a solution. Check that the sides measure 6 ft, $6 + 2 = 8$ ft, and $6 + 4 = 10$ ft.

ANSWERS TO TEST YOUR WORD POWER

1. B; *Example:* $x^2 - 5x - 14$ factors as $(x - 7)(x + 2)$.
2. D; *Example:* The factored form of $x^2 - 5x - 14$ is $(x - 7)(x + 2)$.
3. C; *Example:* The greatest common factor of $8x^2$, $22xy$, and $16x^3y^2$ is $2x$.
4. A; *Example:* $a^2 + 2a + 1$ is a perfect square trinomial; its factored form is $(a + 1)^2$.
5. B; *Examples:* $y^2 - 3y + 2 = 0, x^2 - 9 = 0, 2m^2 = 6m + 8$
6. C; *Example:* See the triangle included in the Quick Review above for **Section 14.7.**

Math in the Media

PUTTING THE BRAKES ON FOLLOWING TOO CLOSELY

Drivers often underestimate stopping distance. Overall **stopping distance** is the sum of **thinking distance** (how far the car travels once you realize you have to brake) and **braking distance** (how far the car travels after you apply the brakes).

The data in the table represents three distinct relationships. The **input** is speed in miles per hour for all three relationships. The **output** is thinking distance in feet for the first relationship, braking distance in feet for the second relationship, and overall stopping distance in feet for the third relationship.

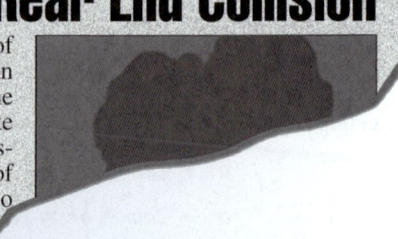

Two Die in Rear End Collision

Two elderly occupants of a 2003 Cadillac Sedan DeVille were killed at the intersection of Interstate 12 and LA Hwy 59 yesterday when the driver of an 18-wheeler failed to come to a stop.

---

1. **(a)** In the relationship between thinking distance and speed, *output* (y) is numerically the same as *input* (x). Write the equation that expresses this.

   (b) When speed is doubled from 20 mph to 40 mph, how does thinking distance change? Does this hold true if a 30 mph speed is doubled?

   (c) Is the equation linear or quadratic? Explain.

Speed	Thinking Distance	Braking Distance	Overall Stopping Distance
20 mph	20 ft	20 ft	40 ft
30 mph	30 ft	45 ft	75 ft
40 mph	40 ft	80 ft	120 ft
50 mph	50 ft	125 ft	175 ft
60 mph	60 ft	180 ft	240 ft

Source: British School of Motoring.

2. **(a)** The relationship between braking distance and speed is given by $y = \frac{1}{20}x^2$. Show that this equation corresponds to the table values for speeds of 20 mph and 40 mph.

   (b) When speed is doubled from 20 mph to 40 mph, how does braking distance change? Does this hold true if a 30 mph speed is doubled?

   (c) Is the equation linear or quadratic? Explain.

3. **(a)** Use the equations from Problems 1 and 2 to write the equation that expresses the relationship between overall stopping distance and speed.

   (b) Is the equation linear or quadratic? Explain.

   (c) A *rule of thumb* for calculating overall stopping distance is to take speed in *tens* of miles per hour, divide by 2, add 1, and multiply the result by the speed in miles per hour. For 40 mph, $4 \div 2 + 1 = 3$; $3 \times 40 = 120$ ft. Show why this rule works.

Chapter 14 ▶▶▶ Review Exercises

[14.1] *Factor out the greatest common factor or factor by grouping.*

1. $15t + 45$

2. $60z^3 + 30z$

3. $44x^3 + 55x^2$

4. $100m^2n^3 - 50m^3n^4 + 150m^2n^2$

5. $2xy - 8y + 3x - 12$

6. $6y^2 + 9y + 4xy + 6x$

[14.2] *Factor completely.*

7. $x^2 + 10x + 21$

8. $y^2 - 13y + 40$

9. $q^2 + 6q - 27$

10. $r^2 - r - 56$

11. $r^2 - 4rs - 96s^2$

12. $p^2 + 2pq - 120q^2$

13. $-8p^3 + 24p^2 + 80p$

14. $3x^4 + 30x^3 + 48x^2$

15. $m^2 - 3mn - 18n^2$

16. $y^2 - 8yz + 15z^2$

17. $p^7 - p^6q - 2p^5q^2$

18. $-3r^5 + 6r^4s + 45r^3s^2$

19. $x^2 + x + 1$

20. $3x^2 + 6x + 6$

[14.3–14.4]

21. To begin factoring $6r^2 - 5r - 6$, what are the possible first terms of the two binomial factors, if we consider only positive integer coefficients?

22. What is the first step you would use to factor $2z^3 + 9z^2 - 5z$?

Factor completely.

23. $2k^2 - 5k + 2$

24. $3r^2 + 11r - 4$

25. $6r^2 - 5r - 6$

26. $10z^2 - 3z - 1$

27. $5t^2 - 11t + 12$

28. $24x^5 - 20x^4 + 4x^3$

29. $-6x^2 + 3x + 30$

30. $10r^3s + 17r^2s^2 + 6rs^3$

31. $-30y^3 - 5y^2 + 10y$

32. $4z^2 - 5z + 7$

33. $-3m^3n + 19m^2n + 40mn$

34. $14a^2 - 27ab - 20b^2$

[14.5]

35. Which one of the following is a difference of squares?

   A. $32x^2 - 1$ **B.** $4x^2y^2 - 25z^2$

   C. $x^2 + 36$ **D.** $25y^3 - 1$

36. Which one of the following is a perfect square trinomial?

   A. $x^2 + x + 1$ **B.** $y^2 - 4y + 9$

   C. $4x^2 + 10x + 25$ **D.** $x^2 - 20x + 100$

Factor completely. In Exercises 42 and 48, use fractions.

37. $n^2 - 64$

38. $25b^2 - 121$

39. $49y^2 - 25w^2$

40. $144p^2 - 36q^2$

41. $x^2 + 100$

42. $x^2 - \dfrac{49}{100}$

43. $z^2 + 10z + 25$

44. $r^2 - 12r + 36$

45. $9t^2 - 42t + 49$

46. $16m^2 + 40mn + 25n^2$

47. $54x^3 - 72x^2 + 24x$

48. $x^2 + \dfrac{1}{3}x + \dfrac{1}{36}$

[14.6] *Solve each equation, and check the solutions.*

49. $(4t + 3)(t - 1) = 0$

50. $(x + 7)(x - 4)(x + 3) = 0$

51. $x(2x - 5) = 0$

52. $z^2 + 4z + 3 = 0$

53. $m^2 - 5m + 4 = 0$

54. $x^2 = -15 + 8x$

55. $3z^2 - 11z - 20 = 0$

56. $81t^2 - 64 = 0$

57. $y^2 = 8y$

58. $n(n - 5) = 6$

59. $t^2 - 14t + 49 = 0$

60. $t^2 = 12(t - 3)$

61. $(5z + 2)(z^2 + 3z + 2) = 0$

62. $x^2 = 9$

63. $49x^3 - 9x = 0$

64. $(2r + 1)(12r^2 + 5r - 3) = 0$

65. $25w^2 - 90w + 81 = 0$

66. $r(r - 7) = 30$

[14.7] *Solve each problem.*

67. The length of a rug is 6 ft more than the width. The area is 40 ft². Find the length and width of the rug.

68. The surface area (*SA*) of a box is given by

$$SA = 2lw + 2lh + 2wh.$$

A treasure chest from a sunken galleon has dimensions as shown in the figure. Its surface area is 650 ft². Find its width.

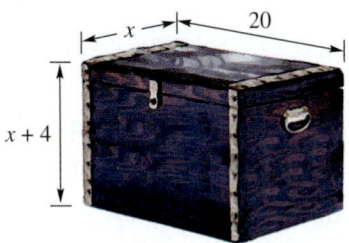

69. The length of a rectangle is three times the width. If the width were increased by 3 m while the length remained the same, the new rectangle would have an area of 30 m². Find the length and width of the original rectangle.

70. The volume of a rectangular box is 120 m³. The width of the box is 4 m, and the height is 1 m less than the length. Find the length and height of the box.

71. The product of two consecutive integers is 29 more than their sum. What are the integers?

72. Two cars left an intersection at the same time. One traveled west, and the other traveled 14 mi less, but to the south. How far apart were they then, if the distance between them was 16 mi more than the distance traveled south?

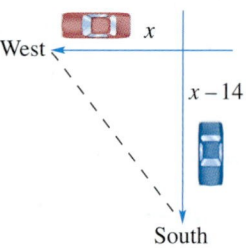

If an object is projected upward with an initial velocity of 128 ft per sec, its height h in feet after t seconds is

$$h = 128t - 16t^2.$$

Find the height of the object after each period of time.

73. 1 sec

74. 2 sec

75. 4 sec

76. For the object described above, when does it return to the ground?

77. Annual revenue in billions of dollars for eBay is shown in the table.

Year	Annual Revenue (in billions of dollars)
2002	1.21
2003	2.17
2004	3.27
2005	4.55
2006	5.77

Source: eBay.

Using the data, we developed the quadratic equation

$$y = 0.05x^2 + 0.95x + 1.19$$

to model eBay revenues y in year x, where $x = 0$ represents 2002, $x = 1$ represents 2003, and so on.

(a) Use the model to find eBay revenue (to the nearest hundredth) in 2005. How does your answer compare to the actual data from the table?

(b) Use the model to estimate annual revenue (to the nearest hundredth) for eBay in 2007.

> > > **Mixed Review Exercises**

78. Which of the following is *not* factored completely?

    A. $3(7t)$ **B.** $3x(7t + 4)$ **C.** $(3 + x)(7t + 4)$ **D.** $3(7t + 4) + x(7t + 4)$

79. Although $(2x + 8)(3x - 4) = 6x^2 + 16x - 32$ is a true statement, the polynomial is not factored completely. Explain why and give the complete factored form.

Factor completely.

80. $z^2 - 11zx + 10x^2$

81. $3k^2 + 11k + 10$

82. $15m^2 + 20mp - 12m - 16p$

83. $y^4 - 625$

84. $6m^3 - 21m^2 - 45m$

85. $24ab^3c^2 - 56a^2bc^3 + 72a^2b^2c$

86. $25a^2 + 15ab + 9b^2$

87. $12x^2yz^3 + 12xy^2z - 30x^3y^2z^4$

88. $2a^5 - 8a^4 - 24a^3$

89. $-12r^2 - 8rq + 15q^2$

90. $100a^2 - 9$

91. $49t^2 + 56t + 16$

Solve.

92. $t(t - 7) = 0$

93. $x^2 + 3x = 10$

94. $25x^2 + 20x + 4 = 0$

Solve each problem.

95. A lot is shaped like a right triangle. The hypotenuse is 3 m longer than the longer leg. The longer leg is 6 m longer than twice the length of the shorter leg. Find the lengths of the sides of the lot.

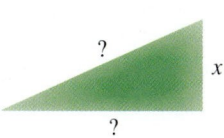

96. A pyramid has a rectangular base with a length that is 2 m more than the width. The height of the pyramid is 6 m, and its volume is 48 m^3. Find the length and width of the base.

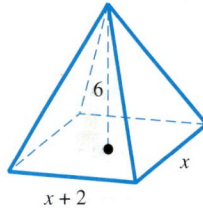

97. The product of the lesser two of three consecutive integers is equal to 23 plus the greatest. Find the integers.

98. If an object is dropped, the distance d in feet it falls in t seconds (disregarding air resistance) is given by the quadratic equation

$$d = 16t^2.$$

Find the distance an object would fall in the following times.

(a) 4 sec   **(b)** 8 sec

99. The floor plan for a house is a rectangle with length 7 m more than its width. The area is 170 m². Find the width and length of the house.

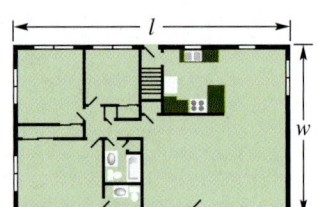

100. The triangular sail of a schooner has an area of 30 m². The height of the sail is 4 m more than the base. Find the length of the base of the sail.

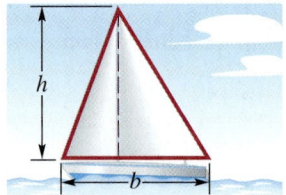

101. The numbers of alternative-fueled vehicles, in thousands, in use for the years 2001–2005 are given in the table.

Year	Number (in thousands)
2001	425
2002	471
2003	534
2004	565
2005	592

Source: Energy Information Administration.

Chevrolet Volt Concept

Using the data, we developed the quadratic equation

$$y = -5x^2 + 72.8x + 354$$

to model the number of vehicles y in year x. Here we used $x = 1$ for 2001, $x = 2$ for 2002, and so on.

(a) What estimate for 2006 is given by the equation?

(b) Why might the estimate for 2006 be unreliable?

1. Which one of the following is the correct, completely factored form of $2x^2 - 2x - 24$?

   A. $(2x + 6)(x - 4)$ **B.** $(x + 3)(2x - 8)$

   C. $2(x + 4)(x - 3)$ **D.** $2(x + 3)(x - 4)$

1. _____

Factor each polynomial completely.

2. $12x^2 - 30x$

2. _____

3. $2m^3n^2 + 3m^3n - 5m^2n^2$

3. _____

4. $2ax - 2bx + ay - by$

4. _____

5. $x^2 - 9x + 14$

5. _____

6. $2x^2 + x - 3$

6. _____

7. $6x^2 - 19x - 7$

7. _____

8. $3x^2 - 12x - 15$

8. _____

9. $10z^2 - 17z + 3$

9. _____

10. $t^2 + 6t + 10$

10. _____

11. $x^2 + \dfrac{1}{36}$

11. _____

12. $y^2 - 49$

12. _____

13. $81a^2 - 121b^2$

13. _____

14. $x^2 + 16x + 64$

14. _____

15. $4x^2 - 28xy + 49y^2$

15. _____

16. $-2x^2 - 4x - 2$

16. _____

17. $6t^4 + 3t^3 - 108t^2$

17. _____

18. $4r^2 + 10rt + 25t^2$

18. _____

19. _____

20. _____

21. _____

22. _____

23. _____

24. _____

25. _____

26. _____

27. _____

28. _____

29. _____

30. _____

19. $4t^3 + 32t^2 + 64t$ **20.** $x^4 - 81$

Solve each equation.

21. $(x + 3)(x - 9) = 0$ **22.** $2r^2 - 13r + 6 = 0$

23. $25x^2 - 4 = 0$ **24.** $x(x - 20) = -100$

25. $t^2 = 3t$ **26.** $(s + 8)(6s^2 + 13s - 5) = 0$

Solve each problem.

27. The length of a rectangular flower bed is 3 ft less than twice its width. The area of the bed is 54 ft². Find the dimensions of the flower bed.

28. Find two consecutive integers such that the square of the sum of the two integers is 11 more than the lesser integer.

29. A carpenter needs to cut a brace to support a wall stud, as shown in the figure. The brace should be 7 ft less than three times the length of the stud. If the brace will be anchored on the floor 15 ft away from the stud, how long should the brace be?

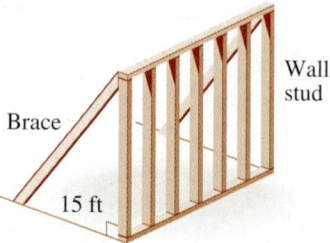

Brace
Wall stud
15 ft

30. The number of Americans using broadband Internet access at home from 2001 through 2006 can be approximated by the quadratic equation

$$y = 2.36x^2 + 4.81x + 13.81,$$

where $x = 0$ represents 2001, $x = 1$ represents 2002, and so on, and y is number in millions. (*Source:* Pew Internet Project, Nielsen/Net Ratings.) Use the model to estimate the number of Americans using broadband Internet at home in 2004. Round your answer to the nearest million.

15

Rational Expressions and Applications

In 2006, Earth's temperature was within 1.8°F of its highest level in 12,000 years. This temperature increase is causing ocean levels to rise as ice fields in Greenland and elsewhere melt. To demonstrate the effects that such global warming is having, British adventurer and endurance swimmer Lewis Gordon Pugh swam at the North Pole in July 2007 in waters that were completely frozen 10 years ago. The swim, in a water hole where polar ice had melted, was in 28.8°F waters, the coldest ever endured by a human. Pugh, who has also swum in the waters of Antarctica, hopes his efforts will inspire world leaders to take climate change seriously. (*Source:* www.breitbart.com, IPCC.)

In Exercise 11 of **Section 15.7,** we use a *rational expression* to determine the rate at which Pugh swam at the North Pole.

15.1 ▶▶▶ The Fundamental Property of Rational Expressions

The quotient of two integers (with denominator not 0), such as $\frac{2}{3}$ or $-\frac{3}{4}$, is called a *rational number*. In the same way, the quotient of two polynomials with denominator not equal to 0 is called a *rational expression*.

> **Rational Expression**
>
> A **rational expression** is an expression of the form $\frac{P}{Q}$, where P and Q are polynomials, with $Q \neq 0$.

Examples of rational expressions include

$$\frac{-6x}{x^3 + 8}, \quad \frac{9x}{y + 3}, \quad \text{and} \quad \frac{2m^3}{8}. \qquad \text{Rational expressions}$$

Our work with rational expressions will require much of what we learned in **Chapters 5 and 6** on polynomials and factoring, as well as the rules for fractions from **Chapter R.**

OBJECTIVE 1 **Find the values of the variable for which a rational expression is undefined.** In the definition of a rational expression $\frac{P}{Q}$, Q cannot equal 0. *The denominator of a rational expression cannot equal 0 because division by 0 is undefined.*

For instance, in the rational expression

$$\frac{8x^2}{x - 3}, \leftarrow \text{Denominator cannot equal 0.}$$

the variable x can take on any value except 3. When x is 3, the denominator becomes $3 - 3 = \mathbf{0}$, making the expression undefined. Thus, x cannot equal 3. We can indicate this restriction by writing $x \neq 3$.

To determine the values for which a rational expression is undefined, use the following procedure.

> **Determining When a Rational Expression Is Undefined**
>
> **Step 1** Set the denominator of the rational expression equal to 0.
>
> **Step 2** Solve this equation.
>
> **Step 3** The solutions of the equation are the values that make the rational expression undefined.

EXAMPLE 1 **Finding Values That Make Rational Expressions Undefined**

Find any values of the variable for which each rational expression is undefined.

(a) $\dfrac{p + 5}{3p + 2}$

Remember that the *numerator* may be any number; we must find any value of p that makes the *denominator* equal to 0 since division by 0 is undefined.

Continued on Next Page

Step 1 Set the denominator equal to 0.

$$3p + 2 = 0$$

Step 2 Solve this equation.

$$3p = -2$$

$$p = -\frac{2}{3}$$

Step 3 The given expression is undefined for $-\frac{2}{3}$, since substituting $-\frac{2}{3}$ for p makes the denominator 0. Thus, $p \neq -\frac{2}{3}$.

(b) $\dfrac{9m^2}{m^2 - 5m + 6}$

$$m^2 - 5m + 6 = 0 \qquad \text{Set the denominator equal to 0.}$$

$$(m - 2)(m - 3) = 0 \qquad \text{Factor.}$$

$$m - 2 = 0 \quad \text{or} \quad m - 3 = 0 \qquad \text{Zero-factor property}$$

$$m = 2 \quad \text{or} \qquad m = 3 \qquad \text{Solve.}$$

The original expression is undefined for 2 and 3, so $m \neq 2$ and $m \neq 3$.

(c) $\dfrac{2r}{r^2 + 1}$

This denominator will not equal 0 for any value of r because r^2 is always greater than or equal to 0, and adding 1 makes the sum greater than 0. Thus, there are no values for which this rational expression is undefined.

———————————— *Work Problem* ⬛1⬛ *at the Side.* ▶

⬛**OBJECTIVE**⬛ **2** **Find the numerical value of a rational expression.**
We use substitution to evaluate a rational expression for a given value of the variable.

┌─ **EXAMPLE 2** **Evaluating Rational Expressions**

Find the numerical value of $\dfrac{3x + 6}{2x - 4}$ for each value of x.

(a) $x = 1$

$$\frac{3x + 6}{2x - 4}$$

$$= \frac{3(1) + 6}{2(1) - 4} \qquad \text{Let } x = 1.$$

$$= \frac{9}{-2}$$

$$= -\frac{9}{2} \qquad \frac{a}{-b} = -\frac{a}{b}$$

(b) $x = 0$

$$\frac{3x + 6}{2x - 4}$$

$$= \frac{3(0) + 6}{2(0) - 4} \qquad \text{Let } x = 0.$$

$$= \frac{6}{-4}$$

$$= -\frac{3}{2} \qquad \text{Lowest terms}$$

Continued on Next Page

---

1 Find any values of the variable for which each rational expression is undefined. Write answers with $\neq$.

(a) $\dfrac{x + 2}{x - 5}$

(b) $\dfrac{3r}{r^2 + 6r + 8}$

(c) $\dfrac{-5m}{m^2 + 4}$

2 Find the numerical value of each rational expression when $x = -3$, $x = 0$, and $x = 3$.

(a) $\dfrac{x}{2x + 1}$

(c) $x = 2$

$$\dfrac{3x + 6}{2x - 4}$$

$$= \dfrac{3(2) + 6}{2(2) - 4} \qquad \text{Let } x = 2.$$

$$= \dfrac{12}{\mathbf{0}}$$

Substituting 2 for x makes the denominator 0, so the expression is undefined when $x = 2$.

(d) $x = -2$

$$\dfrac{3x + 6}{2x - 4}$$

$$= \dfrac{3(-2) + 6}{2(-2) - 4} \qquad \text{Let } x = -2.$$

$$= \dfrac{0}{-8}$$

$$= 0 \qquad \qquad \dfrac{0}{b} = 0$$

Note

The numerator of a rational expression may be any real number. If the numerator equals 0 and the denominator does *not* equal 0, then the rational expression equals 0. See Example 2(d).

◀ Work Problem **2** at the Side.

OBJECTIVE 3 Write rational expressions in lowest terms. A fraction such as $\frac{2}{3}$ is said to be in *lowest terms*. How can "lowest terms" be defined? We use the idea of greatest common factor for this definition, which applies to all rational expressions.

(b) $\dfrac{2x + 6}{x - 3}$

Lowest Terms

A rational expression $\frac{P}{Q}$ $(Q \neq 0)$ is in **lowest terms** if the greatest common factor of its numerator and denominator is 1.

The properties of rational numbers also apply to rational expressions. We use the **fundamental property of rational expressions** to write a rational expression in lowest terms.

Fundamental Property of Rational Expressions

If $\frac{P}{Q}$ $(Q \neq 0)$ is a rational expression and if K represents any polynomial, where $K \neq 0$, then

$$\dfrac{PK}{QK} = \dfrac{P}{Q}.$$

This property is based on the identity property of multiplication, which says that multiplying by 1 leaves any number unchanged.

$$\dfrac{PK}{QK} = \dfrac{P}{Q} \cdot \dfrac{K}{K} = \dfrac{P}{Q} \cdot 1 = \dfrac{P}{Q}$$

The next example shows how to write both a rational number and a rational expression in lowest terms. Notice the similarity in the procedures. In both cases, we factor and then divide out the greatest common factor.

ANSWERS

2. (a) $\dfrac{3}{5}$; 0; $\dfrac{3}{7}$ **(b)** 0; -2; undefined

---

EXAMPLE 3 **Writing in Lowest Terms**

Write each expression in lowest terms.

(a) $\dfrac{30}{72}$

Begin by factoring.

$$= \frac{2 \cdot 3 \cdot 5}{2 \cdot 2 \cdot 2 \cdot 3 \cdot 3}$$

(b) $\dfrac{14k^2}{2k^3}$

Write k^2 as $k \cdot k$ and k^3 as $k \cdot k \cdot k$.

$$= \frac{2 \cdot 7 \cdot k \cdot k}{2 \cdot k \cdot k \cdot k}$$

Group any factors common to the numerator and denominator.

$$= \frac{5 \cdot (2 \cdot 3)}{2 \cdot 2 \cdot 3 \cdot (2 \cdot 3)}$$

$$= \frac{7 \, (2 \cdot k \cdot k)}{k \, (2 \cdot k \cdot k)}$$

Use the fundamental property.

$$= \frac{5}{2 \cdot 2 \cdot 3}, \quad \text{or} \quad \frac{5}{12}$$

$$= \frac{7}{k}$$

Work Problem **3** *at the Side.* ▶

Writing a Rational Expression in Lowest Terms

Step 1 **Factor** the numerator and denominator completely.

Step 2 **Use the fundamental property** to divide out any common factors.

EXAMPLE 4 **Writing in Lowest Terms**

Write each rational expression in lowest terms.

(a) $\dfrac{3x - 12}{5x - 20}$

$$= \frac{3 \, (x - 4)}{5 \, (x - 4)} \qquad \text{Factor. (Step 1)}$$

$$= \frac{3}{5} \qquad \text{Fundamental property (Step 2)}$$

The given expression is equal to $\frac{3}{5}$ for all values of x, where $x \neq 4$ (since the denominator of the original rational expression is 0 when x is 4).

(b) $\dfrac{m^2 + 2m - 8}{2m^2 - m - 6}$

$$= \frac{(m + 4) \, (m - 2)}{(2m + 3) \, (m - 2)} \qquad \text{Factor. (Step 1)}$$

$$= \frac{m + 4}{2m + 3} \qquad \text{Fundamental property (Step 2)}$$

Here $m \neq -\frac{3}{2}$ and $m \neq 2$, since the denominator of the original expression is 0 for these values of m.

---

3 Write each expression in lowest terms.

(a) $\dfrac{15}{40}$

(b) $\dfrac{5x^4}{15x^2}$

(c) $\dfrac{6p^3}{2p^2}$

---

4 Write each rational
expression in lowest terms.

(a) $\dfrac{4y + 2}{6y + 3}$

(b) $\dfrac{8p + 8q}{5p + 5q}$

(c) $\dfrac{x^2 + 4x + 4}{4x + 8}$

(d) $\dfrac{a^2 - b^2}{a^2 + 2ab + b^2}$

From now on, we write statements of equality of rational expressions with the understanding that they apply only to those real numbers that make neither denominator equal to 0.

> **CAUTION**
> *Rational expressions cannot be written in lowest terms until after the numerator and denominator have been factored. Only common factors can be divided out, not common terms.* For example,
>
> $$\frac{6x + 9}{4x + 6} = \frac{3\,(2x + 3)}{2\,(2x + 3)} = \frac{3}{2} \qquad\Big|\qquad \frac{6 + x}{4x} \;\leftarrow\; \text{Numerator cannot be factored.}$$
>
> ↑ Divide out the common factor.   This expression is already in lowest terms.

◀ Work Problem **4** at the Side.

EXAMPLE 5   **Writing in Lowest Terms (Factors Are Opposites)**

Write $\dfrac{x - y}{y - x}$ in lowest terms.

At first glance, there might not seem to be any way in which $x - y$ and $y - x$ can be factored to get a common factor. However, $y - x$ can be factored as

$$y - x$$

> Be careful with signs.

$$= -1\,(-y + x) \qquad \text{Factor out } -1.$$

$$= -1\,(x - y) \qquad \text{Commutative property}$$

With this result in mind, we simplify as follows.

$$\frac{x - y}{y - x}$$

$$= \frac{1\,(x - y)}{-1\,(x - y)} \qquad y - x = -1\,(x - y) \text{ from above.}$$

$$= \frac{1}{-1}, \quad \text{or} \quad -1 \qquad \text{Fundamental property}$$

◀ Work Problem **5** at the Side.

5 Write $\dfrac{x - y}{y - x}$ from Example 5 in lowest terms by factoring -1 from the numerator. How does the result compare to the result in Example 5?

In Example 5, notice that $y - x$ is the **opposite** (or **additive inverse**) of $x - y$. A general rule for this situation follows.

> If the numerator and the denominator of a rational expression are opposites, such as in $\dfrac{x - y}{y - x}$, then the rational expression is equal to -1.

> **CAUTION**
> Although x and y appear in both the numerator and denominator in Example 5, we cannot use the fundamental property right away because they are *terms*, not *factors*. **Terms are added, while factors are multiplied.**

ANSWERS

4. (a) $\dfrac{2}{3}$ (b) $\dfrac{8}{5}$ (c) $\dfrac{x + 2}{4}$ (d) $\dfrac{a - b}{a + b}$

5. The result is -1, the same as in Example 5.

EXAMPLE 6 **Writing in Lowest Terms (Factors Are Opposites)**

Write each rational expression in lowest terms.

(a) $\dfrac{2 - m}{m - 2}$

Since $2 - m$ and $m - 2$ (or $-2 + m$) are opposites, this expression equals -1.

(b) $\dfrac{4x^2 - 9}{6 - 4x}$

$$= \dfrac{(2x + 3)(2x - 3)}{2(3 - 2x)} \qquad \text{Factor the numerator and denominator.}$$

$$= \dfrac{(2x + 3)(2x - 3)}{2(-1)(2x - 3)} \qquad \text{Write } 3 - 2x \text{ as } -1(2x - 3).$$

$$= \dfrac{2x + 3}{2(-1)} \qquad \text{Fundamental property}$$

$$= \dfrac{2x + 3}{-2}, \quad \text{or} \quad -\dfrac{2x + 3}{2} \qquad \dfrac{a}{-b} = -\dfrac{a}{b}$$

(c) $\dfrac{3 + r}{3 - r}$

The quantity $3 - r$ *is not* the opposite of $3 + r$. This rational expression is already in lowest terms.

Work Problem **6** *at the Side.* ▶

OBJECTIVE **4** **Recognize equivalent forms of rational expressions.** When working with rational expressions, it is important to be able to recognize equivalent forms of an expression. For example, the common fraction $-\frac{5}{6}$ can also be written as $\frac{-5}{6}$ and as $\frac{5}{-6}$.

Consider the final rational expression from Example 6(b),

$$-\dfrac{2x + 3}{2}.$$

The $-$ sign representing the factor -1 is in front of the expression, even with the fraction bar. As with the common fraction $-\frac{5}{6}$, the factor -1 may instead be placed in the numerator or in the denominator. Some other equivalent forms of this rational expression are

Use parentheses.

$$\dfrac{-(2x + 3)}{2} \quad \text{and} \quad \dfrac{2x + 3}{-2}.$$

> Multiply *each* term in the parentheses by -1.

By the distributive property,

$$\dfrac{-(2x + 3)}{2} \quad \text{can also be written} \quad \dfrac{-2x - 3}{2}.$$

> **CAUTION**
> $\frac{-2x + 3}{2}$ is *not* an equivalent form of $\frac{-(2x + 3)}{2}$. The sign preceding 3 in the numerator of $\frac{-2x + 3}{2}$ should be $-$ rather than $+$. ***Be careful to apply the distributive property correctly.***

6 Write each rational expression in lowest terms.

(a) $\dfrac{5 - y}{y - 5}$

(b) $\dfrac{m - n}{n - m}$

(c) $\dfrac{25x^2 - 16}{12 - 15x}$

(d) $\dfrac{9 - k}{9 + k}$

7 Decide whether each rational expression is equivalent to

$$-\frac{2x - 6}{x + 3}.$$

(a) $\dfrac{-(2x - 6)}{x + 3}$

(b) $\dfrac{-2x + 6}{x + 3}$

(c) $\dfrac{-2x - 6}{x + 3}$

(d) $\dfrac{2x - 6}{-(x + 3)}$

(e) $\dfrac{2x - 6}{-x - 3}$

(f) $\dfrac{2x - 6}{x - 3}$

ΕXAMPLE 7   **Writing Equivalent Forms of a Rational Expression**

Write four equivalent forms of the rational expression

$$-\frac{3x + 2}{x - 6}.$$

If we apply the negative sign to the numerator, we obtain the equivalent forms

$$\frac{-(3x + 2)}{x - 6}, \quad \text{and, by the distributive property,} \quad \frac{-3x - 2}{x - 6}.$$

If we apply the negative sign to the denominator of the given rational expression, we obtain the equivalent forms

$$\frac{3x + 2}{-(x - 6)} \quad \text{and, by distributing again,} \quad \frac{3x + 2}{-x + 6}.$$

CAUTION

Recall that $-\frac{5}{6} \neq \frac{-5}{-6}$. Thus, in Example 7, it would be incorrect to distribute the negative sign in $-\frac{3x + 2}{x - 6}$ to *both* the numerator *and* the denominator. (Doing this would actually lead to the *opposite* of the original expression.)

◀ *Work Problem* **7** *at the Side.*

15.1 ▶▶▶ Exercises

1. Fill in each blank with the correct response.

(a) The rational expression $\dfrac{x + 5}{x - 3}$ is undefined when $x =$ _____, and is equal to 0 when $x =$ _____.

(b) The rational expression $\dfrac{p - q}{q - p}$ is undefined when $p =$ _____, and in all other cases when written in lowest terms is equal to _____.

2. Make the correct choice for each blank.

(a) $\dfrac{4 - r^2}{4 + r^2}$ _____(is/is not) equal to -1.

(b) $\dfrac{5 + 2x}{3 - x}$ and $\dfrac{-5 - 2x}{x - 3}$ _____(are/are not) equivalent rational expressions.

3. Define *rational expression* in your own words, and give an example.

4. Why can't the denominator of a rational expression equal 0?

Find any value(s) of the variable for which each rational expression is undefined. Write answers with ≠. See Example 1.

5. $\dfrac{2}{5y}$

6. $\dfrac{7}{3z}$

7. $\dfrac{x + 1}{x + 6}$

8. $\dfrac{m + 2}{m + 5}$

9. $\dfrac{4x^2}{3x - 5}$

10. $\dfrac{2x^3}{3x - 4}$

11. $\dfrac{m + 2}{m^2 + m - 6}$

12. $\dfrac{r - 5}{r^2 - 5r + 4}$

13. $\dfrac{x^2 - 3x}{4}$

14. $\dfrac{x^2 - 4x}{6}$

15. $\dfrac{3x}{x^2 + 2}$

16. $\dfrac{4q}{q^2 + 9}$

Find the numerical value of each rational expression when (a) $x = 2$ and (b) $x = -3$. See Example 2.

17. $\dfrac{5x - 2}{4x}$

18. $\dfrac{3x + 1}{5x}$

19. $\dfrac{x^2 - 4}{2x + 1}$

20. $\dfrac{2x^2 - 4x}{3x - 1}$

21. $\dfrac{(-3x)^2}{4x + 12}$

22. $\dfrac{(-2x)^3}{3x + 9}$

23. $\dfrac{5x + 2}{2x^2 + 11x + 12}$

24. $\dfrac{7 - 3x}{3x^2 - 7x + 2}$

Write each rational expression in lowest terms. See Examples 3 and 4.

25. $\dfrac{18r^3}{6r}$

26. $\dfrac{27p^2}{3p}$

27. $\dfrac{4(y - 2)}{10(y - 2)}$

28. $\dfrac{15(m-1)}{9(m-1)}$

29. $\dfrac{(x+1)(x-1)}{(x+1)^2}$

30. $\dfrac{(t+5)(t-3)}{(t-1)(t+5)}$

31. $\dfrac{7m+14}{5m+10}$

32. $\dfrac{8z-24}{4z-12}$

33. $\dfrac{m^2-n^2}{m+n}$

34. $\dfrac{a^2-b^2}{a-b}$

35. $\dfrac{12m^2-3}{8m-4}$

36. $\dfrac{20p^2-45}{6p-9}$

37. $\dfrac{3m^2-3m}{5m-5}$

38. $\dfrac{6t^2-6t}{2t-2}$

39. $\dfrac{9r^2-4s^2}{9r+6s}$

40. $\dfrac{16x^2-9y^2}{12x-9y}$

41. $\dfrac{2x^2-3x-5}{2x^2-7x+5}$

42. $\dfrac{3x^2+8x+4}{3x^2-4x-4}$

43. $\dfrac{zw+4z-3w-12}{zw+4z+5w+20}$

44. $\dfrac{km+4k+4m+16}{km+4k+5m+20}$

45. $\dfrac{ac-ad+bc-bd}{ac-ad-bc+bd}$

46. Which rational expression can be simplified?

A. $\dfrac{x^2+2}{x^2}$ **B.** $\dfrac{x^2+2}{2}$ **C.** $\dfrac{x^2+y^2}{y^2}$ **D.** $\dfrac{x^2-5x}{x}$

Write each rational expression in lowest terms. See Examples 5 and 6.

47. $\dfrac{6-t}{t-6}$

48. $\dfrac{2-k}{k-2}$

49. $\dfrac{m^2-1}{1-m}$

50. $\dfrac{a^2-b^2}{b-a}$

51. $\dfrac{q^2-4q}{4q-q^2}$

52. $\dfrac{z^2-5z}{5z-z^2}$

Write four equivalent forms for each rational expression. See Example 7.

53. $-\dfrac{x+4}{x-3}$

54. $-\dfrac{x+6}{x-1}$

55. $-\dfrac{2x-3}{x+3}$

56. $-\dfrac{5x-6}{x+4}$

57. $-\dfrac{3x-1}{5x-6}$

58. $-\dfrac{2x-9}{7x-1}$

59. The area of the rectangle is represented by x^4+10x^2+21. What is the width?
$\left(\textit{Hint: Use } w=\frac{A}{l}.\right)$

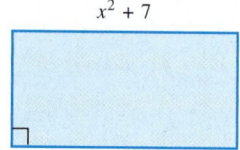

x^2+7

60. The volume of the box is represented by
$$(x^2+8x+15)(x+4).$$
Find the polynomial that represents the area of the bottom of the box.

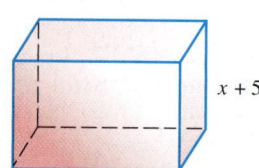

$x+5$

15.2 ▶▶▶ Multiplying and Dividing Rational Expressions

OBJECTIVE 1 Multiply rational expressions. The product of two fractions is found by multiplying the numerators and multiplying the denominators. Rational expressions are multiplied in the same way.

> **Multiplying Rational Expressions**
> The product of the rational expressions $\frac{P}{Q}$ and $\frac{R}{S}$ is shown below.
> $$\frac{P}{Q} \cdot \frac{R}{S} = \frac{PR}{QS}$$
> In words: To multiply rational expressions, multiply the numerators and multiply the denominators.

OBJECTIVES

1 Multiply rational expressions.

2 Find reciprocals.

3 Divide rational expressions.

EXAMPLE 1   Multiplying Rational Expressions

Multiply. Write each answer in lowest terms.

(a) $\dfrac{3}{10} \cdot \dfrac{5}{9}$

(b) $\dfrac{6}{x} \cdot \dfrac{x^2}{12}$

Indicate the product of the numerators and the product of the denominators.

$= \dfrac{3 \cdot 5}{10 \cdot 9}$ $\quad$ $= \dfrac{6 \cdot x^2}{x \cdot 12}$

Leave the products in factored form because common factors are needed to write the product in lowest terms. Factor the numerator and denominator to further identify any common factors. Then use the fundamental property to write each product in lowest terms.

$= \dfrac{3 \cdot 5}{2 \cdot 5 \cdot 3 \cdot 3}$ $\quad$ $= \dfrac{6 \cdot x \cdot x}{2 \cdot 6 \cdot x}$

$= \dfrac{1}{6}$ $\quad$ $= \dfrac{x}{2}$

Work Problem **1** *at the Side.* ▶

EXAMPLE 2   Multiplying Rational Expressions

Multiply. Write the answer in lowest terms.

$\dfrac{x + y}{2x} \cdot \dfrac{x^2}{(x + y)^2}$

$= \dfrac{(x + y)x^2}{2x(x + y)^2}$ Multiply numerators.
   Multiply denominators.

$= \dfrac{(x + y)\, x \cdot x}{2x(x + y)(x + y)}$ Factor; identify common factors.

$= \dfrac{x}{2(x + y)}$ $\dfrac{(x + y)x}{x(x + y)} = 1$; lowest terms

Work Problem **2** *at the Side.* ▶

---

1 Multiply. Write each answer in lowest terms.

(a) $\dfrac{2}{7} \cdot \dfrac{5}{10}$

(b) $\dfrac{3m^2}{2} \cdot \dfrac{10}{m}$

(c) $\dfrac{8p^2q}{3} \cdot \dfrac{9}{q^2p}$

2 Multiply. Write each answer in lowest terms.

(a) $\dfrac{a + b}{5} \cdot \dfrac{30}{2(a + b)}$

(b) $\dfrac{3(p - q)}{q^2} \cdot \dfrac{q}{2(p - q)^2}$

ANSWERS

1. (a) $\dfrac{1}{7}$ (b) $15m$ (c) $\dfrac{24p}{q}$

2. (a) 3 (b) $\dfrac{3}{2q(p - q)}$

3 Multiply. Write each answer in lowest terms.

(a)

$$\frac{x^2 + 7x + 10}{3x + 6} \cdot \frac{6x - 6}{x^2 + 2x - 15}$$

(b)

$$\frac{m^2 + 4m - 5}{m + 5} \cdot \frac{m^2 + 8m + 15}{m - 1}$$

4 Find the reciprocal of each rational expression.

(a) $\dfrac{5}{8}$

(b) $\dfrac{6b^5}{3r^2b}$

(c) $\dfrac{t^2 - 4t}{t^2 + 2t - 3}$

EXAMPLE 3 **Multiplying Rational Expressions**

Multiply. Write the answer in lowest terms.

$$\frac{x^2 + 3x}{x^2 - 3x - 4} \cdot \frac{x^2 - 5x + 4}{x^2 + 2x - 3}$$

$$= \frac{(x^2 + 3x)(x^2 - 5x + 4)}{(x^2 - 3x - 4)(x^2 + 2x - 3)} \qquad \text{\color{blue}Definition of multiplication}$$

$$= \frac{x(x + 3)(x - 4)(x - 1)}{(x - 4)(x + 1)(x + 3)(x - 1)} \qquad \text{\color{blue}Factor.}$$

$$= \frac{x}{x + 1} \qquad \text{\color{blue}Divide out the common factors.}$$

The quotients $\frac{x + 3}{x + 3}, \frac{x - 4}{x - 4},$ and $\frac{x - 1}{x - 1}$ are all equal to 1, justifying the final product $\frac{x}{x + 1}$.

◀ *Work Problem* **3** *at the Side.*

OBJECTIVE **2** **Find reciprocals.**   If the product of two rational expressions is 1, the rational expressions are called **reciprocals** (or **multiplicative inverses**) of each other. The reciprocal of a rational expression is found by interchanging the numerator and the denominator. For example,

$$\frac{2x - 1}{x - 5} \quad \text{has reciprocal} \quad \frac{x - 5}{2x - 1}.$$

EXAMPLE 4 **Finding Reciprocals of Rational Expressions**

Find the reciprocal of each rational expression.

(a) $\dfrac{4p^3}{9q}$ has reciprocal $\dfrac{9q}{4p^3}.$ {\color{blue}Interchange the numerator and denominator.}

(b) $\dfrac{k^2 - 9}{k^2 - k - 20}$ has reciprocal $\dfrac{k^2 - k - 20}{k^2 - 9}.$

◀ *Work Problem* **4** *at the Side.*

OBJECTIVE **3** **Divide rational expressions.**   Suppose we have $\frac{7}{8}$ gal of milk and want to find how many quarts we have. Since 1 qt is $\frac{1}{4}$ gal, we ask, "How many $\frac{1}{4}$s are there in $\frac{7}{8}$?" This would be interpreted as

$$\frac{7}{8} \div \frac{1}{4}, \quad \text{or} \quad \frac{\dfrac{7}{8}}{\dfrac{1}{4}} \quad \text{\color{blue}← The fraction bar means division.}$$

The fundamental property of rational expressions discussed earlier can be applied to rational number values of P, Q, and K.

$$\frac{P}{Q} = \frac{P \cdot K}{Q \cdot K} = \frac{\dfrac{7}{8} \cdot 4}{\dfrac{1}{4} \cdot 4} = \frac{\dfrac{7}{8} \cdot 4}{1} = \frac{7}{8} \cdot \frac{4}{1}. \qquad \text{\color{blue}Let } P = \frac{7}{8}, Q = \frac{1}{4}, \text{ and } K = 4 \text{ (the reciprocal of } Q\text{).}$$

So, to divide $\frac{7}{8}$ by $\frac{1}{4}$, we multiply $\frac{7}{8}$ by the reciprocal of $\frac{1}{4}$, namely $\frac{4}{1}$, or 4. Since $\frac{7}{8}(4) = \frac{7}{2}$, there are $\frac{7}{2}$ qt, or $3\frac{1}{2}$ qt, in $\frac{7}{8}$ gal.

ANSWERS

3. **(a)** $\dfrac{2(x - 1)}{x - 3}$ **(b)** $(m + 5)(m + 3)$

4. **(a)** $\dfrac{8}{5}$ **(b)** $\dfrac{3r^2b}{6b^5}$ **(c)** $\dfrac{t^2 + 2t - 3}{t^2 - 4t}$

The preceding discussion illustrates dividing common fractions. Division of rational expressions is defined in the same way.

> ### Dividing Rational Expressions
> If $\frac{P}{Q}$ and $\frac{R}{S}$ are any two rational expressions, with $\frac{R}{S} \neq 0$, then
> $$\frac{P}{Q} \div \frac{R}{S} = \frac{P}{Q} \cdot \frac{S}{R} = \frac{PS}{QR}$$
> In words: To divide one rational expression by another rational expression, multiply the first rational expression (dividend) by the reciprocal of the second rational expression (divisor).

The next example shows the division of two rational numbers and the division of two rational expressions.

EXAMPLE 5   Dividing Rational Expressions

Divide. Write each answer in lowest terms.

(a) $\dfrac{5}{8} \div \dfrac{7}{16}$ **(b)** $\dfrac{y}{y+3} \div \dfrac{4y}{y+5}$

Multiply the first expression by the reciprocal of the second.

$= \dfrac{5}{8} \cdot \dfrac{16}{7}$ ← Reciprocal of $\frac{7}{16}$

$= \dfrac{5 \cdot 16}{8 \cdot 7}$

$= \dfrac{5 \cdot \mathbf{8} \cdot 2}{\mathbf{8} \cdot 7}$

$= \dfrac{10}{7}$

$= \dfrac{y}{y+3} \cdot \dfrac{y+5}{4y}$ ← Reciprocal of $\frac{4y}{y+5}$

$= \dfrac{y(y+5)}{(y+3)(4y)}$

$= \dfrac{y+5}{4(y+3)}$

Work Problem **5** *at the Side.* ▶

EXAMPLE 6   Dividing Rational Expressions

Divide. Write the answer in lowest terms.

$$\frac{(3m)^2}{(2p)^3} \div \frac{6m^3}{16p^2}$$

$(3m)^2 = 3^2 m^2;$
$(2p)^3 = 2^3 p^3$

$= \dfrac{(3m)^2}{(2p)^3} \cdot \dfrac{16p^2}{6m^3}$ Multiply by the reciprocal.

$= \dfrac{9m^2}{8p^3} \cdot \dfrac{16p^2}{6m^3}$ Power rule for exponents

$= \dfrac{9 \cdot 16m^2 p^2}{8 \cdot 6p^3 m^3}$ Multiply numerators.
                Multiply denominators.

$= \dfrac{3}{mp}$ Lowest terms

Work Problem **6** *at the Side.* ▶

---

5 Divide. Write each answer in lowest terms.

(a) $\dfrac{3}{4} \div \dfrac{5}{16}$

(b) $\dfrac{r}{r-1} \div \dfrac{3r}{r+4}$

(c) $\dfrac{6x-4}{3} \div \dfrac{15x-10}{9}$

6 Divide. Write each answer in lowest terms.

(a) $\dfrac{5a^2 b}{2} \div \dfrac{10ab^2}{8}$

(b) $\dfrac{(3t)^2}{w} \div \dfrac{3t^2}{5w^4}$

ANSWERS

5. (a) $\dfrac{12}{5}$ **(b)** $\dfrac{r+4}{3(r-1)}$ **(c)** $\dfrac{6}{5}$

6. (a) $\dfrac{2a}{b}$ **(b)** $15w^3$

7 Divide. Write each answer in lowest terms.

(a)

$$\frac{y^2 + 4y + 3}{y + 3} \div \frac{y^2 - 4y - 5}{y - 3}$$

(b) $\dfrac{4x(x + 3)}{2x + 1} \div \dfrac{-x^2(x + 3)}{4x^2 - 1}$

8 Divide. Write each answer in lowest terms.

(a) $\dfrac{ab - a^2}{a^2 - 1} \div \dfrac{a - b}{a - 1}$

(b) $\dfrac{x^2 - 9}{2x + 6} \div \dfrac{9 - x^2}{4x - 12}$

EXAMPLE 7 **Dividing Rational Expressions**

Divide. Write the answer in lowest terms.

$$\frac{x^2 - 4}{(x + 3)(x - 2)} \div \frac{(x + 2)(x + 3)}{-2x}$$

$$= \frac{x^2 - 4}{(x + 3)(x - 2)} \cdot \frac{-2x}{(x + 2)(x + 3)} \qquad \text{Multiply by the reciprocal.}$$

$$= \frac{-2x(x^2 - 4)}{(x + 3)(x - 2)(x + 2)(x + 3)} \qquad \begin{array}{l}\text{Multiply numerators.}\\ \text{Multiply denominators.}\end{array}$$

$$= \frac{-2x(x + 2)(x - 2)}{(x + 3)(x - 2)(x + 2)(x + 3)} \qquad \text{Factor the numerator.}$$

$$= \frac{-2x}{(x + 3)^2}, \quad \text{or} \quad -\frac{2x}{(x + 3)^2} \qquad \text{Lowest terms; } \frac{-a}{b} = -\frac{a}{b}$$

◀ *Work Problem* **7** *at the Side.*

EXAMPLE 8 **Dividing Rational Expressions (Factors Are Opposites)**

Divide. Write the answer in lowest terms.

$$\frac{m^2 - 4}{m^2 - 1} \div \frac{2m^2 + 4m}{1 - m}$$

$$= \frac{m^2 - 4}{m^2 - 1} \cdot \frac{1 - m}{2m^2 + 4m} \qquad \text{Multiply by the reciprocal.}$$

$$= \frac{(m^2 - 4)(1 - m)}{(m^2 - 1)(2m^2 + 4m)} \qquad \begin{array}{l}\text{Multiply numerators.}\\ \text{Multiply denominators.}\end{array}$$

$$= \frac{(m + 2)(m - 2)(1 - m)}{(m + 1)(m - 1)(2m)(m + 2)} \qquad \begin{array}{l}\text{Factor; } 1 - m \text{ and } m - 1 \text{ are}\\ \text{opposites.}\end{array}$$

$$= \frac{-1(m - 2)}{2m(m + 1)} \qquad \text{From Section 15.1, } \frac{1 - m}{m - 1} = -1.$$

$$= \frac{-m + 2}{2m(m + 1)}, \quad \text{or} \quad \frac{2 - m}{2m(m + 1)} \qquad \text{Distribute } -1 \text{ in the numerator.}$$

◀ *Work Problem* **8** *at the Side.*

In summary, follow these steps to multiply or divide rational expressions.

Multiplying or Dividing Rational Expressions

Step 1 **Note the operation.** If the operation is division, use the definition of division to rewrite as multiplication.

Step 2 **Multiply** numerators and multiply denominators.

Step 3 **Factor** all numerators and denominators completely.

Step 4 **Write in lowest terms** using the fundamental property.

Steps 2 and 3 may be interchanged based on personal preference.

ANSWERS

7. **(a)** $\dfrac{y - 3}{y - 5}$ **(b)** $-\dfrac{4(2x - 1)}{x}$

8. **(a)** $\dfrac{-a}{a + 1}$ **(b)** $\dfrac{-2x + 6}{3 + x}$

15.2 ▶▶▶ Exercises

1. Match each multiplication problem in Column I with the correct product in Column II.

	I		II
(a)	$\dfrac{5x^3}{10x^4} \cdot \dfrac{10x^7}{2x}$	**A.**	$\dfrac{2}{5x^5}$
(b)	$\dfrac{10x^4}{5x^3} \cdot \dfrac{10x^7}{2x}$	**B.**	$\dfrac{5x^5}{2}$
(c)	$\dfrac{5x^3}{10x^4} \cdot \dfrac{2x}{10x^7}$	**C.**	$\dfrac{1}{10x^7}$
(d)	$\dfrac{10x^4}{5x^3} \cdot \dfrac{2x}{10x^7}$	**D.**	$10x^7$

2. Match each division problem in Column I with the correct quotient in Column II.

	I		II
(a)	$\dfrac{5x^3}{10x^4} \div \dfrac{10x^7}{2x}$	**A.**	$\dfrac{5x^5}{2}$
(b)	$\dfrac{10x^4}{5x^3} \div \dfrac{10x^7}{2x}$	**B.**	$10x^7$
(c)	$\dfrac{5x^3}{10x^4} \div \dfrac{2x}{10x^7}$	**C.**	$\dfrac{2}{5x^5}$
(d)	$\dfrac{10x^4}{5x^3} \div \dfrac{2x}{10x^7}$	**D.**	$\dfrac{1}{10x^7}$

Multiply. Write each answer in lowest terms. See Examples 1 and 2.

3. $\dfrac{10m^2}{7} \cdot \dfrac{14}{15m}$

4. $\dfrac{36z^3}{6z} \cdot \dfrac{28}{z^2}$

5. $\dfrac{16y^4}{18y^5} \cdot \dfrac{15y^5}{y^2}$

6. $\dfrac{20x^5}{-2x^2} \cdot \dfrac{8x^4}{35x^3}$

7. $\dfrac{2(c+d)}{3} \cdot \dfrac{18}{6(c+d)^2}$

8. $\dfrac{4(y-2)}{x} \cdot \dfrac{3x}{6(y-2)^2}$

Find the reciprocal of each rational expression. See Example 4.

9. $\dfrac{3p^3}{16q}$

10. $\dfrac{6x^4}{9y^2}$

11. $\dfrac{r^2+rp}{7}$

12. $\dfrac{16}{9a^2+36a}$

13. $\dfrac{z^2+7z+12}{z^2-9}$

14. $\dfrac{p^2-4p+3}{p^2-3p}$

Divide. Write each answer in lowest terms. See Examples 5 and 6.

15. $\dfrac{9z^4}{3z^5} \div \dfrac{3z^2}{5z^3}$

16. $\dfrac{35q^8}{9q^5} \div \dfrac{25q^6}{10q^5}$

17. $\dfrac{4t^4}{2t^5} \div \dfrac{(2t)^3}{-6}$

18. $\dfrac{-12a^6}{3a^2} \div \dfrac{(2a)^3}{27a}$

19. $\dfrac{3}{2y-6} \div \dfrac{6}{y-3}$

20. $\dfrac{4m+16}{10} \div \dfrac{3m+12}{18}$

21. Explain in your own words how to multiply rational expressions.

22. Explain in your own words how to divide rational expressions.

Multiply or divide. Write each answer in lowest terms. See Examples 3, 7, and 8.

23. $\dfrac{5x - 15}{3x + 9} \cdot \dfrac{4x + 12}{6x - 18}$

24. $\dfrac{8r + 16}{24r - 24} \cdot \dfrac{6r - 6}{3r + 6}$

25. $\dfrac{2 - t}{8} \div \dfrac{t - 2}{6}$

26. $\dfrac{4}{m - 2} \div \dfrac{16}{2 - m}$

27. $\dfrac{5 - 4x}{5 + 4x} \cdot \dfrac{4x + 5}{4x - 5}$

28. $\dfrac{5 - x}{5 + x} \cdot \dfrac{x + 5}{x - 5}$

29. $\dfrac{6(m - 2)^2}{5(m + 4)^2} \cdot \dfrac{15(m + 4)}{2(2 - m)}$

30. $\dfrac{7(q - 1)}{3(q + 1)^2} \cdot \dfrac{6(q + 1)}{3(1 - q)^2}$

31. $\dfrac{p^2 + 4p - 5}{p^2 + 7p + 10} \div \dfrac{p - 1}{p + 4}$

32. $\dfrac{z^2 - 3z + 2}{z^2 + 4z + 3} \div \dfrac{z - 1}{z + 1}$

33. $\dfrac{2k^2 - k - 1}{2k^2 + 5k + 3} \div \dfrac{4k^2 - 1}{2k^2 + k - 3}$

34. $\dfrac{2m^2 - 5m - 12}{m^2 + m - 20} \div \dfrac{4m^2 - 9}{m^2 + 4m - 5}$

35. $\dfrac{2k^2 + 3k - 2}{6k^2 - 7k + 2} \cdot \dfrac{4k^2 - 5k + 1}{k^2 + k - 2}$

36. $\dfrac{2m^2 - 5m - 12}{m^2 - 10m + 24} \div \dfrac{4m^2 - 9}{m^2 - 9m + 18}$

37. $\dfrac{m^2 + 2mp - 3p^2}{m^2 - 3mp + 2p^2} \div \dfrac{m^2 + 4mp + 3p^2}{m^2 + 2mp - 8p^2}$

38. $\dfrac{r^2 + rs - 12s^2}{r^2 - rs - 20s^2} \div \dfrac{r^2 - 2rs - 3s^2}{r^2 + rs - 30s^2}$

39. $\left(\dfrac{x^2 + 10x + 25}{x^2 + 10x} \cdot \dfrac{10x}{x^2 + 15x + 50} \right) \div \dfrac{x + 5}{x + 10}$

40. $\left(\dfrac{m^2 - 12m + 32}{8m} \cdot \dfrac{m^2 - 8m}{m^2 - 8m + 16} \right) \div \dfrac{m - 8}{m - 4}$

41. If the rational expression $\dfrac{5x^2 y^3}{2pq}$ represents the area of a rectangle and $\dfrac{2xy}{p}$ represents the length, what rational expression represents the width?

    ☐ Width

    Length $= \dfrac{2xy}{p}$

    The area is $\dfrac{5x^2 y^3}{2pq}$.

42. Consider the division problem $\dfrac{x - 6}{x + 4} \div \dfrac{x + 7}{x + 5}$.

We know that division by 0 is undefined, so the restrictions on x are $x \neq -4$, $x \neq -5$, and $x \neq -7$. Why is the last restriction needed?

15.3 ▶▶▶ Least Common Denominators

OBJECTIVE 1 Find the least common denominator for a list of fractions. Just as with common fractions, adding or subtracting rational expressions (to be discussed in the next section) often requires a **least common denominator (LCD),** the simplest expression that is divisible by all denominators. For example, the least common denominator for $\frac{2}{9}$ and $\frac{5}{12}$ is 36, because 36 is the smallest positive number divisible by both 9 and 12.

We can often find least common denominators by inspection. For example, the LCD for $\frac{1}{6}$ and $\frac{2}{3m}$ is $6m$. In other cases, we find the LCD by a procedure similar to that used in **Section 14.1** for finding the greatest common factor.

OBJECTIVES

1 **Find the least common denominator for a list of fractions.**

2 **Write equivalent rational expressions.**

Finding the Least Common Denominator (LCD)

Step 1   **Factor** each denominator into prime factors.

Step 2   **List each different denominator factor** the *greatest* number of times it appears in any of the denominators.

Step 3   **Multiply** the denominator factors from Step 2 to get the LCD.

When each denominator is factored into prime factors, every prime factor must be a factor of the least common denominator.

In Example 1, we find the LCD for both numerical and algebraic denominators.

EXAMPLE 1   **Finding Least Common Denominators**

Find the LCD for each pair of fractions.

(a) $\dfrac{1}{24}, \dfrac{7}{15}$ **(b)** $\dfrac{1}{8x}, \dfrac{3}{10x}$

Step 1   Write each denominator in factored form with numerical coefficients in prime factored form.

$$24 = 2 \cdot 2 \cdot 2 \cdot 3 = 2^3 \cdot 3 \qquad 8x = 2 \cdot 2 \cdot 2 \cdot x = 2^3 \cdot x$$
$$15 = 3 \cdot 5 \qquad\qquad\qquad 10x = 2 \cdot 5 \cdot x$$

Step 2   We find the LCD by taking each different factor the *greatest* number of times it appears as a factor in any of the denominators.

The factor 2 appears three times in one product and not at all in the other, so the greatest number of times 2 appears is three. The greatest number of times both 3 and 5 appear is one.

Here, 2 appears three times in one product and once in the other, so the greatest number of times 2 appears is three. The greatest number of times 5 appears is one, and the greatest number of times x appears in either product is one.

Step 3   LCD $= 2 \cdot 2 \cdot 2 \cdot 3 \cdot 5$
$= 2^3 \cdot 3 \cdot 5$
$= 120$

LCD $= 2 \cdot 2 \cdot 2 \cdot 5 \cdot x$
$= 2^3 \cdot 5 \cdot x$
$= 40x$

Work Problem **1** *at the Side.* ▶

1 Find the LCD for each pair of fractions.

(a) $\dfrac{7}{10}, \dfrac{1}{25}$

(b) $\dfrac{7}{20p}, \dfrac{11}{30p}$

(c) $\dfrac{4}{5x}, \dfrac{12}{10x}$

ANSWERS

1. **(a)** 50   **(b)** $60p$ **(c)** $10x$

2 Find the LCD.

(a) $\dfrac{4}{16m^3n}, \dfrac{5}{9m^5}$

(b) $\dfrac{3}{25a^2}, \dfrac{2}{10a^3b}$

EXAMPLE 2 Finding the LCD

Find the LCD for $\dfrac{5}{6r^2}$ and $\dfrac{3}{4r^3}$.

Step 1 Factor each denominator.

$$6r^2 = 2 \cdot \mathbf{3} \cdot r^2$$
$$4r^3 = \mathbf{2^2} \cdot r^3$$

Step 2 The greatest number of times 2 appears is two, the greatest number of times 3 appears is one, and the greatest number of times r appears is three; therefore,

Step 3 $\qquad\qquad$ LCD $= \mathbf{2^2} \cdot \mathbf{3} \cdot \mathbf{r^3} = 12r^3$.

◀ *Work Problem* **2** *at the Side.*

CAUTION

When finding the LCD, use each factor the *greatest* number of times it appears in any *single* denominator, not the *total* number of times it appears. For instance, the greatest number of times r appears as a factor in one denominator in Example 2 is 3, *not* 5.

---

3 Find the LCD for the fractions in each list.

(a) $\dfrac{7}{3a}, \dfrac{11}{a^2 - 4a}$

(b)
$\dfrac{2m}{m^2 - 3m + 2}, \dfrac{5m - 3}{m^2 + 3m - 10},$
$\dfrac{4m + 7}{m^2 + 4m - 5}$

(c) $\dfrac{6}{x - 4}, \dfrac{3x - 1}{4 - x}$

EXAMPLE 3 Finding LCDs

Find the LCD for the fractions in each list.

(a) $\dfrac{6}{5m}, \dfrac{4}{m^2 - 3m}$

$\left. \begin{array}{l} 5m = \mathbf{5} \cdot \mathbf{m} \\ m^2 - 3m = \mathbf{m}\,(\mathbf{m - 3}) \end{array} \right\}$ Factor each denominator.

Use each different factor the greatest number of times it appears.

$$\text{LCD} = \mathbf{5} \cdot \mathbf{m} \cdot (\mathbf{m - 3}) = 5m\,(m - 3)$$

> Be sure to include m as a factor in the LCD.

Because m is not a *factor* of $m - 3$, both factors, m and $m - 3$, must appear in the LCD.

(b) $\dfrac{1}{r^2 - 4r - 5}, \dfrac{3}{r^2 - r - 20}, \dfrac{1}{r^2 - 10r + 25}$

$\left. \begin{array}{l} r^2 - 4r - 5 = (r - 5)(\mathbf{r + 1}) \\ r^2 - r - 20 = (r - 5)(\mathbf{r + 4}) \\ r^2 - 10r + 25 = (\mathbf{r - 5})^2 \end{array} \right\}$ Factor each denominator.

Use each different factor the greatest number of times it appears as a factor.

$$\text{LCD} = (r - 5)^2 (r + 1)(r + 4)$$

(c) $\dfrac{1}{q - 5}, \dfrac{3}{5 - q}$

The expressions $q - 5$ and $5 - q$ are opposites of each other because

$$-(q - 5) = -q + 5 = 5 - q.$$

Therefore, either $q - 5$ or $5 - q$ can be used as the LCD.

◀ *Work Problem* **3** *at the Side.*

OBJECTIVE 2 Write equivalent rational expressions. Once the LCD has been found, the next step in preparing to add or subtract two rational expressions is to use the fundamental property to write equivalent rational expressions. We use the following steps.

Writing a Rational Expression with a Specified Denominator

Step 1 **Factor** both denominators.

Step 2 **Decide what factor(s) the denominator must be multiplied by** in order to equal the specified denominator.

Step 3 **Multiply** the rational expression by that factor divided by itself. (That is, multiply by 1.)

EXAMPLE 4 **Writing Equivalent Rational Expressions**

Write each rational expression as an equivalent expression with the indicated denominator.

(a) $\dfrac{3}{8} = \dfrac{?}{40}$

(b) $\dfrac{9k}{25} = \dfrac{?}{50k}$

Step 1 For each example, first factor the denominator on the right. Then compare the denominator on the left with the one on the right to decide what factors are missing.

$$\dfrac{3}{8} = \dfrac{?}{\mathbf{5 \cdot 8}}$$

$$\dfrac{9k}{25} = \dfrac{?}{25 \cdot \mathbf{2k}}$$

Step 2 A factor of 5 is missing.

Factors of 2 and k are missing.

Step 3 Multiply $\frac{3}{8}$ by $\frac{5}{5}$.

Multiply $\frac{9k}{25}$ by $\frac{2k}{2k}$.

$$\dfrac{3}{8} = \dfrac{3}{8} \cdot \dfrac{\mathbf{5}}{\mathbf{5}} = \dfrac{\mathbf{15}}{40}$$

$$\dfrac{9k}{25} = \dfrac{9k}{25} \cdot \dfrac{\mathbf{2k}}{\mathbf{2k}} = \dfrac{\mathbf{18k^2}}{50k}$$

$\frac{5}{5} = 1$

$\frac{2k}{2k} = 1$

Work Problem **4** *at the Side.* ▶

EXAMPLE 5 **Writing Equivalent Rational Expressions**

Write each rational expression as an equivalent expression with the indicated denominator.

(a)
$$\dfrac{8}{3x + 1} = \dfrac{?}{12x + 4}$$

$$\dfrac{8}{3x + 1} = \dfrac{?}{\mathbf{4(3x + 1)}}$$ Factor the denominator on the right.

The missing factor is 4, so multiply the fraction on the left by $\frac{4}{4}$.

$$\dfrac{8}{3x + 1} \cdot \dfrac{\mathbf{4}}{\mathbf{4}} = \dfrac{\mathbf{32}}{12x + 4}$$ Fundamental property

Continued on Next Page

---

4 Write each rational expression as an equivalent expression with the indicated denominator.

(a) $\dfrac{3}{4} = \dfrac{?}{36}$

(b) $\dfrac{7k}{5} = \dfrac{?}{30p}$

5 Write each rational expression as an equivalent expression with the indicated denominator.

(a) $\dfrac{9}{2a + 5} = \dfrac{?}{6a + 15}$

(b) $\dfrac{12p}{p^2 + 8p} = \dfrac{?}{p^3 + 4p^2 - 32p}$

Factor the denominator in each rational expression.

$$\frac{12p}{p(p + 8)} = \frac{?}{p(p + 8)(p - 4)}$$

$p^3 + 4p^2 - 32p$
$= p(p^2 + 4p - 32)$
$= p(p + 8)(p - 4)$

The factor $p - 4$ is missing, so multiply $\dfrac{12p}{p(p + 8)}$ by $\dfrac{p - 4}{p - 4}$.

$$\frac{12p}{p^2 + 8p} = \frac{12p}{p(p + 8)} \cdot \frac{\boldsymbol{p - 4}}{\boldsymbol{p - 4}} \quad \text{Fundamental property}$$

$$= \frac{12p(p - 4)}{p(p + 8)(p - 4)} \quad \begin{array}{l}\text{Multiply numerators.}\\[4pt]\text{Multiply denominators.}\end{array}$$

$$= \frac{12p^2 - 48p}{p^3 + 4p^2 - 32p} \quad \text{Multiply the factors.}$$

> **Note**
>
> In the next section we add and subtract rational expressions, which sometimes requires the steps illustrated in Examples 4 and 5. While it may be beneficial to leave the equivalent expression in factored form, we multiplied out the factors in the numerator and the denominator in Example 5(b),
>
> $$\frac{12p(p - 4)}{p(p + 8)(p - 4)},$$
>
> in order to give the answer,
>
> $$\frac{12p^2 - 48p}{p^3 + 4p^2 - 32p},$$
>
> in the same form as the original problem.

◀ **Work Problem 5** at the Side.

(b) $\dfrac{5k + 1}{k^2 + 2k} = \dfrac{?}{k^3 + k^2 - 2k}$

ANSWERS

5. (a) $\dfrac{27}{6a + 15}$

(b) $\dfrac{(5k + 1)(k - 1)}{k^3 + k^2 - 2k}$, or $\dfrac{5k^2 - 4k - 1}{k^3 + k^2 - 2k}$

15.3 ▶▶▶ **Exercises**

FOR EXTRA HELP

 MyMathLab

Math XL PRACTICE

 WATCH

 DOWNLOAD

 READ

 REVIEW

Choose the correct response in Exercises 1–4.

1. Suppose that the greatest common factor of a and b is 1. Then the least common denominator for $\frac{1}{a}$ and $\frac{1}{b}$ is

A. a **B.** b **C.** ab **D.** 1.

2. If a is a factor of b, then the least common denominator for $\frac{1}{a}$ and $\frac{1}{b}$ is

A. a **B.** b **C.** ab **D.** 1.

3. The least common denominator for $\frac{11}{20}$ and $\frac{1}{2}$ is

A. 40 **B.** 2 **C.** 20 **D.** none of these.

4. Suppose that we wish to write the rational expression $\dfrac{1}{(x-4)^2(y-3)}$ with denominator $(x-4)^3(y-3)^2$. We must multiply both the numerator and the denominator by

A. $(x-4)(y-3)$ **B.** $(x-4)^2$

C. $x-4$ **D.** $(x-4)^2(y-3)$.

Find the least common denominator for the fractions in each list. See Examples 1 and 2.

5. $\dfrac{2}{15}, \dfrac{3}{10}, \dfrac{7}{30}$

6. $\dfrac{5}{24}, \dfrac{7}{12}, \dfrac{9}{28}$

7. $\dfrac{3}{x^4}, \dfrac{5}{x^7}$

8. $\dfrac{2}{y^5}, \dfrac{3}{y^6}$

9. $\dfrac{5}{36q}, \dfrac{17}{24q}$

10. $\dfrac{4}{30p}, \dfrac{9}{50p}$

11. $\dfrac{6}{21r^3}, \dfrac{8}{12r^5}$

12. $\dfrac{9}{35t^2}, \dfrac{5}{49t^6}$

13. If the denominators of two fractions in prime factored form are $2^3 \cdot 3$ and $2^2 \cdot 5$, what is the factored form of their LCD?

14. Suppose two rational expressions have denominators $(t+4)^3(t-3)$ and $(t+4)^2(t+8)$. Find the factored form of their LCD. What is the similarity between the answers for this problem and for Exercise 13?

15. If two denominators have greatest common factor equal to 1, how can you easily find their least common denominator?

16. Suppose two fractions have denominators a^k and a^r, where k and r are natural numbers, with $k > r$. What is their least common denominator?

Find the least common denominator for the fractions in each list. See Examples 1–3.

17. $\dfrac{9}{28m^2}, \dfrac{3}{12m-20}$

18. $\dfrac{15}{27a^3}, \dfrac{8}{9a-45}$

19. $\dfrac{7}{5b-10}, \dfrac{11}{6b-12}$

20. $\dfrac{3}{7x^2+21x}, \dfrac{1}{5x^2+15x}$

21. $\dfrac{5}{c-d}, \dfrac{8}{d-c}$

22. $\dfrac{4}{y-x}, \dfrac{7}{x-y}$

23. $\dfrac{3}{k^2 + 5k}, \dfrac{2}{k^2 + 3k - 10}$

24. $\dfrac{1}{z^2 - 4z}, \dfrac{4}{z^2 - 3z - 4}$

25. $\dfrac{5}{p^2 + 8p + 15}, \dfrac{3}{p^2 - 3p - 18}, \dfrac{2}{p^2 - p - 30}$

26. $\dfrac{10}{y^2 - 10y + 21}, \dfrac{2}{y^2 - 2y - 3}, \dfrac{5}{y^2 - 6y - 7}$

Write each rational expression as an equivalent expression with the indicated denominator.
See Examples 4 and 5.

27. $\dfrac{4}{11} = \dfrac{?}{55}$

28. $\dfrac{6}{7} = \dfrac{?}{42}$

29. $\dfrac{-5}{k} = \dfrac{?}{9k}$

30. $\dfrac{-3}{q} = \dfrac{?}{6q}$

31. $\dfrac{13}{40y} = \dfrac{?}{80y^3}$

32. $\dfrac{5}{27p} = \dfrac{?}{108p^4}$

33. $\dfrac{5t^2}{6r} = \dfrac{?}{42r^4}$

34. $\dfrac{8y^2}{3x} = \dfrac{?}{30x^3}$

35. $\dfrac{5}{2(m + 3)} = \dfrac{?}{8(m + 3)}$

36. $\dfrac{7}{4(y - 1)} = \dfrac{?}{16(y - 1)}$

37. $\dfrac{-4t}{3t - 6} = \dfrac{?}{12 - 6t}$

38. $\dfrac{-7k}{5k - 20} = \dfrac{?}{40 - 10k}$

39. $\dfrac{14}{z^2 - 3z} = \dfrac{?}{z(z - 3)(z - 2)}$

40. $\dfrac{12}{x(x + 4)} = \dfrac{?}{x(x + 4)(x - 9)}$

41. $\dfrac{2(b - 1)}{b^2 + b} = \dfrac{?}{b^3 + 3b^2 + 2b}$

42. $\dfrac{3(c + 2)}{c(c - 1)} = \dfrac{?}{c^3 - 5c^2 + 4c}$

15.4 ▷▷▷ Adding and Subtracting Rational Expressions

To add and subtract rational expressions, we find least common denominators and write equivalent fractions with the LCD.

OBJECTIVE 1 **Add rational expressions having the same denominator.** We find the sum of two such rational expressions with the same procedure that we used in **Section 4.4** for adding two fractions having the same denominator.

> **Adding Rational Expressions (Same Denominator)**
>
> If $\frac{P}{Q}$ and $\frac{R}{Q}$ ($Q \neq 0$) are rational expressions, then
>
> $$\frac{P}{Q} + \frac{R}{Q} = \frac{P + R}{Q}.$$
>
> In words: To add rational expressions with the same denominator, add the numerators and keep the same denominator.

EXAMPLE 1 **Adding Rational Expressions (Same Denominator)**

Add. Write each answer in lowest terms.

(a) $\dfrac{4}{9} + \dfrac{2}{9}$ **(b)** $\dfrac{3x}{x+1} + \dfrac{3}{x+1}$

The denominators are the same, so the sum is found by adding the two numerators and keeping the same (common) denominator.

$= \dfrac{4+2}{9}$ Add.

$= \dfrac{6}{9}$

$= \dfrac{2 \cdot 3}{3 \cdot 3}$ Factor.

$= \dfrac{2}{3}$ Lowest terms

$= \dfrac{3x+3}{x+1}$ Add.

$= \dfrac{3(x+1)}{x+1}$ Factor.

$= 3$ Lowest terms

Work Problem **1** *at the Side.* ▶

OBJECTIVE 2 **Add rational expressions having different denominators.** As in **Section 4.4**, we use the following steps to add two rational expressions having different denominators.

> **Adding Rational Expressions (Different Denominators)**
>
> **Step 1** **Find the least common denominator (LCD).**
>
> **Step 2** **Write each rational expression** as an equivalent rational expression with the LCD as the denominator.
>
> **Step 3** **Add** the numerators to get the numerator of the sum. The LCD is the denominator of the sum.
>
> **Step 4** **Write in lowest terms** using the fundamental property.

OBJECTIVES

1 Add rational expressions having the same denominator.

2 Add rational expressions having different denominators.

3 Subtract rational expressions.

---

1 Add. Write each answer in lowest terms.

(a) $\dfrac{7}{15} + \dfrac{3}{15}$

(b) $\dfrac{3}{y+4} + \dfrac{2}{y+4}$

(c) $\dfrac{x}{x+y} + \dfrac{1}{x+y}$

(d) $\dfrac{a}{a+b} + \dfrac{b}{a+b}$

(e) $\dfrac{x^2}{x+1} + \dfrac{x}{x+1}$

ANSWERS

1. **(a)** $\dfrac{2}{3}$ **(b)** $\dfrac{5}{y+4}$ **(c)** $\dfrac{x+1}{x+y}$
   (d) 1 **(e)** x

2 Add. Write each answer in lowest terms.

(a) $\dfrac{1}{10} + \dfrac{1}{15}$

> **EXAMPLE 2**   **Adding Rational Expressions (Different Denominators)**

Add. Write each answer in lowest terms.

(a) $\dfrac{1}{12} + \dfrac{7}{15}$ **(b)** $\dfrac{2}{3y} + \dfrac{1}{4y}$

Step 1   First find the LCD, using the methods of the previous section.

$$12 = 2 \cdot 2 \cdot 3 = 2^2 \cdot 3 \qquad\qquad 3y = 3 \cdot y$$
$$15 = 3 \cdot 5 \qquad\qquad\qquad\quad 4y = 2 \cdot 2 \cdot y = 2^2 \cdot y$$
$$\text{LCD} = 2^2 \cdot 3 \cdot 5 = 60 \qquad \text{LCD} = 2^2 \cdot 3 \cdot y = 12y$$

Step 2   Now write each rational expression as an equivalent expression with the LCD (either 60 or $12y$) as the denominator.

$$\frac{1}{12} + \frac{7}{15} = \frac{1\,(5)}{12\,(5)} + \frac{7\,(4)}{15\,(4)} \qquad \frac{2}{3y} + \frac{1}{4y} = \frac{2\,(4)}{3y\,(4)} + \frac{1\,(3)}{4y\,(3)}$$

$$= \frac{5}{60} + \frac{28}{60} \qquad\qquad\qquad = \frac{8}{12y} + \frac{3}{12y}$$

Step 3   Add the numerators. The LCD is the denominator.

Step 4   Write in lowest terms if necessary.

$$= \frac{5 + 28}{60} \qquad\qquad\qquad\qquad = \frac{8 + 3}{12y}$$

$$= \frac{33}{60}, \ \text{ or } \ \frac{11}{20} \qquad\qquad = \frac{11}{12y}$$

◀ Work Problem **2** at the Side.

(b) $\dfrac{6}{5x} + \dfrac{9}{2x}$

> **EXAMPLE 3**   **Adding Rational Expressions**

Add. Write the answer in lowest terms.

$$\frac{2x}{x^2 - 1} + \frac{-1}{x + 1}$$

Step 1   Since the denominators are different, find the LCD.

$$\left. \begin{array}{l} x^2 - 1 = (x + 1)(x - 1) \\[4pt] x + 1 \text{ is prime.} \end{array} \right\} \text{ The LCD is } (x + 1)(x - 1).$$

Step 2   Write each rational expression as an equivalent expression with the LCD as the denominator.

$$\frac{2x}{x^2 - 1} + \frac{-1}{x + 1} \qquad\qquad \text{The LCD is } (x + 1)(x - 1).$$

(c) $\dfrac{m}{3n} + \dfrac{2}{7n}$

$$= \frac{2x}{(x + 1)(x - 1)} + \frac{-1\,(x - 1)}{(x + 1)(x - 1)} \qquad \begin{array}{l}\text{Multiply the second} \\ \text{fraction by } \frac{x-1}{x-1}.\end{array}$$

$$= \frac{2x}{(x + 1)(x - 1)} + \frac{-x + 1}{(x + 1)(x - 1)} \qquad \text{Distributive property}$$

Step 3 $= \dfrac{2x - x + 1}{(x + 1)(x - 1)}$ Add numerators; keep the same denominator.

ANSWERS

2. **(a)** $\dfrac{1}{6}$ **(b)** $\dfrac{57}{10x}$ **(c)** $\dfrac{7m + 6}{21n}$

$$= \frac{x + 1}{(x + 1)(x - 1)} \qquad \text{Combine like terms in the numerator.}$$

Step 4

$$= \frac{1(x + 1)}{(x + 1)(x - 1)} \qquad \text{Identity property for multiplication}$$

> Remember to write 1 in the numerator.

$$= \frac{1}{x - 1} \qquad \text{Divide out the common factors.}$$

Work Problem **3** *at the Side.* ▶

EXAMPLE 4 Adding Rational Expressions

Add. Write the answer in lowest terms.

$$\frac{2x}{x^2 + 5x + 6} + \frac{x + 1}{x^2 + 2x - 3}$$

$$= \frac{2x}{(x + 2)(x + 3)} + \frac{x + 1}{(x + 3)(x - 1)} \qquad \text{Factor the denominators.}$$

$$= \frac{2x(x - 1)}{(x + 2)(x + 3)(x - 1)} + \frac{(x + 1)(x + 2)}{(x + 2)(x + 3)(x - 1)} \qquad \text{The LCD is } (x + 2) \cdot (x + 3)(x - 1).$$

$$= \frac{2x(x - 1) + (x + 1)(x + 2)}{(x + 2)(x + 3)(x - 1)} \qquad \begin{array}{l}\text{Add numerators;} \\ \text{keep the same} \\ \text{denominator.}\end{array}$$

$$= \frac{2x^2 - 2x + x^2 + 3x + 2}{(x + 2)(x + 3)(x - 1)} \qquad \text{Multiply.}$$

$$= \frac{3x^2 + x + 2}{(x + 2)(x + 3)(x - 1)} \qquad \begin{array}{l}\text{Combine like} \\ \text{terms.}\end{array}$$

It is usually more convenient to leave the denominator in factored form. The numerator cannot be factored here, so the expression is in lowest terms.

Work Problem **4** *at the Side.* ▶

EXAMPLE 5 Adding Rational Expressions with Denominators That Are Opposites

Add. Write the answer in lowest terms.

$$\frac{y}{y - 2} + \frac{8}{2 - y} \qquad \text{The denominators are opposites.}$$

$$= \frac{y}{y - 2} + \frac{8(-1)}{(2 - y)(-1)} \qquad \begin{array}{l}\text{Multiply } \frac{8}{2 - y} \text{ by } \frac{-1}{-1} \text{ to get a} \\ \text{common denominator.}\end{array}$$

$$= \frac{y}{y - 2} + \frac{-8}{-2 + y} \qquad \text{Distributive property}$$

$$= \frac{y}{y - 2} + \frac{-8}{y - 2} \qquad \text{Rewrite } -2 + y \text{ as } y - 2.$$

$$= \frac{y - 8}{y - 2} \qquad \begin{array}{l}\text{Add numerators;} \\ \text{keep the same denominator.}\end{array}$$

If we had chosen to use $2 - y$ as the common denominator, the final answer would be in the form $\frac{8 - y}{2 - y}$, which is equivalent to $\frac{y - 8}{y - 2}$.

Work Problem **5** *at the Side.* ▶

---

3 Add. Write each answer in lowest terms.

(a) $\dfrac{2p}{3p + 3} + \dfrac{5p}{2p + 2}$

(b) $\dfrac{4}{y^2 - 1} + \dfrac{6}{y + 1}$

(c) $\dfrac{-2}{p + 1} + \dfrac{4p}{p^2 - 1}$

4 Add. Write each answer in lowest terms.

(a) $\dfrac{2k}{k^2 - 5k + 4} + \dfrac{3}{k^2 - 1}$

(b)
$$\dfrac{4m}{m^2 + 3m + 2} + \dfrac{2m - 1}{m^2 + 6m + 5}$$

5 Add. Write the answer in lowest terms.

$$\dfrac{m}{2m - 3n} + \dfrac{n}{3n - 2m}$$

ANSWERS

3. (a) $\dfrac{19p}{6(p + 1)}$ (b) $\dfrac{2(3y - 1)}{(y + 1)(y - 1)}$

 (c) $\dfrac{2}{p - 1}$

4. (a) $\dfrac{(2k - 3)(k + 4)}{(k - 4)(k - 1)(k + 1)}$

 (b) $\dfrac{6m^2 + 23m - 2}{(m + 2)(m + 1)(m + 5)}$

5. $\dfrac{m - n}{2m - 3n}$, or $\dfrac{n - m}{3n - 2m}$

6 Subtract. Write each answer in lowest terms.

(a) $\dfrac{3}{m^2} - \dfrac{2}{m^2}$

OBJECTIVE 3 Subtract rational expressions.   To subtract rational expressions having the same denominator, use the following rule.

> **Subtracting Rational Expressions (Same Denominator)**
>
> If $\dfrac{P}{Q}$ and $\dfrac{R}{Q}$ ($Q \neq 0$) are rational expressions, then
>
> $$\frac{P}{Q} - \frac{R}{Q} = \frac{P - R}{Q}.$$
>
> In words: To subtract rational expressions with the same denominator, subtract the numerators and keep the same denominator.

EXAMPLE 6 **Subtracting Rational Expressions (Same Denominator)**

Subtract. Write the answer in lowest terms.

$$\frac{2m}{m - 1} - \frac{m + 3}{m - 1}$$
Use parentheses around the quantity being subtracted.

$$= \frac{2m - (m + 3)}{m - 1}$$
Subtract numerators; keep the same denominator.

Be careful with signs.
$$= \frac{2m - m - 3}{m - 1}$$
Distributive property

$$= \frac{m - 3}{m - 1}$$
Combine like terms.

(b) $\dfrac{x}{2x + 3} - \dfrac{3x + 4}{2x + 3}$

> **CAUTION**
>
> Sign errors often occur in problems like the one in Example 6. The numerator of the fraction being subtracted must be treated as a single quantity. *Be sure to use parentheses after the subtraction sign.*

◀ Work Problem **6** at the Side.

In the remaining examples, we subtract rational expressions having different denominators.

EXAMPLE 7 **Subtracting Rational Expressions (Different Denominators)**

Subtract. Write the answer in lowest terms.

$$\frac{9}{x - 2} - \frac{3}{x}$$
The LCD is $x(x - 2)$.

$$= \frac{9x}{x(x - 2)} - \frac{3(x - 2)}{x(x - 2)}$$
Rewrite each expression with the LCD.

$$= \frac{9x - 3(x - 2)}{x(x - 2)}$$
Subtract numerators; keep the same denominator.

Continued on Next Page

ANSWERS

6. (a) $\dfrac{1}{m^2}$ (b) $\dfrac{-2(x + 2)}{2x + 3}$

Be careful here.

$$= \frac{9x - 3x + 6}{x(x-2)} \qquad \text{Distributive property}$$

$$= \frac{6x + 6}{x(x-2)}, \quad \text{or} \quad \frac{6(x+1)}{x(x-2)} \qquad \text{Combine like terms; factor the numerator.}$$

Note

We factored the final numerator in Example 7 to get $\frac{6(x+1)}{x(x-2)}$. The fundamental property does not apply, however, since there are no common factors to divide out. The answer is in lowest terms.

Work Problem **7** *at the Side.* ▶

EXAMPLE 8 **Subtracting Rational Expressions with Denominators That Are Opposites**

Subtract. Write the answer in lowest terms.

$$\frac{3x}{x-5} - \frac{2x-25}{5-x} \qquad \begin{array}{l}\text{The denominators are opposites.} \\ \text{We choose } x-5 \text{ as the common} \\ \text{denominator.}\end{array}$$

$$= \frac{3x}{x-5} - \frac{(2x-25)(-1)}{(5-x)(-1)} \qquad \begin{array}{l}\text{Multiply } \frac{2x-25}{5-x} \text{ by } \frac{-1}{-1} \text{ to get a} \\ \text{common denominator.}\end{array}$$

$$= \frac{3x}{x-5} - \frac{-2x+25}{x-5} \qquad (5-x)(-1) = -5+x = x-5$$

$$= \frac{3x - (-2x+25)}{x-5} \qquad \begin{array}{l}\text{Use parentheses.} \\ \text{Subtract numerators.}\end{array}$$

$$= \frac{3x + 2x - 25}{x-5} \qquad \text{Distributive property}$$

$$= \frac{5x - 25}{x-5} \qquad \text{Combine like terms.}$$

$$= \frac{5(x-5)}{x-5} \qquad \text{Factor.}$$

$$= 5 \qquad \text{Divide out the common factor.}$$

Work Problem **8** *at the Side.* ▶

EXAMPLE 9 **Subtracting Rational Expressions**

Subtract. Write the answer in lowest terms.

$$\frac{6x}{x^2 - 2x + 1} - \frac{1}{x^2 - 1}$$

Begin by factoring the denominators.

$$x^2 - 2x + 1 = (x-1)^2 \quad \text{and} \quad x^2 - 1 = (x-1)(x+1)$$

From the factored denominators, we identify the LCD, $(x-1)^2(x+1)$. *We use the factor $x-1$ twice* because it appears twice in the first denominator.

Continued on Next Page

---

7 Subtract. Write each answer in lowest terms.

(a) $\dfrac{1}{k+4} - \dfrac{2}{k}$

(b) $\dfrac{6}{a+2} - \dfrac{1}{a-3}$

8 Subtract. Write each answer in lowest terms.

(a) $\dfrac{5}{x-1} - \dfrac{3x}{1-x}$

(b) $\dfrac{2y}{y-2} - \dfrac{1+y}{2-y}$

ANSWERS

7. (a) $\dfrac{-k-8}{k(k+4)}$ (b) $\dfrac{5(a-4)}{(a+2)(a-3)}$

8. (a) $\dfrac{5+3x}{x-1}$, or $\dfrac{-5-3x}{1-x}$

  (b) $\dfrac{3y+1}{y-2}$, or $\dfrac{-3y-1}{2-y}$

9 Subtract. Write each answer in lowest terms.

(a) $\dfrac{4y}{y^2 - 1} - \dfrac{5}{y^2 + 2y + 1}$

(b) $\dfrac{3r}{r^2 - 5r} - \dfrac{4}{r^2 - 10r + 25}$

$\dfrac{6x}{(x - 1)^2} - \dfrac{1}{(x - 1)(x + 1)}$ — The LCD is $(x - 1)^2 (x + 1)$.

$= \dfrac{6x(x + 1)}{(x - 1)^2 (x + 1)} - \dfrac{1(x - 1)}{(x - 1)(x - 1)(x + 1)}$ — Fundamental property

$= \dfrac{6x(x + 1) - 1(x - 1)}{(x - 1)^2 (x + 1)}$ — Subtract numerators.

$= \dfrac{6x^2 + 6x - x + 1}{(x - 1)^2 (x + 1)}$ — Distributive property

$= \dfrac{6x^2 + 5x + 1}{(x - 1)^2 (x + 1)}$, or $\dfrac{(2x + 1)(3x + 1)}{(x - 1)^2 (x + 1)}$ — Combine like terms; factor the numerator.

Verify that the final answer is in lowest terms.

◀ Work Problem **9** at the Side.

EXAMPLE 10 **Subtracting Rational Expressions**

Subtract. Write the answer in lowest terms.

$$\frac{q}{q^2 - 4q - 5} - \frac{3}{2q^2 - 13q + 15}$$

To find the LCD, factor each denominator.

$$q^2 - 4q - 5 = (q + 1)(q - 5)$$
$$2q^2 - 13q + 15 = (q - 5)(2q - 3)$$

The LCD here is $(q + 1)(q - 5)(2q - 3)$. Write each rational expression with the LCD, using the fundamental property.

$\dfrac{q}{(q + 1)(q - 5)} - \dfrac{3}{(q - 5)(2q - 3)}$ — The LCD is $(q + 1) \cdot (q - 5)(2q - 3)$.

$= \dfrac{q(2q - 3)}{(q + 1)(q - 5)(2q - 3)} - \dfrac{3(q + 1)}{(q + 1)(q - 5)(2q - 3)}$

$= \dfrac{q(2q - 3) - 3(q + 1)}{(q + 1)(q - 5)(2q - 3)}$ — Subtract numerators.

$= \dfrac{2q^2 - 3q - 3q - 3}{(q + 1)(q - 5)(2q - 3)}$ — Distributive property

$= \dfrac{2q^2 - 6q - 3}{(q + 1)(q - 5)(2q - 3)}$ — Combine like terms.

Verify that the final answer is in lowest terms.

◀ Work Problem **10** at the Side.

10 Subtract. Write each answer in lowest terms.

(a) $\dfrac{2}{p^2 - 5p + 4} - \dfrac{3}{p^2 - 1}$

(b) $\dfrac{q}{2q^2 + 5q - 3} - \dfrac{3q + 4}{3q^2 + 10q + 3}$

ANSWERS

9. (a) $\dfrac{4y^2 - y + 5}{(y + 1)^2 (y - 1)}$

(b) $\dfrac{3r - 19}{(r - 5)^2}$

10. (a) $\dfrac{14 - p}{(p - 4)(p - 1)(p + 1)}$

(b) $\dfrac{-3q^2 - 4q + 4}{(2q - 1)(q + 3)(3q + 1)}$

15.4 ▶▶▶ Exercises

Match the expression in Column I with the correct sum or difference in Column II.

I

1. $\dfrac{x}{x+6} + \dfrac{6}{x+6}$

2. $\dfrac{2x}{x-6} - \dfrac{12}{x-6}$

3. $\dfrac{6}{x-6} - \dfrac{x}{x-6}$

4. $\dfrac{6}{x+6} - \dfrac{x}{x+6}$

5. $\dfrac{x}{x+6} - \dfrac{6}{x+6}$

6. $\dfrac{1}{x} + \dfrac{1}{6}$

7. $\dfrac{1}{6} - \dfrac{1}{x}$

8. $\dfrac{1}{6x} - \dfrac{1}{6x}$

II

A. 2

B. $\dfrac{x-6}{x+6}$

C. -1

D. $\dfrac{6+x}{6x}$

E. 1

F. 0

G. $\dfrac{x-6}{6x}$

H. $\dfrac{6-x}{x+6}$

Note: When adding and subtracting rational expressions, several different equivalent forms of the answer often exist. If your answer does not look exactly like the one given in the back of the book, check to see whether you have written an equivalent form.

Add or subtract. Write each answer in lowest terms. See Examples 1 and 6.

9. $\dfrac{4}{m} + \dfrac{7}{m}$

10. $\dfrac{5}{p} + \dfrac{11}{p}$

11. $\dfrac{a+b}{2} - \dfrac{a-b}{2}$

12. $\dfrac{x-y}{2} - \dfrac{x+y}{2}$

13. $\dfrac{5}{y+4} - \dfrac{1}{y+4}$

14. $\dfrac{4}{y+3} - \dfrac{1}{y+3}$

15. $\dfrac{5m}{m+1} - \dfrac{1+4m}{m+1}$

16. $\dfrac{4x}{x+2} - \dfrac{2+3x}{x+2}$

17. $\dfrac{x^2}{x+5} + \dfrac{5x}{x+5}$

18. $\dfrac{t^2}{t-3} + \dfrac{-3t}{t-3}$

19. $\dfrac{y^2-3y}{y+3} + \dfrac{-18}{y+3}$

20. $\dfrac{r^2-8r}{r-5} + \dfrac{15}{r-5}$

21. Explain with an example how to add or subtract rational expressions with the same denominator.

22. Explain with an example how to add or subtract rational expressions with different denominators.

Add or subtract. Write each answer in lowest terms. See Examples 2, 3, 4, and 7.

23. $\dfrac{z}{5} + \dfrac{1}{3}$

24. $\dfrac{p}{8} + \dfrac{3}{5}$

25. $\dfrac{5}{7} - \dfrac{r}{2}$

26. $\dfrac{10}{9} - \dfrac{z}{3}$

27. $-\dfrac{3}{4} - \dfrac{1}{2x}$

28. $-\dfrac{5}{8} - \dfrac{3}{2a}$

29. $\dfrac{3}{5x} + \dfrac{9}{4x}$

30. $\dfrac{3}{2x} + \dfrac{4}{7x}$

31. $\dfrac{x+1}{6} + \dfrac{3x+3}{9}$

32. $\dfrac{2x-6}{4} + \dfrac{x+5}{6}$

33. $\dfrac{x+3}{3x} + \dfrac{2x+2}{4x}$

34. $\dfrac{x+2}{5x} + \dfrac{6x+3}{3x}$

35. $\dfrac{2}{x+3} + \dfrac{1}{x}$

36. $\dfrac{3}{x-4} + \dfrac{2}{x}$

37. $\dfrac{1}{k+5} - \dfrac{2}{k}$

38. $\dfrac{3}{m+1} - \dfrac{4}{m}$

39. $\dfrac{x}{x-2} + \dfrac{-8}{x^2-4}$

40. $\dfrac{2x}{x-1} + \dfrac{-4}{x^2-1}$

41. $\dfrac{x}{x-2} + \dfrac{4}{x+2}$

42. $\dfrac{2x}{x-1} + \dfrac{3}{x+1}$

43. $\dfrac{t}{t+2} + \dfrac{5-t}{t} - \dfrac{4}{t^2+2t}$

44. $\dfrac{2p}{p-3} + \dfrac{2+p}{p} - \dfrac{-6}{p^2-3p}$

45. What are the two possible LCDs that could be used for the sum

$$\frac{10}{m-2} + \frac{5}{2-m}?$$

46. If one form of the correct answer to a sum or difference of rational expressions is $\frac{4}{k-3}$, what would be an alternative form of the answer if the denominator is $3-k$?

Add or subtract. Write each answer in lowest terms. See Examples 5 and 8.

47. $\dfrac{4}{x-5} + \dfrac{6}{5-x}$

48. $\dfrac{10}{m-2} + \dfrac{5}{2-m}$

49. $\dfrac{-1}{1-y} + \dfrac{3-4y}{y-1}$

50. $\dfrac{-4}{p-3} - \dfrac{p+1}{3-p}$

51. $\dfrac{2}{x-y^2} + \dfrac{7}{y^2-x}$

52. $\dfrac{-8}{p-q^2} + \dfrac{3}{q^2-p}$

53. $\dfrac{x}{5x-3y} - \dfrac{y}{3y-5x}$

54. $\dfrac{t}{8t-9s} - \dfrac{s}{9s-8t}$

55. $\dfrac{3}{4p-5} + \dfrac{9}{5-4p}$

56. $\dfrac{8}{3-7y} - \dfrac{2}{7y-3}$

*In each subtraction problem, the rational expression that follows the subtraction sign has a numerator with more than one term. **Be careful with signs** and find each difference. See Examples 6–10.*

57. $\dfrac{2m}{m-n} - \dfrac{5m+n}{2m-2n}$

58. $\dfrac{5p}{p-q} - \dfrac{3p+1}{4p-4q}$

59. $\dfrac{5}{x^2 - 9} - \dfrac{x + 2}{x^2 + 4x + 3}$

60. $\dfrac{1}{a^2 - 1} - \dfrac{a - 1}{a^2 + 3a - 4}$

61. $\dfrac{2q + 1}{3q^2 + 10q - 8} - \dfrac{3q + 5}{2q^2 + 5q - 12}$

62. $\dfrac{4y - 1}{2y^2 + 5y - 3} - \dfrac{y + 3}{6y^2 + y - 2}$

Perform the indicated operations. See Examples 1–10.

63. $\dfrac{4}{r^2 - r} + \dfrac{6}{r^2 + 2r} - \dfrac{1}{r^2 + r - 2}$

64. $\dfrac{6}{k^2 + 3k} - \dfrac{1}{k^2 - k} + \dfrac{2}{k^2 + 2k - 3}$

65. $\dfrac{x + 3y}{x^2 + 2xy + y^2} + \dfrac{x - y}{x^2 + 4xy + 3y^2}$

66. $\dfrac{m}{m^2 - 1} + \dfrac{m - 1}{m^2 + 2m + 1}$

67. $\dfrac{r + y}{18r^2 + 9ry - 2y^2} + \dfrac{3r - y}{36r^2 - y^2}$

68. $\dfrac{2x - z}{2x^2 + xz - 10z^2} - \dfrac{x + z}{x^2 - 4z^2}$

69. Refer to the rectangle in the figure.

  (a) Find an expression that represents its perimeter. Give the simplified form.

  (b) Find an expression that represents its area. Give the simplified form.

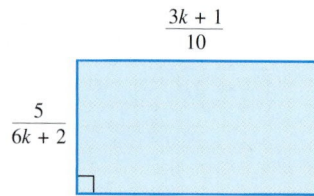

70. Refer to the triangle in the figure. Find an expression that represents its perimeter.

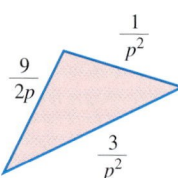

15.5 ▶▶▶ Complex Fractions

The quotient of two mixed numbers in arithmetic, such as $2\frac{1}{2} \div 3\frac{1}{4}$, can be written as a fraction:

$$2\frac{1}{2} \div 3\frac{1}{4} = \frac{2\frac{1}{2}}{3\frac{1}{4}} = \frac{2 + \frac{1}{2}}{3 + \frac{1}{4}}.$$

The last expression is the quotient of expressions that involve fractions. In algebra, some rational expressions also have fractions in the numerator, or denominator, or both.

OBJECTIVES

1 Simplify a complex fraction by writing it as a division problem (Method 1).

2 Simplify a complex fraction by multiplying numerator and denominator by the least common denominator (Method 2).

> **Complex Fraction**
>
> A rational expression with one or more fractions in the numerator, or denominator, or both, is called a **complex fraction.**

Examples of complex fractions include

$$\frac{2 + \frac{1}{2}}{3 + \frac{1}{4}}, \quad \frac{\frac{3x^2 - 5x}{6x^2}}{2x - \frac{1}{x}}, \quad \text{and} \quad \frac{3 + x}{5 - \frac{2}{x}}. \quad \text{Complex fractions}$$

The parts of a complex fraction are named as follows.

$$\left.\begin{array}{c}\dfrac{2}{p} - \dfrac{1}{q}\end{array}\right\} \leftarrow \text{Numerator of complex fraction}$$
$$\qquad\qquad \leftarrow \text{Main fraction bar}$$
$$\left.\begin{array}{c}\dfrac{3}{p} + \dfrac{5}{q}\end{array}\right\} \leftarrow \text{Denominator of complex fraction}$$

OBJECTIVE 1 Simplify a complex fraction by writing it as a division problem (Method 1). Since the main fraction bar represents division in a complex fraction, one method of simplifying a complex fraction involves division. This is the method introduced in **Section 4.6.**

> **Method 1 for Simplifying a Complex Fraction**
>
> **Step 1** Write both the numerator and denominator as single fractions.
>
> **Step 2** Change the complex fraction to a division problem.
>
> **Step 3** Perform the indicated division.

Once again, the first example shows complex fractions from both arithmetic and algebra.

1 Simplify each complex fraction using Method 1.

(a) $\dfrac{\dfrac{2}{5} + \dfrac{1}{4}}{\dfrac{1}{2} + \dfrac{1}{3}}$

(b) $\dfrac{6 + \dfrac{1}{x}}{5 - \dfrac{2}{x}}$

(c) $\dfrac{9 - \dfrac{4}{p}}{\dfrac{2}{p} + 1}$

2 Simplify each complex fraction using Method 1.

(a) $\dfrac{\dfrac{rs^2}{t}}{\dfrac{r^2s}{t^2}}$

(b) $\dfrac{\dfrac{m^2n^3}{p}}{\dfrac{m^4n}{p^2}}$

ANSWERS

1. (a) $\dfrac{39}{50}$ (b) $\dfrac{6x + 1}{5x - 2}$ (c) $\dfrac{9p - 4}{2 + p}$

2. (a) $\dfrac{st}{r}$ (b) $\dfrac{n^2p}{m^2}$

ΕXAMPLE 1 **Simplifying Complex Fractions (Method 1)**

Simplify each complex fraction.

(a) $\dfrac{\dfrac{2}{3} + \dfrac{5}{9}}{\dfrac{1}{4} + \dfrac{1}{12}}$

(b) $\dfrac{6 + \dfrac{3}{x}}{\dfrac{x}{4} + \dfrac{1}{8}}$

Step 1 First, write each numerator as a single fraction.

$$\frac{2}{3} + \frac{5}{9} = \frac{2(3)}{3(3)} + \frac{5}{9}$$
$$= \frac{6}{9} + \frac{5}{9} = \frac{11}{9}$$

$$6 + \frac{3}{x} = \frac{6}{1} + \frac{3}{x}$$
$$= \frac{6x}{x} + \frac{3}{x} = \frac{6x + 3}{x}$$

Do the same thing with each denominator.

$$\frac{1}{4} + \frac{1}{12} = \frac{1(3)}{4(3)} + \frac{1}{12}$$
$$= \frac{3}{12} + \frac{1}{12} = \frac{4}{12}$$

$$\frac{x}{4} + \frac{1}{8} = \frac{x(2)}{4(2)} + \frac{1}{8}$$
$$= \frac{2x}{8} + \frac{1}{8} = \frac{2x + 1}{8}$$

Step 2 Write the equivalent complex fraction as a division problem.

$$\dfrac{\dfrac{11}{9}}{\dfrac{4}{12}} = \frac{11}{9} \div \frac{4}{12}$$

$$\dfrac{\dfrac{6x + 3}{x}}{\dfrac{2x + 1}{8}} = \frac{6x + 3}{x} \div \frac{2x + 1}{8}$$

Step 3 Now use the definition of division and multiply by the reciprocal. Then write in lowest terms using the fundamental property.

$$= \frac{11}{9} \cdot \frac{12}{4}$$
$$= \frac{11 \cdot 3 \cdot 4}{3 \cdot 3 \cdot 4}$$
$$= \frac{11}{3}$$

$$= \frac{6x + 3}{x} \cdot \frac{8}{2x + 1}$$
$$= \frac{3(2x + 1)}{x} \cdot \frac{8}{2x + 1}$$
$$= \frac{24}{x}$$

◀ Work Problem **1** at the Side.

ΕXAMPLE 2 **Simplifying a Complex Fraction (Method 1)**

Simplify the complex fraction.

$\dfrac{\dfrac{xp}{q^3}}{\dfrac{p^2}{qx^2}}$ The numerator and denominator are single fractions, so use the definition of division and then the fundamental property.

$$\frac{xp}{q^3} \div \frac{p^2}{qx^2}$$
$$= \frac{xp}{q^3} \cdot \frac{qx^2}{p^2}$$
$$= \frac{x^3}{q^2p}$$

◀ Work Problem **2** at the Side.

EXAMPLE 3 **Simplifying a Complex Fraction (Method 1)**

Simplify the complex fraction.

$$\dfrac{\dfrac{3}{x+2}-4}{\dfrac{2}{x+2}+1}$$

$$=\dfrac{\dfrac{3}{x+2}-\dfrac{4(x+2)}{x+2}}{\dfrac{2}{x+2}+\dfrac{1(x+2)}{x+2}}$$ Write both second terms with a denominator of $x+2$.

$$=\dfrac{\dfrac{3-4(x+2)}{x+2}}{\dfrac{2+1(x+2)}{x+2}}$$ Subtract in the numerator.

   Add in the denominator.

$$=\dfrac{\dfrac{3-4x-8}{x+2}}{\dfrac{2+x+2}{x+2}}$$ Be careful with signs.

   Distributive property

$$=\dfrac{\dfrac{-5-4x}{x+2}}{\dfrac{4+x}{x+2}}$$ Combine like terms.

$$=\dfrac{-5-4x}{x+2}\cdot\dfrac{x+2}{4+x}$$ Multiply by the reciprocal.

$$=\dfrac{-5-4x}{4+x}$$ Lowest terms

CAUTION

$$\dfrac{\dfrac{a}{b}+\dfrac{c}{d}}{\dfrac{e}{f}+\dfrac{g}{h}}\quad\textit{does not equal}\quad\left(\dfrac{a}{b}+\dfrac{c}{d}\right)\cdot\left(\dfrac{f}{e}+\dfrac{h}{g}\right).$$

Work Problem **3** *at the Side.* ▶

OBJECTIVE **2** **Simplify a complex fraction by multiplying numerator and denominator by the least common denominator (Method 2).** Since any expression can be multiplied by a form of 1 to get an equivalent expression, we can multiply both the numerator and the denominator of a complex fraction by the same nonzero expression to get an equivalent complex fraction. If we choose the expression to be the LCD of all the fractions within the complex fraction, the result will no longer be complex. This is Method 2.

3 Simplify using Method 1.

$$\dfrac{\dfrac{2}{x-1}+\dfrac{1}{x+1}}{\dfrac{3}{x-1}-\dfrac{4}{x+1}}$$

ANSWER

3. $\dfrac{3x+1}{-x+7}$

4 Simplify each complex fraction using Method 2.

(a) $\dfrac{\dfrac{2}{3} - \dfrac{1}{4}}{\dfrac{4}{9} + \dfrac{1}{2}}$

(b) $\dfrac{2 - \dfrac{6}{a}}{3 + \dfrac{4}{a}}$

(c) $\dfrac{\dfrac{p}{5 - p}}{\dfrac{4p}{2p + 1}}$

> **Method 2 for Simplifying a Complex Fraction**
>
> **Step 1** Find the LCD of all fractions within the complex fraction.
>
> **Step 2** Multiply both the numerator and the denominator of the complex fraction by this LCD using the distributive property as necessary. Write in lowest terms.

In the next example, Method 2 is used to simplify the complex fractions from Example 1.

EXAMPLE 4 Simplifying Complex Fractions (Method 2)

Simplify each complex fraction.

(a) $\dfrac{\dfrac{2}{3} + \dfrac{5}{9}}{\dfrac{1}{4} + \dfrac{1}{12}}$

(b) $\dfrac{6 + \dfrac{3}{x}}{\dfrac{x}{4} + \dfrac{1}{8}}$

Step 1 Find the LCD for all denominators in the complex fraction.

The LCD for 3, 9, 4, and 12 is 36. | The LCD for x, 4, and 8 is $8x$.

Step 2 Multiply the numerator and denominator of the complex fraction by the LCD.

$\dfrac{\dfrac{2}{3} + \dfrac{5}{9}}{\dfrac{1}{4} + \dfrac{1}{12}}$

$= \dfrac{36\left(\dfrac{2}{3} + \dfrac{5}{9}\right)}{36\left(\dfrac{1}{4} + \dfrac{1}{12}\right)}$

Multiply each term by 36.

$= \dfrac{36\left(\dfrac{2}{3}\right) + 36\left(\dfrac{5}{9}\right)}{36\left(\dfrac{1}{4}\right) + 36\left(\dfrac{1}{12}\right)}$

$= \dfrac{24 + 20}{9 + 3}$

$= \dfrac{44}{12}$

$= \dfrac{4 \cdot 11}{4 \cdot 3}$

$= \dfrac{11}{3}$

$\dfrac{6 + \dfrac{3}{x}}{\dfrac{x}{4} + \dfrac{1}{8}}$

$= \dfrac{8x\left(6 + \dfrac{3}{x}\right)}{8x\left(\dfrac{x}{4} + \dfrac{1}{8}\right)}$

Multiply each term by $8x$.

$= \dfrac{8x(6) + 8x\left(\dfrac{3}{x}\right)}{8x\left(\dfrac{x}{4}\right) + 8x\left(\dfrac{1}{8}\right)}$

$= \dfrac{48x + 24}{2x^2 + x}$

$= \dfrac{24(2x + 1)}{x(2x + 1)}$

$= \dfrac{24}{x}$

◀ **Work Problem 4** at the Side.

ANSWERS

4. (a) $\dfrac{15}{34}$ (b) $\dfrac{2a - 6}{3a + 4}$ (c) $\dfrac{2p + 1}{4(5 - p)}$

EXAMPLE 5 Simplifying a Complex Fraction (Method 2)

Simplify the complex fraction.

$$\dfrac{\dfrac{3}{5m} - \dfrac{2}{m^2}}{\dfrac{9}{2m} + \dfrac{3}{4m^2}}$$ The LCD for $5m$, m^2, $2m$, and $4m^2$ is $20m^2$.

$$= \dfrac{20m^2\left(\dfrac{3}{5m} - \dfrac{2}{m^2}\right)}{20m^2\left(\dfrac{9}{2m} + \dfrac{3}{4m^2}\right)}$$ Multiply numerator and denominator by $20m^2$.

$$= \dfrac{20m^2\left(\dfrac{3}{5m}\right) - 20m^2\left(\dfrac{2}{m^2}\right)}{20m^2\left(\dfrac{9}{2m}\right) + 20m^2\left(\dfrac{3}{4m^2}\right)}$$ Distributive property

$$= \dfrac{12m - 40}{90m + 15}$$ Multiply and simplify.

Work Problem **5** *at the Side.* ▶

Some students use Method 1 for problems like Example 2, which is the quotient of two fractions, and Method 2 for problems like Examples 1, 3, 4, and 5, which have sums or differences in the numerators or denominators.

EXAMPLE 6 Simplifying Complex Fractions

Simplify each complex fraction. Use either method.

(a) $\dfrac{\dfrac{1}{y} + \dfrac{2}{y+2}}{\dfrac{4}{y} - \dfrac{3}{y+2}}$ We use Method 2, since there are sums and differences in the numerator and denominator.

$$= \dfrac{\left(\dfrac{1}{y} + \dfrac{2}{y+2}\right)y(y+2)}{\left(\dfrac{4}{y} - \dfrac{3}{y+2}\right)y(y+2)}$$ Multiply numerator and denominator by the LCD, $y(y+2)$.

$$= \dfrac{\left(\dfrac{1}{y}\right)y(y+2) + \left(\dfrac{2}{y+2}\right)y(y+2)}{\left(\dfrac{4}{y}\right)y(y+2) - \left(\dfrac{3}{y+2}\right)y(y+2)}$$ Distributive property

$$= \dfrac{1(y+2) + 2y}{4(y+2) - 3y}$$ Fundamental property

$$= \dfrac{y + 2 + 2y}{4y + 8 - 3y}$$ Distributive property

$$= \dfrac{3y + 2}{y + 8}$$ Combine like terms.

Continued on Next Page

---

5 Simplify using Method 2.

$$\dfrac{\dfrac{2}{5x} - \dfrac{3}{x^2}}{\dfrac{7}{4x} + \dfrac{1}{2x^2}}$$

ANSWER

5. $\dfrac{8x - 60}{35x + 10}$

6 Simplify each complex fraction. Use either method.

(a) $\dfrac{\dfrac{1}{x} + \dfrac{2}{x-1}}{\dfrac{2}{x} - \dfrac{4}{x-1}}$

(b) $\dfrac{1 - \dfrac{2}{x} - \dfrac{15}{x^2}}{1 + \dfrac{5}{x} + \dfrac{6}{x^2}}$

(c) $\dfrac{\dfrac{2x+3}{x-4}}{\dfrac{4x^2-9}{x^2-16}}$

(b) $\dfrac{1 - \dfrac{2}{x} - \dfrac{3}{x^2}}{1 - \dfrac{5}{x} + \dfrac{6}{x^2}}$ There are sums and differences in the numerator and denominator, so we use Method 2.

$= \dfrac{\left(1 - \dfrac{2}{x} - \dfrac{3}{x^2}\right)x^2}{\left(1 - \dfrac{5}{x} + \dfrac{6}{x^2}\right)x^2}$ Multiply numerator and denominator by the LCD, x^2.

$= \dfrac{x^2 - 2x - 3}{x^2 - 5x + 6}$ Distributive property

$= \dfrac{(x-3)(x+1)}{(x-3)(x-2)}$ Factor.

$= \dfrac{x+1}{x-2}$ Divide out the common factor.

You may wish to verify that in this example, like the others, *either* method can be used to simplify the complex fractions.

(c) $\dfrac{\dfrac{x+2}{x-3}}{\dfrac{x^2-4}{x^2-9}}$ This is a quotient of two rational expressions, so we use Method 1.

$= \dfrac{x+2}{x-3} \div \dfrac{x^2-4}{x^2-9}$ Write as a division problem.

$= \dfrac{x+2}{x-3} \cdot \dfrac{x^2-9}{x^2-4}$ Multiply by the reciprocal.

$= \dfrac{(x+2)(x+3)(x-3)}{(x-3)(x+2)(x-2)}$ Multiply; factor.

$= \dfrac{x+3}{x-2}$ Divide out the common factors.

 ◀ *Work Problem* **6** *at the Side.*

15.5 ▶▶▶ Exercises

Note: In many problems involving complex fractions, several different equivalent forms of the answer often exist. If your answer does not look exactly like the one given in the back of the book, check to see whether you have written an equivalent form.

1. Consider the complex fraction $\dfrac{\frac{1}{2} - \frac{1}{3}}{\frac{5}{6} - \frac{1}{12}}$. Answer each part, outlining Method 1 for simplifying this complex fraction.

   (a) To combine the terms in the numerator, we must find the LCD of $\frac{1}{2}$ and $\frac{1}{3}$. What is this LCD? Determine the simplified form of the numerator of the complex fraction.

   (b) To combine the terms in the denominator, we must find the LCD of $\frac{5}{6}$ and $\frac{1}{12}$. What is this LCD? Determine the simplified form of the denominator of the complex fraction.

   (c) Now use the results from parts (a) and (b) to write the complex fraction as a division problem using the symbol $\div$.

   (d) Perform the operation from part (c) to obtain the final simplification.

2. Consider the same complex fraction given in Exercise 1, $\dfrac{\frac{1}{2} - \frac{1}{3}}{\frac{5}{6} - \frac{1}{12}}$. Answer each part, outlining Method 2 for simplifying this complex fraction.

   (a) We must determine the LCD of all the fractions within the complex fraction. What is this LCD?

   (b) Multiply every term in the complex fraction by the LCD found in part (a), but at this time do not combine the terms in the numerator and the denominator.

   (c) Now combine the terms from part (b) to obtain the simplified form of the complex fraction.

Simplify each complex fraction. Use either method. See Examples 1–6.

3. $\dfrac{-\frac{4}{3}}{\frac{2}{9}}$

4. $\dfrac{-\frac{5}{6}}{\frac{5}{4}}$

5. $\dfrac{\frac{p}{q^2}}{\frac{p^2}{q}}$

6. $\dfrac{\frac{a}{x}}{\frac{a^2}{2x}}$

7. $\dfrac{\frac{x}{y^2}}{\frac{x^2}{y}}$

8. $\dfrac{\frac{p^4}{r}}{\frac{p^2}{r^2}}$

9. $\dfrac{\frac{4a^4b^3}{3a}}{\frac{2ab^4}{b^2}}$

10. $\dfrac{\frac{2r^4t^2}{3t}}{\frac{5r^2t^5}{3r}}$

11. $\dfrac{\dfrac{m+2}{3}}{\dfrac{m-4}{m}}$

12. $\dfrac{\dfrac{q-5}{q}}{\dfrac{q+5}{3}}$

13. $\dfrac{\dfrac{2}{x}-3}{\dfrac{2-3x}{2}}$

14. $\dfrac{6+\dfrac{2}{r}}{\dfrac{3r+1}{4}}$

15. $\dfrac{\dfrac{1}{x}+x}{\dfrac{x^2+1}{8}}$

16. $\dfrac{\dfrac{3}{m}-m}{\dfrac{3-m^2}{4}}$

17. $\dfrac{a-\dfrac{5}{a}}{a+\dfrac{1}{a}}$

18. $\dfrac{q+\dfrac{1}{q}}{q+\dfrac{4}{q}}$

19. $\dfrac{\dfrac{1}{2}+\dfrac{1}{p}}{\dfrac{2}{3}+\dfrac{1}{p}}$

20. $\dfrac{\dfrac{3}{4}-\dfrac{1}{r}}{\dfrac{1}{5}+\dfrac{1}{r}}$

21. $\dfrac{\dfrac{2}{p^2}-\dfrac{3}{5p}}{\dfrac{4}{p}+\dfrac{1}{4p}}$

22. $\dfrac{\dfrac{2}{m^2}-\dfrac{3}{m}}{\dfrac{2}{5m^2}+\dfrac{1}{3m}}$

23. $\dfrac{\dfrac{t}{t+2}}{\dfrac{4}{t^2-4}}$

24. $\dfrac{\dfrac{m}{m+1}}{\dfrac{3}{m^2-1}}$

25. $\dfrac{\dfrac{1}{k+1}-1}{\dfrac{1}{k+1}+1}$

26. $\dfrac{\dfrac{2}{p-1}+2}{\dfrac{3}{p-1}-2}$

27. $\dfrac{2+\dfrac{1}{x}-\dfrac{28}{x^2}}{3+\dfrac{13}{x}+\dfrac{4}{x^2}}$

28. $\dfrac{4-\dfrac{11}{x}-\dfrac{3}{x^2}}{2-\dfrac{1}{x}-\dfrac{15}{x^2}}$

29. $\dfrac{\dfrac{1}{m-1}+\dfrac{2}{m+2}}{\dfrac{2}{m+2}-\dfrac{1}{m-3}}$

30. $\dfrac{\dfrac{5}{r+3}-\dfrac{1}{r-1}}{\dfrac{2}{r+2}+\dfrac{3}{r+3}}$

31. $2-\dfrac{2}{2+\dfrac{2}{2+2}}$

32. $3-\dfrac{2}{4+\dfrac{2}{4-2}}$

15.6 ▶▶▶ Solving Equations with Rational Expressions

OBJECTIVES

1. **Distinguish between operations with rational expressions and equations with terms that are rational expressions.**

2. **Solve equations with rational expressions.**

3. **Solve a formula for a specified variable.**

OBJECTIVE 1 Distinguish between operations with rational expressions and equations with terms that are rational expressions. Before solving equations with rational expressions, you must understand the difference between sums and differences of terms with rational coefficients, or rational *expressions*, and *equations* with terms that are rational expressions. *Sums and differences lead to expressions to simplify, while equations are solved.*

EXAMPLE 1 Distinguishing between Expressions and Equations

Identify each of the following as an *expression* or an *equation*. Then simplify the expression or solve the equation.

(a) $\dfrac{3}{4}x - \dfrac{2}{3}x$ This is a difference of two terms. It represents an *expression* to simplify since there is no equals sign.

$= \dfrac{3 \cdot 3}{3 \cdot 4}x - \dfrac{4 \cdot 2}{4 \cdot 3}x$ The LCD is 12. Write each coefficient with this LCD.

$= \dfrac{9}{12}x - \dfrac{8}{12}x$ Multiply.

$= \dfrac{1}{12}x$ Combine like terms, using the distributive property: $\dfrac{9}{12}x - \dfrac{8}{12}x = \left(\dfrac{9}{12} - \dfrac{8}{12}\right)x.$

(b) $\dfrac{3}{4}x - \dfrac{2}{3}x = \dfrac{1}{2}$ Because of the equals sign, this is an *equation* to be solved.

$12\left(\dfrac{3}{4}x - \dfrac{2}{3}x\right) = 12\left(\dfrac{1}{2}\right)$ Use the multiplication property of equality to clear fractions. Multiply by 12, the LCD.

$12\left(\dfrac{3}{4}x\right) - 12\left(\dfrac{2}{3}x\right) = 12\left(\dfrac{1}{2}\right)$ Distributive property

$9x - 8x = 6$ Multiply.

$x = 6$ Combine like terms.

Check $\dfrac{3}{4}x - \dfrac{2}{3}x = \dfrac{1}{2}$ Original equation

$\dfrac{3}{4}(6) - \dfrac{2}{3}(6) \stackrel{?}{=} \dfrac{1}{2}$ Let $x = 6$.

$\dfrac{9}{2} - 4 \stackrel{?}{=} \dfrac{1}{2}$ Multiply.

$\dfrac{1}{2} = \dfrac{1}{2}$ True

Since a true statement results, $\{6\}$ is the solution set of the equation.

The ideas of Example 1 can be summarized as follows.

1 Identify each as an *expression* or an *equation*. Then simplify the expression or solve the equation.

(a) $\dfrac{2x}{3} - \dfrac{4x}{9} = 2$

(b) $\dfrac{2x}{3} - \dfrac{4x}{9}$

2 Solve each equation, and check your solutions.

(a) $\dfrac{x}{5} + 3 = \dfrac{3}{5}$

(b) $\dfrac{x}{2} - \dfrac{x}{3} = \dfrac{5}{6}$

> **Uses of the LCD**
>
> When adding or subtracting rational expressions, keep the LCD throughout the simplification. (See Example 1(a).)
>
> When solving an equation with terms that are rational expressions, multiply each side by the LCD so that denominators are eliminated. (See Example 1(b).)

◀ **Work Problem** **1** **at the Side.**

OBJECTIVE **2** **Solve equations with rational expressions.** When an equation involves fractions as in Example 1(b), we use the multiplication property of equality to clear it of fractions. Choose as multiplier the LCD of all denominators in the fractions of the equation.

EXAMPLE 2 **Solving an Equation with Rational Expressions**

Solve $\dfrac{x}{3} + \dfrac{x}{4} = 10 + x$. Check the solution.

$$12\left(\dfrac{x}{3} + \dfrac{x}{4}\right) = 12(10 + x) \qquad \text{Multiply by the LCD, 12, to clear fractions.}$$

$$12\left(\dfrac{x}{3}\right) + 12\left(\dfrac{x}{4}\right) = 12(10) + 12x \qquad \text{Distributive property}$$

$$4x + 3x = 120 + 12x \qquad \text{Multiply.}$$

$$7x = 120 + 12x \qquad \text{Combine like terms.}$$

$$-5x = 120 \qquad \text{Subtract } 12x.$$

$$x = -24 \qquad \text{Divide by } -5.$$

Check
$$\dfrac{x}{3} + \dfrac{x}{4} = 10 + x \qquad \text{Original equation}$$

$$\dfrac{-24}{3} + \dfrac{-24}{4} \stackrel{?}{=} 10 - 24 \qquad \text{Let } x = -24.$$

$$-8 - 6 \stackrel{?}{=} -14$$

$$-14 = -14 \qquad \text{True}$$

The solution set is $\{-24\}$.

◀ **Work Problem** **2** **at the Side.**

> **CAUTION**
> The use of the LCD here is different than in **Section 15.5.** Here, we use the multiplication property of equality to multiply each side of an *equation* by the least common multiple of the denominators, which is the LCD. Earlier, we used the fundamental property to multiply a *fraction* by another fraction that had the LCD as both its numerator and denominator. Be careful not to confuse these two methods.

EXAMPLE 3 Solving an Equation with Rational Expressions

Solve $\dfrac{p}{2} - \dfrac{p-1}{3} = 1$.

$$6\left(\dfrac{p}{2} - \dfrac{p-1}{3}\right) = 6 \cdot 1 \qquad \text{Multiply by the LCD, 6.}$$

$$6\left(\dfrac{p}{2}\right) - 6\left(\dfrac{p-1}{3}\right) = 6 \qquad \text{Distributive property}$$

$$3p - 2(p-1) = 6 \qquad \boxed{\text{Use parentheses around } p-1 \text{ to avoid errors.}}$$

$$3p - 2(p) - 2(-1) = 6 \qquad \text{Distributive property}$$

$\boxed{\text{Be careful with signs.}}$

$$3p - 2p + 2 = 6 \qquad \text{Multiply.}$$

$$p + 2 = 6 \qquad \text{Combine like terms.}$$

$$p = 4 \qquad \text{Subtract 2.}$$

Check that $\{4\}$ is the solution set by replacing p with 4 in the original equation.

Work Problem **3** *at the Side.* ▶

Recall from **Section 15.1** that the denominator of a rational expression cannot equal 0, since division by 0 is undefined. *Therefore, when solving an equation with rational expressions that have variables in the denominator, the solution cannot be a number that makes the denominator equal 0.*

EXAMPLE 4 Solving an Equation with Rational Expressions

Solve $\dfrac{x}{x-2} = \dfrac{2}{x-2} + 2$. Check the proposed solution.

$$\dfrac{x}{x-2} = \dfrac{2}{x-2} + 2 \qquad \begin{array}{l} x \neq 2, \text{ since both denominators} \\ \text{equal 0 if } x \text{ is 2.} \end{array}$$

$$(x-2)\left(\dfrac{x}{x-2}\right) = (x-2)\left(\dfrac{2}{x-2} + 2\right) \qquad \begin{array}{l}\text{Multiply each side} \\ \text{by the LCD, } x-2.\end{array}$$

$$(x-2)\left(\dfrac{x}{x-2}\right) = (x-2)\left(\dfrac{2}{x-2}\right) + (x-2)(2) \qquad \text{Distributive property}$$

$$x = 2 + 2x - 4 \qquad \text{Simplify.}$$

$$x = -2 + 2x \qquad \text{Combine like terms.}$$

$$-x = -2 \qquad \text{Subtract } 2x.$$

$$x = 2 \qquad \text{Divide by } -1.$$

Replacing x with 2, however, causes denominators to equal 0.

Check

$$\dfrac{x}{x-2} = \dfrac{2}{x-2} + 2 \qquad \text{Original equation}$$

$$\dfrac{2}{2-2} \stackrel{?}{=} \dfrac{2}{2-2} + 2 \qquad \text{Let } x = 2.$$

$\boxed{\text{Division by 0 is undefined.}}$

$$\dfrac{2}{0} \stackrel{?}{=} \dfrac{2}{0} + 2$$

Thus, the proposed solution 2 must be rejected, and the solution set is $\varnothing$.

Work Problem **4** *at the Side.* ▶

---

3 Solve each equation, and check your solutions.

(a) $\dfrac{k}{6} - \dfrac{k+1}{4} = -\dfrac{1}{2}$

(b) $\dfrac{2m-3}{5} - \dfrac{m}{3} = -\dfrac{6}{5}$

4 Solve the equation, and check the proposed solution.

$$1 - \dfrac{2}{x+1} = \dfrac{2x}{x+1}$$

ANSWERS

3. (a) $\{3\}$ **(b)** $\{-9\}$

4. $\varnothing$ (When the equation is solved, -1 is a proposed solution. However, because $x = -1$ leads to a 0 denominator in the original equation, there is no solution.)

5 Solve each equation, and check the proposed solutions.

(a) $\dfrac{4}{x^2 - 3x} = \dfrac{1}{x^2 - 9}$

While it is always a good idea to check solutions to guard against arithmetic and algebraic errors, *it is essential to check proposed solutions when variables appear in denominators in the original equation.* Some students like to determine which numbers cannot be solutions *before* solving the equation, as we did in Example 4.

The steps used to solve an equation with rational expressions follow.

> **Solving an Equation with Rational Expressions**
>
> *Step 1* **Multiply each side of the equation by the LCD.** (This clears the equation of fractions.)
>
> *Step 2* **Solve** the resulting equation for proposed solutions.
>
> *Step 3* **Check** each proposed solution by substituting it in the original equation. Reject any that cause a denominator to equal 0.

EXAMPLE 5 **Solving an Equation with Rational Expressions**

Solve $\dfrac{2}{x^2 - x} = \dfrac{1}{x^2 - 1}$. Check the proposed solution.

Step 1 Factor the denominators to find the LCD.

$$\dfrac{2}{x(x-1)} = \dfrac{1}{(x+1)(x-1)} \qquad \text{The LCD is } x(x+1)(x-1).$$

Notice that 0, −1, and 1 cannot be solutions of this equation. Multiply each side of the equation by $x(x+1)(x-1)$.

(b) $\dfrac{2}{p^2 - 2p} = \dfrac{3}{p^2 - p}$

$$x(x+1)(x-1)\,\dfrac{2}{x(x-1)} = x(x+1)(x-1)\,\dfrac{1}{(x+1)(x-1)}$$

Multiply by the LCD.

Step 2

$$2(x+1) = x \qquad \text{Divide out the common factors.}$$
$$2x + 2 = x \qquad \text{Distributive property}$$
$$x + 2 = 0 \qquad \text{Subtract } x.$$
$$x = -2 \qquad \text{Subtract 2.}$$

Step 3 The proposed solution is −2, which does not make any denominator equal 0.

Check

$$\dfrac{2}{x^2 - x} = \dfrac{1}{x^2 - 1} \qquad \text{Original equation}$$

$$\dfrac{2}{(-2)^2 - (-2)} \overset{?}{=} \dfrac{1}{(-2)^2 - 1} \qquad \text{Let } x = -2.$$

$$\dfrac{2}{4 + 2} \overset{?}{=} \dfrac{1}{4 - 1} \qquad \text{Apply the exponents.}$$

$$\dfrac{1}{3} = \dfrac{1}{3} \qquad \text{True}$$

The solution set is {−2}.

◀ *Work Problem* **5** *at the Side.*

EXAMPLE 6 **Solving an Equation with Rational Expressions**

Solve $\dfrac{2m}{m^2 - 4} + \dfrac{1}{m - 2} = \dfrac{2}{m + 2}$.

Factor the first denominator on the left.

$$\dfrac{2m}{(m + 2)(m - 2)} + \dfrac{1}{m - 2} = \dfrac{2}{m + 2} \quad \text{The LCD is } (m + 2)(m - 2).$$

Notice that -2 and 2 cannot be solutions of the equation.

$$(m + 2)(m - 2)\left(\dfrac{2m}{(m + 2)(m - 2)} + \dfrac{1}{m - 2}\right)$$

$$= (m + 2)(m - 2)\dfrac{2}{m + 2} \quad \text{Multiply by the LCD.}$$

$$(m + 2)(m - 2)\dfrac{2m}{(m + 2)(m - 2)} + (m + 2)(m - 2)\dfrac{1}{m - 2}$$

$$= (m + 2)(m - 2)\dfrac{2}{m + 2} \quad \text{Distributive property}$$

$2m + m + 2 = 2(m - 2)$ Divide out the common factors.

$3m + 2 = 2m - 4$ Combine like terms; distributive property

$m + 2 = -4$ Subtract $2m$.

$m = -6$ Subtract 2.

Check to see that $\{-6\}$ is the solution set of the given equation.

——— *Work Problem* **6** *at the Side.* ▶

EXAMPLE 7 **Solving an Equation with Rational Expressions**

Solve $\dfrac{1}{x - 1} + \dfrac{1}{2} = \dfrac{2}{x^2 - 1}$.

The denominator $x^2 - 1$ factors as $(x + 1)(x - 1)$. Multiply each side by the LCD, $2(x + 1)(x - 1)$. Notice that -1 and 1 cannot be solutions.

$$2(x + 1)(x - 1)\left(\dfrac{1}{x - 1} + \dfrac{1}{2}\right) = 2(x + 1)(x - 1)\dfrac{2}{(x + 1)(x - 1)}$$

$$2(x + 1)(x - 1)\dfrac{1}{x - 1} + 2(x + 1)(x - 1)\dfrac{1}{2}$$

$$= 2(x + 1)(x - 1)\dfrac{2}{(x + 1)(x - 1)}$$

$2(x + 1) + (x + 1)(x - 1) = 2(2)$ Divide out the common factors.

$2x + 2 + x^2 - 1 = 4$ Distributive property; multiply.

$x^2 + 2x + 1 = 4$ Combine like terms.

$x^2 + 2x - 3 = 0$ Subtract 4. **[Write in standard form.]**

$(x + 3)(x - 1) = 0$ Factor.

$x + 3 = 0 \quad \text{or} \quad x - 1 = 0$ Zero-factor property

$x = -3 \quad \text{or} \quad x = 1$ Solve for x.

——— **Continued on Next Page**

---

6 Solve each equation, and check the proposed solutions.

(a)

$$\dfrac{2p}{p^2 - 1} = \dfrac{2}{p + 1} - \dfrac{1}{p - 1}$$

(b)

$$\dfrac{8r}{4r^2 - 1} = \dfrac{3}{2r + 1} + \dfrac{3}{2r - 1}$$

ANSWERS

6. **(a)** $\{-3\}$ **(b)** $\{0\}$

7 Solve the equation, and check the proposed solutions.

$$\frac{2}{3x + 1} - \frac{1}{x} = \frac{-6x}{3x + 1}$$

8 Solve each equation, and check the proposed solutions.

(a) $\dfrac{1}{x - 2} + \dfrac{1}{5} = \dfrac{2}{5(x^2 - 4)}$

(b) $\dfrac{6}{5a + 10} - \dfrac{1}{a - 5}$

$= \dfrac{4}{a^2 - 3a - 10}$

Proposed solutions are -3 and 1. However, 1 makes an original denominator equal 0, so 1 is *not* a solution. Check that -3 is a solution.

Check

$$\frac{1}{x - 1} + \frac{1}{2} = \frac{2}{x^2 - 1} \qquad \text{Original equation}$$

$$\frac{1}{-3 - 1} + \frac{1}{2} \stackrel{?}{=} \frac{2}{(-3)^2 - 1} \qquad \text{Let } x = -3.$$

$$\frac{1}{-4} + \frac{1}{2} \stackrel{?}{=} \frac{2}{9 - 1} \qquad \text{Simplify.}$$

$$\frac{1}{4} \stackrel{?}{=} \frac{2}{8}$$

$$\frac{1}{4} = \frac{1}{4} \qquad \text{True}$$

The check shows that $\{-3\}$ is the solution set.

◀ *Work Problem* **7** *at the Side.*

EXAMPLE 8   **Solving an Equation with Rational Expressions**

Solve $\dfrac{1}{k^2 + 4k + 3} + \dfrac{1}{2k + 2} = \dfrac{3}{4k + 12}$.

$$\frac{1}{k^2 + 4k + 3} + \frac{1}{2k + 2} = \frac{3}{4k + 12}$$

$$\frac{1}{(k + 1)(k + 3)} + \frac{1}{2(k + 1)} = \frac{3}{4(k + 3)} \qquad \begin{array}{l}\text{Factor each denominator.} \\ \text{The LCD is } 4(k + 1)(k + 3).\end{array}$$

Notice that -1 and -3 cannot be solutions of this equation, since replacing k with -1 or -3 causes denominators to equal 0.

$$4(k + 1)(k + 3)\left(\frac{1}{(k + 1)(k + 3)} + \frac{1}{2(k + 1)}\right)$$

$$= 4(k + 1)(k + 3)\frac{3}{4(k + 3)} \qquad \text{Multiply by the LCD.}$$

$$4(k + 1)(k + 3)\frac{1}{(k + 1)(k + 3)} + 2 \cdot 2(k + 1)(k + 3)\frac{1}{2(k + 1)}$$

$$= 4(k + 1)(k + 3)\frac{3}{4(k + 3)} \qquad \text{Distributive property}$$

$$4 + 2(k + 3) = 3(k + 1) \qquad \text{Simplify.}$$

$$4 + 2k + 6 = 3k + 3 \qquad \text{Distributive property}$$

$$2k + 10 = 3k + 3 \qquad \text{Combine like terms.}$$

$$10 = k + 3 \qquad \text{Subtract } 2k.$$

$$7 = k \qquad \text{Subtract } 3.$$

The proposed solution, 7, does not make an original denominator equal 0. A check shows that the algebra is correct, so $\{7\}$ is the solution set.

◀ *Work Problem* **8** *at the Side.*

ANSWERS

7. $\left\{\dfrac{1}{2}\right\}$

8. (a) $\{-4, -1\}$ (b) $\{60\}$

OBJECTIVE 3 Solve a formula for a specified variable. Solving a formula for a specified variable was first discussed in **Section 10.3**. *Remember to treat the variable for which you are solving as if it were the only variable, and all others as if they were constants.*

EXAMPLE 9 Solving for Specified Variables

Solve each formula for the specified variable.

(a) $a = \dfrac{v - w}{t}$ for v

> Our goal is to isolate v.

$$a = \dfrac{v - w}{t}$$

$$at = v - w \qquad \text{Multiply by } t.$$

$$at + w = v, \quad \text{or} \quad v = at + w \qquad \text{Add } w.$$

Check Substitute $at + w$ for v in the original equation. The final result will be the identity $a = a$, indicating that the result obtained is correct.

(b) $F = \dfrac{k}{d - D}$ for d

$$F = \dfrac{k}{d - D}$$

> Our goal is to isolate d.

$$F(d - D) = \dfrac{k}{d - D}(d - D) \qquad \begin{array}{l}\text{Multiply by } d - D \\ \text{to clear the fraction.}\end{array}$$

$$F(d - D) = k \qquad \text{Simplify.}$$

$$Fd - FD = k \qquad \text{Distributive property}$$

$$Fd = k + FD \qquad \text{Add } FD.$$

$$d = \dfrac{k + FD}{F} \qquad \text{Divide by } F.$$

If desired, we can write an equivalent form of this answer by expressing the right side as the sum of two fractions.

$$d = \dfrac{k + FD}{F} \qquad \text{Answer from above}$$

$$d = \dfrac{k}{F} + \dfrac{FD}{F} \qquad \begin{array}{l}\text{Definition of addition of} \\ \text{fractions: } \frac{a + b}{c} = \frac{a}{c} + \frac{b}{c}\end{array}$$

$$d = \dfrac{k}{F} + D \qquad \begin{array}{l}\text{Divide out the common} \\ \text{factor from } \frac{FD}{F}.\end{array}$$

Either answer is correct:

$$d = \dfrac{k + FD}{F}, \quad \text{or} \quad d = \dfrac{k}{F} + D.$$

Work Problem **9** *at the Side.* ▶

---

9 Solve each formula for the specified variable.

(a) $r = \dfrac{A - p}{pt}$ for A

(b) $z = \dfrac{x}{x + y}$ for y

---

10 Solve $\dfrac{2}{x} = \dfrac{1}{y} + \dfrac{1}{z}$ for z.

EXAMPLE 10 Solving for a Specified Variable

Solve the formula $\dfrac{1}{a} = \dfrac{1}{b} + \dfrac{1}{c}$ for c.

$$\frac{1}{a} = \frac{1}{b} + \frac{1}{c}$$

> Goal: Isolate c, the specified variable.

$$abc\left(\frac{1}{a}\right) = abc\left(\frac{1}{b} + \frac{1}{c}\right) \qquad \text{Multiply by the LCD, } abc.$$

$$abc\left(\frac{1}{a}\right) = abc\left(\frac{1}{b}\right) + abc\left(\frac{1}{c}\right) \qquad \text{Distributive property}$$

$$bc = ac + ab \qquad \text{Simplify.}$$

$$bc - ac = ab \qquad \text{Subtract } ac \text{ to get both terms with } c \text{ on the same side.}$$

> Pay careful attention here.

$$c(b - a) = ab \qquad \text{Factor out } c.$$

$$c = \frac{ab}{b - a} \qquad \text{Divide by } b - a.$$

CAUTION

Students often have trouble in the step that involves factoring out the variable for which they are solving. In Example 10, we needed to get both terms with c on the same side of the equation. This allowed us to factor out c on the left, and then isolate it by dividing each side by $b - a$.

$$bc = ac + ab \qquad \text{From Example 10}$$

$$bc - ac = ab \qquad \text{Get } c\text{-terms on the same side.}$$

$$c(b - a) = ab \qquad \text{Factor out } c.$$

$$c = \frac{ab}{b - a} \qquad \text{Divide by } b - a.$$

When solving an equation for a specified variable, be sure that the specified variable appears alone on only one side of the equals sign in the final equation.

◀ *Work Problem* **10** *at the Side.*

15.6 ▶▶▶ Exercises

Identify each as an expression *or an* equation. *Then simplify the expression or solve the equation. See Example 1.*

1. $\dfrac{7}{8}x + \dfrac{1}{5}x$

2. $\dfrac{4}{7}x + \dfrac{3}{5}x$

3. $\dfrac{7}{8}x + \dfrac{1}{5}x = 1$

4. $\dfrac{4}{7}x + \dfrac{3}{5}x = 1$

5. $\dfrac{3}{5}y - \dfrac{7}{10}y$

6. $\dfrac{3}{5}y - \dfrac{7}{10}y = 1$

7. Explain how the LCD is used in a different way when adding and subtracting rational expressions compared to solving equations with rational expressions.

8. If we multiply each side of the equation $\dfrac{6}{x+5} = \dfrac{6}{x+5}$ by $x + 5$, we get $6 = 6$. Are all real numbers solutions of this equation? Explain.

Solve each equation, and check your solutions. See Examples 2 and 3.

9. $\dfrac{2}{3}x + \dfrac{1}{2}x = -7$

10. $\dfrac{1}{4}x - \dfrac{1}{3}x = 1$

11. $\dfrac{3x}{5} - 6 = x$

12. $\dfrac{5t}{4} + t = 9$

13. $\dfrac{4m}{7} + m = 11$

14. $a - \dfrac{3a}{2} = 1$

15. $\dfrac{z-1}{4} = \dfrac{z+3}{3}$

16. $\dfrac{r-5}{2} = \dfrac{r+2}{3}$

17. $\dfrac{3p+6}{8} = \dfrac{3p-3}{16}$

18. $\dfrac{2z+1}{5} = \dfrac{7z+5}{15}$

19. $\dfrac{2x+3}{-6} = \dfrac{3}{2}$

20. $\dfrac{4x+3}{6} = \dfrac{5}{2}$

21. $\dfrac{q+2}{3} + \dfrac{q-5}{5} = \dfrac{7}{3}$

22. $\dfrac{b+7}{8} - \dfrac{b-2}{3} = \dfrac{4}{3}$

23. $\dfrac{t}{6} + \dfrac{4}{3} = \dfrac{t-2}{3}$

24. $\dfrac{x}{2} = \dfrac{5}{4} + \dfrac{x-1}{4}$

25. $\dfrac{3m}{5} - \dfrac{3m-2}{4} = \dfrac{1}{5}$

26. $\dfrac{8p}{5} = \dfrac{3p-4}{2} + \dfrac{5}{2}$

27. What values of x would have to be rejected as possible solutions of the equation $\dfrac{1}{x-4} = \dfrac{3}{2x}$?

28. What is wrong with the following problem? "Solve $\dfrac{2}{3x} + \dfrac{1}{5x}$."

Solve each equation, and check your solutions. See Examples 4–8.

29. $\dfrac{5-2x}{x} = \dfrac{1}{4}$

30. $\dfrac{2x+3}{x} = \dfrac{3}{2}$

◑ 31. $\dfrac{k}{k-4} - 5 = \dfrac{4}{k-4}$

32. $\dfrac{-5}{a+5} = \dfrac{a}{a+5} + 2$

33. $\dfrac{3}{x-1} + \dfrac{2}{4x-4} = \dfrac{7}{4}$

34. $\dfrac{2}{p+3} + \dfrac{3}{8} = \dfrac{5}{4p+12}$

35. $\dfrac{x}{3x+3} = \dfrac{2x-3}{x+1} - \dfrac{2x}{3x+3}$

36. $\dfrac{2k+3}{k+1} - \dfrac{3k}{2k+2} = \dfrac{-2k}{2k+2}$

37. $\dfrac{2}{m} = \dfrac{m}{5m+12}$

38. $\dfrac{x}{4-x} = \dfrac{2}{x}$

39. $\dfrac{5x}{14x+3} = \dfrac{1}{x}$

40. $\dfrac{m}{8m+3} = \dfrac{1}{3m}$

41. $\dfrac{2}{z - 1} - \dfrac{5}{4} = \dfrac{-1}{z + 1}$

42. $\dfrac{5}{p - 2} = 7 - \dfrac{10}{p + 2}$

43. $\dfrac{4}{x^2 - 3x} = \dfrac{1}{x^2 - 9}$

44. $\dfrac{2}{t^2 - 4} = \dfrac{3}{t^2 - 2t}$

45. $\dfrac{-2}{z + 5} + \dfrac{3}{z - 5} = \dfrac{20}{z^2 - 25}$

46. $\dfrac{3}{r + 3} - \dfrac{2}{r - 3} = \dfrac{-12}{r^2 - 9}$

47. $\dfrac{1}{x + 4} + \dfrac{x}{x - 4} = \dfrac{-8}{x^2 - 16}$

48. $\dfrac{x}{x - 3} + \dfrac{4}{x + 3} = \dfrac{18}{x^2 - 9}$

49. $\dfrac{4}{3x + 6} - \dfrac{3}{x + 3} = \dfrac{8}{x^2 + 5x + 6}$

50. $\dfrac{-13}{t^2 + 6t + 8} + \dfrac{4}{t + 2} = \dfrac{3}{2t + 8}$

51. $\dfrac{3x}{x^2 + 5x + 6} = \dfrac{5x}{x^2 + 2x - 3} - \dfrac{2}{x^2 + x - 2}$

52. $\dfrac{x + 4}{x^2 - 3x + 2} - \dfrac{5}{x^2 - 4x + 3} = \dfrac{x - 4}{x^2 - 5x + 6}$

53. If you are solving a formula for the letter k, and your steps lead to the equation $kr - mr = km$, what would be your next step?

54. If you are solving a formula for the letter k, and your steps lead to the equation $kr - km = mr$, what would be your next step?

Solve each formula for the specified variable. See Example 9.

55. $m = \dfrac{kF}{a}$ for F

56. $I = \dfrac{kE}{R}$ for E

57. $m = \dfrac{kF}{a}$ for a

58. $I = \dfrac{kE}{R}$ for R

59. $m = \dfrac{y - b}{x}$ for y

60. $y = \dfrac{C - Ax}{B}$ for C

61. $I = \dfrac{E}{R + r}$ for R

62. $I = \dfrac{E}{R + r}$ for r

63. $h = \dfrac{2A}{B + b}$ for b

64. $h = \dfrac{2A}{B + b}$ for B

65. $d = \dfrac{2S}{n(a + L)}$ for a

66. $d = \dfrac{2S}{n(a + L)}$ for L

Solve each equation for the specified variable. See Example 10.

67. $\dfrac{2}{r} + \dfrac{3}{s} + \dfrac{1}{t} = 1$ for t

68. $\dfrac{5}{p} + \dfrac{2}{q} + \dfrac{3}{r} = 1$ for r

69. $\dfrac{1}{a} - \dfrac{1}{b} - \dfrac{1}{c} = 2$ for c

70. $\dfrac{-1}{x} + \dfrac{1}{y} + \dfrac{1}{z} = 4$ for y

71. $9x + \dfrac{3}{z} = \dfrac{5}{y}$ for z

72. $-3t - \dfrac{4}{p} = \dfrac{6}{s}$ for p

Summary Exercises on Rational Expressions and Equations

A common student error is to confuse an equation, such as $\frac{1}{x} + \frac{1}{x-2} = \frac{3}{4}$, with an expression, such as $\frac{1}{x} + \frac{1}{x-2}$. Look for the equals sign to distinguish between them. Equations are solved for a numerical answer, while problems involving any of the four operations result in simplified expressions.

Add: $\dfrac{1}{x} + \dfrac{1}{x-2}$

$$= \dfrac{1(x-2)}{x(x-2)} + \dfrac{x(1)}{x(x-2)}$$ Write with a common denominator.

$$= \dfrac{x-2+x}{x(x-2)}$$ Add numerators; keep the same denominator.

$$= \dfrac{2x-2}{x(x-2)}$$ Combine like terms.

Subtract: $\dfrac{1}{x} - \dfrac{1}{x-2}$

$$= \dfrac{1(x-2)}{x(x-2)} - \dfrac{x(1)}{x(x-2)}$$ Write with a common denominator.

$$= \dfrac{x-2-x}{x(x-2)}$$ Subtract numerators; keep the same denominator.

$$= \dfrac{-2}{x(x-2)}$$ Combine like terms.

Multiply: $\dfrac{1}{x} \cdot \dfrac{1}{x-2}$

$$= \dfrac{1}{x(x-2)}$$ Multiply numerators and multiply denominators.

Divide: $\dfrac{1}{x} \div \dfrac{1}{x-2}$

$$= \dfrac{1}{x} \cdot \dfrac{x-2}{1}$$ Multiply by the reciprocal of the divisor.

$$= \dfrac{x-2}{x}$$

By contrast, consider the *equation*

$$\frac{1}{x} + \frac{1}{x-2} = \frac{3}{4}.$$

Neither 0 nor 2 can be a solution of this equation, since each will cause a denominator to equal 0.

$$4x(x-2)\left(\frac{1}{x} + \frac{1}{x-2}\right) = 4x(x-2)\frac{3}{4}$$ Multiply each side by the LCD, $4x(x-2)$, to clear fractions.

$$4x(x-2)\frac{1}{x} + 4x(x-2)\frac{1}{x-2} = 4x(x-2)\frac{3}{4}$$ Distributive property

$$4(x-2) + 4x = 3x(x-2)$$ Simplify.

$$4x - 8 + 4x = 3x^2 - 6x$$ Distributive property

$$3x^2 - 14x + 8 = 0$$ Standard form

$$(3x-2)(x-4) = 0$$ Factor.

$$3x - 2 = 0 \quad \text{or} \quad x - 4 = 0$$ Zero-factor property

$$x = \frac{2}{3} \quad \text{or} \quad x = 4$$ Solve for x.

The proposed solutions are $\frac{2}{3}$ and 4; neither makes a denominator equal 0. Check by substituting each proposed solution into the original equation to confirm that the solution set is $\left\{\frac{2}{3}, 4\right\}$.

> **Points to Remember when Working with Rational Expressions and Equations**
>
> 1. When simplifying rational expressions, the fundamental property is applied only after numerators and denominators have been *factored*.
> 2. When adding and subtracting rational expressions, the common denominator must be kept throughout the problem and in the final result.
> 3. When simplifying rational expressions, always check to see if the answer is in lowest terms; if it is not, use the fundamental property.
> 4. When solving equations with rational expressions, the LCD is used to clear the equation of fractions. Multiply each side by the LCD. (Notice how this use differs from that of the LCD in Point 2.)
> 5. When solving equations with rational expressions, reject any proposed solution that causes an original denominator to equal 0.

For each exercise, indicate "expression" if an expression is to be simplified or "equation" if an equation is to be solved. Then simplify the expression or solve the equation.

1. $\dfrac{4}{p} + \dfrac{6}{p}$

2. $\dfrac{x^3 y^2}{x^2 y^4} \cdot \dfrac{y^5}{x^4}$

3. $\dfrac{1}{x^2 + x - 2} \div \dfrac{4x^2}{2x - 2}$

4. $\dfrac{8}{m - 5} = 2$

5. $\dfrac{2x^2 + x - 6}{2x^2 - 9x + 9} \cdot \dfrac{x^2 - 2x - 3}{x^2 - 1}$

6. $\dfrac{2}{k^2 - 4k} + \dfrac{3}{k^2 - 16}$

7. $\dfrac{x - 4}{5} = \dfrac{x + 3}{6}$

8. $\dfrac{3t^2 - t}{6t^2 + 15t} \div \dfrac{6t^2 + t - 1}{2t^2 - 5t - 25}$

9. $\dfrac{4}{p + 2} + \dfrac{1}{3p + 6}$

10. $\dfrac{1}{x} + \dfrac{1}{x - 3} = -\dfrac{5}{4}$

11. $\dfrac{3}{t - 1} + \dfrac{1}{t} = \dfrac{7}{2}$

12. $\dfrac{6}{x} - \dfrac{2}{3x}$

13. $\dfrac{5}{4z} - \dfrac{2}{3z}$

14. $\dfrac{k + 2}{3} = \dfrac{2k - 1}{5}$

15. $\dfrac{1}{m^2 + 5m + 6} + \dfrac{2}{m^2 + 4m + 3}$

16. $\dfrac{2k^2 - 3k}{20k^2 - 5k} \div \dfrac{2k^2 - 5k + 3}{4k^2 + 11k - 3}$

17. $\dfrac{2}{x + 1} + \dfrac{5}{x - 1} = \dfrac{10}{x^2 - 1}$

18. $\dfrac{x}{x - 2} + \dfrac{3}{x + 2} = \dfrac{8}{x^2 - 4}$

15.7 ▶▶▶ Applications of Rational Expressions

In **Section 15.6** we solved equations with rational expressions; now we can solve applications that involve this type of equation. The six-step problem-solving method still applies.

OBJECTIVE **1** Solve problems about numbers.

EXAMPLE 1 Solving a Problem about an Unknown Number

If the same number is added to both the numerator and the denominator of the fraction $\frac{2}{5}$, the result is equivalent to $\frac{2}{3}$. Find the number.

Step 1 **Read** the problem carefully. We are trying to find a number.

Step 2 **Assign a variable.**

Let x = the number added to the numerator and the denominator.

Step 3 **Write an equation.** The fraction

$$\frac{2+x}{5+x}$$

represents the result of adding the same number x to both the numerator and the denominator. Since this result is equivalent to $\frac{2}{3}$, the equation is

$$\frac{2+x}{5+x} = \frac{2}{3}.$$

Step 4 **Solve** this equation.

$$3(5+x)\frac{2+x}{5+x} = 3(5+x)\frac{2}{3} \qquad \text{Multiply by the LCD, } 3(5+x).$$

$$3(2+x) = 2(5+x) \qquad \text{Divide out the common factors.}$$

$$6 + 3x = 10 + 2x \qquad \text{Distributive property}$$

$$x = 4 \qquad \text{Subtract } 2x; \text{ subtract } 6.$$

Step 5 **State the answer.** The number is 4.

Step 6 **Check** the solution in the words of the original problem. If 4 is added to both the numerator and the denominator of $\frac{2}{5}$, the result is $\frac{6}{9}$, which is equivalent to $\frac{2}{3}$, as required.

Work Problem **1** *at the Side.* ▶

OBJECTIVE **2** Solve problems about distance, rate, and time.

If an automobile travels at an average rate of 65 mph for 2 hr, then it travels $65 \times 2 = 130$ mi. Recall from **Section 12.4** that this is an example of the basic relationship between distance, rate, and time given by the formula $d = rt$. By solving, in turn, for r and t in the formula, we obtain two other equivalent forms of the formula. The three forms are given below.

Distance, Rate, and Time Relationship

$$d = rt \qquad r = \frac{d}{t} \qquad t = \frac{d}{r}$$

OBJECTIVES

1 Solve problems about numbers.

2 Solve problems about distance, rate, and time.

3 Solve problems about work.

1 Solve each problem.

(a) A certain number is added to the numerator and subtracted from the denominator of $\frac{5}{8}$. The new fraction equals the reciprocal of $\frac{5}{8}$. Find the number.

(b) The denominator of a fraction is 1 more than the numerator. If 6 is added to the numerator and subtracted from the denominator, the result is $\frac{15}{4}$. Find the original fraction.

The next example illustrates the uses of these formulas.

2 Solve each problem.

(a) The world record in the men's 100-m dash was set by Justin Gatlin of the United States in 2006. He ran it in 9.77 sec. What was his speed in meters per second, to the nearest hundredth? (*Source: Guinness World Records.*)

(b) The world record for the women's 3000-m run was set by Julnara Samitova of Russia in 2005. Her speed was 5.539 m per sec. To the nearest second, what was her time? (*Source: Guinness World Records.*)

(c) A small plane flew from Chicago to St. Louis averaging 145 mph. The trip took 2 hr. What is the distance between Chicago and St. Louis?

EXAMPLE 2 Finding Distance, Rate, or Time

(a) The speed of sound is 1088 ft per sec at sea level at 32°F. In 5 sec, under these conditions, sound travels

$$1088 \times 5 = 5440 \text{ ft.}$$
$$\text{Rate} \times \text{Time} = \text{Distance}$$

Here, we found distance, given rate and time, using $d = rt$.

(b) The winner of the first Indianapolis 500 race (in 1911) was Ray Harroun, driving a Marmon Wasp at an average speed of 74.59 mph. (*Source: Universal Almanac.*)
To complete the 500 mi, it took him

$$\text{Distance} \rightarrow \frac{500}{74.59} \approx 6.70 \text{ hr (rounded)}. \leftarrow \text{Time}$$
$$\text{Rate} \rightarrow$$

Here, we found time, given distance and rate, using $t = \frac{d}{r}$. To convert the decimal 0.70 hr to minutes, multiply by 60 to get $0.70(60) = 42$. It took Harroun about 6 hr, 42 min, to complete the race.

(c) At the 2004 Olympic Games in Athens, Greece, Chinese swimmer Luo Xuejuan set an Olympic record of 66.64 sec in the women's 100-m breaststroke swimming event. (*Source: World Almanac and Book of Facts.*)
Her rate was

$$\text{Rate} = \frac{\text{Distance} \rightarrow}{\text{Time} \rightarrow} \frac{100}{66.64} \approx 1.50 \text{ m per sec (rounded)}.$$

Here, we found rate, given distance and time, using $r = \frac{d}{t}$.

◀ *Work Problem* **2** *at the Side.*

Problem-Solving Hint

Many applied problems use the formulas just discussed. The next two examples show how to solve typical applications of the formula $d = rt$.

A helpful strategy for solving such problems is to *first make a sketch* showing what is happening in the problem. *Then make a table* using the information given, along with the unknown quantities. The table will help organize the information, and the sketch will help set up the equation.

EXAMPLE 3 Solving a Problem about Distance, Rate, and Time

Two cars leave Baton Rouge, Louisiana, at the same time and travel east on Interstate 10. One travels at a constant speed of 55 mph and the other travels at a constant speed of 63 mph. In how many hours will the distance between them be 24 mi?

Step 1 **Read** the problem. We are trying to find the time when the distance between the cars will be 24 mi.

─── **Continued on Next Page**

Step 2 **Assign a variable.** Since we are looking for time,

> let t = the number of hours until the distance between them is 24 mi.

The sketch in Figure 1 shows what is happening in the problem.

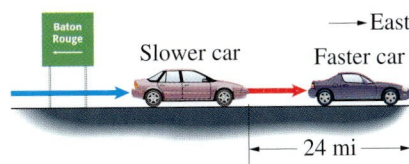

Figure 1

Now, construct a table like the one that follows. Fill in the information given in the problem, and use t for the time traveled by each car. Multiply rate by time to get the expressions for distances traveled.

	Rate	× Time	= Distance	
Faster Car	63	t	$63t$	
Slower Car	55	t	$55t$	

Difference is 24 mi.

The quantities $63t$ and $55t$ represent the two distances. Refer to Figure 1, and notice that the *difference* between the larger distance and the smaller distance is 24 mi.

Step 3 **Write an equation.** Subtract the smaller distance from the larger distance.

$$63t - 55t = 24$$

Step 4 **Solve** the equation.

$$63t - 55t = 24$$
$$8t = 24 \qquad \text{Combine like terms.}$$
$$t = 3 \qquad \text{Divide by 8.}$$

Step 5 **State the answer.** It will take the cars 3 hr to be 24 mi apart.

Step 6 **Check.** After 3 hr the faster car will have traveled

$$63 \times 3 = 189 \text{ mi,}$$

and the slower car will have traveled

$$55 \times 3 = 165 \text{ mi.}$$

Since $189 - 165 = 24$, the conditions of the problem are satisfied.

Problem-Solving Hint

In motion problems like the one in Example 3, once we have filled in two pieces of information in each row of the table, we can automatically fill in the third piece of information, using the appropriate form of the formula relating distance, rate, and time. Then we set up the equation based on our sketch and the information in the table.

Work Problem **3** *at the Side.* ▶

---

3 Solve each problem.

(a) From a point on a straight road, Lupe and Maria ride bicycles in opposite directions. Lupe rides 10 mph and Maria rides 12 mph. In how many hours will they be 55 mi apart?

(b) At a given hour, two steamboats leave a city in the same direction on a straight canal. One travels at 18 mph, and the other travels at 25 mph. In how many hours will the boats be 35 mi apart?

ANSWERS

3. **(a)** $2\frac{1}{2}$ hr **(b)** 5 hr

> **EXAMPLE 4** **Solving a Problem about Distance, Rate, and Time**

The Tickfaw River has a current of 3 mph. A motorboat takes as long to go 12 mi downstream as to go 8 mi upstream. What is the speed of the boat in still water?

Step 1 **Read** the problem. We want the speed of the boat in still water.

Step 2 **Assign a variable.** Let $x =$ the speed of the boat in still water.

Because the current pushes the boat when the boat is going downstream, the speed of the boat downstream will be the *sum* of the speed of the boat and the speed of the current, $(x + 3)$ mph. Because the current slows down the boat when the boat is going upstream, the boat's speed upstream is given by the *difference* between the speed of the boat in still water and the speed of the current, $(x - 3)$ mph. See Figure 2.

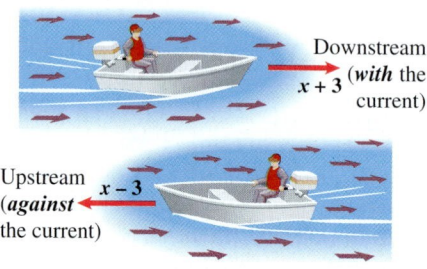

Downstream
$x + 3$ (*with* the current)

Upstream
(*against* the current) $x - 3$

Figure 2

This information is summarized in the following table.

	d	r	t
Downstream	12	$x + 3$	
Upstream	8	$x - 3$	

Fill in the column representing time by using the formula $t = \frac{d}{r}$. Then the time downstream is the distance divided by the rate, or

$$t = \frac{d}{r} = \frac{12}{x + 3},$$ Time downstream

and the time upstream is also the distance divided by the rate, or

$$t = \frac{d}{r} = \frac{8}{x - 3}.$$ Time upstream

The completed table follows.

	d	r	t
Downstream	12	$x + 3$	$\dfrac{12}{x + 3}$
Upstream	8	$x - 3$	$\dfrac{8}{x - 3}$

Times are equal.

Step 3 **Write an equation.** According to the original problem, the time downstream equals the time upstream. The two times from the table must therefore be equal, giving the equation

$$\frac{12}{x + 3} = \frac{8}{x - 3}.$$

Continued on Next Page

Step 4 **Solve.** Begin by multiplying each side by the LCD, $(x + 3)(x - 3)$.

$$\frac{12}{x + 3} = \frac{8}{x - 3}$$

$$(x + 3)(x - 3)\frac{12}{x + 3} = (x + 3)(x - 3)\frac{8}{x - 3}$$

$$12(x - 3) = 8(x + 3) \qquad \text{Divide out the common factors.}$$

$$12x - 36 = 8x + 24 \qquad \text{Distributive property}$$

$$4x = 60 \qquad \text{Subtract } 8x; \text{ add } 36.$$

$$x = 15 \qquad \text{Divide by 4.}$$

Step 5 **State the answer.** The speed of the boat in still water is 15 mph.

Step 6 **Check.** First find the speed of the boat downstream, which is $15 + 3 = 18$ mph. Traveling 12 mi would take

$$t = \frac{d}{r} = \frac{12}{18} = \frac{2}{3} \text{ hr.}$$

The speed of the boat upstream is $15 - 3 = 12$ mph, and traveling 8 mi would take

$$t = \frac{d}{r} = \frac{8}{12} = \frac{2}{3} \text{ hr.}$$

The time upstream equals the time downstream, as required.

Work Problem **4** *at the Side.* ▶

OBJECTIVE **3** **Solve problems about work.** Suppose that you can mow your lawn in 4 hr. Then after 1 hr, you will have mowed $\frac{1}{4}$ of the lawn. After 2 hr, you will have mowed $\frac{2}{4}$, or $\frac{1}{2}$, of the lawn, and so on. This idea is generalized as follows.

> **Rate of Work**
>
> If a job can be completed in t units of time, then the rate of work is
>
> $$\frac{1}{t} \text{ job per unit of time.}$$

> **Problem-Solving Hint**
>
> The relationship between problems involving work and problems involving distance is a very close one. Recall that the formula $d = rt$ says that distance traveled is equal to rate of travel multiplied by time traveled. Similarly, the fractional part of a job accomplished is equal to the rate of work multiplied by the time worked. In the lawn mowing example, after 3 hr, the fractional part of the job done is shown below.
>
> $$\underbrace{\frac{1}{4}}_{\substack{\text{Rate of} \\ \text{work}}} \cdot \underbrace{3}_{\substack{\text{Time} \\ \text{worked}}} = \underbrace{\frac{3}{4}}_{\substack{\text{Fractional part} \\ \text{of job done}}}$$

After 4 hr, $\frac{1}{4}(4) = 1$ whole job has been done.

4 Solve each problem.

(a) A boat can go 10 mi against the current in the same time it can go 30 mi with the current. The current is flowing at 4 mph. Find the speed of the boat with no current.

(b) An airplane, maintaining a constant airspeed, takes as long to go 450 mi with the wind as it does to go 375 mi against the wind. If the wind is blowing at 15 mph, what is the speed of the plane?

> **EXAMPLE 5** **Solving a Problem about Work Rates**

With spraying equipment, Mateo can paint the woodwork in a small house in 8 hr. His assistant, Chet, needs 14 hr to complete the same job painting by hand. If both Mateo and Chet work together, how long will it take them to paint the woodwork?

Step 1 **Read** the problem again. We are looking for time working together.

Step 2 **Assign a variable.** Let $x =$ the number of hours it will take for Mateo and Chet to paint the woodwork, working together.

Begin by making a table. Based on the previous discussion, Mateo's rate alone is $\frac{1}{8}$ job per hour, and Chet's rate is $\frac{1}{14}$ job per hour.

	Rate	Time Working Together	Fractional Part of the Job Done When Working Together
Mateo	$\frac{1}{8}$	x	$\frac{1}{8}x$
Chet	$\frac{1}{14}$	x	$\frac{1}{14}x$

Sum is 1 whole job.

Step 3 **Write an equation.** Together Mateo and Chet complete 1 whole job. We must add the fractional parts and set the sum equal to 1.

$$\underbrace{\frac{1}{8}x}_{\substack{\text{Fractional part} \\ \text{done by Mateo}}} + \underbrace{\frac{1}{14}x}_{\substack{\text{Fractional part} \\ \text{done by Chet}}} = \underbrace{1}_{\text{1 whole job}}$$

Step 4 **Solve.**

$$56\left(\frac{1}{8}x + \frac{1}{14}x\right) = 56(1) \qquad \text{Multiply by the LCD, 56.}$$

$$56\left(\frac{1}{8}x\right) + 56\left(\frac{1}{14}x\right) = 56(1) \qquad \text{Distributive property}$$

$$7x + 4x = 56$$

$$11x = 56 \qquad \text{Combine like terms.}$$

$$x = \frac{56}{11} \qquad \text{Divide by 11.}$$

Step 5 **State the answer.** Working together, Mateo and Chet can paint the woodwork in $\frac{56}{11}$ hr, or $5\frac{1}{11}$ hr.

Step 6 **Check.** Substitute $\frac{56}{11}$ for x in the equation from Step 3.

$$\frac{1}{8}x + \frac{1}{14}x = 1 \qquad \text{Equation from Step 3}$$

$$\frac{1}{8}\left(\frac{56}{11}\right) + \frac{1}{14}\left(\frac{56}{11}\right) \overset{?}{=} 1 \qquad \text{Let } x = \frac{56}{11}.$$

$$\frac{7}{11} + \frac{4}{11} = 1 \qquad \text{True}$$

Our answer, $\frac{56}{11}$ hr, or $5\frac{1}{11}$ hr, seems correct. See the Problem-Solving Hint on the next page for additional strategies for checking.

Problem-Solving Hint

A common error students make when solving a work problem like that in Example 5 is to add the two times, 8 hr and 14 hr, to get an answer of 22 hr. We reason, however, that x, the time it will take Mateo and Chet working together, must be *less than* 8 hr, since Mateo can complete the job by himself in 8 hr.

Another common error students make is to try to split the job in half between the two workers so that Mateo would work $\frac{1}{2}(8)$, or 4 hr, and Chet would work $\frac{1}{2}(14)$, or 7 hr. In this case, Mateo finishes 3 hr before Chet and they have not worked together to get the entire job done as quickly as possible. If Mateo, when he finishes, helps Chet, the job should actually be completed in a time between 4 hr and 7 hr.

Based on this reasoning, does our answer of $5\frac{1}{11}$ hr in Example 5 hold up?

Note

An alternative approach in work problems is to consider the part of the job that can be done in 1 hr. For instance, in Example 5 Mateo can do the entire job in 8 hr, and Chet can do it in 14 hr. Thus, their work rates, as we saw in Example 5, are $\frac{1}{8}$ and $\frac{1}{14}$, respectively. Since it takes them x hours to complete the job when working together, in 1 hr they can paint $\frac{1}{x}$ of the woodwork. The amount painted by Mateo in 1 hr plus the amount painted by Chet in 1 hr must equal the amount they can do together. This leads to the equation

$$\underset{\text{Amount by Mateo} \rightarrow}{\frac{1}{8}} + \underset{\overset{\uparrow}{\text{Amount by Chet}}}{\frac{1}{14}} = \underset{\leftarrow \text{ Amount together}}{\frac{1}{x}}.$$

Compare this with the equation in Step 3 of Example 5. Multiplying each side by $56x$ leads to

$$7x + 4x = 56,$$

the same equation found in the third line of Step 4 in the example. The same solution results.

Work Problem **5** *at the Side.* ▶

5 Solve each problem.

(a) Michael can paint a room, working alone, in 8 hr. Lindsay can paint the same room, working alone, in 6 hr. How long will it take them if they work together?

(b) Roberto can detail his Camaro in 2 hr working alone. His brother Marco can do the job in 3 hr working alone. How long would it take them if they worked together?

Math in the Media

TO PLAY BASEBALL, YOU MUST *WORK* AT IT!

In the 1994 movie *Little Big League,* the young Billy Heywood inherits the Minnesota Twins baseball team and becomes its manager. He leads the team to the Division Championship and then to the playoffs. But before the final playoff game, the biggest game of the year, he can't keep his mind on his job because a homework problem is giving him trouble.

If Joe can paint a house in 3 hours, and Sam can paint the same house in 5 hours, how long does it take for them to do it together?

With the help of one of his players, he is able to solve the problem, and the team goes on to victory.

1. Use the method described in Example 5 of **Section 15.7** to solve this problem.

2. Before the player was able to solve the problem correctly, Billy got "help" from some of the other players. The incorrect answers they gave him were

   (a) 15 hr   **(b)** 8 hr   **(c)** 4 hr.

   Explain the faulty reasoning behind each of these incorrect answers.

3. The player who gave Billy the correct answer solved the problem as follows:

   Using the simple formula a times b over a plus b, we get our answer of one and seven-eighths.

   Show that if it takes one person a hours to complete one job and another b hours to complete the same job, then the expression stated by the player,

   $$\frac{a \cdot b}{a + b}$$

   actually does give the number of hours it would take them to do the job together. (*Hint:* Refer to Example 5 and use a and b rather than 8 and 14. Then solve the resulting formula for x.)

Use Steps 2 and 3 of the six-step problem solving method to set up the equation you would use to solve each problem. (Remember that Step 1 is to read the problem carefully.) Do not actually solve the equation. See Example 1.

1. The numerator of the fraction $\frac{5}{6}$ is increased by an amount so that the value of the resulting fraction is equivalent to $\frac{13}{3}$. By what amount was the numerator increased?

 (a) Let $x = $ _____. (*Step 2*)

 (b) Write an expression for "the numerator of the fraction $\frac{5}{6}$ is increased by an amount."

 (c) Set up an equation to solve the problem.

  (*Step 3*)

2. If the same number is added to the numerator and subtracted from the denominator of $\frac{23}{12}$, the resulting fraction is equivalent to $\frac{3}{2}$. What is the number?

 (a) Let $x = $ _____. (*Step 2*)

 (b) Write an expression for "a number is added to the numerator of $\frac{23}{12}$." Then write an expression for "the same number is subtracted from the denominator of $\frac{23}{12}$."

 (c) Set up an equation to solve the problem.

  (*Step 3*)

Solve each problem. See Example 1.

3. In a certain fraction, the denominator is 4 less than the numerator. If 3 is added to both the numerator and the denominator, the resulting fraction is equivalent to $\frac{3}{2}$. What was the original fraction?

4. In a certain fraction, the denominator is 6 more than the numerator. If 3 is added to both the numerator and the denominator, the resulting fraction is equivalent to $\frac{5}{7}$. What was the original fraction (*not* written in lowest terms)?

5. The denominator of a certain fraction is three times the numerator. If 2 is added to the numerator and subtracted from the denominator, the resulting fraction is equivalent to 1. What was the original fraction (*not* written in lowest terms)?

6. The numerator of a certain fraction is four times the denominator. If 6 is added to both the numerator and the denominator, the resulting fraction is equivalent to 2. What was the original fraction (*not* written in lowest terms)?

7. One-sixth of a number is 5 more than the same number. What is the number?

8. One-third of a number is 2 more than one-sixth of the same number. What is the number?

9. A quantity, its $\frac{3}{4}$, its $\frac{1}{2}$, and its $\frac{1}{3}$, added together, become 93. What is the quantity? (*Source: Rhind Mathematical Papyrus.*)

10. A quantity, its $\frac{2}{3}$, its $\frac{1}{2}$, and its $\frac{1}{7}$, added together, become 33. What is the quantity? (*Source: Rhind Mathematical Papyrus.*)

Solve each problem. See Example 2.

11. In July 2007, British explorer and endurance swimmer Lewis Gordon Pugh became the first person to swim at the North Pole. To highlight climate change, he swam 0.6 mi in 18.833 min in waters created by melted sea ice. What was his rate (to three decimal places)? (*Source: The Gazette,* July 16, 2007.)

12. In the 2004 Summer Olympics in Athens, Greece, Jody Henry of Australia won the women's 100-m freestyle swimming event. Her rate was 1.854 m per sec. What was her time (to the nearest hundredth of a second)? (*Source: World Almanac and Book of Facts.*)

13. The winner of the 2007 Daytona 500 (mile) race was Kevin Harvick who drove his Chevrolet to victory with a rate of 149.335 mph. What was his time (to the nearest thousandth of an hour)? (*Source:* www.daytona500.com)

14. In 2008, Scott Dixon drove his Target Chip Ganassi Racing car to victory in the Indianapolis 500 (mile) race. His rate was 143.567 mph. What was his time (to the nearest thousandth of an hour)? (*Source:* www.indy500.com)

15. In the 2006 Winter Olympics in Torino, Italy, Svetlana Zhurova of Russia won the 500-m speed skating event for women. Her time was 76.57 sec. What was her rate (to three decimal places)? (*Source:* www.espn.com)

16. The winner of the women's 1500-m race in the 2004 Olympics was Kelly Holmes of Great Britain with a time of 3.965 min. What was her rate (to three decimal places)? (*Source:* www.olympics.com)

Set up the equation you would use to solve each problem. Do not actually solve the equation. See Examples 3 and 4.

17. Luvenia can row 4 mph in still water. She takes as long to row 8 mi upstream as 24 mi downstream. How fast is the current? (Let x = speed of the current.)

	d	r	t
Upstream	8	$4 - x$	
Downstream	24	$4 + x$	

18. Julio flew his airplane 500 mi against the wind in the same time it took him to fly it 600 mi with the wind. If the speed of the wind was 10 mph, what was the average speed of his plane in still air? (Let x = speed of the plane in still air.)

	d	r	t
Against the Wind	500	$x - 10$	
With the Wind	600	$x + 10$	

Solve each problem. See Examples 3 and 4.

19. If a migrating hawk travels *m* mph in still air, what is its rate when it flies into a steady headwind of 5 mph? What is its rate with a tailwind of 5 mph?

20. Suppose Stephanie walks *D* miles at *R* mph in the same time that Wally walks *d* miles at *r* mph. Give an equation relating *D, R, d,* and *r.*

21. A boat can go 20 mi against a current in the same time that it can go 60 mi with the current. The current is 4 mph. Find the speed of the boat in still water.

22. A plane flies 350 mi with the wind in the same time that it can fly 310 mi against the wind. The plane has a still-air speed of 165 mph. Find the speed of the wind.

23. The sanderling is a small shorebird about 6.5 in. long, with a thin, dark bill and a wide, white wing stripe. If a sanderling can fly 30 mi with the wind in the same time it can fly 18 mi against the wind when the wind speed is 8 mph, what is the speed of the bird in still air? (*Source:* U.S. Geological Survey.)

24. Airplanes usually fly faster from west to east than from east to west because the prevailing winds go from west to east. The air distance between Chicago and London is about 4000 mi, while the air distance between New York and London is about 3500 mi. If a jet can fly eastbound from Chicago to London in the same time it can fly westbound from London to New York in a 35-mph wind, what is the speed of the plane in still air? (*Source: Encyclopaedia Britannica.*)

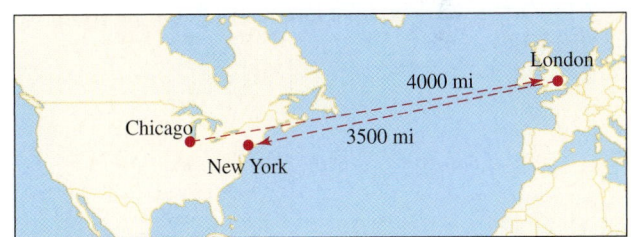

25. Perian Herring's boat goes 12 mph. Find the rate of the current of the river if she can go 6 mi upstream in the same amount of time she can go 10 mi downstream.

26. Sarah Sponholz can travel 8 mi upstream in the same time it takes her to go 12 mi downstream. Her boat goes 15 mph in still water. What is the rate of the current?

Set up the equation you would use to solve each problem. Do not actually solve the equation. See Example 5.

27. Edwin Bedford can tune up his Chevy in 2 hr working alone. His son, Beau, can do the job in 3 hr working alone. How long would it take them if they worked together? (Let *t* represent the time working together.)

	r	*t*	*w*
Edwin		*t*	
Beau		*t*	

28. Working alone, Helio can paint a room in 8 hr. Julianne can paint the same room working alone in 6 hr. How long will it take them if they work together? (Let *t* represent the time working together.)

	r	*t*	*w*
Helio		*t*	
Julianne		*t*	

Solve each problem. See Example 5.

29. Ms. Tseng, a high school mathematics teacher, gave a test on perimeter, area, and volume to her geometry classes. Working alone, it would take her 4 hr to grade the tests. Her student teacher, Jonah Schmidt, would take 6 hr to grade the same tests. How long would it take them to grade these tests if they work together?

30. Geraldo and Luisa Hernandez operate a small laundry. Luisa, working alone, can clean a day's laundry in 9 hr. Geraldo can clean a day's laundry in 8 hr. How long would it take them if they work together?

31. Todd's copier can do a printing job in 7 hr. Scott's copier can do the same job in 12 hr. How long would it take to do the job using both copiers?

32. A pump can pump the water out of a flooded basement in 10 hr. A smaller pump takes 12 hr. How long would it take to pump the water from the basement using both pumps?

33. Hilda can paint a room in 6 hr. Working together with Brenda, they can paint the room in $3\frac{3}{4}$ hr. How long would it take Brenda to paint the room by herself?

34. Grant can completely mess up his room in 15 min. If his cousin Wade helps him, they can completely mess up the room in $8\frac{4}{7}$ min. How long would it take Wade to mess up the room by himself?

35. An inlet pipe can fill a swimming pool in 9 hr, and an outlet pipe can empty the pool in 12 hr. Through an error, both pipes are left open. How long will it take to fill the pool?

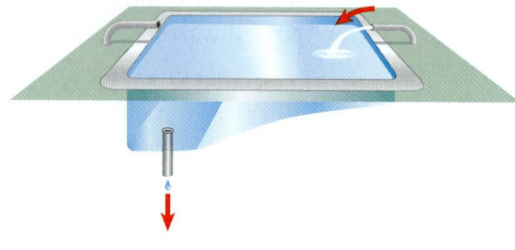

36. One pipe can fill a swimming pool in 6 hr, and another pipe can do it in 9 hr. How long will it take the two pipes working together to fill the pool $\frac{3}{4}$ full?

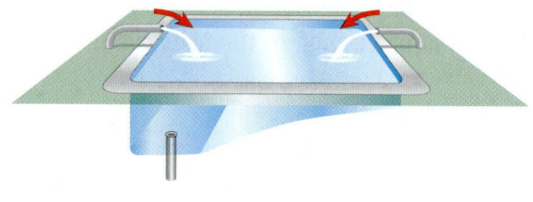

37. Refer to Exercise 35. Assume the error was discovered after both pipes had been running for 3 hr, and the outlet pipe was then closed. How much more time would then be required to fill the pool? (*Hint:* Consider how much of the job had been done when the error was discovered.)

38. A cold water faucet can fill a sink in 12 min, and a hot water faucet can fill it in 15 min. The drain can empty the sink in 25 min. If both faucets are on and the drain is open, how long will it take to fill the sink?

15.8 ▶▶▶ Variation

OBJECTIVE 1 Solve direct variation problems. Suppose that gasoline costs $4.50 per gal. Then 1 gal costs $4.50, 2 gal cost $2\,(\$4.50) = \$9.00$, 3 gal cost $3\,(\$4.50) = \$13.50$, and so on. Each time, the total cost is obtained by multiplying the number of gallons by the price per gallon. In general, if k equals the price per gallon and x equals the number of gallons, then the total cost y is equal to kx. Notice that

> As the *number of gallons **increases,*** the *total cost **increases.***

The reverse is also true:

> As the *number of gallons **decreases,*** the *total cost **decreases.***

The preceding discussion is an example of *variation*. **Two variables vary directly if one is a constant multiple of the other.**

> **Direct Variation**
>
> **y varies directly as x** if there exists a constant k such that
>
> $$y = kx.$$

Also, y is said to be *proportional to x*. The constant k in the equation for direct variation is a numerical value, such as 4.50 in the gasoline price discussion. This value is called the **constant of variation.**

EXAMPLE 1 Using Direct Variation

Suppose y varies directly as x, and $y = 20$ when $x = 4$. Find y when $x = 9$.

Since y varies directly as x, there is a constant k such that $y = kx$. We know that $y = 20$ when $x = 4$. Substituting these values into $y = kx$ and solving for k gives

$$y = kx \qquad \text{Equation for direct variation}$$
$$20 = k \cdot 4 \qquad \text{Substitute the given values.}$$
$$k = 5. \longleftarrow \text{Constant of variation}$$

Since $y = kx$ and $k = 5$,

$$y = 5x. \qquad \text{Let } k = 5.$$

Therefore, when $x = 9$,

$$y = 5x = 5 \cdot 9 = 45. \qquad \text{Let } x = 9.$$

Thus, $y = 45$ when $x = 9$.

Work Problem ①︎ *at the Side.* ▶

OBJECTIVE 2 Solve inverse variation problems. In direct variation, where $k > 0$, as x increases, y increases. Similarly, as x decreases, y decreases. Another type of variation is *inverse variation*. With inverse variation, where $k > 0$,

> As one variable ***increases,*** the other variable ***decreases.***

OBJECTIVES

① **Solve direct variation problems.**

② **Solve inverse variation problems.**

① Solve each problem.

(a) If z varies directly as t, and $z = 11$ when $t = 4$, find z when $t = 32$.

(b) The circumference of a circle varies directly as the radius. A circle with a radius of 7 cm has a circumference of 43.96 cm. Find the circumference if the radius is 11 cm.

ANSWERS

1. (a) 88 **(b)** 69.08 cm

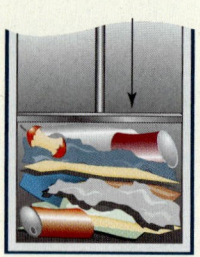

As pressure increases, volume decreases.

Figure 3

For example, in a closed space, volume decreases as pressure increases, as illustrated by a trash compactor. See Figure 3. As the compactor presses down, the pressure on the trash increases; in turn, the trash occupies a smaller space.

> **Inverse Variation**
>
> **y varies inversely as x** if there exists a constant k such that
>
> $$y = \frac{k}{x}.$$

2 Solve the problem.

Suppose z varies inversely as t, and $z = 8$ when $t = 2$. Find z when $t = 32$.

EXAMPLE 2 Using Inverse Variation

Suppose y varies inversely as x, and $y = 3$ when $x = 8$. Find y when $x = 6$.

Since y varies inversely as x, there is a constant k such that $y = \frac{k}{x}$. We know that $y = 3$ when $x = 8$, so we can find k.

$$y = \frac{k}{x} \qquad \text{Equation for inverse variation}$$

$$3 = \frac{k}{8} \qquad \text{Substitute the given values.}$$

$$k = 24 \qquad \text{Solve for } k.$$

Since $y = \frac{24}{x}$, we let $x = 6$ and solve for y.

$$y = \frac{24}{x} = \frac{24}{6} = 4$$

Therefore, when $x = 6$, $y = 4$.

◀ Work Problem **2** at the Side.

3 Solve the problem.

The current in a simple electrical circuit varies inversely as the resistance. If the current is 80 amps when the resistance is 10 ohms, find the current if the resistance is 16 ohms.

EXAMPLE 3 Using Inverse Variation

In the manufacturing of a certain medical syringe, the cost of producing the syringe varies inversely as the number produced. If 10,000 syringes are produced, the cost is \$2 per unit. Find the cost per unit to produce 25,000 syringes.

$$\text{Let} \quad x = \text{the number of syringes produced}$$
$$\text{and} \quad c = \text{the cost per unit.}$$

Since c varies inversely as x, there is a constant k such that

$$c = \frac{k}{x} \qquad \text{Equation for inverse variation}$$

$$2 = \frac{k}{10,000} \qquad \text{Substitute the given values.}$$

$$20,000 = k. \qquad \text{Multiply by 10,000.}$$

Since $c = \frac{k}{x}$,

$$c = \frac{20,000}{25,000} = 0.80. \qquad \text{Let } k = 20,000 \text{ and } x = 25,000.$$

The cost per unit to make 25,000 syringes is \$0.80.

◀ Work Problem **3** at the Side.

1. (a) If the constant of variation is positive and y varies directly as x, then as

x increases, y _____.
(increases/decreases)

(b) If the constant of variation is positive and y varies inversely as x, then as

x increases, y _____.
(increases/decreases)

2. Bill Veeck was the owner of several major league baseball teams in the 1950s and 1960s. He was known to often sit in the stands and enjoy games with his paying customers. Here is a quote attributed to him:

"I have discovered in 20 years of moving around a ballpark, that the knowledge of the game is usually in inverse proportion to the price of the seats."

Explain in your own words the meaning of this statement. (To prove his point, Veeck once allowed the fans to vote on managerial decisions.)

Solve each problem involving direct variation. See Example 1.

⊕ 3. If z varies directly as x, and $z = 30$ when $x = 8$, find z when $x = 4$.

4. If y varies directly as x, and $x = 27$ when $y = 6$, find x when $y = 2$.

5. If d varies directly as r, and $d = 200$ when $r = 40$, find d when $r = 60$.

6. If d varies directly as t, and $d = 150$ when $t = 3$, find d when $t = 5$.

Solve each problem involving inverse variation. See Example 2.

⊕ 7. If z varies inversely as x, and $z = 50$ when $x = 2$, find z when $x = 25$.

8. If x varies inversely as y, and $x = 3$ when $y = 8$, find y when $x = 4$.

9. If m varies inversely as r, and $m = 12$ when $r = 8$, find m when $r = 16$.

10. If p varies inversely as q, and $p = 7$ when $q = 6$, find p when $q = 2$.

Solve each variation problem. See Examples 1–3.

11. For a given base, the area of a triangle varies directly as its height. Find the area of a triangle with a height of 6 in., if the area is 10 in.^{2} when the height is 4 in.

12. The interest on an investment varies directly as the rate of interest. If the interest is $48 when the interest rate is 5%, find the interest when the rate is 4.2%.

13. Hooke's law for an elastic spring states that the distance a spring stretches varies directly with the force applied. If a force of 75 lb stretches a certain spring 16 in., how much will a force of 200 lb stretch the spring?

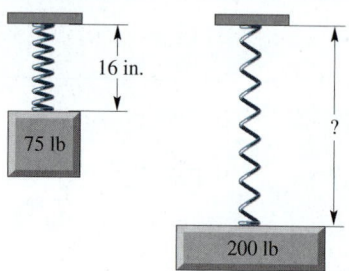

14. The pressure exerted by water at a given point varies directly with the depth of the point beneath the surface of the water. Water exerts 4.34 lb per in.^{2} for every 10 ft traveled below the water's surface. What is the pressure exerted on a scuba diver at 20 ft?

15. For a constant area, the length of a rectangle varies inversely as the width. The length of a rectangle is 27 ft when the width is 10 ft. Find the width of a rectangle with the same area if the length is 18 ft.

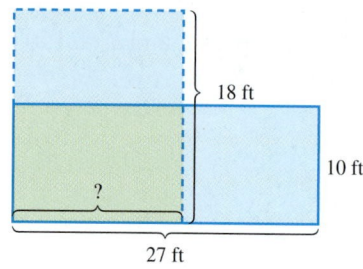

16. Over a specified distance, speed varies inversely with time. If a Dodge Viper on a test track goes a certain distance in one-half minute at 160 mph, what speed is needed to go the same distance in three-fourths minute?

17. If the temperature is constant, the pressure of a gas in a container varies inversely as the volume of the container. If the pressure is 10 lb per ft^2 in a container with volume 3 ft^3, what is the pressure in a container with volume 1.5 ft^3?

18. The current in a simple electrical circuit varies inversely as the resistance. If the current is 20 amps when the resistance is 5 ohms, find the current when the resistance is 8 ohms.

19. In the inversion of raw sugar, the rate of change of the amount of raw sugar varies directly as the amount of raw sugar remaining. The rate is 200 kg per hr when there are 800 kg left. What is the rate of change per hour when only 100 kg are left?

20. The force required to compress a spring varies directly as the change in the length of the spring. If a force of 12 lb is required to compress a certain spring 3 in., how much force is required to compress the spring 5 in.?

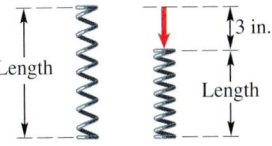

*Use personal experience or intuition to determine whether the situation suggests direct or inverse variation.**

21. The rate and the distance traveled by a pickup truck in 3 hr

22. The number of different lottery tickets you buy and your probability of winning that lottery

23. The number of days from now until December 25 and the magnitude of the frenzy of Christmas shopping

24. Your age and the probability that you believe in Santa Claus

25. The amount of gasoline that you pump and the amount of empty space left in your tank

26. The surface area of a balloon and its diameter

27. The amount of pressure put on the accelerator of a car and the speed of the car

28. The number of days until the end of the baseball season and the number of home runs that Alex Rodriguez has

*The authors thank Linda Kodama of Kapi'olani Community College for suggesting the inclusion of exercises of this type.

*Recall from **Section 6.6** that two triangles are **similar** if they have the same shape (but not necessarily the same size). Similar triangles have side lengths that are proportional—that is, vary directly. The figure shows two similar triangles. Notice that the ratios of the corresponding sides are all equal to $\frac{3}{2}$.*

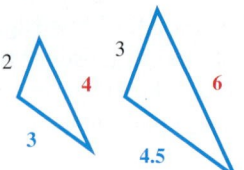

$$\frac{3}{2} = \frac{3}{2} \qquad \frac{4.5}{3} = \frac{3}{2} \qquad \frac{6}{4} = \frac{3}{2}$$

If we know that two triangles are similar, we can set up a direct variation equation to solve for the length of an unknown side.

Find the length x, given that the pair of triangles are similar.

29.

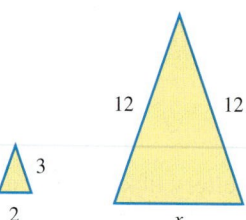

30.

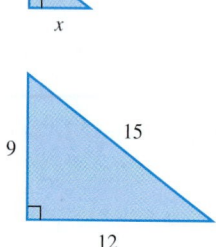

31.

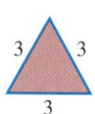

32.

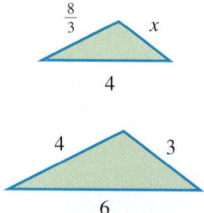

33.

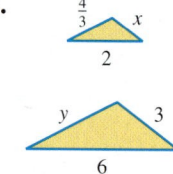

34.

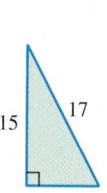

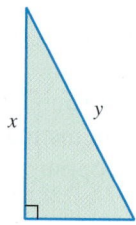

Use similar triangles and direct variation to solve each problem. (Source: Guinness World Records.)

35. One of the tallest candles ever constructed was exhibited at the 1897 Stockholm Exhibition. If it cast a shadow 5 ft long at the same time a vertical pole 32 ft high cast a shadow 2 ft long, how tall was the candle?

36. An enlarged version of the chair used by George Washington at the Constitutional Convention casts a shadow 18 ft long at the same time a vertical pole 12 ft high casts a shadow 4 ft long. How tall is the chair?

Chapter 15 ▶▶▶ Summary

▶ Key Terms

15.1 rational expression   The quotient of two polynomials with denominator not 0 is called a rational expression.

lowest terms   A rational expression is written in lowest terms if the greatest common factor of its numerator and denominator is 1.

15.3 least common denominator (LCD)   The simplest expression that is divisible by all denominators is called the least common denominator.

15.5 complex fraction   A rational expression with one or more fractions in the numerator, denominator, or both, is called a complex fraction.

15.8 direct variation y varies directly as x if there is a constant k such that $y = kx$.

constant of variation   In the equation $y = kx$ or $y = \frac{k}{x}$, the number k is called the constant of variation.

inverse variation y varies inversely as x if there is a constant k such that $y = \frac{k}{x}$.

▶ Test Your Word Power

See how well you have learned the vocabulary in this chapter. Answers, with examples, follow the Quick Review.

1. A **rational expression** is
   A. an algebraic expression made up of a term or the sum of a finite number of terms with real coefficients and whole number exponents
   B. a polynomial equation of degree 2
   C. an expression with one or more fractions in the numerator, denominator, or both
   D. the quotient of two polynomials with denominator not 0.

2. A **complex fraction** is
   A. an algebraic expression made up of a term or the sum of a finite number of terms with real coefficients and whole number exponents
   B. a polynomial equation of degree 2

   C. a rational expression with one or more fractions in the numerator, denominator, or both
   D. the quotient of two polynomials with denominator not 0.

3. In a given set of fractions, the **least common denominator** is
   A. the simplest denominator of all the denominators
   B. the simplest expression that is divisible by all the denominators
   C. the greatest integer that evenly divides the numerator and denominator of all the fractions
   D. the greatest denominator of all the denominators.

4. If two positive quantities x and y are in **direct variation,** and the constant of variation is positive, then
   A. as x increases, y decreases
   B. as x increases, y increases
   C. as x increases, y remains constant
   D. as x decreases, y remains constant.

5. If two positive quantities x and y are in **inverse variation,** and the constant of variation is positive, then
   A. as x increases, y decreases
   B. as x increases, y increases
   C. as x increases, y remains constant
   D. as x decreases, y remains constant.

▶ Quick Review

Concepts

Examples

15.1 The Fundamental Property of Rational Expressions

To find the value(s) for which a rational expression is undefined, set the denominator equal to 0 and solve the equation.

Find the values for which $\dfrac{x-4}{x^2-16}$ is undefined.

$$x^2 - 16 = 0$$
$$(x-4)(x+4) = 0 \qquad \text{Factor.}$$
$$x - 4 = 0 \quad \text{or} \quad x + 4 = 0 \qquad \text{Zero-factor property}$$
$$x = 4 \quad \text{or} \qquad x = -4 \qquad \text{Solve for } x.$$

The rational expression is undefined for 4 and -4, so $x \neq 4$ and $x \neq -4$.

Writing a Rational Expression in Lowest Terms

Step 1 Factor the numerator and denominator.

Step 2 Use the fundamental property to divide out common factors from the numerator and denominator.

Write the expression in lowest terms.

$$\frac{x^2 - 1}{(x-1)^2}$$
$$= \frac{(x-1)(x+1)}{(x-1)(x-1)}$$
$$= \frac{x+1}{x-1}$$

There are often several different equivalent forms of a rational expression.

To find four equivalent forms of $-\dfrac{x-1}{x+2}$, distribute the

$-$ sign in the numerator to get $\dfrac{-(x-1)}{x+2}$, or $\dfrac{-x+1}{x+2}$; do so

in the denominator to get $\dfrac{x-1}{-(x+2)}$, or $\dfrac{x-1}{-x-2}$.

15.2 Multiplying and Dividing Rational Expressions

Multiplying or Dividing Rational Expressions

Step 1 Note the operation. If the operation is division, use the definition of division to rewrite as multiplication.

Step 2 Multiply numerators and multiply denominators.

Step 3 Factor numerators and denominators completely.

Step 4 Write in lowest terms, using the fundamental property.

Steps 2 and 3 may be interchanged based on personal preference.

Multiply. $\dfrac{3x+9}{x-5} \cdot \dfrac{x^2-3x-10}{x^2-9}$

$$= \frac{(3x+9)(x^2-3x-10)}{(x-5)(x^2-9)}$$
$$= \frac{3(x+3)(x-5)(x+2)}{(x-5)(x+3)(x-3)}$$
$$= \frac{3(x+2)}{x-3}$$

Divide. $\dfrac{2x+1}{x+5} \div \dfrac{6x^2-x-2}{x^2-25}$

$$= \frac{(2x+1)(x^2-25)}{(x+5)(6x^2-x-2)}$$
$$= \frac{(2x+1)(x+5)(x-5)}{(x+5)(2x+1)(3x-2)}$$
$$= \frac{x-5}{3x-2}$$

Concepts	Examples

15.3 Least Common Denominators

Finding the LCD

Find the LCD for $\dfrac{3}{k^2 - 8k + 16}$ and $\dfrac{1}{4k^2 - 16k}$.

Step 1 Factor each denominator into prime factors.

Step 2 List each different factor the greatest number of times it appears.

$$k^2 - 8k + 16 = (k - 4)^2 \;\;\} \;\; \text{Factor each}$$
$$4k^2 - 16k = 4k(k - 4) \;\;\} \;\; \text{denominator.}$$

Step 3 Multiply the factors from Step 2 to get the LCD.

$$\text{LCD} = (k - 4)^2 \cdot 4 \cdot k$$
$$= 4k(k - 4)^2$$

Writing Equivalent Rational Expressions

Find the numerator: $\dfrac{5}{2z^2 - 6z} = \dfrac{?}{4z^3 - 12z^2}$.

Step 1 Factor both denominators.

$$\frac{5}{2z(z - 3)} = \frac{?}{4z^2(z - 3)}$$

Step 2 Decide what factors the denominator must be multiplied by to equal the specified denominator.

$2z(z - 3)$ must be multiplied by $2z$.

Step 3 Multiply the rational expression by that factor divided by itself. (That is, multiply by 1.)

$$\frac{5}{2z(z - 3)} \cdot \frac{2z}{2z} = \frac{10z}{4z^2(z - 3)} = \frac{10z}{4z^3 - 12z^2}$$

15.4 Adding and Subtracting Rational Expressions

Adding Rational Expressions

Add. $\dfrac{2}{3m + 6} + \dfrac{m}{m^2 - 4}$

Step 1 Find the LCD.

$$3m + 6 = 3(m + 2) \;\;\} \;\; \text{The LCD is}$$
$$m^2 - 4 = (m + 2)(m - 2) \;\;\} \;\; 3(m + 2)(m - 2).$$

Step 2 Rewrite each rational expression with the LCD as denominator.

$$= \frac{2(m - 2)}{3(m + 2)(m - 2)} + \frac{3m}{3(m + 2)(m - 2)}$$

Step 3 Add the numerators to get the numerator of the sum. The LCD is the denominator of the sum.

$$= \frac{2m - 4 + 3m}{3(m + 2)(m - 2)}$$

Step 4 Write in lowest terms.

$$= \frac{5m - 4}{3(m + 2)(m - 2)}$$

Subtracting Rational Expressions

Subtract. $\dfrac{6}{k + 4} - \dfrac{2}{k}$ The LCD is $k(k + 4)$.

Follow the same steps as for addition, but subtract in Step 3.

$$= \frac{6k}{(k + 4)k} - \frac{2(k + 4)}{k(k + 4)}$$

$$= \frac{6k - 2(k + 4)}{k(k + 4)}$$

$$= \frac{6k - 2k - 8}{k(k + 4)}$$

$$= \frac{4k - 8}{k(k + 4)}, \;\; \text{or} \;\; \frac{4(k - 2)}{k(k + 4)}$$

Concepts	Examples

15.5 Complex Fractions

Simplifying Complex Fractions

Simplify.

Method 1 Simplify the numerator and denominator separately. Then divide the simplified numerator by the simplified denominator.

Method 1

$$\frac{\frac{1}{a} - a}{1 - a} = \frac{\frac{1}{a} - \frac{a^2}{a}}{1 - a} = \frac{\frac{1 - a^2}{a}}{1 - a}$$

$$= \frac{1 - a^2}{a} \div (1 - a)$$

$$= \frac{1 - a^2}{a} \cdot \frac{1}{1 - a}$$

$$= \frac{(1 - a)(1 + a)}{a(1 - a)}$$

$$= \frac{1 + a}{a}$$

Method 2 Multiply the numerator and denominator of the complex fraction by the LCD of all the denominators in the complex fraction. Write in lowest terms.

Method 2

$$\frac{\frac{1}{a} - a}{1 - a} = \frac{\left(\frac{1}{a} - a\right)a}{(1 - a)a} = \frac{\frac{a}{a} - a^2}{(1 - a)a}$$

$$= \frac{1 - a^2}{(1 - a)a} = \frac{(1 + a)(1 - a)}{(1 - a)a}$$

$$= \frac{1 + a}{a}$$

15.6 Solving Equations with Rational Expressions

Solving Equations with Rational Expressions

Solve $\dfrac{x}{x - 3} + \dfrac{4}{x + 3} = \dfrac{18}{x^2 - 9}$.

Step 1 Multiply each side of the equation by the LCD. (This clears the equation of fractions.)

$$\frac{x}{x - 3} + \frac{4}{x + 3} = \frac{18}{(x - 3)(x + 3)} \qquad \text{Factor.}$$

The LCD is $(x - 3)(x + 3)$. Note that 3 and -3 cannot be solutions, as they cause a denominator to equal 0.

$$(x - 3)(x + 3)\left(\frac{x}{x - 3} + \frac{4}{x + 3}\right)$$

$$= (x - 3)(x + 3)\frac{18}{(x - 3)(x + 3)} \qquad \begin{array}{l}\text{Multiply by}\\ \text{the LCD.}\end{array}$$

Step 2 Solve the resulting equation.

$$x(x + 3) + 4(x - 3) = 18 \qquad \text{Distributive property}$$

$$x^2 + 3x + 4x - 12 = 18 \qquad \text{Distributive property}$$

$$x^2 + 7x - 30 = 0 \qquad \text{Standard form}$$

$$(x - 3)(x + 10) = 0 \qquad \text{Factor.}$$

$$x - 3 = 0 \quad \text{or} \quad x + 10 = 0$$
$$\text{Zero-factor property}$$

Step 3 Check each proposed solution by substituting it in the original equation. Reject any value that causes an original denominator to equal 0.

Reject $\longrightarrow x = 3 \quad$ or $\quad x = -10 \qquad$ Solve for x.

Since 3 causes denominators to equal 0, the only solution is -10. Thus, $\{-10\}$ is the solution set.

Concepts	Examples

(15.7) Applications of Rational Expressions

Solving Problems about Distance, Rate, and Time
Use the six-step method.

Step 1 **Read** the problem carefully.

Step 2 **Assign a variable.** Use a table to identify distance, rate, and time. Solve $d = rt$ for the unknown quantity in the table.

On a trip from Sacramento to Monterey, Marge traveled at an average speed of 60 mph. The return trip, at an average speed of 64 mph, took $\frac{1}{4}$ hr less. How far did she travel between the two cities?

Let $x = $ the unknown distance.

	d	r	$t = \dfrac{d}{r}$
Going	x	60	$\dfrac{x}{60}$
Returning	x	64	$\dfrac{x}{64}$

Step 3 **Write an equation.** From the wording in the problem, decide the relationship between the quantities. Use those expressions to write an equation.

Since the time for the return trip was $\frac{1}{4}$ hr less, the time going equals the time returning plus $\frac{1}{4}$.

$$\frac{x}{60} = \frac{x}{64} + \frac{1}{4}$$

Step 4 **Solve** the equation.

$16x = 15x + 240$ Multiply by the LCD, 960.

$x = 240$ Subtract $15x$.

Step 5 **State the answer.**

She traveled 240 mi.

Step 6 **Check** the solution.

The trip there took $\frac{240}{60} = 4$ hr, while the return trip took $\frac{240}{64} = 3\frac{3}{4}$ hr, which is $\frac{1}{4}$ hr less time. The solution checks.

Solving Problems about Work

Step 1 **Read** the problem carefully.

Step 2 **Assign a variable.** State what the variable represents. Put the information from the problem in a table. If a job is done in t units of time, then the rate is $\frac{1}{t}$.

It takes the regular mail carrier 6 hr to cover her route. A substitute takes 8 hr to cover the same route. How long would it take them to cover the route together?

Let $x = $ the number of hours to cover the route together.

The rate of the regular carrier is $\frac{1}{6}$ job per hour; the rate of the substitute is $\frac{1}{8}$ job per hour. Multiply rate by time to get the fractional part of the job done.

	Rate	Time	Part of the Job Done
Regular	$\dfrac{1}{6}$	x	$\dfrac{1}{6}x$
Substitute	$\dfrac{1}{8}$	x	$\dfrac{1}{8}x$

Step 3 **Write an equation.** The sum of the fractional parts should equal 1 (whole job).

The equation is $\dfrac{1}{6}x + \dfrac{1}{8}x = 1$.

Step 4 **Solve** the equation.

The solution of the equation is $\frac{24}{7}$, or $3\frac{3}{7}$. The solution checks, because $\frac{1}{6}\left(\frac{24}{7}\right) + \frac{1}{8}\left(\frac{24}{7}\right) = 1$ is true.

Steps 5 and 6 **State the answer** and **check** the solution.

It would take them $3\frac{3}{7}$ hr to cover the route together.

Concepts	Examples

15.8 Variation

Solving Variation Problems

Step 1 Write the variation equation. Use

$$y = kx \qquad \text{Direct variation}$$

$$\text{or} \quad y = \frac{k}{x}. \qquad \text{Inverse variation}$$

Step 2 Find k by substituting the given values of x and y into the equation.

Step 3 Write the equation with the value of k from Step 2 and the given value of x or y. Solve for the remaining variable.

If y varies inversely as x, and $y = 4$ when $x = 9$, find y when $x = 6$.

$$y = \frac{k}{x} \qquad \text{Equation for inverse variation}$$

$$4 = \frac{k}{9} \qquad \text{Substitute given values.}$$

$$k = 36 \qquad \text{Solve for } k.$$

$$y = \frac{36}{x} \qquad k = 36$$

$$y = \frac{36}{6} \qquad \text{Let } x = 6.$$

$$y = 6$$

ANSWERS TO TEST YOUR WORD POWER

1. D; *Examples:* $-\dfrac{3}{4y}$, $\dfrac{5x^3}{x+2}$, $\dfrac{a+3}{a^2-4a-5}$

2. C; *Examples:* $\dfrac{\frac{2}{3}}{\frac{4}{7}}$, $\dfrac{x-\frac{1}{y}}{x+\frac{1}{y}}$, $\dfrac{2}{a+1}$ $\dfrac{}{a^2-1}$

3. B; *Examples:* The least common denominator of $\dfrac{1}{2}, \dfrac{1}{3}$, and $\dfrac{1}{4}$ is 12. The least common denominator of $\dfrac{1}{x}$ and $\dfrac{1}{x+1}$ is $x(x+1)$.

4. B; *Example:* The equation $y = 3x$ represents direct variation. When $x = 2$, $y = 6$. If x increases to 3, then y increases to $3(3) = 9$.

5. A; *Example:* The equation $y = \dfrac{3}{x}$ represents inverse variation. When $x = 1$, $y = 3$. If x increases to 2, then y decreases to $\dfrac{3}{2}$, or $1\dfrac{1}{2}$.

Chapter 15 ►►► Review Exercises

[15.1] *Find the value(s) of the variable for which each rational expression is undefined.*
Write answers with ≠.

1. $\dfrac{4}{x-3}$

2. $\dfrac{x+3}{2x}$

3. $\dfrac{m-2}{m^2-2m-3}$

4. $\dfrac{2k+1}{3k^2+17k+10}$

*Find the numerical value of each rational expression when **(a)** $x = -2$ and **(b)** $x = 4$.*

5. $\dfrac{x^2}{x-5}$

6. $\dfrac{4x-3}{5x+2}$

7. $\dfrac{3x}{x^2-4}$

8. $\dfrac{x-1}{x+2}$

Write each rational expression in lowest terms.

9. $\dfrac{5a^3b^3}{15a^4b^2}$

10. $\dfrac{m-4}{4-m}$

11. $\dfrac{4x^2-9}{6-4x}$

12. $\dfrac{4p^2+8pq-5q^2}{10p^2-3pq-q^2}$

Write four equivalent expressions for each fraction.

13. $-\dfrac{4x-9}{2x+3}$

14. $-\dfrac{8-3x}{3-6x}$

[15.2] *Multiply or divide. Write each answer in lowest terms.*

15. $\dfrac{8x^2}{12x^5}\cdot\dfrac{6x^4}{2x}$

16. $\dfrac{9m^2}{(3m)^4}\div\dfrac{6m^5}{36m}$

17. $\dfrac{x-3}{4}\cdot\dfrac{5}{2x-6}$

18. $\dfrac{2r+3}{r-4}\cdot\dfrac{r^2-16}{6r+9}$

19. $\dfrac{3q+3}{5-6q}\div\dfrac{4q+4}{2(5-6q)}$

20. $\dfrac{y^2-6y+8}{y^2+3y-18}\div\dfrac{y-4}{y+6}$

21. $\dfrac{2p^2+13p+20}{p^2+p-12}\cdot\dfrac{p^2+2p-15}{2p^2+7p+5}$

22. $\dfrac{3z^2+5z-2}{9z^2-1}\cdot\dfrac{9z^2+6z+1}{z^2+5z+6}$

[15.3] *Find the least common denominator for the fractions in each list.*

23. $\dfrac{1}{8}, \dfrac{5}{12}, \dfrac{7}{32}$

24. $\dfrac{4}{9y}, \dfrac{7}{12y^2}, \dfrac{5}{27y^4}$

25. $\dfrac{1}{m^2 + 2m}, \dfrac{4}{m^2 + 7m + 10}$

26. $\dfrac{3}{x^2 + 4x + 3}, \dfrac{5}{x^2 + 5x + 4}, \dfrac{2}{x^2 + 7x + 12}$

Write each rational expression as an equivalent expression with the indicated denominator.

27. $\dfrac{5}{8} = \dfrac{?}{56}$

28. $\dfrac{10}{k} = \dfrac{?}{4k}$

29. $\dfrac{3}{2a^3} = \dfrac{?}{10a^4}$

30. $\dfrac{9}{x - 3} = \dfrac{?}{18 - 6x}$

31. $\dfrac{-3y}{2y - 10} = \dfrac{?}{50 - 10y}$

32. $\dfrac{4b}{b^2 + 2b - 3} = \dfrac{?}{(b + 3)(b - 1)(b + 2)}$

[15.4] *Add or subtract. Write each answer in lowest terms.*

33. $\dfrac{10}{x} + \dfrac{5}{x}$

34. $\dfrac{6}{3p} - \dfrac{12}{3p}$

35. $\dfrac{9}{k} - \dfrac{5}{k - 5}$

36. $\dfrac{4}{y} + \dfrac{7}{7 + y}$

37. $\dfrac{m}{3} - \dfrac{2 + 5m}{6}$

38. $\dfrac{12}{x^2} - \dfrac{3}{4x}$

39. $\dfrac{5}{a - 2b} + \dfrac{2}{a + 2b}$

40. $\dfrac{4}{k^2 - 9} - \dfrac{k + 3}{3k - 9}$

41. $\dfrac{8}{z^2 + 6z} - \dfrac{3}{z^2 + 4z - 12}$

42. $\dfrac{11}{2p - p^2} - \dfrac{2}{p^2 - 5p + 6}$

[15.5] *Simplify each complex fraction.*

43. $\dfrac{\dfrac{a^4}{b^2}}{\dfrac{a^3}{b}}$

44. $\dfrac{\dfrac{y - 3}{y}}{\dfrac{y + 3}{4y}}$

45. $\dfrac{\dfrac{3m + 2}{m}}{\dfrac{2m - 5}{6m}}$

46. $\dfrac{\dfrac{1}{p} - \dfrac{1}{q}}{\dfrac{1}{q - p}}$

47. $\dfrac{x + \dfrac{1}{w}}{x - \dfrac{1}{w}}$

48. $\dfrac{\dfrac{1}{r + t} - 1}{\dfrac{1}{r + t} + 1}$

[15.6] *Solve each equation. Check your solutions.*

49. $\dfrac{k}{5} - \dfrac{2}{3} = \dfrac{1}{2}$

50. $\dfrac{4-z}{z} + \dfrac{3}{2} = \dfrac{-4}{z}$

51. $\dfrac{x}{2} - \dfrac{x-3}{7} = -1$

52. $\dfrac{3y-1}{y-2} = \dfrac{5}{y-2} + 1$

53. $\dfrac{3}{m-2} + \dfrac{1}{m-1} = \dfrac{7}{m^2 - 3m + 2}$

Solve for the specified variable.

54. $m = \dfrac{Ry}{t}$ for t

55. $x = \dfrac{3y-5}{4}$ for y

56. $\dfrac{1}{r} - \dfrac{1}{s} = \dfrac{1}{t}$ for t

[15.7] *Solve each problem.*

57. In a certain fraction, the denominator is 5 less than the numerator. If 5 is added to both the numerator and the denominator, the resulting fraction is equivalent to $\frac{5}{4}$. Find the original fraction (*not* written in lowest terms).

58. The denominator of a certain fraction is six times the numerator. If 3 is added to the numerator and subtracted from the denominator, the resulting fraction is equivalent to $\frac{2}{5}$. Find the original fraction (*not* written in lowest terms).

59. On June 24, 2007, Dario Franchitti won the first Iowa Corn Indy 250. He drove a Dallara-Honda the 218.75 mi distance, with an average speed of 123.896 mph. What was his time (to the nearest thousandth of an hour)? (*Source:* www.iowaspeedway.com)

60. In the 2006 Winter Olympics in Torino, Italy, Chad Hedrick of the United States won the men's 5000-m speed skating event in 6.245 min. What was his rate (to three decimal places)? (*Source: World Almanac and Book of Facts.*)

61. Zachary and Samuel are brothers who share a bedroom. By himself, Zachary can completely mess up their room in 20 min, while it would take Samuel only 12 min to do the same thing. How long would it take them to mess up the room together?

62. A man can plant his garden in 5 hr, working alone. His daughter can do the same job in 8 hr. How long would it take them if they worked together?

[15.8] *Solve each problem.*

63. If y varies directly as x, and $x = 12$ when $y = 5$, find x when $y = 3$.

64. If a parallelogram has a fixed area, the height varies inversely as the base. A parallelogram has a height of 8 cm and a base of 12 cm. Find the height if the base is changed to 24 cm.

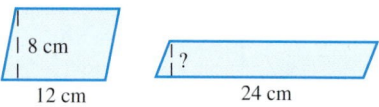

Mixed Review Exercises

Perform the indicated operations.

65. $\dfrac{4}{m - 1} - \dfrac{3}{m + 1}$

66. $\dfrac{8p^5}{5} \div \dfrac{2p^3}{10}$

67. $\dfrac{r - 3}{8} \div \dfrac{3r - 9}{4}$

68. $\dfrac{\dfrac{5}{x} - 1}{\dfrac{5 - x}{3x}}$

69. $\dfrac{4}{z^2 - 2z + 1} - \dfrac{3}{z^2 - 1}$

Solve.

70. $a = \dfrac{v - w}{t}$ for v

71. $\dfrac{2}{z} - \dfrac{z}{z + 3} = \dfrac{1}{z + 3}$

72. $\dfrac{2x}{x^2 - 16} - \dfrac{2}{x - 4} = \dfrac{4}{x + 4}$

73. Anne Kelly flew her plane 400 km with the wind in the same time it took her to go 200 km against the wind. The speed of the wind is 50 km per hr. Find the speed of the plane in still air.

74. At a given hour, two steamboats leave a city in the same direction on a straight canal. One travels at 18 mph, and the other travels at 25 mph. In how many hours will the boats be 70 mi apart?

75. In rectangles of constant area, length and width vary inversely. When the length is 24, the width is 2. What is the width when the length is 12?

76. The longer the term of your subscription to *ESPN: The Magazine*, the less you will have to pay per year. Is this an example of direct or inverse variation?

Chapter 15 ▶▶▶ Test

 Use the Chapter Test Prep Video CD to see fully worked-out solutions to any of the exercises you want to review.

1. Find any values for which $\dfrac{3x - 1}{x^2 - 2x - 8}$ is undefined. Write your answer with $\neq$.

1. _____

2. Find the numerical value of $\dfrac{6r + 1}{2r^2 - 3r - 20}$ when

(a) $r = -2$ and **(b)** $r = 4$.

2. (a) _____

(b) _____

3. Write four rational expressions equivalent to $-\dfrac{6x - 5}{2x + 3}$.

3. _____

Write each rational expression in lowest terms.

4. $\dfrac{-15x^6y^4}{5x^4y}$

5. $\dfrac{6a^2 + a - 2}{2a^2 - 3a + 1}$

4. _____

5. _____

Multiply or divide. Write each answer in lowest terms.

6. $\dfrac{5(d - 2)}{9} \div \dfrac{3(d - 2)}{5}$

7. $\dfrac{6k^2 - k - 2}{8k^2 + 10k + 3} \cdot \dfrac{4k^2 + 7k + 3}{3k^2 + 5k + 2}$

6. _____

7. _____

8. $\dfrac{4a^2 + 9a + 2}{3a^2 + 11a + 10} \div \dfrac{4a^2 + 17a + 4}{3a^2 + 2a - 5}$

8. _____

Find the least common denominator for each list of fractions.

9. $\dfrac{-3}{10p^2}, \dfrac{21}{25p^3}, \dfrac{-7}{30p^5}$

10. $\dfrac{r + 1}{2r^2 + 7r + 6}, \dfrac{-2r + 1}{2r^2 - 7r - 15}$

9. _____

10. _____

Write each rational expression as an equivalent expression with the indicated denominator.

11. $\dfrac{15}{4p} = \dfrac{?}{64p^3}$

12. $\dfrac{3}{6m - 12} = \dfrac{?}{42m - 84}$

11. _____

12. _____

Add or subtract. Write each answer in lowest terms.

13. $\dfrac{4x + 2}{x + 5} + \dfrac{-2x + 8}{x + 5}$

14. $\dfrac{-4}{y + 2} + \dfrac{6}{5y + 10}$

13. _____

14. _____

15. $\dfrac{x + 1}{3 - x} - \dfrac{x^2}{x - 3}$

16. $\dfrac{3}{2m^2 - 9m - 5} - \dfrac{m + 1}{2m^2 - m - 1}$

15. _____

16. _____

Simplify each complex fraction.

17. _____

17. $\dfrac{\dfrac{2p}{k^2}}{\dfrac{3p^2}{k^3}}$

18. $\dfrac{\dfrac{1}{x+3} - 1}{1 + \dfrac{1}{x+3}}$

18. _____

Solve each equation.

19. _____

19. $\dfrac{2}{x-1} - \dfrac{2}{3} = \dfrac{-1}{x+1}$

20. $\dfrac{2x}{x-3} + \dfrac{1}{x+3} = \dfrac{-6}{x^2-9}$

20. _____

21. _____

21. Solve the formula $F = \dfrac{k}{d-D}$ for D.

Solve each problem.

22. _____

22. If the same number is added to the numerator and subtracted from the denominator of $\frac{5}{6}$, the resulting fraction is equivalent to $\frac{1}{10}$. What is the number?

23. _____

23. A boat goes 7 mph in still water. It takes as long to go 20 mi upstream as 50 mi downstream. Find the speed of the current.

24. _____

24. A man can paint a room in his house, working alone, in 5 hr. His wife can do the job in 4 hr. How long will it take them to paint the room if they work together?

25. _____

25. If x varies directly as y, and $x = 12$ when $y = 4$, find x when $y = 9$.

26. _____

26. Under certain conditions, the length of time that it takes for fruit to ripen during the growing season varies inversely as the average maximum temperature during the season. If it takes 25 days for fruit to ripen with an average maximum temperature of 80°F, find the number of days it would take at 75°F. Round your answer to the nearest whole number.

Roots and Radicals

The London Eye opened on New Year's Eve in 1999. This unique structure features 32 observation capsules and has a diameter of 135 meters. Located on the bank of the Thames River, it faces the Houses of Parliament and is the fourth-tallest structure in London. (*Source:* www.londoneye.com)

The formula

$$\text{sight distance} = 111.7\sqrt{\text{height of structure in kilometers}}$$

involves *radicals*, the subject of this chapter, and can be used to determine how far one can see (in kilometers) from the top of a structure on a clear day. In Exercise 55 of **Section 16.6,** we use this formula to determine the truth of the claim that passengers on the London Eye can see Windsor Castle, 25 miles away.

16.1 ▸▸▸ Evaluating Roots

OBJECTIVES

1 Find square roots.

2 Decide whether a given root is rational, irrational, or not a real number.

3 Find decimal approximations for irrational square roots.

4 Use the Pythagorean formula.

5 Find cube, fourth, and other roots.

In **Section 1.8,** we discussed the idea of the *square* of a number. Recall that squaring a number means multiplying the number by itself.

$$\text{If } a = 8, \quad \text{then} \quad a^2 = 8 \cdot 8 = 64.$$

$$\text{If } a = -4, \quad \text{then} \quad a^2 = (-4)(-4) = 16.$$

$$\text{If } a = -\frac{1}{2}, \quad \text{then} \quad a^2 = \left(-\frac{1}{2}\right)\left(-\frac{1}{2}\right) = \frac{1}{4}.$$

In this chapter, we consider the opposite process.

$$\text{If } a^2 = 64, \quad \text{then} \quad a = \mathbf{?}$$

$$\text{If } a^2 = 16, \quad \text{then} \quad a = \mathbf{?}$$

$$\text{If } a^2 = \frac{1}{4}, \quad \text{then} \quad a = \mathbf{?}$$

OBJECTIVE 1 Find square roots. To find a in the three preceding statements, we must find a number that when multiplied by itself results in the given number. The number a is called a **square root** of the number a^2.

---

1 Find all square roots.

(a) 100

(b) 25

(c) 36

(d) $\dfrac{25}{36}$

Early radical symbol

> **EXAMPLE 1**   **Finding All Square Roots of a Number**
>
> Find all square roots of 49.
>
>   To find a square root of 49, think of a number that when multiplied by itself gives 49. One square root is 7 because $7 \cdot 7 = 49$. Another square root of 49 is -7 because $(-7)(-7) = 49$. The number 49 has two square roots, 7 and -7; one is positive, and one is negative.

◀ *Work Problem* **1** *at the Side.*

The **positive** or **principal square root** of a number is written with the symbol $\sqrt{}$. For example, the positive square root of 121 is 11, written

$$\sqrt{121} = \mathbf{11}.$$

The symbol $-\sqrt{}$ is used for the **negative square root** of a number. For example, the negative square root of 121 is -11, written

$$-\sqrt{121} = \mathbf{-11}.$$

The symbol $\sqrt{}$, called a **radical sign**, always represents the positive square root (except that $\sqrt{0} = 0$). The number inside the radical sign is called the **radicand**, and the entire expression, radical sign and radicand, is called a **radical.**

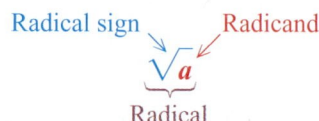

An algebraic expression containing a radical is called a **radical expression.**

    Radicals have a long mathematical history. The radical sign $\sqrt{}$ has been used since sixteenth-century Germany and was probably derived from the letter R. The radical symbol in the margin comes from the Latin word for root, *radix*. It was first used by Leonardo da Pisa (Fibonnaci) in 1220.

We summarize our discussion of square roots as follows.

Square Roots of a

If a is a positive real number, then

$\sqrt{a}$ is the positive or principal square root of a,

and $-\sqrt{a}$ is the negative square root of a.

For nonnegative a,

$$\sqrt{a} \cdot \sqrt{a} = \left(\sqrt{a}\right)^2 = a \quad \text{and} \quad -\sqrt{a} \cdot \left(-\sqrt{a}\right) = \left(-\sqrt{a}\right)^2 = a.$$

Also, $\sqrt{0} = 0$.

▦ **Calculator Tip**   Recall from **Section 5.8** that most calculators have a square root key, usually labeled ⎡√x⎤, that allows us to find the square root of a number. On some models, the square root key must be used in conjunction with the key marked ⎡INV⎤ or ⎡2nd⎤.

EXAMPLE 2   **Finding Square Roots**

Find each square root.

(a) $\sqrt{144}$

The radical $\sqrt{144}$ represents the positive or principal square root of 144. Think of a positive number whose square is 144.

$$12^2 = 144, \quad \text{so} \quad \sqrt{144} = 12.$$

(b) $-\sqrt{1024}$

This symbol represents the negative square root of 1024. A calculator with a square root key can be used to find $\sqrt{1024} = 32$. Then, $-\sqrt{1024} = -32$.

(c) $\sqrt{\dfrac{4}{9}} = \dfrac{2}{3}$ **(d)** $-\sqrt{\dfrac{16}{49}} = -\dfrac{4}{7}$ **(e)** $\sqrt{0.81} = 0.9$

Work Problem ⎡**2**⎤ *at the Side.* ▶

As noted above, when the square root of a positive real number is squared, the result is that positive real number. $\left(\text{Also, } \left(\sqrt{0}\right)^2 = 0.\right)$

EXAMPLE 3   **Squaring Radical Expressions**

Find the *square* of each radical expression.

(a) $\sqrt{13}$

$\left(\sqrt{13}\right)^2 = 13$ Definition of square root

(b) $-\sqrt{29}$

$\left(-\sqrt{29}\right)^2 = 29$ The square of a *negative* number is positive.

(c) $\sqrt{p^2 + 1}$

$\left(\sqrt{p^2 + 1}\right)^2 = p^2 + 1$

Work Problem ⎡**3**⎤ *at the Side.* ▶

⎡**2**⎤ Find each square root.

(a) $\sqrt{16}$

(b) $-\sqrt{169}$

(c) $-\sqrt{225}$

(d) $\sqrt{729}$

(e) $\sqrt{\dfrac{36}{25}}$

(f) $\sqrt{0.49}$

⎡**3**⎤ Find the *square* of each radical expression.

(a) $\sqrt{41}$

(b) $-\sqrt{39}$

(c) $\sqrt{2x^2 + 3}$

ANSWERS

2. **(a)** 4   **(b)** -13 **(c)** -15
   (d) 27   **(e)** $\dfrac{6}{5}$ **(f)** 0.7

3. **(a)** 41   **(b)** 39   **(c)** $2x^2 + 3$

4 Tell whether each square root is *rational, irrational,* or *not a real number.*

(a) $\sqrt{9}$

(b) $\sqrt{7}$

(c) $\sqrt{\dfrac{4}{9}}$

(d) $\sqrt{72}$

(e) $\sqrt{-43}$

OBJECTIVE **2** **Decide whether a given root is rational, irrational, or not a real number.**   All numbers with square roots that are rational are called **perfect squares.**

Perfect squares		Rational square roots
25		$\sqrt{25} = 5$
144	are perfect squares since	$\sqrt{144} = 12$
$\dfrac{4}{9}$		$\sqrt{\dfrac{4}{9}} = \dfrac{2}{3}$

A number that is not a perfect square has a square root that is not a rational number. For example, $\sqrt{5}$ is not a rational number because it cannot be written as the ratio of two integers. Its decimal equivalent (or approximation) neither terminates nor repeats. However, $\sqrt{5}$ is a real number and corresponds to a point on the number line. As mentioned in **Section 10.1,** a real number that is not rational is called an **irrational number.** The number $\sqrt{5}$ is irrational. Many square roots of integers are irrational.

> If a is a *positive* real number that is *not* a perfect square, then
> $$\sqrt{a} \text{ is irrational.}$$

Not every number has a real number square root. For example, there is no real number that can be squared to obtain -36. (The square of a real number can never be negative.) Because of this, $\sqrt{-36}$ *is not a real number.*

> If a is a *negative* real number, then $\sqrt{a}$ is *not* a real number.

CAUTION
Be careful not to confuse $\sqrt{-36}$ and $-\sqrt{36}$. $\sqrt{-36}$ is not a real number since there is no real number that can be squared to obtain -36. However, $-\sqrt{36}$ is the negative square root of 36, which is -6.

EXAMPLE 4   **Identifying Types of Square Roots**

Tell whether each square root is *rational, irrational,* or *not a real number.*

(a) $\sqrt{17}$
   Because 17 is not a perfect square, $\sqrt{17}$ is irrational.

(b) $\sqrt{64}$
   The number 64 is a perfect square, 8^2, so $\sqrt{64} = 8$ is a rational number.

(c) $\sqrt{-25}$
   There is no real number whose square is -25. Therefore, $\sqrt{-25}$ is not a real number.

◀ Work Problem **4** at the Side.

ANSWERS

4. (a) rational   (b) irrational   (c) rational
   (d) irrational   (e) not a real number

> **Note**
>
> Not all irrational numbers are square roots of integers. For example, π (approximately 3.14159) is an irrational number that is not a square root of any integer.

OBJECTIVE 3 Find decimal approximations for irrational square roots. Even if a number is irrational, a decimal that approximates the number can be found using a calculator, as we did in **Section 5.8.** For example, if we use a calculator to find $\sqrt{10}$, the display might show 3.16227766, which is only an *approximation* of $\sqrt{10}$, not an exact rational value.

EXAMPLE 5 **Approximating Irrational Square Roots**

Find a decimal approximation for each square root. Round answers to the nearest thousandth.

(a) $\sqrt{11}$
Using the square root key of a calculator gives $3.31662479 \approx 3.317$, where $\approx$ means "is approximately equal to."

(b) $\sqrt{39} \approx 6.245$ Use a calculator.   **(c)** $-\sqrt{740} \approx -27.203$

Work Problem **5** *at the Side.* ▶

OBJECTIVE 4 Use the Pythagorean formula. Many applications of square roots use the Pythagorean formula. Recall from **Section 14.7** that by this formula if c is the length of the hypotenuse of a right triangle, and a and b are the lengths of the two legs, as shown in Figure 1, then

$$a^2 + b^2 = c^2$$

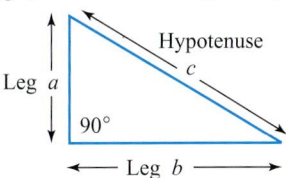

Figure 1

EXAMPLE 6 **Using the Pythagorean Formula**

Find the length of the unknown side of each right triangle with sides a, b, and c, where c is the hypotenuse.

(a) $a = 3, b = 4$
Use the Pythagorean formula to find c^2 first.

$$c^2 = a^2 + b^2$$
$$c^2 = 3^2 + 4^2 \quad \text{Let } a = 3 \text{ and } b = 4.$$
$$c^2 = 9 + 16 \quad \text{Square.}$$
$$c^2 = 25 \quad \text{Add.}$$

Since the length of a side of a triangle must be a positive number, find the positive square root of 25 to get c.

$$c = \sqrt{25} = 5$$

Continued on Next Page

---

5 Find a decimal approximation for each square root. Round answers to the nearest thousandth.

(a) $\sqrt{28}$

(b) $\sqrt{63}$

(c) $-\sqrt{190}$

(d) $\sqrt{1000}$

ANSWERS

5. (a) 5.292   **(b)** 7.937   **(c)** -13.784
     (d) 31.623

6 Find the length of the unknown side in each right triangle. Give any decimal approximations to the nearest thousandth.

(a) $a = 7, b = 24$

(b) $c = 9, b = 5$

Substitute the given values in the Pythagorean formula. Then solve for a^2.

$$c^2 = a^2 + b^2$$
$$9^2 = a^2 + 5^2 \qquad \text{Let } c = 9 \text{ and } b = 5.$$
$$81 = a^2 + 25 \qquad \text{Square.}$$
$$56 = a^2 \qquad \text{Subtract 25.}$$

Use a calculator to find the positive square root of 56 to get a.

$$a = \sqrt{56} \approx 7.483$$

> **CAUTION**
>
> Be careful not to make the common mistake of thinking that $\sqrt{a^2 + b^2}$ equals $a + b$. As Example 6(a) shows, $\sqrt{9 + 16} = \sqrt{25} = 5$. However, $\sqrt{9} + \sqrt{16} = 3 + 4 = 7$. Since $5 \neq 7$, in general,
>
> $$\sqrt{a^2 + b^2} \neq a + b.$$

◀ *Work Problem* **6** *at the Side.*

(b) $c = 15, b = 13$

The Pythagorean formula can be used to solve applied problems that involve right triangles. Use the same six problem-solving steps that we have been using throughout the text.

EXAMPLE 7 **Using the Pythagorean Formula to Solve an Application**

A ladder 10 ft long leans against a wall. The foot of the ladder is 6 ft from the base of the wall. How high up the wall does the top of the ladder rest?

Step 1 **Read** the problem again.

Step 2 **Assign a variable.** As shown in Figure 2, a right triangle is formed with the ladder as the hypotenuse. Let a represent the height of the top of the ladder when measured straight down to the ground.

(c)

8 ⟋ 11 / ?

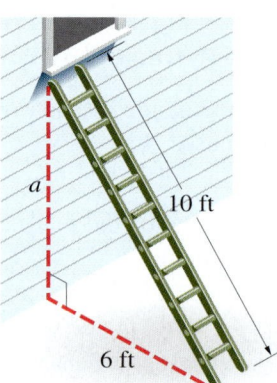

a

10 ft

6 ft

Figure 2

Continued on Next Page

Step 3   **Write an equation** using the Pythagorean formula.

> Substitute carefully.

$$c^2 = a^2 + b^2$$

$$\mathbf{10}^2 = a^2 + \mathbf{6}^2 \qquad \text{Let } c = 10 \text{ and } b = 6.$$

Step 4   **Solve.**

$$100 = a^2 + 36 \qquad \text{Square.}$$

$$64 = a^2 \qquad \text{Subtract 36.}$$

$$\sqrt{64} = a$$

$$a = 8 \qquad \sqrt{64} = 8$$

Choose the positive square root of 64 since *a* represents a length.

Step 5   **State the answer.** The top of the ladder rests 8 ft up the wall.

Step 6   **Check.** From Figure 2, we see that we must have

$$8^2 + 6^2 \stackrel{?}{=} 10^2$$

$$64 + 36 = 100. \qquad \text{True}$$

The check confirms that the top of the ladder rests 8 ft up the wall.

—————————————— *Work Problem* **7** *at the Side.* ▶

OBJECTIVE 5 Find cube, fourth, and other roots.   Finding the square root of a number is the inverse (reverse) of squaring a number. In a similar way, there are inverses to finding the cube of a number, or finding the fourth or higher power of a number. These inverses are the **cube root,** written $\sqrt[3]{a}$, and the **fourth root,** written $\sqrt[4]{a}$. Similar symbols are used for higher roots. In general, we have the following.

$\sqrt[n]{a}$

The *n*th root of *a* is written: $\sqrt[n]{a}$

In $\sqrt[n]{a}$, the number *n* is the **index,** or **order,** of the radical.

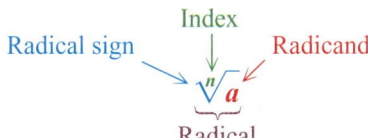

Index

Radical sign   Radicand

Radical

We could write $\sqrt[2]{a}$ instead of $\sqrt{a}$, but the simpler symbol $\sqrt{a}$ is customary since the square root is the most commonly used root.

🖩 **Calculator Tip**   A calculator that has a key marked $\boxed{\sqrt[x]{y}}$, $\boxed{x^y}$, or $\boxed{y^x}$ (again perhaps in conjunction with the $\boxed{\text{INV}}$ or $\boxed{\text{2nd}}$ key) can be used to find other roots.

When working with cube roots or fourth roots, it is helpful to memorize the first few *perfect cubes* ($1^3 = 1$, $2^3 = 8$, $3^3 = 27$, and so on) and the first few *perfect fourth powers* ($1^4 = 1$, $2^4 = 16$, $3^4 = 81$, and so on).

Work Problem **8** *at the Side.* ▶

---

7 A rectangle has dimensions 5 ft by 12 ft. Find the length of its diagonal.

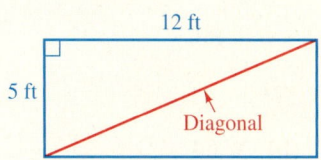

12 ft

5 ft

Diagonal

(Note that the diagonal divides the rectangle into two right triangles with itself as the hypotenuse.)

8 Complete the following list of perfect cubes and perfect fourth powers.

Perfect Cubes	Perfect Fourth Powers
$1^3 = 1$	$1^4 = 1$
$2^3 = 8$	$2^4 = 16$
$3^3 = 27$	$3^4 = 81$
$4^3 = \underline{\ \ \ }$	$4^4 = \underline{\ \ \ }$
$5^3 = \underline{\ \ \ }$	$5^4 = \underline{\ \ \ }$
$6^3 = \underline{\ \ \ }$	$6^4 = \underline{\ \ \ }$
$7^3 = \underline{\ \ \ }$	$7^4 = \underline{\ \ \ }$
$8^3 = \underline{\ \ \ }$	$8^4 = \underline{\ \ \ }$
$9^3 = \underline{\ \ \ }$	$9^4 = \underline{\ \ \ }$
$10^3 = \underline{\ \ \ }$	$10^4 = \underline{\ \ \ }$

ANSWERS

7. 13 ft
8. Perfect cubes: 64; 125; 216; 343; 512; 729; 1000
Perfect fourth powers: 256; 625; 1296; 2401; 4096; 6561; 10,000

9 Find each cube root.

(a) $\sqrt[3]{27}$

(b) $\sqrt[3]{64}$

(c) $\sqrt[3]{-125}$

EXAMPLE 8   Finding Cube Roots

Find each cube root.

(a) $\sqrt[3]{8}$

Look for a number that can be cubed to give 8. Because $2^3 = 8$, $\sqrt[3]{8} = 2$.

(b) $\sqrt[3]{-8} = -2$ because $(-2)^3 = -8$.

(c) $\sqrt[3]{216} = 6$ because $6^3 = 216$.

Notice in Example 8(b) that we can find the cube root of a negative number. (Contrast this with the square root of a negative number, which is not real.) In fact, the cube root of a positive number is positive, and the cube root of a negative number is negative. **There is only one real number cube root for each real number.**

◀ *Work Problem* **9** *at the Side.*

When a radical has an **even index** (square root, fourth root, and so on), **the radicand must be nonnegative** to yield a real number root. Also,

$$\sqrt{a}, \sqrt[4]{a}, \sqrt[6]{a}, \text{ and so on are positive (principal) roots;}$$

$$-\sqrt{a}, -\sqrt[4]{a}, -\sqrt[6]{a}, \text{ and so on are negative roots.}$$

---

10 Find each root.

(a) $\sqrt[4]{81}$

(b) $\sqrt[4]{-81}$

(c) $-\sqrt[4]{81}$

(d) $\sqrt[5]{243}$

(e) $\sqrt[5]{-243}$

EXAMPLE 9   Finding Other Roots

Find each root.

(a) $\sqrt[4]{16} = 2$ because 2 is positive and $2^4 = 16$.

(b) $-\sqrt[4]{16}$

From part (a), $\sqrt[4]{16} = 2$, so the negative root is $-\sqrt[4]{16} = -2$.

(c) $\sqrt[4]{-16}$

For a real number fourth root, the radicand must be nonnegative. There is no real number that equals $\sqrt[4]{-16}$.

(d) $-\sqrt[5]{32}$

First find $\sqrt[5]{32}$. Because 2 is the number whose fifth power is 32, $\sqrt[5]{32} = 2$. Since $\sqrt[5]{32} = 2$, it follows that

$$-\sqrt[5]{32} = -2.$$

(e) $\sqrt[5]{-32}$

Because $(-2)^5 = -32$, $\sqrt[5]{-32} = -2$.

◀ *Work Problem* **10** *at the Side.*

ANSWERS

9. (a) 3 (b) 4 (c) −5
10. (a) 3 (b) not a real number
    (c) −3 (d) 3 (e) −3

Decide whether each statement is true *or* false. *If* false, *tell why.*

1. Every positive number has two real square roots.

2. A negative number has negative square roots.

3. Every nonnegative number has two real square roots.

4. The positive square root of a positive number is its principal square root.

5. The cube root of every real number has the same sign as the number itself.

6. Every positive number has three real cube roots.

Find all square roots of each number. See Example 1.

7. 9
8. 16
9. 64
10. 100
11. 169

12. 225
13. $\dfrac{25}{196}$
14. $\dfrac{81}{400}$
15. 900
16. 1600

Find each square root. See Examples 2 and 4(c).

17. $\sqrt{1}$
18. $\sqrt{4}$
19. $\sqrt{49}$
20. $\sqrt{81}$
21. $-\sqrt{256}$

22. $-\sqrt{196}$
23. $-\sqrt{\dfrac{144}{121}}$
24. $-\sqrt{\dfrac{49}{36}}$
25. $\sqrt{0.64}$
26. $\sqrt{0.16}$

27. $\sqrt{-121}$
28. $\sqrt{-64}$
29. $-\sqrt{-49}$
30. $-\sqrt{-100}$

Find the square of each radical expression. See Example 3.

31. $\sqrt{100}$
32. $\sqrt{36}$
33. $-\sqrt{19}$
34. $-\sqrt{99}$

35. $\sqrt{\dfrac{2}{3}}$
36. $\sqrt{\dfrac{5}{7}}$
37. $\sqrt{3x^2 + 4}$
38. $\sqrt{9y^2 + 3}$

What must be true about the value of a for each statement in Exercises 39–42 to be true?

39. $\sqrt{a}$ represents a positive number.

40. $-\sqrt{a}$ represents a negative number.

41. $\sqrt{a}$ is not a real number.

42. $-\sqrt{a}$ is not a real number.

Write rational, irrational, or not a real number for each number. If a number is rational, give its exact value. If a number is irrational, give a decimal approximation to the nearest thousandth. Use a calculator as necessary. See Examples 4 and 5.

43. $\sqrt{25}$

44. $\sqrt{169}$

45. $\sqrt{29}$

46. $\sqrt{33}$

47. $-\sqrt{64}$

48. $-\sqrt{81}$

49. $-\sqrt{300}$

50. $-\sqrt{500}$

51. $\sqrt{-29}$

52. $\sqrt{-47}$

53. $\sqrt{1200}$

54. $\sqrt{1500}$

Work Exercises 55 and 56 without using a calculator.

55. Choose the best estimate for the length and width (in meters) of this rectangle.

   A. 11 by 6   **B.** 11 by 7   **C.** 10 by 7   **D.** 10 by 6

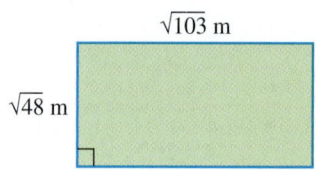

$\sqrt{103}$ m

$\sqrt{48}$ m

56. Choose the best estimate for the base and height (in feet) of this triangle.

   A. $b = 8, h = 5$ **B.** $b = 8, h = 4$
   C. $b = 9, h = 5$ **D.** $b = 9, h = 4$

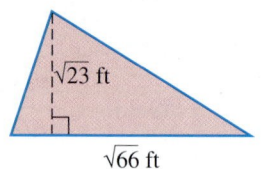

$\sqrt{23}$ ft

$\sqrt{66}$ ft

Find the length of the unknown side of each right triangle with sides a, b, and c, where c is the hypotenuse. See Figure 1 and Example 6. Give any decimal approximations to the nearest thousandth.

57. $a = 8, b = 15$

58. $a = 24, b = 10$

59. $a = 6, c = 10$

60. $b = 12, c = 13$

61. $a = 11, b = 4$

62. $a = 13, b = 9$

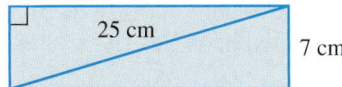 *Solve each problem. See Example 7.*

63. The diagonal of a rectangle measures 25 cm. The width of the rectangle is 7 cm. Find the length of the rectangle.

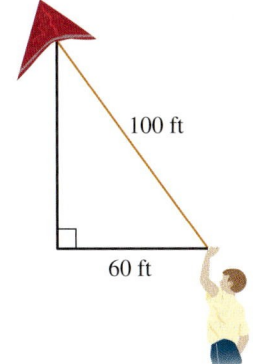

25 cm

7 cm

64. The length of a rectangle is 40 m, and the width is 9 m. Find the measure of the diagonal of the rectangle.

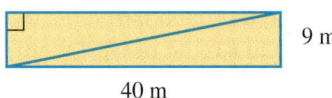

9 m

40 m

65. Tyler is flying a kite on 100 ft of string. How high is it above his hand (vertically) if the horizontal distance between Tyler and the kite is 60 ft?

100 ft

60 ft

66. A guy wire is attached to the mast of a short-wave transmitting antenna. It is attached 96 ft above ground level. If the wire is staked to the ground 72 ft from the base of the mast, how long is the wire?

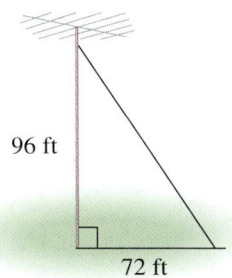

96 ft

72 ft

67. A surveyor measured the distances shown in the figure. Find the distance across the lake between points R and S.

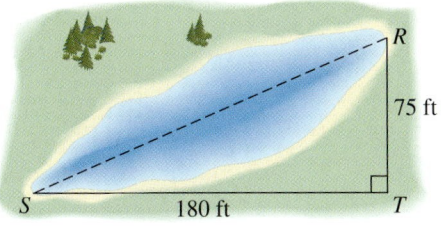

R

75 ft

S 180 ft T

68. A boat is being pulled toward a dock with a rope attached at water level. When the boat is 24 ft from the dock, 30 ft of rope is extended. What is the height of the dock above the water?

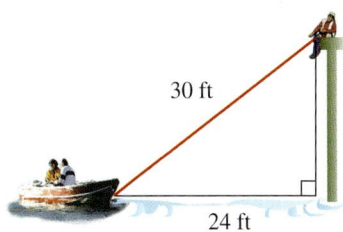

30 ft

24 ft

69. A surveyor wants to find the height of a building. At a point 110.0 ft from the base of the building he sights to the top of the building and finds the distance to be 193.0 ft. How high is the building (to the nearest tenth)?

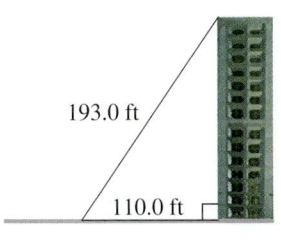

193.0 ft

110.0 ft

70. Two towns are separated by a dense forest. To go from Town B to Town A, it is necessary to travel due west for 19.0 mi, then turn due north and travel for 14.0 mi. How far apart are the towns (to the nearest tenth)?

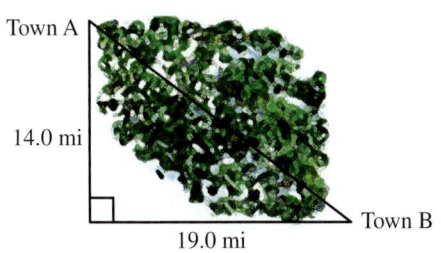

Town A

14.0 mi

Town B

19.0 mi

71. Following Hurricane Katrina, thousands of pine trees in southeastern Louisiana formed right triangles as shown in the photo. Suppose that, for a small such tree, the vertical distance from the base of the broken tree to the point of the break is 4.5 ft. The length of the broken part is 12.0 ft. How far along the ground (to the nearest tenth) is it from the base of the tree to the point where the broken part touches the ground?

72. One of the authors of this text purchased a new rear-projection Toshiba 51H84 television. A television set is "sized" according to the diagonal measurement of the viewing screen. The author purchased a 51-in. TV, so the TV measures 51 in. from one corner of the viewing screen diagonally to the other corner. The viewing screen is 44.5 in. wide. Find the height of the viewing screen (to the nearest tenth).

73. What is the value of x (to the nearest thousandth) in the figure?

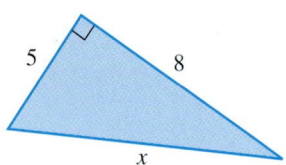

74. What is the value of y (to the nearest thousandth) in the figure?

Find each root. See Examples 8 and 9.

75. $\sqrt[3]{1}$

76. $\sqrt[3]{729}$

77. $\sqrt[3]{125}$

78. $\sqrt[3]{1000}$

79. $\sqrt[3]{-27}$

80. $\sqrt[3]{-64}$

81. $\sqrt[3]{-216}$

82. $\sqrt[3]{-343}$

83. $-\sqrt[3]{-8}$

84. $-\sqrt[3]{-216}$

85. $\sqrt[4]{256}$

86. $\sqrt[4]{625}$

87. $\sqrt[4]{1296}$

88. $\sqrt[4]{10,000}$

89. $\sqrt[4]{-1}$

90. $\sqrt[4]{-625}$

91. $-\sqrt[4]{625}$

92. $-\sqrt[4]{256}$

93. $\sqrt[5]{-1024}$

94. $\sqrt[5]{-100,000}$

16.2 ▶▶▶ Multiplying, Dividing, and Simplifying Radicals

OBJECTIVE 1 Multiply square root radicals. We now develop several rules for finding products and quotients of radicals. Notice that

$$\sqrt{4} \cdot \sqrt{9} = 2 \cdot 3 = 6 \quad \text{and} \quad \sqrt{4 \cdot 9} = \sqrt{36} = 6$$

showing that

$$\sqrt{4} \cdot \sqrt{9} = \sqrt{4 \cdot 9}$$

This result is a particular case of the **product rule for radicals.**

Product Rule for Radicals

For nonnegative real numbers a and b,

$$\sqrt{a} \cdot \sqrt{b} = \sqrt{a \cdot b} \quad \text{and} \quad \sqrt{a \cdot b} = \sqrt{a} \cdot \sqrt{b}$$

In words, the product of two square roots is the square root of the product, and the square root of a product is the product of the two square roots.

EXAMPLE 1 Using the Product Rule to Multiply Radicals

Use the product rule for radicals to find each product.

(a) $\sqrt{2} \cdot \sqrt{3}$ **(b)** $\sqrt{7} \cdot \sqrt{5}$ **(c)** $\sqrt{11} \cdot \sqrt{a}$

$\quad = \sqrt{2 \cdot 3}$ $= \sqrt{35}$ $= \sqrt{11a}$ Assume

$\quad = \sqrt{6}$ $a \geq 0$.

Work Problem **1** *at the Side.* ▶

OBJECTIVE 2 Simplify radicals using the product rule. *A square root radical is simplified when no perfect square factor other than 1 remains under the radical sign.* This is accomplished by using the product rule.

EXAMPLE 2 Using the Product Rule to Simplify Radicals

Simplify each radical.

(a) $\sqrt{20}$

Because 20 has a perfect square factor of 4, we can write

$$\sqrt{20}$$
$$= \sqrt{4 \cdot 5} \qquad \text{4 is a perfect square.}$$
$$= \sqrt{4} \cdot \sqrt{5} \qquad \text{Product rule}$$
$$= 2\sqrt{5}. \qquad \sqrt{4} = 2$$

Thus, $\sqrt{20} = 2\sqrt{5}$. Because 5 has no perfect square factor (other than 1), $2\sqrt{5}$ is called the **simplified form** of $\sqrt{20}$. Note that $2\sqrt{5}$ represents a product, where the factors are 2 and $\sqrt{5}$.

We could also factor 20 into prime factors and look for pairs of like factors. Each pair of like factors produces one factor outside the radical. Thus,

$$\sqrt{20} = \sqrt{2 \cdot 2 \cdot 5} = 2\sqrt{5}.$$

Continued on Next Page

OBJECTIVES

1. **Multiply square root radicals.**

2. **Simplify radicals using the product rule.**

3. **Simplify radicals using the quotient rule.**

4. **Simplify radicals involving variables.**

5. **Simplify other roots.**

---

1 Use the product rule for radicals to find each product.

(a) $\sqrt{6} \cdot \sqrt{11}$

(b) $\sqrt{2} \cdot \sqrt{5}$

(c) $\sqrt{10} \cdot \sqrt{r}, \quad r \geq 0$

ANSWERS

1. (a) $\sqrt{66}$ **(b)** $\sqrt{10}$ **(c)** $\sqrt{10r}$

2 Simplify each radical.

(a) $\sqrt{8}$

(b) $\sqrt{27}$

(c) $\sqrt{50}$

(d) $\sqrt{60}$

(e) $\sqrt{30}$

(b) $\sqrt{72}$

Look for the *largest* perfect square factor of 72. This number is 36, so

$$\sqrt{72}$$
$$= \sqrt{36 \cdot 2} \qquad \text{36 is a perfect square.}$$
$$= \sqrt{36} \cdot \sqrt{2} \qquad \text{Product rule}$$
$$= 6\sqrt{2}. \qquad \sqrt{36} = 6$$

We could also factor 72 into its prime factors and look for pairs of like factors.

$$\sqrt{72} = \sqrt{2 \cdot 2 \cdot 2 \cdot 3 \cdot 3} = 2 \cdot 3 \cdot \sqrt{2} = 6\sqrt{2}$$

In either case, we obtain $6\sqrt{2}$ as the simplified form of $\sqrt{72}$. However, our work is simpler if we begin with the largest perfect square factor.

(c) $\sqrt{300}$
$$= \sqrt{100 \cdot 3} \qquad \text{100 is a perfect square.}$$
$$= \sqrt{100} \cdot \sqrt{3} \qquad \text{Product rule}$$
$$= 10\sqrt{3} \qquad \sqrt{100} = 10$$

(d) $\sqrt{15}$

The number 15 has no perfect square factors (except 1), so $\sqrt{15}$ cannot be simplified further.

◀ *Work Problem* **2** *at the Side.*

EXAMPLE 3 **Multiplying and Simplifying Radicals**

Find each product and simplify.

(a) $\sqrt{9} \cdot \sqrt{75}$
$$= 3\sqrt{75} \qquad \sqrt{9} = 3$$
$$= 3\sqrt{25 \cdot 3} \qquad \text{Factor; 25 is a perfect square.}$$
$$= 3\sqrt{25} \cdot \sqrt{3} \qquad \text{Product rule}$$
$$= 3 \cdot 5 \cdot \sqrt{3} \qquad \sqrt{25} = 5$$
$$= 15\sqrt{3} \qquad \text{Multiply.}$$

Notice that we could have used the product rule to get $\sqrt{9} \cdot \sqrt{75} = \sqrt{675}$, and then simplified. However, the product rule as used here allows us to obtain the final answer without using a large number like 675.

(b) $\sqrt{8} \cdot \sqrt{12}$
$$= \sqrt{8 \cdot 12} \qquad \text{Product rule}$$
$$= \sqrt{4 \cdot 2 \cdot 4 \cdot 3} \qquad \text{Factor; 4 is a perfect square.}$$
$$= \sqrt{4} \cdot \sqrt{4} \cdot \sqrt{2 \cdot 3} \qquad \text{Commutative property; product rule}$$
$$= 2 \cdot 2 \cdot \sqrt{6} \qquad \sqrt{4} = 2$$
$$= 4\sqrt{6} \qquad \text{Multiply.}$$

Continued on Next Page

ANSWERS

2. (a) $2\sqrt{2}$ (b) $3\sqrt{3}$ (c) $5\sqrt{2}$
   (d) $2\sqrt{15}$ (e) cannot be simplified further

(c) $2\sqrt{3} \cdot 3\sqrt{6}$

$= 2 \cdot 3 \cdot \sqrt{3 \cdot 6}$ Commutative property; product rule

$= 6\sqrt{18}$ Multiply.

$= 6\sqrt{9 \cdot 2}$ Factor; 9 is a perfect square.

$= 6\sqrt{9} \cdot \sqrt{2}$ Product rule

$= 6 \cdot 3 \cdot \sqrt{2}$ $\sqrt{9} = 3$

$= 18\sqrt{2}$ Multiply.

Note

We could also simplify Example 3(b) as follows.

$$\sqrt{8} \cdot \sqrt{12}$$

$= \sqrt{4 \cdot 2} \cdot \sqrt{4 \cdot 3}$ Factor.

$= 2\sqrt{2} \cdot 2\sqrt{3}$ $\sqrt{4} = 2$

$= 2 \cdot 2 \cdot \sqrt{2} \cdot \sqrt{3}$ Commutative property

$= 4\sqrt{6}$ Same result

There is often more than one way to find such a product.

Work Problem ③ *at the Side.* ▶

OBJECTIVE 3 Simplify radicals using the quotient rule.   The **quotient rule for radicals** is very similar to the product rule.

Quotient Rule for Radicals

If a and b are nonnegative real numbers and $b \neq 0$, then

$$\sqrt{\frac{a}{b}} = \frac{\sqrt{a}}{\sqrt{b}} \quad \text{and} \quad \frac{\sqrt{a}}{\sqrt{b}} = \sqrt{\frac{a}{b}}.$$

In words, the square root of a quotient is the quotient of the two square roots, and the quotient of two square roots is the square root of the quotient.

EXAMPLE 4 Using the Quotient Rule to Simplify Radicals

Use the quotient rule to simplify each radical.

(a) $\sqrt{\dfrac{25}{9}}$

$= \dfrac{\sqrt{25}}{\sqrt{9}}$

$= \dfrac{5}{3}$

(b) $\dfrac{\sqrt{288}}{\sqrt{2}}$

$= \sqrt{\dfrac{288}{2}}$

$= \sqrt{144}$

$= 12$

(c) $\sqrt{\dfrac{3}{4}}$

$= \dfrac{\sqrt{3}}{\sqrt{4}}$

$= \dfrac{\sqrt{3}}{2}$

Work Problem ④ *at the Side.* ▶

---

③ Find each product and simplify.

(a) $\sqrt{3} \cdot \sqrt{15}$

(b) $\sqrt{10} \cdot \sqrt{50}$

(c) $\sqrt{12} \cdot \sqrt{2}$

(d) $\sqrt{7} \cdot \sqrt{14}$

(e) $3\sqrt{5} \cdot 4\sqrt{10}$

④ Use the quotient rule to simplify each radical.

(a) $\sqrt{\dfrac{81}{16}}$

(b) $\dfrac{\sqrt{192}}{\sqrt{3}}$

(c) $\sqrt{\dfrac{10}{49}}$

ANSWERS

3. **(a)** $3\sqrt{5}$ **(b)** $10\sqrt{5}$ **(c)** $2\sqrt{6}$
   (d) $7\sqrt{2}$ **(e)** $60\sqrt{2}$

4. **(a)** $\dfrac{9}{4}$ **(b)** 8 **(c)** $\dfrac{\sqrt{10}}{7}$

5 Simplify $\dfrac{8\sqrt{50}}{4\sqrt{5}}$.

6 Simplify.

(a) $\sqrt{\dfrac{5}{6}} \cdot \sqrt{120}$

(b) $\sqrt{\dfrac{3}{8}} \cdot \sqrt{\dfrac{7}{2}}$

EXAMPLE 5 Using the Quotient Rule to Divide Radicals

Simplify.

$$\frac{27\sqrt{15}}{9\sqrt{3}}$$

$$= \frac{27}{9} \cdot \frac{\sqrt{15}}{\sqrt{3}} \qquad \text{Multiplication of fractions}$$

$$= \frac{27}{9} \cdot \sqrt{\frac{15}{3}} \qquad \text{Quotient rule}$$

$$= 3\sqrt{5} \qquad \text{Divide.}$$

◀ *Work Problem* **5** *at the Side.*

EXAMPLE 6 Using Both the Product and Quotient Rules

Simplify.

$$\sqrt{\frac{3}{5}} \cdot \sqrt{\frac{1}{5}}$$

$$= \sqrt{\frac{3}{5} \cdot \frac{1}{5}} \qquad \text{Product rule}$$

$$= \sqrt{\frac{3}{25}} \qquad \text{Multiply fractions.}$$

$$= \frac{\sqrt{3}}{\sqrt{25}} \qquad \text{Quotient rule}$$

$$= \frac{\sqrt{3}}{5} \qquad \sqrt{25} = 5$$

◀ *Work Problem* **6** *at the Side.*

OBJECTIVE 4 Simplify radicals involving variables. Simplifying radicals with variable radicands, such as $\sqrt{x^2}$, requires careful analysis. If x represents a nonnegative number, then $\sqrt{x^2} = x$. If x represents a negative number, then $\sqrt{x^2} = -x$, the *opposite* of x (which is positive). For example,

$$\sqrt{5^2} = 5, \qquad \text{but} \qquad \sqrt{(-5)^2} = \sqrt{25} = 5, \quad \text{the } \textit{opposite} \text{ of } -5.$$

This means that the square root of a squared number is always nonnegative. We can use absolute value to express this.

$\sqrt{a^2}$

For any real number a, $\qquad \sqrt{a^2} = |a|$.

The product and quotient rules apply when variables appear under the radical sign, as long as the variables represent only *nonnegative* real numbers. ***To avoid negative radicands, variables under radical signs are assumed to be nonnegative in this text.*** Therefore, absolute value bars are not necessary, since for $x \geq 0$, $|x| = x$.

EXAMPLE 7 **Simplifying Radicals Involving Variables**

Simplify each radical. Assume that all variables represent nonnegative real numbers.

(a) $\sqrt{x^4} = x^2$ since $(x^2)^2 = x^4$.

(b) $\sqrt{25m^6}$

$= \sqrt{25} \cdot \sqrt{m^6}$ Product rule

$= 5m^3$ $(m^3)^2 = m^6$

(c) $\sqrt{8p^{10}}$

$= \sqrt{4 \cdot 2 \cdot p^{10}}$ Factor; 4 is a perfect square.

$= \sqrt{4} \cdot \sqrt{2} \cdot \sqrt{p^{10}}$ Product rule

$= 2 \cdot \sqrt{2} \cdot p^5$ $(p^5)^2 = p^{10}$

$= 2p^5\sqrt{2}$

(d) $\sqrt{r^9}$

$= \sqrt{r^8 \cdot r}$

$= \sqrt{r^8} \cdot \sqrt{r}$ Product rule

$= r^4\sqrt{r}$ $(r^4)^2 = r^8$

(e) $\sqrt{\dfrac{5}{x^2}}$

$= \dfrac{\sqrt{5}}{\sqrt{x^2}}$ Quotient rule

$= \dfrac{\sqrt{5}}{x}$ $x \neq 0$

Note

A quick way to find the square root of a variable raised to an even power is to divide the exponent by the index, 2. For example:

$$\sqrt{x^6} = x^3 \quad \text{and} \quad \sqrt{x^{10}} = x^5$$

$$6 \div 2 = 3 \qquad\qquad 10 \div 2 = 5$$

Work Problem **7** *at the Side.* ▶

OBJECTIVE 5 Simplify other roots.   The product and quotient rules for radicals also work for other roots. To simplify cube roots, look for factors that are *perfect cubes*. A **perfect cube** is a number with a rational cube root. For example, $\sqrt[3]{64} = 4$, and because 4 is a rational number, 64 is a perfect cube. Other roots are handled in a similar manner.

Properties of Radicals

For all real numbers where the indicated roots exist,

$$\sqrt[n]{a} \cdot \sqrt[n]{b} = \sqrt[n]{ab} \quad \text{and} \quad \frac{\sqrt[n]{a}}{\sqrt[n]{b}} = \sqrt[n]{\frac{a}{b}} \quad (b \neq 0).$$

7 Simplify each radical. Assume that all variables represent nonnegative real numbers.

(a) $\sqrt{x^8}$

(b) $\sqrt{36y^6}$

(c) $\sqrt{100p^{12}}$

(d) $\sqrt{12z^2}$

(e) $\sqrt{a^5}$

(f) $\sqrt{\dfrac{10}{n^4}}, \quad n \neq 0$

ANSWERS

7. **(a)** x^4 **(b)** $6y^3$ **(c)** $10p^6$ **(d)** $2z\sqrt{3}$

(e) $a^2\sqrt{a}$ **(f)** $\dfrac{\sqrt{10}}{n^2}$

8 Simplify each radical.

(a) $\sqrt[3]{108}$

(b) $\sqrt[4]{160}$

(c) $\sqrt[4]{\dfrac{16}{625}}$

9 Simplify each radical.

(a) $\sqrt[3]{z^9}$

(b) $\sqrt[3]{8x^6}$

(c) $\sqrt[3]{54t^5}$

(d) $\sqrt[3]{\dfrac{a^{15}}{64}}$

EXAMPLE 8 **Simplifying Other Roots**

Simplify each radical.

(a) $\sqrt[3]{32}$

Remember to write the root index 3 in each radical.

$= \sqrt[3]{8 \cdot 4}$ Factor; 8 is a perfect cube.

$= \sqrt[3]{8} \cdot \sqrt[3]{4}$ Product rule

$= 2\sqrt[3]{4}$

(b) $\sqrt[4]{32}$

Remember to write the root index 4 in each radical.

$= \sqrt[4]{16 \cdot 2}$ Factor; 16 is a perfect fourth power.

$= \sqrt[4]{16} \cdot \sqrt[4]{2}$ Product rule

$= 2\sqrt[4]{2}$

(c) $\sqrt[3]{\dfrac{27}{125}}$

$= \dfrac{\sqrt[3]{27}}{\sqrt[3]{125}}$ Quotient rule

$= \dfrac{3}{5}$

◀ *Work Problem* **8** *at the Side.*

Other roots of radicals involving variables can also be simplified. To simplify cube roots with variables, use the fact that for any real number a,

$$\sqrt[3]{a^3} = a.$$

This is true whether a is positive or negative.

EXAMPLE 9 **Simplifying Cube Roots Involving Variables**

Simplify each radical.

(a) $\sqrt[3]{m^6}$

$= m^2$ $(m^2)^3 = m^6$

(b) $\sqrt[3]{27x^{12}}$

$= \sqrt[3]{27} \cdot \sqrt[3]{x^{12}}$ Product rule

$= 3x^4$ $3^3 = 27;$ $(x^4)^3 = x^{12}$

(c) $\sqrt[3]{32a^4}$

$= \sqrt[3]{8a^3 \cdot 4a}$ Factor; 8 is a perfect cube.

$= \sqrt[3]{8a^3} \cdot \sqrt[3]{4a}$ Product rule

$= 2a\sqrt[3]{4a}$ $(2a)^3 = 8a^3$

(d) $\sqrt[3]{\dfrac{y^3}{125}}$

$= \dfrac{\sqrt[3]{y^3}}{\sqrt[3]{125}}$ Quotient rule

$= \dfrac{y}{5}$

◀ *Work Problem* **9** *at the Side.*

16.2 ▶▶▶ Exercises

Decide whether each statement is true *or* false. *If* false, *show why.*

1. $\sqrt{(-6)^2} = -6$

2. $\sqrt[3]{(-6)^3} = -6$

Use the product rule for radicals to find each product. See Example 1.

3. $\sqrt{3} \cdot \sqrt{5}$

4. $\sqrt{3} \cdot \sqrt{7}$

5. $\sqrt{2} \cdot \sqrt{11}$

6. $\sqrt{2} \cdot \sqrt{15}$

7. $\sqrt{6} \cdot \sqrt{7}$

8. $\sqrt{5} \cdot \sqrt{6}$

9. $\sqrt{13} \cdot \sqrt{r}, r \geq 0$

10. $\sqrt{19} \cdot \sqrt{k}, k \geq 0$

11. Which one of the following radicals is simplified? See Example 2.

   A. $\sqrt{47}$ **B.** $\sqrt{45}$ **C.** $\sqrt{48}$ **D.** $\sqrt{44}$

12. If p is a prime number, is $\sqrt{p}$ in simplified form? Explain your answer.

Simplify each radical. See Example 2.

13. $\sqrt{45}$

14. $\sqrt{200}$

15. $\sqrt{24}$

16. $\sqrt{44}$

17. $\sqrt{90}$

18. $\sqrt{56}$

19. $\sqrt{75}$

20. $\sqrt{18}$

21. $\sqrt{125}$

22. $\sqrt{80}$

23. $\sqrt{145}$

24. $\sqrt{110}$

25. $\sqrt{160}$

26. $\sqrt{128}$

27. $-\sqrt{700}$

28. $-\sqrt{600}$

Find each product and simplify. See Example 3.

29. $\sqrt{3} \cdot \sqrt{18}$ **30.** $\sqrt{3} \cdot \sqrt{21}$ **31.** $\sqrt{12} \cdot \sqrt{48}$ **32.** $\sqrt{50} \cdot \sqrt{72}$

33. $\sqrt{12} \cdot \sqrt{30}$ **34.** $\sqrt{30} \cdot \sqrt{24}$ **35.** $2\sqrt{10} \cdot 3\sqrt{2}$

36. $5\sqrt{6} \cdot 2\sqrt{10}$ **37.** $5\sqrt{3} \cdot 2\sqrt{15}$ **38.** $4\sqrt{6} \cdot 3\sqrt{2}$

39. Simplify the product $\sqrt{8} \cdot \sqrt{32}$ in two ways. First, multiply 8 by 32 and simplify the square root of this product. Second, simplify $\sqrt{8}$, simplify $\sqrt{32}$, and then multiply. How do the answers compare? Make a conjecture (an educated guess) about whether the correct answer can always be obtained using either method when simplifying a product such as this.

40. Simplify the radical $\sqrt{288}$ in two ways. First, factor 288 as $144 \cdot 2$ and then simplify. Second, factor 288 as $48 \cdot 6$ and then simplify. How do the answers compare? Make a conjecture concerning the quickest way to simplify such a radical.

Simplify each radical expression. See Examples 4–6.

41. $\sqrt{\dfrac{16}{225}}$ **42.** $\sqrt{\dfrac{9}{100}}$ **43.** $\sqrt{\dfrac{7}{16}}$ **44.** $\sqrt{\dfrac{13}{25}}$

45. $\dfrac{\sqrt{75}}{\sqrt{3}}$ **46.** $\dfrac{\sqrt{200}}{\sqrt{2}}$ **47.** $\sqrt{\dfrac{5}{2}} \cdot \sqrt{\dfrac{125}{8}}$

48. $\sqrt{\dfrac{8}{3}} \cdot \sqrt{\dfrac{512}{27}}$ **49.** $\dfrac{30\sqrt{10}}{5\sqrt{2}}$ **50.** $\dfrac{50\sqrt{20}}{2\sqrt{10}}$

Simplify each radical. Assume that all variables represent nonnegative real numbers.
See Example 7.

51. $\sqrt{m^2}$

52. $\sqrt{k^2}$

53. $\sqrt{y^4}$

54. $\sqrt{s^4}$

55. $\sqrt{36z^2}$

56. $\sqrt{49n^2}$

57. $\sqrt{400x^6}$

58. $\sqrt{900y^8}$

59. $\sqrt{18x^8}$

60. $\sqrt{20r^{10}}$

61. $\sqrt{45c^{14}}$

62. $\sqrt{50d^{20}}$

63. $\sqrt{z^5}$

64. $\sqrt{y^3}$

65. $\sqrt{a^{13}}$

66. $\sqrt{p^{17}}$

67. $\sqrt{64x^7}$

68. $\sqrt{25t^{11}}$

69. $\sqrt{x^6y^{12}}$

70. $\sqrt{a^8b^{10}}$

71. $\sqrt{81m^4n^2}$

72. $\sqrt{100c^4d^6}$

73. $\sqrt{\dfrac{7}{x^{10}}}, \quad x \neq 0$

74. $\sqrt{\dfrac{14}{z^{12}}}, \quad z \neq 0$

75. $\sqrt{\dfrac{y^4}{100}}$

76. $\sqrt{\dfrac{w^8}{144}}$

77. $\sqrt{\dfrac{x^6}{y^8}}, \quad y \neq 0$

78. $\sqrt{\dfrac{a^4}{b^6}}, \quad b \neq 0$

Simplify each radical. See Example 8.

79. $\sqrt[3]{40}$

80. $\sqrt[3]{48}$

81. $\sqrt[3]{54}$

82. $\sqrt[3]{135}$

83. $\sqrt[3]{128}$

84. $\sqrt[3]{192}$

85. $\sqrt[4]{80}$

86. $\sqrt[4]{243}$

87. $\sqrt[3]{\dfrac{8}{27}}$

88. $\sqrt[3]{\dfrac{64}{125}}$

89. $\sqrt[3]{-\dfrac{216}{125}}$

90. $\sqrt[3]{-\dfrac{1}{64}}$

Simplify each radical. See Example 9.

91. $\sqrt[3]{p^3}$

92. $\sqrt[3]{w^3}$

93. $\sqrt[3]{x^9}$

94. $\sqrt[3]{y^{18}}$

95. $\sqrt[3]{64z^6}$

96. $\sqrt[3]{125a^{15}}$

97. $\sqrt[3]{343a^9b^3}$

98. $\sqrt[3]{216m^3n^6}$

99. $\sqrt[3]{16t^5}$

100. $\sqrt[3]{24x^4}$

101. $\sqrt[3]{\dfrac{m^{12}}{8}}$

102. $\sqrt[3]{\dfrac{n^9}{27}}$

The volume of a cube is found with the formula $V = s^3$, where s is the length of an edge of the cube. Use this information in Exercises 103 and 104.

103. A container in the shape of a cube has a volume of 216 cm³. What is the depth of the container?

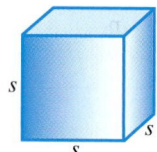

104. A cube-shaped box must be constructed to contain 128 ft³. What should the dimensions (height, width, and length) of the box be?

The volume of a sphere is found with the formula $V = \frac{4}{3}\pi r^3$, where r is the length of the radius of the sphere. Use this information in Exercises 105 and 106.

105. A ball in the shape of a sphere has a volume of 288π in.³. What is the radius of the ball?

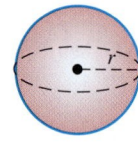

106. Suppose that the volume of the ball described in Exercise 105 is multiplied by 8. How is the radius affected?

Work Exercises 107 and 108 without using a calculator.

107. Choose the best estimate for the area (in square inches) of this rectangle.

    A. 45 **B.** 72 **C.** 80 **D.** 90

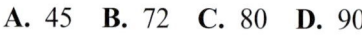

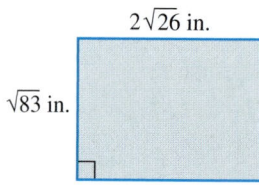

108. Choose the best estimate for the area (in square feet) of the triangle.

    A. 20 **B.** 40 **C.** 60 **D.** 80

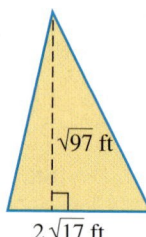

16.3 ▶▶▶ Adding and Subtracting Radicals

OBJECTIVES

1 Add and subtract radicals.

2 Simplify radical sums and differences.

3 Simplify more complicated radical expressions.

OBJECTIVE 1 Add and subtract radicals. We add or subtract radicals by using the distributive property. For example,

$$8\sqrt{3} + 6\sqrt{3} \qquad\qquad 2\sqrt{11} - 7\sqrt{11}$$
$$= (8+6)\sqrt{3} \qquad\qquad = (2-7)\sqrt{11}$$
$$= 14\sqrt{3} \qquad\qquad = -5\sqrt{11}.$$

Only **like radicals**—those that are *multiples of the same root of the same number*—can be combined. In the examples above, $8\sqrt{3}$ and $6\sqrt{3}$ are like radicals, as are $2\sqrt{11}$ and $-7\sqrt{11}$. Below are examples of **unlike radicals.**

$$2\sqrt{5} \text{ and } 2\sqrt{3} \qquad \text{Radicands are different.}$$

as well as $\qquad 2\sqrt{3}$ and $2\sqrt[3]{3} \qquad$ Indexes are different.

Work Problem **1** *at the Side.* ▶

EXAMPLE 1 Adding and Subtracting Like Radicals

Add or subtract, as indicated.

(a) $3\sqrt{6} + 5\sqrt{6}$
$= (3+5)\sqrt{6}$
$= 8\sqrt{6}$

(b) $5\sqrt{10} - 7\sqrt{10}$
$= (5-7)\sqrt{10}$
$= -2\sqrt{10}$

(c) $\sqrt{7} + 2\sqrt{7}$
$= \mathbf{1}\sqrt{7} + 2\sqrt{7}$
$= (\mathbf{1}+2)\sqrt{7}$
$= 3\sqrt{7}$

(d) $\sqrt{5} + \sqrt{5}$
$= 1\sqrt{5} + 1\sqrt{5}$
$= (1+1)\sqrt{5}$
$= 2\sqrt{5}$

(e) $\sqrt{3} + \sqrt{7}$ cannot be added using the distributive property.

Work Problem **2** *at the Side.* ▶

OBJECTIVE 2 Simplify radical sums and differences.

EXAMPLE 2 Simplifying Radicals to Add or Subtract

Add or subtract, as indicated.

(a) $3\sqrt{2} + \sqrt{8}$
$= 3\sqrt{2} + \sqrt{4\cdot2}$ Factor.
$= 3\sqrt{2} + \sqrt{4}\cdot\sqrt{2}$ Product rule
$= 3\sqrt{2} + 2\sqrt{2}$ $\sqrt{4}=2$
$= 5\sqrt{2}$ Add like radicals.

(b) $\sqrt{18} - \sqrt{27}$
$= \sqrt{9\cdot2} - \sqrt{9\cdot3}$ Factor.
$= \sqrt{9}\cdot\sqrt{2} - \sqrt{9}\cdot\sqrt{3}$ Product rule
$= 3\sqrt{2} - 3\sqrt{3}$ $\sqrt{9}=3$

These terms cannot be combined.

Continued on Next Page

1 Indicate whether the radicals in each pair are *like* or *unlike*.

(a) $5\sqrt{6}$ and $4\sqrt{6}$

(b) $2\sqrt{3}$ and $3\sqrt{2}$

(c) $\sqrt{10}$ and $\sqrt[3]{10}$

(d) $7\sqrt{2x}$ and $8\sqrt{2x}$

(e) $\sqrt{3y}$ and $\sqrt{6y}$

2 Add or subtract, as indicated.

(a) $8\sqrt{5} + 2\sqrt{5}$

(b) $-4\sqrt{3} + 9\sqrt{3}$

(c) $12\sqrt{11} - 3\sqrt{11}$

(d) $\sqrt{15} + \sqrt{15}$

(e) $2\sqrt{7} + 2\sqrt{10}$

ANSWERS
1. (a) like (b) unlike (c) unlike (d) like (e) unlike
2. (a) $10\sqrt{5}$ (b) $5\sqrt{3}$ (c) $9\sqrt{11}$ (d) $2\sqrt{15}$ (e) cannot be added

3 Add or subtract, as indicated.

(a) $\sqrt{8} + 4\sqrt{2}$

(b) $\sqrt{27} + \sqrt{12}$

(c) $5\sqrt{200} - 6\sqrt{18}$

4 Simplify each radical expression. Assume that all variables represent non-negative real numbers.

(a) $\sqrt{7} \cdot \sqrt{21} + 2\sqrt{27}$

(b) $\sqrt{3r} \cdot \sqrt{6} + \sqrt{8r}$

(c) $y\sqrt{72} - \sqrt{18y^2}$

(d) $\sqrt[3]{81x^4} + 5\sqrt[3]{24x^4}$

(c) $2\sqrt{12} + 3\sqrt{75}$

$$= 2\left(\sqrt{4} \cdot \sqrt{3}\right) + 3\left(\sqrt{25} \cdot \sqrt{3}\right) \qquad \text{Product rule}$$
$$= 2\left(2\sqrt{3}\right) + 3\left(5\sqrt{3}\right) \qquad \sqrt{4} = 2;\ \sqrt{25} = 5$$
$$= 4\sqrt{3} + 15\sqrt{3} \qquad \text{Multiply.}$$
$$= 19\sqrt{3} \qquad \text{Add like radicals.}$$

◀ *Work Problem* **3** *at the Side.*

OBJECTIVE 3 Simplify more complicated radical expressions.

EXAMPLE 3 **Simplifying Radical Expressions**

Simplify each radical expression. Assume that all variables represent non-negative real numbers.

(a) $\sqrt{5} \cdot \sqrt{15} + 4\sqrt{3}$

$$= \sqrt{5 \cdot 15} + 4\sqrt{3} \qquad \text{Product rule}$$
$$= \sqrt{75} + 4\sqrt{3} \qquad \text{Multiply.}$$
$$= \sqrt{25 \cdot 3} + 4\sqrt{3} \qquad \text{Factor; 25 is a perfect square.}$$
$$= \sqrt{25} \cdot \sqrt{3} + 4\sqrt{3} \qquad \text{Product rule}$$
$$= 5\sqrt{3} + 4\sqrt{3} \qquad \sqrt{25} = 5$$
$$= 9\sqrt{3} \qquad \text{Add like radicals.}$$

(b) $\sqrt{2} \cdot \sqrt{6k} + \sqrt{27k}$

$$= \sqrt{12k} + \sqrt{27k} \qquad \text{Product rule}$$
$$= \sqrt{4} \cdot \sqrt{3k} + \sqrt{9} \cdot \sqrt{3k} \qquad \text{Factor; product rule}$$
$$= 2\sqrt{3k} + 3\sqrt{3k} \qquad \sqrt{4} = 2;\ \sqrt{9} = 3$$
$$= 5\sqrt{3k} \qquad \text{Add like radicals.}$$

(c) $3x\sqrt{50} + \sqrt{2x^2}$

$$= 3x\sqrt{25 \cdot 2} + \sqrt{x^2 \cdot 2} \qquad \text{Factor.}$$
$$= 3x\sqrt{25} \cdot \sqrt{2} + \sqrt{x^2} \cdot \sqrt{2} \qquad \text{Product rule}$$
$$= 3x \cdot 5\sqrt{2} + x\sqrt{2} \qquad \sqrt{25} = 5;\ \sqrt{x^2} = x$$
$$= 15x\sqrt{2} + x\sqrt{2} \qquad \text{Multiply.}$$
$$= 16x\sqrt{2} \qquad \text{Add like radicals.}$$

(d) $2\sqrt[3]{32m^3} - \sqrt[3]{108m^3}$

$$= 2\sqrt[3]{(8m^3)4} - \sqrt[3]{(27m^3)4} \qquad \text{Factor.}$$
$$= 2(2m)\sqrt[3]{4} - 3m\sqrt[3]{4} \qquad \sqrt[3]{8m^3} = 2m;\ \sqrt[3]{27m^3} = 3m$$
$$= 4m\sqrt[3]{4} - 3m\sqrt[3]{4} \qquad \text{Multiply.}$$
$$= m\sqrt[3]{4} \qquad \text{Subtract like radicals.}$$

◀ *Work Problem* **4** *at the Side.*

16.3 ▶▶▶ **Exercises**

Fill in each blank with the correct response.

1. $5\sqrt{2} + 6\sqrt{2} = (5+6)\sqrt{2} = 11\sqrt{2}$ is an example of the _____ property.

2. The radicals $\sqrt[4]{3xy^3}$ and $-6\sqrt[4]{3xy^3}$ are examples of like radicals because both radicals have the same root index, _____, and the same radicand, _____ .

3. $\sqrt{5} + 5\sqrt{3}$ cannot be simplified because the _____ are different.

4. $4\sqrt[3]{2} + 3\sqrt{2}$ cannot be simplified because the _____ are different.

Simplify and add or subtract wherever possible. See Examples 1 and 2.

5. $14\sqrt{7} - 19\sqrt{7}$

6. $16\sqrt{2} - 18\sqrt{2}$

7. $\sqrt{17} + 4\sqrt{17}$

8. $5\sqrt{19} + \sqrt{19}$

9. $6\sqrt{7} - \sqrt{7}$

10. $11\sqrt{14} - \sqrt{14}$

11. $\sqrt{45} + 4\sqrt{20}$

12. $\sqrt{24} + 6\sqrt{54}$

13. $5\sqrt{72} - 3\sqrt{50}$

14. $6\sqrt{18} - 5\sqrt{32}$

15. $-5\sqrt{32} + 2\sqrt{98}$

16. $-4\sqrt{75} + 3\sqrt{12}$

17. $5\sqrt{7} - 3\sqrt{28} + 6\sqrt{63}$

18. $3\sqrt{11} + 5\sqrt{44} - 8\sqrt{99}$

19. $2\sqrt{8} - 5\sqrt{32} - 2\sqrt{48}$

20. $5\sqrt{72} - 3\sqrt{48} + 4\sqrt{128}$

21. $4\sqrt{50} + 3\sqrt{12} - 5\sqrt{45}$

22. $6\sqrt{18} + 2\sqrt{48} + 6\sqrt{28}$

23. $\frac{1}{4}\sqrt{288} + \frac{1}{6}\sqrt{72}$

24. $\frac{2}{3}\sqrt{27} + \frac{3}{4}\sqrt{48}$

Find the perimeter of each figure.

25.

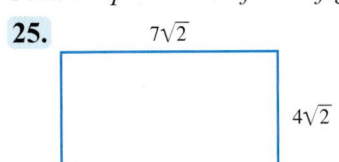

26.

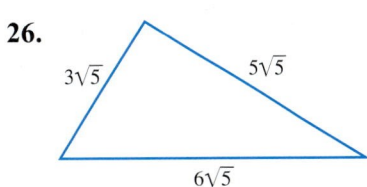

Perform the indicated operations. Assume that all variables represent nonnegative real numbers. See Example 3.

27. $\sqrt{6} \cdot \sqrt{2} + 9\sqrt{3}$

28. $4\sqrt{15} \cdot \sqrt{3} + 4\sqrt{5}$

29. $\sqrt{9x} + \sqrt{49x} - \sqrt{25x}$

30. $\sqrt{4a} - \sqrt{16a} + \sqrt{100a}$

31. $\sqrt{6x^2} + x\sqrt{24}$

32. $\sqrt{75x^2} + x\sqrt{108}$

33. $3\sqrt{8x^2} - 4x\sqrt{2} - x\sqrt{8}$

34. $\sqrt{2b^2} + 3b\sqrt{18} - b\sqrt{200}$

35. $-8\sqrt{32k} + 6\sqrt{8k}$

36. $4\sqrt{12x} + 2\sqrt{27x}$

37. $2\sqrt{125x^2z} + 8x\sqrt{80z}$

38. $\sqrt{48x^2y} + 5x\sqrt{27y}$

39. $4\sqrt[3]{16} - 3\sqrt[3]{54}$

40. $5\sqrt[3]{128} + 3\sqrt[3]{250}$

41. $6\sqrt[3]{8p^2} - 2\sqrt[3]{27p^2}$

42. $8k\sqrt[3]{54k} + 6\sqrt[3]{16k^4}$

43. $5\sqrt[4]{m^3} + 8\sqrt[4]{16m^3}$

44. $5\sqrt[4]{m^5} + 3\sqrt[4]{81m^5}$

Relating Concepts (Exercises 45–48) For Individual or Group Work

Adding and subtracting like radicals is no different than adding and subtracting like terms.
Work Exercises 45–48 in order.

45. Combine like terms: $5x^2y + 3x^2y - 14x^2y$.

46. Combine like terms: $5(p - 2q)^2(a + b) + 3(p - 2q)^2(a + b) - 14(p - 2q)^2(a + b)$.

47. Combine like radicals: $5a^2\sqrt{xy} + 3a^2\sqrt{xy} - 14a^2\sqrt{xy}$.

48. Compare your answers in Exercises 45–47. How are they alike? How are they different?

16.4 ▶▶▶ Rationalizing the Denominator

OBJECTIVE 1 Rationalize denominators with square roots. Although calculators now make it fairly easy to divide by a radical in an expression such as $\frac{1}{\sqrt{2}}$, it is sometimes easier to work with radical expressions if the denominators do not contain any radicals. For example, the radical in the denominator of $\frac{1}{\sqrt{2}}$ can be eliminated by multiplying the numerator and denominator by $\sqrt{2}$, since $\sqrt{2} \cdot \sqrt{2} = \sqrt{4} = 2$.

$$\frac{1}{\sqrt{2}} = \frac{1 \cdot \sqrt{2}}{\sqrt{2} \cdot \sqrt{2}} = \frac{\sqrt{2}}{2} \qquad \text{Multiply by } \tfrac{\sqrt{2}}{\sqrt{2}} = 1.$$

This process of changing the denominator from a radical (irrational number) to a rational number is called **rationalizing the denominator.** *The value of the radical expression is not changed; only the form is changed, because the expression has been multiplied by 1 in the form of $\frac{\sqrt{2}}{\sqrt{2}}$.*

EXAMPLE 1 Rationalizing Denominators

Rationalize each denominator.

(a) $\dfrac{9}{\sqrt{6}}$

$$= \frac{9 \cdot \sqrt{6}}{\sqrt{6} \cdot \sqrt{6}} \qquad \text{Multiply by } \tfrac{\sqrt{6}}{\sqrt{6}} = 1.$$

$$= \frac{9\sqrt{6}}{6} \qquad \sqrt{6} \cdot \sqrt{6} = \sqrt{36} = 6$$

$$= \frac{3\sqrt{6}}{2} \qquad \text{Lowest terms}$$

(b) $\dfrac{12}{\sqrt{8}}$

The denominator could be rationalized by multiplying by $\sqrt{8}$. However, first simplifying the denominator is more direct.

$$\frac{12}{\sqrt{8}}$$

$$= \frac{12}{2\sqrt{2}} \qquad \sqrt{8} = \sqrt{4} \cdot \sqrt{2} = 2\sqrt{2}$$

$$= \frac{12 \cdot \sqrt{2}}{2\sqrt{2} \cdot \sqrt{2}} \qquad \text{Multiply by } \tfrac{\sqrt{2}}{\sqrt{2}} = 1.$$

$$= \frac{12 \cdot \sqrt{2}}{2 \cdot 2} \qquad \sqrt{2} \cdot \sqrt{2} = \sqrt{4} = 2$$

$$= \frac{12\sqrt{2}}{4} \qquad \text{Multiply.}$$

$$= 3\sqrt{2} \qquad \text{Lowest terms}$$

1 Rationalize each denominator.

(a) $\dfrac{3}{\sqrt{5}}$

(b) $\dfrac{-6}{\sqrt{11}}$

(c) $-\dfrac{\sqrt{7}}{\sqrt{2}}$

(d) $\dfrac{20}{\sqrt{18}}$

2 Simplify.

(a) $\sqrt{\dfrac{16}{11}}$

(b) $\sqrt{\dfrac{5}{18}}$

(c) $\sqrt{\dfrac{8}{32}}$

> **Note**
>
> In Example 1(b), we could also have rationalized the original denominator $\sqrt{8}$ by multiplying by $\sqrt{2}$, since $\sqrt{8} \cdot \sqrt{2} = \sqrt{16} = 4$.
>
> $$\dfrac{12}{\sqrt{8}} = \dfrac{12 \cdot \sqrt{2}}{\sqrt{8} \cdot \sqrt{2}} = \dfrac{12\sqrt{2}}{\sqrt{16}} = \dfrac{12\sqrt{2}}{4} = 3\sqrt{2}$$
>
> Both approaches are correct.

◀ *Work Problem* **1** *at the Side.*

OBJECTIVE 2 Write radicals in simplified form.   A radical is considered to be in simplified form if the following three conditions are met.

> **Conditions for Simplified Form of a Radical**
>
> 1. The radicand contains no factor (except 1) that is a perfect square (when dealing with square roots), a perfect cube (when dealing with cube roots), and so on.
> 2. The radicand has no fractions.
> 3. No denominator contains a radical.

EXAMPLE 2 **Simplifying a Radical**

Simplify.

$$\sqrt{\dfrac{27}{5}}$$

$$= \dfrac{\sqrt{27}}{\sqrt{5}} \qquad \text{Quotient rule}$$

$$= \dfrac{\sqrt{27} \cdot \sqrt{5}}{\sqrt{5} \cdot \sqrt{5}} \qquad \text{Rationalize the denominator.}$$

$$= \dfrac{\sqrt{27} \cdot \sqrt{5}}{5} \qquad \sqrt{5} \cdot \sqrt{5} = \sqrt{25} = 5$$

$$= \dfrac{\sqrt{9 \cdot 3} \cdot \sqrt{5}}{5} \qquad \text{Factor.}$$

$$= \dfrac{\sqrt{9} \cdot \sqrt{3} \cdot \sqrt{5}}{5} \qquad \text{Product rule}$$

$$= \dfrac{3 \cdot \sqrt{3} \cdot \sqrt{5}}{5} \qquad \sqrt{9} = 3$$

$$= \dfrac{3\sqrt{15}}{5} \qquad \text{Product rule}$$

◀ *Work Problem* **2** *at the Side.*

ANSWERS

1. (a) $\dfrac{3\sqrt{5}}{5}$ (b) $\dfrac{-6\sqrt{11}}{11}$

   (c) $-\dfrac{\sqrt{14}}{2}$ (d) $\dfrac{10\sqrt{2}}{3}$

2. (a) $\dfrac{4\sqrt{11}}{11}$ (b) $\dfrac{\sqrt{10}}{6}$ (c) $\dfrac{1}{2}$

EXAMPLE 3 **Simplifying a Product of Radicals**

Simplify.

$$\sqrt{\frac{5}{8}} \cdot \sqrt{\frac{1}{6}}$$

$$= \sqrt{\frac{5}{8} \cdot \frac{1}{6}} \qquad \text{Product rule}$$

$$= \sqrt{\frac{5}{48}} \qquad \text{Multiply fractions.}$$

$$= \frac{\sqrt{5}}{\sqrt{48}} \qquad \text{Quotient rule}$$

$$= \frac{\sqrt{5}}{\sqrt{16} \cdot \sqrt{3}} \qquad \text{Product rule}$$

$$= \frac{\sqrt{5}}{4\sqrt{3}} \qquad \sqrt{16} = 4$$

$$= \frac{\sqrt{5} \cdot \sqrt{3}}{4\sqrt{3} \cdot \sqrt{3}} \qquad \text{Rationalize the denominator.}$$

$$= \frac{\sqrt{15}}{4 \cdot 3} \qquad \text{Product rule; } \sqrt{3} \cdot \sqrt{3} = 3$$

$$= \frac{\sqrt{15}}{12} \qquad \text{Multiply.}$$

Work Problem **3** *at the Side.* ▶

3 Simplify.

(a) $\sqrt{\frac{1}{2}} \cdot \sqrt{\frac{5}{6}}$

(b) $\sqrt{\frac{1}{10}} \cdot \sqrt{20}$

(c) $\sqrt{\frac{5}{8}} \cdot \sqrt{\frac{24}{10}}$

4 Simplify. Assume that all variables represent positive real numbers.

(a) $\dfrac{\sqrt{5p}}{\sqrt{q}}$

(b) $\sqrt{\dfrac{5r^2 t^2}{7}}$

EXAMPLE 4 **Simplifying Quotients Involving Radicals**

Simplify. Assume that x and y represent positive real numbers.

(a) $\dfrac{\sqrt{4x}}{\sqrt{y}}$

$$= \frac{\sqrt{4x} \cdot \sqrt{y}}{\sqrt{y} \cdot \sqrt{y}} \qquad \text{Rationalize the denominator.}$$

$$= \frac{\sqrt{4xy}}{y} \qquad \text{Product rule; } \sqrt{y} \cdot \sqrt{y} = y$$

$$= \frac{2\sqrt{xy}}{y} \qquad \sqrt{4} = 2$$

(b) $\sqrt{\dfrac{2x^2 y}{3}}$

$$= \frac{\sqrt{2x^2 y}}{\sqrt{3}} \qquad \text{Quotient rule}$$

$$= \frac{\sqrt{2x^2 y} \cdot \sqrt{3}}{\sqrt{3} \cdot \sqrt{3}} \qquad \text{Rationalize the denominator.}$$

$$= \frac{\sqrt{6x^2 y}}{3} \qquad \text{Product rule; } \sqrt{3} \cdot \sqrt{3} = 3$$

$$= \frac{\sqrt{x^2}\sqrt{6y}}{3} \qquad \text{Product rule}$$

$$= \frac{x\sqrt{6y}}{3} \qquad \sqrt{x^2} = x, \text{ since } x > 0.$$

Work Problem **4** *at the Side.* ▶

ANSWERS

3. (a) $\dfrac{\sqrt{15}}{6}$ (b) $\sqrt{2}$ (c) $\dfrac{\sqrt{6}}{2}$

4. (a) $\dfrac{\sqrt{5pq}}{q}$ (b) $\dfrac{rt\sqrt{35}}{7}$

5 Rationalize each denominator.

(a) $\sqrt[3]{\dfrac{5}{7}}$

OBJECTIVE **3** **Rationalize denominators with cube roots.** A denominator with a cube root is rationalized by changing the radicand in the denominator to a perfect cube, as shown in the next example.

EXAMPLE 5 Rationalizing Denominators with Cube Roots

Rationalize each denominator.

(a) $\sqrt[3]{\dfrac{3}{2}}$

First write the expression as a quotient of radicals. Then multiply the numerator and denominator by the appropriate number of factors of 2 to make the radicand in the denominator a perfect cube. This will eliminate the radical in the denominator. Here, multiply by $\sqrt[3]{2^2}$.

$$\sqrt[3]{\dfrac{3}{2}} = \dfrac{\sqrt[3]{3}}{\sqrt[3]{2}} = \dfrac{\sqrt[3]{3} \cdot \sqrt[3]{2^2}}{\sqrt[3]{2} \cdot \sqrt[3]{2^2}} = \dfrac{\sqrt[3]{3 \cdot 2^2}}{\sqrt[3]{2^3}} = \dfrac{\sqrt[3]{12}}{2} \qquad \sqrt[3]{2^3} = \sqrt[3]{8} = 2$$

Be careful not to multiply by $\sqrt[3]{2}$ here.

Denominator is a perfect cube.

(b) $\dfrac{\sqrt[3]{5}}{\sqrt[3]{9}}$

(b) $\dfrac{\sqrt[3]{3}}{\sqrt[3]{4}}$

Since $\sqrt[3]{4} \cdot \sqrt[3]{2} = \sqrt[3]{2^2} \cdot \sqrt[3]{2} = \sqrt[3]{2^3} = 2$, multiply the numerator and denominator by $\sqrt[3]{2}$.

$$\dfrac{\sqrt[3]{3}}{\sqrt[3]{4}} = \dfrac{\sqrt[3]{3} \cdot \sqrt[3]{2}}{\sqrt[3]{2^2} \cdot \sqrt[3]{2}} = \dfrac{\sqrt[3]{6}}{\sqrt[3]{2^3}} = \dfrac{\sqrt[3]{6}}{2}$$

(c) $\dfrac{\sqrt[3]{2}}{\sqrt[3]{3x^2}}, \quad x \neq 0$

Multiply the numerator and denominator by the appropriate number of factors of 3 and of x to get a perfect cube in the radicand of the denominator. Here, multiply by $\sqrt[3]{3^2 x}$ (that is, $\sqrt[3]{9x}$) since $\sqrt[3]{3x^2} \cdot \sqrt[3]{3^2 x} = \sqrt[3]{(3x)^3} = 3x$.

(c) $\dfrac{\sqrt[3]{4}}{\sqrt[3]{25y}}, \quad y \neq 0$

$$\dfrac{\sqrt[3]{2}}{\sqrt[3]{3x^2}} = \dfrac{\sqrt[3]{2} \cdot \sqrt[3]{3^2 x}}{\sqrt[3]{3x^2} \cdot \sqrt[3]{3^2 x}} = \dfrac{\sqrt[3]{18x}}{\sqrt[3]{(3x)^3}} = \dfrac{\sqrt[3]{18x}}{3x}$$

Be careful not to multiply by $\sqrt[3]{3x^2}$ here.

Denominator is a perfect cube.

CAUTION

A common error in a problem like the one in Example 5(a) is to multiply by $\sqrt[3]{2}$ instead of $\sqrt[3]{2^2}$. Doing this would give a denominator of $\sqrt[3]{2} \cdot \sqrt[3]{2} = \sqrt[3]{4}$. Because 4 is not a perfect cube, the denominator is still not rationalized.

◀ *Work Problem* **5** *at the Side.*

16.4 ▶▶▶ Exercises

Rationalize each denominator. See Examples 1 and 2.

1. $\dfrac{8}{\sqrt{2}}$

2. $\dfrac{12}{\sqrt{3}}$

3. $\dfrac{-\sqrt{11}}{\sqrt{3}}$

4. $\dfrac{-\sqrt{13}}{\sqrt{5}}$

5. $\dfrac{7\sqrt{3}}{\sqrt{5}}$

6. $\dfrac{4\sqrt{6}}{\sqrt{5}}$

7. $\dfrac{24\sqrt{10}}{16\sqrt{3}}$

8. $\dfrac{18\sqrt{15}}{12\sqrt{2}}$

9. $\dfrac{16}{\sqrt{27}}$

10. $\dfrac{24}{\sqrt{18}}$

11. $\dfrac{-3}{\sqrt{50}}$

12. $\dfrac{-5}{\sqrt{75}}$

13. $\dfrac{63}{\sqrt{45}}$

14. $\dfrac{27}{\sqrt{32}}$

15. $\dfrac{\sqrt{24}}{\sqrt{8}}$

16. $\dfrac{\sqrt{36}}{\sqrt{18}}$

17. $\sqrt{\dfrac{1}{2}}$

18. $\sqrt{\dfrac{1}{3}}$

19. $\sqrt{\dfrac{13}{5}}$

20. $\sqrt{\dfrac{17}{11}}$

21. When we rationalize the denominator in an expression such as $\dfrac{4}{\sqrt{3}}$, we multiply both the numerator and denominator by $\sqrt{3}$. By what number are we actually multiplying the given expression, and what property of real numbers justifies the fact that our result is equal to the given expression?

22. In Example 1(a), we show algebraically that $\dfrac{9}{\sqrt{6}}$ is equal to $\dfrac{3\sqrt{6}}{2}$. Support this result numerically by finding the decimal approximation of $\dfrac{9}{\sqrt{6}}$ on your calculator, and then finding the decimal approximation of $\dfrac{3\sqrt{6}}{2}$. What do you notice?

Simplify each product of radicals. See Example 3.

23. $\sqrt{\dfrac{7}{13}} \cdot \sqrt{\dfrac{13}{3}}$

24. $\sqrt{\dfrac{19}{20}} \cdot \sqrt{\dfrac{20}{3}}$

25. $\sqrt{\dfrac{21}{7}} \cdot \sqrt{\dfrac{21}{8}}$

26. $\sqrt{\dfrac{5}{8}} \cdot \sqrt{\dfrac{5}{6}}$

27. $\sqrt{\dfrac{1}{12}} \cdot \sqrt{\dfrac{1}{3}}$

28. $\sqrt{\dfrac{1}{8}} \cdot \sqrt{\dfrac{1}{2}}$

29. $\sqrt{\dfrac{2}{9}} \cdot \sqrt{\dfrac{9}{2}}$

30. $\sqrt{\dfrac{4}{3}} \cdot \sqrt{\dfrac{3}{4}}$

Simplify each radical. Assume that all variables represent positive real numbers.
See Example 4.

31. $\dfrac{\sqrt{7}}{\sqrt{x}}$

32. $\dfrac{\sqrt{19}}{\sqrt{y}}$

33. $\dfrac{\sqrt{4x^3}}{\sqrt{y}}$

34. $\dfrac{\sqrt{9t^3}}{\sqrt{s}}$

35. $\sqrt{\dfrac{5x^3z}{6}}$

36. $\sqrt{\dfrac{3st^3}{5}}$

37. $\sqrt{\dfrac{9a^2r^5}{7t}}$

38. $\sqrt{\dfrac{16x^3y^2}{13z}}$

39. Which one of the following would be an appropriate choice for multiplying the numerator and denominator of $\dfrac{\sqrt[3]{2}}{\sqrt[3]{5}}$ by in order to rationalize the denominator?

    A. $\sqrt[3]{5}$ **B.** $\sqrt[3]{25}$ **C.** $\sqrt[3]{2}$ **D.** $\sqrt[3]{4}$

40. In Example 5(b), we multiplied the numerator and denominator of $\dfrac{\sqrt[3]{3}}{\sqrt[3]{4}}$ by $\sqrt[3]{2}$ to rationalize the denominator. Suppose we had chosen to multiply by $\sqrt[3]{16}$ instead. Would we have obtained the correct answer after all simplifications were done?

Rationalize each denominator. Assume that variables in the denominator represent nonzero real numbers. See Example 5.

41. $\sqrt[3]{\dfrac{5}{9}}$

42. $\sqrt[3]{\dfrac{2}{5}}$

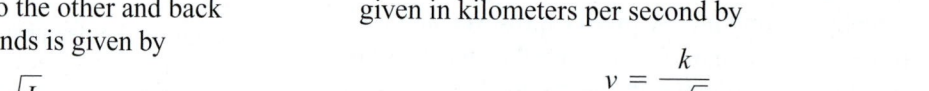 **43.** $\dfrac{\sqrt[3]{4}}{\sqrt[3]{7}}$

44. $\dfrac{\sqrt[3]{5}}{\sqrt[3]{10}}$

45. $\sqrt[3]{\dfrac{3}{4y^2}}$

46. $\sqrt[3]{\dfrac{3}{25x^2}}$

47. $\dfrac{\sqrt[3]{7m}}{\sqrt[3]{36n}}$

48. $\dfrac{\sqrt[3]{11p}}{\sqrt[3]{49q}}$

*In Exercises 49 and 50, (**a**) give the answer as a simplified radical and (**b**) use a calculator to give the answer correct to the nearest thousandth.*

49. The period p of a pendulum is the time it takes for it to swing from one extreme to the other and back again. The value of p in seconds is given by

$$p = k \cdot \sqrt{\dfrac{L}{g}}$$

where L is the length of the pendulum, g is the acceleration due to gravity, and k is a constant. Find the period when $k = 6$, $L = 9$ ft, and $g = 32$ ft per sec per sec.

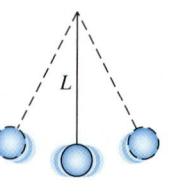

50. The velocity v of a meteorite approaching Earth is given in kilometers per second by

$$v = \dfrac{k}{\sqrt{d}}$$

where d is its distance from the center of Earth and k is a constant. What is the velocity of a meteorite that is 6000 km away from the center of Earth, if $k = 450$?

16.5 ▶▶▶ More Simplifying and Operations with Radicals

We now present a set of guidelines to simplify radical expressions.

Simplifying Radical Expressions

1. If a radical represents a rational number, use that rational number in place of the radical.

   Examples: $\sqrt{49} = 7$; $\sqrt{\dfrac{169}{9}} = \dfrac{13}{3}$

2. If a radical expression contains products of radicals, use the product rule for radicals, $\sqrt[n]{x} \cdot \sqrt[n]{y} = \sqrt[n]{xy}$, to get a single radical.

   Examples: $\sqrt{3} \cdot \sqrt{2} = \sqrt{6}$; $\sqrt[3]{5} \cdot \sqrt[3]{x} = \sqrt[3]{5x}$

3. If a radicand has a factor that is a perfect square, express the radical as the product of the positive square root of the perfect square and the remaining radical factor. A similar statement applies to higher roots.

   Examples: $\sqrt{20} = \sqrt{4 \cdot 5} = \sqrt{4} \cdot \sqrt{5} = 2\sqrt{5}$;
   $\sqrt[3]{16} = \sqrt[3]{8 \cdot 2} = \sqrt[3]{8} \cdot \sqrt[3]{2} = 2\sqrt[3]{2}$

4. If a radical expression contains sums or differences of radicals, use the distributive property to combine like radicals.

   Examples: $3\sqrt{2} + 4\sqrt{2} = 7\sqrt{2}$;
   $3\sqrt{2} + 4\sqrt{3}$ cannot be simplified further.

5. Rationalize any denominator containing a radical.

   Examples: $\dfrac{5}{\sqrt{3}} = \dfrac{5 \cdot \sqrt{3}}{\sqrt{3} \cdot \sqrt{3}} = \dfrac{5\sqrt{3}}{3}$;

   $\sqrt[3]{\dfrac{1}{4}} = \dfrac{\sqrt[3]{1}}{\sqrt[3]{4}} = \dfrac{\sqrt[3]{1} \cdot \sqrt[3]{2}}{\sqrt[3]{4} \cdot \sqrt[3]{2}} = \dfrac{\sqrt[3]{2}}{\sqrt[3]{8}} = \dfrac{\sqrt[3]{2}}{2}$

OBJECTIVE 1 Simplify products of radical expressions.

EXAMPLE 1 **Multiplying Radical Expressions**

Find each product, and simplify.

(a) $\sqrt{5}\left(\sqrt{8} - \sqrt{32}\right)$ ⟶ Simplify inside the parentheses.

$= \sqrt{5}\left(2\sqrt{2} - 4\sqrt{2}\right)$ $\sqrt{8} = 2\sqrt{2}; \sqrt{32} = 4\sqrt{2}$

$= \sqrt{5}\left(-2\sqrt{2}\right)$ Subtract like radicals.

$= -2\sqrt{5 \cdot 2}$ Product rule

$= -2\sqrt{10}$ Multiply.

Continued on Next Page

1 Find each product, and simplify.

(a) $\sqrt{7}\left(\sqrt{2} + \sqrt{5}\right)$

(b) $\sqrt{2}\left(\sqrt{8} + \sqrt{20}\right)$

(c) $\left(\sqrt{2} + 5\sqrt{3}\right)\left(\sqrt{3} - 2\sqrt{2}\right)$

(d) $\left(\sqrt{2} - \sqrt{5}\right)\left(\sqrt{10} + \sqrt{2}\right)$

(b) $\left(\sqrt{3} + 2\sqrt{5}\right)\left(\sqrt{3} - 4\sqrt{5}\right)$ ⟵ Use the FOIL method to multiply.

$$= \underbrace{\sqrt{3}\left(\sqrt{3}\right)}_{\text{First}} + \underbrace{\sqrt{3}\left(-4\sqrt{5}\right)}_{\text{Outer}} + \underbrace{2\sqrt{5}\left(\sqrt{3}\right)}_{\text{Inner}} + \underbrace{2\sqrt{5}\left(-4\sqrt{5}\right)}_{\text{Last}}$$

$= 3 - 4\sqrt{15} + 2\sqrt{15} - 8 \cdot 5$ Product rule

[This does *not* equal $-39\sqrt{15}$.] $= 3 - 2\sqrt{15} - 40$ Add like radicals; multiply.

$= -37 - 2\sqrt{15}$ Combine like terms.

(c) $\left(\sqrt{3} + \sqrt{21}\right)\left(\sqrt{3} - \sqrt{7}\right)$

$$= \sqrt{3}\left(\sqrt{3}\right) + \sqrt{3}\left(-\sqrt{7}\right) + \sqrt{21}\left(\sqrt{3}\right) + \sqrt{21}\left(-\sqrt{7}\right)$$
FOIL

$= 3 - \sqrt{21} + \sqrt{63} - \sqrt{147}$ Product rule

$= 3 - \sqrt{21} + \sqrt{9} \cdot \sqrt{7} - \sqrt{49} \cdot \sqrt{3}$ Factor; 9 and 49 are perfect squares.

$= 3 - \sqrt{21} + 3\sqrt{7} - 7\sqrt{3}$ $\sqrt{9} = 3; \sqrt{49} = 7$

Since there are no like radicals, no terms can be combined.

◀ *Work Problem* **1** *at the Side.*

Example 2 uses the rules for the square of a binomial from **Section 13.4,** $(a + b)^2 = a^2 + 2ab + b^2$ and $(a - b)^2 = a^2 - 2ab + b^2.$

EXAMPLE 2   **Using Special Products with Radicals**

Find each product.

(a) $\left(\sqrt{10} - 7\right)^2$ $(a - b)^2 = a^2 - 2ab + b^2$

$= \left(\sqrt{10}\right)^2 - 2\left(\sqrt{10}\right)(7) + 7^2$ Let $a = \sqrt{10}$ and $b = 7.$

[Do *not* try to combine further here.] $= 10 - 14\sqrt{10} + 49$ $\left(\sqrt{10}\right)^2 = 10; 7^2 = 49$

$= 59 - 14\sqrt{10}$ Combine like terms.

(b) $\left(2\sqrt{3} + 4\right)^2$ $(a + b)^2 = a^2 + 2ab + b^2$

$= \left(2\sqrt{3}\right)^2 + 2\left(2\sqrt{3}\right)(4) + 4^2$ $a = 2\sqrt{3}; b = 4$

$= 12 + 16\sqrt{3} + 16$ $\left(2\sqrt{3}\right)^2 = 4 \cdot 3 = 12$

$= 28 + 16\sqrt{3}$ ⟵ [Do *not* try to combine further here.]

(c) $\left(5 - \sqrt{x}\right)^2$

$= 5^2 - 2(5)\left(\sqrt{x}\right) + \left(\sqrt{x}\right)^2$

$= 25 - 10\sqrt{x} + x, \quad x \geq 0$

◀ *Work Problem* **2** *at the Side.*

2 Find each product. Simplify the answers.

(a) $\left(\sqrt{5} - 3\right)^2$

(b) $\left(4\sqrt{2} + 5\right)^2$

(c) $\left(6 + \sqrt{m}\right)^2, \quad m \geq 0$

ANSWERS

1. (a) $\sqrt{14} + \sqrt{35}$
(b) $4 + 2\sqrt{10}$
(c) $11 - 9\sqrt{6}$
(d) $2\sqrt{5} + 2 - 5\sqrt{2} - \sqrt{10}$

2. (a) $14 - 6\sqrt{5}$ (b) $57 + 40\sqrt{2}$
(c) $36 + 12\sqrt{m} + m$

CAUTION
Only like radicals can be combined. In Examples 2(a) and (b),
$$59 - 14\sqrt{10} \neq 45\sqrt{10} \quad \text{and} \quad 28 + 16\sqrt{3} \neq 44\sqrt{3}.$$

Example 3 uses the rule for the product of the sum and difference of two terms, $(a + b)(a - b) = a^2 - b^2$.

EXAMPLE 3 Using a Special Product with Radicals

Find each product.

(a) $\left(4 + \sqrt{3}\right)\left(4 - \sqrt{3}\right)$ $(a + b)(a - b) = a^2 - b^2$

$= 4^2 - \left(\sqrt{3}\right)^2$ Let $a = 4$ and $b = \sqrt{3}$.

$= 16 - 3$ $4^2 = 16; \left(\sqrt{3}\right)^2 = 3$

$= 13$

(b) $\left(\sqrt{x} - \sqrt{6}\right)\left(\sqrt{x} + \sqrt{6}\right)$

$= \left(\sqrt{x}\right)^2 - \left(\sqrt{6}\right)^2$

$= x - 6, \quad x \geq 0$ $\left(\sqrt{x}\right)^2 = x; \left(\sqrt{6}\right)^2 = 6$

Work Problem **3** *at the Side.* ▶

The pairs of expressions being multiplied in Example 3, $4 + \sqrt{3}$ and $4 - \sqrt{3}$, and $\sqrt{x} - \sqrt{6}$ and $\sqrt{x} + \sqrt{6}$, are **conjugates** of each other. Recall from **Section 5.4** that the expressions $a + b$ and $a - b$ are conjugates.

OBJECTIVE 2 **Use conjugates to rationalize denominators of radical expressions.** To rationalize the denominator in a quotient such as

$$\frac{2}{4 - \sqrt{3}},$$

we multiply the numerator and denominator by $4 + \sqrt{3}$ to obtain

$$\frac{2\left(4 + \sqrt{3}\right)}{\left(4 - \sqrt{3}\right)\left(4 + \sqrt{3}\right)}, \quad \text{or} \quad \frac{2\left(4 + \sqrt{3}\right)}{13}.$$

EXAMPLE 4 Using Conjugates to Rationalize Denominators

Simplify by rationalizing each denominator.

(a) $\dfrac{5}{3 + \sqrt{5}}$

$= \dfrac{5\left(3 - \sqrt{5}\right)}{\left(3 + \sqrt{5}\right)\left(3 - \sqrt{5}\right)}$ Multiply the numerator and denominator by the conjugate of the denominator.

$= \dfrac{5\left(3 - \sqrt{5}\right)}{3^2 - \left(\sqrt{5}\right)^2}$ $(a + b)(a - b) = a^2 - b^2$

$= \dfrac{5\left(3 - \sqrt{5}\right)}{9 - 5}$ $3^2 = 9; \left(\sqrt{5}\right)^2 = 5$

$= \dfrac{5\left(3 - \sqrt{5}\right)}{4}$ Subtract.

Continued on Next Page

---

3 Find each product. Simplify the answers.

(a) $\left(3 + \sqrt{5}\right)\left(3 - \sqrt{5}\right)$

(b) $\left(\sqrt{3} - 2\right)\left(\sqrt{3} + 2\right)$

(c)

$\left(\sqrt{5} + \sqrt{3}\right)\left(\sqrt{5} - \sqrt{3}\right)$

(d)

$\left(\sqrt{10} - \sqrt{y}\right)\left(\sqrt{10} + \sqrt{y}\right),$
$y \geq 0$

4 Rationalize each denominator.

(a) $\dfrac{5}{4 + \sqrt{2}}$

(b) $\dfrac{\sqrt{5} + 3}{2 - \sqrt{5}}$

(c) $\dfrac{7}{5 - \sqrt{x}}$

5 Write each quotient in lowest terms.

(a) $\dfrac{5\sqrt{3} - 15}{10}$

(b) $\dfrac{12 + 8\sqrt{5}}{16}$

(b) $\dfrac{6 + \sqrt{2}}{\sqrt{2} - 5}$

$= \dfrac{(6 + \sqrt{2})(\sqrt{2} + 5)}{(\sqrt{2} - 5)(\sqrt{2} + 5)}$ Multiply the numerator and denominator by the conjugate of the denominator.

$= \dfrac{6\sqrt{2} + 30 + 2 + 5\sqrt{2}}{2 - 25}$ FOIL; $(a + b)(a - b) = a^2 - b^2$

$= \dfrac{11\sqrt{2} + 32}{-23}$ Combine like terms.

$= \dfrac{-11\sqrt{2} - 32}{23}$ $\dfrac{a}{-b} = \dfrac{-a}{b}$

(c) $\dfrac{4}{3 - \sqrt{x}}$

$= \dfrac{4(3 + \sqrt{x})}{(3 - \sqrt{x})(3 + \sqrt{x})}$ Multiply by $\dfrac{3 + \sqrt{x}}{3 + \sqrt{x}} = 1.$

We assume here that $x \geq 0$ and $x \neq 9.$

$= \dfrac{4(3 + \sqrt{x})}{9 - x}$ $3^2 = 9$; $(\sqrt{x})^2 = x$

◀ Work Problem **4** at the Side.

OBJECTIVE **3** **Write radical expressions with quotients in lowest terms.**

EXAMPLE 5 **Writing a Radical Quotient in Lowest Terms**

Write $\dfrac{3\sqrt{3} + 9}{12}$ in lowest terms.

$\dfrac{3\sqrt{3} + 9}{12}$ Don't simplify yet!

$= \dfrac{3(\sqrt{3} + 3)}{3(4)}$ Factor first.

$= 1 \cdot \dfrac{\sqrt{3} + 3}{4}$ Now divide out the common factor; $\dfrac{3}{3} = 1$

$= \dfrac{\sqrt{3} + 3}{4}$ Lowest terms

◀ Work Problem **5** at the side.

CAUTION

An expression like the one in Example 5 can only be simplified by factoring a common factor from the denominator and *each* term of the numerator. For example, *first factor*

$\dfrac{4 + 8\sqrt{5}}{4}$ as $\dfrac{4(1 + 2\sqrt{5})}{4}$ to get $1 + 2\sqrt{5}.$

16.5 ▶▶▶ Exercises

In Exercises 1–4, perform the operations mentally, and write the answers without doing intermediate steps.

1. $\sqrt{49} + \sqrt{36}$ 　　　 **2.** $\sqrt{100} - \sqrt{81}$ 　　　 **3.** $\sqrt{2} \cdot \sqrt{8}$ 　　　 **4.** $\sqrt{8} \cdot \sqrt{8}$

Simplify each expression. Use the five guidelines given in this section. Assume that all variables represent nonnegative real numbers. See Examples 1–3.

5. $\sqrt{5}\left(\sqrt{3} - \sqrt{7}\right)$ 　　　 **6.** $\sqrt{7}\left(\sqrt{10} + \sqrt{3}\right)$ 　　　 **7.** $2\sqrt{5}\left(\sqrt{2} + 3\sqrt{5}\right)$

8. $3\sqrt{7}\left(2\sqrt{7} + 4\sqrt{5}\right)$ 　　　 **9.** $3\sqrt{14} \cdot \sqrt{2} - \sqrt{28}$ 　　　 **10.** $7\sqrt{6} \cdot \sqrt{3} - 2\sqrt{18}$

11. $\left(2\sqrt{6} + 3\right)\left(3\sqrt{6} + 7\right)$ 　　 **12.** $\left(4\sqrt{5} - 2\right)\left(2\sqrt{5} - 4\right)$ 　　 **13.** $\left(5\sqrt{7} - 2\sqrt{3}\right)\left(3\sqrt{7} + 4\sqrt{3}\right)$

14. $\left(2\sqrt{10} + 5\sqrt{2}\right)\left(3\sqrt{10} - 3\sqrt{2}\right)$ 　 **15.** $\left(8 - \sqrt{7}\right)^2$ 　　　 **16.** $\left(6 - \sqrt{11}\right)^2$

17. $\left(2\sqrt{7} + 3\right)^2$ 　　　 **18.** $\left(4\sqrt{5} + 5\right)^2$ 　　　 **19.** $\left(\sqrt{a} + 1\right)^2$

20. $\left(\sqrt{y} + 4\right)^2$ 　　　 **21.** $\left(5 - \sqrt{2}\right)\left(5 + \sqrt{2}\right)$ 　　 **22.** $\left(3 - \sqrt{5}\right)\left(3 + \sqrt{5}\right)$

23. $\left(\sqrt{8} - \sqrt{7}\right)\left(\sqrt{8} + \sqrt{7}\right)$ **24.** $\left(\sqrt{12} - \sqrt{11}\right)\left(\sqrt{12} + \sqrt{11}\right)$ **25.** $\left(\sqrt{y} - \sqrt{10}\right)\left(\sqrt{y} + \sqrt{10}\right)$

26. $\left(\sqrt{t} - \sqrt{13}\right)\left(\sqrt{t} + \sqrt{13}\right)$ **27.** $\left(\sqrt{2} + \sqrt{3}\right)\left(\sqrt{6} - \sqrt{2}\right)$ **28.** $\left(\sqrt{3} + \sqrt{5}\right)\left(\sqrt{15} - \sqrt{5}\right)$

29. $\left(\sqrt{10} - \sqrt{5}\right)\left(\sqrt{5} + \sqrt{20}\right)$ **30.** $\left(\sqrt{6} - \sqrt{3}\right)\left(\sqrt{3} + \sqrt{18}\right)$ **31.** $\left(\sqrt{5} + \sqrt{30}\right)\left(\sqrt{6} + \sqrt{3}\right)$

32. $\left(\sqrt{10} - \sqrt{20}\right)\left(\sqrt{2} - \sqrt{5}\right)$ **33.** $\left(\sqrt{5} - \sqrt{10}\right)\left(\sqrt{x} - \sqrt{2}\right)$ **34.** $\left(\sqrt{x} + \sqrt{6}\right)\left(\sqrt{10} + \sqrt{3}\right)$

35. In Example 1(b), the original expression simplifies to $-37 - 2\sqrt{15}$. Students often try to simplify such expressions by combining -37 and -2 to get $-39\sqrt{15}$, which is incorrect. Explain why.

36. If you try to rationalize the denominator of $\dfrac{2}{4 + \sqrt{3}}$ by multiplying the numerator and denominator by $4 + \sqrt{3}$, what problem arises? What should you multiply by?

Rationalize each denominator. Write quotients in lowest terms. Assume that all variables represent nonnegative real numbers. See Examples 4 and 5.

37. $\dfrac{1}{3 + \sqrt{2}}$ **38.** $\dfrac{1}{4 - \sqrt{3}}$ **39.** $\dfrac{14}{2 - \sqrt{11}}$ **40.** $\dfrac{19}{5 - \sqrt{6}}$

41. $\dfrac{\sqrt{2}}{2 - \sqrt{2}}$

42. $\dfrac{\sqrt{7}}{7 - \sqrt{7}}$

43. $\dfrac{\sqrt{5}}{\sqrt{2} + \sqrt{3}}$

44. $\dfrac{\sqrt{3}}{\sqrt{2} + \sqrt{3}}$

45. $\dfrac{\sqrt{5} + 2}{2 - \sqrt{3}}$

46. $\dfrac{\sqrt{7} + 3}{4 - \sqrt{5}}$

47. $\dfrac{12}{\sqrt{x} + 1}$

48. $\dfrac{10}{\sqrt{x} - 4}$

49. $\dfrac{3}{7 - \sqrt{x}}$

50. $\dfrac{1}{6 + \sqrt{z}}$

Write each quotient in lowest terms. See Example 5.

51. $\dfrac{6\sqrt{11} - 12}{6}$

52. $\dfrac{12\sqrt{5} - 24}{12}$

53. $\dfrac{2\sqrt{3} + 10}{16}$

54. $\dfrac{4\sqrt{6} + 24}{20}$

55. $\dfrac{12 - \sqrt{40}}{4}$

56. $\dfrac{9 - \sqrt{72}}{12}$

Relating Concepts (Exercises 57–62) For Individual or Group Work

Work Exercises 57–62 in order, to see why a common student error is indeed an error.

57. Use the distributive property to write $6(5 + 3x)$ as a sum.

58. Your answer in Exercise 57 should be $30 + 18x$. Why can't we combine these two terms to get $48x$?

59. Repeat Exercise 14 from earlier in this exercise set.

60. Your answer in Exercise 59 should be $30 + 18\sqrt{5}$. Many students will, in error, try to combine these terms to get $48\sqrt{5}$. Why is this wrong?

61. Write the expression similar to $30 + 18x$ that simplifies to $48x$. Then write the expression similar to $30 + 18\sqrt{5}$ that simplifies to $48\sqrt{5}$.

62. Write a short paragraph explaining the similarities between combining like terms and combining like radicals.

Solve each problem.

63. The radius of the circular top or bottom of a tin can with a surface area S and a height h is given by

$$r = \frac{-h + \sqrt{h^2 + 0.64S}}{2}.$$

What radius should be used to make a can with a height of 12 in. and a surface area of 400 in.²?

64. If an investment of P dollars grows to A dollars in 2 yr, the annual rate of return on the investment is given by

$$r = \frac{\sqrt{A} - \sqrt{P}}{\sqrt{P}}.$$

Rationalize the denominator. Then find the annual rate of return r (as a percent) if \$50,000 increases to \$58,320.

Summary Exercises on Operations with Radicals

Perform all indicated operations and express each answer in simplest form. Assume that all variables represent positive real numbers.

1. $5\sqrt{10} - 8\sqrt{10}$

2. $\sqrt{5}\left(\sqrt{5} - \sqrt{3}\right)$

3. $\left(1 + \sqrt{3}\right)\left(2 - \sqrt{6}\right)$

4. $\sqrt{98} - \sqrt{72} + \sqrt{50}$

5. $\left(3\sqrt{5} - 2\sqrt{7}\right)^2$

6. $\dfrac{3}{\sqrt{6}}$

7. $\sqrt[3]{16t^2} - \sqrt[3]{54t^2} + \sqrt[3]{128t^2}$

8. $\dfrac{8}{\sqrt{7} - \sqrt{5}}$

9. $\dfrac{1 + \sqrt{2}}{1 - \sqrt{2}}$

10. $\left(1 + \sqrt[3]{3}\right)\left(1 - \sqrt[3]{3} + \sqrt[3]{9}\right)$

11. $\left(\sqrt{3} + 6\right)\left(\sqrt{3} - 6\right)$

12. $\dfrac{1}{\sqrt{t} + \sqrt{3}}$

13. $\sqrt[3]{8x^3y^5z^6}$

14. $\dfrac{12}{\sqrt[3]{9}}$

15. $\dfrac{5}{\sqrt{6} - 1}$

16. $\sqrt{\dfrac{2}{3x}}$

17. $\dfrac{6\sqrt{3}}{5\sqrt{12}}$

18. $\dfrac{8\sqrt{50}}{2\sqrt{25}}$

19. $\dfrac{-4}{\sqrt[3]{4}}$

20. $\dfrac{\sqrt{6} - \sqrt{5}}{\sqrt{6} + \sqrt{5}}$

21. $\sqrt{75x} - \sqrt{12x}$

22. $\left(5 + 3\sqrt{3}\right)^2$

23. $\left(\sqrt{7} - \sqrt{6}\right)\left(\sqrt{7} + \sqrt{6}\right)$

24. $\sqrt[3]{\dfrac{16}{81}}$

25. $x\sqrt[4]{x^5} - 3\sqrt[4]{x^9} + x^2\sqrt[4]{x}$

26. $\sqrt{6} + \sqrt{6}$

27. $\sqrt{14} + \sqrt{17}$

28. $9\sqrt{24} - 2\sqrt{54} + 3\sqrt{20}$

29. $\sqrt{\dfrac{3}{4}} \cdot \sqrt{\dfrac{1}{5}}$

30. $\dfrac{5}{\sqrt{5}}$

31. $\sqrt[3]{24} + 6\sqrt[3]{81}$

32. $\dfrac{8}{4 - \sqrt{x}}$

33. $\sqrt[3]{4}\left(\sqrt[3]{2} - 3\right)$

34. $\sqrt{32x} - \sqrt{18x}$

35. $\sqrt{\dfrac{5}{8}}$

36. $\left(7 + \sqrt{x}\right)^2$

37. A biologist has shown that the number of different plant species S on a Galápagos Island is related to the area of the island, A (in square miles), by this formula.

$$S = 28.6\sqrt[3]{A}$$

How many plant species (to the nearest whole number) would exist on such an island with the following areas?

(a) 8 mi²   **(b)** 27,000 mi²

16.6 ▷▷▷ Solving Equations with Radicals

A **radical equation** is an equation with a variable in the radicand, such as

$$\sqrt{x+1} = 3 \quad \text{or} \quad 3\sqrt{x} = \sqrt{8x+9}. \qquad \text{Radical equations}$$

OBJECTIVE 1 **Solve radical equations having square root radicals.** The addition and multiplication properties of equality are not enough to solve radical equations. We need a new property, called the *squaring property*.

> ### Squaring Property of Equality
> If each side of a given equation is squared, all solutions of the original equation are *among* the solutions of the squared equation.

> **CAUTION**
> Be very careful with the squaring property: Using this property can give a new equation with *more* solutions than the original equation. For example, starting with the equation $x = 4$ and squaring each side gives
> $$x^2 = 4^2, \quad \text{or} \quad x^2 = 16.$$
> This last equation, $x^2 = 16$, has *two* solutions, 4 or -4, while the original equation, $x = 4$, has only *one* solution, 4. Because of this possibility, checking is more than just a guard against algebraic errors when solving an equation with radicals. It is an essential part of the solution process. ***All proposed solutions from the squared equation must be checked in the original equation.***

EXAMPLE 1 **Using the Squaring Property of Equality**

Solve $\sqrt{p+1} = 3$.

Use the squaring property of equality to square each side of the equation.

$$\sqrt{p+1} = 3$$
$$\left(\sqrt{p+1}\right)^2 = 3^2$$
$$p + 1 = 9 \qquad \left(\sqrt{p+1}\right)^2 = p+1$$
$$\boldsymbol{p = 8} \qquad \text{Subtract 1.}$$

Now check this proposed solution in the original equation.

Check $\qquad \sqrt{p+1} = 3 \qquad$ Original equation

> A check is essential.

$$\sqrt{8+1} \overset{?}{=} 3 \qquad \text{Let } p = 8.$$
$$\sqrt{9} \overset{?}{=} 3$$
$$3 = 3 \qquad \text{True}$$

Because this statement is true, $\{8\}$ is the solution set of $\sqrt{p+1} = 3$. In this case the equation obtained by squaring had just one solution, which also satisfied the original equation.

Work Problem ① *at the Side.* ▶

OBJECTIVES

1. Solve radical equations having square root radicals.

2. Identify equations with no solutions.

3. Solve equations by squaring a binomial.

4. Solve problems using formulas that involve radicals.

---

① Solve each equation. Be sure to check your solutions.

(a) $\sqrt{k} = 3$

(b) $\sqrt{x-2} = 4$

(c) $\sqrt{9-t} = 4$

ANSWERS

1. (a) $\{9\}$ **(b)** $\{18\}$ **(c)** $\{-7\}$

2 Solve each equation.

(a) $\sqrt{3x + 9} = 2\sqrt{x}$

(b) $5\sqrt{x} = \sqrt{20x + 5}$

EXAMPLE 2 Using the Squaring Property with a Radical on Each Side

Solve $3\sqrt{x} = \sqrt{x + 8}$.

$$3\sqrt{x} = \sqrt{x + 8}$$

$$\left(3\sqrt{x}\right)^2 = \left(\sqrt{x + 8}\right)^2 \quad \text{Squaring property}$$

$$3^2\left(\sqrt{x}\right)^2 = \left(\sqrt{x + 8}\right)^2 \quad (ab)^2 = a^2b^2$$

> Be careful here.

$$9x = x + 8 \quad \left(\sqrt{x}\right)^2 = x; \left(\sqrt{x + 8}\right)^2 = x + 8$$

$$8x = 8 \quad \text{Subtract } x.$$

$$x = 1 \quad \text{Divide by 8.}$$

Check

$$3\sqrt{x} = \sqrt{x + 8} \quad \text{Original equation}$$

$$3\sqrt{1} \overset{?}{=} \sqrt{1 + 8} \quad \text{Let } x = 1.$$

$$3(1) \overset{?}{=} \sqrt{9}$$

> This is *not* the solution

$$3 = 3 \quad \text{True}$$

The solution set of $3\sqrt{x} = \sqrt{x + 8}$ is $\{1\}$.

◀ **Work Problem** **2** at the Side.

OBJECTIVE **2** **Identify equations with no solutions.** Not all radical equations have solutions, as shown in Examples 3 and 4.

---

3 Solve $\sqrt{x} = -4$.

EXAMPLE 3 Using the Squaring Property When One Side Is Negative

Solve $\sqrt{x} = -3$.

$$\sqrt{x} = -3$$

$$\left(\sqrt{x}\right)^2 = (-3)^2 \quad \text{Squaring property}$$

$$x = 9 \longleftarrow \text{Proposed solution}$$

Check

$$\sqrt{x} = -3 \quad \text{Original equation}$$

$$\sqrt{9} \overset{?}{=} -3 \quad \text{Let } x = 9.$$

$$3 = -3 \quad \text{False}$$

Because the statement $3 = -3$ is false, the number 9 is *not* a solution of the given equation and is said to be an **extraneous solution**; it must be discarded. In fact, $\sqrt{x} = -3$ has no solution. The solution set is $\emptyset$.

Note

Because $\sqrt{x}$ represents the *principal* or *nonnegative* square root of x in Example 3, we might have seen immediately that there is no solution.

◀ **Work Problem** **3** at the Side.

We use the following steps when solving an equation with radicals.

Solving a Radical Equation

Step 1   **Isolate a radical.** Arrange the terms so that a radical is alone on one side of the equation.

Step 2   **Square each side.**

Step 3   **Combine like terms.**

Step 4   **Repeat Steps 1–3,** if there is still a term with a radical.

Step 5   **Solve the equation.** Find all proposed solutions.

Step 6   **Check all proposed solutions** in the original equation.

> **4** Solve $x = \sqrt{x^2 - 4x - 16}$.

EXAMPLE 4 **Using the Squaring Property with a Quadratic Expression**

Solve $x = \sqrt{x^2 + 5x + 10}$.

Step 1   The radical is already isolated on the right side of the equation.

Step 2   Square each side.

$$x^2 = \left(\sqrt{x^2 + 5x + 10}\right)^2 \qquad \text{Squaring property}$$

$$x^2 = x^2 + 5x + 10 \qquad \left(\sqrt{x^2 + 5x + 10}\right)^2 = x^2 + 5x + 10$$

Step 3 $\quad 0 = 5x + 10 \qquad \text{Subtract } x^2.$

Step 4   This step is not needed.

Step 5 $\quad -10 = 5x \qquad \text{Subtract 10.}$

$\quad\quad -2 = x \qquad \text{Divide by 5.}$

Step 6   **Check**

$$x = \sqrt{x^2 + 5x + 10}$$

$$-2 \stackrel{?}{=} \sqrt{(-2)^2 + 5(-2) + 10} \qquad \text{Let } x = -2.$$

The principal square root of a quantity *cannot* be negative.

$$-2 \stackrel{?}{=} \sqrt{4 - 10 + 10} \qquad \text{Multiply.}$$

$$-2 = 2 \qquad \text{False}$$

Since substituting -2 for x leads to a false result, the equation has no solution, and the solution set is $\emptyset$.

Work Problem **4** *at the Side.* ▶

> **5** Square each expression.
>
> **(a)** $w - 5$
>
> **(b)** $2k - 5$
>
> **(c)** $3m - 2p$

OBJECTIVE **3** **Solve equations by squaring a binomial.** The next examples use the following rules from **Section 13.4.**

$$(a + b)^2 = a^2 + 2ab + b^2 \quad \text{and} \quad (a - b)^2 = a^2 - 2ab + b^2$$

By the second rule, for example,

$$(x - 3)^2$$

$$= x^2 - 2x(3) + 3^2$$

$$= x^2 - 6x + 9.$$

Work Problem **5** *at the Side.* ▶

ANSWERS

4. $\emptyset$

5. **(a)** $w^2 - 10w + 25$ **(b)** $4k^2 - 20k + 25$
   (c) $9m^2 - 12mp + 4p^2$

6 Solve each equation.

(a) $\sqrt{6w + 6} = w + 1$

EXAMPLE 5 **Using the Squaring Property When One Side Has Two Terms**

Solve $\sqrt{2x - 3} = x - 3$.

Square each side, using the rule $(a - b)^2 = a^2 - 2ab + b^2$ to square the binomial on the right side of the equation.

$$\left(\sqrt{2x - 3}\right)^2 = (x - 3)^2$$

> Remember the middle term when squaring.

$$2x - 3 = x^2 - 6x + 9$$

This equation is quadratic because of the x^2-term. To solve it, as shown in **Section 14.6,** we must write the equation in standard form.

$$x^2 - 8x + 12 = 0 \qquad \text{Subtract } 2x; \text{ add } 3; \text{ standard form}$$
$$(x - 6)(x - 2) = 0 \qquad \text{Factor.}$$
$$x - 6 = 0 \quad \text{or} \quad x - 2 = 0 \qquad \text{Zero-factor property}$$
$$x = 6 \quad \text{or} \qquad x = 2 \qquad \text{Solve.}$$

Check *both* of these proposed solutions in the original equation.

Check   If $x = 6$, then

$$\sqrt{2x - 3} = x - 3$$
$$\sqrt{2(6) - 3} \stackrel{?}{=} 6 - 3$$
$$\sqrt{12 - 3} \stackrel{?}{=} 3$$
$$\sqrt{9} \stackrel{?}{=} 3$$
$$3 = 3. \qquad \text{True}$$

If $x = 2$, then

$$\sqrt{2x - 3} = x - 3$$
$$\sqrt{2(2) - 3} \stackrel{?}{=} 2 - 3$$
$$\sqrt{4 - 3} \stackrel{?}{=} -1$$
$$\sqrt{1} \stackrel{?}{=} -1$$
$$1 = -1. \qquad \text{False}$$

Only 6 is a valid solution. (2 is extraneous.) The solution set is $\{6\}$.

◄ *Work Problem* **6** *at the Side.*

(b) $2u - 1 = \sqrt{10u + 9}$

In Example 6, we must isolate the radical on one side of the equation *before* squaring each side. This is Step 1 given in the steps for solving a radical equation. If we forget to do this and square each side, we actually get a more complicated equation that still contains a radical.

EXAMPLE 6 **Rewriting before Using the Squaring Property**

Solve $3\sqrt{x} - 1 = 2x$.

$$3\sqrt{x} - 1 = 2x$$

> This is a key step.

$$3\sqrt{x} = 2x + 1 \qquad \text{Add 1 to isolate the radical.}$$
$$\left(3\sqrt{x}\right)^2 = (2x + 1)^2 \qquad \text{Square each side.}$$
$$9x = 4x^2 + 4x + 1 \qquad (a + b)^2 = a^2 + 2ab + b^2$$
$$4x^2 - 5x + 1 = 0 \qquad \text{Subtract } 9x; \text{ standard form}$$
$$(4x - 1)(x - 1) = 0 \qquad \text{Factor.}$$
$$4x - 1 = 0 \quad \text{or} \quad x - 1 = 0 \qquad \text{Zero-factor property}$$
$$x = \frac{1}{4} \quad \text{or} \qquad x = 1 \qquad \text{Solve.}$$

Continued on Next Page

ANSWERS

6. (a) $\{-1, 5\}$ **(b)** $\{4\}$

Check   If $x = \frac{1}{4}$, then

$$3\sqrt{x} - 1 = 2x$$

$$3\sqrt{\frac{1}{4}} - 1 \stackrel{?}{=} 2\left(\frac{1}{4}\right)$$

$$3\left(\frac{1}{2}\right) - 1 \stackrel{?}{=} \frac{1}{2}$$

$$\frac{1}{2} = \frac{1}{2}. \qquad \text{True}$$

If $x = 1$, then

$$3\sqrt{x} - 1 = 2x$$

$$3\sqrt{1} - 1 \stackrel{?}{=} 2(1)$$

$$3 - 1 \stackrel{?}{=} 2$$

$$2 = 2. \qquad \text{True}$$

Both proposed solutions check, so the solution set is $\left\{\frac{1}{4}, 1\right\}$.

> **CAUTION**
> Errors often occur when each side of an equation is squared. For instance, in Example 6 when each side of
>
> $$3\sqrt{x} = 2x + 1$$
>
> is squared, *the entire binomial on the right must be squared* to get $4x^2 + 4x + 1$. It would be *incorrect* to square the $2x$ and the 1 separately to get $4x^2 + 1$.

Work Problem **7** *at the Side.* ▶

Some radical equations require squaring twice, as in the next example.

EXAMPLE 7   Using the Squaring Property Twice

Solve $\sqrt{21 + x} = 3 + \sqrt{x}$.

$$\sqrt{21 + x} = 3 + \sqrt{x}$$

$$\left(\sqrt{21 + x}\right)^2 = \left(3 + \sqrt{x}\right)^2 \qquad \text{Square each side.}$$

$$21 + x = 9 + 6\sqrt{x} + x \qquad \boxed{\text{Be careful here.}}$$

$$12 = 6\sqrt{x} \qquad \text{Subtract 9; subtract } x.$$

$$2 = \sqrt{x} \qquad \text{Divide by 6.}$$

$$2^2 = \left(\sqrt{x}\right)^2 \qquad \text{Square each side again.}$$

$$4 = x$$

Check   If $x = 4$, then

$$\sqrt{21 + x} = 3 + \sqrt{x} \qquad \text{Original equation}$$

$$\sqrt{21 + 4} \stackrel{?}{=} 3 + \sqrt{4}$$

$$5 = 5. \qquad \text{True}$$

The solution set is $\{4\}$.

Work Problem **8** *at the Side.* ▶

---

7 Solve each equation.

(a) $\sqrt{x} - 3 = x - 15$

(b) $\sqrt{z + 5} + 2 = z + 5$

8 Solve each equation.

(a) $\sqrt{p + 1} - \sqrt{p - 4} = 1$

(b) $\sqrt{2x + 1} + \sqrt{x + 4} = 3$

ANSWERS
7. (a) $\{16\}$ (b) $\{-1\}$
8. (a) $\{8\}$ (b) $\{0\}$

9 Find the area (to the nearest tenth) of a triangle with sides of lengths 12 cm, 15 cm, and 21 cm.

OBJECTIVE **4** **Solve problems using formulas that involve radicals.** In earlier chapters we worked with a variety of formulas, including some to find area. Many useful formulas involve radicals.

The most common formula for the area of a triangle is $A = \frac{1}{2}bh$, where b is the length of the base and h is the height. What if the height is not known? What if we know only the lengths of the sides? Another formula, known as **Heron's formula,** allows us to calculate the area of a triangle if we know the lengths of the sides a, b, and c. Using the lengths of the sides, we must first find the **semiperimeter s,** which is one-half the perimeter.

$$s = \frac{1}{2}(a + b + c) \qquad \text{Semiperimeter}$$

The area $\mathcal{A}$ is given by the formula

$$\mathcal{A} = \sqrt{s(s-a)(s-b)(s-c)}. \qquad \text{Heron's formula}$$

For example, the familiar 3–4–5 right triangle has area $\mathcal{A} = \frac{1}{2}(3)(4) = 6$ square units, calculated using the familiar formula. Using Heron's formula, $s = \frac{1}{2}(3 + 4 + 5) = 6$, and

$$\mathcal{A} = \sqrt{6(6-3)(6-4)(6-5)}$$
$$\mathcal{A} = \sqrt{6 \cdot 3 \cdot 2 \cdot 1}$$
$$\mathcal{A} = \sqrt{36}$$
$$\mathcal{A} = 6.$$

The area is again 6 square units, as expected.

EXAMPLE 8 **Using Heron's Formula to Find the Area of a Triangle**

The sides of a triangular garden plot have lengths 20 ft, 34 ft, and 42 ft. See Figure 3. Find its area.

First find the semiperimeter of the triangle.

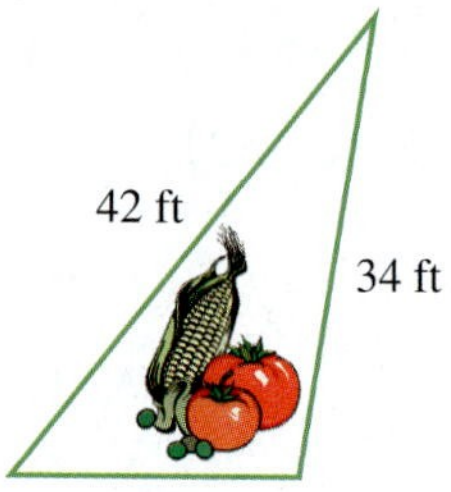

$$s = \frac{1}{2}(a + b + c)$$

$$s = \frac{1}{2}(20 + 34 + 42) \qquad \text{Let } a = 20, b = 34, \text{ and } c = 42.$$

$$s = \frac{1}{2}(96) \qquad \text{Add.}$$

$$s = 48 \qquad \text{Multiply.}$$

Now use Heron's formula to find the area.

$$\mathcal{A} = \sqrt{s(s-a)(s-b)(s-c)}$$

$$\mathcal{A} = \sqrt{48(48-20)(48-34)(48-42)} \qquad \text{Substitute.}$$

$$\mathcal{A} = \sqrt{48(28)(14)(6)} \qquad \text{Subtract.}$$

$$\mathcal{A} = \sqrt{112{,}896} \qquad \text{Multiply.}$$

$$\mathcal{A} = 336 \qquad \text{Use a calculator.}$$

The area of the garden plot is 336 ft².

42 ft

34 ft

20 ft

Figure 3

ANSWER

9. 88.2 cm²

◀ *Work Problem* **9** *at the Side.*

16.6 ▶▶▶ Exercises

Solve each equation. See Examples 1–4.

1. $\sqrt{x} = 7$

2. $\sqrt{k} = 10$

3. $\sqrt{t + 2} = 3$

4. $\sqrt{x + 7} = 5$

5. $\sqrt{r - 4} = 9$

6. $\sqrt{k - 12} = 3$

7. $\sqrt{4 - t} = 7$

8. $\sqrt{9 - s} = 5$

9. $\sqrt{2t + 3} = 0$

10. $\sqrt{5x - 4} = 0$

11. $\sqrt{3x - 8} = -2$

12. $\sqrt{6x + 4} = -3$

13. $\sqrt{w} - 4 = 7$

14. $\sqrt{t} + 3 = 10$

15. $\sqrt{10x - 8} = 3\sqrt{x}$

16. $\sqrt{17t - 4} = 4\sqrt{t}$

17. $5\sqrt{x} = \sqrt{10x + 15}$

18. $4\sqrt{z} = \sqrt{20z - 16}$

19. $\sqrt{3x - 5} = \sqrt{2x + 1}$

20. $\sqrt{5x + 2} = \sqrt{3x + 8}$

21. $k = \sqrt{k^2 - 5k - 15}$

22. $s = \sqrt{s^2 - 2s - 6}$

23. $7x = \sqrt{49x^2 + 2x - 10}$

24. $6x = \sqrt{36x^2 + 5x - 5}$

25. Consider the following "solution"? **WHAT WENT WRONG?** Give the correct solution set.

$$-\sqrt{x - 1} = -4$$
$$-(x - 1) = 16 \qquad \text{Square each side.}$$
$$-x + 1 = 16 \qquad \text{Distributive property}$$
$$-x = 15 \qquad \text{Subtract 1.}$$
$$x = -15 \qquad \text{Multiply by } -1.$$

26. The first step in solving the equation

$$\sqrt{2x + 1} = x - 7$$

is to square each side of the equation. Errors often occur in solving equations such as this one when the right side of the equation is squared incorrectly. What is the square of the right side?

Solve each equation. See Examples 5 and 6.

27. $\sqrt{2x + 1} = x - 7$

28. $\sqrt{3x + 3} = x - 5$

29. $\sqrt{3k + 10} + 5 = 2k$

30. $\sqrt{4t + 13} + 1 = 2t$

31. $\sqrt{5x + 1} - 1 = x$

32. $\sqrt{x + 1} - x = 1$

33. $\sqrt{6t + 7} + 3 = t + 5$

34. $\sqrt{10x + 24} = x + 4$

35. $x - 4 - \sqrt{2x} = 0$

36. $x - 3 - \sqrt{4x} = 0$

37. $\sqrt{x + 6} = 2x$

38. $\sqrt{k + 12} = k$

Solve each equation. See Example 7.

39. $\sqrt{x + 1} - \sqrt{x - 4} = 1$

40. $\sqrt{2x + 3} + \sqrt{x + 1} = 1$

41. $\sqrt{x} = \sqrt{x - 5} + 1$

42. $\sqrt{2x} = \sqrt{x + 7} - 1$

43. $\sqrt{3x + 4} - \sqrt{2x - 4} = 2$

44. $\sqrt{1 - x} + \sqrt{x + 9} = 4$

45. $\sqrt{2x + 11} + \sqrt{x + 6} = 2$

46. $\sqrt{x + 9} + \sqrt{x + 16} = 7$

Solve each problem.

47. The square root of the sum of a number and 4 is 5. Find the number.

48. A certain number is the same as the square root of the product of 8 and the number. Find the number.

49. Three times the square root of 2 equals the square root of the sum of some number and 10. Find the number.

50. The negative square root of a number equals that number decreased by 2. Find the number.

Use Heron's formula to solve each problem. See Example 8.

51. A surveyor has measured the lengths of the sides of a triangular plot of land as 180 ft, 200 ft, and 240 ft. Find the area of the plot. (Round to the nearest whole number.)

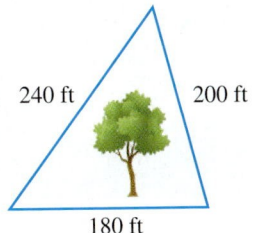

240 ft   200 ft

180 ft

52. A triangular indoor play space in a day care center has sides of length 30 ft, 45 ft, and 65 ft. How much carpeting would be needed to cover this space? (Round to the nearest whole number.)

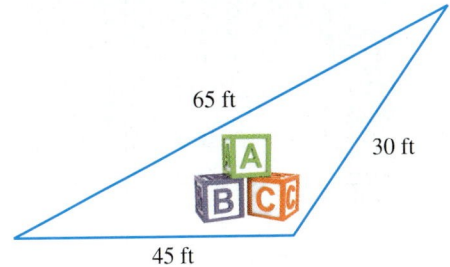

65 ft   30 ft

45 ft

Solve each problem. Give answers to the nearest tenth.

53. To estimate the speed at which a car was traveling at the time of an accident, a police officer drives the car involved in the accident under conditions similar to those during which the accident took place and then skids to a stop. If the car is driven at 30 mph, then the speed at the time of the accident is given by

$$s = 30\sqrt{\dfrac{a}{p}},$$

where a is the length of the skid marks left at the time of the accident and p is the length of the skid marks in the police test. Find s for the following values of a and p.

(a) $a = 862$ ft; $p = 156$ ft

(b) $a = 382$ ft; $p = 96$ ft

(c) $a = 84$ ft; $p = 26$ ft

54. A formula for calculating the distance, d, one can see from an airplane to the horizon on a clear day is

$$d = 1.22\sqrt{x},$$

where x is the altitude of the plane in feet and d is given in miles.

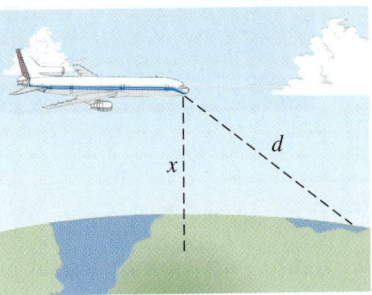

How far can one see to the horizon in a plane flying at the following altitudes?

(a) 15,000 ft

(b) 18,000 ft

(c) 24,000 ft

On a clear day, the maximum distance in kilometers that you can see from a structure is given by the formula

$$\text{sight distance} = 111.7\sqrt{\text{height of structure in kilometers}}.$$

Use this formula and the conversion equations 1 ft ≈ 0.3048 m and 1 km ≈ 0.621371 mi as necessary to solve each problem. Round your answers to the nearest mile. (Source: A Sourcebook of Applications of School Mathematics, NCTM, 1980.)

55. The London Eye is a unique structure that features 32 observation capsules. It is the fourth-tallest structure in London, with a diameter of 135 m. (*Source:* www.londoneye.com) Does the formula justify the claim that on a clear day passengers on the London Eye can see Windsor Castle, 25 mi away?

56. The Empire State Building in New York City is 1250 ft high. (The antenna reaches to 1454 ft.) The observation deck, located on the 102nd floor, is at a height of 1050 ft. (*Source:* www.esbnyc.com) How far could you see on a clear day from the observation deck?

57. The twin Petronas Towers in Kuala Lumpur, Malaysia, were built in 1998. Both towers are 1483 ft high (including the spires). How far would one of the builders have been able to see on a clear day from the top of a spire?

58. The Khufu Pyramid in Giza was built in about 2566 B.C. to a height, at that time, of 482 ft. It is now only about 450 ft high. How far would one of the original builders of the pyramid have been able to see from the top of the pyramid?

Relating Concepts (Exercises 59–64) For Individual or Group Work

Consider the figure, and **work Exercises 59–64 in order.**

59. The lengths of the sides of the entire triangle are 7, 7, and 12. Find the semiperimeter s.

60. Now use Heron's formula to find the area of the entire triangle. Write it as a simplified radical.

61. Find the value of h by using the Pythagorean formula.

62. Find the area of each of the congruent right triangles forming the entire triangle by using the formula $A = \frac{1}{2}bh$.

63. Double your result from Exercise 62 to determine the area of the entire triangle.

64. How do your answers in Exercises 60 and 63 compare?

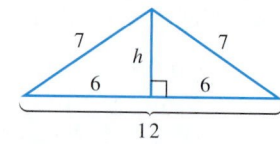

Chapter 16 ▶▶▶ Summary

▶ Key Terms

16.1

square root	A number a is a square root of b if $a^2 = b$.
principal square root	The positive square root of a number is its principal square root.
radicand	The number or expression inside a radical sign is called the radicand.
radical	A radical sign with a radicand is called a radical.
radical expression	An algebraic expression containing a radical is called a radical expression.
perfect square	A number with a rational square root is called a perfect square.
irrational number	A real number that is not rational is called an irrational number.
cube root	A number a is a cube root of b if $a^3 = b$.
index (order)	In a radical of the form $\sqrt[n]{a}$, the number n is the index or order.

Radical sign · Index · $\sqrt[n]{a}$ ← Radicand · Radical

16.2 **perfect cube**   A number with a rational cube root is called a perfect cube.

16.3 **like radicals**   Like radicals are multiples of the same root of the same number.

16.4 **rationalizing the denominator**   The process of changing the denominator of a fraction from a radical (irrational number) to an expression not involving a radical is called rationalizing the denominator.

16.5 **conjugate**   The conjugate of $a + b$ is $a - b$.

16.6 **radical equation**   An equation with a variable in the radicand is a radical equation.
      extraneous solution   A proposed solution that does not satisfy the given equation is an extraneous solution.

▶ New Symbols

$\sqrt{}$ radical sign $\approx$ is approximately equal to $\sqrt[3]{a}$ cube root of a $\sqrt[n]{a}$ nth root of a

▶ Test Your Word Power

See how well you have learned the vocabulary in this chapter. Answers, with examples, follow the Quick Review.

1. A **square root** of a number is
   - A. the number raised to the second power
   - B. the number under a radical sign
   - C. a number that when multiplied by itself gives the original number
   - D. the inverse of the number.

2. A **radicand** is
   - A. the index of a radical
   - B. the number or expression inside the radical sign
   - C. the positive root of a number
   - D. the radical sign.

3. A **radical** is
   - A. a symbol that indicates the nth root
   - B. an algebraic expression containing a square root
   - C. the positive nth root of a number
   - D. a radical sign and the number or expression inside it.

4. The **principal root** of a positive number with even index n is
   - A. the positive nth root of the number
   - B. the negative nth root of the number

   - C. the square root of the number
   - D. the cube root of the number.

5. An **irrational number** is
   - A. the quotient of two integers, with denominator not 0
   - B. a decimal number that neither terminates nor repeats
   - C. the principal square root of a number
   - D. a nonreal number.

(continued)

▶ **Test Your Word Power** (continued)

6. The **Pythagorean formula** states that, in a right triangle,
   A. the sum of the measures of the angles is 180°
   B. the sum of the lengths of the two shorter sides equals the length of the longest side
   C. the longest side is opposite the right angle
   D. the square of the length of the longest side equals the sum of the squares of the lengths of the two shorter sides.

7. **Like radicals** are
   A. radicals in simplest form
   B. algebraic expressions containing radicals

    C. multiples of the same root of the same number
    D. radicals with the same index.

8. The **conjugate** of $a + b$ is
   A. $a - b$
   B. $a \cdot b$
   C. $a \div b$
   D. $(a + b)^2$.

9. **Rationalizing the denominator** is the process of
   A. eliminating fractions from a radical expression
   B. changing the denominator of a fraction from a radical expression to an expression not involving a radical

    C. clearing a radical expression of radicals
    D. multiplying radical expressions.

10. An **extraneous solution** is a value
   A. that makes an equation false and must be discarded
   B. that makes an equation true
   C. that makes an expression equal 0
   D. that checks in the original equation.

▶ **Quick Review**

Concepts	Examples

16.1 Evaluating Roots

If a is a positive real number, then

    $\sqrt{a}$ is the positive or principal square root of a;

    $-\sqrt{a}$ is the negative square root of a; $\sqrt{0} = 0$.

If a is a negative real number, then $\sqrt{a}$ is not a real number.

$$\sqrt{49} = 7$$
$$-\sqrt{81} = -9$$
$$\sqrt{-25} \text{ is not a real number.}$$

If a is a positive rational number, then $\sqrt{a}$ is rational if a is a perfect square. $\sqrt{a}$ is irrational if a is not a perfect square.

$\sqrt{\dfrac{4}{9}}$ and $\sqrt{16}$ are rational. $\sqrt{\dfrac{2}{3}}$ and $\sqrt{21}$ are irrational.

Every real number has exactly one real cube root.

$$\sqrt[3]{27} = 3; \quad \sqrt[3]{-8} = -2$$

Pythagorean Formula

If c is the length of the longest side (hypotenuse) of a right triangle and a and b are the lengths of the shorter sides (legs), then

$$a^2 + b^2 = c^2.$$

Find b for the triangle in the figure.

$$10^2 + b^2 = \left(2\sqrt{61}\right)^2$$
$$100 + b^2 = 4(61)$$
$$100 + b^2 = 244$$
$$b^2 = 144$$
$$b = 12$$

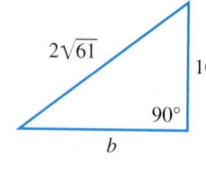

Concepts	Examples

16.2 Multiplying, Dividing, and Simplifying Radicals

Product Rule for Radicals

For nonnegative real numbers a and b,

$$\sqrt{a} \cdot \sqrt{b} = \sqrt{ab} \quad \text{and} \quad \sqrt{ab} = \sqrt{a} \cdot \sqrt{b}$$

$$\sqrt{5} \cdot \sqrt{7} = \sqrt{5 \cdot 7} = \sqrt{35}$$

$$\sqrt{48} = \sqrt{16 \cdot 3} = \sqrt{16} \cdot \sqrt{3} = 4\sqrt{3}$$

Quotient Rule for Radicals

If a and b are nonnegative real numbers and $b \neq 0$, then

$$\sqrt{\frac{a}{b}} = \frac{\sqrt{a}}{\sqrt{b}} \quad \text{and} \quad \frac{\sqrt{a}}{\sqrt{b}} = \sqrt{\frac{a}{b}}$$

If all indicated roots are real, then

$$\sqrt[n]{a} \cdot \sqrt[n]{b} = \sqrt[n]{ab} \quad \text{and} \quad \frac{\sqrt[n]{a}}{\sqrt[n]{b}} = \sqrt[n]{\frac{a}{b}} \quad (b \neq 0).$$

$$\sqrt{\frac{25}{64}} = \frac{\sqrt{25}}{\sqrt{64}} = \frac{5}{8}; \quad \frac{\sqrt{8}}{\sqrt{2}} = \sqrt{\frac{8}{2}} = \sqrt{4} = 2$$

$$\sqrt[3]{5} \cdot \sqrt[3]{3} = \sqrt[3]{15}; \quad \frac{\sqrt[4]{12}}{\sqrt[4]{4}} = \sqrt[4]{\frac{12}{4}} = \sqrt[4]{3}$$

16.3 Adding and Subtracting Radicals

Add and subtract like radicals by using the distributive property. *Only like radicals can be combined in this way.*

$$2\sqrt{5} + 4\sqrt{5}$$
$$= (2 + 4)\sqrt{5}$$
$$= 6\sqrt{5}$$

$$\sqrt{8} - \sqrt{32}$$
$$= 2\sqrt{2} - 4\sqrt{2}$$
$$= -2\sqrt{2}$$

16.4 Rationalizing the Denominator

The denominator of a radical expression can be rationalized by multiplying both the numerator and denominator by a number that will eliminate the radical from the denominator. In some cases, the number will be the conjugate of the denominator.

$$\frac{2}{\sqrt{3}} = \frac{2 \cdot \sqrt{3}}{\sqrt{3} \cdot \sqrt{3}} = \frac{2\sqrt{3}}{3}$$

$$\sqrt[3]{\frac{5}{121}} = \frac{\sqrt[3]{5} \cdot \sqrt[3]{11}}{\sqrt[3]{11^2} \cdot \sqrt[3]{11}} = \frac{\sqrt[3]{55}}{11}$$

16.5 More Simplifying and Operations with Radicals

When appropriate, use the rules for adding and multiplying polynomials to simplify radical expressions.

$$\sqrt{6}\left(\sqrt{5} - \sqrt{7}\right) = \sqrt{30} - \sqrt{42}$$

$$\left(\sqrt{3} + 1\right)\left(\sqrt{3} - 2\right)$$
$$= 3 - 2\sqrt{3} + \sqrt{3} - 2 \quad \text{FOIL}$$
$$= 1 - \sqrt{3} \quad \text{Combine like terms.}$$

These formulas are useful when simplifying radical expressions.

$$(a + b)^2 = a^2 + 2ab + b^2$$
$$(a - b)^2 = a^2 - 2ab + b^2$$
$$(a + b)(a - b) = a^2 - b^2$$

$$\left(\sqrt{13} - \sqrt{2}\right)^2$$
$$= \left(\sqrt{13}\right)^2 - 2\left(\sqrt{13}\right)\left(\sqrt{2}\right) + \left(\sqrt{2}\right)^2$$
$$= 13 - 2\sqrt{26} + 2$$
$$= 15 - 2\sqrt{26}$$

$$\left(\sqrt{5} + \sqrt{3}\right)\left(\sqrt{5} - \sqrt{3}\right)$$
$$= 5 - 3 = 2$$

(continued)

Concepts	Examples

16.5 **More Simplifying and Operations with Radicals** *(continued)*

Any denominators with radicals should be rationalized.

$$\frac{3}{\sqrt{6}} = \frac{3 \cdot \sqrt{6}}{\sqrt{6} \cdot \sqrt{6}} = \frac{3\sqrt{6}}{6} = \frac{\sqrt{6}}{2}$$

If a radical expression contains two terms in the denominator and at least one of those terms is a square root radical, multiply both the numerator and denominator by the conjugate of the denominator.

$$\frac{6}{\sqrt{7} - \sqrt{2}}$$

$$= \frac{6(\sqrt{7} + \sqrt{2})}{(\sqrt{7} - \sqrt{2})(\sqrt{7} + \sqrt{2})}$$

$$= \frac{6(\sqrt{7} + \sqrt{2})}{7 - 2} \qquad \text{Multiply.}$$

$$= \frac{6(\sqrt{7} + \sqrt{2})}{5} \qquad \text{Subtract.}$$

16.6 **Solving Equations with Radicals**

Solving a Radical Equation

Step 1 Isolate a radical.

Step 2 Square each side. (By the squaring property of equality, all solutions of the original equation are *among* the solutions of the squared equation.)

Step 3 Combine like terms.

Step 4 If there is still a term with a radical, repeat Steps 1–3.

Step 5 Solve the equation for proposed solutions.

Step 6 Check all proposed solutions from Step 5 in the original equation.

Solve $\sqrt{2x - 3} + x = 3$

$$\sqrt{2x - 3} = 3 - x \qquad \text{Isolate the radical.}$$

$$\left(\sqrt{2x - 3}\right)^2 = (3 - x)^2 \qquad \text{Square each side.}$$

$$2x - 3 = 9 - 6x + x^2$$

$$0 = x^2 - 8x + 12 \qquad \text{Standard form}$$

$$0 = (x - 2)(x - 6) \qquad \text{Factor.}$$

$$x - 2 = 0 \quad \text{or} \quad x - 6 = 0 \qquad \text{Zero-factor property}$$

$$x = 2 \quad \text{or} \qquad x = 6 \qquad \text{Solve.}$$

A check is essential here. Verify that 2 is the only solution. (6 is extraneous.) The solution set is $\{2\}$.

ANSWERS TO TEST YOUR WORD POWER

1. C; *Examples:* 6 is a square root of 36 since $6^2 = 6 \cdot 6 = 36$; -6 is also a square root of 36.

2. B; *Example:* In $\sqrt{3xy}$, $3xy$ is the radicand.   **3.** D; *Examples:* $\sqrt{144}$, $\sqrt{4xy^2}$, and $\sqrt{4 + t^2}$

4. A; *Examples:* $\sqrt{36} = 6$, $\sqrt[4]{81} = 3$, and $\sqrt[6]{64} = 2$ **5.** B; *Examples:* π, $\sqrt{2}$, $-\sqrt{5}$

6. D; *Example:* In a right triangle where $a = 6$, $b = 8$, and $c = 10$, $6^2 + 8^2 = 10^2$.

7. C; *Examples:* $\sqrt{7}$ and $3\sqrt{7}$ are like radicals; so are $2\sqrt[3]{6k}$ and $5\sqrt[3]{6k}$.

8. A; *Example:* The conjugate of $\sqrt{3} + 1$ is $\sqrt{3} - 1$.

9. B; *Example:* To rationalize the denominator of $\dfrac{5}{\sqrt{3} + 1}$, multiply the numerator and denominator by $\sqrt{3} - 1$

(the conjugate of the denominator) to get $\dfrac{5(\sqrt{3} - 1)}{2}$.

10. A; *Example:* The proposed solution 2 is extraneous when $\sqrt{5q - 1} + 3 = 0$ is solved using the method of **Section 16.6**.

Chapter 16 ▶▶▶ Review Exercises

[16.1] *Find all square roots of each number.*

1. 49   **2.** 81   **3.** 196   **4.** 121   **5.** 256   **6.** 729

Find each root.

7. $\sqrt{16}$ **8.** $-\sqrt{0.36}$ **9.** $\sqrt[3]{-512}$ **10.** $\sqrt[4]{81}$

11. $\sqrt{-8100}$ **12.** $-\sqrt{4225}$ **13.** $\sqrt{\dfrac{144}{169}}$ **14.** $-\sqrt{\dfrac{100}{81}}$

15. Find the value of x.

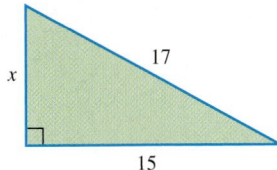

16. An HP f1905 computer monitor has viewing screen dimensions as shown in the figure. Find the diagonal measure of the viewing screen to the nearest tenth. (*Source:* Author's computer.)

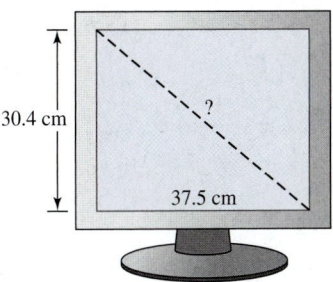

Write rational, irrational, *or* not a real number *for each number. If a number is rational, give its exact value. If a number is irrational, give a decimal approximation for the number. Round approximations to the nearest thousandth.*

17. $\sqrt{73}$ **18.** $\sqrt{169}$ **19.** $-\sqrt{625}$ **20.** $\sqrt{-19}$

[16.2] *Simplify each expression.*

21. $\sqrt{48}$ **22.** $-\sqrt{288}$ **23.** $\sqrt[3]{16}$ **24.** $\sqrt[3]{375}$

25. $\sqrt{12} \cdot \sqrt{27}$ **26.** $\sqrt{32} \cdot \sqrt{48}$ **27.** $-\sqrt{\dfrac{121}{400}}$ **28.** $\sqrt{\dfrac{7}{169}}$

29. $\sqrt{\dfrac{1}{6}} \cdot \sqrt{\dfrac{5}{6}}$

30. $\sqrt{\dfrac{2}{5}} \cdot \sqrt{\dfrac{2}{45}}$

31. $\dfrac{3\sqrt{10}}{\sqrt{5}}$

32. $\dfrac{8\sqrt{150}}{4\sqrt{75}}$

Simplify each expression. Assume that all variables represent nonnegative real numbers.

33. $\sqrt{r^{18}}$

34. $\sqrt{x^{10}y^{16}}$

35. $\sqrt{162x^9}$

36. $\sqrt{\dfrac{36}{p^2}}, \quad p \neq 0$

37. $\sqrt{a^{15}b^{21}}$

38. $\sqrt{121x^6y^{10}}$

39. $\sqrt[3]{y^6}$

40. $\sqrt[3]{216x^{15}}$

[16.3] *Simplify and combine terms where possible.*

41. $7\sqrt{11} + \sqrt{11}$

42. $3\sqrt{2} + 6\sqrt{2}$

43. $3\sqrt{75} + 2\sqrt{27}$

44. $4\sqrt{12} + \sqrt{48}$

45. $4\sqrt{24} - 3\sqrt{54} + \sqrt{6}$

46. $2\sqrt{7} - 4\sqrt{28} + 3\sqrt{63}$

47. $\dfrac{2}{5}\sqrt{75} + \dfrac{3}{4}\sqrt{160}$

48. $\dfrac{1}{3}\sqrt{18} + \dfrac{1}{4}\sqrt{32}$

49. $\sqrt{15} \cdot \sqrt{2} + 5\sqrt{30}$

Simplify each expression. Assume that all variables represent nonnegative real numbers.

50. $\sqrt{4x} + \sqrt{36x} - \sqrt{9x}$

51. $\sqrt{16p} + 3\sqrt{p} - \sqrt{49p}$

52. $3k\sqrt{8k^2n} + 5k^2\sqrt{2n}$

[16.4] *Perform the indicated operations, and write all answers in simplest form. Rationalize all denominators. Assume that all variables represent nonnegative real numbers.*

53. $\dfrac{10}{\sqrt{3}}$

54. $\dfrac{8\sqrt{2}}{\sqrt{5}}$

55. $\dfrac{12}{\sqrt{24}}$

56. $\sqrt{\dfrac{2}{5}}$

57. $\sqrt{\dfrac{5}{14}} \cdot \sqrt{28}$

58. $\sqrt{\dfrac{2}{7}} \cdot \sqrt{\dfrac{1}{3}}$

59. $\sqrt{\dfrac{r^2}{16x}}, \quad x \neq 0$

60. $\sqrt[3]{\dfrac{1}{3}}$

Solve each problem.

61. The radius r of a cone in terms of its volume V is given by this formula.

$$r = \sqrt{\dfrac{3V}{\pi h}}$$

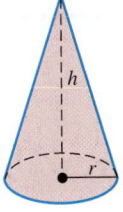

Rationalize the denominator of the radical expression.

62. The radius r of a sphere in terms of its surface area S is given by this formula.

$$r = \sqrt{\dfrac{S}{4\pi}}$$

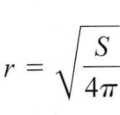

Rationalize the denominator of the radical expression.

[16.5] *Simplify each expression.*

63. $-\sqrt{3}\left(\sqrt{5} + \sqrt{27}\right)$

64. $3\sqrt{2}\left(\sqrt{3} + 2\sqrt{2}\right)$

65. $\left(2\sqrt{3} - 4\right)\left(5\sqrt{3} + 2\right)$

66. $\left(\sqrt{7} + 2\sqrt{6}\right)\left(\sqrt{12} - \sqrt{2}\right)$

67. $\left(2\sqrt{3} + 5\right)\left(2\sqrt{3} - 5\right)$

68. $\left(\sqrt{x} + 2\right)^2, \quad x \geq 0$

Rationalize each denominator.

69. $\dfrac{1}{2 + \sqrt{5}}$

70. $\dfrac{3}{1 + \sqrt{x}}, \quad x \geq 0, x \neq 1$

71. $\dfrac{\sqrt{5} - 1}{\sqrt{2} + 3}$

Write each quotient in lowest terms.

72. $\dfrac{15 + 10\sqrt{6}}{15}$

73. $\dfrac{3 + 9\sqrt{7}}{12}$

74. $\dfrac{6 + \sqrt{192}}{2}$

[16.6] *Solve each equation.*

75. $\sqrt{x} + 5 = 0$

76. $\sqrt{k+1} = 7$

77. $\sqrt{5t+4} = 3\sqrt{t}$

78. $\sqrt{2p+3} = \sqrt{5p-3}$

79. $\sqrt{4x+1} = x - 1$

80. $\sqrt{13+4t} = t + 4$

81. $\sqrt{2-x} + 3 = x + 7$

82. $\sqrt{x} - x + 2 = 0$

83. $\sqrt{x+2} - \sqrt{x-3} = 1$

▷▷▷ Mixed Review Exercises

Simplify each expression if possible. Assume that all variables represent nonnegative real numbers.

84. $2\sqrt{27} + 3\sqrt{75} - \sqrt{300}$

85. $\dfrac{1}{5 + \sqrt{2}}$

86. $\sqrt{\dfrac{1}{3}} \cdot \sqrt{\dfrac{24}{5}}$

87. $\sqrt[3]{54a^7b^{10}}$

88. $-\sqrt{5}\left(\sqrt{2} + \sqrt{75}\right)$

89. $\sqrt{\dfrac{16r^3}{3r}}, \quad s \neq 0$

90. $\dfrac{12 + 6\sqrt{13}}{12}$

91. $\left(\sqrt{5} - \sqrt{2}\right)^2$

92. $\left(6\sqrt{7} + 2\right)\left(4\sqrt{7} - 1\right)$

Solve each equation.

93. $\sqrt{x+2} = x - 4$

94. $\sqrt{k} + 3 = 0$

95. $\sqrt{1+3t} - t = -3$

96. Match each radical in Column I with the equivalent choice in Column II.
Choices may be used once, more than once, or not at all.

I		II	
(a) $\sqrt{64}$	**(b)** $-\sqrt{64}$	**A.** 4	**B.** 8
(c) $\sqrt{-64}$	**(d)** $\sqrt[3]{64}$	**C.** -4	**D.** Not a real number
(e) $\sqrt[3]{-64}$	**(f)** $-\sqrt[3]{-64}$	**E.** 16	**F.** -8

Chapter 16 ▶▶▶ Test

Use the Chapter Test Prep Video CD to see fully worked-out solutions to any of the exercises you want to review.

1. Find all square roots of 400.

1. _____

2. Consider $\sqrt{142}$.

(a) Determine whether it is rational or irrational.

(b) Find a decimal approximation to the nearest thousandth.

2. (a) _____
(b) _____

3. If $\sqrt{a}$ is not a real number, then what kind of number must a be?

3. _____

Simplify where possible. Assume that all variables represent nonnegative real numbers.

4. $\sqrt[3]{216}$

4. _____

5. $-\sqrt{54}$

5. _____

6. $\sqrt{\dfrac{128}{25}}$

6. _____

7. $\sqrt[3]{32}$

7. _____

8. $\dfrac{20\sqrt{18}}{5\sqrt{3}}$

8. _____

9. $3\sqrt{28} + \sqrt{63}$

9. _____

10. $3\sqrt{27x} - 4\sqrt{48x} + 2\sqrt{3x}$

10. _____

11. $\sqrt{32x^2y^3}$

11. _____

12. $(6 - \sqrt{5})(6 + \sqrt{5})$

12. _____

13. $(2 - \sqrt{7})(3\sqrt{2} + 1)$

13. _____

14. $(\sqrt{5} + \sqrt{6})^2$

14. _____

Solve each problem.

15. (a) _____

    (b) _____

15. Find the measure of the unknown leg of this right triangle.

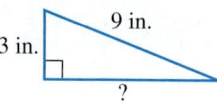

    (a) Give its length in simplified radical form.

    (b) Round the answer to the nearest thousandth.

16. _____

16. In electronics, the impedance Z of an alternating current circuit is given by the formula

$$Z = \sqrt{R^2 + X^2},$$

where R is the resistance and X is the reactance, both in ohms. Find the value of the impedance Z if $R = 40$ ohms and $X = 30$ ohms.

Rationalize each denominator.

17. _____

17. $\dfrac{5\sqrt{2}}{\sqrt{7}}$

18. $\sqrt{\dfrac{2}{3x}}, \quad x \neq 0$

18. _____

19. _____

19. $\dfrac{-2}{\sqrt[3]{4}}$

20. $\dfrac{-3}{4 - \sqrt{3}}$

20. _____

21. _____

21. Write in lowest terms. $\dfrac{\sqrt{12} + 3\sqrt{128}}{6}$

Solve each equation.

22. _____

22. $\sqrt{p} + 4 = 0$

23. $\sqrt{x + 1} = 5 - x$

23. _____

24. _____

24. $3\sqrt{x} - 2 = x$

25. $\sqrt{x + 7} - \sqrt{x} = 1$

25. _____

26. _____

26. Consider the following "solution." ***WHAT WENT WRONG?*** Give the correct solution set.

$$\sqrt{2x + 1} + 5 = 0$$

$$\sqrt{2x + 1} = -5 \qquad \text{Subtract 5.}$$

$$2x + 1 = 25 \qquad \text{Square each side.}$$

$$2x = 24 \qquad \text{Subtract 1.}$$

$$x = 12 \qquad \text{Divide by 2.}$$

The solution set is $\{12\}$.

17

Quadratic Equations

Recreational fishing is big business in the United States. In 2006, nearly 40 million anglers generated over $45.3 billion in retail sales and had a $125.0 billion impact on the U.S. economy. The sportfishing industry created employment for over one million people. If sportfishing were a corporation, it would rank 47th on the 2007 Fortune 500 list of largest American companies based on total sales, ahead of such global giants as Time Warner, IBM, and Microsoft. More Americans fish than play golf and tennis combined. (*Source:* American Sportfishing Association.)

In Example 5 of **Section 17.1**, we apply the *square root property*, a topic of this chapter, to a formula for calculating the length of a bass.

17.1 ▷▷▷ Solving Quadratic Equations by the Square Root Property

OBJECTIVES

1 Solve equations of the form $x^2 = k$, where $k > 0$.

2 Solve equations of the form $(ax + b)^2 = k$, where $k > 0$.

3 Use formulas involving squared variables.

A **quadratic equation** is an equation that can be written in the form

$$ax^2 + bx + c = 0 \qquad \text{Standard form}$$

for real numbers a, b, and c, with $a \neq 0$. As we saw in **Section 14.6**, we can solve a quadratic equation such as $x^2 + 4x + 3 = 0$ by first factoring and then applying the zero-factor property.

$$x^2 + 4x + 3 = 0$$

$$(x + 3)(x + 1) = 0 \qquad \text{Factor.}$$

$$x + 3 = 0 \quad \text{or} \quad x + 1 = 0 \qquad \text{Zero-factor property}$$

$$x = -3 \quad \text{or} \qquad x = -1 \qquad \text{Solve each equation.}$$

The solution set is $\{-3, -1\}$.

◀ *Work Problem* **1** *at the Side.*

1 Solve each equation by factoring.

(a) $x^2 - 2x - 15 = 0$

(b) $2x^2 - 3x + 1 = 0$

OBJECTIVE 1 Solve equations of the form $x^2 = k$, where $k > 0$.
We can solve an equation such as $x^2 = 9$ by factoring as follows.

$$x^2 = 9$$

$$x^2 - 9 = 0 \qquad \text{Subtract 9.}$$

$$(x + 3)(x - 3) = 0 \qquad \text{Factor.}$$

$$x + 3 = 0 \quad \text{or} \quad x - 3 = 0 \qquad \text{Zero-factor property}$$

$$x = -3 \quad \text{or} \qquad x = 3 \qquad \text{Solve each equation.}$$

We might also solve $x^2 = 9$ by noticing that x must be a number whose square is 9. Thus, $x = \sqrt{9} = 3$ or $x = -\sqrt{9} = -3$. This approach is generalized as the **square root property.**

> **Square Root Property**
> If k is a positive number and if $x^2 = k$, then
> $$x = \sqrt{k} \quad \text{or} \quad x = -\sqrt{k}$$

EXAMPLE 1 **Solving Quadratic Equations of the Form $x^2 = k$**

Solve each equation. Write radicals in simplified form.

(a) $x^2 = 16$

By the square root property, if $x^2 = 16$, then

$$x = \sqrt{16} = 4 \quad \text{or} \quad x = -\sqrt{16} = -4.$$

An abbreviation for $x = 4$ or $x = -4$ is $x = \pm 4$ (read "positive or negative 4"). Check each solution by substituting it for x in the original equation. The solution set is $\{-4, 4\}$, or $\{\pm 4\}$.

(b) $z^2 = 5$

The solutions are $z = \sqrt{5}$ or $z = -\sqrt{5}$, so the solution set is $\{-\sqrt{5}, \sqrt{5}\}$, or $\{\pm\sqrt{5}\}$.

Continued on Next Page

ANSWERS

1. **(a)** $\{-3, 5\}$

   (b) $\left\{\frac{1}{2}, 1\right\}$

(c)
$$5m^2 - 32 = 8$$
$$5m^2 = 40 \qquad \text{Add 32.}$$
$$m^2 = 8 \qquad \text{Divide by 5.}$$

> Don't stop here. Simplify the radicals.

$$m = \sqrt{8} \quad \text{or} \quad m = -\sqrt{8} \qquad \text{Square root property}$$
$$m = 2\sqrt{2} \quad \text{or} \quad m = -2\sqrt{2} \qquad \sqrt{8} = \sqrt{4}\cdot\sqrt{2} = 2\sqrt{2}$$

The solution set is $\left\{-2\sqrt{2}, 2\sqrt{2}\right\}$, or $\left\{\pm 2\sqrt{2}\right\}$.

(d) $x^2 = -4$

Because -4 is a negative number and because the square of a real number cannot be negative, ***there is no real number solution*** of this equation. (In this book, we are concerned with finding only *real number* solutions. To use the square root property to find both real number solutions, k must be positive.) The solution set is $\varnothing$.

Work Problem **2** *at the Side.* ▶

OBJECTIVE **2** **Solve equations of the form $(ax + b)^2 = k$, where $k > 0$.** In each equation in Example 1, the exponent 2 had a single variable as its base. We can extend the square root property to solve equations in which the base is a binomial, as shown in the next example.

EXAMPLE 2 **Solving Quadratic Equations of the Form $(x + b)^2 = k$**

Solve each equation.

(a) > Use $x - 3$ as the base.
$$(x - 3)^2 = 16$$
$$x - 3 = \sqrt{16} \quad \text{or} \quad x - 3 = -\sqrt{16} \qquad \text{Square root property}$$
$$x - 3 = 4 \qquad \text{or} \quad x - 3 = -4 \qquad \sqrt{16} = 4$$
$$x = 7 \qquad \text{or} \qquad x = -1 \qquad \text{Add 3.}$$

Check Substitute each solution in the original equation.

$$(x - 3)^2 = 16 \qquad\qquad (x - 3)^2 = 16$$
$$(7 - 3)^2 \stackrel{?}{=} 16 \quad \text{Let } x = 7. \qquad (-1 - 3)^2 \stackrel{?}{=} 16 \quad \text{Let } x = -1.$$
$$4^2 \stackrel{?}{=} 16 \qquad\qquad (-4)^2 \stackrel{?}{=} 16$$
$$16 = 16 \quad \text{True} \qquad\qquad 16 = 16 \quad \text{True}$$

The solutions are 7 and -1, and the solution set is $\{-1, 7\}$.

(b)
$$(x + 1)^2 = 6$$
$$x + 1 = \sqrt{6} \qquad \text{or} \quad x + 1 = -\sqrt{6} \qquad \text{Square root property}$$
$$x = -1 + \sqrt{6} \quad \text{or} \qquad x = -1 - \sqrt{6} \qquad \text{Add } -1.$$

Check
$$\left(-1 + \sqrt{6} + 1\right)^2 = \left(\sqrt{6}\right)^2 = 6;$$
$$\left(-1 - \sqrt{6} + 1\right)^2 = \left(-\sqrt{6}\right)^2 = 6.$$

The solution set is $\left\{-1 + \sqrt{6}, -1 - \sqrt{6}\right\}$, or $\left\{-1 \pm \sqrt{6}\right\}$.

Work Problem **3** *at the Side.* ▶

---

2 Solve each equation. Write radicals in simplified form.

(a) $x^2 = 49$

(b) $x^2 = 11$

(c) $2x^2 + 8 = 32$

(d) $x^2 = -9$

3 Solve each equation.

(a) $(x + 2)^2 = 36$

(b) $(x - 4)^2 = 3$

ANSWERS

2. (a) $\{-7, 7\}$ **(b)** $\left\{-\sqrt{11}, \sqrt{11}\right\}$
   (c) $\left\{-2\sqrt{3}, 2\sqrt{3}\right\}$ **(d)** $\varnothing$

3. (a) $\{-8, 4\}$ **(b)** $\left\{4 + \sqrt{3}, 4 - \sqrt{3}\right\}$

4 Solve $(2x - 5)^2 = 18$.

5 Solve each equation.

(a) $(5x + 1)^2 = 7$

(b) $(7x - 1)^2 = -1$

6 Use the formula in Example 5 to approximate the length of a bass weighing 2.80 lb and having a girth of 11 in.

EXAMPLE 3 Solving a Quadratic Equation of the Form $(ax + b)^2 = k$

Solve $(3r - 2)^2 = 27$.

$$3r - 2 = \sqrt{27} \quad \text{or} \quad 3r - 2 = -\sqrt{27} \qquad \text{Square root property}$$

$$\sqrt{27} = \sqrt{9} \cdot \sqrt{3}$$
$$= 3\sqrt{3}$$

$$3r - 2 = 3\sqrt{3} \quad \text{or} \quad 3r - 2 = -3\sqrt{3}$$

$$3r = 2 + 3\sqrt{3} \quad \text{or} \quad 3r = 2 - 3\sqrt{3} \qquad \text{Add 2.}$$

$$r = \frac{2 + 3\sqrt{3}}{3} \quad \text{or} \quad r = \frac{2 - 3\sqrt{3}}{3} \qquad \text{Divide by 3.}$$

The solution set is $\left\{\dfrac{2 + 3\sqrt{3}}{3}, \dfrac{2 - 3\sqrt{3}}{3}\right\}$.

◀ Work Problem **4** at the Side.

> **CAUTION**
> The solutions in Example 3 are fractions that cannot be simplified. Note that 3 is *not* a common factor in the numerator.

EXAMPLE 4 Recognizing When There Is No Real Solution

Solve $(x + 3)^2 = -9$.

Because the square root of -9 is not a real number, there is no real number solution for this equation. The solution set is $\emptyset$.

◀ Work Problem **5** at the Side.

OBJECTIVE 3 Use formulas involving squared variables.

EXAMPLE 5 Finding the Length of a Bass

We can approximate the weight of a bass, in pounds, given its length L and its girth (distance around) g, where both are measured in inches, using this formula.

$$w = \frac{L^2 g}{1200}$$

Approximate the length of a bass weighing 2.20 lb and having a girth of 10 in. (*Source: Sacramento Bee.*)

$$w = \frac{L^2 g}{1200} \qquad \text{Given formula}$$

$$2.20 = \frac{L^2 \cdot 10}{1200} \qquad w = 2.20,\ g = 10$$

$$2640 = 10L^2 \qquad \text{Multiply by 1200.}$$

$$L^2 = 264 \qquad \text{Divide by 10; interchange the sides.}$$

$$L = \sqrt{264} \quad \text{or} \quad L = -\sqrt{264} \qquad \text{Square root property}$$

A calculator shows that $\sqrt{264} \approx 16.25$, so the length of the bass is a little more than 16 in. (We discard the negative solution $-\sqrt{264} \approx -16.25$, since L represents length.)

◀ Work Problem **6** at the Side.

17.1 ▶▶▶ Exercises

Decide whether each statement is true *or* false. *If* false*, tell why.*

1. If k is a prime number, then $x^2 = k$ has two irrational solutions.

2. If k is a positive perfect square, then $x^2 = k$ has two rational solutions.

3. If k is a positive integer, then $x^2 = k$ must have two rational solutions.

4. If $-10 < k < 0$, then $x^2 = k$ has no real solution.

5. If $-10 < k < 10$, then $x^2 = k$ has no real solution.

6. If k is an integer greater than 24 and less than 26, then $x^2 = k$ has two solutions, -5 and 5.

Solve each equation by using the square root property. Write all radicals in simplest form.
See Example 1.

7. $x^2 = 81$

8. $x^2 = 121$

9. $k^2 = 14$

10. $m^2 = 22$

11. $t^2 = 48$

12. $x^2 = 54$

13. $x^2 = \dfrac{25}{4}$

14. $m^2 = \dfrac{36}{121}$

15. $x^2 = -100$

16. $x^2 = -64$

17. $z^2 = 2.25$

18. $w^2 = 56.25$

19. $r^2 - 3 = 0$ **20.** $x^2 - 13 = 0$ **21.** $7x^2 = 4$

22. $3x^2 = 10$ **23.** $4x^2 - 72 = 0$ **24.** $5z^2 - 200 = 0$

25. $3x^2 - 8 = 64$ **26.** $2x^2 + 7 = 61$ **27.** $5x^2 + 4 = 8$ **28.** $4x^2 - 3 = 7$

Solve each equation by using the square root property. Express all radicals in simplest form. See Examples 2–4.

29. $(x - 3)^2 = 25$ **30.** $(x - 7)^2 = 16$ **31.** $(x + 5)^2 = -13$

32. $(x + 2)^2 = -17$ **33.** $(x - 8)^2 = 27$ **34.** $(x - 5)^2 = 40$

35. $(3x + 2)^2 = 49$ **36.** $(5x + 3)^2 = 36$ **37.** $(4x - 3)^2 = 9$

38. $(7x - 5)^2 = 25$

39. $(5 - 2x)^2 = 30$

40. $(3 - 2x)^2 = 70$

41. $(3x + 1)^2 = 18$

42. $(5x + 6)^2 = 75$

43. $\left(\dfrac{1}{2}x + 5\right)^2 = 12$

44. $\left(\dfrac{1}{3}x + 4\right)^2 = 27$

45. $(4x - 1)^2 - 48 = 0$

46. $(2x - 5)^2 - 180 = 0$

47. Johnny solved the equation in Exercise 39 and wrote his answer as $\left\{ \dfrac{5 + \sqrt{30}}{2}, \dfrac{5 - \sqrt{30}}{2} \right\}$.

Terry solved the same equation and wrote her answer as $\left\{ \dfrac{-5 + \sqrt{30}}{-2}, \dfrac{-5 - \sqrt{30}}{-2} \right\}$.

The teacher gave them both full credit. Explain why both students were correct, although their answers seem to differ.

48. In the solutions found in Example 3 of this section, why is it not valid to simplify by dividing out the 3s in the numerators and denominators?

Solve each problem. See Example 5.

49. One expert at marksmanship can hold a silver dollar at forehead level, drop it, draw his gun, and shoot the coin as it passes waist level. The distance traveled by a falling object is given by

$$d = 16t^2,$$

where d is the distance (in feet) the object falls in t seconds. If the coin falls about 4 ft, use the formula to estimate the time that elapses between the dropping of the coin and the shot.

50. The illumination produced by a light source depends on the distance from the source. For a particular light source, this relationship can be expressed as

$$d^2 = \frac{4050}{I},$$

where d is the distance from the source (in feet) and I is the amount of illumination in foot-candles. How far from the source is the illumination equal to 50 foot-candles?

51. The area A of a circle with radius r is given by the formula

$$A = \pi r^2.$$

If a circle has area 81π in.^{2}, what is its radius?

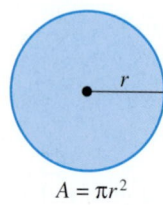

$A = \pi r^2$

52. The surface area S of a sphere with radius r is given by the formula

$$S = 4\pi r^2.$$

If a sphere has surface area 36π ft^2, what is its radius?

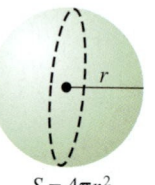

$S = 4\pi r^2$

The amount A that P dollars invested at an annual rate of interest r will grow to in 2 yr is $A = P(1 + r)^2$.

53. At what interest rate will $100 grow to $110.25 in 2 yr?

54. At what interest rate will $500 grow to $572.45 in 2 yr?

55. At what interest rate will $200 grow to $208.08 in 2 yr?

17.2 ▶▶▶ Solving Quadratic Equations by Completing the Square

OBJECTIVE 1 Solve quadratic equations by completing the square when the coefficient of the second-degree term is 1.
The methods we have studied so far are not enough to solve this equation.

$$x^2 + 6x + 7 = 0$$

If we could write the equation in the form $(x + 3)^2$ equals a constant, we could solve it with the square root property discussed in **Section 17.1**. To do that, we need to have a perfect square trinomial on one side of the equation.

Recall from **Section 14.5** that a perfect square trinomial has the form

$$x^2 + 2kx + k^2 \quad \text{or} \quad x^2 - 2kx + k^2,$$

where k represents a positive number.

OBJECTIVES

1 Solve quadratic equations by completing the square when the coefficient of the second-degree term is 1.

2 Solve quadratic equations by completing the square when the coefficient of the second-degree term is not 1.

3 Simplify the terms of an equation before solving.

4 Solve applied problems that require quadratic equations.

EXAMPLE 1 Creating Perfect Square Trinomials

Complete each trinomial so that it is a perfect square. Then factor the trinomial.

(a) $x^2 + 8x +$ _____

The perfect square trinomial will have the form $x^2 + 2kx + k^2$. Thus, the middle term, $8x$, must equal $2kx$.

$$8x = 2kx \quad \longleftarrow \text{Solve this equation for } k.$$
$$4 = k \qquad \text{Divide each side by } 2x.$$

Therefore, $k = 4$ and $k^2 = 4^2 = 16$. The required perfect square trinomial is

$$x^2 + 8x + 16, \quad \text{which factors as} \quad (x + 4)^2.$$

(b) $x^2 - 18x +$ _____

Here the perfect square trinomial will have the form $x^2 - 2kx + k^2$. The middle term, $-18x$, must equal $-2kx$.

$$-18x = -2kx \quad \longleftarrow \text{Solve this equation for } k.$$
$$9 = k \qquad \text{Divide each side by } -2x.$$

Thus, $k = 9$ and $k^2 = 9^2 = 81$. The required perfect square trinomial is

$$x^2 - 18x + 81, \quad \text{which factors as} \quad (x - 9)^2.$$

Work Problem **1** *at the Side.* ▶

1 Complete each trinomial so that it is a perfect square. Then factor the trinomial.

(a) $x^2 + 12x +$ _____

(b) $x^2 - 14x +$ _____

(c) $x^2 - 2x +$ _____

EXAMPLE 2 Rewriting an Equation to Use the Square Root Property

Solve $x^2 + 6x + 7 = 0$.

$$x^2 + 6x = -7 \qquad \text{Subtract 7 from each side.}$$

To solve this equation with the square root property, the quantity on the left side, $x^2 + 6x$, must be written as a perfect square trinomial in the form $x^2 + 2kx + k^2$.

$$x^2 + 6x +$$ _____

Here, $2kx = 6x$, so $k = 3$ and $k^2 = 9$. The required perfect square trinomial is

$$x^2 + 6x + 9, \quad \text{which factors as} \quad (x + 3)^2.$$

Continued on Next Page

ANSWERS

1. **(a)** $36; (x + 6)^2$
   (b) $49; (x - 7)^2$
   (c) $1; (x - 1)^2$

2 Solve $x^2 - 4x - 1 = 0$.

Therefore, if we add 9 to *each* side of $x^2 + 6x = -7$, the equation will have a perfect square trinomial on the left side, as needed.

> This is a key step.

$$x^2 + 6x + 9 = -7 + 9 \qquad \text{Add 9.}$$
$$(x + 3)^2 = 2 \qquad \text{Factor; add.}$$

Now use the square root property to complete the solution.

$$x + 3 = \sqrt{2} \qquad \text{or} \quad x + 3 = -\sqrt{2}$$
$$x = -3 + \sqrt{2} \quad \text{or} \qquad x = -3 - \sqrt{2}$$

Check by substituting $-3 + \sqrt{2}$ and $-3 - \sqrt{2}$ for x in the original equation. The solution set is $\{-3 + \sqrt{2}, -3 - \sqrt{2}\}$.

◀ Work Problem **2** at the Side.

The process of changing the form of the equation in Example 2 from

$$x^2 + 6x + 7 = 0 \quad \text{to} \quad (x + 3)^2 = 2$$

3 Solve each equation by completing the square.

(a) $x^2 + 4x = 1$

is called **completing the square.** Completing the square changes only the form of the equation. To see this, multiply out the left side of $(x + 3)^2 = 2$. Then write the equation in standard form to get $x^2 + 6x + 7 = 0$.

Look again at the original equation in Example 2,

$$x^2 + 6x + 7 = 0.$$

If we take half of 6, the coefficient of x, and square it, we get 9.

$$\frac{1}{2} \cdot 6 = 3 \quad \text{and} \quad 3^2 = 9$$

Coefficient of x \qquad Quantity added to each side

To complete the square in Example 2, we added **9** to each side.

(b) $z^2 + 6z - 3 = 0$

EXAMPLE 3 Completing the Square to Solve a Quadratic Equation

Complete the square to solve $x^2 - 8x = 5$.

To complete the square on $x^2 - 8x$, take half of -8, the coefficient of x, and square it.

$$\frac{1}{2}(-8) = -4 \quad \text{and} \quad (-4)^2 = 16$$

Coefficient of x

Add the result, **16**, to each side of the equation.

$$x^2 - 8x = 5 \qquad \text{Given equation}$$
$$x^2 - 8x + 16 = 5 + 16 \qquad \text{Add 16.}$$
$$(x - 4)^2 = 21 \qquad \text{Factor; add.}$$
$$x - 4 = \sqrt{21} \qquad \text{or} \quad x - 4 = -\sqrt{21} \qquad \text{Square root property}$$
$$x = 4 + \sqrt{21} \quad \text{or} \qquad x = 4 - \sqrt{21} \qquad \text{Add 4.}$$

A check indicates that the solution set is $\{4 + \sqrt{21}, 4 - \sqrt{21}\}$.

◀ Work Problem **3** at the Side.

ANSWERS

2. $\{2 + \sqrt{5}, 2 - \sqrt{5}\}$

3. (a) $\{-2 + \sqrt{5}, -2 - \sqrt{5}\}$

(b) $\{-3 + 2\sqrt{3}, -3 - 2\sqrt{3}\}$

OBJECTIVE **2** **Solve quadratic equations by completing the square when the coefficient of the second-degree term is not 1.**
If a quadratic equation has the form

$$ax^2 + bx + c = 0, \quad \text{where } a \neq 1,$$

then to obtain 1 as the coefficient of x^2, we first divide each side of the equation by a.

EXAMPLE 4 **Solving a Quadratic Equation by Completing the Square**

Solve $4x^2 + 16x = 9$.
   Before completing the square, the coefficient of x^2 must be 1, not 4. We get 1 as the coefficient of x^2 here by dividing each side by 4.

$$4x^2 + 16x = 9$$

The coefficient of x^2 must be 1. $\longrightarrow$ $x^2 + 4x = \dfrac{9}{4}$ Divide by 4.

Next, we begin to complete the square by taking half the coefficient of x, and squaring it:

$$\frac{1}{2}(4) = 2 \quad \text{and} \quad 2^2 = 4.$$

We add the result, **4**, to each side of the equation.

$$x^2 + 4x + 4 = \frac{9}{4} + 4 \qquad \text{Add 4.}$$

$$(x + 2)^2 = \frac{25}{4} \qquad \text{Factor; } \tfrac{9}{4} + 4 = \tfrac{9}{4} + \tfrac{16}{4} = \tfrac{25}{4}.$$

$$x + 2 = \sqrt{\frac{25}{4}} \quad \text{or} \quad x + 2 = -\sqrt{\frac{25}{4}} \qquad \text{Square root property}$$

$$x + 2 = \frac{5}{2} \quad \text{or} \quad x + 2 = -\frac{5}{2} \qquad \text{Take square roots.}$$

$$x = -2 + \frac{5}{2} \quad \text{or} \quad x = -2 - \frac{5}{2} \qquad \text{Subtract 2.}$$

$$x = \frac{1}{2} \quad \text{or} \quad x = -\frac{9}{2} \qquad -2 = -\tfrac{4}{2}$$

Check

$$4x^2 + 16x = 9 \qquad\qquad 4x^2 + 16x = 9$$

$$4\left(\frac{1}{2}\right)^2 + 16\left(\frac{1}{2}\right) \stackrel{?}{=} 9 \qquad 4\left(-\frac{9}{2}\right)^2 + 16\left(-\frac{9}{2}\right) \stackrel{?}{=} 9$$

$$4\left(\frac{1}{4}\right) + 8 \stackrel{?}{=} 9 \qquad\qquad 4\left(\frac{81}{4}\right) - 72 \stackrel{?}{=} 9$$

$$1 + 8 \stackrel{?}{=} 9 \qquad\qquad 81 - 72 \stackrel{?}{=} 9$$

$$9 = 9 \quad \text{True} \qquad\qquad 9 = 9 \quad \text{True}$$

The two solutions, $\frac{1}{2}$ and $-\frac{9}{2}$, check, so the solution set is $\left\{-\frac{9}{2}, \frac{1}{2}\right\}$.

Work Problem **4** *at the Side.* ▶

4 Solve each equation by completing the square.

   (a) $9x^2 + 18x = -5$

   (b) $4t^2 - 24t + 11 = 0$

ANSWERS

4. **(a)** $\left\{-\frac{1}{3}, -\frac{5}{3}\right\}$ **(b)** $\left\{\frac{11}{2}, \frac{1}{2}\right\}$

The steps used to solve a quadratic equation $ax^2 + bx + c = 0$ by completing the square are summarized here.

⑤ Solve each equation by completing the square.

(a) $3x^2 + 5x - 2 = 0$

> **Solving a Quadratic Equation by Completing the Square**
>
> **Step 1** **Be sure the second-degree term has coefficient 1.** If the coefficient of the second-degree term is 1, proceed to Step 2. If it is not 1, but some other nonzero number a, divide each side of the equation by a.
>
> **Step 2** **Write in correct form.** Make sure that all terms with variables are on one side of the equals sign and that all constant terms are on the other side.
>
> **Step 3** **Complete the square.** Take half the coefficient of the first-degree term, and square it. Add the square to each side of the equation. Factor the variable side, and simplify on the other side.
>
> **Step 4** **Solve** the equation by using the square root property.

EXAMPLE 5 **Solving a Quadratic Equation by Completing the Square**

Solve $2x^2 - 7x - 9 = 0$.

Step 1 Get 1 as the coefficient of the x^2-term.

$$x^2 - \frac{7}{2}x - \frac{9}{2} = 0 \qquad \text{Divide by 2.}$$

(b) $2x^2 - 4x - 1 = 0$

Step 2 Add $\frac{9}{2}$ to each side to get the variable terms on the left and the constant on the right.

$$x^2 - \frac{7}{2}x = \frac{9}{2} \qquad \text{Add } \tfrac{9}{2}.$$

Step 3 To complete the square, take half the coefficient of x and square it: $\left[\frac{1}{2}\left(-\frac{7}{2}\right)\right]^2 = \left(-\frac{7}{4}\right)^2 = \frac{49}{16}$.

$$x^2 - \frac{7}{2}x + \frac{49}{16} = \frac{9}{2} + \frac{49}{16} \qquad \boxed{\text{Be sure to add } \tfrac{49}{16} \text{ to } each \text{ side.}}$$

$$\left(x - \frac{7}{4}\right)^2 = \frac{121}{16} \qquad \text{Factor; } \tfrac{9}{2} + \tfrac{49}{16} = \tfrac{72}{16} + \tfrac{49}{16} = \tfrac{121}{16}.$$

Step 4 Solve by using the square root property.

$$x - \frac{7}{4} = \sqrt{\frac{121}{16}} \quad \text{or} \quad x - \frac{7}{4} = -\sqrt{\frac{121}{16}} \qquad \text{Square root property}$$

$$x = \frac{7}{4} + \frac{11}{4} \quad \text{or} \quad x = \frac{7}{4} - \frac{11}{4} \qquad \text{Add } \tfrac{7}{4}; \sqrt{\tfrac{121}{16}} = \tfrac{11}{4}.$$

$$x = \frac{18}{4} = \frac{9}{2} \quad \text{or} \quad x = -\frac{4}{4} = -1 \qquad \text{Simplify.}$$

A check confirms that the solution set is $\left\{-1, \frac{9}{2}\right\}$.

◀ *Work Problem* **⑤** *at the Side.*

EXAMPLE 6 **Solving a Quadratic Equation by Completing the Square**

Solve $4p^2 + 8p + 5 = 0$.

$$4p^2 + 8p + 5 = 0$$

> The coefficient of the second-degree term must be 1.

$$p^2 + 2p + \frac{5}{4} = 0 \qquad \text{Divide by 4.}$$

$$p^2 + 2p = -\frac{5}{4} \qquad \text{Subtract } \tfrac{5}{4}.$$

The coefficient of p is 2. Take half of 2; square the result: $\left[\frac{1}{2}(2)\right]^2 = 1^2 = 1$. Add this result to each side. Then write the left side as a perfect square.

$$p^2 + 2p + 1 = -\frac{5}{4} + 1 \qquad \text{Add 1.}$$

$$(p + 1)^2 = -\frac{1}{4} \qquad \text{Factor; add.}$$

We cannot use the square root property to solve this equation, because the square root of $-\frac{1}{4}$ is not a real number. This equation has no real number solution.* The solution set is $\emptyset$.

─────── *Work Problem* **6** *at the Side.* ▶

OBJECTIVE **3** **Simplify the terms of an equation before solving.**

EXAMPLE 7 **Simplifying before Completing the Square**

Solve $(x + 3)(x - 1) = 2$.

$$(x + 3)(x - 1) = 2 \qquad \text{Given equation}$$

$$x^2 + 2x - 3 = 2 \qquad \text{Multiply using the FOIL method.}$$

$$x^2 + 2x = 5 \qquad \text{Add 3.}$$

$$x^2 + 2x + 1 = 5 + 1 \qquad \text{Complete the square—add } [\tfrac{1}{2}(2)]^2 = 1.$$

$$(x + 1)^2 = 6 \qquad \text{Factor on the left; add on the right.}$$

$$x + 1 = \sqrt{6} \qquad \text{or} \quad x + 1 = -\sqrt{6} \qquad \text{Square root property}$$

$$x = -1 + \sqrt{6} \quad \text{or} \qquad x = -1 - \sqrt{6} \qquad \text{Add } -1.$$

The solution set is $\{-1 + \sqrt{6}, -1 - \sqrt{6}\}$.

─────── *Work Problem* **7** *at the Side.* ▶

> **Note**
> The solutions $-1 \pm \sqrt{6}$ given in Example 7 are *exact*. In applications, decimal solutions are more appropriate. Using the square root key of a calculator, $\sqrt{6} \approx 2.449$. Approximating the two solutions gives
> $$x \approx 1.449 \quad \text{and} \quad x \approx -3.449.$$

---

*The equation in Example 6 has no solution over the *real number system*. In the **complex number system,** however, this equation does have solutions. The complex numbers include numbers whose squares are negative. These numbers are discussed in intermediate and college algebra courses.

6 Solve $5x^2 + 3x + 1 = 0$ by completing the square.

7 Solve each equation.

(a) $r(r - 3) = -1$

(b) $(x + 2)(x + 1) = 5$

ANSWERS

6. $\emptyset$

7. (a) $\left\{\dfrac{3 + \sqrt{5}}{2}, \dfrac{3 - \sqrt{5}}{2}\right\}$

(b) $\left\{\dfrac{-3 + \sqrt{21}}{2}, \dfrac{-3 - \sqrt{21}}{2}\right\}$

8 Suppose a ball is projected upward from ground level with an initial velocity of 128 ft per sec. Its height at time t (in seconds) is given by

$$s = -16t^2 + 128t,$$

where s is in feet. At what times will it be 48 ft above the ground? Give answers to the nearest tenth.

OBJECTIVE **4** **Solve applied problems that require quadratic equations.**   There are many practical applications of quadratic equations. The next example illustrates an application from physics.

EXAMPLE 8   **Solving a Velocity Problem**

If a ball is projected into the air from ground level with an initial velocity of 64 ft per sec, its altitude (height) s in feet in t seconds is given by the formula

$$s = -16t^2 + 64t.$$

How long will it take the ball to be 48 ft above the ground?

Since s represents the height, we substitute **48** for s in the formula and then solve this equation for time t by completing the square.

$48 = -16t^2 + 64t$	Let $s = 48$.
$-3 = t^2 - 4t$	Divide by -16.
$t^2 - 4t = -3$	Interchange the sides.
$t^2 - 4t + 4 = -3 + 4$	Add $\left[\frac{1}{2}(-4)\right]^2 = 4$.
$(t - 2)^2 = 1$	Factor; add.
$t - 2 = 1$ or $t - 2 = -1$	Square root property
$t = 3$ or $t = 1$	Add 2.

The ball reaches a height of 48 ft twice, once on the way up and again on the way down. It takes 1 sec to reach 48 ft on the way up, and then after 3 sec, the ball reaches 48 ft again on the way down.

◀ *Work Problem* **8** *at the Side.*

ANSWER

8. 0.4 sec and 7.6 sec

Complete each trinomial so that it is a perfect square. Then factor the trinomial.
See Example 1.

1. $x^2 + 10x +$ _____

2. $x^2 + 16x +$ _____

3. $x^2 + 2x +$ _____

4. $m^2 - 2m +$ _____

5. $p^2 - 5p +$ _____

6. $x^2 + 3x +$ _____

7. Which step is an appropriate way to begin solving the quadratic equation

$$2x^2 - 4x = 9$$

by completing the square?

A. Add 4 to each side of the equation.

B. Factor the left side as $2x(x - 2)$.

C. Factor the left side as $x(2x - 4)$.

D. Divide each side by 2.

8. In Example 3 of **Section 14.6,** we solved the quadratic equation

$$4p^2 - 26p + 40 = 0$$

by factoring. If we were to solve by completing the square, would we get the same solutions, $\frac{5}{2}$ and 4?

Solve each equation by completing the square. See Examples 2 and 3.

9. $x^2 - 4x = -3$

10. $x^2 - 2x = 8$

11. $x^2 + 5x + 6 = 0$

12. $x^2 + 6x + 5 = 0$

13. $x^2 + 2x - 5 = 0$

14. $x^2 + 4x + 1 = 0$

15. $x^2 - 8x = -4$

16. $m^2 - 4m = 14$

17. $t^2 + 6t + 9 = 0$

18. $k^2 - 8k + 16 = 0$

19. $x^2 + x - 1 = 0$

20. $x^2 + x - 3 = 0$

Solve each equation by completing the square. See Examples 4–7.

21. $4x^2 + 4x - 3 = 0$

22. $9x^2 + 3x - 2 = 0$

23. $2x^2 - 4x = 5$

24. $2x^2 - 6x = 3$

25. $2p^2 - 2p + 3 = 0$

26. $3q^2 - 3q + 4 = 0$

27. $3k^2 + 7k = 4$

28. $2k^2 + 5k = 1$

29. $(x + 3)(x - 1) = 5$

30. $(y - 8)(y + 2) = 24$

31. $(r - 3)(r - 5) = 2$

32. $(k - 1)(k - 7) = 1$

33. $-x^2 + 2x = -5$

34. $-r^2 + 3r = -2$

Solve each problem. See Example 8.

35. If an object is projected upward from ground level on Earth with an initial velocity of 96 ft per sec, its altitude (height) s in feet in t seconds is given by the formula $s = -16t^2 + 96t$. At what times will the object be 80 ft above the ground?

36. At what times will the object described in Exercise 35 be 100 ft above the ground? Round your answers to the nearest tenth.

37. If an object is projected upward on the surface of Mars from ground level with an initial velocity of 104 ft per sec, its altitude (height) s in feet in t seconds is given by the formula $s = -13t^2 + 104t$. At what times will the object be 195 ft above the surface?

38. After how many seconds will the object in Exercise 37 return to the surface? (*Hint:* When it returns to the surface, $s = 0$.)

39. A farmer has a rectangular cattle pen with perimeter 350 ft and area 7500 ft². What are the dimensions of the pen? (*Hint:* Use the figure to set up the equation.)

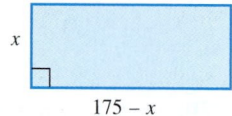

40. The base of a triangle measures 1 m more than three times the height of the triangle. Its area is 15 m². Find the lengths of the base and the height.

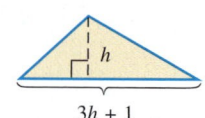

17.3 ▶▶▶ Solving Quadratic Equations by the Quadratic Formula

We can solve any quadratic equation by completing the square, but the method can be tedious. In this section we complete the square on the general quadratic equation

$$ax^2 + bx + c = 0, \qquad a \neq 0, \qquad \text{Standard form}$$

to obtain the *quadratic formula*, which gives the solution(s) of any quadratic equation.

> **Note**
>
> In $ax^2 + bx + c = 0$, there is a restriction that a is not zero. If it were, the equation would be linear, not quadratic.

OBJECTIVE 1 Identify the values of a, b, and c in a quadratic equation. To solve a quadratic equation by this new method, we must first identify the values of a, b, and c in the standard form.

EXAMPLE 1 Identifying Values of a, b, and c in Quadratic Equations

Identify the values of the variables a, b, and c in each quadratic equation $ax^2 + bx + c = 0$.

(a) $2x^2 + 3x - 5 = 0$ — *This equation is in standard form $ax^2 + bx + c = 0$.*

In this example, $a = 2$, $b = 3$, and $c = -5$.

(b) $-x^2 + 2 = 6x$

First write the equation in standard form $ax^2 + bx + c = 0$.

$$-x^2 + 2 = 6x$$

$-x^2$ means $-1x^2$.

$$-x^2 - 6x + 2 = 0 \qquad \text{Subtract } 6x.$$

Here, $a = -1$, $b = -6$, and $c = 2$.

(c) $5x^2 - 12 = 0$

The x-term is missing, so write the equation as

$$5x^2 + 0x - 12 = 0.$$

Then $a = 5$, $b = 0$, and $c = -12$.

(d) $-4x^2 = -x$

In $ax^2 + bx + c = 0$ form, this equation is written $-4x^2 + x = 0$. The constant term c is missing, so $a = -4$, $b = 1$, and $c = 0$.

(e) $(2x - 7)(x + 4) = -23$

$$2x^2 + x - 28 = -23 \qquad \text{Multiply using the FOIL method.}$$

$$2x^2 + x - 5 = 0 \qquad \text{Add 23; standard form}$$

Now, identify the required values: $a = 2$, $b = 1$, and $c = -5$.

——————— *Work Problem* **1** *at the Side.* ▶

OBJECTIVES

1. Identify the values of a, b, and c in a quadratic equation.

2. Use the quadratic formula to solve quadratic equations.

3. Solve quadratic equations with only one solution.

4. Solve quadratic equations with fractions.

---

1 Identify the values of the variables a, b, and c in each quadratic equation $ax^2 + bx + c = 0$.

(a) $5x^2 + 2x - 1 = 0$

(b) $3x^2 = x - 2$

(c) $9x^2 - 13 = 0$

(d) $-x^2 + x = 0$

(e) $(3x + 2)(x - 1) = 8$

ANSWERS

1. (a) $a = 5, b = 2, c = -1$
   (b) $a = 3, b = -1, c = 2$
   (c) $a = 9, b = 0, c = -13$
   (d) $a = -1, b = 1, c = 0$
   (e) $a = 3, b = -1, c = -10$

OBJECTIVE **2** **Use the quadratic formula to solve quadratic equations.** To develop the quadratic formula, we complete the square on $ax^2 + bx + c = 0$ ($a > 0$). For comparison, we also show the corresponding steps for solving $2x^2 + x - 5 = 0$ (from Example 1(e)).

Step 1 Transform so that the coefficient of x^2 is equal to 1.

$$2x^2 + x - 5 = 0 \qquad\qquad ax^2 + bx + c = 0 \quad \text{Standard form}$$

$$x^2 + \frac{1}{2}x - \frac{5}{2} = 0 \quad \text{Divide by 2.} \qquad x^2 + \frac{b}{a}x + \frac{c}{a} = 0 \quad \text{Divide by } a.$$

Step 2 Write so that the variable terms with x are alone on the left side.

$$x^2 + \frac{1}{2}x = \frac{5}{2} \quad \text{Add } \tfrac{5}{2}. \qquad x^2 + \frac{b}{a}x = -\frac{c}{a} \quad \text{Subtract } \tfrac{c}{a}.$$

Step 3 Add the square of half the coefficient of x to each side, factor the left side, and combine terms on the right.

$$x^2 + \frac{1}{2}x + \frac{1}{16} = \frac{5}{2} + \frac{1}{16} \quad \text{Add } \tfrac{1}{16}. \qquad x^2 + \frac{b}{a}x + \frac{b^2}{4a^2} = -\frac{c}{a} + \frac{b^2}{4a^2} \quad \text{Add } \tfrac{b^2}{4a^2}.$$

$$\left(x + \frac{1}{4}\right)^2 = \frac{41}{16} \quad \begin{array}{l}\text{Factor;}\\\text{add on}\\\text{right.}\end{array} \qquad \left(x + \frac{b}{2a}\right)^2 = \frac{b^2 - 4ac}{4a^2} \quad \begin{array}{l}\text{Factor;}\\\text{add on}\\\text{right.}\end{array}$$

Step 4 Use the square root property to complete the solution.

$$x + \frac{1}{4} = \pm\sqrt{\frac{41}{16}} \qquad\qquad x + \frac{b}{2a} = \pm\sqrt{\frac{b^2 - 4ac}{4a^2}}$$

$$x + \frac{1}{4} = \pm\frac{\sqrt{41}}{4} \qquad\qquad x + \frac{b}{2a} = \pm\frac{\sqrt{b^2 - 4ac}}{2a}$$

$$x = -\frac{1}{4} \pm \frac{\sqrt{41}}{4} \qquad\qquad x = -\frac{b}{2a} \pm \frac{\sqrt{b^2 - 4ac}}{2a}$$

$$x = \frac{-1 \pm \sqrt{41}}{4} \qquad\qquad x = \frac{-b \pm \sqrt{b^2 - 4ac}}{2a}$$

The final result on the right is called the **quadratic formula.** (It is also valid for $a < 0$.) *It is a key result that should be memorized.* Notice that there are two values: one for the $+$ sign and one for the $-$ sign.

Quadratic Formula

The solutions of the quadratic equation $ax^2 + bx + c = 0$, $a \neq 0$, are

$$x = \frac{-b + \sqrt{b^2 - 4ac}}{2a} \quad \text{and} \quad x = \frac{-b - \sqrt{b^2 - 4ac}}{2a},$$

or in compact form, $\quad x = \dfrac{-b \pm \sqrt{b^2 - 4ac}}{2a}.$

CAUTION

Notice in the quadratic formula that the fraction bar is under $-b$ as well as the radical. *Be sure to find the values of $-b \pm \sqrt{b^2 - 4ac}$ first, and then divide those results by the value of 2a.*

EXAMPLE 2 Solving a Quadratic Equation by the Quadratic Formula

Solve $2x^2 - 7x - 9 = 0$. (This equation was solved by completing the square in **Section 17.2**, Example 5.)

In this equation, $a = 2$, $b = -7$, and $c = -9$.

$$x = \frac{-b \pm \sqrt{b^2 - 4ac}}{2a} \qquad \text{Quadratic formula}$$

$$x = \frac{-(-7) \pm \sqrt{(-7)^2 - 4(2)(-9)}}{2(2)} \qquad \begin{array}{l}\text{Substitute } a = 2, b = -7,\\ \text{and } c = -9.\end{array}$$

$$x = \frac{7 \pm \sqrt{49 + 72}}{4} \qquad \text{Simplify.}$$

$$x = \frac{7 \pm \sqrt{121}}{4}$$

$$x = \frac{7 \pm 11}{4} \qquad \sqrt{121} = 11$$

$$x = \frac{7 + 11}{4} = \frac{18}{4} = \frac{9}{2} \quad \text{or} \quad x = \frac{7 - 11}{4} = \frac{-4}{4} = -1 \qquad \begin{array}{l}\text{Find } both\\ \text{solutions.}\end{array}$$

Check each solution in the original equation. The solution set is $\left\{-1, \frac{9}{2}\right\}$.

Work Problem **2** *at the Side.* ▶

EXAMPLE 3 Rewriting a Quadratic Equation before Solving

Solve $x^2 = 2x + 1$.

Write the equation in standard form as $x^2 - 2x - 1 = 0$.

$$x = \frac{-b \pm \sqrt{b^2 - 4ac}}{2a} \qquad \text{Quadratic formula}$$

$$x = \frac{-(-2) \pm \sqrt{(-2)^2 - 4(1)(-1)}}{2(1)} \qquad \begin{array}{l}\text{Substitute } a = 1, b = -2,\\ \text{and } c = -1.\end{array}$$

$$x = \frac{2 \pm \sqrt{8}}{2} \qquad \text{Simplify.}$$

$$x = \frac{2 \pm 2\sqrt{2}}{2} \qquad \sqrt{8} = \sqrt{4} \cdot \sqrt{2} = 2\sqrt{2}$$

$$x = \frac{2\left(1 \pm \sqrt{2}\right)}{2} \qquad \boxed{\text{Factor first. Then divide out the common factor.}}$$

$$x = 1 \pm \sqrt{2}$$

The solution set is $\left\{1 + \sqrt{2}, 1 - \sqrt{2}\right\}$.

Work Problem **3** *at the Side.* ▶

2 Solve each equation by using the quadratic formula.

(a) $2x^2 + 3x - 5 = 0$

(b) $6x^2 + x - 1 = 0$

3 Solve $x^2 + 1 = -8x$.

ANSWERS

2. (a) $\left\{1, -\frac{5}{2}\right\}$ **(b)** $\left\{-\frac{1}{2}, \frac{1}{3}\right\}$

3. $\left\{-4 + \sqrt{15}, -4 - \sqrt{15}\right\}$

4 Solve $9x^2 - 12x + 4 = 0$.

EXAMPLE 4 Solving a Quadratic Equation with One Solution

Solve $4x^2 + 25 = 20x$.

Write the equation in standard form as

$$4x^2 - 20x + 25 = 0. \quad \text{Subtract } 20x.$$

Here, $a = 4$, $b = -20$, and $c = 25$. By the quadratic formula,

$$x = \frac{-(-20) \pm \sqrt{(-20)^2 - 4(4)(25)}}{2(4)} = \frac{20 \pm 0}{8} = \frac{5}{2}.$$

In this case, $b^2 - 4ac = 0$, and the trinomial $4x^2 - 20x + 25$ is a perfect square. There is just one solution in the solution set $\left\{\frac{5}{2}\right\}$.

◄ *Work Problem* **4** *at the Side.*

5 Solve.

(a) $x^2 - \frac{4}{3}x + \frac{2}{3} = 0$

Note

The single solution of the equation in Example 4 is a rational number. If all solutions of a quadratic equation are rational, the equation can be solved by factoring as well.

OBJECTIVE **4** Solve quadratic equations with fractions.

EXAMPLE 5 Solving a Quadratic Equation with Fractions

Solve $\frac{1}{10}t^2 = \frac{2}{5}t - \frac{1}{2}$.

$$\frac{1}{10}t^2 = \frac{2}{5}t - \frac{1}{2}$$

$$\mathbf{10}\left(\frac{1}{10}t^2\right) = \mathbf{10}\left(\frac{2}{5}t - \frac{1}{2}\right) \quad \text{Clear fractions; multiply by the LCD, 10.}$$

(b) $x^2 - \frac{9}{5}x = \frac{2}{5}$

$$\mathbf{10}\left(\frac{1}{10}t^2\right) = \mathbf{10}\left(\frac{2}{5}t\right) - \mathbf{10}\left(\frac{1}{2}\right) \quad \text{Distributive property}$$

$$t^2 = 4t - 5 \quad \text{Multiply.}$$

$$t^2 - 4t + 5 = 0 \quad \text{Standard form}$$

Identify $a = 1$, $b = -4$, and $c = 5$. By the quadratic formula,

$$t = \frac{-(-4) \pm \sqrt{(-4)^2 - 4(1)(5)}}{2(1)} \quad \text{Substitute in the quadratic formula.}$$

$$t = \frac{4 \pm \sqrt{16 - 20}}{2} \quad \text{Simplify.}$$

$$t = \frac{4 \pm \sqrt{-4}}{2} \quad \text{Stop here.}$$

Because $\sqrt{-4}$ does not represent a real number, the solution set is $\emptyset$.

◄ *Work Problem* **5** *at the Side.*

17.3 ▶▶▶ Exercises

Write each equation in standard form $ax^2 + bx + c = 0$, if necessary. Then identify the values of a, b, and c. Do not actually solve the equation. See Example 1.

1. $4x^2 + 5x - 9 = 0$

$a = \underline{\hspace{1cm}}, b = \underline{\hspace{1cm}}, c = \underline{\hspace{1cm}}$

2. $8x^2 + 3x - 4 = 0$

$a = \underline{\hspace{1cm}}, b = \underline{\hspace{1cm}}, c = \underline{\hspace{1cm}}$

3. $3x^2 = 4x + 2$

$a = \underline{\hspace{1cm}}, b = \underline{\hspace{1cm}}, c = \underline{\hspace{1cm}}$

4. $5x^2 = 3x - 6$

$a = \underline{\hspace{1cm}}, b = \underline{\hspace{1cm}}, c = \underline{\hspace{1cm}}$

5. $3x^2 = -7x$

$a = \underline{\hspace{1cm}}, b = \underline{\hspace{1cm}}, c = \underline{\hspace{1cm}}$

6. $9x^2 = 8x$

$a = \underline{\hspace{1cm}}, b = \underline{\hspace{1cm}}, c = \underline{\hspace{1cm}}$

Use the quadratic formula to solve each equation. Write all radicals in simplified form, and write all answers in lowest terms. See Examples 2–4.

7. $k^2 + 12k - 13 = 0$

8. $r^2 - 8r - 9 = 0$

9. $p^2 - 4p + 4 = 0$

10. $9x^2 + 6x + 1 = 0$

11. $2x^2 = 5 + 3x$

12. $2z^2 = 30 + 7z$

◐ 13. $2x^2 + 12x = -5$

14. $5m^2 + m = 1$

15. $6x^2 + 6x = 0$

16. $4n^2 - 12n = 0$

17. $-2x^2 = -3x + 2$

18. $-x^2 = -5x + 20$

19. $3x^2 + 5x + 1 = 0$

20. $6x^2 - 6x + 1 = 0$

21. $7x^2 = 12x$

22. $9r^2 = 11r$

23. $x^2 - 24 = 0$

24. $z^2 - 96 = 0$

25. $25x^2 - 4 = 0$

26. $16x^2 - 9 = 0$

◐ 27. $3x^2 - 2x + 5 = 10x + 1$

28. $4x^2 - x + 4 = x + 7$

29. $2x^2 + x + 5 = 0$

30. $3x^2 + 2x + 8 = 0$

31. If we apply the quadratic formula and find that the value of $b^2 - 4ac$ is negative, what can we conclude?

32. A student writes the quadratic formula as $x = -b \pm \dfrac{\sqrt{b^2 - 4ac}}{2a}$. Is this correct? If not, explain the error, and give the correct formula.

Use the quadratic formula to solve each equation. See Example 5.

33. $\dfrac{3}{2}k^2 - k - \dfrac{4}{3} = 0$

34. $\dfrac{2}{5}x^2 - \dfrac{3}{5}x - 1 = 0$

35. $\dfrac{1}{2}x^2 + \dfrac{1}{6}x = 1$

36. $\dfrac{2}{3}t^2 - \dfrac{4}{9}t = \dfrac{1}{3}$

37. $\dfrac{3}{8}x^2 - x + \dfrac{17}{24} = 0$

38. $\dfrac{1}{3}x^2 + \dfrac{8}{9}x + \dfrac{7}{9} = 0$

39. $0.6x - 0.4x^2 = -1$

40. $0.5x^2 = x + 0.5$

41. $0.25x^2 = -1.5x - 1$

Solve each problem.

42. An astronaut on the moon throws a baseball upward. The altitude (height) h of the ball, in feet, x seconds after he throws it, is given by the equation

$$h = -2.7x^2 + 30x + 6.5.$$

At what times is the ball 12 ft above the moon's surface? Give answer(s) to the nearest tenth.

43. A frog is sitting on a stump 3 ft above the ground. He hops off the stump and lands on the ground 4 ft away. During his leap, his height h is given by the equation

$$h = -0.5x^2 + 1.25x + 3,$$

where x is the distance in feet from the base of the stump, and h is in feet. How far was the frog from the base of the stump when he was 1.25 ft above the ground?

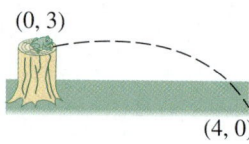

(0, 3)

(4, 0)

44. An old Babylonian problem asks for the length of the side of a square, given that the area of the square minus the length of a side is 870. Find the length of the side. (*Source:* Eves, H., *An Introduction to the History of Mathematics.*)

45. A rule for estimating the number of board feet of lumber that can be cut from a log depends on the diameter of the log. To find the diameter d required to get 9 board feet of lumber, we use the equation

$$\left(\dfrac{d - 4}{4}\right)^2 = 9.$$

Solve this equation for d. Are both answers reasonable?

Summary Exercises on Quadratic Equations

Four algebraic methods have now been introduced for solving quadratic equations written in the form $ax^2 + bx + c = 0$. The following chart shows some advantages and some disadvantages of each method.

Method	Advantages	Disadvantages
1. **Factoring**	It is usually the fastest method.	Not all equations can be solved by factoring. Some factorable polynomials are difficult to factor.
2. **Square root property**	It is the simplest method for solving equations of the form $(ax + b)^2 = $ a number.	Few equations are given in this form.
3. **Completing the square**	It can always be used. (Also, the procedure is useful in other areas of mathematics.)	It requires more steps than other methods.
4. **Quadratic formula**	It can always be used.	It is more difficult than factoring because of the $\sqrt{b^2 - 4ac}$ expression.

Solve each quadratic equation by the method of your choice.

1. $x^2 = 36$

2. $x^2 + 3x = -1$

3. $x^2 - \dfrac{100}{81} = 0$

4. $81t^2 = 49$

5. $z^2 - 4z + 3 = 0$

6. $w^2 + 3w + 2 = 0$

7. $z(z - 9) = -20$

8. $x^2 + 3x - 2 = 0$

9. $(3k - 2)^2 = 9$

10. $(2s - 1)^2 = 10$

11. $(x + 6)^2 = 121$

12. $(5k + 1)^2 = 36$

13. $(3r - 7)^2 = 24$

14. $(7p - 1)^2 = 32$

15. $(5x - 8)^2 = -6$

16. $2t^2 + 1 = t$

17. $-2x^2 = -3x - 2$

18. $-2x^2 + x = -1$

19. $8z^2 = 15 + 2z$

20. $3k^2 = 3 - 8k$

21. $0 = -x^2 + 2x + 1$

22. $3x^2 + 5x = -1$

23. $5x^2 - 22x = -8$

24. $x(x + 6) + 4 = 0$

25. $(x + 2)(x + 1) = 10$

26. $16x^2 + 40x + 25 = 0$

27. $4x^2 = -1 + 5x$

28. $2p^2 = 2p + 1$

29. $3x(3x + 4) = 7$

30. $5x - 1 + 4x^2 = 0$

31. $\dfrac{x^2}{2} + \dfrac{7x}{4} + \dfrac{11}{8} = 0$

32. $t(15t + 58) = -48$

33. $9k^2 = 16(3k + 4)$

34. $\dfrac{1}{5}x^2 + x + 1 = 0$

35. $x^2 - x + 3 = 0$

36. $4x^2 - 11x + 8 = -2$

37. $-3x^2 + 4x = -4$

38. $z^2 - \dfrac{5}{12}z = \dfrac{1}{6}$

39. $5k^2 + 19k = 2k + 12$

40. $\dfrac{1}{2}x^2 - x = \dfrac{15}{2}$

41. $x^2 - \dfrac{4}{15} = -\dfrac{4}{15}x$

42. $(x + 2)(x - 4) = 16$

17.4 ▶▶▶ Graphing Quadratic Equations

OBJECTIVE 1 Graph quadratic equations. In **Chapter 11**, we saw that the graph of a linear equation in two variables is a straight line that represents all the solutions of the equation. In this section, we graph quadratic equations in two variables, of the form

$$y = ax^2 + bx + c.$$

The simplest quadratic equation, $y = x^2$ (or $y = 1x^2 + 0x + 0$), can be graphed in much the same way that straight lines were graphed, by finding ordered pairs that satisfy the equation.

EXAMPLE 1 Graphing a Quadratic Equation

Graph $y = x^2$.

Select several values for x; then find the corresponding y-values. For example, selecting $x = 2$ and substituting in $y = x^2$ gives

$$y = 2^2 = 4,$$

and so the point $(2, 4)$ is on the graph of $y = x^2$. (*Recall that in an ordered pair such as* **(2, 4)**, *the x-value comes first and the y-value second.*)

Work Problem **1** *at the Side.* ▶

If we plot the points from Problem 1 at the side on a coordinate system and draw a smooth curve through them, we obtain the graph in Figure 1. A table of values is shown with the graph.

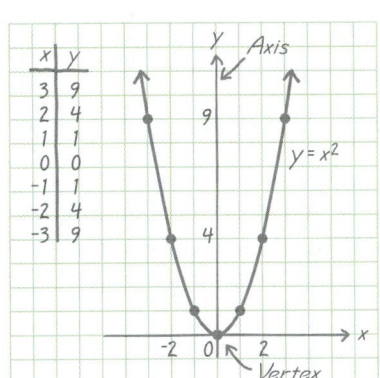

Figure 1

Work Problem **2** *at the Side.* ▶

The curve in Figure 1 is called a **parabola.** The point $(0, 0)$, the *lowest* point on this graph, is called the **vertex** of the parabola. The vertical line through the vertex (the y-axis here) is called the **axis** of the parabola. The axis of a parabola is a **line of symmetry** for the graph, because if the graph is folded on this line, the two halves will coincide.

Every equation of the form $y = ax^2 + bx + c$, with $a \neq 0$, has a graph that is a parabola. Because of its many useful properties, the parabola occurs frequently in real-life applications. For example, the cross sections of radar, spotlight, and telescope reflectors form parabolas.

OBJECTIVES

1 Graph quadratic equations.

2 Find the vertex of a parabola.

---

1 Complete the table of values for $y = x^2$.

x	y
3	
2	4
1	
0	
−1	
−2	
−3	

2 Graph $y = \frac{1}{2}x^2$ by first completing the table of values.

x	y
−2	
−1	
0	
1	
2	

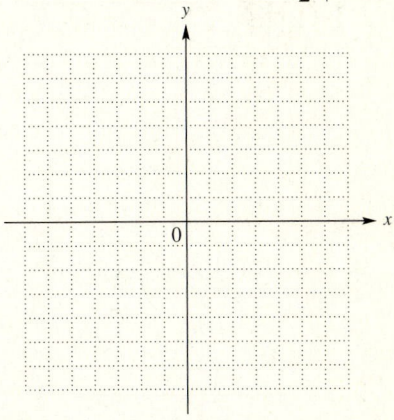

ANSWERS

1. See the table beside Figure 1.

2.

x	y
−2	2
−1	$\frac{1}{2}$
0	0
1	$\frac{1}{2}$
2	2

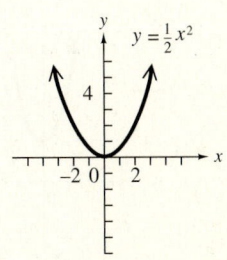

3 Complete each ordered pair for $y = -x^2 + 3$.

$(-2, \_\_)$, $(-1, \_\_)$,

$(1, \_\_)$, $(2, \_\_)$

4 Graph each equation, and identify each vertex.

(a) $y = -x^2 - 3$

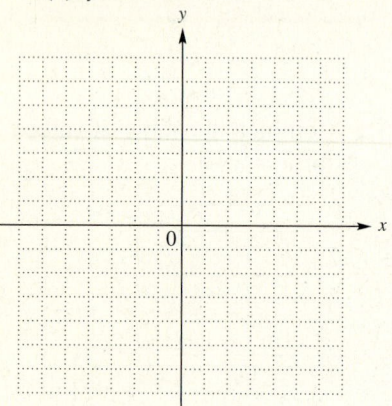

(b) $y = x^2 + 3$

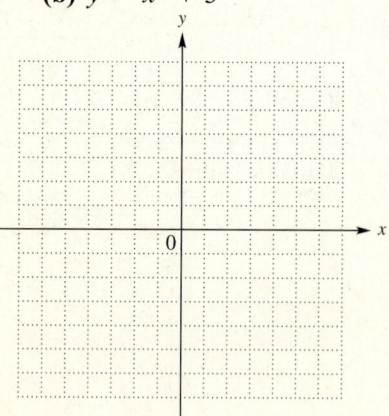

ANSWERS

3. $(-2, -1), (-1, 2), (1, 2), (2, -1)$

4. **(a)**

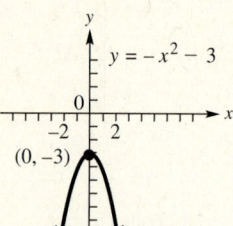

(b)

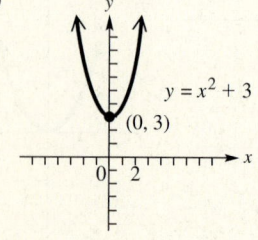

EXAMPLE 2 **Graphing a Parabola by Plotting Points**

Graph $y = -x^2 + 3$.

Find several ordered pairs. Let $x = 0$ to find the y-intercept:

$$y = -x^2 + 3 = -0^2 + 3 = 3.$$

This gives the ordered pair $(0, 3)$.

◀ *Work Problem* **3** *at the Side.*

The ordered pair $(0, 3)$ and the ordered pairs from Problem 3 at the side are listed in the table shown with Figure 2. Plot these points and connect them with a smooth curve as shown in Figure 2. The vertex of this parabola is $(0, 3)$. The graph opens downward because x^2 has a negative coefficient, so the vertex is the *highest* point of the graph.

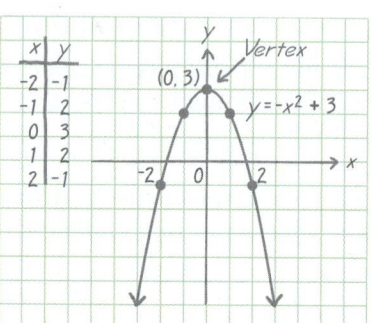

Figure 2

◀ *Work Problem* **4** *at the Side.*

OBJECTIVE 2 Find the vertex of a parabola. The vertex is the most important point to locate when graphing a quadratic equation.

EXAMPLE 3 **Finding the Vertex to Graph a Parabola**

Graph $y = x^2 - 2x - 3$.

We want to find the vertex of the graph. ***Because of its symmetry, if a parabola has two x-intercepts, the x-value of the vertex is exactly halfway between them.*** Therefore, we begin by finding the x-intercepts. We let $y = 0$ in the equation, and solve for x.

$$0 = x^2 - 2x - 3$$

$$0 = (x + 1)(x - 3) \qquad \text{Factor.}$$

$$x + 1 = 0 \quad \text{or} \quad x - 3 = 0 \qquad \text{Zero-factor property}$$

$$x = -1 \quad \text{or} \qquad x = 3$$

There are two x-intercepts, $(-1, 0)$ and $(3, 0)$. Since the x-value of the vertex is halfway between the x-values of the two x-intercepts, it is half their sum.

$$x = \frac{1}{2}(-1 + 3) = \frac{1}{2}(2) = 1 \longleftarrow x\text{-value of the vertex}$$

We find the corresponding y-value by substituting 1 for x in the equation.

$$y = 1^2 - 2(1) - 3 = -4 \longleftarrow y\text{-value of the vertex}$$

The vertex is $(1, -4)$. The axis is the line $x = 1$.

Continued on Next Page

To find the *y*-intercept, substitute 0 for *x* in the equation.

$$y = 0^2 - 2(0) - 3 = -3 \qquad \text{Let } x = 0.$$

The *y*-intercept is $(0, -3)$.

We plot the three intercepts and the vertex, and find additional ordered pairs as needed. For example, if $x = 2$, then

$$y = 2^2 - 2(2) - 3 = -3,$$

leading to the ordered pair $(2, -3)$. A table that includes the ordered pairs we found is shown with the graph in Figure 3.

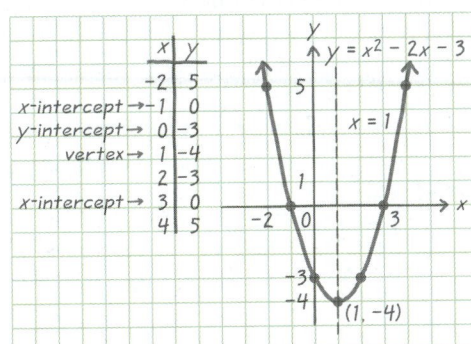

Figure 3

Work Problem **5** *at the Side.* ▶

We can generalize from Example 3. The *x*-coordinates of the *x*-intercepts for $y = ax^2 + bx + c$, by the quadratic formula, are

$$x = \frac{-b + \sqrt{b^2 - 4ac}}{2a} \quad \text{and} \quad x = \frac{-b - \sqrt{b^2 - 4ac}}{2a}.$$

Thus, we can find the *x*-value of the vertex as shown below.

$$x = \frac{1}{2}\left(\frac{-b + \sqrt{b^2 - 4ac}}{2a} + \frac{-b - \sqrt{b^2 - 4ac}}{2a}\right)$$

$$x = \frac{1}{2}\left(\frac{-b + \sqrt{b^2 - 4ac} - b - \sqrt{b^2 - 4ac}}{2a}\right)$$

$$x = \frac{1}{2}\left(\frac{-2b}{2a}\right) \qquad \text{Combine like terms.}$$

$$x = -\frac{b}{2a} \qquad \text{Multiply; lowest terms}$$

For the equation in Example 3, $y = x^2 - 2x - 3$, $a = 1$, and $b = -2$. Thus, the *x*-value of the vertex is

$$x = -\frac{b}{2a} = -\frac{-2}{2(1)} = 1,$$

which is the same *x*-value for the vertex we found in Example 3. (The *x*-value of the vertex is $x = -\frac{b}{2a}$, even if the graph has no *x*-intercepts.)

5 Graph $y = x^2 + 2x - 8$.

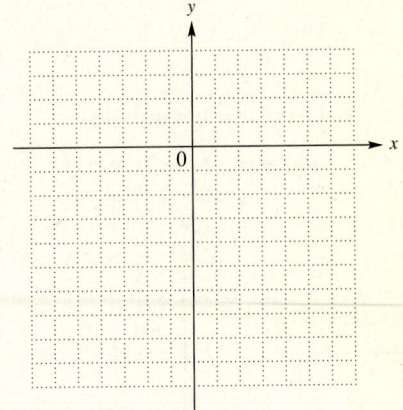

ANSWER

5.

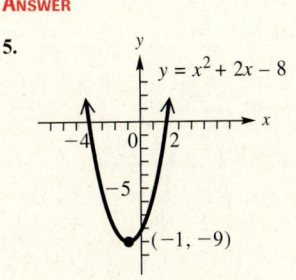

6 Complete each ordered pair for $y = x^2 - 4x + 1$.

$(5,\underline{})$, $(4,\underline{})$, $(-1,\underline{})$

7 Graph each parabola.

(a) $y = x^2 - 3x - 3$

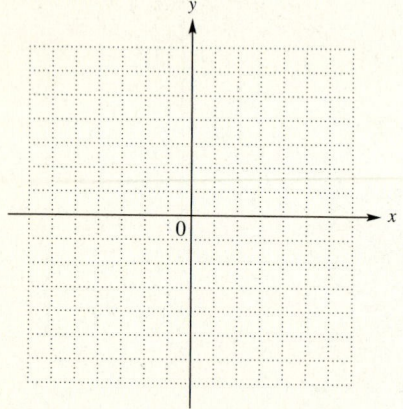

(b) $y = -x^2 + 2x + 4$

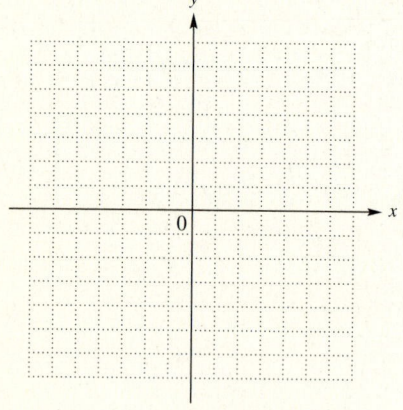

ANSWERS

6. $(5, 6), (4, 1), (-1, 6)$

7. (a)

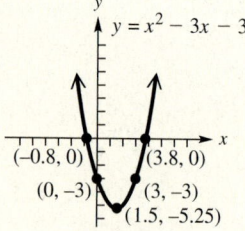

(b)

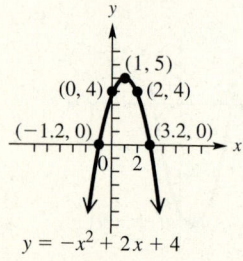

EXAMPLE 4 **Graphing a Parabola**

Graph $y = x^2 - 4x + 1$.

Find the vertex. Here, $a = 1$ and $b = -4$, so the x-value of the vertex is

$$x = -\frac{b}{2a} = -\frac{-4}{2(1)} = 2.$$

The y-value is

$$y = 2^2 - 4(2) + 1 = -3,$$

so the vertex is $(2, -3)$. The axis is the line $x = 2$.

Now find the intercepts. Let $x = 0$ in $y = x^2 - 4x + 1$.

$$y = 0^2 - 4(0) + 1 = 1$$

The y-intercept is $(0, 1)$. Let $y = 0$ to get the x-intercepts. If $y = 0$, the equation becomes $0 = x^2 - 4x + 1$, which cannot be solved by factoring. Use the quadratic formula to solve for x.

$$x = \frac{-(-4) \pm \sqrt{(-4)^2 - 4(1)(1)}}{2(1)}$$ Let $a = 1, b = -4, c = 1$ in the quadratic formula.

$$x = \frac{4 \pm \sqrt{12}}{2}$$ Simplify.

$$x = \frac{4 \pm 2\sqrt{3}}{2}$$ $\sqrt{12} = \sqrt{4} \cdot \sqrt{3} = 2\sqrt{3}$

$$x = \frac{2\left(2 \pm \sqrt{3}\right)}{2}$$ Factor.

$$x = 2 \pm \sqrt{3}$$ Divide out 2.

Use a calculator to find that the x-intercepts are $(3.7, 0)$ and $(0.3, 0)$, to the nearest tenth.

◀ *Work Problem* **6** *at the Side.*

Plot the intercepts, vertex, and the points found in Problem 6. Connect these points with a smooth curve. The graph is shown in Figure 4.

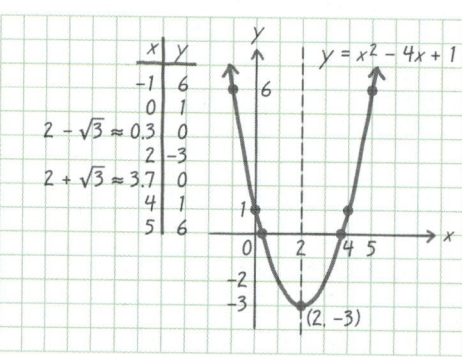

Figure 4

◀ *Work Problem* **7** *at the Side.*

17.4 ▶▶▶ **Exercises**

1. In your own words, explain what is meant by the vertex of a parabola.

2. In your own words, explain what is meant by the line of symmetry of a parabola that opens upward or downward.

Graph each equation. Give the coordinates of the vertex in each case. See Examples 1–4.

3. $y = 2x^2$

4. $y = 3x^2$

5. $y = x^2 - 4$

  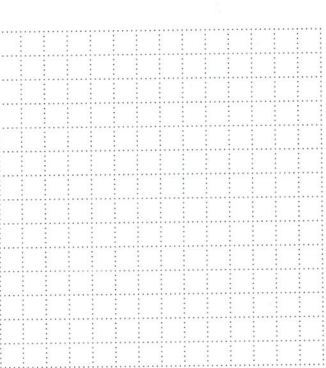

6. $y = x^2 - 6$

7. $y = -x^2 + 2$

8. $y = -x^2 + 4$

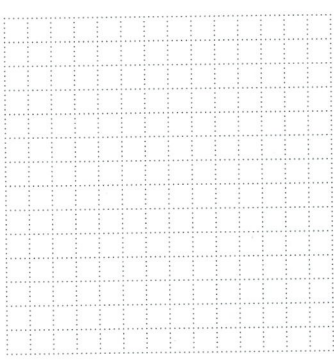 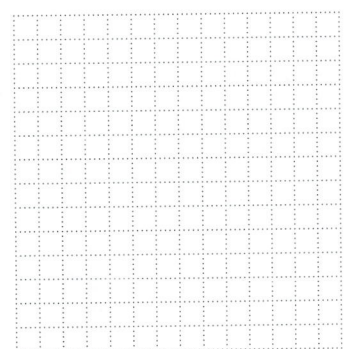 

9. $y = (x + 3)^2$

10. $y = (x - 4)^2$

11. $y = x^2 + 2x + 3$

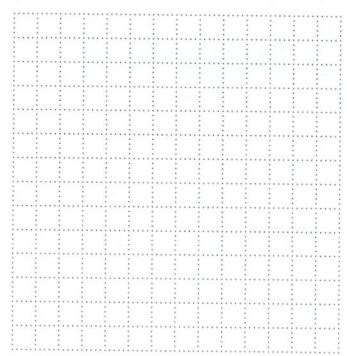

12. $y = x^2 - 4x + 3$

13. $y = -x^2 + 6x - 5$

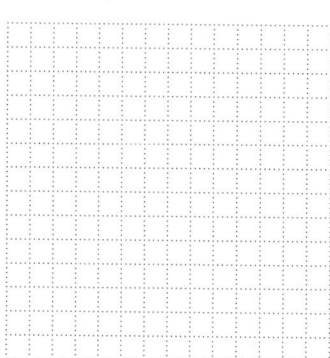

14. $y = -x^2 - 4x - 3$

15. Based on your work in Exercises 3–14, what seems to be the direction in which the parabola $y = ax^2 + bx + c$ opens if $a > 0$? if $a < 0$?

16. See Exercises 10–12. How many real solutions does a quadratic equation have if its corresponding graph has

(a) no x-intercepts

(b) one x-intercept

(c) two x-intercepts?

Solve each problem.

17. The U.S. Naval Research Laboratory designed a giant radio telescope that had a diameter of 300 ft and maximum depth of 44 ft. The graph on the right below describes a cross section of this telescope. Find the equation of this parabola. (*Source: Structure Technology for Large Radio and Radar Telescope Systems*, The MIT Press.)

18. Suppose the telescope in Exercise 17 had a diameter of 400 ft and maximum depth of 50 ft. Find the equation of this parabola.

17.5 ▷▷▷ Introduction to Functions

If gasoline costs $5.00 per gal and you buy **1** gal, then you must pay $5.00(**1**) = $5.00. If you buy **2** gal, your cost is $5.00(**2**) = $10.00; for **3** gal, your cost is $5.00(**3**) = $15.00, and so on. Generalizing, if *x* represents the number of gallons, then the cost is $5.00*x*. If we let *y* represent the cost, then the equation

$$y = 5.00x$$

relates the number of gallons, *x*, to the cost in dollars, *y*. The set of ordered pairs (*x, y*) that satisfy this equation forms a *relation*.

OBJECTIVE 1 Understand the definition of a relation.   In an ordered pair (*x, y*), *x* and *y* are called the **components** of the ordered pair. Any set of ordered pairs is called a **relation**. The set of all first components in the ordered pairs of a relation is the **domain** of the relation, and the set of all second components in the ordered pairs is the **range** of the relation.

EXAMPLE 1 Using Ordered Pairs to Define Relations

(a) {(0, 1), (2, 5), (3, 8), (4, 2)}
This relation has domain {0, 2, 3, 4} and range {1, 2, 5, 8}. The correspondence between the elements of the domain and the elements of the range is shown in Figure 5.

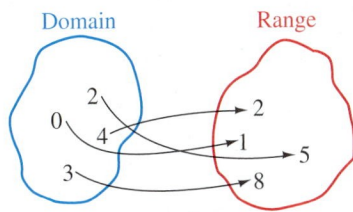

Figure 5

(b) {(**3, 5**), (**3, 6**), (**3, 7**), (**3, 8**)}
The relation has domain {**3**} and range {**5, 6, 7, 8**}.

Work Problem **1** *at the Side.* ▶

OBJECTIVE 2 Understand the definition of a function.   We now investigate an important type of relation, called a *function*.

> **Function**
>
> A **function** is a set of ordered pairs in which each distinct first component corresponds to exactly one second component.

By this definition, the relation in Example 1(a) is a function. The relation in Example 1(b) is *not* a function, because the first component, 3, corresponds to more than one second component. Notice, however, that if the components of the ordered pairs in Example 1(b) were interchanged, giving the relation

$$\{(5, 3), (6, 3), (7, 3), (8, 3)\},$$

then the relation *would* be a function. In that case, each domain element (first component) corresponds to *exactly one* range element (second component).

OBJECTIVES

1 Understand the definition of a relation.

2 Understand the definition of a function.

3 Decide whether an equation defines a function.

4 Use function notation.

5 Apply the function concept in an application.

1 Give the domain and the range of each relation.

(a) {(5, 10), (15, 20), (25, 30), (35, 40)}

(b) {(1, 4), (2, 4), (3, 4)}

ANSWERS

1. **(a)** domain: {5, 15, 25, 35};
    range: {10, 20, 30, 40}
  (b) domain: {1, 2, 3};
    range: {4}

2 Decide whether each relation is a function.

(a) $\{(-2, 8), (-1, 1), (0, 0), (1, 1), (2, 8)\}$

(b) $\{(5, 2), (5, 1), (5, 0)\}$

EXAMPLE 2 **Determining Whether Relations Are Functions**

Determine whether each relation is a function.

(a) $\{(-2, 4), (-1, 1), (0, 0), (1, 1), (2, 4)\}$

Notice that each first component appears once and only once. Because of this, the relation is a function.

(b) $\{(\mathbf{9}, 3), (\mathbf{9}, -3), (4, 2)\}$

The first component 9 appears in two ordered pairs, and corresponds to two different second components. Therefore, this relation is not a function.

◀ *Work Problem* **2** *at the Side.*

The simple relations given in Examples 1 and 2 were defined by listing the ordered pairs or by showing the correspondence with a figure. Most useful functions have an infinite number of ordered pairs and are usually defined with equations that tell how to get the second components, given the first. We have been using equations with x and y as the variables, where x represents the first component (input) and y the second component (output) in the ordered pairs.

Here are some everyday examples of functions.

1. The **cost y** in dollars charged by an express mail company is a function of the **weight x** in pounds determined by the equation $y = 1.5(x - 1) + 9$.

2. In one state, the sales tax is 6% of the price of an item. The **tax y** on a particular item is a function of the **price x**, because $y = 0.06x$.

3. The **distance d** traveled by a car moving at a constant speed of 45 mph is a function of the **time t**. Thus, $d = 45t$.

The function concept can be illustrated by an input-output "machine," as seen in Figure 6. It shows how the express mail company equation $y = 1.5(x - 1) + 9$ provides an output (the cost, represented by y) for a given input (the weight in pounds, given by x).

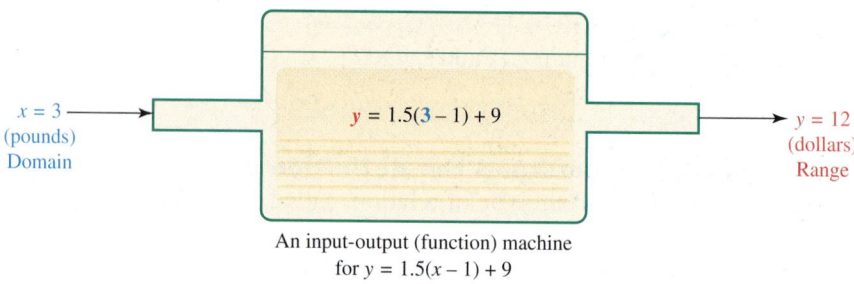

$x = 3$
(pounds)
Domain

$y = 1.5(3 - 1) + 9$

$y = 12$
(dollars)
Range

An input-output (function) machine
for $y = 1.5(x - 1) + 9$

Figure 6

OBJECTIVE 3 **Decide whether an equation defines a function.** Given the graph of an equation, the definition of a function can be used to decide whether or not the graph represents a function. By the definition of a function, each x-value must lead to exactly one y-value. In Figure 7(a) on the next page, the indicated x-value leads to two y-values, so this graph is not the graph of a function. A vertical line can be drawn that intersects this graph in more than one point.

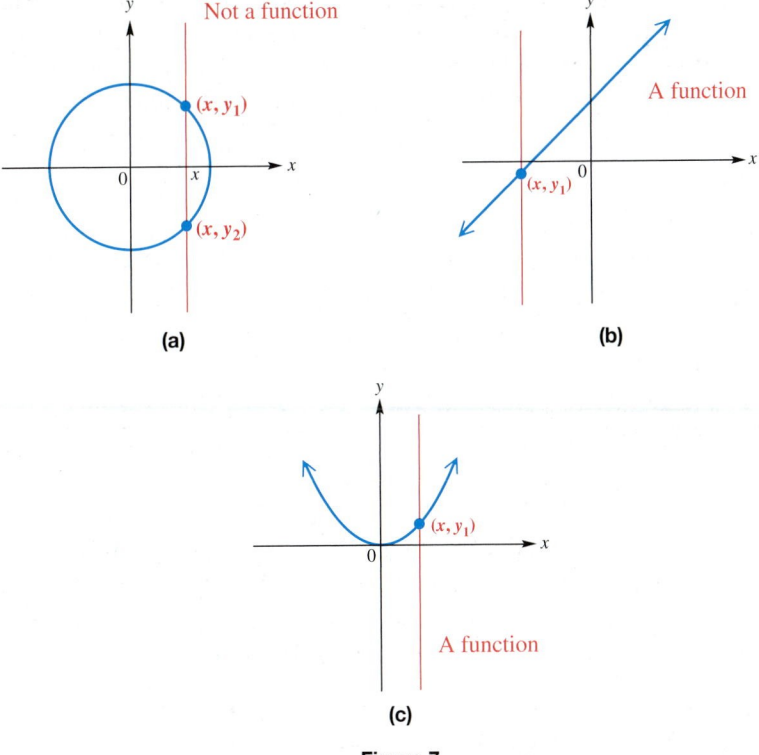

Figure 7

On the other hand, in Figures 7(b) and 7(c) any vertical line will intersect each graph in no more than one point. Because of this, the graphs in Figures 7(b) and 7(c) are graphs of functions. This idea leads to the **vertical line test** for a function.

> **Vertical Line Test**
>
> If a vertical line intersects a graph in more than one point, the graph is not the graph of a function.

As Figure 7(b) suggests, any nonvertical line is the graph of a function. For this reason, *any linear equation of the form $y = mx + b$ defines a function.* (Recall that a vertical line has undefined slope.) Also, any vertical parabola, as in Figure 7(c), is the graph of a function, so *any quadratic equation of the form $y = ax^2 + bx + c$ ($a \neq 0$) defines a function.*

EXAMPLE 3 **Deciding Whether Relations Define Functions**

Decide whether each relation graphed or defined is a function.

(a)

Because there are two ordered pairs with first component -4, as shown in red, this is not the graph of a function.

Continued on Next Page

3 Decide whether each relation graphed or defined is a function.

(a)

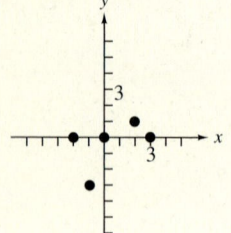

(b)

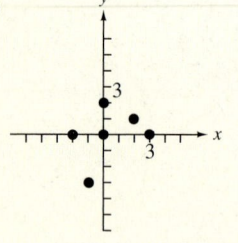

(c)

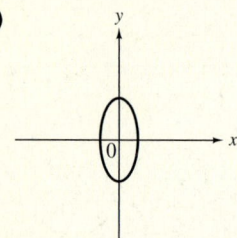

(d)

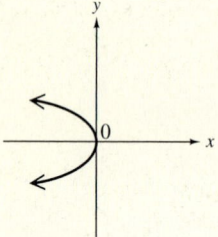

(e) $y = 3$

(b)

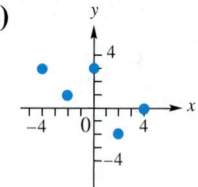

Every first component is paired with one and only one second component, and as a result, no vertical line intersects the graph in more than one point. Therefore, this is the graph of a function.

(c) $y = 2x - 9$

This linear equation is in the form $y = mx + b$. Since the graph of this equation is a line that is not vertical, the equation defines a function.

(d)

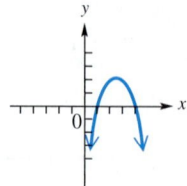

(e)

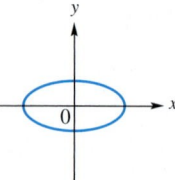

Use the vertical line test. Any vertical line intersects the graph of a vertical parabola just once, so this is the graph of a function.

The vertical line test shows that this graph is not the graph of a function; a vertical line could intersect the graph twice.

(f) $x = 4$

The graph of $x = 4$ is a vertical line, so the equation does *not* define a function. (Every ordered pair has x-value 4.)

◀ *Work Problem* **3** *at the Side.*

OBJECTIVE 4 Use function notation. The letters f, g, and h are commonly used to name functions. For example, the function with defining equation $y = 3x + 5$ may be written

$$f(x) = 3x + 5,$$

where $f(x)$ is read "f of x." The notation $f(x)$ is another way of writing y in a function. For the function defined by $f(x) = 3x + 5$, if $x = 7$ then

$$f(7) = 3 \cdot 7 + 5 \qquad \text{Let } x = 7.$$
$$f(7) = 21 + 5 \qquad \text{Multiply.}$$
$$f(7) = 26. \qquad \text{Add.}$$

Read this result, $f(7) = 26$, as "f of 7 equals 26." The notation $f(7)$ means the value of y when x is 7. The statement $f(7) = 26$ says that the value of y is 26 when x is 7. It also indicates that the point $(7, 26)$ lies on the graph of f.

To find $f(-3)$ for $f(x) = 3x + 5$, substitute -3 for x.

Use parentheses to avoid errors.

$$f(-3) = 3(-3) + 5 \qquad \text{Let } x = -3.$$
$$f(-3) = -9 + 5$$
$$f(-3) = -4$$

CAUTION

The notation $f(x)$ does *not* mean f times x. **The symbol $f(x)$ means the value of the function f when evaluated for x. It represents the y-value that corresponds to x.**

ANSWERS

3. **(a)** function **(b)** not a function
(c) not a function **(d)** not a function
(e) function

> **Function Notation**
> In the notation $f(x)$,
>
> $\qquad f \qquad$ is the name of the function,
>
> $\qquad x \qquad$ is the domain value,
>
> and $\qquad f(x) \qquad$ is the range value y for the domain value x.

EXAMPLE 4 **Using Function Notation**

For the function defined by $f(x) = x^2 - 3$, find the following.

(a) $f(4)$

Substitute 4 for x.

$$f(x) = x^2 - 3$$
$$f(4) = 4^2 - 3 \qquad \text{Let } x = 4.$$

Think: $4^2 = 4 \cdot 4$

$$= 16 - 3 \qquad \text{Apply the exponent.}$$
$$f(4) = 13 \qquad \text{Subtract.}$$

(b) $f(0) = 0^2 - 3$
$f(0) = 0 - 3$
$f(0) = -3$

(c) $f(-3) = (-3)^2 - 3$
$f(-3) = 9 - 3$
$f(-3) = 6$

Work Problem **4** *at the Side.* ▶

OBJECTIVE 5 Apply the function concept in an application.
Because a function assigns to each element in its domain exactly one element in its range, the function concept is used in real-data applications where two quantities are related. Our final example discusses such an application using a table of values.

EXAMPLE 5 **Applying the Function Concept to Population**

Asian-American populations (in millions) are shown in the table.

ASIAN-AMERICAN POPULATION

Year	Population (in millions)
1996	9.7
2000	11.2
2004	13.1
2006	14.9

Source: U.S. Census Bureau.

(a) Use the table to write a set of ordered pairs that defines a function f.

If we choose the years as the domain elements and the populations as the range elements, the information in the table can be written as a set of four ordered pairs. In set notation, the function f is shown below.

$$f = \{(1996, 9.7), (2000, 11.2), (2004, 13.1), (2006, 14.9)\}$$

Continued on Next Page

---

4 For $f(x) = 6x - 2$, find each function value.

(a) $f(-1)$

(b) $f(0)$

(c) $f(1)$

5 The numbers of U.S. children (in millions) educated at home for selected years are given in the table.

School Year	Number of Children
1997	1.1
1999	1.3
2001	1.7
2003	2.2

Source: National Home Education Research Institute, Salem, OR.

(a) Write a set of ordered pairs that defines a function f for the data.

(b) Give the domain and range of f.

(c) Find $f(1999)$.

(d) For what x-value does $f(x)$ equal 1.7 million?

(b) What is the domain of f? What is the range?

We repeat the table and the set of ordered pairs from part (a).

ASIAN-AMERICAN POPULATION

Year	Population (in millions)
1996	9.7
2000	11.2
2004	13.1
2006	14.9

Source: U.S. Census Bureau.

$$f = \{(1996, 9.7), (2000, 11.2), (2004, 13.1), (2006, 14.9)\}$$

The domain is the set of years, or x-values.

$$\{1996, 2000, 2004, 2006\}$$

The range is the set of populations, in millions, or y-values.

$$\{9.7, 11.2, 13.1, 14.9\}$$

(c) Find $f(1996)$ and $f(2004)$.

We refer to the table or the ordered pairs repeated in part (b) to find that

$$f(\mathbf{1996}) = 9.7 \text{ million} \quad \text{and} \quad f(\mathbf{2004}) = 13.1 \text{ million.}$$

(d) For what x-value does $f(x)$ equal 14.9 million? 11.2 million?

We use the table or the ordered pairs found in part (b) to determine

$$f(\mathbf{2006}) = 14.9 \text{ million} \quad \text{and} \quad f(\mathbf{2000}) = 11.2 \text{ million.}$$

◀ *Work Problem* **5** *at the Side.*

ANSWERS

5. **(a)** $f = \{(1997, 1.1), (1999, 1.3),$
$(2001, 1.7), (2003, 2.2)\}$
   (b) domain: $\{1997, 1999, 2001, 2003\}$;
   range: $\{1.1, 1.3, 1.7, 2.2\}$
   (c) 1.3 million
   (d) 2001

17.5 ▶▶▶ **Exercises**

FOR
EXTRA
HELP

  MyMathLab

 Math XL
PRACTICE

WATCH

DOWNLOAD

READ

REVIEW

Complete the following table for the function defined by $f(x) = x + 2$.

x	x + 2	f(x)	(x, y)
0	2	2	(0, 2)
1. 1			
2. 2			
3. 3			
4. 4			

5. Describe the graph of function f in Exercises 1–4 if the domain is $\{0, 1, 2, 3, 4\}$.

6. Describe the graph of function f in Exercises 1–4 if the domain is the set of all real numbers.

Determine whether each relation is a function. Give the domain and the range in Exercises 7–12. See Examples 1–3.

7. $\{(-4, 3), (-2, 1), (0, 5), (-2, -8)\}$

8. $\{(3, 7), (1, 4), (0, -2), (-1, -1), (-2, 5)\}$

9.

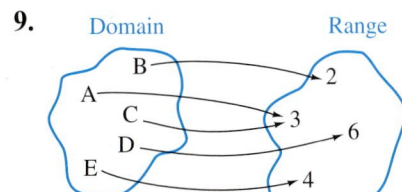

10.

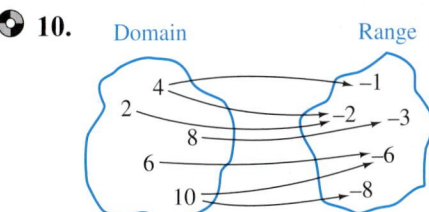

11.

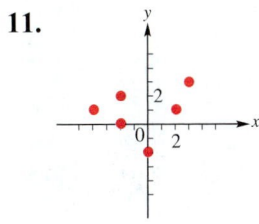

12.

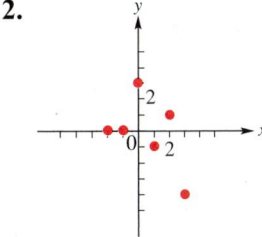

13.

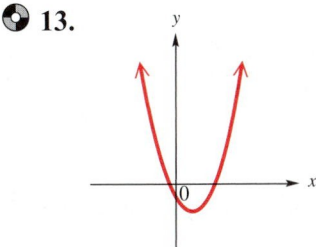

14.

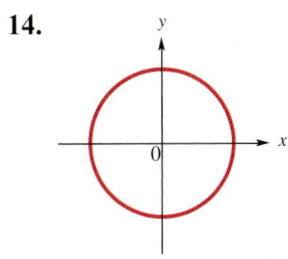

15.

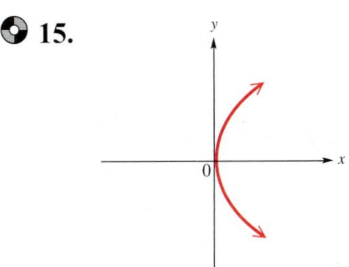

16.

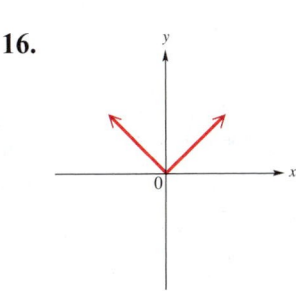

Decide whether each equation defines y as a function of x. (Remember that to be a function, every value of x must give one and only one value of y.) See Example 3.

17. $y = 5x + 3$ **18.** $y = -7x + 12$ **19.** $x = |y|$ **20.** $x = y^2$

Relating Concepts (Exercises 21–24) For Individual or Group Work

*A function defined by $f(x) = 3x - 4$, called a **linear function** because its graph is a straight line, can be graphed by replacing $f(x)$ with y and then using the methods described earlier. Let us assume that some function is written in the form $f(x) = mx + b$, for particular values of m and b. **Work Exercises 21–24 in order.***

21. Since $f(2) = 4$, name the coordinates of one point on the line.

22. Since $f(-1) = -4$, name the coordinates of another point on the line.

23. Use the results of Exercises 21 and 22 to find the slope of the line.

24. Use the slope-intercept form of the equation of a line to write the function in the form $f(x) = mx + b$.

---

*For each function f, find **(a)** $f(2)$, **(b)** $f(0)$, and **(c)** $f(-3)$. See Example 4.*

25. $f(x) = 4x + 3$ **26.** $f(x) = -3x + 5$ **27.** $f(x) = x^2 - x + 2$

28. $f(x) = x^3 + x$ **29.** $f(x) = |x|$ **30.** $f(x) = |x + 7|$

The number of U.S. foreign-born residents has grown by more than 43% since 1990. The graph shows the number of such residents (in millions) for selected years. Use the information in the graph for Exercises 31–35. See Example 5.

31. Write the information in the graph as a set of ordered pairs. Does this set define a function?

32. Suppose that g is the name given to this relation. Give the domain and range of g.

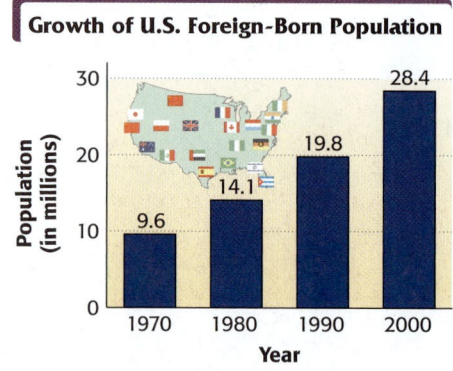

Growth of U.S. Foreign-Born Population

Source: U.S. Census Bureau.

33. Find $g(1980)$ and $g(1990)$.

34. For what value of x does $g(x) = 28.4$ (million)?

35. Suppose $g(2002) = 30.3$ (million). What does this tell you in the context of the application?

Chapter 17 ▶▶▶ Summary

▶ Key Terms

17.4 **parabola** The graph of the quadratic equation $y = ax^2 + bx + c$ is called a parabola.

vertex The vertex of a parabola that opens upward or downward is the lowest or highest point on the graph.

axis The axis of a parabola that opens upward or downward is a vertical line through the vertex.

line of symmetry If a graph is folded on its line of symmetry, the two sides coincide.

17.5 **components** In an ordered pair (x, y), x and y are the components.

relation Any set of ordered pairs is called a relation.

domain The set of all first components in the ordered pairs of a relation is the domain of the relation.

range The set of all second components in the ordered pairs of a relation is the range of the relation.

function A function is a set of ordered pairs in which each first component corresponds to exactly one second component.

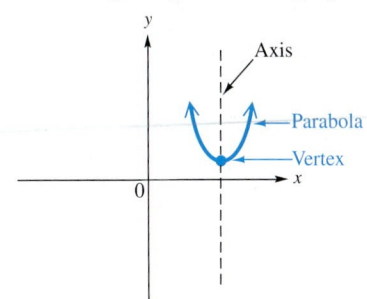

▶ New Symbols

$\pm$ positive or negative (plus or minus)

$f(x)$ function f of x

▶ Test Your Word Power

See how well you have learned the vocabulary in this chapter. Answers, with examples, follow the Quick Review.

1. A **parabola** is the graph of
  A. any equation in two variables
  B. a linear equation
  C. an equation of degree three
  D. a quadratic equation in two variables.

2. The **vertex** of a parabola is
  A. the point where the graph intersects the y-axis
  B. the point where the graph intersects the x-axis
  C. the lowest point on a parabola that opens up or the highest point on a parabola that opens down
  D. the origin.

3. A **relation** is
  A. any set of ordered pairs
  B. a set of ordered pairs in which each distinct first component corresponds to exactly one second component

C. two sets of ordered pairs that are related
  D. a graph of ordered pairs.

4. The **domain** of a relation is
  A. the set of all x- and y-values in the ordered pairs of the relation
  B. the difference between the components in an ordered pair of the relation
  C. the set of all first components in the ordered pairs of the relation
  D. the set of all second components in the ordered pairs of the relation.

5. The **range** of a relation is
  A. the set of all x- and y-values in the ordered pairs of the relation
  B. the difference between the components in an ordered pair of the relation

C. the set of all first components in the ordered pairs of the relation
  D. the set of all second components in the ordered pairs of the relation.

6. A **function** is
  A. any set of ordered pairs
  B. a set of ordered pairs in which each distinct first component corresponds to exactly one second component
  C. two sets of ordered pairs that are related
  D. a graph of ordered pairs.

▶ **Quick Review**

Concepts	Examples

17.1 Solving Quadratic Equations by the Square Root Property

Square Root Property of Equations
If k is positive, and if $x^2 = k$, then
$$x = \sqrt{k} \quad \text{or} \quad x = -\sqrt{k}.$$

The solution set $\{-\sqrt{k}, \sqrt{k}\}$ can also be written $\{\pm\sqrt{k}\}$.

Solve $(2x + 1)^2 = 5$.

$$2x + 1 = \sqrt{5} \qquad \text{or} \quad 2x + 1 = -\sqrt{5}$$
$$2x = -1 + \sqrt{5} \quad \text{or} \qquad 2x = -1 - \sqrt{5}$$
$$x = \frac{-1 + \sqrt{5}}{2} \quad \text{or} \qquad x = \frac{-1 - \sqrt{5}}{2}$$

Solution set: $\left\{\dfrac{-1 + \sqrt{5}}{2}, \dfrac{-1 - \sqrt{5}}{2}\right\}$

17.2 Solving Quadratic Equations by Completing the Square

Solving a Quadratic Equation by Completing the Square

Step 1 If the coefficient of the second-degree term is 1, go to Step 2. If it is not 1, divide each side of the equation by this coefficient.

Step 2 Make sure that all variable terms are on one side of the equation and all constant terms are on the other.

Step 3 Take half the coefficient of x, square it, and add the square to each side of the equation. Factor the variable side, and combine terms on the other side.

Step 4 Use the square root property to solve the equation.

Solve $2x^2 + 4x - 1 = 0$.

$$x^2 + 2x - \frac{1}{2} = 0 \qquad \text{Divide by 2.}$$

$$x^2 + 2x = \frac{1}{2} \qquad \text{Add } \tfrac{1}{2}.$$

$$x^2 + 2x + 1 = \frac{1}{2} + 1 \qquad \left[\tfrac{1}{2}(2)\right]^2 = 1^2 = 1$$

$$(x + 1)^2 = \frac{3}{2} \qquad \text{Factor; add.}$$

$$x + 1 = \sqrt{\frac{3}{2}} \qquad \text{or} \quad x + 1 = -\sqrt{\frac{3}{2}}$$

$$x + 1 = \frac{\sqrt{3} \cdot \sqrt{2}}{\sqrt{2} \cdot \sqrt{2}} \quad \text{or} \quad x + 1 = -\frac{\sqrt{3} \cdot \sqrt{2}}{\sqrt{2} \cdot \sqrt{2}}$$

$$x + 1 = \frac{\sqrt{6}}{2} \qquad \text{or} \quad x + 1 = -\frac{\sqrt{6}}{2}$$

$$x = -1 + \frac{\sqrt{6}}{2} \quad \text{or} \qquad x = -1 - \frac{\sqrt{6}}{2}$$

$$x = \frac{-2}{2} + \frac{\sqrt{6}}{2} \quad \text{or} \qquad x = \frac{-2}{2} - \frac{\sqrt{6}}{2}$$

$$x = \frac{-2 + \sqrt{6}}{2} \quad \text{or} \qquad x = \frac{-2 - \sqrt{6}}{2}$$

Solution set: $\left\{\dfrac{-2 + \sqrt{6}}{2}, \dfrac{-2 - \sqrt{6}}{2}\right\}$

Concepts	Examples

17.3 Solving Quadratic Equations by the Quadratic Formula

Quadratic Formula

The solutions of $ax^2 + bx + c = 0 \, (a \neq 0)$ are

$$x = \frac{-b \pm \sqrt{b^2 - 4ac}}{2a}.$$

Solve $3x^2 - 4x - 2 = 0$.

$$x = \frac{-(-4) \pm \sqrt{(-4)^2 - 4(3)(-2)}}{2(3)}$$

$$x = \frac{4 \pm \sqrt{40}}{6} \qquad \text{Simplify.}$$

$$x = \frac{4 \pm 2\sqrt{10}}{6} \qquad \sqrt{40} = \sqrt{4} \cdot \sqrt{10} = 2\sqrt{10}$$

$$x = \frac{2(2 \pm \sqrt{10})}{2(3)} \qquad \text{Factor.}$$

$$x = \frac{2 \pm \sqrt{10}}{3} \qquad \text{Divide out 2.}$$

Solution set: $\left\{ \dfrac{2 + \sqrt{10}}{3}, \dfrac{2 - \sqrt{10}}{3} \right\}$

17.4 Graphing Quadratic Equations

To graph $y = ax^2 + bx + c$:

Step 1 Find the vertex. Use $x = -\frac{b}{2a}$ and find y by substituting this value for x in the equation.

Graph $y = 2x^2 - 5x - 3$.

$$x = -\frac{b}{2a} = -\frac{-5}{2(2)} = \frac{5}{4} \qquad a = 2, b = -5$$

$$y = 2\left(\frac{5}{4}\right)^2 - 5\left(\frac{5}{4}\right) - 3$$

$$y = 2\left(\frac{25}{16}\right) - \frac{25}{4} - 3$$

$$y = \frac{25}{8} - \frac{50}{8} - \frac{24}{8}$$

$$y = -\frac{49}{8}$$

The vertex is $\left(\frac{5}{4}, -\frac{49}{8}\right)$.

Step 2 Let $x = 0$ to find the y-intercept.

$$y = 2(0)^2 - 5(0) - 3 = -3$$

The y-intercept is $(0, -3)$.

Step 3 Let $y = 0$ to find the x-intercepts (if they exist).

$$0 = 2x^2 - 5x - 3 \qquad \text{Let } y = 0.$$

$$0 = (2x + 1)(x - 3)$$

$$2x + 1 = 0 \qquad \text{or} \quad x - 3 = 0$$

$$2x = -1 \qquad \text{or} \qquad x = 3$$

$$x = -\frac{1}{2}$$

The x-intercepts are $\left(-\frac{1}{2}, 0\right)$ and $(3, 0)$.

(continued)

Concepts	Examples

17.4 Graphing Quadratic Equations (continued)

Step 4 Plot the intercepts and the vertex.

Step 5 Find and plot additional ordered pairs near the vertex and intercepts as needed.

x	y
$-\frac{1}{2}$	0
0	-3
1	-6
$\frac{5}{4}$	$-\frac{49}{8}$
2	-5
3	0

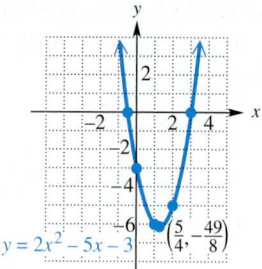

$y = 2x^2 - 5x - 3$ $\left(\frac{5}{4}, -\frac{49}{8}\right)$

17.5 Introduction to Functions

Vertical Line Test

If a vertical line intersects a graph in more than one point, the graph is not the graph of a function.

By the vertical line test, the graph shown is not the graph of a function.

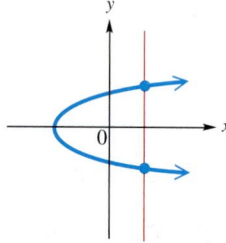

The set of all first components in the ordered pairs of a relation is the **domain** of the relation, and the set of all second components in the ordered pairs is the **range** of the relation.

The function

$$\{(10, 5), (20, 15), (30, 25)\}$$

has domain $\{10, 20, 30\}$ and range $\{5, 15, 25\}$.

To find $f(x)$ for a specific value of x, replace x by that value in the expression for the function f.

If $f(x) = 2x + 7$, find $f(3)$.

$$f(3) = 2(3) + 7$$
$$f(3) = 13$$

ANSWERS TO TEST YOUR WORD POWER

1. D; *Examples*: See Figures 1–4 in **Section 17.4.**
2. C; *Example*: The graph of $y = (x + 3)^2$ has vertex $(-3, 0)$, which is the lowest point on the graph.
3. A; *Example*: $\{(0, 2), (2, 4), (3, 6), (-1, 3)\}$
4. C; *Example*: The domain in the relation given in Answer 3 is the set of x-values $\{0, 2, 3, -1\}$.
5. D; *Example*: The range of the relation given in Answer 3 is the set of y-values $\{2, 4, 6, 3\}$.
6. B; *Example*: The relation given in Answer 3 is a function since each x-value corresponds to exactly one y-value.

Chapter 17 ▶▶▶ Review Exercises

[17.1] *Solve each equation by using the square root property. Express all radicals in simplest form.*

1. $y^2 = 144$

2. $x^2 = 37$

3. $m^2 = 128$

4. $(k + 2)^2 = 25$

5. $(r - 3)^2 = 10$

6. $(2p + 1)^2 = 14$

7. $(3k + 2)^2 = -3$

8. $(3x + 5)^2 = 0$

[17.2] *Solve each equation by completing the square.*

9. $m^2 + 6m + 5 = 0$

10. $p^2 + 4p = 7$

11. $-x^2 + 5 = 2x$

12. $2x^2 - 3 = -8x$

13. $4(x^2 + 7x) + 29 = -20$

14. $(4x + 1)(x - 1) = -7$

Solve each problem.

15. If an object is projected upward on Earth from a height of 50 ft, with an initial velocity of 32 ft per sec, then its height h after t seconds is given by $h = -16t^2 + 32t + 50$, where h is in feet. At what time(s) will it be at a height of 30 ft?

16. Find the lengths of the three sides of the right triangle shown.

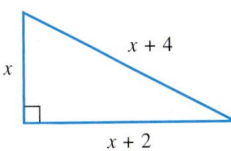

17. What must be added to $x^2 + 3x$ to make it a perfect square?

[17.3]

18. Consider the equation $x^2 - 9 = 0$.

    (a) Solve the equation by factoring.

    (b) Solve the equation by the square root property.

    (c) Solve the equation by the quadratic formula.

    (d) Compare your answers. If a quadratic equation can be solved by both factoring and the quadratic formula, should we always get the same results? Explain.

Solve each equation by using the quadratic formula.

19. $-4x^2 - 2x + 7 = 0$ **20.** $2x^2 + 8 = 4x + 11$ **21.** $x(5x - 1) = 1$

22. $\dfrac{1}{4}x^2 = 2 - \dfrac{3}{4}x$ **23.** $\dfrac{1}{2}x^2 + 3x = 5$

24. A student writes the quadratic formula as $x = -b \pm \sqrt{\dfrac{b^2 - 4ac}{2a}}$. Is this correct? If not, explain the error, and give the correct formula.

[17.4] *Sketch the graph of each equation. Identify each vertex.*

25. $y = -3x^2$ **26.** $y = -x^2 + 5$ **27.** $y = x^2 - 2x + 1$

  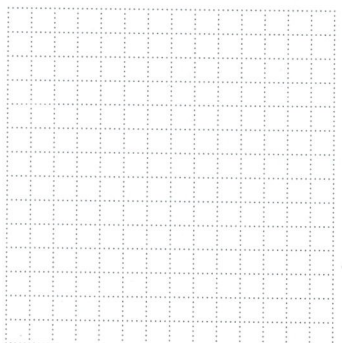

28. $y = -x^2 + 2x + 3$

29. $y = x^2 + 4x + 2$

30. $y = (x + 4)^2$

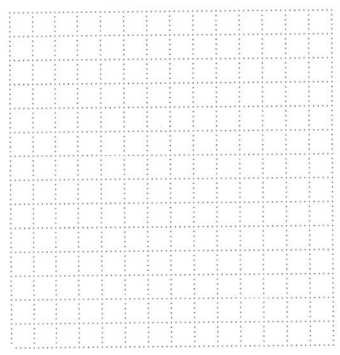

31. Refer to Exercise 17 in **Section 17.4.** Suppose that a telescope has a diameter of 200 ft and a maximum depth of 30 ft. Find the equation for a cross section of the parabolic dish.

[17.5] *Decide whether each relation is or is not a function. In Exercises 32 and 33, give the domain and the range.*

32. $\{(-2, 4), (0, 8), (2, 5), (2, 3)\}$

33. $\{(8, 3), (7, 4), (6, 5), (5, 6), (4, 7)\}$

34.

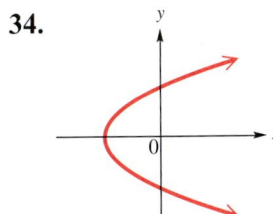

35.

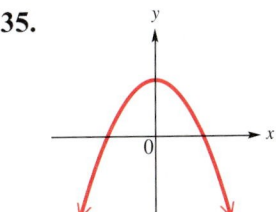

36. $2x + 3y = 12$

37. $y = x^2$

38. $x = 2|y|$

Find (a) $f(2)$ and (b) $f(-1)$.

39. $f(x) = 3x + 2$

40. $f(x) = 2x^2 - 1$

41. $f(x) = |x + 3|$

Becky and Brad are the owners of Cole's Baseball Cards. They have found that the price y, in dollars, of a particular Cal Ripken baseball card depends on the demand x, in hundreds, for the card, according to the function defined by

$$y = -x^2 + 12x - 26.$$

42. What demand produces a price of $6 for the card?

43. Find the vertex of the parabola $y = -x^2 + 12x - 26$. **44.** Give the demand and price that correspond to the vertex.

▶▶▶ **Mixed Review Exercises**

Solve by any method.

45. $(2t - 1)(t + 1) = 54$

46. $(2p + 1)^2 = 100$

47. $(k + 2)(k - 1) = 3$

48. $6t^2 + 7t - 3 = 0$

49. $2x^2 + 3x + 2 = x^2 - 2x$

50. $x^2 + 2x + 5 = 7$

51. $m^2 - 4m + 10 = 0$

52. $k^2 - 9k + 10 = 0$

53. $(5x + 6)^2 = 0$

54. $\dfrac{1}{2}r^2 = \dfrac{7}{2} - r$

55. $x^2 + 4x = 1$

56. $7x^2 - 8 = 5x^2 + 8$

Chapter 17 ▶▶▶ Test

Use the Chapter Test Prep Video CD to see fully worked-out
solutions to any of the exercises you want to review.

Solve by using the square root property.

1. $x^2 = 39$

2. $(x + 3)^2 = 64$

3. $(4x + 3)^2 = 24$

Solve by completing the square.

4. $x^2 - 4x = 6$

5. $2x^2 + 12x - 3 = 0$

Solve by the quadratic formula.

6. $2x^2 + 5x - 3 = 0$

7. $3w^2 + 2 = 6w$

8. $4x^2 + 8x + 11 = 0$

9. $t^2 - \dfrac{5}{3}t + \dfrac{1}{3} = 0$

Solve by the method of your choice.

10. $p^2 - 2p - 1 = 0$

11. $(2x + 1)^2 = 18$

12. $(x - 5)(2x - 1) = 1$

13. $t^2 + 25 = 10t$

1. _____

2. _____

3. _____

4. _____

5. _____

6. _____

7. _____

8. _____

9. _____

10. _____

11. _____

12. _____

13. _____

Solve each problem.

14. _____

14. If an object is projected into the air from ground level with an initial velocity of 64 ft per sec, its altitude (height) s in feet after t seconds is given by the formula $s = -16t^2 + 64t$. At what time(s) will the object be at a height of 64 ft?

15. _____

15. Find the lengths of the three sides of the right triangle.

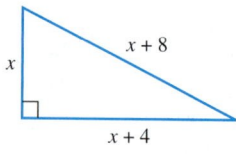

Sketch the graph of each equation. Identify each vertex.

16. _____

16. $y = (x - 3)^2$

17. _____

17. $y = -x^2 - 2x - 4$

18. (a) _____

18. Decide whether each relation represents a function. If it does, give the domain and the range.

(b) _____

   (a) $\{(2, 3), (2, 4), (2, 5)\}$ (b) $\{(0, 2), (1, 2), (2, 2)\}$

19. _____

19. Use the vertical line test to determine whether the graph at the right is that of a function.

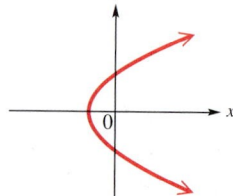

20. _____

20. If $f(x) = 3x + 7$, find $f(-2)$.

Whole Numbers Computation: Pretest

This test will check your skills in doing whole numbers computation, using paper and pencil. Each part of the test is keyed to a section in the Review Chapter, which follows this test. Based on your test results, work the appropriate section(s) in the Review Chapter *before you start Chapter 1.*

Adding Whole Numbers   *(Do not use a calculator.)*

1. 368 + 22	**2.** 7093 + 6073	**3.** 85 + 2968	**4.** 57,208 915 + 59,387

5. 714 + 3728 + 9 + 683,775

1. _____
2. _____
3. _____
4. _____
5. _____

Subtracting Whole Numbers   *(Do not use a calculator.)*

1. 426 − 76	**2.** 3358 − 2729	**3.** 30,602 − 5708

4. 4006 − 97   **5.** 679,420 − 88,033

1. _____
2. _____
3. _____
4. _____
5. _____

Multiplying Whole Numbers   *(Do not use a calculator.)*

1. 3 × 3 × 0 × 6   **2.** 3841
× 7   **3.** (520)(3000)

1. _____
2. _____
3. _____

(continued)

Do not use a calculator; show your work.

4. _____

4.   71
     $\times\,26$

5. Multiply 359 and 48.

6. 853×609

5. _____

6. _____

Dividing Whole Numbers *(Do not use a calculator; show your work.)*

1. _____

1. $3\overline{)69}$

2. $12 \div 0$

3. $\dfrac{25,036}{4}$

4. $7\overline{)5655}$

2. _____

3. _____

4. _____

5. _____

5. $52\overline{)1768}$

6. $45,000 \div 900$

7. $38\overline{)2300}$

8. $83\overline{)44,799}$

6. _____

7. _____

8. _____

Now check your answers on page A–37 in the Answers section at the back of the book. Record the number of problems you worked correctly in each part of the test.

Adding Whole Numbers: _____ correct out of 5.

   If you got 0, 1, or 2 correct, work **Section R.1** in the Review Chapter.

Subtracting Whole Numbers: _____ correct out of 5.

   If you got 0, 1, or 2 correct, work **Section R.2** in the Review Chapter.

Multiplying Whole Numbers: _____ correct out of 6.

   If you got 0, 1, 2, or 3 correct, work **Section R.3** in the Review Chapter.

Dividing Whole Numbers: _____ correct out of 8.

   If you got 0, 1, or 2 correct, work **Sections R.4** and **R.5** in the Review Chapter.

   If you got 3 or 4 correct, work **Section R.5** in the Review Chapter.

Whole Numbers Review

R

> **Note**
> Use the **Whole Numbers Computation: Pretest** on the previous two pages to decide which sections you need to work in this chapter.

R.1 ▶▶▶ Adding Whole Numbers

There are 4 triangles at the left and 2 at the right. In all, there are 6 triangles.

The process of finding the total is called *addition*. Here 4 and 2 were added to get 6. Addition is written with a $+$ sign, as shown below.

$$4 + 2 = 6$$

OBJECTIVE 1 Add two or three single-digit numbers. In addition problems, the numbers being added are called **addends,** and the resulting answer is called the **sum** or **total.**

$$
\begin{array}{r}
4 \quad \leftarrow \text{Addend} \\
+\,2 \quad \leftarrow \text{Addend} \\
\hline
6 \quad \leftarrow \text{Sum (answer)}
\end{array}
$$

Addition problems can also be written horizontally as shown below.

$$4 \quad + \quad 2 \quad = \quad 6$$

Addend   Addend   Sum

OBJECTIVES

1. Add two or three single-digit numbers.
2. Add more than two numbers.
3. Add when regrouping is not required.
4. Add with regrouping.
5. Use addition to solve application problems.
6. Check the sum in addition.

1 Add. Then use the commutative property to write another addition problem and find the sum.

(a) 3 + 4

(b) 9 + 9

(c) 7 + 8

(d) 6 + 9

2 Find each sum.

(a)   5
       4
       6
       9
    + 2

(b)   7
       5
       1
       2
    + 6

(c)   9
       2
       1
       3
    + 4

(d)   3
       8
       6
       4
    + 8

ANSWERS

1. **(a)** 7; 4 + 3 = 7 **(b)** 18; no change
   (c) 15; 8 + 7 = 15 **(d)** 15; 9 + 6 = 15
2. **(a)** 26 **(b)** 21 **(c)** 19 **(d)** 29

> **Commutative Property of Addition**
>
> The **commutative property of addition** states that changing the *order* of the addends in an addition problem does *not* change the sum.

For example, the sum of 4 + 2 is the same as the sum of 2 + 4. Both sums are 6. This allows the addition of the same numbers in a different order.

EXAMPLE 1 **Adding Two Single-Digit Numbers**

Add. Then use the commutative property to write another addition problem and find the sum.

(a) 6 + 2 = **8**   and   2 + 6 = **8**   **(b)** 5 + 9 = **14**   and   9 + 5 = **14**

(c) 8 + 8 = **16**   (No change occurs when commutative property is used.)

◀ *Work Problem* **1** *at the Side.*

> **Associative Property of Addition**
>
> By the **associative property of addition,** changing the *grouping* of addends does *not* change the sum.

For example, the sum of 3 + 5 + 6 may be found in several ways.

$$(3 + 5) + 6 = 8 + 6 \ = \textbf{14}$$ Parentheses tell you to add 3 + 5 first.

$$3 + (5 + 6) = 3 + 11 = \textbf{14}$$ Parentheses tell you to add 5 + 6 first.

Either method gives a sum of 14 because of the associative property of addition.

OBJECTIVE **2** **Add more than two numbers.**   To add several numbers, first write them in a column. Add the first number to the second. Add this sum to the third number; continue until all the numbers are used.

EXAMPLE 2 **Adding More Than Two Numbers**

Find the sum of 2, 5, 6, 1, and 4.

   2 ⎤
   5 ⎦ 2 + 5 = 7
  Ⓖ 7 + 6 = 13
  ① 13 + 1 = 14
+ ④ 14 + 4 = 18
 18 ← Sum

> **Note**
>
> You may also add numbers by starting at the bottom of a column. Adding down from the top or adding up from the bottom will give the same sum.

◀ *Work Problem* **2** *at the Side.*

OBJECTIVE **3** **Add when regrouping is not required.**   If numbers have two or more digits, first you must arrange the numbers in columns so that the ones digits are in the same column, tens are in the same column, hundreds are in the same column, and so on. Next, you add column by column, starting at the right with the ones column.

EXAMPLE 3 **Adding without Regrouping**

Add: 511 + 23 + 154 + 10

First line up the numbers in columns, with the ones column at the right.

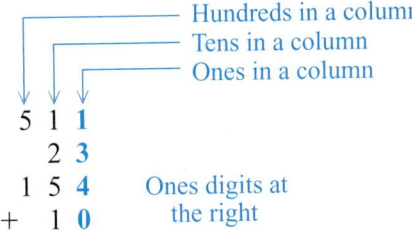

```
      5 1 1
        2 3
      1 5 4
    +   1 0
```

Ones digits at the right

Now start at the right and add the ones digits. Add the tens digits next, and finally, add the hundreds digits.

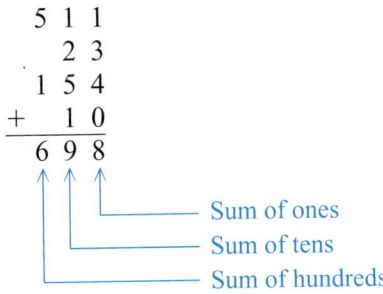

The sum of the four numbers is 698.

Work Problem **3** *at the Side.* ▶

OBJECTIVE 4 Add with regrouping. If the sum of the digits in a column is more than 9, use **regrouping** (sometimes called *carrying*).

EXAMPLE 4 **Adding with Regrouping**

Find the sum of 47 and 29.

Add the digits in the ones column.

```
    47
  + 29
```
7 ones + 9 ones = 16 ones

Regroup 16 ones as 1 ten and 6 ones. Write 6 ones in the ones column and write 1 ten in the tens column.

Write 1 ten in the tens column.

```
    1
   47      7 ones + 9 ones = 16 ones    Regroup 16 ones as
  + 29                                  1 ten and 6 ones.
    6
```
Write 6 ones in the ones column.

Add the digits in the tens column, including the regrouped 1.

```
    1
   47     Remember
  + 29    to add the 1
   76     regrouped ten.
```
1 ten + 4 tens + 2 tens = 7 tens

Work Problem **4** *at the Side.* ▶

3 Add.

(a)
```
    25
  + 73
```

(b)
```
    364
  + 532
```

(c)
```
    42,305
  + 11,563
```

4 Find each sum using regrouping.

(a)
```
    69
  + 26
```

(b)
```
    76
  + 18
```

(c)
```
    56
  + 37
```

(d)
```
    34
  + 49
```

ANSWERS

3. (a) 98   (b) 896   (c) 53,868
4. (a) 95   (b) 94   (c) 93   (d) 83

5 Find each sum, regrouping when necessary.

(a)
```
   481
    79
    38
 + 395
```

(b)
```
  4271
   372
  8976
 + 162
```

(c)
```
    57
     4
   392
   804
    51
 +  27
```

(d)
```
  7821
   435
    72
   305
 + 1693
```

(e)
```
  15,829
     765
      78
      15
       9
       7
 + 13,179
```

EXAMPLE 5 **Adding with Regrouping**

Add: $324 + 7855 + 23 + 7 + 86$

Step 1 Add the digits in the ones column.

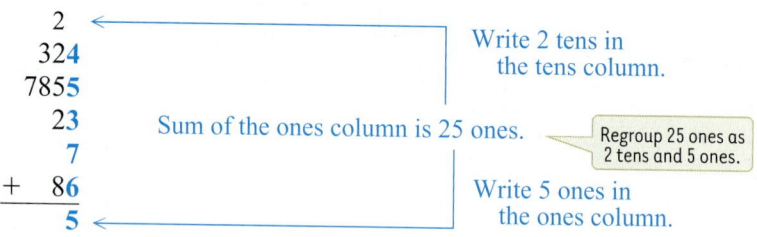

Sum of the ones column is 25 ones.

Regroup 25 ones as 2 tens and 5 ones.

Write 2 tens in the tens column.

Write 5 ones in the ones column.

Step 2 Add the digits in the tens column, including the regrouped 2.

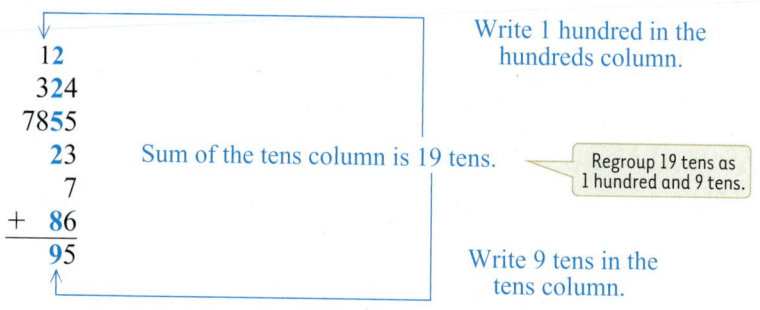

Sum of the tens column is 19 tens.

Regroup 19 tens as 1 hundred and 9 tens.

Write 1 hundred in the hundreds column.

Write 9 tens in the tens column.

Step 3 Add the hundreds column, including the regrouped 1.

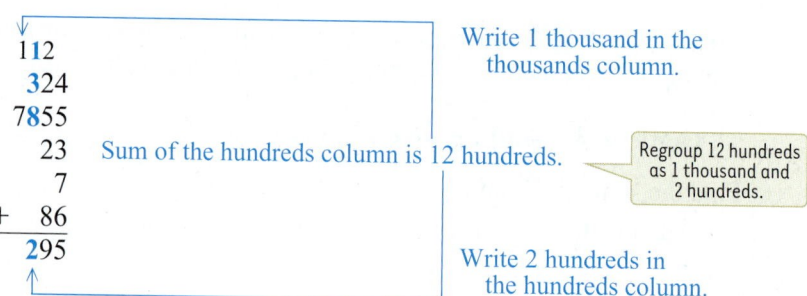

Sum of the hundreds column is 12 hundreds.

Regroup 12 hundreds as 1 thousand and 2 hundreds.

Write 1 thousand in the thousands column.

Write 2 hundreds in the hundreds column.

Step 4 Add the thousands column, including the regrouped 1.

```
  112
  324
 7855
   23
    7
 +  86
 8295
```

Sum of the thousands column is 8 thousands.

Thus, $324 + 7855 + 23 + 7 + 86 = 8295$.

◀ *Work Problem* **5** *at the Side.*

ANSWERS

5. **(a)** 993   **(b)** 13,781   **(c)** 1335
   (d) 10,326   **(e)** 29,882

OBJECTIVE ▮**5**▮ **Use addition to solve application problems.**

EXAMPLE 6 **Applying Addition Skills**

On this map, the distance in miles from one town to another is written along-side the road. Find the shortest route from Altamonte Springs to Clear Lake.

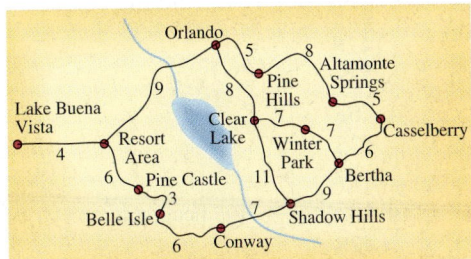

Approach   Add the mileage along various routes from Altamonte Springs to Clear Lake. Then select the shortest route.

Solution   One way from Altamonte Springs to Clear Lake is through Orlando. Add the mileage numbers along this route.

8	Altamonte Springs to Pine Hills
5	Pine Hills to Orlando
+ 8	Orlando to Clear Lake
21 →	miles from Altamonte Springs to Clear Lake, going through Orlando

Another way is through Bertha and Winter Park. Add the mileage numbers along this route.

5	Altamonte Springs to Casselberry
6	Casselberry to Bertha
7	Bertha to Winter Park
+ 7	Winter Park to Clear Lake
25 →	miles from Altamonte Springs to Clear Lake through Bertha and Winter Park

The shortest route from Altamonte Springs to Clear Lake is 21 miles through Orlando.

Work Problem ⑥ *at the Side.* ▶

EXAMPLE 7 **Applying Addition Skills**

Using the map in Example 6 above, find the total mileage from Shadow Hills to Casselberry to Orlando and back to Shadow Hills through Clear Lake.

Approach   Add the mileage from Shadow Hills to Casselberry to Orlando and back to Shadow Hills to find the total.

Solution   Use the numbers from the map.

9	Shadow Hills to Bertha
6	Bertha to Casselberry
5	Casselberry to Altamonte Springs
8	Altamonte Springs to Pine Hills
5	Pine Hills to Orlando
8	Orlando to Clear Lake
+ 11	Clear Lake to Shadow Hills
52 →	miles from Shadow Hills to Casselberry to Orlando and back to Shadow Hills

Work Problem ⑦ *at the Side.* ▶

⑥ Use the map at the left to find the shortest route from Conway to Pine Hills.

⑦ The road is closed between Orlando and Clear Lake, so this route cannot be used. Find the next shortest route from Orlando to Clear Lake.

8 Check each addition. If the sum is incorrect, find the correct sum.

(a)
$$
\begin{array}{r}
32 \\
8 \\
5 \\
+\ 14 \\
\hline
59
\end{array}
$$

(b)
$$
\begin{array}{r}
872 \\
539 \\
46 \\
+\ 152 \\
\hline
1609
\end{array}
$$

(c)
$$
\begin{array}{r}
79 \\
218 \\
7 \\
+\ 639 \\
\hline
953
\end{array}
$$

(d)
$$
\begin{array}{r}
21{,}892 \\
11{,}746 \\
+\ 43{,}925 \\
\hline
79{,}563
\end{array}
$$

OBJECTIVE **6** **Check the sum in addition.** Checking the answer is an important part of problem solving. A common method for checking addition is to re-add from the bottom to top. This is an application of the commutative and associative properties of addition.

EXAMPLE 8 **Checking Addition**

Check each sum. If the sum is incorrect, find the correct sum.

(a) Add down.
$$
\begin{array}{r}
\mathbf{1428} \\
738 \\
63 \\
125 \\
17 \\
+\ 485 \\
\hline
\mathbf{1428}
\end{array}
$$
To check, add up.

Adding down and adding up should give the same sum. In this case, the answers agree, so the sum is probably correct.

(b)
$$
\begin{array}{rr}
 & \mathbf{1033} \\
785 & 785 \\
63 & 63 \\
+\ 185 & +\ 185 \\
\hline
1033 & \mathbf{1033}
\end{array}
$$
Correct, because both answers are the same

To check, add up.

(c)
$$
\begin{array}{rr}
 & \mathbf{2454} \\
635 & 635 \\
73 & 73 \\
831 & 831 \\
+\ 915 & +\ 915 \\
\hline
2444 & \mathbf{2444}
\end{array}
$$
Error, because answers are different

To check, add up.

Re-add to find that the correct sum is 2454.

◀ Work Problem **8** at the Side.

R.1 ▶▶▶ **Exercises**

FOR
EXTRA
HELP

Add. See Examples 2 and 3.

1. **(a)**
```
   5
   7
   6
 + 5
```
(b)
```
   9
   2
   1
   3
 + 4
```
2. **(a)**
```
   2
   9
   5
 + 1
```
(b)
```
   3
   8
   6
   4
 + 8
```

3. **(a)** 3213 + 5715

(b) 38,204 + 21,020

4. **(a)** 6344 + 1655

(b) 63,251 + 36,305

Find each sum, regrouping when necessary. See Examples 4 and 5.

5.
```
   67
 + 83
```
6.
```
   78
 + 36
```
7.
```
  746
 + 905
```
8.
```
  621
 + 359
```
9.
```
  798
 + 206
```

10.
```
  172
 + 156
```
11.
```
  7968
 + 1285
```
12.
```
  1768
 + 8275
```
13.
```
  7896
 + 3728
```
14.
```
  9382
 + 7586
```

15.
```
  3705
  3916
 + 9037
```
16.
```
  6629
  6076
 + 8218
```
17.
```
    32
 + 4977
```
18.
```
   402
 + 9938
```
19.
```
  3077
     8
 + 421
```

20.
```
    56
  7721
 + 172
```
21.
```
  9056
    78
  6089
 + 731
```
22.
```
  4022
   709
  8621
 +  37
```
23.
```
    18
   708
  9286
 + 636
```
24.
```
  1708
   321
    61
 + 8926
```

Check each sum by adding from bottom to top. If an answer is incorrect, find the correct sum. See Example 8.

25.
```
  ----
   179
   214
 + 376
  ----
   759
```
26.
```
  ----
    17
   296
   713
 +  94
  ----
  1220
```
27.
```
  ----
  4713
    28
   615
 +  64
  ----
  5420
```
28.
```
  ----
  6 215
   744
    36
 + 4 284
  ----
  11,279
```

Using the map below, find the shortest route between each pair of cities. See Examples 6 and 7.

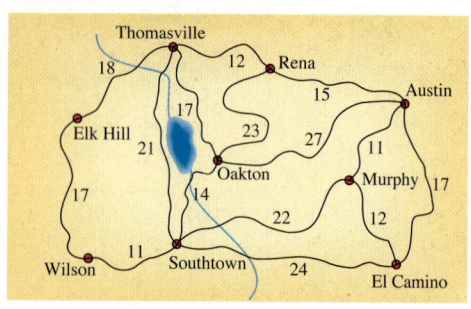

29. Southtown and Rena

30. Elk Hill and Oakton

31. Thomasville and Murphy

32. Austin and Wilson

Solve each application problem.

33. The sale price for a basic auto tune-up is $99, a tire rotation is $24, and an oil change is $29. Find the total cost for all the services.

34. Jane Lim bought an 11-piece set of golf clubs for $120, a dozen golf balls for $9, golf shoes for $45, and a golf glove for $12. How much did she spend in all? (*Source:* Sportmart.)

35. There are 413 women and 286 men on the sales staff. How many people are on the sales staff?

36. One department in an office building has 283 employees while another department has 218 employees. How many employees are in the two departments?

37. According to the latest census estimates, the two states with the highest populations are California with 36,457,549 people and Texas with 23,507,783 people. How many people live in those two states? (*Source:* U.S. Census Bureau.)

38. The two states with the smallest populations are Wyoming with 515,004 people and Vermont with 623,908 people. What is the total population of the two states? (*Source:* U.S. Census Bureau.)

Find the perimeter of (total distance around) each figure.

39.

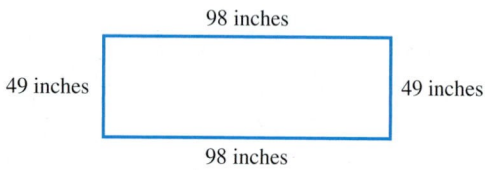

98 inches
49 inches   49 inches
98 inches

40.

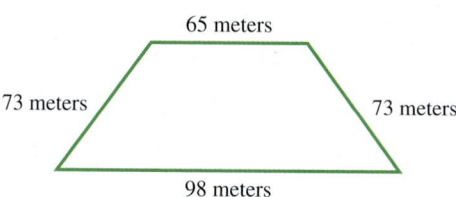

65 meters
73 meters   73 meters
98 meters

41.

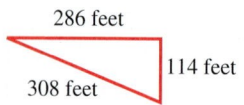

286 feet
114 feet
308 feet

42.

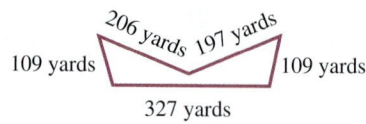

206 yards   197 yards
109 yards   109 yards
327 yards

R.2 ▶▶▶ Subtracting Whole Numbers

Suppose you have \$18, and you spend \$15 for food. You then have \$3 left. There are two different ways of looking at these numbers.

As an addition problem:

$$\$15 \quad + \quad \$3 \quad = \quad \$18$$

Amount spent   Amount left   Original amount

As a subtraction problem:

$$\$18 \quad - \quad \$15 \quad = \quad \$3$$

Original amount   Subtraction symbol   Amount spent   Amount left

OBJECTIVE 1 Change addition problems to subtraction problems or the reverse. As this example shows, an addition problem can be changed to a subtraction problem and a subtraction problem can be changed to an addition problem.

EXAMPLE 1 **Changing Addition Problems to Subtraction**

Change each addition problem to a subtraction problem.

(a) $4 + 1 = 5$

Two subtraction problems are possible, as shown below.

$$5 - 1 = 4 \quad \text{or} \quad 5 - 4 = 1$$

These figures show each subtraction problem.

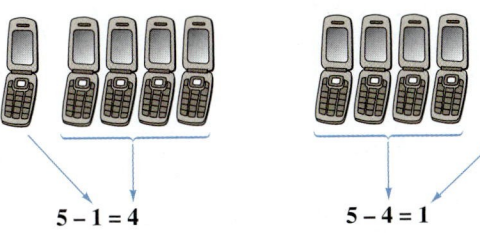

$$5 - 1 = 4 \qquad\qquad 5 - 4 = 1$$

(b) $10 + 17 = 27$

$$27 - 17 = 10 \quad \text{or} \quad 27 - 10 = 17$$

Work Problem **1** *at the Side.* ▶

EXAMPLE 2 **Changing Subtraction Problems to Addition**

Change each subtraction problem to an addition problem.

(a) $8 - 3 = 5$

$$8 = 3 + 5$$

It is also correct to write $8 = 5 + 3$ (using the commutative property).

Continued on Next Page

OBJECTIVES

1 Change addition problems to subtraction problems or the reverse.

2 Identify the minuend, subtrahend, and difference.

3 Subtract when no regrouping is needed.

4 Use addition to check subtraction answers.

5 Subtract with regrouping.

6 Use subtraction to solve application problems.

---

1 Write two subtraction problems for each addition problem.

(a) $4 + 3 = 7$

(b) $6 + 5 = 11$

(c) $150 + 220 = 370$

(d) $623 + 55 = 678$

ANSWERS

1. (a) $7 - 3 = 4$ or $7 - 4 = 3$
(b) $11 - 5 = 6$ or $11 - 6 = 5$
(c) $370 - 220 = 150$ or $370 - 150 = 220$
(d) $678 - 55 = 623$ or $678 - 623 = 55$

2 Write an addition problem for each subtraction problem.

(a) $5 - 3 = 2$

(b) $8 - 3 = 5$

(c) $21 - 15 = 6$

(d) $58 - 42 = 16$

3 Subtract.

(a) $\begin{array}{r} 56 \\ -\ 31 \\ \hline \end{array}$

(b) $\begin{array}{r} 38 \\ -\ 14 \\ \hline \end{array}$

(c) $\begin{array}{r} 378 \\ -\ 235 \\ \hline \end{array}$

(d) $\begin{array}{r} 3927 \\ -\ 2614 \\ \hline \end{array}$

(e) $\begin{array}{r} 5464 \\ -\ 324 \\ \hline \end{array}$

(b) $19 - 14 = 5$
$19 = 14 + 5$

(c) $290 - 130 = 160$
$290 = 130 + 160$

◀ Work Problem **2** at the Side.

OBJECTIVE **2** **Identify the minuend, subtrahend, and difference.** In subtraction, as in addition, the numbers in a problem have names. For example, in the problem $8 - 5 = 3$, the number 8 is the **minuend,** 5 is the **subtrahend,** and 3 is the **difference** or answer.

$$\underset{\underset{\text{Minuend}}{\uparrow}}{8} \quad - \quad \underset{\underset{\text{Subtrahend}}{\uparrow}}{5} \quad = 3 \leftarrow \text{Difference (answer)}$$

$\begin{array}{r} 8 \\ -\ 5 \\ \hline 3 \end{array}$ $\begin{array}{l} \leftarrow \text{Minuend} \\ \leftarrow \text{Subtrahend} \\ \leftarrow \text{Difference} \end{array}$

OBJECTIVE **3** **Subtract when no regrouping is needed.** Subtract two numbers by lining up the numbers in columns so that the digits in the ones place are in the same column. Next, subtract by columns, starting at the right with the ones column.

EXAMPLE 3 **Subtracting Two Numbers without Regrouping**

(a) Ones digits are lined up in the same column.

$\begin{array}{r} 53 \\ -\ 21 \\ \hline 32 \end{array}$

Subtract in the ones column first.

3 ones − 1 one = 2 ones
5 tens − 2 tens = 3 tens

(b) Ones digits are lined up.

$\begin{array}{r} 385 \\ -\ 165 \\ \hline 220 \end{array}$ ← 5 ones − 5 ones = 0 ones

8 tens − 6 tens = 2 tens
3 hundreds − 1 hundred = 2 hundreds

(c) $\begin{array}{r} 9837 \\ -\ 210 \\ \hline 9627 \end{array}$ ← 7 ones − 0 ones = 7 ones

3 tens − 1 ten = 2 tens
8 hundreds − 2 hundreds = 6 hundreds
9 thousands − 0 thousands = 9 thousands

◀ Work Problem **3** at the Side.

OBJECTIVE **4** **Use addition to check subtraction answers.** You can check $8 - 3 = 5$ by *adding* 3 and 5.

$3 + 5 = 8$ so $8 - 3 = 5$ is correct.

EXAMPLE 4 **Checking Subtraction**

Use addition to check each answer. If incorrect, find the correct answer.

(a)
```
   89
 − 47
 ----
   42
```

Rewrite as an addition problem.

Subtraction problem
```
   89
 − 47
 ----
   42
 ----
   89
```
Addition problem
```
   47
 + 42
 ----
   89
```

Because $47 + 42 = 89$, the subtraction was done correctly.

(b) $72 − 41 = 21$

Rewrite as an addition problem.

Does $72 = 41 + 21$? ⟵ 41 + 21 = 62, *not* 72, so there is an error!

But, $41 + 21 = 62$, *not* 72, so the subtraction was done *incorrectly*. Rework the original subtraction to get the correct answer of 31. Then, $41 + 31 = 72$.

(c)
```
   374    ⟵ Match
 − 141
 -----
   233
```
$141 + 233 = 374$

The answer checks.

Work Problem **4** *at the Side.* ▶

OBJECTIVE 5 Subtract with regrouping. If a digit in the minuend is *less* than the one directly below it, **regrouping** is necessary (sometimes called *borrowing*).

EXAMPLE 5 **Subtracting with Regrouping**

Subtract 19 from 57.

Watch the wording!
Subtract 19 **from** 57 is
```
   57              19
 − 19    not      −57
```

Rewrite the problem in vertical format.
```
   5 7
 − 1 9
```

In the ones column, 7 is *less* than 9, so, in order to subtract, you must regroup 1 ten as 10 ones.

5 tens − 1 ten = 4 tens ⟶ 4 17 ⟵ 1 ten = 10 ones, and 10 ones + 7 ones = 17 ones
```
   5̸ 7̸
 − 1 9
```

Now subtract 17 ones minus 9 ones in the ones column.
Then subtract 4 tens minus 1 ten in the tens column.
```
   4 17
   5̸ 7̸
 − 1 9
 ------
   3 8   ⟵ Difference
```

Thus, $57 − 19 = 38$. Check by adding 19 and 38. You should get 57.

Work Problem **5** *at the Side.* ▶

4 Use addition to check each answer. If incorrect, find the correct answer.

(a)
```
   65       Check
 − 23
 ----
   42
```

(b)
```
   46       Check
 − 32
 ----
   24
```

(c)
```
   374      Check
 − 251
 -----
   113
```

(d)
```
   7531     Check
 − 4301
 ------
   3230
```

5 Find each difference.

(a)
```
   67
 − 38
```

(b)
```
   97
 − 29
```

(c)
```
   31
 − 17
```

(d)
```
   863
 −  47
```

(e)
```
   762
 − 157
```

ANSWERS

4. (a) $23 + 42 = 65$; correct
 (b) $32 + 24 ≠ 46$; incorrect; should be 14
 (c) $251 + 113 ≠ 374$; incorrect; should be 123
 (d) $4301 + 3230 = 7531$; correct
5. (a) 29 **(b)** 68 **(c)** 14 **(d)** 816 **(e)** 605

6 Subtract.

(a) 354
     − 82

(b) 457
     − 68

(c) 874
     − 486

(d) 1437
     − 988

(e) 8739
     − 3892

EXAMPLE 6 **Subtracting with Regrouping**

Subtract, regrouping when necessary.

(a) 7856
     − 137

Regroup 1 ten as 10 ones. —— 10 ones + 6 ones = 16 ones

```
    4 16
7 8 5 6
−   1 3 7
7 7 1 9  ← Difference
```

(b) 635
     − 546

Regroup 1 ten as 10 ones. ┌ 10 ones + 5 ones = 15 ones

```
  2 15
6 3 5     Need to regroup
−5 4 6    further because
    9     2 is less than 4
          in tens column.
```

Regroup 1 hundred as 10 tens.
10 tens + 2 tens = 12 tens

```
5 12 15
6  3  5
−5  4  6
   8  9  ← Difference
```

(c) 412
     − 225

```
  0 12
4 1 2     Need to regroup
−2 2 5    further because
    7     0 is less than 2
          in tens column.
```

```
3 10 12
4  1  2
−2  2  5
1  8  7  ← Difference
```

◄ *Work Problem* **6** *at the Side.*

Sometimes a minuend has zeros in some of the positions. In such cases, regrouping may be a little more complicated than what we have shown so far.

EXAMPLE 7 **Regrouping with Zeros**

Subtract.

         4607
       − 3168

There are no tens that can be regrouped into ones. So you must first regroup 1 hundred as 10 tens.

Regroup 1 hundred as 10 tens. —— Write 10 tens.

```
   5 10
4  6  0  7
−3  1  6  8
```

Now regroup 1 ten as 10 ones.

Regroup 1 ten as 10 ones; 10 tens − 1 ten = 9 tens
10 ones + 7 ones = 17 ones

```
      9
   5 10 17
4  6  0  7
−3  1  6  8
         9
```

Continued on Next Page

Complete the problem.

$$
\begin{array}{r}
\overset{9}{} \\
5\ \overset{\cancel{10}}{}\ 17 \\
4\ \cancel{6}\ \cancel{0}\ \cancel{7} \\
-\ 3\ 1\ 6\ 8 \\
\hline
1\ 4\ 3\ 9 \leftarrow \text{Difference}
\end{array}
$$

Check by adding 1439 and 3168; you should get 4607.

──────────── Work Problem **7** at the Side. ▶

EXAMPLE 8 **Regrouping with Zeros**

Find each difference.

(a)   708
        − 149

Write 10 tens. ──────┐   ┌─── Regroup 1 ten as 10 ones.

Regroup 1 hundred ──→ 6 $\cancel{10}$ 18 ←─── 10 ones + 8 ones = 18 ones
as 10 tens.

$$
\begin{array}{r}
\cancel{7}\ \cancel{0}\ \cancel{8} \\
-\ 1\ 4\ 9 \\
\hline
5\ 5\ 9
\end{array}
$$

(b)   380
        − 276

Regroup 1 ten ────┐   ┌── Write 10 ones.
as 10 tens.

$$
\begin{array}{r}
7\ 10 \\
3\ \cancel{8}\ \cancel{0} \\
-\ 2\ 7\ 6 \\
\hline
1\ 0\ 4
\end{array}
$$

(c)   9000
        − 6999

$$
\begin{array}{r}
9\ \ 9 \\
8\ \cancel{10}\ \cancel{10}\ 10 \\
\cancel{9}\ \cancel{0}\ \cancel{0}\ \cancel{0} \\
-\ 6\ 9\ 9\ 9 \\
\hline
2\ 0\ 0\ 1
\end{array}
$$

──────────── Work Problem **8** at the Side. ▶

Recall that an answer to a subtraction problem can be checked by adding.

EXAMPLE 9 **Checking Subtraction**

Use addition to check each answer.

Check

(a)   **613** ←─ Matches   275
        − 275   Correct   + 338
        ─────   ─────
         338   **613**

Continued on Next Page

---

7 Find each difference.

(a)   308
         − 285

(b)   206
         − 148

(c)   5073
        − 1632

8 Find each difference.

(a)   405
         − 267

(b)   370
         − 163

(c)   1570
         − 983

(d)   7001
        − 5193

(e)   4000
        − 1782

9 Use addition to check each answer. If an answer is incorrect, find the correct answer.

(a)
$$\begin{array}{r} 425 \\ -\ 368 \\ \hline 57 \end{array}$$ **Check**

(b)
$$\begin{array}{r} 670 \\ -\ 439 \\ \hline 241 \end{array}$$ **Check**

(c)
$$\begin{array}{r} 14{,}726 \\ -\ 8\ 839 \\ \hline 5\ 887 \end{array}$$ **Check**

10 Use the table from Example 10 at the right.

(a) How many fewer deliveries did Ms. Lopez make on Tuesday than on Friday?

(b) What was the difference in the number of deliveries on Tuesday from the number on Wednesday?

(c) How many more deliveries did she make on Thursday than on Saturday?

Check

(b)
$$\begin{array}{r} \textbf{1915} \\ -\ 1635 \\ \hline 280 \end{array}$$
Matches Correct
$$\begin{array}{r} 1635 \\ +\ 280 \\ \hline \textbf{1915} \end{array}$$

Check

(c)
$$\begin{array}{r} \textbf{15,803} \\ -\ 7\ 325 \\ \hline 8\ 578 \end{array}$$
Does not match Error
$$\begin{array}{r} 7\ 325 \\ +\ 8\ 578 \\ \hline \textbf{15,903} \end{array}$$

Rework the original problem to get the correct answer, 8478. Then, 7325 + 8478 *does* equal 15,803.

◀ Work Problem **9** at the Side.

OBJECTIVE 6 Use subtraction to solve application problems.

EXAMPLE 10 Applying Subtraction Skills

Diana Lopez drives a delivery truck. Using the table below, decide how many more deliveries were made by Ms. Lopez on Monday than on Thursday.

PACKAGE DELIVERY (LOPEZ)

Day	Number of Deliveries
Monday	137
Tuesday	126
Wednesday	119
Thursday	89
Friday	147
Saturday	0

Ms. Lopez made 137 deliveries on Monday, but had only 89 deliveries on Thursday. Find how many more deliveries were made on Monday than on Thursday by subtracting 89 from 137.

$$\begin{array}{r} 137 \\ -\ 89 \\ \hline 48 \end{array}$$ ← Deliveries on Monday
← Deliveries on Thursday
← More deliveries on Monday

Ms. Lopez made 48 more deliveries on Monday than she made on Thursday.

◀ Work Problem **10** at the Side.

ANSWERS

9. (a) 368 + 57 = 425; correct
   (b) 439 + 241 ≠ 670; incorrect; should be 231
   (c) 8839 + 5887 = 14,726; correct
10. (a) 21 fewer deliveries
    (b) 7 fewer deliveries on Wednesday
    (c) 89 more deliveries

R.2 ▶▶▶ Exercises

FOR
EXTRA
HELP

MyMathLab
 Math XL
PRACTICE

WATCH

DOWNLOAD

READ

REVIEW

Use addition to check each subtraction. If an answer is incorrect, find the correct answer.
See Examples 3 and 4.

1.	89	**2.**	47	**3.**	382	**4.**	838
	− 27		− 35		− 261		− 516
	63		13		131		322

Find each difference, regrouping when necessary. See Examples 5–8.

5.	36	**6.**	97	**7.**	83	**8.**	65	**9.**	45
	− 28		− 39		− 58		− 28		− 29

10.	93	**11.**	719	**12.**	916	**13.**	771	**14.**	973
	− 37		− 658		− 618		− 252		− 788

15.	9861	**16.**	6171	**17.**	9988	**18.**	3576	**19.**	38,335
	− 684		− 1182		− 2399		− 1658		− 29,476

20.	61,278	**21.**	40	**22.**	80	**23.**	60	**24.**	70
	− 3 559		− 37		− 73		− 37		− 27

25.	6020	**26.**	7050	**27.**	8503	**28.**	16,004	**29.**	80,705
	− 4078		− 6045		− 2816		− 5 087		− 61,667

30.	72,000	**31.**	66,000	**32.**	77,000	**33.**	20,080	**34.**	80,056
	− 44,234		− 444		− 308		− 96		− 69

Use addition to check each subtraction. If an answer is incorrect, find the correct answer.
See Example 9.

35.	3070	**36.**	1439	**37.**	27,600	**38.**	34,021
	− 576		− 1169		− 807		− 33,708
	2596		270		26,793		727

Solve each application problem. See Example 10.

39. Swimming laps burns 255 calories in 30 minutes. Hiking burns 185 calories in 30 minutes. How many fewer calories are burned in 30 minutes by hiking than by swimming?

40. Lynn Couch had $553 in her checking account. She wrote a check for $308 for school fees. How much is left in her account?

41. An airplane was carrying 254 passengers. When it landed in Atlanta, 183 passengers got off and 109 passengers got on. How many passengers were on the plane then?

42. On Tuesday, 5822 people went to a soccer game, and on Friday, 7994 people went to a soccer game. How many more people went to the game on Friday? What was the total attendance at the two games?

This table shows the median yearly earnings for various occupations. Use the table to answer Exercises 43 and 44.

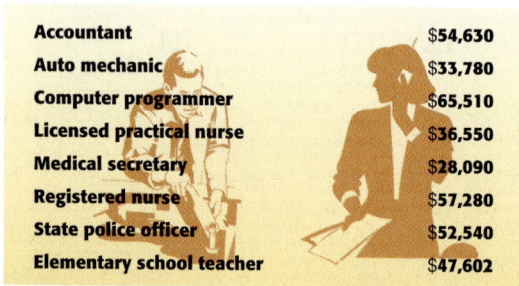

ALL WALKS OF LIFE

Median yearly earnings for selected occupations; starting salaries are lower

Accountant	$54,630
Auto mechanic	$33,780
Computer programmer	$65,510
Licensed practical nurse	$36,550
Medical secretary	$28,090
Registered nurse	$57,280
State police officer	$52,540
Elementary school teacher	$47,602

Source: U.S. Department of Labor *Occupational Outlook Handbook.*

43. (a) Identify the occupations with the highest and lowest earnings.

(b) What is the difference in the yearly earnings for these two occupations?

44. (a) How much less are the yearly earnings of an auto mechanic than a state police officer

(b) How much more are the yearly earnings of a registered nurse than a licensed practical nurse?

45. Downtown Toronto's skyline is dominated by the CN Tower, which rises 1831 feet. The Sears Tower in Chicago is 1451 feet high. Find the difference in height between the two structures. (*Source: World Almanac.*)

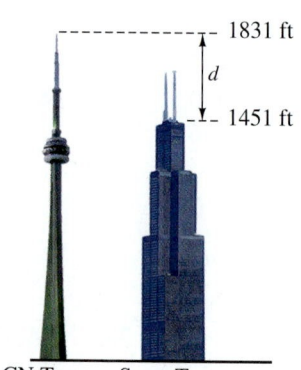

1831 ft

d

1451 ft

CN Tower   Sears Tower

46. The fastest animal in the world, the peregrine falcon, dives at 185 miles per hour. A Boeing 747 cruises at 580 miles per hour. How much faster does the plane cruise than the falcon dives? (*Source: Top 10 of Everything.*)

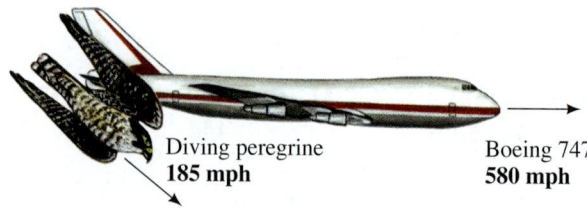

Diving peregrine
185 mph

Boeing 747
580 mph

R.3 ▶▶▶ Multiplying Whole Numbers

Suppose we want to know the total number of computers in a computer lab. The computers are arranged in three rows with four computers in each row. Adding the number 4 a total of 3 times gives 12.

$$4 + 4 + 4 = 12$$

This result is illustrated below. There are 4 computers in each row.

$$
\begin{array}{l}
4 \\
4 \\
+\ 4 \\
\hline
12
\end{array}
$$

3 rows

OBJECTIVE 1 Identify the parts of a multiplication problem.
Multiplication is a shortcut for repeated addition. The numbers being multiplied are called **factors.** The answer is called the **product.** For example, the product of 3 and 4 can be written with the symbol $\times$, a raised dot, parentheses, or, in computer work, an asterisk.

$$
\begin{array}{l}
3 \leftarrow \text{Factor (also called } \textit{multiplicand}\,) \\
\underline{\times\ \ 4} \leftarrow \text{Factor (also called } \textit{multiplier}) \\
12 \leftarrow \text{Product (answer)}
\end{array}
$$

$$3 \times 4 = 12 \qquad 3 \cdot 4 = 12 \qquad (3)(4) = 12 \qquad 3 * 4 = 12$$

Raised dot   In computer work

Work Problem **1** *at the Side.* ▶

Commutative Property of Multiplication

By the **commutative property of multiplication,** changing the *order* of two factors does *not* change the product.

For example: $3 \times 5 = 15$ and $5 \times 3 = 15$

Both products are 15.

CAUTION

Remember, addition also has a commutative property. For example, $4 + 2$ has the same sum as $2 + 4$. Subtraction, however, is **not** commutative.

EXAMPLE 1 Multiplying Two Numbers

Multiply. (Remember, a raised dot means to multiply.) Do the work mentally.

(a) $3 \times 4 = $ **12**   By the commutative property, $4 \times 3 = 12$ also.

(b) $6 \cdot 0 = $ **0**   The product of any number and 0 is 0; if you give no money to each of 6 relatives, you give no money.

(c) $(4)(8) = $ **32**   By the commutative property, $(8)(4) = 32$ also.

Work Problem **2** *at the Side.* ▶

---

1 Identify the factors and the product in each multiplication problem.

(a) $3 \times 6 = 18$

(b) $32 = 8 \times 4$

(c) $5 \cdot 7 = 35$

(d) $(3)(9) = 27$

2 Multiply. Do the work mentally. Then use the commutative property to write another multiplication problem and find the product.

(a) 4×7

(b) 0×9

(c) $8 \cdot 6$

(d) $5 \cdot 5$

(e) $(3)(8)$

ANSWERS

1. **(a)** factors: 3, 6; product: 18
   (b) factors: 8, 4; product: 32
   (c) factors: 5, 7; product: 35
   (d) factors: 3, 9; product: 27
2. **(a)** 28; $7 \times 4 = 28$ **(b)** 0; $9 \times 0 = 0$
   (c) 48; $6 \cdot 8 = 48$ **(d)** 25; no change
   (e) 24; $(8)(3) = 24$

3 Find each product.

(a) $2 \times 3 \times 4$

(b) $6 \cdot 1 \cdot 5$

(c) $(8)(3)(0)$

(d) $3 \times 3 \times 7$

(e) $4 \cdot 2 \cdot 8$

(f) $(2)(2)(9)$

OBJECTIVE 2 Do chain multiplications. Some multiplications involve more than two factors.

> **Associative Property of Multiplication**
>
> By the **associative property of multiplication**, changing the *grouping* of factors does *not* change the product.

EXAMPLE 2 Multiplying Three Numbers

Multiply $2 \times 3 \times 5$.

$$(2 \times 3) \times 5 \qquad \text{Parentheses tell what to do first.}$$
$$6 \quad \times 5 = 30 \leftarrow \text{Product}$$

Also,

$$2 \times (3 \times 5)$$
$$2 \times \quad 15 \quad = 30 \leftarrow \text{Product}$$

By the associative property, either grouping results in the same product.

> **Calculator Tip** The calculator approach to Example 2 above uses chain calculations. Notice that you can enter *all* the factors before pressing the ⊜ key.
>
> $$2 \; \text{⊗} \; 3 \; \text{⊗} \; 5 \; \text{⊜} \; 30$$

◀ *Work Problem* **3** *at the Side.*

OBJECTIVE 3 Multiply by single-digit numbers. Regrouping may be needed in multiplication problems with larger factors.

EXAMPLE 3 Regrouping with Multiplication

Find each product.

(a) $\begin{array}{r} 53 \\ \times\ 4 \\ \hline \end{array}$

Start by multiplying in the ones column. Multiply 4 times 3 ones.

$$\begin{array}{r} 1 \leftarrow \\ 53 \\ \times\ 4 \\ \hline 2 \leftarrow \end{array} \qquad 4 \times 3 \text{ ones} = \textbf{12 ones}$$

Write 1 ten in the tens column.
Write 2 ones in the ones column.

Next, multiply 4 times 5 tens.

$$\begin{array}{r} 1 \\ 53 \\ \times\ 4 \\ \hline 2 \end{array} \qquad 4 \times 5 \text{ tens} = 20 \text{ tens}$$

Add the 1 ten that was written at the top of the tens column.

$$\begin{array}{r} 1 \\ 53 \\ \times\ 4 \\ \hline 212 \end{array} \qquad 20 \text{ tens} + 1 \text{ ten} = \textbf{21 tens}$$

ANSWERS

3. **(a)** 24   **(b)** 30   **(c)** 0   **(d)** 63
   (e) 64   **(f)** 36

Continued on Next Page

(b)   724
$$\times\ \ 5$$

Work as shown below.

$$
\begin{array}{r}
\text{\small 12}\\
724\\
\times\ \ \ 5\\
\hline
3620
\end{array}
$$

← 5×4 ones $= 20$ ones; write 0 ones; write 2 tens in the tens column.

5×2 tens $= 10$ tens; add the 2 regrouped tens to get 12 tens; write 2 tens; write 1 hundred in the hundreds column.

5×7 hundreds $= 35$ hundreds; add the 1 regrouped hundred to get 36 hundreds.

Work Problem (**4**) *at the Side.* ▶

OBJECTIVE **4** **Multiply quickly by numbers ending in zeros.**
The product of two whole number factors is also called a **multiple** of either factor. For example, since $4 \cdot 2 = 8$, the number 8 is a multiple of 4, and 8 is also a multiple of 2. Multiples of 10 are very useful when multiplying. A *multiple of 10* is a whole number that ends in 0, such as 10, 20, or 30; 100, 200, or 300; 1000, 2000, or 3000; and so on. There is a short way to multiply by multiples of 10. Look at the following examples.

$$26 \times 1 = 26$$

$$26 \times 10 = 260$$

$$26 \times 100 = 2600$$

$$26 \times 1000 = 26{,}000$$

Do you see a pattern in the multiplications? These examples suggest the following rule.

> **Multiplying by Multiples of 10**
> To multiply a whole number by 10, by 100, or by 1000, attach one, two, or three zeros to the right of the whole number.

EXAMPLE 4   **Multiplying by Multiples of 10**

Multiply.

(a) $59 \times 10 = 590$
            Attach 0

(b) $74 \times 100 = 7400$
            Attach 00

(c) $803 \times 1000 = 803{,}000$
            Attach 000

Work Problem (**5**) *at the Side.* ▶

You can also find the product of other multiples of ten by attaching zeros.

---

4 Find each product.

(a)   52
$$\times\ 5$$

(b)   79
$$\times\ 0$$

(c)   862
$$\times\ 9$$

(d)   2831
$$\times\ \ \ 7$$

(e)   4714
$$\times\ \ \ 8$$

5 Multiply by attaching zeros.

(a) 45×10

(b) 102×100

(c) 571×1000

(d) 3625×100

(e) 69×1000

ANSWERS

4. **(a)** 260 **(b)** 0 **(c)** 7758 **(d)** 19,817
   (e) 37,712
5. **(a)** 450 **(b)** 10,200 **(c)** 571,000
   (d) 362,500 **(e)** 69,000

6 Find each product by attaching zeros.

(a) 14×50

(b) $(68)(400)$

(c) $\begin{array}{r} 180 \\ \times\ 30 \\ \hline \end{array}$

(d) $\begin{array}{r} 6100 \\ \times\ \ 90 \\ \hline \end{array}$

(e) $\begin{array}{r} 800 \\ \times\ 200 \\ \hline \end{array}$

(f) $(5000)(700)$

(g) $(9)(20{,}000)$

EXAMPLE 5 **Multiplying with Other Multiples of 10**

Find each product.

(a) 75×3000

Multiply 75 by 3, and then attach three zeros.

$$\begin{array}{r} 75 \\ \times\ \ 3 \\ \hline 225 \end{array} \qquad 75 \times 3000 = 225{,}000$$

(b) 150×70

Multiply 15 by 7, and then attach two zeros.

$$\begin{array}{r} 15 \\ \times\ \ 7 \\ \hline 105 \end{array} \qquad 150 \times 70 = 10{,}500$$

◀ *Work Problem* **6** *at the Side.*

OBJECTIVE **5** **Multiply by numbers having more than one digit.**
The next example shows multiplication when both factors have more than one digit.

EXAMPLE 6 **Multiplying with More Than One Digit**

Multiply.

(a) $(46)(23)$

Rewrite the problem in vertical format. Then start by multiplying 46 by 3.

$$\begin{array}{r} 1 \\ 46 \\ \times\ \ 3 \\ \hline 138 \end{array} \leftarrow 46 \times 3 = 138$$

Next, multiply 46 by 20.

$$\begin{array}{r} 1 \\ 46 \\ \times\ 20 \\ \hline 138 \\ 920 \end{array} \leftarrow 46 \times 20 = 920$$

Add the results.

$$\begin{array}{r} 46 \\ \times\ \ 23 \\ \hline 138 \leftarrow 46 \times 3 \\ +\ 920 \leftarrow 46 \times 20 \\ \hline 1058 \leftarrow \text{Add to find the product.} \end{array}$$

Both 138 and 920 are called *partial products*. To save time, the 0 in 920 is usually not written.

$$\begin{array}{r} 46 \\ \times\ \ 23 \\ \hline 138 \\ 92 \\ \hline 1058 \end{array}$$

0 not written. Be very careful to place the 2 in the tens column.

Continued on Next Page

(b)
```
      2 3 3
  ×   1 3 2
  ─────────
      4 6 6
    6 9 9        Tens lined up
  2 3 3          Hundreds lined up
  ─────────
  3 0,7 5 6  ← Product
```

(c)
```
      5 3 8
  ×     4 6
```

First multiply by 6.

Regrouping is needed here.
```
        2 4
      5 3 8
  ×     4 6
  ─────────
      3 2 2 8
```

Now multiply by 4, being careful to line up the tens.
```
        1 3
        2 4
      5 3 8
  ×     4 6
  ─────────
      3 2 2 8   ┐
      2 1 5 2   ┘─ Add the partial products.
  ─────────
    2 4,7 4 8
```

Work Problem **7** *at the Side.* ▶

When 0 appears in the multiplier, be sure to move the partial product to the left to account for the position held by the 0.

EXAMPLE 7 **Multiplication with Zeros**

Find each product.

(a)
```
      1 3 7
  ×   3 0 6
  ─────────
      8 2 2
    0 0 0        Tens lined up
  4 1 1          Hundreds lined up
  ─────────
  4 1,9 2 2
```

(b)
```
      1 4 0 6              1 4 0 6
  ×   2 0 0 1          ×   2 0 0 1
  ───────────          ───────────
      1 4 0 6              1 4 0 6
    0 0 0 0  ← 0 to line up tens
  0 0 0 0    ← 0 to line up hundreds   2 8 1 2 0 0
  2 8 1 2
  ───────────          ───────────
  2,8 1 3,4 0 6        2,8 1 3,4 0 6
```

Zeros are written so that you start writing the partial product 2812 in the *thousands* column.

CAUTION
In Example 7(b) above, in the solution on the right, be careful to insert zeros so that thousands are lined up in the thousands column.

Continued on Next Page

---

7 Find each product.

(a)
```
      3 8
  ×   1 5
```

(b)
```
      3 1
  ×   4 3
```

(c)
```
      6 7
  ×   5 9
```

(d)
```
     2 3 4
  ×   7 3
```

(e)
```
     8 3 5
  × 1 8 9
```

8 Find each product.

(a)   28
    × 60

(b)   817
    × 30

(c)   481
    × 206

(d)   3526
    × 6002

◀ *Work Problem* **8** *at the Side.*

OBJECTIVE **6** **Use multiplication to solve application problems.**

EXAMPLE 8 **Applying Multiplication Skills**

Find the total cost of 24 months of Internet service at $59 per month.

Approach   To find the total cost, multiply the cost for one month of service ($59) by the number of months (24).

Solution   Multiply $59 by 24.

$$
\begin{array}{r}
59 \\
\times\ 24 \\
\hline
236 \\
118 \\
\hline
1416
\end{array}
$$

The total cost of 24 months of Internet service is $1416.

⊞ **Calculator Tip** If you are using a calculator for Example 8 above, press the following keys.

59 ⊗ 24 ⊜ **1416**

◀ *Work Problem* **9** *at the Side.*

9 Find the total cost of these items.

(a) 36 months of cable TV costing $79 each month

(b) 15 laptop computers priced at $1090 each

(c) 60 months of car payments at $389 per month

R.3 ▶▶▶ **Exercises**

Find each product. Try to do the work mentally. See Examples 1 and 2.

1. $3 \times 1 \times 3$ **2.** $2 \times 8 \times 2$ **3.** $9 \times 1 \times 7$ **4.** $2 \times 4 \times 5$ **5.** $9 \cdot 5 \cdot 0$

6. $6 \cdot 0 \cdot 8$ **7.** $(4)(1)(6)$ **8.** $(1)(5)(7)$ **9.** $(2)(3)(6)$ **10.** $(4)(1)(9)$

Find each product. See Examples 3–7.

11. $\begin{array}{r} 35 \\ \times\ 7 \\ \hline \end{array}$ **12.** $\begin{array}{r} 76 \\ \times\ 9 \\ \hline \end{array}$ **13.** $\begin{array}{r} 28 \\ \times\ 6 \\ \hline \end{array}$ **14.** $\begin{array}{r} 83 \\ \times\ 5 \\ \hline \end{array}$ **15.** $\begin{array}{r} 3182 \\ \times\ 6 \\ \hline \end{array}$

16. $\begin{array}{r} 7326 \\ \times\ 5 \\ \hline \end{array}$ **17.** $\begin{array}{r} 36{,}921 \\ \times\ 7 \\ \hline \end{array}$ **18.** $\begin{array}{r} 28{,}116 \\ \times\ 4 \\ \hline \end{array}$ **19.** $\begin{array}{r} 125 \\ \times\ 100 \\ \hline \end{array}$ **20.** $\begin{array}{r} 246 \\ \times\ 100 \\ \hline \end{array}$

21. $\begin{array}{r} 1485 \\ \times\ 30 \\ \hline \end{array}$ **22.** $\begin{array}{r} 8522 \\ \times\ 50 \\ \hline \end{array}$ **23.** $\begin{array}{r} 900 \\ \times\ 300 \\ \hline \end{array}$ **24.** $\begin{array}{r} 400 \\ \times\ 700 \\ \hline \end{array}$ **25.** $\begin{array}{r} 43{,}000 \\ \times\ 2000 \\ \hline \end{array}$

26. $\begin{array}{r} 11{,}000 \\ \times\ 9000 \\ \hline \end{array}$ **27.** $\begin{array}{r} 68 \\ \times\ 22 \\ \hline \end{array}$ **28.** $\begin{array}{r} 82 \\ \times\ 32 \\ \hline \end{array}$ **29.** $\begin{array}{r} 83 \\ \times\ 45 \\ \hline \end{array}$ **30.** $(43)(27)$

31. $(32)(475)$ **32.** $(67)(218)$ **33.** $(729)(45)$ **34.** $(681)(47)$ **35.** $\begin{array}{r} 538 \\ \times\ 342 \\ \hline \end{array}$

36. $\begin{array}{r} 3228 \\ \times\ 751 \\ \hline \end{array}$ **37.** $\begin{array}{r} 8162 \\ \times\ 407 \\ \hline \end{array}$ **38.** $\begin{array}{r} 528 \\ \times\ 106 \\ \hline \end{array}$ **39.** $\begin{array}{r} 6310 \\ \times\ 3008 \\ \hline \end{array}$ **40.** $\begin{array}{r} 3533 \\ \times\ 5001 \\ \hline \end{array}$

Solve each application problem. See Example 8.

41. Giant kelp plants in the ocean can grow 18 inches each day. How much could kelp grow in two weeks? How much could it grow in a 30-day month? (*Source:* Natural Bridges State Park, CA.)

42. A hospital has 20 bottles of thyroid medication, with each bottle containing 2500 tablets. How many of these tablets does the hospital have in all?

43. There are 12 tomato plants to a flat. If a garden center has 48 flats, find the total number of tomato plants.

44. A hummingbird's wings beat about 65 times per second, a chickadee's wings about 27 times per second. How many times does each bird's wings beat in one minute? (*Source: Birder's Handbook.*)

45. A new Prius hybrid automobile gets 45 miles per gallon on the highway. How many highway miles can it travel on 11 gallons of gas?

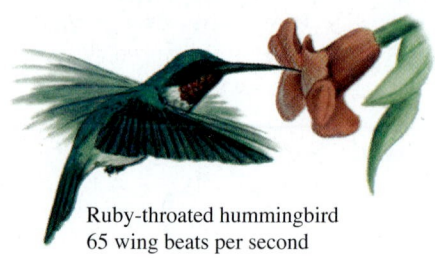

46. Find the total cost of 16 gallons of paint at $27 per gallon.

Ruby-throated hummingbird
65 wing beats per second

Use addition, subtraction, or multiplication to solve each problem.

47. The distance from Reno, Nevada, to the Atlantic Ocean is 2695 miles, while the distance from Reno to the Pacific Ocean is 255 miles. How much farther is it to the Atlantic Ocean than it is to the Pacific Ocean? If you make three round trips from Reno to the Atlantic Ocean, how many frequent flier miles will you earn?

48. The largest living land mammal is the African elephant, and the largest mammal of all time is the blue whale. An African bull elephant may weigh 15,225 pounds and a blue whale may weigh 12 to 25 times that amount. Find the range of weights for the blue whale and the difference between the lightest and heaviest. (*Source: Big Book of Knowledge.*)

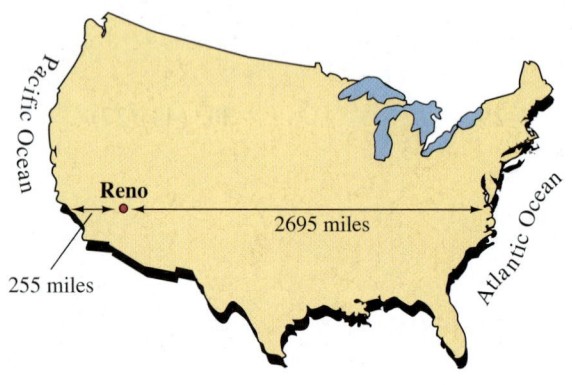

49. A high-fat meal contains 1406 calories, while a low-fat meal contains 348 calories. How many more calories are in seven high-fat meals than in seven low-fat meals?

50. Dannie Sanchez bought four tires at $110 each, two seat covers at $49 each, and six socket wrenches at $3 each. Sales tax was $43. Find the total amount that he spent.

R.4 ▶▶▶ Dividing Whole Numbers

Suppose $12 is to be divided into 3 equal parts. Each part would be $4, as shown here.

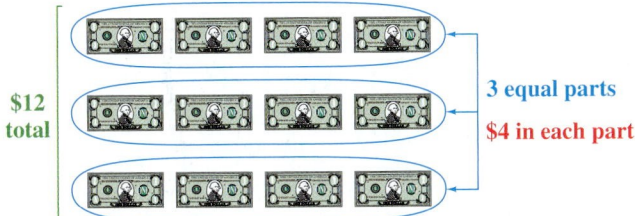

$12 total

3 equal parts
$4 in each part

OBJECTIVES

1 Write division problems in three ways.

2 Identify the parts of a division problem.

3 Divide 0 by a number.

4 Recognize that division by 0 is undefined.

5 Divide a number by itself.

6 Use short division.

7 Use multiplication to check quotients.

8 Use tests for divisibility.

OBJECTIVE 1 Write division problems in three ways. Just as $3 \cdot 4$, 3×4, and $(3)(4)$ are different ways of writing the multiplication of 3 and 4, there are several ways to write 12 divided by 3.

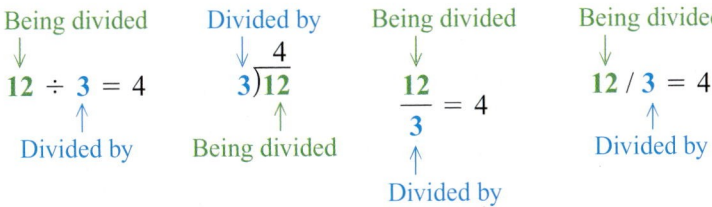

Being divided
$$12 \div 3 = 4$$
Divided by

Divided by
$$3\overline{)12}^{\,4}$$
Being divided

Being divided
$$\frac{12}{3} = 4$$
Divided by

Being divided
$$12\,/\,3 = 4$$
Divided by

We will use three division symbols, $\div$, $\overline{)}$, and $—$. In algebra the bar, $—$, is frequently used. In computer science the slash, $/$, is used.

EXAMPLE 1 Using Division Symbols

Rewrite each division using two other symbols.

(a) $20 \div 4 = 5$

This division can also be written $4\overline{)20}^{\,5}$ or $\dfrac{20}{4} = 5$

(b) $\dfrac{18}{6} = 3$ can also be written $18 \div 6 = 3$ or $6\overline{)18}$

(c) $5\overline{)40}^{\,8}$ can also be written $40 \div 5 = 8$ or $\dfrac{40}{5} = 8$

Work Problem **1** *at the Side.* ▶

OBJECTIVE 2 Identify the parts of a division problem. In division, the number being divided is the **dividend,** the number you are dividing by is the **divisor,** and the answer is the **quotient.**

$$\text{dividend} \div \text{divisor} = \text{quotient}$$

$$\text{divisor}\overline{)\text{dividend}}^{\,\text{quotient}} \qquad \frac{\text{dividend}}{\text{divisor}} = \text{quotient}$$

EXAMPLE 2 Identifying the Parts in a Division Problem

Identify the dividend, divisor, and quotient.

(a) $35 \div 7 = 5$

$$35 \div 7 = 5 \leftarrow \text{Quotient}$$
Dividend   Divisor

Continued on Next Page

1 Rewrite each division using two other symbols.

(a) $48 \div 6 = 8$

(b) $24 \div 6 = 4$

(c) $9\overline{)36}^{\,4}$

(d) $\dfrac{42}{6} = 7$

ANSWERS

1. **(a)** $6\overline{)48}^{\,8}$ and $\dfrac{48}{6} = 8$

   (b) $6\overline{)24}^{\,4}$ and $\dfrac{24}{6} = 4$

   (c) $36 \div 9 = 4$ and $\dfrac{36}{9} = 4$

   (d) $6\overline{)42}^{\,7}$ and $42 \div 6 = 7$

2 Identify the dividend, divisor, and quotient.

(a) $10 \div 2 = 5$

(b) $6 = 30 \div 5$

(c) $\dfrac{28}{7} = 4$

(d) $2\overline{)36}^{\,18}$

3 Divide.

(a) $0 \div 9$

(b) $\dfrac{0}{36}$

(c) $57\overline{)0}$

4 Write each division problem as a multiplication problem.

(a) $6\overline{)18}^{\,3}$

(b) $\dfrac{28}{4} = 7$

(c) $48 \div 8 = 6$

(b) $\dfrac{100}{20} = 5$

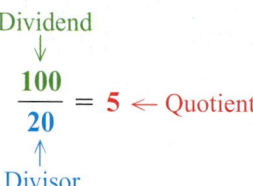

$$\dfrac{\mathbf{100}}{\mathbf{20}} = \mathbf{5} \leftarrow \text{Quotient}$$
$$\uparrow \text{ Divisor}$$

(c) $12\overline{)72}^{\,6}$

$$\mathbf{6} \leftarrow \text{Quotient}$$
$$\mathbf{12}\overline{)72} \leftarrow \text{Dividend}$$
$$\uparrow \text{ Divisor}$$

◀ *Work Problem* **2** *at the Side.*

OBJECTIVE 3 Divide 0 by a number. If no money, or $0, is divided equally among five people, each person gets $0. There is a general rule for dividing 0.

> **Dividing 0**
>
> When 0 is divided by any other number (except 0), the quotient is 0.

EXAMPLE 3 Dividing 0 by a Number

Divide.

(a) $0 \div 12 = \mathbf{0}$

(b) $0 \div 1728 = \mathbf{0}$

(c) $\dfrac{0}{375} = \mathbf{0}$

(d) $129\overline{)0}^{\,\mathbf{0}}$

◀ *Work Problem* **3** *at the Side.*

Recall that a subtraction such as $8 - 3 = 5$ can be written as the addition $8 = 3 + 5$. In a similar way, any division can be written as a multiplication. For example, $12 \div 3 = 4$ can be written as

$$3 \times 4 = 12 \quad \text{or, by the commutative property,} \quad 4 \times 3 = 12$$

EXAMPLE 4 Changing Division to Multiplication

Change each division to multiplication.

(a) $\dfrac{20}{4} = 5$ becomes $4 \cdot 5 = 20$

(b) $8\overline{)48}^{\,6}$ becomes $8 \cdot 6 = 48$

(c) $72 \div 9 = 8$ becomes $9 \cdot 8 = 72$

◀ *Work Problem* **4** *at the Side.*

OBJECTIVE **4** **Recognize that division by 0 is undefined.**
Division by 0 cannot be done. To see why, try to find the answer to this division.

$$9 \div 0 = ?$$

As we have just seen, any division problem can be changed to a multiplication problem.

$$\text{divisor} \cdot \text{quotient} = \text{dividend}$$

If you convert the problem $9 \div 0 = ?$ to its multiplication counterpart, it reads as follows.

$$0 \cdot ? = 9$$

You already know that 0 times any number must always equal 0. Try any number you like to replace the **?** and you'll always get 0 instead of 9. Therefore, the division problem $9 \div 0$ *cannot* be done. Mathematicians say it is *undefined* and have agreed never to divide by 0. However, $0 \div 9$ *can* be done. Check by rewriting it as a multiplication problem.

$$0 \div 9 = 0 \quad \text{because} \quad 9 \cdot 0 = 0 \quad \text{is true.}$$

> **Dividing by 0**
> Dividing by 0 cannot be done. We say that division by 0 is **undefined.** It is impossible to compute an answer.

EXAMPLE 5 **Dividing by 0 Is Undefined**

All of these divisions are *undefined.*

(a) $\dfrac{6}{0}$ is undefined. *It is **not possible** to divide by 0. Write undefined as the answer (**not** 0).*

(b) $0\overline{)8}$ is undefined.

(c) $18 \div 0$ is undefined.

> **Division Involving 0**
> $$\frac{0}{\text{nonzero number}} = 0 \quad \textbf{but} \quad \frac{\text{number}}{0} \text{ is } \textbf{undefined.}$$

> **CAUTION**
> When 0 is the divisor in a problem, write **undefined.**
> **Never divide by 0.**

——— Work Problem **5** at the Side. ▶

Calculator Tip Try these two problems on your calculator. Jot down your answers.

9 ÷ 0 = _____   0 ÷ 9 = _____

When you try to divide by 0, the calculator cannot do it, so it shows the word "Error" in the display, or the letters "ERR" or "E" (for "error").

5 Find the quotient whenever possible.

(a) $\dfrac{8}{0}$

(b) $\dfrac{0}{8}$

(c) $0\overline{)32}$

(d) $32\overline{)0}$

(e) $100 \div 0$

(f) $0 \div 100$

ANSWERS
5. (a) undefined (b) 0 (c) undefined
 (d) 0 (e) undefined (f) 0

6 Divide.

(a) $5 \div 5$

(b) $14\overline{)14}$

(c) $\dfrac{37}{37}$

OBJECTIVE 5 Divide a number by itself.   What happens when a number is divided by itself? For example, $4 \div 4$ or $97 \div 97$?

> **Dividing a Number by Itself**
> When a nonzero number is divided by itself, the quotient is 1.

EXAMPLE 6   Dividing a Nonzero Number by Itself

Divide.

(a) $16 \div 16 = $ **1**

(b) $32\overline{)32}$ with **1** on top

(c) $\dfrac{57}{57} = $ **1**

◀ Work Problem **6** at the Side.

OBJECTIVE 6 Use short division.   Short division is a quick method of dividing a number by a one-digit divisor.

7 Divide.

(a) $2\overline{)18}$

EXAMPLE 7   Using Short Division

Divide $2\overline{)86}$.

First, divide 8 by 2.

$$2\overline{)86} \to 4 \qquad \dfrac{8}{2} = 4$$

Next, divide 6 by 2.

$$2\overline{)86} \to 43 \qquad \dfrac{6}{2} = 3$$

(b) $3\overline{)39}$

The quotient is 43.

◀ Work Problem **7** at the Side.

When two numbers do *not* divide evenly, the leftover portion is called the **remainder.** The remainder is always *less* than the divisor.

(c) $4\overline{)88}$

EXAMPLE 8   Using Short Division with a Remainder

Divide 147 by 4.

Rewrite the problem. $4\overline{)147}$

Because 1 cannot be divided by 4, divide 14 by 4.

(d) $2\overline{)462}$

$$4\overline{)14\,{}^2 7} \to 3 \qquad \dfrac{14}{4} = 3 \text{ with 2 left over}$$

Continued on Next Page

ANSWERS

6. all 1
7. **(a)** 9 **(b)** 13 **(c)** 22 **(d)** 231

Next, divide 27 by 4. The final number left over is the remainder. Write the remainder to the side. "R" stands for remainder.

$$\frac{3\ 6\ \mathbf{R3}}{4)14^27} \qquad \frac{27}{4} = 6 \text{ with 3 left over}$$

The quotient is 36 **R3**.

Work Problem **8** *at the Side.* ▶

EXAMPLE 9   Writing Zeroes in the Quotient

Divide 1439 by 7.

Rewrite the problem. Then, divide 14 by 7.

$$\frac{2}{7)1439} \qquad \frac{14}{7} = 2 \text{ with 0 left over}$$

No need to write 0

Next, divide 3 by 7. But 7 will not go into 3 even one time, so write a 0 in the quotient.

Be sure to write the 0 in the quotient.

$$\frac{2\,0}{7)143^39} \qquad \frac{3}{7} = 0 \text{ with 3 left over}$$

Finally, divide 39 by 7.

$$\frac{2\,0\ 5\ \mathbf{R4}}{7)143^39} \qquad \frac{39}{7} = 5 \text{ with 4 left over}$$

The quotient is 205 **R4**.

Work Problem **9** *at the Side.* ▶

OBJECTIVE 7 Use multiplication to check quotients. Check the answer to a division problem as follows.

> **Checking Division**
>
> $$(\text{divisor} \times \text{quotient}) + \text{remainder} = \text{dividend}$$
>
> Parentheses tell you what to do first. In this case, multiply the divisor by the quotient first and then add the remainder.

EXAMPLE 10   Using Multiplication to Check Division

Check each quotient.

(a) $\dfrac{91\ \mathbf{R3}}{5)458}$

$$(\text{divisor} \times \text{quotient}) + \text{remainder} = \text{dividend}$$
$$(5 \quad \times \quad 91) \quad + \quad 3$$
$$455 \qquad + \quad 3 \quad = \mathbf{458}$$

Matches original dividend so the division was done correctly

Continued on Next Page

---

8 Divide.

(a) $2)\overline{225}$

(b) $3)\overline{275}$

(c) $4)\overline{538}$

(d) $\dfrac{819}{5}$

9 Divide.

(a) $4)\overline{837}$

(b) $\dfrac{747}{7}$

(c) $5)\overline{4538}$

(d) $8)\overline{2440}$

ANSWERS

8. (a) 112 **R1**   (b) 91 **R2**   (c) 134 **R2**
   (d) 163 **R4**
9. (a) 209 **R1**   (b) 106 **R5**   (c) 907 **R3**
   (d) 305

10 Check each division. If a quotient is incorrect, find the correct quotient.

(a) $3\overline{)115}$ $\overset{38 \text{ R}1}{}$

(b) $8\overline{)743}$ $\overset{92 \text{ R}2}{}$

(c) $4\overline{)1312}$ $\overset{328}{}$

(d) $5\overline{)2033}$ $\overset{46 \text{ R}3}{}$

(b) $6\overline{)1258}$ $\overset{29 \text{ R}4}{}$

$$(\text{divisor} \times \text{quotient}) + \text{remainder} = \text{dividend}$$
$$(6 \quad \times \quad 29) \quad + \quad 4$$
$$174 \quad + \quad 4 = \mathbf{178}$$

Does **not** match original dividend of 1258

The quotient does *not* check. Rework the original problem to get the correct quotient, 209 **R**4. Then, to check, $(6 \times 209) + 4 = 1254 + 4 = 1258$, the original dividend.

CAUTION
A common error is forgetting to add the remainder. Be sure to add any remainder when checking a division problem.

◄ *Work Problem* **10** *at the Side.*

OBJECTIVE 8 Use tests for divisibility.   It is often important to know whether a number is divisible by another number. You will find this useful in **Chapter 4** when working with fractions.

Divisibility
A whole number is *divisible* by another whole number if the remainder is 0.

There are some quick tests you can use to decide whether one number is divisible by another.

Tests for Divisibility

A number is divisible by

2	if it ends in 0, 2, 4, 6, or 8.
3	if the sum of its digits is divisible by 3.
4	if the last two digits make a number that is divisible by 4.
5	if it ends in 0 or 5.
6	if it is divisible by both 2 and 3.
8	if the last three digits make a number that is divisible by 8.
9	if the sum of its digits is divisible by 9.
10	if it ends in 0.

The most commonly used divisibility tests are those for 2, 3, 5, and 10.

ANSWERS
10. **(a)** correct
    (b) incorrect; should be 92 **R**7
    (c) correct
    (d) incorrect; should be 406 **R**3

Divisibility by 2

A number is divisible by **2** if the number ends in 0, 2, 4, 6, or 8.

EXAMPLE 11 **Testing for Divisibility by 2**

Which numbers are divisible by 2?

(a) 986

↑
└───── Ends in 6

Because the number ends in 6, which is in the list (0, 2, 4, 6, or 8), the number 986 is divisible by 2.

(b) 3255 is *not* divisible by 2.

↑
└───── Ends in 5, and *not* in 0, 2, 4, 6, or 8

Work Problem **11** *at the Side.* ▶

Divisibility by 3

A number is divisible by **3** if the sum of its digits is divisible by **3.**

EXAMPLE 12 **Testing for Divisibility by 3**

Which numbers are divisible by 3?

(a) 4251

Add the digits.

4 2 5 1

$4 + 2 + 5 + 1 = 12$

> Is 12 divisible by 3? Yes. So the *original* number of 4251 is *also* divisible by 3.

Because 12 is divisible by 3, the number 4251 is divisible by 3.

(b) 29,806

Add the digits.

2 9 8 0 6

$2 + 9 + 8 + 0 + 6 = 25$

> Is 25 divisible by 3? No. So the original number of 29,806 is *not* divisible by 3.

Because 25 is *not* divisible by 3, the number 29,806 is *not* divisible by 3.

CAUTION

Be careful when testing for divisibility by *adding the digits*. This method works only when testing for divisibility by 3 or by 9.

Work Problem **12** *at the Side.* ▶

---

11 Which numbers are divisible by 2?

(a) 612

(b) 315

(c) 2714

(d) 36,000

12 Which numbers are divisible by 3?

(a) 836

(b) 7545

(c) 242,913

(d) 102,484

ANSWERS
11. (a), (c), (d)
12. (b) and (c)

13 Which numbers are divisible by 5?

(a) 160

(b) 635

(c) 3381

(d) 108,605

Divisibility by 5 and by 10

A number is divisible by **5** if it ends in 0 or 5.

A number is divisible by **10** if it ends in 0.

EXAMPLE 13 Determining Divisibility by 5

Which numbers are divisible by 5?

(a) 12,900   ends in 0, so it is divisible by 5.

(b) 4325   ends in 5, so it is divisible by 5.

(c) 392   ends in 2, so it is *not* divisible by 5.

◀ *Work Problem* **13** *at the Side.*

EXAMPLE 14 Determining Divisibility by 10

Which numbers are divisible by 10?

(a) 700 and 9140   end in 0, so both numbers are divisible by 10.

(b) 355 and 18,743   do *not* end in 0, so these numbers are *not* divisible by 10.

◀ *Work Problem* **14** *at the Side.*

14 Which numbers are divisible by 10?

(a) 290

(b) 218

(c) 2020

(d) 11,670

ANSWERS

13. (a), (b), (d)

14. (a), (c), (d)

R.4 ▶▶▶ Exercises

Divide. Then rewrite each division using two other division symbols. See Examples 1, 3, 5, and 6.

1. $\dfrac{12}{12}$ 　　　　 **2.** $\dfrac{9}{0}$ 　　　　 **3.** $24 \div 0$ 　　　 **4.** $4 \div 4$ 　　　 **5.** $\dfrac{0}{4}$

6. $0 \div 8$ 　　　 **7.** $0 \div 12$ 　　　 **8.** $\dfrac{0}{7}$ 　　　 **9.** $0\overline{)21}$ 　　　 **10.** $2 \div 0$

Find each quotient using short division. See Examples 7–9. Also, in Exercises 11–14, identify the dividend, the divisor, and the quotient. See Example 2.

11. $4\overline{)84}$ 　　　 **12.** $2\overline{)66}$ 　　　 **13.** $9\overline{)324}$ 　　　 **14.** $8\overline{)176}$

15. $6\overline{)9125}$ 　　　 **16.** $9\overline{)8371}$ 　　　 **17.** $6\overline{)1854}$ 　　　 **18.** $8\overline{)856}$

19. $4024 \div 4$ 　　　 **20.** $16{,}024 \div 8$ 　　　 **21.** $15{,}019 \div 3$ 　　　 **22.** $32{,}013 \div 8$

23. $\dfrac{26{,}684}{4}$ 　　　 **24.** $\dfrac{16{,}398}{9}$ 　　　 **25.** $\dfrac{74{,}751}{6}$ 　　　 **26.** $\dfrac{72{,}543}{5}$

27. $\dfrac{71{,}776}{7}$ 　　　 **28.** $\dfrac{77{,}621}{3}$ 　　　 **29.** $\dfrac{128{,}645}{7}$ 　　　 **30.** $\dfrac{172{,}255}{4}$

Check each quotient. If a quotient is incorrect, find the correct quotient. See Example 10.

31. $7\overline{)4692}$ 　 **67 R2**

32. $9\overline{)5974}$ 　 **663 R5**

33. $6\overline{)21{,}409}$ 　 **3 568 R2**

34. $4\overline{)103{,}516}$ 　 **25,879**

35. $6\overline{)18{,}023}$ 　 **3 003 R5**

36. $8\overline{)33{,}664}$ 　 **4 208**

37. $6\overline{)69{,}140}$ 　 **11,523 R2**

38. $3\overline{)82{,}598}$ 　 **27,532 R1**

Solve each application problem.

39. Kaci Salmon, a supervisor at Albany Electric, earns $184 for an 8-hour shift. The workers she supervises each earn $112 or $152 for an 8-hour shift. What are the hourly wages for Kaci and for her workers?

40. In a hospital weight loss program, Patient A lost 72 pounds in six months, Patient B lost 81 pounds in nine months, and Patient C lost 91 pounds in seven months. On average, how much did each patient lose each month?

41. Six identical delivery vans for Rosita's Bakery cost a total of $99,600. Find the cost of each van.

42. Ted Slauson, coordinator of Toys for Tots, has collected 2628 toys. If his group gives four toys to each child, how many children will receive the gifts?

A college theater group is looking at two different budgets to cover the cost of props, costumes, and programs for the spring play. They are not sure whether to charge $5, $7, or $9 per ticket. The play's director organized the information in the table below. Use the table for Exercises 43–46.

Budget Amoiunt	Number of $5 Tickets	Number of $7 Tickets	Number of $9 Tickets
$1890			
$2205			

43. For the $1890 budget, find the number of tickets that would need to be sold at $5 each; at $7 each; at $9 each. Write your answers in the table at the left.

44. For the $2205 budget, find the number of tickets that would need to be sold at $5 each; at $7 each; at $9 each. Write your answers in the table at the left.

45. If the director thinks that 300 tickets can be sold, what is the lowest ticket price that will cover the $1890 budget? How much extra money will there be?

46. If the director thinks that 300 tickets can be sold, which ticket price will cover the $2205 budget? How much money will be left over?

47. Circle the numbers that are

  (a) divisible by 2

    358   2047   190   85

  (b) divisible by 3

    736   10,404   5603   78

  (c) divisible by 5.

    53   13,740   985   5506

48. Circle the numbers that are

  (a) divisible by 2

    443   3500   256   74

  (b) divisible by 3

    26,001   9316   92   840

  (c) divisible by 5.

    5051   95   30,652   710

Put a ✓ mark in the blank if the number at the left is divisible by the number at the top of each column. Use the divisibility tests from Examples 11–14.

	2	3	5	10
49. 30	___	___	___	___
51. 184	___	___	___	___
53. 445	___	___	___	___
55. 903	___	___	___	___
57. 5166	___	___	___	___
59. 21,763	___	___	___	___

	2	3	5	10
50. 25	___	___	___	___
52. 192	___	___	___	___
54. 897	___	___	___	___
56. 500	___	___	___	___
58. 8302	___	___	___	___
60. 32,472	___	___	___	___

R.5 ▷▷▷   Long Division

Long division is used to divide by a number with more than one digit.

OBJECTIVE 1   Use long division.   In long division, estimate the various numbers by using a *trial divisor* to get a *trial quotient*.

EXAMPLE 1   Using a Trial Divisor and a Trial Quotient

Divide: $42\overline{)3066}$

Because 42 is closer to 40 than to 50, use 4 as a trial divisor.

$$42$$
↑ —— Trial divisor is 4.

Try to divide the first digit of the dividend by 4. Because 3 cannot be divided by 4, use the first *two* digits, 30.

$$\frac{30}{4} = \mathbf{7} \text{ with remainder 2}$$
↓
$$\mathbf{7} \leftarrow \text{Trial quotient is 7.}$$
$$42\overline{)3066}$$

> Be careful to write the 7 over the 6, because $\frac{306}{42}$ is about 7.

Multiply 7 and 42 to get 294; then subtract 294 from 306.

$$
\begin{array}{r}
7 \\
42\overline{)3066} \\
294 \quad \leftarrow 7 \times 42 = 294 \\
\overline{12} \quad \leftarrow 306 - 294 = 12
\end{array}
$$

Bring down the 6 at the right.

$$
\begin{array}{r}
7 \\
42\overline{)306\mathbf{6}} \\
294\downarrow \\
\overline{12\mathbf{6}} \quad \leftarrow \text{6 brought down}
\end{array}
$$

Use the trial divisor, 4.

$$
\begin{array}{r}
7\mathbf{3} \leftarrow \\
42\overline{)3066} \\
294 \\
126 \quad \leftarrow \text{First two digits of 126} \rightarrow \frac{12}{4} = 3 \\
126 \quad \leftarrow 3 \times 42 = 126 \\
\overline{0}
\end{array}
$$

Check the quotient by multiplying 42 and 73. The product should be 3066, which matches the original dividend.

> **CAUTION**
> The first digit in the answer in long division must be placed in the proper position over the dividend.

Work Problem 1 *at the Side.* ▶

OBJECTIVES

1 Use long division.

2 Divide by multiples of 10.

3 Use multiplication to check quotients.

1 Divide.

(a) $64\overline{)4608}$ *Hint:* 64 is closer to 60 than 70, so use 6 as the trial divisor.

(b) $32\overline{)1792}$ Use ____ as the trial divisor.

(c) $51\overline{)2295}$ Use ____ as the trial divisor.

(d) $\dfrac{6391}{83}$ Use ____ as the trial divisor.

ANSWERS

1. (a) 72   (b) Use 3; 56   (c) Use 5; 45
   (d) Use 8; 77

2 Divide.

(a) $56\overline{)2352}$

(b) $38\overline{)1599}$

(c) $65\overline{)5416}$

(d) $89\overline{)6649}$

EXAMPLE 2 **Dividing to Find a Trial Quotient**

Divide: $58\overline{)2730}$

Use 6 as a trial divisor, because 58 is closer to 60 than to 50.

First two digits → $\dfrac{27}{6}$ = **4** with remainder 3
of dividend

Be careful to write the 4 over the 3.

$$
\begin{array}{r}
4 \quad \leftarrow \text{Trial quotient is 4.}\\
58\overline{)2730}\\
232 \quad \leftarrow 4 \times 58 = 232\\
\overline{41} \quad \leftarrow 273 - 232 = 41 \text{ (smaller than 58,}\\
\text{the divisor)}
\end{array}
$$

Bring down the 0.

$$
\begin{array}{r}
4\\
58\overline{)2730}\\
232\downarrow\\
410 \quad \leftarrow 0 \text{ brought down}
\end{array}
$$

$$
\begin{array}{r}
46\\
58\overline{)2730}\\
232\\
\overline{410}\\
348 \quad \leftarrow 6 \times 58 = 348\\
\overline{62} \quad \leftarrow \text{Greater than 58}
\end{array}
$$

First two digits of 410 $\dfrac{41}{6}$ = **6** with remainder 5

The remainder, 62, is *greater than the divisor,* 58, so **7** should be used in the quotient instead of **6**.

$$
\begin{array}{r}
47 \text{ R4} \leftarrow\\
58\overline{)2730}\\
232\\
\overline{410}\\
406 \quad \leftarrow 7 \times 58 = 406\\
\overline{4} \quad \leftarrow 410 - 406 = 4
\end{array}
$$

◄ Work Problem **2** at the Side.

Sometimes it is necessary to write a 0 in the quotient.

EXAMPLE 3 **Writing Zeros in the Quotient**

Divide: $42\overline{)8734}$

Start as above in Example 2.

$$
\begin{array}{r}
2\\
42\overline{)8734}\\
84 \quad \leftarrow 2 \times 42 = 84\\
\overline{3} \quad \leftarrow 87 - 84 = 3
\end{array}
$$

Bring down the 3.

$$
\begin{array}{r}
2\\
42\overline{)8734}\\
84\downarrow\\
33 \quad \leftarrow 3 \text{ brought down}
\end{array}
$$

Continued on Next Page

Since 33 cannot be divided by 42, write a 0 in the quotient as a placeholder.

$$\begin{array}{r} 2\mathbf{0} \\ 42\overline{)8734} \\ 84 \\ \hline 33 \end{array}$$

Write a 0 in the quotient.

Bring down the final digit, the 4.

$$\begin{array}{r} 20 \\ 42\overline{)873\mathbf{4}} \\ 84 \downarrow \\ \hline 334 \end{array}$$ ← 4 brought down

Complete the problem.

$$\begin{array}{r} 207 \;\mathbf{R}40 \\ 42\overline{)8734} \\ 84 \\ \hline 334 \\ 294 \\ \hline 40 \end{array}$$
← 7 × 42 = 294
← 334 − 294 = 40

The quotient is 207 **R**40.

> **CAUTION**
> There must be a digit in the quotient (answer) above *every* digit in the dividend *once the answer has begun.* Notice that in Example 3 above, a 0 was used to assure an answer digit above every digit in the dividend.

Work Problem **3** *at the Side.* ▶

OBJECTIVE 2 Divide by multiples of 10. When the divisor and dividend both contain zeros at the far right, recall that these numbers are multiples of 10. There is a short way to divide multiples of 10. Look at the following examples.

$$26{,}000 \div 1 = 26{,}000$$
$$26{,}000 \div 10 = 2600$$
$$26{,}000 \div 100 = 260$$
$$26{,}000 \div 1000 = 26$$

Do you see a pattern in these divisions using multiples of 10? These examples suggest the following rule.

> **Dividing by Multiples of 10**
> Divide a whole number by 10, by 100, or by 1000 by dropping one, two, or three zeros from the whole number.

EXAMPLE 4 **Dividing by Multiples of 10**

Divide.

— 0 in divisor

(a) 6<u>0</u> ÷ 1<u>0</u> = 6

 — Drop 0 from dividend.

Continued on Next Page

3 Divide.

(a) 24$\overline{)3127}$

(b) 52$\overline{)10{,}660}$

(c) 39$\overline{)15{,}933}$

(d) 78$\overline{)23{,}462}$

4 Divide by dropping zeros.

(a) $50 \div 10$

(b) $1800 \div 100$

(c) $305,000 \div 1000$

5 Drop zeros and then divide.

(a) $60\overline{)7200}$

(b) $130\overline{)131,040}$

(c) $2600\overline{)195,000}$

6 Check each division. If the quotient is incorrect, find the correct quotient.

(a)
$$
\begin{array}{r}
43 \\
18\overline{)774} \\
72 \\
\hline
54 \\
54 \\
\hline
0
\end{array}
$$

(b)
$$
\begin{array}{r}
42\ \mathbf{R}178 \\
426\overline{)19,170} \\
1\ 704 \\
\hline
1\ 130 \\
952 \\
\hline
178
\end{array}
$$

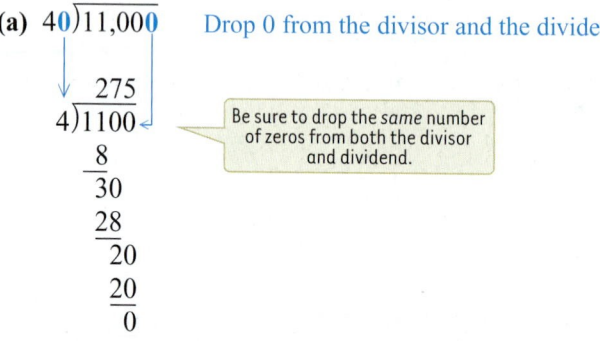

(b) $3\underline{500} \div 1\underline{00} = 35$ — 00 in divisor   — Drop 00 from dividend.

(c) $915,\underline{000} \div 1\underline{000} = 915$ — 000 in divisor   — Drop 000 from dividend.

◀ *Work Problem* **4** *at the Side.*

EXAMPLE 5   **Dividing by Multiples of 10**

Divide by using the shortcut of dropping zeros.

(a) $4\underline{0}\overline{)11,00\underline{0}}$ Drop 0 from the divisor and the dividend.

$$
\begin{array}{r}
275 \\
4\overline{)1100} \\
8 \\
\hline
30 \\
28 \\
\hline
20 \\
20 \\
\hline
0
\end{array}
$$

Be sure to drop the *same* number of zeros from both the divisor and dividend.

(b) $35\underline{00}\overline{)31,5\underline{00}}$ Drop two zeros from the divisor and the dividend.

$$
\begin{array}{r}
9 \\
35\overline{)315} \\
315 \\
\hline
0
\end{array}
$$

◀ *Work Problem* **5** *at the Side.*

OBJECTIVE **3** **Use multiplication to check quotients.**   Quotients in long division can be checked just as quotients in short division were checked. Multiply the quotient and divisor, then add any remainder. The result should match the original dividend.

EXAMPLE 6   **Checking Division**

Check the quotient.

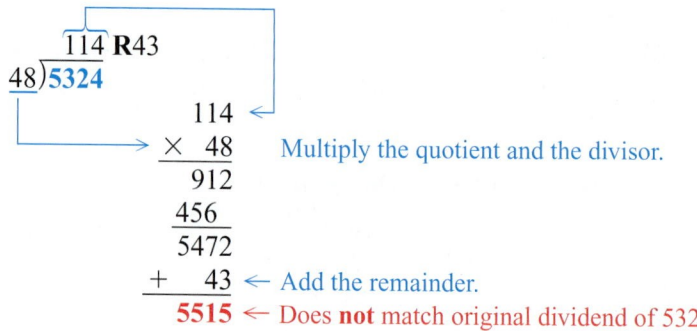

$$
\begin{array}{r}
114\ \mathbf{R}43 \\
48\overline{)5324}
\end{array}
$$

$$
\begin{array}{r}
114 \\
\times\ 48 \\
\hline
912 \\
456\ \ \\
\hline
5472 \\
+\ \ \ 43 \\
\hline
\mathbf{5515}
\end{array}
$$

Multiply the quotient and the divisor.

← Add the remainder.

← Does **not** match original dividend of 5324

The quotient does *not* check. Rework the original problem to get 110 **R**44.

◀ *Work Problem* **6** *at the Side.*

R.5 ▶▶▶ Exercises

*First, indicate what number to use as the trial divisor. Then use the trial divisor to find the first digit in the quotient and write it in the correct position. Do **not** finish the division.*

1. $24\overline{)768}$

Use _____ as the trial divisor.

2. $35\overline{)805}$

Use _____ as the trial divisor.

3. $18\overline{)4500}$

Use _____ as the trial divisor.

4. $28\overline{)3500}$

Use _____ as the trial divisor.

5. $86\overline{)10,327}$

Use _____ as the trial divisor.

6. $51\overline{)24,026}$

Use _____ as the trial divisor.

7. $52\overline{)38,025}$

Use _____ as the trial divisor.

8. $63\overline{)34,400}$

Use _____ as the trial divisor.

9. $77\overline{)249,826}$

Use _____ as the trial divisor.

10. $92\overline{)247,892}$

Use _____ as the trial divisor.

11. $420\overline{)470,800}$

Use _____ as the trial divisor.

12. $190\overline{)901,050}$

Use _____ as the trial divisor.

Find each quotient by using long division. See Examples 1–3. In Exercises 21–24, drop zeros before dividing. See Examples 4 and 5.

13. $29\overline{)1859}$

14. $58\overline{)2204}$

15. $47\overline{)11,121}$

16. $83\overline{)39,692}$

17. $26\overline{)62,583}$

18. $28\overline{)84,249}$

19. $63\overline{)78,072}$

20. $55\overline{)43,223}$

21. $150\overline{)499,760}$

22. $720\overline{)52,560}$

23. $400\overline{)340,000}$

24. $900\overline{)153,000}$

Check each division. If a quotient is incorrect, find the correct quotient. See Example 6.

25. $56\overline{)5943}$ 106 **R**17

26. $87\overline{)3254}$ 37 **R**37

27. $600\overline{)394,800}$ 658 **R**9

28. $300\overline{)139,100}$ 463 **R**200

29. $410\overline{)25,420}$ 62 **R**3

30. $760\overline{)132,600}$ 174 **R**360

31. $72\overline{)32,465}$ 450 **R**65

32. $47\overline{)9570}$ 23 **R**29

Solve each application problem by using addition, subtraction, multiplication, or division.

33. In 1900, the average workweek was 59 hours. Today it is 38 hours. How many more hours were worked each year in 1900 than today? Assume 50 weeks of work per year. (*Source:* Reiman Publications.)

34. The U.S. Government Printing Office uses 255,000 pounds of ink each year. If it does an equal amount of printing on each of 200 workdays in a year, find the weight of the ink used each day.

35. While planning a vacation, a travel agent found hotel prices ranging from $69 per night to $475 per night. What is the difference in cost between the most expensive and least expensive rooms for a five-night stay?

36. Two divorced parents share their child's education costs, which amount to $3718 per year. If one parent pays $1880 each year, find the amount paid by the other parent over five years.

37. Judy Martinez has a 36-month loan. The total debt, including interest, is $11,088. What is her monthly payment?

38. A consultant charged $36,300 for studying the environmental impact of a new shopping mall. If the consultant worked 220 hours, find the rate charged per hour.

39. Clarence Hanks can assemble 42 circuits in one hour. How many circuits can he assemble in a 5-day workweek of 8 hours per day?

40. Tuition at the community college is $57 per credit. There is also a $2 per credit technology fee and a one-time registration fee of $38. How much will Stephanie, a new student, pay to register for 14 credits?

41. A youth soccer association brought in $7588 from fund-raising projects. Expenses of $838 were paid first, with the balance of the money divided evenly among the 18 teams. How much did each team receive?

42. Feather Farms Egg Ranch collected 3545 eggs in the morning and 2575 eggs in the afternoon. If the eggs are packed in flats containing 30 eggs each, find the number of flats needed for packing.

ANSWERS TO SELECTED EXERCISES

Chapter 1

Section 1.1 (pages 5–8)

1. 15; 0; 83,001 **3.** 7; 362,049 **5.** hundreds **7.** hundred-thousands
9. ten-millions **11.** hundred-billions **13.** ten-trillions, hundred-billions,
millions, hundred-thousands, ones **15.** eight thousand, four hundred
twenty-one **17.** forty-six thousand, two hundred five **19.** three million,
sixty-four thousand, eight hundred one **21.** eight hundred forty million,
one hundred eleven thousand, three **23.** fifty-one billion, six million, eight
hundred eighty-eight thousand, three hundred twenty-one **25.** three trillion,
seven hundred twelve million **27.** 46,805 **29.** 5,600,082
31. 271,900,000 **33.** 12,417,625,310 **35.** 600,000,071,000,400
37. three thousand, one hundred fifty-one **39.** 101,280,000
41. one hundred seventy-three million, five hundred twenty-three thousand,
seven hundred **43.** 55,000,800 **45.** six million, four hundred thousand
every day; two billion, three hundred thirty-six million in one year
47. 4,200,000,000 **49.** largest: 97,651,100; ninety-seven million, six
hundred fifty-one thousand, one hundred; smallest: 10,015,679; ten million,
fifteen thousand, six hundred seventy-nine **50.** Answers will vary.
51. sixty-fours; thirty-twos; sixteens; eights **(a)** 101 **(b)** 1010 **(c)** 1111
52. (a) Answers will vary but should mention that the location or place in
which a digit is written gives it a different value. **(b)** 8 = VIII;
38 = XXXVIII; 275 = CCLXXV; 3322 = MMMCCCXXII
(c) The Roman system is *not* a place value system because no matter
what place it's in, M = 1000, C = 100, etc. One disadvantage is that it
takes much more space to write many large numbers; another is that there
is no symbol for zero.

Section 1.2 (pages 15–16)

1. $^+$29,035 feet or 29,035 feet **3.** $^-$128.6 degrees **5.** $^-$18 yards

7. $^+$\$100 or \$100 **9.** $^-6\frac{1}{2}$ pounds

11.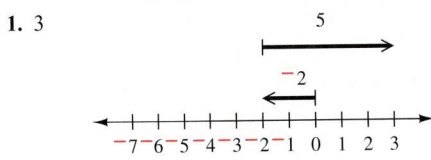

13.

15.

17. > **19.** < **21.** < **23.** > **25.** < **27.** > **29.** >
31. < **33.** 15 **35.** 3 **37.** 0 **39.** 200 **41.** 75 **43.** 8042
45.

46. $^-$1.5, $^-$1, 0, 0.5 **47.** A: may be at risk; B: above normal;
C: normal; D: normal **48. (a)** The patient would think the
interpretation was "above normal" and wouldn't get treatment.
(b) Patient D's score of 0; zero is neither positive nor negative.

Section 1.3 (pages 25–28)

1. 3

3. $^-$7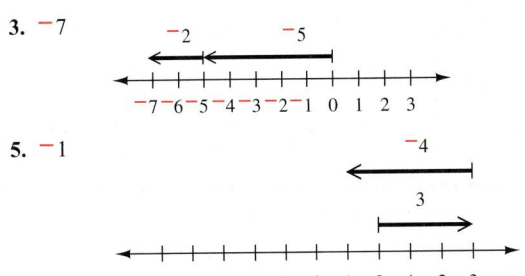

5. $^-$1

7. (a) $^-$10 **(b)** 10 **9. (a)** 12 **(b)** $^-$12 **11. (a)** $^-$50 **(b)** 50
13. (a) 158 **(b)** $^-$158 **15.** The absolute values are the same in each
pair of answers, so the only difference in the sums is the common sign.
17. (a) 2 **(b)** $^-$2 **19. (a)** $^-$7 **(b)** 7 **21. (a)** $^-$5 **(b)** 5
23. (a) 150 **(b)** $^-$150 **25.** Each pair of answers differs only in the sign
of the answer. This occurs because the signs of the addends are reversed.
27. $^-$3 **29.** 7 **31.** $^-$7 **33.** 1 **35.** $^-$8 **37.** $^-$20 **39.** $^-$17
41. $^-$22 **43.** $^-$19 **45.** 6 **47.** $^-$5 **49.** 0 **51.** 5 **53.** $^-$32
55. 13 + $^-$17 = $^-$4 yards **57.** $^-$62 + 50 = $^-$\$12
59. $^-$88 + 35 = $^-$\$53 **61.** Jeff: $^-$20 + 75 + $^-$55 = 0 points;
Terry: 42 + $^-$15 + 20 = 47 points **63.** $^-$2 **65.** 2 **67.** $^-$5 + $^-$18;
$^-$23 **69.** 15 + $^-$4; 11 **71.** 6 + ($^-$14 + 14); 6 + 0 = 6
73. ($^-$14 + $^-$6) + $^-$7; $^-$20 + $^-$7 = $^-$27 **75.** Answers will vary.
Some possibilities are: $^-$6 + 0 = $^-$6; 10 + 0 = 10; 0 + 3 = 3
77. $^-$4116 **79.** 8686 **81.** $^-$96,077

Section 1.4 (pages 31–32)

1. $^-$6; 6 + $^-$6 = 0 **3.** 13; $^-$13 + 13 = 0 **5.** 0; 0 + 0 = 0
7. 14 **9.** $^-$2 **11.** $^-$12 **13.** $^-$25 **15.** $^-$23 **17.** 5 **19.** 20
21. 11 **23.** $^-$60 **25.** 0 **27.** 0 **29.** $^-$6 **31. (a)** 8 **(b)** $^-$2
(c) 2 **(d)** $^-$8 **33. (a)** $^-$3 **(b)** 11 **(c)** $^-$11 **(d)** 3 **35.** $^-$6
37. $^-$5 **39.** 3 **41.** $^-$10 **43.** 12 **45.** $^-$5 **47. (a)** 21 °F;
30 $-$ 21 = 9 degrees difference **(b)** 0 °F; 15 $-$ 0 = 15 degrees
difference **(c)** $^-$17 °F; 5 $-$ $^-$17 = 22 degrees difference
(d) $^-$41 °F; $^-$10 $-$ $^-$41 = 31 degrees difference **49.** The student
forgot to change 6 to its opposite, $^-$6. It should be $^-$6 + $^-$6 = $^-$12.
51. $^-$11 **53.** $^-$5 **55.** $^-$10 **57.** Answers on left: $^-$8; 8. On right:
$^-$1; 1 Subtraction is *not* commutative; the absolute value of the answer is
the same, but the sign changes. **58.** Subtracting 0 from a number does
not change the number. For example $^-$5 $-$ 0 = $^-$5. But subtracting a
number from 0 *does* change the number to its opposite. For example,
0 $-$ $^-$5 = 5.

Section 1.5 (pages 39–42)

1. 630 **3.** $^-$1080 **5.** 7900 **7.** $^-$86,800 **9.** 42,500 **11.** $^-$6000
13. $^-$78,000 **15.** 6000 **17.** 600,000 **19.** $^-$9,000,000
21. 140,000,000 **23.** 20,000,000,000 **25.** 9,000,000,000
27. 30,000 miles **29.** $^-$60 degrees **31.** \$10,000 **33.** 60,000,000
Americans **35.** $^-$300 feet **37.** 700,000 people in Alaska; 40,000,000
people in California **39.** Answers will vary but should mention looking
only at the second digit, rounding first digit up when second digit is 5 or
more, leaving first digit unchanged when second digit is 4 or less. Examples will vary. **41.** *Estimate:* $^-$40 + 90 = 50; *Exact:* $^-$42 + 89 = 47
43. *Estimate:* 20 + $^-$100 = $^-$80; *Exact:* 16 + $^-$97 = $^-$81
45. *Estimate:* $^-$300 + $^-$400 = $^-$700; *Exact:* $^-$273 + $^-$399 = $^-$672
47. *Estimate:* 3000 + 7000 = 10,000; *Exact:* 3081 + 6826 = 9907
49. *Estimate:* 20 + $^-$80 = $^-$60; *Exact:* 23 $-$ 81 = 23 + $^-$81 = $^-$58

51. *Estimate:* $^-40 + ^-40 = ^-80$; *Exact:* $^-39 - 39 =$ $^-39 + ^-39 = ^-78$ **53.** *Estimate:* $^-100 + 30 + ^+70 = 0$; *Exact:* $^-106 + 34 - ^-72 = ^-106 + 34 + ^+72 = 0$ **55.** *Estimate:* $80{,}000 - 50{,}000 = \$30{,}000$; *Exact:* $78{,}650 - 52{,}882 = \$25{,}768$ **57.** *Estimate:* $2000 - 800 - 300 - 400 - 200 - 200 = \$100$; *Exact:* $2120 - 845 - 325 - 365 - 182 - 240 = \$163$ **59.** *Estimate:* $^-100 + 40 + 50 = ^-10$ degrees; *Exact:* $^-102 + 37 + 52 = ^-13$ degrees **61.** *Estimate:* $400 + 100 = 500$ doors and windows; *Exact:* $412 + 147 = 559$ doors and windows

Section 1.6 (pages 49–52)

1. (a) 63 **(b)** 63 **(c)** $^-63$ **(d)** $^-63$ **3. (a)** $^-56$ **(b)** $^-56$ **(c)** 56 **(d)** 56 **5.** $^-35$ **7.** $^-45$ **9.** $^-18$ **11.** $^-50$ **13.** $^-40$ **15.** $^-56$ **17.** 32 **19.** 77 **21.** 0 **23.** 133 **25.** 13 **27.** 0 **29.** 48 **31.** $^-56$ **33.** $^-160$ **35.** 5 **37.** $^-3$ **39.** $^-1$ **41.** 0 **43.** 5 **45.** $^-4$ **47.** Commutative property: changing the *order* of the factors does not change the product.   Associative property: changing the *grouping* of the factors does not change the product. Examples will vary. **49.** Examples will vary. Some possibilities are: **(a)** $6 \cdot ^-1 = ^-6; 2 \cdot ^-1 = ^-2; 15 \cdot ^-1 = ^-15$ **(b)** $^-6 \cdot ^-1 = 6; ^-2 \cdot ^-1 = 2; ^-15 \cdot ^-1 = 15$ The result of multiplying any nonzero number times $^-1$ is the number with the opposite sign. **50.** The products are $4, ^-8, 16, ^-32$. The absolute value doubles each time and the sign changes. The next three products are $64, ^-128$, and 256. **51.** $= 9 \cdot ^-3 + 9 \cdot 5$ Both results are 18. **53.** $= 8 \cdot 25$ Both products are 200. **55.** $= (^-3 \cdot 2) \cdot 5$ Both products are $^-30$. **57.** *Estimate:* $300 \cdot 50 = \$15{,}000$; *Exact:* $324 \cdot 52 = \$16{,}848$ **59.** *Estimate:* $^-10{,}000 \cdot 10 = ^-\$100{,}000$; *Exact:* $^-9950 \cdot 12 = ^-\$119{,}400$ **61.** *Estimate:* $200 \cdot 10 = \$2000$; *Exact:* $182 \cdot 13 = \$2366$ **63.** *Estimate:* $20 \cdot 400 = 8000$ hours; *Exact:* $24 \cdot 365 = 8760$ hours **65.** $^-512$ **67.** 0 **69.** $^-355{,}299$ **71.** $\$247$ **73.** $^-22$ degrees **75.** 772 points

Section 1.7 (pages 59–62)

1. (a) 7 **(b)** 7 **(c)** $^-7$ **(d)** $^-7$ **3. (a)** $^-7$ **(b)** 7 **(c)** $^-7$ **(d)** 7 **5. (a)** 1 **(b)** 35 **(c)** $^-13$ **(d)** 1 **7. (a)** 0 **(b)** undefined **(c)** undefined **(d)** 0 **9.** $^-4$ **11.** $^-3$ **13.** 6 **15.** $^-11$ **17.** undefined **19.** $^-14$ **21.** 10 **23.** 4 **25.** $^-1$ **27.** 0 **29.** 191 **31.** $^-499$ **33.** 2 **35.** $^-4$ **37.** 40 **39.** $^-48$ **41.** 5 **43.** 0 **45.** $2 \div 1 = 2$ but $1 \div 2 = 0.5$, so division is not commutative. **46.** $(12 \div 6) \div 2 = 2 \div 2 = 1; 12 \div (6 \div 2) = 12 \div 3 = 4$; different quotients. Division is not associative. **47.** Similar: If the signs match, the result is positive. If the signs are different, the result is negative. Different: Multiplication is commutative, division is not. You can multiply by 0, but dividing by 0 is undefined. **48.** Examples will vary. The properties are: Any nonzero number divided by itself is 1. Any number divided by 1 is the number. Division by 0 is undefined. Zero divided by any other number (except 0) is 0. **49. (a)** $\dfrac{^-6}{^-1} = 6; \dfrac{^-2}{^-1} = 2; \dfrac{^-15}{^-1} = 15$ **(b)** $\dfrac{6}{^-1} = ^-6; \dfrac{2}{^-1} = ^-2; \dfrac{15}{^-1} = ^-15$ When dividing by $^-1$, change the sign of the dividend to its opposite to get the quotient. **50.** Division is not commutative. $\dfrac{0}{^-3} = 0$ because $0 \cdot ^-3 = 0$. But $\dfrac{^-3}{0}$ is undefined because when $\dfrac{^-3}{0} = ?$ is rewritten as $? \cdot 0 = ^-3$, no number can replace ? and make a true statement. **51.** *Estimate:* $^-40{,}000 \div 20 = ^-2000$ feet; *Exact:* $^-35{,}836 \div 17 = ^-2108$ feet **53.** *Estimate:* $^-200 + 500 = \$300$; *Exact:* $^-238 + 450 = \$212$ **55.** *Estimate:* $400 - 100 = 300$ days; *Exact:* $365 - 106 = 259$ days **57.** *Estimate:* $^-700 \cdot 40 = ^-28{,}000$ feet; *Exact:* $^-730 \cdot 37 = ^-27{,}010$ feet **59.** *Estimate:* $300 \div 5 = 60$ miles; *Exact:* $315 \div 5 = 63$ miles **61.** Average score of 168 **63.** The back shows 520 grams, which is 10 grams more than the front. **65.** $^-\$15$

67. 16 hours, with 40 minutes left over **69.** 33 rooms, with space for 2 people unused. **71.** $^-10$ **73.** undefined **75.** 31.70979198 rounds to 32 years.

Summary Exercises on Operations with Integers (pages 63–64)

1. $^-6$ **2.** 0 **3.** $^-7$ **4.** $^-7$ **5.** 63 **6.** $^-1$ **7.** $^-56$ **8.** $^-22$ **9.** 12 **10.** 8 **11.** $^-13$ **12.** 0 **13.** 0 **14.** $^-17$ **15.** $^-48$ **16.** $^-10$ **17.** $^-50$ **18.** undefined **19.** $^-14$ **20.** $^-6$ **21.** 0 **22.** 16 **23.** $^-30$ **24.** undefined **25.** 48 **26.** $^-19$ **27.** 2 **28.** $^-20$ **29.** 0 **30.** 16 **31.** $^-3$ **32.** $^-36$ **33.** 6 **34.** $^-7$ **35.** $^-5$ **36.** $^-31$ **37.** $^-2$ **38.** $^-32$ **39.** $^-5$ **40.** $^-4$ **41.** $^-9732$ **42.** 100 **43.** 4 **44.** $^-343$ **45.** 5 **46.** $^-6$ **47.** $^-10$ **48.** $^-5$ **49. (a)** The quotient is 0. **(b)** The product is 0. **(c)** The quotient is 1. **50. (a)** The quotient is $^-1$. *Adding* $15 + ^-15$ would give a *sum* of 0. **(b)** Dividing by zero cannot be done; the correct answer is "undefined." **(c)** Work with two numbers at a time: $^-10 \div ^-2$ is 5; then $5 \div ^-5$ is $^-1$, *not* positive 1.

Section 1.8 (pages 71–74)

1. $4 \cdot 4 \cdot 4$; 4 cubed or 4 to the third power **3.** 2^7; 128; 2 to the seventh power **5.** 5^4; 625; 5 to the fourth power **7.** 7^2; $7 \cdot 7$; 49 **9.** 10^1; 10; 10 **11. (a)** 10 **(b)** 100 **(c)** 1000 **(d)** 10,000 **13. (a)** 4 **(b)** 16 **(c)** 64 **(d)** 256 **15.** 9,765,625 **17.** 4096 **19.** 4 **21.** 25 **23.** $^-64$ **25.** 81 **27.** $^-1000$ **29.** 1 **31.** 108 **33.** 200 **35.** $^-750$ **37.** $^-32$ **39. (a)** The answers are $4, ^-8, 16, ^-32, 64, ^-128, 256, ^-512$. When a negative number is raised to an even power, the answer is positive; when raised to an odd power, the answer is negative. **(b)** negative; positive **41.** $^-6$ **43.** 0 **45.** $^-39$ **47.** 16 **49.** 23 **51.** $^-43$ **53.** 7 **55.** $^-3$ **57.** 0 **59.** $^-38$ **61.** 41 **63.** $^-2$ **65.** 13 **67.** 126 **69.** 8 **71.** $\dfrac{27}{^-3} = ^-9$ **73.** $\dfrac{^-48}{^-4} = 12$ **75.** $\dfrac{^-60}{^-1} = 60$ **77.** $^-4050$ **79.** 7 **81.** $\dfrac{27}{0}$ is undefined.

Chapter 1 Review Exercises (pages 85–88)

1. 86, 0, 35,600 **2.** eight hundred six **3.** three hundred nineteen thousand, twelve **4.** sixty million, three thousand, two hundred **5.** fifteen trillion, seven hundred forty-nine billion, six **6.** 504,100 **7.** 620,080,000 **8.** 99,007,000,356
9.

10. $>$ **11.** $<$ **12.** $>$ **13.** $<$ **14.** 5 **15.** 9 **16.** 0 **17.** 125 **18.** $^-1$ **19.** $^-13$ **20.** $^-3$ **21.** 0 **22.** 1 **23.** $^-24$ **24.** $^-7$ **25.** 3 **26.** 0 **27.** $^-17$ **28.** 5; $^-5 + 5 = 0$ **29.** $^-18$; $18 + ^-18 = 0$ **30.** $^-7$ **31.** 17 **32.** $^-16$ **33.** 13 **34.** 18 **35.** $^-22$ **36.** 0 **37.** $^-20$ **38.** $^-1$ **39.** $^-3$ **40.** 14 **41.** 15 **42.** $^-16$ **43.** 3 **44.** $^-8$ **45.** 0 **46.** 210 **47.** 59,000 **48.** 85,000,000 **49.** $^-3000$ **50.** $^-7,060,000$ **51.** 400,000 **52.** $^-200$ pounds **53.** $^-1000$ feet **54.** 400,000,000 directories **55.** 9,000,000,000 people **56.** $^-54$ **57.** 56 **58.** $^-100$ **59.** 0 **60.** 24 **61.** 17 **62.** $^-48$ **63.** 125 **64.** $^-36$ **65.** 50 **66.** $^-72$ **67.** 9 **68.** $^-7$ **69.** undefined **70.** 5 **71.** $^-18$ **72.** 0 **73.** 15 **74.** $^-1$ **75.** $^-5$ **76.** 18 **77.** 0 **78.** 156 days and 2 extra hours **79.** 10,000 **80.** 32 **81.** 27 **82.** 16 **83.** $^-125$ **84.** 8 **85.** 324 **86.** $^-200$ **87.** $^-25$ **88.** $^-2$ **89.** 10 **90.** $^-28$ **91.** $\dfrac{8}{^-8} = ^-1$ **92.** $\dfrac{11}{0}$ is undefined. **93.** associative property of addition **94.** commutative property of multiplication **95.** addition property of 0 **96.** multiplication property of 0 **97.** distributive property **98.** associative property of multiplication **99.** *Estimate:* $\$10{,}000 \cdot 200 = \$2{,}000{,}000$; *Exact:* $\$11{,}900 \cdot 192 = \$2{,}284{,}800$ **100.** *Estimate:* $\$200 + \$400 - \$700 = ^-\$100$; *Exact:* $\$185 + \$428 - \$706 = ^-\$93$ **101.** *Estimate:* $900 \div 20 = 45$ miles; *Exact:* $880 \div 22 = 40$ miles

102. *Estimate:* ($40 • 20) + ($90 • 10) = $1700; *Exact:* ($39 • 19) + ($85 • 12) = $1761   **103.** $^-$$700, $^-$$100, $700, $900, $0, $700
104. January; April   **105.** $2100   **106.** $^-$$1850

Chapter 1 Test (pages 89–90)

1. twenty million, eight thousand, three hundred seven   **2.** 30,000,700,005
3. $\longleftarrow$ + + ● + ●● + + + $\longrightarrow$
$^-3\ ^-2\ ^-1\ 0\ 1\ 2\ 3$

4. $>$; $<$ **5.** 10; 14   **6.** $^-6$ **7.** $^-5$ **8.** 7   **9.** $^-40$ **10.** 10
11. 64   **12.** $^-50$ **13.** undefined   **14.** $^-60$ **15.** $^-5$ **16.** $^-45$
17. 6   **18.** 25   **19.** 0   **20.** 9   **21.** 128   **22.** $^-2$ **23.** 8
24. $^-16$ **25.** An exponent shows how many times to use a factor
in repeated multiplication. Examples will vary. Some possibilities
are $(2)^4 = 2 • 2 • 2 • 2 = 16$ and $(^-3)^2 = (^-3)(^-3) = 9$.
26. Commutative property: changing the *order* of addends does not change
the sum. Associative property: changing the *grouping* of
addends does not change the sum. Examples will vary.   **27.** 900
28. 36,420,000,000   **29.** 350,000   **30.** *Estimate:* $200 + $300 +
$^-$$500 = $0; *Exact:* $184 + $293 + $^-$$506 = $^-$$29
31. *Estimate:* $^-1000 ÷ 10 = ^-100$ yards; *Exact:* $^-1140 ÷ 12 =$
$^-95$ yards   **32.** *Estimate:* $30(200 - 100) = 3000$ calories;
Exact: $31(220 - 110) = 3410$ calories   **33.** $^-10 - ^-100 =$
$^-10 + ^+100 = 90$ degrees difference   **34.** 27 cartons because
26 cartons would leave 28 books unpacked

Chapter 2 Understanding Variables and Solving Equations

Section 2.1 (pages 101–104)

1. c is the variable; 4 is the constant.   **3.** m is the variable; $^-3$ is the constant.   **5.** h is the variable; 5 is the coefficient.   **7.** c is the variable; 2 is
the coefficient; 10 is the constant.   **9.** x and y are variables.   **11.** g is the
variable; $^-6$ is the coefficient; 9 is the constant   **13.** (a) $654 + 10$ is 664
robes.   (b) $208 + 10$ is 218 robes.   (c) $95 + 10$ is 105 robes.
15. (a) $3 • 11$ inches is 33 inches.   (b) $3 • 3$ feet is 9 feet.
17. (a) $3 • 12 - 5$ is 31 brushes.   (b) $3 • 16 - 5$ is 43 brushes.
19. (a) $\frac{332}{4}$ is 83 points.   (b) $\frac{637}{7}$ is 91 points.   **21.** $12 + 12 + 12 + 12$
is 48, $4 • 12$ is 48; $0 + 0 + 0 + 0$ is 0, $4 • 0$ is 0; $^-5 + ^-5 + ^-5 +$
$^-5 = ^-20$, $4 • ^-5$ is $^-20$ **23.** $^-2(^-4) + 5$ is $8 + 5$, or 13;
$^-2(^-6) + ^-2$ is $12 + ^-2$, or 10; $^-2(0) + ^-8$ is $0 + ^-8$, or $^-8$
25. A variable is a letter that represents the part of a rule that varies or
changes depending on the situation. An expression expresses, or tells, the
rule for doing something. For example, $c + 5$ is an expression, and c is the
variable.   **27.** $b • 1 = b$ or $1 • b = b$ **29.** $\frac{b}{0}$ is undefined   or
$b ÷ 0$ is undefined.   **31.** $c • c • c • c • c • c$
33. $x • x • x • x • y • y • y$ **35.** $^-3 • a • a • a • b$ **37.** $9 • x • y • y$
39. $^-2 • c • c • c • c • c • d$ **41.** $a • a • a • b • c • c$ **43.** 16
45. $^-24$ **47.** $^-18$ **49.** $^-128$ **51.** $^-18,432$ **53.** 311,040   **55.** 56
57. $\frac{36}{0}$ is undefined.   **59.** (a) 3 miles   (b) 2 miles   (c) 1 mile
60. (a) $\frac{1}{2}$ mile; take half of the distance for 5 seconds   (b) $7\frac{1}{2}$ seconds;
find the number halfway between 5 seconds and 10 seconds
(c) $12\frac{1}{2}$ seconds; find the number halfway between 10 seconds and
15 seconds

Section 2.2 (pages 115–118)

1. $2b^2$ and b^2; The coefficients are 2 and 1.   **3.** ^-xy and $2xy$;
The coefficients are $^-1$ and 2.   **5.** 7, 3, and $^-4$; The like terms are
constants.   **7.** $12r$ **9.** $6x^2$ **11.** ^-4p **13.** $^-3a^3$ **15.** 0
17. xy **19.** $6t^4$ **21.** $4y^2$ **23.** ^-8x **25.** $12a + 4b$ **27.** $7rs + 14$

29. $a + 2ab^2$ **31.** $^-2x + 2y$ **33.** $7b^2$ **35.** cannot be simplified
37. $^-15r + 5s + t$ **39.** $30a$ **41.** $^-8x^2$ **43.** $^-20y^3$ **45.** $18cd$
47. $21a^2bc$ **49.** $12w$ **51.** $6b + 36$ **53.** $7x - 7$ **55.** $21t + 3$
57. $^-10r - 6$ **59.** $^-9k - 36$ **61.** $50m - 300$ **63.** $8y + 16$
65. $6a^2 + 3$ **67.** $9m - 34$ **69.** $^-25$ **71.** $24x$ **73.** $5n + 13$
75. $11p - 1$ **77.** A simplified expression still has variables, but is written in a simpler way. When evaluating an expression, the variables are all
replaced by specific numbers and the final result is a numerical answer.
79. Like terms have matching variable parts, that is, matching letters and
exponents. The coefficients do not have to match. Examples will vary.
81. Keep the variable part unchanged when combining like terms. The
correct answer is $5x + 8$.   **83.** $^-2y + 9$ **85.** 0   **87.** ^-9x

Summary Exercises on Variables and Expressions (pages 121–122)

1. m is the variable; $^-10$ is the constant.   **2.** c and d are the variables;
$^-8$ is the coefficient.   **3.** x is the variable; 4 is the coefficient; 6 is the
constant.   **4.** (a) 32 yards   (b) 120 inches   **5.** (a) $13,080
(b) $22,342   **6.** $a • d • d • d • d$ **7.** $b • b • b • c • d$
8. $^-7 • a • b • b • b • b • c • c$ **9.** 625   **10.** 0   **11.** 0   **12.** 60
13. $^-8$ **14.** 120   **15.** $^-216$ **16.** $^-800$ **17.** 120,960   **18.** $24b$
19. $7x + 7$ **20.** $^-8c - 32$ **21.** 0   **22.** $12c^2d$ **23.** ^-6f
24. $6w + 8$ **25.** $^-2a - 6b$ **26.** $50x^3y^2$ **27.** $5r^3 + 3r^2 + 2r$
28. $7h^2$ **29.** $^-9$ **30.** $^-32y + 25$ **31.** $36x - 10$ **32.** $19n + 1$
33. (a) Forgot to multiply $6 • 2$; correct answer is $6n + 12$.
(b) Two negative factors give a *positive* product; correct answer is $20a$.
(c) Keep the variable part unchanged; correct answer is $5y - 10$.
34. In the last step, do not change the sign of the first term; keep ^-7x as
^-7x. The correct answer is $^-7x - 9$.

Section 2.3 (pages 129–134)

1. 58 is the solution.   **3.** $^-16$ is the solution.   **5.** $^-12$ is the solution.

7. $p = 4$ **Check** $\underbrace{4 + 5}_{9} = 9$
$9 = 9$

9. $r = 10$ **Check** $8 = \underbrace{10 - 2}_{8}$
$8 = 8$

11. $n = ^-8$ **Check** $^-5 = \underset{\downarrow}{n} + 3$
$^-5 = \underbrace{^-8 + 3}_{^-5}$
$^-5 = ^-5$

13. $k = 18$ **Check** $^-4 + \underset{\downarrow}{k} = 14$
$\underbrace{^-4 + 18}_{14} = 14$
$14 = 14$

15. $y = 6$ **Check** $\underset{\downarrow}{y} - 6 = 0$
$\underbrace{6 - 6}_{0} = 0$
$0 = 0$

17. $r = ^-6$ **Check** $7 = \underset{\downarrow}{r} + 13$
$7 = \underbrace{^-6 + 13}_{7}$
$7 = 7$

19. $x = 11$ **Check** $\underset{\downarrow}{x} - 12 = ^-1$
$\underbrace{11 + ^-12}_{^-1} = ^-1$
$^-1 = ^-1$

21. $t = ^-3$ **Check** $^-5 = ^-2 + \underset{\downarrow}{t}$
$^-5 = \underbrace{^-2 + ^-3}_{^-5}$
$^-5 = ^-5$

23.

$$\overbrace{{}^-2 + {}^-5} = 3$$

Does not balance $\quad {}^-7 \neq 3$

The correct solution is 8. $\quad \overbrace{8 - 5} = 3$

Balances $\quad 3 = 3$

25. $\quad 7 + x = {}^-11$

$$7 + \overbrace{{}^-18} = {}^-11$$

Balances $\quad {}^-11 = {}^-11$

${}^-18$ is the correct solution.

27. $\quad {}^-10 = {}^-10 + b$

$$^-10 = \overbrace{{}^-10 + 10}$$

Does not balance $\quad {}^-10 \neq 0$

The correct solution is 0. $\quad {}^-10 = \overbrace{{}^-10 + 0}$

Balances $\quad {}^-10 = {}^-10$

29. $c = 6$; see *Student's Solutions Manual* for **Checks** of odd-numbered Exercises 29–37; Exercise 39 is in the Solutions section in the back of this book. **31.** $y = 5$ **33.** $b = {}^-30$ **35.** $t = 0$ **37.** $z = {}^-7$ **39.** $w = 3$ **41.** $x = {}^-10$ **43.** $a = 0$ **45.** $y = {}^-25$ **47.** $x = 15$ **49.** $k = 113$ **51.** $b = 18$ **53.** $r = {}^-5$ **55.** $n = {}^-105$ **57.** $h = {}^-5$ **59.** No, the solution is ${}^-14$, the number used to replace x in the original equation. **61.** $g = 295$ graduates **63.** $c = 55$ chirps **65.** $p = \$110$ per month in winter **67.** $m = {}^-19$ **69.** $x = 2$ **71. (a)** Equations will vary. Some possibilities are $n - 1 = {}^-3$ and $8 = x + 10$. **(b)** Equations will vary. Some possibilities are $y + 6 = 6$ and ${}^-5 = {}^-5 + b$. **72. (a)** $x = \dfrac{1}{2}$ **(b)** $y = \dfrac{5}{4}$ **(c)** $n = \$0.85$ **(d)** Equations will vary.

Section 2.4 (pages 139–142)

1. $z = 2$ **Check** $\overbrace{6 \cdot 2} = 12$
$\qquad\qquad\qquad 12 = 12$

3. $r = 4$ **Check** $48 = 12r$
$\qquad\qquad\qquad 48 = \overbrace{12 \cdot 4}$
$\qquad\qquad\qquad 48 = 48$

5. $y = 0$ **Check** $3y = 0$
$\qquad\qquad\qquad \overbrace{3 \cdot 0} = 0$
$\qquad\qquad\qquad 0 = 0$

7. $k = {}^-10$ **Check** ${}^-7k = 70$
$\qquad\qquad\qquad \overbrace{{}^-7 \cdot {}^-10} = 70$
$\qquad\qquad\qquad 70 = 70$

9. $r = 6$ **Check** ${}^-54 = {}^-9r$
$\qquad\qquad\qquad {}^-54 = \overbrace{{}^-9 \cdot 6}$
$\qquad\qquad\qquad {}^-54 = {}^-54$

11. $b = {}^-5$ **Check** ${}^-25 = 5b$
$\qquad\qquad\qquad {}^-25 = \overbrace{5 \cdot {}^-5}$
$\qquad\qquad\qquad {}^-25 = {}^-25$

13. $r = 3$ **Check** $\overbrace{2 \cdot 3} = 6$
$\qquad\qquad\qquad 6 = 6$

15. $p = {}^-3$ **Check** ${}^-12 = 5p - p$
$\qquad\qquad\qquad {}^-12 = \overbrace{5 \cdot {}^-3} - {}^-3$
$\qquad\qquad\qquad {}^-12 = \overbrace{{}^-15 + {}^+3}$
$\qquad\qquad\qquad {}^-12 = {}^-12$

17. $a = {}^-5$ **19.** $x = {}^-10$ **21.** $w = 0$ **23.** $t = 3$ **25.** $t = 0$ **27.** $m = 9$ **29.** $y = {}^-1$ **31.** $z = {}^-5$ **33.** $p = {}^-2$ **35.** $k = 7$

37. $b = {}^-3$ **39.** $x = {}^-32$ **41.** $w = 2$ **43.** $n = 50$ **45.** $p = {}^-10$ **47.** Each solution is the opposite of the number in the equation. So the rule is: When you change the sign of the variable from negative to positive, then change the number in the equation to its opposite. In ${}^-x = 5$, the opposite of 5 is ${}^-5$, so $x = {}^-5$. **49.** Divide by the coefficient of x, which is 3, *not* by the opposite of 3. The correct solution is 5. **51.** $s = 15$ ft **53.** $s = 24$ meters **55.** $y = 27$ **57.** $x = 1$

Section 2.5 (pages 147–152)

1. $p = 1$ **Check** $7(1) + 5 = 12$
$\qquad\qquad\qquad \overbrace{7} + 5 = 12$
$\qquad\qquad\qquad \overbrace{12} = 12$
$\qquad\qquad\qquad 12 = 12$

3. $y = 1$ **Check** $2 = 8y - 6$
$\qquad\qquad\qquad 2 = 8(1) - 6$
$\qquad\qquad\qquad 2 = \overbrace{8 - 6}$
$\qquad\qquad\qquad 2 = 2$

5. $m = 0$ **Check** ${}^-3m + 1 = 1$
$\qquad\qquad\qquad {}^-3(0) + 1 = 1$
$\qquad\qquad\qquad \overbrace{0} + 1 = 1$
$\qquad\qquad\qquad \overbrace{1} = 1$

7. $a = {}^-2$ **Check** $28 = {}^-9a + 10$
$\qquad\qquad\qquad 28 = {}^-9({}^-2) + 10$
$\qquad\qquad\qquad 28 = \overbrace{18} + 10$
$\qquad\qquad\qquad 28 = 28$

9. $x = {}^-4$ **Check** ${}^-5x - 4 = 16$
$\qquad\qquad\qquad {}^-5({}^-4) - 4 = 16$
$\qquad\qquad\qquad \overbrace{20} - 4 = 16$
$\qquad\qquad\qquad \overbrace{16} = 16$

11. $p = 4; 4 = p$ **Check** $6(4) - 2 = 4(4) + 6$
$\qquad\qquad\qquad \overbrace{24} - 2 = \overbrace{16} + 6$
$\qquad\qquad\qquad 22 = 22$

13. $k = {}^-2; {}^-2 = k$ **Check** ${}^-2k - 6 = 6k + 10$
$\qquad\qquad\qquad {}^-2({}^-2) - 6 = 6({}^-2) + 10$
$\qquad\qquad\qquad \overbrace{4} + {}^-6 = \overbrace{{}^-12} + 10$
$\qquad\qquad\qquad {}^-2 = {}^-2$

15. $a = 5; 5 = a$ **Check** ${}^-18 + 7a = 2a + 7$
$\qquad\qquad\qquad {}^-18 + 7(5) = 2(5) + 7$
$\qquad\qquad\qquad \overbrace{{}^-18 + 35} = \overbrace{10 + 7}$
$\qquad\qquad\qquad 17 = 17$

17. $w = 6$ **19.** $y = {}^-9$ **21.** $t = {}^-5$ **23.** $x = 0$ **25.** $h = 1$ **27.** $y = {}^-2$ **29.** $m = {}^-3$ **31.** $w = 2$ **33.** $x = 5$ **35.** $a = 3$ **37.** $b = {}^-3$ **39.** $k = 4$ **41.** $c = 0$ **43.** $y = {}^-5$ **45.** $n = 21$ **47.** $c = 30$ **49.** $p = {}^-2$ **51.** $b = {}^-2$ **53.** The series of steps may vary. One possibility is:

$${}^-2t - 10 = 3t + 5 \qquad \text{Change subtraction to adding the opposite.}$$

$$\begin{array}{r} {}^-2t + {}^-10 = 3t + 5 \\ \underline{2t \qquad\qquad 2t} \end{array} \qquad \text{Add } 2t \text{ to both sides (addition property).}$$

$$\begin{array}{r} 0 + {}^-10 = 5t + 5 \\ \underline{{}^-5 \qquad\quad {}^-5} \end{array} \qquad \text{Add } {}^-5 \text{ to both sides (addition property).}$$

$$\dfrac{{}^-15}{5} = \dfrac{5t}{5} \qquad \text{Divide both sides by 5 (division property).}$$

$${}^-3 = t$$

55. Check $\quad -8 + 4(3) = 2(3) + 2$

$$-8 + \underbrace{12}_{4} = \underbrace{6 + 2}_{8}$$

$$4 \neq 8$$

The check does not balance, so 3 is not the correct solution. The student added $-2a$ to -8 on the left side, instead of adding $-2a$ to $4a$. The correct solution is 5. **57. (a)** It must be negative. **(b)** The sum of x and a positive number is negative, so x must be negative.
58. (a) It must be positive. **(b)** The sum of d and a negative number is positive, so d must be positive. **59. (a)** It must be positive; when the signs are different, the product is negative. **(b)** The product of n and a negative number is negative, so n must be positive. **60. (a)** It must be negative also; when the signs match, the product is positive. **(b)** The product of y and a negative number is positive, so y must be negative.

Chapter 2 Review Exercises (pages 159–160)

1. (a) Variable is k; coefficient is 4; constant is -3. **(b)** $-9y + 20$
2. (a) 70 test tubes **(b)** 106 test tubes **3. (a)** $x \cdot x \cdot y \cdot y \cdot y \cdot y$
(b) $5 \cdot a \cdot b \cdot b \cdot b$ **4. (a)** 9 **(b)** -27 **(c)** -128 **(d)** 720
5. $ab^2 + 3ab$ **6.** $-4x + 2y - 7$ **7.** $16g^3$ **8.** $12r^2t$ **9.** $5k + 10$
10. $-6b - 8$ **11.** $6y$ **12.** $20x + 2$ **13.** Expressions will vary.
One possibility is $6a^3 + a^2 + 3a - 6$.
14. $n = -11$ **Check** $16 + \quad n = 5$

$$16 + -11 = 5$$
$$5 = 5$$

15. $a = 4$ **Check** $-4 + 2 = 2a - 6 - 4$

$$-4 + 2 = 2(4) - 6 - 4$$
$$-2 = \underbrace{8 + -6 + -4}$$
$$-2 = -2$$

16. $m = -8$ **17.** $k = 10$ **18.** $t = 0$ **19.** $p = -6$ **20.** $r = 2$
21. $h = -12$ **22.** $w = 4$ **23.** $c = -2$ **24.** $n = 15$ employees
25. $a = -5$ **26.** $p = 10$ **27.** $y = 5$ **28.** $m = 3$ **29.** $x = 9$
30. $b = 7$ **31.** $z = -3$ **32.** $n = 4$ **33.** $t = 0$ **34.** $d = 5$
35. $b = -2$

Chapter 2 Test (pages 161–162)

1. -7 is the coefficient; w is the variable; 6 is the constant.
2. Buy 177 hot dogs. **3.** $x \cdot x \cdot x \cdot x \cdot x \cdot y \cdot y \cdot y$
4. $4 \cdot a \cdot b \cdot b \cdot b \cdot b$ **5.** -200 **6.** $-4w^3$ **7.** 0 **8.** c
9. cannot be simplified **10.** $-40b^2$ **11.** $15k$ **12.** $21t + 28$
13. $-4a - 24$ **14.** $6x - 15$ **15.** $-9b + c + 6$
16. $x = 5$ **Check** $-4 = x - 9$

$$-4 = 5 - 9$$
$$-4 = -4$$

17. $w = -11$ **Check** $-7w = 77$

$$-7(-11) = 77$$
$$77 = 77$$

18. $p = -14$ **Check** $-p = 14$

$$-1(-14) = 14$$
$$14 = 14$$

19. $a = 3$ **Check** $-15 = -3(a + 2)$

$$-15 = -3(3 + 2)$$
$$-15 = -3(5)$$
$$-15 = -15$$

20. $n = -8$ **21.** $m = 15$ **22.** $x = -1$ **23.** $m = 2$
24. $b = 54$ **25.** $c = 0$ **26.** Equations will vary. Two
possibilities are $x - 5 = -9$ and $-24 = 6y$.

Chapter 3 Solving Application Problems

Section 3.1 (pages 171–174)

1. $P = 36$ cm **3.** $P = 100$ in. **5.** $P = 4$ miles **7.** $P = 88$ mm
9. $s = 30$ ft **11.** $s = 1$ mm **13.** $s = 23$ yards **15.** $s = 2$ ft
17. $P = 28$ yd **19.** $P = 70$ cm **21.** $P = 72$ ft **23.** $P = 26$ in.
25. $l = 9$ cm **Check** 9 cm $+ 9$ cm $+ 6$ cm $+ 6$ cm $= 30$ cm
27. $w = 1$ mile **Check** 4 mi $+ 4$ mi $+ 1$ mi $+ 1$ mi $= 10$ mi
29. $w = 2$ ft **Check** 6 ft $+ 6$ ft $+ 2$ ft $+ 2$ ft $= 16$ ft
31. $l = 2$ m **Check** 2 m $+ 2$ m $+ 1$ m $+ 1$ m $= 6$ m
33. $P = 208$ m **35.** $P = 320$ ft **37.** $P = 54$ mm **39.** $P = 48$ ft
41. $P = 78$ in. **43.** $P = 125$ m **45.** $? = 40$ cm **47.** $? = 12$ in.
49. (a) Sketches will vary. **(b)** Formula for perimeter of an
equilateral triangle is $P = 3s$, where s is the length of one side.
(c) The formula will *not* work for other kinds of triangles because
the sides will have different lengths. **51. (a)** 140 miles
(b) 350 miles **(c)** 560 miles **52. (a)** 70 miles
(b) 175 miles **(c)** 280 miles **(d)** The rate is half of 70 miles
per hour, so the distance will be half as far; divide each result in
Exercise 51 by 2. **53. (a)** 50 hours **(b)** 60 hours **(c)** 150 hours
54. (a) 61 miles per hour **(b)** 57 miles per hour
(c) 65 miles per hour

Section 3.2 (pages 181–184)

1. $A = 77$ ft^2 **3.** $A = 100$ m^2 **5.** $A = 775$ mm^2 **7.** $A = 36$ in.^{2}
9. $A = 105$ cm^2 **11.** $A = 72$ ft^2 **13.** $A = 625$ mi^2
15. $A = 1$ m^2 **17.** $l = 6$ ft **Check** $A = 6$ ft $\cdot 3$ ft; $A = 18$ ft^2
19. $w = 80$ yd **Check** $A = 90$ yd $\cdot 80$ yd; $A = 7200$ yd^2
21. $l = 14$ in. **Check** $A = 14$ in. $\cdot 11$ in.; $A = 154$ in.^{2}
23. $s = 6$ m **25.** $s = 2$ ft **27.** $h = 20$ cm
Check $A = 25$ cm $\cdot 20$ cm; $A = 500$ cm^2 **29.** $b = 17$ in.
Check $A = 17$ in. $\cdot 13$ in. ; $A = 221$ in.^{2} **31.** $h = 1$ m
Check $A = 9$ m $\cdot 1$ m; $A = 9$ m^2 **33.** Height is not part of
perimeter. Square units are used for area, not perimeter.
$P = 25$ cm $+ 25$ cm $+ 25$ cm $+ 25$ cm; $P = 100$ cm
35. Square; $P = 180$ in.; $A = 2025$ in.^{2} **37.** Rectangle; $P = 96$ ft;
$A = 351$ ft^2 **39.** Parallelogram; $P = 60$ cm; $A = 180$ cm^2
41. $P = 48$ m; $A = 144$ m^2 **43.** \$108 **45.** \$725 **47.** 53 yd
49. $P = 52$ ft; $A = 169$ ft^2; 21 ft^2 for each camper (rounded)
51.

5 ft	4 ft	3 ft
▭ 1 ft	▭ 2 ft	▭ 3 ft

52. (a) 5 ft by 1 ft has area of 5 ft^2; 4 ft by 2 ft has area of 8 ft^2; 3 ft by 3 ft
has area of 9 ft^2 **(b)** The square plot 3 ft by 3 ft has the greatest area.
53.

7 ft	6 ft	5 ft	4 ft
▭ 1 ft	▭ 2 ft	▭ 3 ft	▭ 4 ft

54. (a) 7 ft^2, 12 ft^2, 15 ft^2, 16 ft^2 **(b)** Square plots have the greatest area.

Summary Exercises on Perimeter and Area (pages 185–186)

1. Rectangle; $P = 32$ m; $A = 39$ m^2
2. Square; $P = 104$ ft; $A = 676$ ft^2
3. Parallelogram; $P = 34$ yd; $A = 56$ yd^2
4. Rectangle; $P = 36$ cm; $A = 80$ cm^2
5. Square; $P = 36$ in.; $A = 81$ in.^{2}
6. Parallelogram; $P = 30$ m; $A = 45$ m^2
7. Rectangle; $P = 26$ ft; $A = 36$ ft^2
8. Parallelogram; $P = 188$ ft; $A = 2100$ ft^2

9. $w = 1$ ft

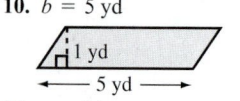

Check 7 ft + 7 ft + 1 ft + 1 ft = 16 ft

10. $b = 5$ yd

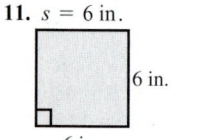

Check $A = 5$ yd • 1 yd
$A = 5$ yd^2

11. $s = 6$ in.

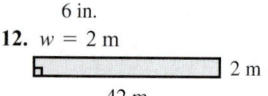

Check $A = 6$ in. • 6 in.
$A = 36$ in.^{2}

12. $w = 2$ m

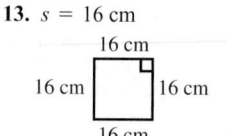

Check $A = 42$ m • 2 m
$A = 84$ m^2

13. $s = 16$ cm

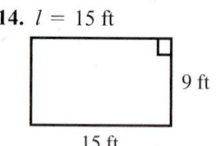

Check $P = 16$ cm + 16 cm + 16 cm + 16 cm = 64 cm

14. $l = 15$ ft

Check $P = 15$ ft + 15 ft + 9 ft + 9 ft = 48 ft

15. 504 meters of fencing   **16.** 13,000 yd^2 − 5000 yd^2 = 8000 yd^2
17. 9 inches   **18.** 176 cm of braid   **19.** $7(15$ ft$^2) = 105$ ft^2 of fabric;
$7(16$ ft$) = 112$ ft of binding   **20.** 48 ft

Section 3.3 (pages 193–196)

1. $14 + x$ or $x + 14$ **3.** $^-5 + x$ or $x + ^-5$ **5.** $20 - x$
7. $x - 9$ **9.** $x - 4$ **11.** ^-6x **13.** $2x$ **15.** $\frac{x}{2}$ **17.** $2x + 8$ or
$8 + 2x$ **19.** $7x - 10$ **21.** $2x + x$ or $x + 2x$
23. $4n - 2 = 26; n = 7$ **Check** $4 • 7 - 2$ does equal 26
$\underbrace{28}\;\; -2$
26
25. $2n + n = ^-15$ or $n + 2n = ^-15; n = ^-5$
Check $2 • ^-5 + ^-5$ does equal $^-15$
$\underbrace{^-10}\; + \;^-5$
$^-15$
27. $5n + 12 = 7n; n = 6$ **Check** $5 • 6 + 12 = 7 • 6$
$\underbrace{30}\; + 12 = \;42$
$42 = 42$
29. $30 - 3n = 2 + n; n = 7$ **Check** $30 - 3 • 7 = 2 + 7$
$\underbrace{30 - 21} = \underbrace{9}$
$9 = 9$
31. Let w be Ricardo's original weight. $w + 15 - 28 + 5 = 177$.
He weighed 185 pounds originally.   **33.** Let c be the number of
cookies the children ate. $18 - c + 36 = 49$. Her children ate
5 cookies.   **35.** Let p be the number of pens in each box.
$6p - 32 - 35 = 5$. There were 12 pens in each box.   **37.** Let d be
each member's dues. $14d + 340 - 575 = ^-25$. Each member
paid $15.   **39.** Let a be Tamu's age. $4a - 75 = a$. Tamu is
25 years old.   **41.** Let m be the amount Brenda spent. $2m - 3 = 81$.
Brenda spent $42.   **43.** Let b be the number of pieces in each bag.
$5b - 3 • 48 = b$. There were 36 pieces of candy in each bag.

45. Let d be the daily amount for an infant. $2d - 4 = 18$. An infant
should receive 11 mg of iron.

Section 3.4 (pages 201–202)

1. a is my age; $a + 9$ is my sister's age. $a + a + 9 = 51$. I am 21;
my sister is 30.   **3.** m is husband's earnings; $m + 1500$ is Lien's.
$m + m + 1500 = 37,500$. Husband earned $18,000; Lien earned
$19,500.   **5.** m is printer's cost; $5m$ is computer's cost;
$5m + m = \$1320$. Printer cost $220; computer cost $1100.
7. Shorter piece is x; longer piece is $x + 10$. So, $x + x + 10 = 78$.
Shorter piece is 34 cm; longer piece is 44 cm.   **9.** Longer piece is x;
shorter piece is $x - 7$. So, $x + x - 7 = 31$. Longer piece is 19 ft;
shorter piece is 12 ft.   **11.** s is number of Senators; $5s - 65$ is
number of Representatives. $s + 5s - 65 = 535$. 100 Senators;
435 Representatives   **13.** First part is x; second part is $x + 25$; third part
is $x + 25$. So, $x + x + x + 25 = 706$. First part is 227 m; second part
is 227 m; third part is 252 m.   **15.** Length is 19 yd.   **17.** Length is
12 ft; width is 6 ft.   **19.** Length is 13 in.; width is 5 in.
21. $P = 52$ in.; $A = 168$ in.^{2}

Chapter 3 Review Exercises (pages 209–210)

1. Square; $P = 112$ cm   **2.** Rectangle; $P = 22$ mi
3. Parallelogram; $P = 42$ yd   **4.** $P = 140$ m   **5.** 12 ft $= 4s$;
$s = 3$ ft   **6.** 128 yd $= 2l + 2(31$ yd$); l = 33$ yd
7. 72 in. $= 2(21$ in.$) + 2w; w = 15$ in.   **8.** $A = 40$ ft^2
9. $A = 625$ m^2 **10.** $A = 208$ yd^2 **11.** 126 ft$^2 = 14$ ft • w;
$w = 9$ ft   **12.** 88 cm$^2 = 11$ cm • $h; h = 8$ cm   **13.** 100 mi$^2 = s • s$;
$s = 10$ mi   **14.** $57 - x$ **15.** $15 + 2x$ or $2x + 15$ **16.** ^-9x
17. $4n + 6 = ^-30; n = ^-9$ **18.** $10 - 2n = 4 + n; n = 2$
19. m is money originally in account. $m - 600 + 750 + 75 = 309$.
$84 was originally in Grace's account.   **20.** c is number of candles
in each box. $4c - 25 = 23$. There were 12 candles in each box.
21. p is Reggie's prize money; $p + 300$ is Donald's prize money.
$p + p + 300 = 1000$. Reggie gets $350; Donald gets $650.
22. w is the width; $2w$ is the length. $84 = 2(2w) + 2(w)$. The width
is 14 cm; the length is 28 cm.   **23. (a)** 36 ft $= 4s; s = 9$ ft
(b) $A = 9$ ft • 9 ft; $A = 81$ ft^2 **24.** Rectangles will vary. Two
possibilities are: $A = 7$ ft • 3 ft, $A = 21$ ft^2; $A = 6$ ft • 4 ft, $A = 24$ ft^2
25. Let f be the fencing for the garden. $36 - f + 20 = 41$.
15 ft of fencing was used on the garden.   **26.** w is the width; $w + 2$ is
the length. $36 = 2(w + 2) + 2 • w$. Width is 8 ft; length is 10 ft.

Chapter 3 Test (pages 211–212)

1. $P = 262$ m   **2.** $P = 40$ in.   **3.** $P = 12$ miles   **4.** $P = 12$ ft
5. $P = 110$ cm   **6.** $A = 486$ mm^2 **7.** $A = 140$ cm^2
8. $A = 3740$ mi^2 **9.** $A = 36$ m^2 **10.** 12 ft $= 4s; s = 3$ ft
11. 34 ft $= 2l + 2(6$ ft$); l = 11$ ft   **12.** 65 in.$^2 = 13$ in. • $h; h = 5$ in.
13. 12 cm$^2 = 4$ cm • $w; w = 3$ cm   **14.** 16 ft$^2 = s^2; s = 4$ ft
15. Answers will vary; one possibility: Linear units like ft are used
to measure length, width, height, and perimeter. Area is measured in
square units like ft^2 (squares that measure 1 ft on each side).
16. $4n + 40 = 0; n = ^-10$ **17.** $7n - 23 = n + 7; n = 5$ **18.** Let m
represent money spent on groceries. $43 - m + 16 = 44$. Son spent $15.
19. Let d represent daughter's age. $39 = 5d + 4$. Daughter is 7 years old.
20. Let p be length of one piece; $p + 4$ be length of second piece.
$p + p + 4 = 118$. One piece is 57 cm; second piece is 61 cm.   **21.** Let w
be width; $4w$ be length. $420 = 2(4w) + 2(w)$. Sketches may vary; length
is 168 ft; width is 42 ft.   **22.** Let h be Marcella's hours; $h - 3$ be Tim's
hours. $h + h - 3 = 19$. Marcella worked 11 hours; Tim worked 8 hours.

Chapter 4 Rational Numbers: Positive and Negative Fractions

Section 4.1 (pages 223–226)

1. $\frac{5}{8}, \frac{3}{8}$ **3.** $\frac{2}{3}, \frac{1}{3}$ **5.** $\frac{3}{2}, \frac{1}{2}$ **7.** $\frac{11}{6}, \frac{1}{6}$ **9.** $\frac{2}{11}, \frac{3}{11}, \frac{4}{11}$ **11.** $\frac{13}{71}, \frac{58}{71}$

13. (a) $\frac{6}{20}$; **(b)** $\frac{19}{20}$ **15.** N: 3; D: 4; 4 equal parts **17.** N: 12; D: 7; 7 equal parts **19.** Proper: $\frac{1}{3}, \frac{5}{8}, \frac{7}{16}$; Improper: $\frac{8}{5}, \frac{6}{6}, \frac{12}{2}$

21.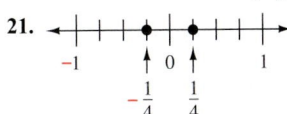

23.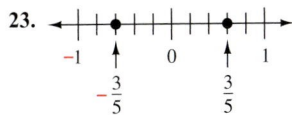

25.
(number line from -1 to 1 with marks at $-\frac{7}{8}$ and $\frac{7}{8}$)

27. $-\frac{3}{4}$ pound **29.** $\frac{3}{10}$ mile **31.** $\frac{2}{5}$ **33.** 0 **35. (a)** $\frac{12}{24}$ **(b)** $\frac{8}{24}$ **(c)** $\frac{16}{24}$ **(d)** $\frac{6}{24}$ **(e)** $\frac{18}{24}$ **(f)** $\frac{4}{24}$ **(g)** $\frac{20}{24}$ **(h)** $\frac{3}{24}$ **(i)** $\frac{9}{24}$ **(j)** $\frac{15}{24}$

37. (a) $-\frac{1}{3}$ **(b)** $-\frac{2}{3}$ **(c)** $-\frac{2}{3}$ **(d)** $-\frac{1}{3}$ **(e)** $-\frac{2}{3}$ **(f)** Some possibilities are: $-\frac{4}{12} = -\frac{1}{3}$; $-\frac{8}{24} = -\frac{1}{3}$; $-\frac{20}{30} = -\frac{2}{3}$; $-\frac{24}{36} = -\frac{2}{3}$.

39. (a) $\frac{1467}{3912}$ **(b)** Divide 3912 by 8 to get 489; multiply 3 by 489 to get 1467. **40. (a)** $\frac{4256}{5472}$ **(b)** Divide 5472 by 9 to get 608; multiply 7 by 608 to get 4256. **41. (a)** $-\frac{1}{5}$ **(b)** Divide 3485 by 2, by 3, and by 5 to see that dividing by 5 gives 697. Or divide 3485 by 697 to get 5. **42. (a)** $-\frac{1}{6}$ **(b)** Divide 4902 by 4, by 6, and by 8 to see that dividing by 6 gives 817. Or divide 4902 by 817 to get 6. **43.** You cannot do it if you want the numerator to be a whole number, because 5 does not divide into 18 evenly. You could use multiples of 5 as the denominator, such as 10, 15, 20, etc. **45.** 10 **47.** -1 **49.** -6 **51.** 3 **53.** 2 **55.** -9 **57.** 1 **59.** -8

61. $\frac{2}{5}$ is unshaded. **63.** $\frac{5}{8}$ is unshaded.

65. One possibility is shown. ◯ ▨ □ □ □ □ □ △ ▲ △

67. One possibility is shown. (! !)! ! !, . . . ? ? ?

Section 4.2 (pages 235–238)

1. (a) yes **(b)** No, 5 is a common factor. **(c)** yes **(d)** yes **(e)** No, 3 is a common factor. **(f)** No, 7 is a common factor. **3. (a)** $\frac{2}{3}$ **(b)** $\frac{2}{3}$ **(c)** $-\frac{1}{4}$ **(d)** $-\frac{1}{2}$ **(e)** $\frac{8}{9}$ **(f)** $-\frac{2}{5}$ **5.** comp. prime comp. neither prime prime comp. comp. **7.** $2 \cdot 3$ **9.** $2 \cdot 2 \cdot 5$ **11.** $5 \cdot 5$ **13.** $2 \cdot 2 \cdot 3 \cdot 3$ **15. (a)** $2 \cdot 2 \cdot 11$ **(b)** $2 \cdot 2 \cdot 2 \cdot 11$ **17. (a)** $3 \cdot 5 \cdot 5$ **(b)** $2 \cdot 2 \cdot 17$ **(c)** $3 \cdot 3 \cdot 3 \cdot 7$

19. $\dfrac{\cancel{2} \cdot \cancel{2} \cdot \cancel{2}}{\cancel{2} \cdot \cancel{2} \cdot \cancel{2} \cdot 2} = \dfrac{1}{2}$ **21.** $\dfrac{\cancel{2} \cdot \cancel{2} \cdot \cancel{2} \cdot \cancel{2} \cdot 2}{\cancel{2} \cdot \cancel{2} \cdot \cancel{2} \cdot \cancel{2} \cdot 3} = \dfrac{2}{3}$ **23.** $\dfrac{2 \cdot \cancel{7}}{3 \cdot \cancel{7}} = \dfrac{2}{3}$

25. $\dfrac{\cancel{2} \cdot 2 \cdot \cancel{3} \cdot 3}{\cancel{2} \cdot \cancel{3} \cdot 7} = \dfrac{6}{7}$ **27.** $\dfrac{2 \cdot 5 \cdot 5}{3 \cdot 3 \cdot 7}$ already in lowest terms

29. $\dfrac{\cancel{3} \cdot \cancel{3} \cdot 3}{\cancel{3} \cdot \cancel{3} \cdot 5} = \dfrac{3}{5}$ **31.** $\dfrac{\cancel{2} \cdot 2 \cdot \cancel{3}}{\cancel{2} \cdot \cancel{3} \cdot 3} = \dfrac{2}{3}$ **33.** $\dfrac{\cancel{5} \cdot 7}{2 \cdot 2 \cdot 2 \cdot \cancel{5}} = \dfrac{7}{8}$

35. $\dfrac{\cancel{2} \cdot \cancel{3} \cdot \cancel{3} \cdot \cancel{5}}{\cancel{2} \cdot 2 \cdot \cancel{3} \cdot \cancel{3} \cdot \cancel{5}} = \dfrac{1}{2}$ **37.** $\dfrac{2 \cdot \cancel{3} \cdot \cancel{5} \cdot \cancel{7}}{3 \cdot \cancel{3} \cdot \cancel{5} \cdot \cancel{7}} = \dfrac{2}{3}$

39. $\dfrac{\cancel{3} \cdot \cancel{11} \cdot 13}{\cancel{3} \cdot 3 \cdot 5 \cdot \cancel{11}} = \dfrac{13}{15}$ **41. (a)** $\dfrac{1}{4}$ **(b)** $\dfrac{1}{2}$ **(c)** $\dfrac{1}{10}$ **(d)** $\dfrac{60}{60} = 1$

43. (a) $\dfrac{1}{3}$ **(b)** $\dfrac{1}{6}$ **(c)** $\dfrac{1}{2}$ **45. (a)** $\dfrac{5}{24}$ **(b)** $\dfrac{2}{3}$ **(c)** $\dfrac{1}{8}$ **47. (a)** The result of dividing 3 by 3 is 1, so 1 should be written above and below all the slashes. The numerator is $1 \cdot 1$, so the correct answer is $\frac{1}{4}$. **(b)** You must divide numerator and denominator by the *same* number. The fraction is already in lowest terms because 9 and 16 have no common factor besides 1. **49.** $\dfrac{2c}{5}$ **51.** $\dfrac{4}{7}$ **53.** $\dfrac{6r}{5s}$ **55.** $\dfrac{1}{7n^2}$ **57.** already in lowest terms **59.** $\dfrac{7}{9y}$ **61.** $\dfrac{7k}{2}$ **63.** $\dfrac{1}{3}$ **65.** $\dfrac{c}{d}$ **67.** $\dfrac{6ab}{c}$ **69.** already in lowest terms **71.** $3eg^2$

Section 4.3 (pages 247–250)

1. $-\dfrac{3}{16}$ **3.** $\dfrac{9}{10}$ **5.** $\dfrac{1}{2}$ **7.** -6 **9.** 36 **11.** $\dfrac{15}{4y}$ **13.** $\dfrac{1}{2}$ **15.** $\dfrac{6}{5}$ **17.** -9 **19.** $-\dfrac{1}{6}$ **21.** $\dfrac{11}{15d}$ **23.** b **25. (a)** Forgot to write 1s in numerator when dividing out common factors. Answer is $\frac{1}{6}$. **(b)** Used reciprocal of $\frac{2}{3}$ in multiplication, but the reciprocal is used only in division. Correct answer is $\frac{16}{3}$. **27. (a)** Forgot to use reciprocal of $\frac{4}{1}$; correct answer is $\frac{1}{6}$. **(b)** Used reciprocal of $\frac{5}{6}$ instead of reciprocal of $\frac{10}{9}$; correct answer is $\frac{3}{4}$. **29.** Rewrite division as multiplication. Leave the first number (dividend) the same. Change the second number (divisor) to its reciprocal by "flipping" it. Then multiply. **31.** $\dfrac{4}{15}$ **33.** $-\dfrac{9}{32}$ **35.** 21 **37.** 15 **39.** undefined **41.** $-\dfrac{55}{12}$ **43.** $8b$ **45.** $\dfrac{3}{5d}$ **47.** $\dfrac{2x^2}{w}$ **49.** $\dfrac{3}{10}$ yd^2 **51.** 80 dispensers **53.** earn \$9300; borrow \$3100 **55.** 9 trips **57.** About 78 infield players **59.** 3200 times **61. (a)** \$58,000 **(b)** \$11,600 **63.** \$25,375 **65.** 7 million horses **67.** 154 million dogs and cats

Section 4.4 (pages 259–262)

1. $\dfrac{7}{8}$ **3.** $-\dfrac{1}{2}$ **5.** $\dfrac{1}{2}$ **7.** $-\dfrac{9}{40}$ **9.** $-\dfrac{13}{24}$ **11.** $-\dfrac{3}{5}$ **13.** $-\dfrac{7}{18}$ **15.** $\dfrac{8}{7}$ **17.** $-\dfrac{3}{8}$ **19.** $\dfrac{3 + 5c}{15}$ **21.** $\dfrac{10 - m}{2m}$ **23.** $\dfrac{8}{b^2}$ **25.** $\dfrac{bc + 21}{7b}$ **27.** $\dfrac{-4 - cd}{c^2}$ **29.** $-\dfrac{44}{105}$ **31.** You cannot add or subtract until all the fractional pieces are the same size. **33. (a)** $\dfrac{15}{20} + \dfrac{8}{20} = \dfrac{23}{20}$ Cannot add fractions with unlike denominators; use 20 as the LCD. **(b)** When rewriting fractions with 18 as the denominator, you must multiply denominator and numerator by the same number. The correct answer is $\dfrac{15}{18} - \dfrac{8}{18} = \dfrac{7}{18}$.

35. (a) $\frac{1}{12}; \frac{1}{12}$ Addition is commutative. **(b)** $\frac{1}{3}; -\frac{1}{3}$ Subtraction is *not* commutative. **(c)** $-\frac{3}{5}; -\frac{3}{5}$ Multiplication is commutative. **(d)** $6; \frac{1}{6}$ Division is *not* commutative. **36. (a)** $0; 0$; The sum of a number and its opposite is 0. **(b)** $1; 1$; When a nonzero number is divided by itself, the quotient is 1. **(c)** $\frac{5}{6}; -\frac{17}{20}$; Multiplying by 1 leaves a number unchanged. **(d)** $1; 1$; A number times its reciprocal is 1. **37.** $\frac{47}{60}$ in. **39.** $\frac{23}{24}$ cubic yard **41.** $\frac{3}{4}$ acre **43.** $\frac{23}{50}$ of workers **45.** $\frac{4}{25}$ of workers **47.** $\frac{7}{24}$ of the day; 7 hours **49.** $\frac{1}{8}$ of the day **51.** $\frac{5}{16}$ inch **53.** $\frac{1}{12}$ mile

Section 4.5 (pages 275–278)

1.
$-2\frac{1}{3}$ $2\frac{1}{3}$

3.
$-\frac{3}{2}$ $\frac{3}{2}$

5. $\frac{9}{2}$ **7.** $-\frac{8}{5}$ **9.** $\frac{19}{8}$ **11.** $-\frac{57}{10}$ **13.** $\frac{161}{15}$ **15.** $4\frac{1}{3}$ **17.** $-2\frac{1}{2}$ **19.** $3\frac{2}{3}$ **21.** $-5\frac{2}{3}$ **23.** $11\frac{3}{4}$ **25.** $7\frac{7}{8}; 2 \cdot 4 = 8$ **27.** $1\frac{5}{21}; 3 \div 3 = 1$ **29.** $5\frac{1}{2}; 4 + 2 = 6$ **31.** $3\frac{2}{3}; 4 - 1 = 3$ **33.** $\frac{17}{18}; 6 \div 6 = 1$ **35.** $6\frac{1}{5}; 8 - 2 = 6$ **37.** $P = 7$ in.; $A = 3\frac{1}{16}$ in.^{2} **39.** $P = 19\frac{1}{2}$ yd; $A = 21\frac{1}{8}$ yd^2 **41.** $13 + 9 = 22$ ft; $21\frac{1}{6}$ ft **43.** $2 \cdot 6 = 12$ ounces; $9\frac{5}{8}$ ounces **45.** $4 - 2 = 2$ miles; $2\frac{3}{10}$ miles **47.** $4 \cdot 5 = 20$ yd; $18\frac{3}{4}$ yd **49.** $\frac{5}{8}$ mile; **51.** $6\frac{3}{8}$ miles **53.** $21\frac{3}{8}$ in. **55.** $24 + 35 + 24 + 35 = 118$ in.; $116\frac{1}{2}$ in. **57.** $25,730 \div 10 = 2573$ anchors; 2480 anchors **59.** $(4 \cdot 22) + (5 \cdot 23) + (3 \cdot 24) + (12 \cdot 1) = 287$ in.; $280\frac{1}{8}$ in.

Summary Exercises on Fractions (pages 279–280)

1. (a) $\frac{3}{8}; \frac{5}{8}$ **(b)** $\frac{4}{5}; \frac{1}{5}$ **2.**
$-\frac{2}{3}$ $\frac{2}{3}$
3. (a) 24 **(b)** 4 **4. (a)** 1 **(b)** -4 **(c)** 9 **5. (a)** $2 \cdot 2 \cdot 2 \cdot 3 \cdot 3$ **(b)** $3 \cdot 5 \cdot 7$ **6. (a)** $\frac{4}{5}$ **(b)** $\frac{7}{8}$ **7.** $\frac{1}{2}$ **8.** $-\frac{5}{24}$ **9.** $\frac{17}{16}$ **10.** $\frac{5}{6}$ **11.** $-\frac{2}{15}$ **12.** $-\frac{3}{8}$ **13.** 56 **14.** $\frac{11}{24}$ **15.** $-\frac{7}{6}$ **16.** $-\frac{19}{12}$ **17.** $\frac{25}{12}$ **18.** 35 **19.** $7\frac{7}{12}; 5 + 3 = 8$ **20.** $11\frac{3}{7}; 2 \cdot 5 = 10$ **21.** $3\frac{3}{10}; 6 - 3 = 3$ **22.** $\frac{16}{35}; 2 \div 4 = \frac{2}{4}$ or $\frac{1}{2}$ **23.** $4; 5 \div 1 = 5$ **24.** $2\frac{2}{3}; 3 - 1 = 2$ **25. (a)** $2\frac{1}{16}$ in. **(b)** $\frac{5}{16}$ in. **26.** $P = 3\frac{1}{2}$ in.; $A = \frac{49}{64}$ in.^{2} **27.** 12 batches **28.** 6¢ **29.** Not sure, 225 adults; Real, 675 adults; Imaginary, 600 adults **30.** diameter $= \frac{1}{4}$ in.; $P = 4\frac{1}{2}$ in. **31.** 23 bottles **32.** 11 lots

Section 4.6 (pages 285–288)

1. $\frac{9}{16}$ **3.** $\frac{8}{125}$ **5.** $-\frac{1}{27}$ **7.** $\frac{1}{32}$ **9.** $\frac{49}{100}$ **11.** $\frac{36}{25}$ or $1\frac{11}{25}$ **13.** $\frac{12}{25}$ **15.** $\frac{1}{100}$ **17.** $-\frac{3}{2}$ or $-1\frac{1}{2}$ **19. (a)** The answers are $\frac{1}{4}, -\frac{1}{8}, \frac{1}{16}, -\frac{1}{32}, \frac{1}{64}, -\frac{1}{128}, \frac{1}{256}, -\frac{1}{512}$. **(b)** When a negative number is raised to an even power, the answer is positive. When a negative number is raised to an odd power, the answer is negative. **20. (a)** Ask yourself, "What number, times itself, is 4?" This is the numerator. Then ask, "What number, times itself, is 9?" This is the denominator. The number under the ketchup is either $\frac{2}{3}$ or $-\frac{2}{3}$. **(b)** The number under the ketchup is $-\frac{1}{3}$. because $\left(-\frac{1}{3}\right)\left(-\frac{1}{3}\right)\left(-\frac{1}{3}\right) = -\frac{1}{27}$. **(c)** Either $\frac{1}{2}$ or $-\frac{1}{2}$. **(d)** No real number works, because both $\left(\frac{3}{4}\right)^2$ and $\left(-\frac{3}{4}\right)^2$ give a *positive* result. **(e)** Either $\frac{1}{3}$ or $-\frac{1}{3}$ inside one set of parentheses and $\frac{1}{2}$ or $-\frac{1}{2}$ inside the other. **21.** -4 **23.** $\frac{5}{16}$ **25.** $\frac{1}{3}$ **27.** $\frac{1}{6}$ **29.** $-\frac{17}{24}$ **31.** $-\frac{4}{27}$ **33.** $\frac{1}{36}$ **35.** $\frac{9}{64}$ in.^{2} **37.** $\frac{19}{10}$ or $1\frac{9}{10}$ miles **39.** 4 **41.** $-\frac{25}{2}$ or $-12\frac{1}{2}$ **43.** $\frac{1}{14}$ **45.** $\frac{5}{18}$ **47.** -8 **49.** $\frac{9}{100}$

Section 4.7 (pages 295–298)

1. $a = 30$ **Check** $\frac{1}{3}(30) = 10$
$10 = 10$

3. $b = -24$ **Check** $-20 = \frac{5}{6}(-24)$
$-20 = -20$

5. $c = 6$ **Check** $-\frac{7}{2}(6) = -21$
$-21 = -21$

7. $m = \frac{3}{4}$ **Check** $\frac{9}{16} = \frac{3}{4}\left(\frac{3}{4}\right)$
$\frac{9}{16} = \frac{9}{16}$

9. $d = -\frac{6}{5}$ **Check** $\frac{3}{10} = -\frac{1}{4}\left(-\frac{6}{5}\right)$
$\frac{3}{10} = \frac{3}{10}$

11. $n = 12$ **Check** $\frac{1}{6}(12) + 7 = 9$
$2 + 7 = 9$
$9 = 9$

13. $r = -9$ **Check** $-10 = \frac{5}{3}(-9) + 5$
$-10 = -15 + 5$
$-10 = -10$

15. $x = 24$ **Check** $\frac{3}{8}(24) - 9 = 0$
$9 - 9 = 0$
$0 = 0$

17. $y = 45$ **19.** $n = -18$ **21.** $x = \frac{1}{12}$ **23.** $b = -\frac{1}{8}$

25. (a) $\frac{1}{6}(18) + 1 = -2$

$3 + 1 = -2$

$4 \neq -2$

Does *not* balance; correct solution is -18.

(b) $-\frac{3}{2} = \frac{9}{4}\left(-\frac{2}{3}\right)$

$-\frac{3}{2} = -\frac{3}{2}$

Balances, so $-\frac{2}{3}$ is correct solution.

27. Some possibilities are: $\frac{1}{2}x = 4$; $-\frac{1}{4}a = -2$; $\frac{3}{4}b = 6$.

29. Let a be the man's age.

$109 = 100 + \frac{a}{2}$

The man is 18 years old.

31. Let a be the woman's age.

$122 = 100 + \frac{a}{2}$

The woman is 44 years old.

33. Let p be the penny size. $\frac{p}{4} + \frac{1}{2} = 3$. The penny size is 10.

35. Let p be the penny size. $\frac{p}{4} + \frac{1}{2} = \frac{5}{2}$. The penny size is 8. **37.** Let h be the man's height. $\frac{11}{2}h - 220 = 209$. The man is 78 in. tall. **39.** Let h be the woman's height. $\frac{11}{2}h - 220 = 132$. The woman is 64 in. tall.

Section 4.8 (pages 305–306)

1. $P = 202$ m; $A = 1914$ m^2 **3.** $P = 5$ ft; $A = \frac{27}{32}$ ft^2

5. $P = 26\frac{1}{4}$ yd; $A = 30\frac{3}{4}$ yd^2 **7.** $P = \frac{454}{15}$ yd or $30\frac{4}{15}$ yd;

$A = \frac{115}{3}$ yd^2 or $38\frac{1}{3}$ yd^2 **9.** $A = 1716$ m^2 **11.** $A = \frac{63}{8}$ ft^2 or $7\frac{7}{8}$ ft^2

13. 132 m of curbing; 726 m^2 of sod **15.** Rectangular solid; $V = 528$ cm^3

17. Rectangular solid or cube; $V = 15\frac{5}{8}$ in.^{3} **19.** Pyramid; $V = 800$ cm^3

21. $V = 106\frac{2}{3}$ ft^3 **23.** $V = 18$ in.^{3} **25.** $V = 651{,}775$ m^3

Chapter 4 Review Exercises (pages 315–316)

1. $\frac{2}{5}; \frac{1}{5}$ **2.** $\frac{3}{10}; \frac{7}{10}$ **3.**

$-\frac{1}{2} \qquad 1\frac{1}{2}$

$-3 \ -2 \ -1 \ \ 0 \ \ 1 \ \ 2 \ \ 3$

4. (a) -4 **(b)** 8 **(c)** -1 **5.** $\frac{7}{8}$ **6.** $\frac{3}{5}$ **7.** already in lowest terms

8. $\frac{3x}{8}$ **9.** $\frac{1}{5b}$ **10.** $\frac{4n}{7m^2}$ **11.** $\frac{1}{16}$ **12.** -12 **13.** $\frac{8}{27}$ **14.** $\frac{1}{6x}$ **15.** $2a^2$

16. $\frac{6}{7k}$ **17.** $\frac{5}{24}$ **18.** $-\frac{2}{15}$ **19.** $\frac{19}{6}$ or $3\frac{1}{6}$ **20.** $\frac{3}{2}$ or $1\frac{1}{2}$ **21.** $\frac{4n+15}{20}$

22. $\frac{3y-70}{10y}$ **23.** $1\frac{5}{13}; 2 \div 2 = 1$ **24.** $2\frac{1}{2}; 7 - 5 = 2$ **25.** $4\frac{1}{20};$

$2 + 2 = 4$ **26.** $-\frac{27}{64}$ **27.** $\frac{1}{36}$ **28.** $-\frac{4}{5}$ **29.** $\frac{7}{6}$ or $1\frac{1}{6}$ **30.** 10

31. $-\frac{4}{27}$ **32.** $w = 20$ **33.** $r = -15$ **34.** $x = \frac{1}{2}$ **35.** $A = 14$ ft^2

36. Rectangular solid; $V = 32\frac{1}{2}$ in.^{3} **37.** Pyramid; $V = 93\frac{1}{3}$ yd^3

38. $\frac{1}{4}$ pound; $7\frac{1}{2}$ pounds **39.** $\frac{5}{6}$ hour; $10\frac{11}{12}$ hours **40.** 12 preschoolers, 40 toddlers, 8 infants **41.** $P = 2\frac{1}{10}$ miles; $A = \frac{9}{40}$ mi^2

Chapter 4 Test (pages 317–318)

1. $\frac{5}{6}; \frac{1}{6}$ **2.**

$-\frac{2}{3} \qquad\qquad 2\frac{1}{3}$

$-3 \ -2 \ -1 \ \ 0 \ \ 1 \ \ 2 \ \ 3$

3. $\frac{1}{4}$

4. already in lowest terms **5.** $\frac{2a^2}{3b}$ **6.** $\frac{13}{15}$ **7.** -2 **8.** $-\frac{7}{40}$ **9.** 14

10. $-\frac{2}{27}$ **11.** $\frac{25}{8}$ or $3\frac{1}{8}$ **12.** $\frac{4}{9}$ **13.** $\frac{9}{16}$ **14.** $\frac{4}{7y}$ **15.** $\frac{24-n}{4n}$

16. $\frac{10+3a}{15}$ **17.** $\frac{1}{18b}$ **18.** $-\frac{1}{18}$ **19.** $-\frac{31}{30}$ or $-1\frac{1}{30}$

20. $5 \div 1 = 5; \frac{64}{15}$ or $4\frac{4}{15}$ **21.** $3 - 2 = 1; \frac{3}{2}$ or $1\frac{1}{2}$ **22.** $d = 35$

23. $t = -\frac{15}{7}$ or $-2\frac{1}{7}$ **24.** $b = 8$ **25.** $x = -15$ **26.** $A = 52$ m^2

27. $A = \frac{117}{2}$ yd^2 or $58\frac{1}{2}$ yd^2 **28.** Rectangular solid; $V = 6480$ m^3

29. Pyramid; $V = 16$ yd^3 **30.** $14\frac{3}{4}$ hours; $1\frac{5}{6}$ hours

31. $\frac{7}{2}$ days or $3\frac{1}{2}$ days **32.** 7392 students work

Chapter 5 Rational Numbers: Positive and Negative Decimals

Section 5.1 (pages 327–330)

1. 7; 0; 4 **3.** 5; 1; 8 **5.** 4; 7; 0 **7.** 1; 6; 3 **9.** 1; 8; 9 **11.** 6; 2; 1

13. 410.25 **15.** 6.5432 **17.** 5406.045 **19.** $\frac{7}{10}$ **21.** $13\frac{2}{5}$ **23.** $\frac{7}{20}$

25. $\frac{33}{50}$ **27.** $10\frac{17}{100}$ **29.** $\frac{3}{50}$ **31.** $\frac{41}{200}$ **33.** $5\frac{1}{500}$ **35.** $\frac{343}{500}$

37. five tenths **39.** seventy-eight hundredths **41.** one hundred five thousandths **43.** twelve and four hundredths **45.** one and seventy-five thousandths **47.** 6.7 **49.** 0.32 **51.** 420.008 **53.** 0.0703 **55.** 75.030

57. Anne should not say "and" because that denotes a decimal point.

59. ten thousandths inch; $\frac{10}{1000} = \frac{1}{100}$ inch **61.** 12 pounds **63.** 3-C

65. 4-A **67.** one and six hundred two thousandths centimeters

69. millionths, ten-millionths, hundred-millionths, billionths; these match the words on the left side of the chart with "ths" attached.

70. First place to the left of the decimal point is ones. "Oneths" would mean a fraction with a denominator of 1, which would equal 1 or more. Anything that is 1 or more is to the *left* of the decimal point.

71. seventy-two million four hundred thirty-six thousand nine hundred fifty-five hundred-millionths **72.** six hundred seventy-eight thousand five hundred fifty-four billionths **73.** eight thousand six and five hundred thousand one millionths **74.** twenty thousand sixty and five hundred five millionths **75.** 0.0302040 **76.** 9,876,543,210.100200300

Section 5.2 (pages 337–338)

1. 16.9 **3.** 0.956 **5.** 0.80 **7.** 3.661 **9.** 794.0 **11.** 0.0980

13. 49 **15.** 9.09 **17.** 82.0002 **19.** $0.82 **21.** $1.22 **23.** $0.50

25. $48,650 **27.** $310 **29.** $849 **31.** $500 **33.** $1.00 **35.** $1000

37. (a) 322 miles per hour **(b)** 107 miles per hour **39. (a)** 186.0 miles per hour **(b)** 763.0 miles per hour **41.** Rounds to $0 (zero dollars) because $0.499 is closer to $0 than to $1. **42.** Round $0.499 to the nearest cent to get $0.50. Guideline: Round amounts less than $1.00 to the nearest cent instead of the nearest dollar. **43.** Rounds to $0.00 (zero cents) because $0.0015 is closer to $0.00 than to $0.01. **44.** Both round to $0.60. Rounding to the nearest thousandth (tenth of a cent) would allow you to identify $0.597 as less than $0.601.

Section 5.3 (pages 345–348)

1. 17.72 **3.** 11.98 **5.** 115.861 **7.** 59.323 **9.** 6 should be written 6.00; sum is 46.22 **11.** $0.3000 = \dfrac{3000 \div 1000}{10,000 \div 1000} = \dfrac{3}{10} = 0.3$ **13.** 89.7
15. 0.109 **17.** 0.91 **19.** 6.661 **21.** The student subtracted in the wrong order; 15.32 should be on top; correct answer is 7.87 **23. (a)** 24.75 in. **(b)** 3.95 in. **25. (a)** 62.27 in. **(b)** 0.39 in. **27.** 23.013 **29.** −45.75 **31.** −6.69 **33.** −6.99 **35.** −4.279 **37.** −0.0035 **39.** 5.37 **41.** *Estimate:* $20 − 10 = 10$; *Exact:* 6.275 **43.** *Estimate:* $−7 + 1 = −6$; *Exact:* −5.8 **45.** *Estimate:* $−40 + (−200) = −240$; *Exact:* −237.571 **47.** *Estimate:* $8 + (+50) = 58$; *Exact:* 59.23 **49.** *Estimate:* $30 − 20 = 10$ million people; *Exact:* 12.1 million people **51.** *Estimate:* $200 + 100 + 90 + 50 + 30 + 20 + 20 = 510$ million people; *Exact:* 527.0 million people **53.** *Estimate:* $2 + 2 + 2 = 6$ m; *Exact:* 6.09 m, which is 0.31 m less than the rhino's height. **55.** *Estimate:* $20 − $9 = 11; *Exact:* $10.88 **57.** *Estimate:* $5 − 5 = 0; *Exact:* $0.30 **59.** *Estimate:* $19 + 2 + 2 + 10 + 2 = 35; *Exact:* $35.25 **61.** $1939.36 **63.** $3.97 **65.** $b = 1.39$ cm **67.** $q = 7.943$ ft

Section 5.4 (pages 353–356)

1. 0.1344 **3.** −159.10 **5.** 15.444 **7.** $34,500.20 **9.** −43.2 **11.** 0.432 **13.** 0.0432 **15.** 0.00432 **17.** 0.0000312 **19.** 0.000009 **21.** 59.6; 4.76; 7226; 32; 803.5; 9. Multiplying by 10, decimal point moves one place to the right; by 100, two places to the right; by 1000, three places to the right. **22.** 5.96; 0.0476; 6.5; 0.32; 8.035; 52.3. Multiplying by 0.1, decimal point moves one place to the left; by 0.01, two places to the left; by 0.001, three places to the left. **23.** *Estimate:* $40 \times 5 = 200$; *Exact:* 190.08 **25.** *Estimate:* $40 \times 40 = 1600$; *Exact:* 1558.2 **27.** *Estimate:* $7 \times 5 = 35$; *Exact:* 30.038 **29.** *Estimate:* $3 \times 7 = 21$; *Exact:* 19.24165 **31.** unreasonable; $289.00 **33.** reasonable **35.** unreasonable; $4.19 **37.** unreasonable; 9.5 pounds **39.** $945.87 (rounded) **41.** $2.45 (rounded) **43.** $81.61 (rounded) **45.** $20,265 **47. (a)** Area before 1929 ≈ 23.2 in.^{2}; Area today ≈ 16.0 in.^{2} **(b)** 7.2 in.^{2} **49. (a)** 0.43 inch **(b)** 4.3 inches **51.** $984.04; $2207.80 **53.** $76.50 **55.** $4.09 (rounded) **57.** $129.25 **59. (a)** $70.05 **(b)** $25.80

Section 5.5 (pages 365–368)

1. −3.9 **3.** 0.47 **5.** 400.2 **7.** 36 **9.** 0.06 **11.** 6000 **13.** 60 **15.** 0.0006 **17.** 25.3 **19.** 516.67 (rounded) **21.** −26.756 (rounded) **23.** 10,082.647 (rounded) **25.** 0.377; 0.0886; 40.65; 0.91; 3.019; 662.57 **(a)** Dividing by 10, decimal point moves one place to the left; by 100, two places to the left; by 1000, three places to the left. **(b)** The decimal point moved to the *right* when multiplying by 10, by 100, or by 1000; here it moves to the *left* when dividing by 10, by 100, or by 1000. **26.** 402; 3.39; 460; 71; 157.7; 8730 **(a)** Dividing by 0.1, decimal point moves one place to the right; by 0.01, two places to the right; by 0.001, three places to the right. **(b)** The decimal point moved to the *left* when multiplying by 0.1, 0.01, or 0.001; here it moves to the *right* when dividing by 0.1, 0.01, or 0.001. **27.** unreasonable; *Estimate:* $40 \div 8 = 5$; $8\overline{)37.8}$ gives 4.725 **29.** reasonable; *Estimate:* $50 \div 50 = 1$ **31.** unreasonable; *Estimate:* $300 \div 5 = 60$; $5.1\overline{)307.02}$ gives 60.2 **33.** unreasonable; *Estimate:* $9 \div 1 = 9$; $1.25\overline{)9.30}$ gives 7.44

35. $4.00 (rounded) **37.** $67.08 **39.** $0.30 **41.** $11.92 per hour **43.** 21.2 miles per gallon (rounded) **45.** 7.37 meters (rounded) **47.** 0.08 meter **49.** 22.49 meters **51.** 14.25 **53.** 3.8 **55.** −16.155 **57.** 3.714 **59.** $0.03 (rounded) **61. (a)** 1,583,333 pieces (rounded) **(b)** 26,389 pieces (rounded) **(c)** 440 pieces (rounded) **63.** 100,000 box tops **65.** 2632 box tops (rounded)

Summary Exercises on Decimals (pages 369–370)

1. $\dfrac{4}{5}$ **2.** $6\dfrac{1}{250}$ **3.** $\dfrac{7}{20}$ **4.** ninety-four and five tenths **5.** two and three ten-thousandths **6.** seven hundred six thousandths **7.** 0.05 **8.** 0.0309 **9.** 10.7 **10.** 6.19 **11.** 1.0 **12.** 0.420 **13.** $0.89 **14.** $3.00 **15.** $100 **16.** −0.945 **17.** 49.6199 **18.** −50 **19.** 15.03 **20.** 0.00488 **21.** −2.15 **22.** 9.055 **23.** 18.4009 **24.** −6.995 **25.** −808.9 **26.** 2.12 **27.** 0.04 **28.** $P = 52.1$ in. **29.** $P = 9.735$ meters **30.** $1.80 **31.** $87.28 **32. (a)** 169 ft^2 **(b)** $0.75 (rounded) **33.** 144 ft^2; $0.69 (rounded) **34. (a)** 9 days (rounded) **(b)** 15.6 pounds (rounded) **35.** Average weight of food eaten by bee is 0.32 ounce. **36.** 0.112 ounce

Section 5.6 (pages 375–378)

1. 0.5 **3.** 0.75 **5.** 0.3 **7.** 0.9 **9.** 0.6 **11.** 0.875 **13.** 2.25 **15.** 14.7 **17.** 3.625 **19.** 6.333 (rounded) **21.** 0.833 (rounded) **23.** 1.889 (rounded) **25. (a)** A proper fraction like $\dfrac{5}{9}$ is less than 1, so $\dfrac{5}{9}$ cannot be equivalent to a decimal number that is greater than 1. **(b)** $\dfrac{5}{9}$ means $5 \div 9$ or $9\overline{)5}$ so correct answer is 0.556 (rounded). This makes sense because both the fraction and decimal are less than 1. **26. (a)** $2.035 = 2\dfrac{35}{1000} = 2\dfrac{7}{200}$, not $2\dfrac{7}{20}$ **(b)** Adding the whole number part gives $2 + 0.35$, which is 2.35 not 2.035. To check, $2.35 = 2\dfrac{35}{100} = 2\dfrac{7}{20}$ but $2.035 = 2\dfrac{35}{1000} = 2\dfrac{7}{200}$ **27.** Just add the whole number part to 0.375. So $1\dfrac{3}{8} = 1.375$; $3\dfrac{3}{8} = 3.375$; $295\dfrac{3}{8} = 295.375$ **28.** It works only when the fraction part has a one-digit numerator and a denominator of 10, or a two-digit numerator and a denominator of 100, and so on. **29.** $\dfrac{2}{5}$ **31.** $\dfrac{5}{8}$ **33.** $\dfrac{7}{20}$ **35.** 0.35 **37.** $\dfrac{1}{25}$ **39.** $\dfrac{3}{20}$ **41.** 0.2 **43.** $\dfrac{9}{100}$ **45.** shorter; 0.72 inch **47.** too much; 0.005 gram **49.** 0.9991 cm, 1.0007 cm **51.** more; 0.05 inch **53. (a)** < **(b)** = **(c)** > **(d)** < **55.** 0.5399, 0.54, 0.5455 **57.** 5.0079, 5.79, 5.8, 5.804 **59.** 0.6009, 0.609, 0.628, 0.62812 **61.** 2.8902, 3.88, 4.876, 5.8751 **63.** 0.006, 0.043, $\dfrac{1}{20}$, 0.051 **65.** 0.37, $\dfrac{3}{8}$, $\dfrac{2}{5}$, 0.4001 **67. (a)** red box; green box **(b)** 0.01 inch **69.** 1.4 in. (rounded) **71.** 0.3 in. (rounded) **73.** 0.4 in. (rounded)

Section 5.7 (pages 383–384)

1. 69.8 (rounded) **3.** $39,622 **5.** $58.24 **7.** 6.1 (rounded) **9.** 2.60 **11. (a)** 2.80 **(b)** 2.93 (rounded) **(c)** 3.13 (rounded) **13.** 15 messages **15.** 516 students **17.** 48.5 pounds of shrimp **19.** 4142 miles (rounded) **21.** 4050 miles **23.** 8 samples **25.** 68 and 74 years (bimodal) **27.** no mode **29.** −10 degrees; 4 degrees; 14 degrees warmer

Section 5.8 (pages 389–392)

1. 4 **3.** 8 **5.** 3.317 (rounded) **7.** 2.236 (rounded) **9.** 8.544 (rounded) **11.** 10.050 (rounded) **13.** 19 **15.** 31.623 (rounded) **17.** 30 is about halfway between 25 and 36, so $\sqrt{30}$ should be about halfway between 5 and 6, or about 5.5. Using a calculator, $\sqrt{30} \approx 5.477$ Similarly, $\sqrt{26}$ should be a little more than $\sqrt{25}$;

by calculator $\sqrt{26} \approx 5.099$ And $\sqrt{35}$ should be a little less than $\sqrt{36}$; by calculator $\sqrt{35} \approx 5.916$ **19.** $\sqrt{1521} = 39$ ft
21. $\sqrt{289} = 17$ in.   **23.** $\sqrt{144} = 12$ mm   **25.** $\sqrt{73} \approx 8.5$ in.
27. $\sqrt{65} \approx 8.1$ yd   **29.** $\sqrt{195} \approx 14.0$ cm   **31.** $\sqrt{7.94} \approx 2.8$ m
33. $\sqrt{65.01} \approx 8.1$ cm   **35.** $\sqrt{292.32} \approx 17.1$ km
37. hypotenuse $= \sqrt{65} \approx 8.1$ ft   **39.** leg $= \sqrt{360,000} = 600$ m
41. hypotenuse $= \sqrt{48.5} \approx 7.0$ ft   **43.** leg $= \sqrt{135} \approx 11.6$ ft.
45. The student used the formula for finding the hypotenuse but the unknown side is a leg, so use leg $= \sqrt{(20)^2 - (13)^2}$. Also, the final answer should be m, not m². Correct answer is $\sqrt{231} \approx 15.2$ m.   **47.** $\sqrt{16,200} \approx 127.3$ ft
48. (a)   **Second**   **(b)** $\sqrt{7200} \approx 84.9$ ft

49. The distance from third to first is the same as the distance from home to second because the baseball diamond is a square.
50. (a) The side length is less than 60 ft.   **(b)** $80^2 = 6400$; $6400 \div 2 = 3200$; $\sqrt{3200} \approx 56.6$ ft

Section 5.9 (pages 397–398)

Please see *Solutions* section for sample *checks* of some odd-numbered exercises. **1.** $h = 4.47$ **3.** $n = -1.4$ **5.** $b = 0.008$
7. $a = 0.29$ **9.** $p = -120$ **11.** $t = 0.7$ **13.** $x = -0.82$
15. $z = 0$ **17.** $c = 0.45$ **19.** $w = 28$ **21.** $p = -40.5$
23. Let d be the adult dose. $0.3\,d = 9$; The adult dose is 30 milligrams.
25. Let d be the number of days. $65.95d + 12 = 275.80$; The saw was rented for 4 days.   **27.** $0.7(220 - a) = 140$; The person is 20 years old.   **28.** $0.7(220 - a) = 126$; The person is 40 years old.
29. $0.7(220 - a) = 134$; $a \approx 28.57$, which rounds to 29. The person is about 29 years old.   **30.** $0.7(220 - a) = 117$; $a \approx 52.86$, which rounds to 53. The person is about 53 years old.

Section 5.10 (pages 407–410)

1. $d = 18$ mm   **3.** $r = 0.35$ km   **5.** $C \approx 69.1$ ft; $A \approx 379.9$ ft²
7. $C \approx 8.2$ m; $A \approx 5.3$ m²   **9.** $C \approx 47.1$ cm, $A \approx 176.6$ cm²
11. $C \approx 23.6$ ft; $A \approx 44.2$ ft²   **13.** $C \approx 27.2$ km; $A \approx 58.7$ km²
15. $A \approx 76.9$ in.²   **17.** $A \approx 57$ cm²   **19.** $A \approx 7850$ yd²
21. $C \approx 91.4$ in.; Bonus: 693 revolutions/mile (rounded)
23. $A \approx 70,650$ mi²

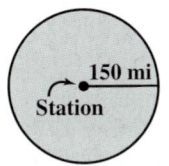

25. watch: $C \approx 3.1$ in.; $A \approx 0.8$ in.²; wall clock: $C \approx 18.8$ in.; $A \approx 28.3$ in.²

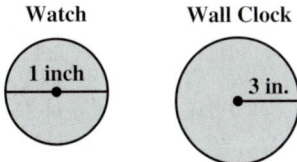

27. $A \approx 78.5$ mi² $-$ 12.6 mi²; difference ≈ 65.9 mi²

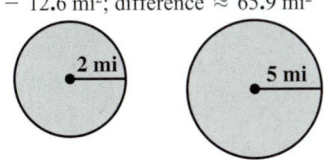

29. (a) $d \approx 45.9$ cm   **(b)** Divide the circumference by π (144 cm $\div$ 3.14).
31. $C \approx 14.6$ ft   **33.** $A \approx 197.8$ cm²   **35.** $V \approx 471$ ft³; $SA \approx 345.4$ ft²
37. $V = 1617$ m³; $SA = 849.4$ m²   **39.** $V \approx 763.0$ in.³; $SA \approx 678.2$ in.²
41. $V = 5550$ mm³; $SA = 2150$ mm²   **43.** Student should use radius of 3.5 cm instead of diameter of 7 cm in the formula; units for volume are cm³, not cm². Correct answer is $V \approx 192.3$ cm³   **45.** $V \approx 3925$ ft³
47. $SA = 163.6$ in.²

Chapter 5 Review Exercises (pages 419–424)

1. 0; 5   **2.** 0; 6   **3.** 8; 9   **4.** 5; 9   **5.** 7; 6   **6.** $\frac{1}{2}$ **7.** $\frac{3}{4}$ **8.** $4\frac{1}{20}$
9. $\frac{7}{8}$ **10.** $\frac{27}{1000}$ **11.** $27\frac{4}{5}$ **12.** eight tenths   **13.** four hundred and twenty-nine hundredths   **14.** twelve and seven thousandths   **15.** three hundred six ten-thousandths   **16.** 8.3   **17.** 0.205   **18.** 70.0066
19. 0.30   **20.** 275.6   **21.** 72.79   **22.** 0.160   **23.** 0.091   **24.** 1.0
25. \$15.83   **26.** \$0.70   **27.** \$17,625.79   **28.** \$350   **29.** \$130
30. \$100   **31.** \$29   **32.** -5.67 **33.** -0.03 **34.** 5.879   **35.** -6.435
36. *Estimate:* 80 million $-$ 50 million $=$ 30 million; *Exact:* 31.4 million people   **37.** *Estimate:* \$300 $-$ \$200 $-$ \$40 $=$ \$60; *Exact:* \$45.80
38. *Estimate:* \$2 $+$ \$5 $+$ \$20 $=$ \$27; \$30 $-$ \$27 $=$ \$3; *Exact:* \$4.14
39. *Estimate:* 2 $+$ 4 $+$ 5 $=$ 11 kilometers; *Exact:* 11.55 kilometers
40. *Estimate:* 6 $\times$ 4 $=$ 24; *Exact:* 22.7106   **41.** *Estimate:* 40 $\times$ 3 $=$ 120; *Exact:* 141.57   **42.** 0.0112   **43.** -0.000355 **44.** reasonable; *Estimate:* 700 $\div$ 10 $=$ 70   **45.** unreasonable;
Estimate: 30 $\div$ 3 $=$ 10; $2.8\overline{)26.6}^{\,9.5}$ **46.** 14.467 (rounded)   **47.** 1200
48. -0.4 **49.** \$708 (rounded)   **50.** \$2.99 (rounded)   **51.** 133 shares (rounded)   **52.** \$3.47 (rounded)   **53.** -4.715 **54.** 10.15   **55.** 3.8
56. 0.64   **57.** 1.875   **58.** 0.111 (rounded)   **59.** 3.6008, 3.68, 3.806
60. 0.209, 0.2102, 0.215, 0.22   **61.** $\frac{1}{8}, \frac{3}{20}, 0.159, 0.17$ **62.** mean: 28.5 digital cameras; median: 19.5 digital cameras   **63.** mean: 44 claims; median: 39 claims   **64.** \$51.05   **65.** Store J: \$107 and \$160 (bimodal); Store K: \$119   **66.** $\sqrt{289} = 17$ in.   **67.** $\sqrt{49} = 7$ cm
68. $\sqrt{104} \approx 10.2$ cm   **69.** $\sqrt{52} \approx 7.2$ in.   **70.** $\sqrt{6.53} \approx 2.6$ m
71. $\sqrt{71.75} \approx 8.5$ km   **72.** $b = 0.25$ **73.** $x = 0$ **74.** $n = -13$
75. $a = -1.5$ **76.** $y = -16.2$ **77.** $d = 137.8$ m   **78.** $r = 1\frac{1}{2}$ in. or 1.5 in.   **79.** $C \approx 6.3$ cm; $A \approx 3.1$ cm²   **80.** $C \approx 109.3$ m; $A \approx 950.7$ m²   **81.** $C \approx 37.7$ in.; $A \approx 113.0$ in.²   **82.** $V \approx 549.5$ cm³; $SA \approx 376.8$ cm²   **83.** $V = 1808.6$ m³; $SA \approx 1205.8$ m²
84. $V \approx 7.9$ ft³; $SA \approx 25.5$ ft²   **85.** 404.865   **86.** -254.8
87. 3583.261 (rounded)   **88.** 29.0898   **89.** 0.03066   **90.** 9.4
91. -15.065 **92.** -9.04 **93.** -15.74 **94.** 8.19   **95.** 0.928   **96.** 35
97. -41.859 **98.** 0.3   **99.** \$3.00 (rounded)   **100.** \$2.17 (rounded)
101. \$35.96   **102.** \$199.71   **103.** \$78.50   **104.** $y = -0.7$
105. $x = 30$ **106.** 15.7 ft long rubber strip; $A \approx 19.6$ ft²
107. mean is 67.6; median is 82   **108.** 20 miles   **109.** $V \approx 49.5$ in.³

Chapter 5 Test (pages 425–426)

1. $18\frac{2}{5}$ **2.** $\frac{3}{40}$ **3.** sixty and seven thousandths   **4.** two hundred eight ten-thousandths   **5.** *Estimate:* 8 $+$ 80 $+$ 40 $=$ 128; *Exact:* 129.2028   **6.** *Estimate:* -6 (1) $= -6$; *Exact:* -6.948
7. *Estimate:* $-80 - 4 = -84$; *Exact:* -82.702
8. *Estimate:* $-20 \div (-5) = 4$; *Exact:* 4.175   **9.** 669.004
10. 480   **11.** 0.000042   **12.** \$1.79 per foot (rounded)
13. Davida, by 0.441 minute   **14.** \$5.35 (rounded)   **15.** $y = -6.15$
16. $x = -0.35$ **17.** $a = 10.9$ **18.** $n = 2.85$ **19.** $0.44, \frac{9}{20}$, 0.4506, 0.451   **20.** 32.09   **21.** 75 books   **22.** 103° and 104° (bimodal)
23. \$11.25   **24.** \$50.55   **25.** $\sqrt{85} \approx 9.2$ cm
26. $\sqrt{279} \approx 16.7$ ft   **27.** $r = 12.5$ in.   **28.** $C \approx 5.7$ km
29. $A \approx 206.0$ cm²   **30.** $V \approx 5086.8$ ft³   **31.** $SA \approx 2599.9$ ft²

Chapter 6 Ratio, Proportion, and Line/Angle/Triangle Relationships

Section 6.1 (pages 433–434)

1. $\dfrac{8}{9}$ **3.** $\dfrac{2}{1}$ **5.** $\dfrac{1}{3}$ **7.** $\dfrac{9}{7}$ **9.** $\dfrac{5}{6}$ **11.** $\dfrac{8}{5}$ **13.** $\dfrac{1}{12}$ **15.** $\dfrac{4}{1}$

17. $\dfrac{1}{2}$ **19.** $\dfrac{36}{1}$ **21.** Answers will vary. One possibility is stocking cards of various types in the same ratios as those in the table. **23.** *White Christmas* to *It's Now or Never*; *White Christmas* to *I Will Always Love You*; *Candle in the Wind* to *I Want to Hold Your Hand* **25.** $\dfrac{6}{1}$ **27.** $\dfrac{38}{17}$ **29.** $\dfrac{1}{4}$ **31.** $\dfrac{34}{35}$

Section 6.2 (pages 439–442)

1. $\dfrac{5 \text{ cups}}{3 \text{ people}}$ **3.** $\dfrac{3 \text{ feet}}{7 \text{ seconds}}$ **5.** $\dfrac{1 \text{ person}}{2 \text{ dresses}}$ **7.** $\dfrac{5 \text{ letters}}{1 \text{ minute}}$ **9.** $\dfrac{21 \text{ dollars}}{2 \text{ visits}}$

11. $\dfrac{18 \text{ miles}}{1 \text{ gallon}}$ **13.** $12 per hour or $12/hour **15.** 5 eggs per chicken or 5 eggs/chicken **17.** 1.25 pounds/person **19.** $103.30/day **21.** 325.9; 21.0 (rounded) **23.** 338.6; 20.9 (rounded) **25.** 4 ounces for $3.65, about $0.913/ounce **27.** 14 ounces for $2.89, about $0.206/ounce **29.** 18 ounces for $1.79, about $0.099/ounce **31.** Answers will vary. For example, you might choose Brand B because you like more chicken, so the cost per chicken chunk may actually be the same or less than Brand A. **33.** 1.75 pounds/week **35.** $12.26/hour **37. (a)** Radiant $0.44; IDT $0.25; Access $0.235 **(b)** Radiant $0.088/min, IDT $0.05/min, Access $0.047/min; Access America is the best buy. **39.** For a 15-minute call: Radiant $0.036/min; IDT $0.031/min; Access $0.047/min; IDT is the best buy. For a 20-minute call: Radiant $0.030/min; IDT $0.029/min; Access $0.047/min; IDT is the best buy. **41.** 0.11 second/meter; 9.$\overline{09}$ or 9.1 meters/second (rounded) **43.** One battery for $1.79; like getting 3 batteries so $1.79 ÷ 3 ≈ $0.597 per battery **45.** Brand P with the 50¢ coupon is the best buy. ($3.39 − $0.50 = $2.89, $2.89 ÷ 16.5 ounces ≈ $0.175 per ounce) **47.** $2.92 (rounded) per month for Verizon, T-Mobile, and Nextel; $3 per month for Sprint. **48.** Average weekdays per month is about 21.7; Verizon ≈ 18 min/weekday; T-Mobile ≈ 28 min/weekday; Nextel and Sprint ≈ 23 min/weekday. **49.** Round to hundredths (nearest cent) to see that T-Mobile is the best buy at $0.07 per "anytime minute"; Verizon ≈ $0.16; Sprint ≈ $0.12; Nextel ≈ $0.10 **50.** Verizon ≈ $1.65 per "anytime minute"; T-Mobile ≈ $1.57; Nextel ≈ $1.63; Sprint = $1.48

Section 6.3 (pages 451–454)

1. $\dfrac{\$9}{12 \text{ cans}} = \dfrac{\$18}{24 \text{ cans}}$ **3.** $\dfrac{200 \text{ adults}}{450 \text{ children}} = \dfrac{4 \text{ adults}}{9 \text{ children}}$ **5.** $\dfrac{120}{150} = \dfrac{8}{10}$

7. $\dfrac{2.2}{3.3} = \dfrac{3.2}{4.8}$ **9.** $\dfrac{3}{5} = \dfrac{3}{5}$; true **11.** $\dfrac{5}{8} = \dfrac{5}{8}$; true **13.** $\dfrac{3}{4} \neq \dfrac{2}{3}$; false

15. 54 = 54; true **17.** 336 ≠ 320; false **19.** 2880 ≠ 2970; false

21. 28 = 28; true **23.** $42\dfrac{1}{2} \neq 45$; false **25.** 66 = 66; true

27. $\dfrac{17 \text{ hits}}{50 \text{ at bats}} = \dfrac{153 \text{ hits}}{450 \text{ at bats}}$ $\begin{array}{l} 50 \cdot 153 = 7650 \\ 17 \cdot 450 = 7650 \end{array}$ Cross products are equal so the proportion is *true*; they hit equally well. **29.** $x = 4$ **31.** $x = 2$ **33.** $x = 88$ **35.** $x = 91$ **37.** $x = 5$ **39.** $x = 10$ **41.** $x \approx 24.44$ (rounded) **43.** $x = 50.4$ **45.** $x \approx 17.64$ (rounded) **47.** $x = 1$

49. $x = 3\dfrac{1}{2}$ **51.** $x = 0.2$ or $x = \dfrac{1}{5}$ **53.** $x = 0.005$ or $x = \dfrac{1}{200}$

55. Find the cross products: 20 ≠ 30, so the proportion is false.

$\dfrac{6\frac{2}{3}}{4} = \dfrac{5}{3}$ or $\dfrac{10}{6} = \dfrac{5}{3}$ or $\dfrac{10}{4} = \dfrac{7.5}{3}$ or $\dfrac{10}{4} = \dfrac{5}{2}$

56. Find the cross products: 192 ≠ 180, so the proportion is false.

$\dfrac{6.4}{8} = \dfrac{24}{30}$ or $\dfrac{6}{7.5} = \dfrac{24}{30}$ or $\dfrac{6}{8} = \dfrac{22.5}{30}$ or $\dfrac{6}{8} = \dfrac{24}{32}$

Summary Exercises on Ratios, Rates, and Proportions (pages 455–456)

1. $\dfrac{6}{5}$ **2.** $\dfrac{2}{7}$ **3.** $\dfrac{2}{1}$ **4.** $\dfrac{13}{11}$ **5.** Comparing the violin to piano, guitar, organ, clarinet and drums gives ratios of $\dfrac{1}{11}, \dfrac{1}{10}, \dfrac{1}{3}, \dfrac{1}{2},$ and $\dfrac{2}{3}$, respectively.

6. (a) guitar to clarinet **(b)** organ to drums, or clarinet to violin **7.** 2.1 points/min; 0.5 min/point **8.** 1.6 points/min; 0.6 min/point **9.** $16.32/hour; $24.48/hour for overtime **10.** $0.50/channel; $0.39/channel; $0.32/channel **11.** 12 ounces for $7.24, about $0.60 per ounce **12.** Brand P with the $2 coupon is the best buy at $0.57 per pound. **13.** $\dfrac{4}{3} = \dfrac{4}{3}$ or $924 = 924$; true **14.** $2.0125 \neq 2.07$; false

15. $68\dfrac{1}{4} = 68\dfrac{1}{4}$; true **16.** $x = 28$ **17.** $x = 3.2$ **18.** $x = 182$ **19.** $x \approx 3.64$ (rounded) **20.** $x \approx 0.93$ (rounded) **21.** $x = 1.56$ **22.** $x \approx 0.05$ (rounded) **23.** $x = 1$ **24.** $x = \dfrac{3}{4}$

Section 6.4 (pages 461–464)

1. 22.5 hours **3.** $7.20 **5.** 42 pounds **7.** $403.68 **9.** 10 ounces (rounded) **11.** 5 quarts **13.** 14 ft, 10 ft **15.** 14 ft, 8 ft **17.** 96 pieces of chicken; 33.6 pounds lasagna; 10.8 pounds deli meats; $5\dfrac{3}{5}$ pounds cheese; 7.2 dozen (about 86) buns; 14.4 pounds potato salad. **19.** 2065 students (reasonable); about 4214 students with incorrect setup (only 2950 students in the group) **21.** about 79 people (reasonable); about 714 people with incorrect setup (only 238 people attended) **23.** 110,838,000 households (reasonable); about 115,408,163 households with incorrect setup (only 113,100,000 U.S. households) **25.** 625 stocks **27.** 4.06 meters (rounded) **29.** 311 calories (rounded) **31.** 10.53 meters (rounded) **33.** You cannot solve this problem using a proportion because the ratio of age to weight is not constant. As Jim's age increases, his weight may decrease, stay the same, or increase. **35.** 5050 students use cream.

37. 120 calories and 12 grams of fiber **39.** $1\dfrac{3}{4}$ cups water, 3 tablespoons margarine, $\dfrac{3}{4}$ cup milk, 2 cups flakes **40.** $5\dfrac{1}{4}$ cups water, 9 tablespoons margarine, $2\dfrac{1}{4}$ cups milk, 6 cups flakes **41.** $\dfrac{7}{8}$ cup water, $1\dfrac{1}{2}$ tablespoons margarine, $\dfrac{3}{8}$ cup milk, 1 cup flakes **42.** $2\dfrac{5}{8}$ cups water, $4\dfrac{1}{2}$ tablespoons margarine, $1\dfrac{1}{8}$ cups milk, 3 cups flakes

Section 6.5 (pages 475–478)

1. line named $\overleftrightarrow{CD}$ or $\overleftrightarrow{DC}$ **3.** line segment named $\overline{GF}$ or $\overline{FG}$ **5.** ray named $\overrightarrow{PQ}$ **7.** perpendicular **9.** parallel **11.** intersecting **13.** $\angle AOS$ or $\angle SOA$ **15.** $\angle CRT$ or $\angle TRC$ **17.** $\angle AQC$ or $\angle CQA$ **19.** right (90°) **21.** acute **23.** straight (180°) **25.** $\angle EOD$ and $\angle COD$; $\angle AOB$ and $\angle BOC$ **27.** $\angle HNE$ and $\angle ENF$; $\angle ACB$ and $\angle KOL$ **29.** 50° **31.** 4° **33.** 50° **35.** 90° **37.** $\angle SON \cong \angle TOM$; $\angle TOS \cong \angle MON$ **39.** $\angle GOH$ measures 63°; $\angle EOF$ measures 37°; $\angle AOC$ and $\angle GOF$ both measure 80°. **41.** True, because $\overrightarrow{UQ}$ is perpendicular to $\overrightarrow{ST}$. **42.** True, because they form a 90° angle, as indicated by the small red square. **43.** False; the angles have the same measure (both are 180°). **44.** False; $\overleftrightarrow{ST}$ and $\overleftrightarrow{PR}$ are parallel. **45.** False; $\overleftrightarrow{QU}$ and $\overleftrightarrow{TS}$ are perpendicular. **46.** True, because both angles are formed by perpendicular lines, so they both measure 90°. **47.** corresponding angles; $\angle 1$ and $\angle 8$, $\angle 2$ and $\angle 5$, $\angle 3$ and $\angle 6$, $\angle 4$ and $\angle 7$; alternate interior angles; $\angle 4$ and $\angle 5$, $\angle 3$ and $\angle 8$. **49.** $\angle 2$, $\angle 4$, $\angle 6$, $\angle 8$ all measure 130°; $\angle 1$, $\angle 3$, $\angle 5$, $\angle 7$ all measure 50°. **51.** $\angle 6$, $\angle 1$, $\angle 3$, $\angle 8$ all measure 47°; $\angle 5$, $\angle 2$, $\angle 7$, $\angle 4$ all measure 133°. **53.** $\angle 6$, $\angle 8$, $\angle 4$, $\angle 2$ all measure 114°; $\angle 7$, $\angle 5$, $\angle 3$, $\angle 1$ all measure 66°. **55.** $\angle 1 \cong \angle 3$, both are 138°; $\angle 2 \cong \angle ABC$, both are 42°.

Section 6.6 (pages 485–488)

1. $\angle 1$ and $\angle 4$; $\angle 2$ and $\angle 5$; $\angle 3$ and $\angle 6$; $\overline{AB}$ and $\overline{DE}$; $\overline{BC}$ and $\overline{EF}$; $\overline{AC}$ and $\overline{DF}$. **3.** $\angle 1$ and $\angle 6$; $\angle 2$ and $\angle 4$; $\angle 3$ and $\angle 5$; $\overline{ST}$ and $\overline{YW}$; $\overline{TU}$ and $\overline{WX}$; $\overline{SU}$ and $\overline{YX}$ **5.** $\angle 1$ and $\angle 6$; $\angle 2$ and $\angle 5$; $\angle 3$ and $\angle 4$; $\overline{LM}$ and $\overline{TS}$; $\overline{LN}$ and $\overline{TR}$; $\overline{MN}$ and $\overline{SR}$ **7.** SAS **9.** SSS **11.** ASA **13.** use SAS: $BC = CE$, $\angle ABC \cong \angle DCE$, $BA = CD$ **14.** use SSS: $WP = YP$, $ZP = XP$, $WZ = YX$ **15.** use SAS: $PS = RS$, $m\angle QSP = m\angle QSR = 90°$, $QS = QS$ (common side) **16.** use SAS: $LM = OM$, $PM = NM$, $\angle LMP \cong \angle OMN$ (vertical angles) **17.** $a = 6$ cm; $b = 15$ cm **19.** $a = 5$ mm; $b = 3$ mm **21.** $a = 24$ in.; $b = 20$ in. **23.** $x = 24.8$ m; Perimeter = 72.8 m; $y = 15$ m; Perimeter = 54.6 m **25.** Perimeter = 8 cm + 8 cm + 8 cm = 24 cm; Area $\approx$ $(0.5)(8 \text{ cm})(6.9 \text{ cm}) \approx 27.6$ cm^2; The area is approximate because the height was rounded to the nearest tenth. **27.** $h = 24$ ft **29.** One dictionary definition is "resembling, but not identical." Examples of similar objects are sets of different size pots or measuring cups; small and large size cans of beans; child's tennis shoe and adult's tennis shoe. **31.** $x = 50$ m **33.** $n = 110$ m

Chapter 6 Review Exercises (pages 497–502)

1. $\dfrac{3}{4}$ **2.** $\dfrac{4}{1}$ **3.** great white shark to whale shark; whale shark to blue whale **4.** $\dfrac{2}{1}$ **5.** $\dfrac{2}{3}$ **6.** $\dfrac{5}{2}$ **7.** $\dfrac{1}{6}$ **8.** $\dfrac{3}{1}$ **9.** $\dfrac{3}{8}$ **10.** $\dfrac{4}{3}$ **11.** $\dfrac{1}{9}$ **12.** $\dfrac{10}{7}$ **13.** $\dfrac{7}{5}$ **14.** $\dfrac{5}{6}$ **15.** $\dfrac{\$11}{1 \text{ dozen}}$ **16.** $\dfrac{12 \text{ children}}{5 \text{ families}}$ **17.** 0.2 page/minute or $\dfrac{1}{5}$ page/minute; 5 minutes/page **18.** \$8/hour; 0.125 hour/dollar or $\dfrac{1}{8}$ hour/dollar **19.** 8 ounces for \$4.98, about \$0.623/ounce **20.** 17.6 pounds for \$18.69 − \$1 coupon, about \$1.005/pound **21.** $\dfrac{3}{5} = \dfrac{3}{5}$ or $90 = 90$; true **22.** $\dfrac{1}{8} \neq \dfrac{1}{4}$ or $432 \neq 216$; false **23.** $\dfrac{47}{10} \neq \dfrac{49}{10}$ or $980 \neq 940$; false **24.** $4.8 = 4.8$; true **25.** $14 = 14$; true **26.** $x = 1575$ **27.** $x = 20$ **28.** $x = 400$ **29.** $x = 12.5$ **30.** $x \approx 14.67$ (rounded) **31.** $x \approx 8.17$ (rounded) **32.** $x = 50.4$ **33.** $x \approx 0.57$ (rounded) **34.** $x \approx 2.47$ (rounded) **35.** 27 cats **36.** 46 hits **37.** \$15.63 (rounded) **38.** 3299 students (rounded) **39.** 68 ft **40.** $27\dfrac{1}{2}$ hours or 27.5 hours **41.** 511 calories (rounded) **42.** 14.7 milligrams **43.** line segment named $\overline{AB}$ or $\overline{BA}$ **44.** line named $\overleftrightarrow{CD}$ or $\overleftrightarrow{DC}$ **45.** ray named $\overrightarrow{OP}$ **46.** parallel **47.** perpendicular **48.** intersecting **49.** acute **50.** obtuse **51.** straight; 180° **52.** right; 90° **53.** (a) 10° (b) 45° (c) 83° **54.** (a) 25° (b) 90° (c) 147° **55.** $\angle 1$ and $\angle 4$ measure 30°; $\angle 3$ and $\angle 6$ measure 90°; $\angle 5$ measures 60° **56.** $\angle 8$, $\angle 3$, $\angle 6$, $\angle 1$ all measure 160°; $\angle 4$, $\angle 7$, $\angle 2$, $\angle 5$ all measure 20°. **57.** SSS **58.** SAS **59.** ASA **60.** $y = 30$ ft; $x = 34$ ft; $P = 104$ ft **61.** $y = 7.5$ m; $x = 9$ m; $P = 22.5$ m **62.** $x = 12$ mm; $y = 7.5$ mm; $P = 38$ mm **63.** $x = 105$ **64.** $x = 0$ **65.** $x = 128$ **66.** $x \approx 23.08$ (rounded) **67.** $x = 6.5$ **68.** $x \approx 117.36$ (rounded) **69.** $\dfrac{8}{5}$ **70.** $\dfrac{33}{80}$ **71.** $\dfrac{15}{4}$ **72.** $\dfrac{4}{1}$ **73.** $\dfrac{4}{5}$ **74.** $\dfrac{37}{7}$ **75.** $\dfrac{3}{8}$ **76.** $\dfrac{1}{12}$ **77.** $\dfrac{45}{13}$ **78.** 24,900 fans (rounded) **79.** $\dfrac{8}{3}$ **80.** 75 ft for \$1.99 − \$0.50 coupon, about \$0.020/ft **81.** 21 ft long; 15 ft wide **82.** (a) 1400 milligrams (b) 100 milligrams **83.** 21 points (rounded) **84.** $\dfrac{1}{2}$ or 0.5 teaspoon **85.** parallel lines **86.** line segment **87.** acute angle **88.** intersecting lines **89.** right angle; 90° **90.** ray **91.** straight angle; 180° **92.** obtuse angle **93.** perpendicular lines **94.** (a) The car turned around in a complete circle. (b) The governor took the opposite view, for example, having once opposed taxes but now supporting them.

95. (a) No; because obtuse angles are $>90°$, their sum would be $>180°$. (b) Yes; because acute angles are $<90°$, their sum could equal 90°. **96.** $\angle 1$ measures 45°; $\angle 3$ and $\angle 6$ measure 35°; $\angle 4$ measures 55°; $\angle 5$ measures 90° **97.** $\angle 5$, $\angle 2$, $\angle 7$, $\angle 4$ all measure 75°; $\angle 6$, $\angle 1$, $\angle 3$, $\angle 8$ all measure 105°.

Chapter 6 Test (pages 503–504)

1. $\dfrac{\$1}{5 \text{ minutes}}$ **2.** $\dfrac{9}{2}$ **3.** $\dfrac{15}{4}$ **4.** 16 ounces for \$1.89 − \$0.50 coupon, about \$0.087/ounce **5.** $x = 25$ **6.** $x \approx 2.67$ (rounded) **7.** $x = 325$ **8.** $x = 10\dfrac{1}{2}$ **9.** 24 orders **10.** 87 students (rounded) **11.** 23.8 grams (rounded) **12.** 60 feet **13.** (e) **14.** (a); 90° **15.** (d) **16.** (g); 180° **17.** Parallel lines are lines in the same plane that never intersect. Perpendicular lines intersect to form a right angle. Sketches will vary. **18.** 9° **19.** 160° **20.** $m\angle 1 = 50°$, $m\angle 2 = 35°$, $m\angle 3 = 95°$, $m\angle 5 = 35°$ **21.** Measures of $\angle 1$, $\angle 3$, $\angle 5$, and $\angle 7$ are all 65°; measures of $\angle 2$, $\angle 4$, $\angle 6$, and $\angle 8$ are all 115°. **22.** ASA (Angle–Side–Angle) **23.** SAS (Side–Angle–Side) **24.** $y = 12$ cm; $z = 6$ cm **25.** In the larger triangle, $x = 12$ mm, $P = 46.8$ mm; in the smaller triangle, $y = 14$ mm, $P = 39$ mm

Chapter 7 Percent

Section 7.1 (pages 515–520)

1. 0.25 **3.** 0.30 or 0.3 **5.** 0.06 **7.** 1.40 or 1.4 **9.** 0.078 **11.** 1.00 or 1 **13.** 0.005 **15.** 0.0035 **17.** 50% **19.** 62% **21.** 3% **23.** 12.5% **25.** 62.9% **27.** 200% **29.** 260% **31.** 3.12% **33.** $\dfrac{1}{5}$ **35.** $\dfrac{1}{2}$ **37.** $\dfrac{11}{20}$ **39.** $\dfrac{3}{8}$ **41.** $\dfrac{1}{16}$ **43.** $\dfrac{1}{6}$ **45.** $1\dfrac{3}{10}$ **47.** $2\dfrac{1}{2}$ **49.** 25% **51.** 30% **53.** 60% **55.** 37% **57.** $37\dfrac{1}{2}$% or 37.5% **59.** 5% **61.** exactly $55\dfrac{5}{9}$%, or 55.6% (rounded) **63.** exactly $14\dfrac{2}{7}$%, or 14.3% (rounded) **65.** 0.08 **67.** 0.42 **69.** 3.5% **71.** 200% **73.** $\dfrac{95}{100}$ or 95% shaded; $\dfrac{5}{100}$ or 5% unshaded **75.** $\dfrac{3}{10}$ or 30% shaded; $\dfrac{7}{10}$ or 70% unshaded **77.** $\dfrac{3}{4}$ or 75% shaded; $\dfrac{1}{4}$ or 25% unshaded **79.** 0.01; 1% **81.** $\dfrac{1}{5}$; 20% **83.** $\dfrac{3}{10}$; 0.3 **85.** 0.5; 50% **87.** $\dfrac{9}{10}$; 0.9 **89.** $1\dfrac{1}{2}$; 150% **91.** 0.08; $\dfrac{2}{25}$ **93.** 5%; 0.05; $\dfrac{1}{20}$ **95.** $\dfrac{1}{4}$; 0.25; 25% **97.** $\dfrac{3}{8}$; 0.375; 37.5% **99.** $\dfrac{1}{7}$; exactly $14\dfrac{2}{7}$%, or 14.3% (rounded) **101.** (a) The student forgot to move the decimal point in 0.35 two places to the right. So $\dfrac{7}{20} = 35\%$. (b) The student did the division in the wrong order. Enter $16 \div 25$ to get 0.64 and then move the decimal point two places to the right. So $\dfrac{16}{25} = 0.64 = 64\%$. **103.** (a) \$78 (b) \$39 **105.** (a) 15 inches (b) $7\dfrac{1}{2}$ inches **107.** (a) 2.8 miles (b) 1.4 miles **109.** 20 children **111.** 60 credits **113.** (a) \$142.50 (b) 50% (c) \$142.50 **115.** (a) 4100 students (b) 50% **117.** (a) 35 problems (b) 0 problems **119.** 50% means 50 out of 100 parts. That's half of the number. A shortcut for finding 50% of a number is to divide the number by 2. Examples will vary.

Section 7.2 (pages 527–528)

1. (a) 10% **(b)** 3000 runners **(c)** unknown **(d)** $\dfrac{10}{100} = \dfrac{n}{3000}$;

300 runners **3. (a)** 4% **(b)** 120 ft **(c)** unknown **(d)** $\dfrac{4}{100} = \dfrac{n}{120}$;

4.8 feet **5. (a)** unknown **(b)** 32 pizzas **(c)** 16 pizzas

(d) $\dfrac{p}{100} = \dfrac{16}{32}$; 50% **7. (a)** unknown **(b)** 200 calories

(c) 16 calories **(d)** $\dfrac{p}{100} = \dfrac{16}{200}$; 8% **9. (a)** 90% **(b)** unknown

(c) 495 students **(d)** $\dfrac{90}{100} = \dfrac{495}{n}$; 550 students **11. (a)** $12\dfrac{1}{2}$%

(b) unknown **(c)** \$3.50 **(d)** $\dfrac{12.5}{100} = \dfrac{3.50}{n}$; \$28 **13.** $\dfrac{250}{100} = \dfrac{n}{7}$;

17.5 hours **15.** $\dfrac{p}{100} = \dfrac{32}{172}$; 18.6% (rounded)

17. $\dfrac{110}{100} = \dfrac{748}{n}$; 680 books **19.** $\dfrac{14.7}{100} = \dfrac{n}{274}$; \$40.28 (rounded)

21. $\dfrac{p}{100} = \dfrac{105}{54}$; 194.4% (rounded) **23.** $\dfrac{4}{100} = \dfrac{0.33}{n}$; \$8.25

25. 150% of \$30 cannot be *less* than \$30; 25% of \$16 cannot be *greater*

than \$16. **27.** The correct proportion is $\dfrac{p}{100} = \dfrac{14}{8}$. The answer should

be labeled with the % symbol. Correct answer is 175%.

Section 7.3 (pages 535–538)

1. 1500 patients **3.** \$15 **5.** 4.5 pounds **7.** \$7.00 **9.** 52 students
11. 870 phones **13.** 4.75 hours **15. (a)** 10% means $\dfrac{10}{100}$ or $\dfrac{1}{10}$.

The denominator tells you to divide the whole by 10. The shortcut for
dividing by 10 is to move the decimal point one place to the left.
(b) Once you find 10% of a number, multiply the result by 2 for 20%
and by 3 for 30%. **17.** 231 programs **19.** 50% **21.** 680 circuits
23. 1080 people **25.** 845 species **27.** \$20.80 **29.** 76%
31. 125% **33.** 4700 employees **35.** 83.2 quarts **37.** 2%
39. 700 tablets **41.** 325 salads **43.** 1.5% **45.** 5 gallons
47. 1029.2 meters **49. (a)** Multiply 0.2 by 100% to change it from a
decimal to a percent. So, 0.20 = 20%. **(b)** The correct equation is

$50 = p \cdot 20$, so the solution is 250%. **51. (a)** $\dfrac{1}{3} \cdot 162 = n$; the solution

is \$54. **(b)** $(0.333333333)(162) = n$; depending upon how your
calculator rounds numbers, the solution is either \$54 or \$53.99999995.
(c) There is no difference or the difference is insignificant.

52. (a) $22 = \dfrac{2}{3} \cdot n$; the solution is 33 cans. **(b)** $22 = (0.666666667)(n)$;

depending upon how your calculator rounds numbers, the solution is
either 33 cans or 32.99999998 cans. **(c)** There is no difference or the
difference is insignificant.

Summary Exercises on Percent (pages 539–540)

1. (a) 0.03; 3% **(b)** $\dfrac{3}{10}$; 0.3 **(c)** $\dfrac{3}{8}$; 37.5% **(d)** $1\dfrac{3}{5}$; 1.6

(e) 0.0625; 6.25% **(f)** $\dfrac{1}{20}$; 0.05 **(g)** 2; 200% **(h)** 0.8; 80%

(i) $\dfrac{9}{125}$; 7.2% **2. (a)** 3.5 ft **(b)** 19 miles **(c)** 105 cows

(d) \$0.08 **(e)** 500 women **(f)** \$45 **(g)** \$87.50 **(h)** 12 pounds

(i) 95 students **3.** $12\dfrac{1}{2}$% or 12.5% **4.** 75 DVDs **5.** \$0.53 (rounded)

6. 175% **7.** 5.9 pounds (rounded) **8.** 600 hours **9.** 50%
10. 500 camp sites **11.** 98 golf balls **12.** 2.7% (rounded)
13. 2300 apartments **14.** 0.63 ounce **15.** 145% **16.** 244 voters
(rounded) **17.** 0.021 inch **18.** 400% **19.** 8% **20.** \$0.68
21. Invitations; \$840 **22.** \$2604 **23. (a)** \$3830.40
(b) \$4166.40 **24. (a)** \$14,784 **(b)** \$89.60

Section 7.4 (pages 547–548)

1. \$37.80 **3. (a)** 10% **(b)** 5% **(c)** 2% **(d)** 1% **5. (a)** 101.6
pounds (rounded) **(b)** 10.1 pounds (rounded) **7.** 13.1% female;
86.9% male (both rounded) **9.** 300 million people (rounded)
11. 138% **13.** 40 problems **15.** 1340 shots (rounded) **17.** 23.7 miles
per gallon (rounded) **19.** March; 24.5 million cans, or 24,500,000 cans
21. 52.5 million cans (January); 38.5 million cans (February) **23.** 64.8%
(rounded) **25.** 9.1% (rounded) **27.** 40% **29.** 678% (rounded)
31. No. 100% is the entire price, so a decrease of 100% would take
the price down to 0. Therefore, 100% is the maximum possible
decrease in the price of something. **33.** George ate more than
65 grams, so the percent must be >100%. Use $p \cdot 65 = 78$ to get
120%. **34.** The team won more than half the games, so the percent
must be >50%. Correct solution is 0.72 = 72%. **35.** The brain could
not weigh 375 pounds, which is more than the person weighs.

$2\dfrac{1}{2}$%, = 2.5% = 0.025, so $(0.025)(150) = n$ and $n = 3.75$ pounds

36. If 80% were absent, then only 20% made it to class. $800 - 640 = 160$
students, or use $(0.20)(800) = n$.

Section 7.5 (pages 557–561)

1. \$6; \$106 **3.** 3%; \$70.04 **5.** \$29.28 (rounded); \$395.26 **7.** \$0.12

(rounded); \$2.22 **9.** $4\dfrac{1}{2}$%; \$13,167 **11.** \$3 + \$1.50 = \$4.50; \$4.83

(rounded); 2(\$3) = \$6; \$6.43 (rounded) **13.** \$8 + \$4 = \$12; \$11.75
(rounded); 2(\$8) = \$16; \$15.67 (rounded) **15.** \$1 + \$0.50 = \$1.50;
\$1.43 (rounded); 2(\$1) = \$2; \$1.91 **17.** \$15; \$85 **19.** 30%; \$126
21. \$4.38 (rounded); \$13.12 **23.** \$3.79 (rounded); \$34.09 **25.** \$42;
\$342 **27.** \$33.30; \$773.30 **29.** \$213.75; \$1713.75 **31.** \$919.67
(rounded); \$18,719.67 **33.** \$1170 **35.** \$7978.13 (rounded)
37. \$26.61 (rounded) **39.** 5% **41.** \$74.25 **43.** \$25.13 **45.** \$216.00
(rounded); \$983.99 **47.** \$2016.38 (rounded) **49.** \$21 (rounded)
51. \$92.38 **53.** \$230.12 **55. (a)** \$18.43 (rounded to nearest cent)
(b) When calculating the discount, the *whole* is \$18.50. But when calcu-
lating the sales tax, the *whole* is only \$17.39 (the discounted price).
56. (a) \$396.05 (rounded to nearest cent). **(b)** 7.53% sales tax
(rounded to the nearest hundredth) would give a final cost of \$398.01.

Chapter 7 Review Exercises (pages 569–572)

1. 0.25 **2.** 1.8 **3.** 0.125 **4.** 0.07 **5.** 265% **6.** 2% **7.** 30%

8. 0.2% **9.** $\dfrac{3}{25}$ **10.** $\dfrac{3}{8}$ **11.** $2\dfrac{1}{2}$ **12.** $\dfrac{1}{20}$ **13.** 75% **14.** 62.5% or

$62\dfrac{1}{2}$% **15.** 325% **16.** 6% **17.** 0.125 **18.** 12.5% **19.** $\dfrac{3}{20}$

20. 15% **21.** $1\dfrac{4}{5}$ **22.** 1.8 **23.** \$46 **24.** \$23 **25.** 9 hours

26. $4\dfrac{1}{2}$ hours **27.** 242 meters **28.** 17,000 cases **29.** 27 cellular

phones **30.** 870 reference books **31.** 9.5% (rounded) **32.** 225%
33. \$2.60 (rounded) **34.** 80 days **35.** 4% **36.** \$575 **37.** 20 people
38. 175% **39. (a)** 3000 patients **(b)** 31.5% decrease (rounded)
40. (a) \$364; **(b)** 224% **41. (a)** 80% **(b)** 6.3% increase (rounded)

42. 10.4% (rounded) **43.** \$0.11 (rounded); \$2.90 **44.** $7\dfrac{1}{2}$%;

\$838.50 **45.** \$4 + \$2 = \$6; \$6.41 (rounded); 2(\$4) = \$8; \$8.55
(rounded) **46.** \$0.80 + \$0.40 = \$1.20; \$1.21 (rounded);
2(\$0.80) = \$1.60; \$1.61 **47.** \$3.75; \$33.75 **48.** 25%; \$189

49. \$68.25; \$418.25 **50.** \$183.60; \$1713.60 **51.** $\dfrac{1}{3}$; $33\dfrac{1}{3}$% (exact)

or 33.3% (rounded) **52.** cards; 68%; 0.68; $\dfrac{17}{25}$ **53.** 277 adults

(rounded) **54.** cards, 188 adults (rounded); sci-fi/simulation,
102 adults (rounded) **55.** 48.4% (rounded) **56.** 7161 dogs (rounded)
57. 36.4% (rounded) **58.** 180.3% increase (rounded) **59.** 38.1%
(rounded) **60.** 265 animals (rounded)

Chapter 7 Test (pages 573–574)

1. 0.75 **2.** 60% **3.** 180% **4.** 7.5% or $7\frac{1}{2}$% **5.** 3.00 or 3

6. 0.02 **7.** $\frac{5}{8}$ **8.** $2\frac{2}{5}$ **9.** 5% **10.** 87.5% or $87\frac{1}{2}$% **11.** 175%

12. 320 laptops **13.** 400% **14.** $19,500 **15.** $8466.75
16. 34% increase (rounded) **17.** To find 50% of a number, divide the number by 2. To find 25% of a number, divide the number by 4. Examples will vary. **18.** Round $31.94 to $30. Then 10% of $30 is $3 and 5% of $30 is half of $3 or $1.50, so a 15% tip estimate is $3 + $1.50 = $4.50. A 20% tip estimate is 2($3) = $6. **19.** Exact tip is $4.79 (rounded). Each person pays $12.24 (rounded). **20.** $3.84; $44.16 **21.** $41.39 (rounded); $188.56 **22.** $815.66 ÷ 6 ≈ $135.94 **23.** $1650; $6650 **24.** $51.60; $911.60

Chapter 8 Measurement

Section 8.1 (pages 585–588)

1. 3; 12 **3.** 8; 2 **5.** 5280; 3 **7.** 2000; 16 **9.** 60; 60 **11. (a)** 2;
(b) 240 **13. (a)** $\frac{1}{2}$ or 0.5 **(b)** 78 **15.** 14,000 to 16,000 lb
17. 27 **19.** 112 **21.** 10 **23.** $1\frac{1}{2}$ or 1.5 **25.** $\frac{1}{4}$ or 0.25
27. $1\frac{1}{2}$ or 1.5 **29.** $2\frac{1}{2}$ or 2.5 **31.** snowmobile/ATV; person walking **33.** 5000 **35.** 17 **37.** 4 to 8 in. **39.** 216 **41.** 28
43. 518,400 **45.** 48,000 **47. (a)** pound/ounces **(b)** quarts/pints or pints/cups **(c)** minutes/hours or seconds/minutes **(d)** feet/inches
(e) pounds/tons **(f)** days/weeks **49.** 174,240 **51.** 800
53. 0.75 or $\frac{3}{4}$ **55.** $1.83 (rounded) **57.** $140 **59. (a)** 1056 sec;
(b) 17.6 min **61. (a)** $12\frac{1}{2}$ qt **(b)** 4 containers, because you can't buy part of a container **63. (a)** 0.8 mi (rounded) **(b)** 13,031 ft **(c)** 2.5 mi (rounded) **64. (a)** 183 yd (rounded) **(b)** 6600 in. **(c)** 0.1 mi (rounded)
(d) 9 ft (rounded) **65. (a)** $1\frac{1}{2}$ to $2\frac{1}{2}$ ft, or 1.5 to 2.5 ft **(b)** 8400 months
(c) 140 more years, for a total of 840 years in all **66. (a)** 283 mi (rounded)
(b) 377 mi (rounded)

Section 8.2 (pages 595–596)

1. 1000; 1000 **3.** $\frac{1}{1000}$ or 0.001; $\frac{1}{1000}$ or 0.001 **5.** $\frac{1}{100}$ or 0.01;
$\frac{1}{100}$ or 0.01 **7.** Answers will vary; about 8 to 10 cm. **9.** Answers will vary; about 20 to 25 mm. **11.** cm **13.** m **15.** km **17.** mm
19. cm **21.** m **23.** Some possible answers are: track and field events, metric auto parts, and lead refills for mechanical pencils. **25.** 700 cm
27. 0.040 m or 0.04 m **29.** 9400 m **31.** 5.09 m **33.** 40 cm
35. 910 mm **37.** less; 18 cm or 0.18 m
39. 5 mm = 0.5 cm
  ← 1 mm = 0.1 cm
41. 0.018 km **43.** 164 cm; 1640 mm **45.** 0.0000056 km

Section 8.3 (pages 603–606)

1. mL **3.** L **5.** kg **7.** g **9.** mL **11.** mg **13.** L **15.** kg
17. unreasonable; too much **19.** unreasonable; too much **21.** reasonable
23. reasonable **25.** Some capacity examples are 2 L bottles of soda and shampoo bottles marked in mL; weight examples are grams of fat listed on food packages and vitamin doses in milligrams. **27.** Unit for your answer (g) is in numerator; unit being changed (kg) is in denominator so it will divide out. The unit fraction is $\frac{1000\ \text{g}}{1\ \text{kg}}$. **29.** 15,000 mL **31.** 3 L
33. 0.925 L **35.** 0.008 L **37.** 4150 mL **39.** 8 kg **41.** 5200 g
43. 850 mg **45.** 30 g **47.** 0.598 g **49.** 0.06 L **51.** 0.003 kg

53. 990 mL **55.** mm **57.** mL **59.** cm **61.** mg **63.** 0.3 L
65. 1340 g **67.** 0.9 L **69.** 3 kg to 4 kg **71.** greater; 5 mg or 0.005 g
73. 200 nickels **75. (a)** 1,000,000
(b) $\dfrac{3.5\ \text{Mm}}{1} \cdot \dfrac{1{,}000{,}000\ \text{m}}{1\ \text{Mm}} = 3{,}500{,}000\ \text{m}$
76. (a) 1,000,000,000
(b) $\dfrac{2500\ \text{m}}{1} \cdot \dfrac{1\ \text{Gm}}{1{,}000{,}000{,}000\ \text{m}} = 0.0000025\ \text{Gm}$
77. (a) 1,000,000,000,000 **(b)** 1000; 1,000,000
78. 1,000,000; 1,000,000,000; $2^{20} = 1{,}048{,}576$; $2^{30} = 1{,}073{,}741{,}824$

Summary Exercises on U.S. Customary and Metric Units (pages 607–608)

1. Length: inch, foot, yard, mile; Weight: ounce, pound, ton; Capacity: fluid ounce, cup, pint, quart, gallon. **2.** Length: millimeter, centimeter, meter, kilometer; Weight: milligram, gram, kilogram; Capacity: milliliter, liter **3. (a)** ft **(b)** yd **(c)** 5280 **4. (a)** min **(b)** 60 **(c)** 24
5. (a) 8 **(b)** gal **(c)** 2 **6. (a)** lb **(b)** 2000 **(c)** 16 **7.** mL **8.** m
9. kg **10.** mg **11.** cm **12.** mm **13.** L **14.** g **15.** 0.45 m
16. 45 sec **17.** 600 mL **18.** 8000 mg **19.** 30 cm **20.** $3\frac{3}{4}$ or 3.75 ft
21. 0.050 or 0.05 L **22.** $4\frac{1}{2}$ or 4.5 gal **23.** 7280 g **24.** 36 oz
25. 0.009 kg **26.** 180 in. **27.** 272 cm; 2720 mm **28.** 2.64 m; 0.00264 km **29.** 0.04 m; 4 cm; 40 mm **30.** 14 cm; 0.14 m; 140 mm
31. $4.65 (rounded) **32.** 10.3 ft (rounded)

Section 8.4 (pages 611–612)

1. $0.83 (rounded) **3.** 89.5 kg **5.** 71 beats (rounded) **7.** 180 cm; $0.02/cm (rounded) **9.** 5.03 m **11.** 4 bottles; 175 mL or 0.175 L
13. 1.89 g **15. (a)** 10 km **(b)** 600 km **(c)** 36,000 km
17. 215 g; 4.3 g; 4300 mg **18.** 330 g; 0.33 g; 330 mg
19. 1550 g; 1500 g; 3 g **20.** 55 g; 50 g; 0.5 g

Section 8.5 (pages 617–620)

1. 21.8 yd **3.** 262.4 ft **5.** 4.8 m **7.** 5.3 oz **9.** 111.6 kg
11. 30.3 qt **13. (a)** about 0.2 oz **(b)** probably not **15.** about 31.8 L
17. about 1.3 cm **19.** 3.5 kg ≈ 7.7 lb so the baby is heavy enough. But 53 cm ≈ 20.7 in., so the baby is not long enough to be in the carrier.
21. −8 °C **23.** 40 °C **25.** 150 °C **27.** 16 °C (rounded) **29.** −20 °C
31. 46 °F (rounded) **33.** 23 °F **35.** 58 °C (rounded); −89 °C (rounded)
37. 10 °C and 41 °C (rounded) **39. (a)** pleasant weather, above freezing but not hot **(b)** 75 °F to 39 °F (rounded) **(c)** Answers will vary. In Minnesota, it's 0 °C to −40 °C; in California, 24 °C to 0 °C.
41. about 1.8 m **42.** about 5.0 kg **43.** about 3040.2 km
44. 38 hr 23 min **45.** about 300 m **46.** less than 3790 mL
47. about 59.4 mL **48.** about 1.6%

Chapter 8 Review Exercises (pages 627–630)

1. 16 **2.** 3 **3.** 2000 **4.** 4 **5.** 60 **6.** 8 **7.** 60 **8.** 5280
9. 12 **10.** 48 **11.** 3 **12.** 4 **13.** $\frac{3}{4}$ or 0.75 **14.** $2\frac{1}{2}$ or 2.5
15. 28 **16.** 78 **17.** 112 **18.** 345,600 **19. (a)** $4153\frac{1}{3}$ yd
(exact) or 4153.3 yd (rounded) **(b)** 2.4 mi (rounded)
20. $2465.20 **21.** mm **22.** cm **23.** km **24.** m **25.** cm
26. mm **27.** 500 cm **28.** 8500 m **29.** 8.5 cm **30.** 3.7 m
31. 0.07 km **32.** 930 mm **33.** mL **34.** L **35.** g **36.** kg
37. L **38.** mL **39.** mg **40.** g **41.** 5 L **42.** 8000 mL
43. 4580 mg **44.** 700 g **45.** 0.006 g **46.** 0.035 L **47.** 31.5 L
48. 357 g (rounded) **49.** 87.25 kg **50.** $1.42 (rounded)
51. 6.5 yd (rounded) **52.** 11.7 in. (rounded) **53.** 67.0 mi (rounded) **54.** 1288 km **55.** 21.9 L (rounded)
56. 44.0 qt (rounded) **57.** 0 °C **58.** 100 °C **59.** 37 °C

60. 20 °C **61.** 25 °C **62.** −15 °C **63.** 28 °F (rounded)
64. 120 °F (rounded) **65.** L **66.** kg **67.** cm **68.** mL
69. mm **70.** km **71.** g **72.** m **73.** mL **74.** g **75.** mg
76. L **77.** 105 mm **78.** $\frac{3}{4}$ hr or 0.75 hr **79.** $7\frac{1}{2}$ ft or 7.5 ft
80. 130 cm **81.** 77 °F **82.** 14 qt **83.** 0.7 g **84.** 810 mL
85. 80 oz **86.** 60,000 g **87.** 1800 mL **88.** 30 °C **89.** 36 cm
90. 0.055 L **91.** 1.26 m **92.** 18 tons **93.** 113 g; 177 °C
(both rounded) **94.** 178.0 lb; 6.0 ft (both rounded) **95.** 13.2 m
96. 22.0 ft **97.** 3.6 m **98.** 39.8 in. **99.** 484,000 lb
100. 19.0 to 22.7 L **101.** **(a)** 242 tons **(b)** 144 in.
102. **(a)** 1.02 m **(b)** 670 cm

Chapter 8 Test (pages 631–632)

1. 36 qt **2.** 15 yd **3.** 2.25 hr or $2\frac{1}{4}$ hr **4.** 0.75 ft or $\frac{3}{4}$ ft **5.** 56 oz
6. 7200 min **7.** kg **8.** km **9.** mL **10.** g **11.** cm **12.** mm
13. L **14.** cm **15.** 2.5 m **16.** 4600 m **17.** 0.5 cm **18.** 0.325 g
19. 16,000 mL **20.** 400 g **21.** 1055 cm **22.** 0.095 L **23.** 3.2 ft
(rounded) **24.** **(a)** 2310 mg **(b)** 500 mg or 0.5 g more **25.** 95 °C
26. 0 °C **27.** 1.8 m **28.** 56.3 kg (rounded) **29.** 13 gal
30. 5.0 mi (rounded) **31.** 23 °C (rounded) **32.** 10 °F (rounded)
33. 6 m ≈ 6.54 yd; $26.03 (rounded) **34.** Possible answers: Use same
system as rest of the world; easier system for children to learn; less use of
fractional numbers; compete internationally.

Chapter 9 Graphs

Section 9.1 (pages 637–640)

1. **(a)** 31,419 points **(b)** Michael Jordan **3.** **(a)** Michael Jordan
(b) Allen Iverson **5.** 11,468 points **7.** West, 27.0; Pettit, 26.4;
O'Neal, 25.9; round to nearest tenth. **9.** The asterisks next to O'Neal's
and Iverson's names mean that they are still actively playing in the NBA.
11. **(a)** 255 calories **(b)** moderate jogging **13.** **(a)** moderate jogging,
aerobic dance, racquetball **(b)** moderate jogging, moderate bicycling,
aerobic dance, racquetball, tennis **15.** 370 calories **17.** **(a)** 366 calories
(rounded) **(b)** 239 calories **19.** about 39 calories **21.** **(a)** 80 million
or 80,000,000 **(b)** 35 million or 35,000,000 **23.** 165 million or
165,000,000 **25.** 5 million or 5,000,000 **27.** 305 million or 305,000,000
29. Answers will vary. One possibility: choose Southwest because it has
the best on-time performance. **30.** Answers will vary. Possibilities include
planning more time between each flight, or doing some or all of your
business via conference calls or e-mail. **31.** Answers will vary. One
possibility: choose Airtran because it has the fewest luggage problems.
32. Answers will vary. Possibilities include buying heavy-duty luggage
or shipping the golf clubs via a delivery service. **33.** Answers will vary.
Possibilities include a lot of bad weather, maintenance problems, new
computer system. **34.** Answers will vary. Possibilities include availability of nonstop flights, convenience of departure times, type and size of
aircraft, availability of low-cost fares.

Section 9.2 (pages 645–650)

1. **(a)** $48,000 **(b)** carpentry, $18,000 **3.** **(a)** $\frac{\$18,000}{\$48,000} = \frac{3}{8}$
(b) $\frac{\$15,000}{\$3000} = \frac{5}{1}$ **5.** **(a)** $\frac{\$4800}{\$48,000} = \frac{1}{10}$ **(b)** $\frac{\$57,600}{\$48,000} = \frac{6}{5}$
7. **(a)** don't know **(b)** quicker **9.** $\frac{720}{6000} = \frac{3}{25}$ **11.** $\frac{1020}{1200} = \frac{17}{20}$
13. $\frac{1740}{180} = \frac{29}{3}$ **15.** 140 people **17.** raids the buffet table; 16 people
19. 64 fewer people **21.** 160 people **23.** mustard; 960 people
25. 64 more people **27.** First, find the percent of the total that is
represented by each item. Next, multiply the percent by 360° to find the
size of each sector. Finally, use a protractor to draw each sector.

29. **(a)** 90° **(b)** 20% **(c)** 10%; 36° **(d)** 10%; 36° **(e)** 15%; 54°
(f) 5%; 18° **(g)** 15%; 54° **(h)** See circle graph below.

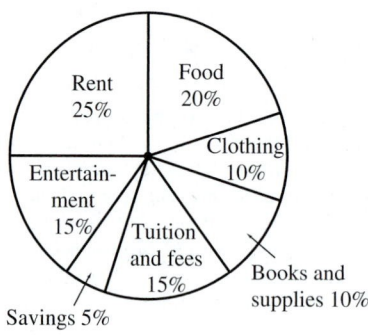

31. **(a)** $200,000 **(b)** 6.25%; 20%; 30%; 25%; 18.75%
(c) 22.5°; 72°; 108°; 90°; 67.5°
(d) See circle graph below.

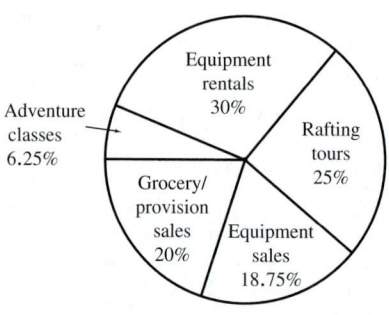

Section 9.3 (pages 655–658)

1. can shop during off hours; 74% **3.** 57%; 342 people **5.** $\frac{1}{2}$, compare
products more easily; nearly $\frac{3}{4}$, shop during off hours **7.** May; 10,000
unemployed **9.** 1500 workers **11.** 2500 workers; 45% increase (rounded)
13. 150,000 gallons **15.** 2004; 250,000 gallons **17.** 550,000 gallons;
367% increase (rounded) **19.** 24.1 million or 24,100,000 PCs
21. 185.6 million or 185,600,000 more PCs **23.** 82.3 million or
82,300,000 PCs; 132% increase (rounded) **25.** **(a)** 3,000,000 DVDs
(b) 1,500,000 DVDs **27.** **(a)** 2,500,000 DVDs **(b)** 3,000,000 DVDs
29. Answers will vary. Possibilities include: Both stores had decreased
sales from 2004 to 2005 and increased sales from 2006 to 2008; Store B
had lower sales than Store A in 2004–05 but higher sales than Store A in
2006–2008. **31.** 2006; $25,000 **33.** 2005, 29% (rounded); 2006,
20%; 2007, 17% (rounded); 2008, 38% (rounded) **35.** Answers will
vary. Possibilities include: The decrease in sales may have resulted from
poor service or greater competition; the increase in sales may have been a
result of more advertising or better service. **37.** Shipments have increased at a rapid rate since 1985. **38.** Answers will vary. Some possibilities are lower prices; more uses and applications for students, home
use, and businesses; improved technology. **39.** **(a)** 104% (rounded)
(b) 159% (rounded) **(c)** 132% (rounded) **(d)** 37% (rounded)
(e) 46% (rounded) **40.** Since 2000, the percent of increase for each
5-year period has been much lower than in earlier periods. **41.** Answers
will vary. Some possibilities are: More people will already own a computer
and not want to buy another; some new invention will replace computers.
42. **(a)** no change; 0% **(b)** 167% increase (rounded) **43.** **(a)** Answers
will vary; perhaps 3,500,000 DVDs in 2009. **(b)** Answers will vary;
perhaps 4,500,000 DVDs in 2009. **(c)** Answers will vary; One possibility
is predicting a continuing increase based on the increase from 2006 to 2008.
44. **(a)** Most people will probably pick Store B because of its greater
sales and more consistent upward trend. **(b)** Answers will vary. Some
possibilities include: age and physical condition of the store, annual
expenses, sales of other products, annual profit.

Section 9.4 (pages 663–664)

1.

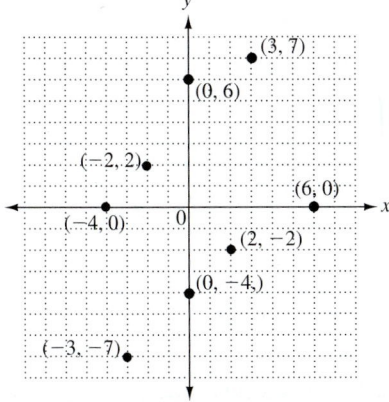

3.

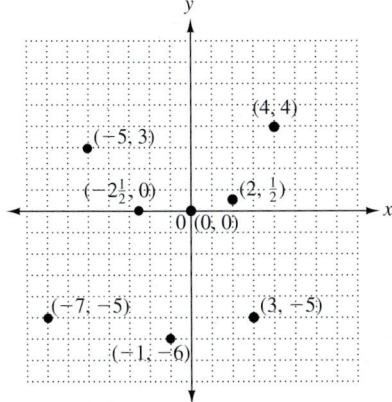

5. A is $(3, 4)$; B is $(5, -5)$; C is $(-4, -2)$; D is approximately $\left(4, \dfrac{1}{2}\right)$; E is $(0, -7)$; F is $(-5, 5)$; G is $(-2, 0)$; H is $(0,0)$.   **7.** III, none, IV, II
9. (a) any positive number   **(b)** any negative number   **(c)** 0   **(d)** any negative number   **(e)** any positive number   **11.** Starting at the origin, move left or right along the x-axis to the number a; then move up if b is positive or move down if b is negative.

Section 9.5 (pages 673–678)

1. 4; $(0, 4)$; 3; $(1, 3)$; 2; $(2, 2)$;
All points on the line are solutions.

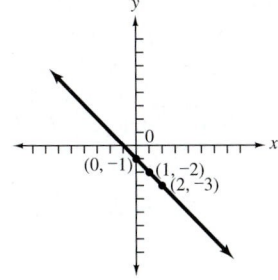

3. -1; $(0, -1)$; -2; $(1, -2)$; -3; $(2, -3)$;
All points on the line are solutions.

5. 4; -1; -6; 99

7.

x	y	(x, y)
1	-1	$(1, -1)$
2	0	$(2, 0)$
3	1	$(3, 1)$

9.

x	y	(x, y)
0	2	$(0, 2)$
-1	1	$(-1, 1)$
-2	0	$(-2, 0)$

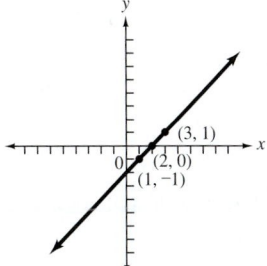 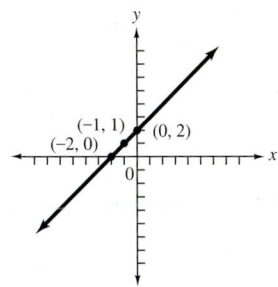

11.

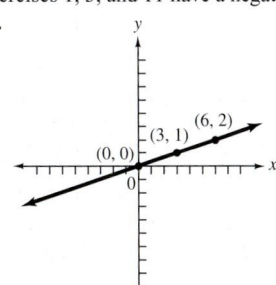

x	y	(x, y)
0	0	$(0, 0)$
1	-3	$(1, -3)$
2	-6	$(2, -6)$

13. The lines in Exercises 7 and 9 have a positive slope. The lines in Exercises 1, 3, and 11 have a negative slope.

15.

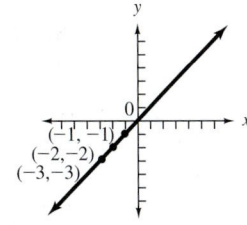

x	y	(x, y)
0	0	$(0, 0)$
3	1	$(3, 1)$
6	2	$(6, 2)$

17.

x	y	(x, y)
-1	-1	$(-1, -1)$
-2	-2	$(-2, -2)$
-3	-3	$(-3, -3)$

19.

x	y	(x, y)
0	3	$(0, 3)$
1	1	$(1, 1)$
2	-1	$(2, -1)$

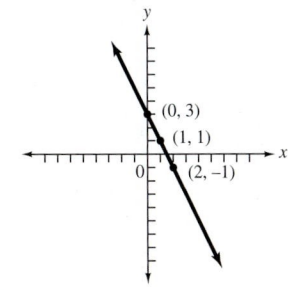

21.

23.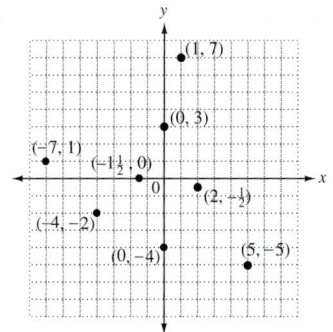

25.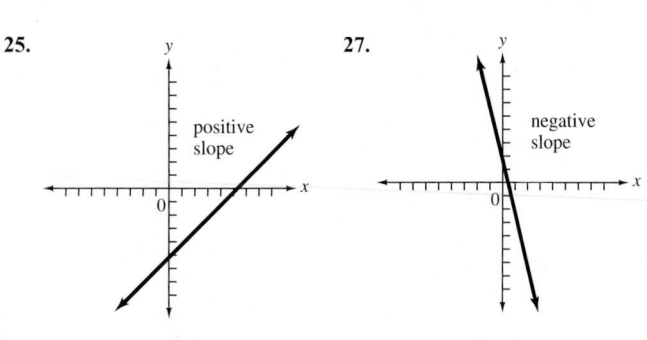

27.

38.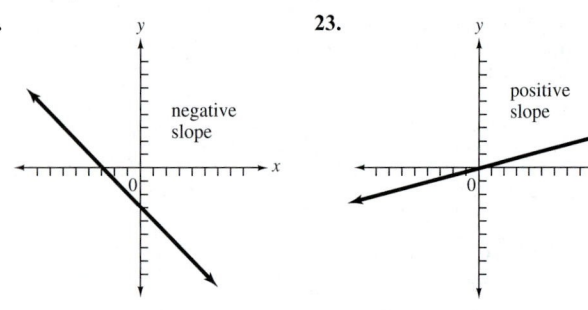

39. A is $(0, 6)$; B is approximately $\left(-2, 2\frac{1}{2}\right)$; C is $(0, 0)$; D is $(-6, -6)$; E is $(4, 3)$; F is approximately $\left(3\frac{1}{2}, 0\right)$; G is $(2, -4)$.

40.
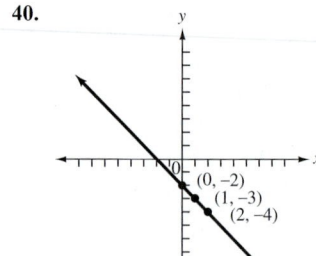

x	y	(x, y)
0	-2	$(0, -2)$
1	-3	$(1, -3)$
2	-4	$(2, -4)$

The graph of $x + y = -2$ has a *negative* slope.

Chapter 9 Review Exercises (pages 685–690)

1. (a) gymnastics **(b)** volleyball **2. (a)** cross country **(b)** volleyball
3. (a) 4653 men **(b)** 13,284 women **4.** men's, 14.7 (rounded); women's, 13.9 (rounded) **5. (a)** 100 inches **(b)** 15 inches **6. (a)** 50 inches **(b)** 55 inches **7. (a)** 35 inches **(b)** 45 inches **8.** 95 inches difference between Juneau and Memphis **9. (a)** lodging; $560 **(b)** food; $400
(c) $1700 **10.** $\frac{560}{400} = \frac{7}{5}$ **11.** $\frac{300}{1700} = \frac{3}{17}$ **12.** $\frac{280}{1700} = \frac{14}{85}$
13. $\frac{300}{160} = \frac{15}{8}$ **14.** painting and wall papering; 63% **15.** construction work; 33% **16.** 43%; 147 homeowners (rounded) **17.** 184 homeowners (rounded) **18. (a)** interior decorating **(b)** construction work and window treatments **19.** Answers will vary. Possibilities include: painting and wall papering are easier to do, take less time, or cost less than construction work. **20.** March; 8,000,000 acre-feet **21.** June; 2,000,000 acre-feet **22.** 5,000,000 acre-feet **23.** 6,000,000 acre-feet **24.** 3,000,000 acre-feet; 37.5% decrease **25.** 3,000,000 acre-feet; 60% decrease **26.** $50,000,000 **27.** $20,000,000 **28.** $20,000,000 **29.** $40,000,000 **30.** Sales decreased for two years and then moved up slightly. Answers will vary. Perhaps there is less new construction, remodeling, and home improvement in the area near Center A. Or, better product selection and service may have reversed the decline in sales. **31.** Sales are increasing. Answers will vary. New construction may have increased in the area near Center B, or greater advertising may attract more attention. **32.** 36° **33.** 35%; 126° **34.** 20% 72° **35.** 25% 90° **36.** $2240; 10% **37.** See graph below.

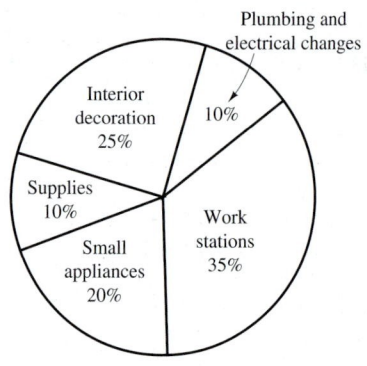

Plumbing and electrical changes 10%; Interior decoration 25%; Supplies 10%; Small appliances 20%; Work stations 35%

There are many other solutions because all points on the line are solutions. Some possibilities are $(-1, -1)$ and $(-2, 0)$.

41.
x	y	(x, y)
0	3	$(0, 3)$
1	4	$(1, 4)$
2	5	$(2, 5)$

42.
x	y	(x, y)
-1	4	$(-1, 4)$
0	0	$(0, 0)$
1	-4	$(1, -4)$

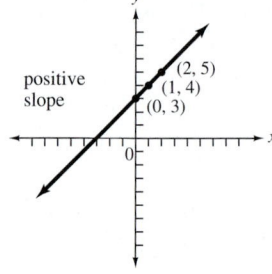

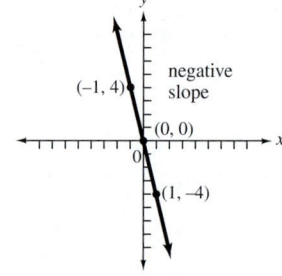

All points on the line are solutions. Some possibilities are $(-1, 2)$ and $(-2, 1)$.

All points on the line are solutions. Some possibilities are $\left(-\frac{1}{2}, 2\right)$ and $\left(\frac{1}{2}, -2\right)$.

Chapter 9 Test (pages 691–694)

1. (a) sardines **(b)** cream cheese **2.** 166 calories (rounded) **3.** 259 mg (rounded) **4. (a)** birds and fish are tied for the greatest number **(b)** 75 bird species and 75 fish species **5.** 55 species **6.** 270 species **7.** television; $812,000 **8.** miscellaneous; $84,000 **9.** $308,000 **10.** $644,000 **11.** 2007; $4000 **12.** $5000; 38% increase (rounded) **13.** 2007; explanations will vary. Some possibilities are: laid off from work, changed jobs, was ill, cut down on hours worked. **14.** 5500 students; 3000 students **15.** College B; 1000 students **16.** Explanations will vary. For example, College B may have added new courses or lowered tuition or added child care. **17.** 35%; 126° **18.** 5%; 18° **19.** 20%; 72° **20.** 30%; 108° **21.** $48,000; 10%

22.

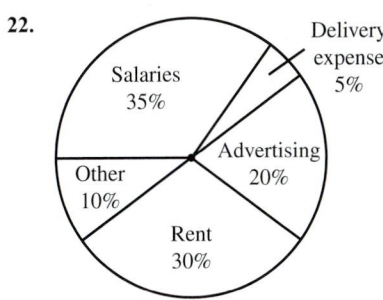

- Salaries 35%
- Delivery expense 5%
- Advertising 20%
- Other 10%
- Rent 30%

23–26.

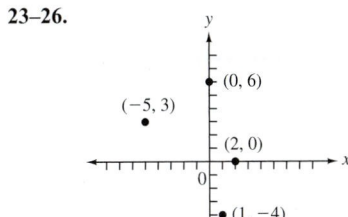

Points: $(0, 6)$, $(-5, 3)$, $(2, 0)$, $(1, -4)$

27. $(0, 0)$; no quadrant **28.** $(-5, -4)$; quadrant III
29. $(3, 3)$; quadrant I **30.** $(-2, 4)$; quadrant II
31.

x	y	(x, y)
0	-4	$(0, -4)$
1	-3	$(1, -3)$
2	-2	$(2, -2)$

Points: $(2, -2)$, $(1, -3)$, $(0, -4)$

32. **(a)** Answers will vary; all points on the line are solutions. Some possibilities are $(3, -1)$ and $(4, 0)$. **(b)** Positive slope

Chapter 10 Real Numbers, Equations, and Inequalities

Section 10.1 (pages 703–704)

1. rational because it terminates **3.** irrational because decimal form never ends or repeats in a fixed block **5.** rational because it's the quotient of integers **7.** rational because digits repeat in a fixed block
9. irrational because decimal form never ends or repeats in a fixed block **11.** **(a)** $3, 7$ **(b)** $0, 3, 7$ **(c)** $-9, 0, 3, 7$
(d) $-9, -1\frac{1}{4}, -\frac{3}{5}, 0, 3, 5.9, 7$ **(e)** $-\sqrt{7}, \sqrt{5}$
(f) All are real numbers.
13. "75 hundredths does not equal three-fourths." False; $0.75 = \frac{3}{4}$.
15. "Negative 4 is less than or equal to negative 5." False; $-4 > -5$ and $-4 \neq -5$. **17.** "Zero is greater than or equal to zero." True, because $0 = 0$. **19.** $19 > 12$ **21.** $\frac{1}{2} \leq \frac{4}{5}$ **23.** $-17 \leq -17$; True
25. $-3 \geq 0$; False **27.** $-6 \neq 6$; True **29.** $45 \leq 40$; False
31. $1 \geq 1$; True **33.** $0 \neq$ undefined; True **35.** **(a)** *Titanic* and *The Dark Knight* **(b)** *Titanic*, *The Dark Knight*, and *Star Wars*
36. **(a)** *The Phantom Menace* **(b)** *E.T.* and *The Phantom Menace*
37. Answers will vary. One possibility is: gross receipts ≤ 436.5 million dollars. **38.** Answers will vary. One possibility is: gross receipts ≥ 436.5 million dollars. **39.** $-4t - 5m$ **41.** $5c + 4d$

43. $-6h + n$ **45.** $3q - 5r + 8s$ **47.** $-19p + 16$
49. $-4y + 22$ **51.** $y^2 + y - 12$ **53.** $6b^2 - 6b + 8$
55. $a + 1$ **57.** $24k - 4$

Section 10.2 (pages 711–714)

1. $m = -1$ **3.** $p = 5$ **5.** $r = 1$ **7.** $x = -\frac{5}{3}$ **9.** $x = -1$
11. $y = 7$ **13.** $p = -4$ **15.** $b = 13$ **17.** $p = 18$ **19.** $x = 12$
21. no solution **23.** all real numbers **25.** no solution
27. no solution **29.** all real numbers **31.** all real numbers
33. No, it is incorrect to divide both sides by a variable. If $-3x$ is added to both sides, the equation becomes $4x = 0$, so $x = 0$ is the correct solution. **35.** $t = 5$ **37.** $x = 0$ **39.** $k = -\frac{7}{5}$
41. $x = -2$ **43.** all real numbers **45.** $x = 120$ **47.** $x = 6$
49. $x = 15,000$ **51.** no solution **53.** $x = 4$ **55.** $y = 0$
57. $x = 20$ **59.** all real numbers **61.** $s = -\frac{13}{8}$

Section 10.3 (pages 719–720)

1. $l = 6$ **3.** $h = 14$ **5.** $c = 5$ **7.** $r = 40$ **9.** $t = 7$
11. $r \approx 1.3$ **13.** $w = 8$ **15.** **(a)** width $= 35$ inches
(b) area $= 1785$ square inches **17.** $d \approx 630$ ft **19.** $r = \frac{d}{t}$
21. $l = \frac{A}{w}$ **23.** $a = P - b - c$ **25.** $p = \frac{I}{rt}$ **27.** $b = \frac{2A}{h}$
29. $r = \frac{A - p}{pt}$ **31.** $h = \frac{V}{\pi r^2}$ **33.** $C = \frac{5}{9}(F - 32)$ or
$C = \frac{5F - 160}{9}$ **34.** **(a)** $P - 2l = 2w$ **(b)** $\frac{P - 2l}{2} = w$
35. **(a)** $\frac{P}{2} = l + w$ **(b)** $\frac{P}{2} - l = w$
36. **(a)** Multiplicative identity property **(b)** A number divided by 1 is equal to itself. **(c)** Multiplication of fractions **(d)** Subtraction of fractions

Section 10.4 (pages 727–730)

1. Use an open circle if the symbol is $>$ or $<$. Use a closed circle if the symbol is $\geq$ or $\leq$. **3.** Every real number less than 2 is a solution. There are an infinite number of solutions, so it is not possible to list them all.
5. (number line, closed circle at 4)
7. (number line, open circle at -3)
9. (number line, closed circles at 8 and 10)
11. (number line, open circle at 0, closed circle at 10)
13. It would imply that $3 < -2$, which is false.
15. $z \geq 1$ (number line, closed circle at 1)
17. $k \geq 5$ (number line, closed circle at 5)
19. $n < -11$ (number line, open circle at -11)
21. It must be reversed when multiplying or dividing by a negative number.
23. His method is incorrect. He divided *by* the positive number 6. The sign of the number divided *into* does not matter.
25. $x < 6$ (number line, open circle at 6)

A N S W E R S

27. $y \ge -10$ — -10

29. $t < -3$ — -3

31. $x \le 0$ — 0

33. $r > 20$ — 20

35. $x \ge -3$ — -3

37. $r \ge -5$ — -5

39. $x < 1$ — 1

41. $x \le 0$ — 0

43. $x \ge 4$ — 4

45. $p < 32$ — 32

47. $x \ge \dfrac{5}{12}$ — $\frac{5}{12}$; 0

49. $k > -21$ — -21

51. all numbers less than 3   **53.** 79 or more   **55.** It is never more than 86 degrees Fahrenheit.   **57.** 32 or greater   **59.** at least \$275

61. 15 minutes   **63.** — 0 4

64. — 0 4

65. — 0 4

66. It is the set of all real numbers. — 0 4

67. The graph would be the set of all real numbers.

Chapter 10 Review Exercises (pages 735–738)

1. rational because it terminates   **2.** rational because it repeats in a fixed block   **3.** rational because $\sqrt{144} = 12$ **4.** irrational because decimal form does not terminate or repeat in a fixed block.

5. True because $\dfrac{2}{3}$ does not equal $\dfrac{6}{10}$.   **6.** True because $-\dfrac{5}{8} = -\dfrac{5}{8}$.

7. False because $-8 < 8$ and $-8 \ne 8$ **8.** False because $-75 < -50$ and $-75 \ne -50$ **9.** 1308   **10.** 90   **11.** 11   **12.** 0
13. $-10x + 20$ **14.** $5 - 3p$ **15.** $17c + 6$ **16.** $-10 + 10y$
17. $-17 - 14r$ **18.** $-2 + 3r$ **19.** $-19k + 54$ **20.** $t - 4$

21. $x = -6$ **22.** $t = \dfrac{3}{2}$ **23.** $r = 20$ **24.** $x = -\dfrac{61}{2}$ **25.** $y = 15$

26. $r = 0$ **27.** no solution   **28.** $x = 20$ **29.** $h = 11$

30. $r = 4.75$ **31.** $w = \dfrac{V}{lh}$ **32.** $h = \dfrac{2A}{b + B}$ **33.** 70.5 feet

34. 8 feet   **35.** — -4

36. — 7

37. — -5 6

38. — $\frac{1}{2}$; 0

39. $y \ge -3$ — -3

40. $t < 2$ — 2

41. $x \ge 3$ — 3

42. $k \ge 46$ — 46

43. $x < -5$ — -5

44. $w < -37$ — -37

45. 94 or more   **46.** all numbers less than or equal to $-\dfrac{1}{3}$

47. $y = 7$ **48.** $r = \dfrac{I}{pt}$ **49.** $x < 2$ **50.** $k = -9$ **51.** $x = 70$

52. $y = \dfrac{13}{4}$ **53.** no solution   **54.** all real numbers   **55.** $a = -1$

56. $a = P - b - c$ **57.** $y = -12$ **58.** $z = -\dfrac{3}{5}$ **59.** 26 inches

60. $20\dfrac{1}{2}$ inches or 20.5 inches   **61.** 450 miles or more

62. 84 or more   **63.** 1.5 meters (rounded)   **64.** 11 feet

Chapter 10 Test (pages 739–740)

1. irrational because the decimal form never ends or repeats in a fixed block   **2.** rational because it repeats in a fixed block

3. False because $-6 > -8$ and $-6 \ne -8$ **4.** False because $\dfrac{3}{4}$ in decimal form is 0.75 and $0.75 = 0.75$ **5.** True because $1.3 > 0.95$ or $1\dfrac{3}{10} > \dfrac{95}{100}$ **6.** -136 **7.** 2   **8.** $5a - 14$ **9.** $-9x^2 - 6x - 8$

10. $x = -6$ **11.** $y = \dfrac{13}{4}$ **12.** $x = -10.8$ **13.** no solution

14. $x = -\dfrac{21}{2}$ **15.** $x = 30$ **16.** all real numbers

17. $p = \$2500$ **18.** 19.9 ft (rounded)   **19.** $r = \dfrac{d}{2}$ **20.** $h = \dfrac{2A}{b}$

21. $r = \dfrac{A - p}{pt}$ **22.** $x < 11$ — 11

23. $x \le 4$ — 4 **24.** 81 or more

Chapter 11 Graphs of Linear Equations and Inequalities in Two Variables

Section 11.1 (pages 751–756)

1. 2005, 2006, 2007   **3.** 2001: about 165 billion lb; 2007: about 185 billion lb   **5.** from 2000 to 2005; about \$0.85   **7.** The price of a gallon of gas was decreasing.   **9.** does; do not   **11.** y
13. 6   **15.** yes   **17.** yes   **19.** no   **21.** yes   **23.** no
25. No. For two ordered pairs (x, y) to be equal, the x-values must be equal and the y-values must be equal. Here we have $4 \ne -1$ and $-1 \ne 4$.

27. 11   **29.** $-\dfrac{7}{2}$ **31.** -4 **33.** -5 **35.** 4; 6; -6; (0, 4); (6, 0); (-6, 8)

37. 3; -5; -15; (0, 3); (-5, 0); (-15, -6)   **39.** -9; -9; -9
41. -6; -6; -6 **43.** 8; 8; 8   **45.** (2, 4); I   **47.** (-5, 4); II
49. (3, 0); no quadrant   **51.** negative; negative   **53.** positive; negative
55. If $xy < 0$, then either $x < 0$ and $y > 0$ or $x > 0$ and $y < 0$. If $x < 0$ and $y > 0$, then the point lies in quadrant II. If $x > 0$ and $y < 0$, then the point lies in quadrant IV.

57.–68.

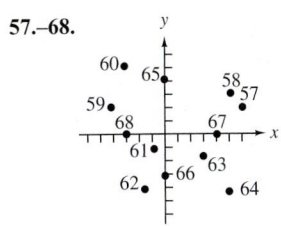

69. $-3; 6; -2; 4$

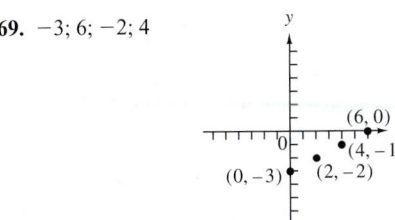

71. $-3; 4; -6; -\dfrac{4}{3}$

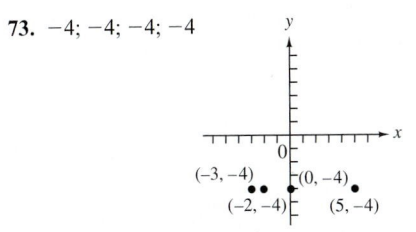

73. $-4; -4; -4; -4$

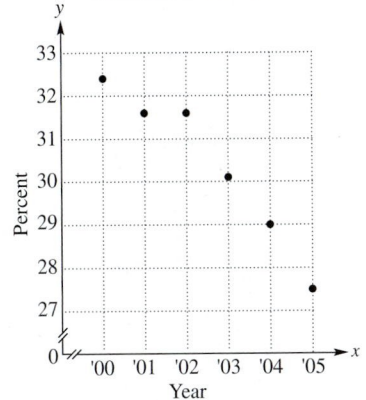

75. The points in each graph appear to lie on a straight line.

77. (a) $(5, 45)$ **(b)** $(6, 50)$ **79. (a)** $(2000, 32.4)$, $(2001, 31.6)$, $(2002, 31.6)$, $(2003, 30.1)$, $(2004, 29.0)$, $(2005, 27.5)$

(b) $(2007, 27.1)$ means that 27.1 percent of 2-year college students in 2007 received a degree within 3 years.

(c)   2-YEAR COLLEGE STUDENTS COMPLETING A DEGREE WITHIN 3 YEARS

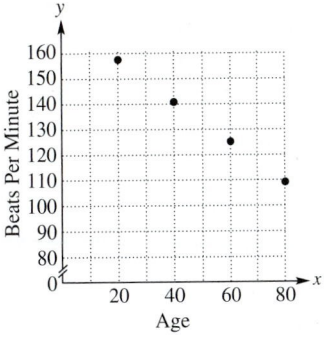

(d) With the exception of the point for 2002, the points lie approximately on a straight line. Rates at which 2-year college students complete a degree within 3 years are generally decreasing.

81. (a) 157, 141, 125, 109 **(b)** (20, 157), (40, 141), (60, 125), (80, 109)

(c)   TARGET HEART RATE ZONE (Upper Limit)

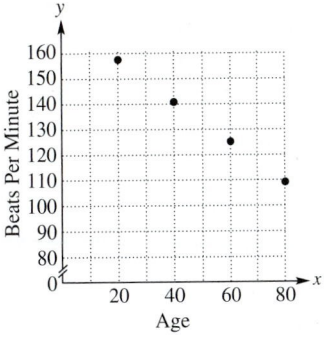

The points lie in a linear pattern.

Section 11.2 (pages 765–770)

1. $5; 5; 3$

3. $1; 3; -1$

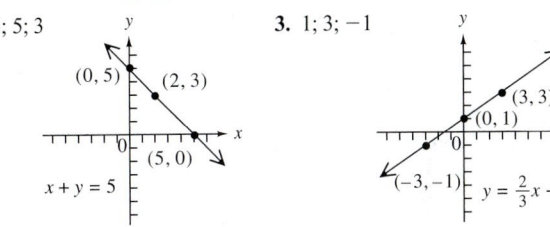

5. $-6; -2; -5$

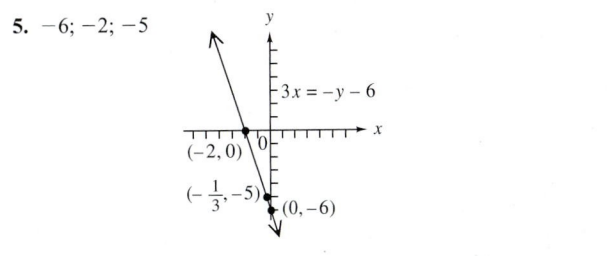

7. (a) A **(b)** C **(c)** D **(d)** B **9.** $(12, 0); (0, -8)$ **11.** $(0, 0); (0, 0)$

13.

15.

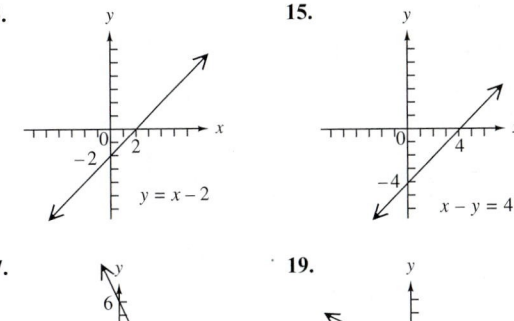

17.

19.

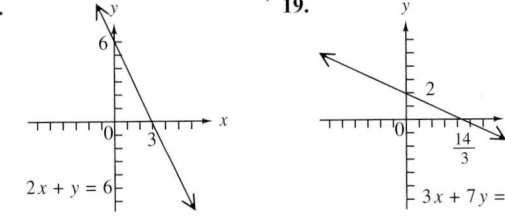

21.

23.

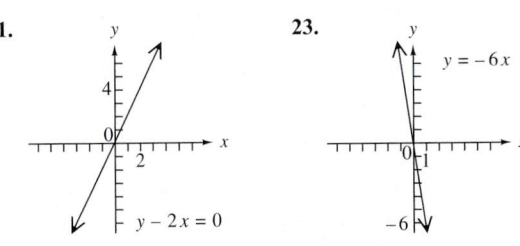

25. 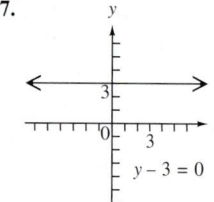   **27.**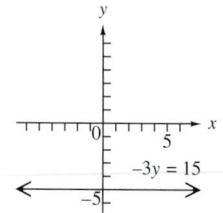

29.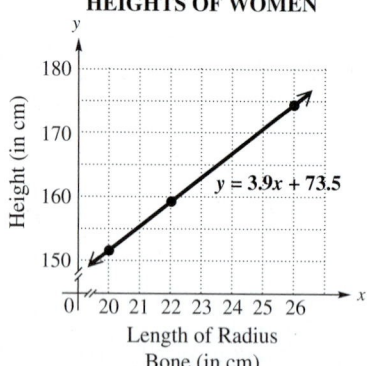

In Exercises 31 and 33, descriptions may vary.
31. The graph is a line with x-intercept $(-3, 0)$ and y-intercept $(0, 9)$.
33. The graph is a horizontal line with y-intercept $(0, -2)$.
35. Choose a value *other than* 0 for either x or y. For example, if $x = -5$, $y = 4$.
37. (a) 151.5 cm, 159.3 cm, 174.9 cm   **(b)** (20, 151.5), (22, 159.3), (26, 174.9)
(c)

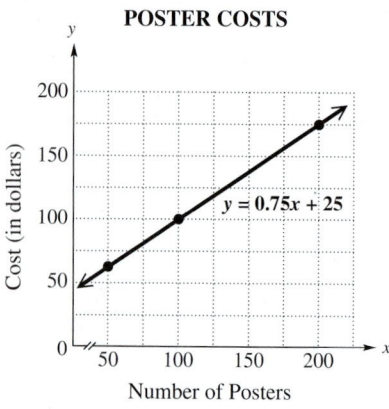

HEIGHTS OF WOMEN

(d) 24 cm; 24 cm
39. (a) \$62.50; \$100   **(b)** 200   **(c)** (50, 62.50), (100, 100), (200, 175)
(d)

POSTER COSTS

41. (a) \$30,000   **(b)** \$15,000   **(c)** \$5000   **(d)** After 5 yr, the SUV has a value of \$5000.   **43. (a)** 1990: 24.1 lb; 2000: 29.5 lb; 2005: 32.2 lb
(b) 1990: 25 lb; 2000: 30 lb; 2005: 32 lb   **(c)** The values are quite close.

Section 11.3 (pages 779–782)

1. $\frac{3}{2}$ **3.** $-\frac{7}{4}$ **5.** 0   **7.** Rise is the vertical change between two different points on a line. Run is the horizontal change between two different points on a line.

9.–12. Answers will vary.

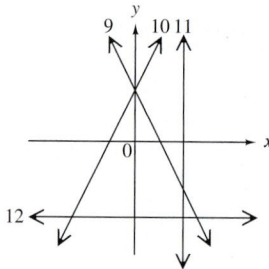

13. (a) falls from left to right   **(b)** horizontal   **(c)** vertical   **(d)** rises from left to right   **15.** Because he found the difference $3 - 5 = -2$ in the numerator, he should have subtracted in the same order in the denominator to get $-1 - 2 = -3$. The correct slope is $\frac{-2}{-3} = \frac{2}{3}$.

17. $\frac{5}{4}$ **19.** $\frac{3}{2}$ **21.** -3 **23.** 0   **25.** undefined   **27.** $-\frac{1}{2}$ **29.** 5

31. $\frac{1}{4}$ **33.** $\frac{3}{2}$ **35.** 0   **37.** undefined   **39.** 1   **41. (a)** negative

(b) 0   **43. (a)** positive   **(b)** negative   **45. (a)** 0   **(b)** negative

47. $\frac{4}{3}; \frac{4}{3}$; parallel   **49.** $\frac{5}{3}; \frac{3}{5}$; neither   **51.** $\frac{3}{5}; -\frac{5}{3}$; perpendicular

53. $\frac{8}{27}$ **55.** 232 thousand, or 232,000   **56.** positive; increased

57. 232,000 students   **58.** -0.95 **59.** negative; decreased
60. 0.95 student per computer

Section 11.4 (pages 791–796)

1. (a) D   **(b)** C   **(c)** B   **(d)** A   **3.** $y = 3x - 3$

5. $y = -x + 3$ **7.** $y = -\frac{1}{2}x + 2$ **9.** $y = 4x - 3$ **11.** $y = 3$

13. (a) C   **(b)** B   **(c)** A   **(d)** D

15. 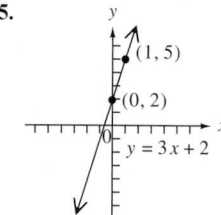   **17.**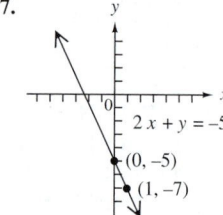

19. 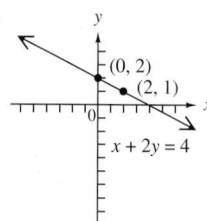   **21.** $y = \frac{1}{2}x + 4$

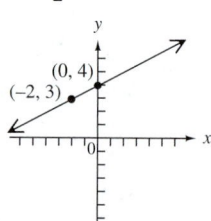

23. $y = -\frac{2}{5}x - \frac{23}{5}$ **25.** $y = 2$

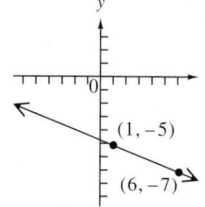   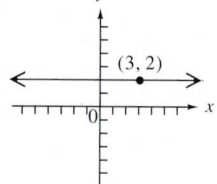

27. $x = 3$ (no slope-intercept form)   **29.** $y = \frac{2}{3}x$

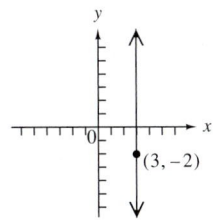

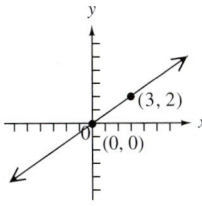

31. $y = 2x - 7$ **33.** $y = -2x - 4$ **35.** $y = \frac{2}{3}x + \frac{19}{3}$

37. $y = x - 3$ **39.** $y = -\frac{5}{7}x - \frac{54}{7}$ **41.** $y = -\frac{2}{3}x - 2$

43. $x = 3$ (no slope-intercept form)   **45.** $y = \frac{1}{3}x + \frac{4}{3}$

47. $y = \frac{3}{4}x - \frac{9}{2}$ **49.** $y = -2x - 3$ **51.** $(0, 32); (100, 212)$ **52.** $\frac{9}{5}$

53. $F - 32 = \frac{9}{5}(C - 0)$ **54.** $F = \frac{9}{5}C + 32$ **55.** $C = \frac{5}{9}(F - 32)$

56. $86°$ **57.** $10°$ **58.** $-40°$ **59.** **(a)** $400   **(b)** $0.25
(c) $y = 0.25x + 400$ **(d)** $425   **(e)** $1500   **61.** **(a)** $(1, 1909), (2, 2079)$,
$(3, 2182), (4, 2272), (5, 2361)$
(b) yes

**AVERAGE ANNUAL COSTS AT
2-YEAR COLLEGES**

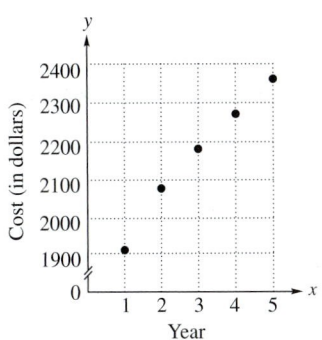

(c) $y = 94x + 1891$ **(d)** $2455

Summary Exercises on Linear Equations and Graphs (pages 797–798)

1. $-3; (0, -6)$ **2.** $-2; (0, -4)$ **3.** $-4; (0, -3)$

4. $-5; (0, -8)$ **5.** $\frac{3}{2}; (0, 6)$ **6.** $\frac{5}{3}; (0, 5)$

7.
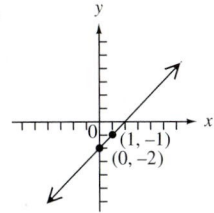

8.
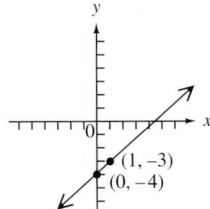

9.
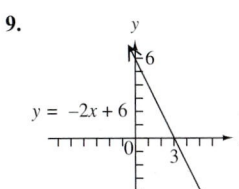

10.
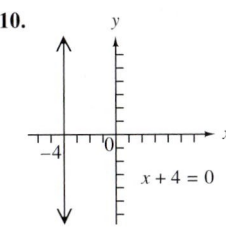

11.
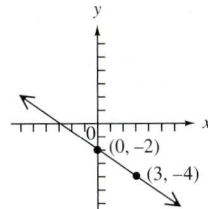

12.
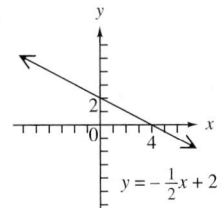

13.
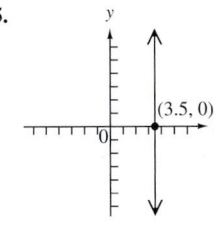

14.
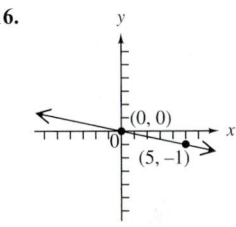

15.
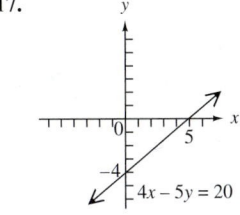

16.
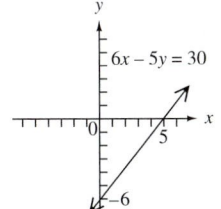

17.
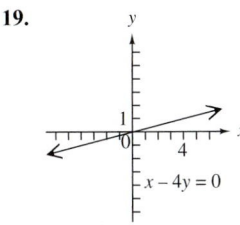

18.
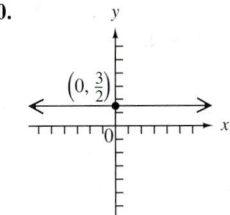

19.
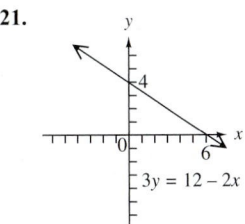

20.
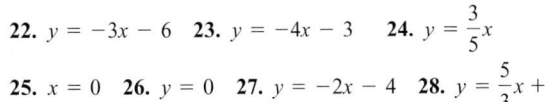

21.
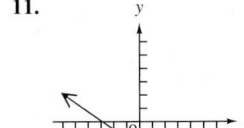

22. $y = -3x - 6$ **23.** $y = -4x - 3$ **24.** $y = \frac{3}{5}x$

25. $x = 0$ **26.** $y = 0$ **27.** $y = -2x - 4$ **28.** $y = \frac{5}{3}x + 5$

Section 11.5 (pages 803–806)

1. false; The point $(4, 0)$ lies on the boundary line $3x - 4y = 12$, which is *not* part of the graph because the symbol $<$ does not involve equality.
3. true   **5.** $>, >$ **7.** $\leq$

ANSWERS

9.

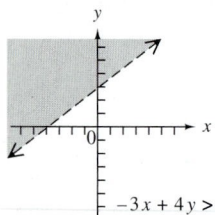

$x + y \geq 4$

11.

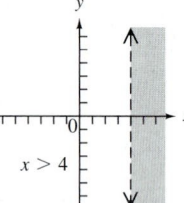

$x + 2y \geq 7$

13.

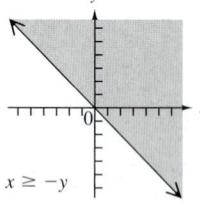

$-3x + 4y > 12$

15.

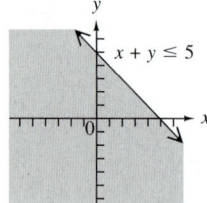

$x > 4$

17.

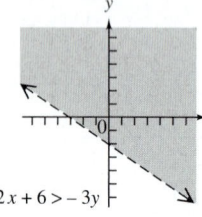

$x \geq -y$

19. A test point cannot lie on the boundary line. It must lie on one side of the boundary.

21.

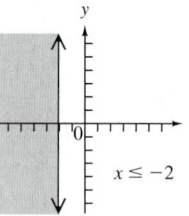

$x + y \leq 5$

23.

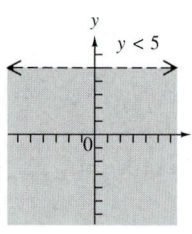

$x + 2y < 4$

25.

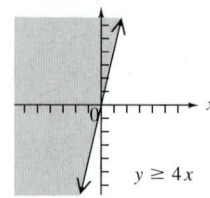

$2x + 6 > -3y$

27.

$y \geq 2x + 1$

29.

$x \leq -2$

31.

$y < 5$

33.

$y \geq 4x$

35. Every point in quadrant IV has a positive x-value and a negative y-value. Substituting into $y > x$ would imply that a negative number is greater than a positive number, which is always false. Thus, the graph of $y > x$ cannot lie in quadrant IV.

37. (a)

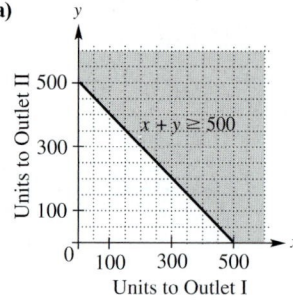

$x + y \geq 500$

(b) $(500, 0)$ and $(200, 400)$; Other answers are possible.

Chapter 11 Review Exercises (pages 811–816)

1. $(2001, 51.8)$, $(2002, 51.8)$, $(2003, 52.8)$, $(2004, 51.3)$, $(2005, 51.6)$, $(2006, 52.5)$, $(2007, 51.4)$ **2.** In the year 2006, 52.5% of first-year college students at two-year public institutions returned for a second year.

3. 2004; 1.5%   **4.** 2003; 1.0%   **5.** $-1; 2; 1$ **6.** $2; \dfrac{3}{2}; \dfrac{14}{3}$

7. $0; \dfrac{8}{3}; -9$ **8.** $7; 7; 7$ **9.** yes   **10.** no   **11.** yes

12.–15.

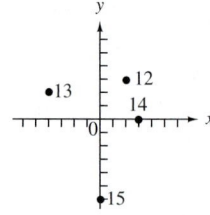

16. x is positive in quadrants I and IV; y is negative in quadrants III and IV. Thus, if x is positive and y is negative, (x, y) must lie in quadrant IV.   **17.** In the ordered pair $(k, 0)$, the y-value is 0, so the point lies on the x-axis. In the ordered pair $(0, k)$, the x-value is 0, so the point lies on the y-axis.   **18.** II   **19.** III   **20.** no quadrant

21. $\left(-\dfrac{5}{2}, 0\right); (0, 5)$ **22.** $\left(-\dfrac{7}{2}, 0\right); (0, -7)$ **23.** $\left(\dfrac{8}{3}, 0\right); (0, 4)$

24.

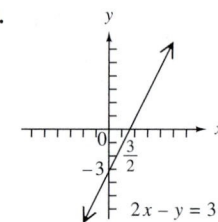

$2x - y = 3$

25.

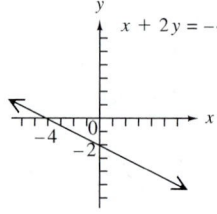

$x + 2y = -4$

26.

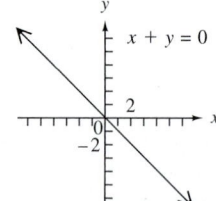

$x + y = 0$

27. $-\dfrac{1}{2}$ **28.** $-\dfrac{2}{3}$ **29.** 0   **30.** undefined   **31.** 3   **32.** $\dfrac{2}{3}$ **33.** $\dfrac{3}{2}$

34. $-\dfrac{1}{3}$ **35.** undefined   **36.** 0   **37.** $\dfrac{3}{2}$ **38. (a)** 2   **(b)** $\dfrac{1}{3}$ **39.** parallel

40. perpendicular   **41.** neither   **42.** 0   **43.** $y = -x + \dfrac{2}{3}$

44. $y = -\dfrac{1}{3}x + 1$ **45.** $y = x - 7$ **46.** $y = \dfrac{2}{3}x + \dfrac{14}{3}$

ANSWERS

47. $y = -\frac{3}{4}x - \frac{1}{4}$ **48.** $y = -\frac{1}{4}x + \frac{3}{2}$ **49.** $y = 1$ **50.** $x = \frac{1}{3}$

51. (a) $y = -\frac{1}{3}x + 5$ **(b)** slope: $-\frac{1}{3}$; y-intercept: $(0, 5)$

(c)

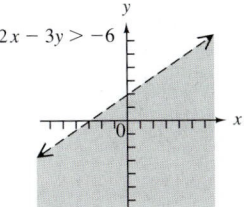

52. (a) B **(b)** D **(c)** A **(d)** C

53.

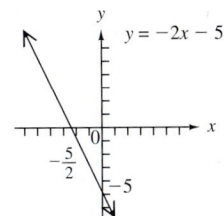

54.

55.

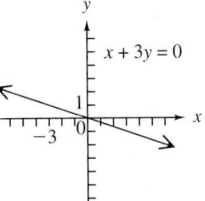

56. A **57.** C, D **58.** A, B, D **59.** D **60.** C **61.** B

62. $\left(-\frac{5}{2}, 0\right)$; $(0, -5)$; -2

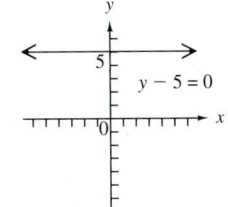

63. $(0, 0)$; $(0, 0)$; $-\frac{1}{3}$

64. no x-intercept; $(0, 5)$; 0

65. $y = -\frac{1}{4}x - \frac{5}{4}$ **66.** $y = -3x + 30$ **67.** $y = -\frac{4}{7}x - \frac{23}{7}$

68.

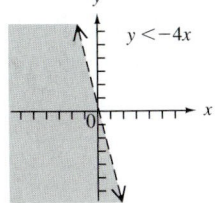

69.

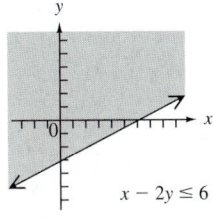

70. 2.5% **71.** Since the graph rises from left to right, the slope is positive.
72. $(2002, 41.2), (2007, 43.7)$ **73.** $y = 0.5x - 959.8$ **74.** 0.5; yes
75. 41.7, 42.2, 42.7, 43.2 **76.** 44.2%; No. The equation is based on data only for 2002 through 2007.

Chapter 11 Test (pages 817–820)

1. between 1998 and 1999, 1999 and 2000, and 2003 and 2004
2. The unemployment rate was increasing.
3. 2003: 6.0%; 2004: 5.5%; decline: 0.5%
4. x-intercept: $(2, 0)$; y-intercept: $(0, 6)$

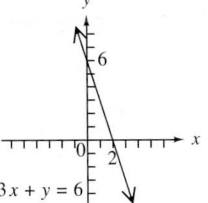

5. x-intercept: $(0, 0)$; y-intercept: $(0, 0)$

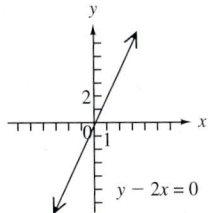

6. x-intercept: $(-3, 0)$; y-intercept: none

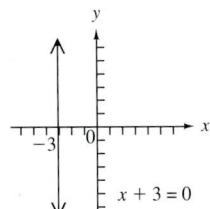

7. x-intercept: none; y-intercept: $(0, 1)$

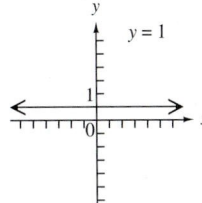

8. x-intercept: $(4, 0)$; y-intercept: $(0, -4)$

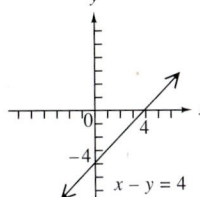

9. $-\frac{8}{3}$ **10.** -2 **11.** undefined **12.** $\frac{5}{2}$ **13.** 0 **14.** $y = 2x + 6$

15. $y = \frac{5}{2}x - 4$ **16.** $y = -9x + 12$ **17.** $y = -\frac{3}{2}x + \frac{9}{2}$

18.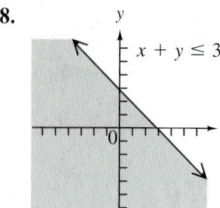
$x + y \leq 3$

19.
$3x - y > 0$

20. The slope is negative since sales are decreasing. **21.** (0, 209), (7, 160); -7 **22.** $y = -7x + 209$ **23.** 174 thousand; The equation gives a good approximation of the actual sales. **24.** In 2007, worldwide snowmobile sales were 160 thousand.

Chapter 12 Systems of Linear Equations and Inequalities

Section 12.1 (pages 827–830)

1. B, because the ordered pair must be in quadrant II. **3.** There is no way that the sum of two numbers can be both 2 and 4 at the same time. **5.** no **7.** yes **9.** yes **11.** no
We show the graphs here only for Exercises 13–17.
13. $\{(4, 2)\}$ **15.** $\{(0, 4)\}$

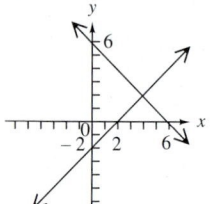

17. $\{(4, -1)\}$

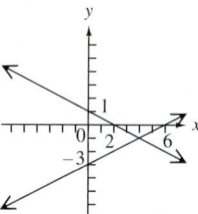

19. $\{(1, 3)\}$ **21.** $\{(0, 2)\}$ **23.** $\emptyset$ (inconsistent system)
25. $\{(x, y) \mid 2x - y = 4\}$ (dependent equations) **27.** $\{(4, -3)\}$
29. $\emptyset$ (inconsistent system) **31. (a)** neither **(b)** intersecting lines
(c) one solution **33. (a)** dependent **(b)** one line **(c)** infinite number of solutions **35. (a)** inconsistent **(b)** parallel lines **(c)** no solution
37. 1980–2000 **39. (a)** 1997–2002 **(b)** 2001 **(c)** 2002 **(d)** (2004, 70) (The *y*-value is approximate.) **(e)** During the period 1997–2004, debit card use went from least popular to most popular of the three methods depicted.

Section 12.2 (pages 837–838)

1. The *y*-value must also be determined. The solution set is $\{(3, 0)\}$.
3. $\{(3, 9)\}$ **5.** $\{(7, 3)\}$ **7.** $\{(-2, 4)\}$ **9.** $\{(-4, 8)\}$ **11.** $\{(3, -2)\}$
13. $\{(x, y) \mid 3x - y = 5\}$ **15.** $\left\{ \left(\frac{1}{3}, -\frac{1}{2} \right) \right\}$ **17.** $\emptyset$
19. $\{(x, y) \mid 3x - 4y = 2\}$ **21.** $\{(2, -3)\}$ **23.** $\{(10, -12)\}$
25. $\{(-4, 2)\}$ **27.** $\{(7, -3)\}$ **29.** $\{(20, 30)\}$ **30.** To find the total cost, multiply the number of bicycles (x) by the cost per bicycle (400 dollars) and add the fixed cost (5000 dollars). Thus, $y_1 = 400x + 5000$ gives this total cost (in dollars). **31.** $y_2 = 600x$ **32.** $y_1 = 400x + 5000, y_2 = 600x$; solution set: $\{(25, 15,000)\}$ **33.** 25; 15,000; 15,000

Section 12.3 (pages 843–846)

1. true **3.** true **5.** $\{(-1, 3)\}$ **7.** $\{(-1, -3)\}$ **9.** $\{(-2, 3)\}$
11. $\left\{ \left(\frac{1}{2}, 4 \right) \right\}$ **13.** $\{(3, -6)\}$ **15.** $\{(7, 4)\}$ **17.** $\{(0, 4)\}$
19. $\{(-4, 0)\}$ **21.** $\{(0, 0)\}$ **23.** $\emptyset$ **25.** $\{(x, y) \mid x - 3y = -4\}$
27. $\{(2, 9)\}$ **29.** $\{(-6, 5)\}$ **31.** $\left\{ \left(-\frac{6}{5}, \frac{4}{5} \right) \right\}$ **33.** $\left\{ \left(\frac{1}{8}, -\frac{5}{6} \right) \right\}$
35. $\{(11, 15)\}$ **37.** $\emptyset$ **39.** $\{(x, y) \mid 2x + y = 0\}$ **41.** $1339 = 1996a + b$
42. $1536 = 2004a + b$ **43.** $1996a + b = 1339, 2004a + b = 1536$; solution set: $\{(24.625, -47,812.5)\}$ **44.** $y = 24.625x - 47,812.5$
45. 1486.8 (million); This is quite a bit less than the actual figure.
46. Since the data do not lie in a perfectly straight line, the quantity obtained from an equation determined in this way will probably be "off" a bit. We cannot put too much faith in models such as this one, because not all sets of data points are linear in nature.

Summary Exercises on Solving Systems of Linear Equations (pages 847–848)

1. (a) Use substitution since the second equation is solved for *y*.
(b) Use elimination since the coefficients of the *y*-terms are opposites.
(c) Use elimination since the equations are in standard form with no coefficients of 1 or -1. Solving by substitution would involve fractions.
2. The system on the right is easier to solve by substitution because the second equation is already solved for *y*. **3. (a)** $\{(1, 4)\}$ **(b)** $\{(1, 4)\}$
(c) Answers will vary. **4. (a)** $\{(-5, 2)\}$ **(b)** $\{(-5, 2)\}$ **(c)** Answers will vary. **5.** $\{(2, 6)\}$ **6.** $\{(-3, 2)\}$ **7.** $\left\{ \left(\frac{1}{3}, \frac{1}{2} \right) \right\}$ **8.** $\emptyset$ **9.** $\{(3, 0)\}$
10. $\left\{ \left(\frac{3}{2}, -\frac{3}{2} \right) \right\}$ **11.** $\{(x, y) \mid 3x + y = 7\}$ **12.** $\{(9, 4)\}$
13. $\left\{ \left(-\frac{5}{7}, -\frac{2}{7} \right) \right\}$ **14.** $\{(4, -5)\}$ **15.** $\emptyset$ **16.** $\{(-4, 6)\}$
17. $\left\{ \left(\frac{19}{3}, -5 \right) \right\}$ **18.** $\left\{ \left(\frac{22}{13}, -\frac{23}{13} \right) \right\}$ **19.** $\{(-12, -60)\}$
20. $\{(2, -4)\}$ **21.** $\{(18, -12)\}$ **22.** $\{(-2, 1)\}$ **23.** $\left\{ \left(13, -\frac{7}{5} \right) \right\}$
24. $\{(10, -9)\}$ **25.** $\{(0.04, 0.9)\}$

Section 12.4 (pages 855–860)

1. D **3.** B **5.** D **7.** C **9.** the second number; $x - y = 48$; The two numbers are 73 and 25. **11.** *The Phantom of the Opera*: 8197; *Cats*: 7485
13. *Spider-Man 3*: \$336.5 million; *Shrek the Third*: \$322.7 million
15. Terminal Tower: 708 ft; Key Tower: 950 ft **17. (a)** 45 units
(b) Do not produce; the product will lead to a loss. **19.** 46 ones;
28 tens **21.** 2 DVDs of *Night at the Museum* and 5 Linkin Park CDs
23. \$2500 at 4%; \$5000 at 5% **25.** The Police: \$107; Van Halen: \$115
27. 80 L of 40% solution; 40 L of 70% solution **29.** 30 lb at \$6 per lb;
60 lb at \$3 per lb **31.** nuts: 40 lb; raisins: 20 lb **33.** bicycle:
13.5 mph; car: 49.3 mph **35.** car leaving Cincinnati: 55 mph;
car leaving Toledo: 70 mph **37.** Roberto: 17.5 mph; Juana: 12.5 mph
39. boat: 10 mph; current: 2 mph **41.** plane: 470 mph; wind: 30 mph

Section 12.5 (pages 865–866)

1. C   **3.** B

5.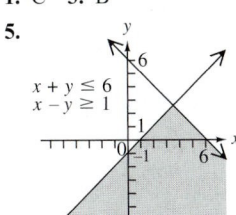

7.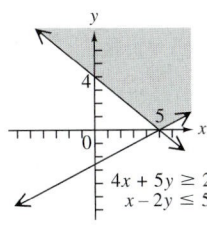

9.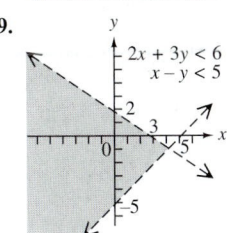

11.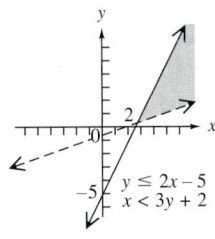

13.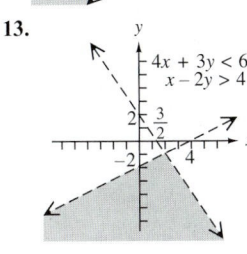

15.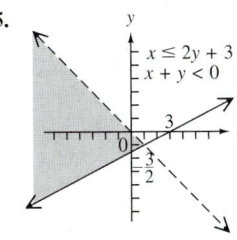

17.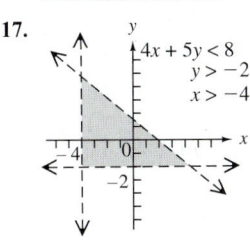

19.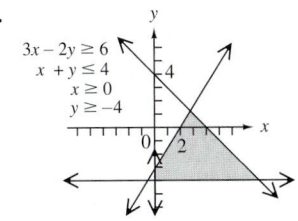

Chapter 12 Review Exercises (pages 871–874)

1. yes   **2.** no   **3.** $\{(3, 1)\}$ **4.** $\{(0, -2)\}$ **5.** $\{(x, y) \mid x - 2y = 2\}$
6. $\emptyset$ **7.** It is not a solution of the system because it is not also a solution of the second equation, $2x + y = 4$.   **8.** $\{(2, 1)\}$ **9.** $\{(3, 5)\}$
10. $\{(6, 4)\}$ **11.** $\emptyset$ **12.** $\{(7, 1)\}$ **13.** $\{(-5, -2)\}$ **14.** $\{(-4, 3)\}$
15. $\{(x, y) \mid 3x - 4y = 9\}$ **16.** (a) 2 (b) 9   **17.** $\{(9, 2)\}$
18. $\left\{\left(\dfrac{10}{7}, -\dfrac{9}{7}\right)\right\}$ **19.** $\{(8, 9)\}$ **20.** $\{(2, 1)\}$ **21.** $\{(6, -4)\}$
22. $\{(-8, 5)\}$ **23.** Subway: 20,755; McDonald's: 13,774
24. *AARP The Magazine:* 23.4 million; *Reader's Digest:* 10.1 million
25. length: 27 m; width: 18 m   **26.** 13 twenties; 7 tens   **27.** 25 lb of
$1.30 candy; 75 lb of $0.90 candy   **28.** plane: 250 mph; wind: 20 mph
29. $7000 at 3%; $11,000 at 4%   **30.** 60 L of 40% solution; 30 L of 70%
solution

31.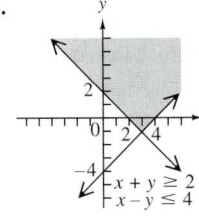

32.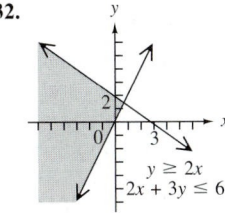

33.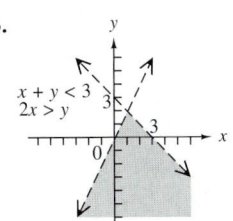

34. B   **35.** B   **36.** $\{(2, 0)\}$ **37.** $\{(-4, 15)\}$ **38.** $\emptyset$

39.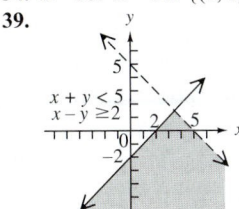

40.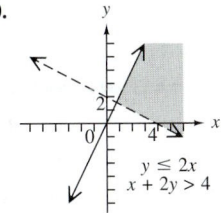

41.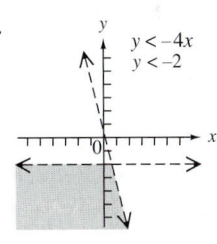

42. 8 in., 8 in., and 13 in.   **43.** Giants: 17; Patriots: 14
44. (a) years 0–6   (b) year 6; about $650

Chapter 12 Test (pages 875–876)

1. $\{(2, -3)\}$ **2.** It has no solution.   **3.** $\{(1, -6)\}$ **4.** $\{(-35, 35)\}$
5. $\{(5, 6)\}$ **6.** $\{(-1, 3)\}$ **7.** $\{(0, 0)\}$ **8.** $\emptyset$ **9.** $\{(x, y) \mid 3x - y = 6\}$
10. $\{(12, -4)\}$ **11.** Memphis and Atlanta: 394 mi; Minneapolis and
Houston: 1176 mi   **12.** Statue of Liberty: 3.6 million; National World
War II Memorial: 5.4 million   **13.** 20 L of 15% solution; 30 L of 40%
solution   **14.** slower car: 45 mph; faster car: 60 mph

15.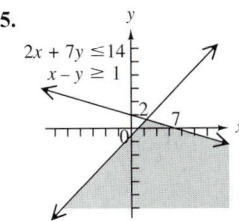

16.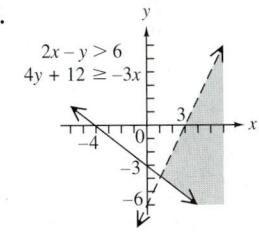

Chapter 13   Exponents and Polynomials

Section 13.1 (pages 883–886)

1. 7; 5   **3.** 8   **5.** 26   **7.** 1; 6   **9.** 1; 1   **11.** 1; $\dfrac{1}{5}$ **13.** 2; $-19, -1$
15. 3; 1, $-8, \dfrac{2}{3}$ **17.** $2m^5$ **19.** $-r^5$ **21.** $\dfrac{2}{3}x^4$ **23.** cannot be simplified;
$0.2m^5 - 0.5m^2$ **25.** $-5x^5$ **27.** $5p^9 + 4p^7$ **29.** $-2y^2$
31. already simplified; 4; binomial   **33.** already simplified;
$6m^5 + 5m^4 - 7m^3 - 3m^2$; 5; none of these   **35.** $x^4 + \dfrac{1}{3}x^2 - 4$; 4;
trinomial   **37.** 7; 0; monomial   **39.** $1.5x^2 - 0.5x$; 2; binomial
41. (a) -1 (b) 5   **43.** (a) 19 (b) -2 **45.** (a) 36 (b) -12
47. (a) -124 (b) 5   **48.** 5; 175   **49.** 87 ft; (1, 87)   **50.** $16.00
51. $27   **52.** 2.5; 130   **53.** $5m^2 + 3m$ **55.** $4x^4 - 4x^2$
57. $\dfrac{7}{6}x^2 - \dfrac{2}{15}x + \dfrac{5}{6}$ **59.** $12m^3 - 13m^2 + 6m + 11$
61. $2.9x^3 - 3.5x^2 - 1.5x - 9$ **63.** $8r^2 + 5r - 12$

65. $5m^2 - 14m + 6$ **67.** $4x^3 + 2x^2 + 5x$
69. $-18y^5 + 7y^4 + 5y^3 + 3y^2 + y$ **71.** $-2m^3 + 7m^2 + 8m - 9$
73. $-11x^2 - 3x - 3$ **75.** $2x^2 + 8x$ **77.** $8x^2 + 8x + 6$
79. $8t^2 + 8t + 13$ **81.** $13a^2b - 7a^2 - b$ **83.** $c^4d - 5c^2d^2 + d^2$
85. $12m^3n - 11m^2n^2 - 4mn^2$ **87. (a)** $23y + 5t$ **(b)** $25°, 67°, 88°$

Section 13.2 (pages 893–894)

1. 1   **3.** false   **5.** false   **7.** t^7 **9.** $\left(\dfrac{1}{2}\right)^5$ **11.** $(-8p)^2$ **13.** The
expression $(-3)^4$ means $(-3)(-3)(-3)(-3) = 81$, while -3^4 means
$-(3 \cdot 3 \cdot 3 \cdot 3) = -81$.   **15.** base: 3; exponent: 5; 243   **17.** base: -3;
exponent: 5; -243 **19.** base: $-6x$; exponent: 4   **21.** base: x; exponent: 4
23. The product rule does not apply to $5^2 + 5^3$ because it is a *sum*, not
a product. $5^2 + 5^3 = 25 + 125 = 150$ **25.** 5^8 **27.** 4^{12} **29.** $(-7)^9$
31. t^{24} **33.** $-56r^7$ **35.** $42p^{10}$ **37.** The product rule does not apply.
39. The product rule does not apply.   **41.** 4^6 **43.** t^{20} **45.** $343r^3$
47. 5^{12} **49.** -8^{15} **51.** $5^5x^5y^5$ **53.** $8q^3r^3$ **55.** $\dfrac{1}{8}$ **57.** $\dfrac{a^3}{b^3}$ **59.** $\dfrac{9^8}{5^8}$
61. $-8x^6y^3$ **63.** $9a^6b^4$ **65.** $\dfrac{5^5}{2^5}$ **67.** $\dfrac{9^5}{8^3}$ **69.** $2^{12}x^{12}$ **71.** -6^5p^5
73. $6^5x^{10}y^{15}$ **75.** x^{21} **77.** $4w^4x^{26}y^7$ **79.** $-r^{18}s^{17}$ **81.** $\dfrac{125a^6b^{15}}{c^{18}}$
83. $25m^6p^{14}q^5$ **85.** $16x^{10}y^{16}z^{10}$ **87.** $30x^7$

Section 13.3 (pages 899–900)

1. $x^2 + 7x + 12$ **3.** $2x^3 + 7x^2 + 7x + 2$ **5.** distributive
7. $-6m^2 - 4m$ **9.** $6p - \dfrac{9}{2}p^2 + 9p^4$ **11.** $6y^5 + 4y^6 + 10y^9$
13. $12x^3 + 26x^2 + 10x + 1$ **15.** $6r^3 + 5r^2 - 12r + 4$
17. $20m^4 - m^3 - 8m^2 - 17m - 15$ **19.** $5x^4 - 13x^3 + 20x^2 + 7x + 5$
21. $m^2 + 12m + 35$ **23.** $n^2 + n - 6$ **25.** $8r^2 - 10r - 3$ **27.** $9x^2 - 4$
29. $9q^2 + 6q + 1$ **31.** $15xy - 40x + 21y - 56$ **33.** $6t^2 + 23st + 20s^2$
35. $-0.3t^2 + 0.22t + 0.24$ **37.** $x^2 - \dfrac{5}{12}x - \dfrac{1}{6}$ **39.** $\dfrac{15}{16}r^2 - \dfrac{1}{4}r - 2r^2$
41. $2x^3 + x^2 - 15x$ **43.** $6y^5 - 21y^4 - 45y^3$ **44.** $(30x + 60)$ yd^2
45. $30x + 60 = 600; \{18\}$ **46. (a)** 10 yd by 60 yd   **(b)** 140 yd
47. $2100   **48.** $1260   **49.** The answers are $x^2 - 16$, $y^2 - 4$, and
$r^2 - 49$. Each product is the difference of the square of the first term and
the square of the last term of the binomials.

Section 13.4 (pages 905–906)

1. (a) $4x^2$ **(b)** $12x$ **(c)** 9   **(d)** $4x^2 + 12x + 9$ **3.** $p^2 + 4p + 4$
5. $z^2 - 10z + 25$ **7.** $16x^2 - 24x + 9$ **9.** $4p^2 + 20pq + 25q^2$
11. $0.64t^2 + 1.12ts + 0.49s^2$ **13.** $25x^2 + 4xy + \dfrac{4}{25}y^2$
15. $9t^3 - 6t^2 + t$ **17.** $-16r^2 + 16r - 4$ **19. (a)** $49x^2$ **(b)** 0
(c) $-9y^2$ **(d)** $49x^2 - 9y^2$; Because 0 is the identity element for
addition, it is not necessary to write "$+ 0$."   **21.** $q^2 - 4$ **23.** $4w^2 - 25$
25. $100x^2 - 9y^2$ **27.** $4x^4 - 25$ **29.** $49x^2 - \dfrac{9}{49}$ **31.** $9p^3 - 49p$
33. $(a + b)^2$ **34.** a^2 **35.** $2ab$ **36.** b^2 **37.** $a^2 + 2ab + b^2$
38. They both represent the area of the entire large square.
39. 1225   **40.** $30^2 + 2(30)(5) + 5^2$ **41.** 1225   **42.** They are
equal.   **43.** $m^3 - 15m^2 + 75m - 125$ **45.** $y^3 + 6y^2 + 12y + 8$
47. $8a^3 + 12a^2 + 6a + 1$
49. $81r^4 - 216r^3t + 216r^2t^2 - 96rt^3 + 16t^4$
51. $3x^5 - 27x^4 + 81x^3 - 81x^2$
53. $-8x^6y - 32x^5y^2 - 48x^4y^3 - 32x^3y^4 - 8x^2y^5$ **55.** 512 cu. units

Section 13.5 (pages 915–916)

1. negative   **3.** negative   **5.** positive   **7.** 0   **9.** 1   **11.** 1   **13.** -1
15. 0   **17.** 0   **19.** 2   **21.** $\dfrac{1}{64}$ **23.** 16   **25.** $\dfrac{49}{36}$ **27.** $\dfrac{1}{81}$ **29.** $\dfrac{8}{15}$
31. $-\dfrac{7}{18}$ **33.** 1   **34.** $\dfrac{5^2}{5^2}$ **35.** $5^{2-2} = 5^0$ **36.** $5^0 = 1$; This supports the
definition of a 0 exponent.   **37.** $\dfrac{1}{9}$ **39.** $\dfrac{1}{6^5}$, or $\dfrac{1}{7776}$ **41.** 216   **43.** $2r^4$
45. $\dfrac{25}{64}$ **47.** $\dfrac{p^5}{q^8}$ **49.** r^9 **51.** $\dfrac{x^5}{6}$ **53.** $3y^2$ **55.** x^3 **57.** $\dfrac{yz^2}{4x^3}$
59. $a + b$ **61.** 343   **63.** $\dfrac{1}{x^2}$ **65.** $\dfrac{64x}{9}$ **67.** $\dfrac{x^2z^4}{y^2}$ **69.** $6x$
71. $\dfrac{1}{m^{10}n^5}$ **73.** $\dfrac{5}{16x^5}$ **75.** $\dfrac{36q^2}{m^4p^2}$

Summary Exercises on the Rules for Exponents (pages 917–918)

1. $\dfrac{6^{12}x^{24}}{5^{12}}$ **2.** $\dfrac{r^6s^{12}}{729t^6}$ **3.** $100,000x^7y^{14}$ **4.** $-128a^{10}b^{15}c^4$ **5.** $\dfrac{729w^3x^9}{y^{12}}$
6. $\dfrac{x^4y^6}{16}$ **7.** c^{22} **8.** $\dfrac{1}{k^4t^{12}}$ **9.** $\dfrac{11}{30}$ **10.** $y^{12}z^3$ **11.** $\dfrac{x^6}{y^5}$ **12.** 0   **13.** $\dfrac{1}{z^2}$
14. $\dfrac{9}{r^2s^2t^{10}}$ **15.** $\dfrac{300x^3}{y^3}$ **16.** $\dfrac{3}{5x^6}$ **17.** x^8 **18.** $\dfrac{y^{11}}{x^{11}}$ **19.** $\dfrac{a^6}{b^4}$ **20.** $6ab$
21. $\dfrac{61}{900}$ **22.** 1   **23.** $\dfrac{343a^6b^9}{8}$ **24.** 1   **25.** -1 **26.** 0   **27.** $\dfrac{27y^{18}}{4x^8}$
28. $\dfrac{1}{a^8b^{12}c^{16}}$ **29.** $\dfrac{x^{15}}{216z^9}$ **30.** $\dfrac{q}{8p^6r^3}$ **31.** x^6y^6 **32.** 0   **33.** $\dfrac{343}{x^{15}}$
34. $\dfrac{9}{x^6}$ **35.** $5p^{10}q^9$ **36.** $\dfrac{7}{24}$ **37.** $\dfrac{r^{14}t}{2s^2}$ **38.** 1   **39.** $8p^{10}q$ **40.** $\dfrac{1}{mn^3p^3}$
41. -1 **42.** $\dfrac{3}{40}$ **43.** Using the product rule, simplify as follows:
$(10^2)^3 = 10^{2 \cdot 3} = 10^6 = 1,000,000$.   **44.** The negative sign is not part of
the base: $-5^4 = -1(5)^4 = -1 \cdot 625 = -625$.

Section 13.6 (pages 921–922)

1. $6x^2 + 8$; 2; $3x^2 + 4$ **3.** $3x^2 + 4$; 2 (These may be reversed.); $6x^2 + 8$
5. To use the method of this section, the divisor must be just one term.
This is true of the first problem, but not the second.   **7.** $30x^3 - 10x + 5$
9. $-4m^3 + 2m^2 - 1$ **11.** $4t^4 - 2t^2 + 2t$ **13.** $a^4 - a + \dfrac{2}{a}$
15. $-2x^3 + \dfrac{2x^2}{3} - x$ **17.** $-9x^2 + 5x + 1$ **19.** $\dfrac{4x^2}{3} + x - \dfrac{2}{3x}$
21. $9r^3 - 12r^2 - 2r + 1 - \dfrac{2}{3r}$ **23.** $-m^2 + 3m - \dfrac{4}{m}$
25. $\dfrac{12}{x} - \dfrac{6}{x^2} + \dfrac{14}{x^3} - \dfrac{10}{x^4}$ **27.** $-4b^2 + 3ab - \dfrac{5}{a}$ **29.** $6x - 2 + \dfrac{1}{x}$
31. $15x^5 - 35x^4 + 35x^3$ **33.** 1423
34. $(1 \times 10^3) + (4 \times 10^2) + (2 \times 10^1) + (3 \times 10^0)$
35. $x^3 + 4x^2 + 2x + 3$ **36.** They are similar in that the coefficients of
the powers of ten are equal to the coefficients of the powers of x. They are
different in that one is a number while the other is a polynomial. They are
equal if $x = 10$.

Section 13.7 (pages 927–928)

1. The divisor is $2x + 5$; the quotient is $2x^3 - 4x^2 + 3x + 2$.   **3.** Divide
$12m^2$ by $2m$ to get $6m$.   **5.** $x + 2$ **7.** $2y - 5$ **9.** $p - 4 + \dfrac{44}{p + 6}$

11. $r - 5$ **13.** $2a - 14 + \frac{74}{2a+3}$ **15.** $4x^2 - 7x + 3$ **17.** $3y^2 - 2y + 2$ **19.** $2x^2 - 2x + 3 + \frac{-1}{x+1}$ **21.** $3k - 4 + \frac{2}{k^2-2}$ **23.** $x^2 + 1$ **25.** $x^2 + 1$ **27.** $2p^2 - 5p + 4 + \frac{6}{3p^2+1}$ **29.** $x^3 + 6x - 7$ **31.** $2x^2 + \frac{3}{5}x + \frac{1}{5}$ **33.** $(x^2 + x - 3)$ units **35.** 33 **36.** 33 **37.** They are the same. **38.** The answers should agree.

Section 13.8 (pages 933–934)

1. 6.1309×10^9; 5.8689×10^9 **3.** in scientific notation **5.** not in scientific notation; 5.6×10^6 **7.** not in scientific notation; 4×10^{-3} **9.** not in scientific notation; 8×10^1 **11.** A number is written in scientific notation if it is the product of a number whose absolute value is between 1 and 10 (inclusive of 1) and a power of 10. **13.** 5.876×10^9 **15.** 8.235×10^4 **17.** 7×10^{-6} **19.** -2.03×10^{-3} **21.** 750,000 **23.** 5,677,000,000,000 **25.** 1,000,000,000,000 **27.** -6.21 **29.** 0.00078 **31.** 0.000000005134 **33.** 6×10^{11}; 600,000,000,000 **35.** 1.5×10^7; 15,000,000 **37.** 8×10^{-3}; 0.008 **39.** 2.4×10^2; 240 **41.** 6.3×10^{-2}; 0.063 **43.** 6.426×10^4; 64,260 **45.** 3×10^{-4}; 0.0003 **47.** 4×10^1; 40 **49.** 1.3×10^{-5}; 0.000013 **51.** 5×10^2; 500 **53.** 2.6×10^{-3}; 0.0026 **55.** 7.205×10^{-6}; 0.000007205 **57.** 1.5×10^{17} mi **59.** \$3554

Chapter 13 Review Exercises (pages 939–942)

1. $22m^2$; degree 2; monomial **2.** $p^3 - p^2 + 4p + 2$; degree 3; none of these **3.** already in descending powers; degree 5; none of these **4.** $-8y^5 - 7y^4 + 9y$; degree 5; trinomial **5.** $-5a^3 + 4a^2$ **6.** $2r^3 - 3r^2 + 9r$ **7.** $11y^2 - 10y + 9$ **8.** $-13k^4 - 15k^2 - 4k - 6$ **9.** $10m^3 - 6m^2 - 3$ **10.** $-y^2 - 4y + 26$ **11.** $10p^2 - 3p - 11$ **12.** $7r^4 - 4r^3 - 1$ **13.** 4^{11} **14.** -5^{11} **15.** $-72x^7$ **16.** $10x^{14}$ **17.** 19^5x^5 **18.** -4^7y^7 **19.** $5p^4t^4$ **20.** $\frac{7^6}{5^6}$ **21.** $27x^6y^9$ **22.** t^{42} **23.** $36x^{16}y^4z^{16}$ **24.** $\frac{8m^9n^3}{p^6}$ **25.** $125x^6$ **26.** The product rule for exponents does not apply here because we want the sum of 7^2 and 7^4, not their product. **27.** $10x^2 + 70x$ **28.** $-6p^5 + 15p^4$ **29.** $6r^3 + 8r^2 - 17r + 6$ **30.** $8y^3 + 27$ **31.** $5p^5 - 2p^4 - 3p^3 + 25p^2 + 15p$ **32.** $x^2 + 3x - 18$ **33.** $6k^2 - 9k - 6$ **34.** $12p^2 - 48pq + 21q^2$ **35.** $2m^4 + 5m^3 - 16m^2 - 28m + 9$ **36.** $a^2 + 8a + 16$ **37.** $9p^2 - 12p + 4$ **38.** $4r^2 + 20rs + 25s^2$ **39.** $r^3 + 6r^2 + 12r + 8$ **40.** $8x^3 - 12x^2 + 6x - 1$ **41.** $4z^2 - 49$ **42.** $36m^2 - 25$ **43.** $25a^2 - 36b^2$ **44.** $4x^4 - 25$ **45.** three; two **46.** $(a+b)^2 = (a+b)(a+b) = a^2 + 2ab + b^2$. The term $2ab$ is not in $a^2 + b^2$. **47.** 2 **48.** $\frac{1}{32}$ **49.** $\frac{25}{36}$ **50.** $-\frac{3}{16}$ **51.** 36 **52.** x^2 **53.** $\frac{1}{p^{12}}$ **54.** r^4 **55.** 2^8 **56.** $\frac{1}{9^6}$ **57.** 5^8 **58.** $\frac{1}{8^{12}}$ **59.** $\frac{1}{m^2}$ **60.** y^7 **61.** r^{13} **62.** $25m^6$ **63.** $\frac{y^{12}}{8}$ **64.** $\frac{1}{a^3b^5}$ **65.** $72r^5$ **66.** $\frac{8n^{10}}{3m^{13}}$ **67.** $\frac{5y^2}{3}$ **68.** $-2x^2y$ **69.** $-y^3 + 2y - 3$ **70.** $p - 3 + \frac{5}{2p}$ **71.** $-x^9 + 2x^8 - 4x^3 + 7x$ **72.** $-2m^2n + mn^2 + \frac{6n^3}{5}$ **73.** $2r + 7$ **74.** $4m + 3 + \frac{5}{3m-5}$ **75.** $2a + 1 + \frac{-8a+12}{5a^2-3}$ **76.** $k^2 + 2k + 4 + \frac{-2k-12}{2k^2+1}$ **77.** 4.8×10^7 **78.** 2.8988×10^{10} **79.** 6.5×10^{-5} **80.** 8.24×10^{-8} **81.** 24,000 **82.** 78,300,000 **83.** 0.000000897 **84.** 0.00000000000995 **85.** 8×10^2; 800 **86.** 4×10^6; 4,000,000 **87.** 2.5×10^{-2}; 0.025 **88.** 1×10^{-2}; 0.01 **89.** 2.796×10^{10} calculations; 1.6776×10^{12} calculations **90.** about 3.3 **91.** 0 **92.** $\frac{243}{p^3}$ **93.** $\frac{1}{49}$ **94.** $49 - 28k + 4k^2$ **95.** $y^2 + 5y + 1$ **96.** $\frac{1296r^8s^4}{625}$ **97.** $-8m^7 - 10m^6 - 6m^5$ **98.** 32 **99.** $5xy^3 - \frac{8y^2}{5} + 3x^2y$ **100.** $\frac{r^2}{6}$ **101.** $8x^3 + 12x^2y + 6xy^2 + y^3$ **102.** $\frac{3}{4}$ **103.** $a^3 - 2a^2 - 7a + 2$ **104.** $8y^3 - 9y^2 + 5$ **105.** $10r^2 + 21r - 10$ **106.** $144a^2 - 1$ **107.** $2x^2 + x - 6$ **108.** $20x^4 + 8x^2$; $25x^8 + 20x^6 + 4x^4$

Chapter 13 Test (pages 943–944)

1. $4t^4 + t^3 - 6t^2 - t$ **2.** $-2y^2 - 9y + 17$ **3.** $-12t^2 + 5t + 8$ **4.** -32 **5.** $\frac{216}{m^6}$ **6.** $-27x^5 + 18x^4 - 6x^3 + 3x^2$ **7.** $2r^3 + r^2 - 16r + 15$ **8.** $t^2 - 5t - 24$ **9.** $8x^2 + 2xy - 3y^2$ **10.** $25x^2 - 20xy + 4y^2$ **11.** $100v^2 - 9w^2$ **12.** $x^3 + 3x^2 + 3x + 1$ **13.** $12x + 36$; $9x^2 + 54x + 81$ **14.** $\frac{1}{625}$ **15.** 2 **16.** $\frac{7}{12}$ **17.** 8^5 **18.** x^2y^6 **19.** $4y^2 - 3y + 2 + \frac{5}{y}$ **20.** $-3xy^2 + 2x^3y^2 + 4y^2$ **21.** $2x + 9$ **22.** $3x^2 + 6x + 11 + \frac{26}{x-2}$ **23.** **(a)** 3.44×10^{11} **(b)** 5.57×10^{-6} **24.** **(a)** 29,600,000 **(b)** 0.0000000607 **25.** 5.89×10^{15} mi

Chapter 14 Factoring and Applications

Section 14.1 (pages 953–954)

1. 4 **3.** 4 **5.** 6 **7.** 1 **9.** 8 **11.** $10x^3$ **13.** xy^2 **15.** 6 **17.** $3m^2$ **19.** $2z^4$ **21.** $2mn^4$ **23.** $y + 2$ **25.** $a - 2$ **27.** $2 + 3xy$ **29.** $x(x - 4)$ **31.** $3t(2t + 5)$ **33.** $\frac{1}{4}d(d - 3)$ **35.** $-6x^2(2x + 1)$ **37.** $5y^6(13y^4 + 7)$ **39.** no common factor (except 1) **41.** $8m^2n^2(n + 3)$ **43.** $-2x(2x^2 - 5x + 3)$ **45.** $13y^2(y^6 + 2y^2 - 3)$ **47.** $9qp^3(5q^3p^2 + 4p^3 + 9q)$ **49.** $(x + 2)(c + d)$ **51.** $(2a + b)(a^2 - b)$ **53.** $(p + 4)(q - 1)$ **55.** $(5 + n)(m + 4)$ **57.** $(2y - 7)(3x + 4)$ **59.** $(y + 3)(3x + 1)$ **61.** $(z + 2)(7z - a)$ **63.** $(3r + 2y)(6r - x)$ **65.** $(w + 1)(w^2 + 9)$ **67.** $(a + 2)(3a^2 - 2)$ **69.** $(4m - p^2)(4m^2 - p)$ **71.** $(y + 3)(y + x)$ **73.** $(z - 2)(2z - 3w)$ **75.** commutative property **76.** $2x(y - 4) - 3(y - 4)$ **77.** No, because it is not a product. It is the difference between $2x(y - 4)$ and $3(y - 4)$. **78.** $(2x - 3)(y - 4)$; yes

Section 14.2 (pages 959–960)

1. a and b must have different signs. **3.** A prime polynomial is one that cannot be factored using only integers in the factors. **5.** 1 and 12, -1 and -12, 2 and 6, -2 and -6, 3 and 4, -3 and -4; The pair with a sum of 7 is 3 and 4. **7.** 1 and -24, -1 and 24, 2 and -12, -2 and 12, 3 and -8, -3 and 8, 4 and -6, -4 and 6; The pair with a sum of -5 is 3 and -8. **9.** C **11.** $x + 11$ **13.** $x - 8$ **15.** $y - 5$ **17.** $x + 11$ **19.** $y - 9$

21. $(y + 8)(y + 1)$ **23.** $(b + 3)(b + 5)$ **25.** $(m + 5)(m - 4)$
27. $(x + 8)(x - 5)$ **29.** $(y - 5)(y - 3)$ **31.** $(z - 8)(z - 7)$
33. $(r - 6)(r + 5)$ **35.** $(a - 12)(a + 4)$ **37.** prime
39. $(r + 2a)(r + a)$ **41.** $(x + y)(x + 3y)$ **43.** $(t + 2z)(t - 3z)$
45. $(v - 5w)(v - 6w)$ **47.** $4(x + 5)(x - 2)$ **49.** $2t(t + 1)(t + 3)$
51. $-2x^4(x - 3)(x + 7)$ **53.** $a^3(a + 4b)(a - b)$
55. $mn(m - 6n)(m - 4n)$ **57.** The factored form $(2x + 4)(x - 3)$
is incorrect because $2x + 4$ has a common factor of 2, which must be
factored out for the trinomial to be *completely* factored.

Section 14.3 (pages 963–964)

1. $(m + 6)(m + 2)$ **3.** $(a + 5)(a - 2)$ **5.** $(2t + 1)(5t + 2)$
7. $(3z - 2)(5z - 3)$ **9.** $(2s - t)(4s + 3t)$ **11.** $(3a + 2b)(5a + 4b)$
13. B **15.** (a) 2; 12; 24; 11 (b) 3; 8 (Order is irrelevant.)
(c) $3m$; $8m$ (d) $2m^2 + 3m + 8m + 12$ (e) $(2m + 3)(m + 4)$
(f) $(2m + 3)(m + 4) = 2m^2 + 11m + 12$ **17.** $(2x + 1)(x + 3)$
19. $(4r - 3)(r + 1)$ **21.** $(4m + 1)(2m - 3)$ **23.** $(3m + 1)(7m + 2)$
25. $(2b + 1)(3b + 2)$ **27.** $(4y - 3)(3y - 1)$ **29.** $3(4x - 1)(2x - 3)$
31. $2m(m - 4)(m + 5)$ **33.** $-4z^3(z - 1)(8z + 3)$
35. $(3p + 4q)(4p - 3q)$ **37.** $(3a - 5b)(2a + b)$ **39.** $(5 - x)(1 - x)$
41. The student stopped too soon. He needs to factor out the common
factor $4x - 1$ to get $(4x - 1)(4x - 5)$ as the correct answer.

Section 14.4 (pages 969–970)

1. B **3.** A **5.** A **7.** $2a + 5b$ **9.** $x^2 + 3x - 4$; $x + 4$, $x - 1$, or
$x - 1$, $x + 4$ **11.** $2z^2 - 5z - 3$; $2z + 1$, $z - 3$, or $z - 3$, $2z + 1$
13. The binomial $2x - 6$ cannot be a factor because it has a common
factor of 2, but the polynomial does not. **15.** $(3a + 7)(a + 1)$
17. $(2y + 3)(y + 2)$ **19.** $(3m - 1)(5m + 2)$ **21.** $(3s - 1)(4s + 5)$
23. $(5m - 4)(2m - 3)$ **25.** $(4w - 1)(2w - 3)$
27. $(4y + 1)(5y - 11)$ **29.** prime **31.** $2(5x + 3)(2x + 1)$
33. $-q(5m + 2)(8m - 3)$ **35.** $3n^2(5n - 3)(n - 2)$
37. $-y^2(5x - 4)(3x + 1)$ **39.** $(5a + 3b)(a - 2b)$
41. $(4s + 5t)(3s - t)$ **43.** $m^4n(3m + 2n)(2m + n)$
45. $-1(x + 7)(x - 3)$ **47.** $-1(3x + 4)(x - 1)$
49. $-1(a + 2b)(2a + b)$ **51.** $5 \cdot 7$ **52.** $(-5)(-7)$
53. The product of $3x - 4$ and $2x - 1$ is $6x^2 - 11x + 4$.
54. The product of $4 - 3x$ and $1 - 2x$ is $6x^2 - 11x + 4$.
55. The factors in Exercise 53 are the opposites of the factors
in Exercise 54. **56.** $(3 - 7t)(5 - 2t)$

Section 14.5 (pages 975–976)

1. 1; 4; 9; 16; 25; 36; 49; 64; 81; 100; 121; 144; 169; 196; 225; 256;
289; 324; 361; 400 **3.** 2 **5.** $(y + 5)(y - 5)$ **7.** $\left(p + \frac{1}{3}\right)\left(p - \frac{1}{3}\right)$
9. prime **11.** $(3r + 2)(3r - 2)$ **13.** $\left(2m + \frac{3}{5}\right)\left(2m - \frac{3}{5}\right)$
15. $4(3x + 2)(3x - 2)$ **17.** $(14p + 15)(14p - 15)$
19. $(4r + 5a)(4r - 5a)$ **21.** prime **23.** $(p^2 + 7)(p^2 - 7)$
25. $(x^2 + 1)(x + 1)(x - 1)$ **27.** $(p^2 + 16)(p + 4)(p - 4)$
29. The teacher was justified, because it was not factored *completely*;
$x^2 - 9$ can be factored as $(x + 3)(x - 3)$. The complete factored form
is $(x^2 + 9)(x + 3)(x - 3)$. **31.** No, it is not a perfect square since the

middle term would have to be $30y$. **33.** $(w + 1)^2$ **35.** $(x - 4)^2$
37. $\left(t + \frac{1}{2}\right)^2$ **39.** $(x - 0.5)^2$ **41.** $2(x + 6)^2$ **43.** $(4x - 5)^2$
45. $(7x - 2y)^2$ **47.** $(8x + 3y)^2$ **49.** $-2h(5h - 2y)^2$
51. $(2x + 3)(5x - 2)$ **52.** $5x - 2$ **53.** Yes. We saw in Exercise 51
that $(2x + 3)(5x - 2)$ is the factored form of $10x^2 + 11x - 6$.
54. The quotient is $x^2 + x + 1$, so $x^3 - 1$ factors as $(x - 1)(x^2 + x + 1)$.

Summary Exercises on Factoring (pages 977–980)

1. F **2.** G **3.** A **4.** B **5.** D **6.** H **7.** C **8.** E **9.** H **10.** D
11. $8m^3(4m^6 + 2m^2 + 3)$ **12.** $2(m + 3)(m - 8)$
13. $7k(2k + 5)(k - 2)$ **14.** prime **15.** $(6z + 1)(z + 5)$
16. $(m + n)(m - 4n)$ **17.** $(7z + 4y)(7z - 4y)$
18. $10nr(10nr + 3r^2 - 5n)$ **19.** $4x(4x + 5)$ **20.** $(4 + m)(5 + 3n)$
21. $(5y - 6z)(2y + z)$ **22.** $(y^2 + 9)(y + 3)(y - 3)$
23. $(m - 3)(m + 5)$ **24.** $(2y + 1)(3y - 4)$ **25.** $8z(4z - 1)(z + 2)$
26. $(p - 1.2)^2$ **27.** $(z - 6)^2$ **28.** $(3m + 8)(3m - 8)$
29. $(y - 6k)(y + 2k)$ **30.** $(4z - 1)^2$ **31.** $6(y - 2)(y + 1)$
32. $\left(x + \frac{1}{4}\right)^2$ **33.** $(p - 6)(p - 11)$ **34.** $(a + 8)(a + 9)$ **35.** prime
36. $3(6m - 1)^2$ **37.** $(z + 2a)(z - 5a)$ **38.** $(2a + 1)(a^2 - 7)$
39. $(2k - 3)^2$ **40.** $(a - 7b)(a + 4b)$ **41.** $(4r + 3m)^2$
42. $(3k - 2)(k + 2)$ **43.** prime **44.** $(a^2 + 25)(a + 5)(a - 5)$
45. $4(2k - 3)^2$ **46.** $(4k + 1)(2k - 3)$ **47.** $6y^4(3y + 4)(2y - 5)$
48. $5z(z - 2)(z - 7)$ **49.** $(8p - 1)(p + 3)$ **50.** $(4k - 3h)(2k + h)$
51. $6(3m + 2z)(3m - 2z)$ **52.** $(2k - 5z)^2$ **53.** $2(3a - 1)(a + 2)$
54. $(3h - 2g)(5h + 7g)$ **55.** $7(2a + 3b)(2a - 3b)$
56. $(5z - 6)(2z + 1)$ **57.** $5m^2(5m - 13n)(5m - 3n)$
58. $(3y - 1)(3y + 5)$ **59.** $(3u + 11v)^2$ **60.** prime
61. $9p^8(3p + 7)(p - 4)$ **62.** $5(2m - 3)(m + 4)$ **63.** $(2 - q)(2 - 3p)$
64. $\left(k + \frac{8}{11}\right)\left(k - \frac{8}{11}\right)$ **65.** $4(4p + 5m)(4p - 5m)$
66. $(m + 4)(m^2 - 6)$ **67.** $(10a + 9y)(10a - 9y)$
68. $(8a - b)(a + 3b)$ **69.** $(a + 4)^2$ **70.** $(2y + 5)(2y - 5)$ **71.** prime
72. $-3x(x + 2y)(x - 2y)$ **73.** $(5a - 7b)^2$ **74.** $8(t^2 + 1)(t + 1)(t - 1)$
75. $-4(x - 3y)^2$ **76.** $25(2a + b)(2a - b)$ **77.** $-2(x - 9)(x - 4)$
78. $(m + 3)(2m - 5n)$ **79.** $2(2x + 5)(3x - 2)$ **80.** $(y^2 + 5)(y^4 - 3)$
81. $(y + 0.8)(y - 0.8)$ **82.** $6p(2p + 1)(p - 5)$

Section 14.6 (pages 987–988)

1. $\{-5, 2\}$ **3.** $\left\{3, \frac{7}{2}\right\}$ **5.** $\left\{-\frac{5}{6}, 0\right\}$ **7.** $\left\{0, \frac{4}{3}\right\}$ **9.** $\left\{-\frac{1}{2}, \frac{1}{6}\right\}$
11. $\{9\}$ **13.** Set each *variable* factor equal to 0, to get $2x = 0$ or
$3x - 4 = 0$. The solution set is $\left\{0, \frac{4}{3}\right\}$. **15.** $\{-2, -1\}$ **17.** $\{1, 2\}$
19. $\{-8, 3\}$ **21.** $\{-1, 3\}$ **23.** $\{-2, -1\}$ **25.** $\{-4\}$ **27.** $\left\{-2, \frac{1}{3}\right\}$
29. $\left\{-\frac{4}{3}, \frac{1}{2}\right\}$ **31.** $\left\{-\frac{2}{3}\right\}$ **33.** $\{-3, 3\}$ **35.** $\left\{-\frac{7}{4}, \frac{7}{4}\right\}$ **37.** $\{-11, 11\}$
39. $\{0, 7\}$ **41.** $\left\{0, \frac{1}{2}\right\}$ **43.** $\{2, 5\}$ **45.** $\left\{-4, \frac{1}{2}\right\}$ **47.** $\left\{-12, \frac{11}{2}\right\}$
49. $\{-2, 0, 2\}$ **51.** $\left\{-\frac{7}{3}, 0, \frac{7}{3}\right\}$ **53.** $\left\{-\frac{5}{2}, \frac{1}{3}, 5\right\}$ **55.** $\left\{-\frac{7}{2}, -3, 1\right\}$
57. (a) 64; 144; 4; 6 (b) No time has elapsed, so the object hasn't fallen
(been released) yet.

Section 14.7 (pages 995–1000)

1. Read; variable; equation; Solve; answer; Check; original

3. *Step 3:* $45 = (2x + 1)(x + 1)$; *Step 4:* $x = 4$ or $x = -\frac{11}{2}$;

Step 5: base: 9 units; height: 5 units; *Step 6:* $9 \cdot 5 = 45$

5. *Step 3:* $192 = 4x(x + 2)$; *Step 4:* $x = 6$ or $x = -8$; *Step 5:* length: 8 units; width: 6 units; *Step 6:* $8 \cdot 6 \cdot 4 = 192$ **7.** length: 14 cm; width: 12 cm **9.** length: 15 in.; width: 12 in. **11.** height: 13 in.; width: 10 in.

13. mirror: 7 ft; painting: 9 ft **15.** 20, 21 **17.** $-3, -2$ or 4, 5

19. $-3, -1$ or 7, 9 **21.** $-2, 0, 2$ or 6, 8, 10 **23.** 12 cm **25.** 12 mi

27. 8 ft **29. (a)** 1 sec **(b)** $\frac{1}{2}$ sec and $1\frac{1}{2}$ sec **(c)** 3 sec

(d) The negative solution, -1, does not make sense since t represents time, which cannot be negative. **31. (a)** 46 million; The result using the model is a little more than 44 million, the actual number for 1996.

(b) 14 **(c)** 184 million; The result is a little more than 182 million, the actual number for 2004. **(d)** 318 million **32.** \$58.6 billion; 16%

33. 2003: \$522.1 billion; 2004: \$610.8 billion; 2005: \$699.5 billion.

34. The answer using the linear equation is close to the actual data for 2004, but not for the other years. **35.** 2003: \$503.9 billion; 2004: \$601.8 billion; 2005: \$718.2 billion **36.** The answers in Exercise 35 are fairly close to the actual data. The quadratic equation models the data better.

37. $(1, 365.1), (2, 423.7), (3, 496.9), (4, 612.1), (5, 714.4)$

38.

U.S. TRADE DEFICIT
(Goods and Services)

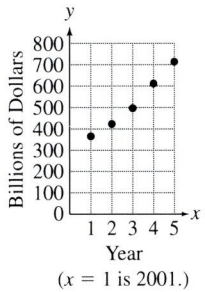

$(x = 1$ is 2001.)

39. \$853.1 billion **40. (a)** The actual deficit is quite a bit less than the estimate. **(b)** No, data for later years might not follow the same pattern.

Chapter 14 Review Exercises (pages 1005–1010)

1. $15(t + 3)$ **2.** $30z(2z^2 + 1)$ **3.** $11x^2(4x + 5)$

4. $50m^2n^2(2n - mn^2 + 3)$ **5.** $(x - 4)(2y + 3)$ **6.** $(2y + 3)(3y + 2x)$

7. $(x + 3)(x + 7)$ **8.** $(y - 5)(y - 8)$ **9.** $(q + 9)(q - 3)$

10. $(r - 8)(r + 7)$ **11.** $(r + 8s)(r - 12s)$ **12.** $(p + 12q)(p - 10q)$

13. $-8p(p + 2)(p - 5)$ **14.** $3x^2(x + 2)(x + 8)$

15. $(m + 3n)(m - 6n)$ **16.** $(y - 3z)(y - 5z)$ **17.** $p^5(p - 2q)(p + q)$

18. $-3r^3(r + 3s)(r - 5s)$ **19.** prime **20.** $3(x^2 + 2x + 2)$

21. r and $6r$, $2r$ and $3r$ **22.** Factor out z. **23.** $(2k - 1)(k - 2)$

24. $(3r - 1)(r + 4)$ **25.** $(3r + 2)(2r - 3)$ **26.** $(5z + 1)(2z - 1)$

27. prime **28.** $4x^3(3x - 1)(2x - 1)$ **29.** $-3(x + 2)(2x - 5)$

30. $rs(5r + 6s)(2r + s)$ **31.** $-5y(3y + 2)(2y - 1)$ **32.** prime

33. $-mn(3m + 5)(m - 8)$ **34.** $(2a - 5b)(7a + 4b)$ **35.** B **36.** D

37. $(n + 8)(n - 8)$ **38.** $(5b + 11)(5b - 11)$ **39.** $(7y + 5w)(7y - 5w)$

40. $36(2p + q)(2p - q)$ **41.** prime **42.** $\left(x + \frac{7}{10}\right)\left(x - \frac{7}{10}\right)$

43. $(z + 5)^2$ **44.** $(r - 6)^2$ **45.** $(3t - 7)^2$ **46.** $(4m + 5n)^2$

47. $6x(3x - 2)^2$ **48.** $\left(x + \frac{1}{6}\right)^2$ **49.** $\left\{-\frac{3}{4}, 1\right\}$ **50.** $\{-7, -3, 4\}$

51. $\left\{0, \frac{5}{2}\right\}$ **52.** $\{-3, -1\}$ **53.** $\{1, 4\}$ **54.** $\{3, 5\}$ **55.** $\left\{-\frac{4}{3}, 5\right\}$

56. $\left\{-\frac{8}{9}, \frac{8}{9}\right\}$ **57.** $\{0, 8\}$ **58.** $\{-1, 6\}$ **59.** $\{7\}$ **60.** $\{6\}$

61. $\left\{-\frac{2}{5}, -2, -1\right\}$ **62.** $\{-3, 3\}$ **63.** $\left\{-\frac{3}{7}, 0, \frac{3}{7}\right\}$ **64.** $\left\{-\frac{3}{4}, -\frac{1}{2}, \frac{1}{3}\right\}$

65. $\left\{\frac{9}{5}\right\}$ **66.** $\{-3, 10\}$ **67.** length: 10 ft; width: 4 ft **68.** 5 ft

69. length: 6 m; width: 2 m **70.** length: 6 m; height: 5 m **71.** 6, 7 or $-5, -4$ **72.** 26 mi **73.** 112 ft **74.** 192 ft **75.** 256 ft **76.** after 8 sec

77. (a) \$4.49 billion; The answer using the model is very close.

(b) \$7.19 billion **78.** D **79.** The factor $2x + 8$ has a common factor of 2. The complete factored form is $2(x + 4)(3x - 4)$.

80. $(z - x)(z - 10x)$ **81.** $(3k + 5)(k + 2)$ **82.** $(3m + 4p)(5m - 4)$

83. $(y^2 + 25)(y + 5)(y - 5)$ **84.** $3m(2m + 3)(m - 5)$

85. $8abc(3b^2c - 7ac^2 + 9ab)$ **86.** prime

87. $6xyz(2xz^2 + 2y - 5x^2yz^3)$ **88.** $2a^3(a + 2)(a - 6)$

89. $-(2r + 3q)(6r - 5q)$ **90.** $(10a + 3)(10a - 3)$ **91.** $(7t + 4)^2$

92. $\{0, 7\}$ **93.** $\{-5, 2\}$ **94.** $\left\{-\frac{2}{5}\right\}$ **95.** 15 m, 36 m, 39 m

96. length: 6 m; width: 4 m **97.** $-5, -4, -3$ or 5, 6, 7

98. (a) 256 ft **(b)** 1024 ft **99.** width: 10 m; length: 17 m **100.** 6 m

101. (a) 611 thousand vehicles **(b)** The estimate may be unreliable because the conditions that prevailed in the years 2001–2005 may have changed, causing either a greater increase or a greater decrease in the numbers of alternative-fueled vehicles.

Chapter 14 Test (pages 1011–1012)

1. D **2.** $6x(2x - 5)$ **3.** $m^2n(2mn + 3m - 5n)$ **4.** $(2x + y)(a - b)$

5. $(x - 7)(x - 2)$ **6.** $(2x + 3)(x - 1)$ **7.** $(3x + 1)(2x - 7)$

8. $3(x + 1)(x - 5)$ **9.** $(5z - 1)(2z - 3)$ **10.** prime **11.** prime

12. $(y + 7)(y - 7)$ **13.** $(9a + 11b)(9a - 11b)$ **14.** $(x + 8)^2$

15. $(2x - 7y)^2$ **16.** $-2(x + 1)^2$ **17.** $3t^2(2t + 9)(t - 4)$ **18.** prime

19. $4t(t + 4)^2$ **20.** $(x^2 + 9)(x + 3)(x - 3)$ **21.** $\{-3, 9\}$ **22.** $\left\{\frac{1}{2}, 6\right\}$

23. $\left\{-\frac{2}{5}, \frac{2}{5}\right\}$ **24.** $\{10\}$ **25.** $\{0, 3\}$ **26.** $\left\{-8, -\frac{5}{2}, \frac{1}{3}\right\}$

27. 6 ft by 9 ft **28.** $-2, -1$ **29.** 17 ft **30.** 49 million

Chapter 15 Rational Expressions and Applications

Section 15.1 (pages 1021–1022)

1. (a) $3; -5$ **(b)** $q; -1$ **3.** A rational expression is a quotient of polynomials, such as $\frac{x + 3}{x^2 - 4}$. **5.** $y \neq 0$ **7.** $x \neq -6$ **9.** $x \neq \frac{5}{3}$

11. $m \neq -3, m \neq 2$ **13.** never undefined **15.** never undefined

17. (a) 1 **(b)** $\frac{17}{12}$ **19. (a)** 0 **(b)** -1 **21. (a)** $\frac{9}{5}$ **(b)** undefined

23. (a) $\frac{2}{7}$ **(b)** $\frac{13}{3}$ **25.** $3r^2$ **27.** $\frac{2}{5}$ **29.** $\frac{x - 1}{x + 1}$ **31.** $\frac{7}{5}$ **33.** $m - n$

35. $\dfrac{3(2m+1)}{4}$ 37. $\dfrac{3m}{5}$ 39. $\dfrac{3r-2s}{3}$ 41. $\dfrac{x+1}{x-1}$ 43. $\dfrac{z-3}{z+5}$

45. $\dfrac{a+b}{a-b}$ 47. -1 49. $-(m+1)$ 51. -1

Answers may vary in Exercises 53–57.

53. $\dfrac{-(x+4)}{x-3}, \dfrac{-x-4}{x-3}, \dfrac{x+4}{-(x-3)}, \dfrac{x+4}{-x+3}$

55. $\dfrac{-(2x-3)}{x+3}, \dfrac{-2x+3}{x+3}, \dfrac{2x-3}{-(x+3)}, \dfrac{2x-3}{-x-3}$

57. $\dfrac{-(3x-1)}{5x-6}, \dfrac{-3x+1}{5x-6}, \dfrac{3x-1}{-(5x-6)}, \dfrac{3x-1}{-5x+6}$ 59. x^2+3

Section 15.2 (pages 1027–1028)

1. (a) B (b) D (c) C (d) A 3. $\dfrac{4m}{3}$ 5. $\dfrac{40y^2}{3}$ 7. $\dfrac{2}{c+d}$

9. $\dfrac{16q}{3p^3}$ 11. $\dfrac{7}{r^2+rp}$ 13. $\dfrac{z^2-9}{z^2+7z+12}$ 15. 5 17. $-\dfrac{3}{2t^4}$ 19. $\dfrac{1}{4}$

21. To multiply two rational expressions, multiply the numerators and multiply the denominators. Write the answer in lowest terms.

23. $\dfrac{10}{9}$ 25. $-\dfrac{3}{4}$ 27. -1 29. $\dfrac{9(m-2)}{-(m+4)}$, or $\dfrac{-9(m-2)}{m+4}$ 31. $\dfrac{p+4}{p+2}$

33. $\dfrac{(k-1)^2}{(k+1)(2k-1)}$ 35. $\dfrac{4k-1}{3k-2}$ 37. $\dfrac{m+4p}{m+p}$ 39. $\dfrac{10}{x+10}$ 41. $\dfrac{5xy^2}{4q}$

Section 15.3 (pages 1033–1034)

1. C 3. C 5. 30 7. x^7 9. $72q$ 11. $84r^5$ 13. $2^3 \cdot 3 \cdot 5$
15. The least common denominator is their product. 17. $28m^2(3m-5)$
19. $30(b-2)$ 21. $c-d$ or $d-c$ 23. $k(k+5)(k-2)$

25. $(p+3)(p+5)(p-6)$ 27. $\dfrac{20}{55}$ 29. $\dfrac{-45}{9k}$ 31. $\dfrac{26y^2}{80y^3}$

33. $\dfrac{35t^2r^3}{42r^4}$ 35. $\dfrac{20}{8(m+3)}$ 37. $\dfrac{8t}{12-6t}$ 39. $\dfrac{14(z-2)}{z(z-3)(z-2)}$

41. $\dfrac{2(b-1)(b+2)}{b^3+3b^2+2b}$

Section 15.4 (pages 1041–1044)

1. E 3. C 5. B 7. G 9. $\dfrac{11}{m}$ 11. b 13. $\dfrac{4}{y+4}$ 15. $\dfrac{m-1}{m+1}$

17. x 19. $y-6$ 21. Combine the numerators and keep the same denominator. For example, $\dfrac{3x+2}{x-6} + \dfrac{-2x-8}{x-6} = \dfrac{x-6}{x-6}$. Then write in lowest terms: $\dfrac{x-6}{x-6} = 1$. 23. $\dfrac{3z+5}{15}$ 25. $\dfrac{10-7r}{14}$ 27. $\dfrac{-3x-2}{4x}$

29. $\dfrac{57}{20x}$ 31. $\dfrac{x+1}{2}$ 33. $\dfrac{5x+9}{6x}$ 35. $\dfrac{3x+3}{x(x+3)}$ 37. $\dfrac{-k-10}{k(k+5)}$

39. $\dfrac{x+4}{x+2}$ 41. $\dfrac{x^2+6x-8}{(x-2)(x+2)}$ 43. $\dfrac{3}{t}$ 45. $m-2$ or $2-m$

47. $\dfrac{-2}{x-5}$, or $\dfrac{2}{5-x}$ 49. -4 51. $\dfrac{-5}{x-y^2}$, or $\dfrac{5}{y^2-x}$

53. $\dfrac{x+y}{5x-3y}$, or $\dfrac{-x-y}{3y-5x}$ 55. $\dfrac{-6}{4p-5}$, or $\dfrac{6}{5-4p}$ 57. $\dfrac{-(m+n)}{2(m-n)}$

59. $\dfrac{-x^2+6x+11}{(x+3)(x-3)(x+1)}$ 61. $\dfrac{-5q^2-13q+7}{(3q-2)(q+4)(2q-3)}$

63. $\dfrac{9r+2}{r(r+2)(r-1)}$ 65. $\dfrac{2x^2+6xy+8y^2}{(x+y)(x+y)(x+3y)}$, or $\dfrac{2x^2+6xy+8y^2}{(x+y)^2(x+3y)}$

67. $\dfrac{15r^2+10ry-y^2}{(3r+2y)(6r-y)(6r+y)}$ 69. (a) $\dfrac{9k^2+6k+26}{5(3k+1)}$ (b) $\dfrac{1}{4}$

Section 15.5 (pages 1051–1052)

1. (a) $6; \dfrac{1}{6}$ (b) $12; \dfrac{3}{4}$ (c) $\dfrac{1}{6} \div \dfrac{3}{4}$ (d) $\dfrac{2}{9}$ 3. -6 5. $\dfrac{1}{pq}$ 7. $\dfrac{1}{xy}$

9. $\dfrac{2a^2b}{3}$ 11. $\dfrac{m(m+2)}{3(m-4)}$ 13. $\dfrac{2}{x}$ 15. $\dfrac{8}{x}$ 17. $\dfrac{a^2-5}{a^2+1}$ 19. $\dfrac{3(p+2)}{2(2p+3)}$

21. $\dfrac{40-12p}{85p}$ 23. $\dfrac{t(t-2)}{4}$ 25. $\dfrac{-k}{2+k}$ 27. $\dfrac{2x-7}{3x+1}$

29. $\dfrac{3m(m-3)}{(m-1)(m-8)}$ 31. $\dfrac{6}{5}$

Section 15.6 (pages 1061–1064)

1. expression; $\dfrac{43}{40}x$ 3. equation; $\left\{\dfrac{40}{43}\right\}$ 5. expression; $-\dfrac{1}{10}y$

7. When solving an equation, we multiply each side by the LCD, which eliminates all denominators. When adding or subtracting fractions, we multiply by a form of 1 so that the result has the LCD in the denominator. The denominators are not eliminated. 9. $\{-6\}$ 11. $\{-15\}$ 13. $\{7\}$

15. $\{-15\}$ 17. $\{-5\}$ 19. $\{-6\}$ 21. $\{5\}$ 23. $\{12\}$ 25. $\{2\}$

27. 0 and 4 29. $\left\{\dfrac{20}{9}\right\}$ 31. $\emptyset$ 33. $\{3\}$ 35. $\{3\}$ 37. $\{-2, 12\}$

39. $\left\{-\dfrac{1}{5}, 3\right\}$ 41. $\left\{-\dfrac{3}{5}, 3\right\}$ 43. $\{-4\}$ 45. $\emptyset$ 47. $\{-1\}$

49. $\{-6\}$ 51. $\left\{-6, \dfrac{1}{2}\right\}$ 53. Transform the equation so that the terms with k are on one side and the remaining term is on the other.

55. $F = \dfrac{ma}{k}$ 57. $a = \dfrac{kF}{m}$ 59. $y = mx + b$ 61. $R = \dfrac{E-Ir}{I}$, or $R = \dfrac{E}{I} - r$ 63. $b = \dfrac{2A-hB}{h}$, or $b = \dfrac{2A}{h} - B$ 65. $a = \dfrac{2S-dnL}{dn}$, or $a = \dfrac{2S}{dn} - L$ 67. $t = \dfrac{rs}{rs-2s-3r}$, or $t = \dfrac{-rs}{-rs+2s+3r}$ 69. $c = \dfrac{ab}{b-a-2ab}$, or $c = \dfrac{-ab}{-b+a+2ab}$

71. $z = \dfrac{3y}{5-9xy}$, or $z = \dfrac{-3y}{9xy-5}$

Summary Exercises on Rational Expressions and Equations (pages 1065–1066)

1. expression; $\dfrac{10}{p}$ 2. expression; $\dfrac{y^3}{x^3}$ 3. expression; $\dfrac{1}{2x^2(x+2)}$

4. equation; $\{9\}$ 5. expression; $\dfrac{x+2}{x-1}$ 6. expression; $\dfrac{5k+8}{k(k-4)(k+4)}$

7. equation; $\{39\}$ 8. expression; $\dfrac{t-5}{3(2t+1)}$ 9. expression; $\dfrac{13}{3(p+2)}$

10. equation; $\left\{-1, \dfrac{12}{5}\right\}$ 11. equation; $\left\{\dfrac{1}{7}, 2\right\}$ 12. expression; $\dfrac{16}{3x}$

13. expression; $\dfrac{7}{12z}$ 14. equation; $\{13\}$

15. expression; $\dfrac{3m+5}{(m+2)(m+3)(m+1)}$ 16. expression; $\dfrac{k+3}{5(k-1)}$

17. equation; $\emptyset$ 18. equation; $\{-7\}$

Section 15.7 (pages 1075–1078)

1. (a) the amount (b) $5+x$ (c) $\dfrac{5+x}{6} = \dfrac{13}{3}$ 3. $\dfrac{9}{5}$ 5. $\dfrac{2}{6}$ 7. -6

9. 36 11. 0.032 mi per min 13. 3.348 hr 15. 6.530 m per sec

17. $\dfrac{8}{4-x} = \dfrac{24}{4+x}$ **19.** into a headwind: $(m-5)$ mph; with a tailwind: $(m+5)$ mph **21.** 8 mph **23.** 32 mph **25.** 3 mph **27.** $\dfrac{1}{2}t + \dfrac{1}{3}t = 1$, or $\dfrac{1}{2} + \dfrac{1}{3} = \dfrac{1}{t}$ **29.** $2\dfrac{2}{5}$ hr **31.** $4\dfrac{8}{19}$ hr **33.** 10 hr **35.** 36 hr **37.** $8\dfrac{1}{4}$ hr

Section 15.8 (pages 1081–1084)

1. (a) increases **(b)** decreases **3.** 15 **5.** 300 **7.** 4 **9.** 6 **11.** 15 in.^{2} **13.** $42\dfrac{2}{3}$ in. **15.** 15 ft **17.** 20 lb per ft^2 **19.** 25 kg per hr **21.** direct **23.** inverse **25.** inverse **27.** direct **29.** 8 **31.** 2 **33.** $x=1; y=4$ **35.** 80 ft

Chapter 15 Review Exercises (pages 1091–1094)

1. $x \neq 3$ **2.** $x \neq 0$ **3.** $m \neq -1, m \neq 3$ **4.** $k \neq -5, k \neq -\dfrac{2}{3}$

5. (a) $-\dfrac{4}{7}$ **(b)** -16 **6. (a)** $\dfrac{11}{8}$ **(b)** $\dfrac{13}{22}$ **7. (a)** undefined **(b)** 1

8. (a) undefined **(b)** $\dfrac{1}{2}$ **9.** $\dfrac{b}{3a}$ **10.** -1 **11.** $\dfrac{-(2x+3)}{2}$

12. $\dfrac{2p+5q}{5p+q}$

Answers may vary in Exercises 13 and 14.

13. $\dfrac{-(4x-9)}{2x+3}, \dfrac{-4x+9}{2x+3}, \dfrac{4x-9}{-(2x+3)}, \dfrac{4x-9}{-2x-3}$

14. $\dfrac{-(8-3x)}{3-6x}, \dfrac{-8+3x}{3-6x}, \dfrac{8-3x}{-(3-6x)}, \dfrac{8-3x}{-3+6x}$ **15.** 2 **16.** $\dfrac{2}{3m^6}$

17. $\dfrac{5}{8}$ **18.** $\dfrac{r+4}{3}$ **19.** $\dfrac{3}{2}$ **20.** $\dfrac{y-2}{y-3}$ **21.** $\dfrac{p+5}{p+1}$ **22.** $\dfrac{3z+1}{z+3}$

23. 96 **24.** $108y^4$ **25.** $m(m+2)(m+5)$ **26.** $(x+3)(x+1)(x+4)$

27. $\dfrac{35}{56}$ **28.** $\dfrac{40}{4k}$ **29.** $\dfrac{15a}{10a^4}$ **30.** $\dfrac{-54}{18-6x}$ **31.** $\dfrac{15y}{50-10y}$

32. $\dfrac{4b(b+2)}{(b+3)(b-1)(b+2)}$ **33.** $\dfrac{15}{x}$ **34.** $-\dfrac{2}{p}$ **35.** $\dfrac{4k-45}{k(k-5)}$

36. $\dfrac{28+11y}{y(7+y)}$ **37.** $\dfrac{-2-3m}{6}$ **38.** $\dfrac{3(16-x)}{4x^2}$ **39.** $\dfrac{7a+6b}{(a-2b)(a+2b)}$

40. $\dfrac{-k^2-6k+3}{3(k+3)(k-3)}$ **41.** $\dfrac{5z-16}{z(z+6)(z-2)}$ **42.** $\dfrac{-13p+33}{p(p-2)(p-3)}$

43. $\dfrac{a}{b}$ **44.** $\dfrac{4(y-3)}{y+3}$ **45.** $\dfrac{6(3m+2)}{2m-5}$ **46.** $\dfrac{(q-p)^2}{pq}$ **47.** $\dfrac{xw+1}{xw-1}$

48. $\dfrac{1-r-t}{1+r+t}$ **49.** $\left\{\dfrac{35}{6}\right\}$ **50.** $\{-16\}$ **51.** $\{-4\}$ **52.** $\varnothing$ **53.** $\{3\}$

54. $t = \dfrac{Ry}{m}$ **55.** $y = \dfrac{4x+5}{3}$ **56.** $t = \dfrac{rs}{s-r}$ **57.** $\dfrac{20}{15}$ **58.** $\dfrac{3}{18}$

59. 1.766 hr **60.** 800.641 m per min **61.** $7\dfrac{1}{2}$ min **62.** $3\dfrac{1}{13}$ hr **63.** $\dfrac{36}{5}$

64. 4 cm **65.** $\dfrac{m+7}{(m-1)(m+1)}$ **66.** $8p^2$ **67.** $\dfrac{1}{6}$ **68.** 3

69. $\dfrac{z+7}{(z+1)(z-1)^2}$ **70.** $v = at + w$ **71.** $\{-2,3\}$ **72.** $\{2\}$

73. 150 km per hr **74.** 10 hr **75.** 4 **76.** inverse

Chapter 15 Test (pages 1095–1096)

1. $x \neq -2, x \neq 4$ **2. (a)** $\dfrac{11}{6}$ **(b)** undefined **3.** (Answers may vary.)

$\dfrac{-(6x-5)}{2x+3}, \dfrac{-6x+5}{2x+3}, \dfrac{6x-5}{-(2x+3)}, \dfrac{6x-5}{-2x-3}$ **4.** $-3x^2y^3$ **5.** $\dfrac{3a+2}{a-1}$

6. $\dfrac{25}{27}$ **7.** $\dfrac{3k-2}{3k+2}$ **8.** $\dfrac{a-1}{a+4}$ **9.** $150p^5$ **10.** $(2r+3)(r+2)(r-5)$

11. $\dfrac{240p^2}{64p^3}$ **12.** $\dfrac{21}{42m-84}$ **13.** 2 **14.** $\dfrac{-14}{5(y+2)}$

15. $\dfrac{x^2+x+1}{3-x}$, or $\dfrac{-x^2-x-1}{x-3}$ **16.** $\dfrac{-m^2+7m+2}{(2m+1)(m-5)(m-1)}$

17. $\dfrac{2k}{3p}$ **18.** $\dfrac{-2-x}{4+x}$ **19.** $\left\{-\dfrac{1}{2}, 5\right\}$ **20.** $\left\{-\dfrac{1}{2}\right\}$

21. $D = \dfrac{dF-k}{F}$, or $D = d - \dfrac{k}{F}$ **22.** -4 **23.** 3 mph **24.** $2\dfrac{2}{9}$ hr

25. 27 **26.** 27 days

Chapter 16 Roots and Radicals

Section 16.1 (pages 1105–1108)

1. true **3.** false; Zero has only one square root. **5.** true

7. $-3, 3$ **9.** $-8, 8$ **11.** $-13, 13$ **13.** $-\dfrac{5}{14}, \dfrac{5}{14}$ **15.** $-30, 30$

17. 1 **19.** 7 **21.** -16 **23.** $-\dfrac{12}{11}$ **25.** 0.8 **27.** not a real number

29. not a real number **31.** 100 **33.** 19 **35.** $\dfrac{2}{3}$ **37.** $3x^2+4$

39. a must be positive. **41.** a must be negative. **43.** rational; 5

45. irrational; 5.385 **47.** rational; -8 **49.** irrational; -17.321

51. not a real number **53.** irrational; 34.641 **55.** C **57.** $c=17$

59. $b=8$ **61.** $c=11.705$ **63.** 24 cm **65.** 80 ft **67.** 195 ft

69. 158.6 ft **71.** 11.1 ft **73.** 9.434 **75.** 1 **77.** 5 **79.** -3

81. -6 **83.** 2 **85.** 4 **87.** 6 **89.** not a real number

91. -5 **93.** -4

Section 16.2 (pages 1115–1118)

1. false; $\sqrt{(-6)^2} = \sqrt{36} = 6$ **3.** $\sqrt{15}$ **5.** $\sqrt{22}$ **7.** $\sqrt{42}$

9. $\sqrt{13r}$ **11.** A **13.** $3\sqrt{5}$ **15.** $2\sqrt{6}$ **17.** $3\sqrt{10}$ **19.** $5\sqrt{3}$

21. $5\sqrt{5}$ **23.** cannot be simplified **25.** $4\sqrt{10}$ **27.** $-10\sqrt{7}$

29. $3\sqrt{6}$ **31.** 24 **33.** $6\sqrt{10}$ **35.** $12\sqrt{5}$ **37.** $30\sqrt{5}$

39. $\sqrt{8} \cdot \sqrt{32} = \sqrt{8 \cdot 32} = \sqrt{256} = 16$. Also, $\sqrt{8} = 2\sqrt{2}$ and $\sqrt{32} = 4\sqrt{2}$, so $\sqrt{8} \cdot \sqrt{32} = 2\sqrt{2} \cdot 4\sqrt{2} = 8 \cdot 2 = 16$. Both methods give the same answer, and the correct answer can always be obtained using either method. **41.** $\dfrac{4}{15}$ **43.** $\dfrac{\sqrt{7}}{4}$

45. 5 **47.** $\dfrac{25}{4}$ **49.** $6\sqrt{5}$ **51.** m **53.** y^2 **55.** $6z$ **57.** $20x^3$

59. $3x^4\sqrt{2}$ **61.** $3c^7\sqrt{5}$ **63.** $z^2\sqrt{z}$ **65.** $a^6\sqrt{a}$ **67.** $8x^3\sqrt{x}$

69. x^3y^6 **71.** $9m^2n$ **73.** $\dfrac{\sqrt{7}}{x^5}$ **75.** $\dfrac{y^2}{10}$ **77.** $\dfrac{x^3}{y^4}$ **79.** $2\sqrt[3]{5}$

81. $3\sqrt[3]{2}$ **83.** $4\sqrt[3]{2}$ **85.** $2\sqrt[4]{5}$ **87.** $\dfrac{2}{3}$ **89.** $-\dfrac{6}{5}$ **91.** p

93. x^3 **95.** $4z^2$ **97.** $7a^3b$ **99.** $2t\sqrt[3]{2t^2}$ **101.** $\dfrac{m^4}{2}$ **103.** 6 cm

105. 6 in. **107.** D

Section 16.3 (pages 1121–1122)

1. distributive **3.** radicands **5.** $-5\sqrt{7}$ **7.** $5\sqrt{17}$ **9.** $5\sqrt{7}$

11. $11\sqrt{5}$ **13.** $15\sqrt{2}$ **15.** $-6\sqrt{2}$ **17.** $17\sqrt{7}$

19. $-16\sqrt{2} - 8\sqrt{3}$ **21.** $20\sqrt{2} + 6\sqrt{3} - 15\sqrt{5}$ **23.** $4\sqrt{2}$

25. $22\sqrt{2}$ **27.** $11\sqrt{3}$ **29.** $5\sqrt{x}$ **31.** $3x\sqrt{6}$ **33.** 0

35. $-20\sqrt{2k}$ **37.** $42x\sqrt{5z}$ **39.** $-\sqrt[3]{2}$ **41.** $6\sqrt[3]{p^2}$ **43.** $21\sqrt[4]{m^3}$
45. $-6x^2y$ **46.** $-6(p-2q)^2(a+b)$ **47.** $-6a^2\sqrt{xy}$
48. The answers are alike because the numerical coefficient of the three answers is the same: -6. Also, the first variable factor is raised to the second power, and the second variable factor is raised to the first power. The answers are different because the variables are different: x and y, then $p-2q$ and $a+b$, and then a and $\sqrt{xy}$.

Section 16.4 (pages 1127–1128)

1. $4\sqrt{2}$ **3.** $\dfrac{-\sqrt{33}}{3}$ **5.** $\dfrac{7\sqrt{15}}{5}$ **7.** $\dfrac{\sqrt{30}}{2}$ **9.** $\dfrac{16\sqrt{3}}{9}$

11. $\dfrac{-3\sqrt{2}}{10}$ **13.** $\dfrac{21\sqrt{5}}{5}$ **15.** $\sqrt{3}$ **17.** $\dfrac{\sqrt{2}}{2}$ **19.** $\dfrac{\sqrt{65}}{5}$

21. We are actually multiplying by 1. The identity property of multiplication justifies our result. **23.** $\dfrac{\sqrt{21}}{3}$ **25.** $\dfrac{3\sqrt{14}}{4}$ **27.** $\dfrac{1}{6}$

29. 1 **31.** $\dfrac{\sqrt{7x}}{x}$ **33.** $\dfrac{2x\sqrt{xy}}{y}$ **35.** $\dfrac{x\sqrt{30xz}}{6}$ **37.** $\dfrac{3ar^2\sqrt{7rt}}{7t}$

39. B **41.** $\dfrac{\sqrt[3]{15}}{3}$ **43.** $\dfrac{\sqrt[3]{196}}{7}$ **45.** $\dfrac{\sqrt[3]{6y}}{2y}$ **47.** $\dfrac{\sqrt[3]{42mn^2}}{6n}$

49. (a) $\dfrac{9\sqrt{2}}{4}$ sec **(b)** 3.182 sec

Section 16.5 (pages 1133–1136)

1. 13 **3.** 4 **5.** $\sqrt{15}-\sqrt{35}$ **7.** $2\sqrt{10}+30$ **9.** $4\sqrt{7}$
11. $57+23\sqrt{6}$ **13.** $81+14\sqrt{21}$ **15.** $71-16\sqrt{7}$
17. $37+12\sqrt{7}$ **19.** $a+2\sqrt{a}+1$ **21.** 23 **23.** 1
25. $y-10$ **27.** $2\sqrt{3}-2+3\sqrt{2}-\sqrt{6}$ **29.** $15\sqrt{2}-15$
31. $\sqrt{30}+\sqrt{15}+6\sqrt{5}+3\sqrt{10}$
33. $\sqrt{5x}-\sqrt{10}-\sqrt{10x}+2\sqrt{5}$ **35.** Because multiplication must be performed before addition, it is incorrect to add -37 and -2. Only like radicals can be combined. **37.** $\dfrac{3-\sqrt{2}}{7}$ **39.** $-4-2\sqrt{11}$
41. $1+\sqrt{2}$ **43.** $-\sqrt{10}+\sqrt{15}$ **45.** $2\sqrt{5}+\sqrt{15}+4+2\sqrt{3}$
47. $\dfrac{12(\sqrt{x}-1)}{x-1}$ **49.** $\dfrac{3(7+\sqrt{x})}{49-x}$ **51.** $\sqrt{11}-2$ **53.** $\dfrac{\sqrt{3}+5}{8}$
55. $\dfrac{6-\sqrt{10}}{2}$ **57.** $30+18x$ **58.** They are not like terms.
59. $30+18\sqrt{5}$ **60.** They are not like radicals. **61.** Make the first term $30x$, so that $30x+18x=48x$; make the first term $30\sqrt{5}$, so that $30\sqrt{5}+18\sqrt{5}=48\sqrt{5}$. **62.** Both like terms and like radicals are combined by adding their numerical coefficients. The variables in like terms are replaced by radicals in like radicals. **63.** 4 in.

Summary Exercises on Operations with Radicals (pages 1137–1138)

1. $-3\sqrt{10}$ **2.** $5-\sqrt{15}$ **3.** $2-\sqrt{6}+2\sqrt{3}-3\sqrt{2}$
4. $6\sqrt{2}$ **5.** $73-12\sqrt{35}$ **6.** $\dfrac{\sqrt{6}}{2}$ **7.** $3\sqrt[3]{2t^2}$
8. $4\sqrt{7}+4\sqrt{5}$ **9.** $-3-2\sqrt{2}$ **10.** 4 **11.** -33
12. $\dfrac{\sqrt{t}-\sqrt{3}}{t-3}$ **13.** $2xyz^2\sqrt[3]{y^2}$ **14.** $4\sqrt[3]{3}$ **15.** $\sqrt{6}+1$
16. $\dfrac{\sqrt{6x}}{3x}$ **17.** $\dfrac{3}{5}$ **18.** $4\sqrt{2}$ **19.** $-2\sqrt[3]{2}$ **20.** $11-2\sqrt{30}$
21. $3\sqrt{3x}$ **22.** $52+30\sqrt{3}$ **23.** 1 **24.** $\dfrac{2\sqrt[3]{18}}{9}$ **25.** $-x^2\sqrt[4]{x}$

26. $2\sqrt{6}$ **27.** cannot be simplified further **28.** $12\sqrt{6}+6\sqrt{5}$
29. $\dfrac{\sqrt{15}}{10}$ **30.** $\sqrt{5}$ **31.** $20\sqrt[3]{3}$ **32.** $\dfrac{8(4+\sqrt{x})}{16-x}$
33. $2-3\sqrt[3]{4}$ **34.** $\sqrt{2x}$ **35.** $\dfrac{\sqrt{10}}{4}$ **36.** $49+14\sqrt{x}+x$
37. (a) 57 species **(b)** 858 species

Section 16.6 (pages 1145–1148)

1. $\{49\}$ **3.** $\{7\}$ **5.** $\{85\}$ **7.** $\{-45\}$ **9.** $\left\{-\dfrac{3}{2}\right\}$ **11.** $\varnothing$
13. $\{121\}$ **15.** $\{8\}$ **17.** $\{1\}$ **19.** $\{6\}$ **21.** $\varnothing$ **23.** $\{5\}$
25. When the left side is squared, the result should be $x-1$, not $-(x-1)$. The correct solution set is $\{17\}$.
27. $\{12\}$ **29.** $\{5\}$ **31.** $\{0,3\}$ **33.** $\{-1,3\}$ **35.** $\{8\}$ **37.** $\{4\}$
39. $\{8\}$ **41.** $\{9\}$ **43.** $\{4,20\}$ **45.** $\{-5\}$ **47.** 21 **49.** 8
51. $17{,}616\ \text{ft}^2$ **53. (a)** 70.5 mph **(b)** 59.8 mph **(c)** 53.9 mph
55. yes; 26 mi **57.** 47 mi **59.** $s=13$ units **60.** $6\sqrt{13}$ sq. units
61. $h=\sqrt{13}$ units **62.** $3\sqrt{13}$ sq. units **63.** $6\sqrt{13}$ sq. units
64. They are both $6\sqrt{13}$.

Chapter 16 Review Exercises (pages 1153–1156)

1. $-7,7$ **2.** $-9,9$ **3.** $-14,14$ **4.** $-11,11$ **5.** $-16,16$
6. $-27,27$ **7.** 4 **8.** -0.6 **9.** -8 **10.** 3 **11.** not a real
number **12.** -65 **13.** $\dfrac{12}{13}$ **14.** $-\dfrac{10}{9}$ **15.** 8 **16.** 48.3 cm
17. irrational; 8.544 **18.** rational; 13 **19.** rational; -25
20. not a real number **21.** $4\sqrt{3}$ **22.** $-12\sqrt{2}$ **23.** $2\sqrt[3]{2}$
24. $5\sqrt[3]{3}$ **25.** 18 **26.** $16\sqrt{6}$ **27.** $-\dfrac{11}{20}$ **28.** $\dfrac{\sqrt{7}}{13}$ **29.** $\dfrac{\sqrt{5}}{6}$
30. $\dfrac{2}{15}$ **31.** $3\sqrt{2}$ **32.** $2\sqrt{2}$ **33.** r^9 **34.** x^5y^8 **35.** $9x^4\sqrt{2x}$
36. $\dfrac{6}{p}$ **37.** $a^7b^{10}\sqrt{ab}$ **38.** $11x^3y^5$ **39.** y^2 **40.** $6x^5$ **41.** $8\sqrt{11}$
42. $9\sqrt{2}$ **43.** $21\sqrt{3}$ **44.** $12\sqrt{3}$ **45.** 0 **46.** $3\sqrt{7}$
47. $2\sqrt{3}+3\sqrt{10}$ **48.** $2\sqrt{2}$ **49.** $6\sqrt{30}$ **50.** $5\sqrt{x}$ **51.** 0
52. $11k^2\sqrt{2n}$ **53.** $\dfrac{10\sqrt{3}}{3}$ **54.** $\dfrac{8\sqrt{10}}{5}$ **55.** $\sqrt{6}$ **56.** $\dfrac{\sqrt{10}}{5}$
57. $\sqrt{10}$ **58.** $\dfrac{\sqrt{42}}{21}$ **59.** $\dfrac{r\sqrt{x}}{4x}$ **60.** $\dfrac{\sqrt[3]{9}}{3}$ **61.** $r=\dfrac{\sqrt{3V\pi h}}{\pi h}$
62. $r=\dfrac{\sqrt{S\pi}}{2\pi}$ **63.** $-\sqrt{15}-9$ **64.** $3\sqrt{6}+12$ **65.** $22-16\sqrt{3}$
66. $2\sqrt{21}-\sqrt{14}+12\sqrt{2}-4\sqrt{3}$ **67.** -13 **68.** $x+4\sqrt{x}+4$
69. $-2+\sqrt{5}$ **70.** $\dfrac{3(1-\sqrt{x})}{1-x}$ **71.** $\dfrac{-\sqrt{10}+3\sqrt{5}+\sqrt{2}-3}{7}$
72. $\dfrac{3+2\sqrt{6}}{3}$ **73.** $\dfrac{1+3\sqrt{7}}{4}$ **74.** $3+4\sqrt{3}$ **75.** $\varnothing$ **76.** $\{48\}$
77. $\{1\}$ **78.** $\{2\}$ **79.** $\{6\}$ **80.** $\{-3,-1\}$ **81.** $\{-2\}$ **82.** $\{4\}$
83. $\{7\}$ **84.** $11\sqrt{3}$ **85.** $\dfrac{5-\sqrt{2}}{23}$ **86.** $\dfrac{2\sqrt{10}}{5}$ **87.** $3a^2b^3\sqrt[3]{2ab}$
88. $-\sqrt{10}-5\sqrt{15}$ **89.** $\dfrac{4r\sqrt{3rs}}{3s}$ **90.** $\dfrac{2+\sqrt{13}}{2}$ **91.** $7-2\sqrt{10}$
92. $166+2\sqrt{7}$ **93.** $\{7\}$ **94.** $\varnothing$ **95.** $\{8\}$ **96. (a)** B **(b)** F
(c) D **(d)** A **(e)** C **(f)** A

Chapter 16 Test (pages 1157–1158)

1. $-20, 20$ **2. (a)** irrational **(b)** 11.916 **3.** a must be negative.

4. 6 **5.** $-3\sqrt{6}$ **6.** $\dfrac{8\sqrt{2}}{5}$ **7.** $2\sqrt[3]{4}$ **8.** $4\sqrt{6}$ **9.** $9\sqrt{7}$

10. $-5\sqrt{3x}$ **11.** $4xy\sqrt{2y}$ **12.** 31

13. $6\sqrt{2} + 2 - 3\sqrt{14} - \sqrt{7}$ **14.** $11 + 2\sqrt{30}$

15. (a) $6\sqrt{2}$ in. **(b)** 8.485 in. **16.** 50 ohms **17.** $\dfrac{5\sqrt{14}}{7}$

18. $\dfrac{\sqrt{6x}}{3x}$ **19.** $-\sqrt[3]{2}$ **20.** $\dfrac{-3\left(4 + \sqrt{3}\right)}{13}$

21. $\dfrac{\sqrt{3} + 12\sqrt{2}}{3}$ **22.** $\emptyset$ **23.** $\{3\}$ **24.** $\{1, 4\}$ **25.** $\{9\}$

26. 12 is not a solution. A check shows that it does not satisfy the original equation. The solution set is $\emptyset$.

Chapter 17   Quadratic Equations

Section 17.1 (pages 1163–1166)

1. true **3.** false; If k is a positive integer that is not a perfect square, then the solutions will be irrational. **5.** false; For values of k that satisfy $0 \le k < 10$, there are real solutions. **7.** $\{-9, 9\}$

9. $\{-\sqrt{14}, \sqrt{14}\}$ **11.** $\{-4\sqrt{3}, 4\sqrt{3}\}$ **13.** $\left\{-\dfrac{5}{2}, \dfrac{5}{2}\right\}$

15. $\emptyset$ **17.** $\{-1.5, 1.5\}$ **19.** $\{-\sqrt{3}, \sqrt{3}\}$ **21.** $\left\{-\dfrac{2\sqrt{7}}{7}, \dfrac{2\sqrt{7}}{7}\right\}$

23. $\{-3\sqrt{2}, 3\sqrt{2}\}$ **25.** $\{-2\sqrt{6}, 2\sqrt{6}\}$ **27.** $\left\{-\dfrac{2\sqrt{5}}{5}, \dfrac{2\sqrt{5}}{5}\right\}$

29. $\{-2, 8\}$ **31.** $\emptyset$ **33.** $\{8 + 3\sqrt{3}, 8 - 3\sqrt{3}\}$ **35.** $\left\{-3, \dfrac{5}{3}\right\}$

37. $\left\{0, \dfrac{3}{2}\right\}$ **39.** $\left\{\dfrac{5 + \sqrt{30}}{2}, \dfrac{5 - \sqrt{30}}{2}\right\}$

41. $\left\{\dfrac{-1 + 3\sqrt{2}}{3}, \dfrac{-1 - 3\sqrt{2}}{3}\right\}$ **43.** $\{-10 + 4\sqrt{3}, -10 - 4\sqrt{3}\}$

45. $\left\{\dfrac{1 + 4\sqrt{3}}{4}, \dfrac{1 - 4\sqrt{3}}{4}\right\}$ **47.** The answers are equivalent. If the answer of either student is multiplied by $\dfrac{-1}{-1}$, it will look like the answer of the other student. **49.** about $\dfrac{1}{2}$ sec **51.** 9 in. **53.** 5% **55.** 2%

Section 17.2 (pages 1173–1174)

1. $25; (x + 5)^2$ **3.** $1; (x + 1)^2$ **5.** $\dfrac{25}{4}; \left(p - \dfrac{5}{2}\right)^2$ **7.** D

9. $\{1, 3\}$ **11.** $\{-3, -2\}$ **13.** $\{-1 + \sqrt{6}, -1 - \sqrt{6}\}$

15. $\{4 + 2\sqrt{3}, 4 - 2\sqrt{3}\}$ **17.** $\{-3\}$

19. $\left\{\dfrac{-1 + \sqrt{5}}{2}, \dfrac{-1 - \sqrt{5}}{2}\right\}$ **21.** $\left\{-\dfrac{3}{2}, \dfrac{1}{2}\right\}$

23. $\left\{\dfrac{2 + \sqrt{14}}{2}, \dfrac{2 - \sqrt{14}}{2}\right\}$ **25.** $\emptyset$

27. $\left\{\dfrac{-7 + \sqrt{97}}{6}, \dfrac{-7 - \sqrt{97}}{6}\right\}$ **29.** $\{-4, 2\}$

31. $\{4 + \sqrt{3}, 4 - \sqrt{3}\}$ **33.** $\{1 + \sqrt{6}, 1 - \sqrt{6}\}$

35. 1 sec and 5 sec **37.** 3 sec and 5 sec **39.** 75 ft by 100 ft

Section 17.3 (pages 1179–1180)

1. $4; 5; -9$ **3.** $3; -4; -2$ **5.** $3; 7; 0$ **7.** $\{-13, 1\}$ **9.** $\{2\}$

11. $\left\{-1, \dfrac{5}{2}\right\}$ **13.** $\left\{\dfrac{-6 + \sqrt{26}}{2}, \dfrac{-6 - \sqrt{26}}{2}\right\}$ **15.** $\{-1, 0\}$

17. $\emptyset$ **19.** $\left\{\dfrac{-5 + \sqrt{13}}{6}, \dfrac{-5 - \sqrt{13}}{6}\right\}$ **21.** $\left\{0, \dfrac{12}{7}\right\}$

23. $\{-2\sqrt{6}, 2\sqrt{6}\}$ **25.** $\left\{-\dfrac{2}{5}, \dfrac{2}{5}\right\}$ **27.** $\left\{\dfrac{6 + 2\sqrt{6}}{3}, \dfrac{6 - 2\sqrt{6}}{3}\right\}$

29. $\emptyset$ **31.** There is no real number solution, so the solution set is $\emptyset$.

33. $\left\{-\dfrac{2}{3}, \dfrac{4}{3}\right\}$ **35.** $\left\{\dfrac{-1 + \sqrt{73}}{6}, \dfrac{-1 - \sqrt{73}}{6}\right\}$ **37.** $\emptyset$

39. $\left\{-1, \dfrac{5}{2}\right\}$ **41.** $\{-3 + \sqrt{5}, -3 - \sqrt{5}\}$ **43.** 3.5 ft

45. $-8, 16$; Only 16 board feet is a reasonable answer.

Summary Exercises on Quadratic Equations (pages 1181–1182)

1. $\{-6, 6\}$ **2.** $\left\{\dfrac{-3 + \sqrt{5}}{2}, \dfrac{-3 - \sqrt{5}}{2}\right\}$ **3.** $\left\{-\dfrac{10}{9}, \dfrac{10}{9}\right\}$

4. $\left\{-\dfrac{7}{9}, \dfrac{7}{9}\right\}$ **5.** $\{1, 3\}$ **6.** $\{-2, -1\}$ **7.** $\{4, 5\}$

8. $\left\{\dfrac{-3 + \sqrt{17}}{2}, \dfrac{-3 - \sqrt{17}}{2}\right\}$ **9.** $\left\{-\dfrac{1}{3}, \dfrac{5}{3}\right\}$

10. $\left\{\dfrac{1 + \sqrt{10}}{2}, \dfrac{1 - \sqrt{10}}{2}\right\}$ **11.** $\{-17, 5\}$ **12.** $\left\{-\dfrac{7}{5}, 1\right\}$

13. $\left\{\dfrac{7 + 2\sqrt{6}}{3}, \dfrac{7 - 2\sqrt{6}}{3}\right\}$ **14.** $\left\{\dfrac{1 + 4\sqrt{2}}{7}, \dfrac{1 - 4\sqrt{2}}{7}\right\}$

15. $\emptyset$ **16.** $\emptyset$ **17.** $\left\{-\dfrac{1}{2}, 2\right\}$ **18.** $\left\{-\dfrac{1}{2}, 1\right\}$ **19.** $\left\{-\dfrac{5}{4}, \dfrac{3}{2}\right\}$

20. $\left\{-3, \dfrac{1}{3}\right\}$ **21.** $\{1 + \sqrt{2}, 1 - \sqrt{2}\}$

22. $\left\{\dfrac{-5 + \sqrt{13}}{6}, \dfrac{-5 - \sqrt{13}}{6}\right\}$ **23.** $\left\{\dfrac{2}{5}, 4\right\}$

24. $\{-3 + \sqrt{5}, -3 - \sqrt{5}\}$ **25.** $\left\{\dfrac{-3 + \sqrt{41}}{2}, \dfrac{-3 - \sqrt{41}}{2}\right\}$

26. $\left\{-\dfrac{5}{4}\right\}$ **27.** $\left\{\dfrac{1}{4}, 1\right\}$ **28.** $\left\{\dfrac{1 + \sqrt{3}}{2}, \dfrac{1 - \sqrt{3}}{2}\right\}$

29. $\left\{\dfrac{-2 + \sqrt{11}}{3}, \dfrac{-2 - \sqrt{11}}{3}\right\}$ **30.** $\left\{\dfrac{-5 + \sqrt{41}}{8}, \dfrac{-5 - \sqrt{41}}{8}\right\}$

31. $\left\{\dfrac{-7 + \sqrt{5}}{4}, \dfrac{-7 - \sqrt{5}}{4}\right\}$ **32.** $\left\{-\dfrac{8}{3}, -\dfrac{6}{5}\right\}$

33. $\left\{\dfrac{8 + 8\sqrt{2}}{3}, \dfrac{8 - 8\sqrt{2}}{3}\right\}$ **34.** $\left\{\dfrac{-5 + \sqrt{5}}{2}, \dfrac{-5 - \sqrt{5}}{2}\right\}$

35. ∅ **36.** ∅ **37.** $\left\{-\dfrac{2}{3}, 2\right\}$ **38.** $\left\{-\dfrac{1}{4}, \dfrac{2}{3}\right\}$ **39.** $\left\{-4, \dfrac{3}{5}\right\}$

40. $\{-3, 5\}$ **41.** $\left\{-\dfrac{2}{3}, \dfrac{2}{5}\right\}$ **42.** $\{-4, 6\}$

Section 17.4 (pages 1187–1188)

1. The vertex of a parabola is the lowest or highest point on the graph.
3. $(0, 0)$ **5.** $(0, -4)$

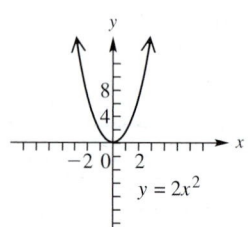 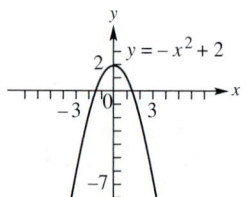

7. $(0, 2)$ **9.** $(-3, 0)$

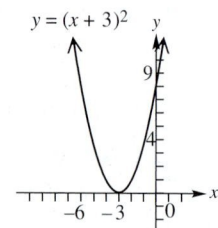

11. $(-1, 2)$ **13.** $(3, 4)$

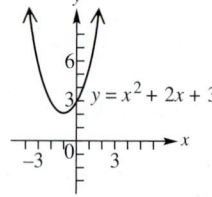 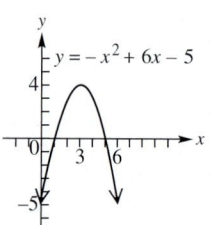

15. If $a > 0$, it opens upward, and if $a < 0$, it opens downward.
17. $y = \dfrac{11}{5625} x^2$

Section 17.5 (pages 1195–1196)

1. $3; 3; (1, 3)$ **3.** $5; 5; (3, 5)$ **5.** The graph consists of the five points
$(0, 2), (1, 3), (2, 4), (3, 5),$ and $(4, 6).$ **7.** not a function;
domain: $\{-4, -2, 0\}$; range: $\{3, 1, 5, -8\}$ **9.** function;
domain: $\{A, B, C, D, E\}$; range: $\{2, 3, 6, 4\}$ **11.** not a function;
domain: $\{-4, -2, 0, 2, 3\}$; range: $\{-2, 0, 1, 2, 3\}$ **13.** function
15. not a function   **17.** function   **19.** not a function   **21.** $(2, 4)$
22. $(-1, -4)$ **23.** $\dfrac{8}{3}$ **24.** $f(x) = \dfrac{8}{3} x - \dfrac{4}{3}$ **25. (a)** 11   **(b)** 3
(c) -9 **27. (a)** 4   **(b)** 2   **(c)** 14   **29. (a)** 2   **(b)** 0   **(c)** 3
31. $\{(1970, 9.6), (1980, 14.1), (1990, 19.8), (2000, 28.4)\}$; yes
33. $g(1980) = 14.1$ (million); $g(1990) = 19.8$ (million)
35. For the year 2002, the function gives 30.3 million foreign-born
residents in the United States.

Chapter 17 Review Exercises (pages 1201–1204)

1. $\{-12, 12\}$ **2.** $\{-\sqrt{37}, \sqrt{37}\}$ **3.** $\{-8\sqrt{2}, 8\sqrt{2}\}$
4. $\{-7, 3\}$ **5.** $\{3 + \sqrt{10}, 3 - \sqrt{10}\}$
6. $\left\{\dfrac{-1 + \sqrt{14}}{2}, \dfrac{-1 - \sqrt{14}}{2}\right\}$ **7.** ∅   **8.** $\left\{-\dfrac{5}{3}\right\}$ **9.** $\{-5, -1\}$
10. $\{-2 + \sqrt{11}, -2 - \sqrt{11}\}$ **11.** $\{-1 + \sqrt{6}, -1 - \sqrt{6}\}$
12. $\left\{\dfrac{-4 + \sqrt{22}}{2}, \dfrac{-4 - \sqrt{22}}{2}\right\}$ **13.** $\left\{-\dfrac{7}{2}\right\}$ **14.** ∅   **15.** 2.5 sec
16. 6, 8, 10   **17.** $\left(\dfrac{3}{2}\right)^2$, or $\dfrac{9}{4}$ **18. (a)** $\{-3, 3\}$ **(b)** $\{-3, 3\}$
(c) $\{-3, 3\}$ **(d)** We will always get the same results, no matter which
method of solution is used.
19. $\left\{\dfrac{-1 + \sqrt{29}}{4}, \dfrac{-1 - \sqrt{29}}{4}\right\}$ **20.** $\left\{\dfrac{2 + \sqrt{10}}{2}, \dfrac{2 - \sqrt{10}}{2}\right\}$
21. $\left\{\dfrac{1 + \sqrt{21}}{10}, \dfrac{1 - \sqrt{21}}{10}\right\}$ **22.** $\left\{\dfrac{-3 + \sqrt{41}}{2}, \dfrac{-3 - \sqrt{41}}{2}\right\}$
23. $\{-3 + \sqrt{19}, -3 - \sqrt{19}\}$ **24.** No, because the fraction bar should
be under both $-b$ and $\sqrt{b^2 - 4ac}$. The term $2a$ should not be in the
radicand. The correct formula is $x = \dfrac{-b \pm \sqrt{b^2 - 4ac}}{2a}$.
25. $(0, 0)$ **26.** $(0, 5)$

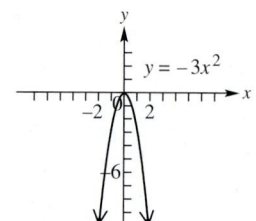 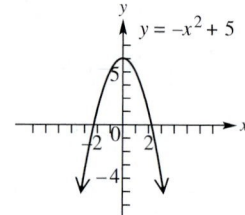

27. $(1, 0)$ **28.** $(1, 4)$

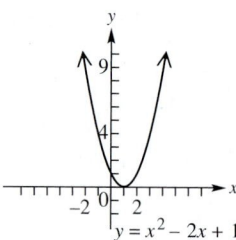 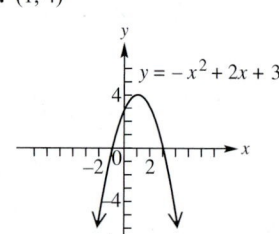

29. $(-2, -2)$ **30.** $(-4, 0)$

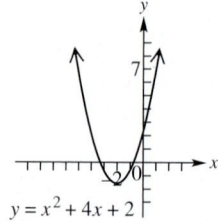

31. $y = \dfrac{3}{1000} x^2$ **32.** not a function; domain: $\{-2, 0, 2\}$;
range: $\{4, 8, 5, 3\}$ **33.** function; domain: $\{8, 7, 6, 5, 4\}$;
range: $\{3, 4, 5, 6, 7\}$ **34.** not a function   **35.** function

36. function **37.** function **38.** not a function **39. (a)** 8
(b) −1 **40. (a)** 7 **(b)** 1 **41. (a)** 5 **(b)** 2 **42.** 400 or 800

43. (6, 10) **44.** demand: 600; price: $10 **45.** $\left\{-\dfrac{11}{2}, 5\right\}$

46. $\left\{-\dfrac{11}{2}, \dfrac{9}{2}\right\}$ **47.** $\left\{\dfrac{-1+\sqrt{21}}{2}, \dfrac{-1-\sqrt{21}}{2}\right\}$ **48.** $\left\{-\dfrac{3}{2}, \dfrac{1}{3}\right\}$

49. $\left\{\dfrac{-5+\sqrt{17}}{2}, \dfrac{-5-\sqrt{17}}{2}\right\}$ **50.** $\left\{-1+\sqrt{3}, -1-\sqrt{3}\right\}$

51. ∅ **52.** $\left\{\dfrac{9+\sqrt{41}}{2}, \dfrac{9-\sqrt{41}}{2}\right\}$ **53.** $\left\{-\dfrac{6}{5}\right\}$

54. $\left\{-1+2\sqrt{2}, -1-2\sqrt{2}\right\}$ **55.** $\left\{-2+\sqrt{5}, -2-\sqrt{5}\right\}$

56. $\left\{-2\sqrt{2}, 2\sqrt{2}\right\}$

Chapter 17 Test (pages 1205–1206)

1. $\left\{-\sqrt{39}, \sqrt{39}\right\}$ **2.** $\{-11, 5\}$ **3.** $\left\{\dfrac{-3+2\sqrt{6}}{4}, \dfrac{-3-2\sqrt{6}}{4}\right\}$

4. $\left\{2+\sqrt{10}, 2-\sqrt{10}\right\}$ **5.** $\left\{\dfrac{-6+\sqrt{42}}{2}, \dfrac{-6-\sqrt{42}}{2}\right\}$

6. $\left\{-3, \dfrac{1}{2}\right\}$ **7.** $\left\{\dfrac{3+\sqrt{3}}{3}, \dfrac{3-\sqrt{3}}{3}\right\}$ **8.** ∅

9. $\left\{\dfrac{5+\sqrt{13}}{6}, \dfrac{5-\sqrt{13}}{6}\right\}$ **10.** $\left\{1+\sqrt{2}, 1-\sqrt{2}\right\}$

11. $\left\{\dfrac{-1+3\sqrt{2}}{2}, \dfrac{-1-3\sqrt{2}}{2}\right\}$ **12.** $\left\{\dfrac{11+\sqrt{89}}{4}, \dfrac{11-\sqrt{89}}{4}\right\}$

13. $\{5\}$ **14.** 2 sec **15.** 12, 16, 20
16. vertex: (3, 0) **17.** vertex: (−1, −3)

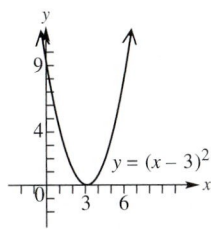
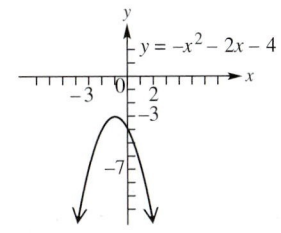

18. (a) not a function **(b)** function; domain: $\{0, 1, 2\}$; range: $\{2\}$
19. not a function **20.** 1

Whole Numbers Computation: Pretest

(pages 1207–1208)
Adding Whole Numbers
1. 390 **2.** 13,166 **3.** 3053 **4.** 117,510 **5.** 688,226

Subtracting Whole Numbers
1. 350 **2.** 629 **3.** 24,894 **4.** 3909 **5.** 591,387

Multiplying Whole Numbers
1. 0 **2.** 26,887 **3.** 1,560,000 **4.** 1846 **5.** 17,232 **6.** 519,477

Dividing Whole Numbers
1. 23 **2.** undefined **3.** 6259 **4.** 807 **R**6 **5.** 34 **6.** 50
7. 60 **R**20 **8.** 539 **R**62

Chapter R

Section R.1 (pages 1215–1216)
1. (a) 23 **(b)** 19 **3. (a)** 8928 **(b)** 59,224 **5.** 150 **7.** 1651
9. 1004 **11.** 9253 **13.** 11,624 **15.** 16,658 **17.** 5009 **19.** 3506
21. 15,954 **23.** 10,648 **25.** incorrect; should be 769 **27.** correct
29. 33 miles **31.** 38 miles **33.** $152 **35.** 699 people
37. 59,965,332 people **39.** 294 inches **41.** 708 feet

Section R.2 (pages 1223–1224)
1. incorrect; should be 62 **3.** incorrect; should be 121 **5.** 8 **7.** 25
9. 16 **11.** 61 **13.** 519 **15.** 9177 **17.** 7589 **19.** 8859 **21.** 3
23. 23 **25.** 1942 **27.** 5687 **29.** 19,038 **31.** 65,556 **33.** 19,984
35. incorrect; should be 2494 **37.** correct **39.** 70 calories **41.** 180
passengers **43. (a)** Highest is computer programmer; lowest is medical
secretary. **(b)** $37,420 **45.** 380 feet

Section R.3 (pages 1231–1232)
1. 9 **3.** 63 **5.** 0 **7.** 24 **9.** 36 **11.** 245 **13.** 168
15. 19,092 **17.** 258,447 **19.** 12,500 **21.** 44,550 **23.** 270,000
25. 86,000,000 **27.** 1496 **29.** 3735 **31.** 15,200 **33.** 32,805
35. 183,996 **37.** 3,321,934 **39.** 18,980,480 **41.** 252 inches;
540 inches **43.** 576 plants **45.** 495 miles **47.** 2440 miles;
16,170 miles **49.** 7406 calories

Section R.4 (pages 1241–1242)
1. 1; $12\overline{)12}$; $12 \div 12$ **3.** undefined; $\dfrac{24}{0}$; $0\overline{)24}$ **5.** 0; $4\overline{)0}$; $0 \div 4$

7. 0; $\dfrac{0}{12}$; $12\overline{)0}$ **9.** undefined; $\dfrac{21}{0}$; $21 \div 0$ **11.** 21; dividend: 84;
divisor: 4; quotient: 21. **13.** 36; dividend: 324; divisor: 9;
quotient: 36 **15.** 1520 **R**5 **17.** 309 **19.** 1006 **21.** 5006 **R**1
23. 6671 **25.** 12,458 **R**3 **27.** 10,253 **R**5 **29.** 18,377 **R**6
31. incorrect; should be 670 **R**2 **33.** incorrect; should be 3568 **R**1
35. correct **37.** correct **39.** $23; $14; $19 **41.** $16,600
43. 378 for $5 tickets; 270 for $7 tickets; 210 for $9 tickets **45.** $7; $210
47. (a) 358; 190 **(b)** 10,404; 78 **(c)** 13,740; 985 **49.** 2, 3, 5, 10
51. 2 **53.** 5 **55.** 3 **57.** 2, 3 **59.** none of the above

Section R.5 (pages 1247–1248)
1. $24\overline{)768}$ $\overset{3}{}$ use 2 as the trial divisor. **3.** $18\overline{)4500}$ $\overset{2}{}$ use 2 as the trial divisor.
5. $86\overline{)10,327}$ $\overset{1}{}$ use 9 as the trial divisor. **7.** $52\overline{)38,025}$ $\overset{7}{}$ use 5 as the trial divisor.
9. $77\overline{)249,826}$ $\overset{3}{}$ use 8 as the trial divisor. **11.** $420\overline{)470,800}$ $\overset{1}{}$ use 4 as the trial divisor. **13.** 64 **R**3
15. 236 **R**29 **17.** 2407 **R**1 **19.** 1239 **R**15 **21.** 3331 **R**110 **23.** 850
25. incorrect; should be 106 **R**7 **27.** incorrect; should be 658
29. incorrect; should be 62 **31.** correct **33.** 1050 hours **35.** $2030
37. $308 **39.** 1680 circuits **41.** $375

Chapter 1 Introduction to Algebra: Integers

Section 1.1 (pages 5–8)

19. To write 3,064.801 in words start at the left: three million, sixty-four thousand, eight hundred one. Do **not** write "eight hundred *and* one at the end. Use "and" *only* when there is a decimal point in the number.

35. Six hundred *trillion,* seventy-one million, four hundred
The first group name is *trillion,* so you need to fill *five groups* of three digits: trillions, billions, millions, thousands, and ones.

$$\underline{6\,0\,0},\underline{0\,0\,0},\underline{0\,7\,1},\underline{0\,0\,0},\underline{4\,0\,0}$$

There are no billions or thousands, so fill these groups with zeros.

43. Fifty-five *million,* eight hundred
The first group name is *millions,* so you need to fill *three groups* of three digits: millions, thousands, and ones.

$$\underline{0\,5\,5},\underline{0\,0\,0},\underline{8\,0\,0}$$

There are no thousands, so fill the thousands group with zeros. When writing the number, you can omit the leading 0 in the millions group.
The number is: 55,000,800

Section 1.2 (pages 15–16)

23. $^-3$ is to the *right* of $^-6$ on the number line, so $^-3$ is *greater than* $^-6$. Write $^-3 > ^-6$.

25. $^-10$ is to the *left* of $^-2$ on the number line, so $^-10$ is *less than* $^-2$. Write $^-10 < ^-2$.

41. $|^-75| = 75$ because the distance between 0 and $^-75$ on the number line is 75 spaces.

Section 1.3 (pages 25–28)

39. $^-17 + 0 = ^-17$
Adding zero to any number leaves the number unchanged.

49. Add from left to right.

$$^-3 + ^-11 + 14 \quad \text{First add } ^-3 + ^-11.$$
$$= ^-14 + 14 \quad \text{Add } ^-14 + 14.$$
$$= 0$$

53. Add from left to right.

$$^-7 + 28 + ^-56 + 3 \quad \text{First add } ^-7 + 28.$$
$$= 21 + ^-56 + 3 \quad \text{Add } 21 + ^-56.$$
$$= ^-35 + 3 \quad \text{Add } ^-35 + 3.$$
$$= ^-32$$

61. First write Jeff's scores as a sum of integers. Then add from left to right.

$$^-20 + 75 + ^-55 \quad \text{First add } ^-20 + 75.$$
$$= 55 + ^-55 \quad \text{Add } 55 + ^-55.$$
$$= 0 \text{ points}$$

First write Terry's scores as a sum of integers. Then add from left to right.

$$42 + ^-15 + 20 \quad \text{First add } 42 + ^-15.$$
$$= 27 + 20 \quad \text{Add } 27 + 20.$$
$$= 47 \text{ points}$$

81. When entering $^-99,000$ on your calculator, be sure to use the *negative* key as opposed to the *subtraction* key.

$$86 + ^-99,000 + 0 + 2837 = ^-96,077$$

Section 1.4 (pages 31–32)

11. $\quad 7 - 19 \quad$ Change subtraction to addition. Change 19 to $^-19$.
$$= 7 + ^-19$$
$$= ^-12$$

17. $\quad ^-3 - ^-8 \quad$ Change subtraction to addition. Change $^-8$ to $^+8$.
$$= ^-3 + ^+8$$
$$= 5$$

23. $\quad ^-30 - 30 \quad$ Change subtraction to addition. Change 30 to $^-30$
$$= ^-30 + ^-30$$
$$= ^-60$$

39. $\quad 3 - ^-3 - 10 - ^-7 \quad$ Change all subtractions to additions.
Change $^-3$ to $^+3$, 10 to $^-10$, and $^-7$ to $^+7$.
$$= 3 + ^+3 + ^-10 + ^+7 \quad \text{Add from left to right. First add } 3 + 3.$$
$$= 6 + ^-10 + ^+7 \quad \text{Then add } 6 + ^-10.$$
$$= ^-4 + ^+7 \quad \text{Finally, add } ^-4 + 7.$$
$$= 3$$

53. $\quad 0 - |^-7 + 2| \quad$ Simplify the sum inside the absolute value bars first:
$^-7 + 2$ is $^-5$.
$$= 0 - |^-5| \quad |^-5| \text{ is 5 because the distance from 0 to } ^-5 \text{ on the number line is 5 spaces.}$$
$$= 0 - 5 \quad \text{Change subtraction to addition. Change 5 to } ^-5.$$
$$= 0 + ^-5 \quad \text{Add.}$$
$$= ^-5$$

Section 1.5 (pages 39–42)

13. $^-7\underline{8},499 \approx ^-78,000$ (nearest thousand)
The next digit is 4 or less, so the thousands place remains 8. Change 4, 9, and 9 to 0.

23. $19,9\underline{5}1,880,500 \approx 20,000,000,000$ (nearest hundred-million)
The next digit is 5 or more, so the hundred-millions place changes from 9 to 10. Write 0 and regroup 1 to billions place. In the billions place, $9 + $ regrouped 1 is 10. Write the 0 and regroup 1 to ten-billions.

31. $\$9\underline{9}42 \approx \$10,000$
Round to the nearest thousand since the leftmost digit is in the thousands place.
Next digit is 5 or more, so the thousands place changes from 9 to 10. Write 0 and regroup 1 to the ten-thousands place. Change 9, 4, and 2 to 0. $9942 is closer to $10,000 than to $9000.

43. $16 + ^-97$
Because the leftmost digit is in the tens place, round 16 and $^-97$ to the nearest ten.
$\underline{1}6$ is closer to 20 than 10.
$^-\underline{9}7$ is closer to $^-100$ than $^-90$.
Estimate: $20 + ^-100 = ^-80$
Exact: $16 + ^-97 = ^-81$

53. $\quad ^-106 + 34 - ^-72 \quad$ Change $34 - ^-72$ to adding the opposite: $34 + ^+72$
$$^-106 + 34 + ^+72$$
Because the leftmost digit is in the hundreds place, round $^-106$ to the nearest hundred.
$^-1\underline{0}6 \approx ^-100$
Because the leftmost digit is in the tens place, round 34 and 72 to the nearest ten.
$3\underline{4} \approx 30$ and $7\underline{2} \approx 70$
Estimate: $^-100 + 30 + ^+70 = 0$
Exact: $^-106 + 34 - ^-72$
$\quad = ^-106 + 34 + ^+72 = 0$

Section 1.6 (pages 49–52)

3. (a) $7(^-8) = ^-56 \quad$ Factors have *different* signs, so the product is *negative.*

(b) $^-7(8) = ^-56 \quad$ Factors have *different* signs, so the product is *negative.*

(c) $7(8) = 56 \quad$ Factors have the *same* sign, so the product is *positive.*

(d) $^-7(^-8) = 56 \quad$ Factors have the *same* sign, so the product is *positive.*

27. $(0)(^-25) = 0$ Use the multiplication property of 0, which says that multiplying any number by 0 gives a product of 0.

31. Multiply from left to right.

$(^-4)(^-2)(^-7)$ Multiply $^-4 \cdot {}^-2$ first. The factors have the *same* sign, so the product is *positive*.

$= 8(^-7)$ Then multiply $8 \cdot {}^-7$. The factors have *different* signs, so the product is *negative*.

$= {}^-56$

39. Because $^-17$ is a *negative* product, the factors must have *different* signs.

$^-17 = 17(^-1)$

45. Because $^-5$ and $^-2$ have the same sign, their product of 10 is positive. The final product of $^-40$ is negative, so the missing factor must be negative.

$(^-4)(^-5)(^-2) = {}^-40$

$(^-4)(10) = {}^-40$

51. $9(^-3 + 5)$ rewritten using the distributive property is

$9 \cdot {}^-3 + 9 \cdot 5$

To check:

$9(^-3 + 5) = 9 \cdot {}^-3 + 9 \cdot 5$

$9(2) = {}^-27 + 45$

$18 = 18$

69. $|6 - 7| \cdot {}^-355{,}299$ Work inside the absolute value bars first.

$= |6 + {}^-7| \cdot {}^-355{,}299$ Change subtraction to addition; change 7 to $^-7$. Then add $6 + {}^-7$ to get $^-1$.

$= |{}^-1| \cdot {}^-355{,}299$ $^-1$ is 1 space from 0, so $|{}^-1|$ is 1

$= 1 \cdot {}^-355{,}299$ Multiplication property of 1

$= {}^-355{,}299$

73. The temperature drops 3 degrees ($^-3$ degrees) for every 1000 feet climbed into the air. An altitude of 24,000 feet would require 24 increases of 1000 feet each.

$^-3 \cdot 24 = {}^-72$ degrees

To find the temperature at an altitude of 24,000 feet, add the change in temperature ($^-72$) to the temperature on the ground (50).

The temperature at 24,000 feet is $50 + {}^-72 = {}^-22$ degrees.

Section 1.7 (pages 59–62)

1. (a) $14 \div 2 = 7$

(*Same* signs, quotient is *positive*)

(b) $^-14 \div {}^-2 = 7$

(*Same* signs, quotient is *positive*)

(c) $14 \div {}^-2 = {}^-7$

(*Different* signs, quotient is *negative*)

(d) $^-14 \div 2 = {}^-7$

(*Different* signs, quotient is *negative*)

7. (a) $\dfrac{0}{50} = 0$; zero divided by any nonzero number is 0.

(b) $\dfrac{50}{0}$ is undefined; division by zero is undefined.

(c) $\dfrac{-11}{0}$ is undefined; division by zero is undefined.

(d) $\dfrac{0}{-11} = 0$; zero divided by any nonzero number is 0.

25. $\dfrac{-18}{18} = {}^-1$ (*Different* signs, quotient is *negative*)

29. $\dfrac{-573}{-3} = 191$ (*Same* signs, quotient is *positive*)

35. $^-64 \div {}^-8 \div {}^-2$ No parentheses, so start at the left: $^-64 \div {}^-8$ is 8. The signs are the *same* so the quotient is *positive*.

$= 8 \div {}^-2$ $8 \div {}^-2$ is $^-4$. The signs are *different* so the quotient is *negative*.

$= {}^-4$

41. $^-5 \div {}^-5(^-10) \div {}^-2$ Start at the left: $^-5 \div {}^-5$ is 1. The signs are the *same* so the quotient is *positive*.

$= 1(^-10) \div {}^-2$ Next, $1(^-10)$ is $^-10$. The signs are *different* so the product is *negative*.

$= {}^-10 \div {}^-2$ Finally, $^-10 \div {}^-2$ is 5. The signs are the *same* so the quotient is *positive*.

$= 5$

57. Descending implies a negative, $^-730$ feet each minute. Use front end rounding:

$^-730$ rounds to $^-700$ and 37 rounds to 40. Use multiplication to find how far the plane descended.

Estimate: $^-700 \cdot 40 = {}^-28{,}000$ feet

Exact: $^-730 \cdot 37 = {}^-27{,}010$ feet

The exact answer of $^-27{,}010$ feet is close to the estimate of $^-28{,}000$ feet.

69. Use division to find the number of rooms.

$$\dfrac{163 \text{ people}}{5 \text{ people per room}}$$

```
      3 2
   5 )1 6 3
      1 5
      1 3
      1 0
        3
```

32 rooms will be full, and that leaves 3 people. So 33 rooms are needed. One room will have only 3 people in it.

73. $^-6(^-8) \div (^-5 - {}^-5)$ Work inside parentheses.

$= {}^-6(^-8) \div (^-5 + {}^+5)$ Change subtraction to addition and $^-5$ to $^+5$. Then add $^-5 + 5$ to get 0.

$= {}^-6 \cdot {}^-8 \div 0$ Multiply $^-6 \cdot {}^-8$ to get 48.

$= 48 \div 0$ Division by zero is **undefined.**

Undefined

Section 1.8 (pages 71–74)

13. (a) $4^1 = 4$

(b) $4^2 = 4 \cdot 4 = 16$

(c) $4^3 = 4 \cdot 4 \cdot 4$

$= 16 \cdot 4$

$= 64$

(d) $4^4 = 4 \cdot 4 \cdot 4 \cdot 4$ Multiply two factors at a time.

$= 16 \cdot 4 \cdot 4$

$= 64 \cdot 4$

$= 256$

21. $(^-5)^2 = (^-5)(^-5)$ *Same* signs.

$= 25$ Product is *positive*.

27. $(^-10)^3 = (^-10)(^-10)(^-10)$

$= 100(^-10)$

$= {}^-1000$

35. $6^1(^-5)^3$ Apply the exponents first. 6^1 is 6 and $(^-5)^3$ is $(^-5)(^-5)(^-5)$ or $^-125$

$= 6(^-125)$

$= {}^-750$

51. $^-7 + 6(8 - 14)$ Work inside parentheses first.

$= {}^-7 + 6(8 + {}^-14)$ Change subtraction to adding the opposite, so $8 + {}^-14$ is $^-6$.

$= {}^-7 + 6(^-6)$ Multiply $6(^-6)$ to get $^-36$.

$= {}^-7 + {}^-36$ Add last.

$= {}^-43$

59. $2 - {}^-5(^-2)^3$ Apply the exponent: $(^-2)^3$ is $^-8$.

$= 2 - {}^-5(^-8)$ Multiply $^-5(^-8)$ to get 40.

$= 2 - 40$ Change $2 - 40$ to $2 + {}^-40$.

$= 2 + {}^-40$ Add last.

$= {}^-38$

67. $4(3^2) + 7(3 + 9) - {}^-6$ Work inside parentheses: $3 + 9$ is 12.

$= 4(3^2) + 7(12) - {}^-6$ Apply exponent: 3^2 is 9.

$= 4(9) + 7(12) - {}^-6$ Multiply $4(9)$ to get 36.

$= 36 + 7(12) - {}^-6$ Multiply $7(12)$ to get 84.

$= 36 + 84 - {}^-6$ Change $84 - {}^-6$ to $84 + {}^+6$.

$= 36 + 84 + 6$ Add $36 + 84$.

$= 120 + 6$ Add $120 + 6$.

$= 126$

75. $\dfrac{2^3 \cdot ({}^-2 - 5) + 4({}^-1)}{4 + 5({}^-6 \cdot 2) + (5 \cdot 11)}$

First do the work in the numerator.

$2^3 \cdot ({}^-2 - 5) + 4({}^-1)$ Work inside parentheses: Change ${}^-2 - 5$ to ${}^-2 + {}^-5$ to get ${}^-7$.

$= 2^3 \cdot ({}^-7) + 4({}^-1)$ Apply the exponent: 2^3 is 8.

$= 8 \cdot ({}^-7) + 4({}^-1)$ Multiply $8({}^-7)$ to get ${}^-56$.

$= {}^-56 + 4({}^-1)$ Multiply $4({}^-1)$ to get ${}^-4$.

$= {}^-56 + {}^-4$ Add last.

$= {}^-60$

Next, do the work in the denominator.

$4 + 5({}^-6 \cdot 2) + (5 \cdot 11)$ Work inside parentheses: ${}^-6 \cdot 2$ is ${}^-12$.

$= 4 + 5({}^-12) + (5 \cdot 11)$ Work inside parentheses: $5 \cdot 11$ is 55.

$= 4 + 5({}^-12) + 55$ Multiply $5({}^-12)$ to get ${}^-60$.

$= 4 + {}^-60 + 55$ Add from left to right.

$= {}^-56 + 55$

$= {}^-1$

The last step is the division:

$\dfrac{{}^-60}{{}^-1} = 60$

Chapter 2 Understanding Variables and Solving Equations

Section 2.1 (pages 101–104)

17. Expression (rule) for ordering brushes:

$3c - 5$

(a) Evaluate the expression when c, the class size, is 12.

$3c - 5$ Replace c with 12.

$\underbrace{3 \cdot 12}_{36} - 5$ Multiply before subtracting.

$36 - 5$

31 brushes must be ordered.

(b) Evaluate the expression when c, the class size, is 16.

$3c - 5$ Replace c with 16.

$\underbrace{3 \cdot 16}_{48} - 5$ Multiply before subtracting.

$48 - 5$

43 brushes must be ordered.

---

37. $9xy^2$ can be written as $9 \cdot x \cdot y \cdot y$. The exponent 2 applies only to the base y and means that y is used as a factor two times.

51. Evaluate $r^2 s^5 t^3$ when r is ${}^-3$, s is 2, and t is ${}^-4$, using a calculator.

$r^2 s^5 t^3$ Replace r with ${}^-3$, s with 2, and t with ${}^-4$.

$({}^-3)^2 (2)^5 ({}^-4)^3$ Use the y^x key for each exponent.

$(9)(32)({}^-64)$ Multiply from left to right.

$(288)({}^-64)$

${}^-18{,}432$

57. Evaluate $\dfrac{z^2}{{}^-3y + z}$ when z is ${}^-6$ and y is ${}^-2$.

$\dfrac{z^2}{{}^-3y + z}$ Replace z with ${}^-6$ and y with ${}^-2$.

$\dfrac{({}^-6)^2}{{}^-3({}^-2) + {}^-6}$ Follow the order of operations.

$\dfrac{36}{0}$ Numerator: $({}^-6)^2$ is ${}^-6 \cdot {}^-6$ or 36. Denominator: ${}^-3({}^-2) + {}^-6$ is $6 + {}^-6$ or 0.

Undefined   Division by 0 is undefined.

Section 2.2 (pages 115–118)

15. $\underbrace{c - c}_{0}$ Any number minus itself is 0.

23. ${}^-x - 6x - x$ These are like terms. Rewrite ${}^-x$ as ${}^-1x$ and x as $1x$.

${}^-1x - 6x - 1x$ Change both subtractions to adding the opposite.

${}^-1x + {}^-6x + {}^-1x$ Add the coefficients.

$\dfrac{({}^-1 + {}^-6 + {}^-1)x}{{}^-8x}$ The variable part, x, stays the same.

35. ${}^-x^3 + 3x - 3x^2 + 2$

There are no like terms. The expression cannot be simplified.

57. ${}^-2(5r + 3)$ Use the distributive property.

${}^-2 \cdot 5r + {}^-2 \cdot 3$

${}^-10r + {}^-2 \cdot 3$ Change addition to subtracting the opposite.

${}^-10r + {}^-6$

${}^-10r - 6$

69. ${}^-5(k + 5) + 5k$ Use the distributive property.

${}^-5 \cdot k + {}^-5 \cdot 5 + 5k$

${}^-5k + {}^-5 \cdot 5 + 5k$ Multiply ${}^-5 \cdot 5$ to get ${}^-25$.

${}^-5k + {}^-25 + 5k$

${}^-5k + 5k + {}^-25$ Rewrite using the commutative property.

$\underbrace{({}^-5 + 5)k}_{} + {}^-25$ Add the coefficients of like terms.

$\underbrace{0k}_{} + {}^-25$ Zero times any number is 0.

$\underbrace{0 + {}^-25}_{{}^-25}$ Zero added to any number is the number.

73. $5 + 2(3n + 4) - n$ Use the distributive property.

$5 + 2 \cdot 3n + 2 \cdot 4 - n$

$5 + 2 \cdot 3n + 2 \cdot 4 - 1n$ Rewrite n as $1n$.

$5 + 6n + 2 \cdot 4 - 1n$ Multiply $2 \cdot 3n$ to get $6n$.

$5 + 6n + 8 - 1n$ Multiply $2 \cdot 4$ to get 8.

$5 + 6n + 8 + {}^-1n$ Change subtraction to adding the opposite.

$6n + {}^-1n + 5 + 8$ Rewrite using the commutative property.

$5n + 5 + 8$ Add the coefficients of like terms; $6 + {}^-1$ is 5.

$5n + 13$ Add $5 + 8$ to get 13.

85. ${}^-10 + 4({}^-3b + 3) + 2(6b - 1)$

Use the distributive property.

${}^-10 + 4 \cdot {}^-3b + 4 \cdot 3 + 2 \cdot 6b - 2 \cdot 1$

Do all the multiplication.

${}^-10 + {}^-12b + 12 + 12b - 2$

Change subtraction to adding the opposite.

${}^-10 + {}^-12b + 12 + 12b + {}^-2$

Group like terms and add the coefficients.

$\underbrace{{}^-12b + 12b}_{0b} + \underbrace{{}^-10 + 12 + {}^-2}_{0}$

Any number times 0 is 0, so $0b$ is 0.

Simplified expression: 0

Section 2.3 (pages 129–134)

21. To solve ${}^-5 = {}^-2 + t$, we want to get the variable, t, by itself on the right side of the equal sign. To do that, we add the opposite of ${}^-2$, which is 2. Then ${}^-2 + 2$ will be 0.

${}^-5 = {}^-2 + t$ Add the opposite of ${}^-2$, which is 2, to both sides.

$\dfrac{2}{{}^-3} = \dfrac{2}{0 + t}$ Adding 0 to t leaves t unchanged.

${}^-3 = t$ The solution is ${}^-3$.

Check ${}^-5 = {}^-2 + t$ Replace t with ${}^-3$.

${}^-5 = {}^-2 + {}^-3$

${}^-5 = {}^-5$ Balances

27. The given solution is 10.

Check ${}^-10 = {}^-10 + b$ Replace b with 10.

${}^-10 = {}^-10 + 10$

${}^-10 \neq 0$ Does **not** balance; 10 is **not** the correct solution.

Find the correct solution.

${}^-10 = {}^-10 + b$ Add the opposite of ${}^-10$, which is 10, to both sides. Then ${}^-10 + 10$ is 0.

$\dfrac{10}{0} = \dfrac{10}{0 + b}$ Adding 0 to b leaves b unchanged.

$0 = b$ The solution is 0.

(continued)

Check $^-10 = {}^-10 + b$ Replace b with 0.

$$^-10 = {}^-10 + 0$$
$$^-10 = {}^-10 \quad \text{Balances}$$

The correct solution is 0.

39. $^-5w + 2 + 6w = {}^-4 + 9$ Rearrange and combine like terms.

$\underbrace{^-5w + 6w} + 2 = \underbrace{^-4 + 9}$ On the left side, combine $^-5w + 6w$.

$1w \quad\; + 2 = \quad 5$ On the right side, Add $^-4 + 9$.

$\qquad\qquad\qquad\qquad$ Add $^-2$ to

$\dfrac{\quad\; -2 \quad\; -2 \quad}{1w \quad\; + 0 =\quad 3}$ both sides.

$\qquad\qquad\qquad\qquad$ $1w$ is equivalent to w.

$\qquad\qquad w = 3$ The solution is 3.

Check

$^-5w + 2 + 6w = {}^-4 + 9$ Replace w with 3.

$^-5 \cdot 3 + 2 + 6 \cdot 3 = {}^-4 + 9$ Multiply.

$^-15 + 2 + 18 = {}^-4 + 9$ Add.

$\qquad\qquad\quad 5 = 5$ Balances

57. $^-9 + 9 = \; 5 \; + h$ Combine like terms.

$\qquad 0 = \; 5 \; + h$

$\dfrac{\quad -5 \quad\; -5 \quad}{\quad -5 =\; 0 \; + h}$ Add $^-5$ to both sides.

$\qquad -5 = h$ The solution is $^-5$.

69. $^-6x + 2x + 6 + 5x$

$= |0 - 9| - |{}^-6 + 5|$

On the left side, rearrange the terms and add like terms.

$^-6x + 2x + 5x + 6 = |0 - 9| - |{}^-6 + 5|$

$x + 6 = |0 - 9| - |{}^-6 + 5|$

On the right side, work inside each set of absolute value bars:

$|0 - 9|$ is $|0 + {}^-9|$ or $|{}^-9|$ or 9

$|{}^-6 + 5|$ is $|{}^-1|$ or 1

So the equation becomes:

$x + 6 = 9 - 1$

Change subtraction to adding the opposite.

$x + 6 = \; 9 + {}^-1$ Add $9 + {}^-1$ to get 8.

$x + 6 = \; 8$

$\dfrac{\quad -6 \quad\; -6 \quad}{x + 0 = \quad 2}$ Add $^-6$ to both sides.

$\qquad x = \quad 2$ The solution is 2.

Section 2.4 (pages 139–142)

15. $^-12 = 5p - p$ Change subtraction to adding the opposite.

$^-12 = 5p + {}^-p$ Rewrite ^-p as ^-1p.

$^-12 = 5p + {}^-1p$ Combine like terms.

$^-12 = 4p$

$\dfrac{^-12}{4} = \dfrac{4p}{4}$ Divide both sides by 4.

$^-3 \; = p$ The solution is $^-3$.

Check

$^-12 = 5p - p$ Replace p with $^-3$.

$^-12 = 5 \cdot {}^-3 - {}^-3$ Multiply.

$^-12 = {}^-15 - {}^-3$ Change subtraction to adding the opposite.

$^-12 = {}^-15 + {}^+3$

$^-12 = {}^-12$ Balances

29. $100 - 96 = 31y - 35y$ Change subtractions to adding the opposite.

$100 + {}^-96 = 31y + {}^-35y$

$4 = {}^-4y$ Combine like terms.

$\dfrac{4}{^-4} = \dfrac{^-4y}{^-4}$ Divide both sides by $^-4$.

$^-1 = y$ The solution is $^-1$.

35. $^-2({}^-4k) = 56$ Use the associative property.

$({}^-2 \cdot {}^-4) \cdot k = 56$

$8k = 56$ Multiply $^-2 \cdot {}^-4$ to get 8.

$\dfrac{8k}{8} = \dfrac{56}{8}$ Divide both sides by 8.

$k = 7$ The solution is 7.

43. $^-n = {}^-50$ Rewrite ^-n as ^-1n.

$^-1n = {}^-50$

$\dfrac{^-1n}{^-1} = \dfrac{^-50}{^-1}$ Divide both sides by $^-1$.

$n = 50$ The solution is 50.

57. $^-37(14x) + 28(21x)$

$= |72 - 72| + |{}^-166 + 96|$

On the right side, simplify within the absolute value bars.

$^-37(14x) + 28(21x) = |0| + |{}^-70|$

Find the absolute values.

$^-37(14x) + 28(21x) = 0 + 70$

Use the associative property to multiply on the left side.

$({}^-37 \cdot 14) \cdot x + (28 \cdot 21) \cdot x = 0 + 70$

$^-518x + 588x = 0 + 70$

$^-518x + 588x = 70$

Combine like terms.

$70x = 70$

Divide both sides by the coefficient, 70, to get x by itself.

$\dfrac{70x}{70} = \dfrac{70}{70}$

$x = 1$

The solution is 1.

Section 2.5 (pages 147–152)

7. $28 = {}^-9a + 10$ To get ^-9a by itself, add $^-10$ to both sides.

$\dfrac{\quad -10 \qquad\qquad -10 \quad}{18 = {}^-9a + 0}$

$18 = {}^-9a$

$\dfrac{18}{^-9} = \dfrac{^-9a}{^-9}$ Divide both sides by $^-9$.

$^-2 = a$ The solution is $^-2$.

Check $28 = {}^-9a + 10$ Use the original equation and replace a with $^-2$.

$28 = {}^-9({}^-2) + 10$

$28 = 18 + 10$

$28 = 28$ Balances

15. Solve by keeping the variable on the *left* side.

$^-18 + 7a = 2a + 7$

$\dfrac{\quad -2a \qquad\;\; -2a \quad}{^-18 + 5a = 0 + 7}$ Add ^-2a to both sides.

$^-18 + 5a = 7$

$\dfrac{\;\; 18 \qquad\qquad 18 \quad}{0 + 5a = 25}$ Add 18 to both sides.

$5a = 25$

$\dfrac{5a}{5} = \dfrac{25}{5}$ Divide both sides by 5.

$a = 5$ The solution is 5.

Solve by keeping the variable on the *right* side.

$^-18 + 7a = 2a + 7$

$\dfrac{\quad -7a \qquad\qquad -7a \quad}{^-18 + 0 = {}^-5a + 7}$ Add ^-7a to both sides.

$^-18 = {}^-5a + 7$

$\dfrac{\quad -7 \qquad\qquad\;\; -7 \quad}{^-25 = {}^-5a + 0}$ Add $^-7$ to both sides.

$^-25 = {}^-5a$

$\dfrac{^-25}{^-5} = \dfrac{^-5a}{^-5}$ Divide both sides by $^-5$.

$5 = a$ The solution is 5.

Check $^-18 + 7a = 2a + 7$

$^-18 + 7(5) = 2(5) + 7$

$^-18 + 35 = 10 + 7$

$17 = 17$ Balances

27. $0 = {}^-2(y + 2)$ Use the distributive property.

$0 = {}^-2y + {}^-4$

$\dfrac{\;\; 4 \qquad\qquad\quad 4 \quad}{4 = {}^-2y + 0}$ Add 4 to both sides.

$4 = {}^-2y$

$\dfrac{4}{^-2} = \dfrac{^-2y}{^-2}$ Divide both sides by $^-2$.

$^-2 = y$ The solution is $^-2$.

37. $7 - 5b = 28 + 2b$ Change subtraction to adding the opposite.

$7 + {}^-5b = 28 + 2b$

$\dfrac{\quad 5b \qquad\qquad 5b \quad}{7 + 0 = 28 + 7b}$ Add $5b$ to both sides.

$7 = 28 + 7b$

$\dfrac{\quad -28 \qquad -28 \quad}{^-21 = 0 + 7b}$ Add $^-28$ to both sides.

$^-21 = 7b$

$\dfrac{^-21}{7} = \dfrac{7b}{7}$ Divide both sides by 7.

$^-3 = b$ The solution is $^-3$.

41. $10(c - 6) + 4 = 2 + c - 58$

Use the distributive property.

$10c - 60 + 4 = 2 + c - 58$

Rewrite subtractions as adding the opposite.

$10c + {}^-60 + 4 = 2 + c + {}^-58$

Combine like terms.

$$10c + {}^-56 = 1c + {}^-56$$

$$\frac{{}^-1c \qquad\qquad {}^-1c}{9c + {}^-56 = 0 + {}^-56} \quad \begin{array}{l}\text{Add } {}^-1c \text{ to}\\ \text{both sides.}\end{array}$$

$$9c + {}^-56 = {}^-56$$

$$\frac{56 \qquad 56}{9c + 0 = 0} \quad \begin{array}{l}\text{Add 56 to}\\ \text{both sides.}\end{array}$$

$$\frac{9c}{9} = \frac{0}{9} \quad \begin{array}{l}\text{Divide } both \text{ sides}\\ \text{by 9.}\end{array}$$

$$c = 0 \quad \text{The solution is 0.}$$

49. $\quad {}^-5(2p + 2) - 7 = 3(2p + 5)$

Use the distributive property on both sides.

$$\quad {}^-10p + {}^-10 - 7 = 6p + 15$$

Write subtraction as adding the opposite.

$$\quad {}^-10p + {}^-10 + {}^-7 = 6p + 15$$

Combine like terms.

$$\quad {}^-10p + {}^-17 = 6p + 15$$

$$\frac{{}^-6p \qquad\qquad {}^-6p}{{}^-16p + {}^-17 = 0 + 15} \quad \begin{array}{l}\text{Add } {}^-6p \text{ to}\\ \text{both sides.}\end{array}$$

$$\quad {}^-16p + {}^-17 = 15$$

$$\frac{17 \qquad 17}{{}^-16p + 0 = 32} \quad \begin{array}{l}\text{Add 17 to}\\ \text{both sides.}\end{array}$$

$$\quad {}^-16p = 32$$

$$\frac{{}^-16p}{{}^-16} = \frac{32}{{}^-16} \quad \begin{array}{l}\text{Divide both}\\ \text{sides by } {}^-16.\end{array}$$

$$p = {}^-2$$

The solution is ${}^-2$.

Chapter 3 Solving Application Problems

Section 3.1 (pages 171–174)

13. The perimeter of a square parking lot is 92 yards (yd). Use the formula for perimeter of a square, $P = 4s$.

$$P = 4s$$

$$92 \text{ yd} = 4s \quad \text{Replace } P \text{ with 92 yd.}$$

$$\frac{92 \text{ yd}}{4} = \frac{4s}{4} \quad \text{Divide both sides by 4.}$$

$$23 \text{ yd} = s$$

The length of one side of the parking lot is 23 yards.

25. Use the formula for the perimeter of a rectangle, $P = 2l + 2w$. The value of P is 30 cm, and the value of w is 6 cm.

$$P = 2l + 2w \quad \begin{array}{l}\text{Replace } P\\ \text{with 30 cm}\\ \text{and } w \text{ with}\\ \text{6 cm}\end{array}$$

$$30 \text{ cm} = 2l + 2 \cdot 6 \text{ cm} \quad \begin{array}{l}\text{Multiply}\\ \text{on the right}\\ \text{side.}\end{array}$$

$$30 \text{ cm} = 2l + 12 \text{ cm}$$

$$\frac{{}^-12 \text{ cm} \qquad\qquad {}^-12 \text{ cm}}{18 \text{ cm} = 2l + 0} \quad \begin{array}{l}\text{Add } {}^-12 \text{ cm}\\ \text{to both sides.}\end{array}$$

$$\frac{18 \text{ cm}}{2} = \frac{2l}{2} \quad \begin{array}{l}\text{Divide both}\\ \text{sides by 2.}\end{array}$$

$$9 \text{ cm} = l$$

The length is 9 cm.

Check

9 cm
6 cm [rectangle] 6 cm
9 cm

$$P = 9 \text{ cm} + 9 \text{ cm} + 6 \text{ cm} + 6 \text{ cm}$$

$$P = 30 \text{ cm}$$

30 cm matches the original perimeter, so 9 cm is the correct length.

35. Add all four sides to find the perimeter. Because the figure is a parallelogram, opposite sides have the same length.

$$P = 100 \text{ ft} + 60 \text{ ft} + 100 \text{ ft} + 60 \text{ ft}$$

$$P = 320 \text{ ft}$$

The perimeter is 320 ft.

45. $\quad P = 10 \text{ cm} + 10 \text{ cm} + 30 \text{ cm}$
$$\qquad + 25 \text{ cm} + \text{?}$$

$$P = 75 \text{ cm} + \text{?}$$

Because the perimeter is 115 cm, replace P with 115 cm.

$$115 \text{ cm} = 75 \text{ cm} + \text{?}$$

$$\frac{{}^-75 \text{ cm} \qquad {}^-75 \text{ cm}}{40 \text{ cm} = 0 \quad + \text{?}} \quad \begin{array}{l}\text{Add } {}^-75 \text{ cm to}\\ \text{both sides.}\end{array}$$

$$40 \text{ cm} = \text{?}$$

The length of the unknown side is 40 cm.

Section 3.2 (pages 181–184)

5. The figure is a parallelogram.

Use the formula $A = bh$.

Turn your book sideways to identify that the height is 25 mm and the base is 31 mm. Replace b with 31 mm and h with 25 mm in the formula.

$$A = 31 \text{ mm} \cdot 25 \text{ mm}$$

$$A = 775 \text{ mm}^2$$

The area is 775 mm^2

Be careful to use mm^2 in your answer (**not** mm) because this is *area*.

21. Use the formula for the area of a rectangle, $A = lw$. The value of A is 154 in.^{2} and the value of w is 11 in.

$$A = l \cdot w \quad \begin{array}{l}\text{Replace } A \text{ with}\\ 154 \text{ in.}^2, \text{ and}\\ w \text{ with 11 in.}\end{array}$$

$$154 \text{ in.}^2 = l \cdot 11 \text{ in.}$$

$$\frac{154 \text{ in.} \cdot \text{in.}}{11 \text{ in.}} = \frac{l \cdot 11 \text{ in.}}{11 \text{ in.}} \quad \begin{array}{l}\text{Divide both}\\ \text{sides by 11 in.}\end{array}$$

$$14 \text{ in.} = l$$

The length of the photo is 14 in.

Check

14 in.
11 in. [Photo $A = 154$ in.^{2}] 11 in.
14 in.

$$A = l \cdot w$$

$$A = 14 \text{ in.} \cdot 11 \text{ in.}$$

$$A = 154 \text{ in.}^2 \quad \begin{array}{l}\text{Matches area given}\\ \text{in the problem.}\end{array}$$

29. Use the area formula for a parallelogram, $A = bh$. Replace A with 221 in.^{2} and h with 13 in.

$$A = b \cdot h$$

$$221 \text{ in.}^2 = b \cdot 13 \text{ in.}$$

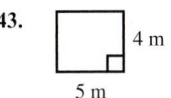

$$\frac{221 \text{ in.} \cdot \text{in.}}{13 \text{ in.}} = \frac{b \cdot 13 \text{ in.}}{13 \text{ in.}} \quad \begin{array}{l}\text{Divide both}\\ \text{sides by}\\ \text{13 in.}\end{array}$$

$$17 \text{ in.} = b$$

The base is 17 in.

Check

[parallelogram with 13 in. height and 17 in. base]

$$A = b \cdot h$$

$$A = 17 \text{ in.} \cdot 13 \text{ in.}$$

$$A = 221 \text{ in.}^2 \leftarrow \text{Matches}$$

43.

[rectangle 5 m by 4 m]

Tyra is decorating the top edges of her walls, so find the *perimeter* of her ceiling.

$$P = 2 \cdot l + 2 \cdot w \quad \begin{array}{l}\text{Perimeter of a}\\ \text{rectangle.}\end{array}$$

$$P = 2 \cdot 5 \text{ m} + 2 \cdot 4 \text{ m} \quad \begin{array}{l}\text{Replace } l \text{ with}\\ \text{5 m and } w\\ \text{with 4 m.}\end{array}$$

$$P = 10 \text{ m} + 8\text{m}$$

$$P = 18 \text{ m}$$

The strip costs $6 per meter. To find the cost of 18 meters, multiply $6(18) to get $108.

To have the top edges of her walls decorated, Tyra will spend $108.

49. The floor of the Coleman tent is a square.

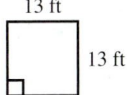

13 ft [square] 13 ft

To find the perimeter, use the formula for the perimeter of a square, $P = 4s$.

$$P = 4s \quad \begin{array}{l}\text{Formula for perimeter of}\\ \text{a square.}\end{array}$$

$$P = 4 \cdot 13 \text{ ft} \quad \text{Substitute 13 ft for } s.$$

$$P = 52 \text{ ft}$$

The perimeter of the tent is 52 ft.

To find the area, use the formula for the area of a square, $A = s^2$.

$$A = s^2 \quad \begin{array}{l}\text{Formula for area of}\\ \text{a square}\end{array}$$

$$A = s \cdot s$$

$$A = 13 \text{ ft} \cdot 13 \text{ ft} \quad \begin{array}{l}\text{Multiply } 13 \cdot 13 \text{ to}\\ \text{get 169.}\end{array}$$

$$A = 169 \text{ ft}^2 \quad \begin{array}{l}\text{Multiply ft} \cdot \text{ft to}\\ \text{get ft}^2.\end{array}$$

The area of the tent is 169 ft^2.

Because the tent sleeps 8 campers, divide the area of the floor, 169 ft^2, by 8.

$$\frac{169 \text{ ft}^2}{8} \approx 21 \text{ ft}^2$$

There is about 21 ft^2 of floor space for each camper.

Section 3.3 (pages 193–196)

29. Let n represent the unknown number.

30	subtract	3 times a number	is	2 plus	the number
↓	↓	↓	↓ ↓	↓	↓
30	−	$3n$	= 2	+	n

$$30 - 3n = 2 + n$$

$$\underline{\quad +3n \qquad\qquad +3n\quad}$$ Add $3n$ to both sides.

$$30 + 0n = 2 + 4n$$

$$\underline{\quad -2 \qquad\qquad -2\quad}$$ Add $^-2$ to both sides.

$$28 \qquad\quad = 0 + 4n$$

$$28 = 4n$$

$$\frac{28}{4} = \frac{4n}{4}$$ Divide both sides by 4.

$$7 = n$$

The number is 7.

Check Three times 7 is 21. When 21 is subtracted from 30, the result is 9; 2 plus 7 is also 9. *True*

33. *Step 1*

The problem is about the number of cookies in the cookie jar.

Unknown: number of cookies the children ate

Known: 18 cookies at the start, three dozen cookies put in the jar, 49 cookies at the end

Step 2(a)

Let c represent the number of cookies the children ate.

Step 3

Cookies at the start	−	Cookies eaten	+	Cookies added	=	Ended up with 49 cookies
↓		↓		↓		↓
18	−	c	+	36	=	49

Step 4

$$18 - c + 36 = 49$$ Write the understood $1c$. Change subtraction to adding the opposite.

$$54 + (^-1c) = 49$$

$$\underline{\quad -54 \qquad\qquad -54\quad}$$ Add $^-54$ to both sides.

$$0 + (^-1c) = ^-5$$

$$\frac{^-1c}{^-1} = \frac{^-5}{^-1}$$ Divide both sides by $^-1$.

$$c = 5$$

Step 5 The children ate 5 cookies.

Step 6

There were 18 cookies at the start. The children ate 5 cookies, so $18 - 5 = 13$ cookies in the jar.

Three dozen cookies were added. There are 12 cookies in a dozen, so 36 cookies were added. $13 + 36 = 49$ cookies, which matches the original problem.

The correct answer is 5 cookies.

37. *Step 1*

The problem is about the bank account of a music club.

Unknown: the dues paid by each member

Known: 14 members, earned $340, spent $575, and account is overdrawn by $25

Step 2(a)

Let d represent the dues paid by each member.

Step 3

Total dues paid	+ $340	− $575	=	Overdrawn bank account
↓	↓	↓		↓
14 · d	+ 340	− 575	=	$^-25$

Step 4

$$14d + 340 - 575 = ^-25$$

$$14d + 340 + (^-575) = ^-25$$ Add on the left.

$$14d + (^-235) = ^-25$$

$$\underline{\qquad\quad +235 \qquad +235\quad}$$ Add 235 to both sides.

$$14d + 0 = 210$$

$$\frac{14d}{14} = \frac{210}{14}$$ Divide by 14.

$$d = 15$$

Step 5 Each member paid $15 in dues.

Step 6

14 members paid dues of $15 each.
$14(\$15) = \$210$ in the bank account.
Earned $340, so
$\$210 + \$340 = \$550$ in the account.
Spent $575, so
$\$550 - \$575 = \$550 + ^-575 = ^-\$25$ in the account.

The account was overdrawn by $25, which matches the original problem.

The correct answer is $15 dues.

41. *Step 1*

The problem is about spending money for clothes.

Unknown: amount spent by Brenda

Known: Consuelo spent $3 less than twice the amount that Brenda spent, Consuelo spent $81.

Step 2(a)

Let m represent the amount of money that Brenda spent.

Step 3

Consuelo spent	=	$3 less than twice the amount that Brenda spent
↓		↓
81	=	$2m - 3$

Step 4

$$81 = 2m - 3$$ Change subtraction to adding the opposite.

$$81 = 2m + (^-3)$$

$$\underline{\,+3 \qquad\qquad +3\,}$$ Add 3 to both sides.

$$84 = 2m + 0$$

$$\frac{84}{2} = \frac{2m}{2}$$ Divide both sides by 2.

$$42 = m$$

Step 5 Brenda spent $42 on clothes.

Step 6

Consuelo spent $3 less than twice the amount that Brenda spent, or

$2 \cdot \$42 + (^-\$3) = \$84 - (\$3) = \$81$, which matches the original problem.

The correct answer is $42.

Section 3.4 (pages 201–202)

5. *Step 1*

The problem is about the price of a computer and the price of a printer.

Unknowns: the computer's price; the printer's price

Known: The computer's price is five times the printer's price. The total paid for both items is $1320.

Step 2(b)

You know the least about the printer's price, so let m represent the printer's price. The price of the computer is 5 times as much, so it can be represented by $5 \cdot m$, or $5m$.

Step 3

The total price for both items is $1320.

Printer price	+	Computer price	=	Total price
↓		↓		↓
m	+	$5m$	=	1320

Step 4

$$m + 5m = 1320$$ Combine like terms.

$$\frac{6m}{6} = \frac{1320}{6}$$ Divide both sides by 6.

$$m = 220$$

Step 5

The printer's price is m, so the printer cost $220. The computer's price is $5m$, so the computer cost $5 \cdot \$220 = \$1100$.

Step 6

Five times $220 is $1100 and the sum of $220 and $1100 is $1320, so the solution checks.

13. *Step 1*

The problem is about cutting a fence into three parts.

Unknowns: Length of first part; length of second part; length of third part

Known: Two parts are of equal length. The third part is 25 m longer than each of the other parts. The entire fence is 706 m long.

Step 2(b)

Let x represent the length of the first part. Because the second part is the same length as the first, use another x for the second part. Then the third part is 25 m longer, so its length is $x + 25$.

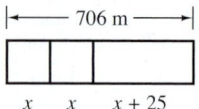

Step 3

The lengths of the three parts add up to 706.

$$x + x + (x + 25) = 706$$

Step 4

$$x + x + x + 25 = 706$$ Combine like terms.

$$3x + 25 = 706$$

$$\underline{\qquad -25 \qquad -25\quad}$$ Add $^-25$ to both sides.

$$3x + 0 = 681$$

$$\frac{3x}{3} = \frac{681}{3}$$ Divide both sides by 3.

$$x = 227$$

Step 5

x is the length of the first part and the second part, so each of those parts is 227 m long. The third part is $x + 25$ or $227 + 25 = 252$ m in length.

The three parts are 227 m, 227 m, and 252 m.

Step 6

Since $252 - 227 = 25$ and $227 + 227 + 252 = 706$, the solution checks.

19. Step 1

The problem is about the length and the width of a jewelry box.

Unknowns: The length of the box; width of the box

Known: The length is 3 in. more than twice the width. The perimeter is 36 in.

Step 2(b)

You know the least about the width, so let w represent the width. The length is 3 in. more than twice the width, so $2w + 3$ represents the length.

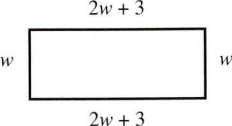

Step 3

Use the formula for the perimeter of a rectangle, $P = 2l + 2w$, or $2l + 2w = P$.

$2 \cdot (2w + 3) + 2 \cdot w = 36$

Step 4

$2(2w + 3) + 2w = 36$ — Use the distributive property.

$4w + 6 + 2w = 36$ — Combine like terms.

$6w + 6 = 36$

$\underline{ -6 \quad -6}$ — Add $^-6$ to both sides.

$\dfrac{6w}{6} = \dfrac{30}{6}$ — Divide both sides by 6.

$w = 5$

Step 5

w is the width, so the width is 5 in.
The length is $2w + 5$, or $2(5) + 3 = 13$ in.

Step 6

The length of 13 in. is 3 in. longer than twice the width of 5 in.

$P = 2l + 2w$

$P = 2 \cdot 13 + 2 \cdot 5$

$P = 26 + 10$

$P = 36$ in. $\rightarrow$ Matches perimeter given in the problem

Chapter 4 Rational Numbers: Positive and Negative Fractions

Section 4.1 (pages 223–226)

7. An area equal to 11 of the $\frac{1}{6}$ parts is shaded: $\frac{11}{6}$

An area equal to 1 of the $\frac{1}{6}$ parts is unshaded: $\frac{1}{6}$

13. (a) Under the circle, it says, "out of every 20 women surveyed." In the circle, it shows 6 women choosing "at work." So 6 out of every 20 women would like flowers delivered at work.

The fraction is $\frac{6}{20}$.

(b) $13 + 6 = 19$ out of every 20 women would like flowers delivered either at home or at work: $\frac{19}{20}$

37. (a) $-\dfrac{2}{6} = -\dfrac{2 \div 2}{6 \div 2} = -\dfrac{1}{3}$

(b) $-\dfrac{4}{6} = -\dfrac{4 \div 2}{6 \div 2} = -\dfrac{2}{3}$

(c) $-\dfrac{12}{18} = -\dfrac{12 \div 6}{18 \div 6} = -\dfrac{2}{3}$

(d) $-\dfrac{6}{18} = -\dfrac{6 \div 6}{18 \div 6} = -\dfrac{1}{3}$

(e) $-\dfrac{200}{300} = -\dfrac{200 \div 100}{300 \div 100} = -\dfrac{2}{3}$

(f) Some possibilities are:

$-\dfrac{1}{3} = -\dfrac{1 \cdot 4}{3 \cdot 4} = -\dfrac{4}{12}$;

$-\dfrac{1}{3} = -\dfrac{1 \cdot 8}{3 \cdot 8} = -\dfrac{8}{24}$;

$-\dfrac{2}{3} = -\dfrac{2 \cdot 10}{3 \cdot 10} = -\dfrac{20}{30}$;

$-\dfrac{2}{3} = -\dfrac{2 \cdot 12}{3 \cdot 12} = -\dfrac{24}{36}$.

57. $\dfrac{150}{150} = 150 \div 150 = 1$

59. $-\dfrac{32}{4} = -32 \div 4 = -8$

Section 4.2 (pages 235–238)

17. (a)

```
       75
      /  \
    (3)   25
         /  \
       (5)  (5)
```

The prime factorization of 75 is $3 \cdot 5 \cdot 5$.

(b)

```
       68
      /  \
    (2)   34
         /  \
       (2)  (17)
```

The prime factorization of 68 is $2 \cdot 2 \cdot 17$.

(c) $3\overline{)189}$ ← 189 is not divisible by 2 (first prime) so use 3 (next prime); quotient is 63

$3\overline{)63}$ ← Divide 63 by 3; quotient is 21

$3\overline{)21}$ ← Divide 21 by 3; quotient is 7

$7\overline{)7}$ ← Divide 7 by 7

1 ← Quotient is 1

The prime factorization of 189 is $3 \cdot 3 \cdot 3 \cdot 7$.

35. The prime factorization of 90 is $2 \cdot 3 \cdot 3 \cdot 5$. The prime factorization of 180 is $2 \cdot 2 \cdot 3 \cdot 3 \cdot 5$.

$\dfrac{90}{180} = \dfrac{\cancel{2} \cdot \cancel{3} \cdot \cancel{3} \cdot \cancel{5}}{\cancel{2} \cdot 2 \cdot \cancel{3} \cdot \cancel{3} \cdot \cancel{5}} = \dfrac{1}{2}$

41. 60 minutes in an hour means 60 parts in the whole.

(a) $\dfrac{15}{60} = \dfrac{\cancel{3} \cdot \cancel{5}}{2 \cdot 2 \cdot \cancel{3} \cdot \cancel{5}} = \dfrac{1}{4}$ of an hour

(b) $\dfrac{30}{60} = \dfrac{\cancel{2} \cdot \cancel{3} \cdot \cancel{5}}{2 \cdot \cancel{2} \cdot \cancel{3} \cdot \cancel{5}} = \dfrac{1}{2}$ of an hour

(c) $\dfrac{6}{60} = \dfrac{\cancel{2} \cdot \cancel{3}}{\cancel{2} \cdot 2 \cdot \cancel{3} \cdot 5} = \dfrac{1}{10}$ of an hour

(d) $\dfrac{60}{60} = 60 \div 60 = 1$ hour

57. $\dfrac{9x^2}{16y^2} = \dfrac{3 \cdot 3 \cdot x \cdot x}{2 \cdot 2 \cdot 2 \cdot 2 \cdot y \cdot y}$

There are no common factors, so the fraction is already in lowest terms.

67. $\dfrac{210ab^3c}{35b^2c^2}$

$= \dfrac{2 \cdot 3 \cdot \cancel{5} \cdot \cancel{7} \cdot a \cdot \cancel{b} \cdot \cancel{b} \cdot b \cdot \cancel{c}}{\cancel{5} \cdot \cancel{7} \cdot \cancel{b} \cdot \cancel{b} \cdot \cancel{c} \cdot c}$

$= \dfrac{6ab}{c}$

Section 4.3 (pages 247–250)

9. $\dfrac{4}{9}$ of $81 = \dfrac{4}{9} \cdot \dfrac{81}{1} = \dfrac{4 \cdot \cancel{9} \cdot 9}{\cancel{9} \cdot 1} = \dfrac{36}{1} = 36$

11. $\left(\dfrac{3x}{4}\right)\left(\dfrac{5}{xy}\right) = \dfrac{\overbrace{3 \cdot x} \cdot 5}{2 \cdot 2 \cdot \underbrace{x \cdot y}}$

$3x$ means $3 \cdot x$

$= \dfrac{3 \cdot \cancel{x} \cdot 5}{2 \cdot 2 \cdot \cancel{x} \cdot y} = \dfrac{15}{4y}$

19. Signs do not match, so the quotient will be negative.

$-\dfrac{2}{3} \div 4 = -\dfrac{2}{3} \div \dfrac{4}{1} = -\dfrac{2}{3} \cdot \dfrac{1}{4}$

Reciprocals

$= -\dfrac{\cancel{2} \cdot 1}{3 \cdot \cancel{2} \cdot 2} = -\dfrac{1}{6}$

23. $\dfrac{ab^2}{c} \div \dfrac{ab}{c} = \dfrac{ab^2}{c} \cdot \dfrac{c}{ab} = \dfrac{\cancel{a} \cdot \cancel{b} \cdot b \cdot \cancel{c}}{\cancel{c} \cdot \cancel{a} \cdot \cancel{b}}$

Reciprocals

$= \dfrac{b}{1} = b$

37. Signs match, so the quotient will be positive.

$$-9 \div \left(-\frac{3}{5}\right) = -\frac{9}{1} \cdot \left(-\frac{5}{3}\right)$$

Reciprocals

$$= \frac{\overset{1}{\cancel{3}} \cdot 3 \cdot 5}{1 \cdot \cancel{3}} = \frac{15}{1} = 15$$

43. $\dfrac{4}{7}$ of $14b = \dfrac{4}{7} \cdot 14b = \dfrac{4}{7} \cdot \dfrac{14b}{1}$

$$= \frac{2 \cdot 2 \cdot 2 \cdot \overset{1}{\cancel{7}} \cdot b}{\underset{1}{\cancel{7}} \cdot 1} = \frac{8b}{1} = 8b$$

53. Todd must earn $\dfrac{3}{4}$ of the $12,400 cost

$$\frac{3}{4} \cdot 12,400 = \frac{3}{4} \cdot \frac{12,400}{1} = \frac{3 \cdot \overset{1}{\cancel{4}} \cdot 3100}{\underset{1}{\cancel{4}} \cdot 1}$$

$$= \frac{9300}{1} = \$9300 \text{ earned}$$

Todd must borrow the rest:
$12,400 − $9300 = $3100

67. Dogs: $175 \cdot \dfrac{2}{5} = \dfrac{175}{1} \cdot \dfrac{2}{5}$

$$= \frac{\overset{1}{\cancel{5}} \cdot 35 \cdot 2}{1 \cdot \underset{1}{\cancel{5}}} = 70$$

Cats: $175 \cdot \dfrac{12}{25} = \dfrac{175}{1} \cdot \dfrac{12}{25}$

$$= \frac{7 \cdot \overset{1}{\cancel{25}} \cdot 12}{\underset{1}{\cancel{25}}}$$

$$= 84$$

$70 + 84 = 154$ million U.S. pets that are dogs or cats.

Section 4.4 (pages 259–262)

7. $\dfrac{3}{8} - \dfrac{3}{5}$

Step 1   The LCD is $8 \cdot 5 = 40$.

Step 2

$$\frac{3}{8} = \frac{3 \cdot 5}{8 \cdot 5} = \frac{15}{40} \quad \text{and} \quad \frac{3}{5} = \frac{3 \cdot 8}{5 \cdot 8} = \frac{24}{40}$$

Step 3   Subtract the numerators. Write the difference over the common denominator.

$$\frac{3}{8} - \frac{3}{5} = \frac{15}{40} - \frac{24}{40}$$

$15 + (-24)$ is -9

$$= \frac{15 - 24}{40} = \frac{-9}{40} \quad \text{or} \quad -\frac{9}{40}$$

Step 4 $-\dfrac{9}{40}$ is already in lowest terms.

15. $2 - \dfrac{6}{7} = \dfrac{2}{1} - \dfrac{6}{7}$

Step 1   The LCD is 7.

Step 2

$$2 = \frac{2 \cdot 7}{1 \cdot 7} = \frac{14}{7} \quad \text{and} \quad \frac{6}{7} \text{ has the LCD.}$$

Step 3

$$\frac{2}{1} - \frac{6}{7} = \frac{14}{7} - \frac{6}{7} = \frac{14 - 6}{7} = \frac{8}{7}$$

Step 4 $\dfrac{8}{7}$ is already in lowest terms.

21. $\dfrac{5}{m} - \dfrac{1}{2}$

Step 1   The LCD is $2 \cdot m$ or $2m$.

Step 2

$$\frac{5}{m} = \frac{5 \cdot 2}{m \cdot 2} = \frac{10}{2m}, \quad \frac{1}{2} = \frac{1 \cdot m}{2 \cdot m} = \frac{1m}{2m}$$

Step 3

$$\frac{5}{m} - \frac{1}{2} = \frac{10}{2m} - \frac{1m}{2m}$$

$$= \frac{10 - 1m}{2m} \quad \text{or} \quad \frac{10 - m}{2m}$$

Step 4   Already in lowest terms.

27. $-\dfrac{4}{c^2} - \dfrac{d}{c}$

Step 1   The LCD is c^2.

Step 2

$$-\frac{4}{c^2} \text{ has the LCD} \quad \text{and} \quad \frac{d}{c} = \frac{d \cdot c}{c \cdot c} = \frac{cd}{c^2}$$

Step 3

$$-\frac{4}{c^2} - \frac{d}{c} = -\frac{4}{c^2} - \frac{cd}{c^2} = \frac{-4 - cd}{c^2}$$

Step 4   Already in lowest terms.

47. $\dfrac{1}{8}$ of the day is spent in class. $\dfrac{1}{6}$ of the day is spent in study. Add to find the total fraction.

$$\frac{1}{8} + \frac{1}{6} = \frac{3}{24} + \frac{4}{24} = \frac{3 + 4}{24} = \frac{7}{24}$$

To find the number of hours, multiply the fraction by 24 hours in a day.

$$\frac{7}{24} \cdot 24 = \frac{7}{24} \cdot \frac{24}{1} = \frac{7 \cdot \cancel{24}}{\cancel{24}} = 7$$

$\dfrac{7}{24}$ of the day (or 7 hours) was spent in class and study.

53. If the total perimeter is $\dfrac{7}{8}$ mile, use subtraction to find the length of the fourth side. The LCD of the fractions is 24.

Length

$$= \frac{7}{8} - \frac{1}{4} - \frac{1}{6} - \frac{3}{8}$$

Change subtractions to adding the opposite.

$$= \frac{7}{8} + \left(-\frac{1}{4}\right) + \left(-\frac{1}{6}\right) + \left(-\frac{3}{8}\right)$$

$$= \frac{21}{24} + \left(-\frac{6}{24}\right) + \left(-\frac{4}{24}\right) + \left(-\frac{9}{24}\right)$$

$$= \frac{21 + (-6) + (-4) + (-9)}{24}$$

Add in the numerator; keep the common denominator.

$$= \frac{2}{24} = \frac{1}{12} \text{ mile}$$

The fourth side is $\dfrac{1}{12}$ mile long.

Section 4.5 (pages 275–278)

13. $10\dfrac{11}{15}$

Step 1:

$15 \cdot 10 = 150; \quad 150 + 11 = 161$

Step 2: $10\dfrac{11}{15} = \dfrac{161}{15}$ Keep the same denominator.

21. Divide 51 by 9.

$$\begin{array}{r} 5 \\ 9\overline{\smash{)}5\ 1} \\ \underline{4\ 5} \\ 6 \end{array}$$

Keep the negative sign.

so $-\dfrac{51}{9} = -5\dfrac{6}{9} = -5\dfrac{2}{3}$ ← Lowest terms

29. $3\dfrac{2}{3}$ rounds to 4   and $1\dfrac{5}{6}$ rounds to 2.

Estimate: $4 + 2 = 6$

To find the exact answer first rewrite each mixed number as an equivalent improper fraction: $3\dfrac{2}{3}$ is $\dfrac{11}{3}$ and $1\dfrac{5}{6}$ is $\dfrac{11}{6}$. Then write $\dfrac{11}{3}$ as an equivalent fraction with a denominator of 6, the LCD for $\dfrac{11}{3}$ and $\dfrac{11}{6}$.

Exact: $3\dfrac{2}{3} + 1\dfrac{5}{6} = \dfrac{11}{3} + \dfrac{11}{6} = \dfrac{22}{6} + \dfrac{11}{6}$

$$= \frac{22 + 11}{6} = \frac{33}{6}$$

$$= 5\frac{3}{6} = 5\frac{1}{2}$$

33. $5\dfrac{2}{3}$ rounds to 6.   *Estimate:* $6 \div 6 = 1$

To find the exact answer, first rewrite the mixed number and the whole number as improper fractions: $5\dfrac{2}{3}$ is $\dfrac{17}{3}$ and 6 is $\dfrac{6}{1}$. Then rewrite the division problem as multiplying by the reciprocal of $\dfrac{6}{1}$.

Exact:

$$5\frac{2}{3} \div 6 = \frac{17}{3} \div \frac{6}{1} = \frac{17}{3} \cdot \frac{1}{6} = \frac{17}{18}$$

Reciprocals

35. $1\dfrac{4}{5}$ rounds to 2.   *Estimate:* $8 - 2 = 6$

To find the exact answer, first rewrite 8 as $\dfrac{8}{1}$ and $1\dfrac{4}{5}$ as the improper fraction $\dfrac{9}{5}$. Then rewrite $\dfrac{8}{1}$ as an equivalent fraction with a denominator of 5, the LCD for $\dfrac{8}{1}$ and $\dfrac{9}{5}$.

Exact: $8 - 1\frac{4}{5} = \frac{8}{1} - \frac{9}{5} = \frac{40}{5} - \frac{9}{5}$

$= \frac{40 - 9}{5} = \frac{31}{5} = 6\frac{1}{5}$

37. The figure is a square.

$P = 4s = 4 \cdot 1\frac{3}{4} = \frac{4}{1} \cdot \frac{7}{4} = \frac{\overset{1}{\cancel{4}} \cdot 7}{1 \cdot \underset{1}{\cancel{4}}}$

$= \frac{7}{1} = 7$ in.

$A = s \cdot s = 1\frac{3}{4} \cdot 1\frac{3}{4} = \frac{7}{4} \cdot \frac{7}{4} = \frac{49}{16}$

$= 3\frac{1}{16}$ in.²

Section 4.6 (pages 285–288)

17. $\left(-\frac{3}{2}\right)^3\left(-\frac{2}{3}\right)^2$ ⎤ Three factors of $-\frac{3}{2}$

$= \left(-\frac{3}{2}\right)\left(-\frac{3}{2}\right)\left(-\frac{3}{2}\right)\left(-\frac{2}{3}\right)\left(-\frac{2}{3}\right)$ — Two factors of $-\frac{2}{3}$

$= -\frac{\overset{1}{\cancel{3}} \cdot \overset{1}{\cancel{3}} \cdot 3 \cdot \overset{1}{\cancel{2}} \cdot \overset{1}{\cancel{2}}}{\underset{1}{\cancel{2}} \cdot \underset{1}{\cancel{2}} \cdot 2 \cdot \underset{1}{\cancel{3}} \cdot \underset{1}{\cancel{3}}}$ Divide out all common factors

$= -\frac{3}{2}$ or $-1\frac{1}{2}$

21. $\frac{1}{5} - 6\left(\frac{7}{10}\right)$

There is no work to be done inside the parentheses. There are no exponents, so after rewriting 6 as $\frac{6}{1}$, start with Step 3, multiplying and dividing.

$= \frac{1}{5} - \underbrace{\frac{6}{1}\left(\frac{7}{10}\right)}$ Multiply.

$= \frac{1}{5} - \underbrace{\frac{42}{10}}$ Subtract; LCD is 10.

$= \frac{2}{10} - \underbrace{\frac{42}{10}}$

$= \frac{2 - 42}{10}$

$= \frac{-40}{10}$ Write $\frac{-40}{10}$ in lowest terms:

$\frac{-4 \cdot \overset{1}{\cancel{10}}}{\underset{1}{\cancel{10}}} = \frac{-4}{1} = -4$

$= -4$ Lowest terms

25. $-\frac{3}{10} \div \frac{3}{5}\left(-\frac{2}{3}\right)$

There is no work to be done inside the parentheses. There are no exponents, so start with Step 3, multiplying and dividing. Change division to multiplying by the reciprocal.

$= -\frac{3}{10} \cdot \frac{5}{3}\left(-\frac{2}{3}\right)$ Divide out common factors and then multiply.

$= -\frac{\overset{1}{\cancel{3}} \cdot \overset{1}{\cancel{3}}}{2 \cdot \overset{1}{\cancel{3}} \cdot \underset{1}{\cancel{3}}}\left(-\frac{2}{3}\right)$

$= -\frac{1}{2}\left(-\frac{2}{3}\right)$ Multiply; matching signs so the product is positive.

$= \frac{1 \cdot \overset{1}{\cancel{2}}}{\underset{1}{\cancel{2}} \cdot 3}$

$= \frac{1}{3}$

39. $\dfrac{-\frac{7}{9}}{-\frac{7}{36}}$ Rewrite using the $\div$ symbol for division.

$= -\frac{7}{9} \div \left(-\frac{7}{36}\right) = -\frac{7}{9} \cdot \left(-\frac{36}{7}\right)$ Reciprocals

$= \frac{\overset{1}{\cancel{7}} \cdot \overset{1}{\cancel{9}} \cdot 4}{\underset{1}{\cancel{9}} \cdot \underset{1}{\cancel{7}}}$ Divide out common factors.

$= \frac{4}{1} = 4$

Section 4.7 (pages 295–298)

15. $\frac{3}{8}x - 9 = 0$ Add 9 to both sides to get the variable term by itself.

$\underline{9 9}$

$\frac{3}{8}x + 0 = 9$ Multiply both sides by $\frac{8}{3}$ (the reciprocal of $\frac{3}{8}$.)

$\frac{\overset{1}{\cancel{8}}}{\underset{1}{\cancel{3}}} \cdot \frac{\overset{1}{\cancel{3}}}{\underset{1}{\cancel{8}}}x = \frac{8}{\underset{1}{\cancel{3}}} \cdot \frac{\overset{3}{\cancel{9}}}{1}$

$x = 24$

Check $\frac{3}{8}(24) - 9 = 0$ The solution is 24. Replace x with 24 in the original equation.

$9 - 9 = 0$

$0 = 0$ Balances

When x is 24, the equation balances, so 24 is the correct solution (**not** 0).

23. $\frac{3}{10} = -4b - \frac{1}{5}$ Add $\frac{1}{5}$ to both sides.

$\underline{\frac{1}{5}\frac{1}{5}}$

$\frac{3}{10} + \frac{1}{5} = -4b + 0$ Rewrite $\frac{1}{5}$ as an equivalent fraction with a denominator of 10, the LCD for $\frac{3}{10}$ and $\frac{1}{5}$.

$\frac{3}{10} + \frac{2}{10} = -4b$ Add on the left side.

$\frac{5}{10} = -4b$

$-\frac{1}{4} \cdot \frac{5}{10} = -\frac{1}{4}(-4b)$ Multiply both sides by $-\frac{1}{4}$, the reciprocal of -4.

$-\frac{5}{40} = b$ Write the fraction in lowest terms.

$-\frac{\overset{1}{\cancel{5}}}{\underset{1}{\cancel{5}} \cdot 8} = b$

$-\frac{1}{8} = b$

35. Let p be the penny size.

$\frac{p}{4} + \frac{1}{2} = \frac{5}{2}$ Add $-\frac{1}{2}$ both sides.

$\underline{-\frac{1}{2}-\frac{1}{2}}$

$\frac{p}{4} + 0 = 2$ Rewrite $\frac{p}{4}$ as $\frac{1}{4}p$.

$\frac{1}{4}p = 2$

$\frac{\overset{1}{\cancel{4}}}{1}\left(\frac{1}{\underset{1}{\cancel{4}}}p\right) = \frac{4}{1}(2)$ Multiply both sides by $\frac{4}{1}$, the reciprocal of $\frac{1}{4}$.

$p = 8$

The penny size is 8.

37. Let h be the man's height.

$\frac{11}{2}h - 220 = $ Recommended weight

$\frac{11}{2}h - 220 = 209$ Add 220 to both sides.

$\underline{220220}$

$\frac{11}{2}h + 0 = 429$

$\frac{\overset{1}{\cancel{2}}}{\underset{1}{\cancel{11}}}\left(\frac{\overset{1}{\cancel{11}}}{\underset{1}{\cancel{2}}}h\right) = \frac{2}{11}(429)$ Multiply both sides by $\frac{2}{11}$, the reciprocal of $\frac{11}{2}$.

$h = \frac{2 \cdot \overset{1}{\cancel{11}} \cdot 39}{\underset{1}{\cancel{11}}}$ Write the fraction in lowest terms.

$h = 78$

The man is 78 inches tall.

Section 4.8 (pages 305–306)

7. $P = 12\frac{3}{5}$ yd $+ 7\frac{2}{3}$ yd $+ 10$ yd

Write mixed numbers as improper fractions.

$P = \frac{63}{5}$ yd $+ \frac{23}{3}$ yd $+ 10$ yd

Rewrite fractions as equivalent fractions with a denominator of 15, the LCD

$P = \frac{189}{15}$ yd $+ \frac{115}{15}$ yd $+ \frac{150}{15}$ yd Add.

$P = \frac{454}{15}$ yd or $30\frac{4}{15}$ yd

$A = \frac{1}{2}bh$ Replace b with 10 yd and h with $7\frac{2}{3}$ yd.

$A = \frac{1}{2}(10 \text{ yd})\left(7\frac{2}{3}\text{ yd}\right)$

Multiply $\frac{1}{2}(10 \text{ yd})$ to get 5 yd.

$A = (5 \text{ yd})\left(7\dfrac{2}{3}\text{ yd}\right)$

$A = \dfrac{5\text{ yd}}{1} \cdot \dfrac{23\text{ yd}}{3}$

$A = \dfrac{5\text{ yd} \cdot 23\text{ yd}}{3}$

$A = \dfrac{115}{3}\text{ yd}^2 \text{ or } 38\dfrac{1}{3}\text{ yd}^2$

9. The *entire* figure is a rectangle.

$A = lw$

$A = 37\text{ m} \cdot 52\text{ m} = 1924\text{ m}^2$

The unshaded portion is a triangle.

$A = \dfrac{1}{2}bh$

$A = \dfrac{1}{2} \cdot \dfrac{52\text{ m}}{1} \cdot \dfrac{8\text{ m}}{1}$

$A = \dfrac{1 \cdot \overset{1}{\cancel{2}} \cdot 26\text{ m} \cdot 8\text{ m}}{\underset{1}{\cancel{2}} \cdot 1 \cdot 1}$

$A = 208\text{ m}^2$

Subtract to find the area of the shaded portion.

$A = 1924\text{ m}^2 - 208\text{ m}^2$

$= 1716\text{ m}^2$

The shaded area is 1716 m^2.

13. Amount of curbing to go "around" the space implies perimeter.

$P = 33\text{ m} + 55\text{ m} + 44\text{ m}$

$P = 132\text{ m}$

132 m of curbing is needed.

Amount of sod to "cover" the space implies area.

The figure is a triangle.

$A = \dfrac{1}{2}bh$

$A = \dfrac{1}{2} \cdot 44\text{ m} \cdot 33\text{ m}$

$A = \dfrac{1}{2} \cdot \dfrac{44\text{ m}}{1} \cdot \dfrac{33\text{ m}}{1}$

$A = \dfrac{1 \cdot \overset{1}{\cancel{2}} \cdot 22\text{ m} \cdot 33\text{ m}}{\underset{1}{\cancel{2}} \cdot 1 \cdot 1}$

$A = 726\text{ m}^2$

726 m^2 of sod is needed.

19. The figure is a pyramid.

First find the area of the rectangular base.

$B = l \cdot w$

$= 8\text{ cm} \cdot 15\text{ cm}$

$= 120\text{ cm}^2$

Now find the volume of the pyramid.

$V = \dfrac{B \cdot h}{3}$

$= \dfrac{\overset{40}{\cancel{120}}\text{ cm}^2 \cdot 20\text{ cm}}{\underset{1}{\cancel{3}}}$

$= \dfrac{800\text{ cm}^3}{1} = 800\text{ cm}^3$

Chapter 5 Rational Numbers: Positive and Negative Decimals

Section 5.1 (pages 327–330)

1. hundredths / thousandths / tens / ones / tenths

7 0.4 8 9

17. thousands / hundreds / tens / ones / tenths / hundredths / thousandths

5 4 0 6.0 4 5

31. $0.205 = \dfrac{205}{1000} = \dfrac{205 \div 5}{1000 \div 5} = \dfrac{41}{200}$

45. 1.075 is one and seventy-five thousandths.

5 is in thousandths place

1.075

one **and** seventy-five **thousandths**

53. Seven hundred three ten-thousandths: fill *four* decimal places for ten-thousandths.

$\dfrac{703}{10,000} = 0.0703$

ten-thousandths place

59. 8-pound test line has a diameter of 0.010 inch which is read "ten thousandths inch."

$0.010 = \dfrac{10}{1000} = \dfrac{10 \div 10}{1000 \div 10} = \dfrac{1}{100}\text{ inch}$

Section 5.2 (pages 337–338)

9. 793.988 to the nearest tenth

Draw a cut-off line after the tenths place:

793.9 | 88

The first digit cut is 8, which is *5 or more*, so round up by adding 1 tenth to the part you are keeping.

$\begin{array}{r} 793.9 \\ + \quad 0.1 \\ \hline 794.0 \end{array}$

$793.988 \approx 794.0$

You must write the zero in the tenths place to show that the number was rounded to the nearest tenth.

27. Round \$310.08 to the nearest dollar.

Draw a cut-off line: \$310. | 08

The first digit cut is 0, which is *4 or less*, so the part you keep stays the same.

Union dues: \$310.08 rounded to the nearest dollar is \$310

(Do **not** write \$310.00 as the answer.)

35. \$999.73 to the nearest dollar

Draw a cut-off line: \$999. | 73

The first digit cut is 7, which is *5 or more*, so round up by adding \$1.

$\begin{array}{r} \$999 \\ + \quad 1 \\ \hline \$1000 \end{array}$

Answer: \$999.73 rounded to the nearest dollar is \$1000

Section 5.3 (pages 345–348)

19. Rewrite 8.339 from 15 in vertical format. Watch the *order* of the numbers.

Line up decimal points.

$\begin{array}{r} 15.000 \\ -8.339 \\ \hline 6.661 \end{array}$ ← Write a decimal point and three zeros.

← Subtract as usual.

37. $-1.7035 - (5 - 6.7)$

Work inside parentheses first. Rewrite the subtraction as adding the opposite.

$5 - 6.7 = 5 + (-6.7) = -1.7$

Now the problem becomes:

$-1.7035 - (-1.7)$

Rewrite subtraction as adding the opposite. Because -1.7035 has the larger absolute value and is negative, the answer will be negative.

$= -1.7035 + 1.7$

$= -0.0035$

43. Using front end rounding, we can estimate $-6.5 + 0.7$ as $-7 + 1 = -6$, so -5.8 is the most reasonable answer.

53. Add the heights of the three basketball players.

Estimate:		*Exact:*
2	Rounds to	1.83
2	Rounds to	2.16
+ 2	Rounds to	+ 2.10
6 meters		6.09 meters

To find the difference, subtract 6.09 meters from the rhino's height of 6.4 meters.

$\begin{array}{r} 6.40 \\ -6.09 \\ \hline 0.31 \end{array}$

The basketball players' combined height of 6.09 meters is 0.31 meter less than the rhino's height of 6.4 meters.

Section 5.4 (pages 353–356)

11. $(7.2)(0.06) = 0.432$

The factors have the *same sign*, so the product is *positive*.

7.2 has *1 decimal place.* } Product has

0.06 has *2 decimal places.* } *3 decimal places.*

19. $(-0.003)^2$ means $(-0.003)(-0.003)$

$\begin{array}{r} 0.003 \\ \times \quad 0.003 \\ \hline 0.000009 \end{array}$ ← *3 decimal places*

← *3 decimal places*

← *6 decimal places*

Write five zeros so you can count over 6 decimal places in the answer. The factors have the *same sign*, so the product is *positive*.

35. A gallon of milk for \$419 is *unreasonable*; it is too much. Write a decimal point in \$419 to get a reasonable answer of \$4.19

43. Multiply the number of gallons that Michelle pumped into her SUV by the price per gallon.

$\begin{array}{r} 20.510 \\ \times \quad \$3.979 \\ \hline \$81.609290 \end{array}$

Round \$81.609290 to the nearest cent (nearest hundredth). Michelle paid \$81.61 for the gas.

47. (a) To find the area of the bill before 1929, multiply length by width. Use a calculator

$$(7.4218)(3.125) = 23.193125$$

Rounding to the nearest tenth gives an area of 23.2 in.^{2}

To find the area after 1929, multiply length by width. $(6.14)(2.61) = 16.0254$

Rounding to the nearest tenth gives an area of 16.0 in.^{2}

(b) Subtract to find the difference in the areas. $23.2 - 16.0 = 7.2$

The difference in the rounded areas is 7.2 in.^{2}

59. (a) Multiply to find the cost for 3 short-sleeved, solid-color shirts. Each shirt is $14.75.

$$\begin{array}{r} \$14.75 \\ \times\quad 3 \\ \hline \$44.25 \end{array}$$

Based on this subtotal, shipping is $5.95. Next, multiply to find the cost of 3 monograms.

$$\begin{array}{r} \$4.95 \\ \times\quad 3 \\ \hline \$14.85 \end{array}$$

Add these amounts, plus $5.00 for a gift box.

$$\begin{array}{ll} \$44.25 & \leftarrow \text{Shirts} \\ 5.95 & \leftarrow \text{Shipping} \\ 14.85 & \leftarrow \text{Monograms} \\ +\ 5.00 & \leftarrow \text{Gift box} \\ \hline \$70.05 \end{array}$$

The total cost is $70.05.

(b) Subtract the cost of the shirts to find the difference.

$$\begin{array}{r} \$70.05 \\ -\ 44.25 \\ \hline \$25.80 \end{array}$$

The monograms, gift box, and shipping added $25.80 to the cost of the gift.

Section 5.5 (pages 365–368)

7. The numbers have the same sign, so the quotient is positive.

$$\begin{array}{r} 3\,6. \\ 1.5\,\overline{\smash{)}\,5\,4\,0.} \\ \underline{4\,5} \\ 9\,0 \\ \underline{9\,0} \\ 0 \end{array}$$

Move decimal point in divisor and dividend 1 place; write one 0 in the dividend. Then line up the decimal point in the quotient.

21. First consider $240.8 \div 9$. Write extra zeros in the dividend so that you can continue dividing.

$$\begin{array}{r} 2\,6.7\,5\,5\,5 \\ 9\,\overline{\smash{)}\,2\,4\,0.8\,0\,0\,0} \\ \underline{1\,8} \\ 6\,0 \\ \underline{5\,4} \\ 6\,8 \\ \underline{6\,3} \\ 5\,0 \\ \underline{4\,5} \\ 5\,0 \\ \underline{4\,5} \\ 5\,0 \\ \underline{4\,5} \\ 5 \end{array}$$

Rounding the quotient to the nearest thousandth gives 26.756. The quotient is *negative* because the divisor and the dividend have *different* signs.

$$-240.8 \div 9 \approx -26.756$$

31. Does $307.02 \div 5.1 = 6.2$?

Estimate: $300 \div 5 = 60$

A quotient of 6.2 is *unreasonable;* it is too small compared to an estimate of 60.

$$\begin{array}{r} 6\,0.2 \\ 5.1\,\overline{\smash{)}\,3\,0\,7\,0.2} \\ \underline{3\,0\,6} \\ 1\,0 \\ \underline{0} \\ 1\,0\,2 \\ \underline{1\,0\,2} \\ 0 \end{array}$$

The correct quotient is 60.2, which is close to the estimate of 60.

55. $-8.68 - \underbrace{4.6(10.4)}\ \div\ 6.4$ Multiply.

$\quad = -8.68 - \underbrace{47.84 \div 6.4}$ Divide.

$\quad = \underbrace{-8.68 -\quad 7.475}$ Subtract.

$\quad =\qquad -16.155$

65. Divide 100,000 box tops (the answer from Exercise 63) by 38 weeks.

$$\begin{array}{r} 2\,6\,3\,1.5 \\ 38\,\overline{\smash{)}\,1\,0\,0,0\,0\,0.0} \\ \underline{7\,6} \\ 2\,4\,0 \\ \underline{2\,2\,8} \\ 1\,2\,0 \\ \underline{1\,1\,4} \\ 6\,0 \\ \underline{3\,8} \\ 2\,2\,0 \\ \underline{1\,9\,0} \\ 3\,0 \end{array}$$

2631.5 rounds to 2632

The school needs to collect 2632 box tops (rounded) during each of the 38 weeks.

Section 5.6 (pages 375–378)

17. $3\dfrac{5}{8} = \dfrac{29}{8}$

Divide 29 by 8.

$$\begin{array}{r} 3.6\,2\,5 \\ 8\,\overline{\smash{)}\,2\,9.0\,0\,0} \\ \underline{2\,4} \\ 5\,0 \\ \underline{4\,8} \\ 2\,0 \\ \underline{1\,6} \\ 4\,0 \\ \underline{4\,0} \\ 0 \end{array}$$

$3\dfrac{5}{8} = 3.625$

33. $0.35 = \dfrac{35}{100} = \dfrac{35 \div 5}{100 \div 5} = \dfrac{7}{20}$

49. Write zeros so that all the numbers have four decimal places.

$1.0100 > 1.0020$ *unacceptable*

$0.9991 > 0.9980$ and

$0.9991 < 1.0020$ *acceptable*

$1.0007 > 0.9980$ and

$1.0007 < 1.0020$ *acceptable*

$0.9900 < 0.9980$ *unacceptable*

The lengths of 0.9991 cm and 1.0007 cm are acceptable.

65. First write $\dfrac{3}{8}$ and $\dfrac{2}{5}$ as decimals.

$$\begin{array}{r} 0.3\,7\,5 \\ 8\,\overline{\smash{)}\,3.0\,0\,0} \\ \underline{2\,4} \\ 6\,0 \\ \underline{5\,6} \\ 4\,0 \\ \underline{4\,0} \\ 0 \end{array} \qquad \begin{array}{r} 0.4 \\ 5\,\overline{\smash{)}\,2.0} \\ \underline{2\,0} \\ 0 \end{array}$$

So, $\dfrac{3}{8} = 0.375$ and $\dfrac{2}{5} = 0.4$. Now write zeros so that all the numbers have four decimal places.

$\dfrac{3}{8} = 0.3750$

$\dfrac{2}{5} = 0.4000 \leftarrow$ second greatest

$0.37 = 0.3700 \leftarrow$ least

$0.4001 = 0.4001 \leftarrow$ greatest

From least to greatest: $0.37, \dfrac{3}{8}, \dfrac{2}{5}, 0.4001$

Section 5.7 (pages 383–384)

1. mean $= \dfrac{\text{sum of all values}}{\text{number of values}}$

$= \dfrac{92 + 51 + 59 + 86 + 68 + 73 + 49 + 80}{8}$

$= \dfrac{558}{8} = 69.75$ (rounds to 69.8)

The mean (average) final exam score was 69.8 (rounded).

7.

Quiz Score	Frequency	Product
3	4	$3 \cdot 4 = 12$
5	2	$5 \cdot 2 = 10$
6	5	$6 \cdot 5 = 30$
8	5	$8 \cdot 5 = 40$
9	2	$9 \cdot 2 = 18$
Totals	18	110

weighted mean $= \dfrac{\text{sum of products}}{\text{total number of quizzes}}$

$= \dfrac{110}{18} = 6.\overline{1}$ (rounds to 6.1)

The mean (average) quiz score was 6.1 (rounded).

9.

Course	Credits	Grade	Product (Credits · Grade)
Biology	4	B (=3)	$4 \cdot 3 = 12$
Biology Lab	2	A (=4)	$2 \cdot 4 = 8$
Mathematics	5	C (=2)	$5 \cdot 2 = 10$
Health	1	F (=0)	$1 \cdot 0 = 0$
Psychology	3	B (=3)	$3 \cdot 3 = 9$
Totals	15		39

GPA $= \dfrac{\text{sum of products}}{\text{total number of credits}}$

$= \dfrac{39}{15} = 2.60$

17. First, arrange the numbers in order from least to greatest.

34, 40, 40, 47, <u>48, 49</u>, 51, 56, 95, 96
         Middle two numbers

The list has 10 numbers so there is no single middle number. The middle *two* numbers are 48 and 49, so the median is the average of those two numbers.

$$\frac{48 + 49}{2} = 48.5 \text{ pounds of shrimp.}$$

25. <u>74</u>, <u>68</u>, <u>68</u>, <u>68</u>, 75, 75, <u>74</u>, <u>74</u>, 70, 77

Both 68 years and 74 years occur three times, which is more often than any of the other values, so each is a mode. This list is *bimodal* (has two modes).

29. For Barrow, Alaska

$$\text{mean} = \frac{-2 + (-11) + (-13) + (-18) + (-15) + (-2)}{6}$$

$$= \frac{-61}{6} \approx -10\,°F$$

For Fairbanks, Alaska

$$\text{mean} = \frac{3 + (-7) + (-10) + (-4) + 11 + 31}{6}$$

$$= \frac{24}{6} = 4\,°F$$

Find the difference: $4 - (-10) = 14$

Fairbanks' mean is 14 degrees warmer than Barrow's mean.

Section 5.8 (pages 389–392)

25. The unknown length is the side opposite the right angle, which is the hypotenuse. Use the formula for finding the hypotenuse.

The legs are 8 in. and 3 in.

$$\text{hypotenuse} = \sqrt{(\text{leg})^2 + (\text{leg})^2}$$
         Legs are 8 and 3

$$= \sqrt{(8)^2 + (3)^2}$$
     (8)(8) is 64 and (3)(3) is 9

$$= \sqrt{64 + 9}$$
Use a calculator to find $\sqrt{73}$.

$$= \sqrt{73}$$
Round to the nearest tenth.

$$\approx 8.5 \text{ in.}$$

The hypotenuse is approximately 8.5 in. long.

35. The hypotenuse is 21.6 km long. One leg is 13.2 km long.

$$\text{leg} = \sqrt{(\text{hypotenuse})^2 - (\text{leg})^2}$$
Hypotenuse is 21.6; one leg is 13.2.

$$= \sqrt{(21.6)^2 - (13.2)^2}$$
     (21.6)(21.6) is 466.56 and
     (13.2)(13.2) is 174.24.

$$= \sqrt{466.56 - 174.24}$$
Use a calculator to find $\sqrt{292.32}$.

$$= \sqrt{292.32}$$

Round to the nearest tenth.

$$\approx 17.1 \text{ km}$$

The length of the leg is approximately 17.1 km.

41. The diagonal brace is the hypotenuse of a right triangle with legs that have lengths of 6.5 ft and 2.5 ft.

$$\text{hypotenuse} = \sqrt{(\text{leg})^2 + (\text{leg})^2}$$
      Legs are 6.5 and 2.5

$$= \sqrt{(6.5)^2 + (2.5)^2}$$

     (6.5)(6.5) is 42.25 and
     (2.5)(2.5) is 6.25.

$$= \sqrt{42.25 + 6.25}$$
Use a calculator to find $\sqrt{48.5}$.

$$= \sqrt{48.5}$$
Round to the nearest tenth.

$$\approx 7.0 \text{ ft}$$

The diagonal brace is about 7.0 ft long.

43.

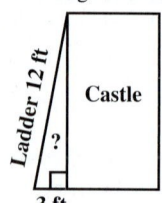

The length of the ladder is the hypotenuse (12 ft) and the length of one leg is (3 ft). Find the length of the other leg.

$$\text{leg} = \sqrt{(\text{hypotenuse})^2 - (\text{leg})^2}$$
Hypotenuse is 12; one leg is 3.

$$= \sqrt{(12)^2 - (3)^2}$$
     (12)(12) is 144 and (3)(3) is 9.

$$= \sqrt{144 - 9}$$
     Use a calculator to find $\sqrt{135}$.

$$= \sqrt{135}$$
     Round to the nearest tenth.

$$\approx 11.6 \text{ ft}$$

The ladder will reach about 11.6 ft high on the building.

Section 5.9 (pages 397–398)

3. $-20.6 + n = -22$ *Add 20.6 to both sides.*

$$\frac{+20.6 \quad\quad = +20.6}{0 + n = -1.4}$$

$$n = -1.4$$

Check

$$-20.6 + n = -22$$
       ↓   *Original equation*

$$-20.6 + (-1.4) = -22$$
     Replace n with -1.4.

$$-22 = -22$$
     Balances

When n is replaced with -1.4, the equation balances, so -1.4 is the correct solution (**not** -22).

11. $-3.3t = -2.31$

$$\frac{-3.3 \cdot t}{-3.3} = \frac{-2.31}{-3.3}$$ *Divide both sides by* -3.3.

On the left side, divide out the common factor of -3.3.

$$\frac{\overset{1}{\cancel{-3.3}} \cdot t}{\underset{1}{\cancel{-3.3}}} = 0.7$$ *On the right side,*
     $-2.31 \div -3.3$ is 0.7.

$$t = 0.7$$

Check $-3.3t = -2.31$
       ↓   *Original equation*

$$-3.3(0.7) = -2.31$$
     Replace t with 0.7.

$$-2.31 = -2.31$$
Balances, so 0.7 is the correct solution.

19. $0.8w - 0.4 = -6 + w$ *Add 0.4 to*
$$\frac{+ 0.4 \quad\quad +0.4}{0.8w + \quad 0 = -5.6 + w}$$ *both sides.*

             Write w
             as 1w.

$$0.8w = -5.6 + 1w$$

$$0.8w = -5.6 + 1w$$ *Add* $-1w$
$$\frac{-1w \quad\quad\quad -1w}{-0.2w = -5.6 + 0}$$ *to both sides*

$$\frac{\overset{1}{\cancel{-0.2}}\,w}{\underset{1}{\cancel{-0.2}}} = \frac{-5.6}{-0.2}$$ *Divide both sides*
      by -0.2

$$w = 28$$

23. *Step 1*

The problem is about doses of a medication.

Unknown: the adult dose

Known: child's dose is 0.3 times the adult dose; child's dose is 9 mg

Step 2   Let d be the adult dose.

Step 3

adult dose multiplied by 0.3	is	child's dose
↓	↓	↓
$0.3d$	$=$	9

Step 4

$$0.3d = 9$$

$$\frac{\overset{1}{\cancel{0.3}}d}{\underset{1}{\cancel{0.3}}} = \frac{9}{0.3}$$ *Divide both sides by 0.3.*

$$d = 30$$

Step 5 The adult dose is 30 milligrams.

Step 6

$$0.3(30 \text{ milligrams}) = 9 \text{ milligrams}$$
(Matches amount given in problem.)

Section 5.10 (pages 407–410)

11. First, find the circumference.

The diameter is $7\frac{1}{2}$ ft, so use the formula with d in it. Write $7\frac{1}{2}$ ft in decimal form as 7.5 ft.

$$C = \pi \cdot d$$

$$\approx 3.14 \cdot 7.5 \text{ ft}$$

$$\approx 23.6 \text{ ft}$$

To find the area, first find the radius by dividing the diameter by 2.

$$r = \frac{7.5 \text{ ft}}{2} = 3.75 \text{ ft}$$

$$A = \pi \cdot r \cdot r$$

$$\approx 3.14 \cdot 3.75 \text{ ft} \cdot 3.75 \text{ ft}$$

$$\approx 44.2 \text{ ft}^2$$

17. First, find the area of a whole circle with a radius of 10 ft.

$$A = \pi \cdot r \cdot r$$

$$\approx 3.14 \cdot 10 \text{ cm} \cdot 10 \text{ cm}$$

$$= 314 \text{ cm}^2$$

Divide the area of the whole circle by 2 to find the area of the semicircle.

$$\frac{314 \text{ cm}^2}{2} = 157 \text{ cm}^2$$

Then find the area of the triangle.

$$A = \frac{1}{2} \cdot b \cdot h$$

$$= \frac{1}{2} \cdot 20 \text{ cm} \cdot 10 \text{ cm}$$

$$= 100 \text{ cm}^2$$

Finally, subtract the area of the triangle from the area of the semicircle.

The shaded area is about

$$157 \text{ cm}^2 - 100 \text{ cm}^2 = 57 \text{ cm}^2$$

21. The distance that a point on the tire tread moves in one complete turn is the same as the circumference of the tire.

$$C = \pi \cdot d$$
$$\approx 3.14 \cdot 29.10 \text{ in.}$$
$$\approx 91.4 \text{ in.}$$

Bonus question:

$$\frac{1 \text{ revolution}}{91.4 \text{ inches}} \cdot \frac{12 \text{ inches}}{1 \text{ foot}} \cdot \frac{5280 \text{ feet}}{1 \text{ mile}}$$
$$\approx 693 \text{ revolutions/mile}$$

23. If the station can be heard for 150 miles in all directions, its signal covers a circular area.

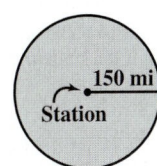

$$A = \pi \cdot r \cdot r$$
$$\approx 3.14 \cdot 150 \text{ mi} \cdot 150 \text{ mi}$$
$$= 70{,}650 \text{ mi}^2$$

There are about 70,650 mi² in the broadcast area.

29. (a)

$C = 144 \text{ cm}$	
$C = \pi \cdot d$	
$144 \text{ cm} = \pi \cdot d$	Replace C with 144 cm.
$144 \text{ cm} \approx 3.14 \cdot d$	Replace π with 3.14.
$\dfrac{144 \text{ cm}}{3.14} \approx \dfrac{3.14 \cdot d}{3.14}$	Divide both sides by 3.14.
$45.9 \approx d$	

The diameter is about 45.9 cm.

(b) Divide the circumference by π (144 cm ÷ 3.14).

39. First, find the radius of the cylinder.

$$r = \frac{d}{2} = \frac{18 \text{ in.}}{2} = 9 \text{ in.}$$

Then find the volume.

$$V = \pi \cdot r \cdot r \cdot h$$
$$\approx 3.14 \cdot 9 \text{ in.} \cdot 9 \text{ in.} \cdot 3 \text{ in.}$$
$$\approx 763.0 \text{ in.}^3$$

Now find the surface area

$$SA = (2 \cdot \pi \cdot r \cdot h)$$
$$+ (2 \cdot \pi \cdot r \cdot r)$$
$$\approx (2 \cdot 3.14 \cdot 9 \text{ in.} \cdot 3 \text{ in.})$$
$$+ (2 \cdot 3.14 \cdot 9 \text{ in.} \cdot 9 \text{ in.})$$
$$\approx 169.56 + 508.68$$
$$\approx 678.2 \text{ in.}^2$$

47. Use the formula for the surface area of a rectangular solid.

$$SA = 2lw + 2lh + 2wh$$
$$= (2 \cdot 5.5 \text{ in.} \cdot 2.8 \text{ in.})$$
$$+ (2 \cdot 5.5 \text{ in.} \cdot 8 \text{ in.})$$
$$+ (2 \cdot 2.8 \text{ in.} \cdot 8 \text{ in.})$$
$$= 30.8 + 88 + 44.8$$
$$= 163.6 \text{ in.}^2$$

The amount of cardboard needed is 163.6 in.²

Chapter 6 Ratio, Proportion, and Line/Angle/Triangle Relationships

Section 6.1 (pages 433–434)

3. $100 to $50

The ratio is $\dfrac{100}{50}$. The common units, $, divide out and are not written. Write this ratio in lowest terms by dividing the numerator and the denominator by 50.

$$\frac{\$100}{\$50} = \frac{100}{50} = \frac{100 \div 50}{50 \div 50} = \frac{2}{1}$$

9. $1\dfrac{1}{4}$ to $1\dfrac{1}{2}$

Write the ratio as $\dfrac{1\frac{1}{4}}{1\frac{1}{2}}$. Then write $1\dfrac{1}{4}$ and $1\dfrac{1}{2}$ as improper fractions. Rewrite as a division problem in horizontal format using the ÷ symbol. Then multiply by the reciprocal of the divisor.

$$\frac{1\frac{1}{4}}{1\frac{1}{2}} = \frac{\frac{5}{4}}{\frac{3}{2}} = \frac{5}{4} \div \frac{3}{2} = \frac{5}{4} \cdot \frac{\overset{1}{2}}{3} = \frac{5}{6}$$

25.

$$\frac{\text{longest side}}{\text{shortest side}} = \frac{1.8 \text{ meters}}{0.3 \text{ meters}} = \frac{1.8}{0.3}$$
$$= \frac{(1.8)(10)}{(0.3)(10)} = \frac{18}{3}$$
$$= \frac{18 \div 3}{3 \div 3} = \frac{6}{1}$$

The ratio of the length of the longest side to the length of the shortest side is $\dfrac{6}{1}$.

29. Subtract to find the increase in price.
$$\$12.50 - \$10 = \$2.50$$

$$\frac{\text{Increase in price} \rightarrow \$2.50}{\text{Original price} \rightarrow \$10} = \frac{2.50}{10}$$
$$= \frac{(2.50)(10)}{(10)(10)} = \frac{25}{100}$$
$$= \frac{25 \div 25}{100 \div 25} = \frac{1}{4}$$

The ratio of the increase in price to the original price is $\dfrac{1}{4}$.

Section 6.2 (pages 439–442)

17. Write the rate as a fraction: $\dfrac{7.5 \text{ pounds}}{6 \text{ people}}$

Then divide 7.5 by 6 to find the unit rate.

$$6\overline{)7.5} \;\; \frac{1.25}{}$$

$$\frac{7.5 \text{ pounds} \div 6}{6 \text{ people} \div 6} = \frac{1.25 \text{ pounds}}{1 \text{ person}}$$

The unit rate is 1.25 pounds/person.

23. Subtract to find the miles traveled.
$$28{,}396.7 - 28{,}058.1 = 338.6$$

Then set up the rate. Divide to find the unit rate and round to the nearest tenth.

$$\frac{\text{miles} \rightarrow}{\frac{\text{per}}{\text{gallon}} \rightarrow} \frac{338.6 \text{ miles}}{16.2 \text{ gallons}} \approx 20.90 \approx 20.9 \text{ miles per gallon}$$

27.

Size	Cost per Unit	
12 ounces	$\dfrac{\$2.49}{12 \text{ ounces}}$	$\approx \$0.208$
14 ounces	$\dfrac{\$2.89}{14 \text{ ounces}}$	$\approx \$0.206$
18 ounces	$\dfrac{\$3.96}{18 \text{ ounces}}$	$= \$0.22$

The lowest cost per ounce is $0.206, so the best buy is 14 ounces for $2.89.

43. One battery for $1.79 that lasts 3 times longer is like getting 3 batteries for $1.79. So $1.79 ÷ 3 ≈ $0.597 per battery.

An eight-pack of AA batteries for $4.99 is $4.99 ÷ 8 ≈ $0.624 per battery.

The better buy is the one battery package.

Section 6.3 (pages 451–454)

13. $\dfrac{150}{200} = \dfrac{200}{300}$

Write each ratio in lowest terms.

$$\frac{150 \div 50}{200 \div 50} = \frac{3}{4} \quad \text{and} \quad \frac{200 \div 100}{300 \div 100} = \frac{2}{3}$$

Because $\dfrac{3}{4}$ is *not* equivalent to $\dfrac{2}{3}$, the proportion is *false*.

23. $\dfrac{18}{15} = \dfrac{2\frac{5}{6}}{2\frac{1}{2}}$

Cross products:

$$18 \cdot 2\frac{1}{2} = \frac{\overset{9}{\cancel{18}}}{1} \cdot \frac{5}{\underset{1}{\cancel{2}}} = 45$$

$$15 \cdot 2\frac{5}{6} = \frac{\overset{5}{\cancel{15}}}{1} \cdot \frac{17}{\underset{2}{\cancel{6}}} = \frac{85}{2}$$
$$= 42\frac{1}{2}$$

The cross products are *unequal*, so the proportion is *false*.

35.

$$\begin{array}{c} x \cdot 18 \\ 42 \cdot 39 \end{array} \quad \text{Cross products}$$

$$x \cdot 18 = 42 \cdot 39 \quad \text{Show that cross products are equivalent.}$$

$$\frac{x \cdot \overset{1}{\cancel{18}}}{\underset{1}{\cancel{18}}} = \frac{1638}{18} \quad \text{Divide both sides by 18.}$$

$$x = 91$$

Check

$$42 \cdot 39 = 1638 \text{ and } 91 \cdot 18 = 1638$$

Cross products are equal, so 91 is the correct solution (***not*** 1638).

41. $\dfrac{99}{55} = \dfrac{44}{x}$ or $\dfrac{9}{5} = \dfrac{44}{x}$ Write $\dfrac{99}{55}$ in lowest terms as $\dfrac{9}{5}$

$$9 \cdot x = 5 \cdot 44 \quad \text{Show that cross products are equivalent.}$$

(continued)

$$\frac{\overset{1}{\cancel{9}} \cdot x}{\underset{1}{\cancel{9}}} = \frac{220}{9}$$

Divide both sides by 9.

$$x = \frac{220}{9}$$

$$x \approx 24.44$$

Round to the nearest hundredth.

Check

$9 \cdot 24.44 = 219.96$ and $5 \cdot 44 = 220$

The cross products are slightly different because you rounded the value of x. However, they are close enough to see that the problem was done correctly and 24.44 is the approximate solution.

49.
$$\frac{2\frac{1}{3}}{1\frac{1}{2}} = \frac{x}{2\frac{1}{4}}$$

$1\frac{1}{2} \cdot x = 2\frac{1}{3} \cdot 2\frac{1}{4}$ Show that cross products are equivalent.

$\frac{3}{2} \cdot x = \frac{7}{\underset{1}{\cancel{3}}} \cdot \frac{\overset{3}{\cancel{9}}}{4}$ Write mixed numbers as improper fractions

$\frac{3}{2} \cdot x = \frac{21}{4}$

$\dfrac{\frac{\overset{1}{\cancel{3}}}{\cancel{2}} \cdot x}{\frac{\cancel{3}}{\cancel{2}}} = \dfrac{\frac{21}{4}}{\frac{3}{2}}$ Divide both sides by $\frac{3}{2}$.

$x = \frac{21}{4} \div \frac{3}{2} = \frac{\overset{7}{\cancel{21}}}{\underset{2}{\cancel{4}}} \cdot \frac{\overset{1}{\cancel{2}}}{\underset{1}{\cancel{3}}} = \frac{7}{2} = 3\frac{1}{2}$

53.
$$\frac{x}{\frac{3}{50}} = \frac{0.15}{1\frac{4}{5}}$$

Change to decimals: $\frac{3}{50}$ is $3 \div 50 = 0.06$

and $1\frac{4}{5}$ is $\frac{9}{5}$ or $9 \div 5 = 1.8$

$$\frac{x}{0.06} = \frac{0.15}{1.8}$$

$x \cdot 1.8 = 0.15(0.06)$ Show that cross products are equivalent.

$\dfrac{x \cdot \overset{1}{\cancel{1.8}}}{\underset{1}{\cancel{1.8}}} = \dfrac{0.009}{1.8}$ Divide both sides by 1.8

$$x = 0.005$$

Change to fractions: 0.15 is $\frac{15}{100}$ or

$$\frac{15 \div 5}{100 \div 5} = \frac{3}{20}$$

$$\frac{x}{\frac{3}{50}} = \frac{\frac{3}{20}}{1\frac{4}{5}}$$

$1\frac{4}{5} \cdot x = \frac{3}{50} \cdot \frac{3}{20}$ Show that cross products are equivalent.

$$\dfrac{1\frac{\overset{1}{\cancel{4}}}{5} \cdot x}{1\frac{\cancel{4}}{5}} = \dfrac{\frac{9}{1000}}{1\frac{4}{5}}$$

Divide both sides by $1\frac{4}{5}$.

$$x = \frac{9}{1000} \div 1\frac{4}{5} = \frac{\overset{1}{\cancel{9}}}{\underset{200}{\cancel{1000}}} \cdot \frac{\overset{1}{\cancel{5}}}{\cancel{9}} = \frac{1}{200}$$

Check

$$0.005 = \frac{5}{1000} \text{ or } \frac{5 \div 5}{1000 \div 5}$$

$$= \frac{1}{200} \leftarrow \text{Matches}$$

The two answers are equivalent.

Section 6.4 (pages 461–464)

9. Step 1 The problem is about how many ounces of rice noodles are needed for a given number of servings.

Unknown: ounces needed for 12 servings
Known: 6 ounces needed for 7 servings

Step 2 Let x be the number of ounces needed for 12 servings.

Step 3
$$\frac{6 \text{ ounces}}{7 \text{ servings}} = \frac{x \text{ ounces}}{12 \text{ servings}}$$

Step 4
$7 \cdot x = 6 \cdot 12$ Cross products

$\dfrac{\overset{1}{\cancel{7}} \cdot x}{\underset{1}{\cancel{7}}} = \dfrac{72}{7}$ Divide both sides by 7.

$$x = 10\frac{2}{7} \approx 10$$

Step 5 You need about 10 ounces for 12 servings.

Step 6 You are making 12 servings, a little less than twice 7 servings. So you need a little less than twice 6 ounces of noodles; an exact answer of 10 ounces is reasonable.

15. The length of the dining area is the same as the length of the kitchen, which you found to be 14 feet in Exercise 13.

Find the width of the dining area. First find its width on the floor plan.

$4.5 - 2.5$ inches $= 2$ inches on the floor plan

$$\frac{1 \text{ inch}}{4 \text{ feet}} = \frac{2 \text{ inches}}{x \text{ feet}}$$
$$1 \cdot x = 4 \cdot 2$$
$$x = 8$$

The dining area is 8 feet wide.

19. Step 1 The problem is about the number of students who will need to take a refresher math course.

Unknown: how many entering students need to take a refresher math course.

Known: 7 out of 10 entering students need to take a refresher math course; 2950 entering students

Step 2 Let x be the number of entering students who need to take a refresher math course.

Step 3
$$\frac{7 \text{ need refresher}}{10 \text{ entering}} = \frac{x \text{ need refresher}}{2950 \text{ entering}}$$

Step 4
$$\frac{7}{10} = \frac{x}{2950}$$

$$10 \cdot x = 7 \cdot 2950$$

$$\dfrac{\overset{1}{\cancel{10}} \cdot x}{\underset{1}{\cancel{10}}} = \dfrac{20{,}650}{10}$$

$$x = 2065$$

Step 5 2065 entering students will probably need a refresher math course.

Step 6 This is a reasonable answer because it's more than half the students, but not all the students.

Incorrect setup
$$\frac{10 \text{ entering}}{7 \text{ need refresher}} = \frac{x \text{ need refresher}}{2950 \text{ entering}}$$

$$7 \cdot x = 10 \cdot 2950$$

$$\dfrac{\overset{1}{\cancel{7}} \cdot x}{\underset{1}{\cancel{7}}} = \dfrac{29{,}500}{7}$$

$$x \approx 4214$$

The *incorrect* setup gives an *unreasonable* estimate of 4214 entering students; there are only 2950 entering students.

31. Coretta's shadow $\rightarrow 1.05$ meters
Coretta's height $\rightarrow 1.68$ meters

$$= \frac{6.58 \text{ meters}}{x} \begin{array}{l} \leftarrow \text{tree shadow} \\ \leftarrow \text{tree height} \end{array}$$

$1.05 \cdot x = (1.68)(6.58)$ Cross products.

$\dfrac{\overset{1}{\cancel{1.05}}}{\underset{1}{\cancel{1.05}}} = \dfrac{11.0544}{1.05}$ Divide both sides by 1.05

$x = 10.528 \approx 10.53$ Round to hundredths.

The height of the tree is about 10.53 meters.

37. First find the number of calories in a $\frac{1}{2}$-cup serving of bran cereal.

$$\frac{\frac{1}{3} \text{ cup}}{80 \text{ calories}} = \frac{\frac{1}{2} \text{ cup}}{x \text{ calories}}$$

$\frac{1}{3} \cdot x = 80 \cdot \frac{1}{2}$ Cross products

$\dfrac{\frac{\overset{1}{\cancel{1}}}{\cancel{3}} \cdot x}{\frac{\cancel{1}}{\cancel{3}}} = \dfrac{40}{\frac{1}{3}}$ Divide both sides by $\frac{1}{3}$.

$x = 40 \div \frac{1}{3}$ or $\frac{40}{1} \cdot \frac{3}{1}$

$x = 120$ calories

Then find the number of grams of fiber in a $\frac{1}{2}$-cup serving of bran cereal.

$$\frac{\frac{1}{3}\text{ cup}}{8\text{ grams of fiber}} = \frac{\frac{1}{2}\text{ cup}}{x\text{ grams of fiber}}$$

$$\frac{1}{3} \cdot x = 8 \cdot \frac{1}{2} \quad \text{\color{blue}Cross products}$$

$$\frac{\frac{1}{\cancel{3}} \cdot x}{\frac{\cancel{1}}{3}} = \frac{4}{\frac{1}{3}} \quad \text{\color{blue}Divide both sides by } \frac{1}{3}.$$

$$x = 4 \div \frac{1}{3} \text{ or } \frac{4}{1} \cdot \frac{3}{1}$$

$$x = 12\text{ grams of fiber}$$

A $\frac{1}{2}$-cup serving of bran cereal provides 120 calories and 12 grams of fiber.

Section 6.5 (pages 475–478)

5. The figure starts at point P and goes on forever in one direction, so it is a ray named $\overrightarrow{PQ}$.

7. The lines are *perpendicular* because they intersect at right angles. The small red square indicates the right angle (90°).

23. Two rays in a straight line pointing in opposite directions measure 180°. An angle that measures 180° is called a *straight angle*.

31. Find the complement of 86° by subtracting.
$$90° - 86° = 4° \leftarrow \text{Complement}$$

35. Find the supplement of 90° by subtracting.
$$180° - 90° = 90° \leftarrow \text{Supplement}$$

37. $\angle SON \cong \angle TOM$ because they are vertical angles.
$\angle TOS \cong \angle MON$ because they are vertical angles.

39. Because $\angle COE$ and $\angle GOH$ are vertical angles, they are also congruent. This means they have the same measure. $\angle COE$ measures 63°, so $\angle GOH$ also measures 63°.

The sum of the measures of $\angle COE$, $\angle AOC$, and $\angle AOH$ equals 180°. Therefore, $\angle AOC$ measures $180° - (63° + 37°) = 180° - 100° = 80°$.

Since $\angle AOC$ and $\angle GOF$ are vertical angles, they are congruent, so $\angle GOF$ also measures 80°.

Since $\angle AOH$ and $\angle EOF$ are vertical angles, they are congruent, so $\angle EOF$ measures 37°.

51. $\angle 6 \cong \angle 1$ (vertical angles), so the measure of $\angle 1$ is also 47°.

$\angle 6 \cong \angle 8$ (corresponding angles), so the measure of $\angle 8$ is also 47°.

$\angle 6 \cong \angle 3$ (alternate interior angles), so the measure of $\angle 3$ is also 47°.

Notice that the exterior sides of $\angle 6$ and $\angle 5$ form a straight angle of 180°. Therefore, $\angle 6$ and $\angle 5$ are supplementary angles and the sum of their measures is 180°. If $\angle 6$ is 47° then $\angle 5$ must be 133° because $180° - 47° = 133°$. So the measure of $\angle 5$ is 133°.

$\angle 5 \cong \angle 2$ (vertical angles), so the measure of $\angle 2$ is also 133°.

$\angle 5 \cong \angle 7$ (corresponding angles), so the measure of $\angle 7$ is also 133°.

$\angle 7 \cong \angle 4$ (vertical angles), so the measure of $\angle 4$ is also 133°.

Section 6.6 (pages 485–488)

3. Notice that *rotating* $\triangle STU$ makes it possible to slide it on top of $\triangle WXY$ so the triangles match.

The corresponding angles are:
$\angle 1$ and $\angle 6$; $\angle 2$ and $\angle 4$; $\angle 3$ and $\angle 5$

The corresponding sides are:
$\overline{ST}$ and $\overline{YW}$; $\overline{TU}$ and $\overline{WX}$; $\overline{SU}$ and $\overline{YX}$

11. On both triangles, two corresponding angles and the side that connects them measure the same, so the Angle-Side-Angle (ASA) method can be used to prove that the triangles are congruent.

19. You want to find the values of a and b in the smaller triangle. Notice that the side that is 6 mm long in the smaller triangle corresponds to the side that is 12 mm long in the larger triangle. Set up a ratio of corresponding sides.

$$\frac{6\text{ mm}}{12\text{ mm}} = \frac{6}{12} = \frac{1}{2} \text{ in lowest terms}$$

Write a proportion to find a.

$$\frac{a}{10} = \frac{1}{2}$$

$$a \cdot 2 = 10 \cdot 1 \quad \text{\color{blue}Cross products are equivalent.}$$

$$\frac{a \cdot \cancel{2}}{\cancel{2}} = \frac{10}{2} \quad \text{\color{blue}Divide both sides by 2.}$$

$$a = 5\text{ mm}$$

Write a proportion to find b.

$$\frac{b}{2} = \frac{1}{2}$$

$$b \cdot 2 = 6 \cdot 1 \quad \text{\color{blue}Cross products are equivalent.}$$

$$\frac{b \cdot \cancel{2}}{\cancel{2}} = \frac{6}{2} \quad \text{\color{blue}Divide both sides by 2.}$$

$$b = 3\text{ mm}$$

25. Since triangles CDE and FGH are similar and all of the sides of CDE are the same length, all the sides of FGH are also the same length. Therefore, each missing side of triangle FGH is 8 cm.

Now add the lengths of all three sides to find the perimeter of triangle FGH.

$$P = 8\text{ cm} + 8\text{ cm} + 8\text{ cm}$$
$$P = 24\text{ cm}$$

Set up a ratio of corresponding sides to find the height (h) of triangle FGH.

$$\frac{10.4}{12} = \frac{h}{8}$$

$$12 \cdot h = 8 \cdot 10.4 \quad \text{\color{blue}Cross products are equivalent.}$$

$$\frac{\cancel{12} \cdot h}{\cancel{12}} = \frac{83.2}{12} \quad \text{\color{blue}Divide both sides by 12.}$$

$$h \approx 6.9\text{ cm} \quad \text{\color{blue}Rounded}$$

Area of triangle FGH
$$= 0.5 \cdot b \cdot h$$
$$\approx (0.5)(8\text{ cm})(6.9\text{ cm})$$
$$\approx 27.6\text{ cm}^2$$

33. The side 50 m long in the smaller triangle corresponds to the side with length n in the larger triangle. Write a proportion to find n.

$$\frac{50}{n} = \frac{100}{100 + 120} = \frac{100}{220}$$

Write $\frac{100}{220}$ in lowest terms as $\frac{5}{11}$.

$$\frac{50}{n} = \frac{5}{11}$$

$$5 \cdot n = 50 \cdot 11 \quad \text{\color{blue}Cross products}$$

$$\frac{\cancel{5} \cdot n}{\cancel{5}} = \frac{550}{5} \quad \text{\color{blue}Divide both sides by 5.}$$

$$n = 110\text{ m}$$

The length of the lake is 110 m.

Chapter 7 Percent

Section 7.1 (pages 515–520)

7. $140\% = 140.\% = 1.40$ or 1.4

Drop the percent sign and move the decimal point two places to the left.

13. $0.5\% = 00.5\% = 0.005$

0 is attached so the decimal point can be moved two places to the left. Drop the percent symbol.

17. $0.5 = 0.50 = 50\%$

    Decimal point *not* written with whole number percents

0 is attached so the decimal point can be moved two places to the right. Attach a percent sign.

27. $2 = 2.00 = 200\%$

Two zeros are attached so the decimal point can be moved two places to the right. Attach a percent sign.

41. $6.25\% = \frac{6.25}{100} = \frac{(6.25)(100)}{(100)(100)}$

$$= \frac{625 \div 625}{10,000 \div 625} = \frac{1}{16}$$

Drop the % symbol, and write 6.25 over 100. Multiply the numerator and denominator by 100 to get whole numbers. Then write the fraction in lowest terms.

43. $16\frac{2}{3}\% = \frac{16\frac{2}{3}}{100} = \frac{\frac{50}{3}}{100}$

$$= \frac{50}{3} \div \frac{100}{1} = \frac{\cancel{50}}{3} \cdot \frac{1}{\cancel{100}} = \frac{1}{6}$$

Drop the % symbol and write $16\frac{2}{3}$ over 100.

Rewrite $16\frac{2}{3}$ as the improper fraction $\frac{50}{3}$.

Rewrite the complex fraction using the $\div$ symbol for division. Then follow the steps for dividing fractions.

57. Multiply $\frac{3}{8}$ by 100%.

$$\frac{3}{8} \cdot 100\% = \frac{3}{8} \cdot \frac{100}{1}\%$$

$$= \frac{3 \cdot \overset{1}{\cancel{4}} \cdot 25}{2 \cdot \cancel{4} \cdot 1}\% = \frac{75}{2}\%$$

$$= 37\frac{1}{2}\% \text{ or } 37.5\%$$

61. Multiply $\frac{5}{9}$ by 100%.

$$\frac{5}{9} \cdot 100\% = \frac{5}{9} \cdot \frac{100}{1}\% = \frac{500}{9}\%$$

$$= 55\frac{5}{9}\% \text{ (exactly), or}$$
$$55.6\% \text{ (rounded)}$$

89. $1.5 = \frac{150}{100} = \frac{150 \div 50}{100 \div 50}$

$$= \frac{3}{2} = 1\frac{1}{2} \leftarrow \text{Mixed number}$$

$$1.50 = 150\% \leftarrow \text{Percent}$$

99. There are 4 canines and 28 total teeth.

$$\frac{4}{28} = \frac{4 \div 4}{28 \div 4} = \frac{1}{7} \leftarrow \text{Fraction}$$

$$\frac{1}{7} = \frac{1}{7} \cdot 100\% = \frac{100}{7}\%$$

$$= 14\frac{2}{7}\% \text{ (exactly), or } 14.3\% \text{ (rounded)}$$

113. (a) 50% of $285 is half of $285 or
$285 ÷ 2 = $142.50

(b) John will have to pay the remaining
50% of the tuition (100% − 50% =
50%).

(c) Since John has to pay 50% (half) of
the tuition, he will pay $285 ÷ 2 =
$142.50

Section 7.2 (pages 527–528)

13. $\underline{250\%}$ of $\underline{\text{7 hours}}$ is $\underline{\text{how long?}}$

Percent   Whole   Part
     (follows   (unknown)
     of)

$$\text{Percent} \rightarrow \frac{250}{100} = \frac{n}{7} \begin{matrix} \leftarrow \text{Part (unknown)} \\ \leftarrow \text{Whole} \end{matrix}$$
$$\text{Always } 100 \rightarrow$$

Step 1 $\frac{250}{100} = \frac{n}{7}$

Step 2 $100 \cdot n = 250 \cdot 7$
$$100 \cdot n = 1750$$

Step 3 $\frac{\overset{1}{\cancel{100}} \cdot n}{\underset{1}{\cancel{100}}} = \frac{1750}{100}$

$$n = 17.5$$

250% of 7 hours is 17.5 hours.

21. $\underline{105}$ employees is $\underline{\text{what percent}}$ of $\underline{54}$ employees?

Part   Percent   Whole
   (unknown)   (follows *of*)

$$\text{Percent (unknown)} \rightarrow \frac{p}{100} = \frac{105}{54} \begin{matrix} \leftarrow \text{Part} \\ \leftarrow \text{Whole} \end{matrix}$$
$$\text{Always } 100 \rightarrow$$

Step 1 $\frac{p}{100} = \frac{105}{54}$

Step 2 $p \cdot 54 = 100 \cdot 105$
$$p \cdot 54 = 10{,}500$$

Step 3 $\frac{p \cdot \overset{1}{\cancel{54}}}{\underset{1}{\cancel{54}}} = \frac{10{,}500}{54}$

$$p \approx 194.4 \quad \text{\textcolor{blue}{Round to nearest tenth.}}$$

105 employees is 194.4% (rounded)
of 54 employees.

23. $\underline{\$0.33}$ is $\underline{4\%}$ of $\underline{\text{what amount?}}$

Part   Percent   Whole (unknown)
         (follows *of*)

$$\text{Percent} \rightarrow \frac{4}{100} = \frac{0.33}{n} \begin{matrix} \leftarrow \text{Part} \\ \leftarrow \text{Whole} \\ \text{(unknown)} \end{matrix}$$
$$\text{Always } 100 \rightarrow$$

Step 1 $\frac{4}{100} = \frac{0.33}{n}$

Step 2 $4 \cdot n = 100 \cdot 0.33$
$$4 \cdot n = 33$$

Step 3 $\frac{\overset{1}{\cancel{4}} \cdot n}{\underset{1}{\cancel{4}}} = \frac{33}{4}$

$$n = 8.25$$

$0.33 is 4% of $8.25.

Section 7.3 (pages 535–538)

5. To find 10% of 45 pounds, divide 45 by 10.
To divide by 10, move the decimal point
one place to the left.

10% of 45. = 4.5 pounds.

Choose 4.5 pounds.

23. Write $12\frac{1}{2}\%$ as 12.5%. Then move the
decimal point two places to the left.

$\underline{12.5\%}$ of $\underline{\text{what number of people}}$ is $\underline{135\text{ people}}$

$$0.125 \quad \cdot \quad n \quad = \quad 135$$

$$\frac{(\overset{1}{\cancel{0.125}})(n)}{\underset{1}{\cancel{0.125}}} = \frac{135}{0.125}$$

$$n = 1080$$

$12\frac{1}{2}\%$ of 1080 people is 135 people.

31. $\underline{\text{What percent}}$ of $\underline{\$264}$ is $\underline{\$330?}$

$$p \quad \cdot \quad 264 \quad = \quad 330$$

$$\frac{p \cdot \overset{1}{\cancel{264}}}{\underset{1}{\cancel{264}}} = \frac{330}{264}$$

$$p = 1.25 = 125\%$$

125% of $264 is $330.

37. $\underline{\$1.48}$ is $\underline{\text{what percent}}$ of $\underline{\$74?}$

$$1.48 \quad = \quad p \quad \cdot \quad 74$$

$$\frac{1.48}{74} = \frac{(p)(\overset{1}{\cancel{74}})}{\underset{1}{\cancel{74}}}$$

$$0.02 = p$$
$$2\% = p$$

$1.48 is 2% of $74.

45. $\underline{225.\%}$ of $\underline{\text{what number of gallons}}$ is $\underline{11.25\text{ gallons?}}$

$$2.25 \quad \cdot \quad n \quad = \quad 11.25$$

$$\frac{(\overset{1}{\cancel{2.25}})(n)}{\underset{1}{\cancel{2.25}}} = \frac{11.25}{2.25}$$

$$n = 5$$

225% of 5 gallons is 11.25 gallons.

Section 7.4 (pages 547–548)

11. Step 1 The problem is about raising
money for scholarships.

Unknown: the percent of goal raised

Known: the goal was $50,000; $69,000
was actually raised.

Step 2 Let p be the percent of the goal that
was raised.

Step 3 Use the percent equation.

percent • whole = part

$$p \cdot 50{,}000 = 69{,}000$$

Step 4

$$\frac{p \cdot \overset{1}{\cancel{50{,}000}}}{\underset{1}{\cancel{50{,}000}}} = \frac{69{,}000}{50{,}000} \quad \textcolor{blue}{\text{Divide both sides by 50,000.}}$$

$$p = 1.38$$

Multiply the solution by 100% to change it
from a decimal to a percent.

$$1.38 = 138\%$$

Step 5 The society raised 138% of their goal.

Step 6 The solution 138% makes sense because the amount raised was *greater* than the
goal, so the percent should be *greater* than
100%.

15. Step 1 The problem is about shots tried by
a basketball player.

Unknown: the number of shots tried

Known: 638 shots were made; 638 is
47.6% of the shots tried

Step 2 Let n be the number of shots tried.

Step 3 The percent is given in the problem:
47.6%. The keyword "of" appears right after
47.6%, so you can use the phrase "47.6%
of the shots he tried" to help you write one
side of the equation.

$\underline{47.6\%}$ *of* $\underline{\text{the shots he tried}}$

$$0.476 \quad \cdot \quad n = 638$$

Step 4

$$\frac{\overset{1}{(0.\cancel{476})}(n)}{\underset{1}{0.\cancel{476}}} = \frac{638}{0.476}$$ Divide both sides by 0.476

$$n \approx 1340$$

Step 5 He tried 1340 shots (rounded).

Step 6 The solution of 1340 shots tried makes sense because 50% of 1340 shots is 670 shots, which is close to the number given in the problem (638 shots).

17. **Step 1** This problem is about mileage an SUV gets with new tires.

Unknown: the amount of increase in mileage

Known: 15% better mileage than old mileage of 20.6 mpg

Step 2 Let n be the amount of increase in mileage with the new tires.

Step 3

Percent • Original = Amount of
mileage increase

$$\underbrace{0.15 \quad • \quad 20.6}_{3.09} = n$$
$$= n$$

Step 4 The new tires should increase her mileage by 3.09 miles per gallon.

Step 5 Her new mileage should be $20.6 + 3.09 = 23.69 \approx 23.7$ miles per gallon (rounded)

Step 6 The solution of 23.7 miles per gallon is reasonable because $23.7 - 20.6 = 3.1$, which is about 15% of 20.6 (3 is 15% of 20).

23. **Step 1** This problem is about finding the percent decrease in the cost of a Model T car.

Unknown: the percent of decrease

Known: $825 was the original cost; $290 was the new cost

Step 2 Let p be the percent decrease.

Step 3 First subtract to find the *amount* of decrease. Then write an equation.

Amount of
decrease $= \$825 - \$290 = \$535$

percent of original = amount of
cost decrease

$$p \quad • \quad \$825 \quad = \quad \$535$$

Step 4

$$\frac{p • \overset{1}{\cancel{825}}}{\underset{1}{\cancel{825}}} = \frac{535}{825}$$ Divide both sides by 825.

$$p \approx 0.648 = 64.8\%$$

Step 5 The percent of decrease was 64.8% (rounded).

Step 6 The solution of 64.8% is reasonable because the amount of decrease was greater than half the original value and less than three-quarters of the original valve. So, the percent should be between 50% and 75%.

29. **Step 1** This problem is about the percent increase in the number of species threatened with extinction in the United States.

Unknown: the percent of increase

Known: there were 78 species on the list in 1967; there were 607 species on the list in 2007

Step 2 Let p be the percent of increase

Step 3 First subtract to find the *amount* of increase $607 - 78 = 529$

percent of original = amount of
number increase

$$p \quad • \quad 78 \quad = \quad 529$$

Step 4

$$\frac{p • \overset{1}{\cancel{78}}}{\underset{1}{\cancel{78}}} = \frac{529}{78}$$ Divide both sides by 78.

$$p \approx 6.78 = 678\%$$

Step 5 The percent of increase is 678% (rounded).

Step 6 The solution of 678% is reasonable because the amount of increase is more than six times as much as the original number (6 times is 600%).

Section 7.5 (pages 557–561)

9. Cost of item is $12,600

Tax rate is unknown (p)

Sales tax is $567

tax • cost of = sales
rate item tax

$$p \quad • \quad 12{,}600 \quad = \quad 567$$

$$\frac{(p)(\overset{1}{\cancel{12{,}600}})}{\underset{1}{\cancel{12{,}600}}} = \frac{567}{12{,}600}$$ Divide both sides by 12,600.

$$p = 0.045 = 4.5\%$$

The tax rate is 4.5% or $4\frac{1}{2}\%$.

Total cost $= \$12{,}600 + \$567 = \$13{,}167$

11. The bill of $32.17 rounds to $30.

Estimate of 15% tip:

10% of $30. is $3 and 5% of $30 is half of $3, or $1.50. An estimate is $3 + \$1.50 = \$4.50$.

Exact 15% tip:

Percent • whole = part

$$(0.15)(32.17) = \$4.8255 \approx \$4.83$$ (rounded)

Estimate of 20% tip:

10% of $30. is $3 and 20% is 2 times $3, or $6. An estimate is $6.

Exact 20% tip:

Percent • whole = part

$$(0.20)(32.17) = \$6.434 \approx \$6.43$$ (rounded)

21. Original price is $17.50

Rate of discount is 25.% = 0.25

rate of discount • original price = amount of discount

$$(0.25)(17.50) = n$$
$$\$4.375 = n$$

The amount of discount is $4.375 \approx \$4.38$.

Then, original price − amount of discount = sale price

$$\$17.50 - \$4.38 = \$13.12$$

The sale price is $13.12.

31. $17,800 at $7\frac{3}{4}\%$ for 8 months

The principal is $17,800. The rate is $7\frac{3}{4}\%$ or 7.75%. Then $07.75\% = 0.0775$ as a decimal.

The time is 8 months, which is $\frac{8}{12}$ of a year.

Use the formula $I = prt$.

$$I = p • r • t$$

$$= \underbrace{(17{,}800)(0.0775)}\left(\frac{8}{12}\right)$$

$$= (1379.5) \qquad \left(\frac{8}{12}\right)$$

$$= \left(\frac{1379.5}{1}\right)\left(\frac{8}{12}\right)$$

$$= \frac{11{,}036}{12} \approx 919.67$$

The interest is $919.67.

amount due = principal + interest
$$= \$17{,}800 + \$919.67$$
$$= \$18{,}719.67$$

The total amount due is $18,719.67.

37. Cost of item is $24.99

Tax rate is $6\frac{1}{2}\%$ or $06.5\% = 0.065$

tax rate • cost of = sales
item tax

$$0.065 • 24.99 = n$$
$$(0.065)(24.99) = n$$
$$1.62 \approx n$$

The sales tax (rounded) is $1.62.

Total cost $= \$24.99 + \$1.62 = \$26.61$

The total cost of the headset is $26.61.

53. **Camcorder:** original price of $287.95; 65% off

Write 65.% as 0.65 in decimal form.

amount of
discount $= (0.65)(\$287.95) \approx \$187.17$

sale
price $= \$287.95 - \$187.17 = \$100.78$

sales tax $= 0.06(\$100.78) \approx \$6.05$

Total $= \$100.78 + \$6.05 = \$106.83$

Jeans: 2 pairs at $48 each; 45% off; write 45.% as 0.45

amount of
discount $= (0.45)(\$48) = \$21.60$

sale price $= \$48 - \$21.60 = \$26.40$

sales tax: no tax on clothing

Total for 2 pairs $= 2(\$26.40) = \$52.80$

Ring: original price of $95; 30% off; write 30.% as 0.30

amount of
discount $= (0.30)(\$95) = \$28.50$

sale price $= \$95 - \$28.50 = \$66.50$

sales tax $= (0.06)(\$66.50) = \$3.99$

Total $= \$66.50 + \$3.99 = \$70.49$

Total Bill $= \$106.83 + \$52.80 + \$70.49 = \$230.12$

SOLUTIONS

Chapter 8 Measurement

Section 8.1 (pages 585–588)

25. 3 in. to feet

$$\frac{\overset{1}{\cancel{3}} \text{ in.}}{1} \cdot \frac{1 \text{ ft}}{\underset{4}{\cancel{12}} \text{ in.}} = \frac{1}{4} \text{ ft or } 1 \div 4 = 0.25 \text{ ft}$$

35. $4\frac{1}{4}$ gal to quarts

$$\frac{4\frac{1}{4} \text{ gal}}{1} \cdot \frac{4 \text{ qt}}{1 \text{ gal}} = \frac{17}{\cancel{4}} \cdot \frac{\overset{1}{\cancel{4}}}{1} \text{ qt} = 17 \text{ qt}$$

43. 6 days to seconds

$$\frac{6 \text{ days}}{1} \cdot \frac{24 \text{ hr}}{1 \text{ day}} \cdot \frac{60 \text{ min}}{1 \text{ hr}} \cdot \frac{60 \text{ sec}}{1 \text{ min}}$$
$$= 6 \cdot 24 \cdot 60 \cdot 60 \text{ sec}$$
$$= 518{,}400 \text{ sec}$$

55. *Step 1*
The problem asks for the price per pound of strawberries.

Step 2
Convert ounces to pounds. Then divide the cost by the number of pounds.

Step 3
To estimate, round $2.29 to $2. Then, there are 16 oz in a pound, so 20 oz is a little more than 1 lb. Thus, $2 ÷ 1 = $2 per pound as our estimate.

Step 4
Use a unit fraction to convert 20 oz to pounds.

$$\frac{\overset{5}{\cancel{20}} \text{ oz}}{1} \cdot \frac{1 \text{ lb}}{\underset{4}{\cancel{16}} \text{ oz}} = \frac{5}{4} \text{ lb} = 1.25 \text{ lb}$$

$$\begin{array}{c} \text{Cost} \to \\ \text{Per} \to \\ \text{Pound} \to \end{array} \frac{\$2.29}{1.25 \text{ lb}} = 1.832 \approx 1.83$$

Step 5
The strawberries are $1.83 per pound (to the nearest cent).

Step 6
The exact answer, $1.83, is close to our estimate of $2.

61. (a) Find the total number of cups per week. Then convert cups to quarts.

$$\frac{2}{3} \cdot 15 \cdot 5 = \frac{2}{\cancel{3}} \cdot \frac{\overset{5}{\cancel{15}}}{1} \cdot \frac{5}{1} = 50 \text{ cups}$$

$$\frac{\overset{25}{\cancel{50}} \text{ cups}}{1} \cdot \frac{1 \text{ pt}}{\underset{1}{\cancel{2}} \text{ cups}} \cdot \frac{1 \text{ qt}}{2 \text{ pt}}$$
$$= \frac{25}{2} \text{ qt} = 12\frac{1}{2} \text{ qt}$$

The center needs $12\frac{1}{2}$ qt of milk per week.

(b) Convert quarts to gallons.

$$\frac{12\frac{1}{2} \text{ qt}}{1} \cdot \frac{1 \text{ gal}}{4 \text{ qt}}$$
$$= \frac{25}{2} \cdot \frac{1}{4} \text{ gal} = 3.125 \text{ gal}$$

The center should order 4 gallon containers, because you can't buy part of a container.

Section 8.2 (pages 595–596)

19. A paper clip is about 3 <u>cm</u> long. Both meters and kilometers would be much too long, and 3 mm is a very tiny length (see ruler in Section 8.2).

25. 7 m to cm

$$\frac{7 \text{ m}}{1} \cdot \frac{100 \text{ cm}}{1 \text{ m}} = 7 \cdot 100 \text{ cm} = 700 \text{ cm}$$

33. 400 mm to cm
From mm to cm is *one* place to the *left* on the conversion line, so move the decimal point *one* place to the *left* also.
40 0. mm = 40.0 cm or 40 cm

37. 82 cm to m
From cm to m is *two* places to the *left* on the conversion line, so move the decimal point *two* places to the *left* also.
82. cm = 0.82 m

0.82 m is less than 1 m, so 82 cm is **less than** 1 m. The difference in length is
1 m − 0.82 m = 0.18 m or
100 cm − 82 cm = 18 cm.

45. 5.6 mm to km

$$\frac{5.6 \text{ mm}}{1} \cdot \frac{1 \text{ m}}{1000 \text{ mm}} \cdot \frac{1 \text{ km}}{1000 \text{ m}}$$
$$= \frac{5.6}{1{,}000{,}000} \text{ km}$$
$$= 0.0000056 \text{ km}$$

Section 8.3 (pages 603–606)

7. Lori caught a small sunfish weighing 150 g. (Weight is measured in mg, g, or kg.)

35. 8 mL to L

$$\frac{8 \text{ mL}}{1} \cdot \frac{1 \text{ L}}{1000 \text{ mL}} = \frac{8}{1000} \text{ L} = 0.008 \text{ L}$$

41. 5.2 kg to g
From kg to g is *three* places to the *right* on the conversion line.
5.200 kg = 5200. g or 5200 g

67. Convert 900 mL to L.

$$\frac{900 \text{ mL}}{1} \cdot \frac{1 \text{ L}}{1000 \text{ mL}} = \frac{900}{1000} \text{ L} = 0.9 \text{ L}$$

On average, we breathe in and out roughly 0.9 L of air every 10 seconds.

73. Convert 1 kg to g.

$$\frac{1 \text{ kg}}{1} \cdot \frac{1000 \text{ g}}{1 \text{ kg}} = 1000 \text{ g}$$

Divide the 1000 g by the weight of 1 nickel.

$$\frac{1000 \text{ g}}{5 \text{ g}} = 200$$

There are 200 nickels in 1 kg of nickels.

Section 8.4 (pages 611–612)

7. Multiply to find the total length of pencil lead in mm, then convert mm to cm.
60 mm • 30 = 1800 mm of pencil lead

$$\frac{\overset{180}{\cancel{1800}} \text{ mm}}{1} \cdot \frac{1 \text{ cm}}{\underset{1}{\cancel{10}} \text{ mm}} = 180 \text{ cm}$$

The total length of pencil lead is 180 cm.
Divide to find the cost per cm.

$$\frac{\$3.29}{180 \text{ cm}} \approx \$0.0183/\text{cm} \approx \$0.02/\text{cm}$$

The cost is $0.02/cm (rounded).

13. Three cups a day for one week is
3 • 7 = 21 cups in one week.

$$\frac{21 \text{ cups}}{1} \cdot \frac{90 \text{ mg}}{1 \text{ cup}} = 1890 \text{ mg}$$

Convert 1890 mg to g.

$$\frac{1890 \text{ mg}}{1} \cdot \frac{1 \text{ g}}{1000 \text{ mg}} = 1.89 \text{ g}$$

Agnete consumes 1.89 g of caffeine in one week.

Section 8.5 (pages 617–620)

3. 80 m to feet

$$\frac{80 \text{ m}}{1} \cdot \frac{3.28 \text{ ft}}{1 \text{ m}} \approx 262.4 \text{ ft}$$

17. Convert 0.5 in. to centimeters.

$$\frac{0.5 \text{ in.}}{1} \cdot \frac{2.54 \text{ cm}}{1 \text{ in.}} = 1.27 \text{ cm} \approx 1.3 \text{ cm}$$

The dwarf gobie is about 1.3 cm long.

29. −4 °F

$$C = \frac{5(-4 - 32)}{9} = \frac{5(-36)}{9}$$
$$= \frac{-180}{9} = -20$$
$$-4 \text{ °C} = -20 \text{ °F}$$

39. (a) Since the comfort range of the boots is from 24 °C to 4 °C, you would wear these boots in pleasant weather—above freezing, but not hot.

(b) Change 24 °C to Fahrenheit.

$$F = \frac{9C}{5} + 32 = \frac{9(24)}{5} + 32$$
$$= \frac{216}{5} + 32$$
$$= 43.2 + 32 = 75.2$$

Thus, 24 °C ≈ 75 °F.
Change 4 °C to Fahrenheit.

$$F = \frac{9C}{5} + 32 = \frac{9(4)}{5} + 32$$
$$= \frac{36}{5} + 32$$
$$= 7.2 + 32 = 39.2$$

Thus, 4 °C ≈ 39 °F.

The boots are designed for Fahrenheit temperatures of about 75 °F to about 39 °F.

(c) The range of metric temperatures in January would depend on where you live. In Minnesota, it's 0 °C to −40 °C, and in California it's 24 °C to 0 °C.

47. Fuel left after landing

$$\frac{2 \text{ fl oz}}{1} \cdot \frac{1 \text{ qt}}{32 \text{ fl oz}} \cdot \frac{0.95 \text{ L}}{1 \text{ qt}} \cdot \frac{1000 \text{ mL}}{1 \text{ L}}$$
$$= \frac{950}{16} \text{ mL} = 59.375 \text{ mL} \approx 59.4 \text{ mL}$$

Chapter 9 Graphs

Section 9.1 (pages 637–640)

7. To find the average number of points scored per game, divide the number in the points column by the number in the games column for each of the three players.

   Round answers to the nearest tenth to match other numbers in that column in the table.

   Jerry West: $\dfrac{25{,}192}{932} \approx 27.0$

   Bob Pettit: $\dfrac{20{,}880}{792} \approx 26.4$

   Shaquille O'Neal: $\dfrac{25{,}454}{981} \approx 25.9$

17. **(a)** $\dfrac{322 \text{ calories}}{110 \text{ pounds}} = \dfrac{x \text{ calories}}{125 \text{ pounds}}$
    Find the cross products.
    $110 \cdot x = 322 \cdot 125$
    Divide both sides by 110.

    $\dfrac{\overset{1}{\cancel{110}}x}{\underset{1}{\cancel{110}}} = \dfrac{40{,}250}{110}$

    $x \approx 366$

    The person would burn approximately 366 calories.

    (b) $\dfrac{210 \text{ calories}}{110 \text{ pounds}} = \dfrac{x \text{ calories}}{125 \text{ pounds}}$
    Find the cross products.
    $110 \cdot x = 210 \cdot 125$
    Divide both sides by 110.

    $\dfrac{\overset{1}{\cancel{110}}x}{\underset{1}{\cancel{110}}} = \dfrac{26{,}250}{110}$

    $x \approx 239$

    The person would burn approximately 239 calories.

27. There are 30.5 symbols in all.
    $30.5 (10 \text{ million}) = 305 \text{ million}$
    The total number is 305 million or 305,000,000.

Section 9.2 (pages 645–650)

5. **(a)**

   $\dfrac{\text{floor covering} + \text{painting} + \text{window coverings}}{\text{total remodeling budget}}$

   $= \dfrac{\$1200 + \$2400 + \$1200}{\$48{,}000}$

   $= \dfrac{\$4800}{\$48{,}000} = \dfrac{1}{10}$

   (b) $\dfrac{\text{actual total}}{\text{budget total}} = \dfrac{\$57{,}600}{\$48{,}000} = \dfrac{6}{5}$

19. The percent for "runs and hides" is 35%.

    percent $\cdot$ whole $=$ part
    $\downarrow \qquad \downarrow \qquad \downarrow$
    35% $\cdot$ 400 $= n$
    (0.35) (400) $= n$
    140 $= n$

    The number of people who said "runs and hides" is 140.

The number of people who said "watches from a safe perch" is 76 (from Exercise 16).

$140 - 76 = 64 \quad$ so $\quad$ 64 fewer people said," watches from a safe perch," than said, "runs and hides."

31. **(a)** Total sales $= \$12{,}500 + \$40{,}000$
    $+ \$60{,}000 + \$50{,}000 + \$37{,}500$
    $= \$200{,}000$

    (b) Adventure classes $= \$12{,}500$

    percent of total $= \dfrac{12{,}500}{200{,}000} = 0.0625$
    $= 6.25\%$

    Grocery/provision sales $= \$40{,}000$

    percent of total $= \dfrac{40{,}000}{200{,}000} = 0.2 = 20\%$

    Equipment rentals $= \$60{,}000$

    percent of total $= \dfrac{60{,}000}{200{,}000} = 0.3 = 30\%$

    Rafting tours $= \$50{,}000$

    percent of total $= \dfrac{50{,}000}{200{,}000} = 0.25 = 25\%$

    Equipment sales $= \$37{,}500$

    percent of total $= \dfrac{37{,}500}{200{,}000} = 0.1875$
    $= 18.75\%$

    (c) Adventure classes:
    number of degrees $= (0.0625)(360°)$
    $= 22.5°$

    Grocery/provision sales:
    number of degrees $= (0.2)(360°) = 72°$

    Equipment rentals:
    number of degrees $= (0.3)(360°) = 108°$

    Rafting tours:
    number of degrees $= (0.25)(360°) = 90°$

    Equipment sales:
    number of degrees $= (0.1875)(360°)$
    $= 67.5°$

    (d)

Section 9.3 (pages 655–658)

3. 57% say they find better prices on-line.

   percent $\cdot$ whole $=$ part
   $\downarrow \qquad \downarrow \qquad \downarrow$
   57% $\cdot$ 600 $= n$
   (0.57) (600) $= n$
   342 $= n$

   342 people gave this answer.

11. The number of unemployed workers increased from 5500 in February of 2007 to 8000 in April of 2007. The amount of increase was $8000 - 5500 = 2500$ workers.

    $\begin{array}{ccccc} \text{percent of} & \text{February 2007} & = & \text{Amount} \\ \downarrow & \downarrow \text{ unemployed workers} & \downarrow & \text{of increase} \\ p & \cdot \quad 5500 & = & 2500 \end{array}$

    $\dfrac{p \cdot \overset{1}{\cancel{5500}}}{\underset{1}{\cancel{5500}}} = \dfrac{2500}{5500}$

    $p = \dfrac{5}{11}$

    $p \approx 0.45 = 45\%$

    The percent of increase was 45% (rounded).

23. The amount of increase in shipments from 1995 to 2000 was 144.6 million $-$ 62.3 million $= 82.3$ million or 82,300,000 PCs.

    $\begin{array}{ccccc} \text{percent of} & 1995 & = & \text{amount} \\ \downarrow & \downarrow \text{ shipments} & \downarrow & \text{of increase} \\ p & \cdot \quad 62.3 & = & 82.3 \end{array}$

    $\dfrac{p \cdot \overset{1}{\cancel{62.3}}}{\underset{1}{\cancel{62.3}}} = \dfrac{82.3}{62.3}$

    $p \approx 1.32 = 132\%$

    The percent of increase was 132% (rounded).

33. **For 2005:** Profit is $10,000, Sales are $35,000

    $\begin{array}{ccccc} \text{profit} & \text{is} & \text{what percent} & \text{of} & \text{sales} \\ \downarrow & \downarrow & \downarrow & & \downarrow \quad \downarrow \\ 10{,}000 & = & p & \cdot & 35{,}000 \end{array}$

    $\dfrac{10{,}000}{35{,}000} = \dfrac{p \cdot \overset{1}{\cancel{35{,}000}}}{\underset{1}{\cancel{35{,}000}}}$

    $0.29 \approx p$
    $29\% \approx p$

    For 2006: Profit is $5,000, Sales are $25,000

    $\dfrac{\$5{,}000}{\$25{,}000} = p$

    $p = 0.20 = 20\%$

    For 2007: Profit is $5,000, Sales are $30,000

    $\dfrac{\$5{,}000}{\$30{,}000} = p$

    $p \approx 0.17 = 17\%$

    For 2008: Profit is $15,000, Sales are $40,000

    $\dfrac{\$15{,}000}{\$40{,}000} = p$

    $p \approx 0.38 = 38\%$

Section 9.4 (pages 663–664)

7. For $(-3, -7)$ the pattern is $(-, -)$, so the point is in Quadrant III.

   The point corresponding to $(0, 4)$ is on the y-axis, so it is not in any quadrant.

   For $(10, -16)$ the pattern is $(+, -)$, so the point is in Quadrant IV.
   For $(-9, 5)$ the pattern is $(-, +)$, so the point is in Quadrant II.

9. (a) Any *positive* number, because points in Quadrant II have the pattern $(-, +)$.

(b) Any *negative* number, because points in Quadrant IV have the pattern $(+, -)$.

(c) 0, because points that are not in any quadrant have a zero as one of the numbers in the ordered pair.

(d) Any *negative* number, because points in Quadrant III have the pattern $(-, -)$.

(e) Any *positive* number, because points in Quadrant I have the pattern $(+, +)$.

Section 9.5 (pages 673–678)

3. $x + y = -1$

To complete the table, replace x in the equation with each value listed for x.

When x is 0: $\quad 0 + y = -1 \quad$ so y must be -1.

This gives the ordered pair $(0, -1)$.

When x is 1: $\quad 1 + y = -1 \quad$ so y must be -2.

This gives the ordered pair $(1, -2)$.

When x is 2: $\quad 2 + y = -1 \quad$ so y must be -3.

This gives the ordered pair $(2, -3)$.

x	y	Ordered Pair (x, y)
0	-1	$(0, -1)$
1	-2	$(1, -2)$
2	-3	$(2, -3)$

Plot $(0, -1)$ and $(1, -2)$ and $(2, -3)$ on the grid. Draw a line connecting the points and extend it in both directions.

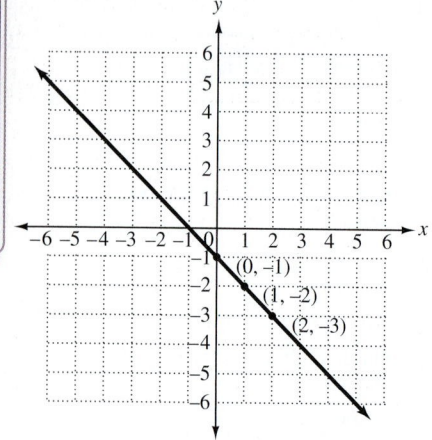

All points on the line are solutions so there are many possibilities.

Two other possible solutions are $(-1, 0)$ and $(-2, 1)$.

To check that $(-1, 0)$ and $(-2, 1)$ are solutions, substitute into the original equation.

19. $y = -2x + 3$

x	$y = -2 \cdot x + 3$	(x, y)
0	$-2(0) + 3$ is 3	$(0, 3)$
1	$-2(1) + 3$ is 1	$(1, 1)$
2	$-2(2) + 3$ is -1	$(2, -1)$

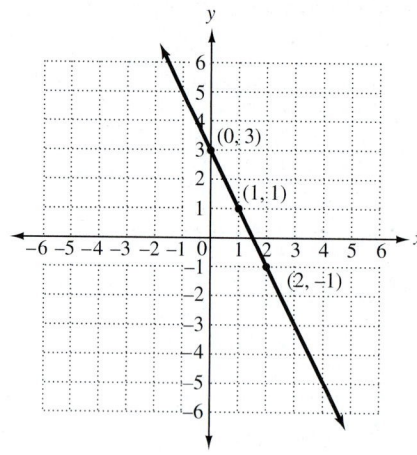

23. $y = \dfrac{1}{4}x$

One possible table that uses 0 and multiples of 4 as values of x is shown.

x	$y = \dfrac{1}{4} \cdot x$	(x, y)
0	$\dfrac{1}{4}(0)$ is 0	$(0, 0)$
4	$\dfrac{1}{4}(4)$ is 1	$(4, 1)$
8	$\dfrac{1}{4}(8)$ is 2	$(8, 2)$

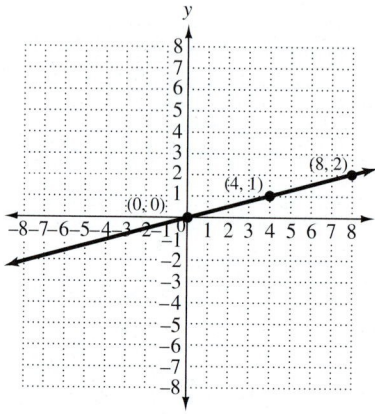

The line has a positive slope (it rises from left to right).

Chapter 10 Real Numbers, Equations, and Inequalities

Section 10.1 (pages 703–704)

9. (a) $\sqrt{2} \approx 1.41421356$ so it is irrational because the decimal value never ends or repeats in a fixed block.

29. $3^2[(-10 \div 5) + 7] \le 40$
$$9[-2 + 7] \le 40$$
$$9[5] \le 40$$
$$45 \le 40 \quad \textit{False}$$

45. $-(-3q + 5r - 8s)$
$$-1(-3q + 5r - 8s)$$
$$(-1 \cdot -3q) + (-1 \cdot 5r) - (-1 \cdot 8s)$$
$$3q + (-5r) - (-8s)$$
$$3q - 5r + 8s$$

55. $-3(-a + 1) - (2a - 4)$
$$(-3 \cdot -a) + (-3 \cdot 1) - 1(2a - 4)$$
$$(3a) \quad + \quad (-3) \quad - 2a + 4$$
$$3a - 3 - 2a + 4 \quad \text{Combine terms}$$
$$a + 1$$

Section 10.2 (pages 711–714)

7. *Step 1*
$$x + 3 = -(2x + 2)$$
$$x + 3 = -1(2x + 2)$$
$$x + 3 = -2x - 2 \quad \text{Distributive property}$$

Step 3
$$x + 3 = -2x - 2$$
$$\underline{+2x} \qquad \underline{+2x} \qquad \text{Add } 2x$$
$$3x + 3 = \qquad -2$$
$$\underline{-3} \qquad \underline{-3} \qquad \text{Subtract 3}$$
$$3x = \qquad -5$$

Step 4
$$\frac{3x}{3} = \frac{-5}{3} \qquad \text{Divide by 3}$$
$$x = -\frac{5}{3}$$

Step 5

Check by replacing x in the *original* equation with $-\frac{5}{3}$.
$$\left(-\frac{5}{3}\right) + 3 = -\left(2 \cdot -\frac{5}{3} + 2\right)$$
$$-\frac{5}{3} + \frac{9}{3} = -1\left(-\frac{10}{3} + \frac{6}{3}\right)$$
$$\frac{4}{3} = -1\left(-\frac{4}{3}\right)$$
$$\frac{4}{3} = \frac{4}{3} \qquad \text{Balances, so } -\frac{5}{3} \text{ is correct solution.}$$

31. $10(-2x + 1) = -14(x + 2) + 38 - 6x$
$$-20x + 10 = -14x - 28 + 38 - 6x$$
$$-20x + 10 = -20x + 10$$
$$\underline{+20x} \qquad \underline{+20x} \qquad \text{Add } 20x$$
$$\frac{}{10} = \frac{}{10}$$

Since $10 = 10$ is a *true* statement, all *real numbers* are solutions of the equation.

39. $\dfrac{2}{3}k - \left(k + \dfrac{1}{4}\right) = \dfrac{1}{12}(k + 4)$

The least common denominator of all the fractions in the equation is 12, so multiply both sides by 12 and solve for k. Use the distributive property.
$$12\left[\frac{2}{3}k - \left(k + \frac{1}{4}\right)\right] = 12\left[\frac{1}{12}(k + 4)\right]$$
$$12\left(\frac{2}{3}k\right) - 12\left(k + \frac{1}{4}\right) = 12\left(\frac{1}{12}k + \frac{4}{12}\right)$$

$$8k - 12k - 12\left(\frac{1}{4}\right) = 1k + 4$$

$$8k - 12k - 3 = k + 4$$

$$-4k - 3 = k + 4$$

$$\frac{-k}{-5k - 3} = \frac{-k}{4} \qquad \text{Subtract } k$$

$$\frac{+3}{-5k} = \frac{+3}{7} \qquad \text{Add 3}$$

$$\frac{-5k}{-5} = \frac{7}{-5} \qquad \text{Divide by } -5$$

$$k = -\frac{7}{5}$$

The solution is $-\frac{7}{5}$.

47. To clear the equation of decimals, we multiply both sides by 100. To multiply by 100, move the decimal point two places to the right.

$$1.00x + 0.05(12 - x) = 0.10(63)$$

$$100x + 5(12 - x) = 10(63)$$

$$100x + 60 - 5x = 630$$

$$95x + 60 = 630$$

$$\frac{-60}{95x} = \frac{-60}{570} \qquad \begin{array}{l}\text{Subtract} \\ 60\end{array}$$

$$\frac{95x}{95} = \frac{570}{95} \qquad \begin{array}{l}\text{Divide} \\ \text{by 95}\end{array}$$

$$x = 6$$

The solution is 6.

Section 10.3 (pages 719–720)

17. The circumference 1978 feet. You want to find the diameter, so use the circumference of a circle formula that involves diameter (d).

$$C = \pi d$$

$$1978 \approx 3.14d$$

$$\frac{1978}{3.14} \approx \frac{3.14d}{3.14} \qquad \text{Divide by 3.14}$$

$$629.9 \approx d$$

Round 629.9 to 630, so the dome has a diameter of 630 ft, to the nearest whole foot.

27. $A = \frac{1}{2}bh$ for b

$$2(A) = 2\left(\frac{1}{2}bh\right) \qquad \text{Multiply by 2}$$

$$2A = bh$$

$$\frac{2A}{h} = \frac{bh}{h} \qquad \text{Divide by } h$$

$$\frac{2A}{h} = b \quad \text{or} \quad b = \frac{2A}{h}$$

29. $A = p + prt$ for r

$$A = p + prt$$

$$\frac{-p}{A - p} = \frac{-p}{prt} \qquad \text{Subtract } p$$

$$\frac{A - p}{pt} = \frac{prt}{pt} \qquad \text{Divide by } pt$$

$$\frac{A - p}{pt} = r \quad \text{or} \quad r = \frac{A - p}{pt}$$

Section 10.4 (pages 727–730)

11. $0 < y \le 10$

Place an open circle at 0 (because 0 is not part of the graph) and a solid circle at 10 (because 10 is part of the graph); then draw a line segment between the two circles.

35. $-0.02x \le 0.06$ Multiply by 100

$$\frac{-2x}{-2} \le \frac{6}{-2} \qquad \begin{array}{l}\text{Divide by } -2; \text{ reverse} \\ \text{symbol from } \le \text{ to } \ge\end{array}$$

$$x \ge -3$$

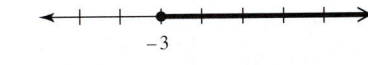

45. $\frac{2}{3}(p + 3) > \frac{5}{6}(p - 4)$

$$6\left(\frac{2}{3}\right)(p + 3) > 6\left(\frac{5}{6}\right)(p - 4) \qquad \begin{array}{l}\text{Multiply} \\ \text{by 6, the} \\ \text{LCD}\end{array}$$

$$4(p + 3) > 5(p - 4)$$

$$4p + 12 > 5p - 20$$

$$\frac{-12}{4p} > \frac{-12}{5p - 32} \qquad \begin{array}{l}\text{Subtract} \\ 12\end{array}$$

$$\frac{-5p}{-1p} > \frac{-5p}{-32} \qquad \begin{array}{l}\text{Subtract} \\ 5p\end{array}$$

$$\frac{-1p}{-1} > \frac{-32}{-1}$$

Divide by -1; reverse symbol from $>$ to $<$

$$p < 32$$

53. Let x be the score on the fifth test.

The average of the five tests	is at least	80

$$\frac{89 + 78 + 73 + 81 + x}{5} \ge 80$$

Multiply by 5

$$5\left(\frac{89 + 78 + 73 + 81 + x}{5}\right) \ge 5(80)$$

$$89 + 78 + 73 + 81 + x \ge 400$$

$$321 + x \ge 400$$

Subtract 321

$$x \ge 79$$

Twylene must receive a score of 79 or more so that her average is at least 80.

Chapter 11 Graphs of Linear Equations and Inequalities in Two Variables

Section 11.1 (pages 751–756)

23. Is $(5, -6)$ a solution of the equation $x = -6$?

Since y does not appear in the equation, we just substitute 5 for x.

$$x = -6$$

$$5 \overset{?}{=} -6 \qquad \text{Let } x = 5.$$

The result is false, so $(5, -6)$ is not a solution of the equation $x = -6$.

43. The given equation $x - 8 = 0$ may be written as $x = 8$. For any value of y, the value of x will always be 8. The completed table of values follows.

x	y
8	8
8	3
8	0

Section 11.2 (pages 765–770)

29. $-3y = 15$

$$y = \frac{15}{-3}, \text{ or } -5 \qquad \text{Divide by } -3.$$

For any value of x, the value of y is -5. Three ordered pairs are $(-2, -5)$, $(0, -5)$, and $(1, -5)$. Plot these points and draw a line through them. The graph is a horizontal line.

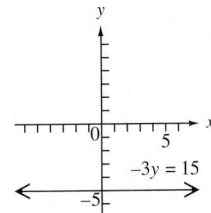

Section 11.3 (pages 779–782)

27. Use the slope formula with

$$\left(-\frac{7}{5}, \frac{3}{10}\right) = (x_1, y_1)$$

and

$$\left(\frac{1}{5}, -\frac{1}{2}\right) = (x_2, y_2).$$

$$\text{slope } m = \frac{\text{change in } y}{\text{change in } x}$$

$$= \frac{y_2 - y_1}{x_2 - x_1}$$

$$= \frac{-\frac{1}{2} - \frac{3}{10}}{\frac{1}{5} - \left(-\frac{7}{5}\right)} \qquad \text{Substitute.}$$

$$= \frac{-\frac{5}{10} - \frac{3}{10}}{\frac{1}{5} + \frac{7}{5}} \qquad \begin{array}{l}\text{Get a common} \\ \text{denominator in} \\ \text{the numerator;} \\ \text{simplify the} \\ \text{denominator.}\end{array}$$

(continued)

SOLUTIONS

$$= \dfrac{-\frac{8}{10}}{\frac{8}{5}}$$ Subtract in the numerator; add in the denominator.

$$= -\dfrac{8}{10} \div \dfrac{8}{5}$$ $\frac{a}{b} = a \div b$

$$= -\dfrac{8}{10} \cdot \dfrac{5}{8}$$ Multiply by the reciprocal.

$$= -\dfrac{1}{2}$$ Multiply; lowest terms

53. The slope (or grade) of the hill is the ratio of the rise to the run, or the ratio of the vertical change to the horizontal change. Since the rise is 32 and the run is 108, the slope is

$$\dfrac{32}{108} = \dfrac{8 \cdot 4}{27 \cdot 4} = \dfrac{8}{27}.$$ Lowest terms

Section 11.4 (pages 791–796)

3. The rise is 3 and the run is 1, so the slope m is given by

$$m = \dfrac{\text{rise}}{\text{run}} = \dfrac{3}{1} = 3.$$

The y-intercept is $(0, -3)$, so $b = -3$. The equation of the line, written in slope-intercept form $y = mx + b$, is

$$y = 3x - 3.\quad m = 3, b = -3$$

45. $\left(\dfrac{1}{2}, \dfrac{3}{2}\right), \left(-\dfrac{1}{4}, \dfrac{5}{4}\right)$

First, find the slope of the line.

$$m = \dfrac{\frac{5}{4} - \frac{3}{2}}{-\frac{1}{4} - \frac{1}{2}} = \dfrac{\frac{5}{4} - \frac{6}{4}}{-\frac{1}{4} - \frac{2}{4}} = \dfrac{-\frac{1}{4}}{-\frac{3}{4}}$$

$$= -\dfrac{1}{4} \div \left(-\dfrac{3}{4}\right) = -\dfrac{1}{4}\left(-\dfrac{4}{3}\right) = \dfrac{1}{3}$$

Now use the point $\left(\dfrac{1}{2}, \dfrac{3}{2}\right)$ for (x_1, y_1) and $m = \dfrac{1}{3}$ in the point-slope form.

$$y - y_1 = m(x - x_1)$$

$$y - \dfrac{3}{2} = \dfrac{1}{3}\left(x - \dfrac{1}{2}\right)$$ Substitute.

$$y - \dfrac{3}{2} = \dfrac{1}{3}x - \dfrac{1}{6}$$ Distributive property

$$y = \dfrac{1}{3}x - \dfrac{1}{6} + \dfrac{9}{6}$$ Add $\frac{3}{2} = \frac{9}{6}$.

$$y = \dfrac{1}{3}x + \dfrac{4}{3}$$ Combine like terms; $\frac{8}{6} = \frac{4}{3}$

47. Solve the given equation for y.

$$3x = 4y + 5$$

$$3x - 4y = 5$$ Subtract $4y$.

$$-4y = -3x + 5$$ Subtract $3x$.

$$y = \dfrac{3}{4}x - \dfrac{5}{4}$$ Divide by -4.

The slope is $\dfrac{3}{4}$. A line parallel to this line has the same slope. Use the point-slope form with $m = \dfrac{3}{4}$ and $(x_1, y_1) = (2, -3)$.

$$y - y_1 = m(x - x_1)$$

$$y - (-3) = \dfrac{3}{4}(x - 2)$$ Substitute.

$$y + 3 = \dfrac{3}{4}x - \dfrac{3}{2}$$

$$y = \dfrac{3}{4}x - \dfrac{3}{2} - \dfrac{6}{2}$$ Subtract $3 = \frac{6}{2}$.

$$y = \dfrac{3}{4}x - \dfrac{9}{2}$$ Combine like terms.

Chapter 12 Systems of Linear Equations and Inequalities

Section 12.1 (pages 827–830)

21. $2x - 3y = -6$

$$y = -3x + 2$$

To graph $2x - 3y = -6$, find the intercepts.

$$2x - 3(0) = -6\quad \text{Let } y = 0.$$

$$2x = -6$$

$$x = -3$$

The x-intercept is $(-3, 0)$.

$$2(0) - 3y = -6\quad \text{Let } x = 0.$$

$$-3y = -6$$

$$y = 2$$

The y-intercept is $(0, 2)$.

Plot the intercepts, $(-3, 0)$ and $(0, 2)$, and draw the line through them.

To graph the second line, start by plotting the y-intercept, $(0, 2)$. From this point, go 3 units down and 1 unit to the right (because the slope is -3) to reach the point $(1, -1)$. Draw the line through $(0, 2)$ and $(1, -1)$.

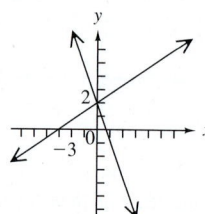

The lines intersect at their common y-intercept, $(0, 2)$.

Solution set: $\{(0, 2)\}$

25. $4x - 2y = 8$ (1)

$$2x = y + 4\quad (2)$$

Graph the line $4x - 2y = 8$ using its intercepts, $(2, 0)$ and $(0, -4)$.

Graph the equation $2x = y + 4$ using its intercepts, $(2, 0)$ and $(0, -4)$. Since both equations have the same intercepts, they are equations of the same line.

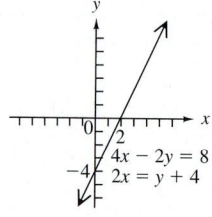

There are an infinite number of solutions, so we use set-builder notation to write the solution set. Rewrite equation (2) in standard form.

$$2x - y = 4$$

The solution set is

$$\{(x, y) \mid 2x - y = 4\}.$$

The given system consists of dependent equations.

Section 12.2 (pages 837–838)

15. $6x - 8y = 6$ (1)

$$2y = -2 + 3x\quad (2)$$

Solve equation (2) for y.

$$2y = -2 + 3x$$

$$y = \dfrac{3x - 2}{2}\quad (3)$$

Substitute $\dfrac{3x - 2}{2}$ for y in equation (1).

$$6x - 8y = 6\quad (1)$$

$$6x - 8\left(\dfrac{3x - 2}{2}\right) = 6$$

$$6x - 4(3x - 2) = 6$$

$$6x - 12x + 8 = 6$$

$$-6x + 8 = 6$$

$$-6x = -2$$

$$x = \dfrac{-2}{-6}, \quad \text{or} \quad \dfrac{1}{3}$$

To find y, let $x = \dfrac{1}{3}$ in equation (3).

$$y = \dfrac{3x - 2}{2} = \dfrac{3\left(\frac{1}{3}\right) - 2}{2} = \dfrac{1 - 2}{2} = -\dfrac{1}{2}$$

Solution set: $\left\{\left(\dfrac{1}{3}, -\dfrac{1}{2}\right)\right\}$

19. $12x - 16y = 8$ (1)

$$3x = 4y + 2\quad (2)$$

Solve equation (2) for x.

$$3x = 4y + 2$$

$$x = \dfrac{4y + 2}{3}\quad (3)$$

Substitute $\dfrac{4y + 2}{3}$ for x in equation (1).

$$12x - 16y = 8\quad (1)$$

$$12\left(\dfrac{4y + 2}{3}\right) - 16y = 8$$

$$4(4y + 2) - 16y = 8$$

$$16y + 8 - 16y = 8$$

$$8 = 8\quad \text{True}$$

This true result means that every solution of one equation is also a solution of the other, so the system has an infinite number of solutions. To write the solution set, rewrite equation (2) in standard form.

$$3x - 4y = 2$$

Solution set: $\{(x, y) \mid 3x - 4y = 2\}$

25. $\dfrac{x}{5} + 2y = \dfrac{16}{5}$ (1)

$$\dfrac{3x}{5} + \dfrac{y}{2} = -\dfrac{7}{5}\quad (2)$$

Multiply each side of equation (1) by 5.

$$5\left(\frac{x}{5} + 2y\right) = 5\left(\frac{16}{5}\right)$$
$$x + 10y = 16 \qquad (3)$$

Multiply each side of equation (2) by 10.

$$10\left(\frac{3x}{5} + \frac{y}{2}\right) = 10\left(-\frac{7}{5}\right)$$
$$6x + 5y = -14 \qquad (4)$$

We now have the simplified system

$$x + 10y = 16 \qquad (3)$$
$$6x + 5y = -14. \qquad (4)$$

Solve equation (3) for x.

$$x = 16 - 10y \qquad (5)$$

Substitute $16 - 10y$ for x in equation (4).

$$6x + 5y = -14 \qquad (4)$$
$$6(16 - 10y) + 5y = -14$$
$$96 - 60y + 5y = -14$$
$$-55y = -110$$
$$y = 2$$

To find x, let $y = 2$ in equation (5).

$$x = 16 - 10y \qquad (5)$$
$$x = 16 - 10(2)$$
$$x = -4$$

Solution set: $\{(-4, 2)\}$

Section 12.3 (pages 843–846)

25.
$$-x + 3y = 4 \qquad (1)$$
$$-2x + 6y = 8 \qquad (2)$$

Multiply equation (1) by -2 and add the result to equation (2).

$$\begin{array}{ll} 2x - 6y = -8 & (3) \\ -2x + 6y = 8 & (2) \\ \hline 0 = 0 & \text{True} \end{array}$$

The true statement indicates that the equations of the original system are dependent. To obtain an equation in standard form, with $A > 0$ in $Ax + By = C$, multiply equation (1) by -1.

$$x - 3y = -4$$

Solution set: $\{(x, y) \mid x - 3y = -4\}$

39.
$$6x + 3y = 0 \qquad (1)$$
$$-18x - 9y = 0 \qquad (2)$$

Multiply equation (1) by 3 and add the result to equation (2).

$$\begin{array}{ll} 18x + 9y = 0 & (3) \\ -18x - 9y = 0 & (2) \\ \hline 0 = 0 & \text{True} \end{array}$$

This true result, $0 = 0$, means that the system has an infinite number of solutions. To obtain an equation for the solution set in which the coefficients have greatest common factor 1, divide both sides of equation (1) by 3 to obtain the equivalent equation

$$2x + y = 0.$$

Solution set: $\{(x, y) \mid 2x + y = 0\}$

Section 12.4 (pages 855–860)

33. Step 1

Read the problem again.

Step 2

Assign variables.

Let x = the average speed of the bicycle;
y = the average speed of the car.

	r	t	d
Bicycle	x	7.5	$7.5x$
Car	y	7.5	$7.5y$

↑
Use the formula
$d = rt$.

Step 3

The total distance traveled by the bicycle and the car is 471 mi, so

$$7.5x + 7.5y = 471 \qquad (1)$$

The car traveled 35.8 mi faster than the bicycle, so

$$y = x + 35.8 \qquad (2)$$

We now have a system of equations.

$$7.5x + 7.5y = 471 \qquad (1)$$
$$y = x + 35.8 \qquad (2)$$

Step 4

Because equation (2) is already solved for y, the substitution method is a good choice. Substitute $x + 35.8$ for y in equation (1).

$$7.5x + 7.5y = 471 \quad (1)$$
$$7.5x + 7.5(x + 35.8) = 471$$
$$ \text{Let } y = x + 35.8$$
$$7.5x + 7.5x + 268.5 = 471$$
$$ \text{Distributive property}$$
$$15x + 268.5 = 471$$
$$ \text{Combine like terms.}$$
$$15x = 202.5$$
$$ \text{Subtract 268.5}$$
$$x = 13.5$$
$$ \text{Divide by 15}$$

To find the value of y substitute 13.5 for x in equation (2).

$$y = x + 35.8 \qquad (2)$$
$$y = 13.5 + 35.8 \quad \text{Let } x = 13.5$$
$$y = 49.3$$

Step 5

The average speed of the bicycle is 13.5 mph, and the average speed of the car is 49.3 mph.

Step 6

In 7.5 hr, the total distance traveled by the bicycle and the car is

$$7.5(13.5) + 7.5(49.3) = 471 \text{ mi.}$$

Since $49.3 - 13.5 = 35.8$, the car traveled 35.8 mph faster than the bicycle, as stated.

Section 12.5 (pages 865–866)

15. $x \le 2y + 3$

$x + y < 0$

Graph $x = 2y + 3$ as a solid line through $(3, 0)$ and $(5, 1)$. Using $(0, 0)$ as a test point will result in the true statement $0 \le 3$, so shade the region containing the origin.

Graph $x + y = 0$ as a dashed line through $(0, 0)$ and $(1, -1)$. Using $(1, 0)$ as a test point will result in the false statement $1 < 0$, so shade the region *not* containing $(1, 0)$.

The solution set of the system is the intersection of the two shaded regions. It includes the portion of the line $x = 2y + 3$ that bounds the region but not the portion of the line $x + y = 0$.

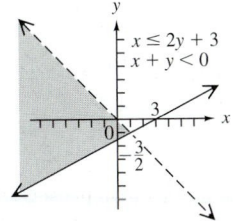

$$x \le 2y + 3$$
$$x + y < 0$$

19. $3x - 2y \ge 6$

$x + y \le 4$

$x \ge 0$

$y \ge -4$

Graph $3x - 2y = 6$, $x + y = 4$, $x = 0$, and $y = -4$ as solid lines. All four inequalities are true for $(2, -2)$. Shade the region bounded by the four lines, which contains the test point $(2, -2)$. The solution set is the shaded region.

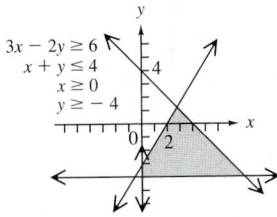

$$3x - 2y \ge 6$$
$$x + y \le 4$$
$$x \ge 0$$
$$y \ge -4$$

Chapter 13 Exponents and Polynomials

Section 13.1 (pages 883–886)

21. $\dfrac{1}{2}x^4 + \dfrac{1}{6}x^4$

$$= \left(\frac{1}{2} + \frac{1}{6}\right)x^4 \quad \text{Distributive property}$$
$$= \left(\frac{3}{6} + \frac{1}{6}\right)x^4 \quad \begin{array}{l}\text{Write fractions with a}\\\text{common denominator.}\end{array}$$
$$= \frac{4}{6}x^4 \quad \text{Add fractions.}$$
$$= \frac{2}{3}x^4 \quad \text{Lowest terms}$$

37. $0.8x^4 - 0.3x^4 - 0.5x^4 + 7$

$$= (0.8 - 0.3 - 0.5)x^4 + 7$$
$$= 0x^4 + 7$$
$$= 7$$

Since 7 can be written as $7x^0$, the degree of the polynomial is 0. The simplified polynomial has one term, so it is a monomial.

57. Add.

$$\frac{2}{3}x^2 + \frac{1}{5}x + \frac{1}{6}$$
$$\frac{1}{2}x^2 - \frac{1}{3}x + \frac{2}{3}$$

Rewrite so that the fractions in each column have a common denominator; then add column by column.

(continued)

$$\frac{4}{6}x^2 + \frac{3}{15}x + \frac{1}{6}$$
$$\frac{3}{6}x^2 - \frac{5}{15}x + \frac{4}{6}$$
$$\overline{\frac{7}{6}x^2 - \frac{2}{15}x + \frac{5}{6}}$$

75. Use the formula for the perimeter of a square, $P = 4s$, with $s = \frac{1}{2}x^2 + 2x$.

$$P = 4s$$
$$P = 4\left(\frac{1}{2}x^2 + 2x\right) \qquad \text{Substitute.}$$
$$P = 4\left(\frac{1}{2}x^2\right) + 4(2x) \qquad \text{Distributive property}$$
$$P = 2x^2 + 8x \qquad \text{Multiply.}$$

Section 13.2 (pages 893–894)

47. $(-5^2)^6$

$$= (-1 \cdot 5^2)^6 \qquad \text{Write the factor } -1.$$
$$= (-1)^6 \cdot (5^2)^6 \qquad \text{Power rule (b)}$$
$$= 1 \cdot 5^{2\cdot6} \qquad \text{Power rule (a)}$$
$$= 5^{12}$$

81. $\left(\dfrac{5a^2b^5}{c^6}\right)^3, \quad c \neq 0$

$$= \frac{(5a^2b^5)^3}{(c^6)^3} \qquad \text{Power rule (c)}$$
$$= \frac{5^3(a^2)^3(b^5)^3}{(c^6)^3} \qquad \text{Power rule (b)}$$
$$= \frac{125a^6b^{15}}{c^{18}} \qquad \text{Power rule (a)}$$

83. $(-5m^3p^4q)^2(p^2q)^3$

$$= (-1 \cdot 5m^3p^4q)^2(p^2q)^3 \qquad \text{Write the factor } -1.$$
$$= (-1)^2 \cdot 5^2 \cdot (m^3)^2 \cdot (p^4)^2$$
$$\quad \cdot q^2 \cdot (p^2)^3 \cdot q^3 \qquad \text{Power rule (b)}$$
$$= 1 \cdot 25 \cdot m^6p^8q^2p^6q^3 \qquad \text{Power rule (a)}$$
$$= 25m^6p^{8+6}q^{2+3} \qquad \text{Product rule}$$
$$= 25m^6p^{14}q^5$$

Section 13.3 (pages 899–900)

37. $\left(x - \dfrac{2}{3}\right)\left(x + \dfrac{1}{4}\right)$

$$\qquad\quad \mathbf{F} \qquad \mathbf{O} \qquad \mathbf{I} \qquad \mathbf{L}$$
$$= x(x) + x\left(\frac{1}{4}\right) + \left(-\frac{2}{3}\right)x + \left(-\frac{2}{3}\right)\frac{1}{4}$$
$$= x^2 + \frac{1}{4}x - \frac{2}{3}x - \frac{1}{6}$$
$$= x^2 + \left(\frac{3}{12}x - \frac{8}{12}x\right) - \frac{1}{6}$$
$$= x^2 - \frac{5}{12}x - \frac{1}{6}$$

Section 13.4 (pages 905–906)

11. $(0.8t + 0.7s)^2$
$$= (0.8t)^2 + 2(0.8t)(0.7s) + (0.7s)^2$$
$$\qquad\qquad (a + b)^2 = a^2 + 2ab + b^2$$
$$= 0.64t^2 + 1.12ts + 0.49s^2$$

17. $-(4r - 2)^2$
First square the binomial.
$$(4r - 2)^2$$
$$= (4r)^2 - 2(4r)(2) + 2^2$$
$$= 16r^2 - 16r + 4$$
Now multiply by -1.
$$-1(16r^2 - 16r + 4)$$
$$= -16r^2 + 16r - 4$$

27. $(2x^2 - 5)(2x^2 + 5)$
$$= (2x^2)^2 - 5^2 \quad (a+b)(a-b) = a^2 - b^2$$
$$= 4x^4 - 25 \quad (2x^2)^2 = 2^2x^4 = 4x^4$$

55. Use the formula for the volume of a rectangular solid, $V = lwh$. Here, $V = (x + 2)(x + 2)(x + 2)$ or $V = (x + 2)^3$. If $x = 6$, then
$$V = (6 + 2)^3 = 8^3 = 512.$$
The volume of the cube is 512 cu. units.

Section 13.5 (pages 915–916)

15. $(-2)^0 - 2^0$
$$= 1 - 1 \quad a^0 = 1$$
$$= 0$$

31. $-2^{-1} + 3^{-2}$
$$= -(2^{-1}) + 3^{-2} \quad \text{In } -2^{-1}, \text{ the base is 2.}$$
$$= -\frac{1}{2^1} + \frac{1}{3^2} \quad a^{-n} = \frac{1}{a^n}$$
$$= -\frac{1}{2} + \frac{1}{9} \quad \text{Apply the exponents.}$$
$$= -\frac{9}{18} + \frac{2}{18} \quad \begin{array}{l}\text{Write fractions}\\\text{with a common}\\\text{denominator.}\end{array}$$
$$= -\frac{7}{18} \quad \text{Add.}$$

71. $\dfrac{(m^7n)^{-2}}{m^{-4}n^3}$

$$= \frac{(m^7)^{-2}n^{-2}}{m^{-4}n^3} \quad \text{Power rule (b)}$$
$$= \frac{m^{7(-2)}n^{-2}}{m^{-4}n^3} \quad \text{Power rule (a)}$$
$$= \frac{m^{-14}n^{-2}}{m^{-4}n^3}$$
$$= m^{-14-(-4)}n^{-2-3} \quad \text{Quotient rule}$$
$$= m^{-10}n^{-5}$$
$$= \frac{1}{m^{10}n^5} \quad a^{-n} = \frac{1}{a^n}$$

Section 13.6 (pages 921–922)

19. $\dfrac{-3x^3 - 4x^4 + 2x}{-3x^2}$

$$= \frac{-4x^4 - 3x^3 + 2x}{-3x^2} \quad \begin{array}{l}\text{Write in}\\\text{descending}\\\text{powers.}\end{array}$$
$$= \frac{-4x^4}{-3x^2} - \frac{3x^3}{-3x^2} + \frac{2x}{-3x^2} \quad \begin{array}{l}\text{Divide}\\\text{each term}\\\text{by } -3x^2.\end{array}$$
$$= \frac{4x^2}{3} + x - \frac{2}{3x} \quad \begin{array}{l}\text{Quotient rule; be}\\\text{careful with signs.}\end{array}$$

Notice how the third term is written with x in the denominator, which is **not** the same as $-\frac{2}{3}x$ or $-\frac{2x}{3}$. In $-\frac{2}{3x}$, we are *dividing* by x; in $-\frac{2}{3}x$ and $-\frac{2x}{3}$, we are *multiplying* by x. Applying the quotient rule to the term $\frac{2x}{-3x^2}$ gives

$$\frac{2x}{-3x^2} = \frac{2x^1}{-3x^2} = -\frac{2}{3}x^{1-2} = -\frac{2}{3}x^{-1}$$
$$= -\frac{2}{3}\left(\frac{1}{x}\right) = -\frac{2}{3x}.$$

$\left(\frac{4}{3}x^2\right.$ is an acceptable form for the first term, $\frac{4x^2}{3}$. Why?$\left.\right)$

29. Use the formula for the area of a rectangle, $A = lw$, with $A = 12x^2 - 4x + 2$ and $w = 2x$.

$$A = lw$$
$$12x^2 - 4x + 2 = l(2x) \quad \begin{array}{l}\text{Substitute for}\\ A \text{ and } w.\end{array}$$
$$\frac{12x^2 - 4x + 2}{2x} = l \quad \begin{array}{l}\text{Divide each}\\\text{side by } 2x.\end{array}$$
$$\frac{12x^2}{2x} - \frac{4x}{2x} + \frac{2}{2x} = l \quad \begin{array}{l}\text{Divide each}\\\text{term by } 2x.\end{array}$$
$$6x - 2 + \frac{1}{x} = l \quad \begin{array}{l}\frac{2}{2x} = \frac{2}{2} \cdot \frac{1}{x}\\ = 1 \cdot \frac{1}{x} = \frac{1}{x}\end{array}$$

$6x - 2 + \dfrac{1}{x}$ represents the length of the rectangle.

Section 13.7 (pages 927–928)

23. $(x^4 - x^2 - 2) \div (x^2 - 2)$

Use 0 as the coefficient of the missing x^3- and x-terms in the dividend and the missing x-term in the divisor.

$$\begin{array}{r}
x^2 \qquad\qquad + 1 \\
x^2 + \mathbf{0x} - 2 \overline{)x^4 + \mathbf{0x^3} - \; x^2 + \mathbf{0x} - 2} \\
\underline{x^4 + 0x^3 - 2x^2} \\
x^2 + 0x - 2 \\
\underline{x^2 + 0x - 2} \\
0
\end{array}$$

The remainder is 0. The answer is the quotient, $x^2 + 1$.

29. $\dfrac{2x^5 + x^4 + 11x^3 - 8x^2 - 13x + 7}{2x^2 + x - 1}$

$$2x^2 + x - 1\overline{)2x^5 + x^4 + 11x^3 - 8x^2 - 13x + 7}$$

with quotient $x^3 + 6x - 7$:

$$\begin{array}{r}
x^3 \qquad\quad + 6x - 7 \\
2x^5 + x^4 - x^3 \\
\hline
12x^3 - 8x^2 - 13x \\
12x^3 + 6x^2 - 6x \\
\hline
-14x^2 - 7x + 7 \\
-14x^2 - 7x + 7 \\
\hline
0
\end{array}$$

The remainder is 0. The answer is the quotient, $x^3 + 6x - 7$.

Section 13.8 (pages 933–934)

51. $\dfrac{4 \times 10^5}{8 \times 10^2}$

$= \dfrac{4}{8} \times \dfrac{10^5}{10^2}$

$= 0.5 \times 10^3$
      Divide; quotient rule

$= (5 \times 10^{-1}) \times 10^3$
      Write 0.5 in scientific notation.

$= 5 \times (10^{-1} \times 10^3)$
      Associative property

$= 5 \times 10^2$
      Product rule

$= 500$
      Write without exponents.

55. $\dfrac{(1.65 \times 10^8)(5.24 \times 10^{-2})}{(6 \times 10^4)(2 \times 10^7)}$

$= \dfrac{1.65 \times 5.24}{6 \times 2} \times \dfrac{10^8 \times 10^{-2}}{10^4 \times 10^7}$
      Associative and commutative properties

$= 0.7205 \times \dfrac{10^6}{10^{11}}$ Product rule

$= 0.7205 \times 10^{-5}$ Quotient rule

$= (7.205 \times 10^{-1}) \times 10^{-5}$
      Write 0.7205 in scientific notation.

$= 7.205 \times (10^{-1} \times 10^{-5})$
      Associative property

$= 7.205 \times 10^{-6}$ Product rule

$= 0.000007205$
      Write without exponents.

Chapter 14 Factoring and Applications

Section 14.1 (pages 953–954)

63. $18r^2 + 12ry - 3xr - 2xy$

$= (18r^2 + 12ry) + (-3xr - 2xy)$
      Group the terms.

$= 6r(3r + 2y) - x(3r + 2y)$
      Factor each group.

$= (3r + 2y)(6r - x)$
      Factor out the common factor, $3r + 2y$.

71. $y^2 + 3x + 3y + xy$

$= y^2 + 3y + xy + 3x$
      Rearrange terms.

$= (y^2 + 3y) + (xy + 3x)$
      Group the terms.

$= y(y + 3) + x(y + 3)$
      Factor each group.

$= (y + 3)(y + x)$
      Factor out the common factor, $y + 3$.

Section 14.2 (pages 959–960)

35. $a^2 - 8a - 48$

Find two integers whose product is -48 and whose sum is -8. Since c is negative, one integer must be positive and one must be negative.

Factors of -48	Sums of Factors
$-1, 48$	47
$1, -48$	-47
$-2, 24$	22
$2, -24$	-22
$-3, 16$	13
$3, -16$	-13
$-4, 12$	8
$4, -12$	-8 ←
$-6, 8$	2
$6, -8$	-2

Thus,

$a^2 - 8a - 48$ factors as $(a + 4)(a - 12)$.

51. $-2x^6 - 8x^5 + 42x^4$

First, factor out the negative common factor, $-2x^4$.

$$-2x^6 - 8x^5 + 42x^4$$
$$= -2x^4(x^2 + 4x - 21)$$

Now factor $x^2 + 4x - 21$.

Factors of -21	Sums of Factors
$1, -21$	-20
$-1, 21$	20
$3, -7$	-4
$-3, 7$	4 ←

Thus,

$x^2 + 4x - 21$ factors as $(x - 3)(x + 7)$.

The complete factored form is

$$-2x^6 - 8x^5 + 42x^4$$
$$= -2x^4(x - 3)(x + 7).$$

55. $m^3n - 10m^2n^2 + 24mn^3$

First, factor out the GCF, mn.

$$m^3n - 10m^2n^2 + 24mn^3$$
$$= mn(m^2 - 10mn + 24n^2)$$

The expressions $-6n$ and $-4n$ have a product of $24n^2$ and a sum of $-10n$. The complete factored form is

$$m^3n - 10m^2n^2 + 24mn^3$$
$$= mn(m - 6n)(m - 4n).$$

Section 14.3 (pages 963–964)

33. $-32z^5 + 20z^4 + 12z^3$

First factor out the negative common factor, $-4z^3$.

$$-32z^5 + 20z^4 + 12z^3$$
$$= -4z^3(8z^2 - 5z - 3)$$

To factor $8z^2 - 5z - 3$, find two integers whose product is $8(-3) = -24$ and whose sum is -5. These integers are -8 and 3. Now rewrite the given trinomial and factor it.

$$-32z^5 + 20z^4 + 12z^3$$
$$= -4z^3(8z^2 - 5z - 3)$$
$$= -4z^3(8z^2 - 8z + 3z - 3)$$
   $-5z = -8z + 3z$
$$= -4z^3[(8z^2 - 8z) + (3z - 3)]$$
   Group the terms.
$$= -4z^3[8z(z - 1) + 3(z - 1)]$$
   Factor each group.
$$= -4z^3[(z - 1)(8z + 3)]$$
   Factor out common factor, $z - 1$.
$$= -4z^3(z - 1)(8z + 3)$$

39. $5 - 6x + x^2$

Find two integers whose product is $5(1) = 5$ and whose sum is -6. The integers are -1 and -5.

$$5 - 6x + x^2$$
$$= 5 - x - 5x + x^2$$
$$= (5 - x) + (-5x + x^2)$$
   Group the terms.
$$= 1(5 - x) - x(5 - x)$$
   Factor each group.
$$= (5 - x)(1 - x)$$
   Factor out the common factor, $5 - x$.

Section 14.4 (pages 969–978)

43. $6m^6n + 7m^5n^2 + 2m^4n^3$

Factor out the GCF, m^4n.

$$6m^6n + 7m^5n^2 + 2m^4n^3$$
$$= m^4n(6m^2 + 7mn + 2n^2)$$

Now factor $6m^2 + 7mn + 2n^2$ by trial and error. Possible factors of $6m^2$ are $6m$ and m or $3m$ and $2m$. Possible factors of $2n^2$ are $2n$ and n.

$$(3m + 2n)(2m + n)$$
$$= 6m^2 + 7mn + 2n^2 \qquad \text{Correct}$$

The complete factored form is

$$6m^6n + 7m^5n^2 + 2m^4n^3$$
$$= m^4n(3m + 2n)(2m + n).$$

Section 14.5 (pages 975–976)

13. $4m^2 - \dfrac{9}{25}$

Because $4m^2 = (2m)^2$ and $\dfrac{9}{25} = \left(\dfrac{3}{5}\right)^2$,

$4m^2 - \dfrac{9}{25}$ is a difference

of squares.

$$4m^2 - \dfrac{9}{25}$$
$$= (2m)^2 - \left(\dfrac{3}{5}\right)^2$$
$$= \left(2m + \dfrac{3}{5}\right)\left(2m - \dfrac{3}{5}\right)$$

39. $x^2 - 1.0x + 0.25$

The first and last terms are perfect squares, x^2 and $(-0.5)^2$. The trinomial is a perfect square, since the middle term is

$$2 \cdot x \cdot (-0.5) = -1.0x.$$

Therefore,

$$x^2 - 1.0x + 0.25$$
$$= (x)^2 - 2(x)(0.5) + (0.5)^2$$
$$= (x - 0.5)^2.$$

Section 14.6 (pages 987–988)

51. $9y^3 - 49y = 0$

To factor the polynomial, begin by factoring out the greatest common factor.

$$y(9y^2 - 49) = 0$$

Now factor $9y^2 - 49$ as the difference of two squares.

$$y(3y + 7)(3y - 7) = 0$$

Set each of the three factors equal to 0 and solve.

$$y = 0 \quad \text{or} \quad 3y + 7 = 0 \quad \text{or} \quad 3y - 7 = 0$$
$$\qquad\qquad 3y = -7 \quad \text{or} \qquad 3y = 7$$
$$\qquad\qquad y = -\frac{7}{3} \quad \text{or} \qquad y = \frac{7}{3}$$

Solution set: $\left\{ -\frac{7}{3}, 0, \frac{7}{3} \right\}$

53. $(2r + 5)(3r^2 - 16r + 5) = 0$

$$(2r + 5)(3r - 1)(r - 5) = 0$$

       Factor $3r^2 - 16r + 5$.

Set each of the three factors equal to 0, and solve the resulting equations.

$$2r + 5 = 0 \quad \text{or} \quad 3r - 1 = 0 \quad \text{or} \quad r - 5 = 0$$
$$2r = -5 \quad \text{or} \qquad 3r = 1 \quad \text{or} \qquad r = 5$$
$$r = -\frac{5}{2} \quad \text{or} \qquad r = \frac{1}{3}$$

Solution set: $\left\{ -\frac{5}{2}, \frac{1}{3}, 5 \right\}$

Section 14.7 (pages 995–1000)

13. Let $\quad x =$ the length of a side of the square painting.

Then $x - 2 =$ the length of a side of the square mirror.

Since the formula for the area of a square is $A = s^2$, the area of the painting is x^2, and the area of the mirror is $(x - 2)^2$. The difference between their areas is 32, so

$$x^2 - (x - 2)^2 = 32$$
$$x^2 - (x^2 - 4x + 4) = 32$$
$$x^2 - x^2 + 4x - 4 = 32$$
$$4x - 4 = 32$$
$$4x = 36$$
$$x = 9.$$

The length of a side of the painting is 9 ft. The length of a side of the mirror is

$$9 - 2 = 7 \text{ ft}.$$

As a check,

$$9^2 - 7^2 = 81 - 49 = 32, \quad \text{as required.}$$

---

21. Let $\quad x =$ the least even integer.

Then $x + 2 =$ the next even integer and $x + 4 =$ the third even integer.

$$x^2 + (x + 2)^2 = (x + 4)^2$$
$$x^2 + x^2 + 4x + 4 = x^2 + 8x + 16$$

       Square the binomials.

$$2x^2 + 4x + 4 = x^2 + 8x + 16$$

       Combine like terms.

$$x^2 - 4x - 12 = 0$$

       Standard form

$$(x + 2)(x - 6) = 0$$

       Factor.

$$x + 2 = 0 \quad \text{or} \quad x - 6 = 0$$

       Zero-factor property

$$x = -2 \quad \text{or} \qquad x = 6$$

If $x = -2$, then $x + 2 = 0$ and $x + 4 = 2$.
If $x = 6$, then $x + 2 = 8$ and $x + 4 = 10$.

The integers are -2, 0, and 2, or 6, 8, and 10.

Check

$(-2)^2 + 0^2 \overset{?}{=} 2^2$	$6^2 + 8^2 \overset{?}{=} 10^2$
$4 + 0 \overset{?}{=} 4$	$36 + 64 \overset{?}{=} 100$
$4 = 4$	$100 = 100$
True	True

Chapter 15 Rational Expressions and Applications

Section 15.1 (pages 1021–1022)

21. (a) $\dfrac{(-3x)^2}{4x + 12}$

$$= \frac{(-3 \cdot 2)^2}{4(2) + 12} \quad \text{Let } x = 2.$$
$$= \frac{(-6)^2}{8 + 12}$$
$$= \frac{36}{20}$$
$$= \frac{9}{5}$$

(b) $\dfrac{(-3x)^2}{4x + 12}$

$$= \frac{[-3(-3)]^2}{4(-3) + 12} \quad \text{Let } x = -3.$$
$$= \frac{9^2}{-12 + 12}, \quad \text{or} \quad \frac{81}{0}$$

Since substituting -3 for x makes the denominator 0, the given rational expression is *undefined* when $x = -3$.

35. $\dfrac{12m^2 - 3}{8m - 4}$

$$= \frac{3(4m^2 - 1)}{4(2m - 1)} \quad \text{Factor out the common factors.}$$
$$= \frac{3(2m + 1)(2m - 1)}{4(2m - 1)} \quad \text{Factor again in the numerator.}$$
$$= \frac{3(2m + 1)}{4} \quad \text{Divide out the common factors.}$$

---

43. We factor the numerator and denominator by grouping in this rational expression.

$$\frac{zw + 4z - 3w - 12}{zw + 4z + 5w + 20}$$

$$= \frac{(zw + 4z) - (3w + 12)}{(zw + 4z) + (5w + 20)} \quad \text{Group the terms.}$$

$$= \frac{z(w + 4) - 3(w + 4)}{z(w + 4) + 5(w + 4)} \quad \text{Factor each group.}$$

$$= \frac{(w + 4)(z - 3)}{(w + 4)(z + 5)} \quad \text{Factor by grouping.}$$

$$= \frac{z - 3}{z + 5} \quad \text{Divide out the common factor.}$$

59. $lw = A$ Formula for the area of a rectangle

$$w = \frac{A}{l} \quad \text{Divide by } l.$$

$$w = \frac{x^4 + 10x^2 + 21}{x^2 + 7} \quad \text{Substitute for } A \text{ and } l.$$

$$w = \frac{(x^2 + 7)(x^2 + 3)}{x^2 + 7} \quad \text{Factor.}$$

$$w = x^2 + 3 \quad \text{Divide out the common factor.}$$

Note: If it is not apparent that A,

$$x^4 + 10x^2 + 21,$$

factors as $(x^2 + 7)(x^2 + 3)$, we can use "long division" to find the quotient $\dfrac{A}{l}$.

Remember to insert zeros for the coefficients of the missing terms.

$$\begin{array}{r}
x^2 \qquad\quad + 3 \\
x^2 + 0x + 7 \overline{)x^4 + 0x^3 + 10x^2 + 0x + 21} \\
\underline{x^4 + 0x^3 + 7x^2} \qquad\qquad \\
3x^2 + 0x + 21 \\
\underline{3x^2 + 0x + 21} \\
0
\end{array}$$

The width of the rectangle is $x^2 + 3$.

Section 15.2 (pages 1027–1028)

29. $\dfrac{6(m - 2)^2}{5(m + 4)^2} \cdot \dfrac{15(m + 4)}{2(2 - m)}$

$$= \frac{6 \cdot 15(m - 2)^2(m + 4)}{5 \cdot 2(m + 4)^2(2 - m)} \quad \text{Multiply.}$$

$$= \frac{2 \cdot 3 \cdot 3 \cdot 5(m - 2)(m - 2)(m + 4)}{5 \cdot 2(m + 4)(m + 4)(-1)(m - 2)} \quad \text{Factor.}$$

$$= \frac{3 \cdot 3(m - 2)}{(m + 4)(-1)} \quad \text{Divide out the common factors.}$$

$$= \frac{9(m - 2)}{-(m + 4)}, \quad \text{or} \quad \frac{-9(m - 2)}{m + 4}$$

37. $\dfrac{m^2 + 2mp - 3p^2}{m^2 - 3mp + 2p^2} \div \dfrac{m^2 + 4mp + 3p^2}{m^2 + 2mp - 8p^2}$

$$= \frac{m^2 + 2mp - 3p^2}{m^2 - 3mp + 2p^2} \cdot \frac{m^2 + 2mp - 8p^2}{m^2 + 4mp + 3p^2}$$

       Multiply by the reciprocal.

$$= \frac{(m + 3p)(m - p)(m + 4p)(m - 2p)}{(m - 2p)(m - p)(m + 3p)(m + p)}$$
Multiply numerators and multiply denominators; factor.

$$= \frac{m + 4p}{m + p}$$
Divide out the common factors.

41. The formula for the area of a rectangle is $A = lw$. Solving for w gives $w = \dfrac{A}{l}$, which can be written as $w = A \div l$, since the fraction bar indicates division. Substituting the given expressions for A and l gives

$$\underset{\underset{\downarrow}{A}}{} \quad \underset{\underset{\downarrow}{l}}{}$$

$$\frac{5x^2 y^3}{2pq} \div \frac{2xy}{p}$$

$$= \frac{5x^2 y^3}{2pq} \cdot \frac{p}{2xy} \quad \text{Multiply by the reciprocal.}$$

$$= \frac{5xy^2}{4q} \quad \text{Multiply; divide out the common factors.}$$

as the width.

Section 15.3 (pages 1033–1034)

37. $\dfrac{-4t}{3t - 6} = \dfrac{?}{12 - 6t}$

Factor each denominator. (On the right, rewrite $12 - 6t$ as $-6t + 12$, and factor.)

$$\frac{-4t}{3(t - 2)} = \frac{?}{-6(t - 2)}$$

The missing factor is -2, so multiply the fraction on the left by $\dfrac{-2}{-2}$.

$$\frac{-4t}{3(t - 2)} \cdot \frac{-2}{-2} = \frac{8t}{-6(t - 2)}, \text{ or } \frac{8t}{12 - 6t}$$

Section 15.4 (pages 1041–1044)

31. $\dfrac{x + 1}{6} + \dfrac{3x + 3}{9}$

First, write the second fraction in lowest terms.

$$\frac{3x + 3}{9} \text{ is } \frac{3(x + 1)}{9}, \text{ or } \frac{x + 1}{3}.$$

The problem becomes

$$\frac{x + 1}{6} + \frac{x + 1}{3}.$$

The LCD is 6. Thus,

$$\frac{x + 1}{6} + \frac{x + 1}{3}$$

$$= \frac{x + 1}{6} + \frac{x + 1}{3} \cdot \frac{2}{2} \quad \text{Get a common denominator.}$$

$$= \frac{x + 1}{6} + \frac{2(x + 1)}{6}$$

$$= \frac{x + 1 + 2x + 2}{6} \quad \text{Add numerators; distributive property}$$

$$= \frac{3x + 3}{6} \quad \text{Combine like terms.}$$

$$= \frac{3(x + 1)}{3 \cdot 2} \quad \text{Factor.}$$

$$= \frac{x + 1}{2}. \quad \text{Lowest terms}$$

43. $\dfrac{t}{t + 2} + \dfrac{5 - t}{t} - \dfrac{4}{t^2 + 2t}$

$$= \frac{t}{t + 2} + \frac{5 - t}{t} - \frac{4}{t(t + 2)} \quad \text{Factor.}$$

$$= \frac{t}{t + 2} \cdot \frac{t}{t} + \frac{5 - t}{t} \cdot \frac{t + 2}{t + 2}$$

$$- \frac{4}{t(t + 2)} \quad \text{LCD} = t(t + 2)$$

$$= \frac{t \cdot t + (5 - t)(t + 2) - 4}{t(t + 2)}$$
Write using the LCD.

$$= \frac{t^2 + 5t + 10 - t^2 - 2t - 4}{t(t + 2)}$$
Multiply in the numerator.

$$= \frac{3t + 6}{t(t + 2)} \quad \text{Combine like terms.}$$

$$= \frac{3(t + 2)}{t(t + 2)} \quad \text{Factor.}$$

$$= \frac{3}{t} \quad \text{Lowest terms}$$

61. $\dfrac{2q + 1}{3q^2 + 10q - 8} - \dfrac{3q + 5}{2q^2 + 5q - 12}$

$$= \frac{2q + 1}{(3q - 2)(q + 4)} - \frac{3q + 5}{(2q - 3)(q + 4)}$$
Factor.

$$= \frac{(2q + 1)(2q - 3)}{(3q - 2)(q + 4)(2q - 3)}$$

$$- \frac{(3q + 5)(3q - 2)}{(2q - 3)(q + 4)(3q - 2)}$$
LCD $= (3q - 2)(q + 4)(2q - 3)$

$$= \frac{(2q + 1)(2q - 3) - (3q + 5)(3q - 2)}{(3q - 2)(q + 4)(2q - 3)}$$
Subtract numerators.

$$= \frac{(4q^2 - 4q - 3) - (9q^2 + 9q - 10)}{(3q - 2)(q + 4)(2q - 3)}$$
Multiply in the numerator.

$$= \frac{4q^2 - 4q - 3 - 9q^2 - 9q + 10}{(3q - 2)(q + 4)(2q - 3)}$$
Distributive property

$$= \frac{-5q^2 - 13q + 7}{(3q - 2)(q + 4)(2q - 3)}$$
Combine like terms.

---

65. $\dfrac{x + 3y}{x^2 + 2xy + y^2} + \dfrac{x - y}{x^2 + 4xy + 3y^2}$

$$= \frac{x + 3y}{(x + y)(x + y)} + \frac{x - y}{(x + 3y)(x + y)}$$

$$= \frac{(x + 3y)(x + 3y)}{(x + y)(x + y)(x + 3y)}$$

$$+ \frac{(x - y)(x + y)}{(x + 3y)(x + y)(x + y)}$$
LCD $= (x + y)(x + y)(x + 3y)$

$$= \frac{(x^2 + 6xy + 9y^2) + (x^2 - y^2)}{(x + y)(x + y)(x + 3y)}$$

$$= \frac{2x^2 + 6xy + 8y^2}{(x + y)(x + y)(x + 3y)},$$

$$\text{or } \frac{2x^2 + 6xy + 8y^2}{(x + y)^2(x + 3y)}$$

There is no need to factor out 2 in the numerator, since the denominator does not have 2 as a factor. The answer is in lowest terms.

Section 15.5 (pages 1051–1052)

15. $\dfrac{\dfrac{1}{x} + x}{\dfrac{x^2 + 1}{8}}$ Use Method 2.

$$= \frac{8x\left(\dfrac{1}{x} + x\right)}{8x\left(\dfrac{x^2 + 1}{8}\right)} \quad \text{LCD} = 8x$$

$$= \frac{8 + 8x^2}{x(x^2 + 1)} \quad \text{Distributive property}$$

$$= \frac{8(1 + x^2)}{x(x^2 + 1)} \quad \text{Factor.}$$

$$= \frac{8}{x} \quad \text{Lowest terms}$$

29. $\dfrac{\dfrac{1}{m - 1} + \dfrac{2}{m + 2}}{\dfrac{2}{m + 2} - \dfrac{1}{m - 3}}$ Use Method 2.

$$= \frac{(m - 1)(m + 2)(m - 3)\left(\dfrac{1}{m - 1} + \dfrac{2}{m + 2}\right)}{(m - 1)(m + 2)(m - 3)\left(\dfrac{2}{m + 2} - \dfrac{1}{m - 3}\right)}$$
LCD $= (m - 1)(m + 2)(m - 3)$

$$= \frac{(m + 2)(m - 3) + 2(m - 1)(m - 3)}{2(m - 1)(m - 3) - (m - 1)(m + 2)}$$
Distributive property

$$= \frac{(m - 3)[(m + 2) + 2(m - 1)]}{(m - 1)[2(m - 3) - (m + 2)]}$$
Factor out $m - 3$ in the numerator and $m - 1$ in the denominator.

$$= \frac{(m - 3)[m + 2 + 2m - 2]}{(m - 1)[2m - 6 - m - 2]}$$
Distributive property

$$= \frac{3m(m - 3)}{(m - 1)(m - 8)} \quad \text{Combine like terms.}$$

SOLUTIONS

31. $2 - \dfrac{2}{2 + \dfrac{2}{2+2}}$

$= 2 - \dfrac{2}{2 + \dfrac{2}{4}}$ Simplify the denominator.

$= 2 - \dfrac{2}{\dfrac{5}{2}}$ $2 + \dfrac{2}{4} = \dfrac{4}{2} + \dfrac{1}{2} = \dfrac{5}{2}$

$= 2 - \left(2 \cdot \dfrac{2}{5}\right)$ $\dfrac{2}{\dfrac{5}{2}} = 2 \div \dfrac{5}{2} = 2 \cdot \dfrac{2}{5}$

$= 2 - \dfrac{4}{5}$ Multiply.

$= \dfrac{10}{5} - \dfrac{4}{5}$ $2 = \dfrac{10}{5}$

$= \dfrac{6}{5}$ Subtract.

Section 15.6 (pages 1061–1064)

11. $\dfrac{3x}{5} - 6 = x$

$5\left(\dfrac{3x}{5} - 6\right) = 5(x)$ Multiply by the LCD, 5.

$5\left(\dfrac{3x}{5}\right) - 5(6) = 5x$ Distributive property

$3x - 30 = 5x$

$-30 = 2x$ Subtract $3x$.

$-15 = x$ Divide by 2.

Check by substituting -15 for x in the original equation.
Solution set: $\{-15\}$

21. $\dfrac{q+2}{3} + \dfrac{q-5}{5} = \dfrac{7}{3}$

$15\left(\dfrac{q+2}{3} + \dfrac{q-5}{5}\right) = 15\left(\dfrac{7}{3}\right)$
    Multiply by the LCD, 15.

$15\left(\dfrac{q+2}{3}\right) + 15\left(\dfrac{q-5}{5}\right) = 5 \cdot 7$

$5(q+2) + 3(q-5) = 35$

$5q + 10 + 3q - 15 = 35$

$8q - 5 = 35$

$8q = 40$

$q = 5$

A check confirms that 5 is the solution.
Solution set: $\{5\}$

37. $\dfrac{2}{m} = \dfrac{m}{5m+12}$

$m(5m+12)\left(\dfrac{2}{m}\right) = m(5m+12)\left(\dfrac{m}{5m+12}\right)$

    Multiply by the LCD, $m(5m+12)$.

$(5m+12)(2) = m(m)$ Divide out the common factors.

$10m + 24 = m^2$ Multiply.

$m^2 - 10m - 24 = 0$ Standard form

$(m+2)(m-12) = 0$ Factor.

$m + 2 = 0$ or $m - 12 = 0$
    Zero-factor property

$m = -2$ or $m = 12$

The proposed solutions are -2 and 12.

Check $\dfrac{2}{-2} \overset{?}{=} \dfrac{-2}{5(-2)+12}$ Let $m = -2$.

$-1 = \dfrac{-2}{2}$ True

Check $\dfrac{2}{12} \overset{?}{=} \dfrac{12}{5(12)+12}$ Let $m = 12$.

$\dfrac{1}{6} = \dfrac{12}{72}$ True

Solution set: $\{-2, 12\}$

65. $d = \dfrac{2S}{n(a+L)}$ for a

We need to isolate a on one side of the equation.

$d = \dfrac{2S}{n(a+L)}$

$d \cdot n(a+L) = \dfrac{2S}{n(a+L)} \cdot n(a+L)$
    Multiply by $n(a+L)$.

$dn(a+L) = 2S$

$dna + dnL = 2S$ Distributive property

$dna = 2S - dnL$ Subtract dnL.

$a = \dfrac{2S - dnL}{dn}$,   Divide by dn.

or $a = \dfrac{2S}{dn} - L$

71. $9x + \dfrac{3}{z} = \dfrac{5}{y}$ for z

$yz\left(9x + \dfrac{3}{z}\right) = yz\left(\dfrac{5}{y}\right)$ Multiply by the LCD, yz.

$yz(9x) + yz\left(\dfrac{3}{z}\right) = yz\left(\dfrac{5}{y}\right)$ Distributive property

$9xyz + 3y = 5z$

$9xyz - 5z = -3y$ Get the z-terms on one side.

$z(9xy - 5) = -3y$ Factor out z.

$z = \dfrac{-3y}{9xy - 5}$,   Divide by $9xy - 5$.

or $z = \dfrac{3y}{5 - 9xy}$

Section 15.7 (pages 1075–1078)

25. Let x represent the rate of the current of the river. Then $(12 - x)$ is the rate upstream (against the current) and $(12 + x)$ is the rate downstream (with the current).

Use $t = \dfrac{d}{r}$ to complete the table.

	d	r	t
Upstream	6	$12 - x$	$\dfrac{6}{12-x}$
Downstream	10	$12 + x$	$\dfrac{10}{12+x}$

Since the times are equal, we get the following equation.

$\dfrac{6}{12-x} = \dfrac{10}{12+x}$

$(12+x)(12-x)\dfrac{6}{12-x}$
$= (12+x)(12-x)\dfrac{10}{12+x}$

Multiply by the LCD, $(12+x)(12-x)$.

$6(12+x) = 10(12-x)$

$72 + 6x = 120 - 10x$

$16x = 48$

$x = 3$

The rate of the current of the river is 3 mph.

33. Let x = the number of hours it would take Brenda to paint the room by herself.

	Rate	Time Working Together	Fractional Part of the Job Done When Working Together
Hilda	$\dfrac{1}{6}$	$3\dfrac{3}{4} = \dfrac{15}{4}$	$\dfrac{1}{6} \cdot \dfrac{15}{4} = \dfrac{5}{8}$
Brenda	$\dfrac{1}{x}$	$3\dfrac{3}{4} = \dfrac{15}{4}$	$\dfrac{1}{x} \cdot \dfrac{15}{4} = \dfrac{15}{4x}$

Since together Hilda and Brenda complete 1 whole job, we must add their individual fractional parts and set the sum equal to 1.

$\dfrac{5}{8} + \dfrac{15}{4x} = 1$

$8x\left(\dfrac{5}{8} + \dfrac{15}{4x}\right) = 8x(1)$ Multiply by the LCD, $8x$.

$8x\left(\dfrac{5}{8}\right) + 8x\left(\dfrac{15}{4x}\right) = 8x$

$5x + 30 = 8x$

$30 = 3x$

$10 = x$

It will take Brenda 10 hr to paint the room by herself.

35. Let x = the number of hours to fill the pool with both pipes left open.

	Rate	Time to Fill the Pool	Part Done by Each Pipe
Inlet pipe	$\dfrac{1}{9}$	x	$\dfrac{1}{9}x$
Outlet pipe	$\dfrac{1}{12}$	x	$\dfrac{1}{12}x$

Part done by inlet pipe	−	Part done by outlet pipe	=	Full pool
↓		↓		↓
$\dfrac{1}{9}x$	−	$\dfrac{1}{12}x$	=	1

$$36\left(\frac{1}{9}x - \frac{1}{12}x\right) = 36(1) \qquad \text{Multiply by the LCD, 36.}$$

$$36\left(\frac{1}{9}x\right) - 36\left(\frac{1}{12}x\right) = 36$$

$$4x - 3x = 36$$

$$x = 36$$

It will take 36 hr to fill the pool.

Section 15.8 (pages 1081–1084)

11. For a given base, the area A of a triangle varies directly as its height h, so there is a constant k such that $A = kh$. Find the value of k.

$$10 = k(4) \qquad \text{Let } A = 10, \ h = 4.$$

$$k = \frac{10}{4} = 2.5 \qquad \text{Solve for } k.$$

When $k = 2.5$, $A = kh$ becomes

$$A = 2.5h$$

Now find A when $h = 6$.

$$A = 2.5(6) = 15$$

When the height of the triangle is 6 in., the area of the triangle is 15 in.^{2}

21. For a constant time of 3 hr, if the rate of the pickup truck *increases,* then the distance traveled *increases.* Thus, the variation between the quantities is *direct.*

29. Ratios of corresponding sides are equal, so since $\dfrac{12}{3} = 4$, we must have

$$\frac{x}{2} = 4, \text{ and thus, } x = 8.$$

Chapter 16 Roots and Radicals

Section 16.1 (pages 1105–1108)

55. $\sqrt{103} \approx \sqrt{100} = 10$ Length

$\sqrt{48} \approx \sqrt{49} = 7$ Width

The best estimate for the length and width, in meters, of the rectangle is 10 by 7, choice C.

71. Let $a = 4.5$ and $c = 12.0$. Use the Pythagorean formula.

$$c^2 = a^2 + b^2$$

$$(12.0)^2 = (4.5)^2 + b^2 \qquad \text{Substitute carefully.}$$

$$144 = 20.25 + b^2$$

$$123.75 = b^2$$

$$b = \sqrt{123.75}$$

$$b \approx 11.1 \qquad \text{Use a calculator.}$$

The distance from the base of the tree to the point where the broken part touches the ground is 11.1 ft (to the nearest tenth).

---

83. $-\sqrt[3]{-8} = -(-2) = 2$

Section 16.2 (pages 1115–1118)

37. $5\sqrt{3} \cdot 2\sqrt{15}$

$= 5 \cdot 2 \cdot \sqrt{3 \cdot 15}$ Commutative property; product rule

$= 10\sqrt{45}$ Multiply.

$= 10\sqrt{9 \cdot 5}$ Factor; 9 is a perfect square.

$= 10\sqrt{9} \cdot \sqrt{5}$ Product rule

$= 10 \cdot 3 \cdot \sqrt{5}$ $\sqrt{9} = 3$

$= 30\sqrt{5}$ Multiply.

105. Use the formula for the volume of a sphere.

$$V = \frac{4}{3}\pi r^3$$

Let $V = 288\pi$ and solve for r.

$$288\pi = \frac{4}{3}\pi r^3 \qquad V = 288\pi$$

$$\frac{3}{4}(288\pi) = \frac{3}{4}\left(\frac{4}{3}\pi r^3\right) \qquad \text{Multiply by } \frac{3}{4}.$$

$$216\pi = \pi r^3 \qquad \text{Simplify.}$$

$$216 = r^3 \qquad \text{Divide by } \pi.$$

$$\sqrt[3]{216} = r \qquad \text{Take cube roots.}$$

$$6 = r$$

The radius is 6 in.

107. $2\sqrt{26} \approx 2\sqrt{25} = 2 \cdot 5 = 10$ Length

$\sqrt{83} \approx \sqrt{81} = 9$ Width

Using 10 and 9 as estimates for the length and the width of the rectangle gives us

$$10 \cdot 9 = 90$$

as an estimate for the area, in square inches. Choice D is the best estimate.

Section 16.3 (pages 1121–1122)

19. $2\sqrt{8} - 5\sqrt{32} - 2\sqrt{48}$

$= 2\left(\sqrt{4} \cdot \sqrt{2}\right) - 5\left(\sqrt{16} \cdot \sqrt{2}\right)$
$\quad - 2\left(\sqrt{16} \cdot \sqrt{3}\right)$ Product rule

$= 2\left(2\sqrt{2}\right) - 5\left(4\sqrt{2}\right) - 2\left(4\sqrt{3}\right)$
$\qquad\qquad\qquad \sqrt{4} = 2; \ \sqrt{16} = 4$

$= 4\sqrt{2} - 20\sqrt{2} - 8\sqrt{3}$ Multiply.

$= -16\sqrt{2} - 8\sqrt{3}$ Subtract like radicals.

23. $\dfrac{1}{4}\sqrt{288} + \dfrac{1}{6}\sqrt{72}$

$= \dfrac{1}{4}\sqrt{144 \cdot 2} + \dfrac{1}{6}\sqrt{36 \cdot 2}$

$= \dfrac{1}{4}\left(\sqrt{144} \cdot \sqrt{2}\right) + \dfrac{1}{6}\left(\sqrt{36} \cdot \sqrt{2}\right)$
$\qquad\qquad\qquad\qquad\qquad \text{Product rule}$

$= \dfrac{1}{4}\left(12\sqrt{2}\right) + \dfrac{1}{6}\left(6\sqrt{2}\right)$
$\qquad\qquad \sqrt{144} = 12; \ \sqrt{36} = 6$

$= 3\sqrt{2} + 1\sqrt{2}$ Multiply.

$= 4\sqrt{2}$ Add like radicals.

25. Use the formula for the perimeter of a rectangle.

$$P = 2l + 2w$$

$$P = 2\left(7\sqrt{2}\right) + 2\left(4\sqrt{2}\right)$$

$$P = 14\sqrt{2} + 8\sqrt{2}$$

$$P = 22\sqrt{2}$$

43. $5\sqrt[4]{m^3} + 8\sqrt[4]{16m^3}$

$= 5\sqrt[4]{m^3} + 8\sqrt[4]{16 \cdot m^3}$

$= 5\sqrt[4]{m^3} + 8\sqrt[4]{16}\,\sqrt[4]{m^3}$ Product rule

$= 5\sqrt[4]{m^3} + 8 \cdot 2 \cdot \sqrt[4]{m^3}$ $\sqrt[4]{16} = 2$

$= 5\sqrt[4]{m^3} + 16\sqrt[4]{m^3}$ Multiply.

$= 21\sqrt[4]{m^3}$ Add like radicals.

Section 16.4 (pages 1127–1128)

37. $\sqrt{\dfrac{9a^2r^5}{7t}}$

$= \dfrac{\sqrt{9a^2r^5}}{\sqrt{7t}}$ Quotient rule

$= \dfrac{\sqrt{9a^2r^4 \cdot r}}{\sqrt{7t}}$ Factor.

$= \dfrac{\sqrt{9a^2r^4} \cdot \sqrt{r}}{\sqrt{7t}}$ Product rule

$= \dfrac{3ar^2\sqrt{r}}{\sqrt{7t}}$ $\sqrt{9} = 3; \ \sqrt{a^2} = a,$ since $a > 0; \ \sqrt{r^4} = r^2$

$= \dfrac{3ar^2\sqrt{r} \cdot \sqrt{7t}}{\sqrt{7t} \cdot \sqrt{7t}}$ Rationalize the denominator.

$= \dfrac{3ar^2\sqrt{7rt}}{7t}$ Product rule; $\sqrt{7t} \cdot \sqrt{7t} = 7t$

41. $\sqrt[3]{\dfrac{5}{9}}$

$= \dfrac{\sqrt[3]{5}}{\sqrt[3]{9}}$ Quotient rule

$= \dfrac{\sqrt[3]{5} \cdot \sqrt[3]{3}}{\sqrt[3]{9} \cdot \sqrt[3]{3}}$ Multiply numerator and denominator by $\sqrt[3]{3}$ to make radical in denominator a perfect cube.

$= \dfrac{\sqrt[3]{15}}{\sqrt[3]{27}}$ Product rule

$= \dfrac{\sqrt[3]{15}}{3}$ $\sqrt[3]{27} = 3$

49. (a) $p = k \cdot \sqrt{\dfrac{L}{g}}$

$p = 6 \cdot \sqrt{\dfrac{9}{32}}$ Let $k = 6$, $L = 9$, $g = 32$.

$p = \dfrac{6\sqrt{9}}{\sqrt{32}}$

$p = \dfrac{6 \cdot 3}{\sqrt{16 \cdot 2}}$

$p = \dfrac{18}{4\sqrt{2}}$

$p = \dfrac{9}{2\sqrt{2}}$

$p = \dfrac{9 \cdot \sqrt{2}}{2\sqrt{2} \cdot \sqrt{2}}$ Rationalize the denominator.

$p = \dfrac{9\sqrt{2}}{4}$

The period of the pendulum is $\dfrac{9\sqrt{2}}{4}$ sec.

(b) Using a calculator, $\dfrac{9\sqrt{2}}{4} \approx 3.182$ sec.

Section 16.5 (pages 1133–1136)

13. $\left(5\sqrt{7} - 2\sqrt{3}\right)\left(3\sqrt{7} + 4\sqrt{3}\right)$

$= 5\sqrt{7}\left(3\sqrt{7}\right) + 5\sqrt{7}\left(4\sqrt{3}\right)$

$\quad - 2\sqrt{3}\left(3\sqrt{7}\right) - 2\sqrt{3}\left(4\sqrt{3}\right)$
   FOIL

$= 15 \cdot 7 + 20\sqrt{21} - 6\sqrt{21} - 8 \cdot 3$
   Multiply.

$= 105 + 14\sqrt{21} - 24$ Multiply numbers; add like radicals.

$= 81 + 14\sqrt{21}$ Subtract.

43. $\dfrac{\sqrt{5}}{\sqrt{2} + \sqrt{3}}$

$= \dfrac{\sqrt{5}\left(\sqrt{2} - \sqrt{3}\right)}{\left(\sqrt{2} + \sqrt{3}\right)\left(\sqrt{2} - \sqrt{3}\right)}$

Multiply numerator and denominator by the conjugate of the denominator.

$= \dfrac{\sqrt{5} \cdot \sqrt{2} - \sqrt{5} \cdot \sqrt{3}}{\left(\sqrt{2}\right)^2 - \left(\sqrt{3}\right)^2}$

Distributive property; $(a + b)(a - b) = a^2 - b^2$

$= \dfrac{\sqrt{10} - \sqrt{15}}{2 - 3}$ Product rule; $\left(\sqrt{a}\right)^2 = a$

$= \dfrac{\sqrt{10} - \sqrt{15}}{-1}$

$= -\sqrt{10} + \sqrt{15}$

55. $\dfrac{12 - \sqrt{40}}{4}$

$= \dfrac{12 - \sqrt{4} \cdot \sqrt{10}}{4}$ Product rule

$= \dfrac{12 - 2\sqrt{10}}{4}$ $\sqrt{4} = 2$

$= \dfrac{2\left(6 - \sqrt{10}\right)}{2(2)}$ Factor numerator and denominator.

$= \dfrac{6 - \sqrt{10}}{2}$ Lowest terms

Section 16.6 (pages 1145–1148)

19. $\sqrt{3x - 5} = \sqrt{2x + 1}$

$\left(\sqrt{3x - 5}\right)^2 = \left(\sqrt{2x + 1}\right)^2$
   Squaring property

$3x - 5 = 2x + 1$ $\left(\sqrt{a}\right)^2 = a$

$x = 6$ Subtract $2x$; add 5.

Check $\sqrt{3x - 5} = \sqrt{2x + 1}$

$\sqrt{3(6) - 5} \stackrel{?}{=} \sqrt{2(6) + 1}$

   Let $x = 6$.

$\sqrt{13} = \sqrt{13}$ True

Solution set: $\{6\}$

29. $\sqrt{3k + 10} + 5 = 2k$

$\sqrt{3k + 10} = 2k - 5$ Subtract 5.

$\left(\sqrt{3k + 10}\right)^2 = (2k - 5)^2$ Squaring property

$3k + 10 = 4k^2 - 20k + 25$

$\left(\sqrt{a}\right)^2 = a$; square the binomial.

$4k^2 - 23k + 15 = 0$ Standard form

$(4k - 3)(k - 5) = 0$ Factor.

$k = \dfrac{3}{4}$ or $k = 5$ Zero-factor property

Checking both proposed solutions in the original equation will show that 5 is a solution, but $\dfrac{3}{4}$ is not.

Solution set: $\{5\}$

45.

$\sqrt{2x + 11} + \sqrt{x + 6} = 2$

$\sqrt{2x + 11} = 2 - \sqrt{x + 6}$
   Isolate a radical.

$\left(\sqrt{2x + 11}\right)^2 = \left(2 - \sqrt{x + 6}\right)^2$
   Square both sides.

$2x + 11 = 4 - 4\sqrt{x + 6} + x + 6$

$\left(\sqrt{a}\right)^2 = a^2$; square the binomial.

$x + 1 = -4\sqrt{x + 6}$
   Isolate a radical.

$(x + 1)^2 = \left(-4\sqrt{x + 6}\right)^2$
   Square both sides again.

$x^2 + 2x + 1 = 16(x + 6)$

Square the binomial; $\left(a\sqrt{b}\right)^2 = a^2 b$

$x^2 + 2x + 1 = 16x + 96$
   Distributive property

$x^2 - 14x - 95 = 0$ Standard form

$(x + 5)(x - 19) = 0$ Factor.

$x = -5$ or $x = 19$ Zero-factor property

Checking both proposed solutions in the original equation will show that -5 is a solution, but 19 is not.

Solution set: $\{-5\}$

49. Let $x =$ the unknown number.

Write an equation.

$3\sqrt{2} = \sqrt{x + 10}$

To solve this equation, square both sides and then solve the resulting equation.

$\left(3\sqrt{2}\right)^2 = \left(\sqrt{x + 10}\right)^2$

   Square both sides.

$18 = x + 10$

$\left(3\sqrt{2}\right)^2 = 9 \cdot 2 = 18$

$8 = x$ Subtract 10.

Check   Since $\sqrt{8 + 10} = \sqrt{18}$, and $\sqrt{18} = 3\sqrt{2}$, the number is 8.

Chapter 17 Quadratic Equations

Section 17.1 (pages 1163–1166)

21. $7x^2 = 4$

$x^2 = \dfrac{4}{7}$ Divide by 7.

$x = \sqrt{\dfrac{4}{7}}$ or $x = -\sqrt{\dfrac{4}{7}}$

   Square root property

$x = \dfrac{\sqrt{4}}{\sqrt{7}} \cdot \dfrac{\sqrt{7}}{\sqrt{7}}$ or $x = -\dfrac{\sqrt{4}}{\sqrt{7}} \cdot \dfrac{\sqrt{7}}{\sqrt{7}}$

   Rationalize denominators.

$x = \dfrac{2\sqrt{7}}{7}$ or $x = -\dfrac{2\sqrt{7}}{7}$

Solution set: $\left\{ -\dfrac{2\sqrt{7}}{7}, \dfrac{2\sqrt{7}}{7} \right\}$

43. $\left(\dfrac{1}{2}x + 5\right)^2 = 12$

$\dfrac{1}{2}x + 5 = \sqrt{12}$ or $\dfrac{1}{2}x + 5 = -\sqrt{12}$

$\dfrac{1}{2}x = -5 + \sqrt{12}$ or $\dfrac{1}{2}x = -5 - \sqrt{12}$
   Subtract 5.

$\dfrac{1}{2}x = -5 + 2\sqrt{3}$ or $\dfrac{1}{2}x = -5 - 2\sqrt{3}$

   $\sqrt{12} = \sqrt{4} \cdot \sqrt{3} = 2\sqrt{3}$

$x = -10 + 4\sqrt{3}$ or $x = -10 - 4\sqrt{3}$
   Multiply each term by 2.

Solution set: $\left\{ -10 + 4\sqrt{3},\, -10 - 4\sqrt{3} \right\}$

53.
$$A = P(1 + r)^2$$
$$110.25 = 100(1 + r)^2$$
Let $A = 110.25, P = 100$.
$$1.1025 = (1 + r)^2$$
Divide by 100.
$$(1 + r)^2 = 1.1025$$
Interchange sides.
$$1 + r = \sqrt{1.1025} \quad \text{or} \quad 1 + r = -\sqrt{1.1025}$$
Square root property
$$r = -1 + 1.05 \text{ or} \quad r = -1 - 1.05$$
Subtract 1; use a calculator.
$$r = 0.05 \quad \text{or} \quad r = -2.05$$

Reject the solution -2.05 since the rate of interest cannot be negative. The rate is 0.05, or 5%.

Section 17.2 (pages 1173–1174)

17. $t^2 + 6t + 9 = 0$

The left side of this equation is already a perfect square.

$$(t + 3)^2 = 0 \qquad \text{Factor.}$$
$$t + 3 = 0 \qquad \text{Square root property}$$
$$t = -3 \qquad \text{Subtract 3.}$$

A check verifies that the solution is -3.
Solution set: $\{-3\}$

19. $x^2 + x - 1 = 0$
$$x^2 + x = 1 \quad \text{Add 1.}$$

Take half of 1, the coefficient of x, square it, and add the result to each side.

$$x^2 + x + \frac{1}{4} = 1 + \frac{1}{4}$$
$$\text{Add } \left[\frac{1}{2}(1)\right]^2 = \frac{1}{4}.$$
$$\left(x + \frac{1}{2}\right)^2 = \frac{5}{4}$$
Factor; add.

$$x + \frac{1}{2} = \sqrt{\frac{5}{4}} \quad \text{or} \quad x + \frac{1}{2} = -\sqrt{\frac{5}{4}}$$

$$x + \frac{1}{2} = \frac{\sqrt{5}}{2} \quad \text{or} \quad x + \frac{1}{2} = -\frac{\sqrt{5}}{2}$$

$$x = -\frac{1}{2} + \frac{\sqrt{5}}{2} \quad \text{or} \quad x = -\frac{1}{2} - \frac{\sqrt{5}}{2}$$

$$x = \frac{-1 + \sqrt{5}}{2} \quad \text{or} \quad x = \frac{-1 - \sqrt{5}}{2}$$

Solution set: $\left\{\dfrac{-1 + \sqrt{5}}{2}, \dfrac{-1 - \sqrt{5}}{2}\right\}$

23.
$$2x^2 - 4x = 5$$
$$x^2 - 2x = \frac{5}{2} \quad \text{Divide by 2.}$$
$$x^2 - 2x + 1 = \frac{5}{2} + 1$$
$$\text{Add } \left[\frac{1}{2}(-2)\right]^2 = 1.$$
$$(x - 1)^2 = \frac{7}{2} \quad \text{Factor; add.}$$

$$x - 1 = \sqrt{\frac{7}{2}} \quad \text{or} \quad x - 1 = -\sqrt{\frac{7}{2}}$$

$$x - 1 = \frac{\sqrt{7}}{\sqrt{2}} \quad \text{or} \quad x - 1 = -\frac{\sqrt{7}}{\sqrt{2}}$$

$$x - 1 = \frac{\sqrt{14}}{2} \quad \text{or} \quad x - 1 = -\frac{\sqrt{14}}{2}$$

Rationalize denominators by multiplying by $\dfrac{\sqrt{2}}{\sqrt{2}}$.

$$x = 1 + \frac{\sqrt{14}}{2} \quad \text{or} \quad x = 1 - \frac{\sqrt{14}}{2}$$

$$x = \frac{2}{2} + \frac{\sqrt{14}}{2} \quad \text{or} \quad x = \frac{2}{2} - \frac{\sqrt{14}}{2}$$

$$x = \frac{2 + \sqrt{14}}{2} \quad \text{or} \quad x = \frac{2 - \sqrt{14}}{2}$$

Solution set: $\left\{\dfrac{2 + \sqrt{14}}{2}, \dfrac{2 - \sqrt{14}}{2}\right\}$

Section 17.3 (pages 1179–1180)

21. $7x^2 = 12x$

Write the equation in standard form.
$$7x^2 - 12x = 0$$

Substitute $a = 7$, $b = -12$, and $c = 0$ into the quadratic formula.

$$x = \frac{-b \pm \sqrt{b^2 - 4ac}}{2a}$$

$$x = \frac{-(-12) \pm \sqrt{(-12)^2 - 4(7)(0)}}{2(7)}$$

$$x = \frac{12 \pm \sqrt{144 - 0}}{14}$$

$$x = \frac{12 \pm 12}{14}$$

$$x = \frac{12 + 12}{14} = \frac{24}{14} = \frac{12}{7} \quad \text{or}$$

$$x = \frac{12 - 12}{14} = \frac{0}{14} = 0$$

Solution set: $\left\{0, \dfrac{12}{7}\right\}$

27. $3x^2 - 2x + 5 = 10x + 1$

Write the equation in standard form.
$$3x^2 - 12x + 4 = 0$$

Substitute $a = 3$, $b = -12$, and $c = 4$ into the quadratic formula.

$$x = \frac{-b \pm \sqrt{b^2 - 4ac}}{2a}$$

$$x = \frac{-(-12) \pm \sqrt{(-12)^2 - 4(3)(4)}}{2(3)}$$

$$x = \frac{12 \pm \sqrt{144 - 48}}{6}$$

$$x = \frac{12 \pm \sqrt{96}}{6}$$

$$x = \frac{12 \pm 4\sqrt{6}}{6}$$

$$\sqrt{96} = \sqrt{16} \cdot \sqrt{6} = 4\sqrt{6}$$

$$x = \frac{2(6 \pm 2\sqrt{6})}{2 \cdot 3} \qquad \text{Factor.}$$

$$x = \frac{6 \pm 2\sqrt{6}}{3} \qquad \text{Divide out 2.}$$

Solution set: $\left\{\dfrac{6 + 2\sqrt{6}}{3}, \dfrac{6 - 2\sqrt{6}}{3}\right\}$

39. $0.6x - 0.4x^2 = -1$

To eliminate the decimals, multiply each side by 10.

$$6x - 4x^2 = -10$$

Write this equation in standard form.

$$4x^2 - 6x - 10 = 0$$

Divide each side by 2 so that we can work with smaller coefficients in the quadratic formula.

$$2x^2 - 3x - 5 = 0$$

Use the quadratic formula with $a = 2$, $b = -3$, and $c = -5$.

$$x = \frac{-(-3) \pm \sqrt{(-3)^2 - 4(2)(-5)}}{2(2)}$$

$$x = \frac{3 \pm \sqrt{9 + 40}}{4}$$

$$x = \frac{3 \pm \sqrt{49}}{4}$$

$$x = \frac{3 \pm 7}{4}$$

$$x = \frac{3 + 7}{4} = \frac{10}{4} = \frac{5}{2} \quad \text{or}$$

$$x = \frac{3 - 7}{4} = \frac{-4}{4} = -1$$

Solution set: $\left\{-1, \dfrac{5}{2}\right\}$

Section 17.4 (pages 1187–1188)

17. Because the vertex is at the origin, an equation of the parabola is of the form
$$y = ax^2.$$

As shown in the figure, one point on the graph has coordinates (150, 44).

$$y = ax^2 \qquad \text{General equation}$$
$$44 = a(150)^2 \qquad \text{Let } x = 150, y = 44.$$
$$44 = 22,500a \qquad \text{Apply the exponent.}$$
$$a = \frac{44}{22,500} \qquad \text{Solve for } a.$$
$$a = \frac{4 \cdot 11}{4 \cdot 5625}, \quad \text{or} \quad \frac{11}{5625} \qquad \text{Lowest terms}$$

Thus, an equation of the parabola is

$$y = \frac{11}{5625}x^2.$$

▶▶▶ INDEX

STUDY SKILLS

Feeling a little uncertain of your ability to learn and remember mathematics? As you work through this textbook, watch for the special study skills pages. You'll find out how your brain learns and remembers best, and you'll learn brain-friendly ways to use your textbook, take lecture notes, do homework, make study cards and mind maps, prepare for and take tests, and much more. It's your ticket to success in mathematics! (See the Table of Contents in the front of the book for study skills titles and locations.)

FORMULAS

(*See inside back cover for geometry formulas.*)

Percent proportion: $\dfrac{\text{part}}{\text{whole}} = \dfrac{\text{percent}}{100}$

Percent equation: $\text{part} = \text{percent} \cdot \text{whole}$

Simple interest: $\text{Interest} = \text{principal} \cdot \text{rate} \cdot \text{time}$

Celsius to Fahrenheit: $F = \dfrac{9C}{5} + 32$

Fahrenheit to Celsius: $C = \dfrac{5(F - 32)}{9}$

$\text{Mean} = \dfrac{\text{sum of all values}}{\text{number of values}}$

ORDER OF OPERATIONS

1. Do all operations inside *parentheses* or *other grouping symbols*.
2. Simplify any expressions with *exponents* and find any *square roots*.
3. *Multiply* and *divide*, proceeding from left to right.
4. *Add* and *subtract*, proceeding from left to right.

Example 1

$6^2 - 8 \div 2 + (7 + 3)$	Work inside parentheses.
$6^2 - 8 \div 2 + 10$	Simplify exponents.
$36 - 8 \div 2 + 10$	Divide.
$36 - 4 + 10$	Add and subtract from left to right.
$32 + 10$	
42	

Example 2

$4 + 10 \div 2(6) - 3(8 \div 4)$	Work inside parentheses.
$4 + 10 \div 2(6) - 3(2)$	Multiply and divide from left to right.
$4 + 5(6) - 3(2)$	
$4 + 30 - 3(2)$	
$4 + 30 - 6$	Add and subtract from left to right.
$34 - 6$	
28	

A phrase to help you remember the order of operations:

Please	**Excuse**	**My**	**Dear**	**Aunt**	**Sally**
Parentheses;	**Exponents;**	**Multiply and Divide;**		**Add and Subtract**	

FRACTION-DECIMAL-PERCENT EQUIVALENTS

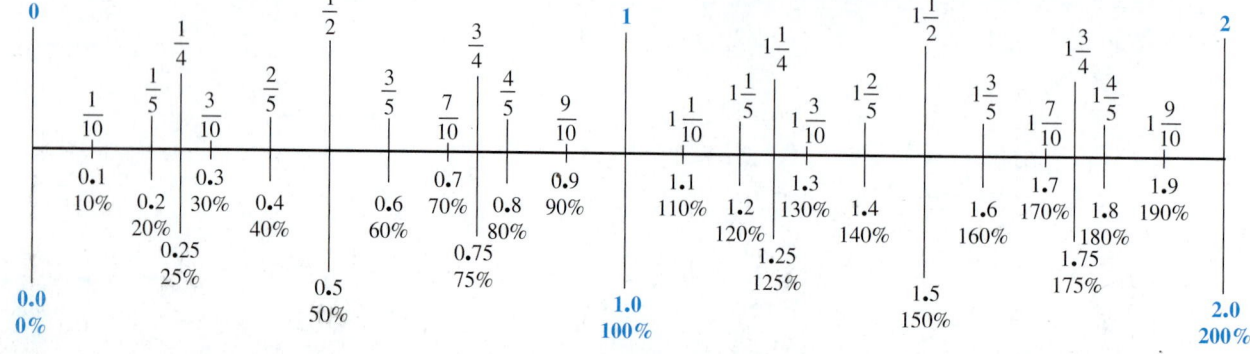